ALBRIGHT'S CHEMICAL ENGINEERING HANDBOOK

ALBRIGHT'S CHEMICAL ENGINEERING HANDBOOK

Edited by
Lyle F. Albright
*Purdue University, West Lafayette
Indiana, USA*

CRC Press is an imprint of the
Taylor & Francis Group, an **informa** business

CRC Press
Taylor & Francis Group
6000 Broken Sound Parkway NW, Suite 300
Boca Raton, FL 33487-2742

© 2009 by Taylor & Francis Group, LLC
CRC Press is an imprint of Taylor & Francis Group, an Informa business

No claim to original U.S. Government works
Printed in the United States of America on acid-free paper
10 9 8 7 6 5 4 3 2 1

International Standard Book Number-13: 978-0-8247-5362-7 (Hardcover)

This book contains information obtained from authentic and highly regarded sources. Reasonable efforts have been made to publish reliable data and information, but the author and publisher cannot assume responsibility for the validity of all materials or the consequences of their use. The authors and publishers have attempted to trace the copyright holders of all material reproduced in this publication and apologize to copyright holders if permission to publish in this form has not been obtained. If any copyright material has not been acknowledged please write and let us know so we may rectify in any future reprint.

Except as permitted under U.S. Copyright Law, no part of this book may be reprinted, reproduced, transmitted, or utilized in any form by any electronic, mechanical, or other means, now known or hereafter invented, including photocopying, microfilming, and recording, or in any information storage or retrieval system, without written permission from the publishers.

For permission to photocopy or use material electronically from this work, please access www.copyright.com (http://www.copyright.com/) or contact the Copyright Clearance Center, Inc. (CCC), 222 Rosewood Drive, Danvers, MA 01923, 978-750-8400. CCC is a not-for-profit organization that provides licenses and registration for a variety of users. For organizations that have been granted a photocopy license by the CCC, a separate system of payment has been arranged.

Trademark Notice: Product or corporate names may be trademarks or registered trademarks, and are used only for identification and explanation without intent to infringe.

Library of Congress Cataloging-in-Publication Data

Albright's chemical engineering handbook / editor Lyle Albright.
   p. cm.
  Includes bibliographical references and index.
  ISBN 978-0-8247-5362-7 (alk. paper)
  1. Chemical engineering--Handbooks, manuals, etc. I. Albright, Lyle Frederick, 1921- II. Title.

TP151.A565 2008
660--dc22   2007020174

Visit the Taylor & Francis Web site at
http://www.taylorandfrancis.com

and the CRC Press Web site at
http://www.crcpress.com

Table of Contents

Preface ix
The Editor xi
Contributors xiii

Chapter 1 Physical and Chemical Properties 1
Allan H. Harvey

Chapter 2 Mathematics in Chemical Engineering 35
Sinh Trinh, Nandkishor Nere, and Doraiswami Ramkrishna

Chapter 3 Engineering Statistics 199
Daniel W. Siderius

Chapter 4 Thermodynamics of Fluid Phase and Chemical Equilibria 255
Kwang-Chu Chao, David S. Corti, and Richard G. Mallinson

Chapter 5 Fluid Flow 393
Ron Darby

Chapter 6 Heat Transfer 479
Kenneth J. Bell

Chapter 7 Radiation Heat Transfer 567
Z. M. Zhang and David P. DeWitt

Chapter 8 Mass Transfer 591
James R. Fair

Chapter 9 Industrial Mixing Technology 615
Douglas E. Leng, Sanjeev S. Katti, and Victor Atiemo-Obeng

Chapter 10 Liquid-Liquid Extraction 709
D. William Tedder

Chapter 11 Chemical Reaction Engineering 737
J. B. Joshi and L. K. Doraiswamy

Chapter 12 Distillation 969
James R. Fair

Chapter 13 Absorption and Stripping 1073
James R. Fair

Chapter 14 Adsorption 1119
Kent S. Knaebel

Chapter 15 Process Control..1173
James B. Riggs, William J. Korchinski, and Arkan Kayihan

Chapter 16 Conceptual Process Design, Process Improvement, and Troubleshooting 1267
Donald R. Woods, Andrew N. Hrymak, and James R. Couger

Chapter 17 Chemical Process Safety 1437
Richard W. Prugh

Chapter 18 Environmental Engineering: A Review of Issues, Regulations, and Resources ...1485
Bradly P. Carpenter, Douglas E. Watson, and Brooks C. Carpenter

Chapter 19 Biochemical Engineering 1501
James M. Lee

Chapter 20 Measuring Physical Properties 1531
Lyle F. Albright

Chapter 21 Selecting Materials of Construction (Steels and Other Metals)............................1539
David A. Hansen

Chapter 22 Solid/Liquid Separation..1597
Frank M. Tiller, Wenping Li, and Wu Chen

Chapter 23 Drying: Principles and Practice 1667
Arun S. Mujumdar

Table of Contents

Chapter 24 Dry Screening of Granular and Powder Materials 1717
A. J. DeCenso and Nash McCauley

Chapter 25 Conveying of Bulk Solids 1729
Fred Thomson

Chapter 26 Principles and Applications of Electrochemical Engineering 1737
Peter N. Pintauro

Chapter 27 Patents and Intellectual Property..1831
M. Henry Heines

Chapter 28 Communication..1841
F. S. Oreovicz

Chapter 29 Ethical Concerns of Engineers...1859
Lyle F. Albright

Appendix: Conversion Factors 1867

Index 1881

Table of Contents

Chapter 24: Dry Screening of Granular and Powder Materials 1717
J. Freeman, J. Nash McCauley

Chapter 25: Conveying of Bulk Solids 1779
Reza Baradaran

Chapter 26: Principles and Applications of Biomechanical Engineering 1777
Jorge M. Fuentes

Chapter 27: Process and Reactive Dyestuffs 1831
(illegible)

Chapter 28: Granular Flow 1841
K.S. Chowdary

Chapter 29: Transportation and Logistics 1855
Eric L. Allston

Appendix: Conversion Factors 1867

Index 1881

Preface

This handbook was written to provide a thorough discussion of the most important topics of interest to engineers and scientists in chemically oriented fields. The expected readers will vary from students in the university to those employed in industry, academia, and the government. Because the engineering disciplines are broad and complex, and growing more so, a wide variety of subjects needed to be covered in the 29 chapters of this handbook. The first 27 chapters are technical in nature; the last two chapters are not. Because technical personnel need to communicate their ideas with others, one chapter focuses on communication approaches. Ethics has also become a key issue, especially in the last several years, and this is covered in the final chapter. As industry becomes increasingly internationalized, ethical concerns will likely continue to grow, because standards often vary in different countries.

Let me share some of the thoughts that I had as I planned and organized this handbook. First, a chapter in this handbook should differ from one to be expected in a textbook. In a handbook, each chapter should be succinct, providing basic information (including case examples) and indicating where additional information can be found. The topics selected for the various chapters in this handbook were chosen with the advice and counsel of individuals whose opinions I respect. Some overlap of material on specific examples is sometimes found in two or more chapters. For example, the determination and prediction of chemical and physical properties is discussed in both Chapter 1 (Physical and Chemical Properties) and Chapter 4 (Thermodynamics). As editor, I permitted and even encouraged some overlap when the authors were reaching their conclusions from different perspectives. But I tried to be certain that the different authors were each aware of this overlap so that they could handle it to the best advantage of everyone involved.

Second, there have been major advances in technical information in the last few years. It was therefore imperative that these advances be reported and discussed as needed. These include fundamentals, new approaches, and improved applications. Much better mathematical and statistical models are now available. Computers have become of ever-increasing importance, leading to much improved research, plant design, plant operations, and so forth. Several groups currently market important computer models, and these are reported here. In some cases, free information can be found on the Internet. For example, the National Institute of Standards and Technology (NIST) has made available a large statistics handbook at no charge. Chapter 3 of the current handbook emphasizes the applications of statistics to chemically oriented problems.

Third, the selection of an author for a specific chapter was often made after receiving the advice of others. For all chapters, I outlined my thoughts on the expected emphasis that I hoped to be presented throughout the handbook. In all cases, the authors were given the chance to modify my suggestions. As a result, even better manuscripts were received. Several authors later added one or more coauthors, whom I welcomed. In my opinion, this handbook is blessed with expert authors.

I hope that this handbook will promote better engineering and plant operations.

The Editor

Professor Lyle F. Albright, emeritus professor of chemical engineering at Purdue University, is proud that he was able to assemble 43 distinguished authors for the 29 chapters of this handbook. These authors have been associated with 13 universities in the United States, three universities in other countries, and industrial companies, government laboratories, and consultants. Professor Albright says his education increased greatly from reading the manuscripts of this book. This handbook emphasizes established fundamentals plus newer developments. Although no single handbook can provide all the necessary details, this one provides many important approaches for both students and the professional engineer.

In preparing this handbook, Professor Albright called on his 65 plus years in industry, academia, and consulting. He first served as a shift supervisor from 1939 to 1941 at Dow Chemical in a large semi-works plant that produced a highly purified butadiene. Several years later, he learned that this butadiene was employed to help develop a process to produce synthetic rubber, which helped to keep the Allied armies mobile during World War II. He was also employed by E.I. DuPont de Nemours, Inc., at the Hanford Engineering Works from 1944 to 1946 as part of the Manhattan Project.

After obtaining his Ph.D. in chemical engineering at the University of Michigan, he joined Colgate-Palmolive Co. in Jersey City. His academic career includes the University of Oklahoma (1951–1955) and Purdue University (1955–present). Sabbaticals were at the University of Texas (summer, 1952) and Texas A&M University (all of 1985). For the last 50 years, he has been an active consultant in the following areas: production of high-quality alkylates in refineries, ethylene and propylene production, nitration, partial hydrogenation of vegetable oils, and pulping of wood. His consulting work emphasizes the need to understand the fundamentals of the process in order to improve plant operations. That theme is carried over to this handbook.

Contributors

Lyle F. Albright
School of Chemical Engineering
Purdue University
West Lafayette, Indiana

Victor Atiemo-Obeng
The Dow Chemical Company
Midland, Michigan

Kenneth J. Bell
Chemical Engineering School
Oklahoma State University
Stillwater, Oklahoma

Bradly Carpenter
Greenfield Environmental, Inc.
Maple Grove, Minnesota

Brooks C. Carpenter
Greenfield Environmental, Inc.
Maple Grove, Minnesota

Kwang-Chu Chao
Purdue University (retired)
Fremont, California

Wu Chen
The Dow Chemical Company
Freeport, Texas

David S. Corti
School of Chemical Engineering
Purdue University
West Lafayette, Indiana

James R. Couper
Fayetteville, Arkansas

Ronald Darby
Department of Chemical Engineering
Texas A&M University
College Station, Texas

A. J. DeCenso
Formerly of Rotex, Inc.
Cincinnati, Ohio

David P. DeWitt (deceased)
Edgewater, Maryland

L. K. Doraiswamy
Chemical Engineering Department
Iowa State University
Ames, Iowa

James Fair
Chemical Engineering Department
University of Texas
Austin, Texas

David A. Hansen
Retired from Fluor Daniel
Montgomery, Texas

Allan H. Harvey
Physical and Chemical Properties Division
National Institute of Standards and Technology
Boulder, Colorado

M. Henry Heines
Townsend and Townsend and Crew
San Francisco, California

Andrew N. Hrymak
Chemical Engineering Department
McMaster University
Hamilton, Ontario, Canada

J. B. Joshi
Department of Chemical Technology
University of Mumbai
Mumbai, India

Sanjeev S. Katti
The Dow Chemical Company
Midland, Michigan

Arkan Kayihan
Expedia, Inc.
Seattle, Washington

Kent S. Knaebel
Adsorption Research, Inc.
Dublin, Ohio

William J. Korchinski
Advanced Industrial Modeling, Inc.
Santa Barbara, California

James M. Lee
Chemical Engineering Department
and Division of Bioengineering Environmental Systems
Washington State University
Pullman, Washington

Douglas E. Leng
Leng Associates
Midland, Michigan

Wenping Li
Chemical Engineering Department
University of Houston
Houston, Texas

Richard G. Mallinson
School of Chemical Engineering and Materials Science
University of Oklahoma
Norman, Oklahoma

Nash McCauley
Retired from Rotex, Inc.
Cincinnati, Ohio

Arun S. Mujumdar
Mechanical Engineering Department
National University of Singapore
Singapore, Indonesia

Nandkishor Nere
School of Chemical Engineering
Purdue University
West Layfayette, Indiana

Frank Oreovicz
Purdue University (retired)
West Lafayette, Indiana

Peter N. Pintauro
Department of Chemical and Biomolecular Engineering
Vanderbilt University
Nashville, Tennessee

Richard W. Prugh
Chilworth Technology, Inc.
Monmouth Junction, New Jersey

Doraiswami Ramkrishna
School of Chemical Engineering
Purdue University
West Lafayette, Indiana

James B. Riggs
Chemical Engineering Department
Texas Tech
Lubbock, Texas

Daniel W. Siderius
Department of Chemistry
Washington University
St. Louis, Missouri

D. William Tedder
School of Chemical Engineering
Georgia Institute of Technology
Atlanta, Georgia

Fred Thomson (deceased)
Landenberg, Pennsylvania

Frank M. Tiller (deceased)
Chemical Engineering Department
University of Houston
Houston, Texas

Sinh Trinh
Rentech, Inc.
Denver, Colorado

Douglas E. Watson
Greenfield Environmental, Inc.
Downers Grove, Illinois

Contributors

Donald R. Woods
Chemical Engineering Department
McMasters University
Hamilton, Ontario, Canada

Z. M. Zhang
Woodruff School of Mechanical Engineering
Georgia Institute of Technology
Atlanta, Georgia

Z. M. Zhang
Woodruff School of Mechanical Engineering
Georgia Institute of Technology
Atlanta, Georgia

1 Physical and Chemical Properties

Allan H. Harvey

CONTENTS

1.1 Introduction 2
1.2 Thermodynamic Properties of Pure Fluids 3
    1.2.1 Importance of Pure-Fluid Properties 3
    1.2.2 Relative Importance of Different Properties 3
    1.2.3 Water and Steam 3
    1.2.4 Pure Fluids with Reference-Quality Data 5
    1.2.5 Pure Fluids with Moderate Amounts of Data 5
    1.2.6 Pure Fluids with Little or No Data 7
    1.2.7 Ideal-Gas Properties 8
    1.2.8 Critical Constants and Acentric Factors for Pure Fluids 8
1.3. Thermodynamic Properties of Single-Phase Mixtures 8
    1.3.1 Density 8
    1.3.2 Caloric Properties 10
1.4. Phase Equilibria for Mixtures 10
    1.4.1 Types of Phase-Equilibrium Calculations 10
    1.4.2 Equation-of-State Methods 11
    1.4.3 Activity-Coefficient Methods 12
    1.4.4 Choosing a Method 13
    1.4.5 Sources of Data 14
1.5. Transport Properties 14
    1.5.1 Kinetic Theory for Transport Properties 14
    1.5.2 Viscosity 15
    1.5.3 Thermal Conductivity 16
    1.5.4 Diffusivity 17
1.6. Aqueous Electrolyte Solutions 17
    1.6.1 Vapor-Liquid Equilibria and Activity Coefficients 17
    1.6.2 Density and Enthalpy 18
    1.6.3 Transport Properties 19
1.7. Properties for Chemical Reaction Equilibria 20
1.8. Measurement of Fluid Thermophysical Properties 20
    1.8.1 When Experiments Are Necessary 20
    1.8.2 General Considerations 21
    1.8.3 Density 22
    1.8.4 Heat Capacity and Caloric Properties 22
    1.8.5 Pure-Component Vapor Pressure 23
    1.8.6 Mixture Vapor-Liquid Equilibria 24

	1.8.7 Liquid-Liquid Equilibria	25
	1.8.8 Viscosity	25
	1.8.9 Thermal Conductivity	26
	1.8.10 Electrolyte Solutions	27
1.9.	Overview of Major Data Sources	27
	1.9.1 Introductory Comments	27
	1.9.2 NIST (Including TRC)	28
	1.9.3 DIPPR	28
	1.9.4 DECHEMA	29
	1.9.5 DDB	29
	1.9.6 NEL	29
	1.9.7 Landolt-Börnstein	29
	1.9.8 Beilstein	29
	1.9.9 Gmelin	30
	1.9.10 Process Simulation Software	30
Acknowledgments		30
References		30

1.1 INTRODUCTION

No single handbook could tabulate more than a small fraction of the physical and chemical property data needed by engineers. Therefore, this chapter does not contain extensive tables of data, but instead points readers to reliable sources of data and to methods for extrapolation, estimation, or measurement of data.

We cannot emphasize enough the importance of *quality* of data. Much data—whether in handbooks, on the Internet, or in scientific journals—are inaccurate or simply wrong. This can be due to problems with experiments, errors in processing measurements, misuse of extrapolation or estimation techniques, or something as simple as a copying error. For the responsible engineer, the goal is not just to "get a number," but to get a *reliable* number. Obtaining reliable physical and chemical property data requires *evaluation* of data. This involves expert evaluation of experimental techniques (including sample purity), consistency tests, comparisons among multiple data sets and multiple measurements for the same substance, and other factors such as trends within chemical families. It is preferable to use sources where the data are evaluated and where some indication of their quality is given.

A related issue is *uncertainty*. A datum has little value if one does not know whether it is uncertain by 1% or 100%. Ideally, all data would have a quantitative uncertainty given, which could be propagated into engineering design calculations. In practice, we often have to settle for approximate or qualitative estimates of uncertainties, but the more that can be said about uncertainty, the better.

Data sources listed here range from those that are free, to data available at low cost (for example in a single book or inexpensive database), to databases that may cost thousands of dollars. Of course, engineers want to save money, but often "you get what you pay for." While one can sometimes take advantage of free products from government agencies or academic groups, reliable data often cost money, because data collection and evaluation require skilled labor. The engineer who uses free data (perhaps from a Web search) of unknown quality as the basis for a multimillion-dollar design is being foolish if more trustworthy data could be obtained for a reasonable price.

As process simulation programs become routine tools, many engineers treat their thermodynamic calculations as a "black box" without giving thought to the underlying data or models. To their credit, developers of process simulators have spent much effort to validate both data and models. However, it is unwise to put blind trust in numbers merely because they are produced by a computer; the best software can still produce nonsense if inappropriate thermodynamic methods and data are used. This chapter should provide resources to help the reader make informed judgments about which models and data to choose in process simulation, and to supplement simulation software in those cases where necessary data are missing.

Before proceeding, we mention sources for a few areas not covered in this chapter. Basic chemical thermodynamics is the subject of Chapter 4. For polymers and their solutions, the *Polymer Handbook* [1] is an indispensable source, and more on polymer thermophysical properties may be found in two books from AIChE's DIPPR project [2, 3]. The estimation of properties of mixtures described by distillation curves (typically petroleum fractions), or of the pseudocomponents derived from such curves, is covered in the *API Technical Data Book* [4]. Many molecular data, such as dipole moments and spectroscopic constants, are tabulated in the *NIST Chemistry Webbook* [5].

1.2 THERMODYNAMIC PROPERTIES OF PURE FLUIDS

1.2.1 Importance of Pure-Fluid Properties

While chemical engineers most often deal with mixtures, knowledge of pure-fluid properties is indispensable for at least three reasons. First, many processes make use of fluids that are close enough to pure that pure-fluid properties suffice. Many industrial uses of water and steam fall into this category. Second, pure-fluid behavior can often be used as a surrogate for a mixture of similar compounds; for example, sometimes it is desirable to model hydrocarbon mixtures as a small number (perhaps even one) of known components. Third, and perhaps most important, most models for thermophysical properties of mixtures employ properties of the pure components as a starting point; the underlying pure-component data must be accurate in order to describe the mixture.

1.2.2 Relative Importance of Different Properties

While the relative importance of properties depends on the context, it is often the case in chemical engineering that the vapor pressure is the most important pure-fluid property. This is because of the prevalence of separation operations that depend on vapor-liquid equilibria, and also because of the importance of volatility for safety and environmental concerns.

Because energy usage and heat transfer are important in process operations, caloric properties (enthalpy, heat capacity) can be considered the second most important area. The enthalpy of vaporization is particularly important in vapor-liquid separations.

Volumetric properties (density and, to a lesser extent, derivative properties such as isothermal compressibility) enter into many process calculations. However, it is usually relatively easy to measure and/or predict fluid densities with sufficient accuracy for most purposes.

Other properties of interest include surface tension and transport properties (viscosity, thermal conductivity, diffusivity). Transport properties will be covered in a later section.

1.2.3 Water and Steam

Water, in pure or nearly pure form, is widely used both as a process stream and (as cooling water or steam) as a heat-transfer fluid. Because of its importance, international standards exist for its thermophysical properties. These standards are set by the International Association for the Properties of Water and Steam (IAPWS; see www.iapws.org).

In the past, engineers used "steam tables" to obtain properties of water and steam. While such books still exist [6, 7], it is now usually more convenient to use software that implements IAPWS property standards [8]. Table 1.1 reports the thermodynamic properties of water for saturated liquid and vapor. Such short tables are useful for quick reference; design calculations usually require more extensive tables or software.

TABLE 1.1
Thermodynamic Properties of Saturated Water and Steam as a Function of Temperature

T, °C	p, MPa	Density, kg/m³		Enthalpy, kJ/kg		Entropy, kJ/(kg·K)	
		ρ_L	ρ_V	h_L	h_V	s_L	s_V
0.01	0.000 612	999.79	0.004 855	0.00	2500.9	0.0000	9.1555
5	0.000 873	999.92	0.006 802	21.02	2510.1	0.0763	9.0248
10	0.001 228	999.65	0.009 407	42.02	2519.2	0.1511	8.8998
15	0.001 706	999.06	0.012 841	62.98	2528.3	0.2245	8.7803
20	0.002 339	998.16	0.017 314	83.91	2537.4	0.2965	8.6660
25	0.003 170	997.00	0.023 075	104.83	2546.5	0.3672	8.5566
30	0.004 247	995.61	0.030 415	125.73	2555.5	0.4368	8.4520
35	0.005 629	993.99	0.039 674	146.63	2564.5	0.5051	8.3517
40	0.007 385	992.18	0.051 242	167.53	2573.5	0.5724	8.2555
45	0.009 595	990.17	0.065 565	188.43	2582.4	0.6386	8.1633
50	0.012 352	988.00	0.083 147	209.34	2591.3	0.7038	8.0748
55	0.015 762	985.66	0.104 56	230.26	2600.1	0.7680	7.9898
60	0.019 946	983.16	0.130 43	251.18	2608.8	0.8313	7.9081
65	0.025 042	980.52	0.161 46	272.12	2617.5	0.8937	7.8296
70	0.031 201	977.73	0.198 43	293.07	2626.1	0.9551	7.7540
75	0.038 595	974.81	0.242 19	314.03	2634.6	1.0158	7.6812
80	0.047 414	971.77	0.293 67	335.01	2643.0	1.0756	7.6111
85	0.057 867	968.59	0.353 88	356.01	2651.3	1.1346	7.5434
90	0.070 182	965.30	0.423 90	377.04	2659.5	1.1929	7.4781
95	0.084 608	961.88	0.504 91	398.09	2667.6	1.2504	7.4151
100	0.101 42	958.35	0.598 17	419.17	2675.6	1.3072	7.3541
110	0.143 38	950.95	0.826 93	461.42	2691.1	1.4188	7.2381
120	0.198 67	943.11	1.1221	503.81	2705.9	1.5279	7.1291
130	0.270 28	934.83	1.4970	546.38	2720.1	1.6346	7.0264
140	0.361 54	926.13	1.9667	589.16	2733.4	1.7392	6.9293
150	0.476 16	917.01	2.5481	632.18	2745.9	1.8418	6.8371
160	0.618 23	907.45	3.2596	675.47	2757.4	1.9426	6.7491
170	0.792 19	897.45	4.1222	719.08	2767.9	2.0417	6.6650
180	1.0028	887.00	5.1588	763.05	2777.2	2.1392	6.5840
190	1.2552	876.08	6.3954	807.43	2785.3	2.2355	6.5059
200	1.5549	864.66	7.8610	852.27	2792.0	2.3305	6.4302
210	1.9077	852.72	9.5885	897.63	2797.3	2.4245	6.3563
220	2.3196	840.22	11.615	943.58	2800.9	2.5177	6.2840
230	2.7971	827.12	13.985	990.19	2802.9	2.6101	6.2128
240	3.3469	813.37	16.749	1037.6	2803.0	2.7020	6.1423
250	3.9762	798.89	19.967	1085.8	2800.9	2.7935	6.0721
260	4.6923	783.63	23.712	1135.0	2796.6	2.8849	6.0016
270	5.5030	767.46	28.073	1185.3	2789.7	2.9765	5.9304
280	6.4166	750.28	33.165	1236.9	2779.9	3.0685	5.8579
290	7.4418	731.91	39.132	1290.0	2766.7	3.1612	5.7834
300	8.5879	712.14	46.168	1345.0	2749.6	3.2552	5.7059
310	9.8651	690.67	54.541	1402.2	2727.9	3.3510	5.6244
320	11.284	667.09	64.638	1462.2	2700.6	3.4494	5.5372
330	12.858	640.77	77.050	1525.9	2666.0	3.5518	5.4422
340	14.601	610.67	92.759	1594.5	2621.8	3.6601	5.3356
350	16.529	574.71	113.61	1670.9	2563.6	3.7784	5.2110

TABLE 1.1 *(Continued)*
Thermodynamic Properties of Saturated Water and Steam as a Function of Temperature

T, °C	p, MPa	Density, kg/m³		Enthalpy, kJ/kg		Entropy, kJ/(kg·K)	
		ρ_L	ρ_V	h_L	h_V	s_L	s_V
360	18.666	527.59	143.90	1761.7	2481.5	3.9167	5.0536
370	21.044	451.43	201.84	1890.7	2334.5	4.1112	4.8012
T_c	22.064	322.00	322.00	2084.3	2084.3	4.4070	4.4070

Note: Critical temperature T_c = 373.946°C.

Source: Data generated from A. H. Harvey, A. P. Peskin, and S. A. Klein, *NIST/ASME Steam Properties*, NIST Standard Reference Database 10, Version 2.2, National Institute of Standards and Technology, Gaithersburg, MD, 2000. Available on-line at http://www.nist.gov/srd/nist10.htm.

1.2.4 Pure Fluids with Reference-Quality Data

Some chemicals have received particular attention because of their industrial importance. These include the major components of air, common light hydrocarbons, and common refrigerants, for which accurate measurements of a variety of properties have been made over wide ranges of temperature and pressure. This has allowed the development of comprehensive, reference-quality equations of state (EOS) that can be used to calculate thermodynamic properties of the pure fluids essentially within the uncertainty of the underlying data. When a fluid has such an EOS available, that is the preferred data source.

Reference-quality EOS typically have a complicated functional form, with many parameters. However, software is available [9, 10] that implements these EOS for many fluids; for example, the database from NIST [9] incorporates formulations for approximately 80 pure fluids. A subset of this information is available in the *NIST Chemistry Webbook* [5].

1.2.5 Pure Fluids with Moderate Amounts of Data

Many other substances have insufficient data for a reference-quality EOS. There may be a few vapor-pressure data, and perhaps some other measurements such as liquid density or heat capacity. In such cases, there are two basic approaches.

The first approach is simple correlations of properties. This is most convenient when the property of interest is a function of only one variable, such as the vapor pressure (a function of temperature only). Such correlations can also be useful for properties of liquids (as long as they are not close to the critical point), since most liquid-phase properties are relatively insensitive to pressure and can be approximately represented as a function of temperature only. Several databases [11–14] contain correlations for pure-fluid properties as a function of temperature for many common substances, and vapor-pressure correlations for many substances are in some additional sources [5, 15]. It is important to be aware of the range of conditions in which a correlation has been fitted, as extrapolation can lead to significant errors.

If sufficient data exist but have not been correlated, it is of course possible to fit the data yourself. In such cases, one should choose an equation with a physical basis and/or a reliable record of correlating the property in question for other fluids. The functional forms used in the databases mentioned in the previous paragraph can provide guidance in this regard. In addition to the databases mentioned in the previous paragraph, sources with extensive pure-fluid data exist for vapor pressures

[16, 17], liquid densities [18], liquid heat capacities [19, 20], liquid heat capacities at 298.15 K [21], and enthalpies of vaporization [22].

For the vapor pressure, the Antoine equation is widely used:

$$\ln\left(\frac{p^{sat}}{p_0}\right) = A - \frac{B}{T+C} \quad (1.1)$$

where p^{sat} is the vapor pressure; p_0 is a reference pressure (typically 1 in the units of pressure being used); T is the absolute temperature in kelvins, and A, B, and C are parameters fitted to the data. Sometimes Equation (1.1) is written with a base-10 logarithm, and sometimes with the denominator written as $(T + C - 273.15)$, so when using reported Antoine parameters, one must be careful to use the correct equation format. Because the Antoine equation has a physical basis (it is derived from the Clausius-Clapeyron equation), it can be extrapolated for small distances in temperature, as long as one stays well below the critical temperature. Extended versions of the Antoine equation, with more parameters, are sometimes used to cover a wider temperature range.

The Wagner equation for vapor pressure is capable of covering a wide range of temperatures:

$$\ln\left(\frac{p^{sat}}{p_c}\right) = \frac{1}{T_r}(a\tau + b\tau^{1.5} + c\tau^3 + d\tau^6) \quad (1.2a)$$

where $T_r = T/T_c$; T_c and p_c are the critical temperature and pressure of the fluid, respectively; $\tau = 1 - T_r$; and a, b, c, and d are adjustable parameters. Because the Wagner equation is constrained to give the correct critical pressure, it is better suited for extrapolation to high temperatures. Unlike the Antoine equation, it requires reliable values of T_c and p_c. Sometimes a slightly different Wagner form gives better results:

$$\ln\left(\frac{p^{sat}}{p_c}\right) = \frac{1}{T_r}(a\tau + b\tau^{1.5} + c\tau^{2.5} + d\tau^5) \quad (1.2b)$$

Use of the Wagner equations requires relatively extensive and internally consistent data; they may produce unphysical slopes of the vapor-pressure curve if fitted to inconsistent data. Neither the Wagner nor the Antoine equation should be extrapolated far below the temperature range in which it was fitted. Vapor-pressure data at low temperatures (where the vapor pressures are small) are scarce. Often, better estimates of the vapor pressure at low temperatures may be obtained from extrapolation techniques that make use of heat-capacity data, as discussed in *The Properties of Gases and Liquids* [15].

The second approach for calculating thermodynamic properties when some data are available is to use an equation of state (see Chapter 4, "Thermodynamics of Fluid Phase and Chemical Equilibria"). The most popular EOS are cubic equations, especially the Soave-Redlich-Kwong (SRK) and Peng-Robinson (PR) equations. The parameters in these equations were originally computed from the critical parameters and the acentric factor. However, it is also possible to fit the EOS parameters directly to experimental data; it has become common practice to fit parameters in advanced versions of these EOS to vapor-pressure data. Such fitting is generally necessary to get a good representation of the vapor pressure for polar fluids. Twu et al. [23] provide an overview of modern cubic EOS technology.

Because software for cubic EOS calculations is widely available, it is tempting to use these EOS for everything without considering that choice. This is unwise, because these methods have limitations. The common EOS forms like SRK and PR were optimized primarily for hydrocarbons;

without refinement, they are less reliable for polar fluids. Cubic EOS are inaccurate near the critical point. Also, their prediction of liquid densities and heat capacities tends to be inaccurate unless additional modifications are made.

The two approaches presented in this section need not be mutually exclusive; one can use a correlation for some properties and an equation of state for others. Because of the poor liquid density predictions of many cubic equations of state, it is common to use semiempirical correlations for liquid density (such as the Rackett equation [15]) while using an EOS for vapor-liquid equilibria.

1.2.6 PURE FLUIDS WITH LITTLE OR NO DATA

Often, engineers must estimate thermophysical properties for compounds where few, if any, measurements exist. A comprehensive source for such estimation techniques is the book *The Properties of Gases and Liquids* [15], which should be consulted by anybody who is serious about property estimation. We will restrict ourselves here to more general comments.

Many estimation techniques are *corresponding-states* methods, where the properties of the target fluid are scaled to the properties of a well-known fluid. Scaling factors are often based on the critical temperature and pressure and the acentric factor (which in many cases must be estimated). It is essential to recognize that most correlations were developed for certain classes of fluids, making it dangerous to use them for fluids that are very different from those used to develop the correlation. For example, a correlation developed for nonpolar hydrocarbons should not be applied to a polar fluid such as ammonia or methanol.

Other estimation techniques are categorized as *group-contribution* methods, where a property (or a parameter in a property model) is estimated based on the chemical structure of a compound, with each piece of the molecule contributing to the total. These methods depend on the existence of data on enough compounds to be able to regress contributions due to different groups. Group-contribution methods should also not be used for systems significantly different from those used to develop the method.

Sometimes the results from estimation techniques can be calibrated if only one or two data points are known. For example, if only a single vapor-pressure datum is known and the estimation method produces a vapor pressure that is low by 30% at that point, one can adjust the predicted vapor pressure at other temperatures by 30% to produce an improved estimate. Such estimates become more uncertain the farther one gets from the experimental datum.

One may also extrapolate from limited data if more extensive data exist for a similar compound. Most properties for similar compounds follow similar curves when plotted on appropriate coordinates (for example, $\ln(p^{sat})$ versus $1/T$ for vapor pressure). When all else fails, one can extrapolate a single data point on such a plot by drawing a curve through the point that is parallel to that for a similar compound where data exist. One must be able to judge what is a similar compound; for organic compounds, the addition of methyl groups or changes in branching typically do not qualitatively alter the fluid thermodynamic properties, whereas the addition or subtraction of functional groups (-OH, -Cl, -NO, etc.) is more significant. Data for similar compounds may also be used to validate purely predictive methods; a method that agrees with reliable data for 2-pentanol can be used with more confidence for 2-hexanol.

Predictions based on as little as one data point are much more reliable than those using no data at all. Sometimes, a single data point can be found in a source other than the usual property databases. For example, the normal boiling point (NBP) provides a point on the vapor-pressure curve. The NBP (along with the room-temperature liquid density, usually reported as specific gravity) is often reported in chemical supply catalogs, although the uncertainty is usually unknown. Such single points may also sometimes be found in chemistry handbooks such as *Beilstein* [24] or *Gmelin* [25] (see Sections 9.8 and 9.9), in the *CRC Handbook of Chemistry and Physics* [26], or even in the literature article where the compound is first reported.

An additional alternative is to measure the needed data. The difficulty and cost will depend on the fluid and on the property and conditions of interest. Experiments may well be needed if an important project is at stake, especially if you do not have high confidence in the available estimation methods. Measurement of fluid properties is discussed in Section 1.8 of this chapter.

1.2.7 IDEAL-GAS PROPERTIES

For gases at relatively low pressures, the ideal-gas law is often sufficient to calculate the density. The ideal-gas heat capacity (and its integrated form, the ideal-gas enthalpy) are used not only for calculating the properties of gases assumed to be ideal, but also as a starting point in many calculations of the heat capacity and enthalpy of nonideal fluids.

Ideal-gas heat capacities may be estimated from gas-phase heat-capacity measurements, but more often they are calculated from statistical mechanics based on molecular information, usually obtained by spectroscopy [27–29]. The results of such calculations have been tabulated for many molecules [5, 11–15, 28–30]. Predictive methods also exist based on molecular structure; the leading methods are reviewed in *The Properties of Gases & Liquids* [15]. Calculations from one of these methods (that of Benson) are available in the *NIST Chemistry Webbook* [5].

1.2.8 CRITICAL CONSTANTS AND ACENTRIC FACTORS FOR PURE FLUIDS

The values of temperature, pressure, and density at the critical point, and the acentric factor (which characterizes the shape of the vapor-pressure curve) are seldom of direct interest. However, they are used in many correlations and equations of state.

The critical temperature and pressure have been measured for a few hundred fluids; these are mostly relatively small molecules because most larger molecules with high critical temperatures begin to decompose before the critical temperature is reached. The critical density is less often measured but is seldom used in correlations. Extensive tabulations of critical parameters exist [5, 11–15, 31]; in some cases these sources supplement measured values with those obtained from estimation techniques. Some sources [11, 12, 14, 15] also contain values for the acentric factor. It is also possible to calculate the acentric factor from its definition (see Chapter 4, "Thermodynamics") from the vapor pressure and values of the critical temperature and pressure.

Techniques exist to estimate the critical parameters and acentric factor from molecular structure; these are reviewed in *The Properties of Gases and Liquids* [15]. Since these techniques depend on the regression of experimental data to assign contributions to individual groups within the molecule, they are reliable only when the functional groups in the molecule for which the prediction is made are present in similar molecules in the dataset used to develop the correlation.

1.3 THERMODYNAMIC PROPERTIES OF SINGLE-PHASE MIXTURES

1.3.1 DENSITY

Three major approaches are used to predict the density of a multicomponent vapor or liquid. Many methods may be enhanced if some mixture data are available.

One approach is to use a mixture equation of state (EOS). This could be one of the cubic EOS mentioned in Section 1.2.5, but their poor performance for pure-fluid densities also carries over to mixtures. More sophisticated mixture EOS are available that make use of the reference-quality equations of state described in Section 1.2.4. If such an EOS exists for each component in a mixture, such an approach can produce good densities. A computer database is available [9] that implements this approach for common refrigerants and light hydrocarbons.

Mixture EOS models generally contain adjustable binary parameters that can improve the predictions if some experimental mixture data are available. Often, these models have been optimized to reproduce vapor-liquid equilibria rather than densities, although such parameters still usually improve the density prediction compared to omitting them completely. If the components in a mixture are chemically similar, often the binary parameters can be set to zero without significant loss of accuracy. For pairs where experimental data are not available, the binary interaction parameter for a similar pair may provide a reasonable estimate.

For gases at low and moderate pressures, it is often preferable to use the virial expansion that provides successive corrections to the ideal-gas law (see Chapter 4, "Thermodynamics"). The first correction (called the second virial coefficient, B) has been derived from volumetric data for many pure fluids. Higher virial coefficients are much less well known, as are the cross-coefficients for interactions between unlike molecules. Estimation techniques for second (and to a lesser extent third) virial coefficients exist [15] and work reasonably well for many fluids, especially organic compounds of low polarity. Dymond et al. have compiled extensive experimental data for virial coefficients [32].

A second approach is the mixture corresponding-states approach. These are similar to the corresponding-states correlations for pure fluids mentioned in Section 1.2.6, but with the addition of "mixing rules" to obtain effective mixture parameters (for example, critical properties) to insert into the correlations. These methods are thoroughly discussed in *The Properties of Gases and Liquids* [15]. These models can also contain binary interaction parameters, and the comments given above for such parameters in mixture EOS models apply here as well.

The third approach, used primarily for liquid mixtures, requires reliable pure-component density values (see Section 1.2). One starts with an "ideal" mixture volume, defined by

$$v^{id} = \sum_i x_i v_i \tag{1.3}$$

where v is the volume per mole, the sum is taken over all components in the mixture, x_i is the mole fraction of component i, and all volumes are evaluated at the same temperature and pressure. For chemically similar compounds, the ideal mixture volume as defined by Equation (1.3) usually produces reliable mixture volumes (and therefore densities, since the molar density is the reciprocal of the molar volume) if the pure-component values are known accurately.

Data are sometimes available for the excess volume of mixtures, particularly binary mixtures. These may be used to improve upon Equation (1.3). The excess volume is defined by

$$v^E = v - v^{id} = v - \sum_i x_i v_i \tag{1.4}$$

The simplest representation of excess volume is a quadratic composition dependence for v^E, where the single parameter for each binary pair may be evaluated from one mixture density. Under the assumption that the interaction between a pair of components is not affected by the presence of a third component, the mixture excess volume is

$$v^E = \sum_i \sum_j x_i x_j A_{ij} \tag{1.5}$$

If good volumetric data are available for all the binary pairs in a mixture (except perhaps for trace components), Equation (1.5) can provide good liquid densities. For pairs of components that are chemically dissimilar, higher-order terms beyond A_{ij} may be necessary.

The most significant limitation of Equation (1.5) is that all the pure-component volumes must be at the same temperature and pressure. If the pure components are not all in the same phase at this temperature and pressure, the definition of an "ideal" mixture volume in Equation (1.3) loses its usefulness. In practice, Equation (1.5) is useful only for liquid mixtures where all the components are liquids in their pure state at the condition of interest. Equation (1.5) also tends to be less useful in mixtures containing dissimilar components and/or many different functional groups. More parameters are usually needed in such systems, and the implicit assumption that the pair interactions are unaffected by other components in the mixture is less likely to be true.

To use Equation (1.5), reliable data are needed for v^E in binary mixtures. These data are also useful in equation-of-state or corresponding-states approaches if it is desired to fit a binary parameter to improve the performance. References to many binary excess-volume data (but not the data themselves) may be found in the series of books by Wisniak and Tamir [33], and data may be found in the Dortmund Data Bank [13] and a volume of the Landolt-Börnstein series [34].

1.3.2 Caloric Properties

The three approaches mentioned in the previous section may also be used to describe caloric properties (enthalpy, entropy, heat capacity) of mixtures. The same considerations mentioned earlier are also true for the application of mixture equations of state and corresponding-states methods for the prediction of caloric properties.

For the third approach mentioned in Section 1.3.1 (modeling a liquid solution in terms of deviation from ideality), one can use Equations (1.3) to (1.5), substituting the excess enthalpy h^E for the excess volume. As with density, such an approach is useful primarily for mixtures of liquids, and for cases where the interactions in solution are not too complex. Strong interactions in solution, such as hydrogen bonding, have a greater effect on h^E than on v^E. h^E data can be found through the bibliography by Wisniak and Tamir [33] and in some additional databases and compilations [13, 31, 35].

The excess enthalpy of a liquid mixture can be rigorously related to the excess Gibbs energy g^E of the solution; models for g^E are typically used for calculating phase equilibria with activity coefficients. The relationship is

$$h^E = -RT^2 \left[\frac{\partial (g^E/RT)}{\partial T} \right]_{p,x} \tag{1.6}$$

where R is the molar gas constant. Equation (1.6) seldom provides a successful approach in practice, probably because the temperature dependence of g^E models is often not very accurate (especially if the parameters have been fitted at only a single temperature). However, it provides a reasonable approximation if good phase-equilibrium data exist over a range of temperatures. Conversely, if measurements of h^E are available, Equation (1.6) may be used to find the temperature dependence of g^E (and therefore the temperature dependence of activity coefficients).

1.4 PHASE EQUILIBRIA FOR MIXTURES

1.4.1 Types of Phase-Equilibrium Calculations

The basic principles of phase equilibria are discussed in the chapter covering thermodynamics (Chapter 4). In general, two or more phases are in equilibrium when they have the same temperature, pressure, and fugacity (or, equivalently, chemical potential) for each species.

While an enormous variety of phase equilibria exist, if we restrict ourselves to fluids, we need consider only the possible presence of a vapor and one or more liquid phases. The case most commonly encountered in chemical engineering (for example, in distillation) is vapor-liquid equilibrium (VLE). Multiple liquid phases (such as oil and water) can be in equilibrium, so one

Physical and Chemical Properties

can have liquid-liquid equilibrium (LLE), VLLE, LLLE, etc. Chemical engineers rarely deal with more than two simultaneous liquid phases. Phase equilibria involving solids are outside the scope of this chapter; thermodynamic modeling of solid solubility in liquids is covered in some standard texts [15, 36].

The principles and algorithms for calculating fluid-phase equilibria are discussed in many textbooks [36–40]. Here, we focus on methods and data requirements for calculating the component fugacities in a phase as a function of temperature, pressure, and composition; this is the key element in all phase-equilibrium calculations.

1.4.2 Equation-of-State Methods

As discussed in the thermodynamics chapter (Chapter 4), an equation of state (EOS) can be used to calculate the fugacities of all components in a mixture. This approach finds widespread use in the chemical and petroleum refining industries; cubic equations of state are used most often, particularly the Soave–Redlich–Kwong (SRK) and Peng–Robinson (PR) equations.

A prerequisite for mixture EOS calculations is reliable EOS parameters for the pure components. As discussed in Section 1.2.5, these may be obtained in a generalized way from critical constants and the acentric factor, or they may be fitted to data for the specific fluid. For an accurate representation of mixture phase equilibria, the EOS must produce accurate vapor pressures for the pure components.

Once the pure-component EOS parameters are established, the next piece of the EOS model is the *mixing rules*. A mixing rule is an algorithm for determining a parameter for the mixture from the composition, the pure-component parameter values, and perhaps other data. The simplest mixing rule would be a linear mole-fraction average of the pure-component values; this rule is typically used for the "b" parameter (associated with size) in the SRK or PR equations. Phase equilibria in these EOS are much more sensitive to the mixing rule for the "a" parameter (associated with intermolecular energy); the simplest common mixing rule for this parameter is

$$a_{mix} = \sum_i \sum_j x_i x_j a_{ij} \tag{1.7}$$

where the sums extend over all components and values where $i = j$ are the pure-component values.

Equation (1.7) requires a_{ij} for unlike pairs. The method of calculating a_{ij} from a_{ii} and a_{jj} is called a *combining rule*. It is in the combining rule that a *binary interaction parameter* is typically introduced:

$$a_{ij} = (a_{ii} a_{jj})^{1/2} (1 - k_{ij}) \tag{1.8}$$

where the parameter k_{ij} (which should be small unless the mixture is highly nonideal) is fitted to data for the binary mixture of components i and j. Mixture phase equilibria are sensitive to the k_{ij}, which may be temperature dependent.

Simple mixing rules such as Equation (1.7) are usually adequate for mixtures of nonpolar fluids. For mixtures containing polar compounds, more complicated mixing rules have been devised, often containing more than one adjustable parameter per binary. These methods are beyond the scope of this chapter; more details may be found elsewhere [15, 22, 41, 42]. For some complex mixtures, particularly those containing polymers or associating components, better results can be obtained from a molecular-based approach such as the statistical associating fluid theory (SAFT). Advanced models like SAFT require more effort to implement, but often they can be accessed as a part of commercial process simulation software or other software packages.

A German academic group has provided a software package called PE [43] that may be used to perform phase-equilibrium and density calculations with many different equations of state and mixing rules. Parameters are built in for a limited number of components and mixtures, but the software also has the capability to fit parameters to pure-component and mixture data.

It is also possible to calculate mixture phase equilibria based on the reference-quality equations of state described in Section 1.2.4. In this approach, the Helmholtz energy is written as the sum of pure-component contributions (given by the reference-quality EOS for each pure component) plus a residual term. The residual term contains one or more adjustable parameters for each binary. Software implementing this approach is available [9].

For some important classes of compounds, methods have been developed to predict EOS parameters, including binary interaction parameters, from molecular structure. The PSRK method [44] has found significant use, and a promising new method is known as VTPR [45].

1.4.3 Activity-Coefficient Methods

The use of activity coefficients for phase equilibria is based on writing the fugacity f_i of each component in the liquid phase as the product of an ideal term and a correction (activity coefficient γ_i) for nonideality:

$$f_i^L = \gamma_i x_i f_i^{0L} \tag{1.9}$$

where the standard-state fugacity f_i^{0L} is the fugacity of pure component i at the temperature and pressure of interest. Fugacity f_i^{0L} is related to the pure-component vapor pressure p_i^{sat} by

$$f_i^{0L} = p_i^{sat} \phi_i^{sat} \exp\left[\int_{p_i^{sat}}^{p} \frac{v_i^L}{RT} dp\right] \tag{1.10}$$

where ϕ_i^{sat} is the fugacity coefficient of component i at saturation, v_i^L is the liquid molar volume of pure component i, and the exponential is the Poynting factor, which accounts for the effect of pressure on liquid fugacity. Fugacity coefficient ϕ_i^{sat} can be assumed to be unity in many cases (exceptions would be if p_i^{sat} is significantly above atmospheric pressure or if component i associates strongly in the vapor phase [i.e., carboxylic acids, HF]). The Poynting factor can also be taken as unity at pressures near atmospheric; for higher pressures, it may be simplified by assuming that v_i^L is independent of pressure.

The activity coefficients γ_i are typically computed from a model for the excess Gibbs energy g^E, as described in the thermodynamics chapter (Chapter 4). The most popular are the Wilson, NRTL, and Uniquac models, described in detail in many places [15, 36–40]. They contain two or three adjustable (and possibly temperature-dependent) parameters per binary. One cannot predict which model will be best for a given system; however, the Wilson equation is incapable of describing LLE.

For binary pairs where no data exist to which to fit parameters in activity-coefficient models, group-contribution methods have been developed to estimate these parameters based on molecular structure. The leading method, UNIFAC [46], usually provides reasonable estimates for mixtures of organic compounds.

A recent alternative to group-contribution activity-coefficient estimation methods is based on interactions between surface charge distributions (determined by quantum-mechanical calculations) of molecules in solution. The solvation model used for the charge-distribution calculation is known as COSMO; the most widely used method based on this technique is called COSMO-RS [47].

Physical and Chemical Properties

Equation (1.9) produces liquid-phase fugacities, but vapor-phase fugacities are also required for phase-equilibrium calculations. At low pressures, one can usually assume ideal-gas behavior for the vapor, in which case the fugacity of each component is equal to its partial pressure. At higher pressures, the virial expansion or another equation of state may be used. Correction for vapor-phase nonideality is essential in systems with strong vapor-phase associations, such as those containing carboxylic acids. The leading such method is that of Hayden and O'Connell [48]. If vapor-phase fugacity corrections are to be used in phase-equilibrium calculations, they must be included in the same manner when fitting parameters in the activity-coefficient model.

For systems containing dissolved supercritical gases, Equation (1.10) cannot be used directly because the pure-component vapor pressure p_i^{sat} is undefined for the gaseous component. If the solubility of the gas is small, Henry's law may be used to describe the fugacity of the dissolved gas. The Henry's constant k_H of a solute in a solvent is defined by

$$k_H = \lim_{x_2 \to 0} \frac{f_2}{x_2} \qquad (1.11)$$

where subscript 2 designates the solute. Equation (1.11), without the infinite-dilution limit, may be used to describe solute fugacity f_2 when the pressure is relatively low and x_2 is small (less than 0.01 for typical systems). Values of Henry's constant k_H for solute-solvent pairs may be derived from experimental solubility data. If the solvent is a mixture of liquids, methods exist [15, 36] for estimating k_H in the mixture from its value in each component of the solvent.

A prerequisite for successful use of activity-coefficient models is accurate knowledge of the vapor pressure of each pure component (see Section 1.2); this is the dominant factor in Equation (1.10) and provides the pure-component endpoints that anchor the mixture calculations. The second most important factor is the binary interaction parameters, which must be fitted to reliable binary data or estimated with a reliable method. Some activity-coefficient models also require molar volumes of the pure components. Additional parameters such as critical constants and acentric factors may be required if vapor-phase fugacity corrections are used.

1.4.4 Choosing a Method

Whether to use an EOS or an activity-coefficient method is sometimes just a matter of convenience (for example, which method is easily available in software), but there are some general principles to consider. Perhaps the most important is that activity-coefficient methods require the vapor pressure of each component at the temperature of interest; this renders these methods unsuitable for systems containing supercritical gases (unless the gas can be treated with Henry's law, as mentioned above). This makes EOS methods preferable for mixtures containing light components, such as methane-containing mixtures in hydrocarbon processing. Equations of state typically have difficulty representing highly nonideal systems, such as mixtures of polar and nonpolar fluids. Activity-coefficient models (or EOS with sophisticated mixing rules that incorporate a g^E model) are normally preferred for such systems.

An important consideration when using either method for VLE is the importance of correct prediction of the pure-component vapor pressures. For activity-coefficient models, this requires the direct use of a correlation for p_i^{sat}; for EOS models, it may require (especially for polar fluids) fitting the $a(T)$ term in the EOS to vapor-pressure data.

For LLE, the equilibrium is dominated by the activity coefficients. While VLE for simple systems can be approximated by setting activity coefficients equal to unity (Raoult's law), LLE always requires binary interaction parameters. In liquid-liquid extraction systems, parameters based on binary data alone may be insufficient for accurate design; a few experimental ternary data for LLE tie-lines often provide significant improvement.

When using binary interaction parameters, it is best to use parameters fitted to binary data at or near the temperature of interest. If data are available at multiple temperatures, it is possible to include limited temperature dependence. It is also possible to fit parameters for a binary pair to ternary data, but only if parameters for the other two binary pairs in the ternary system are already known. Binary interaction parameters are not interchangeable between methods, so it is important to use exactly the same EOS or activity-coefficient model in both data regression and phase-equilibrium calculations.

Finally, VLLE calculations can sometimes be simplified in systems containing water and hydrocarbons. Because the solubility of hydrocarbons in water is very small, simplified calculations can be made by assuming a pure liquid water phase. Methods exist [4] to estimate the amount of water present in the vapor and dissolved in the liquid hydrocarbon phase. Such a simplification could not be performed if the amount of hydrocarbon in the water were important (for example, if wastewater contamination were a key design variable), but it is often adequate for calculations in petroleum refining.

1.4.5 Sources of Data

Both EOS and activity-coefficient methods require binary interaction parameters. In process simulation software, the necessary parameters may already be built into a data bank. Sometimes, parameters for the system of interest may be found in the literature. If not, however, the parameters must be fitted to mixture data.

The Dortmund Data Bank (DDB) [13] contains a large amount of mixture VLE and LLE data. A large collection of printed data is the DECHEMA Chemistry Data Series [31]; another printed source is the *International Data Series: Selected Data on Mixtures* [49]. Knowledge of an azeotrope (where the coexisting vapor and liquid have the same composition) can be an important piece of VLE data; azeotropic data may be found in the DDB [13] and in a printed compilation [50]. Extensive data for solubility, primarily of gases in various solvents, are in the *IUPAC Solubility Data Series* [51]; some data from this series are now available on the Internet [52]. Solubility data for organic compounds in water are collected in the AQUASOL database [53]; a large subset of these data is available in book form [54]. Some Henry's constants of solutes in water are available in the *NIST Chemistry Webbook* [5], and high-quality correlations have been produced [55] for the Henry's constants of common gases in water over a wide range of temperatures. The compilation of Linke and Seidell [56], while old, is still a valuable source of data for solubilities of inorganic compounds (including salts and other solids) in various liquids.

1.5 TRANSPORT PROPERTIES

1.5.1 Kinetic Theory for Transport Properties

For simple approximations to intermolecular interactions, the kinetic theory of gases has been well developed for the computation of transport properties at low densities. Theory and theory-based correlations are reviewed in references [15] and [57]. If the molecules are modeled as hard spheres of diameter σ and molar mass M, kinetic theory gives the following relations for the viscosity η, thermal conductivity λ, and diffusivity D of dilute gases:

$$\eta = C_\eta \frac{T^{1/2} M^{1/2}}{\sigma^2} \tag{1.12}$$

$$\lambda = C_\lambda \frac{T^{1/2}}{M^{1/2} \sigma^2} \tag{1.13}$$

$$D = C_D \frac{T^{1/2}}{M^{1/2}\rho\sigma^2} \tag{1.14}$$

where T is the absolute temperature, ρ is the density, and the multiplying constants can be computed by the theory at varying levels of approximation.

Unfortunately, real molecules differ significantly from hard spheres, so Equation (1.12) to (1.14) are not directly useful for real fluids. Additional correction factors can be added to these equations for fairly realistic spherically symmetric interactions; these can represent nonpolar fluids that are roughly spherical, such as the noble gases and CH_4. However, most molecules of interest are far from spherical, and kinetic theory is still intractable for molecular interactions that are not spherically symmetric. Therefore, the direct applicability of kinetic theory for calculating transport properties of real fluids is limited. However, kinetic theory plays an important role in guiding the functional form of semiempirical correlations such as those discussed below.

1.5.2 Viscosity

At low and moderate pressures, the viscosity of a gas is nearly independent of pressure and can be correlated for engineering purposes as a function of temperature only. Equations have been proposed based on kinetic theory and on corresponding-states principles; these are reviewed in *The Properties of Gases and Liquids* [15], which also includes methods for extending the calculations to higher pressures. Most methods contain molecular parameters that may be fitted to data where available. If data are not available, the parameters can be estimated from better-known quantities such as the critical parameters, acentric factor, and dipole moment. The predictive accuracy for gas viscosities is typically within 5%, at least for the sorts of small- and medium-sized, mostly organic, molecules used to develop the correlations.

For gas mixtures, the available methods are similar [15]. Some interpolate between pure-component viscosity values, while others utilize mixing rules to produce mixture correlation parameters from properties of each component. Predictive accuracy is somewhat worse than for pure components, but still usually within 10%. Simplistic combinations of pure-component viscosities (such as a linear mole-fraction average) can be quite inaccurate for gas mixtures, especially if the components differ greatly in polarity or in molar mass.

For liquid viscosity, theory is lacking and correlations are largely empirical. The main variation is with temperature; the effect of pressure is small for liquids well removed from the critical point. It is common to correlate the viscosity (or sometimes the kinematic viscosity, $\nu = \eta/\rho$) with a logarithmic dependence in reciprocal temperature:

$$\ln \eta = A + \frac{B}{T} \tag{1.15}$$

Equation (1.15) can accurately represent the viscosity only in limited temperature ranges of tens of Kelvins. For wider temperature ranges (particularly at low temperatures), an additional constant may be added, putting the correlation in the form of the Antoine equation, Equation (1.1), which in the context of viscosity is called the Vogel–Tammann–Fulcher equation. These correlations often extrapolate poorly, especially at temperatures above about $0.7T_c$. Predictive methods, as described in *The Properties of Gases and Liquids* [15], typically make use of group contributions and some thermodynamic information (critical properties, etc.) to estimate parameters in a correlating equation. These predictions can be subject to large errors on occasion, but often are accurate to within 15%.

For liquid mixtures, the available methods interpolate between pure-component viscosities, which may be known from experiment or estimated with a predictive method. Unfortunately, there is no clearly best method for interpolation. The most commonly used form is

$$\ln \eta_{\text{mix}} = \sum_i x_i \ln \eta_i + \frac{1}{2} \sum_i \sum_j x_i x_j G_{ij} \tag{1.16}$$

where the sums extend over all species. The G_{ij} parameter may be fitted to binary data where available or else set to zero; structure-based estimation techniques for G_{ij} have also been developed. *The Properties of Gases and Liquids* [15] discusses the merits of this and similar mixing rules. All obtain reasonable results (within about 10% if the pure-component viscosities are accurately known) for most classes of fluids *if* all the mixture components are liquids well below their critical temperatures. Such mixing rules are notoriously unreliable for mixtures of light and heavy components, such as crude oil containing dissolved methane.

The methods in the preceding paragraphs are effective for gases at moderately low densities and for dense liquids well below the critical temperature. For intermediate densities (high-temperature liquids, compressed gases, supercritical fluids), a corresponding-states approach is preferred. In this approach, the properties of the fluid are mapped onto those of a well-known reference fluid such as propane. The mapping parameters may depend on the fluid's critical properties, acentric factor, and/or vapor-pressure curve; one-fluid mixing rules for these quantities are used to map mixtures onto the reference fluid. The most commonly used such approach, SUPERTRAPP [58], has been implemented in NIST databases [9, 59].

For a few fluids (water, air components, light hydrocarbons, common refrigerants), extensive viscosity data exist and have been fitted to comprehensive equations. These reference-quality correlations are available in the pure-fluid property databases mentioned in Section 1.2.4.

For completeness, we mention that the viscosity of a fluid diverges to infinity in the limit as the critical point is approached. The effects of this viscosity divergence are confined to such a tiny region around the critical point that it can usually be ignored.

Viscosity data for pure components are available in several places [11–14, 60–64], and some collections of mixture data (mostly for binaries) also exist [31, 63, 65]. Some additional data references are cited in *The Properties of Gases and Liquids* [15].

1.5.3 THERMAL CONDUCTIVITY

Kinetic theory is useful for vapor-phase thermal conductivities, but a complication (compared to the viscosity) is that molecules can store thermal energy in internal modes. Almost always, an approach due to Eucken is used in which the correlated quantity is $(\lambda M/\eta C_V)$, where λ is the thermal conductivity, M is the molar mass, η is the viscosity (computed as described in Section 1.5.2), and C_V is the constant-volume heat capacity that contains contributions from molecular rotation and vibration. *The Properties of Gases and Liquids* [15] reviews several methods for estimating this factor, and also a group-contribution corresponding-states method that is useful for predictions for some classes of organic compounds. Methods exist [15] for extending the low-density results to somewhat higher pressures.

As with the viscosity, simplistic mole-fraction averaging of pure-component values is not advisable for gas-phase thermal conductivities. The most common method for predicting mixture values has the form

$$\lambda_{\text{mix}} = \sum_i \frac{y_i \lambda_i}{\sum_j y_j A_{ij}} \tag{1.17}$$

where various methods exist [15] for estimating A_{ij} (which is unity if $i = j$). A corresponding-states approach, similar to that for pure components, can also be used.

Physical and Chemical Properties

Correlations and estimation methods for liquid thermal conductivities are mostly empirical. At low temperatures (below or near the normal boiling point), the thermal conductivity is roughly linear in temperature over short temperature ranges; more complex temperature functions are required to cover a larger range of temperatures. Predictive methods are summarized in *The Properties of Gases and Liquids* [15].

For the thermal conductivity of liquid mixtures well below the critical point of each component, a linear mass-fraction average of the pure-component values is often a reasonable approximation. Such an average usually somewhat overpredicts the mixture value, and more complex mixing rules have been proposed [15] that give better quantitative results.

For high-temperature liquids, compressed gases, and other systems at intermediate densities, a corresponding-states treatment is preferable. The SUPERTRAPP model for thermal conductivity is similar to that described for viscosity in Section 1.5.2, and is available in the same NIST databases [9, 59].

Reference-quality correlations exist for the thermal conductivity of a few well-measured fluids such as water. These are available in the pure-fluid databases mentioned in Section 1.2.4.

An additional complication in describing the thermal conductivity at intermediate densities is its divergence to infinity in the critical region. Unlike viscosity, the divergence of the thermal conductivity manifests itself in a wide region around the critical point (contributing more than 1% for densities roughly within 50% of ρ_c at temperatures up to roughly $1.4T_c$). Calculations of thermal conductivity in this region require consideration of the near-critical divergence, either explicitly or by a corresponding-states approach where the formulation for the reference fluid has the near-critical contribution built in. A recent engineering-oriented correlation [66] incorporates this divergence.

There are collections of thermal conductivity data for pure components [11–14, 67] and mixtures [31, 65]. Some additional data references are cited in *The Properties of Gases and Liquids* [15].

1.5.4 Diffusivity

Diffusion coefficients are important for mass-transfer operations (see Chapter 8, "Mass Transfer"). There are several differently defined diffusion coefficients (self-diffusion coefficient, interdiffusion [or mutual diffusion] coefficient, intradiffusion [or tracer diffusion] coefficient); this can be a source of confusion. These are delineated in standard references [15, 68, 69].

For gases at low and moderate pressures, correlations and predictive methods are based on kinetic theory [15]. Because of the similarity in molecular mechanisms between viscosity and diffusivity, molecular parameters derived from viscosity data may often be successfully used to predict gas-phase diffusivities.

In liquids, predictive methods for diffusivity are typically semiempirical, relating the diffusivity of a solute at infinite dilution to the solvent viscosity, the molar volumes of the components, and sometimes other quantities [15]. For finite concentrations, the manner in which the diffusion coefficients pass from one infinite-dilution limit to the other is sometimes complex, and the models that exist [15] typically have a parameter that must be fitted to data.

Several sources of experimental diffusivity data are mentioned in the corresponding chapter of *The Properties of Gases and Liquids* [15].

1.6 AQUEOUS ELECTROLYTE SOLUTIONS

1.6.1 Vapor–Liquid Equilibria and Activity Coefficients

For systems with electrolytes dissolved in water, the methods discussed in previous sections are usually not appropriate, due to the presence of charged species. We will focus on aqueous electrolytes, but most methods discussed may be applied to electrolytes dissolved in other solvents. The modeling of electrolytes in mixed solvents is especially difficult and is beyond the scope of this chapter. The thermodynamics of electrolyte solutions is discussed in most physical chemistry textbooks. The key quantities are the osmotic coefficient (which describes the effect of the dissolved electrolyte on the vapor pressure of the solvent) and the activity coefficient (which is related to the chemical potential of the solute).

Correlations for the activity and osmotic coefficients of aqueous electrolytes begin with the theoretically rigorous Debye-Hückel limiting law for dilute solutions. Because the Debye-Hückel theory is accurate only at very low concentrations, it is supplemented by semiempirical terms. The most widely used are the Pitzer model and the electrolyte-NRTL model, both of which may be extended (perhaps with additional parameters) to systems containing multiple salts. A thorough description of the Pitzer model and a listing of parameters is in the monograph edited by Pitzer [70]. Sources for the electrolyte-NRTL model include the original journal article [71] and the book by Zemaitis et al. [72].

Several sources exist [72–74] containing evaluated data for the activity and/or osmotic coefficients of single electrolytes in water at 25°C. Data at other temperatures, or for mixed salts, are more scarce, but some compilations and databases exist [13, 74–76].

If data are lacking on the system of interest, it is often a fair approximation (especially at low and moderate concentrations) to use a "model-substance" approach. The behavior of an electrolyte is assumed to be similar to that of a known electrolyte of the same charge type. For example, NaCl is a model substance for 1:1 salts. This approach is particularly useful in estimating the temperature dependence of activity and osmotic coefficients; when these coefficients are known only at 25°C, the model-substance approach may be used to estimate the effect of temperature.

1.6.2 Density and Enthalpy

The methods described in Section 1.3 are generally unsuitable for electrolyte solutions. Electrolytes cannot be simply incorporated into equations of state, and simple mixing rules or corresponding-states approaches do not work because the properties of the pure electrolyte have little relationship to the contributions their ions make to mixture properties.

Typically, density and enthalpy are modeled by starting with a definition of ideal mixing in which each solute contributes as it does at infinite dilution:

$$v^{id} = x_1 v_1 + \sum_i x_i v_i^{\circ} \qquad (1.18)$$

where subscript 1 denotes the solvent, the sum includes all the solutes, and v_i° is the *standard-state volume* of solute i, which is the infinite-dilution limit of the solute's partial molar volume. Additional multiplicative factors appear in the equations if, as is commonly the case for electrolytes, concentration is expressed in molality (moles per kilogram of solvent) instead of mole fraction. An analogous equation applies for the enthalpy:

$$h^{id} = x_1 h_1 + \sum_i x_i h_i^{\circ} \qquad (1.19)$$

For dilute solutions, Equation (1.18) and Equation (1.19) are sufficient to describe the density and enthalpy. For more concentrated solutions, corrections must be applied. The corrections to the density are obtained from the pressure derivative of an activity-coefficient expression, while those for enthalpy are obtained from a temperature derivative. Details are given by Zemaitis et al. [72] and Pitzer [70]; the latter also tabulates parameters for temperature dependence in the Pitzer activity-

Physical and Chemical Properties

coefficient model. Since it is rare to have activity-coefficient data over a wide range of temperature or pressure, the usual approach for applying these expressions is to treat the parameters for temperature dependence or pressure dependence as adjustable and fit them to data for enthalpies and densities, respectively.

Some fairly well-validated estimation techniques exist for the standard-state properties of electrolytes in water. The one in widest use is the Helgeson–Kirkham–Flowers (HKF) correlation, as implemented in software known as SUPCRT92. Versions of the software may be found with a Web search, and a site [77] provides access to the most current set of parameters.

Experimental data for densities and enthalpies of electrolyte solutions are found in some compilations [75, 78] and in the ELDAR database [76]. The book by Zaytsev and Aseyev [74] contains extensive tables based on smoothed experimental results; some caution is needed with such tables because smoothing procedures can introduce artifacts. The *CRC Handbook of Chemistry and Physics* [26] also contains density data for many electrolytes in water at 20°C.

1.6.3 Transport Properties

The viscosity of aqueous electrolyte solutions is typically modeled as the relative viscosity η_r:

$$\eta_r = \eta/\eta_0 \tag{1.20}$$

where η_0 is the viscosity of pure water at the same temperature and pressure (which may be obtained from the sources mentioned in Section 1.2.3). At low concentrations (up to 0.1 molal and sometimes as much as 1 molal), the viscosity is described by the Jones-Dole equation:

$$\eta_r = 1 + AI^{1/2} + \sum_i c_i B_i \tag{1.21}$$

where c_i is the concentration (molality) of ionic species i, I is the ionic strength defined by

$$I = 0.5 \sum_i c_i z_i^2$$

z_i is the charge on species i, coefficient A depends on various solute and solvent properties, and the coefficients B_i are specific to the individual ions. Parameters for the Jones-Dole equation at room temperature are tabulated by Marcus [79]. A semiempirical extension of the Jones-Dole equation to higher concentrations, and also a method for extrapolating room-temperature parameters to higher temperatures, are described by Lencka et al. [80]. Jiang and Sandler [81] have developed a different method, based on liquid-state theory, that also appears promising for correlation and limited prediction of electrolyte solution viscosities.

The thermal conductivity of aqueous electrolyte solutions is typically described at room temperature by the following empirical equation:

$$\lambda = \lambda_0 + \sum_i \alpha_i c_i \tag{1.22}$$

where λ_0 is the thermal conductivity of pure water and α_i is a parameter specific to ion i. Values of α_i for common ions are tabulated by McLaughlin [82], who also gives an extension of the method to other temperatures.

Experimental data for the transport properties of aqueous electrolytes are found in several sources [29, 75, 76]. Tables based on smoothed data are in the book by Zaytsev and Aseyev [74]. Values of the viscosity for many electrolyte solutions at 20°C are tabulated in the *CRC Handbook* [26]. While now somewhat dated, the book by Horvath [83] discusses additional correlation methods, data sources, and parameters for transport properties of electrolyte solutions, including material on diffusion coefficients and electrical conductivity.

1.7 PROPERTIES FOR CHEMICAL REACTION EQUILIBRIA

The thermodynamic equilibrium constant for a chemical reaction is a function of temperature only and is related to the standard-state Gibbs energy change for the reaction. The standard state for each component is typically that of the pure substance at the temperature of interest and 0.1 MPa. The standard-state Gibbs energy for each component is taken as that corresponding to the formation of each compound from its elements. Standard-state Gibbs energies are typically tabulated at 298.15 K. Values of the standard-state enthalpy of formation and the heat capacity are used to calculate the standard-state Gibbs energy (and thence the thermodynamic equilibrium constant) at other temperatures.

For many species, thermochemical properties (Gibbs energy and enthalpy of formation, heat capacity) can be found in the *NIST Chemistry Webbook* [5]. A good written source, especially for small molecules, is the *JANAF Tables* [30]. The *NBS Tables of Chemical Thermodynamic Properties* [84] provides thermochemical properties at 298.15 K for many species, including ions in aqueous solution. Many pure-component thermodynamic databases include enthalpy and Gibbs energy of formation for the compounds included [11–14]. For vapor species, ideal-gas properties are often adequate; these may be obtained from the sources described in Section 1.2.7.

It is also possible to estimate standard-state thermodynamic properties from molecular structures, particularly for gas-phase species. Group-contribution methods for such estimations are reviewed in *The Properties of Gases and Liquids* [15]. The most widely used method is that of Benson, for which software is available [85]; more limited calculations are available in the *NIST Chemistry Webbook* [5].

For relatively small molecules, it is becoming routine to calculate thermochemical properties in the ideal-gas state with computational quantum mechanics. For organic species with fewer than about 10 nonhydrogen atoms, the methods are sufficiently well developed that they can rival or even surpass the accuracy that can be obtained from experiment.

Reaction equilibria in solution (acid-base neutralization, etc.) significantly affect phase equilibria and other properties mentioned in previous sections, rendering those calculations much more difficult. In many aqueous electrolyte systems, properly accounting for the speciation in solution is the most important part of the problem.

1.8 MEASUREMENT OF FLUID THERMOPHYSICAL PROPERTIES

1.8.1 WHEN EXPERIMENTS ARE NECESSARY

When data are needed, the option of measurement should always be considered. Even for those without experimental facilities, there is the option of contracting the work. Experiments may be more time-consuming than computerized estimation techniques, but if the data are of sufficient importance, the effort may be a good investment.

When deciding whether experiments are needed when data are lacking, one must weigh the merits of measurement versus estimation. One important factor is the reliability of available estimation methods. If a method has been demonstrated to be reliable for systems very similar to the one of interest, then measurement may be unnecessary. On the other hand, if the only available estimation method is of dubious reliability and/or was not developed for similar substances, experiments may be needed.

A second important factor is the difficulty of the experiment. Some measurements, such as liquid densities at ambient temperatures, provide accurate data easily, but in other cases experiments may be difficult or infeasible. Complications that would argue against experiments include high temperatures and/or pressures (or very low temperatures or pressures requiring cryogenic or vacuum equipment); chemicals that are unstable, corrosive, or toxic; and chemicals that are not available in sufficient purity or quantity. Some properties, such as high-pressure phase equilibria and the thermal conductivity of polar fluids, are more difficult to measure accurately. A few laboratories do have capabilities for more difficult measurements, so the option of contracting with such a lab for measurements may be considered.

A third important factor is the degree of accuracy to which the information is required. Often, estimation techniques provide reasonable "rough" answers, while experimental measurements (if done well) reduce the uncertainty significantly. It may be preferable to use estimates for screening calculations and even preliminary design, and to make measurements only for components and mixtures that are of vital importance in the final design. Often, several measurement techniques exist for a property, and more accurate results can be obtained at greater expense and effort. For some applications, a simple, inexpensive measurement may be adequate, even for final design. For others, more painstaking measurements may be needed to obtain the required accuracy.

In the remainder of this section, we will briefly discuss the measurement of fluid thermophysical properties and the level of accuracy that can be obtained for a given property. More detailed information may be found in a series of books from the International Union of Pure and Applied Chemistry (IUPAC) [69, 86–90].

1.8.2 GENERAL CONSIDERATIONS

An essential consideration in experimental work is *safety*. This includes issues of toxicity, flammability, etc., for all chemicals involved, and also hazards created by interactions within mixtures. Safety must be considered at all stages in the design and construction of experiments.

Another important factor is *experience*. While some instruments may be used successfully with little training, in many cases considerable skill is needed to measure data safely and accurately. Especially for more precise and specialized measurements, there is no substitute for hands-on experience with the apparatus. Companies that do not maintain experimental expertise may need to turn to outside consultants or contractors when the need for data arises.

The control and measurement of *temperature* and *pressure* are essential for accurate experimental work, especially at temperatures and pressures differing significantly from ambient conditions. Sometimes, a 0.1 K deviation in temperature can change a property by an amount larger than the precision of the measuring apparatus. Two of the IUPAC books mentioned above [87, 89] discuss the measurement of temperature and pressure. In most cases, the farther one goes from ambient temperatures and pressures, the larger the experimental uncertainty and/or the more complex and expensive is the apparatus required to maintain accuracy.

Purity may be an important factor. Purification procedures are discussed in reference [91]. The fluid being measured may be available in different grades; the purity (and expense) required will depend on the accuracy needed and the property being measured. Some properties (density, heat capacity) are in most cases not significantly affected by small amounts of impurities. The vapor pressure and vapor-liquid equilibria are quite sensitive to impurities, particularly if the impurity is much more volatile (e.g., dissolved air) than the fluids being measured.

Many experimental apparatus require *calibration* with substances whose properties are accurately known. Water (see Section 1.2.3) is the most common calibration fluid for liquid-phase properties, but other fluids such as toluene are sometimes used. Vapor-phase properties are often calibrated with helium, argon, nitrogen, or air. An IUPAC book [92] describes recommended reference materials for different property measurements. Many reference materials for calibrations are available through the NIST Standard Reference Materials program [http://www.nist.gov/srm/] and other national metrology laboratories.

1.8.3 Density

For liquids at ambient conditions, one can obtain a reasonably accurate density simply by weighing a known volume of the liquid. For rough work, one can use something as simple as a volumetric flask; for more precise work, calibrated volumes (known as pycnometers) are used. One can easily obtain 1% accuracy, and careful pycnometry can obtain 0.1%.

More accurate density measurements may be made by instruments that take advantage of the principle of Archimedes, where the apparent weight of an object immersed in a fluid is diminished by that of the fluid displaced. In the simplest version of this experiment, a "sinker" of known mass and volume is immersed in the fluid while suspended by a wire from an analytical balance. More sophisticated versions may use a magnet to suspend the sinker or measure the difference between two sinkers of similar mass and surface area but different volume. With care and good control of temperature and pressure, such instruments can achieve uncertainties of 0.02% or lower for both vapor and liquid densities.

Another method of measuring density relies on the change in the resonant frequency of a tube (often U-shaped) when it is filled with a fluid. Vibrating-tube densimeters are commercially available; they can be a convenient measuring tool in many circumstances. These instruments must be calibrated (usually with water if liquids are being measured, although for liquids whose density is significantly different from water a calibration fluid with a similar density is preferable) at the temperature and pressure of interest. While the precision of these instruments is often better than 0.01%, very careful calibration and temperature control are required to obtain better than 0.1% uncertainty in practice.

Another alternative, particularly for high-pressure measurements, is the isochoric method. A cell of known volume is filled with the fluid, and then the pressure is measured as the temperature of the cell is changed. This provides data along curves of approximately constant density, known as isochors. Corrections are made for the expansion of the vessel due to temperature and pressure. This method is most useful for supercritical fluids and other situations where the fluid is fairly compressible. Uncertainties with this method can be on the order of 0.1%, but are often higher at elevated temperatures and pressures.

Expansion methods are often used for measuring gas densities. In these methods, a sample is expanded from a small volume to a larger volume (where the ratio of volumes is accurately known), holding the temperature constant and measuring the pressure ratio. Typically, multiple expansions are used (a successive expansion technique known as the Burnett method is popular), with the final state being at a pressure sufficiently low that the density is accurately known by other means (such as correction of the ideal-gas law by the second virial coefficient). The Burnett expansion method may achieve uncertainties in density as low as 0.01%.

1.8.4 Heat Capacity and Caloric Properties

The measurement of heat capacity and related quantities is known as *calorimetry*. Most often the constant-pressure heat capacity C_p is measured; some instruments measure the constant-volume heat capacity C_V. Often, what is actually measured is not the derivatives C_p and C_V but an energy change divided by a small but finite temperature change. In some cases, the original "enthalpy increment" data may be more useful than the approximate heat capacities derived from them. In addition to the IUPAC books referenced in Section 1.8.1, the monograph of Hemminger and Höhne [93] has extensive information about calorimetry.

One category of calorimetric measurements is *batch* or *static calorimetry*, where, in the simplest implementation, the sample is contained in a vessel, a measured amount of energy is added (usually electrically), and the temperature change is measured. Corrections are required for the heat capacity of the calorimeter itself; sometimes parallel measurements are made with identical calorimeters (one full, one empty) to determine this correction. For liquids, this method usually yields not C_p but C_σ, where the path of the derivative is along the vapor-liquid saturation curve. This quantity is negligibly different from C_p as long as the amount of vapor in the cell is small and the experiment is conducted far from the critical point. For vapors and supercritical fluids, batch calorimeters most naturally yield C_V. Batch calorimetry can achieve accuracy of about 1% (or even 0.1% for some research instruments); the uncertainty typically increases at high temperatures and pressures.

Flow calorimetry has become the leading method for fluids and fluid mixtures. In a flow calorimeter, energy is added (or removed) at a known rate while fluid is flowing through the instrument at a known rate. Once steady state is reached, measurement of the temperature before and after the heating section yields the heat capacity C_p. Sometimes, to cancel out sources of error, measurements are made of the difference between a fluid and a well-known reference such as water. Flow instruments may also measure heats of mixing; the instrument can have two inputs and one output and measure either the temperature change upon mixing or the rate of energy input (or removal) to keep the temperature at its initial value. Commercial instruments are available for flow calorimetry on liquids near room temperature; several laboratories have the capability for flow calorimetry at more extreme conditions. The accuracy of flow calorimetry depends on minimizing heat leaks and careful control and measurement of flow rate and temperature; uncertainties in measured heat capacities or heats of mixing are typically on the order of 1%.

The other common category of calorimetry is *differential* methods, in which the thermal behavior of the substance being measured is compared to that of a reference sample whose behavior is known. In *differential scanning calorimetry* (DSC), the instrument measures the difference in power needed to maintain the samples at the same temperature. In *differential thermal analysis* (DTA), the samples are heated in a furnace whose temperature is continuously changed (usually linearly), and the temperature difference between the sample and the reference sample as a function of time can yield thermodynamic information. DSC and DTA are most commonly used for determining the temperature of a phase transition, particularly for transitions involving solids. In addition, DSC experiments can yield values for the enthalpy of a phase transition or the heat capacity. Commercial DSC and DTA instruments are available.

1.8.5 PURE-COMPONENT VAPOR PRESSURE

In the measurement of vapor pressure, it is essential that the sample be thoroughly degassed; even a small amount of dissolved air or other volatile component will distort the measurement. Degassing usually involves either distillation of the sample or placing the sample (sometimes after freezing) under vacuum.

The most straightforward method for vapor-pressure measurement is the *static* method, in which the pressure of the vapor above a pure liquid is measured directly with a manometer, pressure gauge, or pressure transducer. All parts of the apparatus must be maintained at a temperature at least as high as that of the sample in order to avoid condensation. Static techniques may be used at high temperatures and pressures with appropriate apparatus construction, but they become difficult at low vapor pressures due to the difficulty of pressure measurement and the effects of impurities. With good equipment and procedures, the accuracy of static vapor pressure measurements can be on the order of 0.1%.

Another conceptually simple method is *ebulliometry*, where the liquid is boiled under total reflux at a fixed pressure and the boiling temperature is measured. Because of the difficulty of accurately measuring the temperature of a boiling liquid, ebulliometers typically measure the temperature of the condensing vapor. An advantage to ebulliometry is that volatile impurities are purged as the system approaches steady state, so the need for degassing is reduced. The attainable accuracy of ebulliometry can exceed that for static methods, but it is more difficult to extend ebulliometry to high pressures.

Sometimes *comparative* measurements are used, where a substance whose vapor pressure is well known (such as water) serves as a reference. This can be done either by measuring the temperatures of the sample and the reference fluid boiling under the same pressure (comparative ebulliometry), or by measuring the pressure difference between the sample and reference fluid at a common temperature. Comparative measurements can be relatively fast and convenient if a suitable reference fluid is available. With good temperature measurement and an accurately known reference fluid, the accuracy of comparative ebulliometry can be better than 0.1%.

In the *gas saturation* method, a carrier gas such as helium or air is bubbled slowly through the sample, coming to equilibrium with it. The material exiting the apparatus is collected, and the amount of sample dissolved can be measured at the end of the experiment. Knowledge of the number of moles carried away, the number of moles of carrier gas used, and the pressure of equilibration allows calculation of the partial pressure of the sample at equilibrium, which is approximately its vapor pressure. Care must be taken to ensure that all of the carrier gas is equilibrated with the sample and to prevent premature condensation before the solute is collected. This method is typically applied for vapor pressures below about 3 kPa, and may achieve 1% accuracy with careful use.

For very low vapor pressures (below about 10 Pa), the vapor pressure can be measured by the *Knudsen* method, in which the rate of escape of vapor through a small hole is measured. The kinetic theory of gases (with correction factors for hole geometry, etc.) is used to relate the rate of effusion to the vapor pressure. The attainable accuracy of such methods is around 1% to 10% at best, but it is often the only experimental option for substances of very low volatility.

1.8.6 MIXTURE VAPOR-LIQUID EQUILIBRIA

The experimental measurement of vapor-liquid equilibria in mixtures is a huge subject; space considerations dictate that we only briefly discuss the major categories of techniques. More information may be found in some books [87, 90, 94].

The *analytic* method involves confining a mixture in a two-phase state, measuring its temperature and pressure, and analyzing samples from each phase to obtain the equilibrium compositions. This is challenging in practice, largely due to the difficulty of sampling. Samples must be small to avoid disturbing the equilibrium, and they must be extracted without changing their composition. Often, internal stirring or mixing is needed to obtain equilibrium in a reasonable period of time. This approach is widely used and can provide accurate data if done carefully. Sometimes only the liquid composition is measured; while this provides less information, it may suffice if the vapor phase is predominantly a single component whose solubility in the liquid is the quantity of most interest. Sometimes measurement of the total pressure and the liquid composition are combined with approximate thermodynamic analysis of the vapor (for example, with the virial equation) to yield a complete picture of the vapor-liquid equilibrium.

In some cases, it is possible to measure the composition of one or both phases without withdrawing a sample. Techniques for *in situ* composition measurements include various forms of spectroscopy and measurements of other properties such as the refractive index.

A related *dynamic* method makes use of a recirculating still to produce equilibrated samples of vapor and liquid. This method is often used at low and moderate pressures; it is particularly suited for mixtures where the volatilities of the components are similar.

A dynamic technique known as *differential ebulliometry* can measure the effect of small amounts of solute on a solvent. The difference in temperature is measured between a pure fluid and a solution, both of which are boiling at the same pressure. Differential ebulliometry is a common technique for determining infinite-dilution activity coefficients, particularly when the solute is less volatile than the solvent.

The *synthetic* method is often used to determine phase boundaries (bubble points or dew points) at high pressure and/or temperature. In this method, the cell is filled with a mixture of a known composition in a single-phase region. Then, the temperature or the volume is changed until a second phase begins to form; in this manner the composition, temperature, and pressure at the bubble or dew point are known. The formation of the second phase may be detected visually or through measurement of some property such as heat capacity or density that exhibits a discontinuity or change in slope at a phase transition. A disadvantage of this technique is that the composition of the coexisting phase cannot be determined, but it is often used in situations where sampling is impractical.

While gas solubilities in a liquid may be measured by the analytic or synthetic techniques mentioned above, specialized methods have been developed for gas solubility measurements, especially near ambient conditions. The typical method is to place a known amount of gas in contact with a known amount of degassed liquid, allow the system to equilibrate, and then measure the amount of gas remaining (often by knowing the vapor-phase volume and measuring its pressure). The difference between initial and final amounts of gas is the amount dissolved in the liquid. It is common for such methods to achieve 1% accuracy in the solubility, and the best instruments can achieve 0.1%.

The solubility of relatively nonvolatile components (including solids) in a gas may be measured by the *gas saturation* method mentioned in Section 1.8.5. The gas is passed through the solute, and the amount of solute collected by a given amount of gas is measured. This method is commonly used to measure solubilities in supercritical fluids, where the effects of pressure and vapor-phase nonideality make the partial pressure of the solute in the vapor much larger than its vapor pressure. The major challenges are analysis of the solute collected, avoiding condensation of the solute before it is collected, and ensuring that all the gas is saturated with the solute.

1.8.7 Liquid-Liquid Equilibria

Liquid–liquid phase equilibria are most often measured by the analytic method mentioned above, although the synthetic method is used on occasion. Issues of sampling and equilibration are again important. If the densities of the liquid phases are similar, it can be difficult to achieve separation. Entrainment of one phase in the other is a particular problem in aqueous/organic systems, where emulsions can form. For nonpolar solutes like hydrocarbons in water, the solubility is so small that analysis becomes difficult, and even a tiny amount of contamination with the hydrocarbon phase cannot be tolerated. Reviews exist [87, 90, 95] describing apparatus for measuring liquid-liquid equilibria at near-ambient pressures.

1.8.8 Viscosity

The emphasis of this section is on Newtonian fluids whose viscosity is independent of shear rate. Most methods are discussed in an IUPAC volume [69]. Because most of these methods actually measure the kinematic viscosity (ratio of viscosity to density), knowledge of the density of the fluid is required to obtain the viscosity itself.

Perhaps the most familiar technique is the capillary-flow method. The working principle is the Hagen-Poiseuille relationship between the flow rate through a tube of fixed diameter, the pressure drop, and the viscosity. In practice, because the capillary diameter appears to the fourth power in the working equation and is difficult to determine accurately, capillary viscometers are usually calibrated with reference fluids such as water or reference oils that are available from viscometer manufacturers and some national laboratories.

There are two types of capillary viscometers. In gravitational instruments, gravity drives vertical flow through a capillary and a timer is used to measure the flow rate. For liquids with low vapor pressures, an open-ended viscometer is suitable; this design has been well studied, and commercial instruments are available. Accuracy of 1% or better can be achieved. For more volatile liquids, sealed instruments may be used. A different analysis must be applied to obtain correct results with a sealed capillary viscometer [96]. Sealed capillary viscometers have not yet been commercialized, and the uncertainty is currently 2% at best.

A second type of capillary viscometer generates a constant volumetric flow rate with a pumping device, and the viscosity is determined by measuring the pressure drop. Such viscometers have been used mostly in research, but some commercial instruments are available. These viscometers may be used over a wide range of conditions for both gases and liquids, and in some cases the uncertainty can be reduced to less than 0.1%.

Another simple method is to measure the falling or rolling of a sphere through a liquid; this motion is described by Stokes's law under certain simplifying assumptions. This method is commonly used for viscous liquids, with an accuracy on the order of 3%. It can also be used for liquids at high pressures, with some loss of accuracy.

A more sophisticated technique involves measuring the damping of the oscillations of a body immersed in a fluid. The oscillating body may be a sphere, a thin disk, or a cylinder. Research instruments of this type can attain better than 0.5% accuracy (0.05% in the best cases) for both vapors and liquids over a wide range of conditions. In a variant known as the oscillating-cup method, the fluid is contained inside a hollow body; this method is more often used for high-temperature measurements on fluids such as molten metals or molten salts.

Three techniques related to oscillating-body methods can be mentioned. One involves measuring the torsional oscillations of a piezoelectric crystal immersed in a fluid; it can achieve accuracies of about 2% and has been used in a variety of conditions. A second involves measuring the drag on a rotating cylinder magnetically suspended in the fluid; it can achieve uncertainties as low as 0.15% in the dilute-gas region and 0.4% for higher densities. A third technique involves measuring the damping of a vibrating wire due to shear and to its inertia; this can yield simultaneous measurements of density and viscosity, with uncertainties in viscosity near 4% if both properties are being measured, or near 0.5% if the density is already accurately known. The first two techniques have been implemented in commercially available viscometers.

Finally, there are industrial rheological instruments in common use that allow determination of viscosities for liquids to within a few percent, which is sufficient for many purposes. These involve measuring the torque required to maintain a given velocity for a fluid confined in the annulus between two concentric cylinders, one of which is rotated (*Couette* viscometer), or a fluid confined between a flat plate and a rotating cone (*cone-and-plate* viscometer).

1.8.9 Thermal Conductivity

The traditional way to measure thermal conductivity is with *steady-state* instruments, in which a measured heat flux is compared to a temperature difference between surfaces. Most often the geometry is coaxial cylinders, a thin wire inside a cylinder, or parallel plates. In such instruments, eliminating convection currents is crucial; many old data taken with steady-state instruments are unreliable because of convection. Multiple experiments at different heat fluxes are often performed to verify the absence of convection. With good design and operation, such instruments may achieve accuracy in the 1% to 3% range.

A newer method is the *transient hot-wire* method, where an electric current is passed through a metal wire immersed in the fluid. The resistance of the wire is affected by its temperature, which in turn is affected by the dissipation of heat from the wire's surface, which depends on the thermal conductivity of the fluid. These instruments require sophisticated data analysis, but that is no longer an obstacle with the ready availability of personal computers. The absence of convection is relatively easy to verify. The best research instruments can achieve an accuracy of better than 1%. Measurements on conducting fluids (such as polar liquids) are more difficult because of the need to electrically insulate the wire. Other geometries, such as needle-shaped cylinders and thin strips, are also sometimes used for transient measurements.

Commercial instruments exist for cylindrical probes for transient measurements and for steady-state parallel-plate measurements. These generally require calibration with reference fluids of known thermal conductivity.

1.8.10 ELECTROLYTE SOLUTIONS

Most of the techniques mentioned above can be applied to electrolyte solutions in a straightforward manner. It may be necessary to use different apparatus materials due to the corrosive nature of some electrolyte systems, especially at high temperatures. In addition, special techniques exist to measure activity and osmotic coefficients in electrolyte solutions. These methods are discussed in more detail in reference [70].

In the *isopiestic* method, the vapor pressure of a solution is determined by equilibrating it with that of a reference solution (aqueous NaCl is often used) whose vapor pressure as a function of salt concentration and temperature is well known. This method can provide osmotic coefficients accurate to better than 1% over a wide range of conditions, but it is accurate only for electrolyte molalities above approximately 0.1 mol·kg^{-1}. The necessary equilibration may take several days, and specialized apparatus is required for measurements significantly above room temperature.

Potentiometry may also be used to determine activity coefficients of electrolytes; the measured e.m.f. of an electrochemical cell is related to the activities of the ions. These measurements can yield very accurate values near room temperature for systems where reversible and reproducible electrodes have been developed. Potentiometry at high temperatures is much more difficult; this is an area of active research.

1.9 OVERVIEW OF MAJOR DATA SOURCES

1.9.1 INTRODUCTORY COMMENTS

Table 1.2 lists major data sources by property for both pure components and mixtures. It is also useful to organize the information by database; that is the purpose of this section. We will describe a number of major databases and compilations and how they may be accessed. It is impossible for such a list to be complete; we have tried to list the most comprehensive sources we are aware of. Sources that are confined to one or two properties can be found in the section corresponding to that property (and in Table 1.2). Failure to mention a particular data source should not be taken as an indication that the source is not valuable.

TABLE 1.2
Major Sources of Thermophysical Property Data for Pure Fluids and Mixtures

Property	Pure-Component Data Sources	Mixture Data Sources
Vapor pressure	[5, 11, 12, 13, 14, 15, 16, 17]	N/A
Caloric properties	[11, 12, 13, 14, 19, 20, 21, 22]	[13, 31, 33, 35]
Density	[11, 12, 13, 14, 15, 18]	[13, 33, 34]
Ideal-gas properties	[5, 11, 12, 13, 14, 15, 28, 29, 30]	N/A
Viscosity	[11, 12, 13, 14, 60, 61, 62, 63, 64]	[31, 63, 65]
Thermal conductivity	[11, 12, 13, 14, 67]	[31, 65]
Mixture-phase equilibria	N/A	[13, 31, 49, 50, 51, 52]

Sources: Numbered as listed in References section.

1.9.2 NIST (Including TRC)

One of the mandates of the U.S. National Institute of Standards and Technology (NIST) is to provide reliable data to meet the needs of industry, including the physical and chemical property data that are the subject of this chapter.

One of the main sources of data from the NIST is the Standard Reference Database program [http://www.nist.gov/srd/]. Computer databases are available for the thermophysical properties of a variety of industrially important fluids and fluid mixtures, including water, air, cryogenic fluids, common refrigerants (including refrigerant blends), and natural gas [8,9]. These are available as stand-alone programs or as source code for integration into other software.

The *NIST Chemistry Webbook* [5] provides information on a large number of chemical compounds. This includes thermophysical property information (a subset of that available in the Standard Reference Databases) for several important pure fluids. Structural information is available for a large number of compounds, and for many of these data are given for vapor pressure, heats of formation and phase change, and/or ideal-gas heat capacity.

The NIST's Thermodynamics Research Center (TRC) has a large collection of pure-fluid thermodynamic and transport properties; tables of recommended values and correlations exist both in paper form and in a computer database [12]. The TRC has also produced books with comprehensive compilations for organic compounds (sometimes also available as software) for vapor pressure [17], liquid density [18], and ideal-gas heat capacity [29], in addition to a compilation on virial coefficients [32]. Their major archival database of experimental pure-component and mixture data is called Source [97]; it is currently available only to members of their consortium. Some data for mixtures of organic compounds are published in the periodical *Selected Data on Mixtures* [49]. More information is at http://trc.nist.gov.

Other sources of data from the NIST include the *JANAF Tables* for standard-state properties of small molecules [30], *The NBS Tables of Chemical Thermodynamic Properties* for species (including aqueous ions) at 298.15 K [84], and the *Journal of Physical and Chemical Reference Data* (jointly published by NIST and the American Institute of Physics).

1.9.3 DIPPR

The Design Institute for Physical Properties (DIPPR) of the American Institute of Chemical Engineers (AIChE) sponsors a variety of projects to provide data for the chemical industries. The DIPPR's Project 801 produces a database containing pure-component thermodynamic and transport properties for over 1900 industrially important chemicals. Recommended values are given for a number of single-valued properties (such as critical parameters, normal boiling point, standard-state Gibbs energy, and enthalpy of formation), and correlations are given for 15 temperature-dependent properties (such as vapor pressure, heat of vaporization, and transport properties for the saturated liquid and vapor). Evaluated experimental data are used where possible, but sometimes estimation techniques are used to fill in missing values so that a complete set of recommended and evaluated data is given for each compound included in the database. The original data used to develop the recommended values are also available. DIPPR 801 data are available in both book and electronic form [11]. The most recent data are restricted to the project's industrial sponsors; the version available to the general public lags behind by about two years.

DIPPR Project 911 compiles and evaluates data for environmental and health purposes. This includes data on flammability, toxicity, biological oxygen demand, and related properties, and data for air/water and octanol/water partitioning. A discontinued project compiled several volumes of data for transport properties of mixtures [65]. Information about the DIPPR, including other projects relevant to physical and chemical properties, may be found at http://www.aiche.org/dippr/.

1.9.4 DECHEMA

Beginning in the 1970s, the German chemical engineering society DECHEMA has produced the *Chemistry Data Series*, containing thermophysical property data for fluids and fluid mixtures. Many volumes are devoted to mixture phase equilibria, and other volumes cover critical properties, heats of mixing, infinite-dilution activity coefficients, mixture transport properties, electrolyte solutions, polymer solutions, and solubilities. A complete listing is available [31]. Many of these (and additional) data are available in their DETHERM database [http://www.dechema.de/detherm.html], which can also be accessed through the STN system of the Chemical Abstracts Service [http://www.cas.org/stn.html].

1.9.5 DDB

Some of the compilers of the *DECHEMA Chemistry Data Series* expanded their work into a comprehensive data bank for thermophysical property data. The Dortmund Data Bank (DDB) historically concentrated on mixture data, and it contains tens of thousands of datasets both for mixture phase equilibria (including electrolytes) and for excess properties of mixing. In recent years, a large amount of pure-component property data has also been added.

More information about the DDB, including a list of distributors, may be found at their Web site [13]. One notable feature is a free lookup program that allows one to search the database and see whether or not the DDB contains data for a particular system. Much of the data in the DDB is also available through the DETHERM database mentioned in Section 1.9.4.

1.9.6 NEL

In Great Britain, the National Engineering Laboratory (NEL, formerly a government agency but now privatized) has produced a database for thermodynamic and transport properties. PPDS contains correlations for properties of a large number of pure components; these are based on evaluated experimental data where possible but also include some estimated properties. For mixtures, the database contains binary interaction parameters fitted to data for use with common equation-of-state and liquid-activity methods for calculating phase equilibria. Information is available at their Web site [14].

1.9.7 Landolt-Börnstein

Since the late 1800s, the Landolt-Börnstein series, published in Germany, has provided important critical compilations of data for physics and related fields. Several volumes of the physical chemistry group have already been cited [17, 18, 32, 34]. Additional compilations include data on static dielectric constants and vapor-liquid surface tension. Further information on the Landolt-Börnstein series may be found at http://www.landolt-boernstein.com. Most of their current volumes may also be accessed on-line; information about on-line access may be found at the Web site.

1.9.8 Beilstein

Beilstein's Handbook of Organic Chemistry, greatly updated and expanded since its original German publication in the late 1800s, has long been the standard reference for information on organic chemistry. While its focus is on chemical aspects (structure, reactions, spectra, analysis), thermodynamic property information (boiling points, critical properties, thermochemical data) is also reported for some compounds, with references to the original literature. Because *Beilstein* contains millions of compounds, it is a good place to search for information on organic chemicals that are not common enough to be included in other databases.

While many technical reference libraries have *Beilstein* in print form, it is more convenient to search for information in the electronic form of the database. In addition, the printed series is no longer being updated. The electronic database allows a variety of searches on compounds and properties. Access to the electronic database is offered through CrossFire Beilstein [see http://www.mdl.com] or through the STN database system of the Chemical Abstracts Service [http://www.cas.org/stn.html]. Some details given in the print version (such as experimental methods) were not carried over to the electronic version, so in some cases it may be necessary to consult the print version or the literature citation for further information.

1.9.9 GMELIN

Gmelin's Handbook of Inorganic and Organometallic Chemistry, first published in German in the 1700s, is to inorganic and organometallic chemistry what *Beilstein* is to organic chemistry. Like *Beilstein*, it is more focused on chemical aspects, but contains some thermodynamic property information. Also like *Beilstein*, much of the data (without some supplementary information) has been carried over into an electronic database that continues to be updated. Access to the electronic database is offered through CrossFire Gmelin [see http://www.mdl.com] or the STN database system of the Chemical Abstracts Service [http://www.cas.org/stn.html].

1.9.10 PROCESS SIMULATION SOFTWARE

In chemical process design, it is common to obtain property data via process simulation software. This is sometimes a good option; the major simulation packages incorporate data from NIST, DIPPR, and other reliable sources. Caution is in order, however, especially for temperature-dependent properties and for mixture quantities such as binary interaction parameters in activity-coefficient models or equations of state. The danger is that the parameters in the simulation software may be fitted at conditions other than those of interest, and often it is difficult to tell where the parameters came from or under what conditions they are applicable. It is important to verify that the parameters in the software are appropriate for the conditions in which they are being used; that might involve a query to the vendor or doing some sample calculations to validate them against data.

ACKNOWLEDGMENTS

This chapter is a contribution of the National Institute of Standards and Technology, not subject to copyright in the United States. The identification of certain commercial products in this chapter neither constitutes nor implies recommendation or endorsement by the U.S. government or the National Institute of Standards and Technology. Material for some sections was supplied by Dr. Arno Laesecke of NIST (Section 1.8.8) and Dr. Song Yu of Columbia University (Sections 1.9.8 and 1.9.9). Several members of the NIST Experimental Properties of Fluids Group provided helpful input for Section 1.8.

REFERENCES

1. Brandrup, J., E. H. Immergut, and E. A. Grulke, eds., *Polymer Handbook*, 4th ed., John Wiley & Sons, New York, 1999.
2. Danner, R. P., and M. S. High, *Handbook of Polymer Solution Thermodynamics*, AIChE, New York, 1993.
3. Caruthers, J. M. (ed.), *Handbook of Diffusion and Thermal Properties of Polymers and Polymer Solutions*, AIChE, New York, 1998.
4. American Petroleum Institute, *Technical Data Book: Petroleum Refining*, 6th ed., API, Washington, DC, 1997; available on-line at http://www.epcon.com.

5. Linstrom, P. J., and W. G. Mallard, eds., *NIST Chemistry Webbook*, NIST Standard Reference Database 69, National Institute of Standards and Technology, Gaithersburg, MD;. http://webbook.nist.gov.
6. Wagner, W., and H.-J. Kretzschmar, *International Steam Tables*, 2nd ed., Springer-Verlag, Berlin, 2006.
7. Parry, W. T., J. C. Bellows, J. S. Gallagher, and A. H. Harvey, *ASME International Steam Tables for Industrial Use*, ASME Press, New York, 2000.
8. Harvey, A. H., A. P. Peskin, and S. A. Klein, *NIST/ASME Steam Properties*, NIST Standard Reference Database 10, Version 2.2, National Institute of Standards and Technology, Gaithersburg, MD, 2000. http://www.nist.gov/srd/nist10.htm.
9. Lemmon, E. W., M. O. McLinden, and M. L. Huber, *NIST Reference Fluid Thermodynamic and Transport Properties—REFPROP*, NIST Standard Reference Database 23, Version 7.0, National Institute of Standards and Technology, Gaithersburg, MD, 2002. http://www.nist.gov/srd/nist23.htm.
10. Wagner, W., C. Bonsen, and U. Overhoff, FLUIDCAL, Software to Calculate Thermodynamic and Transport Properties for Pure Substances, Version 2004, Lehrstuhl für Thermodynamik, Ruhr-Universität Bochum, Bochum, Germany, 2004. http://www.ruhr-uni-bochum.de/thermo/.
11. Design Institute for Physical Properties, DIPPR Project 801: Evaluated Process Design Data. http://www.aiche.org/dippr/ and http://dippr.byu.edu. Data from DIPPR 801 have also been issued in printed form at various times; the most recent is R. L. Rowley, W. V. Wilding, J. L. Oscarson, Y. Yang, R. J. Rowley, T. E. Daubert, and R. P. Danner (eds.), *Physical and Thermodynamic Properties of Pure Chemicals: DIPPR*, Taylor & Francis, London, 2003.
12. Thermodynamics Research Center, *TRC Thermodynamic Tables—Hydrocarbons* and *TRC Thermodynamic Tables—Nonhydrocarbons*. Both are loose-leaf multivolume publications with supplements available on a subscription basis; see http://trc.nist.gov. A software version of this database is: X. Yan, Q. Dong, X. Hong, and M. Frenkel, NIST/TRC Table Database, NIST Standard Reference Database 85, Version 1.5, National Institute of Standards and Technology, Gaithersburg, MD, 2001. http://trc.nist.gov/databases/Webtable.htm.
13. Dortmund Data Bank (DDB). http://www.ddbst.de.
14. PPDS Database. http://www.ppds.co.uk.
15. Poling, B. E., J. M. Prausnitz, and J. P. O'Connell, *The Properties of Gases and Liquids*, 5th ed., McGraw-Hill, New York, 2001.
16. Boublik, T., V. Fried, and E. Hala, *The Vapour Pressures of Pure Substances*, 2nd ed., Elsevier, Amsterdam, 1984.
17. *Vapor Pressure of Chemicals* (subvols. A, B, and C), in *Landolt-Börnstein: Numerical Data and Functional Relationships in Science and Technology—New Series*, vol. 4/20, Springer-Verlag, Berlin, 1999–2001. Some of these data are available in a NIST Standard Reference Database at http://www.nist.gov/srd/nist87.htm.
18. *Thermodynamic Properties of Organic Compounds and Their Mixtures* (subvols. B–J), in *Landolt-Börnstein: Numerical Data and Functional Relationships in Science and Technology—New Series*, vol. 4/8, Springer-Verlag, Berlin, 1996–2003.
19. Zábranskà, M., V. Rùzicka, V. Majer, and E. S. Domalski, *Heat Capacity of Liquids: Critical Review and Recommended Values*, American Institute of Physics, Melville, NY, 1996.
20. Zábranskà, M., V. Rùzicka, and E. S. Domalski, *J. Phys. Chem. Ref. Data* 30: 1199 (2001).
21. Domalski, E. S., and E. D. Hearing, *J. Phys. Chem. Ref. Data* 25: 1 (1996).
22. Majer, V., and V. Svoboda, *Enthalpies of Vaporization of Organic Compounds, a Critical Review and Data Compilation*, Blackwell Scientific, Oxford, 1985.
23. Twu, C. H., W. D. Sim, and V. Tassone, *Chem. Eng. Prog.* 98 (11): 58 (Nov. 2002).
24. *Beilstein's Handbook of Organic Chemistry*.
25. *Gmelin's Handbook of Inorganic and Organometallic Chemistry*.
26. Lide, D. R. (ed.), *CRC Handbook of Chemistry and Physics*, 86th ed., CRC Press, Boca Raton, FL, 2005.
27. Rowley, R. L., *Statistical Mechanics for Thermophysical Property Calculations*, Prentice Hall, Englewood Cliffs, NJ, 1994.
28. Stull, D. R., E. F. Westrum, and G. C. Sinke, *The Chemical Thermodynamics of Organic Compounds*, Wiley, New York, 1969.
29. Frenkel, M., G. J. Kabo, K. N. Marsh, G. N. Roganov, and R. C. Wilhoit, *Thermodynamics of Organic Compounds in the Gas State*, vols. 1 and 2, Thermodynamics Research Center, College Station, TX, 1994 (now distributed by CRC Press); available at http://www.nist.gov/srd/nist88.htm.

30. Chase, M. W., Jr., ed., *NIST-JANAF Thermochemical Tables*, 4th ed., American Institute of Physics, Melville, NY, 1998.
31. Society for Chemical Engineering and Biotechnology, *DECHEMA Chemistry Data Series*;. http://www.dechema.de/CDS-design-1.html.
32. Dymond, J. D., K. N. Marsh, and R. C. Wilhoit, *Virial Coefficients of Pure Gases and Mixtures* (subvols. A and B), in *Landolt-Börnstein: Numerical Data and Functional Relationships in Science and Technology—New Series*, vol. 4/21, Springer-Verlag, Berlin, 2002–2003.
33. Wisniak, J., and A. Tamir, *Mixing and Excess Thermodynamic Properties: A Literature Source Book*, Elsevier, Amsterdam, 1978; see also *Supplement 1*, Elsevier, Amsterdam, 1982; *Supplement 2*, Elsevier, Amsterdam, 1986.
34. *Densities of Liquid Systems* (subvols. A and B), in *Landolt-Börnstein: Numerical Data and Functional Relationships in Science and Technology—New Series*, vol. 4/1, Springer-Verlag, Berlin, 1974–1977.
35. Christensen, J. J., R. W. Hanks, and R. M. Izatt, *Handbook of Heats of Mixing*, Wiley, New York, 1982; J. J. Christensen, R. L. Rowley, and R. M. Izatt, *Handbook of Heats of Mixing (Supplementary Volume)*, Wiley, New York, 1988.
36. Prausnitz, J. M., R. N. Lichtenthaler, and E. Gomes de Azevedo, *Molecular Thermodynamics of Fluid-Phase Equilibria*, 3rd ed., Prentice Hall, Upper Saddle River, NJ, 1999.
37. Walas, S., *Phase Equilibria in Chemical Engineering*, Butterworth, Boston, 1985.
38. Sandler, S. I., *Chemical and Engineering Thermodynamics*, 3rd ed., Wiley, New York, 1998.
39. Elliott, J. R., and C. T. Lira, *Introductory Chemical Engineering Thermodynamics*, Prentice Hall, Upper Saddle River, NJ, 1999.
40. Smith, J. M., H. C. Van Ness, and M. M. Abbott, *Introduction to Chemical Engineering Thermodynamics*, 6th ed., McGraw-Hill, New York, 2001.
41. Anderko, A., in *Equations of State for Fluids and Fluid Mixtures* (J. V. Sengers, R. F. Kayser, C. J. Peters, and H. J. White, Jr., eds.), 75–126, Elsevier, Amsterdam, 2000.
42. Sandler, S. I., and H. Orbey, in *Equations of State for Fluids and Fluid Mixtures* (J. V. Sengers, R. F. Kayser, C. J. Peters, and H. J. White, Jr., eds.), 321–357, Elsevier, Amsterdam, 2000.
43. Thermische Verfahrenstechnik; available on-line at http://www.tu-harburg.de/vt2/pe2000/.
44. Horstmann, S., K. Fischer, and J. Gmehling, *Fluid Phase Equil.* 167: 173 (2000), and references therein.
45. Ahlers, J., and J. Gmehling, *Ind. Eng. Chem. Res.* 41: 5890 (2002).
46. Wittig, R., J. Lohmann, and J. Gmehling, *Ind. Eng. Chem. Res.* 42: 183 (2003), and references therein.
47. Klamt, A., and F. Eckert, *Fluid Phase Equil.* 172: 43 (2000); F. Eckert and A. Klamt, *Ind. Eng. Chem. Res.* 40: 2371 (2001); O. Spuhl and W. Arlt, *Ind. Eng. Chem. Res.* 43: 852 (2004); see also http://www.cosmologic.de.
48. Hayden, J. G., and J. P. O'Connell, *Ind. Eng. Chem. Process Des. Dev.* 14: 209 (1975).
49. International Data Series, Selected Data on Mixtures, Series A: Thermodynamic Properties of Non-Reacting Binary Systems of Organic Substances. Quarterly publication with annual index, 1973–present; see http://trc.nist.gov/IDS/ids.htm.
50. Gmehling, J., J. Menke, J. Krafczyk, and K. Fischer, *Azeotropic Data*, 3 vols., Wiley-VCH, Weinheim, 2004.
51. *IUPAC Solubility Data Series*. Over 80 volumes published beginning in 1979; a listing is given at http://www.iupac.org/publications/sds/volumes.html. Beginning with vol. 66 (1998), new volumes are published in the *Journal of Physical and Chemical Reference Data*.
52. IUPAC-NIST Solubility Database, http://srdata.nist.gov/solubility/.
53. AQUASOL Database; http://www.pharmacy.arizona.edu/outreach/aquasol/.
54. Yalkowsky, S. H., and Y. He, *Handbook of Aqueous Solubility Data*, CRC Press, Boca Raton, FL, 2003.
55. Fernández-Prini, R., J. L. Alvarez, and A. H. Harvey, *J. Phys. Chem. Ref. Data* 32: 903 (2003).
56. Linke, W. F., and A. Seidell, *Solubilities: Inorganic and Metal-Organic Compounds*, American Chemical Society, Washington, DC, vol. 1, 1958; vol. 2, 1965.
57. Millat, J., J. H. Dymond, and C. A. Nieto de Castro (eds.), *Transport Properties of Fluids: Their Correlation, Prediction and Estimation*, Cambridge University Press, Cambridge, 1996.
58. Huber, M. L., and H. J. M. Hanley, in *Transport Properties of Fluids: Their Correlation, Prediction and Estimation* (J. Millat, J. H. Dymond, and C. A. Nieto de Castro, eds.), pp. 283–295, Cambridge University Press, Cambridge, 1996.

59. Huber, M. L., NIST Thermophysical Properties of Hydrocarbon Mixtures Database (SUPERTRAPP), NIST Standard Reference Database 4, Version 3.1, National Institute of Standards and Technology, Gaithersburg, MD, 2003. http://www.nist.gov/srd/nist4.htm.
60. Stephan, K., and K. Lucas, *Viscosity of Dense Fluids*, Plenum Press, New York, 1979.
61. Viswanath, D. S., and G. Natarajan, *Data Book on the Viscosity of Liquids*, Hemisphere, New York, 1989.
62. Vargaftik, N. B., Y. K. Vinogradov, and V. S. Yargin, *Handbook of Physical Properties of Liquids and Gases: Pure Substances and Mixtures*, 3rd ed., Begell, New York, 1996.
63. Wohlfarth, C., and B. Wohlfarth, *Viscosity of Pure Organic Liquids and Binary Liquid Mixtures: Pure Organometallic and Organononmetallic Liquids and Binary Liquid Mixtures* (subvol. A), in *Landolt-Börnstein: Numerical Data and Functional Relationships in Science and Technology—New Series*, vol. 4/18, Springer-Verlag, Berlin, 2001.
64. Wohlfarth, C., and B. Wohlfarth, *Viscosity of Pure Organic Liquids and Binary Liquid Mixtures: Pure Organic Liquids* (subvol. B), in *Landolt-Börnstein: Numerical Data and Functional Relationships in Science and Technology—New Series*, vol. 4/18, Springer-Verlag, Berlin, 2002.
65. Gammon, B. E., K. N. Marsh, and A. K. R. Dewan, *Transport Properties and Related Thermodynamic Data of Binary Mixtures* (Parts 1–5), AIChE, New York, 1993–1998.
66. Mathias, P. M., V. S. Parekh, and E. J. Miller, *Ind. Eng. Chem. Res.* 41: 989 (2002).
67. Vargaftik, N. B., L. P. Filippov, A. A. Tarzimanov, and E. E. Totskii, *Handbook of Thermal Conductivity of Liquids and Gases*, CRC Press, Boca Raton, FL, 1994.
68. Cussler, E. L., *Diffusion: Mass Transfer in Fluid Systems*, 2nd ed., Cambridge University Press, Cambridge, 1997.
69. Wakeham, W. A., A. Nagashima, and J. V. Sengers, eds., *Measurement of the Transport Properties of Fluids*, vol. 3 of *Experimental Thermodynamics*, Blackwell Scientific, Oxford, 1991.
70. Pitzer, K. S., ed., *Activity Coefficients in Electrolyte Solutions*, 2nd ed., CRC Press, Boca Raton, FL, 1991.
71. Chen, C.-C., H. I. Britt, J. F. Boston, and L. B. Evans, *AIChE J.* 28: 588 (1982); C.-C. Chen and L. B. Evans, *AIChE J.* 32: 444 (1986).
72. Zemaitis, J. F., Jr., D. M. Clark, M. Rafal, and N. C. Scrivner, *Handbook of Aqueous Electrolyte Solutions*, AIChE, New York, 1986.
73. Hamer, W. J., and Y.-C. Wu, *J. Phys. Chem. Ref. Data* 1: 1047 (1972); B. R. Staples and R. L. Nuttall, *J. Phys. Chem. Ref. Data* 6: 385 (1977); R. N. Goldberg and R. L. Nuttall, *J. Phys. Chem. Ref. Data* 7: 263 (1978); R. N. Goldberg, R. L. Nuttall, and B. R. Staples, *J. Phys. Chem. Ref. Data* 8: 923 (1979); R. N. Goldberg, *J. Phys. Chem. Ref. Data* 8: 1005 (1979); R. N. Goldberg, *J. Phys. Chem. Ref. Data* 10: 1 (1981); R. N. Goldberg, *J. Phys. Chem. Ref. Data* 10: 671 (1981); B. R. Staples, *J. Phys. Chem. Ref. Data* 10: 765 (1981); B. R. Staples, *J. Phys. Chem. Ref. Data* 10: 779 (1981).
74. Zaytsev, I. D., and G. G. Aseyev, *Properties of Aqueous Solutions of Electrolytes*, CRC Press, Boca Raton, FL, 1992.
75. Lobo, V. M. M., *Handbook of Electrolyte Solutions* (Parts A and B), Elsevier, Amsterdam, 1990.
76. ELDAR Database; available on-line at http://www.uni-regensburg.de/Fakulta-eten/nat_Fak_IV/Physikalische_Chemie/Kunz/eldar/eldhp.html; also available through DECHEMA's DETHERM database at http://www.dechema.de/detherm-design-1.html.
77. SUPCRT database updates; available on-line at http://geopig.asu.edu/supcrt_data_base_updates.html.
78. Söhnel, O., and P. Novotny, *Densities of Aqueous Solutions of Inorganic Substances*, Elsevier, Amsterdam, 1985.
79. Marcus, Y., *Ion Properties*, Marcel Dekker, New York, 1997.
80. Lencka, M. M., A. Anderko, S. J. Sanders, and R. D. Young, *Int. J. Thermophys.* 19: 367 (1998).
81. Jiang, J., and S. I. Sandler, *Ind. Eng. Chem. Res.* 42: 6267 (2003).
82. McLaughlin, E., *Chem. Rev.* 64: 389 (1964).
83. Horvath, A. L., *Handbook of Aqueous Electrolyte Solutions*, Ellis Horwood, Chichester, 1985.
84. Wagman, D. D., W. H. Evans, V. B. Parker, R. H. Schumm, I. Halow, S. M. Bailey, K. L. Churney, and R. L. Nuttall, "The NBS Tables of Chemical Thermodynamic Properties," *J. Phys. Chem. Ref. Data* 11 (supp. 2): 1982.
85. ASTM, *Chemical Thermodynamic and Energy Release Evaluation—CHETAH*, computer program; available on-line at http://www.chetah.usouthal.edu.

86. McCullough, J. P., and D. W. Scott, eds., *Calorimetry of Non-reacting Systems*, vol. 1 of *Experimental Thermodynamics*, Butterworths, London, 1968.
87. Le Neindre, B., and B. Vodar, eds., *Experimental Thermodynamics of Non-reacting Fluids*, vol. 2 of *Experimental Thermodynamics*, Butterworths, London, 1975.
88. Marsh, K. N., and P. A. G. O'Hare, eds., *Solution Calorimetry*, vol. 4 of *Experimental Thermodynamics*, Blackwell Scientific, Oxford, 1994.
89. Goodwin, A. R. H., K. N. Marsh, and W. A. Wakeham, eds., *Measurement of the Thermodynamic Properties of Single Phases*, vol. 6 of *Experimental Thermodynamics*, Elsevier, Amsterdam, 2003.
90. Weir, R. D., and T. W. de Loos, eds., *Measurement of the Thermodynamic Properties of Multiple Phases*, vol. 7 of *Experimental Thermodynamics*, Elsevier, Amsterdam, 2005.
91. Armarego, W. L. F., and C. L. L. Chai, *Purification of Laboratory Chemicals*, 5th ed., Butterworth-Heinemann, Amsterdam, 2003.
92. Marsh, K. N., ed., *Recommended Reference Materials for the Realization of Physicochemical Properties*, Blackwell Scientific, New York, 1987.
93. Hemminger, W., and G Höhne, *Calorimetry: Fundamentals and Practice*, Verlag Chemie, Weinheim, 1984.
94. Raal, J. D., and A. L. Mühlbauer, *Phase Equilibria: Measurement and Computation*, Taylor & Francis, Bristol, PA, 1998.
95. Novák, J. P., J. Matou, and J. Pick, *Liquid-Liquid Equilibria*, vol. 7 of *Studies in Modern Thermodynamics*, Elsevier, Amsterdam, 1987.
96. Laesecke, A., T. O. D. Luddecke, R. F. Hafer, and D. J. Morris, *Int. J. Thermophys.* 20: 401 (1999).
97. Frenkel, M., Q. Dong, R. C. Wilhoit, and K. R. Hall, *Int. J. Thermophys.* 22: 215 (2001).

2 Mathematics in Chemical Engineering

Sinh Trinh, Nandkishor Nere, and Doraiswami Ramkrishna

CONTENTS

2.1 Introduction 39
2.2 Equations Encountered in Chemical Engineering 40
    2.2.1 Processes Governed by Difference Equations 42
        2.2.1.1 Scalar Difference Equations 42
        2.2.1.2 Vector Difference Equations 43
    2.2.2 Processes Governed by Linear Equations..44
        2.2.2.1 Steady-State Continuous Countercurrent Staged Extraction.....................44
        2.2.2.2 Steady-State First-Order Reactions in a Stirred Tank Reactor 45
    2.2.3 Processes Governed by Nonlinear Equations..45
        2.2.3.1 Concentration of a Species in a Chemical Reaction at Equilibrium 45
        2.2.3.2 Vapor-Liquid Equilibria 46
        2.2.3.3 Steady State of a Continuous Stirred-Tank Reactor 47
    2.2.4 Processes Governed by Ordinary Differential Equations..47
        2.2.4.1 Dynamics of a Continuous Stirred Tank Reactor......................................47
        2.2.4.2 Steady State of a Tubular Reactor 48
    2.2.5 Processes Governed by Partial Differential Equations..48
        2.2.5.1 Dynamics of a Tubular Reactor 48
        2.2.5.2 Dynamics of Chromatography...49
    2.2.6 Processes Governed by Integral Equations..50
        2.2.6.1 Particle Size Distribution in Continuous Comminution Process..............50
        2.2.6.2 Determination of Pore Size Distribution in Porous Media.......................50
        2.2.6.3 Integral Equations Resulting Out of the Boundary-Value Problems
                with Mixed Derivative Boundary Condition 51
    2.2.7 Processes Governed by Integrodifferential Equations 52
    2.2.8 Processes Governed by Stochastic Differential Equations......................................52
2.3 The Number System...53
    2.3.1 The Real Number System 53
        2.3.1.1 Powers and Roots...53
        2.3.1.2 Logarithm 54
    2.3.2 Sequences and Series 54
    2.3.3 Tests for Convergence of Sequence and Series 55
        2.3.3.1 Integral Test..55
        2.3.3.2 Ratio Test 56
        2.3.3.3 Root Test 56
        2.3.3.4 Comparison Test...56
        2.3.3.5 Limit Comparison Test 57
    2.3.4 Taylor Series 57

		2.3.5	Binomial Theorem 58
		2.3.6	Inequalities 58
			2.3.6.1 Algebraic Inequalities 58
			2.3.6.2 Integral Inequalities 59
	2.4	Differential and Integral Calculus 60
		2.4.1	Functions, Limits, and Continuity 60
		2.4.2	The Derivative 61
			2.4.2.1 An Example Application of the Derivatives 62
		2.4.3	The Mean Value Theorem 62
			2.4.3.1 An Example Application of the Mean Value Theorem 62
		2.4.4	L'Hôspital's Rule 63
		2.4.5	Implicit Function Theorem 64
		2.4.6	The Integral 64
			2.4.6.1 Improper Integrals 65
			2.4.6.2 An Example Application of the Integrals 66
	2.5	Vector Analysis 66
		2.5.1	Vector Algebra 66
		2.5.2	Vector Calculus 68
		2.5.3	Orthogonal Curvilinear Coordinate Systems 69
			2.5.3.1 Scale Factors and Metric Tensors 69
			2.5.3.2 Differential Operators in Curvilinear Coordinate System 70
			2.5.3.3 Circular Cylindrical Coordinates 71
			2.5.3.4 Spherical Coordinates 71
			2.5.3.5 Elliptic Cylindrical Coordinates 72
			2.5.3.6 Prolate Spheroidal Coordinates 72
			2.5.3.7 Oblate Spheroidal Coordinates 73
			2.5.3.8 Parabolic Cylinder Coordinates 73
			2.5.3.9 Parabolic Coordinates 74
			2.5.3.10 Conical Coordinates 74
			2.5.3.11 Ellipsoidal Coordinates 75
			2.5.3.12 Paraboloidal Coordinates 75
			2.5.3.13 Bispherical Coordinates 76
			2.5.3.14 Toroidal Coordinates 76
		2.5.4	Vector Integral Theorems 76
		2.5.5	Gradients of Sum and Product 77
	2.6	Dimensional Analysis 78
		2.6.1	Theory of Dimensional Analysis 78
		2.6.2	Applications of Dimensional Analysis 79
	2.7	Algebraic Equations 81
		2.7.1	System of Linear Equations 81
			2.7.1.1 Solution of Linear Equations 83
		2.7.2	Nonlinear Equations 85
			2.7.2.1 Polynomial Equations 85
			2.7.2.2 Numerical Solutions of Nonlinear Equations 88
	2.8	Difference Equations 92
		2.8.1	Method of Solution for Homogeneous Equations 92
		2.8.2	Method of Solution for Inhomogeneous Equations 92
		2.8.3	Numerical Solutions to Ordinary Differential Equations 94
			2.8.3.1 Explicit Methods 96
			2.8.3.2 Implicit Methods 99

2.9	Ordinary Differential Equations	101
	2.9.1 Linear First-Order Differential Equation	101
	2.9.2 Nonlinear First-Order Differential Equation	101
	2.9.2.1 Autonomous Nonlinear Equation	101
	2.9.2.2 Implicit Equation	101
	2.9.2.3 Lagrange Equation	102
	2.9.2.4 Separable Equation	103
	2.9.2.5 Bernoulli Equation	103
	2.9.2.6 Exact Differential Equations	104
	2.9.2.7 Homogeneous Equations	104
	2.9.2.8 Ricatti Equations	105
	2.9.3 Second-Order Differential Equations	106
	2.9.3.1 Homogeneous Linear Equations with Constant Coefficients	106
	2.9.3.2 Inhomogeneous Linear Differential Equations with Constant Coefficients	107
	2.9.3.3 Green's Function	108
	2.9.3.4 Green's Function by Eigenfunction (Mercer's) Expansions	110
	2.9.4 Linear Higher-Order Differential Equations	111
	2.9.4.1 Homogeneous Equation	111
	2.9.4.2 Inhomogeneous Equation	112
	2.9.4.3 Application of Higher Order Equations	112
	2.9.4.4 Finite Element Method	114
2.10	Partial Differential Equations	115
	2.10.1 First-Order Partial Differential Equations	115
	2.10.2 Classification of Second-Order Equations	118
	2.10.3 Parabolic Equations	119
	2.10.3.1 Separation of Variables	119
	2.10.3.2 Inhomogeneous Equation, Duhamel's Principle	120
	2.10.3.3 Inhomogeneous Boundary Conditions	122
	2.10.3.4 Similarity Solutions	122
	2.10.3.5 Inhomogeneous Equation in Infinite Domain	123
	2.10.4 Hyperbolic Equations	124
	2.10.4.1 d'Alembert's Solution	124
	2.10.4.2 Separation of Variables	126
	2.10.5 Elliptic Equations	128
	2.10.5.1 Poisson Integral Formula	129
	2.10.5.2 Green's Function	130
	2.10.6 Computational Fluid Mechanics	131
2.11	Integral Equations	131
	2.11.1 Volterra Integral Equations	131
	2.11.2 Methods of Solution for Volterra Equations	132
	2.11.2.1 Kernel Is Only a Function of Dependent Variable	132
	2.11.2.2 Separable Kernels	133
	2.11.2.3 Degenerate (Finite Rank) Kernels	133
	2.11.2.4 Difference Kernels	133
	2.11.2.5 Method of Resolvent Kernels	134
	2.11.2.6 The Generalized Abel Integral Equation	135
	2.11.2.7 Numerical Solution of Volterra Integral Equations of the Second Kind	135
	2.11.3 Fredholm Integral Equations	136

   2.11.4 Methods of Solution for Fredholm Equations 136
     2.11.4.1 Degenerate Kernels 136
     2.11.4.2 Method of Fredholm Resolvent Kernels 137
     2.11.4.3 Method of Iterated Kernels 138
     2.11.4.4 Symmetric Kernels 139
     2.11.4.5 Numerical Solution of Nonhomogeneous Fredholm Equation
        of the Second Kind 139
     2.11.4.6 Solution of Ill-Posed Fredholm Equations of the First Kind 140
     2.11.4.7 Method of Regularization 142
2.12. Complex Variables 143
  2.12.1 Properties of Complex Numbers 143
  2.12.2 Analytic Functions 144
     2.12.2.1 Elementary Functions 147
     2.12.2.2 Multivalued Functions 147
     2.12.2.3 Logarithmic Functions 148
     2.12.2.4 Cauchy Integral Formula 148
     2.12.2.5 Laurent Series 149
  2.12.3 Residue Theorem 149
     2.12.3.1 Calculus of Residues 149
     2.12.3.2 Application of Complex Integration 150
  2.12.4 Argument Principle and Rouché Theorem 151
  2.12.5 Conformal Mapping 152
2.13 Integral Transforms 155
  2.13.1 Laplace Transform 156
     2.13.1.1 Convolution Property 156
     2.13.1.2 Application of Laplace Transform 156
  2.13.2 Fourier Transform 157
     2.13.2.1 Convolution Property 157
     2.13.2.2 Application of Fourier Transform 158
     2.13.2.3 Application of Fourier Sine Transform 159
     2.13.2.4 Application of Fourier Cosine Transform 160
  2.13.3 Mellin Transform 160
     2.13.3.1 Convolution Property 160
     2.13.3.2 Application of Mellin Transform 161
  2.13.4 Hankel Transform 162
     2.13.4.1 Property of Hankel Transform 162
     2.13.4.2 Application of Hankel Transform 162
2.14 Calculus of Variations 163
  2.14.1 Euler-Lagrange Differential Equation 163
  2.14.2 Euler-Lagrange Equations for a Functional of n-Dependent Variables 164
  2.14.3 Euler-Lagrange Equations for a Functional Involving n-Order Derivative 165
  2.14.4 Euler-Lagrange Equations for a Functional of Two Independent Variables 165
  2.14.5 Application of Calculus of Variations 165
2.15 Stochastic Differential Equations 166
  2.15.1 Differential Chapman-Kolmogorov Equation 167
  2.15.2 Connection between the Fokker-Planck Equation and Stochastic
     Differential Equation 167
  2.15.3 Îto Stochastic Integral 168
     2.15.3.1 One-Dimensional Îto Formula 169
     2.15.3.2 An Example Application of a Stochastic Differential Equation 169
2.16 Asymptotic Approximations and Expansions 170

2.17 Steady-State Multiplicity and Stability...173
    2.17.1 Steady-State Multiplicity of a Chemical Reactor...173
        2.17.1.1 Steady-State Multiplicity of CSTR...173
        2.17.1.2 Steady-State Multiplicity of a Tubular Reactor 174
    2.17.2 Analysis of Multiplicity by Singularity Theory 176
        2.17.2.1 Application of Singularity Theory to the Analysis of CSTR 177
    2.17.3 Stability of Steady-State Solution...179
        2.17.3.1 Linear Stability Analysis..179
        2.17.3.2 Method of Lyapunov's Function..180
Appendix 2.A: Mathematical Software 182
Appendix 2.B: Dirac Delta Function..184
Appendix 2.C: Tables of Integral Transforms 185
References 191

2.1 INTRODUCTION

Chemical engineering is concerned with the study of systems in which matter undergoes changes in composition, energy, and morphological structure brought about by physicochemical processes spanning a wide spectrum of spatio-temporal scales. The methodology of chemical engineering consists of analysis and synthesis.

1. Analysis comprises mainly
   Observation aided by all available tools of scientific instrumentation
   Investigation based on the use of physical conservation principles together with phenomenological equations of material behavior within the framework of continuum and statistical mechanics, chemical kinetics, and thermodynamics
2. Synthesis comprises
   Design and control of processes using the methods of optimization, control theory, and artificial intelligence
   Design of products of chemical origin using the methods of molecular theory, artificial intelligence, and practical experience so as to satisfy specified physicochemical properties

Because of the complexity of chemical engineering systems, the use of theory is often admixed with economic use of empiricism founded on the methods of dimensional analysis. With increasingly available software for symbolic manipulations and numerical computations, it is not difficult to obtain solutions to mathematical problems. As a result, a large amount of tedious and detailed calculation is often relegated to such software programs, which may be insensitive to specific oddities of the physical process. Consequently, the software can only provide correct solutions if the problems are correctly formulated.

The correct formulation of the physical problem means that the resulting mathematical problem be well-posed. Well-posedness requires that a unique solution exists and that it be continuous with respect to small perturbations in the parameters of the governing equations, which are due to uncertainties in measurements and fluctuations in operations. These requirements are not always satisfied in reality.

From the engineering standpoint, existence of a solution is relegated to finding it either analytically or computationally. In the latter case, care should be exercised to ensure that results of computation are bona fide solutions to the problems, since it has been documented (Skufca, 2004) that numerical instability in a well-tested initial-value problem solver, Runge-Kutta-Felberg, can induce chaos under certain conditions.

Uniqueness of a solution is not assured for a certain class of problems, particularly nonlinear steady-state problems. Thus one frequently encounters multiple solutions, requiring strategies for finding them.

When multiple steady states exist, the stability of these solutions and the startup condition must be contemplated to ensure that the desired steady state is attained.

Continuity with respect to small perturbations in the governing equations is more difficult to ascertain. It can arise in problems where the mathematical equations are structurally unstable. Typical examples of these problems are systems undergoing bifurcation. In structurally unstable systems, small perturbations in system parameters lead to large changes in the solution. For example, slight perturbations can alter the stability characteristics of the solutions. In addition to these structurally unstable systems, there is a class of engineering problems that is sensitive with respect to small errors due to, perhaps, uncertainties in measurements or other fluctuations in the surrounding environments. These problems arise in determining unknown functions of interest from experimental data. They are known broadly as inverse problems.

A particular class of problems whose solutions are sensitive to initial conditions is known as chaotic problems. The phenomenon of chaos has been observed in fluid mechanics, chemical reactions, and biological systems (Cvitanocic, 1987). A special feature of these systems is unpredictability. The chaotic solutions are so sensitive to initial conditions that two systems with minute differences in their initial states can eventually diverge from each other. Thus their long-term dynamics are unpredictable.

The foregoing issues are clearly important, since understanding them can affect the engineering designs and operations, but they have not received adequate attention from practicing engineers. In this chapter, various mathematical tools for investigation and computation of chemical engineering problems are introduced to help the practicing engineer to adapt to the modern approach to the design and operation of chemical engineering equipment. The special features of this chapter are to bring these issues to the attention of the readers and address some of them in sections where they are encountered using the techniques currently available in the literature. Special focus is placed on (a) simple analyses of equations to ensure that numerical solutions are reliable and (b) the formulation of a class of ill-posed problems, the inverse problems.

The specific objectives of this chapter are to

1. Sensitize the reader to the issues of well-posedness of a mathematical problem
2. Provide concise coverage of solution techniques in linear and nonlinear problems and provide directions for more detailed treatment
3. Introduce the engineer briefly to symbolic and numerical software for solving mathematical problems of interest. It should be noted that we have not cited literature in the text, however, pertinent references are included in the bibliography.

The organization of this chapter preserves the hierarchy of topics generally followed in the applied mathematics literature. However, the reader is encouraged to refer to Table 2.1 for general guidelines on what chemical engineering activity would lead to each of the itemized sections.

2.2 EQUATIONS ENCOUNTERED IN CHEMICAL ENGINEERING

The mathematical equations arising in chemical engineering are obtained by applying the conservation principles of mass, energy, and momentum to chemical reaction systems. These equations express the accumulation of mass, energy, or momentum in the system as a result of the combined effects of net fluxes across the boundary and the consumption or production of these quantities in the system. These equations can be classified as algebraic equations, ordinary differential equations, partial differential equations, integral equations, and integrodifferential equations. The reader is referred to various free Web sources (e.g., http://en.wikipedia.org/wiki/Trigonometry, etc.) for the details pertaining to the trigonometry that has not been covered in this chapter.

Algebraic equations arise in systems that are independent of space and time variations, and ordinary differential equations govern systems that either vary with space or vary with time. In

TABLE 2.1
Chemical Engineering Activities Related to Mathematics Employed and Relevant Sections in Chapter 2

Chemical Engineering Activity	Mathematics Employed	Section
1. Primer to mathematics	Number system	2.3
	Differential and integral calculus	2.4
	Vector analysis	2.5
	Complex variables	2.12
	Integral transforms	2.13, Appendix 2.C
	Calculus of variations	2.14
	Linear algebra	2.7
2. Material and energy balances (filtration, drying, crystallization, extraction, distillation, and absorption)	Algebraic equations	2.7
	Ordinary differential equations	2.9
	Partial differential equations	2.7, 2.9, 2.10, 2.11
	Algebraic equations	2.7
	Difference equations	2.8
3. Reactors and reactor analysis (gas-, liquid-, and solid-phase reactors; packed bed; solid catalyst; fluid liquid bed; liquid liquid bed; ion exchange; etc.)	Algebraic equations	2.7
	Ordinary differential equations	2.9
	Partial differential equations	2.10
	Integrodifferential equations	2.7, 2.9, 2.10, 2.11
	Stability analysis	2.17
	Multiplicity of steady states	2.17
	Calculus of variations	2.14
	Stochastic differential equations	2.15
	Asymptotic approximations and expansions	2.16
4. Fluid flow	Ordinary differential equations	2.9
	Partial differential equations	2.10
	Integral equations	2.7, 2.11
	Integrodifferential equations	2.7, 2.9, 2.10, 2.11
5. Scale-up	Dimensional analysis	2.6
6. Thermodynamics	Differential and integral calculus	2.4
	Algebraic equations	2.7
	Ordinary differential equations	2.9
	Partial differential equations	2.10

other words, algebraic equations govern steady-state systems that are spatially uniform. Examples are steady-state equations for a continuous stirred-tank reactor (CSTR).

Differential equations are equations that contain the derivatives of the unknown functions. They must be supplemented with auxiliary conditions to completely specify a problem. Auxiliary conditions must be prescribed at one or more points in the domain of the independent variables representing the boundary of the domain interface between different regions, and so on. Those equations with prescribed conditions at one point are called initial-value problems, and those with prescribed conditions on the boundary of the domain are appropriately called boundary-value problems. Initial-value problems generally govern the dynamics of the systems, while boundary-value problems describe the systems in steady state.

Ordinary differential equations govern systems that vary either with time or space, but not both. Examples are equations that govern the dynamics of a CSTR or the steady state of tubular reactors. Both the dynamics of a CSTR and the steady state of a plug-flow reactor are governed by first-order ordinary differential equations with prescribed initial conditions. The steady-state tubular reactors with axial dispersion are governed by a second-order differential equation with the boundary conditions specified at the entrance and exit of the reactor (boundary-value problems). The existence and uniqueness of a solution for an initial-value problem are guaranteed by a general theorem. Partial differential equations arise in steady-state systems with variations in more than one spatial coordinate. Unsteady-state systems may include temporal variations. Transient equations for tubular reactors are examples of partial differential equations.

Integral equations are equations that contain the integral of the unknown functions. There are two types of integral equations: Volterra and Fredholm integral equations. Fredholm equations feature integrals with *fixed* limits, while Volterra equations have integrals in which the limits of integration are the independent variables. These equations can be obtained by direct formulation or by the reduction of differential equations. Volterra equations are reduced from initial-value problems, and Fredholm equations from boundary-value problems.

Integrodifferential equations involve both derivatives and integrals of the unknown functions. These equations arise in radiative transport problems in weakly absorbing media and naturally arise in the modeling of particulate systems. For more information, see the monograph by Ramkrishna (2000).

There are many systems that can fluctuate randomly in space and time and cannot be described by deterministic equations. For example, Brownian motion of small particles occurs randomly because of random collisions with molecules of the medium in which the particles are suspended. It is useful to model such systems with what are known as stochastic differential equations. Stochastic differential equations feature "noise" terms representing the behavior of random elements in the system. Other examples of stochastic behavior arise in chemical reaction systems involving a small number of molecules, such as in a living cell or in the formation of particles in emulsion drops, and so on. A useful reference on stochastic methods is Gardiner (2003).

The different equations encountered in mathematical modeling can be further classified as linear and nonlinear equations. Linear equations arise in systems where the unknowns in the equations are present in the first power. Linear equations enjoy the principle of superposition, i.e., the sum of the solutions is also a solution of the equations. Linearity allows the original problem to be partitioned into simpler component problems that can be solved separately and superimposed to obtain the solution to the original problem.

In what follows, we present representative chemical engineering applications arising in each of the foregoing classes of equations so as to familiarize the reader with their scope. Introductory discussion is deferred to the individual sections covering each topic.

2.2.1 Processes Governed by Difference Equations

We may classify the difference equations encountered in chemical engineering applications into two types, namely, scalar and vector difference equations. The unknowns in difference equations are functions of discrete independent variables.

2.2.1.1 Scalar Difference Equations

Dissolution of a species: We consider the dynamics of dissolution of a solid mounted on a revolving assembly to show that the mathematical description of its dissolution gives rise to a simple scalar difference equation.

Let x_0 = initial concentration of the dissolved solid in the solution; x_s = solubility of the solid species in a solvent; x_n = concentration of dissolved species at time $t_n = (\Delta t)n$, where n stands for the number of discrete time steps, (Δt). For dissolution to occur, we must of course assume that $x_0 < x_s$. The rate law for the concentration of dissolved solid in the solution follows the equation

$$\frac{x_{n+1} - x_n}{\Delta t} = K_{La}(x_s - x_n) \qquad (2.1)$$

where K_{La} is the overall mass transfer coefficient (s^{-1}), which can be determined experimentally for a given speed of rotation and a fixed area of exposure. This equation can be put in the form

$$y_{n+1} = y_n(1-\lambda) + \lambda \tag{2.2}$$

where

$$y_n = \frac{x_n}{x_s}, \quad \lambda = K_{La}\Delta t \tag{2.3}$$

and

$$0 < \lambda \ll 1 \tag{2.4}$$

Thus Equation (2.2) represents the dissolution process in terms of a scalar difference equation that has a straightforward solution.

2.2.1.2 Vector Difference Equations

We present here how vector difference equations arise in one of the most classical chemical engineering operations, i.e., distillation. Methods to solve the difference equations are presented in subsection 2.8.

Multicomponent rectification: Consider a multicomponent mixture of m-species continuously fed into a distillation column (Acrivos and Amundson, 1955; Amundson, 1966; Ramkrishna and Amundson, 1985). Let $x_{n,i}$ and $y_{n,i}$ be the compositions of the ith species on the nth plate for the liquid phase and the vapor phase, respectively. Based on constant molal overflow of liquid with a downflow rate L and a vapor upward flow rate V, the steady-state mass balance for the ith species in the rectifying section above the nth plate leads to the equation

$$y_{n+1,i} = \frac{R}{R+1}x_{n,i} + \frac{x_{0,i}}{R+1}, \quad i = 1,2,\ldots,m \tag{2.5}$$

where $R \equiv L/(V-L)$ is the reflux ratio and $x_{0,i}$ is the known mole fraction of the ith species in the product. The standard assumption of equilibrium between the liquid and the vapor leaving the plate is expressed by a relationship of the form

$$x_{n,i} = \frac{p_i y_{n,i}}{\sum_{j=1}^{m} p_j y_{n,j}}, \quad p_i > 0 \tag{2.6}$$

where p_i ($i = 1,2,3,\ldots,m$) are constants representing relative volatilities. The relationship satisfies the requirement that the mole fractions must be summed to unity. To solve for the nth plate composition, it is necessary to calculate the composition on all the intermediate plates.

The following procedure is useful to calculate the composition just above the feed plate. Let $\{X_{n,i}\}$ be defined iteratively as

$$X_{n,i} = Rp_i X_{n-1,i} + p_i \sum_{j=1}^{m} X_{n-1,j} x_{0,j} \tag{2.7}$$

Also, we have

$$X_{0,i} = p_i, \quad i = 1, 2, \ldots, m \tag{2.8}$$

If $X_{n,i}$ are known, the original variables $x_{n,i}$ are then recovered from

$$x_{n,i} = \frac{X_{n,i} x_{0,i}}{\sum_{j=1}^{m} X_{n,j} x_{0,j}} \tag{2.9}$$

To solve for $X_{n,i}$, it is convenient to define the vector $\mathbf{X}_n = (X_{n,1}, X_{n,2}, \ldots, X_{n,m})$ and the matrix operator A, whose elements are given by

$$a_{ij} \equiv R p_i \delta_{ij} + p_i x_{0,j} \tag{2.10}$$

$\mathbf{X}_n$ is then satisfied by a difference equation of the form

$$\mathbf{X}_n = \mathbf{A} \mathbf{X}_{n-1} \tag{2.11}$$

The solution of Equation (2.11) is given by

$$\mathbf{X}_n = \mathbf{A}^n \mathbf{X}_0 \tag{2.12}$$

where $\mathbf{X}_0 = (p_0, p_1, \ldots, p_m)$. The solution of the vector difference Equation (2.12) is expressed in terms of the eigenvalues of the eigenvectors of the matrix $\mathbf{A}$ (for details, see Amundson, 1966 or Ramkrishna and Amundson, 1985). Similarly, one can obtain a vector difference equation for the stripping section.

2.2.2 Processes Governed by Linear Equations

2.2.2.1 Steady-State Continuous Countercurrent Staged Extraction

Consider a sequence of n equilibrium extraction stages, which are countercurrently fed with the extracting solvent phase entering stage 1, and the raffinate stream entering stage n (Ramkrishna and Amundson, 1985). The flow rates of the solute-free extract and raffinate are denoted as S and R, respectively. The solute concentration in the jth stage measured in mole ratios is Y_j in the extract phase and X_j in raffinate. The equilibrium relationship between Y and X is assumed to be $Y = KX$. The steady-state mass balance is given by the following equations:

$$\text{Stage 1: } SY_0 - (SK + R)X_1 + RX_2 = 0 \tag{2.13}$$

$$\text{Stages 2, 3, \ldots, } n-1: SKX_{j-1} - (SK + R)X_j + RX_{j+1} = 0, \; j = 2, 3, \ldots, n-1 \tag{2.14}$$

$$\text{Stage } n: SKX_{n-1} - (SK + R)X_n + RX_{n+1} = 0 \tag{2.15}$$

which are clearly the linear equations that can be concisely written as

$$Ax = b$$

where

$$A = \begin{bmatrix} \beta_1 & \gamma_1 & 0 & 0 & \cdots & 0 & 0 \\ \alpha_2 & \beta_2 & \gamma_2 & 0 & \cdots & 0 & 0 \\ \vdots & & & \cdots & & & \vdots \\ 0 & 0 & \cdots & 0 & \alpha_{n-1} & \beta_{n-1} & \gamma_{n-1} \\ 0 & 0 & \cdots & 0 & 0 & \alpha_n & \beta_n \end{bmatrix} \quad (2.16)$$

and

$$\mathbf{x} = (X_1, X_2, \ldots, X_{n+1})^T \quad \mathbf{b} = (-SY_0, 0, \ldots, -RX_{n+1})^T \quad \alpha_j = SK \quad \beta_j = -(SK + R) \quad \gamma_j = R$$

2.2.2.2 Steady-State First-Order Reactions in a Stirred-Tank Reactor

Consider the sequence of reversible first-order reactions (Ramkrishna and Amundson, 1985)

$$A_1 \underset{k_1'}{\overset{k_1}{\rightleftarrows}} A_2 \underset{k_2'}{\overset{k_2}{\rightleftarrows}} \cdots \underset{k_{n-1}'}{\overset{k_{n-1}}{\rightleftarrows}} A_n$$

in the continuous stirred-tank reactor. Let F be the volumetric flow rate of the feed, V be the volume of the reactor, and $\mathbf{x}_f$ and $\mathbf{x} = (x_1, \ldots, x_n)$ are the vectors comprising the mole fractions of species A_i in the feed and exit streams, respectively. The steady-state mass balance can be concisely represented by the set of linear equations as

$$\mathbf{Kx} + \frac{F}{V}(\mathbf{x}_f - \mathbf{x}) = 0 \quad (2.17)$$

where the rate constants, i.e., the elements k_{ij} of the matrix $\mathbf{K}$, are given by

$$k_{1,1} = -k_1, \quad k_{1,2} = k_1', \quad k_{n-1,n} = k_{n-1}, \quad k_{n,n} = -k_{n-1}'$$

$$k_{ij} = k_{i-1}\delta_{i-1,j} - (k_{i-1}' + k_i)\delta_{i,j} + k_i'\delta_{i,j-1}$$

2.2.3 Processes Governed by Nonlinear Equations

2.2.3.1 Concentration of a Species in a Chemical Reaction at Equilibrium

Consider the following equilibrium chemical reaction wherein the reactants are present in stoichiometric proportion:

$$N_2 + 3H_2 \rightleftarrows 2NH_3$$

Let the total pressure of the gases be P_t, and let x be the equilibrium partial pressure of NH_3 whose value is desired at equilibrium:

$$P_{N_2} + P_{H_2} + P_{NH_3} = P_t$$

By stoichiometry $P_{H_2} = 3P_{N_2}$. Hence

$$P_{N_2} = \frac{1}{4}(P_t - x), \quad P_{H_2} = \frac{3}{4}(P_t - x)$$

The equilibrium constant for the chemical reaction is given by

$$\frac{P_{NH_3}^2}{P_{N_2} P_{H_2}^3} = K$$

Substituting the partial pressure in terms of equilibrium partial pressure gives

$$\frac{256 x^2}{27(P_t - x)^4} = K$$

or

$$x = \frac{3}{16}\sqrt{3K}(P_t - x)^2$$

that is a quadratic equation in x.

2.2.3.2 Vapor-Liquid Equilibria

Consider one mole per hour of a stream consisting of n volatile liquids with known compositions, and $x_{f,i}$ to be continuously separated into vapor and liquid streams at a given temperature and pressure (Reklaitis, 1983). It is desired to determine the steady-state flow rates of the vapor stream and of the liquid stream and their compositions. Let K_i be the vapor-liquid equilibrium constants, $K_i = y_i/x_i$, where x_i and y_i are the liquid and vapor fractions, respectively. K_i is calculated from Raoult's law, $K_i = p_i(T)/P$, where p_i is the vapor pressure obtained from the Antoine equation. The flow rate of vapor stream, V, is obtained by solving the following nonlinear equation resulting out of the material balance on each of the species:

$$\sum_{i=1}^{n} \frac{x_{f,i}(1 - K_i)}{1 + V(K_i - 1)} = 0 \tag{2.18}$$

The flow rate of the liquid stream is

$$L = 1 - V$$

The mole fractions are obtained from

$$x_i = \frac{x_{f,i}}{1 + V(K_i - 1)} \tag{2.19}$$

and the mole fractions in the vapor phase are given by

$$y_i = K_i x_i$$

2.2.3.3 Steady State of a Continuous Stirred-Tank Reactor

Consider a general chemical reaction occurring in a single phase and transforming a particular reactant into products in a CSTR having volume V, with an inlet volumetric flow rate F of the feed at temperature T_f containing reactant at concentration C_f. The reactor is cooled by a heat exchanger whose heat-transfer coefficient and surface area are U and A, respectively. The temperature of the coolant is maintained at T_c. Let C_p be the capacity of the reaction mixture, ρ be the mixture density, $r(C,T)$ be the reaction rate per unit volume of the reactor, and $-\Delta H$ be the heat of reaction. The steady-state mass and energy balances in the CSTR are given by the following equations:

$$F(C_f - C) - Vr(C,T) = 0 \qquad (2.20)$$

$$\rho C_p F(T_f - T) - UA(T - T_c) + (-\Delta H)Vr(C,T) = 0 \qquad (2.21)$$

The above system of nonlinear equations can be combined into a single equation by eliminating the reaction term to obtain T in terms of C:

$$T = \frac{F(-\Delta H)(C_f - C)}{\rho C_p F + UA} + \frac{\rho C_p F T_f + UAT_c}{\rho C_p F + UA} \qquad (2.22)$$

Substituting T into the mass balance equation yields a nonlinear equation in C.

2.2.4 Processes Governed by Ordinary Differential Equations

2.2.4.1 Dynamics of a Continuous Stirred-Tank Reactor

We consider here the same continuous reactor of Section 2.2.3.3 under unsteady-state conditions. Assuming that all temperature dependent parameters remain constant at some average values, the mass and energy balances in the CSTR are governed by the following equations:

$$\frac{dC_A}{dt} = \frac{F}{V}(C_f - C_A) - r(C,T) \qquad (2.23)$$

$$\frac{dT}{dt} = \frac{F}{V}(T_f - T) - \frac{UA}{\rho C_p V}(T - T_c) + \frac{(-\Delta H)}{\rho C_p} r(C,T) \qquad (2.24)$$

$$C(0) = C_0$$

$$T(0) = T_0$$

The above equations constitute a system of first-order nonlinear differential equations whose steady-state versions are given in the previous section.

2.2.4.2 Steady State of a Tubular Reactor

Consider the reaction in Section 2.2.3.3 occurring in a homogeneous tubular reactor of length l, cross-sectional area A, and the perimeter P. The reactant is fed into the reactor with velocity v, which is assumed to be uniform over the tube cross section. The constant mass-dispersion coefficient and the thermal conductivity are denoted by D_m and k, respectively. Heat exchange takes place through the wall of the reactor maintained at the coolant temperature of T_c with the heat-transfer coefficient U. We assume here no radial variation of temperature, i.e., $T \equiv T(z)$. The mass and energy balance are given by the following differential equations:

$$D_m \frac{d^2C}{dz^2} - v\frac{dC}{dz} - r(C,T) = 0, \quad 0 < z < l \tag{2.25}$$

$$k\frac{d^2T}{dz^2} - v\rho C_p \frac{dT}{dz} - \frac{UP}{A}(T-T_c) + (-\Delta H)r(C,T) = 0, \quad 0 \leq z < l \tag{2.26}$$

The boundary conditions are

$$-D_m \frac{dC}{dz} = v(C_f - C), \quad z = 0$$

$$\frac{dC}{dz} = 0, \quad z = l$$

$$-k\frac{dT}{dz} = \rho C_p v(T_f - T), \quad z = 0$$

$$\frac{dT}{dz} = 0, \quad z = l$$

For an adiabatic reactor, i.e., $U = 0$, T can be solved in terms of C, and these equations are combined into one nonlinear differential equation to be solved for C.

2.2.5 Processes Governed by Partial Differential Equations

2.2.5.1 Dynamics of a Tubular Reactor

We reflect on the more involved case of a tubular reactor described in Section 2.2.4.2, where we will consider the radial variation of temperature in addition to its axial variation. Thus $T \equiv T(r,z)$. Furthermore, we also assume that the diffusion coefficient D_m is constant. Consequently, the mass and energy balance are given by the following partial differential equations:

$$\frac{DC}{Dt} = D_m \nabla^2 C - r(C,T) = 0, \quad 0 < z < l, \; 0 < r < R \tag{2.27}$$

$$\rho C_p \frac{DT}{Dt} = k\nabla^2 T + (-\Delta H)r(c,T) = 0, \quad 0 < z < l, \; 0 < r < R \tag{2.28}$$

where

$$\nabla^2 = \frac{1}{r}\frac{\partial}{\partial r}\left(r\frac{\partial}{\partial r}\right) + \frac{\partial^2}{\partial z^2}$$

$$\frac{D}{Dt} = \frac{\partial}{\partial t} + \boldsymbol{v}\cdot\nabla$$

$$\boldsymbol{v} = v_r\boldsymbol{\delta}_r + v_z\boldsymbol{\delta}_z$$

and

$$\nabla = \boldsymbol{\delta}_r\frac{\partial}{\partial r} + \boldsymbol{\delta}_z\frac{\partial}{\partial z}$$

The boundary conditions are given by the following equations:

$$-D_m\frac{\partial C}{\partial z} = v_z(C_f - C),\ z = 0$$

$$\frac{\partial C}{\partial z} = 0,\ z = l$$

$$\frac{\partial C}{\partial r} = 0,\ r = R$$

$$-k\frac{\partial T}{\partial z} = \rho C_p v_z(T_f - T),\ \text{at } z = 0$$

$$\frac{\partial T}{\partial z} = 0,\ z = l$$

$$-k\frac{\partial T}{\partial r} = U(T - T_w),\ r = R$$

where C_f and T_f are the concentration of a species in a feed and the temperature of the feed, respectively, while T_w is the wall temperature.

2.2.5.2 Dynamics of Chromatography

Chromatographic separation is achieved by selective adsorption of chemical species in a packed column. Consider a one-dimensional, isothermal chromatographic column, which is fed with a mixture of m-species at axial velocity v, uniform across the cross-section. Let c_i be the moles per unit volume concentration of species i in liquid phase, and n_i be the concentration of the species in the moles per unit volume. The local mass balances for each of the species result in the following set of first-order partial differential equations (Rhee et al., 1986):

$$\varepsilon\frac{\partial c_i}{\partial t} + \varepsilon\frac{\partial v c_i}{\partial z} + (1-\varepsilon)\frac{\partial n_i}{\partial t} = 0 \qquad (2.29)$$

where ε is the porosity. The concentrations of species i in the liquid phase and particle phase are assumed to be in equilibrium, which is given by a relationship (Langmuir isotherms) such as

$$n_i = f_i(c_i)$$

Differentiating this relationship gives

$$\frac{\partial n_i}{\partial t} = f'(c_i)\frac{\partial c_i}{\partial t} \tag{2.30}$$

Substituting the above equation in the mass balances yields the partial differential equation for the concentration dynamics:

$$\frac{\partial c_i}{\partial t}[\varepsilon+(1-\varepsilon)f'(c_i)]+\varepsilon\frac{\partial vc_i}{\partial z}=0 \tag{2.31}$$

2.2.6 Processes Governed by Integral Equations

2.2.6.1 Particle Size Distribution in Continuous Comminution Process

Consider a continuous comminution process in which the feed consists of large particles and the smaller comminuted particles are being removed at the same mass flow rate as the feed. Following Ramkrishna (2000), the population balance equation may be written as

$$\frac{\partial n(x,t)}{\partial t} = \frac{n_f(x)-n(x,t)}{\theta} + \int_x^\infty v(y)g(y)P(x|y)n(y,t)dy - g(x)n(x,t) \tag{2.32}$$

where $n(x,t)$ is the number density (assumed to be spatially uniform), $n_f(x,t)$ is the number density in the feed and θ is the mean residence time. The comminution process is described by the breakage frequency $g(x)$ and the size distribution of particles of size x formed from the breakage of a particle size y is defined by $P(x|y)$. The steady state integro-differential equation may be written in terms of the so-called "breakage kernel," $K(x,y) \equiv v(y)g(y)P(x|y)$ as follows.

$$g(x)n_s(x) = \frac{n_f(x)-n_s(x)}{\theta} + \int_x^\infty K(x,y)n_s(y)dy \tag{2.33}$$

where $n_s(x)$ is the steady-state exit number density.

2.2.6.2 Determination of Pore Size Distribution in Porous Media

The pore size distribution in a porous medium is determined by measuring the counterdiffusion rates of two gases across the medium in a Wicke-Kallenbach experiment (Brown and Travis, 1983). The experiments are carried out without a temperature or pressure gradient across the pellet.

The thickness of the porous medium is L. The mole fractions of a gaseous species A at the near face and a distant face at L are denoted by y_{A0} and y_{AL}, respectively. Define α as a function of molecular weights of the gases A and B as follows:

$$\alpha = \frac{M_A+M_B}{M_B}$$

Mathematics in Chemical Engineering

For diffusion through a cylindrical pore, an equation describing the dependence of diffusion rate on the porosity (ε), temperature (T), pressure (P), and pore size (r) is given as

$$N_A(P,T) = \frac{\varepsilon D_{AB}^0 P_0}{RTL\alpha} \int_{r_{min}}^{r_{max}} \left\{ \ln\left[\frac{1-\alpha y_{AL} + (3D_{AB}^0 P_0)/[(2rP)(8RT/\pi M_A)^{1/2}]}{1-\alpha y_{A0} + (3D_{AB}^0 P_0)/[(2rP)(8RT/\pi M_A)^{1/2}]}\right]\right\} \frac{f(r)}{\tau(r)} dr \quad (2.34)$$

where $f(r)$ and $\tau(r)$ are the pore size distribution and the tortuosity factor, respectively. D_{AB}^0 is the binary diffusion coefficient at the atmospheric pressure, P_0. Notwithstanding the fact that pressure and pore radius may have different ranges, we may loosely define Equation (2.34) as a Fredholm equation of the first kind, where the unknown function to be determined is $f(r)/\tau(r)$, given the diffusion measurements of N_A at a fixed temperature and various pressures.

2.2.6.3 Integral Equations Resulting Out of the Boundary-Value Problems with Mixed Derivative Boundary Condition

While descriptions of some of the chemical engineering systems may directly yield integral equations as illustrated in the above examples, for some systems one can transform the governing differential equations into integral equations with computational advantage for their solution. An example of such a transformation follows.

Consider a two-dimensional slab of length l (along x direction) and width b (along y direction) with its insulated ends consisting of heat generation defined by $f(x,y)$, cooled by the liquid flowing along its boundary (Ramkrishna and Amundson, 1979). Let k, h, P, t, C_p, and W be the thermal conductivity of the slab, heat transfer coefficient through the wall, heat transfer area per unit length of the slab at the cooled surface, coolant temperature, specific heat of the coolant, and coolant flow rate, respectively. The boundary-value problem describing the two-dimensional variation of slab temperature, $T(x,y)$, can be completely described as follows:

$$\frac{\partial^2 T}{\partial x^2} + \frac{\partial^2 T}{\partial x^2} = -f(x,y), \quad 0 < y < b, \quad 0 < x < l \quad (2.35)$$

subjected to the boundary conditions given by

$$y = 0, \quad \frac{\partial T}{\partial y} = 0, \quad 0 < x < l \quad (2.36)$$

$$y = b, \quad -k\frac{\partial T}{\partial y} = h(T-t) = \frac{WC_p}{P}\frac{dt}{dx}, \quad 0 < x < l \quad (2.37)$$

$$x = 0, \quad x = l, \quad \frac{\partial T}{\partial x} = 0, \quad 0 < y < b \quad (2.38)$$

The solution to Equation (2.35), i.e., the temperature distribution in the slab, can be written in terms of Green's function as follows:

$$T(x,y) = \int_0^1 d\xi \int_0^b d\eta\, G(x,y,\xi,\eta) f(\xi,\eta) + \frac{h}{k}\int_0^1 d\xi\, t(\xi) G(x,y,\xi,b) \quad (2.39)$$

The boundary condition in Equation (2.37) represents a mixed derivative boundary condition, which can be written in the form

$$\frac{dt}{dx} = \beta[T(x,b) - t], \quad t(0) = t_0, \quad \beta \equiv \frac{hP}{WC_p} \qquad (2.40)$$

and can be rewritten as

$$t(x) = t_0 e^{-\beta x} + \beta \int_0^x T(\xi, b) e^{-\beta(x-\xi)} d\xi \qquad (2.41)$$

Substitution for $T(\xi,b)$ from Equation (2.39) results in the Fredholm integral equation:

$$t(x) = g(x) + \lambda \int_0^1 K(x,\xi) t(\xi) d\xi \qquad (2.42)$$

where the expressions for the functions $g(x)$ and $K(x,\xi)$ can be readily obtained from Equations (2.41) and (2.39). Thus, solution of the integral Equation (2.42) completely specifies the solution to the boundary-value problem of Equation (2.39).

2.2.7 Processes Governed by Integrodifferential Equations

Most chemical systems involve multiple phases. The description of the interaction between them leads to both the differential as well as integral terms. The *population balance equations* form an important class of integrodifferential equations.

The population balance equation is a framework for the modeling of particulate systems. These include dispersions involving solid particles, liquid drops, and gas bubbles spanning a variety of systems of chemical engineering interest. The detailed derivation of the population balance equation and its applications can be found in Ramkrishna (1985, 2000). Publications pioneering the general application of population balance are by Hulburt and Katz (1964), Randolph and Larson (1964), and Frederickson et al. (1967).

Let $f(\mathbf{x},\mathbf{r},t)$ be the number density of particles with the particle state, $\mathbf{x}$, called as internal coordinate at a spatial location $\mathbf{r}$ at time t. Let $\dot{\mathbf{X}}(\mathbf{x},\mathbf{r},\mathbf{Y},t)$ and $\dot{\mathbf{R}}(\mathbf{x},\mathbf{r},\mathbf{Y},t)$ be the velocities along the internal and external coordinates, respectively, at time t in an environment of state $\mathbf{Y}$. Let $P(\mathbf{x},\mathbf{r},\mathbf{Y},t)$ and $D(\mathbf{x},\mathbf{r},\mathbf{Y},t)$ be the rates of production and destruction of particles. Then, following Ramkrishna (2000), the population balance equation for the number density is written as

$$\frac{\partial f}{\partial t} + \nabla \cdot \dot{\mathbf{X}} f + \frac{1}{V_0} \int_{A_i} f \dot{\mathbf{R}} \cdot d\mathbf{A} + \frac{1}{V_0} \int_{A_o} f \dot{\mathbf{R}} \cdot d\mathbf{A} = P - D$$

where V_0 is the system volume, and A_i and A_o are the areas of inlet and outlet, respectively. The production and destruction terms most commonly resulting out of aggregation and breakage involve integral terms imparting population balance equations as a class of the integrodifferential equations.

2.2.8 Processes Governed by Stochastic Differential Equations

In industrial operations, there often exist random fluctuations in the operating conditions from a variety of sources. Aris and Amundson (1958) show how fluctuations in the outputs of a continuous reactor caused by specified fluctuations in the input variables can be estimated by linearization.

For dynamic effects, stochastic differential equations pose a reasonable alternative to model such situations. A stochastic differential equation is written in the following form:

$$\frac{dX_t}{dt} = a(t, X_t) + b(t, X_t)\xi(t) \qquad (2.43)$$

where $\xi(t)$ is the random fluctuations. The alternative to Equation (2.43) is

$$X_t = X_0 + \int_0^t a(s, X_s)ds + \int_0^t b(s, X_a)\xi(s)ds \qquad (2.44)$$

The fluctuations are often modeled as a Wiener process, i.e., $\xi(t)dt = dW(t)$ (Gardiner, 2003); the function is known as *white* noise. For its precise mathematical definition and that of the Wiener process, the reader is referred to Gardiner (2003). Then Equation (2.44) is replaced by

$$X_t = X_0 + \int_0^t a(s, X_s)ds + \int_0^t b(s, X_s)dW(s) \qquad (2.45)$$

Equation (2.45) is called the Îto stochastic equation.

For information on how stochastic differential equations can be used to analyze the effect of the fluctuations in operating conditions, see Rao et al. (1974a). The actual method of solving nonlinear stochastic equations can be found in Rao et al. (1974b). Algorithms simpler than Rao et al. (1974b) are reported elsewhere (e.g., Pardoux and Talay, 1985).

2.3 THE NUMBER SYSTEM

A prerequisite for engineering calculations is the understanding of the number system. The reader may refer to the free on-line Web source http://mathworld.wolfram.com/topics/NumberTheory.html for the details.

2.3.1 THE REAL NUMBER SYSTEM

Integers are whole numbers spanning both positive and negative numbers including zero, while natural numbers consist of only positive numbers excluding zero. In addition to integers, rational numbers or fractional numbers are the ratios of integers. Irrational numbers are the numbers not representable by a ratio of integers. An example of an irrational number is π, the ratio of circumference of a circle to its diameter. The real number system consists of rational and irrational numbers (integers are special cases of rational numbers). Apart from addition, subtraction, multiplication, and division, operations on real numbers involving powers, roots, and logarithms are necessary for solving equations governing the engineering systems.

2.3.1.1 Powers and Roots

The nth power of a real number is defined as

$$a^n = \underbrace{a \cdot a \cdot a \cdots a}_{n \text{ times}}$$

where a is called the base and n is called the exponent. Powers with negative exponents are defined as

$$a^{-n} = \frac{1}{a^n}, \quad a \neq 0$$

The nth root of a real number is defined as

$$x = a^{1/n} = \sqrt[n]{a} \text{ if and only if } x^n = a$$

Properties of powers and roots are as follows:

$$a^n a^m = a^{m+n}; \quad \frac{a^n}{a^m} = a^{n-m}; \quad (ab)^n = a^n b^n; \quad \left(\frac{a}{b}\right)^n = \frac{a^n}{b^n}$$

$$a^{n/m} = \sqrt[m]{a^n} = (\sqrt[m]{a})^n; \quad \sqrt[m]{\sqrt[n]{a}} = \sqrt[mn]{a}; \quad \sqrt[n]{ab} = \sqrt[n]{a}\sqrt[n]{b}$$

$$\sqrt[n]{\frac{a}{b}} = \frac{\sqrt[n]{a}}{\sqrt[n]{b}} \quad (b \neq 0)$$

2.3.1.2 Logarithm

The logarithm to the base $c > 0$,

$$x = \log_c a$$

is defined by

$$c^x = a \quad \text{or} \quad c^{\log_c a} = a$$

Properties of logarithms:

$$\log_c a = 1; \quad \log_c c^p = p$$

$$\log_c(ab) = \log_c a + \log_c b; \quad \log_c\left(\frac{a}{b}\right) = \log_c a - \log_c b$$

$$\log_c(a^n) = n \log_c a; \quad \log_c(\sqrt[n]{a}) = \frac{1}{n} \log_c a$$

2.3.2 Sequences and Series

A sequence is a function that maps integers to a subset of real numbers. A sequence can be viewed as a set of numbers indexed by integers. A finite sequence terminates after a finite number of terms. The following is an example of a finite sequence:

$$\{a_i\}_{i=0}^{5} = \{2i+1\}_{i=0}^{5} = \{1, 3, 5, 7, 9, 11\}$$

Mathematics in Chemical Engineering

A sequence that extends indefinitely is an infinite sequence. For example, $\{a_i\}_{i=1}^{\infty} = \{2i+1\}_{i=0}^{\infty}$ is a sequence of odd integers that is an example of a divergent sequence. The following is an example of a real sequence:

$$\{a_i\}_{i=1}^{\infty} = \left\{\frac{1}{i}\right\}_{i=1}^{\infty}$$

If an infinite sequence approaches a definite number, then the sequence is convergent. Thus the real sequence represented above is a convergent sequence that converges to zero.

A sequence is called increasing (decreasing) if $a_{n+1} > a_n (a_{n+1} < a_n)$ and is called nondecreasing (nonincreasing) if $a_{n+1} \geq a_n (a_{n+1} \leq a_n)$. A sequence is called strictly monotone if it is either increasing or decreasing. A sequence is bounded if there is a number M such that $|a_n| \leq M$. A bounded strictly monotone sequence converges.

A series is the sum of a sequence. Thus, given a sequence, one can form a series by summing the sequence. A finite sequence generates a finite series, and an infinite sequence generates an infinite series.

From a series, one can form an infinite sequence by defining the partial sum as follows:

$$A_n = \sum_{i=1}^{n} a_i$$

The sequence formed from this series is $A_1, A_2, A_3, \ldots$, where

$$A_2 + a_1 + a_2, \quad A_3 = a_1 + a_2 + a_3, \quad A_4 = a_1 + a_2 + a_3 + a_4 \ldots$$

However, the criteria for convergence of a sequence and of a series are different. A convergent sequence may not generate a convergent series. A sequence formed by partial sums of a convergent series is convergent.

2.3.3 Tests for Convergence of Sequence and Series

We list here some of the tests that can be used to determine the convergence of series without having to evaluate the series:

$$\sum_{n=1}^{\infty} a_n$$

2.3.3.1 Integral Test

The series converges if and only if the integral

$$\int_{1}^{\infty} f(x)dx$$

converges, where $f(k) = a_k \geq 0$. For example, the harmonic series is divergent because

$$\sum_{n=1}^{\infty} \frac{1}{n}$$

is divergent because

$$\int_1^{\infty} \frac{1}{x} dx = \infty$$

2.3.3.2 Ratio Test

$$\rho = \lim_{n \to \infty} \frac{a_{n+1}}{a_n}$$

The series is convergent if $\rho < 1$; divergent if $\rho > 1$. It is indeterminate if $\rho = 1$. For example, the series

$$\sum_{n=0}^{\infty} \frac{1}{n!}$$

is convergent because

$$\lim_{n \to \infty} \frac{a_{n+1}}{a_n} = \lim_{n \to \infty} \frac{1}{n+1} = 0 < 1$$

2.3.3.3 Root Test

$$\rho = \lim_{n \to \infty} \sqrt[n]{a_n}$$

The series is convergent if $\rho < 1$; divergent if $\rho > 1$. It is indeterminate if $\rho = 1$. For example, the series

$$\sum_{n=1}^{\infty} \frac{1}{n^n}$$

is convergent because

$$\lim_{n \to \infty} \sqrt[n]{a_n} = \lim_{n \to \infty} \frac{1}{n} = 0 < 1$$

2.3.3.4 Comparison Test

If $0 < a_n < (>) b_n$ and the series

Mathematics in Chemical Engineering

$$\sum_{n=1}^{\infty} b_n$$

is convergent (divergent), then the series

$$\sum_{n=1}^{\infty} a_n$$

is convergent (divergent).

For example, the series

$$\sum_{n=1}^{\infty} \frac{1}{n^2+n}$$

is convergent because

$$\frac{1}{n^2+n} < \frac{1}{n^2}$$

and the latter series is convergent by integral test. Thus the former series is convergent.

2.3.3.5 Limit Comparison Test

$$\rho = \lim_{n \to \infty} \frac{a_n}{b_n}$$

If $0 < \rho < \infty$, then both the series converge or diverge. For example, the series

$$\sum_{n=1}^{\infty} \frac{n}{n^2+n}$$

is divergent because the harmonic series is divergent, and by the limit comparison test, both the series diverge.

2.3.4 Taylor Series

If $f(x)$ is analytic within and on a circle C centered at x_0 of radius R, then the Taylor series representation of $f(x)$ is given by

$$f(x) = \sum_{n=0}^{\infty} \frac{f^{(n)}(x_0)}{n!}(x-x_0)^n$$

2.3.5 Binomial Theorem

For any positive integer n,

$$(a+b)^n = a^n + na^{n-1}b + \frac{n(n-1)}{2!}a^{n-2}b^2 + \frac{n(n-1)(n-2)}{3!}a^{n-2}b^3 + \cdots +$$

$$= \sum_{m=0}^{n} C_m^n a^{n-m} b^m$$

where

$$C_m^n = \frac{n!}{(n-m)!m!} = C_{n-m}^n$$

$$n! = n(n-1)(n-2)\cdots 1, \quad 0! = 1$$

C_m^n satisfies the following identity:

$$2^n = 1 + C_1^n + C_2^n + \cdots + C_n^n$$

that is obtained by letting $a = b = 1$. The expansion of binomial terminates when n is an integer. A similar expansion holds for $b = 1$ and noninteger n,

$$(1+a)^n = 1 + na + \frac{n(n-1)}{2!}a^2 + \frac{n(n-1)(n-2)}{3!}a^3 + \cdots +$$

2.3.6 Inequalities

We list here some of the important inequalities followed by an example showing the use of one of the inequalities in the context of a chemical reaction. Let $\{a_i\}_{i=1}^n$ and $\{b_i\}_{i=1}^n$ be sequences of real numbers and $p, q \geq 1$.

2.3.6.1 Algebraic Inequalities

$$\text{Triangular inequality:} \quad \left|\sum_{i=1}^{n} a_i\right| \leq \sum_{i=1}^{n} |b_i| \tag{2.46}$$

$$\text{Hölder's inequality:} \quad \sum_{i=1}^{n} |a_i b_i| \leq \left(\sum_{i=1}^{n} |a_i|^p\right)^{1/p} \left(\sum_{i=1}^{n} |b_i|^q\right)^{1/q}, \quad \frac{1}{p} + \frac{1}{q} = 1 \tag{2.47}$$

For $p = q = 2$, Holder's inequality is known as the Cauchy–Schwartz inequality.

Minkowski's inequality:
$$\left(\sum_{i=1}^{n}|a_i+b_i|^p\right)^{1/p} \leq \left(\sum_{i=1}^{n}|a_i|^p\right)^{1/p} + \left(\sum_{i=1}^{n}|a_i|^p\right)^{1/p} \quad (2.48)$$

Arithmatic-geometric means inequality:
$$\frac{1}{n}\sum_{i=1}^{n}a_i \geq \left(\prod_{i=1}^{n}a_i\right)^{1/n}, \quad a_i \geq 0 \quad (2.49)$$

Chebyshev's inequality:
$$\frac{1}{n}\sum_{i=1}^{n}a_i b_i \geq \left(\frac{1}{n}\sum_{i=1}^{n}a_i\right)\left(\frac{1}{n}\sum_{i=1}^{n}b_i\right), \quad a_1 \geq \cdots \geq a_m, b_1 \geq \cdots \geq b_m \quad (2.50)$$

2.3.6.2 Integral Inequalities

Modulus inequality:
$$\left|\int_a^b f(x)dx\right| \leq \int_a^b |f(x)|dx \quad (2.51)$$

Hölder inequality:
$$\int_a^b |f(x)g(x)|dx \leq \left(\int_a^b |f(x)|^p dx\right)^{1/p} \left(\int_a^b |g(x)|^q\right)^{1/q} \quad (2.52)$$

For $p = q = 2$,

Arithmetic-geometric inequality:
$$\frac{1}{b-a}\int_a^b f(x)dx \geq \exp\left[\frac{1}{b-a}\int_a^b \ln f(x)dx\right], \quad f(x) \geq 0 \quad (2.53)$$

$$\left(\frac{1}{b-a}\int_a^b f(x)dx\right)\left(\frac{1}{b-a}\int_a^b g(x)dx\right) \leq \frac{1}{b-a}\int_a^b f(x)g(x)dx, \quad f'(x), \; g'(x) \geq 0 \quad (2.54)$$

Example 2.1: Application of Inequalities

Consider a reactant undergoing an nth-order reaction with rate kc^n in dispersed-phase droplets. The concentration c is assumed to vary between 0 and c_0. If the distribution of solute concentration is denoted $f(c)$, which is nonnegative and such that $\int_0^{c_0} f(c)dc = 1$, the average concentration in the drop phase is then

$$\bar{c} = \int_0^{c_0} f(c)dc$$

The reaction rate in the drop phase is then

$$\overline{kc^n} = k\int_0^{c_0} c^n f(c)dc$$

On the other hand, the reaction rate in terms of the average concentration is given by

$$k\bar{c}^n = k\left[\int_0^{c_0} cf(c)dc\right]^n$$

It is easy to show that the rate of reaction as described by the lumped dispersed phase analysis underestimates the actual rate by using the Hölder's inequality. Let $g(c) = cf^{1/p}$ and $h(c) = f^{1/q}$. Thus,

$$\int_0^{c_0} c[f(c)]^{1/p+1/q} dc \leq \left(\int_0^{c_0} c^p f(c)\right)^{1/p} \left(\int_0^{c_0} f(c)dc\right)^{1/q}$$

The second integral on the right-hand side of the inequality is unity by definition. Hence, with $p = n$ and $q = 1/(n-1)$,

$$k\bar{c}^n \leq k\overline{c^n}, \quad n > 1$$

2.4 DIFFERENTIAL AND INTEGRAL CALCULUS

2.4.1 Functions, Limits, and Continuity

A function $y = f(x)$ is a rule that assigns a unique value of y for each of the given values of x in the definition of a domain of f, and x is said to be an independent variable and y the dependent variable. There may exist more than one value of x for a given value of y. The set of values of x is called the domain of a function, and the corresponding set of y is called the range of a function.

A function $f(x)$ approaches a limit y_0 as x approaches x_0; for any given $\varepsilon > 0$, then there exists a $\delta > 0$ such that $|f(x) - y_0| < \varepsilon$ for $|x - x_0| < \delta$. Symbolically,

$$\lim_{x \to x_0} f(x) = y_0$$

The limits follow simple arithmetic operations, e.g., the sum of the limits is the limit of the sums, etc. A function is said to be continuous at x_0 if

$$\lim_{x \to x_0} f(x) = f(x_0)$$

For example,

$$f(x) = \begin{cases} \dfrac{1}{x}, & x > 0 \\ 0, & x = 0 \end{cases}$$

is not continuous at $x = 0$ because

$$\lim_{x \to 0} f(x) = \infty \neq f(0) = 0$$

2.4.2 THE DERIVATIVE

The derivative is the rate of change of the function with respect to its argument. It follows that if the derivative is positive, then the function is increasing, and if the derivative is negative, then the function is decreasing.

A function is said to be differentiable at $x = a$ if

$$f'(a) \equiv \frac{df(a)}{dx} \equiv \lim_{x \to a} \frac{f(x) - f(a)}{x - a}$$

exists. The derivative of $f(x)$ at $x = a$ is the instantaneous rate of change of $f(x)$ with respect to x at $x = a$. Graphically, it is the slope of a tangent line to the curve $f(x)$ at the point $(a, f(a))$.

Following are the rules of differentiation for the differentiable functions $f(x)$ and $g(x)$:

$$\frac{dc}{dx} = 0, \quad c \neq f(x)$$

$$\frac{d[cf(x)]}{dx} = c\frac{df(x)}{dx}, \quad c = \text{constant}$$

$$\frac{d}{dx}[f(x) \pm g(x)] = \frac{df(x)}{dx} \pm \frac{dg(x)}{dx}$$

$$\frac{d}{dx}f(x)g(x) = f(x)\frac{dg(x)}{dx} + g(x)\frac{f(x)}{dx}$$

$$\frac{d}{dx}\frac{f(x)}{g(x)} = \frac{f(x)\frac{dg(x)}{dx} - g(x)\frac{df(x)}{dx}}{[g(x)]^2}$$

$$\frac{d}{dx}f(g(x)) = \frac{df(g)}{dg}\frac{dg(x)}{dx}$$

$$\frac{d^n}{dx^n}f(x)g(x) = \sum_{m=0}^{n} C_m^n \frac{d^{n-m}}{dx^{n-m}}f(x)\frac{d^m}{dx^m}g(x)$$

$$\frac{d}{dx}[f(x)]^n = n[f(x)]^{n-1}\frac{df(x)}{dx}$$

A higher-order derivative of a function is defined recursively in terms of the lower-order derivative.

The second-order derivative is the concavity of the function at x. If the second derivative of a function at x_0 is negative, then the function is concave down at x_0. The function is concave up at x_0 if its second derivative is positive. If the second derivative vanishes at x_0, then it is called as an inflection point, where the function essentially changes its concavity.

When the derivative is zero, the function reaches its local extremum. Whether this local extremum is a maximum or minimum depends on additional information. If, in addition, the second derivative is negative, then the function attains maximum at x_0; if the second derivative is positive at the local extremum, then it is a minimum.

2.4.2.1 An Example Application of the Derivatives

The concentration of the intermediate product, B, at any time, t, as a result of the following reactions in series

$$A \xrightarrow{k_1} B \xrightarrow{k_2} C$$

with first-order reaction kinetics with respect to A and B and with the initial concentration of the reactant, A, equal to C_{A0}, is given by

$$C_B(t) = \frac{k_1 C_{A0}}{k_2 - k_1}[e^{-k_1 t} - e^{-k_2 t}]$$

To find out the space-time, τ, that maximizes the concentration of B, the derivative of C_B with respect to time t is set equal to zero at $t = \tau$ to yield

$$\left.\frac{dC_B}{dt}\right|_{t=\tau} = 0 \Rightarrow \tau = \frac{\ln k_2 - \ln k_1}{k_2 - k_1}$$

$$\left.\frac{d^2 C_B}{dt^2}\right|_{t=\tau} < 0, \text{ at } t = \tau$$

Hence, the maximum concentration of the species B, $C_{B,m}$ is given by

$$C_{B,m} = \frac{k_1 C_{A0}}{k_2 - k_1}\left[\left(\frac{k_1}{k_2}\right)^{k_1/k_2 - k_1} - \left(\frac{k_1}{k_2}\right)^{k_2/k_2 - k_1}\right]$$

2.4.3 The Mean Value Theorem

Let $f(x)$ and $g(x)$ be the differentiable functions on $[a,b]$. If $g(b) \neq g(a)$ and $g'(x) \neq 0$, then there exists $c \in (a,b)$ such that

$$\frac{f'(c)}{g'(c)} = \frac{f(b) - f(a)}{g(b) - g(a)}$$

In particular, if $g(x) = x$, then

$$f'(c) = \frac{f(b) - f(a)}{b - a}$$

which is the familiar form of the mean value theorem.

2.4.3.1 An Example Application of the Mean Value Theorem

To account for the effects of transport resistance in a catalyst pellet, effectiveness factors (defined as the ratio between the observed reaction rate to that obtained when neither concentration nor temperature gradients exist between the ambient phase and any point in the catalyst pellet) are used in modeling chemical reactors. It is very important that the effectiveness factors do not change the nature of the reactions. One of the properties that the effectiveness factors must possess is positive definiteness. This property ensures that the transport resistance does not change the direction of a reaction. Here we illustrate the use of the mean value theorem to show that the overall effectiveness factor is positive.

Reactions in series were discussed in Section 2.4.2.1. For reactions occurring in a spherical catalyst pellet with the pore radius R, the concentration of the intermediate product, B, is proportional to the overall effectiveness factor given by

$$\bar{\eta} = \frac{\eta_2 \phi_2^2 - \eta_1 \phi_1^2}{\phi_2^2 - \phi_1^2}$$

where effectiveness factors for each of the reactions are given by

$$\eta_i = \frac{3\phi_i \coth(3\phi_i) - 1}{3\phi_i^2} \quad \text{and the Thiele moduli as,} \quad \phi_i = \frac{R}{3}\sqrt{\frac{a_p k_i}{D_B}}, \quad i = 1, 2$$

where D_B is the diffusivity of the product B. Thus,

$$\bar{\eta} = \frac{\phi_2 \coth(3\phi_2) - \phi_1 \coth(3\phi_1)}{\phi_2^2 - \phi_1^2}$$

Setting $f(c) \equiv \sqrt{c} \coth(3\sqrt{c})$, $g(c) \equiv c$ and applying the mean value theorem,

$$\bar{\eta} = \frac{f'(c)}{g'(c)}, \quad \phi_1^2 < c < \phi_2^2$$

that ensures that the overall effectiveness factor is positive if

$$f(x) = x \coth(3x)$$

is positive. It can be readily shown that $f'(x)$ is indeed positive for all positive x.

2.4.4 L'Hôspital's Rule

The limit of a function as its independent variable approaches a value is indeterminate if it has one of the following forms:

$$\frac{0}{0}, \frac{\infty}{\infty}, 0 \cdot \infty, 0^0, \infty^0, 1^\infty, \infty - \infty \tag{2.55}$$

L'Hôspital's rule is applicable to the first two forms, namely, 0/0 or ∞/∞. The rest of the forms can be rewritten in terms of the first two.

For limits that yield 0/0 or ∞/∞, L'Hôspital's rule states

$$\lim_{x \to x_0} \frac{f(x)}{g(x)} = \lim_{x \to x_0} \frac{f'(x)}{g'(x)}$$

since $\infty - 1/0$, $0 \cdot \infty$ can be stated as the first two forms. Limit of the form $\infty - \infty$ is equivalent to $1 - \infty/\infty$.

Logarithms of the limits can be used to find the limits involving exponents using L'Hôspital's rule.

Example 2.2: Application of L'Hôspital's Rule

$$\lim_{x \to \infty} \left(\cos \frac{1}{x} \right)^x$$

Let

$$y = \left(\cos \frac{1}{x} \right)^x \Rightarrow \ln y = x \ln \cos \frac{1}{x}$$

$$\lim_{x \to \infty} \ln y = \lim_{x \to \infty} x \ln \cos \frac{1}{x} = \lim_{x \to \infty} \frac{\ln \cos \frac{1}{x}}{\frac{1}{x}} = -\lim_{x \to \infty} \frac{\sin \frac{1}{x}}{\cos \frac{1}{x}} = 0$$

$$\lim_{x \to \infty} \left(\cos \frac{1}{x} \right)^x = 1$$

2.4.5 Implicit Function Theorem

Given a function $f(x,y)$ such that

$$f(x_0, y_0) = 0, \quad \text{and} \quad \left. \frac{\partial f}{\partial y} \right|_{(x_0, y_0)} \neq 0$$

then there exists a neighborhood $B(x_0)$ and a unique function $g(x)$ such that $y = g(x)$ for $x \in B(x_0)$, $f(x, g(x)) = 0$.

2.4.6 The Integral

Given a bounded function $f(x)$ defined on a bounded interval $[a,b]$, the definite Riemann integral of $f(x)$ is defined as

$$\int_a^b f(x)dx = \lim_{\|\Delta x_i\|} \sum_{i=1}^{n} f(\xi_i) \Delta x_i$$

where

$$x_{i-1} < \xi_i < x_i, \quad \|\Delta x_i\| = \max_{i=1}^{n}(x_i - x_{i-1}), \quad a = x_0 < x_1 < \cdots < x_n = b$$

Mathematics in Chemical Engineering

The term $f(x)$ is known as the integrand; a and b are the limits of integration. The interpretation of a definite integral is the area bounded by the function $f(x)$ and the x-axis between the lines $x = a$ and $x = b$.

The integral possesses the following properties:

$$\int_a^b f(x)dx = -\int_b^a f(x)dx$$

$$\int_a^c f(x)dx = \int_a^b f(x)dx + \int_b^c f(x)dx$$

$$\int_a^b [f(x)+g(x)]dx = \int_a^b f(x)dx + \int_a^b g(x)dx$$

Rules of integration:

1. Integration by parts,

$$\int_a^b f(x)g'(x)dx = f(x)g(x)\Big|_{x=a}^{x=b} - \int_a^b g(x)f'(x)dx$$

2. If $x = h(t)$, then $dx = h'(t)dt$ and

$$\int_a^b f(x)dx = \int_{h^{-1}(a)}^{h^{-1}(b)} f[h(t)]dt$$

3. Leibnitz's rule: If $p(y)$ and $q(y)$ are continuously differentiable functions of y,

$$\frac{\partial}{\partial y}\int_{p(y)}^{q(y)} f(x,y)dx = \frac{dq}{dy}f(q,y) - \frac{dp}{dy}f(p,y) + \int_{p(y)}^{q(y)} \frac{\partial}{\partial y}f(x,y)dx$$

The fundamental theorem of calculus: Given a continuous function $f(x)$ on the bounded interval $[0,x]$, then

$$\frac{d}{dx}\int_0^x f(\xi)d\xi = f(x)$$

2.4.6.1 Improper Integrals

There are two types of improper integrals. One type of improper integral is that where the integrand becomes infinite at one of the limits of integration. The other type is that where one or both of the limits of integration are infinite. The improper integral is defined as

$$\int_a^\infty f(x)dx = \lim_{b\to\infty}\int_a^b f(x)dx$$

If the limit exists, then the improper integral converges; otherwise, the improper integral diverges. Given $f(x)$ on $[a,b]$ such that $f(a) = \infty$, then the improper integral is defined as

$$\int_a^b f(x)dx = \lim_{x \to a} \int_x^b f(\xi)d\xi$$

The improper integrals are said to be convergent if and only if the limits are convergent.

The Riemann integral is convergent for a bounded function with a finite number of discontinuities. It is possible to construct a pathological function that has an infinite number of discontinuities. Although such a function is bounded, its Riemann integral does not exist. The Lebesgue integral is designed to overcome such a deficiency.

2.4.6.2 An Example Application of the Integrals

We reproduce here some useful formulae for the determination of area under the curve, the arc length, volume of solid revolution, and the surface area that are the results of the application of the integrals. Let $f(x)$ be a real valued function defined on an interval (a,b). Let A be the area bounded by the function, the x-axis, and the lines $x = a$ and $x = b$. Then

$$A = \int_a^b f(x)dx \tag{2.56}$$

and the arclength, S, of the curve defined by the function is

$$S = \int_a^b \sqrt{1+\left(\frac{df}{dx}\right)^2}\, dx \tag{2.57}$$

The volume of the solid generated by revolving the curve $f(x)$ about the x-axis is

$$\pi \int_a^b [f(x)]^2 dx$$

and the surface area of the solid body is given by

$$2\pi \int_a^b f(x)\sqrt{1+\left(\frac{df}{dx}\right)^2}\, dx \tag{2.58}$$

2.5 VECTOR ANALYSIS

2.5.1 VECTOR ALGEBRA

A vector $\mathbf{v}$ in n-dimensional space is an array consisting of n scalar quantities, i.e., $\mathbf{v} = [v_1 \; v_2 \; \ldots \; v_n]^T$. In three-dimensional space, a vector, $\mathbf{v}$, can be represented geometrically as directed line segments with directions and magnitudes. The terms v_x, v_y, and v_z are the components of the vector $\mathbf{v}$ along the x-, y-, and z-axes, respectively. The magnitude of $\mathbf{v}$, $|\mathbf{v}|$, is $\sqrt{v_x^2 + v_y^2 + v_z^2}$, which measures the length of the vector or the distance from the origin to the tip of the vector. The angles α, β, and γ between the vector $\mathbf{v}$ and basis vectors δ_x, δ_y, and δ_y of the coordinate axes x, y, and z, respectively, are calculated from the direction cosines of the vector,

Mathematics in Chemical Engineering

$$\cos\alpha = \frac{v_x}{\sqrt{v_x^2 + v_y^2 + v_z^2}} \quad \cos\beta = \frac{v_y}{\sqrt{v_x^2 + v_y^2 + v_z^2}} \quad \cos\gamma = \frac{v_z}{\sqrt{v_x^2 + v_y^2 + v_z^2}} \quad (2.59)$$

From these operations, it is clear that negating a vector is reversing its direction. It is possible to resolve a vector into its components, and vice versa, i.e.,

$$\mathbf{v} = [v_x \quad v_y \quad v_z]^T \quad \text{or} \quad \mathbf{v} = v_x \boldsymbol{\delta}_x + v_y \boldsymbol{\delta}_y + v_z \boldsymbol{\delta}_z$$

Addition of vectors is carried out componentwise. Given two three-dimensional vectors **a** and **b**,

$$\mathbf{a} \pm \mathbf{b} = [a_1 \pm b_1 \quad a_2 \pm b_2 \quad a_3 \pm b_3]$$

it follows that addition of vectors is commutative and associative:

$$\mathbf{a} + \mathbf{b} = \mathbf{b} + \mathbf{a} \quad (2.60)$$

$$(\mathbf{a} + \mathbf{b}) + \mathbf{c} = \mathbf{a} + (\mathbf{b} + \mathbf{c}) \quad (2.61)$$

Scalar product: The scalar product of two three-dimensional vectors **a** and **b** is defined as

$$\mathbf{a} \cdot \mathbf{b} = |\mathbf{a}||\mathbf{b}| \cos(\mathbf{a},\mathbf{b})$$

where $\cos(\mathbf{a},\mathbf{b})$ is the cosine of the angle between the vectors **a** and **b**. It is a componentwise operation,

$$\mathbf{a} \cdot \mathbf{b} = \sum_{i=1}^{3} a_i b_i$$

or it can be simply written as $a_i b_i$ using Einstein's convention of summation over repeated indices. It follows that scalar products are commutative and distributive.

Vector product: The vector product of two three-dimensional vectors **a** and **b** is

$$\mathbf{a} \times \mathbf{b} = \begin{vmatrix} \boldsymbol{\delta}_x & \boldsymbol{\delta}_y & \boldsymbol{\delta}_z \\ a_x & a_y & a_z \\ b_x & b_y & b_z \end{vmatrix}$$

Componentwise operation is defined as

$$\mathbf{a} \times \mathbf{b} = \sum_{i=1}^{3}\sum_{j=1}^{3}\sum_{k=1}^{3} \varepsilon_{ijk} a_i b_j \boldsymbol{\delta}_k$$

or it can be simply represented by $\varepsilon_{ijk} a_i b_j \boldsymbol{\delta}_k$, where

$$\varepsilon_{123} = \varepsilon_{231} = \varepsilon_{312} = 1 \tag{2.62}$$

$$\varepsilon_{132} = \varepsilon_{321} = \varepsilon_{213} = -1 \tag{2.63}$$

$$\varepsilon_{ijk} = \varepsilon_{ijj} = \varepsilon_{iji} = 0 \tag{2.64}$$

Scalar triple product:

$$(\mathbf{a} \times \mathbf{b}) \cdot \mathbf{c} = \begin{vmatrix} a_x & a_y & a_z \\ b_x & b_y & b_z \\ c_x & c_y & c_z \end{vmatrix}$$

Vector triple product:

$$\mathbf{a} \times (\mathbf{b} \times \mathbf{c}) = (\mathbf{a} \cdot \mathbf{c})\mathbf{b} - (\mathbf{a} \cdot \mathbf{b})\mathbf{c}$$

Quadruple scalar product:

$$(\mathbf{a} \times \mathbf{b}) \cdot (\mathbf{c} \times \mathbf{d}) = (\mathbf{a} \cdot \mathbf{c})(\mathbf{b} \cdot \mathbf{d}) - (\mathbf{a} \cdot \mathbf{d})(\mathbf{b} \cdot \mathbf{c})$$

Quadruple vector product:

$$(\mathbf{a} \times \mathbf{b}) \times (\mathbf{c} \times \mathbf{d}) = [(\mathbf{a} \times \mathbf{b}) \cdot \mathbf{d}]\mathbf{c} - [(\mathbf{a} \times \mathbf{b}) \cdot \mathbf{c}]\mathbf{d}$$

2.5.2 Vector Calculus

We list here some of the useful vector operations in a differential setting that are generally encountered in chemical engineering:

$$\frac{d\mathbf{a}}{dt} = \frac{da_x}{dt}\boldsymbol{\delta}_x + \frac{da_y}{dt}\boldsymbol{\delta}_y + \frac{da_z}{dt}\boldsymbol{\delta}_z$$

$$\frac{d}{dt}(\mathbf{a}+\mathbf{b}) = \frac{d\mathbf{a}}{dt} + \frac{d\mathbf{b}}{dt}$$

$$\frac{d}{dt}(\phi\mathbf{a}) = \phi\frac{d\mathbf{a}}{dt} + \mathbf{a}\frac{d\phi}{dt}$$

$$\frac{d}{dt}(\mathbf{a}\cdot\mathbf{b}) = \frac{d\mathbf{a}}{dt}\cdot\mathbf{b} + \mathbf{a}\cdot\frac{d\mathbf{b}}{dt}$$

$$\frac{d}{dt}(\mathbf{a}\times\mathbf{b}) = \frac{d\mathbf{a}}{dt}\times\mathbf{b} + \mathbf{a}\times\frac{d\mathbf{b}}{dt}$$

$$\frac{d}{dt}(\mathbf{a}\times\mathbf{b}\cdot\mathbf{c}) = \frac{d\mathbf{a}}{dt}\times\mathbf{b}\cdot\mathbf{c} + \mathbf{a}\times\frac{d\mathbf{b}}{dt}\cdot\mathbf{c} + \mathbf{a}\times\mathbf{b}\cdot\frac{d\mathbf{c}}{dt}$$

$$\frac{d}{dt}(\mathbf{a}\times\mathbf{b}\times\mathbf{c}) = \frac{d\mathbf{a}}{dt}\times(\mathbf{b}\times\mathbf{c}) + \mathbf{a}\times\left(\frac{d\mathbf{b}}{dt}\times\mathbf{c}\right) + \mathbf{a}\times\left(\mathbf{b}\times\frac{d\mathbf{c}}{dt}\right)$$

2.5.3 Orthogonal Curvilinear Coordinate Systems

2.5.3.1 Scale Factors and Metric Tensors

Given a transformation of coordinate system from (x_1,x_2,x_3) to an orthogonal curvilinear coordinate system (ξ_1,ξ_2,ξ_3),

$$x_1 \equiv x_1(\xi_1,\xi_2,\xi_3) \quad x_2 \equiv x_2(\xi_1,\xi_2,\xi_3) \quad x_3 \equiv x_3(\xi_1,\xi_2,\xi_3)$$

we have

$$dx_i = \sum_{j=1}^{3} \frac{\partial x_i}{\partial \xi_j} d\xi_j$$

and

$$dl^2 = \sum_{i=1}^{3} dx_i^2 = \sum_{i=1}^{3}\sum_{j=1}^{3} g_{ij} d\xi_i d\xi_j$$

where the elements g_{ij} are given by

$$g_{ij} = \sum_{k=1}^{3} \frac{\partial x_k}{\partial \xi_i} \frac{\partial x_k}{\partial \xi_j} = \left[\sum_{k=1}^{3} \frac{\partial \xi_i}{\partial x_k} \frac{\partial \xi_i}{\partial x_k}\right]$$

provided the Jacobian of the transformation

$$J = \begin{vmatrix} \dfrac{\partial x_1}{\partial \xi_1} & \dfrac{\partial x_1}{\partial \xi_2} & \dfrac{\partial x_1}{\partial \xi_3} \\ \dfrac{\partial x_2}{\partial \xi_1} & \dfrac{\partial x_2}{\partial \xi_2} & \dfrac{\partial x_2}{\partial \xi_3} \\ \dfrac{\partial x_3}{\partial \xi_1} & \dfrac{\partial x_3}{\partial \xi_2} & \dfrac{\partial x_3}{\partial \xi_3} \end{vmatrix}$$

does not vanish.

Let the scale factors be defined by

$$h_1 = \sqrt{g_{11}} \quad h_2 = \sqrt{g_{22}} \quad h_3 = \sqrt{g_{33}}$$

then the differential volume element

$$dV = h_1 h_2 h_3 d\xi_1 d\xi_2 d\xi_3$$

and the elements of the area ds_i on the surfaces u_i = constant, for $i = 1,2,3$, are

$$ds_1 = h_2 h_3 d\xi_1 d\xi_3; \quad ds_2 = h_1 h_3 d\xi_1 d\xi_3; \quad ds_3 = h_1 h^2 d\xi^2$$

2.5.3.2 Differential Operators in Curvilinear Coordinate System

We present here the expressions for various differential vector operators that can be of great utility for transforming the equations in different coordinate systems. Given a curvilinear coordinate system with basis vectors $(h_1 \delta_1 \quad h_2 \delta_2 \quad h_3 \delta_3)$,

$$\frac{\partial \delta_1}{\partial \xi_1} = -\frac{\delta_2}{h_2}\frac{\partial h_1}{\partial \xi_2} - \frac{\delta_3}{h_3}\frac{\partial h_1}{\partial \xi_3} \qquad \frac{\partial \delta_1}{\partial \xi_2} = \frac{\delta_2}{h_1}\frac{\partial h_2}{\partial \xi_1} \qquad \frac{\partial \delta_1}{\partial \xi_3} = \frac{\delta_3}{h}\frac{\partial h_3}{\partial \xi_1}$$

$$\frac{\partial \delta_2}{\partial \xi_2} = -\frac{\delta_1}{h_1}\frac{\partial h_2}{\partial \xi_1} - \frac{\delta_3}{h_3}\frac{\partial h_2}{\partial \xi_3} \qquad \frac{\partial \delta_2}{\partial \xi_1} = \frac{\delta_1}{h_2}\frac{\partial h_1}{\partial \xi_2} \qquad \frac{\partial \delta_2}{\partial \xi_3} = \frac{\delta_3}{h_2}\frac{\partial h_3}{\partial \xi_2}$$

$$\frac{\partial \delta_3}{\partial \xi_3} = -\frac{\delta_1}{h_1}\frac{\partial h_3}{\partial \xi_1} - \frac{\delta_2}{h_2}\frac{\partial h_3}{\partial \xi_2} \qquad \frac{\partial \delta_3}{\partial \xi_1} = \frac{\delta_1}{h_3}\frac{\partial h_1}{\partial \xi_3} \qquad \frac{\partial \delta_3}{\partial \xi_2} = \frac{\delta_2}{h_3}\frac{\partial h_2}{\partial \xi_3}$$

let

$$\mathbf{f} = f_1 \delta_1 + f_2 \delta_2 + f_3 \delta_3$$

$$\nabla \Phi = \delta_1 \frac{\partial \Phi}{\partial \xi_1} + \delta_2 \frac{\partial \Phi}{\partial \xi_2} + \delta_3 \frac{\partial \Phi}{\partial \xi_3}$$

$$\nabla \cdot \mathbf{f} = \frac{1}{h_1 h_2 h_3} \left(\frac{\partial}{\partial \xi_1}(h_2 h_3 f_1) + \frac{\partial}{\partial \xi_2}(h_1 h_3 f_2) + \frac{\partial}{\partial \xi_3}(h_1 h_2 f_3) \right)$$

$$\nabla \times \mathbf{f} = \frac{1}{h_1 h_2 h_3} \begin{vmatrix} h_1 \delta_1 & h_2 \delta_2 & h_3 \delta_3 \\ \frac{\partial}{\partial \xi_3} & \frac{\partial}{\partial \xi_3} & \frac{\partial}{\partial i_3} \\ h_1 f_1 & h_2 f_2 & h_3 f_3 \end{vmatrix}$$

$$\nabla^2 = \frac{1}{h_1 h_2 h_3} \left(\frac{\partial}{\partial \xi_1}\left(\frac{h_2 h_3}{h_1}\frac{\partial}{\partial \xi_3}\right) + \frac{\partial}{\partial \xi_2}\left(\frac{h_1 h_3}{h_2}\frac{\partial}{\partial \xi_2}\right) + \frac{\partial}{\partial \xi_3}\left(\frac{h_1 h_2}{h_3}\frac{\partial}{\partial \xi_3}\right) \right)$$

$$\nabla(\nabla \cdot \mathbf{f}) = \sum_{i=1}^{3} \frac{1}{h_i}\frac{\partial}{\partial \xi_i}\left\{\frac{1}{h_1 h_2 h_3}\left[\frac{\partial f_1 h_2 h_3}{\partial \xi_1} + \frac{\partial f_2 h_1 h_3}{\partial \xi_2} + \frac{\partial f_3 h_1 h_2}{\partial \xi_3}\right]\right\} \delta_i$$

Mathematics in Chemical Engineering

$$\nabla \times (\nabla \times \mathbf{f}) = \frac{\boldsymbol{\delta}_1}{h_2 h_3} \left\{ \frac{\partial}{\partial \xi_2} \left[\frac{h_3}{h_1 h_2} \left(\frac{f_2 h_2}{\partial \xi_1} - \frac{f_1 h_1}{\partial \xi_2} \right) \right] - \frac{\partial}{\partial \xi_3} \left[a c h_2 h_1 h_3 \left(\frac{f_1 h_1}{\partial \xi_3} - \frac{f_3 h_3}{\partial \xi_2} \right) \right] \right\}$$

$$+ \frac{\boldsymbol{\delta}_2}{h_1 h_3} \left\{ \frac{\partial}{\partial \xi_3} \left[\frac{h_1}{h_2 h_3} \left(\frac{\partial f_3 h_3}{\partial \xi_2} - \frac{\partial f_2 h_2}{\partial \xi_3} \right) \right] - \frac{\partial}{\partial \xi_1} \left[\frac{h_3}{h_1 h_2} left \left(\frac{\partial f_2 h_2}{\partial \xi_3} - \frac{\partial f_1 h_1}{\partial \xi_2} \right) \right] \right\}$$

$$+ \frac{\boldsymbol{\delta}_3}{h_1 h_2} \left\{ \frac{\partial}{\partial \xi_1} \left[\frac{h_2}{h_1 h_3} \left(\frac{\partial f_1 h_1}{\partial \xi_3} - \frac{\partial f_3 h_3}{\partial \xi_1} \right) \right] - \frac{\partial}{\partial \xi_2} \left[\frac{h_1}{h_2 h_3} \left(\frac{\partial f_3 h_3}{\partial \xi_2} - \frac{\partial f_2 h_2}{\partial \xi_3} \right) \right] \right\}$$

$$\nabla^2 \mathbf{f} = \nabla(\nabla \cdot \mathbf{f}) - \nabla \times (\nabla \times \mathbf{f})$$

In the following discussion, we list the coordinate transformations followed by the scaling factors for a variety of coordinate systems.

2.5.3.3 Circular Cylindrical Coordinates

$$\xi_1 = r, \ \xi_2 = \theta, \ \xi_3 = z, \ x = r\cos\theta, \ y = r\cos\theta, \ z = z$$

$$h_1 = 1 \quad h_2 = r \quad h_3 = 1 \tag{2.65}$$

$$\nabla \Phi = \frac{\partial \Phi}{\partial r} \boldsymbol{\delta}_r + \frac{1}{r}\frac{\partial \Phi}{\partial \theta} \boldsymbol{\delta}_\theta + \frac{\partial \Phi}{\partial z} \boldsymbol{\delta}_z$$

$$\nabla \cdot \mathbf{f} = \frac{1}{r}\frac{\partial r f_r}{\partial r} + \frac{1}{r}\frac{\partial f_\theta}{\partial \theta} + \frac{\partial f_z}{\partial z}$$

$$\nabla \times \mathbf{f} = \frac{1}{r} \begin{vmatrix} \boldsymbol{\delta}_r & r\boldsymbol{\delta}_\theta & \boldsymbol{\delta}_z \\ \frac{\partial}{\partial r} & \frac{\partial}{\partial \theta} & \frac{\partial}{\partial z} \\ f_r & f_\theta & f_z \end{vmatrix}$$

$$\nabla^2 = \frac{1}{r}\frac{\partial}{\partial r}\left(r\frac{\partial}{\partial r}\right) + \frac{1}{r^2}\frac{\partial^2}{\partial \theta^2} + \frac{\partial^2}{\partial z^2}$$

2.5.3.4 Spherical Coordinates

$$\xi_1 = r, \ \xi_2 = \theta, \ \xi_3 = \phi, \ x = r\sin\theta\cos\phi, \ y = r\sin\theta\sin\phi, \ z = r\cos\theta$$

$$h_1 = 1 \quad h_2 = r \quad h_3 = r\sin\theta \tag{2.66}$$

$$\nabla \Phi = \frac{\partial \Phi}{\partial r}\delta_r + \frac{1}{r}\frac{\partial \Phi}{\partial \theta}\delta_\theta + \frac{1}{r\sin\theta}\frac{\partial \Phi}{\partial \phi}\delta_\phi$$

$$\nabla \cdot \mathbf{f} = \frac{1}{r^2}\frac{\partial}{\partial r}(r^2 f_r) + \frac{1}{r\sin\theta}\frac{\partial}{\partial \theta}(f_\theta \sin\theta) + \frac{1}{r\sin\theta}\frac{\partial f_\phi}{\partial \phi}$$

$$\nabla \times \mathbf{f} = \frac{1}{r^2 \sin\theta}\begin{vmatrix} \delta_r & r\delta_\theta & r\sin\theta\,\delta_\phi \\ \frac{\partial}{\partial r} & \frac{\partial}{\partial \theta} & \frac{\partial}{\partial \phi} \\ f_r & rf_\theta & r\sin\theta\,f_\phi \end{vmatrix}$$

$$\nabla^2 = \frac{1}{r^2}\frac{\partial}{\partial r}\left(r^2\frac{\partial}{\partial r}\right) + \frac{1}{r^2 \sin\theta}\frac{\partial}{\partial \theta}\left(\sin\theta\frac{\partial}{\partial \theta}\right) + \frac{1}{r^2 \sin^2\theta}\frac{\partial^2}{\partial \phi^2}$$

2.5.3.5 Elliptic Cylindrical Coordinates

$$x = a\cosh\xi_1 \cos\xi_2, \quad y = a\sinh\xi_1 \sin\xi_2$$

where ξ_1 = constant is a confocal ellipse described by

$$\left(\frac{x}{a\cosh\xi_1}\right)^2 + \left(\frac{y}{a\sinh\xi_1}\right)^2 = 1$$

and ξ_2 = constant is a confocal hyperbola described by

$$\left(\frac{x}{a\cos\xi_2}\right)^2 - \left(\frac{y}{a\sin\xi_2}\right)^2 = 1$$

$$h_1 = h_2 = a\sqrt{\cosh^2\xi_1 - \cos^2\xi_2} \quad h_3 = 1$$

2.5.3.6 Prolate Spheroidal Coordinates

$$x = a\cosh\xi_1 \cos\xi_2, \quad y = a\sinh\xi_1 \sin\xi_2 \sin\xi_3, \quad z = a\sinh\xi_1 \sin\xi_2 \cos\xi_3$$

where ξ_1 = constant is a prolate spheroid described by

$$\left(\frac{x}{a\cosh\xi_1}\right)^2 + \left(\frac{y}{a\sinh\xi_1}\right)^2 + \left(\frac{z}{a\sinh\xi_1}\right)^2 = 1$$

ξ_2 = constant is a hyperboloid two sheets described by

$$\left(\frac{x}{a\cos\xi_2}\right)^2 + \left(\frac{y}{a\sin\xi_2}\right)^2 - \left(\frac{z}{a\sin\xi_2}\right)^2 = 1$$

and ξ_3 = constant is a half-plane described by

$$\tan\xi_3 = \frac{y}{z}$$

$$h_1 = h_2 = a\sqrt{\sinh^2\xi_1 + \sin^2\xi_2} \quad h_3 = a\sinh\xi_1 \sin\xi_2$$

2.5.3.7 Oblate Spheroidal Coordinates

$$x = a\cosh\xi_1 \sin\xi_2 \sin\xi_3, \quad y = a\sinh\xi_1 \cos\xi_2, \quad z = a\cosh\xi_1 \sin\xi_2 \cos\xi_3$$

where ξ_1 = constant is an oblate spheroid described by

$$\left(\frac{x}{a\cosh\xi_1}\right)^2 + \left(\frac{y}{a\sinh\xi_1}\right)^2 + \left(\frac{z}{a\cosh\xi_1}\right)^2 = 1$$

ξ_2 = constant is a hyperboloid one sheet described by

$$\left(\frac{x}{a\sin\xi_2}\right)^2 - \left(\frac{y}{a\cos\xi_2}\right)^2 + \left(\frac{z}{a\sin\xi_2}\right)^2 = 1$$

and ξ_3 = constant is a half plane described by

$$\tan\xi_3 = \frac{x}{z}$$

$$h_1 = h_2 = a\sqrt{\cosh^2\xi_1 - \sin^2\xi_2} \quad h_3 = a\cosh\xi_1 \sin\xi_2$$

2.5.3.8 Parabolic Cylinder Coordinates

$$x = \frac{1}{2}(\xi_2^2 - \xi_1^2), \quad y = \xi_1\xi_2$$

where ξ_1 = constant is a parabolic cylinder described by

$$y^2 = \xi_1^2(\xi_1^2 + 2x)$$

and ξ_2 = constant is a parabolic cylinder described by

$$y^2 = \xi_1^2(\xi_1^2 - 2x)$$

$$h_1 = h_2 = \sqrt{\xi_1^2 + \xi_2^2} \quad h_3 = 1$$

2.5.3.9 Parabolic Coordinates

$$x = \frac{1}{2}(\xi_2^2 - \xi_1^2), \quad y = \xi_1\xi_2 \sin\xi_3, \quad z = \xi_1\xi_2 \cos\xi_3$$

where ξ_1 = constant is a paraboloid of revolution described by

$$y^2 + z^2 = \xi_1^2(\xi_1^2 + 2x)$$

and ξ_2 = constant is a paraboloid of revolution described by

$$y^2 + z^2 = \xi_1^2(\xi_1^2 - 2x)$$

$$\tan\xi_3 = \frac{y}{z}$$

$$h_1 = h_2 = \sqrt{\xi_1^2 + \xi_2^2} \quad h_3 = \xi_1\xi_2$$

2.5.3.10 Conical Coordinates

$$x^2 = \left(\frac{\xi_1\xi_2\xi_3}{bc}\right)^2 \quad y^2 = \frac{\xi_1^2(b^2 - \xi_2^2)(\xi_3^2 - b^2)}{b^2(c^2 - b^2)} \quad z^2 = \frac{\xi_1^2(c^2 - \xi_2^2)(c^2 - \xi_3^2)}{b^2(c^2 - b^2)}$$

$$0 < \xi_2^2 < b^2 < \xi_3^2 < c^2$$

where ξ_1 = constant is a sphere described by

$$x^2 + y^2 + z^2 = r^2 = \xi_1^2$$

ξ_2 = constant is an elliptic cone described by

$$\frac{x^2}{\xi_2^2} - \frac{y^2}{b^2 - \xi_2^2} - \frac{z^2}{c^2 - \xi_2^2} = 0$$

and ξ_3 = constant is an elliptic cone described by

$$\frac{x^2}{\xi_3^2} - \frac{y^2}{b^2 - \xi_3^2} - \frac{z^2}{c^2 - \xi_3^2} = 0$$

$$h_1 = 1 \quad h_2 = \sqrt{\frac{\xi_1^2(\xi_3^2 - \xi_2^2)}{(b^2 - \xi_2^2)(c^2 - \xi_2^2)}}, \quad h_3 = \sqrt{\frac{\xi_1^2(\xi_3^2 - \xi_2^2)}{(\xi_3^2 - b^2)(c^2 - \xi_3^2)}}$$

2.5.3.11 Ellipsoidal Coordinates

$$x^2 = \left(\frac{\xi_1 \xi_2 \xi_3}{bc}\right)^2 \quad y^2 = \frac{(\xi_1^2 - a^2)(\xi_2^2 - a^2)(a^2 - \xi_3^2)}{a^2(b^2 - a^2)} \quad z^2 = \frac{(\xi_1^2 - b^2)(b^2 - \xi_2^2)(b^2 - \xi_3^2)}{b^2(b^2 - a^2)}$$

$$0 \leq \xi_3^2 < a^2 < \xi_2^2 < b^2 < \xi_1^2 < \infty$$

where ξ_1 = constant is an ellipsoid described by

$$\left(\frac{x}{\xi_1}\right)^2 + \frac{y^2}{\xi_1^2 - a^2} - \frac{z^2}{\xi_1^2 - b^2} = 1$$

ξ_2 = constant is a hyperbola one sheet described by

$$\left(\frac{x}{\xi_2}\right)^2 + \frac{y^2}{\xi_2^2 - a^2} - \frac{z^2}{b^2 - \xi_2^2} = 1$$

ξ_3 = constant is a hyperbola two sheets described by

$$\left(\frac{x}{\xi_3}\right)^2 - \frac{y^2}{a^2 - \xi_3^2} - \frac{z^2}{b^2 - \xi_3^2} = 1$$

$$h_1 = \left[\frac{(\xi_1^2 - \xi_2^2)(\xi_1^2 - \xi_3^2)}{(\xi_1^2 - a^2)(\xi_1^2 - b^2)}\right]^{1/2} \quad h_2 = \left[\frac{(\xi_1^2 - \xi_3^2)(\xi_1^2 - \xi_2^2)}{(\xi_2^2 - a^2)(b^2 - \xi_2^2)}\right]^{1/2} \quad h_3 = \left[\frac{(\xi_1^2 - \xi_3^2)(\xi_2^2 - \xi_3^2)}{(a^2 - \xi_3^2)(b^2 - \xi_3^2)}\right]^{1/2}$$

2.5.3.12 Paraboloidal Coordinates

$$x^2 = 4\frac{(\xi_2 - a)}{a - b}(a - \xi_1)(a - \xi_3), \quad y^2 = 4\frac{(\xi_2 - b)}{a - b}(b - \xi_1)(\xi_2 - b), \quad z = \xi_1 + \xi_2 + \xi_3 - a - b$$

$$0 < \xi_1 < b < \xi_3 < a < \xi_2$$

where ξ_1 = constant is an elliptic paraboloid described by

$$\frac{x^2}{a - \xi_1} + \frac{y^2}{b - \xi_1} = 4(z - \xi_1)$$

ξ_2 = constant is an elliptic paraboloid described by

$$\frac{x^2}{a - \xi_2} + \frac{y^2}{b - \xi_2} = 4(z - \xi_2)$$

ξ_3 = constant is a hyperbolic paraboloid described by

$$\frac{x^2}{a-\xi_3} - \frac{y^2}{\xi_3-b} = 4(z-\xi_3)$$

$$h_1 = \left[\frac{(\xi_2-\xi_1)(\xi_2-\xi_3)}{(\xi_2-a)(\xi_2-b)}\right]^{1/2} \quad h_2 = \left[\frac{(\xi_2-\xi_1)(\xi_3-\xi_1)}{(a-\xi_1)(b-\xi_1)}\right]^{1/2} \quad h_3 = \left[\frac{(\xi_3-\xi_1)(\xi_2-\xi_1)}{(a-\xi_3)(\xi_3-b)}\right]^{1/2}$$

2.5.3.13 Bispherical Coordinates

$$x = a\xi_3 \frac{\sqrt{1-\xi_2^2}}{\xi_1-\xi_2}; \quad y = a\frac{\sqrt{(1-\xi_2^2)(1-\xi_3^2)}}{\xi_1-\xi_2}; \quad z = a\frac{\sqrt{\xi_1^2-1}}{\xi_1-\xi_2}$$

$$h_1 = \frac{a}{(\xi_1-\xi_2)\sqrt{\xi_1^2-1}}; \quad h_2 = \frac{a}{(\xi_1-\xi_2)\sqrt{1-\xi_2^2}}; \quad h_3 = \frac{a}{\xi_1-\xi_2}\sqrt{\frac{1-\xi_2^2}{1-\xi_3^2}}$$

2.5.3.14 Toroidal Coordinates

$$x = a\frac{\sqrt{\xi_1^2-1}}{\xi_1-\xi_2}; \quad y = a\frac{\sqrt{(\xi_1^2-1)(1-\xi_3^2)}}{\xi_1-\xi_2}; \quad z = a\frac{\sqrt{1-\xi_2^2}}{\xi_1-\xi_2}$$

$$h_1 = \frac{a}{(\xi_1-\xi_2)\sqrt{\xi_1^2-1}}; \quad h_2 = \frac{a}{(\xi_1-\xi_2)\sqrt{1-\xi_2^2}}; \quad h_3 = \frac{a}{\xi_1-\xi_2}\sqrt{\frac{\xi_2^2-1}{1-\xi_3^2}}$$

2.5.4 Vector Integral Theorems

Let V be a volume bounded by a closed surface S, and let $\mathbf{f}$ be a continuously differentiable vector field in V and S. If $d\mathbf{s}$ is an outward normal vector to the differential area, then we have the following equations as per various useful theorems.

Gauss's divergence theorem:

$$\iiint_V \nabla \cdot \mathbf{f}\,dv = \iint_S \mathbf{f}\cdot d\mathbf{s}$$

Gradient theorem:

$$\iiint_V \nabla f\,dv = \iint_S f\,d\mathbf{s}$$

Curl theorem:

$$\iiint_V \nabla \times \mathbf{f}\,dv = -\iint_S \mathbf{f}\times d\mathbf{s}$$

Stokes's theorem: Let S be a surface bounded by a closed curve C and $d\mathbf{r}$ be the tangent vector. Then

$$\iint_S (\nabla \times \mathbf{f}) \cdot d\mathbf{s} = \int_c \mathbf{f} \cdot d\mathbf{r}$$

Generalized form of Gauss's divergence theorem:

$$\iiint_V (\mathbf{a} \cdot \nabla)\mathbf{f} + \mathbf{f}(\nabla \cdot \mathbf{a})dv = \iint_S \mathbf{f}(\mathbf{a} \cdot d\mathbf{s})$$

Green's identities:

$$\iiint_V [\nabla\psi \cdot \nabla\phi + \psi\nabla^2\phi]dv = \iint_S \psi\nabla\phi \cdot d\mathbf{s}$$

$$\iiint_V [\psi\nabla^2\phi - \phi\nabla^2\psi]dv = \iint_S [\psi\nabla\phi - \phi\nabla\psi] \cdot d\mathbf{s}$$

Green's theorem: Let R be a region in a plane bounded by a closed curve C and P, and let Q be any two continuously differentiable functions in R and on C. Then

$$\iint_R \left[\frac{\partial Q}{\partial \xi_1} - \frac{\partial P}{\partial \xi_2}\right]d\xi_1 d\xi_2 = \oint_l [Qd\xi_2 + Pd\xi_1]$$

Reynolds's transport theorem: Let f be a continuous function of space and time defined throughout the volume $V(t)$ bounded by a closed surface $S(t)$ moving with velocity $\mathbf{v}$. Then

$$\frac{D}{Dt}\iiint_{V(t)} f dv = \iiint_{V(t)} \frac{\partial f}{\partial t}dv + \iint_{S(t)} f\mathbf{v} \cdot d\mathbf{s} = \iiint_{V(t)} \left(\frac{Df}{Dt} + f\nabla \cdot \mathbf{v}\right)dv$$

where

$$\frac{D}{Dt} \equiv \frac{\partial}{\partial t} + \mathbf{v} \cdot \nabla$$

2.5.5 Gradients of Sum and Product

We list here some of the important formulae involving the gradients of the sum and the product:

$$\nabla(\phi + \psi) = \nabla\phi + \nabla\psi$$

$$(\mathbf{a} \cdot \nabla)(\mathbf{f} + \mathbf{g}) = (\mathbf{a} \cdot \nabla)\mathbf{f} + (\mathbf{a} \cdot \nabla)\mathbf{g}$$

$$\nabla \cdot (\mathbf{f} + \mathbf{g}) = \nabla \cdot \mathbf{f} + \nabla \cdot \mathbf{g}$$

$$\nabla \times (\mathbf{f} + \mathbf{g}) = \nabla \times \mathbf{f} + \nabla \times \mathbf{g}$$

$$\nabla(\phi\psi) = \psi\nabla\phi + \phi\nabla\psi$$

$$\nabla(\mathbf{f}\cdot\mathbf{g}) = (\mathbf{f}\cdot\nabla)\mathbf{g} + (\mathbf{g}\cdot\nabla)\mathbf{f} + \mathbf{f}\times(\nabla\times\mathbf{g}) + \mathbf{g}\times(\nabla\times\mathbf{f})$$

$$\nabla\cdot(\phi\cdot\mathbf{f}) = \phi(\nabla\cdot\mathbf{f}) + \mathbf{f}\cdot\nabla\phi$$

$$(\mathbf{a}\cdot\nabla)(\phi\mathbf{f}) = \phi[(\mathbf{a}\cdot\nabla)\mathbf{f}] + \mathbf{f}[(\mathbf{a}\cdot\nabla)\phi]$$

$$(\mathbf{a}\cdot\nabla)(\mathbf{f}\times\mathbf{g}) = \mathbf{f}\times[(\mathbf{a}\cdot\nabla)\mathbf{g}] + [(\mathbf{a}\cdot\nabla)\mathbf{f}]\times\mathbf{g}$$

$$\nabla\cdot(\phi\mathbf{f}) = \mathbf{f}\cdot\nabla\phi + \phi\nabla\cdot\mathbf{f}$$

$$\nabla\cdot(\mathbf{f}\times\mathbf{g}) = \mathbf{g}\cdot(\nabla\times\mathbf{f}) - \mathbf{f}\cdot(\nabla\times\mathbf{g})$$

$$\nabla\times(\phi\mathbf{f}) = \nabla\phi\times\mathbf{f} + \phi\nabla\times\mathbf{f}$$

$$\nabla\times(\mathbf{f}\times\mathbf{g}) = (\mathbf{g}\cdot\nabla)\mathbf{f} - (\mathbf{f}\cdot\nabla)\mathbf{g} + \mathbf{f}\nabla\cdot\mathbf{g} - \mathbf{g}\nabla\cdot\mathbf{f}$$

2.6 DIMENSIONAL ANALYSIS

Dimensional homogeneity is fundamental to equations relating variables in the description of natural processes. The recognition of this basic attribute is the substance of dimensional analysis, which results in the reduction of relevant parameters to the essential minimum. Physical quantities comprise combinations of one or more basic dimensions. Table 2.2 shows some commonly used physical quantities in engineering expressed in terms of basic dimensions (mass, M; length, L; time, T; temperature, θ).

2.6.1 THEORY OF DIMENSIONAL ANALYSIS

Consider a specific process that involves n physical quantities comprising m basic dimensions among them. The dimension of each physical quantity is defined as the product of each of the basic dimensions raised to an appropriate exponent. For example, velocity can be expressed as LT^{-1}. Each physical quantity is then viewed as a vector of elements representing the respective exponents. Thus in the *MLT* system involving mass (M)-length (L)-time (T) as basic dimensions, velocity would be represented by the row vector [0 1 −1]. By lining up the vectors representing the different physical quantities (in columns), we obtain the *dimensional matrix*. A dimensionless group is a multiplicative combination of physical quantities with no net dimension and has the zero vector as its representative. Thus the formation of a dimensionless group can be identified with expressing the zero vector as a linear combination of vectors representing physical quantities in the group. From linear algebra, it follows that the number of independent dimensionless groups that can be formed is $n - r$, where r is the rank of the dimensional matrix, which is known as Buckingham's π-theorem. If the basic dimensions are independent, then $r = m$. Once the dimensionless groups $\pi_1, \pi_2, \ldots, \pi_{n-r}$ are determined, the physical process is described abstractly by an equation of the form $f(\pi_1, \pi_2, \ldots, \pi_{n-r}) = 0$. A couple of examples follow.

Mathematics in Chemical Engineering

TABLE 2.2
Dimensions of Commonly Used Physical Quantities in Engineering

Physical Quantity	Dimension
Mass	M
Length	L
Time	T
Density	ML^{-3}
Curvature	L^{-1}
Moment of inertia	ML^2
Momentum	MLT^{-1}
Viscosity	$M(LT)^{-1}$
Kinematic viscosity	L^2T^{-1}
Frequency	T^{-1}
Acceleration	LT^{-2}
Force	MLT^{-2}
Pressure	$M(LT^2)^{-1}$
Angular acceleration	T^{-2}
Surface tension	MT^{-2}
Energy, Torque	ML^2T^{-2}
Power	ML^2T^{-3}
Temperature	θ
Heat capacity per unit mass	$L^2(T^2\theta)^{-1}$
Heat capacity per unit volume	$M(T^2L\theta)^{-1}$
Temperature gradient	$(L\theta)^{-1}$
Conductivity	$ML(T^3\theta)^{-1}$

2.6.2 Applications of Dimensional Analysis

Example 2.3
Here we illustrate the use of dimensional analysis to determine the relationship between the drag force, F, on a smooth sphere and the affecting physical quantities, namely, the diameter of the sphere, D, the fluid velocity, V, density, ρ, and viscosity, μ. It is assumed that $f(D,V,F,\rho,\mu)=0$.

The dimensional matrix is formed as follows:

	D	V	F	ρ	μ
M	0	0	1	1	1
L	1	1	1	-3	-1
T	0	-1	-2	0	-1

To make the groups of all the physical quantities dimensionless, it is convenient to form the products as $D^{k_1}V^{k_2}F^{k_3}\rho^{k_4}\mu^{k_5}$. From the matrix, the following equations are formed:

$$k_3 + k_4 + k_5 = 0$$
$$k_1 + k_2 + k_3 - 3k_4 - k_5 = 0 \qquad (2.67)$$
$$-k_2 - 2k_3 - k_5 = 0$$

According to Buckingham's π-theorem, there exist two dimensionless groups. Thus there are two linearly independent solutions to Equation (2.67), and these are $k_1 = k_2 = -2$, $k_3 = 1$, $k_4 = -1$, $k_5 = 0$ and $k_1 = k_2 = k_4 = 1$, $k_3 = 0$, $k_5 = -1$. The resulting dimensionless groups are given by

$$\pi_1 = \frac{F}{\rho V^2 D^2} \quad \pi_2 = \frac{VD\rho}{\mu}$$

Experimental data are, however, correlated as $C_D = (8/\pi)\pi_1$, known as the drag coefficient, while π_2 is the well-known Reynolds number.

Example 2.4

The objective of this example is to show the utility of dimensional analysis in the interpretation of experimental data. Bose and coworkers (Barenblett, 1996) performed experiments with various fluids, namely, water, chloroform, bromoform, and mercury. The fluids were allowed to flow through a pipe in the turbulent regime. The times τ required for the fluids of density ρ and viscosity μ to attain a certain specified total discharge (volume) of Q under an imposed pressure drop P were measured. A graph of log P as a function of log$(1/\tau)$ showed four separate curves representing four different fluids. The same data can be plotted in terms of the new dimensionless groups, showing the collapse of all the data on a single curve through dimensional analysis, as was first shown by Von Kármán (1957):

The dimensional matrix may be written as

	P	τ	Q	ρ	μ
M	1	0	0	1	1
L	-1	0	3	-3	-1
T	-2	1	0	0	-1

The dimensionless groups are formed using $P^{k_1}\tau^{k_2}Q^{k_3}\rho^{k_4}\mu^{k_5}$:

$$k_1 + k_4 + k_5 = 0$$
$$-k_1 + 3k_3 - 3k_4 - k_5 = 0 \quad (2.68)$$
$$-2k_1 + k_2 - k_5 = 0$$

The two linearly independent solutions are $k_1 = 0$, $k_2 = -1$, $k_3 = 2/3$, $k_4 = 1$, $k_5 = -1$ and $k_1 = k_2 = 1$, $k_5 = -1$, $k_3 = k_4 = 0$. The two dimensionless groups are

$$\frac{\tau P}{\mu} \quad \text{and} \quad \frac{\rho Q^{2/3}}{\tau \mu}$$

The functional relation is

$$f\left(\frac{\tau P}{\mu}, \frac{\rho Q^{2/3}}{\tau \mu}\right)$$

Rearranging,

$$P = \frac{\mu}{\tau} g \left(\frac{\rho Q^{2/3}}{\tau \mu} \right) \quad (2.69)$$

When the experimental data were plotted using $\log(\tau P/m)$ as a function of $\log(\rho Q^{2/3}/\tau\mu)$, a single curve was indeed obtained for all the fluids. Thus the functional relationship in Equation (2.69) can be seen to be verified experimentally.

Software: Dimensional Analysis Toolbox in MatLab, written by Steffen Brückner, may be available free of cost for academic use. It may be obtained from http://www.sbrs.net/ along with the usage instruction. The user is referred to the license policy prior to use.

2.7 ALGEBRAIC EQUATIONS

We present here the solution methods for algebraic equations that may further be classified into linear and nonlinear equations.

2.7.1 System of Linear Equations

The subject of linear equations is best described in terms of concepts associated with linear algebra and matrix theory. The reader is referred to Amundson (1966) for details. We present here only the basic definitions and results that are important for the solution of linear algebraic equations.

Consider m equations in n unknowns $x_1, x_2, \ldots, x_n$ given by

$$\begin{aligned} a_{11}x_1 + a_{12}x_2 + \cdots + a_{1n}x_n &= b_1 \\ a_{21}x_1 + a_{22}x_2 + \cdots + a_{2n}x_n &= b_2 \\ &\vdots \\ a_{m1}x_1 + a_{m2}x_2 + \cdots + a_{mn}x_n &= b_m \end{aligned} \quad (2.70)$$

It can be compactly represented by vector-matrix notation as

$$\mathbf{Ax} = \mathbf{b} \quad (2.71)$$

where $\mathbf{A}$ is a rectangular matrix array of (known) coefficients, $\mathbf{x}$ is a column vector of the n unknowns, and $\mathbf{b}$ is a column vector of known constants, as shown below:

$$\mathbf{A} = \begin{bmatrix} a_{11} & a_{12} & \cdots & a_{1n} \\ a_{21} & a_{22} & \cdots & a_{2n} \\ \vdots & \vdots & \ddots & \vdots \\ a_{m1} & a_{m2} & \cdots & a_{mn} \end{bmatrix}, \mathbf{x} = \begin{bmatrix} x_1 \\ x_2 \\ \vdots \\ x_n \end{bmatrix}, \text{ and } \mathbf{b} = \begin{bmatrix} b_1 \\ b_2 \\ \vdots \\ b_m \end{bmatrix} \quad (2.72)$$

When the number of equations is less than the number of unknowns ($m < n$), at least $m - n$ of the unknowns must be assumed arbitrarily in terms of which the remaining unknowns can be calculated. When $m = n$, the number of equations is equal to the number of unknowns, a situation generally viewed to be sufficient for solving for all the unknowns. However, this issue is connected with situations when solutions exist in general. When $m > n$, the number of equations exceeds the number of unknowns, raising concerns about the consistency of the set of equations. The question of when a set of equations can be solved for the unknowns regardless of the relative values of m and n is answered elegantly by matrix theory. A few preliminary definitions are necessary before the necessary and sufficient conditions for the existence of the solution to a set of algebraic equations can be understood.

Vectors: The collection of all vectors containing n elements satisfying basic axioms (see, for example, Ramkrishna and Amundson (1985)) is called a linear space denoted by $\Re^n$. The basic axioms define scalar multiplication, vector addition (from which evolves the concept of linear combination), and a null vector that has all elements zero. Thus a linear combination of vectors $\mathbf{x_j}, j = 1,2,\ldots,k$, is expressed as the vector $\Sigma_{j=1}^{k} \alpha_j \mathbf{x_j}$, where the α_j's are numbers. If the α_j's are all zero, the linear combination gives the zero vector, 0. Linear spaces contain (linear) subspaces that are subsets in which linear combinations are contained within the set. Linear subspaces are also linear spaces.

Linear dependence: A set of vectors such as $\mathbf{x_i}, i = 1,2,\ldots,k$, is said to be linearly dependent if it is possible to find scalars not all zeroes such that the linear combination gives the zero vector. If this is not possible, the set is said to be linearly dependent. No vector in a linearly independent set can be expressed as a linear combination of the remaining vectors.

The maximum number of linearly independent vectors in a linear space or subspace is said to be the dimension of the space. Such a linearly independent set of vectors is said to be a *basis* for that space, by which it is meant that any arbitrary vector in the space can be expressed as a linear combination of the basis set.

Matrices: We consider the matrix $\mathbf{A}$ defined in Equation (2.72). If it has only one column ($n = 1$), it is the same as a column vector. If it has only one row, then it is a row vector. Thus vectors may be regarded as matrices themselves. The following properties are of interest.

Rank of a matrix: The rank of matrix $\mathbf{A}$ is the largest order of square array whose determinant is nonzero. Clearly, the rank of the matrix above cannot exceed the minimum of m and n. The definition of a determinant can be found in Amundson (1966). The matrix $\mathbf{A}$ may be regarded as a *transformation* of vectors in $\Re^n$ into a range of vectors, denoted $R(\mathbf{A})$, a subspace of $\Re^m$. The rank of $\mathbf{A}$ may also be seen to be the dimension of $R(\mathbf{A})$. When $m = n$, $\mathbf{A}$ is said to be a *square* matrix of order n. If its rank is less than n, it is said to be *singular*. Clearly, the determinant of the singular matrix is zero.

A square matrix that has only nonzero elements along its diagonal is called a diagonal matrix. If the diagonal matrix has all its diagonal elements equal to unity, it is called the identity matrix, denoted by $\mathbf{I}$. The diagonal matrix $\mathbf{D}$ and the identity matrix $\mathbf{I}$ are shown below:

$$\mathbf{D} = \begin{bmatrix} d_{11} & 0 & \cdots & 0 \\ 0 & d_{22} & \cdots & 0 \\ \vdots & \vdots & \ddots & \vdots \\ 0 & 0 & \cdots & d_{nn} \end{bmatrix}, \quad \mathbf{I} = \begin{bmatrix} 1 & 0 & \cdots & 0 \\ 0 & 1 & \cdots & 0 \\ \vdots & \vdots & \ddots & \vdots \\ 0 & 0 & \cdots & 1 \end{bmatrix}$$

Matrix addition/subtraction: Two matrices $\mathbf{A}$ and $\mathbf{B}$ are said to be compatible for addition/subtraction if both the matrices have the same numbers of rows and columns. The addition/subtraction of two matrices is carried out elementwise and simply follows

$$C = A \pm B \Rightarrow c_{ij} = a_{ij} \pm b_{ij} \quad (i=1,\ldots,n;\ j=1,\ldots,n)$$

Matrix multiplication: Two matrices, **A** and **B**, are said to be conformable for multiplication in the order **AB** if **A** has the same number of columns as **B** has rows. The multiplication is defined by

$$C = AB \Rightarrow c_{ik} = \sum_{i=1}^{n} a_{ij} b_{jk}$$

Thus square matrices can always be multiplied in any order. Also, we have **IA** = **A** for any matrix **A**, which also implies that an identity matrix raised to any exponent also gives **I**. A nonsingular matrix has an *inverse* matrix, denoted by A^{-1}, with the property $A^{-1}A = AA^{-1} = I$.

LU decomposition: For a square matrix **A** of order n, given that the determinants of the matrices A_p ($p = 1,2,\ldots,n-1$) formed by the elements at the intersection of the first p rows and columns of **A** are nonzeroes, then there exists a unique lower triangular matrix **L** and a unique upper triangular matrix **U** such that

$$A = LU$$

2.7.1.1 Solution of Linear Equations

We will first address the question of when the set of linear equations in Equation (2.71) has a solution. Homogeneous equations always have a solution. This is also evident from the fact that **x** = **0** is always a solution. Whether nontrivial solutions also exist depends on the rank of the matrix **A**. If r is the rank of the matrix **A**, then there are $(n-r)$ linearly independent solutions obtained by assuming arbitrary values for suitably selected $(n-r)$ variables and solving for the remaining r variables. One way to do this is as follows. Assume that the equations are so arranged that the rth-order upper left-hand matrix is nonsingular. Set *one* of the $(n-r)$ variables $x_{r+1}, x_{r+2},\ldots x_n$ to be unity and all others equal to zero. Solve for the variables $x_1, x_2,\ldots x_r$. This generates one of the linearly independent solutions. The others can be generated similarly. Clearly, the nontrivial solution exists when $n = r$.

Equation (2.71) has a unique solution when its homogeneous version has only the zero solution. Since the latter occurs only when $r = n$, we conclude that, for the case of $m = n$, i.e., the number of equations is the same as the number of unknowns, a unique solution occurs regardless of **b** when **A** is nonsingular. When $m \neq n$ and $r = n$, a unique solution exists for **Ax** = **b** as long as the ranks of matrices **A** and [**A**|**b**] are the same, where the *augmented* matrix [**A**|**b**] is defined by

$$[A\,|\,b] = \begin{bmatrix} a_{11} & a_{12} & \cdots & a_{1n} & b_1 \\ a_{21} & a_{22} & \cdots & a_{2n} & b_2 \\ \vdots & & & & \\ a_{m1} & a_{m2} & \cdots & a_{mn} & b_m \end{bmatrix}$$

For $r < n$, the solution for **Ax** = **b** exists when **b** satisfies the condition

$$b^T z_j = 0 \quad j = 1, 2,\ldots,(m-r) \tag{2.73}$$

where z_j are the $(m - r)$ linearly independent solutions of the homogeneous equation $\mathbf{b}^T\mathbf{z_j} = \mathbf{0}$. In this case, the solution to the inhomogeneous Equation (2.71) is not unique, since a particular solution to the inhomogeneous equation plus an arbitrary nontrivial solution of its homogeneous version is also a solution to the inhomogeneous Equation (2.71).

It should be noted that the calculation of an inverse matrix requires considerable computational time, and hence various methods have been proposed for the solution of a set of linear equations that take advantage of the structure of the matrix $\mathbf{A}$. The methods may be divided into two classes, namely, direct and iterative methods.

Cramer's rule, LU decomposition, and Gauss elimination methods form the class of *direct methods*, whereas Jacobi, Gauss-Seidel, and successive overrelaxation methods constitute the class of *iterative methods*. Among the direct methods, the Gauss elimination method, especially its partial pivoting variant, is the preferred method for large problems when sufficient computer memory is available for matrix storage. It relies on the transformation of the coefficient matrix into an upper triangular matrix problem and solving it by back substitution.

Iterative methods are sometimes used due to ease of computer coding and lesser computational storage requirements. The Jacobi method is the simplest iterative method but has slower convergence in comparison with the Gauss-Seidel method. In the Gauss-Seidel method, the $(k+1)$th iteration of the value of the unknown x_i is given by

$$x_i^{(k+1)} = \left(-\sum_{j=1}^{i-1} a_{ij} x_j^{(k+1)} - \sum_{j=i+1}^{i-1} a_{ij} x_j^{(k)} + b_i \right) \bigg/ a_{ii}, \quad i = 1, 2, \ldots, m$$

The successive overrelaxation method is a variant of the Gauss-Seidel method, wherein the $(k+1)$th iteration is a weighted average of the Gauss Seidel kth and $(k+1)$th estimates x_i and $x_i^{(k+1)}$, respectively. The reader is referred to Jensen and Jeffreys (1977) for a detailed account on the matrices and solution methods.

Generalized inverse (pseudoinverse): To obtain an approximate solution of an overdetermined system of linear equations, i.e., when the number of equations is greater than the number of unknowns $(m > n)$, a vector $\mathbf{x}$ is sought to minimize the square of residuals, $\mathbf{r}^T\mathbf{r}$, where

$$\mathbf{r} = \mathbf{Ax} - \mathbf{b}$$

The condition for minimization can be written as

$$d\mathbf{r}^T\mathbf{r} = 0$$

than implies

$$\mathbf{A}^T\mathbf{Ax} = \mathbf{A}^T\mathbf{b}$$

The matrix $\mathbf{B} = \mathbf{A}^T\mathbf{A}$ is symmetric and positive definite. If its inverse exists, then the least square solution is given by

$$\mathbf{x} = \mathbf{B}^{-1}\mathbf{A}^T\mathbf{b} = (\mathbf{A}^T\mathbf{A})^{-1}\mathbf{A}^T\mathbf{b}$$

The matrix $(\mathbf{A}^T\mathbf{A})^{-1}\mathbf{A}^T$ is called a pseudoinverse of $\mathbf{A}$, which becomes the actual inverse $\mathbf{A}^{-1}$ for $m = n$.

2.7.2 Nonlinear Equations

In what follows, we discuss the solution methods for polynomial equations, followed by the most commonly used iterative solution methods.

2.7.2.1 Polynomial Equations

A polynomial of degree n, $P_n(x)$, is represented as

$$P_n(x) = \sum_{k=0}^{n} a_k x^k = a_n x^n + a_{n-1} x^{n-1} + \cdots + a_1 x + a_0$$

where $a_n \neq 0$. A corresponding polynomial equation is written as

$$\sum_{k=0}^{n} a_k x^k = a_n x^n + a_{n-1} x^{n-1} + \cdots + a_1 x + a_0 = 0$$

A number r is called a root of the polynomial if

$$P_n(r) = 0$$

The fundamental theorem of algebra states that a polynomial of degree n has n number of roots, although some of the roots may be complex numbers. If a root of a polynomial of degree n is known to be, say r, then a polynomial of degree $n - 1$ is given by

$$P_{n-1}(x) = \frac{P_n(x)}{x - r}$$

In what follows, we provide a list of some of the important rules and formulae pertaining to polynomial equations.

Routh-Hurwitz criterion: The number of roots with positive real parts of a real polynomial equation is the number of sign changes in the following sequence:

$$\{D_1, \ D_1 D_2, \ D_2 D_3, \ \ldots, \ D_{n-1} D_{n-2}, \ a_0\}$$

where

$$D_1 = a_1 \quad D_2 = \begin{vmatrix} a_1 & a_0 \\ a_3 & a_2 \end{vmatrix} \quad D_3 = \begin{vmatrix} a_1 & a_0 & 0 \\ a_3 & a_2 & a_1 \\ a_5 & a_4 & a_3 \end{vmatrix} \quad D_4 = \begin{vmatrix} a_1 & a_0 & 0 & 0 \\ a_3 & a_2 & a_1 & a_0 \\ a_5 & a_4 & a_3 & a_2 \\ a_7 & a_6 & a_5 & a_4 \end{vmatrix} \cdots$$

$$D_n = \begin{vmatrix} a_1 & a_0 & 0 & 0 & \cdots & 0 \\ a_3 & a_2 & a_1 & 0 & \cdots & 0 \\ \cdots & \cdots & \cdots & \cdots & \cdots & \cdots \\ a_{2n-1} & a_{2n-2} & a_{2n-3} & a_{2n-4} & \cdots & a_n \end{vmatrix}$$

Descartes's rule of signs: The number of positive real roots of a real polynomial equation is either equal to the number of sign changes in the sequence of coefficients $\{a_n, a_{n-1}, \ldots, a_0\}$, or it is less by an even number. Application of this rule to $P(-x)$ gives the number of negative real roots.

Budan's rule of signs: The number of zeroes in the interval (a,b) is either equal to $V(a) - V(b)$ or less by an even number, where $V(x)$ is the number of sign changes in the sequence

$$\{P_n(x), P'_n(x), P''_n(x), \ldots, P_n^n(x)\}$$

Bounds on real roots: If the first k coefficients $(a_n, a_{n-1}, \ldots, a_{k-1})$ of a real polynomial are nonnegative, then all the real roots are less than U where

$$U = 1 + \sqrt[k]{\frac{M}{a_n}}$$

$$M = \max_{0 \le j \le k} |a_j|, \quad a_j < 0$$

Application of this bound to $P_n(-x)$ gives the lower bound L on the roots. Thus, all the real roots of a real polynomial are located in the interval (L, U).

Polynomials of degree $n \le 4$ can be solved by the following formulae.

Quadratic formula: The roots of a polynomial of degree $n = 2$ are given by the quadratic formula

$$x_1 = \frac{-a_1 + \sqrt{D}}{2a_2}$$

$$x_2 = \frac{-a_1 - \sqrt{D}}{2a_2}$$

where

$$D = a_1^2 - 4a_2 a_0$$

If $D > 0$, then there are two distinct real roots. If $D < 0$, then there are two distinct complex roots. If $D = 0$, then the repeated roots are given by

$$x_1 = x_2 = \frac{-a_1}{2a_2}$$

Cardano's formula: The roots of a polynomial of degree $n = 3$ are given by the cubic formula. The number of real roots of a cubic equation depends on the sign of the polynomial discriminant

$$D = \left(\frac{p}{3}\right)^3 + \left(\frac{q}{2}\right)^2$$

where

$$p = \left(\frac{3a_1 - a_2^2}{3}\right)$$

and

$$q = \left(\frac{9a_1 - a_2 27 a_0 - 2a_2^3}{27}\right)$$

If $D > 0$, there is one real root and two complex roots.
If $D < 0$, there are three real roots.
If $D = 0$, there is one real root and a real double root or a triple real root.

For $D \geq 0$, the roots are given by

$$x_1 = u_1 + u_2 - \frac{a_2}{3a_3} \qquad (2.74)$$

$$x_2 = m_1 u_1 + m_2 u_2 - \frac{a_2}{3a_3} \qquad (2.75)$$

$$x_3 = m_2 u_1 + m_1 u_2 - \frac{a_2}{3a_3} \qquad (2.76)$$

where

$$u_{1,2} = \sqrt[3]{-\frac{q}{2} \pm \sqrt{D}} \qquad m_{1,2} = \frac{1}{2}\left(-1 \pm i\sqrt{3}\right)$$

For $D < 0$, let

$$\rho = \sqrt{-\left(\frac{p}{3}\right)^3} \quad \text{and} \quad \cos\phi = -\frac{q}{2\rho}$$

Then the roots are given by

$$x_1 = 2\sqrt[3]{\rho}\cos\left(\frac{\phi}{3}\right) - \frac{a_2}{3a_3} \qquad (2.77)$$

$$x_2 = 2\sqrt[3]{\rho}\cos\left(\frac{\phi}{3} + \frac{2\pi}{3}\right) - \frac{a_2}{3a_3} \qquad (2.78)$$

$$x_3 = 2\sqrt[3]{\rho}\cos\left(\frac{\phi}{3} + \frac{4\pi}{3}\right) - \frac{a_2}{3a_3} \qquad (2.79)$$

Quartic formula: For $n = 4$, the roots are obtained by solving the cubic resolvent equation (Harris and Stocker, 1998):

$$8y^3 - 4\frac{a_2}{a_4}y^2 + \frac{2a_3a_1 - 8a_0}{a_4}y + 4\frac{a_2a_0 - a_3^2 a_0 - a_1^2}{a_4} = 0$$

The roots to the quartic equation are obtained by solving

$$x^2 + \frac{a_3 + a_4 D}{2a_4} + \left(y + \frac{a_3 y - a_1}{a_4 D}\right) = 0$$

where

$$D = \pm\sqrt{8y + \frac{a_3^2}{a_4} - 4\frac{a_2}{a_4}}$$

If $a_3 = a_1 = 0$, then the quartic equation is reduced to a quadratic equation by substituting $y = x^2$.

2.7.2.2 Numerical Solutions of Nonlinear Equations

We discuss here the use of two most prominent iterative methods of solutions of nonlinear equations through illustrations that are relevant to chemical engineers.

2.7.2.2.1 Successive Substitution

The steady-state mass and energy balances in a CSTR are given by Equations (2.20) and (2.21), respectively, which are reproduced here for ready reference:

$$F(C_f - C) - Vr(C,T) = 0 \tag{2.80}$$

$$\rho F C_p (T_f - T) - UA(T - T_c) + (-\Delta H)Vr(C,T) = 0 \tag{2.81}$$

These equations are written in nondimensionalized form as

$$1 - u - DaR(u,u) = 0 \tag{2.82}$$

$$1 - v + \beta DaR(u,u) = 0 \tag{2.83}$$

with the following dimensionless parameters (Morbidelli et al., 1986):

$$u = \frac{C}{C_f} \quad v = \frac{T}{T_m} \quad Da = \frac{Vr(C_f, T_m)}{FC_f} \quad \delta = \frac{UA}{F\rho C_p}$$

$$\beta = \frac{(-\Delta H)C_f}{\rho C_p T_m (1+\delta)} \quad R = \frac{r(C,T)}{r(C_f, T_m)} \quad T_m = \frac{T_f + \delta T_c}{1+\delta}$$

These can be combined by eliminating the reaction term to express the relationship between the dimensionless concentration and the dimensionless temperature as

$$u = \frac{1 + \beta - v}{\beta} \tag{2.84}$$

to give the dimensionless energy equation,

$$1 - v + \beta DaR\left(\frac{1+\beta-v}{v}, v\right) = 0 \tag{2.85}$$

The equation is rewritten as

$$v = 1 + \beta DaR\left(\frac{1+\beta-v}{v},\right) \tag{2.86}$$

so that it exhibits the form $x = g(x)$. This equation is solved by the following nonlinear recurrence equation (e.g., Reklaitis, 1983):

$$x_{n+1} = g(x_n) \tag{2.87}$$

Given an initial guess x_0, $g(x_0)$ is evaluated to give x_1. The process is repeated to obtain a sequence $\{x_0, x_1, \ldots, \}$ that may converge to the solution. The condition for the convergence of a sequence is given by

$$\left|g'(x)\right|_{x=x_\infty} < 1$$

where x_∞ is the solution. Thus, the condition ensures that the iteration converges to a stable solution. In general, the system of n equations in n unknowns is written in vector form as

$$\mathbf{f}(\mathbf{x}) = \mathbf{0} \qquad (2.88)$$

which can be solved by iterating the following equation:

$$\mathbf{x}_{n+1} = \mathbf{g}(\mathbf{x}_n)$$

The criterion for the convergence of such a sequence is that the absolute eigenvalues of the following Jacobian matrix evaluated at the root are less than 1:

$$\mathbf{Jg}(\mathbf{x})\bigg|_{\mathbf{x}=\mathbf{x}_\infty} = \begin{bmatrix} \dfrac{\partial g_1}{\partial x_1} & \dfrac{\partial g_1}{\partial x_2} & \cdots & \dfrac{\partial g_1}{\partial x_n} \\ \dfrac{\partial g_2}{\partial x_1} & \vdots & \cdots & \dfrac{\partial g_2}{\partial x_n} \\ \dfrac{\partial g_n}{\partial x_1} & \dfrac{\partial g_n}{\partial x_2} & \cdots & \dfrac{\partial g_n}{\partial x_n} \end{bmatrix}_{\mathbf{x}=\mathbf{x}_\infty}$$

The convergence condition stated above is difficult to ascertain, since it requires eigenvalues of the Jacobian to be evaluated at the root. A relatively loose condition is to ensure that the eigenvalues of the Jacobian are less than unity in the interval where the expected root is located. Using this condition and the application of the Gershgorin theorem (Bell, 1965), the condition for the convergence of iterations can be ensured for the function $g(x)$.

Example 2.5: Application of the Convergence Condition

Consider first-order reactions represented by

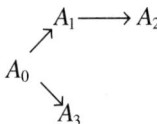

The first-order rate constants k_1 and k_3 associated with the reactions producing A_1 and A_3, respectively, are determined by measuring the rate of consumption of the reactant A_0 and the production of A_3. The rate constant, k_2, for the reaction producing A_2 is determined by solving a nonlinear equation that is derived as follows.

The concentrations of A_0, A_1, and A_3 as functions of time are obtained by integrating the following equations:

$$\frac{dA_0}{dt} = -(k_1 + k_3)A_0 \qquad (2.89)$$

Mathematics in Chemical Engineering

$$\frac{dA_1}{dt} = k_1 A_0 - k_2 A_1 \tag{2.90}$$

$$\frac{dA_3}{dt} = k_3 A_0 \tag{2.91}$$

$$A_0(0) = A_i \tag{2.92}$$

$$A_1(0) = A_2(0) = A_3(0) = 0 \tag{2.93}$$

to give

$$A_0 = A_i e^{-(k_1+k_3)t} \tag{2.94}$$

$$A_1 = \frac{k_1 A_i (e^{-(k_1+k_3)t} - e^{-k_2 t})}{k_2 - k_1 - k_3} \tag{2.95}$$

$$A_3 = \frac{k_3 A_i (1 - e^{-(k_1+k_3)t})}{k_1 + k_3} \tag{2.96}$$

Determination of k_2 involves the measurement of the concentration of A_1. Because A_1 is an intermediate product, the maximum concentration of A_1 occurs at t_{max} when

$$(k_1 + k_3) e^{-(k_1+k_3)t_{max}} = k_2 e^{-k_2 t_{max}} \tag{2.97}$$

In Equation (2.97), k_1, k_3, and t_{max} are known from experimental measurements. The unknown k_2 is obtained by solving this equation.

This equation admits two solutions. One solution occurs at the time less than t_{max} and the other at time greater than t_{max}. When the method of successive substitution is applied to the equation

$$k_2 = (k_1 + k_3) e^{k_2 - (k_1+k_3)t_{max}}$$

the condition for convergence for this equation is satisfied when $k_2 < 1/t_{max}$. The second solution cannot be obtained from this equation. By rewriting Equation (2.97) as

$$k_2 = (k_1 + k_3) + \frac{\ln[k_2/(k_1+k_3)]}{t_{max}}$$

the second solution with $k_2 > 1/t_{max}$ can now be obtained, since the condition for convergence is satisfied.

2.7.2.2.2 Newton-Raphson Iteration

The convergence of the successive substitution method is slow, i.e., it may require many iterations for the sequence to converge. The Newton-Raphson method has a faster rate of convergence, which is given as follows:

$$\mathbf{Jf}(\mathbf{x}_n)[\mathbf{x}_{n+1} - \mathbf{x}_n] = \mathbf{f}(\mathbf{x}_n) \tag{2.98}$$

where

$$\mathbf{Jf}(\mathbf{x}) = \begin{bmatrix} \dfrac{\partial f_1}{\partial x_1} & \dfrac{\partial f_1}{\partial x_2} & \cdots & \dfrac{\partial f_1}{\partial x_n} \\ \dfrac{\partial f_2}{\partial x_1} & \vdots & \cdots & \dfrac{\partial f_2}{\partial x_n} \\ \dfrac{\partial f_n}{\partial x_1} & \dfrac{\partial f_n}{\partial x_2} & \cdots & \dfrac{\partial f_n}{\partial x_n} \end{bmatrix}$$

The sequence $\mathbf{x}_{n+1}$ is generated by solving the system of linear equations defined by Equation (2.98). This sequence converges faster but requires a more accurate initial guess for x_0.

2.8 DIFFERENCE EQUATIONS

Difference equations are discrete analogues of differential equations. In this section, we present the methods of solving linear difference equations.

2.8.1 Method of Solution for Homogeneous Equations

Given a homogeneous kth-order linear difference equation

$$a_k y_{n+k} + a_{k-1} y_{n+k-1} + \cdots + a_0 y_n = 0 \tag{2.99}$$

the general solution of Equation (2.99) is determined by the solutions of the characteristic equation

$$a_k r^k + a_{k-1} r^{n+k-1} + \cdots + a_0 = 0 \tag{2.100}$$

If $r_1, r_2, \ldots, r_k$ are distinct roots of the characteristic equation, then the general solution is given by

$$y_n = \sum_{m=1}^{k} C_m r_m^n \tag{2.101}$$

where C_m's are constants to be determined from the boundary conditions.

If $r_1, r_2, \ldots, r_{m-p}$ are real but not distinct, and if r_p is a root of p-multiplicity, then the general solution is

$$y_n = \sum_{m=1}^{m} \sum_{k_j=0}^{k_{p-1}} C_{m,k_j} r_{m-p}^n k^m r_p^n \tag{2.102}$$

2.8.2 Method of Solution for Inhomogeneous Equations

Given the following inhomogeneous linear equations:

TABLE 2.3
Functional Form of Particular Solutions for y_p Depending on p_n

p_n	y_p
a^n	Aa^n
k^m	$\sum_{j=0}^{m} A_j k^j$
$\sin(bn)$ or $\cos(bn)$	$A\sin(bn) + B\cos(bn)$
$a^n \sin(nb)$ or $\cos(bn)$	$a^n[A\sin(bn) + B\cos(bn)]$

$$a_k y_{n+k} + a_{k-1} y_{n+k-1} + \cdots + a_0 y_n = p_n \quad (2.103)$$

the general solution of Equation (2.103) is given by

$$y_n = y_h + y_p$$

where y_h satisfies the homogeneous Equation (2.99) and y_p is a particular solution that depends on the functional form of p_n. The method of undetermined coefficients is used to solve for the particular solution y_p. This method assumes that the particular solution y_p depends on the form of p_n, as shown in Table 2.3.

Example: Application of the Solution Method
Consider the difference equation given by

$$8y_{n+2} - 6y_{n+1} + y_n = 2^n$$

The characteristic equation is

$$8r^2 - 6r + 1 = 0$$

The roots are $r_1 = 0.25$ and $r_2 = 0.5$. The particular solution has the functional form

$$y_p = A2^n$$

$$8y_{p,n+2} - 6y_{n+1} + y_n = 8A2^{n+2} - 6A2^{n+1} + A2^n = 2^n(32A - 12A + A) = 2^n$$

$$11A = 1 \quad \text{or} \quad A = \frac{1}{11}$$

The general solution is

$$y_n = C_1 \left(\frac{1}{4}\right)^n + C_2 \left(\frac{1}{2}\right)^n + \frac{1}{11} 2^n$$

The constants C_1 and C_2 can be calculated through the application of boundary conditions, enabling the complete specification of the solution.

2.8.3 Numerical Solutions to Ordinary Differential Equations

In this subsection, some commonly used numerical schemes that involve difference equations to solve ordinary differential equations are presented along with their stability characteristics. Simple examples to illustrate the effects of step size on the convergence of numerical methods are shown.

A simple discretization of the first-order linear differential equation

$$\frac{dy}{dz} = -\lambda y, \quad a \leq x \leq b, \quad y(a) = y_0$$

by a straightforward discretization using Euler's rule yields the following difference equation:

$$y_{n+1} = y_n - h\lambda y_n = (1 - h\lambda)y_n$$

where $h = (b - a)/n$. The solution is given by

$$y_n = (1 - h\lambda)^n y_0$$

The solution of the differential equation approaches zero as x approaches infinity. The solution of the difference equation clearly depends on h, and it approaches zero only if $h\lambda < 1$. For $h\lambda > 1$, the solution of the difference equation oscillates between very large positive and negative values for large n. Thus, the solution of the difference equation converges only for values h that satisfy $h\lambda < 1$.

For a nonlinear equation,

$$\frac{dy}{dx} = f(x, y), \quad a \leq x \leq b, \quad y(a) = y_0$$

the necessary and sufficient condition through convergence of Euler's rule is (Lambert, 1973)

$$\sup_{x \in (a,b); y \in (-\infty, \infty)} \left(h \left| \frac{\partial f}{\partial y} \right| \right) < 1$$

This example illustrates that the step size, h, plays a crucial role in the convergence of the numerical solution of a differential equation. Depending on the equation to be solved and the method of discretization, h can be adjusted to obtain convergence of a numerical solution.

A general method of solving a differential equation is to divide the interval of integration into n subintervals and then approximate the derivative by a one-step difference quotient, as in the Euler method or k-step difference quotient:

$$\frac{dy}{dx} = \frac{1}{h} \sum_{i=0}^{k} \alpha_i y_i \qquad (2.104)$$

where α_i's are constants with values ensuring the limit of the right-hand side of Equation (2.104) to the derivative of y at $x = 0$. Such an approximation introduces errors in computation. It also introduces

Mathematics in Chemical Engineering

an extraneous parameter h. The example shows that the growth or decay of the error is related to this parameter. Hence the strategy is to use the step length to control the growth or decay of the error. To derive conditions for controlling the error, a few notations and definitions are introduced.

The order of a method is the integer p such that the difference between the term on the right-hand side of Equation (2.104) and y' is proportional to h^{p+1}. The solution of the difference equation converges to that of the differential equation if the difference between the two solutions vanishes as the step size, h, approaches zero. A difference scheme is said to be zero-stable if the numerical solution of the equation for $f(x,y) = 0$ is zero.

In general, for $f(x,y) \neq 0$, the solution of the differential equation is approximated by the solution of the following difference equation:

$$\sum_{i=0}^{k} \alpha_i y_i = h \sum_{i=0}^{k} \beta_i f_i \qquad (2.105)$$

where α_k's and β_k's are the constants. If $\beta_k = 0$, the method is explicit; if $\beta_k \neq 0$, the method is implicit. The k-step methods require $k - 1$ starting values before they can be used to calculate the solution at the k-step. Explicit methods are easier to implement but are generally less stable. For nonlinear equations, implicit methods require iterative solutions, since the values of the unknown function at subsequent steps are not known.

The stability of a method leads to the condition that the modulus of the root, r, of the equation below should be less than unity:

$$\sum_{i=0}^{k} (\alpha_i - \bar{h}\beta_i) r^i = 0 \qquad (2.106)$$

where $\bar{h} \equiv h \frac{\partial f}{\partial y}$. Methods that satisfy this condition are absolutely stable. It is shown (Lambert, 1973) that for every consistent, zero-stable method, one root of Equation (2.106) is greater than unity in magnitude for positive and small $\bar{h}$. However, for positive $\bar{h}$, the solution of the differential equation also grows. Thus, for the solutions to be meaningful, it is required that the rate of growth of the error be less than that of the solution. If r_1 is the root with the largest magnitude, then the condition for convergence is

$$|r_i| < |r_1| \quad i = 2, 3, \ldots \qquad (2.107)$$

Methods that satisfy the conditions of Equation (2.107) are said to be relatively stable.

Often, the step size is chosen to increase the accuracy of the numerical solution. More importantly, the step size controls not only the error, but also the stability of the numerical scheme. The following problem is selected to illustrate the stability characteristics of the numerical method:

$$\frac{dy}{dx} = -50y^2 \quad y(0) = 1 \qquad (2.108)$$

The theoretical solution is

$$y_t(x) = \frac{1}{1 + 50x} \qquad (2.109)$$

TABLE 2.4
Theoretical and Numerical Solutions of Equation (2.108) by Four-Stage Runge-Kutta Method (y_{rk})

x	y_t, Equation (2.109)	y_{rk}, $h = 0.02$	y_{rk}, $h = 0.04$
0	1	1	1
0.04	0.3333	0.326900	−0.3333
0.08	0.2	0.197676	−0.958119
0.12	0.142857	0.141669	−578.994

The interval of absolute stability for four-stage Runge-Kutta is (−2.78,0). If the numerical solution is to converge to the theoretical solution, the step size h must be selected such that $-2.78 < \bar{h} < 0$. Because $dy/dx < 0$, the solution is a monotonically decreasing function of x, the maximum value of y is $y(0) = 1$, and the minimum of $\bar{h} = -100h < 2.78$. For $h = 0.01$ and 0.02, $\bar{h} = -1, -2$, respectively, which remain in the interval of absolute stability. For $h = 0.04$, $\bar{h} = -4$, which falls outside of the interval of absolute stability. Hence, the numerical solution diverges for $h = 0.04$.

The extension of a single equation to a system is accomplished with vector-dependent variables. In principle, the stability condition is still valid with the replacement of $\bar{h}$ by $h\rho$, where ρ is the maximum among all the absolute eigenvalues of the Jacobian of the vector $\mathbf{f}(x,\mathbf{y})$.

The reader is referred to Lambert (1973) for an excellent treatise on the computational methods for ordinary differential equations.

In what follows, we list the explicit and implicit methods for ready reference. The differences will be obvious through the formulations.

2.8.3.1 Explicit Methods

2.8.3.1.1 Runge-Kutta Methods

There exists a family of Runge-Kutta methods that differs by the number of functional evaluations of $f(x_r, y_r)$. A method that requires R functional evaluations is known as the R-stage Runge-Kutta method.

Two-stage Runge-Kutta method
1. Modified Euler method:

$$y_{n+1} - y_n = hf\left(x_n + \frac{1}{2}h, y_n + \frac{1}{2}hf(x_n, y_n)\right)$$

   Interval of absolute stability: (−2,0)
2. Improved Euler method:

$$y_{n+1} - y_n = \frac{1}{2}h[f(x_n, y_n) + f(x_n + h, y_n + hf(x_n, y_n))]$$

   Interval of absolute stability: (−2,0)

Three-stage Runge-Kutta method
1. Heun's third-order formula:

$$y_{n+1} - y_n = \frac{h}{4}(k_1 + 3k_3) \tag{2.110}$$

$$k_1 = f(x_n, y_n) \tag{2.111}$$

$$k_2 = f\left(x_n + \frac{1}{3}h, y_n + \frac{1}{3}hk_1\right) \tag{2.112}$$

$$k_3 = f\left(x_n + \frac{2}{3}h, y_n + \frac{2}{3}hk_2\right) \tag{2.113}$$

Interval of absolute stability: $(-2.51, 0)$

2. Kutta's third-order rule

$$y_{n+1} - y_n = \frac{h}{6}(k_1 + 4k_2 + k_3) \tag{2.114}$$

$$k_1 = f(x_n, y_n) \tag{2.115}$$

$$k_2 = f\left(x_n + \frac{1}{2}h, y_n + \frac{1}{2}hk_1\right) \tag{2.116}$$

$$k_3 = f(x_n + h, y_n - hk_1 + 2hk_2) \tag{2.117}$$

Interval of absolute stability: $(-2.51, 0)$

Four-stage Fourth-order Runge-Kutta method
1.

$$y_{n+1} - y_n = \frac{h}{6}(k_1 + 2k_2 + 2k_3 + k_4) \tag{2.118}$$

$$k_1 = f(x_n, y_n) \tag{2.119}$$

$$k_2 = f\left(x_n + \frac{1}{2}h, y_n + \frac{1}{2}hk_1\right) \tag{2.120}$$

$$k_3 = f\left(x_n + \frac{1}{2}h, y_n + \frac{1}{2}hk_2\right) \tag{2.121}$$

$$k_4 = f(x_n + h, y_n + hk_3) \tag{2.122}$$

Interval of absolute stability: $(-2.78, 0)$

2.

$$y_{n+1} - y_n = \frac{h}{8}(k_1 + 3k_2 + 3k_3 + k_4) \qquad (2.123)$$

$$k_1 = f(x_n, y_n) \qquad (2.124)$$

$$k_2 = f\left(x_n + \frac{1}{3}h, y_n + \frac{1}{3}hk_1\right) \qquad (2.125)$$

$$k_3 = f\left(x_n + \frac{2}{3}h, y_n - \frac{1}{3}hk_1 + hk_2\right) \qquad (2.126)$$

$$k_4 = f(x_n + h, y_n + hk_1 - hk_2 + hk_3) \qquad (2.127)$$

Interval of absolute stability: $(-2.78, 0)$

Fifth-order six-stage Runge-Kutta method

$$y_{n+1} - y_n = \frac{h}{90}(7k_1 + 32k_3 + 12k_4 + 32k_5 + 7k_6) \qquad (2.128)$$

$$k_1 = f(x_n, y_n) \qquad (2.129)$$

$$k_2 = f\left(x_n + \frac{1}{2}h, y_n + \frac{1}{2}hk_1\right) \qquad (2.130)$$

$$k_3 = f\left(x_n + \frac{1}{4}h, y_n + \frac{1}{16}h(3k_1 + k_2)\right) \qquad (2.131)$$

$$k_4 = f\left(x_n + \frac{1}{2}h, y_n + \frac{1}{2}hk_3\right) \qquad (2.132)$$

$$k_5 = f\left(x_n + \frac{3}{4}h, y_n + \frac{3}{16}h(-k_2 + 2k_3 + 3k_4)\right) \qquad (2.133)$$

$$k_6 = f\left(x_n + hy_n + \frac{1}{7}h(k_1 + 4k_2 + 6k_3 - 12k_4 + 8k_5)\right) \qquad (2.134)$$

Interval of absolute stability: $(-5.7, 0)$

2.8.3.1.2 Adams-Bashforth Family of Methods

For an in-depth discussion of the Adams-Bashforth family of methods, see Johnson and Riess (1982).

One-step Adams-Bashforth or Euler's Rule

$$y_{n+1} - y_n = h f_n \tag{2.135}$$

Interval of absolute stability: $(-2,0)$

Two-step Adams-Bashforth

$$y_{n+2} - y_{n+1} = \frac{h}{2}(3 f_{n+1} - f_n) \tag{2.136}$$

Interval of absolute stability: $(-1,0)$

Three-step Adams-Bashforth

$$y_{n+3} - y_{n+2} = \frac{h}{12}(23 f_{n+2} - 16 f_{n+1} + 5 f_n) \tag{2.137}$$

Interval of absolute stability: $(-6/11,0)$

Four-step Adams-Bashforth

$$y_{n+4} - y_{n+3} = \frac{h}{24}(55 f_{n+3} - 59 f_{n+2} + 37 f_{n+1} - 9 f_n) \tag{2.138}$$

Interval of absolute stability: $(-3/10,0)$

2.8.3.2 Implicit Methods

2.8.3.2.1 Trapezoidal, Simpson, and Runge-Kutta

Trapezoidal rule

$$y_{n+1} - y_n = \frac{h}{2}(f_{n+1} + f_n) \tag{2.139}$$

Interval of absolute stability: $(-\infty,0)$

Simpson's rule

$$y_{n+2} - y_n = \frac{h}{3}(f_{n+2} + 4 f_{n+1} + f_n) \tag{2.140}$$

Interval of relative stability: $(0,\infty)$

Fourth-order two-stage implicit Runge-Kutta method

$$y_{n+1} - y_n = \frac{h}{2}(k_1 + k_2) \qquad (2.141)$$

$$k_1 = f\left(x_n + \left(\frac{1}{2} + \frac{\sqrt{3}}{6}\right)h, y_n + \frac{1}{4}hk_1 + \left(\frac{1}{4} + \frac{\sqrt{3}}{6}\right)hk_2\right) \qquad (2.142)$$

$$k_2 = f\left(x_n + \left(\frac{1}{2} - \frac{\sqrt{3}}{6}\right)h, y_n + \left(\frac{1}{4} - \frac{\sqrt{3}}{6}\right)hk_1 + \frac{1}{4}hk_2\right) \qquad (2.143)$$

Interval of absolute stability: $(-\infty, 0)$.

2.8.3.2.2 Adams-Moulton Family of Methods

One-step Adams-Moulton, backward Euler's rule

$$y_{n+1} - y_n = hf_{n+1} \qquad (2.144)$$

Interval of absolute stability: $(-\infty, 0)$

Two-step Adams-Moulton, trapezoidal rule

$$y_{n+2} - y_{n+1} = \frac{h}{2}(f_{n+2} + f_{n+1}) \qquad (2.145)$$

Interval of absolute stability: $(-\infty, 0)$

Three-step Adams-Moulton

$$y_{n+3} - y_{n+2} = \frac{h}{12}(5f_{n+2} + 8f_{n+1} - f_{n+1}) \qquad (2.146)$$

Interval of absolute stability: $(-6, 0)$

Four-step Adams-Moulton

$$y_{n+4} - y_{n+3} = \frac{h}{24}(9f_{n+4} + 19f_{n+3} - 5f_{n+2} + f_{n+1}) \qquad (2.147)$$

Interval of absolute stability: $(-3, 0)$

Software: The method chosen of course depends on the nature of the equation. The analysis presented earlier may be of great help in selecting the right kind of step size balancing the stability and convergence.

2.9 ORDINARY DIFFERENTIAL EQUATIONS

In this section, formulae for the solution of selected differential equations are presented. The advantage of having the solution expressed in terms of the formula is that the solution is known without having to know the values of the parameters. Sometimes, the solution is given as an implicit nonlinear algebraic equation. In such cases, the advantage of analytical formulae over numerical computation diminishes.

2.9.1 Linear First-Order Differential Equation

A linear differential equation is an equation in which the unknown function appears in linear form, i.e., with the power of 1. The general linear first-order differential equation is

$$\frac{dy}{dx} + p(x)y(x) = q(x)$$

The general solution to this differential equation subjected to the condition

$$y(0) = y_0$$

is given by

$$y(x) = y_0 \exp\left[-\int_0^x p(t)dt\right] + \int_0^x q(t)\exp\left[\int_x^t p(s)ds\right]dt \qquad (2.148)$$

2.9.2 Nonlinear First-Order Differential Equation

2.9.2.1 Autonomous Nonlinear Equation

An autonomous equation is an equation in which the independent variable does not appear explicitly. The general form is

$$\frac{dy}{dx} = F(y), \quad y(0) = y_0$$

The general solution is given by the implicit function

$$\int_{y_0}^{y} \frac{dy}{F(y)} = x$$

2.9.2.2 Implicit Equation

An implicit differential equation is an equation in which the dependent variable is expressed as a function of its derivative:

$$y = f\left(\frac{dy}{dx}\right) \qquad (2.149)$$

$$\text{Let } p = \frac{dy}{dx} \quad \text{thus} \quad dy = pdx$$

From Equation (2.149),

$$dy = f'(p)dp = pdx$$

The solution is given by

$$y = f(p), \quad x = \int \frac{f'(p)dp}{p} + c \qquad (2.150)$$

$$x = f\left(\frac{dy}{dx}\right)$$

$$\text{Let } p = \frac{dy}{dx} \quad \text{thus} \quad dy = pdx$$

From Equation (2.150),

$$dx = f'(p)dp$$

$$dy = pdx = pf'(p)dp$$

The solution is given by

$$y = \int pf'(p)dp + c, \quad x = f(p)$$

2.9.2.3 Lagrange Equation

$$y = xf\left(\frac{dy}{dx}\right) + g\left(\frac{dy}{dx}\right)$$

$$\text{Let } p = \frac{dy}{dx} \quad \text{thus} \quad dy = pdx$$

$$pdx = f(p)dx + xf'(p)dp + g'(p)dp$$

Dividing throughout by dp gives

Mathematics in Chemical Engineering

$$\frac{dx}{dp}[p-f(p)] = xf'(p) + g'(p)$$

that can be solved to obtain the solution $x(p)$, which in conjunction with the following equation forms the complete solution.

$$y = xf(p) + g(p) = \int^p \frac{[f'(p) + g'(p)]dp}{p - f(p)}$$

2.9.2.4 Separable Equation

$$\frac{dy}{dx} = f(x)g(y)$$

The solution to the equation with $y(0) = y_0$ is given by

$$\int_{y_0}^{y} \frac{dy}{g(y)} = \int_0^x f(x)dx$$

2.9.2.5 Bernoulli Equation

$$\frac{dy}{dx} + P(x)y = Q(x)y^n, \quad y(0) = y_0$$

is transformed into a linear equation by the substitution

$$u(x) = y(x)^{1-n}$$

to get the equation for $u(x)$ as

$$\frac{1}{(1-n)}\frac{du}{dx} + P(x)u = Q(x), \quad y(0) = y_0$$

The solution is given by

$$y(x) = \left\{ y_0^{1-n} \exp\left[(n-1)\int_0^x P(t)dt\right] + (1-n)\int_0^x \exp\left[(1-n)\int_x^t P(s)ds\right]Q(t)dt \right\}^{1/1-n}$$

Example 2.6

Consider the following isothermal parallel reactions in a constant-volume batch reactor with the reaction order of 1 for the first reaction and order n for the second reaction:

$$A \xrightarrow{k_1} B$$

$$nA \xrightarrow{k_n} C$$

The mass balance for the species A is given by

$$\frac{dA}{dt} = -k_1 A - n k_n A^n, \quad A(0) = A_0$$

The solution is given by

$$A(t) = \left[\left(A_0^{1-n} + \frac{nk_n}{k_1} \right) e^{(n-1)k_1 t} - \frac{nk_n}{k_1} \right]^{1/(1-n)}, \quad n > 1$$

2.9.2.6 Exact Differential Equations

$$N(x,y)\frac{dy}{dx} + M(x,y) = 0, \quad y(0) = y_0$$

This is often written in the form

$$M(x,y)dx + N(x,y)dy = 0$$

The equation is said to be exact if

$$\frac{\partial M}{\partial y} = \frac{\partial N}{\partial x}$$

Then the solution is given by the implicit function, $\psi(x,y)$, where

$$\psi(x,y) = 0 \int_0^x M(t,y)dt + \int_{y_0}^y \left[N(x,s) - \int_0^x \frac{\partial M(t,s)}{\partial s} dt \right] ds$$

2.9.2.7 Homogeneous Equations

A nonlinear differential equation

$$\frac{dy}{dx} = f(x,y)$$

is said to be homogeneous of degree 0 if

$$f(tx,ty) = f(x,y) = tF\left(\frac{y}{x}\right)$$

The transformation $y = xv$ converts the equation to

$$x\frac{dv}{dx} = F(v) - v$$

that then can be solved by the solution method for a separable equation.

2.9.2.8 Ricatti Equations

$$\frac{dy}{dx} = P(x)y^2 + Q(x)y + R(x), \quad y(0) = y_0 \qquad (2.151)$$

Ricatti equations can be solved if P, Q, and R are constant. The solution is then

$$x = \int_{y_0}^{y} \frac{dz}{Pz^2 + Qz + R}$$

In general, if a particular solution $y_1(x)$ to Equation (2.151) is known, then letting $y(x) = z(x) + y_1(x)$ transforms Equation (2.151) to the Bernoulli equation:

$$\frac{dz}{dx} = P(x)z^2 + (2P(x)y_1 + Q(x))z$$

For the dynamics of a CSTR of volume V fed with the reactant concentration, C_f, with an nth-order chemical reaction,

$$\frac{dC}{dt} = \frac{F}{V}(C_f - C) - kC^n$$

$$C(0) = C_0$$

The solution is given by

$$t = \int_{C_0}^{C} \frac{dC}{F(C_f - C)/V - kC^n} \qquad (2.152)$$

For $n = 2$, the integral in Equation (2.152) can be evaluated to give

$$-\omega t = \ln\left[\frac{(C - C_+)(C_0 - C_-)}{(C - C_-)(C_0 - C_+)}\right]$$

that can be solved for C as

$$C = \frac{C_+(C_0 - C_-) - C_-(C_0 - C_+)e^{-\omega t}}{(C_0 - C_-) - (C_0 - C_+)e^{-\omega t}}$$

where

$$C_\pm = \frac{F}{2kV}\left(1 \pm \sqrt{1 + 4C_f kV/F}\right) \quad \text{and} \quad \omega = \frac{F}{V}\sqrt{1 + 4C_f kV/F}$$

2.9.3 SECOND-ORDER DIFFERENTIAL EQUATIONS

2.9.3.1 Homogeneous Linear Equations with Constant Coefficients

The mass balance for a first-order reaction in a tubular reactor with a flow velocity of v and the concentration of reactant in feed, C_f, undergoing a reaction with the rate constant of k is governed by the following second-order linear equation with constant coefficients:

$$D\frac{d^2C}{dx^2} - v\frac{dC}{dx} - kC = 0 \qquad (2.153)$$

subjected to boundary conditions:

$$-D\frac{dC}{dx} = v(C_f - C), \quad x = 0$$

$$\frac{dC}{dx} = 0, \quad x = L$$

where L is the reactor length and D is the diffusion coefficient. The solutions are determined by the roots of the characteristic equation,

$$D\lambda^2 = v\lambda - k = 0$$

that is given by

$$\lambda_{1,2} = \frac{v \pm \sqrt{v^2 + 4Dk}}{2D}$$

$$C(x) = A\exp(\lambda_1 x) + B\exp(\lambda_2 x)$$

Equivalently,

$$C(x) = \exp\left(\frac{v}{2D}x\right)[A\cosh(\gamma x) + B\sinh(\gamma x)] \qquad (2.154)$$

where

$$\gamma = \frac{\sqrt{v^2 + 4Dk}}{2D}$$

$$A = \frac{vC_f(v\sinh(\gamma L) + 2D\gamma\cosh(\gamma L))}{(v^2 + 2Dk)\sinh(\gamma L) + 2Dv\gamma\cosh(\gamma L)}$$

$$B = -\frac{vC_f(v\cosh(\gamma L) + 2D\gamma\sinh(\gamma L))}{(v^2 + 2Dk)\sinh(\gamma L) + 2Dv\gamma\cosh(\gamma L)}$$

2.9.3.2 Inhomogeneous Linear Differential Equations with Constant Coefficients

Consider an inhomogeneous linear differential equation with constant coefficients given by

$$D\frac{d^2C}{dx^2} - v\frac{dC}{dx} - kC = S(x) \qquad (2.155)$$

The solution of Equation (2.155) is given by

$$C(x) = Ay_1(x) + By_2(x) + C_p(x)$$

where $C_p(x)$ is obtained by the method of variation of parameters, as follows. Let

$$C_p(x) = \alpha(x)y_1(x) + \beta(x)y_2(x)$$

where $y_1(x)$ and $y_2(x)$ are the two linearly independent solutions given in the previous section as

$$y_1(x) = \exp\left(\frac{v}{2D}x\right)\cosh(\gamma x), \quad y_2(x) = \exp\left(\frac{v}{2D}x\right)\sinh(\gamma x)$$

where

$$\gamma = \frac{\sqrt{v^2 + 4Dk}}{2D}$$

$$C'_p(x) = \alpha'(x)y_1(x) + \beta'(x)y_2(x) + \alpha(x)y'_1(x) + \beta(x)y'_2(x)$$

For $C'(x)$ to assume the same functional form as if α and β were constants, it is required that

$$\alpha'(x)y_1(x) + \beta'(x)y_2(x) = 0$$

The second derivative is then

$$C''_p(x) = \alpha'(x)y'_1(x) + \alpha(x)y''_1(x) + \beta'(x)y'_2(x) + \beta(x)y''_2(x)$$

Substituting $C''_p(x)$ and $C'_p(x)$ into the nonhomogeneous equation, one obtains a system of first-order differential equations for $\alpha(x)$ and $\beta(x)$:

$$\alpha'(x)y_1(x) + \beta'(x)y_2(x) = 0$$

$$\alpha'(x)y'_1(x) + \beta'(x)y'_2(x) = S(x)$$

The solution is given by

$$\alpha(x) = -\int^x \frac{S(\xi)}{y_1(\xi)y_2'(\xi) - y_1'(\xi)y_2(\xi)} y_2(\xi)d\xi$$

$$\beta(x) = \int^x \frac{S(\xi)}{y_1(\xi)y_2'(\xi) - y_1'(\xi)y_2(\xi)} y_1(\xi)d\xi$$

Substituting $\alpha(x)$ and $\beta(x)$ in $C_p(x)$ gives the particular solution

$$C_p(x) = \int^x \frac{y_1(t)y_2(x) - y_1(x)y_2(t)}{y_1(t)y_2'(t) - y_1'(t)y_2(t)} S(t)dt$$

and the general solution

$$C(x) = Ay_1 + By_2 + \int^x \frac{y_1(t)y_2(x) - y_1(x)y_2(t)}{y_1(t)y_2'(t) - y_1'(t)y_2(t)} S(t)dt$$

where A and B are constants that can be determined on the specification of boundary conditions.

2.9.3.3 Green's Function

The method of Green's function expresses the solution of the inhomogeneous boundary-value problem as the integral representation of the inhomogeneous function (Friedman, 1956; Stakgold, 1979). Given the boundary-value problem,

$$a_2 \frac{d^2y}{dx^2} + a_1 \frac{dy}{dx} + a_0 y = g(x)$$

$$\alpha_1 \frac{dy}{dx} + \beta_1 y = 0, \quad x = 0 \qquad (2.156)$$

$$\alpha_2 \frac{dy}{dx} + \beta_2 y = 0, \quad x = 1$$

The above boundary-value problem is transformed to the following self-adjoint forms:

$$\frac{d}{dx}\left(p(x)\frac{dy}{dx}\right) + q(x)y = f(x)$$

$$B_1(y) = \alpha_1 \frac{dy}{dx} + \beta_1 y = 0, \quad x = 0 \qquad (2.157)$$

$$B_2(y) = \alpha_2 \frac{dy}{dx} + \beta_2 y = 0, \quad x = 1$$

$$p(x) = \exp\left(\int^x \frac{a_1}{a_2} dx\right), \quad q(x) = \frac{a_0(x)p(x)}{a_2(x)}, \quad f(x) = \frac{p(x)g(x)}{a_2(x)}$$

Green's function for Equation (2.157) is given by

$$G(x,\xi) = \begin{cases} \dfrac{v_1(x)v_2(\xi)}{p(\xi)W(v_1(\xi),v_2(\xi))} & x \le \xi \\ \dfrac{v_1(\xi)v_2(x)}{p(\xi)W(v_1(\xi),v_2(\xi))} & x > \xi \end{cases}$$

where $v_i(x)$ ($i = 1,2$) are the two linearly independent solutions of the homogeneous equation such that $B_i(v_i) = 0$, and $W(u,v)$ is the Wronskian of u and v:

$$W(u,v) = \begin{vmatrix} u & v \\ u' & v' \end{vmatrix}$$

that is nonzero, because u and v are linearly independent solutions.

The $pW(u,v)$ is constant in the interval $[0,1]$ for any pair of functions u and v satisfying the homogeneous version of the differential equation in (2.156) and the homogeneous boundary conditions.

The solution of Equation (2.157) is given by

$$y(x) = \int_0^1 G(x,\xi) g(\xi) d\xi$$

Equation (2.156), with the following nonhomogeneous boundary conditions,

$$\alpha_1 \frac{dy}{dx} + \beta_1 y = \gamma_1, \quad x = 0$$

$$\alpha_2 \frac{dy}{dx} + \beta_2 y = \gamma_2, \quad x = 1$$

is given by

$$y(x) = \frac{\gamma_2 p(1)[\alpha_2 v_2(1) - \beta_2 v_2'(1)]}{p(x)W(v_1(x),v_2(x))} v_1(x) - \frac{\gamma_1 p(0)[\alpha_1 v_1(0) - \beta_1 v_1'(0)]}{p(x)W(v_1(x),v_2(x))} v_2(x) + \int_0^1 G(x,\xi) g(\xi) d\xi$$

Example 2.7
The steady-state mass balance in a tubular reactor with a source term (reactant fed along the length of the reactor) is governed by the following boundary-value problem:

$$D \frac{d^2 C}{dx^2} - v \frac{dC}{dx} - ky = -S(x)$$

$$D \frac{dC}{dx} - vC = -vC_f, \quad x = 0$$

$$\frac{dC}{dx} = 0, \quad x = L$$

The solution is given by

$$C(x) = C_1(x) + \int_0^1 \exp\left(-\frac{v}{D}x\right) G(x,\xi) S(\xi) d\xi$$

where $C_1(x)$ is the solution of the homogeneous equation and $C_0 \neq 0$ is given by Equation (2.154), and the Green's function $G(x,\xi)$ for $C_0 = 0$ is given by

$$G(x,\xi) = \begin{cases} -\dfrac{4D^2 q v_1(x) v_2(\xi)}{4Dqv\cosh(qL) + (4D^2 q^2 + v^2)\sinh(q)} & x < \xi \\ -\dfrac{4D^2 q v_1(\xi) v_2(x)}{4Dqv\cosh(qL) + (4D^2 q^2 + v^2)\sinh(q)} & x \geq \xi \end{cases} \quad (2.158)$$

where the two linearly independent solutions v_1 and v_2 satisfying $B_i(v_i)$ are

$$v_1(x) = \exp\left(\frac{v}{2D}x\right)\left[\cosh(qx) + \frac{vL}{2Dq}\sinh(qx)\right]$$

$$v_2(x) = \exp\left(\frac{v}{2D}x\right)\left[\cosh(q(x-L)) - \frac{vL}{2Dq}\sinh(q(x-L))\right]$$

2.9.3.4 Green's Function by Eigenfunction (Mercer's) Expansions

Consider the Green's function defined by the following differential equations:

$$\frac{d}{dx}\left(p(x)\frac{dG}{dx}\right) + q(x)G - \lambda r(x)G = \delta(x-\xi)$$

$$\alpha_1 \frac{dG(a,\xi;\lambda)}{dx} + \beta_1 G(a,\xi;\lambda) = 0, \quad x = a \quad (2.159)$$

$$\alpha_2 \frac{dG(b,\xi;\lambda)}{dx} + \beta_2 G(b,\xi;\lambda) = 0, \quad x = b$$

where $\delta(x-\xi)$ is the dirac delta function (see Appendix 2.B). Let $y_i(x)$ be the normalized eigenfunctions corresponding to the eigenvalues λ_i of the operator, then we have, $Ly_i = \lambda_i y$ where

$$L \equiv -\frac{1}{r(x)}\left[\frac{d}{dx}\left(p(x)\frac{d}{dx}\right) + r(x)q(x)\right] \quad (2.160)$$

Since L is self-adjoint, the eigenfunctions y_i and y_j, corresponding to different eigenvalues λ_i and λ_j, respectively, are orthogonal, i.e.,

$$\int_a^b y_i(x) y_j(x) r(x) dx = \delta_{i,j} = \begin{cases} 1 & \text{if } i = j \\ 0 & \text{if } i \neq j \end{cases}$$

Let

$$G(x;\xi;\lambda) = \sum_i c_i(\xi;\lambda) y_i(x) \qquad (2.161)$$

where

$$c_i = \int_a^b r(x) y_i(x) G(x,\xi;\lambda) dx$$

Multiplying Equation (2.159) by $y_i(x)$ and integrating over (a,b) gives

$$\int_a^b \left[\frac{d}{dx}\left(p(x)\frac{dG}{dx} \right) + q(x)G - \lambda r(x)G \right] y_i dx = \int_a^b \delta(x-\xi) y_i dx$$

$$= \int_a^b \underbrace{\left[\frac{d}{dx}\left(p(x)\frac{dy_i}{dx} \right) + q(x)y_i \right]}_{\lambda_i r(x) y_i(x)} G dx - \int_a^b \lambda r(x) y_i G dx = \int_a^b \delta(x-\xi) y_i(x) dx$$

$$\lambda_i \int_a^b r(x) y_i(x) G dx - \lambda \int_a^b r(x) y_i(x) G dx = y_i(\xi)$$

$$c_i(\lambda_i - \lambda) = y_i(\xi)$$

$$G(x,\xi;\lambda) = \sum_i \frac{y_i(x) y_i(\xi)}{\lambda_i - \lambda}$$

2.9.4 LINEAR HIGHER-ORDER DIFFERENTIAL EQUATIONS

2.9.4.1 Homogeneous Equation

$$\sum_{i=1}^n a_i \frac{d^i y}{dt^i} = 0$$

The general solution is given by

$$\sum_{i=1}^n a_i \lambda^i = 0$$

where λ_i are the roots of the polynomial. Now (a) if the roots of the polynomial are distinct, then the solution is given by

$$y(x) = \sum_{i=1}^n c_i \exp(\lambda_i x)$$

or (b) if λ_i is a root with multiplicity $k_i (\Sigma_j^m k_j = n)$, then the solution is given by

$$y(x) = \sum_{i=1}^{m} \left(\sum_{k_j=0}^{k_i-1} c_{i,k_j} t^{k_j} \right) \exp(\lambda_i x)$$

2.9.4.2 Inhomogeneous Equation

$$\sum_{i=1}^{n} a_i \frac{d^i y}{dt^i} = f(t)$$

Let $y_i = \exp(\lambda_i x)$, $i = 1,2,\ldots,n$ be the n solutions of the homogeneous equation, i.e., the n solutions for $f(t) = 0$. Let us define the vectors

$$\mathbf{Y}_i = \begin{bmatrix} y_i \\ \dfrac{dy_i}{dx} \\ \cdots \\ \dfrac{d^{n-1} y_i}{dx^{n-1}} \end{bmatrix} \quad \boldsymbol{\delta}_n = \begin{bmatrix} 0 \\ 0 \\ \vdots \\ 1 \end{bmatrix}$$

The Wronskian of these solutions is defined as

$$W(\mathbf{Y}_1, \mathbf{Y}_2, \ldots, \mathbf{Y}_n) = \begin{vmatrix} y_1 & y_2 & \cdots & y_n \\ \dfrac{dy_1}{dx} & \dfrac{dy_2}{dx} & \cdots & \dfrac{dy_n}{dx} \\ \dfrac{d^2 y_1}{dx} & \vdots & \cdots & \dfrac{d^2 y_n}{dx} \\ \vdots & \vdots & \vdots & \vdots \\ \dfrac{d^{n-1} y_1}{dx} & \dfrac{d^{n-1} y_2}{dx} & \cdots & \dfrac{d^{n-1} y_n}{dx} \end{vmatrix}$$

and let

$$W_1 = W(\boldsymbol{\delta}_n, \mathbf{Y}_2, \ldots, \mathbf{Y}_n) \cdots W_n = W(\mathbf{Y}_1, \mathbf{Y}_2, \ldots, \boldsymbol{\delta}_n) \quad \text{and} \quad \boldsymbol{\delta}_n = [0 \quad 0 \quad \cdots \quad 0 \quad 1]^T$$

$$y(x) = y_h(x) + \sum_{i=1}^{n} y_i(x) \int_0^x \frac{W_i}{W(\mathbf{Y}_1, \mathbf{Y}_2, \ldots, \mathbf{Y}_n)} f(t) dt$$

where $y_h(x)$ is the solution of the homogeneous equation.

2.9.4.3 Application of Higher-Order Equations

Consider the sequence of reactions following the first-order kinetics

$$A \xrightarrow{k_1} B \xrightarrow{k_2} C$$

taking place in a tubular reactor. The mass balances for A and B are governed by the following system of boundary-value problems:

$$D_A \frac{d^2 C_A}{dx^2} - v_A \frac{dC_A}{dx} - k_1 C_A = 0 \tag{2.162}$$

$$x = 0, \quad v_A(C_{A0} - C_A) = -D_A \frac{dC_A}{dx}; \quad x = L, \quad \frac{dC_A}{dx} = 0 \tag{2.163}$$

$$D_B \frac{d^2 C_B}{dx^2} - v_B \frac{dC_B}{dx} + k_1 C_A - k_2 C_B = 0 \tag{2.164}$$

$$x = 0, \quad v_B C_B = D_B \frac{dC_B}{dx}; \quad x = L, \frac{dC_B}{dx} = 0 \tag{2.165}$$

where v_A and v_B are the convective velocities of species A and B, respectively, while D_A and D_B indicate the diffusivities of species A and B, respectively. A single fourth-order differential equation can be obtained by eliminating C_A from Equation (2.164), as

$$D_A D_B \frac{d^4 C_B}{dx^4} - (D_A v_B + D_B v_A) \frac{d^3 C_B}{dx^3} \ldots$$
$$- (D_A k_2 - v_A v_B + D_B k_1) \frac{d^2 C_B}{dx^2} + (v_A k_2 + v_B k_1) \frac{dC_B}{dx} + k_1 k_2 C_B = 0 \tag{2.166}$$

The boundary conditions are derived by eliminating C_A from Equation (2.164),

$$x = 0, \quad -D_A D_B \frac{d^3 C_B}{dx^3} + (D_A v_B + D_B v_A) \frac{d^2 C_B}{dx^2} \ldots$$
$$+ (D_A k_2 - v_A v_B) \frac{dC_B}{dx} - v_A k_2 C_B - k_1 v_A C_{A0} = 0 \tag{2.167}$$

$$x = L, \quad D_B \frac{d^3 C_B}{dx^3} - v_B \frac{d^2 C_B}{dx^2} = 0 \tag{2.168}$$

Equation (2.166), along with the two original and the two new boundary conditions given by Equation (2.167) and Equation (2.168), form a boundary-value problem for C_B.

Alternatively, substituting the solution of Equation (2.162), which is given by Equation (2.154), into Equation (2.164) gives an equation for C_B that is solved by the method of Green's function:

$$C_B(x) = k_2 \int_0^L C_A(\xi) G(x, \xi) d\xi$$

where the Green's function is given by Equation (2.158).

2.9.4.4 Finite Element Method

Various formulations of finite element methods have been proposed. For an exhaustive account on finite element methods, the reader is referred to Chen (2005), Donea (2003), Reddy (2005), etc. We present here one of the popular formulations known as the weak formulation of the governing differential equation that, instead of requiring the solution to be twice continuously differentiable, requires that the derivative of the solution be square integrable. We illustrate the weak formulation of the boundary-value problem, Equation (2.157).

To derive the weak formulation, multiplying Equation (2.157) by a test function, $\phi(x)$ (any member of a family of suitably smooth functions), and integrating by parts over an arbitrary element (x_1, x_2) we get

$$-\int_{x_1}^{x_2} a \frac{dy}{dx} \frac{d\phi}{dx} dx + \int_{x_1}^{x_2} qy\phi dx - \int_{x_1}^{x_2} f\phi dx \cdots$$
$$+ \frac{a(x_2)\phi(x_2)}{\alpha_2}[-\beta_2 y(x_2)+\gamma_2] - \frac{a(x_1)\phi(x_1)}{\alpha_1}[-\beta_1 y(x_1)+\gamma_1]$$

Let the approximation solution on the local element $y^{(e)}$ be

$$y^{(e)}(x) = \sum_{j=1}^{n} y_j^{(e)} \psi_j^{(e)}(x)$$

where $y_j^{(e)}$ are the parameters to be determined, and $\psi_j^{(e)}(x)$ are the approximation functions given by

$$\psi_1^{(e)}(x) = \frac{x_{e+1} - x}{x_{e+1} - x_e} \quad \psi_2^{(e)}(x) = \frac{x - x_e}{x_{e+1} - x_e} \quad x_e \leq x \leq x_{e+1}$$

The functions $\psi_j^{(e)}(x)$ are endowed with the following properties:

1. $\psi_i^{(e)}(x_j) = \begin{cases} 0 & \text{if } i \neq j \\ 1 & \text{if } i = j \end{cases} \quad x_1 = x_e \quad x_2 = x_{e+1}$ (2.169)

2. $\sum_{i=1}^{2} \psi_i^{(e)}(x) = 1$ (2.170)

Selecting $\phi = \psi_i$ and substituting the approximate solution into the weak formulation, one obtains the local Galerkin finite element equation,

$$-\sum_{j=1}^{n}\left[\int_{x_1}^{x_2}\left(a\frac{d\psi_i}{dx}\frac{d\psi_j}{dx}+q\psi_i\psi_j\right)dx - \frac{a(x_2)\beta_2}{\alpha_2}\psi_i(x_2)\psi_j(x_2) + \frac{a(x_1)\beta_1}{\alpha_1}\psi_i(x_1)\psi_j(x_1)\right]y_j^{(e)}\cdots$$
$$-\int_{x_1}^{x_2} f\psi_i dx + \frac{a(x_2)\psi_i(x_2)\gamma_2}{\alpha_2} - \frac{a(x_1)\psi_i(x_1)\gamma_1}{\alpha_1}$$

The above equation may be written in matrix form with the following matrix elements:

$$K_{11}^{(e)} = \int_{x_1}^{x_2}\left(a\frac{d\psi_1}{dx}\frac{d\psi_1}{dx}+q\psi_1\psi_1\right)dx+\frac{a(x_1)\beta_1}{\alpha_1} \quad (2.171)$$

$$K_{12}^{(e)} = K_{21}^{(e)} = \int_{x_1}^{x_2}\left(a\frac{d\psi_1}{dx}\frac{d\psi_2}{dx}+q\psi_1\psi_2\right)dx \quad (2.172)$$

$$K_{22}^{(e)} = \int_{x_1}^{x_2}\left(a\frac{d\psi_2}{dx}\frac{d\psi_2}{dx}+q\psi_1\psi_1\right)dx-\frac{a(x_2)\beta_1}{\alpha_2} \quad (2.173)$$

$$F_1^{(e)} = -\int_{x_1}^{x_2} f\psi_1 dx - \frac{a(x_1)\gamma_1}{\alpha_1} \quad (2.174)$$

$$F_2^{(e)} = -\int_{x_1}^{x_2} f\psi_2 dx + \frac{a(x_2)\gamma_2}{\alpha_2} \quad (2.175)$$

2.10 PARTIAL DIFFERENTIAL EQUATIONS

A partial differential equation in a function $z(x,y)$ of two independent variables x and y may be represented in general by the implicit equation

$$F(x,y,z,z_x,z_y,\ldots) = 0$$

where

$$z_x \equiv \frac{\partial z}{\partial x} \text{ and } z_y \equiv \frac{\partial z}{\partial y}$$

are partial derivatives. If only first derivatives are involved, the equation is referred to as a first order partial differential equation. If higher derivatives are involved, one has higher order partial differential equations. If the function F is linear in the dependent variable z and its derivatives, it is referred to as a linear partial differential equation. If it is nonlinear with respect to z, we have a *quasilinear* partial differential equation.

2.10.1 First-Order Partial Differential Equations

We consider a first-order partial differential equation in a function z of two variables x and y. The quasilinear partial differential equation is given by

$$a(x,y,z)\frac{\partial z(x,y)}{\partial x}+b(x,y,z)\frac{\partial z(x,y)}{\partial y}=c(x,y,z) \quad (2.176)$$

that must be solved subject to the specification of z on some initial curve represented parametrically by $x = f(s)$, $y = g(s)$, and $z = h(s)$. The method of characteristics reduces the partial differential equation to a system of ordinary differential equations and proceeds as follows. We define a characteristic direction on the plane $x - y$ by the differential equations

$$\frac{\partial x}{\partial t} = a(x,y,z), \quad \frac{\partial y}{\partial t} = b(x,y,z), \quad \frac{\partial z}{\partial t} = c(x,y,z) \tag{2.177}$$

the third of which follows from substituting the first two into the partial differential Equation (2.176). The ordinary differential equations are solved subject to the initial conditions $x(0,s) = f(s)$, $y(0,s) = g(s)$, $z(0,s) = h(s)$. The ordinary differential Equations (2.177) are solved by the methods used for ordinary differential equations to obtain the solution $x(t,s)$, $y(t,s)$, $z(t,s)$. The entire solution surface is generated by varying the parameter s over the range of the initial curve. If it is desired to get the solution in the form $z = Z(x,y)$, then we must invert (t,s) to obtain x,y. We consider the following example, from Rhee et al. (1986).

Example 2.8

The dynamics of a plug-flow reactor with a fluid velocity, v, and the first-order reaction kinetics is governed by

$$\frac{\partial c}{\partial t} + v\frac{\partial c}{\partial x} = -kc$$

$$c(x,0) = f(x), \quad c(0,t) = g(t)$$

The characteristic equations are

$$\frac{dt}{dx} = 1$$

$$\frac{dx}{ds} = v$$

$$\frac{dc}{ds} = -kc$$

Figure 2.1 shows the space-time domain divided by the line $x = vt$ into two regions, $x > vt$ and $x < vt$. For the characteristic lines in the region $x > vt$, the characteristics intersect the x-axis on which the initial condition is specified. The initial condition is parameterized by ξ:

$$t = 0, \quad x = \xi, \quad c = f(\xi) \quad \text{for } s = 0$$

In the region $x < vt$, feed condition is used:

$$x = 0, \quad t = \eta, \quad c = g(\eta) \quad \text{for } s = 0$$

For $x > vt$, the characteristic line emanating from the condition $t = 0$, $x = \xi$ is $f(\xi)$, and integration from $t = 0$ to $t = s$ gives

$$t = s, \quad x = vs + \xi, \quad c(s,\xi) = f(\xi)\exp(-ks)$$

Eliminating s and ξ from the initial and boundary conditions,

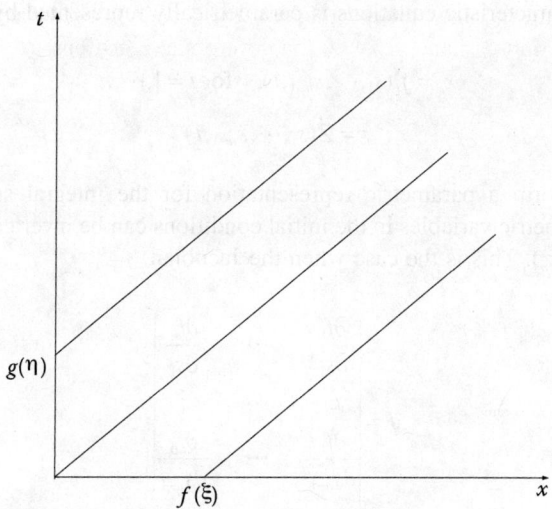

FIGURE 2.1 Characteristic plane for integration of plug-flow reactor.

$$c(x,t) = f(x-vt)\exp(-kt), \quad \text{for } x > vt$$

For $x < vt$, the characteristic line evolving from the condition $x = 0$, $t = \eta$ is represented by $g(\eta)$. Thus,

$$t = s+\eta, \quad x = vs, \quad c(s,\eta) = g(\eta)\exp(-ks)$$

Eliminating s and η in terms of x and t gives

$$c(x,t) = g\left(t - \frac{x}{v}\right)\exp\left(-\frac{kx}{v}\right) \quad \text{for } x < vt$$

The above can be generalized for n variables, as below (John, 1986). Given a first-order partial differential equation in n variables,

$$\sum_{i=1}^{n} a_i(x_1, x_2, \ldots, x_n, z) z_{x_i} = c(x_1, x_2, \ldots, x_n, z)$$

The characteristic curves in $x_1, x_2, \ldots, x_n, z$ are given by

$$\frac{dx_i}{dt} = a_i(x_1, x_2, \cdots, x_n, z) \quad \text{for } i = 1, \cdots, n$$

$$\frac{dz}{dt} = c(x_1, x_2, \cdots, x_n, z)$$

x_i and z satisfy the *initial conditions* in $n - 1$ dimensions given parametrically,

$$x_i = f_i(s_1, \ldots, s_{n-1}, t), \quad \text{for } i = 1, \ldots, n$$

$$z = h(s_1, \ldots, s_{n-1})$$

The solution of the characteristic equations is parametrically represented by

$$x_i = f_i(s_1, \cdots, s_{n-1}, t), \quad \text{for } i = 1, \cdots, n$$

$$z = Z(s_1, \cdots, s_{n-1}, t)$$

These equations form a parametric representation for the integral surface $z = z(x_1,\ldots,x_n)$, provided that the parametric variables in the initial conditions can be inverted to obtain $(s_1,\ldots,s_{n-1},t)$ as functions of $(x_1,\ldots,x_n)$. This is the case when the Jacobian

$$J = \begin{vmatrix} \dfrac{\partial f_1}{\partial s_1} & \cdots & \dfrac{\partial f_n}{\partial s_1} \\ \vdots & \cdots & \vdots \\ \dfrac{\partial f_1}{\partial s_{n-1}} & \cdots & \dfrac{\partial f_n}{\partial s_{n-1}} \\ a_1 & \cdots & a_n \end{vmatrix}$$

does not vanish.

2.10.2 Classification of Second-Order Equations

Second-order equations are classified into different types, depending on the coefficients of the second-order derivatives. The auxiliary conditions to determine a unique solution depend on the type of equation. Auxiliary conditions specified incorrectly with the type of equation lead to an ill-posed problem that may not reflect physical reality.

The general second-order partial differential equation in n-variables is written as

$$\sum_{i=1}^{n}\sum_{k=1}^{n} a_{ik}(x_1, x_2, \ldots x_n)\frac{\partial^2 \Phi}{\partial x_i \partial x_k} + B\frac{\partial^2 \Phi}{\partial t^2} + C\left(x_1,\ldots,x_n; \Phi; \frac{\partial \Phi}{\partial x_1},\ldots,\frac{\partial \Phi}{\partial x_n}\right) = 0 \qquad (2.178)$$

The equation is elliptic if and only if $B = 0$ and is hyperbolic if and only if $B < 0$. If $B = 0$ and the matrix $[a_{ik}]$ is nonnegative definite, the equation is parabolic.

For hyperbolic and parabolic equations, the conditions on $(x_1,x_2,\ldots,x_n)$ are prescribed on the boundary of the region, and the conditions on t must be prescribed at $t = 0$. These equations are often referred to as evolution equations. These equations describe the time evolution of a process from a given initial configuration.

As a consequence of the evolutionary nature of the hyperbolic and parabolic equations, boundary conditions are specified on the boundary of the spatial variables, and initial conditions are specified on the time variable, t.

Elliptic equations describe steady-state or equilibrium processes. For such equations, all auxiliary conditions must be prescribed on the boundary of the region of interest. Initial conditions given to elliptic equations lead to ill-posed problems. Solutions to these ill-posed problems exhibit sensitivity to the initial data. Small changes to the initial data cause large changes in the solution.

There are three kinds of boundary conditions for elliptic equations. If the values of the unknown function are prescribed on the boundary, then the problem is called the Dirichlet problem. If the derivatives of the unknown function are prescribed on the boundary, then it is called the Neumann problem. If a linear combination of the function values and the derivatives is specified, then it is called the Robin problem.

The methods for solving a second-order partial differential equation are separation of variables, similarity variable, Laplace transform, Fourier transform, and Hankel transform. Each of the methods is applicable to all three types of equations. However, the geometry and boundary conditions largely determine the proper application of the method of solution.

The solutions for one-dimensional parabolic and hyperbolic problems and two-dimensional elliptic problems are presented in this section. The technique for solving one-dimensional problems can be applied to solve higher-dimensional problems. An elaborate account of the physics and mathematics of the partial differential equation can be found in textbooks and monographs on partial differential equations (see, for example, Evans, 1998; and Guenther and Lee, 1996).

2.10.3 Parabolic Equations

A prototype of the parabolic equation is the diffusion equation. Diffusion equations are commonly encountered in chemical engineering processes. In the following subsections, the methods of solving one-dimensional diffusion equations are presented. These methods can be easily extended to higher-dimensional problems.

2.10.3.1 Separation of Variables

Consider a one-dimensional diffusion equation of the following form:

$$\mathcal{L}u = \left(\frac{\partial}{\partial t} - \frac{\partial^2}{\partial x^2}\right) u = 0 \quad 0 < x < 1$$

$$u(x,0) = u_0(x), \quad u(0,t) = u(1,t) = 0$$

The method of separation of variables is applicable for linear homogeneous equations, i.e., $\mathcal{L}(u + v) = \mathcal{L}u + \mathcal{L}v$, $\mathcal{L}(\alpha u) = \alpha \mathcal{L}u$, in a finite domain. The method of separation of variables assumes that

$$u(x,t) = U(x)T(t)$$

Substituting into the equation gives

$$\frac{T'(t)}{T(t)} = \frac{X''(x)}{X(x)}$$

$$T(0) = u_0(x), \quad X(0) = X(1) = 0$$

that shows that the ratios of the unknown functions must be constant:

$$\frac{T'}{T} = -\lambda^2 = \frac{X''}{X}$$

Thus,

$$X(x) = A\sin(\lambda x) + B\cos(\lambda x)$$

$$X(0) = 0 \Rightarrow B = 0, \quad X(1) = 0 \Rightarrow \lambda = n\pi$$

Because the equation is linear, the principle of superposition applies, i.e., the sum of linearly independent solutions is also a solution. Thus a general solution is

$$u(x,t) = \sum_{n=1}^{\infty} A_n \sin(n\pi x)\exp(-n^2\pi^2 t)$$

The expression for the coefficient A_n can be obtained as follows. The initial condition gives

$$u(x,0) = u_0(x) = \sum_{n=1}^{\infty} A_n \sin(n\pi x)$$

Multiplying the above equation by $\sin(m\pi x)$ and integrating over $(0,1)$ gives

$$A_n = 2\int_0^1 u_0(x)\sin(n\pi x)dx$$

The answer is obtained by making use of the orthogonal property,

$$2\int_0^1 \sin(n\pi x)\sin(m\pi x)dx = \delta_{m,n}, \quad \text{where } \delta_{m,n} \text{ is the Kronecker delta}$$

The method of separation of variables is applicable for higher-order and higher-dimensional equations, but it requires that the equation and boundary conditions be homogeneous in a finite domain.

2.10.3.2 Inhomogeneous Equation, Duhamel's Principle

Duhamel's principle helps to represent solutions to inhomogeneous initial boundary-value problems in terms of the solutions to homogeneous problems. The solution of the inhomogeneous equation,

$$\frac{\partial u}{\partial t} = \frac{\partial u^2}{\partial x^2} + h(x,t) \tag{2.179}$$

subjected to boundary conditions

$$u(0,t) = u(1,t) = u(x,0) = 0$$

is given by

$$u(x,t) = \int_0^t v(x,t-\tau;\tau)d\tau \tag{2.180}$$

where $v(x,t;\tau)$ is the solution of the following homogeneous equation:

$$\frac{\partial v}{\partial t} = \frac{\partial^2 v}{\partial x^2}$$

$$v(0,t;\tau) = v(1,t;\tau) = 0$$

$$v(x,0;\tau) = h(x,\tau)$$

the solution of which clearly follows:

$$v(x,t,\tau) = 2\sum_{n=1}^{\infty} \sin(n\pi x) \int_0^1 h(x',\tau)\sin(n\pi x')dx' e^{-n^2\pi^2 t}$$

Duhamel's principle can be confirmed from the fact that it yields the nonhomogeneous equation starting from the solution itself, as follows:

$$\frac{\partial u}{\partial t} = v(x,0,t) + \int_0^t \frac{\partial u}{\partial t}(x,t-\tau,\tau)d\tau$$

$$\frac{\partial u}{\partial t} = h(x,t) + \int_0^t \frac{\partial u}{\partial t}(x,t-\tau,\tau)d\tau$$

$$\frac{\partial^2 u}{\partial x^2} = \int_0^t \frac{\partial^2 u}{\partial x^2}(x,t-\tau,\tau)d\tau = \int_0^t \frac{\partial u}{\partial t}(x,t-\tau,\tau)d\tau$$

$$\frac{\partial^2 u}{\partial x^2} = \frac{\partial u}{\partial t} - h(x,t)$$

that is the nonhomogeneous equation.

Use of Equation (2.180) gives

$$u(x,t) = \int_0^t 2\sum_{n=1}^{\infty} \sin(n\pi x) \int_0^1 h(x',\tau)\sin(n\pi x')e^{-n^2\pi^2(t-\tau)}dx'd\tau$$

Thus the solution of the inhomogeneous equation

$$\frac{\partial u}{\partial t} = \frac{\partial^2 u}{\partial x^2} + h(x,t)$$

$$u(0,t) = u(1,t) = 0$$

$$u(x,0) = u_0(x)$$

is obtained by the superposition principle (i.e., a linear combination of solutions to the linear system is also a solution to the same linear system):

$$u(x,t) = \sum_{n=1}^{\infty} A_n \sin(n\pi x)\exp(-n^2\pi^2 t) + \int_0^t \sum_{n=1}^{\infty} B_n(\tau)\sin(n\pi x)\exp[-n^2\pi^2(t-\tau)]d\tau$$

where

$$A_n = 2\int_0^1 u_0(x)\sin(n\pi x)dx$$

$$B_n(\tau) = 2\int_0^1 h(x,\tau)\sin(n\pi x)dx$$

2.10.3.3 Inhomogeneous Boundary Conditions

Generally, the initial boundary-value problem given by

$$\frac{\partial u}{\partial t} = \frac{\partial^2 u}{\partial x^2}$$

$$u(x,0) = u_0(x), \quad u(0,t) = a(t), \quad u(1,t) = b(t)$$

is solved by letting $u(x,t) = v(x,t) + w(x,t)$ such that $v(0,t) = v(1,t) = 0$. Thus, $w(0,t) = a(t)$, $w(1,t) = b(t)$. The simplest function $w(x,t)$ that satisfies the boundary conditions is

$$w(x,t) = [b(t) - a(t)]x + a(t)$$

Substitution for $u(x,t)$ into the equation gives

$$\frac{\partial v}{\partial t} = \frac{\partial^2 v}{\partial x^2} + [a'(t) - b'(t)]x - a'(t)$$

$$v(0,t) = v(1,t) = 0$$

$$v(x,0) = u_0(x) + [a(0) - b(0)]x - a(0)$$

The solution of the transformed equation can then be easily obtained using Duhamel's principle with $h(x,t) = [a'(t) - b'(t)]x - a'(t)$.

The methods presented here can be neatly integrated into the subject of linear operators (Ramkrishna and Amundson, 1985), extending considerably the variety of problems that can be solved. For the solution using the method of Fourier transforms, the reader is referred to Varma and Morbidelli (1997).

2.10.3.4 Similarity Solutions

The equation suggests a similarity solution when there is no characteristic length scale, i.e., the domain is either a semi-infinite or infinite region. Consider the diffusion equation in a positive real line:

$$\frac{\partial u}{\partial t} = \frac{\partial^2 u}{\partial x^2}, \quad 0 < x < \infty$$

$$u(x,0) = 0, \quad u(0,t) = u_0(t)$$

Let the transformed variable be $\eta = y/(2\sqrt{t})$ and hence $u(x,t) = \Theta(t)F(\eta)$. Such a scaling behavior is typical of diffusion processes:

$$\frac{\partial u}{\partial x} = \frac{\partial u}{\partial \eta}\frac{\partial \eta}{\partial x} = \Theta F'(\eta)\frac{1}{2\sqrt{t}}; \quad \frac{\partial^2 u}{\partial x^2} = \Theta F''(\eta)\frac{1}{4t}$$

and

$$\frac{\partial u}{\partial t} = \Theta' F + \Theta F' \frac{\partial \eta}{\partial t} = \Theta' F + \Theta F' \left(-\frac{\eta}{2t}\right)$$

Substituting the transformed variables into the diffusion equation yields

$$\Theta(F'' + 2\eta F') = 4t\Theta' F$$

If $\Theta(t) = ct^n$, then the equation becomes

$$F'' + 2\eta F' - 4nF = 0, \quad F(0) = 1, \quad F(\infty) = 0$$

that is a linear second-order ordinary differential equation. The solution $F_n(\theta)$, depending on the exponent n, is given by

$$F_n(\theta) = \frac{2}{\sqrt{\pi}} \int_\theta^\infty \left(1 - \frac{\theta^2}{s^2}\right)^n \exp(-s^2)\,ds$$

For each $n = 1, 2, \ldots$, $F_n(\eta)$ also satisfies the original diffusion equation. By the superposition principle for linear equations,

$$u(x,t) = \sum_{n=0}^\infty a_n t^n F_n\left(\frac{x}{2\sqrt{t}}\right)$$

is a solution. Applying the remaining condition at $x = 0$,

$$u(0,t) = u_0(t) = \sum_{n=0}^\infty a_n t^n$$

Assuming that $u_0(t)$ is analytic, a_n can be evaluated using power series expansion.

2.10.3.5 Inhomogeneous Equation in Infinite Domain

Consider the temperature distribution in an infinite rod governed with the heat source described by the equation

$$\frac{\partial u}{\partial t} = \frac{\partial^2 u}{\partial x^2} + h(x,t), \quad -\infty < x < \infty$$

$$u(x,0) = u_0(x)$$

The problem is solved by superimposing the solutions

$$u = v_1 + v_2$$

where $v_1(x,t)$ is the solution of

$$\frac{\partial v_1}{\partial t} = \frac{\partial^2 v_1}{\partial x^2}, \quad -\infty < x < \infty$$

$$v_1(x,0) = u_0(x)$$

and $v_2(x,t)$ satisfies

$$\frac{\partial v_2}{\partial t} = \frac{\partial^2 v_2}{\partial x^2} + h(x,t), \quad -\infty < x < \infty$$

$$v_2(x,0) = 0$$

For the inhomogeneous equation,

$$\frac{\partial v}{\partial t} = \frac{\partial^2 v}{\partial x^2} + h(x,t), \quad 0 < x < \infty$$

$$v(0,t) = a(t)$$

$$v(x,0) = u_0(x)$$

the solution is given with the use of Duhamel's principle (Section 2.10.3.2):

$$v(x,t) = u_h(x,t) + \int_0^t \int_0^\infty [k(x-s,t-\tau) - k(x+s,t-\tau)] h(s,\tau) ds d\tau$$

where $u_h(x,t)$ is the solution of the homogeneous equation, and $k(x,t)$ is the fundamental solution.

2.10.4 Hyperbolic Equations

A prototype of a hyperbolic equation is the wave equation. The wave equation governs many physical phenomena such as the propagation of sound waves, water waves, vibration of a membrane in a two-dimensional setting, and vibration of a string in a one-dimensional setting.

2.10.4.1 d'Alembert's Solution

The one-dimensional wave equation in an infinite domain can be solved by a change of independent variables to reduce the wave equation that can be integrated to produce d'Alembert's solution. Consider a one-dimensional wave equation,

$$\frac{\partial^2 u}{\partial t^2} = \frac{\partial^2 u}{\partial x^2}, \quad -\infty < x < \infty$$

$$u(x,0) = f(x) \tag{2.181}$$

$$\frac{\partial u(x,0)}{\partial t} = g(x)$$

Let $\xi = x - t$ and $\eta = x + t$. Then the equation is transformed to

$$\frac{\partial^2 u}{\partial \xi \partial \eta} = 0$$

The general solution is given by

$$u(x,t) = F(x-t) + G(x+t)$$

Applying the initial conditions,

$$F(x) + T(x) = f(x)$$
$$-F'(x) + G'(x) = g(x)$$

Solving the above system for F and G gives

$$F(x-t) = \frac{1}{2}f(x-t) - \frac{1}{2}\int_0^{x-t} g(s)ds$$

$$G(x+t) = \frac{1}{2}f(x+t) + \frac{1}{2}\int_0^{x+t} g(s)ds$$

Therefore

$$u(x,t) = \frac{1}{2}[f(x-t) + f(x+t)] + \frac{1}{2}\int_{x-t}^{x+t} g(s)ds$$

For an inhomogeneous linear equation,

$$\frac{\partial^2 u}{\partial t^2} = \frac{\partial^2 u}{\partial x^2} + h$$

$$u(x,0) = f(x)$$

$$\frac{\partial u(x,0)}{\partial t} = g(x)$$

The solution is obtained by defining

$$u(x,t) = v(x,t) + w(x,t)$$

where $v(x,t)$ is the solution of Equation (2.181) and $w(x,t)$ satisfies

$$\frac{\partial^2 w}{\partial t^2} = \frac{\partial^2 w}{\partial x^2}, \quad -\infty < x < \infty$$

$$w(x,0) = 0$$

$$\frac{\partial w(x,0)}{\partial t} = 0$$

Defining $\xi = x - t$ and $\eta = x + t$ transforms the above equation to

$$\frac{\partial^2 u}{\partial \xi \partial \eta} = -\frac{1}{4}h\left(\frac{1}{2}(\xi+\eta), -\frac{1}{2}(\xi-\eta)\right)$$

$$w(s,s) = \frac{\partial w(s,s)}{\partial \xi} = \frac{\partial w(s,s)}{\partial \eta} = 0$$

the solution of which is

$$w(x,t) = \frac{1}{2}\int_0^t \int_{x-t}^{x+t} h(y,z)dydz$$

Therefore, the solution of the inhomogeneous wave equation is given by

$$u(x,t) = \frac{1}{2}[f(x-t) + f(x+t)] + \frac{1}{2}\int_{x-t}^{x+t} g(s)ds + \frac{1}{2}\int_0^t \int_{x-t+\tau}^{x+t+\tau} h(y,z)dydz$$

2.10.4.2 Separation of Variables

Consider a one-dimensional inhomogeneous equation

$$\frac{\partial^2 u}{\partial t^2} = \frac{\partial^2 u}{\partial x^2} + h$$

subjected to the boundary conditions

$$u(0,t) = a(t)$$
$$u(1,t) = b(t)$$
$$u(x,0) = f(x)$$
$$\frac{\partial u(x,0)}{\partial t} = g(x)$$

Let $u(x,t) = v(x,t) + x[b(t) - a(t)] + a(t)$, where $v(x,t)$ satisfies

$$\frac{\partial^2 v}{\partial t^2} = \frac{\partial^2 v}{\partial x^2} + h - x[b''(t) - a''(t)] - a''(t)$$

$$v(0,t) = v(1,t) = 0$$

$$v(x,0) = f(x) - x[b(0) - a(0)] - a(0)$$

$$\frac{\partial v(x,0)}{\partial t} = g(x) - x[b'(0) - a'(0)] - a'(0)$$

The problem for $v(x,t)$ is solved by defining

$$v(x,t) = v_1(x,t) + v_2(x,t)$$

where $v_1(x,t)$ is the solution of

$$\frac{\partial^2 v_1}{\partial t^2} = \frac{\partial^2 v_1}{\partial x^2}$$

$$v_1(0,t) = v_1(1,t) = 0$$

$$v_1(x,0) = f(x) - x[b(0) - a(0)] - a(0)$$

$$\frac{\partial v_1(x,0)}{\partial t} = g(x) - x[b'(0) - a'(0)] - a'(0)$$

that is solved by letting $v_1(x,t) = X(x)T(t)$. The equation becomes

$$X'' = -\lambda^2 X$$

$$X(0) = X(1) = 0$$

$$X(x) = C_1 \sin(\lambda x) + C_2 \cos(\lambda x)$$

$$X(0) = 0 = C_2$$

$$X(1) = 0 \Rightarrow \lambda = n\pi$$

$$T'' = -\lambda^2 T$$

$$T_n(t) = A_n \sin(n\pi t) + B_n \cos(n\pi t)$$

The general solution is

$$v_1(x,t) = \sum_{n=1}^{\infty} [A_n \sin(n\pi t) + B_n \cos(n\pi t)] \sin(n\pi x)$$

where

$$B_n = 2 \int_0^1 \{f(x) - x[b(0) - a(0)] - a(0)\} \sin(n\pi x) dx$$

$$A_n = \frac{2}{n\pi} \int_0^1 \{g(x) - x[b'(0) - a'(0)] - a'(0)\} \sin(n\pi x) dx$$

and $v_2(x,t)$ is the solution of

$$\frac{\partial^2 v_2}{\partial t^2} = \frac{\partial^2 v_2}{\partial x^2} + h - x[b''(t) - a''(t)] - a''(t)$$

$$v_2(0,t) = v_2(1,t) = v_2(x,0) = 0$$

$$\frac{\partial v_2(x,0)}{\partial t} = 0$$

which is solved by Duhamel's principle (Section 2.10.3.2), which states that the solution of the inhomogeneous equation is given by

$$v_2(x,t) = \int_0^t w(x, t-\tau, \tau) d\tau$$

where $w(x,t)$ satisfies

$$\frac{\partial^2 w}{\partial t^2} = \frac{\partial^2 w}{\partial x^2}$$

$$w(0,t,\tau) = w(1,t,\tau) = w(x,0,\tau) = 0$$

$$\frac{\partial w(x,0,\tau)}{\partial t} = h(x,\tau) - x[b''(\tau) - a''(\tau)] - a''(\tau)$$

The solution is given by

$$w(x,t,\tau) = \sum_{1}^{\infty} C_n(\tau) \sin(n\pi t) \sin(n\pi x)$$

where

$$C_n(\tau) = \frac{2}{n\pi} \int_0^1 \{h(x,\tau) - x[b''(\tau) - a''(\tau)] - a''(\tau)\} \sin(n\pi x) dx$$

Thus the solution of the inhomogeneous equation is given by

$$u(x,t) = x[b(t) - a(t)] + a(t) + \sum_{n=1}^{\infty} [A_n \sin(n\pi t) + B_n \cos(n\pi t)] \sin(n\pi x) +$$

$$\int_0^t \sum_{n=1}^{\infty} C_n(\tau) \sin(n\pi(t-\tau)) \sin(n\pi x) d\tau$$

2.10.5 Elliptic Equations

A prototype of the elliptic equation is the Laplace equation. The Laplace equation describes the steady-state distribution of energy or matter inside the domain given the distribution on the boundary. The dynamics of a given process may or may not come to a steady state. If the dynamics does not lead to an equilibrium, then the solution does not exist. Consider the steady-state temperature distribution in a unit cube governed by the following Poisson's equation:

$$\nabla^2 u = f(x,y,z), \quad (x,y,z) \in \mathcal{D}$$

where $\mathcal{D}$ is the domain of the region. The general boundary condition is written as

$$-\alpha \frac{\partial u}{\partial n} + Bu = \gamma(x,y,z) \in \beta$$

When $\alpha = 0$, the boundary condition is called the Dirichlet boundary condition; when $\beta = 0$, it is called the Neumann boundary condition. Otherwise, it is known as the Robin boundary condition. When $\gamma \neq 0$, the boundary condition is homogeneous. Otherwise, it is inhomogeneous. When $\alpha = 0$ and $\beta \neq 0$, u on the boundary is known and is described by γ. When $\alpha \neq 0$ and $\beta = 0$, the derivative is described on the boundary, and u on the boundary is the unknown that must be solved. When both $\alpha \neq 0$ and $\beta \neq 0$, both the u and its derivative are unknown.

The elliptic equation describes a steady-state or an equilibrium process within the region. Physically, steady state is not attained unless the net rate of generation is balanced with the net flux into the region. The physics is manifested in the existence of the solution. Thus, the elliptic equation does not admit a solution unless the condition for the existence of a steady state is satisfied.

For $\beta = 0$, the condition for the existence of a solution of the Poisson equation is

$$\int_B \gamma dS = \int_D f(x,y,z) dx dy dz$$

The physical interpretation of the above condition is that, at steady state, the generation must be balanced by the efflux.

2.10.5.1 Poisson Integral Formula

Consider the Laplace equation

$$\nabla^2 u = \frac{1}{r}\frac{\partial}{\partial r}\left(r\frac{\partial u}{\partial r}\right) + \frac{1}{r^2}\frac{\partial^2 u}{\partial \theta^2} = 0, \quad 0 \leq r < 1, \quad 0 \leq \theta \leq 2\pi$$

$$u(1,\theta) = f(\theta)$$

Assume that $u(r,\theta) = R(r)\Theta(\theta)$. Substituting into the equation and rearranging, we get

$$\frac{r^2 R'' + rR'}{R} = -\frac{\Theta''}{\Theta} = \lambda^2$$

The solution is given by

$$\Theta(\theta) = A\cos(\lambda\theta) + B\sin(\lambda\theta)$$

To determine the value of λ, the condition of periodicity is imposed:

$$\Theta(\theta + 2\pi) = \Theta(\theta)$$

Thus $\lambda = n$, $n = 0, 2, 4, \ldots$. The solution for R is

$$R(r) = Cr^n + Dr^{-n}$$

To ensure that the solution is finite at $r = 0$, $D = 0$. Thus, by the superposition principle, the general solution is given by

$$u(r,\theta) = \frac{a_0}{2} + \sum_{n=1}^{\infty} r^n (a_n \cos(n\theta) + b_n \sin(n\theta))$$

where

$$a_n = \frac{1}{\pi} \int_0^{2\pi} f(\theta)\cos(n\theta)d\theta, \quad b_n = \frac{1}{\pi} \int_0^{2\pi} f(\theta)\sin(n\theta)d\theta$$

Substituting a_n and b_n into the solution, we get

$$u(r,\theta) = \frac{1}{\pi} \int_0^{2\pi} \left\{ \frac{1}{2} + \sum_{n=1}^{\infty} r^n [\cos(n\theta)\cos(n\phi) + \sin(n\theta)\sin(n\phi)] \right\} f(\phi)d\phi$$

$$= \frac{1}{\pi} \int_0^{2\pi} \left\{ \frac{1}{2} + \sum_{n=1}^{\infty} r^n \cos[n(\phi - \theta)] \right\} f(\phi)d\phi$$

$$= \frac{1-r^2}{2\pi} \int_0^{2\pi} \frac{f(\phi)d\phi}{1 - 2r\cos(\phi - \theta) + r^2}$$

that is obtained by summing the series in the braces.

The solution of the Laplace equation in a unit sphere with $f(\theta,\phi)$ prescribed on the surface of the sphere is given by

$$u(r,\theta,\phi) = \frac{1}{4\pi} \int_0^{2\pi} \int_0^{\pi} \frac{(1-r^2)f(\theta',\phi')}{(1 - 2r\cos\Theta + r^2)^{3/2}} \sin\phi' d\phi' d\theta'$$

where

$$\cos\Theta = \cos\phi\cos\phi' + \sin\phi\sin\phi'\cos(\theta - \theta')$$

2.10.5.2 Green's Function

The solution of the Poisson equation

$$\nabla^2 u = h, \quad 0 \leq r \leq 1, \quad 0 \leq \theta \leq 2\pi$$

$$u(1,\theta) = 0$$

is given by

$$u(r,\theta) = \int_0^1 \int_0^{2\pi} G(r,\theta;\rho,\phi)h(\rho,\phi)\rho d\rho d\phi$$

Mathematics in Chemical Engineering

where the Green's function is given by

$$G(r,\theta;\rho,\phi) = \frac{1}{4\pi}\log[1+r^2\rho^2 - 2r\rho\cos(\theta-\phi)] - \frac{1}{4\pi}\log[r^2+\rho^2 - 2r\rho\cos(\theta-\phi)]$$

Generally, the equation

$$\nabla^2 u = h, \quad 0 \le r < 1, \quad 0 \le \theta \le 2\pi$$

$$u(1,\theta) = f(\theta)$$

is solved by letting $u = v + w$, where $v(r,\theta)$ is solved by the Poisson integral formula and $w(r,\theta)$ is solved by the Green's function.

The Fourier transform can be used to solve the Dirichlet problem in the infinite domain, and the Fourier sine transform can be used in the semi-infinite domain. The Fourier cosine transform is appropriate for the Neumann problem in the semi-infinite domain.

2.10.6 Computational Fluid Mechanics

The governing equations of fluid flows are seldom analytically solved. Often, the analytical solutions involve restricted assumptions and approximations. Hence, many engineering problems involving fluid flows must be solved numerically.

Software: *In view of the rigor involved in numerical solution, many commercial software packages have been developed that serve the purpose of computational fluid dynamics (CFD). Appendix 2.A gives a listing of sources for various commercial as well as free CFD codes. These CFD codes may be broadly categorized into either finite volume method based or finite element method based. For a detailed account of computational methods, see the books by Patankar (1980), Ferziger and Peric (2002), Ranade (2002), Chen (2005), Reddy (2005), and so forth.*

2.11 INTEGRAL EQUATIONS

Mathematical models of physical processes yield differential equations when changes brought about in the system arise from purely local effects. Thus diffusion, convection, etc., contribute to local changes alone. However, there are many processes in which changes are brought about by cumulative contributions from afar. Examples are radiative heat transport; transport of mass, momentum, and energy between particles and the continuous phase in a dispersed phase system; and so on. Integral equations arise from the modeling of such systems. The methods for solving integral equations are presented in the following subsections with illustrative examples.

There are two classes of integral equations, Fredholm and Volterra. In the following subsections, the methods for solving each class of integral equations are presented.

2.11.1 Volterra Integral Equations

Volterra integral equations are of the following forms:

$$\int_a^x K(x,y)u(y)dy = f(x) \qquad (2.182)$$

$$u(x) - \lambda \int_a^x K(x,y)u(y)dy = f(x) \qquad (2.183)$$

$$p(x)u(x) - \lambda \int_a^x K(x,y)u(y)dy = f(x) \qquad (2.184)$$

The first equation is known as a Volterra equation of the first kind, and the last two are of the second and third kinds, respectively. An integral equation of the third kind can be transformed into an integral equation of the second kind if $p(x) \neq 0$.

Differentiating the integral equation of the first kind with respect to the independent variable x yields (under suitable continuity and differentiability conditions on various functions in the integral equation)

$$\frac{df}{dx} = \lambda \int_0^x \frac{\partial K(x,y)}{\partial x} u(y)dy + \lambda K(x,x)u(x) \qquad (2.185)$$

If $K(x,x) \neq 0$, then the Volterra equation of the first kind can be converted to a Volterra equation of the second kind,

$$u(x) + \int_0^x H(x,t)u(t)dt = g(x)$$

where

$$g(x) \equiv \frac{1}{K(x,x)} \frac{df}{dx}, \quad H(x,t) \equiv \frac{1}{K(x,x)} \frac{\partial K(x,t)}{\partial x} \qquad (2.186)$$

Volterra equations are classified as regular and singular integral equations. If $K(x,y) < \infty$, the equations are called regular; otherwise, they are defined as singular. If $f(x) = 0$, the equations are termed homogeneous.

Each type of equation is further categorized into linear and nonlinear equations based on whether or not the unknowns in the equations appear as a first power. The following is an example of a nonlinear Volterra equation:

$$u(x) = f(x) + \int_0^x K(x,y)F(u(y))dy \qquad (2.187)$$

Other forms of nonlinear integral equations also arise in applications.

2.11.2 Methods of Solution for Volterra Equations

The monograph by Linz (1985) provides a good theoretical and practical treatment of Volterra integral equations. In what follows, we list the solutions in accordance with the kernel type.

2.11.2.1 Kernel Is Only a Function of Dependent Variable

For $K(x,y) = K(y)$,

$$u(x) = f(x) + \int_0^x K(y)u(y)dy \qquad (2.188)$$

is equivalent to

$$\frac{du}{dx} = \frac{df}{dx} + K(x)u(x), \quad \text{with } u(0) = f(0) \qquad (2.189)$$

Equation (2.189) is a first-order linear differential equation. The solution is given by

$$u(x) = f(0)\exp\left[\int_0^x K(y)dy\right] + \int_0^x \frac{df}{dt}\exp\left[\int_t^x K(y)dy\right]dt \qquad (2.190)$$

2.11.2.2 Separable Kernels

For a separable kernel, $K(x,y) = P(x)Q(y)$,

$$u(x) = f(x) + P(x)\int_0^x Q(y)u(y)dy \qquad (2.191)$$

The solution is given by

$$u(x) = f(x) + P(x)\int_0^x Q(y)f(y)\exp\left[\int_y^x P(s)Q(s)ds\right]dy \qquad (2.192)$$

2.11.2.3 Degenerate (Finite Rank) Kernels

$$u(x) = f(x) + \sum_{i=1}^n \int_0^x P_i(x)Q_i(y)u(y)dy \qquad (2.193)$$

The solution is

$$u(x) = f(x) + \sum_{i=1}^n P_i(x)v_i(x) \qquad (2.194)$$

where $y_i(x)$ is the solution of

$$\frac{dv_i(x)}{dx} = Q_i(x)\left[f(x) + \sum_{j=1}^n P_j(x)v_j(x)\right], \quad v_i(x) = 0 \quad i = 1,2,3,\ldots,n \qquad (2.195)$$

2.11.2.4 Difference Kernels

When the kernel $K(x,y)$ depends only on the difference $(x - y)$, it is termed a *difference kernel*:

$$u(x) = f(x) + \lambda \int_0^x K(x-y)u(y)dy \qquad (2.196)$$

The equation is solved by Laplace transform with the help of the convolution theorem:

$$\bar{u}(s) = \frac{\bar{f}(s)}{1 - \lambda \bar{K}(s)}, \quad \lambda \bar{K} \neq 1$$

where $\bar{u}(s)$, $\bar{f}(s)$, and $\bar{K}(s)$ are the Laplace transforms of $u(x)$, $f(x)$, and $K(x)$, respectively. Inverting the transform,

$$u(x) = \mathcal{L}^{-1}\left\{\frac{\bar{f}(s)}{1 - \lambda \bar{K}(s)}\right\}$$

The inversion of the right hand side above would of course depend on the specific form of K(s). The method of resolvent kernels in the section below can also be applied here.

$$u(x) = f(x) + \lambda \int_0^x K(x,y)u(y)dy$$

2.11.2.5 Method of Resolvent Kernels

$$u(x) = f(x) + \lambda \int_0^x K(x,y)f(y)dy \qquad (2.197)$$

The solution of a Volterra equation is given in terms of a resolvent kernel,

$$u(x) = f(x) + \lambda \int_0^x \Gamma(x,y;\lambda)f(y)dy \qquad (2.198)$$

where the resolvent kernel is given in terms of the Neumann series as

$$\Gamma(x,y;\lambda) = \sum_{i=0}^{\infty} \lambda^i K_{i+1}(x,y) \qquad (2.199)$$

$$K_1(x,y) = K(x,y) \qquad (2.200)$$

$$K_2(x,y) = \int_y^x K(x,s)K(s,y)ds \qquad (2.201)$$

$$\vdots \qquad (2.202)$$

$$K_{i+1}(x,y) = \int_y^x K(x,s)K_i(s,y)ds \qquad (2.203)$$

2.11.2.6 The Generalized Abel Integral Equation

$$f(x) = \int_0^x \frac{u(y)}{(x-y)^\alpha} dy, \quad 0 < \alpha < 1 \qquad (2.204)$$

Following Tricomi (1957), the solution is given by

$$u(x) = \frac{\sin\alpha\pi}{\pi} \frac{d}{dx} \int_0^x \frac{f(y)}{(x-y)^{1-\alpha}} dy = \frac{\sin(\alpha\pi)}{\pi} \left[\frac{f(0)}{x^{1-\alpha}} + \int_0^x \frac{f'(y)}{(x-y)^{1-\alpha}} dy \right] \qquad (2.205)$$

2.11.2.7 Numerical Solution of Volterra Integral Equations of the Second Kind

In this section, a simple method for the numerical solution of Volterra equations of the second kind is presented. An excellent treatment for the equation of the second kind can be found in Linz (1985).
Consider the following linear Volterra equation:

$$u(x) = f(x) + \int_a^x K(x, y)u(y) dy \qquad (2.206)$$

The simplest numerical method to solve Equation (2.206) is to subdivide the interval (a,x) into n equal subintervals and evaluate the integral by summing the areas of the trapezoids in each of the subintervals. Let $\Delta x = (x - a)/n$, $n \geq 1$. Then

$$\int_a^x K(x,y)u(y)dy = \Delta x \left[\frac{1}{2}(K(x,a)u(a) + K(x,x)u(x)) + \sum_{i=1}^{n-1} K(x, y_i)u(y_i) \right]$$

The solution $u(x)$ is then obtained at the discrete points $x_i = a + i\Delta x$. The integral Equation (2.206) is approximated by the following system of equations:

$$u(a) = f(a)$$

$$u(x_i) = f(x_i) + \Delta x \left[\frac{1}{2}(K(x_i,a)u(a) + K(x_i,x_n)u(x_n)) + \sum_{j=1}^{i-1} K(x_i, y_j)u(y_j) \right] \qquad (2.207)$$

$$i = 1, 2, \ldots, n, \quad y_j \leq x_i$$

In Equation (2.207), it is assumed that if the final index of the summation is less than the initial index, then the summation is zero. The system Equation (2.207) constitutes a system of $n + 1$ linear equations in $n + 1$ unknowns u_i, $i = 0, 1, 2, \ldots, n$, which can be easily solved by back substitution.
Nonlinear Volterra equations are solved by iterative methods. Two iterative methods are presented here. One is Newton's method, and the other is the method of successive substitution.
Consider the following nonlinear Volterra equation:

$$u(x) = f(x) + \int_a^x K(x,y,u(y))dy \qquad (2.208)$$

Equation (2.209) is linearized by Newton's method to give

$$u_{k+1}(x) = f(x) + \int_a^x \left\{ K(x,y,u_k(y)) + \frac{\partial K}{\partial u}(x,y,u_k)u_{k+1}(y) - \frac{\partial K}{\partial u}(x,y,u_k(y))u_k(y) \right\} dy \qquad (2.209)$$

Successive substitution is based on the following equation:

$$u_{k+1}(x) = f(x) + \int_a^x K(x,y,u_k(y))dy \qquad (2.210)$$

The trapezoidal rule is used to convert Equation (2.210) to a set of algebraic equations.

Jerry (1985) discusses the solution to the nonlinear Volterra equations. The existence of linear equations is discussed in Section 2.7.1.

2.11.3 Fredholm Integral Equations

The equations of the form

$$\int_a^b K(x,y)u(y)dy = f(x) \qquad (2.211)$$

$$u(x) - \lambda \int_a^b K(x,y)u(y)dy = f(x) \qquad (2.212)$$

$$a(x)u(x) - \lambda \int_a^b K(x,y)u(y)dy = f(x) \qquad (2.213)$$

are called the Fredholm integral equations of the first, second, and third kinds, respectively. Every boundary-value problem can be reduced to a Fredholm equation via the Green's function. Fredholm equations can be further classified into regular and singular integral equations. If the kernel of the equation is finite everywhere in the domain of the integration, the equation is regular. Otherwise, it is singular.

2.11.4 Methods of Solution for Fredholm Equations

We list here the solutions to Fredholm integral equations in accordance with the type of kernel.

2.11.4.1 Degenerate Kernels

The kernel $K(x,y)$ is said be a degenerate kernel if it can be expressed as $\Sigma_{i=1}^n a_i(x)b_i(y)$:

$$u(x) = f(x) + \lambda \sum_{i=1}^n \int_a^b a_i(x)b_i(y)u(y)dy \qquad (2.214)$$

$$= f(x) + \lambda \sum_{i=1}^{n} a_i(x) \int_a^b b_i(y) u(y) dy \qquad (2.215)$$

$$u(x) = f(x) + \lambda \sum_{i=1}^{n} c_i a_i(x) \qquad (2.216)$$

where

$$c_k = \int_a^b b_k(y) u(y) dy \qquad (2.217)$$

Substituting Equation (2.216) in Equation (2.217) gives a linear system of equation for c_k:

$$c_k = f_k + \lambda \sum_{i=1}^{n} a_{ki} c_i, \quad k = 1, 2, \ldots, n$$

where

$$f_k = \int_a^b b_k(x) f(x) dx \qquad (2.218)$$

$$a_{ki} = \int_a^b b_k(x) a_i(x) dx \qquad (2.219)$$

If the determinant of the matrix $\mathbf{I} - \lambda \mathbf{A}$ is zero, then either the integral equation has no solution or it has infinitely many solution.

2.11.4.2 Method of Fredholm Resolvent Kernels

The solutions of

$$u(x) = f(x) + \lambda \int_a^b K(x, y) u(y) dy$$

is given by

$$u(x) = f(x) + \lambda \int_a^b \Gamma(x, y; \lambda) f(y) dy$$

where

$$\Gamma(x, y; \lambda) = \frac{D(x, y; \lambda)}{D(\lambda)}, \quad D(\lambda) \neq 0$$

The functions $\Gamma(x,y;\lambda)$, $D(x,y;\lambda)$, and $D(\lambda)$ are the Fredholm resolvents of the kernel, Fredholm minor, and Fredholm determinant, respectively. The $D(x,y;\lambda)$ is defined as

$$D(x,y;\lambda) = \sum_{n=0}^{\infty} \frac{(-\lambda)^n}{n!} B_n(x,y)$$

where

$$B_n(x,y) = C_n K(x,y) - n \int_a^b K(x,s) B_{n-1}(s,y) ds, \quad B_0(x,y) = K(x,y)$$

where

$$C_n = \int_a^b B_{n-1}(y,y) dy, \quad n = 1, 2, \ldots, \quad C_0 = 1$$

and

$$D(\lambda) = \sum_{n=0}^{\infty} \frac{(-\lambda)^n}{n!} C_n, \quad C_0 = 1$$

2.11.4.3 Method of Iterated Kernels

The solution of

$$u(x) = f(x) + \lambda \int_a^b K(x,y) u(y) dy$$

is given by

$$u(x) = f(x) + \lambda \int_a^b \Gamma(x,y;\lambda) f(y) dy$$

The iterated kernel $\Gamma(x,y;\lambda)$ is defined as

$$\Gamma(x,y;\lambda) = \sum_{i=1}^{\infty} \lambda^{i-1} K_i(x,y) \qquad (2.220)$$

where

$$K_1(x,y) \equiv K(x,y), \quad K_i(x,y) = \int_a^b K(x,s) K_{i-1}(s,y) ds$$

The series defined in Equation (2.220) is convergent for $\max|\lambda B_n| < 1$, where

$$B = \sqrt{\int_a^b \int_a^b K^2(x,y)dxdy}$$

2.11.4.4 Symmetric Kernels

If the kernel is symmetric, i.e., $K(x,y) = K(y,x)$, then the eigenvalues λ_i of the integral operator, defined by

$$u_i(x) = \int_a^b K(x,y)u(y)dy$$

are real, and the corresponding eigenvectors u_i are orthogonal, i.e.,

$$\int_a^b u_i(x)u_j(x)dx = 0, \quad \text{for } i \neq j$$

For square-integrable kernels

$$\int_a^b \int_a^b K^2(x,y)dxdy < \infty$$

Following Jerry (1985), the solution of the nonhomogeneous Fredholm Equation (2.210) is given by

$$u(x) = f(x) + \lambda \sum_{i=1}^{\infty} \frac{a_i u_i(x)}{\lambda_i - \lambda}$$

where

$$a_i = \int_a^b f(x)u_i(x)dx$$

and

$$u_i(x) = \lambda_i \int_a^b K(x,y)u_i(y)dy$$

2.11.4.5 Numerical Solution of Nonhomogeneous Fredholm Equation of the Second Kind

In this subsection, a simple trapezoidal rule is used to derive a numerical scheme to solve Fredholm equations of the second kind. The use of a Gaussian quadrature to approximate the integral is discussed by Press et al. (1992).

Consider the following Fredholm equation:

$$u(x) = f(x) + \int_a^b K(x,y)u(y)dy \qquad (2.221)$$

The equation is solved at discrete points $\{x_i\}$, and the integral is approximated by the trapezoidal rule to give the following system of equations:

$$u(x_i) = f(x_i) + \Delta x \left\{ \frac{1}{2}[K(x_i,a)u(a) + K(x_i,b)u(b)] + \sum_{j=1}^{n-1} K(x_i,y_j)u(y_j) \right\} \qquad (2.222)$$

$$i = 0, 1, \ldots n$$

where $x_0 = a$, $x_n = b$, $\Delta x = (b-a)/n$. Equation (2.223) represents a system of $n+1$ equations in $n+1$ unknowns $\{u_i\}_{i=0}^n$.

For the nonlinear equation,

$$u(x) = f(x) + \int_a^b K(x,y,u(y))dy \qquad (2.223)$$

The simplest way to solve Equation (2.223) is by successive substitution:

$$u_{k+1}(x) = f(x) + \int_a^b K(x,y,u_k(y))dy \qquad (2.224)$$

Equation (2.224) is solved by discretizing the integral using the trapezoidal rule and iterating the resulting system of algebraic equations until convergence criteria are met.

The second method of solution is by linearization using Newton's method to give

$$u_{k+1}(x) = f(x) + \int_a^b \left\{ K(x,y,u_k(y)) + \frac{\partial K}{\partial u}(x,y,u_k(y))u_{k+1}(y) - \frac{\partial K}{\partial u}(x,y,u_k(y))u_k(y) \right\} dy \qquad (2.225)$$

The trapezoidal rule is used to discretize Equation (2.225) to form a system of linear equations that is solved iteratively for a given closer initial guess of $u_0(x)$.

2.11.4.6 Solution of Ill-Posed Fredholm Equations of the First Kind

It is seen in Section 2.2.6 that Fredholm equations of the first kind arise from inverse problem formulation, i.e., the determination of model parameters from experimental measurements. Inverse problems are generally ill posed. They lack the one or more of the three properties required for reliable model predictions:

1. Existence of solution
2. Uniqueness of solution
3. Stability with respect to small perturbations in model parameters

In this subsection, the ill-posed nature of Fredholm integral equations of the first kind is demonstrated with an example equation. A technique known as *regularization* to overcome these issues is presented. The essential idea for regularization is to approximate the integral equation of the first kind with the integral equation of the second kind, which is a well-posed problem (see Wing, 1991).

To demonstrate the ill-posed nature, consider the following equation with degenerate kernel:

$$f(x) = \int_a^b \sum_{i=1}^n \alpha_i(x)\beta_i(y)u(y)dy$$

$$= \sum_{i=1}^n \alpha_i(x)\int_a^b \beta_i(y)u(y)dy = \sum_{i=1}^n c_i\alpha_i(x)$$

(2.226)

where

$$c_i = \int_a^b \beta_i(y)u(y)dy$$

For a given solution $u(x)$, $u_1(x) = u(x) + cg(x)$ is also a solution for arbitrary constant c, provided that $g(x)$ satisfies

$$\int_a^b \beta_i(x)g(x)dx = 0, \quad i = 1, 2, \cdots, n$$

To solve Equation (2.226), assume that

$$u(y) = \sum_{i=1}^n u_i\beta_i(y) \tag{2.227}$$

Substitution of Equation (2.227) in Equation (2.226) gives

$$\sum_{i=1}^n c_i\alpha_i(x) = \sum_{i=1}^n \alpha_i(x)\int_a^b \left\{\sum_{j=1}^n u_i\beta_i(y)\beta_j(y)(y)\right\}dy$$

$$= \sum_{i=1}^n \alpha_i(x)\sum_{j=1}^n u_j \int_a^b \beta_i(y)\beta_j(y)dy \tag{2.228}$$

$$= \sum_{i=1}^n \alpha_i(x)\sum_{j=1}^n a_{ij}u_j$$

where

$$a_{ij} = \int_a^b \beta_i(y)\beta_j(y)dy$$

For linearly independent functions $\alpha_i(x)$, Equation (2.230) is a system of linear equations,

$$Au = c$$

to be solved for u_i, which is used in the solution Equation (2.227).

2.11.4.7 Method of Regularization

This section presents a simplified treatment of the technique of Tikhonov regularization. The idea of regularization is to convert a original ill-posed problem (which means that the error in the solution is magnified by errors in the input data) into a well-posed problem for which the error in the solution is under control. More specifically, consider the solution of the following Fredholm integral equation of the first kind.

$$\int_a^b K(x,y)u(y)dy = f(x), \text{ i.e., } Au = f$$

where A is clearly a linear integral operator. By direct solution of the above equation we imply the minimization of the norm of the residual $Au-f$. This is ill-posed because small errors in f produce large errors in the solution u. To regularize this we ask to minimize

$$J(u) = \|Au - f\| + \alpha\|u\| \tag{2.229}$$

where $\alpha > 0$. This formulation is a well-posed problem because it prevents excessive detail in the solution. It is shown (Kress, 1989) that the minimizer u_α of Equation (2.231) is the unique solution of the equation

$$\alpha u + A^*Au = A^*f \tag{2.230}$$

where

$$A^*Au = \int_a^b K(y,x)K(x,y)u(y)dy$$

and

$$A^*f = \int_a^b K(y,x)f(y)dy$$

Thus, the ill-posed problem of a Fredholm equation of the first kind is approximated by the well-posed problem of the second kind.

The error of approximation is represented by the functional $J(u)$, which depends on the regularization parameter α. There are two components in this error. One component involves $\|Au - f\|$, which is due to the solution of the ill-posed problem, while the other component involves regularization $\alpha\|u\|$. The relative importance of each term in this functional is depicted in Figure 2.2. The strategy for choosing α is to minimize the functional $J(u)$. The criterion is called the L criterion in view of the shape of the function $J(u)$.

For the inverse problems of chemical engineering interest, the reader is referred to Ramkrishna and coworkers (Sathyagal et al. (1985), Muralidhar and Ramkrishna (1989), Wright and Ramkrishna (1992), etc.).

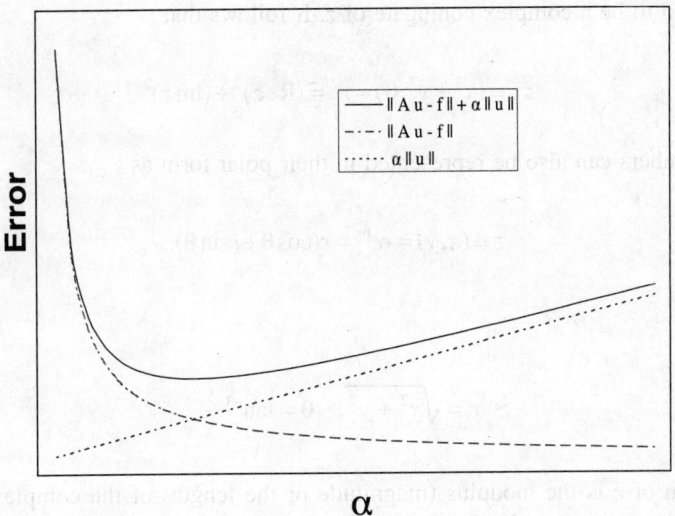

FIGURE 2.2 Total error of approximation as represented by the functional $J(u)$.

2.12 COMPLEX VARIABLES

Complex numbers owe their origin to the quest for the square root of a negative number. Thus the so-called imaginary number $i \equiv \sqrt{(-1)}$ is a fundamental element of complex numbers, written as $z = x + iy$, in which x is the *real* part and y is the *imaginary* part. Although real numbers quantify physical quantities, complex numbers provide very convenient representations of many physical phenomena. In quantum mechanics, the wave function is a complex function. Two-dimensional, incompressible, irrotational flows are represented by a complex flow potential, $w = \varphi + i\psi$, with φ, the velocity potential, as the real part, and ψ, the stream function, as the imaginary part.

2.12.1 Properties of Complex Numbers

A complex variable, $z = x + iy$, is alternatively represented as $z \equiv (x,y)$. Two complex numbers are equal if and only if their corresponding real and imaginary parts are equal. The addition, multiplication, and division of two complex numbers are given by

$$(x_1, y_1) \pm (x_2, y_2) = (x_1 \pm x_2, y_1 \pm y_2)$$

$$(x_1, y_1)(x_2, y_2) = (x_1 x_2 - y_1 y_2, y_1 y_2 + x_1 y_2)$$

and

$$\frac{(x_1, y_1)}{(x_2, y_2)} = \frac{(x_1 x_2 + y_1 y_2, x_2 y_1 - x_1 y_2)}{x_2^2 + y_2^2}, \quad \text{for } x_2^2 + y_2^2 \neq 0$$

The real number $\sqrt{x^2 + y^2}$ is called the modulus of a complex number, z. Unlike real numbers, complex numbers cannot be ordered (such as greater than or less than).

A special operation in a complex number system is

$$\overline{z} = x - iy = (x, -y)$$

The term $\bar{z}$ is said to be a complex conjugate of z. It follows that

$$z\bar{z} \equiv (x^2 + y^2, 0) = r^2 = (\text{Re } z)^2 + (\text{Im } z)^2$$

Complex numbers can also be represented in their polar form as

$$z \equiv (x, y) = re^{i\theta} = r(\cos\theta + i\sin\theta)$$

where

$$r = \sqrt{x^2 + y^2}, \quad \theta = \tan^{-1}\frac{y}{x}$$

The interpretation of r is the modulus (magnitude or the length) of the complex number, and θ (phase angle) is the argument of z, $\theta = \arg(z)$. Geometrically

$$z = re^{i\theta} = re^{i(\theta + 2n\pi)}$$

Thus $\arg(z)$ as a function should be restricted, such as

$$-\pi < \arg(z) < \pi$$

Such restriction defines the principal value of $\arg(z)$, which is denoted as $\text{Arg}(z)$:

$$\text{Arg}(z_1 z_2) = \text{Arg}(z_1) + \text{Arg}(z_2)$$

$$\text{Arg}\left(\frac{z_1}{z_2}\right) = \text{Arg}(z_1) + \text{Arg}(z_2)$$

The nth root of unity is defined as

$$z^n = 1 \Rightarrow z = \cos\left(\frac{2k\pi}{n}\right) + i\sin\left(\frac{2k\pi}{n}\right), \quad k = 1, 2, \cdots, n-1$$

2.12.2 Analytic Functions

A complex function $w = f(z)$ assigns a value of w in the complex plane to each value of z in the subset of the complex plane D. The set of complex numbers, D, is called the domain of definition of f. When the domain of definition is not specified, then it is understood that the function takes on all values of z for which f is defined. For example, the domain of $f(z) = 1/z$ is all of z such that $z \neq 0$.

Functions of a complex variable often encounter indefiniteness from being multivalued. To make a function single-valued, a function of a complex variable also includes its domain of definition.

Given a function of a complex variable $f(z)$, it is always possible to write

$$f(z) = f(x + iy) = u(x, y) + iv(x, y)$$

where $u(x,y)$ and $v(x,y)$ are real-valued functions or real variables (x,y). For example,

$$f(z) = \frac{1}{z} = \frac{1}{x+iy} = \frac{x-iy}{(x+iy)(x-iy)} = \frac{x}{x^2+y^2} - i\frac{y}{x^2+y^2}$$

Thus

$$u(x,y) = \frac{x}{x^2+y^2}, \quad v(x,y) = \frac{-y}{x^2+y^2}$$

Given complex variables $z = (x,y)$, $z_0 = (x_0,y_0)$ and a function $w = f(z)$, the limit is defined as

$$\lim_{z \to z_0} f(z) = w_0 \Rightarrow \text{for a given } \varepsilon \text{ there exists a } \delta \text{ such that } |f(z) - w_0| < \varepsilon \text{ whenever } |z - z_0| < \delta$$

To be consistent, it is required that the limit of a function $f(z)$ as z tends to z_0 approaches in an arbitrary manner, not just in a particular direction. The limit is written as

$$\lim_{z \to z_0} f(z) = \lim_{(x,y) \to (x_0,y_0)} u(x,y) + i \lim_{(x,y) \to (x_0,y_0)} v(x,y)$$

A function is said to be continuous at z_0 if

$$\lim_{z \to z_0} f(z) = f(z_0)$$

For example, the function

$$f(z) = \begin{cases} \dfrac{1}{z}, & z \neq 0 \\ 0 & z = 0 \end{cases}$$

is not continuous at $z_0 = 0$ because

$$\lim_{z \to 0} \neq f(0)$$

The derivative of a function $f(z)$ at z_0 is defined as

$$f'(z_0) = \lim_{z \to z_0} \frac{f(z) - f(z_0)}{z - z_0}$$

It follows that the rules of differentiation for real functions also hold for complex variables, i.e.,

The addition rule:

$$\frac{d}{dz}(f(z)\pm g(z))=\frac{df}{dz}\pm\frac{dg}{dz}$$

The product rule:

$$\frac{d}{dz}(f(z)g(z))=f(z)\frac{dg}{dz}+g(z)\frac{dy}{dz}$$

The quotient rule:

$$\frac{d}{dz}\frac{f(z)}{g(z)}=\frac{g(z)\frac{df}{dz}-f(z)\frac{dg}{dz}}{g^2(z)}$$

The derivative of a function of a complex variable $f(x + iy) = u(x,y) + iv(x,y)$ exists at $z_0 = (x_0,y_0)$ if and only if the following equations are satisfied:

$$\frac{\partial u}{\partial x}=\frac{\partial v}{\partial y} \quad \text{and} \quad \frac{\partial v}{\partial x}=-\frac{\partial u}{\partial y} \qquad (2.231)$$

Equations (2.233) are known as Cauchy–Riemann equations. In polar coordinates, the $f(r_0 e^{i\theta}) = u(r_0,\theta_0) + iv(r_0,\theta_0)$ is differentiable at (r_0,θ_0) if and only if

$$\frac{\partial u}{\partial r}=\frac{1}{r}\frac{\partial v}{\partial \theta} \quad \text{and} \quad \frac{\partial v}{\partial r}=-\frac{1}{r}\frac{\partial u}{\partial \theta} \qquad (2.232)$$

It follows that the real and imaginary parts of a differentiable function are not independent, i.e., given $u(x,y)$, $v(x,y)$ can be constructed with an arbitrary constant. Given a differentiable function $f(z)$, the real and imaginary parts of f satisfy the Laplace equation:

$$\frac{\partial^2 u}{\partial x^2}+\frac{\partial^2 u}{\partial y^2}=0$$

$$\frac{\partial^2 v}{\partial x^2}+\frac{\partial^2 v}{\partial y^2}=0$$

The $u(x,y)$ of a differentiable function $f(z)$ is also called a *harmonic function*, whereas $v(x,y)$ is called a conjugate harmonic function.

It also follows that if $f(z)$ is analytic, then the real and imaginary parts satisfy the Cauchy-Riemann equations, and it can be represented by a Taylor series in the neighborhood of z_0. A complex-valued function that is analytic in the whole complex plane is called an *entire function*. If a complex-valued function fails to be analytic at z_0 but is analytic at every other point in the neighborhood of z_0, then z_0 is said to be an isolated singular point of f. For example, 0 is an isolated singular point of $f(z) = 1/z$.

2.12.2.1 Elementary Functions

In accord with the definition of an entire function (analytic at all finite points of the complex plane), a polynomial

$$p(z) = \sum_{m=0}^{n} a_m x^m$$

is an entire function. The exponential, sine, and cosines are entire functions. They are represented by Taylor series expansion:

$$\sin(z) = \frac{e^{iz} - e^{-iz}}{2i} \quad \text{and} \quad \cos(z) = \frac{e^{iz} + e^{-iz}}{2}$$

Define

$$\sinh(z) = \frac{e^z - e^{-z}}{2}$$

$$\cosh(z) = \frac{e^z + e^{-z}}{2}$$

It follows that

$$\sin(iz) = i\sinh(z), \quad \sinh(iz) = i\sin(z) \quad \text{and}$$

$$\cosh(iz) = \cos(z), \quad \cos(iz) = \cosh(z)$$

$$\cosh^2(z) - \sinh^2(z) = 1$$

$$|\sin(z)|^2 = \sin^2 x + \sinh^2 y, \quad |\sinh(z)|^2 = \sinh^2 x + \sin^2 y \quad \text{and}$$

$$|\cos(z)|^2 = \cos^2 x + \cosh^2 y, \quad |\cosh(z)|^2 = \cosh^2 x + \cos^2 y$$

2.12.2.2 Multivalued Functions

Some of the functions that are useful in practical applications are not uniquely defined. Examples of such functions are nth roots of a complex variable for $n > 1$. For a specific example, consider the case when $n = 2$:

$$w = \sqrt{z} = re^{1/2(\theta + 2k\pi)i}$$

where k is an integer. With no restriction on k, $w = 1$ is the value corresponding to infinitely many values of k. Such behavior is not acceptable according to the definition of a function. To avoid this, restriction is placed on the values of k such that $k < n$. Geometrically, this restriction means that the radius vector from the origin to z is not allowed to circulate around the origin more than once. The semi-infinite ray extending from the origin is known as a branch cut. The origin is known as a branch point.

2.12.2.3 Logarithmic Functions

Given the entire function

$$z = re^{i\theta}$$

the function log z is

$$\log z = \log r + i\theta$$

where $r = |z|$ and $q = \arg(z)$. Because

$$re^{i\theta} = re^{i(\theta + 2k\pi)}$$

$$\theta = \arg(z) = \arg(z) + 2k\pi$$

is not a single-valued function. The situation is similar to that in the previous section. Defining the principal value of the logarithmic function,

$$\text{Log } z = \text{Log } r + i\Theta, \quad (r > 0, -\pi < \Theta \leq \pi)$$

The mapping $w = \text{Log } z$ is a single-valued and continuous function in its domain of definition, which is the set of all nonzero complex numbers; its range is the strip $-\pi < Im(w) \leq \Theta$. Thus the function Log z is analytic.

2.12.2.4 Cauchy Integral Formula

Consider a curve C parameterized by t, $z(t) = x(t) + iy(t)$, $a \leq t \leq b$ and a complex function $f(x + iy) = u(x,y) + iv(x,y)$. The contour integral of $f(z)$ along the curve z is

$$\int_C f(z)dz = \int_a^b (ux' - vy')dt + i\int_a^b (vx' + vy')dt$$

$$= \int_C udx - vdy + i\int_C vdx + udy$$

If $f(z)$ is analytic everywhere within and on a simple closed contour C, taken in the positive sense (i.e., traversing the curve, C, in counterclockwise direction), and if z_0 is any point interior to C, then

$$f(z_0) = \frac{1}{2\pi i} \int_C \frac{f(z)dz}{z - z_0} \tag{2.233}$$

A simple curve is a closed curve that does not intersect itself. Equation (2.235) is called the Cauchy integral formula. As a consequence of the Cauchy integral formula, we can write

$$\frac{d^n f(z_0)}{dz^n} = \frac{n!}{2\pi i} \int_C \frac{f(z)dz}{(z - z_0)^{n+1}}$$

2.12.2.5 Laurent Series

Let the function f be analytic throughout an annular domain $R_1 < |z - z_0| < R_2$, and let C denote any positive oriented simple closed contour around z_0 and lying in that domain. Then at each point z in the domain, $f(z)$ has the series representation

$$f(z) = \sum_{n=0}^{\infty} a_n (z-z_0)^n + \sum_{n=1}^{\infty} \frac{b_n}{(z-z_0)^n}$$

where

$$a_n = \frac{1}{2\pi i} \int_{C_1} \frac{f(s)ds}{(s-z_0)^{n+1}} \quad (n = 0,1,2,3,\ldots)$$

$$b_n = \frac{1}{2\pi i} \int_{C_2} \frac{f(s)ds}{(s-z_0)^{-n+1}} \quad (n = 1,2,3,4,\ldots)$$

is called the Laurent series.

The complex number that is the coefficient of $1/(z - z_0)$ (i.e., b_1 of the Laurent series) is known as the residue of f at the isolated singular point z_0. The principal part of f is the portion of the series involving negative exponents. If $b_n = 0$ for all $n > m$, then f is said to have a pole of order m. If $m = 1$, it is called a simple pole. If $m = \infty$, f is said to have an essential singularity. If $m = 0$, f is said to have a removable singularity.

2.12.3 Residue Theorem

Let $f(z)$ be analytic inside and on a simple closed contour C, except for a finite number of singular points, $z_1, z_2, \ldots, z_m$ inside C. Then

$$\int_{C_k} f(z)dz = 2\pi i \sum_{n=1}^{m} \operatorname{Res} f(z); \quad \text{for } z = z_k \quad (k = 1,2,3,\ldots,n)$$

2.12.3.1 Calculus of Residues

If f has a pole of order m at z_0, then

$$\phi(z) = (z-z_0)^m f(z) = b_m + b_{m-1}(z-z_0) + b_{m-2}(z-z_0)^2 +$$

$$\cdots + b_1(z-z_0)^{m-1} + \sum_{n=0}^{\infty} a_n (z-z_0)^{n+m}$$

where $b_m \neq 0$; hence the point z_0 is a removable singular point of the function ϕ. It is seen that

$$b_m = \phi(z_0)$$

From the Taylor expansion of $\phi(z)$,

$$b_1 = \frac{1}{(m-1)!} \frac{d^{m-1}\phi(z_0)}{dz^{m-1}}$$

If $m = 1$, then

$$b_1 = \phi(z_0) = \lim_{z \to z_0}(z - z_0)f(z)$$

Consequently,

$$f(z) = \frac{\phi(z_0)}{(z-z_0)^m} + \frac{\phi'(z_0)}{(z-z_0)^{m-1}} + \cdots + \frac{\phi^{(m-1)}(z_0)}{(m-1)!} \frac{1}{z-z_0} + \sum_{n=m}^{\infty} \frac{\phi^{(n)}(z_0)}{n!}(z-z_0)^{n-m}$$

For example, if $q(z_0) = 0$, then

$$\phi(z) = \frac{p(z_0) + p'(z_0)(z-z_0) + \cdots}{q''(z_0)/2! + q'''(z-z_0)/3! + \cdots}$$

and

$$b_1 = \phi'(z_0) = 2\frac{p'(z_0)}{q''(z_0)} - \frac{2}{3}\frac{p(z_0)q'''(z_0)}{[q''(z_0)]^2}$$

2.12.3.2 Application of Complex Integration

One of the many applications of the theory of complex variables is the application of the residue theorem to evaluate definite real integrals. Another is to use conformal mapping to solve boundary-value problems involving harmonic functions. The residue theorem is also very useful in evaluating integrals resulting from solutions of differential equations by the method of integral transforms.

Evaluation of definite real integrals using the residue theorem requires proper specification of contours. For some problems, contours can be as simple as a semicircle in the upper half of the complex plane. Problems utilizing this contour are those with integration limits extending over the real line, i.e., $(-\infty,\infty)$, and the integrand being a rational function. We now illustrate the use of complex integration through an example (Fisher, 1999) as follows.

Example 2.9
Evaluate

$$\int_0^{\infty} \frac{\log x \, dx}{x^2 + b^2}$$

There are two simple poles, ib and $-ib$, on the imaginary axis. An appropriate contour for this integral is shown in Figure 2.3.

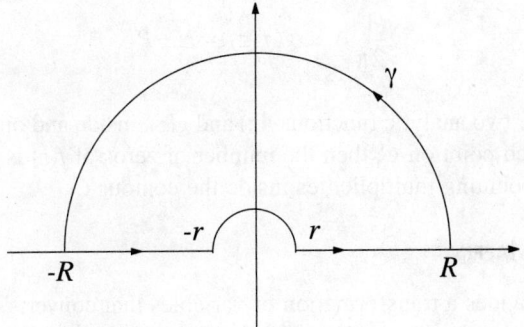

FIGURE 2.3 Contour for evaluation integral of Example 2.9.

$$\int_\gamma \frac{\log z}{z^2+b^2} dz = \int_0^\pi \frac{\log x}{x^2+b^2} dx + iR \int_0^\pi \frac{(\log R + i\theta)e^{i\theta}}{R^2 e^{2i\theta}+b^2} d\theta$$

$$+ \int_{-R}^{-r} \frac{\log|x| + \pi i}{(|x|e^{i\pi})^2 + b^2} dx + ir \int_\pi^0 \frac{(\log r + i\theta)e^{i\theta}}{r^2 e^{2i\theta}+b^2} d\theta$$

As $r \to 0$ and $R \to \infty$, the two integrals over the two semicircles vanish and the remaining integrals become

$$\int_0^\infty \frac{\log x}{x^2+b^2} dx + \int_{-\infty}^0 \frac{\log|x|+i\pi}{x^2+b^2} dx = \int_{-\infty}^\infty \frac{\log|x|}{x^2+b^2} dx + i\pi \int_{-\infty}^0 \frac{dx}{x^2+b^2}$$

$$= \int_{-\infty}^\infty \frac{\log|x|}{x^2+b^2} dx + \frac{i\pi}{2b}$$

The residue at $z = b^{i\pi/2}$ of the integrand is

$$\frac{\log z}{z^2+b^2} = \frac{\log z}{(z+ib)(z-ib)}\bigg|_{z=be^{i\pi/2}} = \frac{\log b + i\pi/2}{2ib}$$

Applying the residue theorem to the contour integral gives

$$\int_\gamma \frac{\log z}{z^2+b^2} = \frac{\pi}{b}\left[\log b + \frac{i\pi}{2}\right] = 2\int_{-\infty}^\infty \frac{\log|x|}{x^2+b^2} dx + \frac{i\pi^2}{2b}$$

that after simplifying gives

$$\int_0^\infty \frac{\log x}{x^2+b^2} dx = \frac{\pi}{2b} \log b$$

2.12.4 Argument Principle and Rouché Theorem

Let f be the meromorphic (i.e., it is analytic throughout the domain except possibly for poles) in the domain interior to a positively oriented simple closed contour C with N number of zeros and P number of poles in the interior of C. Then the *argument principle* states that

$$\frac{1}{2\pi}\Delta_C \arg(f(z)) = N - P$$

Rouché theorem: For two analytic functions $f(z)$ and $g(z)$, inside and on a simple closed contour C, if $|f(z)| > |g(z)|$ at each point on C, then the number of zeros of $f(z)$ is the same as the number of zeros of $f(z) + g(z)$, counting multiplicities inside the contour C.

2.12.5 Conformal Mapping

Conformal mapping provides a transformation of variables that converts one mathematical problem into another. Its importance lies in the greater ease with which the transformed problem may be solved than the original problem. Applications arise in steady-state conduction or diffusion problems.

Let C_1 and C_2 be two smooth curves in the z plane passing through z_0, and let α be the angle between these two curves measured in the counterclockwise direction. Let Γ_1 and Γ_2 be the images of $w = f(x)$ of C_1 and C_2 in the w plane, respectively. If the angle between the curves Γ_1 and Γ_2 measured in the counterclockwise direction is α, then the mapping $w = f(z)$ is said to be conformal.

A mapping is said to be conformal at a point z_0 if it preserves the magnitude and sense of the angle between any two smooth arcs through the point z_0. Thus, at each point z where a function $f(z)$ is analytic and $f'(z) \neq 0$, the mapping $w = f(z)$ is conformal.

Theorem 1: Suppose that the analytic function $f(z) = u(x,y) + iv(x,y)$ maps a domain D_z in the z plane onto a domain D_w in the w plane. If $h(u,v)$ is a harmonic function defined on D_w, then the function

$$H(x,y) = h[u(x,y), v(x,y)]$$

is harmonic in D_z.

Theorem 2: Suppose that the analytic function $f(z) = u(x,y) + iv(x,y)$ maps an arc C in the z plane onto an arc Γ in the w plane. Let $f(z)$ be conformal on C and let a function $h(u,v)$ be differentiable on Γ. If the function $h(u,v)$ satisfies either of the conditions

$$h = c \quad \text{or} \quad \frac{dh}{dn} = 0$$

along Γ, then the function $H(x,y) = h[u(x,y), v(x,y)]$ satisfies the corresponding condition, $H = c$ or $dH/dN = 0$ along C. Here dH/dN denotes the derivative normal to C.

Theorems 1 and 2 are useful for the construction of analytic functions satisfying given boundary conditions. We will now illustrate some of the applications taken from Brown and Churchill (1996).

Example 2.10: The Dirichlet Boundary-Value Problem

Consider a thin semi-infinite plate defined by $y \geq 0$ whose faces are insulated and whose edge $y = 0$ is kept at temperature zero except for the segment $-1 < x < 1$. The segment $(-1 < x < 1)$ is kept at temperature unity. The steady-state temperature distribution $T(x,y)$ is given by the Laplace equation

$$\nabla^2 T = 0$$

for $-\infty < x < \infty$, $y \geq 0$. Then

$$T(x,0) = \begin{cases} 1 & \text{when} \quad |x| < 1 \\ 0 & \text{when} \quad |x| > 1 \end{cases}$$

Also, $|T(x,y)| < M$, where M is some positive constant. The above problem defined in the $(x - y)$ plane may be solved using conformal mapping to transform it into a problem in a suitably defined $(u - v)$ plane.

The region in the $u - v$ plane will be the image of the half-plane under a transformation $w = f(z)$ that is analytic in the domain $y > 0$ and is conformal along the boundary $y = 0$, except at the points $(1,0)$ and $(-1,0)$, where it is undefined. The transformation is given by

$$w = \log\left|\frac{z-1}{z+1}\right| + i \arg \frac{z-1}{z+1}$$

To see this transformation, let

$$z = 1 + r_1 \exp(i\theta_1) \quad \text{and} \quad z = -1 + r_2 \exp(i\theta_2)$$

Then

$$w = \log \frac{r_1}{r_2} + i(\theta_1 - \theta_2)$$

The segment $|x| < 1$, corresponding to the angle $\theta_1 - \theta_2 = \pi$, is mapped to the boundary $v = \pi$ of the strip. The remainder of the x-axis, $|x| > 1$, is mapped to the boundary $v = 0$ of the strip. The condition $T(x,0)$ for $|x| > 1$ is transformed to $T(u,\pi) = 0$ and $T(x,0) = 0$, and for $|x| > 1$ it is transformed to $T(u,0) = 0$. A bounded harmonic function of u and v, which is zero on edge $v = 0$ of the strip and unity on the edge $v = \pi$, is

$$T(u,v) = \frac{v}{\pi}$$

We have

$$\frac{z-1}{z+1} = \frac{x-iy}{x+iy} = \frac{x^2 + y^2 - 1 + 2iy}{(x+1)^2 + y^2}$$

and, by definition,

$$\arg z = \tan^{-1} \frac{y}{x}$$

Thus, using the transformation w expressed in terms of the x,y coordinate, we find that

$$v = \arg \frac{(z-1)(\bar{z}+1)}{(z+1)(z\mp 1)} = \arg \frac{x^2 + y^2 - 1 + 2iy}{(x+1)^2 + y^2} = \tan^{-1} \frac{2y}{x^2 + y^2 + 1}$$

The solution of the Laplace equation is then expressed as

$$T(x,y) = \frac{1}{\pi} \tan^{-1}\left(\frac{2y}{x^2+y^2-1}\right)$$

Example 2.11: The Neumann Boundary-Value Problem
Find the harmonic function in the first quadrant $x,y > 0$ satisfying

$$\begin{cases} \dfrac{\partial T(x,0)}{\partial y} = 0 & \text{when} \quad 0 < x < 1 \\ T(x,0) = 0 & \text{when} \quad x > 1 \end{cases}$$

$$T(0,y) = 0$$

The function

$$z = \sin w = \sin u \cosh v + i \cos u \sinh v = x + iy$$

maps the first quadrant onto the infinite strip,

$$0 \le u \le \frac{\pi}{2}, \quad v = 0 \text{ onto } 0 < x < 1$$

$$u = \frac{\pi}{2}, \quad v \ge 0 \text{ onto } v \ge 1$$

$$u = 0, \quad v \ge 0 \text{ onto } v \ge 1$$

The boundary conditions become

$$T(0,v) = 0$$

$$T(1,v) = 1$$

$$\frac{\partial T(u,0)}{\partial v} = 0, \quad 0 < u < \frac{\pi}{2}$$

The harmonic function that satisfies the boundary conditions is

$$T(u,v) = \frac{2}{\pi} u$$

The mapping implies that

$$\frac{x^2}{\sin^2 u} - \frac{y^2}{\cos^2 u} = 1$$

which is an equation of a hyperbola with vertices at (±sin u,0) and foci at (±1,0). By the definition of a hyperbola,

$$\sqrt{(x+1)^2+y^2} - \sqrt{(x-1)^2+y^2} = 2\sin u$$

Substitution into $T(x,y)$ gives

$$T(x,y) = \frac{2}{\pi}\sin^{-1}\left(\frac{1}{2}\sqrt{(x+1)^2+y^2} - \sqrt{(x-1)^2+y^2}\right)$$

The reader is referred to Fisher (1999) for more applications.

2.13 INTEGRAL TRANSFORMS

The general integral transform of a function $f(t)$ is defined as follows:

$$F(\lambda) = \int_a^b K(\lambda,t)f(t)dt \qquad (2.234)$$

Integral transforms are distinguished by the way the kernel, $K(\lambda,t)$, and the integration limits are defined. The kernels for different transforms are listed along with the integration limits in Table 2.5, where J_v is the Bessel function of the first kind. Generally, the Laplace transform is used to solve initial-value problems, while the Fourier transform is used to solve boundary-value problems where the domain of the independent variable used is the real line $(-\infty,\infty)$. Fourier sine and cosine transforms are most useful for problems over the half-line domain $(0,\infty)$. The Fourier sine transform is more appropriate for the Dirichlet boundary-value problems. On the other hand, the Fourier cosine transform is suitable for the Neumann boundary-value problems. Other transforms are used in connection with the geometry of the problem under consideration.

**TABLE 2.5
Kernels for Different Transforms and Integration Limits**

Transform	(a,b)	K(t,λ)
Laplace	$(0,\infty)$	$e^{-\lambda t}$
Fourier	$(-\infty,\infty)$	$\frac{1}{\sqrt{2\pi}}e^{-i\lambda t}$
Fourier cosine	$(0,\infty)$	$\sqrt{\frac{2}{\pi}}\cos(\lambda t)$
Fourier sine	$(0,\infty)$	$\sqrt{\frac{2}{\pi}}\sin(\lambda t)$
Mellin	$(0,\infty)$	$t^{\lambda-1}$
Hankel	$(0,\infty)$	$tJ_v(\lambda t),\ v \geq -\frac{1}{2}$

Integral transforms can be used to solve ordinary differential equations by converting them to algebraic equations. In what follows, the convolution properties of the different transforms have been listed, followed by the methods of integral transform used to solve (a) one-dimensional diffusion equations in the infinite and semi-infinite domains and (b) Laplace equations in the cylindrical geometries.

2.13.1 Laplace Transform

The Laplace transform of $f(t)$ is denoted as $\mathcal{L}f(t)$, and the inversion is given by

$$f(t) = \frac{1}{2\pi i} \int_{\gamma-i\infty}^{\gamma+i\infty} e^{\lambda t} F(\lambda) d\lambda \qquad (2.235)$$

where γ is chosen such that all singularities of the integrand lie to the left of γ.

2.13.1.1 Convolution Property

If $F(\lambda)$ and $G(\lambda)$ are the Laplace transforms of $f(t)$ and $g(t)$, respectively, then

$$\begin{aligned}
\mathcal{L}\left[\int_0^t f(s)g(t-s)ds\right] &= \int_0^\infty e^{-\lambda t} \int_0^t f(s)g(t-s)ds\,dt \\
&= \int_0^\infty f(s)ds \int_s^\infty f(t-s)e^{-\lambda t} dt \\
&= \int_0^\infty f(s)e^{-\lambda s} ds \int_0^\infty g(y)e^{-\lambda y} dy \\
\mathcal{L}\left[\int_0^t f(s)g(t-s)ds\right] &= F(\lambda)G(\lambda)
\end{aligned} \qquad (2.236)$$

2.13.1.2 Application of Laplace Transform

The temperature distribution in a semi-infinite bar that is heated at one end is described by the solution of the following problem:

$$\frac{\partial u}{\partial t} = \frac{\partial^2 u}{\partial x^2} \qquad (2.237)$$

subjected to the boundary conditions

$$-\frac{\partial u(0,t)}{\partial x} = q(t), \quad u(x,0) = 0$$

Define

$$U(x,\lambda) = \int_0^\infty u(x,t)e^{-\lambda t} dt$$

and

$$Q(\lambda) = -\int_0^\infty q(t)e^{-\lambda t}dt$$

Multiply Equation (2.239) by $e^{-\lambda t}$ and integrate over $(0,\infty)$,

$$\frac{d^2U}{dx^2} - \lambda U = 0$$

$$-\frac{dU(0)}{dx} = Q$$

the solution of which is given by

$$U = \frac{Q}{\sqrt{\lambda}} e^{-\sqrt{\lambda}x}$$

Let $F(\lambda) = Q$ and $G(\lambda) = \exp(-\sqrt{\lambda}x)/\sqrt{\lambda}$. From the table of Laplace transforms (Appendix 2.C),

$$\mathcal{L}^{-1}G(\lambda) = K(x,t) = \frac{e^{-x^2/(4t)}}{\sqrt{\pi t}}$$

The application of convolution theorem yields

$$u(x,t) = \int_0^t q(\tau)K(x,t-\tau)d\tau$$

For more examples using inversion integrals, the reader is referred to Churchill et al. (1976).

2.13.2 Fourier Transform

The Fourier transform of $f(t)$ is denoted as $\mathcal{F}f(t) = F(\lambda)$, and the inversion is given by

$$f(t) = \frac{1}{\sqrt{2\pi}} \int_{-\infty}^\infty e^{-i\lambda t} F(\lambda) d\lambda \qquad (2.238)$$

2.13.2.1 Convolution Property

If $F(\lambda)$ and $G(\lambda)$ are the Fourier transforms of $f(t)$ and $g(t)$, respectively, then

$$\mathcal{F}\left[\int_{-\infty}^\infty f(t-s)g(s)ds\right] = \frac{1}{2\pi} \int_{-\infty}^\infty e^{i\lambda t} \int_{-\infty}^\infty f(t-s)g(s)ds\,dt$$

$$= \frac{1}{\sqrt{2\pi}} \int_{-\infty}^\infty g(s)ds \int_{-\infty}^\infty e^{i\lambda t} f(t-s)dt \qquad (2.239)$$

$$= \frac{1}{\sqrt{2\pi}} \int_{-\infty}^\infty e^{i\lambda s} g(s)ds \int_{-\infty}^\infty e^{i\lambda y} f(y)dy$$

$$\mathcal{F}\left[\int_{-\infty}^\infty f(t-s)g(s)ds\right] = F(\lambda)G(\lambda)$$

2.13.2.2 Application of Fourier Transform

The fundamental solution of the heat equation, i.e., the Dirichlet problem in infinite domain, $(-\infty,\infty)$, is obtained by solving

$$\frac{\partial u}{\partial t} = \frac{\partial^2 u}{\partial x^2}, \quad -\infty < x < \infty \qquad (2.240)$$

subjected to the boundary conditions

$$u(x,0) = u_0(x)$$

Define

$$U(\lambda,t) = \frac{1}{\sqrt{2\pi}} \int_{-\infty}^{\infty} u(x,t)\exp(i\lambda x)dx$$

Multiplying Equation (2.242) by $e^{i\lambda x}$ and integrating over $(-\infty,\infty)$ yields

$$\frac{dU}{dt} = -\lambda^2 U$$

$$U(\lambda,0) = U_0(\lambda)$$

where

$$U_0(\lambda) = \int_{-\infty}^{\infty} u_0(x)\exp(i\lambda x)dx$$

The solution is given by

$$U(\lambda,t) = U_0(\lambda)\exp(-\lambda^2 t)$$

Inverting the Fourier transform with the use of the convolution theorem yields

$$u(x,t) = \int_{-\infty}^{\infty} k(x-y,t)u_0(y)dy$$

where the function

$$k(x,t) = \mathcal{F}^{-1}[\exp(-\lambda^2 t - i\lambda x)] = \frac{1}{\sqrt{4\pi t}}\exp\left[\frac{-x^2}{4t}\right] \qquad (2.241)$$

is called the fundamental solution of the diffusion equation.

Mathematics in Chemical Engineering

2.13.2.3 Application of Fourier Sine Transform

The temperature distribution in a semi-infinite rod with one end of the rod kept at the prescribed temperature is described by the solution of the following initial-boundary-value problem:

$$\frac{\partial u}{\partial t} = \frac{\partial^2 u}{\partial x^2}, \quad 0 < x < \infty \qquad (2.242)$$

subjected to the boundary conditions

$$u(0,t) = a(t)$$

$$u(x,0) = u_0(x)$$

Define

$$U(\lambda,t) = \sqrt{\frac{2}{\pi}} \int_0^\infty \sin(\lambda x) u(x,t) dx$$

$$U(\lambda,0) = U_0(\lambda) = \sqrt{\frac{2}{\pi}} \int_0^\infty \sin(\lambda x) u_0(x) dx$$

The transformed equation becomes

$$\frac{dU}{dt} + \lambda^2 U = \lambda \sqrt{\frac{2}{\pi}} a(t)$$

the solution of which is

$$U(\lambda,t) = U_0(\lambda) e^{-\lambda^2 t} + \lambda \sqrt{\frac{2}{\pi}} \int_0^t e^{-\lambda^2(t-\tau)} a(\tau) d\tau$$

Inversion yields

$$u(x,t) = \sqrt{\frac{2}{\pi}} \int_0^\infty U(\lambda,t) \sin(\lambda x) d\lambda$$

$$= \sqrt{\frac{2}{\pi}} \int_0^\infty U_0(\lambda) \sin(\lambda x) e^{-\lambda^2 t} d\lambda + \frac{2}{\pi} \int_0^\infty \int_0^t \lambda e^{-\lambda^2(t-\tau)} a(\tau) \sin(\lambda x) d\tau d\lambda$$

$$= \frac{2}{\pi} \int_0^\infty \int_0^\infty [e^{-\lambda^2 t} \sin(\lambda s) \sin(\lambda x)] u_0(s) d\lambda ds + \frac{2}{\pi} \int_0^t \int_0^\infty \lambda e^{-\lambda^2(t-\tau)} \sin(\lambda x) a(\tau) d\lambda d\tau$$

Using the trigonometric property, $2 \sin(A) \sin(B) = \cos(A - B) - \cos(A + B)$, it follows that

$$u(x,t) = \frac{1}{\pi}\int_0^\infty \int_0^\infty e^{-\lambda^2 t}[\cos(\lambda(x-s)) - \cos(\lambda(x+s))]u_0(s)d\lambda ds$$

$$+ \frac{2}{\pi}\int_0^t \int_0^\infty \lambda e^{-\lambda^2(t-\tau)} \sin(\lambda x) a(\tau) d\lambda d\tau$$

The definition of $k(x,t)$ [Equation (2.243)] can be used to arrive at the following:

$$u(x,t) = \int_0^\infty [k(x-s,t) - k(x+s,t)]u_0(s)ds - 2\int_0^t \frac{\partial k(x,t-\tau)}{\partial x} a(\tau)d\tau$$

where $k(x,t)$ is given by Equation (2.243).

2.13.2.4 Application of Fourier Cosine Transform

The temperature distribution in a semi-infinite rod with prescribed heat flux applied at one end can be obtained as follows. The temperature distribution in the rod is governed by the following equation:

$$\frac{\partial u}{\partial t} = \frac{\partial^2 u}{\partial x^2} + h(x,t), \quad 0 < x < \infty \tag{2.243}$$

subjected to the boundary conditions

$$-\frac{\partial u(0,t)}{\partial x} = a(t)$$

$$u(x,0) = u_0(x)$$

Following the procedure similar to that described in the previous subsection, the solution obtained by the Fourier cosine transform is given as

$$u(x,t) = \int_0^\infty [k(x-s,t) + k(x+s,t)]u_0(s)ds - 2\int_0^t k(x,t-\tau)a(\tau)d\tau +$$

$$\int_0^t \int_0^\infty [k(x-s,t-\tau) + k(x+s,t-\tau)]h(s,\tau)dsd\tau$$

where the $k(x,t)$ is defined by Equation (2.243)

2.13.3 MELLIN TRANSFORM

2.13.3.1 Convolution Property

Let $F(\lambda)$ and $G(\lambda)$ be the Mellin transforms of $f(x)$ and $g(x)$, respectively. Then

$$\mathcal{M}\left[\int_0^\infty f\left(\frac{x}{y}\right)\frac{g(y)}{y}dy\right] = F(\lambda)G(\lambda) \tag{2.244}$$

2.13.3.2 Application of Mellin Transform

The steady-state temperature distribution in a wedge of radius a with the angle between the edges equal to 2α (i.e., spanning from $-\alpha$ to α) is governed by the following Dirichlet boundary-value problem in terms of polar coordinates (r,θ):

$$\frac{\partial^2 u}{\partial r^2} + \frac{1}{r}\frac{\partial u}{\partial r} + \frac{1}{r^2}\frac{\partial^2 u}{\partial \theta^2} = 0 \qquad (2.245)$$

subjected to the boundary conditions

$$u(r,\pm\alpha) = \begin{cases} 1 & 0 \le r < a \\ 0 & r > a \end{cases}$$

Define

$$U(\lambda,\theta) = \int_0^\infty u(r,\theta) r^{\lambda-1} dr$$

Multiply Equation (2.247) by $r^{\lambda+1}$ and integrate over $(0,\infty)$ to get

$$\frac{d^2}{d\theta^2}\int_0^\infty u r^{\lambda-1} dr + \int_0^\infty \frac{\partial^2 u}{\partial r^2} r^{\lambda+1} dr + \int_0^\infty \frac{\partial u}{\partial r} r^\lambda dr = 0$$

Applying integration by parts to the last two terms yields

$$\frac{d^2 U}{d\theta^2} + \lambda^2 U = 0$$

$$U(\lambda,\pm\alpha) = \frac{a^\lambda}{\lambda}$$

The solution is given by

$$U(\lambda,\theta) = \frac{a^\lambda \cos(\lambda\theta)}{\lambda \cos(\lambda\alpha)}$$

Following Davies (2002), the inversion gives the required temperature distribution as

$$u(r,\theta) = \begin{cases} 1 - \dfrac{1}{\pi}\arctan\left[\dfrac{2(ar)^{\pi/(2\alpha)}\cos(\pi\theta/(2\alpha))}{a^{\pi/\alpha} - r^{\pi/\alpha}}\right] & 0 \le r < a \\ \dfrac{1}{\pi}\dfrac{2(ar)^{\pi/(2\alpha)}\cos(\pi\theta/(2\alpha))}{a^{\pi/\alpha} - r^{\pi/\alpha}} & r > a \end{cases}$$

2.13.4 Hankel Transform

2.13.4.1 Property of Hankel Transform

Let $\mathcal{B}_k$ be the Bessel differential operator of order k, i.e.,

$$\mathcal{B}_k \equiv \frac{d^2}{dr^2} + \frac{1}{r}\frac{d}{dr} - \left(\frac{k}{r}\right)^2$$

If the Hankel transform of $f(r)$ is denoted as $\mathcal{H}_k f(r) = F_k(\lambda)$, then (Zauderer, 1983; Poularikas, 2000)

$$\mathcal{H}_k[\mathcal{B}_k f(r)] = -\lambda^2 \mathcal{H}_k[f(r)] + [rf'(r)J_k(\lambda r) - \lambda r f J'_k(\lambda r)]\big|_{r=0}^{r=\infty}$$

The second term on the right-hand side is specified using the boundary conditions. The inverse Hankel transform is given by

$$f(r) = \int_0^\infty \lambda F_k(\lambda) J_k(\lambda r)\, d\lambda$$

2.13.4.2 Application of Hankel Transform

The Hankel transform may be used to solve numerous boundary-value problems in a relatively straightforward way, using various properties of Bessel functions. We present here the solution to one of the boundary-value problems (Davies, 2002) of interest.

The axisymmetric steady-state temperature distribution in an infinite cylinder with one of its ends maintained at constant temperature is described by the solution of

$$\frac{\partial^2 u}{\partial r^2} + \frac{1}{r}\frac{\partial u}{\partial r} + \frac{\partial^2 u}{\partial z^2} = 0 \qquad (2.246)$$

subjected to the boundary conditions

$$u(r,0) = u_0(r)$$

$$\lim_{z \to \infty} u(r,z) = 0$$

Define

$$U(\lambda, z) = \int_0^\infty u(r,z) J_0(r,\lambda) r\, dr$$

In addition to the stated boundary conditions, it is assumed that

$$\lim_{r \to \infty}\left[r\frac{\partial u}{\partial r} J_0(\lambda r)\right] = \lim_{r \to \infty} r u J_0(\lambda r) = 0 \qquad (2.247)$$

Multiply Equation (2.248) by $rJ_0(r\lambda)$ and integrate over $(0,\infty)$ to get

$$\frac{d^2}{dz^2}\int_0^\infty uJ_0(r\lambda)rdr + \int_0^\infty \left(\frac{\partial^2 u}{\partial r^2}+\frac{1}{r}\frac{\partial u}{\partial r}\right)J_0(r\lambda)rdr$$

The second integral is the Hankel transform of the Bessel operator of zero order. Using the property of the Hankel transform, Equation (2.248) is transformed with the conditions stated by Equation (2.249) to

$$\frac{d^2U}{dz^2}-\lambda^2 U=0$$

$$U(\lambda,0)=U_0(\lambda)=\int_0^\infty u_0(r)J_0(\lambda r)rdr$$

$$\lim_{z\to\infty} U(\lambda,z)=0$$

$$U(\lambda,z)=U_0(\lambda)e^{-z\lambda}$$

Inversion of the transform gives

$$u(r,z)=\int_0^\infty U_0(\lambda)e^{-z\lambda}J_0(r\lambda)\lambda d\lambda$$

The inverse transform can be easily obtained from the table for Hankel transforms in Appendix 2.C.

2.14 CALCULUS OF VARIATIONS

The calculus of variations is concerned primarily with determining maxima and minima of quantities that depend on functions. It can be effectively used for optimization problems (Denn, 1969). We present here a brief overview of the variational calculus followed by a simple illustrative example.

2.14.1 EULER-LAGRANGE DIFFERENTIAL EQUATION

Consider the following integral:

$$\mathcal{F}(y)=\int_a^b F(y,y',x)dx$$

that evaluates to a real number for a given function of $F(y(x), y'(x))$. The objective of calculus of variations is to find the unknown function $y(x)$ that optimizes the integral $\mathcal{F}(y)$. The integral $\mathcal{F}(y)$ is called the *functional*, which assigns a real number for a given function y.

To seek an optimal function, it is necessary to find the stationary value of the functional about the optimal solution y:

$$\delta\mathcal{F}=\mathcal{F}(y+h)-\mathcal{F}(y)=\frac{d}{dt}\mathcal{F}(y+th)\Big|_{t=0}$$

The stationary value is obtained by solving $\delta\mathcal{F}=0$, where $h(x)$ is an arbitrary continuously differentiable function:

$$\delta \mathcal{F} = \int_a^b \left(\frac{\partial F}{\partial y} h + \frac{\partial F}{\partial y'} h' \right) dx = 0$$

Integrating the second term by parts yields

$$\delta \mathcal{F} = \int_a^b \left(\frac{\partial F}{\partial y} - \frac{d}{dx}\frac{\partial F}{\partial y'} \right) h\,dx + F_{y'}h \Big|_{x=a}^{x=b} = 0 \qquad (2.248)$$

There are two possibilities:

Case 1: The variational equation is subjected to the fixed end-point conditions

$$y(a) = y_a, \quad y(b) = y_b$$

that translate to

$$h(a) = h(b) = 0$$

Equation (2.250) becomes

$$\delta \mathcal{F} = \int_a^b \left(\frac{\partial F}{\partial y} - \frac{d}{dx}\frac{\partial F}{\partial y'} \right) h\,dx = 0 \qquad (2.249)$$

Because $h(x)$ is an arbitrary continuously differentiable function, the variational equation of the *functional* requires

$$F_y - \frac{d}{dx} F_{y'} = 0 \qquad (2.250)$$

that is known as a *Euler-Lagrange* equation establishing a necessary condition for $y(x)$ to be *optimal*.

Case 2: The variational equation is subjected to the conditions

$$F_{y'}\Big|_{x=a} = F_{y'}\Big|_{x=b} = 0$$

These conditions are known as the natural boundary conditions. In either case, the optimal function satisfies the Euler-Lagrange equation.

2.14.2 Euler-Lagrange Equations for a Functional of n-Dependent Variables

An approach similar to the one adopted in the previous subsection can be used to derive the Euler-Lagrange equations for a functional that possesses n-dependent variables $y_1, y_2, \ldots y_n$ to give the following system of equations:

…

$$F_{y_i} - \frac{d}{dx}F_{y'_i} = 0, \quad i=1,2,\ldots,n$$

2.14.3 Euler-Lagrange Equations for a Functional Involving nth-Order Derivative

The Euler-Lagrange equation for a functional that contains the derivatives of nth order is

$$F_y - \frac{d}{dx}F_{y'} + \cdots + (-1)^n \frac{d^n}{dx^n}F_{y^n} = 0$$

where the value of the function and its $(n-1)$th derivatives at the end points are specified as

$$y(a) = y_a, \; y'(a) = y_a^{(2)}, \; \ldots, \; y^{(n-1)}(a) = y_a^{(n-1)}$$

$$y(b) = y_b, \; y'(b) = y_b^{(2)}, \; \ldots, \; y^{(n-1)}(b) = y_b^{(n-1)}$$

2.14.4 Euler-Lagrange Equations for a Functional of Two Independent Variables

The Euler-Lagrange equation for a functional of two independent variables is (Gel'fand and Fomin, 2000)

$$F_y - \frac{d}{dx_1}F_{y_{x_1}} - \frac{d}{dx_2}F_{y_{x_2}} = 0$$

2.14.5 Application of Calculus of Variations

Example 2.12
The area of the surface generated by rotating the curve $y(x)$ joining two given points (a,b) about the x-axis is

$$2\pi \int_a^b y\sqrt{1+y'^2}\,dx$$

The curve that minimizes the surface area is given by the following Euler-Lagrange equation (Gel'fand and Fomin, 2000):

$$F_y - \frac{d}{dx}F_{y'} = 0$$

that is independent of x. The general solution can be obtained as follows. Multiplication by y' after expanding the derivative yields the following condition:

$$F - y'F_{y'} = c_1$$

where c_1 is a constant. Using

$$F(y, y', x) = y\sqrt{1+y'^2}, \quad F_{y'} = \frac{yy'}{\sqrt{1+y'^2}}$$

$$y\sqrt{1+y'^2} - \frac{yy'^2}{\sqrt{1+y'^2}} = c_1$$

Solving for y' gives

$$y' = \sqrt{c_1 y^2 - 1}$$

the solution of which is given by

$$y = c_1 \cosh\left(\frac{x+c_2}{c_1}\right)$$

which is an equation describing a catenary curve, where c_2 is another constant. The constants c_1 and c_2 can be determined from the condition that the curve representing the solution should pass through the two (prescribed) end points.

2.15 STOCHASTIC DIFFERENTIAL EQUATIONS

Processes in which a certain time-dependent random variable $X(t)$ exists are called *stochastic processes*. Loosely, these are the processes that evolve probabilistically in time. We can measure values $x_1, x_2, x_3, \ldots$, etc., of $X(t)$ at times $t_1, t_2, t_3, \ldots$, and we assume that a set of joint probability densities, $p(x_1, t_1; x_2, t_2; x_3, t_3; \ldots)$, exists that describes the system completely. For a treatise on stochastic processes, the reader is referred to Gardiner (2003). The applications related to chemical engineering can be followed from Rao et al. (1974a and 1974b).

The conditional probability densities can be defined in terms of the joint probability density functions as

$$p(x_1, t_1; x_2, t_2; \ldots | y_1, \tau_1; y_2, \tau_2; \ldots) = \frac{p(x_1, t_1; x_2, t_2; \ldots; y_1, \tau_1; y_2, \tau_2; \ldots)}{p(y_1, \tau_1; y_2, \tau_2; \ldots)} \tag{2.251}$$

The definitions in Equation (2.253) are valid independently of the ordering of the times. However, it is usual to consider only times that increase from right to left, i.e.,

$$t_1 \geq t_2 \geq t_3 \cdots \geq \tau_1 \geq \tau_2 \ldots \tag{2.252}$$

Stochastic differential equations then obviously describe the evolution of stochastic processes in terms of a dependent random variable $X(t)$.

Conditional probabilities can be seen as predictions of the future values of $X(t)$ (i.e., $x_1, x_2, \ldots$ at times $t_1, t_2, \ldots$) given the knowledge of the past values ($y_1, y_2, \ldots$ at times $\tau_1, \tau_2, \ldots$). A stochastic process is said to be completely independent if

$$p(x_1, t_1; x_2, t_2; \ldots | y_1, \tau_1; y_2, \tau_2; \ldots) = \Pi_i p(x_i, t_i)$$

Mathematics in Chemical Engineering

that means that the value of $\mathbf{X}$ at time t is completely independent of its values in the past (or future).

If a stochastic process evolves in time satisfying the ordering given by Equation (2.254) and the conditional probability is determined by the knowledge of the most recent condition, then such a process is called a *Markov* process. We have

$$p(\mathbf{x}_1,t_1;\mathbf{x}_2,t_2;\dots|\mathbf{y}_1,\tau_1;\mathbf{y}_2,\tau_2;\dots) = p(\mathbf{x}_1,t_1;\mathbf{x}_2,t_2;\dots;\mathbf{y}_1,\tau_1) \qquad (2.253)$$

2.15.1 Differential Chapman-Kolmogorov Equation

The equation that governs the conditional probability function is the differential Chapman-Kolmogorov equation:

$$\frac{\partial}{\partial t} p(z,t|y,t') = -\sum_i \frac{\partial}{\partial z_i}[A_i(z,t)p(z,t|y,t')] + \sum_{i,j} \frac{1}{2}\frac{\partial^2}{\partial z_i \partial z_j}[B_{i,j}(z,t)p(z,t|y,t')]$$

$$+ \int dx [W(z|x,t)p(x,t|y,t') - W(x|z,t)p(z,t|y,t')]$$

where

$$W(x|z,t) = \lim_{\Delta t \to 0} p(x, t+\Delta t|z,t)/\Delta t$$

For the quantities $W(x|z,t)$ equal to zero, the differential Chapman-Kolmogorov equation takes the form of the Fokker-Planck equation:

$$\frac{\partial}{\partial t} p(z,t|y,t') = -\sum_i [A_i(z,t)p(z,t|y,t')] + \sum_{i,j} \frac{1}{2}\frac{\partial^2}{\partial z_i \partial z_j}[B_{i,j}(z,t)p(z,t|y,t')]$$

and the corresponding process is known as a diffusion process. The vector $A(z,t)$ is known as the drift vector and the matrix $B(z,t)$ as the diffusion matrix. The derivation of the equation requires that the diffusion matrix be symmetric and positive definite. The property of the diffusion process is that the sample path $y(t)$ is a continuous function of time, but it is nowhere differentiable.

2.15.2 Connection between the Fokker-Planck Equation and Stochastic Differential Equation

Given a general stochastic differential equation

$$\frac{dx}{dt} = a(z,t) + b(x,t)\xi(t) \qquad (2.254)$$

or, equivalently, the stochastic integral equation

$$x - x(0) = \int_0^t a(x(t'),t')dt' + \int_0^t b[x(t'),t']dW(t') \qquad (2.255)$$

the conditional probability density function for the sample path is the Fokker-Planck equation,

$$\frac{\partial}{\partial t}p(x,t|x_0,t_0) = -\frac{\partial}{\partial x}[a(x,t)p(x,t|x_0,t_0)] + \frac{1}{2}\frac{\partial^2}{\partial x^2}[b(x,t)^2 p(x,t|x_0,t_0)]$$

For a stochastic differential equation for n variables,

$$d\mathbf{x} = \mathbf{A}(\mathbf{x},t)dt + \mathbf{B}(\mathbf{x},t)d\mathbf{W}(t)$$

where $\mathbf{W}(t)$ is an n-variable Wiener process, the Fokker-Planck equation for the conditional probability density $p \equiv p(\mathbf{x},t|\mathbf{x}_0,t_0)$ is given by

$$\frac{\partial p}{\partial t} = -\sum_i \frac{\partial}{\partial x_i}[A_i(\mathbf{x},t)p] + \sum_{i,j} \frac{1}{2}\frac{\partial^2}{\partial x_i \partial x_j}\{[\mathbf{B}(\mathbf{x},t)\mathbf{B}^T(\mathbf{x},t)]_{i,j}\,p\}$$

2.15.3 Îto Stochastic Integral

The solution of a stochastic differential equation is expressed in terms of the integral with respect to a sample function W(t). The stochastic integral

$$\int_{t_0}^{t} G(t')dW(t')$$

is obtained by dividing the interval $[t_0, t]$ into n subintervals such that

$$t_0 \leq t_1 \leq \cdots \leq t_{n-1} \leq t$$

where the intermediate points τ_i satisfy

$$t_{i-1} \leq \tau_i \leq t_i$$

The Îto stochastic integral is defined as a limit of the partial sum

$$\int_{t_0}^{t} G(t')dW(t') = \text{ms-lim}_{n\to\infty} \sum_{i=1}^{n} G(t_{i-1})[W(t_i) - W(t_{i-1})]$$

where ms-lim is the mean square limit defined as

$$\text{ms-lim}_{n\to\infty} X_n = X \Leftrightarrow \lim_{n\to\infty} \int p(\omega)[X_n(\omega) - X(\omega)]d\omega = 0$$

The Îto integration formula, which is presented in the next subsection, is useful for the purpose of evaluating a stochastic integral.

2.15.3.1 One-Dimensional Îto Formula

Let $X(t)$ be an Îto process given by (Oksendal, 2000)

$$dX = udt + vdW(t)$$

Let $g(t,x) \in C^2([0,\infty) \times \mathbb{R})$. Then

$$Y(t) = g(t,x)$$

is again an Îto process, and

$$dY = \frac{\partial g}{\partial t}(t,X)dt + \frac{\partial g}{\partial x}(t,X)dX + \frac{1}{2}\frac{\partial^2 g}{\partial x^2}(t,X) \cdot (dX)^2$$

that is called the Îto formula, where $(dX)^2 = (dX) \cdot (dX)$ is evaluated using the following rules:

$$dt \cdot dt = dt \cdot dW = dW \cdot dt = 0, \quad dW \cdot dW = dt$$

The following example shows the application of the Îto integration formula to evaluate a stochastic integral.

Example 2.13
Evaluate the integral

$$\int_0^t W(t')dt'$$

Choose $X = W$ and $g(t,x) = 1/2x^2$. Then

$$Y = g(t,W) = \frac{1}{2}W^2 \Rightarrow dY = d\left(\frac{1}{2}W^2\right)$$

By the Îto formula,

$$dY = \frac{\partial g}{\partial t}dt + \frac{\partial g}{\partial x}dW + \frac{1}{2}\frac{\partial^2 g}{\partial x^2}(dW)^2 = W(t)dW(t) + \frac{1}{2}(dW)^2 = W(t)dW(t) + \frac{1}{2}dt$$

Thus,

$$\int_0^t d\left(\frac{1}{2}W^2\right) = \frac{1}{2}W^2(t) = \int_0^t W(t')dW(t') + \frac{1}{2}t$$

2.15.3.2 An Example Application of a Stochastic Differential Equation

The concentration of a species undergoing a first-order chemical reaction with the rate constant of k with noise may be described by the following stochastic differential equation:

$$\frac{dC}{dt} = -kC + C\alpha\xi(t) \quad t=0, \quad C=C_0$$

$$dC = -kCdt + C\alpha\xi(t)dt$$

2.16 ASYMPTOTIC APPROXIMATIONS AND EXPANSIONS

Many problems in transport and chemical reaction engineering are nonlinear and cannot be solved analytically. A powerful approach to solve such problems lies in the method of matched asymptotic expansions that often provide analytical expressions for the solution. The method is based on an expansion whose convergence is based on concepts somewhat different from that usually understood. An example is considered below to clarify the nature of such expansions. Such expansions can be used in the solution of nonlinear equations for limiting values of parameters associated with the problem. Several examples are available in the chemical engineering literature (Leal, 1992, 2007; Deen, 1998; Varma and Morbidelli, 1997).

Example 2.15
A detailed treatment of this example is given by Erdelyi (1956). Let

$$S(x) \sum_{n=0}^{\infty} (-1)^n n! x^n \qquad (2.256)$$

The foregoing sum is definitely divergent because of $n!$. However, let us proceed without regard for the divergence to invoke the well-known result from the theory of gamma functions,

$$n! = \int_0^\infty e^{-t} t^n dt \qquad (2.257)$$

Using Equation (2.260) in Equation (2.59), one has

$$S(x) = \sum_{n=0}^{\infty} (-1)^n \int_0^\infty e^{-t} x^n t^n dt \qquad (2.258)$$

Disregarding conditions required for interchanging the summation and integration, one obtains

$$S(x) = \int_0^\infty e^{-t} \sum_{n=0}^{\infty} (-1)^n (xt)^n dt \qquad (2.259)$$

The alternating series within the integrand on the right-hand side of Equation (2.258) can now be summed exactly to write from Equation (2.259)

$$S(x) = \int_0^\infty \frac{e^{-t}}{1+xt} dt \qquad (2.260)$$

Note that the right-hand side is a very well-behaved function for any real value of x. It was not legitimately obtained because the interchange of summation and integration cannot be performed under the circumstances. Note also that this is not a case of an irrational procedure producing a valid result. The result is not valid because the left-hand side simply does not exist.

The issue in question is whether the series in Equation (2.256) can still approximate the function in Equation (2.260) in some way. To entertain this possibility, consider the evaluation of the left-hand side up to, say, m terms defined by

$$S_m(x) \equiv \sum_{n=0}^{m} (-1)^n n! x^n \qquad (2.261)$$

Compute the foregoing sum for $m = 0,1,2,3,\ldots$ for $x = 0.1, 0.01, 0.001$ and compare it with the evaluation of the right-hand side of Equation (2.263), which is obtained by the following equation:

$$S(x) = \frac{1}{x} e^{1/x} E_1\left(\frac{1}{x}\right) \qquad (2.262)$$

where

$$E_1(x) = \int_x^{\infty} \frac{e^{-t}}{t} dt \qquad (2.263)$$

On the other hand, for each value of x, compare the values obtained from Equation (2.261) with increasing m. The results are presented below.

x	$S(x)$	$S_1(x)$	$S_2(x)$	$S_{25}(x)$
0.1	0.915633	0.90000	0.920000	−0.19594
0.01	0.990194	0.99000	0.990200	0.990194
0.001	0.999002	0.99900	0.999002	0.999002

The feature most worthy of notice is that while the summation with an increase in the number of terms (i.e., m) does not improve the approximation to the function, the approximation improves as the value of x itself is increased. At $x = 0.1$, the divergence has already manifested with 10 terms. Although at the lower values of x (0.01, 0.001) the expansion appears to converge in the usual sense, with sufficient increase in the number of terms, one expects divergence to set in. This is then the idea behind an asymptotic expansion of a function. Here the parameter used for conveying the idea of asymptotic expansion was x. In what follows, it is more convenient to use ε as the parameter.

Consider the *asymptotic expansion* of the function $T(\mathbf{x},\varepsilon)$ with ε as a parameter. In writing the expansion, two alternative order relations, denoted O and o, are generally used. Thus we set

$$T(\mathbf{x},\varepsilon) = f_0(\varepsilon)T_0(\mathbf{x}) + f_1(\varepsilon)T_1(\mathbf{x}) + f_2(\varepsilon)T_2(\mathbf{x}) + O(f_3(\varepsilon)) \qquad (2.264)$$

where the coefficients $f_n(\varepsilon)$ are called gauge functions. The neglected term $O(f_3(\varepsilon))$ simply means that it behaves as $f_3(\varepsilon)$ as $\varepsilon \to 0$. An alternative notation for the foregoing situation is $o(f_3(\varepsilon))$, which means that the (magnitude of the) largest neglected term is smaller than $|f_2(\varepsilon)|$ as $\varepsilon \to 0$. This notation is more useful when the identity of $f_3(\varepsilon)$ is unknown. Now the expansion for $T(\mathbf{x},\varepsilon)$ is not *necessarily convergent* in the usual sense of

$$\lim_{N\to\infty}\left|T(\mathbf{x},\varepsilon)-\sum_{n=1}^{N}f_n(\varepsilon)T_n(\mathbf{x})\right|=0 \qquad (2.265)$$

but a necessary condition for asymptotic convergence is

$$\lim_{\varepsilon\to 0}\left|T(\mathbf{x},\varepsilon)-\sum_{n=1}^{N}f_n(\varepsilon)T_n(\mathbf{x})\right|=0, \quad \text{fixed } N \qquad (2.266)$$

This definition is consistent with the example shown earlier. A necessary condition for asymptotic convergence is given by

$$\lim_{\varepsilon\to 0}\frac{f_{n+1}(\varepsilon)}{f_n(\varepsilon)} \quad \text{for all } n \qquad (2.267)$$

Suppose the expansion of $T(\mathbf{x},\varepsilon)$ is desired in a domain Ω. If the asymptotic convergence occurs for all $\mathbf{x}$ in Ω, the expansion is said to be a *regular asymptotic expansion*.

Generally, the occurrence of the regular asymptotic expansions is in finite (bounded) domains. In other situations, the convergence of a particular expansion may occur only in a subdomain. For example, the entire domain may be made up of two separate subdomains, Ω^+ separated from another denoted Ω^-, with an intersection $\partial\Omega$ separating the two domains. This is also called the *overlap region*. Thus, we may have two separate expansions of the function $T(\mathbf{x},\varepsilon)$ as follows:

$$T(\mathbf{x},\varepsilon)=\sum_{n=0}^{N^+}f_n^+(\varepsilon)T_n^+(\mathbf{x}), \quad \mathbf{x}\in\Omega^+ \qquad (2.268)$$

$$T(\mathbf{x},\varepsilon)=\sum_{n=0}^{N^-}f_n^-(\varepsilon)T_n^-(\mathbf{x}), \quad \mathbf{x}\in\Omega^- \qquad (2.269)$$

The expansions in the individual subdomains are known as *singular expansions*. To determine the expansion in the entire domain, it is necessary to match the two expansions in the overlap region $\partial\Omega$. There are two strategies that are possible. One is referred to as a "matching" strategy, requiring the same functional forms for the gauge functions from either side in the overlap region. Thus in this case one has

$$N^+ = N^-, \quad f_n^+(\varepsilon)=f_n^-(\varepsilon), \quad T_n^+(\mathbf{x})=T_n^-(\mathbf{x}), \quad \mathbf{x}\in\partial\Omega$$

A less rigorous strategy is satisfied by numerically equating the two solutions by "patching" as follows:

$$\sum_{n=0}^{N^+}f_n^+(\varepsilon)T_n^+(\mathbf{x})=\sum_{n=0}^{N^-}f_n^-(\varepsilon)T_n^-(\mathbf{x}), \quad \mathbf{x}\in\partial\Omega$$

Generally, $T(x,\varepsilon)$ satisfies a nonlinear boundary-value problem. We obtain a singular perturbation problem when, upon letting $\varepsilon \to 0$, the order of the differential equation is reduced.

2.17 STEADY-STATE MULTIPLICITY AND STABILITY

Nonlinear equations may admit no real solutions or multiple real solutions. For example, the quadratic equation can have no real solutions or two real solutions. Thus, it is important to know whether a given equation governing the behavior of an engineering system can admit more than one solution, since it is related to the issue of operation and performance of the system. In this subsection, criteria for the existence of multiple steady-state solutions to the governing equations of a CSTR and tubular reactors and, subsequently, the stability of these multiple steady states are presented.

2.17.1 STEADY-STATE MULTIPLICITY OF A CHEMICAL REACTOR

The analysis of steady-state multiplicity of a nonisothermal chemical reactor is complicated due to the number of parameters involved and the exponential nonlinearity in the temperature dependence of the kinetic function. In this and the next subsection, the results of the analysis of steady-state multiplicity are presented. The derivations of these results are detailed in a review chapter by Morbidelli et al. (1986).

2.17.1.1 Steady-State Multiplicity of a CSTR

The dimensionless steady-state equations of mass and energy balance for a CSTR are reproduced here:

$$1 - u - Da \, r(u,v) = 0 \qquad (2.270)$$

$$1 - v + \beta Da \, r(u,v) = 0 \qquad (2.271)$$

For the nondimensionalization, the reader is referred to Section 2.7. From the physical arguments, the dimensionless concentration u is bounded between 0 and 1, i.e., $0 \leq u \leq 1$. The relationship between the dimensionless concentration u and dimensionless temperature v can be obtained as

$$u = \frac{1+\beta-v}{\beta} \qquad (2.272)$$

that shows the corresponding bounds on the dimensionless temperature:

$$1 \leq v \leq 1+\beta \quad \text{for } \beta > 0$$

$$1 - |\beta| \leq v \leq 1 \quad \text{for } \beta < 0$$

Thus substitution of Equation (2.272) into Equation (2.271) yields

$$1 - v + \beta Da \, r\left(\frac{1+\beta-v}{v}, v\right) \qquad (2.273)$$

For nth-order kinetics, Equation (2.273) is written as

$$F(v) \equiv \frac{(1+\beta-v)^n}{v-1} \exp\left[\gamma\left(1-\frac{1}{v}\right)\right] = \frac{\beta^{n-1}}{Da} \qquad (2.274)$$

Because $F(1) = \infty$, $F(1 + \beta) = 0$, and $F(v) > 0$ for $1 \leq v \leq 1 + \beta$, Equation (2.292) admits a unique solution if $F(v)$ is monotonically decreasing, i.e.,

$$\frac{dF}{dv} \leq 0 \quad \text{for} \quad 1 \leq v \leq 1+\beta \qquad (2.275)$$

If the condition in Equation (2.275) is violated, then uniqueness is assured if and only if

$$G(\bar{Z}) \geq \gamma \quad \text{for} \quad 0 < Z < 1 \qquad (2.276)$$

where

$$G(Z;n,\beta) \equiv \frac{(1+\beta Z)^2 [(1+Z(n-1)]}{\beta Z(1-Z)}, \quad Z = \frac{v-1}{\beta}$$

and $z \equiv \bar{Z}$ is the one and only root in the interval $(0,1)$ of the following equation:

$$\beta(n-1)Z^3 - (n-1)(2\beta+1)Z^2 - (2+\beta)Z + 1 = 0 \qquad (2.277)$$

that is obtained from $G'(Z) = 0$. If the condition in Equation (2.276) is violated, then multiple solutions exist for $Da_* < Da < Da^*$, where

$$Da_* = \frac{\beta^{n-1}}{F(v_+)} \quad \text{and} \quad Da^* = \frac{\beta^{n-1}}{F(v_-)}$$

where $v_\pm = 1 + \beta Z_\pm$ and $Z_\pm$ are the roots of $G(Z) = \gamma$.

2.17.1.2 Steady-State Multiplicity of a Tubular Reactor

The analysis of multiplicity for a tubular reactor is involved because its mass and energy balances are governed by nonlinear boundary-value problems. The uniqueness conditions for a tubular reactor are more conservative than those for a CSTR. The exact bounds for the uniqueness require numerical solutions.

The dimensionless mass and energy balance in a tubular reactor are (Morbidelli et al. 1986)

$$\frac{1}{Pe_m}\frac{d^2u}{dx^2} - \frac{du}{dx} - Da r(u,v) = 0 \qquad (2.278)$$

$$\frac{1}{Pe_t}\frac{d^2v}{dx^2} - \frac{dv}{dx} - \delta(v - v_c) + \beta Da r(u,v) = 0 \qquad (2.279)$$

with the boundary conditions

$$\frac{du}{dx} = Pe_m(u-1), \quad \frac{dv}{dx} = Pe_t(v-1), \quad \text{at } x = 0 \qquad (2.280)$$

$$\frac{du}{dx} = 0, \quad \frac{dv}{dx} = 0, \quad \text{at } x = 1 \qquad (2.281)$$

where the dimensionless variables are defined as follows:

$$u = \frac{C}{C_f} \quad v = \frac{T}{T_f} \quad v_c = \frac{T_c}{T_f} \quad x = \frac{z}{L} \quad Da = \frac{Lr(C_f, T_f)}{VC_f} \quad d = \frac{UPL}{V\rho A C_p}$$

$$\beta = \frac{(-\Delta H)C_f}{\rho C_p T_f} \quad R = \frac{r(C,T)}{r(C_f, T_f)} \quad Pe_m = \frac{VL}{D_m} \quad Pe_t = \frac{VL\rho C_p}{D_t} \quad \gamma = \frac{E}{RT_f}$$

Case 1: $Pe_m = Pe_t = Pe$, $\delta = 0$
A *sufficient* condition for *uniqueness* is

$$\frac{dF}{dv} \leq 0 \quad \text{for} \quad 1 \leq v \leq 1+\beta \qquad (2.282)$$

where

$$F(v) = \frac{r\left(\frac{1+\beta-v}{\beta}, v\right)}{v-1}$$

When the condition in Equation (2.282) is violated, the following condition ensures uniqueness:

$$Da \leq \frac{\mu}{\beta r\left(\frac{1+\beta-v}{\beta}, v\right)} \quad \text{for} \quad 1 \leq v \leq 1+\beta$$

where $p = \lambda_1/Pe + Pe/4$ and λ_1 is the smallest root of the equation

$$\tan\sqrt{\lambda} = \frac{Pe\sqrt{\lambda}}{\lambda - Pe^2/4}, \quad 0 < \lambda_1 < \pi^2$$

Case 2: $Pe_m = Pe_t = Pe$, $\delta > 0$ (Morbidelli et al., 1986; Varma and Amundson, 1972)
A *sufficient* condition is given by

$$Da \leq \frac{\mu + \delta}{\beta R'\left(\frac{1+\beta-v}{\beta}, v\right) + [\delta(a_1 - a_2)/2\mu]} \quad \text{for } 1 \leq v \leq 1+\beta$$

where μ is the same parameter as defined in the previous case and

$$a_1 = \sup \frac{\partial r}{\partial u} > 0, \quad a_2 = \inf \frac{\partial r}{\partial u} \geq 0$$

Case 3: $Pe_m \neq Pe_p, \delta \neq 0$

$$Da \leq \frac{-\beta\mu_m a_3 + (\mu_t + \delta)a_2 + \sqrt{\Delta}}{a_3 \beta[2a_2 + a_1 \exp(|Pe_t - Pe_m|)]}$$

where

$$\Delta = \beta\mu_m a_3 + (\mu_t + \delta)^2 + 2\beta\mu_m(\mu_m + \delta)a_1 a_3 \exp(|Pe_t - Pe_m|), \quad a_3 = \sup \frac{\partial r}{\partial v} \qquad (2.283)$$

where $\mu_i = \lambda_1/Pe_i + Pe_i/4$, $i = m, t$ and λ_1 is the same as previously defined.

The exact bounds for multiplicity can be determined by computationally solving the boundary-value problems with the use of the singularity theory.

2.17.2 Analysis of Multiplicity by Singularity Theory

Techniques based on the implicit function theorem have been used to predict the existence of multiple solutions in a CSTR (Chang and Calo, 1979). An extension of catastrophe theory known as singularity theory has also been effectively used to determine the conditions for the existence of multiple solutions in a CSTR and a tubular reactor (Balakotaiah and Luss, 1981, 1982; Witmer et al., 1986). In this subsection, the technique of singularity to find the maximum number of solutions of a single mathematical equation and its application to analysis of the multiplicity of a CSTR are presented (Luss, 1986; Balakotaiah at al., 1985). The details of singularity theory can be found in Golubitsky and Schaeffer (1985).

For the first-order reaction, the steady-state equations for mass and energy balance in a CSTR can be combined into a single equation represented as

$$f(x,\mathbf{p}) = 0 \qquad (2.284)$$

where $\mathbf{p}$ is a vector of physical quantities. A singular point $(x_s, \mathbf{p}_s)$ of codimension k is defined as

$$f(x_s, \mathbf{p}_s) = \frac{df(x_s, \mathbf{p}_s)}{dx} = \cdots = \frac{d^k f(x_s, \mathbf{p}_s)}{dx^k} = 0, \quad \frac{d^{k+1} f(x_s, \mathbf{p}_s)}{dx^{k+1}} \neq 0 \qquad (2.285)$$

The application of catastrophe theory (Hofp, 1960) predicts that in the neighborhood of a singular point of codimension k, the qualitative features represented by Equation (2.284) are similar to that of the polynomial,

$$y^{k+1} - \sum_{i=0}^{k-1} \alpha_i y^i \qquad (2.286)$$

In the neighborhood of such a singular point, Equation (2.284) admits $k+1$ solutions if

$$\text{rank}[b_{i,j}] = k$$

where the elements $b_{i,j}$ are the coefficients of the expansion,

$$\frac{\partial f}{\partial p_i}(y, \boldsymbol{p}_s) = \sum_{j=1}^{k} \frac{b_{i,j}}{j!}(y-y_s)^{j-1} + O(|y-y_s|^k)$$

Thus, the singular points of Equation (2.284) characterized by Equation (2.285) allow one to analyze the multiplicity of Equation (2.284) from the knowledge of the polynomial Equation (2.286).

2.17.2.1 Application of Singularity Theory to the Analysis of CSTR

The mass and energy balance equations for a CSTR can be combined into a single equation,

$$x - (1-x)Da \exp\left[\frac{Bx + \alpha y_c Da}{1 + \alpha Da + (\alpha Da + Bx/\gamma)}\right] = F(x, \boldsymbol{p}) = 0 \qquad (2.287)$$

where the dimensionless variables are defined as

$$x = \frac{C_f - C_a}{C_f} \quad y_c = \frac{E}{RT_f}\left(\frac{T_c - T_f}{T_f}\right) \quad B = \frac{E}{RT_f}\frac{(-\Delta H)C_f}{\rho C_p T_f}$$

$$\gamma = \frac{E}{RT_f} \quad Da = \frac{V}{F}\exp\left[\frac{E}{RT_f}\right] \quad \alpha = \frac{hA}{\rho C_p V}\exp\left[\frac{-E}{RT_f}\right]$$

and

$$\boldsymbol{p} = (Da, B, \alpha, y_c, \gamma)$$

For adiabatic ($\alpha = 0$) CSTR

$$F(x, Da, \boldsymbol{p}_s) = \frac{\partial F}{\partial x}(x, Da, \boldsymbol{p}_s) = 0 \qquad (2.288)$$

The solutions of Equation (2.288) give

$$x_\pm = \frac{1 - \frac{2}{\gamma} \pm \sqrt{1 - \frac{4}{B}(1+B/\gamma)}}{2(1+B/\gamma^2)}$$

$$Da = \frac{x_\pm}{1 - x_\pm}\exp\left(\frac{-Bx_\pm}{1 + Bx_\pm/\gamma}\right)$$

Multiplicity exists if and only if

$$B > \frac{4}{1-4/\gamma} > 0 \tag{2.289}$$

(i) Adiabatic CSTR, hysteresis variety

In addition to the condition for multiplicity in Equation (2.289), the condition for the existence of hysteresis variety is defined by

$$\frac{\partial^2 F}{\partial x^2}(x, Da, p_s) = 0$$

that gives the solution

$$\alpha = \frac{1-x_h}{x_h}\left[\frac{4Bx_h(1-x_h)}{\gamma^2(1-2x_h)}\right]\exp\left\{\gamma - \frac{2}{1-2x_h}\right\}$$

where

$$x_h = \frac{\frac{4B}{\gamma}\left(1+\frac{y_c}{\gamma}\right) + 4y_c - B \pm \sqrt{D}}{2\left[4y_c - 2B + \frac{4B}{\gamma}\left(1+\frac{y_c}{\gamma}\right)\right]}$$

and

$$D = \left[\frac{4B}{\gamma}\left(1+\frac{y_c}{\gamma}\right)\right]^2 - 4y_c\left[4y_c - 2B + \frac{4B}{\gamma}\left(1+\frac{y_c}{\gamma}\right)\right]$$

(ii) Isola variety

$$F(x, Da, p_s) = \frac{\partial F}{\partial x}(x, Da, p_s) = \frac{\partial F}{\partial Da}(x, Da, p_s) = 0 \tag{2.290}$$

The condition in Equation (2.290) gives parametric representations by the equations

$$x_b^5 B^3/\gamma^2 - x_b^3[1-x_b+y_c x_b/\gamma^2 - 2/\gamma]B^2 + x_b[1+y_c(1-x_b)-2x_b y_c/\gamma]B - y_c = 0$$

$$\alpha = \frac{B(1-x_b)^2}{Bx_b^2 - y_c}\exp\left\{\frac{Bx_b^2}{1+Bx_b^2/\gamma}\right\}$$

where the parameter $0 < x_b < 1$.

(iii) Pitch-fork singular point

$$F(x, Da, p_s) = \frac{\partial F}{\partial x}(x, Da, p_s) = \frac{\partial^2 F}{\partial x^2}(x, Da, p_s) = \frac{\partial F}{\partial Da}(x, Da, p_s) = 0 \quad (2.291)$$

and

$$\frac{\partial^3 F}{\partial x^3} \frac{\partial^2 F}{\partial x \partial Da} < 0$$

Solving Equation (2.291) gives

$$B = \frac{\gamma^2 y_c (1-2x)^2}{x[4(1-x)(\gamma+y_c) - \gamma^2(1-2x)]} \quad \alpha Da = \frac{Bx(1-x)}{Bx^2 - y_c} \quad \text{and} \quad \gamma < \frac{8}{3}$$

where x is the root of

$$x^3 + A_1 x^2 + A_2 x + A_3 = 0$$

with

$$A_1 = (\gamma^2 - 2\gamma - 4y_c)/2y_c$$
$$A_2 = [(\gamma^2 + y_c)(4\gamma + 4y_c - \gamma^2) - (\gamma^2 - 2\gamma - 2y_c)^2]/4\gamma^2 y_c$$
$$A_3 = (\gamma - 2)[\gamma(\gamma - 4) - 4y_c]/8\gamma y_c$$

and Da is the solution of Equation (2.305) for the corresponding α, Da, B, and x.

These conditions establish the boundary of the region in the space $(Da, B, \alpha, y_c, \gamma)$. The multiplicity patterns in the interior of each region are established by solving the equation for one set of parameters in the region.

2.17.3 Stability of Steady-State Solution

The existence of multiple solutions does not ensure that these solutions are physically attainable. In order for these solutions to be physically attainable, they must be stable. The linear stability analysis presented here provides the necessary conditions for the stability. The method of Lyapunov's function can also be used to assess the stability and the magnitude of the permissible perturbations so that the reactor returns to the steady state. In the case of linear stability analysis, the eigenvalues of a differential operator determine the stability. An excellent account of the stability analysis of chemical reactors can be found in Perlmutter (1972).

2.17.3.1 Linear Stability Analysis

Linear stability analysis is carried out on dynamical equations linearized about the steady-state solution (u_s, v_s). The steady-state solution is stable if the eigenvalues of the system of linearized equations are negative, unstable if those are positive, and indeterminate otherwise.

The dynamics of the CSTR is governed by the following dimensionless equations:

$$\frac{du}{dt} = 1 - u - Da u^n e^{-\gamma/v} \qquad (2.292)$$

$$\frac{dv}{dt} = (1+\delta)(1 - v + \beta Da u^n e^{-\gamma/v}) \qquad (2.293)$$

where the parameters are given by

$$u = \frac{C}{C_f} \quad \delta = \frac{UA}{F\rho C_p} \quad T_m = \frac{T_f + \delta T_c}{1+\delta}$$

$$v = \frac{T}{T_m} \quad \gamma = \frac{E}{RT_m} \quad Da = \frac{kVC_f^{n-1}}{F} \quad \beta = \frac{(-\Delta H)C_f}{\rho C_p T_m}$$

The linearized equations about the steady-state solution (u_s, v_s) can be written as

$$\frac{du}{dt} = -[1 + Dae^{-\gamma/v_s}]u - \frac{Da\gamma u_s}{v_s^2} e^{-\gamma/v_s} v \qquad (2.294)$$

$$\frac{dv}{dt} = (1+\delta)\left\{\beta Dae^{-\gamma/v_s} u - \left[1 - \frac{\beta Da\gamma u_s}{v_s^2} e^{-\gamma/v_s}\right]v\right\} \qquad (2.295)$$

The eigenvalues of the system of Equations (2.294) and (2.295) are negative if and only if

$$1 + Dae^{-\gamma/v_s} - \beta Da\gamma u_s/v_s^2 e^{-\gamma/v_s} > 0$$

and

$$2 + \delta + Dae^{-\gamma/v_s} - (1+\delta)(\beta Da\gamma u_s/v_s^2 e^{-\gamma/v_s}) >$$

which are the necessary conditions for the steady-state solution to be stable.

2.17.3.2 Method of Lyapunov's Function

The method of Lyapunov's function consists of constructing a function $V(x,y)$ with the following properties:

(i) $\qquad V(x, y) > 0 \quad \text{for } x, y \neq 0$

(ii) $\qquad \dfrac{dV(x, y)}{dt} = \nabla V \cdot f < 0 \quad \text{where } \dfrac{dx}{dt} = f$

For the steady-state solution (u_s, v_s) of a CSTR, the dynamical equations in terms of deviation variables (Denn, 1975),

$$x = u - u_s, \quad y = v - v_s$$

are written as

$$\frac{dx}{dt} = -(1 + Dae^{-\gamma/v}) - Dau_s\Delta(v, v_s)y \quad (2.296)$$

$$\frac{dy}{dt} = (1+\delta)[\beta Dae^{-\gamma/v}x - (1 - \beta Da\Delta(v, v_s))y] \quad (2.297)$$

where

$$\Delta(v, v_s) = \frac{e^{-\gamma/v} - e^{-\gamma/v_s}}{v - v_s}$$

To construct a Lyapunov's function, consider $V(x,y) = (x^2 + y^2)/2$:

$$\frac{dV(x,y)}{dt} = x\{-[1 + Dae^{-\gamma/v}]x - Dau_s\Delta(v, v_s)y)\} \cdots$$

$$+ y(1+\delta)\{-[1 - \beta Dau_s\Delta(v, v_s)]y + Da\beta e^{-\gamma/v}x\}$$

$$= -\{(1 + Dae^{-\gamma/v})x^2 + Da[u_s\Delta(v, v_s) - (1+\delta)\beta e^{-\gamma/v}]xy + (1+\delta)[1 - \beta Dau_s\Delta(v, v_s)]y^2\}$$

$\dot{V}(x,y)$ is negative definite if the terms in the braces are positive definite. Thus $\dot{V}(x,y)$ is negative definite if and only if

$$1 + Dae^{-\gamma/v} > 0$$

$$4(1+\delta)(1 + Dae^{-\gamma/v})[1 - \beta Dau_s\Delta(v, v_s)] - Da^2\{u_s\Delta(v, v_s) - (1+\delta)\beta e^{-\gamma/v}\}^2 > 0$$

The first inequality is true for power law kinetics. The second inequality holds only for certain values of v. Let v_c be the value of y such that $|v_c - v_s|$ is the minimum and for which the second inequality is violated. Then, the stability is ensured for all (u,v) such that

$$(u - u_s)^2 + (v - v_s)^2 < (v_c - v_s)^2$$

Other types of Lyapunov's functions may also be constructed.

For the general treatment in regard to Lyapunov's function, the reader may see Khalil (1996).

APPENDIX 2.A: MATHEMATICAL SOFTWARE

The development of software over the decades has enabled engineers to perform increasingly complex calculations based on the solution of a variety of mathematical equations. While analytic solutions can be well handled by symbolic mathematical software, the system of equations requiring numerical solution can be tackled using a variety of software packages. Furthermore, one can recognize that some software packages are designed to address specific classes of mathematical problems and methods of solutions. The purpose of this brief appendix is to make the reader aware of such tools along with their specific utilities and their sources.

Table 2.A.1 lists various software packages along with their utility in reference to the relevant mathematics in chemical engineering. It should be noted that the table lists only the tools that the authors are aware of and that it is *not complete* by any means. Furthermore, the mathematical utility of the software packages represents only a *fraction* of their many capabilities.

TABLE 2.A.1
Software Packages Relevant to Mathematical Calculations in Chemical Engineering

Software	Mathematical Utility	Source
1. Aspen HYSYS and Aspen Plus	Process simulation and optimization	www.aspentech.com
2. Auto 2000	Continuation and bifurcation problems	www.acm.caltech.edu
3. CFD Solvers	Computational fluid dynamics	
CFX and FLUENT		www.ansys.com
FLOW-3D		www.flow3d.com
PHOENIX		www.cham.co.uk
SPIKE, TINA, FINS		www.fluidgravity.co.uk
List of free CFD software		www.icemcfd.com
4. Comsol	Finite element analysis, free equation-based modeling, general periodic boundary conditions, evaluation of material, energy balances	www.comsol.com
5. DIVA	Nonlinear equations, discrete algebraic equations, continuation and stability analysis, parameter analysis, and parameter estimation	www.mpi-magdeburg.mpg.de (see also Mangold et al., 2000)
6. gPROMS	Nonlinear equations, steady-state and dynamic simulations, steady-state and dynamic optimization, and parameter estimation	www.psenterprise.com
7. JACOBIAN	Ordinary differential equations, differential algebraic equations, partial differential equations, discrete/continuous dynamic simulation, sensitivity analysis, optimization, and parameter estimation	www.numericatech.com
8. Mathematica	Analytical integration and differentiation, algebraic and differential equations, and optimization problems	www.wolfram.com
9. Mathcad	Symbolic mathematics, calculus, algebra, differential equations, statistics, geometry, and transforms	www.mathsoft.com
10. Maple	Symbolic mathematics, calculus, algebra, differential equations, statistics, geometry, and transforms	www.maplesoft.com

Continued

TABLE 2.A.1 *(Continued)*
Software Packages Relevant to Mathematical Calculations in Chemical Engineering

Software	Mathematical Utility	Source
11. MatLab	Analytical integration and differentiation, linear algebra, statistics, optimization, numerical integration, Fourier analysis, filtering, ordinary differential equations, partial differential equations, and matrix manipulations	www.mathworks.com
12. Maxima	Manipulation of symbolic expressions and numerics (e.g., differentiation; integration; Taylor series; Laplace transforms; ordinary differential equations; systems of linear equations, polynomials, and sets; vectors; matrices; and tensors)	www.maxima.sourceforge.net
13. Octave	Linear and nonlinear problems	www.gnu.org
14. Parsival	Description and simulation of dispersed systems, and Integrodifferential equations	www.cit-wulkow.de
15. XPP/XPPAUT	Ordinary differential equations, difference and delay equations, functional equations, boundary-value problems, and stochastic equations	www.math.pitt.edu

APPENDIX 2.B: DIRAC DELTA FUNCTION

The Dirac delta or Dirac's delta, introduced by the theoretical physicist Paul Dirac is an interesting function with signficant utility, which is formally denoted by δ and defined as follows:

$$\delta(x) = \begin{cases} \infty & x = 0 \\ 0 & x \neq 0 \end{cases}$$

that is constrained to satisfy

$$\int_{-\infty}^{\infty} \delta(x) dx = 1$$

Thus, the function may be visualized as the one representing an infinitely sharp peak bounding unit area.

Some of the important properties are listed below:

1. $\int_{-\infty}^{\infty} f(x)\delta(x-a)dx = f(a)$

2. $\int_{-\infty}^{\infty} \delta(\alpha x)dx = \frac{1}{|\alpha|}$ for a scalar α

3. $\int_{-\infty}^{\infty} f(t)\delta(t-\tau)dt = f(\tau)$

4. $\int_{-\infty}^{\infty} f(x)\delta^n(x)dx = -\int_{-\infty}^{\infty} \frac{df}{dx}\delta^{n-1}(x)dx$

5. $\int_{-1}^{1} \delta\left(\frac{1}{x}\right)dx = 0$

6. $\delta(x) = \int_{-\infty}^{\infty} e^{-2\pi i \lambda x} dx$

Mathematics in Chemical Engineering

APPENDIX 2.C: TABLES OF INTEGRAL TRANSFORMS

We provide here six tables of various integral transforms for ready reference. The extensive listing can be obtained from Gradshteyn and Ryzhik (1983).

TABLE 2.C.1
Laplace Transforms

$$F(\lambda) = \int_0^\infty e^{-\lambda x} f(x)\,dx \qquad f(x) = \frac{1}{\sqrt{2\pi i}} \int_{\gamma-i\infty}^{\gamma+i\infty} F(\lambda) e^{\lambda x}\,dx$$

$F(\lambda)$	$f(x)$
$\dfrac{1}{\lambda^{n+1}}$	$\dfrac{x^n}{n!},\ n \ge 0$
$\dfrac{a}{\lambda^2 + a^2}$	$\sin(ax)$
$\dfrac{\lambda}{\lambda^2 + a^2}$	$\cos(ax)$
$\dfrac{\lambda}{(\lambda+a)^2}$	$e^{-ax}(1-ax)^2$
$\dfrac{\lambda}{\lambda^4 + 4a^4}$	$\dfrac{1}{2a^2}\sin(ax)\sinh(ax)$
$\dfrac{\lambda^3}{\lambda^4 + 4a^4}$	$\cos(ax)\cosh(ax)$
$\dfrac{\lambda^3}{\lambda^4 - a^4}$	$\dfrac{1}{2}[\cos(ax) + \cosh(as)]$
$\dfrac{\lambda^2}{\lambda^4 - a^4}$	$\dfrac{1}{2a}[\sin(ax) + \sinh(ax)]$
$\dfrac{\lambda}{(\lambda^2 + a^2)^2}$	$\dfrac{x\sin(ax)}{2a}$
$\tan^{-1}\dfrac{a}{\lambda}$	$\dfrac{\sin(ax)}{x}$
$\dfrac{a}{\lambda^2 - a^2}$	$\sinh(ax),\ \lambda > |a|$
$\dfrac{\lambda}{\lambda^2 - a^2}$	$\cosh(ax),\ \lambda > |a|$
$\dfrac{\lambda}{(\lambda^2 + a^2)^2}$	$\dfrac{x}{2a}\sin(ax)$
$\dfrac{1}{(\lambda^2 + a^2)^2}$	$\dfrac{1}{2a^3}[\sin(ax) - ax\cos(ax)]$
$e^{-a\sqrt{\lambda}}$	$\dfrac{a}{2\sqrt{\pi x^3}}e^{-a^2/(4x)}$
$\dfrac{e^{-a\sqrt{\lambda}}}{\sqrt{\lambda}}$	$\dfrac{1}{\sqrt{\pi x}}e^{-a^2/(4x)}$

Continued

TABLE 2.C.1 *(Continued)*
Laplace Transforms

$$F(\lambda) = \int_0^\infty e^{-\lambda x} f(x)dx \qquad f(x) = \frac{1}{\sqrt{2\pi i}} \int_{\gamma-i\infty}^{\gamma+i\infty} F(\lambda)e^{\lambda x} dx$$

$\dfrac{e^{-a\sqrt{\lambda}}}{\lambda}$	$\dfrac{2}{\sqrt{\pi}} \int_{a/\sqrt{x}}^\infty e^{-y^2} dy = \text{erfc}\left(\dfrac{a}{2\sqrt{x}}\right)$
$(-1)^n \dfrac{d^n F}{d\lambda^n}$	$x^n f(x)$
$\dfrac{F(\lambda)}{\lambda}$	$\int_0^x f(t)dt$
$F(\lambda - a)$	$e^{ax} f(x)$
$\lambda F(\lambda) - F(0)$	$\dfrac{df}{dx}$

TABLE 2.C.2
Fourier Transforms

$$F(\lambda) = \frac{1}{\sqrt{2\pi}} \int_{-\infty}^\infty f(x)e^{i\lambda x} d\lambda \qquad f(x) = \frac{1}{\sqrt{2\pi}} \int_{-\infty}^\infty F(\lambda)e^{-i\lambda x} d\lambda$$

$\sqrt{2\pi}\delta(\lambda)$	1
$(\pi/2)^{1/2} i\ \text{sing}(\lambda)$	$1/x$
$1/\sqrt{2\pi}$	$\delta(x)$
$1/|\lambda|$	$1/|x|$
$(2/\pi)^{1/2}\Gamma(1-a)\sin(\tfrac{1}{a}\pi)/|\lambda|^{1-a}$	$1/|x|^a$
$(2/\pi)^{1/2}\delta(1+a)$	e^{iax}
$a(2/\pi)^{1/2}/(a^2+\lambda^2)$	$e^{-a|x|},\ a>0$
$2ai\lambda(2/\pi)^{1/2}/(a^2+\lambda^2)^2,\ \lambda>0$	$xe^{-a|x|},\ a>0$
$(2/\pi)^{1/2}(a^2-\lambda^2)/((a^2+\lambda^2)^2),\ \lambda>0$	$|x|e^{-a|x|},\ a>0$
$[a+(a^2+\lambda^2)^{1/2}]^{1/2}/(a^2+\lambda^2)^{1/2},\ \lambda>0$	$e^{-a|x|}/|x|^{1/2},\ a>0$
$e^{-\lambda^2/4a^2}/(a\sqrt{2})$	$e^{-a^2 x^2}$
$(2/\pi)^{1/2} e^{-a|\lambda|}/a$	$1/(a^2+x^2),\ \text{Re}\,a>0$
$-i(2/\pi)^{1/2}\lambda e^{-a|\lambda|}/2a$	$x/(a^2+x^2),\ \text{Re}\,a>0$
$\dfrac{1}{(2a)^{1/2}}\sin\left(\dfrac{\lambda^2}{4a}+\dfrac{\pi}{4}\right)$	$\sin(ax^2)$

TABLE 2.C.2 (Continued)
Fourier Transforms

$$F(\lambda) = \frac{1}{\sqrt{2\pi}} \int_{-\infty}^{\infty} f(x) e^{i\lambda x} d\lambda \qquad f(x) = \frac{1}{\sqrt{2\pi}} \int_{-\infty}^{\infty} F(\lambda) e^{-i\lambda x} d\lambda$$

$F(\lambda)$	$f(x)$
$\dfrac{1}{(2a)^{1/2}} \cos\left(\dfrac{\lambda^2}{4a} - \dfrac{\pi}{4}\right)$	$\cos(ax^2)$
$\dfrac{(\pi/2)^{1/2} \sin(\pi a/b)}{b[\cosh(\pi\lambda/b) + \cos(\pi a/b)]}$	$\dfrac{\sinh(ax)}{\cosh(bx)},\ 0 < a < b$
$\dfrac{i(\pi/2)^{1/2} \sinh(\pi\lambda/b)}{b[\cosh(\pi\lambda/b) + \cos(\pi a/b)]}$	$\dfrac{\cosh(ax)}{\sinh(bx)},\ 0 < a < b$
$\begin{cases}(\pi/2)^{1/2} & \text{for } |\lambda| < a \\ 0 & \text{for } |\lambda| > a\end{cases}$	$\sin(ax)/x$
$(2/\pi^3)^{1/2} e^{\pi\lambda}/(1 + e^{\pi\lambda})^2$	$x/\sinh x$
$(2/\pi)^{1/2} (-i\lambda)^{-(1+v)} v!$	x^v sign $xv < -a$ but not integer
$(2\pi)^{1/2} \cos[\pi/2(1+v)]/\Gamma(1+v)^{1+v}$	$|x|^v\ v < -1$ but not integer
$\dfrac{i \operatorname{sign} \lambda (2\pi)^{1/2} \sin[\pi/2(1+v)]\Gamma(v+1)}{|\lambda|^{(1+v)}}$	$|x|^v$ sign $xv < -1$ but not integer
$(\pi/2)^{1/2} \cot(\pi a - i\lambda\pi)/(a - i\lambda)$	$e^{-ax} \ln|1 - e^{-x}|,\ -1 < \operatorname{Re} a < 0$
$(\pi/2)^{1/2} \csc(\pi a - i\lambda\pi)/(a - i\lambda)$	$e^{-ax} \ln(1 - e^{-x}),\ -1 < \operatorname{Re} a < 0$
$F(\lambda)/(i\lambda)$	$\int_0^x f(t)\,dt$
$F(\lambda - a)$	$e^{iax} f(x)$
$(-i\lambda)^n F(\lambda)$	$\dfrac{d^n f}{dx^n}$

TABLE 2.C.3
Fourier Sine Transforms

$$F(\lambda) = \sqrt{\frac{2}{\pi}} \int_0^\infty f(x)\sin(\lambda x)dx \qquad f(x) = \sqrt{\frac{2}{\pi}} \int_0^\infty F(\lambda)\sin(\lambda x)d\lambda$$

$F(\lambda)$	$f(x)$
$1/x$	$(\pi/2)^{1/2}\,\text{sign}\,\lambda,\quad \lambda>0$
$(2/\pi)^{1/2}\lambda^{v-1}\Gamma(1-v)\cos(v\pi/2),\quad \lambda>0$	$x^{-v},\quad 0<\text{Re}\,v<1$
$\dfrac{1}{(2\pi)^{1/2}}\ln\left\|\dfrac{\lambda+a}{\lambda-a}\right\|,\quad \lambda>0$	$\dfrac{\sin(ax)}{x},\quad a>0$
$\begin{cases}\sqrt{\dfrac{\pi}{2}}\lambda & \text{for } \lambda<a \\ a\sqrt{\dfrac{\pi}{2}} & \text{for } \lambda>a\end{cases}$	$\dfrac{\sin(ax)}{x^2}$
$a\sqrt{\dfrac{\pi}{2}}\lambda^{-1/2}J_1(2a\lambda^{1/2}),\quad \lambda>0$	$\sin\dfrac{a^2}{x},\quad a>0$
$\sqrt{\dfrac{\pi}{2}}\lambda^{-1/2}Y_0(2a\lambda^{1/2})+K_0(2a\lambda^{1/2}),\quad \lambda>0$	$x^{-1}\sin\dfrac{a^2}{x},\quad a>0$
$\sqrt{\dfrac{\pi}{2}}a^{-1}\lambda^{1/2}J_1(2a\lambda^{1/2}),\quad \lambda>0$	$x^{-2}\sin\dfrac{a^2}{x},\quad a>0$
$\sqrt{\dfrac{\pi}{2}}e^{-a\lambda},\quad \lambda>0$	$\dfrac{x}{a^2+x^2},\quad a>0$
$2\pi^{-1/2}a^{-1}\lambda e^{-a\lambda},\quad \lambda>0$	$\dfrac{x}{(a^2+x^2)^2}$
$2\pi^{-1/2}a^{-1}\lambda e^{-a\lambda},\quad \lambda>0$	$\dfrac{x}{(a^2+x^2)^2}$
$\pi/2^{1/2}a^{-2}(1-e^{-a\lambda}),\quad \text{Re}\,\lambda>0$	$x^{-1}(a^2+x^2)^{-1}\,\text{Re}\,a>0$
$2/\pi^{1/2}\lambda/(a^2+\lambda^2),\quad \text{Re}\,\lambda>0$	$e^{-ax},\quad \text{Re}\,a>0$
$2/\pi^{1/2}\,2a\lambda/(a^2+\lambda^2)^2,\quad \lambda>0$	$xe^{-ax},\quad \text{Re}\,a>0$
$2/\pi^{1/2}\tan^{-1}\dfrac{\lambda}{a},\quad \lambda>0$	$x^{-1}e^{-ax},\quad \text{Re}\,a>0$
$(2/\pi)^{1/2}\Gamma(v)(a^2+\lambda^2)^{-v/2}\sin\left[v\tan^{-1}\dfrac{\lambda}{a}\right],\quad \lambda>0$	$x^{-v-1}e^{-ax},\quad \text{Re}\,a>0,\quad \text{Re}\,v>0$
$(2/\pi)^{1/2}a^{-1}\tanh\dfrac{1}{2a}\pi\lambda,\quad \lambda>0$	$\csc(ax),\quad \text{Re}\,a>0$
$(2/\pi)^{1/2}a^{-1}\dfrac{1}{\tan(\pi\lambda/a)}-\lambda,\quad \lambda>0$	$\dfrac{1}{\tanh(ax)},\quad \text{Re}\,a>0$

TABLE 2.C.4
Fourier Cosine Transforms

$$F(\lambda) = \sqrt{\frac{2}{\pi}} \int_0^\infty f(x)\cos(\lambda x)dx \qquad f(x) = \sqrt{\frac{2}{\pi}} \int_0^\infty F(\lambda)\cos(\lambda x)d\lambda$$

$F(\lambda)$	$f(x)$
$\dfrac{(2/\pi)^{1/2}\sec\left(\frac{1}{2}v\pi\right)\lambda^{v-1}}{\Gamma(v)},\ \lambda>0$	$x^{-v},\ 0<\operatorname{Re}v<1$
$\dfrac{(2/\pi)^{1/2}e^{-a\lambda}}{a},\ \lambda>0$	$(x^2+a^2)^{-1},\ \operatorname{Re}a>0$
$\dfrac{(2/\pi)^{1/2}(1+a\lambda)e^{-a\lambda}}{2a^3},\ \lambda>0$	$(x^2+a^2)^{-2},\ \operatorname{Re}a>0$
$\sqrt{2}\left(\dfrac{\lambda}{2a}\right)^v \dfrac{K_v(a\lambda)}{\Gamma\left(1+\frac{1}{2}\right)},\ \lambda>0$	$(x^2+a^2)^{-v-1/2},\ \operatorname{Re}a>0,\ \operatorname{Re}v>-1/2$
$\dfrac{(2/\pi)^{1/2}}{(a^2+\lambda^2)},\ \lambda>0$	$e^{-a\lambda},\ \operatorname{Re}a>0$
$\dfrac{(2/\pi)^{1/2}(a^2-\lambda^2)}{(a^2+\lambda^2)^2},\ \lambda>0$	$xe^{-a\lambda},\ \operatorname{Re}a>0$
$(2/\pi)^{1/2}\dfrac{\Gamma(v)\cos\left[v\tan^{-1}\left(\frac{\lambda}{a}\right)\right]}{(a^2+\lambda^2)^{v/2}},\ \lambda>0$	$x^{v-1}e^{-ax},\ \operatorname{Re}a>0,\ \operatorname{Re}v>0$
$\dfrac{e^{-\lambda^2/4a^2}}{\|a\|\sqrt{2}},\ \lambda>0$	$e^{-a^2x^2},\ \operatorname{Re}a>0$
$\dfrac{\tan^{-1}\left(\frac{2}{\lambda^2}\right)}{\sqrt{2\pi}},\ \lambda>0$	$x^{-1}e^{-x}\sin x$
$\dfrac{1}{2\sqrt{a}}\left[\cos\left(\dfrac{\lambda^2}{4a}\right)-\sin\left(\dfrac{\lambda^2}{2a}\right)\right],\ \lambda>0$	$\sin(ax^2)$
$\left(\dfrac{\pi}{2}\right)^{1/2}\dfrac{\sin(\pi a/b)}{b[\cosh(\pi\lambda/b)+\cos(\pi/b)]},\ \lambda>0$	$\dfrac{\sinh(ax)}{\sinh(bx)}$
$\dfrac{(2\pi)^{1/2}\cos(\pi a/2b)\cosh(\pi\lambda/2b)}{b[\cosh(\pi\lambda/b)+\cos(\pi/b)]},\ \lambda>0$	$\dfrac{\cosh(ax)}{\cosh(bx)}$

TABLE 2.C.5
Mellin Transforms

$$F(\lambda) = \int_0^\infty x^{\lambda-1} f(x)\, dx \qquad f(x) = \frac{1}{2\pi i} \int_{\gamma-i\delta}^{\gamma+i\delta} F(\lambda) x^{-\lambda}\, dx$$

$\Gamma(\lambda)\cos\dfrac{\pi}{2}\lambda, \quad 0 < \mathrm{Re}\,\lambda < 1$	$\cos x$
$\Gamma(\lambda)\sin\dfrac{\pi}{2}\lambda, \quad 0 < \mathrm{Re}\,\lambda < 1$	$\sin x$
$\Gamma(\lambda), \quad \mathrm{Re}\,\lambda > 0$	e^{-x}
$\dfrac{\Gamma(a-\lambda)\Gamma(\lambda)}{\Gamma(a)}, \quad 0 < \mathrm{Re}\,\lambda < \mathrm{Re}\,a$	$(1+x)^{-a}$
$\dfrac{\pi}{\sin\lambda\pi}, \quad 0 < \mathrm{Re}\,\lambda < 1$	$(1+x)^{-1}$
$\dfrac{\Gamma(\lambda) e^{i\lambda\pi/2}}{\alpha^\lambda}, \quad 0 < \mathrm{Re}\,\lambda < 1$	e^{iax}
$\dfrac{\pi}{\lambda\sin\lambda\pi}, \quad -1 < \mathrm{Re}\,\lambda < 0$	$\mathrm{Log}(1+x)$
$\dfrac{\Gamma(\lambda)}{2a^{\lambda/2}}$	$e^{-ax^2}, \quad a > 0$
$\dfrac{\pi\cos\lambda\theta}{\sin\lambda\pi}, \quad \lambda \ne k\pi, \quad k \in \mathbb{Z}$	$\dfrac{1 + x\cos\theta}{1 + 2x\cos\theta + x^2}$
$\dfrac{\pi\sin\lambda\theta}{\sin\lambda\pi}, \quad \lambda \ne k\pi, \quad k \in \mathbb{Z}$	$\dfrac{x\sin\theta}{1 + 2x\cos\theta + x^2}$
$\dfrac{dF(\lambda)}{d\lambda}$	$(\log x) f(x)$
$F(\lambda + a)$	$x^a f(x), \quad a \in \mathbb{C}$
$a^{-\lambda} F(\lambda)$	$f(ax), \quad a > 0$
$\dfrac{1}{a} F\!\left(\dfrac{\lambda}{a}\right)$	$f(x^a), \quad a > 0$
$-\dfrac{1}{\lambda} F(\lambda + 1)$	$\displaystyle\int_0^x f(t)\, dt$
$\dfrac{1}{\lambda} F(\lambda + 1)$	$\displaystyle\int_x^\infty f(t)\, dt$

TABLE 2.C.6
Hankel Transforms

$$f(\lambda) = \int_0^\infty J_v(x\lambda) x F(x) dx \qquad F(x) = \int_0^\infty f(\lambda) J_v(x\lambda) \lambda d\lambda$$

$f(\lambda)$		$F(x)$	
$\dfrac{2^\mu a^{v+\mu+1} \Gamma(v+1) J_{v+\mu+1}(a\lambda)}{\lambda^{\mu+1}}$,	$v > -1$	$x^v (a^2 - x^2)^\mu Y(a-x)$,	$a > 0$, $\mu > -1$
$\dfrac{2^s \Gamma((v+s+1)/2)}{\lambda^{s+1} \Gamma((v-s+1)/2)}$		x^{s-1}	$-v-1 < s < v+1$
$\dfrac{2^v \lambda^v \Gamma(v+1/2)}{\sqrt{\pi}(a^2+\lambda^2)^{v+1/2}}$,	$v > 0$	$x^{v-1} e^{-ax}$,	$a > 0$
$\dfrac{2^{v+1} a \lambda^v \Gamma(v+3/2)}{\sqrt{\pi}(a^2+\lambda^2)^{v+3/2}}$,	$v > 0$	$x^v e^{-ax}$,	$a > 0$
$\dfrac{\lambda}{\sqrt{a^2+\lambda^2}(a+\sqrt{a^2+\lambda^2})^v}$,	$v > -1$	$x^{-1} e^{-ax}$,	$a > 0$
$\dfrac{\lambda^v (a + v\sqrt{a^2+\lambda^2})}{(a^2+\lambda^2)^{3/2}(a+\sqrt{a^2+\lambda^2})^2}$,	$v > -2$	e^{-ax},	$a > 0$
$(a^2/2)^{v+1} \lambda^v e^{-a^2\lambda^2/4}$,	$v > -1$	$x^v e^{-x^2/a^2}$,	$a \neq 0$
$\begin{cases}(a^2-\lambda^2)^{-1/2} & 0 < \lambda < a \\ 0 & \lambda > a\end{cases}$	$v = 0$	$\dfrac{\sin(ax)}{x}$,	$a > 0$

REFERENCES

Abramowitz, M., and Stegun, I., *Handbook of Mathematical Functions with Formulas, Graphs, and Mathematical Tables*, Mineola, NY: Dover Publications, 1972.

Acrivos, A., and Amundson, N. R., "On the Steady State Fractionation of Multicomponent and Complex Mixtures in an Ideal Cascade, Part 3: Discussion of the Numerical Method of Calculation," *Chem. Eng. Sci.* 4 (3) (1955): 141–148.

Amann, H., *Ordinary Differential Equations: An Introduction to Nonlinear Analysis*, Berlin: Walter de Gruyter, 1990.

Ames, W. F., *Nonlinear Partial Differential Equations in Engineering*, New York: Academic Press, 1965.

Amundson, N. R., *Mathematical Methods in Chemical Engineering: Matrices and Their Application*, Englewood Cliffs, NJ: Prentice-Hall, 1966.

Anderson, R. L., and Ibragimov, N. H., *Lie-Backlund Transformations in Applications*, Philadelphia: Society for Industrial and Applied Mathematics, 1979.

Anger, G., *Inverse Problems in Differential Equations*, New York: Plenum Press, 1990.

Antimirov, M. Ya., Kolyshkin, A. A., and Vaillancourt, R., *Complex Variables*, San Diego: Academic Press, 1998.

Aris, R., and Amundson, N. R., "Statistical Analysis of a Reactor, Linear Theory," *Chem. Eng. Sci.* 9 (1958): 250–262.

Aris, R., *Vectors, Tensors, and The Basic Equations of Fluid Mechanics*, Mineola, NY: Dover Publications, 1962.

Aris, R., *Introduction to the Analysis of Chemical Reactors*, Englewood Cliffs, NJ: Prentice-Hall, 1965.

Arnold, L., *Stochastic Differential Equations*, New York: Wiley-Interscience, 1974.

Arnold, V. I., *Ordinary Differential Equations*, trans. R. A. Silverman, Cambridge, MA: MIT Press, 1985.

Arnold, V. I., *Mathematical Methods of Classical Mechanics*, trans. K. Vogtmann and A. Weinstein, Berlin: Springer-Verlag, 1978.
Balakotaiah, V., and Luss, D. "Analysis of the Multiplicity Patterns of a CSTR," *Chem. Eng. Comm.* 13 (1981): 111–132.
Balakotaiah, V., and Luss, D. "Analysis of the Multiplicity Patterns of a CSTR," *Chem. Eng. Comm.* 15 (1982): 185–189.
Balakotaiah, V., Luss, D., and Keyfitz, B., "Steady State Multiplicity Analysis of Lumped-Parameter Systems Described by a Set of Algebraic Equations," *Chem. Eng. Comm.* 36 (1982): 121–147.
Balakotaiah, V., and Luss, D. "Steady-State Multiplicity Features of Chemical Reacting Systems," *Chem. Eng. Edu.* 24 (1986): 12–56.
Banks, H. T., and Kunisch, K., *Estimation Techniques for Distributed Parameter Systems*, Boston: Birkhauser, 1989.
Barenblatt, G. I., *Scaling Phenomena in Fluid Mechanics*, Cambridge, UK: Cambridge University Press, 1994.
Barenblatt, G. I., *Scaling, Self-Similarity and Intermediate Asymptotics*, Cambridge, UK: Cambridge University Press, 1996.
Bell, H. E., "Gerschgorin's Theorem and the Zeros of Polynomials," *Am. Math. Monthly* 72 (1965): 292–295.
Bender, C. M., and Orszag, S. A., *Advanced Mathematical Methods for Scientists and Engineers*, New York: McGraw-Hill, 1978.
Bleistein, N., and Handelsman, R. A., *Asymptotic Expansions of Integrals*, Mineola, NY: Dover Publications, 1986.
Bluman, G. W., and Kumei, S., *Symmetries and Differential Equations*, Berlin: Springer-Verlag, 1989.
Borisenko, A. I., and Tarapov, I. E., *Vector and Tensor Analysis with Applications*, Mineola, NY: Dover Publications, 1979.
Boyce, W. E., and DiPrima, R. E., *Elementary Differential Equations and Boundary Value Problems*, 7th ed., New York: John Wiley and Sons, 2000.
Brown, L. F., and Travis, B. J., "Using Diffusion Measurements To Determine Pore-Size Distributions in Porous Materials," *Chem. Eng. Sci.* 38 (6) (1983): 843–847.
Carslaw, H. S., *An Introduction to the Theory of Fourier's Series and Integrals*, 3rd rev. ed., Mineola, NY: Dover Publications, 1950.
Chen, Z., *Finite Element Methods and Their Applications*, Berlin: Springer-Verlag, 2005.
Churchchill, R. V., Brown, J. W., and Verhey, R. F., *Complex Variables and Applications*, 3rd ed., New York: McGraw-Hill, 1976.
Cloud, M. J., and Drachman, B. C., *Inequalities with Applications to Engineering*, Berlin: Springer-Verlag, 1998.
Cochran, J. A., *Applied Mathematics, Principles, Techniques, and Applications*, Belmont, CA: Wadsworth, International Group, 1982.
Coddington, E. A., and Levinson, N., *Theory of Ordinary Differential Equations*, New York: McGraw-Hill, 1955.
Collatz, L., *Functional Analysis and Numerical Mathematics*, trans. H. Oser, New York: Academic Press, 1996.
Collins, E. R., *Mathematical Methods for Physicists and Engineers*, Mineola, NY: Dover Publications, 1999.
Colton, D., and Kress, R., *Inverse Acoustic and Electromagnetic Scattering Theory*, Berlin: Springer-Verlag, 1998.
Corduneanu, C., *Integral Equations and Applications*, Cambridge: Cambridge University Press, 1991.
Corduneanu, C., *Principles of Differential and Integral Equations*, London: Chelsea Publishing, 1977.
Corduneanu, C., and Sandberg, I. W., *Volterra Equations and Applications*, Amsterdam: Gordon and Breach, 2000.
Corduneanu, C., *Integral Equations and Applications*, Cambridge: Cambridge University Press, 1991.
Dautray, R., and Lions, J., *Physical Origin and Classical Methods*, vol. 1 of *Mathematical Analysis and Numerical Methods for Science and Technology*, Berlin: Springer-Verlag, 1985.
Dautray, R., and Lions, J., *Functional and Variational Methods*, vol. 2 of *Mathematical Analysis and Numerical Methods for Science and Technology*, Berlin: Springer-Verlag, 1985.
Dautray, R., and Lions, J., *Spectral Theory and Applications*, vol. 3 of *Mathematical Analysis and Numerical Methods for Science and Technology*, Berlin: Springer-Verlag, 1985.
Dautray, R., and Lions, J., *Integral Equations and Numerical Methods*, vol. 4 of *Mathematical Analysis and Numerical Methods for Science and Technology*, Berlin: Springer-Verlag, 1985.
Dautray, R., and Lions, J., *Evolution Problems II: The Navier-Stokes and Transport Equations in Numerical Methods*, vol. 5 of *Mathematical Analysis and Numerical Methods for Science and Technology*, Berlin: Springer-Verlag, 1985.

Dautray, R., and Lions, J., *Evolution Problems II*, vol. 6 of *Mathematical Analysis and Numerical Methods for Science and Technology*, Berlin: Springer-Verlag, 1985.
Davies, B., *Integral Transforms and Their Applications*, 3rd ed., Berlin: Springer-Verlag, 2002.
Davis, E. B., *Spectral Theory and Differential Operators*, Cambridge, UK: Cambridge University Press, 1995.
Debnath, L., *Nonlinear Partial Differential Equations for Scientists and Engineers*, Boston: Birkhauser, 1997.
Deen, W. M., *Analysis of Transport Phenomena*, Oxford: Oxford University Press, 1998.
Denn, M., *Optimization by Variational Methods*, New York: McGraw-Hill, 1969.
Denn, M., *Stability of Reaction and Transport Processes*, Englewood Cliffs, NJ: Prentice-Hall, 1975.
Derrick, W. R., *Complex Analysis and Application*, 2nd ed., Belmont, CA: Wadsworth International Group, 1984.
Dettman, J. W., *Applied Complex Variables*, Mineola, NY: Dover Publications, 1984.
DiBenedetto, E., *Partial Differential Equations*, Boston: Birkhauser (1995).
Donea, J., and Huerta, A., *Finite Element Methods for Flow Problems*, New York: John Wiley & Sons, 2003.
Drazin, P. G., *Nonlinear Systems*, Cambridge, UK: Cambridge University Press, 1992.
Duffy, D. G., *Transform Methods for Solving Partial Differential Equations*, Boca Raton, FL: CRC Press, 1994.
Edwards, R. E., *Functional Analysis, Theory and Applications*, Mineola, NY: Dover Publications, 1995.
Erd´elyi, A., *Asymptotic Expansions*, Mineola, NY: Dover Publications, 1956.
Ferziger, J. H., and Peric, M., *Computational Methods for Fluid Dynamics*, 3rd ed., Berlin: Springer, 2002.
Fisher, S. D., *Complex Variables*, 2nd ed., Mineola, NY: Dover Publications, 1999.
Flanigan, F. J., *Complex Variables, Harmonic and Analytic Functions*, Mineola, NY: Dover Publications, 1983.
Folland, G. B., *Introduction to Partial Differential Equations*, Princeton, NJ: Princeton University Press, 1995.
Fowler, A. C., *Mathematical Models in the Applied Sciences*, Cambridge, UK: Cambridge University Press, 1997.
Ford, W. B., *Studies on Divergent Series and Summability* and *The Asymptotic Developments of Functions Defined by MacLaurin Series*, London: Chelsea Publishing, 1960.
Franklin, J. N., *Matrix Theory*, Mineola, NY: Dover Publications, 2000.
Fredrickson, A. G., Ramkrishna, D., and Tsuchiya, H. M., "Statistics and Dynamics of Procaryotic Cell Populations," *Mathematical Biosciences* 1 (1967): 327–374.
Friedman, B., *Principle and Techniques of Applied Mathematics*, New York: John Wiley & Sons, 1956.
Gardiner, C. W., *Handbook of Stochastic Methods*, 2nd ed., Berlin: Springer-Verlag, 2003.
Gel'fand, I. M., and Fomin, S. V., *Calculus of Variations*, trans. R. A. Silverman, Mineola, NY: Dover Publications, 2000.
Gel'fand, I. M., *Lectures on Linear Algebra*, trans. A. Shenitzer, Mineola, NY: Dover Publications Inc. (1989).
Glasko, V. B., *Inverse Problems of Mathematical Physics*, trans. A. Bincer, Melville, NY: American Institute of Physics, 1984.
Glendining, P., *Stability, Instability and Chaos: An Introduction to the Theory of Nonlinear Differential Equations*, Cambridge, UK: Cambridge University Press, 1994.
Goldberg, S., *Unbounded Linear Operators, Theory and Applications*, Mineola, NY: Dover Publications, 1996.
Golubitsky, M., and Schaeffer, D. G., *Singularities and Groups in Bifurcation Theory*, vol. 1, Berlin: Springer-Verlag, 1985.
Golubitsky, M., and Schaeffer, D. G., *Singularities and Groups in Bifurcation Theory*, vol. 2, Berlin: Springer-Verlag, 1988.
Gradshteyn, I. S., and Ryzhik, I. M., *Table of Integrals, Series, and Products*, corrected and enlarged ed., New York: Academic Press, 1983.
Groetsch, C. W., *Inverse Problems in the Mathematical Sciences*, Moscow: MIR Publishers, 1983.
Guenther, R. B., and Lee, J. W., *Partial Differential Equations of Mathematical Physics and Integral Equations*, Mineola, NY: Dover Publications, 1996.
Hahn, W., *Theory and Application of Liapunov's Direct Method*, Englewood Cliffs, NJ: Prentice-Hall, 1963.
Harris, J. W., and Stocker, H., *Handbook of Mathematics and Computational Science*, Berlin, Springer-Verlag, 1998.
Hartman, P., *Ordinary Differential Equations*, 2nd ed., Boston: Birkhauser (1982).
Hildebrand, F. B., *Methods of Applied Mathematics*, 2nd ed., Mineola, NY: Dover Publications, 1992.
Hildebrandt, S., and Leis, R., eds., *Partial Differential Equations and Calculus of Variations*, Berlin: Springer-Verlag, 1988.
Hochstadt, H., *Differential Equations: A Modern Approach*, Mineola, NY: Dover Publications, 1964.

Hochstadt, H., *Integral Equations*, New York: John Wiley, 1973.
Holmes, M. H., *Introduction to Perturbations Methods*, Berlin: Springer-Verlag, 1995.
Hopf, H., "The work of R. Thom," in Proceedings International Congress Mathematics. 1958, New York, 1960.
Hulburt, H. M., and Katz, S.,"Some Problems in Particle Technology: A Statistical Mechanical Formulation", *Chem. Eng. Sci.* 19 (8) (1964): 555–574.
Isaacson, E., and Keller, H. B., *Analysis of Numerical Methods*, New York: John Wiley & Sons, 1966.
Isakov, V., *Inverse Problems for Partial Differential Equations*, Berlin: Springer-Verlag, 1998.
Iooss, G., and Joseph, D. D., *Elementary Stability and Bifurcation Theory*, Berlin: Springer-Verlag, 1990.
Jensen, V. G., and Jeffreys, G. V., *Mathematical Methods in Chemical Engineering*, New York: Academic Press, 1977.
Jerri, A. J., *Introduction to Integral Equations with Applications*, New York: Marcel Dekker, 1985.
John, F., *Partial Differential Equations*, 4th ed., Berlin: Springer-Verlag, 1982.
Johnson, L. W., and Riess, R. D., *Numerical Analysis*, 2nd ed., Reading, MA: Addison Wesley, 1982.
Jordan, D. W., and Smith, P., *Nonlinear Ordinary Differential Equations*, 2nd ed., Oxford: Oxford University Press, 1989.
Kahn, P. J., *Introduction to Linear Algebra*, New York: Harper & Row, 1967.
Karman, T. von, *Aerodynamics*, Ithaca, NY: Cornell University Press, 1957.
Keener, J. P., *Principles of Applied Mathematics, Transformation and Approximation*, Reading, MA: Addison Wesley, 1988.
Keller, H. B., *Numerical Methods for Two-Point Boundary Value Problems*, Mineola, NY: Dover Publications, 1992.
Kevorkian, J., and Cole, J. D., *Multiple Scale and Singular Perturbation Methods*, Berlin: Springer-Verlag, 1996.
Khalil, H. K., *Nonlinear Systems*, 2nd ed., Englewood Cliffs, NJ: Prentice-Hall, 1996.
Kirsch, A., *An Introduction to the Mathematical Theory of Inverse Problems*, Berlin: Springer-Verlag, 1996.
Kondo, J., *Integral Equations*, Oxford: Oxford University Press, 1991.
Kress, R., *Linear Integral Equations*, Berlin: Springer-Verlag, 1989.
Kubo, R., *Stochastic Processes in Chemical Physics*, ed. K. E. Shuler, New York: Wiley-Interscience, 1969.
Lakin, W. D., and Sanchez, D. A., *Topics in Ordinary Differential Equations*, Mineola, NY: Dover Publications, 1982.
Lakshmikantham, V., Leela, S., and Martynyuk, A. A., *Stability Analysis of Nonlinear Systems*, New York: Marcel Dekker, 1989.
Lamb, G. L., Jr., *Introductory Applications of Partial Differential Equations*, New York: John Wiley & Sons, 1995.
Lambert, J. D., *Computational Methods in Ordinary Differential Equations*, New York: John Wiley & Sons, 1973.
Lawrence, P., *Differential Equations and Dynamical Systems*, Berlin: Springer-Verlag, 1991.
Leal, G., *Advanced Transport Phenomena: Fluid Mechanics and Convective Transport Processes,* New York: Cambridge University Press.
LePage, W. R., *Complex Variables and the Laplace Transform for Engineers*, Mineola, NY: Dover Publications, 1961.
Leal, G., *Laminar Flow and Convective Transport Processes: Scaling Principles and Asymptotic Analysis*, Boston: Butterworth-Heinemann, 1992.
Lightstone, A. H., *Fundamentals of Linear Algebra*, New York: Appleton-Century-Crofts, 1969.
Linz, P., *Analytical and Numerical Methods for Volterra Equations*, Philadelphia: SIAM (1985).
Lord, R., "On the Viscosity of Argon as Affected by Temperature," *Proc. R. Soc. London* 66 (1899–1900): 68–74.
Lord, R., "The Principle of Similitude," *Nature* 95 (1915): 66–68.
Mangold, M., Kienle, A., Gilles, E. D., and Mohl, K. D., "Nonlinear Computation in DIVA Methods and Applications," *Chem. Eng. Sci.* 55 (2): 441–454 (2000).
Markushevich, A. I., *The Theory of Analytic Functions: A Brief Course*, Moscow: MIR Publisher, 1983.
Marsden, J., *Basic Complex Analysis*, New York: W.H. Freeman, 1973.
Menzel, D., *Mathematical Physics*, Mineola, NY: Dover Publications, 1953.
Mikhailov, V. P., *Partial Differential Equations*, Moscow: MIR Publisher, 1983.
Moler, C., *Numerical Computing with MATLAB*, Philadelphia: SIAM, 2004.

Morbidelli, M., Varma, A., and Aris, R., "Reactor Steady-State Multiplicity and Stability," in *Chemical Reaction and Reactor Engineering*, 973–1055, New York: Marcel Dekker, 1986.
Morse, P. M., and Feshback, H., *Methods of Theoretical Physics, Part I*, New York: McGraw-Hill, 1953.
Morse, P. M., and Feshback, H., *Methods of Theoretical Physics, Part II*, New York: McGraw-Hill, 1953.
Muralidhar, R., and Ramkrishna, D., "Inverse Problems of Agglomeration Kinetics, 2: Binary Clustering Coefficients from Self-Preserving Spectra," *J. Colloid Interface Sci.* 131 (2) (1989): 503–513.
Muralidhar, R., Ramkrishna, D., Das, P. K., and Kumar, R., "Coalescence of Rigid Droplets in a Stirred Dispersion, 2: Band-Limited Force Fluctuations, *Chem. Eng. Sci.* 43 (7) (1988): 1559–1568.
Murdock, J. A., *Perturbations, Theory and Methods*, New York: John Wiley & Sons, 1991.
Muskhelishvili, N. I., *Singular Integral Equations*, Mineola, NY: Dover Publications, 1991.
Myvskis, A. D., translated from Russian by V. M. Volosov and I. G. Volosova, *Advanced Mathematics for Engineers*, Moscow: MIR Publishers, 1975.
Naimark, M. A., *Linear Differential Operators, Part I: Elementary Theory of Linear Differential Operators*, New York: Ungar Publishing, 1967.
Naimark, M. A., *Linear Differential Operators, Part II: Linear Differential Operators in Hilbert Space*, New York: Ungar Publishing, 1968.
Nayfeh, A. H., *Perturbation Methods*, New York: John Wiley & Sons, 1973.
Nayfeh, A. H., *Problems in Perturbation*, New York: John Wiley & Sons, 1985.
Noble, B., *Methods Based on The Wiener-Hopf Technique for the Solution of Partial Differential Equations*, London: Chelsea Publishing, 1988.
Oksendal, B., *Stochastic Differential Equations, an Introduction with Applications*, 5th ed., Berlin: Springer-Verlag, 2000.
Pardoux, E., and Talay, D., "Discretization and Simulation of Stochastic Differential-Equations," *Acta Applicandae Mathematicae* 3 (1): 23–47 (1985).
Patankar, S. V., *Numerical Heat Transfer and Fluid Flow*, New York: McGraw-Hill, 1980.
Pavlou, S., and Costas, G. V., "Optimal Catalyst Activity Profile in Pellets with Shell-Progressive Poisoning: The Case of Fast Linear Kinetics," *Chem. Eng. Sci.* 45 (3) (1990): 695–703.
Pearson, C. E., ed., *Handbook of Applied Mathematics*, New York: Van Nostrand Reinhold, 1983.
Pelmutter, D., *Stability of Chemical Reactors*, Englewod Cliffs, NJ: Prentice-Hall, 1972.
Petkov, V., and Lazarov, R., eds., *Integral Equations and Inverse Problems*, International Conference on Integral Equations and Inverse Problems (Varna, Bulgaria, 1989), Harlow, Essex, England: Longman Scientific & Technical, and New York: Wiley, 1991.
Petrovski, I. G., *Ordinary Differential Equations*, Mineola, NY: Dover Publications, 1966.
Petrovski, I. G., *Partial Differential Equations*, Mineola, NY: Dover Publications, 1991.
Piccinini, L. C., Stampacchia, G., and Vidossich, G., *Ordinary Differential Equations in R^n, Problems and Methods*, Berlin: Springer-Verlag, 1984.
Pipkin, A. C., *A Course on Integral Equations*, Berlin: Springer-Verlag, 1991.
Polyanin, A. D., Zhurov, A. I., and Vyazmin, A. V., "Generalized Separation of Variables in Nonlinear Heat and Mass Transfer Equations," *J. Non-Equillib. Thermodyn.* 25 (2000): 252–267.
Porter, D., and Stirling, D. S. G., *Integral Equations: A Practical Treatment from Spectral Theory to Applications*, Cambridge, UK: Cambridge University Press, 1990.
Poularikas, A., *The Transform and Application Handbook*, 2nd ed., Boca Raton, FL: CRC Press, 2000.
Prasolov, V. V., *Problems and Theorems in Linear Algebra*, Providence, RI: American Mathematical Society, 1991.
Press, W. H., Flannery, B. P., Teukolsky, S. A., and Vetterling, W. T., *Numerical Recipes*, Cambridge, UK: Cambridge University Press, 1992.
Protter, M. H., and Weinberger, H. F., *Maximum Principles in Differential Equations*, Berlin: Springer-Verlag, 1984.
Ramkrishna, D., "The Status of Population Balances," *Rev. Chem. Eng.* 3 (1985): 49–95.
Ramkrishna, D., *Population Balances: Theory and Applications to Particulate Systems in Engineering*, New York: Academic Press, 2000.
Ramkrishna, D., and Amundson, N. R. "Boundary Value Problems in Transport with Mixed or Oblique Derivative Boundary Conditions, I," *Chem. Eng. Sci.* 34 (1979): 301–308.
Ramkrishna, D., and Amundson, N. R., *Linear Operator Methods in Chemical Engineering, with Applications to Transport and Chemical Reaction Systems*, Englewod Cliffs, NJ: Prentice-Hall, 1985.

Ranade, V. V., *Computational Flow Modeling for Chemical Reactor Engineering*, New York: Academic Press, 2002.

Randolph, A. D., and Larson, M. A., *Theory of Particulate Processes*, New York: Academic Press, 1971.

Rao, N. J., Borwanker, J. D., and Ramkrishna, D., "Numerical Solution of Ito Integral Equations," *SIAM Journal on Control*, 12 (1974): 124–139.

Rao, N. J., Ramkrishna, D., and Borwanker, J. D., "Nonlinear Stochastic Simulation of Stirred Tank Reactors," *Chem. Eng. Sci.* 29 (1974): 1193–1204.

Rashevsky, N., *Mathematical Biophysics: Physico-Mathematical Foundations of Biology*, vol. 1, Mineola, NY: Dover Publications, 1960.

Reddy, J. N., *An Introduction to the Finite Element Method*, New York: McGraw-Hill, 2005.

Reklaitis, G. V., *Introduction to Material and Energy Balances*, New York: John Wiley & Sons, 1983.

Rhee, H., Aris, R., and Amundson, N. R., *Theory and Application of Single Equations*, vol. 1 of *First-Order Partial Differential Equations*, Mineola, NY: Dover Publications, 2001.

Rhee, H., Aris, R., and Amundson, N. R., *Theory and Application of Hyperbolic Systems of Quasi-linear Equations*, vol. 2 of *First-Order Partial Differential Equations*, Mineola, NY: Dover Publications, 2001.

Richtmyer, R. D., *Principles of Advanced Mathematical Physics*, vol. 1, Berlin: Springer-Verlag, 1985.

Richtmyer, R. D., *Principles of Advanced Mathematical Physics*, vol. 2, Berlin: Springer-Verlag, 1985.

Robinson, J., *Infinite-Dimensional Dynamical Systems, from Basic Concepts to Actual Calculations*, Cambridge, UK: Cambridge University Press, 2001.

Rubinstein, I., and Rubinstein, L., *Partial Differential Equations in Classical Mathematical Physics*, Cambridge, UK: Cambridge University Press, 1998.

Sagan, H., *Boundary and Eigenvalue Problems in Mathematical Physics*, Mineola, NY: Dover Publications, 1961.

Sagan, H., *Introduction to the Calculus of Variations*, Mineola, NY: Dover Publications, 1993.

Sathyagal, A. N., Ramkrishna, D., and Narsimhan, G., "Solution of Inverse Problems in Population Balances, II: Particle Break-Up," *Computers Chem. Eng.* 19 (4): 437 (1995).

Schneider, H., and Barker, G. P., *Matrices and Linear Algebra*, Mineola, NY: Dover Publications, 1989.

Shilov, G. E., *Elementary Functional Analysis*, Mineola, NY: Dover Publications, 1995.

Shilov, G. E., *Linear Algebra*, Mineola, NY: Dover Publications, 1971.

Silverman, R. A., *Introductory Complex Analysis*, Mineola, NY: Dover Publications, 1972.

Sirovich, L., *Introduction to Applied Mathematics*, Berlin: Springer-Verlag, 1988.

Skufca, J. D., "Analysis Still Matters: A Surprising Instance of Failure of Runge-Kutta-Felberg ODE Solvers," *Siam Review* 46 (2004): 729.

Smith, D. R., *Variation Methods in Optimization*, Mineola, NY: Dover Publications, 1998.

Smoller, J., *Shock Waves and Reaction-Diffusion Equations*, 2nd ed., Berlin: Springer-Verlag, 1994.

Sneddon, I. N., *Fourier Transforms*, Mineola, NY: Dover Publications, 1995.

Sneddon, I. N., *The Use of Integral Transforms*, New York: McGraw-Hill, 1972.

Sobolev, S. L., *Partial Differential Equations of Mathematical Physics*, Mineola, NY: Dover Publications, 1989.

Sperb, R., *Maximum Principles and Their Applications*, New York: Academic Press, 1981.

Stakgold, I.,*Green's Functions and Boundary Value Problems*, New York: John Wiley, 1979.

Starzhinskii, V. M., *Applied Methods in the Theory of Nonlinear Oscillations*, Moscow: MIR Publishers, 1980.

Taylor, G. I., "The Formation of a Blast Wave by a Very Intense Explosion, I: Theoretical Discussion," *Proc. Roy. Soc. A*. 201 (1950a): 159–174.

Taylor, G. I., "The Formation of a Blast Wave by a Very Intense Explosion, II: The Atomic Explosion of 1945," *Proc. Roy. Soc. A*. 201 (1950b): 175–186.

Temam, R., *Infinite-Dimensional Dynamical Systems in Mechanics and Physics*, Berlin: Springer-Verlag, 1997.

Titchmarsh, E. C., *The Theory of Functions*, Oxford: Oxford University Press, 1988.

Titchmarsh, E. C., *Introduction to the Theory of Fourier Integrals*, Londn: Chelsea Publishing, 1986.

Trujillo, D. M., and Busby, H. R., *Practical Inverse Analysis on Engineering*, Boca Raton, FL: CRC Press, 1997.

Varma, A., and Amundsom, N. R., "Some Problems Concerning the Non-Adiabatic Tubular Reactor," *Can. J. Chem. Eng.* 50: 470 (1972).

Varma, A., and Morbidelli, M., *Mathematical Methods in Chemical Engineering*, New York, Oxford: Oxford University Press, 1997.

Varma, A., Morbidelli, M., and Wu, H., *Parametric Sensitivity in Chemical Systems*, Cambridge, UK: Cambridge University Press, 1999.
Vasil'eva, A. B., Butuzov, V. F., and Kalachev, L. V., *The Boundary Function Method for Singular Perturbation Problems*, Philadelphia: Society for Industrial and Applied Mathematics, 1995.
Vladimirov, V. S., *Equations of Mathematical Physics*, Moscow: MIR Publishers, 1984.
Wallace, P. R., *Mathematical Analysis of Physical Problems*, Mineola, NY: Dover Publications, 1972.
Wasow, W., *Asymptotics Expansions for Ordinary Differential Equations*, Mineola, NY: Dover Publications, 1987.
Weinberger, H. F., *A First Course in Partial Differential Equations with Complex Varaibles and Transform Methods*, Mineola, NY: Dover Publications, 1995.
Weinstock, R., *Calculus of Variations with Applications to Physics & Engineering*, Mineola, NY: Dover Publications, 1974.
Wilf, Herbert S., *Mathematics for the Physical Sciences*, Mineola, NY: Dover Publications, 1978.
Williams, W. E., "The Reduction of Boundary Value Problems to Fredholm Integral Equations of the Second Kind," *Zeitschrift fur Angewandte Mathematik und Physik*, 13 (2) (1962): 133–152.
Witmer, G., Balakotaiah, V., and Luss, D., "Finding Singular Points of Two Point Boundary Value Problems," *J. Compu. Phys.*, 65 (1986): 244–250.
Wolfgang, W., *Differential and Integral Inequalities*, Berlin: Spring-Verlag, 1970.
Wright, H., and Ramkrishna, D., "Solutions of Inverse Problems in Population Balances, I: Aggregation Kinetics," *Computers Chem. Eng.* 16 (12): 1019 (1992).
Yosida, K., *Lectures on Differential and Integral Equations*, Mineola, NY: Dover Publications, 1991.
Zauderer, E., *Partial Differential Equations of Applied Mathematics*, New York: John Wiley, 1983.
Zemanian, A. H., *Distribution Theory and Transform Analysis, an Introduction to Generalized Functions, with Applications*, Mineola, NY: Dover Publications, 1987.
Zemanian, A. H., *Generalized Integral Transformations*, Mineola, NY: Dover Publications, 1987.
Zwillinger, D., *Handbook of Differential Equations*, New York, Academic Press, 1989.

Varma, A., Morbidelli, M., and Wu, H., *Parametric Sensitivity in Chemical Systems*, Cambridge, UK: Cambridge University Press, 1999.

Vanden Eijnden, A.B., Bourgeois, T., and Lauckner, L. Vs, *The Boundary Function Method in Singular Perturbation Problems*, Philadelphia: Society for Industrial and Applied Mathematics, 1995.

Vladimirov, V.S., *Equations of Mathematical Physics*, Moscow: MIR Publishers, 1984.

Wallace, P.R., *Mathematical Analysis of Physical Problems*, Mineola, NY: Dover Publications, 1972.

Weaver, W., *Asymptotic Expansions for Ordinary Differential Equations*, Mineola, NY: Dover Publications, 1987.

Weinberger, H. F., *A First Course in Partial Differential Equations with Complex Variables and Transform Methods*, Mineola, NY: Dover Publications, 1995.

Weinstock, R., *Calculus of Variations, with Applications to Physics & Engineering*, Mineola, NY: Dover Publications, 1974.

Willett, Herbert S., *Mathematics for the Physical Sciences*, Mineola, NY: Dover Publications, 1978.

Williams, W.E., "The Singular Boundary Value Problems in Predicting the pH Evolution of the Second Kind," *Zeit. Angew. Mathematik und Physik* 37(5) (1986): 737–742.

Wang, G., Daniellidis, V., and Lin, S.H., *Finite-element Solution of Two-Point Boundary Value Problems*, J. Comp. Phys. 15 (1984): 231-239.

Wolfgang, W., *Differential and Integral Equations*, Berlin: Springer-Verlag, 1970.

Wygant, E. and Ramakrishna, D., "Solutions of Inverse Problems in Population Balances. I. Aggregation Kinetics," *Computers Chem. Eng.* 16 (12) (1992): 1081.

Yndurain, F., *The Problem of Differential Equations*, Mineola, NY: Dover Publications, 1961.

Zauderer, E., *Partial Differential Equations of Applied Mathematics*, New York: John Wiley, 1983.

Zemanian, A. H., *Distribution Theory and Transform Analysis, an Introduction to Generalized Functions, with Applications*, Mineola, NY: Dover Publications, 1987.

Zemanian, A. H., *Generalized Integral Transformations*, Mineola, NY: Dover Publications, 1987.

Zwillinger, D., *Handbook of Differential Equations*, New York: Academic Press, 1989.

3 Engineering Statistics

Daniel W. Siderius

CONTENTS

3.1 Introduction 200
    3.1.1 Data Types 200
    3.1.2 Random Variables 200
    3.1.3 Probability Distributions 201
        3.1.3.1 Discrete Probability Distributions 201
        3.1.3.2 Continuous Probability Density Distributions 201
    3.1.4 Cumulative Probability Distribution Functions 202
    3.1.5 Characteristic Parameters of a Probability Distribution 202
    3.1.6 Statistics of Small Sets of Data 203
3.2 Probability Distributions 203
    3.2.1 Binomial Distribution—Discrete Variable 204
    3.2.2 Geometric Distribution—Discrete Variable 205
    3.2.3 Poisson Distribution—Discrete Variable 205
    3.2.4 Normal or Gaussian Distribution—Continuous Variable 206
    3.2.5 t Distribution of Sample Means—Continuous Variable 207
    3.2.6 Chi-Square Distribution for the Sample Variance—Continuous Variable 210
    3.2.7 F Distribution for the Ratio of Sample Variances—Continuous Variable 211
3.3 Confidence Intervals of Population Parameters 212
    3.3.1 Confidence Interval of a Mean: Population Variance Known 213
    3.3.2 Confidence Interval of a Mean: Population Variance Unknown 225
    3.3.3 Confidence Interval of the Difference of Two Means 226
    3.3.4 Confidence Interval for the Variance 228
    3.3.5 Confidence Interval for the Ratio of Two Variances 229
    3.3.6 Summary of Confidence Intervals 230
3.4 Statistical Hypothesis Testing 231
3.5 Least Squares Regression 233
    3.5.1 Simple Linear Least Squares Regression 234
        3.5.1.1 Basic Algorithm 234
        3.5.1.2 Estimation of Uncertainty 237
    3.5.2 Generalized Multiple Linear Regression 239
        3.5.2.1 Basic Algorithm 240
        3.5.2.2 Estimation of Uncertainty 241
    3.5.3 Analysis of Variance 243
    3.5.4 Nonlinear Regression 245
3.6 Statistical Analysis of Error Propagation 245
3.7 Design of Experiments 247
    3.7.1 Design Matrices 247
    3.7.2 Design for One-Factor Experiment 248
        3.7.2.1 $\hat{y} = \beta_0 = \overline{y}$ 248

        3.7.2.2 $\hat{y} = \beta_1 x$ 248
        3.7.2.3 $\hat{y} = \beta_0 + \beta_1 x$ 248
  3.7.3   Design for Several Factors 248
  3.7.4   Factorial Design with More Levels 251
  3.7.5   Blocks and New Duplicates 252
  3.7.6   Computer-Aided Experimental Design 252
Appendix 253
Acknowledgment 254
References 254

3.1 INTRODUCTION

Statistics finds important applications by engineers, scientists, and industrial managers when planning and executing research, plant design and operation, marketing and sales programs. Improved operations, better and more uniform products, increased safety, and additional profitability often result. Experimental data contain errors or uncertainties for a variety of factors.

In analyzing a chemical process, it is important to first establish threshold knowledge of the key operating variables, or at least most of them. Such variables affect the operating costs, amount and quality of products formed, and of course profitability. Statistics are often employed to develop mathematical models for all of the above. Plant operators frequently need to be notified, because all of the following likely vary with time: sales of products, availability and cost of feed streams, character of desired product (varies with seasons, such as type of gasoline), etc. Statistics can indicate how to maximize profits.

Numerous resources are available that describe how statistics can be best employed. These include literature references (see list of references located at the end of this chapter), software packages, and the *Handbook of Statistical Methods* prepared by the National Institute of Standards and Testing [1]. The latter consists of about 3500 pages and contains coverage of topics of technical importance, including many examples. Its Internet address is http://www.itl.nist.gov/div898/handbook/index.htm. On the Internet, it is listed as *NIST/SEMATECH e-Handbook of Statistical Methods*. It is free to all and is highly recommended. The appendix at the end of this chapter provides an outline of the topics covered in the handbook. Suitable mathematical software packages are also available from several vendors, including (most commonly used) Microsoft Excel, Mathematica, MatLab®, and JMP. Several examples in this chapter refer to Microsoft Excel, because it simplifies many statistical tasks.

3.1.1 Data Types

Data are of two categories: discrete and continuous. Each type has its own statistical parameters, probability models, and statistical tools. Some statistical tests are applicable to only one type. Data are discrete if there is a finite set of possible outcomes. A data set is continuous if the data may, at least theoretically, assume any value over an interval. Large sets of discrete data are often modeled as continuous, because statistical tools for continuous data are typically simpler.

3.1.2 Random Variables

A variable is random if it is associated with an experiment (a test in the broadest sense, such as throwing dice, measuring temperature, collecting poll data, etc.) in which the outcome may take on values according to its data type. For example, the outcome of throwing two six-sided dice and taking their sum is an integer between 2 and 12, one of several discrete outcomes. Measuring temperature at some point in a room, however, takes on a random value depending on the temperature profile of the room, since the random values are between the upper and lower temperatures in the room.

Engineering Statistics

3.1.3 PROBABILITY DISTRIBUTIONS

The statistical description of an experiment is essential to defining the probability distribution of the data. A mathematical function that connects a particular outcome of an experiment to the chance or probability that the particular outcome will occur is needed. The form of a probability distribution will depend on the type of random variable. Regardless of the type of data, the sum of the probabilities of all possible outcomes adds to unity, i.e., 1.0.

3.1.3.1 Discrete Probability Distributions

For discrete random variables, the probability distribution can often be determined using mathematical intuition, as all experiments are characterized by a fixed set of outcomes. For example, consider an experiment in which a six-sided die is thrown. The variable x denotes the number on the die face, and $P(x)$ is the probability distribution, i.e., the chance of observing x on the face following a throw. Since this random variable is discrete, if the die is fair, then the probability of any possible value is equal and is given by $1/n$, where n is the number of sides, since the sum of all possible outcomes must be unity. The distribution is, therefore, called uniform. Hence, for $n = 6$, $P(x) = 1/6$, where $x = 1, 2, 3, 4, 5,$ or 6.

Example 3.1

Determine the probability distribution $P(x)$ in which two six-sided dice are thrown simultaneously, where x represents the sum of faces of the dice.

Solution

The sum for this experiment may be $x = 2, 3, 4, \ldots, 11,$ or 12. With two dice, there may be multiple outcomes that result in the particular x; the probability is not uniform. Therefore, the probability of observing x is simply the number of outcomes resulting in x divided by the number of possible outcomes (totaling 36):

x	Number of Outcomes	$P(x)$
2	1	1/36
3	2	1/18
4	3	1/12
5	4	1/9
6	5	5/36
7	6	1/6
8	5	5/36
9	4	1/9
10	3	1/12
11	2	1/18
12	1	1/36

This example demonstrates the simple method by which a discrete probability distribution can be determined.

3.1.3.2 Continuous Probability Density Distributions

For a continuous variable, it is improper to speak of the probability of a *particular* outcome occurring due to the infinite number of possible outcomes. Hence, the probability of a particular outcome is zero, and is more appropriately described by a probability density distribution within a small window. For example, consider an unbiased random number generator that generates an infinite number of real numbers between 0 and 3.0. Let θ be the random number generated and define $f(\theta)d\theta$ as the probability density distribution. The relative probability of observing a particular number is equal to that of any other number (both being between 0 and 3). Because the sum (or in the case of a continuous probability density distribution, the integral) of all probabilities is equal to unity, one can write

$$\int_0^3 f(\theta)d\theta = 1 \tag{3.1}$$

Therefore, since the distribution is uniform, $f(\theta) = 1/3$. Such distributions provide an evaluation of reproducibility for data, but not accuracy.

3.1.4 Cumulative Probability Distribution Functions

A cumulative probability distribution function characterizes a set of outcomes between an upper and a lower bound. For a discrete distribution function, the associated distribution is usually denoted as $F(a \leq x \leq b)$, where a and b are the lower and upper bounds, respectively. The new function F is determined by summing the probability of independent outcomes that result in x between a and b. The distribution of a continuous probability density distribution can be written in the same manner and is determined by modifying the bounds of the integral appropriately. These properties of cumulative distribution functions are illustrated in the following example.

Example 3.2

When throwing two six-sided dice and taking their sum, determine the probability of observing a sum between 7 and 10, inclusive.

Solution

Probabilities are summed for face sums of 7, 8, 9, and 10 as follows:

$$F(7 \leq x \leq 10) = \sum_{i=7}^{10} P(i) = \frac{1}{6} + \frac{5}{36} + \frac{1}{9} + \frac{1}{12} = \frac{1}{2} \tag{3.2}$$

Example 3.3

For the random number generator mentioned in Section 3.1.3.2 (which generates a number between 0 and 3, such that $f(x) = 1/3$), determine the probability of observing a random number between 0.4 and 1.7.

Solution

The probability density distribution is integrated between 0.4 and 1.7:

$$F(0.4 \leq x \leq 1.7) = \int_{0.4}^{1.7} f(\theta)d\theta = \int_{0.4}^{1.7} \frac{1}{3} d\theta = 0.4333 \tag{3.3}$$

The bounds of a cumulative distribution might be $-\infty$ or ∞ if those bounds are mathematically meaningful to the probability model. For a lower bound $a = -\infty$, the cumulative distribution is interpreted as the probability of observing a value smaller than b. The positive infinity bound is exactly opposite.

3.1.5 Characteristic Parameters of a Probability Distribution

Probability distributions are characterized by several terms: the *mean* is the expected value of a population (denoted as μ) and the *variance* is the spread, or width, of a distribution (denoted as

Engineering Statistics

σ^2) for discrete and continuous random variables. The standard deviation of a probability distribution is the square root of the variance and is denoted as σ. The units of the variance are those of the mean squared, and the standard deviation has the same units as the mean. Other parameters relating to probability distributions—skewness, kurtosis, and higher-order moments—are discussed in the literature [2–4].

3.1.6 Statistics of Small Sets of Data

Often it is not feasible to collect a large set of data, and a sample of the population is taken that (hopefully) closely approximates the probability distribution. This sample has its own statistics that are analogous, though not identical, to the population parameters. Like the population parameters mentioned before, the most important sample statistics are the sample mean and sample variance, denoted $\bar{x}$ and s^2, respectively.

For a set of n data points, the ith observation is denoted x_i. The sample mean (arithmetic average), $\bar{x}$, and sample variance, s^2, are defined [2] as follows:

$$\bar{x} = \frac{\sum_{i=1}^{n} x_i}{n} \tag{3.4}$$

$$s^2 = \frac{\sum_{i=1}^{n} (x_i - \bar{x})^2}{n-1} \tag{3.5}$$

The $n-1$ in the denominator of the sample variance equals the number of degrees of freedom. For a data set with a known sample mean, knowledge of only $n-1$ observations is necessary to determine the nth observation, constraining the problem by one degree of freedom. The sample standard deviation, (s), like the population counterpart, σ, is the square root of the sample variance.

The sample mean and sample variance are usually not equal to the population mean and population variance, although they are similar in meaning. These sample statistics help approximate the population parameters. Statistical tests and assumptions of the expected distribution of sample data provide intervals that should confidently contain the population parameters, as discussed in Section 3.3.

3.2 PROBABILITY DISTRIBUTIONS

Probability distributions are often employed in statistical analyses. When only success or failure can occur (two discrete outcomes), the following distributions can be employed: binomial distribution, geometric distribution, and Poisson distribution. In some laboratory or industrial trials, the results are simply a success or a failure. Sections 3.3 and 3.4 discuss the preferred statistical tests and the distribution suited for each test. For example, when dealing with a sample mean, statistical tests require the t distribution or normal (Gaussian) distribution, depending on the context. Statistical tests related to the variability of products or processes generally use the chi-square and F distributions. As the variability of a product or process becomes of more interest (e.g., Six-Sigma analysis, in which one goal is reduced product variability), statistical tests applicable to measures of data variability become increasingly useful. Other probability distributions, such as the Weibull distribution, log-normal distribution, beta distribution, etc., are less commonly used [2–4].

TABLE 3.1
Definition of Population Statistical Parameters

	Discrete Random Variable	Continuous Random Variable
Population mean	$\mu = \sum_i x_i P(x_i)$	$\mu = \int_{-\infty}^{\infty} x f(x) dx$
Population variance	$\sigma^2 = \sum_i (x_i - \mu)^2 P(x_i)$	$\sigma^2 = \int_{-\infty}^{\infty} (x - \mu)^2 f(x) dx$

Sources: Montgomery, Douglas C., and Runger, George C. 2003. *Applied Statistics and Probability for Engineers*, 3rd ed., New York: John Wiley & Sons. Metcalfe, Andrew V., *Statistics in Engineering*, 1994. London: Chapman & Hall. Fraser, D.A.S, *Probability and Statistics: Theory and Applications*, 1976, North Scituate, MA: Duxbury Press.

3.2.1 BINOMIAL DISTRIBUTION—DISCRETE VARIABLE

The binomial distribution describes the probability of x successes in n trials when only success or failure occurs. This distribution assumes that each trial is independent and that the probability of a success in each trial is constant. This probability is denoted p. The probability function for a binomial distribution [2] is

$$P(x,n) = \binom{n}{x} p^x (1-p)^{n-x} \tag{3.6}$$

$$\binom{n}{x} = \frac{n!}{x!(n-x)!} \tag{3.7}$$

Equation (3.6) can be derived using intuition. The probability of observing x successes and $n - x$ failures is $p^x(1 - p)^{n-x}$, and

$$\binom{n}{x}$$

is the binomial coefficient that equals the number of arrangements of subsets of size x given a total number of elements n. "(The binomial coefficient also equals the redundancy of the experiment.)" Using the appropriate equations in Table 3.1 along with the expression for $P(x,n)$ given in Equation (3.6), the mean value of x is $\mu = np$ and the variance of x is $\sigma^2 = np(1 - p)$.

Example 3.4

Suppose that a six-sided die is thrown 20 times. What is the probability of a particular number turning up 6 times? (A success is defined as that particular number turning up, and a failure is *any other* number turning up.)

Solution

The probability of any side is $p = 1/6$, $x = 6$, and $n = 20$. Using Equation (3.6),

Engineering Statistics

$$P(6,20) = \binom{20}{6}\left(\frac{1}{6}\right)^6\left(\frac{5}{6}\right)^{14} = 0.0647$$

3.2.2 Geometric Distribution—Discrete Variable

The geometric distribution indicates the probability of conducting x trials to obtain a success in an experiment in which there are only two possible outcomes. Like the binomial distribution, this is another Bernoulli process. Each trial is assumed to be independent, and the probability of observing a success is constant over all trials, denoted p. The probability distribution for the geometric distribution [2] is

$$P(x) = (1-p)^{x-1} p \tag{3.8}$$

Similar to the binomial distribution, Equation (3.8) can also be derived intuitively. Since there is only one set of experimental trials that results in $x - 1$ failures followed by one success, the distribution has no binomial coefficient or redundancy. Consequently, the probability of the first success occurring in the xth trial is the probability of $x - 1$ failures multiplied by the probability of one success.

The expected value of the geometric distribution is $\mu = 1/p$, and the variance is $\sigma^2 = (1-p)/p^2$ [2].

Example 3.5

For a six-sided die, what is the probability of a chosen face turning up for the first time on the fourth throw?

Solution

For each throw of the die, the probability of a given face showing is $p = 1/6$. Since we desire the first success to be on the fourth throw, $x = 4$. Hence, the solution is given by

$$P(4) = \left(\frac{5}{6}\right)^{4-1}\left(\frac{1}{6}\right) = 0.0965$$

3.2.3 Poisson Distribution—Discrete Variable

The Poisson distribution describes the probability of a discrete number of events occurring within a fixed interval, given that the probability of the event occurring is independent of the size of the interval. For example, suppose that a manufacturing defect occurs with an average rate of occurrence p and the products are manufactured over an interval n. The expected number of defects is clearly np. The Poisson distribution is one way of describing the probability that x defects will occur in an interval n. This distribution is a generalization of the binomial distribution to an infinite number of trials. The mathematical form of the Poisson distribution [2] is

$$P(x) = \frac{e^{-\lambda}\lambda^x}{x!} \tag{3.9}$$

where x is the number of actual events and $\lambda = np$ is the *expected* number of events in the interval n.

The expected value of the Poisson distribution and the variance both equal λ.

Example 3.6

During continuous paper production, a tear occurs on average once every 500 feet. What is the probability that five tears will occur in 1000 feet?

Solution

The average rate of tears is $p = 1/500$ (ft^{-1}) and the interval size is $n = 1000$ ft. Therefore, the *expected* number of tears is

$$\lambda = 1000 \text{ ft} \cdot \left(\frac{1}{500 \text{ ft}}\right) = 2$$

Using Equation (3.9), the probability of $x = 5$ tears occurring in 1000 feet of product is

$$P(5) = \frac{e^{-2} 2^5}{5!} = 0.0361$$

3.2.4 Normal or Gaussian Distribution—Continuous Variable

The normal or Gaussian distribution is commonly employed for large data sets, since they generally resemble a characteristic bell-shaped curve. Although the shape of each distribution may be qualitatively different, all bell curves collapse to the same distribution when scaled appropriately. As discussed in Section 3.3, the normal distribution also describes the distribution of sample means when the population variance is known. The analytic form of the normal distribution [2] is

$$f(x) = \frac{1}{\sqrt{2\pi\sigma^2}} \exp\left(\frac{-(x-\mu)^2}{2\sigma^2}\right) \tag{3.10}$$

The mean and standard deviation of the normal distribution are μ and σ, respectively. Since the normal distribution is designed for continuous data, the cumulative distribution function is more practical than the probability density function. For a particular data population, the cumulative distribution [2] is as follows:

$$F(x) = \frac{1}{\sqrt{2\pi\sigma^2}} \int_{-\infty}^{x} \exp\left(\frac{-(x'-\mu)^2}{2\sigma^2}\right) dx' = \frac{1}{2}\left(1 + \text{erf}\left(\frac{x-\mu}{\sqrt{2\sigma^2}}\right)\right) \tag{3.11}$$

The error function, *erf(y)*, is introduced because the normal distribution cannot be integrated analytically. The error function is found in many mathematical tables and software packages [5–8]. Summarizing, $F(x)$ is the probability of an observation being less than or equal to x. Conversely, the probability of an observation exceeding x is $1 - F(x)$. From the properties of integrals, it is straightforward to show that the probability of an observation being between a and b, $F(a \leq x \leq b)$, can be expressed as $F(b) - F(a)$. Values of $F(x)$ are also available in the literature or suitable software. For example, the probability of an observation falling within one standard deviation of the mean, $F(\mu - \sigma \leq x \leq \mu + \sigma)$, is 0.683. The probability of an observation being within two standard deviations is 0.955, and that within three is 0.997, a progression often referred to as the 68-95-99 rule [2].

The normal distribution is reduced to the *standard* normal distribution by introducing the transformation variable z, defined [2] as follows:

$$z = \frac{x - \mu}{\sigma} \tag{3.12}$$

Making this substitution into Equation (3.6) or (3.7) reduces the generic normal distribution to one with mean 0 and standard deviation 1, collapsing all possible normal distributions onto a standard curve. Tabulated values of the cumulative distribution function F are usually presented in terms of the transformation variable z. Sample values of $F(z)$ are presented in Table 3.2. Microsoft Excel contains an intrinsic function, NORMSDIST, that produces the cumulative probability for a standard normal variable z given as its argument. A companion function, NORMSINV, outputs the z value for a given $F(z)$. The Microsoft Excel manual or the electronic help files [5, 6] provide command syntax and usage examples.

Example 3.7
If the fuel mileage of a particular vehicle is normally distributed with a mean of 23.2 miles per gallon (mpg) and a standard deviation of 1.75 mpg,

1. What is the probability that the fuel mileage will fall in the manufacturer-stated range of 21 to 25 mpg?
2. What is the probability that the fuel mileage will exceed 25 mpg?

Solution

Solution 1
1. Using Equation (3.12), the z values for $x = 21$ and $x = 25$ are
   $z_{low} = (21 - 23.2)/1.75 = -1.26$
   $z_{high} = (25 - 23.2)/1.75 = 1.03$
2. Using Table 3.2, the cumulative distribution values for these z are 0.89617 (or 0.896) and 0.84849 (or 0.848). Hence,
   $F(z_{low}) = 1 - F(-z_{low}) = 1 - 0.896 = 0.104$
   $F(z_{high}) = 0.848$
   The probability that the fuel mileage is less than 21 mpg is 0.104, and the probability that the mileage is less than 25 mpg is 0.848. Therefore, the probability of the fuel mileage being between 21 and 25 mpg is

$$F(21 \leq x \leq 25) = F(-1.26 \leq z \leq 1.03) = 0.848 - 0.104 = 0.744$$

Solution 2
The probability that it exceeds 25 mpg is $1 - 0.848 = 0.152$.

3.2.5 *t* DISTRIBUTION OF SAMPLE MEANS—CONTINUOUS VARIABLE

With only a few data points, the resulting sample mean and variance are not equal to the true population parameters. The *t* distribution [9] then describes the distribution of sample means, where the test statistic is

$$t = \frac{\bar{x} - \mu}{s/\sqrt{n}} \tag{3.13}$$

This distribution is often used to compute confidence intervals or perform hypothesis tests on the sample mean. Writing under the pseudonym *Student*, W. S. Gosset showed that the *t*-test statistic

TABLE 3.2
Cumulative Probability, α, of the Standard Normal Distribution as a Function of the Standard Variable z. The left column indicates z to the tenths digit and the top row indicates z to the hundredths. For example, the α value for z = 1.38 is 0.91621.

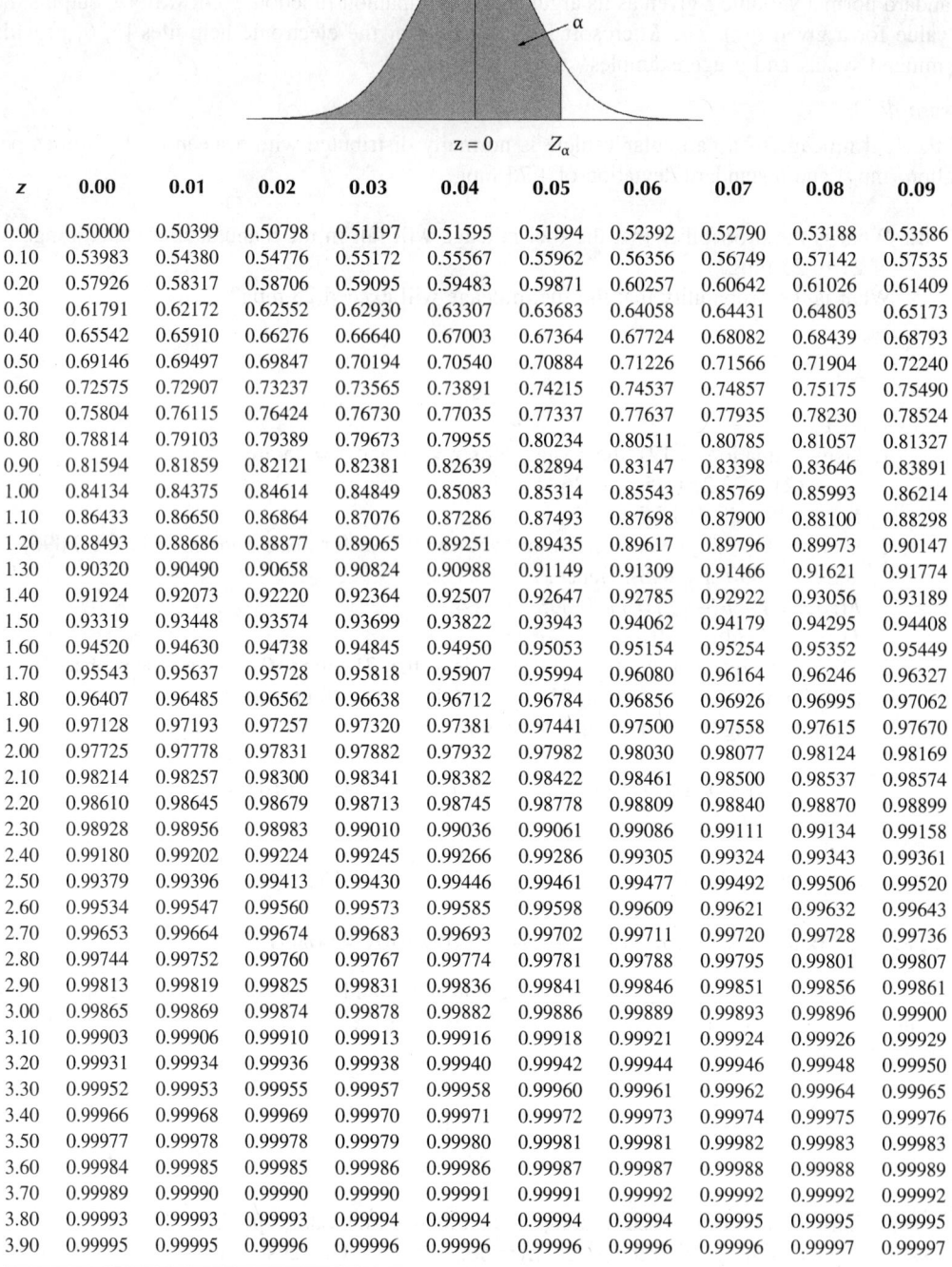

z	0.00	0.01	0.02	0.03	0.04	0.05	0.06	0.07	0.08	0.09
0.00	0.50000	0.50399	0.50798	0.51197	0.51595	0.51994	0.52392	0.52790	0.53188	0.53586
0.10	0.53983	0.54380	0.54776	0.55172	0.55567	0.55962	0.56356	0.56749	0.57142	0.57535
0.20	0.57926	0.58317	0.58706	0.59095	0.59483	0.59871	0.60257	0.60642	0.61026	0.61409
0.30	0.61791	0.62172	0.62552	0.62930	0.63307	0.63683	0.64058	0.64431	0.64803	0.65173
0.40	0.65542	0.65910	0.66276	0.66640	0.67003	0.67364	0.67724	0.68082	0.68439	0.68793
0.50	0.69146	0.69497	0.69847	0.70194	0.70540	0.70884	0.71226	0.71566	0.71904	0.72240
0.60	0.72575	0.72907	0.73237	0.73565	0.73891	0.74215	0.74537	0.74857	0.75175	0.75490
0.70	0.75804	0.76115	0.76424	0.76730	0.77035	0.77337	0.77637	0.77935	0.78230	0.78524
0.80	0.78814	0.79103	0.79389	0.79673	0.79955	0.80234	0.80511	0.80785	0.81057	0.81327
0.90	0.81594	0.81859	0.82121	0.82381	0.82639	0.82894	0.83147	0.83398	0.83646	0.83891
1.00	0.84134	0.84375	0.84614	0.84849	0.85083	0.85314	0.85543	0.85769	0.85993	0.86214
1.10	0.86433	0.86650	0.86864	0.87076	0.87286	0.87493	0.87698	0.87900	0.88100	0.88298
1.20	0.88493	0.88686	0.88877	0.89065	0.89251	0.89435	0.89617	0.89796	0.89973	0.90147
1.30	0.90320	0.90490	0.90658	0.90824	0.90988	0.91149	0.91309	0.91466	0.91621	0.91774
1.40	0.91924	0.92073	0.92220	0.92364	0.92507	0.92647	0.92785	0.92922	0.93056	0.93189
1.50	0.93319	0.93448	0.93574	0.93699	0.93822	0.93943	0.94062	0.94179	0.94295	0.94408
1.60	0.94520	0.94630	0.94738	0.94845	0.94950	0.95053	0.95154	0.95254	0.95352	0.95449
1.70	0.95543	0.95637	0.95728	0.95818	0.95907	0.95994	0.96080	0.96164	0.96246	0.96327
1.80	0.96407	0.96485	0.96562	0.96638	0.96712	0.96784	0.96856	0.96926	0.96995	0.97062
1.90	0.97128	0.97193	0.97257	0.97320	0.97381	0.97441	0.97500	0.97558	0.97615	0.97670
2.00	0.97725	0.97778	0.97831	0.97882	0.97932	0.97982	0.98030	0.98077	0.98124	0.98169
2.10	0.98214	0.98257	0.98300	0.98341	0.98382	0.98422	0.98461	0.98500	0.98537	0.98574
2.20	0.98610	0.98645	0.98679	0.98713	0.98745	0.98778	0.98809	0.98840	0.98870	0.98899
2.30	0.98928	0.98956	0.98983	0.99010	0.99036	0.99061	0.99086	0.99111	0.99134	0.99158
2.40	0.99180	0.99202	0.99224	0.99245	0.99266	0.99286	0.99305	0.99324	0.99343	0.99361
2.50	0.99379	0.99396	0.99413	0.99430	0.99446	0.99461	0.99477	0.99492	0.99506	0.99520
2.60	0.99534	0.99547	0.99560	0.99573	0.99585	0.99598	0.99609	0.99621	0.99632	0.99643
2.70	0.99653	0.99664	0.99674	0.99683	0.99693	0.99702	0.99711	0.99720	0.99728	0.99736
2.80	0.99744	0.99752	0.99760	0.99767	0.99774	0.99781	0.99788	0.99795	0.99801	0.99807
2.90	0.99813	0.99819	0.99825	0.99831	0.99836	0.99841	0.99846	0.99851	0.99856	0.99861
3.00	0.99865	0.99869	0.99874	0.99878	0.99882	0.99886	0.99889	0.99893	0.99896	0.99900
3.10	0.99903	0.99906	0.99910	0.99913	0.99916	0.99918	0.99921	0.99924	0.99926	0.99929
3.20	0.99931	0.99934	0.99936	0.99938	0.99940	0.99942	0.99944	0.99946	0.99948	0.99950
3.30	0.99952	0.99953	0.99955	0.99957	0.99958	0.99960	0.99961	0.99962	0.99964	0.99965
3.40	0.99966	0.99968	0.99969	0.99970	0.99971	0.99972	0.99973	0.99974	0.99975	0.99976
3.50	0.99977	0.99978	0.99978	0.99979	0.99980	0.99981	0.99981	0.99982	0.99983	0.99983
3.60	0.99984	0.99985	0.99985	0.99986	0.99986	0.99987	0.99987	0.99988	0.99988	0.99989
3.70	0.99989	0.99990	0.99990	0.99990	0.99991	0.99991	0.99992	0.99992	0.99992	0.99992
3.80	0.99993	0.99993	0.99993	0.99994	0.99994	0.99994	0.99994	0.99995	0.99995	0.99995
3.90	0.99995	0.99995	0.99996	0.99996	0.99996	0.99996	0.99996	0.99996	0.99997	0.99997

is distributed according to a complicated expression involving gamma functions. When $n \to \infty$, the t distribution becomes identical to the normal distribution, and a data set with n greater than about 25 is effectively normal. Qualitatively, the t distribution is similar to the normal distribution, $f(t)$ being symmetric about its maximum value at $t = 0$. The cumulative distribution function for the variable $t_{\alpha,\nu}$ is listed in Table 3.3 for given values of ν and α, $\nu = n - 1$ representing the number

---

TABLE 3.3
***t* Variable for the *t* Distribution as a Function of Cumulative Probability α and Degrees of Freedom ν**

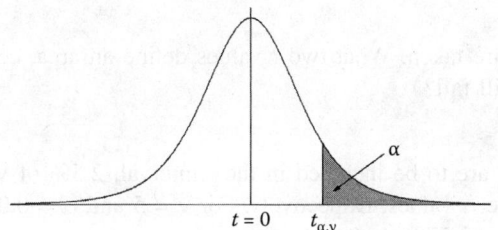

α ν	0.45	0.3	0.2	0.15	0.1	0.05	0.025	0.01	0.005	0.0025	0.001	0.0005
1	0.1584	0.7265	1.3764	1.9626	3.0777	6.3138	12.7062	31.8205	63.6567	127.3213	318.3088	636.6192
2	0.1421	0.6172	1.0607	1.3862	1.8856	2.9200	4.3027	6.9646	9.9248	14.0890	22.3271	31.5991
3	0.1366	0.5844	0.9785	1.2498	1.6377	2.3534	3.1824	4.5407	5.8409	7.4533	10.2145	12.9240
4	0.1338	0.5686	0.9410	1.1896	1.5332	2.1318	2.7764	3.7469	4.6041	5.5976	7.1732	8.6103
5	0.1322	0.5594	0.9195	1.1558	1.4759	2.0150	2.5706	3.3649	4.0321	4.7733	5.8934	6.8688
6	0.1311	0.5534	0.9057	1.1342	1.4398	1.9432	2.4469	3.1427	3.7074	4.3168	5.2076	5.9588
7	0.1303	0.5491	0.8960	1.1192	1.4149	1.8946	2.3646	2.9980	3.4995	4.0293	4.7853	5.4079
8	0.1297	0.5459	0.8889	1.1081	1.3968	1.8595	2.3060	2.8965	3.3554	3.8325	4.5008	5.0413
9	0.1293	0.5435	0.8834	1.0997	1.3830	1.8331	2.2622	2.8214	3.2498	3.6897	4.2968	4.7809
10	0.1289	0.5415	0.8791	1.0931	1.3722	1.8125	2.2281	2.7638	3.1693	3.5814	4.1437	4.5869
11	0.1286	0.5399	0.8755	1.0877	1.3634	1.7959	2.2010	2.7181	3.1058	3.4966	4.0247	4.4370
12	0.1283	0.5386	0.8726	1.0832	1.3562	1.7823	2.1788	2.6810	3.0545	3.4284	3.9296	4.3178
13	0.1281	0.5375	0.8702	1.0795	1.3502	1.7709	2.1604	2.6503	3.0123	3.3725	3.8520	4.2208
14	0.1280	0.5366	0.8681	1.0763	1.3450	1.7613	2.1448	2.6245	2.9768	3.3257	3.7874	4.1405
15	0.1278	0.5357	0.8662	1.0735	1.3406	1.7531	2.1314	2.6025	2.9467	3.2860	3.7328	4.0728
16	0.1277	0.5350	0.8647	1.0711	1.3368	1.7459	2.1199	2.5835	2.9208	3.2520	3.6862	4.0150
17	0.1276	0.5344	0.8633	1.0690	1.3334	1.7396	2.1098	2.5669	2.8982	3.2224	3.6458	3.9651
18	0.1274	0.5338	0.8620	1.0672	1.3304	1.7341	2.1009	2.5524	2.8784	3.1966	3.6105	3.9216
19	0.1274	0.5333	0.8610	1.0655	1.3277	1.7291	2.0930	2.5395	2.8609	3.1737	3.5794	3.8834
20	0.1273	0.5329	0.8600	1.0640	1.3253	1.7247	2.0860	2.5280	2.8453	3.1534	3.5518	3.8495
21	0.1272	0.5325	0.8591	1.0627	1.3232	1.7207	2.0796	2.5176	2.8314	3.1352	3.5272	3.8193
22	0.1271	0.5321	0.8583	1.0614	1.3212	1.7171	2.0739	2.5083	2.8188	3.1188	3.5050	3.7921
23	0.1271	0.5317	0.8575	1.0603	1.3195	1.7139	2.0687	2.4999	2.8073	3.1040	3.4850	3.7676
24	0.1270	0.5314	0.8569	1.0593	1.3178	1.7109	2.0639	2.4922	2.7969	3.0905	3.4668	3.7454
25	0.1269	0.5312	0.8562	1.0584	1.3163	1.7081	2.0595	2.4851	2.7874	3.0782	3.4502	3.7251
30	0.1267	0.5300	0.8538	1.0547	1.3104	1.6973	2.0423	2.4573	2.7500	3.0298	3.3852	3.6460
35	0.1266	0.5292	0.8520	1.0520	1.3062	1.6896	2.0301	2.4377	2.7238	2.9960	3.3400	3.5911
40	0.1265	0.5286	0.8507	1.0500	1.3031	1.6839	2.0211	2.4233	2.7045	2.9712	3.3069	3.5510
45	0.1264	0.5281	0.8497	1.0485	1.3006	1.6794	2.0141	2.4121	2.6896	2.9521	3.2815	3.5203
50	0.1263	0.5278	0.8489	1.0473	1.2987	1.6759	2.0086	2.4033	2.6778	2.9370	3.2614	3.4960
100	0.1260	0.5261	0.8452	1.0418	1.2901	1.6602	1.9840	2.3642	2.6259	2.8707	3.1737	3.3905
∞	0.1257	0.5244	0.8416	1.0364	1.2816	1.6449	1.9600	2.3263	2.5758	2.8070	3.0902	3.2905

of degrees of freedom and α being the percentage of the distribution lying above the particular t. The Microsoft Excel functions TDIST and TINV can be used to calculate t values or cumulative probabilities [5, 6]. The t distribution is most useful in determining the confidence interval of a measured sample mean and testing hypotheses, as described in Sections 3.3 and 3.4.

Example 3.8
Ten measurements are taken. Above what t value will 10% of the measurements fall?

Solution

As $n = 10$, $v = n - 1 = 9$. Because 10% of the distribution lies above the t value, $\alpha = 0.1$. Then, reading from Table 3.3, for $v = 9$ and $\alpha = 0.1$, $t = 1.3830$. Therefore $P(t > 1.3830) = 0.1$.

Example 3.9
Now six measurements are taken. What two t values define an area, centered at $t = 0$, in which 95% of measurements will fall?

Solution

If 95% of measurements are to be included in the t interval, 2.5% of values will fall above and below the upper and lower t values, respectively. For $v = 5$ and $\alpha = 0.025$, the t value is 2.5706. Therefore $F(-2.5706 < t < 2.5706) = 0.95$.

3.2.6 CHI-SQUARE DISTRIBUTION FOR THE SAMPLE VARIANCE— CONTINUOUS VARIABLE

As in the t distribution that describes the distribution of sample means, a specific probability distribution is necessary to describe the distribution of the variance of samples. The chi-square distribution [2] describes the probability distribution of the sample variance for a sample of size n, where the test statistic χ^2 is

$$\chi^2 = \frac{(n-1)s^2}{\sigma^2} \tag{3.14}$$

The chi-square distribution is used to perform statistical tests on the sample variance. It is highly asymmetric for small values of n, but becomes more symmetric and similar to a normal distribution as n becomes large, such as 20 or 30. The cumulative distribution function of the chi-square distribution is listed in Table 3.4 as a function of v and α, where $v = n - 1$ is the number of degrees of freedom and α is the percentage of the distribution above the particular χ^2. Microsoft Excel has built-in functions, CHIDIST and CHIINV, that compute a chi-square distribution [5, 6].

Example 3.10
Twenty measurements are taken. Above what χ^2 value will 10% of the measurements fall?

Solution

Because $n = 20$, $v = n - 1 = 19$. From Table 3.4, for $v = 19$ and $\alpha = 0.1$, $\chi^2 = 27.2036$. Therefore, $F(\chi^2 > 27.2036) = 0.1$.

Example 3.11
Ten measurements are taken. Define the areas for which 10% of measurements will have lower χ^2 and 10% will have higher χ^2.

Solution

The two χ^2 values to be found are those for $n = 10$ ($v = 9$), such that 10% of the distribution lies above the higher χ^2 value and 10% of the distribution lies below the smaller χ^2 value. The higher

TABLE 3.4
χ^2 Variable for Chi-Square Distribution as a Function of Cumulative Probability α and Degrees of Freedom ν

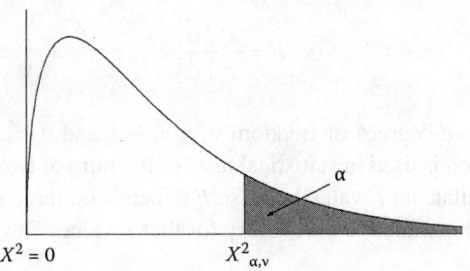

α												
ν	0.995	0.99	0.95	0.9	0.75	0.5	0.2	0.1	0.05	0.1	0.005	0.001
1	0.0000	0.0002	0.0039	0.0158	0.1015	0.4549	1.6424	2.7055	3.8415	2.7055	7.8794	10.8276
2	0.0100	0.0201	0.1026	0.2107	0.5754	1.3863	3.2189	4.6052	5.9915	4.6052	10.5966	13.8155
3	0.0717	0.1148	0.3518	0.5844	1.2125	2.3660	4.6416	6.2514	7.8147	6.2514	12.8382	16.2662
4	0.2070	0.2971	0.7107	1.0636	1.9226	3.3567	5.9886	7.7794	9.4877	7.7794	14.8603	18.4668
5	0.4117	0.5543	1.1455	1.6103	2.6746	4.3515	7.2893	9.2364	11.0705	9.2364	16.7496	20.5150
6	0.6757	0.8721	1.6354	2.2041	3.4546	5.3481	8.5581	10.6446	12.5916	10.6446	18.5476	22.4577
7	0.9893	1.2390	2.1673	2.8331	4.2549	6.3458	9.8032	12.0170	14.0671	12.0170	20.2777	24.3219
8	1.3444	1.6465	2.7326	3.4895	5.0706	7.3441	11.0301	13.3616	15.5073	13.3616	21.9550	26.1245
9	1.7349	2.0879	3.3251	4.1682	5.8988	8.3428	12.2421	14.6837	16.9190	14.6837	23.5894	27.8772
10	2.1559	2.5582	3.9403	4.8652	6.7372	9.3418	13.4420	15.9872	18.3070	15.9872	25.1882	29.5883
11	2.6032	3.0535	4.5748	5.5778	7.5841	10.3410	14.6314	17.2750	19.6751	17.2750	26.7568	31.2641
12	3.0738	3.5706	5.2260	6.3038	8.4384	11.3403	15.8120	18.5493	21.0261	18.5493	28.2995	32.9095
13	3.5650	4.1069	5.8919	7.0415	9.2991	12.3398	16.9848	19.8119	22.3620	19.8119	29.8195	34.5282
14	4.0747	4.6604	6.5706	7.7895	10.1653	13.3393	18.1508	21.0641	23.6848	21.0641	31.3193	36.1233
15	4.6009	5.2293	7.2609	8.5468	11.0365	14.3389	19.3107	22.3071	24.9958	22.3071	32.8013	37.6973
16	5.1422	5.8122	7.9616	9.3122	11.9122	15.3385	20.4651	23.5418	26.2962	23.5418	34.2672	39.2524
17	5.6972	6.4078	8.6718	10.0852	12.7919	16.3382	21.6146	24.7690	27.5871	24.7690	35.7185	40.7902
18	6.2648	7.0149	9.3905	10.8649	13.6753	17.3379	22.7595	25.9894	28.8693	25.9894	37.1565	42.3124
19	6.8440	7.6327	10.1170	11.6509	14.5620	18.3377	23.9004	27.2036	30.1435	27.2036	38.5823	43.8202
20	7.4338	8.2604	10.8508	12.4426	15.4518	19.3374	25.0375	28.4120	31.4104	28.4120	39.9968	45.3147
25	10.5197	11.5240	14.6114	16.4734	19.9393	24.3366	30.6752	34.3816	37.6525	34.3816	46.9279	52.6197
30	13.7867	14.9535	18.4927	20.5992	24.4776	29.3360	36.2502	40.2560	43.7730	40.2560	53.6720	59.7031
40	20.7065	22.1643	26.5093	29.0505	33.6603	39.3353	47.2685	51.8051	55.7585	51.8051	66.7660	73.4020
50	27.9907	29.7067	34.7643	37.6886	42.9421	49.3349	58.1638	63.1671	67.5048	63.1671	79.4900	86.6608
60	35.5345	37.4849	43.1880	46.4589	52.2938	59.3347	68.9721	74.3970	79.0819	74.3970	91.9517	99.6072
70	43.2752	45.4417	51.7393	55.3289	61.6983	69.3345	79.7147	85.5270	90.5312	85.5270	104.2149	112.3169
80	51.1719	53.5401	60.3915	64.2778	71.1445	79.3343	90.4053	96.5782	101.8795	96.5782	116.3211	124.8392
90	59.1963	61.7541	69.1260	73.2911	80.6247	89.3342	101.0537	107.5650	113.1453	107.5650	128.2989	137.2084
100	67.3276	70.0649	77.9295	82.3581	90.1332	99.3341	111.6667	118.4980	124.3421	118.4980	140.1695	149.4493

χ^2 is the value having ν = 9 and α = 0.1, which, from Table 3.4, is χ^2 = 14.6837. The lower value is defined such that 10% of χ^2 values fall below or, equivalently, 90% fall above. This χ^2 value is that having ν = 9 and α = 0.9, which is 4.1682. These two χ^2 values describe a region containing 80% of χ^2 values for the given n. Therefore, one can write F(χ^2 < 4.1682) = 0.1, F(4.1682 < χ^2 < 14.6837) = 0.8, and F(χ^2 > 14.6837) = 0.1.

3.2.7 F Distribution for the Ratio of Sample Variances— Continuous Variable

A question that may arise is whether the variability (the variance) of a measured variable changes from sample to sample. Significant changes in the variability may indicate a manufacturing problem.

Changes in variability are often quantified by examining the ratio of the variance of two samples, where a statistically significant (see Section 3.3) ratio not equal to unity indicates a change. The proper distribution to describe this ratio is the F distribution [2], which describes the ratio of two chi-square variables (e.g., a ratio of variances). The test statistic for this distribution is

$$F = \frac{s_1^2/\sigma_1^2}{s_2^2/\sigma_2^2} \qquad (3.15)$$

where samples 1 and 2 have degrees of freedom $v_1 = n_1 - 1$ and $v_2 = n_2 - 1$, respectively. Owing to its form, the F distribution is used in statistical tests of the ratio of two sample variances. Multiple tables are required to calculate an F value because F depends on three parameters: the two degrees of freedom and the desired cumulative probability for that F value. The number of tables is reduced by the relationship [2]

$$F_{1-\alpha, v_1, v_2} = \frac{1}{F_{\alpha, v_2, v_1}} \qquad (3.16)$$

where $F_{1-\alpha, v_1, v_2}$ is the F value for a cumulative probability of $1 - \alpha$ and degrees of freedom v_1 and v_2 in the numerator and denominator, respectively. The F distribution is highly asymmetric, but becomes normal as the number of degrees of freedom approaches infinity. Tables 3.5 through 3.9 display the F values for cumulative probabilities (α) of 0.01, 0.025, 0.05, 0.1, and 0.25, respectively. The Microsoft Excel functions FDIST and FINV provide F values and cumulative probabilities [5, 6]. Since the F distribution depends on three variables, these Excel functions are simpler to use than reading from tables.

Example 3.12
Five measurements are taken in sample 1 and seven measurements are taken in sample 2. Above what F value will 10% of the measurements fall?

Solution
Solving with a confidence level of 90% ($\alpha = 0.1$), with $v_1 = 5 - 1 = 4$ and $v_2 = 7 - 1 = 6$, and reading from Table 3.8, $F_{0.1, 4, 6} = 3.1808$. Therefore, 10% of F values fall above 3.1808 for the given degrees of freedom.

Example 3.13
For the same two sample sizes, what two F values define an interval for which 5% of measurements fall above the higher F value (95% fall above this F) and 5% fall below the lower F value?

Solution
We solve such that 5% of measurements fall below the lower F value and 95% fall above this F, 5% of measurements fall above the higher F value. Therefore, the two F values for this interval are $F_{0.95, 4, 6}$ and $F_{0.05, 4, 6}$. The lower F value is not listed in one of the tables, so the relationship shown in Equation (3.16) is used to replace $F_{0.95, 4, 6}$ with $1/F_{0.05, 6, 4}$. Reading from Table 3.7, $F_{0.05, 4, 6} = 4.5337$ and $F_{0.05, 6, 4} = 6.1631$. Then, $F_{0.95, 4, 6} = 1/F_{0.05, 6, 4} = 0.1623$. Therefore, 90% of F measurements fall between 0.1623 and 4.5337 for the specified samples.

3.3 CONFIDENCE INTERVALS OF POPULATION PARAMETERS

When collecting experimental data, experimental measurements are *sample* parameters and do not represent the true *population* parameters exactly. Hence, based on the experimental measurements of sample parameters, in what interval does the population parameter lie? Using the probability

Engineering Statistics

distributions discussed in Section 3.2, one can estimate many population parameters. When population parameters are known, tests can be made as to some predetermined interval of confidence. Formalized hypothesis testing is discussed in Section 3.4. Both confidence intervals and hypothesis testing require a preselected *confidence level*. For most statistical analyses, the preferred confidence level is 95%. Selecting a lower confidence level increases the margin of error in statistical analysis, and vice versa.

Confidence intervals of population parameters are determined from sample parameters as follows:

1. A confidence level of the interval indicates the probability that the true population parameter lies within that interval. The confidence level is often written as $100(1 - \alpha)\%$, where $100\alpha\%$ is the probability that the population parameter lies outside the interval. With a confidence level of 95%, $\alpha = 0.05$ (so that $1.0 - 0.05$ equals 0.95).
2. The interval of the test statistic is determined from the confidence level. For a generic test statistic y, there is an interval

$$y_l \leq y \leq y_h \tag{3.17}$$

such that $100(1 - \alpha)\%$ of y values fall between y_l and y_h. Equivalently,

$$F(y_l \leq y \leq y_h) = 1 - \alpha \tag{3.18}$$

If the interval defines a region below which the population parameter lies, then y_l is replaced with $-\infty$. Likewise, y_h is replaced with ∞ to determine the minimum value of the population parameter. Both situations are termed "one-sided confidence intervals," as the interval is bounded on only one side. Otherwise, the interval is "two-sided" and the interval has bounds on two sides. Distributions, y_l and y_h are chosen to be centered about $y = 0$, so that $y_l = -y_h$.
3. The test statistic is replaced by its definition in terms of the population and sample parameters.
4. The inequality defining the confidence interval is manipulated algebraically to yield the confidence interval of the population parameter.

The following subsections describe how confidence intervals are constructed using the probability distributions discussed in Section 3.2. Confidence intervals can also be constructed with other probability distributions than those discussed here (see references [1–4]).

3.3.1 Confidence Interval of a Mean: Population Variance Known

According to the central limit theorem, the probability density distribution of $\bar{x}$ for a sample of size n will be given by a standard normal distribution with the test statistic defined [2] as

$$z = \frac{\bar{x} - \mu}{\sigma/\sqrt{n}} \tag{3.19}$$

For lower and upper z values z_l and z_h, respectively, the confidence interval for μ with confidence level $100(1 - \alpha)\%$ is defined by the following inequalities, which follow from the replacement of y by z, y_l by z_l, and y_h by z_h in Equation (3.17):

$$z_l \leq \frac{\bar{x} - \mu}{\sigma/\sqrt{n}} \leq z_h \tag{3.20}$$

TABLE 3.5
F Variable for F Distribution as a Function of Degrees of Freedom v_1 and v_2 for Cumulative Probability $\alpha = 0.01$

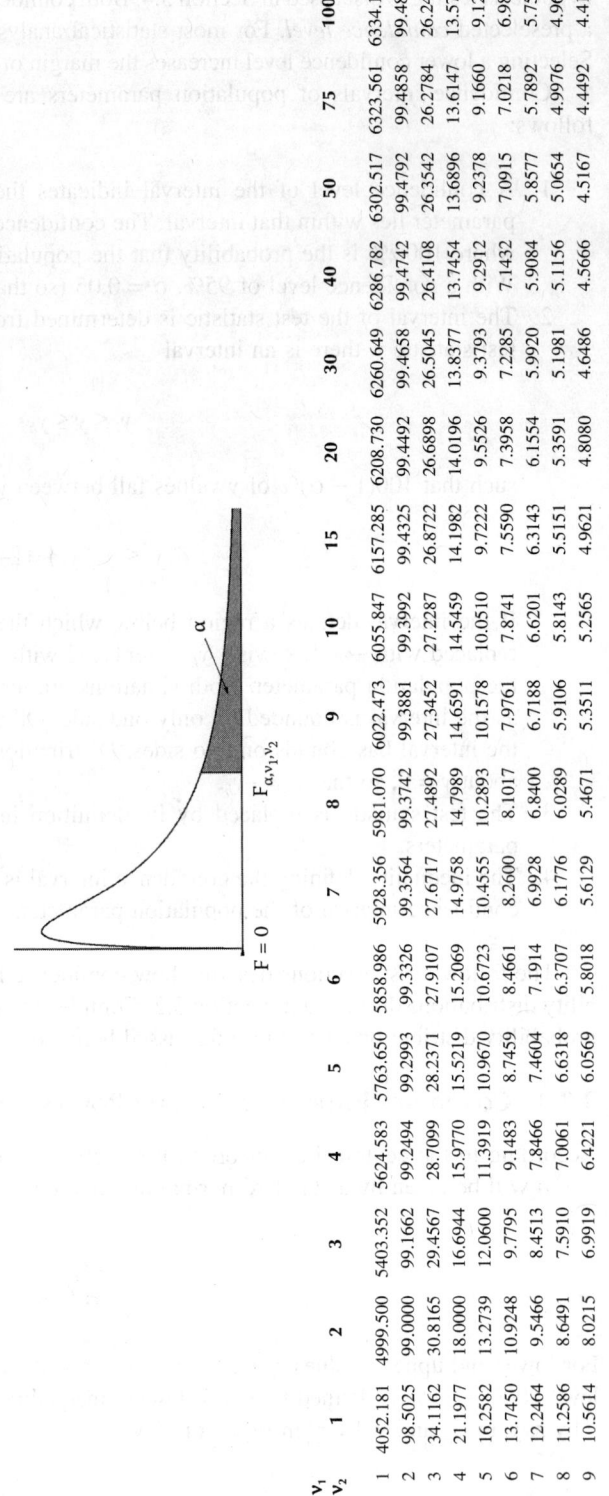

v_1 \ v_2	1	2	3	4	5	6	7	8	9	10	15	20	30	40	50	75	100
1	4052.181	4999.500	5403.352	5624.583	5763.650	5858.986	5928.356	5981.070	6022.473	6055.847	6157.285	6208.730	6260.649	6286.782	6302.517	6323.561	6334.110
2	98.5025	99.0000	99.1662	99.2494	99.2993	99.3326	99.3564	99.3742	99.3881	99.3992	99.4325	99.4492	99.4658	99.4742	99.4792	99.4858	99.4892
3	34.1162	30.8165	29.4567	28.7099	28.2371	27.9107	27.6717	27.4892	27.3452	27.2287	26.8722	26.6898	26.5045	26.4108	26.3542	26.2784	26.2402
4	21.1977	18.0000	16.6944	15.9770	15.5219	15.2069	14.9758	14.7989	14.6591	14.5459	14.1982	14.0196	13.8377	13.7454	13.6896	13.6147	13.5770
5	16.2582	13.2739	12.0600	11.3919	10.9670	10.6723	10.4555	10.2893	10.1578	10.0510	9.7222	9.5526	9.3793	9.2912	9.2378	9.1660	9.1299
6	13.7450	10.9248	9.7795	9.1483	8.7459	8.4661	8.2600	8.1017	7.9761	7.8741	7.5590	7.3958	7.2285	7.1432	7.0915	7.0218	6.9867
7	12.2464	9.5466	8.4513	7.8466	7.4604	7.1914	6.9928	6.8400	6.7188	6.6201	6.3143	6.1554	5.9920	5.9084	5.8577	5.7892	5.7547
8	11.2586	8.6491	7.5910	7.0061	6.6318	6.3707	6.1776	6.0289	5.9106	5.8143	5.5151	5.3591	5.1981	5.1156	5.0654	4.9976	4.9633
9	10.5614	8.0215	6.9919	6.4221	6.0569	5.8018	5.6129	5.4671	5.3511	5.2565	4.9621	4.8080	4.6486	4.5666	4.5167	4.4492	4.4150

Engineering Statistics

10	10.0443	7.5594	6.5523	5.9943	5.6363	5.3858	5.2001	5.0567	4.9424	4.8491	4.5581	4.4054	4.2469	4.1653	4.1155	4.0479	4.0137
11	9.6460	7.2057	6.2167	5.6683	5.3160	5.0692	4.8861	4.7445	4.6315	4.5393	4.2509	4.0990	3.9411	3.8596	3.8097	3.7421	3.7077
12	9.3302	6.9266	5.9525	5.4120	5.0643	4.8206	4.6395	4.4994	4.3875	4.2961	4.0096	3.8584	3.7008	3.6192	3.5692	3.5014	3.4668
13	9.0738	6.7010	5.7394	5.2053	4.8616	4.6204	4.4410	4.3021	4.1911	4.1003	3.8154	3.6646	3.5070	3.4253	3.3752	3.3070	3.2723
14	8.8616	6.5149	5.5639	5.0354	4.6950	4.4558	4.2779	4.1399	4.0297	3.9394	3.6557	3.5052	3.3476	3.2656	3.2153	3.1468	3.1118
15	8.6831	6.3589	5.4170	4.8932	4.5556	4.3183	4.1415	4.0045	3.8948	3.8049	3.5222	3.3719	3.2141	3.1319	3.0814	3.0124	2.9772
16	8.5310	6.2262	5.2922	4.7726	4.4374	4.2016	4.0259	3.8896	3.7804	3.6909	3.4089	3.2587	3.1007	3.0182	2.9675	2.8981	2.8627
17	8.3997	6.1121	5.1850	4.6690	4.3359	4.1015	3.9267	3.7910	3.6822	3.5931	3.3117	3.1615	3.0032	2.9205	2.8694	2.7996	2.7639
18	8.2854	6.0129	5.0919	4.5790	4.2479	4.0146	3.8406	3.7054	3.5971	3.5082	3.2273	3.0771	2.9185	2.8354	2.7841	2.7139	2.6779
19	8.1849	5.9259	5.0103	4.5003	4.1708	3.9386	3.7653	3.6305	3.5225	3.4338	3.1533	3.0031	2.8442	2.7608	2.7093	2.6386	2.6023
20	8.0960	5.8489	4.9382	4.4307	4.1027	3.8714	3.6987	3.5644	3.4567	3.3682	3.0880	2.9377	2.7785	2.6947	2.6430	2.5718	2.5353
25	7.7698	5.5680	4.6755	4.1774	3.8550	3.6272	3.4568	3.3239	3.2172	3.1294	2.8502	2.6993	2.5383	2.4530	2.3999	2.3267	2.2888
30	7.5625	5.3903	4.5097	4.0179	3.6990	3.4735	3.3045	3.1726	3.0665	2.9791	2.7002	2.5487	2.3860	2.2992	2.2450	2.1698	2.1307
40	7.3141	5.1785	4.3126	3.8283	3.5138	3.2910	3.1238	2.9930	2.8876	2.8005	2.5216	2.3689	2.2034	2.1142	2.0581	1.9795	1.9383
50	7.1706	5.0566	4.1993	3.7195	3.4077	3.1864	3.0202	2.8900	2.7850	2.6981	2.4190	2.2652	2.0976	2.0066	1.9490	1.8677	1.8248
60	7.0771	4.9774	4.1259	3.6490	3.3389	3.1187	2.9530	2.8233	2.7185	2.6318	2.3523	2.1978	2.0285	1.9360	1.8772	1.7937	1.7493
70	7.0114	4.9219	4.0744	3.5996	3.2907	3.0712	2.9060	2.7765	2.6719	2.5852	2.3055	2.1504	1.9797	1.8861	1.8263	1.7410	1.6954
80	6.9627	4.8807	4.0363	3.5631	3.2550	3.0361	2.8713	2.7420	2.6374	2.5508	2.2709	2.1153	1.9435	1.8489	1.7883	1.7015	1.6548
90	6.9251	4.8491	4.0070	3.5350	3.2276	3.0091	2.8445	2.7154	2.6109	2.5243	2.2442	2.0882	1.9155	1.8201	1.7588	1.6707	1.6231
100	6.8953	4.8239	3.9837	3.5127	3.2059	2.9877	2.8233	2.6943	2.5898	2.5033	2.2230	2.0666	1.8933	1.7972	1.7353	1.6461	1.5977

TABLE 3.6
F Variable for F Distribution as a Function of Degrees of Freedom v_1 and v_2 for Cumulative Probability $\alpha = 0.025$

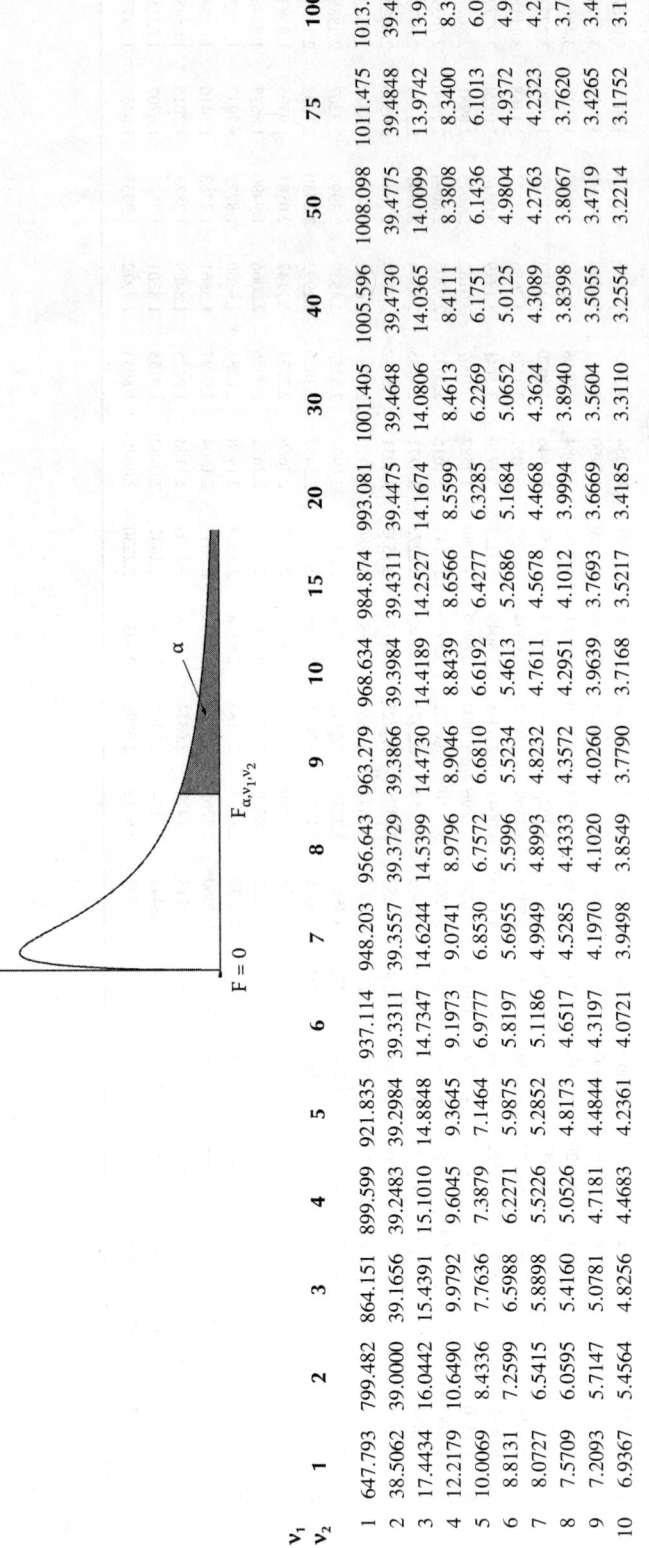

v_1 v_2	1	2	3	4	5	6	7	8	9	10	15	20	30	40	50	75	100
1	647.793	799.482	864.151	899.599	921.835	937.114	948.203	956.643	963.279	968.634	984.874	993.081	1001.405	1005.596	1008.098	1011.475	1013.163
2	38.5062	39.0000	39.1656	39.2483	39.2984	39.3311	39.3557	39.3729	39.3866	39.3984	39.4311	39.4475	39.4648	39.4730	39.4775	39.4848	39.4875
3	17.4434	16.0442	15.4391	15.1010	14.8848	14.7347	14.6244	14.5399	14.4730	14.4189	14.2527	14.1674	14.0806	14.0365	14.0099	13.9742	13.9562
4	12.2179	10.6490	9.9792	9.6045	9.3645	9.1973	9.0741	8.9796	8.9046	8.8439	8.6566	8.5599	8.4613	8.4111	8.3808	8.3400	8.3195
5	10.0069	8.4336	7.7636	7.3879	7.1464	6.9777	6.8530	6.7572	6.6810	6.6192	6.4277	6.3285	6.2269	6.1751	6.1436	6.1013	6.0800
6	8.8131	7.2599	6.5988	6.2271	5.9875	5.8197	5.6955	5.5996	5.5234	5.4613	5.2686	5.1684	5.0652	5.0125	4.9804	4.9372	4.9154
7	8.0727	6.5415	5.8898	5.5226	5.2852	5.1186	4.9949	4.8993	4.8232	4.7611	4.5678	4.4668	4.3624	4.3089	4.2763	4.2323	4.2101
8	7.5709	6.0595	5.4160	5.0526	4.8173	4.6517	4.5285	4.4333	4.3572	4.2951	4.1012	3.9994	3.8940	3.8398	3.8067	3.7620	3.7393
9	7.2093	5.7147	5.0781	4.7181	4.4844	4.3197	4.1970	4.1020	4.0260	3.9639	3.7693	3.6669	3.5604	3.5055	3.4719	3.4265	3.4034
10	6.9367	5.4564	4.8256	4.4683	4.2361	4.0721	3.9498	3.8549	3.7790	3.7168	3.5217	3.4185	3.3110	3.2554	3.2214	3.1752	3.1517

n													
11	6.7241	5.2559	4.6300	4.2751	4.0440	3.8806	3.7586	3.6638	3.5879	3.5257	3.3299	3.2261	3.1176	3.0613	3.0268	2.9800	2.9561
12	6.5538	5.0959	4.4742	4.1212	3.8911	3.7283	3.6065	3.5118	3.4358	3.3735	3.1772	3.0728	2.9633	2.9063	2.8714	2.8238	2.7996
13	6.4143	4.9653	4.3472	3.9959	3.7667	3.6043	3.4827	3.3880	3.3120	3.2497	3.0527	2.9477	2.8373	2.7797	2.7443	2.6961	2.6715
14	6.2979	4.8567	4.2417	3.8919	3.6634	3.5014	3.3799	3.2853	3.2093	3.1469	2.9493	2.8437	2.7324	2.6742	2.6384	2.5895	2.5646
15	6.1995	4.7650	4.1528	3.8043	3.5764	3.4147	3.2934	3.1987	3.1227	3.0602	2.8621	2.7559	2.6437	2.5850	2.5488	2.4993	2.4739
16	6.1151	4.6867	4.0768	3.7294	3.5021	3.3406	3.2194	3.1248	3.0488	2.9862	2.7875	2.6808	2.5678	2.5085	2.4719	2.4218	2.3961
17	6.0420	4.6189	4.0112	3.6648	3.4379	3.2767	3.1556	3.0610	2.9849	2.9222	2.7230	2.6158	2.5020	2.4422	2.4053	2.3545	2.3285
18	5.9781	4.5597	3.9539	3.6083	3.3820	3.2209	3.0999	3.0053	2.9291	2.8664	2.6667	2.5590	2.4445	2.3842	2.3468	2.2956	2.2692
19	5.9216	4.5075	3.9034	3.5587	3.3327	3.1718	3.0509	2.9563	2.8801	2.8172	2.6171	2.5089	2.3937	2.3329	2.2952	2.2434	2.2167
20	5.8715	4.4612	3.8587	3.5147	3.2891	3.1283	3.0074	2.9128	2.8365	2.7737	2.5731	2.4645	2.3486	2.2873	2.2493	2.1969	2.1699
25	5.6864	4.2909	3.6943	3.3530	3.1287	2.9685	2.8478	2.7531	2.6766	2.6135	2.4110	2.3005	2.1816	2.1183	2.0787	2.0239	1.9955
30	5.5675	4.1821	3.5893	3.2499	3.0265	2.8667	2.7460	2.6513	2.5746	2.5112	2.3072	2.1952	2.0739	2.0089	1.9681	1.9112	1.8816
40	5.4239	4.0510	3.4633	3.1261	2.9037	2.7444	2.6238	2.5289	2.4519	2.3882	2.1819	2.0677	1.9429	1.8752	1.8324	1.7722	1.7405
50	5.3403	3.9749	3.3902	3.0544	2.8326	2.6736	2.5530	2.4579	2.3808	2.3168	2.1090	1.9933	1.8659	1.7963	1.7520	1.6892	1.6558
60	5.2856	3.9253	3.3425	3.0077	2.7863	2.6274	2.5068	2.4117	2.3344	2.2702	2.0613	1.9445	1.8152	1.7440	1.6985	1.6337	1.5990
70	5.2470	3.8903	3.3090	2.9748	2.7537	2.5949	2.4743	2.3791	2.3017	2.2374	2.0277	1.9100	1.7792	1.7069	1.6604	1.5939	1.5581
80	5.2183	3.8643	3.2841	2.9504	2.7295	2.5708	2.4502	2.3549	2.2775	2.2130	2.0026	1.8843	1.7523	1.6790	1.6318	1.5639	1.5271
90	5.1962	3.8443	3.2649	2.9315	2.7109	2.5522	2.4316	2.3363	2.2588	2.1942	1.9833	1.8644	1.7315	1.6574	1.6095	1.5404	1.5028
100	5.1786	3.8284	3.2496	2.9166	2.6961	2.5374	2.4168	2.3215	2.2439	2.1793	1.9679	1.8486	1.7148	1.6401	1.5917	1.5215	1.4833

TABLE 3.7
F Variable for F Distribution as a Function of Degrees of Freedom v_1 and v_2 for cumulative probability $\alpha = 0.05$

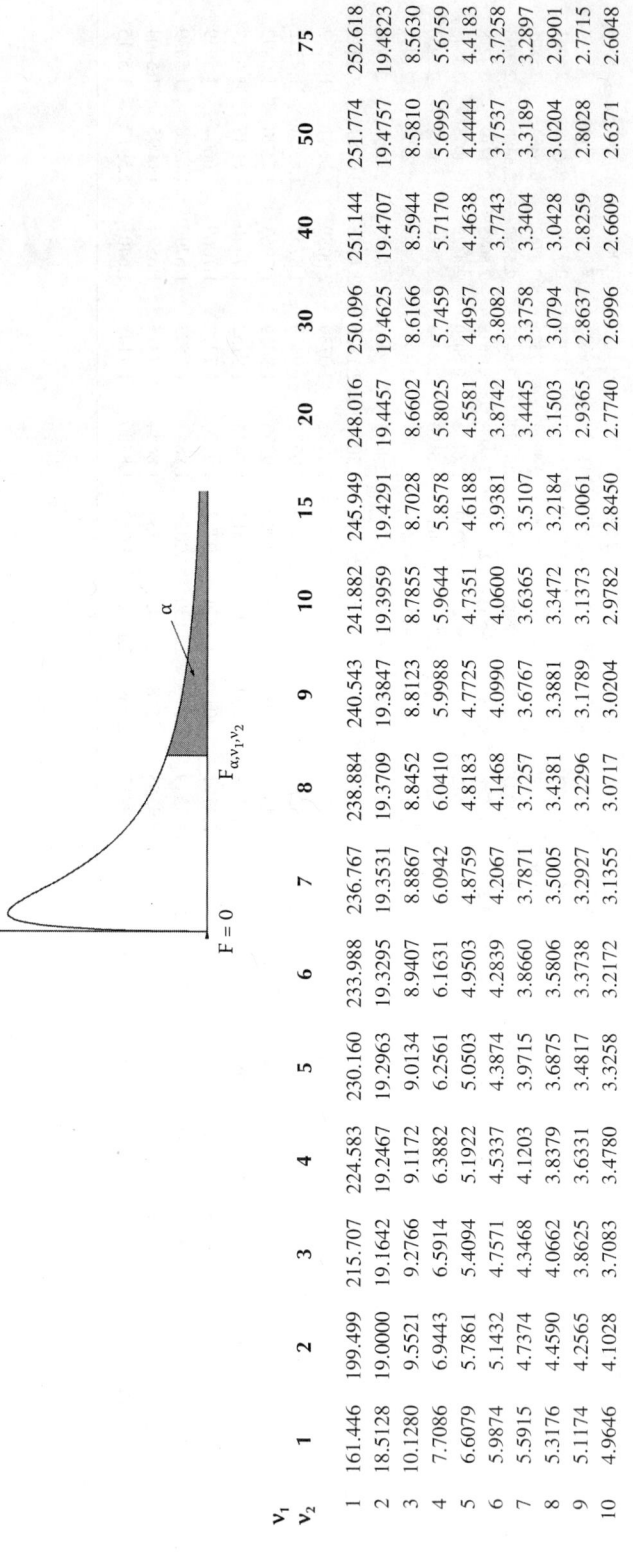

v_1 \ v_2	1	2	3	4	5	6	7	8	9	10	15	20	30	40	50	75	100
1	161.446	199.499	215.707	224.583	230.160	233.988	236.767	238.884	240.543	241.882	245.949	248.016	250.096	251.144	251.774	252.618	253.043
2	18.5128	19.0000	19.1642	19.2467	19.2963	19.3295	19.3531	19.3709	19.3847	19.3959	19.4291	19.4457	19.4625	19.4707	19.4757	19.4823	19.4857
3	10.1280	9.5521	9.2766	9.1172	9.0134	8.9407	8.8867	8.8452	8.8123	8.7855	8.7028	8.6602	8.6166	8.5944	8.5810	8.5630	8.5539
4	7.7086	6.9443	6.5914	6.3882	6.2561	6.1631	6.0942	6.0410	5.9988	5.9644	5.8578	5.8025	5.7459	5.7170	5.6995	5.6759	5.6640
5	6.6079	5.7861	5.4094	5.1922	5.0503	4.9503	4.8759	4.8183	4.7725	4.7351	4.6188	4.5581	4.4957	4.4638	4.4444	4.4183	4.4051
6	5.9874	5.1432	4.7571	4.5337	4.3874	4.2839	4.2067	4.1468	4.0990	4.0600	3.9381	3.8742	3.8082	3.7743	3.7537	3.7258	3.7117
7	5.5915	4.7374	4.3468	4.1203	3.9715	3.8660	3.7871	3.7257	3.6767	3.6365	3.5107	3.4445	3.3758	3.3404	3.3189	3.2897	3.2749
8	5.3176	4.4590	4.0662	3.8379	3.6875	3.5806	3.5005	3.4381	3.3881	3.3472	3.2184	3.1503	3.0794	3.0428	3.0204	2.9901	2.9747
9	5.1174	4.2565	3.8625	3.6331	3.4817	3.3738	3.2927	3.2296	3.1789	3.1373	3.0061	2.9365	2.8637	2.8259	2.8028	2.7715	2.7556
10	4.9646	4.1028	3.7083	3.4780	3.3258	3.2172	3.1355	3.0717	3.0204	2.9782	2.8450	2.7740	2.6996	2.6609	2.6371	2.6048	2.5884

11	4.8443	3.9823	3.5874	3.3567	3.2039	3.0946	3.0123	2.9480	2.8962	2.8536	2.7186	2.6464	2.5705	2.5309	2.5066	2.4734	2.4566
12	4.7472	3.8853	3.4903	3.2592	3.1059	2.9961	2.9134	2.8486	2.7964	2.7534	2.6169	2.5436	2.4663	2.4259	2.4010	2.3671	2.3498
13	4.6672	3.8056	3.4105	3.1791	3.0254	2.9153	2.8321	2.7669	2.7144	2.6710	2.5331	2.4589	2.3803	2.3392	2.3138	2.2791	2.2614
14	4.6001	3.7389	3.3439	3.1122	2.9582	2.8477	2.7642	2.6987	2.6458	2.6022	2.4630	2.3879	2.3082	2.2663	2.2405	2.2051	2.1870
15	4.5431	3.6823	3.2874	3.0556	2.9013	2.7905	2.7066	2.6408	2.5876	2.5437	2.4034	2.3275	2.2468	2.2043	2.1780	2.1419	2.1234
16	4.4940	3.6337	3.2389	3.0069	2.8524	2.7413	2.6572	2.5911	2.5377	2.4935	2.3522	2.2756	2.1938	2.1507	2.1240	2.0873	2.0685
17	4.4513	3.5915	3.1968	2.9647	2.8100	2.6987	2.6143	2.5480	2.4943	2.4499	2.3077	2.2304	2.1477	2.1040	2.0769	2.0396	2.0204
18	4.4139	3.5546	3.1599	2.9277	2.7729	2.6613	2.5767	2.5102	2.4563	2.4117	2.2686	2.1906	2.1071	2.0629	2.0354	1.9975	1.9780
19	4.3808	3.5219	3.1274	2.8951	2.7401	2.6283	2.5435	2.4768	2.4227	2.3779	2.2341	2.1555	2.0712	2.0264	1.9986	1.9601	1.9403
20	4.3513	3.4928	3.0984	2.8661	2.7109	2.5990	2.5140	2.4471	2.3928	2.3479	2.2033	2.1242	2.0391	1.9938	1.9656	1.9267	1.9066
25	4.2417	3.3852	2.9912	2.7587	2.6030	2.4904	2.4047	2.3371	2.2821	2.2365	2.0889	2.0075	1.9192	1.8718	1.8421	1.8008	1.7794
30	4.1709	3.3158	2.9223	2.6896	2.5336	2.4205	2.3343	2.2662	2.2107	2.1646	2.0148	1.9317	1.8409	1.7918	1.7609	1.7176	1.6950
40	4.0847	3.2317	2.8387	2.6060	2.4495	2.3359	2.2490	2.1802	2.1240	2.0773	1.9245	1.8389	1.7444	1.6928	1.6600	1.6137	1.5892
50	4.0343	3.1826	2.7900	2.5572	2.4004	2.2864	2.1992	2.1299	2.0733	2.0261	1.8714	1.7841	1.6872	1.6337	1.5995	1.5508	1.5249
60	4.0012	3.1504	2.7581	2.5252	2.3683	2.2541	2.1665	2.0970	2.0401	1.9926	1.8364	1.7480	1.6491	1.5943	1.5590	1.5085	1.4814
70	3.9778	3.1277	2.7355	2.5027	2.3456	2.2312	2.1435	2.0737	2.0166	1.9689	1.8117	1.7223	1.6220	1.5661	1.5300	1.4779	1.4498
80	3.9604	3.1108	2.7188	2.4859	2.3287	2.2142	2.1263	2.0564	1.9991	1.9512	1.7932	1.7032	1.6017	1.5449	1.5081	1.4548	1.4259
90	3.9469	3.0977	2.7058	2.4729	2.3157	2.2011	2.1131	2.0430	1.9856	1.9376	1.7789	1.6883	1.5859	1.5284	1.4910	1.4366	1.4070
100	3.9362	3.0873	2.6955	2.4626	2.3053	2.1906	2.1025	2.0323	1.9748	1.9267	1.7675	1.6764	1.5733	1.5151	1.4772	1.4220	1.3917

TABLE 3.8
F Variable for F Distribution as a Function of Degrees of Freedom v_1 and v_2 for Cumulative Probability $\alpha = 0.1$

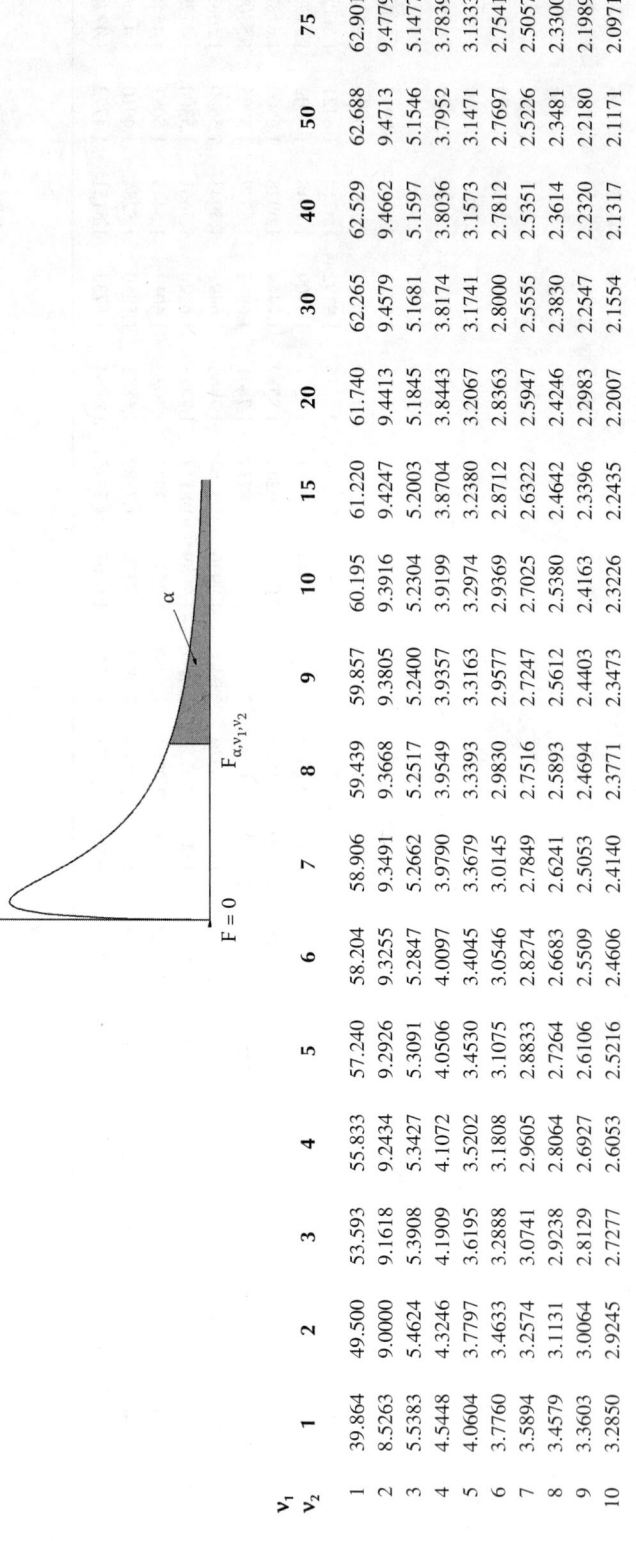

v_1 v_2	1	2	3	4	5	6	7	8	9	10	15	20	30	40	50	75	100
1	39.864	49.500	53.593	55.833	57.240	58.204	58.906	59.439	59.857	60.195	61.220	61.740	62.265	62.529	62.688	62.901	63.007
2	8.5263	9.0000	9.1618	9.2434	9.2926	9.3255	9.3491	9.3668	9.3805	9.3916	9.4247	9.4413	9.4579	9.4662	9.4713	9.4779	9.4813
3	5.5383	5.4624	5.3908	5.3427	5.3091	5.2847	5.2662	5.2517	5.2400	5.2304	5.2003	5.1845	5.1681	5.1597	5.1546	5.1477	5.1443
4	4.5448	4.3246	4.1909	4.1072	4.0506	4.0097	3.9790	3.9549	3.9357	3.9199	3.8704	3.8443	3.8174	3.8036	3.7952	3.7839	3.7782
5	4.0604	3.7797	3.6195	3.5202	3.4530	3.4045	3.3679	3.3393	3.3163	3.2974	3.2380	3.2067	3.1741	3.1573	3.1471	3.1333	3.1263
6	3.7760	3.4633	3.2888	3.1808	3.1075	3.0546	3.0145	2.9830	2.9577	2.9369	2.8712	2.8363	2.8000	2.7812	2.7697	2.7541	2.7463
7	3.5894	3.2574	3.0741	2.9605	2.8833	2.8274	2.7849	2.7516	2.7247	2.7025	2.6322	2.5947	2.5555	2.5351	2.5226	2.5057	2.4971
8	3.4579	3.1131	2.9238	2.8064	2.7264	2.6683	2.6241	2.5893	2.5612	2.5380	2.4642	2.4246	2.3830	2.3614	2.3481	2.3300	2.3208
9	3.3603	3.0064	2.8129	2.6927	2.6106	2.5509	2.5053	2.4694	2.4403	2.4163	2.3396	2.2983	2.2547	2.2320	2.2180	2.1989	2.1892
10	3.2850	2.9245	2.7277	2.6053	2.5216	2.4606	2.4140	2.3771	2.3473	2.3226	2.2435	2.2007	2.1554	2.1317	2.1171	2.0971	2.0869

11	3.2252	2.8595	2.6602	2.5362	2.4512	2.3891	2.3416	2.3040	2.2735	2.2482	2.1671	2.1230	2.0762	2.0516	2.0364	2.0157	2.0050
12	3.1766	2.8068	2.6055	2.4801	2.3940	2.3310	2.2828	2.2446	2.2135	2.1878	2.1049	2.0597	2.0115	1.9861	1.9704	1.9489	1.9379
13	3.1362	2.7632	2.5603	2.4337	2.3467	2.2830	2.2341	2.1953	2.1638	2.1376	2.0532	2.0070	1.9576	1.9315	1.9153	1.8931	1.8817
14	3.1022	2.7265	2.5222	2.3947	2.3069	2.2426	2.1931	2.1539	2.1220	2.0954	2.0095	1.9625	1.9119	1.8852	1.8686	1.8457	1.8340
15	3.0732	2.6952	2.4898	2.3614	2.2730	2.2081	2.1582	2.1185	2.0862	2.0593	1.9722	1.9243	1.8728	1.8454	1.8284	1.8049	1.7929
16	3.0481	2.6682	2.4618	2.3327	2.2438	2.1783	2.1280	2.0880	2.0553	2.0281	1.9399	1.8913	1.8388	1.8108	1.7934	1.7694	1.7570
17	3.0262	2.6446	2.4374	2.3077	2.2183	2.1524	2.1017	2.0613	2.0284	2.0009	1.9117	1.8624	1.8090	1.7805	1.7628	1.7382	1.7255
18	3.0070	2.6239	2.4160	2.2858	2.1958	2.1296	2.0785	2.0379	2.0047	1.9770	1.8868	1.8368	1.7827	1.7537	1.7356	1.7106	1.6976
19	2.9899	2.6056	2.3970	2.2663	2.1760	2.1094	2.0580	2.0171	1.9836	1.9557	1.8647	1.8142	1.7592	1.7298	1.7114	1.6858	1.6726
20	2.9747	2.5893	2.3801	2.2489	2.1582	2.0913	2.0397	1.9985	1.9649	1.9367	1.8449	1.7938	1.7382	1.7083	1.6896	1.6636	1.6501
25	2.9177	2.5283	2.3170	2.1842	2.0922	2.0241	1.9714	1.9292	1.8947	1.8658	1.7708	1.7175	1.6589	1.6272	1.6072	1.5792	1.5645
30	2.8807	2.4887	2.2761	2.1422	2.0492	1.9803	1.9269	1.8841	1.8490	1.8195	1.7223	1.6673	1.6065	1.5732	1.5522	1.5225	1.5069
40	2.8353	2.4404	2.2261	2.0909	1.9968	1.9269	1.8725	1.8289	1.7929	1.7627	1.6624	1.6052	1.5411	1.5056	1.4830	1.4507	1.4336
50	2.8087	2.4120	2.1967	2.0608	1.9660	1.8954	1.8405	1.7963	1.7598	1.7291	1.6269	1.5681	1.5018	1.4648	1.4409	1.4068	1.3885
60	2.7911	2.3933	2.1774	2.0410	1.9457	1.8747	1.8194	1.7748	1.7380	1.7070	1.6034	1.5435	1.4755	1.4373	1.4126	1.3769	1.3576
70	2.7786	2.3800	2.1637	2.0269	1.9313	1.8600	1.8044	1.7596	1.7225	1.6913	1.5866	1.5259	1.4567	1.4176	1.3922	1.3552	1.3352
80	2.7693	2.3702	2.1535	2.0165	1.9206	1.8491	1.7933	1.7483	1.7110	1.6796	1.5741	1.5128	1.4426	1.4027	1.3767	1.3388	1.3180
90	2.7621	2.3625	2.1457	2.0084	1.9123	1.8406	1.7846	1.7395	1.7021	1.6705	1.5644	1.5025	1.4315	1.3911	1.3646	1.3258	1.3044
100	2.7564	2.3564	2.1394	2.0019	1.9057	1.8339	1.7778	1.7324	1.6949	1.6632	1.5566	1.4943	1.4227	1.3817	1.3548	1.3153	1.2934

TABLE 3.9
F Variable for F Distribution as a Function of Degrees of Freedom v_1 and v_2 for Cumulative Probability $\alpha = 0.25$

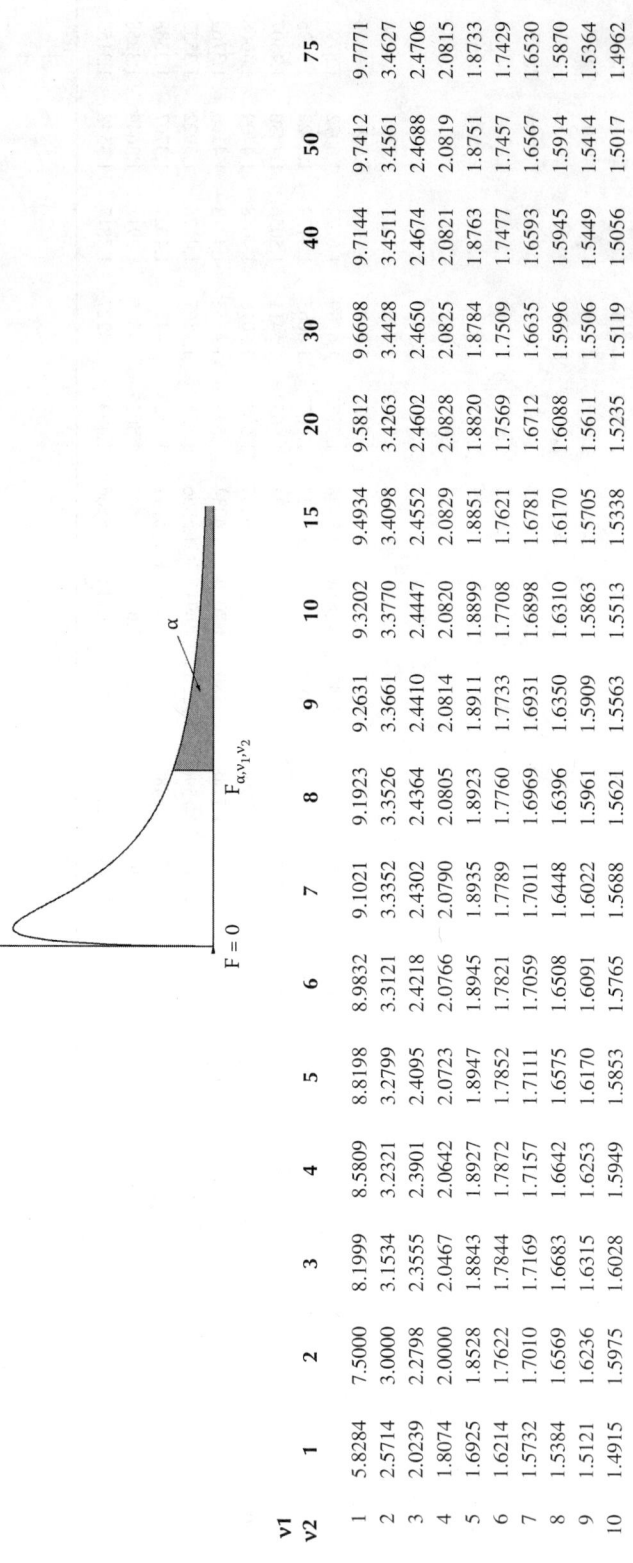

v1\v2	1	2	3	4	5	6	7	8	9	10	15	20	30	40	50	75	100
1	5.8284	7.5000	8.1999	8.5809	8.8198	8.9832	9.1021	9.1923	9.2631	9.3202	9.4934	9.5812	9.6698	9.7144	9.7412	9.7771	9.7951
2	2.5714	3.0000	3.1534	3.2321	3.2799	3.3121	3.3352	3.3526	3.3661	3.3770	3.4098	3.4263	3.4428	3.4511	3.4561	3.4627	3.4661
3	2.0239	2.2798	2.3555	2.3901	2.4095	2.4218	2.4302	2.4364	2.4410	2.4447	2.4552	2.4602	2.4650	2.4674	2.4688	2.4706	2.4715
4	1.8074	2.0000	2.0467	2.0642	2.0723	2.0766	2.0790	2.0805	2.0814	2.0820	2.0829	2.0828	2.0825	2.0821	2.0819	2.0815	2.0813
5	1.6925	1.8528	1.8843	1.8927	1.8947	1.8945	1.8935	1.8923	1.8911	1.8899	1.8851	1.8820	1.8784	1.8763	1.8751	1.8733	1.8724
6	1.6214	1.7622	1.7844	1.7872	1.7852	1.7821	1.7789	1.7760	1.7733	1.7708	1.7621	1.7569	1.7509	1.7477	1.7457	1.7429	1.7414
7	1.5732	1.7010	1.7169	1.7157	1.7111	1.7059	1.7011	1.6969	1.6931	1.6898	1.6781	1.6712	1.6635	1.6593	1.6567	1.6530	1.6511
8	1.5384	1.6569	1.6683	1.6642	1.6575	1.6508	1.6448	1.6396	1.6350	1.6310	1.6170	1.6088	1.5996	1.5945	1.5914	1.5870	1.5848
9	1.5121	1.6236	1.6315	1.6253	1.6170	1.6091	1.6022	1.5961	1.5909	1.5863	1.5705	1.5611	1.5506	1.5449	1.5414	1.5364	1.5338
10	1.4915	1.5975	1.6028	1.5949	1.5853	1.5765	1.5688	1.5621	1.5563	1.5513	1.5338	1.5235	1.5119	1.5056	1.5017	1.4962	1.4933

11	1.4749	1.5767	1.5798	1.5704	1.5598	1.5502	1.5418	1.5346	1.5284	1.5229	1.5041	1.4930	1.4805	1.4737	1.4694	1.4634	1.4603
12	1.4613	1.5595	1.5609	1.5504	1.5389	1.5286	1.5197	1.5120	1.5054	1.4996	1.4796	1.4678	1.4544	1.4471	1.4425	1.4361	1.4327
13	1.4500	1.5452	1.5451	1.5336	1.5214	1.5105	1.5011	1.4931	1.4861	1.4801	1.4590	1.4465	1.4324	1.4247	1.4198	1.4129	1.4094
14	1.4403	1.5331	1.5317	1.5194	1.5066	1.4952	1.4854	1.4770	1.4697	1.4634	1.4414	1.4284	1.4136	1.4055	1.4003	1.3931	1.3893
15	1.4321	1.5227	1.5202	1.5071	1.4938	1.4820	1.4718	1.4631	1.4556	1.4491	1.4263	1.4127	1.3973	1.3888	1.3834	1.3758	1.3718
16	1.4249	1.5137	1.5103	1.4965	1.4827	1.4705	1.4601	1.4511	1.4433	1.4366	1.4131	1.3990	1.3830	1.3742	1.3685	1.3606	1.3565
17	1.4186	1.5057	1.5015	1.4872	1.4730	1.4605	1.4497	1.4405	1.4325	1.4256	1.4014	1.3869	1.3704	1.3613	1.3554	1.3472	1.3429
18	1.4130	1.4988	1.4938	1.4790	1.4644	1.4516	1.4406	1.4312	1.4230	1.4159	1.3911	1.3762	1.3592	1.3497	1.3437	1.3352	1.3307
19	1.4081	1.4925	1.4870	1.4717	1.4568	1.4437	1.4325	1.4228	1.4145	1.4073	1.3819	1.3666	1.3492	1.3394	1.3332	1.3244	1.3198
20	1.4037	1.4870	1.4808	1.4652	1.4500	1.4366	1.4252	1.4153	1.4069	1.3995	1.3736	1.3580	1.3401	1.3301	1.3237	1.3146	1.3099
25	1.3870	1.4661	1.4577	1.4406	1.4242	1.4099	1.3977	1.3871	1.3781	1.3701	1.3422	1.3252	1.3056	1.2945	1.2873	1.2771	1.2717
30	1.3761	1.4524	1.4426	1.4244	1.4073	1.3923	1.3795	1.3685	1.3590	1.3507	1.3213	1.3033	1.2823	1.2703	1.2626	1.2514	1.2455
40	1.3626	1.4355	1.4239	1.4045	1.3863	1.3706	1.3571	1.3455	1.3354	1.3266	1.2952	1.2758	1.2529	1.2397	1.2310	1.2184	1.2116
50	1.3546	1.4255	1.4128	1.3927	1.3739	1.3576	1.3437	1.3317	1.3213	1.3122	1.2795	1.2592	1.2350	1.2208	1.2115	1.1979	1.1904
60	1.3493	1.4188	1.4055	1.3848	1.3657	1.3491	1.3348	1.3226	1.3119	1.3026	1.2690	1.2481	1.2229	1.2081	1.1983	1.1838	1.1757
70	1.3455	1.4141	1.4002	1.3793	1.3598	1.3430	1.3285	1.3161	1.3052	1.2958	1.2616	1.2401	1.2142	1.1989	1.1886	1.1734	1.1650
80	1.3427	1.4106	1.3964	1.3751	1.3554	1.3384	1.3238	1.3112	1.3002	1.2906	1.2559	1.2341	1.2076	1.1919	1.1813	1.1656	1.1567
90	1.3406	1.4079	1.3933	1.3719	1.3520	1.3349	1.3201	1.3074	1.2964	1.2866	1.2516	1.2294	1.2024	1.1864	1.1756	1.1593	1.1502
100	1.3388	1.4057	1.3909	1.3693	1.3493	1.3321	1.3172	1.3044	1.2933	1.2835	1.2481	1.2256	1.1983	1.1819	1.1709	1.1543	1.1449

$$\bar{x} - z_h \frac{\sigma}{\sqrt{n}} \leq \mu \leq \bar{x} - z_l \frac{\sigma}{\sqrt{n}} \qquad (3.21)$$

(Equation (3.21) is obtained by rearranging Equation (3.20), which follows from applying Equation (3.17) to the standard normal variable z.) If, for instance, the confidence interval is chosen to be symmetric about μ (as is usually done), z_h is normally written as $z_{\alpha/2}$ to denote that $\alpha/2$ percentage points fall above z_h. Symmetry would also provide $z_l = -z_h = z_{\alpha/2}$. The following example illustrates the process of determining the confidence interval for a population mean.

Example 3.14

For a filtration experiment, the variance of the solids collection rate is known to be $\sigma^2 = 4.50$. After three experimental runs, the collection rate was determined to have a sample mean $\bar{x} = 25.50$. Determine the two-sided confidence interval of the population mean μ for a confidence level of 95%.

Solution

1. With a two-sided confidence interval, 2.5% of the distribution is above and 2.5% is below the upper and lower bounds, respectively. Therefore, $\alpha = 0.05$.
2. Using Table 3.2, the z value is 1.96 and -1.96 for the upper and lower bounds, because $F(z \leq 1.96) = 0.975$. Because the normal distribution is symmetric, the lower bound is $z_l = -1.96$. (One could confirm this by showing that $F(z \leq -1.96) = 0.025$.)
3. Using Equation (3.7),

$$-1.96 \leq \frac{25.5 - \mu}{\sqrt{4.5}/\sqrt{3}} \leq 1.96$$

4. The 95% confidence interval for μ is calculated as

$$23.1 \leq \mu \leq 27.9$$

Since the confidence interval depends on $1/\sqrt{n}$, it becomes narrower as n increases, as the next example shows.

Example 3.15

When seven additional experimental runs are completed, n increases from 3 to 10, and the sample mean becomes $\bar{x} = 25.76$. In this case, σ, z_l, and z_h are unchanged. Determine the new 95% confidence interval for μ.

Solution

One can proceed directly to step 3. Entering $\bar{x} = 25.76$ and $n = 10$ into Equation (3.20),

$$-1.96 \leq \frac{25.76 - \mu}{\sqrt{4.5}/\sqrt{10}} \leq 1.96$$

Solving yields a slightly smaller interval:

$$24.5 \leq \mu \leq 27.1$$

Example 3.16

For a confidence level of 80% ($\alpha = 0.2$) and with $\bar{x} = 25.76$ and $n = 10$, determine the confidence interval for μ.

Engineering Statistics

Solution

From Table 3.2, the z values defining the interval are $z_l = -1.28$ and $z_h = 1.28$. (These values were obtained by linear interpolation.) Therefore, Equation (3.20) becomes

$$-1.282 \le \frac{25.76 - \mu}{\sqrt{4.5}/\sqrt{10}} \le 1.282$$

The confidence interval for μ has hence narrowed as indicated next:

$$24.9 \le \mu \le 26.6$$

3.3.2 Confidence Interval of a Mean: Population Variance Unknown

When the population variance σ^2 is not known, the confidence interval for a population mean cannot be determined using a standard normal distribution, but it can be determined with the t distribution. As discussed in Section 3.2.5, one defines the test statistic t in terms of $\bar{x}$, μ, n, and the sample variance s^2 (see Equation (3.13)). By replacing the generic test statistic y in Equation (3.17) with t, the interval for μ is defined [2] by

$$t_l \le \frac{\bar{x} - \mu}{s/\sqrt{n}} \le t_h \qquad (3.22)$$

or, equivalently,

$$\bar{x} - t_h \frac{s}{\sqrt{n}} \le \mu \le \bar{x} - t_l \frac{s}{\sqrt{n}} \qquad (3.23)$$

Since the t value depends on $\nu = n - 1$, the number of degrees of freedom, and because the t distribution is symmetric about $t = 0$, t_l is usually written as $t_{1-\alpha/2,\nu}$ and, likewise, t_h is often written as $t_{\alpha/2,\nu}$. As in the previous section, the confidence level is $100(1 - \alpha)\%$. The t distribution is highly sensitive to the number of degrees of freedom or, equivalently, n. For samples sizes larger than about 25, though, the t distribution is almost identical to the normal distribution and will not change significantly if n increases. For large samples, one could replace the t variable with z (Equation (3.19)) and just assume that $s^2 = \sigma^2$. Sometimes the ratio $s/\sqrt{n}$ is referred to as the standard error (*SE*) and is reported rather than the standard deviation or variance. Especially for least squares regression parameters, the standard error is easier to compute than the variance and, as such, replaces $s/\sqrt{n}$ for testing a hypothesis or in the calculation of confidence intervals. The following examples indicate how a confidence interval for μ can be generated using the t distribution.

Example 3.17

In Example 3.14, the sample of size $n = 3$ had a mean of $\bar{x} = 25.50$ and a sample variance of $s^2 = 4.8$ (slightly larger than the presumed population variance used in the previous examples). Determine the two-sided confidence interval for μ for a confidence level of 95%.

Solution

1. For a confidence level of 95%, $\alpha = 0.05$.
2. Reading from Table 3.3, $F(t \le 4.303) = 0.975$ for $\nu = 3 - 1 = 2$. Since the t distribution is symmetric, $F(-4.303 \le t \le 4.303) = 0.95$. Therefore, the lower bound is $t_{1-\alpha/2,n} = -4.303$ and the upper bound is $t_{\alpha/2,\nu} = 4.303$.

3. Equation (3.22) becomes

$$-4.303 \leq \frac{25.5 - \mu}{\sqrt{4.8}/\sqrt{3}} \leq 4.303$$

4. The 95% confidence interval for μ is, then,

$$20.06 \leq \mu \leq 30.94$$

This interval is larger when defined by a t statistic. Some difference is due to the sample variance being slightly larger than the population variance in the previous example, but the major difference is due to the t distribution being much wider than the standard normal distribution when the sample size is small. The following example demonstrates the effect of sample size on confidence intervals.

Example 3.18
If the sample size for the solids collection experiment had been increased to $n = 10$, but the sample mean and sample variance were unchanged, what would be the 95% confidence interval for μ?

Solution
The lower and upper t values are $t_{0.975,9} = -2.262$ and $t_{0.025,9} = 2.262$ (as listed in Table 3.3). Entering the parameters into Equation (3.23), as shown below, there is a smaller interval as compared with the case of three samples:

$$25.50 - 2.262 \frac{\sqrt{4.8}}{\sqrt{10}} \leq \mu \leq 25.50 + 2.262 \frac{\sqrt{4.8}}{\sqrt{10}}$$

and the 95% confidence interval is, then,

$$23.93 \leq \mu \leq 27.07$$

3.3.3 Confidence Interval of the Difference of Two Means

Occasionally, an experiment should determine if two populations have different population means. For example, following process modification, has the product's specification(s) changed? Measuring the confidence interval for the difference in the population means is a statistical test that can be employed. The t distribution is widely used in this case [2]. Table 3.10 shows the equations to be used. Because the population variances of the two samples may differ, two sets of equations are necessary.

The variables $\bar{x}_i$, s_i^2, and n_i in Table 3.10 are the sample mean, sample variance, and number of data points of the ith sample, respectively. In the calculation of ν in the case of unequal population variances, the result should be rounded to the nearest integer. The t distribution can approximate the difference of sample means for populations with unequal population. The t variable is obtained via Table 3.3. Table 3.10 is also important; one should not assume that the population variances are identical without evidence. When in doubt, assume that the populations have differing variances, as this may provide a wider confidence interval (and a larger margin of error in drawing conclusions).

The most important conclusion is to determine whether there is any difference in population means. Depending on the chosen confidence level, the resulting confidence interval may contain the value zero. For such a case, at the given level of confidence, the populations do not have statistically different means, despite any difference in the sample means or variances. Conversely, if zero is not contained in the confidence interval, then the population means differ at that confidence level.

**TABLE 3.10
Statistical Parameters for the Difference in Population Mean of Two Samples**

Equal Population Variance $\sigma_1^2 = \sigma_2^2$	Unequal Population Variance $\sigma_1^2 \neq \sigma_2^2$
$t = \dfrac{(\bar{x}_1 - \bar{x}_2) - (\mu_1 - \mu_2)}{s_p\sqrt{\dfrac{1}{n_1} + \dfrac{1}{n_2}}}$	$t = \dfrac{(\bar{x}_1 - \bar{x}_2) - (\mu_1 - \mu_2)}{\sqrt{\dfrac{s_1^2}{n_1} + \dfrac{s_2^2}{n_2}}}$
$s_p^2 = \dfrac{(n_1 - 1)s_1^2 + (n_2 - 1)s_2^2}{n_1 + n_2 - 2}$	...
$\nu = n_1 + n_2 - 2$	$\nu = \dfrac{\left(\dfrac{s_1^2}{n_1} + \dfrac{s_2^2}{n_2}\right)^2}{\dfrac{(s_1^2/n_1)^2}{n_1 - 1} + \dfrac{(s_2^2/n_2)^2}{n_2 - 1}}$

Source: Montgomery, Douglas C., and Runger, George C., *Applied Statistics and Probability for Engineers*, 3rd ed., John Wiley & Sons, New York (2003).

Example 3.19

This example is adapted from the *NIST/SEMATECH e-Handbook of Statistical Methods*, Section 7.3.1 [1]. Two different procedures are employed to assemble a device. For procedure 1, the mean assembly time for a sample of $n_1 = 11$ is 36.1 s with standard deviation 4.91 s. Procedure 2 has mean assembly time 32.2 s with standard deviation 2.54 s for a sample of $n_2 = 9$. Are the means statistically different at a confidence level of 95% if the populations are assumed to have identical variance?

Solution

The degrees of freedom and pooled variance are calculated before computing the interval. Since the populations have equal variance, the appropriate equations are those in the left side of Table 3.10:

$$s_p^2 = \frac{(n_1 - 1)s_1^2 + (n_2 - 1)s_2^2}{n_1 + n_2 - 2} = \frac{(11-1)\cdot 4.91^2 + (9-1)\cdot 2.54^2}{11 + 9 - 2} = 16.3$$

$$\nu = n_1 + n_2 - 2 = 11 + 9 - 2 = 18$$

1. For a confidence level of 95%, $\alpha = 0.05$.
2. With $\nu = 18$ and $\alpha = 0.05$, the lower and upper t values (from Table 3.3) are -2.1009 and 2.1009, respectively.
3. The t variable is defined by the left side of Table 3.10. Entering this t into Equation (3.17) and yields the confidence interval

$$t_{0.975,18} \leq \frac{(\bar{x}_1 - \bar{x}_2) - (\mu_1 - \mu_2)}{s_p\sqrt{\dfrac{1}{n_1} + \dfrac{1}{n_2}}} \leq t_{0.025,18}$$

4. Entering known values and rearranging,

$$0.0922 \text{ s} \leq \mu_1 - \mu_2 \leq 7.71 \text{ s}$$

Since zero is not included in the confidence interval, the mean assembly time of procedure 2 is less than that of procedure 1.

Example 3.20
Repeat Example 3.19 if it is not known whether the populations have equal variances.

Solution
Since the population variances are not equal, the degrees of freedom are calculated using the appropriate equation from the right side of Table 3.10 as follows:

$$\nu = \frac{\left(\dfrac{s_1^2}{n_1} + \dfrac{s_2^2}{n_2}\right)^2}{\dfrac{(s_1^2/n_1)^2}{n_1 - 1} + \dfrac{(s_2^2/n_2)^2}{n_2 - 1}} = \frac{\left(\dfrac{4.91^2}{11} + \dfrac{2.54^2}{9}\right)^2}{\dfrac{(4.91^2/11)^2}{11 - 1} + \dfrac{(2.54^2/9)^2}{9 - 1}} = 15.53 \approx 16$$

(Note that ν is rounded to the nearest integer, as noninteger degrees of freedom are not sensible.) The upper and lower t values (from Table 3.3) are 2.1120 and −2.1120, respectively. Entering the t variable in the right side of Table 3.10 into Equation (3.17), the interval is defined by

$$t_{0.975,16} \leq \frac{(x_1 - x_2) - (\mu_1 - \mu_2)}{\sqrt{\dfrac{s_1^2}{n_1} + \dfrac{s_2^2}{n_2}}} \leq t_{0.025,16}$$

Entering the known values into the above equation yields

$$0.285 \text{ s} \leq \mu_1 - \mu_2 \leq 7.52 \text{ s}$$

Once again, zero is not included in the interval and, as such, the two procedures have differing mean assembly times. Interestingly, the interval generated without the assumption of equal population variances is smaller than the previous case, owing mainly to the large difference between the two sample standard deviations.

3.3.4 Confidence Interval for the Variance

As discussed in Section 3.2.6, the variance of a sample can be calculated using the chi-square distribution. Computing the confidence interval for the population variance from the sample statistic is straightforward using the procedure laid out in Section 3.3 with the chi-square test statistic shown in Equation (3.14). Since the chi-square distribution is not symmetric, the upper and lower bounds are not simply opposite in sign, as they are in the normal or t distributions.

Example 3.21
For the solids collection experiment in Examples 3.17 and 3.18, the variance of the collection rate of a sample of size $n = 10$ was stated to be 4.8. Determine the 90% confidence interval of the population variance.

Engineering Statistics

Solution

1. For a confidence level of 90%, $\alpha = 0.1$.
2. From Table 3.4, the lower and upper bounds of the interval are $\chi_l^2 = 0.3.3251$ and $\chi_h^2 = 16.919$, respectively, for $\nu = n - 1 = 9$.
3. The interval is defined by

$$\chi_l^2 \leq \frac{(n-1)s^2}{\sigma^2} \leq \chi_h^2$$

4. Entering the known values,

$$\frac{(n-1)s^2}{\chi_h^2} \leq \sigma^2 \leq \frac{(n-1)s^2}{\chi_l^2}$$

the population variance is calculated to be between 2.6 and 13.

3.3.5 Confidence Interval for the Ratio of Two Variances

The confidence interval for the ratio of two variances is determined using the F distribution, as discussed in Section 3.2.7. Again, the procedure discussed in Section 3.3 is followed, using the F variable defined in Equation (3.15). Also, the F distribution is asymmetric and, as such, will have upper and lower bounds on the F variable that are not simply opposite in sign. When a confidence interval for the ratio of two variances contains unity (1.0), then, subject to the selected level of confidence, the two variances are not different.

Example 3.22

For the two procedures previously analyzed, the standard deviation of the assembly time for procedure 1 was found to be 4.91 s for a sample of size $n_1 = 11$. Procedure 2 had standard deviation 2.54 s for $n_2 = 9$. Determine the 95% confidence interval for the ratio of the population variances, σ_1^2/σ_2^2.

Solution

Following the procedure in Section 3.3, using the F variable and its definition:

1. For a confidence level of 95%, $\alpha = 0.05$.
2. The lower F variable can be denoted $F_{1-\alpha/2,\nu_1,\nu_2}$ and the upper $F_{\alpha/2,\nu_1,\nu_2}$. For these sample sizes, $\nu_1 = 10$ and $\nu_2 = 8$. Reading from Table 3.6 (or using the Microsoft Excel function FINV), $F_{0.025,10,8} = 4.2951$, and utilizing the symmetry rule of Section 3.2.7, $F_{0.975,10,8} = 1/F_{0.025,8,10}$. Again, from Table 3.6, one has $F_{0.975,10,8} = 1/3.8549 = 0.2594$.
3. Replacing the generic variable y with F,

$$F_{1-\alpha/2,\nu_1,\nu_2} \leq \frac{s_1^2/\sigma_1^2}{s_2^2/\sigma_2^2} \leq F_{\alpha/2,\nu_1,\nu_2}$$

4. Rearranging and entering the known values,

$$\frac{s_1^2/s_2^2}{F_{\alpha/2,v_1,v_2}} \leq \frac{\sigma_1^2}{\sigma_2^2} \leq \frac{s_1^2/s_2^2}{F_{1-\alpha/2,v_1,v_2}}$$

$$0.870 \leq \frac{\sigma_1^2}{\sigma_2^2} \leq 14.4$$

Hence, the ratio of population variances falls between 0.870 and 14.4. Since this interval contains unity, the ratio may be unity at a 95% confidence level and, hence, the two population variances do not differ based on this test.

3.3.6 SUMMARY OF CONFIDENCE INTERVALS

The confidence intervals discussed in this subsection are shown in Table 3.11 [2] in two-sided form. These intervals can be converted to one-sided form by removing the appropriate inequality and replacing the remaining $(1 - \alpha/2)$ or $\alpha/2$ term with $(1 - \alpha)$ or α. For example, the generic two-sided confidence interval $y_{\alpha/2} \leq y \leq y_{1-\alpha/2}$ is replaced with $y \leq y_{1-\alpha}$ to define a one-sided interval with an upper bound.

TABLE 3.11
Summary of Several Two-Sided Confidence Intervals

Interval Type	Sample Parameter	Two-Sided Interval with Confidence Level $100(1 - \alpha)\%$
Population mean (μ), variance (σ^2) known	$\bar{x}$	$\bar{x} - z_{\alpha/2}\frac{\sigma}{\sqrt{n}} \leq \mu \leq \bar{x} - z_{1-\alpha/2}\frac{\sigma}{\sqrt{n}}$
Population mean (μ), variance (σ^2) unknown	$\bar{x}$	$\bar{x} - t_{\alpha/2,n-1}\frac{s}{\sqrt{n}} \leq \mu \leq \bar{x} - t_{1-\alpha/2,n-1}\frac{s}{\sqrt{n}}$
Difference between population means $(\mu_1 - \mu_2)$ for $\sigma_1^2 = \sigma_2^2$	$\bar{x}_1 - \bar{x}_2$	$\bar{x}_1 - \bar{x}_2 - t_{\alpha/2,n_1+n_2-2} s_p\sqrt{\frac{1}{n_1}+\frac{1}{n_2}} \leq \mu_1 - \mu_2 \leq \bar{x}_1 - \bar{x}_2 - t_{1-\alpha/2,n_1+n_2-2} s_p\sqrt{\frac{1}{n_1}+\frac{1}{n_2}}$ where $s_p = \sqrt{\frac{(n_1-1)s_1^2 + (n_2-1)s_2^2}{n_1+n_2-2}}$
Difference between population means $(\mu_1 - \mu_2)$ for $\sigma_1^2 \neq \sigma_2^2$	$\bar{x}_1 - \bar{x}_2$	$\bar{x}_1 - \bar{x}_2 - t_{\alpha/2,v}\sqrt{\frac{s_1^2}{n_1}+\frac{s_2^2}{n_2}} \leq \mu_1 - \mu_2 \leq \bar{x}_1 - \bar{x}_2 - t_{1-\alpha/2,v}\sqrt{\frac{s_1^2}{n_1}+\frac{s_2^2}{n_2}}$ where $v = \frac{\left(\frac{s_1^2}{n_1}+\frac{s_2^2}{n_2}\right)^2}{\frac{(s_1^2/n_1)^2}{n_1-1}+\frac{(s_2^2/n_2)^2}{n_2-1}}$
Population variance (σ^2)	s^2	$\frac{(n-1)s^2}{\chi^2_{\alpha/2,n-1}} \leq \sigma^2 \leq \frac{(n-1)s^2}{\chi^2_{1-\alpha/2,n-1}}$
Ratio of population variances (σ_1^2/σ_2^2)	s_1^2/s_2^2	$\frac{s_1^2/s_2^2}{F_{\alpha/2,n_1-1,n_2-1}} \leq \frac{\sigma_1^2}{\sigma_2^2} \leq \frac{s_1^2/s_2^2}{F_{1-\alpha/2,n_1-1,n_2-1}}$

Source: Montgomery, Douglas C., and Runger, George C., *Applied Statistics and Probability for Engineers*, 3rd ed., John Wiley & Sons, New York (2003).

Engineering Statistics

3.4 STATISTICAL HYPOTHESIS TESTING

Statistical tests are available for certain populations using available information; the analyses can be performed at a given confidence level by the methods explained. When an improved mathematical model is available, better conclusions can be realized.

Statisticians often use the following terms: A **null hypothesis** is a statement proposed for test and is identified as H_0. The **alternative hypothesis**, H_1, is the competing statement regarding the data that is true if the null hypothesis is false. If the null hypothesis is correct, the **acceptance region** indicates the correct interval of desired confidence level, or $100(1 - \alpha)\%$, and the **critical region** is the interval(s) in which the null hypothesis is false.

To perform a hypothesis test:

1. Identify the parameter of interest such as the population mean, μ, or the variance, σ^2.
2. State the null hypothesis, H_0. Then identify the alternative hypothesis, H_1. (The alternative hypothesis can be two sided or one sided.)
   Examples:
   a. H_0: The mean reaction rate is 50 mol/l · s; $\mu = 50$ mol/l · s.
      H_1: The mean reaction rate is not 50 mol/l · s (two sided); $\mu \neq 50$ mol/l · s, or
      H_1: The mean reaction rate is less than 50 mol/l · s (one sided); $\mu < 50$ mol/l · s.
   b. H_0: The variance exceeds 10 cm²; $\sigma^2 > 10$ cm².
      H_1: The variances is less than 10 cm²; $\sigma^2 \leq 10$ cm².
3. Select a significance level for α. (It is critical that the significance level be chosen *before* performing the statistical test so that the statistical test does not bias the selection of a significance level, which could result in incorrect conclusions.) For example, $\alpha = 0.05$ (confidence level is 95%) or $\alpha = 0.01$ (confidence level is 99%).
4. Identify the test statistic and associated probability distribution appropriate to the hypothesis.
   Examples:
      Mean with unknown variance: t variable and t distribution
      Two variances: F variable and F distribution
5. Determine the critical region(s) in terms of the test statistic. For example, one-sided t-test: $t \leq t_{\alpha,\nu}$, or two-sided F-test: $F < F_{\alpha/2,\nu_1,\nu_2}$ and $F > F_{1-\alpha/2,\nu_1,\nu_2}$.
6. Calculate the test statistic for the sample data.
7. Accept or reject the null hypothesis H_0 based on the calculated test statistic and the bounds of the critical region.

A new hypothesis may be proposed and the procedure repeated if the null hypothesis is rejected, but the new hypothesis is biased, since it is based on previously tested information. This procedure is utilized in the following examples.

Example 3.23

For a decomposition reaction at constant volume and temperature (see Example 3.26 in Section 3.5.1.1), the concentration of species A follows a first-order exponential decay given by

$$C_A(t) = C_{A0} \exp(-kt)$$

where C_A is the concentration of species A in moles per liter, t is time in seconds, C_{A0} is the concentration of species A at $t = 0$, and k is the rate constant for the particular temperature with units of inverse seconds. In Example 3.26, k is estimated by linear regression to equal 0.503 h⁻¹ with a standard error of 0.022 h⁻¹ with a sample size of $n = 11$. The rate constant was thought to be 0.530 h⁻¹ for the particular temperature. Does a hypothesis test support this value of k at a confidence level of 95%?

Solution

We must determine whether 0.530 s⁻¹ is an acceptable value for the rate constant.

1. The value of interest is the mean value of the rate constant. For this example, the mean value will be denoted μ and is equal to 0.530 h⁻¹. The estimate of this mean is denoted $\bar{x}$ and is given by the measured value, 0.503 h⁻¹. The standard error, $SE(\bar{x})$, is 0.022 h⁻¹.
2. The two hypotheses:
   H_0: $k = 0.530$ h⁻¹ (the given value is correct).
   H_1: $k \neq 0.530$ h⁻¹ (the given value is incorrect).
3. Since the stated confidence level is to be 95%, the significance level is $\alpha = 0.05$.
4. For a mean with unknown population variance, the proper test statistic is the t variable, defined for this case [2] as

$$t = \frac{\bar{x} - \mu}{SE(\bar{x})}$$

5. For a linear regression, the number of degrees of freedom is $\nu = n - 2$ because two degrees of freedom are taken by the fitting parameters (see Section 3.5). Thus, $\nu = 9$. The critical region will be the upper and lower tails of the t distribution that have cumulative probability of 0.025 each. Thus, reading from Table 3.3, the lower t value is $t_{11,0.975} = -2.26$ and the upper t value is $t_{11,0.025} = 2.26$. Thus, the critical region is defined by

$$t < -2.26 \quad \text{and} \quad t > 2.26$$

Since this hypothesis test is two sided, there are two regions in which the null hypothesis could be rejected.
6. The t variable is

$$t = \frac{\bar{x} - \mu}{SE(\bar{x})} = \frac{0.503\ h^{-1} - 0.530\ h^{-1}}{0.022\ h^{-1}} = -1.23$$

Since the t variable is neither greater than 2.26 nor less than −2.26, and therefore, falls outside the critical region, the null hypothesis should be accepted. The alternative hypothesis is rejected. At a confidence level of 95%, one may state that the mean rate constant k is 0.530 s⁻¹.

Example 3.24

During a polymerization reaction, the viscosity of a reactor effluent is measured periodically to determine the reaction progress. The measurement is repeated so that $n = 6$. If the true variance exceeds 0.8 cSt² (cSt = centistokes), the measurements are repeated to achieve better precision. For this sample, the mean viscosity was determined to be 106.4 cSt with variance 1.9 cSt². At a confidence level of 95%, does the variance exceed the desired variability?

Solution

1. The value of interest is the population variance σ^2, which is estimated for this sample of $n = 6$ with sample variance $s^2 = 1.9$ cSt².
2. The hypotheses are
   H_0: $\sigma^2 \leq 0.8$ cSt².
   H_1: $\sigma^2 > 0.8$ cSt².
3. For a 95% confidence level, the significance level is $\alpha = 0.05$.

4. Using chi-square statistics, the test statistic [2] is

$$\chi^2 = \frac{(n-1)s^2}{\sigma^2}$$

5. The number of degrees of freedom for this sample is $\nu = n - 1 = 5$. For this one-sided test, this hypothesis should be rejected if the chi-square test statistic lies in the upper tail of the chi-square distribution for $\nu = 4$ with cumulative probability equal to 0.05. Reading from Table 3.4, this tail begins at $\chi^2_{0.05,5} = 11.1$. The critical region is therefore $\chi^2 > 11.1$.
6. For this sample, the chi-square test statistic is

$$\chi^2 = \frac{(n-1)s^2}{\sigma^2} = \frac{6(5-1)\cdot 1.9 \text{ cSt}^2}{0.8 \text{ cSt}^2} = 11.9$$

7. This chi-square value does not lie in the critical region. Therefore, at a confidence level of 95%, the variance does exceed 0.8 cSt². Therefore, the null hypothesis is rejected and the alternative hypothesis is accepted. The measurements should be repeated in order to reduce the measurement variability.

The previous two examples (3.23 and 3.24) implement the hypothesis-testing procedure for two straightforward statistical tests. Hypothesis testing may also be extended to the difference between the means of two samples or the ratio of two variances. The relevant test statistics for these calculations were presented in Section 3.3 of this chapter. The NIST/SEMATECH handbook [1] discusses other hypothesis tests in its Section 7, particularly Subsections 7.2 through 7.4. The textbook *Applied Statistics and Probability for Engineers*, by Montgomery and Runger [2], also presents various tests.

3.5 LEAST SQUARES REGRESSION

Regression analysis is often employed to fit experimental data to a mathematical model. The purpose may be to determine physical properties or constants (e.g., rate constants, transport coefficients), to discriminate between proposed models, to interpolate or extrapolate data, etc. The model should provide estimates of the uncertainty in calculations from the resulting model and, if possible, make use of available error in the data. An initial model (or models) may be empirical, but with advanced knowledge of reactors, distillation columns, other separation devices, heat exchangers, etc., more sophisticated and fundamental models can be employed. As a starting point, a linear equation with a single independent variable may be initially chosen. Of importance, is the mathematical model linear? In general, a function, f, of a set of adjustable parameters, $\{\beta_j\}$, is linear if a derivative of that function with respect to any adjustable parameter is not itself a function of any other adjustable parameter, that is,

$$\frac{\partial^2 f}{\partial \beta_j \partial \beta_k} = 0 \qquad (3.24)$$

Some functions do not satisfy the test of Equation (3.24), but by suitable manipulation they can become linear and are termed *intrinsically* linear. Linear regression can be generalized to a straightforward analytic procedure, while nonlinear regression generally has no unique solution.

For both regression types, one begins with a set of independent variables $\mathbf{X}_i$ (bold denoting vector notation); a measured, dependent response y_i; and a regression function $\hat{y}(\mathbf{X}_i)$ with adjustable parameters $\{\beta_j\}$. Each response may be assumed to be distributed normally with variance σ_i^2. If one makes the assumption that each measurement is independent, then the joint probability of the set of n data points [2] is

$$P = \prod_{i=1}^{n}\left[\frac{1}{\sqrt{2\sigma_i^2}} \exp\left(\frac{-[y_i - \hat{y}(\mathbf{X}_i)]^2}{2\sigma_i^2}\right)\right] \qquad (3.25)$$

The parameters that produce the best model maximize P. One can alternatively maximize $\ln P$ instead of P and still arrive at the maximum probability. Therefore, Equation (3.25) is transformed to

$$\ln P = \ln\left(\prod_{i=1}^{n}\frac{1}{\sqrt{2\sigma_i^2}}\right) - \frac{1}{2}\sum_{i=1}^{n}\frac{[y_i - \hat{y}(\mathbf{X}_i)]^2}{\sigma_i^2} \qquad (3.26)$$

Since the first term on the right-hand side does not depend on the adjustable parameters, maximizing P is equivalent to minimizing the summation in Equation (3.26). The optimal solution occurs when, for all β_j,

$$\frac{\partial}{\partial \beta_j}\left(\sum_{i=1}^{n}\frac{[y_i - \hat{y}(\mathbf{X}_i)]^2}{\sigma_i^2}\right) = 0 \qquad (3.27)$$

One usually does not have values or reliable estimates of σ_i^2. Hence, the assumption that the variance of each y_i is uniform, i.e., $\sigma_i = \sigma$, is made so that Equation (3.27) reduces to

$$\frac{\partial}{\partial \beta_j}\left(\sum_{i=1}^{n}[y_i - \hat{y}(\mathbf{X}_i)]^2\right) = 0 \qquad (3.28)$$

When reliable σ_i are available, the adjustable parameters may be obtained from Equation (3.27), but the derivation is more involved than the simpler case of uniform variance. More information regarding the full derivation is provided by Press et al. [10]. The remainder of this section focuses on regression analyses with the assumption of constant variance, with separate discussions of simple linear, generalized multiple linear, and nonlinear regression.

3.5.1 SIMPLE LINEAR LEAST SQUARES REGRESSION

Linear least squares regression is the most common method of fitting a response that is a function of a single independent variable. Many nonlinear functions may be transformed to simple linear functions, extending the capabilities of the simplest regression algorithm.

3.5.1.1 Basic Algorithm

For linear least squares, the following model is entered into the equations of the previous subsection:

$$\hat{y}(x) = \beta_0 + \beta_1 x \qquad (3.29)$$

With this chosen model, the adjustable parameters are obtained by simultaneously solving the two equations that result from Equation (3.28). If the chosen model is intrinsically linear, it must first be transformed to the form of Equation (3.29) before proceeding. Defining the following terms (SS denotes "sums of squares") simplifies further analysis [10]:

$$SS_{xx} = \sum_{i=1}^{n}(x_i - \bar{x})^2 = \sum_{i=1}^{n} x_i^2 - n \cdot \bar{x}^2 \quad (3.30)$$

$$SS_{yy} = \sum_{i=1}^{n}(y_i - \bar{y})^2 = \sum_{i=1}^{n} y_i^2 - n \cdot \bar{y}^2 \quad (3.31)$$

$$SS_{xy} = \sum_{i=1}^{n}(x_i - \bar{x})(y_i - \bar{y}) = \sum_{i=1}^{n} x_i y_i - n \cdot \bar{x} \cdot \bar{y} \quad (3.32)$$

Using these terms, the values of the adjustable parameters that maximize P are

$$\beta_1 = \frac{SS_{xy}}{SS_{xx}} \quad (3.33)$$

$$\beta_0 = \bar{y} - \beta_1 \bar{x} = \bar{y} - \frac{SS_{xy}}{SS_{xx}} \bar{x} \quad (3.34)$$

Example 3.25
Consider the general mathematical model for a first-order exponential reaction as follows:

$$C_A(t) = C_{A0} \exp(-kt) \quad (3.35)$$

where C_A is the concentration of species A, k is the rate constant, and t is time. C_{A0} denotes the initial concentration. Transform this model to an intrinsically linear equation, with which one could determine k using linear regression.

Solution
Taking a logarithm of Equation (3.35) linearizes the equation:

$$\ln C_A(t) = \ln C_{A0} - kt \quad (3.36)$$

With Equation (3.29), $y = \ln C_A(t)$, $x = t$, $\beta_0 = \ln C_{A0}$, and $\beta_1 = -k$. One could use experimental measurements of C_A versus t to estimate the initial concentration and decay constant.

Example 3.26
For the decomposition of A according to the chemical reaction

$$A \rightarrow 2B$$

the rate of reaction at constant volume and temperature follows a first-order exponential decay given by

$$C_A(t) = C_{A0} \exp(-kt)$$

and the variables are as identified in Example 3.25. The table below lists C_A as a function of t that was measured experimentally. Determine k via a linear regression.

t (h)	$C_A(t)$ (mol/l)	$y_i = \ln C_A(t)$
0.0	7.96	2.07
1.0	4.95	1.60
2.0	3.67	1.30
3.0	2.10	0.742
4.0	1.73	0.548
5.0	0.637	−0.451
6.0	0.566	−0.569
7.0	0.224	−1.50
8.0	0.202	−1.60
9.0	0.0688	−2.68
10.0	0.0670	−2.70

Solution

The above equation is identical in form to Equation (3.35). Therefore, its intrinsically linear form is

$$\ln C_A(t) = \ln C_{A0} - kt$$

As in Example 3.25, let $y = C_A$ and $x = t$. Again, $\beta_0 = \ln C_{A0}$ and $\beta_1 = -k$. The new y values are listed in the table, and the x values are identical to those in the t column. First, the average values of x and y are $\bar{x} = 5.0$ h and $\bar{y} = -0.273$ and $n = 11$. Then, following Equations (3.30)–(3.32), the sums of squares are

$$SS_{xx} = \sum_{i=1}^{n} x_i^2 - n \cdot \bar{x}^2 = 110 \text{ h}^2$$

$$SS_{yy} = \sum_{i=1}^{n} y_i^2 - n \cdot \bar{y}^2 = 28.2$$

$$SS_{xy} = \sum_{i=1}^{n} x_i y_i - n \cdot \bar{x} \cdot \bar{y} = -55.3 \text{ h}$$

The regression coefficients, following Equations (3.33) and (3.34), are

$$\beta_1 = \frac{SS_{xy}}{SS_{xx}} = -0.503 \text{ h}^{-1}$$

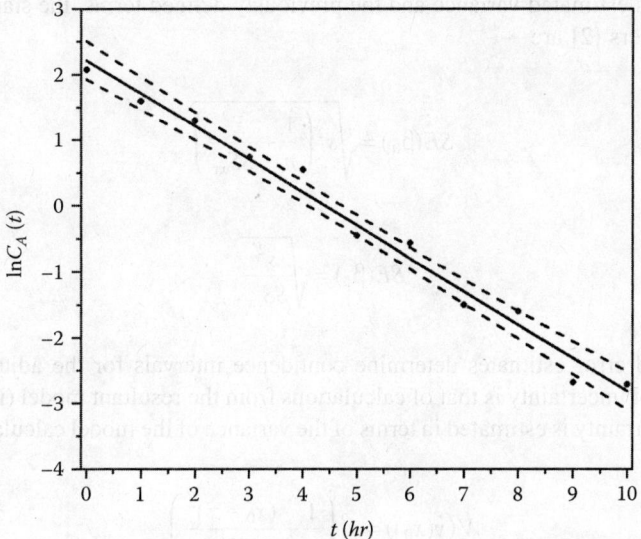

FIGURE 3.1 Linear regression for Example 3.25. Filled circles represent the experimental data and the linear model is shown with the solid line. The dashed lines mark the bounds of the 95% confidence interval on predictions for the linear model.

$$\beta_0 = \bar{y} - \frac{SS_{xy}}{SS_{xx}}\bar{x} = 2.219$$

Therefore, the reaction rate constant is

$$k = -\beta_1 = 0.503 \text{ h}^{-1}$$

Figure 3.1 indicates that the linear equation tested provides a good correlation of the experimental data.

3.5.1.2 Estimation of Uncertainty

Uncertainty in the linear regression is estimated by determining the standard errors of adjustable parameters and predictions from the linear model. With a reliable estimate of the variance of the response variable, σ^2, the known value is used for uncertainty predictions. Otherwise, the variance is estimated in terms of the sum of square residuals. The residual at each measurement is defined as

$$e_i = y_i - \hat{y}(x_i) = y_i - \beta_0 - \beta_1 x_i \tag{3.37}$$

In terms of the sum of the square residuals, the variance of the model can be estimated [2] by

$$s^2 = \frac{\sum_{i=1}^{n} e_i^2}{n-2} = \frac{SS_{yy} - \frac{(SS_{yy})^2}{SS_{xx}}}{n-2} \tag{3.38}$$

In terms of this estimated variance and the previously defined terms, the standard errors of the adjustable parameters [2] are

$$SE(\beta_0) = \sqrt{s^2\left(\frac{1}{n} + \frac{\bar{x}^2}{SS_{xx}}\right)} \tag{3.39}$$

$$SE(\beta_1) = \sqrt{\frac{s^2}{SS_{xx}}} \tag{3.40}$$

These standard error estimates determine confidence intervals for the adjustable parameters. Another measure of uncertainty is that of calculations from the resultant model (i.e., the uncertainty of $\hat{y}(x)$). This uncertainty is estimated in terms of the variance of the model calculations [2], given by

$$V(\hat{y}(x_0)) = s^2\left(\frac{1}{n} + \frac{(x_0 - \bar{x})^2}{SS_{xx}}\right) \tag{3.41}$$

where x_0 is the value of the dependent variable at which the model prediction and associated variance are calculated. Note that this variance depends on the distance between x_0 from the mean x value. The variance is wider for points far from the mean $\bar{x}$. This variance allows for the determination of a confidence interval of the true value of y in terms of the model parameters [2]

$$\beta_0 + \beta_1 x_0 - t_{\alpha/2, n-2}\sqrt{s^2\left(\frac{1}{n} + \frac{(x_0 - \bar{x})^2}{SS_{xx}}\right)} \le y \le \beta_0 + \beta_1 x_0 + t_{\alpha/2, n-2}\sqrt{s^2\left(\frac{1}{n} + \frac{(x_0 - \bar{x})^2}{SS_{xx}}\right)} \tag{3.42}$$

where α denotes the significance level.

The accuracy of the regression is often measured by the coefficient of determination (R^2) defined [2] as follows:

$$R^2 = 1 - \frac{\sum_{i=1}^{n} e_i^2}{SS_{yy}} \tag{3.43}$$

Mathematically, R^2 measures the proportion of the error in $\hat{y}(x)$ arising from the regression. When R^2 is nearly equal to unity, the model predicts the experimental data points well. It does not, however, mean that the model represents the true data trend. To evaluate the model, examine the plot comparing the predicted and experimental values.

Example 3.27

For the concentration-versus-time data given in Example 3.26, determine the standard errors of the regression parameters β_0 and β_1 and the R^2 value. Plot the 95% confidence intervals for calculations from the model.

Solution

Entering the known values from Example 3.25 into Equations (3.38)–(3.40) and (3.43),

$$s^2 = \frac{\sum_{i=1}^{n}(y_i - \beta_0 - \beta_1 x_i)^2}{n-2} = 0.0515$$

$$SE(\beta_0) = \sqrt{s^2\left(\frac{1}{n} + \frac{\bar{x}^2}{SS_{xx}}\right)} = 0.128$$

$$SE(\beta_1) = \sqrt{\frac{s^2}{SS_{xx}}} = 0.022 \text{ h}^{-1}$$

$$R^2 = 1 - \frac{\sum_{i=1}^{n} e_i^2}{SS_{yy}} = 0.984$$

Because $\beta_1 = -k$, the standard error of the reaction constant is, then

$$SE(k) = SE(\beta_1) = 0.022 \text{ h}^{-1}$$

The upper and lower bounds of the 95% confidence interval must be computed for several values of x_0 to provide the 95% confidence interval that surrounds the regression line. As an example, for $x_0 = t = 1$ h, the upper bound is

$$\beta_0 + \beta_1 x_0 - t_{\alpha/2, n-2} \sqrt{s^2 \left(\frac{1}{n} + \frac{(x_0 - \bar{x})^2}{S_{xx}}\right)}$$

$$= 2.22 - 0.503(\text{h}^{-1}) \cdot 1\text{ h} - 2.262 \sqrt{0.0515 \left(\frac{1}{11} + \frac{(1\text{ h} - 5\text{ h})^2}{110 \text{ h}^2}\right)}$$

$$= 1.966$$

This calculation (and the lower-bound calculation) is repeated for a number of data points and plotted in Figure 3.1 as the dashed lines. For this data set, the 95% confidence interval is not narrow; it is wider at its ends than at the center, as expected from Equation (3.41).

3.5.2 GENERALIZED MULTIPLE LINEAR REGRESSION

Linear regression is limited in application. For more complex models, multiple independent variables or even combinations of several independent variables are needed. These "multiple linear" functions are particularly useful for fitting experimental data to empirical models that have no physical basis. Generalized multiple linear regression is usually accomplished via matrix algebra (see [11–13] or a linear algebra text employing matrix algebra).

3.5.2.1 Basic Algorithm

The model consists of a single measured response y_i and k independent variables x_i [2]. The independent variables may be any chosen independent variable or any linear function of the independent variables. For example, if pressure and temperature are independent variables P and T, respectively, then the set of independent variables x_i may be a combination of P, T, $P \cdot T$, P^2, T^2, etc. The model for y_i may be

$$\hat{y}(x_i) = \beta_0 + \beta_1 x_{i1} + \beta_2 x_{i2} + \ldots + \beta_k x_{ik} \tag{3.44}$$

with $\{\beta_j\}$ being the set of $(k+1)$ adjustable parameters. (The $(k+1)$th parameter is the constant term in the model.) Solving for an optimal $\{\beta_j\}$ is accomplished more easily using a vector-matrix technique. The following vectors and matrix are defined:

$$\hat{y} = \begin{bmatrix} \hat{y}(x_1) \\ \hat{y}(x_2) \\ \vdots \\ \hat{y}(x_n) \end{bmatrix} \quad X = \begin{bmatrix} 1 & x_{11} & x_{12} & \cdots & x_{1k} \\ 1 & x_{21} & x_{22} & \cdots & x_{2k} \\ \vdots & \vdots & \vdots & \ddots & \vdots \\ 1 & x_{n1} & x_{n2} & \cdots & x_{nk} \end{bmatrix} \quad \beta = \begin{bmatrix} \beta_0 \\ \beta_1 \\ \vdots \\ \beta_k \end{bmatrix} \quad y = \begin{bmatrix} y_1 \\ y_2 \\ \vdots \\ y_n \end{bmatrix} \quad x_i = \begin{bmatrix} 1 \\ x_{i1} \\ x_{i2} \\ \vdots \\ x_{ik} \end{bmatrix} \tag{3.45}$$

For n data points, Equation (3.44) may be rewritten as

$$\hat{y} = X\beta \tag{3.46}$$

The residual vector is written as

$$e = y - \hat{y} \tag{3.47}$$

Tedious manipulation indicates that the square error, $|e|^2$, is minimized when the set of adjustable parameters [2] is as follows:

$$\beta = (X^T X)^{-1} X^T y \tag{3.48}$$

where the superscript T denotes the transpose vector or matrix and the superscript -1 indicates the inverse matrix.

Example 3.28

For specific heat (C_p)-vs.-temperature (T) data, the following model is often employed:

$$\hat{C}_p(T) = \beta_0 + \beta_1 T + \beta_2 T^2$$

Determine its coefficients for the data in the following table:

T (K)	C_p (J/mol·K)
200	34.96
300	35.94
400	37.32
500	38.21
600	39.46

Solution

The independent "variables" are $x_1 = T$ and $x_2 = T^2$, and the dependent response is $y = C_p$. The $\mathbf{y}$ vector and $\mathbf{X}$ matrix are, then,

$$\mathbf{y} = \begin{bmatrix} 34.96 \\ 35.94 \\ 37.32 \\ 38.21 \\ 39.46 \end{bmatrix} \frac{\text{J}}{\text{mol} \cdot \text{K}} \quad \text{and} \quad \mathbf{X} = \begin{bmatrix} 1 & 200\text{ K} & 40,000\text{ K}^2 \\ 1 & 300\text{ K} & 90,000\text{ K}^2 \\ 1 & 400\text{ K} & 160,000\text{ K}^2 \\ 1 & 500\text{ K} & 250,000\text{ K}^2 \\ 1 & 600\text{ K} & 360,000\text{ K}^2 \end{bmatrix}$$

Substituting $\mathbf{y}$ and $\mathbf{X}$ into Equation (3.48), the optimal set of parameters is found to be

$$\beta_0 = 32.7 \frac{\text{J}}{\text{mol} \cdot \text{K}}$$

$$\beta_1 = 1.01 \times 10^{-2} \frac{\text{J}}{\text{mol} \cdot \text{K}^2}$$

$$\beta_2 = 3.57 \times 10^{-7} \frac{\text{J}}{\text{mol} \cdot \text{K}^3}$$

3.5.2.2 Estimation of Uncertainty

The uncertainty of a multiple linear model can be estimated with a reliable estimate of the variance of y; a σ^2 value can also be used for uncertainty predictions. Otherwise, one requires an approximation based on the sum of square residuals. The residual vector for a multiple linear model was previously given in Equation (3.47). In terms of $\mathbf{e}$ and other previously defined quantities, the variance may be estimated [2] by

$$s^2 = \frac{\mathbf{e}^T \mathbf{e}}{n-k-1} = \frac{\mathbf{y}^T \mathbf{y} - \boldsymbol{\beta}^T \mathbf{X}^T \mathbf{y}}{n-k-1} = \frac{\sum_{i=1}^{n}(y_i - \hat{y}(\mathbf{x}_i))^2}{n-k-1} \tag{3.49}$$

All other uncertainty quantities are given in terms of the covariance matrix $\mathbf{C}$, defined [2] by

$$\mathbf{C} = (\mathbf{X}^T \mathbf{X})^{-1} = \begin{bmatrix} C_{00} & C_{01} & \cdots & C_{0k} \\ C_{10} & C_{11} & \cdots & C_{1k} \\ \vdots & \vdots & \ddots & \vdots \\ C_{k0} & C_{k1} & \cdots & C_{kk} \end{bmatrix} \tag{3.50}$$

which is a symmetric matrix, i.e., $C_{ij} = C_{ji}$. In terms of the elements of $\mathbf{C}$, the estimated standard error of the jth adjustable parameter [2] is

$$SE(\beta_j) = \sqrt{s^2 C_{jj}} \tag{3.51}$$

and the covariance of any two adjustable parameters [2] is

$$\text{cov}(\beta_i, \beta_j) = s^2 C_{ij} \quad i \neq j \qquad (3.52)$$

Predictions from the model [2] have variance

$$V(\hat{y}(\mathbf{x}_0)) = s^2 \mathbf{x}_0^T \mathbf{C} \mathbf{x}_0 \qquad (3.53)$$

The confidence interval for predictions from the model is

$$\hat{y}(\mathbf{x}_0) - t_{\alpha/2, n-k-1}\sqrt{s^2 \mathbf{x}_0^T \mathbf{C} \mathbf{x}_0} \leq y \leq \hat{y}(\mathbf{x}_0) + t_{\alpha/2, n-k-1}\sqrt{s^2 \mathbf{x}_0^T \mathbf{C} \mathbf{x}_0} \qquad (3.54)$$

Model adequacy may again be measured using R^2 (coefficient of determination), which is given by Equation (3.43). The interpretation of R^2 for multiple linear regression is identical to that used in the simple linear version of least squares.

Example 3.29

Referring to Example 3.28, where C_p was fit to a polynomial function of T, determine the standard error of each regressor β_j and the variance of the model's prediction for $T = 200$ K.

Solution

To determine the standard error of each β_j, one first needs the covariance matrix and s^2. Entering the $\mathbf{X}$ matrix determined in Example 3.28 into Equation (3.50), the covariance matrix $\mathbf{C}$ is

$$\mathbf{C} = (\mathbf{X}^T \mathbf{X})^{-1} = \begin{bmatrix} 15.8 & -8.40 \times 10^{-2} \text{K}^{-1} & 1.00 \times 10^{-4} \text{K}^{-2} \\ -8.40 \times 10^{-2} \text{K}^{-1} & 4.67 \times 10^{-4} \text{K}^{-2} & -5.71 \times 10^{-7} \text{K}^{-3} \\ 1.00 \times 10^{-4} \text{K}^{-2} & -5.71 \times 10^{-7} \text{K}^{-3} & 7.14 \times 10^{-10} \text{K}^{-4} \end{bmatrix}$$

Following Equation (3.49), the variance estimator is

$$s^2 = \frac{\mathbf{y}^T \mathbf{y} - \boldsymbol{\beta}^T \mathbf{X}^T \mathbf{y}}{n - k - 1} = \frac{0.0434 \left(\dfrac{\text{J}}{\text{mol} \cdot \text{K}}\right)^2}{5 - 2 - 1} = 0.0217 \left(\frac{\text{J}}{\text{mol} \cdot \text{K}}\right)^2$$

Therefore, via Equation (3.51), the standard errors of the adjustable parameters are

$$SE(\beta_0) = \sqrt{s^2 C_{00}} = 0.586 \frac{\text{J}}{\text{mol} \cdot \text{K}}$$

$$SE(\beta_1) = \sqrt{s^2 C_{11}} = 3.18 \times 10^{-3} \frac{\text{J}}{\text{mol} \cdot \text{K}^2}$$

$$SE(\beta_2) = \sqrt{s^2 C_{22}} = 3.94 \times 10^{-6} \frac{\text{J}}{\text{mol} \cdot \text{K}^3}$$

For the data at $T = 200$ K, the $\mathbf{x}_0$ vector (a column vector identical to the first row of $\mathbf{X}$; see Equation (3.45)) is

$$\mathbf{x}_0 = \begin{bmatrix} 1 \\ 200 \text{ K} \\ 40{,}000 \text{ K}^2 \end{bmatrix}$$

Then, by Equation (3.53), the variance of the model prediction at $T = 200$ K is

$$V(\hat{y}(\mathbf{x}_0)) = s^2 \mathbf{x}_0^T \mathbf{C} \mathbf{x}_0 = 0.886 \left(\frac{\text{J}}{\text{mol} \cdot \text{K}} \right)^2$$

The variance estimation could be repeated for multiple $\mathbf{x}_0$ points to determine confidence intervals for predictions from the model.

3.5.3 Analysis of Variance

Analysis of variance (ANOVA) is a method of testing a regression. With a regression model, a simplified hypothesis test examines the forms of uncertainty; some uncertainty arises from the regression, while some uncertainty is unexplained by the model. One determines whether the regression model is significant by comparing the significance of each uncertainty source. First, the following error terms [2] are defined:

$$SS_T = \mathbf{y}^T \mathbf{y} - n(\bar{y})^2 \tag{3.55}$$

$$SS_R = \hat{\mathbf{y}}^T \mathbf{y} - n(\bar{y})^2 \tag{3.56}$$

$$SS_E = \mathbf{y}^T \mathbf{y} - \hat{\mathbf{y}}^T \mathbf{y} \tag{3.57}$$

where SS_T denotes the *total corrected sum of squares*, while SS_R and SS_E denote the *regression sum of squares* and *error sum of squares*, respectively. It is straightforward to show that $SS_T = SS_R + SS_E$. Second, the number of degrees of freedom of each sum-of-squares term is determined. The number of degrees of freedom of SS_T is equal to $n - 1$. SS_R and SS_E have degrees of freedom k (k being the number of regressors) and $n - k - 1$, respectively. Third, the *mean square* terms, MS_R and MS_E, are calculated by dividing the SS_R and SS_E terms by their respective number of degrees of freedom [2]:

$$MS_R = \frac{SS_R}{k} \tag{3.58}$$

$$MS_E = \frac{SS_E}{n - k - 1} \tag{3.59}$$

Lastly, the ratio MS_R/MS_E is computed and denoted F_0. Since F_0 is a ratio of two variance terms, it is distributed according to an F distribution with degrees of freedom $v_1 = k$ and $v_2 = n - k - 1$. Thus, if that ratio exceeds a predetermined critical F value, the regression is significant. Equivalently, if the cumulative probability that the F variable exceeds the given ratio (usually

TABLE 3.12
Table Form Summary of ANOVA Procedure

Source of Variance	Degrees of Freedom	Sum of Squares	Mean Square	F_0	P
Regression	k	SS_R	$MS_R = SS_R/k$	MS_R/MS_E	$P(F > F_0)$
Error	$n-k-1$	SS_E	$MS_E = SS_E/(n-k-1)$	...	...
Total	$n-1$	SS_T	...	...	...

Source: Montgomery, Douglas C., and Runger, George C., *Applied Statistics and Probability for Engineers*, 3rd ed., John Wiley & Sons, New York (2003).

denoted P) is smaller than the significance level α, the regression is significant. The ANOVA procedure is usually presented as a table, such as the summary given in Table 3.12.

Example 3.30
With a significance level of α = 0.05 (95% confidence level), use an ANOVA table to determine whether the multiple linear regression model determined in Example 3.28 is significant.

Solution
For the multiple linear regression in Example 3.28, there were five data points and three regressors for n = 5 and k = 2. From the model and basis data in that example, the following sums of squares were computed via Equations (3.55 through 3.57):

For example,

$$SS_T = \mathbf{y}^T\mathbf{y} - n(\bar{y})^2 = [34.96 \quad 35.94 \quad 37.32 \quad 38.21 \quad 39.46] \begin{bmatrix} 34.96 \\ 35.94 \\ 37.32 \\ 38.21 \\ 39.46 \end{bmatrix} \left(\frac{J}{mol \cdot K}\right)^2$$

$$- 5\left(\frac{34.96 + 35.94 + 37.32 + 38.21 + 39.46}{5} \frac{J}{mol \cdot K}\right)^2 = 12.744 \left(\frac{J}{mol \cdot K}\right)^2$$

Then,

$$SS_T = 12.744 \left(\frac{J}{mol \cdot K}\right)^2$$

$$SS_R = 12.701 \left(\frac{J}{mol \cdot K}\right)^2$$

$$SS_E = 0.043 \left(\frac{J}{mol \cdot K}\right)^2$$

Computing the mean square terms and entering the data into an ANOVA table yields

Source of Variance	Degrees of Freedom	Sum of Squares	Mean Square	F_0	P
Regression	2	12.744	6.351	292.67	3.41×10^{-3}
Error	2	0.043	0.0217	...	...
Total	4	12.701	...	...	...

Since the P value of the ANOVA test does not exceed the chosen significance level $\alpha = 0.05$, the regression is deemed significant. For this example, the FDIST function in Microsoft Excel [5, 6] was used to determine the P value, 3.41×10^{-3}, of the ANOVA table (see Table 3.12).

3.5.4 Nonlinear Regression

Since no general algorithm or procedure is available for least squares regression when the mathematical fitting function is nonlinear, one begins with Equation (3.26) and assumes that $\sigma_i = \sigma$ (uniform variance); the model probability will be maximized when the sum of square error (e.g., Equation (3.57)) is minimized. Completing the regression then becomes a problem of functional minimization. Algorithms for functional minimization in literature sources include the quasi-Newton method, the Levenburg-Marquardt algorithm, or the method of steepest descent [10]. Software such as JMP (produced by SAS) or the Solver feature of Microsoft Excel may be helpful. One starts with an estimate (ideally based on intuition) and repeats the calculations until some solution criterion is satisfied. Common to all the techniques, though, is the fact that functional minimization techniques are not guaranteed to produce the *global* minimum of the objective function. The solution produced by such an algorithm could very well be a *local* minimum in the parameter space, unless the entire parameter space has been thoroughly searched. If multiple local minima exist, then the algorithm could arrive at any minimum, depending on the starting guess. Although nonlinear regression is difficult to implement, it remains useful.

3.6 STATISTICAL ANALYSIS OF ERROR PROPAGATION

Often, the quantity of interest in an experiment is not measured directly, but is computed via a mathematical equation or model. For example, the overall heat-transfer coefficient (say, U) of a specific heat exchanger might be determined indirectly by measuring the inlet and outlet temperatures. Repeated experiments provide an estimate of the variance of U, but this variance does not account for possible experimental errors (e.g., the thermocouple errors in the temperature measurements). The error in each measurement accumulates in the overall error of a calculated quantity in a manner known as *propagation of error*. Measurement error can arise from random variability or instrument sensitivity. There is, however, a mathematical approach to deal with these errors.

Suppose that a number of variables, $(x_1, x_2, \ldots, x_n)$, are measured and used to calculate a quantity y according to the relationship

$$y(x_1, x_2, \ldots, x_n) = a_1 x_1 + a_2 x_2 + \cdots + a_n x_n = \sum_{i=1}^{n} a_i x_i \qquad (3.60)$$

For this mathematical model, the variance of the computed value of y is given by Barry [14]

$$\sigma^2(y) = \sum_{i=1}^{n} a_i^2 \sigma^2(x_i) + 2\sum_{i=1}^{n}\sum_{j=i+1}^{n} a_i a_j Cov(x_i, x_j) \tag{3.61}$$

where $\sigma^2(x_i)$ is the variance of the ith measured variable, and $Cov(x_i, x_j)$ is the covariance of the i–j pair of measurements. The covariance term in Equation (3.61) allows for the measured variables $(x_1, x_2, \ldots, x_n)$ to be correlated. The measured variables are often uncorrelated, which allows for removing the second summation term in Equation (3.61). If the measured variables are indeed correlated, the covariance relationship between two variables may be unknown, seemingly leaving Equation (3.61) useless. One may, however, simplify the problem by assuming that the covariance terms are zero, reducing Equation (3.61) to

$$\sigma^2(y) = \sum_{i=1}^{n} a_i^2 \sigma^2(x_i) \tag{3.62}$$

Thus, for a computed value that is linearly dependent on all measured variables, the variance of the computed value can be predicted in a straightforward fashion from the variance of the individual measured variables.

Since the relationship between a computed quantity and measured variables is often nonlinear, an approximate method is needed to calculate the variance of the computed quantity. Suppose that a quantity y is given by a nonlinear function and that the individual measured quantities $(x_1, x_2, \ldots, x_n)$ are independent. (The assumption of independence removes the need for covariance terms.) Then the variance of y may be approximated [15] as

$$\sigma^2(y) = \left(\frac{\partial y}{\partial x_1}\bigg|_{\{x_j\}}\right)^2 \sigma^2(x_1) + \left(\frac{\partial y}{\partial x_2}\bigg|_{\{x_j\}}\right)^2 \sigma^2(x_2) + \cdots + \left(\frac{\partial y}{\partial x_n}\bigg|_{\{x_j\}}\right)^2 \sigma^2(x_n) = \sum_{i=1}^{n}\left(\frac{\partial y}{\partial x_i}\bigg|_{\{x_j\}}\right)^2 \sigma^2(x_i) \tag{3.63}$$

where $\{x_j\}$ indicates that the partial derivatives of y are evaluated at the current set of measured variables $(x_1, x_2, \ldots, x_n)$. (Equation (3.63) reduces to Equation (3.62) for the case of a linear function. Despite this similarity, Equation (3.63) is still an approximation [15].)

With Equations (3.62) and (3.63), one can determine estimates of the variance of a calculated value from the variance of measured variables. The individual variance terms, $\sigma^2(x_i)$, might not be known, however, and one must make estimates. Through repeated experiments, $\sigma^2(x_i)$ could be estimated by $s^2(x_i)$, the sample variance of variable x_i. If one has information about the sensitivity of a measuring device, this previous knowledge could be input as a variance term.

Example 3.31

For a fully developed laminar flow in a pipe, the Hagen-Poiseuille equation is used to calculate the pressure drop per unit length ($\Delta P/L$) at a given volumetric flow rate $\dot{V}$, pipe diameter D, and fluid viscosity μ:

$$\left(\frac{\Delta P}{L}\right) = \frac{\pi D^4}{128\mu \dot{V}} \tag{3.64}$$

Engineering Statistics

From this equation, determine an expression for the variance of $(\Delta P/L)$ in terms of the variances of D, $\dot{V}$, and μ.

Solution

In terms of Equation (3.63), $y = (\Delta P/L)$, and the x variables are $x_1 = D$, $x_2 = \dot{V}$, and $x_3 = \mu$. Therefore, Equation (3.63) becomes

$$\sigma^2\left(\frac{\Delta P}{L}\right) = \left(\frac{\partial}{\partial D}\left(\frac{\Delta P}{L}\right)\bigg|_{\{D,\dot{V},\mu\}}\right)^2 \sigma^2(D) + \left(\frac{\partial}{\partial \dot{V}}\left(\frac{\Delta P}{L}\right)\bigg|_{\{D,\dot{V},\mu\}}\right)^2 \sigma^2(\dot{V}) + \left(\frac{\partial}{\partial \mu}\left(\frac{\Delta P}{L}\right)\bigg|_{\{D,\dot{V},\mu\}}\right)^2 \sigma^2(\mu)$$

For the Hagen-Poiseuille equation (Equation (3.64)), the above becomes

$$\sigma^2\left(\frac{\Delta P}{L}\right) = \left(\frac{4\pi D^3}{128\mu \dot{V}}\right)^2 \sigma^2(D) + \left(\frac{-\pi D^4}{128\mu \dot{V}^2}\right)^2 \sigma^2(\dot{V}) + \left(\frac{-\pi D^4}{128\mu^2 \dot{V}}\right)^2 \sigma^2(\mu)$$

With knowledge of the variances of the measured variables, one can compute the variance of the pressure drop per unit length. The computed variance can then be used to determine confidence intervals or perform hypothesis tests.

3.7 DESIGN OF EXPERIMENTS

Statistics has an increasing role in the design of experiments and equipment. One first evaluates available information relative to proposed experiments or the equipment. The process being considered could be chemical reactions or reactors; separation processes, including distillation, absorption, stripping, adsorption, etc.; and heat exchangers, etc.

To employ statistics, the first step is to develop a mathematical model for the unit or process. With a well-established process that has been operating for a considerable period of time, a mathematical model that is highly fundamental based on well-determined steps or reactions can often be developed. For example, Hougen and Watson [16] many years ago showed how the numerous steps in catalytic reactions can be modeled in a relatively fundamental manner. For distillation units, relatively fundamental models can generally be developed to indicate how changes of the following affect operations: reflux ratio, process conditions, cost of feed streams, product and feed compositions, demand for product, etc.

With a new process or product, empirical models are often employed in the initial stages. The objective of a research program is to obtain key information as rapidly and cheaply as possible.

3.7.1 Design Matrices

A designed experiment provides a plan of the runs to be performed and the order in which they will be executed. The operating conditions of each run are specified, and replicates are included. The proposed design lists the levels selected for each factor in a given run. It is called the *design matrix* **D**, and is often similar to the **X** matrix of linear regression (see Section 3.5); it does not include the columns of ones for the pure constant term, nor does it include the high-order columns for squares and other functions of the factors. The design matrix is composed of a column vector for each run and is identical to the corresponding column of the **X** matrix for $\beta_j x_j$ terms. It retains this form even if a direct proportionality term is not included in the model. For example, if x enters the model only as $\ln x$ (a logarithm), the design matrix contains a column of x's rather than the $\ln x$.

If the design matrix is in a coded form (i.e., the values of the x_j terms are scaled or changed to a simpler number), the coding equations must be designated.

3.7.2 Design for One-Factor Experiment

The preferred choice of conditions for experiments depends on the model. Some designs for several simple models are discussed next. The experiments are hopefully conducted to minimize the random error of the response, and σ^2 is therefore estimated by s^2.

3.7.2.1 $\hat{y} = \beta_0 = \bar{y}$

For this constant-value model, the only consideration is how many times to replicate, since the conditions are constant. The selection of n can be logically determined from the confidence interval equation. The width of this interval is proportional to $t_{\alpha/2,\nu}$ (the t variable for chosen confidence level and number of runs) and $s/\sqrt{n}$, the standard error of the average. More runs reduce the confidence interval in two ways. Increasing n decreases the standard error, since it is inversely proportional to the square root of n. Also, the value of $t_{\alpha/2,\nu}$ decreases with increasing n as the number of degrees of freedom, $\nu = n - 1$, increases. After selecting n to obtain a desired confidence interval, it is a trial-and-error calculation, since $t_{\alpha/2,\nu}$ depends upon n and is obtained from Table 3.3. Of course, an estimate of the experimental error is needed. An estimate is normally based on knowledge of the process to obtain a preliminary value. After a few runs are completed, an experimental value of s^2 is calculated, and a revised value can be determined.

3.7.2.2 $\hat{y} = \beta_1 x$

For this proportionality model, first determine the confidence interval of the slope, which is a fixed value that is calculated by

$$V(\beta_1) \propto \frac{1}{\sum_{i=1}^{n} x^2} \tag{3.65}$$

since the matrix **C** is 1×1 for this case. Sufficient runs are made to obtain reliable values to minimize the confidence interval. At $x = 1$, the confidence interval on the response is equal to the confidence interval on β_1.

3.7.2.3 $\hat{y} = \beta_0 + \beta_1 x$

The general straight-line model is often used as a starting point for model development. The confidence interval often widens at the ends (see Section 3.5). More levels of the factor act to reduce the confidence interval. Equally spaced levels are often tested. The number of levels depends on the complexity of the model or on the number of additional terms to be considered. Enough levels should be provided for left-out terms. The number of levels should exceed the number of parameters expected. Even if a straight line is likely, more than three levels should be considered, plus some replication.

3.7.3 Design for Several Factors

With several factors to study, the size of the experimental design becomes large, time consuming, and expensive. Two levels are often selected for an exploratory design.

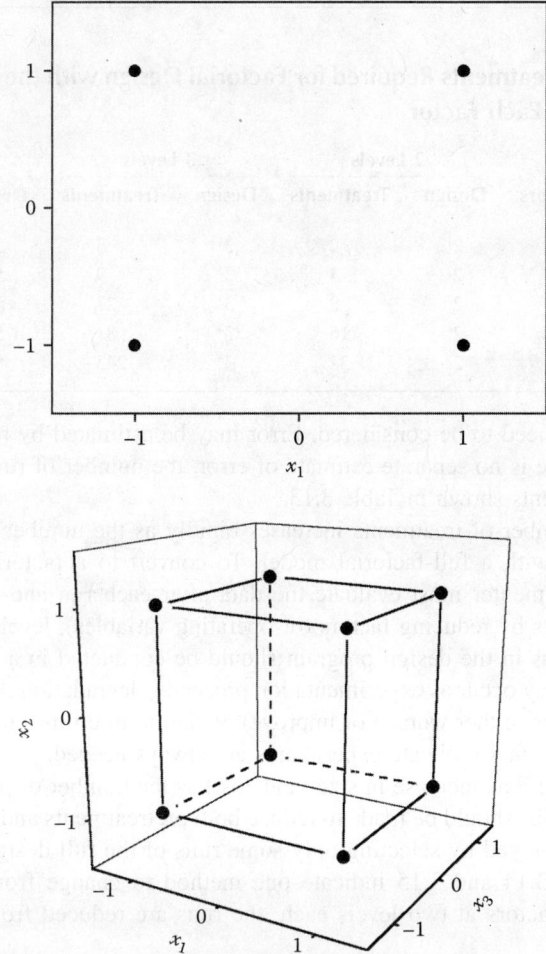

FIGURE 3.2 Two-level factorial design for (a) two factors and (b) three factors.

The arrangement of runs for these designs is illustrated for a full-factorial design. In a complete factorial, all combinations of all levels of the factors are tested. For a two-level factorial design with two factors, there are four treatments, as seen in Figure 3.2(a). The levels of each factor, often called *low* and *high*, are coded as $x = -1$ and $x = +1$, respectively. (Since only 1's enter the matrix when there are two levels of each factor, a short form of the x matrix may show only the − and + signs and omits the 1's.) All combinations of these two-level factors are displayed in *factor space* in Figure 3.2(a). Figure 3.2(b) shows the corresponding factorial design for three factors at two levels each, i.e., a total of eight experiments. The number of treatments required by a factorial design is given by the product of the number of levels l for each factor, 1 to k, or

$$\text{Number of Treatments} = l_1 \times l_2 \times l_3 \times \cdots \times l_k \tag{3.66}$$

For two-level factorial designs, the number of treatments (or experiments) simplifies to 2^k, where k is the number of factors or variables considered. Using only two levels greatly reduces the number of experiments required. Table 3.13 compares required treatments for a factorial design with several levels of each factor. The number of treatments increases both with the number of factors and the number of levels per factor. Five levels of more than two factors is usually prohibitive.

TABLE 3.13
Number of Treatments Required for Factorial Design with the Same Number of Levels for Each Factor

Number of Factors	2 Levels Design	2 Levels Treatments	3 Levels Design	3 Levels Treatments	5 Levels Design	5 Levels Treatments
1	2^1	2	3^1	3	5^1	5
2	2^2	4	3^2	9	5^2	25
3	2^3	8	3^3	27	5^3	125
4	2^4	16	3^4	81	5^4	625
5	2^5	32	3^5	243	5^5	3125

Some replicates also need to be considered. Error may be estimated by replicating only a few of the treatments. If there is no separate estimate of error, the number of runs required must exceed the number of treatments shown in Table 3.13.

As noted, the number of treatments increases rapidly as the number of factors and levels of treatments increases with a full-factorial model. To convert to a factorial design having fewer treatments, the experimenter must evaluate the data after each run and seek how to reduce the number of experiments by reducing factors (or operating variables), levels to be investigated, etc. As a general rule, runs in the design program should be conducted in a random order, since the following problems may occur as experimentation proceeds: degradation of the feed stocks, change in operating techniques (either worsen or improve), variation in utilities used (e.g., temperature of cooling water), etc. Further replicate experiments are always needed.

Experimental programs increase in size (and cost) as the number of pretreatments and factors increase. Considerations should be made to reduce both pretreatments and factors. Partial-factorial designs are often employed by selecting only some runs of the full design and designating them as −1 or +1. Tables 3.14 and 3.15 indicate one method to change from full factorial to half-factorial. With four factors at two levels each, the runs are reduced from 16 to 8 by removing

TABLE 3.14
Full-Factorial Design for Four Factors with Two Levels Each: 2^4 Design

Treatment	I	A	B	C	D	AB	AC	AD	BC	BD	CD	ABC	ABD	ACD	BCD	ABCD
(1)	+1	−1	−1	−1	−1	+1	+1	+1	+1	+1	+1	−1	−1	−1	−1	+1
(2)	+1	−1	−1	−1	+1	+1	+1	−1	+1	−1	−1	−1	+1	+1	+1	−1
(3)	+1	−1	−1	+1	−1	+1	−1	+1	−1	+1	−1	+1	−1	+1	+1	−1
(4)	+1	−1	−1	+1	+1	+1	−1	−1	−1	−1	+1	+1	+1	−1	−1	+1
(5)	+1	−1	+1	−1	−1	−1	+1	+1	−1	−1	+1	+1	+1	−1	+1	−1
(6)	+1	−1	+1	−1	+1	−1	+1	−1	−1	+1	−1	+1	−1	+1	−1	+1
(7)	+1	−1	+1	+1	−1	−1	−1	+1	+1	−1	−1	−1	+1	+1	−1	+1
(8)	+1	−1	+1	+1	+1	−1	−1	−1	+1	+1	+1	−1	−1	−1	+1	−1
(9)	+1	+1	−1	−1	−1	−1	−1	−1	+1	+1	+1	+1	+1	+1	−1	−1
(10)	+1	+1	−1	−1	+1	−1	−1	+1	+1	−1	−1	+1	−1	−1	+1	+1
(11)	+1	+1	−1	+1	−1	−1	+1	−1	−1	+1	−1	−1	+1	−1	+1	+1
(12)	+1	+1	−1	+1	+1	−1	+1	+1	−1	−1	+1	−1	−1	+1	−1	−1
(13)	+1	+1	+1	−1	−1	+1	−1	−1	−1	−1	+1	−1	−1	+1	+1	+1
(14)	+1	+1	+1	−1	+1	+1	−1	+1	−1	+1	−1	−1	+1	−1	−1	−1
(15)	+1	+1	+1	+1	−1	+1	+1	−1	+1	−1	−1	+1	−1	−1	−1	−1
(16)	+1	+1	+1	+1	+1	+1	+1	+1	+1	+1	+1	+1	+1	+1	+1	+1

TABLE 3.15
Example of Half-Factorial Design of Table 3.14 via the D = ABC Alias

Treatment	I	A	B	C	D	AB	AC	AD	BC	BD	CD	ABC	ABD	ACD	BCD	ABCD
(1)	+1	−1	−1	−1	−1	+1	+1	+1								
(4)	+1	−1	−1	+1	+1	+1	−1	−1								
(6)	+1	−1	+1	−1	+1	−1	+1	−1								
(7)	+1	−1	+1	+1	−1	−1	−1	+1								
(10)	+1	+1	−1	−1	+1	−1	−1	+1								
(11)	+1	+1	−1	+1	−1	−1	+1	−1								
(13)	+1	+1	+1	−1	−1	+1	−1	−1								
(16)	+1	+1	+1	+1	+1	+1	+1	+1								

treatments according to the alias **D = ABC**. Treatments in Table 3.14 with this alias (and the equivalent aliases **AD = BC, BD = AC, CD = AB, A = BCD, B = ACD, C = ABC,** and **I = ABCD**) are selected for the half-factorial design; all other treatments are removed. Fewer interactions of factors are investigated, but when interesting results are obtained, more runs are added and, hence, investigated. Montgomery and Runger [2] thoroughly discuss the method of aliasing for reducing factorial designs.

3.7.4 FACTORIAL DESIGN WITH MORE LEVELS

Attempts should always be made to use theoretical mathematical models, but generally only empirical ones need to be employed in the initial investigation. When a factor causes curvature of the results, at least three levels must be investigated. Initially, it is likely that the factor(s) causing the curvature is unknown. The center point at $x_1 = 0$ and $x_2 = 0$ is often a good one to investigate first, since it does not alter the orthogonality. Replicates are often run at the center, and they are interspersed with other treatments. (Complete orthogonality is maintained when runs at the zero level in all factors are added to the **X** matrix.)

Although a central point offers a measure of curvature, it cannot distinguish which factor is responsible for the curvature. Additional levels of each of the factors are needed to provide sufficient treatments to assess the model. A quadratic form of the model containing all possible terms to the second order is convenient and can fit many response behaviors:

$$\hat{y} = \beta_0 + \beta_1 x + \beta_2 x_2 + \beta_{12} x_1 x_2 + \beta_{11} x_1^2 + \beta_{22} x_2^2 \quad (3.67)$$

Added terms are often planned in a *star design*, with high- and low-level treatments in each factor out from the center point. Figure 3.3(a) shows one design with treatments of the star design in solid. With two factors, these can be two scaled-units from the center, and the result is a 3 × 3 factorial design rotated 45°. With three or more factors, the *composite design* of the two-level factorial and the star radiating from the center has fewer treatments than a three-level factorial (see Figure 3.3(b)). Sufficient treatments are offered for fitting the full quadratic model extension of Equation (3.67) in more factors. The spacing of the treatments of the star design can be varied, but typically these are between 1 and 2 units from the center in the coded scale.

An important extra advantage of this composite design is that the level of the star-design treatments can be individually adjusted to provide five levels for any or all factors. Extending the range of a factor beyond the −1 to +1 levels aids in determining curvature and can offer advantages in narrowing the confidence interval on the response.

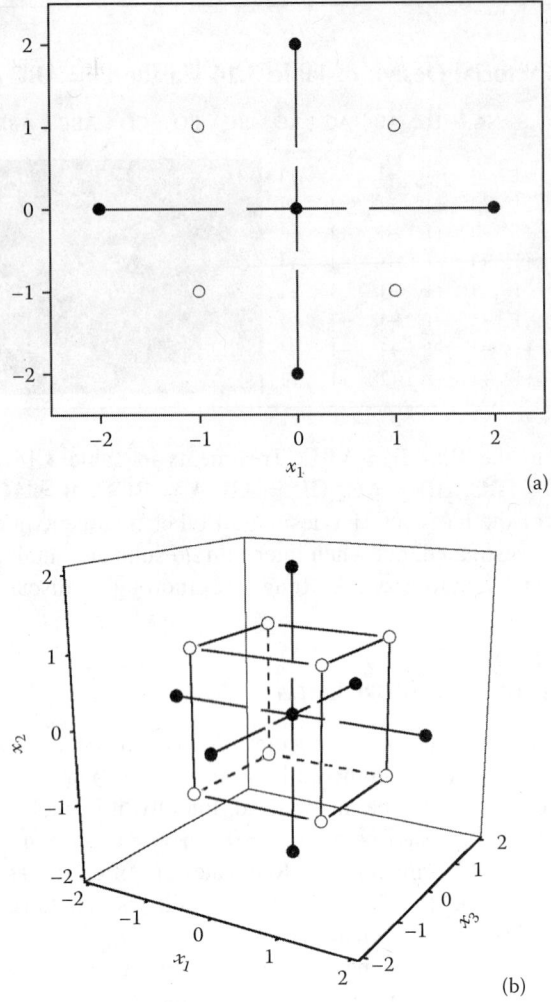

FIGURE 3.3 Composite designs, two-level factorial plus star design, for (a) two factors and (b) three factors.

3.7.5 BLOCKS AND NEW DUPLICATES

Sometimes a block of runs is performed to estimate reproducibility. For example, one block (or group) of runs is made with one feedstock or one reactor system. In these cases, partial factorial runs may act as different blocks, or replicates are performed.

3.7.6 COMPUTER-AIDED EXPERIMENTAL DESIGN

Several computer software packages are helpful in designing experiments. A current commonly used software is JMP, produced by SAS [17]. Typically, design-of-experiments software will ask the user to input the number of factors and levels and the size of the design, e.g., 1/2 or 1/4 factorials or even smaller. Afterward, the software self-selects those terms to alias and then will usually display the model appropriate to the factorial design. The software may also randomize the order of experiments or enter the experiments into blocks if the user desires. JMP is particularly useful, as it contains both the experimental-design software and multiple linear regression algorithms, and both are designed to work in tandem, providing a platform for experimental design and follow-up mathematical modeling.

APPENDIX

As indicated in the introduction to this chapter, the freely available *NIST/SEMATECH e-Handbook of Statistical Methods* (available online at http://www.itl.nist.gov/div898/handbook/index.htm) is recommended as a resource for readers. This large handbook provides valuable information to engineers and scientists on the application of statistics to real-world problems. The main headings of its Table of Contents are as follows:

1. Exploratory Data Analysis
    a. Introduction
    b. Assumptions
    c. Techniques
    d. Case Studies
2. Measurement Process Characterization
    a. Characterization
    b. Control
    c. Calibration
    d. Gauge R&R Studies
    e. Uncertainty Analysis
    f. Case Studies
3. Production Process Characterization
    a. Introduction
    b. Assumptions
    c. Data Collection
    d. Data Analysis
    e. Case Studies
4. Process Modeling
    a. Introduction
    b. Assumptions
    c. Data Collection
    d. Data Analysis
    e. Interpretation & Use
    f. Case Studies
5. Process Improvement
    a. Introduction
    b. Assumptions
    c. Choosing an Experiment Design
    d. Analysis of DOE Data
    e. Advanced Topics
    f. Case Studies
6. Process or Product Monitoring and Control
    a. Introduction
    b. Test Product for Acceptability
    c. Univariate and Multivariate Control Charts
    d. Time Series Models
    e. Tutorials
    f. Case Studies
7. Product and Process Comparisons
    a. Introduction
    b. Comparisons: One Process

  c. Comparisons: Two Processes
  d. Comparisons: Three + Processes
 8. Assessing Product Reliability
  a. Introduction
  b. Assumptions/Prerequisites
  c. Reliability Data Collection
  d. Reliability Data Analysis

ACKNOWLEDGMENT

The author is grateful to the late Robert Eckert for notes that assisted in the compilation of this handbook chapter.

REFERENCES

1. National Institute of Standards and Testing, Gaithersburg, MD, *NIST/SEMATECH e-Handbook of Statistical Methods*; available online at http://www.itl.nist.gov/div898/handbook/index.htm; accessed 22 May 2007.
2. Montgomery, Douglas C., and Runger, George C. 2003. *Applied Statistics and Probability for Engineers*, 3rd ed. New York: John Wiley & Sons.
3. Metcalfe, Andrew V. 1994. *Statistics in Engineering*. London: Chapman & Hall.
4. Fraser, D.A.S. 1976. *Probability and Statistics: Theory and Applications*. North Scituate, MA: Duxbury Press.
5. Microsoft Excel electronic help files 2007.
6. Microsoft Excel online assistance; available online at http://office.microsoft.com/en-us/assistance/CH790018021033.aspx; accessed 22 May 2007.
7. Mathematica electronic help files 2007.
8. MATLAB electronic help files 2007.
9. Gosset, William S., "The Probable Error of a Mean," *Biometrika*, 6: 1–25 (1908).
10. Press, William H., Flannery, Brian P., Teukolsky, Saul A., and Vetterling, William T. 1992. *Numerical Recipes in FORTRAN*, 2nd ed. Cambridge: Cambridge University Press.
11. Amundson, Neal R. 1966. *Mathematical Methods in Chemical Engineering*. Englewood Cliffs, NJ: Prentice-Hall.
12. Frazer, Robert A., Duncan, William J., and Collar, Arthur R. 1960. *Elementary Matrices and Some Applications to Dynamics and Differential Equations*. New York: Cambridge University Press.
13. Lay, David C. 2005. *Linear Algebra and Its Applications*. Reading, MA: Addison Wesley.
14. Barry, Austin B. 1978. *Errors in Practical Measurement in Science, Engineering, and Technology*. New York: John Wiley & Sons.
15. Bragg, Gordon M. 1974. *Principles of Experimentation and Measurement*. Englewood Cliffs, NJ: Prentice-Hall.
16. Hougen, Olaf. A., and Watson, Kenneth. M. 1947. *Chemical Process Principles*, vol. 3, 943–58. New York: John Wiley & Sons.
17. SAS JMP™ electronic help files 2007.

4 Thermodynamics of Fluid Phase and Chemical Equilibria

Kwang-Chu Chao, David S. Corti, and Richard G. Mallinson

CONTENTS

- 4.1 Principles of Thermodynamics 256
  - 4.1.1 Introduction 256
  - 4.1.2 Temperature and the Ideal Gas 257
  - 4.1.3 First Law of Thermodynamics and Internal Energy 258
    - 4.1.3.1 Open Systems 261
  - 4.1.4 Second Law of Thermodynamics and Entropy 264
  - 4.1.5 Equilibrium Energy Functions 268
    - 4.1.5.1 Equilibrium Criterion I 268
    - 4.1.5.2 Equilibrium Criterion II 269
    - 4.1.5.3 Equilibrium Criterion III 271
    - 4.1.5.4 Equilibrium Criterion IV 273
    - 4.1.5.5 Equilibrium Criterion V 275
  - 4.1.6 Open System and Chemical Potential 277
  - 4.1.7 Partial Molar Quantities and the Gibbs-Duhem Relation 280
- 4.2 Volumetric and Thermodynamic Properties 284
  - 4.2.1 Pressure–Volume–Temperature Relationship 284
  - 4.2.2 Principle of Corresponding States 287
  - 4.2.3 Phase Rule 290
  - 4.2.4 Phase Behavior of Mixtures 291
    - 4.2.4.1 Gas-Liquid Equilibrium 291
    - 4.2.4.2 Liquid-Liquid Equilibrium 293
    - 4.2.4.3 Gas-Gas Equilibrium 294
  - 4.2.5 Equations of State 295
    - 4.2.5.1 Van der Waals-Type Equations 295
    - 4.2.5.2 Perturbation Equations 299
    - 4.2.5.3 The Virial and Extended Virial Equations 309
    - 4.2.5.4 Extended Virial Equations 312
  - 4.2.6 Energy Functions of Ideal Gases and Mixtures 313
  - 4.2.7 Residual Functions and Energy Functions from Equations of State 317
    - 4.2.7.1 Helmholtz Energy 317
    - 4.2.7.2 Gibbs Energy 318
    - 4.2.7.3 Entropy 319
    - 4.2.7.4 Internal Energy 319
    - 4.2.7.5 Enthalpy 320
    - 4.2.7.6 Some General Comments 320
  - 4.2.8 Fugacity 321
- References 324

4.3 Liquid Solutions 325
    4.3.1 Ideal and Real Solutions 325
    4.3.2 Ideal and Excess Solution Properties...328
    4.3.3 Activity-Coefficient Models 329
        4.3.3.1 Redlich–Kister Equation 330
        4.3.3.2 Van Laar Equation 330
        4.3.3.3 Regular Solutions...332
        4.3.3.4 Flory–Huggins Equation 334
        4.3.3.5 Wilson's Local-Composition Equation..336
        4.3.3.6 Nonrandom Two-Liquids (NRTL) Equation 338
        4.3.3.7 The Complete Local-Composition Equation...339
        4.3.3.8 UNIQUAC Equation 341
    4.3.4 Group Contribution Methods 343
        4.3.4.1 Four Postulates of Group Solution 343
        4.3.4.2 Analytical Solution of Groups...345
    4.3.5 Gibbs Energy Models of Liquid Solutions...347
References 350
4.4 Fluid-Phase Equilibria..351
    4.4.1 Vapor-Liquid Equilibrium in a Single-Component Fluid..351
    4.4.2 Vapor-Liquid Equilibrium of Ideal Mixtures..355
    4.4.3 Vapor-Liquid Equilibrium by γ-ϕ Models 358
        4.4.3.1 Low-Pressure Models...358
        4.4.3.2 A High-Pressure Model 362
    4.4.4 Vapor-Liquid Equilibrium by φ-φ Models..364
        4.4.4.1 φ-φ Models—φ from vdW Mixing Rules...364
        4.4.4.2 Free-Energy-Matching Mixing Rules..367
    4.4.5 Liquid-Liquid Equilibrium Models..367
    4.4.6 Gas-Solid Equilibrium Models 372
References 374
4.5 Chemical Reaction Equilibria 375
    4.5.1 Chemical Reaction Equilibrium 375
    4.5.2 Equilibrium Constants 376
    4.5.3 Phase Rule for Chemically Reacting Species..384
    4.5.4 Open Systems with Reaction 385
    4.5.5 Stoichiometric Formulation..388
References 391
Data References 392

4.1 PRINCIPLES OF THERMODYNAMICS

4.1.1 Introduction

Thermodynamics has played an increasingly important role over the years in the engineering analysis and synthesis of chemical processes. There has been an expansion of this role with the success of vapor-pressure-fitted van der Waals equations of state and group contribution methods. By far the major part of computer time in chemical process design is now devoted to thermodynamic calculations. This chapter presents the thermodynamic methods needed to perform the analysis and synthesis, to start from basic principles, to describe the phenomena, to develop the methods, and to illustrate applications.

In a brief review of the terminology of thermodynamics, the portion of the universe that is studied is called a *system*. The rest of the universe that interacts with the system is the *surroundings*.

The system is *open* if it exchanges mass with its surroundings. A *closed* system cannot exchange mass with its surroundings. The system is *adiabatic* if it does not exchange heat with the surroundings. A system is *isolated* if it is precluded from exchanging mass or energy with its surroundings.

A system at *equilibrium* does not change with time, and no transport flux of momentum, heat, or mass occurs. An equilibrium system is characterized by a *reversible* response to differential changes in external forces. A small increase of applied pressure produces a compression effect on the fluid; a small decrease of pressure produces an expansion effect. For a system at pressure p, either reversible compression or reversible expansion takes place, depending on the sign of dp. Similarly, with a differential increase in temperature of the surroundings from its equilibrium temperature, heat is absorbed by the equilibrium system; conversely, heat is rejected by the system with an infinitesimal decrease in temperature of the surroundings. Thus reversible exchange of heat with a system at T can be either heat input from a source at $T + |dT|$ or output to a sink at $-|dT|$. A reversible process is an ideal process taking place at near zero speed in an equilibrium system. Equilibrium and reversibility go hand in hand.

A system can be composed of several phases. A *phase* is a homogeneous part of a system. The phases of a system at equilibrium are also at equilibrium. An *extensive property* of a system is the sum of the parts of the system. Thus volume is extensive. The extensive property of an equilibrium phase is directly proportional to its mass m. An *intensive property*, like density, temperature, or pressure, is independent of extension and assumes a uniform value in an equilibrium phase.

The *state* of an equilibrium phase is specified by its chemical composition and a relatively few intensive properties, generally two, e.g., temperature and pressure, temperature and mass density, or density and pressure. The number and kind of intensive properties required are determined by experience. To specify the state of a nonequilibrium system generally requires more than the values of a few intensive properties. Thus a system undergoing heat transfer may require a mathematical function to describe its temperature distribution and thus its state.

State properties are functions of the state and are independent of the previous history of the system; state properties are reproducible and have no memory. The density of water at a fixed state is the same any time it is measured; the density of water is a state property. However, the hardness of a specimen of steel is not a state property, inasmuch as it assumes different values at the same state, depending on its history of heat treatment.

A change of state is called a *process*. A system goes through an *adiabatic* process if no heat exchange occurs during the change of state. An *isothermal* process takes place at a constant temperature, and an *isobaric* process occurs at constant pressure.

4.1.2 TEMPERATURE AND THE IDEAL GAS

Temperature is a measure of the degree of hotness of an object. Objects at various degrees of hotness can be assigned numbers, one to each object, in order of increasing hotness. This sequence of numbers represents temperature. Two objects in thermal contact eventually reach a state of the same hotness. They are at thermal equilibrium and at the same temperature. Two objects that are separately at thermal equilibrium with a third object are also at the same temperature.

A temperature scale can be set up by reference to an observable effect of temperature on an object, for example, thermal expansion, thermoelectric effect, or intensity of radiation of energy. Thus the thermal expansion effect can be observed by partially filling a glass capillary with mercury. The height of the meniscus of the mercury rises or falls as the hotness rises or falls. The capillary tube can be marked with numbers as a linear progression of distance to indicate temperature. Many devices can be devised for the measurement of temperature. The fundamental problem is to find a temperature scale that is independent of any arbitrarily selected measuring device employing an arbitrarily selected medium.

The answer is the ideal-gas temperature scale defined as follows: Measure the volume V of a fixed quantity of gas or gas mixture at a sequence of low pressures p at the temperature of the object to be determined; form the pV products and extrapolate to determine a zero-pressure limiting value; call it β. Repeat this experiment at the steam point to determine a benchmark value β_S. Repeat again at the ice point to determine another benchmark value β_I. The temperature of the object on the ideal-gas temperature scale in kelvins, K, is given by

$$T = 100\beta / (\beta_S - \beta_I) \tag{4.1}$$

In Equation (4.1), the steam-point temperature and the ice-point temperature are defined to differ by 100°K. The ideal-gas temperature is independent of the species and quantity of gas and of the apparatus used for the measurement. Substituting β_I for β in Equation (4.1), the ice point has been found to be 273.15° and, similarly, the steam point to be 373.15°K.

In terms of ideal-gas temperature, it follows from Equation (4.1) that all gases are described by

$$\lim_{p \to 0} pv = \kappa T \tag{4.2}$$

where κ is a constant in the equation but depends on the quantity of gas. The quantity assumes a universal constant value for 1 mole of gas by Avogadro's law, which states that equal volumes of gas of any chemical species measured at the same temperature and pressure contain equal numbers of molecules at ideal-gas conditions,

$$\lim_{p \to 0} pv = RT \tag{4.3}$$

where v denotes molar volume and R is a universal constant determined to be 8.314510 J/(mol·K).

For real gases, the pV product attains the limiting value RT at zero pressure, where intermolecular potential energy vanishes. The extrapolation to zero pressure frees the pV product from the effect of intermolecular forces. An ideal gas is defined as one that is free of intermolecular potential energy at finite pressures. For the ideal gas, the pv product equals RT at all pressures, and is called the ideal-gas equation,

$$pv = RT \tag{4.4}$$

This equation is an approximation of real gas behavior at high temperatures or low pressures. The approximation remains good at lower temperatures at progressively lower pressures. Where the degree of approximation is acceptable, the equation is in common use for its simplicity.

A real gas or real liquid is said to be in the ideal-gas state when its intermolecular potential is ignored. The properties of the fluid in the ideal-gas state can be calculated with ease using Equation (4.4). One simple universal equation applies to all substances, requiring no substance-specific parameters. However, for most real states, the ideal-gas equation is inadequate, and real-fluid properties are obtained by adding to the ideal-gas equation the contribution of intermolecular potential in the form of deviation functions, also called residual functions. A major objective of Section 4.2 is to derive the deviation functions from the equation of state of the substance. Because the ideal-gas properties are known, to find the deviation function is as good as finding the state function of a real substance. In this way the ideal-gas equation is used universally in all equation-of-state calculations of thermodynamic functions.

4.1.3 First Law of Thermodynamics and Internal Energy

The first law of thermodynamics states that energy can be converted from one form to another or transported from one system to another, but the sum of the energies involved—kinetic energy, potential energy, electric energy, internal energy (which is a state function), and other forms—is always conserved and remains constant. Energy is neither destroyed nor created. Thus the change in the total energy of a system including mechanical energy and internal energy is equal to the flow of energy into the system in the form of heat or work. Let U denote internal energy and E denote mechanical energy, including potential energy, kinetic energy, and the like. Let Q denote heat, the flow of energy due to a temperature difference. Let W be the work, or mechanical energy in transition. By the first law,

$$dU + dE = \delta Q + \delta W \tag{4.5}$$

We follow the convention of assigning a positive sign to Q or W for flow into the system. A negative value for δQ designates heat loss from the system; a negative value of δW denotes work done by the system. With the use of δ rather than d, the differential δQ or δW indicates a differential quantity that is not the differential of a state function and, therefore, is not an exact differential. Equation (4.5) brings out the essence of the first law by using dU for the differential of U.

Applying Equation (4.5) to the surroundings by changing the signs of the flow terms,

$$dU_{sur} + dE_{sur} = -\delta Q - \delta W \tag{4.6}$$

Adding Equation (4.5) and Equation (4.6),

$$d(U+E) + d(U+E)_{sur} = 0 \tag{4.7}$$

Integrating,

$$(U+E) + (U+E)_{sur} = C \tag{4.8}$$

where C is a constant showing that the total energy of the system and surroundings is always conserved. Energy can be converted from one form to another, but it cannot be created from nothing, nor annihilated into nonexistence. The first law is a generalization of the principle of conservation of mechanical energy that applies to mechanical systems in the absence of dissipative forces. By introducing internal energy as a state function, the first law turns the principle of conservation of energy into a general principle.

Being of a molecular nature, internal energy is made up of:

- Kinetic energy of motion of molecules
- Potential energy of interaction of molecules
- Energy of rotation of atoms and groups of atoms within molecules
- Energy of vibration of the chemical bonds
- Chemical energy of bonding of atoms
- Excitation of electrons
- Energy of the atomic nuclei

Since not all of these energies have been completely elucidated, the absolute value of internal energy is not known, nor is it of urgent concern, as it is the *change* of internal energy that accompanies a process that is of interest, rather than the absolute value. When values of internal energy need to be tabulated or reported, they are based on a convenient datum state.

Of the various forms of molecular energy, ideal gas is unique in lacking potential energy of molecular interaction; the molecules do not appreciably attract or repulse one another. Since the potential energy of the molecules is the only form of molecular energy that is volume dependent, the internal energy of the ideal gas is independent of pressure or density, but is a function of temperature only:

$$(\partial U / \partial p)_T = 0 \tag{4.9}$$

$$(\partial U / \partial \rho)_T = 0 \tag{4.10}$$

While the intermolecular forces are absent, the internal structure of the molecule and the attendant energies remain unaltered from the real gas. The internal energy, heat capacity, and related functions retain their specific values for each substance in the ideal-gas state.

For the phenomena addressed in this chapter, changes in the mechanical energy of a substance are insignificant compared with changes in the internal energy; mechanical energy contents will not be included in the first law equation in the rest of the chapter. We will simply write

$$dU = \delta Q + \delta W \tag{4.11}$$

A change of volume is always associated with the phenomena addressed in this chapter; we will consider volume change as the only means by which work is exchanged with the surroundings, $\delta W = -pdV$, the sign convention being for the work to be positive when done on the system. Thus

$$dU = \delta Q - pdV \tag{4.12}$$

As a state function, how does U change with state? Let us express dU in terms of the state variables and V,

$$dU = (\partial U / \partial V)_T dV + (\partial U / \partial T)_V dT \tag{4.13}$$

Combining the two equations, we express the heat exchange in terms of changes of state,

$$\delta Q = [(\partial U / \partial V)_T + p]dV + (\partial U / \partial T)_V dT \tag{4.14}$$

Let C_V denote heat capacity at constant volume defined by

$$(\delta Q = C_v dT)_v \tag{4.15}$$

Comparison of Equation (4.14) and Equation (4.15) shows that

$$C_V = (\partial U / \partial T)_V \tag{4.16}$$

At constant volume, the change of U with a change of from state 1 to 2 is given by

$$U_2 - U_1 = \int_1^2 C_v dT \tag{4.17}$$

C_V can be measured calorimetrically or, for the ideal gas, calculated from spectroscopic data. For a real gas or liquid, it can be obtained by combining the ideal-gas value with equation-of-state calculations to be discussed in Section 4.2.7, where the isothermal variation of U with V will also be discussed.

4.1.3.1 Open Systems

A closed system has been considered to define internal energy and to develop the fundamental differential form of internal energy. When internal energy is added to a system by transfer of mass from the surroundings into the system, the system is said to be open. The energy conservation equation, Equation (4.11), is extended to open systems by considering the addition of a differential number of moles dn_{in} in an input stream to the system and the leaving of a differential number of moles dn_{out} in a stream out of the system. By the principle of conservation of energy—adding the differential internal energy and the associated differential pv work of propelling the streams into the system to Equation (4.11) for any number of input streams and likewise subtracting the leaving streams—we obtain the differential change dU of the system

$$dU = \delta Q + \delta W + \sum_{in} u_{in} dn_{in} + \sum_{in} p_{in} v_{in} dn_n - \sum_{out} u_{out} dn_{out} - \sum_{out} p_{out} v_{out} dn_{out} \qquad (4.18)$$

By explicitly expressing the pv work associated with the input and output flows in separate terms so as to exclude them from the work term W, the meaning of W in Equation (4.11) remains unchanged as it reappears in Equation (4.18).

We now define a new energy function, called the enthalpy, H, by

$$H = U + pV \qquad (4.19)$$

Accordingly, Equation (4.18) can be rewritten as

$$dU = \delta Q + \delta W + \sum_{in} h_{in} dn_{in} - \sum_{out} h_{out} dn_{out} \qquad (4.20)$$

Reexpressed in time rate of change,

$$\dot{U} = \dot{Q} + \dot{W} + \sum_{in} h_{in} \dot{n}_{in} - \sum_{out} h_{out} d\dot{n}_{out} \qquad (4.21)$$

at steady conditions, with all flows at a constant rate and the system unchanging in time, $\dot{U}$. Rearranging the equation, we obtain

$$\sum_{out} \dot{H}_{out} = \sum_{in} \dot{H}_{in} + \dot{Q} + \dot{W} \qquad (4.22)$$

This equation, known as enthalpy balance, is extensively used in engineering analyses of processes. The system under consideration is then the whole chemical plant or process. Thus the enthalpy balance displays the sources and destinations of energy in the plant or the process. Chemical reactions commonly produce large enthalpy changes and heat effects. Often, the enthalpy balance of chemical processes is dominated by the thermal chemistry of the process. In the following, several examples illustrate the application of the first law of open systems.

Example 4.1: Throttle

A throttle is a device (such as a partly closed valve or porous plug) that reduces the pressure of a flowing fluid without any shaft work or significant acceleration of the fluid (i.e., no appreciable change in the kinetic energy of the fluid). With a single inlet and outlet stream, and assuming that these streams and the heat transfer are negligible, Equation (4.22) reduces to

$$h_{out} = h_{in} \qquad (4.23)$$

A throttle does not change the enthalpy of the fluid; throttling is an isenthalpic process. For a given input state and a specified outlet pressure, one finds the outlet temperature by conducting a one-dimensional search for a temperature at which the enthalpy is equal to the input enthalpy. An enthalpy chart provides a convenient means for the search. Another method of solution is to apply the Joule-Thompson coefficient, μ_{JT}, defined by

$$\mu_{JT} = \left(\frac{\partial T}{\partial P}\right)_h \qquad (4.24)$$

The sign of the above partial derivative determines whether a fluid expanded through a throttle experiences a temperature drop. For an ideal gas, $\mu_{JT} = 0$, so that there is no change for a given pressure drop. For real gases, at high enough temperatures (depending upon the substance of interest), $\mu_{JT} < 0$, so that the temperature of the gas increases upon expansion. Below some inversion temperature, μ_{JT} becomes positive, so that the gas now cools upon expansion through a throttle. For example, this inversion temperature is about 1319 for CO_2 and about 40 for He. Typically, another inversion temperature exists at even lower temperatures, where μ_{JT} again becomes negative. Also, there is usually a single inversion pressure, where above that pressure μ_{JT} is always negative. For example, this inversion pressure is 884 bar for CO_2 and around 27.5 bar for He. Typically, a throttle is used to cool a gas upon expansion. If the temperature drop is large enough, the gas can be partially liquefied.

Example 4.2: Heat Exchangers

Consider a heat exchanger operating with two streams, A and B, with steady flow rates of $\dot{n}_A$ and $\dot{n}_B$. Since the volume of the heat exchanger is constant and no shaft work is performed, $\dot{W} = 0$. Also, assuming that all transfer of heat occurs between the two streams inside the heat exchanger, so that no heat is lost to the surroundings, then $\dot{Q} = 0$. With kinetic and potential energy changes being small, the energy balance reduces to

$$\Delta h_A \dot{n}_A + \Delta h_B \dot{n}_B = 0 \qquad (4.25)$$

where $\Delta h_i = h_{i,out} - h_{i,in}$. If the incoming and outgoing pressure of a stream is the same, then Δh_i can be calculated from

$$\Delta h_i = \int_{T_{i,in}}^{T_{i,out}} C_{p,i} dT$$

Example 4.3: Venting of an Insulated Tank

Consider a rigid (constant-volume), well-insulated (adiabatic) tank containing a gas at some initial high pressure. Then a valve is opened and the gas is slowly vented through a pipe. Assuming that the changes in kinetic energy are small (as compared with the enthalpy fluxes and internal energy changes of the system), Equation (4.20) reduces to

$$dU = -h_{out}\delta n_{out} = h_{out}dN$$

where N denotes the number of moles in the tank. In general, the enthalpy of the exiting stream, h_{out}, may not have the same properties of the gas remaining in the tank. If we now assume that the venting is slow and the contents of the tank are well mixed, we can replace h_{out} with the molar enthalpy of the contents of the entire tank, h. The internal energy of the gas in the tank varies because both mass is removed and its properties (such as and p) change; the first law can be written as

$$dU = d(Nu) = Ndu + udN = hdN$$

or

$$\frac{du}{h-u} = \frac{dN}{N} \tag{4.26}$$

where u is the molar internal energy. The above expression is a general relation describing how the molar internal energy of the contents of the tank varies with a change in the mass, or moles, of the system.

If the gas in the tank is ideal, where $h\, u = Pv = RT$ and $du = C_v dT$, then Equation (4.26) is given by

$$\frac{C_v}{R}\frac{dT}{T} = \frac{dN}{N}$$

which upon integration yields (assuming a constant heat capacity)

$$\left(\frac{N_f}{N_i}\right) = \left(\frac{T_f}{T_i}\right)^{C_v/R} \tag{4.27}$$

The above relation can be rewritten in terms of the pressure of the tank. Noting that the volume of the tank is fixed, and that for an ideal gas $PV = NRT$, Equation (4.27) can be rewritten as

$$\left(\frac{P_f}{P_i}\right) = \left(\frac{T_f}{T_i}\right)^{C_p/R} \tag{4.28}$$

where, for an ideal gas, $C_P = C_v + R$.

Example 4.4: Filling of an Insulated Tank from a Constant-Temperature and -Pressure Source

Consider a rigid (constant-volume), well-insulated (adiabatic) tank that is initially evacuated ($N_i = P_i = 0$). The tank is to be filled with a gas from some external source, assumed to be large enough that the temperature and pressure of the gas entering the tank do not vary throughout the process. Gas enters the tank until the pressures of the tank and the external source are the same. Neglecting heat transfer and changes in kinetic energy, Equation (4.20) now reduces to

$$dU = h_{in}\delta n_{in} = h_{in}dN$$

where we have also made use of the mass balance equation. In this example $dN > 0$. Since the enthalpy of the incoming stream, h_{in}, is constant, we may integrate both sides of the above equation to obtain

$$U_f - U_i = h_{in}(N_f - N_i)$$

Because $N_i = 0$, $U_i = 0$, so that the above reduces to

$$U_f = N_f u_f = h_{in} N_f$$

or

$$u_f = h_{in} = u_{in} + p_{in} v_{in} \tag{4.29}$$

The first term on the far right side represents the molar internal energy that entered the tank, while the second term represents the work done by the incoming stream as it enters the tank.

For an ideal gas, $pv = RT$ and $du = c_v dT$, so that Equation (4.29) becomes (by using an average heat capacity)

$$u_f - u_{in} = c_v(T_f - T_{in}) = p_{in} v_{in} = RT_{in}$$

or

$$T_f = \frac{c_p}{c_v} T_{in} \tag{4.30}$$

Because $C_p > C_v$, then $T_f > T_{in}$. This problem clearly illustrates one of the differences between open and closed systems and of the importance of the work due to flow (which, in this example, is responsible for the increase of the final temperature above that of the external source). As an example, air, modeled as a pure-component ideal gas, is well represented by $c_v = (5/2)R$ and $c_p = (7/2)R$, so that $T_f = (7/5)T_{in}$, a 40% temperature rise above T_{in}.

4.1.4 Second Law of Thermodynamics and Entropy

The second law according to Planck is widely cited: *It is impossible to construct a machine operating in cycles that will convert heat into work without producing any other changes in the surroundings.* Another form of the second law is: *It is impossible to operate any process solely to transfer heat from a lower temperature to a higher one.* It can be shown that both are equivalent and both are deducible from the following statement:

There exists an extensive state function called entropy S that is defined by

$$dS = \delta Q / \theta \tag{4.31}$$

where δQ is the differential heat absorbed reversibly from the surroundings and θ is the thermodynamic temperature of the system. The thermodynamic temperature is an absolute temperature that can never be less than zero, and is otherwise unspecified. It will be shown shortly that it is identical to the ideal-gas temperature.

Thermodynamics of Fluid Phase and Chemical Equilibria

The entropy change made by a process, whether reversible or irreversible, of a closed system is given by

$$dS = d_eS + d_iS \tag{4.32}$$

where $d_eS = \delta Q/\theta$ denotes entropy change due to transfer of heat into the system and θ is the temperature where the heat transfer takes place; d_iS is the entropy generated due to other processes taking place within the system,

$$d_iS > 0 \quad \text{for irreversible and possible processes} \tag{4.33}$$

$$d_iS = 0 \quad \text{for reversible and possible processes} \tag{4.34}$$

$$d_iS < 0 \quad \text{never} \tag{4.35}$$

The second law gives a direction to changes: change proceeds in the direction that increases entropy, but never to decrease entropy. The fundamental criterion is established for the completion of irreversible changes and the attainment of equilibrium, thus providing the basis for the study of equilibrium phenomena.

The temperature θ of the second law, called the thermodynamic temperature, will be shown to be identical to the ideal-gas temperature T. To proceed, we consider the Carnot cycle to produce mechanical work from heat. The Carnot cycle is an idealized reversible cycle by which a working fluid is confined in a cylinder-piston device to (a) absorb heat from a high-temperature source while producing mechanical work and (b) reject heat to a low-temperature sink while being compressed toward the original state. The successive steps of a Carnot cycle are as follows:

1. Isothermal expansion at a high temperature θ_1, absorbing heat Q_1. The change of entropy of the working fluid $\Delta S = Q_1/\theta_1$ by Equation (4.31).
2. Adiabatic expansion to cool down to a low temperature θ_2, absorbing no heat. The change of entropy equals zero.
3. Isothermal compression at θ_2, absorbing heat Q_2 (a negative quantity from the sign convention). The entropy change $\Delta S = Q_2/\theta_2$.
4. Adiabatic compression designed to bring the system back to the original state at θ_1. The entropy change equals zero.

Let $-W$ denote the total work done by the system in going through one complete cycle. The first law requires

$$\Delta U = Q_1 + Q_2 + W \tag{4.36}$$

Since U is a state function, its change for the cycle is zero, $\Delta U = 0$, and

$$Q_1 + Q_2 = -W \tag{4.37}$$

Since entropy is a state function, its change is also zero. Adding up the entropy changes for all four steps of the cycle,

$$Q_1/\theta_1 + Q_2/\theta_2 = 0 \qquad (4.38)$$

The cycle is characterized by the absorption of heat at only one temperature and by the rejection of heat at only one temperature, such that the effect of temperature on the quantity of heat absorbed or rejected can be revealed. By Equation (4.38), the effect is a simple proportionality,

$$Q_1/\theta_1 = -Q_2/\theta_2 \qquad (4.39)$$

The efficiency of the cycle is defined to be the fraction of the heat absorbed at the high temperature that is converted into work, $(-W)/Q_1$. By combining Equation (4.37) and Equation (4.38), the efficiency of the Carnot cycle is obtained,

$$(-W)/Q_1 = (\theta_1 - \theta_2)/\theta_1 \qquad (4.40)$$

The theoretical efficiency of the ideal reversible cycle is seen to depend only on the temperatures of the heat source and sink, and is independent of the working fluid confined in the piston-cylinder device.

Let us use an ideal gas in a Carnot cycle and find the efficiency of the cycle by using ideal-gas properties in enumerating the changes in the four steps of the cycle. Let us designate the intial state of step 1 with the subscript A, the initial state of step 2 with B, and so on. The high temperature at step 1, which is θ_1 on the thermodynamic temperature scale, will be T_1 on the ideal-gas temperature scale. The low temperature of step 3 will be T_2, corresponding to θ_2. The work and heat terms of a step will be designated with subscripts 1, 2, 3, or 4.

1. Isothermal expansion at a high temperature T_1:

$$U_B - U_A = 0$$

$$W_1 = -\int p dV = -nRT_1 \ln(V_B/V_A)$$

$$Q_1 = -W_1$$

2. Adiabatic expansion to cool down from T_1 to T_2:

$$Q_2 = 0$$

$$U_C - U_B = W_2$$

3. Isothermal compression at a low temperature T_2:

$$U_D - U_C = 0$$

$$W_3 = -\int p dV = -nRT_2 \ln(V_D/V_C)$$

$$Q_3 = -W_3$$

Thermodynamics of Fluid Phase and Chemical Equilibria

4. Adiabatic compression to heat up to the high temperature T_1:

$$Q_4 = 0$$

$$U_A - U_D = W_4$$

Since $U_A = U_B$ and $U_C = U_D$, $W_2 = -W_4$. These two work terms cancel when the work terms of the cycle are summed, leaving the total work done on the working fluid to be

$$W = W_1 + W_3 \tag{4.41}$$

Forming the efficiency by dividing $-W$ with Q_1,

$$-W/Q_1 = [T_1 \ln (V_B/V_A) + T_2 \ln (V_D/V_C)]/[T_1 \ln (V_B/V_A)] \tag{4.42}$$

The volumes V_B and V_C are related because B and C are on an adiabatic line,

$$nRT \ln (V_B/V_C) = C_V \ln (T_2/T_1) \tag{4.43}$$

Similarly V_A and V_D are related to the same temperature ratio; the two volume ratios are equal, $V_B/V_C = V_A/V_D$. Upon rearranging,

$$V_B/V_A = V_C/V_D \tag{4.44}$$

The efficiency of the ideal-gas Carnot cycle is found upon substituting Equation (4.43) into Equation (4.42),

$$(-W)/Q_1 = (T_1 - T_2)/T_1 \tag{4.45}$$

The ideal-gas Carnot cycle is a particular case of the Carnot cycle; its efficiency must be given by the general formula of Carnot cycles. Comparison of Equation (4.45) with Equation (4.40) shows

$$\theta_1/\theta_2 = \theta_1/\theta_2 \tag{4.46}$$

$$\theta_1/\theta_1 = \theta_2/\theta_2 \tag{4.47}$$

Since the two temperatures 1 and 2 are arbitrary, Equation (4.47) reveals that the two temperature scales are proportional,

$$\theta/T = k \tag{4.48}$$

where k is a constant. For convenience, k is set to be equal to 1 to give

$$\theta = T \tag{4.49}$$

The thermodynamic temperature is identical to the ideal-gas temperature. Both will be referred to as the absolute temperature. From here on we will use the absolute temperature for the calculation of entropy change according to $dS = \delta q/T$.

The efficiency of the Carnot cycle is the highest among all machines for the production of work from heat working between any two given temperatures. When reversible and operating between the same high temperature T_1 and low temperature T_2, it *absorbs* heat Q_2 at T_2 and *rejects* heat Q_1 at T_1 while *consuming* work W. The ratio W/Q_1 remains the same as before. Suppose an irreversible machine exists that has a higher efficiency than the Carnot cycle. Let it absorb heat Q_1' at T_1 and produce the same work W as consumed by the reversed Carnot. Then, as supposed,

$$W/Q1' > W/Q1 \tag{4.50}$$

$$Q1' < Q1 \tag{4.51}$$

If the reversed Carnot cycle were coupled to the irreversible machine and driven by it, there would be a net amount of heat $Q_1 - Q_1' > 0$ flowing into the reservoir from this coupled machine. There being no net work production, this amount of heat is obtained from the low-temperature reservoir at T_2. The coupled machine would be pumping heat from a lower temperature to a higher temperature while producing no other changes in the surroundings. But such a result is impossible by the second law.

The efficiencies of all reversible machines are equal. For, if any one should have a different efficiency, the machine with the higher efficiency can be operated to drive the one with lower efficiency in reverse. The combined machine would be capable of pumping heat from a lower temperature to a higher temperature without producing any other changes in the surroundings, hence violating the second law.

The Carnot cycle reveals a fundamental difference of heat from mechanical energies. Whereas the conversion of mechanical energy to heat can be readily completed, the conversion of heat to work is incomplete, even under ideal conditions. Only part of the heat can be converted, and the rest has to be rejected to a lower temperature sink. In thermodynamics, work denotes all forms of energy that are interconvertible without any theoretical limit on the degree of convertibility. Thus potential energy, kinetic energy, electrical energy, surface energy, etc., belong to the category of work.

4.1.5 Equilibrium Energy Functions

Consider the change of entropy of a system that is isolated and cannot exchange heat or work with the surroundings. Not being able to exchange work means a constant volume, $dV = 0$. Following from the further restriction of no heat exchange, the first law requires no change of internal energy, $dU = 0$. By the second law, there can be no entropy change due to heat exchange with the surroundings. But entropy can increase due to irreversible processes within the system,

$$(dS > 0)_{U,V} \tag{4.52}$$

The increase persists as long as irreversible processes take place until entropy reaches a maximum

$$(dS = 0)_{U,V} \tag{4.53}$$

and the system attains an equilibrium state. Equilibrium criteria are described in the following subsections.

4.1.5.1 Equilibrium Criterion I

At fixed U and V, a system seeks equilibrium by increasing its entropy until equilibrium is attained as S attains a maximum.

Figure 4.1 shows an S-V-U space. The equilibrium states are represented by a surface in this space. The space on the concave side of the equilibrium surface is taken up by nonequilibrium

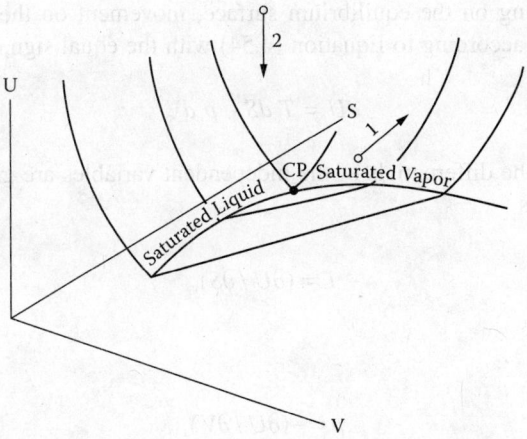

FIGURE 4.1 [S-V-U] space.

states, such as a system at a nonuniform temperature distribution or at varying compositions from one region to the next. A point in this space represents an isolated system that is being propelled by irreversible processes to move up in S toward the equilibrium surface, as indicated by arrow 1. The ascent in S comes to a stop upon reaching the equilibrium surface. The final point on the surface represents the equilibrium state specified by the variables U and V. The space on the other side of the surface cannot be reached.

Alternatively, an equilibrium state in the S-V-U space can be specified by the variables S and V. In the first-law equation, Equation (4.11),

$$dU = \delta Q + \delta W$$

we replace δQ with $\delta Q \leq TdS$, in which, by the second law, the equal sign (=) applies for reversible processes, while the less-than sign (<) applies for irreversible processes. At the same time, replace δW with $-pdV$. There follows

$$dU \leq TdS - pdV \tag{4.54}$$

At fixed values of S and V,

$$(dU \leq 0)_{S,V} \tag{4.55}$$

In Figure 4.1, change of the system at constant S and V is illustrated by arrow 2 pointing in the direction of descending U as a result of irreversible processes taking place in the system. The descent in U comes to a stop upon reaching the equilibrium surface. The point of landing on the surface is the equilibrium state at the specified S and V. There,

$$(dU = 0)_{S,V} \tag{4.56}$$

Equation (4.56) states another criterion of equilibrium: equilibrium criterion II.

4.1.5.2 Equilibrium Criterion II

At fixed S and V, a system seeks equilibrium by decreasing internal energy until the equilibrium state is attained as U reaches a minimum.

Subsequent to landing on the equilibrium surface, movement on the surface is achieved by varying S or V, or both, according to Equation (4.54) with the equal sign,

$$dU = T\,dS - p\,dV \tag{4.57}$$

The coefficients of the differentials of the independent variables are given by the geometry of the $U(S,V)$ surface. Thus

$$T = (\partial U / \partial S)_v \tag{4.58}$$

and

$$p = -(\partial U / \partial V)_s \tag{4.59}$$

where S and V are used as state variables to describe a process, and U is the state function that expresses the changes in the system in a most useful way. To illustrate, the work that can be extracted from a fluid will be found for a reversible adiabatic nonflow process. The differential work done by a fluid in a nonflow process is given by

$$-\delta W = p\,dV \tag{4.60}$$

In a reversible adiabatic process, $\delta Q = 0$ and $dS = 0$; the process is isentropic. Comparison with Equation (4.57) shows

$$(\delta U = dW)_s \tag{4.61}$$

Integration gives

$$(-\Delta U = -W)_S \tag{4.62}$$

The decrease in U is the measure of reversible nonflow work that can be extracted from a fluid upon isentropic expansion. The state function U, upon being measured or calculated for various states, becomes useful for engineering design of work-production processes that expand the fluid from one state to another that are tabulated or calculated.

Equation (4.57) indicates the fundamental differential relation of equilibrium U to the variables S and V. The simplicity and directness of the relation indicate that U is best suited for the exploration of phenomena expressed in the independent variables $[S,V]$. Equilibrium U is said to be the natural, or canonical, energy function of its natural variables S and V. We set out to find energy functions that are natural to other variables. The variables of greatest general usefulness are the sets $[S,V]$, $[S,P]$, $[T,V]$, and $[T,p]$. Having had the fundamental differential equation and natural energy function for the first set of variables $[S,V]$, our method is to change variables one at a time, starting from the known, to successively find new ones.

Switching to the variables $[S,P]$, we find enthalpy to be just the natural energy function. Enthalpy has been defined in Equation (4.19) as

$$H = U + pV \tag{4.63}$$

The differential of H is

$$dH = dU + p\,dV + V\,dp \tag{4.64}$$

Substitution of Equation (4.54) into Equation (4.64) gives

$$dH \leq T\,dS + V\,dp \tag{4.65}$$

In Equation (4.65) we have changed variables to $[S,p]$ by Legendre transformation.

The less-than-or-equal-to sign ($\leq$) in Equation (4.65) applies to two processes: The "equal" portion of the sign refers to reversible processes, while the less-than sign applies to irreversible processes. At fixed $[S,p]$, Equation (4.65) reduces to

$$(dH)_{S,p} < 0 \tag{4.66}$$

Enthalpy continues to decrease, with irreversible processes taking place in the system until equilibrium is attained as the enthalpy reaches a minimum,

$$(dH)_{S,p} = 0 \tag{4.67}$$

Here we arrive at the third criterion of equilibrium.

4.1.5.3 Equilibrium Criterion III

At fixed S and p, a system seeks equilibrium by decreasing enthalpy until the equilibrium state is attained at which enthalpy is a minimum.

In $[H,S,p]$ space, equilibrium states are represented by a surface. Space above the surface is occupied by nonequilibrium states; space below the surface is unattainable. Motion on the surface is propelled by changes dS or dp, or both, according to Equation (4.65) with the equal sign,

$$dH = T\,dS + V\,dp \tag{4.68}$$

Equation (4.68) shows that the natural energy function of the set of variables $[S,p]$ is H and that the natural variables of H are S and p.

The geometry of the equilibrium surface is determined by the coefficients of Equation (4.68),

$$(\partial H / \partial S)_p = T \tag{4.69}$$

and

$$(\partial H / \partial p)_s = V \tag{4.70}$$

An immediate application of H as the natural energy function of S and p is its use as a measure of the work of expansion at reversible adiabatic flow conditions. The differential work of expansion at flow is given by $-V\,dp$. Comparison with Equation (4.68) shows that

$$(-dH = -dW)_s \tag{4.71}$$

Integrating,

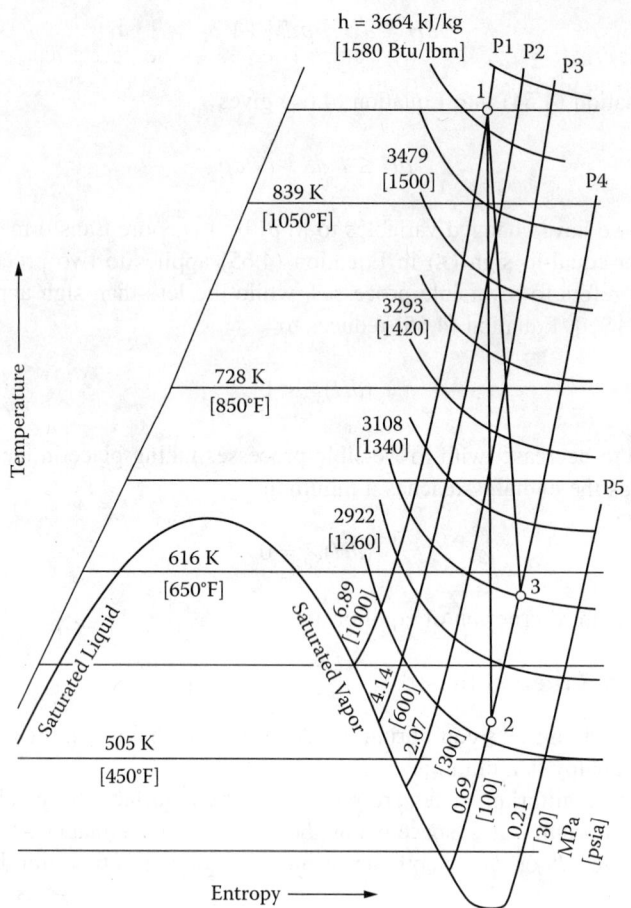

FIGURE 4.2 T-S chart for steam. (From Keenan, J. H., Keyes, F. G., Hill, P. G., and Moore, J. G. *Steam Table, Thermodynamic Properties of Water, Including Vapor, Liquid, and Solid Phases*. New York: John Wiley, 1969. With permission.)

$$(-\Delta H = -W)_s \tag{4.72}$$

The decrease of enthalphy is the measure of reversible work that can be extracted from a flowing fluid upon expansion.

Example 4.5: Work of Expansion of Steam

Steam at 6.89 MPa and 849° (1150°F) is expanded adiabatically to 0.69 MPa in a turbine. How much reversible flow work can be extracted from 1 kg of steam?

Solution

The reversible adiabatic flow work done by the steam is given by $(H_1 - H_2)$. From Figure 4.2, H_1 = 3687 kJ/kg and H_2 = 2961 kJ/kg. Therefore, $-W$ = 3687 − 2961 = 726 kJ/kg.

If the expansion is adiabatic and irreversible, entropy increases during the expansion. Point 3 shows a possible final state. The enthalpy of the final state would be greater than the H_2 given above, and the work produced, being equal to the decrease in H, would be reduced.

Example 4.6: Enthalpy as a Measure of Heat Effect

Enthalpy is useful as a measure of the heat effect of constant-pressure processes. By the first law, $dU = \delta q - p\, dV$ when pV work is the only work term. At a condition of constant pressure, $p\, dV =$

Thermodynamics of Fluid Phase and Chemical Equilibria

$d(pV)$. Moving the pV term to the same side with U and combining, $d(U + pV) = \delta q$, or $dH = \delta q$. Integrating, we obtain $\Delta H = Q$. Enthalpy change is equal to the heat effect at constant pressure. The calculation of heat effect by using enthalpy is illustrated in Example, Section I..G.

Heat capacity at constant pressure C_p is defined to be $(\delta q/dT)_p$. In view of the above discussion, we obtain $C_p = (dH/dT)_p$.

The natural variables S and V of U are transformed to T and V by Legendre transformation of U to the energy function A, called the Helmholtz free energy, or Helmholtz energy. By definition,

$$A = U - TS \tag{4.73}$$

Differentiation gives

$$dA = dU - T\,dS - S\,dT \tag{4.74}$$

From Equation (4.54)

$$dU - T\,dS \leq -p\,dV \tag{4.75}$$

Substitution of Equation (4.75) into Equation (4.74) gives

$$dA \leq -S\,dT - p\,dV \tag{4.76}$$

The equal sign applies to reversible processes and the less-than sign to irreversible processes.

It follows at constant T and V that

$$(dA \leq 0)_{T,V} \tag{4.77}$$

Helmholtz energy decreases as irreversible processes take place in the system at constant T and V, until the irreversible processes are exhausted and the system attains equilibrium at which

$$(dA = 0)_{T,V} \tag{4.78}$$

The criterion of equlibrium is obtained.

4.1.5.4 Equilibrium Criterion IV

At constant T and V, a system seeks equilibrium by decreasing A until equilibrium is attained at a state of minimum A.

Subsequent to attaining equilibrium, changes in A are propelled by dT or dV, or both, according to Equation (4.76) with the equal sign

$$dA = -S\,dT - p\,dV \tag{4.79}$$

Helmholtz energy is displayed as the natural energy function of its natural variables T and V. The geometry of the equilibrium surface is expressed by the coefficients of the differentials,

$$(\partial A / \partial T)_v = -S \tag{4.80}$$

and

$$(\partial A / \partial V)_T = -p \tag{4.81}$$

Equation (4.80) and Equation (4.81) are useful for calculating the Helmholtz energy at various states.

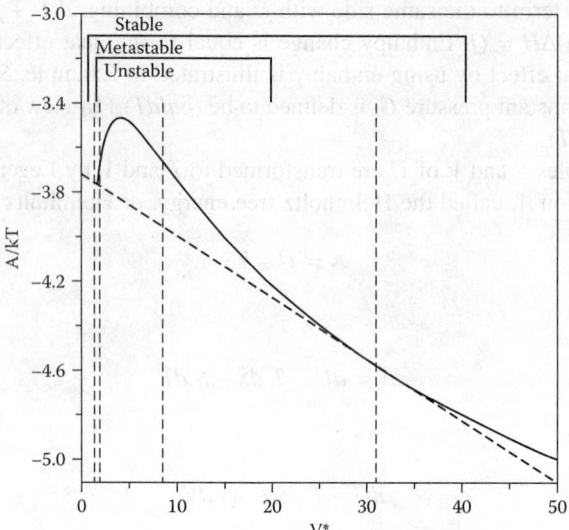

FIGURE 4.3 Helmholtz energy of the Lennard-Jones fluid at a condensation temperature $T^* = 1$ (From Watson, B. S., and Chao, K. C. *J. Chem. Phys.* 96: 9046, 1992. With permission.).

Figure 4.3 shows the molar Helmholtz energy of a fluid at a constant T in the condensation range. The curve represents the A of equilibrium states as a function of volume. A point above the curve represents a nonequilibrium state. The natural tendency is for the point to move down until it settles on the curve.

Example 4.7: Mechanical Instability

An equilibrium state of a homogeneous phase system is not stable if its Helmholtz energy is lowered by separating into two equilibrium phases of different densities. Thus in Figure 4.3 the equilibrium state at point *a* splits into two phases *d* and *e*, the saturation states. Points *b* and *c* show a possible intermediate state in separation. Line *d-e* is tangent to the equilibrium curve at both *d* and *e*. The Helmholtz energy reaches a minimum at point *f* on the line *d-e* at the conclusion of the separation. The slope of the tangent is the vapor pressure.

The Helmholtz energy isotherm of a sub-critical fluid is made up of five regions: (1) Unstable region in which

$$(\partial^2 A / \partial V^2)_T < 0 \tag{4.82}$$

Unstable states do not exist in practice. No homogeneous fluid phase can be found in this density range. (2) Meta-stable region—There are two meta stable regions, one on either side of the unstable region. The second derivative is greater than zero in the metastable region. The metastable region is bounded on one side by the spinodal point where the second derivative of A is zero, and by the saturation point on the other side. Because its equilibrium A stands above the common tangent *d-e*. A metastable state naturally decomposes into saturated states like *d* and *e*, but can be obtained experimentally with care to exdlude nucleation. Super-heated liquid and supersaturated vapor are metastable. (3) Stable region—The second derivative of A with respect to v is positive in this region. There are two stable regions; each is bounded by a saturation point. Vapor phase exists at densities below the saturation value, and liquid at densities above.

Finally, we look for the natural energy function for the variables T and p. The Helmholtz energy of the natural variables T and V offers a starting point. We define the state function Gibbs free energy G by

$$G = A + pV \tag{4.83}$$

Differentiation of Equation (4.83) gives

$$dG = dA + p\,dV + V\,dp \tag{4.84}$$

Using Equation (4.75) to eliminate dA from Equation (4.84) gives

$$dG \leq -S\,dT - V\,dp \tag{4.85}$$

The equal sign applies for reversible processes, while the less-than sign applies for irreversible processes. At constant T and p, if not at equilibrium, the system seeks to decrease its Gibbs energy,

$$(dG \leq 0)_{T,p} \tag{4.86}$$

As the system eventually reaches equilibrium, G attains a minimum,

$$(dG = 0)_{T,p} \tag{4.87}$$

Equation (4.86) and Equation (4.87) express a criterion of equilibium.

4.1.5.5 Equilibrium Criterion V

At constant T and p, a system seeks equilibrium by decreasing G until equilibrium is attained at a state of minimum G.

Equilibrium states in the space $[G,T,p]$ are represented by a surface. A change of G on the equilibrium surface is given by Equation (4.85) with the equal sign,

$$dG = -S\,dT + V\,dp \tag{4.88}$$

The equation displays the Gibbs energy as the natural energy function of its natural variables T and p. The coefficients of the differentials in this equation reflect the geometry of the equilibrium surface,

$$(\partial G / \partial T)_p = -S \tag{4.89}$$

$$(\partial G / \partial p)_T = V \tag{4.90}$$

The equations are useful for finding the change of G as T or p is changed; the equations are also useful for deriving the change of functions that will be defined based on G.

Figure 4.4 shows the Gibbs energy of a binary mixture at a constant T and p with the datum state of either component selected to be the pure fluid at the given T and p. The equilibrium states are on the curve. The nonequilibrium states are above the curve and seek to descend to the equilibrium curve. For instance, to form a mixture of composition m from the pure fluids, the initial state of the two fluids upon being put together before molecular mixing takes place is m with $G = 0$. Molecular mixing irreversibly lowers the G of the mixture until mixing is complete and G reaches a minimum at point n on the equilibrium curve.

Example 4.8: Diffusional Instability

Not all of the equilibrium states in Figure 4.4 are stable; some will seek to decompose into two equilibrium phases of different compositions. Take point c. The shape of the curve at c is such that,

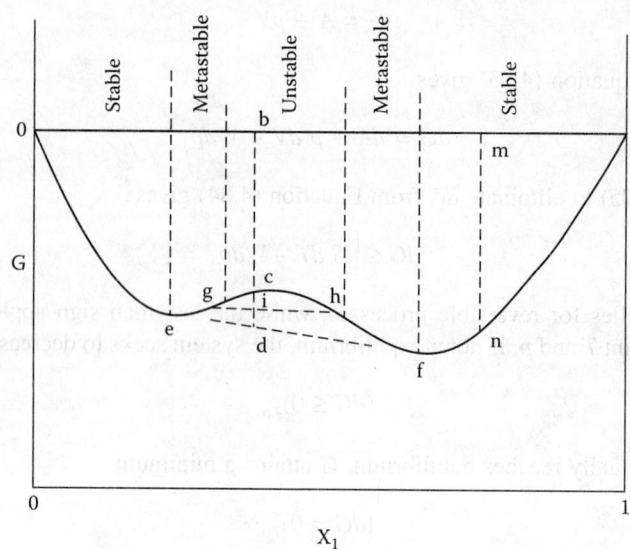

FIGURE 4.4 Gibbs energy of a partially miscible solution at a constant [T,p].

should a small difference in composition develop in any small part of the fluid due to molecular diffusional fluctuation, the difference will be promoted by a decrease in G. The difference grows, and two parts of the solution will be formed to become more and more different until they eventually separate into two phases e and f. Points g and h show an intermediate stage of the separation process. The Gibbs energies of g and h add up to a total G of the system at i, which is lowered from c. The separated phases come to a stop at e and f, yielding a minimal total G at d. Points e and f are the two immiscible mixtures at equilibrium.

The sufficient condition of diffusional instability is

$$(\partial^2 G / \partial x^2)_{T,p} < 0 \tag{4.91}$$

The equation applies to the equilibrium states in the unstable region of x. To the outside of e and f, we find the stable states for which the second derivative of G with respect to x is greater than 0. The states between the unstable region and e or f, despite having a positive value for the second derivative of G with respect to x, seek to decompose into the two phases at e and f to lower their Gibbs energy. These states are metastable.

The state variables T and p are particularly suited for systems at heterogeneous phase equilibrium because equilibrium phases coexist at uniform T and p. As the natural energy function of [T,p], the Gibbs energy and derived functions will be used to describe phase equilibrium.

The state functions U, H, A, and G are called energy functions. Each is associated with a pair of natural variables. To map the energy functions of a substance over a range of states by experimental measurement, by correlation of data, or by theoretical calculations, it is advantageous to develop *one* energy function as a function of its natural variables. Then all other energy functions can be obtained by elementary mathematical operations. The procedure is illustrated in the following, with Helmholtz free energy A having been worked out as a function of [T,V]. From the defining differential equation of A, Equation (4.79), the partial derivatives have been identified in Equation (4.80) and Equation (4.81): $S = -(\partial A/\partial T)_V$, $p = -(\partial A/\partial V)_T$. Substitution in the defining equations of the energy functions leads successively to

$$G = A + pV = A[T,V] - V(\partial A / \partial V)_T \tag{4.92}$$

Thermodynamics of Fluid Phase and Chemical Equilibria

$$U = A + TS = A[T,V] - T(\partial A / \partial T)_V \tag{4.93}$$

$$H = A + TS + pV = A[T,V] - T(\partial A / \partial T)_V + V(\partial A / \partial V)_T \tag{4.94}$$

In this manner, all energy functions other than A are obtained, all in terms of state variables $[T,V]$ in the region in which A is known.

4.1.6 Open System and Chemical Potential

An open system can exchange mass with its surroundings, and mole numbers of the chemical substances are the new state variables required for the specification of the system. Consider the Gibbs energy of an open system. In addition to its natural state variables T and p, there are the state variables, $n_1, n_2, \ldots, n_i, \ldots, n_v$, where n_i stands for mole number of species i, and n is the number of chemical species in the system,

$$G = G[T, p, n_1, \ldots] \tag{4.95}$$

A differential change of G is expressed by

$$dG = (\partial G / \partial T)_{p,ni} dT + (\partial G / \partial p)_{T,ni} dp + \sum_i (\partial G / \partial n_i)_{T,p,nj} dn_i \tag{4.96}$$

Since the mole numbers are constants in a closed system, the first two partial derivatives in Equation (4.96) are identical to the respective partial derivatives for the system if closed. Reference to Equation (4.88) shows

$$(\partial G / \partial T)_{p,ni} = -S \tag{4.97}$$

$$(\partial G / \partial p)_{T,ni} = V \tag{4.98}$$

Substituting back into Equation (4.96) gives

$$dG = -S\,dT + V\,dP + \sum_i \mu_i dn_i \tag{4.99}$$

where we have defined

$$\mu_i = (\partial G / \partial n_i)_{T,p,nj} \tag{4.100}$$

The constancy of T and p in the partial derivative identifies μ_i as the partial molar Gibbs energy. Because of its key importance in phase equilibrium and chemical equilibrium, μ is given the name chemical potential.

The other energy functions are related to G by

$$G = A + pV = H - TS = U + pV - TS \tag{4.101}$$

Differentiation of Equation (4.101) and combination with Equation (4.99) gives

$$dU = T\,dS - p\,dV + \sum_i \mu_i\,dn_i \tag{4.102}$$

$$dH = T\,dS + V\,dp + \sum_i \mu_i\,dn_i \tag{4.103}$$

$$dA = -S\,dT - p\,dV + \sum_i \mu_i\,dn_i \tag{4.104}$$

$$dG = -S\,dT + V\,dp + \sum_i \mu_i\,dn_i \tag{4.105}$$

These four equations complete the Gibbs fundamental forms. It follows from these equations that μ_i is also the partial mole number derivative of U, H, or A in addition to that of G,

$$\mu_i = (\partial U / \partial n_i)_{S,V,nj} = (\partial H / \partial n_i)_{S,p,nj} = (\partial A / \partial n_i)_{T,V,nj} \tag{4.106}$$

However, the partial differentiations here are not at constant T and p; consequently, μ_i is not partial molar U, nor partial molar H, nor partial molar A. The conditions of the partial differentiation for U and H cannot be physically interpreted, making the partial derivatives of U and H of little interest for the description of real systems. But the partial differentiation of A is at a physically meaningful condition that makes the partial derivative useful for the calculation of chemical potential.

Chemical potential plays a fundamental role in phase and chemical equilibria. Consider a heterogeneous system at a constant T and p consisting of a number of phases that are open and are at equilibrium with respect to one another. The chemical potential, like any intensive property, must assume a uniform value in any phase because each phase is at equilibrium. The chemical potential is the same value in different phases at equilibrium, so that diffusional mass transfer does not take place between the phases. Consider the transport of dn_i, moles of i from phase α of higher μ_i to phase β of lower μ_i. The change of G is

$$dG = [(\partial G_\beta / \partial n_i)_{T,p,nj} - (\partial G_\alpha / \partial n_i)_{T,p,nj}]dn_i \tag{4.107}$$

$$= [\mu_{\beta,i} - \mu_{\beta,i}]dn_i \tag{4.108}$$

showing that $dG < 0$. By equilibrium criterion V, a system at constant T and p seeks equilibrium by decreasing G. Mass transfer then occurs, and the system is not at equilibrium. For the phases to be at equilibrium, $dG = 0$, it is necessary that $\mu_{\alpha,1} = \mu_{\beta,1}$. Therefore, the condition of phase equilibrium is expressed by

$$T_\alpha = T_\beta = \ldots \quad \text{for all phases} \tag{4.109}$$

$$p_\alpha = p_\beta = \ldots \quad \text{for all phases} \tag{4.110}$$

Thermodynamics of Fluid Phase and Chemical Equilibria

$$\mu_{\alpha,1} = \mu_{\beta,1} = \ldots \quad \text{for all phases} \tag{4.111}$$

$$\mu_{\alpha,2} = \mu_{\beta,2} = \ldots \quad \text{for all phases} \tag{4.112}$$

for all chemical species.

These equations are the basis for phase equilibrium. To develop methods of calculation of phase equilibrium on this basis, efforts will be made in two stages:

1. Transforming chemical potential to a more convenient quantity, called fugacity, as the measure of equilibrium; and learning how fugacity depends on the species and state of the mixture. These are the subjects of Sections 4.2 and 4.3.
2. Searching for the state at which the fugacities are matched in between the phases while satisfying the specified conditions of temperature, pressure, moles, enthalpy, or other factors. This is the subject of Section 4.4.

Gibbs developed the fundamental criterion of chemical equilibrium in terms of chemical potentials. Consider a chemical reaction in the general form

$$\sum_i \nu_i M_i = 0 \tag{4.113}$$

where ν_i denotes the stoichiometric coefficient of species i in the reaction, and M_i is the molecular formula. By usual convention, ν_i is positive for the products and negative for the reactants. The stoichiometry of the reaction is expressed by

$$dn_i / \nu_i = d\xi \quad \text{for all } i \tag{4.114}$$

where ξ is the degree of advancement or extent of the reaction. All mole number changes can be expressed in terms of ξ, from Equation (4.114),

$$dn_i = \nu_i d\xi \quad \text{for all } i \tag{4.115}$$

For the reaction at constant T and V to be at equilibrium, by criterion V, Helmholtz energy must be at a minimum, $dA = 0$,

$$dA = -SdT - pdV + \sum_i \mu_i dn_i \tag{4.116}$$

$$= \left(\sum_i \mu_i \nu_i\right) d\xi \tag{4.117}$$

$$= 0 \tag{4.118}$$

Since $d\xi$ is arbitrary, Equation (4.117) requires

$$\sum_i \mu_i \nu_i = 0 \qquad (4.119)$$

Hence the sum of chemical potentials of the reactants must balance the sum of the products.

By considering the reaction system at constant T and p, and equilibrium criterion V, $dG = 0$, one arrives at the same conclusion that Equation (4.119) is the condition of chemical reaction equilibrium.

Equation (4.119) is the fundamental equation of chemical equilibrium. To make progress toward a useful description of chemical equilibrium, efforts are required in two stages:

1. Replace chemical potential with fugacity and find out how fugacity depends on the species and the state of the mixture. This the subject of Sections 4.2 and 4.3.
2. Search for the state at which the fugacities satisfy the criterion of chemical equilibrium while meeting the specified conditions of temperature, pressure, moles, enthalpy, or other factors. This is the subject of Section 4.5.

4.1.7 PARTIAL MOLAR QUANTITIES AND THE GIBBS-DUHEM RELATION

The fundamental differential form for Gibbs energy can be integrated to express Gibbs energy as a sum of contributions by the components. From Equation (4.99),

$$dG = -SdT + Vdp + \sum_i (\partial G/\partial n_i)_{T,p,nj} dn_i \qquad (4.120)$$

upon integration at constant T, p, and composition, and thus constant μ_i,

$$G = \sum_i (\partial G/\partial n_i)_{T,p,nj} n_i = \sum_i \mu_i n_i \qquad (4.121)$$

By Equation (4.121), the partial molar Gibbs energy of component i represents the contribution of i to the total Gibbs energy of the mixture.

An extensive property Y in general can be similarly represented in terms of contributions by the components. Let Y be a state function of T, p, and the mole numbers n_i: $Y(T,p,n_1,n_2,...)$, where n_i denotes a mole number. The differential of Y at constant p is

$$dY = \sum_i (\partial Y/\partial n_i)_{T,p,nj} dn_i \qquad (4.122)$$

This equation is integrated at constant T, p, and composition, and thus at constant $(\partial Y/\partial n_i)_{T,p,nj}$, in a homogeneous phase to give

$$Y = \sum_i (\partial Y/\partial n_i)_{T,p,nj} n_i = \sum_i Y_i n_i \qquad (4.123)$$

where $Y_i = (\partial/\partial n_i)_{T,p,nj}$ stands for the partial molar Y of i in a homogeneous phase. Thus, for a phase,

$$U = \sum_i U_i n_i \qquad (4.124)$$

$$H = \sum_i H_i n_i \qquad (4.125)$$

$$A = \sum_i A_i n_i \qquad (4.126)$$

$$V = \sum_i V_i n_i \qquad (4.127)$$

An extensive property of a mixture is the sum of the mole number times the partial molar quantity of the components for a phase. A component in mixture exists in a partial state, assuming partial property values; it contributes to the property of the mixture by its partial value.

The partial molar quantities are not independent, but are mutually related and related to T and p by the Gibbs–Duhem equation. Starting with Gibbs energy, we differentiate Equation (4.121) to obtain

$$dG = \sum_i \mu_i dn_i + \sum_i n_i d\nu_i \qquad (4.128)$$

Comparing this equation with dG of Equation (4.105), we have, for a homogeneous phase,

$$\sum_i n_i d\mu_i = -S\,dT + V\,dp \qquad (4.129)$$

For enthalpy, we differentiate H as a state function of T, p, and n_i, $i = 1,2,\ldots$,

$$dH = (dH/dT)_{p,n_i} dT + (dH/dp)_{T,n_i} dp + \sum_i H_i dn_i \qquad (4.130)$$

We next diffferentiate Equation (4.125) to give

$$dH = \sum_i H_i dn_i + \sum_i n_i dH_i \qquad (4.131)$$

Comparing the two equations, we have

$$\sum_i n_i dH_i = (dH/dT)_{p,n_i} dT + (dH/dp)_{T,n_i} dp \qquad (4.132)$$

At constant T and p, expressing only the effect of composition change,

$$\sum_i n_i dH_i = 0 \qquad (4.133)$$

The method of derivation for H is generally useful for extensive properties to relate the effect of intensive properties on partial molar properties.

Example 4.9: Heat Generated by Adding a Stream to a Tank

In a batch-feed operation, 100 mol of pure liquid is added at the rate of 2 mol/min to a tank of liquid solution initially containing 100 mol of pure A. The tank is stirred, and its contents are assumed to be thoroughly mixed and uniform in composition at all times. The tank is jacketed to provide for heat removal to keep the tank contents at a constant temperature of 300°K. The pressure is constant at 1 atm.

Find the rate of evolution of heat of dilution as a function of time from $t = 0$ to $t = 50$ min in units of kJ/min. Find the total heat evolved.

Given:

Partial Molar Enthalpy at 300

Mol % of	H_B, kJ/mol
0	–6,240
10	–17,556
20	–25,916
30	–31,768
40	–38,038
50	–43,890
60	–46,816
100	–56,430

No information is given about the partial enthalpy of A.

Solution

For the mixing process at constant pressure, the heat release is equal to the decrease in enthalpy upon mixing according to Example I.E-2. We need to find the enthalpy change of mixing. For the addition of dn moles of into the body of mixture containing n_A moles of A and n_B moles of B, the initial state enthalpy is

$$H_i = (n_A H_A + n_B H_B) + H_B^0 dn_B$$

where the superscript 0 denotes pure liquid and subscripts A or B denote a partial molar quantity of A or B, respectively. The final state enthalpy is

$$H_f = (n_A H_A + n_B H_B) + d(n_A H_A + n_B H_B)$$

The differential change of enthalpy upon mixing is

$$dH = H_f - H_i = d(n_A H_A + n_B H_B) - H_B^0 dn_B$$

$$= (H_A dn_A + n_A dH_A + n_B dH_B + H_B dn_B) - H_B^0 dn_B$$

Thermodynamics of Fluid Phase and Chemical Equilibria

The first term in the last equation vanishes because n_A is unchanged. The second and third terms add up to zero according to the Gibbs-Duhem equation. There is left only

$$dH = (H_B - H_B^0)\, dn_B$$

that says that only the change of state of the differential moles of B added needs to be considered. The rate of heat production is

$$Q = (H_B - H_B^0) dn_B/dt$$

Substituting the table values of H_B and H_B^0, we find a large quantity of heat is produced, like the mixing of sulfuric acid into water. The numerical calculation is left as an exercise for the reader.

The total heat evolved is given by integrating the differential heat evolution,

$$Q = \int_0^{} (H_B - H_B^0) dn_B$$

Even with the addition of a finite quantity of B changing the composition from pure A to 50% A, the total enthalpy change does not involve changes in the partial enthalpy of A. To verify the formula for Q, we find the total change of H with the mixing of 50 mol A and 50 mol B:

$$Q = H_{\text{total mix}} - (n_A H_A^0 + n_B H_B^0)$$

$$= (n_A H_A + n_B H_B) - (n_A H_A^0 + n_B H_B^0)$$

Let us develop the integral formula

$$Q = \int_0^{} (H_B - H_B^0) dn_B$$

Because H_B^0 is constant,

$$= \int_0^{} H_B dn_B - H_B^0 n_B$$

By partial integration,

$$= (H_B n_B)\big|_0 - \int_0^{} n_B dH_B - H_B^0 n_B$$

Using Gibbs–Duhem equation for the integrand,

$$= H_B n_B + \int n_A dH_A - H_B^0 n_B$$

n_A is constant:

$$= H_B n_B + n_A H_A - n_A H_A^0 - H_B^0 n_B$$

Thus the integral is verified to agree with the total enthalpy calculation. Although only H_B is used in the integral method of calculation, information about H_A and the molar of the mixture are implied in H_B and can be explicitly revealed by integrating H_B to yield H_A, and subsequently adding the two partials to obtain the molar of the mixture.

For 1 mole of a binary mixture, in general, by the Gibbs-Duhem equation at constant and p,

$$x_A \, dH_A + xB \, dH_B = 0$$

Transporting the second term to the other side and integrating,

$$H_A - H_A^0 = -\int_0 (x_B / x_A)(dH_B / dx_B) dx_B$$

Having found H_A, the total mixture is obtained by summing,

$$= x_A H_A + x_B H_B$$

The enthalpy of the mixture becomes completely known as soon as the partial molar enthalpy of one component is determined.

4.2 VOLUMETRIC AND THERMODYNAMIC PROPERTIES

4.2.1 Pressure–Volume–Temperature Relationship

A substance can exist as a gas, liquid, or solid. The gas state exists at low pressure. Within a range of temperature, depending on the substance, as the pressure is increased at a constant temperature, the gas condenses into a liquid. At temperatures above the range, the gas stays unchanged in phase state to very high pressures. At temperatures below the range, the gas condenses upon compression directly into a solid.

Figure 4.5 shows the pressure variation with volume at various constant temperatures for a typical substance. The isotherms are ordered: the higher the temperature, the higher is the isotherm; the isotherms do not intersect. At T_1, the highest temperature shown, the substance is a gas at all pressures, and the ideal-gas law is often a good approximation. This isotherm approximates a hyperbola.

At a lower temperature T_2, the gas state exists at low pressures. As the pressure is increased to point 5 in Figure 4.5, the vapor becomes saturated and begins to condense. With further reduction of volume, more condensation occurs while the pressure stays unchanged. The intensive properties of the vapor and liquid phases remain constant until condensation is complete at point 6. At higher pressures, a compressed liquid is found that changes volume only slightly upon further compression.

At temperature T_3 lower than T_2, the low-pressure gas becomes condensed as it is compressed, just like at T_2. Upon being sufficiently compressed, the liquid begins to solidify at point 9. The pressure stays constant as volume is reduced while solidification progresses along the line segment 9 to 10 until solidification is complete at point 10. Only solid exists at higher pressure in the p–V space beyond the line f-g-h.

At temperature T_4 lower than T_3, the low-pressure gas condenses directly into a solid at a usually low pressure called the *sublimation pressure*. The condensation proceeds as volume is reduced at constant pressure until the condensation is complete at point 12. Further reduction of volume of the solid requires application of high pressure. The gas states are found in the area marked G, generally at low pressure and high temperature. Liquids are found in the area marked L, and solid states in S.

Thermodynamics of Fluid Phase and Chemical Equilibria

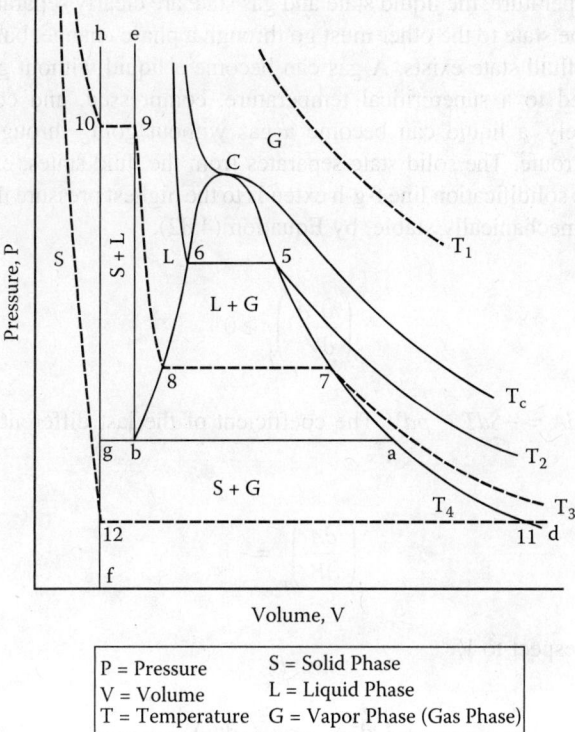

FIGURE 4.5 *p-V* diagram for pure component.

The saturated vapor states are on line c-a-d. On segment c-a, the saturated vapor condenses to a liquid "dew" as the first liquid drops appear. On segment a-d, the saturated vapor condenses into a solid "frost." The saturated liquid states are on line c-b, where the last vapor bubbles are disappearing. On line b-c, the liquid is saturated and begins forming a solid. On line g-h, the solid is imminently melting into a liquid; on line f-g, it is subliming to a vapor. Two phases coexist in three areas in Figure 4.5: liquid and gas labeled, L+G; solid and gas, labeled S+G; and solid and liquid, labeled S+L. Three phases, gas, liquid, and solid, coexist on one line: a-b-g. Temperature and pressure remain constant at the triple point, while volume may vary.

The two saturation lines—c-a representing saturated gas and c-b representing saturated liquid—meet at c, which is the highest point of the L+G two-phase area. Point c is the gas-liquid *critical state*. As temperature is increased from below to approach the critical, the flat gas-liquid coexistence segment of the isotherm shrinks to the point c at the critical temperature. The critical point is a point of inflection of the isotherm. At c,

$$\left(\frac{\partial p}{\partial V}\right)_T = 0 \tag{4.134}$$

and

$$\left(\frac{\partial^2 p}{\partial V^2}\right)_T = 0 \tag{4.135}$$

Below the critical temperature, the liquid state and gas state are clearly separated in the L+G region. The transition from one state to the other must go through a phase change, but at temperature above the critical, only one fluid state exists. A gas can become a liquid without going through a phase transition when heated to a supercritical temperature, compressed, and cooled to a subcritical temperature. Conversely, a liquid can become a gas without going through a phase change by following the reverse route. The solid state separates from the fluid states, either gas or liquid, by a phase transition. The solidification line f-g-h extends to the highest pressure that has been observed.

For a state to be mechanically stable, by Equation (4.82),

$$\left(\frac{\partial^2 A}{\partial V^2}\right)_T \geq 0 \tag{4.136}$$

By Equation (4.79), $dA = -SdT - pdV$. The coefficient of the last differential in this equation is identified as

$$\left(\frac{\partial A}{\partial V}\right)_T = -p \tag{4.137}$$

Differentiating with respect to V,

$$\left(\frac{\partial^2 A}{\partial V^2}\right)_T = -\left(\frac{\partial p}{\partial V}\right)_T \tag{4.138}$$

Comparing Equation (4.127) and Equation (4.138) shows that mechanical stability requires

$$\left(\frac{\partial p}{\partial V}\right)_T \leq 0 \tag{4.139}$$

The isotherms must have a negative or zero slope. In Figure 4.5 the slope is negative for single-phase states and zero for the two-phase states.

Figure 4.6 shows the pVT relationship in the pT plane with lines of constant volume called isochores. The isochore v_1 is a high-density line; it starts from the two-phase coexistence line of the S+L regions to enter into the liquid region, rapidly gaining pressure as temperature is increased. The isochores v_2 and v_3 are at successively larger volume or lower density; both start from the two-phase coexistence line of L+G, which is the *vapor pressure curve*, to enter into the liquid region. The slope of the isochore drops successively from v_1 to v_3. The isochores v_4 and v_5, with successively larger volume and both being larger than the critical, likewise start from the L+G coexistence line, but go into the gas region. The isochore v_6, starting from the two-phase line S+G, becomes a sublimed vapor as p and T increase; its volume is the largest shown.

Isochores with $v < v_c$ are above the L+G two-phase coexistence line; isochores with $v > v_c$ are below the line. The isochore with $v = v_c$ coincides with the L+G coexistence line over the entire range from the triple point t to the critical point c. Beyond the critical point it is a continuous extension of the coexistence line. The isochores are remarkable by their near linearity. The curvature, generally small, is the largest at states close to the two-phase line.

The S+G two-phase coexistence line, or the *sublimation pressure curve*, f-t, starts from the triple-point t to proceed downward toward the left. The S+L two-phase coexistence line, t-h, is the *melting point curve*. Typically the melting temperature is higher at elevated pressure, as shown in

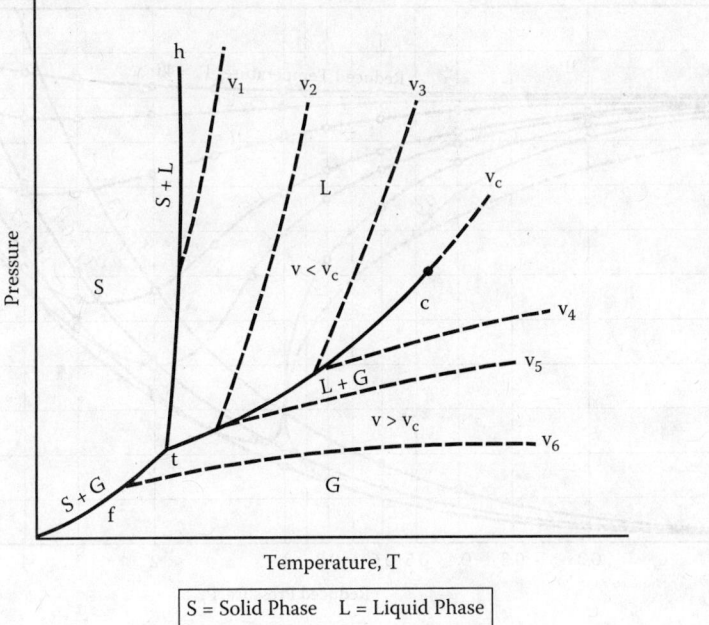

FIGURE 4.6 T diagram for pure component.

Figure 4.6. An important exception is found with water; its melting temperature is lowered as pressure is raised. The melting curve extends without limit in the direction of high pressure. All three two-phase coexistence curves—the S+G, S+L, and L+G—meet at the triple point where the three phases S, L, and G coexist.

4.2.2 Principle of Corresponding States

The experimentally observed similarity of the pressure-volume-temperature behavior of many fluids can be represented by the principle of corresponding states (PCS), according to which various substances behave in the same way when expressed on suitable reduced scales. Thus the pVT surfaces of different substances superimpose upon reduction with appropriate scale factors. The reducing scale factors commonly employed are the properties of the fluid at a singular point such as the critical point. On the reduced scale, one general pVT relationship is followed by a number of substances, i.e., p_r is the same function of T_r and v_r, where the subscript r denotes a reduced dimensionless quantity and the subscript c the quantity at the critical point:

$$P_r = p/p_c \tag{4.140}$$

$$T_r = T/T_c \tag{4.141}$$

$$v_r = v/v_c \tag{4.142}$$

For practical calculations, the pVT relationship is conveniently expressed in a compressibility-factor z chart, where $z \equiv pv/RT$. Figure 4.7 shows z as a function of p_r at various T_r. This figure well represents the simple fluids Ar, Kr, and Xe, but is at best an approximate for other fluids.

For an ideal gas, $z = 1$. All real-fluid isotherms approach $z = 1$ as $p_r \to 0$. The high T_r isotherms are close to being horizontal and stay close to the line $z = 1$ at all values of p_r. This is the behavior of common gases such as air, hydrogen, and helium at ambient and higher temperatures. As T_r

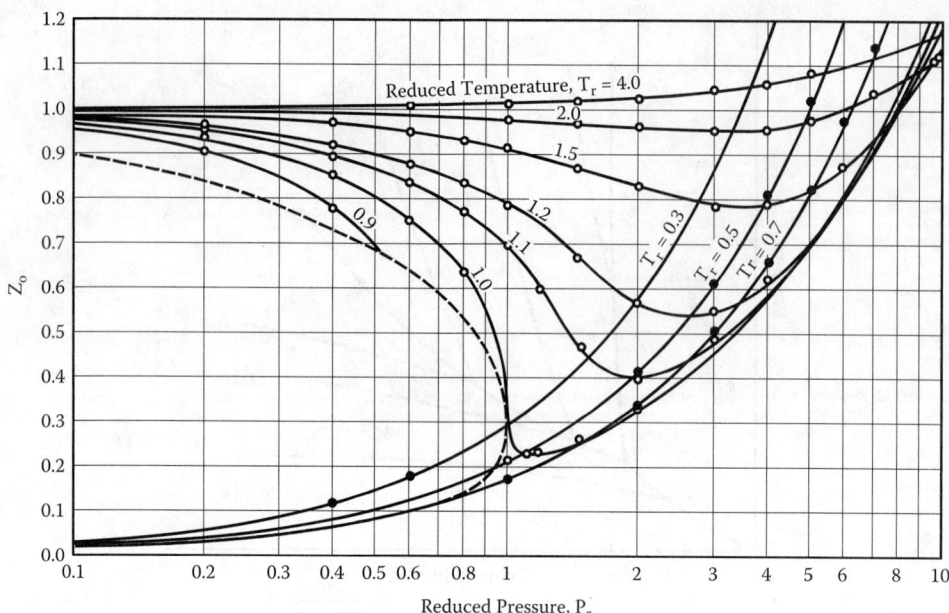

FIGURE 4.7 z_o for simple fluids. (From Lee, B. I., and Kesler, M. *AIChE J.* 21: 510–527, 1975. With permission.)

decreases, the isotherms become lower. At the Boyle point, the isotherm becomes horizontal at $p_r = 0$, which defines the Boyle point. Below the Boyle point, the isotherms approach $p_r = 0$ with a negative slope. Until the temperature is lowered to $T_r = 1$, the isotherms are continuous from $p_r = 0$ to high pressure. Below $T_r = 1$, the isotherms extends smoothly from $p_r = 0$ only to the point of saturation of the vapor, where it descends vertically at a constant pressure until it reaches the saturated liquid state. Increase of pressure beyond the vapor pressure sends the isotherm up steadily due to the small compressibility of the liquid, eventually to large values of z.

To extend the PCS correlation to fluids other than the simple fluids, it is observed that deviation from the PCS correlation of the simple fluids occurs in a systematic way, depending on the shape, size, and chemical nature of the molecule. A prominent example is the reduced vapor pressure p_{rs} as a function of the reduced temperature T_r. Although it is almost exactly the same function for the three simple fluids, the reduced-vapor-pressure lines of other substances are lower with more elongated or more polar molecules. Pitzer and Brewer [1] adopt the lowering of the reduced-vapor-pressure curve of a substance from that of the simple fluids as the basis for defining a parameter to extend the PCS correlation to substances other than simple fluids. At the reduced temperature of 0.7, the reduced vapor pressure of simple fluids equals 0.1 and $-\log p_{rs} = 1$. Taking the difference at $T_r = 0.7$ between the $(-\log p_{rs})$ of a substance and that of the simple fluids, Pitzer and Brewer defined the acentric factor

$$\omega \equiv (-\log p_{rs}) - 1 \qquad (4.143)$$

The minus value of $\log p_{rs}$ is used to make it a positive number. The lower the reduced-vapor-pressure curve of a substance, the larger is the positive value $-\log p_{rs}$ and, hence, the larger ω for the substance. The acentric factor is zero for the simple fluids and increases as the molecules are more nonspherical or more polar.

With the use of ω as the third parameter (in addition to T_c and p_c), the pVT relationship of nonpolar and weakly polar substances, called normal fluids, is correlated on the basis of PCS by extending the z correlation of simple fluids according to

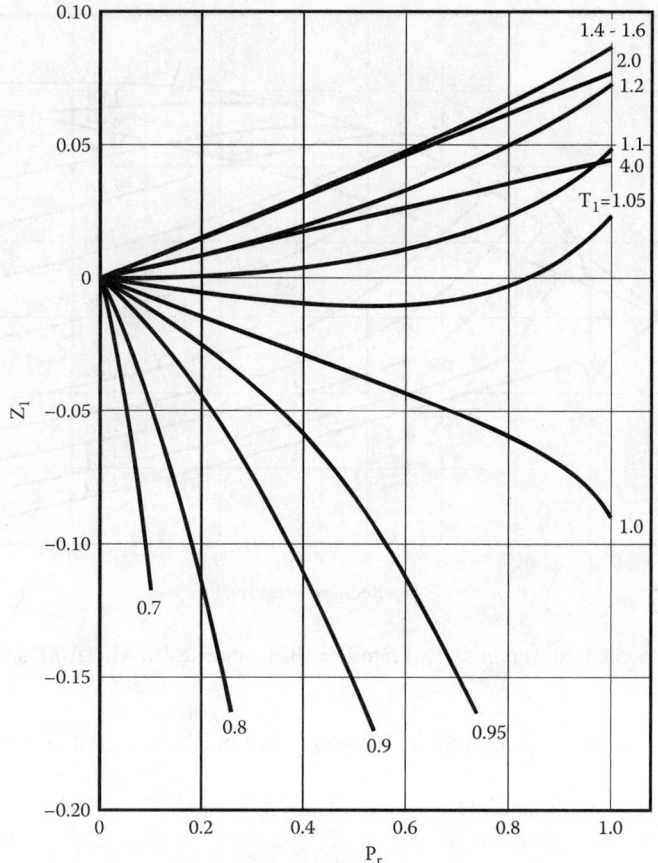

FIGURE 4.8 z_1 for normal fluids at $p_r < 1$. (From Lee, B. I., and Kesler, M. *AIChE J.* 21: 510–527, 1975. With permission.)

$$z[T_r,p_r] = z_0[T_r,p_r] + \omega z_1[T_r,p_r] \qquad (4.144)$$

where z_0 is the generalized compressibility factor of simple fluids, and z_1 is the generalized correction function for normal fluids The square brackets indicate the independent variables. Figure 4.7 presents z_0; Figure 4.8, z_1 for $p_r < 1$; and Figure 4.9, z_1 for $p_r > 1$.

Normal fluids are hydrocarbons, ethers, esters, and many organic halides. Strongly hydrogen-bonded substances like water, acetic acid, and hydrogen fluoride are not normal fluids; neither are quantum fluids like helium and hydrogen. Lee and Kesler [2] developed an equation to represent the three-parameter generalized correlation. The equation is presented in Section 4.2.5, entitled "Equations of State." Figures 4.7, 4.8, and 4.9 are based on the Lee and Kesler equation.

Example 4.10
Find the Density of Compressed Methane at 300 and 68 atm

Solution
From Chapter 1 ("Physical and Chemical Properties"), we obtain for methane:

$$T_c = 190.7°K, \ p_c = 45.8 \text{ atm}, \ \omega = 0.013.$$

The reduced variables are found: $T_r = 300/190.7 = 1.573$ and $p_r = 68/45.8 = 1.49$. From Figure 4.8, $z_0 = 0.89$. From Figure 4.9, $z_1 = 0.14$. By Equation (4.144), $z = 0.89 + (0.013)(0.14) = 0.892$ and

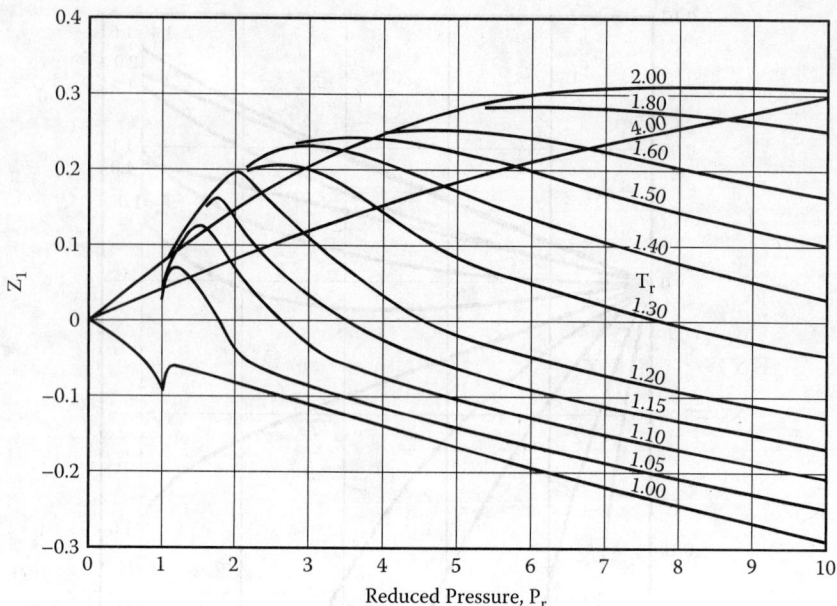

FIGURE 4.9 z_1 for normal fluids at $p_r > 1$. (From Lee, B. I., and Kesler, M. *AIChE J.* 21: 510–527, 1975. With permission..

$$v = zRT/p = (0.892)(82.0)(300)/68 = 323 \text{ cm}^3/\text{mol} = 20.2 \text{ cm}^3/\text{gm}$$

For methane, ω is small, and the correction from z_0 is small.

4.2.3 PHASE RULE

The phase rule due to J. Willard Gibbs gives the number of intensive variables that can be specified to completely fix the intensive state of a system at heterogeneous phase equilibrium. The number of intensive variables that can be specified is called the degree of freedom F. Experience tells us that for a given material existing as a one-phase gas, liquid, or solid, it requires two intensive variables to be specified, for example T and p, or T and ρ, to fix the intensive state of the system; thus $F = 2$. But for a pure substance, for instance steam and liquid water, to coexist in two equilibrium phases, only one intensive variable can be specified, for example T or p. By fixing the temperature, the pressure is fixed at the vapor pressure, and all intensive properties of the coexisting vapor or liquid, such as the density of the vapor, density of the liquid, etc., are all fixed; thus $F = 1$.

To find the degree of freedom in general for a system consisting of C components coexisting in P phases, Gibbs considered the conditions of the phases in coexistence, Equations (4.145) to (4.147). There are in total $(2 + C)(P - 1)$ equations. For each phase there are $2 + (C - 1)$ independent variables, e.g., p, and $(C - 1)$ compositions variables, adding up for the P phases to a total number of $P(2 + C - 1)$. How many independent variables can be specified for the equations to have a solution? The answer is: the number of variables minus the number of equations,

$$F = P(2 + C - 1) - (2 + C)(P - 1) \tag{4.145}$$

or

$$F = C - P + 2 \tag{4.146}$$

Thermodynamics of Fluid Phase and Chemical Equilibria

Equation (4.146) expresses the *Gibbs phase rule* in the absence of chemical reactions, surface effects, or membrane partitions. It is the most used form of the phase rule.

For a chemically reacting system at reaction equilibrium, the chemical potentials of the reactants are related by the equilibrium equations, Equation (4.119), there being one such equation for each independent reaction. The number of equations governing the chemical potentials is increased by the number of independent reactions R, correspondingly reducing the degree of freedom by R. The phase rule for reacting systems is, therefore,

$$F = C - P + 2 - R \tag{4.147}$$

The phase behavior of a one-component system will illustrate the phase rule. In the pT diagram of Figure 4.6, the single-phase states are in the areas labeled G, L, or S, representing gas, liquid, or solid, respectively. By Equation (4.146), $F = C - P + 2 = 1 - 1 + 2 = 2$. There are two degrees of freedom, e.g., T and p, or T and ρ. Specifying two intensive variables completely fixed the intensive state of the phase.

There is one degree of freedom on a two-phase coexistence line, L+G, S+L, or S+G in Figure 4.6. Specifying T or p of both phases, or ρ of one of the phases, completely fixed the intensive states of both phases at the triple point (point t), where the three phases (G, L, S) are in coexistence; then $F = C - P + 2$ with zero degrees of freedom.

4.2.4 Phase Behavior of Mixtures

The phase behavior of single-component systems has been discussed as part of the pVT relationship presented in Section 4.2.1. Examining the phase behavior of mixtures, we observe that, with mixtures, phase behavior remains one facet of the pVT relationship. But a new phenomenon is encountered with mixtures: phases at equilibrium are generally of different compositions. These mixtures show a great variety of phase behavior that can often be exploited to make separations. We examine in broad terms the qualitative features of the phase behavior of binary mixtures of various types. Experience has shown a wealth of phenomena displayed by binary mixtures.

We start with gas-liquid equilibrium (gle). For simple mixtures of similar components, gas-liquid coexistence states are observed at a wide range of T and p that are bounded at the high-temperature side by the critical states and at the low-temperature side by the formation of solids. In mixtures of large difference in molecular attractive forces, liquid immiscibility takes place. On the high-temperature side, liquid-liquid equilibrium (lle) either merges into gas-liquid equilibrium (gle) or becomes bounded by separate critical states. With extremely dissimilar molecules, the liquid-liquid coexistence phenomenon persists without limit to high temperature, where it turns into gas-gas equilibrium (gge).

4.2.4.1 Gas-Liquid Equilibrium

A simple case of gas-liquid equilibrium is shown in Figure 4.10. The dashed lines on the planes of $x_B = 0$ and $x_B = 1$ are the vapor pressure curves of the components A and B. The curves end up at the critical points C_A and C_B. Two-phase border loops are shown at T_1 and T_2. The upper branch of a loop is the *bubble-point curve*, upon which are the states of imminent formation of gas bubbles. Above the bubble point, only liquid exists. The lower branch is the *dew-point curve*, or the states of imminent formation of a liquid dew. Below the dew point, only gas exists. Enclosed within the loop are the states at heterogeneous phase coexistence. A pair of coexisting phases at equilibrium are located at the end points of a horizontal line segment in the loop. To illustrate, one such segment in the loop T_1 is shown to have one end point 3 on the dew-point curve and the other end point f on the bubble-point curve. A mixture at point g is at a heterogeneous coexistence state, and is made

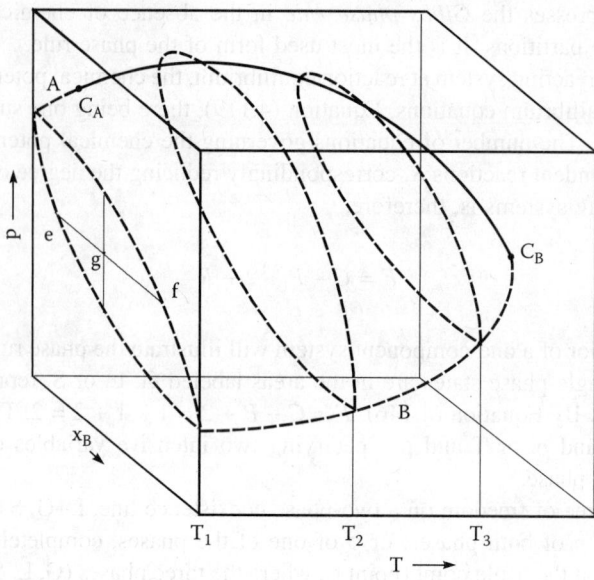

FIGURE 4.10 Gas-liquid equilibria of a binary system.

up of a phase at e and another phase at f, both phases being at different compositions. The total composition of the two phases is represented by g.

More than one phase border loop exists at one temperature for some mixtures of components of greater disparity. Figure 4.11 shows the isothermal gle of n-heptane + ethanol at 313° and 333°K. Two connected loops occur at each temperature. Both branches of the lower curves are dew-point states; the upper curve, bubble-point states. The dew state and bubble state on the same branch of the saturation loop and at the same T and p are at phase equilibrium. To illustrate, two phases are at equilibrium at 313°K and 20 kPa and also at 333° and 50 kPa. All four pairs are shown with dashed line segments.

At the connection point of the two saturation loops, the gas and the liquid at equilibrium are of the same composition but different densities. The mixture of equal gas and liquid composition is called an azeotrope. The azeotrope composition changes with p and T. It is a minimum boiling azeotrope when boiling occurs at a temperature less than those of the two components. Maximum boiling azeotropes are also found, though not as often.

To examine gle at elevated pressures we return to Figure 4.10. At T_1, which is below the critical temperature of both A and B, the bubble- and dew-point curves both start at the low-pressure end at the vapor pressure of B, which is the less volatile component. At higher pressure, the two curves diverge and finally both end at the vapor pressure of A, the more volatile component. At T_2, which is between the critical temperatures of A and B, the bubble- and dew-point curves at the low-pressure end still start at the vapor pressure of B, but at the higher-pressure side they meet at the mixture critical state. The critical points of mixtures of varying composition form the *mixture critical loci*, a space curve that connects the critical states of A and B.

Unusual behavior is observed even for simple mixtures in the near critical region. Depending on the relative location of the critical state, the vaporization-condensation process in the near critical region known as *retrograde condensation* occurs, and two dew points are found on the constant-pressure line. Between the dew points is a point of maximum condensation. Thus heating the fluid causes a liquid to appear, to increase to a maximum, then to decrease, and finally to disappear. Cooling the fluid along the same line produces the opposite sequence of events. The phenomenon is also often observed upon expanding or compressing the fluid at a constant temperature; it is called *retrograde condensation of the first kind*. In contrast, *Retrograde condensation of the second*

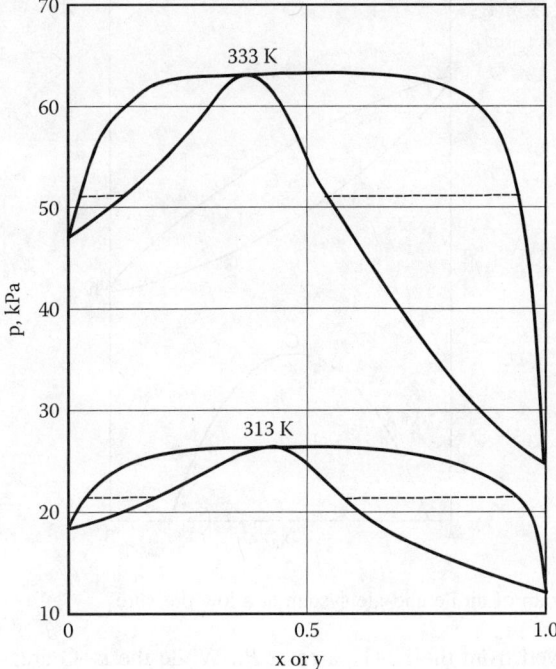

FIGURE 4.11 A maximum-pressure azeotrope.

kind is when a vapor phase first appears, increases to maximum, and then decreases. The retrograde phenomenon is employed in natural gas engineering for the recovery of condensables from a high-pressure natural gas. The maximum recovery is obtained by separating the phases after a suitable expansion from the high underground pressure.

The critical loci of mixture systems are of much interest because they define the boundary in pT space for light alkanes. They are approximately parabolic, connecting the critical points of the pure components and rising to a point of maximum pressure in between. This simple behavior is observed for binary mixtures of light n-paraffins, including methane and a C_2–C_5 paraffin, ethane and C_3–C_{18} n-paraffin, and propane and some other n-paraffins. For mixtures of greater difference in molecular sizes or attractive forces, the critical loci are more complex due to the formation of solids or immiscible liquids that terminate the critical locus.

4.2.4.2 Liquid-Liquid Equilibrium

Many liquids are only partially miscible. Two liquid phases are formed when the two pure liquids are brought together in some proportions. Figure 4.12 shows the T–X diagram of a mixture at low pressures (and lower temperatures). According to the phase rule, for a binary system the degree of freedom F is equal to $(4 - P)$, where P is the number of phases. If we consider two liquid phases coexisting, $F = 2$. At a fixed pressure P_1, we can fix the temperature to completely determine the intensive state of the two-phase system. Thus the saturated liquid compositions are functions of T only. If component A is the less volatile and B the more volatile, the boiling point of A at P_1 is T_g; the boiling point of B is T_h; and $T_g > T_h$. The boiling point of the two-liquid heterogeneous mixture is T_i and is a fixed point at P_1 with $F = 0$, as three phases, two liquids, and a gas are in coexistence.

The figure shows that at P_1, lle can exist at $T < T_i$; at $T > T_i$, lle cannot exist, but is replaced by gle until $T = T_g$. Three phases can coexist at T_i for mixtures with total composition in the range $x_B > x > x_A$. At $T > T_g$, only gas can be found. For the phase behavior of this mixture at higher pressures, the L+G area of P_2 is evolved from the combination of L_A+G and L_B+G areas at P_1. The

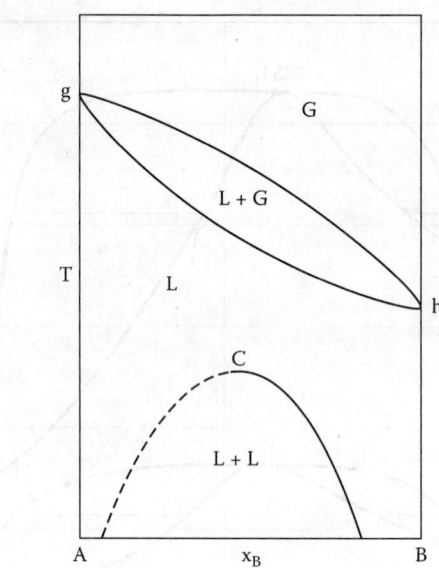

FIGURE 4.12 T–X diagram of an lle and gle system at a low pressure.

L+L area at P_2 is evolved from the L_A+L_B area at P_1. While the L+G areas and the L+L area are contiguous in P_1, the L+G and L+L areas are separated at P_2. Since the L+G area at P_2 touches both coordinates $x = 0$ and $x = 1$, P_2 is below the critical pressures of both A and B. At P_2, T_g is the boiling point of A and T_h is that of B. A liquid-liquid critical point appears at P_2, located at the peak of the L+L region. Since the two-phase region is located below it, this critical temperature is called an upper critical solution temperature, UCST. This temperature changes with T, p, and x to form a curve similar to the gl critical loci. The vapor pressure curves of A and B are similar to the gl critical loci. These curves terminate at the critical points at high p and T; the gle space is bounded by the critical loci. For some systems, even solidification of the heavier component occurs. As an example, Figure 4.13 shows such solidification and the formation of a quaternary point for mixtures of methane and n-heptane at high pressures.

4.2.4.3 Gas-Gas Equilibrium

Gas-gas equilibrium refers to the fluid-phase equilibrium that exists at high temperature and very high pressures above the critical points of the components and that persists to ever higher temperature and pressure apparently without limit. The occurrence of gge in mixtures follows the breakup of the gle critical locus and the merging of one branch of the broken gle critical loci with the UCST. The other branch of the critical locus, unhinged from the critical point of the light component, becomes free to move up to ever higher temperature and pressure. As an example, helium-nitrogen mixtures show such behavior. These are two branches of the saturation loop that intersect on the vapor-pressure curve of nitrogen. Above the T_c of helium, there is no intersection on the other composition axis. At a higher temperature T_2, the loop develops a pinch, indicating a maximum mutual solubility at the pinch pressure. The pinch tightens with increasing temperature until at T_3 the pinch is complete, and the saturation loop divides into two branches with a critical state located on each. The locus of critical states at the lower branches is connected to the critical point of nitrogen. The upper branch with its critical state proceeds to ever higher temperatures. The saturation loops of the upper branch are all open-ended at the top and remain open to the highest pressures explored. The openness is displayed by many systems of weak interaction between the unlike components, notably water + hydrocarbons.

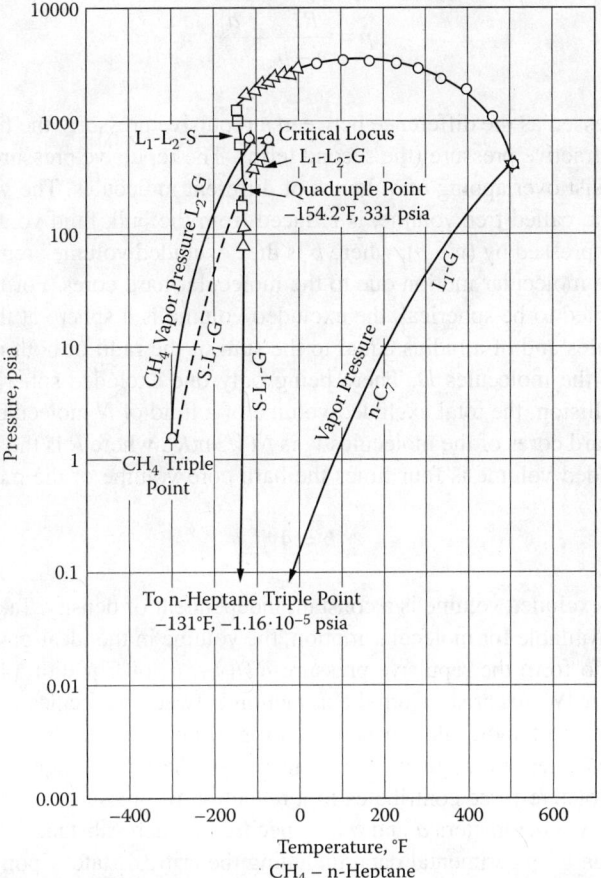

FIGURE 4.13 pT diagram of methane + n-heptane system. (From Chang, H. L., Hurt, L. J., and Kobayashi, R. *AIChE J.* 12: 1212, 1966. With permission.)

4.2.5 Equations of State

An equation of state (eos) is an algebraic expression of the pressure-volume-temperature relationship of a substance. A good equation of state faithfully represents the isotherms, the isochores, and the generalized correlation previously presented. The algebraic form makes it convenient to derive formulas for the energy functions and other thermodynamic properties. The derived formulas together with the equation of state are useful for calculating properties of pure fluids and mixtures, for example, density of gas or liquid, vapor pressure, vapor-liquid equilibrium (vle), liquid-liquid equilibrium (lle), enthalpy, and entropy. Here we review some of the more important eos's. A more extensive review of equations of state can be found in Walas [3] and Sandler et al. [4]. Most belong to one of three types:

1. Van der Waals equations
2. Perturbation equations
3. Virial and extended virial equations

4.2.5.1 Van der Waals-Type Equations

The van der Waals (vdW) equation of state (eos) is as follows:

$$p = \frac{RT}{v-b} - \frac{a}{v^2} \qquad (4.148)$$

Pressure is expressed as the difference between a repulsive pressure (the first term on the right-hand side) and an attractive pressure (the second term). The repulsive pressure is due to the forces of molecules that resist overlapping of volumes of different molecules. The volume open and free for molecular motion, called free volume, is reduced from the bulk fluid volume. In the vdW eos, the free volume is expressed by $(v - b)$, where b is the "excluded volume" representing the volume that is unavailable to molecular motion due to the molecular hard cores. For the collision between two molecules assumed to be spherical, the excluded volume is a sphere at the center of either of the colliding molecules and of a radius equal to the sum of the radii of both molecules in contact, i.e., the diameter of the molecules D. There being only one excluded spherical volume for each two molecules in collision, the total excluded volume of a fluid of N molecules is $(N/2)(4/3)(\pi D)^3$. The volume of the hard cores of the molecules v_b is $N(4/3)\pi R^3$, where R is the radius of a molecule. Therefore, the excluded volume is four times the hard core volume of the molecules:

$$b = 4v_b \qquad (4.149)$$

In the vdW eos, the excluded volume is a constant independent of density. Taking out the excluded volume as being unavailable for molecular motion, the volume in the ideal-gas equation is replaced by the free volume to form the repulsive pressure $RT/(v - b)$ of Equation (4.148). The attractive pressure $-a/v^2$ in the vdW eos arises from the attraction between molecules in proximity. Since the frequency of encounter of molecules in pairs is proportional to the concentration of molecules squared, the attractive pressure is given by the inverse of the volume squared. The negative sign indicates that the attractive force contributes to a reduction of pressure.

In the vdW eos, two parameters a and b are specific for each substance. These are determined by fitting the equation to experimental data, including the critical state. Upon applying the critical conditions, Equation (4.134) and Equation (4.135), to the eos and solving for a and b from the two resulting equations, we obtain

$$a = 27R^2T_c^2 / (64p_c) \qquad (4.150)$$

$$b = RT_c / (8p_c) \qquad (4.151)$$

The vdW eos is reducible to a dimensionless form when Equation (4.150) and Equation (4.151) are substituted into Equation (4.149). The reduction gives

$$p_r = \frac{8T_r}{3v_r - 1} - \frac{3}{v_r^2} \qquad (4.152)$$

The vdW eos is the first to describe the phenomena of condensation/vaporization, vapor pressure, and critical state. Although only qualitative, it often gives more reliable answers than many complex equations.

With mixtures, the parameters a and b need to represent the effect of molecular encounters, two molecules at a time. For pure fluids, encounters are between molecules of the same kind. In a mixture, the probability of finding a molecule i at a position is x_i; the probability of finding a molecule j at the proximity is x_j. The probability of finding both i and j in the proximity is the product of the two independent probabilities, $x_i x_j$, assuming no correlation. This is the probability of an encounter of species i and j, which is the probability of a_{ij} being produced, where a_{ij} is the a of ij interaction. The a of the mixture is the sum of the a_{ij}'s produced by binary encounters of all species,

$$a = \sum_i \sum_j x_i x_j a_{ij} \tag{4.153}$$

where the summation is over all molecular specifics in the mixture. Similarly, we have for b of the mixture,

$$b = \sum_i \sum_j x_i x_j b_{ij} \tag{4.154}$$

A parameter with indices that are identical is just the pure-component parameter, i.e., $a_{ii} \equiv a_i$ and $b_{ii} \equiv b_i$. The a_{ij} with $i \neq j$ is a cross-interaction parameter, and is found only in mixtures containing i and j, but never in a pure substance. Since a expresses the effect of intermolecular attractive energy, for similar molecules i and j, a_{ij} is approximated by the geometric mean of a_{ii} and a_{jj} in analogy to the London dispersion energy. But the geometric mean becomes worse as the i and j molecules become more dissimilar. It is generally useful to represent a_{ij} by

$$a_{ij} = (1 - k_{ij})\sqrt{a_i a_j} \tag{4.155}$$

where k_{ij} is an adjustable mixture parameter that is nearly vanishing for similar substances i and j but becomes significantly different from zero for highly dissimilar molecules. Values of k_{ij} can be determined by referring to mixture data and have been reported for many pairs of substances [5].

Assuming molecular hard cores to be spheres, we obtain the cross-interaction parameter b_{ij} of Equation (4.154),

$$b_{ij}^{1/3} = (b_i^{1/3} + b_j^{1/3})/2 \tag{4.156}$$

But the simpler formula is widely adopted as an approximation:

$$b_{ij} = (b_i + b_j)/2 \tag{4.157}$$

Substituting Equation (4.157) into Equation (4.154) leads to the linear mixing rule for b:

$$b = \sum_i x_i b_i \tag{4.158}$$

The mixing rules described above for vdW eos's, known as van der Waals one-fluid mixing rules, or simply van der Waals mixing rules, apply as well to the *vdW-type* cubic equations described next.

The van der Waals one-fluid mixing rules work well, yielding good results for most mixtures of nonpolar and weakly polar substances. However, for strongly polar molecules, associating molecules, and molecules with specific interaction, the vdW mixing rules are unreliable. The free-energy-matching mixing rules have been developed from theory of solutions for molecules that strongly correlate. Refer to Section 4.3.5 for discussion.

The fundamental assumption of the vdW eos is that the intermolecular force is separated into repulsive and attractive pressures. The separate representation of the repulsive and attractive forces is maintained in the vdW-type eos and the perturbation equations, but is dropped in the virial equations and the extended virial equations.

Redlich and Kwong (RK) [6] changed the vdW eos as follows:

$$p = \frac{RT}{v-b} - \frac{a}{T^{0.5}v(v+b)} \quad (4.159)$$

When the constants a and b are fitted to the critical temperature and pressure, they are given by

$$a = 0.42747R\ 2T_c^{2.5} / p_c \quad (4.160)$$

$$b = 0.08664\ RT_c / p_c \quad (4.161)$$

One change made by RK is to introduce the factor $1/T^{0.5}$ into the attractive pressure, in essence making the vdW attractive-pressure parameter α a decreasing function of temperature. The second change is to replace the v^2 with $v(v + b)$, opening the possibility of replacing v^2 with a second-degree polynomial of v. The repulsive pressure is left unchanged. Thus the RK equation gives improved representation of the gas state, including the second virial coefficient, the fugacity of vapors up to the vapor pressure, and fugacity of gases up to high pressures. But it is of no help for liquid states.

Wilson [7] correlated the vapor pressure of normal fluids to obtain a modified RK eos as follows:

$$p = \frac{RT}{v-b} - \alpha \frac{a_c}{v(v+b)} \quad (4.162)$$

where a_c and b are constants determined at the critical state. The new factor is introduced as a function of reduced temperature,

$$a = T_r[1 - (1.57 + 1.62\omega)(1/T_r - 1)] \quad (4.163)$$

where ω is Pitzer's acentric factor, as defined in Equation (4.143). Equation (4.163) is obtained by fitting α to vapor pressure of various substances at various temperatures and correlating the fitted values. Although Wilson's equation does not accurately fit vapor pressure, it is the key to the success of eos for the accurate representation of the chemical potential of liquids and, in consequence, other energy functions of liquids and vapor-liquid equilibrium.

Soave [8] retained the form of the Wilson eos, Equation (4.162), but developed a new function to fit vapor pressure,

$$\alpha = [1 + (0.480 + 1.574\omega - 0.176\omega^2)(1 - T_r^{0.5})]^2 \quad (4.164)$$

The equation, referred as the SRK eos, gives quantitative fitting of vapor pressure, good representation of the fugacity of liquid, and improved representation of the energy functions of liquid for normal fluids, although liquid density is not well represented. The equation was the first to be widely used for both the gas and liquid phases and, hence, for gas-liquid equilibrium in engineering calculations.

Thermodynamics of Fluid Phase and Chemical Equilibria

In the Peng and Robinson (PR) [9] eos, the volume dependence of the attractive pressure is changed as follows:

$$p = \frac{RT}{v-b} - \alpha \frac{a_c}{v(v+b) = b(v+b)} \tag{4.165}$$

The a and b are fitted to critical temperature and pressure to give

$$a_c = 0.45724\, R^2 T_c^2 / p_c \tag{4.166}$$

$$b = 0.07780\, RT_c / p_c \tag{4.167}$$

The functional form of Soave's α is retained, but the fitting coefficients of the terms are adjusted,

$$\alpha = [1 + (0.37464 + 1.54226\omega - 0.26992\omega^2)(1 - T_r^{0.5})]^2 \tag{4.168}$$

These changes result in improvement in the liquid molar volume while retaining the good representation of fugacity and energy functions of the liquid state for normal fluids.

The SRK and PR equations follow the principle of corresponding states in the three-parameter form; only the commonly available critical properties T_c, p_c, and ω are required to apply the equation to a substance. The simple vdW mixing rules work well with these equations. Hence they are widely used for the calculation of vapor-liquid equilibrium in mixtures.

The PR eos has been extended by Stryjek and Vera [10] to polar substances by introducing a substance-specific parameter. The modification also gives improved fitting of the vapor pressure of normal fluids. The function of the PRSV equation is given by

$$\alpha = [1 + \kappa(1 - T_r^{0.5})]^2 \tag{4.169}$$

with

$$\kappa = \kappa_0 + \kappa_1(1 + T_r^{0.5})(0.7 - T_r) \tag{4.170}$$

$$\kappa_0 = 0.378893 + 1.4897153\omega - 0.17131848\omega^2 + 0.0196554\omega^3 \tag{4.171}$$

with κ_1 being an adjustable parameter characteristic of each substance. Table 4.1 reports the value of κ_1 for 26 substances of common interest. For water and alcohols, the value of κ_1 applies from low temperatures up to the critical point. For all other compounds, slightly better results are obtained using $\kappa_1 = 0$ for reduced temperatures above 0.7. This table also shows the low deviation of the predicted vapor pressures.

4.2.5.2 Perturbation Equations

Equations of state developed from perturbation theory are composed of a reference fluid equation to which are added perturbation terms. The reference fluid expresses the high repulsive energy that determines the molecular correlation and hence the structure of the fluid. The attractive energies that are relatively weak are treated as perturbations. Perturbation theory opens the door to the separate development of statistical theories for various intermolecular energies: hard-sphere-fluid

TABLE 4.1
PRSV Equation-of-State Parameters

	κ_1	ω	P_s, % deviation
Inorganic			
Carbon dioxide	0.04285	0.225	0.544
Ammonia	0.001	0.2517	0.12
Water	−0.06635	0.3438	0.29
Hydrocarbons			
Methane	−0.00159	0.01045	0.458
Ethane	0.02669	0.09781	0.472
Propane	0.03136	0.15416	0.782
n-Butane	0.03443	0.20096	0.545
n-Pentane	0.03946	0.25143	0.783
n-Hexane	0.05104	0.30075	1.106
n-Heptane	0.04648	0.35022	0.885
n-Octane	0.04464	0.39822	0.546
n-Decane	0.0451	0.49052	0.99
Cyclonehexane	0.07023	0.20877	0.363
Benzene	0.07019	0.20929	0.541
Toluene	0.03849	0.26323	0.363
Ethylbenzene	0.03994	0.3027	0.4
p-Xylene	0.01277	0.32141	0.584
Other organics			
Methanol	−0.16816	0.56533	0.915
Ethanol	−0.03374	0.64439	0.949
1-Propanol	0.21419	0.62013	0.196
Acetone	−0.00888	0.30667	0.435
Butanone	0.05717	0.18909	0.72
Dimethyl ether	0.05717	0.18909	0.72
Acetonitrile	−0.13991	0.3371	5.632
Acetic acid	−0.19724	0.4594	0.379
Furfural	−0.03471	0.39983	3.067

pressure, hard-sphere-chain pressure, pressures of fluids of various hard bodies, square-well attractive pressure, Lennard-Jones attractive pressure, intramolecular pressure, rotational pressure, association pressure, etc. The appropriate pressures are summed to form the equation of state of the fluid.

A commonly used reference fluid is the hard-sphere fluid expressed by the Carnahan-Starling (CS) equation of state:

$$z = \frac{1+\eta+\eta^2-\eta^3}{(1-\eta)^3} \qquad (4.172)$$

where z denotes the compressibility factor pv/RT; $\eta = \pi\rho\sigma^3/6$ is a reduced dimensionless density, also referred to as the packing fraction, expressing the fraction of volume that is packed by the spheres; ρ is the density of molecules n/v; and σ is the diameter of the hard sphere. This equation is based on the fluid structure derived from correlation theory and is verified by molecular simulation. It takes the place of the repulsive pressure $RT/(v − b)$ of the vdW eos, which is an oversimplification in counting excluded volume as due to all molecules in binary collisions. In liquids and dense gases, clusters of more than two molecules are abundant, and the excluded volumes of the pairs in the clusters overlap; the excluded volume in vdW eos is an overcount; and the vdW repulsive pressure is excessively high.

The double-series attractive pressure due to Alder et al. [13] is in common use. The attractive energy of square-well molecules is simulated with the method of molecular dynamics and fitted to a double series in T and ρ. Integration of U/T^2 gives A, and differentiation with v gives the attractive pressure, here expressed as a compressibility factor,

$$z_{att} = \sum_{n=1}^{4}\sum_{m=1}^{9} mD_{nm}(u/kT)^n/(v^0/v)^m \qquad (4.173)$$

where u denotes the depth of the energy well, v^0 is the close-packed volume of the hard core spheres equal to $N\sigma^3/\sqrt{2}$, and σ is the diameter of the hard sphere.

Chen and Kreglewski [11] developed a perturbation equation based on Boublik's [12] hard-convex-body equation as the reference fluid equation. The repulsive pressure is Alder et al.'s double-series polynomial with modified coefficients fitted to experimental argon data. The Boublik–Alder–Chen–Kreglewski (BACK) eos is

$$z = z_{rep} + z_{att} \qquad (4.174)$$

$$z_{rep} = 1 + \frac{(3\alpha+1)y + (3\alpha^2 - 3\alpha - 2)y^2 + (1-\alpha^2)y^3}{(1-y)^3} \qquad (4.175)$$

where z_{att} is given by Equation (4.173) with Chen arid Kreglewski's [11] polynomial coefficients. The variables of the equation are: $y = 0.74048 \, v^0/v$ is the packing fraction; α is a nonsphericity parameter; v^0 is the close-packed volume; and u is an interaction energy parameter. Both v^0 and u are temperature dependent:

$$v^0 = v^{00}[1 - C\exp(-3u^0/kT)] \qquad (4.176)$$

$$u = u^0(1+\eta/kT) \qquad (4.177)$$

The five substance-specific parameters—u^0/k, v^{00}, α, η/k, C—and the attractive-pressure polynomial coefficients can be looked up in the literature [11]. The BACK eos is capable of high-accuracy representation of pVT and was used to prepare American Petroleum Institute Project 44 tables of thermodynamic properties.

In the perturbed hard-chain theory (PHCT) of Beret and Prausnitz [14] and Donohue and Prausnitz [15], the reference fluid is modeled as chains of tangential hard spheres. Since the fluid is still composed of hard spheres, albeit bonded, the CS eos is applied with modification to account for the bonding. The hard-sphere-chain equation is

$$z = \frac{1 + (4c-3)r\eta + (3-2c)(r\eta)^2 - (r\eta)^3}{(1-r\eta)^3} \qquad (4.178)$$

One modification is to replace the packing fraction η of Equation (4.172) with $r\eta$ to account for the r spheres in one chain molecule. The other modification is to revise the terms in the numerator of Equation (4.172) in light of the increased degrees of freedom of the chain of spheres from that of the single sphere. Let the number of external degrees of freedom of a chain molecule be denoted by $3c$. These are transformed from the $3r$ degrees of translational motion of the r spheres that make up the chain. In general, $3c$ is smaller than $3r$ but larger than 3 and consists in the three translational modes and some rotational modes. Any vibrational modes are of high energy and independent of the density of the fluid; they are not included in the $3c$ degrees of external motion. The rotational modes are of lower energy. Though not fully free, some are density dependent. They need to be addressed in equations of state of polyatomic molecular fluids. Prigogine [16] expressed these rotational modes as equivalent translational degrees of freedom. Flory et al. [17] constructed a cell theory for the equivalent translational modes that is widely used in polymer equations of state and polymer solution theory. Beret and Prausnitz [14] and Donohue and Prausnitz [15] reinterpreted the terms in the numerator of the CS equation in light of the added equivalent translational degrees of freedom of the chain molecule. The result is the new terms in the numerator of Equation (4.178). These terms revert to their forms in Equation (4.172) upon setting $3c = 3$.

The attractive pressure of the chain-of-spheres model is the sum of the double-series attractive pressures of the c external degrees of freedom,

$$z = -c \sum_n \sum_m \frac{m A_{nm}}{\tilde{v}^m \tau^n} \tag{4.179}$$

where the reduced variables are $\tau = ckT/(q)$ and $\tilde{v} = v\sqrt{2}/(Nr\sigma^3)$ and the variable ε is the characteristic intermolecular potential energy of a molecule per unit surface area, εq is the surface area of a molecule, and q is the characteristic attractive energy of a molecule. The constants A_{nm} have been adjusted to fit data for alkanes.

The PHCT pressure is the sum of the hard-sphere-chain pressure of Equation (4.172) and the attractive pressure of the chain of spheres of Equation (4.179),

$$z_{PHCT} = z_R + z_{att} \tag{4.180}$$

Three adjustable parameters are specific for each substance: εq, $r\sigma^3$, and c. The parameters are fitted to pVT and vapor pressure data. The PHCT is applicable to a very wide domain: for simple molecules where $c \to 1$, it reduces to the perturbed Carnahan-Starling eos; at low density, it gives a reasonable second virial coefficient and approaches ideal gas at very low density; for high densities, PHCT is essentially equal to the Prigogine–Flory–Patterson theory of polymer liquids.

In the chain-of-rotators (COR) equation proposed by Chien, Greenkorn, and Chao [17], a chain molecule is modeled as joined spheres in which adjacent covalent bonds can rotate. Rotation takes place because the chain is not straight; any two adjacent bonds are joined at an angle like the radials from the center of a pyramid to the four corners. Chien et al. found the pressure of rotation by comparing the pressure of Boublik and Nezbeda's hard dumbbell fluid [18] with the pressure of Carnahan and Starling's hard spheres. The rotational pressure found is

$$z_{rot} = c \left(\frac{e-1}{2} \right) \left(\frac{3\eta + 3e\eta^2 - (e+1)\eta^3}{(1-\eta)^3} \right) \tag{4.181}$$

where η is a reduced density defined to be v_h/v, also referred to as the packing fraction; v_h is the hard core volume; c is the rotational degrees of freedom of a molecule; and e is the eccentricity parameter of the model dumbbell rotator, defined by

$$e = \frac{(1+L)(2+L)}{2+3L-L^3} \quad \text{for } L \leq 1 \tag{4.182}$$

where $L = l/\sigma$, with l being the center-to-center distance and σ representing the diameter of the spheres. Chien et al. [17] took a two-carbon segment as the model rotor. Substituting for ethane gives a value of 1.078 for e. The probability of rotation-caused collision is significant in liquids and dense gases, and the rotational pressure is significant. The probability of collision is small in dilute gas, and the rotational pressure drops rapidly with decreasing density down to z values much lower than the values that the CS hard spheres would ever attain. The z of rotation approaches 0 instead of 1 at zero density. There can be any number of rotators in a molecule without affecting the fluid from approaching ideal-gas behavior at zero density.

The COR eos is made up of three terms: the CS hard-sphere eos, the rotational pressure of Equation (4.181), and Alder et al.'s double-series attractive pressure modified for nonspherical molecules,

$$z = \frac{1+\eta+\eta^2-\eta^3}{(1-\eta)^3} + c\left(\frac{3-1}{2}\right)\left(\frac{3\eta+3e\eta^2-(e+1)\eta^3}{(1-\eta)^3}\right) \\ + \left[1+\frac{c}{2}(B_0+B_1/\tilde{T}+B_2\tilde{T})\right]\sum_n\sum_m \frac{mA_{nm}}{\tilde{T}^n\tilde{v}^m} \tag{4.183}$$

where the reduced temperature $\tilde{T} = T/T^*$ (T^* being a substance-specific characteristic temperature) and the reduced volume $\tilde{v} = v/v_0$ (v_0 being the closest packed volume and also a substance-specific value). Since $\eta = v_h/v$ and $v_h = (\pi\sqrt{2}/6)v_0 = 0.740478$ (the value of v_0 based on the assumption of hard spheres), $\eta = 0.74048/\tilde{v}$. Table 4.2 presents parameters for several common substances, comparisons of predicted values versus experimental vapor pressure, and density of saturated liquid data.

The usefulness of equations of state can be extended with the method of group contribution. By this method, the eos parameters of molecules are synthesized from those of the constituent

TABLE 4.2
COR Parameters and Comparison of Calculated Data for Some Saturated Fluid Properties

Substance	T (K)	c	v_0 (cm³/mol)	Vapor Pressure (% AAD)	Saturated v_{liq} (% AAD)
Methane	151.71	0	21.192	0.79	0.04
Ethane	225.44	2	30.52	1.98	0.53
Propane	263.57	3.2	41.51	1.57	0.21
n-Butane	293.25	4.4	52.24	0.83	0.45
n-Pentane	315.61	5.6	63.1	0.52	0.34
n-Octane	356.55	9.6	95.68	0.56	0.69
Benzene	384.44	4.8	52.25	1.34	0.14
Toluene	395.2	6	63.4	1.26	0.27
CO_2	205.52	5.2	19	0.45	1.14
H_2S	278.82	1.7	20.5	0.57	0.71

Note: AAD = average absolute deviation.

groups of the molecule. Once the group parameters are determined, properties of the molecular fluids can be calculated. Since many molecules are composed of a relatively small number of groups, the group contribution method can be used for many molecular fluids.

To make it suitable for group contribution, Pults et al. [19] simplified the COR eos by replacing the double-series attractive pressure with the Redlich-Kwong attractive pressure in Equation (4.159). The COR group contribution (CORGC) eos is

$$z = \frac{1+\eta+\eta^2-\eta^3}{(1-\eta)^3} + c\left(\frac{e-1}{2}\right)\left(\frac{3\eta+3e\eta^2-(e+1)\eta^3}{(1-\eta)^3}\right) - \frac{a(T)}{RT[v+b(T)]} \quad (4.184)$$

where y is the reduced density equal to $b(T)/(4v)$; $b(T)$ is the covolume, a function of T; c is the degree of rotational freedom; e is a constant = 1.078; and $a(T)$ is a T-dependent attractive-pressure parameter.

The eos parameters a, b, and c are made up of group contributions according to the following equations:

$$a(T) = \sum_i^{N_c} \sum_j^{N_c} x_i x_j \sum_m^{N_i} \sum_n^{N_j} v_{im} v_{jn} q_m q_n q_{mn}(T) \quad (4.185)$$

$$b_i(T) = \sum_m^{N_i} v_{im} b_m(T) \quad (4.186)$$

$$b(T) = \sum_i^{N_c} x_i b_i(T) \quad (4.187)$$

$$c_i = \sum_m^{N_i} v_{im} c_m \quad (4.188)$$

$$c = \sum_i^{N_c} x_i c_i \quad (4.189)$$

In these equations, v_{im} is the number of groups m in molecule i. The index N_i is the number of group species in molecule i, and N_c represents the number of components in the mixture. The coordination number q_m is a normalized surface area of group m based on a value of $q_m = 10$ for methane.

Equation (4.185) combines the interactions of groups to give the total interaction of the fluid. It is a generalization of the van der Waals mixing rule to include groups (or atoms) as well as molecules as entities that interact. To count the interaction of groups is in accord with the nature of polyatomic molecules. Equations (4.185) to (4.189) are given in their general forms for pure fluids and mixtures. For pure fluids, set $i = j = 1$, $N_i = N_j = 1$, and $x_i = x_j = 1$. The equations simplify to van der Waals mixing rule for a mixture of monatomic molecules, with $v_{im} = 1$ for all i and all n.

Equation (4.186) sums up the excluded volume of the constituent groups to form the excluded volume of the molecule. The linear sum is justified because the excluded volume of a group is defined as $(4v_h)$, where v_h stands for the hard core volume. Equation (4.187) linearly combines the

TABLE 4.3
CORGC Parameters for Groups

Group	b_m^* (cm³/mol)	T_m^* (K)	c_m	q_m	$\hat{a}_{mm} \times 10^{-4}$ (cm⁶bar/mol²)	$\hat{T}_{mm}$ (K)
al-CH₃	59.155	316.36	1.00	7.31	7.2608	275.28
al-CH₂-al	43.954	1062.7	0.73	4.66	7.7334	482.99
>CH-	47.264	2118.6	0.30	2.69	−24.037	579.74
>C<	50.953	4462.3	0.00	0.72	−1967.1	298.81
(CH)$_{ar}$	34.696	681.97	0.30	3.45	12.140	578.26
al-C$_{ar}$	34.074	7431.3	0.20	1.03	27.123	1273.5
ar-CH₃	41.656	422.33	1.00	7.31	4.3832	498.28
ar-CH₂-	35.034	2707.2	0.73	4.66	3.2194	1297.0
ri-C$_{ar}$	27.064	8415.8	0.20	0.724	151.11	1715.6
(CH₂)$_{cyc}$	38.041[a] 40.869[b]	652.27	0.73	4.66	7.9113	566.35
-(CH)$_{cyc}$	31.025[a] 35.410[b]	4969.3	0.50	1.97	2.4801	906.93
=CH₂	59.116	282.22	0.85	6.41	6.9675	306.11
=CH-	39.673	920.29	0.70	3.72	10.091	817.28
H₂	18.006	100*10⁴	0.00	4.76	0.7068	464.77
CH₄	81.067	225.16	0.00	10.0	4.8616	258.86
C₂H₆	114.95	375.86	2.00	14.6	6.1357	372.85
CO₂	83.169	319.49	2.50	11.1	9.3356	278.74
H₂S	83.411	418.94	0.0	10.7	10.304	399.49
C₂H₄	118.23	282.22	1.70	12.8	7.4862	294.97

[a] Value in a five-membered naphthenic ring.
[b] Value in a six-membered naphthenic ring
( )$_{ar}$ member of an aromatic ring; ( )$_{cyc}$ member of a naphthenic ring; -al attached to an aliphatic group; -ar attached to an aromatic ring; -ri attached to a ring group (either aromatic or naphthenic).

excluded volumes of the molecules to form the excluded volume of the mixture by the van der Waals mixing rule. Similarly, the summing of the constituent number of rotors of the groups to form the number of rotors of the molecule and the linear combination of the number of rotors of the molecules to form the total number of rotors of the mixture are carried out in Equation (4.188) and Equation (4.189).

The group properties a and b are functions of temperature as follows:

$$a_{mn} = \hat{a}_{mn} \left(\frac{T}{\hat{T}_{mn}}\right)^{-0.18135} \exp\left(\frac{-T}{\hat{T}_{mn}}\right) \tag{4.190}$$

$$b_m = b_m^* \exp(-T/T_m^*) \tag{4.191}$$

Six parameters describe the contributions of a group m, and these are $\hat{a}_{mm}, \hat{T}_{mm}, b_m^*, T_m^*, c_m$, and q_m. Table 4.3 presents the values for 21 groups, including small molecules: hydrogen, methane, ethane, H_2S, and CO_2 are counted as independent groups.

The interaction between unlike groups m and n, appearing as a_{mn}, requires two additional parameters, $\hat{a}_{mm}$ and $\hat{T}_{mm}$, for its description. Unlike interaction parameters have been reported by Pults [20].

To apply to high polymers, Sy-Siong-Kiao, Caruthers, and Chao [21] extend the COR eos to obtain the polymer chain-of-rotator (PCOR) equation,

$$\frac{pv_m}{RT} = c_m \left(\frac{e-1}{2}\right)\frac{3\eta + 3e\eta^2 - (e+1)\eta^3}{(1-\eta)^3} - \frac{a_{mm}}{RT(v_m + b_m)} \qquad (4.192)$$

The equation addresses the molar volume of a segment of the polymer v_m, the subscript m indicating a segmental molar quantity. The parameters of the equation are segmental parameters: thus b_m is the excluded volume of a segment; $\eta = b_m/(4v_m)$ is the density of the fluid expressed in a reduced dimensionless form, also called the packing fraction; and a_{mm} is the attractive-pressure parameter of the segment. The eccentricity parameter of the rotators is e, equal to 1.078 for the model rotator. Results of data reduction indicate the attractive parameter a_{mm} to vary weakly and linearly with temperature,

$$a_m = a_{m0} + a_{m1}T \qquad (4.193)$$

The parameter b_m does not change with temperature, being a constant value for each polymer. The temperature independence of b_m in the range of temperature of the data indicates low kinetic energy of the giant polymer molecules.

Comparing the PCOR with the COR eos, Equation (4.183), the translational pressure in the COR eos is missing in the PCOR eos. The long-chain polymer molecules are so entangled and entwined that the molecules make only translational rotations.

The PCOR eos describes only a liquid state. In agreement with experimental observation, a high polymer does not vaporize or exert a vapor pressure as pressure approaches zero. The equation does give an accurate quantitative representation of the pVT relationship of high-polymer melts. Average absolute deviation of the equation from the experimental specific volumes is 0.1% or less for the complete range of the experimental data covering a pressure range of 0 to 100 or 200 MPa for all polymers studied.

By generalizing Flory's dimer theory, Hall and coworkers [22, 23] obtained the GF-D equation of state, which is as follows:

$$Z_m = Z_1 + \frac{v_{e,m} - v_{e,1}}{v_{e,2} - v_{e,1}}(Z_2 - Z_1) \qquad (4.194)$$

where the compressibility factor of the chain molecule is expressed as an extrapolation of the monomer and dimer values linearly with respect to the excluded volume. The chain molecule is modeled as an m-mer of tangential hard spheres with an attractive potential. The incremental compressibility factor with the addition of a sphere to the chain is a constant equal to the increment at the formation of the dimer from the monomer. The compressibility factor for the monomer or dimer fluid is given by

$$Z_i = Z_i^{rep} + Z_i^{att} \quad i = 1 \text{ or } 2 \qquad (4.195)$$

For the hard core reference fluids, the following monomer and dimer equations apply:

$$Z_1^{rep} = \frac{1 + \eta + \eta^2 - \eta^3}{(1-\eta)^3} \qquad (4.196)$$

$$Z_2^{rep} = \frac{1 + 2.45696\eta + 4.10386\eta^2 - 3.75503\eta^3}{(1-\eta)^3} \tag{4.197}$$

The attractive pressures of the monomer and dimer are, respectively,

$$Z_1^{att} = \frac{-288.15745}{(T^*)^2} \eta^2 \frac{1 - 5.4019\eta}{(1 + 6.75237\eta)^4} \tag{4.198}$$

$$+ \frac{-9.5}{T^*} \eta \frac{1 - 1.3086\eta - 5.72921\eta^2 + 9.50043\eta^3 - 2.3751\eta^4}{(1-\eta)^4} \tag{4.199}$$

with

$$Z_2^{att} = \frac{-492.36296}{(T^*)^2} \eta^2 \frac{2 - 12.31907\eta}{(1 + 8.26765\eta)^4}$$

$$+ \frac{-12.00332}{T^*} \eta \frac{1 + 1.56564\eta - 15.12892\eta^2 + 19.14672\eta^3 - 4.78668\eta^4}{(1-\eta)^4} \tag{4.200}$$

$$\eta = mv_h / v$$
$$T^* = kT / \varepsilon \tag{4.201}$$

where m is the number of segments per molecule, v_h is the molar volume of the hard spheres, and ε is the energy of interaction. These three factors are the pure-component properties in the GF-D eos that can be regressed from experimental pure-component saturated properties.

The excluded volumes of a monomer, dimer, and trimer are given, respectively, by

$$v_{e,1} = (4/3)\pi\sigma^3 \tag{4.202}$$

$$v_{e,2} = (9/4)\pi\sigma^3 \tag{4.203}$$

$$v_{e,3} = 9.82605\sigma^3 \tag{4.204}$$

where σ is the hard-sphere diameter. For interpolation at m between 1 and 2, the formula below is used,

$$v_{e,m} = (1.064343 + 3.246773m - 0.122327m^2)\sigma^3 \tag{4.205}$$

For $m > 2$, the excluded volume is given by the following:

$$v_{e,m} = v_{e,2} + (m - 2)(v_{e,3} - v_{e,2}) \tag{4.206}$$

The GF-D has been fitted with high accuracy to over 100 substances by Wu and Chen [24]. The parameters v^0 and $/k$ are temperature dependent:

$$v^0 = v^{00} + 0.005T \qquad (4.207)$$

$$\varepsilon/k = \varepsilon^0/k - 0.004T \qquad (4.208)$$

The temperature dependences of v^0 and ε/k do not apply to alkanols and water; these quantities are kept constant independent of T.

The statistical-associated fluid theory (SAFT) of Chapman et al. [25, 26] is based on the perturbation theory of Wertheim [27]. The model molecule is a chain of hard spheres that is perturbed with a dispersion attractive potential and association potential. The residual Helmholtz energy a^R of the fluid is given by the sum of the Helmholtz energies of: the initially free hard spheres; bonding the hard spheres to form a chain; the dispersion attractive potential; and the association potential,

$$a^R = a_{hs} + a_{chain} + a_{disp} + a_{assoc} \qquad (4.209)$$

Wertheim found a_{chain} and a_{assoc} in his perturbation theory. Upon differentiating the Helmholtz energy with respect to volume, the equation of state is obtained,

$$z = z^{id} + z_{hs} + z_{chain} + z_{disp} + z_{assoc} \qquad (4.210)$$

where the ideal-gas $z^{id} = 1$, and

$$z_{hs} = r \frac{4\eta - 2\eta^2}{(1-\eta)^3} \qquad (4.211)$$

$$z_{chain} = (1-r) \frac{5\eta - 2\eta^2}{(1-\eta)(2-\eta)} \qquad (4.212)$$

$$z_{disp} = r \sum_n \sum_m m D_{nm} \left(\frac{u}{kT}\right)^n \left(\frac{\eta}{\eta_{cp}}\right)^m \qquad (4.213)$$

$$z_{assoc} = \rho \sum_S \left(\frac{1}{X^S} - \frac{1}{2}\right) \left(\frac{\partial X^S}{\partial \rho}\right) \qquad (4.214)$$

where r is the number of spheres in a chain; η is the packing fraction $= r\rho_A(\pi/6)d^3$; ρ is the molar density of the chain molecules; d is the temperature-dependent diameter of the spheres given by Equation (4.176) in terms of the 0°K value σ, which is related to the eos parameter v_0, the closest packed volume of the spheres at 0°K, $N_A\sigma^3 = \sqrt{2}\, v_0$; and $\eta_{cp} = \pi\sqrt{2}/6$. By Chen and Kreglewski, the energy well depth u is temperature dependent and is given by Equation (4.177).

Associating molecules are bonded at association sites. There can be a number of association sites on a molecule. In Equation (4.214), X^S denotes the mole fraction of site S that is not bonded and is given by

$$X^S = \left\{1 + N_A \sum_Y \rho X^Y \frac{2-\eta}{2(1-\eta)^3} \left[\exp\left(\frac{\varepsilon^{SY}}{kT}\right) - 1\right]\right\}^{-1} \qquad (4.215)$$

Here the summation with respect to Y is over all different kinds of sites in the fluid, and ε^{SY} and κ^{SY} are the characteristic energy and volume, respectively, for association between sites S and Y.

Three adjustable parameters characterize the chain of spheres of the molecule: r, the number of spheres making up a chain; v_0, the molar volume of the spheres at closest packing; and ε, the temperature-independent depth of the energy well of sphere-to-sphere interaction. To characterize the association, two more parameters are required for each pair of association: ε^{SY} and κ^{SY}. Huang and Radosz [28] reported SAFT parameter values for many substances.

The SAFT eos has been applied successfully to pure fluids and fluid mixtures of small, large, nonassociating, and associating molecules. The representation of pVT is generally quite good. The strength of the SAFT is in phase equilibrium of associating substances. Liquid-liquid equilibrium is found predominantly in mixtures of associating substances, and SAFT is superior for liquid-liquid equilibrium.

4.2.5.3 The Virial and Extended Virial Equations

A virial equation is a power-series expansion of the compressibility factor isotherm in molar density from the zero-density ideal gas, $z = 1$ at $\rho = 0$, which is the common point of intersection of all z isotherms,

$$z = 1 + B\rho + C\rho^2 + D\rho^3 + \ldots \quad (4.216)$$

The coefficients B, C, D,—known as the second, third, and fourth virial coefficients—are functions of temperature that are specific for a substance. The equation was originally used by Kammerlingh-Onnes and coworkers for fitting experimental pVT data on compressed gases. By adding terms to the equation as needed, the fitting can be extended to higher densities. It was later discovered that the virial coefficients are given by molecular theory as integrals of intermolecular potential in molecular clusters of various sizes: the second virial coefficient by clusters of two molecules, the third virial coefficient by triplets, and so on. Molecular theory teaches that the virial equation applies to mixtures in the same form as Equation (4.216) and that the mixture virial coefficients are combinations of the pure-component coefficients and cross-interaction coefficients,

$$B = \sum_i \sum_j x_i x_j B_{ij} \quad (4.217)$$

$$C = \sum_i \sum_j \sum_k x_i x_j x_k X_{ijk} \quad (4.218)$$

where the subscript denotes molecular species; x_i denotes mole fraction of species i; $B_{ii} \equiv B_i$ of pure i; and B_{ij} denotes the cross interaction of i and j interaction. Similarly, $C_{iii} \equiv C_i$ of pure i, and C_{ijk} denotes the C of the ijk interaction. A cross-interaction coefficient is also a function of temperature specific to the indicated species.

The virial equation is truncated in application. Figure 4.14 shows the comparison of B-virial equation and B + C virial equation with generalized correlation of Pitzer [1] at $T_r = 0.8$, 1.0, and 1.5. The uppermost curve at each temperature is the generalized z. The virial equation curves are based on virial coefficients from the same source. Close approximation of real gas behavior is obtained by the B-virial equation at pressures up to the vapor pressure at $T_r = 0.8$. At lower temperatures approximation is as good or better, as the existent pressure range of the gas phase is reduced. On the higher temperature side, at $T_r = 1$, we see the B equation holds up till p_r reaches to about 0.5; and at $T_r = 1.5$, the B equation stays close to the generalized correlation up to about

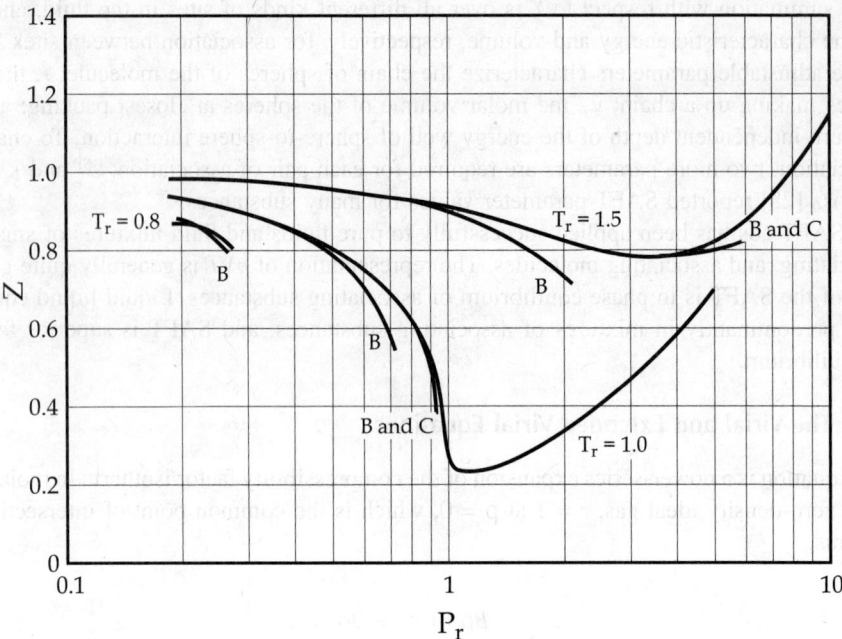

FIGURE 4.14 Comparison of virial equation with a generalized correlation.

the critical pressure. Enhanced by its simple form, and availability of the second virial coefficient, the B-virial equation is convenient and widely used for the vapor phase in low pressure and medium pressure vapor-liquid equilibrium work.

For vapor-liquid equilibrium calculations up to moderate pressures, the B equation is suitable and convenient for the vapor phase for its applicability and simple form. Formulas have been derived from statistical theory for the calculation of virial coefficients, including B, from intermolecular potential energy functions, but intermolecular energy functions are hardly known quantitatively for real molecules. B is found for practical calculations by correlating experimental B values. Pitzer [1] correlated B of normal fluids in a generalized form with acentric factor ω as the third parameter,

$$\frac{Bp_c}{RT_c} = B_0[T_r] + \omega B_1[T_r] \tag{4.219}$$

where $T_r \equiv T/T_c$ and where B_0 (of simple fluids) and B_1 generalize the correlation to normal fluids. Tsonopoulos [29] revised Pitzer's correlation and extended the correlation to polar and hydrogen-bonded fluids using B_0 and B_1 functions as follows:

$$\frac{Bp_c}{RT_c} = B_0[T_r] + \overline{\omega} B_1[T_r] + B_2[T_r] \tag{4.220}$$

$$B_0 = 0.1445 - 0.330/T_r - 0.1385/T_r^2 - 0.0121/T_r^3 - 0.000607/T_r^8 \tag{4.221}$$

$$B_1 = 0.0637 + 0.331/T_r^2 - 0.423/T_r^3 - 0.008/T_r^8 \tag{4.222}$$

$$B_2 = a/T_r^6 - b/T_r^8 \tag{4.223}$$

The B_2 function expresses the polarity and hydrogen-bond contributions. The coefficient a in the last equation above is a function of the reduced dipole moment μ_R:

$$a = -2.140 \times 10^{-4} \mu_R - 4.308 \times 10^{-21} \mu_R^8 \tag{4.224}$$

The reduced dipole moment is defined by

$$\mu_R = \frac{10^5 \mu^2 p_c}{T_c^2} 0.9869 \tag{4.225}$$

The units in the equation are: debye for μ, bar for p_c, and K for T_c. Equation (4.224) applies to a variety of nonhydrogen-bonded polar fluids, including ketones, acetaldehyde, acetonitrile, and ethers. Better results are obtained with specific equations. For ketones use

$$a = -2.0483 * 10^{-4} \mu_R \tag{4.226}$$

For ethers use

$$a = -\exp[-12.63147 + 2.09681 \ln \mu_R] \tag{4.227}$$

For water, $a = 0.0279$, and for aliphatic alcohols, $a = 0.0878$.

The coefficient b in Equation (4.223) is zero for nonhydrogen-bonded polar fluids. For hydrogen-bonded fluids,

$$b = 9.08 \times 10^{-3} + 6.957 \times 10^{-4} \mu_R \tag{4.228}$$

For water, $b = 0.0229$.

For strongly associated fluids, the virial equation is not useful for the description of nonideal-gas behavior. The theory of association is a more suitable approach; see Equation (4.214).

Combining rules are needed to estimate the cross second virial coefficients B_{ij} that are required in applying the virial equation to mixtures. Equation (4.220) is used for this purpose by treating the cross-interaction quantities as though they are properties of a real substance that follow the principle of corresponding states:

$$T_{cij} = (T_{ci}T_{cj})^{1/2} - \Delta T_{cij} \tag{4.229}$$

$$\omega_{ij} = (\omega_{ii} + \omega_{jj})/2 \tag{4.230}$$

$$z_{cij} = 0.291 - 0.08\omega_{ij} \tag{4.231}$$

$$v_{cij} = \frac{1}{8}(v_{cii}^{1/3} + v_{cjj}^{1/3})^3 \tag{4.232}$$

$$p_{cij} = z_{cij} RT_{cij} / v_{cij} \tag{4.233}$$

For the polar and hydrogen-bond interaction terms, the following rules are suggested: polar-polar interaction can be approximated with the geometric mean; two hydrogen bonds interact with full force; hydrogen bond interacts with dipoles with reduced force; and nonpolar molecules do not interact with poles or hydrogen bonds.

4.2.5.4 Extended Virial Equations

The extended virial equations are made up of a truncated virial series followed by a closure term or terms. In the Benedict-Webb-Rubin (BWR) Equation (4.177), the closure term is an exponential,

$$\begin{aligned} P &= RT\rho + (B_0 RT - A_0 - C_0/T^2)\rho^2 + (bRT - a)\rho^3 + \alpha a \rho^6 \\ &\quad + c\rho^3(1+\gamma\rho^2)(1/T^2)\exp(-\gamma\rho^2) \end{aligned} \tag{4.234}$$

The eight constants required have been reported for numerous hydrocarbons.

The Lee-Kesler eos [2] is an extended BWR equation in reduced variables, as follows:

$$z = 1 + \frac{B}{v_r} + \frac{C}{v_r^2} + \frac{D}{v_r^5} + \left(\frac{c_4}{T_r^3 v_r^2}\right)(\beta + \tilde{a}/v_r^2)\exp(-\tilde{a}/v_r^2) \tag{4.235}$$

$$B = b_1 - \frac{b_2}{T_r} - \frac{b_3}{T_r^2} - \frac{b_4}{T_r^3} \tag{4.236}$$

$$C = c_1 - \frac{c_2}{T_r} + \frac{c_3}{T_r^3} \tag{4.237}$$

$$D = d_1 + \frac{d_2}{T_r} \tag{4.238}$$

where $v_r = p_c v/(RT_c)$ and $T_r = T/T_c$.

Using this equation, Lee and Kesler developed a generalized correlation of the pVT relationship of normal fluids in the following form:

$$z = z_0 + (\omega/\omega_r)(z_0 - z_r) \tag{4.239}$$

The subscript 0 denotes simple fluids with acentric factor $\omega = 0$, and subscript r denotes a reference fluid with acentric factor $\omega_r = 0.3978$ from its origin in n-octane. The Lee-Kesler equation constants for z_0 and z_r are presented in Table 4.4. The compressibility factors in Equation (4.239)—z, z_0, z_r—are at the same $[T_r, p_r]$. This correlation is a three-parameter generalized correlation that improves the Pitzer correlation to a wider range of states. The lower temperature bound of the correlation is extended from $T_r = 0.8$ to 0.3.

Extended virial equations of many terms and constants have been developed for the highly accurate representation of experimental data. Some are developed specifically for standard tables of density and derived thermodynamic functions such as entropy and enthalpy. Bender [31] extended the virial equation to a 20-constant equation to represent argon, oxygen, methane, hydrogen, ethene,

TABLE 4.4
Lee-Kesler Equation of State Constants

Constant	Simple fluids	Reference fluids
b_1	0.1181193	0.2026579
b_2	0.265728	0.331511
b_3	0.154790	0.027655
b_4	0.030323	0.203488
c_1	0.0236744	0.0313385
c_2	0.0186984	0.0503618
c_3	0.0	0.016901
c_4	0.042724	0.041577
$d_1 * 10^4$	0.155488	0.48736
$d_2 * 10^4$	0.623689	0.0740336
β	0.65392	1.226
γ	0.060167	0.03754

and propene. Jacobsen and Stewart [32] further extended the virial equation to 32 constants to fit nitrogen. The steam table equation of Keenan, Keyes, Hill, and Moore employs 54 constants [33].

4.2.6 Energy Functions of Ideal Gases and Mixtures

A useful way to find the value of energy functions of real fluids is to calculate it from a suitable equation of state. The calculation gives the deviation of the desired property from its ideal-gas value, called the residual function or deviation function. The energy function is obtained upon adding the residual function and the ideal-gas function. In this subsection we develop the ideal-gas energy functions; in the next subsection we derive the residual functions and sum up with the ideal-gas value.

The ideal gas is defined by

$$p = nRT/V \quad (4.240)$$

and the internal energy, being independent of volume or pressure, is then a function of temperature only,

$$U^* = U^*[T] \quad (4.241)$$

The asterisk (*) indicates an ideal-gas quantity, and the square brackets denote the independent variable or variables. The ideal gas U is given in standard references for a large number of chemical substances.

The enthalpy is defined by

$$H = U + pV \quad (4.242)$$

Forming the pV product from Equation (4.240) and substituting into Equation (4.242),

$$H^* = U^* + nRT \quad (4.243)$$

which shows that H is also independent of p or V, but varies with T,

$$H^* = H^*[T] \tag{4.244}$$

The ideal-gas enthalpy can be found in standard references for a large number of chemical substances.

Ideal-gas free energies A and G are, however, p or V dependent. Consider Helmholtz energy first. From the fundamental differential Equation (4.79),

$$dA = -SdT - pdV \tag{4.245}$$

At constant T, $dT = 0$, we substitute Equation (4.240) into Equation (4.245) and integrate from the volume V^0 at the standard pressure p^0 of either 1 atm or 1 bar to the system volume V,

$$A^* - A_0 = nRT \ln(V_0/V) \tag{4.246}$$

Replacing the standard-state volume in Equation (4.246) with $(RT/1)$ we obtain

$$A^* = A^0 + nRT \ln \frac{RT}{V(1)} \tag{4.247}$$

The factor 1 standing for the standard pressure is often left out of the equation, but is kept here as a reminder that the units of the other factors in RT/V must be consistent with the units of the standard pressure. The ideal-gas A^0 or G^0 is tabulated in standard references.

To find the Gibbs energy of an ideal gas, start from the fundamental differential Equation (4.88):

$$dG = -SdT + Vdp \tag{4.248}$$

At constant $T(dT = 0)$, substitute Equation (4.240) into Equation (4.248) and integrate from a standard pressure $p^0 = 1$ atm or 1 bar to p to obtain

$$G^* - G_0 = nRT \ln(p/p_0) \tag{4.249}$$

If pressure is in units of atmospheres or bars in Equation (4.249), to agree with the units of the standard state, p^0 can be omitted from the equation, leaving

$$G^* = G_0 + nRT \ln p \tag{4.250}$$

The ideal-gas standard-state G^0 can be looked up from standard references for a large number of substances.

To find the entropy of an ideal gas, we combine the enthalpy of Equation (4.244) with the Gibbs energy of Equation (4.250),

$$S^* = (H^* - G^*)/T \tag{4.251}$$

$$= S^0 - nR \ln p \tag{4.252}$$

To find the energy functions of ideal-gas mixtures, we start with the ideal-gas equation of state, Equation (4.240), recognizing that the number of moles n in the equation includes the moles of all species in the mixture, i.e., $n = n_1 + n_2 + \cdots$. To show the mole numbers of the various species, we write the ideal-gas eos as follows:

$$p = (n_1 + n_2 + \cdots)RT/V \qquad (4.253)$$

Expanding into several terms,

$$p = n_1 RT/V + n_2 RT/V + \cdots \qquad (4.254)$$

where $n_1 RT/V$ is just the pressure of the n_1 moles of species 1 at the V and T of the mixture. It is the partial pressure p_1 of species 1,

$$p_1 = n_1 RT/V \qquad (4.255)$$

$$= (n_1/n)(nRT/V) \qquad (4.256)$$

$$= y_1 p \qquad (4.257)$$

Equation (4.257) shows that the partial pressure of species 1 is also given by $y_1 p$. Similar equations hold for each component. Equation (4.253) becomes

$$p = p_1 + p_2 + \cdots \qquad (4.258)$$

The pressure of an ideal-gas mixture is the sum of the pressures of the components, each being pure at state $[n_i, T, V]$ or $[T, p_i]$.

Since the molecules of an ideal gas do not appreciably attract one another and do not take up any appreciable volume, each component in a mixture behaves like it is existing by itself, leading to the additive partial-pressure phenomenon. J. W. Gibbs stated the principle of independent action: The energy function of an ideal-gas mixture at $[T,p]$ is the sum of the energy functions of the components, each in its pure state at $[T,p_i]$,

$$w^*[T,p] = \sum y_i w_i^*[T, p_i] \qquad (4.259)$$

where w^* denotes an energy function per mole of mixture, and w_i^* is the molar value of pure i.

Accordingly, the internal energy of an ideal-gas mixture is made up of the pure-component internal energies,

$$u^*[T,p] = \sum y_i u_i^*[T, p_i] \qquad (4.260)$$

Since the U of an ideal gas is independent of pressure, the pressure in the brackets [] can be dropped,

$$u^*[T] = \sum y_i u_i^*[T] \qquad (4.261)$$

Similarly, the molar enthalpy of an ideal-gas mixture can be obtained by summing the pure-component enthalpies,

$$h^*[T] = \sum y_i h_i^*[T] \qquad (4.262)$$

The Helmholtz energy of an ideal-gas mixture is, following the principle of independent action,

$$a^*[T,p] = \sum y_i a_i^*[T,p_i] \qquad (4.263)$$

It is often useful to relate the mixture a^* to the standard value a_i^0 of the components at the standard pressure of 1 atm or 1 bar so it can be looked up in standard references. By substituting $p_i = py_i$ and making use of Equation (4.247), we obtain

$$a_i^*[T,p_i] = a_i^0 + RT \ln y_i + RT \ln p \qquad (4.264)$$

Substituting into Equation (4.263) gives

$$a^*[T,p] = \sum y_i a_i^0 + RT \sum y_i \ln y_i + RT \ln p \qquad (4.265)$$

Relating the mixture to the pure components all at the same p, we rewrite Equation (4.265) as follows:

$$a^*[T,p] = \sum y_i a_i^*[T,p] + RT \sum y_i \ln y_i \qquad (4.266)$$

The Gibbs energy of an ideal-gas mixture is obtained by adding RT to the Helmholtz energy equations. Employing standard values of the pure components and Equation (4.265),

$$g^*[T,p] = \sum y_i g_i^0 + RT \ln p + RT \sum y_i \ln y_i \qquad (4.267)$$

To relate the mixture g to the pure-component values at the same pressure, we have from Equation (4.266)

$$g^*[T,p] = \sum y_i g_i^0[T,p] + RT \sum y_i \ln y_i \qquad (4.268)$$

The entropy of an ideal-gas mixture is obtained by combining the internal energy of Equation (4.261) and the Helmholtz energy of Equation (4.265):

$$s^* = (u^* - a^*)/T \qquad (4.269)$$

$$= \sum y_i s_i^0 - R \ln p - R \sum y_i \ln y_i \qquad (4.270)$$

Equation (4.270) is rewritten upon referring to Equation (4.252) to relate the mixture entropy to the pure gases at the same pressure,

$$s^* = \sum y_i s_i^* - R \sum y_i \ln y_i \qquad (4.271)$$

Thermodynamics of Fluid Phase and Chemical Equilibria

4.2.7 Residual Functions and Energy Functions from Equations of State

A residual function is defined to be the difference between the property of a real fluid and that of the fluid as an ideal gas at the same state. The real-fluid property is obtained by adding the residual function to the ideal-gas property. The calculation of residual function from an equation of state follows the identity

$$W - W^* = \int_{\infty}^{V}\left[\left(\frac{\partial W}{\partial V}\right)_T - \left(\frac{\partial W^*}{\partial V}\right)_T\right]dV \qquad (4.272)$$

where W denotes the thermodynamic function and the asterisk (*) denotes the ideal gas. The derivative $(\partial W/\partial V)_T$ is obtained from the equation of state, while the ideal gas $(\partial W^*/\partial V)_T$ is obtained from the formulas presented earlier for the four energy functions and entropy. Other thermodynamic functions, when needed, can be derived from the formulas for the energy functions and entropy presented here.

4.2.7.1 Helmholtz Energy

We will begin the derivation with Helmholtz energy, as it is the natural energy function for the independent variables T and V of equations of state. By the fundamental differential equation for A, Equation (4.81)

$$\left(\frac{\partial a}{\partial v}\right)_T = -p \qquad (4.273)$$

The p in this equation is the equation of state. For an ideal gas,

$$\left(\frac{\partial a^*}{\partial v}\right)_T = -\frac{RT}{v} \qquad (4.274)$$

By Equation (4.272), we obtain the residual Helmholtz energy

$$a - a^* = \int_{\infty}^{v}\left(-p + \frac{RT}{v}\right)dv \qquad (4.275)$$

Using Equation (4.247) for a^*, we obtain

$$a^0 = \int_{\infty}^{v}\left(-p + \frac{RT}{v}\right)dv + RT\ln\frac{RT}{v} + a^0 \qquad (4.276)$$

The standard state a^0 for a mixture is combined from the components according to

$$a^0 = \sum y_i a_i^0 + RT\sum y_i \ln y_i \qquad (4.277)$$

Example 4.11
Find the Helmholtz Energy of Peng-Robinson eos

Solution
The PR eos is Equation (4.165),

$$p = \frac{RT}{v-b} - \frac{\alpha a_c}{v(v+b)+b(v-b)} \qquad (4.278)$$

where α is a function of T, and a_c and b are constants.

Substituting Equation (4.278) into Equation (4.275) and then integrating gives the residual Helmholtz energy by the PR eos,

$$a - a^* = RT \ln \frac{v}{v-b} + \frac{\alpha a_c}{2\sqrt{2}b} \ln \frac{v+(1-\sqrt{2})b}{v+(1+\sqrt{2})b} \qquad (4.279)$$

Replacing a^* with a^0 by using Equation (4.276), we obtain the Helmholtz energy,

$$a = RT \ln \frac{RT}{v-b} + \frac{\alpha a_c}{2\sqrt{2}b} \ln \frac{v+(1-\sqrt{2})b}{v+(1+\sqrt{2})b} + a^0 \qquad (4.280)$$

4.2.7.2 Gibbs Energy

To obtain Gibbs energy from eos, we begin by forming the residual function

$$g - g^* = (a - a^*) + pv - RT \qquad (4.281)$$

Expressing $(a - a^*)$ in terms of an eos by Equation (4.275), we obtain

$$g - g^* = \int_{\infty}^{v} \left(-p + \frac{RT}{v}\right) dv + pv - RT \qquad (4.282)$$

Using Equation (4.250) to replace g^* with g^0,

$$g = \int_{\infty}^{v} \left(-p + \frac{RT}{v}\right) dv + pv - RT + RT \ln \frac{RT}{v} + g^0 \qquad (4.283)$$

Example 4.12: Gibbs Energy by PR eos
The integral in Equation (4.282) for the PR eos will be taken from Equation (4.279). The pv term in Equation (4.282) will be formed by using the PR eos presented in Equation (4.278). It follows that

$$g - g^* = RT \ln \frac{v}{v-b} + \frac{\alpha a_c}{2\sqrt{2}b} \ln \frac{v+(1-\sqrt{2})b}{v+(1+\sqrt{2})b}$$
$$+ \frac{RTv}{v-b} - \frac{\alpha a_c v}{v(v+b)+b(v-b)} - RT \qquad (4.284)$$

To obtain g, the terms of Equation (4.284) are used in Equation (4.283),

$$g = RT \ln \frac{v}{v-b} + \frac{\alpha a_c}{2\sqrt{2}b} \ln \frac{v+(1-\sqrt{2})b}{v+(1+\sqrt{2})b}$$

$$+ \frac{RT_v}{v-b} - \frac{\alpha a_c v}{v(v+b)+b(v-b)} - RT + RT \ln \frac{RT}{v} + g^0 \qquad (4.285)$$

4.2.7.3 Entropy

The fundamental differential form of dU shows

$$s = -\left(\frac{\partial a}{\partial T}\right)_v \qquad (4.286)$$

Accordingly, by differentiating Equation (4.275), we have the residual entropy in terms of eos,

$$s - s^* = \int_\infty^v \left[\left(\frac{\partial p}{\partial T}\right)_v - \frac{R}{v}\right] dv \qquad (4.287)$$

To obtain the entropy relative to the standard state, we substitute Equation (4.252) into Equation (4.287),

$$s = \int_\infty^v \left[\left(\frac{\partial p}{\partial T}\right)_v - \frac{R}{v}\right] dv + R \ln \frac{v}{RT} + s^0 \qquad (4.288)$$

4.2.7.4 Internal Energy

To obtain internal energy from an eos, we first obtain its residual function by combining the residual Helmholtz energy and residual entropy,

$$u - u^* = (a - a^*) + T(s - s^*) \qquad (4.289)$$

Substituting Equation (4.275) for $(a - a^*)$ and Equation (4.287) for $(s - s^*)$ in Equation (4.289),

$$u - u^* = \int_\infty^v \left[T\left(\frac{\partial p}{\partial T}\right)_v - p\right] dv \qquad (4.290)$$

In the ideal-gas state, $u^* = u^0$. We rewrite Equation (4.290) to give the value of u itself,

$$u = \int_\infty^v \left[T\left(\frac{\partial p}{\partial T}\right)_v - p\right] dv + u^0 \qquad (4.291)$$

Example 4.13: Internal Energy by Peng-Robinson eos
The PR eos is differentiated to give

$$\left(\frac{\partial p}{\partial T}\right)_v = \frac{R}{v-b} - \frac{a_c \alpha'}{v^2 + 2bv - b^2} \tag{4.292}$$

where α' stands for $d\alpha/dT$ obtained from Equation (4.162),

$$\alpha' = -2[1 + (0.37465 + 1.5422\omega - 0.26992\omega)(1 - T_r^{0.5})]$$
$$(0.37464 + 1.5422\omega - 0.26992\omega^2)/(T_r^{0.5} T_c) \tag{4.293}$$

Substituting p of Equation (4.159) and its derivative Equation (4.292) into Equation (4.291) and integrating, we obtain internal energy by PR eos,

$$u = \frac{a_c(\alpha - T\alpha')}{2\sqrt{2}b} \ln \frac{v + (1-\sqrt{2})b}{v + (1+\sqrt{2})b} + u^0 \tag{4.294}$$

The internal energy deviation from ideal gas $(u - u^0)$ given in this equation is entirely due to the attractive-force term of the eos. The hard core makes no contribution to the internal energy, as the hard core pressure vanishes in the combination $T(\partial p/\partial T)_v - p$. The hard core in a van der Waals-type eos does not contribute to internal energy. However, in some van der Waals-type eos's, the hard core has a slight temperature dependence, and that does make a weak contribution to internal energy.

4.2.7.5 Enthalpy

To obtain enthalpy from eos we add pv to the internal energy. From Equation (4.290), we obtain the general formula for residual enthalpy from eos,

$$h - h^* = \int_{\infty}^{v} \left[T\left(\frac{\partial p}{\partial T}\right)_v - p \right] dv + pv - RT \tag{4.295}$$

For an ideal gas, enthalpy is independent of v but is a function of T only, and $h^* = h^0$. Equation (4.295) is rewritten to give h itself,

$$h = \int_{\infty}^{v} \left[T\left(\frac{\partial p}{\partial T}\right)_v - p \right] dv + pv - RT + h^0 \tag{4.296}$$

Example 4.14: Enthalpy by Peng-Robinson eos
By adding $(pv - RT)$ to Equation (4.294), we obtain enthalpy by PR eos,

$$h = \frac{a_c(\alpha - T\alpha')}{2\sqrt{2}b} \ln \frac{v + (1-\sqrt{2})b}{v + (1+\sqrt{2})b} + \frac{RTb}{v-b} - \frac{v\alpha a_c}{v^2 + 2bv - b^2} + h^0 \tag{4.297}$$

4.2.7.6 Some General Comments

The independent variables in the formulas for thermodynamic functions derived from eos are $[T,v]$. To use the formula for a state specified in $[T,p]$, the volume at the given state must first be found by solving the eos. The formula is equally useful for a pure fluid or a fluid mixture. However, the equation-of-state parameters and standard-state quantities for a mixture, though no different in symbols from those of a pure fluid, are combinations of the pure-component values and are functions of composition. In spite of this difference, they are used the same way in the formulas of this section.

The formulas presented here are valid for fluids at a homogeneous one-phase equilibrium state. They are not to be directly applied to a fluid at an unstable state, for ordinary interest is not on the unstable fluid as a homogeneous phase, but on the saturated phases that separate from the unstable fluid. Separate calculations on the separated phases need to be performed with the eos-derived formulas for the individual saturated phases, and summed if desired. The calculations to find the saturated equilibrium phases are the subject of Section 4.4.

For isothermal processes, the change in enthalpy is equal to the change in residual enthalpy, for the only difference between enthalpy and residual enthalpy is h^0, which is canceled when forming the isothermal change. Similarly, an isothermal change in internal energy is equal to the change in residual internal energy. It is more convenient to use the residual quantity in both instances.

For nonisothermal processes, the change of energy functions contains a change of standard state. The standard-state values need to be known. The ideal-gas standard-state values are given in standard tables. These have to be looked up or, alternatively, heat capacity data are employed instead of the standard-state values. Consider the change of enthalpy; the difference is found by integrating the heat capcity,

$$h_2^0 - h_1^0 = \int_1^2 c_p^0 dT \tag{4.298}$$

Similarly for internal energy and entropy,

$$u_2^0 - u_1^0 = \int_1^2 (c_p^0 - R) dT \tag{4.299}$$

$$s_2^0 - s_1^0 = \int_1^2 \left(\frac{c_p^0}{T}\right) dt \tag{4.300}$$

4.2.8 FUGACITY

Fugacity is a transformation of chemical potential for convenient use in place of chemical potential for both phase and chemical equilibrium. The transformation is suggested from observing the chemical potential in an ideal-gas mixture,

$$\mu_i^* = g_i^0 + RT \ln(p_i / 1) \tag{4.301}$$

which is obtained by differentiation of Equation (4.267). By this equation, the chemical potential varies monotonously with p_i at a given temperature, going up or down as p_i goes up or down. Specifying a value for μ_i^* gives a value of p_i. Partial pressure can be used in place of chemical potential as the driving force for chemical reaction or phase change for ideal gases.

Fugacity, f, in general is defined by the transformation equation,

$$RT \ln(f_i / 1) = \mu_i - g_i^0 \tag{4.302}$$

The factor 1 in the equation is the standard-state pressure of 1 atm or 1 bar, and will be left out in subsequent equations for simplicity, with the understanding that f has the dimension of pressure in units of either atmospheres or bars, depending on the convention for g_i^0. It follows from the two equations above that, for an ideal-gas mixture, $f_i = p_i$, reducing for a pure ideal gas to $f = p$.

Upon substitution of μ_i from the defining equation of fugacity, Equation (4.302), in the phase-equilibrium equation, Equation (4.111) et seq., canceling out g_i^0 and RT, and exponentiating, we obtain the fugacity equality condition of heterogeneous phase equilibrium,

$$f_{\alpha i} = f_{\beta i} \tag{4.303}$$

for all components i and all phases α and β. The fugacity of any component must assume the same value in all phases in a heterogeneous system at equilibrium. These equations will be used in calculations in place of the chemical potential equality conditions of Equation (4.111) et seq. in the following sections of this chapter.

Why is fugacity so convenient that it is to be used in place of chemical potential in calculations? The reasons are:

1. Fugacity is freed from the ideal-gas standard state g^0. It is completely determined by the properties of the fluid at the temperature of interest.
2. Low-pressure gas states can be conveniently represented by f, for $f_i = p_i$, for an ideal gas. It follows for a real gas at a low or moderate pressure that $f_i \approx p_i$. The fugacity of a dilute component in a liquid is equal to the fugacity of the component at a small partial pressure in a gas mixture at equilibrium with the liquid, again $f_i \approx p_i$. Since the partial pressure of a gas is a well-behaved mathematical quantity, the fugacity is also well behaved at small partial pressures in a gas or at a small concentration in liquids. In contrast, for a dilute component, as its $p_i \to 0$ in an ideal-gas mixture, $\mu_i \to -\infty$ by Equation (4.301). It follows that $\mu_i \to -\infty$ for a dilute component in a real gas or in a liquid. The limit of $-\infty$ is ill-behaved and is avoided with the replacement of μ_i by f_i, which simply approaches zero.

To obtain a formula to calculate fugacity from an equation of state, we appeal to A, as its natural variables, T and V, are the independent variables of eos. We start with Equation (4.106) and carry out the indicated partial differentiation on Equation (4.276) to obtain

$$\mu_i = \int_{\infty}^{v} \left[\frac{RT}{v} - \left(\frac{\partial p}{\partial n_i} \right)_{T,V,nj} \right] dv + RT \ln \frac{RT/1}{v} + RT \ln y_i + g_i^0 \tag{4.304}$$

Substitution of this μ_i into Equation (4.302) gives

$$RT \ln(f_i/1) = \int_{\infty}^{v} \left[\frac{RT}{v} - \left(\frac{\partial p}{\partial n_i} \right)_{T,V,nj} \right] dv + RT \ln \frac{y_i RT}{v} \tag{4.305}$$

The ratio of fugacity to partial pressure, $f_i/(y_i p)$, called the *fugacity coefficient*, ϕ_i, is in common use as a measure of departure of a real fluid f_i from its ideal-gas value. For a pure fluid, the fugacity coefficient is simply $\phi \equiv f/p$. For an ideal-gas mixture, $\phi_i = 1$ for all i; for a pure ideal gas, $\phi = 1$. For a real fluid, by rearranging Equation (4.305),

$$RT \ln \phi_i = \int_\infty^v \left[\frac{RT}{v} - \left(\frac{\partial p}{\partial n_i} \right)_{T,V,nj} \right] dv - RT \ln z \qquad (4.306)$$

Equation (4.306) is well behaved even as $p_i \to 0$; it is the preferred equation for calculating fugacity. For a pure fluid, the equation reduces to

$$RT \ln \phi = \int_\infty^v \left(\frac{RT}{v} - p \right) dv - RT \ln z + RT(z-1) \qquad (4.307)$$

Although generally used for pure fluids, Equation (4.307) is sometimes also used to find the fugacity of a mixture, $\phi_m \equiv f_m/p$. The fugacity of a mixture is just an altered representation of the Gibbs energy of a mixture, $RT \ln f_m = g - g^0$.

Example 4.15: Fugacity Equation from the Redlich-Kwong (RK) eos
The RK eos, Equation (4.159), will be slightly altered as follows:

$$p = RT/(v-b) - a/[v(v+b)] \qquad (4.308)$$

where the parameter a is a function of temperature. The temperature dependence will not show itself explicitly in this example because fugacity is completely determined by the property at the temperature of interest. Since the Wilson and the Soave eos's differ from the RK only in the temperature dependence of a, the fugacity equation is obtained for the other two eos's with the proper selection of the temperature function for a.

The parameters a and b in Equation (4.308) for mixtures are functions of composition. For the present example, we adopt the vdW one-fluid mixing rules,

$$a = \sum_i \sum_j y_i y_j a_{ij} \qquad (4.309)$$

$$b = \sum_i y_i b_i \qquad (4.310)$$

where y_i denotes the mole fraction of component i.

Substituting the a and b combination rules in Equation (4.308) and replacing y_i with the mole numbers n_i, we obtain

$$p = \frac{nRT}{V - \sum n_i b_i} - \frac{n^2 \sum_i \sum_j n_i n_j a_{ij}}{V\left(V + \sum n_i b_i\right)} \qquad (4.311)$$

where $n = \Sigma n_i$, $n_i = ny_i$, and $V = nv$.

Taking the partial derivative $(\partial p/\partial n_i)_{T,p,nj}$, converting the n's back to the y's, and performing the integration indicated in Equation (4.306), we obtain the fugacity coefficient,

$$\ln \phi_k = \ln \frac{v}{v-b} + \frac{b_k}{v-b} - \frac{2\sum_i y_i a_{ik}}{RTb} \ln \frac{v}{v-b} + \frac{ab_k}{RTb^2}\left(\ln \frac{v}{v+b} - \frac{v}{v+b}\right) - \ln z \qquad (4.312)$$

REFERENCES

1. Pitzer, K. S., and Brewer, L. *Revision of Thermodynamics*, by G. N. Lewis and M. Randall. New York: McGraw-Hill, 1961.
2. Lee, B. I., and Kesler, M. *AIChE J.* 21: 510–527, 1975.
3. Walas, S. M. *Phase Equilibrium in Chemical Engineering*. Stoneham, MA: Butterworth, 1985, 3–102.
4. Sandler, S. I., Orbey, H., and Lee, B. I. In *Models for Thermodynamic and Phase Equilibria Calculations*. Edited by S. I. Sandler. New York: Marcel Dekker, 1994, 87–186.
5. Han, S. J., Lin, H. M., and Chao, K. C. *Chem. Eng. Sci.* 43: 2327–2367, 1988.
6. Redlich, O., and Kwong, J. N. S. *Chem. Rev.* 44: 233–244, 1949.
7. Wilson, G. M. *Adv. Cryogen. Eng.* 9: 168–174, 1964.
8. Soave, G. *Chem. Eng. Sci.* 27: 1197–1203, 1972.
9. Peng, D. Y., and Robinson, D. B. *Ind. Eng. Chem. Fund.* 15: 59–64, 1976.
10. Stryjek, R., and Vera, J. H. *Canadian J. Chem. Eng.* 64: 323–333, 1986.
11. Chen, S. S., and Kreglewski, A. *Ber. Bunsenges. Phys. Chem.* 81: 1049, 1977.
12. Boublik, T. *J. Chem. Phys.* 63: 4084, 1975.
13. Alder, B. J., Young, D. A., and Mark, M. A. *J. Chem. Phys.* 56: 3013, 1972.
14. Beret, S., and Prausnitz, J. M. *AIChE J.* 21: 1123, 1975.
15. Donohue, M. D., and Prausnitz, J. M. *AIChE J.* 24: 849, 1978.
16. Prigogine, I. *The Molecular Theory of Solutions*. New York: North Holland, 1957.
17. Chien, C. H., Greenkorn, R. A., and Chao, K. C. *AIChE J.* 29: 560–571, 1983.
18. Boublik, T., and Nezbeda, I. *Chem. Phys. Lett.* 46: 315, 1977.
19. Pults, J. D., Greenkom, R. A., and Chao, K. C. *Chem. Eng. Sci.* 44: 2553, 1989.
20. Pults, J. D., Greenkorn, R. A., and Chao, K. C. *Fluid Phase Equilibria* 51: 147, 1989.
21. Sy-Siong-Kiao, R., Caruthers, J. M., and Chao, K. C. *Ind. Eng. Chem. Res.* 35: 1446–1455, 1996.
22. Honnell, K. G., and Hall, C. K. *J. Chem. Phys.* 90: 1841–1855, 1989.
23. Yithiraj, A., and Hall, C. K. *J. Chem. Phys.* 95: 8494–8506, 1991.
24. Wu, C. S., and Chen, Y. P. *Fluid Phase Equilibria* 101: 3–26, 1994.
25. Chapman, W. G., Gubbins, K. E., Jackson, G., and Radosz, M. *Fluid Phase Equilibria* 52: 31, 1989.
26. Chapman, W. G., Gubbins, K. E., Jackson, G., and Radosz, M. *Ind. Eng. Chem. Res.* 29: 1709, 1990.
27. Wertheim, M. S. *J. Chem. Phys.* 87: 127, 1987.
28. Huang, S. H., and Radosz, M. *Ind. Eng. Chem. Res.* 29: 2284, 1990; 30: 1944, 1991.
29. Tsonopoulos, C. *AIChE J.* 20: 263, 1974; 21: 827, 1975; 24, 1112, 1978.
30. Benedict, M., Webb, G. B., and Rubin, L. C. *J. Chem. Phys.* 8: 334, 1940.
31. Bender, E. *Cryogenics* 15: 667, 1975.
32. Jacobsen, R. T., and Stewart, R. *Br. J. Phys. Chem. Ref. Data* 2: 757, 1973.
33. Keenan, J. H., Keyes, F. G., Hill, P. G., and Moore, J. G. *Steam Table, Thermodynamic Properties of Water, Including Vapor, Liquid, and Solid Phases*. New York: John Wiley, 1969.

4.3 LIQUID SOLUTIONS

4.3.1 IDEAL AND REAL SOLUTIONS

The fugacity of liquid solutions is basic to phase equilibrium and chemical equilibrium of liquids. Raoult observed that the equilibrium partial pressure of a component of a liquid solution with like substances is given by

$$p_i = x_i p_i^S \quad \text{for } i = 1, 2, \ldots, v \tag{4.313}$$

where v denotes the number of components in the solution; p_i denotes the partial pressure of i in the coexisting equilibrium vapor mixture, py_i; x_i is the mole fraction of i in solution; and p_i^S is the vapor pressure of i at the temperature of the solution. Raoult's law is useful for its simplicity and generality. One needs to know only the vapor pressure. As pure liquid i is diluted to form the solution, its equilibrium partial pressure is reduced from the vapor pressure by a factor equal to the mole fraction x_i, reflecting a purely dilution effect, while molecular interaction remains unchanged given that all solutes are highly alike. Though very useful, Raoult's law is restricted to low pressures. The essence of the law does not find full expression in the form of Raoult's law.

The ideal-solution law is a more general statement of Raoult's law in terms of fugacity. In solution of highly alike substances,

$$f_i = x_i f_i^P \tag{4.314}$$

where f_i is the fugacity of component i in solution, and f_i^P is the fugacity of pure liquid i at the T and p of the solution, called the standard state of the component. The standard-state fugacity is the vapor pressure, p_i^S, transformed by

$$f_i^P = p_i^S \varphi_i^S \exp\{v_L^S(p - p_i^S)/RT\} \tag{4.315}$$

where φ_i^S is the fugacity coefficient of the saturated vapor; v_L^S is the molar volume of saturated liquid i; and the exponential factor, called the Poynting correction factor, expresses the incremental fugacity due to compression from pressure p_i^S to the pressure of the solution. The standard-state fugacity is also given in the form of a generalized correlation of fugacity of pure liquids.

The ideal-solution law, Equation (4.314), reduces to Raoult's law, Equation (4.313), under the simplifying conditions:

1. Saturated vapor of i being highly dilute, $\varphi_i^S = 1$
2. The Poynting correction factor being nearly equal to 1
3. The equilibrium vapor behaves as an ideal-gas mixture, so the fugacity coefficient φ_i of component i in the equilibrium vapor equals 1.

Expressed in fugacity, the ideal-solution law is not restricted to low pressure, as it is in Raoult's law.

A solution is ideal if it satisfies the ideal-solution law. No real solution is rigorously ideal, but solutions of similar substances approach ideal-solution behavior as the similarity increases. Solutions of xylene isomers, for example, deviate from ideal-solution law by about 1% at the maximum. Close members of the same homologous series are often assumed to be ideal. It is not unusual to calculate mixtures of paraffin hydrocarbons with the ideal-solution equation. Ideal-solution law is the basis for ideal K values often used in industry. However, ideal-solution law is of great value in another way, and that is to provide a basis for introducing a correction factor, known as the activity coefficient.

The activity coefficient is the ratio of the fugacity of a component in a real solution to its ideal-solution value,

$$\gamma_i = \frac{f_i}{x_i f_i^p} \tag{4.316}$$

For ideal solutions, $\gamma_i = 1$ for all i at all compositions. The more different γ is from 1, the more nonideal is the solution. According to Equation (4.316), the fugacity of a component in a real solution is given by

$$f_i = \gamma_i x_i f_i^p \tag{4.317}$$

The activity coefficient expresses the nonideal-solution behavior of fugacity. The formal development of models for the activity coefficient in solution thermodynamics follows.

Both the ideal-solution Equation (4.314) and the real-solution Equation (4.317) reveal the motivation to use vapor pressure as the basis to find fugacity of components in liquids. This method of using vapor pressure and activity coefficient was practically the sole method for the quantitative calculation of fugacities of liquid solutions until the 1970s. Then, van der Waals-type equations of state were employed to calculate liquid fugacities with good accuracy using Equation (4.307) to perform the integration across the condensation range. Subsequently, some extended virial equations and the perturbation equations were used. Different methods are suitable for various classes of solutions or in different pressure ranges. The activity-coefficient method is suitable for highly nonideal solutions at low pressure and is also suitable for high pressure when the activity coefficient is incorporated in an equation of state. The activity-coefficient method is not readily applied to liquid solutions of light gases.

The fugacities f_i or activity coefficients γ_i of a liquid solution are measured in vapor-liquid equilibrium experiments. In commonly employed methods, the liquid solution is brought in contact and kept in contact with a vapor mixture of the same components until equilibrium is attained between the phases. A sample of the vapor is then withdrawn and analyzed to determine its mole fractions y_i, $i = 1,2,\ldots$. Similarly for the liquid sample, the mole fractions $x_i = 1,2,\ldots$ are determined. Together with the measured p, an experimental point of vapor-liquid equilibrium is given by

$$T, p, x_i \ (i = 1,2,\ldots) \quad \text{and} \quad y_i \ (i = 1,2,\ldots)$$

The fugacities in the gas mixtures are obtained from the gas-phase data: using an equation of state, one can calculate the fugacity coefficient ϕ_i from the experimental y_i and p and then form the product to give

$$f_{iv} = p y_i \phi_i \tag{4.318}$$

A simple equation of state is sufficient to find ϕ_i for the present purposes because the gas phase is usually at a low pressure. The virial equation truncated after B is useful, and correlation of B is found in Section 4.2.5.

By Equation (4.303), the fugacity of a component in the liquid is equal to that of the component in the equilibrium vapor. In light of Equation (4.318), the experimental fugacities of the components in the liquid are determined as

$$f_{iL} = p y_i \phi_i \tag{4.319}$$

From the liquid fugacities thus obtained, the experimental activity coefficients follow from Equation (4.315),

$$\gamma_i = \frac{p y_i \phi_i}{x_i f_i^P} \tag{4.320}$$

Experimental activity-coefficient data are required for the equations to be made useful. To determine the mixture-specific parameters in the activity-coefficient equations, the equation is fitted to the experimental data. A large volume of data and many parameter values for various activity-coefficient equations have been reported. Chapter 1 ("Physical and Chemical Properties") gives sources for this information. The Dechema Chemistry Data Series also presents experimental data and correlations by activity-coefficient equations in many volumes published over several decades.

For mixtures of ethanol + water, the activity coefficient of either component approaches 1 as its mole fraction approaches 1. In this solution, and generally for positively deviating solutions, the activity coefficient of a component steadily increases as its mole fraction decreases. The activity coefficient attains the highest value at infinite dilution. In this mixture, the infinite dilution value often reaches approximately 6 to 8. The activity coefficient rises to much higher values in solutions of a polar/associated substance + a nonpolar one; coefficients of hundreds or higher are encountered in a phase-separated liquid-liquid system, such as water plus hydrocarbons. On the other hand, in solutions of nonpolar substances, such as hydrocarbons, a 20% deviation from the ideal solution earns the designation of definitely nonideal; however, mixtures of homologues like the aliphatics are more ideal. Long-chain molecules make a different class. When small molecules are mixed with these giants, their activity coefficients are as low as 0.01. But mixing one polymer with another polymer often does not result in a homogeneous solution.

The activity coefficients of the various components in a solution are related by the Gibbs-Duhem equation. Converting μ_i to f_i in Equation (4.128) gives the Gibbs-Duhem equation for fugacity at constant $[T,p]$,

$$\sum_k n_k d \ln f_k = 0 \tag{4.321}$$

Substituting Equation (4.315) for f_i results in the Gibbs-Duhem equation for activity coefficient at constant $[T,p]$,

$$\sum_k n_k d \ln \gamma_k = 0 \tag{4.322}$$

For a binary solution, the interdependence of the activity coefficients is determined by rearranging Equation (4.322) as follows:

$$x_A \frac{\partial \ln \gamma_A}{\partial x_A} = x_B \frac{\partial \ln \gamma_B}{\partial x_B} \tag{4.323}$$

The ratio of the slopes of the $\ln \gamma$ curves is equal to the inverse of the ratio of the x's, and the slope of a $\ln \gamma_i$ curve approaches zero as x_i approaches 1. The departure from ideal-solution behavior for a component occurs when it is diluted; the predominant component behaves as if it were in an ideal solution. In a dilute solution, where the solute follows Henry's law, the solvent follows the ideal-solution law.

4.3.2 IDEAL AND EXCESS SOLUTION PROPERTIES

Activity coefficient is a function of the state of a mixture. An activity-coefficient equation is required to calculate the fugacities of real solutions. The interrelationship of the activity coefficients through the Gibbs-Duhem equation implies that the activity-coefficient equations of all components are derivatives of a common thermodynamic function. Since the activity coefficient is an expression of the nonideal behavior of a component, a thermodynamic function is needed to express the nonideality of the total solution and then to obtain from it the activity-coefficient equation.

With the fugacity equation of a real solution, Equation (4.317), taking the logarithm of the equation and multiplying by RT gives

$$RT \ln f_i = RT \ln f_i^P + RT \ln x_i + RT \ln \gamma_i \tag{4.324}$$

Transforming the fugacities f_i, and f_i^P into chemical potentials by using Equation (4.302) gives

$$\mu_i = g_i^P + RT \ln x_i + RT \ln \gamma_i \tag{4.325}$$

where g_i^P is the molar Gibbs energy of pure liquid i at the $[T,p]$ of the solution.

Forming the molar Gibbs energy from the partial values of Equation (4.325),

$$g = \sum_i x_i g_i^P + RT \sum_i x_i \ln x_i + RT \sum_i x_i \ln \gamma_i \tag{4.326}$$

The molar Gibbs energy of an ideal solution is obtained by setting $\gamma_i = 1$ for all i in Equation (4.326),

$$g^{\text{Id}} = \sum_i x_i g_i^P + RT \sum_i x_i \ln x_i \tag{4.327}$$

Even for an ideal solution, the Gibbs energy is not equal to the sum of the pure-liquid constituents.

The change of Gibbs energy upon mixing is the difference between the Gibbs energy of solution and the sum of its pure-liquid constituents at the same $[T,p]$. For 1 mole of solution,

$$g^M \equiv g - \sum_i x_i g_i^P \tag{4.328}$$

Thus, according to Equation (4.326),

$$g^M = RT \sum_i x_i \ln x_i + RT \sum_i x_i \ln \gamma_i \tag{4.329}$$

This is the change in Gibbs energy when the pure liquids are mixed to form 1 mole of solution. For the formation of 1 mole of ideal solution with $\gamma_i = 1$, Equation (4.329) simplifies to

Thermodynamics of Fluid Phase and Chemical Equilibria

$$g^{M,Id} = RT \sum_i x_i \ln x_i \qquad (4.330)$$

Since $\ln x_i$ is always < 0, $g^{M,Id} < 0$, and Gibbs energy is decreased in forming an ideal solution.

The excess Gibbs energy of mixing is defined as the excess of g^M of a real solution above that of the ideal solution,

$$g^E = g^M - g^{M,Id} \qquad (4.331)$$

In view of Equation (4.329) and Equation (4.330), for 1 mole of solution,

$$g^E = RT \sum_i x_i \ln \gamma_i \qquad (4.332)$$

The excess Gibbs energy is the sum of the nonidealities of the components and is the measure of the departure of the total solution from ideal-solution behavior. Hence it is the focal point for theoretical formulation. To facilitate the formulation, g^E is conveniently separated into parts,

$$g^E = u^E - Ts^E + pv^E \qquad (4.333)$$

The excess internal energy, u^E, reflects the varied interaction energies of the different molecules of a solution. The excess entropy reflects the orderliness/randomness of the molecular structure. The excess volume is usually left out as being negligible.

The activity-coefficient equation is obtained from g^E by taking the partial molar derivative of (ng^E) and setting to zero the sum of the derivatives of $\ln \gamma_k$ on account of the Gibbs-Duhem equation, leaving

$$g_i^E = RT \ln \gamma_i \quad i = 1, 2, \ldots, v \qquad (4.334)$$

where g_i^E stands for partial molar excess Gibbs energy and v is the number of components.

Activity coefficients of all components in a solution are obtained from one formulation of g.

4.3.3 Activity-Coefficient Models

Some commonly used activity coefficient equations are presented here. The equations contain parameters that reflect molecular interaction and must be known to use the equation. The parameters are determined by fitting the equation to experimental activity coefficient data. For multi-component mixtures, which are the main interest in engineering applications, the number of equation parameters can be large. The required experimental data to determine the parameters can be excessive.

Simplification of the equation and reduction of data requirement have been achieved in newer equations by addressing binary molecular interactions only. Interactions of clusters of greater than two molecules are not explicitly addressed, but their constituent molecular pairs are included as binary interactions. In this way only binary interaction parameters appear in all activity coefficient equations—for binary solutions as well as for multi-component solutions. By fitting binary solution data all solution parameters can be obtained, for multi-component solutions as well. The need for experimental data on multi-component solutions is eliminated, as all required parameters can be determined by fitting binary solution data.

Experience has shown that the binary interaction equations are at least as good as any larger cluster interaction equation in fitting data; usually they are superior. Their convenience and economy can be over-riding. The binary interaction equations have become the preferred equations to use. The other types of equations are included in this section only because they are still in the literature.

4.3.3.1 Redlich–Kister Equation

Redlich and Kister (RK) [1] employ a parabola $B_{12}x_1x_2$ as the basic form for the g^E of a binary solution. A power series in $(x_1 - x_2)$ is added to to express the asymmetry, if any, of the g^E function,

$$\frac{g_{12}^E}{RT} = x_1x_2[B_{12} + C_{12}(x_1 - x_2) + D_{12}(x_1 - x_2)^2] \tag{4.335}$$

Binary parameters B, C, and D are to be determined by fitting activity-coefficient data. Progressively more parameters are used, depending on the complexity of the function, but D is generally the highest employed. Differentiation of Equation (4.335) gives the activity coefficients

$$\ln \gamma_1 = (B + 3C + 5D)x_2^2 - 4(C + 4D)x_2^3 + 12Dx_2^4$$
$$\ln \gamma_2 = (B - 3C + 5D)x_1^2 + 4(C - 4D)x_1^3 + 12Dx_1^4 \tag{4.336}$$

The RK equation is strictly data fitting with a power series. The judiciously selected function gives a good representation of experimental data for many nonpolar, polar, and associated systems with activity coefficients either greater than or smaller than 1. It is easy to use. But for three-component and higher solutions, the RK equation becomes complex and unwieldy; they are not presented here.

4.3.3.2 Van Laar Equation

Van Laar [2] developed an activity-coefficient equation for liquid solutions based on the vdW eos. This equation has found wide use. Based on the observation that strongly nonideal solutions were associated with large heats of mixing, it was conjectured that nonideality was due to heat of mixing only. Neglecting s^E,

$$g^E = h^E \tag{4.337}$$

For liquids, pv^E is small, and excess enthalpy is assumed equal to u^E,

$$g^E = u^E \tag{4.338}$$

Van Laar [2] modified the van der Waals (vdW) equation to reduce the internal energy by Equation (4.291):

$$u = -\frac{a}{v} + u^0 \tag{4.339}$$

Forming u^E from the energies of the liquids, and setting the u^E to be g^E,

$$g^E = -\frac{a_m}{v_m} + \sum_i x_i \frac{a_i}{v_i} \tag{4.340}$$

where the subscript m denotes the mixture and subscripts $i, j, \ldots$ are the pure liquids. The energies of the ideal gases, u^0, disappear from Equation (4.340) because $u^M = 0$ for the ideal gases.

Van Laar set the liquid molar volumes v in Equation (4.340) to be b as an estimate:

$$g^E = -\frac{a_m}{b_m} + \sum_i x_i \frac{a_i}{b_i} \tag{4.341}$$

To set v to be equal to b is highly unrealistic. However, the b as well as a will be treated as adjusted parameters in van Laar's activity-coefficient equation to fit experimental data. Van Laar also followed van der Waals mixing rules for a_m and b_m (cf. Equation (4.180) and Equation (4.181)):

$$a_m = \sum_i x_i x_j a_{ij} \tag{4.342}$$

$$b_m = \sum_i x_i b_i \tag{4.343}$$

with $a_{ii} = a_i$ and $a_{ij} = \sqrt{a_i a_j}$.

Substitution of Equation (4.342) and Equation (4.343) into Equation (4.341) and algebraic rearrangement gives, for binary solutions,

$$g^E = \frac{x_1 x_2 b_1 b_2}{x_1 b_1 + x_2 b_2}\left(\frac{\sqrt{a_1}}{b_1} - \frac{\sqrt{a_2}}{b_2}\right)^2 \tag{4.344}$$

Taking the partial molar derivatives gives

$$RT \ln \gamma_1 = \frac{A_{12}}{\left(1 + \frac{A_{12}}{A_{21}} \frac{x_1}{x_2}\right)^2} \tag{4.345}$$

$$RT \ln \gamma_2 = \frac{A_{21}}{\left(1 + \frac{A_{21}}{A_{12}} \frac{x_2}{x_1}\right)^2} \tag{4.346}$$

where

$$A_{12} = b_1 \left(\frac{\sqrt{a_1}}{b_1} - \frac{\sqrt{a_2}}{b_2}\right)^2 \tag{4.347}$$

$$A_{21} = b_2 \left(\frac{\sqrt{a_2}}{b_2} - \frac{\sqrt{a_1}}{b_1} \right)^2 \qquad (4.348)$$

Equations (4.345) and (4.346) are the van Laar activity-coefficient equations that are used for fitting data by adjusting the parameters A_{21} and A_{12}. Although these parameters are derived from pure-component parameters, as shown in Equations (4.347) and (4.348), they are, nevertheless, considered mixture-specific binary interaction parameters and are thus indicated with subscripts because they are determined by fitting binary mixture data. Equations (4.347) and (4.348), while showing the source of derivation of the parameters, are not used for their determination.

From the assumed $s^E = 0$, the van Laar equation gives $(RT \ln \gamma)$ as a constant, independent of temperature. The van Laar equation gives a good representation of the activity coefficient of many mixtures, including highly nonideal solutions. However, because it always gives activity coefficient greater than 1, it is not to be used for negatively deviating solutions with $g^E < 0$ and $\gamma < 1$.

By following the same procedure, the activity coefficient in a ternary solution is obtained,

$$RT \ln \gamma_1 = \left[x_2^2 A_{12} \left(\frac{A_{21}}{A_{12}} \right)^2 + x_3^2 A_{13} \left(\frac{A_{31}}{A_{13}} \right)^2 + x_2 x_3 \frac{A_{21} A_{31}}{A_{12} A_{13}} \left(A_{12} + A_{13} - A_{32} \frac{A_{13}}{A_{31}} \right) \right] \Bigg/ \left(x_1 x_2 \frac{A_{21}}{A_{12}} + x_3 \frac{A_{31}}{A_{13}} \right)^2 \qquad (4.349)$$

To obtain the activity coefficient of component 2 or 3, the subscripts are shifted in the order 1-2-3-1.

The three-component equation is complex, and the complexity increases rapidly with the number of components of the solution. Hence, the van Laar equation is rarely used for solutions containing more than two components.

4.3.3.3 Regular Solutions

Hildebrand et al. [3] define a regular solution as one with negligible excess entropy that is composed of nonpolar or slightly polar compounds that do not chemically associate or hydrogen bond. Like van Laar's solutions, the nonideality of regular solutions is an energy effect. Study of solubility data led to the observation that the energy effect is more precisely an energy density effect. The energy density is defined for a pure liquid by

$$a = \frac{u^* - u}{v} \qquad (4.350)$$

The residual energy or the molecular interaction energy $(u^* - u)$ is obtained from the heat of vaporization minus RT and plus any vapor nonideality correction.

Scatchard [4] suggests that the molar energy of a solution be given by

$$(u^* - u)_m = v_m \sum_{i>} \sum_j \phi_i \phi_j a_{ij} \qquad (4.351)$$

where ϕ denotes a volume fraction defined by

$$\phi_i = \frac{x_i v_i^P}{\sum_k x_k v_k^P} \qquad (4.352)$$

The superscript P denotes a pure liquid. The molar volume of the solution v_m is given by

$$v_m = \sum_i x_i v_i^P \qquad (4.353)$$

ignoring any change of volume with mixing.

The excess energy of the solution is obtained by combining the energy of the solution from Equation (4.351) and the energies of the pure liquids from Equation (4.350), and canceling the ideal-gas energies,

$$u^E = \sum_i \phi_i v_i a_i - \sum_{i>} \sum_j v_m \phi_i \phi_j a_{ij} \qquad (4.354)$$

With the assumed zero s^E and neglecting v^E, the u^E of Equation (4.354) is set to be g^E. Combining terms of the summation gives

$$g^E = v_m \sum_{i>} \sum_j \phi_i \phi_j (a_i - a_{ij}) \qquad (4.355)$$

Taking partial molar derivative of g^E gives

$$RT \ln \gamma_k = v_k \left(a_k - 2 \sum_j \phi_j a_{kj} + \sum_i \sum_j \phi_i f_j a_{ij} \right) \qquad (4.356)$$

Introducing the assumption $a_{ij} = (a_i a_j)^{1/2}$ simplifies Equation (4.356) to

$$RT \ln \gamma_k = v_k (\delta_k - \delta_m)^2 \qquad (4.357)$$

where δ is the solubility parameter defined to be the square root of the energy density,

$$\delta_k = a_k^{1/2} \qquad (4.358)$$

For a solution,

$$\delta_m = \sum_i \phi_i \delta_i \qquad (4.359)$$

For a binary solution, Equation (4.357) reduces to

$$RT \ln \gamma_A = v_A \phi_B^2 (\delta_B - \delta_A)^2 \qquad (4.360)$$

$$RT \ln \gamma_B = v_B \phi_A^2 (\delta_A - \delta_B)^2 \qquad (4.361)$$

The v and δ are pure-component properties. There are no mixture parameters to adjust for data fitting in Equation (4.357), Equation (4.360), and Equation (4.361). The regular solution equation is predictive based on pure-component liquid properties. It is generally useful for a first-order estimate for nonpolar, nonassociating, nonhydrogen bonding solutions. $RT \ln \gamma_k$ of a regular solution by Equation (4.357) is not dependent on T,

$$\left(\frac{\partial g^E}{\partial T} \right)_{p,x} = -s^E = 0 \qquad (4.362)$$

The regular solution equation is useful for solutions of normal fluids encountered in the petroleum and gas industries up to high pressure. The regular solution equation is useful for the calculation of solubility of gases. Since the temperature of usual interest is greatly above the critical temperature of the light gases, a hypothetical standard state has to be determined for the regular solution equation to be useful for the light gases. This and other topics related to gas solubility will be found in Section 4.4.

4.3.3.4 Flory–Huggins Equation

The dissolution of polymers into a solvent produces little heat effect for polymers and solvent of similar chemical structure. However, the solvents in these solutions are much less volatile than in an ideal solution with activity coefficients much lower than 1.

Solutions with no heat of mixing are called athermal, and $h^E = 0$. The nonideality of an athermal solution is due to excess entropy. Flory and Huggins independently found the excess entropy of polymer solutions by counting the number of ways by which space is occupied by molecules of different sizes. The more ways to occupy space, the greater is the entropy. According to statistical mechanics, entropy is determined by the number of micromolecular states, $S = k \ln \Omega$, where k is the Boltzmann constant equal to the universal ideal-gas constant per molecule R/N_A; and for molecular solutions, Ω is the number of states for the molecules to occupy space. For the purpose of counting the states, the elemental segments of the polymer molecular chain are assumed to be located on a regular lattice structure like that of a crystal, with each segment taking up one lattice point. Suppose the polymer molecules are placed on the lattice structure one at a time so that the elemental segments of the molecule take up contiguous sites. The number of ways to place the ith polymer molecule on the lattice, ω_i, is $(M_0 - M_{i-1})[z(1 - M_{i-1}/M_0)]^{m-1}$, where the factors are:

- $(M_0 - M_{i-1})$ is the number of ways to place the first segment of the ith molecule on the lattice, with M_0 being the total number of lattice sites and M_{i-1} being the number of lattice sites that are occupied after $i - 1$ molecules have been placed.
- z is the coordination number of the lattice, i.e., the number of closest neighbors of a site. For face-centered and hexagonal close-packed lattices, $z = 12$, and for body-centered packing, $z = 8$. Thus a number from 8 to 10 would be a reasonable value to assign to the hypothetical lattice of a liquid. Although used in the counting, the value of z does not make a difference to the entropy.

$(1 - M_{i-1}/M_0)$ represents the fraction of unoccupied sites. Upon being multiplied by z, the product represents the average number of vacant neighbor sites available for the placement of a segment of the ith molecule.

m is the number of segments of a polymer molecule.

The number of ways the molecules occupy space, Ω, is the continuous product of the ω_i's of all the polymer molecules. Substituting Ω in the entropy equation and converting the lattice sites into molecular core volumes, the entropy of mixing of a polymer solution is found to be

$$s^M = -R \sum_i x_i \ln \phi_i \qquad (4.363)$$

where ϕ_i is the volume fraction of component i in solution given by

$$\phi_i = \frac{x_i v_i}{\sum_j x_j v_j} \qquad (4.364)$$

and v_i is the molar hard core volume of the linear molecule.

The excess entropy following from Equation (4.363) is

$$s^E = -R \sum x_i \ln(\phi_i / x_i) \qquad (4.365)$$

For a solution of molecules of the same size—$\phi_i = x_i$ by Equation (4.364), and $s^E = 0$—the solution reverts to an ideal solution.

There being no difference in the interaction energy of the molecules under consideration, $u^E = 0$, and $g^E = -Ts^E$,

$$g^E = -RT \sum x_i \ln(\phi_i / x_i) \qquad (4.366)$$

Taking the partial molar derivatives gives

$$\ln \gamma_i = \ln \frac{\phi_i}{x_i} + 1 - \frac{\phi_i}{x_i} \qquad (4.367)$$

This is the Flory-Huggins (FH) equation for athermal solutions [5, 6]. The equation formally appears symmetric to all components of a mixture, but in fact it describes the great asymmetry between the polymer and the small solvent in solution. Consider the solution of a small solvent and a polymer that is r times the volume of the solvent. The solvent may well be the monomer, here designated as component 1, and the polymer may be an r-mer as component 2. Setting v_2/v_1 in Equation (4.367) gives

$$\ln a_1 = \ln \phi_1 + \left(1 - \frac{1}{r}\right) \phi_2 \qquad (4.368)$$

$$\ln a_2 = \ln \phi_2 + (1-r)\phi_1 \qquad (4.369)$$

where a_i is the activity $\equiv x_i \gamma_i \equiv f_i / f_i^P$. The equations are, in fact, asymmetric with respect to solvent and polymer.

It is remarkable that no empirical mixture parameters and no experimental data are required to use the equation. The only parameters in the Flory-Huggins equation are the hard core volumes v_i, which are a pure-component property, and the atomic or group contribution values are found in standard compilations. Since the v_i's are significant in the FH equation only in terms of their ratios, pure-liquid molar volumes are often used for v_i in place of hard core volumes. For solutions of polymers of the same chemical formula, molecular masses are legitimate substitutes for v_i, for the same reason. Thus the volume fractions ϕ_i can be substituted by mass fractions w_i. Either volume fraction or mass fraction is directly related to laboratory data. To avoid mole fractions, the activity a_i from Equations (4.368) and (4.369) can be used to calculate f_i by $f_i \equiv a_i f_i^P$.

A surprisingly simple but elegant statistical mechanical result is represented by Equations (4.367), (4.368), and (4.369). Though originally developed for polymer solutions, it is also applicable to solutions of small molecules to account for molecular size differences. It has been made part of some activity-coefficient equations, as will be discussed in the following.

Polymer solutions are not athermal when the solvent and the polymer are of different chemical structure. To account for the effect of energy of mixing, Flory suggested, from counting the change of contacts of the different molecular species upon mixing, an explanation for the excess energy as follows:

$$u^E = RT \frac{v_m}{v_1} \chi \phi_1 \phi_2 \qquad (4.370)$$

where v is molar liquid volume and (v_m/v_1) is a measure of the total molecular contacts. The exchange energy, $\chi \equiv \chi_{12} - (\chi_{11} + \chi_{22})/2$, expressing the energy of formation of a 1–2 contact at the expense of a 1–1 contact and a 2–2 contact, is a mixture-energy parameter for each solvent-polymer combination at a fixed temperature, and is a constant independent of molecular mass for homopolymers. This constant is zero for solutions of a polymer in its monomer.

By adding the u^E of Equation (4.370) with the s^E of Equation (4.365) to form g^E, and by differentiating the g^E, the activity coefficient is obtained,

$$\ln a_1 = \chi \phi_2^2 + \ln \phi_1 + \left(1 - \frac{1}{r}\right)\phi_2 \qquad (4.371)$$

$$\ln a_2 = \div \chi_1^2 + \ln \phi_2 + (1-r)\phi_1 \qquad (4.372)$$

where subscript 1 denotes solvent and 2 is a polymer. These are Flory's χ equations, in which the χ parameter is adjusted for data fitting. Although supposed to be constant, χ has been found to vary with concentration and with polymer molecular mass to varying extent, depending on the mixture system. The χ equation is not as generally useful as the Flory-Huggins equation.

4.3.3.5 Wilson's Local-Composition Equation

Wilson [7] postulated that the nonideality of a solution is caused by a realignment of molecules at the microscopic neighborhood due to the different attractive energies of the various species. Certain molecules are strongly attracted to the central molecule preferentially. The local composition about the central molecule becomes altered from the bulk composition of the solution. Wilson postulates the local composition to determine the entropy of the central molecule.

The probability of a molecule j being at the immediate neighborhood of a central molecule i is x_j, if there is no correlation between the i and j molecules. In case the two are correlated, the probability is x_j times the total correlation. The total correlation can be found by solving an integral equation, given the interaction potential energy function. It is, however, impractical to build the total correlation into a theory of excess functions. Wilson replaced the total correlation with the simpler direct correlation, which is part of the total correlation, to express the probability of finding a j molecule in the microscopic neighborhood of a central i molecule by $x_j \exp(-g_{ji}/RT)$, where g_{ji} is the energy of the i-j interaction and the Boltzmann exponential factor is the direct correlation.

The local mole fraction of j about i is obtained upon normalization,

$$x_{ji} = \frac{x_j \exp(-g_{ji}/RT)}{\sum_k x_k \exp(-g_{ki}/RT)} \tag{4.373}$$

where x_j denotes the bulk mole fraction of j and x_{ji} represents the mole fraction of j in the neighborhood of i.

The local mole fraction is converted to the local volume fraction as follows:

$$\phi_{ji} = \frac{x_j v_j \exp(-g_{ji}/RT)}{\sum_k x_k v_k \exp(-g_{ki}/RT)} \tag{4.374}$$

where v_i is the volume of i. Wilson postulated that the local volume fraction of i in its own neighborhood determines its entropy in the same way that volume fraction determines the entropy in the Flory-Huggins theory of solution of molecules of different sizes (cf. Equation (4.363)). The entropy of mixing of the solution is therefore given by

$$s^M = -R \sum_i x_i \ln \phi_{ii} \tag{4.375}$$

Ignoring any excess energy contribution, Wilson's excess Gibbs energy is $-Ts^E$ as follows:

$$g^E = RT \sum_i x_i \ln(\phi_{ii}/x_i) \tag{4.376}$$

Substitution of ϕ_{ii} by Equation (4.374) into Equation (4.376) and rearrangement gives

$$g^E = -RT \sum_i x_i \ln\left(\sum_i x_k \Lambda_{ik}\right) \tag{4.377}$$

where we have defined the interaction parameter

$$\Lambda_{ik} \equiv \frac{v_k}{v_i} \exp \frac{-(g_{ik} - g_{ii})}{RT} \tag{4.378}$$

In general, $g_{ik} = g_{ki}$, $\Lambda_{12} \neq \Lambda_{21}$, and $\Lambda_{ii} = 1$.

The activity coefficient is obtained by taking partial molar derivatives of g^E,

$$\ln \gamma_i = 1 - \ln\left(\sum_j x_j \ln \Lambda_{ij}\right) - \sum_k \left(\frac{x_k \Lambda_{ki}}{\sum_j x_j \Lambda_{kj}}\right) \tag{4.379}$$

For a binary solution, Equation (4.379) reduces to

$$\ln \gamma_1 = -\ln(x_1 + \Lambda_{12} x_2) + x_2 \left(\frac{\Lambda_{12}}{x_1 + \Lambda_{12} x_2} - \frac{\Lambda_{21}}{\Lambda_{21} x_1 + x_2}\right) \tag{4.380}$$

$$\ln \gamma_2 = -\ln(x_2 + \Lambda_{21} x_1) - x_1 \left(\frac{\Lambda_{12}}{x_1 + \Lambda_{12} x_2} - \frac{\Lambda_{21}}{\Lambda_{21} x_1 + x_2}\right) \tag{4.381}$$

Two adjustable parameters, Λ_{12} and Λ_{21}, need to be determined by fitting binary-solution data to Equation (4.380) and Equation (4.381). Parameters obtained from binary solutions are useful and sufficient in Equation (4.379) for multicomponent solutions, since no higher interaction parameters are required in the multicomponent equation.

The Wilson equation is widely used for many nonpolar, polar, and associated solutions in vapor-liquid equilibrium systems. It is often best for hydrogen-bonded substances. For multicomponent solutions, it makes effective use of binary-solution parameters to give good results, but it cannot predict the liquid immiscibility phenomena.

4.3.3.6 Nonrandom Two-Liquids (NRTL) Equation

To extend the activity-coefficient equation to partially miscible solutions, Renon and Prausnitz [8] introduced a factor to the exponential energy term in Wilson's equation. With $\alpha < 1$, the effect is to suppress the preferential attraction of molecules to the central molecule. The local mole fraction of component 2 about component 1 in a binary solution is given by

$$x_{21} = \frac{x_2 \exp[-\alpha_{12}(g_{21} - g_{11})/RT]}{x_1 + x_2 \exp[-\alpha_{12}(g_{21} - g_{11})/RT]} \tag{4.382}$$

and the local mole fraction of 1 about 1 is

$$x_{11} = \frac{x_1}{x_1 + x_2 \exp[-\alpha_{12}(g_{21} - g_{11})/RT]} \tag{4.383}$$

The factor α is assumed to be a constant for a binary system, $\alpha_{11} = \alpha_{22} = \alpha_{12} = \alpha_{21}$. The partial contribution of component 1 to the excess Gibbs energy of the solution is assumed to be $g^{(1)} - g_p^1$, where

$$g^{(1)} = x_{11} g_{11} + x_{21} g_{21} \tag{4.384}$$

and g_p^1 is the g of 1 in the pure-liquid state, set to be g_{11}. Renon and Prausnitz [8] changed Wilson's g, which is energy to Gibbs energy. Summing the contributions of both 1 and 2,

$$g^E = x_1(g^{(1)} - g_{11}) + x_2(g^{(2)} - g_{22})$$ (4.385)

The activity coefficient obtained from Equation (4.385) is

$$\ln \gamma_1 = x_2^2 \left[\tau_{21} \frac{\exp(-2\alpha_{12}\tau_{21})}{[x_1 + x_2 \exp(-\alpha_{12}\tau_{21})]^2} + \tau_{12} \frac{\exp(-\alpha_{12}\tau_{12})}{[x_2 + x_1 \exp(-\alpha_{12}\tau_{21})]^2} \right]$$ (4.386)

$$\ln \gamma_2 = x_1^2 \left[\tau_{12} \frac{\exp(-2\alpha_{12}\tau_{12})}{[x_2 + x_1 \exp(-\alpha_{12}\tau_{12})]^2} + \tau_{21} \frac{\exp(-\alpha_{12}\tau_{21})}{[x_1 + x_2 \exp(-\alpha_{12}\tau_{21})]^2} \right]$$ (4.387)

where

$$\tau_{12} = \frac{g_{12} - g_{22}}{RT}$$ (4.388)

$$\tau_{21} = \frac{g_{21} - g_{11}}{RT}$$ (4.389)

with

$$g_{12} = g_{21}$$ (4.390)

The general equation for a multicomponent solution is

$$\ln \gamma_i = \frac{\sum_j x_j \tau_{ji} G_{ji}}{\sum_k x_k G_{ki}} + \sum_j \frac{x_j G_{ij}}{\sum_k G_{kj} x_k} \left(\tau_{ij} - \frac{\sum_m x_m \tau_{mj} G_{mj}}{\sum_k G_{kj} x_j} \right)$$ (4.391)

where $G_{ij} = \exp(-\alpha_{ij}\tau_{ij})$.

The NRTL equations—Equation (4.386), Equation (4.387), and Equation (4.391)—contain three parameters for each binary system that are adjusted to fit data. Experience indicates that α varies in the range 0.20 to 0.47. Where experimental data are scarce, the value of α can be set by referring to known values of similar mixtures. Renon and Prausnitz [8] recommended values for broad classes of mixtures. A typical value of α is 0.3. The NRTL equation offers advantages over the Wilson equation for strongly nonideal mixtures, and especially for partially miscible systems.

4.3.3.7 The Complete Local-Composition Equation

Wilson's account for nonideality is all entropic. Only s^E contributes to g^E, while u^E is ignored. Hence Wilson's local-composition equation is incapable of describing immiscibility phenomena. Wang and Chao [9] modify the Wilson equation by restoring the missing u^E to g^E, thus completing the formation of g^E. By counting the contacts between the various molecular species based on the local mole fractions, they expressed the excess energy as follows:

$$u^E = \frac{1}{2}\sum_i x_i \sum_j z x_{ji}(g_{ji} - g_{ii}) \tag{4.392}$$

where g is energy as in Wilson's equation, and z is the coordination number, i.e., the number of closest neighbor molecules.

Adding Equation (4.392) to Wilson's $-Ts^E$ gives the complete g^E,

$$g^E = \frac{z}{2}\sum_i x_i \sum_j x_{ji}(g_{ji} - g_{ii}) + RT \sum_i x_i \ln(\phi_{ii}/x_i) \tag{4.393}$$

The activity coefficient obtained from Equation (4.393) is

$$\ln \gamma_i = \frac{1}{RT}\left(\frac{z}{2}\right)\left[\sum_j x_{ji}(g_{ji} - g_{ii}) + \sum_j x_j \sum_k x_{kj}\left(\frac{x_{ij}}{x_i}\right)(g_{ij} - g_{kj})\right]$$
$$+ 1 - \ln\left(\sum_j x_j \Lambda_{ij}\right) - \sum_k \left(\frac{x_k \Lambda_{ki}}{\sum_j x_j \Lambda_{kj}}\right) \tag{4.394}$$

where Λ_{ij} is defined as

$$\Lambda_{ij} = \frac{v_j}{v_i}\exp\frac{-(g_{ji} - g_{ii})}{RT} \tag{4.395}$$

For binary mixtures, Equation (4.394) simplifies to

$$\ln \gamma_1 = \frac{1}{RT}\left(\frac{z}{2}\right)\left[x_{21}^2(g_{21} - g_{11}) + x_2 x_{22}\frac{x_{12}}{x_1}(g_{21} - g_{22})\right]$$
$$- \ln(x_1 + \Lambda_{12}x_2) + x_2\left(\frac{\Lambda_{12}}{x_1 + x_2\Lambda_{12}} - \frac{\Lambda_{21}}{x_2 + x_1\Lambda_{21}}\right) \tag{4.396}$$

$$\ln \gamma_2 = \frac{1}{RT}\left(\frac{z}{2}\right)\left[x_{21}^2(g_{12} - g_{22}) + x_1 x_{11}\frac{x_{21}}{x_2}(g_{21} - g_{11})\right]$$
$$- \ln(x_2 + \Lambda_{21}x_1) + x_1\left(\frac{\Lambda_{12}}{x_1 + x_2\Lambda_{12}} - \frac{\Lambda_{21}}{x_2 + x_1\Lambda_{21}}\right) \tag{4.397}$$

The effective volume v is generally set to be the pure-liquid molar volume. The coordination number z has been found to be 6 by fitting the equation to a number of solutions. Two adjustable parameters $(g_{21}-g_{11})$ and $(g_{12}-g_{22})$ are to be determined by fitting experimental data on the binary solution. The binary-solution parameters are useful for multicomponent solutions in Equation (4.394).

Comparison with experimental data shows that the complete local-composition equation preserves the quality of Wilson's equation in describing vapor-liquid equilibrium of completely miscible systems. There are no more than slight differences between the complete equation and Wilson's equation in the fitting of data. But the complete local-composition (CLC) equation extends Wilson's local-composition equation to partially miscible solutions. Good predictions of the coexistent liquid compositions of ternary mixtures based on the binary parameters have been found for water + ethyl acetate + ethanol, for water + methyl acetate + acetone, and for water + acrylonitrile + acetonitrile.

4.3.3.8 UNIQUAC Equation

In the UNIQUAC equation (UNIversal QUAsi Chemical), Abrams and Prausnitz [10] and Maurer and Prausnitz [11] account for the congregation of molecules toward the strongly attractive by using the quasichemical equilibrium theory of Guggenheim, in which two interacting molecules are compared with a quasimolecule. The quasimolecule AB formed from unlike molecules A and B is postulated as being at association/dissociation equilibrium with the quasimolecules AA and BB,

$$AA + BB \leftrightarrow 2AB \tag{4.398}$$

The numbers of quasimolecules, i.e., the numbers of interacting pairs, are governed by the equilibrium equation

$$\frac{N_{AB}^2}{N_{AA}N_{BB}} = \frac{1}{4}\exp\left(-\frac{2w_{AB}}{RT}\right) \tag{4.399}$$

where the exchange energy, w_{AB}, defined by $\varepsilon_{AB} - (\varepsilon_{AA} + \varepsilon_{BB})/2$, is the energy of "dissociation/association." There is one equilibrium equation for each unlike pair like AB.

To satisfy the equilibrium equations, the number of contacts AB is given by

$$N_{AB} = x_A x_B \exp\left(\frac{-\varepsilon_{AB}}{RT}\right) \tag{4.400}$$

and similarly for all unlike and like pairs, i.e., for A = 1,2,…,ν; B = 1,2,…,ν. Thus the local contact numbers in Wilson's local-composition equation are recovered similarly to correlation. However, in UNIQUAC, Equation (4.400) is used to form the excess energy without normalization.

An extended Flory and Huggins equation is added in the UNIQUAC equation to account for the entropy effects of molecular size differences. The UNIQUAC g^E is made up of two parts,

$$g^E = g^E_{\text{combinatorial}} + g^E_{\text{residual}} \tag{4.401}$$

The component parts are given by

$$\frac{g^E_{\text{combinatorial}}}{RT} = \sum_i x_i \ln\frac{\phi_i^*}{x_i} + \frac{z}{2}\sum_i q_i x_i \ln\frac{\theta_i}{\phi_i^*} \tag{4.402}$$

$$\frac{g^E_{\text{residual}}}{RT} = \sum_i q_i' x_i \ln\left(\sum_j \theta_j' \tau_{ji}\right) \tag{4.403}$$

where z, the coordination number, is set equal to 10. The segment fraction ϕ^* and area fractions θ and θ' are given by

$$\phi_i^* = \frac{r_i x_i}{\sum_j r_j x_j} \tag{4.404}$$

$$\theta_i = \frac{q_i x_i}{\sum_j q_j x_j} \tag{4.405}$$

$$\theta_i' = \frac{q_i' x_i}{\sum_j q_j' x_j} \tag{4.406}$$

Parameter r measures the number of segments of a molecule for the term v in the Flory-Huggins equation. Parameters q_i and q' are surface areas that are interchangeable for all except strongly hydrogen-bonded water and alcohols. Parameters r, q, and q' are pure-component molecular structure parameters. The combinatorial g^E is dependent only on pure-component parameters. The residual g^E depends additionally on binary interaction parameters τ_{ij} and τ_{ji},

$$\tau_{ij} = \exp\left(-\frac{a_{ij}}{T}\right) \tag{4.407}$$

$$\tau_{ji} = \exp\left(-\frac{a_{ji}}{T}\right) \tag{4.408}$$

where $a_{ij} = \Delta u_{ij}/R$ is an energy of interaction expressed in Kelvins. The terms a_{ij} and a_{ji} are the adjustable parameters for fitting binary-solution data.

The UNIQUAC activity-coefficient equation is given by

$$\ln \gamma_i = \ln \frac{\phi_i^*}{x_i} + \frac{z}{2} q_i \ln \frac{\theta_i}{\phi_i^*} + \ell_i - \frac{\phi_i^*}{x_i} \sum_j x_j \ell_j - q_i' \ln\left(\sum_j \theta_j' \tau_{ji}\right)$$
$$+ q_i' - q_i' \sum_j \frac{\theta_j' \tau_{ij}}{\sum_k \theta_k' \tau_{kj}} \tag{4.409}$$

where

$$\ell_j = \frac{z}{2}(r_j - q_j) - (r_j - 1) \tag{4.410}$$

For a binary solution, the equations are simplified to the following:

$$\ln \gamma_1 = \ln \frac{\Phi_1^*}{x_1} + \frac{2}{z} q_1 \ln \frac{q_1}{\Phi_1^*} + \Phi_2^* \left(\ell_1 - \frac{r_1}{r_2} \right)$$
$$- q_1' \ln(\theta_1' + \theta_2' q_1') + \theta_2' q_1' \left(\frac{\tau_{21}}{\theta_1' + \theta_1' \tau_{21}'} - \frac{\tau_{12}}{\theta_2' + \theta_1' \tau_{12}} \right) \tag{4.411}$$

$$\ln \gamma_2 = \ln \frac{\Phi_2^*}{x_2} + \frac{z}{2} q_2 \ln \frac{\theta_2}{\Phi_2^*} \left(\ell_2 - \frac{r_2}{r_1} \ell_1 \right)$$
$$- q_2' \ln(\theta_2' + \theta_1' \tau_{12}) + \theta_1' q_2' \left(\frac{\tau_{12}}{\theta_2' + \theta_1' \tau_{12}} - \frac{\tau_{21}}{\theta_1' + \theta_2' \tau_{21}} \right) \tag{4.412}$$

The UNIQUAC equation, based on the name UNIversal QUAsi Chemical, is applicable to liquid solutions of hydrocarbons, alcohols, nitriles, ketones, aldehydes, organic acids, and water. Partially miscible solutions are represented. The two interaction parameters are determined by fitting binary-solution data, and the equations are useful for binary as well as multicomponent solutions.

An important advantage of the UNIQUAC equation is its applicability to the solution of groups. A group is a structural part of a molecule, such as the methyl group, the methylene, the hydroxyl, the nitrile, etc. By viewing a solution of molecules as a solution of the existent groups, many molecular solutions can be described with a relatively small number of adjustable parameters. The descriptive power of the equation becomes greatly magnified. The UNIQUAC equation is renamed the UNIFAC (universal functional activity coefficient) equation when applied to solutions of groups, as discussed next.

4.3.4 Group Contribution Methods

A small number of groups make up a great variety of molecules. This fact has motivated studies of group contribution to excess properties of solutions. By viewing a solution of molecules as a solution of the groups, many solutions can be described.

To be useful as the basis for conceptually reducing solutions of molecules to solutions of groups, a group must have a distinct chemical structure, a distinct geometric configuration, and a distinct force field. It must possess all of these qualities independent of the molecule in which it finds itself. Take the methylene group. It has a distinctive chemical structure, $-CH_2$, and a distinctive geometric configuration, with the two hydrogen atoms occupying two apexes of the C-tetrahedron and with the two bonds extending toward the remaining two apexes. Methylene groups in many molecules satisfy these two conditions. The force field is determined by the electronic configuration. An indication of the electronic configuration is the localized electric charges within the group. From *ab initio* quantum mechanical calculation of whole molecules, Wu and Sandler [12] found the charge on a methylene in paraffin molecules to be +0.002e, which is practically neutral; but in a methylene adjacent to a hydroxyl group in an ethanol molecule, the charge is 0.333e. The two methylenes are different and require different interaction parameter values. Using quantum calculations, they have identified a large number of distinct groups.

4.3.4.1 Four Postulates of Group Solution

Wilson and Deal [13] suggest four postulates to describe a solution of groups for the representation of molecular solutions as follows:

1. The partial molar G^E, or simply $\ln \gamma_i$, of a molecular species is made up of two parts: one due to the size differences, called $\ln \gamma_i^s$, and the other due to interaction of the groups, called $\ln \gamma_i^g$,

$$\ln \gamma_i = \ln \gamma_i^s + \ln \gamma_i^g \qquad (4.413)$$

The subscript i denotes a molecule.

2. The contribution due to size differences of the molecules is given by the Flory-Huggins athermal solution equation. The constituent atoms (other than hydrogen) m_i are used as a measure of molecular size:

$$\ln \gamma_i^s = \ln \frac{m_i}{\sum_j x_j m_j} + 1 - \frac{m_i}{\sum_j x_j m_j} \qquad (4.414)$$

3. A group i has an activity coefficient Γ_i that is a function of group fractions X_k, $k = 1, 2,\ldots,v$, and temperature. The group fraction is the mole fraction of the group in the group solution that is converted from the mole fraction x of the molecules that make up the solution:

$$X_k = \frac{\sum_j v_{kj} x_j}{\sum_l \sum_j v_{lj} x_j} \qquad (4.415)$$

where X_k is the group fraction of group k, x_j is the mole fraction of molecules j, and v_{kj} is the number of k groups in a j molecule. The activity of a group can be represented by a function F:

$$\ln \Gamma_k = F_k(X_1, X_2, X_3,\ldots; T) \qquad (4.416)$$

where F can be one of the activity-coefficient equations such as the Wilson equation or the UNIQUAC equation.

4. The group interaction contribution to the activity coefficient of molecule j, $\ln \gamma_j^g$, is the sum of $\ln(\Gamma_k / \Gamma_{kj}^*)$ of all groups k in molecule j,

$$\ln \gamma_j^g = \sum_k v_{kj} (\ln \Gamma_k - \ln \Gamma_{kj}^*) \qquad (4.417)$$

where Γ_k is given by Equation (4.416) and Γ_{kj}^* is the activity coefficient of group k in pure liquid j, and is given by the same function F of Equation (4.416):

$$\ln \Gamma_{kj}^* = F_k(X_{1,j}^*, X_{2,j}^*,\ldots; T) \qquad (4.418)$$

The group fraction $X_{1,j}^*$ is that of group 1 in pure liquid j, and similarly for the other groups. The subtraction of $\Gamma_{1,j}^*$ in Equation (4.417) ensures that the standard state of the pure liquid is satisfied; γ_j^g always approaches 1 as $x_j \to 1$ in any solution.

4.3.4.2 Analytical Solution of Groups

In the analytical solution of groups (ASOG) method of Derr and Deal [14], the four postulates of Wilson and Deal are implemented. The activity coefficient due to molecular size differences is given by the Flory-Huggins equation, Equation (4.414), and the group activity coefficient is given by the Wilson equation,

$$\ln \Gamma_i = 1 - \ln \sum_j X_j a_{ij} - \sum_k \frac{X_k a_{ki}}{\sum_j X_j a_{kj}} \qquad (4.419)$$

Equation (4.419) is identical to Equation (4.379) except for the replacement of mole fraction x by group fraction X, and molecular-interaction parameters Λ by group-interaction parameters a.

Kojima and Tochigi [15] have reported 143 sets of group-group interaction parameters a_{ij} for Equation (4.419) for 31 groups. In their work, the group count n_{kj}, the number of k groups in a j molecule, is adjusted from the molecular formula for use in Equation (4.415) and Equation (4.417):

1. Methyl and methylene are assumed to be the same and are lumped together as methylene.
2. Methyne, CH, is assumed to be CH = 0.8 CH_2.
3. Similarly, carbonium C in alkanes is C = 0.5 CH_2.
4. Water is counted as 1.6 OH groups; v H_2O = 1.6.

The count of groups in the Flory-Huggins equation is, however, not altered. In Equation (4.414), m_i remains the number of nonhydrogen atoms of molecule i.

Benzene and n-hexane have activity coefficients up to about 1.5 in the dilute range. For hydrocarbon mixtures, this is a rather large activity coefficient. Ethanol and n-heptane, however, have activity coefficients up to about 10 in the dilute range. Large activity coefficients occur for mixtures of very different components, ethanol being highly polar and n-heptane being nonpolar. The activity coefficients of acetone + methanol are only as high as about 2, which is a moderate value for these highly polar substances if compared with their activity coefficients in solutions with nonpolar substances. Comparison with experimental data shows that all three mixtures are well represented by the ASOG model with Kojima and Tochigi's parameters.

The UNIFAC (universal functional activity coefficient) method [16] is similar to the ASOG method and is based on the four postulates of Wilson and Deal [13] regarding solution of groups. In UNIFAC, the activity coefficient is made of two parts,

$$\ln \gamma_i = \ln \gamma_i^c + \ln \gamma_i^R \qquad (4.420)$$

The superscripts c and R denote, respectively, the combinatorial and residual contributions. The term γ_i^c is given by

$$\ln \gamma_i^c = (\ln \Phi_i / x_i + 1 - \Phi_i / x_i) - \frac{1}{2} z q_i (\ln \Phi_i / \theta_i + 1 - \Phi_i / \theta_i) \qquad (4.421)$$

where x is the mole fraction and z the coordination number, or the number of closest neighbors assumed to be 10:

$$\Phi_i = x_i r_i \Big/ \sum_j x_j r_j \qquad (4.422)$$

The core volume r_i is given by

$$r_i = \sum_k v_{ki} R_k \qquad (4.423)$$

with v_{ki} being the number of group k in molecule i, and R_k being the volume of group k.
   The surface area θ_i is given by

$$\theta_i = x_i q_i \Big/ \sum_j x_j q_j \qquad (4.424)$$

The surface area fraction q_i is the sum of the group values:

$$q_i = \sum_k v_{ki} Q_k \qquad (4.425)$$

where Q_k is the surface area of group k.
   The residual activity coefficient is the sum of group activity coefficients Γ_k,

$$\ln \gamma_i^R = \sum_k v_{ki} \ln \Gamma_k - \ln \Gamma_k^{(i)} \qquad (4.426)$$

where the summation of all groups in molecule i and $\Gamma_k^{(i)}$ is the Γ_k in a group solution that is purely molecules of i.
   The group Γ_k is given by the UNIQUAC equation applied to the groups,

$$\ln \Gamma_k = Q_k \left[1 - \ln\left(\sum_m \theta_m \Psi_{mk} \right) - \sum_m \left(\theta_m \Psi_{mk} \Big/ \sum_n \theta_n \Psi_{nm} \right) \right] \qquad (4.427)$$

$$\Psi_{nm} = \exp(-a_{nm}/T) \qquad (4.428)$$

$$\theta_m = Q_m X_m \Big/ \sum_n Q_n X_n \qquad (4.429)$$

The group fraction X is given by

$$X_m = \frac{\sum_j v_{mj} x_j}{\sum_j \sum_n v_{nj} x_j} \qquad (4.430)$$

The $\Gamma_k^{(i)}$ in Equation (4.426) is determined from Equation (4.427) at group fractions $X_k^{(i)}$ of the group solution that is purely molecules of i.
   Group volume R and surface area Q are presented for 50 groups by Hansen et al. [17]. While the group volume R and surface area Q are obtained from analysis of pure substance data, the group-interaction parameters are obtained from correlation of mixture vapor-liquid equilibrium data.

4.3.5 GIBBS ENERGY MODELS OF LIQUID SOLUTIONS

One begins by forming excess Gibbs energy, g^E, from the equation of state, according to Equation (4.283):

$$g = \int_{\infty}^{v}\left(-p + \frac{RT}{v}\right)dv + pv - RT + RT \ln\frac{RT}{v} + g^0 \tag{4.431}$$

where p stands for the equation of state. Excess Gibbs energy, g^E, is formed by substituting the eos of the mixture in this equation, followed by substituting the eos of the pure components and then combining the results at the same $[T,p]$ according to

$$g^E = g - \sum_i x_i g_i^P - RT \sum_i x_i \ln x_i \tag{4.432}$$

where g denotes the molar Gibbs energy of the mixture, and g_i^P is the molar Gibbs energy of pure i.

Using the Peng-Robinson eos, Equation (4.159), for illustration, we obtain from Equation (4.431) and Equation (4.432)

$$\frac{g_{eos}^E}{RT} = \ln\left(\frac{v}{v-b}\right) - \sum_i x_i \ln\left(\frac{v_i^P}{v_i^P - b_i}\right) + \frac{a}{2\sqrt{2}bRT} \times$$

$$\ln\left[\frac{v+b(1-\sqrt{2})}{v+b(1+\sqrt{2})}\right] - \frac{1}{2\sqrt{2}RT}\sum_i x_i(a_i/b_i)\ln\left[\frac{v_i^P + b_i(1-\sqrt{2})}{v_i^P + b_i(1+\sqrt{2})}\right] \tag{4.433}$$

$$+ \frac{p}{RT}\left(v - \sum_i x_i v_i^P\right) - \ln v + \sum_i x_i \ln v_i^P$$

where the superscript P indicates a pure component, and v is the molar volume of the mixture.

The eos parameters a and b for the mixture, so far unspecified by any mixing rules, make their appearance in Equation (4.433). By setting the g^E_{eos} equal to the g^E_{sm} of a solution model, one could solve for a to make g^E_{sm} part of a, thus specifying its composition dependence and making up a mixing rule. By using the a thus obtained in the eos, the solution model becomes incorporated in the eos. To proceed to solve for a, the v's in Equation (4.433) that are so far undetermined must be specified.

The g^E_{sm} solution models fit low-pressure vapor-liquid equilibrium data for many liquid solutions. These models with fitted parameters are the prime interest to be incorporated into equations of state. To set Equation (4.433) to be equal to these g^E_{sm}'s, the v's in the equation are set to the standard-state pure-liquid volumes of a low-pressure vapor-liquid equilibrium mixture. Novenario et al. [19] calculated the saturated liquid volume for a large number of substances at low pressures with the PR eos and expressed the volume as a multiple of b,

$$v_i^P = kb_i \tag{4.434}$$

The pressure of a saturated Peng-Robinson liquid with $k = 1.15$ is between 0 and 2 atm for all the substances tested. Since Gibbs energy of a liquid is insensitive to pressure at low pressures, $v_i^P = 1.15\ b_i$ is adopted as the standard state in PR eos for pure liquids at low-pressure vapor-liquid equilibrium. Similarly, the volume of the liquid mixture is set to be $v = 1.15\ b$. Substitution of Equation (4.434) into Equation (4.433) leads to

$$\frac{g_{\text{eos}}^E}{RT} = \frac{1}{2\sqrt{2}RT}\ln\left(\frac{k+1-\sqrt{2}}{k+1+\sqrt{2}}\right)\left[\frac{a}{b} - \sum_i x_i\left(\frac{a_i}{b_i}\right)\right]$$

$$+ \frac{pk}{RT}\left(b - \sum_i x_i b_i\right) - \ln b + \sum_i x_i \ln b_i \tag{4.435}$$

where $k = 1.15$ for the PR eos.

Upon adopting the additive mixing rule for b,

$$b = \sum_i x_i b_i \tag{4.436}$$

The middle term of Equation (4.435) can be eliminated, and Equation (4.435) becomes

$$\frac{g_{\text{eos}}^E}{RT} = \frac{1}{2\sqrt{2}RT}\ln\left(\frac{k+1-\sqrt{2}}{k+1+\sqrt{2}}\right)\left[\frac{a}{b} - \sum_i x_i\left(\frac{a_i}{b_i}\right)\right]$$

$$- \ln b + \sum_i x_i \ln b_i \tag{4.437}$$

Upon setting the g_{eos}^E of Equation (4.437) to be equal to g_{sm}^E of the solution model, we solve for a/b to obtain

$$\frac{a}{b} = \frac{2\sqrt{2}RT}{\ln\left(\frac{k+1-\sqrt{2}}{k+1+\sqrt{2}}\right)}\left(\frac{g_{\text{sm}}^E}{RT} + \ln b - \sum_i x_i \ln b_i\right) - \sum_i x_i\left(\frac{a_i}{b_i}\right) \tag{4.438}$$

Equation (4.436) and Equation (4.438) make a set of mixing rules for a and b. The fugacity coefficient obtained from the mixing rules is

$$\ln \phi_i = -\ln\left[\frac{p(v-b)}{RT}\right] + \frac{b_i}{b}(z-1)$$

$$+ \frac{a}{2\sqrt{2}bRT}\left[\frac{1}{na}\left(\frac{\partial n^2 a}{\partial n_i}\right) - \frac{b_i}{b}\right]\ln\left[\frac{v+b(1-\sqrt{2})}{v+b(1+\sqrt{2})}\right] \tag{4.439}$$

with

$$\frac{1}{na}\left(\frac{\partial n^2 a}{\partial n_i}\right) = \frac{b_i}{b} + \frac{b}{a}\left\{K\left[-\ln\gamma_i + \ln\left(\frac{b_i}{b}\right) + 1 - \frac{b_i}{b}\right] + \frac{a_i}{b_i}\right\} \tag{4.440}$$

and

$$K = \frac{-2\sqrt{2}RT}{\ln\left(\frac{k+1-\sqrt{2}}{k+1+\sqrt{2}}\right)} \tag{4.441}$$

The Peng–Robinson–Stryjek–Vera (PRSV) eos with the incorporation of UNIFAC activity-coefficient predicts well experimental data of numerous mixtures, including ethanol-water mixtures at 150–350°C and acetone-with mixtures at 100–250°C.

The free-energy-matching method of incorporating a solution model into an eos is not restricted to a cubic eos. The incorporation of a solution model in a noncubic equation will be illustrated in the following with the chain-of-rotators (COR) eos, Equation (4.183). The excess Gibbs energy is obtained by substituting this equation in Equation (4.431) successively for the mixture and the pure components and combining the results according to Equation (4.432). A lengthy equation is obtained, but this is simplified when the molar volumes are set at the value of the standard state $V_i = kb_i$ and $v = kb$. Novenario et al. [19] found that with $k = 0.6$, the fluids are low-pressure saturated liquids by the COR eos. The equation is further simplified with the adoption of linear mixing rules for b and c,

$$b = \sum x_i b_i \qquad (4.442)$$

$$c = \sum x_i c_i \qquad (4.443)$$

leading to

$$g^E_{eos} = \ln(1+1/k)\left[-\left(\frac{a}{b}\right)+\sum_i x_i\left(\frac{a_i}{b_i}\right)\right]+RT\left(\sum_i x_i \ln b_i - \ln b\right) \qquad (4.444)$$

Upon setting the g^E of the low-pressure COR liquids of Equation (4.444) equal to that of solution model g^E_{sm}, we obtain the mixing rule

$$\frac{a}{b} = \frac{RT}{\ln(1+1/k)}\left(\frac{g^E_{sm}}{RT}+\sum_i x_i \ln b_i - \ln b\right)+\sum_i x_i\left(\frac{a_i}{b_i}\right) \qquad (4.445)$$

where $k = 0.6$. Equation (4.442), Equation (4.443), and Equation (4.445) make up the set of *free-energy-matching mixing rules* for the COR eos. The fugacity coefficient for a mixture component i obtained from the COR eos with the mixing rules is

$$\ln \phi_i = \frac{y(4-3y)}{(1-y)^2}+\frac{c_i}{2}(\alpha-1)\left[\frac{(\alpha+4)y-3y^2}{(1-y)^2}+(\alpha+1)\ln(1-y)\right]$$

$$+\left(\frac{b_i}{b}\right)(z-1)-\ln z - \frac{\ln(1+4y)}{RT} \qquad (4.446)$$

$$\left[\frac{a_i}{b_i}+\frac{RT}{\ln(1+1/k)}\left(-\ln \gamma_i + \ln\left(\frac{b_i}{b}\right)-1-\left(\frac{b_i}{b}\right)\right)\right]$$

Mixing rules based on setting g^E_{eos} to be equal to g^E_{sm} are often called free-energy-matching mixing rules. Huron and Vidal [18] pioneered the free-energy-matching method to apply to cubic equations of state. In their method, the molal volumes of fluids that are mixed are set to be equal to b, so that the mixing state is at infinite pressure. As a result, the eos does not fit experimental low-pressure data, unless the solution-model parameters in the incorporated eos are refitted to the data. Huron-Vidal's method cannot utilize the wealth of correlated low-pressure g_{sm}^E and γ, including correlations such as ASOG and UNIFAC. By setting the mixing at a low-pressure liquid state, the method of Novenario et al. [19] succeeded in making use of liquid-solution models with parameters fitted to low-pressure data.

In the method of Wong and Sandler [20, 21], mixing rules are developed for a and b to satisfy two conditions:

1. To equate the excess Helmholtz energy of the eos at infinite pressure to that of a solution model at a low pressure
2. To impart a quadratic composition dependence to the second virial coefficient of the eos

To achieve the first condition, it is found that

$$g^E(T, p = 1 \text{ bar}) \approx a^E(T, p = 1 \text{ bar}) \approx a^E(T, \text{high pressure}) \approx a^E(T, p)$$

Using the Peng-Robinson eos for illustration, the excess Helmholtz energy is found by combining the Helmholtz energy given by Equation (4.280). Upon setting $v = b$ in the excess function to attain infinite pressure,

$$a_{eos}^E = -\frac{a}{b} + \sum_i x_i \frac{a_i}{b_i} \qquad (4.447)$$

where the un-subscripted a and b refer to the mixture. Setting this excess function to be equal to the g_{sm}^E of a solution model and solving for the a/b of the eos,

$$\frac{a}{b} = -g_{sm}^E + \sum_i x_i \frac{a_i}{b_i} \qquad (3.448)$$

This is one of the two equations that are to be solved to give the combining rules for a and b. We turn to the second virial coefficient to find the other equation to be solved for a and b.

$$b = \frac{\sum_i \sum_j x_i x_j \left(b - \frac{a}{RT}\right)_{ij}}{1 + \frac{g_{sm}^E}{RT} - \sum_i x_i \left(\frac{a_i}{b_i RT}\right)_i} \qquad (4.448)$$

Equation (4.447) and Equation (4.448) together make the mixing rules for a and b.

The free-energy-matching method of Wong et al. [20, 21] has been found to have good accuracy with experimental value data for ethanol-water mixtures and for acetone-water mixtures.

REFERENCES

1. Redlich, O., and Kister, A. T. *J. Am. Chem. Soc.* 71: 505, 1949.
2. van Laar, J. J. *Z. Phys. Chem.* 72: 723, 1929.

3. Hildebrand, J. H., Prausnitz, J. M., and Scott, R. L. *Regular and Related Solutions*, Englewood Cliffs, NJ: Prentice Hall, 1970.
4. Scatchard, G. *Chem. Rev.* 8: 321, 1931.
5. Flory, P. J. *Principles of Polymer Chemistry*, Ithaca, NY: Cornell University Press, 1953.
6. Huggins, M. L. *Physical Chemistry of Polymers*, New York: Wiley, 1958.
7. Wilson, G. M. *J. Am. Chem. Soc.* 86: 127, 1964.
8. Renon, H., and Prausnitz, J. M. *AIChE J.* 14: 135, 1968.
9. Wang, W. C., and Chao, K. C. *Chem. Eng. Sci.* 38: 1483–1492, 1983.
10. Abrams, D., and Prausnitz, J. M. *AIChE J.* 21: 116, 1975.
11. Mauer, G., and Prausnitz, J. M. *Fluid Phase Equilibria* 2: 91, 1978.
12. Wu, H. S., and Sandler, S. I. *Ind. Eng. Chem. Res.* 30: 881, 889, 1991.
13. Wilson, G. M., and Deal, C. H. *Ind. Eng. Chem. Fund.* 1: 20, 1962.
14. Derr, E. L., and Deal, C. H. *Inst. Chem. Eng. Symp. Ser., London* 3(32): 40, 1969.
15. Kojima, K., and Tochiji, K. *Prediction of Vapor-Liquid Equilibria by the ASOG Method*, New York: Elsevier Scientific Publishing, 1979.
16. Fredenslund, A., Jones, R. L., and Prausnitz, J. M. *AIChE J.* 21: 1086, 1975.
17. Hansen, H. K., Rasmussen, P., Fredenslund, A., Shiller, M., and Mehling, J. G. *Ind. Eng. Chem. Res.* 30: 2352–2355, 1991.
18. Huron, M. J., and Vidal, J. *Fluid Phase Equilibria* 3: 255–271, 1979.
19. Novenario, C. R., Caruthers, J. M., and Chao, K.C. *Ind. Eng. Chem. Res.* 35: 269–277, 1996.
20. Wong, D. S. H., and Sandler, S. I. *AIChE J.* 38: 671, 1992.
21. Wong, D. S. H., Orbey, H., and Sandler, S. I. *Ind. Eng. Chem. Res.* 31: 2033, 1992.

4.4 FLUID-PHASE EQUILIBRIA

4.4.1 Vapor-Liquid Equilibrium in a Single-Component Fluid

Vapor pressure is the pressure of a single-component fluid at vapor-liquid equilibrium at a given temperature. Conversely, at constant pressure, temperature remains constant as long as both phases coexist. Vapor pressure is the primary variable in engineering operations of evaporation, condensation, power generation, drying, humidification, and a host of others. Normal boiling point (the temperature at which vapor pressure equals 1 atm) and the critical point at the upper end of the vapor pressure curve are among the most commonly used property constants.

The vapor pressure increases rapidly as temperature is increased; an increase of five orders of magnitude is common from the triple point to the critical point. To find the vapor pressure as a function of T, we begin by finding its differential equation. Since the chemical potential of the vapor and that of the liquid are equal for all states on the vapor pressure curve, differential changes of the chemical potentials with differential changes dT and dp on the vapor pressure curve are also equal,

$$dg_v = dg_L \tag{4.449}$$

For a pure substance, the chemical potential is the molar Gibbs energy. The change of g_v due to dT and dp is given by Equation (4.88):

$$dg_v = -s_v dT + v_v dp \tag{4.450}$$

Similarly, for the equilibrium liquid,

$$dg_L = -s_L dT + v_L dp \tag{4.451}$$

Setting the last two equations to be equal according to Equation (4.449) and rearranging gives

$$\frac{dp}{dT} = \frac{\Delta_v s}{\Delta_v v} \qquad (4.452)$$

The symbol Δ_v denotes the change with vaporization. The change of entropy with vaporization is equal to the heat of vaporization divided by T, therefore:

$$\frac{dp}{dT} = \frac{\Delta_v h}{T \Delta_v v} \qquad (4.453)$$

This is the Clapeyron equation, the differential equation for vapor pressure as a function of T.

To integrate the equation at low pressure, we assume: the vapor volume to be given by the ideal-gas equation; the liquid volume to be negligible; and the heat of vaporization to be unchanging with T. We obtain from Equation (4.453) the Clausius-Clapeyron equation,

$$d \ln p = \frac{\Delta_v h}{RT^2} dT \qquad (4.454)$$

Integration gives

$$\ln p = A - \frac{\Delta_v h}{RT} \qquad (4.455)$$

where A is an integration constant. Lumping the parameters, we have the vapor pressure equation

$$\ln p = A - B/T \qquad (4.456)$$

which contains two parameters to be adjusted to fit the data. This equation gives good fitting of data in a moderate temperature range, but becomes only approximate when extended to a large temperature range. By fitting the equation to the critical state and the normal boiling point, we obtain a quick approximation at the interpolation range. The deviation grows quite large at extrapolated temperatures below the normal boiling point.

$$\ln p = \frac{\ln p_c - \ln 1.0}{1/T_b - 1/T_c} \left(\frac{1}{T_b} - \frac{1}{T} \right) \qquad (4.457)$$

Here pressure is atmospheric, and T is absolute temperature. Subscript b denotes boiling point, and c denotes the critical point.

Another approximate equation is obtained by interpolation between the critical point and $T_r = 0.7$ by employing the acentric factor,

$$\log P_r = \frac{7}{3} \left(\frac{1}{0.7} - \frac{1}{T_r} \right)(\omega + 2) - (1 - \omega) \qquad (4.458)$$

A common logarithm is used in this equation. Both Equation (4.456) and Equation (4.458) are useful for making an initial estimate of either T or p in iterative computer calculations for vapor-liquid equilibrium calculations.

Thermodynamics of Fluid Phase and Chemical Equilibria

The Antoine equation is an empirical modification of Equation (4.456) as follows:

$$\ln p = A - B/(T + C) \tag{4.459}$$

With an additional parameter C, the Antoine equation is widely used for vapor pressure data up to about 2 bars. Over the entire temperature range, an equation of greater complexity is required. Several have been obtained by integrating Equation (4.454) based on various temperature functions for $\Delta_v h$ and $\Delta_v v$ and empirical observation. The Frost-Karkwarf-Thodos equation [1–3] is widely used with constants reported [4] for a large number of substances. The equation is

$$\ln p = A + B/T + C \ln T + Dp/T^2 \tag{4.460}$$

When or if p appears on both sides of the equation, the equation has to be solved by trial and error to find the vapor pressure.

Vapor pressure of normal fluids has been correlated in a generalized reduced form by Lee and Kessler [5] with the following:

$$\ln P_r = F^{(0)}[T_r] + \omega F^{(1)}[T_r] \tag{4.461}$$

$$F^{(0)} = 5.92714 - 6.09648/T_r - 1.28862 \ln T_r + 0.169347 T_r^6 \tag{4.462}$$

$$F^{(1)} = 15.2518 - 15.6875/T_r - 13.4721 \ln T_r + 0.43577 T_r^6 \tag{4.463}$$

A vapor pressure function $p^S[T]$ is implicit in an equation of state that covers both the gas and the liquid states. The vapor pressure functions of some equations of state are accurate over a wide temperature range, and with the use of a computer program, the implicit function becomes a convenient source of vapor pressure information; some programs can calculate vapor pressure with suitable data or with slightly expanded program code.

To find the vapor pressure using an equation of state (eos), we begin with vapor-liquid equilibrium for a single-component system:

$$f_V = f_L \tag{4.464}$$

The fugacity by an eos is given by Equation (4.307) as follows:

$$\ln(f/p) = \int_\infty^v \left(\frac{1}{v} - \frac{p}{RT}\right) dv - \ln z + z - 1$$

Taking the difference between the fugacities of the saturated liquid and saturated vapor, we obtain

$$\ln(f_L/f_V) = \frac{1}{RT} \int_{v_L}^{v_V} p\, dv + \frac{p_L v_L - p_V v_V}{RT} \tag{4.465}$$

The subscripts L and V denote the saturated liquid and the saturated vapor, respectively. An isotherm is an S-shaped curve, while an isobar is a horizontal line. At a phase-equilibrium state, $f_L = f_V$ and $p_L = p_V = p^S$, where p^S stands for the vapor pressure. Equation (4.465) rearranges to

$$p^S = \frac{1}{v_V - v_L} \int_{v_L}^{v_V} p\, dv \qquad (4.466)$$

This is the implicit vapor pressure equation. It is solved by repeated substitution. Beginning with an estimated p^S, v_V and v_L are calculated at a given T until they converge adequately.

Example 4.16
Find the implicit vapor pressure equation of the Redlich-Kwong eos. The RK eos, Equation (4.153), will be written in a simplified form,

$$p = \frac{RT}{v - b} - \frac{a_T}{v(v + b)} \qquad (4.467)$$

where a_T is a function of T and is given as $a/T^{0.5}$ for the RK eos. Substituting Equation (4.159) into Equation (4.465) and integrating gives

$$p^S = \frac{1}{v_V - v_L}\left[RT \ln \frac{v_V}{v_L} - \frac{a_T}{b} \ln \frac{v_V(v_L + b)}{v_L(v_V + b)} \right] \qquad (4.468)$$

The saturated volumes v_V and v_L in Equation (4.468) are solved in the RK eos, Equation (4.467), at the vapor pressure, which makes the equation implicit. The equation can be readily solved for the saturated pressure by repeated substitution. Agreement with vapor pressure is qualitative at best.

Many eos's also provide only a qualitative representation of vapor pressure data due in part to the fact that vapor pressure by eos is determined by the integral $\int p\, dv$ of the eos in Equation (4.466) in the condensation range from v_L to v_V. The eos is either unstable or metastable in this range. No experimental data exist for the unstable states. Only scarce data exist for some metastable states. It is impossible to fit the eos to the pv isotherm in this range, as the isotherm hardly exists in the range of interest.

Yet it is desired that the eos give an accurate representation of vapor pressure. The key to achieving this goal is in Equation (4.468), which shows that the p^S by the RK eos is dependent on the attractive-pressure parameter a. By solving Equation (4.468) for a and letting p^S be the experimental vapor pressure, we obtain

$$a_T = \frac{RT \dfrac{v_V - b}{v_L - b} - p^S(v_V - v_L)}{\dfrac{1}{b} \ln \dfrac{v_V(v_L + b)}{v_L(v_V + b)}} \qquad (4.469)$$

This equation is implicit in a_T, inasmuch as the saturated volumes are dependent on a_T, but the equation can be readily solved, for instance, by the method of repeated substitution. By using the solved value of a_T in the eos, Equation (4.162), the vapor pressure calculated by the eos simply reproduces the experimental vapor pressure data. Wilson correlated the a_T's that are fitted to vapor pressure data of a number of normal fluids to obtain Equation (4.157), the Wilson eos. Vapor pressure calculated by the Wilson eos is improved over that of the RK eos, but the accuracy still leaves something to be desired. Soave correlated the vapor pressure, fitting a_T with Equation (4.164). Even better, the Soave eos is useful for the quantitative calculation of vapor pressure. In addition, the Peng-Robinson and the chain-of-rotators eos's provide quantitative calculations of vapor pressure.

Thermodynamics of Fluid Phase and Chemical Equilibria

The success of vapor pressure calculation by an eos is a result of the equality of the calculated fugacities of the saturated vapor and the liquid at the true vapor pressure. The vapor fugacity is generally calculated with good accuracy, as the departure from ideal gas is not great for the gas. Hence the liquid fugacity is likewise accurately calculated by such an eos, which is a major triumph for the eos, for the departure from ideal gas is always quite large for the liquid. Since fugacity is transformed Gibbs energy, the Gibbs energy of the liquid is well represented by the eos. Because entropy is equal to the temperature derivative of Gibbs energy, the entropy of saturated liquid is well represented by the eos. By $h = g + Ts$, the eos achieves good representation of enthalpy of the liquid. Thus, by fitting vapor pressure, the eos attains quantitative fitting of the energy functions and entropy of the saturated liquid.

The initial success of the van der Waals eos to describe the existence of both the gas and the liquid marked a milestone in eos development. The more recent achievement of quantitative fitting of fugacity and energy functions of liquids is another milestone. However, the correlation of liquid pVT data is not improved by vdW-type equations, since adjustment is made of the integral of pdv in the condensation range. Liquid density predictions often deviate; using the Soave eos, detractions are often 10%–20%, depending on the temperature. The lack of fitting of pVT of the liquids points to the deteriorating representation of the energy functions at *compressed* liquid states, as the *incremental* changes of the energy functions from the *saturated* liquid state on are dependent on the pVT of the liquid. In this regard, it is noteworthy that the perturbation equations provide much better fitting of pVT data of liquids. It can be concluded that the vdW-type eos's, in spite of the great success of some of them in the description of vapor-liquid equilibrium (vle), are in general not suitable for high-pressure work: pVT, thermodynamic functions, liquid/liquid equilibrium, and so on. The perturbation eos's are the choice for high-pressure work.

4.4.2 VAPOR-LIQUID EQUILIBRIUM OF IDEAL MIXTURES

An ideal mixture is one in which the ideal-solution law is followed in all phases. For the liquid phase in an ideal mixture, by Equation (4.314), the fugacity of a component i is given by

$$f_i^L = x_i f_i^{PL} \tag{4.470}$$

where the superscript L denotes a liquid, and PL a pure liquid. The fugacity of pure i is taken at the $[T,p]$ of the system. The fugacity of a component in the vapor mixture is similarly given by

$$f_i^V = y_i f_i^{PV} \tag{4.471}$$

where y_i, is the mole fraction of i in the vapor. This equation is called the *Lewis fugacity rule*.

At vapor-liquid equilibrium, the fugacity of i in the liquid is equal to that in the vapor, and it follows, by setting the two equations above to be equal and rearranging, that

$$y_i / x_i = f_i^{PL} / f_i^{PV} \tag{4.472}$$

The ratio of the mole fractions y_i and x_i at equilibrium is defined to be the K value of i, often called the equilibrium ratio or equilibrium constant,

$$K_i \equiv y_i / x_i \tag{4.473}$$

The definition is general for ideal as well as nonideal mixtures, and is a measure of the volatility of a component. A component with $K > 1$ is referred to as a light component, and one with $K < 1$, a heavy component.

The K values in an ideal mixture are given, according to Equation (4.472), by

$$K_i^{\text{Id}} = f_i^{\text{PL}} / f_i^{\text{PV}} \tag{4.474}$$

The pure-component fugacity is a substance-specific function of $[T,p]$. It follows from Equation (4.474) that ideal K values are substance-specific functions of $[T,p]$, but they are also independent of the composition of the mixture. Ideal K values provide an approximate description of mixtures of isomers, mixtures of near-neighbor homologues, and mixtures of isomers of near-neighbor homologues.

K values are useful for vapor-liquid equilibrium (vle) calculations. Five vle calculations are basic in applications:

1. Bubble-point temperature
   Given: liquid composition and pressure
   To find: vapor composition and temperature
2. Bubble-point pressure
   Given: liquid composition and temperature
   To find: vapor composition and pressure
3. Dew-point temperature
   Given: vapor composition and pressure
   To find: liquid composition temperature
4. Dew-point pressure
   Given: vapor composition and temperature
   To find: liquid composition and pressure
5. Flash
   Given: T, p, and overall composition of feed mixture
   To find: fraction of liquid formed, and composition and fraction of vapor formed and composition

The solution of the bubble- and dew-point problems using K values will be illustrated with the bubble-point temperature determination: Start by making an initial estimate of the bubble-point temperature. Using a chosen K-value chart, read K_i for the components of the mixture at the given pressure and the estimated temperature. Calculate the mole fractions of the incipient vapor by using the K values and the given x_i: $y_i = K_i x_i$ for all i. Sum the y_i's and check if the sum is equal to 1,

$$\sum_i K_i x_i = 1 \tag{4.475}$$

If the equation is not satisfied to within a specified small tolerance, the summation is repeated using a new estimated temperature until convergence is obtained The converged T is the bubble-point temperature. The method is a one-dimensional temperature search.

To find the bubble-point pressure, begin by making an estimate of the bubble pressure and carry out a one-dimensional search for the pressure at which $\Sigma_i K_i x_i = 1$ is satisfied to within a tolerance. Make a new chart reading for new K values at each new pressure. The converged pressure is the sought bubble pressure, and the converged $y_i = K_i x_i$ is the mole fraction of the vapor.

The dew-point problems are similarly one-dimensional searches, but the criterion to satisfy becomes the formation of an incipient dew expressed by the equation

Thermodynamics of Fluid Phase and Chemical Equilibria

$$\sum y_i/K_i = 1 \qquad (4.476)$$

The solution of a flash problem requires the satisfaction of material balances in addition to the equilibrium between the vapor and the liquid. Let us take 1 mole of total feed as a basis and denote the number of moles of liquid formed as n_L and the number of moles of vapor formed as n_V. Overall mass balance requires

$$n_L + n_V = 1 \qquad (4.477)$$

The component mass balance requires

$$n_L x_i + n_V y_i = z_i \qquad (4.478)$$

where z_i denotes the given mole fraction i in the overall feed mixture. Equilibrium between the vapor and liquid phases requires

$$y_i = K_i x_i \qquad (4.479)$$

Upon substituting Equation (4.479) into Equation (4.478) and solving for x_i,

$$x_i = \frac{z_i}{n_L + (1 - n_L)K_i} \qquad (4.480)$$

The x_i's must add up to 1,

$$\sum_i \frac{z_i}{n_L + (1 - n_L)K_i} = 1 \qquad (4.481)$$

Multiplying Equation (4.480) by K_i we obtain

$$y_i = \frac{K_i z_i}{n_L + (1 - n_L)K_i} \qquad (4.482)$$

The y_i's must add up to 1,

$$\sum_i \frac{K_i z_i}{n_L + (1 - n_L)K_i} = 1 \qquad (4.483)$$

Either Equation (4.481) or Equation (4.483) can be solved for n_L, with n_V to follow, by searching with a suitable numerical procedure in the interval from 0 to 1. The existence of an extremum within the search interval can be troublesome, though not serious; for example, interval halving will provide a solution. However, a more robust equation for numerical solution is obtained by combining the two equations. Subtracting one equation from the other gives

$$\sum_i \frac{z_i}{n_L + (1-n_L)K_i} - \sum_i \frac{K_i z_i}{n_L + (1-n_L)K_i} = 0 \qquad (4.484)$$

The necessary and sufficient condition for the search for an n_L to succeed is that $\Sigma_i K_i z_i > 1$ and $\Sigma_i z_i/K_i > 1$ so as to ascertain that the specified $[T,p]$ is bracketed by the dew point and the bubble point. When the conditions are satisfied, numerous one-dimensional search procedures can be used. Upon finding n_L and n_V, the x_i's follow from Equation (4.480) and the y_i's from Equation (4.482).

Calculation of vapor-liquid equilibrium states using K values is particularly convenient for ideal mixtures for which the K values are independent of composition, changing only with temperature or pressure or both. While convenient, the K-value method does satisfy all the equilibrium conditions: the component fugacities are equal in both phases for all components in the mixture, and temperature and pressure are equal in both phases. It follows that the K-value method can be used in general for nonideal mixtures as well. The composition dependence of the K values of nonideal mixtures is addressed next.

4.4.3 Vapor-Liquid Equilibrium by γ-ϕ Models

Most mixtures are not ideal. An activity coefficient is then introduced in the liquid solution and a fugacity coefficient in the vapor of the mixture. A γ-ϕ model is made up of a selected activity-coefficient equation for the liquid and a selected equation of state for the vapor plus the associated standard-state equations and parameters.

4.4.3.1 Low-Pressure Models

Low pressure is generally meant to be less than 5 bar. The departure of a low-pressure gas mixture from ideal-gas mixture behavior is generally minor, and a simple equation of state is commonly employed to calculate the ϕ_i. The γ-ϕ model primarily counts on the activity-coefficient model to describe the vle phenomenon.

4.4.3.1.1 Fugacities of the Liquid Components
The fugacities in the liquid are given by

$$f_{iL} = x_i \gamma_i f_{iL}^P \qquad i = 1, 2 \ldots v \qquad (4.485)$$

An activity-coefficient equation is to be selected which, to a large extent, defines the character of the model. Examples will be given later of models of various activity-coefficient equations.

The required standard-state fugacity f_{iL}^P in Equation (4.485) is given by

$$f_L^P = p^S \phi^S \exp\left[\frac{v_L^S (p - p^S)}{RT}\right] \qquad (4.486)$$

The subscript i is left out of Equation (4.486) to address the pure component. Of the three factors on the right-hand side of the equation that make up the standard fugacity, the first one, p^S, is the primary factor. For a nonpolar component at a low pressure, a good approximation ignores the other two factors, thereby simplifying the standard fugacity to be just the vapor pressure. Such a simplification may introduce an overestimate of the standard fugacity by ≈1% for a nonpolar liquid at its normal boiling point. The next higher approximation is to ignore only the exponential factor that accounts for the pressure effect on the liquid fugacity. This factor generally does not differ significantly from 1. For the difference $(p - p^S)$ as much as 5 atm, the exponential factor does not differ more than 2 or 3% from 1. However, the fugacity coefficient of the saturated vapor ϕ^S cannot be ignored for polar components, as it is often significantly smaller than 1. Values as low as ≈ 0.3 have been reported for some carboxylic acids. The fugacity coefficient can be estimated with the truncated second virial equation. The virial equation is discussed in Section 4.2.5.

For normal fluids, an alternative to the use of vapor pressure to find the standard liquid fugacity f_L, is to use a generalized correlation such as the correlation by the acentric factor method,

$$\log v = \log v_0 + \omega \log v_1 \tag{4.487}$$

where v is f/p and ω is the acentric factor. The tabulated liquid values of Curl and Pitzer [6] and Lydersen, Greenkorn, and Hougen [7] are fitted by Chao and Seader [8] with the following functions:

$$\log v_0 = 5.75748 - 3.01761/T_r - 4.98500 T_r + 2.02299 T_r^2$$
$$+ (0.08427 + 0.26667 T_r - 0.31138 T_r^2) p_r \tag{4.488}$$
$$+ (-0.02655 + 0.02883 T_r) p_r^2 - \log p_r$$

$$\log v_1 = -4.23893 + 8.65808 T_r - 1.22060/T_r - 3.15224 T_r^8 \tag{4.489}$$
$$- 0.025(p_r - 0.6)$$

Equation (4.488) and Equation (4.489) are valid in the range $1.3 > T_r > 0.5$ and $p_r < 9.0$. Upon obtaining v from the three equations above, f_L is found by

$$f_L = vp \tag{4.490}$$

4.4.3.1.2 Fugacities of the Gas Components

For the gas mixture, the fugacities are given by

$$f_{iV} = y_i \phi_i p \quad i = 1, 2 \ldots v \tag{4.491}$$

where the fugacity coefficient ϕ_i is to be obtained from a selected eos. Nonpolar mixtures at a low pressure of 1 bar or lower are often assumed to be ideal gas, $\phi_i = 1$ for all i. The ideal-gas assumption is made for even higher pressure gases, depending on the degree of approximation that is acceptable. For nonpolar and polar gases to moderate pressures, the virial eos truncated at the second virial coefficient gives reliable results of ϕ_i. The results may not be as reliable for highly associated substances like carboxylic acids and HF. Using the second virial eos, the fugacity coefficient is given by

$$\ln \phi_i = \frac{2}{v} \sum_j y_j B_{ij} - \ln z \quad i = 1, 2 \ldots, v \tag{4.492}$$

The v and z in this equation are to be solved from the second virial eos,

$$z = 1 + \frac{1}{v}\sum_i\sum_j y_i y_j B_{ij} \qquad (4.493)$$

at the specified $[T,p]$. A correlation of second virial coefficient is found in Section 4.2.5

4.4.3.1.3 Equilibrium Equations and Calculations

Equilibrium between the gas and liquid phases requires the fugacities of component i in both phases to be equal,

$$x_i \gamma_i f_{iL}^P = p y_i \phi_i, \quad i = 1,2\ldots v \qquad (4.494)$$

These are the equilibrium equations of the γ-ϕ model that are to be solved for the unknown variables among T, p, x, y, n_L, or n_V from the known variables simultaneously with the mass balances and/or other applicable constraint equations. A common method of solution of the simultaneous equations is to use K values. From Equation (4.494) the K values are formed,

$$K_i \equiv y_i / x_i = \gamma_i f_{iL}^P / (p \phi_i) \quad i = 1,2,\ldots v \qquad (4.495)$$

As described in Section 4.4.2, the K values are useful for the calculation of the bubble, dew, or flash states. However, the method of calculation described in Section 4.4.2 needs to be modified because the K values formed by Equation (4.495) are not constant at a specified $[T,p]$ but change with composition. The method of repeated substitution is suited to the use of composition-dependent K values: One begins by making an initial estimate for the variables that are being sought, be it p, T, x_i, y_i, n_V, or n_L, while using specified values of the specified variables. The K_i's are computed by Equation (4.495). The bubble, dew, or flash calculations are made as described in Section 4.4.2 to obtain values of the sought variables, be it T, p, x_i, y_i, n_V, or n_L. The calculated variables are generally not the same as the initial estimates. Use the calculated values as the new estimates and repeat the equilibrium calculation until adequate convergence occurs.

The equilibrium calculations are straightforward without iteration if the gas phase is an ideal-gas mixture, as it may be assumed at low pressures for normal fluids. Consider the bubble-point pressure problem: given T and x_i, find p and y_i. From Equation (4.487),

$$y_i p = x_i \gamma_i f_{iL}^P / \phi_i \qquad (4.496)$$

Summing over i and setting Σy_i equal to 1 at the bubble pressure, we obtain the bubble pressure

$$p = \sum_i x_i \gamma_i f_{iL}^P / \phi_i \qquad (4.497)$$

The vapor composition is obtained by dividing Equation (4.496) by Equation (4.497),

$$y_i = (x_i \gamma_i f_{iL}^P / \phi_i) \Big/ \left(\sum_j x_j \gamma_j f_{jL}^P / \phi_j \right) \qquad (4.498)$$

The two equations are the complete solutions of the bubble-point problem, but the equations are implicit because the ϕ_i's are functions of y_i and p, and f_{iL}^P and f_{jL}^P are functions of p. Repeated

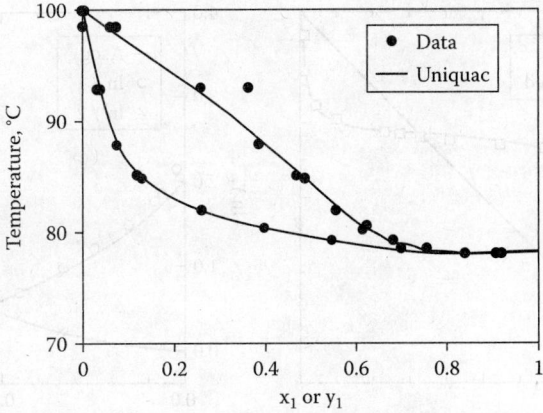

FIGURE 4.15 T-x-y phase diagram for ethanol(1) + water(2) mixtures at 1 atm calculated from UNIQUAC compared with data. (Kojima & Tochigi, New York: Elsevier Scientific Publishing, 1979. With permission.)

substitution leads to solution. The solution of the equations becomes a straightforward matter of substitution for low-pressure mixtures upon assuming ideal gas for the vapor, whereupon $\phi_i = 1$. It is also necessary to leave out the exponential factor from f_{iL}^P so that it is simplified to be $\phi_i^S p_i^S$ or, for nonpolar mixtures, to p_i^S.

For the basic vle problems other than the bubble pressure, there is no special advantage to using equations similar to Equation (4.497) and Equation (4.498) in place of the K-value method, as they do not afford an explicit straightforward solution, but require an iterative solution.

An extensive compilation of experimental vle data is found in the Dechema Chemistry Data Series, which includes thousands of mixture systems. Several eos's do a fair-to-excellent job of correlating equilibrium data and calculating activity coefficients, including Margules, Van Laar, Wilson, NRTL, UNIFAC, and UNIQUAC, as indicated for data at atmospheric pressure. The examples all have minimum boiling azeotropes. UNIQUAC gave the best agreement of predicted versus experimental data values for the following binary mixtures: ethanol–water (Figure 4.15), benzene–cyclohexane (Figure 4.16), and ethanol-in-heptane. ASOG was also found to do an excellent job of correlating the above-mentioned systems. Figure 4.17 shares the results for the ethanol–n-hexane mixture.

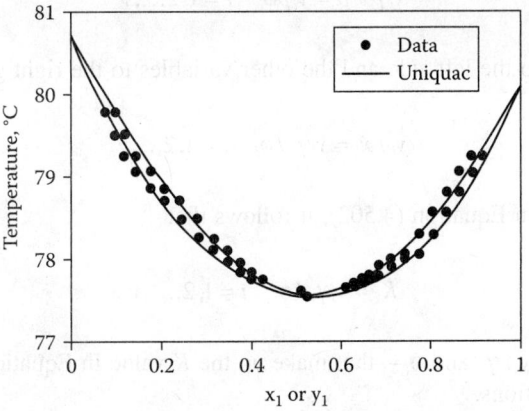

FIGURE 4.16 T-x-y phase diagram for benzene(1) + cyclohexane(2) mixtures at 1 atm calculated from UNIQUAC compared with data. (Kojima, K. & Tochigi, K. Prediction of Vapor-Liquid Equilibrium by the ASOG Method. New York: Elsevier Scientific Publishing Co., 1979. With permission.)

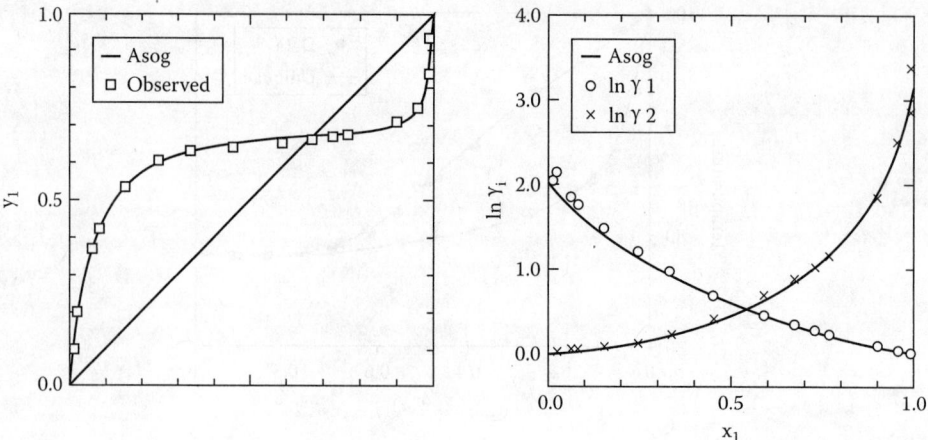

FIGURE 4.17 x-y and ln γ-x diagram for n-hexane (1) + ethanol (2) mixtures at 1 atm calculated from ASOG compared with data. (From Kojima, K., and Tochiji, K. *Prediction of Vapor-Liquid Equilibria by the ASOG Method*, New York: Elsevier Scientific Publishing, 1979. With permission.)

4.4.3.2 A High-Pressure Model

The Chao-Seader correlation [8] is applicable up to high pressures for normal fluids, including hydrocarbons (paraffins, olefins, aromatics, and naphthenes) and inert gases (methane and hydrogen). The correlation is a combination of three parts: (a) pure-liquid fugacity coefficient v_i, (b) liquid-solution activity coefficient γ_i, and (c) gas mixture fugacity coefficient ϕ_i. From these parts, the liquid fugacities and vapor fugacities are obtained:

Liquid fugacities:
$$f_{iL} = x_i \gamma_i v_i p \quad i = 1, 2 \ldots v \tag{4.499}$$

Vapor fugacities:
$$f_{iV} = y_i p \phi_i \quad i = 1, 2 \ldots v \tag{4.500}$$

The vle equations are obtained by setting the liquid and gas fugacities equal for each component,

$$x_i \gamma_i v_i p = y_i p \phi_i \quad i = 1, 2 \ldots, v \tag{4.501}$$

Rearranging x_i and y_i, to the left side and the other variables to the right gives

$$y_i / x_i = v_i \gamma_i / \phi_i \quad i = 1, 2, \ldots v \tag{4.502}$$

Upon setting $K_i \equiv y_i/x_i$ in Equation (4.502), it follows that

$$K_i = v_i \gamma_i / \phi_i \quad i = 1, 2, \ldots, v \tag{4.503}$$

The three factors—v_i, γ_i, and ϕ_L—that make up the K value in Equation (4.503) are described in the following subsections.

4.4.3.2.1 Pure-Liquid Fugacity Coefficient

The liquid fugacity coefficient plays the part of an ideal K value. Indeed, it *is* the K value if the liquid solution is ideal $\gamma_i = 1$ and the gas phase is an ideal-gas mixture $\phi_i = 1$. Equations (4.487) to (4.489) give v as a function of T_r and p_r. The valid range of these equations is $1.3 > T_r > 0.5$ and pressure up to $p_r \approx 9$.

For light gases such as methane and hydrogen, states of common engineering interest are at much higher T_r than the valid range of states for Equations (4.487) to (4.489). A special equation is developed for each gas to fit v values of the hypothetical liquid of the gas. The equation for methane is

$$\log v = 2.43840 - 2.24550/T_r - 0.34084T_r$$
$$+ 0.00212T_r^2 - 0.00223T_r^3 \qquad (4.504)$$
$$+ (0.10486 - 0.03691T_r)p_r - \log p_r$$

The equation is valid at $530 >> 200$ and pressure up to 55 MPa.

For hydrogen,

$$\log v = 0.8669098 + 9.9195929/T_r - 31.4141598/T_r^2$$
$$+ 37.4500656/T_r^3 + (0.0183734 - 0.1460381/T_r$$
$$+ 0.8545416/T_r^2 - 1.1687170/T_r^3)p_r \qquad (4.505)$$
$$- 0.0000691p_r^2 - \log p_r$$

The hydrogen of Equation (4.505) is an update by Jin, Greenkorn, and Chao [9] from the original Chao and Seader correlation. The upper temperature limit is extended by the update to about 730°K. The lower limit remains at 200°K, with pressure up to 25 MPa.

4.4.3.2.2 Liquid-Solution Activity Coefficient γ_i

The activity coefficients are given by the regular solution equation,

$$RT \ln \gamma_i = v_i(\delta_i - \delta_m)^2 \qquad (4.506)$$

The equation is discussed in Section 4.3.3. Two parameters are required for each component: v_i is the pure i liquid molar volume at 298°K, and δ_i is the solubility parameter of component i. The solubility parameter of the mixture δ_m is given by the volumetric average of the components $\delta_m = \Sigma_i x_i \delta_i / \Sigma_i x_i v_i$. Parameters are available for numerous substances. Additional parameters can be readily determined from their definitions, when the need arises, except for light gases.

4.4.3.2.3 Gas Fugacity Coefficient ϕ_i

In the Chao-Seader correlation, the fugacity coefficient in the gas mixture is obtained from the Redlich-Kwong eos and is as follows:

$$\ln \phi_i = \ln \frac{v}{v-b} + \frac{b_i}{v-b} - \frac{2\sum_j y_j a_{ji}}{RTb} \ln \frac{v}{v-b} + \frac{ab_i}{RTb^2}\left(\ln \frac{v}{v+b} - \frac{b}{v+b}\right) - \ln z \qquad (4.507)$$

The molar v in this equation is to be solved from the RK eos,

$$p = \frac{RT}{v-b} - \frac{a}{T^{0.5}v(v+b)} \tag{4.508}$$

The resulting Chao-Seader correlation predicts K values of both methane/n-heptane mixtures and propane/isopentane mixtures down to at least $T_r = 0.5$.

4.4.4 Vapor-Liquid Equilibrium by ϕ-ϕ Models

In a ϕ-ϕ model, one equation of state is used to obtain both the gas-phase fugacity coefficients ϕ_{iV} and the liquid-phase fugacity coefficients ϕ_{iL}. Upon setting the gas and liquid fugacities of a component i to be equal and canceling the pressure on both sides of the equation, we obtain

$$\phi_{iV} y_i = \phi_{iL} x_i \tag{4.509}$$

The K value, defined to be the ratio y_i/x_i, is obtained by rearranging Equation (4.509) as the ratio of the fugacity coefficients,

$$K_i = \phi_{iL} / \phi_{iV} \tag{4.510}$$

While the same eos-derived ϕ equation is used for both ϕ_{iL} and ϕ_{iV}, the input variables to the equation are different: for ϕ_{iL}, the liquid mol fractions x_i and liquid molar volume v_L are the input; for ϕ_{iV}, the gas mol fractions y_i and gas molar volume v_V are the input. Find the liquid volume v_L by solving the eos at specified $[x_i\, p,]$ and pick the small root; find the vapor volume v_V by solving the eos at specified, $[y_i\, p,]$ and pick the large root. The vapor-liquid equilibrium calculations using K values are much the same as in the use of γ-ϕ K values.

4.4.4.1 ϕ-ϕ Models—ϕ from vdW Mixing Rules

In this subsection, we present examples of ϕ-ϕ models in which ϕ is derived from vdW mixing rules by which the parameters a and b of mixtures are combinations of the parameters of the components according to

$$a = \sum_i \sum_j x_i x_j a_{ij} \tag{4.511}$$

$$b = \sum_i \sum_j x_i x_j b_{ij} \tag{4.512}$$

The identical double subscripts refer to a pure component, thus $a_{ii} \equiv a_i$ and $a_{ii} \equiv b_i$. The cross interaction a_{ij} is conventionally expressed as an adjusted geometric mean of the pure-component a's:

$$a_{ij} = (1 - k_{ij})\sqrt{a_i a_j} \tag{4.513}$$

The parameter k_{ij} is obtained by fitting experimental data on binary mixtures of i and j. It is common practice to estimate b_{ij} as the arithmetic mean,

$$b_{ij} = b_i + b_j / 2 \tag{4.514}$$

Substituting Equation (4.514) into Equation (4.512) leads to the linear combination rule for b,

$$b = \sum_i x_i b_i \qquad (4.515)$$

These rules are useful for normal fluids, including nonpolar and weakly polar substances.

Example 4.17: vle by Soave eos

The Soave eos has been presented as Equation (4.162) and Equation (4.164). Using van der Waals mixing rules, the fugacity equation is obtained and given as Equation (4.312). The vapor/liquid equilibrium values were calculated for methane/n-butane mixtures at 310.9°K. Both dew-point and bubble-point curves were in excellent agreement at all vapor pressures up to the critical point. For the hydrogen/propane mixture at 310.9°K, there is excellent agreement of the calculated dew-point curve with data at all pressures from the vapor pressure of propane to the critical point, but agreement is not as good on the bubble-point curve. It was assumed that $k = 0$ in these calculations. Because of the large difference between hydrogen and propane molecules, adjustment of k may improve the calculated values.

To find out the accuracy of vle calculations, Han et al. [10] compared the Soave eos K values ϕ_{iL}/ϕ_{iV} calculated at the experimental equilibrium states with experimental K values, y_i/x_i. The deviation of the eos K value from the experimental is chosen for the comparison for the reason that the K value is the quantity used in the vle calculations; furthermore, the deviation of the K values is identical to the deviation from equality of the calculated liquid fugacity and the vapor fugacity at the experimental equilibrium state, as shown by

$$\frac{K_{i,cal}}{K_{i,exp}} = \frac{\phi_{iL}/\phi_{iV}}{y_i/x_i} = \frac{\phi_{iL} x_i}{\phi_{iV} y_i} = \frac{f_{iL}}{f_{iV}} \qquad (4.516)$$

A cross-interaction coefficient, k, was determined for each binary system by minimizing the sum of the squared relative deviations for the K values as compared with experimental data points. Numerous mixtures were included, such as symmetrical binary mixtures, binary mixtures containing hydrogen, and binary mixtures containing methane. For the numerous mixtures investigated, the average absolute deviation (AAD) often was in the range of ≤5% but occasionally was slightly over 10%. The cross-iteration coefficient, k, was relatively high for mixtures of hydrogen/heavy solvents, but was in the range of 0.01 to 0.7 for mixtures of methane and hydrocarbons.

Requiring no substance-specific parameters other than the critical properties and acentric factor, and using van der Waals mixing rules, the Soave eos gives quantitative results for vle calculations for normal fluids, including hydrocarbons and nonpolar light gases. However, density calculations with this eos show deviations over a wide range of temperatures of 10%–20%. When the vle results are to be reported in moles and mole fraction, the Soave equation finds wide application.

Example 4.18: vle by Peng-Robinson eos

The Peng-Robinson eos has been presented as Equation (4.165). Using van der Waals mixing rules, the fugacity equation is obtained by substituting the eos in Equation (4.306), leading to the following:

$$\ln \phi_i = \frac{b_i}{b}(z-1) \ln\left[\frac{p}{RT}(v-b)\right] - \frac{a}{2\sqrt{2}bRT}\left(\frac{2\sum_k x_k a_{ki}}{a} - \frac{b_i}{b}\right)\ln\left(\frac{v+2.414b}{v-0.414b}\right) \qquad (4.517)$$

The vle phase results for methane and propane mixtures at 213.71°K were calculated using an interaction parameter to 12 of 0.0114; excellent agreement was found with experimental dew points and bubble points at pressures up to the initial point. Similar calculations (with good agreement with experimental data) have also been made for mixtures of methane and haptane and of carbon dioxide and n-butane.

Han et al. [10] made many comparisons of the Peng-Robinson eos K values ϕ_{iL}/ϕ_{iV} calculated with experimental K values y_i/x_i. For the eos K-value calculations, a cross-interaction coefficient k is determined for each binary system. For 20 symmetric binary systems for which the experimental temperature is below the critical of both components, the average absolute deviation (AAD) of the calculated from the experimental was found to be $\approx 2.5\%$. The AAD for the individual systems fall in the range from the smallest value of $\approx 1.0\%$ for ethylene + propylene to the largest value of $\approx 4.3\%$ for propylene + isobutene.

The PR eos gives excellent quantitative representations of the vle of normal fluids, including hydrocarbons and nonpolar and slightly polar light gases, using van der Waals mixing rules. The representation of liquid density is fairly good, making vle results expressed in volume fractions readily useful. The PR eos is widely used in engineering design applications.

Example 4.19: vle Chain-of-Rotators eos

The chain-of-rotators (COR) eos has been presented in Equation (4.183). For mixtures, using van der Waals mixing rules for a and b and linear additive rule for c,

$$c = \sum_i x_i c_i \tag{4.518}$$

and following the method of Equation (4.306) leads to the fugacity coefficient equation

$$\ln \phi_i = \frac{y(4-3y)}{(1-y)^2} + \frac{c_i}{2}(\alpha-1)\left[\frac{(\alpha+4)y-3y^2}{(1-y)^2} + (\alpha+1)\ln(1-y)\right]$$
$$-\frac{\ln(1+4y)}{RT}\left[-b_i\left(\frac{a}{b^2}\right) + \frac{2}{b}\left(\sum_j x_j a_{ij}\right)\right] + \frac{b_i}{b}(z-1) - \ln z \tag{4.519}$$

The parameters a and b in this COR eos are functions of T as follows:

$$a = a_1 \exp(-a_2 T) \tag{4.520}$$

$$b = b_1 \exp(-b_2 T^{1.5}) \tag{4.521}$$

Five parameters—a_1, a_2, b_1, b_2, and c—are required to be known for the eos to apply to a substance. The parameter values have been reported in the literature for numerous substances. For 11 hydrocarbon mixtures, the average AAD was 2.3% (ranging from 0.5 to 5.65%). For 12 methane-containing mixtures, the AAD was 3.92% (ranging from 1.32 to 8.16%). The five parameters required for each substance are a drawback to convenient application. The eos is superior in its representation of density of liquid and compressed fluid states—to within 1% up to the highest pressures.

The chain-of-rotators eos often predicts vle information for ternary systems except near the critical point; a good example is the mixture of methane, n-butane, and decane.

Han et al. [10] compared the COR eos K values ϕ_{iL}/ϕ_{iV} calculated with experimental K values y_i/x_i. For the eos calculations, a cross-interaction coefficient k is determined for each binary by minimizing the sum of the squared relative deviations for the K values for both components in all the reported experimental data points.

4.4.4.2 Free-Energy-Matching Mixing Rules

By matching the excess free energy of an equation of state to that of a solution model, eos parameters for mixtures are obtained from the solution model. The solution model is thus made part of the eos. Incorporating a suitable solution model, the eos becomes applicable to mixtures of highly nonideal polar and associating substances. As part of an eos, the solution model is extended to apply to high pressure. The method of incorporating a solution model into an eos is described in Section 4.3.5.

Example 4.20: Peng-Robinson-Stryjek-Vera (PRSV) eos + UNIFAC
The PR eos has been modified by Stryjek and Vera to extend to polar substances that do not follow the three-parameter principle of corresponding states. The modified eos is fitted to the vapor pressure of polar substances with additional substance-specific parameters. The PRSV equation has been described in Equation (4.163) et seq. The free-energy-matched mixture eos parameters are given in Equations (4.436) and (4.438); the fugacity coefficients are given in Equation (4.439). PRSV eos using the UNIFAC activity coefficient predicts the vle data for both ethanol/water mixtures at 423–623°K and acetone/water mixtures at 373–523°K from low to high pressure.

Example 4.21: COR eos + UNIFAC
Free-energy-matching mixing rules for COR eos parameters were obtained in Equation (4.442), Equation (4.443), and Equation (4.445). The fugacity coefficient for a mixture component obtained from the COR eos with the mixing rules was given in Equation (4.446). Upon substituting the UNIFAC in these equations, fugacities are calculated for both the liquid and gas mixtures to form the vle phase diagram, for example, the vle diagram of ethanol/water mixture as predicted using UNIFAC or the PRSV eos plus UNIFAC.

4.4.5 Liquid-Liquid Equilibrium Models

When the liquid-liquids are highly nonideal, two immiscible liquid phases form. Then only a few activity-coefficient models and equations of state can be used.

Using activity-coefficient models, the equality of component fugacities of the two liquid phases at equilibrium is expressed by

$$x_{i\alpha}\gamma_{i\alpha}f_i^P = x_{i\beta}\gamma_{i\beta}f_i^P \quad i = 1, 2, \ldots, \nu \tag{4.522}$$

where the subscript α denotes one phase and β the other phase, and superscript P denotes the pure-liquid standard state. Since the standard-state fugacity is the same on both sides of the equation, it cancels out, leaving

$$x_{i\alpha}\gamma_{i\alpha} = x_{i\beta}\gamma_{i\beta} \quad i = 1, 2, \ldots, \nu \tag{4.523}$$

To find the equilibrium-phase compositions, it is convenient to form the distribution coefficient and perform flash calculations similar to the vle calculations described in Section 4.4.2. For liquid-liquid equilibrium (lle), the distribution coefficient is defined as

$$c_i = x_{i\alpha} / x_{i\beta} \tag{4.524}$$

By Equation (4.523), c_i is calculated by

TABLE 4.5
Find Liquid-Liquid Equilibrium Phase Compositions in a Three-Component Mixture

1. Initialization: Assign a value for x_1. Make initial estimate of mol fractions of all components in both phases other than x_1.
2. $x_{2\alpha} = x_{2\beta}\gamma_{2\beta}/\gamma_{2\alpha}$
3. $x_{3\alpha} = x_{3\beta}\gamma_{3\beta}/\gamma_{3\alpha}$
4. $x_{2\alpha} = x_{2\alpha}(1 - x_{1\alpha})/(x_{2\alpha} + x_{3\alpha})$
5. $x_{3\alpha} = x_{3\alpha}(1 - x_{1\alpha})/(x_{2\alpha} + x_{3\alpha})$
6. $x_{1\beta} = x_{1\alpha}\gamma_{1\alpha}/\gamma_{2\beta}$
7. $x_{2\beta} = x_{2\alpha}\gamma_{2\alpha}/\gamma_{2\beta}$
8. $x_{3\beta} = x_{3\alpha}\gamma_{3\alpha}/\gamma_{3\beta}$
9. $x_{1\beta} = x_{1\beta}/(x_{1\beta} + x_{2\beta} + x_{3\beta})$
10. $x_{2\beta} = x_{2\beta}/(x_{1\beta} + x_{2\beta} + x_{3\beta})$
11. $x_{3\beta} = x_{3\beta}/(x_{1\beta} + x_{2\beta} + x_{3\beta})$
12. Check convergence: Are all mol fractions unchanged within a specified tolerance from the preceding iteration values? If unchanged, calculation is successfully completed. If changed, repeat iteration by going to step 2.

Source: Adapted from data provided by Sandler, S. I. *Chemical and Engineering Thermodynamics*, 3rd ed., New York: John Wiley & Sons, 1999.

$$c_i = \gamma_{i\alpha} / \gamma_{i\beta} \tag{4.525}$$

from the chosen activity-coefficient model at the phase compositions of either the initial estimate or the current iteration. Using c_i in place of K_i, perform a flash calculation on an estimated total feed composition in between the estimated equilibrium phase compositions. Since the c_i's are composition dependent, it is necessary to iterate until convergence. The flash calculation is repeated with the obtained phase compositions to recalculate the activity coefficients; by Equation (4.525) obtain c_i; then perform a flash calculation to obtain phase compositions and check convergence. Flash calculation is a relatively robust method for lle calculations, but it is still not as robust as it is for vle calculations. Initial estimated compositions are particularly important for success.

Another method to calculate equilibrium-phase compositions makes use of repeated substitutions in the fugacity equality conditions, Equation (4.523), while disregarding any mass balances required in the flash calculations. Suppose T and p are given. The variables are $2v$ mol fractions. There are v equilibrium equations and two equations expressing the sum of x being equal to 1 in either phase, totaling $(v + 2)$ equations. The degree of freedom is the number of variables $2v$ minus the number of equations $(v + 2)$; $F = v - 2$. The method will be illustrated with a three-component mixture, $v = 3$ and $F = 1$. One degree of freedom must be specified; let it be $x_{1\alpha}$. Table 4.5 shows a flow sheet of the method.

Besides being expressed in terms of activity coefficients, the fugacities of a liquid solution can also be calculated from equations of state in the form of a fugacity coefficient ϕ_i. The equality of fugacities of two liquid phases at equilibrium becomes expressed by

$$\phi_{i\alpha} p x_{i\alpha} / \phi_{i\beta} p x_{i\beta} \tag{4.526}$$

Canceling p from both sides of the equation,

$$\phi_{i\alpha} x_{i\alpha} = \phi_{i\beta} x_{i\beta} \tag{4.527}$$

For the calculation of the equilibrium phase compositions, it is convenient to use the distribution coefficient c_i ($\equiv x_{i\alpha}/x_{i\beta}$) to perform a flash calculation. The distribution coefficient for the flash calculation is now formed according to Equation (4.527) by

$$c_i = \phi_{i\beta} / \phi_{i\alpha} \qquad (4.528)$$

Using the c_i's from this equation, the flash method and the method illustrated in Table 4.3 are just as applicable as using c_i from Equation (4.525).

Example 4.22: lle by NRTL

The NRTL activity-coefficient equations have been presented in Section 4.3.3; they are generally preferred in the representation of lle in ternary systems with the use of only binary parameters. With the best estimates of the parameters α_{12}, it is possible to obtain a good ternary lle by using the other two binary parameters fitted to vle data. But a reliable estimate of α_{12} remains elusure. Renon et al. [11] report that reliable representation for ternary systems can only be obtained by fitting all nine binary parameters to extensive ternary data.

Example 4.23: lle by UNIQUAC

For the UNIQUAC equation, there are two adjustable equation parameters for each binary. For the binary that is partially miscible, the best way to determine the two binary parameters is to fit the mutual solubility data. For the completely miscible binaries, useful interaction parameters can be obtained from vle data. However, fitting vle data to within experimental accuracy does not uniquely determine the binary parameters. The choice of a particular set of parameters can have a significant effect on the representation of the ternary lle. For the ternary system of chloroform, water, and acetone at 333°K, for example, the two binary parameters are first determined from mutual solubility data for chloroform and water and then the other binary parameters for the two miscible binaries. Somewhat improved predictions occur by fitting binary parameters to the miscible binaries. Similar predictions have also been found for ternary systems of ethyl acetate, ethanol, and water.

Some predictions have also been obtained with the UNIQUAC equation for termary systems containing two immiscible binaries and one completely miscible binary. The ternary of a mixture of m-hexane, aniline, and methyl cyclopentane is such an example.

Example 4.24: lle by UNIFAC

Less success has been obtained to date in predicting lle as compared with vle data. The absolute mean deviation between experimental and calculated mole fractions for 17 ternary lle test systems is found to be 9.22 mole %. To improve fitting of data, UNIFAC parameters have been redetermined based on lle data, and a new table of interaction parameters has been prepared by Magnussen et al. [12] while keeping the pure-component parameters R and Q unchanged. Using the lle parameters, the fitting of data is improved. The absolute mean deviation between data and calculations is reduced to 1.73 mole % for the same 17 text mixtures.

Table 4.6 is an abridgment of the table showing interaction parameters. Having different tables presents no great difficulty in the storing and using the parameters. But difficulty arises in vlle calculations, such as in extractive distillation where vle and lle coexist. A choice has to be made for either set of parameters.

Example 4.25: lle by UNIQUAC

The UNIQUAC equation has been presented in Section 4.3.3. There are two adjustable equation parameters for each binary. For the binary that is partially miscible, the best way to determine the two binary parameters is to fit the mutual solubility data. For the completely miscible binaries, useful interaction parameters can be obtained from vle data. However, fitting vle data to within experimental accuracy does not uniquely determine the binary parameters. The choice of a particular set of parameters can have a significant effect on the representation of the ternary lle. The following example has been predicted: chloroform-water-acetone at 333°K (using UNIQUAC calculation with binary parameters). Even better results were obtained by fitting the binary parameters of the miscible binaries to ternary lle data. Some examples of good predictions include the following

TABLE 4.6
UNIFAC LLE Group-Interaction Parameters

	1 CH	2 C=C	3 ACH	4 ACCH	5 OH	6 H O	7 ACOH	8 CH CO	9 CHO	10 COOH	11 COOC	12 CH O
1 CH	0	74.54	−114.8	−115.7	644.6	1300	2255	472.6	158.1	139.4	972.4	662.1
2 C=C	292.3	0	340.7	4102	724.4	896	...	343.7	−214.7	1647	−577.5	289.3
3 ACH	156.5	−94.78	0	167	703.9	859.4	1649	593.7	362.3	461.8	6	32.14
4 ACCH	104.4	−269.7	−146.8	0	4000	5695	292.6	916.7	1218	339.1	5688	213.1
5 OH	328.2	470.7	−9.21	1.27	0	28.73	−195.5	67.07	1409	−104	195.6	262.5
6 H O	342.4	220.6	372.8	203.7	−122.4	0	344.5	−171.8	−349.9	−465.7	−6.32	64.42
7 ACOH	−159.8	...	−473.2	−470.4	−63.15	−595.9	0	−825.7	...	...	−898.3	...
8 CH CO	66.56	306.1	−78.31	−73.87	216	634.8	−568	0	−37.36	1247	258.7	5.202
9 CHO	146.1	517	−75.3	223.2	−431.3	623.7	...	128	0	0.75	−245.8	...
10 COOH	1744	−48.32	75.49	147.3	118.4	652.3	...	−101.3	1051	0	−117.6	−96.62
11 COOC	−320.1	485.6	114.8	−170	180.6	385.9	337.3	58.84	1090	1417	0	−235.7
12 CH O	1571	76.44	52.13	65.69	137.1	212.8	...	52.38	...	1402	461.3	0

Thermodynamics of Fluid Phase and Chemical Equilibria

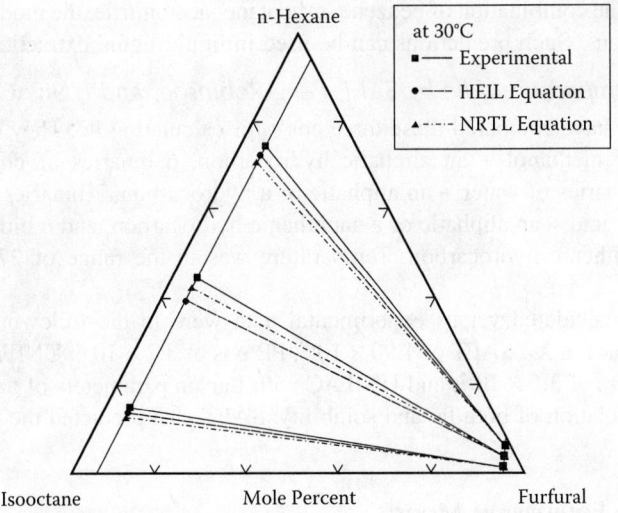

FIGURE 4.18 NRTL-calculated and observed lle in iso-octane (1) + n-hexane (2) + furfural (3) at 30°C. (From J.M. Prausnitz, AIChE J. 14:135, 1968.)

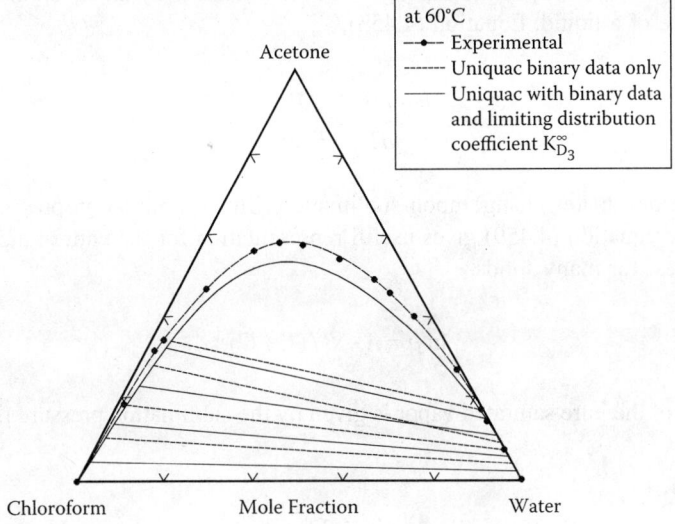

FIGURE 4.19 UNIQUAC-calculated and observed lle in chloroform (1) + water (2) + acetone (3) 60°C. Concentration is mass fractions. (From D. Abrams, J.M. Prausnitz, AIChE J. 21:116, 1975.)

systems: (a) furfural–benzene–isooctane; (b) furfural–n-hexane–isooctone (see Figure 4.18); (c) chloroform-acetone-water (see Figure 4.19); (d) n-hexane–hexane–aniline–methylpentane. An important difference in these systems is the plait point (meeting point of the two immiscible liquid phases); here the activity-coefficient equations are invalid.

Example 4.26: lle by the Complete Local-Composition Model
Addition of an excess energy term to the Hang–Chao–Hilson complete local-composition model, presented earlier, can be used to model liquid-liquid equilibrium. For combinations of water-ethyl acetate-ethanol, predicted values agree well with data at 328°K at low concentrations of ethanol. The predictions are poorer as the ethanol concentration approaches the critical amount. For the combination of water-ethyl acetate-acetonitrile, predictions are better even in the region of critical

concentration. For the combination of benzene–n-heptane–acetonitrile, the model predicts well even at high concentrations. Such predictions can be used in liquid-liquid extractions.

Example 4.27: Comparison of lle by SAFT, Peng-Robinson, and UNIFAC

You and Chen [13] have compared these three eos's for calculating lle. They employed data for 5 binary mixtures of methanol + an aliphatic hydrocarbon, 6 binaries of phenol + an aliphatic hydrocarbon, 12 binaries of water + an aliphatic or a hydrocarbon, 7 binaries of water + an ester, 5 binaries of acetic acid + an aliphatic or a naphthenic hydrocarbon, and 6 binaries of aniline + an aliphatic or a naphthenic hydrocarbon. Temperature was in the range of 273–343°K, with one exception at 236°K.

Predictions of calculated versus experimental data were in the following order of absolute average duration (aad) of X_1: SAFT of 1.90×10^{-2}, PR eos of 3.7×10^{-2}; UNIFAC with Magnussen et al. [12] parameters of 3.5×10^{-2}, and UNIFAC with Larsen parameters of 5.4×10^{-2}. SAFT was also best for extrapolation of both lle and solubility. SAFT also predicted the ternary systems that tested best.

4.4.6 Gas-Solid Equilibrium Models

Gas-solid equilibrium for a single-component system is commonly referred to as sublimation equilibrium. Sublimation pressure, the vapor pressure of a solid, is basic to the modeling of solid-gas equilibrium. Sublimation pressure changes with temperature by an equation similar to that of the vapor pressure of a liquid, Equation (4.453),

$$\frac{dp}{dT} = \frac{\Delta h}{T \Delta v} \qquad (4.529)$$

where the Δ sign designates change upon sublimation. Since sublimation pressure is usually not large, the Antoine equation (4.459) gives useful representation for the entire range of temperature of common interest for many solids,

$$\ln p = A - B/(T+C) \qquad (4.530)$$

The fugacity of the pure saturated vapor is given by the sublimation pressure times the fugacity coefficient,

$$fv = p^S \phi^S \qquad (4.531)$$

The superscript S denotes a saturated state. The fugacity coefficient ϕ can be calculated from an eos or a correlation of a second virial coefficient, but is often assumed to be 1 at the low vapor pressure of many solids.

The fugacity of the solid at sublimation equilibrium is equal to the fugacity of the pure saturated vapor and is therefore also given by Equation (4.531),

$$fs = p^S \phi^S \qquad (4.532)$$

Mixture gas-solid equilibrium is basic to supercritical extraction, the extraction of a solid component from a mixture of solids with a compressed gas at a state above and about its critical point. Another example is deposition of solid carbon dioxide, hydrogen sulfide, or heavy hydrocarbons from natural gas at cryogenic processing conditions. Such deposition often needs to be avoided.

In modeling mixture gas-solid equilibrium, consideration is given to pure solid phase or phases; solid solutions are so rare they will not be considered. The modeling is simplified in two ways: First, modeling the fugacity of a solid is simple; second, there are no equilibrium equations for the components that do not form solids; they remain in the gas phase. Denoting a solid-forming component by i, the equilibrium equation is

$$f_{iS} = f_{iG} \tag{4.533}$$

The subscript S denotes the solid phase; G, the gas phase.

To find the fugacity of the solid, Equation (4.532) is modified by the effect of the system pressure p being different from p_i^S. Thus,

$$f_{iS} = p_i^S \phi_i^S \exp[v_i(p - p_i^S)/RT] \tag{4.534}$$

where v_i denotes the molar volume of solid i. Equation (4.534) is similar to the standard-state fugacity of a liquid-solution component.

The fugacity in the gas phase can be obtained from the fugacity coefficient ϕ_i calculated with a suitable equation of state,

$$f_{iG} = py_i\phi_i \tag{4.535}$$

The formula for the calculation of ϕ_i with an eos can be found in Section 4.2.8.

The equilibrium equation is obtained by setting Equation (4.534) and Equation (4.535) to be equal,

$$py_i\phi_i = p_i^S \phi_i^S \exp[v_i(p - p_i^S)/RT] \tag{4.536}$$

The concentration of i in the equilibrium gas mixture is usually the objective of interest. Solving for y_i from Equation (4.536) gives the mole fraction of i in the gas,

$$y_i = (p_i^S/p)(\phi_i^S/\phi_i)\exp[v_i(p - p_i^S)/RT] \tag{4.537}$$

The first parenthesized factor of the equation is the ideal y_i; it is what y_i would be if the gas mixture were an ideal gas, the pure saturated vapor of i were also an ideal gas, and the Poynting pressure effect were negligible. The actual y_i in a compressed gas is often many times greater than the ideal value, since the factor ϕ_i becomes much smaller than 1. Hence the actual solubility increases up to its real value by a factor of 1000 or more.

A common problem in supercritical extraction is as follows: given the T, p, and i-free composition of the gas, calculate the mole fraction of i and mole fractions of all other components in the gas at equilibrium with pure solid i. The solution is found from Equation (4.537), which gives the required y_i. But there is a snag: y_i must be known to calculate ϕ_i to substitute into Equation (4.537). One makes an initial estimate of y_i and then calculates the y's of other components by normalization, calculates ϕ_i with an equation of state, and substitutes ϕ_i and all required variables into Equation (4.537) to calculate y_i. One compares the calculated y_i with the initial estimate and continues until suitable convergence.

Example 4.28: Supercritical Extraction

In supercritical extraction, a compressed gas at a temperature above and near its critical temperature is used as a solvent to dissolve a solid component. At near-critical states, molecular clusters are formed in abundance; the formation of clusters of small solvent molecules about a large solute molecule effectively contributes to the solvent power of the compressed gas. The fugacity coefficient ϕ_i is reduced, and the solubility y is raised.

An attractive application is the extraction of heat-sensitive biological materials using light gases such as carbon dioxide, ethane, and ethylene as the solvent gas. Their critical temperatures are 304, 305, and 282°K, respectively. Hence, they are often good solvents for supercritical extractions near room temperature. Expansion of the saturated gas from the extractor precipitates the extract for recovery, and the gas is ready to be recycled by compression to the extractor for reuse. Heating and cooling requirements are minimal. Meanwhile, the extract is not exposed to any high-temperature or low-temperature conditions, a vital consideration for heat-sensitive materials, such as many biological products.

In the food industry, supercritical carbon dioxide is used for the decaffeination of coffee, deodorization of plant oils, and extraction of hops and spices. In the tobacco industry, supercritical carbon dioxide is utilized in denicotinizing tobacco. Studies have also been made for the refining of coal by supercritical extraction.

As a general rule, extraction is best when the density of the compressed gas approaches that of the liquid. Often there is a large increase of solubility as the pressure increases at pressures just above the critical pressure. Further increased pressure then results in little increased solubility.

Haselow [14] has compared predictions made with seven eos's for binary mixtures. The four best sets of predictions as expressed as absolute average deviations of 31 binary mixtures four are as follows: Redlich-Kwong, 34; cubic chain-of-rotators, 37; Hans-Coy-Bono-Kwolz-Harling, 38; and Heyen, 40. The other three showed much larger deviations. It can be concluded that predictions involving supercritical fluids are not easy to make with high accuracy because of the high uncertainity in predicting critical properties. Such properties had to be estimated in several systems by group-contribution methods. Furthermore, several eos's, otherwise considered reliable, are not suitable for predicting supercritical information or for solid solutes.

Some success has, however, been obtained in predicting data involving solid phase, e.g., solubility of solid carbon dioxide in compressed air at 143°K (using a truncated second virial eos at pressures up to 40 atm). Good predictions were also realized for existence of solid carbon dioxide in liquid and gaseous mixtures of methane and carbon dioxide. The Peng-Robinson eos was used with success.

REFERENCES

1. Frost, A. A., and Kalkwarf, D. R. *J. Chem. Phys.* 21: 264, 1953.
2. Pasek, G., and Thodos, G. *J. Chem. Eng. Data* 7: 21, 1962.
3. Reynes, E. G., and Thodos, G. *Ind. Eng. Chem. Fund.* 1: 127, 1962.
4. Harlacher, E. A., and Braun, W. G. *Ind. Eng. Chem. Proc. Des. Dev.* 9: 479, 1970.
5. Lee, B. I., and Kesler, M. G. *AIChE J.* 21: 510, 1975.
6. Curl, R. F., and Pitzer, K. S. *Ind. Eng. Chem.* 50: 265, 1958.
7. Lydersen, A. L., Greenkorn, R. A., and Hougen, O. A. Engineering Experimental Station Report no.4, Madison: University of Wisconsin, October 1955.
8. Chao, K. C., and Seader, J. D. *AIChE J.* 7: 598, 1961.
9. Jin, Z. L., Greenkorn, R. A., and Chao, K. C. *AIChE J.* 41: 1602, 1995.
10. Han, S. J., Lin, H. M., and Chao, K. C. *Chem. Eng. Sci.* 43: 2327–2367, 1988.
11. Renon, H., and Prausnitz, J. M. *AIChE J.* 14: 143, 1968.
12. Magnussen, T., Rasmussen, P., and Fredenslund, A. *Ind. Eng. Chem. Proc. Des. Dev.* 20: 331–339, 1981.

13. Yu, M. L., and Chen, Y. P. *Fluid Phase Equilibria* 94: 149–165, 1994.
14. Haselow, J. H., Greenkown, R. A., and Chao, K. C. *Am. Chem. Soc. Symp. Ser.* 300: 156–179, 1986.

4.5 CHEMICAL REACTION EQUILIBRIA

In Section 4.1.6, the fundamental criterion of chemical reaction equilibrium was derived, yielding a corelationship between the chemical potentials of all the components participating in a given chemical reaction. Given that the chemical potentials of the reactants and products are not all independent, the criterion of chemical reaction equilibrium can be used to determine the equilibrium compositions of the reacting species. The development and use of the chemical equilibrium relations are the focus of this section.

4.5.1 Chemical Reaction Equilibrium

As noted previously, a given chemical reaction can be expressed in the following general form:

$$\sum_i v_i M_i = 0 \tag{4.538}$$

where v_i denotes the stoichiometric coefficient of species i participating in the reaction, and M_i is the molecular formula. By convention, v_i is positive for the products and negative for the reactants. The stoichiometry of the reaction can be expressed by

$$dn_i / v_i = d\xi \quad \text{for all } i \tag{4.539}$$

where ξ is the degree of advancement or extent of the reaction. Since the total number of moles of reactants and products is not necessarily fixed as the reaction proceeds, all mole number changes for each species can be expressed in terms of ξ, for which

$$dn_i = v_i d\xi \quad \text{for all } i \tag{4.540}$$

The following occurs at chemical reaction equilibrium:

$$\sum_i \mu_i v_i = 0 \tag{4.541}$$

where μ_i is the chemical potential of species i in the mixture. The sum of the chemical potentials of all species participating in the reaction weighted by their stoichiometric coefficient is zero.

In arriving at Equation (4.541), the derivation implicitly assumes that independent chemical reactions were occurring in the system. In general, what species may be consumed or generated through chemical reaction may not be known *a priori*. A counterpart to Equation (4.541), called the nonstoichiometric formulation, may then be derived. In this case, all possible (if known) components are included, and the minimization of an appropriate thermodynamic potential proceeds by taking into account an atom balance over the given components. The resulting equilibrium compositions are determined without the need to specify any particular reaction pathways. In general, the minimization requires heavy computational effort. The nonstoichiometric approach is not considered here, but the reader can find additional details in the literature.

Even if the actual chemical reaction pathways are not known, a set of linear independent reactions can be generated from some given set of components, thereby allowing Equation (4.541). This approach is known as the stoichiometric formulation and is considered in some detail in Section 4.5.5. We must emphasize that the independent reactions generated by this method do not necessarily describe the actual chemical reactions that occur in the system. These independent reactions relate the mole changes of each species in each reaction to changes in the extent of reaction via a stoichiometric coefficient, i.e., Equation (4.540). Since the final equilibrium state will be independent of the particular pathway taken, the final equilibrium compositions of the given species will still be correctly described by Equation (4.541).

4.5.2 Equilibrium Constants

To begin with, consider a system in which only one independent reaction occurs. The chemical potential of each species i in the mixture can be represented as follows (see Section 4.2.8):

$$\mu_i = g_i^0 + RT \ln \frac{f_i}{f_i^{P,0}} \quad (4.542)$$

where g_i^0 is the molar Gibbs free energy of pure i in its standard state (denoted by the superscript 0) at the given temperature and reference pressure p^0 (typically chosen to be unity in the units of pressure that are chosen, such as 1 bar), f_i is the fugacity of species i in the mixture at the given temperature and pressure p of the mixture, and $f_i^{P,0}$ is the fugacity of pure i in its standard state at and p^0. Combining Equation (4.541) with Equation (4.541) yields

$$\sum_i v_i \left(g_i^0 + RT \ln \frac{f_i}{f_i^{P,0}} \right) = 0 \quad (4.543)$$

or

$$\prod_i \left(\frac{f_i}{f_i^{P,0}} \right)^{v_i} = \exp\left[-\left(\sum_i v_i g_i^0 \bigg/ RT \right) \right] \quad (4.544)$$

The equilibrium compositions of all components participating in the reaction are determined from Equation (4.544).

By convention, the standard-state Gibbs energy change upon reaction, Δg_{rx}^0, is written as

$$\Delta g_{rx}^0 = \sum_i v_i g_i^0 \quad (4.545)$$

with all the components being in their respective standard states. The equilibrium constant, K_a, for the reaction is defined as

$$K_a \equiv \prod_i \left(\frac{f_i}{f_i^{P,0}} \right)^{v_i} \quad (4.546)$$

so that

$$K_a = \exp[-\Delta g_{rx}^0 / RT] \qquad (4.547)$$

Together, Equation (4.545) and Equation (4.546) are equivalent to Equation (4.544). The above procedure can be extended to describe more than one independent reaction. Separate equilibrium constants, along with separate standard-state Gibbs energy changes upon reaction, are introduced for each independent reaction.

Listing Δg_{rx}^0 for all possible reactions is not feasible, but tables are available for a large number of compounds for which the Gibbs energy and enthalpy of formation of the species from their elemental forms have been determined. To obtain Δg_{rx}^0 and the standard-state heat of reaction, Δh_{rx}^0, from these tables at the temperature (and pressure) of interest, one has that

$$\Delta g_{rx}^0 = \sum_i v_i \Delta g_{f,i}^0 \qquad (4.548)$$

and

$$\Delta h_{rx}^0 = \sum_i v_i \Delta h_{f,i}^0 \qquad (4.549)$$

where $\Delta g_{f,i}^0$ and $\Delta h_{f,i}^0$ are the Gibbs energy and enthalpy of formation of the species in their standard states from their elements, respectively. A table of $\Delta g_{f,i}^0$ and $\Delta h_{f,i}^0$, for some widely used substances, is provided at the end of this chapter, as well as a brief discussion of additional sources that provide data on other compounds not listed.

Typically (and by convention), $\Delta g_{f,i}^0$ and $\Delta h_{f,i}^0$ are provided at 298.15°K and a pressure of 1 bar. Also by convention, $\Delta g_{f,i}^0 = \Delta h_{f,i}^0 = 0$ for elements existing in their naturally occurring state at these conditions. For example, hydrogen exists as H_2 (g).

If the element exists in the solid state, information about the crystal form is also necessary. Carbon, for example, often exists as graphite. Other forms of carbon do not yield zero values of $\Delta g_{f,i}^0$ and $\Delta h_{f,i}^0$. In general, the states of aggregation of the components specified in the reactions of interest and listed in the tables for $\Delta g_{f,i}^0$ and $\Delta h_{f,i}^0$ may not be the same.

If Δg_{rx}^0 and Δh_{rx}^0 are needed at another temperature (besides 298.15°K), then the following sets of relations can be used. Temperature variations of Δg_{rx}^0 are calculated from the following Gibbs-Helmholtz relation:

$$d\left(\frac{\Delta g_{rx}^0 / T}{dT}\right) = -\frac{\Delta h_{rx}^0}{T^2} \qquad (4.550)$$

which can be integrated if Δh_{rx}^0 is known as a function of temperature. The value Δh_{rx}^0 itself can be determined from the following expression:

$$\Delta h_{rx}^0(T) = \Delta h_{rx}^0(T_o) + \int_{T_o}^T \Delta C_p^0 dT \qquad (4.551)$$

where

$$\Delta C_p^0 = \sum_i v_i C_{p,i}^0 \qquad (4.552)$$

in which $C_{p,i}^0$ is the isobaric heat capacity of pure i in its standard state at the reference pressure p^0.

Equation (4.550) also allows one to determine how the equilibrium constant varies with temperature. With Equation (4.547), Equation (4.550) implies that

$$\frac{d \ln K_a}{dT} = -\frac{d(\Delta g_{rx}^0 / RT)}{dT} = \frac{\Delta h_{rx}^0}{RT^2} \qquad (4.553)$$

If the reaction is exothermic, or $\Delta h_{rx}^0 < 0$, then K_a decreases with an increase in temperature. Conversely, K_a increases with an increase in temperature when the reaction is endothermic, or $\Delta h_{rx}^0 < 0$.

Substitution of Equation (4.551) into Equation (4.553) yields, upon an integration by parts, the following expression for the standard-state Gibbs energy change upon reaction:

$$\frac{\Delta g_{rx}^0(T)}{RT} = \frac{\Delta g_{rx}^0(T_0) - \Delta h_{rx}^0(T_0)}{RT_0} + \frac{\Delta h_{rx}^0(T_0)}{RT} + \frac{1}{T} \int_{T_0}^{T} \frac{\Delta C_p^0}{R} dT - \int_{T_0}^{T} \frac{\Delta C_p^0}{RT} dT \qquad (4.554)$$

which then can be used to determine K_a at a given temperature. Often, Δh_{rx}^0 does not vary appreciably with temperature. Thus, Equation (4.551) can be integrated with a constant Δh_{rx}^0 to yield

$$\ln K_a(T) = \ln K_a(T_0) = -\frac{\Delta h_{rx}^0}{R}\left(\frac{1}{T} - \frac{1}{T_0}\right) \qquad (4.555)$$

implying that $\ln K_a$ is a linear function of $1/T$. Equation (4.555) provides a useful means of correlating equilibrium constants with temperature. In general, over small temperature changes, $\ln K_a$ is nearly linear to $1/T$ in plots.

The determination of both Δg_{rx}^0 and Δh_{rx}^0 requires that the reference state, or standard state, be known. In general, the choice of the standard state varies. For gases, both real and ideal, the standard state is chosen as the pure materials in an ideal-gas unit fugacity state (that is, at a pressure p^0, which is unity for whatever choice of pressure units is desired; hence, f_i must also be expressed in the same pressure units). Thus, in this case, Δg_{rx}^0 and Δh_{rx}^0 represent changes for the components if they were ideal gases, and not the corresponding changes for the real gases (likewise, in Equations (4.551 and 4.552), $C_{p,i}^0$ represents the isobaric heat capacity of pure i as an ideal gas). With this choice of standard state, the reference fugacities, $f_i^{p,0}$, are always chosen to be unity fugacities, so that, in effect, Δg_{rx}^0 and Δh_{rx}^0 are functions of temperature only. Consequently, K_a is only a function of temperature.

Some tables list the Gibbs energies of formation for states that are different from the ideal-gas unity fugacity state. For example, for substances that are typically solutes, properties at two or more standard states may be given. One standard state is the solute as a pure solid (in the crystalline state) and the other is for the solute in solution at a specified concentration.

In the cases discussed above, most of the standard states are chosen at a fixed reference pressure. Hence, neither K_a nor Δg_{rx}^0 is a function pressure. Yet, if one or more of the standard states for any of the components were chosen to be at the system pressure, p, then both K_a and Δg_{rx}^0 would become

Thermodynamics of Fluid Phase and Chemical Equilibria

functions of this pressure. The pressure change of Δg_{rx}^0, and therefore K_a, can be determined from the following relation:

$$\left(\frac{\partial \Delta g_{rx}^0}{\partial p}\right)_T = \sum_i v_i \left(\frac{\partial g_i^0}{\partial p}\right)_T = -RT\left(\frac{d \ln K_a}{dp}\right)_T \qquad (4.556)$$

If g_i^0 is not a function of p, then the above derivatives all vanish. If it is a function of p, then $(\partial g_i^0 / \partial p) = v_i^0$ is the molar volume of pure i in the chosen standard state. Choosing a standard state that is independent of pressure is certainly convenient, since the above pressure derivatives do not need to be evaluated.

If all species participating in the reaction are in the gas phase with the standard-state fugacities $f_i^{p,0}$ taken as unit fugacities, Equation (4.546) can be expressed as

$$K_a = \prod_i f_i^{v_i} = \left(\prod_i \phi_i^{v_i}\right)\left(\prod_i y_i^{v_i}\right) P^v = K_\phi K_y P^v \qquad (4.557)$$

where

$$v = \sum_i v_i$$

and ϕ_i is the fugacity coefficient of i in the mixture. If the gas is an ideal-gas mixture, then Equation (4.557) reduces to

$$K_a = K_y P^v \quad \text{(ideal-gas mixture)} \qquad (4.558)$$

For liquid mixtures, where (as shown earlier)

$$\frac{f_i}{f_i^{p,0}} = \frac{f_i^p}{f_i^{p,0}} \gamma_i x_i$$

then Equation (4.546) is given by

$$K_a = \left(\prod_i \left(\frac{f_i^p}{f_i^{p,0}}\right)^{v_i}\right)\left(\prod_i \gamma_i^{v_i}\right)\left(\prod_i x_i^{v_i}\right) \qquad (4.559)$$

If the standard state is chosen to be the pure liquid in the same phase at a pressure of 1 bar or its vapor pressure, then each $f_i^p / f_i^{p,0}$ is usually not much different from unity, and so the equilibrium constant reduces to

$$K_a = \left(\prod_i \gamma_i^{v_i}\right)\left(\prod_i x_i^{v_i}\right) = K_\gamma K_x \quad \text{(condensed phase; pressure correction neglected)} \qquad (4.560)$$

The pressure correction for each $f_i^p / f_i^{p,0}$ can, however, be readily evaluated. If the liquid solution is ideal (so that $\gamma_i = 1$), a further simplification results:

$$K_a = K_x \quad \text{(ideal solution; pressure correction neglected)} \tag{4.561}$$

For an equilibrium constant of a heterogeneous chemical reaction where the components are present in separate phases, the standard states of each component may be different. Let there be a reaction between a pure solid, say component 1, and additional components all present in the vapor phase. The equilibrium constant is now expressed as

$$K_a \left(\frac{f_1^p}{f_1^{p,o}} \right)^{v_1} \prod_{i \neq 1} f_i^{v_i}$$

where, in the gas phase, the standard-state fugacities are taken as unit fugacity. With the standard state of the pure solid chosen to be a pressure of 1 bar or the solid's vapor pressure at the given temperature, $f_1^p / f_1^{p,0}$ is often taken as being approximately equal to unity. Hence, the equilibrium constant reduces to

$$K_a = \prod_i f_i^{v_i} \quad \text{(pressure correction of the solid neglected)} \tag{4.562}$$

Further simplification would result if the gaseous components were assumed to behave as an ideal-gas mixture (see Equation [4.558]).

Example 4.29

Calculate the equilibrium mole fractions for the decomposition of nitrogen tetroxide as a result of the reaction N_2O_4 (g) = 2 NO_2 (g) at 25°C and 1 bar. Assume the components form an ideal-gas mixture. Initially there is only 1 mol of N_2O_4.

Solution

The equilibrium relation is given by Equation (4.557), in which

$$K_a = \prod_i f_i^{v_i} = \left(\prod_i \phi_i^{v_i} \right) \left(\prod_i y_i^{v_i} \right) P^v = \frac{y^2 NO_2}{y N_2 O_4} = \exp[-\Delta g_{rx}^o (T = 25°C, P = 1 \text{ bar}) / RT]$$

where the fugacity coefficients are all unity, since the components form an ideal-gas mixture and the standard-state fugacities were chosen as 1 bar. The pressure also cancels out of the equation, since $P = 1$ bar. Using the table provided at the end of this chapter for the Gibbs energy of formation for each component, then, with Equation (4.548),

$$\Delta g_{rx}^0 (T = 25°C, P = 1 \text{ bar}) = 2\Delta g_{f,NO_2}^0 - \Delta g_{f,N_2O_4}^0 = 4.7 \text{ kJ/mol}$$

so that $K_a = 0.15$.

If ξ moles of N_2O_4 react, there will be $1 - \xi$ moles of N_2O_4 remaining. In addition, 2ξ moles of NO_2 would have been produced. Therefore, there will be $(1 - \xi) + (2\xi) = 1 + \xi$ total moles at equilibrium. Hence, the equilibrium mole fractions of each species will be given by

$$y_{N_2O_4} = \frac{1-\xi}{1+\xi} \quad y_{NO_2} = \frac{2\xi}{1+\xi}$$

Now, at equilibrium

$$K_a = \frac{y^2_{NO_2}}{y_{N_2O_4}} \quad \text{or} \quad 0.15 = \frac{4\xi^2}{(1-\xi)(1+\xi)}$$

Solution of the above equation yields

$$\xi = 0.19 \text{ moles} \quad y_{N_2O_4} = 0.68 \quad y_{NO_2} = 0.32$$

In this case, 19% of the initial amount of N_2O_4 was decomposed by the reaction.

If the reaction had proceeded with 1 mol of nitrogen present (as an inert), the equilibrium constant would still have the same value. (Those components not participating in the reaction do not appear in either Equation (4.541) or Equation (4.544).) But with 1 mol of nitrogen still present after the reaction has taken place, the total number of moles at equilibrium would now be $1 + (1 + \xi) = 2 + \xi$, so that

$$y_{N_2O_4} = \frac{1-\xi}{2+\xi} \quad y_{NO_2} = \frac{2\xi}{2+\xi} \quad y_{N_2} = \frac{1}{2+\xi}$$

Thus,

$$K_a = \frac{y^2_{NO_2}}{y_{N_2O_4}} \, 0.15 = \frac{4\xi^2}{(1-\xi)(2+\xi)}$$

where now

$$\xi = 0.25 \text{ moles} \quad y_{N_2O_4} = 0.33 \quad y_{NO_2} = 0.22 \quad y_{N_2} = 0.44$$

In this case, 25% of the initial amount of N_2O_4 was decomposed by the reaction. The inert, though not changing the value of the equilibrium constant, dilutes the mole fractions of the components, thereby increasing the extent of the reaction.

The above N_2O_4 decomposition is an example of a homogeneous reaction, in that all the components participating in the reaction are found in the same phase. Homogeneous reactions cannot go to completion (in that one of the reacting species is completely consumed). In some cases, however, the equilibrium constant may be so large, or so small, that the reaction does not essentially proceed at all, or goes essentially to completion (with negligible error). In contrast, heterogeneous reactions, in which the components participating in the reaction are found in two or more phases, can go to completion. A qualification is needed, however, for heterogeneous reactions. If some of the reactants and products both appear in the same phase, then the reaction (albeit heterogeneous) cannot go to completion.

As a final example, let us repeat the above decomposition reaction, with the inert, but no longer assuming ideal-gas behavior. Because, by convention, the standard state for components, whether real or ideal, in the gas phase is the unit fugacity of the pure component as an ideal gas, the equilibrium constant is the same as before, $K_a = 0.15$. One again begins with Equation (4.557),

$$K_a = \prod_i f_i^{v_i} = \left(\prod_i \phi_i^{v_i}\right)\left(\prod_i y_i^{v_i}\right)P^v = K_\phi K_y P^v$$

but the above relation cannot be further simplified. With $P = 1$ bar, we have that

$$0.15 = \frac{y^2_{NO_2}}{y_{N_2O_4}} \frac{\phi^2_{NO_2}}{\phi_{N_2O_4}}$$

whereas before

$$y_{N_2O_4} = \frac{1-\xi}{2+\xi} \quad y_{NO_2} = \frac{2\xi}{2+\xi} \quad y_{N_2} = \frac{1}{2+\xi}$$

In this nonideal example, the fugacity coefficients are also functions of temperature, pressure (which is 1 bar), and mole fractions. Numerical solution is required for this case, in addition to an expression for the fugacity coefficients of the components of the mixture (see Section 4.2.8).

Example 4.30

Now consider the following two gas-phase reactions with the given equilibrium constants:

$$A + B = C \quad K_{a1} = 1.5$$

$$C + D = 2E \quad K_{a2} = 2.5$$

The reactions occur at 2 bar, and initially there are 2 moles of A, 1 mole of B, and 3 moles of D. Assume the components form an ideal-gas mixture.

Solution

Begin by defining two extents of reaction, ξ_1 and ξ_2. At equilibrium, there will be $2\xi_1$ moles of A, $1\xi_1$ moles of B, $\xi_1 - \xi_2$ moles of C, $3 - \xi_2$ moles of D, and $2\xi_2$ moles of E. Thus, the total number of moles of all species at equilibrium will be $6 - \xi_1$ moles. Therefore,

$$K_{a1} = \frac{y_C}{y_A y_B P} \quad \text{or} \quad 1.5 = \frac{(\xi_1 - \xi_2)(6 - \xi_1)}{2(2 - \xi_1)(1 - \xi_1)}$$

$$K_{a2} = \frac{y_E^2}{y_C y_D} \quad \text{or} \quad 2.5 = \frac{4\xi_2^2}{(\xi_1 - \xi_2)(3 - \xi_2)}$$

Numerical solution of the above two expressions yields $\xi_1 = 0.73$ moles and $\xi_2 = 0.54$ moles. Hence, at equilibrium

$$y_A = 0.24 \quad y_B = 0.050 \quad y_C = 0.036 \quad y_D = 0.47 \quad y_E = 0.21$$

Example 4.31

Consider the production of carbon black, C(s), from methane in a reactor maintained at a pressure of 1 bar and a given temperature T:

$$CH_4(g) = C(s) + 2H_2(g)$$

With 1 mole of CH_4 charged to the reactor, determine the equilibrium compositions of methane and hydrogen, and the fraction of methane charged that has reacted. Assume methane and hydrogen form an ideal-gas mixture.

Solution

For the given reaction,

$$K_a = \left(\frac{f_C^P}{f_C^{0,o}}\right)\frac{f_{H_2}^2}{f_{CH_4}} = \frac{y_{H_2}^2}{y_{CH_4}}P = \frac{y_{H_2}^2}{y_{CH_4}} = \frac{(1-y_{CH_4})^2}{y_{CH_4}}$$

where, for the ideal-gas components, the standard-state fugacities were chosen as 1 bar, and the ratio of the fugacity of the pure solid to the fugacity of the pure solid in its standard state is equal to approximately unity. The value of K_a is determined from the pure-component standard states of all the species. Solving the above equation yields the one physical root of

$$y_{CH_4} = \left(1 + \frac{K_a}{2}\right) - \sqrt{K_a\left(1 + \frac{K_a}{4}\right)}$$

As expected, $y_{CH_4} \to 1$ as $K_a \to 0$ and $y_{CH_4} \to 0$ as $K_a \to \infty$. Since K_a is between these limits, y_{CH_4} falls between zero and unity. Hence, the above heterogeneous reaction does not go to completion. Despite the appearance of a second phase (the solid), both methane (reactant) and hydrogen (product) appear in the same gaseous phase, preventing y_{CH_4} (or y_{H_2}) from equaling zero or unity.

With 1 mole of methane charged to the reactor, the final mole fractions of methane and hydrogen are given by

$$y_{CH_4} = \frac{1-\xi}{1+\xi} \quad y_{H_2} = \frac{2\xi}{1+\xi}$$

Hence, the above equilibrium relation can be rewritten in terms of the extent of reaction

$$K_a = \frac{4\xi^2}{(1-\xi)(1+\xi)}$$

which can be rearranged to yield

$$\xi = \sqrt{\frac{K_a}{4+K_a}}$$

Consistent with the trend for the equilibrium composition of methane, the above relation indicates that $\xi \to 0$ as $K_a \to 0$ and $\xi \to 1$ mole as $K_a \to \infty$. Since the initial amount of methane charged to the reactor is 1 mole, the above equation also describes the fraction of methane that has reacted (as well as providing the number of moles of solid carbon that was produced by the reaction).

4.5.3 PHASE RULE FOR CHEMICALLY REACTING SPECIES

The Gibbs phase rule is modified in the following way when both phase and chemical equilibria need to be satisfied:

$$F = n = 2 - \pi - r$$

where n is the number of components, π is the number of phases, and r is the number of independent reactions. Each reaction (via Equation (4.541)) provides another constraint on the chemical potentials of the species participating in the reaction, serving to decrease the number of independently variable intensive properties of the system.

As an example of the application of the phase rule with reaction, consider a mixture of four components in which only two participate in a reaction. If all components are present in the gas phase, then $F = 4 + 2 - 1 - 1 = 4$. Hence, only four intensive variables are independently variable (e.g., T, P, y_1, and y_2; or P, y_1, y_2, and y_3) if the vapor is ideal. For example, the independent reaction yields the following form of the equilibrium constant:

$$K_a(T) = \frac{y_2}{y_1}$$

and not all sets of four intensive variables can be chosen to describe the system. The above reaction-equilibrium relation reveals that T, y_1, and y_2 are not independently variable. Once any two of these three variables are chosen, the value of the other is fixed by the reaction-equilibrium relation. Thus, from the set of four intensive variables denoting the degrees of freedom of the system, T, y_1, and y_2 cannot all be chosen. At least two others, such as p and y_3, must be chosen to generate a complete set of four intensive variables (e.g., T, y_1, P, and y_3).

If the gas phase is not ideal, the restriction on T, y_1, and y_2 is removed. Fugacity coefficients would now appear in the reaction-equilibrium relations, which themselves are functions of T, p, and the mole fractions of the components (say y_1, y_2, and y_3). Hence, the reaction-equilibrium relation no longer provides a constraint on just T, y_1, and y_2, although it still provides a constraint on, p, and all the $(n-1)$ mole fractions. T, y_1, and y_2 can again be chosen, in addition to a fourth intensive variable (such as p or y_3), to generate a set of four intensive variables denoting the degrees of freedom of the system.

As another example, consider the heterogeneous reaction A (g) + B (g) = C (s). The phase rule states that $F = 3 + 2 - 2 - 1 = 2$, so that only two intensive variables are independently variable. Since C is present as a pure solid, one of those independent variables cannot be a mole fraction of C. Only the mole fraction of A, or B, in the vapor phase is relevant. The appropriate reaction-equilibrium relationship is

$$K_a(T) = \frac{f_C^P / f_C^{P,o}}{(f_A / f_A^{P,o})(f_A / f_A^{P,o})}$$

Now, the left-hand side of the above expression is only a function of T (the standard state of the pure solid is chosen as $P = 1$ bar or the vapor pressure, and so is only a function of T), while the right-hand side is only a function of T, p, and y_A (or y_B). With the degrees of freedom set at two, any two of these intensive variables can be chosen without overspecifying the system.

Thermodynamics of Fluid Phase and Chemical Equilibria

If we add, however, an additional requirement that A and B are to be fed into a reactor in the stoichiometric ratio 1:1, then as the reaction proceeds (again A (g) + B (g) = C (s)), the mole fractions of A and B will both be fixed at 0.5. (Since the product is a solid, and not a vapor, the chosen stoichiometric feed ratio ensures that $y_A = y_B = 0.5$.) Consequently, another constraint has been applied to the system, thereby decreasing the degrees of freedom of the system. Since $y_A = 0.5$ at all times, T and p are no longer independently variable. For this particular choice of the stoichiometric feed, the degrees of freedom of the system have been reduced to 1 (unity).

This previous example may, at first sight, appear to be an apparent exception to the Gibbs phase rule. The derivation of the Gibbs phase rule, however, does not explicitly account for the additional constraints generated by the case of stoichiometric feeds. The Gibbs phase rule can, of course, be modified as above to handle the additional constraints placed on the system (assuming the experimenter is aware of them). In general, care should be exercised when analyzing the allowed degrees of freedom of a system with both phase and chemical equilibria.

4.5.4 Open Systems with Reaction

Let us now consider an open system with chemical reaction to illustrate how the previously discussed concepts are incorporated. Without loss of generality, let us consider a steady-state flow process with a reactor having a single inlet and outlet stream. The open-system energy balance then becomes

$$0 = \dot{Q} + \dot{W} + H_{in}\dot{n}_{in} - H_{out}\dot{n}_{out} \qquad (4.563)$$

where $\dot{n}_{in}, \dot{n}_{out}$ represent molar flow rates and are not necessarily the same due to the possible change of the total molar flow rate due to reaction.

Let us also consider the case in which the reactor is adiabatic and the exiting stream has reached chemical equilibrium (i.e., the kinetics of the reaction is sufficiently fast so that the components reach equilibrium before exiting the reactor). For this adiabatic case, we are interested in computing the exiting temperature of the outlet stream, which is commonly referred to as the *adiabatic reaction temperature*. The outlet, or adiabatic reaction, temperature is determined from both the energy balance and the relations for chemical equilibrium.

In the adiabatic limit with no shaft work generated or done on the reactor, the energy balance reduces to

$$H_{in}\dot{n}_{in} = H_{out}\dot{n}_{out} \qquad (4.564)$$

Since the incoming and outgoing streams are mixtures, the molar enthalpy of the streams can be expressed in terms of the partial molar enthalpies of each component,

$$H_{in} = \sum_i y_{in,i} \overline{H}_{in,i} \quad \text{and} \quad H_{in} = \sum_i y_{out,i} \overline{H}_{out,i} \qquad (4.565)$$

where, for example, $\overline{H}_{in,i}$ is evaluated at the temperature, pressure, and composition of the inlet stream. Now, the mole balance for each component participating in the reaction is given by

$$\dot{n}_{out,i} = \dot{n}_{in,i} + v_i \dot{\xi} \qquad (4.566)$$

where $\dot{\xi}$ is the steady-state rate of the extent of reaction. The above mole balance implies that

$$\dot{n}_{out} = \sum_i \dot{n}_{out,i} = \dot{n}_{in} + \dot{\xi} \sum_i \nu_i \qquad (4.567)$$

with the energy balance being rewritten as

$$H_{in}\dot{n}_{in} = \dot{n}_{in}\sum_i y_{in,i}\bar{H}_{in,i} = \dot{n}_{out}\sum_i y_{out,i}\bar{H}_{out,i} = H_{out}\dot{n}_{out} \qquad (4.568)$$

which yields, in combination with the mole balance, Equation (4.567), the following upon rearrangement:

$$\sum_i \dot{n}_{in,i}(\bar{H}_{out,i} - \bar{H}_{in,i}) = -\dot{\xi}\sum_i \nu_i \bar{H}_{out,i} = -\dot{\xi}\Delta\bar{H}_{rxn,out} \qquad (4.569)$$

where $\Delta\bar{H}_{rxn,out}$ is the heat of reaction based on the outlet conditions and is calculated not from the enthalpies of the pure components, but from their partial molar enthalpies (so that enthalpy of mixing effects are included in this term).

Equation (4.569) is a general result for the steady-state adiabatic reactor with a single inlet and outlet stream. For convenience, let us assume that we have an ideal-gas mixture, so that $\bar{H}_i = H_i(T)$, where H_i is the pure-component molar enthalpy of species i. Therefore, Equation (4.569) becomes

$$\sum_i \dot{n}_{in,i}(H_i(T_{out}) - H_i(T_{in})) = -\dot{\xi}\Delta h_{rxn}(T_{out}) \qquad (4.570)$$

where $\Delta h_{rxn}(T_{out})$ is now the heat of reaction (as defined by Equation (4.549)) at the exiting temperature T_{out}. The left side of Equation (4.570) can be rewritten as an integral over the pure-component isobaric heat capacities from T_{in} to T_{out}, while the heat of reaction, if known at some reference temperature T_0, can be rewritten using Equation (4.551), so that

$$\sum_i \dot{n}_{in,i} \int_{T_{in}}^{T_{out}} C_{p,i} dT = -\dot{\xi}\Delta h_{rxn}^0(T_0) - \dot{\xi}\sum_i \nu_i \int_{T_0}^{T_{out}} C_{p,i} dT \qquad (4.571)$$

or

$$\dot{\xi} = \frac{-\sum_i \dot{n}_{in,i} \int_{T_{in}}^{T_{out}} C_{p,i} dT}{\Delta h_{rxn}^0(T_0) - \sum_i \nu_i \int_{T_0}^{T_{out}} C_{p,i} dT} \qquad (4.572)$$

with T_{in} known, and $\dot{\xi}$ is just a function of T_{out}.

Since the exiting stream is at chemical equilibrium, we are able to use the equilibrium relations discussed in Section 4.5.2. With the exiting temperature equal to T_{out}, the equilibrium constant for the ideal-gas mixture at the outlet is equal to

Thermodynamics of Fluid Phase and Chemical Equilibria

$$K_a(T_{out}) = \left(\prod_i y_{out,i}^{v_i}\right) P_{out}^v \qquad (4.573)$$

The exiting compositions are related to $\dot{\xi}$ in the following manner:

$$y_{out,i} = \frac{\dot{n}_{out,i}}{\dot{n}_{out}} = \frac{\dot{n}_{in,i} + v_i \dot{\xi}}{\dot{n}_{in} + \dot{\xi}\sum_i v_i} \qquad (4.574)$$

Since the pressure of the reactor is typically known, along with the molar flow rates of the components of the inlet stream, Equations (4.572) and (4.573) provide two equations with two unknowns, which can be solved for $\dot{\xi}$ and T_{out}. In other words, the energy balance and the chemical equilibrium relations generate a complete set of equations to determine the adiabatic reaction temperature. Whether T_{out} is greater than or less than T_{in} depends upon whether the reaction is exothermic or endothermic.

Example 4.32
Determine the outlet temperature and equilibrium composition of ammonia synthesized in an adiabatic reactor from the following reaction:

$$N_2 + 3H_2 = 2NH_3$$

N_2 and H_2 are fed into the reactor at 25°C with molar flow rates of 1 mol/s and 3 mol/s, respectively. The reactor is held at a pressure of 300 bar, and the components form an ideal-gas mixture.

Solution
From Equation (4.574), the equilibrium compositions are

$$y_{out,N_2} = \frac{1-\dot{\xi}}{4-2\dot{\xi}} \quad y_{out,H_2} = \frac{3-3\dot{\xi}}{4-2\dot{\xi}} \quad y_{out,NH_3} = \frac{2\dot{\xi}}{4-2\dot{\xi}}$$

so that Equation (4.573) is given by

$$K_a(T_{out}) = \frac{4\dot{\xi}^2(4-2\dot{\xi})^2}{(1-\dot{\xi})(3-3\dot{\xi})^3 P_{out}^2} = \frac{16\dot{\xi}^2(2-\dot{\xi})^2}{27(1-\dot{\xi})^4 P_{out}^2} \qquad (4.575)$$

or, equivalently,

$$\frac{P_{out}\sqrt{27 K_a(T_{out})}}{4} = \frac{\dot{\xi}(2-\dot{\xi})}{(1-\dot{\xi})^2}$$

We can solve the above relation for $\dot{\xi}$, retaining the one root for which the mole fractions are positive (i.e., $0 \le \dot{\xi} \le 1$),

$$\dot{\xi} = 1 - \sqrt{\frac{1}{1+\Gamma(T_{out})}}$$

where

$$\Gamma(T_{out}) = \frac{P_{out}\sqrt{27 H_a(T_{out})}}{4}$$

The value of K_a at T_{out} can be determined from Equation (4.554). For $T_0 = 298.15°$ and a unit ideal-gas fugacity of $P = 1$ bar, $\Delta g^0_{f,N_2} = 0$, $\Delta g^0_{f,H_2} = 0$, and $\Delta g^0_{f,NH_3} = -16.5$ kJ/mol, so that $\Delta g^0_{rx}(T_0) = -33$ kJ/mol, and $\Delta h^0_{f,N_2} = 0$, $\Delta h^0_{f,H_2} = 0$, and $\Delta h^0_{f,NH_3} = -46.1$ kJ/mol, so that $\Delta h^0_{rx}(T_0) = -92.2$ kJ/mol. (These values were taken from Table 4.7.) Therefore, $\ln K_a (T = 298.15 \text{ K}) = 13.3$. Isobaric heat capacities of the components are found in Smith et al. (2001).

Using the given data, one finds from Equation (4.554) that, for example, $\ln K_a(T = 800°K) = -11.6$. Since the heat of reaction is negative, K_a decreases with an increase in T_{out}. Thus, Equation (4.575) indicates that $\dot{\xi}$ decreases with an increase in T_{out}. For $P_{out} = 300$ bar, $\dot{\xi}$ ranges from $\dot{\xi} = 0.998$ mol/s at $T_{out} = 298.15$ to $\dot{\xi} = 0.0955$ mol/s at $T_{out} = 1000°K$.

Now, $\dot{\xi}$ is also related to T_{out} via the energy balance relation, Equation (4.572). Since the heat of reaction is negative and the reactor is adiabatic, Equation (4.572) reveals that an increase in $\dot{\xi}$ results in an increase in T_{out}. For example, with $T_{in} = 298.15° = 25°C$, $T_{out} = 300°$ corresponds to $\dot{\xi} = 0.00232$ mol/s, while $T_{out} = 800°$ corresponds to $\dot{\xi} = 0.753$ mol/s.

When plotted as $\dot{\xi}$ versus T_{out}, Equation (4.572) shows $\dot{\xi}$ increasing with an increase in T_{out}, while Equation (4.573) shows $\dot{\xi}$ decreasing with an increase in T_{out}. The point at which Equation (4.572) and Equation (4.575) intersect determines the adiabatic outlet temperature and the extent of reaction. For $T_{in} = 298.15$, we find that $T_{out} = 688$ and $\dot{\xi} = 0.563$ mol/s. Thus, the outlet compositions are equal to $y_{out,N_2} = 0.152$, $y_{out,H_2} = 0.456$, and $y_{out,NH_3} = 0.392$.

4.5.5 Stoichiometric Formulation

As stated in Section 4.5.1, even if the actual chemical reaction pathways are not known, linear independent reactions can be generated from the given components, thereby allowing one to determine equilibrium compositions via Equation (4.541). This approach, known as the stoichiometric formulation, is discussed below. Although the independent reactions generated by this method do not necessarily describe the chemical reactions that occur in the system, they enable one, however, to relate the mole changes of each species in each reaction to changes in the extent of reaction via a stoichiometric coefficient. Since the final equilibrium state is independent of the particular pathway taken, the final equilibrium compositions of the given species are still correctly described by the chemical equilibrium condition in Equation (4.541).

The stoichiometric formulation begins with the introduction of a matrix **D** in which each element d_{ij} denotes the number of atoms of type i in component j. Each row of the matrix therefore represents an element found in each component, while the columns represent a given component present in the system. Let there be m total elements and n' total components. Note that components found in two or more phases, e.g., H_2O (l) and H_2O (g), are counted as distinct components. Thus, if there are up to n components in each of p phases, then $n \leq n' \leq np$.

Example 4.33

Given CO_2, H_2, C, and H_2O in the gaseous phase and H_2O in a separate liquid phase, generate the corresponding matrix **D**.

TABLE 4.7
Standard Enthalpies and Gibbs Energies of Formation (kJ/mol) at 298.15° for 1 Mole of Each Substance from Its Elements

Chemical Species		State	Δh_f^0	Δg_f^0
Paraffins				
Methane	CH_4	(g)	−74.5	−50.5
Ethane	C_2H_6	(g)	−83.8	−31.9
Propane	C_3H_8	(g)	−104.7	−24.3
n-Butane	C_4H_{10}	(g)	−125.8	−16.6
1-Alkenes				
Ethylene	C_2H_4	(g)	52.5	68.5
Propylene	C_3H_6	(g)	19.7	62.2
1-Butene	C_4H_8	(g)	1.2	70.3
Miscellaneous organics				
Acetylene	C_2H_2	(g)	227.5	210.0
Benzene	C_6H_6	(l)	49.1	124.5
Toluene	C_7H_8	(l)	12.2	113.6
Cyclohexane	C_6H_{12}	(l)	−156.2	26.9
Ethanol	C_2H_6O	(l)	−277.7	−174.8
Methanol	CH_4O	(l)	238.7	−166.3
Inorganic compounds				
Ammonia	NH_3	(g)	−46.1	−16.5
Carbon dioxide	CO_2	(g)	−393.5	−394.4
Carbon monoxide	CO	(g)	−110.5	−137.2
Nitrogen dioxide	NO_2	(g)	33.2	51.3
Nitrogen tetroxide	N_2O_4	(g)	9.2	97.9
Water	H_2O	(g)	−241.8	−228.6
Water	H_2O	(l)	−285.8	−237.1

Note: All standard states are at 25°C. The standard state for a gas (g) is the pure ideal gas at 1 bar. For liquids (l), the standard state is the substance as a liquid at 25°C and 1 bar.

Source: Adapted from data provided by Sandler, S. I. *Chemical and Engineering Thermodynamics*, 3rd ed., New York: John Wiley & Sons, 1999.

Solution
The matrix **D** is given by

$$\mathbf{D} = \begin{array}{c|ccccc} & CO_2 & H_2 & C & H_2O(g) & H_2O(l) \\ \hline C & 1 & 0 & 1 & 0 & 0 \\ H & 0 & 2 & 0 & 2 & 2 \\ O & 2 & 0 & 0 & 1 & 1 \end{array}$$

Once the matrix **D** is determined, the smallest number of linear independent reactions can be generated from the known set of components. Let the rank of the matrix **D** be denoted by c. With the number of components given as n', the number of independent reactions is equal to $n' - c$. The stoichiometric coefficient for component j participating in reaction r ($r = 1,...,n' - c$) can be obtained from the following vector equation:

$$\mathbf{Dv}_r = 0 \qquad (4.576)$$

where $\mathbf{v}_r$ is a column vector of stoichiometric coefficients $v_{j,r}$ in which the stoichiometric coefficient of component j in the independent reaction r is $v_{j,r}$, where $r = 1,...,n' - c$.

Example 4.34

Determine the number of independent reactions, and suggest some possible reaction pathways for CO_2, H_2, C, CO, and H_2O.

Solution

The matrix $\mathbf{D}$ is given by

$$\mathbf{D} = \begin{array}{c|ccccc} & CO_2 & H_2 & C & CO & H_2O \\ \hline C & 1 & 0 & 1 & 0 & 0 \\ H & 0 & 2 & 0 & 0 & 2 \\ O & 2 & 0 & 0 & 1 & 1 \end{array}$$

When the matrix is reduced to echelon form by Gauss-Jordan elimination, the rank of the matrix can be shown to be equal to 3. With $n' = 5$, the number of independent reactions is $5 - 3 = 2$. Equation (4.575) requires that, for each of the two independent reactions,

$$\begin{pmatrix} d_{11} & d_{12} & d_{13} & d_{14} & d_{15} \\ d_{21} & d_{22} & d_{23} & d_{24} & d_{25} \\ d_{31} & d_{32} & d_{33} & d_{34} & d_{35} \end{pmatrix} \times \begin{pmatrix} v_1 \\ v_2 \\ v_3 \\ v_4 \\ v_5 \end{pmatrix} = \begin{pmatrix} 0 \\ 0 \\ 0 \end{pmatrix}$$

With the given elements of the matrix $\mathbf{D}$, the above vector equation reduces to

$$\begin{aligned} v_1 \quad\quad\quad + v_3 + v_4 \quad\quad\quad &= 0 \\ +2v_2 \quad\quad\quad + 2v_5 &= 0 \\ 2v_1 \quad\quad\quad + v_4 + v_5 &= 0 \end{aligned}$$

These relations describe a set of three equations with five unknowns. Hence, we are free to choose the values of any two coefficients. For example, let us choose $v_1 = 1$ and $v_2 = 0$. Therefore, $v_3 = 1$, $v_4 = -2$, and $v_5 = 0$. Thus, one independent reaction would be

$$CO_2 + C - 2CO = 0$$

when $v_1 = 0$ and $v_2 = 1$, so that $v_3 = -1$, $v_4 = 1$, and $v_5 = -1$, and the second independent reaction would be

$$H_2 - C + CO - H_2O = 0$$

The above reactions represent two possible independent reactions. Other initial choices for two out of the five stoichiometric coefficients may generate different reactions, but these new reactions can always be generated by linear combinations of the two reactions given above.

Repeating the above problem, but with H_2O in both the liquid and vapor phases (note that we have implicitly considered in the previous example components in different phases, since the component C would most likely be found in the solid phase), then

$$\mathbf{D} = \begin{array}{c|cccccc} & CO_2 & H_2 & C & CO & H_2O(g) & H_2O(l) \\ \hline C & 1 & 0 & 1 & 1 & 0 & 0 \\ H & 0 & 2 & 0 & 0 & 2 & 2 \\ O & 2 & 0 & 0 & 1 & 1 & 1 \end{array}$$

The rank of this matrix is again three. With $n' = 6$, the number of independent reactions is now $6 - 3 = 3$. With the above given elements of the matrix $\mathbf{D}$, the vector equation for the stoichiometric coefficients reduces to

$$v_1 + v_3 + v_4 = 0$$
$$ +2v_2 +2v_5 +2v_6 = 0$$
$$2v_1 + v_4 + v_5 + v_6 = 0$$

The above relations describe a set of three equations with six unknowns. Hence, we are free to choose the values of any three coefficients. For example, let us choose $v_1 = 1$ and $v_2 = v_5 = 0$. Therefore, $v_3 = 1$, $v_4 = -2$, and $v_6 = 0$. Thus, one independent reaction is the same as before,

$$CO_2 + C - 2CO = 0$$

Now choose $v_1 = v_6 = 0$ and $v_2 = 1$, so that $v_3 = -1$, $v_4 = 1$, and $v_5 = -1$. The second independent reaction is again equal to

$$H_2 - C + CO - H_2O\;(g) = 0$$

Finally, choose $v_1 = v_2 = 0$ and $v_5 = 1$, so $v_3 = 0$, $v_4 = 0$, and $v_6 = -1$. Thus, the third independent reaction is

$$H_2O\;(g) - H_2O\;(l) = 0$$

This reaction is simply a restatement of the phase-equilibrium condition of equality of chemical potentials of a given component in the phases in which it is found.

REFERENCES

Elliot, J. R., and Lira, C. T. *Introductory Chemical Engineering Thermodynamics*, Upper Saddle River, NJ: Prentice Hall, 1999.
Levine, I. N. *Physical Chemistry*, 3rd ed., New York: McGraw-Hill, 1988.
Sandler, S. I. *Chemical and Engineering Thermodynamics*, 3rd ed., New York: John Wiley & Sons, 1999.

Smith, J. M., Van Ness, H. C., and Abbott, M. M. *Introduction to Chemical Engineering Thermodynamics*, 6th ed., New York: McGraw-Hill, 2001.

Tester, J. W., and Modell, M. *Thermodynamics and Its Applications*, 3rd ed., Upper Saddle River, NJ: Prentice Hall, 1997.

DATA REFERENCES

Where to find thermophysical property data?

See Chapter 1 of this handbook plus several of the books listed in the references for this chapter for comprehensive tables of energies of formation. Two other widely referenced sources from which enthalpies and free energies of formation have been compiled are:

"TRC Thermodynamic Tables—Hydrocarbons," Thermodynamic Research Center, Texas A&M Univ. System, College Station, TX.

"The NBS Tables of Chemical Thermodynamic Properties," *Physical and Chemical Reference Data* 11 (suppl. 2): 1982.

When thermodynamic property data cannot be found, one may use group-contribution methods to estimate the enthalpy and Gibbs energy of formation. These methods are discussed in detail in

Reid, R. C., Prausnitz, J. M., and Poling, B. E., *The Properties of Gases and Liquids*, 4th ed., New York: McGraw-Hill, 1987.

5 Fluid Flow

Ron Darby

CONTENTS

- 5.1 Introduction 394
- 5.2 Fluid Properties 394
  - 5.2.1 Classification of Material/Fluid Properties 395
  - 5.2.2 Measurement of Viscosity 398
    - 5.2.2.1 Cup and Bob (Couette) Viscometer 398
    - 5.2.2.2 Tube Flow (Poiseuille) Viscometer 398
  - 5.2.3 Viscous Fluid Models 399
- 5.3 Conservation Principles 404
  - 5.3.1 Conservation of Mass 404
  - 5.3.2 Conservation of Energy 405
    - 5.3.2.1 Macroscopic Energy Balance 405
  - 5.3.3 Conservation of Momentum 407
- 5.4 Fluid Statics 408
  - 5.4.1 The Basic Equation of Fluid Statics 408
    - 5.4.1.1 Constant Density 408
    - 5.4.1.2 Ideal Gas 409
    - 5.4.1.3 The Standard Atmosphere 410
  - 5.4.2 Moving Systems 410
  - 5.4.3 Static Forces on Solid Boundaries 411
- 5.5 Pipe Flow 419
  - 5.5.1 Flow Regines 419
  - 5.5.2 Pressure-Flow Relations for Pipe Flows 419
  - 5.5.3 Newtonian Fluids 419
    - 5.5.3.1 Laminar Flow 419
    - 5.5.3.2 Turbulent Flow 420
    - 5.5.3.3 Rough Pipe 420
    - 5.5.3.4 All Flow Regimes 420
    - 5.5.3.5 Water in Sch 40 Pipe 422
  - 5.5.4 Power Law Fluids 422
    - 5.5.4.1 Laminar Flow 422
    - 5.5.4.2 All Flow Regimes 426
  - 5.5.5 Bingham Plastic Fluids 426
    - 5.5.5.1 Laminar Flow 427
    - 5.5.5.2 All Flow Regimes 428
  - 5.5.6 Fitting Losses 428
    - 5.5.6.1 2-K (Hooper) Method 429
    - 5.5.6.2 3-K Method 429
    - 5.5.6.3 Non-Newtonian Fluids 429
  - 5.5.7 Pipe Flow Analysis 431
  - 5.5.8 Economical Diameter 433

	5.5.9 Noncircular Conduits	435
	5.5.10 Turbulent Drag Reduction	437
	5.5.11 Compressible Flows	438
	5.5.11.1 Isothermal Pipe Flow	440
	5.5.11.2 Adiabatic Flow	441
	5.5.11.3 Choked Flow	441
	5.5.11.4 Isentropic Nozzle Flow	442
	5.5.11.5 Supersonic Flow	442
5.6	Pumps and Compressors	443
	5.6.1 Pump Types	443
	5.6.2 Centrifugal Pump Characteristics	444
	5.6.2.1 Required Head	446
	5.6.2.2 Composite Curves	446
	5.6.2.3 Cavitation and NPSH	447
	5.6.2.4 Specific Speed	449
	5.6.3 Compressors	449
	5.6.3.1 Isothermal, Isentropic, and Polytropic Operations	451
	5.6.3.2 Staged Operation	452
5.7	Flow Measurement and Control	453
	5.7.1 Pitot Tube	453
	5.7.2 Venturi and Nozzle	454
	5.7.3 Orifice Meter	455
	5.7.3.1 Incompressible Flow	455
	5.7.3.2 Compressible Flow	460
	5.7.3.3 Loss Coefficient	462
	5.7.3.4 Applications	462
5.8	Control Valves	462
	5.8.1 Valve Characteristics	464
	5.8.2 Valve Sizing Equations—Incompressible Flows	464
	5.8.2.1 Matching Valve Trim and System Characteristics	465
	5.8.3 Compressible Flows	468
	5.8.4 Viscosity Correction	473
References		476

5.1 INTRODUCTION

The flow behavior of fluids is crucial to virtually every aspect of the chemical process and related industries. The most common application involves the transportation of fluids through piping systems containing fittings, valves, etc., by means of a driving force such as that provided by a pump, static elevation change, or some other source of pressure. While this is indeed important, an understanding of fluid flow is the key to various other operations such as packed and fluidized beds, fluid-solid separations, safety relief systems, flow metering and control, etc. These flows may be either laminar or turbulent, and the fluids may be either incompressible Newtonian or non-Newtonian liquids, compressible gases, or complex two-phase flows involving gas-liquid, solid-liquid, or solid-gas combinations. Various aspects of each of these will be considered in this chapter.

5.2 FLUID PROPERTIES

The properties of a fluid that influence flow behavior are its density and viscosity. Fluids may be classified as incompressible or compressible, depending upon whether the density is constant or a

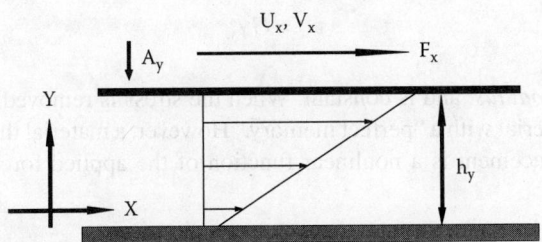

FIGURE 5.1 Simple shear.

function of pressure, and as Newtonian or non-Newtonian, depending upon whether the viscosity is constant or a function of flow conditions. Liquids can usually be considered to be incompressible, but may be either Newtonian or non-Newtonian. Gases are compressible, but are invariably Newtonian. Both liquids (Newtonian and non-Newtonian) and gases will be considered.

5.2.1 Classification of Material/Fluid Properties

The flow behavior of a fluid is determined by the fluid properties and how it responds to the forces exerted on or within the fluid. The forces, F_j (and the corresponding stress components, $\sigma_{ij} = F_j/A_i$), acting at any point within a fluid, may be either isotropic (i.e., uniform in all directions, such as pressure, P) or anisotropic (such as shear stresses, τ_{ij}, which are directionally dependent):

$$\sigma_{ij} = -P\delta_{ij} + \tau_{ij} \tag{5.1}$$

where δ_{ij} is the "unit tensor" (or Kronecker delta). The pressure P is a negative (compressive) stress, since tensile stresses are defined as positive. In laminar flows, the shear stresses τ_{ij} depend on the relative motion (deformation) within the fluid and the *rheological properties** as defined by a *constitutive equation*, which is a unique property of the fluid structure. In turbulent flows, these stress components are dominated by the inertial momentum flux associated with turbulent eddy motion and the fluid density. These shear stresses are zero in any fluid in a state of rest or uniform motion.

The rheological nature of a material can be defined by a relation between components of stress and deformation in *simple shear*, as illustrated in Figure 5.1.

The shear stress (τ_{yx}) and the shear strain (γ_{yx}) are given by

$$\tau_{yx} = F_x / A_y \tag{5.2}$$

$$\gamma_{yx} = U_x / h_y = \frac{du_x}{dy} \tag{5.3}$$

A material that is a *rigid solid* does not deform no matter how much force is applied, i.e.,

$$\gamma_{yx} = 0 \tag{5.4}$$

However, if the material is a *linear elastic (Hookean) solid*,

---

* Rheology is the study of the deformation and flow behavior of materials, both fluids and solids.

$$\tau_{yx} = G\gamma_{yx} \tag{5.5}$$

where G is the *shear modulus*, and is constant. When the stress is removed, the strain also goes to zero, representing a material with a "perfect memory." However, a material that exhibits a "memory," but for which the displacement is a nonlinear function of the applied force, is a *nonlinear (non-Hookean) elastic solid*:

$$G = \tau_{yx} / \gamma_{yx} = fn\ (\tau\ \text{or}\ \gamma) \tag{5.6}$$

G is still the shear modulus, but it is no longer a constant; it is a *function* of the displacement gradient (strain) (γ_{yx}), or of the magnitude of the applied stress (τ_{yx}), i.e., $G(\gamma)$ or $G(\tau)$. The particular form of the function depends upon the specific nature (i.e., constitution) of the material.

If the material is a fluid with negligible resistance to deformation (e.g., a low-pressure gas), the governing equation is

$$\tau_{yx} = 0 \tag{5.7}$$

This is called an *inviscid (Pascalian) fluid*. However, if the fluid molecules exhibit a significant attraction, and the resistance to flow is directly proportional to the relative motion, the material is a *Newtonian fluid*:

$$\tau_{yx} = \mu \dot{\gamma}_{yx} \tag{5.8}$$

where $\dot{\gamma}_{yx}$ is the *rate of shear strain* or, for short, *shear rate*:

$$\dot{\gamma}_{yx} = \frac{d\gamma_{yx}}{dt} = \frac{dv_x}{dy} = \frac{V_x}{h_y} \tag{5.9}$$

Here μ is the fluid *viscosity*, defined as $\mu = d\tau_{yx}/d\dot{\gamma}_{yx}$. If the shear stress and shear rate are not proportional, the fluid is *non-Newtonian*, in which case the viscosity is not constant but is a *function* of either the shear rate or shear stress:

$$\eta = \frac{\tau_{yx}}{\dot{\gamma}_{yx}} = fn\ (\tau\ \text{or}\ \dot{\gamma}) \tag{5.10}$$

Most common fluids of simple structure are Newtonian (i.e., water, air, glycerine, oils, etc.). However, fluids with complex structures (i.e., high polymer melts or solutions, suspensions, emulsions, foams, etc.) are generally non-Newtonian. Examples of non-Newtonian behavior include mud, paint, ink, mayonnaise, shaving cream, polymer melts and solutions, toothpaste, etc. Many "two-phase" systems (e.g., suspensions, emulsions, foams, etc.) are purely viscous fluids and do not exhibit significant elastic or "memory" properties. However, many high polymer fluids (e.g., melts and solutions) are *viscoelastic* and exhibit both elastic (memory) as well as nonlinear viscous (flow) properties. A classification of material behavior is summarized in Table 5.1 (in which the subscripts have been omitted for simplicity). Only purely viscous Newtonian and non-Newtonian fluids are considered here. The properties and flow behavior of viscoelastic fluids are the subject of numerous books and papers (e.g., Darby, 1976; Bird et al., 1987).

TABLE 5.1
Classification of Materials

Rigid Solid (Euclidian)	Linear Elastic Solid (Hookean)	Nonlinear Elastic Solid (Non-Hookean)	Visco-Elastic Fluids and Solids (Non-Linear)	Nonlinear Viscous Fluid (Non-Newtonian)	Linear Viscous Fluid (Newtonian)	Inviscid Fluid (Pascalian)
$\gamma = 0$	$\tau = G\gamma$	$\tau = fn(\gamma)$	$\tau = n(\gamma, \dot{\gamma}, \ldots)$	$\tau = fn(\dot{\gamma})$	$\tau = \mu\dot{\gamma}$	$\tau = 0$
	or	or		or	or	
	$G = \tau/\gamma$	$G = \tau/\gamma$		$\eta = \tau/\dot{\gamma}$	$\mu = \tau/\dot{\gamma}$	
	Shear Modulus (Constant)	Modulus Function of γ or τ		Viscosity Function of $\dot{\gamma}$ or τ	Viscosity (Constant)	

⟵ Purely Elastic Solids   Purely Viscous Fluids ⟶

Elastic Deformations Store Energy   Viscous Deformations Dissipate Energy

Source: Darby, R. *Chemical Engineering Fluid Mechanics*, 2nd ed., New York: Marcel-Dekker, 2001.

5.2.2 Measurement of Viscosity

Two common methods for measuring viscosity are the cup and bob (Couette) and the tube flow (Poiseuille) viscometers.

5.2.2.1 Cup and Bob (Couette) Viscometer

This viscometer consists of two concentric cylinders, the outer "cup" and the inner "bob," with the test fluid in the annular gap. The radius of the bob is R_i, that of the cup is R_o, and the length of surface in contact with the sample is L. One of the cylinders (preferably the cup) is rotated at various controlled angular velocities (Ω), and the resulting torque (T) is measured. The shear stress is

$$\tau_{r\theta} = \frac{T}{2\pi r^2 L} = \tau \tag{5.11}$$

Setting $r = R_i$ gives the stress on the bob surface, and $r = R_o$ gives the stress on the cup. If the gap is small (i.e., $(R_o - R_i)/R_i \leq 0.02$), the average shear stress is evaluated at $(R_i + R_o)/2$, and the corresponding average shear rate is given by

$$\dot{\gamma}_{yx} = \frac{dv_\theta}{dr} = \frac{\Omega}{1-\beta} = \dot{\gamma} \tag{5.12}$$

where $\beta = R_i/R_o$. If the gap is not small, the shear rate at the bob is given by

$$\dot{\gamma} = \frac{2\Omega}{n'(1-\beta^{2/n'})} \tag{5.13}$$

which is accurate to 1–2% in most cases, with an extreme error of about 5% in the worst case, for any size gap and any type of fluid (e.g., Darby, 1985). Here n' is the point slope of the log-log plot of T versus Ω:

$$n' = \frac{d(\log T)}{d(\log Q)} \tag{5.14}$$

A series of data points of T versus Ω must be obtained in order to determine n' at each point. The viscosity at each shear rate (or shear stress) is then determined by dividing the shear stress at the bob by the shear rate at the bob for each data point.

5.2.2.2 Tube Flow (Poiseuille) Viscometer

Viscosity can also be determined by measuring the total pressure drop ($\Delta \Phi = \Delta P + \rho g \Delta z$) and flow rate ($Q$) in steady laminar flow through a uniform circular tube of length L and diameter D (this is called *Poiseuille flow*). The shear stress at the tube wall (τ_w) is determined from the measured pressure drop:

$$\tau_w = -\frac{\Delta \Phi}{4L/D} \tag{5.15}$$

and the shear rate at the tube wall ($\dot{\gamma}_w$) is given by

$$\dot{\gamma}_w = \Gamma\left(\frac{3n'+1}{4n'}\right) \tag{5.16}$$

where

$$\Gamma = \frac{32Q}{\pi R} = \frac{8V}{D} \tag{5.17}$$

and

$$n' = \frac{d\log \tau_w}{d\log \Gamma} = \frac{d\log(-\Delta\Phi)}{d\log Q} \tag{5.18}$$

is the point slope of the log-log plot of $\Delta\Phi$ versus Q, evaluated at each data point. This n' is the same as that determined by the cup and bob viscometer for a given fluid.

5.2.3 Viscous Fluid Models

The nature of the dependence of the shear stress (or viscosity) upon shear rate determines the class of viscous fluid behavior (Figures 5.2 and 5.3). The shear stress and shear rate may both be either positive or negative, although (by the usual "mechanics" convention) they always have the same sign.

A *Newtonian fluid* obeys the equation

$$\tau = \mu\dot{\gamma} \tag{5.19}$$

where μ is the viscosity.

If the shear stress–shear rate data appear linear but intersect the shear stress axis at a value of τ_o, the material is called a *Bingham plastic*:

$$\tau = \tau_o + \mu_\infty\dot{\gamma} \quad \text{for } |\tau| \geq \tau_o \tag{5.20}$$

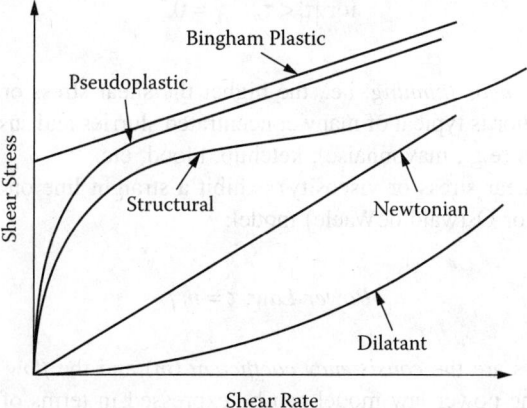

FIGURE 5.2 Shear stress versus shear rate for various fluids.

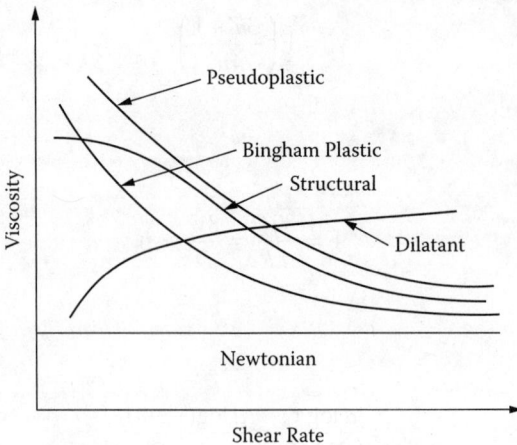

FIGURE 5.3 Viscosity versus. shear rate for fluids in Figure 5.4.

The two rheological properties required—the *yield stress*, τ_o, and the *high-shear-limiting (or plastic) viscosity*, μ_∞—determine the flow behavior of a Bingham plastic. The viscosity function for the Bingham plastic is

$$\eta(\dot{\gamma}) = \frac{\tau_o}{\dot{\gamma}} + \mu_\infty \tag{5.21}$$

or

$$\eta(\tau) = \frac{\mu_\infty}{1 - \tau_o/\tau} \tag{5.22}$$

Since this material will not "flow" unless the shear stress exceeds the yield stress, these equations apply only when $|\tau| > \tau_o$. For smaller values of the shear stress, the material behaves as a rigid solid, i.e.,

$$\text{for } |\tau| < \tau_o \quad \dot{\gamma} = 0 \tag{5.23}$$

The Bingham plastic is *shear thinning*, i.e., the higher the shear stress or shear rate, the lower is the viscosity. This behavior is typical of many concentrated slurries and suspensions, such as muds, paints, foams, emulsions (e.g., mayonnaise), ketchup, blood, etc.

If the data (either shear stress or viscosity) exhibit a straight line on a log-log plot, the fluid follows the *power law* (or Ostwald deWaele) model:

$$\text{Power Law: } \tau = m\dot{\gamma}^n \tag{5.24}$$

The viscosity parameters are the *consistency coefficient* (m) and the *flow index* (n). The apparent viscosity function for the power law model can be expressed in terms of either the shear rate or the shear stress:

$$\eta(\dot{\gamma},\tau) = m\dot{\gamma}^{n-1} = m^{1/n}\tau^{(n-1)/n} \tag{5.25}$$

For $n = 1$, the power law fluid model represents Newtonian behavior; for $n < 1$, the fluid is *shear thinning* (or *pseudoplastic*); and for $n > 1$, the model represents *shear thickening* (or *dilatant*) behavior. Most non-Newtonian fluids are shear thinning, whereas shear-thickening behavior is relatively rare, being observed primarily for some concentrated suspensions of very small particles (e.g., starch suspensions) and some unusual polymeric fluids (e.g., molecules that exhibit shear-induced structure). Although the power law model is very popular for curve-fitting viscosity data over a limited range of shear rate, it is dangerous to use this model to extrapolate beyond the range of the data because (for $n < 1$) it predicts an ever-increasing viscosity as the shear rate decreases, and an ever-decreasing viscosity as the shear rate increases, both of which are physically unrealistic.

A typical viscosity characteristic of many non-Newtonian fluids (e.g., polymeric fluids, flocculated suspensions, colloids, foams, gels, etc.) is illustrated by the curves labeled *structural viscosity* in Figures 5.2 and 5.3. These fluids exhibit Newtonian behavior at very low and very high shear rates, with shear thinning or pseudoplastic behavior at intermediate shear rates. This can often be attributed to a reversible "structure" or network that forms in the "rest" or equilibrium state. When the material is sheared, the structure breaks down, resulting in a shear-dependent (shear thinning) behavior. This type of behavior is exhibited by fluids as diverse as polymer solutions, blood, latex emulsions, paint, mud (sediment), etc. An example of a useful model that represents this type of behavior is the *Carreau* model:

$$\text{Carreau: } \eta(\dot{\gamma}) = \eta_\infty + \frac{\eta_o - \eta_\infty}{[1+(\lambda\dot{\gamma})^2]^p} \tag{5.26}$$

This model contains four rheological parameters: the *low shear limiting viscosity*, η_o; the *high shear-limiting viscosity*, η_∞; a *time constant*, λ; and the *shear-thinning index*, p. This viscosity model can represent a wide variety of materials, although data over a range of six to eight *decades* of shear rate may be required to define the complete shape of the curve (and to determine values for all four parameters). The Carreau model reduces to various other popular models over certain ranges of shear rate (including the Bingham plastic and power law models) as follows:

1. *Low to intermediate shear rate range*: If $\eta_\infty \ll (\eta, \eta_o)$, the Carreau model reduces to a three-parameter model (η_o, λ, and p) called the *Ellis* model, which is equivalent to a power law model with a low shear-limiting viscosity:

$$\text{Ellis: } \eta(\dot{\gamma}) = \frac{\eta_o}{[1+(\lambda\dot{\gamma})^2]^p} \tag{5.27}$$

2. *Intermediate to high shear rate range*: If $\eta_o \gg (\eta, \eta_\infty)$ and also $(\lambda\dot{\gamma})^2 \gg 1$, the model reduces to the equivalent of a power law model with a high shear-limiting viscosity, called the *Sisko* model:

$$\text{Sisko: } \eta(\dot{\gamma}) = \eta_\infty + \frac{\eta_o}{(\lambda\dot{\gamma})^{2p}} \tag{5.28}$$

Although this appears to have four parameters, it is in reality a three-parameter model, since the combination $\eta_o/(\lambda)^{2p}$ is a single parameter, along with p and η_∞. If the value of p in the Sisko model is set equal to 1/2, the result is equivalent to the *Bingham plastic* model:

$$\text{Bingham: } \eta(\dot{\gamma}) = \eta_\infty + \frac{\eta_o}{\lambda \dot{\gamma}} \qquad (5.29)$$

where the yield stress τ_o corresponds to η_o/λ, and η_∞ is the (high shear) limiting viscosity.

3. *Intermediate shear rate*: For $\eta_\infty \ll \eta \ll \eta_o$ and $(\lambda\dot{\gamma})^2 \gg 1$, the Carreau model reduces to the *power law model*:

$$\text{Power Law: } \eta(\dot{\gamma}) = \frac{\eta_o}{(\lambda\dot{\gamma})^{2p}} \qquad (5.30)$$

where the power law parameters m and n are equivalent to the following combination of Carreau parameters:

$$m = \frac{\eta_o}{\lambda^{2p}}, \quad n = 1 - 2p \qquad (5.31)$$

Other models have been proposed to provide a more accurate fit of some detailed viscosity data. Two of these are the *Meter model*, a stress-dependent viscosity model that has the same general characteristics as the Carreau model:

$$\text{Meter: } \eta(\dot{\gamma}) = \eta_\infty + \frac{\eta_o - \eta_\infty}{[1+(\tau/\sigma)^{2a}]} \qquad (5.32)$$

where σ is a characteristic stress parameter and a is the shear-thinning index, and the *Yashuda model*. The Yashuda model (1981) is also similar to the Carreau model, but with one additional parameter (a total of five parameters):

$$\text{Yashuda: } \eta = \eta_\infty + \frac{(\eta_o - \eta_\infty)}{[1+(\lambda\dot{\gamma})^{2a}]^{p/a}} \qquad (5.33)$$

which reduces to the Carreau model for $a = 1$. (This is also sometimes called the *Carreau-Yashuda model*.) This model is particularly useful for representing polymer melt data for broad-molecular-weight polymers for which the zero-shear viscosity is approached very gradually. Each of these last three models reduces to Newtonian behavior at very low and very high shear rates, and to power law behavior at intermediate shear rates.

It should be stressed that the specific model that best represents a given non-Newtonian, and the values of the parameters defined by the model, cannot be predicted accurately from first principles or from more basic information. The specific function and the values of the parameters must be determined from laboratory viscosity measurements.

Example 5.1

The following data were taken in a cup and bob viscometer, having a bob radius of 2.5 cm, a cup radius of 2.75 cm, and a bob length of 6.5 cm. Determine the viscosity of the sample and the nature of the shear dependence of this viscosity.

Fluid Flow

Torque (dyne-cm)	Speed (rpm)
3,000	2
6,000	4
11,700	10
14,500	20
17,800	40

Solution

The viscosity is the shear stress at the bob [Equation (5.11)] divided by the shear rate at the bob [Equation (5.13)]. The value of n' in Equation (5.13) is determined from the *point slope* of the (log T) vs. (log rpm) plot at each data point. The plot is shown in Figure 5.4. In general, if the data do not fall on a straight line on this plot, the point slope (tangent) at each data point must be determined, giving a different value of n' for each data point. The slope can be determined graphically by using numerical techniques, or by averaging the local slopes between successive points. Best results are usually obtained if the data are smoothed before the slopes are computed. The results of the calculations are shown below, where the values of n' are determined by averaging the slopes between successive points, Equation (5.13) gives the shear rate, and Equation (5.11) gives the shear stress:

Shear Stress at Bob (dyn/cm²)	n'	Shear Rate at Bob (1/s)	Viscosity (Poise)
11.8	1	2.41	4.87
23.5	0.864	4.90	4.80
45.8	0.519	13.1	3.49
56.8	0.303	29.6	1.92
69.7	0.296	59.6	1.17

Note: 10 dyn/cm² = 1 Pa, and a viscosity of 10 Poise = 1 Pa·s = 1000 cP

A plot of viscosity vs. shear rate (Figure 5.5) shows that the fluid approaches Newtonian behavior at low shear rates and is shear thinning at higher shear rates, with power law–type behavior. As shown in Figure 5.5, the Ellis model gives a good representation of these data.

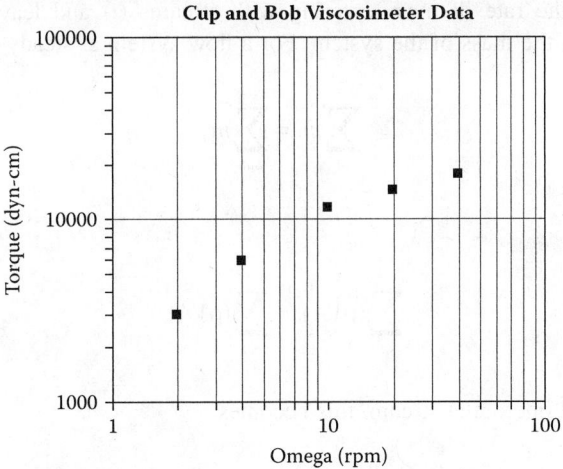

FIGURE 5.4 Cup and bob viscosimeter data.

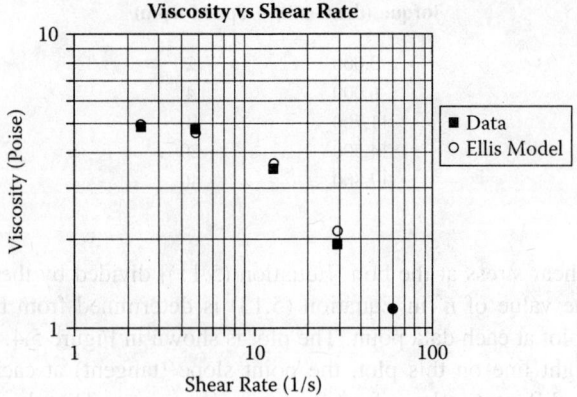

FIGURE 5.5 Viscosity vs. shear rate.

5.3 CONSERVATION PRINCIPLES

The flow behavior of fluids is governed by the basic laws for conservation of mass, energy, and momentum coupled with appropriate expressions for the irreversible rate processes (e.g., friction loss) as a function of fluid properties, flow conditions, geometry, etc. These conservation laws can be expressed in terms of "microscopic" or point values of the variables, or in terms of "macroscopic" or integrated average values of these quantities. In principle, the macroscopic balances can be derived by integration of the microscopic balances. However, unless the "local" microscopic details of the flow field are required, it is often easier and more convenient to start with the macroscopic balance equations.

5.3.1 Conservation of Mass

The macroscopic conservation of mass, or continuity, equation for any system is

$$\sum_{in} \dot{m}_i - \sum_{out} \dot{m}_o = \frac{dm_s}{dt} \tag{5.34}$$

where $\dot{m}_i$ and $\dot{m}_o$ are the rate of mass entering with streams (i) and leaving with streams (o), respectively, and m_s is the mass of the system. For a flow system at steady state,

$$\sum_{in} \dot{m}_i = \sum_{out} \dot{m}_o \tag{5.35}$$

or

$$\sum_{in} (\rho V A)_i = \sum_{out} (\rho V A)_o \tag{5.36}$$

For only one inlet and one outlet stream, this becomes

$$\dot{m}_i = \dot{m}_o \tag{5.37}$$

5.3.2 Conservation of Energy

The most common forms of energy that must be considered in flowing fluids are internal (thermal), kinetic, potential (due to gravity), mechanical (work), and heat. Energy can enter or leave any system either with the streams entering or leaving or through the boundary of the system. The latter includes heat (Q) and "shaft" work (W), so called because it is commonly associated with work done by a pump, compressor, mixer, etc., which is driven by a shaft. The sign conventions for heat (Q) and work (W) are arbitrary, and vary from one reference to another. The common "engineering" conventions are that heat *added* to the system and work that is *generated by* the system (such as a turbine) are *positive*, but work that is *consumed* by the system (e.g., by a pump) is *negative*. This convention is also more consistent with the "driving force" interpretation of the terms in the Bernoulli form of the energy balance, as shown below.

5.3.2.1 Macroscopic Energy Balance

The macroscopic conservation of energy for any system is

$$\sum \left(h + gz + \frac{V^2}{2} \right)_i \dot{m}_i - \sum \left(h + gz + \frac{V^2}{2} \right)_o \dot{m}_o + \dot{Q} - \dot{W} = \frac{d}{dt}\left[\left(u + gz + \frac{V^2}{2} \right)_{sys} m_{sys} \right] \quad (5.38)$$

where $h = u + P/\rho$ is the *enthalpy* per unit mass of fluid. For only one inlet and one exit stream, $\dot{m}_i = \dot{m}_o = \dot{m}$ and, for steady state, this can be written as

$$\Delta h + g\Delta z + \frac{1}{2}\Delta V^2 = q - w \quad (5.39)$$

where $q = \dot{Q}/\dot{m}$, $w = \dot{W}/\dot{m}$ are the heat added to the system and the work done by the system, respectively, per unit mass of fluid. From the definition of enthalpy, and accounting for the irreversible dissipation of energy (i.e., friction loss), this is equivalent to

$$\int_{P_i}^{P_o} \frac{dP}{\rho g} + (z_o - z_i) + \frac{1}{2g}(\alpha_o V_o^2 - \alpha_i V_i^2) + \frac{e_f}{g} + \frac{w}{g} = 0 \quad (5.40)$$

where

$$e_f = (u_o - u_i) - q + \int_{P_i}^{P_o} P d\left(\frac{1}{\rho}\right) \quad (5.41)$$

is the irreversible energy dissipation due to "friction." Equation (5.40) is the "generalized" or "engineering" Bernoulli equation, and is the most useful equation for analysis of flows in conduits. For an incompressible (constant density) fluid, the equation can be written as

$$H_P + H_s + h_w = h_f + H_v \quad (5.42)$$

which is the "head" form of the equation, since all terms have dimensions of length. Here, H_P ($= -\Delta P/\rho g$) is the decrease in "pressure head"; H_z ($= -\Delta z$) is the decrease in "static head"; h_w ($= -w/g$) is the "pump head"; h_f ($= e_f/g$) is the friction "head loss"; and H_v ($= \Delta[\alpha V^2/2g]$) is the increase in "velocity head." For gases, if the pressure change is less than about 20%, the incompressible equation can be applied with reasonable accuracy by assuming the fluid density to be constant at a value equal to the average density in the system.

The α factors in the kinetic energy (or velocity head) term represent a correction factor to account for the deviation from plug flow through the conduit. For a Newtonian fluid in laminar flow in a circular tube, the profile is parabolic and the value of α is 2. For a highly turbulent flow, the profile is much flatter and $\alpha \approx 1.06$ (depending on the Reynolds number), although for practical purposes it is usually assumed that $\alpha = 1$ for turbulent flow.

Example 5.2

The mass flow rate of a hot coal-oil slurry can be determined by injecting a small sidestream of cool oil at a controlled rate and measuring the change in temperature downstream of the injection point. Calculate the flow rate of a slurry that is initially at 146.9°C (300°F) and has a density of 1200 kg/m³ and a specific heat of 0.7 kcal/kg°C (Btu/lb$_m$°F), if a sidestream of oil is injected at a rate of 0.454 kg/s (1 lb$_m$/s). The temperature of the sidestream is 15.6°C (60°F), and it has a density of 800 kg/m³ and a specific heat of 0.6 kcal/kg°C (Btu/lb$_m$°F). The temperature of the slurry downstream of the injection point is 147.8°C (298°F) with no sidestream injected, and 146.1°C (295°F) with the stream injected.

Solution

The steady-state energy balance applied to the slurry and sidestreams, excluding any pump in the line, is

$$\sum_{in}(h+gz+V^2/2)_i \dot{m}_i - \sum_{out}(h+gz+V^2/2)_o \dot{m}_o + \dot{Q} = 0$$

Assuming negligible change in elevation and kinetic energy, this becomes

$$h_{s1}\dot{m}_s + h_o\dot{m}_o - h_{s2}\dot{m}_s - h_{o2}\dot{m}_o + \dot{Q} = 0$$

where the subscripts s = slurry, o = oil, 1 = point upstream of injection point, and 2 = point downstream of injection point. $\dot{Q}$ is the rate of heat lost from the pipe between point 1 and point 2. Assuming $\Delta h = c_p \Delta T$ (i.e., negligible pressure drop), this becomes

$$\dot{m}_s c_{ps}(T_1 - T_2) + \dot{m}_o p_{po}(T_o - T_1) + \dot{Q} = 0$$

For no sidestream injection, $T_2 = T'_2 = 147.8°C$ (298°F):

$$\dot{Q} = \dot{m}_s c_{ps}(T'_2 - T_1)$$

Inserting this into the previous equation and solving for $\dot{m}_s$,

$$\dot{m}_s = \frac{\dot{m}_o c_{po}(T_2 - T_o)}{c_{ps}(T'_2 - T_2)} = \frac{(0.454 \text{ kg/s})(0.6 \text{ kcal/kgC})(146.1-15.6)°C}{(0.7 \text{ kcal/kgC})(147.8-146.1)°C} = 29.9 \text{ kg/s}$$

Fluid Flow

5.3.3 Conservation of Momentum

Because momentum is a vector quantity, the directional properties of each term must be considered. Also, Newton's second law provides an equivalence between force and the rate of momentum. Thus, for a flow system, the conservation of momentum must include the forces that act on the fluid mass of the system as well as the momentum carried into and out of the system.

The macroscopic momentum balance equation is

$$\sum_{On\ Fluid} \vec{F} + \sum_{in} (\dot{m}\vec{V})_i - \sum_{out} (\dot{m}\vec{V})_o = \frac{d}{dt}(m\vec{V})_{sys} \tag{5.43}$$

For three-dimensional space, this equation is represented by three component equations, one for each direction. If there is only one entering and one leaving stream, then $\dot{m}_i = \dot{m}_o = \dot{m}$. For steady state, the momentum equation becomes

$$\sum_{On\ Fluid} \vec{F} = \sum_{out} (\dot{m}\vec{V})_o - \sum_{in} (\dot{m}\vec{V})_i \tag{5.44}$$

Note that $\dot{m}$ is a scalar (i.e., $\dot{m} = \rho \vec{V} \cdot \vec{A}$ is a scalar product) and

$$\sum_{On\ Fluid} \vec{F}$$

represents all of the forces that can act *on* the fluid "*system.*" All of these forces, as well as each velocity component, must be referenced to a common reference frame or coordinate system. All forces due to pressures acting on the fluid in the system should be determined from the *gage* pressure, provided that the atmospheric pressure surrounding the system is uniform.

Example 5.3

A vessel containing water at 15.6°C (60°F) is fitted with a relief valve to keep the pressure in the vessel at or below 791 kPa (100 psig). The valve has a 101.6-mm (4-in.) inlet and is attached to the top of the vessel, and opens when the pressure reaches 791 kPa (100 psig). Determine the force (magnitude and direction) exerted on the valve at maximum flow conditions when the valve is open. The valve discharge is 90° from the direction of the inlet flow through a 152.4-mm (6-in.) diameter exit (see Figure 5.6) to the atmosphere. The loss coefficient for the open valve is 3 (see Section 5.7). Assume a steady state, and neglect gravity forces.

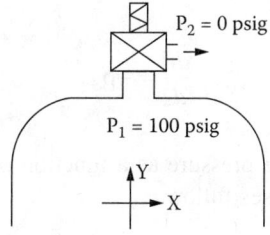

FIGURE 5.6 Force on relief valve.

Solution

The system is the fluid between the tank exit (at 791 kPa) and the valve exit (at 101 kPa). The loss coefficient relates the friction loss in the valve to the velocity leaving the valve, $e_f = K_f V_2^2/2$. Since the tank is much larger than the valve discharge, the velocity (and kinetic energy) in the tank can be neglected, and the flow is highly turbulent (which can be verified), so that $\alpha = 1$. The exit velocity (V_2) is then determined from the Bernoulli equation:

$$V_2 = \left[\frac{2(P_1 - P_2)}{\rho(1 + K_f)}\right]^{1/2} = \left[\frac{2(690 \times 10^3 \text{ Pa})}{(1000 \text{ kg/m}^3)(1+3)}\right]^{1/2} = 18.6 \text{ m/s}$$

and the mass flow rate through the valve is determined from the continuity equation:

$$\dot{m} = \rho V_2 A_2 = (1000 \text{ kg/m}^3)(18.6 \text{ m/s})\frac{\pi}{4}(0.1524 \text{ m})^2 = 339 \text{ kg/s}$$

The x and y components of the force on the valve are determined from the corresponding components of the momentum balance equation:

$$\sum_{\text{On Fluid}} F_x = -(F_x)_{\text{On Valve}} = \dot{m}(V_{x2} - V_{x1}) = (339 \text{ kg/s})(18.6 \text{ m/s}) = 6305 N$$

$$\sum_{\text{On Fluid}} F_y = -(F_y)_{\text{On Valve}} + P_1 A_1 = \dot{m}(V_{2y} + V_{1y}) = 0$$

$$(F_y)_{\text{On Valve}} = (791 \times 10^3 \text{ Pa})\frac{\pi}{4}(0.1016 \text{ m})^2 = 6413 N$$

$$\theta = \tan^{-1}(F_y/F_x) = 45.5°$$

Thus there is a force component in the $-x$ direction of 6300 N and in the $+y$ direction of 5600 N. The net force is $|F| = \sqrt{F_x^2 + F_y^2} = 9000$ N, and it acts in a direction up and to the left at an angle of 45.5° from the horizontal.

5.4 FLUID STATICS

5.4.1 THE BASIC EQUATION OF FLUID STATICS

If there is no *relative* motion within the fluid, the differential form of the energy balance (Bernoulli equation) reduces to

$$\frac{dP}{dz} = -\rho g \qquad (5.45)$$

Integration of this equation gives the pressure as a function of elevation (z) if the variations of g and ρ are known. Various special cases follow.

5.4.1.1 Constant Density

For a constant-density "isochoric" (or incompressible) fluid, under constant gravity,

$$P_1 - P_2 = \rho g(z_2 - z_1) \tag{5.46}$$

or

$$\Phi_1 = \Phi_2 = constant \tag{5.47}$$

where $\Phi = P + \rho g z$ is the sum of the local pressure (P) and static head ($\rho g z$). Φ is a constant at all points within a continuous incompressible fluid. For example, this can be applied to a manometer containing a fluid with density ρ_m attached to a vessel or conduit containing a fluid with density ρ_f to relate the manometer reading (Δh) to the combined pressure/head difference ($\Delta \Phi$):

$$\Delta \Phi = -\Delta \rho g \Delta h \tag{5.48}$$

where $\Delta \Phi = \Phi_2 - \Phi_1$, $D\rho = \rho_m - \rho_f$, Δh = manometer reading, $\Phi = P + \rho g z$, and z is the elevation of the pressure tap. This is the basic *manometer equation* and may be applied to any manometer in any orientation.

5.4.1.2 Ideal Gas

If the fluid is described by the ideal gas law ($\rho = PM/RT$), and the temperature and g are constant over all z, the integral of the basic equation from (P_1, z_1) to (P_2, z_2) gives

$$P_2 = P_1 \exp\left(-\frac{Mg\Delta z}{RT}\right) \tag{5.49}$$

where $\Delta z = z_2 - z_1$, M is the molecular weight, T is absolute temperature, and R is the gas constant.
For an ideal gas under *isentropic* conditions,

$$\frac{P}{\rho^k} = constant = \frac{P_1}{\rho_1^k} \tag{5.50}$$

where $k = c_p/c_v$ is the specific heat ratio for the gas (for an ideal gas, $c_p = c_v + R/M$). If the density is eliminated from this and the ideal gas law, the temperature-pressure relation becomes

$$\frac{T}{T_1} = \left(\frac{P}{P_1}\right)^{(k-1)/k} \tag{5.51}$$

Using this and the ideal gas law to eliminate ρ and T from Equation (5.45), the latter can be integrated to give

$$P_2 = P_1 \left[1 - \left(\frac{k-1}{k}\right)\frac{gM\Delta z}{RT_1}\right]^{k/(k-1)} \tag{5.52}$$

The corresponding temperature-elevation relation is

$$T_2 = T_1 \left[1 - \left(\frac{k-1}{k}\right)\frac{gM\Delta z}{RT_1}\right] \tag{5.53}$$

5.4.1.3 The Standard Atmosphere

The "standard atmosphere" is based on a representative observed average of the temperature distribution in the atmosphere, averaged over the entire earth and over all seasons:

$$\text{For } 0 < z < 11 \text{ km: } \frac{dT}{dz} = -6.5°/\text{km} = -G \tag{5.54}$$

$$\text{For } z > 11 \text{ km: } T = -56.5°C$$

where the average temperature at sea level ($z = 0$) is taken to be 15°C (285 K). Using this expression along with the ideal gas law gives the pressure as a function of elevation:

$$P_2 = P_1 \left[1 - \frac{G\Delta z}{T_0 - Gz_1} \right]^{Mg/RG} \tag{5.55}$$

where $T_0 = 285$ K.

5.4.2 MOVING SYSTEMS

The basic equation of statics also applies to fluids in motion, as long as there are no *relative* motion or velocity *gradients* that give rise to shear stresses. If the motion involves a uniform acceleration, an additional component of the pressure gradient results, as illustrated.

If the fluid accelerates upward with an acceleration of a_z, this is equivalent to an increase in the gravitational acceleration, i.e.,

$$\frac{dP}{dz} = -\rho(g + a_z) \tag{5.56}$$

Likewise, an acceleration in any direction, such as the i direction, will result in a pressure gradient within the fluid in the $-i$ direction of magnitude ρa_i:

$$\frac{\partial P}{\partial x_i} = -\rho a_i \tag{5.57}$$

Example 5.4
Determine the shape of the free surface in an open container that is rotating uniformly at an angular velocity of ω (see Figure 5.7).

Solution
The uniform rotation results in an (inward) radial acceleration equal to $\omega^2 r$. This gives rise to a radial pressure gradient in the fluid, which is in addition to the vertical pressure gradient due to gravity. The total pressure differential within the fluid is thus

$$dP = \left(\frac{\partial P}{\partial z}\right) dz + \left(\frac{\partial P}{\partial r}\right) dr = \rho(-g\,dz + \omega^2 r\,dr)$$

Fluid Flow

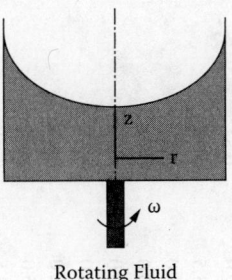

Rotating Fluid

FIGURE 5.7 Rotating fluid.

Since the pressure on the surface is constant,

$$(dP)_s = 0 = -\rho g (dz)_s + \rho \omega^2 (r dr)_s$$

This can be integrated to give

$$z = z_o + \frac{\omega^2 r^2}{2g}$$

which is the equation of a parabola.

5.4.3 Static Forces on Solid Boundaries

The force exerted on a solid boundary by a fluid pressure is given by

$$\vec{F} \int_A P d\vec{A} \tag{5.58}$$

Since both force and area are vectors (whereas pressure is a scalar), the direction of the force is determined by the orientation of the surface on which the pressure acts, i.e., the *projected* area of the surface corresponding to the component of force.

This can be applied, for example, to determine the required pipe wall thickness (t) that will withstand a given pressure (P) for a given pipe size ($D = 2R$) and "working stress" of the metal (σ):

$$t = \frac{RP}{\sigma} \tag{5.59}$$

The dimensionless pipe wall thickness (times 1000) is the *schedule number* of the pipe. This expression is only approximate, as it does not make any allowance for such factors as pipe threads, corrosion, etc. To compensate for these effects, an additional allowance is made in the wall thickness for the working definition of the "schedule thickness," t_s:

$$\text{Schedule Number} = \frac{1000P}{\sigma} = \frac{1750 t_s - 200}{D_o} \tag{5.60}$$

where both t_s and D_o (the pipe outside diameter) are in inches. This expression only applies to pipe larger than about 3 in., since smaller pipes must have a thicker wall just for dimensional integrity and rigidity. Standard pipe dimensions are given in Table 5.2.

TABLE 5.2
Standard Steel Pipe Dimensions and Capacities

Nominal Pipe Size, in.	Outside Diameter, in.	Schedule No.	Wall Thickness, in.	Inside Diameter, in.	Cross-Sectional Area		Circumference, ft, or Surface. ft²/ft of Length		Capacity at 1-ft/s Velocity			Weight of Plain-End Pipe, lb/ft
					Metal. in.²	Flow ft²	Outside	Inside	U.S. gal/min	lb/h Water		
1/4	0.405	10S	0.049	0.307	0.055	0.00051	0.106	0.0804	0.231	115.5		0.19
		40ST, 40S	.068	.269	.072	.00040	.106	.0705	.179	89.5		0.24
		80XS, 80S	.095	.215	.093	.00025	.106	.0563	.113	56.5		0.31
1/4	0.540	10S	.065	.410	.097	.00092	.141	.107	.412	206.5		0.33
		40ST.40S	.088	.364	.125	.00072	.141	.095	.323	161.5		0.42
		80XS, 80S	.119	.302	.157	.00050	.141	.079	.224	112.0		0.54
3/8	0.675	10S	.065	.545	.125	.00162	.177	.143	.727	363.5		0.42
		40ST,40S	.091	.493	.167	.00133	.177	.129	.596	298.0		0.57
		80XS, 80S	.126	.423	.217	.00098	.177	.111	.440	220.0		0.74
1/2	0.840	5S	.065	.710	.158	.00275	.220	.186	1.234	617.0		0.54
		10S	.083	.674	.197	.00248	.220	.176	1.112	556.0		0.67
		40ST,40S	.109	.622	.250	.00211	.220	.163	0.945	472.0		0.85
		80XS,80S	.147	.546	.320	.00163	.220	.143	0.730	365.0		1.09
		160	.188	.464	.385	.00117	.220	.122	0.527	263.5		1.31
		XX	.294	.252	.504	.00035	.220	.066	0.155	77.5		1.71
3/4	1.050	5S	.065	.920	.201	.00461	.275	.241	2.072	1036.0		0.69
		10S	.083	.884	.252	.00426	.275	.231	1.903	951.5		0.86
		40ST, 40S	.113	.824	.333	.00371	.275	.216	1.665	832.5		1.13
		80XS, 80S	.154	.742	.433	.00300	.275	.194	1.345	672.5		1.47
		160	.219	.612	.572	.00204	.275	.160	0.917	458.5		1.94
		XX	.308	.434	.718	.00103	.275	.114	0.461	230.5		2.44
1	1.315	5S	.065	1.185	.255	.00768	.344	.310	3.449	1725		0.87
		10S	.109	1.097	.413	.00656	.344	.287	2.946	1473		1.40
		40ST 40S	.133	1.049	.494	.00600	.344	.275	2.690	1345		1.68
		80XS, 80S	.179	0.957	.639	.00499	.344	.250	2.240	1120		2.17
		160	.250	0.815	.836	.00362	.344	.213	1.625	812.5		2.84
		XX	.358	0.599	1.076	.00196	.344	.157	0.878	439.0		3.66

Nom. Size	OD	Schedule	Wall	ID							
1 1/4	1.660	5S	.065	1.530	0.326	.01277	.435	.401	5.73	2865	1.11
		10S	.109	1.442	0.531	.01134	.435	.378	5.09	2545	1.81
		40ST, 40S	.140	1.380	0.668	.01040	.435	.361	4.57	2285	2.27
		80XS, 80S	.191	1.278	0.881	.00891	.435	.335	3.99	1995	3.00
		160	.250	1.160	1.107	.00734	.435	.304	3.29	1645	3.76
		XX	.382	0.896	1.534	.00438	.435	.235	1.97	985	5.21
1 1/2	1.900	5S	.065	1.770	0.375	.01709	.497	.463	7.67	3835	1.28
		10S	.109	1.682	0.614	.01543	.497	.440	6.94	3465	2.09
		40ST, 40S	.145	1.610	0.800	.01414	.497	.421	6.34	3170	2.72
		80XS, 80S	.200	1.500	1.069	.01225	.497	.393	5.49	2745	3.63
		160	.281	1.338	1.429	.00976	.497	.350	4.38	2190	4.86
		XX	.400	1.100	1.885	.00660	.497	.288	2.96	1480	6.41
2	2.375	5S	.065	2.245	0.472	.02749	.622	.588	12.34	6170	1.61
		10S	.109	2.157	0.776	.02538	.622	.565	11.39	5695	2.64
		40ST, 40S	.154	2.067	1.075	.02330	.622	.541	10.45	5225	3.65
		80ST, 80S	.218	1.939	1.477	.02050	.622	.508	9.20	4600	5.02
		160	.344	1.687	2.195	.01552	.622	.436	6.97	3485	7.46
		XX	.436	1.503	2.656	.01232	.622	.393	5.53	2765	9.03
2 1/2	2.875	5S	.083	2.709	0.728	.04003	.753	.709	17.97	8985	2.48
		10S	.120	2.635	1.039	.03787	.753	.690	17.00	8500	3.53
		40ST, 40S	.203	2.469	1.704	.03322	.753	.647	14.92	7460	5.79
		80XS, 80S	.276	2.323	2.254	.02942	.753	.608	13.20	6600	7.66
		160	.375	2.125	2.945	.02463	.753	.556	11.07	5535	10.01
		XX	.552	1.771	4.028	.01711	.753	.464	7.68	3840	13.69
3	3.500	5S	.083	3.334	0.891	.06063	.916	.873	27.21	13,605	3.03
		10S	.120	3.260	1.274	.05796	.916	.853	26.02	13,010	4.33
		40ST, 40S	.216	3.068	2.228	.05130	.916	.803	23.00	11,500	7.58
		80XS, 80S	.300	2.900	3.016	.04587	.916	.759	20.55	10,275	10.25
		160	.438	2.624	4.213	.03755	.916	.687	16.86	8430	14.32
		XX	.600	2.300	5.466	.02885	.916	.602	12.95	6475	18.58
3 1/2	4.0	5S	.083	3.834	1.021	.08017	1.047	1.004	35.98	17,990	3.48
		10S	.120	3.760	1.463	.07711	1.047	.984	34.61	17,305	4.97
		40ST, 40S	.226	3.548	2.680	.06870	1.047	.929	30.80	15,400	9.11
		80XS 80S	.318	3.364	3.678	.06170	1.047	.881	27.70	13,850	12.5

Continued

TABLE 5.2 (Continued)
Standard Steel Pipe Dimensions and Capacities

Nominal Pipe Size, in.	Outside Diameter, in.	Schedule No.	Wall Thickness, in.	Inside Diameter, in.	Cross-Sectional Area		Circumference, ft, or Surface. ft²/ft of Length		Capacity at 1-ft/s Velocity			Weight of Plain-End Pipe, lb/ft
					Metal. in.²	Flow ft²	Outside	Inside	U.S. gal/min	lb/h Water		
4	4.5	5S	.083	4.334	1.152	.10245	1.178	1.135	46.0	23,000		3.92
		10S	.120	4.260	1.651	.09898	1.178	1.115	44.4	22,200		5.61
		40ST, 40S	.237	4.026	3.17	.08840	1.178	1.054	39.6	19,800		10.79
		80XS, 80S	.337	3.826	4.41	.07986	1.178	1.002	35.8	17,900		14.98
		120	.438	3.624	5.58	.07170	1.178	0.949	32.2	16,100		19.00
		160	.531	3.438	6.62	.06647	1.178	0.900	28.9	14,450		22.51
		XX	.674	3.152	8.10	.05419	1.178	0.825	24.3	12,150		27.54
5	5.563	5S	.109	5.345	1.87	.1558	1.456	1.399	69.9	34,950		6.36
		10S	.134	5.295	2.29	.1529	1.456	1.386	68.6	34,300		7.77
		40ST, 40S	.258	5.047	4.30	.1390	1.456	1.321	62.3	31,150		14.62
		80XS, 80S	.375	4.813	6.11	.1263	1.456	1.260	57.7	28,850		20.76
		120	.500	4.563	7.95	.1136	1.456	1.195	51.0	25,500		27.01
		160	.625	4.313	9.70	.1015	1.456	1.129	45.5	22,750		32.96
		XX	.750	4.063	11.34	.0900	1.456	1.064	40.4	20,200		38.55
6	6.625	5S	.109	6.407	2.23	.2239	1.734	1.677	100.5	50,250		7.60
		10S	.134	6.357	2.73	.2204	1.734	1.664	98.9	49,450		9.29
		40ST, 40S	.280	6.065	5.58	.2006	1.734	1.588	90.0	45,000		18.97
		80XS, 80S	.432	5.761	8.40	.1810	1.734	1.508	81.1	40,550		28.57
		120	.562	5.501	10.70	.1650	1.734	1.440	73.9	36,950		36.39
		160	.719	5.187	13.34	.1467	1.734	1.358	65.9	32,950		45.34
		XX	.864	4.897	15.64	.1308	1.734	1.282	58.7	29,350		53.16
8	8.625	5S	.109	8.407	2.915	.3855	2.258	2.201	173.0	86,500		9.93
		10S	.148	8.329	3.941	.3784	2.258	2.180	169.8	84,900		13.40
		20	.250	8.125	6.578	.3601	2.258	2.127	161.5	80,750		22.36
		30	.277	8.071	7.265	.3553	2.258	2.113	159.4	79,700		24.70
		40ST, 40S	.322	7.981	8.399	.3474	2.258	2.089	155.7	77,850		28.55
		60	.406	7.813	10.48	.3329	2.258	2.045	149.4	74,700		35.64

Fluid Flow

		Schedule									
10	10.75	80XS, 80S	.500	7.625	12.76	.3171	2.258	1.996	142.3	71,150	43.39
		100	.594	7.437	14.89	.3017	2.258	1.947	135.4	67,700	50.95
		120	.719	7.187	17.86	.2817	2.258	1.882	126.4	63,200	60.71
		140	.812	7.001	19.93	.2673	2.258	1.833	120.0	60,000	67.76
		XX	.875	6.875	21.30	.2578	2.258	1.800	115.7	57,850	72.42
		160	.906	6.813	21.97	.2532	2.258	1.784	113.5	56,750	74.69
		5S	.134	10.482	4.47	.5993	2.814	2.744	269.0	134,500	15.19
		10S	.165	10.420	5.49	.5922	2.814	2.728	265.8	132,900	18.65
		20	.250	10.250	8.25	.5731	2.814	2.685	257.0	128,500	28.04
		30	.307	10.136	10.07	.5603	2.814	2.655	252.0	126,000	34.24
		40ST, 40S	.365	10.020	11.91	.5475	2.814	2.620	246.0	123,000	40.48
		80S, 60XS	.500	9.750	16.10	.5185	2.814	2.550	233.0	116,500	54.74
		80	.594	9.562	18.95	.4987	2.814	2.503	223.4	111,700	64.43
		100	.719	9.312	22.66	.4729	2.814	2.438	212.3	106,150	77.03
		120	.844	9.062	26.27	.4479	2.814	2.372	201.0	100,500	89.29
		140, XX	1.000	8.750	30.63	.4176	2.814	2.291	188.0	94,000	104.13
		160	1.125	8.500	34.02	.3941	2.814	2.225	177.0	88,500	115.64
12	12.75	5S	0.156	12.438	6.17	.8438	3.338	3.26	378.7	189,350	20.98
		10S	0.180	12.390	7.11	.8373	3.338	3.24	375.8	187,900	24.17
		20	0.250	12.250	9.82	.8185	3.338	3.21	367.0	183,500	33.38
		30	0.330	12.090	12.88	.7972	3.338	3.17	358.0	179,000	43.77
		ST, 40S	0.375	12.000	14.58	.7854	3.338	3.14	352.5	176,250	49.56
		40	0.406	11.938	15.74	.7773	3.338	3.13	349.0	174,500	53.52
		XS, 80S	0.500	11.750	19.24	.7530	3.338	3.08	338.0	169,000	65.42
		60	0.562	11.626	21.52	.7372	3.338	3.04	331.0	165,500	73.15
		80	0.688	11.374	26.07	.7056	3.338	2.98	316.7	158,350	88.63
		100	0.844	11.062	31.57	.6674	3.338	2.90	299.6	149,800	107.32
		120, XX	1.000	10.750	36.91	.6303	3.338	2.81	283.0	141,500	125.49
		140	1.125	10.500	41.09	.6013	3.338	2.75	270.0	135,000	139.67
		160	1.312	10.126	47.14	.5592	3.338	2.65	251.0	125,500	160.27
14	14	5S	0.156	13.688	6.78	1.0219	3.665	3.58	459	229,500	23.07
		10S	0.188	13.624	8.16	1.0125	3.665	3.57	454	227,000	27.73
		10	0.250	13.500	10.80	0.9940	3.665	3.53	446	223,000	36.71
		20	0.312	13.376	13.42	0.9750	3.665	3.50	438	219,000	45.61
		30, ST	0.375	13.250	16.05	0.9575	3.665	3.47	430	215,000	54.57

Continued

TABLE 5.2 (Continued)
Standard Steel Pipe Dimensions and Capacities

Nominal Pipe Size, in.	Outside Diameter, in.	Schedule No.	Wall Thickness, in.	Inside Diameter, in.	Cross-Sectional Area		Circumference, ft, or Surface, ft²/ft of Length		Capacity at 1-ft/s Velocity		Weight of Plain-End Pipe, lb/ft
					Metal, in.²	Flow ft²	Outside	Inside	U.S. gal/min	lb/h Water	
16	16	40	0.438	13.124	18.66	0.9397	3.665	3.44	422	211,000	63.44
		XS	0.500	13.000	21.21	0.9218	3.665	3.40	414	207,000	72.09
		60	0.594	12.812	25.02	0.8957	3.665	3.35	402	201,000	85.05
		80	0.750	12.500	31.22	0.8522	3.665	3.27	382	191,000	106.13
		100	0.938	12.124	38.49	0.8017	3.665	3.17	360	180,000	130.85
		120	1.094	11.812	44.36	0.7610	3.665	3.09	342	171,000	150.79
		140	1.250	11.500	50.07	0.7213	3.665	3.01	324	162,000	170.21
		160	1.406	11.188	55.63	0.6827	3.665	2.93	306	153,000	189.11
16	16	5S	0.165	15.670	8.21	1.3393	4.189	4.10	601	300,500	27.90
		10S	0.188	15.624	9.34	1.3314	4.189	4.09	598	299,000	31.75
		10	0.250	15.500	12.37	1.3104	4.189	4.06	587	293,500	42.05
		20	0.312	15.376	15.38	1.2985	4.189	4.03	578	289,000	52.27
		30, ST	0.375	15.250	18.41	1.2680	4.189	3.99	568	284,000	62.58
		40, XS	0.500	15.000	24.35	1.2272	4.189	3.93	550	275,000	82.77
		60	0.656	14.688	31.62	1.1766	4.189	3.85	528	264,000	107.50
		80	0.844	14.312	40.19	1.1171	4.189	3.75	501	250,500	136.61
		100	1.031	13.938	48.48	1.0596	4.189	3.65	474	237,000	164.82
		120	1.219	13.562	56.61	1.0032	4.189	3.55	450	225,000	192.43
		140	1.438	13.124	65.79	0.9394	4.189	3.44	422	211,000	223.64
		160	1.594	12.812	72.14	0.8953	4.189	3.35	402	201,000	245.25
18	18	5S	0.165	17.670	9.25	1.7029	4.712	4.63	764	382,000	31.43
		10S	0.188	17.624	10.52	1.6941	4.712	4.61	760	379,400	35.76
		10	0.250	17.500	13.94	1.6703	4.712	4.58	750	375,000	47.39
		20	0.312	17.376	17.34	1.6468	4.712	4.55	739	369,500	58.94
		ST	0.375	17.250	20.76	1.6230	4.712	4.52	728	364,000	70.59

Fluid Flow

	30	0.438	17.124	24.16	1.5993	4.712	4.48	718	359,000	82.15
	XS	0.500	17.000	27.49	1.5763	4.712	4.45	707	353,500	93.45
	40	0.562	16.876	30.79	1.5533	4.712	4.42	697	348,500	104.67
	60	0.750	16.500	40.64	1.4849	4.712	4.32	666	333,000	138.17
	80	0.938	16.124	50.28	1.4180	4.712	4.22	636	318,000	170.92
	100	1.156	15.688	61.17	1.3423	4.712	4.11	602	301,000	207.96
	120	1.375	15.250	71.82	1.2684	4.712	3.99	569	284,500	244.14
	140	1.562	14.876	80.66	1.2070	4.712	3.89	540	270,000	274.22
	160	1.781	14.438	90.75	1.1370	4.712	3.78	510	255,000	308.50
20	5S	0.188	19.624	11.70	2.1004	5.236	5.14	943	471,500	39.78
	10S	0.218	19.564	13.55	2.0878	5.236	5.12	937	467,500	46.06
	10	0.250	19.500	15.51	2.0740	5.236	5.11	930	465,000	52.73
	20, ST	0.375	19.250	23.12	2.0211	5.236	5.04	902	451,000	78.60
	30, XS	0.500	19.000	30.63	1.9689	5.236	4.97	883	441,500	104.13
	40	0.594	18.812	36.21	1.9302	5.236	4.92	866	433,000	123.11
	60	0.812	18.376	48.95	1.8417	5.236	4.81	826	413,000	166.40
	80	1.031	17.938	61.44	1.7550	5.236	4.70	787	393,500	208.87
	100	1.281	17.438	75.33	1.6585	5.236	4.57	744	372,000	256.10
	120	1.500	17.000	87.18	1.5763	5.236	4.45	707	353,500	296.37
	140	1.750	16.500	100.3	1.4849	5.236	4.32	665	332,500	341.09
	160	1.969	16.062	111.5	1.4071	5.236	4.21	632	316,000	397.17
24	5S	0.218	23.564	16.29	3.0285	6.283	6.17	1359	679,500	55.37
	10, 10S	0.250	23.500	18.65	3.012	6.283	6.15	1350	675,000	63.41
	20, ST	0.375	23.250	27.83	2.948	6.283	6.09	1325	662,500	94.62
	XS	0.500	23.000	36.90	2.885	6.283	6.02	1295	642,500	125.49
	30	0.562	22.876	41.39	2.854	6.283	5.99	1281	640,500	140.68
	40	0.688	22.624	50.39	2.792	6.283	5.92	1253	626,500	171.29
	60	0.969	22.062	70.11	2.655	6.283	5.78	1192	596,000	238.35
	80	1.219	21.562	87.24	2.536	6.283	5.64	1138	569,000	296.58
	100	1.531	20.938	108.1	2.391	6.283	5.48	1073	536,500	367.39
	120	1.812	20.376	126.3	2.264	6.283	5.33	1016	508,000	429.39
	140	2.062	19.876	142.1	2.155	6.283	5.20	965	482,500	483.12
	160	2.344	19.312	159.5	2.034	6.283	5.06	913	456,500	52.13

Continued

TABLE 5.2 (Continued)
Standard Steel Pipe Dimensions and Capacities

Nominal Pipe Size, in.	Outside Diameter, in.	Schedule No.	Wall Thickness, in.	Inside Diameter, in.	Cross-Sectional Area		Circumference, ft, or Surface. ft²/ft of Length		Capacity at 1-ft/s Velocity		Weight of Plain-End Pipe, lb/ft
					Metal. in.²	Flow ft²	Outside	Inside	U.S. gal/min	lb/h Water	
30	30	5S	0.250	29.500	23.37	4.746	7.854	7.72	2130	1,065,000	79.43
		10, 10S	0.312	29.376	29.10	4.707	7.854	7.69	2110	1,055,000	98.93
		ST	0.375	29.250	34.90	4.666	7.854	7.66	2094	1,048,000	118.65
		20, XS	0.500	29.000	46.34	4.587	7.854	7.59	2055	1,027,500	157.53
		30	0.625	28.750	57.68	4.508	7.854	7.53	2020	1,010,000	196.08

Note: 5S, 10S, and 40S are extracted from Stainless Steel Pipe, ANSI B36.19–1976, with permission of the publisher, the American Society of Mechanical Engineers, New York. ST = standard wall, XS = extra strong wall, XX = double extra strong wall, and Schedules 10 through 160 are extracted from Wrought-Steel and Wrought-iron Pipe, ANSI B36.10–1975, with permission of the same publisher. Decimal thicknesses for respective pipe sizes represent their nominal or average wall dimensions. Mill tolerances as high as ±12 1/2% are permitted.

Plan-end pipe is produced by a square cut. Pipe is also shipped from the mills threaded, with a threaded coupling on one end, or with the ends beveled for welding, or grooved or sized for patented couplings. Weights per foot for threaded and coupled pipe are slightly greater because of the weight of the coupling, but it is not available larger than 12 in or lighter than Schedule 30 sizes 8 through 12 in, or Schedule 40 sizes 6 in. and smaller.

To convert inches to millimeters, multiply by 25.4; to convert square inches to square millimeters, multiply by 645; to convert feet to meters, multiply by 0.3048; to convert square feet to square meters, multiply by 0.0929; to convert pounds per foot to kilograms per meter, multiply by 1.49; to convert gallons to cubic meters, multiply by 3.7854×10^{-3}; and to convert pounds to kilograms, multiply by 0.4536.

Fluid Flow

5.5 PIPE FLOW

5.5.1 FLOW REGIMES

The flow regime in a pipe is determined by the Reynolds number. For a Newtonian fluid:

$$N_{Re} = \frac{DV\rho}{\mu} = \frac{\rho V^2}{\mu V / D} \tag{5.61}$$

This represents the ratio of the momentum flux carried by the fluid axially along the tube to the viscous momentum flux transported normal to the flow in the radial direction. Laminar flows are dominated by the fluid viscosity and are stable, whereas turbulent flows are dominated by the fluid density inertia and are unstable.

5.5.2 PRESSURE-FLOW RELATIONS FOR PIPE FLOWS

For steady, fully developed one-dimensional flow in a uniform pipe, the engineering Bernoulli and momentum equations provide equivalent interpretations of the friction loss:

$$e_f = K_f \frac{V^2}{2} = \frac{\tau_w}{\rho}\left(\frac{4fL}{D}\right) \tag{5.62}$$

where K_f is the "loss coefficient" and τ_w is the wall shear stress. For pipe, these are related to the dimensionless (Fanning) friction factor (f) by

$$K_f = \frac{4fL}{D}; \quad \tau_w = \frac{f}{2}\rho V^2 \tag{5.63}$$

A loss coefficient can be defined for any element in which energy is dissipated (pipe, fittings, valves, etc.), although the friction factor is defined only for pipe flow. All that is necessary to describe the pressure-flow relation for pipe flows is Bernoulli's equation and a knowledge of the friction factor, which depends upon flow conditions, pipe size, and fluid properties.

5.5.3 NEWTONIAN FLUIDS

5.5.3.1 Laminar Flow

For a Newtonian fluid in laminar flow,

$$f = \frac{4\pi D\mu}{Q\rho} = \frac{16\mu}{DV\rho} = \frac{16}{N_{Re}} \tag{5.64}$$

i.e., only one dimensionless group, fN_{Re}, is required to characterize laminar pipe flow, which has a value of 16. When this is introduced into the Bernoulli equation, the result is

$$Q = -\frac{\pi \Delta \Phi D^4}{128\mu L} \tag{5.65}$$

which is the *Hagen-Poiseuille* equation.

5.5.3.2 Turbulent Flow

For smooth pipe for Reynolds numbers from 4000 to 10^6, the *von Karman* equation provides an implicit relation between the friction factor and the Reynolds number:

$$\frac{1}{\sqrt{2}} = 4.0\log(N_{Re}\sqrt{f}) - 0.04 \qquad (5.66)$$

where $N_{Re}\sqrt{f}$ is

$$N_{Re}\sqrt{f} = \left(\frac{e_f D^3 \rho^2}{2L\mu^2}\right)^{1/2} \qquad (5.67)$$

which is independent of flow rate.

5.5.3.3 Rough Pipe

Colebrook extended the von Karman equation to account for tube wall roughness (ε/D) as follows:

$$\frac{1}{\sqrt{f}} = -4\log\left[\frac{\varepsilon/D}{3.7} + \frac{1.255}{N_{Re}\sqrt{f}}\right] \qquad (5.68)$$

The Colebrook equation is convenient for determining the flow rate from the allowable friction loss (e.g., driving force), tube size, and fluid properties. Published plots of f vs. N_{Re} and ε/D (i.e., the Moody diagram) are usually generated from the Colebrook equation.

The actual size of the roughness elements on any surface will vary with the material, age and usage, deposits, dirt, scale, rust, etc. Typical values for various materials are given in Table 5.3. The most common pipe material—clean, new, commercial steel—has an effective roughness of about 0.0018 in. (0.045 mm). Other surfaces, such as concrete, may vary as much as several orders of magnitude, depending upon the nature of the surface finish.

5.5.3.4 All Flow Regimes

The expressions for the friction factor in both laminar and turbulent flows were combined into a single expression by Churchill (1977) as follows:

$$f = 2\left[\left(\frac{8}{N_{Re}}\right)^{12} + \frac{1}{(A+B)^{3/2}}\right]^{1/12} \qquad (5.69)$$

where

$$A = \left[2.457\ln\left(\frac{1}{\left(\frac{7}{N_{Re}}\right)^{0.9} + \frac{0.27\varepsilon}{D}}\right)\right]^{16} \qquad (5.70)$$

TABLE 5.3
Equivalent Roughness of Various Surfaces

Material	Condition	Roughness Range	Typical Value
Drawn Brass, Copper, Stainless	New	0.01-0.0015 mm (0.0004-0.00006 in.)	0.002 mm (0.00008 in.)
Commercial Steel	New	0.1-0.02 mm (0.004-0.0008 in.)	0.045 mm (0.0018 in.)
	Light Rust	1.0-0.15 mm (0.04-0.006 in.)	0.3 mm (0.015 in.)
	General Rust	3.0-1.0 mm (0.1-0.04 in.)	2.0 mm (0.08 in.)
Iron	Wrought, New	0.045 mm (0.002 in.)	0.045 mm (0.002 in.)
	Cast, New	1.0-0.25 mm (0.04-0.01 in.)	0.30 mm (0.025 in.)
	Galvanized	0.15-0.025 mm (0.006-0.001 in.)	0.15 mm (0.006 in.)
	Asphalt Coated	1.0-0.1 mm (0.04-0.004 in.)	0.15 mm (0.006 in.)
Sheet Metal	Ducts Smooth Joints	0.1-0.02 mm (0.004-0.0008 in.)	0.03 mm (0.0012 in.)
Concrete	Very Smooth	0.18-0.025 mm (0.007-0.001 in.)	0.04 mm (0.0016 in.)
	Wood Floated, Brushed	0.8-0.2 mm (0.03-0.007 in.)	0.3 mm (0.012 in.)
	Rough, Visible Form Marks	2.5-0.8 mm (0.1-0.03 in.)	2.0 mm (0.08 in.)
Wood	Stave, Used	1.0-0.25 mm (0.035-0.01 in.)	0.5 mm (0.02 in.)
Glass or Plastic	Drawn Tubing	0.01-0.0015 mm (0.0004-0.00006 in.)	0.002 mm (0.00008 in.)
Rubber	Smooth Tubing	0.07-0.006 mm (0.003-0.00025 in.)	0.01 mm (0.0004 in.)
	Wire Reinforced	4.0-0.3 mm (0.15-0.01 in.)	1.0 mm (0.04 in.)

$$B = \left(\frac{37,530}{N_{Re}}\right)^{16} \tag{5.71}$$

This equation gives the Fanning friction factor over the entire range of Reynolds numbers within the accuracy of the data used to construct the Moody diagram, including a reasonable estimate for the intermediate or transition region between laminar and turbulent flows. Corresponding expressions for the friction factor for non-Newtonian fluids in pipes are given below for two (two-parameter) models: the power law and Bingham plastic.

Example 5.5

Determine the power required to pump water at 60°F at a rate of 500 gpm through a 6-in. sch 40 pipe, 10 miles long, if the pressure at the entrance and discharge of the line is atmospheric.

Solution

At this temperature, the viscosity of water is 1 cP and the density is 1 g/cc. The actual ID of the sch 40 pipe is 154.1 mm (6.065 in.), and it is assumed to have a roughness of 0.045 mm (0.0018 in.). The power is the work done by the pump per unit mass of fluid ($-w$) times the mass flow rate, $\dot{m}$. From the Bernoulli equation for no change in pressure or elevation, this is

$$-w\dot{m} = e_f \dot{m} = \left(\frac{4fL}{D}\right)\left(\frac{V^2}{2}\right)\rho Q = \frac{32 fL Q^3}{\pi^2 D^5}$$

The friction factor (f) depends on the Reynolds number ($N_{Re} = 4Q\rho/\pi D\mu = 2.61 \times 10^5$) and the relative roughness ($\varepsilon/D = 2.97 \times 10^{-4}$). From the Churchill equation (or the Colebrook equation, by iteration), $f = 0.00432$. Using this value in the above equation gives $-w\dot{m} = 81.3$ kW (109 hp).

5.5.3.5 Water in Sch 40 Pipe

For the special case of water at 60°F in sch 40 steel ($\varepsilon = 0.045$ mm or 0.0018 in.) pipe, the relation between flow rate, velocity, pressure drop, and pipe size is tabulated in Table 5.4. The range of values tabulated covers most of the range that would be expected in practice. Note that the friction loss is tabulated as "pressure drop" in psi per 100 ft of pipe, which is equivalent to (100 $\rho e_f/L$) in English engineering units.

5.5.4 Power Law Fluids

The power law model is very popular for representing the viscosity of a wide variety of non-Newtonian fluids because of its simplicity and versatility. However, care should be exercised in its application; for reliable results, the range of shear stress (or shear rate) expected in the application should not extend beyond the range of the rheological data used to evaluate the model parameters. Both laminar and turbulent pipe flow of highly loaded slurries of fine particles, for example, can often be adequately represented by either of these two models, as shown by Darby et al. (1992).

5.5.4.1 Laminar Flow

The form of the power law model that applies to tube flow is

$$\tau_{rx} = -m\left(-\frac{dv_x}{ddr}\right)^n \tag{5.72}$$

since the velocity gradient (shear rate) and shear stress are negative. This can be integrated to give the flow rate:

$$Q = \pi \left(\frac{\tau_w}{mR}\right)^{1/n} \left(\frac{n}{3n+1}\right) R^{(3n+1)/n} \tag{5.73}$$

which is the power law fluid equivalent of the Hagen-Poiseuille equation. The dimensionless form of this equation is

$$fN_{RePL} = 16 \tag{5.74}$$

Fluid Flow

TABLE 5.4
Flow of Water through Schedule 40 Steel Pipe

Pressure Drop per 100 Feet and Velocity in Schedule 40 Pipe for Water at 60°F

Discharge		1/8"		1/4"		3/8"		1/2"		3/4"		1"		1 1/4"		1 1/2"		2"		2 1/2"		3"		3 1/2"		4"		5"	
Gallons per Minute	Cubic Ft per Second	Velocity Feet per Second	Press. Drop lbf per Sq In.	Velocity Feet per Second	Press. Drop lbf per Sq In.	Velocity Feet per Second	Press. Drop lbf per Sq In.	Velocity Feet per Second	Press. Drop lbf per Sq In.	Velocity Feet per Second	Press. Drop lbf per Sq In.	Velocity Feet per Second	Press. Drop lbf per Sq In.	Velocity Feet per Second	Press. Drop lbf per Sq In.	Velocity Feet per Second	Press. Drop lbf per Sq In.	Velocity Feet per Second	Press. Drop lbf per Sq In.	Velocity Feet per Second	Press. Drop lbf per Sq In.	Velocity Feet per Second	Press. Drop lbf per Sq In.	Velocity Feet per Second	Press. Drop lbf per Sq In.	Velocity Feet per Second	Press. Drop lbf per Sq In.	Velocity Feet per Second	Press. Drop lbf per Sq In.
.2	0.000446	1.13	1.86	0.616	0.359																								
.3	0.000668	1.69	4.22	0.924	0.903	0.504	0.159	0.317	0.061																				
.4	0.000891	2.26	6.98	1.23	1.61	0.672	0.345	0.422	0.086																				
.5	0.00111	2.82	10.5	1.54	2.39	0.840	0.539	0.528	0.167	0.301	0.033																			
.6	0.00134	3.39	14.7	1.85	3.29	1.01	0.751	0.633	0.240	0.361	0.041																			
.8	0.00178	4.52	25.0	2.46	5.44	1.34	1.25	0.844	0.408	0.481	0.102																			
1	0.00223	5.65	37.2	3.08	8.28	1.68	1.85	1.06	0.600	0.602	0.155	0.371	0.048																	
2	0.00446	11.29	134.4	6.16	30.1	3.36	6.58	2.11	2.10	1.20	0.526	0.743	0.164	0.429	0.044															
3	0.00668			9.25	64.1	5.04	13.9	3.17	4.33	1.81	1.09	1.114	0.336	0.644	0.090	0.473	0.043													
4	0.00891			12.33	111.2	6.72	23.9	4.22	7.42	2.41	1.83	1.49	0.565	0.858	0.150	0.630	0.071													
5	0.01114					8.40	36.7	5.28	11.2	3.01	2.75	1.86	0.835	1.071	0.223	0.788	0.104													
6	0.01337					10.08	51.9	6.33	15.8	3.61	3.54	2.23	1.17	1.29	0.309	0.946	0.145	0.574	0.044											
8	0.01761					13.44	91.1	8.45	27.7	4.81	6.60	2.97	1.99	1.72	0.518	1.26	0.241	0.765	0.073											
10	0.02228							10.56	42.4	6.02	9.99	3.71	2.99	2.15	0.774	1.58	0.361	0.956	0.108	0.670	0.046									
15	0.03342									9.03	21.6	5.57	6.36	3.22	1.63	2.37	0.755	1.43	0.224	1.01	0.094									
20	0.04456									12.03	37.8	7.43	10.9	4.29	2.78	3.16	1.28	1.91	0.375	1.34	0.158	0.868	0.056							
25	0.05570											9.28	16.7	5.37	4.22	3.94	1.93	2.39	0.561	1.68	0.234	1.09	0.083	0.812	0.041					
30	0.06684											11.14	23.8	6.44	5.92	4.73		2.87	0.786	2.01	0.327	1.30	0.114	0.974	0.056					
35	0.07798											12.99	32.2	7.51	7.90	5.52	3.64	3.35	1.05	2.35	0.436	1.52	0.151	1.14	0.071	0.882	0.041			
40	0.8912											14.85	41.5	8.59	10.24	6.30	4.65	3.83	1.35	2.68	0.556	1.74	0.192	1.30	0.095	1.01	0.052			
45	0.1003													9.67	12.80	7.09	5.85	4.30	1.67	3.02	0.668	1.95	0.239	1.46	0.117	1.13	0.064			
50	0.1114													10.74	15.66	7.88	7.15	4.78	2.03	3.35	0.839	2.17	0.288	1.62	0.142	1.26	0.076			
60	0.1337													12.89	22.2	9.47		5.74	2.87	4.02	1.18	2.60	0.406	1.95	0.204	1.51	0.107			
70	0.1560															11.05	13.71	6.70	3.84	4.69	1.59	3.04	0.540	2.27	0.261	1.76	0.143	1.12	0.047

Continued

TABLE 5.4 (Continued)
Flow of Water through Schedule 40 Steel Pipe

Pressure Drop per 100 Feet and Velocity in Schedule 40 Pipe for Water at 60°F

Discharge			10"		12"		14"		6"		8"	
Gallons per Minute	Cubic Ft per Second	Velocity Feet per Second	Press. Drop lbf per Sq In.	Velocity Feet per Second	Press. Drop lbf per Sq In.	Velocity Feet per Second	Press. Drop lbf per Sq In.	Velocity Feet per Second	Press. Drop lbf per Sq In.	Velocity Feet per Second	Press. Drop lbf per Sq In.	Velocity Feet per Second	Press. Drop lbf per Sq In.
80	0.1782	7.65	4.97	5.36	2.03	3.47	0.687	2.60	0.334	2.02	0.160	1.28	0.060			12.62	17.59
90	0.2005	8.60	6.20	6.03	2.53	3.91	0.861	2.92	0.416	2.27	0.224	1.44	0.074			14.20	22.0
100	0.2228	9.56	7.59	6.70	3.09	4.34	1.05	3.25	0.509	2.52	0.272	1.60	0.090	1.11	0.036	15.78	26.9
125	0.2785	11.97	11.76	8.38	4.71	5.43	1.61	4.06	0.769	3.15	0.415	2.01	0.135	1.39	0.055	19.72	41.4
150	0.3342	14.36	16.70	10.05	6.69	6.51	2.24	4.87	1.08	3.78	0.680	2.41	0.190	1.67	0.077		
175	0.3899	16.75	22.3	11.73	8.97	7.60	3.00	5.68	1.44	4.41	0.774	2.81	0.253	1.94	0.102		
200	0.4456	19.14	28.8	13.42	11.68	8.68	3.87	6.49	1.85	5.04	0.985	3.21	0.323	2.22	0.130		
225	0.5013			15.09	14.63	9.77	4.83	7.30	2.32	5.67	1.23	3.61	0.401	2.50	0.162		
250	0.557					10.85	5.93	8.12	2.84	6.30	1.46	4.01	0.495	2.78	0.195		
275	0.6127					11.94	7.14	8.93	3.40	6.31	1.79	4.41	0.583	3.05	0.234		
300	0.6684					13.00	8.36	9.74	4.02	7.56	2.11	4.81	0.683	3.33	0.275		
325	0.7241					14.12	9.89	10.53	4.09	8.19	2.47	5.21	0.797	3.61	0.320		
350	0.7798							11.36	5.41	8.82	2.84	5.62	0.919	3.89	0.367		
375	0.8355							12.17	6.18	9.45	3.25	6.02	1.05	4.16	0.416		
400	0.8912							12.98	7.03	10.08	3.68	6.42	1.19	4.44	0.471		
425	0.9469							13.80	7.89	10.71	4.12	6.82	1.33	4.72	0.529		
450	1.003							14.61	8.80	11.34	4.60	7.22	1.48	5.00	0.590		
475	1.059	1.93	0.054							11.97	5.12	7.62	1.64	5.27	0.653		
500	1.114	2.03	0.059							12.60	5.65	8.02	1.81	5.55	0.720		
550	1.225	2.24	0.071							13.85	6.79	8.82	2.17	6.66	0.861		
600	1.337	2.44	0.083							15.12	8.04	9.63	2.55	6.00	1.02		
650	1.448	2.64	0.097									10.43	2.98	7.22	1.18		
700	1.560	2.85	0.112	2.01	0.047							11.23	3.43	7.78	1.35		
750	1.671	3.05	0.127	2.15	0.054							12.03	3.92	8.33	1.55		
800	1.782	3.25	0.143	2.29	0.061							12.83	4.43	8.88	1.75		
850	1.894	3.46	0.160	2.44	0.068	2.02	0.042					13.64	5.00	9.44	1.96		

Fluid Flow

Flow (gpm)	Q (cfs)	10″ v	10″ ΔP	12″ v	12″ ΔP	14″ v	14″ ΔP	16″ v	16″ ΔP	18″ v	18″ ΔP	20″ v	20″ ΔP	24″ v	24″ ΔP
900	2.005	3.66	0.179	2.58	0.075	2.13	0.047	…	…	…	…	…	…	…	…
950	2.117	3.86	0.198	2.72	0.083	2.25	0.052	…	…	…	…	…	…	…	…
1000	2.228	4.07	0.218	2.87	0.091	2.37	0.057	…	…	…	…	…	…	…	…
1100	2.451	4.48	0.260	3.15	0.110	2.61	0.068	…	…	…	…	…	…	…	…
1200	2.674	4.88	0.306	3.44	0.128	2.85	0.080	2.18	0.042	…	…	…	…	…	…
1300	2.896	5.29	0.355	3.73	0.150	3.08	0.093	2.36	0.048	…	…	…	…	…	…
1400	3.119	5.70	0.409	4.01	0.171	1.32	0.107	2.54	0.055	…	…	…	…	…	…
1500	3.342	6.10	0.466	4.30	0.195	3.56	0.122	2.72	0.063	…	…	…	…	…	…
1600	3.565	6.51	0.527	4.59	0.219	3.79	0.138	2.90	0.071	…	…	…	…	…	…
1800	4.010	7.32	0.663	5.16	0.276	4.27	0.172	3.27	0.088	2.58	0.050	…	…	…	…
2000	4.456	8.14	0.808	5.73	0.339	4.74	0.209	3.63	0.107	2.87	0.060	…	…	…	…
2500	5.570	10.17	1.24	7.17	0.515	5.93	0.321	4.54	0.163	3.59	0.091	…	…	…	…
3000	6.684	12.20	1.76	8.10	0.731	7.11	0.451	5.45	0.232	4.30	0.129	3.46	0.075	…	…
3500	7.798	14.24	2.38	10.03	0.982	8.30	0.607	6.35	0.312	5.02	0.173	4.04	0.101	…	…
4000	8.912	16.27	3.08	11.47	1.27	9.48	0.787	7.26	0.401	5.74	0.222	4.62	0.129	3.19	0.052
4500	10.03	18.31	3.87	12.90	1.60	10.67	0.990	8.17	0.503	6.46	0.280	5.20	0.162	3.59	0.065
5000	11.14	20.35	4.71	14.33	1.95	11.85	1.71	9.08	0.617	7.17	0.340	5.77	0.199	3.99	0.079
6000	13.37	24.41	6.74	17.20	2.77	14.23	1.17	10.89	0.877	8.61	0.483	6.93	0.280	4.79	0.111
7000	15.60	28.49	9.11	20.07	3.74	16.60	2.31	12.71	1.18	10.04	0.652	8.08	0.376	5.59	0.150
8000	17.82	…	…	22.93	4.84	18.96	2.99	14.52	1.51	11.47	0.839	9.23	0.488	6.38	0.192
9000	20.05	…	…	25.79	6.09	21.34	3.76	16.34	1.90	12.91	1.05	10.39	0.608	7.18	0.242
10000	22.28	…	…	28.66	7.46	23.71	4.61	18.15	2.34	14.34	1.28	11.54	0.739	7.58	0.294
12000	26.74	…	…	34.40	10.7	28.45	6.59	21.79	3.33	17.21	1.83	13.85	1.06	0.58	0.416
14000	31.19	…	…	…	…	33.19	8.89	25.42	4.49	20.08	2.45	16.16	1.43	11.17	0.562
16000	35.65	…	…	…	…	…	…	29.05	5.83	22.95	3.18	18.47	1.85	12.77	0.723
18000	40.10	…	…	…	…	…	…	32.68	7.31	25.82	4.03	20.77	2.32	14.36	0.907
20000	44.56	…	…	…	…	…	…	36.31	9.03	28.69	28.69	23.08	2.86	15.96	1.12

Note: For pipe lengths other than 100 ft, the pressure drop is proportional to the length. Thus, for 50 ft of pipe, the pressure drop is approximately one half the value given in the table for 300 feet, three times the given value, etc.

Velocity is a function of the cross-sectional flow area; thus, it is constant for a given flow rate and is independent of pipe length.

For calculations for pipe other then Schedule 40, see explanation on next page.

Source: Darby, R., *Chemical Engineering Fluid Mechanics*, 2nd ed., New York: Marcel Dekker, 2001.

where

$$N_{\text{Re}PL} = \frac{8D^n V^{2-n} \rho}{m[2(3n+1)/n]^n} \tag{5.75}$$

is the power law Reynolds number.

5.5.4.2 All Flow Regimes

There are insufficient data in the literature to provide a reliable estimate of the effect of roughness on friction loss for non-Newtonian fluids in turbulent flow. However, the influence of roughness is normally neglected, since the laminar boundary layer thickness for such fluids is typically much larger than for Newtonian fluids (i.e., the flow conditions most often fall in the "hydraulically smooth" range for common pipe materials). An expression by Darby et al. (1992) for f for the power law fluid, which applies to both laminar and turbulent flow, is

$$f = (1-\alpha)f_L + \frac{\alpha}{[f_T^{-8} + f_{Tr}^{-8}]^{1/8}} \tag{5.76}$$

where

$$f_L = \frac{16}{N_{\text{Re}PL}} \tag{5.77}$$

$$f_T = \frac{0.0682 n^{-1/2}}{N_{\text{Re}PL}^{1/(1.87+2.39n)}} \tag{5.78}$$

$$f_{Tr} = 1.79 \times 10^{-4} \exp(-5.24n) N_{\text{Re}PL}^{(0.414+0.757n)} \tag{5.79}$$

and the parameter α is given by

$$\alpha = \frac{1}{(1+4^{-\Delta})} \tag{5.80}$$

where

$$\Delta = N_{\text{Re}PL} - N_{\text{Re}PLc} \tag{5.81}$$

and the "critical" Reynolds number that corresponds to the onset of turbulence is given by

$$N_{\text{Re}PLc} = 2100 + 875(1-n) \tag{5.82}$$

5.5.5 Bingham Plastic Fluids

The Bingham plastic model usually provides a good representation for the viscosity of concentrated slurries, suspensions, sediments, emulsions, foams, etc. Such materials often exhibit a yield stress,

Fluid Flow

which must be exceeded before the material will flow at a significant rate. Other examples include paint, ink, shaving cream, mayonnaise, etc. There are also many fluids that may have a yield stress that is less pronounced, such as blood, fine silt, etc.

A "plastic" is really two materials. At low stresses, below the yield stress (τ_o), it behaves as a solid, whereas above the yield stress it behaves as a fluid. The Bingham model for this material is

$$\text{For } |\tau| < \tau_o: \quad \dot{\gamma} = 0$$
$$\text{For } |\tau| > \tau_o: \quad \tau = \pm\tau_o + \mu_\infty \dot{\gamma} \tag{5.83}$$

Since the shear stress and shear rate can be either positive or negative, the "plus" is used in the former case, and the "minus" for the latter. For tube flow, the "minus" sign should be used.

5.5.5.1 Laminar Flow

There is a finite distance from the center of the pipe over which the stress is always less than the yield stress, so in this region the material behaves as a solid and does not "yield" but moves as a rigid "plug." The radius of this plug (r_o) is

$$r_o = R \frac{\tau_o}{\tau_w} \tag{5.84}$$

Outside of this plug region, the stress exceeds the yield stress and the material will flow. The total flow rate is determined by combining the flow in the "plug" with that in the "fluid" region to give

$$Q = \frac{\pi R^3 \tau_w}{4\mu_\infty} \left[1 - \frac{4}{3}\left(\frac{\tau_o}{\tau_w}\right) + \frac{1}{3}\left(\frac{\tau_o}{\tau_w}\right)^4 \right] \tag{5.85}$$

which is known as the *Buckingham–Reiner* equation. In dimensionless form, it can be written as

$$f_L = \frac{16}{N_{Re}} \left[1 + \frac{N_{He}}{6 N_{Re}} - \frac{1}{3} \frac{N_{He}^4}{f^3 N_{Re}^7} \right] \tag{5.86}$$

where the Reynolds number is given by

$$N_{Re} = \frac{DV\rho}{\mu_\infty} \tag{5.87}$$

and

$$N_{He} = \frac{D^2 \rho \tau_o}{\mu_\infty^2} \tag{5.88}$$

is the *Hedstrom number*. Note that the Bingham plastic reduces to a Newtonian fluid if $\tau_o = N_{He} = 0$, for which $fN_{Re} = 16$. There are actually only two independent dimensionless groups in Equation (5.86), i.e., fN_{Re} and N_{He}/N_{Re} (which is also called the *Bingham number*, $N_{Bi} = [D\tau_o/\mu V]$). The

Buckingham-Reiner equation is implicit in f, so it must be solved by iteration for known values of N_{Re} and N_{He}.

5.5.5.2 All Flow Regimes

For the Bingham plastic fluid, there is no abrupt transition from laminar to turbulent flow, as is observed for Newtonian fluids. Instead, a gradual deviation from purely laminar flow to fully turbulent flow occurs. For turbulent flow, the friction factor can be represented by the empirical expression of Darby and Melson (1982) (as modified by Darby et al., 1992):

$$f_T = \frac{10^a}{N_{Re}^{0.193}} \tag{5.89}$$

where

$$a = -1.41[1 + 0.146\exp(-2.9 \times 10^{-5} N_{He})] \tag{5.90}$$

For all values of N_{Re} and N_{He}, f is given by

$$f = (f_L^m + f_T^m)^{1/m} \tag{5.91}$$

where

$$m = 1.7 + \frac{40{,}000}{N_{Re}} \tag{5.92}$$

and f_L is given by the Buckingham–Reiner equation (5.86).

Example 5.6

Calculate the power required to pump a coal slurry that is 50% (by weight) solids at a rate of 315.5 m³/s (500 gpm) through a 6-in. sch 40 pipeline (ID = 154.1 mm) that is 16.1 km (10 mi) long. The slurry behaves as a Bingham plastic, with a yield stress of 1.16 Pa, a limiting viscosity of 40.6 × 10⁻³ Pa·s, and a density of 1162 kg/m³.

Solution

The same general equation used in Example 5.5 applies, i.e.,

$$-w\dot{m} = e_f \dot{m} = \left(\frac{4fL}{D}\right)\left(\frac{V^2}{2}\right)\rho Q = \frac{32 f L \rho Q^3}{\pi^2 D^5}$$

where now the friction factor depends on both the Reynolds number ($N_{Re} = 4Q\rho/\pi D\mu_\infty = 6.42 \times 10^3$) and the Hedstrom number ($N_{He} = D^2\rho\tau_o/\mu_\infty^2 = 7.78 \times 10^3$). Using Equations (5.90) to (5.93), we find: $a = -1.574$, $m = 7.93$, $f_L = 0.2045$, $f_T = 0.00491$, $f = 0.205$. Thus, the above equation gives $-w\dot{m} = 3860$ kW (5170 hp).

5.5.6 Fitting Losses

The loss coefficient for pipe is related to the Fanning friction factor by

$$K_f = \frac{4fL}{D} \tag{5.93}$$

For other "loss elements" (e.g., fittings, valves, constrictions, expansions, contractions, etc.), a variety of methods exist for determining K_f. These include the "constant K_f" (a rough estimate), the $(L/D)_{eq}$ method (a better, but limited, estimate), the Crane (1991) method (a good estimate for fully turbulent flow), and the 2-K (Hooper, 1981, 1988) and 3-K (Darby, 2001) methods. The latter two methods provide good values over the greatest range of fitting sizes and Reynolds numbers, and are presented below, with the 3-K method being the most accurate.

5.5.6.1 2-K (Hooper) Method

Hooper (1981, 1988) correlated loss coefficients for a variety of fittings as a function of Reynolds number and diameter by the equation

$$e_f = \frac{K_f V^2}{2} \quad \text{where} \quad K_f = \frac{K_1}{N_{Re}} + K_\infty \left(1 + \frac{1}{ID_{in}}\right) \tag{5.94}$$

where ID_{in} is the internal diameter in inches of the pipe that contains the fitting. This correlation is valid over a wide range of Reynolds numbers and a limited range of fitting sizes. Tables 5.5 and 5.6 give values of K_1 and K_∞ for various contractions and expansions and for entrance and exit losses.

5.5.6.2 3-K Method

Although the 2-K method applies over a wide range of Reynolds numbers, the scaling term (e.g., 1/ID) works best for relatively small fittings. A more accurate scaling law based on data for a wider range of valve and fitting sizes from a variety of sources has been given by Darby (1991, 2001) based on data for various valves, tees, and elbows and is represented by the 3-K equation

$$K_f = \frac{K_1}{N_{Re}} + K_i \left(1 + \frac{K_d}{D_{n.in}^{0.3}}\right) \tag{5.95}$$

where $D_{n.in}$ is the pipe nominal diameter in inches. Values of the 3 K's (K_1, K_i, and K_d are given in Table 5.7 for various valves, tees, and elbows. (For comparison, typical values of $(L/D)_{eq}$ for these fittings are also included.) All of the K's are dimensionless except for K_d, which has units of in.$^{0.3}$. If units of millimeters are used for the diameter, the values of K_d in Table 5.7 should be multiplied by 2.64. This method is based on a wide range of observations from different sources (Darby, 2001), and is recommended as providing the best values over the widest range of Reynolds numbers and fitting sizes.

The definition of K_f (i.e., $K_f = 2e_f/V^2$) involves the kinetic energy of the fluid, $V^2/2$. For sections in which the flow area changes (e.g., pipe entrance, exit, expansion, contraction, etc.), the entering and leaving velocities will be different. Since the value of the velocity used with the definition of K_f is arbitrary, it is very important to know which velocity is implied when values of the loss coefficient are used from various sources (e.g., handbooks, manuals, texts, etc.). In most cases it is the largest velocity, through the smallest flow area, but the values in Table 5.6 for contractions and expansions are all used with the upstream velocity.

5.5.6.3 Non-Newtonian Fluids

There are insufficient data in the literature to enable reliable correlation or prediction of friction loss in valves and fittings for non-Newtonian fluids. As a first approximation, however, it may be assumed that a correlation analogous to the 2-K or 3-K method should apply to non-Newtonian

TABLE 5.5
Loss Coefficients for Expansions and Contractions

K_t to be used with upstream velocity head, $V_1^2/2$. $\beta = d/D$

Contraction

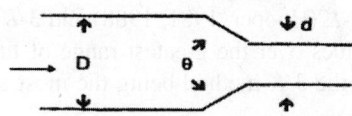

$\theta < 45°$

$N_{Re,1} < 2500$:

$$K_t = 1.6\left[1.2 + \frac{160}{N_{Re,1}}\right]\left[\frac{1}{\beta^4} - 1\right]\sin\frac{\theta}{2}$$

$N_{Re,1} > 2500$:

$$K_t = 1.6[0.6 + 1.92f_1]\left[\frac{1-\beta^2}{\beta^4}\right]\sin\frac{\theta}{2}$$

$\theta > 45°$

$N_{Re,1} < 2500$:

$$K_t = \left[1.2 + \frac{160}{N_{Re,1}}\right]\left[\frac{1}{\beta^4} - 1\right]\left[\sin\frac{\theta}{2}\right]^{1/2}$$

$N_{Re,1} > 2500$:

$$K_t = [0.6 + 1.92f_1]\left[\frac{1-\beta^2}{\beta^4}\right]\left[\sin\frac{\theta}{2}\right]^{1/2}$$

Expansion

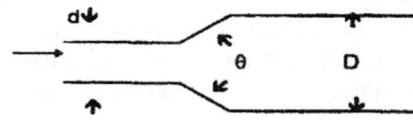

$\theta < 45°$

$N_{Re,1} < 4000$:

$$K_t = 5.2(1-\beta^4)\sin\frac{\theta}{2}$$

$N_{Re,1} > 4000$:

$$K_t = 2.6(1+3.2f_1)(1-\beta^2)^2\sin\frac{\theta}{2}$$

$\theta > 45°$

$N_{Re,1} < 4000$:

$$K_t = 2(1-\beta^4)$$

$N_{Re,1} > 4000$:

$$K_{t1} = (1+3.2f_1)(1-\beta^2)^2$$

Source: Hooper, W. B., "Calculate Head Loss Caused by Change in Pipe Size," *Chem. Eng.*, p. 89, Nov. 7 (1988).

fluids if the (Newtonian) Reynolds number is replaced by a single corresponding dimensionless group that adequately characterizes the influence of the non-Newtonian properties. For the power law model, the appropriate Reynolds number was derived for pipe flow:

$$N_{RePL} = \frac{2^{7-3n}\rho Q^{2-n}}{m\pi^{2-n}D^{4-5n}}\left[\frac{n}{3n+1}\right]^n \tag{5.96}$$

For the Bingham plastic, a corresponding expression for the Reynolds number based on the ratio of inertial to viscous momentum flux is

Fluid Flow

TABLE 5.6
2-K Constants for Entrance and Exit Losses

$$K_1 = K_1/N_{Re} + K_\infty$$

Entrance
Inward projecting (Borda)
$K_1 = 160$, $K_\infty = 1.0$

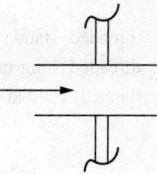

Flush (rounded)
$K_1 = 160$

r/d	K_∞
0.0 (sharp)	0.5
0.02	0.28
0.04	0.24
0.06	0.15
0.10	0.09
0.15 & up	0.04

For pipe exit:
$K_\infty = 1.0$ for all geometries
$K_1 = 0.0$

Orifice:

$$K_\infty = \frac{2.91}{\beta^4}(1-\beta^2)(1-\beta^4) = \frac{(1-\beta^2)(1-\beta^4)}{C_o^2 \beta^4}$$

$\beta = D_o/D_p$
$K_1 = 0.0$

Source: Hooper, W. B., "The 2-K Method Predicts Head Losses in Pipe Fittings," *Chem. Eng.*, p. 97, Aug. 24 (1981).

$$N_{ReBP} = \frac{4Q\rho}{\pi D \mu_\infty (1 + \pi D^3 \tau_o / 24 Q \mu_\infty)} = \frac{N_{Re}}{1 + N_{He}/(6N_{Re})} \tag{5.97}$$

5.5.7 Pipe Flow Analysis

Pipe flow analyses may require the determination of either the driving force, flow rate, or pipe diameter. The procedures for these analyses for Newtonian and non-Newtonian fluids are similar, utilizing the engineering Bernoulli equation and the dimensionless expressions for the Fanning friction factor. For these purposes, the engineering Bernoulli equation can be written as

$$DF = \sum_i e_{fi} + \frac{1}{2}(\alpha_2 V_2^2 - \alpha_1 V_1^2) \tag{5.98}$$

where

$$DF = -\left(\frac{\Delta P}{\rho} + g\Delta z + w\right) \tag{5.99}$$

TABLE 5.7
3-K Constants for Loss Coefficients for Valves and Fittings
$$K_f = K_1/N_{Re} + K_i(1 + K_d/D_{in.}^{0.3})$$

Fitting					$(L/D)_{eq}$	K_1	K_i	K_d
Elbows	90°	threaded, standard		$(r/D = 1)$	30	800	0.14	4.0
		threaded, long radius		$(r/D = 1.5)$	16	800	0.071	4.2
		flanged, welded, bends		$(r/D = 1)$	20	800	0.091	4.0
				$(r/D = 2)$	12	800	0.056	3.9
				$(r/D = 4)$	14	800	0.066	3.9
				$(r/D = 6)$	17	800	0.075	4.2
		mitered	1 weld	(90°)	60	1000	0.27	4.0
			2 welds	(45°)	15	800	0.068	4.1
			3 welds	(30°)	8	800	0.035	4.2
	45°	threaded standard		$(r/D = 1)$	16	500	0.071	4.2
		long radius		$(r/D = 1.5)$		500	0.052	4.0
		mitered						
			1 weld	(45°)	15	500	0.086	4.0
			2 welds	(22.5°)	6	500	0.052	4.0
	180°	threaded, close-return bend		$(r/D = 1)$	50	1000	0.23	4.0
		flanged		$(r/D = 1)$		1000	0.12	4.0
		all		$(r/D = 1.5)$		1000	0.10	4.0
Tees	Through-branch							
	(as elbow)	threaded		$(r/D = 1)$	60	500	0.274	4.0
				$(r/D = 1.5)$		800	0.14	4.0
		flanged		$(r/D = 1)$	20	800	0.28	4.0
		stub-in branch				1000	0.34	4.0
	Run-through	threaded		$(r/D = 1)$	20	200	0.091	4.0
		flanged		$(r/D = 1)$		150	0.05	4.0
		stub-in branch				100	0	0
Valves	Angle valve	45°	full line size, $\beta = 1$		55	950	0.25	4.0
		90°	full line size, $\beta = 1$		150	1000	0.69	4.0
	Globe valve	standard, $\beta = 1$			340	1500	1.70	3.6
	Plug valve	branch flow			90	500	0.41	4.0
		straight through			18	300	0.084	3.9
		three-way (flow through)			30	300	0.14	4.0
	Gate valve	standard, $\beta = 1$			8	300	0.037	3.9
	Ball valve	standard, $\beta = 1$			3	300	0.017	3.5
	Diaphragm	dam type				1000	0.69	4.9
	Swing check	$V_{min} = 35[\rho \text{ (lb}_m/\text{ft}^3)]^{-1/2}$			100	1500	0.46	4.0
	Lift check	$V_{min} = 40[\rho \text{ (lb}_m/\text{ft}^3)]^{-1/2}$			600	2000	2.85	3.8

Source: Darby, R., *Chemical Engineering Fluid Mechanics*, 2nd ed., Marcel Dekker, New York (2001).

is the net "driving force" for moving the fluid, and

$$\sum_i e_{fi} = \frac{1}{2}\sum_i (V^2 K_f)_i = \frac{8Q^2}{\pi^2}\sum_i \left(\frac{K_f}{D^4}\right)_i \quad (5.100)$$

where the summation is over each fitting and segment of pipe (of diameter D_i) in the system. The loss coefficients for the pipe and fittings are given by the Fanning friction factor in the 3-K formula:

Fluid Flow

$$K_{pipe} = \frac{4fL}{D}, \; K_{fit} = \frac{K_1}{N_{Re}} + K_i\left(1 + \frac{K_d}{D_{n,in}^{0.3}}\right) \tag{5.101}$$

In many applications the kinetic energy terms in Equation (5.98) are negligible (or cancel out), although this should be verified for each situation. The loss coefficients are a function of the Reynolds number, which for Newtonian fluids is

$$N_{Re} = \frac{4Q\rho}{\pi D \mu} \tag{5.102}$$

as well as the relative roughness if Newtonian, or the flow index (n) if power law, or the Hedstrom number if Bingham plastic. The Bernoulli equation can thus be written as

$$DF = \frac{8Q^2}{\pi^2}\left[\sum_i \left(\frac{K_f}{D^4}\right)_i + \frac{\alpha_2}{D_2^4} - \frac{\alpha_1}{D_1^4}\right] \tag{5.103}$$

where the α's are the kinetic energy correction factors at the upstream (1) and downstream (2) points ($\alpha = 2$ for laminar flow and $\alpha = 1$ for turbulent flow of a Newtonian fluid). This equation can be solved directly for either DF, or Q, or D provided all the loss coefficients (K_f's) are known. If DF is to be determined, the Reynolds number and hence all of the K_f's can be calculated directly, so the procedure is straightforward. If Q or D are to be determined, it is necessary to start with an estimate of each of the K_f's, use this to calculate Q or D, and use the result to calculate the Reynolds number and hence to verify (or revise) the K_f values. For the unknown diameter case, only one diameter $D = D_i$ can be determined. The procedure is the same for non-Newtonian as for Newtonian fluids.

5.5.8 Economical Diameter

When installing a piping system, both the "best" pipe and the "best" pump must usually be selected to deliver a prescribed flow. The term "best" in this case refers to that combination of pipe and pump that will minimize the total system cost, including both the capital cost of the pipe and pump stations as well as operating (energy) cost, i.e.,

Capital cost of pipe (CCP)
Capital cost of pump stations (CCPS)
Energy cost to power pumps (EC)

Since energy cost is continuous and capital costs are "one time," it is assumed that the capital costs are amortized over a period of Y years, the "economic lifetime" of the pipeline. The reciprocal of this ($X = 1/Y$) is the fraction of the total capital cost charged off per year. The capital cost and the energy cost per year combine to give the total cost. (There are other costs, such as maintenance, etc., but these are minor and will not materially influence the result.)

Data on the cost of typical pipelines of various sizes can be represented by (Darby and Melson, 1982)

$$CCP = aD_{ft}^p L \tag{5.104}$$

where D_{ft} is the pipe ID in feet, and the parameters a and p depend upon the grade of steel in the pipe, as shown in Table 5.8. Likewise, the capital cost of (installed) pump stations (for 370 kW (500 hp) and over) is a linear function of the pump power (see Figure 5.8):

TABLE 5.8
Cost of Pipe (1980 $)

	Pipe Cost Correlation Parameters[b]				
Pipe Grade[a]	ANSI 300#	ANSI 400#	ANSI 600#	ANSI 900#	ANSI 1500#
a	23.1	23.9	30.0	38.1	55.3
p	1.16	1.22	1.31	1.35	1.39

[a] ANSI grades correspond roughly to Schedules 20, 30, 40, 80, and 120, respectively.
[b] Pipe cost: $/ft = $a(ID_{ft})^p$, $/m = 3.28 \times a(0.3048\, ID_m)^p$

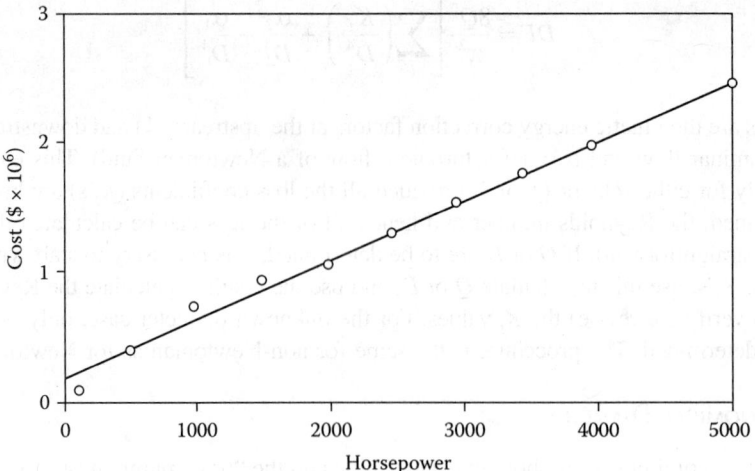

FIGURE 5.8 Cost of pump stations (1980 $).

$$CCPS = A + BHP/\eta_e \tag{5.105}$$

where $A = \$172,800$, $B = \$605/\text{kW}$ ($\$450.8/\text{hp}$) (in 1980 $), and HP/η_e is the power rating of the pump.

The energy cost is determined from the pumping power requirement, as calculated from the Bernoulli equation:

$$HP = -w\dot{m}\left[\frac{2fLV^2}{D} + \frac{\Delta\Phi}{\rho}\right] = \frac{32 fL\dot{m}^3}{\pi^2\rho^2 D^5} + \dot{m}\frac{\Delta\Phi}{\rho} \tag{5.106}$$

and the total energy cost per year is

$$EC = \frac{CHP}{\eta_e} \tag{5.107}$$

where C is the unit energy cost (e.g., \$/hp·yr, \$/kW·h, etc.) and η_e is the pump efficiency. The total annual cost of the pipeline is the sum of the capital and energy costs. The pipe diameter that minimizes this total cost (i.e., D_{ec}) is given by

$$D_{ec} = \left[\left(\frac{B+CY}{ap\eta_e}\right)\left(\frac{160 f \dot{m}^3}{\pi^2 \rho^2}\right)\right]^{1/(p+5)} \quad (5.108)$$

where $Y = 1/X$. Since this equation is implicit (i.e., f depends on D_{ec}), it is best solved by iteration, starting with an assumed value for f, calculating the corresponding value of D_{ec}, and using this to evaluate N_{Re} and hence verify (revise) f. The same procedure can be used for non-Newtonian as for Newtonian fluids. Note that the cost data in Figure 5.8 are circa 1980. These data can be used with an appropriate inflation factor to apply to current costs, or the energy cost as of 1980 can be used with these values, since all of the cost factors in Equation (5.108) appear as a ratio, so that any inflation factor will cancel out when applied to both capital and energy costs.

5.5.9 Noncircular Conduits

Application of the conservation of momentum to the steady flow of a fluid in a uniform conduit with any arbitrary cross-sectional shape results in an expression for the momentum due to wall drag (or the equivalent frictional energy dissipation) of the form

$$e_f = \frac{\tau_w}{\rho}\left(\frac{4L}{D_h}\right) \quad (5.109)$$

where

$$D_h = 4\frac{A}{W_p} \quad (5.110)$$

Here, A is the cross-sectional area of the conduit, and W_p is the "wetted perimeter" (i.e., the length of the solid boundary in the cross section that is wetted by the fluid). For a circular pipe filled with liquid, $D_h = D$, the pipe diameter. Thus, D_h is the geometric parameter that is equivalent to the diameter of a circular pipe and is called the *hydraulic diameter.*

This result can be used to apply the previous equations for circular pipes to conduits of any other shape, by replacing D in the appropriate equation with D_h for the noncircular conduit. This gives excellent results for turbulent flows, for which the boundary layer is generally thin relative to the flow area dimensions, since the wall resistance (i.e., friction loss) is confined to a region very near the wall and, consequently, is not very sensitive to the shape of the cross section.

For laminar flows, the "laminar sublayer" fills the entire cross section. Thus, the effect of wall drag is influenced to a greater extent by the shape of the cross section. For many noncircular conduits with laminar flows, either theoretical analyses (similar to the Hagen-Poiseuille equation) or numerical analyses for the pressure-flow relation have been conducted. The results can be expressed in dimensionless form as

$$fN_{Reh} = K \quad (5.111)$$

similar to the dimensionless form of the Hagen-Poiseuille equation (5.64), for which $K = 16$. Table 5.9 shows the results in terms of the value of K for a variety of noncircular cross sections. It is

TABLE 5.9
Laminar Flow Factors for Noncircular Conduits

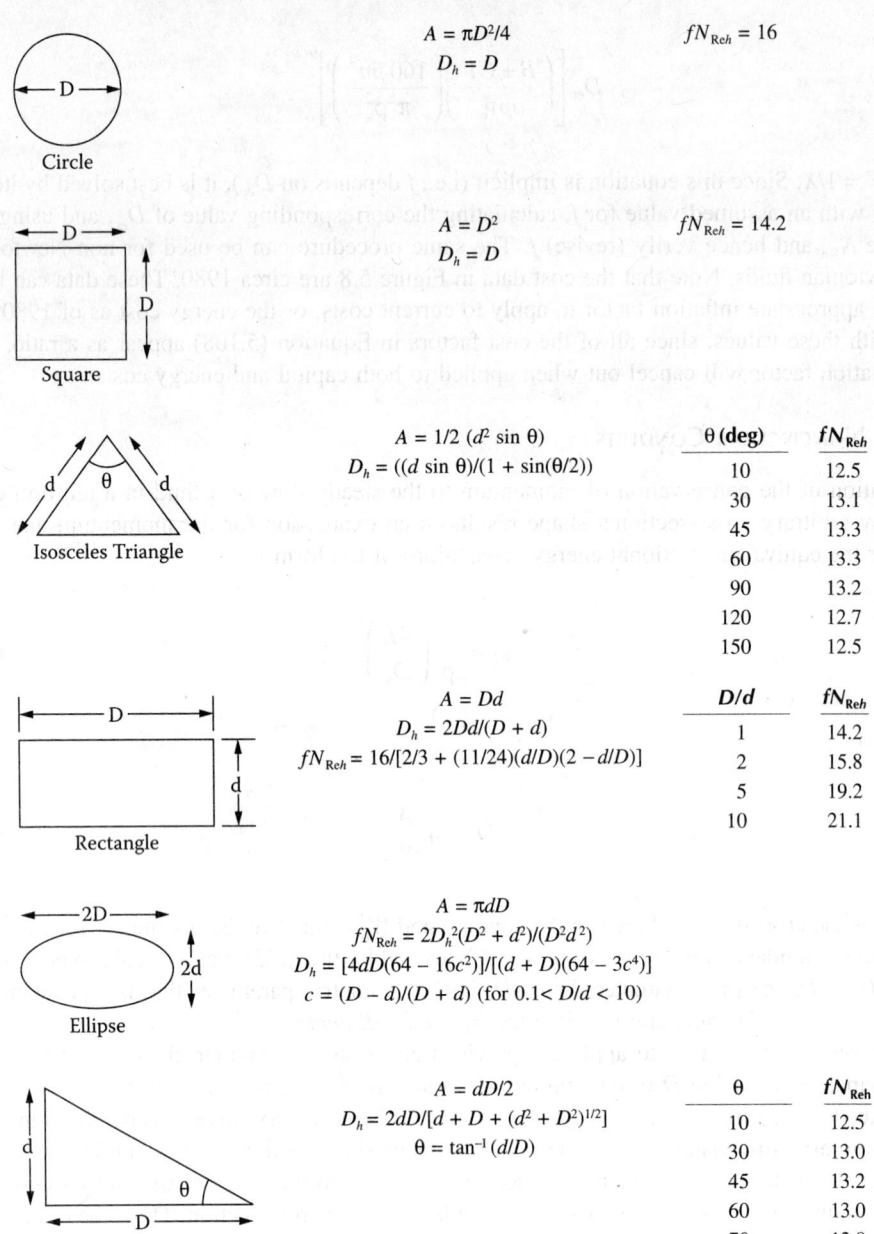

Circle	$A = \pi D^2/4$, $D_h = D$	$fN_{Reh} = 16$	
Square	$A = D^2$, $D_h = D$	$fN_{Reh} = 14.2$	

Isosceles Triangle: $A = 1/2 (d^2 \sin \theta)$, $D_h = ((d \sin \theta)/(1 + \sin(\theta/2)))$

θ (deg)	fN_{Reh}
10	12.5
30	13.1
45	13.3
60	13.3
90	13.2
120	12.7
150	12.5

Rectangle: $A = Dd$, $D_h = 2Dd/(D + d)$, $fN_{Reh} = 16/[2/3 + (11/24)(d/D)(2 - d/D)]$

D/d	fN_{Reh}
1	14.2
2	15.8
5	19.2
10	21.1

Ellipse: $A = \pi dD$, $fN_{Reh} = 2D_h^2(D^2 + d^2)/(D^2 d^2)$, $D_h = [4dD(64 - 16c^2)]/[(d + D)(64 - 3c^4)]$, $c = (D - d)/(D + d)$ (for $0.1 < D/d < 10$)

Right Triangle: $A = dD/2$, $D_h = 2dD/[d + D + (d^2 + D^2)^{1/2}]$, $\theta = \tan^{-1}(d/D)$

θ	fN_{Reh}
10	12.5
30	13.0
45	13.2
60	13.0
70	12.8
90	12.0

important to note that the proper definitions of the fanning friction factor (f) and the hydraulic Reynolds number be used in these expressions, i.e.,

$$f = \frac{e_f}{(4L/D_h)(V^2/2)} \quad (5.112)$$

Fluid Flow

$$N_{Reh} = \frac{D_h V \rho}{\mu} \qquad (5.113)$$

The velocity in these equations can be replaced by Q/A (where A can be expressed in terms of D_h for the geometry of interest). However, *do not* use $V = 4Q/\pi D^2$, since this is for circular pipes only.

For turbulent flows in noncircular conduits, the Moody diagram or Churchill equation can be used with good results, if the relative roughness is taken as ε/D_h.

5.5.10 Turbulent Drag Reduction

A phenomenon known as *turbulent drag reduction* (or the *Toms effect*) occurs in solutions of a variety of very high polymers. As much as 85% *less* energy may be required to pump solutions of certain high polymers at concentrations of 100 ppm or less through a pipe, when compared with the energy required to pump the solvent alone at the same flow rate through the same pipe. Although the mechanism is still under debate, the model of Darby and Chang (1984) as modified by Darby and Pivsa-Art (1991), based on energy storage by the elastic properties of the polymers, shows that, for smooth tubes, the friction factor vs. Reynolds number relation for Newtonian fluids (e.g., the Colebrook or Churchill equation) may also be used for drag-reducing flows, provided that: (a) the Reynolds number is defined using the properties (e.g., viscosity) of the Newtonian solvent, and (b) the Fanning friction factor is modified as follows:

$$f_p = \frac{f_s}{\sqrt{1 + N_{De}^2}} \qquad (5.114)$$

Here, f_s is the Fanning friction factor for the Newtonian solvent, as predicted using the (Newtonian) Reynolds number; f_p is a "generalized" Fanning friction factor that applies to polymer (e.g., drag reducing) solutions as well as Newtonian fluids; and N_{De} is the dimensionless Deborah number, which reflects the fluid viscoelastic properties and accounts for the storage of energy by the elastic deformations (for Newtonian fluids, $N_{De} = 0$). This correlation collapses data for a wide range of polymers at various concentrations in a wide range of tube sizes all on the same line for f_p vs. N_{Re}, which can be represented by the classic Colebrook equation (for Newtonian fluids in smooth tubes).

The complete expression for N_{De} is given by Darby and Pivsa-Art (1991) in terms of the viscoelastic properties of the fluid. These properties are complex, and cannot be predicted accurately from first principles or other basic properties, and must be measured for each polymer solution. However, a simplified expression has been given by Darby and Pivsa-Art (1991), in which these rheological parameters are contained within two "constants," k_1 and k_2, which depend only on the specific polymer solution and its concentration:

$$N_{De} = k_2 \left(\frac{8 \mu_s N_{Res}}{\rho D^2} \right)^{k_1} N_{Res}^{0.34} \qquad (5.115)$$

Values of k_1 and k_2 for various polymer/tube systems are given in Table 5.10. (Values of k_1 and k_2 can be determined for a given polymer solution from laboratory measurements of pressure drop in smooth tubes at two flow rates in the turbulent range.) These values can be used with the model to predict friction loss for that solution at any Reynolds number in any size pipe. If the Colebrook equation for smooth tubes is used, the appropriate generalized expression for the friction factor is

TABLE 5.10
Parameters for k_1 and k_2 for Various Polymer Solutions

Polymer	Conc. (mg/kg)	Diam. (cm)	k_1	k_2 (s^{k1})	Reference
Guar gum	20	1.27	0.05	0.009	Wang (1972)
(Jaguar A-20-D)	50		0.06	0.014	
	200		0.07	0.022	
	500		0.10	0.029	
	1000		0.16	0.028	
Guar gum	30		0.05	0.008	White (1966)
	60		0.06	0.010	
	240		0.08	0.016	
	480		0.11	0.018	
Polyacrylamide	100	0.176	0.093	0.0342	Darby (1991)
(Separan AP-30)	250	to	0.095	0.0293	
Fresh	500	1.021	0.105	0.0244	
(Separan AP-30)	100		0.088	0.0431	
Degraded	250		0.095	0.0360	
	500		0.103	0.0280	
(AP-273)	10	1.090	0.12	0.0420	White (1975)
(PAM E198)	10	0.945	0.21	0.0074	Virk (1970)
	280			0.0078	
(PAA)	300	2.0 & 3.0	0.40	0.0050	Hoffmann (1975)
	700		0.53	0.0049	
(ET-597)	125	0.69,	0.47	0.00037	Astarita (1969)
	250	1.1, &	0.39	0.0013	
	500	2.05	0.30	0.0061	
Hydroxyethyl cellulose	100	2.54	0.10	0.0074	Wang (1972)
(OP-100M)	200		0.16	0.0072	
	500		0.24	0.0068	
	1000		0.35	0.0063	
(HEC)	2860	4.8, 1.1 & 2.05	0.02	0.0310	Savins (1969)
Polyethylene oxide	10	5.08	0.22	0.017	Goren (1967)
(WSR 301)	20		0.21	0.016	
	50		0.19	0.014	
(W205)	10	0.945	0.31	0.0022	Virk (1970)
	105		0.26	0.0080	
Xanthan gum	1000	0.52	0.02	0.046	Bewersdorff (1988)
(Rhodopol 23)					

Source: Darby, R., and Pivsa-Art, S., *Canad. J. Chem. Eng.*, 69, 1395 (1991).

$$f_p = \frac{0.41}{[\ln(N_{\text{Re}s}/7)]^2} \frac{1}{(1+N_{De}^2)^{1/2}} \tag{5.116}$$

5.5.11 Compressible Flows

For gases, the dependence of density upon pressure and temperature has a significant effect on the flow behavior when the change in pressure is significant (e.g., 20% or more) relative to the absolute pressure in the system. At low velocities (relative to the speed of sound), the relative change in pressure and associated effects are often small, and the assumption of constant density (evaluated at an average pressure) is usually reasonable. However, when the gas velocity approaches the speed of sound, the effects of compressibility become the most significant.

Under most conditions, the gas properties can be described adequately by the ideal gas law, i.e.,

$$\rho = \frac{PM}{RT} \tag{5.117}$$

In general, the further the temperature and pressure are from the critical temperature and pressure of the gas (i.e., the higher the temperature and lower the pressure), the better the properties are represented by the ideal gas law.

For a steady flow of an incompressible fluid in a uniform pipe, the only property that varies along the pipe is pressure. However, for a compressible fluid when the pressure varies (i.e., drops), the density also drops, which means that the velocity must increase for a given mass flow. The kinetic energy thus increases, which results in a decrease in the internal energy and the temperature. This process is usually described as adiabatic, or locally isentropic, with the effect of friction loss included separately. A limiting case is the isothermal condition, although special means are usually required to achieve constant temperature. Under *isothermal* conditions for an ideal gas,

$$\frac{P}{\rho} = constant = \frac{P_1}{\rho_1} = \frac{P_2}{\rho_2} \ldots \text{etc.} \tag{5.118}$$

whereas for *isentropic* conditions,

$$\frac{P}{\rho^k} = constant = \frac{P_1}{\rho_1^k} = \frac{P_2}{\rho_2^k} \ldots \text{etc.} \tag{5.119}$$

where $k = c_p/c_v$ is the "isentropic exponent." For an ideal gas, $c_p = c_v + R/M$, and for diatomic gases $k \approx 1.4$, whereas for triatomic and higher gases, $k \approx 1.3$. Table 5.11 lists some properties of various gases, including the isentropic exponents. As the downstream pressure drops, the velocity in the pipe increases until it reaches the speed of sound at the pipe exit. At this point, no further increase in velocity is possible by reducing pressure downstream, since pressure changes are transmitted at the speed of sound, and the flow is said to be "choked." The speed of sound in any medium is given by

$$c = \sqrt{\frac{K}{\rho}} \tag{5.120}$$

where K is the bulk modulus of the material. For gases,

$$K = \rho \left(\frac{\partial P}{\partial \rho}\right)_s = \rho k \left(\frac{\partial P}{\partial \rho}\right)_T \tag{5.121}$$

which for an ideal gas becomes

$$c = \sqrt{\frac{kP}{\rho}} = \sqrt{\frac{kRT}{M}} \tag{5.122}$$

TABLE 5.11
Properties of Gases

Name of Gas	Chemical Formula or Symbol	Approx. Molecular Weight (M)	Weight Density, Pounds per Cubic Foot (π)	Specific Gravity Relative to Air (S_g)	Individual Gas Constant (R)	Specific Heat at Room Temperature (Btu/Lb °F)		Heat Capacity per Cubic Foot		k Equal to C_p/C_v
						C_p^a	C_v^b	C_p	C_v	
Acetylene (ethyne)	C_2H_2	26.0	.0682	0.907	59.4	0.350	0.269	.0239	.0184	1.30
Air	—	29.0	.0752	1.000	53.3	0.241	0.172	.0181	.0129	1.40
Ammonia	NH_3	17.0	.0448	0.596	91.0	0.523	0.396	.0234	.0178	1.32
Argon	A	39.9	.1037	1.379	38.7	0.124	0.074	.0129	.0077	1.67
Butane	C_4H_{10}	58.1	.1554	2.067	26.5	0.395	0.356	.0614	.0553	1.11
Carbon dioxide	CO_2	44.0	.1150	1.529	35.1	0.205	0.158	.0236	.0181	1.30
Carbon monoxide	CO	28.0	.0727	0.967	55.2	0.243	0.173	.0177	.0126	1.40
Chlorine	Cl_2	70.9	.1869	2.486	21.8	0.115	0.086	.0215	.0162	1.33
Ethane	C_2H_6	30.0	.0789	1.049	51.5	0.386	0.316	.0305	.0250	1.22
Ethylene	C_2H_4	28.0	.0733	0.975	55.1	0.400	0.329	.0293	.0240	1.22
Helium	He	4.0	.01039	0.1381	386.3	1.250	0.754	.0130	.0078	1.66
Hydrogen chloride	HCl	36.5	.0954	1.268	42.4	0.191	0.135	.0182	.0129	1.41
Hydrogen	H_2	2.0	.00523	0.0695	766.8	3.420	2.426	.0179	.0127	1.41
Hydrogen sulphide	H_2S	34.1	.0895	1.190	45.2	0.243	0.187	.0217	.0167	1.30
Methane	CH_4	16.0	.0417	0.554	96.4	0.593	0.449	.0247	.0187	1.32
Methyl chloride	CH_3Cl	50.5	.1342	1.785	30.6	0.240	0.200	.0322	.0268	1.20
Natural gas	—	19.5	.0502	0.667	79.1	0.560	0.441	.0281	.0221	1.27
Nitric oxide	NO	30.0	.0780	1.037	51.5	0.231	0.165	.0180	.0129	1.40
Nitrogen	N_2	28.0	.0727	0.967	55.2	0.247	0.176	.0180	.0127	1.41
Nitrous oxide	N_2O	44.0	.1151	1.530	35.1	0.221	0.169	.0254	.0194	1.31
Oxygen	O_2	32.0	.0831	1.105	48.3	0.217	0.155	.0180	.0129	1.40
Propane	C_3H_4	44.1	.1175	1.562	35.0	0.393	0.342	.0462	.0402	1.15
Propene (propylene)	C_3H_6	42.1	.1091	1.451	36.8	0.358	0.314	.0391	.0343	1.14
Sulphur dioxide	SO_2	64.1	.1703	2.264	24.0	0.154	0.122	.0262	.0208	1.26

Note: Weight density values were obtained by multiplying density of air by specific gravity of gas. For values at 60°F, multiply by 1.0154.

Natural gas values are representative only. Exact characteristics require knowledge of specific constituents.

[a] c_p = Specific heat at constant pressure.
[b] c_v = Specific heat al constant volume.

Source: Molecular weight, specific gravity, individual gas constant, and specific heat values were abstracted from, or based on, data in Table 24 of Mark's *Standard Handbook for Mechanical Engineers* (7th edition)

5.5.11.1 Isothermal Pipe Flow

For isothermal flow in a pipe, the mass flux ($G = \dot{m}/A$) can be determined by integrating the differential form of the Bernoulli to give

$$G = \sqrt{P_1 \rho_1} \left[\frac{(1 - P_2^2 / P_1^2)}{4 fL / D - 2 \ln(P_2 / P_1)} \right]^{1/2} \tag{5.123}$$

Neglecting the logarithmic term in the denominator (which comes from the change in kinetic energy of the gas) gives the *Weymouth equation*. Furthermore, if the average density of the gas is used in the Weymouth equation, i.e.,

Fluid Flow

$$\bar{\rho} = \frac{(P_1 + P_2)M}{2RT} \quad \text{or} \quad \frac{M}{2RT} = \frac{\bar{\rho}}{P_1 + P_2} \tag{5.124}$$

the result is identical to the Bernoulli equation for an incompressible fluid in a straight, uniform pipe:

$$G = \left[\frac{\bar{\rho}(P_1 - P_2)}{2fL/D}\right]^{1/2} = \sqrt{P_1\bar{\rho}} \left[\frac{2(1 - P_1/P_2)}{4fL/D}\right]^{1/2} \tag{5.125}$$

The mass flux reaches a maximum (choked flow) when the velocity reaches the speed of sound. At this point, the mass flow rate is independent of any further decrease in the exit pressure but will still vary with changes in the upstream pressure. The equation for G is implicit, since the friction factor depends upon the Reynolds number, which depends on G. However, the Reynolds number under choked flow conditions is typically high enough that fully turbulent flow occurs, in which case the friction factor depends only on the relative pipe roughness, i.e.,

$$f = \frac{0.0625}{\left[\log\left(\frac{3.7}{\varepsilon/D}\right)\right]^2} \tag{5.126}$$

5.5.11.2 Adiabatic Flow

For adiabatic flow, the usual procedure is to assume "locally isentropic" conditions and include the friction loss to account for irreversibilities. In this case, the pressure, temperature, density, and velocity all change along the pipe. For an ideal gas under isentropic conditions,

$$T = T_1 \left(\frac{P}{P_1}\right)^{(k-1)/k} \tag{5.127}$$

and the expression for the mass flux is

$$G = \sqrt{P_1\rho_1} \left[\frac{2\left(\frac{k}{k+1}\right)\left(1 - \left(\frac{P_2}{P_1}\right)^{(k+1)/k}\right)}{\frac{4fL}{D} - \frac{2}{k}\ln\left(\frac{P_2}{P_1}\right)}\right]^{1/2} \tag{5.128}$$

If the system contains fittings as well as straight pipe, the term $(2fL/D)$ is replaced by $\Sigma(K_f/2)$, where ΣK_f represents the sum of all the loss coefficients in the system.

5.5.11.3 Choked Flow

The pressure at which choked flow occurs (P_2^*) is determined by the conditions for sonic flow. For isothermal flow, the critical (choked) mass flux is given by

$$G^* = \sqrt{P_1\rho_1} \left(\frac{P_2^*}{P_1}\right)^{1/2} \tag{5.129}$$

and P_2^* is determined from

$$\Sigma K_f = \left(\frac{P_1}{P_2^*}\right)^2 - 2\ln\left(\frac{P_1}{P_2^*}\right) - 1 \tag{5.130}$$

The pressure at the (inside of the) end of the pipe at which the flow becomes sonic (P_2^*) is a unique function of the upstream pressure (P_1) and the sum of the loss coefficients in the system (K_f). The equation is implicit for P_2^* and can be solved by iteration for given values of K_f and P_1. For adiabatic flow, the corresponding expressions are

$$G^* = \sqrt{P_1\rho_1}\left[k\left(\frac{P_2^*}{P_1}\right)^{(k+1)/k}\right]^{1/2} \tag{5.131}$$

and

$$\Sigma K_f = \frac{2}{(k+1)}\left[\left(\frac{P_1}{P_2^*}\right)^{(k+1)/k} - 1\right] - \frac{2}{k}\ln\left(\frac{P_1}{P_2^*}\right) \tag{5.132}$$

5.5.11.4 Isentropic Nozzle Flow

The mass flux for the adiabatic flow of an ideal gas flowing through a frictionless conduit or a constriction (such as an orifice, valve, etc.) for which $A_1 \gg A_2$ (i.e., $V_1 \ll V_2$) is given by

$$G = \sqrt{P_1\rho_1}\left[\left(\frac{2k}{k-1}\right)\left(\frac{P_2}{P_1}\right)^{2/k}\left\{1 - \left(\frac{P_2}{P_1}\right)^{(k-1)/k}\right\}\right]^{1/2} \tag{5.133}$$

The mass flow is a maximum when

$$\frac{P_2}{P_1} = \frac{P_2^*}{P_1} = \left(\frac{2}{k+1}\right)^{k/(k-1)} \tag{5.134}$$

which, for $k = 1.4$ (e.g., air), has a value of 0.528. Thus, if the downstream pressure is approximately one-half or less of the upstream pressure, the flow will be choked. The mass flow rate under adiabatic conditions is always somewhat greater than that under isothermal conditions, but the difference is normally <20%. In fact, for long piping systems ($L/D > 1000$), the difference is usually less than 5% (e.g., Holland, 1973).

5.5.11.5 Supersonic Flow

The conservation of mass, energy, and momentum equations can be written for the adiabatic flow of an ideal gas, and, when arranged in dimensionless form,

$$G^* = \sqrt{P_1\rho_1}\left(\frac{P_2^*}{P_1}\right)^{1/2} \tag{5.135}$$

Fluid Flow

shows that all flow variables can be expressed in terms of three dimensionless variables: the isentropic ratio, k; the Mach number, $N_{Ma} = V/c$; and the dimensionless pipe length, L/D. An inspection of these equations shows that, for $N_{Ma} < 1$, as the distance down the pipe (dL) increases, V also increases, but P, ρ, and T will decrease. However, for $N_{Ma} > 1$, just the opposite is true, i.e., V and T decrease but P and ρ will increase with distance down the pipe. That is, a flow that is initially subsonic will approach (as a limit) sonic flow as L increases, while an initially supersonic flow will also approach sonic flow. Thus all flows, regardless of their starting conditions, will tend toward the speed of sound as the gas progresses down a uniform pipe. The only way that a subsonic flow can be transformed into a supersonic flow is through a converging-diverging nozzle, where the speed of sound is reached at the nozzle throat. Supersonic flow relations are found in many fluid mechanics books (e.g., Hall, 1951; Shapiro, 1953).

Example 5.7

Oxygen must be fed to a reactor at a constant rate of 4.55 kg/s (10 lb$_m$/s) from a storage tank in which the pressure is 791 kPa (100 psig) at a temperature of 21°C (70°F). The pressure in the reactor fluctuates from 115 to 170 kPa (2 to 10 psig), so you want to insert a choke in the line to maintain a constant flow rate. If the choke is a 0.61-m (2-ft) length of tubing, what should the diameter of the tubing be?

Solution

For the flow to be independent of the downstream pressure, it must be choked. Assuming adiabatic flow, Equation (5.132) gives the total loss coefficient for the system when the downstream pressure is $P_2 \leq P_2^*$, where $P_{2max} = 170$ kPa (10 psig = 24.7 psia). The mass flux under choked conditions (G^*) is given by Equation (5.131), which depends on P_2^*/P_1 (where $P_1 = 791$ kPa) as well as the tube diameter. Thus the solution is iterative, in accordance with the following procedure:

1. Assume a value for (P_2^*/P_1) and solve Equation (5.132) for ΣK_f and Equation (5.126) for G^*.
2. Determine the tube diameter from $A = \dot{m}/G^*$, and $D = \sqrt{4A/\pi}$, where $\dot{m} = 4.55$ kg/s = 10 lb$_m$/s.
3. Calculate $N_{Re} = (\dot{m}/\pi D \mu)$, using $\mu = 0.02$ cP.
4. Assuming $\varepsilon = 0.0455$ mm (0.0018 in.), determine f (from, for example, the Churchill equation) and $K_f = 4fL/D$.
5. If other losses in the line are small relative to that of the tubing, then $K_f = \Sigma K_f$, which was determined in step 1.
6. Repeat steps 1 to 5 until the values of K_f agree. The proper tube diameter is then determined from step 2.

Following this procedure gives: (P_2^*/P_1) = 0.66, $P_2^* = 522$ Pa (75.7 psia), $D = 0.0485$ m (0.159 ft), $G^* = 2470$ kg/s·m² (504 lb$_m$/s·ft²), $N_{Re} = 5.96 \times 10^6$, and $f = 0.00485$.

5.6 PUMPS AND COMPRESSORS

5.6.1 Pump Types

There are a wide variety of pumps, but most can be broadly classified as either positive displacement or centrifugal. Positive-displacement pumps include piston, plunger, diaphragm, screw, gear, progressing cavity pumps, etc., and operate at a fixed volumetric flow rate under varying head or load requirements. Consequently, the discharge line from the pump should never be closed without allowing for recycle around the pump, or damage to the pump could result. Such pumps have limited flow capacity, but are capable of relatively high pressures, and are appropriate for high-pressure requirements, very viscous fluids, and applications where precisely controlled or metered flow rate is required.

Centrifugal pumps operate by the transfer of energy (or angular momentum) from a rotating impeller to the fluid, normally inside a casing. The kinetic energy of the fluid is increased by the angular momentum imparted by the high-speed impeller, and is then converted to pressure energy (or head) in a diverging (volute) area. The developed head depends upon the pump design and the size, shape, and speed of the impeller, and the flow rate is determined by the fluid resistance of the system in which the pump is installed. Centrifugal pumps operate at approximately constant head over a wide range of flow rates (within limits). They can be operated in a "closed off" condition (i.e., closed discharge line), since the liquid will recirculate within the pump without causing damage. However, this condition should be avoided, because energy dissipation within the pump can result in excessive heating of the fluid or the pump, and cavitation or unstable operation could occur with adverse consequences. Centrifugal pumps are appropriate for "ordinary" liquids (i.e., low to moderate viscosity) and low to moderate pressures over a wide variety of flow conditions.

5.6.2 Centrifugal Pump Characteristics

The hydraulic work ($-w$) delivered per unit mass of fluid, the pressure (ΔP) developed by the pump, and the pump head (H_p) are related by

$$-w = \frac{\Delta P}{\rho} = gH_p \tag{5.135}$$

The pump efficiency, η_e, is the hydraulic work delivered to the fluid divided by the work delivered to the pump by the motor ($-w_m$):

$$\eta_e = \frac{-w}{-w_m} \tag{5.136}$$

The efficiency of a pump depends upon the pump and impeller design, the size and speed of the impeller, and the operating flow rate, and it is measured by the pump manufacturer.

When selecting a pump for a particular application, the flow capacity and required pump head must be specified. Although a variety of different pumps might satisfy the given specifications, the best pump is generally the one that requires the least power to operate at the specified conditions. The horsepower required for the pump-driving motor is determined by

$$HP = -w_m \dot{m} = \frac{\Delta P Q}{\eta_e} = \frac{\rho g H_p Q}{\eta_e} \tag{5.137}$$

The motor power is also related to the motor torque (Γ) and speed (ω) by

$$HP = \Gamma \omega = \frac{\rho g H_p Q}{\eta_e} \tag{5.138}$$

The angular momentum balance provides a relation for the required motor torque:

$$\Gamma = \dot{m} \omega R_i^2 = \rho Q \omega R_i^2 \tag{5.139}$$

Fluid Flow

where R_i is the impeller radius. Eliminating Γ from Equations (5.138) and (5.139) shows that

$$H_p = \frac{\eta_e \omega^2 R_i^2}{g} \qquad (5.140)$$

That is, the pump head is determined primarily by the size and speed of the impeller and the pump efficiency, independent of the flow rate and fluid density. This is approximately correct for most centrifugal pumps over a wide range of flow rates. However, there is a limit to the flow that a given pump can handle, and the head will drop off significantly as the flow rate approaches this limit, which is typically near the maximum or "best" efficiency point (BEP). Pump characteristic curves are determined by the pump manufacturer. Figure 5.9 shows a typical set of curves for a given pump with various impeller diameters at a specific motor speed. (To convert H_p from feet to meters, multiply by 0.3048, and to convert Q from gallons per minute (gpm) to cubic meters per second (m³/s), multiply by 6.31×10^{-5}.) Curves for constant efficiency and NPSH (net positive suction head) are also shown on the plot. Operation at conditions on the right branch of the efficiency contours (to the right of the "maximum normal capacity" line) should be avoided, since this could result in unstable operation. Pump curves are normally determined by measurements using water. Although the pump head is independent of fluid density, the power is proportional to the fluid density so that the power (HP) curves are the motor horsepower required to pump water at 16°C (60°F) (multiply the values of HP by 0.746 to convert from horsepower to kilowatts). These must be corrected for density when operating with other fluids or temperatures. The curves labeled "minimum NPSH" refer to the cavitation characteristics of the pump (see below).

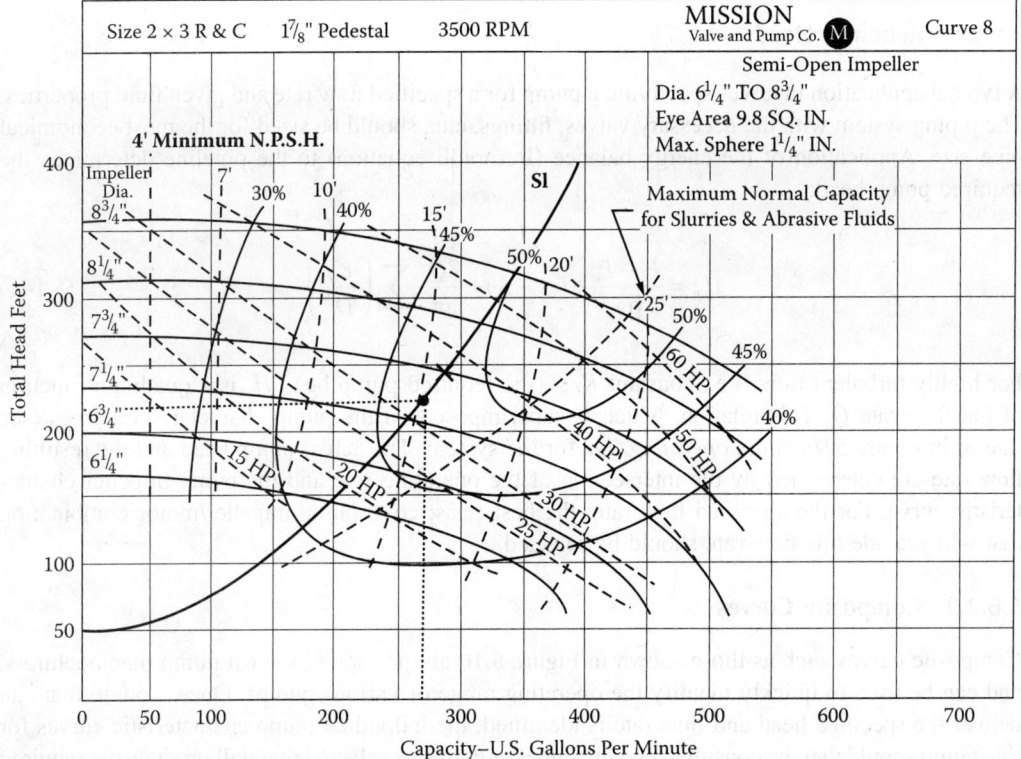

FIGURE 5.9 Typical centrifugal pump characteristic curves. (From TRW Mission Pump Brochure.)

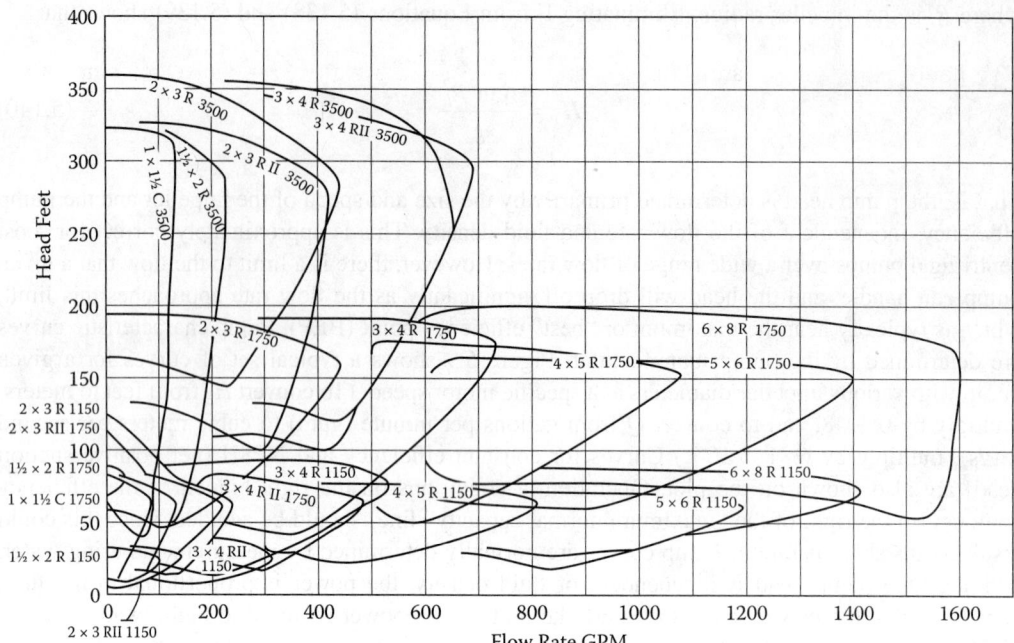

FIGURE 5.10 Typical centrifugal pump composite curves. (From TRW Mission Pump Brochure [manufacturer's catalog].)

5.6.2.1 Required Head

A typical application involves specifying a pump for a specified flow rate and given fluid properties. The piping system with the necessary valves, fittings, etc., should be sized for the most economical pipe size. Application of the energy balance (Bernoulli equation) to the pipeline determines the required pump head:

$$H_p = \frac{P_2 - P_1}{\rho g} + (z_2 - z_1) + \frac{8Q^2}{g\pi^2}\sum_i \left(\frac{K_f}{D^4}\right)_i \tag{5.141}$$

For highly turbulent flow (i.e., constant K_f's), the required pump head H_p is a quadratic function of the flow rate Q. This relation, which is superimposed on the pump characteristic curves (see line Sl in Figure 5.9), is the *operating line* for the system. The actual pump head and the resulting flow rate are determined by the intersection of the operating line and the pump impeller characteristic curve. For the specified flow rate, the best (least cost) pump/impeller/motor combination that will provide this flow rate should be selected.

5.6.2.2 Composite Curves

Composite curves such as those shown in Figure 5.10 are provided by most pump manufacturers, and can be used to quickly identify the operating range of various pumps. Once a pump that can deliver the specified head and flow rate is identified, the individual pump characteristic curves for that pump should then be consulted and the impeller diameter selected that will produce the required head or greater. This can be repeated for various pump, impeller, and speed combinations to determine the combination that results in the least power requirement.

Fluid Flow

5.6.2.3 Cavitation and NPSH

The fluid velocity in and around the pump impeller is much higher than the velocity entering or leaving the pump, and the pressure is lowest where the velocity is highest. If the minimum pressure in the pump is lower than the fluid vapor pressure, then vaporization or *cavitation* will occur. The pump head is the same if it is filled with gas or liquid, but the pump pressure, which is proportional to the fluid density, can be several orders of magnitude lower for a gas than a liquid at the same head (this condition is known as *vapor lock*). Cavitation can result in serious damage to a pump or impeller. To prevent cavitation, it is necessary that the pressure at all points within the pump be greater than the fluid vapor pressure. This will be true if the pump suction pressure is sufficiently higher than the fluid vapor pressure. The difference between the suction pressure and the minimum pressure in the pump depends upon the pump design, impeller size and speed, and flow rate. The minimum suction pressure relative to the fluid vapor pressure at which cavitation will not occur is called the *minimum required NPSH* (net positive suction head). The NPSH values are shown on the pump characteristic curves (e.g., Figure 5.9). The NPSH is almost independent of impeller diameter at low flow rates and increases with flow rate as well as with impeller diameter at higher flow rates. A distinction is sometimes made between the NPSH "required" to prevent cavitation (NPSHR) and the actual head (e.g., pressure) "available" at the pump suction (NPSHA), i.e., H_s. A pump will *not* cavitate if (NPSHA = H_s) > (NPSHR + Vapor Pressure Head), i.e.,

$$H_s = \frac{P_s}{\rho g} \geq NPSH + \frac{P_v}{\rho g} \tag{5.142}$$

where P_v is the liquid vapor pressure. The NPSH at the operating point determines where the pump can be installed in a piping system to avoid cavitation. If the pressure at the upstream entrance to the suction line is P_1, the maximum distance above this point that the pump can be located without cavitating (i.e., the maximum suction lift) is determined by Bernoulli's equation applied from P_1 to P_s:

$$h_{max} = \frac{P_1 - P_v}{\rho g} - NPSH + \frac{V_1^2 - V_s^2}{2g} - \frac{\sum (e_f)_s}{g} \tag{5.143}$$

where V_1 is the velocity entering the suction line, V_s is the velocity at the pump inlet (suction), and $\Sigma(e_f)_s$ is the total friction loss in the suction line from the upstream entrance (point 1) to the pump inlet, including all pipe, fittings, etc. If the maximum suction lift (h_{max}) is negative, the pump must be located below the upstream entrance to the suction line to prevent cavitation. It is best to be conservative when interpreting the suction head required to prevent cavitation. The minimum required NPSH on the pump curves is normally determined using water at 16°C (60°F) with the discharge line fully open. Even though a centrifugal pump will run with a closed discharge line and no bypass, the recirculation within the pump will increase the dissipative heating, which, in turn, increases the minimum required NPSH. This is especially true with high-efficiency pumps, which have close clearances.

Example 5.8

Determine the pump specifications required to transfer water at 30°C (86°F) from an open pond to an open tank located 61 m (200 ft) above the pond, at a rate of 0.0126 m³/s (200 gpm). The pipe is 3 in. sch 40 (ID = 77.93 mm) commercial steel (ε = 0.0457 mm [0.0018 in.]) and is 67 m (220 ft) long, and includes four flanged elbows. There is 3.05 m (10 ft) of horizontal pipe in both the pump suction and discharge lines. If the pump has the characteristics shown in Figure 5.9, determine: (a) the required pump head (in meters); (b) the proper impeller diameter to use; (c) the pump efficiency and NPSH at the operating point; (d) the power of the motor required to drive the pump; and (e) the maximum height above the pond that the pump can be located without cavitating. Fluid properties: $\rho = 996$ kg/m^3, $\mu = 0.798$ cP, $P_v = 31.824$ mm Hg.

Solution

The required pump head is calculated from Equation (5.141), with $P_1 = P_2$:

$$H_p = (z_2 - z_1) + \frac{8Q^2}{g\pi^2}\sum_i\left(\frac{K_f}{D^4}\right)_i$$

where

$$\sum_i K_f = 4K_{el} + \left(\frac{4fL}{D}\right)_{pipe}$$

and K_{el} is determined from the 3-K correlation:

$$K_{el} = \frac{800}{N_{Re}} + 0.091\left(1 + \frac{4.0}{D_{n,in}^{0.3}}\right)$$

The Reynolds number is

$$N_{Re} = \frac{4Q\rho}{\pi D \mu} = \frac{4(0.0126 \text{ m}^3/\text{s})(996 \text{ kg/m}^3)}{\pi(0.07793 \text{ m})(0.000798 \text{ Pa s})} = 2.57 \times 10^5$$

so that $K_{el} = 0.353$ (4 $K_{el} = 1.41$). From the Churchill equation, $f = 0.00475$, which gives

$$K_{pipe} = \frac{4fL}{D} = \frac{4(0.00475)(67 \text{ m})(1000 \text{ mm/m})}{77.93 \text{ mm}} = 16.36$$

Inserting these values into the Bernoulli equation [Equation (5.141)] results in $H_p = 67.3$ m (221 ft). Referring to Figure 5.9, this pump will deliver a head of 220 ft (67.1 m) (which is close enough) at a flow rate of 200 gpm (0.0126 m³/s) using an impeller with a diameter of 7-1/4 in. (184 mm). (Note that if the desired operating head and flow rate intersect at a point between the curves for available impeller diameters, the next larger impeller must be used, which will result in a pump head somewhat higher than necessary. This will result in a flow rate higher than desired, unless a valve is included in the line that can be adjusted to take the excess head as friction loss.)

From Figure 5.9 at the operating point, read $\eta_e = 0.42$ and NPSH = 11 ft (3.35 m). The required motor power is then given by

$$HP = \frac{\rho Q g H_p}{\eta_e} = \frac{(996 \text{ kg/m}^3)(0.0126 \text{ m}^3/\text{s})(9.81 \text{ m/s}^2)(67.3 \text{ m})}{0.42} = 19.8 \text{ kw (26.6 hp)}$$

The maximum suction lift for the pump that can be achieved without cavitation is given by Equation (5.144) as

Fluid Flow

$$h_{\max} = \frac{P_1 - P_v}{\rho g} - NPSH + \frac{V_1^2 - V_s^2}{2g} - \frac{\sum(e_f)_s}{g}$$

where $P_1 = 1$ atm, $V_1 = 0$, $V_s = 2.65$ m/s (8.68 ft/s), and $\Sigma(e_f)_s$ is the total friction loss in the suction line, which is assumed to include two elbows, 3.05 m (10 ft) of horizontal pipe, and $h_{\max}$ m of vertical pipe:

$$\sum(e_f)_s = \left[\frac{4f(L+h_{\max})_s}{D} + 4K_{el}\right]\frac{V_s^2}{2}$$

Note that $h_{\max}$ occurs on both sides of the equation. Solving this equation gives $h_{\max} = 5.55$ m (18.2 ft). That is, the pump will cavitate if it is located more that 5.5 m (18 ft) above the pond. Conservative practice would be to locate the pump closer than this to the pond.

5.6.2.4 Specific Speed

The flow rate, head, and impeller speed at the best efficiency point (BEP) of the pump characteristic are used to define the pump *specific speed*:

$$N_s = \frac{N\sqrt{Q}}{H_P^{3/4}} \quad \text{or} \quad \frac{\text{rpm}\sqrt{\text{gpm}}}{\text{ft}^{3/4}} \tag{5.144}$$

Although N_s is dimensionless, it is common practice to use selected mixed units for the parameters in N_s. In English units, these are N (rpm), Q (gpm), and H_P (ft). For N (rpm), Q (m³/s), and H_P (m), the corresponding value of N_s should be multiplied by 51.6 to give the same value of N_s determined using consistent English units. N_s represents roughly the ratio of the pump capacity to the head at the speed corresponding to the maximum efficient point (BEP), and depends primarily on the design of the pump and impeller. This can vary widely from almost pure radial flow to almost pure axial flow (e.g., a fan). Some examples of various types of impellers are shown in Figure 5.11. Radial-flow impellers have the highest head and lowest flow capacity (low N_s), whereas axial-flow impellers have a high flow capacity and low head characteristic (high N_s). Figure 5.12 indicates the range of flow rates and efficiencies for various impeller designs with corresponding values of N_s. The maximum efficiency occurs for mixed-flow impellers at a value of N_s of about 3000 (English units).

There are conflicting parameters in the proper design of a centrifugal pump. For example, the lower the suction velocity (V_s), the less the tendency to cavitate. This would dictate that the eye of the impeller should be as large as possible to minimize V_s. However, a large impeller eye means a large vane speed at the impeller inlet, which is destabilizing with respect to recirculation. Hence, it is advisable to design the impeller with the smallest eye diameter that is practicable.

5.6.3 Compressors

A compressor is basically a pump for a high-pressure compressible fluid. The compressibility properties of the fluid (gas) are normally significant when the pressure in a system changes by 30% or more. For small pressure changes, a fan or blower is an appropriate pump for a gas. Fan performance can be described using the incompressible flow equations, since the relative pressure change is small. Like pumps, compressors may be either positive displacement or centrifugal, the former being suitable for relatively high pressure ratios and low flow rates, while the latter are

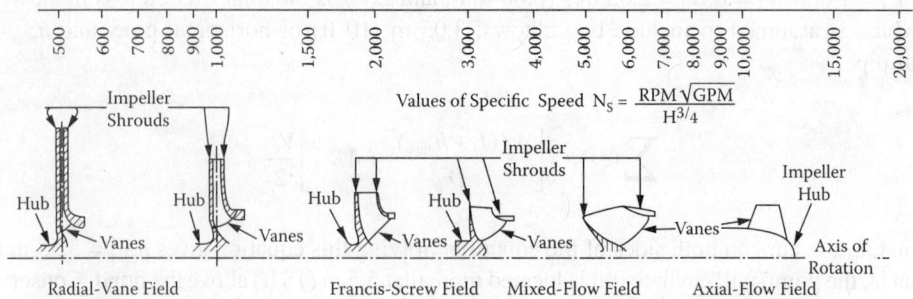

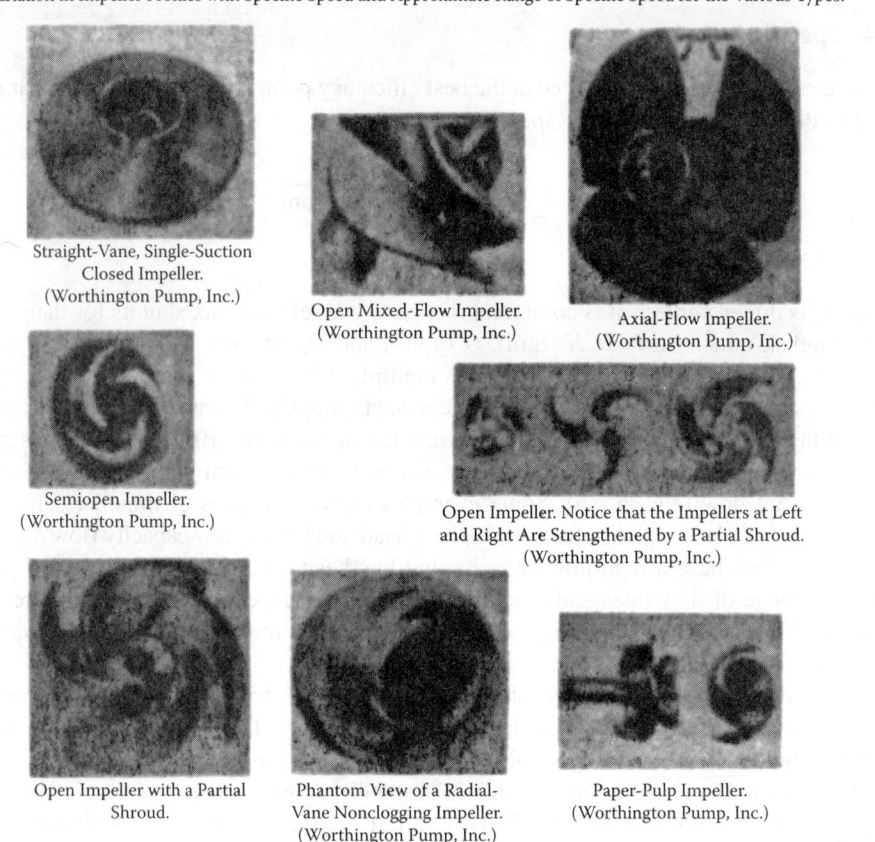

FIGURE 5.11 Impeller designs and specific speed characteristics. (From Karassik et al., 1976.)

Fluid Flow

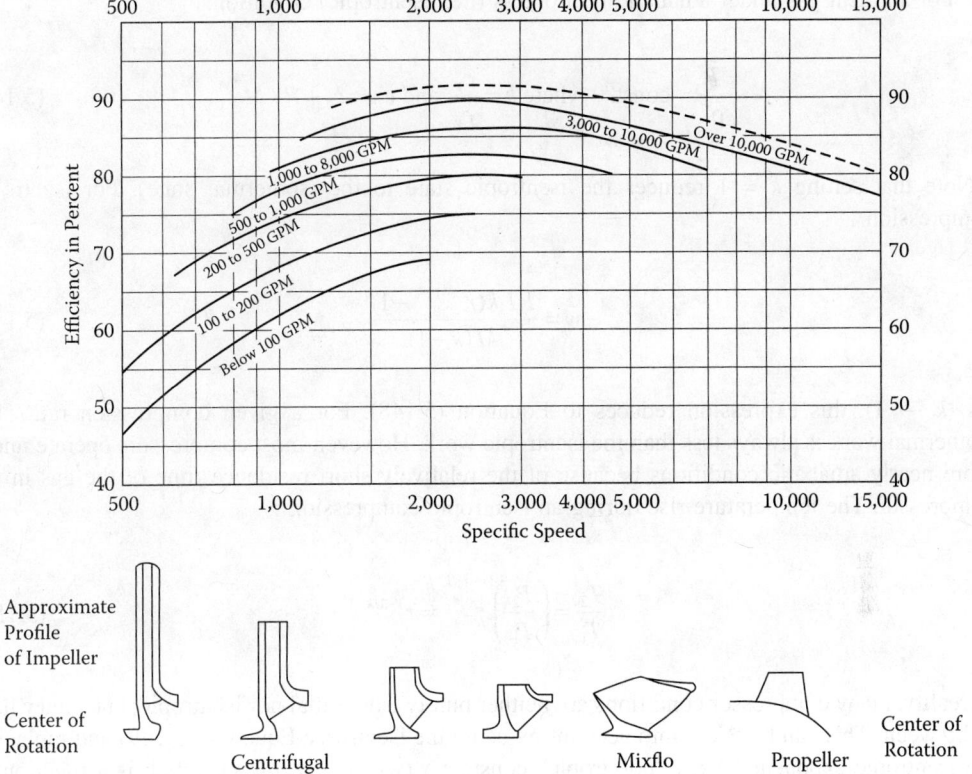

FIGURE 5.12 Correlation between impeller shape, specific speed, and efficiency. (From Drassik et al., 1976.)

more suited for higher flow rates but lower pressure ratios. The governing equations depend on whether the operation is isothermal or adiabatic. Expressions here assume that the ideal gas law applies, although this assumption can be modified by inclusion of a compressibility correction factor, as necessary. For an ideal (frictionless) compressor, the compression work is given by

$$-w = \int_{P_1}^{P_2} \frac{dP}{\rho} \tag{5.145}$$

and, for an ideal gas,

$$\rho = \frac{PM}{RT} \tag{5.146}$$

5.6.3.1 Isothermal, Isentropic, and Polytropic Operations

For isothermal conditions, Equation (5.145) becomes

$$-w = \frac{RT}{M} \ln r \tag{5.147}$$

where $r = P_2/P_1$ is the *compression ratio*.

For an ideal gas under adiabatic frictionless (i.e., isentropic) conditions,

$$\frac{P}{\rho^k} = \text{const}, \quad \text{where } k = \frac{c_p}{c_v} \text{ and } c_p = c_v + R/M \tag{5.148}$$

(Note that setting $k = 1$ reduces the isentropic state to the isothermal state). For isentropic compression,

$$-w = \frac{RT_1 k(r^{(k-1)/k} - 1)}{M(k-1)} \tag{5.149}$$

As ($k \to 1$), this expression reduces to Equation (5.148). For a given compression ratio, the isothermal work is always less than the isentropic work. However, most compressors operate under more nearly adiabatic conditions because of the relatively short residence time of the gas in the compressor. The temperature rise during an isentropic compression is

$$\frac{T_2}{T_1} = \left(\frac{P_2}{P_1}\right)^{(k-1)/k} = r^{k-1/k} \tag{5.150}$$

In reality, many compressor conditions are neither purely isothermal nor isentropic, but somewhere in between. This can be taken into account by using the isentropic Equation (5.149) and replacing the isentropic exponent k by a "polytropic" constant γ (where $1 < \gamma < k$), which is a function of the compressor design as well as the properties of the gas.

5.6.3.2 Staged Operation

Multiple compressor stages can be arranged in series to increase the overall compression ratio. To increase the overall efficiency, the gas is cooled between stages by "interstage coolers." As the number of stages increases, the total compression work for isentropic compression with interstage cooling to the initial temperature (T_1) approaches that of isothermal compression at T_1. The optimum compression ratio for each stage that minimizes the total compression work for any number of stages (n) with interstage cooling to the initial temperature is

$$r = \frac{P_2}{P_1} = \frac{P_3}{P_2} = \cdots = \frac{P_{n+1}}{P_n} = \left(\frac{P_{n+1}}{P_1}\right)^{1/n} \tag{5.151}$$

If there is no interstage cooling, or for interstage cooling to a temperature other than T_1, the optimum compression ratio for each stage (i) is related to the temperature entering that stage (T_i) by

$$T_i \left(\frac{P_{i+1}}{P_i}\right)^{(k-1)/k} = T_i r_i^{(k-1)/k} = \text{constant } t \tag{5.152}$$

The above relations apply to ideal (frictionless) compressors. To account for friction losses, the ideal computed work is divided by the compressor efficiency, η_e, to get the total work that must be supplied to the compressor:

Fluid Flow

$$(-w)_{ideal} = \frac{(-w)_{ideal}}{\eta_e} \tag{5.153}$$

The energy "lost" due to friction is dissipated into thermal energy, which raises the temperature of the gas. This temperature rise is in addition to that due to the isentropic compression, so that the total temperature rise across an adiabatic compressor stage is

$$T_2 = T_1 r^{(k-1)/k} + \left(\frac{1-\eta_e}{\eta_e}\right)\left(\frac{-w_{ideal}}{c_v}\right) \tag{5.154}$$

Centrifugal compressors can operate at speeds from 9,000 to 50,000 rpm, with single-stage compression ratios from 1.2 to 9, and efficiencies from 30 to 85%, depending upon the specific design and application.

5.7 FLOW MEASUREMENT AND CONTROL

Some common methods for measuring flow rate in conduits, namely, the pitot tube, venturi, nozzle, and orifice meters, will be illustrated here. However, there are a great many other devices in use, such as turbine, vane, Coriolis, ultrasonic, and magnetic flow meters. The manufacturer or supplier of these devices should be consulted for their specifications and range of applicability.

5.7.1 PITOT TUBE

The volumetric flow rate (Q) of a fluid through a conduit may be determined by integrating the local ("point") velocity over the cross section of the conduit. If the conduit cross section is circular,

$$Q = \int_0^{\pi R^2} V(r)d(\pi r^2) = 2\pi \int_0^R V(r)r\,dr \tag{5.155}$$

The pitot tube may be used to measure the local velocity $V(r)$ at a given position in the conduit, as illustrated in Figure 5.13. A differential pressure measuring device (e.g., a manometer, transducer, or DP cell) measures the pressure difference between two tubes, which is related directly to the local velocity:

$$V_1 = \sqrt{\frac{2(P_2 - P_1)}{\rho}} \tag{5.156}$$

The pitot-static tube includes an annular tube surrounding the stagnation tube, with holes in the sides through which the static pressure is measured. For this configuration, errors introduced by

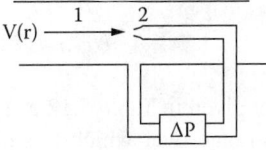

FIGURE 5.13 Pitot tube.

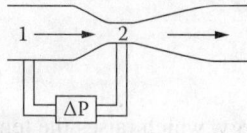

FIGURE 5.14 Venturi meter.

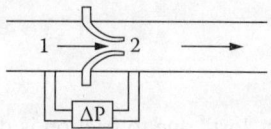

FIGURE 5.15 Nozzle.

the presence of the probe and the stem inserted into the flow are minimized if the probe length is 14 times the probe diameter, and the static pressure holes are six probe diameters from the end of the probe. Also, axial alignment of the probe is important for accurate measurement, and the probe should be located no closer than 100 pipe diameters from any bend or flow constriction. The total flow rate (Q) through the conduit can be determined by measuring a sufficient number of radial points across the conduit to enable accurate evaluation of the integral in Equation (5.155).

The pitot tube requires considerable effort and time to obtain an accurate value of the total flow rate. However, the probe offers minimal resistance to the flow and hence is very efficient from the standpoint of the resulting permanent pressure drop due to friction. It is also the only practical means for determining the flow rate in very large conduits (e.g., smokestacks). Errors increase as the ratio of the probe diameter to pipe diameter increases, and for accurate readings the probe diameter should be no larger than 2% of the tube diameter. This limits accurate application to pipes of about 40-cm diameter or larger.

5.7.2 Venturi and Nozzle

Other devices can be used to determine the flow rate from a single measurement. These are sometimes referred to as *obstruction meters*, since the basic principle involves introducing an "obstruction" (e.g., a constriction) into the flow channel and then measuring the pressure drop across this obstruction, which depends on the flow rate. Two such devices, the venturi meter and the nozzle, are illustrated in Figures 5.14 and 5.15, respectively. In both cases, the pressure drop from a point upstream of the meter to a point in a plane with the minimum flow area (A_2) is related to the velocity V_2 by the Bernoulli equation:

$$V_2 = C_v \sqrt{\frac{-2\Delta P}{\rho(1-\beta^4)}} \qquad (5.157)$$

where C_v is the discharge coefficient for the venturi, which is a function of the conduit Reynolds number, as shown in Figure 5.16. The coefficient accounts for deviation from plug flow as well as irreversible effects. However, since the coefficient is not greatly different from 1.0 at high Reynolds numbers (having a value of about 0.985 at [pipe] Reynolds numbers, N_{ReD}, above about 2×10^5), this indicates that these nonidealities are small. According to Miller (1983), for $N_{ReD} > 4000$, the discharge coefficient for the venturi, as well as for the nozzle and orifice, can be described as a function of N_{ReD} and β by the general equation

$$C = C_\infty + \frac{b}{N_{ReD}^n} \qquad (5.158)$$

where the parameters C_∞, b, and n are given in Table 5.12 as a function of β (the ratio of the throat diameter to the pipe diameter). The range over which this equation applies and its approximate accuracy are given in Table 5.13. The pressure recovery in the venturi is relatively large, so that the net friction loss across the entire meter is a relatively small fraction of the measured pressure drop, as indicated in Figure 5.17.

Fluid Flow

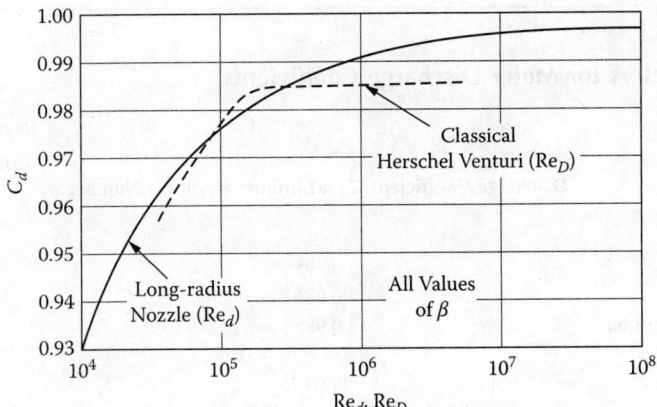

FIGURE 5.16 Venturi and nozzle discharge coefficient versus Reynolds number. (From White, F. M., *Fluid Mechanics*, 3rd ed., McGraw-Hill, New York (1994).)

For the orifice and nozzle, the flow, since the expansion area changes abruptly downstream, is uncontrolled, and considerable eddying occurs downstream. This dissipates more energy, resulting in a significantly higher net pressure loss and lower pressure recovery. The above equations assume that the device is horizontal, i.e., the pressure taps on the pipe are located in the same horizontal plane. Note that the Reynolds number used for the venturi coefficient in Figure 5.16 is based upon the pipe diameter (D), whereas the Reynolds number used for the nozzle coefficient is based upon the nozzle diameter (d) (also note that $N_{ReD} = \beta N_{Red}$). There are various "standard" designs for the nozzle, and the reader should consult the literature for details (e.g., Miller, 1983). The discharge coefficient for nozzles can also be described by Equation (5.158), with the appropriate parameters given in Table 5.12. International standard shapes for the nozzle and venturi are shown in Figure 5.18.

5.7.3 Orifice Meter

The orifice meter is illustrated in Figure 5.19. The flow equation is the same as for the venturi and nozzle, the major difference being the *vena contracta*, which is the contraction of the fluid stream just downstream of the orifice to an area that is approximately 60% of the orifice area. If the pipe diameter is D, the orifice diameter is d, and the diameter of the vena contracta is d_2, the contraction ratio for the vena contracta is defined as $C_c = A_2/A_o = (d_2/d)^2$. For highly turbulent flow, $C_c \approx 0.61$.

5.7.3.1 Incompressible Flow

For incompressible flow, the flow equation becomes

$$\dot{m} = C_o A_o \sqrt{\frac{2\rho \Delta P}{(1-\beta^4)}} \tag{5.159}$$

Figure 5.20 shows the orifice (discharge) coefficient C_o as a function of the orifice Reynolds number (N_{Red}) and $\beta = d/D$. There are a variety of "standard" orifice plate and pressure tap designs (e.g., Miller, 1983). The ASME specifications for the most common concentric square-edged orifices are shown in Figure 5.21. The various pressure tap locations are illustrated in Figure 5.22. Radius taps, for which the location is scaled to the pipe diameter, are the most reliable. Corner taps and flange taps are the most convenient, as they can be installed in the orifice flange and so do not require

TABLE 5.12
Values of Parameters for Meter Discharge Coefficients

Primary Device	Discharge Coefficient C_∞ at Infinite Reynolds Number	Reynolds Number Term Coefficient b	Exponent n
Venturi			
Machined inlet	0.995	0	0
Rough cast inlet	0.984	0	0
Rough welded sheet-iron inlet	0.985	0	0
Universal Venturi Tube[b]	0.9797	0	0
Lo-Loss tube[c]	$1.005 - 0.471\beta + 0.564\beta^2 - 0514\beta^3$	0	0
Nozzle:			
ASME long radius	0.9975	$-6.53\beta^{0.5}$	0.5
ISA	$0.9900 - 0.2262\beta^{4.1}$	$1708 - 8936\beta + 19{,}779\beta^{4.7}$	1.15
Orifice:			
Venturi nozzle (ISA inlet)	$0.9858 - 0.196\beta^{4.5}$	0	0
Corner taps	$0.5959 + 0.0312\beta^{2.1} - 0.184\beta^6$	$91.71\beta^{2.5}$	0.75
Flange taps (D in inches)			
$D \geq 2.3$	$0.5959 + 0.0312\beta^{2.1} - 0.184\beta^6 + 0.09\dfrac{\beta^4}{D*(1-\beta^4)} - 0.0337\dfrac{\beta^3}{D}$	$91.71\beta^{2.5}$	0.75
$2 \leq D \leq 2.3^4$	$0.5959 + 0.0312\beta^{2.1} - 0.184\beta^6 + 0.039\dfrac{\beta^4}{1-\beta^4} - 0.0337\dfrac{\beta^3}{D}$	$91.71\beta^{2.5}$	0.75
Flange taps (D^* in millimeters)			
$D^* \geq 58.4$	$0.5959 + 0.0312\beta^{2.1} - 0.184\beta^6 + 2.286\dfrac{\beta^4}{D*(1-\beta_4)-\beta^4} - 0.856\dfrac{\beta^3}{D*}$	$91.71\beta^{2.5}$	0.75
$50.8 \leq D^* \leq 58.4^d$	$0.5959 + 0.0312\beta^{2.1} - 0.184\beta^6 + 0.039\dfrac{\beta^4}{1-\beta^4} - 0.856\dfrac{\beta^3}{D*}$	$91.71\beta^{2.5}$	0.75
D and $D/2$ taps	$0.5959 + 0.0312\beta^{2.1} - 0.184\beta^6 + 0.039\dfrac{\beta^4}{1-\beta^4} - 0.0158\beta^3$	$91.71\beta^{2.5}$	0.75
2 1/2 D and 8D taps	$0.5959 + 0.461\beta^{2.1} + 0.48\beta^6 + 0.039\dfrac{\beta^4}{1-\beta^4}$	$91.71\beta^{2.5}$	0.75

[a] Detailed Reynolds number, line size, beta ratio, and other limitations are given in Table 5.13.

[b] From BIF CALC-440/441; the manufacturer should be consulted for exact coefficient information.

[c] Derived from the Badger Meter, Inc. Lo-Loss tube coefficient curve; the manufacturer should be consulted for exact coefficient information.

[d] For $1/2 \leq D \leq 1\ 1/2$ in. ($12 \leq D^* \leq 40$ mm), use flow coefficient equation (10.1) or Equation (10.2) given in Chapter 10, (Miller), with $C = \sqrt{1-\beta^4}\,K$.

[e] Source: Stolz (1978).

Source: Miller, R. W., *Flow Measurement Engineering Handbook*, McGraw-Hill, New York (1983).

TABLE 5.13
Range and Accuracy of Flow Meter Equation

Primary Device	Nominal Pipe Diameter D, in mm	Beta Ratio β	Pipe Reynolds Number R_o Range	Coefficient Accuracy, %†
Venturi				
Machined inlet	2–10 (50–250)	0.4–0.75	2×10^5 to 10^6	±1
Rough cast	4–32(100–800)	0.3–0.75	2×10^5 to 10^6	±0.7
Rough-welded sheet-iron inlet	8–48 (200–1500)	0.4–0.7	2×10^5 to 10^6	±1.5
Universal Venturi Tube‡	≥3(≥ 75)	0.2–0.75	$> 7.5 \times 10^4$	±0.5
Lo-Loss‡	3–120(75–3000)	0.35–0.85	1.25×10^5 to 3.5×10^6	±1
Nozzle[i]				
ASME	2–16(50–400)	0.25–0.75	10^4 to 10^7	±2.0
ISA	2–20(50–500)	0.3–0.6	10^5 to 10^6	±0.8
		0.6–0.75	2×10^5 to 10^7	23–0.4
Venturi nozzle	3–20 (75–500)	0.3–0.75	2×10^5 to 2×10^6	$= 1.2 \pm 1.54\beta_4$
Orifice				
Corner, flange, D and $D/2$	2–36 (50–900)¶	0.2–0.6	10^4 to 10^7	±0.6
		0.6–0.75	10^4 to 10^7	± 6
		0.2–0.75	2×10^3 to 10^4	$\pm 0.6 \pm \beta$
2 1/2 D and 8D (Pipetaps)	2–36(50–900)	0.2–0.5	10^4 to 10^7	± 0.8
		0.51–0.7		± 1.6
Eccentric†				
Flange and vena contracta	4 (100)	0.3–0.75	10^4 to 10^6	±2
	6–14 (150–350)	0.3–0.75	10^4 to 10^6	±1.5
Segmental†				
Flange and vena contracta	4–14(150–350)	0.35–0.75	10^4 to 10^6	±2
Quadrant-edged‡				
Flange and corner	1–4(25–100)	0.25–0.6	250 to 6×10^4	±2–±25
Conical entrance‡				
Corner		0.1–0.3	25 to 2×10^4	=2–±25

† ISO 5167 (1980) and ASME *Fluid Meters* (1971) show slightly different values for some devices.
‡ The manufacturer should be consulted for recommendations.
[i] Curves of discharge coefficient versus Reynolds number appear in Figure 5.20.

For $1/2 \leq D \leq 1/2$ in. ($12 \leq D^* \leq 40$ mm), use flow coefficient Equation (10.1) or Equation (10.2) given in Chapter 10 (Miller) with $C = \sqrt{1-\beta^4}\, K$.

Source: Miller, R. W., *Flow Measurement Engineering Handbook*, McGraw-Hill, New York (1983).

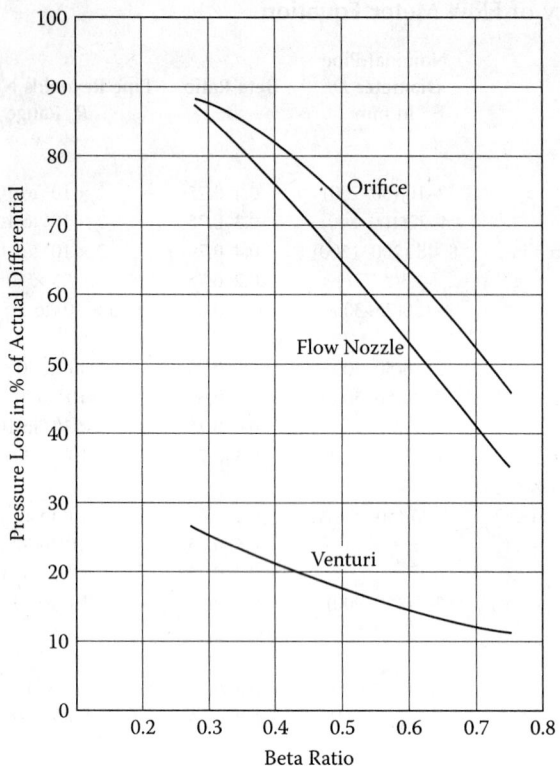

FIGURE 5.17 Unrecovered pressure loss in various meters. (From Miller, R. W., *Flow Measurement Engineering Handbook*, McGraw-Hill, New York (1983).)

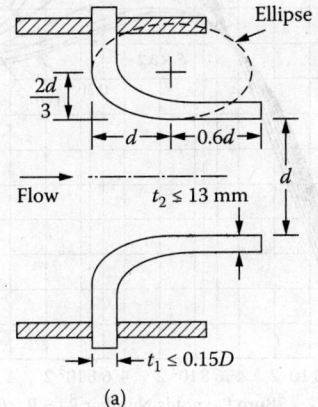

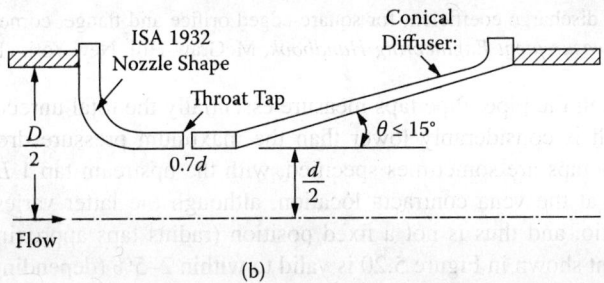

FIGURE 5.18 International standard shapes for nozzle and venturi meters. (From White, F. M., *Fluid Mechanics*, 3rd ed., McGraw-Hill, New York (1994).)

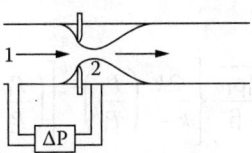

FIGURE 5.19 Orifice meter.

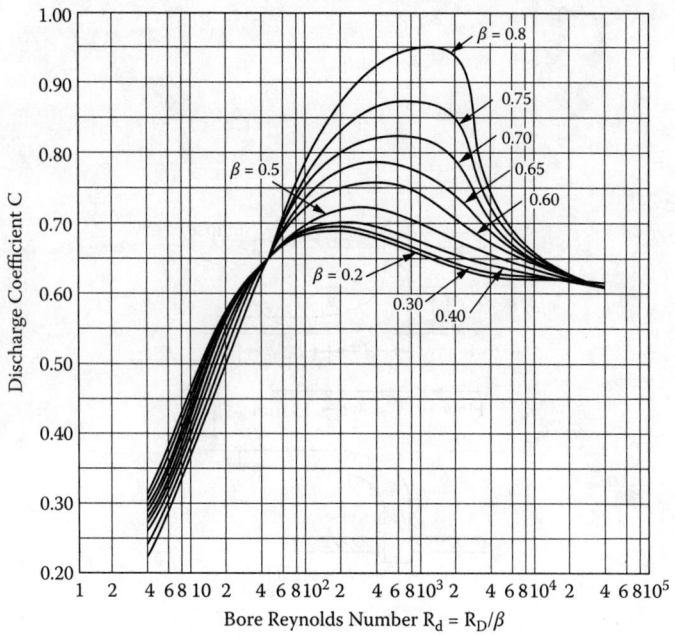

FIGURE 5.20 Orifice discharge coefficient for square-edged orifice and flange, corner, or radius tap. (From Miller, R. W., *Flow Measurement Engineering Handbook*, McGraw-Hill, New York (1983).)

additional taps through the pipe. Pipe taps measure essentially the total unrecovered pressure drop (friction loss), which is considerably lower than the maximum pressure drop across the orifice plate. Vena contracta taps are sometimes specified, with the upstream tap 1 D from the plate and the downstream tap at the vena contracta location, although the latter varies with the Reynolds number and beta ratio, and thus is not a fixed position (radius taps approximate this condition). The orifice coefficient shown in Figure 5.20 is valid to within 2–5% (depending upon the Reynolds number and β) for all pressure tap locations except pipe and vena contracta taps. More accurate values can be calculated from Equation (5.160) for high Reynolds numbers, with the parameter expressions given in Table 5.12 for the specific orifice and pressure tap arrangements.

5.7.3.2 Compressible Flow

For an ideal gas flowing through an orifice under adiabatic conditions, the flow equation is

$$\dot{m} = C_o A_o \sqrt{\frac{P_1 \rho_1}{1-\beta^4}} \left\{ \frac{2k}{k-1} \left(\frac{P_2}{P_1}\right)^{2/k} \left[\left(\frac{P_1}{P_2}\right)^{(k-1)/k} - 1 \right] \right\}^{1/2} \quad (5.160)$$

where the value of C_o is assumed to be the same as for a incompressible flow. The ratio of this equation to the incompressible equation is called the *expansion factor Y*. Thus,

$$\dot{m} = C_o A_o Y \sqrt{\frac{2\rho_1 \Delta P}{(1-\beta^4)}} \quad (5.161)$$

where the density ρ_1 is evaluated at the upstream pressure (P_1). Values of Y are shown as a function of $\Delta P/P_1$ and β for a square-edged orifice, nozzles, and venturi meters for values of $k = c_p/c_v$ of

Fluid Flow

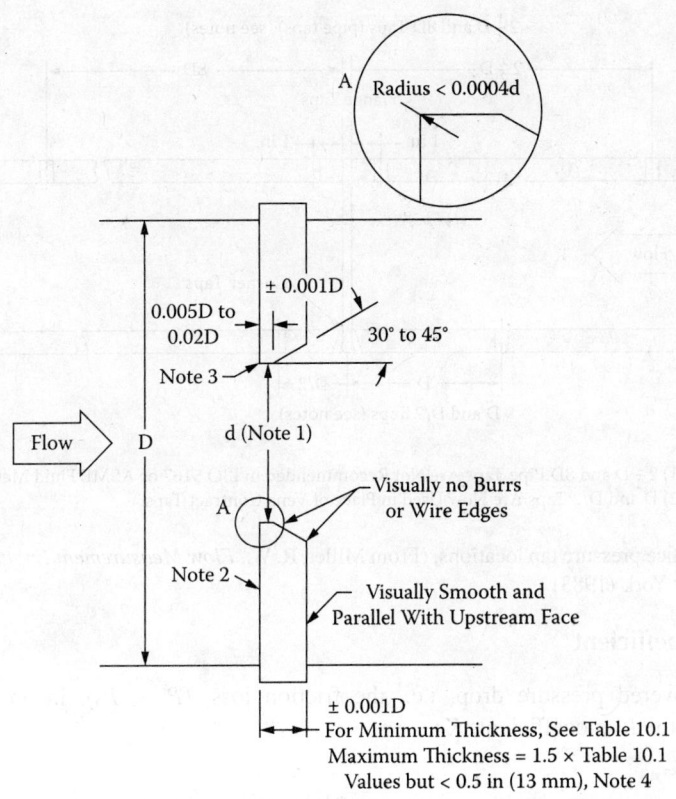

FIGURE 5.21 Concentric square-edged orifice specifications. (From Miller, R. W., *Flow Measurement Engineering Handbook*, McGraw-Hill, New York (1983).)

1.3 and 1.4 in Figure 5.23. The lines in Figure 5.23 for the orifice can be represented by the following equation for radius taps (Miller, 1983):

$$Y = 1 - \frac{\Delta P}{kP_1}(0.41 + 0.35\beta^4) \tag{5.162}$$

and for pipe taps by

$$Y = 1 - \frac{\Delta P}{kP_1}[0.333 + 1.145(\beta^4 + 0.7\beta^5 + 12\beta^{13})] \tag{5.163}$$

The "buttons" at the end of the lines in Figure 5.23 for the nozzle and venturi correspond to critical (choked) flow. Although there are no "buttons" shown on the lines for the orifice, choked flow does occur in orifices as well. However, the choked flow is not as reproducible for orifices because of the variable effects of the vena contracta.

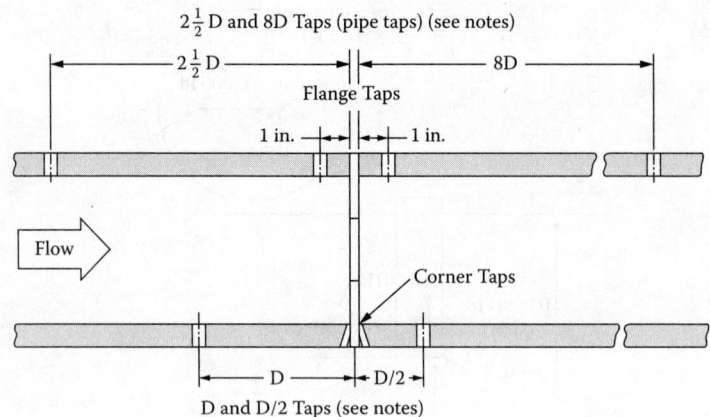

FIGURE 5.22 Orifice pressure tap locations. (From Miller, R. W., *Flow Measurement Engineering Handbook*, McGraw-Hill, New York (1983).)

5.7.3.3 Loss Coefficient

The total unrecovered pressure drop, i.e., the friction loss, $(P_1 - P_3)$, in an orifice meter is characterized by the loss coefficient, K_f:

$$K_f = \frac{(1-\beta^4)(1-\beta^2)}{C_o^2 \beta^4} \tag{5.164}$$

which is to be used with the pipe velocity (V_1). If K_f is based upon the velocity through the orifice (V_o) instead of the pipe velocity, the β^4 term in the denominator should be omitted.

5.7.3.4 Applications

Determining the flow rate for a given pipe/orifice geometry and known pressure drop from Equation (5.162) requires an iterative procedure, since the orifice coefficient C_o depends on the Reynolds number (e.g., Figure 5.20), which cannot be found until the flow rate is known. This procedure is usually simplified by initially assuming $C_o = 0.61$.

If the orifice diameter is to be determined for a specified value or range of flow rate and pressure drop in a given pipe, it is more convenient to rearrange Equation (5.162) for β as follows:

$$\beta = \left[\frac{X}{1+X}\right]^{1/4} \quad \text{where } X = \frac{8}{\rho_1 \Delta P}\left[\frac{\dot{m}}{\pi D^2 Y C_o}\right] \tag{5.165}$$

An iterative procedure for C_o is still required, since $C_o = fn(\beta)$, and this can be simplified as above by initially assuming $C_o = 0.61$.

5.8 CONTROL VALVES

A control valve is a specially designed globe valve that acts as a "variable resistance" in the line to control the flow rate. Closing down on the valve decreases the area between the plug and the

Fluid Flow

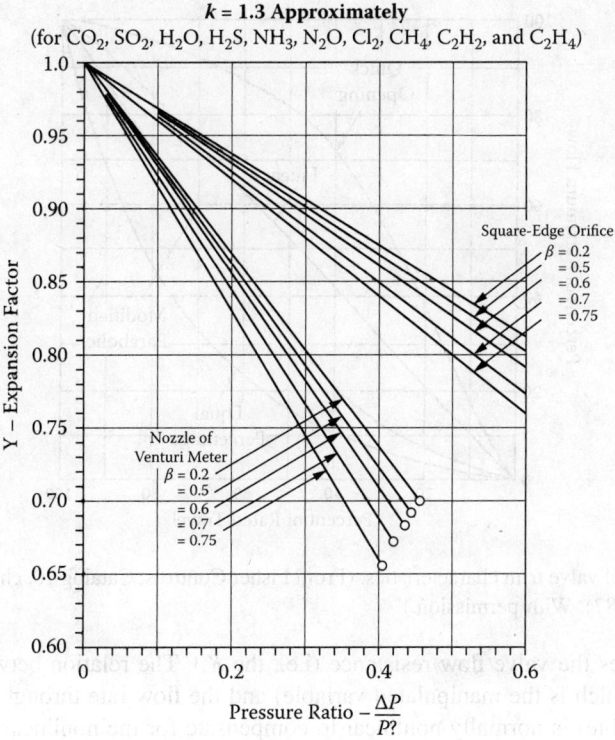

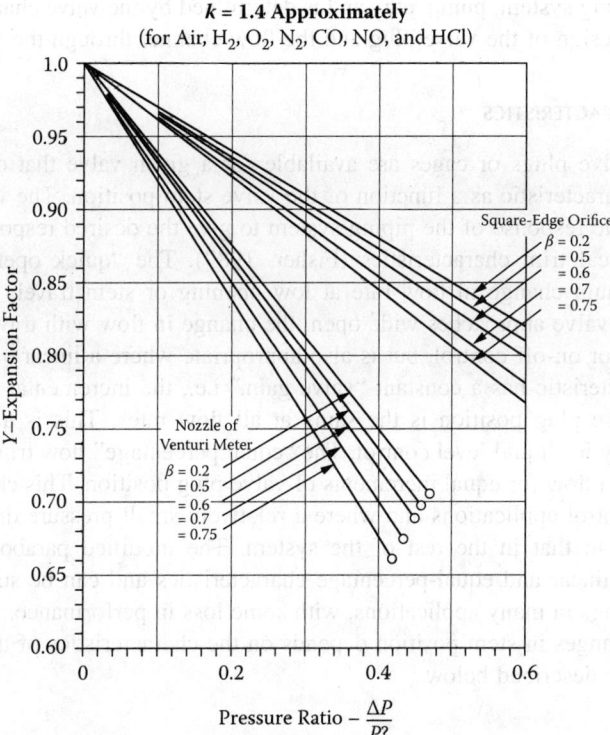

FIGURE 5.23 Expansion factor for square-edged orifice and nozzle or venturi meter; (a) $k = 1.3$, (b) $k = 1.4$. (From Crane Co., "Flow of Fluids through Valves, Fittings, and Pipe," Technical Manual 410, Crane Co., New York (1978).)

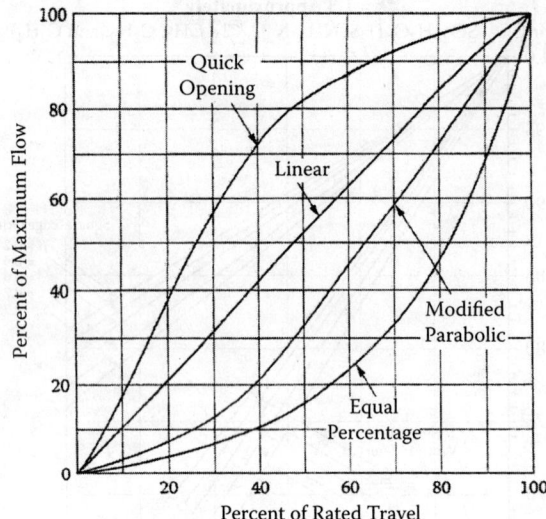

FIGURE 5.24 Control valve trim characteristics. (From Fisher Controls, Catalog 10, chap. 2, Fisher Controls, Marshalltown, IA (1987). With permission.)

seat, which increases the valve flow resistance (i.e., the K_f). The relation between the valve stem or plug position (which is the manipulated variable) and the flow rate through the valve (which is the controlled variable) is normally nonlinear to compensate for the nonlinear pressure-flow characteristics of the piping system, pump, etc., and is determined by the valve characteristic or "trim," i.e., the shape and design of the valve plug and the flow channel through the valve.

5.8.1 Valve Characteristics

Different shaped valve plugs or cages are available for a given valve that determine the valve response or trim characteristic as a function of the valve stem position. The valve trim is chosen to match the dynamic response of the piping system to give the desired response relation. Figure 5.24 illustrates typical trim characteristics (Fisher, 1987). The "quick opening" characteristic provides the maximum change in flow rate at low opening or stem travel, with a fairly linear relationship. As the valve approaches wide open, the change in flow with travel approaches zero. This is best suited for on-off control, but is also appropriate where a linear valve is desired. The "linear" flow characteristic has a constant "valve gain," i.e., the incremental change in flow rate with change in valve plug position is the same at all flow rates. This is a commonly desired property, particularly for liquid level control. The "equal percentage" flow trim provides the same percentage change in flow for equal increments of valve plug position. This characteristic is often used in pressure control applications and where a relatively small pressure drop across the valve is required relative to that in the rest of the system. The modified parabolic characteristic is intermediate to the linear and equal-percentage characteristics and can be substituted for equal-percentage valve plugs in many applications, with some loss in performance. However, the actual flow response to changes in stem position depends on the characteristics of the piping system as well as the valve, as described below.

5.8.2 Valve Sizing Equations—Incompressible Flows

For incompressible fluids, Bernoulli's equation relates the pressure drop across a valve and the flow rate through the valve in terms of the loss coefficient, K_f:

Fluid Flow

$$Q = AV = A\sqrt{\frac{2\Delta P}{\rho K_f}} \tag{5.166}$$

where A is an appropriate flow area, V is the velocity through that area, and $\Delta P = (P_1 - P_2)$ is the pressure drop across the valve. However, in a control valve the area (and hence V) is a variable, and the internal flow geometry can be quite complex. The pressure drop is not the maximum value (which would occur if P_2 were at the vena contracta, as for an orifice meter), but is the *net unrecovered* pressure loss corresponding to a point (P_2), which is far enough downstream that any possible pressure recovery has occurred. Combining the flow area and geometrical factors and the density of the reference fluid with the friction loss coefficient leads to the following equation for incompressible fluids:

$$Q = C_v\sqrt{\frac{\Delta P}{SG_v}} = C_v\sqrt{\rho_w g h_v} = 0.658 \left[\sqrt{\text{psi/ft}}\right] C_v \left[\text{gpm}/\sqrt{\text{psi}}\right] \sqrt{h_v} \text{ [ft]} \tag{5.167}$$

This equation defines the *flow coefficient*, C_v. Here, SG is the fluid specific gravity (relative to water), ρ_w is the density of water, and h_v is the head loss across the valve. Valve C_v's are determined by the manufacturer, and the values are different for each valve and also vary with the valve opening (or stem travel) for a given valve. Although Equation (5.167) is similar to the flow equation for flow meters, the flow coefficient C_v is *not dimensionless*, but has dimensions of $[L^3][L/M]^{1/2}$. More specifically, the "normal engineering" units for C_v are gpm/(psi)$^{1/2}$:

Q = volumetric flow rate (gpm for liquids or scfh for gas or steam)
SG = specific gravity (relative to water for liquids [62.3 lb$_m$/ft^3] or air at 60°F and 1 atm for gases [0.0764 lb$_m$/ft^3])
ρ_1 = density at upstream conditions (lb$_m$/ft^3)
P_1 = upstream pressure (psia)
ΔP = total (net) pressure drop across valve (psi)

Typical flow coefficient values are shown in Table 5.14, in which K_m applies to cavitating and flashing liquids and C_1 applies to critical (choked) compressible flow, as discussed below.

5.8.2.1 Matching Valve Trim and System Characteristics

Selection of the proper valve size and trim to be used for a given application requires matching the valve, piping system, and pump characteristics for the desired response (Darby, 1997). The operating point for a piping system depends upon the pressure-flow behavior of both the system and the pump. The control valve is part of the piping system and acts as a variable resistance in the system (the valve loss coefficient K_f increases as the discharge coefficient C_v decreases as the valve is closed). The operating point for the system is where the pump head (H_p) characteristic intersects the system head requirement (H_s):

$$H_s = \frac{\Delta P}{\rho g} + \Delta z + \frac{Q^2}{g}\left[\frac{8}{\pi}\sum\left(\frac{K_f}{D^4}\right) + \frac{1}{\rho_w C_v^2}\right] \tag{5.168}$$

where the last term is the head loss through the control valve, h_v (Equation (5.167)), and C_v depends upon the valve stem travel, X (see Figure 5.24), i.e.,

$$C_v = C_{v\max} f(X) \tag{5.169}$$

TABLE 5.14
Example Flow Coefficient Values for a Linear Control Valve

Linear Characteristic

Coefficients	Body Size, inch	Port Diameter, inch	Total Travel, inch	\multicolumn{10}{c}{Valve Opening, Percent of Total Travel}	K_m(1) and C_1
				10	20	30	40	50	60	70	80	90	100	
C_v (Liquid)	2 & 3 x 2	1-7/8	1-1/2	1.69	9.45	21.9	33.4	42.7	50.0	55.6	59.6	61.9	63.6	.72
	3 & 4 x 3	2-7/8	2	3.41	25.4	52.6	76.0	96.4	114	127	133	135	136	.91
	4 & 6 x 4	3-5/8	2	6.69	25.1	50.1	77.9	106	134	157	175	185	188	.86
	6 & 8 x 6	5-3/8	3	9.40	63.8	138	212	282	339	373	389	398	405	.81
C_g (Gas)	2 & 3 x 2	1-7/8	1-1/2	60.8	328	729	1110	1400	1600	1710	1780	1810	1840	28.9
	3 & 4 x 3	2-7/8	2	142	839	1760	2540	3240	3680	4320	4490	4540	4570	33.6
	4 & 6 x 4	3-5/8	2	229	791	1530	2350	3250	4190	5090	5850	6360	6580	35.0
	6 & 8 x 6	5-3/8	3	287	1910	4060	6160	8400	10,600	12,300	13,300	13,800	14,100	34.8
C_s (Steam)	2 & 3 x 2	1-7/8	1-1/2	3.04	16.3	36.5	55.5	70.0	80.0	85.5	89.0	90.5	92.0	28.9
	3 & 4 x 3	2-7/8	2	7.10	42.0	88.0	127	162	194	216	225	227	229	33.6
	4 & 6 x 4	3-5/8	2	11.5	39.6	76.5	118	163	210	255	293	318	329	35.0
	6 & 8 x 6	6-3/8	3	14.4	95.5	203	308	420	530	615	665	690	705	34.8

Equal Percentage Characteristic

Coefficients	Body Size, inch	Port Diameter, inch	Total Travel, inch	10	20	30	40	50	60	70	80	90	100	K_m and C_1
C_v (Liquid)	2 & 3 x 2	1-7/8	1-1/8	1.04	1.59	3.52	6.99	12.1	19.7	30.5	40.9	44.9	50.7	.79
	3 & 4 x 3	2-7/8	1-1/2	2.58	5.17	10.80	16.2	28.9	44.9	62.6	82.9	104	117	.91
	4 & 6 x 4	3-5/8	1-1/2	3.44	7.12	13.1	21.8	34.8	54.0	80.4	109	132	154	.71
	6 & 8 x 6	5-3/8	2-1/2	5.27	13.0	22.1	35.3	57	93	141	194	246	308	.64
C_g (Gas)	2 & 3 x 2	1-7/8	1-1/8	41.5	61.2	123	233	401	653	996	1320	1460	1590	31.4
	3 & 4 x 3	2-7/8	1-1/2	88.9	175	381	636	985	1530	2190	2890	3610	4000	34.2
	4 & 6 x 4	3-5/8	1-1/2	134	240	430	700	1080	1650	2450	3440	4210	5140	33.4
	6 & 8 x 6	5-3/8	2-1/2	152	422	673	1020	1710	2730	3990	5490	7350	9220	29.9
C_s (Steam)	2 & 3 x 2	1-7/8	1-1/8	2.08	3.06	6.15	11.7	20.1	32.7	49.8	66.0	73.0	79.5	31.4
	3 & 4 x 3	2-7/8	1-1/2	4.45	8.75	19.1	31.9	49.3	76.5	110	145	181	200	34.2
	4 & 6 x 4	3-5/8	1-1/2	6.70	12.0	21.5	35.0	54.0	82.5	124	172	211	257	33.4
	6 & 8 x 6	5-3/8	2-1/2	7.60	21.1	33.7	51.0	85.5	137	209	275	368	461	29.9

Modified Equal Percentage Characteristic

Coefficients	Body Size, inch	Port Diameter, inch	Total Travel, inch	10	20	30	40	50	60	70	80	90	100	K_m and C_1
C_v (Liquid)	2 & 3 x 2	1-7/8	1-1/2	1.07	2.55	6.87	15.1	28.6	38.3	47.6	53.7	57.3	60.4	.73
	3 & 4 x 3	2-7/8	2	3.08	8.63	18.5	34.3	57.8	84.5	106	123	131	135	.86
	4 & 6 x 4	3-5/8	2	4.49	10.7	21.8	41.2	71.0	107	141	166	183	193	.85
	6 & 8 x 6	5-3/8	3	5.67	16.4	29.3	52.0	92.5	151	217	280	346	380	.75
C_g (Gas)	2 & 3 x 2	1-7/8	1-1/2	43.0	95.9	230	493	874	1260	1530	1660	1720	1800	29.8
	3 & 4 x 3	2-7/8	2	105	295	635	1140	1930	2890	3720	4250	4470	4540	33.6
	4 & 6 x 4	3-5/8	2	172	337	663	1280	2240	3380	4470	5480	6460	6670	34.6
	6 & 8 x 6	5-3/8	3	200	298	894	1520	2620	4330	6270	8210	10,700	12,900	32.9
C_s (Steam)	2 & 3 x 2	1-7/8	1-1/2	2.15	4.80	11.5	24.7	43.7	63.0	76.5	83.0	86.0	90.0	29.8
	3 & 4 x 3	2-7/8	2	5.25	14.8	31.8	57.0	96.5	145	186	213	224	227	33.6
	4 & 6 x 4	3-5/8	2	8.60	16.9	33.2	64.0	112	169	224	274	323	334	34.6
	6 & 8 x 6	5-3/8	3	10.0	24.9	44.7	76.0	131	217	314	411	535	625	32.9

This table lists The K_m values for the C_v coefficients and the C_1 values for the C_g and C_s coefficients at 100% total travel.

A typical pump and system curve is illustrated in Figure 5.25. The effect on the system curve of partially closing the valve is shown. Closing down on the valve (reducing X) decreases the valve C_v and increases the head loss through the valve, h_v. This shifts the system curve upward by an amount h_v at a given flow rate (h_v depends on flow rate). The range of possible flow rates (i.e., the "turndown" ratio) lies between the intersection on the pump curve of the system curve with a "fully open" valve (Q_{max}, corresponding to $C_{v,max}$) and the intersection of the system curve with the (partly) closed valve. The desired operating point should be as close as practical to Q_{max}, since this corresponds to an open valve with minimum flow resistance and is the most efficient region on the pump curve. Thus the flow is controlled by closing down on the valve (i.e., reducing X and C_v, and thus raising h_v). The minimum operating flow rate (Q_{min}) is established by the turndown ratio required for proper control. These limits set the size of the valve (e.g., the required $C_{v,max}$), and the head-flow-rate behavior of the system over the desired flow range determines the proper trim for the valve. The system curve is shifted by the amount h_v in response to closing the valve (i.e., reducing X), where

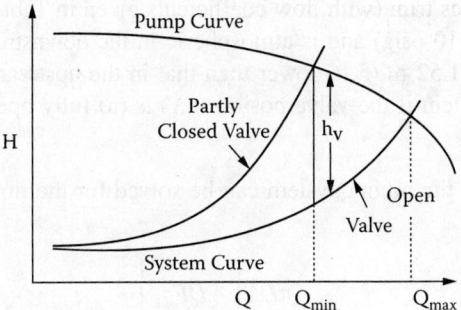

FIGURE 5.25 Effect of control valve on operating point.

$$h_v = \frac{Q^2}{\rho_w g C_{v,\max}^2 f^2(X)} \tag{5.170}$$

where $f(X)$ represents the valve trim characteristic function.

The trim may be chosen to provide as sensitive a response between valve stem travel (X) and flow rate (Q) as possible, or to give a linear relation between X and Q. However, the piping resistance and pump head are nonlinear functions of the flow rate, and the flow through the control valve is a nonlinear function of the valve stem travel, X. Thus, the most appropriate valve trim for a given installation is best determined by evaluating Q as a function of X for various trim characteristics, and choosing the trim that provides the most sensitive or most linear response over the operating range. For a given valve (e.g., $C_{v,\max}$) and a given flow system, this can be done by calculating the valve position (X) for various flow rates (Q) for a given trim function $f(X)$ from the system and valve equations (Equations (5.168) to (5.171)). The trim that gives the most linear (or most sensitive) relation between X and Q is then chosen. This process can be aided by fitting the trim curves by an empirical equation of the form

$$\text{Linear trim: } f(X) = X \tag{5.171}$$

$$\text{Parabolic trim: } f(X) = X^2 \text{ or } X^n \tag{5.172}$$

$$\text{Equal-percentage trim: } f(X) = \frac{\exp(aX^n)-1}{\exp(a)-1} \tag{5.173}$$

$$\text{Quick-opening trim: } f(X) = 1 - [a(1-X) - (a-1)(1-X)^n] \tag{5.174}$$

where a and n are parameters that can be adjusted to give the best fit to the trim curves. It is also often possible to fit the pump characteristic curves by a parabolic equation of the form

$$H_p = H_o + aQ + bQ^2 \tag{5.175}$$

Example 5.9

Two tanks containing water at 15.6°C (60°F) are connected by a piping system that contains 30.5 m (100 ft) of 3-in. sch 40 pipe (ID = 77.9 mm), four flanged elbows, two ball valves, and a control valve with equal-percentages trim (with flow coefficients given in Table 5.14). The pressure in the upstream tank is 170 kPa (10 psig) and is atmospheric in the downstream tank, and the elevation of the downstream tank is 1.52 m (5 ft) lower than that in the upstream tank. Determine the flow rate of the water in the system if the valve position (X) is (a) fully open and (b) half-open.

Solution

The Bernoulli equation for the piping system can be solved for the flow rate (Q) to give

$$Q = \frac{\pi D^2}{2\sqrt{2}} \left[\frac{DF}{\sum K_f} \right]^{1/2}$$

where ΣK_f includes all losses in the pipe, fittings, and the control valve as well as all entrance and exit losses. Although the loss coefficients for the pipe and fittings are dependent upon the Reynolds number (hence on Q), a first estimate can be obtained by assuming fully turbulent flow. In this case, the fitting loss coefficients are determined by the 3-K correlation by omitting the K_m term. This gives (4 K_{el}) = 1.41 and (2 K_{bv}) = 0.132. For pipe in fully turbulent flow:

$$f = \frac{0.0625}{\left[\log\left(\frac{3.7D}{\varepsilon} \right) \right]^2} = 0.00433$$

so that $K_{pipe} = 4fL/D = 6.77$. From Equation (5.167), the loss coefficient for the control valve is related to the valve flow coefficient by

$$(K_f)_{CV} = 1.732 \frac{2A^2}{C_v^2 \rho_w}$$

where A is the pipe cross-section area. The factor 1.732 includes the conversion factors required for C_v in gpm/(psi)$^{1/2}$ (Table 5.14), A in mm^2, and ρ_w in kg/m^3. From Table 5.14, the value of C_v is 28.9 for the valve half-open, and 117 when fully open, giving $(K_f)_{cv}$ of 5.75 and 94.2, respectively. Setting $K_{entr} = 0.5$ and $K_{exit} = 1.0$ gives $\Sigma K_f = 15.6$ for the fully open valve and 104.2 for the half-open valve. Inserting these values into the equation for Q gives $Q = 0.0156$ m^3/s (248 gpm) for the open valve and 0.00606 m^3/s (96 gpm) for the half-open valve.

5.8.3 COMPRESSIBLE FLOWS

The minimum pressure in the valve (P_{vc}) generally occurs at the vena contracta, just downstream of the flow orifice. The pressure then rises downstream to P_2, with the amount of pressure recovery depending upon the valve design. If P_{vc} is less than the fluid vapor pressure (P_v), the liquid will partially vaporize, forming bubbles. If the pressure recovers to a value greater than P_v, these bubbles may collapse suddenly, setting up local shock waves, which can result in considerable damage. The result is *cavitation*, as opposed to *flashing*, which occurs if the recovered pressure remains below P_v and the bubbles do not collapse. After the first vapor cavities form, the flow rate will no longer be proportional to the square root of the pressure difference across the valve due to the decreasing density of the mixture. If sufficient vapor forms, the flow can become choked, at which point the flow rate will be independent of the downstream pressure as long as P_1 remains constant. The critical pressure ratio ($r_c = P_{2c}/P_v$) at which choking will occur is shown in Figure 5.26 for

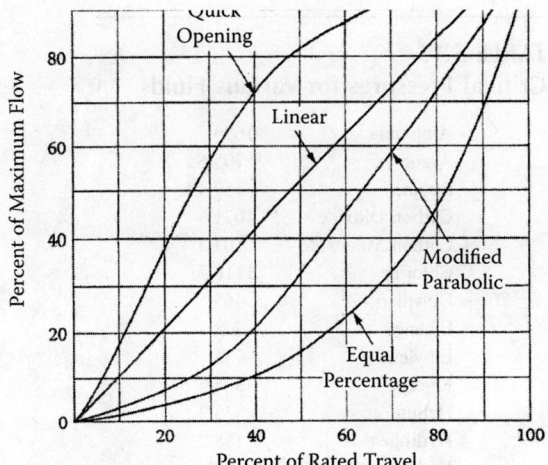

FIGURE 5.26 Critical pressure ratios for water. (From Fisher Controls, Catalog 10, chap. 2, Fisher Controls, Marshalltown, IA (1987).)

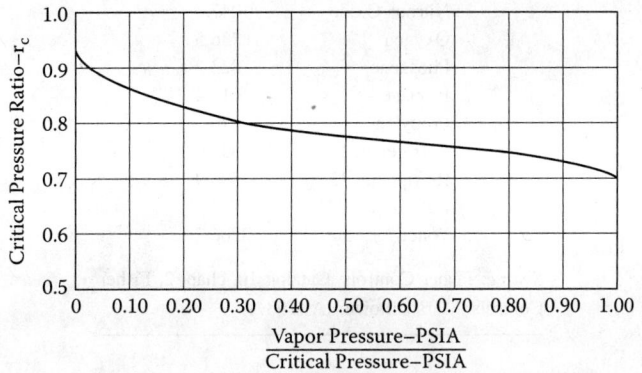

FIGURE 5.27 Critical pressure ratio for cavitating and flashing liquids. (From Fisher Controls, Catalog 10, chap. 2, Fisher Controls, Marshalltown, IA (1987).)

water and Figure 5.27 for other liquids, as a function of the liquid vapor pressure (P_v) relative to the fluid thermodynamic critical pressure (P_c). Table 5.15 lists the critical pressure values for some common fluids. An equation that represents the critical pressure ratio, r_c, with acceptable accuracy (Fisher Controls, 1977) is

$$r_c = 0.96 - 0.28\sqrt{\frac{P_v}{P_c}} \tag{5.176}$$

With r_c known, the allowable pressure drop across the valve at which cavitation occurs is given by

$$\Delta P_c = K_m(P_1 - r_c P_v) \tag{5.177}$$

where K_m is the valve recovery coefficient (which is a function of the valve design). The recovery coefficient is defined as the ratio of the overall net pressure drop ($P_1 - P_2$) to the maximum pressure drop from upstream to the vena contracta ($P_1 - P_{vc}$):

TABLE 5.15
Critical Pressures for Various Fluids

Ammonia	1636
Argon	705.6
Butane	550.4
Carbon Dioxide	1071.6
Carbon Monoxide	507.5
Chlorine	1118.7
Dowtherm A	465
Ethane	708
Ethylene	735
Fluorine	808.5
Helium	33.2
Hydrogen	188.2
Hydrogen Chloride	1198
Isobutane	529.2
Isobutylene	580
Methane	673.3
Nitrogen	492.4
Nitrous Oxide	1047.6
Oxygen	736.5
Phosgene	823.2
Propane	617.4
Propylene	670.3
Refrigerant 11	635
Refrigerant 12	596.9
Refrigerant 22	716
Water	3206.2

Source: Fisher Controls, Catalog 10, chap. 2, Fisher Controls, Marshalltown, IA (1987).

$$K_m = \frac{P_1 - P_2}{P_1 - P_{vc}} \tag{5.178}$$

Values of K_m for the Fisher Controls Example valve are given in the last column of Table 5.14, and representative values for other valves at the fully open condition are given in Table 5.16. If $\Delta P > \Delta P_c$, the value of ΔP_c is used as the pressure drop in the standard liquid sizing equation to determine Q; otherwise, the value of $(P_1 - P_2)$ is used:

$$Q = C_v \sqrt{\Delta P_c / SG} \tag{5.179}$$

The notation used here is that from the Fisher Controls literature (e.g., Fisher Controls, 1990). The ANSI/ISAS 75.01 standard for control valves (e.g., Baumann, 1991; Hutchison, 1971) gives the same equations with the notation $F_L = (K_m)^{1/2}$ and $F_F = r_c$ in place of the factors K_m and r_c.

For relatively low pressure drops, the effect of compressibility is negligible, and the general flow equation (Equation (5.167)) applies. Including conversion factors for flow rate in *scfh* and the density of air at standard conditions (1 atm, 520°R), this equation is

$$Q_{scfh} = 1362 C_v P_1 \sqrt{\frac{\Delta P}{P_1 (SG) T_1}} \tag{5.180}$$

TABLE 5.16
Representative Values of Fully Open K_m for Various Valves

Body Type	K_m
Globe: single port, flow opens	0.70–0.80
Globe: double port	0.70–0.80
Angle: flow closes	
Venturi outlet liner	0.20–0.25
Standard seat ring	0.50–0.60
Angle: flow opens	
Maximum orifice	0.70
Minimum orifice	0.90
Ball Valve:	
V-Notch	0.40
Conventional	0.30
Butterfly valve:	
60° open	0.55
90° open	0.30

Source: Hutchison, J. W., *ISA Handbook of Control Valves*, Instrument Society of America, Research Triangle Park, NC (1971).

The effect of variable density can be accounted for by an expansion factor Y, as is done for flow in pipes and meters, in which case Equation (5.181) can be written as

$$Q_{scfh} = 1362 C_v P_1 Y \sqrt{\frac{X}{(SG)T_1}} \tag{5.181}$$

where

$$X = \frac{\Delta P}{P_1} = \frac{P_1 - P_2}{P_1} \tag{5.182}$$

Deviations from the ideal gas law may be incorporated by multiplying T_1 in Equation (5.180) or (5.182) by the compressibility factor, Z, for the gas. The expansion factor Y depends upon the pressure drop X, the dimensions (clearance) in the valve, the gas-specific heat ratio k, and the Reynolds number (the effect of which is often negligible). The expansion factor for a given valve can be represented, to about ±2%, by the expression (Hutchison, 1971)

$$Y = 1 - \frac{X}{3X_T} \tag{5.183}$$

where X_T depends upon the specific valve, as illustrated in Figure 5.28. The value of X_T can be determined from Figure 5.28 as the value of X where $Y = 2/3 \times 1.4/k$.

When the gas velocity reaches the speed of sound, choked flow occurs and the mass flow rate reaches a maximum. From Equation (5.184), this is equivalent to a maximum in $Y\sqrt{X}$, which occurs at $Y = 0.667$. This corresponds to the terminus of the lines in Figure 5.29, i.e., X_T is the

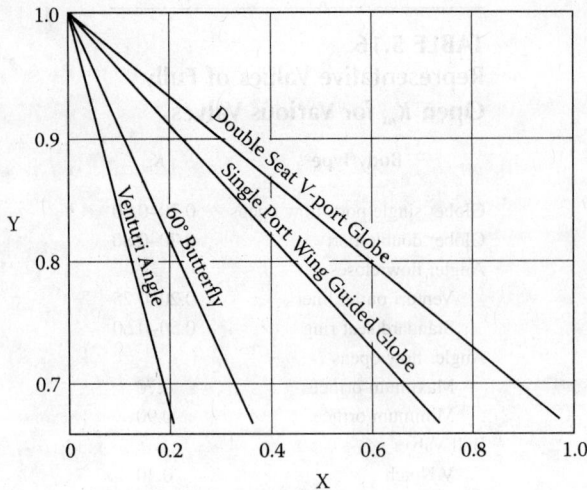

FIGURE 5.28 Expansion factor (Y) as a function of pressure drop ratio (X) for four different types of control valves. (From Hutchison, J. W., *ISA Handbook of Control Valves*, Instrument Society of America, Research Triangle Park, NC (1971).)

pressure ratio across the valve at which choking occurs, and any further increase in X (e.g., ΔP) due to lowering P_2 can have no effect on the flow rate.

The flow coefficient C_v is determined by calibration with water, and it is not entirely satisfactory for predicting the flow rate of compressible fluids under choked flow conditions. This has to do with the fact that different valves exhibit different pressure recovery characteristics with gases and hence will choke at different pressure ratios, which is not significant for liquid flows. For this reason, another flow coefficient, C_g, is often determined by calibration with air under critical flow conditions (Fisher Controls, 1977). The corresponding flow equation for gas flow is

$$Q_{critical} = C_g P_1 \left(\frac{520}{SGT} \right)^{1/2} \tag{5.184}$$

Equation (5.181), which applies at low pressure drops, and Equation (5.185), which applies to critical (choked) flow, have been combined into one general "universal" empirical equation by Fisher (1977), by using a sine function to represent the transition between the limits of both of these states:

$$Q_{scfh} = C_g \sqrt{\frac{520}{(SG)T_1}} P_1 \sin \left[\frac{3417}{C_1} \sqrt{\frac{\Delta P}{P_1}} \right]_{degrees} \tag{5.185}$$

Here, $C_1 = C_g/C_v$ and is determined by measurements on air. Values of C_1 are listed in the last column in Table 5.14 for the valve illustrated. C_1 is also approximately equal to $40\sqrt{X_T}$ (Hutchison, 1971). For steam or vapor at any pressure, a corresponding equation is

$$Q_{lb/hr} = 1.06 C_g \sqrt{P_1 \rho_1} \sin \left[\frac{3417}{C_1} \sqrt{\frac{\Delta P}{P_1}} \right]_{degrees} \tag{5.186}$$

where ρ_1 is the density of the gas at P_1, in lb_m/ft^3. When the argument of the sine term (in brackets) in Equation (5.185) or (5.186) is equal to 90° or more, the flow has reached critical flow conditions (choked) and cannot increase above this value without increasing P_1. Under these conditions, the sine term is equal to unity for this and all larger values of ΔP.

The flow coefficients referred to above are determined by calibration with air. For applications with other gases, the difference between the properties of air and those of the other gas must be considered. The gas density can be incorporated into the equations, but a correction must be made for the specific heat ratio ($c_p/c_v = k$) as well. This can be done by considering the expression for the ideal (isentropic) flow of a gas through a nozzle, which can be written (in "engineering units") as follows:

$$Q_{scfh} = \frac{3.78 \times 10^5 A_2 P_1}{SG\sqrt{RT}} \sqrt{\frac{k}{k-1}\left[\left(\frac{P_2}{P_1}\right)^{2/k} - \left(\frac{P_2}{P_1}\right)^{(k+1)/k}\right]} \quad (5.187)$$

Critical (choked) flow will occur in the nozzle throat when the pressure ratio is

$$r = \frac{P_2}{P_1} = \left(\frac{2}{k+1}\right)^{k/(k-1)} \quad (5.188)$$

Thus, for choked flow, Equation (5.188) becomes

$$Q_{scfh} = \frac{3.78 \times 10^5 A_2 P_1}{SG\sqrt{RT}} \sqrt{\left(\frac{k}{k+1}\right)\left(\frac{2}{k+1}\right)^{2/(k-1)}} \quad (5.189)$$

The quantity in the radical, which is a function only of $k[fn(k)]$, represents the dependence of the flow rate on the gas property. This may be used to define a correction factor, C_2, that can be used as a multiplier to correct the flow rate for air to that for any other gas:

$$C_2 = \frac{fn(k)_{gas}}{fn(k)_{air}} = \frac{\sqrt{\left[\frac{k}{k+1}\right]\left[\frac{2}{k+1}\right]^{2/(k-1)}}}{0.4839} \quad (5.190)$$

A plot of C_2 versus k from Equation (5.190) is shown in Figure 5.29.

5.8.4 Viscosity Correction

A correction for fluid viscosity must be applied to the flow coefficient (C_v) for liquids other than water. This correction factor (F_v) is obtained from Figure 5.30 by the following procedure, depending upon whether the objective is to find the valve size for a given Q and ΔP, to find Q for a given valve and ΔP, or to find ΔP for a given valve and Q.

To determine the valve size for a given Q and ΔP, calculate the required C_v as follows:

$$C_v = \frac{Q}{\sqrt{\Delta P / SG}} \quad (5.191)$$

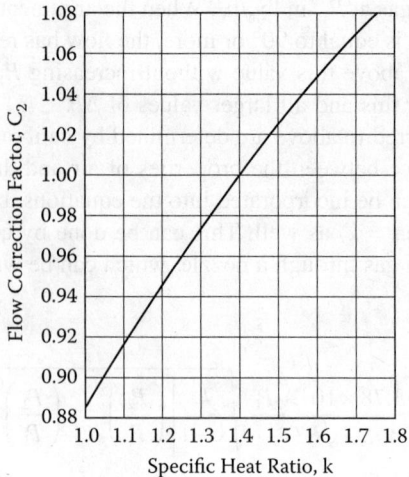

FIGURE 5.29 Correction factor for gas properties.

Then determine the Reynolds number for the valve from

$$N_{Re} = \frac{17250 Q}{\sqrt{C_v} \nu_{cs}} \tag{5.192}$$

where Q is in gpm, ΔP is in psi, and ν_{cs} is the fluid kinematic viscosity (μ/ρ) in centistokes. The viscosity correction factor, F_v, is then read from the middle line on Figure 5.30 and used to calculate a corrected value of C_v as follows:

$$C_{v_c} = C_v F_v \tag{5.193}$$

The proper valve size and percent opening are then found from the table for the valve flow coefficient (e.g., Table 5.14) at the point where the coefficient is equal to or higher than this corrected value.

To predict flow rate for a given valve (i.e., a given C_v) and given ΔP, the maximum flow rate (Q_{max}) is determined as

$$Q_{max} = C_v \sqrt{\Delta P / SG} \tag{5.194}$$

The Reynolds number is then calculated from Equation (5.192), and the viscosity correction factor, F_v, is read from the bottom curve in Figure 5.30. The corrected flow rate is then

$$Q_c = \frac{Q_{max}}{F_v} \tag{5.195}$$

To predict pressure drop for a given valve (C_v) and given flow rate (Q), calculate the Reynolds number as above and read the viscosity correction factor, F_v, from the top line of Figure 5.30. The predicted pressure drop across the valve is then

Fluid Flow

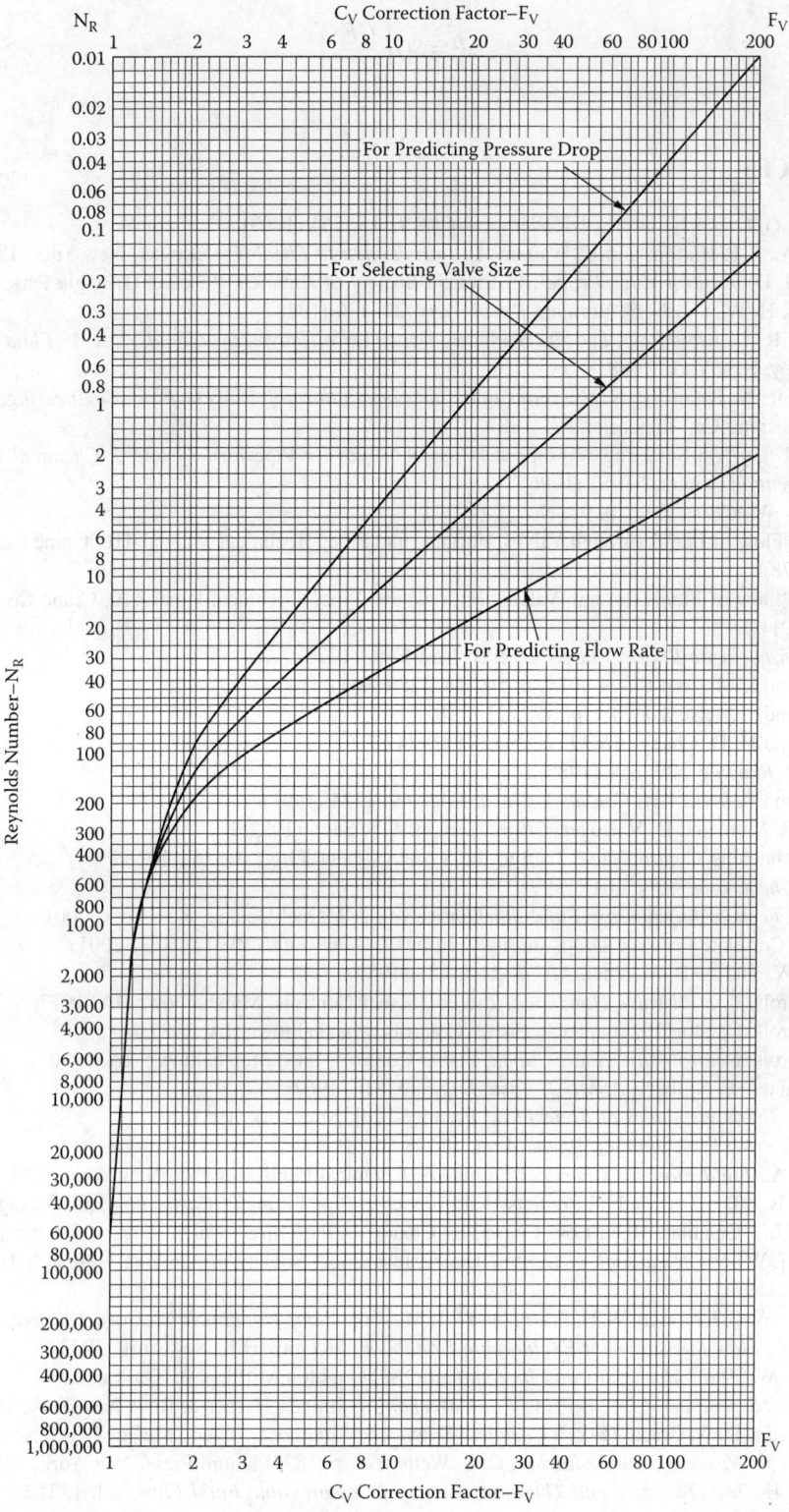

FIGURE 5.30 Viscosity correction factor for C_v. (From Fisher Controls, 1977.)

$$\Delta P = SG\left(\frac{QF_v}{C_v}\right)^2 \qquad (5.196)$$

REFERENCES

Astarita, G., G. Greco Jr., and L. Nicodemo, *AIChE J.*, 15, 564 (1969).
Barnes, H. A., J. F. Hutton, and K. Walters, *An Introduction to Rheology*, Elsevier, New York, 1989.
Baumann, H. D., *Control Valve Primer*, Instrument Society of America, Research Triangle Park, NC, 1991.
Bewersdorff, H. W., and N. S. Berman, *Rheol. Acta*, 27, 130 (1988).
Bird, R. B., R. C. Armstrong, and O. Hassager, *Dynamics of Polymeric Liquids*, vol. 1, *Fluid Mechanics*, Wiley, New York, 1988.
Cheremisinoff, N. P., and P. N. Cheremisinoff, *Instrumentation for Process Flow Engineering*, Technomic Publishing Co., Lancaster, PA, 1987.
Christy, J. R. E., "On Selecting Appropriate Control Valves for Pipework Systems," *Chemical Engineering Education*, winter, 54–57 (1996).
Churchill, S. W., *Chem. Eng.*, p. 91, Nov. 7 (1977).
Crane Co., "Flow of Fluids through Valves, Fittings, and Pipe," Technical Manual 410, Crane Co., New York (1978).
Crane Co., "Flow of Fluids through Valves, Fittings, and Pipe," Technical Paper 410, Crane Co., New York (1991).
Darby, R., *Viscoelastic Fluids*, Marcel Dekker, New York (1976).
Darby, R., and J. Melson, *Chem. Eng.*, p. 59, Dec. 28 (1981).
Darby, R., and J. Melson, *J. Pipelines*, 2, 11 (1982).
Darby, R., and H. D. Chang, *AIChE J.*, 30, 274 (1984).
Darby, R., *J. Rheology*, 29, 359 (1985).
Darby, R., and S. Pivsa-Art, *Canad. J. Chem. Eng.*, 69, 1395 (1991).
Darby, R., R. Mun, and D. V. Boger, *Chem. Eng.*, p. 116, Sept. (1992).
Darby, R., Matching Control Valve Trim to the System, *Chem. Eng.*, 104 (6), 147 (1997).
Darby, R., *Chem. Eng.*, July, 101 (1999).
Darby, R., *Chemical Engineering Fluid Mechanics*, vol. 2, Marcel Dekker, New York, 2001.
Darby, R., "Correlate Pressure Drops through Fittings," *Chem. Eng.*, 108 (4), 127 (2001).
Dodge, D. W., and A. B. Metzner, *AIChE J.*, 5, 189 (1959).
Fisher Controls, *Control Valve Handbook*, 2nd ed., Fisher Controls, Marshalltown, IA, 1977.
Fisher Controls, Catalog 10, chapter 2, Fisher Controls, Marshalltown, IA, 1987.
Fisher Controls, *Control Valve Source Book*, Fisher Controls, Intl., Marshalltown, IA, 1990.
Goren, Y., and J. F. Norbury, *ASME J. Basic Eng.*, 89, 816, 1967.
Hall, N. A., *Thermodynamics of Fluid Flow*, Prentice-Hall, New York, 1951.
Hoffmann, L., and P. Schummer, *Rheol. Acta*, 17, 98, 1978.
Holland, F. A., *Fluid Flow for Chemical Engineers*, Chemical Publishing Co., New York, 1973.
Hooper, W. B., "The 2-K Method Predicts Head Losses in Pipe Fittings," *Chem. Eng.*, p. 97, Aug. 24 (1981).
Hooper, W. B., "Calculate Head Loss Caused by Change in Pipe Size," *Chem. Eng.*, p. 89, Nov. 7 (1988).
Hutchison, J. W., *ISA Handbook of Control Valves*, Instrument Society of America, Research Triangle Park, NC, (1971).
Krassik, I. J., W. C. Krutzsch, W. H. Frazer, and J. P. Messina, *Pump Handbook*, McGraw-Hill, New York, 1976.
Miller, R. W., *Flow Measurement Engineering Handbook*, McGraw-Hill, New York, 1983.
Murdock, J. W., *Fluid Mechanics and Its Applications*, Houghton Mifflin Co., Boston, 1976.
Olson, R. M., *Essentials of Engineering Fluid Mechanics*, 4th ed., Harper & Row, New York, 1980.
Perry, R. H., and D. W. Green, *Perry's Chemical Engineers' Handbook*, 7th ed., McGraw-Hill, New York, 1997.
Savins, J. G., in *Viscous Drag Reduction*, C. S. Wells, Ed., p. 183, Plenum Press, New York, 1969.
Shapiro, A. H., *The Dynamics and Thermodynamics of Compressible Fluid Flow*, vol. 1, The Ronald Press, New York, 1953.
Virk, P. S., and H. Baher, *Chem. Eng. Sci.*, 25, 1183 (1970).

Virk, P. S., *AIChE J.*, 21, 625 (1975).
Wang, C. B., *Ind. Eng. Chem. Fund.*, 11, 566 (1972).
White, A., *J. Mech. Eng. Sci.*, 8, 452 (1966).
White, D. Jr., and R. J. Gordon, *AIChE J.*, 21, 1027 (1975).
White, F. M., *Fluid Mechanics*, 3rd ed., McGraw-Hill, New York, 1994.

6 Heat Transfer

Kenneth J. Bell

CONTENTS

6.1 Introduction 480
6.2 Conduction Heat Transfer 481
    6.2.1 Mechanisms of Conduction and the Basic Equation 481
    6.2.2 One-Dimensional Steady-State Conduction 482
    6.2.3 Thermal Contact Resistance 486
    6.2.4 Extended Surfaces ("Fins") 487
        6.2.4.1 Types of Extended Surface 487
        6.2.4.2 Convective Heat Transfer 488
        6.2.4.3 Heat Transfer in Fins of Constant Cross Section 488
        6.2.4.4 Heat Transfer in Radial (Spiral) Fins 491
        6.2.4.5 Other Extended Surface Geometries 493
    6.2.5 Two- and Three-Dimensional Steady-State Conduction: Shape Factors 493
    6.2.6 Transient Conduction in Simple Solids 497
    6.2.7 Numerical Methods 503
6.3 Single-Phase Convection Heat Transfer 503
    6.3.1 Mechanisms of Convection 503
    6.3.2 Film Coefficient of Heat Transfer 504
    6.3.3 Dimensionless Numbers 504
        6.3.3.1 Reynolds Number, Re 504
        6.3.3.2 Nusselt Number, Nu 505
        6.3.3.3 Stanton Number, St 506
        6.3.3.4 Colburn j-Factor for Heat Transfer, j_H 506
        6.3.3.5 Prandtl Number, Pr 506
        6.3.3.6 Graetz Number, Gz 506
        6.3.3.7 Peclet Number, Pe 506
        6.3.3.8 Grashof Number, Gr 507
        6.3.3.9 The Sieder-Tate Term 507
    6.3.4 Single-Phase Heat-Transfer Correlations for Common Geometries in Forced Convection 507
        6.3.4.1. Inside Round Tubes 507
        6.3.4.2 Internally Enhanced Tubes 510
        6.3.4.3 Inside Annular Channels 510
        6.3.4.4 Flow across a Circular Cylinder 512
        6.3.4.5 Flow across Tube Banks 513
        6.3.4.6 Heat Transfer in Packed and Fluidized Beds 519
    6.3.5 Single-Phase Heat Transfer in Natural Convection 520
        6.3.5.1 Natural Convection from a Vertical Plane Surface 521
        6.3.5.2 Natural Convection from Horizontal Plates 522
        6.3.5.3 Natural Convection from a Horizontal Cylinder 523

| | | 6.3.5.4 | Natural Convection between Two Horizontal Parallel Plates | 523 |

| | 6.3.5.5 | Other Geometries | 523 |

6.4 Condensation and Vaporization Heat Transfer 523
    6.4.1 Mechanisms of Condensation 523
    6.4.2 Design Equations for Filmwise Condensation 524
        6.4.2.1 Condensation on Vertical Plane and Tubular Surfaces 524
        6.4.2.2 Condensation inside a Horizontal Tube 528
        6.4.2.3 Condensation outside Horizontal Tubes and Tube Banks 529
    6.4.3 Special Cases in Condensation 530
        6.4.3.1. Enhanced Surfaces in Condensation 530
        6.4.3.2 Condensation in the Presence of a Noncondensable Gas 530
        6.4.3.3 Condensation of a Multicomponent Vapor 530
        6.4.3.4 Condensation of Superheated Vapor 530
    6.4.4 Mechanisms of Vaporization 531
    6.4.5 Boiling Heat-Transfer Correlations 532
        6.4.5.1 Nucleate Boiling 532
        6.4.5.2 Critical Heat Flux in Pool Boiling 533
        6.4.5.3 Natural and Forced Convection Vaporization 533
    6.4.6 Special Cases in Vaporization 535
        6.4.6.1 Boiling Outside Tube Bundles 535
        6.4.6.2 Enhanced Surfaces in Boiling 536
        6.4.6.3 Subcooled Boiling 536
6.5 Heat Exchangers 536
    6.5.1 Introduction 536
    6.5.2 Types of Heat Exchangers and Their Selection 537
        6.5.2.1 Criteria for Heat-Exchanger Selection 537
        6.5.2.2 Double-Pipe Heat Exchangers 537
        6.5.2.3 Shell-and-Ttube Heat Exchangers 538
        6.5.2.4 Multitube ("Hairpin") Heat Exchangers 545
        6.5.2.5 Gasketed-Plate Heat Exchanger and Related Partially Welded
                Variants 545
        6.5.2.6 Plate-Fin (Matrix) Heat Exchangers 547
        6.5.2.7 Air-Cooled Heat Exchangers 547
        6.5.2.8 Mechanically Aided Heat Exchangers 549
    6.5.3 Principles of Heat-Exchanger Design 550
        6.5.3.1 Heat Transfer between Two Fluids Separated by a Wall 550
        6.5.3.2 The Basic Design Integral 552
        6.5.3.3 The Mean Temperature Difference Concept 552
    6.5.4 Logic of the Heat-Exchanger Design Process 560
    6.5.5 Fouling 562
References 563

6.1 INTRODUCTION

This chapter deals with the transfer of heat in systems likely to be of interest to chemical engineers. These situations include transfer within a single phase (especially between two surfaces of a solid), between a solid surface and a fluid, between fluids separated by a solid surface, and between surfaces of neighboring bodies. The basic heat-transfer processes discussed include conduction, single-phase convection, vaporization, and condensation. These processes are described physically, and representative equations are given for calculating the rate at which heat is transferred.

The fundamental relationships are then extended to the design of heat exchangers, including a description of the most important types of heat exchangers and their areas of application.

The literature on both the fundamental processes and the applications to practical heat-transfer devices is immense, and only the most important references are given here. The *Heat Exchanger Design Handbook* [1] is a particularly useful source, being both comprehensive and giving special attention to design and application of heat exchangers.

6.2 CONDUCTION HEAT TRANSFER

6.2.1 Mechanisms of Conduction and the Basic Equation

Heat in solid, liquid, or gaseous matter is the random kinetic energy of the electrons, atoms, or molecules present; temperature is a measure of the average kinetic energy possessed by the assembly of electrons, atoms, and molecules. In a solid, the atoms or molecules are on the average fixed in a given position, but vibrate about this position, moving more rapidly at higher temperatures. In metallic solids, free electrons (so called because they are not associated with a given atom but can move freely within the boundaries of the body) represent a substantial fraction of the total kinetic energy (or heat content) of the body. The electrons move faster on average in a hot metal than in a colder one.

Conduction in a nonmetallic solid occurs when higher-temperature, more rapidly vibrating atoms (or molecules) transfer some of their kinetic energy to their lower-temperature neighbors, and so on through the extent of the solid. The same mechanism also occurs in a metallic solid, but the energy transport by the free electrons is generally even more important. Since the free electrons are also responsible for the conduction of an electrical current through a metal, a relationship exists between the ability of a metal to conduct heat and to conduct electricity. Details of the conduction process are complicated. Fortunately, for the present purposes, all of the complexities may be bypassed and the problem handled in a very simple way: To a very good approximation, the heat flux, $\dot{q}/A$, in conduction is directly proportional to the temperature gradient, $-(\partial T/\partial x)$, and the constant of proportionality is termed the thermal conductivity, k:

$$\left(\frac{\dot{q}}{A}\right)_x = -k_x \left(\frac{\partial T}{\partial x}\right) \tag{6.1}$$

In Equation (6.1), the terms are subscripted x because certain solids (e.g., wood, single crystals, and some highly oriented materials like pyrolitic graphite) have different thermal conductivities in different directions, and the subscript indicates the direction of heat flow. Most engineering materials, however, may be treated as isotropic (properties independent of direction) and the subscript dropped. The minus (−) sign in Equation (6.1) arises from the fact that heat flows from a high temperature to a lower one, and therefore $(\partial T/\partial x)$ is inherently negative; hence, the thermal conductivity and the heat flux are positive.

Equation (6.1) also applies in principle to gases and liquids. The temperature difference (and therefore the density difference) in the fluid often results in natural convection effects (discussed below), usually substantially enhancing the heat-transfer rate.

The thermal conductivity is experimentally determined, and appropriate values can be found in Part V of [1] or many other sources. Values can be sensitive to composition, (e.g., a small amount of alloying material usually significantly reduces k), to density, and to microstructure. The thermal conductivity is also a function of the temperature and hence varies with position in a conducting body. However, it is usually satisfactory to evaluate the conductivity at the arithmetic mean of the surface temperatures and use that value in the calculations.

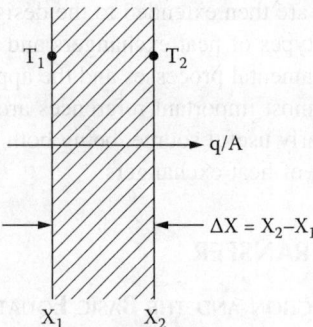

FIGURE 6.1 Conduction through a plane wall.

Equation (6.1) can be generalized to apply to three-dimensional, transient conduction with internal heat sources or sinks, but most chemical engineering applications can be treated as one- or two-dimensional, steady-state cases as described below. More advanced treatments are given in the literature [1–3].

6.2.2 One-Dimensional Steady-State Conduction

Equation (6.1) can be readily integrated for several important cases of one-dimensional steady-state conduction, including the following:

1. Plane slab of thickness Δx, with uniform temperatures on each surface (Figure 6.1). Integrating Equation (6.1) with $T = T_1$ on one surface (at $x = x_1$) and $T = T_2$ on the other surface (at $x = x_2$) gives

$$\frac{\dot{q}}{A} = \frac{k|(T_1 - T_2)|}{(x_2 - x_1)} = \frac{k|(T_1 - T_2)|}{\Delta x} \tag{6.2}$$

The vertical bars on the temperature difference indicate that the absolute value is to be used. It is assumed that the face dimensions Δy and Δz (where $\Delta y \Delta z = A$) either are large compared with Δx, or that the edges of the slab are well-insulated (adiabatic).

2. Multiple plane slabs in series with uniform temperatures on each surface (Figure 6.2). For two slabs of thicknesses Δx_1 and Δx_2 with thermal conductivities k_1 and k_2, and with $T = T_1$ on the front face of the first slab and $T = T_2$ on the rear face of the second slab, Equation (6.2) gives

$$\frac{\dot{q}}{A} = \frac{k_1|(T_1 - T^*)|}{\Delta x_i} = \frac{k_2|(T^* - T_2)|}{\Delta x_2} \tag{6.3}$$

where T^* is the temperature of the contact surface between the two slabs. This assumes no contact resistance between the two slabs; contact resistance is discussed in Section 6.2.3. T^* can be eliminated from Equation (6.3) to give

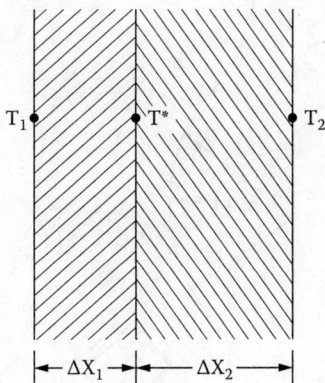

FIGURE 6.2 Conduction through two plane walls in series.

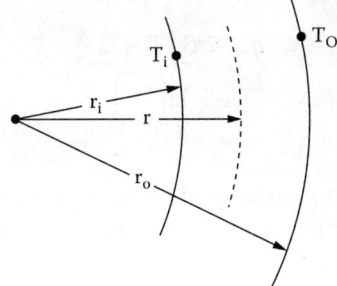

FIGURE 6.3 Conduction through a cylindrical wall.

$$\frac{\dot{q}}{A} = \frac{|(T_1 - T_2)|}{\frac{\Delta x_1}{k_1} + \frac{\Delta x_2}{k_2}} \qquad (6.4)$$

Equation (6.4) can be extended to any number of plane slabs in series. Note that the right-hand side has the form of an overall driving force $(T_1 - T_2)$, divided by the sum of the individual resistances $(\Delta x_i / k_i)$. The interface temperature T^* can be found by substituting $\dot{q}/A$ back into Equation (6.3).

3. Cylindrical tube with wall thickness $(r_o - r_i)$ and uniform temperatures on each surface (Figure 6.3). For this geometry, the area available for conduction increases from the inner wall to the outer according to the relationship $A(r) = 2\pi r L$, where r is the radius from the centerline of the tube and L is the length of the tube. Equation (6.1) may be rewritten

$$\frac{\dot{q}}{L} = -k(2\pi r)\left(\frac{dT}{dr}\right) \qquad (6.5)$$

Integrating this equation with the conditions that $T = T_i$ at $r = r_i$ (inside surface of the tube) and $T = T_o$ at $r = r_o$ (outside surface) gives

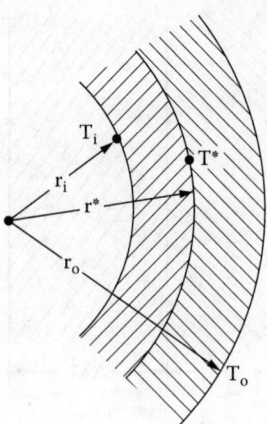

FIGURE 6.4 Conduction through a composite tube wall.

$$\frac{\dot{q}}{L} = \frac{2\pi k |(T_i - T_o)|}{\ln\left(\dfrac{r_o}{r_i}\right)} \tag{6.6}$$

The diametral ratio (D_o / D_i) can be used in the logarithmic term.
The heat flux, i.e., the heat-flow rate per unit area of heat-transfer surface, may be found from

$$\frac{\dot{q}}{A_i} = \frac{\dot{q}}{2\pi r_i L} = \frac{k|(T_i - T_o)|}{r_i \ln(r_o / r_i)} \tag{6.7}$$

for the inside surface area of the tube, and from

$$\frac{\dot{q}}{A_o} = \frac{\dot{q}}{2\pi r_o L} = \frac{k|(T_i - T_o)|}{r_o \ln(r_o / r_i)} \tag{6.8}$$

for the outside surface area. It is necessary to specify the area involved whenever a heat flux is given.

4. Multiple concentric cylinders with uniform temperatures on inside and outside surfaces (Figure 6.4), with zero contact resistance at the interface. Applying Equation (6.6) to each cylinder gives

$$\frac{\dot{q}}{L} = \frac{2\pi k_i |(T_i - T^*)|}{\ln\left(\dfrac{r^*}{r_i}\right)} = \frac{2\pi k_0 |(T^* - T_0)|}{\ln\left(\dfrac{r_0}{r^*}\right)} \tag{6.9}$$

where k_i and k_o are the thermal conductivities of the inner and outer cylinders, respectively. Eliminating T^* gives

Heat Transfer

$$\frac{\dot{q}}{L} = \frac{|(T_i - T_0)|}{\dfrac{\ln(r^*/r_i)}{2\pi k_i} + \dfrac{\ln(r_0/r^*)}{2\pi k_0}} \qquad (6.10)$$

The right-hand side is the overall driving force divided by the sum of the resistances, which can be extended to include additional concentric cylinders. The value of $(\dot{q}/L)$ found from Equation (6.10) can be inserted into Equation (6.9) to find T^*.

Example 6.1

Calculate the rate of heat transfer through a bimetallic tube consisting of an outer tube of low-carbon steel and an inner tube ("liner") of 347 stainless steel. The outer tube has an outside diameter of 31.8 mm (1.25 in.) and an inside diameter of 19.0 mm (0.75 in.). The liner has an outside diameter of 18.9 mm (0.75 in.) and a wall thickness of 20 BWG (= 0.889 mm = 0.035 in.), giving a nominal inside diameter of 17.3 mm (0.680 in.).

The inside wall temperature of the liner is 260°C (500°F) and the outside wall temperature of the outer tube is 100°C (212°F). Assume the liner and the outer tube are in perfect thermal contact.

Solution

Estimate that the average temperature of the outer tube is 150°C (302°F) and that of the liner is 230°C (446°F). (These estimates can be checked later.) Then the thermal conductivities are 49.1 W/m·K (28.4 Btu/h·ft·°F) for the low-carbon steel and 17.5 W/m·K (10.1 Btu/h·ft·°F) for the liner. Substituting into Equation (6.10) using diameters:

$$\frac{\dot{q}}{L} = \frac{260-100}{\dfrac{\ln(31.8/19.0)}{2\pi(49.1)} + \dfrac{\ln(19.0/17.3)}{2\pi(17.5)}} = \frac{160}{1.67\times 10^{-3} + 8.52\times 10^{-4}}$$

$$\frac{\dot{q}}{L} = 6.34\times 10^4 \text{ W/m} \;(= 6.60\times 10^4 \text{ Btu/ft})$$

The interface temperature can be calculated from Equation (6.9), rewritten, and using diameters:

$$T^* = T_i - \frac{(\dot{q}/L)\ln(D^*/D_i)}{2\pi k_i}$$

$$= 260 - \frac{(6.34\times 10^4)\ln(19.0/17.3)}{2\pi(17.5)} = 260 - 54.0 = 206°C \;(= 403°F)$$

giving an average temperature of the stainless steel of 233°C (451°F) and of the low-carbon steel 153°C (307°F). These are close enough to the assumed temperatures and the corresponding thermal conductivities that no further calculations are required.

The heat fluxes on the inside and outside tube surfaces can now be calculated:

$$\frac{\dot{q}}{A_i} = \frac{\dot{q}}{\pi D_i L} = \frac{6.34\times 10^4}{\pi(0.01714)} = 1.18\times 10^6 \text{ W/m}^2 (= 3.74\times 10^5 \text{ Btu/hr ft}^2)$$

for the inside surface area of the tube and

$$\frac{\dot{q}}{A_o} = \frac{\dot{q}}{\pi D_o L} = \frac{6.34 \times 10^4}{\pi(0.0315)} = 6.41 \times 10^5 \text{ W/m}^2 (= 2.03 \times 10^5 \text{ Btu/hr ft}^2)$$

for the outside surface of the tube.

6.2.3 Thermal Contact Resistance

When two solid surfaces, even finely machined ones, are placed in physical contact, there are still gaps between the two surfaces due to the inherent irregularities—roughness—of each surface. These gaps are usually filled with the ambient atmosphere, which usually has a lower effective thermal conductivity than the materials in contact. Heat transfer from one surface to another thus tends to be concentrated at those areas where the surfaces are in direct physical contact, which are also the areas on which the pressure forces act to keep the two surfaces in static equilibrium. The concentration of the heat-flow paths on only a part of the adjoining surfaces increases the local temperature gradients and in effect creates an additional resistance to heat transfer at the interface. This is termed the *thermal contact resistance*, and its possible effect in reducing the rate of heat transfer through composite materials (such as the bimetallic tube described above) must be considered.

Several factors determine the magnitude of the resistance, including

1. The roughnesses of the surfaces (height of surface irregularities)
2. The topology of the surfaces, e.g., whether the surfaces are grooved or randomly bumpy (grainy roughness), etc.
3. Hardness of the surface materials
4. Pressure exerted by one surface on the other
5. Thermal conductivities of the surface materials and the interstitial fluid
6. Other possible thermal transport mechanisms between the two surfaces, including radiation, natural convection, and free molecule transport at low pressures

The most generally applicable predictive method for contact resistance is ascribed to Irvine and Taborek in Section 2.4.6 of the *Heat Exchanger Design Handbook* [1]. Their method requires numerical values of the above properties, which usually do not have high accuracy, and the authors estimate about 25% mean error, with an error spread of a factor of 2 about the actual value. A typical thermal contact resistance is on the order of

$$1 - 2 \times 10^{-4} \frac{\text{m}^2 \text{K}}{\text{W}} \ (6 - 12 \times 10^{-4} \text{ hr ft}^2 \text{ °F/Btu})$$

The contact resistance is also affected by differential thermal expansion when the surfaces are at elevated temperatures or with large temperature differences in the two materials. The problem is aggravated by thermal cycling, especially if one material is heated above its elastic behavior range, and must be considered with finned tubes in air-cooled heat exchangers.

Example 6.2

Recalculate the results of Example 6.1 assuming a thermal contact resistance of 2×10^{-4} m²·K/W at the interface between the low-carbon and stainless steels. Assume that the dimensions are unchanged.

Heat Transfer

Solution

The additional resistance may be added to the denominator of Equation (6.10):

$$\frac{\dot{q}}{L} = \frac{160}{\dfrac{\ln(31.8/19.0)}{2\pi(49.1)} + 2\times 10^{-4} + \dfrac{\ln(19.0/17.3)}{2\pi(17.5)}}$$

$= 5.88 \times 10^4$ W/m (6.12×10^4 Btu/ft), a 7.5% reduction.

6.2.4 Extended Surfaces ("Fins")

6.2.4.1 Types of Extended Surface

It is sometimes advantageous to add heat-transfer area to a tube or other heat-transfer device by using extended surface or fins on the primary surface. These additional areas may be in the form of longitudinal or axial fins (Figure 6.5) or radial fins (Figure 6.6). The fins may be formed by extrusion or upsetting of the surface of a thick-walled tube (integral fins), or wrapped on the surface under tension ("tension-wound"), or welded or brazed to the tube. Studs or pin fins may also be used (Figure 6.7).

Using extended surfaces is addressed in several later sections on specific types and applications of heat exchangers. This section analyzes the efficiency of these surfaces, which is a problem of heat conduction, with the additional consideration of heat transfer to or from the surrounding fluid by convection.

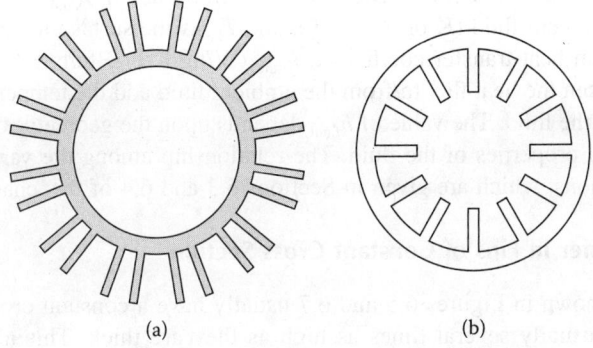

FIGURE 6.5 Longitudinal fins on a tube: (a) external, (b) internal.

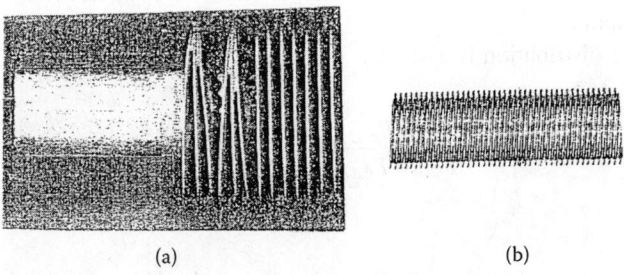

FIGURE 6.6 Radial fins on a tube: (a) high fin, (b) low fin.

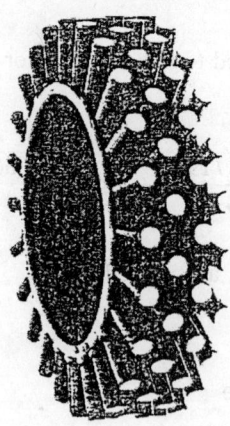

FIGURE 6.7 Studs or pin fins on a tube.

6.2.4.2 Convective Heat Transfer

Convection is the transfer of heat to or from, and within, flowing fluids. Section 6.3 of this chapter provides a more extensive treatment of convective heat transfer. For the analysis of heat-transfer fins, the rate of heat transfer to/from the fin surface from/to the ambient fluid is given by the equation

$$(\dot{q}/A)_{fin} = h_{fluid} \left| (T_{fluid} - T_{fin}) \right| \tag{6.11}$$

where $(\dot{q}/A)_{fin}$ is the heat flux (W/m² or Btu/h·ft²) to or from the fin, T_{fluid} is the bulk (mixed mean) temperature of the ambient fluid (K or °C or °F), and T_{fin} is the surface temperature of the fin (K or °C or °F). The film heat-transfer coefficient, h_{fluid} (W/m²·K or Btu/h·ft²·°F), is the constant of proportionality between the heat flux to/from the ambient fluid and the temperature differential for heat transfer to/from the fluid. The value of h_{fluid} depends upon the geometry of the system and the velocity and physical properties of the fluid. The relationship among the variables is in the form of empirical correlations, which are given in Sections 6.3 and 6.4 of this chapter.

6.2.4.3 Heat Transfer in Fins of Constant Cross Section

Fins such as those shown in Figures 6.5 and 6.7 usually have a constant cross-sectional area for conduction and are usually several times as high as they are thick. This allows the use of the one-dimensional conduction equation to calculate the temperature profile in the fin (Figure 6.8). It is also usually assumed that the film heat-transfer coefficient is uniform over the surface (nonconservative) and that the fin tip is adiabatic (i.e., no heat transfer, which is a slightly conservative assumption).

The temperature distribution is given by

$$\frac{T_{fin} - T_{fluid}}{T_{fin,b} - T_{fluid}} = \frac{\cosh N(X - x)}{\cosh N \, X} \tag{6.12}$$

where

$$N = \sqrt{\frac{2h_{fluid}}{Yk_{fin}}} \tag{6.13}$$

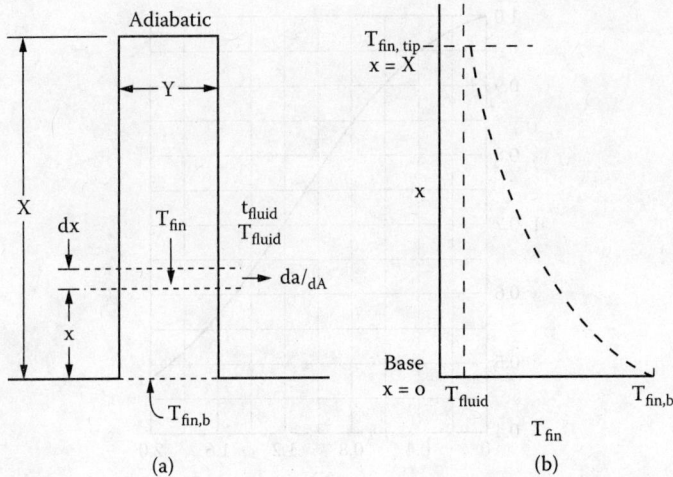

FIGURE 6.8 (a) Schematic for heat transfer in a constant cross-sectional fin, (b) typical temperature profile.

for the longitudinal fin, and

$$N = \sqrt{\frac{4h_{fluid}}{D_{stud}k_{fin}}} \qquad (6.14)$$

for the stud fin.

In these equations, $T_{fin,b}$ is the temperature at the base of the fin, X is the height of the fin, Y is the thickness of the fin, D_{stud} is the diameter of the stud fin, and k_{fin} is the thermal conductivity of the fin material.

The fin efficiency, η_{fin}, is defined as the ratio of the actual amount of heat transferred by the fin to the amount that would be transferred if the fin material had an infinitely high thermal conductivity. For the longitudinal fin,

$$\eta_{fin} = \frac{\tanh\left[X\sqrt{\frac{2h_{fluid}}{Yk_{fin}}}\right]}{X\sqrt{\frac{2h_{fluid}}{Yk_{fin}}}} = \frac{\tanh(XN)}{XN} \qquad (6.15)$$

and for the stud fin,

$$\eta_{fin} = \frac{\tanh\left[X\sqrt{\frac{4h_{fluid}}{D_{stud}k_{fin}}}\right]}{X\sqrt{\frac{4h_{fluid}}{D_{stud}k_{fin}}}} = \frac{\tanh(XN)}{XN} \qquad (6.16)$$

Figure 6.9 shows η_{fin} as a function of XN.

The effective area for heat transfer from a finned surface is the sum of the "prime" area (the tube surface between the fins) and the finned area multiplied by the fin efficiency:

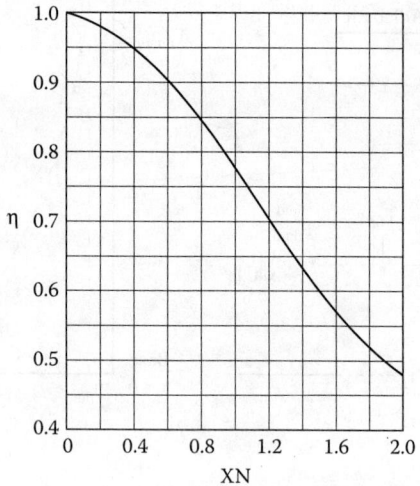

FIGURE 6.9 Fin efficiency for longitudinal and stud fins of constant cross-sectional area.

$$A_{eff} = A_{prime} + A_{fin}\eta_{fin} \tag{6.17}$$

In using the fin efficiency, the seemingly precise analytical solutions actually are premised upon several assumptions of unknown validity. Therefore, the resulting numerical values must be used cautiously and conservatively. And it must be remembered that, when using convective heat-transfer correlations for finned surfaces, the experimental data underlying these correlations were reduced using computed fin efficiencies.

Example 6.3

Twelve longitudinal low-carbon steel fins, each 15.8 mm (0.625 in.) high and 3.18 mm (0.125 in.) thick, are uniformly spaced around the outside surface of a 2-in. Schedule 40 (outside diameter = 6.033 cm = 2.375 in.) low-carbon steel pipe. The outside pipe wall temperature is 150°C (302°F), and it is cooled by a viscous oil flowing along the pipe at an average bulk temperature of 40°C (104°F). The oil has a film heat-transfer coefficient of 120 W/m²·K (21.1 Btu/h·ft²·°F), which is assumed constant over the outside surface of the pipe and the fins. The thermal conductivity of the steel is assumed constant at 49.1 W/m·K (28.4 Btu/h·ft·°F). Find the rate of heat transfer from the finned pipe per unit length and compare this with the heat transfer from the same length of unfinned pipe.

Solution

The heat-transfer area of the fins per meter of finned pipe (neglecting the fin tips) is

$$\frac{A_{fin}}{L} = 12(2)(15.8 \times 10^{-3}) = 0.379 \text{ m}^2/\text{m} \ (= 1.25 \text{ ft}^2/\text{ft})$$

The outside surface area of the pipe, excluding the area covered by the bases of the fins, A_{prime}, is

$$\frac{A_{prime}}{L} = \pi(6.033 \times 10^{-2}) - 12(3.18 \times 10^{-3})$$

$$= 0.1514 \text{ m}^2/\text{m} \ (= 0.497 \text{ ft}^2/\text{ft})$$

Heat Transfer

The fin efficiency is found from Equation (6.15):

$$X\sqrt{\frac{2h_{fluid}}{Yk_{fin}}} = 15.8\times 10^{-3}\sqrt{\frac{2(120)}{3.18\times 10^{-3}(49.1)}} = 0.619$$

Since this quality is dimensionless, the same value would be found for any set of consistent units:

$$\eta_{fin} = \frac{\tanh(0.619)}{0.619} = 0.890$$

The effective area is

$$\frac{A_{eff}}{L} = 0.1514 + 0.379(0.890) = 0.489 \text{ m}^2/\text{m} \; (= 1.603 \text{ ft}^2/\text{ft})$$

The heat transfer from the finned pipe is

$$\frac{\dot{q}}{L} = h_{fluid}\left(\frac{A_{eff}}{L}\right)(T_{pipe} - T_{fluid})$$

$$= 120(0.489)(150 - 40)$$

$$= 6.45 \times 10^3 \text{ W/m} \; (= 6.70 \times 10^3 \text{(Btu/hr)/ft})$$

By comparison, the heat transfer from the plain pipe is

$$\left(\frac{\dot{q}}{L}\right)_{plain} = h_{fluid}(\pi D_o)(T_{pipe} - T_{fluid})$$

$$= 120(\pi)(6.033\times 10^{-2})(150 - 40)$$

$$= 2.50 \times 10^3 \text{ W/m} \; (= 2.60 \times 10^3 \text{ (Btu/hr)/ft})$$

6.2.4.4 Heat Transfer in Radial (Spiral) Fins

The analysis of fin efficiency in radial, or spiral, fins (Figure 6.6) is substantially more complex because of the changing areas for both conduction and convection heat transfer. Again assuming a uniform heat-transfer coefficient, constant fin thickness, and adiabatic fin tip, Gardner [4] obtained the results shown in Figure 6.10, where D_o and D_i are the outside (tip) and inside (root) diameters of the fin, respectively.

Example 6.4

A low-carbon steel tube with a 25.4 mm (1 in.) outside diameter is radially finned with an aluminum (Alloy 1100) strip 15.8 mm (0.625 in.) high and 0.483 mm (0.019 in.) average thickness uniformly spaced at 400 fins/m (10.2 fins/in.). The outside surface of the steel tube is 80°C (176°F), and the tube is cooled by air in cross flow, with a bulk air temperature of 30°C (86°F) with a film heat-transfer coefficient of 70 W/m²K (12.3 Btu/h·ft²·°F), assumed uniform over the entire outside surface

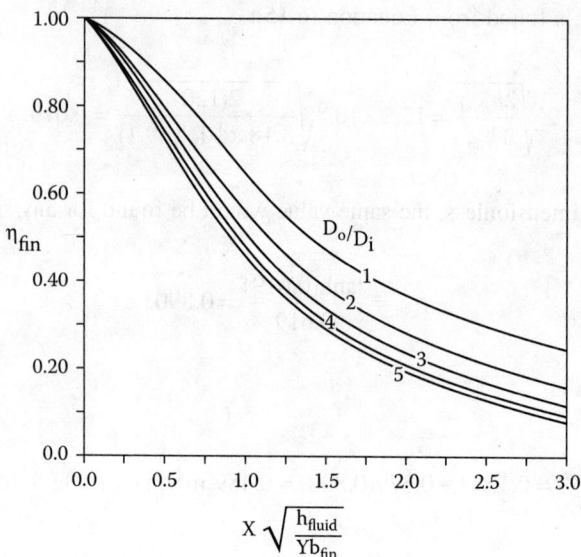

FIGURE 6.10 Fin efficiency for radial fins of constant thickness.

of the tube and fins. Assume negligible contact resistance. The thermal conductivity of Alloy 1100 is 218 W/m·K (126 Btu/h·ft·°F).

Calculate the heat-transfer rate per unit length of tube and compare that result with the unfinned tube under the same conditions.

Solution

The total fin area per unit length of tube is

$$\frac{A_{fin}}{L} = \frac{\pi}{4}(D_{fin}^2 - D_{root}^2)(2)(400)$$

$$= \frac{\pi}{4}[(0.0570)^2 - (0.0254)^2](2)(400)$$

$$= 1.636 \text{ m}^2/\text{m} \ (= 5.38 \text{ ft}^2/\text{ft})$$

The prime tube outside area (that area not covered by the fin bases) is

$$\frac{A_{prime}}{L} = \pi(0.0254)[1 - 400(0.483 \times 10^{-3})]$$

$$= 6.44 \times 10^{-2} \text{ m}^2/\text{m} \ (= 0.211 \text{ ft}^2/\text{ft})$$

The fin efficiency is found from

$$X\sqrt{\frac{h_{fluid}}{Yk_{fin}}} = 15.8 \times 10^{-3}\sqrt{\frac{70}{(0.483 \times 10^{-3})(218)}} = 0.407$$

$$\frac{D_{fin}}{D_{root}} = \frac{25.4 + 2(15.8)}{25.4} = 2.24$$

From Figure 6.10,

$$\eta_{fin} = 0.83$$

Then,

$$\frac{A_{eff}}{L} = 6.44 \times 10^{-2} + 0.83(1.636)$$

$$= 1.42 \text{ m}^2/\text{m} \ (= 4.67 \text{ ft}^2/\text{ft})$$

and, from Equation (6.11),

$$\frac{\dot{q}}{L} = 70(1.42)(80-30)$$

$$= 4.97 \times 10^3 \text{ W/m} \ (= 5.17 \times 10^3 \text{ (Btu/hr)/ft})$$

By comparison, the bare tube without fins has an outside area of

$$\frac{A_0}{L} = \pi D_0 = \pi(25.4 \times 10^{-3})$$

$$= 7.98 \times 10^{-2} \text{ m}^2/\text{m} \ (= 0.2618 \text{ ft}^2/\text{ft})$$

and a heat-transfer rate of

$$\frac{\dot{q}}{L} = 70(7.98 \times 10^{-2})(80-30)$$

$$= 279 \text{ W/m} \ (= 290 \text{ (Btu/hr)/ft})$$

In this case, the finned tube transfers 16.7 times as much heat per unit length as the plain tube.

6.2.4.5 Other Extended Surface Geometries

A wide variety of other geometrical configurations have been proposed for enhancing heat transfer in single-phase, condensing, and vaporizing systems. Extended treatments of many of these geometries are available in the literature [5–7].

6.2.5 TWO- AND THREE-DIMENSIONAL STEADY-STATE CONDUCTION: SHAPE FACTORS

Many conduction problems involving one-, two-, and three-dimensional steady-state heat flows employ a "shape factor" S, defined by Equation (6.18):

$$S = \frac{\dot{q}}{k\Delta T} \qquad (6.18)$$

where $\dot{q}$ is the rate of heat flow (W or Btu/h), k is the thermal conductivity of the conducting medium, and ΔT is the temperature difference between the heat source and the heat sink (both of which are assumed to be isothermal).

The shape factor may be found by one of two primary methods: analytical solution of the fundamental equations, or numerical solution in generalized dimensions presented in simple empirical equations to a sufficient degree of accuracy.

A solution of useful shape factors is given in Figure 6.11, taken from Parker et al. [8].

Example 6.5

A rectangular steel cold box with 4.76 mm (3/16 in.) thick walls has interior dimensions of 1.0 m by 1.25 m by 2.0 m (3.28 ft by 4.10 ft by 6.56 ft). It is insulated on all sides by a 10 cm (3.94 in.) thick layer of rock wool with a thermal conductivity of 0.040 W/m·K (0.023 Btu/h·ft·°F). The interior wall temperature is −20°C (−4°F), and the exterior surface of the insulation is at 40°C (104°F).

Calculate the heat-transfer rate from the surrounding atmosphere to the interior of the box.

Solution

The thermal resistance of the steel wall is negligible compared with that of the insulation. Assume zero contact resistance between the wall and the insulation.

Refer to the last diagram in Figure 6.11:

$$\sum L = 4(1.0) + 4(1.25) + 4(2.0) = 17.0 \text{ m} (= 55.8 \text{ ft})$$

$$\Delta X = 0.10 \text{ m} (= 0.328 \text{ ft})$$

$$A = 2(1.0)(1.25) + 2(1.0)(2) + 2(1.25 \times 2.0) = 11.5 \text{ m}^2 (= 124 \text{ ft}^2)$$

$$S = \frac{11.5}{0.10} + 0.54(17.0) + 1.2(0.1) = 124.3 \text{ m} (= 408 \text{ ft})$$

$$\dot{q} = (124.3 \text{ m})\left(0.040 \frac{\text{W}}{\text{m} \cdot \text{K}}\right)(40 - (-20))°\text{C}$$

$$\dot{q} = 298 \text{ W} (= 1020 \text{ Btu/hr})$$

Note that this calculation does not consider the convective heat transfer on either side of the insulated surface. An average temperature difference between the outside atmosphere and the outer surface could be obtained by using

$$\dot{q} = \bar{h} A_{outside} (T_{atm} - T_{surface})$$

where $\bar{h}$ is an appropriate film heat-transfer coefficient (Section 6.3), $A_{outside}$ is the outside area of the box, and T_{atm} and $T_{surface}$ are the temperatures of the atmosphere and the outside surface of the box, respectively. This assumes that the entire outside area of the box is exposed to the atmosphere and that $\bar{h}$ is constant and uniform.

Shape	Diagram	$S = \dfrac{\dot{q}}{k\,\Delta T}$
Long hollow cylinder of length L.	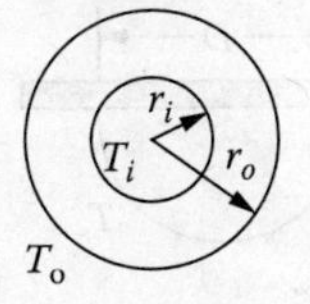	$\dfrac{2\pi L}{\ln \dfrac{r_o}{r_i}}$
Hollow sphere.	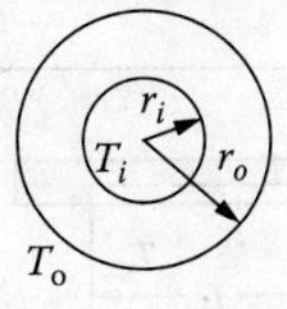	$\dfrac{4\pi r_o r_i}{r_o - r_i}$
Cylinder with square insulation, length L.	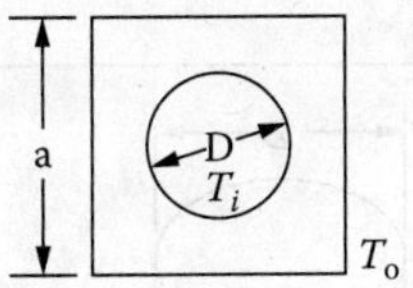	$\dfrac{2\pi L}{\ln\left(1.08\,\dfrac{a}{D}\right)}$
Eccentric parallel cylinders of length L, with eccentricity e.	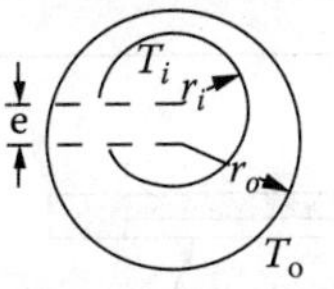	$\dfrac{2\pi L}{\ln\left(\dfrac{\sqrt{(r_o+r_i)^2 - e^2} + \sqrt{(r_o-r_i)^2 - e^2}}{\sqrt{(r_o+r_i)^2 - e^2} - \sqrt{(r_o-r_i)^2 - e^2}}\right)}$
Cylinder of diameter D and length L buried in a semi-infinite medium having a temperature at great distance T_∞. The cylinder is located a distance z from the surface.	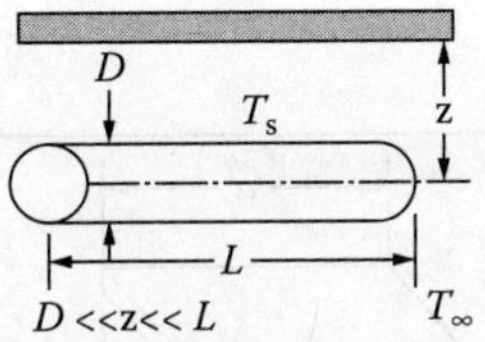	$\dfrac{2\pi L}{\ln\dfrac{2L}{D}\left(1 + \dfrac{\ln(L/2z)}{\ln(2L/D)}\right)}$
A vertical cylinder of length L and diameter D, placed in a semi-infinite medium having an adiabatic surface, and temperature T_∞ at large distance.	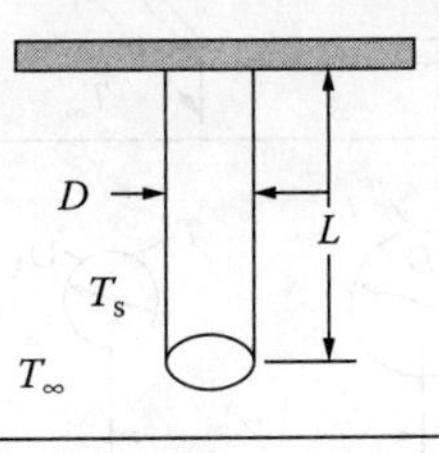	$\dfrac{2\pi L}{\ln\left(\dfrac{4L}{D}\right)}$
Sphere of diameter D, buried at distance z below adiabatic surface in a semi-infinite medium having temperature T_∞ at a large distance.	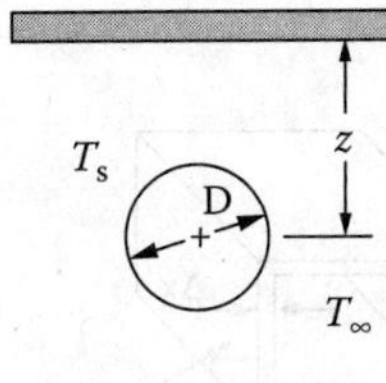	$\dfrac{2\pi D}{1 + \dfrac{D}{4z}}$

FIGURE 6.11 Shape factors for various steady-state conduction geometries. (From Parker, J. D., Boggs, S. H., and Blick, E. F., *Introduction to Fluid Mechanics and Heat Transfer*, Addison-Wesley, Reading, MA, 1969.)

Continued

Shape	Diagram	$S = \dfrac{\dot{q}}{k\,\Delta T}$
Half-sphere submerged into the surface of a semi-infinite medium with otherwise adiabatic surface and temperature T_∞ at a large distance.	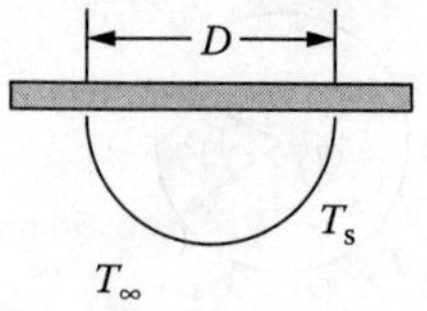	πD
Disk placed on surface of semi-infinite medium with otherwise adiabatic surface and temperature T_∞ at a large distance.	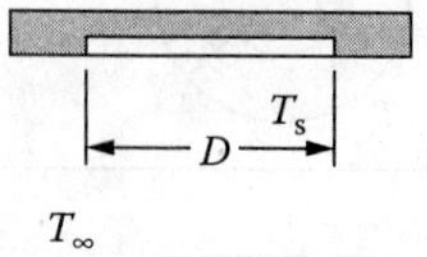	$2D$
Thin disk in infinite medium having temperature T_∞ at a large distance.	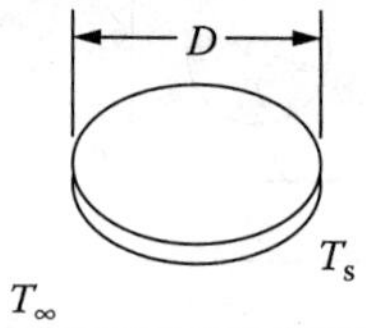	$4D$
Rectangular plate on surface of semi-infinite medium with otherwise adiabatic surface and temperature T_∞ at large distance. Plate has dimensions $a \times b$.	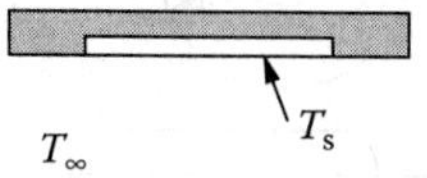	$\dfrac{\pi a}{\ln \dfrac{4a}{b}}$ $a > b$
Thin rectangular plate of dimensions $a \times b$ buried in infinite medium having temperature T_∞ at large distance	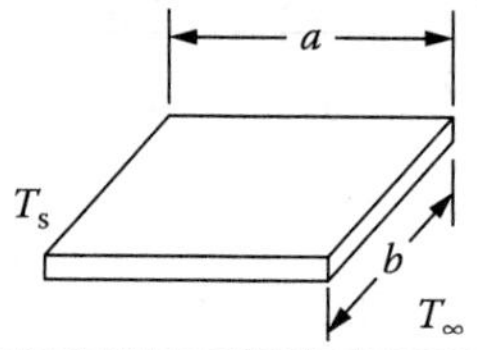	$\dfrac{2\pi a}{\ln \dfrac{4a}{b}}$ $a > b$
Conduction between cylinders of length L a distance Z apart, located in an infinite medium. No heat loss from cylinders to medium is considered.	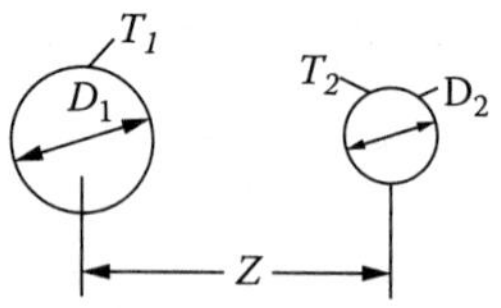	$\dfrac{2\pi L}{\cosh^{-1}\left(\dfrac{4Z^2 - D_1^2 - D_2^2}{2D_1 D_2}\right)}$
Conduction between inside and outside surfaces of a rectangular box having uniform inside and outside surface temperatures. Wall thickness Δx is less than any inside dimensions.	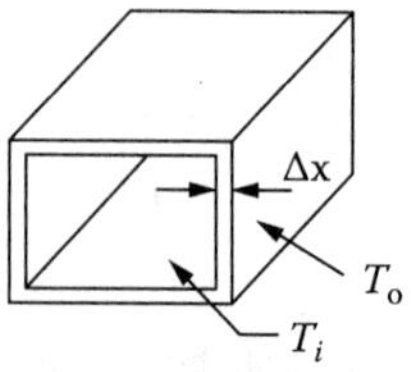	$\dfrac{A}{\Delta x} + 0.54\Sigma L + 1.2\,\Delta x$ ΣL = Sum of all inside lengths, Δx = Thickness of walls, A = Inside surface area.

FIGURE 6.11 Continued.

6.2.6 Transient Conduction in Simple Solids

The analytical solutions for transient conduction in plates, cylinders, and spheres have been obtained by Heisler [9] and the solutions represented graphically for more convenient use. These solutions are for the case of a solid of initially uniform temperature T_0 exposed at time zero to a surrounding fluid medium at a constant temperature T_∞. The surface of the solid is cooled or heated by the fluid with a constant convective heat-transfer coefficient $\bar{h}$. The solid is assumed to have a constant thermal conductivity k_s and a constant thermal diffusivity α, defined as

$$\alpha = \frac{k_s}{\rho c_p} \quad (6.19)$$

where ρ is the density and c_p is the specific heat of the solid.

The charts for the flat plate of infinite length and width are given in Figures 6.12a and 6.12b, charts for infinitely long cylinders in Figures 6.13a and 6.13b, and for the sphere in Figures 6.14a and 6.14b. The first chart in each pair relates the dimensionless temperature at the center of the solid,

$$\frac{T_M - T_\infty}{T_o - T_\infty} \quad (6.20)$$

to the dimensionless time, expressed as the Fourier number

$$Fo = \frac{\alpha t}{L^2} \quad (6.21)$$

for the plane, and

$$Fo = \frac{\alpha t}{R^2} \quad (6.22)$$

for the cylinder and sphere.

In the Fourier number, α is the thermal diffusivity (Equation (6.19)), t is the time, L is the half-thickness of the plane, and R is the radius of the cylinder or sphere.

The reciprocal Biot number ($1/Bo$) is the parameter incorporating the convection. It is given as

$$\frac{1}{Bo} = \frac{k_s}{\bar{h}L} \quad (6.23)$$

for the plane, and

$$\frac{1}{Bo} = \frac{k_s}{\bar{h}R} \quad (6.24)$$

for the cylinder and sphere. The Biot number is a measure of the ability of an object to transfer heat to/from an external fluid compared with its ability to conduct heat internally. *Fo* and *1/Bo* are dimensionless.

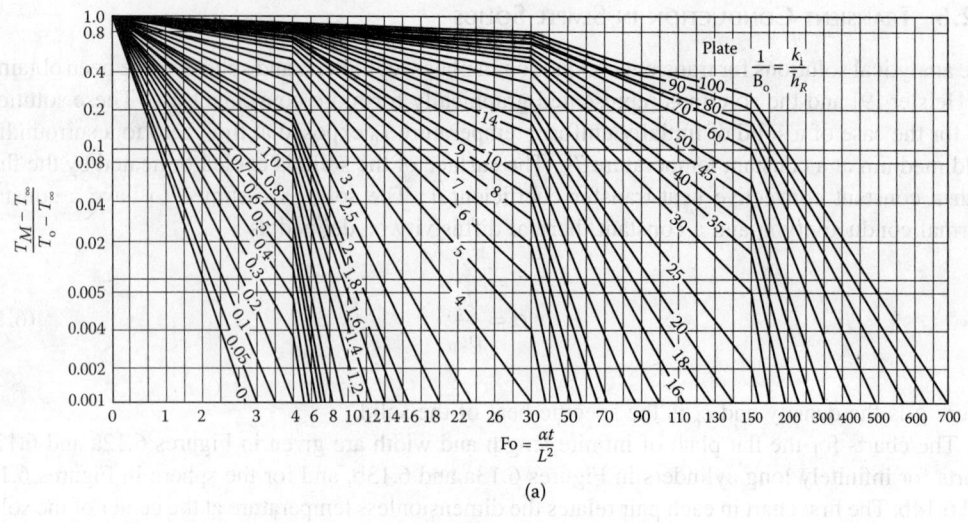

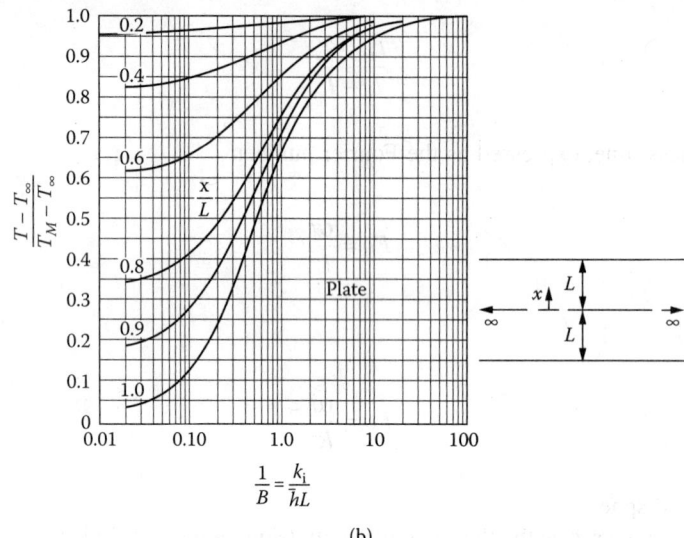

FIGURE 6.12 (a) Variation of the midplane temperature of a plane solid with dimensionless time. (b) Variation of temperature in a plane solid as a function of position.

The second chart in each pair (Figures 6.12b, 6.13b, and 6.14b) relates the temperature at any position in the solid to the temperature at the center at any given time, using as parameters $1/Bo$ and the relative position in the solid between the center and the surface.

A property of these solutions is that they may be combined to give the solution to additional cases. For example, consider the case of a rectangular solid having dimensions X, Y, Z. The temperature at a given position within that solid at a given time $T(x,y,z,t)$ is given by

$$T(x,y,z,t) = T(x,t)T(y,t)\,T(z,t) \tag{6.25}$$

where $T(x,t)$ is the solution found for position x at time t assuming Y and Z to be infinite, $T(y,t)$ the solution for position y at time t assuming X and Z to be infinite, and $T(z,t)$ the solution for position z at time t assuming X and Y to be infinite.

The case for a short cylinder is illustrated in the following example.

Heat Transfer

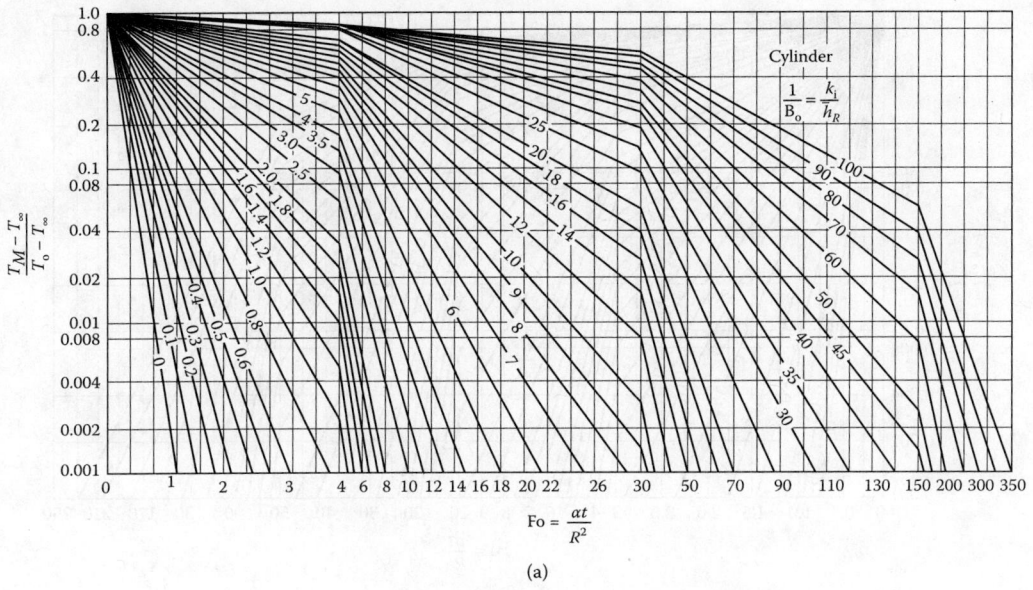

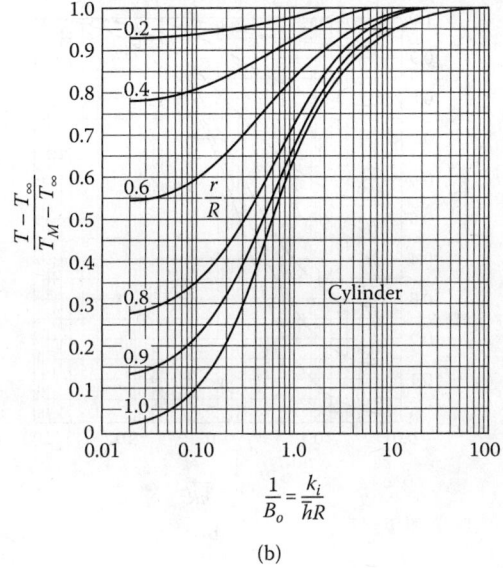

FIGURE 6.13 (a) Variation of the center temperature of a cylinder with dimensionless time. (b) Variation of temperature in a cylinder as a function of position.

Example 6.6

Small cylinders of a plastic are to be heated by air at 65°C (149°F) from a uniform temperature of 20°C (68°F) to a midpoint temperature of 60°C (140°F). The cylinder diameter and length are, respectively, 6.0 mm (0.236 in.) and 25 mm (0.984 in.), and the average convective heat-transfer coefficient between the air and the cylinder is estimated to be 50 W/m²K (8.8 Btu/h·ft²·°F). The properties of the plastic are density 1186 kg/m³ (74 lb$_m$/ft³), thermal conductivity 0.194 W/m·K (0.112 Btu/h·ft·°F), and specific heat 1520 J/kgK (= 0.365 Btu/lb$_m$°F).

Find the heating time necessary to bring the midpoint temperature to 60°C.

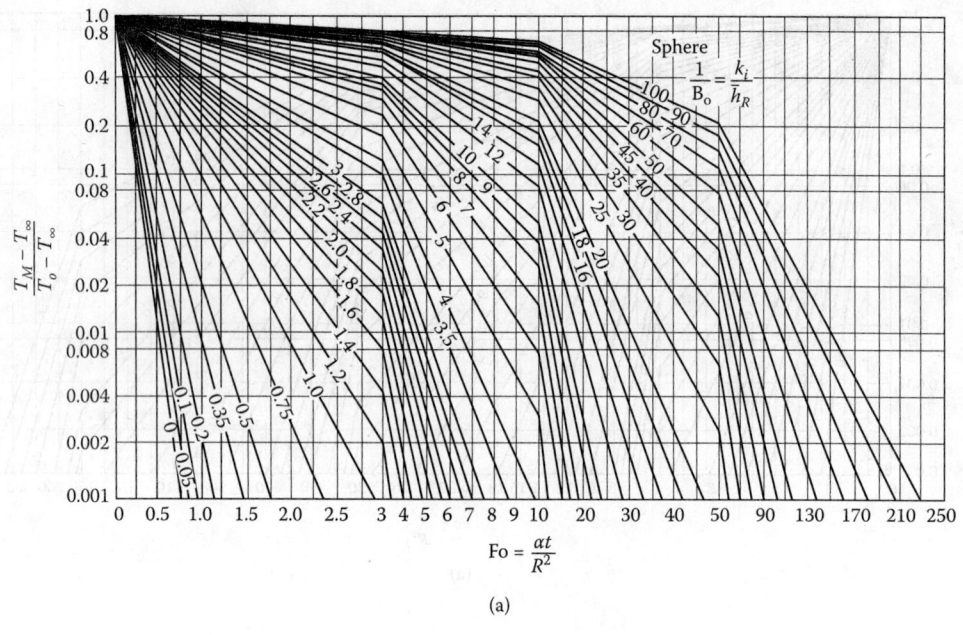

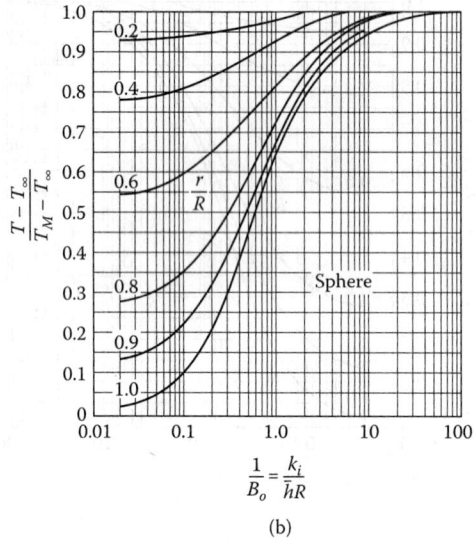

FIGURE 6.14 (a) Variation of the center temperature of a sphere with dimensionless time. (b) Variation of temperature in a sphere as a function of position.

Solution

The small cylinder can be modeled as an infinitely long cylinder of radius 3.0 mm, intersected by plane surfaces of 12.5 mm half-thickness perpendicular to the cylinder axis. (See Figure 6.15.)

The thermal diffusivity is

$$\alpha = \frac{0.194}{1,186(1,520)} = 1.076 \times 10^{-7}\ \frac{m^2}{sec} \left(= 4.15 \times 10^{-3}\ \frac{ft^2}{hr} \right)$$

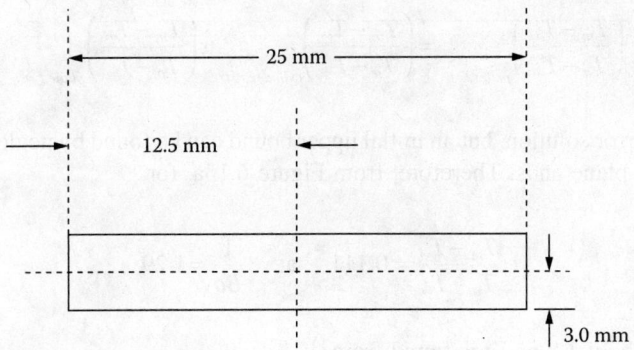

FIGURE 6.15 Cylindrical acrylic pellet analyzed in Example 6.6.

The Fourier number for the cylinder is

$$Fo = \frac{(1.076 \times 10^{-7})t}{(3.0 \times 10^{-3})^2} = \frac{(4.15 \times 10^{-3})t}{(0.118/12)^2(3600)}$$

$$= 1.196 \times 10^{-2} t$$

where t is in seconds.

The Fourier number for the plane is

$$Fo = \frac{1.076 \times 10^{-7} t}{(12.5 \times 10^{-3})^2} = \frac{4.15 \times 10^{-3} t}{(0.492/12)^2(3600)}$$

$$= 6.89 \times 10^{-4} t$$

where t is in seconds.

The reciprocal Biot number for the cylinder is

$$\frac{1}{Bo} = \frac{0.194}{50(3.0 \times 10^{-3})} = \frac{0.112}{8.8(0.118/12)} = 1.29$$

and for the plane it is

$$\frac{1}{Bo} = \frac{0.194}{50(12.5 \times 10^{-3})} = \frac{0.112}{8.8(0.492/12)} = 0.310$$

The dimensionless temperature change for the center of the cylinder is

$$\frac{T_m - T_\infty}{T_o - T_\infty} = \frac{60 - 65}{20 - 65} = \frac{140 - 149}{68 - 149} = 0.111$$

This quantity is the product of the solutions for the cylinder and the plane for the same time t^*:

$$\left(\frac{T_m-T_\infty}{T_o-T_\infty}\right)_{pellet,t=t^*} = \left(\frac{T_m-T_\infty}{T_o-T_\infty}\right)_{cylinder,t=t^*} \times \left(\frac{T_m-T_\infty}{T_o-T_\infty}\right)_{plane,t=t^*}$$

This is a trial-and-error solution, but an initial upper bound can be found by neglecting the relatively minor effect of the plane ends. Therefore, from Figure 6.13a, for

$$\frac{T_m-T_\infty}{T_o-T_\infty} = 0.111 \quad \text{at} \quad \frac{1}{Bo} = 1.29$$

Fo is found to be about 1.8 and t is found from

$$t = \frac{1.8}{1.196 \times 10^{-2}} = 151 \text{ s}$$

For a second estimate, try $t = 120$ s, with $Fo = 1.44$ for the cylinder and $Fo = 0.083$ for the plane from Figure 6.12a. For the cylinder,

$$\left(\frac{T_m-T_\infty}{T_o-T_\infty}\right)_{cylinder} = 0.17$$

and for the plane,

$$\left(\frac{T_m-T_\infty}{T_o-T_\infty}\right)_{plane} = 0.92$$

giving an estimate for

$$\left(\frac{T_m-T_\infty}{T_o-T_\infty}\right)_{pellet} = 0.156$$

A third estimate of $t = 135$ s gives $Fo = 1.61$ for the cylinder, $Fo = 0.093$ for the plane,

$$\left(\frac{T_m-T_\infty}{T_o-T_\infty}\right)_{cylinder} = 0.14, \quad \left(\frac{T_m-T_\infty}{T_o-T_\infty}\right)_{plane} = 0.88, \quad \text{and} \quad \left(\frac{T_m-T_\infty}{T_o-T_\infty}\right)_{pellet} = 0.12$$

This is as close as the charts can be read, and probably as close as the normal uncertainties and approximations in such problems justify, so the required heating time is 135 s.

The temperature on the outside surface of the pellet at the lengthwise center can be found from Figure 6.13b. Reading at

$$\frac{r}{R} = 1.0 \quad \text{and} \quad \frac{1}{Bo} = 1.29$$

gives

$$\left(\frac{T - T_\infty}{T_0 - T_\infty}\right) = 0.73$$

where T is the temperature at the surface. Then,

$$T = 65 + 0.73\,(60 - 65) = 61.4°C$$

6.2.7 Numerical Methods

The previous sections have dealt with a number of important conduction problems, but all had specific geometric limitations and usually idealized assumptions. Numerical methods (which, for all practical purposes, require the use of a computer) can handle a much broader range of problems, including any geometry that can be described in terms of surfaces and intersects, variable physical properties (including temperature and spatial variation), and surface heat-transfer mechanisms, including convection and radiation, internal heat generation, and transient behavior.

The methods used are too specialized and detailed to describe here, but Patankar's book [10] is a good starting point. Commercial software codes are widely available.

6.3 SINGLE-PHASE CONVECTION HEAT TRANSFER

6.3.1 Mechanisms of Convection

Most industrial heat-transfer problems involve the transfer of heat between a solid surface and a flowing fluid. This process is termed *convective heat transfer*, where convection may be defined as the transport of heat from one point to another in a flowing fluid as a result of macroscopic motions of the fluid, the heat being carried as internal energy.

In laminar flow, the heat transfer occurs by conduction from the wall to the adjacent fluid molecules and then from molecule to molecule in the flowing fluid. The fluid immediately at the wall is normally at zero velocity relative to the wall and at the wall temperature. The temperature profile in the flowing fluid is determined by the thermal properties of the fluid (mainly the thermal conductivity) and its velocity profile.

In turbulent flow, the heat is first transferred by conduction through a very thin layer of relatively uneddied flow immediately adjacent to the wall. Further from the wall, small eddies develop in the flow, becoming larger at greater distances from the wall. The effect of the eddies is to rapidly disperse the heat from near the wall to the bulk of the fluid. Hence, the major resistance to heat transfer in turbulent flow is in the laminar region near the wall; the greater the Reynolds number, the thinner this region is, and the greater the rate of heat transfer.

Many flows in practice are neither clearly laminar nor turbulent, but may be termed as "transitional," "developing," or "disturbed." The best known example is the transition region in a tube or pipe, described in greater detail below.

Another distinction among flows is whether the flow is forced by an external means such as a pump (termed *forced convection*) or whether the flow arises as a result of a density difference developed in the fluid circuit as a result of the heat transfer (termed *natural convection* or *thermosiphon action*). Some cases include both mechanisms.

In any case, the fluid temperature at the wall is different from that at the centerline of the flow. When the fluid is being heated, the wall temperature is higher than the centerline temperature, and vice versa for cooling the fluid. The "bulk" or "mixing cup" temperature of the fluid, T, is the temperature that would be measured if the total flow through a cross section were collected over a given period and perfectly mixed. The bulk temperature is intermediate between the wall temperature and the centerline temperature (but usually close to the latter) and is the temperature that is calculated from a heat balance. The difference between the wall, bulk, and centerline temperatures must be considered when dealing with a temperature-sensitive fluid.

These heat-transfer mechanisms have been modeled both analytically and numerically for many flow geometries. Most numerical methods require a computer to obtain an analytical answer, and because of the simplifications and assumptions required to obtain a solution, they are usually more accurate than the empirical methods discussed in this chapter. Good introductions to these methods are provided by Hewitt [1], Patankar [10], and Shah and London [11].

6.3.2 Film Coefficient of Heat Transfer

For the majority of convective heat-transfer cases, the heat flux ($\dot{q}/A$) between the surface and the fluid is nearly proportional to the temperature difference between the bulk fluid and the wall ($T - T_w$). A constant of proportionality h or α is termed the *film coefficient* of heat transfer and is defined by

$$\dot{q}/A = h|(T - T_w)| \tag{6.26}$$

where the bars denote that the absolute value of temperature difference is to be used. The area A used in the equation is usually (but not necessarily; see Section 6.3.3) the surface area across which the heat actually flows; this area should always be carefully defined to avoid confusion. For example, for heat transfer to a fluid inside a plain cylindrical tube, the heat-transfer area is $A_i = \pi d_i L$, where d_i is the inside diameter, L is the tube length over which heat transfer occurs, and h_i is defined as the film heat-transfer coefficient based on (or referenced to) the inside heat-transfer surface of the tube.

The value of h is found from correlations specific to the flow geometry (e.g., inside a tube or across a bank of tubes) and also depends upon the fluid velocity and physical properties. These correlations come primarily from experimental data and are usually expressed as functions of dimensionless numbers to generalize them. The form of the correlation may arise from theory or mechanistic models. In a very few instances, dimensional equations may be the only forms available; they must be used with data given in the stated dimensions and are usually of very limited generality.

Certain heat-transfer processes, notably nucleate boiling, do not follow the proportionality described above. Nevertheless, the concept of a film heat-transfer coefficient is so convenient for practical computation that it is often used in these cases. Special care is required to ensure that the heat flux, the temperature difference, and the coefficient are consistent.

Because the film coefficient correlations are based on experimental data, the calculated coefficients are not more accurate than those data. Further, the conditions in the experiments are often different from those in practical applications. The result is that the best correlations (e.g., the Petukhov-Popov correlation for turbulent flow [17] in long plain round tubes) have an uncertainty of about ±10%, and correlations for more complex cases are significantly less accurate. The consequences of these uncertainties must be considered when applying the calculated results to the design of equipment or interpretation of equipment performance.

6.3.3 Dimensionless Numbers

Heat-transfer-coefficient correlations are usually presented in terms of dimensionless numbers, which are groups of variables that have no net dimensions when evaluated in any consistent system of units. The most important dimensionless numbers for heat transfer are defined next.

6.3.3.1 Reynolds Number, Re

The Reynolds number is defined as

$$\mathrm{Re} = \frac{D\rho V}{\mu} = \frac{DG}{\mu} = \frac{DV}{\nu} \qquad (6.27)$$

where D is a characteristic length of the system, ρ is the density of the flowing fluid, V is a characteristic velocity of the fluid, and μ is the fluid absolute viscosity. $G\,(=\rho V)$ is the mass velocity (more correctly, mass flux) of the fluid, and $\nu = (\mu/\rho)$ is the kinematic viscosity. The exact definitions of D and V depend upon the geometry of the system and must be carefully defined in each case.

For most calculations, the physical properties of the fluid are evaluated at the average bulk temperature of the fluid between the inlet and outlet (except for the wall viscosity μ_w in the Sieder-Tate correction factor below).

The Reynolds number is a measure of the rate of momentum transport in a flowing fluid to the rate of viscous dissipation of momentum to the bounding solid surface. More simply, it is a measure of the intensity of turbulence in the fluid. At low Reynolds numbers, the flow is laminar and typically goes through a transition region before becoming fully turbulent at higher Re. The values of Re characterizing these flow regimes are different for each flow geometry. For a fully developed flow in a plain round tube, the flow is laminar for Re < 2100–2300 and turbulent for Re > 7,000–10,000.

Example 6.7

Calculate the Reynolds number for water at 40°C (104°F) flowing at 2 m/s (6.56 ft/s) inside a plain round tube with 25.4 mm outside diameter by 21.2 mm inside diameter (1 in., 14 BWG tube, inside diameter 0.834 in.).

Solution

At 40°C, water has a density of 992 kg/m³ (61.9 lb$_m$/ft³) and a viscosity of 653×10^{-6} Ns/m² (0.653 cp or 1.58 lb$_m$/ft·h). From Equation (6.27),

$$\mathrm{Re} = \frac{(21.2 \times 10^{-3}\ \mathrm{m})(992\ \mathrm{kg/m^3})(2\ \mathrm{m/s})}{(653 \times 10^{-6}\ \mathrm{Ns/m^2})((\mathrm{kg\ m/s^2})/\mathrm{N})} = 64{,}400$$

or

$$\mathrm{Re} = \frac{(0.834\ \mathrm{in.})(61.9\ \mathrm{b_m/ft^3})(6.56\ \mathrm{ft/s})(3600\ \mathrm{s/hr})}{(12\ \mathrm{in./ft})(1.58\ \mathrm{lb_m/ft\ hr})} = 64{,}400$$

This value is in the turbulent flow regime.

6.3.3.2 Nusselt Number, Nu

The Nusselt number is defined as

$$\mathrm{Nu} = \frac{hD}{k} \qquad (6.28)$$

where h is the film-transfer coefficient, k is the thermal conductivity of the fluid, and D is a characteristic length of the system. The Nusselt number nondimensionalizes the film coefficient in terms of the thermal conductivity and a measure of the distance the heat must penetrate the fluid.

6.3.3.3 Stanton Number, St

An alternative way to represent the heat-transfer coefficient is by the Stanton number, defined as

$$St = \frac{h}{V \rho c_p} \tag{6.29}$$

where c_p is the specific heat of the fluid at constant pressure. The Stanton number is the ratio of the heat-transfer rate to the heat transported in the stream.

6.3.3.4 Colburn j-Factor for Heat Transfer, j_H

The Colburn j-factor is another representation of the heat-transfer coefficient and arises from a boundary layer theory model. It is defined as

$$j_H = St \; Pr^{2/3} \left(\frac{\mu_w}{\mu}\right)^{0.14} \tag{6.30}$$

where St is the Stanton number defined in Section 6.3.3.3, Pr is the Prandtl number defined in Section 6.3.3.5, and $(\mu_w/\mu)^{0.14}$ is the Sieder-Tate term described in Section 6.3.3.9.

6.3.3.5 Prandtl Number, Pr

The Prandtl number is defined as

$$Pr = \frac{c_p \mu}{k} \tag{6.31}$$

where c_p is the specific heat at constant pressure, μ is the absolute viscosity, and k is the thermal conductivity of the fluid. The Prandtl number is the ratio of the momentum diffusivity (μ/ρ) to the thermal diffusivity ($k/\rho c_p$) of the fluid.

6.3.3.6 Graetz Number, Gz

The Graetz number is defined as

$$Gz = \frac{D}{L} \cdot \frac{V \rho c_p D}{k} \tag{6.32}$$

where L is the distance from the start of heating (for heat transfer at a given point) or length of heat-transfer surface (for total amount of heat transferred). The Graetz number is primarily of interest for laminar flow heat transfer with a developing temperature profile.

6.3.3.7 Peclet Number, Pe

The Peclet number is defined as

$$Pe = \frac{V \rho c_p D}{k} \tag{6.33}$$

Heat Transfer

$$= \text{Re}\,\text{Pr} \tag{6.34}$$

The Peclet number reflects the ratio of heat transferred by convection to that transferred by conduction and is most commonly found in applications in laminar flow or with liquid metals.

6.3.3.8 Grashof Number, Gr

The Grashof number is defined as

$$Gr = \frac{g\beta \Delta T L^3}{\nu^2} = \frac{g\beta \rho^2 \Delta T L^3}{\mu^2} \tag{6.35}$$

where g is the gravitational acceleration (9.81 m/s² or 32.2 ft/s² for most applications), β is the thermal expansion coefficient for the fluid, and ΔT is the temperature difference between the bulk fluid and the heat-transfer surface. L is a characteristic dimension that is defined for each geometry. The Grashof number appears in natural convection heat-transfer correlations. The Rayleigh number (Ra) is the product of the Grashof and Prandtl numbers:

$$Ra = \frac{g\beta \rho^2 \Delta T L^3 c_p}{\mu k} \tag{6.36}$$

6.3.3.9 The Sieder-Tate Term

In heat transfer, the fluid at the wall has the same temperature as the wall, different from the bulk temperature. While all fluid properties are to some degree functions of temperature, the viscosity is most strongly affected. If the viscosity at the wall is lower than the bulk value, the boundary layer is thinner than for the isothermal case, and the heat-transfer coefficient is higher than that predicted for the constant property case. Correspondingly, the coefficient is reduced if the viscosity at the wall is higher than the bulk value. The effect is usually small (5 to 10%) but can be much larger with viscous fluids or large temperature differences.

Sieder and Tate [12] suggested a simple but broadly applicable correction factor, $(\mu/\mu_w)^{0.14}$, where μ is the viscosity at the bulk average temperature and μ_w is the viscosity at the corresponding wall temperature. The correction factor should be limited to the range from 0.5 to 2.0. The viscosity of liquids generally decreases with increasing temperature, while gases show the opposite behavior.

6.3.4 SINGLE-PHASE HEAT-TRANSFER CORRELATIONS FOR COMMON GEOMETRIES IN FORCED CONVECTION

A comprehensive collection of single-phase correlations is given by Kakac et al. [13]. A selection of the most important cases is treated in this section.

6.3.4.1. Inside Round Tubes

6.3.4.1.1 Laminar Flow, $\text{Re} = \dfrac{D_i \rho V_i}{\mu} < 2100$

Prediction of the heat-transfer coefficient for this problem is complicated by the fact that the local coefficient is very high at the start of heat transfer and decreases as the conduction into or out of the fluid builds up an adverse temperature gradient. A number of analytical and numerical solutions for this and other constant cross-sectional geometries are given by Shah and London [11], but most of them assume constant physical properties and hence fail to show the effect of natural convection induced by the density change due to heating or cooling. One well-known analytical solution is the so-called Graetz-Nusselt problem (parabolic velocity profile) for fully developed laminar flow in a plain round tube (described by Shah and London [11] and Sieder and Tate [12]). The exact solution is unwieldy and is usually represented by the Hausen Equation (15):

$$\frac{\bar{h}_i D_i}{k} = \left[3.65 + \frac{0.0668 \operatorname{Re} \operatorname{Pr}(D_i/L)}{1+0.04[\operatorname{Re}\operatorname{Pr}(D_i/L)]^{2/3}}\right]\left(\frac{\mu}{\mu_w}\right)^{0.14} \quad (6.37)$$

where $\bar{h}_i$ is the average heat-transfer coefficient over the heat-transfer length of the tube L. The Sieder-Tate term has been added. This equation does not include natural convection effects. The Palen-Taborek equation [16] does consider natural convection:

$$\frac{\bar{h}_i D_i}{k} = 2.5 + 4.55(\operatorname{Re}°)^{0.37}(d_i/L)^{0.37}\operatorname{Pr}^{0.17}\left(\frac{\mu}{\mu_w}\right)^{0.14} \quad (6.38)$$

where

$$\operatorname{Re}° = \operatorname{Re} + 0.8 \operatorname{Gr}^{0.5} \exp(-0.42/\operatorname{Gr}^2) \quad (6.39)$$

The temperature difference ΔT used in the Grashof number is the average temperature difference between the bulk fluid and the wall over the heat-transfer length. This equation applies only to horizontal tubes.

For vertical tubes, the natural convection effect depends upon both the flow direction and whether the fluid is being heated or cooled. This problem is examined in the work of Jakob [14] and in Section 2.5.10 of the *Heat Exchanger Design Handbook* [1].

6.3.4.1.2 Turbulent Flow, $\operatorname{Re} = \dfrac{D_i \rho V_i}{\mu} > 7000$

The Sieder-Tate equation [12],

$$\frac{h_i D_i}{k} = 0.023 \operatorname{Re}^{0.8} \operatorname{Pr}^{1/3}\left(\frac{\mu}{\mu_W}\right)^{0.14} \quad (6.40)$$

is convenient and reliable for Reynolds numbers above about 7000. The Petukhov-Popov equation [17] is generally regarded as the best available for this case:

$$\frac{h_i D_i}{k} = \frac{(f_F/2)\operatorname{Re}\operatorname{Pr}}{1.07 + 12.7(f_F/2)^{1/2}(\operatorname{Pr}^{2/3}-1)}\left(\frac{\mu}{\mu_W}\right)^{0.14} \quad (6.41)$$

where

$$f_F = (3.64 \log_{10} \operatorname{Re} - 3.28)^{-2} \quad (6.42)$$

(The Sieder-Tate term has been added, and f_F is the Fanning friction factor for smooth tubes in turbulent flow.)

The local heat-transfer coefficient near the tube entrance is somewhat higher than these values because of the disturbed flow conditions. The above values are reached after about 40 diameters into the tube.

Example 6.8

Calculate the heat-transfer coefficient and heat flux for water at the conditions given in Example 6.7. The tube wall temperature is 60°C (140°F).

Solution

At 40°C (104°F), water has a specific heat of 4.183 kJ/kg·K (1.00 Btu/lb$_m$) and a thermal conductivity of 0.631 W/m·K (0.365 Btu/h·ft·°F). At 60°C (140°F), water has a viscosity of 467×10^{-6} N·s/m² or 0.467 cP (1.13 lb$_m$/ft·h). Other properties are given in Example 6.7.

The Reynolds number in Example 6.7 was found to be 64,400, so the flow is turbulent, and the use of Equations (6.40)–(6.42) is required.

Solution by the Sieder-Tate equation:
The Prandtl number at 40°C is

$$\text{Pr} = \frac{(4.183 \text{ kJ/kg K})(635 \times 10^{-6} \text{ Ns/m}^2)(1 \text{ kg m/s}^2 \text{ N})(1000 \text{ Ws/kJ})}{(0.631 \text{ W/mK})} = 4.21$$

The Sieder-Tate term is

$$\left(\frac{\mu}{\mu_w}\right)^{0.14} = \left(\frac{653 \times 10^{-6}}{467 \times 10^{-6}}\right)^{0.14} = \left(\frac{1.58}{1.13}\right)^{0.14} = 1.048$$

From Equation (6.40), the Nusselt number is

$$\text{Nu} = \frac{h_i D_i}{k} = 0.023(64,400)^{0.8}(4.21)^{1/3}(1.048) = 274$$

and the film heat-transfer coefficient (based on inside of tube surface area) is

$$h_i = 274\left(\frac{0.631 \text{ W/mK}}{21.2 \times 10^{-3} \text{m}}\right) = 8,150 \text{ W/m}^2\text{K (or 8,150 W/m}^2\text{C), or 1,430 Btu/hr ft}^2 \text{ °F}$$

The corresponding heat flux is

$$q/A_i = h_i(T_W - T_b) = (8,150 \text{ W/m}^2\text{C})(60-40)\text{C} = 163,000 \text{ W/m}^2, \text{ or } 51,500 \text{ Btu/hr ft}^2$$

Solution by the Petukhov-Popov equation:
Using the values for Re, Pr, and $(\mu/\mu_\infty)^{0.14}$ developed above, Equation (6.42) gives

$$f_F = [3.64 \log_{10} 64,400 - 3.28]^{-2} = 0.00494$$

and Equation (6.41) gives

$$Nu = \frac{h_i D_i}{k} = \frac{(0.00494/2)(64,400)(4.21)}{1.07+12.7\left(\frac{0.00494}{2}\right)^{1/2}[(4.21)^{2/3}-1]}(1.048) = 337$$

Then

$$h_i = 337\left(\frac{0.631\,\text{W/mK}}{21.2\times10^{-3}\,\text{m}}\right) = 10{,}020\ \text{W/m}^2\text{K (or 10,020 W/m}^2\text{K), or 1,760 Btu/hr ft}^2\,°\text{F}$$

The coefficients are based on the inside-tube heat-transfer surface. The corresponding fluxes are

$$q/A_i = (10{,}020\ \text{W/m}^2\text{C})(60-40)\text{C} = 200{,}400\ \text{W/m}^2,\ \text{or 63,400 Btu/hr ft}^2$$

The agreement between the two correlations in this case is within about 20%, which is an example of the possible differences in two different but accepted correlations.

6.3.4.1.3 Transition Flow, 2100 < Re < 7000

Prediction of the heat-transfer coefficient in the transition flow regime is uncertain due to the strong effects of entrance conditions and instability of the flow pattern. Gnielinski [18] modified the Petukhov-Popov equation to accommodate the transition region and extend it into the turbulent flow range:

$$\frac{h_i D_i}{k} = \frac{(f_F/2)(\text{Re}-1000)\text{Pr}}{1+12.7(f_F/2)^{1/2}(\text{Pr}^{2/3}-1)}\left[1+\frac{D_i}{L}\right]^{2/3}\left(\frac{\mu}{\mu_w}\right)^{0.14} \quad (6.43)$$

Again, f_F is given by Equation (6.42). The Sieder-Tate term has been added to the equation.

6.3.4.2 Internally Enhanced Tubes

Many techniques—straight, spiraled, and circular fins; twisted tape and wire inserts; and various corrugations—improve the heat-transfer coefficient or increase the heat-transfer area inside tubes. There is always an increased pressure drop as well as increased cost associated with these enhancements. There may also be a change in the fouling characteristics, which may be either a positive or negative factor, depending upon the circumstances. These tubes are ordinarily considered in cases where the inside-tube heat-transfer coefficient is substantially lower than the outside coefficient, since the use of enhanced tubes can only be justified if increased cost and possible operational complications are offset by a savings in the overall size and cost of the heat exchanger. Webb [7] provides an excellent overview of enhanced heat transfer for a wide variety of geometries and operational conditions.

6.3.4.3 Inside Annular Channels

An annular channel is the space between two concentric cylindrical tubes, shown in Figure 6.16. This geometry is used in double-pipe heat exchangers, as described later. The primary interest is on heat transfer on the outside surface of the inner tube. In evaluating the Reynolds, Nusselt, and

Heat Transfer

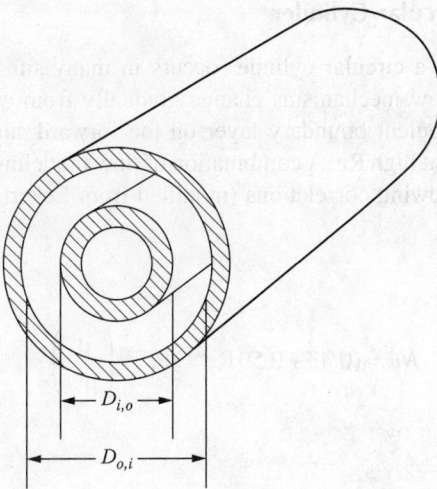

FIGURE 6.16 Annulus channel formed between two concentric round cylinders. $D_{o,i}$ is the inside diameter of the outer cylinder, and $D_{i,o}$ is the outside diameter of the inner cylinder.

Peclet numbers, the characteristic length for an annulus is the diametral clearance between the two cylinders, $(D_{o,i} - D_{i,o})$, where $D_{o,i}$ is the inside diameter of the outer cylinder and $D_{i,o}$ is the outside diameter of the inner cylinder.

6.3.4.3.1 Laminar Flow, Re ≤ 2300

Heat transfer to a laminar flow in an annulus is complicated by the fact that both the velocity and thermal profiles are simultaneously developing near the entrance and, often, over the length of the heated channel. Natural convection may also be a factor. It is usually conservative (i.e., predicted heat-transfer coefficients are lower than those experienced) to use equations for the fully developed flow.

For the usual case of heat transfer on the inner surface of the annulus, Martin [19] recommends, for fully developed flow, the following:

$$\mathrm{Nu} = \frac{\bar{h}(D_{o,i} - D_{i,o})}{k} = 3.66 + 1.2\left(\frac{D_{i,o}}{D_{o,i}}\right)^{-0.8} \tag{6.44}$$

Equations for other cases are given in the *Heat Exchanger Design Handbook* [1].

6.3.4.3.2 Turbulent flow, Re ≥ 7000

Petukhov and Roizen [20] recommend, for heat transfer at the inner surface of the annulus,

$$\frac{\mathrm{Nu}_{\mathrm{annulus}}}{\mathrm{Nu}_{\mathrm{tube}}} = 0.86\left(\frac{D_{i,o}}{D_{o,i}}\right)^{-0.16} \tag{6.45}$$

where $\mathrm{Nu}_{\mathrm{tube}}$ is calculated from Equations (6.41) and (6.42).

6.3.4.3.3 Transition Flow, 2300 < Re < 7000

Calculations in transition flow are very uncertain, and design in this range should be avoided if possible. The *Heat Exchanger Design Handbook* [1] gives an extended procedure involving linear interpolation with Reynolds number between 2,300 and 10,000, giving generally conservative results.

6.3.4.4 Flow across a Circular Cylinder

Heat transfer in flow across a circular cylinder occurs in many situations over a wide range of conditions. The dominant flow mechanisms change gradually from creeping laminar at very low Re to a fully developed turbulent boundary layer on the forward surface of the cylinder with a strongly eddied wake region at high Re. A combination of flow modeling and extensive experimental data have suggested the following correlations (modified from Eckert and Drake [21]):

For $1 \leq Re \leq 1000$:

$$Nu = (0.43 + 0.50\,Re^{0.5})\,Pr^{0.38}\left(\frac{\mu}{\mu_w}\right)^{0.25} \qquad (6.46)$$

For $1000 \leq Re \leq 2 \times 10^5$:

$$Nu = 0.25\,Re^{0.6}\,Pr^{0.38}\left(\frac{\mu}{\mu_w}\right)^{0.25} \qquad (6.47)$$

In these equations, the cylinder diameter D_o is used in Nu and Re, and the undisturbed approach velocity is used in Re.

Example 6.9

Calculate the rate of heat loss by convection from the outer surface of an uninsulated 4-in. Schedule 40 pipe (D_o = 4.500 in. = 0.114 m) exposed to wind blowing at 8.0 m/s (26.3 ft/s) perpendicular to the pipe. The air is at −23°C (−9.4°F), and the pipe surface is at 27°C (80.6°F).

Solution

Air properties at −23°C and 1 atm abs are

Density, ρ = 1.41 kg/m³ (0.088 lb$_m$/ft³)
Viscosity, μ = 1.49 × 10⁻⁵ kg/m·s (0.0360 lb$_m$/ft·h)
Thermal conductivity, k = 0.0223 W/m·K (0.0127 Btu/h·ft·°F)
Specific heat, c_p = 1.00 × 10³ Ws/kgK (0.24 Btu/lb$_m$°F)

The air viscosity at the surface of the pipe (at 27°C) is μ_w = 1.98 × 10⁻⁵ kg/m·s (0.0478 lb$_m$/ft·h).

$$Re = \frac{D_o \rho V}{\mu} = \frac{(0.114\text{ m})(1.41\text{ kg/m}^3)(8.0\text{ m/s})}{(1.49 \times 10^{-5}\text{ kg/ms})} = 86,300$$

$$Pr = \frac{c_p \mu}{k} = \frac{(1.00 \times 10^3\text{ Ws/kg K})(1.49 \times 10^{-5}\text{ kg/ms})}{0.0223\text{ W/m K}} = 0.668$$

From Equation (6.47),

$$\frac{hD_o}{k} = 0.25(86,300)^{0.6}(0.668)^{0.38}\left(\frac{1.49 \times 10^{-5}}{1.98 \times 10^{-5}}\right)^{0.25} = 183$$

Heat Transfer

$$h = 183 \frac{(0.0223 \text{ W/m} \cdot \text{K}))}{0.114 \text{ m}} = 35.6 \text{ W/m}^2\text{K} \, (= 6.27 \text{ Btu/h} \cdot \text{ft}^2 \, °\text{F})$$

The heat flux is

$$Q/A = h(T_{pipe} - T_{air})$$

$$= (35.6 \text{ W/m}^2\text{K})[(27°\text{C}) - (-23°\text{C})]$$

$$= 1780 \text{ W/m}^2 = 564 \text{ Btu/h} \cdot \text{ft}^2$$

and the heat loss per meter of pipe is

$$Q/h = \left(1780 \frac{\text{W}}{\text{m}^2}\right)(\pi)(0.114 \text{ m})$$

$$= 637 \text{ W/m} \, (= 663 \text{ Btu/hr per foot of length})$$

6.3.4.5 Flow across Tube Banks

Banks of tubes are used in shell-and-tube heat exchangers and in air-cooled heat exchangers because they provide a high heat-transfer surface per unit volume of heat exchanger in a mechanically strong, easily manufactured, and reasonably maintainable configuration. Plain tubes are commonly used when the heat-transfer coefficients on each side of the surface are comparable, and circumferentially finned tubes are often used when one coefficient is substantially lower. Low fins (1.5 to 3 mm high) are often used in shell-and-tube exchangers, and high fins (10 to 25 mm high) in air-cooled exchangers.

6.3.4.5.1 Banks of Plain Tubes

The Reynolds number used for characterizing flow is

$$\text{Re} = \frac{D_o \rho V_{max}}{\mu} \qquad (6.48)$$

where D_o is the outside diameter of the tube, V_{max} the maximum velocity attained during flow across the bank (i.e., through the passages where adjacent tubes approach most closely), and ρ and μ are, respectively, the density and viscosity of the fluid at its average bulk temperature in the bank. There is no sharp transition in flow behavior as the Reynolds number increases from the laminar flow region (Re < 100) to the fully turbulent region (Re > 6000). There is, however, a strong dependence on the geometry of the tube bank.

For shell-and-tube exchangers, four different basic geometries of tube layout are used, defined by the angle between the approach-flow direction and the axis connecting the centers of adjacent tubes (Figure 6.17). The 30° and 45° orientations are generally preferred on the basis of heat transfer vs. pressure-drop performance. The 30° configuration provides more heat-transfer area per unit volume than the 45° arrangement and is usually chosen if mechanical cleaning of the outside heat-transfer surface is not required. The 45° layout provides clear lanes through the tube field for cleaning.

The other important geometrical parameter is the pitch ratio, defined as the distance between the centers of adjacent tubes, P_t, divided by the tube outside diameter, D_o. The smaller the pitch

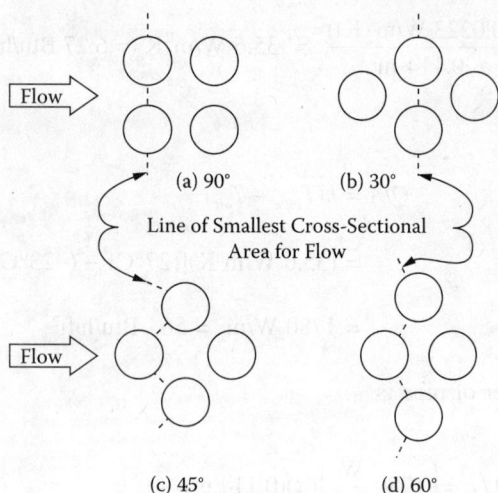

FIGURE 6.17 Typical tube layouts used in shell-and-tube heat exchangers.

ratio, the greater the surface area per unit volume; practical construction considerations set the minimum pitch ratio to about 1.25. As the pitch ratio is increased, the pressure drop for a given flow rate decreases much faster than the heat-transfer coefficient, but requires a relativity modest increase in heat-exchanger size. For handling liquids, the pitch ratio is usually chosen between 1.25 and 1.5; for gases and in vacuum service (including condensing and vaporizing), pitch ratios of 2 or even greater can be considered.

Generalized transfer and pressure-drop correlations for several commonly used ideal (no bypass or leakage flows) tube banks are shown in Figures 6.18, 6.19a, and 6.19b over the usual Reynolds number range. Figure 6.20 gives more specialized and accurate curves over a narrower but critical range.

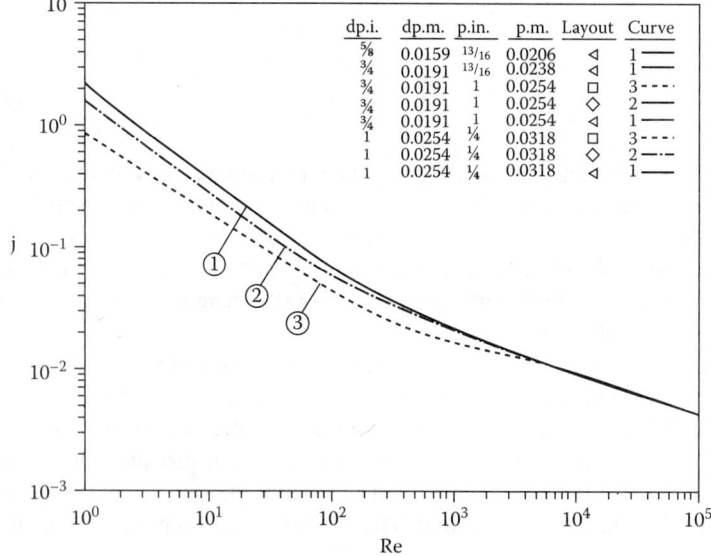

FIGURE 6.18 Correlation of Colburn j-factors for ideal tube banks.

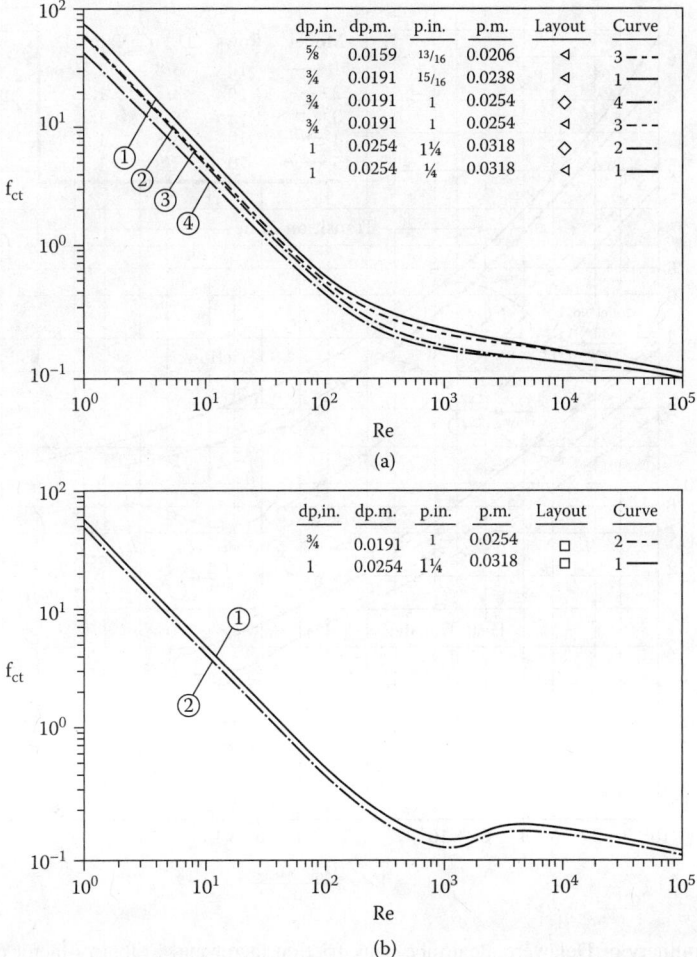

FIGURE 6.19 (a) Correlation of friction factors for ideal tube banks, 30° and 45° layouts. (b) Correlation of friction factors for ideal tube banks, 90° layouts.

This heat-transfer correlation is in terms of the Colburn j-factor, defined as

$$j = \left(\frac{h}{c_p \rho V_{max}}\right) Pr^{2/3} \left(\frac{\mu_w}{\mu}\right)^{0.14} \tag{6.49}$$

where V_{max} is the highest velocity reached by the fluid in passing through the tube bank, i.e., through the smallest cross-sectional area for flow as shown in Figure 6.17. Note that for the 45° and 60° configurations, this velocity is reached between tubes in adjacent rows for pitch ratios in the usual range of interest ($P_t/D_o < 1.71$ for 45° layout). This equation is valid for tube banks with 10 or more rows of tubes in depth. For shallower tube banks, there is some reduction in the average coefficient. Zukauskas et al. [24] present an extended discussion of this phenomenon and a generalized (but less accurate) correlation for a wide range of tube-bank geometry.

The pressure-drop correlation is given in Figures 6.18, 6.19a, 6.19b, and 6.20 in terms of the modified Chilton-Genereaux friction factor f_{ct} as a function of Reynolds number, where

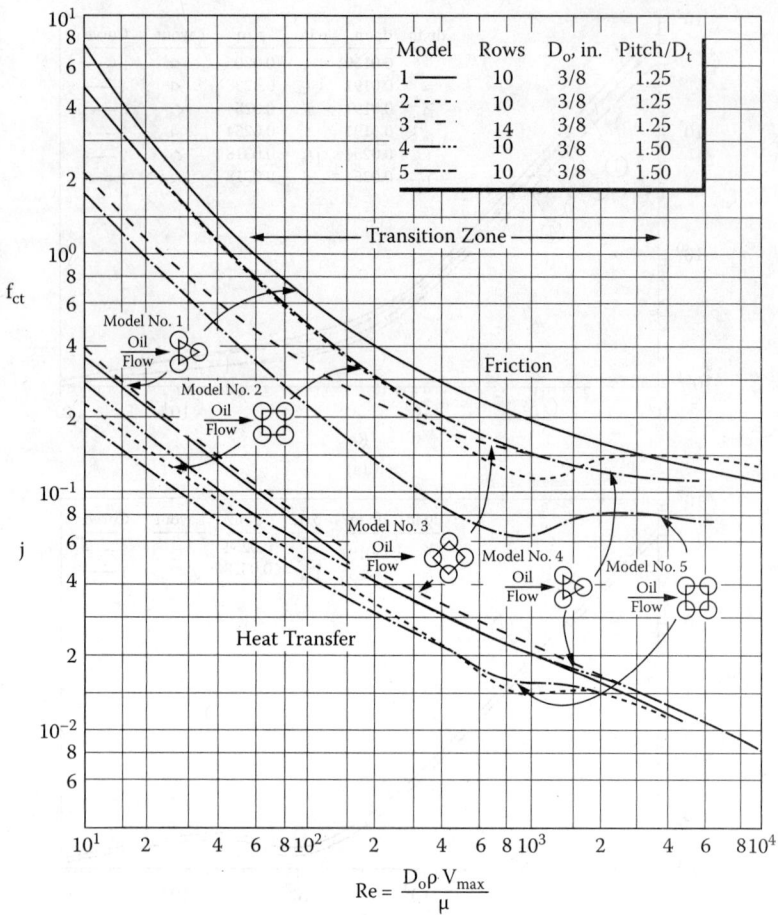

FIGURE 6.20 Summary of Delaware ideal tube-bank friction factor and Colburn j-factor data.

$$f_{ct} = \frac{2\Delta p g_c}{4\rho V_{max}^2 N}\left(\frac{\mu}{\mu_w}\right)^{0.14} \tag{6.50}$$

In this equation, N is the number of major restrictions in the tube bank (i.e., the number of times the flow reaches its maximum velocity in flowing through the tube bank). In the 30° and 90° arrangements, N is equal to the number of tube rows crossed in the bank; for the 45° and 60° layouts, N is one less than the number of rows crossed. The term Δp is the frictional pressure drop across the tube bank.

Example 6.10

Calculate the heat-transfer coefficient and frictional pressure loss for the following case: Water at an average temperature of 20°C (68°F) is flowing at the rate of 50 kg/s (110 lb_m/s) across a tube bank composed of 274 tubes, each 25.4 mm (1.00 in.) in diameter and 1 m long in a 30° layout with a 1.25 pitch ratio, as shown in Figure 6.21.

There are 7 rows with 20 tubes per row and 6 rows with 19 tubes per row. The heat-transfer and pressure-drop characteristics are represented by curve 1 in Figures 6.18 and 6.19a and by model 1 in Figure 6.20. The tube surface temperature is 60°C (140°F).

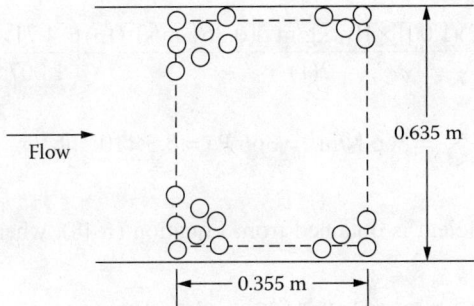

FIGURE 6.21 Flow across the 30°, 1.25 pitch ratio tube bank in Example 6.10.

Solution
Water properties at 20°C are

Density, $\rho = 1.001 \times 10^3$ kg/m³ (= 62.5 lb$_m$/ft³)
Viscosity, $\mu = 1.007 \times 10^{-3}$ kg/ms (= 2.44 lb$_m$/ft hr)
Thermal conductivity, $k = 0.597$ W/m K (= 0.345 Btu/hr ft °F)
Specific heat, $c_p = 4.183 \times 10^3$ Ws/kg K (= 0.999 Btu/lb$_m$ °F)
Water viscosity at 60°C, $\mu_w = 4.71 \times 10^{-4}$ kg/ms (= 1.70 lb$_m$/ft hr)

Free-flow area: Because there are different numbers of tubes in adjacent rows, there are two different free-flow areas and, hence, different calculated heat-transfer coefficients and pressure losses per row. For the rows with 20 tubes, the free-flow area will be smaller and the velocity, heat-transfer coefficient, and pressure drop per row will be higher than those for rows with 19 tubes. We can obtain a reasonable estimate by using an average velocity. For a row with 20 tubes, the minimum free-flow area is

$$S_{min,20} = [0.635 \text{ m} - 20(25.4 \times 10^{-3} \text{ m})](1 \text{ m}) = 0.127 \text{ m}^2 \; (= 1.367 \text{ ft}^2)$$

and the corresponding velocity is

$$V_{max,20} = \frac{(50 \text{ kg/s})}{(1001 \text{ kg/m}^3)(0.127 \text{ m}^2)} = 0.393 \text{ m/s} = 1.29 \text{ ft/s}$$

The corresponding values for a 19-tube row are $S_{min,19} = 0.152$ m² (1.64 ft²) and $V_{max,19} = 0.329$ m/s (1.08 ft/s). The average maximum velocity is 0.361 m/s. (Large clearances between the outermost tubes and the side walls result in significant and ineffective bypass flows and should be avoided in good design.)

Continuing with this average velocity,

$$Re = \frac{(2.54 \times 10^{-2} \text{ m})(1.001 \times 10^3 \text{ kg/m}^3)(0.329 \text{ m/s})}{(1.007 \times 10^{-3} \text{ kg/ms})} = 8310$$

From Figure 6.20, using curve 1, $f_{ct} = 0.12$ and $j = 0.0090$. The pressure drop for $N = 13$ major restrictions (13 rows of a 30° layout) is from Equation (6.49):

$$\Delta p = \frac{4(0.12)(1.001 \times 10^3 \text{ kg/m}^3)(0.361 \text{ m/s})^2(13)}{2(1)} \left(\frac{4.71 \times 10^{-4}}{1.007 \times 10^{-3}} \right)^{0.14}$$

$$= 366 \frac{\text{kg}}{\text{m} \cdot \text{s}^2} = 366 \text{ N/m}^2 = 366 \text{ Pa} = 5.3 \times 10^{-2} \text{ lb}_f/\text{in.}^2$$

The heat-transfer coefficient is obtained from Equation (6.48), where

$$\text{Pr} = \frac{(4.183 \times 10^3 \text{ Ws/kg K})(1.007 \times 10^{-3} \text{ kg/ms})}{(0.597 \text{ W/m K})} = 7.06$$

and

$$h = \frac{0.0090(4.183 \times 10^3 \text{ Ws/kg K})(1.001 \times 10^3 \text{ kg/m}^3)(0.361 \text{ m/s})}{(7.06)^{2/3} \left(\frac{4.71 \times 10^{-4}}{1.007 \times 10^{-3}} \right)^{0.14}}$$

$$= 4.11 \times 10^3 \text{ W/m}^2\text{K} \; (= 724 \text{ Btu/h} \cdot \text{ft}^2 \cdot {}^\circ\text{F})$$

6.3.4.5.2 Banks of Low-Finned Tubes

The limited experimental results of Williams and Katz [25] and Briggs et al. [26] can be used to modify the results for plain tubes to apply to cross flow in banks of low-finned tubes commonly used in shell-and-tube heat exchangers.

The plain-tube j-factor curves can be used for the corresponding low-finned tubes for Reynolds numbers above 1000. At lower Re, the retardation of the flow between the fins reduces the j-factor. It is necessary to calculate V_{max}, taking into account the flow area between the fins. The root diameter of the tube (i.e., the tube diameter at the base of the fins) is used in the Reynolds number. The coefficient thus calculated is based on the entire external tube area including fins, but a fin efficiency (calculated by the method of Section 6.2.4.3) must be applied to the fin area.

For calculating pressure drop, the friction factors given in Figures 6.19a, 6.19b, and 6.20 should be multiplied by a factor of 1.5.

6.3.4.5.3 Banks of High-Finned Tubes

The experimentally based correlation of Briggs and Young [27] is recommended for this case:

$$\frac{h_o D_r}{R} = 0.134 \, \text{Re}^{0.68} \, \text{Pr}^{1/3} \left(\frac{H}{s} \right)^{-0.2} \left(\frac{Y}{s} \right)^{-0.12} \tag{6.51}$$

The heat-transfer coefficient h_o is referenced to the entire outside tube surface, including fins, and the fin efficiency is found from Section 6.2.4.3. D_r is the root diameter of the tube (to the base of the fins), H is the fin height, Y is the fin thickness, and s is the face-to-face spacing between the fins. The Reynolds number Re is defined as

$$\text{Re} = \frac{D_r \rho V_{max}}{\mu} \tag{6.52}$$

Heat Transfer

where V_{max} is calculated at the minimum flow area between tubes, including the flow area between the fins.

Pressure drop is calculated from the correlation of Robinson and Briggs [28]:

$$\Delta p = \frac{f_r N \rho V_{max}^2}{g_c} \quad (6.53)$$

where N is the number of rows of tubes, and f_r is the friction factor found from the following equation:

$$f_r = 18.93 \operatorname{Re}^{-0.316} \left(\frac{P_t}{D_r}\right)^{-0.927} \left(\frac{P_t}{P_1}\right)^{0.52} \quad (6.54)$$

where Re is defined by Equation (6.52), P_t is the distance between centers of adjacent tubes, and P_1 is the distance between centers of adjacent tubes in successive rows, measured along the diagonal. For the usual equilateral triangular tube layout, $(P_t/P_1) = 1$.

In-line banks of high-finned tubes are not usually used in air-cooled heat-transfer equipment because the preferential flow path between the fin tips of adjacent tubes allows flow bypassing and reduces the apparent heat-transfer coefficient. This effect tends to disappear in the deeper tube banks used in convection sections of fired heaters.

6.3.4.6 Heat Transfer in Packed and Fluidized Beds

Packed beds, often composed of catalyst pellets, are used widely in the chemical industries. Fluid flows through the bed and exchanges heat with the bed material. Heat-transfer processes within the bed and between the bed and the container walls are also of concern. Gnielinski ([29], summarized in [1]) presents a comprehensive listing of available data and gives the following equations valid over a Reynolds number range from 100 to 2×10^4 and void fractions $0.26 < \psi < 1$:

$$\operatorname{Nu}_{packed\ bed} = \frac{(h_{packed\ bed}) D_{sphere}}{k_{fluid}} = f_\psi \operatorname{Nu}_{single\ sphere} \quad (6.55)$$

where

$$\operatorname{Nu}_{single\ sphere} = 2 + \sqrt{\operatorname{Nu}_{lam}^2 + \operatorname{Nu}_{turb}^2} \quad (6.56)$$

$$\operatorname{Nu}_{lam} = 0.664 \sqrt{\operatorname{Re}_\psi} \sqrt[3]{\operatorname{Pr}} \quad (6.57)$$

$$\operatorname{Nu}_{turb} = \frac{0.037 \operatorname{Re}_\psi^{0.8} \operatorname{Pr}}{1 + 2.443 \operatorname{Re}_\psi^{-0.1}(\operatorname{Pr}^{2/3} - 1)} \quad (6.58)$$

$$\operatorname{Re}_\psi = \frac{D_{sphere} \rho V_{free}}{\mu} \quad (6.59)$$

$$f_\psi = 1 + 1.5(1 - \psi) \quad (6.60)$$

$$\psi = \frac{V_{total} - V_s}{V_{total}} \qquad (6.61)$$

In these equations, D_{sphere} is the diameter of the spheres; V_{free} is the fluid velocity in the cross section of the empty vessel; ρ and μ are, respectively, the density and absolute (dynamic) viscosity of the fluid; and V_{total} and V_s are, respectively, the volume of the empty vessel and the total volume of the spheres respectively.

The corresponding pressure drop can be estimated by the Ergun Equation [27]:

$$\frac{\Delta p}{L} \frac{D_p}{\rho V_s^2} \left(\frac{\varepsilon^3}{1-\varepsilon} \right) = 150 \frac{1-\varepsilon}{\left(\frac{D_p \rho V_s}{\mu} \right)} + 1.75 \qquad (6.62)$$

where L is the height of the bed and ε is the porosity.

Fluidized beds are widely used in chemical processing, but the heat-transfer and fluid-flow characteristics are too complex to be adequately treated here. Martin (in [1], Section 2.8.4) gives an excellent overview.

6.3.5 Single-Phase Heat Transfer in Natural Convection

In natural convection heat transfer, fluid motion is induced by the density difference between the fluid in contact with the solid heat-transfer surface and the bulk fluid. The density difference is related to the temperature difference between wall and bulk by

$$\Delta \rho = \rho_{wall} - \rho_{bulk} = \beta \bar{\rho} (T_{bulk} - T_{wall}) \qquad (6.63)$$

where β is the thermal expansion coefficient of the fluid at constant pressure, defined by

$$\beta = -\frac{1}{r} \left(\frac{\partial \rho}{\partial T} \right)_p \qquad (6.64)$$

and $\bar{\rho}$ is the mean density,

$$\bar{\rho} = \frac{1}{2}(\rho_{wall} + \rho_{bulk}) \qquad (6.65)$$

For small density differences, β can be assumed constant. The value of β can be found from a table of density vs. temperature (at constant pressure) over the temperature range of interest.

The resulting induced flow may be laminar (usually at small temperature differences and/or viscous fluids) or turbulent, or often in a transition or mixed laminar and turbulent regime. Correlations may be based on analytical or experimental studies, and numerical methods are now available. A major limitation to analytical solutions is that constant viscosity is generally assumed, whereas the variation of viscosity with temperature is likely to have a major effect upon the velocity gradient and the dominant flow regime near the heat-transfer surface.

Other factors that need to be considered in any natural convection calculation include the geometry of the container (if any), flows generated from (or limited by) other surfaces in the vicinity, the interaction with forced flows, and possible radiative heat-transfer interactions. Because natural convection heat-transfer coefficients are generally low, radiation effects should always be considered in any analysis. It should be noted that local convective heat-transfer coefficients often vary with position and can be quite different from the mean value over a specified surface. Typical correlations of mean values for several common geometries are given below; more extensive treatments are given in the *Heat Exchanger Design Handbook* [1], Sections 2.5.7–2.5.10, and in Hewitt et al. [30] and Incropera and DeWitt [31]. Physical properties used in these correlations are to be evaluated at the "film temperature," taken to be the average between the solid surface and bulk fluid temperatures.

6.3.5.1 Natural Convection from a Vertical-Plane Surface

For a vertical-plane heat-transfer surface of height L, the Grashof number is defined as

$$\text{Gr}_L = \frac{gL^3\rho^2\beta|T_{surf} - T_{fluid}|}{\mu^2} \tag{6.66}$$

A local value for any distance x from the leading edge ($0 < x \le L$) is defined by replacing L by x in this equation. The corresponding Rayleigh number is found by multiplying Gr by the Prandtl number,

$$\text{Ra}_{L\text{ or }x} = \text{Gr}_{L\text{ or }x}\,\text{Pr} \tag{6.67}$$

For Ra $< 10^9$, the flow is laminar, and the heat-transfer coefficient varies with distance. For Ra $> 10^9$, the flow is turbulent, and the heat-transfer coefficient is independent of position. The corresponding equations are from the *Heat Exchanger Design Handbook* [1], Section 2.5.7, by Churchill and Chu [32]. These are presented as follows.

1. Laminar, constant temperature surface:
   The local heat-transfer coefficient at x is

$$\text{Nu}_x = \frac{h_x x}{k} = 0.68 + 0.503\text{Ra}_x^{1/4} f_1(\text{Pr}) \tag{6.68}$$

   where

$$f_1(\text{Pr}) = \left[1 + \left(\frac{0.492}{\text{Pr}}\right)^{9/16}\right]^{-4/9} \tag{6.69}$$

   The average coefficient from $x = 0$ to $x = x$ ($x \le L$) is

$$\overline{\text{Nu}_x} = \frac{\overline{h_x} x}{k} = 0.68 + 0.67\text{Ra}_x^{1/4} f_1(\text{Pr}) \tag{6.70}$$

2. Laminar, constant heat-flux surface:
   The local coefficient at x is

$$\text{Nu}_x = \frac{h_x x}{k} = 0.68 + 0.563 R_x^{1/4} f_2(\text{Pr}) \tag{6.71}$$

where

$$f_2(\text{Pr}) = \left[1 + \left(\frac{0.437}{\text{Pr}}\right)^{9/16}\right]^{-4/9} \quad (6.72)$$

The average coefficient from $x = 0$ to $x = x$ ($x \leq L$) is

$$\overline{\text{Nu}_x} = 0.67 \text{Ra}_x^{1/4} f_2(\text{Pr}) \quad (6.73)$$

where ΔT in Ra_x is based on the wall temperature at $x/2$.

3. Turbulent flow, $\text{Ra}_x > 10^9$:
   The heat-transfer coefficient for a turbulent boundary layer is

$$\overline{\text{Nu}_x} = \frac{\overline{h}_x x}{k} = 0.15 \text{Ra}_x^{1/3} f_3(\text{Pr}) \quad (6.74)$$

where

$$f_3(\text{Pr}) = \left[1 + \left(\frac{0.492}{\text{Pr}}\right)\right]^{-16/27}$$

4. Combined laminar and turbulent flow:
   These equations can be combined to give an average coefficient over a vertical length L:

$$\overline{\text{Nu}_L} = 0.68 + 0.67 \text{Ra}_L^{1/4} [f_1(\text{Pr})][1 + 1.6 \times 10^{-8} \text{Ra}_L [f_1(\text{Pr})]^4]^{1/12} \quad (6.75)$$

6.3.5.2 Natural Convection from Horizontal Plates

Heat transfer in this case depends upon the dimensions of the surface and which surface (upper or lower) is heated or cooled. Hewitt et al. [30] recommend the following equations:

1. Upper surface of heated plate or lower surface of cooled plate:

$$\overline{\text{Nu}_L} = \frac{\overline{h}L}{k} = 0.54 \text{Ra}_L^{1/4} \quad \text{for } (10^4 \leq \text{Ra}_L \leq 10^7) \quad (6.76)$$

$$\overline{\text{Nu}_L} = \frac{\overline{h}L}{k} = 0.15 \text{Ra}_L^{1/3} \quad \text{for } (10^7 \leq \text{Ra}_L \leq 10^{11}) \quad (6.77)$$

2. Lower surface of heated plate or upper surface of cooled plate:

$$\overline{\text{Nu}_L} = 0.27 \text{Ra}_L^{1/4} \quad \text{for } (10^5 \leq \text{Ra}_L \leq 10^{10}) \quad (6.78)$$

The characteristic length L used in the Rayleigh and Grashof numbers is defined as

$$L = \frac{A}{P} \qquad (6.79)$$

where A is the plate surface area and P is the perimeter.

6.3.5.3 Natural Convection from a Horizontal Cylinder

For a long horizontal cylinder (i.e., negligible end effects), Churchill and Chu [32] recommend

$$\overline{Nu}_D = \frac{\bar{h}D}{k} = \left\{ 0.60 + \frac{0.387 Ra_D^{1/6}}{\left[1 + \left(\frac{0.559}{Pr}\right)^{9/16}\right]^{8/27}} \right\}^2 \quad \text{for } (Ra_D \leq 10^{12}) \qquad (6.80)$$

The cylinder diameter D is the characteristic length in the Nusselt and Rayleigh numbers, and $\bar{h}$ is the mean heat-transfer coefficient over the entire surface of the cylinder.

6.3.5.4 Natural Convection between Two Horizontal Parallel Plates

For two horizontal parallel plates whose dimensions are large compared with the spacing L between them, and where the bottom plate is hotter than the top plate, Globe and Dropkin [33] recommend

$$\overline{Nu}_L = \frac{\bar{h}L}{k} = 0.069 Ra_L^{1/3} Pr^{0.074} \quad \text{for } (3 \times 10^5 < Ra_L < 7 \times 10^9) \qquad (6.81)$$

6.3.5.5 Other Geometries

Correlations for a variety of other geometries are provided in the *Heat Exchanger Design Handbook* [1] and by Gnielinski [29] and DeWitt and Incropera [31].

6.4 CONDENSATION AND VAPORIZATION HEAT TRANSFER

6.4.1 Mechanisms of Condensation

A vapor or vapor/gas mixture will condense to a liquid on any surface that is even slightly below the saturation temperature (or dew point) of the vapor at the existing pressure. This is true even if the vapor is highly superheated or multicomponent. In the latter case, the dew point decreases as condensation proceeds.

There are four mechanisms of condensation: dropwise, filmwise, direct contact, and homogenous. In dropwise condensation, the vapor initially condenses as multitudes of tiny droplets on preferred sites on the surface. The droplets grow rapidly by conduction through the droplet until gravity or vapor shear overcomes surface tension and the droplets run off the surface. As long as the surface remains nonwetting, the cycle is repeated. Dropwise condensation gives heat-transfer coefficients as high as 60,000 W/m²K (10,000 Btu/h·ft²·°F) for steam. Most surfaces are initially nonwetting, at least for steam, because of residual shop oil or other surface contamination, but become wetting after a short time, typically a few hours. Once the surface is rendered wetting, filmwise condensation dominates, with substantially reduced coefficients. Attempts to maintain a nonwetting surface by introducing promoters such as metallic stearates or by coating the surface with a thin film of Teflon have proven operationally impractical. However, care must be taken when starting up new equipment with condensing steam as a heating medium because the initial high heat-transfer rate can lead to high surface temperatures and excessive fouling rates on the other stream.

When the condensing surface is completely wetted, the condensate film flows under the influence of gravity and vapor shear, and its heat-transfer characteristics can be reasonably predicted from hydrodynamic principles. This is the usual design mode for condensation and is developed in detail in the next section.

Direct-contact condensation occurs when a subcooled liquid is sprayed into the vapor or allowed to flow down a structural surface in a vapor space. Condensation occurs directly on the surface of the coolant. Heat-transfer rates for this equipment (such as barometric condensers) are strongly dependent upon the sprayer characteristics and/or the particular structured surface, and must be obtained from the vendor.

If a body of vapor can be sufficiently subcooled without contact with a surface (by radiation, for example), theory predicts that a liquid phase can form spontaneously. This process is termed *homogenous condensation*, and fog formation is often cited as an example. However, the subcooling required is so great that it is unlikely to occur in any process situation. Fog formation *does* occur in process condensers, but the probable mechanism is condensation on tiny droplets carried over from a previous processing step (a distillation column, for example, where bubble collapse produces large numbers of submicroscopic droplets).

The present discussion is limited to filmwise condensation on a few common geometries. There are several comprehensive and specialized references that can be consulted: the *Heat Exchanger Design Handbook* [1] covers an especially wide range of theory and application, while Hewitt et al. [30] and Incropera and DeWitt [31] present good summaries. Collier and Thome [34] cover convective condensation processes.

6.4.2 Design Equations for Filmwise Condensation

For condensation of a pure vapor, the basic assumptions are that (a) the latent heat of condensation is released at the interface between the vapor and the liquid condensate film and (b) the interface temperature is equal to the saturation temperature of the vapor at the existing pressure. The heat-transfer rate through the condensate film depends upon the thickness and heat-transfer mechanisms of the film, which in turn depend upon the geometry of the condensing surface and the flow regime of the condensate (laminar or turbulent, gravity driven or vapor-shear driven). The presentation and discussion of the correlations is organized in the following sections according to the basic geometry.

6.4.2.1 Condensation on Vertical-Plane and Tubular Surfaces

6.4.2.1.1 Gravity Driven

In the absence of a strong vapor flow, the condensate film on a vertical surface drains under the influence of gravity only. At low flow rates, the flow is laminar, and a theoretical analysis by Nusselt [35] is applicable. The local value of the coefficient at a distance x from the start of condensation is

$$h_x = \left[\frac{k_\ell^3 \rho_\ell (\rho_\ell - \rho_v) \lambda g}{4 \mu_\ell (T_{sat} - T_w) x} \right]^{1/4} \tag{6.82}$$

where k_ℓ, ρ_ℓ, and μ_ℓ are the thermal conductivity, density, and viscosity of the condensate, respectively; ρ_v is the vapor density; λ is the latent heat of condensation; g is the gravitational acceleration (usually 9.81 m/s² or 32.2 ft/s²); and T_{sat} and T_w are the vapor saturation temperature and wall temperature, respectively.

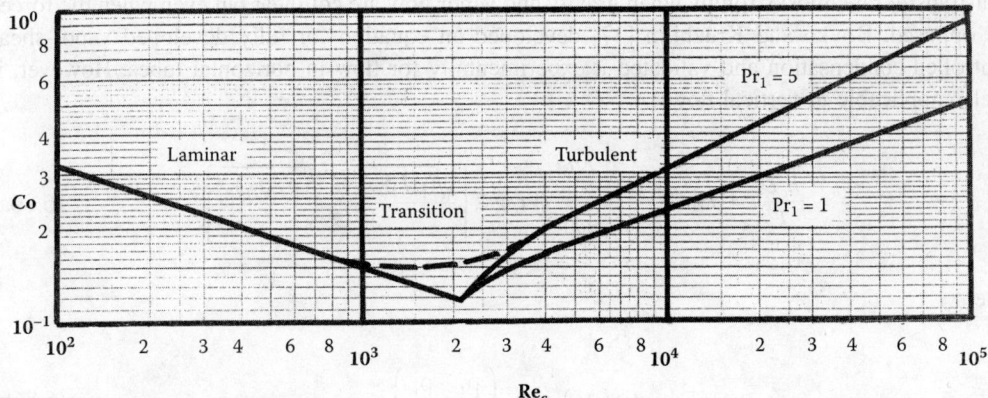

FIGURE 6.22 Condensation number versus Reynolds number for gravity-driven condensation on a vertical surface.

The average coefficient for a condensing surface of length L is obtained by integrating Equation (6.82), assuming an isothermal surface, to obtain

$$h_c = 0.943 \left[\frac{k_\ell^3 \rho_\ell (\rho_\ell - \rho_v) \lambda g}{\mu_\ell (T_{sat} - T_w) L} \right]^{1/4} \quad (6.83)$$

It is usually sufficiently accurate to use an average wall temperature and to evaluate the condensate properties at the arithmetic mean film temperature, $1/2(T_{sat} + T_w)$.

These equations are strictly valid only for a plane vertical surface. However, they can be used for inside or outside vertical tubes with small error because the condensate film is thin compared with the diameter of a typical tube. Because of rippling and other nonidealities, the predicted coefficients are about 10–20% below experimental values.

A condensate Reynolds number can be defined as

$$Re_c = \frac{4\Gamma}{\mu_\ell} \quad (6.84)$$

where Γ is the condensate mass flow rate m_c per unit length of drainage perimeter. For a plane wall, the drainage perimeter is the width of the plane; for a tube, it is the circumference, πD_o or πD_i. At Re_c above about 1000, the condensate becomes turbulent. The transition region shows substantially higher but not very predictable coefficients, compared with the Nusselt solution. The fully turbulent film was analyzed by Colburn [36]. These results are shown in Figure 6.22 in generalized form of the condensation number, Co, defined as

$$Co = h_c \left[\frac{\mu_\ell^2}{k_\ell^3 \rho_\ell (\rho_\ell - \rho_v) g} \right]^{1/3} \quad (6.85)$$

as a function of Re_c.

6.4.2.1.2 Vapor-Shear Driven

If vapor velocities are high enough, the vapor shear on the condensate interface initiates turbulence in the condensate film and causes it to flow off the surface more rapidly. These effects increase the heat-transfer rate, most strongly when gravity and vapor flow are collinear but even when the forces are opposed. Boyko and Kruzhilin [37] developed an equation for fully developed vapor-shear controlled condensation and validated it experimentally for flow in horizontal tubes. However, it is also applicable to vertical flows:

$$\frac{h_c D_i}{k_\ell} = 0.024 \left(\frac{D_i G_T}{\mu_\ell} \right)^{0.8} (Pr_\ell)^{0.43} \left[\frac{\sqrt{(\rho/\rho_m)_i} + \sqrt{(\rho/\rho_m)_o}}{2} \right] \quad (6.86a)$$

where

$$(\rho/\rho_m)_i = 1 + \left(\frac{\rho_\ell - \rho_v}{\rho_v} \right) x_i \quad (6.86b)$$

and

$$(\rho/\rho_m)_o = 1 + \left(\frac{\rho_\ell - \rho_v}{\rho_v} \right) x_o \quad (6.86c)$$

The term h_c is the effective mean heat-transfer coefficient for condensing a stream from an inlet quality (mass fraction vapor) x_i to an outlet quality x_o. For the usual case of totally condensing a saturated vapor stream ($x_i = 1$; $x_o = 0$), the bracketed term in Equation (6.86a) becomes

$$\left[\frac{1 + \sqrt{\rho_\ell/\rho_v}}{2} \right] \quad (6.86d)$$

G_T is the mass velocity (mass flux) of the entering fluid, defined as the mass flow rate in kg/s (lb_m/h) divided by the cross-sectional area of the tube or the flow channel. The subscript ℓ refers to the physical properties of the condensate and v to the vapor. Equation (6.86a) may be used to calculate the local condensing heat-transfer coefficient at any quality x by replacing the bracketed term by

$$\left[\sqrt{1 + \left(\frac{\rho_\ell - \rho_v}{\rho_v} \right) x} \right] \quad (6.86e)$$

6.4.2.1.3 Selection of a Condensing Correlation

Four possibilities for a condensing film on a vertical surface exist: laminar, transition, turbulent gravity driven, and turbulent vapor-shear driven. How does one select the correct one? Consideration of the problem reveals that the flow situation that most likely exists in a given case is also the one that gives the highest coefficient. Therefore, the recommended procedure is to find the coefficients given, respectively, by Equation (6.82), Figure 6.22, and Equation (6.86a, b, c, d, e), and choose the highest one. Caution, however, should be taken because all of the correlations have uncertainties of at least 10–20%.

Example 6.11

Saturated propane vapor at 55°C (131°F) and 1900 kPa abs (275 lb_f/in.² abs) is flowing downward at a rate of 25 kg/h (55.1 lb_m/h) through a vertical 1-in. 14 BWG tube (inside diameter of 0.834

in. = 21.2 mm). The inside surface temperature of the tube is 40°C (104°F). How long a tube is required to completely condense the propane vapor?

Solution

Latent heat of condensation of propane at 55°C (131°F): 271 kJ/kg (116.5 Btu/lb$_m$)
Density of propane vapor at 5°C and 1900 kPa abs: 44.0 kg/m^3 (2.75 lb$_m$/ft^3)
Physical properties of liquid propane at a film temperature of 47.5°C (117.5°F):
  Density: 453 kg/m^3; (28.3 lb$_m$/ft^3)
  Viscosity: 75.6 × 10^{-6} Pa·s = 75.6 × 10^{-6} N·s/m^2; (0.183 lb$_m$/ft·h)
  Thermal conductivity: 82.3 × 10^{-3} W/m·K; (0.0476 Btu/h·ft·°F)
  Specific heat: 3.07 kJ/(kgK); (= 0.733 Btu/lb$_m$°F)
  Prandtl number:

$$\text{Pr}_\ell = \frac{(3.07 \text{ kJ/kg K})(75.6 \times 10^{-6} \text{ N s/m}^2)}{(82.3 \times 10^{-3} \text{ W/m K})}\left(\frac{1 \text{ W}}{1 \text{ J/s}}\right)\left(\frac{10^3 \text{ J}}{1 \text{ kJ}}\right)\left(\frac{1 \text{ kg m/s}^2}{1 \text{ N}}\right) = 2.82$$

The mass velocity, G_T, is

$$G_T = \left(25 \frac{\text{kg}}{\text{hr}}\right) \Big/ \left[\frac{\pi(21.2 \times 10^{-3} \text{ m})^2}{4}\right]\left(3600 \frac{\text{s}}{\text{hr}}\right) = 19.7 \text{ kg/m}^2\text{s} \left(= 1.45 \times 10^4 \frac{\text{lb}_m}{\text{ft}^2\text{hr}}\right)$$

$$\Gamma = \frac{(25 \text{ kg/hr})}{(3600 \text{ s/hr})\pi(21.2 \times 10^{-3} \text{ m})} = 0.104 \text{ kg/ms} (= 252 \text{ lb}_m/\text{ft hr})$$

$$\text{Re}_c = \frac{4(0.104 \text{ kg/ms})}{75.6 \times 10^{-6} \text{ Pa s}}\left(\frac{1 \text{ Pa s}}{1 \text{ kg m s}}\right) = 5,500$$

indicating that condensate film is turbulent.

From Figure 6.22, read Co = 0.15. From Equation (6.85),

$$h_c = 0.15\left\{\frac{(82.3 \times 10^{-3} \text{ W/m K})^3(453 \text{ kg/m}^3)[(453 - 44)\text{kg/m}^3](9.81 \text{ m/s}^2)}{(75.6 \times 10^{-6} \text{ Ns/m}^2)^2}\right\}^{1/3}$$

$$= 4450 \text{ W/m}^2 \text{ K} (= 784 \text{ Btu/h ft}^2)$$

Check the Boyko-Kruzhilin equation: Since the condensation path is from saturated vapor in ($x_i = 1.0$) to saturated liquid out ($x_o = 0$), use Equation (6.86d) for the bracketed term in Equation (6.86a):

$$\frac{h_c D_i}{k_\ell} = 0.024\left[\frac{(21.2 \times 10^{-3} \text{ m})(19.7 \text{ kg/m}^2\text{s})}{(75.6 \times 10^{-6} \text{ N s/m}^2)}\left(\frac{1 \text{ N s}^2}{\text{kg m}}\right)\right]^{0.8}(2.82)^{0.43}\left[\frac{1+\sqrt{453/44}}{2}\right] = 77.8$$

$$h_c = 77.8\left[\frac{82.3 \times 10^{-3} \text{ W/m K}}{21.2 \times 10^{-3} \text{ m}}\right] = 302 \text{ W/m}^2\text{K} (= 53.2 \text{ Btu/h·ft}^2\cdot°\text{F})$$

In this case, the condensation process that gives the highest coefficient is gravity-driven, turbulent film, and the proper coefficient to use is $h_c = 4{,}450 \text{ W/m}^2\text{K}$.

The total heat duty to be removed should include the relatively small contribution of subcooling the condensate from the saturation temperature to the mean film temperature of about 47.5°C:

$$q = (25 \text{ kg/hr})\left[(271 \text{ kJ/kg}) + \left(3.07 \frac{\text{kJ}}{\text{kg K}}\right)(7.5 \text{ K})\right]$$

$$= 7.35 \times 10^3 \text{ kJ/h} = 2.04 \text{ kW} (= 6.97 \times 10^3 \text{ Btu/h})$$

The tube inside surface area required is

$$A_i = \frac{\dot{q}}{h_c \Delta T} = \frac{2.04 \times 10^3 \text{ W}}{(4{,}450 \text{ W/m}^2 \text{ K})(15 \text{ K})} = 0.0306 \text{ m}^2 (= 0.328 \text{ ft}^2)$$

and the length L required is

$$L = \frac{0.0306 \text{ m}^2}{\pi(21.2 \times 10^{-3} \text{ m})} = 0.459 \text{ m} (= 1.51 \text{ ft})$$

6.4.2.2 Condensation inside a Horizontal Tube

6.4.2.2.1 Gravity Driven

At low flow rates, a vapor may be condensed inside a horizontal or nearly horizontal tube under the condition that the condensate film drains down the inside surface of the tube by gravity and in laminar flow into the liquid pool at the bottom of the tube and then out the tube outlet. Kern [38] proposed a simple modification of the Nusselt equation for condensation outside horizontal tubes (see next section) that can be used for this case:

$$h_c = 0.761\left[\frac{k_\ell^3 \rho_\ell (\rho_\ell - \rho_v) g L}{m_c \mu_\ell}\right]^{1/3} \tag{6.87}$$

where m_c is the mass flow rate of the condensate formed in the tube.

6.4.2.2.2 Vapor-Shear Driven

At sufficiently high vapor flow rates, vapor shear causes the condensate film to become turbulent and drain from the tube more rapidly. The Boyko-Kruzhilin equation (6.86a, b, c, d, e) was validated by condensing steam data.

6.4.2.2.3 Selection of a Correlation

Similar to the case for vertical tubes, the correlation (either Equation (6.87) or Equation (6.86a, b, c, d, e)) giving the higher coefficient for a given case most likely represents the actual controlling flow situation. There is a transition region between the two correlations, where the actual coefficient may be 20–30% higher than the prediction of either correlation. In designing for condensing inside "horizontal" tubes, it is important to ensure that the tube is actually slightly inclined downward toward the exit; 1/4 to 1/2 in. per foot of length is usually sufficient for good drainage.

6.4.2.3 Condensation outside Horizontal Tubes and Tube Banks

6.4.2.3.1 Single Horizontal Tubes with Gravity-Driven, Laminar Flow

Nusselt's solution for this case [35] is

$$h_c = 0.725 \left[\frac{k_\ell^3 \rho_\ell (\rho_\ell - \rho_v) \lambda g}{\mu_\ell (T_{sat} - T_w) D} \right]^{1/4} \tag{6.88}$$

which may be put in the equivalent forms

$$h_c = 0.952 \left[\frac{k_\ell^3 \rho_\ell (\rho_\ell - \rho_v) g L}{m_c \mu_\ell} \right]^{1/3} \tag{6.89}$$

where m_c is the mass flow rate of the condensate, and

$$h_c = 1.51 \left[\frac{k_\ell^3 \rho_\ell (\rho_\ell - \rho_v) g}{\mu_\ell^2} \right]^{1/3} \mathrm{Re}_c^{-1/3} \tag{6.90a}$$

where

$$\mathrm{Re}_c = \frac{4 m_c}{\mu_\ell L} \tag{6.90b}$$

and L is the length of the tube.

6.4.2.3.2 Multiple Horizontal Tubes in a Vertical Row

Nusselt [35] also analyzed this problem, assuming that the condensate drained from an upper tube at bottom dead center as a continuous sheet, falling on the next lower tube and flowing down and around that tube in undisturbed laminar flow. Under these highly idealized conditions, the average coefficient for a row of tubes N tubes high is

$$h_{c,N} = h_{c,1} N^{-1/4} \tag{6.91}$$

Because of the flow disturbances implicit in the actual drainage of the condensate, this equation is too conservative, and an alternative procedure is described in Section 6.4.2.3.4 below.

6.4.2.3.3 Effect of High Cross-Flow Vapor Velocity

A high vapor velocity, in cross flow on a cylinder, causes rippling and turbulence in the condensate and results in an increase (compared with Equation (6.89)) in the heat-transfer coefficient beginning at vapor Reynolds numbers (based on maximum vapor velocity flowing around the tube and the tube diameter) above 20,000, rising to as much as a 10-fold increase at Reynolds numbers about 100,000 [39]. Such high velocities are usually unacceptable because of high pressure drop, vibration, and erosion.

6.4.2.3.4 Estimation of Heat Transfer for Flow across a Tube Bundle

Because of the opposing effects (which are individually poorly understood) of the phenomena described in the previous two sections, any calculation of condensing heat-transfer coefficient on a bundle of horizontal tubes (as in a shell-and-tube heat exchanger) is uncertain. An acceptable procedure for most purposes is to ignore both the row-number effect and the vapor-shear enhancement and use the single-horizontal-tube equations (6.88), (6.89), (6.90a), and (6.90b) for the entire tube bundle.

6.4.3 Special Cases in Condensation

6.4.3.1. Enhanced Surfaces in Condensation

Enhanced surfaces can often significantly increase the effective heat-transfer coefficient in condensation, especially if the condensing heat-transfer coefficient is the limiting factor in the overall heat-transfer-coefficient equation. Such enhancements include low fins on horizontal tubes, which increase the heat-transfer area, and fluting on vertical tubes and plane surfaces, which thins the condensate film over part of the surface by surface-tension effects. However, these improvements are limited by condensate retention between the fins and flooding of the drainage paths [7, 34].

6.4.3.2 Condensation in the Presence of a Noncondensable Gas

Condensing vapors often contain a gas that is noncondensable and effectively insoluble in the condensate, such as a small concentration of air in steam. The condensation process may be little affected by the presence of noncondensables at the start of condensation. As the condensation proceeds, the concentration of the noncondensable increases, the dew point decreases (decreasing the temperature difference for heat transfer), and the mass-transfer resistance to the vapor diffusing to the condensing surface increases and may eventually control the rate of condensation. This process was first seriously studied by Colburn and Hougen [40] and more recently by Mueller [41]. A fundamentally sound design procedure requires commercially available computer-based design methods, but a reasonable estimate can be made using the method described next.

6.4.3.3 Condensation of a Multicomponent Vapor

Condensation of a multicomponent vapor, often including noncondensable gases, is a common design problem in the process industries and in energy-producing cycles and air-conditioning and refrigeration systems using mixed working fluids or refrigerants. As the less-volatile components preferentially condense, the dew point continuously falls, and the rate of condensation is controlled by the diffusion and counterdiffusion of the various components to and away from the condensing surface. It is usually assumed that the remaining vapor and the condensate are in thermodynamic equilibrium at every point in the process, but this is only approximately true, even if care is taken to keep the total vapor and condensate streams in intimate contact throughout the condenser. If the condensate is allowed to fall out of contact with the remaining vapor, the condenser acts as a poor fractionator, and the condensing capability is sharply reduced.

The basic fluid dynamics and heat- and mass-transfer processes for multicomponent condensation are poorly understood, and the computation is difficult; available design methods are both heuristic and feasible only for computer solution. The basic model was developed by Silver [42] and put in more general form by Bell and Ghaly [43]. Computer-based design methods that have been validated against experimental data are commercially available.

6.4.3.4 Condensation of Superheated Vapor

A superheated vapor may be condensed directly from the vapor state by contact with a surface that is below the saturation temperature or the dew point of the vapor. The heat-transfer coefficient is predicted by the equations in this section if the saturation temperature (*not* the superheat temperature) is used as the effective temperature for heat transfer, i.e., a desuperheater/condenser may be designed as if all of the heat load (including the sensible heat for desuperheating the vapor to the saturation temperature) were transferred at the same rate as for the saturated vapor as long as the condenser surface temperature and the coolant temperature are below the saturation temperature.

A further consequence is that the superheated vapor will fail to directly condense *only* if the surface temperature is above the saturation temperature, and this will only occur if the local heat flux is *higher* than would be obtained if the vapor were condensing on the surface. Note that this implies, first, that a desuperheater is not necessary to desuperheat a vapor if the only concern is that the heat-transfer rate of a superheated vapor is too low, and second, that care must be taken with superheated vapors such that the local heat-transfer rate and surface temperature are not too high for the coolant stream, causing excessive fouling or film boiling (see next section on vaporization).

6.4.4 MECHANISMS OF VAPORIZATION

Vaporization processes may be divided into "pool boiling," in which the hot surface is immersed in a pool of liquid and the vapor bubbles may flow freely away from the hot surface, driven by the difference in phase densities, and "convective vaporization," where the liquid and vapor flow together along or away from the hot surface, driven either by natural convection of the two-phase mixture (termed *thermosiphon action*) or by an externally forced convection (e.g., by a circulating pump).

Pool boiling may be further divided into several processes, dependent primarily upon the temperature difference between the hot surface temperature and the saturation temperature (boiling temperature) of the liquid. Typical behavior for a light hydrocarbon at its saturation temperature boiling off of a typical metal tube surface is shown in Figure 6.23, in which the heat flux to the liquid is plotted against the temperature difference between the hot surface, T_w, and the saturation temperature, T_{sat}.

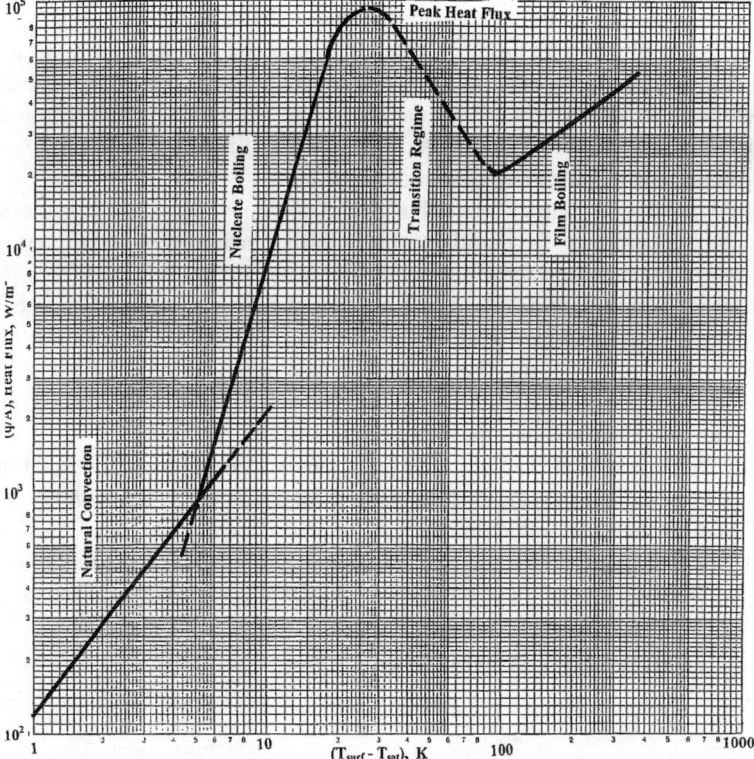

FIGURE 6.23 Pool boiling curve for a light hydrocarbon at atmospheric pressure.

At temperature differences below about 3–5°K (5–9°F), the heat transfer is by natural convection to the liquid; the slightly superheated liquid rises quiescently to the free-liquid surface and vaporizes with no obvious evidence of boiling action. At temperature differences above about 5°K (9°F), vapor bubbles begin to form at specific nucleation sites on the surface, grow, detach, and rise to the free surface, while the process repeats itself regularly. Nucleation sites are usually microscopic pits, scratches, grain boundaries, or other surface irregularities where gas or vapor is trapped to initiate the bubble growth. The smoother the surface, the greater is the superheat required to initiate nucleation and nucleate boiling. Surfaces such as glass may be so devoid of nucleation sites that the liquid may be highly superheated before a random piece of dust sets off a sudden nucleation, desuperheating, and burst of vapor generation. Rough surfaces may nucleate at very small superheats.

Once stable nucleate boiling is established, the heat flux and vapor generation rate are strong functions of the temperature difference. As the surface temperature increases, more nucleation sites are activated, and the bubble generation rate increases at already active sites. Each surface has its own characteristics, and those can change with time as a result of corrosion, fouling, mechanical damage, or even how the equipment was shut down. Consequently, any estimate of nucleate boiling heat-transfer coefficients is approximate.

However, there is an upper limit on the heat flux that can be sustained in nucleate boiling. As the heat flux, and therefore the vapor generation, rises, it is increasingly difficult for sufficient liquid to reach the surface to maintain the vapor generation. At the peak nucleate boiling heat flux, the liquid reaching the surface is equal to the vapor generation rate. This condition, termed the "peak" or "critical" heat flux, can be estimated using equations given in the next section.

If the peak heat flux is exceeded even momentarily and/or locally, a portion of the surface dries out and usually further heats up, and becomes essentially useless for heat transfer. The dry-out process will usually spread over more surface and result in a pronounced decrease in the heat-transfer rate. This is termed the *transition boiling regime*, and extended operation in this regime may result in rapidly increasing fouling and loss of heat-transfer capability, until the next cleaning. Once the transition boiling regime has been entered, only a sharp reduction in the surface temperature, or even complete shutdown, can restore the nucleate boiling regime. If the heat source is hot enough, the surface temperature rises until the surface is covered by a continuous film of vapor and the heat-transfer rate reaches a minimum. This condition is termed the *Leidenfrost point* and marks the beginning of the "film boiling regime."

In film boiling, all heat is transferred by conduction/convection through the vapor and by radiation from the hot surface to the liquid interface. Further increases in surface temperature result in a slow increase in the heat transfer. However, heat transfer in film boiling is poor and may result in severe fouling and even thermal decomposition of the boiling fluid. If the heat source is a constant-flux device, such as an electrical resistance heater, the surface temperature may rise to the melting point and the equipment may be permanently damaged. Film boiling should be avoided. When starting up boiling equipment, it is important to introduce the cold liquid first to prevent initial overheating of the surface into the film boiling regime.

In convective vaporization, the same boiling regimes are encountered, but modified by the net motion of the two-phase fluid past the surface. At low velocities or high heat fluxes, the convection effect is small, and nucleate boiling dominates. At higher velocities, the heat-transfer rate is dominated by the two-phase mixture sweeping across the surface. It is still important to avoid transition and film boiling, but the onset of these phenomena is complicated by many factors. (See [1, 34].)

6.4.5 Boiling Heat-Transfer Correlations

6.4.5.1 Nucleate Boiling

Prediction of nucleate-boiling heat transfer is complicated by the strong dependence of bubble nucleation on boiling surface characteristics that are hard to quantify or even identify and measure.

Heat Transfer

Of the many correlations in the literature, none give consistently accurate predictions. As good as any, and simpler than most, is the Mostinski [44] equation, which is dimensional and must be used with the units specified:

$$h = 0.1011 p_{cr}^{0.69}(q/A)^{0.7}[1.8 p_R^{0.17} + 4 p_R^{1.2} + 10 p_R^{10}] \tag{6.92a}$$

where h is the nucleate-boiling heat-transfer coefficient in W/m²K, p_{cr} is the critical pressure of the boiling fluid in bar abs, (q/A) is the boiling heat flux in W/m², and p_R is the reduced pressure, the pressure of the boiling process divided by the critical pressure of the boiling liquid (p_R is dimensionless).

In U.S. customary units, the Mostinski equation is

$$h = 0.00658 p_{cr}^{0.69}(q/A)^{0.7}[1.8 p_R^{0.17} + 4 p_R^{1.2} + 10 p_R^{10}] \tag{6.92b}$$

Here, h is in Btu/h·ft²·°F, p_{cr} is in lb$_f$/in.² abs, and (q/A) is in Btu/h·ft². In both of these equations,

$$(q/A) = h(T_{surf} - T_{sat}) \tag{6.93}$$

6.4.5.2 Critical Heat Flux in Pool Boiling

The critical heat flux in pool boiling can be predicted reasonably closely by an equation developed by Zuber et al. [45] and Lienhard and Dhir [46], as reported by Hewitt et al. [30].

$$(q/A)_{crit} = 0.149 i_{fg} \rho_v^{1/2}[\sigma g(\rho_\ell - \rho_v)]^{1/4} \tag{6.94}$$

where i_{fg} is the latent heat of vaporization; ρ_ℓ and ρ_v are the densities of the liquid and vapor phases, respectively; σ is the surface tension; and g is the gravitational acceleration. In using U.S. customary units, it is necessary to include $g_c = 4.17 \times 10^8$ lb$_m$·ft/lb$_f$·h² in the bracketed term to get the correct dimensions. Equation (6.94) is valid within about ±15% for most pool boiling geometries (flat plate, cylinder, etc.), but a more complete and accurate table of correction factors is given by Collier and Thome [34].

6.4.5.3 Natural and Forced Convection Vaporization

The present best correlation for convective vaporization upwards in a vertical tube (the usual case) is by Steiner and Taborek (47), also described in detail by Collier and Thome [34]. It combines the nucleate-boiling and convective contributions in a continuous function that satisfies the asymptotic conditions at both pool boiling and full convective flow-dominated extremes. The complete method is too extensive to include here, but the following equations are valid for most boiling cases of interest.

The heat-transfer coefficient for the combined nucleate-boiling and convective vaporization processes, h_{TP}, is related to the nucleate-boiling coefficient h_{NB}, and the convective coefficient, h_c, by

$$h_{TP} = [(h_{NB})^3 + (h_c)^3]^{1/3} = \frac{(q/A)}{T_{surf} - T_{sat}} \tag{6.95a}$$

Either the nucleate-boiling correlation given by Equations (6.92a) and (6.92b) or the extended methods given in the literature [34, 47] can be used for h_{NB}. The two-phase convective coefficient h_c is related to the single-phase liquid coefficient $h_{\ell 0}$ by

$$h_c = h_{\ell 0} F_{TP} \qquad (6.95b)$$

where $h_{\ell o}$ is calculated from Equations (6.41) and (6.42), treating the entire mass flow rate as liquid. The two-phase flow correction factor F_{TP} is given by

$$F_{TP} = [(1-x)^{1.5} + 1.9 x^{0.6} (\rho_\ell / \rho_v)^{0.35}]^{1.1} \qquad (6.95c)$$

Example 6.12

A two-phase mixture of steam and water is flowing upward in a 1-in. OD (25.4 mm) 14 BWG tube (ID = 0.834 in. [21.2 mm]) at 185°C (366°F) and 11.225 bar abs (165 psig). The total mass flow rate is 0.284 kg/s (0.626 lb$_m$/s). The local quality (mass fraction vapor) is 0.05, and the local heat flux is 250 kW/m² (79,250 Btu/h·ft²). Calculate the local heat-transfer coefficient and the local inside tube wall temperature.

Solution

The thermophysical properties required are

Liquid density, ρ_ℓ: 881.67 kg/m³ (54.99 lb$_m$/ft³)
Vapor density, ρ_v: 5.745 kg/m³ (0.3634 lb$_m$/ft³)
Liquid viscosity, μ_ℓ: 145.2 × 10^{-6} kg/m·s (0.351 lb$_m$/ft·h)
Liquid thermal conductivity, k_ℓ: 671.1 × 10^{-3} (0.388 Btu/h·ft·°F)
Liquid Prandtl number, Pr$_\ell$: 0.957
Critical pressure, p_{crit}: 220.55 bar abs (3206 lb$_f$/in.² abs)
Reduced pressure, p_R: 11.225 bar/220.55 bar = 0.0509

First, evaluate h_{TP} from Equation (6.95a), utilizing Equation (6.92a) to calculate h_{NB}, and using Equations (6.41), (6.42), (6.95c), and (6.95b) to calculate h_o, F_{TP}, and h_c, respectively.

Calculation of h_{NB}:

$$h_{NB} = 0.1011(220.55)^{0.69}(250,000)^{0.7} \times [1.8(0.0509)^{0.17} + 4(0.0509)^{1.2} + 10(0.0509)^{10}]$$

$$= 30{,}100 \text{ W/m}^2 \text{ K } (= 5300 \text{ Btu/hr ft}^2 \text{°F})$$

(Note: This is a dimensional equation, so the units are not "calculable" in the usual sense.)
Calculation of h_o:

$$\text{Re} = \frac{D_i \rho_\ell v_{\ell 0}}{\mu_\ell}$$

$$v_{\ell o} = \frac{0.284 \text{ kg/s}}{\left(881.67 \dfrac{\text{kg}}{\text{m}^3}\right) \dfrac{\pi}{4} (21.2 \times 10^{-3} \text{ m})^2} = 0.913 \text{ m/s } (= 2.99 \text{ ft/s})$$

$$\text{Re} = \frac{(21.2 \times 10^{-3} \text{ m})(881.67 \text{ kg/m}^3)(0.913 \text{ m/s})}{(145.2 \times 10^{-6} \text{ kg/m s})} = 118{,}000$$

$$f_F = (3.64 \log_{10} 118{,}000 - 3.28)^{-2} = 0.00434$$

$$\frac{h_{\ell 0} D_i}{k_\ell} = \frac{\left(\dfrac{0.00434}{2}\right)(118{,}000)(0.957)}{1.07 + 12.7\left(\dfrac{0.00434}{2}\right)^{1/2}[(0.957)^{2/3} - 1]} = 233$$

$$h_{\ell o} = 233 \frac{(671.1 \times 10^{-3})}{(21.2 \times 10^{-3})} = 7{,}390 \frac{\text{W}}{\text{m}^2\,\text{K}} \left(= 1300 \frac{\text{Btu}}{\text{hr ft}^2\,°\text{F}}\right)$$

Calculation of F_{TP}:

$$F_{TP} = \left[(1 - 0.05)^{1.5} + 1.9(0.05)^{0.6}\left(\frac{881.67}{5.745}\right)^{0.35}\right]^{1.1} = 3.05$$

$$h_c = 7{,}390(3.05) = 22{,}500 \frac{\text{W}}{\text{m}^2\,\text{K}} \left(= 3960 \frac{\text{Btu}}{\text{hr ft}^2\,°\text{F}}\right)$$

Calculation of h_{TP}:

$$h_{TP} = [(30{,}100)^3 + (22{,}500)^3]^{1/3}$$

$$h_{TP} = 33{,}800 \text{ W/m}^2\,\text{K} \ (= 5950 \text{ Btu/hr ft}^2\,°\text{F})$$

Calculation of $T = T_{wall} - T_{sat}$:

$$\Delta T = \frac{q/A}{h_{TP}} = \frac{250{,}000 \text{ W/m}^2}{33{,}800 \text{ W/m}^2\,\text{K}}$$

$$\Delta T = 7.40°\text{C} \ (= 13.3°\text{F})$$

6.4.6 Special Cases in Vaporization

6.4.6.1 Boiling outside Tube Bundles

Kettle reboilers generate vapor for process purposes by boiling the liquid outside a bundle of horizontal tubes. This process was historically considered to be an example of pool boiling, with the coefficient calculated by a typical nucleate boiling correlation and a maximum heat flux either fixed by experience [48] or by some modification of Equation (6.94) reflecting the size of the tube bundle [49, 50].

Recent studies have demonstrated that the boiling process is actually a convective one, with the vapor generated on the lower tubes creating a rising and growing two-phase flow across the upper tubes. The vapor and liquid separate at the top of the bundle and the clear liquid flows downward around the sides of the bundle to complete the circuit. Computer-based design methods employ this model, and the existing database is discussed by Collier and Thome [34].

6.4.6.2 Enhanced Surfaces in Boiling

A variety of enhancements to the boiling process are commercially available. Low-finned tubes are commonly used with many organic liquids (which tend to give low boiling coefficients) to increase the heat-transfer area available. Other surfaces are designed to increase the availability of nucleation sites or increase the circulation rate of the boiling liquid past the heat-transfer surface by natural convection. These surfaces are particularly effective in initiating and improving the boiling performance at very low temperature differences. However, they generally do little or nothing to increase the critical heat flux. Manufacturers' data and recommended design practices are necessary to fully utilize the advantages of these special surfaces.

6.4.6.3 Subcooled Boiling

Boiling also will occur with a liquid that is below its saturation temperature if the hot surface temperature is sufficiently above the nucleation temperature for the surface. The bubbles thus formed collapse quickly when they move from the surface into the bulk subcooled liquid, resulting in high heat-transfer rates. Moderate subcooling at the entrance to a vaporizer can be assumed to be heated by heat transfer at the same rates as for the saturated boiling process, if the saturation temperature of the fluid at the existing pressure is used in the temperature difference.

6.5 HEAT EXCHANGERS

6.5.1 INTRODUCTION

Heat exchangers vary in heat-transfer surface area from less than 1 m^2 (10 ft^2) to over 10,000 m^2 (100,000 ft^2), in service temperatures from near absolute zero to over 1400°C (2500°F), and in pressures from near full vacuum to over 600 bar (atmospheres). The smaller ones used for routine duties like jacket water and lube oil cooling can often be ordered quickly from standardized designs or assembled from off-the-shelf components. The larger exchangers and those used for process stream heating, cooling, or phase change are usually custom-designed, implying among other things specialized design and fabrication personnel and facilities, and longer delivery times. Those intended for extreme conditions are always custom designs.

This section describes the major types of heat exchangers used in the process industries, their major construction features and options, and their most common applications. Next is a subsection devoted to the basic equations of heat-exchanger design, including the concept of an overall heat-transfer coefficient, the basic design equation (used mainly in computer-based design methods), and the mean temperature difference formulation (used primarily with hand-based design methods). The next subsection describes the basic logic of heat-exchanger design methods, whether by hand or by computer. The last subsection introduces fouling mechanisms and techniques for amelioration.

The *Heat Exchanger Design Handbook* [1] is the most comprehensive reference, but Shah and Sekulic [51] and Kakac and Liu [52] are more handy and cover most areas. Hewitt et al. [30] includes much information on heat exchangers. Qualitative factors in design and application are discussed by Bell [53]. Other, more specialized references are given with the descriptions of the various types.

Most heat exchangers for the process industries are designed by computer programs. There are commercially available design programs for most common heat-exchanger types. Many manufacturers employ their own programs. Many have proprietary programs for their equipment, which vary in accuracy and reliability. All design programs are to some degree heuristic in that they incorporate correlations that have not been tested over the full range of possible process variables. All make assumptions (such as the uniformity of flow distribution among a multitude of parallel tubes, for example) that are not exactly achievable and may be seriously compromised in some situations. The input data are always to some degree uncertain, especially

Heat Transfer

when multicomponent systems are involved. Operational parameters vary under normal conditions and may change significantly with long-term process changes or during startup, shut down, or emergency operations. It is highly recommended that every heat-exchanger design be reviewed by someone knowledgeable in both the equipment and the process application before being accepted.

6.5.2 Types of Heat Exchangers and Their Selection

6.5.2.1 Criteria for Heat-Exchanger Selection

Different heat-exchanger types may be feasible for any given service. The criteria for selecting the best exchanger for a given application are listed below in the usual order of importance:

1. The heat exchanger must perform the required thermal changes on the process streams within the allowed pressure drops.
2. The heat exchanger must withstand service conditions. There are mechanical stresses imposed by the exchanger's weight and those externally imposed by piping stresses, wind and seismic loading, and mechanical handling during initial installation, normal operation, and turnaround. The temperature differences within the exchanger create thermal stresses, which are dealt with by various design features. The heat exchanger must withstand corrosive attack, primarily handled by selection of the materials of construction. Erosion and vibration are controlled by limiting velocities, especially in certain critical areas near the nozzles and wherever the flow changes directions in the heat exchanger. The heat exchanger must also be designed either to minimize the buildup of fouling or to withstand the mechanical effects of fouling as it develops.
3. The heat exchanger must allow mechanical and chemical cleaning of any portions of heat-transfer surface or other vital components that become fouled, and it must allow replacement of the tubes, gaskets, and other short-lived components during the normal lifetime of the exchanger. Maintenance should require minimum downtime, handling difficulties, and labor cost.
4. The heat exchanger must provide operational flexibility, permitting operation over the probable range of conditions without instability, excessive fouling, vibration problems, or freeze-up that might damage the exchanger itself. Changes in process conditions (e.g., change in feed-stream composition, turndown) and in environmental conditions (e.g., daily and seasonal changes in atmospheric temperature) must be considered.
5. Cost is always an important factor. Cost considerations include first cost and installation as well as the cost of lost production. The latter cost (while the heat exchanger is out of service or overloaded) often is a critical criterion in determining the acceptability of a heat exchanger.
6. Other criteria include maximum weight, length, or diameter limitations as well as the use of standard tube sizes or other replaceable components that are carried in inventory.
7. The experience (good or bad) of operating and maintenance personnel with existing heat exchangers should be considered.

6.5.2.2 Double-Pipe Heat Exchangers

A typical double-pipe heat exchanger is illustrated in Figure 6.24. It consists of an inner pipe or tube within a larger diameter outer pipe, with one fluid flowing through the inner pipe and another flowing through the annular space between the inner and outer pipes. The inner pipe may have external or internal longitudinal fins (or both, or other internal enhancement) to change the heat-transfer area ratio. (See Section 6.5.3 for the rationale for the use of finning.) Multiple units may be connected in series to provide the required heat-transfer area. Parallel exchangers may be

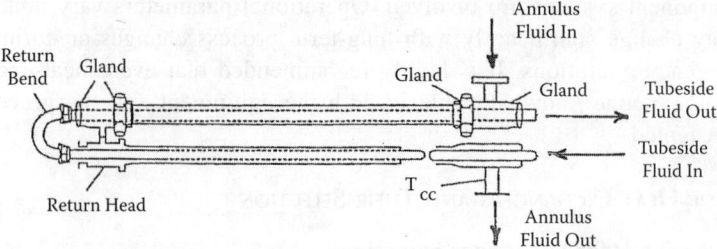

FIGURE 6.24 Basic double-pipe heat exchanger, with arrows showing countercurrent flow.

provided to meet pressure-drop limits. Pure countercurrent, pure cocurrent, and various series- or parallel-flow arrangements may be used.

Double-pipe exchangers are comparatively bulky, heavy, and expensive per unit heat-transfer area and hence are usually limited to units of less than 50 m^2 (500 ft^2), though there is no absolute limit. The advantages of the double-pipe exchanger lie in the flexibility of application and piping arrangement. They can be quickly assembled by plant maintenance crews and are usually cleanable on either side. Uniform and stable flow distribution in each channel of a double-pipe heat exchanger is assured by providing a pump for each parallel section; this can be of particular importance in cooling viscous liquids.

6.5.2.3 Shell-and-Tube Heat Exchangers

Shell-and-tube heat exchangers are the workhorses of the process industries and other applications. The basic configuration, flow across/along a bank of cylindrical tubes, is mechanically strong, provides high heat-transfer surface per unit volume, and allows the use of a variety of special design features to meet many unusual and extreme conditions of chemical processing. The cost per unit area is relatively low, especially in low-carbon steel. The major disadvantage is that they are inflexible once they are constructed, and an exchanger that does not perform satisfactorily in service is usually difficult to modify. Sometimes, performance can be upgraded by replacing the tubes with externally or internally enhanced tubes (e.g., low-finned tubes, twisted wire or tape inserts, etc.).

Most shell-and-tube exchangers are built to the standards of the Tubular Exchanger Manufacturers Association—TEMA Standards [54]—which cover nomenclature, manufacturing standards, recommended good practices, and much else, but do not include thermal design methods. These exchangers generally must conform to the ASME Boiler and Pressure Vessel Code [55] or comparable standards in other countries. Most refinery and chemical process exchangers also satisfy API Standard 660 [56], and power plant exchangers are covered under the various standards of the Heat Exchange Institute [57]. Singh and Soler [58] present mechanical design procedures, and Yokell [59] gives a comprehensive survey of exchanger nomenclature, components, manufacturing techniques, inspection procedures, and maintenance.

The TEMA standards identify three classes of exchanger construction: Class R covers exchangers for the generally severe requirements of the refinery and related processing industries. Class C meets the moderate requirements of commercial and general process applications. Class B meets the special requirements of chemical process service, mainly addressed to the use of special alloys for corrosion resistance. There is relatively little difference between the requirements, and most process-plant exchangers are built to Class R standards.

Figure 6.25 is a schematic diagram of a typical shell-and-tube heat exchanger. The tubes are the basic component of the shell-and-tube exchanger, providing the heat-transfer surface between one fluid flowing inside the tubes and another fluid flowing across the outside of the tubes. The tubes are generally drawn or extruded seamless metal, or welded. Low-carbon or alloy steel, stainless steel, copper, Admiralty, cupronickel, Inconel, aluminum, or titanium tubes are common; other materials may be specified for special applications.

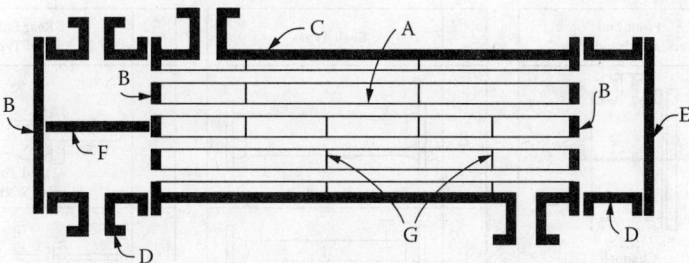

FIGURE 6.25 Diagram of a typical fixed tubesheet heat exchanger, TEMA AEL, with two tube-side passes. (A: tubes, B: tubesheets, C: shell, D: tube-side channels and nozzles, E: channel covers, F: pass dividers, G: baffles)

The tubes may be either bare or with low fins on the outside. Low-fin tubes are used when the fluid on the outside of the tubes (the "shell-side" fluid) has a substantially lower heat-transfer coefficient than the fluid on the inside of the tubes (the "tube-side" fluid). A low-fin surface provides 2 1/2 to 5 times as much heat-transfer area on the outside as the corresponding bare tube, and this area ratio helps to offset the lower heat-transfer coefficient. If low-fin tubes are used, it is necessary to calculate the efficiency by the methods given in Section 6.2.4 of this chapter. Other tube enhancements are available; the thermal/hydraulic characteristics of these devices generally need to be provided by the manufacturer.

The tubes are inserted into drilled holes in the tube sheets ("B" in Figure 6.25) and held in place either by roller expansion into two circumferential grooves cut into each tubesheet hole, or by welding to the tubesheet, or both. A properly expanded tube-to-tubesheet joint is very strong, but may eventually leak; in this case, a seal weld may also be used.

The shell ("C" in Figure 6.25) confines and guides the shell-side fluid through the exchanger. TEMA gives a standard nomenclature to designate the shell configuration and the associated tube-side channels and nozzles, shown here as Figure 6.26. The basic configuration of a shell-and-tube exchanger can be quickly conveyed by three letters chosen from Figure 6.26: The first letter, chosen from the first column, identifies the front head (tube-side inlet); the second letter, from the second column, identifies the shell type (especially the shell-side nozzle positions); the third letter identifies the rear head type. Thus, the exchanger in Figure 6.25 is AEL.

The E shell is the most common configuration; the shell-side fluid enters at one end of the shell and exits at the other. The nozzles may be on opposite sides of the shell (as shown) or on the same side, as piping convenience may dictate. The shell may be horizontal or vertical—especially for condensing (flow downward) or vaporizing (flow upward).

The F shell has a longitudinal baffle on the shell side, which allows for two shell-side passes and countercurrent flow if there are also two tube-side passes. (The advantages and disadvantages of multipass flow are discussed below.) However, it is important to the thermal performance that there be no physical leakage around the side edges of the longitudinal baffle. This can be achieved by welding the longitudinal baffle to the front tubesheet and the shell; in small-diameter shells, this is achieved by cutting the shell in half longitudinally and welding it with the longitudinal baffle between the two halves. Various baffle-sealing devices are also used. Heat transfer also occurs across the baffle by conduction, reducing the efficiency of the exchanger. The effect can be minimized by using a "sandwich" baffle with a layer of insulation. Rozenmann and Taborek [60] discuss the thermal effects of longitudinal baffle leakage. The F shell is almost exclusively used for single-phase flow in the shell.

G and H shells are used primarily for vaporizing the shell-side fluid, especially as horizontal thermosiphon reboilers. Partial longitudinal baffles force the entering liquid to flow across the entire length of the tubes and minimize the possibility of dry-out or stagnant vapor pockets. Leakage

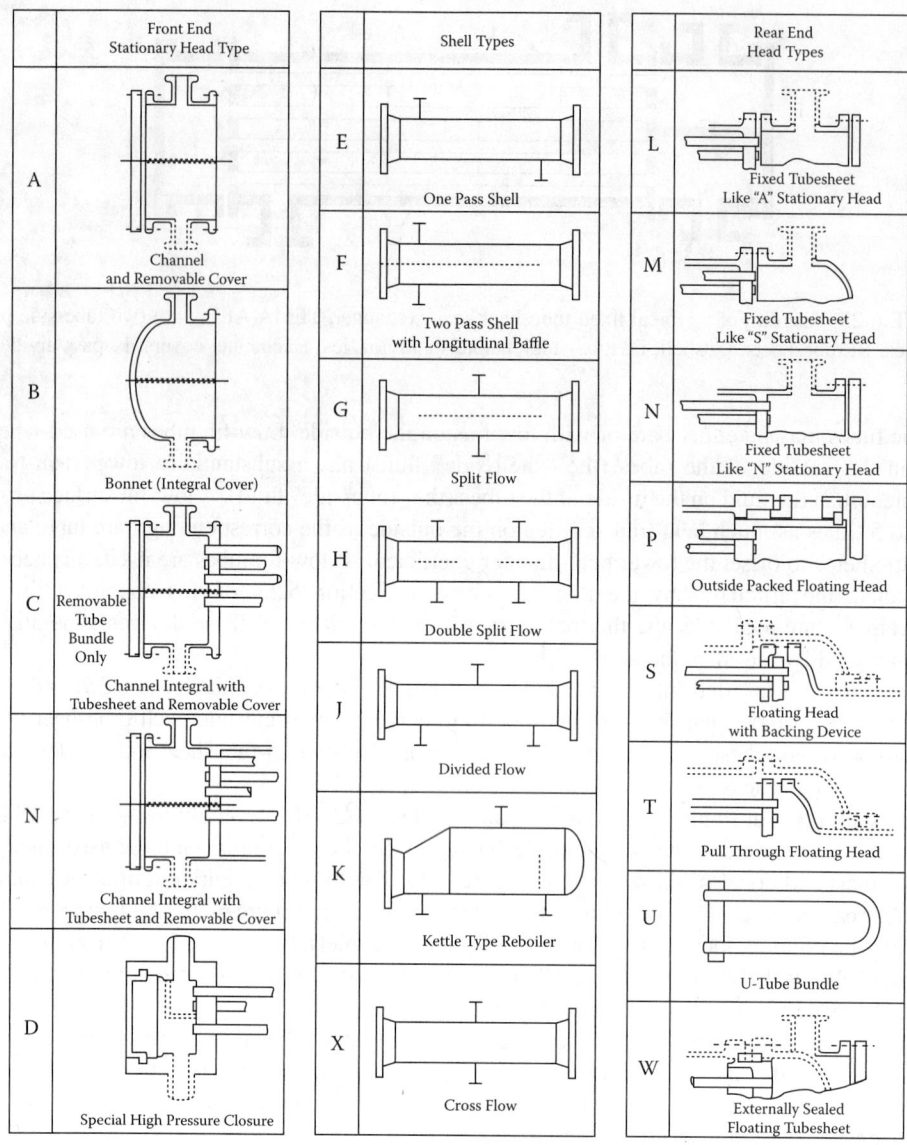

FIGURE 6.26 TEMA shell-and-tube heat exchanger nomenclature. (Reprinted with permission from *Standards of Tubular Exchanger Manufacturers Association*, 9th ed., Tarrytown, NY (2007).)

across the longitudinal baffles is less serious in this case than for the F shell; the baffles should fit closely in the shell and be secured against vibration, but need not be welded to the shell.

The J shell is mainly used for condensation of low-pressure vapor because it splits the flow, reducing velocities and resulting in lower pressure drops. It is also sometimes used for single-phase flow. The J shell can also be used with the two nozzles on top, and two J shells can be mounted nozzle to nozzle in series for long-condensing-range vapors.

The K shell is used for boiling on the shell side ("kettle reboiler") with steam or hot fluid inside a bundle of U-tubes; less commonly, a floating head may be used with straight tubes. The large-diameter shell allows for disengagement of most of the entrained liquid from the vapor; further drying of the vapor can be achieved with mesh demister pads or centrifugal vane separators at the outlet shell-side nozzle. The liquid level is ordinarily maintained at the top of the tube bundle by a weir (as shown in the drawing) or an external liquid-level control device.

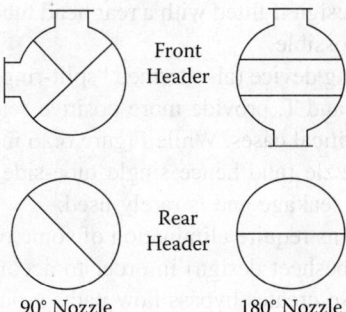

FIGURE 6.27 Possible nozzle and pass-partition plate configuration for four tube-side passes.

The X (or cross flow) shell gives the lowest pressure drop of all the configurations and hence is commonly used as a vacuum condenser. It is essential to secure good vapor distribution across the length of the shell. This can be accomplished in three ways: (a) multiple inlet nozzles from a separate manifold, (b) omitting several rows of tubes and cutting down the support baffles at the top of the tube field, or (c) attaching a large elongated nozzle ("bathtub nozzle") to the top of the shell.

The front-end (stationary head) types are identified by the first letter in the TEMA shorthand nomenclature. The A type is a flanged channel that bolts to the shell flange on one side and has a flanged removable cover on the other. In Figure 6.26, the dashed lines through the center and lower nozzle denote the position of the pass divider (or pass-partition plate) F in Figure 6.25, if two tube passes are used. Arrangements for more tube-side passes are shown in Figure 6.27. The removable cover allows tube inspection, cleaning, or replacement without disturbing the piping to the exchanger. Type B front-end header is the bonnet type, which is often less expensive than the A type but does not allow tube inspection without disconnecting the tube-side plumbing.

Types C and N front headers provide increasingly monolithic construction with fewer gaskets and hence less opportunity for leakage, but also less flexibility of maintenance and greater concern for thermal stress relief. The N type is especially used for nuclear power and hazardous fluid applications.

The D head uses a special closure similar to the breech of an artillery piece, with application to high-pressure feed-water heaters in power plants and some high-pressure chemical processes.

The rear head design is driven by the need to accommodate or relieve the thermal stresses between shell-and-tube bundles due to their different temperatures. If the temperature differences are not too great, a fixed tubesheet rear head, TEMA types L and M (essentially identical to types A and B front heads), can be used. For low shell-side pressures, thermal stress relief can be provided by expansion rolls or bellows in the shell. Tube-side expansion joints can be employed with a floating-head design inside a shell bonnet to provide a single tube-side pass.

If the thermal stresses are too great to be accommodated by a fixed tubesheet design, a U-tube configuration (TEMA U) can often be used. This design eliminates the second tubesheet, and each tube is free to expand or contract independently of the shell or other tubes, and the bundle usually can be easily removed for cleaning or repair. However, countercurrent flow is possible only with an F shell, which usually means that the tube bundle cannot be removed. The U-bend area must be stabilized against vibration, either by rod arrays supporting the tubes or by moving the shell nozzle forward of the U-bend and inserting a full circle tube support at the start of the U-bends. Only the outermost tubes in a U bundle can be individually replaced. It is customary to plug any leaking tubes until the thermal performance is significantly compromised and then remove and replace the entire bundle.

The outside packed and externally sealed floating-head designs, TEMA P and W, rely upon packing between the shell flange and the floating tubesheet to seal the shell-side fluid against leakage to the atmosphere and are largely limited to water or inert gases as the shell-side fluid.

They do allow single-tube pass design if fitted with a rear head tube-side nozzle and flexible piping. Individual tube replacement is possible.

The floating head with backing device (also termed "split-ring floating head") and pull-through floating-head designs, TEMA S and T, provide more positive sealing of the shell-side fluid, with the T design used for the most critical cases. While Figure 6.26 indicates (by the dashed lines) that a rear-head packed tube-side nozzle (and hence single tube-side pass) is possible, this increases the possibility of shell-side fluid leakage and is rarely used.

All of the floating-head designs require elimination of some of the peripheral tubes (compared with the same diameter fixed tubesheet design) in order to accommodate the tubesheet flange or skirt. Omission of these tubes also creates bypass flow paths around the tube bundle, reducing the flow rate and heat-transfer coefficient across the tubes and reducing the effective temperature difference for heat transfer. These bypass lanes can be partially blocked and the thermal effectiveness largely restored by inserting sealing strips—metal strips that are about as wide as the clearance between the shell and the tube bundle and run the length of the exchanger through slots cut into the baffles. The strips are installed symmetrically on both sides of the tube bundle and both sides of the central plane—one pair about every six tube rows in the direction of flow is suggested. Tie rods and spacers located in the bypass lanes also help block the flow but are not as effective as sealing strips.

The essential purpose of the shell-side baffles (item G in Figure 6.25) is to support the tubes against sagging and vibration. A secondary role is to guide the shell-side flow across and through the tube bundle to achieve maximum thermal effectiveness for the allowable pressure drop. The most common configuration is the single segmental baffle, shown in Figure 6.28a. The baffle cut and spacing are design variables chosen (within the limits set by the TEMA standards) to best meet these requirements. Typically, the baffle cut is 15–25% of the shell inside diameter for liquids and 40–45% for gases. Spacing and cut will usually be selected so that the baffle window free-flow area is about 50–100% of the tube bundle free cross-flow area. A tube vibration analysis is used in all questionable cases to verify that damaging levels of vibration will not occur. TEMA also specifies the diametral clearances between tube and baffle hole and between baffle and shell inside diameter; these clearances are necessary for bundle assembly and removal and for tube replacement.

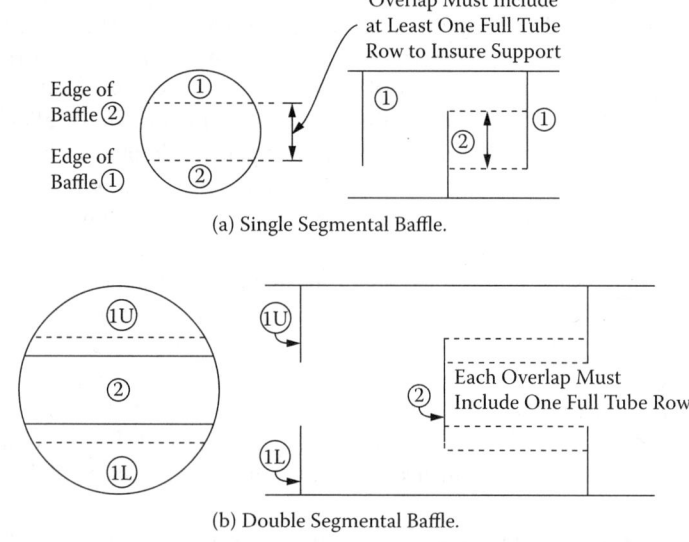

FIGURE 6.28 (a) Single-segmental baffle configuration. (b) Double-segmental baffle configuration.

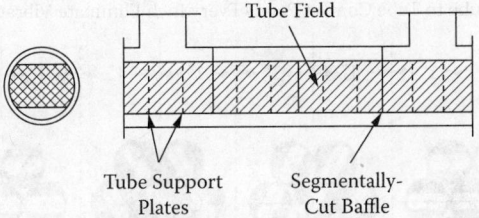

FIGURE 6.29 Diagram of a shell-and-tube heat exchanger with no tubes in the window (NTIW).

FIGURE 6.30 Schematic of the RODbaffle™ configuration for shell-and-tube heat exchangers.

FIGURE 6.31 Schematic of shell-side flow in a helically baffled heat exchanger. (Courtesy, Lummus Technology Heat Transfer, Bloomfield, NJ, USA, with permission.)

Alternative tube support methods include double-segmental baffles (shown in Figure 6.28b), strip baffles, "no tubes in the window" (NTIW) construction (Figure 6.29), RODbaffles™ (Figure 6.30), helical baffles (Figure 6.31), and self-supporting tubes (Figure 6.32). Strip baffles are an extension of the double-segmental baffle concept, further dividing the central baffle into two or more strips with gaps between, and adding a third (or more) strips to the two wing baffles. It is essential that all of the segments/strips overlap their neighbors by at least one full row of tubes to maintain bundle integrity.

The "no tubes in the window" (NTIW) design (Figure 6.29) eliminates all of the tubes that pass through the window areas. These tubes have the longest unsupported span between baffles and are thus the most prone to vibrate. To further secure the remaining tubes, intermediate support plates can be inserted between the baffles. These plates further reduce the unsupported length, but have little effect upon pressure drop or heat transfer.

The RODbaffle design (Figure 6.30) replaces the plate baffles with an array of parallel rods welded to a full circle rod or plate that fits inside the shell. The rods in each grid have a diameter equal to the nominal clearance between the tubes and are spaced so that two rows of tubes fit between two adjacent rods. Four grids form a set, and any number of sets can be used in an exchanger, depending upon length. The second grid in the set is similar to the first, but has the rods offset by one tube diameter. The third and fourth grids are like the first and second, respectively, but rotated 90°. The individual grids are spaced from 100 mm (4 in.) to 300 mm (12 in.) apart, as

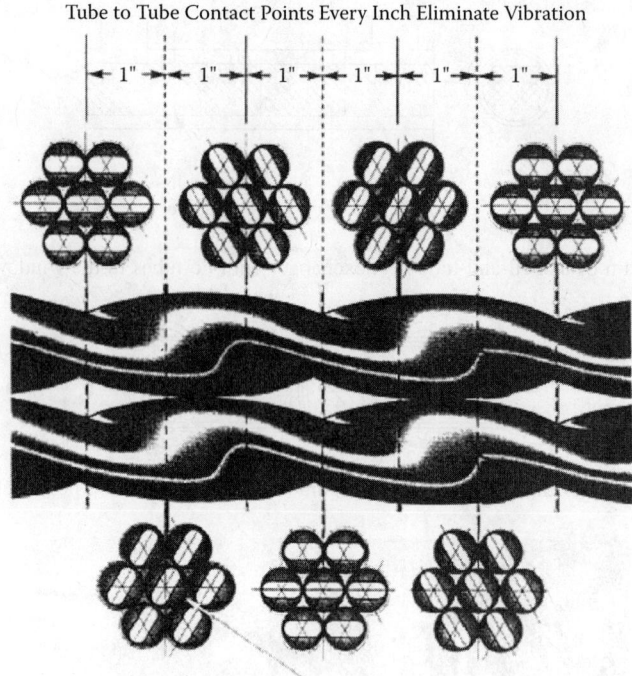

FIGURE 6.32 Diagram of two Twisted Tubes®, showing how tubes are mutually self-supporting. (Courtesy of Koch Heat Transfer Company LP.)

required, to provide support and rigidity, and each tube is supported on all four sides in each set. The sets are held together by welded tie rods. Tube vibration is effectively eliminated, and the nearly longitudinal flow gives significantly reduced pressure drop and a somewhat reduced heat-transfer coefficient than the corresponding conventionally baffled exchanger. Annular distributors (see below) are often specified for RODbaffle exchangers to minimize pressure drop and ensure good distribution.

Helically baffled heat exchangers, illustrated in Figure 6.31, are a recent development. The spiral path through the shell side eliminates or reduces eddy regions and bypass flows, resulting in improved heat transfer for a given pressure drop [61].

Twisted Tube™ construction (Figure 6.32) eliminates the need for baffles altogether, since the tubes can be oriented ("tuned") to contact surrounding tubes periodically along the exchanger. The shell-side flow is longitudinal, and special correlations have been developed for both heat transfer and pressure drop on both sides of the surface. Laminar flow inside the tube is enhanced by the secondary flow induced by the twisted flow path.

Impingement plates are another shell-side construction feature that may be required to prevent tube vibration or erosion. The impingement plate is placed under the shell-side inlet nozzle and should be larger than the nozzle opening, with an escape area off the edges of the plate that is twice the flow area of the nozzle. The impingement plate is attached to the shell or the tie rods. TEMA standards specify the conditions under which an impingement plate is required.

An alternative to the impingement plate is the annular distributor (Figure 6.33), which, through the use of a partial double shell, absorbs the impact of the inlet fluid jet and distributes the fluid more uniformly into the tube bundle. An annular distributor is designed to require a low-pressure

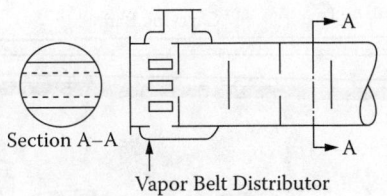

FIGURE 6.33 Annular distributor for a shell-and-tube heat exchanger, with double-segmental baffles.

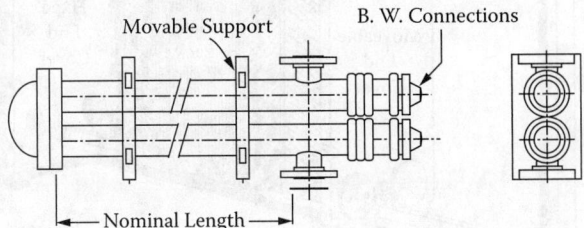

FIGURE 6.34 Schematic of a hairpin heat exchanger. (Reprinted by permission of Koch Heat Transfer LP.)

loss, and is often specified for use with RODbaffle exchangers. It also allows flexibility in the placement of the first baffle. Annular collectors can also be used for the exit flow.

6.5.2.4 Multitube ("Hairpin") Heat Exchangers

The multitube heat exchanger, also termed a "hairpin" heat exchanger, in reference to its prime construction feature, is shown schematically in Figure 6.34. It is intermediate between the double-pipe and the shell-and-tube types, consisting of a bundle of U-tubes arranged in two parallel shells connected by an end fitting accommodating the U-bend. This provides true countercurrent flow. Tube support is required as in shell-and-tube exchangers; the U-bends are active heat-transfer surfaces and must be stabilized against vibration. Sizes range up to 2500 m^2 (25,000 ft^2).

6.5.2.5 Gasketed-Plate Heat Exchanger and Related Partially Welded Variants

Figure 6.35 is a schematic of the basic gasketed-plate heat exchanger, consisting of a pack of embossed thin plates gasketed around the outer edges and the inlet and outlet fluid ports to seal against fluid leakage to the atmosphere and the other fluid. A pair of plates with the gasketing shown as heavy lines is shown in Figure 6.36. The plates are pressed from a wide variety of metals; however, low-carbon steel is not used because it is usually not cost competitive with an equivalent shell-and-tube exchanger. Plastic plates (usually fiber reinforced or graphite loaded for enhanced thermal conductivity) are available for especially corrosive applications. The plates are available in sizes up to about 4 m (14 ft) by 1.2 m (4 ft) wide, and a single large rack can hold up to 500 plates.

The plates can be supplied with some ports closed off to allow for various series- or parallel-flow arrangements for each stream to optimize the pressure-drop utilization and match the temperature profiles of the streams. Plates with special port designs and increased spacing are available for condensing or partial-vaporization services. The plates are embossed with a chevron or staggered rectangular pattern, providing numerous contact areas between adjacent plates to protect against collapse on the low-pressure side. The resulting tortuous flow pattern causes turbulent flow, even at low Reynolds numbers, resulting in high heat-transfer coefficients and pressure gradients.

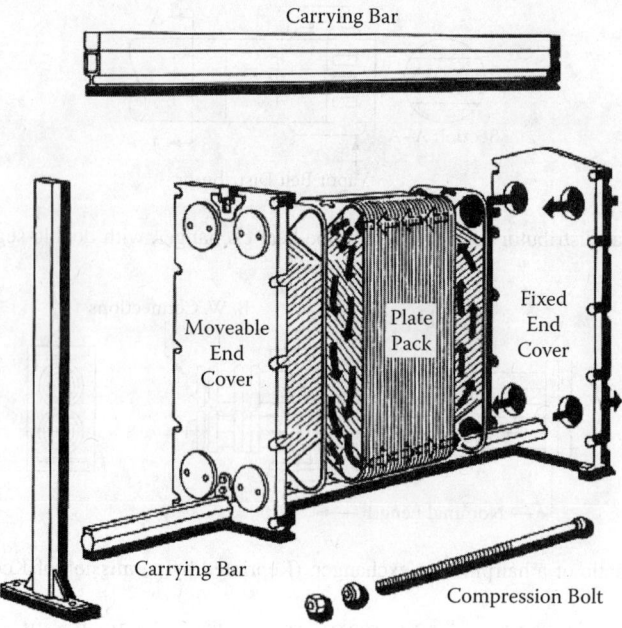

FIGURE 6.35 Exploded diagram of a gasketed-plate heat exchanger. (Reproduced with permission of Alfa Laval Inc.)

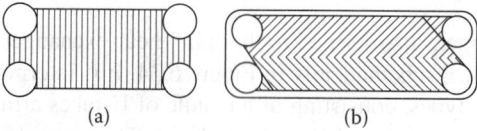

FIGURE 6.36 Two types of plates for the gasketed-plate heat exchanger: (a) parallel-corrugated plate; (b) cross-corrugated plate ("herringbone" or "chevron" pattern).

The gaskets are made from a variety of elastomers; they are confined in grooves around the plate edges and ports, and must be firmly and uniformly compressed to provide an effective seal. These gasket sets can generally be reused, allowing plate exchangers to be regularly disassembled for cleaning or sterilization. The gasket material and construction limit the temperature range and pressures of the application. Small plates can be used up to about 20–25 bar (300–375 psig) at temperatures below about 100°C (210°F), whereas large plates are limited to about 8 bar (120 psig). Maximum temperatures are about 150°C (300°F) with reusable gaskets. Jacketed asbestos gaskets can be used to about 250°C (480°F), but cannot be reused. Limits vary among manufacturers and with the specific designs. Designs with pairs of plates welded together have been manufactured. These designs allow higher pressure operation on the welded channel side, but sacrifice some of the flexibility of disassembly and reassembly. A diffusion-bonded Monobloc™ construction is also available for very high pressure operation, but it cannot be disassembled.

Major manufacturers have design and rating programs that reflect the pressure-drop and heat-transfer characteristics of their specific plates, which can vary widely and are generally proprietary. API [62] provides standards for gasketed-plate exchangers. Where higher alloy construction is required, gasketed-plate exchangers generally have a lower capital cost and are more compact than shell-and-tube exchangers of comparable thermal capacity. Fouling is generally less severe (though fluids containing larger solid particles or fibrous materials tend to plug the fine channels) due to the high turbulence and shear stress developed, and most designs can be disassembled easily for cleaning. Reassembly without leaks can be more difficult unless maintenance crews are specially

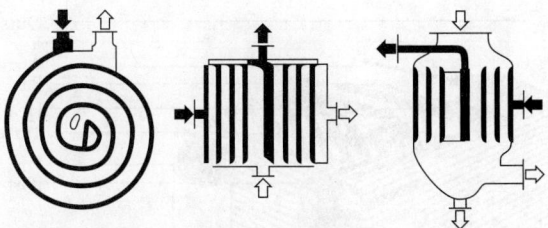

a. Principle of SHE   b. Principle of Type 1   c. Principle of Type 2

FIGURE 6.37 Schematic diagrams of spiral-plate heat exchangers: (a, b) basic construction for two fluids in counterflow, (c) configured for condensation or reboiler service.

trained and equipped. The low temperature and pressure limits are the chief restriction upon the use of gasketed-plate exchangers.

The spiral-plate heat exchanger (Figure 6.37) is closely related to the gasketed plate. One type is constructed by winding two long plates together in a spiral to form open channels for each fluid and then welding the open edges of adjacent plates on each side to form two continuous channels, each open on one edge on opposite sides. The open edges are then closed off by covers on either side of the spiral structure, held in place by peripheral clamps. The plates are spaced by welded stubs or embossed nubs on the plates. Gaskets are used to seal the plate edges against the covers and the atmosphere. Another type has one channel left open for cross flow in that channel and can be used in column internal and external reflux condensers and reboilers.

Spiral-plate exchangers provide high surface densities. The induced secondary flow results in enhanced heat-transfer coefficients, especially for viscous fluids, and reduced sedimentation-type fouling. Spiral-plate exchangers can be easily opened for cleaning by high-velocity water jets.

6.5.2.6 Plate-Fin (Matrix) Heat Exchangers

Plate-fin, or matrix, heat exchangers offer a high volumetric density of heat-transfer surface, on the order of 1000 m^2/m^3 (based on the entire surface in contact with any fluid) and hence are of interest in cryogenic processes requiring extreme insulation and in gas-to-gas applications. The essential construction features are shown in Figure 6.38. The prime heat-transfer surface is the aluminum fin, consisting of thin sheets that have been folded to form arrays of longitudinal fins and brazed to parting sheets separating the fluid channels. Side bars close the channels against the surroundings. The fins, parting sheets, and side bars are stacked in the desired pattern, omitting the side bars where fluid inlets and exits are desired, and the entire assembly is then furnace- or salt-bath brazed. Finally the inlet and exit manifolds are welded to the assembly.

The diagram shows three streams in one exchanger, and units handling up to 12 or 14 different streams have been constructed. Different numbers of parallel channels and different fin geometries can be used for the several fluids. Only nonfouling fluids can be used in plate-fin exchangers due to the fine channels and the inability to disassemble for cleaning. A major concern is achieving uniform flow distribution among the several parallel channels for each fluid, requiring careful design of the distributor section for each channel.

Each manufacturer has specialized design procedures for their equipment. The manufacturers have formed ALPEMA, the Brazed Aluminium Plate-Fin Manufacturers' Association, which has issued *The ALPEMA Standards* [63] for the design, construction, operation, and maintenance of this equipment.

6.5.2.7 Air-Cooled Heat Exchangers

Air-cooled heat exchangers use air as the cooling medium to dissipate low-temperature waste heat, and the two basic configurations are shown in Figure 6.39. Design is dictated by the poor thermal

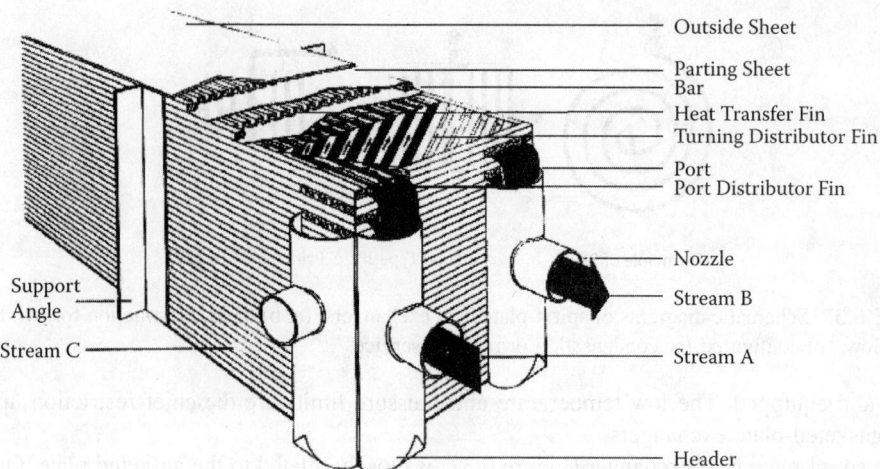

FIGURE 6.38 Cutaway view of a brazed aluminum plate fin (or matrix) heat exchanger configured for three streams. (Courtesy Chart Energy and Chemicals, Inc., La Crosse, WI, a wholly owned subsidiary of Chart Industries, Cleveland, OH.)

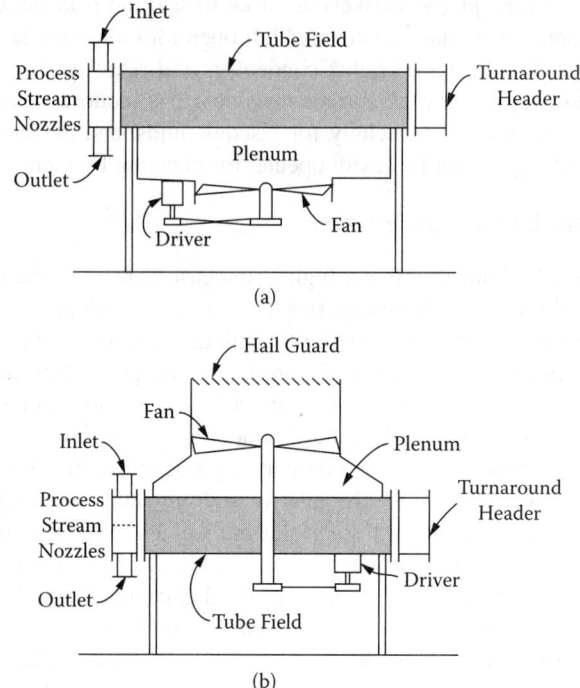

FIGURE 6.39 Schematic diagrams of air-cooled heat exchangers: (a) forced draft, (b) induced draft.

properties of air. The low density of atmospheric air and its low specific heat require the moving of large volumes of air, requiring large single-stage axial flow fans. These fans can only achieve low pressure rises; a typical design pressure drop for the air side is 3/4 in. of water (0.027 psi, or 0.19 kPa). Consequently, the air is limited to low velocities—a face velocity of about 10 ft/s (3 m/s) for a six-row tube bank is typical. The low velocity and low thermal conductivity result in low heat-transfer coefficients and require the use of high-finned tubes (Section 6.3.4.5.3).

The basic heat-transfer surface is a bank of 1 to 20 (but usually 3 to 12) rows of high-finned tubes. The fluids enter the tubes by box headers (pipe headers at higher pressures), with the tubes welded into one wall of the header; internal dividers are used in the headers to create multiple tube passes. The header cover may be bolted to the header, or the header may be welded, with threaded plugs opposite the tube ends to facilitate assembly and tube cleaning. The tubes are slanted downward toward the exit (about 1/4 in. per ft) to facilitate drainage. Occasionally, especially for turbine steam condensation, an A-frame configuration is employed, with the tubes slanted about 30° from the vertical and the fans mounted in forced draft at the base of the A, thereby reducing the plan area required.

Forced- and induced-draft arrangements are used in about equal numbers, depending upon the relative advantages in a particular situation. The forced-draft arrangement locates the fan and driver below the tube bundle, allowing greater rigidity and access for servicing the fan and driver, and creating turbulence in the air flow across the tube field. However, the air flow is somewhat maldistributed, with highest velocities immediately over the fan blade tips and significantly lower velocities in the corners and edges of the layout and over the fan hubs. The air exhaust velocity is essentially the face velocity, and is more subject to recirculation around, underneath, and back into the cooler, reducing the effective temperature difference for heat transfer. The induced-draft configuration produces a higher exhaust velocity (typically about 2 1/2 times the face velocity) and results in a more uniform velocity distribution across the tube field. However, the heat-transfer coefficient for the first two or three rows of tubes is reduced significantly while the turbulent flow is developing [64]. The plenum shroud protects the tubes from hail and from sudden rain storms.

Proper selection and positioning of the fans is crucial to successful operation. Standard practice calls for the fan area to be about 40% of the face area. The fan placement should provide the most uniform air flow through the tube bank possible. A scale drawing of the fan circle over the tube-field layout should be made, especially for air-cooled exchangers with multiple tube-side services, to ensure that all of the tube field is reasonably included in the flow field.

The fans may be driven by electric motors or by turbines. In either case, it is almost always necessary to use either a gear reducer or a belt drive to achieve the proper fan speed; maximum fan tip speed is about 1500 ft/min (7.7 m/s). Where feasible, two fans (or more) should be used with each bay of the exchanger. This allows one fan to be on constant speed, while the other is controlled to achieve the desired process outlet temperature for existing ambient conditions. Both variable and variable-pitch control systems are used. API [65] provides the standards for air-cooled heat exchangers in the process industries.

6.5.2.8 Mechanically Aided Heat Exchangers

Two kinds of heat exchangers require direct application of mechanical energy to the heat-transfer surface to achieve reasonable heat-transfer rates. The first is the stirred-tank reactor with a motor-driven paddle, propeller, or helical flight inside the reactor. A wide variety of heat-transfer surface configurations is available, including tube coils, welded plates, and helical tubing coils. External jackets are also available and may be used independently or in conjunction with the internal surface. Correlations for heat-transfer and power requirements are mostly proprietary to the manufacturer, but some general correlations are available in Volume 3 of the *Heat Exchanger Design Handbook* [1].

Scraped-surface exchangers, illustrated in Figure 6.40, are used for fractional crystallizers, dewaxing coolers, and heating and cooling of extremely viscous or solid-containing liquids. They are specialty items, manufactured by a relatively small number of firms, each of which has its own design methods and correlations. They are expensive and require constant attention and maintenance. The type shown in the figure applies a constant scraping force to the solid surface as fixed by the spring characteristics. Another type—the "hydrodynamic blade"—has the scraper blade

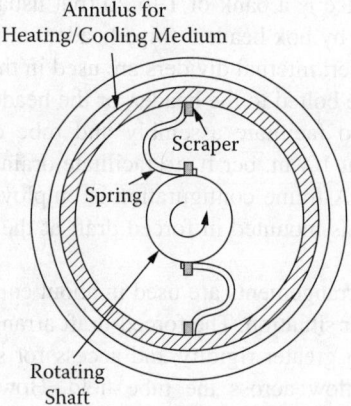

FIGURE 6.40 Cross-sectional view of a scraped-surface heat exchanger with spring-loaded blades.

mounted on a jointed arm, with the blade free to swing outward by centrifugal force as the shaft rotates; the force acting on the scraped surface is thus controlled by the rotational speed.

6.5.3 Principles of Heat-Exchanger Design

6.5.3.1 Heat Transfer between Two Fluids Separated by a Wall

Figure 6.41 shows a cross section of a typical heat-exchanger tube with a hot fluid inside at temperature T and a cold fluid outside at t. (The following argument applies to hot fluid outside and cold inside with appropriate change of sign. It applies whether the respective heat-transfer processes are single-phase convection, or condensation or vaporization, always noting that heat flows from the hot fluid to the cold.) The figure also shows fouling films on both surfaces, which is almost always the practical situation; fouling is discussed in more detail in Section 6.5.5.

Under the reasonable assumptions that the heat-transfer process is at near-steady state (i.e., conditions at any given point do not vary rapidly with time) and that the longitudinal heat-transfer rates along the tube are negligible compared with those radially through the tube wall, the heat-transfer rate, Q (watts or Btu/h), for a tube of length L (m or ft) is given by the following equations:

$$Q = h_i A_i (T - T_{fi}) \tag{6.96a}$$

$$Q = \frac{A_i (T_{fi} - t_{wi})}{R_{fi}} \tag{6.96b}$$

$$Q = \frac{(2\pi L) k_w (t_{wi} - t_{wo})}{\ln(r_o / r_i)} \tag{6.96c}$$

$$Q = \frac{A_o (t_{wo} - t_{fo})}{R_{fo}} \tag{6.96d}$$

$$Q = h_o A_o (t_{fo} - t) \tag{6.96e}$$

Heat Transfer

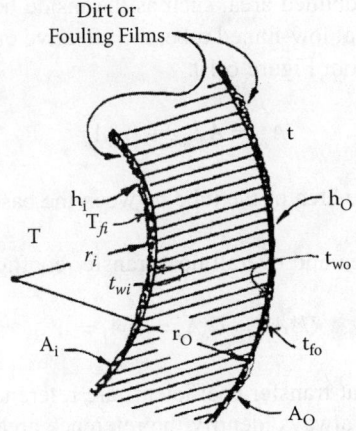

FIGURE 6.41 Cross section of fluid-to-fluid heat transfer through a tube wall.

In these equations, h_i and h_o are the inside and outside film heat-transfer coefficients, A_i and A_o are the inside and outside surface areas of the clean tube, R_{fi} and R_{fo} are the inside and outside surface fouling resistances, T_{fi} is the interface temperature between the fluid and the surface of the fouling deposit inside the tube, t_{fo} is the interface temperature between the fluid and fouling surface on the outside, t_{wi} and t_{wo} are the inside and outside tube wall temperatures, and k_w is the thermal conductivity of the tube material. Each of the terms in the denominator is a resistance to the heat flow, and their sum is the total resistance to heat transfer between the two fluids. These equations can be combined to eliminate the intermediate temperatures and find the heat-transfer rate using only the overall temperature difference $(T - t)$:

$$Q = \frac{T - t}{\dfrac{1}{h_i A_i} + \dfrac{R_{fi}}{A_i} + \dfrac{\ln(r_o / r_i)}{2\pi L k_w} + \dfrac{R_{fo}}{A_o} + \dfrac{1}{h_o A_o}} \tag{6.97}$$

The overall heat-transfer coefficient U^* is based on any convenient reference area A^* and is defined by the equation

$$Q = U^* A^* (T - t) \tag{6.98}$$

Comparing the last two equations gives the overall heat-transfer coefficient in terms of individual resistances:

$$U^* = \frac{1}{\dfrac{A^*}{h_i A_i} + \dfrac{R_{fi} A^*}{A_i} + \dfrac{A^* \ln(r_o / r_i)}{2\pi L k_w} + \dfrac{R_{fo} A^*}{A_o} + \dfrac{A^*}{h_o A_o}} \tag{6.99}$$

Frequently, but not always, the total outside area of the tubes in a heat exchanger, A_o, is chosen as the reference area, and the resulting equation is

$$U = \frac{1}{\dfrac{A_o}{h_i A_i} + \dfrac{R_{fi} A_o}{A_i} + \dfrac{A_o \ln(r_o / r_i)}{2\pi L k_w} + R_{fo} + \dfrac{1}{h_o}} \tag{6.100}$$

Any other convenient, well-defined area, such as the inside heat-transfer area A_i, can be used as the reference area. For a typical low-finned tube, the effective outside area of the tube, A_{eff}, must include the fin efficiency, η_{fin}, from Figure 6.10.

$$A_{eff} = A_{prime} + \eta_{fin}A_{fin} \tag{6.101}$$

where A_{prime} is the outside surface area of the tube between the bases of the fins and A_{fin} is the total surface area of all the fins.

The relationship between area and overall heat-transfer coefficient is

$$U^*A^* = U_oA_o = U_iA_i = \ldots \tag{6.102}$$

Because the values of the heat-transfer coefficients are referenced to the area upon which they are based, care must be taken to always identify the reference area in any situation where there is any possibility of confusion or misunderstanding.

6.5.3.2 The Basic Design Integral

The above equations relate the heat-transfer rate to the local temperature difference $(T - t)$ and the heat-transfer area A through the overall heat-transfer coefficient U. In almost all practical situations, one or both temperatures will vary along the length of the heat exchanger. The change in temperature of each stream is calculated from the enthalpy balance on that stream. To apply Equation (6.98) to the design of a heat exchanger in which the temperature difference is not constant, the equation needs to be written in differential form:

$$dA^* = \frac{dQ}{U^*(T-t)} \tag{6.103}$$

which can then be formally integrated over the entire heat duty of the heat exchanger, Q_T:

$$A^* = \int_0^{Q_T} \frac{dQ_T}{U^*(T-t)} \tag{6.104}$$

This is the basic design equation for heat exchangers having two fluids exchanging heat across a well-defined solid surface. The local fluid temperatures are calculated as a function of the entering temperatures and the amount of heat exchanged up to a given point in the heat exchanger. The individual film heat-transfer coefficients, and hence the overall coefficient, may also be functions of the amount exchanged, especially in condensing and vaporizing services, as discussed in Section 6.4.

Most large heat exchangers are designed using commercially available computer programs to evaluate Equation (6.104) numerically. However, if certain assumptions are made, the equation can be analytically solved to give a simple algebraic solution. This procedure is discussed in the next section.

6.5.3.3 The Mean Temperature Difference Concept

If certain assumptions are made, Equation (6.104) can be analytically integrated for a number of important heat-exchanger configurations and applications. The following set of assumptions is reasonably valid for many cases:

1. All elements of a given stream have the same thermal history. This implies good mixing of each stream at each cross section. Significant flow maldistribution or bypassing violates this assumption.
2. The heat exchanger is at steady state. This assumption is usually satisfied during normal operation.
3. Each stream has a constant specific heat. Isothermal phase transitions (corresponding to an effective specific heat of infinity) satisfy this requirement and, in fact, simplify the solution.
4. The overall heat-transfer coefficient is constant. This usually requires that the individual film coefficients, or at least the controlling one, be essentially constant.
5. The flow is either entirely cocurrent or countercurrent, as shown in Figures 6.42a, 6.42b, 6.43a, and 6.43b. This assumption is relaxed below for many multipass or cross-flow heat exchangers.
6. The heat exchanger is perfectly insulated, i.e., it does not exchange heat with any connections or surroundings.

Applying these assumptions and the appropriate heat balances to Equation (6.104) results in

$$A^* = \frac{Q_T}{U^*(LMTD)} \tag{6.105}$$

where A^* is the total reference heat-transfer area required in the heat exchanger, U^* is the overall heat-transfer coefficient based on A^*, and $LMTD$ is the logarithmic mean temperature difference for the exchanger, given by

$$LMTD = \frac{(T_i - t_i) - (T_o - t_o)}{\ln\frac{(T_i - t_o)}{(T_o - t_i)}} \quad \text{for cocurrent flow} \tag{6.106}$$

and

$$LMTD = \frac{(T_i - t_o) - (T_o - t_i)}{\ln\frac{(T_i - t_o)}{(T_o - t_i)}} \quad \text{for countercurrent flow} \tag{6.107}$$

Typical temperature profiles are shown for cocurrent flow in Figure 6.42b, where the outlet cold fluid temperature can never be greater than the outlet hot fluid temperature, and for countercurrent flow in Figure 6.43b, where the outlet cold fluid temperature can be greater than the outlet hot fluid temperature. It is readily shown that the countercurrent arrangement always transfers as much or more heat in a given heat exchanger than the cocurrent configuration for the same inlet temperatures and flow rates if the assumptions are satisfied. However, the surface temperature in the cocurrent arrangement is more nearly constant, which may be important if a heat-sensitive fluid or temperature-sensitive fouling is involved.

The mean temperature difference formulation may be extended to many heat-exchanger configurations that are neither purely countercurrent nor purely cocurrent. As noted in Section 6.5.2.3, most shell-and-tube exchangers have multiple tube passes, resulting in the tube-side fluid being in countercurrent flow to the shell-side in some passes and cocurrent in others. Typical temperature profiles for a 1-2 shell-and-tube exchanger are shown in Figures 6.44a and

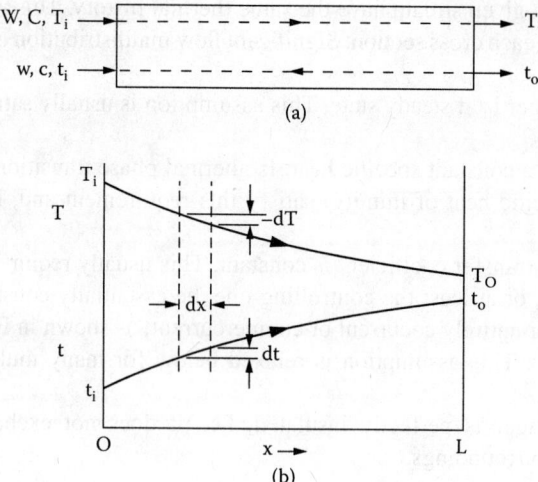

FIGURE 6.42 (a) Cocurrent flow pattern through a heat exchanger. (b) Typical temperature profiles through a cocurrent heat exchanger.

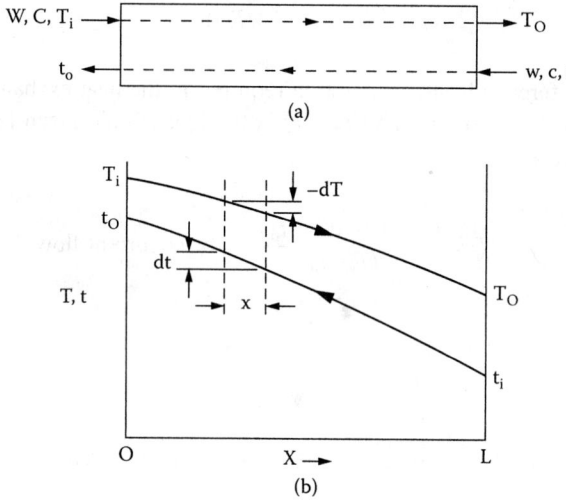

FIGURE 6.43 (a) Countercurrent flow pattern through a heat exchanger. (b) Typical temperature profiles through a countercurrent heat exchanger.

6.44b. Similarly, air-cooled exchangers as described in Section 6.5.2.7 have the air in cross flow to the tubes.

If certain additional assumptions are made, many of these cases can be handled by a modification of Equation (6.105):

$$A^* = \frac{Q_T}{U^*(MTD)} \tag{6.108a}$$

where

$$(MTD) = F(LMTD)_{counter\,current} \tag{6.108b}$$

Heat Transfer

and (MTD) is the effective mean temperature difference. The MTD is found by multiplying the logarithmic mean temperature difference *for countercurrent flow* (Equation (6.107)) by F, the configuration correction factor, which is discussed in detail below.

For this formulation to apply to shell-and-tube exchangers, it is assumed that each tube pass has the same number of tubes and that the tube-side coefficient in each pass is the same. The analysis also requires that there be sufficient baffles that the shell-side flow can be treated as longitudinal to the tubes, rather than as a series of cross-flow sections. Three or four baffles may be sufficient to meet this criterion if the shell-side temperature change is less than the minimum temperature difference between shell-side and tube-side fluids, and eight or more baffles are almost always sufficient for the assumption to be satisfied.

For air-cooled exchangers, it is assumed that each pass has the same number of tubes, that the coefficient is the same for all tubes, and that the passes are arranged countercurrent to the air flow.

The configuration correction factor F can be obtained by analytical or numerical integration of Equations (6.108a) and (6.108b) as a function of two dimensionless ratios, P and R, defined as

$$P = \frac{t_o - t_i}{T_i - t_i}; \quad R = \frac{T_i - T_o}{t_o - t_i} \tag{6.109}$$

The analytical solutions for F can be incorporated into computer design programs, and the numerical solutions can be represented over the practical range by curve fits. For hand use, however, graphical representation is more convenient and revealing. Extensive tables and graphs of F as a function of P and R for most configurations are given in the literature [1, 30, 51, 54, 56, 64, 65]. A few of the most common ones are included here as Figures 6.45 to 6.52.

Figures 6.44a and 6.44b show typical temperature profiles for a 1-2 shell-and-tube exchanger. If the outlet temperatures of the two streams are equal for this configuration, the value of F is usually close to 0.8. It is possible to design for the outlet temperature of the cold stream to exceed

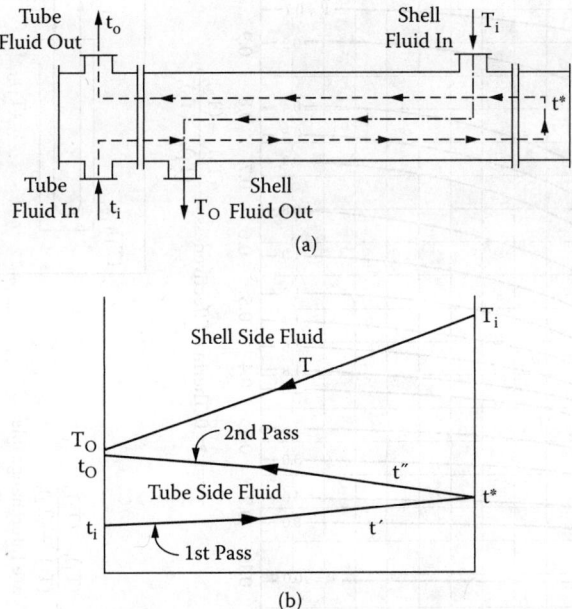

FIGURE 6.44 (a) Nominal flow pattern through a one-shell pass, two-tube pass (1-2) shell-and-tube heat exchanger. (b) Typical temperature profiles through a one-shell pass, two-tube pass (1-2) shell-and-tube heat exchanger.

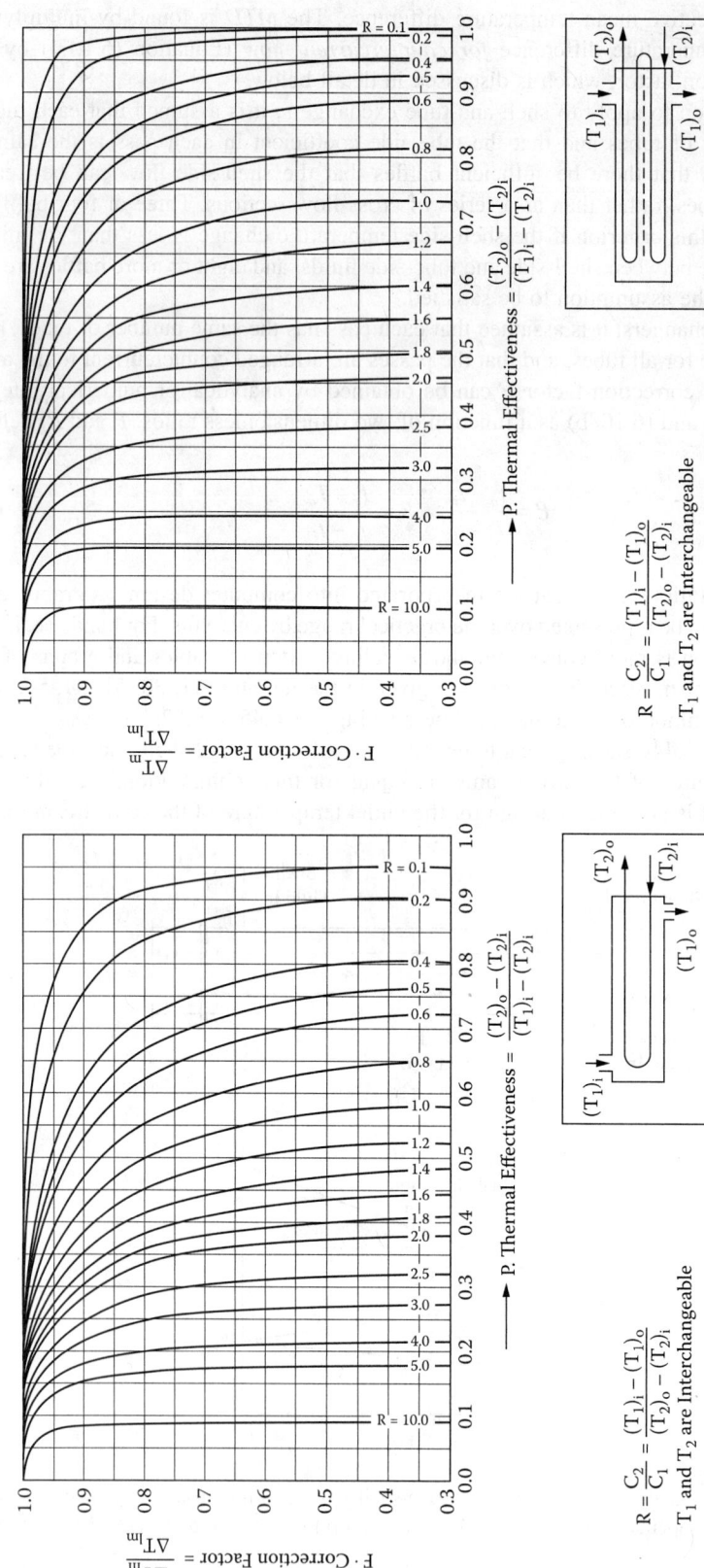

FIGURE 6.46 F chart for two E shells in series with any even number of tube passes or an F shell with four, eight, etc., tube passes. (Adapted from Hewitt, G. F., Ed., *Heat Exchanger Design Handbook—1998*, Begell House, New York (1998).)

FIGURE 6.45 F chart for one E shell with any even number of tube passes. (Adapted from Hewitt, G. F., Ed., *Heat Exchanger Design Handbook—1998*, Begell House, New York (1998).)

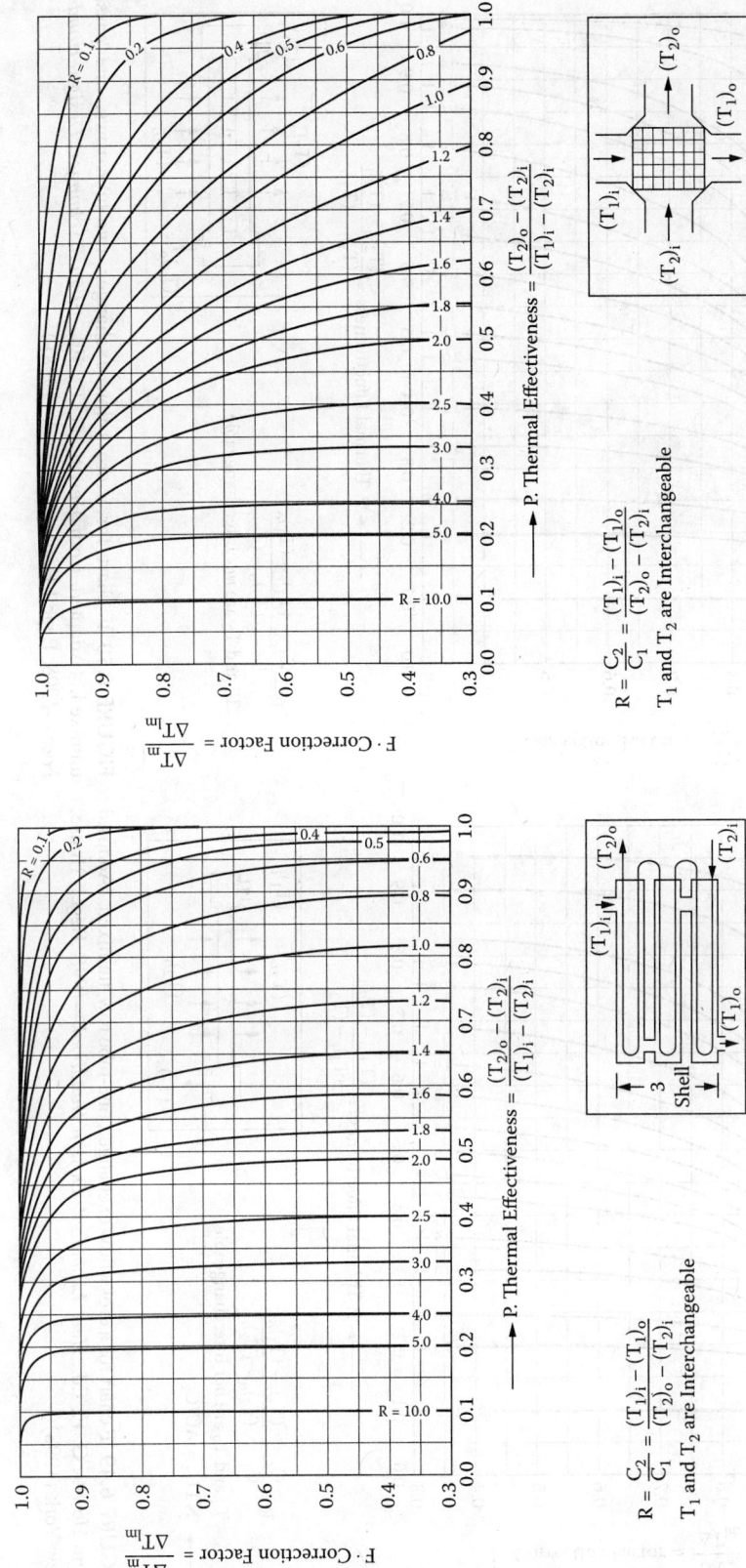

FIGURE 6.47 F chart for three E shells in series, each with any even number of tube passes. (Adapted from Hewitt, G. F., Ed., *Heat Exchanger Design Handbook—1998*, Begell House, New York (1998).)

FIGURE 6.48 F chart for a cross-flow exchanger, both fluids unmixed. (Adapted from Hewitt, G. F., Ed., *Heat Exchanger Design Handbook—1998*, Begell House, New York (1998).)

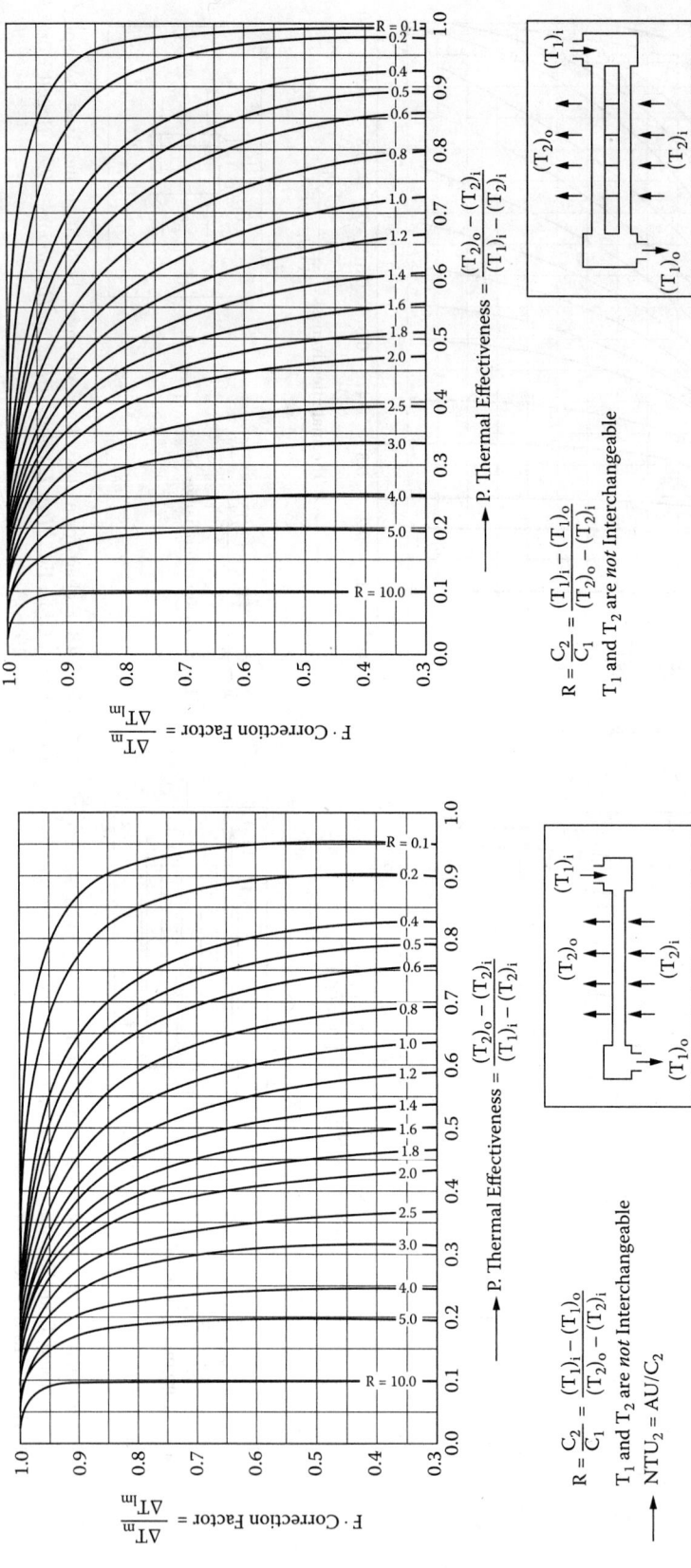

FIGURE 6.49 F chart for a cross-flow exchanger, one-tube row, unmixed. (Adapted from Hewitt, G. F., Ed., *Heat Exchanger Design Handbook—1998*, Begell House, New York (1998).)

FIGURE 6.50 F chart for a cross-flow exchanger, two-tube rows, one pass, unmixed. (Adapted from Hewitt, G. F., Ed., *Heat Exchanger Design Handbook—1998*, Begell House, New York (1998).)

Heat Transfer

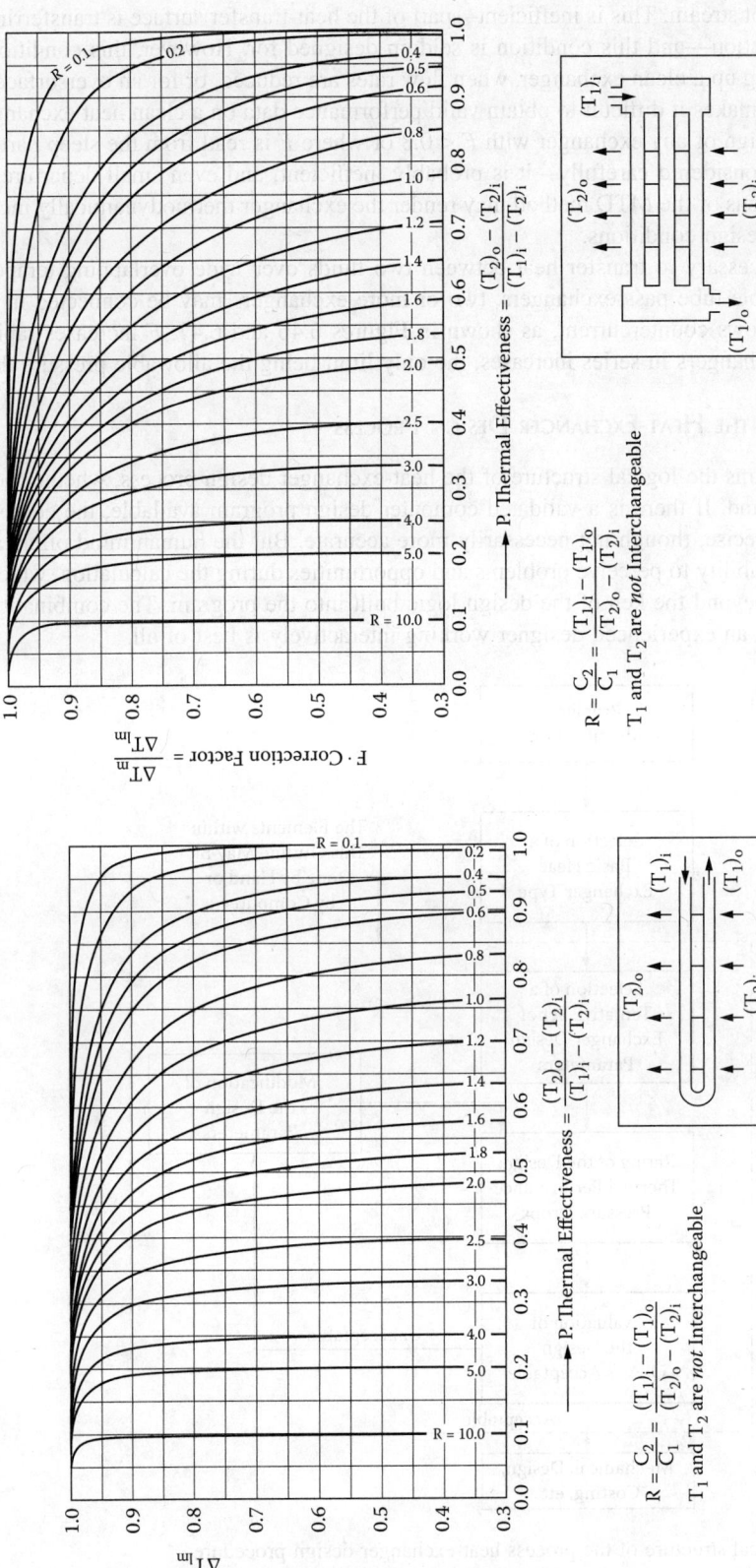

FIGURE 6.52 F chart for a cross-flow exchanger, three-tube rows, one pass, unmixed. (Adapted from Hewitt, G. F., Ed., *Heat Exchanger Design Handbook—1998*, Begell House, New York (1998).)

FIGURE 6.51 F chart for a cross-flow exchanger, two-tube rows, two passes, unmixed between passes. (Adapted from Hewitt, G. F., Ed., *Heat Exchanger Design Handbook—1998*, Begell House, New York (1998).)

the outlet of the hot stream. This is inefficient—part of the heat-transfer surface is transferring heat in the wrong direction—and this condition is seldom designed for. However, this condition may occur when starting up a clean exchanger, when flow rates are reduced, or for an oversurfaced heat exchanger. It also makes it difficult to obtain valid performance data on a clean heat exchanger. As a general rule, design of any exchanger with $F < 0.8$ or where F is read from the steep part of the curve should be considered carefully—it is probably inefficient, and even small departures from the ideal assumptions of the MTD method may render the exchanger thermodynamically incapable of achieving the design conditions.

When it is necessary to transfer heat between two fluids over wide overlapping temperature ranges with multiple tube pass exchangers, two or more exchangers may be connected in series, with the overall flows countercurrent, as shown in Figures 6.46 and 6.47. F increases rapidly as the number of exchangers in series increases, the only limit being the allowable pressure drops.

6.5.4 Logic of the Heat-Exchanger Design Process

Figure 6.53 diagrams the logical structure of the heat-exchanger design process, whether done by computer or by hand. If there is a validated computer design program available, the computer is faster and more precise, though not necessarily more accurate. But the human mind offers advantages, such as the ability to perceive problems and opportunities during the calculation, which may lead to solutions beyond the ken of the design logic built into the program. The combination of a good program and an experienced designer working interactively is best of all.

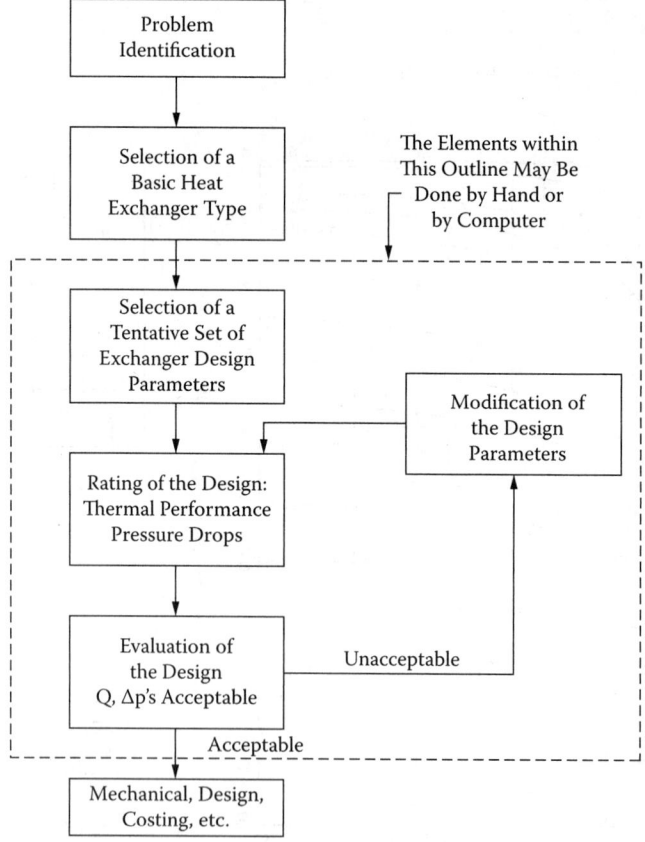

FIGURE 6.53 Logical structure of the process heat-exchanger design procedure.

The first step is to completely and unambiguously define the problem and provide the designer with all of the data required. This will include the flow rates, compositions, temperatures, and pressures of the streams. It is usually necessary to provide the thermodynamic and thermophysical properties over the range of temperatures and pressures to be encountered. Condensing and/or vaporizing curves, including equilibrium compositions, enthalpies, and phase mass fractions as functions of temperature for multicomponent mixtures, are essential. Design fouling resistances should be supplied as available from experience. Any requirements or limitations on diameter, length, weight, piping connections, supports, and construction features to facilitate maintenance must be given.

The next decision—selection of an exchanger type—is usually prefigured in arriving at the data provided above. However, it is well to consider alternative candidates for operational and maintenance advantages as well as for first cost. Criteria to be included are

1. Level of confidence in the design method
2. Level of confidence in the fabrication techniques
3. Level of confidence in plant operating and maintenance personnel
4. Operational flexibility, short and long term

The logic blocks in the outlined area carry out the thermal/hydraulic design of the exchanger. Whether by computer or hand, a starting configuration must be chosen, and there are certain minimum data that must be supplied even for a computer program, e.g., tube diameter, wall thickness, material, TEMA type, maximum shell length and diameter, etc. Most computer programs have a default process whereby most of the unspecified mechanical data are arbitrarily selected. With a computer program, the greatest concern is that the computations will converge to a feasible design from the initial data. For hand design, it is highly desirable to start with a design as close as possible to the final design to minimize computational time. An estimation procedure to provide a starting design for shell-and-tube exchangers is given in Section 3.1.4 of the *Heat Exchanger Design Handbook* [1].

The rating program, Figure 6.54, is the core of the design computations. It contains the subprograms for calculating the geometrical parameters (e.g., flow cross sections and heat-transfer area), the correlations for heat-transfer coefficients and temperature differences, and the correlations for pressure drops.

In the next step, the calculated thermal capability and the pressure drops for the candidate design are compared with the required values. The thermal capability must be satisfied (with any specified safety factor included), and the pressure drops should use as much of the allowable pressure drops as possible without exceeding the limits. Given, however, the uncertainties in the correlations and other design and operational factors, the allowable pressure drops should be chosen and interpreted to reflect this uncertainty. The computer is rigid in its decision regarding the acceptability of the design. The designer must apply the judgment whether a design with a somewhat

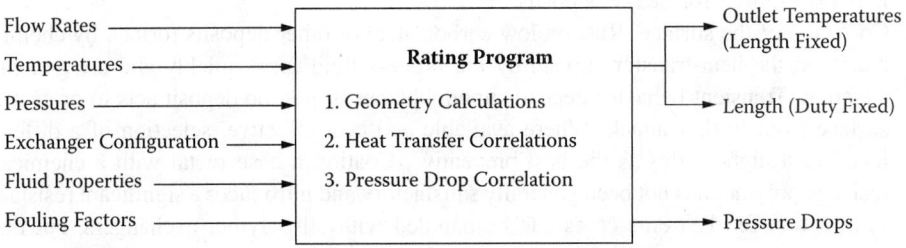

FIGURE 6.54 Rating program: core block of the heat-exchanger design process.

larger predicted pressure than the "allowable" but with superior qualifications in other respects should be accepted.

If any of the rating output is deemed "unacceptable" by the program, the "Design Modification" program takes over, following a very detailed sequence of evaluating what is acceptable and what needs to be changed in the design. The objective function is usually to minimize the surface area required within the pressure-drop limitations, since this usually leads to the minimum first cost. The changes made must satisfy applicable codes and standards, as well as accepting only feasible construction features and commercially available components, e.g., tubes. The number of logical paths through such a program is enormous, and it can never be completely checked out. Writing such a program requires the very highest knowledge and skill. Even so, illogical or suboptimal results may ensue, and the output must be carefully evaluated every time. The prudent designer will consider whether rerunning the program with a different set of input specifications (e.g., changing the TEMA type or the tube layout) may not produce a superior design.

6.5.5 Fouling

Fouling is any unwanted deposit on the heat-transfer surface of a heat exchanger or any deposit that interferes with the desired flow pattern in the heat exchanger. In the first case, fouling increases the resistance to heat transfer, resulting in reduced thermal capability or requiring higher temperature differences or increased velocities to maintain thermal performance. In the second case, pressure drop increases at the same throughput, and a higher fraction of the flow may be forced into bypass areas.

Fouling is the greatest unsolved problem in heat-exchanger design. The most comprehensive book on fouling is by Bott [66], and there are many individual papers, but there are few quantitative and generalizable principles. TEMA Standards [54] provides tables of suggested values of the fouling resistance for a wide variety of services, but the mechanisms are so complex and dependent upon the specifics of each stream, exchanger design, and operating conditions that predictions are very uncertain.

Fouling processes are time dependent, starting at or near zero in new or freshly cleaned exchangers, increasing in time, sometimes to an asymptotic limit and other times increasing without limit until the exchanger is shut down and cleaned. Fouling deposits are often due to a combination of mechanisms, which complicates both the choice of a fouling resistance and the recommended amelioration technique.

There are several fouling mechanisms:

1. Sedimentation: Deposition of silt or other suspended material in the fluid on the heat-transfer surface. These deposits are generally weak and can be reduced or eliminated by increasing the velocity. The fouling resistance often reaches a constant value as long as conditions are unchanged. In this case, additional surface can be provided through a fouling resistance, R_f, in the overall heat-transfer coefficient equation. Cleaning can usually be effected by the use of high-velocity water sprays, though mechanical scraping may be required for heavy deposits.
2. Corrosion of the surface: Rust on low-carbon steel or other deposits formed by chemical attack on the heat-transfer surface by the process fluid form quickly and are generally adherent. Transient behavior depends upon whether or not the deposit acts to protect the surface from further attack. Where available and cost effective, selection of a different metal or a higher alloy is the best preventive. Coating a base metal with a chemically resistant polymer has not been generally satisfactory and introduces a significant resistance to heat transfer. Extreme cases can be handled with all-polymer exchangers, but these have poor heat-transfer characteristics and low strength and temperature limits. Heavy corrosion layers can usually be removed by appropriate chemical treatment or by water jets or mechanical scraping, but this only exposes the cleaned surface to renewed attack.

3. Inverse solubility: Certain substances—notably calcium sulfate (gypsum)—occurring in water supplies are less soluble in hot water than in cold and will crystallize out upon a hot surface. Depending upon the strength of the deposit and the operational strategy (e.g., increasing the hot stream temperature to maintain heat flux), the deposit may reach an asymptote or continue growing (as in the constant heat-flux case). Fouling in cooling-water systems is usually complicated by other fouling mechanisms, notably sedimentation, corrosion, and biological fouling (discussed below). Process streams may show similar behavior to other stream components. Chemical controls to inhibit the deposition or weaken the deposit by destroying the crystalline structure are generally the best approach to controlling this type of fouling. Without such action, these deposits are often strong and usually require high-velocity water jets or mechanical scraping to remove.
4. Freezing: Freezing occurs when the surface temperature is below the freezing point of the fouling substance (which may be a minor constituent, neon in a helium liquefaction process, for example). Freezing can only be avoided by ensuring that the surface temperature is always and everywhere above the freezing temperature of any constituent. This is an example of the proper use of cocurrent flow. The deposit is usually easily removed by periodically raising the temperature and flushing out the melted residue.
5. Polymerization, charring, or chemical reaction: Many substances will form deposits of high-molecular-weight solids when heated too strongly (often exacerbated by the presence of oxygen). The deposits may range from hard and brittle to sticky, and frequent, vigorous mechanical cleaning is often the only choice to control fouling. Designing for lower surface temperatures (e.g., higher velocities, cocurrent flow) is sometimes possible, but selection of heat-exchanger type and configuration to simplify mechanical cleaning is also important.
6. Biological fouling: Water systems are subject to microbiological fouling (growth of algae and bacterial films) and macrobiological fouling (growth of barnacles, mussels, and other large organisms). Microbiological fouling is an organic film structure with a high water content on the heat-transfer surface. Biological fouling can be minimized by use of copper alloy tubing and can be controlled by chlorination and similar chemical treatments, both of which introduce operational and environmental problems. Microbiological fouling often occurs in conjunction with sedimentation, inverse solubility, and/or corrosion fouling, and a water-treatment specialist should be consulted in these cases. Macrobiological fouling occurs most commonly on the tubesheets and header surfaces, where the local velocities are low enough to allow the larvae to gain a foothold. Once established, the organisms can grow large enough to interfere with or entirely block fluid flow to the heat-transfer channels. These organisms are firmly attached and require vigorous mechanical action to knock them off.

The best source of design information on fouling is to find someone who has had to deal with a similar situation in the past and can give a historical perspective on what happened and how it was dealt with. The literature, especially the book by Bott [66], can be a source of case studies and contacts, and this information is fairly freely shared in the profession. The specialty water treatment chemical suppliers have large files of recorded experience in that area, and there is increasing use of chemicals in controlling fouling in process streams as well.

REFERENCES

1. Hewitt, G. F., Ed., *Heat Exchanger Design Handbook—1998*, Begell House, New York, 1998.
2. Arpaci, V. S., *Conduction Heat Transfer*, Addison-Wesley, Reading, MA, 1966.
3. Ozisik, M. N., *Heat Conduction*, John Wiley and Sons, New York, 1980.

4. Gardner, K. A., "Efficiency of Extended Surfaces," *Trans. ASME* 67, 621–631 (Nov. 1945).
5. Kern, D. Q., and Kraus, A. D., *Extended Surface Heat Transfer*, McGraw-Hill, New York, 1972.
6. Kraus, A. D., Azziz, A., Welty, J. R., and Aziz, A., *Extended Surface Heat Transfer*, McGraw-Hill, New York, 2000.
7. Webb, R. L., *Principles of Enhanced Heat Transfer*, John Wiley, New York, 1994.
8. Parker, J. D., Boggs, S. H., and Blick, E. F., *Introduction to Fluid Mechanics and Heat Transfer*, Addison-Wesley, Reading, MA, 1969.
9. Heisler, M. P., "Temperature Charts for Induction and Constant Temperature Heating," *Trans. ASME* 69, 227–236, (1947).
10. Patankar, S. V., *Numerical Heat Transfer and Fluid Flow*, Taylor and Francis, Philadelphia, PA, 1980.
11. Shah, R. K., and London, A. L., *Laminar Flow Forced Convection in Ducts*, *Advances in Heat Transfer*, supp. 1, Academic Press, New York, 1978.
12. Sieder, E.W., and Tate, G.E., "Heat Transfer and Pressure Drop of Liquids in Tubes," *Ind. Eng. Chem.* 28, 1429 (1936).
13. Kakac, S., Shah, R. K., and Aung, W., *Handbook of Single Phase Convective Heat Transfer*, John W. Wiley and Sons, New York, 1987.
14. Jakob, M., *Heat Transfer*, vol. 1, John W. Wiley and Sons, New York, 1949.
15. Hausen, H., "VDIZ. Beih." *Verfahrenstech.* 4, 91 (1943).
16. Palen, J. W., and Taborek, J., "An Improved Heat Transfer Correlation for Laminar Flow of High Prandtl Number Liquids in Horizontal Tubes," *AIChE Symposium Series* 81 (245), 90–96 (1985).
17. Petukhov, B. S., "Heat Transfer and Friction in Turbulent Pipe Flow with Variable Liquid Properties," *Advances in Heat Transfer*, vol. 6, Hartnett, J. P., and Irvine, T. F., Jr., Eds., pp. 504–561, Academic Press, New York, 1970.
18. Gnielinski, V., "New Equations for Heat and Mass Transfer in Turbulent Pipe and Channel Flow," *Int. Chem. Eng.* 16, 359–368 (1976).
19. Martin, H., in Ref. 1, citing Gauler, K., "Wärme–und Stoffübertragung an eine mit bewegte ebene Grenzfläche bei Grenzschichströmung," Dr.-Ing., University of Karlsruhe, Germany (1972).
20. Petukhov B. S., and Roizen, L. I., "An Experimental Investigation of Heat Transfer in a Turbulent Flow of Gas in Tubes of Annular Section," *High Temperature* (USSR) 1, 373–380 (1963).
21. Eckert, E. R. G., and Drake, R. M., Jr., *Analysis of Heat and Mass Transfer*, p. 406, McGraw-Hill, New York, 1972.
22. Bell, K. J., "Process Heat Transfer Notes," Oklahoma State University, Stillwater, OK, 1972.
23. Bergelin, O. P., Brown, G. A., and Doberstein, S. C., *Trans. ASME* 74, 953 (1952).
24. Zukauskas, A., Skrinska, A., Zingzda, J., and Gnielinski, V., "Banks of Plain and Finned Tubes," in *Heat Exchanger Design Handbook—1998*, Hewitt, G. F., Ed., Section 2.5.3, Begell House, New York, 1998.
25. Williams, R. B., and Katz, D. L., "Performance of Finned Tubes in Shell and Tube Heat Exchangers," *Trans. ASME* 74, 1307–1320 (1952).
26. Briggs, D. E., Katz, D. L., and Young, E. H., "How to Design Finned Tube Heat Exchangers," *Chem. Eng. Prog.* 59 (11), 49–59 (1963).
27. Briggs, D. E., and Young, E. H., "Heat Transfer—Houston," *Chem. Eng. Prog. Symposium Series No. 41*, 59, 1 (1963).
28. Robinson, K. K., and Briggs, D. E., "Heat Transfer—Los Angeles," *Chem. Eng. Prog. Symposium Series No. 64*, 62, 177 (1965).
29. Gnielinski, V., "Equations for the Calculation of Heat and Mass Transfer during Flow through Stationary Spherical Packings at Moderate and High Peclet Numbers," *Int. Chem. Eng.* 21, 378–383 (1981).
30. Hewitt, G. F., Shires, G. L., and Bott, T. R., *Process Heat Transfer*, CRC Press, Boca Raton, FL, 1994.
31. Incropera, F. P., and DeWitt, D. P., *Fundamentals of Heat and Mass Transfer*, 4th ed., John Wiley and Sons, New York, 1996.
32. Churchill, S. W., and Chu, H. H. S., "Correlating Equations for Laminar and Turbulent Free Convection from a Horizontal Cylinder," *Int. J. Heat Mass Transfer* 18, 1049 (1975).
33. Globe, S., and Dropkin, D., "Natural Convection Heat Transfer in Liquids Confined between Two Horizontal Plates," *J. Heat Transfer* 81C, 24, (1959).
34. Collier, J. G., and Thome, J. R., *Convective Boiling and Condensation*, 3rd ed., Clarendon Press, Oxford, U.K., 1994.

35. Nusselt, W., "Surface Condensation of Water Vapor, Parts I and II," *VDI* 60, 541 and 569 (1916).
36. Colburn, A. P., "The Calculation of Condensation Where a Portion of the Condensate Layer Is in Turbulent Motion," *Trans. AIChE* 30, 187 (1933–34).
37. Boyko, L. D., and Kruzhilin, G. N., "Heat Transfer and Hydraulic Resistance during Condensation of Steam in a Horizontal Tube and in a Bundle of Tubes," *Int. J. Heat Mass Transfer* 10, 361 (1967).
38. Kern, D. Q., *Process Heat Transfer*, McGraw-Hill, New York, 1950.
39. Diehl, J. E., and Unruh, C. H., ASME Paper 58-HT 20, 1958.
40. Colburn, A. P., and Hougen, O. A., "Design of Cooler-Condensers for Mixtures of Vapors with Non-condensing Gases," *Ind. Eng. Chem.* 26 (11), 117 (1934).
41. Mueller, A. C., "Review of Thermal Design Methods for Multicomponent Condensation," *Heat Transfer Engineering* 20 (4), 6, (1999).
42. Silver, L., "Gas Cooling with Aqueous Condensation," *Trans. I ChemE* 25, 30 (1947).
43. Bell, K. J., and Ghaly, M. A., "An Approximate Generalized Design Method for Multicomponent/Partial Condensers," *AIChE Symposium Series* 69 (131, "Heat Transfer"), 72 (1973).
44. Mostinski, I. L., "Calculation of Heat Transfer and Critical Heat Fluxes in Liquids," *Teploenergetika* 10 (4), 66 (1963).
45. Zuber, N., Tribus, M., and Westwater, J. W., "The Hydrodynamic Crisis in Pool Boiling of Saturated Liquids," in *International Developments in Heat Transfer*, p. 230, ASME, New York, 1963.
46. Lienhard, J. H., and Dhir, V. K., "Peak Boiling Heat Flux from Finite Bodies," *Trans. ASME J. Heat Transfer* 95, 152 (1973).
47. Steiner, D., and Taborek, J., "Flow Boiling Heat Transfer in Vertical Tubes Correlated by an Asymptotic Model," *Heat Transfer Engineering* 13 (2), 43 (1992).
48. Kern, D. Q., *Process Heat Transfer*, McGraw-Hill, New York, 1950.
49. Palen, J. W., and Small, W. M., *Hydrocarbon Processing* 43 (11), 199 (1964).
50. Palen, J. W., Yarden, A., and Taborek, J., "Characteristics of Boiling Outside Large-Scale Horizontal Multi-Tube Bundles," *AIChE Symposium Series* (No. 118, "Heat Transfer—Tulsa"), 68, 50–61 (1972).
51. Shah, R. K., and Sekuli, D. P., *Fundamentals of Heat Exchanger Design*, John Wiley & Sons, Hoboken, NJ, 2003.
52. Kakac, S., and Liu, H., *Heat Exchangers: Selection, Rating, and Thermal Design*, CRC Press, Boca Raton, FL, 1998.
53. Bell, K.J., "Process Heat Exchanger Design: Qualitative Factors in Selection and Application," chap. 2 in *Recent Developments in Chemical Process and Plant Design*, Liu, Y. A., McGee, H. A., Jr., and Epperly, W. R., Eds., John Wiley & Sons, New York, 1987.
54. Tubular Exchanger Manufacturers Association (TEMA), *Standards of Tubular Exchanger Manufacturers Association*, 9th ed., Tarrytown, NY, 2007.
55. American Society of Mechanical Engineers, ASME Boiler and Pressure Vessel Code, Section VIII, Division 1: "Unfired Pressure Vessels," ASME, New York (new editions are published regularly).
56. American Petroleum Institute, API Standard 660, "Shell-and-Tube Heat Exchangers for General Refinery Services," API, Washington, DC, 2003.
57. Heat Exchange Institute, *Standards for Power Plant Heat Exchangers*, 4th ed., Cleveland, OH, 2004. (Other standards for power plant equipment are also available from this source.)
58. Singh, K. P., and Soler, A. I., *Mechanical Design of Heat Exchangers and Pressure Vessel Components*, Arcturus, Cherry Hill, NJ, 1984.
59. Yokell, S., *A Working Guide to Shell and Tube Heat Exchangers*, McGraw-Hill, New York, 1990.
60. Rozenmann, T., and Taborek, J., "The Effect of Leakage through the Longitudinal Baffle in the Performance of Two-Pass Shell Heat Exchangers," *AIChE Symposium Series* (No. 118, "Heat Transfer—Tulsa"), 68, 12–20 (1972).
61. Kral, D., Stehlik, P., Van der Ploeg, H. J., and Master, B. I., "Helical Baffles in Shell-and-Tube Heat Exchangers, Part I: Experimental Verification," *Heat Transfer Engineering* 17 (1), 93–101 (1996).
62. American Petroleum Institute, API Standard 662, "Plate Heat Exchangers for General Refinery Services," API, Washington, DC, 2002.
63. Brazed Aluminium Plate-Fin Heat Exchanger Manufacturers' Association (ALPEMA), *The ALPEMA Standards*, 2nd ed. (2000); available on-line at http://www.alpema.org/stand.htm, accessed 14 June 2007.

64. Kays, W. M., and London, A. L., *Compact Heat Exchangers*, 3rd ed., Krieger Publishing, Malabar, FL, 1998.
65. American Petroleum Institute, API Standard 661, "Air-Cooled Heat Exchangers for General Refinery Service," API, Washington, DC, 2002.
66. Bott, T. R., *Fouling of Heat Exchangers*, Elsevier Science, Amsterdam, 1995.

7 Radiation Heat Transfer

Z. M. Zhang and David P. DeWitt

CONTENTS

7.1 Introduction 567
7.2 Surface Emission 568
    7.2.1 Intensity, Emissive Power, and Irradiation 568
    7.2.2 The Blackbody 569
7.3 Radiative Properties of Solids 570
    7.3.1 Emissivity 570
    7.3.2 Absorptance, Reflectance, and Transmittance 573
    7.3.3 Kirchhoff's Law 574
7.4 Radiative Energy Exchange between Surfaces 575
    7.4.1 Radiosity 575
    7.4.2 The View Factor 576
    7.4.3 Radiation Exchange between Blackbody Surfaces 576
    7.4.4 Enclosure with N Diffuse-Gray Surfaces—The Net Radiation Method 579
    7.4.5 The Network Representation 580
7.5 Gas and Particle Radiation 583
    7.5.1 Radiative Transfer Equation 583
    7.5.2 Radiative Properties of Gases and Particles 585
7.6 Radiation Thermometry 586
References 588

7.1 INTRODUCTION

Radiation heat transfer is important in furnaces, flames, high-temperature materials processing and manufacturing, solar energy utilization, cooling and insulation in space, and cryogenic systems. Typical furnaces include ethylene, vinyl chloride, styrene, and steam generation units. Thermal radiation is relevant to many industrial heating, cooling, and drying processes. Even at room temperature, heat transfer by radiation is often on the same order of magnitude as that by free convection. The main features of radiation that are distinct from conduction and convection are: energy can be transferred without an intervening medium; and energy transfer is not proportional to the temperature difference between surfaces, and is dependent upon the surface radiative properties, which are functions of wavelength and temperature.

Although radiation can travel in a vacuum, it originates from matter. All forms of matter emit radiation through the mechanisms of electronic transitions and lattice vibrations. In most solids and liquids, radiation emitted from the interior is strongly absorbed by adjoining molecules. Therefore, radiation from these materials can be treated as a surface phenomenon. Radiation in gases and some semitransparent solids and liquids, however, must be treated as a volumetric phenomenon.

Radiation travels at the speed of light, which is $c_0 = 2.9979 \times 10^8$ m/s in vacuum. Radiative transport is alternatively viewed as the propagation of a collection of particles, called photons, or

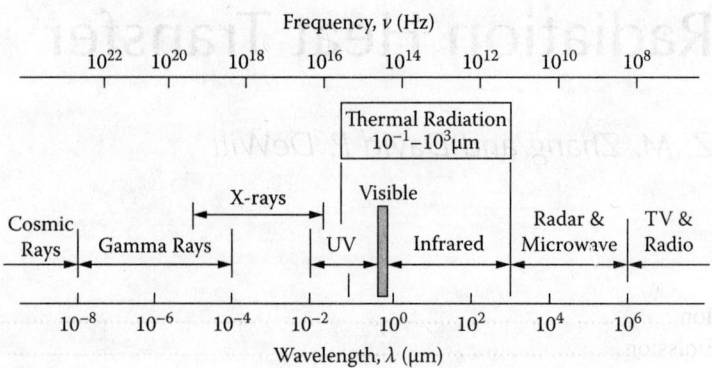

FIGURE 7.1 Thermal radiation: a region of the electromagnetic-wave spectrum.

as electromagnetic waves. In the *particle theory*, the energy of an individual photon is proportional to its frequency,

$$E_{photon} = h\nu \tag{7.1}$$

where $h = 6.6261 \times 10^{-34}$ J·s is Planck's constant, and ν is the frequency [Hz]. The momentum of each photon is $h\nu/c$. The particle theory is important in deriving the spectral distribution of thermal radiation and in the study of gas emission and absorption. The *wave theory* treats radiation as electromagnetic waves. One of the most important characteristics of a wave is its wavelength λ, which is related to its frequency and the speed of light by

$$\lambda \nu = c \tag{7.2}$$

The frequency is independent of the medium in which radiation propagates, but the speed of light and the wavelength are inversely proportional to the refractive index (n) of the medium. This is to say, $\lambda = \lambda_0 / n$ and $c = c_0 / n$. The refractive index of air is about 0.03% greater than 1 (the refractive index of vacuum), and therefore it is often approximated as 1.

Radiation at a single wavelength (or within a very narrow spectral band) is called monochromatic radiation (examples are lasers and high-temperature gas emission), whereas radiation in a broad wavelength region is called polychromatic or broadband radiation (such as that emitted from the sun). The visible spectrum covers the wavelength region from about 0.4 to 0.7 µm, a small region within the very broad electromagnetic spectrum shown in Figure 7.1. Thermal radiation, which is pertinent to heat transfer, refers to the spectral region from approximately 0.1 to 1000 µm, where the emission depends on temperature. It includes part of the ultraviolet region and the entire visible and infrared regions.

7.2 SURFACE EMISSION

7.2.1 INTENSITY, EMISSIVE POWER, AND IRRADIATION

As illustrated in Figure 7.2, the *spectral intensity* of radiation emitted by an element surface dA in the direction (θ, ϕ) is defined as the rate of emitted energy (radiant power) dq per unit wavelength interval about λ, per unit solid angle about this direction, and per unit area of dA projected to the direction (θ, ϕ):

Radiation Heat Transfer

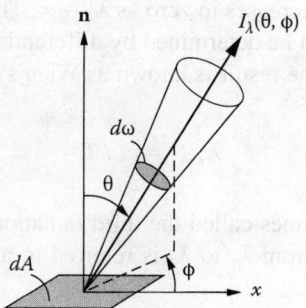

FIGURE 7.2 The definition of spectral intensity exitent from a surface.

$$I_\lambda(\lambda, \theta, \phi) = \frac{dq}{dA \cos\theta \, d\omega \, d\lambda} \tag{7.3}$$

where $d\omega = \sin\theta d\theta d\phi$, and the unit of I_λ is W/m²·sr·μm. The total intensity is the integral of the spectral intensity over all wavelengths.

The *emissive power E* is the total radiant power exitent from the surface (per unit area) toward the hemisphere due to thermal emission, which may be obtained by integrating dq in Equation (7.3) over the hemisphere and over all wavelengths. The spectral emissive power E_λ is the emissive power per unit wavelength interval about λ. If the emitted intensity is the same in all directions, the surface is called a diffuse emitter, for which $E = \pi I$ and $E_\lambda = \pi I_\lambda$.

The radiant power incident on a surface per unit area is called the *irradiation G*. Diffuse irradiation refers to the condition when the incident radiation is uniformly distributed in all directions, i.e., the incoming intensity is independent of the direction.

7.2.2 The Blackbody

The *blackbody* is an ideal surface that absorbs all incident radiation regardless of wavelength and direction. Furthermore, a blackbody is a diffuse emitter and, at any prescribed wavelength and temperature, no surface can emit more energy than a blackbody. The total emissive power of a blackbody depends only on its temperature and is given by the Stefan-Boltzmann law,

$$E_b = \pi I_b = \sigma T^4 \tag{7.4}$$

where $\sigma = 5.6704 \times 10^{-8}$ W/m²·K⁴ is the Stefan-Boltzmann constant, and T is the absolute temperature in kelvin [K]. In a medium of refractive index $n \neq 1$, the right-hand side (RHS) of Equation (7.4) needs to be multiplied by n^2. The spectral emissive power of a blackbody is given by Planck's law or the Planck distribution [1–4]:

$$E_{\lambda,b}(\lambda, T) = \pi I_{\lambda,b}(\lambda, T) = \frac{c_1}{\lambda^5(e^{c_2/\lambda T} - 1)} \tag{7.5}$$

where $c_1 = 3.7418 \times 10^8$ W·μm⁴/m² and $c_2 = 1.4388 \times 10^4$ μm·K are called the first and second radiation constants, respectively. Equation (7.5) can be integrated over all wavelengths to obtain Equation (7.4).

At any specific wavelength, the higher the temperature, the greater the spectral emissive power. At any temperature, the emissive power approaches zero as $\lambda \to 0$, increases with wavelength until it reaches a maximum, and then decreases to zero as $\lambda \to \infty$. The wavelength corresponding to the maximum emissive power λ_{max} can be determined by differentiating Equation (7.5) with respect to λ and setting the result to zero. The result is known as Wien's displacement law:

$$\lambda_{max} = c_3 / T \tag{7.6}$$

where $c_3 = 2897.8$ μm·K is sometimes called the third radiation constant. The fraction of emitted energy in the wavelength region from λ_1 to λ_2 is referred to as the blackbody band fraction and has the form

$$F_{\lambda_1-\lambda_2} = \frac{1}{\sigma T^4} \int_{\lambda_1}^{\lambda_2} E_{\lambda,b} d\lambda = \int_{\lambda_1 T}^{\lambda_2 T} \frac{c_1}{\sigma (\lambda T)^5 (e^{c_2/\lambda T} - 1)} d(\lambda T) \tag{7.7}$$

Figure 7.3a represents the Planck distribution for blackbody spectral emissive power with $E_{\lambda,b} / \sigma T^5$ as a function of λT. The band fraction of emitted energy in the region from 0 to λT is equal to the shaded area, which is expressed as $F_{0 \to \lambda T}$ and shown in Figure 7.3b. About a quarter of the emitted energy is at wavelengths shorter than λ_{max}, and nearly 95% of the emitted energy is distributed between $0.6\lambda_{max}$ and $6\lambda_{max}$. The spectral distribution of solar radiation can be approximated as a blackbody at 5800 K. Therefore, $\lambda_{max} \approx 0.5$ μm, and the spectral region (95%) important for solar radiation is between 0.3 and 3 μm. For radiation from objects near room temperature, $\lambda_{max} \approx 10$ μm and the spectral region of importance (95%) is between 6 and 60 μm.

If $\lambda T \ll c_2$, the RHS of Equation (7.5) can be approximated as $c_1 \lambda^{-5} e^{-c_2/\lambda T}$; this is known as Wien's approximate formula, which yields a relative error of less than 0.7% up to λ_{max}. If $\lambda T \gg c_2$, the RHS of Equation (7.5) is approximately equal to $c_1 T / c_2 \lambda^4$, which is known as the Rayleigh-Jeans formula and is applicable only for very long wavelengths.

Blackbody radiation is achieved in an isothermal enclosure or cavity under thermodynamic equilibrium, as shown in Figure 7.4a. A uniform and isotropic radiation field is formed inside the enclosure. The total or spectral irradiation on any surface inside the enclosure is diffuse and identical to that of the blackbody emissive power. The spectral intensity is the same in all directions and is a function of λ and T given by Planck's law. If there is an aperture with an area much smaller compared with that of the cavity (see Figure 7.4b), λ the radiation field may be assumed unchanged and the outgoing radiation approximates that of blackbody emission. All radiation incident on the aperture is completely absorbed as a consequence of reflection within the enclosure. Blackbody cavities are used for measurements of radiant power and radiative properties, and for calibration of radiation thermometers (RTs) traceable to the International Temperature Scale of 1990 (ITS-90) [5].

7.3 RADIATIVE PROPERTIES OF SOLIDS

7.3.1 EMISSIVITY

The ratio of the spectral emissive power of an object to that of a blackbody at the same temperature defines the spectral emissivity:

$$\varepsilon_\lambda (\lambda, T) \equiv \frac{E_\lambda (\lambda, T)}{E_{\lambda,b} (\lambda, T)} \tag{7.8}$$

The total emissivity is defined as and related to ε_λ by

Radiation Heat Transfer

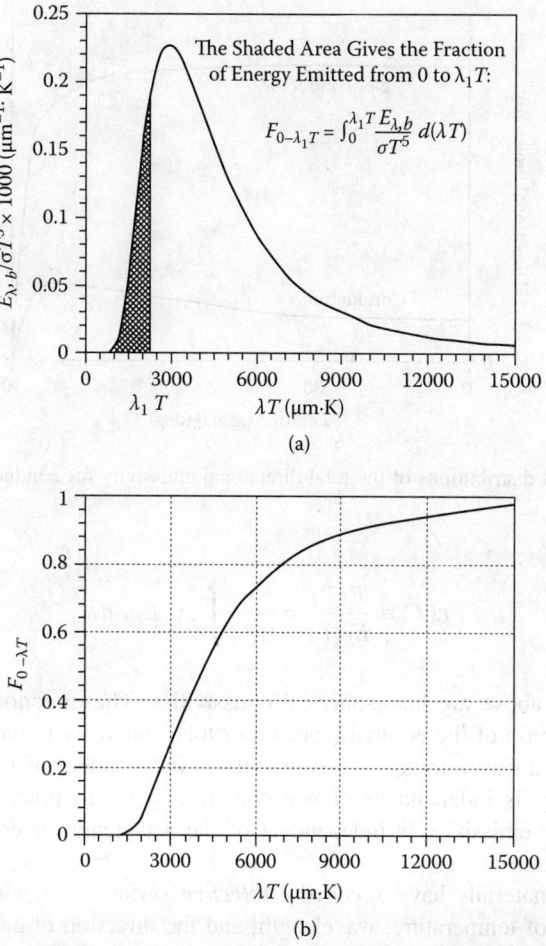

FIGURE 7.3 Blackbody characteristics: (a) Planck's law for spectral emissive power, and (b) band fraction for emission over the range from 0 to λT.

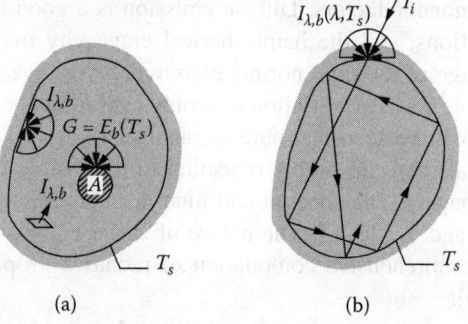

FIGURE 7.4 Blackbody characteristics for isothermal enclosures: (a) intensity is the same in all directions, and irradiation on any surface inside the enclosure is equal to the blackbody emissive power, and (b) emission through a small aperture approximates that of a blackbody, and the cavity acts as a perfect absorber.

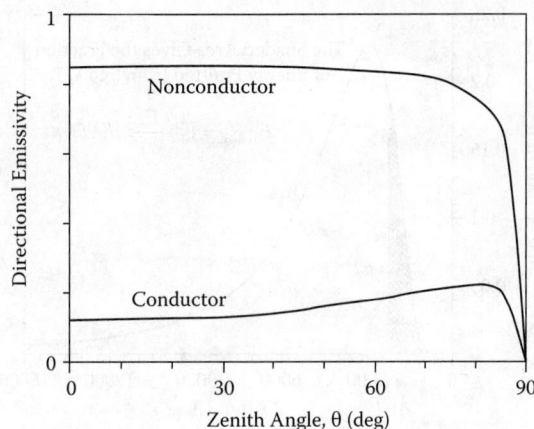

FIGURE 7.5 Directional distributions of the total directional emissivity for conductors (metals) and nonconductors (dielectrics).

$$\varepsilon(T) \equiv \frac{E(T)}{E_b(T)} = \frac{1}{\sigma T^4} \int_0^\infty \varepsilon_\lambda E_{\lambda,b} \, d\lambda \tag{7.9}$$

The ε_λ and ε defined above are hemispherical emissivities. The directional emissivities may be defined based on the ratio of the emitted (spectral or total) intensity to the blackbody (spectral or total) intensity. For a diffuse emitter, the emissivity is independent of the direction. For a *gray surface*, the emissivity is independent of wavelength, $\varepsilon = \varepsilon_\lambda$. In practice, the surface may be considered gray if the emissivity is independent of the wavelength over the spectral regions of irradiation and emission.

Many industrial materials have *spectrally selective* surfaces for which the emissivity is a complicated function of temperature, wavelength, and the direction of emission, and depends on the surface conditions. For perfectly smooth surfaces, the emissivity can be predicted from the electromagnetic wave theory [2–4]. Typical directional distributions of the total, directional emissivity are shown in Figure 7.5. For conductors (metallic materials), the emissivity is nearly constant for $\theta \leq 40°$, increases to a maximum near 80°, and finally drops to zero. For nonconductors, the emissivity is nearly constant for $\theta \leq 70°$ and decreases sharply to zero. The hemispherical emissivity is slightly higher ($\leq 30\%$) than the normal emissivity for conductors and slightly lower ($\leq 5\%$) than the normal emissivity for nonconductors. Diffuse emission is a good first-order approximation in many engineering applications, and the hemispherical emissivity may be assumed equal to the normal emissivity. The range of the total normal emissivities for a variety of materials near room temperature is listed in Table 7.1. The variation is mainly caused by the different types of materials and surface conditions. It can be seen that pure metals usually have a very low emissivity. Some spectrally selective materials that are highly reflecting in the visible may have a high emissivity (such as white paint and paper). The spectral and total normal emissivities for selected materials are shown in Figures 7.6 and 7.7 [1, 6]. The nature of surface finishing and coating can greatly affect the emissivity. A comprehensive compilation of radiative properties of solids is given by Touloukian and DeWitt [6].

Example 7.1

Assuming that ε_λ of stainless steel shown in Figure 7.6 is independent of temperature, how will its total emissivity ε change as the surface temperature increases?

TABLE 7.1
Ranges of the Total Normal Emissivity for Some Common Materials at 25°C

Material	ε_n
Highly polished metals, foils, and films	0.02–0.07
Polished metals	0.04–0.13
Unpolished metals	0.1–0.4
Oxidized metals	0.25–0.68
Oxides and ceramics	0.4–0.8
Graphite	0.76–0.94
Minerals and glasses	0.78–0.95
Vegetation, water, and skin	0.88–0.97
Special paints and anodized finishes	0.90–0.98

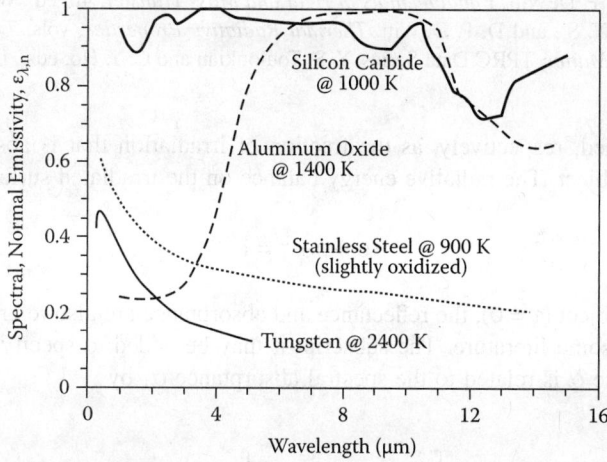

FIGURE 7.6 Spectral normal emissivity of selected materials. (Adapted from Incropera, F. P., and D. P. DeWitt, *Fundamentals of Heat and Mass Transfer*, 4th ed., Wiley, New York, 1996; and from Touloukian, Y. S., and D. P. DeWitt, *Thermal Radiative Properties*, vols. 7, 8, and 9, in *Thermophysical Properties of Matter*, TPRC Data Series, Y. S. Touloukian and C. Y. Ho, eds., IFI Plenum, New York, 1970–1972.)

Solution

As the temperature increases, the Planck distribution of the blackbody spectral emissive power will shift toward the short wavelength, so that more radiant power will be emitted at relatively shorter wavelengths. Using Equation (7.7), it can be estimated that at 400 K, less than 2% emitted power is at $\lambda < 4$ μm; whereas at 1000 K, more than 48% emitted power is at $\lambda < 4$ μm (see Figure 7.3b). Because the spectral emissivity decreases as the wavelength increases, the total emissivity will increase as temperature increases. The same reasoning can be used to explain the trends that the total emissivity of aluminum oxide decreases but that of tungsten increases as the temperature increases, as shown in Figure 7.7.

7.3.2 ABSORPTANCE, REFLECTANCE, AND TRANSMITTANCE

When radiation is incident on an object, a portion of it is absorbed inside the material, another portion is reflected, and the remaining is transmitted. The absorptance α, reflectance ρ, or trans-

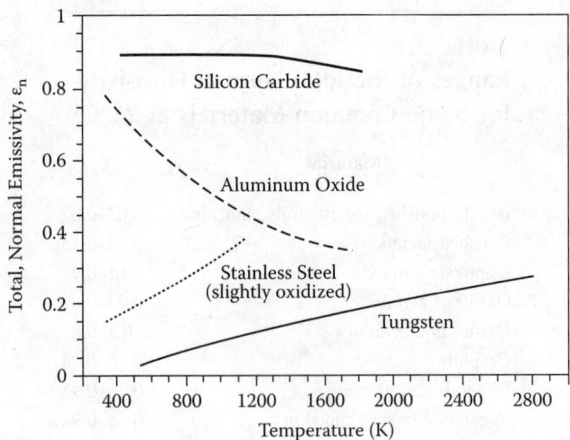

FIGURE 7.7 Temperature dependence of the total normal emissivity of selected materials. (Adapted from Incropera, F. P., and D. P. DeWitt, *Fundamentals of Heat and Mass Transfer*, 4th ed., Wiley, New York, 1996; and from Touloukian, Y. S., and D. P. DeWitt, *Thermal Radiative Properties*, vols. 7, 8, and 9, in *Thermophysical Properties of Matter*, TPRC Data Series, Y. S. Touloukian and C. Y. Ho, eds., IFI Plenum, New York, 1970–1972.)

mittance τ are defined, respectively, as the fraction of irradiation that is absorbed, reflected, or transmitted by the object. The radiative energy balance on the irradiated surface yields

$$\alpha + \rho + \tau = 1 \tag{7.10}$$

For an opaque object ($\tau = 0$), the reflectance and absorptance are also referred to as reflectivity and absorptivity in some literature. The subscript λ may be added to specify spectral properties. The total absorptance α is related to the spectral absorptance α_λ by

$$\alpha = \int_0^\infty \alpha_\lambda G_\lambda(\lambda) d\lambda \bigg/ \int_0^\infty G_\lambda(\lambda) d\lambda \tag{7.11}$$

where $G_\lambda(\lambda)$ represents the spectral distribution of the irradiation. In general, the value of α depends on the spectral and directional distributions of the incident radiation (the radiative environment to which the surface is exposed) and the spectral absorptance. Equation (7.11) also holds for the reflectance or transmittance when α is replaced by ρ or τ.

Sometimes it is important to consider the direction of reflected irradiation exitent from a surface. A property called the bidirectional reflectance distribution function (BRDF) is used to specify the directional distribution of the reflected intensity for a specified direction of incident radiation [2–4]. A specular surface is a mirrorlike surface for which the incidence angle is equal to the reflection angle. For a diffusely reflecting surface, the reflected intensity is the same in all directions, and if perfectly reflective, the BRDF is $1/\pi$ sr^{-1}.

7.3.3 Kirchhoff's Law

Consider an object in a blackbody enclosure. Under thermodynamic equilibrium, the emitted radiant power $\varepsilon A_s E_b$ must equal the absorbed power $\alpha A_s G$, where A_s is the area of the surface. Because $G = E_b$, it follows that

$$\varepsilon = \alpha \tag{7.12}$$

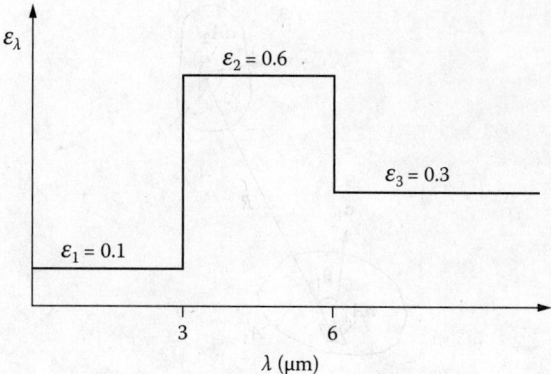

FIGURE 7.8 Example 7.2: spectral emissivity distribution of the spectrally selective surface.

It can also be shown that $\varepsilon_\lambda = \alpha_\lambda$ under the same condition [2]. These equalities between emissivities and absorptivities are known as Kirchhoff's laws. Because the absorptivity depends on the irradiation conditions, Kirchhoff's relations may not hold in general, except for the directional, spectral properties. It can be shown that $\varepsilon_\lambda = \alpha_\lambda$ if a surface is diffuse or diffusely irradiated. Furthermore, $\varepsilon = \alpha$ if the surface is diffuse-gray.

Example 7.2

Consider a hypothetical surface that is opaque and diffuse. At 300 K, its spectral emissivity ε_λ may be approximated with the wavelength dependence shown in Figure 7.8. Determine its total emissivity, ε, and the absorptivity for radiation from the sun, α_S, and a blackbody furnace at 2000 K, α_f.

Solution

The total emissivity can be calculated from the spectral emissivity using the Planck distribution; see Equation (7.9). Using Equation (7.7), it can be shown that at 300 K nearly 98% of the blackbody radiation is at wavelengths longer than 6 μm; therefore, the total emissivity at 300 K is equal to the spectral emissivity beyond 6 μm (see Figure 7.8). Hence, $\varepsilon \approx \varepsilon_3 = 0.3$. Because the surface is diffuse, from Kirchhoff's law, $\alpha_\lambda = \varepsilon_\lambda$. The total absorptivity can be calculated from Equation (7.11). The irradiation from the sun is proportional to the emissive power of a blackbody at 5800 K. Nearly 97% of the solar irradiation is at wavelengths below 3 μm; hence, the solar absorptivity is $\alpha_s \approx \varepsilon_1 = 0.1$ (see Figure 7.8). For a furnace at 2000 K, $G_\lambda \propto E_{\lambda,b}(\lambda, T)$; from Equation (7.11), we have

$$\alpha_f = \frac{1}{\sigma T^4} \int_0^\infty \varepsilon_\lambda E_{\lambda,b} d\lambda = \varepsilon_1 F_{0-\lambda_1 T} + \varepsilon_2 (F_{0-\lambda_2 T} - F_{0-\lambda_1 T}) + \varepsilon_3 (1 - F_{0-\lambda_2 T})$$

$$= 0.1 \times 0.7378 + 0.6 \times (0.9451 - 0.7378) + 0.3 \times (1 - 0.9451) = 0.2146.$$

where Equation (7.7) has been used to compute the band fractions. One can also use Figure 7.3b to obtain a rough estimate. Note that for spectrally selective surfaces (like the one given in this example), in general, $\varepsilon \neq \alpha$.

7.4 RADIATIVE ENERGY EXCHANGE BETWEEN SURFACES

7.4.1 RADIOSITY

The radiation exitent from an opaque surface can include emitted and reflected components. The *radiosity* is the sum of the emissive power and the portion of irradiation that is reflected by the

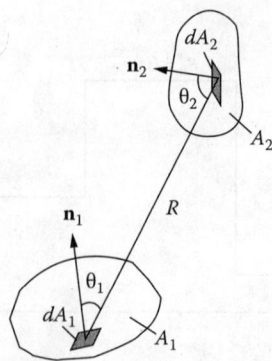

FIGURE 7.9 Definition of the view factor.

surface. The *total* radiosity is $J = \rho G + E$, and the *spectral* radiosity is $J_\lambda = \rho_\lambda G_\lambda + E_\lambda$. Radiosity is a useful concept in calculating radiation exchange between surfaces. Like E and G, J is also a hemispherical property.

7.4.2 THE VIEW FACTOR

Radiative exchange between diffuse-gray and opaque surfaces, including blackbodies, have many engineering applications. In the simplest case, the medium between the surfaces can be assumed *nonparticipating* (no absorption, emission, or scattering during propagation). The *view factor* (also called the configuration factor or shape factor) F_{ij} is defined as the fraction of radiation leaving surface i that is intercepted by surface j. For diffuse-gray surfaces with uniform radiosity, the view factor from A_1 to A_2 (see Figure 7.9) is

$$F_{12} = \frac{1}{A_1} \int_{A_2} \int_{A_1} \frac{\cos\theta_1 \cos\theta_2}{\pi R^2} dA_1 dA_2 \qquad (7.13)$$

The view factor depends only upon the geometric arrangement of the surfaces, and satisfies the *reciprocity relation* $A_i F_{ij} = A_j F_{ji}$. The view factor must be between 0 and 1. In an enclosure consisting of N surfaces, the *summation rule* gives

$$\sum_{j=1}^{N} F_{ij} = 1$$

The term F_{ii} refers to the fraction of radiation that leaves surface i and is directly intercepted by surface i itself. For flat or convex surfaces, $F_{ii} = 0$; and for concave surfaces, F_{ii} is nonzero. Table 7.2 provides equations for calculating the view factors for some common configurations often encountered in practical applications [1, 2, 8]. The view factors for more complicated geometries can be found in [2].

7.4.3 RADIATION EXCHANGE BETWEEN BLACKBODY SURFACES

If all surfaces are blackbodies, the net radiative heat-transfer rate from surface i to j is

$$q_{ij} = E_{bi} A_i F_{ij} - E_{bj} A_j F_{ji} = A_i F_{ij} (\sigma T_i^4 - \sigma T_j^4) \qquad (7.14)$$

TABLE 7.2
View Factors for Some Simple Common Geometries

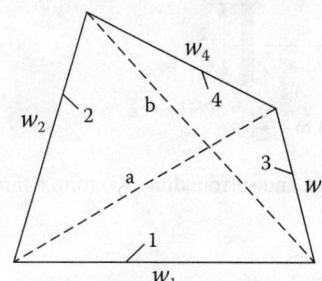

$$F_{12} = \frac{w_1 + w_2 - b}{2w_1}; \quad F_{13} = \frac{w_1 + w_3 - a}{2w_1}$$

$$F_{14} = \frac{a + b - w_2 - w_3}{2w_1}$$

Long Strips (cross-strings method [8])

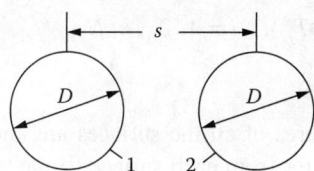

$$F_{12} = \frac{1}{\pi}\left[\sin^{-1}\left(\frac{D}{s}\right) + \sqrt{\left(\frac{s}{D}\right)^2 - 1} - \frac{s}{D}\right]$$

Long Cylinders of the Same Diameter

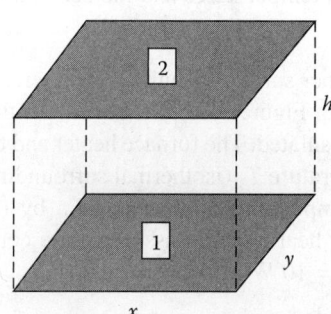

$$F_{12} = \frac{2}{\pi XY}\left[\ln\frac{AB}{\sqrt{1+X^2+Y^2}} + XB\tan^{-1}\frac{X}{B}\right.$$
$$\left. + YA\tan^{-1}\frac{Y}{A} - X\tan^{-1}X - Y\tan^{-1}Y\right]$$

where $X = x/h, Y = y/h, A = \sqrt{1+X^2}$, and $B = \sqrt{1+Y^2}$

Aligned Parallel Rectangles

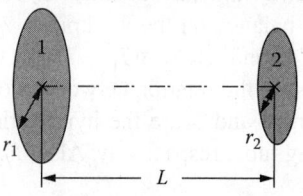

$$S = 1 + \frac{1+(r_2/L)^2}{(r_1/L)^2}$$

$$F_{12} = \frac{1}{2}\left[S - \sqrt{S^2 - 4\left(\frac{r_2}{r_1}\right)^2}\right]$$

Coaxial Parallel Disks

Sources: Incropera, F. P., and D. P. DeWitt, *Fundamentals of Heat and Mass Transfer*, 4th ed., Wiley, New York, 1996; Siegel, R., and J. R. Howell, *Thermal Radiation Heat Transfer*, 3rd ed., Hemisphere, Washington, 1992; and Hottel, H. C., and A. F. Sarofim, *Radiative Transfer*, McGraw-Hill, New York, 1967.

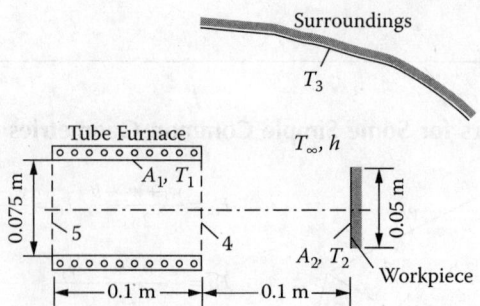

FIGURE 7.10 Example 7.3: tube furnace (1), workpiece (2), and surroundings (3) form a three-black-surface enclosure.

In an enclosure of N blackbody surfaces, the net radiative heat-transfer rate from surface i is

$$q_i = \sum_{j=1}^{N} q_{ij} = \sum_{j=1}^{N} A_i F_{ij} (\sigma T_i^4 - \sigma T_j^4), \quad i = 1, 2, \ldots, N \tag{7.15}$$

Given the geometry of the enclosure, if the temperatures of all the surfaces are known, Equation (7.15) can be used to determine the net radiative transfer from each surface. If the temperatures of k ($1 \leq k \leq N$) surfaces and the net heat transfer from the remaining ($N - k$) surfaces are known, Equation (7.15) may be solved to obtain the unknown temperatures and the net heat-transfer rates.

Example 7.3

An electrically heated tubular furnace with a black inner surface of uniform temperature $T_1 = 1200$ K irradiates a coaxial thin disk workpiece, as shown in Figure 7.10. The surface of the disk facing the furnace is painted black and the back side is well insulated. The furnace heater and the workpiece are placed in a large room with a uniform wall temperature T_3 (isothermal surroundings) of 300 K and air temperature $T_\infty = 293$ K. Determine the temperature of the disk, T_2, by (a) neglecting convective heat transfer and (b) including convective heat transfer, assuming the convective heat-transfer coefficient between the disk and the air is $h = 10$ W/m²·K.

Solution

(a) The black surface of the furnace (1), workpiece (2) and surroundings (3) form a three-surface enclosure. Performing a radiative energy balance on the workpiece, $q_{1-2} - q_{2-3} = 0$, and using the reciprocity relation $A_1 F_{12} = A_2 F_{21}$, find $A_2 F_{21} \sigma (T_1^4 - T_2^4) - A_2 F_{23} \sigma (T_2^4 - T_3^4) = 0$. The view factor F_{21} may be calculated using the relation between two concentric disks (Table 7.2), $F_{21} = F_{24} - F_{25} = 0.086$, where 4 and 5 are the hypothetical circular surfaces at the right end and left end of the heating tube, respectively. Also, $F_{23} = 1 - F_{21}$. The result is $T_2 = 655.6$ K.

(b) The energy balance for surface 2 including convection is $q_{1-2} - q_{2-3} - q_{conv} = 0$, or

$$A_2 F_{21} (\sigma T_1^4 - \sigma T_2^4) - A_2 F_{23} (\sigma T_2^4 - \sigma T_3^4) - A_2 h (T_2 - T_\infty) = 0$$

Solving by iteration, find $T_2 = 601$ K. Although convective heat loss is only about half of the radiative heat loss, it should not be neglected.

7.4.4 Enclosure with N Diffuse-Gray Surfaces—The Net Radiation Method

Consider an enclosure of N diffuse-gray, opaque surfaces, each surface being isothermal and having a uniform radiosity and irradiation. The radiative energy balance for the ith surface gives $q_i = A_i(J_i - G_i)$, where $J_i \equiv E_i + \rho_i G_i = \varepsilon_i E_{bi} + (1-\varepsilon_i)G_i$. Combining these two equations to eliminate G_i, one obtains

$$q_i = \frac{A_i \varepsilon_i}{1 - \varepsilon_i}(E_{bi} - J_i) \tag{7.16}$$

The total radiation reaching surface i from the N-isothermal surfaces in the enclosure (including itself) is

$$A_i G_i = \sum_{j=1}^{N} A_j F_{ji} J_j = \sum_{j=1}^{N} A_i F_{ij} J_j$$

Therefore,

$$q_i = A_i \left(J_i - \sum_{j=1}^{N} F_{ij} J_j \right) = \sum_{j=1}^{N} A_i F_{ij}(J_i - J_j) = \sum_{j=1}^{N} q_{ij} \tag{7.17}$$

where q_{ij} is the net radiative heat-transfer rate from surface i to surface j.

If the surface temperature T_i is specified for surfaces $i = 1, 2, ..., k$ ($1 \leq k \leq N$) and the net radiative heat rate q_i is specified for surfaces $i = k+1, k+2, ..., N$, then Equations (7.16) and (7.17) may be rearranged as the following:

$$\begin{cases} \dfrac{\sigma T_i^4 - J_i}{(1-\varepsilon_i)/\varepsilon_i A_i} = \sum_{j=1}^{N} \dfrac{J_i - J_j}{1/A_i F_{ij}}, & i = 1, 2, 3, ..., k \\[2ex] \sum_{j=1}^{N} \dfrac{J_i - J_j}{1/A_i F_{ij}} = q_i, & i = k+1, k+2, ..., N \end{cases} \tag{7.18}$$

which gives N-linear algebraic equations that can be solved simultaneously for the N-unknown radiosities $J_1, J_2, ..., J_N$. The unknown q_i ($i = 1, 2, ..., k$) and T_i ($i = k+1, k+2, ..., N$) can be calculated using Equation (7.18). The above equations may be rearranged in matrix format and solved by matrix inversion [2, 3].

If the temperature of each surface is specified, the spectral intensity leaving a surface is the sum of the emitted and the reflected intensities, therefore,

$$I_{\lambda,i} = \varepsilon_{\lambda,i} I_{\lambda,b} + (1-\varepsilon_{\lambda,i}) \sum_{j=1}^{N} F_{ij} I_{\lambda,j}, \quad i = 1, 2, ... N \tag{7.19}$$

The N-linear algebraic equations given above can be solved for the spectral intensities $I_{\lambda,i}$. The foregoing treatment, the net-radiation method, is limited to opaque, diffuse-gray surfaces, which can be extended to include some specular surfaces using the *imaging method*. If the irradiation is not uniform, the *integration method* using differential surfaces is often used. In addition, the *Monte Carlo method*, which is a statistical method, can also be used to determine the radiative heat transfer between enclosure surfaces. Detailed discussions of these more complicated can be found in the literature [2–4, 7, 8].

7.4.5 The Network Representation

The radiative energy balances of Equations (7.16) and (7.17) can be represented in a network. Compared with an electric network, E_b and J_i are analogous to the potential, q_i and q_{ij} are analogous to the current, and $(1-\varepsilon_i)/A_i\varepsilon_i$ and $1/A_iF_{ij}$ are analogous to the resistances. The network analogy provides a useful way for visualizing radiation exchange in an enclosure and is a convenient tool for calculating the radiative exchange in an enclosure consisting of two or three surfaces. For an enclosure of two surfaces, using the *network method* as shown in Figure 7.11,

$$q_1 = -q_2 = q_{12} = \frac{\sigma T_1^4 - \sigma T_2^4}{\dfrac{1-\varepsilon_1}{\varepsilon_1 A_1} + \dfrac{1}{A_1 F_{12}} + \dfrac{1-\varepsilon_2}{\varepsilon_2 A_2}} \tag{7.20}$$

If surface 1 is a plane wall exposed to isothermal surroundings as shown in Figure 7.12a, then $F_{12} = 1$ and $A_1/A_2 \ll 1$. The net radiative heat-transfer rate from A_1 to its surroundings is

$$q_{rad} = q_1 = A_1 \varepsilon_1 \sigma (T_1^4 - T_{sur}^4) \tag{7.21}$$

Radiative heat transfer is often coupled with convective heat transfer to form the boundary conditions for conduction problems. A *radiation heat transfer coefficient* may be defined by $q_{rad} = A_1 h_{rad}(T_1 - T_{sur})$, where $h_{rad} = \varepsilon_1 \sigma (T_1^2 + T_{sur}^2)(T_1 + T_{sur})$ is a strong function of the temperatures and emissivity. As a first-order estimate, h_{rad} may be compared with the convection heat transfer

FIGURE 7.11 The radiation network for a two-surface enclosure.

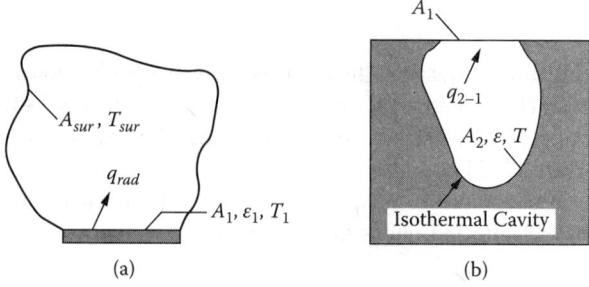

FIGURE 7.12 Applications of net radiation exchange between two surfaces: (a) a small, convex surface and large, isothermal surroundings; (b) an isothermal cavity radiating into cold, black surroundings.

Radiation Heat Transfer

coefficient (h) in a particular problem to see whether radiation heat transfer is significant. For a surface at 310 K, with an emissivity of 0.8 and the surroundings at 298 K, $h_{rad} \approx 5$ W/m²·K, which is comparable with the values of h (2 to 10 W/m²·K) due to natural convection of air in the same temperature range [1].

Another application of the two-surface enclosure is to determine the effective (apparent) emissivity of an isothermal cavity, as shown in Figure 7.12b. The effective emissivity is defined as the ratio of the radiant power from the cavity to that emitted by a blackbody of the same temperature with an area equal to the cavity opening. The radiation leaving surface 2 to the space above the cavity opening (surface 1) q_{2-1} may be obtained from Equation (7.20) by setting surface 1 as a blackbody at 0 K. Hence,

$$\varepsilon_{eff} = \frac{q_{2-1}}{A_1 \sigma T^4} = \frac{\varepsilon}{\varepsilon + (1-\varepsilon)A_1/A_2} = \frac{\varepsilon}{1-(1-\varepsilon)(1-A_1/A_2)} \quad (7.22)$$

As expected, the effective emissivity reaches unity as A_1/A_2 approaches zero. The effective absorptivity α_{eff} is equal to ε_{eff} and may be viewed as the ratio of the radiant energy absorbed by the cavity walls to that incident through the opening. Blackbody cavities have important applications in radiometry and radiation thermometry.

Example 7.4

In a petrochemical furnace as shown in Figure 7.13, a bank of tubes containing the process fluid is heated to $T_2 = 600$ K by placing the tubes between a large hot plate (with area A_1) and a refractory (insulated) wall ($A_3 = A_1$). The hot plate, with an emissivity of $\varepsilon_1 = 0.7$, is maintained at $T_1 = 1200$ K. The emissivity of the tube is $\varepsilon_2 = 0.5$. The diameter of each tube is $D = 50$ mm and the central distance between adjacent tubes is $s = 100$ mm. (a) Determine the heating rate that must be provided to the hot plate per unit surface area. (b) Determine the temperature of the refractory surface. (c) If the tubes are to be heated to $T_2' = 800$ K with the same heat-transfer rate, what is the required hot wall temperature T_1'?

Solution

Neglect convection heat transfer, and assume all surfaces are diffuse, gray, and of infinite extent. Since the bottom wall (A_3) is insulated, the net radiative heat flux at that surface must be zero. This type of surface is called a *reradiating surface*. The radiation network is shown in Figure 7.14. For an area of 1 m², $A_1 = A_3 = 1$ m², the number of tubes is $M = 10$, and the length of each tube is $L = 1$ m. Hence, $A_2 = M\pi DL = 1.57$ m² (the surface area of the tubes). If there were only one tube and the plates were infinitely large, the view factor between the tube and each plate would be 0.5.

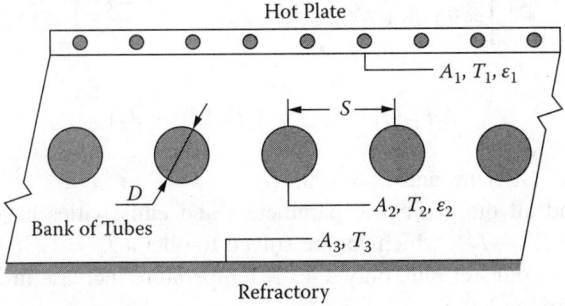

FIGURE 7.13 Example 7.4: hot furnace plate (1), tubes (2), and refractory wall (3) form a three-surface enclosure with a reradiating surface.

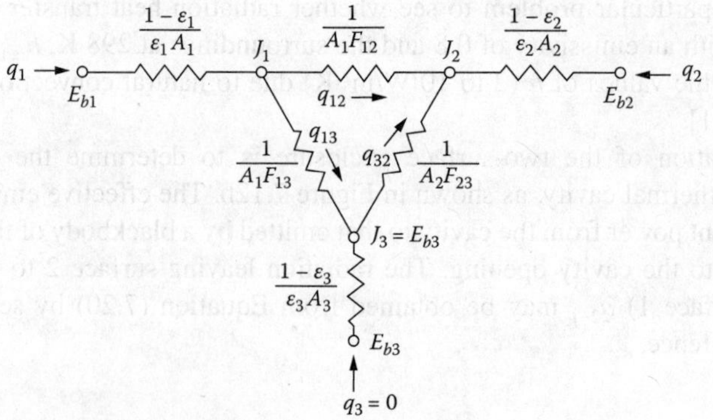

FIGURE 7.14 Network representation for a three-surface enclosure.

The view factor between the tube bank and each plate can be approximated by subtracting the view factor between two tubes from 0.5:

$$F_{21} = F_{23} = \frac{1}{2} - \frac{1}{\pi}\left[\sin^{-1}\frac{D}{S} + \sqrt{\left(\frac{S}{D}\right)^2 - 1} - \frac{S}{D}\right] = 0.4186$$

Therefore, $F_{12} = \frac{A_2}{A_1}F_{12} = 0.6576$ and $F_{13} = 1 - F_{12} = 0.3424$.

(a) $q_1 = q_{12} = -q_2 = \dfrac{\sigma(T_1^4 - T_2^4)}{\dfrac{1-\varepsilon_1}{A_1\varepsilon_1} + \left(A_1F_{12} + \dfrac{1}{1/A_1F_{13} + 1/A_2F_{23}}\right)^{-1} + \dfrac{1-\varepsilon_2}{A_2\varepsilon_2}}$

$$= \frac{5.67\times 10^{-8}(1200^4 - 600^4)}{\dfrac{1-0.7}{0.7} + \left(0.6576 + \dfrac{1}{1/0.3424 + 1/(1.57\times 0.4186)}\right)^{-1} + \dfrac{1-0.5}{1.57\times 0.5}}$$

$$= 50.14 \text{ kW/m}^2$$

(b) $J_1 = E_{b1} - q_1\left(\dfrac{1-\varepsilon_1}{A_1\varepsilon_1}\right) = 96.08 \text{ kW/m}^2$, $J_2 = E_{b2} - q_2\left(\dfrac{1-\varepsilon_2}{A_2\varepsilon_2}\right) = 39.29 \text{ kW/m}^2$

$$A_1F_{13}(J_1 - E_{b3}) = A_2F_{23}(E_{b3} - J_2)$$

yields $E_{b3} = 58.74$ kW/m² and $T_3 = 1009$ K.

(c) Because q_{12} and all the geometric parameters and emissivities have not changed, we have $T_1^4 - T_2^4 = T_1'^4 - T_2'^4$, which can be solved to obtain $T_1' = 1238.6$ K. That is to say, for the same heat transfer rate, only a 40 K temperature increase in T_1 will increase the tube temperature by 200 K.

Note that the emissivity value of the reradiating surface does not affect the results. If surfaces A_1 and A_2 are treated as blackbodies, $\varepsilon_1 = \varepsilon_2 = 1$, then the predicted net radiative heat transfer rate q_{12} would be 97.3 kW/m², which is almost twice the value calculated in part (a).

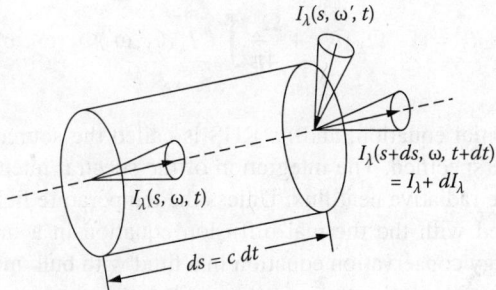

FIGURE 7.15 Radiative transfer in a participating medium.

7.5 GAS AND PARTICLE RADIATION

The above discussion assumes that the medium between surfaces does not participate in the thermal radiative process, which is a good approximation for nonpolar gases, such as Ar, N_2, and O_2. Polar gases, such as carbon dioxide, water vapor, and hydrocarbon gases, emit and absorb radiation in specific wavelength intervals (i.e., bands) over a wide range of temperatures. Gas molecules and aerosol in the atmosphere, pulverized coal in a combustor, and soot in flames can also scatter radiation. Radiative transfer in *participating media* is important in the study of the earth's radiation environment and in combustion processes. A brief summary of the radiative transfer equation and radiative properties of gases and particles is given here.

7.5.1 Radiative Transfer Equation

As shown in Figure 7.15, the spectral intensity in a participating medium, $I_\lambda = I_\lambda(s,\omega,t)$, depends on the location, the coordinate s, its direction (represented by the solid angle ω), and the time t. In a time interval dt, the beam travels from s to $s + ds$ ($ds = c\, dt$), and the intensity is (a) attenuated by absorption and scattering and (b) enhanced by emission and in-scattering. The macroscopic description of the radiation intensity is known as the *radiative transfer equation* (RTE) [4]:

$$\frac{dI_\lambda}{ds} = \frac{1}{c}\frac{\partial I_\lambda}{\partial t} + \frac{\partial I_\lambda}{\partial s} = \kappa_\lambda I_{\lambda,b}(T) - \beta_\lambda I_\lambda + \frac{\sigma_\lambda}{4\pi}\int_{4\pi} I_\lambda(s,\omega',t)\Phi_\lambda(\omega',\omega)d\omega' \quad (7.23)$$

where κ_λ and σ_λ are the absorption and scattering coefficients, respectively; $\beta_\lambda = \kappa_\lambda + \sigma_\lambda$ is the attenuation (or extinction) coefficient; and $\Phi_\lambda(\omega',\omega)$ is the *scattering phase function* ($\Phi_\lambda \equiv 1$ for isotropic scattering). The RHS of Equation (7.23) is composed of three terms: the first one accounts for the contribution of emission (which depends on the local temperature T); the second one is the attenuation by absorption and scattering; and the third one is the contribution of in-scattering from all directions (solid angle 4π) to the direction ω.

In most engineering applications, the transient term is negligible. The *single scattering albedo* is defined as $\Omega_\lambda = \sigma_\lambda / \beta_\lambda$. A dimensionless optical coordinate (sometimes called optical thickness) is often used, which is defined as

$$\zeta_\lambda = \int_0^s \beta_\lambda ds$$

The RTE in terms of the optical coordinate and the single scattering albedo is [2–4]

$$\frac{dI_\lambda}{d\zeta_\lambda} + I_\lambda = (1-\Omega_\lambda)I_{\lambda,b} + \frac{\Omega_\lambda}{4\pi}\int_{4\pi} I_\lambda(\zeta_\lambda,\omega')\Phi_\lambda(\omega',\omega)d\omega' \qquad (7.24)$$

This is an integro-differential equation, and its RHS is called the source function. The boundary conditions also need to be specified. The integration of the spectral intensity over all wavelengths and all directions gives the radiative heat flux. Unless the temperature field is prescribed, Equation (7.24) is generally coupled with the thermal diffusion equation in a macroscopically stationary medium and with the energy conservation equation in a fluid with bulk movement. Under radiation equilibrium (steady state with no heat generation and without conduction or convection), the governing equation (energy equation) prescribes that the divergence of the radiative heat flux is zero.

Analytical solutions of the RTE rarely exist for applications with multidimensional and non-homogeneous media. Approximate models have been developed to deal with special types of problems, including Hottel's *zonal method* [8], the *differential and moment methods* (often using the spherical harmonic approximation), and the *discrete ordinates method*. The statistical model using the Monte Carlo method is often used for complicated geometries and radiative properties. Detailed discussions of these solution methods can be found in the literature [2–4, 7, 8] and are beyond the scope of this chapter. For an isothermal medium without scattering, integrating the RTE yields

$$I_\lambda(\zeta_\lambda) = I_\lambda(0)e^{-\zeta_\lambda} + I_{\lambda,b}(1-e^{-\zeta_\lambda}) \qquad (7.25)$$

where $e^{-\zeta_\lambda}$ and $(1-e^{-\zeta_\lambda})$ may be regarded, respectively, as the transmissivity and emissivity of the gas along the path length 0 to s. Optically thin and optically thick limits refer to the cases when $\zeta_\lambda \ll 1$ (essentially transparent) and $\zeta_\lambda \gg 1$ (essentially opaque).

As an example, consider a sphere of radius R with a black wall of temperature T_w filled with an isothermal gas at T_g, as shown in Figure 7.16. Neglect scattering and assume that the gas has a refractive index of 1 and an absorption coefficient κ that is independent of the wavelength (gray assumption). The spectral intensity at the wall is a function of angle θ. From Equation (7.25), we have $I_\lambda(\theta) = I_{\lambda,b}(T_w)e^{-2\kappa R\cos\theta} + I_{\lambda,b}(T_g)(1-e^{-2\kappa R\cos\theta})$. Since κ is not a function of the wavelength, a similar expression can be written for the total intensity by omitting the subscript λ. The heat flux on the wall can be obtained by integrating the intensity over the hemisphere:

$$q'' = \int_{\phi=0}^{2\pi}\int_{\theta=0}^{\pi/2} I(\theta)\cos\theta\sin\theta d\theta d\phi.$$

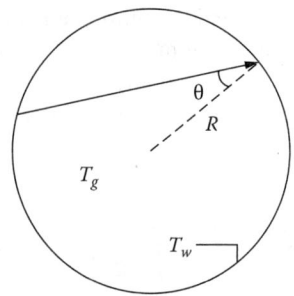

FIGURE 7.16 Example 7.5: an isothermal gas in a sphere.

Radiation Heat Transfer

The result is $q'' = \tau_g E_b(T_w) + \varepsilon_g E_b(T_g)$, where τ_g is regarded as the mean transmissivity, and $\varepsilon_g = 1 - \tau_g$ is the mean emissivity (which is the same as the mean absorptivity $\alpha_g = 1 - \tau_g$). The integration yields $\tau_g = 2(2\kappa R)^{-2}[1 - (1 + 2\kappa R)e^{-2\kappa R}]$. The radiative heat-transfer rate from the gas to the wall is readily calculated from $q = 4\pi R^2 (\varepsilon_g \sigma T_g^4 - \alpha_g \sigma T_w^4)$. In practice, the gas emits and absorbs at particular wavelength bands. Finding the mean total emissivity and absorptivity is important for first-order engineering heat-transfer calculations.

7.5.2 Radiative Properties of Gases and Particles

The molecular diameter of common gases is on the order of nanometers, much smaller than the wavelengths important for thermal radiation. Therefore, scattering by molecular gases is usually negligible in radiative heat transfer. When radiation travels through a cloud of gas, some of the energy may be absorbed. The absorption of radiant energy (or photons) raises the energy levels of individual molecules. Conversely, gas molecules may spontaneously lower their energy levels and emit photons. These changes in energy levels are called *radiative transitions*, which include bound-bound transitions (between nondissociated molecular states), bound-free transitions (between nondissociated and dissociated states), and free-free transitions (between dissociated states).

Bound-free and free-free transitions usually occur at very high temperatures (greater than about 5000 K) and are in the ultraviolet and visible regions. The most important transitions for radiative heat transfer are bound-bound transitions between vibrational energy levels coupled with rotational transitions. The photon energy (or frequency) must be exactly the same as the difference between two energy levels if the photon is to be absorbed or emitted; therefore, the quantization of the energy levels results in discrete spectral lines for absorption and emission. The rotational lines superimposed on a vibrational line give a band of closely spaced spectral lines, called the vibration-rotation spectrum. In addition to the natural line broadening due to Heisenberg's uncertainty principle, collision and Doppler effects are two other reasons for line broadening, although the resulting lines are often still very narrow, on the order of 0.05 cm^{-1}. Wavenumber (in cm^{-1}), which is the inverse of wavelength, is often used to describe the spectral lines. The shape and width of each line depend on the temperature and pressure of the medium. The Lorentz profile is commonly used to define the shape of these lines for most engineering applications. More elaborated discussions on the absorption and emission mechanisms can be found in the literature [2–4]. A detailed treatment on the radiative properties of gases was given by Tien [9].

The large number of quantum states in polyatomic gases can produce spectral lines that are spaced very close to each other; some lines may overlap as a consequence of broadening. The results are called vibration-rotation bands. Band models are therefore developed to facilitate the computation of the total radiative properties. In a narrow-band model, the absorption coefficients of individual lines are added up in a spectral interval of 50 cm^{-1} or so. Narrow-band models are much simpler than the line-by-line calculation, yet still require a large database and significant computational effort. The wide-band models, on the other hand, predict the total absorption and emission over an entire vibrational band. Typical wide-band models are the box model and the exponential wide-band model. Among other gases, water vapor (H_2O) and carbon dioxide (CO_2) are the major species that participate in radiative transfer, both in the atmosphere and in combustion furnaces. The four most important bands for H_2O are centered at wavelengths of 1.38 μm (7250 cm^{-1}), 1.87 μm (5350 cm^{-1}), 2.7 μm (3760 cm^{-1}), and 6.3 μm (1600 cm^{-1}). For CO_2, the major bands are at 2.7 μm (3715 cm^{-1}), 4.3 μm (2349 cm^{-1}), and 15 μm (667 cm^{-1}). For CO, the major absorption band is at 4.7 μm (2143 cm^{-1}). Detailed formulations and parameters for various gas species can be found in the literature [2–4].

For heat-transfer calculations in many engineering applications, only the spectrally integrated (i.e., total) gas emissivity and absorptivity are needed in the calculation. For isothermal gases, the band emissivity can be integrated considering the blackbody intensity distribution at the gas temperature. If two absorbing gases appear together, the integration can be performed separately and the total emissivities should be added and then subtracted by the contribution of the overlapping bands; hence,

$$\varepsilon_{a+b} = \varepsilon_a + \varepsilon_b - \Delta\varepsilon_{ab} \tag{7.26}$$

where subscripts a and b represent two different gases, typically H_2O and CO_2, and $\Delta\varepsilon_{ab}$ is the contribution due to overlapping bands (which is about 2.7 μm for H_2O and CO_2). A similar expression can be obtained for the total absorptance. However, the integration must be weighted over the Planck distribution at the wall temperature. Notice that the total emissivity is a function of the gas temperature, pressure, partial pressure, and path length; and the total absorptivity is also a function of the wall temperature. Most heat transfer textbooks present the total emissivity and absorptivity of H_2O and CO_2 in charts [1, 2], an approach first introduced by Hottel [8]. New charts and equations have also been proposed. A summary can be found in the textbook by Modest [4]. For nonisothermal gases, the zonal method is widely employed for calculating radiative heat transfer in combustion chambers [2, 8].

Small particles are important for radiative transfer in atmosphere and combustion chambers [2–4]. Soot particles ranging from 5 to 100 nm in size are formed during a combustion process and emit thermal radiation (often stronger than the emission from combustion gases) in a continuous spectrum over the infrared and some visible regions. In pulverized-coal combustion furnaces, coal, char, and fly ash particles are important, as are soot particles. Particles can also scatter electromagnetic waves or photons, causing a change in the direction of propagation. In the early 20th century, Mie developed a theory for a spherical particle based on Maxwell's electromagnetic wave equations, known as the *Mie scattering theory*, which can be used to predict the scattering-phase function. In the case when the particle sizes are small compared with the wavelength (Rayleigh limit), the formulation reduces to the simple expression obtained earlier by Rayleigh. In the Rayleigh limit, the scattering efficiency is inversely proportional to the wavelength to the fourth power. This factor explains why the sky is blue and why the sun appears red at sunset. The absorption efficiency, on the other hand, is proportional to the inverse of the wavelength. The scattering phase function is symmetric as far as forward and backward scattering is concerned, and is nearly isotropic [2–4]. Soot particles generally fall in the Rayleigh limit. For spheres whose diameters are much greater than the wavelength, diffraction theory and geometric optics based on specular surface or diffuse surface can be applied [4]. If the scattering by one particle is affected by the presence of other particles, the scattering is called dependent scattering and requires more complicated treatments. Detailed discussions of particle properties and scattering theory can be found from [2–4] and references therein.

7.6 RADIATION THERMOMETRY

Radiation thermometry (or pyrometry) is a subject that involves determining the temperature of an object from measurement of its exitent intensity. A radiation thermometer (RT) is a radiometer calibrated to indicate the temperature of a blackbody, as shown in Figure 7.17a. RTs have been used in high-temperature furnaces, glass processing, steel processing, crystal growth, aluminum manufacturing, and thermal imaging [10]. The application of RTs to high-temperature furnaces is addressed in [11]. A more recent review on surface temperature measurements using optical techniques is given by Zhang [5]. When an RT is used to measure the temperature of a real surface, as shown in Figure 7.17b, two issues arise. The first is the unknown emissivity of the surface and the second is the influence of the surrounding radiation. Various methods have been developed to deal with these problems, including the creation of a blackbody cavity on the surface, the two-color method, and the use of a controlled reference source [5, 10, 11]. The development of optical

Radiation Heat Transfer

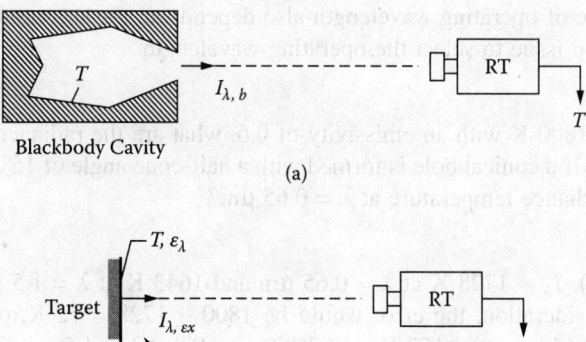

FIGURE 7.17 Radiation thermometry: (a) calibration against a blackbody cavity; (b) measurement of a real surface.

fibers has allowed radiometric temperature measurement for surface locations otherwise inaccessible by imaging radiometers.

The *measurement equation* of a spectral radiation thermometer may be approximated as

$$S_\lambda = C_\lambda L_{\lambda,ex}(\lambda) \tag{7.27}$$

where S_λ is the detector output signal, $L_{\lambda,ex}(\lambda)$ is the *exitent spectral radiance*,* and C_λ is an instrument constant that is independent of the target material and temperature. The *radiance temperature* T_λ is defined by the following equation:

$$L_{\lambda,b}(\lambda, T_\lambda) = L_{\lambda,ex}(\lambda) \tag{7.28}$$

For a freely radiating target (i.e., with cold black surroundings), the exitent spectral radiance is due only to the emission by the target (see Figure 7.17), therefore,

$$L_{\lambda,ex}(\lambda) = L_{\lambda,e}(\lambda, T) = \varepsilon'_\lambda \, L_{\lambda,b}(\lambda, T) \tag{7.29}$$

where ε'_λ is the directional spectral emissivity, and $L_{\lambda,b}(\lambda, T)$ is the blackbody spectral radiance given by Planck's law, Equation (7.5). By combining Equations (7.28) and (7.29) and applying Wien's approximate formula, one obtains the temperature equation that relates the surface temperature to the radiance temperature as follows:

$$\frac{1}{T} = \frac{1}{T_\lambda} + \frac{\lambda}{c_2} \ln \varepsilon'_\lambda \tag{7.30}$$

The effect of the emissivity uncertainty on the temperature accuracy decreases as λ decreases. However, the wavelength at which $L_{\lambda,b}(\lambda, T)$ is a maximum is given by Wien's displacement law.

---

* Although the concept of radiance is the same as the intensity defined in Equation (7.3), it is commonly used in radiation thermometry. Therefore, both the term radiance and its symbol L are retained here.

In practice, the choice of operating wavelength also depends on the surrounding radiation. Therefore, it is an important issue to select the operating wavelength.

Example 7.5

For a surface at $T = 1800$ K with an emissivity of 0.6, what are the radiance temperatures at $\lambda = 0.65$ μm and 1.5 μm? If a conical hole is formed with a half-cone angle of 15°, what is the effective emissivity and the radiance temperature at $\lambda = 0.65$ μm?

Solution

From Equation (7.30), $T_\lambda \approx 1728$ K at $\lambda = 0.65$ μm and 1643 K at $\lambda = 1.5$ μm. If the emissivity is not taken into consideration, the error would be $1800 - 1728 = 72$ K, or 4% for a radiation thermometer at $\lambda = 0.65$ μm, and 157 K or 8.7% for an RT at $\lambda = 1.5$ μm. If a cavity is formed, assuming that the surface is diffuse, the effective emissivity can be approximated by Equation (7.22). Because the half-cone angle is 15°, $A_1/A_2 = \sin 15°$ and $\varepsilon_{eff} = 0.853$. The radiance temperature based on the effective emissivity is $T_{\lambda=0.65\,\mu m} \approx 1777$ K. The error in RT will be reduced to 23 K or 1.3%. The effective emissivity for a specular surface would be even larger in this case, which would result in a smaller difference between T_λ and T.

If the target is not a blackbody and the surrounding radiation is not negligible, $L_{\lambda,ex}(\lambda)$ is the sum of the emitted and reflected spectral radiances:

$$L_{\lambda,ex}(\lambda) = L_{\lambda,e}(\lambda, T) + L_{\lambda,r}(\lambda) \tag{7.31}$$

The reflected spectral radiance depends on the distribution of the surrounding radiation incident on the target and the bidirectional reflectance distribution function of the target, and except for well-controlled environments or in some limited cases, $L_{\lambda,r}(\lambda)$ is difficult to determine.

For flame or gas temperature measurements, the exitent radiance is proportional to $L_{\lambda,b}(1 - e^{-\zeta_\lambda})$, where ζ_λ is the optical thickness. The emission from the surface may transmit through the gas and needs to be taken into consideration [11]. Although RTs can be used to measure temperatures for high-temperature surfaces and flames with uncertainties ranging from a few kelvin to a few tens of kelvin, great care must be taken in the calibration and operation of RTs. Calibration is usually done with high-temperature blackbody furnaces. In the measurement, the field of view and the angle of incidence need to be correctly chosen. The alignment needs to be carefully performed. Background radiation and other radiation sources should be avoided.

REFERENCES

1. Incropera, F. P., and D. P. DeWitt, *Fundamentals of Heat and Mass Transfer*, 4th ed., Wiley, New York, 1996.
2. Siegel, R., and J. R. Howell, *Thermal Radiation Heat Transfer*, 3rd ed., Hemisphere, Washington, 1992.
3. Brewster, M. Q., *Thermal Radiative Transfer and Properties*, Wiley, New York, 1992.
4. Modest, M. F., *Radiative Heat Transfer*, McGraw-Hill, New York, 1993.
5. Zhang, Z. M., in *Annual Review of Heat Transfer*, vol. 11, C. L. Tien, ed., 351–411, Begell House, New York, 2000.
6. Touloukian, Y. S., and D. P. DeWitt, *Thermal Radiative Properties*, vols. 7, 8, and 9, in *Thermophysical Properties of Matter*, TPRC Data Series, Y. S. Touloukian and C. Y. Ho, eds., IFI Plenum, New York, 1970–1972.
7. Sparrow, E. M., and R. D. Cess, *Radiation Heat Transfer*, augmented ed., Hemisphere, Washington, 1981.
8. Hottel, H. C., and A. F. Sarofim, *Radiative Transfer*, McGraw-Hill, New York, 1967.

9. Tien, C. L. in *Advances in Heat Transfer*, vol. 5, T. F. Irvine, Jr., and J. P. Hartnett, eds., 253–324, Academic Press, New York, 1968.
10. DeWitt, D. P., and G. D. Nutter. eds. *Theory and Practice of Radiation Thermometry*, Wiley, New York, 1988.
11. DeWitt, D. P., and L. F. Albright. eds. *Measurement of High Temperatures in Furnaces and Processes*, vol. 82, no. 249, AIChE Symposium Series, American Institute of Chemical Engineers, New York, 1986.

8 Mass Transfer

James R. Fair

CONTENTS

8.1 Introduction...591
8.2 Molecular Diffusion Coefficients..592
    8.2.1 Gases—Binary Mixtures..592
    8.2.2 Liquids—Binary Mixtures..594
    8.2.3 Gases—Multicomponent Mixtures...597
    8.2.4 Liquids—Multicomponent Mixtures...598
    8.2.5 Diffusion through Porous Solids..598
    8.2.6 Gases—Diffusion through Membranes..600
8.3 Mass Transfer at a Phase Boundary...601
    8.3.1 Stagnant-Film Model..602
    8.3.2 Penetration Model...602
    8.3.3 Surface Renewal Model..604
8.4 Mass Transfer across a Phase Boundary..604
    8.4.1 Two-Film Model..604
    8.4.2 Interfacial Area..607
    8.4.3 Volumetric Coefficients, Gas-Liquid..607
    8.4.4 Volumetric Coefficients, Liquid-Liquid..608
    8.4.5 Evaporation of Spills...611
8.5 Summary..612
Nomenclature 612
References 614

8.1 INTRODUCTION

Mass can be transported in various ways, for example by truck, aircraft, or barge. Here the context is quite different; we shall use the term "mass transfer" to characterize the movement of molecules under the influence of a concentration gradient. Of special concern will be molecular movement across a phase boundary. The primary applications of such interphase transfer will involve gas-liquid, vapor-liquid, liquid-liquid, and fluid-solid transfer for absorption/stripping, distillation, extraction, and adsorption. The principles described can easily be extended to other operations such as leaching and solids drying.

Common to all the above example applications is the relationship known as Fick's first law:

$$J_i = -D_i \frac{\partial C_i}{\partial z} \tag{8.1}$$

where

    J_i = a molar mass flux of species i, g-moles/(s − cm^2)

C_i = concentration of species i, g-moles/cm^3
Z = distance, cm
D_i = a proportionality constant, called the *diffusion coefficient* for species i, cm^2/s

Equation (8.1) is basic and states that the rate of diffusion of a molecule, in mixture with others, moves at a rate directly proportional to its concentration gradient and the cross section through which it diffuses, and is inversely proportional to the distance through which it must travel. It implies steady-state diffusion.* The negative sign in Equation (8.1) is simply to make the flux positive, since the concentration gradient is negative. Variations of Equation (8.1) can be used to allow concentration expressed in mole fraction or mass fraction, for example,

$$J_i = -C_m D_i \frac{\partial y_i}{\partial z} \tag{8.1a}$$

where

J_i = molar mass flux of species i, g-moles/(s·cm^2)
C_m = average total concentration, total g-moles/cm^3
z = distance, cm
y_i = mole fraction of species i (g-moles i/total g-moles)
D_i = proportionality constant, the *diffusion coefficient* for species i, cm^2/s

The value of D depends upon the conditions of transport. For the often-used case of diffusion under nonturbulent conditions, *molecular diffusion* prevails, and D is a *molecular diffusion coefficient*.† As such, its value depends on temperature, pressure, relative size of molecules involved, and in some cases, whether all molecules, including i, are polar. If turbulent conditions prevail, we have an *eddy diffusion coefficient*, usually designated by the symbol ε. The models used in this chapter do not involve the turbulent case, largely because it lacks a firm basis for estimation, i.e., degrees of turbulence are not easily evaluated.

Equation (8.1) has its counterparts for heat flux (Fourier's equation) and momentum flux. Evaluation of D_i is relatively simple for binary systems, i.e., species i moving through, say, species j. If i is moving through several components that can be grouped as a single pseudo component, evaluation of D_i is straightforward. However, if all species in the mixture are diffusing at different rates (and in different directions), one must use the Maxwell-Stefan relationships, which are beyond the coverage of the present chapter. Taylor and Krishna [1] provide extensive coverage of multi-component diffusion.

8.2 MOLECULAR DIFFUSION COEFFICIENTS

8.2.1 GASES—BINARY MIXTURES

Diffusion coefficients for some binary gas mixtures have been measured and are reported in various compendia, such as *Perry's Handbook* [2]. Of concern here are the models available for *estimating* the coefficients, or for extrapolating the values of measured coefficients. A number of predictive models have been presented for the case of binary gas mixtures. The models are based on experimental data, where the movement of one component is measured under carefully controlled laminar conditions. A model combining both accuracy and ease of use is due to Fuller et al. [3]:

---

* Fick's second law deals with transient diffusion, not covered in this chapter.
† The terms *diffusion coefficient* and *diffusivity* are synonymous. The former will be used here.

Mass Transfer

$$D_{ij} = \frac{1.0\left(10^{-3}\right)T^{1.75}}{P\left[\left(\Sigma v\right)_i^{1/3} + \left(\Sigma v\right)_j^{1/3}\right]^2} \left(\frac{1}{M_i} + \frac{1}{M_j}\right)^{1/2} \quad (8.2)$$

where
- D_{ij} = gaseous diffusion coefficient, cm²/s
- T = absolute temperature, °K
- M = molecular weight
- P = absolute pressure, atm
- Σ_v = diffusion volume, the sum of atomic volumes in each molecule; values obtained from Table 8.1.

Equation (8.2) shows that the coefficient is inversely proportional to pressure and proportional to the 1.75 power of absolute temperature. It shows that the relative sizes of the molecules are also important. Equation (8.2) is reported to predict values within ±7% for nonpolar and polar molecules. In general, it is derived from the kinetic theory of gases.

For more careful (and more tedious) work, the model of Hirschfelder-Bird-Spotz [4] is recommended. It involves the Lennard-Jones potential between diffusing species, and that potential can be corrected slightly when both species are polar, using the method of Brokaw [5]. For engineering work, Equation (8.2) is normally adequate. According to the kinetic theory, the gas diffusion coefficient is independent of concentration.

TABLE 8.1
Atomic and Molecular Diffusion Volumes for Equation (8.2)

Atomic and Structural Diffusion Volume Increments, v

C	16.5	(Cl)	19.5
H	1.98	(S)	17.0
O	5.48	Aromatic ring	−20.2
(N)	5.69	Heterocyclic ring	−20.2

Diffusion Volumes, Σv, for Simple Molecules

H_2	7.07	CO_2	26.9
D_2	6.70	N_2O	35.9
He	2.88	NH_3	14.9
N_2	17.9	H_2O	12.7
O_2	16.6	(CCl_2F_2)	114.8
Air	20.1	(SF_6)	69.7
Ar	16.1	(Cl_2)	37.7
Kr	22.8	(Br_2)	67.2
CO	18.9	(SO_2)	41.1

Note: Atoms and molecules in parentheses indicate very few data available.

Source: Reid, R. C., Prausnitz, J. M., and Sherwood, T. K., *The Properties of Gases and Liquids,* 3rd ed., 554. New York: McGraw-Hill, 1977.

Example 8.1

Estimate the gaseous diffusion coefficient for methanol diffusing through water vapor at 25°C and 1.0 atm pressure.

Solution

Equation (8.2) is used, with the following values:

$T = 298°K$
$M_i = 32.0$ (methanol)
$M_j = 18.1$ (water)
$P = 1.0$ atm
$\Sigma v_i = 16.5 + 4(1.98) + 5.48 = 29.9$ (Table 8.1)
$\Sigma v_j = 12.7$ (Table 8.1)

By Equation (8.2),

$$D_{ij} = \frac{1.0(10)^{-3}(298)^{1.75}}{1.0\left[(29.9)^{1/3}+(12.7)^{1/3}\right]^2}\left(\frac{1}{32.0}+\frac{1}{18.1}\right)^{0.5} = 0.213 \text{ cm}^2/\text{s}$$

8.2.2 Liquids—Binary Mixtures

The liquid phase is less predictable than the gas phase, and methods for predicting diffusion coefficients do not have the support that gases have from the kinetic theory of gases. The Stokes-Einstein relationship forms the basis for modeling:

$$D_{ij}^L = \frac{kT}{4\pi r_i \mu_j} \tag{8.3}$$

Wilke and Chang [7] modified this relationship to develop a *dimensional* relationship for the dilute case, i.e., low concentrations of i in j:

$$D_{ij}^{LO} = \frac{7.4(10^{-8})(\phi_j M_j)^{1/2} T}{V_i^{0.6} \mu_j} \tag{8.4}$$

where

D_{ij}^{LO} = liquid diffusion coefficient for dilute solute i in solvent j, cm²/s

ϕ_j = correction factor for associated liquid solvent; $\phi = 1.0$ for nonassociated liquids, $= 2.4$ for water, 1.9 for methanol, and 1.5 for ethanol

M_j = molecular weight of solvent j

T = absolute temperature, °K

μ_j = viscosity of solvent, mPa·s (centipoise)

V_i = molar volume of solution at normal boiling point, cm³/g-mole

Values of V_i may be estimated by means of the LeBas parameters given in Table 8.2.

TABLE 8.2
LeBas Additive Volumes for Wilke-Chang Equation (8.4)

	Increment (cm³/g-mole)
Carbon	14.8
Hydrogen	3.7
Oxygen (except where noted below)	7.4
in methyl esters and ethers	9.1
in ethyl esters and ethers	9.9
in higher esters and ethers	11.0
in acids	12.0
joined to S, P, N	8.3
Nitrogen	
doubly bonded	15.6
in primary amines	10.5
in secondary amines	12.0
Bromine	27.0
Chlorine	24.6
Fluorine	8.7
Iodine	37
Sulfur	25.6
Ring	
three-membered	−6.0
four-membered	−8.5
five-membered	−11.5
six-membered	−15.0
naphthalene	−30.0
anthracene	−47.5

Note: The additive volume procedure should not be used for simple molecules. Instead, use the following values:

H_2	14.3	SO_2	44.8	COS	51.5	O_2	25.6
NO	23.6	Cl_2	48.4	N_2	31.2	N_2O	36.4
Br_2	53.2	Air	29.9	NH_3	25.8	I_2	71.5
CO	30.7	H_2O	18.9	CO_2	34.0	H_2S	32.9

For solute-solvent mixtures that are ideal, values of intermediate concentrations can be estimated on the following basis:

$$D_{ij}^L = x_i D_{ji}^{LO} + x_j D_{ij}^{LO} \tag{8.5}$$

Thus, the two infinite-dilution coefficients are calculated, and intermediate values of the coefficient are based on a linear relationship based on mole fractions. A 50-50 molar mixture would have a coefficient of

$$\frac{D_{ij}^{LO} + D_{ji}^{LO}}{2}$$

For non-ideal liquid mixtures, the Vignes [8] relationship should be used:

$$D_{ij}^L = \left(D_{ij}^{LO}\right)^{x_j} \left(D_{ji}^{LO}\right)^{x_i} \left(1 + x_i \frac{d \ln \gamma_i}{d x_i}\right) \tag{8.6}$$

For an ideal-liquid mixture, this equation shows a linear plot of ln D vs. molar concentration. The term γ_i is the liquid-phase activity coefficient, and the term $d \ln\gamma_i/dx_i$ is the slope of the conventional plot of the logarithm of the activity coefficient versus mole fraction (see Figures 12.7 and 12.9 in Chapter 12, or see Chapter 4). Thus, for the value of the liquid diffusion coefficient of a concentrated binary mixture i-j, Equations (8.5) or (8.6) should be used, depending on ideality of the solution.

When water is the solute, diffusing at very low concentration through a solvent, the dimensional relationship by Sitaraman et al. [9] gives a somewhat better fit than Equation (8.4):

$$D_{ij}^{LO} = 16.8(10^{-10})\left(\frac{M_j^{0.5} \lambda_j^{0.333} T}{\mu_j V_i^{0.5} \lambda_i^{0.3}}\right)^{0.93} \tag{8.7}$$

where
λ_i and λ_j = latent heats of vaporization of the solute and solvent, respectively, at their normal boiling points, J/kg
μ_j = viscosity of the solute, cP

and other terms are the same as for Equation (8.4). Special relationships apply when an organic is diffusing through an ionic solution; for such a case, consult Hirschfelder et al. [4] or Reid et al. [6]

Example 8.2

For a liquid solution of 10 mole-% methanol in water, calculate the diffusion coefficients for each component. Temperature is 25°C, pressure is 1.0 atm, viscosity of the methanol is 0.55 cP, and viscosity of water is 0.95 cP.

Solution

Let i be methanol and j be water. The infinite-dilution coefficient for methanol in water is computed using Equation (8.4). Values of terms are

Association parameters ϕ = 2.6 for water and 1.9 for methanol
Molecular weights M = 18.1 for water and 32.0 for methanol
Molecular volumes (Table 8.2): for methanol = 14.8 + 4(3.7) + 7.4 = 37.0; for water = 18.9

1. Methanol in water

$$D_{ij}^{LO} = \frac{7.40(10)^{-8}\left[(2.6)(18.1)\right]^{1/2}(298)}{(37)^{0.6}(0.95)} = 1.824(10)^{-5} \text{ cm}^2/\text{s}$$

2. Water in methanol

$$D_{ji}^{LO} = \frac{7.40(10)^{-8}\left[(1.9)(32.0)\right]^{1/2}(298)}{(18.9)^{0.6}(0.55)} = 5.36(10)^{-5} \text{ cm}^2/\text{s}$$

3. 10% methanol
Because the mixture is nonideal, a relationship between composition and activity coefficient is needed. An estimate, based on van Laar coefficients $A_{ij} = 0.36$ and $A_{ji} = 0.22$,

$$\ln \gamma_i = \frac{0.83}{\left[1 + 1.627\,x_i/x_j\right]^2}$$

from which, for $x_i = 0.10$,

$$\partial \ln \gamma_i / x_i = -2.0275$$

and for $1 + x_i$,

$$(d\ln\gamma_i/dx_i) = 1 + 0.10\,(-2.0275) = 0.7972$$

Finally, by Equation (8.6),

$$D_{ij,x_i=0.10}^{L} = \left[1.824(10)^{-5}\right]^{0.9}\left[5.36(10)^{-5}\right]^{0.1}(0.7972) = 1.624(10)^{-5} \text{ m}^2/\text{s}$$

Note: For a simple molar average of infinite-dilution coefficients,

$$D_{ij}^{L} = 2.19(10)^{-5} \text{ cm}^2/\text{s}$$

8.2.3 Gases—Multicomponent Mixtures

As mentioned earlier, prediction of the coefficient for species i diffusing through a mixture of $j,k,l,\ldots$ depends on the diffusion characteristics of the other components. For the special case where i is diffusing through other components that are not diffusing, the Wilke [10] relationship may be used:

$$D_{i-\text{mixture}} = \frac{1 - y_i}{\sum_{j}^{n} \dfrac{y_j}{D_{ij}}} \tag{8.8}$$

The components $j,k,l,\ldots$ are considered to be a "stagnant barrier mixture."

Equation (8.8) provides a good approximation for dilute cases, even if the barrier mixture is not "stagnant." For more rigorous approaches, the text by Taylor and Krishna[1] should be consulted.

8.2.4 Liquids—Multicomponent Mixtures

If the mixture is ideal, or nearly so, a modification of the Vignes equation (8.6) may be used, considering the mixture $j,k,l,\ldots$ as a pseudo component. Thus, for a ternary ideal mixture i,j,k,

$$D^{LO}_{i-mixture}\,\mu_{ijk} = \left(D^{LO}_{ij}\,\mu_j\right)^{x_j} \left(D^{LO}_{ik}\right)^{x_k} \tag{8.9}$$

For, say, a ternary non-ideal mixture, activity-coefficient data must be available. In such a case, the result of Equation (8.9) can be multiplied by $(1+x_i(\partial \ln \gamma_i / \partial x_i))$ as in the Vignes equation (8.6). One should recognize, however, that activity-coefficient data for mixtures with more than two components are scarce; the best source for such data is the Gmehling series [11].

8.2.5 Diffusion through Porous Solids

For a gas diffusing through the very small pores of an adsorbent or solid catalyst, a form of Fick's equation (Equation (8.1)), called the slab equation, may be used:

$$J_i = -\frac{D_{ki}}{z_o}\left(C_{i2}-C_{i1}\right) = \frac{D_{ki}}{RT}\left(\frac{(p_1-p_2)_i}{z_o}\right) \tag{8.10}$$

where

z_o = length of pore traveled, cm
R = gas constant, 82.057 (atm·cm³)/ (g-moles·°K)
T = absolute temperature, °K
p_i = partial pressure of i, obtained from concentration using the gas law:

$$C_i = \frac{n_i}{V} = \frac{p_i}{RT} \tag{8.11}$$

and D_{ki} is the *Knudsen diffusion coefficient* for species i, so called because, in the very small and crooked pores, the molecules collide more frequently with the pore wall than with each other. Thus, the value of this coefficient depends more on the pore geometry than on the diffusing species:

$$D_{ki} = D_k = 2/3\,r_e\left(\frac{8RT}{\pi M_i}\right)^{0.5} = 9700\,r_e\sqrt{\frac{T}{M_i}} \;\; \text{cm}^2/\text{s} \tag{8.12}$$

with r_e being the effective pore radius, cm.

The pore size may be known, or it can be estimated:

$$r_e = \frac{2\varepsilon}{S\rho_p} \tag{8.13}$$

where

ε = fraction voids in solid particle
S = total surface area of solid, cm²/g
ρ_p = particle density, g/cm³

Mass Transfer

A correction for D_k is needed for the usual case of a pore that is very crooked, called the "tortuosity factor" τ:

$$D_{k,\text{eff}} = D_k \frac{\varepsilon}{\tau} \tag{8.14}$$

This factor has an observed range of 1 to 7. A good average is $\tau = 4$.

For larger pore sizes, molecular diffusion may play a role, and the molecular diffusion and Knudsen coefficients can be combined:

$$\frac{1}{D_{i,\text{eff}}} = \frac{1}{D_i} + \frac{1}{D_{k,\text{eff}}} \tag{8.15}$$

where $D_i = D_{ij}$ = coefficient determined by Equation (8.2) and $D_{k,\text{eff}}$ is obtained from Equation (8.14).

Example 8.3

Propane (mol. wt. = 44) is diffusing through nitrogen in the pore structure of 4 × 6-mesh Sorb-Tech activated carbon. Temperature is 298°K. Assume a tortuosity factor of 4. The carbon has the following properties [12]:

Surface area = 1096 m²/g = 1.096 × 10⁷ cm²/g
Bulk density = 0.49 g/cm³
Internal void fraction = 0.543
Particle density = 1.00 g/cm³

Solution

First, the effective pore radius is estimated by Equation (8.13):

$$r_e = \frac{2(0.543)}{1{,}096(10)^7 (1.00)} = 0.99(10)^{-10} \text{ cm}$$

Next, the Knudsen coefficient is calculated by Equation (8.12):

$$D_{ki} = 9.7(10)^3 \, 0.99(10)^{-10} \sqrt{298/44} = 25.0(10)^{-7} \text{ cm}^2/\text{s}$$

The coefficient is then corrected for tortuosity by Equation (8.14):

$$D_{ki,\text{eff}} = 25.0(10)^{-7} \frac{0.543}{4} = 3.40(10)^{-7} \text{ cm}^2/\text{s}$$

By Equation (8.2), the molecular diffusion coefficient for propane in nitrogen is 0.113 cm²/s. Because $D_i \gg D_{ki,\text{eff}}$, it is clear that Knudsen diffusion controls.

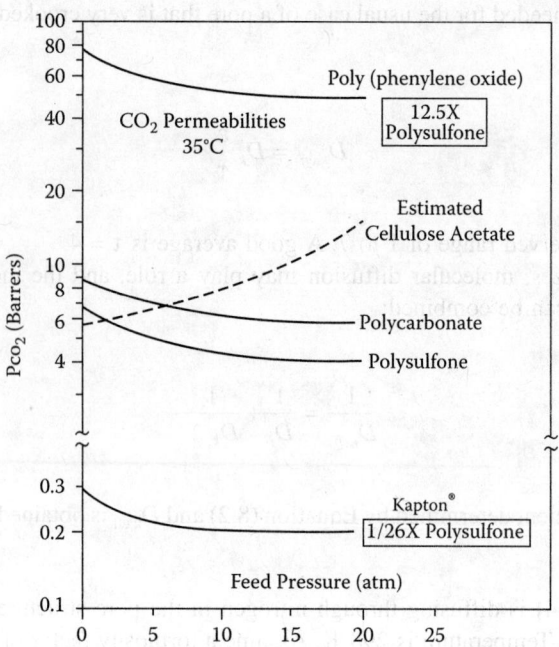

FIGURE 8.1 Permeabilities for carbon dioxide passing through several membrane materials. (From Fair, J. R. *Chem. Proc.* 52 (10): 81 (1989). With permission.)

8.2.6 GASES—DIFFUSION THROUGH MEMBRANES

Again, Fick's law is used, with the length term equal to the membrane thickness. The diffusion coefficient is usually called a *permeability* and is measured by observing the flux J_i for a given membrane material coupled with a known pressure difference on either side of the membrane. For a semipermeable membrane, the permeability P is a function of both a mobility term and a solubility term, i.e., $P = DS$. Because the separating power of a membrane is based on different values of P for each component, it follows that the selectivity term α_{ij} is

$$\alpha_{ij} = \frac{P_i}{P_j} = \frac{D_i}{D_j}\frac{S_i}{S_j} = \left(\frac{\text{mobility}}{\text{selectivity}}\right)\left(\frac{\text{solubility}}{\text{selectivity}}\right) \qquad (8.16)$$

To indicate the influence of the membrane material on a diffusing species, Figure 8.1 shows permeabilities for carbon dioxide moving through different membranes under the influence of a pressure-driving force [13]. The permeability is calculated from measured flux:

$$P_{CO_2} = \frac{\text{Std. cm}^3 \text{ diffusing}}{\text{s} - \text{cm}^2 - \text{cm Hg driving force}}$$

Researchers in membrane technology have adopted the special permeability unit called the Barrer:

$$1.0\,\text{Barrer} = \frac{10^{10}\left(\text{cm}^3 \text{ @ STP}\right)(\text{cm})}{\text{cm}^2 (\text{sec})(\text{cm Hg})}$$

In some cases, two or more different materials are sandwiched to form the membrane. A combined permeability follows the same style as Equation (8.15).

8.3 MASS TRANSFER AT A PHASE BOUNDARY

For molecules approaching a phase boundary (gas-liquid, liquid-liquid, gas-solid, liquid-solid), movement is represented by Fick's law (Equation (8.1)) or, more practically, the slab equation [Equation (8.10)]. Examples of application are

1. A species in a *gas mixture moving to a boundary with a liquid* where, if the species is soluble in the liquid, the species moves across the phase boundary and into the liquid. This is a type of unidirectional mass transfer found in absorption separations.
2. A species in a *liquid mixture moving to a boundary with an immiscible liquid* where, if soluble, the species passes into the immiscible liquid. This is unidirectional mass transfer found in liquid-liquid extraction separations.
3. A species in a *gas mixture moving to a boundary with a solid* where, if appropriate, it moves into the structure of the solid. The solid can be an adsorbent or a catalyst, for example. Diffusion within the pore structure may be a consideration. Transfer may be unidirectional or counterdiffusional, for example, if products are leaving a solid catalyst.
4. A species in a *liquid mixture moving to a boundary with a vapor* and passing countercurrently to another species, leaving the vapor and passing into the liquid. This is often equimolar counterdiffusion, as found in distillation.

For the counterdiffusion steady-state case, as in distillation, the slab equation is

$$J_i = \frac{D_i}{z_o}(C_{i1} - C_{i2}) = \frac{D_i}{RT\,z_o}(p_{i1} - p_{i2}) = \frac{D_i\,P}{RT\,z_o}(y_{i1} - y_{i2}) = \frac{D_i^L\,C_i}{z_o}(x_{i1} - x_{i2}) \quad (8.17)$$

where the partial-pressure driving forces denote a gas and the mole-fraction driving forces denote a gas (y) or liquid (x).

For unidirectional steady-state mass transfer, as in absorption, adsorption, and stripping,

$$J_i = \frac{D_i}{z_o}\left(\frac{C_{i1} - C_{i2}}{C_{jm}}\right) = \frac{D_i\,P}{RT\,z_o}\left(\frac{p_{i1} - p_{i2}}{p_{jm}}\right) = \frac{D_i\,P}{RT\,z_o}\left(\frac{y_{i1} - y_{i2}}{y_{jm}}\right) = \frac{D_i^L\,C_i}{z_o}\left(\frac{x_{i1} - x_{i2}}{x_{jm}}\right) \quad (8.18)$$

where C_{jm}, p_{jm}, y_{jm}, and x_{jm} include all components other than i in the mixture. The differences between Equations (8.17) and (8.18) should be noted carefully. For "lean gas" or "lean liquid" cases, where the concentration of the diffusing component i is very low, the fluxes have about the same value. The difference has to do with the assignment of a stationary diffusional plane (equimolar transfer) or a moving plane (unidirectional transfer) and the phenomenological derivation of the two modes; this is beyond the scope of the present work. Flux designations of J and N are sometimes used to denote the difference. For the derivations, the reader may consult works of Pigford et al. [14], Hines and Maddox [15], or other basic texts on mass transfer.

Given Equations (8.17) and (8.18), the question arises as to how they can be applied to practical operations such as those noted (distillation, absorption, adsorption, etc.). The problem is in assigning a value to the distance z_0. The driving force can be calculated, and the diffusion coefficient can be estimated, as just discussed. The determination of z_0 is the basis for several theories, discussed in the following sections.

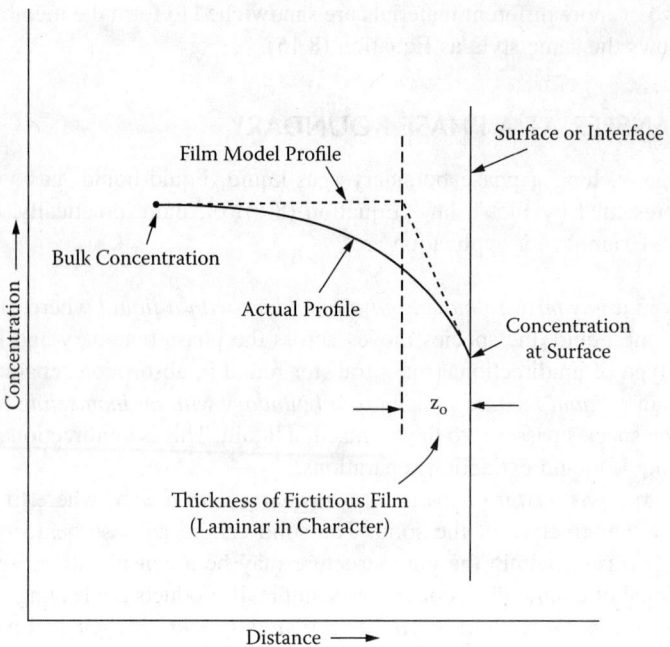

FIGURE 8.2 Elements of the film theory.

8.3.1 Stagnant-Film Model

This model, developed by Whitman [16] in 1923, is still used. The assumption is that adjacent to the interface there exists a thin film that is laminar in character and through which transfer is by molecular diffusion only. In this film, the entire concentration driving force exists, i.e., outside the film, in the bulk fluid, the concentration is constant at some steady state. Flow outside the film may be laminar or turbulent, but inside the film, flow conditions are completely laminar. Figure 8.2 shows the elements of the theory.

It is possible in some cases to measure the thickness of the film, which is akin to the boundary layer in turbulent flow. In most cases, however, this is not practical, if indeed possible. Accordingly, the quotient D/z_0 is simply called a *mass transfer coefficient*. Equations for the different mass transfer coefficients are shown in Table 8.3.

The coefficient is subject to certain variables, not all quantifiable:

1. The coefficient increases or decreases according to the variables affecting the diffusion coefficient, Equations (8.2) or (8.4).
2. The film is subject to the fluid mechanics of the bulk mixture. For example, high bulk fluid velocities can thin the film and thus increase the value of the coefficient.

Even if the flow of the bulk fluid is laminar, there is still considered to be a separate film at the interface.

One must be very careful about the units used. Those shown in Table 8.3 are typical, but many other combinations can be used.

8.3.2 Penetration Model

In searching for a more realistic model than one containing a "fictitous film," Higbie [17] proposed the elimination of the film idea and replacing it with an unsteady-state model that permits turbulent conditions adjacent to the interface. Sometimes called the "Higbie model," it contains the following elements:

TABLE 8.3
Mass Transfer Coefficient Definitions

	Driving Force	Equimolar	Unidirectional	Units	Typical Units	
Gases	p	$k'_g = \dfrac{D_{ij}P}{RTz_o}$	$k_g = \dfrac{D_{ij}P}{RTz_o p_{jm}}$	$\dfrac{\text{moles}}{\text{time-area-pressure}}$	$\dfrac{\text{gm-moles}}{\text{s-cm}^2\text{-atm}}$	
Gases	y	$k'_y = \dfrac{D_{ij}P}{RT z_o}$	$k_y = \dfrac{D_{ij}P}{RTz_o y_{jm}}$	$\dfrac{\text{moles}}{\text{time-area-molefraction}}$	$\dfrac{\text{gm-moles}}{\text{s-cm}^2}$	
Gases	C	$k'_c = \dfrac{D_{ij}}{z_o}$	$k_c = \dfrac{D_{ij}P}{z_o p_{jm}}$	$\dfrac{\text{moles}}{\text{time-area-moles/volume}}$	$\dfrac{\text{gm-moles}}{\text{s-cm-gm}^2\text{ moles/cm}^3} = \dfrac{\text{cm}}{\text{s}}$	
Liquids	C	$k'_L = \dfrac{D_{ij}^L}{z_o}$	$k_L = \dfrac{D_{ij}^L}{z_o x_{jm}}$	$\dfrac{\text{moles}}{\text{time-area-moles/volume}}$	$\dfrac{\text{gm moles}}{\text{s-cm-gm}^2\text{ moles/cm}^3} = \dfrac{\text{cm}}{\text{s}}$	
Liquids	x	$k'_x = \dfrac{D_{ij}^L C_i}{z_o}$	$k_x = \dfrac{D_{ij}^L C_i}{z_o x_{jm}}$	$\dfrac{\text{moles}}{\text{time-area-mole fraction}}$	$\dfrac{\text{gm moles}}{\text{s-cm}^2}$	

- There are no films.
- Eddies bring "packets" of material to the interface, where they are exposed for a short time.
- All exposure times are the same.
- Mass transfer occurs only during exposure.

While Higbie proposed the model only for the liquid phase, it has been extended to include the gas phase. To satisfy Fick's theory, Higbie assumed that, at the time of exposure, the transfer process involved only molecular diffusion or, alternatively, the packets assumed a laminar character as they approached the interface. The net result of Higbie's model provides the flux equation:

$$J_i = 2\left(C_{iB} - C_{i,\text{int}}\right)\sqrt{\dfrac{D_{ij}}{\pi \theta_i}} \tag{8.19}$$

where
J_i = flux of i, g-moles/s·cm² of interface
$C_{i,B}$ = bulk concentration of i, g-moles/cm³
$C_{i,\text{int}}$ = interfacial concentration of i, g-moles/cm³
D_{ij} = molecular diffusion coefficient, cm²/s
π = 3.1416...
θ_i = exposure time of i at the interface, s

Thus, the mass transfer coefficient is $2\sqrt{D_{ij}/\pi\theta_i}$, cm/s.

One apparent advantage of this model is that it predicts a dependence of flux on the one-half power of the diffusion coefficient, which is closer to reality than the first power for the molecular diffusion model. And it is more realistic, in that eddy-type packets are considered. However, it does not have an application advantage, since there is little physical basis for estimating the exposure time, let alone a range of exposure times, just as there is little experimental support for the stagnant film of finite thickness.

8.3.3 Surface Renewal Model

This model, proposed originally by Danckwerts [18], is an extension of the penetration model. Whereas Higbie assumed that all exposure times were the same, Danckwerts provided for a range of times, based on probability theory. After a given exposure, the surface (interface) was renewed, leading to the name of the theory. The elements of the model are

- There are no films.
- Eddies bring packets to the interface, remaining there for a variety of times.
- The surface (interface) is renewed on the basis of fluid flow and geometry.
- Mass transfer occurs only during exposure.

The flux equation by this model is

$$J_i = \left(C_{iB} - C_{i,\text{int}}\right)\sqrt{S D_{ij}} \tag{8.20}$$

where S is the fractional surface renewal rate, s^{-1}; and the mass transfer coefficient is $\sqrt{S D_{ij}}$, cm/s. In reality, the difference between the penetration and surface-renewal models is the factor of 2 in the flux equation. The models are used interchangeably, with little attempt to predict the range of exposure times. Use of these models is described in the following section.

8.4 MASS TRANSFER ACROSS A PHASE BOUNDARY

The practical applications provided here all involve two phases, with molecules transferring between them. Thus, there are two resistances to transfer, plus possibly a third resistance at the interface itself. We have just discussed transfer within a phase and ending at a phase boundary, such as an interface. It is necessary to couple individual phase resistances to characterize the overall transfer process. The first attempt at this, and indeed a lasting one, was presented by Lewis and Whitman [19] as the *two-film theory*. More recently it has been called simply the *two-resistance theory*, eliminating the requirement that transport in each phase be handled by the film concept.

8.4.1 Two-Film Model

To describe this model, let us assume that a species is transferring from a gas phase to a liquid phase, as in gas absorption. Figure 8.3 diagrams the process. If the transfer flux is based on a unit area of interface, we may write the total flux equation as follows:

$$J_i = k_{y,i}\left(y_{ib} - y_{i,\text{int}}\right) = k_{x,i}\left(x_{i,\text{int}} - x_{ib}\right) \tag{8.21}$$

[gas]   [liquid]

where
 J_i = mass flux of i, g-moles/s·cm² interfacial area
 k_x, k_y = mass transfer coefficients for gas and liquid, respectively, defined in Table 8.3
 y_{ib} = mole fraction of i in the bulk gas phase
 $y_{i,\text{int}}$ = mole fraction of i in the gas, at the interface
 $x_{i,\text{int}}$ = mole fraction of i in the liquid, at the interface

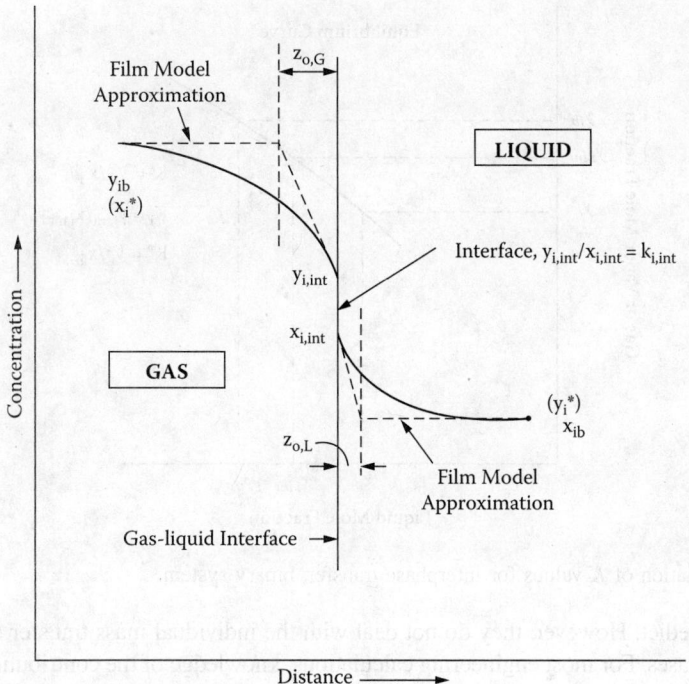

FIGURE 8.3 Two-film theory applied to unidirectional flow, gas to liquid.

x_{ib} = mole fraction of i in the bulk liquid

To make Equation (8.21) tractable for design purposes, it is necessary to make the assumption that *thermodynamic equilibrium exists at the interface*, i.e., $y^*_{i,\text{int}}/x_{i,\text{int}} = K_i$ or $y_{i,\text{int}}/x^*_{i,\text{int}} = K_i$, where K_i is the conventional thermodynamic equilibrium ratio and the asterisks denote an equilibrium composition. Note that the equilibrium designation depends on the concentration units used. For example, if gas concentrations are in partial pressures and liquid compositions are in molar concentrations, a Henry's law coefficient, $H_i = p^*_{i,\text{int}}/C_{i,\text{int}}$, must be used.

It is also expedient to define a fictitious liquid composition in equilibrium with the bulk gas concentration, i.e., $x_{ib}^* = y_{ib}/K'$, and to define a fictitious gas composition in equilibrium with the bulk liquid composition, $y_i^* = K'' x_{ib}$. Because the K value is composition dependent, values of K, K', and K'' will not necessarily be the same. Figure 8.4 illustrates this difference on a y versus x plot.

The assumption of interfacial equilibrium permits definition of overall mass transfer relationships for transferring species i:

$$J_i = K_{ox}(x_i^* - x_{ib}) \tag{8.22a}$$

$$= K_{oy}(y_{ib} - y_i^*) \tag{8.22b}$$

where
K_{ox} = overall coefficient based on liquid concentrations
K_{oy} = overall coefficient based on gas concentrations

These expressions cover the overall concentration change, bulk to bulk, and take into account the total resistance to mass transfer offered by both phases. They are useful in that they eliminate the need for interfacial concentrations, which are rarely available and are difficult if not impossible to

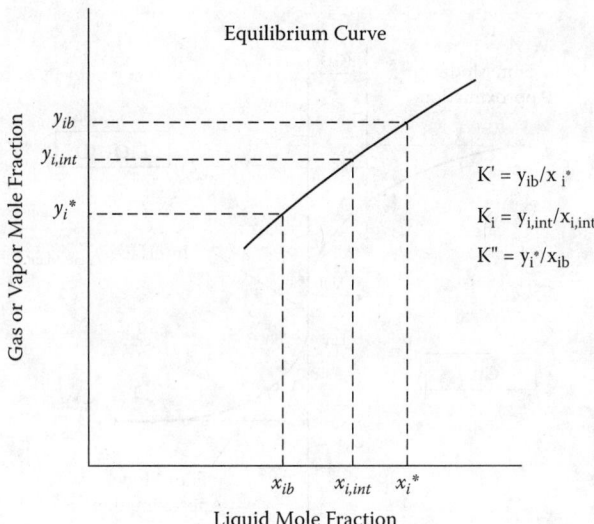

FIGURE 8.4 Variation of K values for interphase transfer, binary system.

measure or to predict. However, they do not deal with the individual mass transfer characteristics of the individual phases. For most engineering calculations, knowledge of the contributions of the phases to mass transfer are needed, obtainable from the following relationship for transferring species i:

$$\frac{1}{K_{ox}} = \frac{1}{K_i k_y} + \frac{1}{k_x} \tag{8.23}$$

$$\frac{1}{K_{oy}} = \frac{1}{k_y} + \frac{K_i}{k_x} \tag{8.24}$$

Equations (8.23) and (8.24) are equivalent and represent a series resistance expression:

$$R_{total} = R_{gas} + R_{liquid}$$

When essentially all of the resistance is on the gas side of the interface, $k_x \gg k_y$, and $K_{oy} \approx k_y$ or $K_{ox} \approx k_y K_i$. A well-known example of pure gas-phase resistance is the humidification of dry air with pure water. Because there is no concentration gradient in the liquid, the interface concentration equals the vapor pressure of water at the temperature under consideration. Thus the only driving force is $p^{vap} - p_b$, where p_b = partial pressure of water in the air, equivalent to the humidity. Many distillation separations have most of the resistance in the vapor phase.

Conversely, when essentially all of the resistance is on the liquid side, $K_{oy} \approx k_g/K_i$ or $K_{ox} \approx k_x$. For test work, this situation is simulated by the stripping of oxygen from water by air. Because of the limited solubility of oxygen in water, the value of K_i is very high, and the first term on the right-hand side of Equation (8.23) is essentially zero. This is the situation for many strippers, where the solute is relatively insoluble in the solvent.

Equations (8.22)–(8.24) and their associated discussion refer to a single diffusing component. Further, they are based only on one set of symbols from those given in Table 8.3. Equations with other symbols are included in Table 8.4.

TABLE 8.4
Gas-Liquid Mass Transfer

$$J_i = k_g \left(p_{ib} - p_{i,\text{int}} \right) = k_x \left(x_{i,\text{int}} - x_{ib} \right)$$

$$\frac{1}{K_{og}} = \frac{1}{k_g} + \frac{H_i}{k_x} \qquad H_i = \text{Henry's law coefficient}, = \frac{p_{ib}}{x_{ib}}$$

$$\text{or} \quad \frac{1}{K_{oL}} = \frac{1}{H_i k_g} + \frac{1}{k_x}$$

$$J_i = k_y \left(y_{ib} - y_{i,\text{int}} \right) = k_c \left(C_{i,\text{int}} - C_i \right)_L$$

$$\frac{1}{K_{oy}} = \frac{1}{k_y} + \frac{H_i'}{k_c} \qquad H_i' = \text{Henry's law coefficient} = \frac{y_{ib}}{C_{ib}}$$

$$\text{or} \quad \frac{1}{K_{oL}} = \frac{1}{k_y H_i'} + \frac{1}{k_c}$$

$$J_i = k_y \left(y_{ib} - y_{i,\text{int}} \right) = k_x \left(x_{i,\text{int}} - x_{ib} \right)$$

$$\frac{1}{K_{oy}} = \frac{1}{k_y} + \frac{K_i}{k_c} \qquad K_i = \text{Equilibrium ratio} = \frac{y_{ib}}{x_{ib}}$$

$$\text{or} \quad \frac{1}{K_{ox}} = \frac{1}{k_y K_i} + \frac{1}{k_x}$$

8.4.2 Interfacial Area

The mass flux is based on a unit cross-sectional area, for example, at the interface. For a few simple contacting geometries the interfacial area may be known, and some of these are discussed later. For most situations, particularly when significant turbulence is present, the area is not known, and methods for its estimation are unreliable. In such cases, the area is coupled with the mass transfer coefficient to form a *volumetric mass transfer coefficient*.

8.4.3 Volumetric Coefficients, Gas-Liquid

When the mass transfer coefficient is combined with the interfacial area per unit volume, a volumetric mass transfer coefficient results with typical units,

$$k_y a \text{ or } k_x a = \frac{\text{gm moles}}{\text{s} - \text{cm}^2 - \text{mole fraction}} \times \frac{\text{cm}^2}{\text{cm}^3} = \frac{\text{gm moles}}{\text{s} - \text{cm}^3 - \text{mole fraction}}$$

The two-film, or two-resistance, model then results:

$$\frac{1}{K_{ox}a} = \frac{1}{K_i k_y a} + \frac{1}{k_x a} \qquad (8.25)$$

$$\frac{1}{K_{oy}a} = \frac{1}{k_y a} + \frac{K_i}{k_x a} \qquad (8.26)$$

where a is the interfacial area per unit volume.

For design purposes, representative expressions for total transfer, considering resistance of both phases, are

Absorption, gas to liquid:

$$N_i = K_{oy,i} a \left(y_i - y_i^* \right) = K_{ox,i} a \left(x_i^* - x_i \right) \qquad (8.27)$$

Stripping, liquid to gas:

$$N_i = K_{oy,i} a \left(y_i^* - y_i \right) = K_{ox,i} a \left(x_i - x_i^* \right) \qquad (8.28)$$

In Equations (8.27 and 8.28), N_i is the total moles/time of i transferred, and a is the effective interfacial area at the phase boundary. Individual phase resistances are handled by Equations (8.25) or (8.26).

Distillation, equimolar counterdiffusion:

$$N_i = K_i' a \left(y_i - y_i^* \right) \qquad (8.29)$$

By convention, distillation calculations are based on the gas (vapor) phase, since usually most of the resistance is in that phase. Note also that since components are diffusing in both directions, the coefficient has a slightly different definition (see Table 8.3 for the individual mass transfer coefficients, which are primed). Further, the equilibrium ratio is replaced by the slope of the equilibrium curve, m. The overall coefficient is, then,

$$K_{oy}' a = \frac{1}{\dfrac{1}{k_{oy}' a} + \dfrac{m}{k_x' a}} \qquad (8.30)$$

For a binary mixture, a typical y^*, x graph is shown in Figure 8.5. It is clear that with a change in location within the distillation column, the slope varies, being relatively large at the bottom. The liquid-side resistance to mass transfer is thus greater at the bottom than at the top, since the $m/k_x' a$ term represents the liquid-phase resistance.

8.4.4 VOLUMETRIC COEFFICIENTS, LIQUID-LIQUID

For two immiscible liquids in contact, a phase boundary occurs, and it is possible to transfer a solute I between them. The analogy between liquid-liquid and gas-liquid systems is apparent, and the equations for gas-liquid mass transfer apply. The phase with the lower density is called the

Mass Transfer

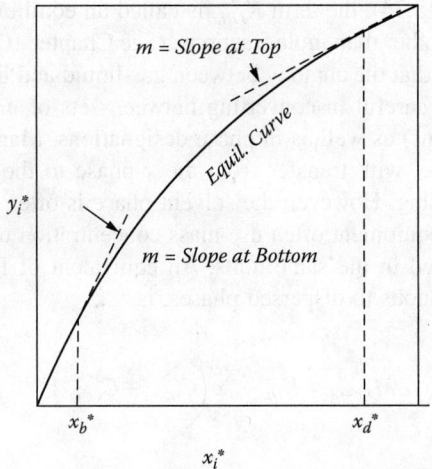

FIGURE 8.5 Representative equilibrium curve for binary distillation.

light phase, and the phase with the higher density is called the *heavy phase*. In design work, especially for liquid-liquid extraction, a distinction can be made between the dispersed phase and the continuous phase and between the extract (solvent) phase and the raffinate phase. These latter distinctions need not enter into the present discussion. For convenience in the analogy, we shall designate the light phase as the y-phase and the heavy phase as the x-phase.

Diffusion coefficients for the phases are calculated by methods given earlier. For the two-film model, equilibrium at the interface is assumed. In some cases, the interfacial area may be estimated, e.g., if a minor phase is dispersed such that discrete drops of a given size may be assumed, and if the fractional holdup of that phase (in the other phase) can be determined, then the use of volumetric coefficients is not necessary. Although such instances are relatively rare, and are related to the type of contactor used, they can be handled in the same fashion as gas-liquid mass transfer.

The liquid-liquid equivalents of Equations (8.27) and (8.28) are

Light phase to heavy phase:

$$N_i = K_{oy,i} a \left(y_i - y_i^* \right) = K_{ox,i} a \left(x_i^* - x_i \right) \tag{8.31}$$

Heavy phase to light phase:

$$N_i = K_{oy,i} a \left(y_i^* - y_i \right) = K_{ox,i} a \left(x_i - x_i^* \right) \tag{8.32}$$

and, with subscripts L and H (light and heavy) substituted for y and x,

$$\frac{1}{K_{oL,i} a} = \frac{1}{K_{L/H,i} k_{L,i} a} + \frac{1}{k_{H,i} a} \tag{8.33}$$

$$\frac{1}{K_{oH,i} a} = \frac{1}{k_{L,i} a} + \frac{K_{L/H,i}}{k_{H,i} a} \tag{8.34}$$

In Equations (8.33) and (8.34), the term $K_{L/H,i}$ is called an equilibrium *distribution coefficient*, and is often carried in terms other than mole fractions. (See Chapter 10 on liquid-liquid extraction.) The point to be made here is that the analogy between gas-liquid and liquid-liquid systems applies. However, one must be very careful in converting between sets of units (concentrations or mass fractions versus mole fractions) as well as in phase designations. Many designers call the y phase the solvent (or extract) phase, with transfer from the x phase to the y phase, with the stripping analogy in gas-liquid contacting. However, the solvent phase is often the light phase.

Designers of extraction equipment often use mass concentration units, which requires that the equilibrium ratio be expressed in the same units. An equivalent of Equations (8.31) and (8.32), using movement from continuous to dispersed phases, is

$$M_{i,c \to d} = K_{OL,cont}\, a \left(C'_{i,cont} - C'^{*}_{i,cont} \right) \qquad (8.35)$$

where
$\quad M$ = g/(s-cm^3) transferred
$\quad a$ = interfacial area, cm^2/cm^3
$\quad K_{OL,cont}$ = overall coefficient for the diffusing species (i.e., solute), cm/s
$\quad C'$ = mass concentration in bulk continuous phase, g/cm^3
$\quad C'^{*}$ = continuous phase concentration in equilibrium with bulk dispersed phase, i.e., $C'^{*} = m_{cd}\, C'_{disp}$
$\quad m_{cd}$ = equilibrium distribution coefficient, g/g

Example 8.4 has been included to help clarify the extraction adaptation of the gas-liquid relationships.

Example 8.4

Acetone is being removed from water by contacting with toluene in a packed extraction column. Water is the continuous [heavy] phase, and toluene is the dispersed [light] phase. Calculate the overall mass transfer coefficient for this process.

Data:
$\quad$ Continuous phase: density = 0.994 g/cm^3, viscosity = 0.92 cP, molecular weight ≈ 18 (lean solution)
$\quad$ Dispersed phase: density = 0.860 g/cm^3, viscosity = 0.54 cP, molecular weight ≈ 92 (lean solution)
$\quad$ Molecular weight of solute = 58
$\quad$ Distribution coefficient (equivalent to equilibrium ratio) = C_{disp}/C_{cont} = 0.67

Solution

Seibert and Fair [20] have found, for the system in question and for the particular packing type and flow rates, effective interfacial area = a = 1.46 cm^2/cm^3, and the following phase mass transfer coefficients:

$k_{L,cont}$ = 4.20 × 10^{-3} cm/s
$k_{L,disp}$ = 9.81 × 10^{-3} cm/s

Because concentrations are used, the equilibrium ratio is also in terms of concentrations. Thus,

$$K_{oL,cont} = \frac{1}{k_{L,cont}} + \frac{1}{m_{DC} k_{L,disp}} = \frac{1}{4.20 \times 10^{-3}} + \frac{1}{0.67(9.81 \times 10^{-3})} = 2.56 \times 10^{-3} \text{ cm/s}$$

$$K_{oL,cont} a = (1.46)(2.56 \times 10^{-3}) = 3.73 \times 10^{-3} \text{ s}^{-1}$$

8.4.5 Evaporation of Spills

An important application of mass transfer principles deals with the evaporation of a hazardous liquid after an inadvertent spill. Here, the liquid must vaporize and diffuse through the adjacent air. The slab equation (Equation (8.17)), with partial-pressure driving force, may be used:

$$J_i = \frac{D_{ij}}{RT\, z_o}(p_{i1} - p_{i2}) = \frac{k_{ci}}{RT}\left(p_i^{sat} - p_{i2}\right) \tag{8.36}$$

where k_{ci} is a mass transfer coefficient for the liquid transferring into the air. If the evaporating liquid is relatively pure, there is no liquid-side resistance to mass transfer. The driving force is thus the vapor pressure of the liquid, and the downstream partial pressure can be neglected. The flux J_i is the amount of liquid evaporated per unit time per area of the spill.

Studies of water evaporation have shown the mass transfer coefficient to be a simple function of wind velocity U_j [21]:

$$k_{water,g} = 0.67 U_j^{0.78} \tag{8.37}$$

By film theory, the wind velocity affects the film thickness z_0 in Equation (8.36).

For liquids other than water, the U.S. Environmental Protection Agency recommends the simple ratio [22]:

$$\frac{k_{ci}}{k_{water,gas}} = \left(\frac{M_{water}}{M_i}\right)^{1/3} \tag{8.38}$$

which is an approximation of the diffusion coefficient ratio in Equation (8.36).

A final design equation then results:

$$J_i = \frac{0.67 U_j^{0.78} (18/M_i)^{1/3} p_i^{sat}}{RT} \tag{8.39}$$

which of course can be simplified by combining the constant terms. If the amount and area of the spill are known, Equation (8.39) can be used to determine the time required for essentially complete vaporization.

Example 8.5

A leak of toluene generates a spill area of 3.0 m². The temperature is 32°C, and the average wind velocity is 0.5 m/s. What is the rate of evaporation of the toluene?

Data:
Toluene mol. wt. = 92, and its vapor pressure = 40 mm Hg, or 0.053 atm.

Solution
By Equation (8.39),

$$J_i = \frac{0.67(0.5)^{0.78}(18/92)^{1/3} \, 0.053}{(82.057 \times 305.2)} = 4.79 \times (10^{-7}) \frac{\text{gm} - \text{moles}}{\text{s} - \text{cm}^2}$$

For 3×10^4 cm² and converting to familiar units, rate of evaporation = 0.175 lb/min.

8.5 SUMMARY

This chapter deals with the diffusional transfer of mass to and across a phase boundary. In particular, gas-liquid, gas-solid, and liquid-liquid phase combinations have been considered. Process applications include absorption, stripping, distillation, extraction, adsorption, and the diffusional aspects of chemical reactions on a solid surface. For steady-state transfer operations, the rates of mass transfer can be correlated by variations of Fick's first law, which states that the rate is directly proportional to the concentration driving force and the extent of interfacial area, and inversely proportional to the distance of movement of the mass to the interface.

To make Fick's law tractable for engineering calculations, several theories, or models, have been discussed. Their application usually introduces some empiricism, since distance of diffusion and extent of interfacial area are usually indeterminate. For most calculations, the time-tested film model is appropriate, not because it represents with validity the physical situation, but rather because it is supported by a wealth of experience as well as the ready availability of data, particularly molecular diffusion coefficients. The penetration and surface-renewal models have specialized applications and also depend on empirically derived parameters.

Not all possible applications of mass transfer theory have been discussed, and multicomponent systems have been treated as pseudo binary or ternary systems. To delve deeper, the reader should consult specialized books, some of which are listed in the References section.

NOMENCLATURE

a	interfacial area, s^{-1}
c_m	mean concentration, g-moles/cm³
C	concentration, g-moles/cm³
C_m	average total concentration, g-moles/cm³
D	molecular diffusion coefficient of gas, cm²/s
D^L	molecular diffusion coefficient, liquid phase, cm²/s
D^{LO}	molecular diffusion coefficient of liquid at infinite dilution, cm²/s
H	Henry's law coefficient, p/x or y/C
i	diffusing species i
j	diffusing species j
J	diffusion flux for unidirectional transfer, g-moles/s·cm²
k	diffusing species k
k	Boltzmann constant, Equation (8.3)
K	overall mass transfer coefficient

K	gas-liquid or vapor-liquid equilibrium ratio
l	diffusing species l
m	slope of equilibrium line
M	molecular weight
N	diffusion flux for equimolar transfer, g-moles/cm^2·s
p	partial pressure, atm
P	total (absolute) pressure, atm
r	pore radius, cm
R	gas constant
s	seconds
S	fractional surface renewal, s^{-1}
S	total surface area of solid, cm^2/g (Equation (8.13))
T	absolute temperature, °K
V	molar volume, cm^3/g-mole
x	liquid mole fraction
y	gas or vapor mole fraction
z	diffusion distance, cm
z_0	thickness of film, cm

Greek letters

γ	thermodynamic activity coefficient
ε	internal void fraction of solid
θ	exposure time, s
λ	latent heat of vaporization, J/kg (Equation (8.7))
μ	viscosity, cP
π	3.1416…
ρ	Density, g/cm^3
Σ_v	diffusion volume (Equation (8.2))
τ	tortuosity factor, dimensionless
ϕ	correction factor for liquid association (Equation (8.4))

Subscripts

b	bulk basis
eff	effective
G	gas
i	component i
int	interface basis
L	liquid
x	liquid basis, mole fractions
y	gas basis, mole fractions

Superscripts

*	equilibrium
G	gas
L	liquid

REFERENCES

1. Taylor, R., and Krishna, R. *Multicomponent Mass Transfer.* New York: John Wiley, 1993.
2. Perry, R. H., and Green, D. W. *Perry's Chemical Engineers' Handbook*, 7th ed., 328–332. New York: McGraw-Hill, 1997.
3. Fuller, E. N., Schettler, P. D., and Giddings, J. C. *Ind. Eng. Chem.* 58 (1996): (5) 19.
4. Hirschfelder, J. O., Bird, R. B., and Spotz, E. L. *Chem. Rev.* 44 (1949): 205.
5. Brokaw, R. S. *Ind. Eng. Chem. Proc. Des. Devel.* 8 (1969): 240.
6. Reid, R. C., Prausnitz, J. M., and Sherwood, T. K., *The Properties of Gases and Liquids,* 3rd ed., 554. New York: McGraw-Hill, 1977.
7. Wilke, C. R., and Chang, P. *AIChE J.* 1 (1955): 264.
8. Vignes, A. *Ind. Eng. Chem. Fundam.* 5 (1966): 189.
9. Sitaraman, R., Ibraham, S. H., and Kuloor, N. R. *J. Chem. Eng. Data* 8 (1963): 198.
10. Wilke, C. R. *Chem. Eng. Prog.* 45 (1949): 218.
11. Gmehling, J., Onken, U., and Arlt, W. *Vapor-Liquid Equilibrium Data Collection.* Frankfurt/Main, Germany: Dechema, 1977.
12. Alvarez-Trevit, J. A. "Steam Regeneration of Activated Carbon Adsorbents," Ph.D. diss., The University of Texas at Austin, 1995.
13. Fair, J. R. *Chem. Proc.* 52 (1989): (10) 81.
14. Sherwood, T. K., Pigford, R. L., and Wilke, C. R. *Mass Transfer.* New York: McGraw-Hill, 1975.
15. Hines, A. L., and Maddox, R. N. *Mass Transfer—Fundamentals and Applications*, 20–25. Englewood Cliffs, NJ: Prentice-Hall, 1985.
16. Whitman, W. G., *Chem. Met. Eng.* 29 (1923): 146.
17. Higbie, R., *Trans. AIChE* 31 (1935): 365.
18. Danckwerts, P. V. *Ind. Eng. Chem.* 43 (1951): 1460.
19. Lewis, W. K., and Whitman, W. G. *Ind. Eng. Chem.* 16 (1924): 1215.
20. Seibert, A. F., and Fair, J. R. *Ind. Eng. Chem. Res.* 27 (1988): 470.
21. Mackay, D., and Matsu, R. S. *Can. J. Chem. Eng.* 51(1973): 433.
22. Peress, J. *Chem. Eng. Prog.* 99 (2003): (4) 32.

9 Industrial Mixing Technology

Douglas E. Leng, Sanjeev S. Katti, and Victor Atiemo-Obeng

CONTENTS

9.1	Introduction and Overview	617
	9.1.1 Turbulent, Transitional, and Laminar Mixing	617
	9.1.2 Process Effects	617
	9.1.3 Equipment and Design	618
9.2	Symbols, Dimensionless Groups, and Terms Used in Mixing	619
	9.2.1 Common Symbols Used in Mixing	619
	9.2.2 Common Dimensionless Groups Used in Mixing Correlations	620
	9.2.3 Significance of Commonly Used Mixing Terms	620
9.3	Stirred Tanks: Equipment and Function	623
	9.3.1 Impellers and Their Characteristics	623
	9.3.2 Power	624
	9.3.3 Multiple Impellers	625
	9.3.4 Flow Discharge	626
	9.3.5 Baffles	626
	9.3.6 Vessel Shape	628
	9.3.7 Mechanical Considerations	628
	9.3.8 Bottom Clearance	629
	9.3.9 Vendor's Role	629
9.4	Blending	630
	9.4.1 Introduction and Scope	630
	9.4.2 Turbulent, Transitional, and Laminar Flow Blending	630
	9.4.3 The Nature of Turbulent Flow	632
	9.4.4 Shear Rates	633
	9.4.5 Flow Patterns	633
	9.4.6 Mixing Time—Turbulent and Transitional Flows	635
	9.4.7 Mixing Time—Laminar Flow	639
9.5	Reactive Mixing	639
	9.5.1 Introduction	640
	9.5.2 Fundamental Concepts in Reactive Mixing	640
	9.5.3 Macro-Mixing and Micro-Mixing	642
	9.5.4 Ideal Flow Patterns in Reactors and Residence-Time Distribution (RTD)	643
	9.5.5 Micro-Mixing and Segregation	644
	9.5.6 Micro-Mixing and Selectivity	644
	9.5.7 Micro-Mixing: Practical Implications	647
	9.5.8 Reactive Mixing in Multiphase Systems	647
	9.5.9 Equipment Types and Design Guidelines	649
	9.5.10 Scale-Up	651
	9.5.11 Some Practical Guidelines for Small-Scale (Lab) Experimentation	652

9.6 Solid-Liquid Mixing 653
    9.6.1 Introduction 653
    9.6.2 Settling Solids 653
    9.6.3 Particle Suspension in Stirred Vessels 655
    9.6.4 Just-Suspended Conditions 656
    9.6.5 Floating Solids 657
    9.6.6 Uniform Solids Concentrations 657
    9.6.7 Power Requirements 659
    9.6.8 Equipment 659
    9.6.9 Solids Suspension by Jet Mixing 660
9.7 Gas-Liquid Mixing 660
    9.7.1 Introduction and Scope 660
    9.7.2 Mass Transfer—General 661
    9.7.3 Equipment and Its Function 662
    9.7.4 Impeller Characteristics in Gas-Liquid Mixing 663
    9.7.5 Mass Transfer and Gas Holdup 666
    9.7.6 Gas Residence Time 668
    9.7.7 Scale-Up 669
9.8 Immiscible Liquid-Liquid Mixing 671
    9.8.1 Introduction 671
    9.8.2 Characterization of Immiscible Liquid-Liquid Systems 671
    9.8.3 Drop Sizes 672
    9.8.4 Dispersion of Drops, Laminar Flow, and Low Viscosity 673
    9.8.5 Dispersion of Low-Viscosity (Inviscid) Drops ($\mu < 0.020$ Pa·s) Turbulent Flow 674
    9.8.6 Dispersion of Higher-Viscosity Drops ($\mu > 0.02$ Pa·s) Turbulent Flow 676
    9.8.7 Time for Dispersion 677
    9.8.8 Coalescence of Suspended Drops 677
    9.8.9 Population-Balance Methods 678
    9.8.10 Creating a Dispersion—Maintaining Drop Suspension 678
    9.8.11 Simultaneous Suspension, Dispersion, and Coalescence 680
    9.8.12 Equipment Used for Liquid-Liquid Mixing 681
    9.8.13 Processes 681
9.9 Static In-Line Mixers 682
    9.9.1 Introduction to In-Line Static Mixers 682
    9.9.2 Hydrodynamics and Other Characteristics of Mixing in Static Mixers 683
    9.9.3 Static Mixer Selection and Design Issues 684
    9.9.4 Reynolds Number and Flow Regime 685
    9.9.5 Component Flow and Viscosity Ratios 686
    9.9.6 Variation Coefficient in Blending Applications 687
    9.9.7 Dispersed-Phase Size Distribution in Liquid-Liquid and Gas-Liquid Systems 688
    9.9.8 Interfacial Areas for Gas-Liquid Systems 690
    9.9.9 Mass-Transfer Coefficients for Gas-Liquid Dispersions 690
    9.9.10 Heat-Transfer Enhancement in Laminar Flow Applications 691
    9.9.11 Pressure Drop and Power Requirements 691
    9.9.12 Mixing Efficiency 692
    9.9.13 Injection Considerations and Designs 693
    9.9.14 Pump Selection and Flow Control 693
9.10 Jet Mixing 694
    9.10.1 Introduction 694
    9.10.2 Principles 694

Industrial Mixing Technology 617

    9.10.3 Equipment...694
    9.10.4 Correlations 695
9.11 Heat Transfer in Mixing Equipment...697
    9.11.1 Introduction 697
    9.11.2 Important Heat-Transfer Considerations...698
    9.11.3 Fundamentals of Heat Transfer in Agitated Vessels................................699
    9.11.4 Heat-Transfer Surfaces and Effective Area 700
    9.11.5 Jackets and Other Applied Devices 700
    9.11.6 Internal Pipe Coils..701
    9.11.7 Other Internal Devices 701
    9.11.8 External Auxiliary Devices 701
    9.11.9 Process-Side Heat-Transfer Correlations...702
    9.11.10 Service-Side Heat-Transfer Correlations 703
References..705

9.1 INTRODUCTION AND OVERVIEW

9.1.1 TURBULENT, TRANSITIONAL, AND LAMINAR MIXING

It has been estimated that well over 70% of all mixing is done under turbulent conditions. Turbulent mixing occurs in pipelines, liquid jets, impingement devices, mixing vessels, tee-junctions, static mixers, pumps and orifices, and anywhere turbulence can be created. Eddy interactions lead to turbulent mixing in low-viscosity fluids. Large, high-energy eddies become smaller as a result of viscous-energy dissipation. During this process, regions of segregation are reduced in both size and intensity by the eddy motion. This process is normally fast. In turbine-equipped baffle-stirred vessels, turbulent mixing develops at impeller Reynolds numbers (Re), $(D^2N\rho/\mu) \geq 10^4$. Turbulent mixing is faster than transitional-flow mixing and is orders of magnitude faster than laminar mixing. Virtually all gas-gas mixing occurs under turbulent conditions. Commonly used units of power intensity are kW/m^3 or hp per 1000 gal.

At transitional flow conditions, larger diameter impellers are used to improve mixing times, but they require more power than equivalent mixing rates under turbulent conditions. Baffles improve mixing rates for Re > 300, but have the opposite effect for Re below this point. Helical ribbon-, anchor-, and gate-type impellers are commonly used at the low end of the transitional flow range.

Laminar mixing is a two-stage process: (1) Discrete layers or lumps of fluid are stretched by mechanically induced shear. The more they are stretched, the thinner they become. (2) Diffusion decreases the concentration gradients between adjacent, yet still distinct layers. Laminar mixing is slower, and much more energy demanding, than turbulent or transitional mixing. If complete stretching does not occur throughout the fluid, uniformity will not be achieved. Stirred vessels containing impellers are used for viscosities of up to 50,000 cP. At higher viscosities, equipment used includes extruders (single and twin screw), kneaders, Banbury mixers, and static mixers. Gear pumps are used to move material through static (motionless) mixers. Good mixing demands that all of the fluid be exposed to the stretching process. Power is measured in terms of horsepower/pound.

9.1.2 PROCESS EFFECTS

Mixing enables a chemical reaction or a physical process to be carried out efficiently, thereby achieving a desirable result. Rapid local mixing, created by turbulence, often affects the yield from a set of chemical reactions. Segregation, due to poor mixing, can lead to lower yields and formation of undesirable by-products. Sometimes, simple changes, such as relocating the feed point, can lead to an improved performance. In cases where reaction rates are very fast, such as in reaction injection molding (RIM), impinging jet mixers create the rapid mixing needed for producing high-quality products. Mixing helps promote desirable rates of polymer chain growth for polymerizations. Many exothermic reactions, such as nitration and emulsion polymerization reactions, are operated by continuous reagent/feed addition. Such processes require rapid mixing to avoid local segregation that could lead to undesirable side reactions. Neutralization reactions often form solid precipitates. Conditions of locally high supersaturation, due to inadequate mixing, lead to the formation of hard-to-filter fine particles. This can be prevented by introducing reagents to the impeller region, where local mixing rates are fast.

For two-phase systems, mixing promotes faster mass transfer by creating higher interfacial area due to smaller bubbles or drops. Turbulence also helps reduce the boundary-layer resistance around drop or bubble surfaces, leading to faster mass transfer.

For solid-liquid systems, agitation provides solids suspension, either by maintaining movement at the bottom of a vessel, or by maintaining a desired level of uniformity throughout the vessel.

Agitation is required for blending, where the process objective is to promote homogenization of a fluid to a desired degree of uniformity, often in a specified time. The rate of approach to uniformity is an exponential process. Complete uniformity takes a very long time. In practice, one sets specifications to typically 95 or 99% uniformity, based on the process requirements.

The subjects discussed in this chapter are described more fully in the *Handbook of Industrial Mixing* [1].

9.1.3 Equipment and Design

It is important to match mixing equipment capabilities with process requirements. While it is desirable to have an optimum design and operating conditions for every step in the process sequence, it is seldom practical to do so. For example, specialty and pharmaceutical processes require the use of multipurpose reactors. An important consideration is to understand how less-than-ideal equipment will function in all stages of operation.

Documenting mixing performance data is vital to future troubleshooting. During the life of the equipment, modifications of both processes and equipment are common. For example, increased production requirements could lead to higher process concentrations and viscosity. Under such conditions, mixing may become inadequate, leading to regions of stagnation. Documented performance conditions for the original process are useful for diagnosing how the process responds to new conditions. A simple design or operational change can often meet the new challenge.

Mixing intensities vary greatly throughout a stirred vessel. While turbulent mixing can exist in the impeller region, transitional or laminar-flow conditions can exist elsewhere. Energy dissipation near the impeller is 40–50 times greater than in other regions (see Zhou and Kresta [2]). Common practice introduces feed to the surface of the liquid. While this avoids plugging problems and feed-pipe stagnation, it places the feed in a weakly mixed region.

Computational fluid dynamics or CFD (also known as computational fluid mixing, CFM) was introduced to the chemical process industries in the late 1980s. CFD/CFM is a numerical technique for solving fluid relationships such as conservation, transport, and the Navier-Stokes equations. Commercial CFD software enables one to predict the effects that geometry, feed location, physical properties, and operating conditions have on conditions in the vessel. Typical results predict velocity profiles, rates of energy dissipation, concentrations, and flow streamlines as they would occur in the vessel. This tool enables one to appreciate the good and bad features for each considered design. CFD simulations are based on assumptions. Some are low risk, but some impose high risk. Experimental validation is important particularly for nontrivial applications. Validation is advisable. Published velocity profile data can often help to validate results. At the time of this writing, CFD is weak in its ability to model large-scale turbulence and multiphase flow.

9.2 SYMBOLS, DIMENSIONLESS GROUPS, AND TERMS USED IN MIXING

9.2.1 Common Symbols Used in Mixing

		Units	
Symbol	Significance	Metric	English
a	Interfacial area of drops, bubbles, and solids per unit volume	cm^2/cm^3	ft^2/ft^3
B	Baffle width	m	ft
c_p	Specific heat at constant pressure	J/(kg·°K)	Btu/(lb·°F)
C	Impeller clearance, distance between lower tip of the impeller and the bottom of the vessel	m	ft
c_{B0}	Initial concentration of B	mol/m^3	
d	Diameter of dispersed phase particle, drop, or bubble	m	ft
d_{32}	Sauter mean drop or bubble diameter	m	ft
d_j	Jet diameter	m	ft
D	Impeller diameter (diameter swept out by the impeller)	m	ft
H	Height of liquid measured from the bottom of the tank	m	ft
h	Film heat-transfer coefficient, often accompanied with subscripts i, j, o, (inside, jacket, outside)	J/(s·m^2·°K)	Btu/(h·ft^2·°F)
k	Thermal conductivity	W/(m·°K)	Btu/(ft·°F)
k_2	Specific reaction rate constant	various	
$k_L a$	Volumetric mass transfer coefficient	s^{-1}	
M_i	Mass transfer rate for species (i)	g mole/s	lb mole/s
N	Impeller speed	s^{-1}	rpm
N_{JS}	Impeller speed to just suspend particles (all solids move)	s^{-1}	rpm
n_B	Number of blades on the turbine hub		
P	Power drawn by the impeller	kW	hp
P_G	Power drawn by the impeller under gassing conditions	kW	hp
Q	Volumetric gas flow rate, liquid flow rate	m^3/s	ft^3/s
S	Spacing between impellers	m	ft
T	Tank inside diameter	m	ft
τ_r	Reaction time	s	
τ_m	Mixing time	s	
V	Volume of contents in a vessel	m^3	gal
V_j	Jet velocity	m/s	ft/s
W	Impeller blade width	m	ft
X_S	Selectivity variable for chemical reactions for s		
F	Holdup, often used with subscripts g, s to designate gas, solid, etc.	fraction, sometimes %	
s	Interfacial or surface tension	N/m	lb$_f$/ft
D	Molecular diffusivity	m^2/s	ft^2/s
μ	Dynamic viscosity	Pa·s	cP
u	Kinematic viscosity	m^2/s	ft^2/h
r	Density	kg/m^3	lb/ft^3
H	Henry's law constant		

9.2.2 Common Dimensionless Groups Used in Mixing Correlations

Name	Symbols	Significance
Reynolds number	$Re = D^2N\rho/\mu$	Ratio of inertial to viscous stress; defines turbulent, transitional, and laminar-flow regimes; a correlating parameter for flow-sensitive terms in mixing
Power number	$Np = P/\rho N^3 D^5$	A function of the impeller Reynolds number and impeller type; Np is constant at $Re > 10^4$; increases with decreasing Re in the laminar region
Flow number	$Nq = Q/ND^3$	Ratio of actual impeller flow to an idealized flow
Fourier number	$Fo = \mu\theta/\rho T^2$	Used to correlate mixing times; also referred to as the vessel Reynolds number
Gas flow number	$Fl_G = Q/ND^3$	Ratio of gas to idealized liquid flow
Dämköhler number	$Da = \tau_M/\tau_R$	Ratio of mixing time to reaction time
	$Da = k_2 c_{BO}/E$	Half life of a reaction to the rate of engulfment
Capillary number	$Ca = \tau \cdot r_D/\sigma$	Ratio of fluid stress on a drop to its interfacial resistance
Froude number	$Fr = N^2D/g$	Ratio of centrifugal to gravity forces; used for vortex definition and in gas-liquid mixing correlations
Nusselt number	$Nu = hD/k$	Used with Re and Pr to correlate heat-transfer data
Prandtl number	$Pr = c_p\mu/k$	Ratio of convective to conductive heat transfer
Weber number	$We = N^2D^3\rho/\sigma$	Ratio of inertia to surface/interfacial forces for gas-liquid and immiscible L-L systems; used to correlate bubble/drop size

9.2.3 Significance of Commonly Used Mixing Terms

Power Number, Np: The power number, Np, sometimes referred to as Po, is a measure of the relative drag of the impeller. Streamline curved blades, like hydrofoils and retreat-curve impellers, have less drag than flat blades; consequently, their power numbers are lower than those for flat-blade impellers. Power numbers of some of the more popular impellers are given in Table 9.1. The calculation of power from impeller diameter, speed, and liquid density is given by Equation (9.1).

Flow Number, Nq: The magnitude of the flow number is a measure of an impeller's ability to produce flow. The larger the flow number, the greater is the flow. The total impeller flow consists of the direct discharge flow plus entrained flow. Most reported flow numbers include both flows. Equation (9.3) shows the use of Nq in calculating the total discharge flow.

Mixing time, θ_M: Mixing time, θ_M, is the time it takes to mix initially segregated materials to a specified degree of uniformity. For example, it takes 60% longer to mix to 99% uniformity than to mix to only 95%. Mixing times can be calculated from the methods described in Section 9.4. For certain reactions, it is important to have mixing times shorter than reaction times. The Dämköhler number, Da, is the ratio of mixing to reaction times and is further described in Section 9.5.

Impeller Reynolds number, Re, N_{Re}, and vessel Reynolds number, Re: The impeller and vessel Reynolds numbers are the ratios of inertia to viscous forces. They are indicators of flow conditions: turbulent, laminar, or transitional. They are used to correlate other quantities such as the power number, Figure 9.1, and the inside heat-transfer coefficients shown by Equation (9.85).

Micro-Mixing: Micro-mixing is the smallest scale of mixing. In terms of dimensions, it is at or below the Kolmogoroff microscale that can be calculated by Equation (9.9). Micro-

TABLE 9.1
Characteristics of Commonly Used Impellers

Model	Flow Pattern	No. of Blades	Features	Np	Nq	Supplier
HE-3	Highly axial	3	Efficient turbulent flow, blending, and suspending of solids; efficient pumping impeller in gas-liquid systems	0.23–0.33	0.54	Chemineer
CD-6/BT-6	Highly radial	6	Effective gas dispersing impeller; BT-6 handles higher gas flows	2.7–3.7	0.72	Chemineer
Maxflow T	Between high-efficiency HE-3 and 45° PBT	3, 4, or 6	High solidity reduces cavitation tendency in gas-handling applications; good pumping in the presence of gas	0.4–0.9	...	Chemineer
Mark II	Axial	3 or 6	High solidity reduces cavitation tendency; highly effective blending in side-entering mode	0.22–0.75 depends on pitch, D/T, and no. of blades	...	Chemineer
R-100	Highly radial	5 or 6	Commonly used for gassing; prone to cavitate; also used for making dispersions and emulsions	5.75 depends on pitch and D/T	0.8	Lightnin
D-6						Chemineer
A-200 P-4, P-6	Mixed flow	4	Good for viscous mixing, general blending operations; more shear than the hydrofoils	1.0–1.3	0.72–0.82	Lightnin Chemineer
A-310	Highly axial	3	High pumping, efficient, available in many materials of construction, side-entering version is the A-312	0.30	0.56	Lightnin
A-315	Highly axial	4	High solidity makes this a good gas-dispersing impeller; high pumping circulates gas-liquid mixture throughout vessel	0.75	0.73	Lightnin
A-320	Highly axial	3	Good for high-viscosity applications, often interchanging with A-200, more axial than the A-200, but somewhat dependent on Reynolds number	0.65	0.68	Lightnin
R-510	Weak radial	6	A bar turbine that produces high shear for dispersing and shredding operations; similar in use to the R-500, bars are mounted on a disc with alternate blade on top and bottom of disc	0.61	...	Lightnin

Continued

TABLE 9.1 (Continued)
Characteristics of Commonly Used Impellers

Model	Flow Pattern	No. of Blades	Features	Np	Nq	Supplier
R-500	Weak radial	many	Very high shear, effective in dissolving polymers, forming solutions; removes boundary-layer resistance	0.17–0.3	...	Lightnin
Curved Blade turbine	Highly radial	6	Lower shear rate than straight blade; not commonly used today	3.91	0.82	Various
Propeller 1.5/1.0 Pitch	Strong axial	3	Efficient, costly, heavy compared with HE-3 and A-310	0.87/0.35	0.77/0.44	Various
Curved Blade Turbine	Radial	4	Good pumping due to broad blades; a glassed steel impeller	1.7–1.4	0.54	Pfaudler
Retreat-curve impeller (RCI)	Radial	3	Used with glass-lined equipment; usually at bottom of vessel and with large D/T ratios; effective general-purpose impeller; can be used for limited gas dispersion, with good baffling	0.5–0.3	0.24	Pfaudler DeDietrich Tycon
Vertical-blade turbine	Radial	2 + 2	Similar to vertical pitch flat-blade design; glassed steel, overlapping hub design	2.7	0.65	Pfaudler
Turbofoil	Axial	4	A glassed hydrofoil, similar to HE-3 or A-310, except has four blades	0.4–0.3	0.48	Pfaudler

Industrial Mixing Technology

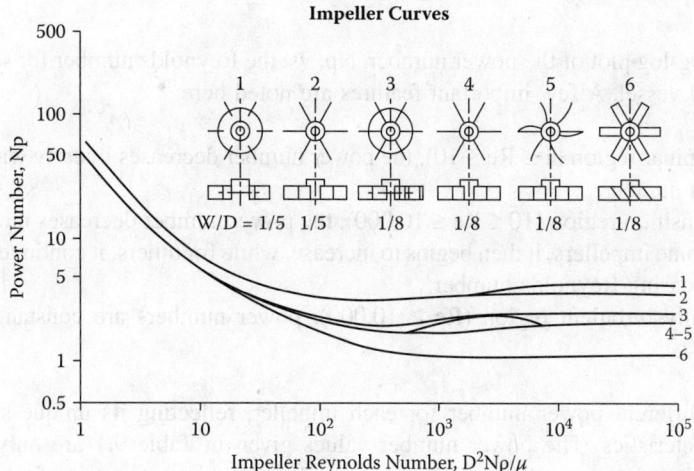

FIGURE 9.1 Reynolds number vs. power number for six turbine impellers. (Modified from Rushton et al., 1980. With permission.)

mixing is the scale involved with kinetically controlled chemical reactions. Effective micro-mixing usually requires high-energy input.

Meso-Mixing: Meso-mixing is a term used to describe the intermediate scale of mixing between micro-mixing and macro-mixing. More specifically, it is the turbulent exchange between turbulent impeller flow and the surrounding fluid. Refer to Figure 9.9.

Macro-Mixing: Macro-mixing is distributive mixing caused by large-scale flows. It is analogous to convective mixing. The rapid blending in a stirred vessel is due to macro-mixing.

9.3 STIRRED TANKS: EQUIPMENT AND FUNCTION

9.3.1 IMPELLERS AND THEIR CHARACTERISTICS

This section describes mixer geometry: impellers (types listed in Figure 9.3), the number required, their characteristics, importance of location, vessel shape, and effects of baffles. An example is given to show a basic calculation of mixer power and flow.

Impellers available in the pre-1960 era would have been limited to four- and six-blade disc turbines (also known as radial-flow turbines or RFT or Rushton turbines), the four- and six-blade 45° pitch blade turbines (PBT), the four- and six-blade flat-blade turbines (FBT), and the three-blade retreat-curve impellers (RCI).

While these older impeller types are still used, numerous hydrofoil designs are now available. These contoured blade hydrofoils, originally developed by R. Weetman, require much less power and produce nearly axial flow, two big advantages they have over the older designs. True axial flow is desirable, since it permits more efficient top-to-bottom mixing or turnover. Hydrofoils are also specialized for different applications, such as viscous fluid mixing, gas-liquid mixing, as well as low-viscosity turbulent mixing. They are ideally suited for flow-dominated applications such as multiphase suspension and blending. They work well with other impellers.

Recent designs used for gas-liquid mixing include the Scaba and Chemineer's BT-6 and CD-6 originating from the work of Professors J. M. Smith and A. Bakker. The forward-facing cup-shaped blades delay the onset of cavitation, thus enabling more power to be delivered for dispersion. Other impeller designs include sawtooth blades such as the Lightnin R-500, used to produce high shear rates in low-viscosity fluids; Ekato's INTERMIG design for moderately high-viscosity blending; and the older helical and anchor designs for high-viscosity mixing.

9.3.2 POWER

Figure 9.1 is a log-log plot of the power number, Np, vs. the Reynolds number for several impellers in a fully baffled vessel. A few important features are noted here:

- In the laminar region ($1 \leq Re \leq 10$), the power number decreases linearly with increasing Reynolds number.
- In the transition region ($10 \leq Re \leq 10,000$), the power number decreases more gradually, and for some impellers, it then begins to increase, while for others, it continues to decrease with increasing Reynolds number.
- In the fully turbulent region ($Re \geq 10,000$), power numbers are constant, but design dependent.

There is a different power number for each impeller, reflecting its unique shape and drag-producing characteristics. The power number values given in Table 9.1 are only applicable for turbulent flow. Manufacturers offer a variety of impellers, the characteristics of which are given in Table 9.1.

The power requirement for a single impeller operating in a baffled vessel can be calculated from

$$P = Np \cdot \rho \cdot N^3 D^5 \tag{9.1}$$

In English units, Equation (9.1) becomes

$$HP = Po \cdot \rho \left(lb \cdot ft^{-3}\right) \cdot N^3 \left(s^{-1}\right) \cdot D(ft) / 32.2 \left(ft \cdot s^{-2}\right) \cdot 550 \left(ft - lb \cdot hr^{-1}\right) \tag{9.1a}$$

In metric units, Equation (9.1) becomes

$$kW = Np \cdot \rho \cdot \left(kg \cdot m^{-3}\right) \cdot N^3 \left(s^{-1}\right) \cdot D^5 (m) / 1,000 \tag{9.1b}$$

In practice, the Reynolds number must first be determined to obtain the power number. The impeller Reynolds number is defined for stirred vessels and given by

$$Re = \frac{D^2 N \rho}{\mu} \tag{9.2}$$

Example 9.1

Determine the power needed to agitate a fluid using a Rushton impeller, given the specific gravity of the liquid is 1.0, the tank diameter is 3.0 m, the height of liquid is 3.0 m, the impeller diameter is 1.0 m, the speed is 1.0 s^{-1}, and the liquid viscosity is 1.0 cP.

Solution

Substituting the Reynolds number in Equation (9.2), $Re = (1.0^2)(1)(1,000)/0.001 = 10^6$. Flow is therefore fully turbulent. The power number for a Rushton impeller, the top curve in Figure 9.1, is Np = 5.0. The density = 1000 kg/m³, and the power is

$$P_{Rushton} = 5.0(1000)(1.0^3)(1.0^5)/1000 = 5.0 \text{ kW}$$

If a 45° PBT is selected (line 6, Figure 9.1) at a Re = 10^6, the power number is 1.3. The power is

$$P_{45°PBT} = 1.3(1000)(1.0^3)(1.0^5)/1000 = 1.3 \text{ kW}$$

Zhou and Kresta [2] have determined the energy distribution for different impellers. They reported the dissipation rates in the impeller region to be 38.1% for the A-310, 43.4% for the Rushton turbine, and 70.5% for the 45° PBT of the total energy input to the impeller.

9.3.3 Multiple Impellers

Many applications benefit from the use of two or more impellers. Multiple impellers are commonly used in tall vessels, where $H/T \geq 1.2$. The upper impellers are often hydrofoils, and the selection of the lower turbine depends on the application. For example, gas-liquid and liquid-liquid dispersion applications frequently combine hydrofoils in the upper position and disc turbines in the lower position. The practice of placing several radial discharging disc turbines at equal intervals on the shaft leads to compartmentalization and poor turnover. Two or more axial-flow turbines, such as hydrofoil types, are often used for blending of low-viscosity homogeneous materials. Multiple 45° pitch blade turbines can be used for viscous blending. Low-viscosity blending in square-shaped vessels can normally be done using a single impeller. Multiple impellers are generally not used for solids suspension applications.

Multiple impellers need not be the same size or type. For example, in gas-liquid applications, a good practice is to assign 75% of the energy to gas dispersion and 25% to circulation. Using this power allocation, the assumption of additive power and values of Np enables one to calculate the respective impeller diameters.

Power calculation is more complicated for certain multiple impeller configurations. The power demand is less than additive if one impeller "receives flow" from an adjacent impeller. Frequently, multiple axial or mixed-flow impellers are used to create better flow distribution. The effects of impeller spacing, blade angle, number of blades, the degree of baffling, as well as the effect of Reynolds number on the power number are given by Bates [3] and shown in Figure 9.2. The ordinate is the ratio of two impellers to that of a single flat-blade turbine, and the abscissa is the impeller spacing.

With reference to Figure 9.2, observe that the power demand for multiple radial impellers spaced at one or more impeller diameters apart is additive, but at closer spacing, power is greater than the sum of each impeller. This is due to flow interference. As mentioned earlier, the total power demand for multiple mixed-flow impellers is complex.

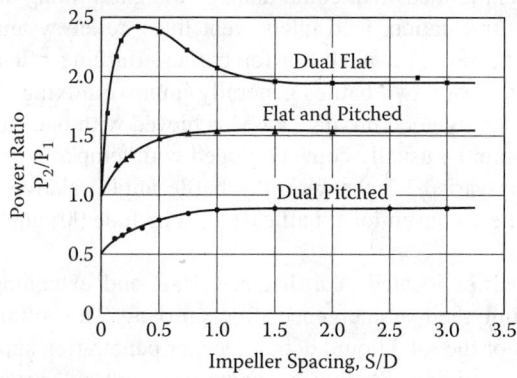

FIGURE 9.2 Effect of dual turbine spacing on power. (Modified from Bates et al., 1963. With permission.)

9.3.4 Flow Discharge

The discharge flow rate from most impellers is expressed in terms of the flow number N_q. Values are given in Table 9.1. The flow number, Nq, is defined by

$$Nq = \frac{Q}{ND^3} \qquad (9.3)$$

where Q represents the total flow (direct + entrained) in m³/s (ft³/s), N is the speed in s⁻¹, and D is the impeller diameter in m (ft).

Flow rate calculations can be used to estimate the mean circulation time, τ_C, of fluid in the vessel. It is the liquid volume divided by the impeller flow rate. A rule of thumb is that the approximate mixing time, τ_M, is three to four times the mean circulation time, τ_C.

The flow rate resulting from complementing dual impellers is not significantly greater than for a single impeller. The performance improvement comes from better flow distribution. The mean circulation time is the time a particle takes to complete a circuit in the vessel. Every vessel has a distribution of circulation times. Multiple impellers reduce the circulation time distribution (see Section 9.4).

Figure 9.3 shows many of the impellers listed in Table 9.1.

9.3.5 Baffles

Baffles are used to convert tangential to axial flow and are required for all stirred-tank turbulent-flow applications. They are optional for transitional-flow conditions ($10 \leq Re \leq 10{,}000$) and should never be used for laminar-flow applications. Within the transitional-flow range, a good rule of thumb is to use some baffling for $300 \leq Re \leq 1000$ but not below 300. Baffles are required for most low-viscosity, multiphase applications.

The standard fully baffled vessel consists of four equally spaced vertical plates, T/12 to T/10 wide, and set away from the wall a distance of T/72 to T/90. Baffle lengths extend from near the liquid surface to the bottom of the straight wall. Baffles reaching the surface create eddies, which are helpful for engulfing addition streams or floating solids. Surface eddies also lead to incorporation of headspace gases, which sometimes need to be prevented. Short baffles located in the impeller region can substitute for full-length designs. Baffles used with glass-lined equipment include the beavertail, finger, "h," "d," and fin baffles. These are shown in Figure 9.4. One or two of these baffles are normally suspended from top openings on the vessel. Care must be taken to avoid excessive drag and fluctuating loads that could damage the glass lining of the support nozzle.

The demand for instrumentation, feed inlets, vent lines, relief systems, and inspection ports places a heavy demand on nozzle availability for baffles. Baffling often becomes secondary to these other process needs. While two baffles generally improve mixing rates by as much as 20% over single baffles, surprisingly good mixing can be achieved with one "h," "d," or fin-type baffle. The baffles shown in Figure 9.4 usually come equipped with temperature-sensing provisions. The degree of baffling can be varied by changing the baffle angle relative to flow, thus creating a variable flow resistance. Less conventional baffle designs include 90° angle-shaped baffles welded to the vessel walls.

Short top-entering baffles located at a distance 2/3R and extending to mid-depth convert angular to axial momentum; their submergence affects mixing. The optimum depth for suspended baffles is from 1/3 to 2/3 of the total liquid depth. Deeper penetration suppresses overall momentum and results in poorer mixing. Shallower penetration fails to convert enough tangential to axial momentum. Applications requiring changing liquid levels or operation at low liquid levels are often equipped with one deep and one shallow baffle. Similar arguments also hold for wall-attached baffles.

Industrial Mixing Technology

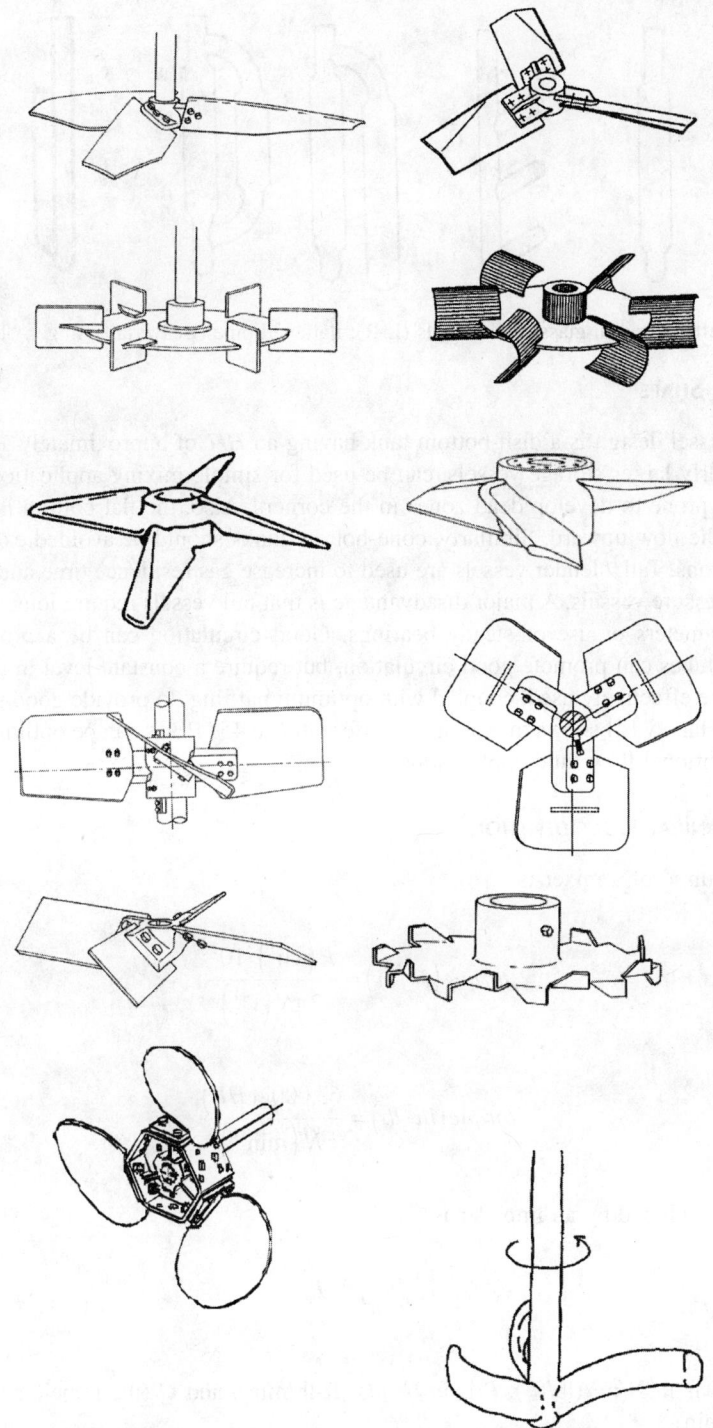

FIGURE 9.3 Commonly used impellers (L-R): (row 1) Lightnin A-310, Chemineer HE-3; (row 2) Rushton (R-100, D-6), Chemineer CD-6; (row 3) Pfaudler Turbofoil T, Lightnin C-102 Mark II; (row 4) Lightnin A-315, Lightnin A-320; (row 5) 45° pitched blade turbine, Lightnin R-500; (row 6) Prochem Max Flow T, Pfaudler 4-blade, curved blade turbine, retreat-curve impeller RCI.

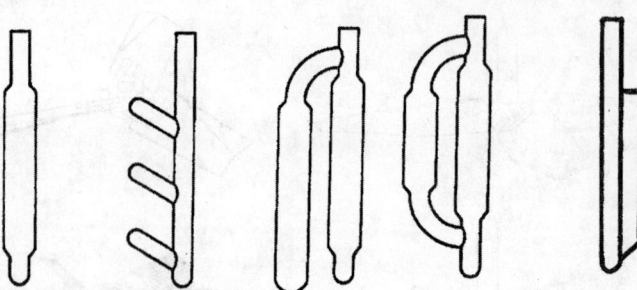

FIGURE 9.4 Baffles used in glass steel vessels (L-R): flattened pipe (Beavertail), finger, "h," "d," and fin.

9.3.6 Vessel Shape

The optimal vessel design is a dish-bottom tank having an *H/T* of approximately 1.2. Flat-bottom tanks, particularly larger storage vessels, can be used for simple mixing applications, but the flat-bottom tank is prone to develop dead zones in the corners. Also, the flat bottom is not conducive to redirecting the flow upward. Similarly, cone-bottom tanks should be avoided except for certain solids applications. Tall, slender vessels are used to increase gas residence time and to reduce wall thickness of pressure vessels. A major disadvantage is that tall vessels require long shafts requiring larger shaft diameters or use of steady bearings. Good circulation can be a problem for these designs. Draft tubes can promote good circulation, but require a constant level in the tank. Single impellers can be effectively used, coupled with optimum baffling, to provide good mixing in tanks up to 2:1 *H/T*. Harvey [4] shows that multiple different-size 45° PBTs can be optimized to improve mixing in transitional flow fluid applications.

9.3.7 Mechanical Considerations

The torque required of a mixer is

$$torque(N \cdot m) = \frac{P(kW) \cdot 10^3}{2\pi N(s^{-1})} \tag{9.4a}$$

$$torque(in \cdot lb) = \frac{63,000 \cdot (HP)}{N(\min^{-1})} \tag{9.4b}$$

The head developed by an impeller is

$$head \propto \frac{P}{Q} \tag{9.5}$$

where the head is in N/m² (lb/ft²), *P* is in N·m/s (ft-lb/min), and *Q* (the impeller discharge flow) is in m³/s (ft³/min).

Impeller size (*D/T* ratio) influences whether flow or turbulence governs the process of mixing. Oldshue [5] shows, for equal process results, that impeller diameter affects power and torque characteristics, as shown in Figure 9.5. The values for power and torque are normalized to the values for an impeller with *D/T* = 0.333.

D/T ratios can be in the range of 0.2 to 0.7, but commonly they are 0.3 to 0.5. At equal power, large impellers rotate more slowly and require large gearboxes compared with smaller impellers.

Industrial Mixing Technology

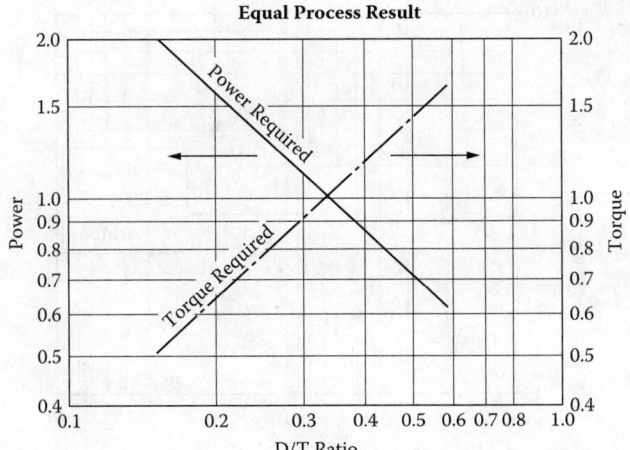

FIGURE 9.5 Typical curve of power decreasing while torque increases with *D/T*. (From Oldshue, J. Y., 1983.)

Large impellers create more flow and relatively less turbulence than small impellers. Small impellers display the opposite effects. There is a minimum cost of mixing (operating vs. capital) associated with the selection of *D/T*. This minimum is usually found at 0.35 to 0.45.

A complete mechanical design should include a calculation of the critical speed. When the operating speed is close to or exceeds the critical shaft speed, vibrations can lead to mechanical instabilities and severe equipment damage. For more information, see the chapter by King [6] in *Mixing in the Process Industries* or the chapter by Dickey and Fasano in the *Handbook of Industrial Mixing* [1].

9.3.8 Bottom Clearance

The clearance, expressed as "C" or as a ratio C/T or C/D, is the distance from the bottom edge of an impeller to the bottom of the vessel. C/D for an impeller in the middle is 0.5. General-purpose vessels used for multiple operations use an impeller placed very close to the bottom of the tank. This enables solids to be handled more efficiently as well as loading and unloading operations. When the impeller is some distance from the bottom, a small impeller "tickler" can help solve mixing problems with settling solids.

Impeller clearance affects total discharge flow and its direction. For example, when axial-flow impellers are placed close to the bottom, they produce a radial discharge. Upward-angled, retreat-curve impellers are always located at the bottom of the vessel, where they produce strong radial flow and produce good circulation while handling level changes and solids effectively.

Impeller clearance also affects power draw, as is shown in Figure 9.6 for six-blade Rushton, PBT, and flat-blade turbines.

9.3.9 The Vendor's Role

In the authors' experience, the vendor can play a valuable and often underappreciated role in supplying broad mixing experience. It is important to establish a trusting yet confidential relationship early on. Competitive vendor proposals serve to confirm designs and cost, but they are only as accurate as the information supplied to them. Industrial in-house mixing technology is important to validate vendor proposals. The vendor is often valuable in proposing new designs and in troubleshooting. Modern equipment suppliers are sophisticated in their knowledge of mixing, often possessing CFD tools and providing in-house experimental facilities for the benefit of their customers.

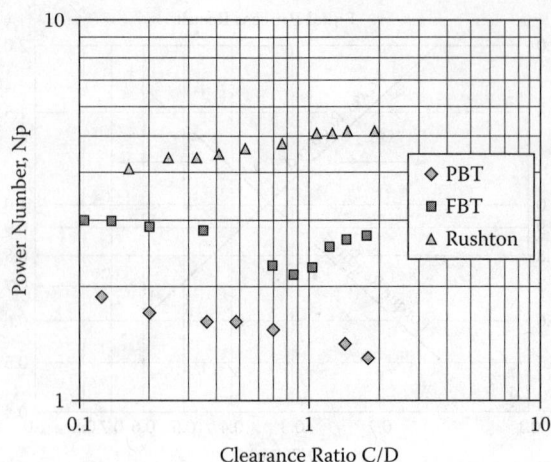

FIGURE 9.6 Effect of clearance on power. (From Bates et al., 1963).

9.4 BLENDING

9.4.1 INTRODUCTION AND SCOPE

Blending is the mixing of different, but miscible, fluids. The topics that are discussed in this section include: fluid properties (rheology), mixing mechanisms, equipment used, and the energy required to mix to a certain degree of uniformity in a given time. This section emphasizes the stirred tank, but blending is also done in static mixers and by jet mixing. These topics are discussed in Sections 9.9 and 9.10 in this chapter. The kinetics of blending is discussed as well as how blend time is related to design, impeller selection, and operation.

The fluid viscosity is one of the primary considerations when selecting an impeller. Turbine-type impellers are used for fluid viscosity of up to 50,000 cP. Smaller diameter turbines are used for low-viscosity fluids; larger ones for higher viscosity applications. Beyond 50,000 cP, the helical ribbon impeller, screw/draft tube designs, gate types, paddles, and anchors are used. While gates, paddles, and anchors are considerably less expensive than the helical ribbon or screw/draft tube designs, they are less able to produce good top-to-bottom flow, essential for good mixing.

9.4.2 TURBULENT, TRANSITIONAL, AND LAMINAR FLOW BLENDING

Blending is all about creating a specified degree of homogeneity from initially segregated materials in a given time. The *degree of mixing* is a term used to describe the quality of the result. For example, θ_{95} refers to the time it takes to reach 95% uniformity, and θ_{99} the time to reach 99% uniformity. Higher degrees of uniformity require longer mixing times. An example is given later in this section.

Turbulent and laminar mixing are quite different. Turbulent mixing takes place by three mechanisms: turbulent eddy motion, bulk or convective flow, and molecular diffusion. Laminar flow has no eddies to assist in mixing. Laminar mixing first depends on creating very thin layers between initially unmixed components, followed by molecular diffusion.

Efficient laminar mixing requires that very thin layers of fluid be well distributed throughout the vessel. Laminar mixing is slow and can take orders of magnitude longer to reach an equivalent state of uniformity than turbulent mixing.

In designing blending systems, it is important to establish whether conditions will be turbulent, transitional, or laminar. Turbulent mixing occurs at impeller Reynolds numbers greater than 10^4.

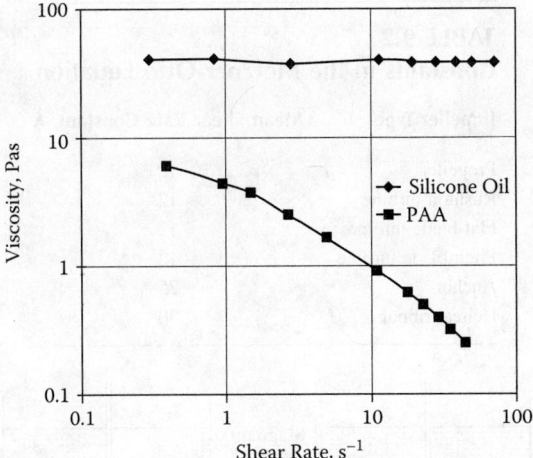

FIGURE 9.7 Typical viscosity shear rate curves for Newtonian silicone fluid and non-Newtonian polyacrylamide. (From Ulbricht et al., 1986.)

Mixing times are normally short. If the Reynolds number lies between 10 and 10^4, conditions are referred to as transitional. Mixing times will be longer than at turbulent conditions, but short compared with laminar flow conditions. At a Reynolds number of <10, conditions are laminar. While stirred tanks using proprietary devices can be used at these conditions, one ought to consider using a motionless mixer.

While many process fluids are Newtonian, some are non-Newtonian (as seen in Figure 9.7). For such cases, it is not sufficient to use a single value for viscosity to determine the impeller Reynolds number. Concentrated slurries are typically non-Newtonian and particle-size dependent. They are frequently shear-thinning. Dilatant behavior seldom occurs. If the system is simply shear thinning, it is usually possible to describe its rheological behavior with a simple power-law relationship:

$$\tau = k\gamma^n \tag{9.6}$$

where n is the flow behavior index and k is the viscosity at a shear rate of 1.0 s^{-1}.

Metzner [8] determined that the mean shear rate, $\bar{\gamma}$ (s^{-1}), for a stirred tank in the transitional to laminar regime can be determined by the Metzner-Otto equation:

$$\bar{\gamma} = KN \tag{9.7}$$

where K values are found in Table 9.2, and N is the speed in s^{-1}.

Shear-thinning fluids can cause a number of problems. Near the impeller zone, where shear rates are high, the apparent viscosity is low and the region is well mixed. Further from the impeller, shear rates are much lower, the apparent viscosity is higher, and the fluid is often stagnant. Caverns, shown in Figure 9.8 (regions of well-mixed fluids in a stagnant matrix), can develop yield-stress fluids. The shape of the cavern depends on the discharge pattern of the impeller.

Cavern formation, shown in Figure 9.8, is very real and highly undesirable. It can be avoided by better distribution of the shear stress (using multiple impellers), thus creating lower local effective viscosity and higher circulation rates throughout the vessel.

TABLE 9.2
Constants in the Metzner-Otto Equation

Impeller Type	Mean Shear Rate Constant, K
Propeller	10
Rushton turbine	12
Flat-blade turbine	12
Pitch-blade turbine	12
Anchor	25
Helical ribbon	30

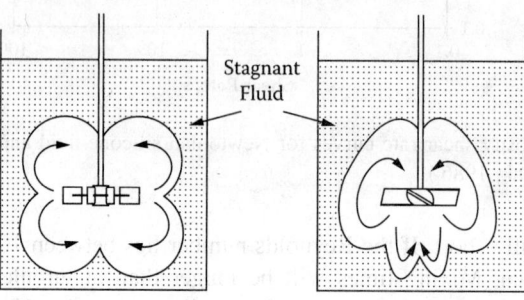

FIGURE 9.8 Shape of cavity in a shear-thinning suspension. (From Wichterle et al., 1981.)

Recent advances in commercial CFD software have enabled this problem to be modeled. The diameter and height of caverns have been correlated:

$$\left(\frac{D_c}{D}\right)^3 = N_P \cdot \left(\frac{\rho \cdot N^2 D^2}{\tau_y \pi^2 \left(1/3 + H_c/D_c\right)}\right) \tag{9.8}$$

where D_C and H_C are the cavern diameter and height and τ_y is the yield stress. The value of H_C/D_C for a Rushton turbine was found to be 0.4 [9].

Example 9.2

Calculate the impeller Reynolds number for a power-law fluid with k = 1.5 poise, n = 0.7, and density = 1000 kg/m^3. Assume the vessel diameter is 2 m, and the impeller, a 45° PBT, is 0.7 m in diameter. The operating speed is 90 rpm. Determine the nature of flow.

Solution

The mean shear rate constant K, for a 45° PBT, from Table 9.2 is 12. The mean shear rate at 90 rpm is 12 (90/60) = 18 s^{-1}. The viscosity is defined as the shear stress/shear rate and is $\mu = k\gamma^{(n-1)}$ for a power-law fluid. For the problem at hand, this becomes 1.5 (18)$^{-0.3}$ = 0.630 poise (0.063 kg/m·s). The Reynolds number is 0.7^2 (90/60) 1000/0.063 = 11,666. The flow is fully turbulent in this example.

9.4.3 THE NATURE OF TURBULENT FLOW

Large eddies are produced when an impeller passes through a low-viscosity liquid. These eddies are close to the impeller tip and are similar in size to the blade width. Figure 9.9 shows the energy $E(k)$ as a continuous function of wave number (k), along with terms commonly used in turbulent mixing. Large eddies depend on how the turbulence is generated. Smaller eddies do not. Eddy size,

Industrial Mixing Technology

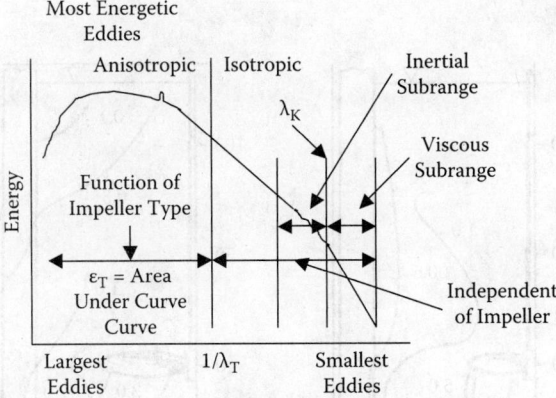

FIGURE 9.9 Turbulence energy spectrum, with λ_T as the length scale.

λ, is inversely proportional to wave number, k. The size and energy content of these eddies decrease with increasing distance from the impeller tip.

The Kolmogoroff scale can be calculated from

$$\lambda_k = \left(\frac{\nu^3}{\varepsilon_T}\right)^{1/4} \quad (9.9)$$

where λ_k is the Kolmogoroff length scale, ε_T is the local energy dissipation rate, and ν is the kinematic viscosity. For a power input of 1 kW/m³ (5 hp/1000 USG), λ_k is about 30 μ. This value is an effective size for many fast chemical reactions.

The energy distribution in a stirred vessel is not uniform. As mentioned earlier, investigators have shown dissipation rates near the impeller to be 10 to 50 times greater than the vessel average.

Figure 9.10 shows energy-dissipation contours for four impellers. The numbers represent fractions of the average energy input. It is important to understand energy distribution because it affects all processes requiring intensive mixing. This includes fast multipath chemical reactions, bubble and drop dispersion, and solids dissolution. Subsequent sections review these topics.

9.4.4 SHEAR RATES

The highest shear rates in a mixing vessel are close to the blades. Local impeller shear rates are proportional to tip speed (πDN) and are impeller specific. Disc turbines produce higher shear rates than propellers or hydrofoils (Figure 9.11). Certain applications, such as gas dispersion, require high shear rates while suspension polymerization requires a low shear rate. The vessel average shear rate, as described by the Metzner and Otto equation (9.7), is dependent on rotational speed, not tip speed.

9.4.5 FLOW PATTERNS

Figure 9.12 shows the idealized flow patterns for radial- and axial-flow turbines. The radial-flow turbine (R-100, D-6, or Rushton) (Figure 9.12, upper left) creates two completely separate circulating zones. Shear is very high in the impeller discharge region, and the impinging discharge flow divides at the wall, forming the two zones. The axial-flow turbine (Figure 9.12, upper right) would include propellers and hydrofoils (A-310, A-315, A-320, HE-3, Maxflow T, etc.). This impeller creates a

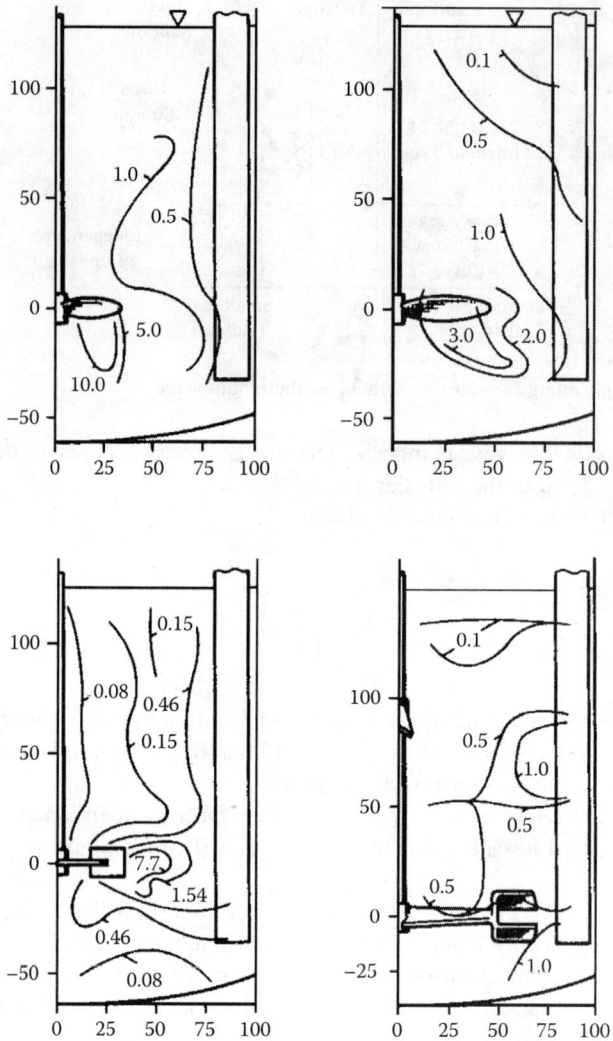

FIGURE 9.10 Energy contours for different impellers (L-R): (top) small propeller, large propeller; (bottom): Rushton turbine, Ekato-MIG. (From Todtenhaupt et al., 1991.)

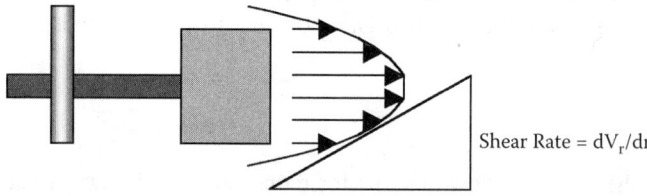

FIGURE 9.11 Typical velocity profile of the discharge flow from a disc turbine.

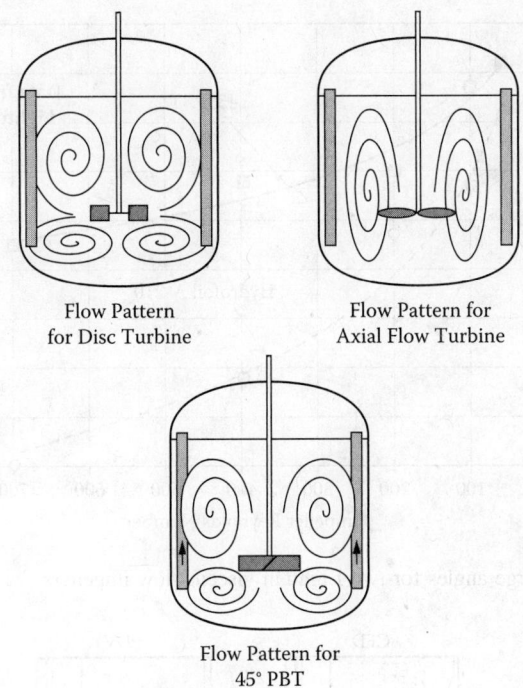

FIGURE 9.12 Typical flow patterns for radial, axial, and 45° pitch blade turbines.

desirable top-to-bottom single-flow loop pattern. The shear rate in the impeller region is much lower than the radial-flow turbine at comparable tip speeds. Flow distribution is good, and there is no flow stagnation anywhere. The 45° PBT (Figure 9.12, bottom) produces a downward and outward flow discharge angle. It also produces fairly high shear. The returning flow from the bottom zone (at the wall) helps to feed the upper zone and create fairly good top-to-bottom turnover, but not as good as for true axial impellers, since there are two distinct circulation regions that develop. Flow in the center region directly below the PBT is upward. This creates a conically shaped, poorly mixed region under the impeller and can occasionally lead to processing problems. In the above discussion, all vessels are fully baffled. Flow pattern information helps determine the best-feed point location.

Viscosity, hence Reynolds number, affects the performance of axial-flow impellers. The discharge angle (measured from the horizontal) decreases with increasing viscosity, causing flow patterns to change. Propellers discharge flow in a similar pattern to the 45° PBT as shown in Figure 9.13, but in a more vertical direction. The flow discharge angle becomes increasingly more radial with decreasing impeller Reynolds number.

Flow discharge angles are important for multi-impeller configurations. If flow from adjacent impellers fails to "communicate," separate circulation zones develop. This is seldom desirable. Harvey [4] shows the effect of multiple impellers on flow patterns in Figure 9.14. CFD analyses are shown on the left, and experimental laser Doppler flow measurements are shown on the right.

Experimental and CFD results are in close agreement. Observe that impeller 2 is able to capture flow from impeller 3 above it. Had impeller 3 been larger, its discharge would have extended beyond the suction region of impeller 2, and flow would be less. Both impeller size and spacing are important variables that affect flow pattern.

9.4.6 Mixing Time—Turbulent and Transitional Flows

Mixing-time considerations are important for analyzing competing chemical reactions, polymerizations, and precipitations. While data are available on the effect of speed on mixing time, design

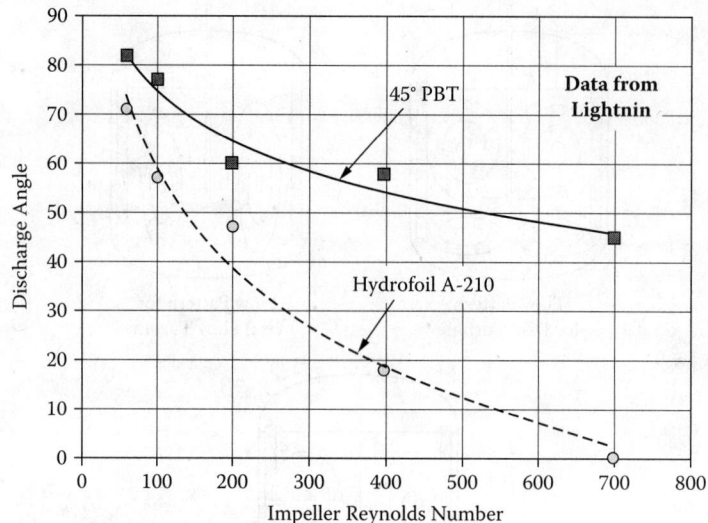

FIGURE 9.13 Flow discharge angles for two Lightnin viscous flow impellers.

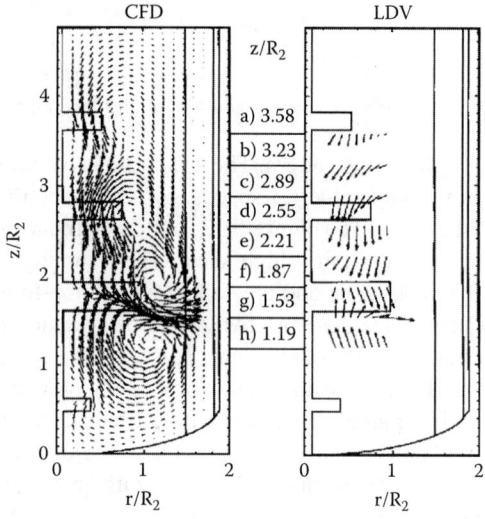

FIGURE 9.14 Experimental and CFD modeling of a series of 45° PBTs for fluids in the transition region. (From Harvey et al., 1997.)

details are usually lacking. For example, the effects of multiple impellers, clearance, spacing, and baffle design on mixing time are not well known, but the effects of speed, impeller diameter, and impeller type are better understood.

Many investigators have found that, under turbulent conditions, mixing is complete after three to four turnovers. This is shown schematically in Figure 9.15.

The mean circulation time (assuming the turbulent flow discharge, Q, from an impeller is $Q = N_q ND^3$, the tank shape is $T = H$, its volume is V_R, and D/T is constant) is

$$\theta_C = \frac{V_R}{Q} \propto \frac{T^3}{N_q ND^3} \qquad (9.10)$$

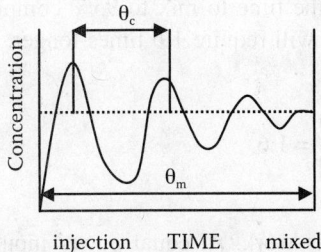 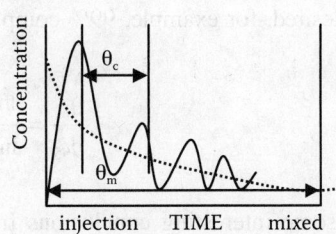

FIGURE 9.15 Typical tracer-time curves for batch and continuous stirred vessels.

For scale-up, if D/T = constant and the same impeller type is selected,

$$\theta_C = \frac{1}{N_q N (D/T)^3} = \text{constant} \cdot \frac{1}{N} \quad (9.11)$$

$$\theta_c N = \text{constant} \quad (9.12)$$

Thus, for scaling turbulent flow cases, the product of mixing time and speed is constant. Industrial scale-up is seldom done using constant mixing time as a criterion, because power demand is usually too high.

This does not apply for transitional flows. Work at BHRG's FMP program reported by Grenville [12] shows that Re, Fo, and Np can be arranged to correlate mixing times for fluids in the turbulent and transitional flow regimes. Data were taken from mixing-time experiments in vessels from 0.3 to 2.67 m in diameter, using different impeller types and sizes. These included a 45° PBT and diameters of (T/3 and T/2), a flat-blade turbine of diameter of (T/3), and a hydrofoil of diameter of (T/2). For Re > 200, power numbers and $N\theta$ were constant, as in Equation (9.12). In the transition region, $N\theta \propto \text{Re}^a$, where $a = -1.0$ for the FBT and PBT, and -0.84 for hydrofoils. The correlations represent a large amount of data and are given by Equations (9.13) to (9.15).

At turbulent conditions:

$$Np^{1/3} Re = \frac{5.2}{F_o} \quad (9.13)$$

where $F_O = \mu\theta/\rho T^2$, and the relative standard deviation is ±10%. F_O is the Fourier number, also referred to as the vessel Reynolds number.

In the transition flow region,

$$Np^{1/3} Re \sqrt{F_o} = 184 \quad (9.14)$$

Here the relative standard deviation is ±17.4%

At the crossover point,

$$Re_{crossover} = \frac{6,370}{Np^{1/3}} \quad (9.15)$$

These equations enable one to calculate θ_{95}, or the time to mix to 95% completion. If more uniformity is desired, for example, 99% complete, it will require 1.6 times longer, as shown by

$$\frac{\theta_{99}}{\theta_{95}} = \frac{\ln(0.01)}{\ln(0.05)} = 1.6 \qquad (9.16)$$

There are some interesting conclusions from this work. At equal power input and impeller diameter D, mixing time is impeller independent. Impellers with a low power number, such as A-310 or HE-3, must operate at higher speeds to attain comparable power to a higher-power-number impeller. Higher speeds mean lower torque and a smaller, less costly design.

Example 9.3

Addition polymerization requires that monomer groups be added to a core polymer containing reactive ends. If the chain ends see low monomer concentration, end groups will be short; but if locally high concentrations are encountered, end groups will be long. Quality considerations demand uniformity of chain length of the end groups. Rapid monomer mixing is essential. Kinetic studies showed the half-life of the addition reaction to be $\cong 5.0$ minutes. The pre-polymer is dissolved in a solvent, the viscosity at the operating conditions is 1000 cP, and fluid behavior is Newtonian. The reactor is 6 ft in diameter × 8 ft high, with top and bottom elliptical heads. The active average volume is 1160 gal. The viscosity remains constant during end-group addition. Design a mixing system to give 99% complete mixing in 1.0 min, or at 20% of the reaction half-life.

Solution

Equations (9.11) to (9.13) must be solved iteratively. An equation solver (or spreadsheet) provides a convenient way to do this. First, a Lightnin A-320 impeller is selected. (A 45° PBT could also have been a reasonable first choice.) Choose $D = 28$ in. so that $D/T \approx 0.4$. The height of liquid is 72 in. and the volume including the elliptical head is 1162 USG. A single impeller is selected, since the H/T is 1.0. Assume a speed; determine the Re and the crossover point for transitional flow (Equation (9.15)) and the mixing time θ_{99}. The trial-and-error solution using a spreadsheet is shown in Table 9.3.

We are mainly interested in numbers in the last (eighth) column. A speed of 145 rpm was selected to give a 99% mixing time (valid for the transitional region) in just under a minute. Had the Reynolds number been higher than values in the sixth column, the correlation for turbulent flow (column 7) would have been used. The power requirement is 2.58 hp. The volumetric based power is 2.22 hp/1000 gal.

TABLE 9.3
Solution to Problem 3

Speed (rpm)	HP	HP/1000 gal	Re	Re Trans	θ_{95} (min) Turb	θ_{95} (min) Trans	θ_{99} (min) Trans
100	0.85	0.73	843	6719	0.38	1.26	2.01
110	1.13	0.97	927	6736	0.34	1.05	1.67
120	1.46	1.26	1011	6782	0.32	0.88	1.41
130	1.86	1.60	1095	6764	0.29	0.76	1.21
140	2.32	2.00	1180	6776	0.27	0.65	1.05
145	2.58	2.22	1222	6781	0.26	0.61	0.98

TABLE 9.4
Constants for Equation (9.17)

Impeller Type	B	k
Helical ribbon	300	36.7
Anchor	180	16.2

9.4.7 Mixing Time—Laminar Flow

Laminar mixing takes place at Re < 10. The process tasks are the same for turbulent and transitional flow. Other important process requirements might include the ingestion of solids and powders. Since viscosities and power input are very high, there is lots of energy to aid in the dispersion of agglomerates. Impellers used include anchor, screw, and gate-type and helical ribbon designs, each having many design variables such as blade width, wall clearance, pitch angle, and single- or double-flight helical ribbon.

Equation (9.17) gives the general form of the power equation described by Tatterson in *Fluid Mixing and Gas Dispersion in Agitated Tanks* [13]:

$$N_P = \frac{C(n)}{Re}$$
$$C(n) = Bk^{n-1}$$
(9.17)

where C is a function of n, the flow behavior index for a power-law fluid; Re is the Reynolds number; and B and k values are given in Table 9.4.

Mixing times depend on geometry, speed, and rheology, in the general form of Equation (9.18)

$$N\theta_M = k_M$$
(9.18)

where k_M is referred to as the mixing or homogenization number.

Straight lines in Figure 9.16 indicate that k_M is constant. Lines for the ribbon are virtually identical, while those for the paddle are quite different. Nagata [14] explains the difference as follows. CMC is a pseudoplastic fluid, and TiO_2, a slurry, is a Bingham plastic. The ribbon impeller produces effective axial flows, but less shearing action. A large axial flow is necessary to maintain particles in a proper suspended state. The paddle creates mostly tangential motion, not conducive for solids suspension, and creates slippage planes in the slurry. Shear stress cannot be transferred to fluid elements effectively, and mixing times are long. We conclude that while paddles and anchors are simple to construct, they are ineffective mixers compared with helical impellers. However, they are helpful in promoting improved heat transfer at the wall. This is discussed in Section 9.11.

9.5 REACTIVE MIXING

This section discusses the mixing of reagents to produce a chemical reaction within a single phase, with a brief introduction to reactive mixing in multiphase systems. The latter is a rather complex subject, and a detailed treatment is beyond the scope of this section.

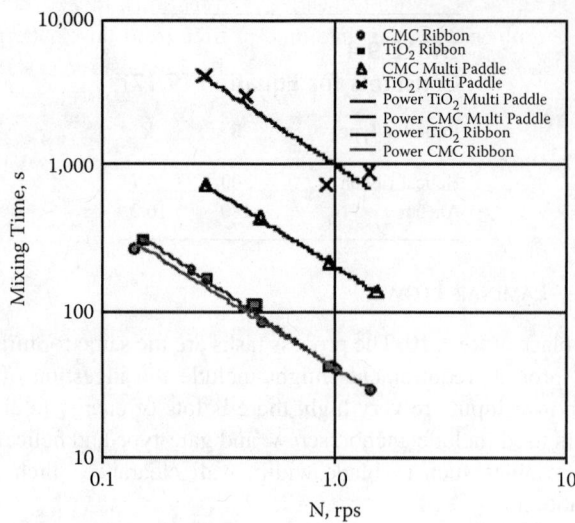

FIGURE 9.16 Comparison of the mixing time for ribbon and multistage paddle in non-Newtonian liquids. (From Nagata, 1975.)

9.5.1 Introduction

The challenges of mixing of fluids during reaction are important for fast reactions in homogeneous systems as well as for all multiphase systems. There are two issues involved: degree of segregation and timing. The degree of segregation refers to whether mixing occurs on the microscopic level (mixing at the molecular level in the limit) or the macroscopic level (mixing of clumps or globs of fluid). The second issue refers to how fast or slow is the change in the degree of segregation from the initial (feed) state to the final (exit) state, i.e., whether the macrofluid (glob of fluid) reaches its ultimate size early or late as it flows through the reactor. Since both macro-mixing and micro-mixing vary with reactor size, the following three undesirable effects are typically observed upon scale-up of mixing sensitive reaction systems:

1. Reduced reaction rates and conversion
2. Reduced selectivity
3. Reduced product quality (increased impurity levels)

Since mixing is very efficient on a laboratory scale, the detrimental effects of mixing observed on larger scales generally come as an unwanted surprise.

9.5.2 Fundamental Concepts in Reactive Mixing

When two miscible reactant fluids A and B are mixed, it is normally assumed that the two streams first form a homogeneous mixture that then reacts. This is true if, and only if, the chemical reaction is slower (say requiring more than 100–400 s) than the mixing time (typically 0.1 s to 50 s in a well-designed turbulent reactor). Mixing time is strongly dependent on scale. When the half-lives of the two processes are similar in magnitude, or when mixing is slower than reaction, mixing and reaction proceed simultaneously, not consecutively. Reaction occurs in localized zones. As the intrinsic reaction rate increases, the sizes of the reaction zones shrink to a plane. Only a small part of the reactor volume is used for reaction. For example, in any proton transfer reactions in aqueous phase (half-life of less than a millionth of a second), the reaction takes place in a very small volume, as may be visually observed in acid-base neutralization reactions. Indeed, acid-base reactions with suitable indicators are often used to determine mixing times and flow patterns.

A single reaction is rarely the case in practice. Usually, the reaction systems are quite complex, with several products, intermediates, and undesired by-products. However, all reaction schemes can be ultimately analyzed in terms of series/parallel reactions.

Parallel reactions:

$$A \rightarrow R \text{ (Product)}$$

$$A \rightarrow S \text{ (By-product)}$$

For reactions in parallel, the concentration level of reactant is the key to the control of product distribution. A high reactant concentration favors the reaction of higher order or the one with the larger reaction rate constant.

Series reactions:

$$A \rightarrow R \rightarrow S$$

For reactions in series, the mixing of fluids of different compositions is the key to formation of intermediate. The maximum possible amount of all intermediates is obtained if fluids of different compositions and at different stages of conversion are not allowed to mix. Thus, a plug-flow reactor will always give a much higher yield of the intermediate R than a CSTR.

Series-parallel reactions:

$$A + B \xrightarrow{k_1} R \text{ (desirable)}$$

$$B + R \xrightarrow{k_2} S \text{ (undesirable)}$$

Series-parallel reactions can be analyzed in terms of their constituent series reaction and constituent parallel reactions. Hence, as far as A, R, and S are concerned, this scheme may be viewed as the series reaction above. On the other hand, this scheme may be viewed as a parallel reaction with respect to B. Thus, the concentration level of B has no effect on the product distribution. When R is the desired product, the best way of contacting A and B is to react A uniformly, while adding B any convenient way.

Once a working reaction scheme is determined, it is possible to determine the optimal contacting for favorable product distribution, as illustrated by Levenspiel [15]. Figures 9.17 and 9.18 from [15] illustrate the contacting patterns for various combinations of high and low concentrations in noncontinuous and continuous flow operations. Here we are concerned about influencing the reactor performance by primarily changing the bulk contacting pattern on a macro scale for relatively slow reactions. We will return to fast reactions and micro-mixing later.

The contacting pattern is not the only issue that determines the batch, semi-batch, or continuous mode of reactor operation. Sometimes, heat transfer and control of exothermic reactions may require

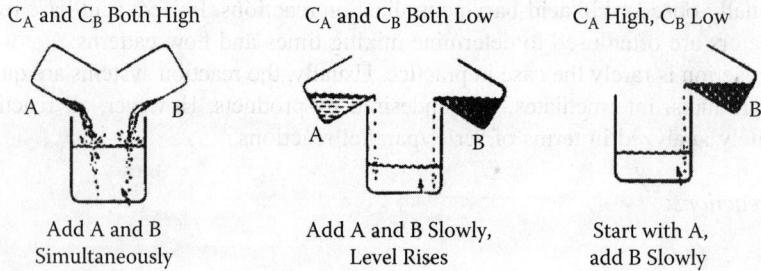

FIGURE 9.17 Contacting patterns for various combinations of high and low concentrations of reactants in noncontinuous operations. (From Levenspiel, 1972.)

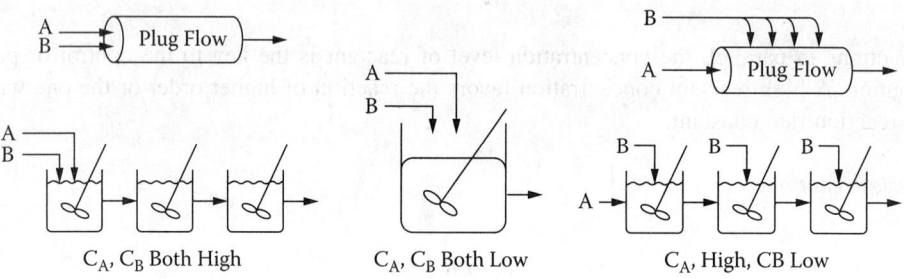

FIGURE 9.18 Contacting patterns for various combinations of high and low concentration of reactants in continuous flow operations. (From Levenspiel, 1972.)

a semi-batch mode of operation. At other times, safety considerations may require very little buildup of a potentially dangerous intermediate.

If kinetics are known for all the reactions (a rare occasion in practice), it is possible to predict the yield and selectivity in ideal reactors or their combinations. Please refer to any reaction engineering textbook for more information on this subject.

9.5.3 Macro-Mixing and Micro-Mixing

Turbulence can be conceptualized as the transfer of kinetic energy from large eddies (on the order of the size of the impeller blade width) to successively smaller eddies. The smaller eddies originally in the inertial sub-range disintegrate further into the viscous sub-range, and eventually all energy is dissipated into heat. This subject is dealt with more fully in the discussion on blending in Section 9.4.

Macro-mixing is mixing that is associated with the length scales larger than the viscous sub-range. Macro-mixing is visible, a result of flow rate, fluid motion, and convection. Flow patterns, circulation cells, and back-mixing regions are macro-mixing phenomena. The residence-time distribution (RTD) is related to macro-mixing.

Micro-mixing is mixing associated at length scales in the viscous sub-range or smaller. Molecular diffusion and turbulence intensity are important micro-mixing parameters. In reality, there is no sharp distinction between these two types of mixing. The names macro- and micro-mixing are conceptualizations of processes that take place simultaneously, and on all scales. Intense micro-mixing without macro-mixing (or vice versa) does not lead to an effective mixing process.

9.5.4 Ideal Flow Patterns in Reactors and Residence-Time Distribution (RTD)

Common practice is to model reactors as some combination of elementary and idealized building blocks, such as

- Continuous stirred-tank reactors (CSTRs)
- Plug-flow reactors (PFRs) with or without axial dispersion
- Dead volumes
- Bypass flows
- Recycle flows
- Inlets and outlets

When can a reactor be considered an ideal CSTR? A rule of thumb is that a stirred tank can be considered ideal when the mixing time is less than about 20% of the mean vessel residence time and, for reacting systems, less than about 5% of the characteristic reaction time.

Residence-time distribution (RTD) is a useful concept in characterizing the macro-mixing in a continuous flow system. However, it does not describe micro-mixing. If the flow pattern in a mixer is known, the residence-time distribution (RTD) can be calculated, and there is no need to measure it. The practice, however, is to measure the RTD and then attempt to deduce the flow pattern from it. An RTD measurement shows the correlation between the concentration of a reactant at two locations of the stirred vessel, typically the inlet and the outlet. An RTD measurement will result in information on:

- Bypass flow rate
- Dead volume
- Active CSTR and PFR volumes

An RTD measurement does not reveal details of the flow patterns within the vessel, or the sequence of the flow through the various regions. For complex geometries, an RTD measurement can be useful in determining the presence of stagnation. There are limitations on when to use RTD measurements to predict the specific flow patterns or the reactor performance. For example, the sequence of a CSTR and PFR cannot be determined from an RTD measurement alone. Swapping locations of a CSTR and a PFR in series will result in the same RTD. However if a non-first-order reaction takes place in a CSTR and a PFR in series, the resulting conversion does depend on sequence. RTD is not sufficient to predict the reactor performance except in the case of first-order reactions. In a reaction scheme with any other order, the reactor performance is determined not only by the length of time that a fluid element spends in the vessel, but also by its surrounding environment during its travel from inlet to outlet.

Quantitative information on the flow pattern can also be determined by direct flow visualization, although this method is limited to optically transparent liquids, sufficient light, suitable flow followers, and adequate observation points. Direct observation provides information about recirculation (back-mixing), short circuits, and dead zones.

Computational fluid dynamics (CFD) has emerged as a very valuable tool in modeling the real flow patterns in chemical reactors. It represents a quantum leap from the idealized reactor models or their modifications, such as the tanks-in-series or axial-dispersion models to account for non-idealities. It has the potential to account for flow and reactions inside a reactor in their entirety. CFD has been used successfully to predict the flow patterns and reactor performance in the case of reactions involving macro-mixing effects.

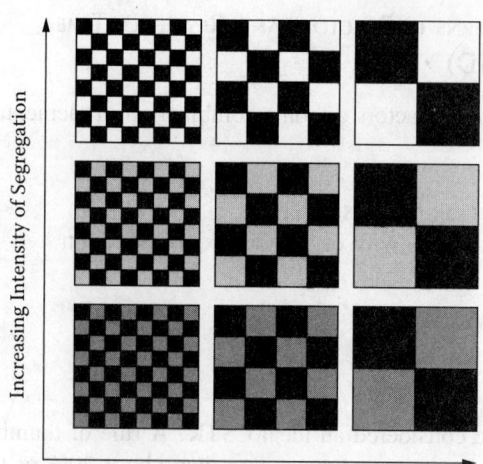

FIGURE 9.19 Scale and intensity of segregation.

9.5.5 MICRO-MIXING AND SEGREGATION

Each feed stream to a reactor usually contains only one reactant, and the large portions of fluid thus introduced must be broken down rapidly so that each reactant can be dispersed to encounter the others and react. The large turbulent eddies break down the fluid elements to successively smaller sizes until a size is reached when the inertial forces are unlikely to reduce it further, and the energy decays due to viscous action. Based on work presented in the literature [16–20], it is possible to make an estimate of the size of agglomerates.

For example, Kolmogoroff's length scale, λ_k, represents the eddy size that is associated with the boundary between the inertial and viscous subranges and is defined by

$$\lambda_k = \left(v^3 / \varepsilon_T \right)^{1/4} \tag{9.19}$$

where ε_T is the local energy dissipation rate expressed as power per unit mass, and v is the kinematic viscosity.

For a typical power input of 1 kW/m³ (P/V = 5 hp/1000 gal) with a waterlike fluid (v of 0.009 cm²/s) in a stirred reactor, λ_k is 30 µm. Note that this is the scale at which molecular diffusion dominates, and a 30-µm (micron) diameter sphere still contains 4×10^{14} molecules of water!

The *scale of segregation* effectively measures the size of unmixed regions. As mixing proceeds, the scale of segregation is reduced, due to liquid motion.

The *intensity of segregation* measures the concentration variations in the mixture. As mixing proceeds, the intensity of segregation is reduced, due to diffusion (Figure 9.19).

Consider two limiting cases to help explain the effect of segregation on a single reaction: Feed streams containing reactants A and B are available, each first as a micro-fluid (free to mix) and then as a macro-fluid (segregation maintained) [15]. Micro-fluids A and B behave in the expected manner and reaction occurs. However, upon mixing of macro-fluids, no reaction takes place because molecules of A cannot contact molecules of B (except at the interface, which has a zero volume in this idealization). These two situations are illustrated in Figure 9.20.

9.5.6 MICRO-MIXING AND SELECTIVITY

If only a single fast (or instantaneous) reaction is involved, relatively slow mixing and state of segregation may cause a change in the apparent reaction rate (reduction in plant capacity), but no

Industrial Mixing Technology

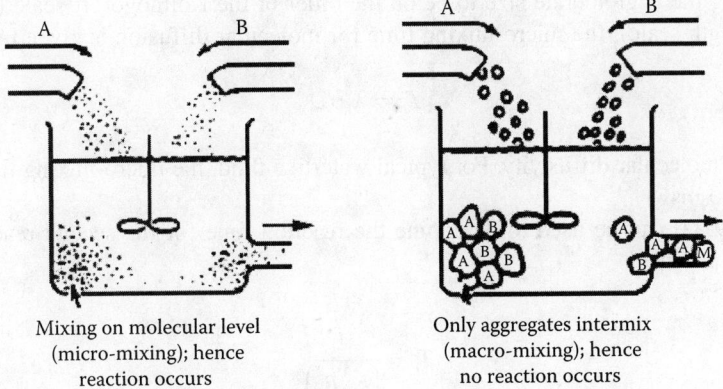

FIGURE 9.20 Differences in behavior of micro-fluids in the reaction of A and B.

effect on product quality (no side reactions). However, this is rarely the case in reality; multiple series and parallel reactions are always involved. Effects not observed on a small scale may appear on large scale, that is, color, yield loss, lower quality, and so forth.

Consider the series-parallel reaction scheme, also referred to as the competitive consecutive reaction scheme, discussed earlier. The desired product R continues to react with the initial reactant B to produce the undesired product S. Usually the reaction conditions (temperature or catalyst) are arranged such that k_1 is much faster than k_2. The selectivity, X_S, is defined as

$$X_S = \frac{2S}{R + 2S} \qquad (9.20)$$

For optimum performance, X_S should be kept as low as possible.

If we let equimolar quantities of A and B react, the following diagram clarifies how mixing can control selectivity under two limiting cases:

Initial State
(Completely Segregated)

```
A    A    A    A
A    A    A    A
B    B    B    B
B    B    B    B
```

Poor Mixing:	Ideal Mixing:
$\tau_M \gg \tau_R$	$\tau_M \ll \tau_R$
First reacts, then mixes	First mixes, then reacts

↓   ↓

```
A  A  A  A        A  B  A  B
R  R  R  R        B  A  B  A
B  B  B  B        A  B  A  B
                  B  A  B  A
```

↓   ↓

```
A  A  A  A        R  R  R  R
S  S  S  S        R  R  R  R
 (Undesirable)     (Desirable)
```

Considering the agglomerate size to be on the order of the Kolmogoroff scale (or some other appropriate length scale), the micro-mixing time for molecular diffusion is given by

$$T_m = \lambda_\kappa^2/D \tag{9.21}$$

where D is the molecular diffusivity. For typical waterlike fluid, the micro-mixing time falls in the range of 1 to 20 ms.

Equation (9.22) can be used to determine the reaction time for the second reaction, forming an undesirable product:

$$T_r = \frac{1}{\left(k_2^* B_o\right)} \tag{9.22}$$

where B_o is the initial or local concentration of B. It is important to remember that the local reaction rate is not given by the bulk concentration, but by the feed or local concentration.

The ratio of time constants for mixing and reaction defines the Dämköhler number or the micro-mixing modulus, a dimensionless number shown in

$$Da = \frac{k_2 \lambda_k^2 B_o}{D} \tag{9.23}$$

The dependence of X_S on the micro-mixing modulus is given in Figure 9.21 at various ratios of the two kinetics constants. The S-shaped curve for high ratio of k_1/k_2 (practically relevant for reasons stated earlier) shows that the selectivity is sensitive to micro-mixing, especially if the Dämköhler number is around 1 or greater.

Bourne and coworkers [21–23] have developed the mathematics and applied their model to predict the effects of stoichiometric ratio, startup of a semi-batch reactor, effect of volumetric feed ratio, batch vs. continuous operation, etc. They have also experimentally demonstrated the use of reaction systems with well-characterized kinetics to determine the level of micro-mixing. Thus chemical reactions can be considered as molecular probes to be used to study segregation. Other

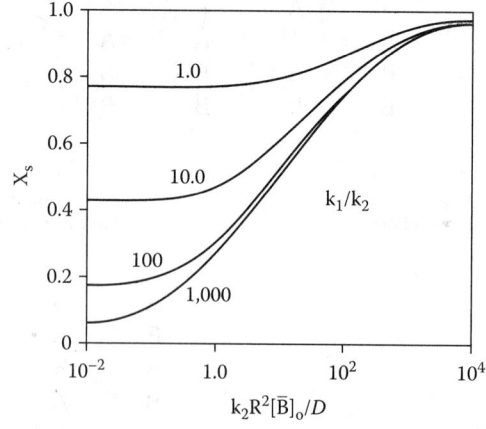

FIGURE 9.21 Influence of k_1/k_2 on product distribution from CSTR; parameter: k_1/k_2 ($a = 1$; $N_{AO} = N_{BO}$; $\tau = 100\ \tau_E$). (From Baldyga et al., 1984.)

techniques such as optical or conductivity probes have limited spatial resolution and cannot give information down to the smallest scales of turbulence.

There are other more advanced micro-mixing models such as the engulfment, deformation, and diffusion (EDD) model [24–26] and the interaction by exchange with the mean (IEM) model [27]. The reader should consult the original papers for details.

9.5.7 Micro-Mixing: Practical Implications

Always suspect micro-mixing to be important when reactions are fast and lab-scale studies show minimal side reactions. If possible, make the quantitative comparison using the appropriate reaction scheme and micro-mixing modulus. Laboratory reactors tend to run at much higher turbulence levels than plant reactors.

When micro-mixing is important, the point of addition of ingredients makes a large difference on selectivity [28]. In such cases, ingredients should be added at the point of highest energy dissipation and highest turbulence. Note that the local energy dissipation rate is not uniform in an agitated vessel (see Figure 9.10) and can vary by more than an order of magnitude in both directions from the average energy dissipation rate calculated from the power per unit volume. X_S is a strong function of the local power dissipation rate, ε_T. In a stirred vessel, a high ε_T region is within and downstream of the impeller tip region. The best addition point is just before the bulk of the flow enters a region with a high local energy dissipation rate, such as the suction of an impeller.

High local energy dissipation is very advantageous for fast competitive and consecutive reactions. In such cases, improvements in micro-mixing will result in reduction of unwanted materials, in improvement of quality, and possibly in increased production capacity.

9.5.8 Reactive Mixing in Multiphase Systems

The physical aspects of gas-liquid, liquid-liquid, solid-liquid, and gas-liquid-solid systems are discussed in the subsequent sections of this chapter. Using the guidelines given there, it is possible to get an estimate of local and average mass-transfer coefficients, interfacial areas, and contacting patterns. This section briefly considers the effect of reactions on mass transfer, a subject treated elsewhere in this handbook and in advanced texts [29, 30]. Note that while we will refer mostly to gas-liquid systems, the same treatment would more or less apply to liquid-liquid and liquid-solid systems. In the case of gas-liquid-solid systems, it may be possible to determine the controlling resistance and simplify the analysis to a two-phase system, as far as the reaction part is concerned.

In gas-liquid reactions carried out in agitated tanks, generally the gas-side resistance is small compared with the liquid-side resistance, and hence the following discussion is limited to only processes where the liquid-side resistance controls. If a pure gas is used, there is no gas-side resistance.

In reactions involving gas-liquid mixing, one of the first things to determine is the controlling process: (bulk) chemical reaction or mass transfer. Experimentally, this is done by checking the effect of agitator speed on rate of reaction, as shown in Figure 9.22.

If the process is mass-transfer controlled, there are three possible regimes:

Film diffusion (regime 2)
Film kinetics (regime 3)
Instantaneous reaction (regime 4)

The volumetric mass-transfer coefficient, $k_L a$, increases as the stirrer speed increases. (Refer to correlations given in Section 9.7 later in this chapter.) If the process is bulk-reaction controlled, then the stirrer speed increases and the volumetric mass-transfer coefficient, $k_L a$, increases (refer to correlations given in Section 9.7). If the process is mass-transfer controlled (regimes 2, 3, 4),

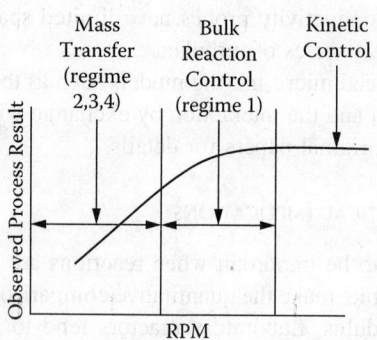

FIGURE 9.22 Effect of agitation speed on observed process rate.

the observed rate of reaction increases. However, a condition may be reached when the observed rate does not increase with increasing the rpm or mixing intensity, indicating the overall process is bulk-reaction controlled (regime 1). For the exact mathematical conditions for the various regimes, please refer to the illustrative example and the reference book [29]. Design speed should be at the minimum required to keep reaction conditions in regime 1.

Certain fast or instantaneous reactions (for example, proton transfer reactions) are always mass-transfer controlled. The enhancement factor, I, accounts for the effect of chemical reactions on mass transfer and is defined as the ratio of the process rate over the mass-transfer rate in absence of a chemical reaction.

- In regime 2, the reaction is fast enough to keep the bulk liquid phase concentration of the gaseous reactant essentially zero but not fast enough to occur substantially in the liquid film. There is no enhancement of mass transfer due to reaction. Diffusion and reaction take place in series fashion.
- In regime 3, the reaction is sufficiently fast to completely consume the gaseous reactant in the liquid film. Diffusion and reaction are occurring simultaneously in a parallel fashion in the liquid film. The true liquid-side mass-transfer coefficient has no effect on the overall process rate.
- In regime 4, the reaction is so fast (virtually instantaneous) that the gaseous reactant (A) and the liquid-phase reactant (B) cannot coexist. Diffusion of A from the interface and diffusion of B from the bulk liquid toward the reaction plane control the overall process.

There are systems that could fall between these regimes, and the division between them is not always clear-cut.

It is important to determine the regime of a particular reaction system, since the equipment choice and the effect of design and operating variables on the process performance depend on the regime. A lack of this fundamental understanding leads to many apparent discrepancies between different scales of operations and sometimes to scale-up failures. For instance, in the lab, the process may be operating in regime 1, while in production scale it could operate in regime 2. In this specific case, on the lab scale, the increasing concentration of the liquid-phase reactant increases the process rate, while on the production scale, there would be a essentially no effect. Alternatively, one may design equipment for minimal mass-transfer requirements based on the confirmation of regime 1 on lab scale only to find much lower process rates. There would be a negligible effect of partial pressure of A if the process operates in regime 4. The effect of temperature is minimal if the system is in regimes 2 and 4. It is substantial in regime 3 (apparent activation energy is half of the true activation energy) and maximum in regime 1 (apparent activation energy is equal to the true activation energy).

Industrial Mixing Technology

9.5.9 Equipment Types and Design Guidelines

For typical reactive mixing systems in a stirred tank, a dual impeller system, Figure 9.23, with a radial-flow impeller at the bottom and an axial flow on the top is recommended [31–33]. We recommend the following ratios as initial starting points for a typical batch with liquid height equal to tank diameter:

- Lower impeller diameter/vessel diameter, $D_l/T = 0.25$
- Upper impeller diameter/vessel diameter, $D_u/T = 0.33$
- Distance between impellers/lower impeller diameter = 1.75
- Clearance of lower impeller from bottom = $0.75 \times D_l$

If a single impeller is desired, we recommend a 45° four-blade PBT with a diameter equal to one-third of the tank diameter located at a clearance of 0.75 times the impeller diameter from the bottom as the starting point.

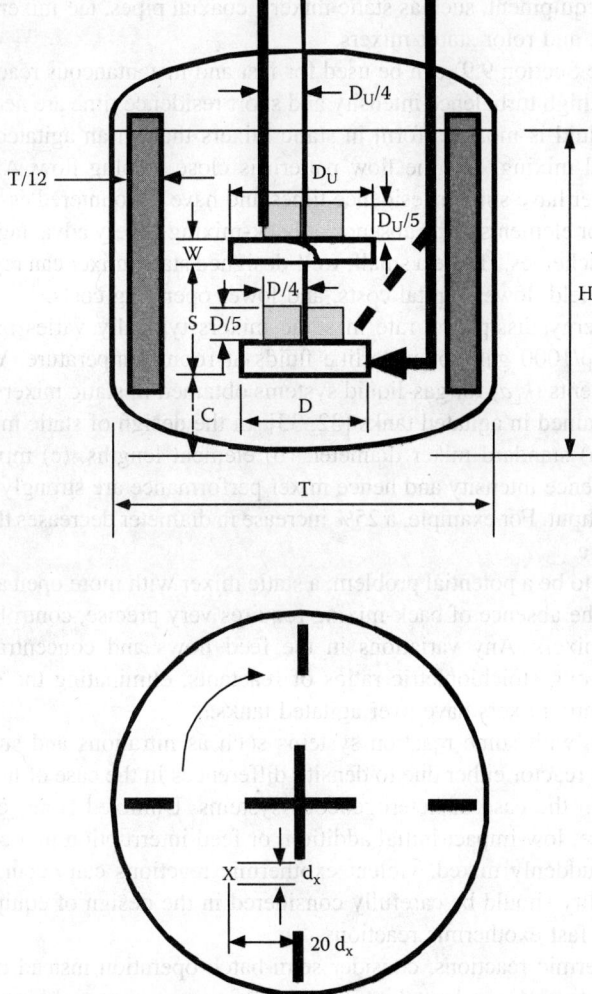

FIGURE 9.23 Feed point preferences for reactive mixing. (From Fasano, J. B., and W. R. Penney, *Chem. Eng. Prog.* 87: 46 (1991). With permission.)

The feed should be introduced to the suction of the impeller, as shown in Figure 9.23 [31], to take full advantage of maximum energy dissipation to promote rapid blending of the feed. Unfortunately, it is the worst possible location for getting back-mixing into the feed pipe with potential subsequent reactions under milder turbulent conditions leading to plugging problems.

To minimize back-mixing into feed pipe and to prevent the feed jet from penetrating through the impeller zone, Fasano and Penney [31] recommend:

- For feed location above impeller: $0.09\, V_t < V_f < 0.12\, V_t$
- For feeding in the plane of impeller: $V_t < V_f < 40$ ft/s

where V_t = impeller tip speed, and V_f = feed jet velocity.

Other devices are used to conduct reactive mixing. Reaction injection-molding systems (RIM) use high-energy opposing jets, each delivering a reactive component such as for polyurethane parts. Reaction rates are fast relative to diffusion. High-velocity impinging jets ensure intensive mixing for RIM applications. A typical micro-mixing Kolmogoroff scale (referred to as striation thickness in reaction injection mixing, RIM, literature) is 10 to 50 μm. Reactive mixing can also be carried out in other types of equipment, such as static mixers, coaxial pipes, tee-mixers, dynamic propeller driven in-line mixers, and rotor-stator mixers.

Static mixers (see Section 9.9) can be used for fast and instantaneous reactions involving low-viscosity fluids when high turbulence intensity and short residence time are needed. The turbulence encountered by the fluid is more uniform in static mixers than in an agitated tank. Static mixers give very good radial mixing, and the flow pattern is close to plug flow. All the fluid elements leaving the static mixer have similar residence times and have encountered essentially the identical environment (neighbor elements). The absence of back-mixing is very advantageous for consecutive competitive reaction schemes. Hence a small, well-designed static mixer can replace a large agitated tank with improved yield, lower capital costs, and lower operating costs.

The turbulent energy dissipation rate in static mixers typically varies between 100 to 1000 kW/m³ (20 to 200 hp/1000 gal) for waterlike fluids at room temperature. Values of volumetric mass-transfer coefficients ($k_L a$) for gas-liquid systems obtained in static mixers are 10 to 100 times higher than those obtained in agitated tanks [32, 33]. In the design of static mixers as reactors, the main variables are (a) standard mixer diameter, (b) element lengths, (c) mixer voidage, and (d) pressure drop. Turbulence intensity and hence mixer performance are strongly dependent on diameter for a fixed throughput. For example, a 25% increase in diameter decreases the energy dissipation rate by a factor of five.

If plugging is apt to be a potential problem, a static mixer with more open area (higher voidage) should be selected. The absence of back-mixing requires very precise, controlled metering of feed streams into static mixers. Any variations in the feed flows and concentrations could lead to uncorrectable and wrong stoichiometric ratios of reactants, eliminating the selectivity and yield improvements that static mixers have over agitated tanks.

In stirred vessels, with some reaction systems such as nitrations and sulfonations, unmixed zones can form in the reactor either due to density differences in the case of homogeneous systems or phase separation in the case of heterogeneous systems. Unmixed zones can result from poor design, agitator failure, low-impact initial addition, or feed interruption in a continuous reactor. If unmixed fluids are suddenly mixed, violent exothermic reactions can occur, leading to a major disaster. This possibility should be carefully considered in the design of equipment and operating procedures involving fast exothermic reactions.

For highly exothermic reactions, consider semi-batch operation instead of a batch operation. The semi-batch operation allows for a better control of temperature and heat removal by limiting the feed rate of the reactants.

If the reaction intermediates are potentially hazardous, their concentration levels should be kept minimal in the reactors. For continuous processes, consider a plug-flow operation instead of a

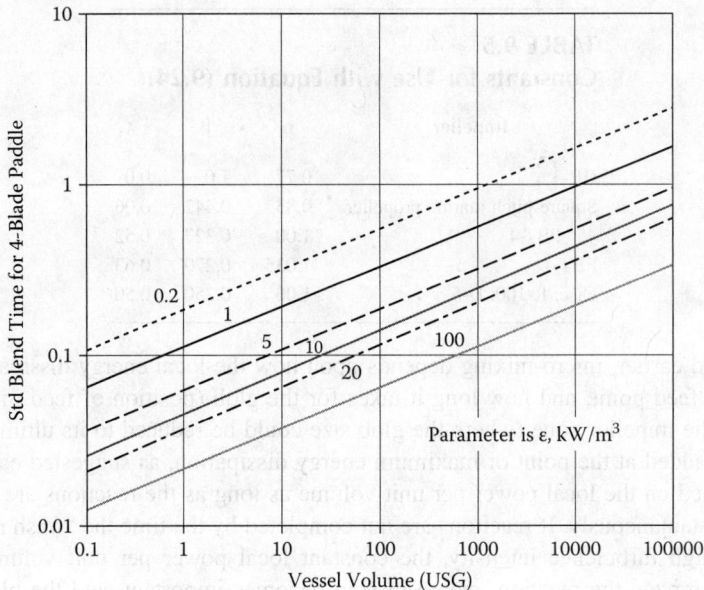

FIGURE 9.24 Blend time comparisons at constant P/V for different scales. (From Fasano et al., 1991.)

CSTR to minimize the inventory of intermediates. A plug-flow reactor would also typically have smaller volumes, hence a smaller inventory of reactants, intermediates, and products than a CSTR for the same throughput. This can be a big safety advantage.

9.5.10 Scale-Up

The intrinsic kinetics of a reaction system is independent of scale. The fundamental chemistry is the same whether on a lab scale or plant scale. What changes with scale is the environment. In the case of multiphase systems, the mass-transfer characteristics change drastically with scale. It is virtually impossible to keep all aspects of the fluid environment equal upon scale-up, as reported by Leng [35]. Specifically, mixing parameters such as mixing time, power per unit volume, turbulence intensity and its distribution, average shear and maximum shear, tip-speed, Reynolds number, and flow distribution, etc., cannot all be kept constant. This necessitates choosing the most essential mixing scale-up variable for the system and compromising and understanding the impact other variables will have on the process.

For example, typical ranges of blend times with scale for a standard four-blade 45° PBT are given in Figure 8 from Fasano and Penney [31].

Figure 9.24 shows that for a typical 1-hp/1000-gal (0.2 kW/m³) power input, the blend time would go up from about 0.1 min in a 1-gal reactor to 0.8 min in a 10,000-gal reactor. Would this type of increase have any detrimental effect on the reaction system? Of course the answer depends on the specific chemistry. Comparative mixing times are given for different impellers. These are shown in the following equation for the correction factor, CF. If the mix time is known for a given impeller system, multiplying this time by the CF gives the mix time for a different system:

$$CF = \alpha \cdot (sg)^{1/3} \left[-\ln(1-U)/4.605 \right] \cdot \left[\frac{T}{Z} \right]^{0.056} \left[\frac{\beta}{D/T} \right]^{X} \tag{9.24}$$

where sg = specific gravity, U = extent of complete mixing, T = vessel diameter, Z = height of straight side, and D = impeller diameter. Constants are given in Table 9.5.

TABLE 9.5
Constants for Use with Equation (9.24)

Impeller	α	β	X
HE-3	0.77	1.0	0.0
Square-pitch marine propeller	0.83	0.442	0.06
45° PBT-4	1.00	0.327	0.52
FBT-4	1.035	0.270	0.63
Disc: R-100, D-6	1.06	0.250	0.50

As discussed earlier, micro-mixing depends upon how the local energy dissipation affects the glob size at the feed point, and how long it takes for the glob (position of feed in the circulation loop) to reach the impeller zone (where the glob size could be reduced to its ultimate size). If the ingredients are added at the point of maximum energy dissipation, as suggested earlier, the scale-up could be based on the local power per unit volume as long as the reactions are extremely fast, i.e., virtually instantaneously. If reactions are not completed by the time the "fresh reactants" leave the region of high turbulence intensity, the constant local power per unit volume may not be adequate. To complete the reaction, macro-mixing becomes important, and the blend time could become the limiting scale-up criterion.

The chemical kinetics are usually not known for many industrial reactions and are often quite difficult to determine. While CFD holds the most promising approach for homogeneous systems when the kinetics are fully known, some experimental work with verification of CFD predictions is required for a successful scale-up. Most frequently, there is no luxury of determining detailed kinetics and doing CFD computations and verification. Hence, the following procedure (modified from Fasano [31]) is recommended:

- Conduct experiments to determine yield, (X_S), varying N, D/T, feed location, and feed time or feed rate. If possible, this should be done at two scales (for example, 1-gal and 25-gal reactors).
- Plot the yield data vs. mixing time, power per unit volume, local power per unit volume, etc., and determine the appropriate scale-up criterion.
- If blend time is the appropriate criterion, use Figure 9.24 to determine the feasibility of using an agitated vessel without a pump-around loop.
- If blend time is excessive on a large scale without requiring impractical power input, use a circulation loop containing static mixers.
- If local power per unit volume (micro-mixing) is the criterion, generally there should be no scale-up difficulty in terms of design. Make sure that macro-mixing does not become limiting upon scale-up.

9.5.11 Some Practical Guidelines for Small-Scale (Lab) Experimentation

To set process conditions successfully in industrial design, it is useful to first determine the optimal conditions and design on a model scale. When applied properly, upscaling and downscaling have immense advantages. One of the important, but often ignored, functions of small-scale research is to learn from "blunders."

The lab reactor should be geometrically similar to a full-scale reactor, with the proper impeller and baffles as appropriate, not round-bottom flasks with magnetic stirrers. The latter ones are not useful to gather scale-up data.

Create a similar environment to that of the large scale. This means using the same concentrations and purity of reactants as on the industrial scale. In some reaction systems, it may be possible to reduce mixing sensitivity by changing temperature.

Run the lab reactor at the same blend time as the large scale and determine/compare the product quality. If you are told "mixing is not important," ask the researcher to run the reaction without any agitation. Mixing often appears to be noncritical at the bench scale because it is not rate limiting and turnover is fast. This usually changes with scale-up. One method to determine the effect of micro-mixing on product quality is to measure purity, color, presence of by-products, etc., as a function of agitator speed. Care must be taken to select very slow speeds capable of simulating conditions in the intended large reactor. For example, the effect of rate of addition of reactants on quality needs to be explored for batch and semi-batch reactions. Fast addition rates can lead to poor quality. This critical "addition time" usually depends on impeller speed (energy dissipation), the number and location of injection points, and reactant concentration for a given system.

The minimum recommended scale ratio for experimentation must not be less than 1/100 to 1/500 of full-scale equipment. If critical mass-transfer-dependent multiphase reactions are involved, the scale-up ratio should be reduced to 1/10 to 1/50.

9.6 SOLID-LIQUID MIXING

9.6.1 INTRODUCTION

Solid-liquid mixing involves the suspension, distribution, and the drawing down of solids by agitation. In addition to vessel geometry, impeller variables include type, diameter, number, speed, and location. Process results include the desired level (quality) of suspension, such as just off-the-bottom, complete uniformity, or any intermediate condition. The slurry properties, density difference (solid/liquid), viscosity, and solids concentration all determine how difficult the task may be. As alternatives to stirred vessels, jets (see Section 9.10) can be used for light-duty suspension. Literature references deal mainly with settling solids as opposed to floating solids. We will try to address both conditions.

Solid-liquid mixing is used extensively throughout the chemical process industries. Examples include the suspension of solid catalysts in a reactor, the dissolution of solids to form a solution, the suspension of solids in crystallization, suspension polymerization, and fermentation. Crystallizers are usually designed to maximize flow, hence solids suspension, and to minimize shear to avoid crystal attrition. A common application is to provide solids-liquid mixing to feed a solids separator such as a centrifuge. Here the task is to provide a uniform solids feed as levels recede in the feed vessel.

Solids concentrations can vary from a few percent to well over 50% in a typical stirred tank. Solids concentration, particle shape, and the viscosity of the suspending phase are the main factors affecting the rheology and settling characteristics of the slurry. Cubic- and spherical-shaped solids tend to form Newtonian slurries, while needle-, oblong-, and plate-shaped solids form thixotropic slurries. Such slurries exhibit yield stresses even at quite low solids concentrations. This can lead to the development of caverns, as shown in Section 9.4. Proper design can usually overcome these stagnation problems.

9.6.2 SETTLING SOLIDS

The settling rate of a solid in a liquid is determined by its drag, a force defined by many parameters, including the shape-dependent drag coefficient. The motion of particle settling is described by the particle Reynolds number, as seen in

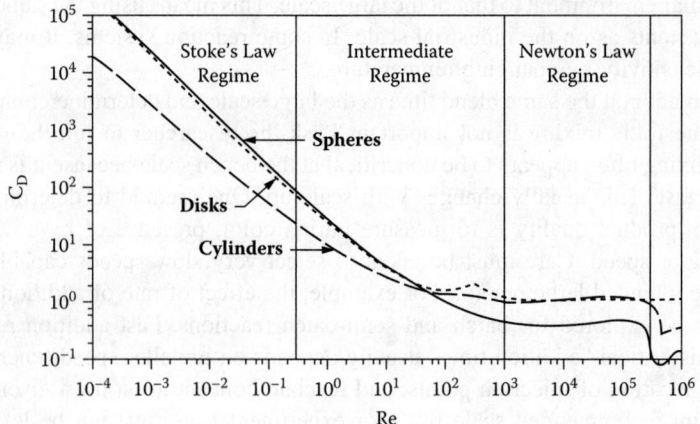

FIGURE 9.25 Dependence of drag coefficient on Reynolds number for spheres, cylinders, and discs.

TABLE 9.6
Expressions for C_D for Spheres in Different Flow Regimes

Region	Reynolds Number	Expression for C_D
Stokes law regime	$Re \leq 0.3$	$C_D = 24/Re_p$
Intermediate regime	$0.3 \leq Re_p \leq 1000$	$C_D = 18.5/Re_p^{3/5}$
Newton's law regime	$1000 \leq Re_p \leq 3.5\ 10^5$	$C_D = 0.445$

$$Re_{particle} = \frac{d_p V_t \rho_l}{\mu_l} \tag{9.25}$$

where d_p is the particle diameter, V_t is the terminal settling velocity, and ρ_l and μ_l are the density and viscosity of the suspending phase. Stokes's law applies when the particle Reynolds number is less than 0.3.

Figure 9.25 shows the effect of particle shape on the drag coefficient over a range of Reynolds numbers.

The following equation enables one to calculate the free settling velocity of a particle:

$$V_t = \left(\frac{4 g_c d_p (\rho_s - \rho_l)}{3 C_D \rho_l} \right)^{1/2} \tag{9.26}$$

The drag coefficient C_D is a variable depending on the flow conditions represented by the particle Reynolds number and is given in Table 9.6 or Figure 9.25.

Equation (9.27) is the corresponding terminal settling velocity expression for the laminar regime or Stokes's law given in Table 9.6:

$$V_t = \frac{g_c d_p (\rho_s - \rho_l)}{18\,\mu} \tag{9.27}$$

Equation (9.28) gives the settling velocity for the turbulent regime:

$$V_t = 1.73 \sqrt{\frac{g_c d_p (\rho_s - \rho_l)}{\rho_l}} \qquad (9.28)$$

When slurry concentrations reach 15–20% solids, a phenomenon known as hindered settling occurs, which reduces settling velocities over those predicted by Equations (9.27 and 9.28). Equation (9.29) is proposed by Maude [36] in an attempt to correct for hindered settling.

$$V_{hs} = V_t (1 - \bar{C})^m \qquad (9.29)$$

where $m = 4.65$ for Stokes's law regime, and 2.33 for Newton's law regime.

The settling velocity correction factor $(1 - \bar{C})^m$ is 0.98 at 1% and 0.89 at 5%, for Newton law regime.

9.6.3 Particle Suspension in Stirred Vessels

The just-suspended state is defined as the condition where no particle remains on the bottom of the vessel (or upper surface of the liquid) for longer than 1 to 2 s. At just-suspended conditions, all solids are in motion, but their concentration in the vessel is not uniform. There is no solid buildup in corners or behind baffles. This condition is ideal for many mass- and heat-transfer operations, including chemical reactions and dissolution of solids. At just-suspended conditions, the slip velocity is high, and this leads to good mass/heat-transfer rates. The precise definition of the just-suspended condition coupled with the ability to observe movement using glass or transparent tank bottoms has enabled consistent data to be collected. These data have helped with the development of reliable, semi-empirical models for predicting the just-suspended speed. Complete suspension refers to nearly complete uniformity. Power requirement for the just-suspended condition is much lower than for complete suspension.

Many agitated waste-treatment ponds operate at conditions shown in Figure 9.26(a). Power is low and solids build up in the corners of the vessel, and that buildup is acceptable. Figure 9.26(b) represents the just-suspended state, an excellent situation for establishing good mass and heat transfer around a particle. Energy levels are low to moderate, and this condition is the most common in industry. Figure 9.26(c) represents uniform solids suspension. This is desirable for certain reactions and feeding separation equipment such as filters, centrifuges, or continuous processing equipment. Particles are fully entrained by the flows, and slip velocities are lower than for the just-suspended case, illustrated in Figure 9.26(b). The relative standard deviation or RSD has become a quantitative tool to describe the degree of uniformity, as is shown later in this section.

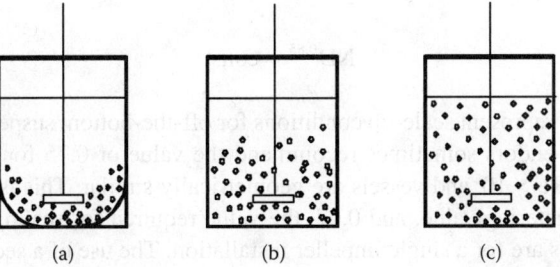

FIGURE 9.26 (a–c). Typical solids suspension conditions.

TABLE 9.7
Power Numbers and *s* Values (Zwietering Equation) for Different Impellers, Clearances (*C/T*), and Impeller Sizes (*D/T*)

Impeller Type	C/T	D/T	Po	s
4–45°-PBT, downflow	1/5	1/3	1.4	5.7
	1/4	1/3		6.2
	1/4	1/2		5.8
	1/4	1/4		7.1
6–45°-PBT, downflow	1/4	1/2	1.6	5.7
	1/4	1/2		6.4
A-310, downflow	1/5	1/3	0.25	7.6
	1/4	1/3		7.9
	1/4	1/3		7.8
	1/4	1/2		6.0
HE-3, downflow	1/4	1/2	0.25	6.2
	1/4	1/3		7.2
Intermig (2 off)	1/6 2/3	7/10	0.7	7.4
6–45°-PBT, upflow	1/4	1/2	1.6	6.9

9.6.4 JUST-SUSPENDED CONDITIONS

The determination of the just-suspended speed, N_{JS}, for off-the-bottom suspension has been extensively studied [37, 38], most recently by the British Hydrodynamics Group (BHRG). All studies support the Zwietering equation given by

$$N_{js} = s \upsilon^{0.1} \left[\frac{g \Delta \rho}{\rho_l} \right]^{0.45} X^{0.13} d_p^{0.20} D^{-0.85} \qquad (9.30)$$

where N_{js} = speed in rps to just-suspended particles, $\Delta \rho$ = difference in density between the solid and liquid phases, ρ_l the density of the liquid phase in kg/m³, X = mass ratio of solids to liquid, υ = kinematic liquid viscosity in m²/s, g = 9.8 m/s², d_p = average particle size in m, D = impeller diameter in m, and s is the geometry factor given below in Table 9.7.

The speed for just suspension can be calculated by means of Equation (9.30) using s values given in Table 9.7. Note that s varies with impeller type, clearance, and *D/T*. Additional values for s may be found in the *Handbook of Industrial Mixing* (chapter 10, 560) [1].

For purposes of scale-up, the Zwietering equation predicts for equal systems that

$$ND^{0.85} = \text{const} \qquad (9.31)$$

Equation (9.31) would represent scale-up conditions for off-the-bottom suspension. Lacking precise process information, vendors sometimes recommend the value of 0.75 for the scale-up exponent for conditions where Re > 10^4 and vessels are geometrically similar. This is an intermediate value between 0.85, the Zwietering value, and 0.67, the value required for scaling solids uniformity.

These relationships are for a single impeller installation. The use of a second impeller has little effect on N_{JS}, but does provide for better solids uniformity.

9.6.5 Floating Solids

The formation of slurries of nonwetting, floating solids occurs in two stages. First, clumps of dry solids are engulfed by surface eddies. Once entrained, gas is separated from the solids as the clumps pass through the impeller. Successful engulfment depends on a locally high velocity interacting with the settled solids. While such flows are easy to create at the bottom of the vessel, they are hard to form at the top. The task is made quite difficult by (a) the flat, deformable upper surface in contrast to a smooth dish-shaped bottom that helps suspend solids assisted by a directed impeller flow; (b) the poor ability of the impeller to direct a discharge stream to engage the solids; and (c) the poor solids wetting that resists engulfment. Floating wetted solids are much easier to form into slurries.

If a surface vortex can be formed, it helps to incorporate floating solids. This can be a central vortex created by full-body swirl, or a localized one such as those formed behind a baffle. In both cases, the tip of the vortex helps feed solids to a region beneath the surface, where stronger flows exist. An impeller submerged a distance T/4 can do a good job of forming a slurry. Here care must be taken not to engulf gas. If gas bubbles attach to the floating solids, it makes suspension forming much more difficult.

Rigid, thixotropic floating solids can be broken up by shear using rakelike arms extending down into the slurry from horizontal arms located above the surface. This design breaks up tendencies to stagnate. The drag produced by the fingers requires little power yet creates the necessary shear stress to break up caking solids.

More power is needed to create suspensions of floating than for settling solids. A 45° PBT placed at a depth of T/4 from the surface in combination with a second impeller placed lower down in the vessel usually works well and should avoid gas entrainment. The placement of baffles is critical. If a central vortex is to be used to incorporate solids, the vessel should be baffle-free in the upper half of the tank. Short, wide baffles suspended from the top of the tank extending to a depth of T/3 are an alternative for initiation of engulfment [39]. A large $D/T = 0.6$, four-blade 45° PBT placed near the bottom of the tank was used in Joosten's work. The minimum speed N_{DF} for just-suspending conditions is given by Equation (9.32):

$$\frac{N_{DF}^2 D}{g} = 3.6 \cdot 10^{-2} \cdot \left(\frac{D}{T}\right)^{-3.65} \left(\frac{\Delta\rho}{\rho_1}\right)^{0.42} \quad (9.32)$$

It takes more energy to reach the critical speed for suspending floating solids than that for settling solids. Scale-up is $ND^{0.5}$ = constant (similar geometry and $Re > 10^4$). The 0.5 value for the exponent suggests that more power is needed for scale-up than for constant P/V.

9.6.6 Uniform Solids Concentrations

Occasionally, processes require solids uniformity in the vessel. Uniformity is more difficult to measure than off-the-bottom suspension. The best acceptable parameter is the relative standard deviation, or RSD. This involves multiple depth sampling and analysis under steady conditions at known positions in the vessel. The following equation gives a definition of the RSD. Examples of this measure of uniformity are given by Mak [40].

$$RSD_j = \frac{1}{C_m}\left[\frac{1}{n-1}\sum_{i=1}^{n}(C_{ij} - C_m)^2\right]^{1/2} \quad (9.33)$$

where C_m is the overall mean concentration, n is the number of sample points, and C_{ij} is the concentration of the ith position, jth speed.

Low values of the RSD mean better uniformity. Dual impellers help establish better uniformity by effectively distributing flow, hence solids, throughout the vessel. Recent CFD studies have shown reasonably good agreement between measured and calculated solids concentration profiles.

Design recommendations for obtaining good solids uniformity include using:

- Dish-shaped bottom heads
- Four baffles T/12 wide, placed T/72 away from the wall, and extending from the lower tangent line to the surface of the liquid
- Impellers: 45° PBT, hydrofoil (HE-3, A-310), or a bottom-located RCI
- A second impeller, if solids uniformity is essential, and $H/T > 1.2$

Mak [41] has shown that at three different scales, the RSD scales up using constant P/V. This is shown in Figure 9.27. Increasing the power (increasing N^3D^2) reduces the relative standard deviation (RSD). The data for three widely different scales were originally plotted against various different scaling criteria; the one shown in Figure 9.27 is for constant P/V. The close agreement for the three scales validates the P/V scaling assumption. This agreement for large-scale vessels is important. It implies that model suspension experiments can safely be conducted in fairly small equipment. At the Dow Chemical Company, the author determined the power needed to form completely uniform suspension of fast-settling solids. Studies were done in three bench-scale vessels of 0.3 m, 0.457 m, and 0.61 m in diameter using process materials. Equivalent solids suspension at different sizes was found at P/V = constant. This result was applied to a (20-ft diameter) 50,000-gal slurry reactor. Sampling at different positions showed solids uniformity had been reached.

A useful rule of thumb for solids suspension is to relate the solids settling velocity to the calculated suspension velocity created by the impeller. When the flow velocity exceeds the solids settling velocity by a factor of 1.5 to 2.0, usually good performance can be expected. As an example, we calculate the speed to completely suspend solids in a 3.05-m (10-ft) vessel equipped with a 1.22-m (4-ft) HE-3 impeller. The solids settling velocity is 3.05 m/min (10 ft/min). Furthermore, it is assumed that downflow occurs in an area of 1.2 D and that the circulation pattern is single loop. The annular upflow area becomes $(\pi/4)(3.048^2 - 1.52^2) = 5.48$ m². The velocity in this region must

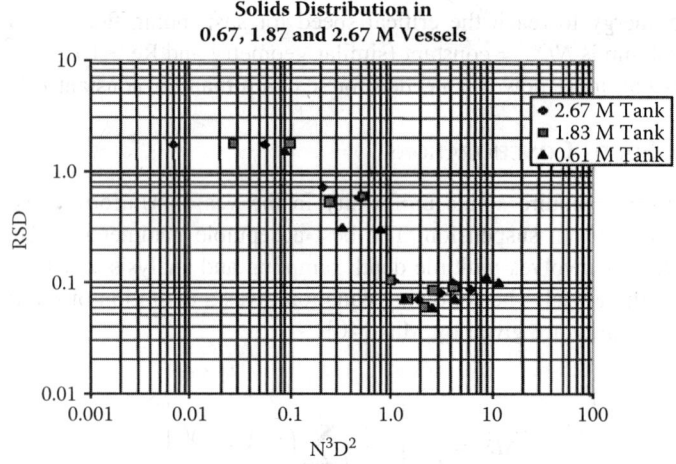

FIGURE 9.27 Power per unit volume as the scale-up criterion for solids distribution. (From Mak et al., 1990.)

TABLE 9.8
Relative Power and Speed Required for Solids Suspension with Solids-Settling Velocities of 5–18.3 m/min

Criterion	Speed Ratio	Power Ratio
On-bottom suspension	1.0	1.0
Off-bottom suspension	1.7	5.0
Complete uniformity	2.9	25.0

be ≥2(3.05) or 6.1 m/min or a volume flow of 33.4 m³/min. This must match the pumping rate of the impeller. Nq for the HE-3 (Table 9.1) is 0.54. Solving $Q = N_q N D^3$ for N gives 136 rpm as the speed required to deliver twice the settling velocity, and 102 rpm for 1.5 times the settling velocity.

9.6.7 Power Requirements

There is a heavy penalty for specifying uniform slurry conditions if that is not really required. Oldshue [42] points out, as is shown in Table 9.8, the differences (right-hand column) in power between conditions of complete uniformity and just suspended.

The calculation of power follows methods described previously, using the estimated weight average density, ρ, in the expression for Reynolds number and power. The viscosity for a Newtonian system is best measured by using a paddle-type rotating bob viscometer. The Metzner-Otto method is commonly used to determine the Re for a non-Newtonian system, where viscosity is a function of shear rate. This method consists of determining the mean shear rate from $\bar{\gamma} = KN$, where N is the stirring speed in rps, K is 10 for a propeller, and $\bar{\gamma}$ is the mean shear rate is in s⁻¹. Viscosity will not influence the calculation of power when Re ≥ 200.

As stated previously, complete solids uniformity requires scale-up by P/V = constant, Re ≥ 10⁴, and geometric similarity, including baffles. Scale-up is given by Equation (9.34) for constant power per unit volume:

$$ND^{2/3} = \text{const} \tag{9.34}$$

9.6.8 Equipment

A variety of impellers can be used for just-suspension purposes. Generally, for settling solids where $\rho_s \geq \rho_l$, a down-pumping axial or mixed-flow impeller is better than radially discharging impellers. One exception is the low-shear retreat-curve impeller (RCI), normally located at the bottom of the vessel. It does a good job of suspension and minimizes crystal breakage. Clearances of $C = T/4$ to $T/6$ are a common practice. A good choice for impeller diameter is $D/T = 1/3$ to $1/2$. Overly large impellers produce complex flows at the bottom of the vessel, with often undesirable results. Multiple impellers are not needed for off-bottom suspension applications [43], since only solid movement at the bottom is required. Any additional impellers add to the power consumption, and improve uniformity, but do not affect bottom movement. With the exception of viscous and non-Newtonian slurries, vessels need to be baffled. Four baffles, $T/10$ to $T/12$ wide, are desirable and are set away from the wall a distance ≈$T/72$. When vessel proportions exceed $H/T = 1.2$, multiple impellers are recommended. A good design might include an upper hydrofoil and a lower PBT.

Example 9.4
Calculate the minimum suspension speed and the power requirement for the following:

Solids $X = 0.30$
$D = 0.51$ m
$dp = 10^{-4}$ m
$v = 10^{-6}$ m²/s
$(g \Delta \rho / \rho) = 9.81 \times 2.34$ m/s²
$s = 3.9$ (for flat-blade turbines, taken to be equivalent to the glassed impeller)

Solution

The basis for the calculation is the Zwietering Equation (Equation (9.29)). Substituting, we get

$$N_{JS} = 3.9 \cdot \left(10^{-6}\right)^{0.1} \left[9.81 \cdot 2.34\right]^{0.45} 0.30^{0.13} \left(10^{-4}\right)^{0.2} 0.51^{-0.85} = 1.92 \, rps$$

Given that the impeller is a glass-coated retreat-curve type, the approximate calculation for the minimum suspension speed is 115 rpm. A recommended design speed would be 130–150 rpm. This would guarantee more than minimum suspension conditions. The power at 150 rpm based on a power number of 0.6 would be 0.323 kW (≈0.5 hp).

9.6.9 Solids Suspension by Jet Mixing

Liquid jets can be used to suspend solids. There are both simple nozzle designs and proprietary designs such as the Aerocleve-Pentech. In the latter, multiple jets radiate outward from the central feeder. Jets function in a similar fashion to impeller streams, so that many of the terms used in the Zwietering equation apply to solids suspended by jets. Effective suspension is created when the jet is centrally placed at a height of about T/6 to T/10 above the bottom of the floor of the tank. Jet solid suspension is designed on the basis of the V_{JS}, the jet velocity required to just suspend particles. Particle movement is the same as defined for the case of the agitated vessel. Research at BHRG for the FMP consortium has developed extensive technology on jet solid suspension for its members. For more information on jet mixing, see Section 9.10.

9.7 GAS-LIQUID MIXING

9.7.1 Introduction and Scope

Gas-liquid mixing is one of the most frequently used operations in the process industries, being involved with waste treatment, fermentation, hydrogenation, oxidation, chlorination, and gas stripping. Throughout this section, liquid is the continuous phase and gas is the dispersed phase. The purpose of gas-liquid mixing is to introduce soluble gases into the liquid phase, where reactions can take place. Mixing affects how quickly this process occurs by creating an interfacial bubble surface area and a suitable bubble distribution throughout the vessel. If the interfacial area controls the rate of a reaction, reactions are referred to as being mass-transfer-controlled. If chemistry is slow and mass transfer is not limiting, reactions are referred to as being kinetically limiting.

Energy provided by agitation creates an interfacial area by dispersing bubbles. High-shear impellers (four- or six-blade flat-blade disc designs) are often used, since they can deliver more power (at a given speed and diameter) to the fluid than other impeller designs. The disc, shown in Figure 9.3, helps collect gas and direct it to the blades, where dispersion occurs. While disc turbines are most commonly used for gas dispersion, any impeller capable of delivering energy to the liquid can be used to disperse gas. This includes the 45° or 60° PBT, the FBT, hydrofoils, and even retreat-curve impellers.

Cavitation commonly occurs in g-l mixing operations, and this reduces the impeller function. Cavitation, shown in Figure 9.28, is a condition where large pockets of gas collect and cling to the

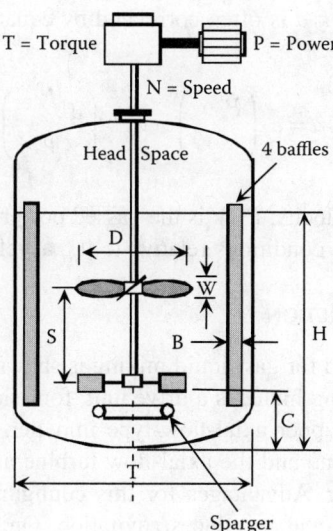

FIGURE 9.28 General design for g-l mixing.

back side of impeller blades. It is associated with high gas flow rates and low impeller speeds. All impellers cavitate, some more easily than others. A seriously cavitating impeller is incapable of dispersing gas or providing good mixing, since adequate power can no longer be transferred to the fluid. It is important to predict the onset of cavitation and its effect on the process. The tendency for an impeller to cavitate is design dependent.

Impellers used for gas-liquid mixing have three tasks:

1. Creation of bubble surface area for mass transfer
2. Attainment of gas holdup
3. Avoidance of impeller flooding

9.7.2 Mass Transfer—General

Equation (9.35) describes the rate of mass transfer, N_i, of species (i) from the gas bubble to the continuous liquid phase. It depends on (a) the total surface area, a, at the gas-liquid interface; (b) the driving force, $(C^*_{Li} - C_{Li})$; and (c) the mass-transfer coefficient, k_L. C^*_{Li} represents the concentration of component (i) in the liquid phase in equilibrium with gas in the bubble, and C_{Li} represents the concentration of component (i) in the bulk of the liquid. Normally, a and k_L are combined into a lumped parameter, $k_L a$, since determination of separate values of a and k_L is difficult. The units of $k_L a$ are in min^{-1} or s^{-1}.

$$N_i = k_L a \left(C^*_{Li} - C_{Li} \right) \tag{9.35}$$

At a given fraction gas holdup, Φ, and bubble diameter, d_{32}, the total surface area of the gas, a, is related as shown by

$$a = 6 \cdot \Phi / d_{32} \tag{9.36}$$

where d_{32} is the Sauter mean bubble diameter (also written as d_{SM}). Note that a in Equations (9.35) and (9.36) is the same variable, the interfacial bubble area per unit volume, and has units of m^2/m^3.

The mass-transfer coefficient $k_L a$ is often correlated by equations such as

$$k_L a = c \cdot \left(P_g/V\right)^{\alpha} \cdot v_{sg}^{\beta} \cdot \left(\mu^*/\mu_{ref}\right)^{\gamma} \qquad (9.37)$$

where V_{sg} is the superficial gas velocity, P_g/V is the gassed power per unit volume, C is a constant, and μ^* is the viscosity at process conditions relative to μ_{ref} at reference conditions.

9.7.3 Equipment and Its Function

A conventional stirred vessel used for gas-liquid mixing is shown in Figure 9.29. A very common configuration for gas-liquid mixing includes a drive unit, four sidewall baffles, a sparger, a lower gas-dispersing impeller, and an upper axial-flow-type impeller. Some operations place the gas-dispersing turbine as the upper one and the axial-flow turbine as the lower one. In this case, the axial-flow turbine is up-pumping. Advantages for this configuration are that it greatly reduces fluctuating loading on the drive, and it reduces cavitation. On the negative side, gas holdup is more difficult to achieve. Various vessel shapes can be used for gas-liquid mixing. Tall vessels, e.g., $H/T = 4/1$ containing multiple impellers, have the advantage of providing longer gas residence times, but it is more difficult for these to provide good turnover. The minimum H/T should be at least 1:1. A common impeller-to-tank diameter ratio for gas dispersion is $D/T = 0.4$, but this can vary from 0.33 to 0.50. The energy input for gas dispersion is high and can vary from 0.5 to 6.0 kW/m³ (2.5 to 30 hp/1000 gal). Many applications fall into 1–2 kW/m³ (5–10 hp/1000 gal) based on gassed conditions. Because of this high energy, other aspects such as uniformity are usually met with ease.

Effective gas dispersion occurs in stages. First, large bubbles form from a sparger or inlet pipe. These are entrained by flow to the impeller blades, where the second stage of dispersion takes place. Smaller bubbles form here. Ring spargers, shown in Figure 9.29, are commonly used for introducing gas to the impeller. They are described more fully later in this section. Pipes can also be used to introduce gas. These can be single or multiple dip-pipe arrangements. Because of their large openings, pipes deliver larger bubbles than ring spargers. These bubbles can bypass the impeller. In pipe sparging, entering gas is sheared away from the tip, creating bubbles by a shedding mechanism.

Direct loading is a term used to describe gas bubbles passing directly from the point of entry to the impeller. *Indirect loading* refers to the condition where bubbles first pass up the tank, then get redirected by entrainment to the impeller for dispersion.

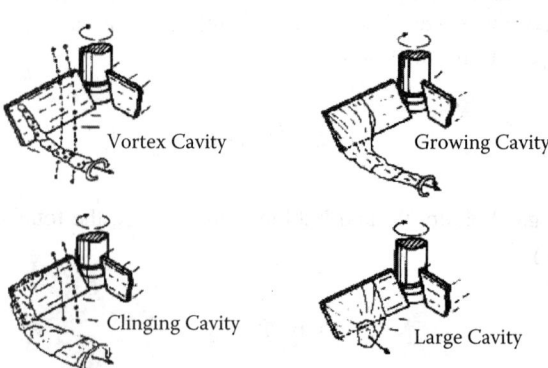

FIGURE 9.29 Development of cavity formation on a 45° PBT.

At constant speed, the impeller power demand decreases with increasing gas flow. This is because (a) the average mixed density decreases and (b) blades cavitate as shown in Figure 9.28. A cavitating Rushton impeller loses up to 60% of its ungassed power draw. Both impeller blade shape and the number of impeller blades affect cavitation. The CD-6, BT-6, and Scaba designs are examples of low-cavitating impellers (see Figure 9.3). Their curved "trailing surfaces" delay the onset of cavitation, allowing more gas to be dispersed.

Baffles must always be used for low-viscosity fluids for gas-liquid mixing. They help convert tangential to axial momentum, producing shear for dispersion and axial flow for good mixing. In the absence of baffles, the tangential swirl forces all bubbles to leave the liquid because of centrifugal forces.

Curved-blade disc turbines are good choices for gas dispersion (see Figure 9.3(a), second row, right). Note that the open cup advances into the liquid. A second impeller, like an axial-flow hydrofoil, helps improve circulation and gas holdup. Multiple disc turbines should be avoided, since they lead to compartmentalization and poorer overall mixing.

Ring spargers are an excellent means for good initial distribution of gas and this avoids an unbalanced loading of the impeller. The relatively small bubbles are easily entrained to the impeller blades. A good design practice is to make the cross-sectional area of the ring two to three times larger than the sum of the area of the holes. The sparger ring diameter ought to be ≈80% of the diameter of the dispersing turbine. Larger ring diameters run the risk of gas bypassing the turbine blades. The ring is located concentrically below the dispersing turbine at a clearance (C) of $0.3 \leq (C/D) \leq 0.5$.

It is often not feasible or easy to use a ring sparger. An alternative is to use single or multiple pipes as spargers. These can be roughly 80% as effective as ring spargers.

Lightnin proposed a four-pipe sparger design to be used with their A-315 four-blade hydrofoil impeller for G-L mixing. (See Figure 9.3(a), fourth row, left.) Gas was introduced through four equally placed pipes spaced 90° apart and discharged close to the bottom of the vessel. There gas was sheared away by tangential liquid flow from the impeller. This design resulted in an indirect loading condition but virtually avoided flooding. The design could create high gas holdup, avoiding cavitation and providing excellent circulation and bubble distribution. The chief drawback was that it was more difficult to create very small bubbles, because of the low power number of the A-315.

Static mixers are also used for continuous gas-liquid operations (see Section 9.9). The orientation of the mixer is important. A vertical orientation with both gas and liquids passing cocurrently downward is desirable. Considerable vendor information is available on gas-liquid dispersion in static mixers.

9.7.4 Impeller Characteristics in Gas-Liquid Mixing

When gas passes through the impeller, several things happen, and their extent depends on the gas flow rate:

- The impeller discharge flow rate is lower.
- The power delivered to the agitator decreases by as much as 60%, depending on the impeller selected.
- Vessel flow patterns become dominated by gas flow rather than impeller flow.
- Regions below the bottom impeller are often void of bubbles.
- Cavitation and finally flooding effects can totally dominate the vessel, leading to very poor gas dispersion and mixing.

Figure 9.28 shows typical cavity formation at the blade tips of a 45° PBT. It is easy to see how these cavities affect dispersion and mixing conditions in the vessel. Cavities reach a steady-state

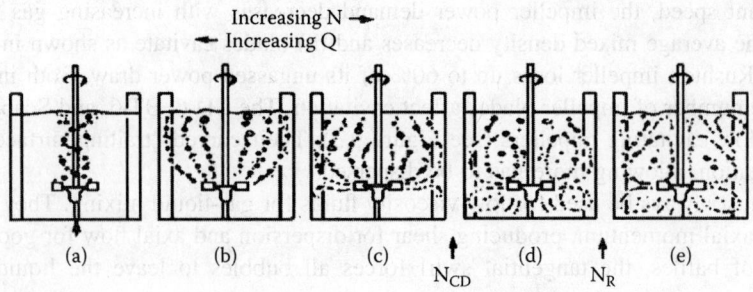

FIGURE 9.30 Changing vessel performance with agitation and gassing rates. (From Nienow et al., 1977.)

size at constant conditions. The sweeping motion of the blades first collects gas bubbles. There they impinge on the blades, creating large cavities (bottom right) because of coalescence. Gas leaves the cavities by shedding from the vortex tip. (See upper right picture in Figure 9.28.)

Figure 9.30 shows the changing bubble distribution patterns in the vessel as gassing and agitation rates change. Figure 9.30(a) shows a flooded impeller. Gas, instead of being dispersed by the impeller, passes centrally up the middle of the tank to the surface. As the agitation rate increases, N_{CD} is reached (shown by Figure 9.30(d)), and a condition known as "complete" dispersion is reached. At still higher speeds, N_R, gas bubbles are entrained back to the impeller. This condition (Figure 9.30(e)) represents excellent circulation and bubble distribution throughout the vessel. Figures 9.30(a–e) show the effects of decreased gas flow at a given rate of agitation. At Figure 9.30(a), while agitation is strong, the gas flow is high, and the impeller cavitates, resulting in gas bypassing the impeller.

The conditions shown by Figure 9.30(a–e) are represented schematically in a regime map, Figure 9.31, where the Froude number is shown as a function of the gas flow number. The Froude number, Fr, is N^2D/g, and the gas flow number, Fl_g, is Q/ND^3. These numbers are calculated for a design and its operation; the values are then placed on the regime map to determine if the design is workable. Such a procedure can help avoid costly scale-up mistakes.

The following are design guidelines for good gas-liquid mixing. Conditions shown by Figure 9.30(a, b) exist for Fr < 0.05. This condition must be avoided. Regime I represents conditions shown by Figure 9.30(e). While dispersion and mixing are excellent, power may be excessive for most applications. Regime II is a practical compromise between performance and energy cost. Most industrial gas-liquid applications will fall into regime II. Regime III is reached by high gassing rates or by low-speed conditions. Regime III is equivalent to conditions shown by Figure 9.30(c). The bubble distribution is not uniform, and holdup is decreased. In some cases, operation in regime III is acceptable, provided that liquid circulation below the impeller is good. The actual location of all lines is a function of the impeller selection. It is essential to determine into which region the system falls, both from a standpoint of design and for troubleshooting. Some suppliers provide charts similar to Figure 9.31 with their literature and proposals.

The following applies to Rushton (D-6) turbines having $(D/T) = 0.40$ [44]:

- Flooding occurs when $Fl_g > \approx 1.2$ Fr (Figure 9.30(a) and the lower-sloped line in Figure 9.31).
- Gas recirculation begins when Fr > ≈ 2.75 $Fl_g^{1/2}$ (Figure 9.30(e) and the upper line in Figure 9.31).
- Vortex and large cavities on upper impellers (multiple impeller systems) develop when $Fl_g Fr \times 0.2$ (Figure 9.30(c))

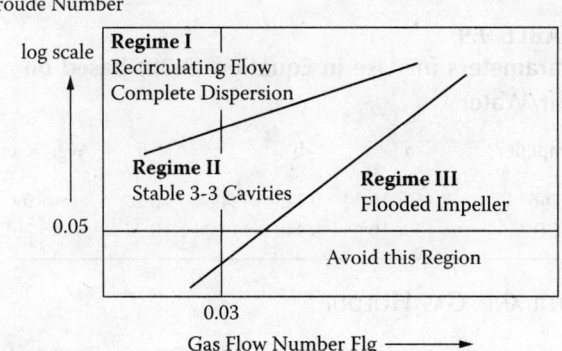

FIGURE 9.31 Flow regimes in gas-liquid systems.

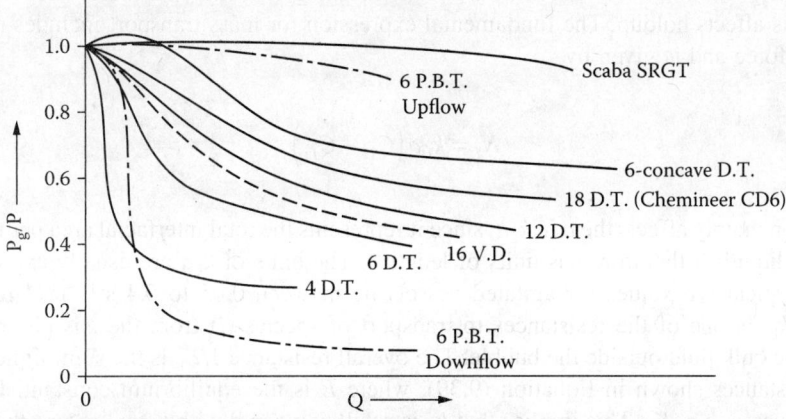

FIGURE 9.32 Impellers used for gassing: the effect of gas on power. (From Middleton, 1995.)

- Vortex and large cavities will begin to develop when $Fl_g \geq 0.03$
- Stable cavities develop when $Fr > 0.05$

Mass transfer and gas holdup depend on gassed power input. It is therefore important to know the power delivered under gassed conditions. Experimental curves such as those shown in Figure 9.32 relate the ratio of gassed to ungassed power, P_G/P_{UG}, to either gas flow rate or the gas flow number, Fl_g. Data in Figure 9.32 show how nine different impellers compare. Note that:

1. All impellers show a decrease in power with increasing gas flow.
2. The Scaba and A-200 or PBT-6 up-pumping turbines are least affected by gas flow.
3. The power ratio for the CD-6 drops to only 0.7, but the Rushton (R-100, DT-6) falls to 0.4.
4. More power is retained when impellers with more blades are used.

Actual power ratios can be calculated using Equation (9.38) (developed by Chemineer) with constants given in Table 9.9. The viscosity μ is in centipoise.

$$P_G/P = 1 - (a \cdot \mu + b) \cdot \tanh(c \cdot Fl_g) \cdot Fr^d \tag{9.38}$$

TABLE 9.9
Parameters for Use in Equation (9.38) Based on Air/Water

Impeller	a	b	c	d	Avg. % error
D-6	-7.15×10^{-4}	0.723	24.54	0.25	7.09
CD-6	-1.15×10^{-4}	0.440	12.08	0.37	5.20

9.7.5 Mass Transfer and Gas Holdup

The general features of mass transfer were previously given in Section 9.7.1. The mass-transfer coefficient, $k_L a$, and gas fraction holdup, Φ, are closely related and depend on the same variables. Equation (9.36) relates area, holdup, and mean bubble size, and the main variables affected by agitation is d_{32} and a. Bubble rise time and thus escaping tendency are a function of the bubble size, and this affects holdup. The fundamental expression for mass transport includes the $k_L a$, and the driving force and is given by

$$N_i = k_L a \left(C_{Li}^* - C_{Li} \right) \tag{9.35}$$

Agitation mainly affects the a in $k_L a$, since it represents the total interfacial area per unit volume of gas plus liquid. It therefore has units of length^{-1}. The units of $k_L a$ are usually expressed in s^{-1} or min^{-1}. Typical $k_L a$ values for agitated vessels lie between 0.05 to 0.4 s^{-1}. The mass-transfer coefficient k_L, is one of the resistances to transport of species (i) from the gas phase inside the bubble to the bulk fluid outside the bubble. The overall resistance $1/K_L$ is the sum of the inside and outside resistances shown in Equation (9.39), where E is the equilibrium constant. In the great majority of cases, $k_L \approx K_L$. This implies that k_G is small compared with k_L and means that the liquid film resistance outside the bubble is controlling:

$$\frac{1}{K_L} = \frac{1}{Ek_G} + \frac{1}{k_L} \tag{9.39}$$

The gas film resistance, k_G, for small bubbles is controlled by diffusion, D_G, as shown by

$$k_G = \frac{2\pi^2 D_G}{3 d_{32}} = 6.6 \frac{D_G}{d_{32}} \tag{9.40}$$

The liquid-side mass-transfer coefficient, $k_L a$ (s^{-1}), is related to the superficial gas velocity, V_{SG} (m/s), and the gassed power per unit volume of liquid, P_G/V, (kW/m^3). The viscosity term μ^*/μ accounts for the effect of process viscosity on the mass-transfer coefficient relative to standard conditions, typically water at 20°C:

$$k_L a = C \cdot \left(\frac{P_g}{V} \right)^\alpha \cdot \left(V_{SG} \right)^\beta \cdot \left(\frac{m^*}{m_{ref}} \right)^\gamma \tag{9.37}$$

TABLE 9.10
Constants for Equation (9.37) Used for the Determination of k_La

Author	System	Impeller Type	D/T	C	α	β	γ
Middleton (43)	Air/water	Various single and multiple	...	0.24	0.7	0.3	...
Middleton (43)	Air/viscous fluids	Various single and multiple	...	0.113	0.56	0.3	−1
Middleton (43)	Air/electrolyte soln.	R-100/D-6	0.5	2.3	0.7	0.6	...
Middleton (43)	Air/water	R-100/D-6	0.5	1.2	0.7	0.6	...
Kar (44)	Air/water	CD-6, D-6	0.4–0.5	0.63	0.52	0.53	...
Kar (44)	Air/water	CD-6 + A-310, D-6 + A-310	0.4–0.5	0.78	0.48	0.57	...
Nocentini (45)	Air/water	D-6 + D-6	...	...	0.62	0.40	−1.17

Table 9.10 gives values of the constants in Equation (9.37). Values of $\alpha = 0.7$ and $\beta = 0.3$ for scale-up are suggested by Middleton [45].

Smith [48] obtained Equation (9.41) for the correlation of $k_L a$ by grouping terms of Equation (9.37) and setting $\alpha = 1$, $\beta = 0.45$, and $\gamma = -0.7$

$$k_L a = 1.25 \cdot 10^{-4} \left(\frac{D}{T}\right)^{2.8} Fr^{0.6} \cdot Re^{0.7} \cdot Fl_g^{0.45} \left(\frac{D}{g}\right)^{-0.5} \tag{9.41}$$

The most important physical property affecting k_L is the liquid viscosity, since it is directly related to the film resistance around bubbles as well as their rise velocity. The Weber number used extensively for immiscible liquid-liquid dispersion (see Section 9.8) is seldom used for gas-liquid systems. The surface tension for many gases and liquids is more constant than the interfacial tension used for liquid-liquid systems.

Gas holdup, Φ, is expressed as a volume fraction and is represented by $V_G/(V_G + V_L)$, where V_G and V_L are the volumes of gas and liquid, respectively, in the vessel. Gas holdup values can vary from 0.01 (1%) to as high as 0.50 (50%) for industrial applications. Holdup depends on bubble size, bubble rise velocity, and bubble coalescence. Gas holdup can be easily measured by observing liquid heights under gassing and nongassing conditions (Figure 9.29):

$$\Phi = \frac{H_{aerated} - H_{unaerated}}{H_{aerated}} \tag{9.42}$$

An expression commonly used for gas holdup is

$$\Phi = K \left(\frac{Pg}{V}\right)^\alpha V_{SG}^\beta \tag{9.43}$$

The mean values from a review [49] are $\alpha = 0.45$ and $\beta = 0.48$. Values listed in Table 9.11 were found by Kar [46] using a ring-type sparger and $C/D = 0.5$, $D/T = 0.4$–0.51. Gas holdup has also been correlated using dimensionless groups. For example, Smith [48] proposes:

$$\Phi = 0.85 \cdot (Re \cdot Fr \cdot Fl_g)^{0.35} \cdot \left(\frac{D}{T}\right)^{1.35} \tag{9.44}$$

TABLE 9.11
Constants for Gas Holdup Equation (9.43)

Author	System	Type	D/T	K	α	β
Kar [46]	Air/water	CD-6	0.4	2.92	0.19	0.88
Kar [46]	Air/water	CD-6	0.5	3.93	0.25	0.92
Kar [46]	Air/water	D-6	0.4	1.15	0.3	0.64
Kar [46]	Air/water	D-6	0.5	2.58	0.39	0.81
Parthasarathy [50]	Air/NaCl/water	D-6	0.4	1.01	0.24	0.67

TABLE 9.12
Constants for Gas Holdup Equation (9.43)

Configuration	Impellers	Deionized Water			Na$_2$SO$_4$ (28 g/l)			PVP (7.4 mPa·s)		
		K	α	β	K	α	β	K	α	β
Rushton	1	0.222	0.215	0.684	0.0069	0.664	0.523	0.0096	0.699	0.483
	2	0.223	0.264	0.699	0.0179	0.520	0.415	0.0361	0.518	0.479
	3	0.278	0.217	0.662	0.0651	0.435	0.548	0.0586	0.536	0.654
A-315	1	1.13	0.231	0.781	0.0481	0.429	0.580	0.206	0.454	0.804
	2	0.321	0.195	0.695	0.114	0.385	0.649	0.312	0.379	0.782
	3	0.328	0.216	0.691	0.130	0.355	0.619	0.286	0.390	0.769
A-310 + R-100	1 R	0.222	0.215	0.684	0.0069	0.664	0.523	0.0096	0.699	0.483
	1 A-310 + 1 R-100	0.151	0.244	0.566	0.0660	0.392	0.514	0.0705	0.425	0.521
	2 A-310 + 1 R-100	0.256	0.244	0.636	0.103	0.403	0.587	0.0832	0.491	0.590

Source: Pinelli, D., M. Nocentini, and F. Magelli, "Hold-Up in Low Viscosity Gas-Liquid Systems Stirred with Multiple Impellers: Comparison of Different Agitators Types and Sets," I. C. Engrs., Eighth European Conference on Mixing, University of Cambridge (*I. Chem. E.* 136, 81–88 [1994]).

Gas holdup studies in tall vessels including three impeller configurations were reported by Pinelli [51] for (a) three equally spaced Rushton turbines, (b) three A-315 turbines, and (c) one Rushton (bottom) and two A-315 turbines. Holdup was correlated using Equation (9.43), with constants shown in Table 9.12. The units for P_G/V are in W/m^3, V in m/s, and holdup in volume fraction gas.

For the case of equally spaced Rushton turbines in deionized water (fast coalescence), all constants are virtually the same. Power was additive, suggesting that the turbines functioned independently. While there is some variation in K for other configurations, values of α and β are nearly constant. The gas flow constant dominates over power. This is not the case for noncoalescing polyvinylpyrolidone (PVP) and salt-containing solutions. Experiments included one impeller with $H/T = 1.0$, two impellers with $H/T = 2.0$, and three impellers with $H/T = 3$. All impellers were equally spaced.

Holdup and bubble size can be rough indications of the quality and intensity of mixing, but less information is revealed than for $k_L a$.

9.7.6 Gas Residence Time

The gas residence time becomes important for sparingly soluble gases and cases where a high gas conversion is desired. The gas residence time depends on viscosity, flow rate, bubble size, and the height of liquid in the vessel. In Figures 9.30(a–c), bubble rise time and liquid height are dominant factors, but in Figures 9.30(d–e), entrained flow is also important. Actually, very little has been

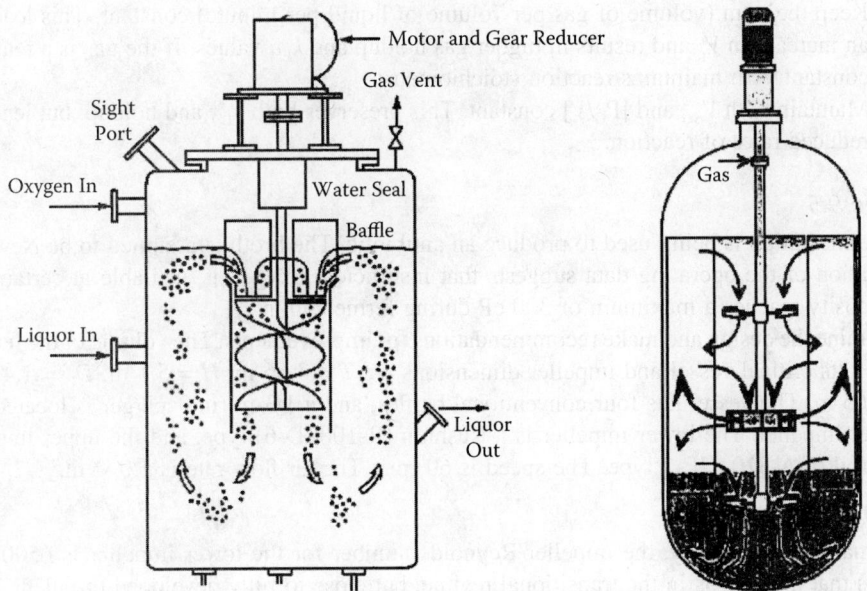

FIGURE 9.33 Circulate headspace reactors: Biazzi/Ekato hydrogenation reactor (right) and Praxair's AGR reactor (left).

published on the residence-time distribution for gas (RTD_G) as a function of design and operating conditions. Still, RTD is important when carrying out chemical reactions, particularly in cases where the gas is sparingly soluble in the liquid phase. The residence time will be a distribution, with larger bubbles having a shorter residence time than smaller bubbles. If visual inspection is possible, a rough order of magnitude is to stop agitation and measure the time it takes for all bubbles to reach the surface.

The gas residence time can be increased by (a) returning the headspace gases back to the gas feed by means of an external compressor; (b) using a self-inducing hollow tube impeller [52, 53]; or (c) using a propriety design configuration, such as Praxair's AGR reactor or Biazzi/Ekato's reactor, both shown in Figure 9.33. These designs are intended to recycle headspace gasses, thereby increasing the gas residence time.

In the Biazzi/Ekato reactor, gas enters the hollow shaft through holes located above the liquid. It is ejected in holes around the eight-blade turbine, where dispersion occurs. The flat-blade turbine has an upper and lower disc to improve suction pressure. Two axial-flow turbines—one above and one below the disperser—provide circulation in the vessel. Closely spaced vertical baffles double as heat exchangers. In the Praxair AGR design (advanced gas reactor), headspace gases are engulfed by a close-fitting high-speed screw operating in a draft tube. A flat-blade dispersing impeller at the bottom of the screw helps disperse the gas. Several design variations are available. The design is reported to give good flow and bubble distribution. It can be retrofitted to existing vessels.

9.7.7 Scale-Up

A scale-up problem that needs to be recognized is that, at small scale, a large quantity of gas enters through the liquid surface. This is not the case at the industrial scale. Humphrey [54] outlines a method for dealing with this scaling problem, and a method to correct for surface aeration.

It is recommended that well-gassed conditions such as those shown in Figures 9.30(d, e) be used for scaling. It is recommended that $k_L a$ and holdup be scaled using Equations (9.37) and (9.43).

Two criteria are often used for gas-liquid scale-up:

1. Keep the vvm (volume of gas per volume of liquid per minute) constant. This leads to an increase in V_g and results in higher gas holdup and $k_L a$ values. If the gas is a reagent, constant vvm maintains reaction stoichiometry.
2. Maintain both V_{SG} and $[P_G/V]$ constant. This preserves both $k_L a$ and holdup, but leads to reduced rates of reaction.

Example 9.5

Batch fermentation is being used to produce an antibiotic. The broth is assumed to be Newtonian. Examination of the operating data suggests that insufficient oxygen is available at certain times. The viscosity reaches a maximum of 300 cP during fermentation.

Examine the design and make recommendations for improvements. The volume of the fermenter is 80 m³. Its critical vessel and impeller dimensions are $T = 3.66$ m, $H = 5.1$ m, $D_L = 1.4$ m, and $D_U = 1.65$ m. The vessel has four conventional baffles, and a 1.3-m ring sparger is located below the lower impeller. The lower impeller is a Rushton (R-100, D-6) type, and the upper impeller is a hydrofoil of A-310, HE-3 type. The speed is 60 rpm. The air flow rate is 1.0 vvm.

Solution

At the maximum viscosity, the impeller Reynolds number for the lower impeller is 6500, which indicates that the flow is in the transitional regime, but close to fully developed turbulent flow. At ungassed conditions, the Rushton impeller ($D = 1.4$ m, Np = 5.75) consumes 30.5 kW (40.98 hp). The upper impeller ($D = 1.65$ m, Np = 0.3) consumes 3.68 kW (4.93 hp). The combined power is 34.2 kW (45.91 hp). The quiescent volume at a height of 5.1 m (200 in.), including the dished head, is 50.5 m³ (13,333 gal). This makes the P/V for the lower R-100 impeller = 0.61 kW/m³ (3.1 hp/1000 gal). The gas flow at 1.0 vvm is 50.5 m³/min (13,333 gal/min), 0.841 m³/s (46.5 ft³/s). The superficial gas velocity, V_{SG}, is 0.125 m/s (0.41 ft/s). The Froude number, $N^2 D/g$, is 0.1423 and the gas flow number, Fl_g, (i.e., Q_g/ND^3), is 0.482.

1. Determine the gassed power input using Equation (9.38). Substituting values, P_G/P_{UG} is 0.688. So $P_G/V = 0.419$ kW/m³ (2.13 hp/1000) gal.
2. Determine bubble dynamics in the vessel. Flooding occurs when $Fl_g > 1.2$ Fr. In this case, 1.2 Fr = 0.171. From this, we conclude that the impeller is flooded. The Froude number is greater than 0.05, so operations (with reference to Figure 9.31) are in regime III, and gas will not reach the lower portions of the vessel. Conditions represented by Figure 9.30(a) or 9.30(b) probably exist. The best option is to increase the speed and then consider increasing the gas flow rate (first the speed).
3. Solving $Fl_g > 1.2$ Fr for N, at $Q = 1.317$ m³/s (46.5 ft³/s), using $D = 1.397$ m (55 in), gives $N > 85$ rpm. This speed would result in a 41% increase in shear rate, which might be unacceptable to the microorganisms. Maintaining the same impellers operating at 85 rpm results in a power draw of 91.44 kW or a P/V of 1.69 kW/m³ (10 hp/1000 gal).
4. If a CD-6 having an Np = 3.67 were selected, the combined impeller power draw at ungassed conditions would be 60.0 kW (80.44 hp). The corresponding Froude number = 0.28, and the gas flow number is 0.341. Thus 1.2 Fr = 0.342, so conditions are just at the onset of flooding. Increasing the gas flow by 20% results in $Q = 1.58$ m³/s (55.8 ft³/s), which would make $Fl_g = 0.409$. This condition would place operations solidly in region II, a more desirable condition.
5. Calculate gassed power, $k_L a$, and holdup at these new conditions. $P_G/P_{UG} = 0.745$, and $P_G = 55.47(0.745) = 41.32 + 4.55 = 45.87$ kW. (For $k_L a$, use only the power draw for the lower turbine.) Use Kar's constants found in Table 9.10, $A = 0.78$, $\alpha = 0.48$, $\beta = 0.57$, and $\gamma = -0.77$. The input variables are $P_G/V = 0.756$ kW/m³ and $V_{SG} = 0.15$ m/s. The calculation for $k_L a$ using Equation (9.37) is 0.00286 s⁻¹. The high viscosity causes this value to be low. At 1 cP, the $k_L a$ would be 0.231 s⁻¹, an acceptable value. Calculations

for gas holdup using Pinelli's correlation give 43.8% for noncoalescing and 23.1% for air-water or coalescing systems. These numbers are very high. Note that the P_G/V is in units of W/m^3 in Pinelli's correlation.

9.8 IMMISCIBLE LIQUID-LIQUID MIXING

9.8.1 Introduction

Immiscible liquid-liquid mixing involves dispersion, suspension, and coalescence of liquid drops in a second liquid phase. Effects of mixing are complex, often poorly understood, and lack industrially usable measuring tools. Scale-up is particularly difficult, since coalescence dispersion and suspension are affected to different degrees by scale. For a more in-depth treatment of this subject, see Chapter 12 by Leng and Calabrese in the *Handbook of Industrial Mixing* [1].

Industrial applications of liquid-liquid mixing include emulsion, suspension polymerization, alkylation, nitration, phase transfer catalysis, solvent extraction, decantation, centrifugation, and electrostatic precipitation. The dynamic behavior of drops is important to the understanding of liquid-liquid mixing. This includes flow direction and intensity as well as the velocity gradients, commonly known as shear. Shear, whether turbulent or laminar, causes drops to deform, disperse, collide, and coalesce. On the other hand, drop suspension depends on bulk flow movement. Whether a drop breaks, or remains intact, depends on stress variation and magnitude. When a drop breaks, there can be few or many daughter drops formed. Higher energies during dispersion lead to the formation of many more daughter drops. Coalescence probability depends on collision, contact time, and film drainage time. Coalescence, dispersion, and suspension processes occur simultaneously but often to different degrees, depending on local flow.

Coalescence is also controlled by the condition of drop surfaces. Surfactants reduce the interfacial tension and help preserve drop stability, therefore affecting drop sizes. Surface-active materials are important in suspension/emulsion polymerization processes.

9.8.2 Characterization of Immiscible Liquid-Liquid Systems

Characterization of immiscible liquid-liquid systems into broad groups based on coalescence rates can sometimes help reduce the complexity that must be considered. Noncoalescing systems can be treated as dispersions only. Examples include dilute dispersions, typically <5% dispersed phase, and stabilized, more-concentrated systems. Many industrial systems are either noncoalescing or very slowly coalescing.

A simple test to characterize relative drop stability is to measure the time it takes to separate a given dispersion into two distinct clear layers. Table 9.13 gives rough guidelines for such a characterization.

TABLE 9.13
System Characterization by Coalescence Rates

Time for Phase Separation	Characterization	Process Implication
< 10 s	Very rapid coalescence	Expect severe scale-up problems
<1 min	Fast coalescence	Scale-up problems exist, but can be managed by the proper selection of equipment; process is apt to operate differently at different scales
2–5 min	Moderate coalescence	Design for coalescence; use broad-blade impellers producing high flow
>5 min	Slow coalescence	Problem can be considered as dispersion only

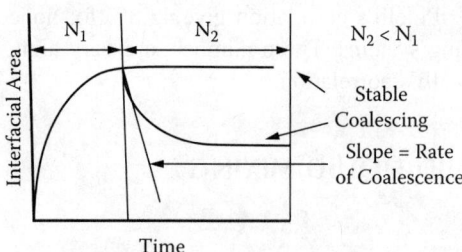

FIGURE 9.34 Dynamic test for coalescence.

TABLE 9.14
Dispersions Classified by Drop Size

Drop Size	Comments	Equipment, Agents Used
<0.5 μm	Stabilized by Brownian motion	Emulsifiers, ultrasonic devices, rotor-stator mixers, impingement mixers, etc.
$0.5 \geq d \geq 3.0$ μm	Marginally stable, can cream and separate	Rotor-stator, impingement mixers, static mixers, emulsifiers
>3.0 μm	Often unstable	Static mixers, in-line mixers, stirred vessels

A more accurate dynamic test for relative drop stability uses agitation and an optical transmission probe. First, a dispersion is created using vigorous mixing, N_1. Agitation is then decreased to N_2. If the interfacial area (indicated by a decrease in optical transmission) decreases, it is due to either coalescence or settling. If settling is the cause, choose a higher value for N_2. If the transmission remains constant after speed change, it suggests that the system is noncoalescing. The probe reading is constant under dynamic equilibrium between dispersion and coalescence. If the system coalesces, the optical transmission will increase, indicating a loss of interfacial area. The probe used is similar to that described by Rodger [55]. The output is schematically illustrated in Figure 9.34. If the optical transmission is recorded and the probe calibrated (e.g., with glass beads), the initial slope of the curve gives the coalescence rate. Another way to classify liquid-liquid systems is by drop size, as shown by Table 9.14. This section focuses on drops larger than ≈5.0 μm.

Initially, dispersion occurs throughout the vessel, but with time, the dispersion region shrinks to that close to the impeller. Experimental data and computational fluid dynamics (CFD) have confirmed that energy dissipation close to the impeller can be ≈40 times greater than the mean for the vessel. Coalescence rates are fastest in regions of gentle flow, where drop contact/rest times are longer. Under stagnant conditions, drops settle, touch each other, and can coalesce to form a condensed layer. Under gentle dynamic conditions, drops collide and remain in contact so that, if sufficient film thinning takes place, coalescence occurs. If the contact time is brief, so that insufficient drainage occurs, drops will separate and no coalescence occurs. Coalescence also results from drop collisions with impeller blades, baffles, and vessel walls.

9.8.3 Drop Sizes

The Sauter mean drop diameter, d_{SM} or d_{32}, defined by Equation (9.45), is most commonly used to characterize drop size because it relates to the volume fraction of the dispersed phase, Φ, and the interfacial area, a. The Sauter mean drop diameter is also known as the volume-to-surface average drop diameter. The interfacial area, a, in Equation (9.45) is also used to deal with mass transfer, such as $k_L a$. Other commonly used terms are d_{50}, d_{90}, and d_{max}. They represent the midsize, the 90th percentile, and the largest size in the drop size distribution, respectively, on a volume basis. The

Industrial Mixing Technology

dispersed-phase fraction, Φ, the Sauter mean drop diameter, d_{SM}, and the interfacial area, a, are related as shown by Equation (9.45).

$$d_{SM} = d_{32} = \frac{6\Phi}{a} = \frac{\sum_{i=1}^{i=n} n_i d_i^3}{\sum_{i=1}^{i=n} n_i d_i^2} \tag{9.45}$$

The maximum surviving drop diameter, d_{max}, is approximately 1.6 times d_{SM}. The Sauter mean drop diameter can be calculated using Equation (9.45) from a population of n drops of different sizes, d_i, d_{i+1}, ..., d_n. The drop size distribution often becomes self-preserving (similar shape distribution) after ≈30 passes through the impeller region.

9.8.4 Dispersion of Drops, Laminar Flow, and Low Viscosity

Single-drop studies have shown that a drop in a rotational shear field will both elongate and undergo internal rotation with increasing shear, until the critical deformation D_{crit} is reached. Beyond this point, breakage occurs. Taylor [56] established the relationship between drop deformation, D, and the magnitude of the rotational shear field. This is shown by Equation (9.46), where $P = \mu_D/\mu_C$:

$$D_{crit} = \frac{L-B}{L+B} = \left(\frac{\bar{G} \cdot d_D \mu_C}{2\gamma}\right) \cdot \left(\frac{1.1875 \cdot P + 1}{P+1}\right) = We \cdot f(P) \tag{9.46}$$

Figure 9.35 [57] and similar curves from Karam and Bellinger [58] show regions for drop breakage. Conditions must be above the curve for dispersion to occur. The burst energy E_B is a function of the variables shown in Figure 9.35.

Leng and Quarderer [59] modified Equation (9.46) by substituting impeller variables (N, D, etc.) for G_B. The resulting equation for dispersion under laminar conditions (boundary layer flow)

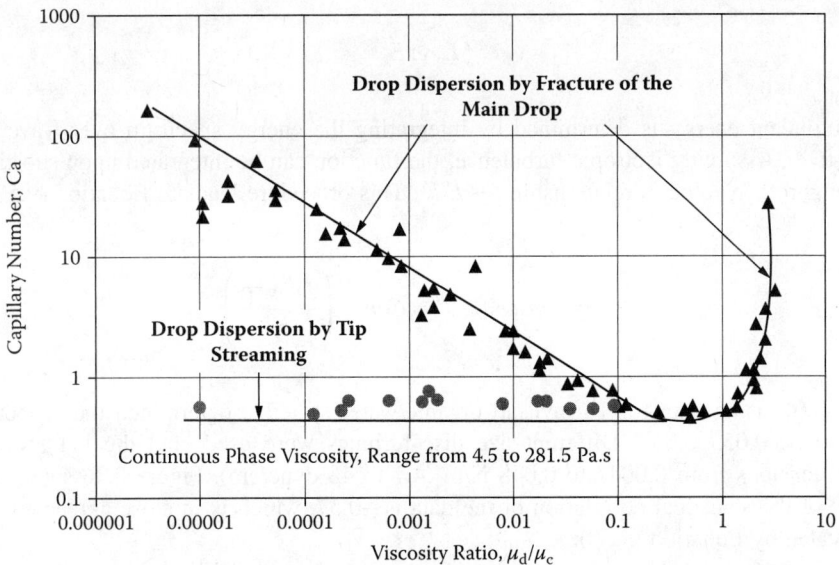

FIGURE 9.35 Critical bursting conditions as a function of the viscosity ratio. (From Grace, 1982.)

for a loop impeller of diameter D and a circular cross-sectional diameter D_C is given by Equation (9.47). A loop impeller consists of rectangular-shaped arms made from pipe. The diameter is usually $D/T = 0.67$, and the height is nearly equal to the depth of liquid in the vessel. The number of arms is commonly two, but four arms have been used successfully. The ratio of V_{tan}/V_{tip} at the tip of the impeller determines the value of k, where V_{tan} is the tangential velocity component at the impeller tip.

$$d_{max} = Const \cdot \gamma \cdot \left(\frac{D_C}{\mu_C \rho_C}\right)^{0.5} \cdot \left[ND(1-k)\right]^{-1.5} \left(\frac{P+1}{1.19P+1}\right) f(P) \qquad (9.47)$$

Equation (9.48) gives the expression for a flat paddle:

$$d_{max} = Const \cdot \gamma \cdot \left(\frac{1}{\mu_C \rho_C}\right)^{0.5} \left[N^{-1.5}D^{-1}(1-k^2)\right]^{-0.75} \cdot \left(\frac{P+1}{1.19P+1}\right) \cdot f(P) \qquad (9.48)$$

Equation (9.48) was tested commercially for scale-up using geometrically similar vessels and for drop sizes ranging from 300 to 1200 μm. Noncoalescing suspension polymerization experiments enabled sizes to be determined for finished beads using screen-analysis measurements. Equation (9.48) was confirmed for dimensionally similar vessels from 8.2 to 15,000 liters. Long dispersion times were used to ensure that complete dispersion had occurred.

9.8.5 Dispersion of Low-Viscosity (Inviscid) Drops ($\mu < 0.020$ Pa·s) Turbulent Flow

The maximum drop size is determined by equating the surface energy, E_S, associated with a drop of diameter d with the equivalent turbulent energy, E_T, associated with a spherical volume of fluid of diameter d shown by Equation (9.49).

$$E_T \geq E_s \qquad (9.49)$$

The turbulent energy is determined by integrating the energy spectrum over wave numbers from $1/d$ to ∞. Assuming isotropic turbulence, the function can be integrated upon substitution of the Kolmogoroff hypothesis relationship for $E(k)$. This procedure leads to Equation (9.50) [60].

$$\frac{d_{max}}{D} = Const \cdot We^{-0.6} = Const \cdot \left(\frac{D^3 N^2 \rho}{\gamma}\right)^{-0.6} \qquad (9.50)$$

Chen [60], using data for 14 inviscid organic/water systems, determined the proportionality constant to be 0.053–0.057. Different-size disc turbines were used, and the L-L systems had interfacial tensions from 0.0047 to 0.048 N/m (4.7 to 48 dyne/cm). Figure 9.36 shows the data. The slope of the statistical correlation of the data is −0.57, which is in close agreement with the −0.6 indicated by Equation (9.50).

Ample literature data show that drop diameters depend on the Weber number. However, much early data were erroneous due to unreported effects of coalescence, the presence of contaminants, and nonsteady-state operating conditions.

Industrial Mixing Technology

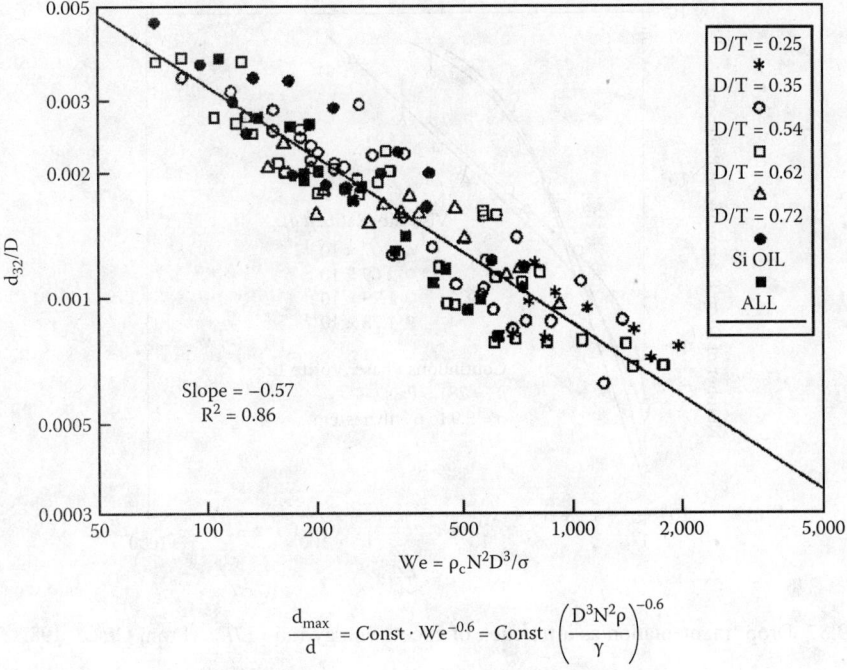

$$\frac{d_{max}}{d} = \text{Const} \cdot We^{-0.6} = \text{Const} \cdot \left(\frac{D^3 N^2 \rho}{\gamma}\right)^{-0.6}$$

FIGURE 9.36 Data of Chen and Middleman for 14 L-L systems ratios. (From Chen et al., 1967.)

Chen's equation applies to low-viscosity fluids and Rushton impellers. It is shown by

$$\frac{d_{32}}{D} = 0.053 \cdot \left(\frac{D^3 N^2 \rho}{\gamma}\right)^{-0.6} \qquad (9.51)$$

When impellers other than the Rushton (D-6 or R-100) are being used, McManamey [61] recommends using Equation (9.52):

$$\frac{d_{32}}{D} = C_1 Np^{-0.4} We^{-0.6} \qquad (9.52)$$

where Np is the power number of the impeller used for dispersion and C_1 is a system constant. It is recommended to first use Equation (9.51) to calculate the drop size, and then ratio the power numbers to arrive at the predicted size for the different impeller types.

Within the equilibrium range (also referred to as the inertia subrange) in turbulent flow, eddy size is related to the kinematic viscosity, and local energy dissipation by the Kolmogoroff hypothesis is given in

$$\lambda_k = \left(\frac{v^3}{\varepsilon}\right)^{1/4} \qquad (9.53)$$

For example, a stirred vessel containing water having a power input of 1.0 kW/m³ (5 hp/1000 USG) has a Kolmogoroff length scale, η, of 30 μm. Drops smaller than this scale are entrained and not further dispersed. Larger drops are broken up by eddy interactions.

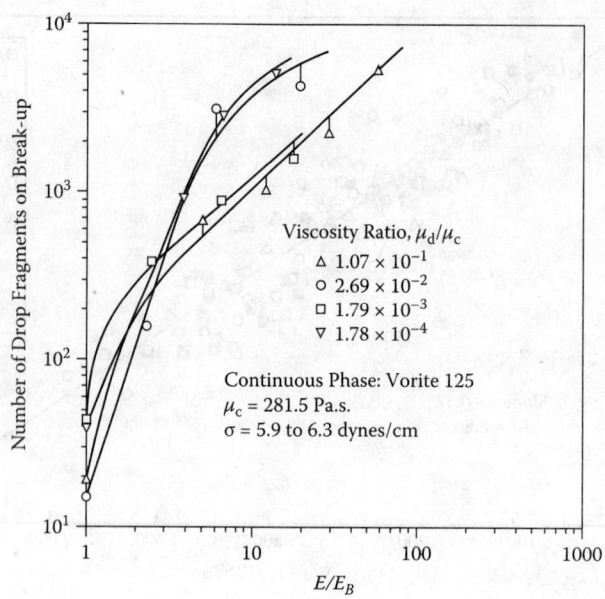

FIGURE 9.37 Drop fragmentation as a function of excess energy ratio E/E_B. (From Grace, 1982.)

Drop size distribution will be addressed briefly. When a drop breaks, there can be few or many fragments. The breadth of the distribution of daughter drops depends on the level of energy intensity at the time of breakage. Figure 9.37 shows that a large number of drops can form from a single drop at high energy. E/E_B is the ratio of local energy E to minimum bursting energy E_B for the drop. The viscosity ratio also affects drop breakage. More fragments form when μ_d/μ_c is small. In applications where the breadth of the size distribution must be minimized, such as in suspension polymerization, it is essential to provide uniform energy distribution.

9.8.6 Dispersion of Higher-Viscosity Drops ($\mu > 0.02$ Pa·s) Turbulent Flow

Calabrese et al. [60] and Wang et al. [63] show that the breakage condition in Equation (9.49) should include a term E_v for dealing with drop viscosity. Equations (9.54) and (9.55) show the result:

$$E_T \geq E_S + E_V \tag{9.54}$$

Equation (9.55) by Calabrese et al. [62] includes viscous terms:

$$\frac{\rho_c \varepsilon^{-2/3} d_{max}^{5/3}}{\gamma} = C_5 \left[1 + C_6 \left(\frac{\rho_c}{\rho_d}\right)^{1/2} \frac{\mu_d \varepsilon^{-1/3} D_{max}^{1/3}}{\gamma} \right] \tag{9.55}$$

When interfacial tension is the main resistance ($E_s \geq E_v$), Equation (9.52) reduces to Equation (9.56), which is equivalent to Equation (9.50):

$$d_{max} = const \cdot \left(\frac{\gamma^{0.6}}{\rho_C^{0.6} \varepsilon^{0.4}} \right) \tag{9.56}$$

Industrial Mixing Technology

When the disperse-phase viscosity provides the main resistance ($E_v \geq E_s$), Equation (9.56) becomes

$$d_{max} = C_8 \left(\rho_c \cdot \rho_d \right)^{-3/8} \cdot \mu_d^{3/4} \cdot \overline{\varepsilon}^{-1/4} \tag{9.57}$$

9.8.7 TIME FOR DISPERSION

The time for equivalent dispersion is scale dependent. A large vessel takes much longer to disperse drops to an equivalent degree of dispersion than a small vessel. If scale-up demands equal dispersion time, the larger vessel will be less completely dispersed than the smaller one. If the desired result is equal drop size, one solution is to increase the speed for the larger vessel and acknowledge that drop sizes will continue to be dispersion-time dependent.

9.8.8 COALESCENCE OF SUSPENDED DROPS

Coalescence is the combining of drops either by collision or at a surface. It is desirable in some applications, while undesirable in others. It aids mass transfer and separation processes, but it is harmful to suspension and emulsion polymerization. Coalescence can occur when drops collide with one another or come to rest on surfaces and interfaces. Collisions between drops can result in either coalescence or rebounding. The impact creates transient forces that act on the colliding drops to thin the film separating them. As this film gets thinner, rupture (coalescence) occurs when the thickness reaches a critical value. If the film thickness does not reach this critical thickness during contact, the drops will depart without coalescing.

The coalescence rate is the product of the collision rate and the coalescence efficiency. The collision rate depends on the drop population and the velocity gradient. The coalescence efficiency depends on film thinning during contact and the time of contact.

Figure 9.38 shows two liquid drops approaching one another. The net force due to the change in momentum caused by the collision is F. Drops are separated from each other by a film of continuous-phase liquid of thickness h.

The leading edges are flattened, since drops are deformable. This deformation creates a parallel disclike geometry representing the thinning liquid acting under F. When a critical thickness is reached, rupture occurs. Equation (9.58) shows the rate of thinning. The reasoning is similar for drops resting at a flat liquid-liquid interface.

$$\frac{dh}{dt} = -\left(\frac{2F}{3\pi\eta R^4} \right) h^3 \tag{9.58}$$

The rate of thinning, dh/dt, depends on the force F, the fourth power of R, the thickness of the film h, and the viscosity η of the continuous phase.

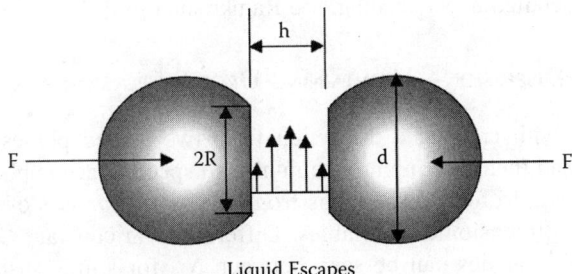

FIGURE 9.38 Film drainage between two colliding drops.

Upon integration and substitution of the initial and final conditions, $h = h_0$ at $t = 0$, and $h = h$ at time t, Equation (9.58) becomes

$$\Delta t = t - t_0 = \frac{3\pi\eta R^4}{4F}\left(\frac{1}{h^2} - \frac{1}{h_o^2}\right) \quad (9.59)$$

The initial distance h_0 is large compared with h, the thickness of the film at time t. The change in time, Δt, is the time it takes to reach a critical thickness for film rupture. Several versions of this equation exist that include internal circulation within the drop, rigid yet deformable interfaces, and complete interface mobility [64, 65].

This relationship has practical implications:

- If Δt, the total drainage, is sufficient such that the critical separation thickness is reached during the contact interval, the drops will coalesce.
- If insufficient drainage occurs during the contact interval, Δt, the drops will not coalesce.
- A low interfacial tension leads to greater flattening of the approaching drop surfaces. This produces a larger disc and makes the thinning process slower. Surfactants lower the interfacial tension, γ, and reduce the likelihood of coalescence. The surfactant may also reduce interfacial mobility, leading to further retardation of drainage rates.
- A higher impact force, F might increase the likelihood of coalescence, but it also creates larger discs, trapping more liquid and making drainage more difficult. Centrifuges and electrostatic coalescers are examples of force-amplification devices; decanters, on the other hand, depend on gravitational force.
- A higher continuous-phase viscosity reduces drainage rates and coalescence probability. For this reason, if two similar volumes of immiscible liquids are dispersed, the fluid with the higher viscosity will normally become the continuous phase.

The presence of solids at the interface usually retards film drainage rates, thereby reducing the probability of coalescence. A few suspension-polymerization processes use solid particles as suspending agents.

9.8.9 Population-Balance Methods

Population-balance analysis has been adapted to both coalescence and dispersion of drops in numerous papers by Calabrese, Ramkrishna, and Tavlarides. The analyses with these tools have led to a considerably better understanding of breakage kernels, breakage rates, coalescence efficiency, and collision rates. However, the description and use of these tools goes beyond the scope of this chapter. For a detailed understanding, see Ramkrishna [66].

9.8.10 Creating a Dispersion—Maintaining Drop Suspension

This subsection deals with creating a dispersion from two settled phases of different density, obtaining uniformity, and then determining the minimum speed to accomplish a dispersion.

The minimum speed, N_m, to form dispersions from two settled phases of different densities is expressed in terms of dimensionless variables. Differences in constant C reflect the effect of impeller clearance. Similarities can be seen between N_{JS} for solids suspension and N_{min} for

TABLE 9.15
Constants for Use in Equation (9.45)

Impeller Type	Clearance	C_1	α_1
1 propeller	H/4	15.32	0.28
1 propeller	3H/4	9.97	0.55
1 propeller	H/2	15.31	0.39
2 propellers	H/4 + 3H/4	5.24	0.92
1 pitch-blade turbine	H/4	6.82	1.05
1 pitch-blade turbine	3H/4	6.2	0.82
1 pitch-blade turbine	H/2	2.99	1.59
2 pitch-blade turbines	H/4 + 3H/4	3.35	0.87
1 flat-blade turbine	H/4	3.18	1.62
1 flat-blade turbine	3H/4	...	...
1 flat-blade turbine	H/2	3.99	0.88
2 flat-blade turbines	H/4 + 3H/4	...	...
1 curved-blade turbine	H/4	3.61	1.46
1 curved-blade turbine	3H/4	...	...
1 curved-blade turbine	H/2	4.71	0.80
2 curved-blade turbines	H/4 + 3H/4	4.29	0.54

drop suspension. Both are functions of density differences, viscosity, and impeller diameter. Skelland et al. [67] showed suspension speeds for equal phase volumes of two immiscible liquids. They also studied the effect of impeller location by placing a single impeller midway in the dense phase ($C = H/4$), at the phase interface ($C = H/2$), and midway in the light phase ($C = 3H/4$). For studies involving dual impellers, the impellers were placed midway in both phases to determine the minimum speed to form a suspension. Table 9.15 gives values for the constants in

$$N_m D^{0.5} / g^{0.5} = C_1 \left(\frac{T}{D}\right)^{\alpha_1} \left(\frac{\mu_c}{\mu_d}\right)^{1/9} \left(\frac{\Delta\rho}{\rho_c}\right)^{0.25} \left(\frac{\sigma}{D^2 \rho_c g}\right)^{0.3} \qquad (9.60)$$

N_m is the stirrer speed in rps required to form a complete suspension. The C constants indicate the importance of impeller location. Single propellers have large constants, thus requiring higher speed. Impellers placed at the L-L interface use a minimum power to form dispersions. Flat-blade, radial discharge impellers do not work efficiently when placed in the light phase. More recent work by Armenante et al. [68] deal with the just-dispersed conditions in vessels equipped with multiple impellers.

Several methods can be used to form L-L dispersions. Placing the impeller in the lower phase will draw the upper phase in by interfacial vortex feeding. If placed in the upper layer, the impeller would normally disperse the lower phase into the light phase. However, the initially formed dispersion can be unstable. As mentioned earlier, the more viscous phase normally becomes the continuous phase. If the less viscous phase is initially the continuous phase, phase inversion is apt to happen. Another method commonly used to form dispersions is to continuously add a second phase while agitating the first phase. This method forces the added fluid to be dispersed. If phase volumes or viscosities are different, phase inversions can occur regardless of how the dispersion was first formed.

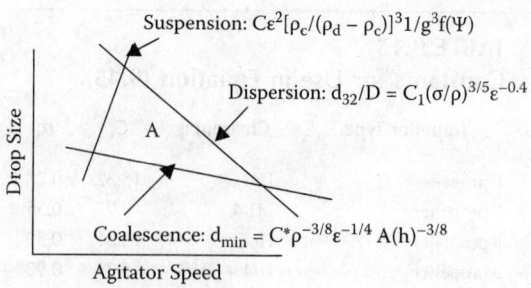

FIGURE 9.39 Dispersion, coalescence, and suspension. (From Church et al., 1961.)

9.8.11 Simultaneous Suspension, Dispersion, and Coalescence

Church [69] describes the simultaneous processes taking place in a stirred vessel. Figure 9.39 shows the result of three processes that form a bounded region. Within this region, equilibrium stability exists. For example, a drop within the region bounded by the three lines can grow by coalescence until it reaches the upper dispersion line. At this point it is dispersed by the local energy. Smaller drops below the coalescence line grow by coalescence until they reach the coalescence line. A drop located to the left of the suspension line will separate. The expressions that describe the lines in Figure 9.39 are given as Equations (9.61 to 9.63).

Dispersion:

$$d_{32}/D = C_1 \cdot (We)^{-0.6} = \left(\frac{\sigma^{3/5}}{\rho_c^{3/5} \cdot \overline{\varepsilon}^{2/5}}\right) \tag{9.61}$$

Coalescence:

$$d_{32} = C_2 N^{-0.75} D^{-0.5} \rho^{-0.375} A(h)^{0.375} \tag{9.62}$$

where $A(h)$ is the adhesive or attractive force and is a function of the separation distance, h.

Suspension:

$$d_{max} = C_3 D^4 N^6 \left(\frac{\Delta\rho}{\rho_c}\right)^3 \left(\frac{1}{g}\right)^3 f(\Psi) \tag{9.63}$$

Regions around the impeller are usually dominated by high turbulence that causes dispersion. Elsewhere in the vessel, turbulence is much lower, and this promotes coalescence. Church [69] refers to this ongoing process as an agitation-stabilized system. This is fine, but the problem comes when one is scaling up a coalescing system. In a small vessel, circulation times are short and the turbulence zones dominate. The reverse is true for a large vessel. Simply keeping P/V constant and the geometry similar does not guarantee comparable drop size with scale-up. As mentioned earlier, an added complication for noncoalescing systems is the time to reach comparable stages of dispersion in different scales. Strongly coalescing systems reach steady state much more quickly than noncoalescing systems.

TABLE 9.16
Common Types of Equipment Used in Immiscible L-L Mixing

Description	Impeller Types	Batch/Continuous	Desired Result	Comments
Stirred tanks, baffles	flat, pitch, and disc type	either	$30 \leq d_{32} \leq 300$ µm	general; mass-transfer operations
Stirred tanks, baffles	retreat curve	either	$30 \leq d_{32} \leq 500$ µm	general; emulsion polymerization
Stirred tanks, no baffles	paddle, loop, special types	batch	$100 \leq d_{32} \leq 1000$ µm	suspension polymerization; suspending agent required
Static/in-line mixers	none	continuous	$10 \leq d_{32} \leq 200$ µm	dispersant or protective colloid needed
Rotor-stator mixers	slotted ring + slotted stator	usually continuous	$1 \leq d_{32} \leq 50$ µm	sparse data for scale-up; need extensive testing
Impingement mixers	none	continuous	$1 \leq d_{32} \leq 50$ µm	sparse data; work with vendors
Homogenizers	valve homogenizers, ultrasonic mixers	continuous	$0.1 \leq d_{32} \leq 10$ µ	sparse data; work with vendors

9.8.12 Equipment Used for Liquid-Liquid Mixing

Liquid-liquid dispersion similar to gas-liquid dispersion depends on power being delivered to the dispersed phase. Many different power sources have been studied, including static mixers, stirred vessels, colloid mills, liquid whistles, ultrasonic devices, and valve homogenizers. At equivalent conditions, Davies [70] found that the maximum surviving drop size correlated with energy supplied and was to a degree independent of the device. A straight-line (log-log) plot represented all drop sizes with energy very well. This does not mean the drop size distribution is independent of the device. Indeed, devices producing very high local energy are likely to produce a much finer drop size distribution (dsd), because of how the drops fracture.

Process requirements, such as particle/drop size or distribution, determine the selection of equipment to be used in immiscible liquid mixing, as shown in Table 9.16. Energy is required for drop dispersion. Commonly used are agitated, baffled, and jacketed vessels, typically with impellers where D/T varies from 0.25 to 0.7. Large, high-flow impellers are used to help deal with strong drop-coalescence tendencies. Particular attention must be paid to mixing in the top and bottom regions of the vessel when significant density differences exist, for example, >200 kg/m³. A second impeller should be used to cope with settling problems.

9.8.13 Processes

A common goal of suspension polymerization is to make uniform-sized drops that polymerize to form uniform size beads. Suspending agents prevent coalescence, thus preserving drops created by agitation. They also reduce the interfacial tension and drop size. Excessively high stirring leads to the creation of unduly small drops/beads. Slow agitation leads to large beads of broad size distribution and can lead to settling of the dispersed phase. Low, uniform shear rates are desirable. Baffling, if used, should be minimal. Impellers include retreat-curve turbines, two- and four-blade paddles, and rectangular-shaped open pipe loops. Impellers having large blade widths promote better circulation while maintaining low shear. Jacketed vessels are used to remove the heat of polymerization.

Dispersion, suspension, and heat removal must be simultaneously controlled in dealing with suspension polymerization. In Figure 9.39, the intersection of suspension and dispersion lines represents the largest maximum drop size that can be produced under existing conditions. Selection of different impellers can affect the position of this intersection. In continuous extraction applications, the goal is to create a large interfacial area, yet not make the drops so small that they are difficult to coalesce.

Other types of equipment such as static (motionless) mixers, rotor-stator mixers, and impingement mixers are used to create dispersions (see Section 9.9).

Example 6

In this example, solvent extraction is to be used to recover a product from a fermentation process. Processing information includes: broth viscosity $\mu = 0.3$ Pa·s (300 cP), interfacial tension $\sigma = 0.003$ N/m (3.0 dynes/cm), bulk density $\rho_C = 1000$ kg/m³ (1.0 g/cm³), vessel volume = 3.54 m³ (750 gal), vessel diameter = 1.524 m (5.0 ft), and $D/T = 0.4$. Laboratory studies showed that acceptable extraction efficiency was obtained with mean drop sizes of 50 μm. Determine the power and speed required to produce 50-μm drops in the 750-gal extractor.

Solution

The solvent will disperse in the broth phase and will be slow to coalesce, since the broth has high viscosity. Equation (9.51) from Chen [60], $d_{max}/D = 0.053(D^3N^2\rho/\sigma)^{-0.6}$, will be used. Substituting $d_{32} = 0.005$ cm, $D = 61$ cm, $\sigma = 3.0$ dyne/cm, and $\rho = 1.0$ g/cm³, then solving for N gives $N = 48$ rpm (0.8 rps).

The Reynolds number, Re = $D^2N\bar{\rho}/\mu$, is 7700 using a volume average viscosity of 38.6 cP. Flow is transitional but nearly turbulent. The power requirement using Np = 5.0, a mean density of 1.0, and impeller speed of 48 rpm, gives

$$hp = Np\,\bar{\rho}\,N^3D^5/17{,}719 = 0.3$$

9.9 STATIC IN-LINE MIXERS

9.9.1 INTRODUCTION TO IN-LINE STATIC MIXERS

A static mixer, also known as a motionless mixer, is a mixing device without any moving parts. It is also known as a pipeline mixer. It creates mixing as fluids pass through stationary geometric elements located inside the pipe. Static mixing elements vary in geometry from ordinary pipe internals such as orifices and baffles to specially designed complex structures. Turbulent pipe flow will ordinarily achieve a reasonably high degree of radial mixing in a length of roughly 100 pipe diameters. Static mixers achieve the same mixing in lengths of three to five pipe diameters.

There are over 30 different designs of static mixer elements. The more popular commercial ones and their suppliers include:

- Chemineer Kenics: KM series and HEV, Figure 9.40(a)
- Sulzer Chemtech: (or Koch) SMX, SMXL, SMV, SMVL, Figure 9.40(b)
- Ross: LPD, LLPD, Figure 9.40(c); ISG, Figure 9.40(d)
- Komax mixer

Static mixers are readily available in a wide variety of materials, including metals, alloys, and plastics in sizes from 1/2 to 24 inches in diameter.

Static mixers are widely used in situations where continuous in-line or pipeline mixing is required. Table 9.17 shows that applications span the entire spectrum from laminar to turbulent flow regimes, and they cover a wide variety of mixing processes involving the blending of gas-gas

Industrial Mixing Technology

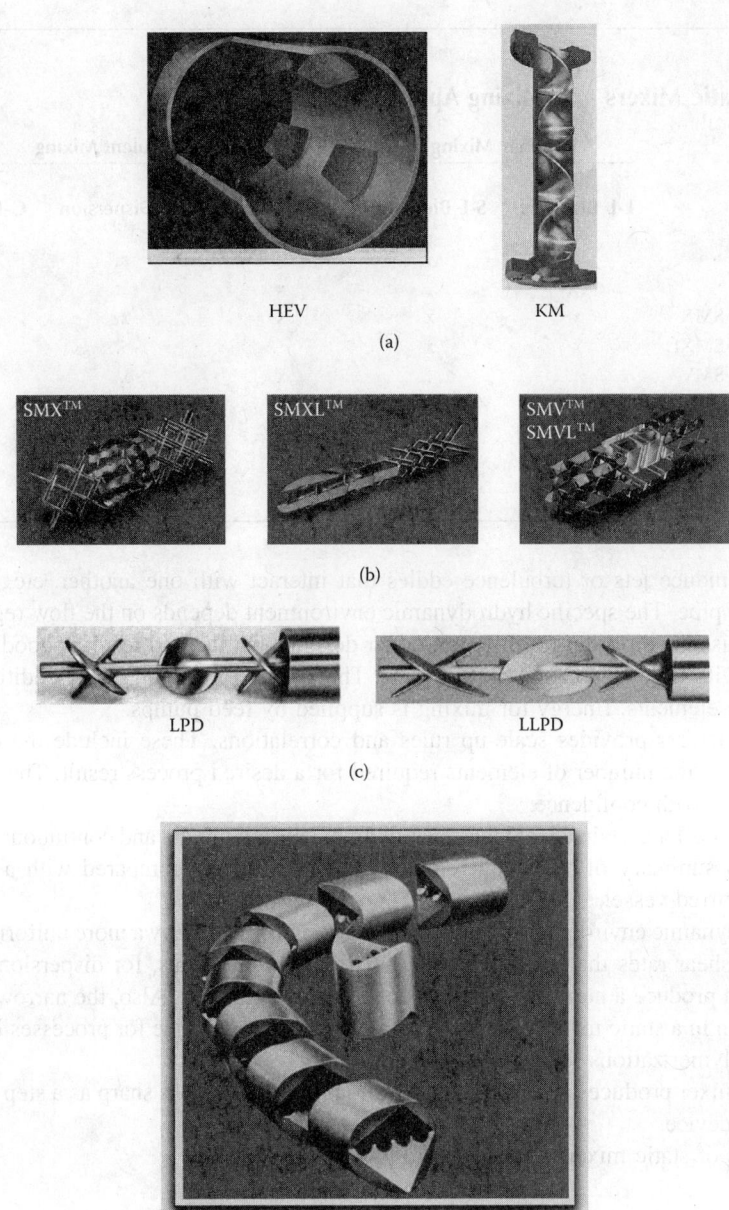

FIGURE 9.40 (a) Chemineer Kenics static mixers. (b) Sulzer or Koch static mixers. (c) Ross static mixers. (d) Ross ISG static mixer. (From Ross Systems and Controls.)

(G-G) or miscible liquid-liquid (L-L) streams as well as dispersions of solid-liquid (S-L), gas-liquid (G-L), and immiscible liquid-liquid (L-L) systems. Static mixers are also used in heat-exchanger tubes to enhance heat transfer under laminar flow conditions.

9.9.2 Hydrodynamics and Other Characteristics of Mixing in Static Mixers

Mixing by static mixing elements results from the hydrodynamic environment created by the fixed elements. These elements cause the fluids to divide and recombine, rotate clockwise and then

TABLE 9.17
Selected Static Mixers and Mixing Applications

	Laminar Mixing		Turbulent Mixing		
Mixer	L-L Blending	S-L Blending	G-G or L-L Blending	L-L Dispersion	G-L Dispersion
Kenics KM	x	...	x	x	x
Kenics HEV	...	...	x	...	...
Sulzer or Koch SMX	x	x	x	x	x
Sulzer or Koch SMXL	x	x	...	...	...
Sulzer or Koch SMV	...	...	x	x	x
Ross LPD or LLPD	x	...	...	x	x
Ross ISG	x	x	...	...	...
Komax mixer	...	...	x	x	x
Komax ultramixer	x	...	...	...	...

anticlockwise, induce jets or turbulence eddies that interact with one another, etc., as the fluids move down the pipe. The specific hydrodynamic environment depends on the flow regime, laminar or turbulent. It is also different for different mixer designs, but the end result is good transverse or radial mixing with little or no axial back-mixing. The energy for mixing is the additional pressure drop due to the elements. Energy for mixing is supplied by feed pumps.

The manufacturer provides scale-up rules and correlations. These include the estimation of pressure drop and the number of elements required for a desired process result. These are reliable and may be used with confidence.

Static mixers offer certain advantages over dynamic in-line mixers and continuous stirred tanks. Table 9.18 is a summary of the characteristics of a static mixer compared with a conventional mechanically stirred vessel.

The hydrodynamic environment in a static mixer is characterized by a more uniform distribution of energy and shear rates than in a conventional agitated tank. Thus, for dispersion processes, a static mixer can produce a narrower drop or bubble size distribution. Also, the narrower residence-time distribution in a static mixer makes it the preferred mixing device for processes involving fast reactions or polymerizations requiring no back-mixing.

The static mixer produces an outlet concentration profile nearly as sharp as a step-change input in a plug-flow device.

Limitations of static mixers include the following:

- Provide little or no axial mixing (but excellent radial mixing)
- Cannot compensate for temporal variations in concentration or flow
- Pressure drop increases with the number of elements
- Certain slurries are prone to plug
- Low residence time and holdup are limitations for many reactions
- Performance is flow sensitive in turbulent applications
- Not easy to adapt to unexpected performance requirements

9.9.3 Static Mixer Selection and Design Issues

The proper selection or design of a static mixer starts with a clear understanding of process objectives and limitations. Table 9.19 gives design information for blending and dispersion applications. The following steps are recommended:

TABLE 9.18
Comparative Characteristics of the Static Mixer and the Continuous Stirred-Tank Mixer

Item	Static Mixer	Continuous Stirred Tank
Capital cost	lower	higher
Operating cost	lower because of lower power	higher
Maintenance cost	lower—no rotating component to wear out	higher—bearings, seals, gears, gearbox oil changes, etc.
Space requirements	smaller	larger
Hazard	volume small, contained, less hazard	large volume, potential for uncontrolled events
Environmental aspects	low leak potential	greater potential for leakage from seals
Hydrodynamic environment	(a) approaches plug-flow radial mixing with minimal back-mixing (b) narrower residence-time distribution (c) more-uniform distribution of energy dissipation and shear rates	(a) totally back-mixed (b) broader residence-time distribution (c) broader distribution of energy dissipation and shear rates
Residence time for reaction	short	long
Applicability	(a) continuous-flow applications (b) all fluid-mixing processes, but mostly used for blending and heat transfer (c) all flow regimes—laminar, transitional, and turbulent (d) wider ranges of temperature, pressure, and chemical environments (e) not suitable for slower reactions	(a) batch, semi-batch, or continuous (b) all fluid-mixing applications (c) more suited to transitional and turbulent flow regimes (d) narrower ranges of temperature, pressure, and chemical environments (e) suitable for slower reactions

1. *Specify the desired process result*: This is the first step to consider. The process listed results listed in Table 9.19 quantify the objective of the mixing process as described below.
2. *Assemble the process information pertinent to the application*: The lists in Table 9.19 are necessary for a complete process definition. They provide insights into the nature of the mixing process. For example, the component fluid phases and their miscibility establish whether one has a blending or dispersion application.
3. *Calculate static mixing process parameters*: The pertinent process parameters, along with the process information, help establish how difficult the mixing is for the process, as discussed below.
4. *Select and design mixer to give the desired process result*: In the selection and design, the pertinent items listed in Table 9.19 are specified. Suppliers of static mixers can do this final step reliably, provided that they have all pertinent information from the process definition steps. It often involves the iterative evaluation of several mixer designs and sizes for the one that best matches the specified process result. The mixer selection is best based on the mixing efficiency. But in retrofit cases, it is often selected on the basis of the design with the lowest pressure drop or shortest length to fit the retrofit design constraints.

9.9.4 Reynolds Number and Flow Regime

The value of the Reynolds number, computed using Equation (9.64), determines whether the flow is turbulent or laminar:

TABLE 9.19
Key Information for Blending and Dispersion in a Static Mixer

Key Information	Blending Applications: Miscible Fluids	Dispersion Applications: Immiscible Fluids
Process information	Number of component flows	Number of component flows
	Component flow rates	Component flow rates
	Component phases	Component phases
	Component flow viscosity	Component flow viscosity
	Component flow density	Component flow density
	Combined flow rate	Combined flow rate
	Combined flow viscosity	Combined flow viscosity
	Combined flow density	Combined flow density
	Allowable pressure drop in mixer	Interfacial tension
	Heat transfer: Yes/No	Allowable pressure drop in mixer
	Reacting flows: Yes/No	Heat transfer: Yes/No
		Reacting flows: Yes/No
		Inter-phase mass transfer: Yes/No
Static mixing process parameters	Reynolds No.	Reynolds No.
	Component flow ratios	Component flow ratios
	Component viscosity ratios	Component viscosity ratios
		Weber No.
Desired process result	Concentration/temperature homogeneity	Concentration/temperature homogeneity
	Variation coefficient	Drop size and distribution
	Heat-transfer rate	Interfacial area
		Mass-transfer coefficient
		Heat-transfer coefficient
Mixer selection and design information	Number of mixers	Number of mixers
	Mixer design	Mixer design
	Feed arrangements	Feed arrangements
	Injector design	Injector design
	Mixer diameter	Mixer diameter
	Mixer length or number of elements	Mixer length or number of elements
	Mixer pressure drop	Mixer pressure drop
	Mixing efficiency	Mixing efficiency

$$\text{Re}' = \frac{\rho V D_h}{\mu} \tag{9.64}$$

where Re' is a dimensionless correlating parameter for static mixer design, and D_h is the hydraulic diameter. The transition from laminar to turbulent flow for pipes occurs at Re = 2100, but in static mixers, the transition point may be as low as 100. Blending under turbulent flow usually requires fewer than five elements. Laminar blending requires many more, depending on the desired uniformity.

9.9.5 COMPONENT FLOW AND VISCOSITY RATIOS

These determine the difficulty of the mixing application, as illustrated in Figure 9.41. The wider the viscosity difference between the component streams, the more difficult is the blending application. The most difficult application is the addition of a small, low-viscosity stream into a large, high-viscosity stream in laminar flow. For example, Tables 9.20(a) and 9.20(b) show effects

Industrial Mixing Technology

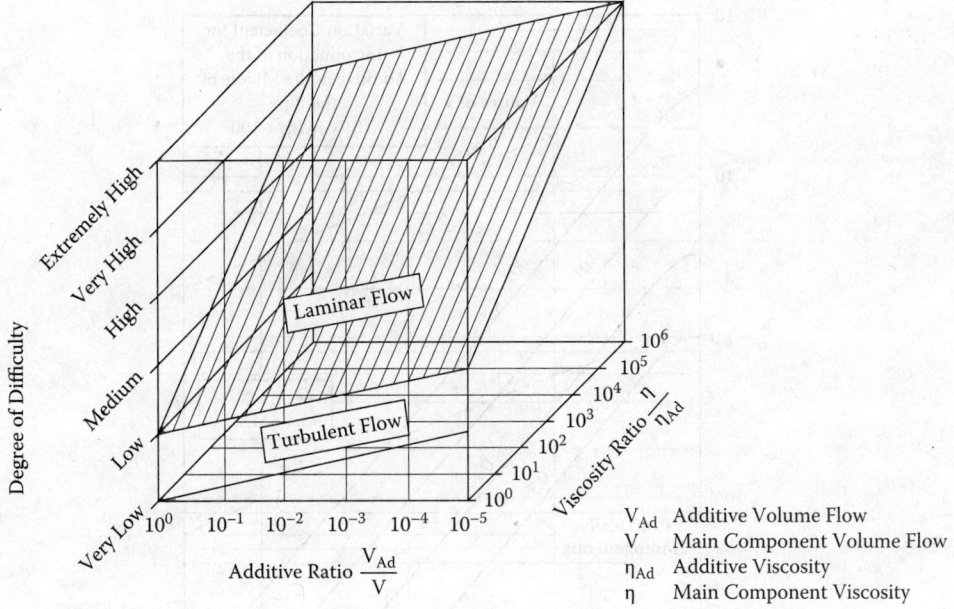

FIGURE 9.41 Additive and flow ratio effects.

TABLE 9.20(A)
Laminar Blending: Effect of Viscosity Ratio on Required Number of Elements

Viscosity Ratio: Primary to Minor Stream	Number of Ross LPD Modules	Number of Ross ISG elements
10^{-1} to 10^3	4–6 elements	10
10^3 to 10^4	6–6 elements	14
10^4 to 10^5	6–8 elements	20

TABLE 9.20(B)
Laminar Blending: Effect of Viscosity Ratio on Required Number of SMX Elements

Flow Ratio: Primary to Minor Stream	L/D @ $\sigma/x =$ 0.05 for Viscosity Ratio of 10^3	L/D @ $\sigma/x =$ 0.05 for Viscosity Ratio of 10^5
10^{-3}	17	20
10^{-2}	15	18
10^{-1}	5	5

on the required number of elements for higher values of the viscosity ratio for Ross and Sulzer SMX, respectively.

9.9.6 Variation Coefficient in Blending Applications

The variation coefficient, $\sigma/\bar{x}$, is used in blending applications to characterize the homogeneity in a static mixer. Equation (9.65) defines $\sigma/\bar{x}$:

$$\frac{\sigma}{\bar{x}} = \frac{\sigma}{\sigma_o} \sqrt{\frac{1-\bar{x}}{\bar{x}}} \tag{9.65}$$

where σ is the standard deviation of point concentration values, x_i, at a given cross section of the mixer; σ_0 is the standard deviation at the mixer inlet; and $\bar{x}$ is the arithmetic mean of x_i values.

Figure 9.42 is a characteristic plot of $\sigma/\bar{x}$ as a function of the length-to-diameter ratio, L/D, for Sulzer SMX and SMXL mixers. It represents what happens when an additive is introduced at

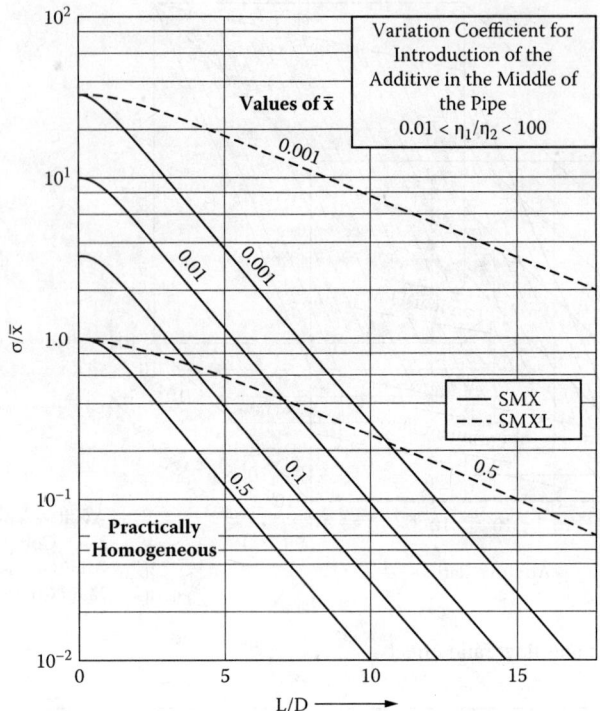

FIGURE 9.42 Variation coefficient as a function of mixer length for Koch SMX and SMXL mixers.

the center of the main flow under laminar conditions in the static mixers. The parameter, the arithmetic mean value, $\bar{x}$, represents the fraction of the additive in the total flow.

Key points in Figure 9.42 are

- Longer mixer lengths are required to reduce the variation coefficient.
- Shorter lengths are required for the SMX than for the SMXL.
- Longer mixers are required as the ratio of the additive decreases; it is easier to blend 50/50 than 99/1.
- The length required is insensitive to the viscosity ratio over a range of $0.01 < \mu_1/\mu_2 < 100$.

Key points in Figure 9.43 are

- In laminar flow, there are big differences in performance between different mixers.
- The top line represents an open pipe.
- Mixing efficiency does not necessarily follow Figure 9.44. Pressure drop needs to be considered.

A coefficient of variation, $\sigma/\bar{x}$, of 0.05 is considered to be a normal level of homogeneity. Less homogeneity may be acceptable if further processing will be done. More homogeneity may be required for critical reactions, particularly when dealing with fast competing chemical reactions.

9.9.7 DISPERSED-PHASE SIZE DISTRIBUTION IN LIQUID-LIQUID AND GAS-LIQUID SYSTEMS

The static mixer creates a hydrodynamic environment where critical viscous shear stresses deform drops, causing them to break. The dispersion process decreases the drop sizes and increases the

Industrial Mixing Technology

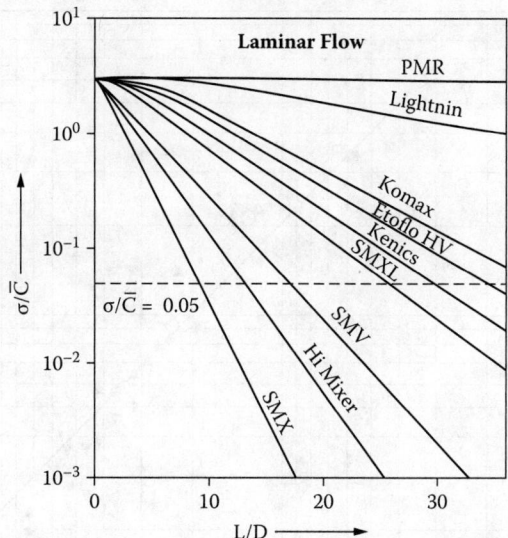

FIGURE 9.43 Comparative mixing performance of various static mixers in laminar flow.

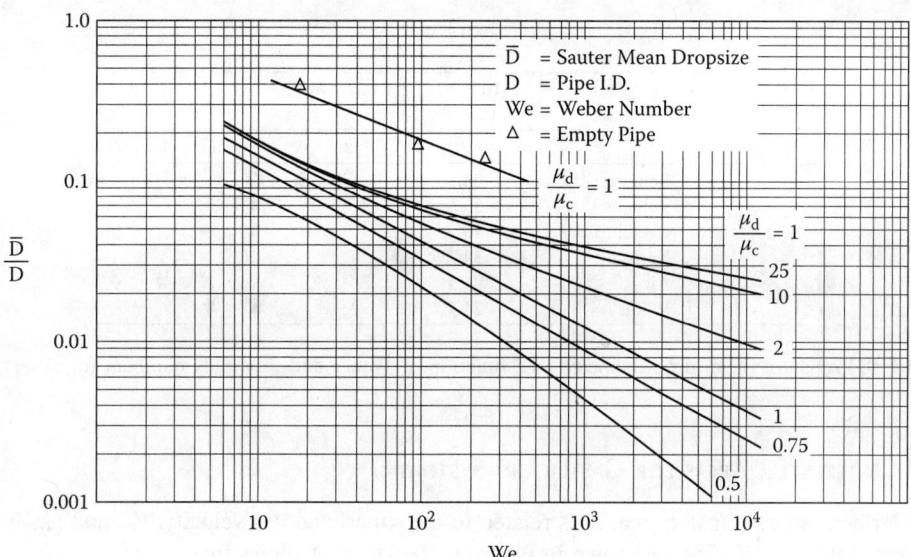

FIGURE 9.44 Sauter mean drop diameter versus Weber number for Kenics KM mixers.

interfacial area between the phases as fluids pass through the mixer until the minimum drop size is attained. The drop size distribution for a noncoalescing system is narrower in a static mixer than in a stirred tank because the distribution of shear forces is more uniform. Typically, 70 to 80% of the volume of droplets will lie within ±20 to 30% of the mean drop diameter, and roughly 12 elements are required to reach the minimum drop diameter [72]. The Sauter mean drop size, d_{32}, is correlated by the Weber and Reynolds numbers, as shown by Figures 9.44 and 9.45, which present drop size d_{32} data for Kenics KM and Sulzer SMV mixers, respectively. It is clear from both figures that d_{32} decreases with increasing Weber number and, from Figure 9.44, d_{32} decreases as the ratio of the dispersed to the continuous liquid phase viscosity decreases.

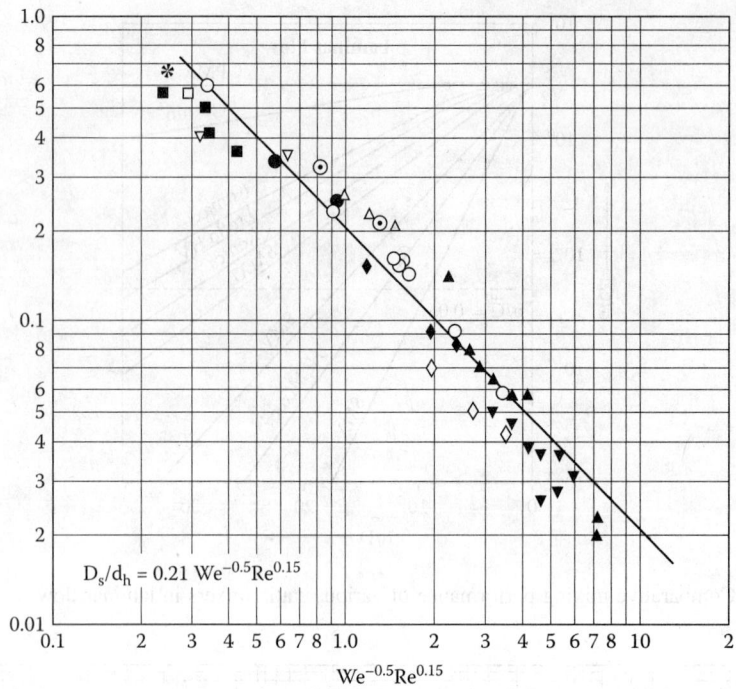

FIGURE 9.45 Sauter mean drop diameter as a function of Weber and Reynolds numbers for Sulzer SMV mixer.

9.9.8 Interfacial Areas for Gas-Liquid Systems

The interfacial mass transfer area, a, is related to the superficial gas velocity, V_G, and gas holdup, Φ_G, for a Sulzer SMV [73], as shown in Figure 9.46. The plot shows that

$$a \propto V_G^{0.85} \phi_G^{0.15} \qquad (9.66)$$

where the proportionality constant varies with the mixer geometry.

9.9.9 Mass-Transfer Coefficients for Gas-Liquid Dispersions

In the same study of gas-liquid mixing in SMV [73], the mass-transfer coefficient, k_L, was found to increase with the average liquid velocity, V_L, as

$$k_L \propto V_L^{1.22} \qquad (9.67)$$

Industrial Mixing Technology

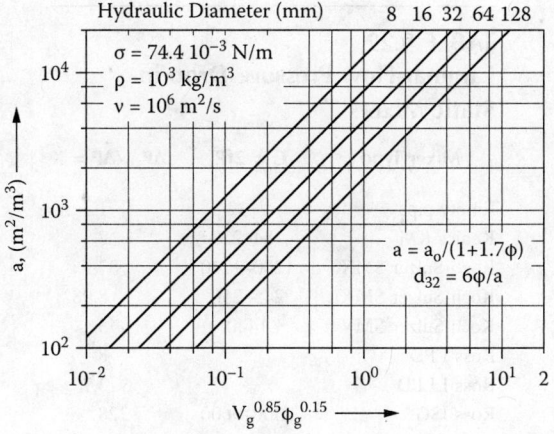

FIGURE 9.46 Interfacial area vs. product of gas velocity and dispersed-phase volume fraction.

TABLE 9.21
Constants for Heat-Transfer Equation (9.69)

Type	α	β
SMX	2.6	0.35
SMXL	0.98	0.38

The Sherwood number in terms of the Schmidt and Reynolds numbers is

$$Sh = 0.0062\, Re^{1.22}\, Sc^{1/3} \tag{9.68}$$

9.9.10 Heat-Transfer Enhancement in Laminar Flow Applications

The heat-transfer coefficient, h, for laminar flow in tubes can be enhanced by a factor of two- to sixfold using static mixer elements in the heat-exchanger tubes. The heat-transfer coefficient is correlated using the Nusselt, Prandtl, and Reynolds numbers. Table 9.21 gives constants for Koch SMX and SMXL mixers for

$$Nu = \alpha \cdot (Re \cdot Pr)^{\beta} \tag{9.69}$$

9.9.11 Pressure Drop and Power Requirements

The pressure drop in a static mixer, ΔP_{SM}, is proportional to the pressure drop in an empty tube of the same length, ΔP, and its accurate prediction is vital to pump sizing:

$$\Delta P_{SM} = K\, \Delta P_{EP} \tag{9.70}$$

K depends on the mixer design and the Reynolds number. The exact values of K can be obtained from supplier literature. Typical values given in terms of the Darcy friction factor are presented in Tables 9.22–9.24. The power required for mixing in static mixers is

TABLE 9.22
Laminar-Flow Pressure Drop in Static Mixers

Mixer Type	$C = 2fR_e$	$\Delta P_{SM}/\Delta P = K$
Empty pipe	32	1
Kenics KM	184–220	7
Koch-Sulzer SMX	1200–1240	37.5
Koch-Sulzer SMXL	245–250	8.98
Koch-Sulzer SMV	1440	45
Ross LPD	...	8
Ross LLPD	...	3.7
Ross ISG	7300–9600	228
Komax	592–620	25

Note: f is the Fanning friction factor. The Darcy friction factor is 4f.

TABLE 9.23
Turbulent Flow Friction Factors

Mixer Type	Darcy's Friction Factor
Empty pipe	0.02
Kenics HEV	0.43
Kenics KM	2–3
Sulzer SMXL	3
Sulzer SMV	6–12
Sulzer SMX	12

TABLE 9.24
Measured Laminar Mixing Efficiency for Selected Static Mixers

Mixer Type	$P_{SM}/P(L/D)$ @ $\sigma/\bar{X} = 0.05$	(L/D) @ $\sigma/\bar{X} = 0.05$	$P_{SM}/P = K$
Sulzer SMXL	130–200	17–26	8
Kenics KM	180–200	25–29	7
Sulzer SMX	320–350	9	39
Komax	730	29–38	25
Sulzer SMV	800	18	45
Ross ISG	1700–3000	6–10	300

$$\text{Power} = Q \Delta P_{SM} \tag{9.71}$$

where Q is the total flow rate through the static mixer.

9.9.12 MIXING EFFICIENCY

The mixing efficiency, ε, of a static mixer has been defined as a normalized product of the pressure drop and the mixing length required to achieve a specified variation coefficient of 0.05:

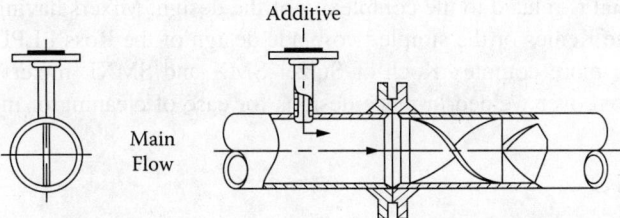

FIGURE 9.47 Simple T-connection.

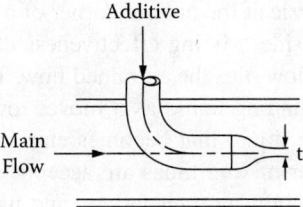

FIGURE 9.48 Coaxial injector design.

$$\varepsilon = \frac{\Delta P_{SM}}{\Delta P}\left(\frac{L}{D}\right) \qquad (9.72)$$

Measured values of ε for several static mixers are ranked in Table 9.24. The corresponding a, for comparison, is shown in Figure 9.46.

9.9.13 Injection Considerations and Designs

A well-designed feed entry system can reduce the number of elements required to achieve a given task. The optimum design depends on the volumetric ratios of flow, component viscosity and density, and the type of operation to be carried out. A simple T-type connection as in Figure 9.47 is normally adequate for two component streams, where both the flow ratio and viscosity ratio (larger to smaller) are ≤10. In these applications, the low-viscosity side stream should be injected at about twice the velocity of the main stream and injected just short of the centerline of the pipe. The same type of T-connection is used to feed high-viscosity side streams (as in a side-arm extruder feeding a main polymer stream), since most other types of injectors have a tendency to plug.

The injector shown in Figure 9.48 is recommended for immiscible liquid-liquid or gas-liquid dispersion or blending in cases where the flow and properties of the component streams vary widely. The additive or minor stream must be introduced at twice the velocity of the main stream and as close to the leading mixer element as practical, but definitely within one pipe diameter upstream of the first element. The injector must discharge at the centerline of the primary stream. An extra element or two may be needed if the injection of the minor stream is not at the centerline.

9.9.14 Pump Selection and Flow Control

It is important to provide adequate control of the flow and concentration of the component steams because static mixers cannot compensate for variations in concentration or flow with time. Centrifugal pumps are used for low-viscosity fluids (viscosity ≤ 500 m·Pa·s). Rotary gear or screw pumps are recommended for higher-viscosity or high-pressure applications. Avoid using pulsating pumps such as reciprocating piston or air-driven diaphragm pumps. Other important criteria include potential for plugging, ease of cleaning, materials of construction, cost, and availability.

Plugging potential is related to the complexity of the design. Mixers having streamlined shape elements, such as the Kenics or the simple two-blade design of the Ross LLPD and LPD, are less apt to plug than the more complex Koch or Sulzer SMX and SMXL mixers. Removable mixer elements are preferred over welded-in-place designs for ease of cleaning or modification.

9.10 JET MIXING

9.10.1 INTRODUCTION

Liquid jets, as opposed to impellers/drives, can provide effective mixing in vessels. A typical configuration is to place a pipe/nozzle at the bottom corner of a vessel and directed upward toward the liquid surface on the opposite side. Mixing effectiveness depends on creating flow. This flow is comprised of direct pumped jet flow plus the entrained flow. The jet stream must be free in order to entrain large quantities of surrounding liquid as it moves toward the upper surface. The longer the path of the jet, the more is the liquid that becomes entrained. Jets in tanks provide effective low-viscosity blending when longer mixing times are acceptable. They can be adapted to various vessel shapes, including rail cars, highway transporters, and flat-bottomed storage tanks. Jets can also be used for light-duty solids suspension, as noted in Section 9.6. Jets should not used for dispersing immiscible liquids, blending viscous fluids, or for any application requiring rapid mixing. Frequently, a transfer pump can serve double duty as a means to provide jet mixing. Mixing times for jet mixers are given by expressions shown in Section 9.10.4. Jet mixing is similar in many ways to mixing using side-entering, wall-mounted propeller mixers.

9.10.2 PRINCIPLES

A free jet moving diagonally through a vessel entrains approximately 15 to 20 times its pumped flow rate. As a first approximation, estimates of mixing times (95%) can be made assuming five turnovers and a 20:1 entrainment ratio. This gives

$$\theta_{approx} = 5V / 20 \cdot (pumping\ rate) \tag{9.73}$$

where V is the volume of liquid in the vessel.

While maximizing jet length is important for flow entrainment, applications requiring changing levels sometimes force compromises to be made. It is advisable to aim the jet at the corner representing the lowest expected operating level. This avoids the jet shooting out from the surface when lower levels are reached. At very low liquid levels, the jet must be aimed horizontally across the floor of the vessel. For this case, entrained flows are a half, due to floor effects, and mixing times will be twice that suggested by Equation (9.73).

It is important to create an efficient flow pattern in the vessel. A typical pattern is shown in Figure 9.49. When the jet is aimed diagonally across the tank, in the top view the pattern is symmetrical, and in both top and front views, two symmetrical flow patterns develop.

9.10.3 EQUIPMENT

Figure 9.49 shows the most common arrangement for a jet-stirred vessel. The jet stream expands at an angle of 15° to 25° due to entrainment and decrease in fluid velocity. The jet velocity at the centerline decreases with distance from the nozzle according to

$$V_z = 6 \cdot \left(\frac{d_J}{z}\right) V_J \tag{9.74}$$

Industrial Mixing Technology

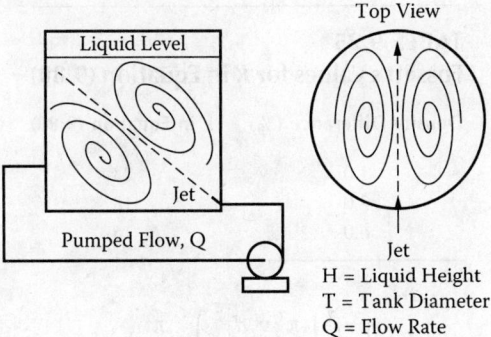

FIGURE 9.49 Typical arrangement for a jet-stirred vessel.

and

$$Q_Z = \alpha \cdot \left(\frac{z}{d_J}\right) \cdot Q_J, \quad (9.75)$$

where $\alpha \cong 3.0$.

9.10.4 Correlations

Researchers at DuPont (J. Gray) developed a useful, but conservative, correlation for jet mixing time, θ_M, given by

$$\theta_M = \frac{30 \cdot V_T \cdot d_J}{Q \cdot (H \cdot T)^{1/2}} \quad (9.76)$$

where d_J is the diameter of a jet required to deliver a flow, Q, the velocity of which is 25–30 ft/s; V is the volume of liquid in the vessel; H is the liquid height; and T is the vessel diameter. Any set of consistent units can be used.

Grenville [74] reported on jet mixing work done by the fluid-mixing processes (FMP) consortium, where Equation (9.77) was derived:

$$\frac{v_J \theta_M}{d_J} = 5.69 \cdot \left(\frac{T}{d_J}\right)^{3/2} \left(\frac{H}{d_J}\right)^{1/2} \quad (9.77)$$

Equation (9.77) is an empirical expression derived by correlating jet-mixing data from vessels ranging in diameter from 0.61 to 3.98 m. The correlation has an RSD of ±8.18%. For the case of $T = H$, Equation (9.77) becomes

$$\frac{v_J \theta_M}{d_J} = 5.69 \cdot \left(\frac{T}{d_J}\right)^2 \quad (9.78)$$

Equation (9.79) gives the power P (watts) required by a jet mixer:

TABLE 9.25
Fossett's Values for K in Equation (9.80)

Density Difference (%)	K in Equation (9.80)
1.0	23
2.0	18
6.0	15

$$P = \frac{\rho v_J^2}{2}\left(\frac{\pi \cdot v_J d_J^2}{4}\right) = \frac{\pi}{8}\rho v^3 d_J^2 \qquad (9.79)$$

where ρ is in kg/m³, d_J is in m, and v_J is in m/s.

An important reference for jet mixing is the work of Fossett [75]. He studied the blending of tetraethyl lead in gasoline stored in very large tanks. Fossett's work included liquids of different densities, ρ_1 and ρ_2. The jet was submerged a distance S from the surface and was aimed at an angle of α (measured from the horizontal). Equation (9.80) gives a relationship for h, the excess head needed by the jet to overcome the density differences:

$$h = \frac{KS}{\sin^2 \alpha}\left(\frac{\rho_2 - \rho_1}{\rho_2}\right) \qquad (9.80)$$

Values of K are experimental and are given in Table 9.25.

Equation (9.81) gives Fossett's equation for mixing time for liquids of equal density:

$$\theta_M = \frac{8T^2}{(Qv_J)^{1/2}} = \frac{0.144T^2}{G^{1/3}h^{1/4}} \qquad (9.81)$$

where G is in gallons per minute, T is in ft, h is the excess head given by Equation (9.80), and θ_M is in hours.

Example 9.7
A 3.658-m (12 ft) diameter by 4.572-m (15 ft) tall tank is to be mixed using a 0.3785-m³/min (100 gpm) pump. The liquid level in the vessel is constant at 3.658m (12 ft), and fluids have equal densities. Determine the mixing time for 95% homogeneity.

Solution
The tank contains 38.44 m³ (10,152 gal) of liquid.

1. As a first approximation, use Equation (9.73) to estimate the mixing time. Direct substitution in Equation (9.73) gives $\theta_M = 25$ min.
2. Using Equation (9.76), first calculate d_J for a velocity of 9.14 m/s (30 ft/s):

$$d_J = \left(\frac{4Q}{\pi v_J}\right)^{1/2} = \left(\frac{4 \cdot 0.3875}{60 \cdot \pi \cdot 9.14}\right)^{1/2} = 0.0296 \text{ m} = 1.17 \text{ in}$$

$$\theta_M = \frac{30 \cdot V \cdot d_J}{Q(H \cdot T)^{1/2}} = \frac{30 \cdot 38.44 \cdot 0.0296}{0.3785 \cdot (3.658 \cdot 3.658)^{1/2}} = 6.74 \text{ min}$$

3. Using Equation (9.78),

$$\theta_M = 5.69 \left(\frac{d_J}{v_J}\right) \cdot \left(\frac{T}{d_J}\right)^2 = 5.69 \left(\frac{0.0269}{9.14}\right) \cdot \left(\frac{3.658}{0.0296}\right)^2 = 281s = 4.7 \text{ min}$$

The first estimate is very conservative and is used to consider if jet mixing is feasible. The 4.7-min figure is probably more accurate.

9.11 HEAT TRANSFER IN MIXING EQUIPMENT

9.11.1 INTRODUCTION

Heat transfer is an important consideration when the fluid motion in the vessel is in the laminar flow regime. It influences the design and operation of agitated process vessels such as reactors, evaporators, and crystallizers. For a review of working relationships, see Dream [76].

Heating and cooling of fluids in these vessels are necessary to:

- Remove the heat of reaction
- Provide uniform temperature in a vessel
- Provide accurate temperature control in a given process

It is important to recognize that agitation improves heat transfer by its effect on the process-side (inside the process vessel) heat-transfer resistance. Usually, the process-side heat-transfer resistance is the controlling resistance. The design challenge is to select and design an agitation system to minimize the process-side heat-transfer resistance while meeting other mixing requirements.

Proximity and nonproximity impellers are the two major designs used in mixing applications. Proximity relates to distance from the vessel wall. Figure 9.50(a) shows a nonproximity impeller typically used for turbulent conditions. Its blades are not close to the vessel wall. "Close proximity" agitators like anchors and helical ribbons, illustrated in Figures 9.50(b, c), are typically used for high-viscosity applications.

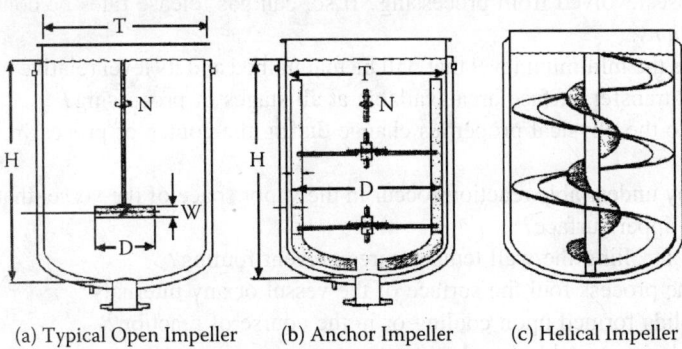

(a) Typical Open Impeller   (b) Anchor Impeller   (c) Helical Impeller

FIGURE 9.50 Types of mixing impellers for heat-transfer applications.

An agitated vessel may be operated in either a continuous, batch, or semi-batch mode. In continuous operations, the typical heat-transfer requirement is to maintain a set process temperature by either adding or removing heat, depending on the chemical reaction involved. In batch operations, the heat-transfer process can have a number of different functions at different stages of the operation. Examples include the:

- Establishment of initial reaction temperature
- Maintenance of a set temperature
- Cooling of a product to a final desired temperature

The heat-transfer coefficient on both the "process" (agitated) and "service" (jacket) side may change dramatically during the course of processing, usually as a result of physical property or chemical changes.

Heat transfer seldom dictates equipment design. Process mixing requirements dictate design for the majority of agitated tank systems. Heat transfer is then a necessary adjunct, and the design objective is to accommodate a suitable means of matching the heat-transfer requirements to other process requirements.

Surface area for process heat transfer is made available by means of jackets, coils, baffles, and plates. When these fail to adequately meet process requirements, pumps and external heat exchangers are commonly used. Under certain conditions, condensers can be designed to remove process heat through the refluxing of a solvent or reactant.

9.11.2 Important Heat-Transfer Considerations

The following are some of the key issues to consider for designing a new system or to troubleshoot an existing one:

Process characteristics:
- Is the process continuous, semi-batch, or batch?
- Is an exothermic reaction involved?
- Is the heat of reaction known? What is the magnitude of heat release?
- Is there a wall temperature limitation? (reactivity, purity, fouling)
- Are internal surface area devices acceptable?
- Is temperature control important?
- Can desirable heat-transfer rates be maintained by controlling the reaction?
- Is corrosivity a problem?
- Are gases evolved from processing? If so, can gas release rates be controlled?

Batch operations:
- What is the minimum level that will be maintained and its level relative to the agitator?
- Is heat-transfer surface area available at all stages of processing?
- How do the physical properties change during the course of processing?

Fouling:
- Can any undesirable reactions occur in the vapor space of the vessel that may deposit on the upper surface?
- Will controlling the wall temperatures prevent fouling?
- Will the process foul the surface of the vessel or any internals?
- Are solids formed upon cooling or in the course of reaction?
- Is the design suitable for cleaning?

Safety:
- Is the heat release due to mixing?

- Is there a choice of heat-transfer media?
- Can temporary power loss create a sudden heat release when power is restored?

9.11.3 Fundamentals of Heat Transfer in Agitated Vessels

Equation (9.82) is the basic heat-transfer equation for heat transfer between two fluids separated by a wall:

$$Q = U_o A \cdot \Delta T \qquad (9.82)$$

where Q = heat flow, kW (Btu/h); U_O = overall heat-transfer coefficient, kW/m²·°K (Btu/h·ft²·°F); A = area for heat transfer, m² (ft²); and ΔT = temperature driving force, °K (°F).

The discussion now turns to the individual resistances comprising U_O and how it is affected by impeller selection and surface area. The overall heat-transfer coefficient for a jacketed vessel can be obtained from the individual resistances by use of Equation (9.83):

$$\frac{1}{U_i} = \frac{1}{h_i} + \frac{x}{k} + \frac{A_i}{h_j A_j} + r_j + r_i \qquad (9.83)$$

where h_i = heat-transfer film coefficient, kW/m²·°K (Btu/h·ft²·°F); x = wall thickness, m (ft); k = thermal conductivity of wall W/m·°K (Btu·ft/h·ft²·°F); A_i = reference area, m² (ft²); A_j = area inside of jacket; r_i = fouling resistance, reference side (inside surface of vessel); r_j = fouling resistance, inside jacket; and U_i = overall heat-transfer coefficient, W/m²·°K (Btu/h·ft²·°F).

These resistances are illustrated in Figure 9.51. The subscript "i" in Equation (9.83) refers to the coefficient at the inside wall of the mixing vessel; the subscript "j" refers to the jacket side. The other terms are the wall resistance and the fouling resistances for either side. A similar equation can be written for an internal coil or other device. In situations where both a jacket and an internal device are used, the overall coefficients for each type of surface should be calculated separately, and the two Q's should be added to obtain the overall heat-transfer capability.

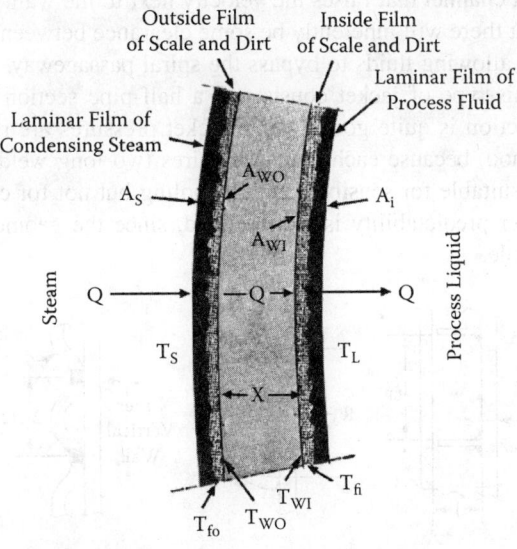

FIGURE 9.51 Heat-transfer resistances.

9.11.4 Heat-Transfer Surfaces and Effective Area

Jackets, internal helical pipe coils, tube baffles, and plate coil baffles are used to provide heat-transfer surface area. The only surface area effective for heat transfer is that portion that is wetted by both service and process fluids. The effective heat-transfer area for some items may be determined as follows:

- Use the total wetted area for plain or spirally baffled jackets.
- The area between the half-pipes is not totally effective for heat transfer when using half-pipe coil jackets, usually fabricated from 2-, 3-, or 4-in. pipes with typical 3/4-in. spacing.
- The total outside wetted area is effective for internal helical coils:

$$A_{co} = \pi \, d_{co} H_c n \left(\pi \, d_c^2 + n^{-2} \right)^{0.6} \tag{9.84}$$

where n = number of coil turns per foot of coil height = $1/p$; H_C = total height of coil, ft; and d_C = centerline diameter of coil helix, ft.

9.11.5 Jackets and Other Applied Devices

Jackets form what amounts to a double wall on the mixing vessel. Several varieties are possible, as illustrated in Figure 9.52. The plain jacket is the simplest construction and therefore the lowest initial cost. It is suitable for condensing heating fluids such as steam, but results in very poor performance using sensible heat-transfer fluids. Large passage areas limit the ability to create good wall velocities.

Installation of "agitating" nozzles is recommended if sensible liquid heating or cooling is to be used with plain jackets. Nozzles produce liquid jets directing the inlet jacket fluid in a spiral fashion into the jacket. This increases the effective velocity and turbulence level. Vendors have information dealing with their performance and installation.

> *Spiral-baffled jacket*: The spiral-plate baffle consists of a spiral strip welded edgewise to the shell. This forms a channel that raises the velocity next to the wall. The largest drawback to this baffle is that there will inherently be some clearance between the edge of the baffle and the tank wall, allowing fluids to bypass the spiral passageway.
>
> *Half-pipe jacket*: This type of jacket consists of a half-pipe section welded to the vessel wall. This construction is quite good if high jacket pressures are required, but it is also an expensive method, because each course requires two long welds along each edge of the cut pipe. It is suitable for sensible heating/cooling but not for condensing/vaporizing fluids. Heat-transfer predictability is well defined, since the geometry is known and no bypassing is possible.

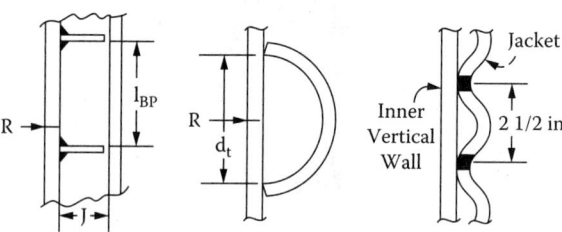

FIGURE 9.52 Jacket designs: spiral, half pipe, dimpled.

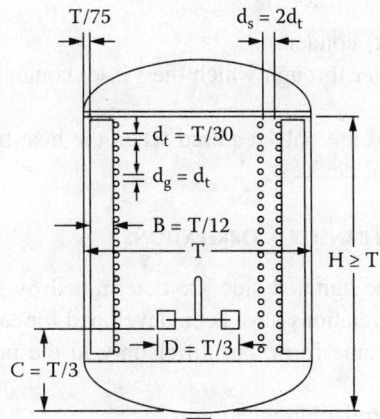

FIGURE 9.53 Recommended geometry for internal pipe coils.

Dimpled jacket: A dimpled jacket consists of an outer shell having regular indentations of the shell material. These "dimples" are intended to promote turbulence by creating high local velocities at the dimple. Heat-transfer information concerning dimple-jacketed vessels are proprietary to the fabricators; there is a little information in the open literature [77].

Other devices: There are other special devices such as clamp-on plate coils, weld-on plate coils, etc., that are often used when an unjacketed vessel needs limited heat-transfer capability.

Internal devices: Internal devices comprise coiled pipe (or tubing), baffles of various types that are also heat-transfer devices, and sometimes even the agitator. All internals interfere with the flow patterns within the vessel and likely lead to the formation of stagnant regions. Consequently, the use of internal devices needs to be carefully considered from the standpoint of harm to good mixing. If fouling is a problem (known or suspected), any internals should be avoided. Internal devices, of any type, increase mechanical complexity and maintenance. It is recommended that internal connections be welded, not flanged, to minimize maintenance.

9.11.6 Internal Pipe Coils

Internal pipe coils consist of one to three helical (concentric) coils of pipe located inside the mixing vessel for sensible heating/cooling. The effectiveness of these coils is directly related to the flow patterns generated by the agitator. Considerable area can be added in this manner, and if all the heat-transfer capability must be provided within the mixing vessel itself, this is an effective means of doing so. Potential difficulties include cleaning, mechanical integrity (the coil must be supported), and installation (both weight and access). The presence of coils and their support structures *always detracts* from mixing performance. The recommended geometry for the use of coils is shown in Figure 9.53.

9.11.7 Other Internal Devices

Tubes and flat hollow plates can be formed into assemblies that act as baffles for the mixing application. Various designs include linking surface areas by "U" bends, often outside the vessel.

9.11.8 External Auxiliary Devices

An external auxiliary heat-transfer device is usually one of the following:

- A standard condenser
- A reflux (or knock-back) condenser
- A sensible heat exchanger through which the vessel contents are circulated

These external auxiliary devices are only required when the heat-transfer requirements cannot be met by use of jackets or internal devices.

9.11.9 Process-Side Heat-Transfer Correlations

Heat-transfer coefficients on the agitated side are determined by the same principles as for any other heat-transfer process. Correlations have been developed for each of the major impeller types. These are all basically of the same form, but differ only in the pre-proportionality constant and values of the exponents.

The general form of the correlating equation is

$$Nu = \frac{h_o T}{k} = K \cdot Re^a \cdot Pr^b \cdot \left(\frac{\mu}{\mu_w}\right)^c \tag{9.85}$$

where K accounts for geometry variations.

Note that Equation (9.85) is basically the same general form as the familiar Dittus-Boelter equation for heat transfer in tubes. The basic heat-transfer mechanism is identical. It is dependent on the flow of fluid next to the heat-transfer surfaces, whether these are the vessel walls or some internals. Differences in the correlations are therefore mainly due to the differences in flow characteristics generated by the different impellers relative to the surface under consideration. This is reflected in the value of K.

These correlations are all of an overall nature and do not take variation in flow due to the flow pattern into account, nor do they consider any local variation in fluid bulk temperature in the vessel. If these things are critical to the application being considered, either experimentation, CFD modeling, or consultation with a vendor will be required.

The process-side heat-transfer coefficient (for flat-blade turbines) for heat transfer to a jacket is based on the work of Brooks [78]. The recommended correlation is

$$Nu = \frac{h_o T}{k} = 0.74 \cdot Re^{0.66} \cdot Pr^{0.33} \cdot \left(\frac{\mu}{\mu_w}\right)^{0.14} \tag{9.86}$$

where Re = impeller Reynolds number; h_o = heat-transfer coefficient at the vessel wall; k = thermal conductivity of the process fluid, W/m·°K (Btu/h·ft·°F); T = tank diameter, m (ft); Nu = Nusseldt number; Pr = Prandtl number; Re = Reynolds number; μ = viscosity of the process fluid, Pa·s (cP); and μ_w = viscosity of the process fluid at the vessel wall, Pa·s (cP).

Other authors have extended the basic correlations to include more details of the impeller geometry, such as blade width, pitch, and number of blades. As long as the process is in the turbulent regime, most of these geometrical variables have little impact on heat transfer, and their use is not recommended until details of an agitation system are selected or in place.

Ackley [79] suggests use of Equation (9.87) for unbaffled retreat-curve (blade) impellers typically used in glass-lined vessels:

$$Nu = \frac{h_o T}{k} = 0.68 \cdot Re^{0.67} \cdot Pr^{0.33} \cdot \left(\frac{\mu}{\mu_w}\right)^{0.14} \tag{9.87}$$

Equation (9.88) is to be used when the vessel is baffled [79]:

$$Nu = \frac{h_o T}{k} = 0.33 \cdot \text{Re}^{0.67} \cdot \text{Pr}^{0.33} \cdot \left(\frac{\mu}{\mu_w}\right)^{0.14} \quad (9.88)$$

For proximity impellers, such as the helical ribbon, for Re < 130, Blazinski [80] gives:

$$Nu = \frac{hD_T}{k} = 0.248 \cdot \text{Re}^{0.5} \cdot \text{Pr}^{0.33} \cdot \left(\frac{\mu}{\mu_w}\right)^{0.14} \cdot \left(\frac{e}{D}\right)^{-0.22} \cdot \left(\frac{i}{D}\right)^{-0.28} \quad (9.89)$$

where e = clearance = $(D_T - D)/2$, m (ft); i = ribbon pitch, m (ft); D = tank diameter, m (ft); D_T = vessel diameter, m (ft).

Oldshue [81] gives Equation (9.90) for the heat-transfer coefficient on the outside coils:

$$\frac{h_{coil} D}{k} = 0.17 \cdot \text{Re}^{0.67} \cdot \text{Pr}^{0.37} \left(\frac{D}{T}\right)^{0.1} \cdot \left(\frac{d}{T}\right)^{0.5} \cdot \left(\frac{\mu}{\mu_s}\right)^m \quad (9.90)$$

where h_O = heat-transfer coefficient for outside of coil; d = tube diameter making up helix, m (ft); and D = impeller diameter, m (ft).

This correlation is valid for tube spacing between two and four tube diameters and for all "practical" tube diameters.

9.11.10 Service-Side Heat-Transfer Correlations

A conservative estimate for the condensing-steam coefficient used in plain jackets is 5.678 kW/m²·°K (1000 Btu/h·ft²·°F). Any organic fluid will have a lower value due mainly to its lower thermal conductivity.

Bondi [82] proposes the following equations for sensible fluid in plain jacket with no agitating nozzles.

Turbulent flow conditions where Re > 10,000,

$$\frac{h_j D_e}{k} = 0.027 \cdot \text{Re}^{0.8} \cdot \text{Pr}^{0.33} \cdot \left(\frac{\mu}{\mu_w}\right)^{0.14} \quad (9.91)$$

In this case, the Reynolds number is defined as

$$\text{Re} = D_e V \rho / \mu$$

where all properties pertain to the jacket fluid

Laminar flow conditions where Re ≤ 10,000,

$$\frac{h_j D_e}{k} = 1.86 \cdot \left[\text{Re} \cdot \text{Pr} \cdot \frac{D_e}{L}\right]^{0.33} \cdot \left(\frac{\mu}{\mu_w}\right)^{0.14} \quad (9.92)$$

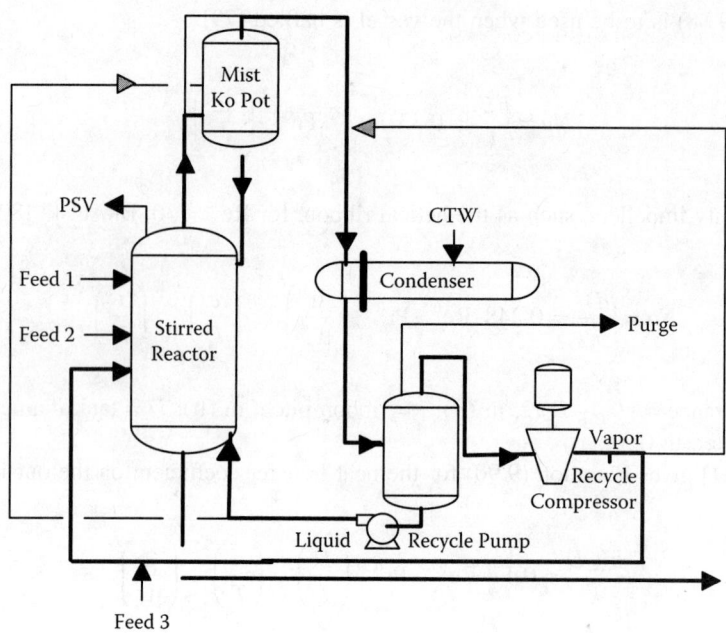

FIGURE 9.54 Reactor system schematic.

These equations require the use of the equivalent diameter, D_e, of the jacket. This is calculated by

$$D_e = \left(\frac{D_{jo}^2 - D_{ji}^2}{D_{ji}} \right)$$

and flow area A_x, by

$$A_x = \pi \cdot \left(\frac{D_{jo}^2 - D_{ji}^2}{4} \right)$$

The flow area is used to calculate the velocity, V, in the Reynolds number term.

Example 9.8

A reactor is operated according to the system schematic shown in Figure 9.54. Boiling of the carrier that is condensed out of the vapor provides most cooling, and both condensate and uncondensed gas are recycled to the reactor. This reactor has a dimpled jacket. Calculate the heat duty of the jacket under the given conditions.

The following system engineering data are provided.

Reactor: 12-ft diameter, 24-ft straight side height, 12-ft liquid level
Impeller: 5-ft diameter flat-blade turbine @ 37 rpm
Production: 10,000 lb/h
Heat of reaction: 1500 Btu/lb
Operating temp.: 85°C (176°F)

Physical properties of vessel contents:
Density: 37.58 lb/ft^3
Viscosity: 0.192 cP
Thermal conductivity: 0.0637 Btu/h·ft°F
Specific heat: 0.6197 Btu/lb°F
Coolant:
Water @ 31°C
Velocity = 2 ft/s
Density: 62.11 lb/ft^3
Viscosity: 0.78 cP
Thermal conductivity: 0.3564 Btu/h·ft·°F
Specific heat: 1.0 Btu/lb·°F

Solution

1. Calculate vessel-side Reynolds number:

$$Re = D^2N\rho/\mu = 5^2\ (37/60)\ 37.58\ (10^4)/6.72\ (0.192) = 4{,}490{,}314$$

2. Calculate the vessel-side heat-transfer coefficient using Equation (9.86):

$$h_o = (0.0637/12)0.74(4{,}490{,}314)^{0.66}(4.522)^{0.33}(1)^{0.14} = 158.8\ \text{Btu/h·ft}^2\text{·°F}$$

3. Calculate the coolant-side from Equation (9.91) using $D_e = 0.66$ in., $A_x = 1.98$ in.^{2} per foot of circumference = 1.98 πD = 0.518 ft^2:

$$F = vA_x = 1.037\ \text{ft}^3/\text{s} = 231{,}808\ \text{lb/h}$$

$$h_j = 579\ \text{Btu/h·ft}^2\text{·°F}$$

4. Calculate the overall coefficient:

$$1/U = 1/h_j + 1/h_i + x/(12)10.0 = 597^{-1} + 158.8^{-1} + 0.5/(12)10.0 = 0.012$$

$$U = 82\ \text{Btu/h·ft}^2\text{·°F}$$

5. Estimate the Q for the jacket: assume the temperature difference is 85 – 31°C and consider only the straight side area:

$$Q = UA\Delta T = 82\ (452)\ 97.2 = 3.6\ \text{MM Btu/h}$$

Note that this is only about 24% of the total heat to be removed of 15 MM Btu/h!

REFERENCES

1. Paul, E. L., V. Atiemo-Obeng, and S. Kresta, *Handbook of Industrial Mixing*, Wiley-Interscience, New York, 2004.
2. Zhou, G., and S. M. Kresta, *Inst. Chem. Engrg.* 74 (Part A): 379 (1996).
3. Bates, R. L., P. L. Fondy, and R. H. Corpstein, *I&EC Proc. Res. & Dev.* 2: 310 (1963).

4. Harvey, A. D., S. P. Wood, and D. E. Leng, *Chem. Eng. Sci.* 52: 1479 (1997).
5. Oldshue, J. Y., *Fluid Mixing Technology*, M.-H. Publications, Chemical Engineering, New York, 1983.
6. King, R., in *Mixing in the Process Industries*, N. Harnby, M. F. Edwards, and A.W. Nienow, eds., 414, Butterworth Heinemann, Oxford, 1992.
7. Ulbricht, J. J., and J. J. Patterson, *Mixing of Liquids by Mechanical Agitation*, J. J. Ulbricht, ed., vol. of *Chemical Engineering: Concepts and Reviews*, Gordon and Breach Science Publishers, New York, London, Paris, Montreux, Tokyo, (1986).
8. Metzner, A. B., and R. E. Otto, *AIChE J.* 3: 3 (1957).
9. Wichterle, K., and O. Wein, *Int. Chem. Eng.* 21: 116 (1981).
10. Elson, T. P., D. J. Cheesman, and A. W. Nienow, *Chem. Eng. Sci.* 41: 2555 (1986).
11. Todtenhaupt, P., E. Todtenhaupt, and W. Muller, *Handbook of Mixing Technology*, EKATO Ruhr-und Mischtechnik. Schopfheim, Germany GmbH, 1991.
12. Grenville, R. K., S. Ruszkowski, and E. Garred, "Blending of Miscible Liquids in the Turbulent and Transitional Regimes," NAMF Mixing XV, Banff, Canada, 1995.
13. Tatterson, G. B., *Fluid Mixing and Gas Dispersion in Agitated Tanks*, McGraw-Hill, New York, 1991.
14. Nagata, S., *Mixing Principles and Applications*, Halstead Press, a division of John Wiley and Sons, New York, 1975.
15. Levenspiel, O., *Chemical Reaction Engineering*, John Wiley, New York, 1972.
16. Kolmogoroff, A. N., *Compt. Rend. Acad. Sci. URSS* 30: 301 (1941).
17. Kolmogoroff, A. N., *Compt. Rend. Acad. Sci. URSS* 32: 16 (1941).
18. Rosensweig, R. E., *AIChE J.* 10: 92 (1964).
19. Batchelor, G. K. J., *Fluid Mech.* 5: 113 (1959).
20. Corrsin, S., *AIChE J.* 10: 87 (1964).
21. Baldyga, J., and J. R. Bourne, *Chem. Eng. Commun.* 28: 231 (1984).
22. Baldyga, J., and J. R. Bourne, *Chem. Eng. Commun.* 28: 243 (1984).
23. Baldyga, J., and J. R. Bourne, *Chem. Eng. Commun.* 28: 259 (1984).
24. Baldyga, J., and J. R. Bourne, *Chem. Eng. J.* 42: 83 (1989).
25. Baldyga, J., and J. R. Bourne, *Chem. Eng. J.* 42: 93 (1989).
26. Baldyga, J., and J. R. Bourne, *Chem. Eng. J.* 45: 25 (1990).
27. David, R., and J. Villermaux, *Chem. Eng. Comm.* 54: 333 (1987).
28. Tosun, G., "An Experimental Study of the Effect of Mixing in $BaSO_4$ Precipitation Reaction," Sixth European Conference on Mixing, Pavia, Italy, 1988.
29. Doraiswamy, L. K., and M. M. Sharma, *Heterogeneous Reactions*, vols. 1 and 2, Wiley Interscience, New York, 1984.
30. Danckwerts, P. V., *Gas-Liquid Reactions*, McGraw-Hill, New York, 1970.
31. Fasano, J. B., and W. R. Penney, *Chem. Eng. Prog.* 87: 46 (1991).
32. Paul, E., *Chem. Ind.* 21: 320 (1990).
33. Paul, E., *Chem. Eng. Sci.* 43: 1773 (1988).
34. Middleton, J. C., "Motionless Mixers as Gas-Liquid Contacting Devices," AIChE 71 Annual Meeting, Miami, 1978.
35. Leng, D. E., *Chem. Eng. Prog.* 87: 23 (1991).
36. Maude, A. D., and R. L. Whitmore, *Br. J. Appl. Phys.* 9: 477 (1958).
37. Zwietering, T. N., *Chem. Eng. Sci.* 8: 244 (1958).
38. Nienow, A. W., *Chem. Eng. Sci.* 23: 1453 (1968).
39. Joosten, G. E. H., J. G. M. Schilder, and A. M. Broere, *Trans. Instn. Chem. Engrs.* 55: 220 (1977).
40. Mak, A. T. C., S. Yang, and G. Ozcan-Taskin, "The Effect of Scale on the Suspension and Distribution of Solids in Stirred Vessels," G.F.d.G.d. Procedes, Mixing IX, Paris, 1997.
41. Mak, A. T. C., and S. W. Ruszkowski, "Scaling-Up of Solids Distribution in Stirred Vessels," I. Chem.E., Mixing IV, University of Bradford (*I. Chem. E.* 121: 379 1990).
42. Oldshue, J. Y., and N. R. Herbst, *A Guide to Fluid Mixing*, Lightnin, Rochester, NY, 1992.
43. Armenante, P. M., H. Yu-Tsang, and L. I. Tong, *Chem. Eng. Sci.* 47: 2865 (1992).
44. Nienow, A.W., D. J. Wilson, and J. C. Middleton, Proc. 2nd European Conf. on Mixing, Cambridge, England, 1977
45. Middleton, J. C., in *Mixing in the Process Industries*, E. A. N. Harnby, ed., 322, Butterworth Heinemann, Oxford, 1995.

46. Kar, K. K., personal communication, 1996.
47. Nocentini, M., D. Fajner, and F. Magelli, *I&EC Res.* 32: 19 (1993).
48. Smith, J. M., "Simple Performance Correlations for Agitated Vessels," North American Mixing Forum (NAMF): Mixing XIII, Banff, Canada, 1991.
49. Rewatkar, V. B., A. J. Deshpande, A. B. Pandit, and J. B. Joshi, *Can. J. Chem. Eng.* 71: 226
50. Parthasarathy, R., and N. Ahmed, "Gas Holdup in Stirred Vessels: Bubble Size and Power Input Effects," M. B. a. G. Fromont, Ed., Seventh European Congress on Mixing, Royal Flemish Society of Engineers, Brugge, 1991.
51. Pinelli, D., M. Nocentini, and F. Magelli, "Hold-Up in Low Viscosity Gas-Liquid Systems Stirred with Multiple Impellers: Comparison of Different Agitators Types and Sets," I. Chem. Engrs., Eighth European Conference on Mixing, University of Cambridge (*I. Chem. E.*136:81 1994).
52. Zlokarnik, M., *Chem. Ing. Tech.* 38: 357 (1966).
53. Joshi, J. B., A. B. Pandit, and M. M. Sharma, *Chem. Eng. Sci.* 37: 813 (1982).
54. Fuchs, R., D. D. Y. Ryu, and A. E. Humphrey, *Ind. Eng. Chem. Proc. Dev.* 10: 190 (1971).
55. Rodger, W. A., V. G. Trice, and J. H. Rushton, *Chem. Eng. Prog.* 52: 515 (1956).
56. Taylor, G. I., *Proc. Roy. Soc.* A146: 501 (1934).
57. Grace, H. P., *Chem. Eng. Commun.* 14: 225 (1982).
58. Karam, H. J., and J. C. Bellinger, *I&EC Fund.* 1: 576 (1968).
59. Leng, D. E., and G. J. Quarderer, *Chem. Eng. Commun.* 14: 177 (1982).
60. Chen, H. T., and S. Middleman, *AIChE Jour.* 13: 989 (1967).
61. McManamey, W. J., *Chem. Eng. Sci.* 34: 345 (1979).
62. Calabrese, R. V., T. P. K. Chang, and P. T. Dang, *AIChE Jour.* 32: 657 (1986).
63. Wang, C. Y., and R. V. Calabrese, *AIChE J.* 32: 667 (1986).
64. Scheele, G. F., and D. E. Leng, *Chem. Eng. Sci.* 26: 1867 (1971).
65. Murdoch, P. G., and D. E. Leng, *Chem. Eng. Sci.* 26: 1881 (1971).
66. Ramkrishna, D., *Population Balances*, John Wiley, New York, 2001
67. Skelland, A.H.P., and R. Seksaria, Ind. Eng. Chem Proc. Des. Dev. 17: 56 (1978)
68. Armenante, P. M., Y. T. Yang, and T. Li, *Chem. Eng. Sci.* 47: 2865 (1992).
69. Church, J. M., and R. Shinnar, *Ind. Eng. Chem.* 53: 479 (1961).
70. Davies, J. T., *Chem. Eng. Sci.* 42: 213–220 (1987).
71. Pahl, M. H., and E. Muschelknautz, *Int. Chem. Eng.* 22: 197 (1982).
72. Kenics, product bulletin, 1986.
73. Grosz-Roel, F., "Gas/Liquid Mass Transfer with Static Mixing Units," 4th European Conference on Mixing, BHRA, Noordwijkerhout, Netherlands, 1982.
74. Grenville, R. K., and J. N. Tilton, *Trans. I. Chem. E.* 74: 390 (1996).
75. Fossett, H., and L. E. Prosser, *Proc. I. Mech. E.* 160: 224 (1949).
76. Dream, R. F., *Chem. Eng.*Jen. 90 (1999).
77. Garvin, J., *CEP* April, 73–75 (2001).
78. Brooks, G., and G. J. Su, *Chem. Eng. Prog.* 55: 54 (1959).
79. Ackley, E. J., *Chem. Eng.* 133 (1960).
80. Blazinski, H, and C. Kuncewicz, *Int. Chem. Eng.* 21: 679 (1981).
81. Oldshue, J. Y., and A. T. Gretton, *Chem. Eng. Prog.* 50: 615 (1954).
82. Bondy, F., and S. Lippa, *Chem. Eng.* 62: (1983).

10 Liquid-Liquid Extraction

D. William Tedder

CONTENTS

10.1 Introduction 709
10.2 Dispersion, Mass Transfer, and Coalescence 712
10.3 Reactive Systems 713
    10.3.1 Lignin Extraction 714
    10.3.2 Simplified TBP Reaction Models 714
    10.3.3 TBP Solvent Cleanup 715
10.4 Distribution Coefficients 716
    10.4.1 Thermodynamic Models 716
    10.4.2 Nonreactive Systems 716
    10.4.3 Reactive Systems 718
    10.4.4 Empirical Distribution Models 720
10.5 Design of Extraction Systems 720
    10.5.1 Phase Diagrams 721
    10.5.2 Countercurrent Extractors 723
    10.5.3 Kremser Equation 725
10.6 Industrial Extraction Equipment 726
    10.6.1 Mixer Settlers 726
    10.6.2 Reciprocating Plate Columns 726
10.7 Internet Sites 729
10.8 Economic Analysis for Vertical Contactors 729
    10.8.1 List of Symbols 731
References 732

10.1 INTRODUCTION

Liquid-liquid extraction (LLE) is widely used in chemical, petroleum refinery, pharmaceutical, mining, and the nuclear industries to separate chemicals in liquid mixtures [1]. Solid-liquid extraction (SLE) is also of importance and is related in many ways. LLE involves two relatively immiscible liquids and several operational events, as summarized next.

1. The two liquids are dispersed in each other. One liquid is dispersed as liquid droplets, and the other forms a more or less continuous liquid phase. At least one of the liquids is a mixture of two or more chemical species, and the objective is to separate it.
2. While the two phases are dispersed, one or more of the chemical species (the solute) transfers across the liquid-liquid interface into the other phase. The phase into which the desired transfer occurs is designated as the *extract*; the phase from which the transfer occurs is designated as the *raffinate*. The amount and rate of mass transfer are of major importance, as discussed below.
3. Later the extract and raffinate are separated, often by gravity separation.

4. Since the amount of transfer (or separation) in a single stage (or single contact) is generally inadequate, multiple contacts are often employed in industrial processes. Generally, countercurrent flow of the two liquids is employed in the overall separation process. With multiple stages, most of the desired component often can be extracted.
5. Frequently, the extract is processed to recover the solute in a more or less pure form. This latter step is sometimes one of the following: distillation, evaporation, freeze drying, crystallization or precipitation, and filtration.

When solvent extraction is employed, two largely immiscible liquids (e.g., an aqueous feed and an organic solvent) are mixed together, allowed to settle, and then separated. During these steps, one or more species that are initially dissolved in the feed liquid transfer to the other liquid phase to form an extract. The feed liquid residual, or raffinate, is thus partly depleted of the solutes, and the original solvent becomes enriched in them. In this way, LLE can be used to separate and purify solutes.

Thermodynamic equilibrium between the two liquids determines the direction of mass transfer and the theoretical amount of compound(s) transferred in a given step. The rate of transfer depends on the level of agitation provided to the dispersion and the interfacial areas between the phases. After the extraction step is completed, separation of phases is (hopefully) rapid. As already indicated, the separated phases are often then sent countercurrent to another extraction unit [2]. Countercurrent cascades of mixers and settlers generally provide the most efficient use of solvent.

Two liquid phases may be contacted in different ways to achieve separation. In the laboratory, the simplest example of LLE involves the use of a separatory funnel where a feed mixture, perhaps an aqueous phase from a reactor, is contacted with several washes of solvent (say, diethyl ether) to extract a solute. If the same aqueous phase is contacted repeatedly with fresh amounts of extracting solvent, such a process is said to be *crosscurrent*.

Industrial applications usually are based on the most efficient methods that involve the countercurrent flow of two liquids across multiple stages. A variety of mechanical devices are available that can be used to achieve such mixing and separate the resulting extract and raffinate and so achieve the desire separation. Once the extraction is complete, almost always both the extract and raffinate need to be treated to recover solvent residuals as well as the desired products.

Liquid-liquid extraction is often advantageous if the solute is heat sensitive and can be recovered by LLE at ambient conditions. It is also advantageous whenever direct separation (e.g., distillation) is too expensive. LLE tends to be advantageous whenever the solvent has a high affinity for the solute and is highly selective for the solute, and the solute is present in very low concentrations and cannot be easily removed by direct separation techniques. Under the right conditions, it can provide high recovery and concentration factors to enable the economic purification of resource materials even when they are found in dilute forms. It is often used to reduce energy consumption by enabling preconcentration of a resource with minimal energy expenditures.

The disadvantages of solvent extraction usually revolve around the solvent cost. Since high solvent losses can easily make an extraction process uneconomical, considerable effort is usually expended to find solvents that are economical and can be easily recovered and recycled.

There are usually many candidate solvents for any particular application. Important factors to consider are (1) the affinity of the solute for the solvent (i.e., its distribution coefficient should be large); (2) the affinity of other species in the mixture for the solvent (i.e., their distribution coefficients should be small); (3) solvent safety considerations (e.g., flammability and toxicity); (4) solvent handling properties such as density, viscosity, and vapor pressure; (5) solvent solubility in the raffinate phase (high solubilities may translate into high solvent losses unless steps are taken to prevent such losses); and (6) solvent cost. In addition, liquid-liquid interfacial tension affects the interfacial area and the rate of mass transfer between the phases.

Safety and health are also often issues for extraction, particularly whenever the potential of fire from using highly flammable solvents is high. On the other hand, LLE is favored in the nuclear industry to separate highly radioactive mixtures because of its relative simplicity and ease to operate safely under remote conditions [3].

Solvent toxicity can also be an issue, but less so as newer green* solvents are developed to replace traditional solvents that are unacceptable because of environmental concerns. There is, for example, growing interest in the use of ionic liquids in LLE, a relatively new and unexplored class of potential solvents.

On the other hand, it is often advantageous to use solvents that are already on a plant site rather than to introduce a new solvent. Such solvents are usually less expensive. Existing storage equipment can often be used in the former case, but introducing a new solvent may require additional storage and handling equipment, with additional costs. Using an existing solvent rather than a new one may avoid new cross-contamination problems that must be addressed with respect to plant safety, operability, and product quality. Solvent extraction can be very advantageous for recovering a high-quality product. On the other hand, its complexity, usually greater than separations by direct methods, is a disadvantage.

LLE may be more complicated to operate than a direct separation system (e.g., distillation), since LLE involves more processing steps (e.g., extraction, solute removal from the extract, solvent recycle, and solvent recovery from the raffinate). The solvent cost, and the additional steps that are required to recycle the solvent, increase the cost of LLE and can make it unattractive.

The complexity of the solvent recovery operation depends largely on the mutual solubility of the two liquid phases. Cleanup is generally simpler and more economical if the two liquid phases are nearly immiscible, but more expensive and complex unit operations (e.g., extractive distillation) may be required if the two phases are partly miscible. Commonly, aqueous raffinates are steam stripped to recover and recycle the solvent.

Solvent recycle from the extract is often more complex than solvent recycle from the raffinate. In the case of metal extraction, the metal solute is typically removed by stripping (washing) the extract to transfer it into a second aqueous phase for subsequent metal recovery. This aqueous phase containing the metal may then be steam stripped similarly to the raffinate from the original extraction step. In the case of organics extraction, the extract is often treated by extractive distillation to recover both the products and the solvent for recycle.

Solvent extraction is used extensively to recover chemicals from natural products. Solvents are used to extract and concentrate natural oils and products in the bioprocessing industries (nutraceutical, food, pharmaceutical, feed, cosmetic, biotechnology) in quantities from grams to metric tons. Biotechnology applications include the recovery of primary and secondary metabolites [4]. Extraction is used to recover vegetable oils and food products. It is used to process a variety of materials including groundnut, mustard seed, soybean, palm kernal, sunflower, rice bran, copra, cottonseed, and minor oil seeds like neem, mahua, watermelon seed, castor seed, and so on.

It is used in the mining industry to recover metals such as copper and nickel. Parasite plants, based on solvent extraction, are used in the phosphate industry to recover by-product uranium from crude phosphoric acid. The uranium concentration in phosphoric acid is very low but, because of the high volume of phosphoric acid that is produced to meet agricultural needs, considerable uranium can be recovered using solvent extraction. In the nuclear industry [5], solvent extraction is used to purify uranium and plutonium [using the plutonium and uranium recovery by extraction (PUREX) process], zirconium from hafnium, and for many other applications. It is also used in environmental applications to clean soil, say, to remove polychlorinated biphenyls (PCBs), dioxins, pesticides, and other hazardous pollutants.

Solvent extraction is widely used in laboratory applications on a very small scale but also on much larger industrial scale. In fact, the variety of materials, both organic and inorganic, that are processed using some form of solvent extraction is staggering. Moreover, the methods in which solvent extraction is applied to vastly different recovery and cleaning operations continues to grow

---

* A "green" process is one that doesn't use toxic or otherwise environmentally unfriendly chemicals.

substantially as new, environmentally acceptable techniques are developed and applied to solve separation problems.

Supercritical or near-critical fluids can be used both for extraction and chromatography. Many chemicals, primarily organic species, can be separated and analyzed using this approach [6], which is particularly useful in the food industry. Substances that are useful as supercritical fluids include carbon dioxide, water, ethane, ethene, propane, xenon, ammonia, nitrous oxide, and a fluoroform. Carbon dioxide is most commonly used, typically at a pressure near 100 bar. The required operating pressure ranges from about 43 bar for propane to 221 bar for water. Sometimes a solvent modifier is added (also called an *entrainer* or *cosolvent*), particularly when carbon dioxide is used.

On a laboratory scale, supercritical fluids may be contacted with a substrate on a once-through basis. Typically, the pressurized fluid is heated and passed through a chamber containing a solid or liquid sample. The desired solute selectively dissolves in the pressurized fluid and is transported out of the chamber and through a throttle into a container where the pressurized fluid evaporates at atmospheric pressure, leaving a solute residue in the container.

Small-scale supercritical extractions may be either dynamic or static. If the fluid is pumped through the sample chamber continuously, then the method is dynamic. If the sample container is filled batchwise with pressurized fluid, then the method is static. In the latter case, the sample container is charged, equilibrated statically, and then discharged through a throttle to collect the residue as before. In either case, the pressurized fluid or solvent is not recycled, and this feature limits the applicability to industrial use, because solvent costs, even with recycle, often determine whether or not extraction is economical.

More important solvent extraction references include the *Handbook of Solvent Extraction*, edited by Lo, Baird, and Hanson [7]. International Solvent Extraction Conferences (ISECs) have been held, and the proceedings, usually two or three volumes from each conference, provide considerable information, particularly of ongoing studies. Akell and King [8] have edited a volume of selected papers taken from ISEC '83, including papers of significant chemical engineering content and which focus on developments in solvent extraction equipment design. Schügerl [4] has written a nomograph on the use of LLE in biotechnology, primarily to recover metabolites. Kulov [9] has edited a book focusing on the kinetics of LLE. This last book also includes chapters on gas-liquid kinetics, which is appropriate since many of our kinetic models for LLE are derived from analogy with gaseous diffusion. Blumberg [10] also provides a book on LLE that may prove helpful. Marcus and Kertes [11] and Marcus, Kertes, and Yanir [12] provide extensive data on extraction systems that are based on solvation and ion exchange reactions.

10.2 DISPERSION, MASS TRANSFER, AND COALESCENCE

Quantitative and (hopefully, at least) qualitative considerations are helpful in characterizing a liquid-liquid system for a potential extraction application. Batch shakeout tests are frequently the easiest way to determine basic feasibility by simply measuring the primary and secondary break times and by analyses to measure the compositions of the equilibrated phases. Such tests are readily conducted by mixing small volumes of each phase in a vial, which is then vigorously agitated and placed on a lab bench to settle. The resulting behavior of the liquid-liquid mixture depends on physical properties and system characteristics. The greater the density difference and interfacial tension between the two liquid phases, for example, the more rapidly the phases tend to separate. More viscous systems separate more slowly.

The primary break time is that length of time required to form a clearly defined interface between the two phases, although both phases may still remain hazy. The physical properties of the liquid-liquid system and the vigor of mixing are of critical importance and usually provide qualitative information that permits rapid screening of large numbers of solvents with minimal effort. The resulting primary break time should be on the order of 1 or 2 min; Treybal [13] suggests

Liquid-Liquid Extraction

this time could be as great as 5 to 10 min. It is helpful if a liquid-liquid system exhibits a primary break time less than 1 min.

As a general rule, greater mass transfer between two phases tends to decrease the rate at which they separate, so longer primary break times may indicate that more mass transfer occurred between the two phases during mixing. On the other hand, liquid-liquid mixtures containing surfactants, even in very small concentrations (e.g., ppm), may form stable emulsions that do not separate at all. In some instances, the vigor of agitation, especially when surfactants are present, dramatically affects the rate of phase separation. Highly energetic mixing, resulting in very small droplets, may form stable emulsions, while less energetic mixing, resulting in larger droplet size distributions, may slowly separate into two distinct phases. The addition of ionic salts may accelerate phase separations.

A primary break time of less than about 30 sec, resulting in two clear phases, probably indicates that relatively little mass transfer has occurred. A primary break time greater than 10 min may be impractical to work with but could be interesting if significant mass transfer has occurred. In the latter case, it may be possible to accelerate the break time, for example, by adjusting the mixture temperature or by dissolving an inextractable salt in the aqueous phase.

The presence of trace concentrations of surfactant impurities (e.g., ppm concentrations of silicates) can complicate matters by yielding stable emulsions, especially if the shakeout test is based on vigorous agitation. When stable emulsions result from vigorous agitation in a bench test, then a contactor that generates a very small drop size distribution (e.g., a high-speed centrifugal contactor) may not be a good choice. On the other hand, the desired separation may occur if the agitation is more gentle and yields only a larger droplet size distribution (e.g., as results from a reciprocating or sieve-plate column). In some cases, special techniques may be required to break an emulsion, such as passing it upflow though a fiber bed that accelerates coalescence of the dispersed phase. If very vigorous agitation yields emulsions with much larger primary break times than for mixtures that are equilibrated using only gentle agitation (but still sufficient to form a liquid-liquid dispersion), then a contacting device that results in larger droplet size distributions is indicated.

The secondary break time is the time needed for the quiescent liquid-liquid mixture to achieve clarity in each of the bulk phases after rapid agitation. After initially breaking, the two liquid phases are usually hazy, especially if significant mass transfer has occurred. This haze is from very small droplets dispersed in each phase. Because of their small size, they move more slowly to the interface. After settling overnight, the two phases may appear clear when examined in normal room lighting, but examination of such liquid phases in a darkened room with a laser beam will likely reveal the continuing presence of droplets in both phases, even after many weeks of quiescent equilibration.

In some cases, the secondary haze can be removed by filtering the phase and allowing the droplets to coalesce. Centrifugation may be employed but is usually more expensive on an industrial scale. Often, some phase clarification is required.

10.3 REACTIVE SYSTEMS

Solvent extraction may be used to separate both organic and inorganic species. Since the affinity of a species for a particular phase is related to its similarity to that phase, oily species are attracted to oily phases and are said to be lipophilic (i.e., oil seeking). Species that are not oily in nature are not attracted to oily phases and are said to by lipophobic (i.e., oil avoiding). Species that are lipophobic are often hydrophilic (i.e., water seeking), and vice versa (i.e., hydrophobic species are often lipophilic).

Sometimes chemical reaction or solvation chemistry is exploited by the extraction process. Often, coordination reactions render a lipophobic species lipophilic by complexation with suitable organic ligands, which cluster around the lipophobic moiety in the complex and enable it to transfer into the solvent phase. The oily organic or solvent phase in such instances usually consists of an inert diluent and an active extractant that reacts with the inorganic species. The extractant solvates the solutes so that they transfer into the organic layer. Chemical reactions may be characterized either as solvation reactions or those involving ion exchange. Solute mass transfer in this instance is from the aqueous to the oily solvent phase.

Similarly, chemical reactions may be used to convert a lipophilic chemical into a hydrophilic, water-seeking species. Solute mass transfer then occurs from the oily solvent phase into the aqueous phase. Pure fluids or mixtures of species that form hydrogen bonds or contain polar moieties are usually highly nonideal. In pure fluids, strong attractive interactions between like molecules may cause molecular aggregation through dipole-dipole and other interactions. In mixtures, specific interactions can occur between molecules of the same species (self-association) or between molecules of different species (solvation).

10.3.1 Lignin Extraction

Relatively large, complex lignin molecules can be extracted by complexation, solvation, and coordination with the correct solvents. The black liquor produced by the Kraft pulping process is an alkaline mixture of lignin-bearing micelles. It is essentially opaque and darkly colored, hence its name. When acidified with dilute sulfuric or hydrochloric acid, the micelles are acidified, and the oily lignin is liberated. It floats to the top of the aqueous phase to form an interfacial precipitate.

A solvent mixture can form an effective solvation system for lignin obtained from black liquor, whereas individual components in the solvent mix are ineffective. Acetone alone is miscible with black liquor, so it cannot be used as a solvent by itself, since it doesn't form a second liquid phase when mixed with water. The solvent 2-ethylhexanol is immiscible with water but does not extract lignin from acidified black liquor. On the other hand, a one-to-one by volume mixture of 2-ethylhexanol and acetone extracts much of the acidified lignin and about half of the total dissolved solids. The resulting aqueous phase is water-white after acidification and extraction by this solvent mixture. The resulting organic layer is ruby red. Presumably, acetone displaces sufficient waters of hydration around lignin to enable the resulting acetone-lignin complex to coordinate with 2-ethylhexanol and transfer into the organic layer. This system dramatically illustrates the power of extraction by solvation and coordination.

10.3.2 Simplified TBP Reaction Models

As used in the tri-*n*-butyl phosphate (TBP), or PUREX, process to recover uranium (U) and plutonium (Pu), the solvent is typically a blend of about 30 vol% TBP and 70 vol% of an aliphatic diluent, say C-12 to C-15 branched paraffins. Spent nuclear fuel[5] is dissolved in nitric acid and then contacted with this TBP solvent. Actinide elements, primarily uranium and plutonium, are coextracted into the organic layer by coordination and solvation with TBP. The extraction of uranium, for example, is thought to occur by the following reaction [14–21]:

$$UO_2^{2+} \cdot 6H_2O(aq) + 2(NO_3)_2(aq) + 2TBP(o) \rightleftharpoons$$
$$UO_2(NO_3)_2 \cdot 2TBP(o) + 6H_2O(aq)$$
(10.1)

indicating that uranyl nitrate hexahydrate in the aqueous phase coordinates with two molecules of TBP in the organic phase to form an anhydrous uranyl nitrate solvate with TBP in the organic phase and leaves six waters of hydration behind in the aqueous phase when it transfers.

The extraction of water by TBP requires a more rigorous model, but the extraction of U and HNO_3 by TBP can be modeled fairly accurately using simplified models based on equilibrium constants and the law of mass action. At least one reaction must be assumed for each species that extracts. In the case of uranium and nitric acid, the following simplified reactions [5] can be assumed:

Liquid-Liquid Extraction

$$UO_2^{2+}(aq) + 2NO_3^-(aq) + 2TBP(o) \rightleftharpoons UO_2(NO_3)_2 \cdot 2TBP(o) \quad (10.2)$$

$$H^+(aq) + NO_3^-(aq) + TBP(o) \rightleftharpoons HNO_3 \cdot TBP(o) \quad (10.3)$$

A simplified extraction model assumes that both liquids are ideal mixtures and neglects any changes in density. Equilibrium concentrations are then estimated by solving the equation set,

$$[UO_2^{2+}(aq)][NO_3^-(aq)]^2[TBP_f(o)]^2 = (1/K_U)[UO_2(NO_3)_2 \cdot 2TBP(o)] \quad (10.4)$$

$$[H^+(aq)][NO_3^-(aq)][TBP_f(o)] = (1/K_H)[HNO_3 \cdot TBP(o)] \quad (10.5)$$

along with the TBP material balance,

$$2[UO_2(NO_3)_2 \cdot 2TBP)_2(o)] + [HNO_3 \cdot TBP(o)] + [TBP_f(o)] = [TBP_o(o)] \quad (10.6)$$

where all quantities in brackets are molar concentrations, $[TBP_o(o)]$ is the stoichiometric TBP concentration before any reaction occurs (about 1 M in 30 vol% TBP), and $[TBP_f(o)]$ is the molar concentration of uncomplexed or "free" TBP in the organic phase. Also K_U and K_H are the approximate equilibrium constants, about 5.5 M^{-4} and 0.145 M^{-2} at 23°C, respectively.

This model is an oversimplification in that it ignores water extraction, the TBP dimer, and TBP concentrations in the aqueous phase, and it assumes ideal mixtures. Nonetheless, it is reasonably effective in predicting extraction behavior, because it correctly models this system using mass action and competition for the uncomplexed or free TBP.

Using this model, the distribution coefficients for uranium is calculated as

$$D_U = \frac{[UO_2(NO_3)_2 \cdot 2TBP(o)]}{[UO_2^{2+}(aq)]} = K_U[NO_3^-(aq)]^2[TBP_f(o)]^2 \quad (10.7)$$

and, for nitric acid extraction,

$$D_H = \frac{[HNO_3 \cdot TBP(o)]}{[H^+(aq)]} = K_H[NO_3^-(aq)][TBP_f(o)] \quad (10.8)$$

Therefore, the uranium extraction has a second-order concentration dependence on the aqueous nitrate and the organic TBP concentrations. Nitric acid extraction has a first-order dependence on the aqueous nitrate and the organic TBP concentrations. The addition of an inextractable nitrate salt (e.g., NaNO$_3$) increases the extraction of both uranium and nitric acid. This model also predicts that uranium extraction is decreased by reducing the total nitrate concentration, which is the procedure for back-extracting uranium, by washing the extract with about 0.1 M nitric acid. Also, uranium and nitric acid compete for free TBP, as indicated in Equations (10.7) and (10.8).

10.3.3 TBP Solvent Cleanup

Reactions of the solvent with species in a feed mixture may also adversely affect an extraction system. Because of the radiation field from fission products in spent nuclear fuel, TBP reacts by radiolysis and acid hydrolysis to form dibutyl and monobutyl phosphoric acids (DBP and MBP, respectively). These latter species are also highly effective in extracting U and Pu, but at conditions that are incompatible with normal PUREX operations. While TBP strips well in the presence of dilute acid, say around 0.01 M HNO_3 or less, DBP and MBP are acidic extractants that complex actinides strongly under these same conditions. For this reason, it becomes difficult, if not impossible, to remove actinides from TBP extract using a dilute acid scrub after MBP and DBP have grown into it as radiolysis products. To maintain solvent viability, it must first be treated with aqueous sodium carbonate wash to remove the degradation products, after which it can be recycled to extract more U and Pu.

Like lignin in alkaline NaOH, DBP and MBP can be removed from TBP solvent by forming micellar structures in alkaline sodium carbonate wash solutions. Actinides that were originally complexed by DBP and MBP in the TBP organic phase prior to carbonate wash are also transferred from the solvent and are loosely incorporated into the micellar structure of the wash, the aqueous raffinate.

Reactive solvent extraction can then be used to recover the actinides from such scrub solutions, not by extracting the actinides but rather by extracting the degradation products [22] into an alcohol solvent layer. Acidification of carbonate scrub washes with nitric acid destroys the carbonate and forms an interfacial crud that floats to the top of the aqueous phase. This crud is similar in behavior to that formed by acidified lignin.

If the same acidification is performed while the scrub solution is actively mixed with 2-ethylhexanol, then a different result is achieved. In this latter case, the degradation products extract into the organic phase rather than forming an interfacial crud. When properly extracted, the resulting aqueous phase is then suitable for recycle back to the PUREX process and subsequent actinide recovery using conventional PUREX chemistry. The 2-ethylhexanol containing the degradation products can be treated using caustic wash to remove them and to enable 2-ethylhexanol solvent recycle.

10.4 DISTRIBUTION COEFFICIENTS

The initial solvent and feed concentrations, the desired final concentrations, and equilibrium behavior determine the direction of mass transfer, the minimum solvent-to-feed ratio, and the minimum theoretical tray requirements. These theoretical trays (or stages) are analogous in many respects to theoretical plates of a distillation column, and absorption (or stripping) columns discussed by Fair in another chapter. For any particular solvent-to-feed ratio, equilibrium relationships and the operating line determine theoretical stage requirements.

Experimental equilibrium data are almost always essential. While theoretical models are available to predict such data, and are very helpful in preliminary design, pilot studies and scaleup should be based on confirmed experimental measurements. Many nomographs providing data are available. Wisniak and Tamir [23–26], Sorensen and Arlt [27–29], and Macedo and Rasmussen [30] provide useful compilations of LLE data. Tiegs [31] provides activity coefficient data at infinite dilution. Although these data are generally quite good, users are advised to at least spot check them for accuracy.

10.4.1 THERMODYNAMIC MODELS

Thermodynamic models predict phase equilibrium. They can be based on solubility considerations alone or on chemical reactions between the two liquid phases.

10.4.2 NONREACTIVE SYSTEMS

When two liquid phases are thoroughly mixed in a closed system at constant temperature and pressure,* the mixture reaches thermodynamic equilibrium. At equilibrium, the fugacities of each species in each liquid phase are equal, and

---

* The pressure dependence is weak except near the critical point.

Liquid-Liquid Extraction

$$\hat{f}_i^\alpha = \hat{f}_i^\beta \tag{10.9}$$

At low to moderate pressures, $\hat{f}_i = \gamma_i x_i P_i^{sat}$ and

$$\gamma_i^\alpha x_i^\alpha P_i^{sat} = \gamma_i^\beta x_i^\beta P_i^{sat} \tag{10.10}$$

where $\hat{f}_i^\alpha$ and $\hat{f}_i^\beta$ are the fugacities of the ith species in the α and β phases, respectively; γ_i^α and γ_i^β are the activity coefficients in the respective phase; and x_i^α and x_i^β are the respective mole fractions. The variable P_i^{sat} is the pure component vapor pressure for the ith species. Since P_i^{sat} appears on both sides of Equation (10.10), it cancels out and, consequently, the pure component vapor pressure has no effect on liquid-liquid phase equilibria. The distribution coefficient, $K_{D_i}^{\alpha/\beta}$, is thus given by

$$K_{D_i}^{\alpha/\beta} = \frac{x_i^\alpha}{x_i^\beta} = \frac{\gamma_i^\beta}{\gamma_i^\alpha} \tag{10.11}$$

and is the basis for using thermodynamic models (e.g., UNIFAC [32, 33] or UNIQUAC*, as discussed elsewhere in this book) to estimate phase equilibrium behavior.

Activity coefficients of solutes at infinite dilution, γ_i^∞, are nearly constant (i.e., by Henry's law). By the Lewis and Randall rule, the activity coefficients of nearly pure solutes approach unity as $x_i^\beta \to 1.0$. In this case, Equation (10.11) is approximately

$$K_{D_i}^{\alpha/\beta} \approx \frac{x_i^\alpha}{1} \approx \frac{1}{\gamma_i^\alpha} \tag{10.12}$$

so the solubility limits x_i^α and x_i^β can be used to estimate γ_i^∞ in either liquid phase.

While solvent volatility at ambient conditions has little effect on liquid-liquid equilibria [because the pure component vapor pressure cancels out in Equation (10.9)], solvent volatility nonetheless can be important in solvent regeneration, particularly if simple or extractive distillation is used to recycle solvent. These techniques are commonly used in solvent regeneration when extraction is applied to the recovery of many organic chemicals (e.g., acetic acid [34] recovery, which typically involves extractive distillation to regenerate the solvent and produce glacial acetic acid). Then the latent heat of vaporization is also an important consideration. Usually, however, it is desirable to have regeneration without solvent distillation, as that approach is often inefficient and expensive.

The distribution ratio, K_{D_i}, in Equation (10.11) is defined as a mole fraction ratio of the ith solute concentration in each liquid phase. Sometimes the distribution ratio, D_i, is defined as a ratio of solute mass fractions. In either case, the distribution ratio usually refers to equilibrium concentrations (either mass or mole fractions, but sometimes molar concentrations) in either phase.

Insofar as the distribution coefficient, K_{D_i} or D_i, is constant† and not a function of composition, then Nernst's law [35] is obeyed, especially for nonreactive, simple systems. The lack of a composition dependence may be taken as an indication that a particular system is not reactive. In such cases, the distribution of solute between the two phases is primarily determined by its solubility in each phase; hence Henry's law and/or the Lewis and Randall rule apply.

---

* Many simulators (e.g., HYSYS or Aspen Plus) incorporate these models, and many others and generally are readily available to users.
† Empirically, D_i is often more nearly constant than K_{D_i}.

10.4.3 Reactive Systems

As mentioned above, chemical theory has been applied in many instances to explain nonideal behavior [36, 37]. Dolezalek [38] was one of the first to assume that hydrogen bonding forms new chemical species in mixtures. Since then, chemical theory has been applied to a wide range of thermodynamic problems [39–42].

Using chemical theory, one must hypothesize reactions to form specific aggregates (i.e., the "true" species) by complexation. This is a disadvantage for at least two reasons. First, there are usually many plausible aggregates. Second, each hypothesized aggregate introduces additional adjustable parameters. Thus, significant experimental data are needed to fit model parameters and to distinguish between alternative equilibria. On the other hand, chemical models can be highly effective while still relatively simple (e.g., the TBP model above), and they can give insight into the nature of the solvent extraction process and the best ways to exploit chemical behavior.

Lattice theories [37] enable one to consider nonspecific physical forces (e.g., molecular dipole moments, induction effects, and London dispersion forces) and have been applied successfully to model nonideality in a wide range of mixtures. Guggenheim [43] was the first to develop a quasichemical theory using lattice models. Wilson [44], Renon and Prausnitz [45], Abrams and Prausnitz [46], and Vera et al. [47] modified it for nonrandom mixtures. Panayiotou and Vera [48, 49] developed expressions for estimating local surfaces and compositions based on quasichemical theory. Kumar et al. [50] revised the equations for multicomponent mixtures. Martinez [51] applied lattice theory to clustering and dissociation. Sayegh and Vera [52] provide a review.

Tri-n-butyl phosphate (TBP) is a reactive extractant that has been widely used in the processing of heavy metal ores and spent fuel elements [53]. It is often mixed with a diluent to improve solvent properties (e.g., to reduce viscosity or to control its extraction power). Such diluents include paraffins such as Amsco 125-82 and n-heptane, benzene, and, to a lesser extent, chloroform and carbon tetrachloride.

Tributyl phosphate extracts metals by solvation and can be considered a reactive solvent, but experimental evidence suggests that it may also dimerize [54]. Of course, adding paraffinic diluents to TBP decreases the degree of TBP self-association by mass action, and diluents such as chloroform or 1-butanol may also form TBP solvates [55]. Phosphorus NMR shift in the system n-heptane and TBP suggests that diluent mass action and TBP dissociation in the dilute TBP range affect phosphorus chemical shift [54–56]. Rytting et al. [57] arrived at a similar conclusion from studying TBP and carbon tetrachloride mixtures in the dilute TBP range. Extraction studies often suggest the existence of complexes of the TBP monomer and dimer in the presence of diluents [58, 59], but virtually all of these investigators have assumed that solvation and association reactions form ideal mixtures in the TBP-diluent phase rather than attempting to evaluate activities.

A more realistic approach attempts to describe each equilibrium in terms of its thermodynamic equilibrium constant, molar volumes of specific aggregates, and a heat of reaction to estimate the chemical contribution to excess enthalpy H^E_{chem}. The first two parameters contribute to the excess Gibbs energy g^E model.

A more general reaction model considers both chemical (specific) and physical (nonspecific) contributions to excess Gibbs energy. Using TBP as an example, one first begins by assuming that the physical and chemical contributions are separable so that

$$g^E = g^E_{chem} + g^E_{phys} \tag{10.13}$$

Then the chemical contribution can be computed from the estimated "true" mole fractions. With the assumption that the mixture is ideal, chemical activities simply equal the true mole fractions of monomeric species, and Prigogine and Defay [60] have shown that, regardless of the solvation or association reactions that may occur, in an ideal true mixture with $g^E_{phys} = 0$,

Liquid-Liquid Extraction

$$\hat{a}_i = \gamma_{i,s} x_{i,s} = z_{i_1} \quad (10.14)$$

where $\hat{a}_i$ and $\gamma_{i,s}$ are the activity and activity coefficient based on the stoichiometric composition $x_{i,s}$, and where z_{i_1} is the true mole fraction of the ith monomer. The reference states for A and B in this case are the pure, unassociated liquids with chemical potentials, pure component Gibbs energies, fugacities, and activities related by $(\mu_i - G_i^o)/RT = \ln \hat{a}_i = \ln \hat{f}_i/f_i^o$.

The extract (or solvent) phase is seldom ideal, however, since the molecules are not uniform in size, and other, nonspecific interactions may occur; but the true mole fractions can still be used to estimate physical contributions to g^E via molecular size and shape difference terms and using regular solution theory.

The stoichiometric mole fractions of diluent and TBP or other reactive extractants (with $i = \{A,B\}$, respectively) are simply the overall mole fractions $x_{i,s}$, molar concentrations $C_{i,s}$, or volume fractions $\phi_{i,s}$.

True mole fractions z_i, molar concentrations C_i, and volume fractions ϕ_i can be computed simply from pure component properties if one assumes that the excess molar volume $V^E = 0$. This assumption can be made initially so as to satisfy component volume and material balances exactly for each mixture. So, using the pure component molar volumes V_i at a given temperature, then $V = \sum_{i=1}^{2} x_{i,s} V_i$, $C_{i,s} = x_{i,s}/V$, and $\phi_{i,s} = V_i C_{i,s}$.

Specific chemical interactions can be considered by assuming appropriate combinations of solvation and self-association reactions. If, for example, the true species are the diluent and TBP monomers z_{A_1} and z_{B_1}, the TBP dimer z_{B_2}, and various combinations of aggregates or complexes (e.g., AB, z_{AB}; AB$_2$; z_{AB_2}; and so forth), then for the jth solvation or self-association reaction in a binary mixture of diluent (A) and TBP (B):

$$n_j A_1 + m_j B_1 \rightleftharpoons A_{n_j} B_{m_j} \quad (10.15)$$

where n_j molecules of A_1 monomer and m_j molecules of B_1, the TBP monomer, are assumed to form a complex. At equilibrium, the chemical potentials of the two monomeric species and the complex are equal, and the product of their activities equals the thermodynamic equilibrium constant $K_{\hat{a}_j}$:

$$K_{\hat{a}_j} = \frac{\hat{a}_{A_{n_j} B_{m_j}}}{\hat{a}_{A_1}^{n_j} \hat{a}_{B_1}^{m_j}} \quad (10.16)$$

Defining volume fraction-based activity coefficients for the jth species γ'_j, where $\hat{a}_j = \gamma'_j \phi_j$, leads to

$$K_{\hat{a}_j} = K_{\phi_j} K_{\gamma'_j} = K_{z_j} K_{\gamma_j} \quad (10.17)$$

Also, $\hat{a}_j = \gamma_j z_j$ for each species, γ_j is the true mole fraction-based activity coefficient for the jth species, and $K_{\hat{a}_j}$ is the thermodynamic equilibrium constant.

Physical interactions may be approximated using regular solution theory, and differences in molecular size and shape by using the estimated true species concentrations to estimate the activity coefficient products K_{γ_j} and $K_{\gamma'_j}$ in each mixture and the reference states.

Given a set of true species compositions, differences in molecular sizes, available surface areas, and cohesive energy densities still exist. The physical effects contributing to nonideality may be divided into three terms:

$$g^E_{phys} = g^E_{ms} + g^E_{sd} + g^E_{rs} \tag{10.18}$$

By equating activities for the stoichiometric species with those for the monomers, one eventually obtains an expression for the stoichiometric activity coefficient:

$$\gamma_{i,s} = \frac{\gamma_{i_1} z_{i_1}}{x_{i,s} \gamma^o_{i_1} z^o_{i_1}} \tag{10.19}$$

in terms of monomeric mole fractions in the true mixture and reference states, the stoichiometric mole fraction, and activity coefficients in the true mixture and reference states.

Additional details describing these models can be found elsewhere [61–63]. Such models describe extraction and enthalpic behavior, and measured shifts in nuclear magnetic resonance.

10.4.4 Empirical Distribution Models

In preliminary design work, it is convenient to correlate distribution coefficients on a mass-fraction basis. An empirical correlation technique that is simple to use and often highly effective is

$$\ln D_i = \beta_0 + \sum_{i=1}^{n} \beta_i X_i + \beta_{n+1} V_m + \beta_{n+2}/T \tag{10.20}$$

where the β_i are empirical parameters; X_i are mass fractions at equilibrium in the mixture; V_m is the stoichiometric volume fraction of the active extractant, which is a solvent blend; and T is the temperature in either K or R. The variable n is the number of components in the mixture. While it is true that D_i may exhibit a composition dependence on any of the species in a mixture (i.e., $\beta_i \neq 0$, $i = 1,2,\ldots n$), it is also commonly true that only a few of the β_i parameters inside the summation term on the right-hand side of Equation (10.20) are statistically nonzero. In fact, the strongest dependence may simply be on the solute mass fraction, while other species can be ignored. In any event, the appropriateness of Equation (10.20) is easily determined by fitting experimental data.

Equation (10.20) is often sufficient to model a liquid-liquid extraction system, especially in the early stages of design. Its simplicity is advantageous, but it has no theoretical basis. As a consequence, users should be careful not to extrapolate it to untested compositions. An additional limitation is the fact that Equation (10.20) does not predict the locus of the mutual solubility curve, and, for $\beta_i > 0$, one might be tempted by this equation to extrapolate to conditions that are actually outside the two-phase region (see Figure 10.1).

Many useful solvents are actually blends of a diluent and a modifier. These blends are often formulated on a volume basis (e.g., 30 vol% TBP in Isopar-M), and for such systems the V_m variable is helpful if the solvent blend can be designed and optimized.

The key feature of Equation (10.20) that makes it particularly useful is the fact that D_i often exhibits a weaker composition dependence than K_{D_i}, which is based on the ratio of mole fractions as shown in Equation (10.11).

10.5 DESIGN OF EXTRACTION SYSTEMS

Liquid-liquid extraction columns may be designed in three different ways: (1) as a collection of equilibrium stages, (2) as a continuous differential contactor with mass transfer, or (3) using purely kinetic models. The first two methods are more commonly used (particularly the first) and, when correctly and carefully performed, they give essentially the same results. The latter method, design

Liquid-Liquid Extraction

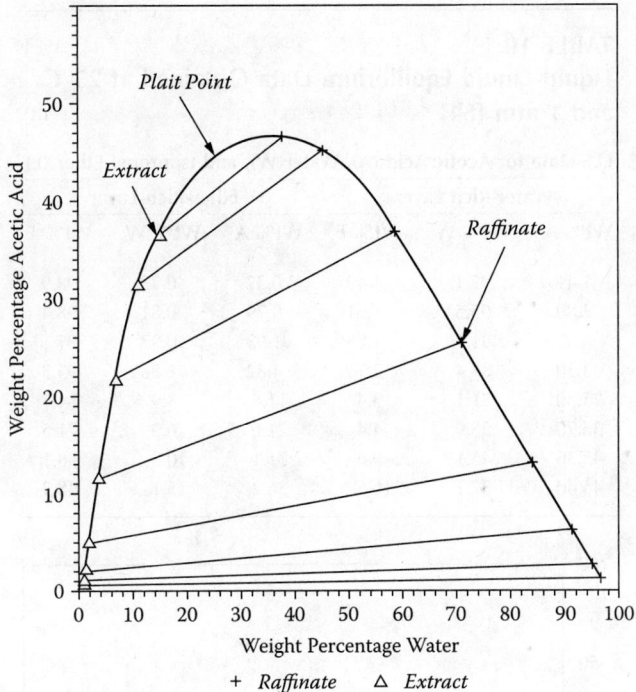

FIGURE 10.1 Liquid-liquid equilibrium data at 25°C for water, acetic acid, and isopropyl ether, showing raffinate and extract compositions at several tie lies and the plait point. At the plait point, both phases dissolve into a single phase with the same composition. The weight percentage of isopropyl ether equals 100 less the weight percentage of water less the weight percentage of acetic acid.

based on mass transfer rates and extraction kinetics, can be important, particularly in high-speed contactors with very short liquid-liquid contact times, but it is less frequently utilized. The first method requires an estimate of stage efficiency. The theoretical stages are divided by the stage efficiency to estimate the actual stages. The second method requires an estimate of the height of column that is equivalent either to a theoretical stage or a transfer unit [2].

10.5.1 Phase Diagrams

Analysis of liquid-liquid behavior and the design of an extraction cascade usually begin with the acquisition of phase equilibrium data. Example data are provided in Table 10.1 for the ternary system, acetic acid, water, and isopropyl ether. Each row in Table 10.1 corresponds to one tie line linking the organic (extract) and aqueous (raffinate) phase compositions at equilibrium. It is convenient to plot such data in an $x - y$ diagram as shown in Figure 10.1. The mutual solubility curve indicates the raffinate and extract compositions, and it defines the limits of the two-phase region. The extract and raffinate merge into a single phase at the plait point.

Any point on Figure 10.1 defines one ternary composition. Points interior to the mutual solubility diagram form two phases as shown in Figure 10.2, where a mixture with composition M in this diagram is shown to separate into two phases with extract composition E and raffinate composition R. The inverse lever rule defines the relative amounts of the two phases. So, given the line segment $\overline{EM}$ and $\overline{MR}$, relative amounts are

$$\frac{\overline{EM}}{\overline{MR}} = \frac{R}{E} = \frac{x_{wm} - x_{we}}{x_{wr} - x_{wm}} = \frac{x_{am} - x_{ae}}{x_{ar} - x_{am}} \tag{10.21}$$

TABLE 10.1
Liquid-Liquid Equilibrium Data Obtained at 25°C and 1 atm [64]

LLE Data for Acetic Acid (A), Water(W), and Isopropyl Ether (E)

Water-Rich Layer			Ether-Rich Layer		
Wt% A	Wt% W	Wt% E	Wt% A	Wt% W	Wt% E
1.41	97.1	1.49	0.37	0.73	98.9
2.89	95.5	1.61	0.79	0.81	98.4
6.42	91.7	1.88	1.93	0.97	97.1
13.30	84.4	2.3	4.82	1.88	93.3
25.50	71.1	3.4	11.4	3.9	84.7
36.70	58.9	4.4	21.6	6.9	71.5
45.30	45.1	9.6	31.1	10.8	58.1
46.40	37.1	16.5	36.2	15.1	48.7

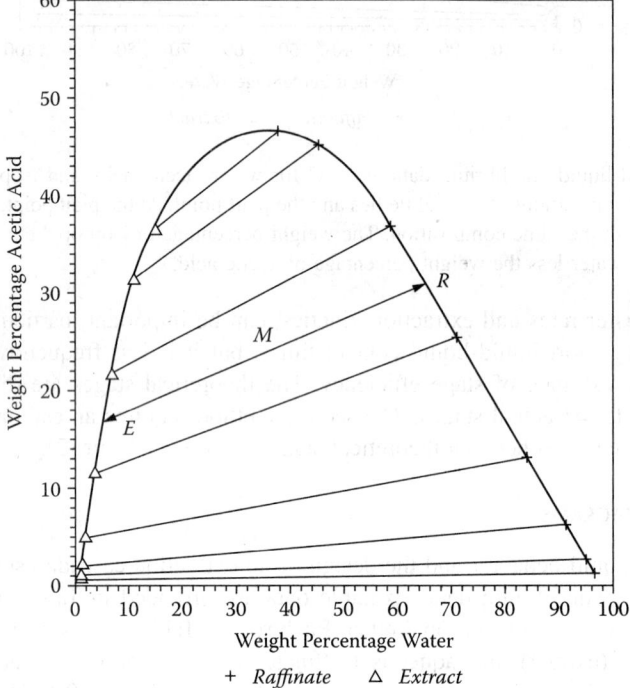

FIGURE 10.2 Liquid-liquid data for the system water, acetic acid, and isopropyl ether. Since mixture M falls inside the mutual solubility curve, it separates into extract and raffinate phases with compositions E and R, respectively. The equilibrium data are at 25°C and 1 atm.

where R and E are the masses of raffinate and extract, respectively, and x_{ij} are mass fractions, with $i = \{w,a\}$ referring to either with water or acetic acid mass fractions and $j = \{e,m,r\}$ referring to either point E, M, or R.

The point M is also known as the mix point, and it defines the overall stoichiometric composition of the mixture (i.e., the composition if only a single phase was formed by mixing), while the extract and raffinate compositions E and R are found from constructing a line parallel to the nearest tie lies that also intersects the mix points and the mutual solubility curve.

Liquid-Liquid Extraction

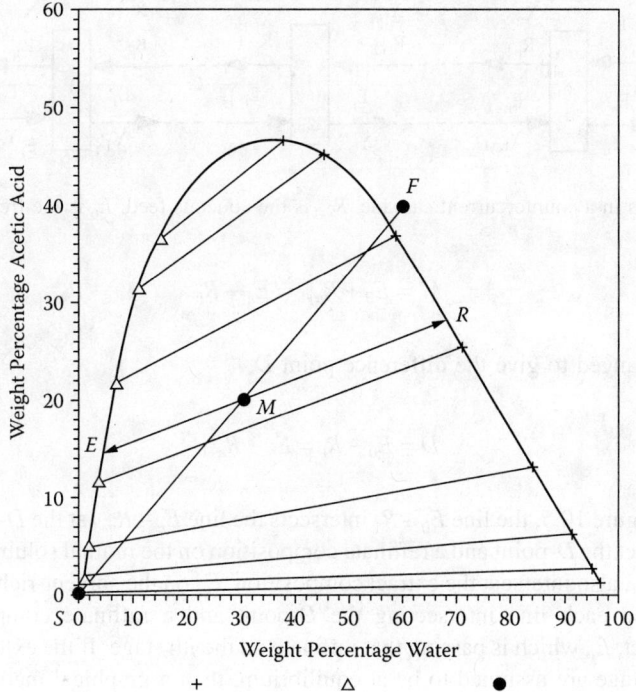

FIGURE 10.3 Liquid-liquid data at 25°C for the system water, acetic acid, and isopropyl ether. Mixing liquids with compositions F and S in equal proportions by mass results in mix point M, which separates into phases with compositions E and R. The equilibrium data are at 25°C and 1 atm.

The compositions of feeds and solvent mixtures may lie outside the mutual solubility curve but, if upon mixing, the overall stoichiometry is such that the M point falls inside the mutual solubility curve, then two liquid phases will form. Mixtures outside of the mutual solubility curve are only a single phase. Thus, Figure 10.2 indicates that the addition of acetic acid to water increases the solubility of isopropyl ether in both the extract and raffinate phases. Aqueous mixtures containing 50 mass% or more of acetic acid are miscible with isopropyl ether in all proportions.

Consider mixing equal amounts of pure isopropyl ether with an aqueous feed that is 40 mass% acetic acid in water. The mix point M then lies at a point equidistant between the feed composition point F and the solvent composition point S. The resulting two phases then have the compositions E and R, as illustrated in Figure 10.3.

10.5.2 Countercurrent Extractors

Cascades of countercurrent extractors, as shown in Figure 10.4, give the highest recovery of solute while requiring the least amount of solvent [65], so they are of the greatest industrial interest. The alternatives, cocurrent and crosscurrent extractions, are of less interest industrially, although the latter is sometimes used in laboratory applications. Steady-state material balances written around the jth stage in Figure 10.4 lead to the equation

$$R_{j-1}x_{j-1} + E_{j+1}y_{j+1} = R_j x_j + E_j y_j \tag{10.22}$$

For a cascade consisting of n stages, R_1 is the feed and E_{n+1} is the fresh solvent added to the cascade, or column, at the opposite end. The mix point, M, can then be defined in terms of

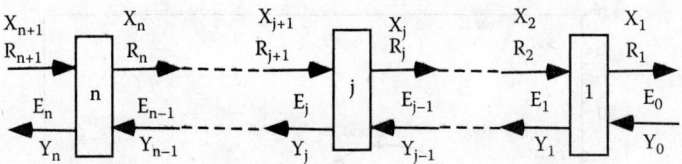

FIGURE 10.4 Stages in a countercurrent cascade. R_{n+1} is the aqueous feed. E_0 is the fresh solvent.

$$M = E_0 + R_{n+1} = E_n + R_1 \tag{10.23}$$

which can be rearranged to give the difference point D,

$$D = E_0 - R_1 = E_n - R_{n+1} \tag{10.24}$$

As shown in Figure 10.5, the line $E_0 - R_1$ intersects the line $E_n - R_{n+1}$ at the D-point. Graphically, all lines that intersect the D-point and a raffinate composition on the mutual solubility curve between R_1 and R_{n+1}, say R_{j+1}, also intersect the extract composition, E_j, on the solvent-rich side of the mutual solubility curve. So each line intersecting the D-point and a raffinate composition, R_{j+1}, also intersects the extract, E_j, which is passing that raffinate on the jth stage. If the extract E_j and raffinate R_j leaving the jth stage are assumed to be at equilibrium, then a graphical method can be used to step off theoretical stages, as shown in Figure 10.6.

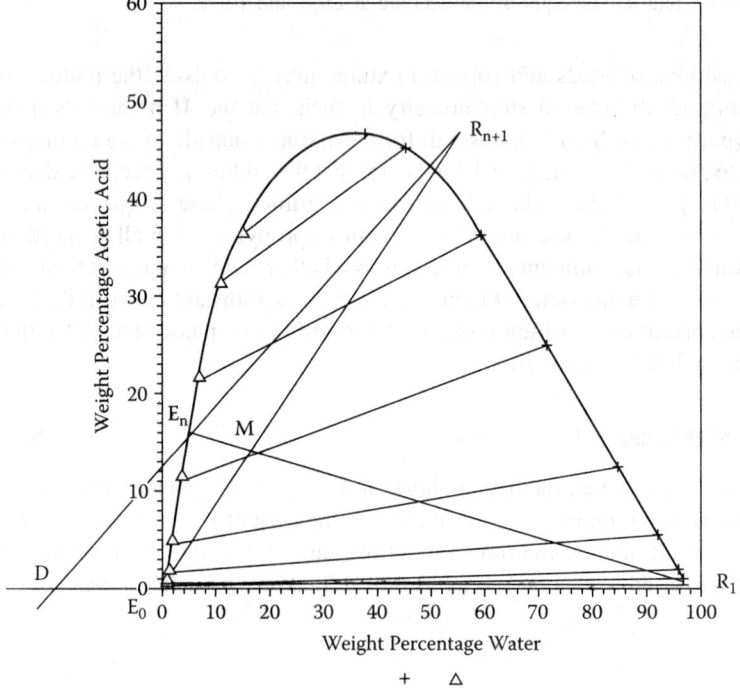

FIGURE 10.5 Liquid-liquid data at 25°C for the system water, acetic acid, and isopropyl ether. Mixing liquids with compositions R_{n+1} and E_0 results in mix point M. Specification of R_1 yields E_n, since the same line also intersects M. The equilibrium data are at 25°C and 1 atm.

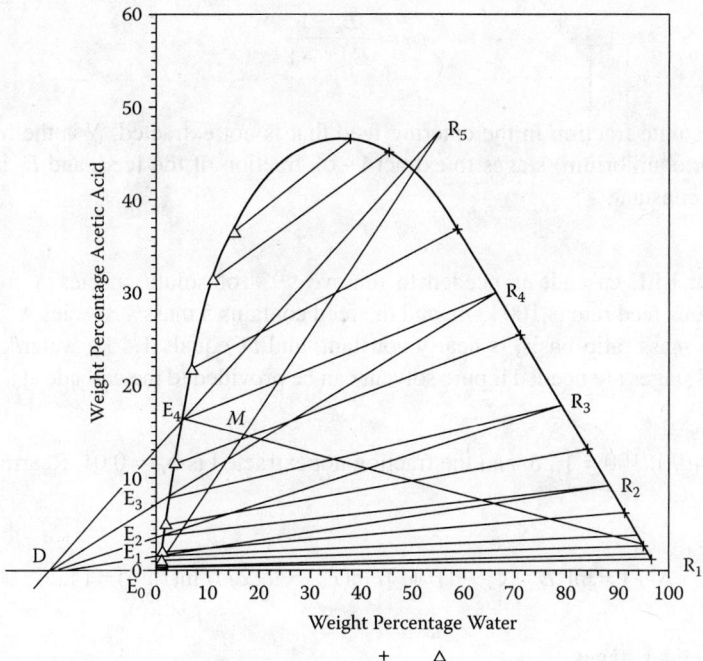

FIGURE 10.6 Liquid-liquid data at 25°C for the system water, acetic acid, and isopropyl ether. Theoretical stages are stepped off by either drawing lines through D to obtain an extract composition or interpolating between nearby tie lines to estimate the extract composition on the next stage. The equilibrium data are at 25°C and 1 atm.

Graphical methods are suitable whenever the two liquid phases are partly miscible as for the water, acetic acid, and isopropyl ether system. If the two liquid phases are nearly immiscible, and their mutual solubility is not affected by the solute concentration, then the Kremser equation discussed below may be applicable.

10.5.3 Kremser Equation

Liquid-liquid systems that obey Nernst's law can be particularly simple to analyze, since the equilibrium relationship is linear in composition. If the two liquid phases are nearly immiscible, and if the solute concentrations are low in either phase, then the operating or material balance line is also nearly linear. Writing a material balance around the jth stage and the raffinate end of the cascade in Figure 10.4 gives the operating line,

$$y_j = \frac{R_{j+1}}{E_j} x_{j+1} + \frac{E_0 y_0 - R_1 x_1}{E_j} \tag{10.25}$$

where E_j and R_j are the extract and raffinate flow rates leaving the jth stage, usually assumed to be in thermodynamic equilibrium. If $E_0 \approx E_1 \approx \ldots E_j$ and $R_1 \approx R_2 \ldots R_{j+1}$, then the subscripts on E and R may be dropped, and the ratio R/E is nearly constant. In this case, the operating line, Equation (10.25), is also linear, and $(Ey_0 - Rx_1)/E$ is constant.

Under these conditions, the extraction factor $E_e = K_{D_i} E/R$ or $E_e = D_i E/R$ is also constant,* and the Kremser [66] group method applies:

---

* E_e is dimensionless. If K_{D_i} is used, then E/R is a molar ratio. If D_i is used, then E/R is a mass ratio.

$$\phi_E = \frac{E_e - 1}{E_e^{N+1} - 1} \qquad (10.26)$$

where ϕ_E is the solute fraction in the entering feed that is not extracted, N is the required number of theoretical (or equilibrium) stages to extract $1 - \phi_E$ fraction of the feed, and E_e is the extraction factor, which is constant.

Example: 10.1

A countercurrent LLE cascade is needed to remove 99% of solute species A from an aqueous phase. The aqueous feed rate is 100 kg/h, and the feed contains 5 mass% species A. The distribution coefficient, on a mass ratio basis, is nearly constant, and D_i equals 1.4 kg water/kg solvent. How many theoretical stages are needed if pure solvent can be provided to the cascade at a rate of 90 kg/h?

Solution

$E_e = D_i E/R = 1.4(90)/100 = 1.26$, and the fraction not extracted is $\phi_E = 0.01$. Rearranging Equation (10.26) gives

$$N + 1 = \ln[(E_e + \phi_E - 1)/\phi_E]/\ln(E_e) = \ln(27)/\ln(1.26) = 14.26$$

$N = 13.26$ theoretical stages.

10.6 INDUSTRIAL EXTRACTION EQUIPMENT

In-depth treatment of extraction equipment can be found in several books. Godfrey and Slater [67] provide more current details, but the treatments provided by Lo, Baird, and Hanson [7] and Skelland and Tedder [68] are also useful.

10.6.1 MIXER SETTLERS

Mixer-settlers are simply tanks configured to permit mixing in one followed by decantation in a second. Each stage consists of a mixer and settler pair (i.e., two tanks). When several mixer-settler stages are configured countercurrently, they constitute a cascade as indicated in Figure 10.4.

The advantages of mixer settlers are (1) simple maintenance and operation, (2) high tolerance for undissolved solids (e.g., as often occurs in the mining industry), (3) high tolerance for interfacial precipitates, (4) simple adjustment of mixing from stage to stage, and (5) easy measurement of concentration profiles throughout the cascade. The disadvantages of mixer-settlers are (1) potentially high inventory and (2) longer residence times than with other types of contactors. This latter disadvantage can be particularly serious if one is processing materials that degrade the solvent (e.g., fission products with a high radiation field).

Mixer-settlers are widely used, particularly in the mining industry. Vendors are available, and they are easily designed. In the laboratory, small-scale, mini-mixer-settlers are available that offer as many as 16 countercurrent stages and are capable of achieving steady state using 500 mL of feed or less.

10.6.2 RECIPROCATING PLATE COLUMNS

In reciprocating plate columns, the height equivalent to a theoretical stage (*HETS*) is a strong function of the reciprocating speed, *Af*, the product of the reciprocation amplitude and frequency. Karr columns actually exhibit optimal values for *Af* where the (*HETS*) is clearly minimized. These

Liquid-Liquid Extraction

minima are a function of both the physical properties of the liquid-liquid system and the column geometry, particularly the column diameter.

The (*HETS*) for reciprocating plate extraction columns has been correlated [69] using

$$\Gamma_L = 24.4 C_{dt}^{-0.075} N_{Re}^{-0.015} N_{We}^{-0.525} N_{Sl}^{-1.15} \left(\frac{U_d}{U_c}\right)^{0.210} \quad (10.27)$$

where $\Gamma_L = HETS/D$, $(Af) < (Af)_{opt}$, and $(Af)_{opt}$ is the reciprocating speed that minimizes the (*HETS*).

At reciprocating speeds greater than $(Af)_{opt}$, the correlation

$$\Gamma_U = 28.8 C_{dt}^{0.651} C_{dr}^{-0.997} N_{Bo}^{0.776} \left(\frac{U_c}{U_d}\right)^{0.582} \quad (10.28)$$

describes the observed behavior, where $\Gamma_U = HETS/D$.

Equations (10.27) and (10.28) can be accurately combined using a method suggested by Churchill[70] to yield an overall correlation:

$$\frac{HETS}{D} = \Gamma_L \left(1 + \left[\frac{\Gamma_U}{\Gamma_L}\right]^{20}\right)^{1/20} \quad (10.29)$$

Figure 10.7 compares experimental data with predictions using this correlation. The optimal values for (*Af*) that minimize (*HETS*) depend strongly on the column diameter, and the (*HETS*) minima are more pronounced for larger-diameter columns. In general, the (*HETS*) increases with column diameter, and it becomes increasingly important to operate at $(Af)_{opt}$ for larger-diameter columns.

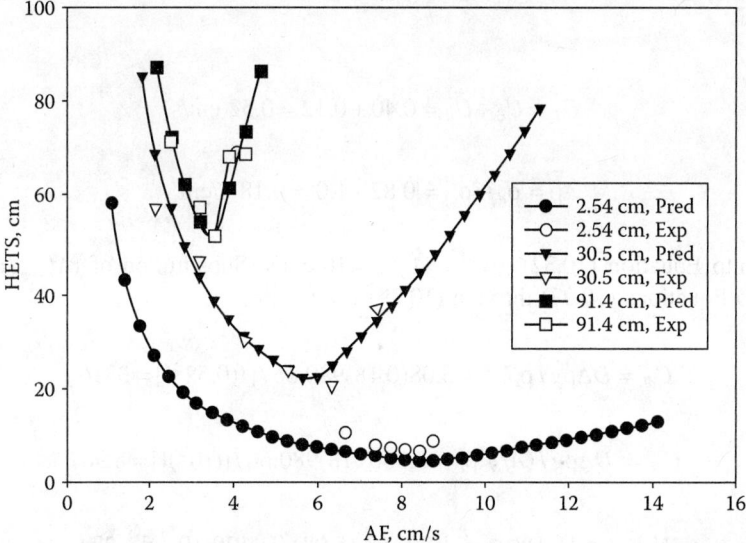

FIGURE 10.7 Predicted *HETS* for reciprocating columns using Equation (10.29). The column diameter affects the optimal reciprocating speed and the sensitivity of the *HETS* to that speed. The larger a column diameter, the more important it is to operate at the optimal reciprocating speed, *Af*.

Rearranging Equation (10.27) gives the following dependencies:

$$HETS \propto D^{0.385}\sigma^{0.525}(Af)^{-1.15}U_T^{0.235}\rho_c^{-0.465}(\Delta\rho g)^{-0.146} \qquad (10.30)$$

for $(Af) < (Af)_{opt}$. Similarly, rearrangement of Equation (10.28) gives these dependencies:

$$HETS \propto D^{0.892}\sigma^{-0.776}(Af)^{+1.99}U_T^{1.30}\rho_c^{+1.65}(\Delta\rho g)^{-0.872} \qquad (10.31)$$

for $(Af) < (Af)_{opt}$. Analysis of these dependencies can be helpful when comparing these correlations with those developed by other investigators.

The minimum ($HETS$) for a column can be estimated from an estimate of $(Af)_{opt}$. The latter quantity may be obtained by differentiating Equation (10.29) and setting $\partial HETS / \partial(Af) = 0$:

$$(Af)_{opt} = \left\{\frac{\Gamma_U}{\Gamma_L}(Af)^{-3.15}\right\}^{-1/3.15}$$

$$(Af)_{opt} = \left\{1.24 D^{0.515}\sigma^{-1.30}U_T^{1.06}\rho_c^{2.12}(\Delta\rho g)^{-0.8}\mu_c^{0.015}\left(\frac{U_c}{U_d}\right)^{0.794}\right\}^{-1/3.15} \qquad (10.32)$$

which predicts $(Af)_{opt} \pm 6\%$.

Example: 10.2

Estimate the minimum $HETS$ for a 2-in diameter reciprocating plate column. The liquid-liquid interfacial tension is 30 dyne/cm. The solvent phase density and viscosity are 0.82 g/cm³ and 1.3 cP, respectively. The aqueous phase density and viscosity are 1 g/cm³ and 1.129 cP, respectively. Water is the continuous phase, and its linear velocity down the column is U_c = 0.40 cm/s. The velocity of the dispersed solvent phase is U_d = 0.12 cm/s.

Solution

$$U_T = U_c + U_d = 0.40 + 0.12 = 0.52 \text{ cm/s}$$

$$\Delta\rho = |\rho_d - \rho_c| = |0.82 - 1.0| = 0.18 \text{ g/cm}^3$$

Substituting into Equation (10.32) gives $(Af)_{opt} = 10$ cm/s. Substitution of $(Af)_{opt}$ and the other properties into Equations (10.27) through (10.29) gives

$$C_{dt} = D\Delta\rho g / \rho_c U_T^2 = 5.08(0.18)980.66 / [1(0.52^2)] = 3316$$

$$C_{dr} = D\Delta\rho g / (Af)^2 \rho_c = 5.08(0.18)980.66 / [(10^2)1] = 8.967$$

$$N_{Re} = \rho_c U_T D / \mu_c = 1.0(0.52)(5.08) / [1.129(10^{-2})] = 234$$

$$N_{We} = D U_T^2 \rho_c / g_c \sigma = 5.08(0.52^2)1.0 / [1.0(30)] = 0.046$$

TABLE 10.2
Internet Sites Providing Solvent Extraction Services

Provider	Service	Internet Address
QVF Process Systems, Inc.	Equipment vendor	www.qvfps.com
Sulzer Chemtech USA, Inc.	Equipment vendor	www.nutter1.com
Varian, Inc.	Hydromatrix extraction cartridges	www.varianinc.com
Visimix Software	Mixing parameter estimates	on-line.visimix.com

$$N_{Sl} = Af/U_T = 10/0.52 = 19.2$$

$$N_{Bo} = D^2 \Delta\rho g / g_c \sigma = 5.08^2 (0.12) 980.66 / 30 = 151.8$$

$$\Gamma_L = 1.613, \ \Gamma_U = 1.636, \text{ and } HETS = 8.5 \text{ cm}.$$

10.7 INTERNET SITES

Several Internet sites that feature solvent extraction information are listed in Table 10.2.

10.8 ECONOMIC ANALYSIS FOR VERTICAL CONTACTORS

The use of solvent extraction technology results in capital expenses to install the equipment and operating expenses. Contact equipment vendors (see Table 10.2) for quotes on specialized solvent extraction equipment. Preliminary capital cost for continuous differential contactors can be obtained by sizing them and then cost estimating them as vertical pressure vessels with internals, e.g., using Guthrie's [71, 72] methods. Mixer-settlers are simply a collection of mixing and settling tanks assembled with appropriate piping. Examples are listed below. Processing highly radioactive systems requires nuclear-grade equipment, which essentially doubles their cost.

Choose materials of construction based on corrosion considerations. Column diameters are determined by specifying linear velocities for the two phases. Column heights are determined by estimating the actual number of stages based on the theoretical stage requirements and average stage efficiency. Internals in pulse columns are very similar to those in distillation towers, especially for sieve trays. Therefore, distillation correlations can be used to estimate FOB purchased and installed costs for continuous differential contactors, if they are assumed to be pulse columns.

From Guthrie,[71] tray FOB purchase costs, E'_i, for sieve trays can be estimated using the correlation

$$E'_i = E_i (F_S + F_m) \tag{10.33}$$

where F_S is 1.0, 1.4, or 2.2 for trays with 24-, 18-, and 12-in spacing, respectively. The tray material factor, F_m, is 0.0, 1.7, or 8.9 for carbon steel, stainless steel, and monel, respectively. The factor E_i is the FOB purchased cost for carbon steel sieve trays, 24-in spacing, either shop or field installed. It is correlated to the column dimensions if shop or field installed using the correlations in Table 10.3. Table 10.4 provides correlations for estimating the FOB purchased equipment costs for vertical process vessels (e.g., extraction columns) and horizontal process vessels (e.g., surge tanks).

TABLE 10.3
Correlations for Estimating fob Purchased Costs for Internals (e.g., Distillation Trays, Absorber Packings, Solvent Extraction Columns, and Others. Referenced to Mid-1968 Dollars with (M&S = 273.1). Correlated Using Information from Guthrie [71]

Item	Correlation
Sieve Trays, 24-in spacing, shop installed[a,b]	$E_i = C_4(HT-10)\left(\dfrac{M\&S}{273.1}\right)$ $C_4 = \begin{cases} 8.665D - 1.218 & D < 4 \text{ ft} \\ 16.67D - 33.35 & 4 \le D \le 6 \text{ ft} \\ 23.33D - 73.31 & 6 < D \le 7 \text{ ft} \\ 25D - 75 & 7 \le D \le 10 \text{ ft} \end{cases}$ D = nominal column diameter, ft HT = nominal column height, ft, actual trays plus 10 ft
Sieve Trays, 24-in spacing, field installed[c]	$E_i = C_4(HT-10)\left(\dfrac{M\&S}{273.1}\right)$ $C_4 = 20D - 60 \quad D > 10 \text{ ft}$ D = nominal column diameter, ft HT = nominal column height, ft, actual trays plus 10 ft

[a] Add 0.08 E_i as indirects for freight, taxes, and insurance.
[b] Add 0.05 E_i to estimate shop labor for installation.
[c] Add 0.58 E_i for field labor and 0.75 E_i for indirects.

Example: 10.3

Estimate the fob purchased cost of a stainless steel pulse column when the Marshall & Swift (M&S) index is 1200. The column is 30 ft high and 2 ft in diameter. Internals are estimated as sieve trays with 1.5-ft spacing.

Solution

From the column dimensions, the bare module cost for the internals, if they are constructed of carbon steel and have 24-in spacing, would be

$$C_4 = 8.665(2) - 1.218 = 16.112$$

$$E_i = 16.112(20)(1200/273.1) = \$1{,}416 \text{ for the internals.}$$

Since the internals have 1.5-ft spacing and are of stainless steel,

$$E_i' = \$1{,}416(1.4 + 1.7) = \$4{,}390 \text{ fob.}$$

For the column, we have $C_3 = 33.143/30 + 0.127 = 1.23177$:

$$E = 2000(2.0/1.23177)^{1.047}(1200/273.1) = \$14{,}598 \text{ if the column is of carbon steel.}$$

Liquid-Liquid Extraction

TABLE 10.4
Correlations for Estimating Base Equipment Costs for Process Vessels. Referenced to Mid-1968 Dollars with (M&S = 273.1). Correlated Using Guthrie's Figures (Low Pressure, Carbon Steel Basis)

Item	Correlation
Vertical process vessels[a]	$E = 2000 \left(\dfrac{D}{C_3}\right)^{1.047} \left(\dfrac{M\&S}{273.1}\right)$
	$C_3 = \begin{cases} 20.867/HT + 0.778 & HT < 15 \text{ ft} \\ 33.143/HT + 0.127 & HT \geq 15 \text{ ft} \end{cases}$
	D = nominal column diameter, ft
	HT = nominal column height, ft, actual trays plus 10 ft
Horizontal process vessels[b]	$E = 1375 \left(\dfrac{D}{C_3}\right)^{0.9747} \left(\dfrac{M\&S}{273.1}\right)$
	$C_3 = \begin{cases} 20.867/HT + 0.778 & HT < 15 \text{ ft} \\ 33.143/HT + 0.127 & HT \geq 15 \text{ ft} \end{cases}$
	HT = nominal horizontal length of vessel, ft
	D = nominal diameter of horizontal vessel, ft

[a] Distillation columns, absorbers, scrubber, strippers, flash drums, etc.
[b] Accumulator drums, surge vessels, decanters, etc.

Because it is stainless, $E' = 3.67E = \$53,575$, and the total fob purchase cost is $E' + E'_i = 53,575 + 4,390 = \$57,965$.

10.8.1 List of Symbols

A	= the amplitude of the reciprocating cycle, measured from the mean position to an extreme, or one-half of the stroke length, cm
(Af)	= the reciprocating speed, cm/s or cm/min
$(Af)_{opt}$	= the reciprocating speed yielding the minimum $HETS$ value, cm/s
$\hat{a}_i$	= chemical activity of species i
$C_{i,s}$	= stoichiometric molar concentration of species i, moles/L, $V^E = 0$
C	= true molar concentration of species i, moles/L, $V^E \neq 0$
C_{dr}	= $D\Delta\rho g/(Af)^2 \rho_c$, a reciprocating drag coefficient
C_{dt}	= $D\Delta\rho g/\rho_c U_T^2$, a column drag coefficient
D	= the column diameter, cm, or difference point in Equation (10.24)
f	= the reciprocating frequency s^{-1}
$\hat{f}_i$	= the fugacity of species i in a liquid phase
g	= gravity acceleration, 980.66 cm/s^2
g^E	= excess Gibbs energy, J/mol
g_c	= gravity constant, 1.0 g-cm/d-s^2
h	= plate spacing in column mass transfer zone, cm
$HETS$	= height equivalent to a theoretical stage, cm
$K_{\hat{a}j}$	= thermodynamic equilibrium constant for complex j

K_{z_j}	= product of true mole fractions for complex j
K_{ϕ_j}	= product of true volume fractions for complex j
$K_{\gamma'_j}$	= product of true volume fraction-based activity coefficients for complex j
K_{γ_j}	= product of true mole fraction-based activity coefficients for complex j
m_j	= diluent coordination number in jth complex
n_j	= solute coordination number in jth complex
N_{Bo}	= a Bond number
N_{RE}	= $\rho_c U_T D / \mu_c$, a Reynolds number
N_{Sl}	= Af / U_T, a Strouhal number
N_{We}	= $DU_T^2 \rho_c / g_c \sigma$, a Weber number
P_i^{sat}	= pure component vapor pressure for species i, mm H_g
R	= universal gas constant, J/mol-K
T	= temperature, K
U_c	= the continuous phase superficial velocity in cm/s
U_d	= the dispersed phase superficial velocity in cm/s
U_T	= net superficial velocity of phases = $U_d + U_c$, cm/s
V_i	= molar liquid volume for species i assuming $V^E = 0$
$\chi_{i,s}$	= stoichiometric mole fraction of species i
z_i	= true mole fraction of species i
z_{i_1}	= true mole fraction of ith monomer
β_j	= empirical parameter
$\Delta\rho$	= $\lvert \rho_d - \rho_c \rvert$, g/ml
Γ_L	= suboptimal $HETS/D$ correlations for $(Af) < (Af)_{opt}$
Γ_U	= the superoptimal $HETS/D$ correlation for $(Af) > (Af)_{opt}$
γ_j	= true mole fraction-based activity coefficient for species j
$\gamma_{i,s}$	= stoichiometric activity coefficient for species i
μ_c	= viscosity of the continuous phase, cP
$\phi_{,i,s}$	= stoichiometric volume fraction of species i
ϕ_i	= volume fraction of species i if $V^E = 0$
ρ_e	= density of the continuous phase, g/ml
ρ_d	= density of the dispersed phase, g/ml
ρ_d	= interfacial tension, d/cm

REFERENCES

1. A. W. Francis, *Handbook for Components in Solvent Extraction*. New York: Gordon & Breach, Science Publishers, 1972.
2. E. J. Henley and J. D. Seader, *Equilibrium-Stage Separation Operations in Chemical Engineering*. New York: John Wiley & Sons, 1981.
3. J. T. Long, *Engineering for Nuclear Fuel Reprocessing*, 2nd ed. La Grange, IL: American Nuclear Society, 1978.
4. K. Schügerl, *Solvent Extraction in Biotechnology*. New York: Springer-Verlag, 1994.
5. M. Benedict, T. H. Pigford, and H. W. Levi, *Nuclear Chemical Engineering*, 2nd ed. New York: McGraw-Hill, 1981.
6. J. R. Williams and A. A. Clifford, eds., *Supercritical Fluid Methods and Protocols*, vol. 13 of *Methods in Biotechnology*. Totowa, NJ: Humana Press, 2000.
7. T. C. Lo, M. H. I. Baird, and C. Hanson, eds., *Handbook of Solvent Extraction*. New York: John Wiley & Sons, 1983.
8. R. B. Akell and C. J. King, eds., *New Developments in Liquid-Liquid Extractors: Selected Papers from ISEC '83*, vol. 80 (238) of *AIChE Symposium Series*. New York: American Institute of Chemical Engineers, 1984.

9. N. N. Kulov, ed., *Liquid-Liquid Systems*. Commack, NY: Nova Science Publishers, Inc., 1996.
10. R. Blumberg, ed., *Liquid-Liquid Extraction*. New York: Academic Press, 1988.
11. Y. Marcus. 1961. Extraction of tracer quantities of uranium (VI) from nitric acid by tri-*n*-butyl phosphate. *J. Phys. Chem.* 65: 1647–1648.
12. Y. Marcus and A. S. Kertes, *Ion Exchange and Solvent Extraction of Metal Complexes*. New York: Wiley Interscience, 1969.
13. R. E. Treybal, *Liquid Extraction*, 2nd ed. New York: McGraw-Hill, 1963.
14. W. Davis, Jr. October 1962. Thermodynamics of extraction of nitric acid by tri-*n*-butyl phosphate hydrocarbon diluent solutions: I. distribution studies with TBP in Amsco 125-82 at intermediate and low acidities. *Nuc. Sc. & Eng.* 14 (2): 159–168.
15. W. Davis, Jr. and H. J. De Bruin. 1964. New activity coefficients of 0–100% nitric acid. *J. Inorg. Nuc. Chem.* 26: 1069–1083.
16. W. Davis, Jr. October 1962. Thermodynamics of extraction of nitric acid by tri-*n*-butyl phosphate-hydrocarbon-diluent solutions: II. densities, molar volumes, and water solubilities of TBP-Amsco 125-82-nitric acid-water solutions. *Nuc. Sc. & Eng.* 14 (2): 169–173.
17. W. Davis, Jr. October 1962. Thermodynamics of extraction of nitric acid by tri-*n*-butyl phosphate hydrocarbon-diluent solutions: III. comparison of literature data. *Nucl. Sci. Eng.* 14 (2): 174–178.
18. W. Davis, Jr., P. S. Lawson, H. J. deBruin, and J. Mrochek. June 1965. Activities of the three components in the system water nitric acid-uranyl nitrate hexahydrate at 25. *J. of Phys. Chem.* 69 (6): 1904–1914.
19. W. Davis, Jr., J. Mrochek, and C. J. Hardy. 1966. The system: tri-*n*-butyl phosphate (TBP)-nitric acid-water I: activities of TBP in equilibrium with aqueous nitric acid and partial molar volumes of the three components in the TBP phase. *J. Inorg. Nucl. Chem.* 28: 2001–2014.
20. W. Davis, Jr. and J. Mrochek. 1967. Activities of tributyl phosphate in tributyl phosphate-uranyl nitrate-water solutions. *Proc. Int. Conf. Solvent Extraction Chemistry*, 283–295.
21. W. Davis, Jr., J. Mrochek, and R. R. Judkins. 1970. Thermodynamics of the two-phase water-uranyl nitrate-tributyl phosphate-Amsco 125-82 system. *J. Inorg. Nucl. Chem.* 32: 1689.
22. E. P. Horwitz, G. W. Mason, C. A. A. Bloomquist, R. A. Leonard, and G. J. Bernstein, The extraction of DBP and MBP from actinides: Application to the recovery of actinides from TBP-sodium carbonate solutions, in J. D. Navratil and W. W. Schulz, eds., *Actinide Separations*, vol. 117 of *ACS Symposium Series*. Washington, DC: American Chemical Society, 1980, 475–498.
23. J. Wisniak and A. Tamir, eds., *Liquid-Liquid Equilibrium and Extraction: A Literature Source Book*. New York: Elsevier, November 1980.
24. J. Wisniak and A. Tamir, eds., *Liquid-Liquid Equilibrium and Extraction: A Literature Source Book: part B*. New York: Elsevier, November 1981.
25. J. Wisniak and A. Tamir, eds., *Liquid-Liquid Equilibrium and Extraction: A Literature Source Book: suppl. 1*. New York: Elsevier, February 1985.
26. J. Wisniak and A. Tamir, eds., *Liquid-Liquid Equilibrium and Extraction: A Literature Source Book: suppl. 2*. New York: Elsevier, February 1987.
27. J. M. Sorensen and W. Arlt, eds., *Liquid-Liquid Equilibrium: Chemistry Data Series, vol. V: part 1, Binary Systems*. Frankfurt: DECHEMA, 1979.
28. J. M. Sorensen and W. Arlt, eds., *Liquid-Liquid Equilibrium: Chemistry Data Series, vol. V: part 2, Ternary Systems*. Frankfurt: DECHEMA, 1980.
29. J. M. Sorensen and W. Arlt, eds., *Liquid-Liquid Equilibrium: Chemistry Data Series, vol. V: part 3, Ternary and Quaternary Systems*. Frankfurt: DECHEMA, 1980.
30. E. A. Macedo and P. Rasmussen, eds., *Liquid-Liquid Equilibrium: Chemistry Data Series, vol. V: suppl. 1*. DECHEMA, Frankfurt, 1987.
31. D. Tiegs., *Activity Coefficients at Infinite Dilution, C1–C9, Chemistry Data Series, vol. IX, part 1*. Frankfurt: DECHEMA, 1986.
32. A. Fredenslund, J. Gmehling, and P. Rasmussen, *Vapor-Liquid Equilibrium using UNIFAC*. Amsterdam: Elsevier, 1977.
33. V. G. Yurkin, M. P. Shapovalov, S. V. Shepeleva, and A. M. Rozen. September 1985. Application of the UNIFAC model to extraction systems. Calculation of mutual phase solubility in trialky phosphate-water systems by the UNIFAC model. *Trans. Zh. Obsh. Khim.*, 55 (9): 1937–1943.

34. C. J. King, Acetic acid extraction, in T. C. Lo, M. H. I. Baird, and C. Hanson, eds., *Handbook of Solvent Extraction*. New York: John Wiley & Sons, Inc., 1983, 567–573.
35. Y. Marcus, Principles of solubility and solutions, in J. Rydberg, C. Musikas, and G. R. Choppin, eds., *Principles and Practices of Solvent Extraction*, chap. 2. New York: Marcel Dekker, 1992, 21–70.
36. W. E. Acree, *Thermodynamic Properties of Nonelectrolyte Solutions*. New York: Academic Press, 1984.
37. J. M. Prausnitz, R. N. Lichtenthaler, and E. G. de Avzevedo, *Molecular Thermodynamics of Fluid-Phase Equilibria*, 2nd ed. Englewood Cliffs, NJ: Prentice Hall, 1986.
38. F. Dolezalek. 1908. Zur Theorie der Binaren Gemische und Korzentrierten Loungen, *Z. Phys. Chem.* 64: 727.
39. Ioannis G. Economou and Marc D. Donohue. December 1991. Chemical, quasi-chemical and perturbation theories for associating fluids. *AIChE J.*, 37 (12): 1875–1894.
40. M. M. Abbott and H. C. Van Ness. September 15, 1992. Thermodynamics of solutions containing reactive species. A guide to fundamentals and applications. *Fluid Phase Equilib.* 77: 53–119.
41. H. R. Rabie and J. H. Vera. April 4, 1995. Chemical theory for ion distribution equilibria in reverse micellar systems. New experimental data for aerosol-OT-isooctane-water-salt systems. *Langmuir* 11: 1162–1169.
42. Scott W. Campbell. November 1994. Chemical theory for mixtures containing any number of alcohols. *Fluid Phase Equilib.* 102: 61–84.
43. E. A. Guggenheim, *Mixtures*. Oxford, UK: Clarendon, 1952.
44. G. M. Wilson. January 1964. Vapor-liquid equilibrium. XI. a new expression for the excess free energy of mixing. *J. Am. Chem. Soc.* 86: 127–130.
45. H. Renon and J. M. Prausnitz. 1968. Local compositions in thermodynamic excess functions for liquid mixtures. *AIChE J.* 14: 135.
46. D. S. Abrams and J. M. Prausnitz. 1975. Statistical thermodynamics of liquid mixtures: A new expression for the excess Gibbs energy of partly or completely miscible systems. *AIChE J.* 21 (1): 116–128.
47. J. H. Vera, S. G. Sayegh, and G. A. Ratcliff. 1977. A quasi lattice-local composition model for the excess Gibbs free energy of liquid mixtures. *Fluid Phase Equilib.* 1: 113.
48. C. Panayiotou and J. H. Vera. 1980. The quasi-chemical approach for nonrandomness in liquid mixtures: Expression for local surfaces and local compositions with an applications to polymer solutions. *Fluid Phase Equilib.* 5: 55.
49. C. Panayiotou and J. H. Vera. 1982. Statistical thermodynamics of γ-mer fluids and their mixtures. *Polymer J.* 14: 681.
50. S. K. Kumar, U. W. Suter, and R. C. Reid. 1987. A statistical mechanics based lattice model equation of state. *Ind. Eng. Chem. Res.* 26: 2532.
51. Gregory M. Martinez. August 1994. Lattice statistics for size, clustering and dissociation with example applications in vapor-liquid equilibria. *Chem. Eng. Sci.* 49 (15): 2423–2435.
52. S. G. Sayegh and J. H. Vera. 1980. Lattice-model expressions for the combinatorial entropy of liquid mixtures: A critical discussion. *Chem. Eng. J.* 19: 1.
53. W. W. Schulz and J. D. Navratil, eds., *Science and Technology of Tri-n-Butyl Phosphate*, vol. 3. Boca Raton, FL: CRC Press, 1986a.
54. K. Choi and D. W. Tedder. 1995. Nuclear magnetic resonances of TBP-diluent mixtures. *Spectrochimica Acta part A* 51: 2301–2305.
55. W. W. Schulz and J. D. Navratil, eds., *Science and Technology of Tri-n-Butyl Phosphate*, vol. 4. Boca Raton, FL: CRC Press, 1986b.
56. K. Choi, *Molecular Interactions in Polar Solvents*. Ph.D. thesis, Atlanta, GA: School of Mechanical Engineering, Georgia Institute of Technology, 1995.
57. J. H. Rytting, A. Goldkamp, and S. Lindenbaum. 1975. Heats of dilution of trialkyl phosphates in isooctane and carbon tetrachloride: Interpretation in terms of self-association. *J. Solution Chem.* 4: 1005.
58. C. R. Blaylock and D. W. Tedder. 1989. Competitive equilibria in the system: Water, nitric acid, tri-n-butyl phosphate and Amsco 125-82. *Sol. Ext. & Ion Exch.* 7 (2): 249–271.
59. A. M. Rozen, L. P. Khorhorina, V. G. Yurkin, and N. M. Novikova. 1963. The reaction of tributyl phosphate (TBP) and TBP-solvate with diluents. *Proc. Acad. Sci. USSR* (transl.) 153: 1387.
60. I. Prigogine and R. Defay, *Chemical Thermodynamics*. London: Longmans Green & Co., 1954.

61. D. W. Tedder, Water extraction, in W. W. Schulz, J. D. Navratil, and A. S. Kertes, eds., *CRC Science and Technology of Tributyl Phosphate: Extraction of Acids*, vol. 4, chap. 3. Boca Raton, FL: CRC Press, 1991, 35 –70.
62. K. Choi and D. W. Tedder. 1996. Molecular interactions in tri-*n*-butyl phosphate-diluent systems. *Ind. Eng. Chem. Res.* 35: 2048–2059.
63. K. Choi and D. W. Tedder. January 1996. Molecular interactions in chloroform diluent systems. *AIChE J.* 43 (1): 196–211.
64. D. F. Othmer, R. E. White, and E. Trueger. October 1941. Liquid-liquid extraction data. *Ind. Eng. Chem.* 23 (10): 1240–1248.
65. E. J. Henley and J. D. Seader, *Equilibrium-Stage Separation Operations in Chemical Engineering*. New York: John Wiley & Sons, 1981.
66. A. Kremser. 1930. *Natl. Petroleum News* 22 (21): 43–49.
67. J. C. Godfrey and M. J. Slater, eds., *Liquid-Liquid Extraction Equipment*. New York: John Wiley & Sons, 1994.
68. A. H. P. Skelland and D. W. Tedder, Extraction–organic chemicals processing, in R.W. Rousseau, ed., *Handbook of Separation Processing Technology*. New York: John Wiley & Sons, 1987.
69. W. Y. Tawfik, A. J. Eckles IV, and D. W. Tedder. 1988. Performance correlations for reciprocating plate extraction columns. *Sol. Ext. & Ion Exch.* 6 (4): 563–584.
70. S. W. Churchill, *The Interpretation and Use of Rate Data: the Rate Concept*. New York: McGraw–Hill, 1974.
71. K. M. Guthrie, Data and techniques for preliminary capital cost estimating, in *Modern Cost Engineering Techniques*. New York: McGraw-Hill, 1970,80 –106.
72. D. W. Tedder, *Preliminary Chemical Process Design and Economics*. Marietta, GA: Hickory Mountain, 2005.

11 Chemical Reaction Engineering

J. B. Joshi and L. K. Doraiswamy

CONTENTS

11.1 Scope and Format of the Chapter 739
    11.1.1 Nomenclature 741
11.2 Reaction and Reactor Fundamentals 741
    11.2.1 Scope 741
    11.2.2 Simple Reactions 741
    11.2.3 Complex (Multiple and Multistep) Reactions 742
    11.2.4 Stoichiometry 742
    11.2.5 Reaction Rates 743
    11.2.6 Extension to Complex (Multiple) Reactions 745
        11.2.6.1 Independent Reactions 749
        11.2.6.2 Reaction Coordinate 749
        11.2.6.3 Some Common Classes of Complex Reactions 749
    11.2.7 Ideal Reactors 750
        11.2.7.1 Batch Reactor 750
        11.2.7.2 Continuous Flow Reactors 751
        11.2.7.3 The Role of Mixing 752
11.3 Reactions with an Interface 753
    11.3.1 Gas-Solid Catalytic Reactions 753
        11.3.1.1 Catalysis by Solids 753
        11.3.1.2 Kinetics of Reactions on Solid Surfaces 756
        11.3.1.3 Role of Diffusion within the Pellet (Internal Diffusion) 759
        11.3.1.4 Role of External Diffusion 763
        11.3.1.5 Relative Roles of Internal and External Diffusion 764
        11.3.1.6 Effects of Various Factors on Catalyst Effectiveness 764
        11.3.1.7 Laboratory Reactors for Accurate Kinetic Data 765
    11.3.2 Gas-Solid Noncatalytic Reactions 770
        11.3.2.1 Modeling of Gas-Solid Reactions 770
        11.3.2.2 Extensions to the Basic Models 781
        11.3.2.3 Models that Account for Structural Variations 782
        11.3.2.4 A General Model That Can Be Reduced to Specific Ones 785
    11.3.3 Fluid-Fluid Reactions 785
        11.3.3.1 Theory of Mass Transfer Accompanied by Irreversible Chemical Reaction 786
        11.3.3.2 Laboratory Reactors for Fluid-Fluid Reactions 788
    11.3.4 Solid-Liquid Reactions 797
    11.3.5 Gas-Liquid-Solid Reactions 797

11.4　Classification of Reactors and Their Description 799
　11.4.1　Introduction 799
　11.4.2　Pressure Energy 800
　　11.4.2.1　Gas-Liquid Reactors 800
　　11.4.2.2　Liquid-Liquid Reactors 812
　　11.4.2.3　Solid-Liquid Reactors 812
　　11.4.2.4　Gas-Solid Catalytic Fixed-Bed Reactors 813
　　11.4.2.5　Gas-Solid Catalytic Fluidized-Bed Reactors 821
　　11.4.2.6　Gas-Solid Noncatalytic Reactors 835
　11.4.3　Kinetic Energy (Stirred Tank Reactors) 839
　　11.4.3.1　Introduction 839
　　11.4.3.2　Blending 841
　　11.4.3.3　Gas Dispersion 841
　　11.4.3.4　Solid Suspension 843
　　11.4.3.5　Heat Transfer 844
　　11.4.3.6　Blending in Gas-Liquid Systems 844
　　11.4.3.7　Blending in Solid-Liquid Systems 845
　　11.4.3.8　Suspension in Gas-Liquid-Solid Systems 845
　　11.4.3.9　Heat Transfer in Gas-Liquid Systems 846
　　11.4.3.10　Dead End Systems 846
　11.4.4　Potential Energy (Film Contactors) 849
　11.4.5　Comparison of Gas-Liquid Reactors 849
11.5　Case Studies 849
　11.5.1　Introduction 849
　Case Study 11.1　Homogeneous Liquid Phase Simple Reaction: $A \rightarrow R$ 852
　Case Study 11.2　Homogeneous Gas-Phase Complex Reaction: Oxidation of
　　　NO Using Ozone 869
　Case Study 11.3　Gas-Solid (Catalytic) Reaction: Modeling of a Complex
　　　Reaction on a Deactivating Catalyst (a Model Scheme) 870
　Case Study 11.4　Gas-Solid (Catalytic) Reaction in Fixed-Bed NINA and
　　　Adiabatic Reactors: Reduction of Nitrobenzene to Aniline 878
　Case Study 11.5　Comparison of Selected Fluid-Bed Models for the Catalytic
　　　Reduction (by Hydrogen) of Nitrobenzene to Aniline 883
　Case Study 11.6　Gas-Solid Noncatalytic Reaction: Development of a Solid
　　　Sorbent for Cleaning Coal Gas Followed by Modeling of the Reaction
　　　and a Conceptual Reactor Design 893
　Case Study 11.7　Gas-Liquid Reaction: Air Oxidation of Sodium Sulfide (a Simple
　　　Reaction with Typical Problems of Gas-Liquid Reactions) 900
　Case Study 11.8　Gas-Liquid Reaction: Absorption of NO_x Gases for the
　　　Manufacture of Nitric Acid (a System with Multiple Complexities) 917
　Case Study 11.9　Gas-Liquid-Solid (Noncatalytic) Reaction: Oxydesulfurization
　　　of Coal in a Slurry Reactor 919
　Case Study 11.10　Gas-Liquid-Solid (Noncatalytic) Reaction: Carbonation of Lime 925
　Case Study 11.11　Gas-Liquid-Solid (Catalytic) Reaction: Hydrogenation of an
　　　Organic Compound 934
　Case Study 11.12　Solid Reaction Followed by a Gas-Solid Reaction: Manufacture
　　　of Methylchlorosilanes 943
Postscript 954
Acknowledgments 956
Nomenclature 956
References 961

11.1 SCOPE AND FORMAT OF THE CHAPTER

Any chemical process can be characterized in terms of three major components: prereaction, reaction, and postreaction. Usually (but not necessarily) the first and last comprise only physical operations such as mixing, heating, vaporization, distillation, crystallization, extraction, absorption, adsorption, and so on. No chemical reaction is involved except in cases where physical separation is facilitated or enhanced by a deliberately imposed reaction (to be distinguished from the reaction of interest). The middle component constitutes the actual chemical reaction occurring in the process, i.e., the reaction of interest. The reaction can be of any type, involving one or more phases, and is performed in reactors of various designs. The design depends on the system used, such as gas, liquid, gas-liquid, liquid-liquid, gas-solid, gas-liquid-solid, and so forth. In the 1930s and 1940s, the analysis and design of reactors was usually referred to as *applied reaction kinetics*. Then, around 1960, the term *chemical reaction engineering* was used (by Octave Levenspiel) to describe this area of chemical engineering.

Chemical reaction engineering (CRE) has grown in over half a century as a highly sophisticated area of chemical engineering. The first attempts were centered round the development of common principles that cut across various industries. As chemical engineers applied these principles to other emerging areas such as biochemical processes, electrochemical processes, photochemical processes, sonochemical processes, and others, new theories had to be developed to accommodate the special needs of different areas. Thus a new trend became discernible where CRE retained its core value and structure but many specialized areas (e.g., biochemical reaction engineering, electrochemical reaction engineering, photochemical reaction engineering, and microreactor engineering) began to spin off with roots in CRE. This inclusiveness of CRE has been its hallmark over the last 30 years.

The present chapter applies the fundamental principles of CRE to specialized examples of practical interest. Many standard texts on CRE start with a detailed presentation of reactors for homogeneous reactions and then extend the treatment to multiphase reactors. In the present chapter, homogeneous reactors constitute only a small part of the treatment and are covered under one of several case studies included. The bulk of the treatment is devoted to multiphase reactions and reactor characteristics. A broad outline of the scope of the chapter is presented in Figure 11.1.

The following format is used. Important physicochemical aspects of CRE (commonly known as reaction analysis) are briefly outlined in Section 11.2, to provide the fundamental framework for the analysis and design of reactors involving one or more phases, referred to generally as single-phase or multiphase reactors, respectively. These include treatment of simple and complex (multiple) reactions, stoichiometry, reaction rates (with a brief reference to the concept of the rate-determining step), and experimental techniques.

For single-phase (homogeneous) reactions, there are no phase boundaries. Mass transfer limitations are not controlling except in very fast reactions. On the other hand, where two or more phases are present, mass or heat transfer across an interface may be the controlling step. Depending on the type of reaction system involved [e.g., gas-liquid, liquid-liquid, gas-solid (catalytic), gas-solid (noncatalytic), and gas-liquid-solid], specific methods of analysis have been developed. Obtaining reliable laboratory data for such heterogeneous reactions is more difficult than for homogeneous reactions. One simple way to collect intrinsic kinetic data, not falsified by transport disguises, is to use the so-called gradientless reactors. All these aspects of *reactions with an interface* are outlined in Section 11.3.

Then, in Section 11.4, a practically oriented survey is presented of several industrial rectors. The treatment begins with a classification of reactors based on the mode of energy input. A full quantitative treatment of this subject is called *reactor analysis* or *modeling*. Such an undertaking is complicated and is not germane to the present effort.

The more important classes of reactors are discussed in Section 11.5 as specific *case studies* of importance. Most case studies include a description of the theory involved, determination of the rate-controlling step, process design (including reactor modeling), estimation of design parameters, and

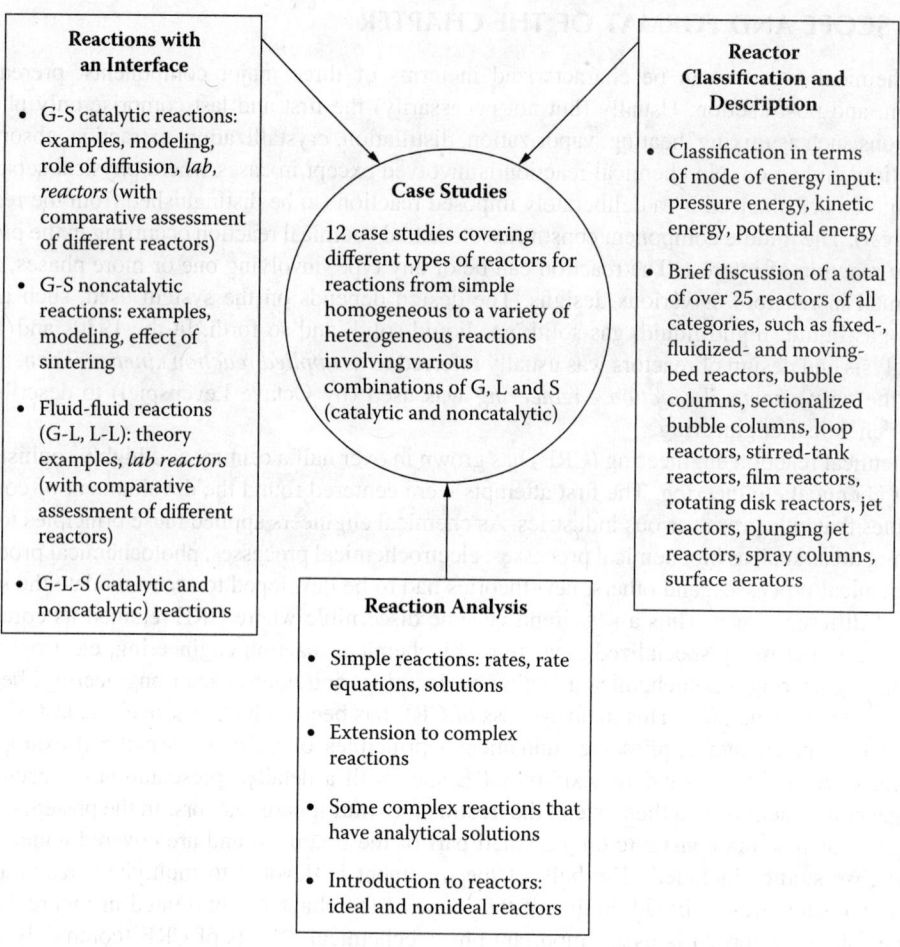

FIGURE 11.1 Scope and format of the chapter.

optimization. A special feature of the present chapter is its treatment of reactors in terms of rigorous case studies following a qualitative description of numerous reactors commonly used in industry.

Sometimes reaction rates can be enhanced by using multifunctional reactors, i.e., reactors in which more than one function (or operation) can be performed. Examples of reactors with such multifunctional capability, or *combo* reactors, are distillation column reactors in which one of the products of a reversible reaction is continuously removed by distillation thus driving the reaction forward; extractive reaction; biphasing; membrane reactors in which separation is accomplished by using a reactor with membrane walls; and simulated moving-bed (SMB) reactors in which reaction is combined with adsorption. Typical industrial applications of multifunctional reactors are esterification of acetic acid to methyl acetate in a distillation column reactor, synthesis of methyl-*ter*-butyl ether (MTBE) in a similar reactor, vitamin K synthesis in a membrane reactor, oxidative coupling of methane to produce ethane and ethylene in a similar reactor, and esterification of acetic acid to ethyl acetate in an SMB reactor. These specialized reactors are increasingly used in industry, mainly because of the obvious reduction in the number of equipment. These reactors are considered by Fair in Chapter 12.

Aspects of CRE not covered in this chapter include computational fluid dynamics (CFD). Total integration of CRE and CFD will generate much more reliable and theoretically sound procedures for reactor design but may not happen for the next 15 to 20 years. Meanwhile, incremental infusions of CFD in CRE will add increasing rigor to existing design procedures and methods for combining

ical Reaction Engineering

complex chemistry with reactor design. Process control, stability, unsteady-state analysis of reactors, and estimation procedures for design parameters are not covered in this chapter. With continuously increasing interaction of CRE with CFD, the present empirical methods of estimation will almost certainly be replaced by more theoretical ones.

11.1.1 Nomenclature

In view of the special format of this chapter, and the need to use a number of symbols (both regular and Greek) and subscripts, especially for the case studies, the following lists have been prepared: (1) a common list for Sections 1 through 4 and (2) individual lists for the case studies. Many items have been defined only in terms of the main symbols, with more detailed specifications through appropriate subscripts listed under the heading *Subscripts*. However, in the interest of clarity, subscripts have also been used in the main lists where necessary.

11.2 REACTION AND REACTOR FUNDAMENTALS

11.2.1 Scope

The design of a chemical reactor always starts with the analysis of the reaction involved, identification of the best system for its commercial operation [homogeneous, gas-solid (catalytic), gas-liquid, gas-liquid-solid (slurry), and so on], and design of reactor for the chosen system. Note that the system is often pre-fixed by the nature of the reaction itself, but it may be possible to alter the original system in several ways, e.g., intentional addition of a liquid phase, use of an immobilized homogeneous catalyst in place of its more common soluble analog, use of a trickle-bed reactor instead of the slurry reactor in which the laboratory experiments might have been performed, and so forth. Depending on the system, the definition of the reaction rate and the units of rate constants will vary. The reaction itself may be simple or complex, and mathematical tools for handling these are necessary. These various aspects of reactions are briefly outlined in this section.

A variety of reactors are used, particularly for multiphase reactions. The more important aspects of these *reactions* are covered in Section 11.3, while Section 11.4 discusses multiphase *reactors*. For an appreciation of these reactors, an understanding of the fundamental types of reactors, called *ideal reactors,* is necessary. Thus, we end this section with a brief presentation of these reactors along with an introduction to the concept of nonideal flow—a feature of many industrial reactors.

11.2.2 Simple Reactions

A simple reaction such as

$$A \to R + S \qquad \text{R11.1}$$

is usually one in which only a single step is involved, defined by a single rate equation. This reaction may, however, actually proceed through several elementary steps such as

$$A + A \to A^* + A \qquad \text{R11.1a}$$

$$A^* + A \to A + A \qquad \text{R11.1b}$$

$$A^* \to R + S \qquad \text{R11.1c}$$

In this set of elementary steps, the intermediate (A^*) is transient and not experimentally detectable. Thus, whereas this mechanistic representation is useful in understanding the nature of the reaction, from the engineering point of view, the overall "single step" represented by R11.1 is adequate. Its mechanistic evolution is largely irrelevant. The rate at which this reaction proceeds is given by

$$rate = (constant)\,[A] \tag{11.1}$$

The concentration of A is raised to a power equal to the stoichiometric coefficient of A in the reaction (unity in this case). We refer to such reactions as *stoichiometric* reactions. If experimental data conform to the empirical form

$$rate = (constant)\,[A]^n \tag{11.2}$$

where n is some constant not equal to the stoichiometric coefficient, then the reaction is considered to be nonstoichiometric. The exponent on the concentration term, whether stoichiometric or nonstoichiometric, is called the *reaction order*. The proportionality constant is typically called the *rate constant*. However, because it is a function of temperature and pressure, the term "constant" is misleading and "coefficient" is a more correct choice (like, for example, mass or heat transfer coefficient and diffusion coefficient). However, "rate constant" is accepted.

11.2.3 Complex (Multiple and Multistep) Reactions

Let us now consider the following reaction in which several ethanolamines (mono-, di-, and tri-) are formed and referred to as a *multiple reaction:*

$$NH_3 + H_2C\underset{O}{-}CH_2 \rightarrow NH_2-(CH_2CH_2OH)$$

$$NH_2(CH_2CH_2OH) + H_2C\underset{O}{-}CH_2 \rightarrow NH=(CH_2CH_2OH)_2 \qquad R11.2$$

$$NH=(CH_2CH_2OH)_2 + H_2C\underset{O}{-}CH_2 \rightarrow N\equiv(CH_2CH_2OH)_3$$

These intermediates are experimentally detectable. The literature refers to such a reaction as a *multistep reaction*. We prefer the term "multiple" and reserve "multistep" to a reaction scheme in which a number of steps, each requiring a different set of operating conditions, are employed to synthesize a compound starting from a given set of raw materials. Thus, the following scheme for synthesizing benzaldehyde ($PhCHO$) starting from toluene ($PhCH_3$) is a *multistep reaction:*

$$PhCH_3 + Cl_2 \rightarrow PhCH_2Cl + HCl$$

$$PhCH_2Cl + NaOH \rightarrow PhCH_2OH + NaCl \qquad R11.3$$

$$4PhCH_2OH + 2HNO_3 \rightarrow 4PhCHO + N_2O + 5H_2O$$

In this chapter, both simple and multiple reactions are considered.

11.2.4 Stoichiometry

Reaction rates (as we shall see later) are always expressed in terms of concentrations of the reacting species or their partial pressures. It is usually convenient to express the concentrations of the various

Chemical Reaction Engineering

components in terms of the conversion of the key component and the initial concentrations of the others. This is done by simple stoichiometric considerations. For the general reaction

$$v_A A + v_B B \rightarrow v_R R + v_s S \quad \text{R11.4}$$

the concentration equations in terms of the conversion X_A are listed in Table 11.1 along with equations for other important terms. Since there can be a change in volume due to a change in number of moles, temperature, pressure, or combinations thereof, the equations also include the effects of these factors.

11.2.5 Reaction Rates

The term *reaction rate* denotes the rate of a reaction in a general way. We now define it specifically as the moles reacting per unit of *reaction space* per unit time. The reaction space can be the empty volume of the reacting space, void volume of the catalyst bed, volume of the catalyst, weight of the catalyst, surface area of the catalyst, or volume of one or all of the phases present in a heterogeneous system. Where time is concerned, its definition depends on the type of reactor used. There are essentially two broad classes of reactors: batch and flow. In a batch reactor, time (t) corresponds to the elapsed time from the start of a run. In a flow reactor, it is given by

$$\bar{t} = \frac{V}{Q_0} = \frac{L}{u} \quad (11.3)$$

that is, the ratio of reactor volume to flow rate (or reactor length to fluid velocity). For a reaction with no volume change, this is also referred to as the *residence time*. The various definitions of the rate and their interconversions are summarized in Table 11.2. A negative sign before the rate represents reactant disappearance, and a positive sign represents product formation.

Equation (11.2) can be expressed as follows:

$$Rate = (kinetic\ term)(potential\ term) \quad (11.4)$$

where the kinetic term is the rate constant representing both the thermal and kinetic effects, and the potential term is the concentration representing the driving force. This form is the well-known power law equation. For reactions involving catalysts, a third term (the adsorption term) is also needed to account for surface catalysis. The exponent on the potential term represents the reaction order. The main feature of the kinetic term k is its temperature dependence, which is normally given by the well-known Arrhenius equation,

$$k = k^\circ e^{-E/R_g T} \quad (11.5)$$

where k° and E are the frequency factor and activation energy, respectively, of the reaction. A slightly more complex form given by

$$k = k^\circ T^m e^{-E/R_g T} \quad (11.6)$$

has also been proposed but is almost never warranted. Several theories for explaining the rate constant have been proposed (see, e.g., Pilling and Seakins, 1995), but they are not considered here.

There are essentially two methods of determining the kinetic parameters k and n: differential and integral. In the differential method, [A] is plotted as a function of time for a batch reactor or

TABLE 11.1
Concentration Equations in Terms of X_A for the General Reaction

$$v_A A + v_B B \rightarrow v_R R + v_s S$$

$$v_A A + v_B B \rightarrow v_R R + v_s S$$

i = B, R, or S
(+ for product, − for reactant)

Batch

$$[i] = \frac{N_i}{V}$$

$$V = V_0 \left(\frac{N_T}{N_{T0}}\right)\left(\frac{P_0}{P}\right)\left(\frac{T}{T_0}\right)$$

$$= V_0 (1 + \varepsilon_A X_A)\left(\frac{P_0}{P}\right)\left(\frac{T}{T_0}\right)$$

$$= V_0 \text{ for constant volume}$$

Flow

$$[i] = \frac{F_i}{Q}$$

$$Q = Q_0 \left(\frac{F_T}{F_{T0}}\right)\left(\frac{P_0}{P}\right)\left(\frac{T}{T_0}\right)$$

$$= Q_0 (1 + \varepsilon_A X_A)\left(\frac{P_0}{P}\right)\left(\frac{T}{T_0}\right)$$

$$= Q_0 \text{ for constant volume}$$

$$\delta_A = \frac{v_S + v_R - v_B - v_A}{v_A}$$

$$\varepsilon_A = y_{A0}\delta_A = \frac{V_{X_A=1} - V_{X_A=0}}{V_{X_A=0}}$$

$$[i] = \frac{y_{i0} p_0}{R_g T_0}, \quad [A] = \frac{y_{A0} p_0}{R_g T_0}$$

$$[A] = \frac{[A]_0 (1 - X_A)}{1 + \varepsilon_A X_A}$$

$$[i] \left\{ \frac{[A]_0 \left(\psi_i \pm \frac{v_i}{v_A} X_A\right)}{1 + \varepsilon_A X_A} \right\} \left(\frac{P}{P_0}\right)\left(\frac{T_0}{T}\right)$$

$$\psi'_i = \frac{[i]_0}{[A]_0} = \frac{F_{i0}}{F_{A0}} = \frac{y_{i0}}{y_{A0}}$$

Note: When there is no volume change, $\varepsilon_A = 0$ and when the system is isothermal and isobaric, $P = P_0$ and $T = T_0$.

TABLE 11.2
Different Definitions of the Reaction Rate

$$\text{Rate} = -\frac{1}{M}\frac{dN_A}{dt} = k[A]^n$$

M	Rate, r			Rate constant, k	
	Symbol	Units		Symbol	Units
Reactor volume	r_A, r_{VA}	mol/m³ reactor s		k_V	(m³ reactor/mole)$^{n-1}$ (1/s)
catalyst volume	r_{vA}	mol/m³ cat s		k_v	(m³ reactor/mol)^n (mol/m³ cat) (1/s)
catalyst weight	r_{wA}	mol/kg cat s		k_w	(m³ reactor/mol)^n (mol/kg cat) (1/s)
catalyst surface	r_{SA}	mol/m² cat s		k_S, k'	(mol/m² cat) (m³ reactor/mol)^n (1/s)
interfacial area	r'_A	mol/m² interface s		k'	(m³/mol)^n (mol/m²) (1/s)
Interconversions:	$k_V = (1 - \varepsilon_b)\, k_v$			ε_b = bulk voidage	
	$k_w = k_v/\rho_c$			ρ_c = density of catalyst, kg/m³	
	$k_V = k'a$			a = interfacial area, m²/m³	

V/Q_0 for a flow reactor, and the rate is determined by differentiating the curve either graphically or analytically. In the integral method, however, different forms of the rate equation are assumed, and the integrated forms are verified against experimental data. Integrated forms for several types of reactions and rate forms are summarized in Table 11.3, and procedures for determining the rate parameters by both differential and integral methods are illustrated Figure 11.2.

A useful method of obtaining the kinetic parameters for a nonunimolecular reaction such as

$$A + B \rightarrow R + S \qquad \text{R11.5}$$

is noteworthy. Let the rate equation be given by

$$-r_A = k_2[A]^m[B]^n \tag{11.7}$$

where k_2 is the second-order rate constant, m and n are the orders with respect to A and B, respectively, and the overall order is $(m + n)$. The reaction will be pseudo-mth order if B is in excess and pseudo-nth order if A is in excess. In fact, a simple experimental way of determining the orders is to first use A in excess (thus ensuring practically no change in its concentration) and determine n, and then use B in excess and determine m.

Another practical feature of reaction analysis is that the rate equation can be written in terms of concentrations as in the equations presented above or in terms of partial pressures. The relationship between the two forms is explained in Table 11.4. It is important to distinguish between the various units and use the correct units in any formulation.

11.2.6 Extension to Complex (Multiple) Reactions

A relatively common industrial situation is one in which a number of components react with one another in more than one reaction, resulting in a complex reaction network. Mathematically, a complex system consisting of N components and M reactions can be represented as

$$\sum_{i=1}^{N} v_{ij} r_j = 0, \qquad j = 1, 2, 3, \ldots M \tag{11.8}$$

where v_{ij} is the stoichiometric coefficient of A_i in the jth reaction. The rate equation for this system may be expressed concisely as

TABLE 11.3
Analytical Solutions (Design Equations) for Simple Reactions in a Batch Reactor (Also Valid for PFR with t Replaced by $\bar{t}$)

Reaction	Rate Equation	Analytical Solution[a]
1. $A \to R$	$-r_A = k[A]$	$kt = \ln\left(\dfrac{1}{1-X_A}\right)$, $\dfrac{[A]}{[A]_0} = e^{-kt}$
2. $2A \to R$	$-r_A = k[A]^2$	$k[A]_0 t = \left[\dfrac{X_A}{1-X_A}\right]$, $\dfrac{[A]}{[A]_0} = \dfrac{1}{1+k[A]_0 t}$
3. $3A \to R$	$-r_A = k[A]^3$	$k[A]_0^2 t = \dfrac{1}{2}\left[\dfrac{1}{(1-X_A)^2} - 1\right]$, $2kt = \dfrac{1}{[A]^2} - \dfrac{1}{[A]_0^2}$
4. $A \to R$	$-r_A = k[A]^n$	$k[A]_0^{n-1} t = \dfrac{1}{(n-1)}\left[(1-X_A)^{1-n} - 1\right]$, $n \neq 1$
5. $A + B \to R$ $\psi_B = 1$	$-r_A = k[A][B]$	$k[A]_0 t = \left[\dfrac{X_A}{(1-X_A)}\right]$, $t = \dfrac{1}{k}\left(\dfrac{1}{[A]} - \dfrac{1}{[A]_0}\right)$
6. $A + B \to R$ $\psi_B \neq 1$	$-r_A = k[A][B]$	$k[A]_0 t = \dfrac{1}{(\psi_B - 1)} \ln\left[\dfrac{\psi_B - X_A}{\psi_B(1-X_A)}\right]$
7. $\upsilon_A A + \upsilon_B B \to R$ $\psi_B = \nu_B/\nu_A$	$-r_A = k[A][B]$	$k[A]_0 t = \dfrac{1}{\psi_B}\left[\dfrac{X_A}{1-X_A}\right]$
8. $A + 2B \to R$ $\psi_B = 2$	$-r_A = k[A][B]^2$	$k[A]_0^2 t = \dfrac{1}{8}\left[\dfrac{1}{(1-X_A)^2} - 1\right]$
9. $A + B \to R$ $\psi_B = 1$	$-r_A = k[A][B]^2$	$k[A]_0^2 t = \dfrac{1}{2}\left[\dfrac{1}{(1-X_A)^2} - 1\right]$
10. $A + 2B \to R$ $\psi_B \neq 2$	$-r_A = k[A][B]^2$	$k[A]_0^2 t = \dfrac{1}{(2-\psi_B)^2}\left[\ln\dfrac{\psi_B - 2X_A}{\psi_B(1-X_A)} + \dfrac{2X_A(2-\psi_B)}{\psi_B(\psi_B - 2X_A)}\right]$
11. $\upsilon_A A + \upsilon_B B \to R$ $\psi_B \neq \upsilon_B/\upsilon_A$	$-r_A = k[A][B]$	$k[A]_0 t = \dfrac{1}{\left[\psi_B - \left(\dfrac{\upsilon_B}{\upsilon_A}\right)\right]} \ln\left[\dfrac{\psi_B - \left(\dfrac{\upsilon_B}{\upsilon_A}\right)X_A}{\psi_B(1-X_A)}\right]$
12. $\upsilon_A A + \upsilon_B B \to R$ $\psi_B = \upsilon_B/\upsilon_A$	$-r_A = k[A]^n[B]^m$	$k[A_0]^{m+n-1} t = \dfrac{1}{\psi_B(m+n-1)}\left[\dfrac{1}{(1-X_A)^{m+n-1}} - 1\right]$

[a] LHS = $k[A_0]^{n-1} t$, where k has the units of an nth [or $(m+n)$]-order reaction, $\left(\dfrac{m^3}{mol}\right)^{n-1} \dfrac{1}{s}$; $\psi_B = \dfrac{[B]_0}{[A]_0}$.

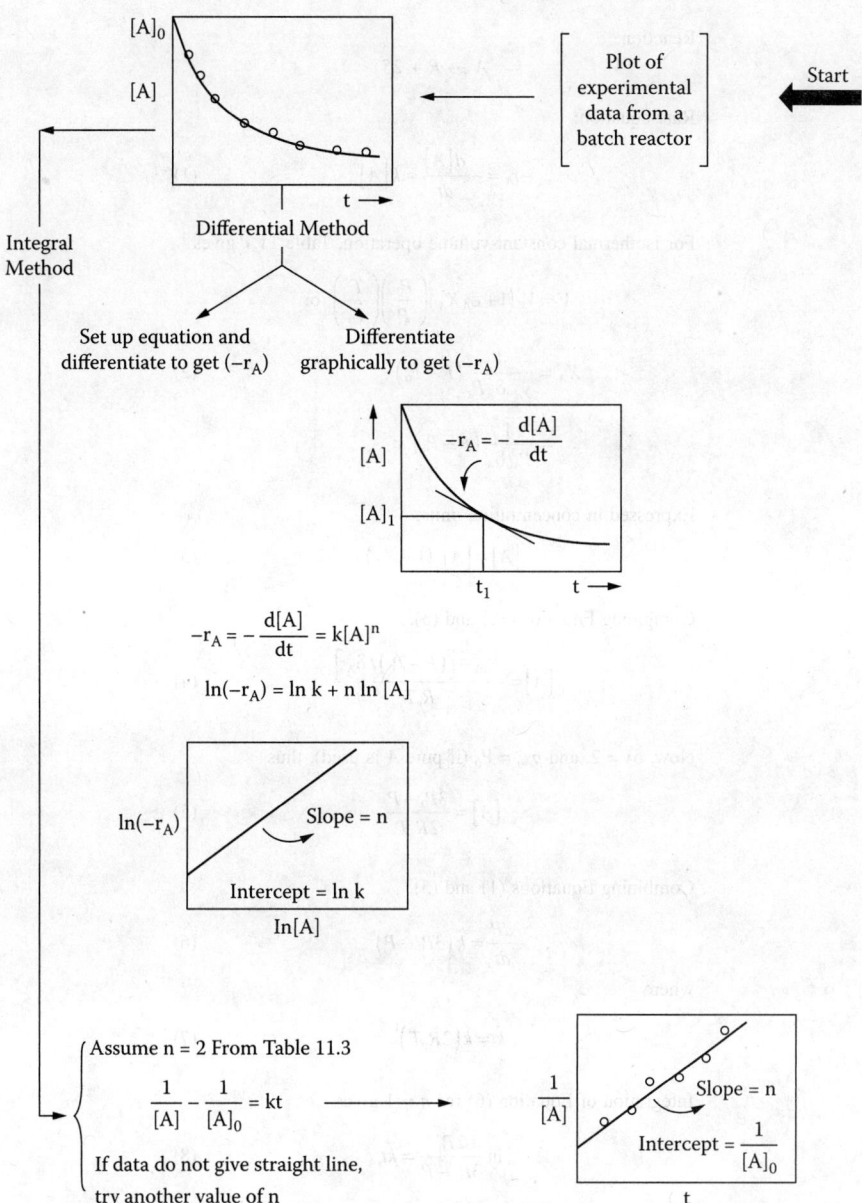

FIGURE 11.2 Determination of reaction rate parameters from differential and integral reactor data.

TABLE 11.4
Rate Equation in Terms of Partial Pressures for an Illustrative Reaction $A \to R + 2S$, and Its Use in Extracting Kinetic Parameters

Reaction:[a]

$$A \to R + 2S$$

Rate equation:

$$-r_A = -\frac{d[A]}{dt} = k[A]^n \quad (1)$$

For isothermal constant-volume operation, Table 11.1 gives

$$V = V_0(1+\varepsilon_A X_A)\left(\frac{P_0}{P}\right)\left(\frac{T}{T_0}\right), \text{ or}$$

$$X_A = \frac{1}{y_{A0}\delta_A P_0}(P-P_0) \quad (2)$$

$$= \frac{1}{P_{A0}\delta_A}(P-P_0)$$

Expressed in concentration units,

$$[A] = [A]_0 (1-X_A) \quad (3)$$

Combining Equations (2) and (3),

$$[A] = \frac{P_{A0} - [(P-P_0)/\delta_A]}{R_g T} \quad (4)$$

Now, $\delta_A = 2$, and $p_{A0} = P_0$ (if pure A is used), thus

$$[A] = \frac{3P_0 - P}{2R_g T} \quad (5)$$

Combining Equations (1) and (5),

$$\frac{dP}{dt} = \breve{k}(3P_0 - P)^n \quad (6)$$

where

$$\breve{k} = k(2R_g T)^{1-n} \quad (7)$$

Integration of Equation (6) for $n = 1$ gives

$$\ln\frac{2P_0}{3P_0 - P} = \breve{k}t \quad (8)$$

Thus, if the reaction is first order, either the differential method [Equation (6)] or the integral method [Equation (8)] can be used. If n – 1, only the differential method can be used.

[a] Classical example: gas-phase decomposition of di-t-butyl peroxide (Peters and Skorpinski, 1965; see also Fogler, 1999; Doraiswamy, 2001).

Chemical Reaction Engineering

$$\frac{d(\mathbf{c}V)}{dt} = \mathbf{\nu r} \tag{11.9}$$

where **c** is a vector ($N \times 1$ matrix) of component concentrations, $\mathbf{\nu}$ is an ($N \times M$) matrix of stoichiometric coefficients, and **r** is a vector ($M \times 1$ matrix) of reaction rates.

11.2.6.1 Independent Reactions

The reaction network as represented by Equation (11.9) refers only to those among all possible reactions that may be regarded as independent. A set of reactions is said to be independent if no reaction from the set can be obtained by algebraic additions of other reactions (as such or in multiples thereof), and each member contains one new species exclusively. For any given set of reactions (including those that are not independent), the number of independent reactions is the rank of the reaction matrix (see Aris 1965; 1969; Nauman 1987; Doraiswamy 2001).

11.2.6.2 Reaction Coordinate

Where complex reactions are concerned, conversion of a reactant is much less important than the actual moles converted in each step. This has the advantage of reducing the number of equations to be solved. Thus, for the reaction

$$A + 2B \rightarrow 2R \qquad \text{R11.6a}$$

$$2C + 3D \rightarrow S \qquad \text{R11.6b}$$

we can write

$$-\frac{N_A - N_{A0}}{1} = -\frac{N_B - N_{B0}}{2} = \frac{N_R - N_{R0}}{2} = \xi_1 \tag{11.10}$$

$$-\frac{N_C - N_{C0}}{2} = -\frac{N_D - N_{D0}}{3} = \frac{N_S - N_{S0}}{1} = \xi_2 \tag{11.11}$$

where ξ_1 and ξ_2 are called the *reaction coordinates* for reactions R11.6a and b, respectively. The compositions of all of the components are written in terms of these coordinates. Then, the following facts emerge: the number of equations is six if written in terms of the rates of disappearance/formation of the individual components, and two if written in terms of the reaction coordinates. This reduction in number is particularly important in networks involving numerous reactions (in the hundreds in petroleum cracking).

11.2.6.3 Some Common Classes of Complex Reactions

Network analysis can be simplified for well known classes of complex reactions such as parallel reactions, series (or consecutive) reactions, independent reactions, series-parallel reactions, triangular reactions, and the so-called Denbigh reaction. These reactions may be represented as follows:

1. Parallel
$A \nearrow^{R}_{\searrow S}$

2. Series
$A \to R \to S$

3. Independent
$A \to R$
$B \to S$

4. Series-Parallel
$A + B \to R$
$B + R \to S$

5. Triangular
(triangle: A, R, S)

6. Denbigh
$A \to R \to S$ with $\to T$, $\to U$

R11.7

Analytical solutions can often be found (see, e.g., Levenspiel 1999; Doraiswamy 2001). These are for batch reactors (also applicable to plug-flow reactors). Solutions for mixed-flow reactors (see Section 11.2.7.2) will be different.

11.2.7 IDEAL REACTORS

11.2.7.1 Batch Reactor

The material balance for any reactor can be written as

$$\text{Input} - \text{output} = \text{accumulation} + \text{disappearance (or formation)} \qquad (11.12)$$

For a batch reactor, the input and output terms are absent, and we have for reactant A

$$-r_A = \frac{dN_A}{dt} \frac{1}{V} \qquad (11.13)$$

Integration with no volume change, with $N_A = N_{A0}(1 - X_A)$, gives the following:

nth order

$$k[A]_0^2 t = \frac{1}{n-1}(1 - X_A)^{1-n}, \; n \neq 1 \qquad (11.14)$$

First order

$$kt = \ln\left(\frac{1}{1 - X_A}\right) \qquad (11.15)$$

The corresponding general equation for a reaction with volume change is

$$t = [A]_0 \int_0^{X_A} \frac{dX_A}{(-r_A)(1 + \varepsilon_A X_A)} \qquad (11.16)$$

Chemical Reaction Engineering

11.2.7.2 Continuous Flow Reactors

A reactor is now considered through which A is continuously passed and the products continuously withdrawn with no accumulation. The simplest case is the plug-flow reactor (PFR) in which elements of the fluid move unidirectionally as plugs, with no feedback from the downstream to the upstream side. Since the composition changes with distance, the material balance will hold only for a finite volume element dV with a conversion of dX_A and gives

$$F_{A0}dX_A = (-r_A)dV \tag{11.17}$$

On integration and assuming no volume change, this gives

$$\frac{V}{F_{A0}} = \int_0^{X_A} \frac{dX_A}{(-r_A)} \tag{11.18}$$

or in terms of residence time as

$$\bar{t} = \frac{V[A]_0}{F_{A0}} = [A]_0 \int_0^{X_A} \frac{dX_A}{(-r_A)} \tag{11.19}$$

Integration of this equation leads to Equation (11.14) for an nth order reaction and Equation (11.15) for a first-order reaction, with $\bar{t}$ replacing t. With volume change, the corresponding equation is (11.16). Thus the batch and plug-flow reactors are exactly identical.

A mixed-flow reactor (MFR), also known as the continuous stirred tank reactor (CSTR), is fully mixed at the molecular level, and the composition of the exiting stream is identical to that within the reactor. In this case, the material balance of Equation (11.12) is applied for the entire reactor and not just for a differential element as in a PFR. No integration is needed; the equation becomes

$$Q_0[A]_0 - (-r_{Ae})V = Q_e[A]_e \tag{11.20}$$

where $(-r_{Ae})$ is the final (exit) rate. Assuming $Q_o = Q$, we get

$$\tau = \frac{[A]_0 X_{Ae}}{-r_{Ae}} \tag{11.21}$$

that for a first-order reaction becomes

$$X_{Ae} = \frac{k\bar{t}}{1 + k\bar{t}} \tag{11.22}$$

The flows in PFR and MFR can be precisely defined by simple mathematical equations, and the batch reactor is simply the batch version of the PFR. A reactor is now considered where the flow is between plug and fully mixed, i.e., a nonideal reactor. Two common examples of such partially mixed reactors are the recycle reactor and the tanks-in-series reactor. In the recycle reactor, part of the outlet from a reactor is recycled at the inlet, thus establishing some mixing between the downstream and the upstream fluids. In the tanks-in-series reactor, several mixed-flow reactors are operated in series. A single MFR is fully mixed, whereas an infinite number of MFRs (or a sufficiently large number) is identical to plug flow. The degree of mixing in the two partially mixed reactors can be controlled by the recycle ratio (R) in one case and the number of tanks (N) in the other. Yet another nonideal reactor is the variable volume reactor, in which only a fraction of the batch reactor contents is emptied. In other words, V^* (= V/V_M, where V_M is the maximum volume) is fixed at a value between 1 (when the reactor functions as an MFR) and 0 (when it functions as a batch reactor). The relationship between the principal parameters of the three reactors is given by

$$V^* = \frac{R}{1+R} = \frac{1}{N} \tag{11.23}$$

11.2.7.3 The Role of Mixing

The concept of mixing is inherent in the recycle and tanks-in-series reactors, since deviations from plug or fully mixed flow are expressed in terms of R and N. Rigorous accounting of mixing must, however, be based on more fundamental concepts of mixing. In many situations, not all elements of a fluid flowing through a reactor leave the reactor at the same time. That is, the residence times for different elements vary. The residence time distribution (RTD) is denoted by a curve that represents the amount of fluid with times between t and $t + dt$ flowing out in the exit stream. When normalized with respect to the total flow (i.e., expressed as fraction of the total flow), this RTD satisfies the condition

$$\int_0^\infty \delta(t)dt = 0 \tag{11.24}$$

where $\delta(t)$ is the distribution function. The mean residence time is then given by

$$\bar{t} = \int_0^\infty t\delta(t)dt = 0 \tag{11.25}$$

For the two ideal cases referred to earlier, those corresponding to plug flow and fully mixed flow, the distribution functions take the following forms:

Plug flow (Dirac delta function):

$$\delta(t) = g(\bar{t}) \tag{11.26}$$

which shows $\delta(t) = 0$ at all times except at $t = \bar{t}$.

Mixed flow (exponential distribution):

$$\delta(t) = \frac{1}{\bar{t}} \exp\left(-\frac{t}{\bar{t}}\right) \tag{11.27}$$

The two distributions are explained in Figure 11.3.

In the equations presented above, the term "fully mixed" represents mixing at the molecular level. There is another mixing condition in which clumps or aggregates of molecules enter the reactor and flow through it without interacting with one another. Within each clump, however, there is complete mixing of molecules, and the residence time of each molecule is the same as that of

Chemical Reaction Engineering

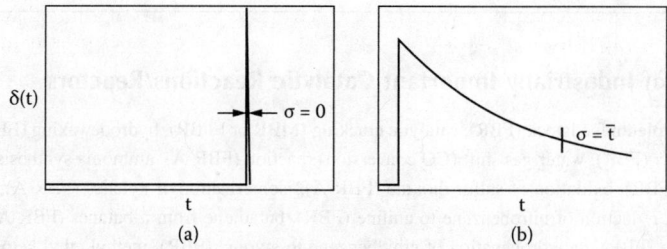

FIGURE 11.3 Plots showing Dirac delta and exponential distributions (for plug-flow and mixed-flow reactors, respectively).

the clump itself. The clumps are fully separated from one another, and this flow is known as *segregated flow*. Assuming that each clump behaves as a batch reactor and that the clumps are exponentially distributed, the following expression for the exit reactant concentration as a function of time can be derived for an *n*th-order reaction:

$$\left(\frac{[\bar{A}]}{[A]_0}\right)_e = \frac{1}{\bar{t}} \int_0^\infty [1 + (n-1)[A]_0^{n-1} kt]^{1/1-n} e^{-t/\bar{t}} dt \tag{11.28}$$

11.3 REACTIONS WITH AN INTERFACE

Many industrially important reactions are characterized by an interface across which heat/mass transfer occurs. They are called *heterogeneous reactions*. They include fluid-fluid reactions, gas-solid catalytic reactions, gas-solid noncatalytic reactions, and solid-solid reactions.

11.3.1 GAS-SOLID CATALYTIC REACTIONS

Over 70% of the chemicals produced today use a catalyst. The total annual purchasing price of catalysts amounts to about $10 billion, and each dollar's worth of catalyst often produces $100 to $1000 worth of products.

Catalysts are generally insoluble solids (heterogeneous catalysts, including bound catalysts), but they can also be soluble (homogeneous catalysts). Table 11.5 lists a few processes employing solid catalysts. The mechanisms for homogeneous catalysts are quite well understood today, and consequently the design of a homogeneous catalyst from first principles is possible. The advantages of homogeneous catalysts can often be transmitted to their heterogeneous counterparts (with an interface), e.g., immobilize a homogeneous catalyst on a solid support. We can realize the advantages of a solid catalyst (e.g., dispensing with the need to recover and purify the catalyst and permitting the use of the popular packed-bed reactor) and of homogeneous catalysts (e.g., their enzyme-like selectivity and greater amenability to design from first principles). Such immobilized catalysts are also discussed here.

11.3.1.1 Catalysis by Solids

Solid catalysts are widely used such as in the sulfuric acid, ammonia, fertilizer, petroleum, and petrochemical industries. The trend is to produce more organic chemicals by catalytic processes (usually employing solid catalysts). These catalysts can be classified in two ways.

1. Those that represent categories that cut across different types of reactions, such as zeolites.
2. Those that are specific to different classes of reactions, such as hydrogenation, oxidation, isomerization, and so on.

TABLE 11.5
Some Examples of Industrially Important Catalytic Reactions/Reactors

Desulfurization of petroleum feedstock (FBR), catalytic cracking (MBR or Fl BR), hydrodewaxing (FBR), steam reforming of methane or naphtha (FBR), water-gas shift (CO conversion) reaction (FBR-A), ammonia synthesis (FBR-A), methanol from synthesis gas (FBR), oxidation of sulfur dioxide (FBR-A), isomerization of xylenes (FBR-A), catalytic reforming of naphtha (FBR-A), reduction of nitrobenzene to aniline (FBR), butadiene from n-butanes (FBR-A), ethylbenzene by alkylation of benzene (FBR), dehydrogenation of ethylbenzene to styrene (FBR), methyl ethyl ketone from sec-butyl alcohol (by dehydrogenation) (FBR), formaldehyde from methanol (FBR), disproportionation of toluene (FBR-A), dehydration of ethanol (FBR-A), dimethylaniline from aniline and methanol (FBR), vinyl chloride from acetone (FBR), vinyl acetate from acetylene and acetic acid (FBR), phosgene from carbon monoxide (FBR), dichloroethane by oxichlorination of ethylene (FBR), oxidation of ethylene to ethylene oxide (FBR), oxidation of benzene to maleic anhydride (FBR), oxidation of toluene to benzaldehyde (FBR), phthalic anhydride from o-xylene (FBR), furane from butadiene (FBR), acrylonitrile by ammoxidation of propylene (Fl BR)

Note: FBR = fixed-bed reactor (usually multitubular), FBR-A = adiabatic fixed-bed reactor, Fl BR = fluidized bed reactor, MBR = moving-bed reactor.

Adapted from Doraiswamy, 2001.

Table 11.6 adapted from Doraiswamy (2001) provides a partial list of these categories. The main features and applications of these two categories of catalysts are discussed in more detail by Doraiswamy (2001). Of these zeolites, TS-1 catalysts and homogeneous catalysts immobilized on solid supports deserve special mention.

Zeolites were first employed in the petroleum industry and then the petrochemicals industry through Mobil's ZSM-5 catalysts. They are crystalline aluminosilicates possessing characteristic pore and cage structures. Chemical transformations have three restrictions imposed by the lattice structure, collectively referred to as *shape selectivity*:

1. Reactant selectivity (where only certain reactant molecules can enter the pores)
2. Product selectivity (where only certain product molecules can leave the pores)
3. Transition state selectivity (where only certain transition state dimensions are viable)

Zeolites function as Brönsted or Lewis acid catalysts or (less frequently) as basic catalysts. Examples of the former are alkylation and isomerization of aromatics, such as the isomerization of xylenes; an example of the latter is the use of cesium zeolite in the synthesis of the key intermediate, 4-methylthiazole, used in the preparation of the anthelmintic, thiabendazole.

TS-1 catalysts are zeolites containing titanium silicate used in the oxidation of organic substrates by hydrogen peroxide. Only those cations that meet certain specific steric requirements can fit into the tetrahedral positions of the zeolite lattice. One such cation is titanium. TS-1 catalysts are very powerful and selective oxidizing catalysts and are potentially among the most useful catalysts developed. For example, phenol can be oxidized to catechol and hydroquinone.

Because homogeneous catalysts lend themselves to *a priori* design (although much still must be done in this direction) and solid catalysts do not, "bottling" of a homogeneous catalyst within a solid provides a good practical means of combining the advantages of the two types of catalysts. There are four classes of such immobilized, bottled, or "heterogenized" homogeneous catalysts, as shown in Table 11.7 (Doraiswamy, 2001) along with important subcategories of each. These catalysts find wide application in the case of enzymes. A particularly important class of heterogenized catalysts is the supported aqueous phase (SAP) catalyst. This is a distinctive class, since here the catalyst is first solubilized in the aqueous phase (which in itself is an important advance), in contrast to the common case where it is soluble in the organic phase and is then heterogenized

TABLE 11.6
Classification of Solid Catalysts in Organic Synthesis

Solid Catalysts			
Principal Types of Catalysts	**Principal Types of Reactions**	**Heterogenized Catalysts**	**Role of Solvent/Additive**
Improved traditional catalysts	Oxidation (Sheldon, 1981; Jorgensen, 1989)	Supported liquid phase catalysts (Rony and Roth 1975; Chan and Rinker 1978; Doraiswamy and Sharma 1984; Arhancet et al. 1990, 1991)	Physical effects (Gilbert and Mercier 1993)
Zeolites (Corma 1991; Perot and Guisnet 1990; Hoelderich and vanBekkum 1991)	Hydrogenation (Johnstone et al. 1985; Kijenski et al. 1988; Kumbhar et al. 1994)	Catalysts bound to polymer or inorganic supports (including zeolites) (Bailey and Langer 1981; Sherrington and Hodge 1988; Clark et al. 1992)	Chemical effects (Gilbert and Mercier 1993)
Asymmetric catalysts (Aitken and Kilenyi 1992; Federsel 1993; Sheldon 1981)	Various other reactions, e.g., alkylation, isomerization, etc.	Immobilized enzyme catalysts (Tanaka et al. 1980)	
Ion-exchange resins (Olah et al. 1986; Chakrabarti and Sharma 1993)			
Bimetallic catalysts (Raab et al. 1993; Dodgson 1993; Uner et al. 1994)			
Heteropolyacids (Misono 1987, 1993)			
Clays (McKillop and Young 1979; Smith 1992; Balogh and Laszlo 1993)			
Solid superacids (Rajadhyaksha and Joshi 1991; Kumbhar et al. 1994)			
Solid bases (Maki et al. 1993; Torok et al. 1993)			
Metallic glasses (Shibata and Masumoto 1987; Molnar et al. 1989)			
Cyclodextrins (Tabushi and Kuroda 1983; Syamala et al. 1986)			
Titanates (Doraiswamy 2001)			

Source: Doraiswamy, L. K., *Prog. Sur. Sci.* 37: 1; *Organic Synthesis Engineering*, New York: Oxford, 2001.

TABLE 11.7
Classification of Methods for Immobilizing (or Heterogenizing) Homogeneous Catalysts

Heterogenized Homogeneous Catalysts

Supported liquid-phase catalysts (SLPC)	Organic solid (polymer)-bound catalysts	Inorganic solid-bound catalysts	Synthesis within porous structure (zeolites)
Supported organic-phase (SOP) catalysts	Modification of preformed polymers	Copolymerization with inorganic monomers	
Supported aqueous-phase (SAP) catalysts	Polymerization of functionalized monomers	Surface bonding with organic oxides	
		Anchoring on functionalized oxides	

Source: Doraiswamy, L. K., 1991; *Organic Synthesis Engineering*, New York: Oxford, 2001.

[supported organic phase (SOP) catalyst]. Water-soluble homogeneous catalysts often have the advantage that the product is returned to the organic phase while the catalyst remains in the aqueous phase, thus minimizing separation problems. This advantage continues in the supported version. A useful application of SAP catalysts is the use of water-soluble homogeneous catalyst (SAP-Ru-BINAP-4SO$_3$Na) supported on controlled pore glass in the enantioselective hydrogenation of 2-(6'-methoxy-2'-naphthyl) acrylic acid to naproxen (Wan and Davis, 1993a, b).

This brief overview of solid catalysts addresses only the nature, class, and catalytic properties of a variety of catalysts. Modeling the kinetics of a reaction occurring on a solid surface is a challenging task and is firmly rooted in the principles of surface science. As this is still an evolving area, empirical shortcuts are often invoked. Furthermore, for solid catalysts, the reactant(s) must first diffuse into the solid, and product(s) must diffuse out of it. Also, the heat evolved or required must be transported between the solid and the fluid bulk. Hence, diffusion accompanied by reaction becomes a major consideration. These *microenvironmental* aspects of solid catalysts are briefly described below.

11.3.1.2 Kinetics of Reactions on Solid Surfaces

In formulating rate equations for homogeneous reactions, concentrations were used to represent the qualitative presence of different components. What is important in catalytic reactions is the "surface concentration" or fractional coverage of the surface by the adsorbed species. Because there is no easy way of directly measuring surface coverage, theoretical equations are used to relate the fractional surface coverage to the corresponding concentration or partial pressure in the homogeneous phase from where the molecules strike the surface and are adsorbed on it. These equations are known as *isotherms*, the most common being the Langmuir isotherm. This isotherm is based on the assumption that the surface behaves ideally. An ideal surface is one in which

1. All active centers on the surface, i.e., centers on which catalysis can occur, are equally active, and the heat of adsorption is independent of surface coverage by previously adsorbed molecules, and
2. There is no interaction between the adsorbed molecules.

For the case of a single molecular species A, the equation is

$$\theta_A = \frac{K_A p_A}{\left(1 + K_A p_A\right)} \quad (11.29)$$

and for a number of competitively adsorbing species, A, B, C, ...,

Chemical Reaction Engineering

$$\theta_A = \frac{K_A p_A}{(1 + K_A p_A + K_B p_B + K_C p_C + \ldots)} \tag{11.30}$$

where K_A, K_B, K_C, ... are adsorption equilibrium constants of A, B, C, ..., atm^{-1}, and p_A, p_B, p_C, ... are the partial pressures of A, B, C, ..., atm. In addition to equations for θ_A, θ_B, θ_C, ..., it is also necessary to have expressions for the fraction of the surface not covered by any molecule (since the rate of adsorption will also depend on this). This fraction is generally represented by θ_v (fraction of vacant sites) and, for a system consisting of molecules A, B, C, ..., is given by

$$\theta_v = \frac{1}{(1 + K_A p_A + K_B p_B + K_C p_C + \ldots)} \tag{11.31}$$

If the different species involved adsorb noncompetitively, each on a different class of sites, then there is no competition for the same sites. Thus, for the noncompetitive adsorption of two species A and B, the expressions for θ would be

$$\theta_A = \frac{K_A p_A}{(1 + K_A p_A)} \tag{11.32}$$

$$\theta_B = \frac{K_B p_B}{(1 + K_B p_B)} \tag{11.33}$$

The equations get more complicated as the ideal surface assumptions are relaxed, but the Langmuir isotherm is adequate for most engineering purposes. Surface nonideality is discussed, among others, by Boudart and Djeja Mariadasseu (1984), Thomas and Thomas (1999), and Doraiswamy (1991).

Methods of formulating rate equations using the Langmuir isotherm have been extensively documented (e.g., Yang and Hougen, 1950; Barnard and Mitchell, 1964; Satterfield, 1970, 1991; Doraiswamy and Sharma, 1984; Froment and Bischoff, 1990; Butt, 1999; Thomas and Thomas, 1999). They all depend on the concept of the rate-determining step.

11.3.1.2.1 The Rate-Determining Step

The rate-determining step is often defined as the slowest of a series of steps that occur. In catalysis, adsorption of reactant, surface reaction, and desorption all occur in series. Because all of these occur at steady state, they should all proceed at the same rate. Therefore, the word "slowest" is a misnomer. The controlling step is really the step that consumes most of the driving force. Even in the case where all of the steps are fast enough to have reached equilibrium, i.e., the steady overall rate = ($rate_{forward} - rate_{reverse}$), there will be a controlling step. This is the step for which the ratio of the two rates is significantly different from 1. For all other steps, the forward and reverse rates are both so high that the ratio tends to be almost unity.

11.3.1.2.2 Formulation of Models for Competitive Adsorption

Consider the reaction

$$A \rightarrow R \tag{R11.8}$$

Assuming the surface reaction to be the controlling step, the rate equation is given by

$$-r_A = k\theta_A \tag{11.34}$$

Incorporating the Langmuir expression for θ_A,

$$-r_A = \frac{kK_A p_A}{(1 + K_A p_A + K_R p_R)} \tag{11.35}$$

where the units of k are the units of the rate. This is a typical so-called Langmuir–Hinshelwood–Hougen–Watson (LHHW) model.

For the reaction

$$A \rightarrow R + S \qquad \text{R11.9}$$

again assume that surface reaction is controlling, but note that two product molecules are formed. Thus, a vacant site must be available next to the one on which A is adsorbed, to accommodate the second molecule. This leads to the equation

$$-r_A = k\theta_A \theta_v \tag{11.36}$$

Substituting for θ_A and θ_v, we get

$$-r_A = \frac{k\, K_A p_A}{(1 + K_A p_A + K_R p_R + K_S p_S)^2} \tag{11.37}$$

For the reaction

$$A + B \rightarrow R + S \qquad \text{R11.10}$$

the rate equation (for surface reaction control) is

$$-r_A = \frac{k\, K_A\, K_B\, p_A\, p_B}{(1 + K_A p_A + K_B p_B + K_R p_R + K_S p_S)^2} \tag{11.38}$$

Where one of the molecules, say A, is dissociated, it can be shown that $K_A P_A$ should be substituted by $\sqrt{K_A p_A}$.

All LHHW equations can be written in a general consolidated form as

$$\text{Rate} = \frac{(\text{kinetic term})(\text{potential term})}{(\text{adsorption term})^n} \tag{11.39}$$

where n is the number of sites involved. Yang and Hougen (1950) have tabulated expressions for each of the three terms for four major types of reactions. In addition to surface reaction control models, they have also considered adsorption and desorption control models. When one of the reactants is not adsorbed, say B in R11.10, then θ_B is replaced simply by p_B and Equation (11.38) is suitably modified. These are the Eley–Rideal models. Selected equations involving different rate-controlling steps, as formulated from Yang and Hougen's tables, were listed by Doraiswamy (2001).

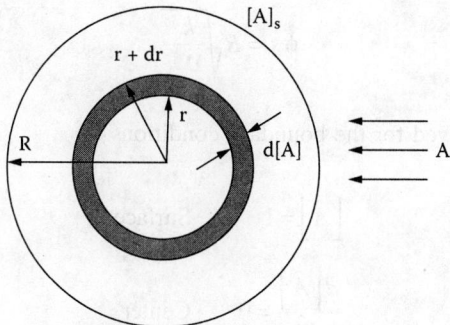

FIGURE 11.4 A differential section of a spherical pellet.

11.3.1.2.3 Formulation of Models for Noncompetitive Adsorption

Consider reaction R11.10 and assume that A and R are competitively adsorbed on one set of surface sites and B and S on another set of sites. A and R do not compete with B and S for the same sites. Then, from equations similar to (11.32) and (11.33), we can write

$$-r_A = \frac{k\, K_A\, K_B\, p_A\, p_B}{(1 + K_A p_A + K_R p_R)(1 + K_B p_B + K_S p_S)} \tag{11.40}$$

11.3.1.2.4 Steps in Selecting an LHHW Model

The first step is to develop equations for all possible controlling steps. They will include, among many possibilities, dissociation of one or more of the reactants, the common single-site mechanism for reactant decomposition, the dual-site mechanism where two sites are involved in the decomposition, and half-site mechanism where two molecules are adsorbed on a single site. Having formulated sufficient models (usually 15 to 20 are even excess of 100), various experimental and sophisticated statistical methods are available for selecting the most probable model, as described in many texts (e.g., Froment and Bischoff, 1990). A comprehensive stepwise procedure (not discussed here) is suggested by Doraiswamy (2001).

11.3.1.3 Role of Diffusion within the Pellet (Internal Diffusion)

For a reaction within a solid catalyst, the reactant must first diffuse into it, leading to a lowering of its concentration in the inner regions of the solid. Clearly, even as A diffuses inward, it is reacting to form the product, but at a progressively diminishing rate as it moves inward. The actual rate constant is the true or intrinsic rate constant multiplied by an *effectiveness factor*, which is a function of the rate constant, diffusion coefficient, and pellet shape and size. Being a codeterminant of the rate, it is an important factor in the analysis and design of catalytic reactors.

11.3.1.3.1 Isothermal Effectiveness Factors

Consider a first-order reaction in a spherical pellet (Figure 11.4). The following continuity equation can be written:

$$\frac{d^2[\hat{A}]}{d\hat{R}} + \frac{2}{\hat{R}} \frac{d[\hat{A}]}{d\hat{R}} = \phi_{s1}^2 [\hat{A}]^n \tag{11.41}$$

where ϕ_{s1} is called the Thiele modulus for a first-order reaction in a spherical pellet and is defined by

$$\phi_{s1} = R\sqrt{\frac{k_v}{D_{eA}}} \tag{11.42}$$

Equation (11.41) can be solved for the boundary conditions

$$\begin{aligned}\left[\hat{A}\right] &= 1 & \text{Surface} \\ \frac{d\left[\hat{A}\right]}{d\hat{R}} &= 0 & \text{Center}\end{aligned} \tag{11.43}$$

to give

$$\left[\hat{A}\right] = \frac{\sinh\left(\phi_{s1}\hat{R}\right)}{\hat{R}\sinh\phi_{s1}} \tag{11.44}$$

An effectiveness factor is now defined as

$$\varepsilon = \frac{\text{actual rate based on average concentration within the pellet}}{\text{rate based on surface concentration throughout the pellet}} \tag{11.45}$$

leading to

$$\begin{aligned}\varepsilon &= \frac{3}{\phi_{s1}}\left(\frac{1}{\tanh\phi_{s1}} - \frac{1}{\phi_1}\right) \\ &= \frac{3}{\phi_{s1}^2}\left(\phi_{s1}\coth\phi_{s1} - 1\right)\end{aligned} \tag{11.46}$$

Similar equations can be derived for a flat plate and a cylinder. Instead of writing different equations for different shapes, the following single continuity equation can be written for a first-order reaction by defining a generalized reduced length parameter ξ (r/R for a sphere or a cylinder and ℓ/L for a flat plate):

$$\frac{d^2\left[\hat{A}\right]}{d\xi^2} + \frac{\beta}{\xi}\frac{d\left[\hat{A}\right]}{d\xi} = \phi_1^2\left[\hat{A}\right] \tag{11.47}$$

where β is a shape factor with values of 2 for a sphere, 1 for a cylinder, and 0 for a flat plate. The Thiele modulus ϕ_1 (with the shape suffix s removed) is a generalized modulus for all shapes for a first-order reaction given by

$$\phi_1 = s\sqrt{\frac{k_v}{D_{eA}}} \tag{11.48}$$

where now s is a new length parameter (sometimes also called the *shape factor*) defined as (Aris 1957)

Chemical Reaction Engineering

$$s = \frac{\text{volume of shape}}{\text{external area of shape}} \quad (11.49)$$

and is equal to $R/3$ for a sphere, $R/2$ for a cylinder, and L for a flat plate (where $2L$ is the plate thickness). The modulus can be further generalized to represent a reaction of any order in a catalyst of any shape by defining it as

$$\phi = s\sqrt{\frac{(n+1)k_v[A]_s^{n-1}}{2\mathbf{D}_{eA}}} \quad (11.50)$$

Note that both the shape and order suffixes have now been removed. A plot of ε vs. ϕ_1 is shown in Figure 11.5.

The Thiele modulus requires knowledge of the true rate constant, which may not always be available. It is more conveniently defined based on the actual measured rate of a reaction. This is called the Weisz modulus (1954) and, for a general shape with s as the diffusion length, is defined as

$$\phi_a = s^2 \frac{(\text{actual rate})}{\mathbf{D}_{eA}[A]_S} = s^2 \left[\frac{(\text{true rate})}{\mathbf{D}_{eA}[A]_S}\right]\varepsilon \quad (11.51)$$

A plot of ε vs. this modulus ϕ_a is included in Figure 11.5.

From Figure 11.5, there are two asymptotic regions—one at low values of the modulus (the kinetic control regime) and the other at high values of the modulus (the diffusion control regime). The intermediate regime corresponds to mixed control, both by reaction and diffusion. The point

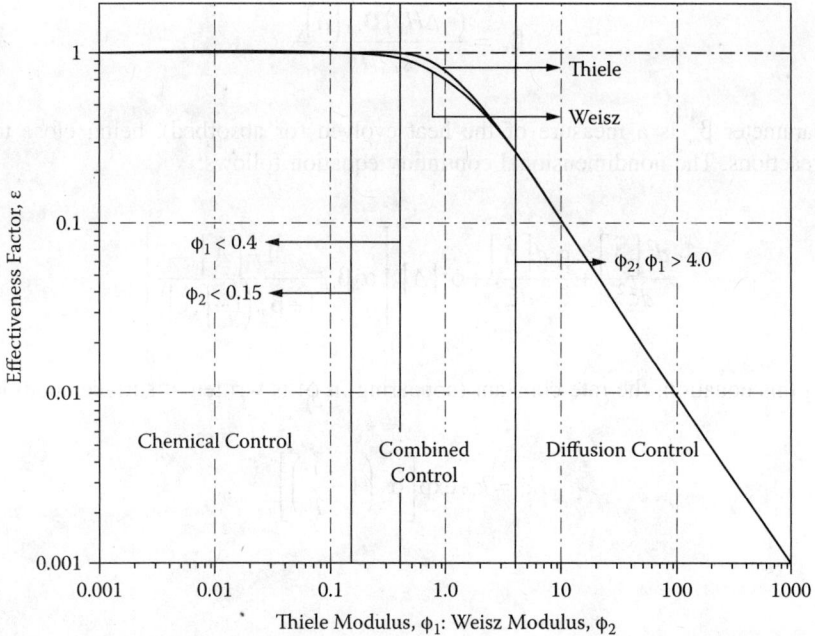

FIGURE 11.5 Effectiveness factor as a function of Thiele modulus and Weisz modulus for a simple isothermal reaction (using the generalized length s) (Doraiswamy, 2001).

of transition from chemical to diffusion control is a practically useful value, since it represents a pellet of optimal dimension. A larger pellet would attract diffusional resistance, whereas a smaller one would involve a higher pressure drop in the reactor. The various regions are clearly marked in the figures. The effectiveness factor concept developed here is based on power law models for the reaction rate. Attempts to extend the analysis to LHHW kinetics (e.g., Roberts and Satterfield, 1965, 1966; Aris, 1975; Satterfield, 1991; Doraiswamy and Sharma, 1984; Doraiswamy, 1991) are not discussed here.

11.3.1.3.2 Nonisothermal Effectiveness Factors

The equations developed above can be extended to nonisothermal situations, i.e., to reactions for which the heat effects cannot be neglected for highly exothermic reactions. The basic equation used here is the mass-energy balance:

$$\begin{pmatrix} \text{Heat generated by reaction of} \\ \text{A in the pellet} \end{pmatrix} = \begin{pmatrix} \text{Heat transferred from} \\ \text{the pellet to the surface} \end{pmatrix}$$

$$\mathbf{D}_{eA}\left([A]_S - [A]\right)\left(-\Delta H_r\right) = k_{T,er}\left(T - T_S\right) \tag{11.52}$$

In dimensionless form, this becomes

$$\hat{T} = \frac{T}{T_S} = 1 + \beta_m \left(1 - [\hat{A}]\right) \tag{11.53}$$

where

$$\beta_m = \frac{(-\Delta H_r)\mathbf{D}_{eA}[A]_S}{k_{T,er}T_S} \tag{11.54}$$

The parameter β_m is a measure of the heat evolved (or absorbed), being close to zero for athermal reactions. The nondimensional continuity equation follows:

$$\frac{d^2[\hat{A}]}{d\xi^2} + \frac{\beta}{\xi}\frac{d[\hat{A}]}{d\xi} + \phi^2[A]^n \left[\alpha_S \beta_m \frac{1-[\hat{A}]}{1+\beta_m\left(1-[\hat{A}]\right)}\right] \tag{11.55}$$

In deriving this equation, the rate constant (appearing in ϕ) is written at surface conditions:

$$k_v = k_{vS} \exp\left[\alpha_S\left(1 - \frac{1}{\hat{T}}\right)\right] \tag{11.56}$$

where

$$\alpha_S = \frac{E}{R_g T_S} \tag{11.57}$$

Chemical Reaction Engineering

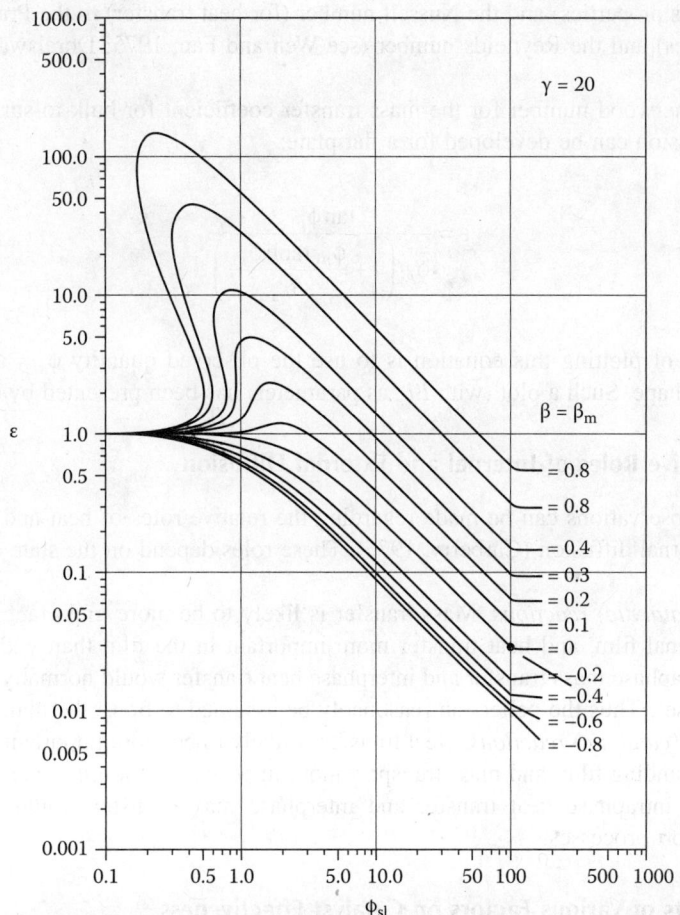

FIGURE 11.6 Effectiveness factor plots for a simple nonisothermal reaction in a spherical pellet (Weisz and Hicks, 1962).

The solution to this equation is graphically displayed in Figure 11.6 (Weisz and Hicks, 1962). β_m is a parameter in the solution. Extension of the analysis to LHHW kinetics introduces another parameter, $K_A p_A$ (see Rajadhyaksha et al., 1976), and an analytical solution is quite straightforward. A striking feature of the plots presented in the figure is that at high values of γ ($= \alpha_s \beta_m$), multiple solutions are possible (see Aris, 1975, for a detailed discussion). This corresponds to the region of very high effectiveness factors. Operating in this region (known as *operating at the edge*) is obviously very beneficial but requires very precise control strategies. In the future, operating at the edge may become common practice.

11.3.1.4 Role of External Diffusion

In the preceding section, it was assumed that the concentration at the surface is known. In other words, the surface concentration can be taken as equal to the concentration in the bulk of the fluid phase. This is true only if there is essentially no mass transfer resistance across the film (or boundary layer) surrounding the pellet. If such a resistance is present, the actual surface concentration will be lower for the reactant and higher for the product than in the fluid bulk. A Sherwood number for mass transfer would then be needed in the analysis. Several empirical correlations have been proposed that relate the Sherwood number (for mass transfer) to the Reynolds number (for flow) and Schmidt number (for mass properties) and the Nusselt number (for heat transfer) to the Prandtl number (for thermal properties) and the Reynolds number (see Wen and Fan, 1975; Doraiswamy and Sharma, 1984).

Using the Sherwood number for the mass transfer coefficient for bulk-to-surface transfer, the following expression can be developed for a flat plate:

$$\varepsilon = \frac{\tan \phi_{p1}}{\phi_{p1}\left(1 + \dfrac{\phi_{p1} \tanh \phi_{p1}}{Bi_m}\right)} \tag{11.58}$$

A practical way of plotting this equation is to use the observed quantity $\phi_a = \varepsilon \phi_1^2$, where ϕ_1 is independent of shape. Such a plot (with Bi_m as parameter) has been presented by Carberry (1976).

11.3.1.5 Relative Roles of Internal and External Diffusion

The following observations can be made regarding the relative roles of heat and mass transfer in internal and external diffusion (Carberry, 1976). These roles depend on the state of the bulk fluid, gas, or liquid.

Gas-solid (catalytic) reactions. Mass transfer is likely to be more important within the pellet than in the external film, and heat transfer more important in the film than within the pellet. In other words, intraphase mass transfer and interphase heat transfer would normally be the dominant transport processes. Thus the pellet can reasonably be assumed to be isothermal.

Liquid-solid (catalytic) reactions. Heat transfer is likely to be more important within the pellet than in the surrounding film, and mass transport more important in the film than within the pellet. In other words, intraphase heat transfer and interphase mass transfer would normally be the dominant transport processes.

11.3.1.6 Effects of Various Factors on Catalyst Effectiveness

The equations and plots presented in the foregoing sections largely pertain to the diffusion of a single component followed by reaction. There are several other situations of industrial importance on which considerable information is available. They include biomolecular reactions in which the diffusion-reaction problem must be extended to two molecular species, reactions in the liquid phase, reactions in zeolites, reactions in immobilized catalysts, and extension to complex reactions (see Aris, 1975; Doraiswamy, 2001). Several factors influence the effectiveness factor, such as pore shape and constriction, particle size distribution, micro-macro pore structure, flow regime (bulk or Knudsen), transverse diffusion, gross external surface area of catalyst (as distinct from the total pore area), and volume change upon reaction. Table 11.8 lists the major effects of all these situations and factors.

11.3.1.6.1 Practical Effects of Diffusion in Complex Reactions

The analysis of complex reactions was briefly reviewed in Section 11.2.2. Diffusion can also greatly influence such complex reactions. In fact, an understanding of the role of diffusion can be incorporated in the design of the reactor system to increase or decrease the rate of certain steps and thus enhance the selectivity of the desired product. Some salient features of the effect of diffusion on some selected complex schemes are outlined in Table 11.9.

11.3.1.6.2 Criteria Based on Measurable Quantities for Eliminating Transport Disguises

In determining the true kinetic parameters of a reaction, all mass and heat transfer effects (or *disguises*, as they are sometimes called) must be eliminated. Several criteria (based on measurable

TABLE 11.8
Practical Effects of Various Factors on Effectiveness Factor in Simple Reactions

	Effect of	Nature of Effect	Reference
1.	Pore size distribution	Radial dispersion has negligible effect in kinetic regime but lowers ε in diffusion regime. Pore length dispersion has less effect.	Schmalzer (1969)
2.	Flow regime	Knudsen diffusion is much more detrimental than bulk diffusion.	Scott (1962), Otani et al. (1965)
3.	External surface	Effect is large for pellets of low porosity. Desirable to coat only the surface with catalyst in such cases.	Carberry and Kulkarni (1973) Varghese et al. (1978)
4.	Nonuniform environment around pellet	Where temperature gradients in a reactor are steep, the environment around a pellet will not be uniform, but is practically uninfluenced by this.	Bischoff (1968) Copelowitz and Aris (1970)
5.	Change in volume	Falls with increase in ϕ and the volume change factor θ defined as $(v_B - 1)X_{AS}$.	Weekman and Gorring (1965)
6.	Negative order kinetics	The usual adverse effect of diffusion turns to advantage	Smith et al. (1975)
7.	Dilution of pellet by inert	Beyond a certain fraction of active catalyst in the pellet, ε increases in the diffusion regime.	Ruckenstein (1970) Vargese and Wolf (1980)
8.	Configurational diffusion (where molecules are of the same size as the pores) mainly in zeolites	Catalysts involving configurational diffusion usually undergo deactivation. The highest initial activity is obtained for pellets with a uniform pore size.	Weisz (1957) Hughes and Mann (1978) Rajagopalan and Luss (1979)
9.	Surface diffusion	Effect is most noticeable at high temperatures and low partial and total pressures. A modified Thiele modulus $\phi_{mod} = \phi_{p1}(1/1 + h_s)$, where $k_s = D_s K_A/D_{bA}$ can be used to prepare the usual effectiveness factor plot.	Barrer (1963) Roberts and McKee (1979) Kammermeyer and Rutz (1959) Krasuk and Smith (1965) Sohn et. al (1970)
10.	Temperature	See Section (11.3.1.3.2).	—
11.	Transverse diffusion	All treatments are restricted to axial diffusion. The effect of transverse diffusion can be significant, but usually is negligible at values of $\ell/R < 50$, and may have to be considered only for very thin slabs.	Bischoff(1965)

quantities) have been formulated for this purpose. Doraiswamy (2001) has tabulated criteria for mass and heat transfer effects both for internal and external transport limitations (see Table 11.10).

11.3.1.7 Laboratory Reactors for Accurate Kinetic Data

In designing various types of commercial reactors (fixed, moving, or fluidized), an equation representing the true intrinsic kinetics of the reaction at hand is required. To obtain in equation free of any diffusional intrusions, the choice is between (1) collecting data without paying too much attention to these disguises and then correcting them through appropriate correction factors like the effectiveness factor, and (2) ensuring that diffusional disguises are absent through the criteria listed in the table. Specially designed reactors can often accomplish this latter objective. Other important factors to be considered in selecting (constructing) a suitable laboratory reactor are ease of sampling and accuracy of measurements using these samples, isothermality of the reactor, accuracy of residence time measurement, the effects of catalyst decay on conversion and selectivity measurements, and construction difficulties and cost. Many of these features associated with a variety of laboratory reactors for reactions involving a solid catalyst have been explicitly discussed by Week-

TABLE 11.9
Effect of Diffusion on Some Industrially Important Classes of Complex Reactions

Reaction[a]	Pore Structure	Intrinsic Selectivity, Yield	Actual Selectivity, Yield	Main Features
1. Independent $A \xrightarrow{1} R$ $B \xrightarrow{2} S$	Monodispersed	$s = \dfrac{k_{v1}}{k_{v2}}$	$s_a = \left(\dfrac{s}{-D}\right)$	Greater diffusional resistance for reaction 2 enhances y_{Ra}
2. Parallel $A \nearrow^{1,m} R$ $\searrow_{2,n} S$	Monodispersed	$s = \dfrac{k_{v1}}{k_{v2}}$ $Y_R = \dfrac{1}{1+p_{nm}}$ where $p_{nm} = \dfrac{k_{v2}}{k_{v1}}[A]_s^{m-1}$	$s_a = s^{1/2}$ $\dfrac{Y_{Ra}}{Y_R} = \dfrac{n+1}{1m-n+1}$ (Roberts 1972; Pawlawski 1961)	Y_{Ra}/Y_R decreases as p_{nm} and ϕ increase

3. Consecutive

A $\xrightarrow{1}$ R $\xrightarrow{2}$ S

Monodispersed:
$$Y_R = \frac{S}{1-s}\left[(1-X_A)^s - (1-X_A)\right]$$
$$s = \frac{k_{v1}}{k_{v2}}, \quad S = \frac{1}{s}$$

Biodispersed:
$$s\frac{k_{v1}}{k_{v2}}, \quad S = \frac{1}{s}$$
$$S = \frac{1}{s^{1/2}}$$

(Carberry 1962; Doraiswamy and Sharma 1984)

Selectivity decreases with increasing complexity of pore structure

4. Parallel-consecutive[b]

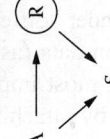

Monodispersed:
$$Y_{RB} = \frac{[R]}{[B]_0}$$
$$Y_{RA} = \frac{[R]}{[A]_0}$$
$$S = \frac{1}{s^{1/4}}$$

Numerical solution for $(Y_{RA})_a$, $(Y_{RB})_a$

$(Y_{RB})_a$ decreases with increase in $[B]_0/[A]_0$; it increases monotonically with X_A for $[B]_0/[A]_0 < 1$ (approx.), beyond which it shows a maximum

---

[a] 1,2 refer to reactions 1 and 2, respectively; m,n are reaction orders.
[b] A second reactant B is assumed to be present for reaction 4.

TABLE 11.10
Criteria for the Absence of Mass and Heat Transfer Effects in Gas-Solid Catalytic Reactions (Doraiswamy, 2001)*

Interphase	Intraparticle	Interphase and Intraparticle
Mass transfer	Isothermal	Isothermal
1. $\eta = 1 - \dfrac{k_a d_p^{1.5}}{11\sqrt{D_b u}}$	1. $\dfrac{d_p^2 r_a}{4 D_{eA} [A]_s} < 1, n = 1$	1. $\alpha_b \beta_{mb} + 0.3 n \alpha_b \left(\dfrac{(-\Delta H)_r r_a d_p}{2 h_{fp} T_b} \right) \leq 0.05 n$
Ruthven (1968)	Weisz and Prater (1954); Weisz (1957)	Mears (1971a); see also Guha and Narsimhan (1972)
		Nonisothermal
2. $\dfrac{r_a d_p}{2[A]_b k_G} < \dfrac{0.15}{n}$	2. $\dfrac{d_p^2 r_a}{4 D_{eA} [A]_s} < |n|, n \neq 0$	2. $\dfrac{r_a d_p^2}{4[A]_b D_{eA}} < \dfrac{1 + 0.33 \alpha_b \dfrac{(-\Delta H_r)(-r_a) d_p}{2 h_{fp} T_b}}{\left| n - \alpha_b \beta_{mb} \right| \left(1 + 0.33 \left(\dfrac{r_a d_p}{2[A]_b k_G} \right) \right)}$
	Stewart and Villadsen (1969); Mears (1971a)	Mears (1971a)

* r_a = observed rate (mol/m³ cat s); k_a = observed rate constant (1/s); α_b = α at bulk conditions; (β_{mb} = β_m at bulk conditions; η = external effectiveness factor; θ = volume change parameter given by $(v_B - 1) y_{As}$, where v_B is the stoichiometric coefficient in the reaction $A \rightarrow V_B B$ and y_{As} is the mole fraction of A at the catalyst surface. See Mears (1971b, 1976) for interparticle criteria (i.e., criteria for the reactor as a whole) such as that for the absence of axial diffusion.

man (1974) and in a more general way by Doraiswamy and Tajbl (1974), Doraiswamy and Sharma (1984), Hofmann (1986), Pratt (1987), and most recently by Doraiswamy (2001).

The internal diffusional effect can be avoided by performing experiments with catalysts of different sizes under otherwise identical conditions and determining that size above which there is a lowering of conversion due to internal diffusional resistance. The external mass transfer effect can be examined by operating the reactor at different feed velocities but at the same residence time (i.e., keeping W/F constant). The velocity beyond which the conversion levels off should be noted and used as the minimum velocity at which to operate an integral reactor. The same method holds for a fully mixed reactor, in which the agitation parameter is the extent of stirring.

Several reactors normally used to obtain kinetic data for a nondecaying catalyst under diffusion-free conditions (as established by tests just described) are shown in Figure 11.7. The integral reactor (operated under plug-flow conditions) is the simplest, but the conversion-residence time data obtained must be differentiated to get the rates (see Figure 11.2). The differential reactor consists of a very small amount of catalyst giving conversions of 1 to 2%. This enables direct determination of the rate corresponding to the average component partial pressures in the bed. Clearly, the analytical methods used for determining the compositions of the inlet and outlet streams must be accurate. Also, since the differential element used corresponds to a small axial section of the integral reactor, it is necessary to use synthetic feeds that will contain all the components of the process stream, including impurities. The main advantage is that the measurements give the rates directly. This advantage can be retained in a CSTR with the additional advantage that integral level conversions (usually in excess of 10%) can be obtained under "differential conditions." In other words, there is no need for differentiating integral conversion data (as in a typical integral reactor). The fully mixed condition can be realized in many ways, most importantly by placing the catalyst in baskets attached to the stirrer (the Carberry reactor), by attaching the catalyst to the pot provided with a suitable internal recirculation system (the Berty reactor), or by attaching the catalyst to the

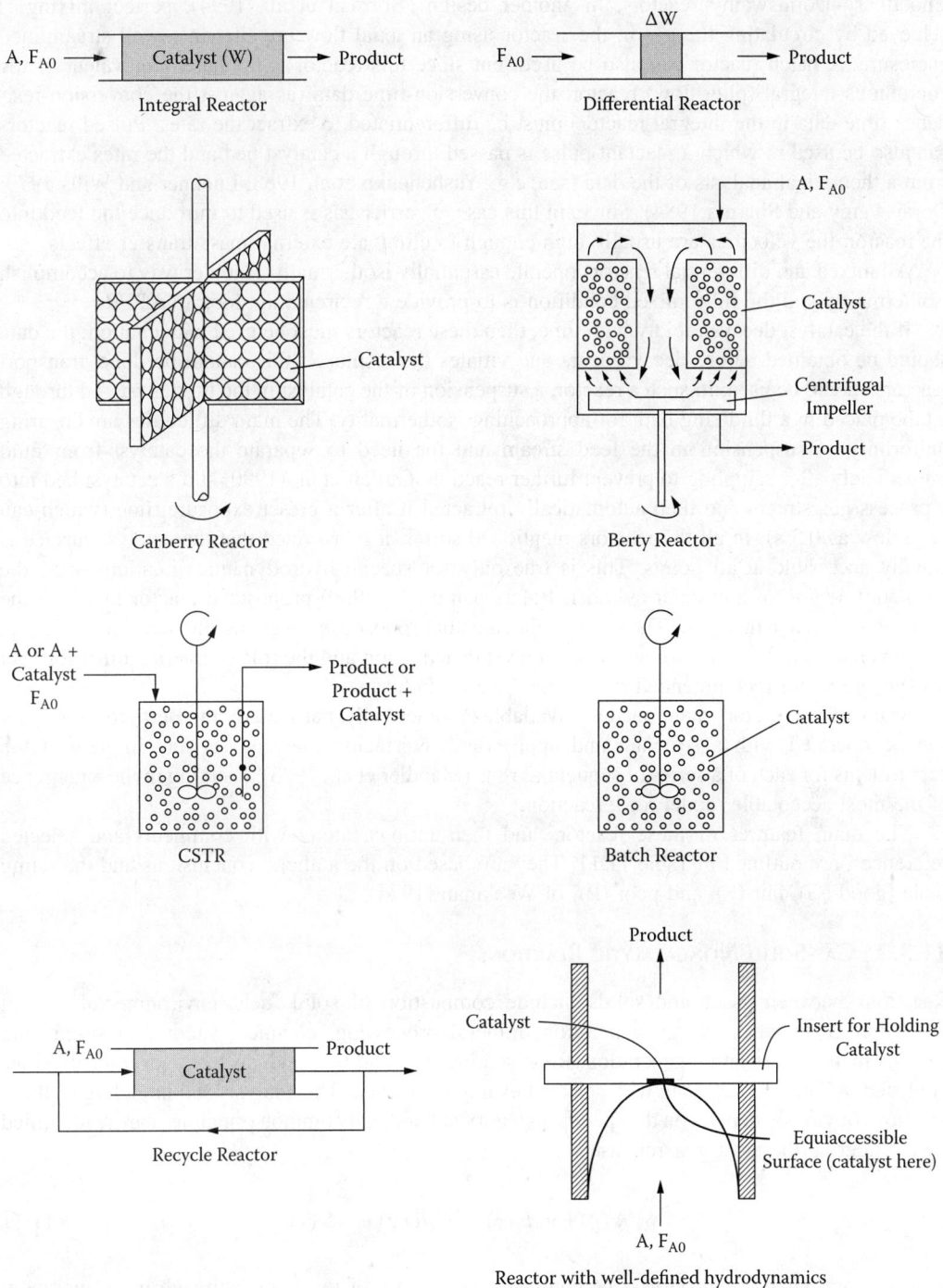

FIGURE 11.7 Some useful laboratory reactors for obtaining precise kinetic data for gas-solid catalytic reactions.

stirrer as in the Carberry reactor but rotating the pot provided with suitable baffles (the Choudhary–Doraiswamy reactor). In another design (Borman et al., 1994), perfect mixing is achieved by circulating the gas in the reactor using an axial flow impeller in a well streamlined enclosure. A batch reactor can also be used, but since this reactor is the batch equivalent of the continuous integral (plug-flow) reactor, the conversion-time data (as against the conversion-residence time data in the integral reactor) must be differentiated to extract the rates. Pulsed reactors can also be used in which a reactant pulse is passed through a catalyst bed and the rates extracted from a theoretical analysis of the data (see, e.g., Yushchenko et al. 1968; Luckner and Wills 1973; Doraiswamy and Sharma 1984). Since, in this case, a carrier gas is used to introduce the feed into the reactor, the velocities are usually high enough to eliminate external mass transfer effects.

All mixed and differential reactors operate essentially isothermally. Another way to accomplish isothermality and the fully mixed condition is to provide a recirculation loop to a PFR.

If the catalyst decays rapidly with time, then these reactors must be used with caution: the data should be obtained before decay occurs and vitiates the results. If this cannot be done, transport reactors should be used. In such a reactor, a suspension of the catalyst in the fluid is passed through a tube placed in a fluidizing bath for approaching isothermality. The main difficulties are ensuring uniformity of suspension in the feed stream and the need to separate the catalyst from fluid immediately after sampling to prevent further reaction. Gullett et al. (1990) slid a catalyst bed into a process gas stream and then automatically retracted it after a preset exposure time (which can be as low as 0.3 s). In all the reactors mentioned so far, it is assumed that the catalyst surface is equally accessible at all points. This is true only for special hydrodynamic situations, e.g., the stagnation region of a circular cylinder. Balaraman et al. (1980) proposed a reactor in which the catalyst is placed in this zone. The single-pellet reactor proposed by Hegedus and Petersen (1973a–c, 1974) is particularly useful for studying catalyst deactivation and the role of internal diffusion, but it is generally not recommended for routine kinetic studies.

Many of the reactors mentioned are available commercially, particularly the Berty reactor. They can be operated with a software and appropriate interfacing that can set and implement the experiments for each of a series of sequential runs (Mandler et al., 1983), resulting in the emergence of the most acceptable model for a reaction.

The main features of these reactors and their ratings, along with comments and selected references, are outlined in Table 11.11. They are based on the authors' conclusions and the rating scale [good (G), fair (F), and poor (P)] of Weekman (1974).

11.3.2 Gas-Solid Noncatalytic Reactions

Reactions between gases and solids include combustion of solid fuels, environmental control (pollution abatement), energy generation, mineral processing, chemical vapor deposition, and catalyst manufacture and regeneration. Representing the solid by s and gas by g, several categories are listed in Table 11.12 along with selected examples of each. The analysis and modeling of these reactions obviously depend on the specific category at hand, but common principles can be identified by considering the most general case,

$$\nu_A A(g) + \nu_B B(s) \rightarrow \nu_R R(g) + \nu_S S(s) \qquad \text{R11.11}$$

Thus our presentation will largely be confined to this class of reactions, although brief references will also be made to other classes, notably $A(g) + B(s) \rightarrow R(g)$, represented by the gasification of coal.

11.3.2.1 Modeling of Gas-Solid Reactions

The first model, the shrinking core model (SCM) or the sharp interface model (SIM), was proposed about half a century ago. Other models also describe the behavior of the solid as it undergoes

TABLE 11.11
Laboratory Reactors for Gas-Solid Catalytic Reactions: Their Principal Features and Ratings

	Main Features	Isothermality	Diffusion-Free Operation	Contact Time Determination	Rate Measurement	Comments	Selected References
1. Integral reactor	A single-tube fixed-bed reactor; temperature variation can be controlled by catalyst dilution; easiest to construct and operate; best heated/cooled in a fluidized sand bath	P[a]-F[c]	F	F	I[d]	Gives a practical "feel" for the scaled up version; dilution desirable but causes a "dilution effect" if greater than experimental error in measuring conversion; use the Sofekun-Rollins-Doraiswamy criterion to evaluate this effect, multiple taps can give multiple conversion data for each run.	Musick et al. (1972); van den Bleek et al. (1969); Sofekun et al. (1994)
2. Differential reactor	Small bed, 1–2 g catalyst; low conversion places heavy demand on accuracy	G[a]-F[b]	G	P-F	D	Use of recycle eliminates diffusional effects; an integral reactor can supply the required partially reacted feed	Pansing and Malloy (1962); Lunde and Kestner (1974)
3. Fully mixed reactor a. Rotating catalyst	Catalyst in basket(s) attached to stirrer (Carberry reactor)	G	G	F-G	D	Direct measurement of bed temperature almost impossible; otherwise an excellent design	Carberry (1964); Tajbl et al. (1966, 1967); Doraiswamy and Sharma (1984)
b. Stirred fluid, stationary catalyst	Fluid stirred with special stirrer (Berty reactor) or by rotating pot with baffles (Choudhary-Doraiswamy reactor)					Bed temperature readily measurable; uniform gas circulation through bed uncertain	Berty (1973); Choudhary and Doraiswamy (1972); Doraiswamy and Tajbl (1974)
c. Continuous stirred tank	Fluid + catalyst passed continuously through a stirred reactor, or just the fluid is passed with a fixed amount of catalyst restrained in the reactor	P-F	G	F-G	D	Difficulty in quickly separating catalyst from fluid in samples can lead to wrong composition values	
d. Recycle reactor	Large fraction of product is recycled	F-G	G	G	D	A recycle to feed ratio >20 ensures full mixing	Perkins and Rase (1958); Butt et al. (1962) see also Carberry (1976); Levenspiel (1999)

Continued

TABLE 11.11 (Continued)
Laboratory Reactors for Gas-Solid Catalytic Reactions: Their Principal Features and Ratings

	Main Features	Isothermality	Diffusion-Free Operation	Contact Time Determination	Rate Measurement	Comments	Selected References
4. Stirred batch reactor	A batch of fluid and catalyst is placed in a stirred reactor and the progress of reaction is followed as a function of time.	G	G	G	I[d]	Generally not recommended for gas-solid reactions	Galiski and Hightower (1970) Richardson and Friedrich (1975)
5. Pulse reactor	A microreactor in which a pulse of feed is introduced; can be integral or differential	G	G	P	I[e]	Good for a rapidly deactivating catalyst; far removed from the "real world"	Hegedus and Petersen (1973a,b,c; 1974)
6. Single-pellet reactor	Used specifically to study deactivation and internal diffusion in a catalyst pellet (Hegaedus-Petersen reactor)	F-G	NA	G	D	Good mainly for studying role of deactivation; can also be used for kinetic studies but not recommended	Balaraman et al. (1980)
7. Reactor with well-defined hydrodynamics	The forward stagnation zone of a circular cylinder used for obtaining an equiaccessible surface which allows accurate accounting of mass transfer (Balaraman-Mashelkar-Doraiswamy reactor)	F-G	NA	F-G	I[e]	Precise accounting of mass transfer effect possible if present	Doraiswamy and Tajbl (1974) Doraiswamy and Sharma (1984) Froment and Bischoff (1990)
8. Nonisothermal reactor (including adiabatic reactor)	Reactor operated without any effort to make it isothermal; can also be operated adiabatically	NA	F	F	I[e]	Industrial conditions are best simulated; adiabatic operation requires only temperature profile measurement and no chemical analysis as in all other reactors.	

*G = good, F = fair, D = direct, I = indirect, a = in general, b = for highly exothermic or endothermic reactions, c= with bed dilution, d = refer Figure 11.2, e = see reference(s), NA = not applicable.

TABLE 11.12
Industrially Important Examples of Different Types of Noncatalytic Gas-Solid Reactions[a]

Type	General Reaction Scheme	Reaction
A	Solid + fluid → solid + fluid	Roasting of zinc ore
		$2ZnS(s) + 3O_2(g) \rightarrow 2ZnO(s) + 2SO_2(g)$
		Production of uranium tetrachloride by chlorination of
		$UO_2(s) + CCl_4(g) \rightarrow UCl_4(s) + CO_2(g)$
		Selective chlorination of iron in ilmenite
		$FeTiO_3(s) + CO(g) + Cl_2(g) \rightarrow FeCl_2(g) + CO_2(g) + TiO_2(s)$
B	Solid + fluid → solid	Nitrogenation of calcium carbide to produce cyanamide
		$CaC_2(s) + N_2(g) \rightarrow CaCN_2(s) + C(s)$
		Rusting reaction of iron
		$2Fe(s) + O_2(g) \rightarrow 2FeO(s)$
		Absorption of SO_2 by dry limestone injection
		$CaO(s) + SO_2(g) \rightarrow CaSO_3(s)$
C	Solid → fluid + solid	Calcination of limestone
		$CaCO_3(s) \rightarrow CaO(s) + CO_2(g)$
		Decomposition of magnesium hydroxide
		$Mg(OH)_2(s) \rightarrow MgO(s) + H_2O(g)$
D	Solid + fluid → fluid	Production of carbon disulfide
		$C(s) + S_2(g) \rightarrow CS_2(g)$
		Chlorination of rutile to titanium tetrachloride
		$TiO_2(s) + 2C(s) + 2Cl_2(g) \rightarrow TiCl_4(g) + 2CO(g)$
		Gasification of carbon
		$C(s) + H_2O(g) \rightarrow CO(g) + H_2(g)$
E	Solid → fluid	Decomposition of ammonium chloride
		$NH_4Cl(s) \rightarrow NH_3(g) + HCl(g)$
		Decomposition of ammonium sulfate
		$(NH_4)_2SO_4(s) \rightarrow 2NH_3(g) + SO_3(g) + H_2O(g)$
F	Fluid → solid + fluid	Mond process for nickel production
		$Ni(CO)_4(g) \rightarrow Ni(s) + 4CO(g)$
		Oxidation of silicon tetrachloride to silicon dioxide
		$SiCl_4(g) + O_2(g) \rightarrow SiO_2(s) + 2Cl_2(g)$
		Burning of titanium tetrachloride to rutile
		$TiCl_4(g) + O_2(g) \rightarrow TiO_2(s) + 2Cl_2(g)$

[a] General reaction: $\nu_A A + \nu_B B = \nu_R R + \nu_s S$.

reaction with a gas (or is reacting by itself to produce a gas). These categories (many not mutually exclusive) include shrinking (or expanding) core, volume reaction, reaction zone, particle-pellet (or grain), grain-micrograin, discrete, computational, and percolation models. The percolation models are based on the statistical physics of disordered media and include such phenomena as aggregation processes, scaling, network modeling of the pore space, discretization, and random walk representation of diffusion processes. For a review of percolation models, see Sahimi et al. 1990. Research on the other models continues because of their usefulness (see e.g., Szekely et al. 1976; Ramachandran and Doraiswamy 1982; Kulkarni and Doraiswamy 1986; Doraiswamy and Kulkarni 1987; Bhatia and Gupta 1993).

The simplest models are those in which the internal structure of a pellet is not considered, and its behavior as a whole is modeled. These are normally called the *macroscopic* or *basic* models. In other models, the behavior of the distinctive elements of a pellet, such as the grain, micrograin, or the pore, constitutes the central feature; such models account for structural changes during reaction. The so-called random pore models are the most common.

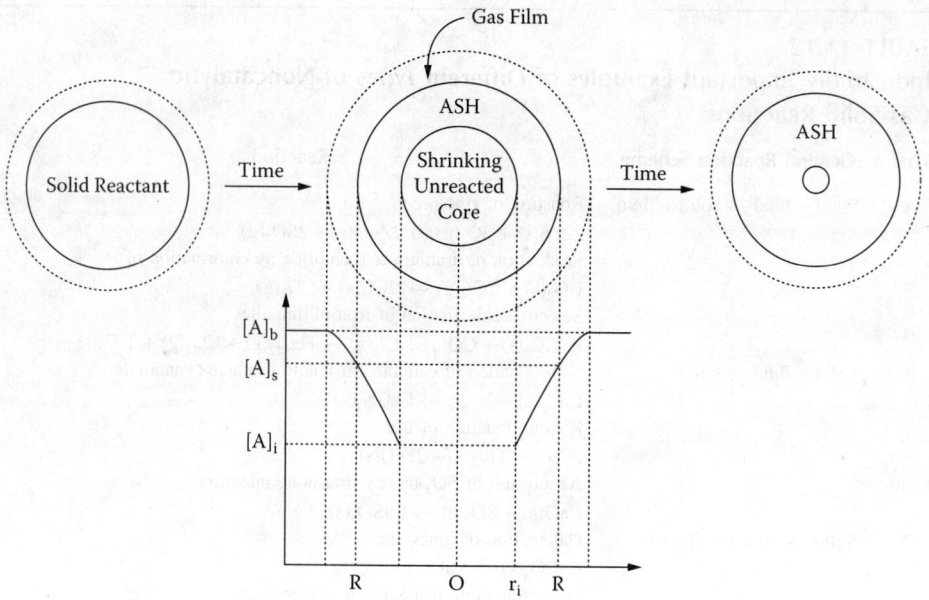

FIGURE 11.8 The shrinking core model (SCM) for gas-solid noncatalytic reactions.

11.3.2.1.1 Shrinking Core Model (SCM)

Figure 11.8 illustrates the basic features of SCM. The gas first diffuses through the film surrounding the pellet and reacts at the interface. As the reaction progresses, the interface moves inward, leaving behind a shell of the exhausted solid (called ash). In effect, the unreacted core shrinks until the entire solid has reacted. This behavior is possible only if the solid is nonporous. Otherwise, the gas will diffuse below the surface, and the reaction will no longer be confined to the interface.

The mathematical analysis of this model is facilitated by the pseudo-steady-state (PSS) assumption, i.e., the interface remains stationary while the mass flux equations are written. This is generally satisfied for gas-solid reactions. The following equations for makers of A diffusing per unit time for a single pellet, $\mathbf{R}_A$ can be written:

Diffusion through gas film:

$$\mathbf{R}_A = 4\pi R^2 k_G \qquad (11.59)$$

Diffusion through ash layer:

$$\mathbf{R}_A = \frac{4\pi R\, r_i \mathbf{D}_{e,As}}{R - r_i} \qquad (11.60)$$

Chemical reaction at the interface:

$$\mathbf{R}_A = 4\pi r_i^2 k_S \qquad (11.61)$$

Noting that conversion in a spherical pellet is related to the ratio of the initial and interface radii by the equation

$$X_B = 1 - \left(\frac{r_i}{R}\right)^3 \tag{11.62}$$

Equations (11.59) through (11.61) are combined to give

$$\mathbf{R}_A = \left(\frac{1}{4\pi R^2 k_G} + \frac{R - r_i}{4\pi R r_i \mathbf{D}_{e,As}} + \frac{1}{4\pi r_i^2 k_s}\right)^{-1} [A]_s \tag{11.63}$$

To express the rate in terms of the solid reactant B, we write the following equation for the rate of movement of the sharp interface:

$$-\frac{d}{dt}\left((4/3)\pi r_i^3 \frac{\rho_B}{M_B}\right) = \frac{\nu_B}{\nu_A} \mathbf{R}_A = \nu \mathbf{R}_A \tag{11.64}$$

Substituting Equation (11.63) for $\mathbf{R}_A$ in Equation (11.64) and integrating, we get an equation of the general form

$$\tau = f_1(X_A) + f_2(X_A) + f_3(X_A) \tag{11.65}$$

or

$$\tau = \tau_f + \tau_a + \tau_c \tag{11.66}$$

where τ_f, τ_a, and τ_c represent, respectively, the times required for complete conversion if film transfer, ash diffusion, or reaction alone were to be the controlling step. The functions f_1, f_2, and f_3, and the various τ_s, assume different forms for different geometries of the pellet and are defined in Table 11.13. The dependence of τ on pellet size varies for different controlling regimes: first-order in R for reaction control, second-order for ash diffusion control, and 1.5- to 2.0-order for film diffusion control.

Improvements and modifications of the above procedure have been suggested. In one method (Tine, 1985; Villa and Quiroga, 1989), by using Laplace transforms of the basic mass balance equations, the diffusion and reaction rate coefficients are integrated into a single coefficient. This is of limited use, because film diffusion is ignored in this development. Sometimes a porous film of a solid product is deposited on the ash layer, adding one more resistance to the overall process. An example is the reduction of ilmenite with hydrogen forming a porous film of iron on the ash (Briggs and Sacco, 1991). Thus, one has to be cautious in routinely applying the additivity of resistances principle for treating combined control as was done in developing Equation (11.65) or (11.66).

A practically useful extension of SCM is when a relatively fragile solid undergoes reaction by SCM, such as a pellet of lime encased in a strong shell. This creates an additional diffusional layer. The development of such a pellet and its modeling are described in Case Study 11.6.

An alternative approach is to express the results in analogy with those for catalytic systems, in terms of an effectiveness factor (Ishida and Wen, 1968). Unlike in catalytic pellets, here the rate changes with time. Hence, the effectiveness factor also changes with time (i.e., with r_i), and the following equation can be derived for a first-order reaction in a sphere:

TABLE 11.13
Time-Conversion Relationships for SIM for Different Particle Geometries[a]

$$\upsilon_A A(g) + \upsilon_B B(s) \rightarrow R(g) + S(s)$$

		Functional forms for		
Controlling Regime	Flat Plate	Cylinder	Sphere	τ
Film diffusion, $f_1(X_B)$	X_B	X_B	X_B	$\dfrac{\rho_B R}{\upsilon k_g [A]_b}$
Ash diffusion, $f_2(X_B)$	X_B^2	$X_B + (1-X_B)\ln(1-X_B)$	$1 - (1-X_B)^{2/3} + 2(1-X_B)$	$\dfrac{\rho_B R^2}{2\upsilon D_{e,As}[A]_b}$
Reaction, $f_3(X_B)$	X_B	$1(1-X_B)^{1/2}$	$1-(1-X_B)^{1/3}$	$\dfrac{\rho_B R}{\upsilon k_s [A]_b^n}$

[a] Conversion $X_B = 1 - (r/R)^s$, where $s = 1, 2,$ and 3 for flat plate, cylinder, and sphere, respectively.

$$\varepsilon = \left[1 + \hat{R}_i Da \left(\frac{1}{Sh} + \frac{1-\hat{R}_i}{\hat{R}_i}\right)\right]^{-1} \tag{11.67}$$

where Sh (same as Bi_m) $= k_G R/\mathbf{D}_{e,As}$ and $Da = k_S R[A]_b^{m-1}/\mathbf{D}_{e,As}$ (with $m = 1$). The equation takes an implicit form for nonfirst-order reactions.

SCM is a phenomenological model that predicts the total conversion of a solid in a finite time and is well suited for many practical systems. However, it cannot account for such features as the leveling off of conversion at a value lower than the total conversion. Most importantly, it is not suitable for porous solids.

11.3.2.1.2 Volume Reaction Model

When the solid is porous, the reaction occurs throughout the pellet, with no sharp interface. If diffusion is assumed to be fast, the gas concentration will be uniform throughout the pellet, leading to the so-called homogeneous model. The rate of reaction can then be simply written as

$$r_A = k_v [A]^m [B]^n \tag{11.68}$$

The general conservation equations for the solid and reactant species for the volume reaction model in dimensionless form are as follows:

$$\nabla^2 [\hat{A}] = \phi^2 [\hat{A}]^m [\hat{B}]^n \tag{11.69}$$

$$-\frac{d[\hat{B}]}{d\hat{t}} = [\hat{A}]^m [\hat{B}]^n \tag{11.70}$$

where

$$\phi = \left[\frac{[A]_b^{m-1}[B]_0^n}{D_{e,As}}\right]^{1/2} \tag{11.71}$$

and

$$\hat{t} = \nu k_s [A]_b^m [B]_0^{n-1} t \tag{11.72}$$

The boundary conditions are

$$\hat{t}=0, \quad \hat{R}=1: \lfloor \hat{B} \rfloor = 1$$

$$\hat{t}>0, \quad \hat{R}=1: \frac{d\lfloor \hat{A} \rfloor}{d\hat{R}} = Sh\left(1-\lfloor \hat{A} \rfloor\right) \quad \text{or} \quad [\hat{A}]=1 \tag{11.73}$$

$$\hat{t}>0, \quad \hat{R}=0: \frac{d\lfloor \hat{A} \rfloor}{d\hat{R}} = 0$$

No analytical solution for this set of equations is possible for arbitrary values of m and n. However, analytical solutions can sometimes be found. For example, for the homogeneous model corresponding to low values of ϕ (and hence uniform concentration of) A throughout the pellet, the solution is

$$X_B = \begin{cases} 1-\exp(-\hat{t}) & \text{for } n=1 \\ \hat{t} & \text{for } n=0 \end{cases} \tag{11.74}$$

A practically important case is when $m = 1$. In this situation, Equations (11.69) and (11.70) can be combined into a single equation by defining a cumulative gas concentration as

$$\psi = \int_0^{\hat{t}} [\hat{A}] d\hat{t} \tag{11.75}$$

The transformed equation for the case of $m = n = 1$ is as follows (del Borghi et al., 1976; Dudukovic and Lamba, 1978; see also Ramachandran and Kulkarni, 1980):

$$\nabla^2 \psi = \phi^2 \left[1-\exp(-\psi)\right] \tag{11.76}$$

with boundary conditions

$$\psi = \hat{t}, \quad \left(\frac{d\psi}{d\hat{R}}\right)_{\hat{R}=0} = 0 \tag{11.77}$$

This transformation is extremely useful for systems with no structural changes and for reactions with power law kinetics.

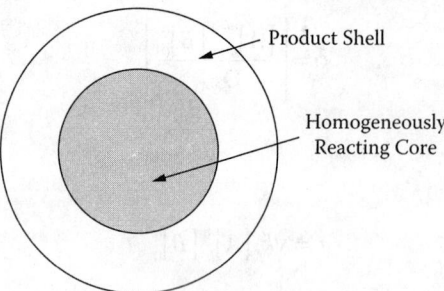

FIGURE 11.9 The two-zone model for gas-solid noncatalytic reactions.

The importance of reaction orders m and n has been examined at length (see, e.g., Doraiswamy and Sharma, 1984). The case of $m = 1$, $n = 0$ (i.e., zero order with respect to the solid reactant) is particularly important, since the gas concentration can drop to zero within the pellet, depending on the value of ϕ. In fact, a critical value given by

$$\phi_{cr} = \frac{6}{2/Sh + 1} \tag{11.78}$$

exists beyond which the concentration of A can fall to zero at some point within the pellet. For $\phi < \phi_{cr}$, the concentration would be finite at all points in the pellet, and Equation (11.75) describes the conversion-time behavior.

11.3.2.1.3 The Zone Models

The homogeneous model behaves in part as a shrinking core model when the reaction-diffusion interaction is such that the outer layers become exhausted, leading to the formation of an ash layer as in SIM. The difference, however, is that the reaction is not topochemical; i.e., it is not confined to the interface but occurs throughout the reactant matrix (core) as in the homogeneous model (Figure 11.9). Ishida and Wen (1968) have derived equations for this so-called two-zone model. A more general model is, however, one in which a reaction zone is sandwiched between the ash layer and the unreacted core (Bowen and Cheng, 1969). The model, sketched in Figure 11.10, is characterized by three stages (Mantri et al., 1976): (1) zone formation starting from the pellet surface until it has reached a thickness determined by the reaction-diffusion interaction for the system, (2) zone travel to the interior leaving a layer of ash as the shell, and (3) zone collapse as it merges with the core (thus becoming a two-zone model), the reaction continuing in the core until the entire solid is exhausted. The experimental results of Prasannan and Doraiswamy (1982) on the oxidation of zinc sulfide reveal three stages of the reaction. The zone width is clearly a function of the Thiels modulus. Their model reveals that, when the zone thickness is zero, it reduces to SIM, and when it is of the pellet dimension, it reduces to the homogeneous model. In an extension of this model to reversible reactions, Khan (1999) postulates two reaction zones.

11.3.2.1.4 The Particle-Pellet or Grain Models

Although such models are based on the granular structure of the pellet, the first version did not account for changes in the grain size. Hence, they can broadly be considered as spanning the macroscopic models in which pore evolution is ignored and the structural models in which the progress of reaction is explicitly related to pore evolution with time (i.e., to structural changes). The basic feature of these models, sketched in Figure 11.11, is that the grains constituting a pellet are spherical and of the same size, that each grain reacts according to SIM, and that the size of the grain does not change with reaction (thereby implying no voidage change with reaction and hence no pore evolution).

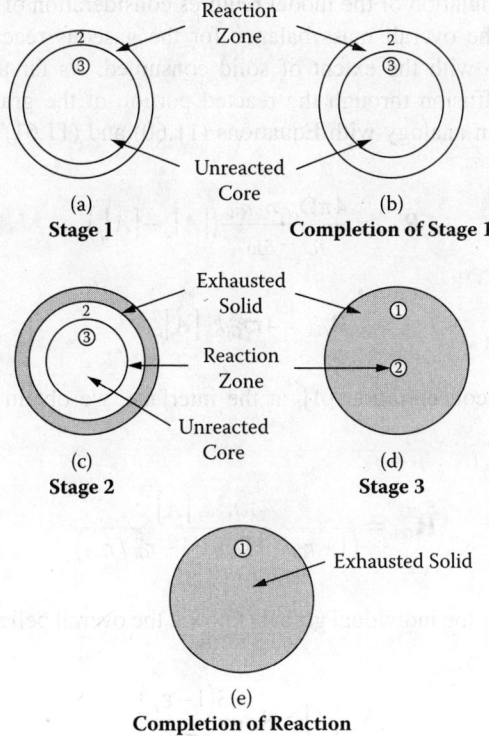

FIGURE 11.10 The three-zone model for gas-solid noncatalytic reactions.

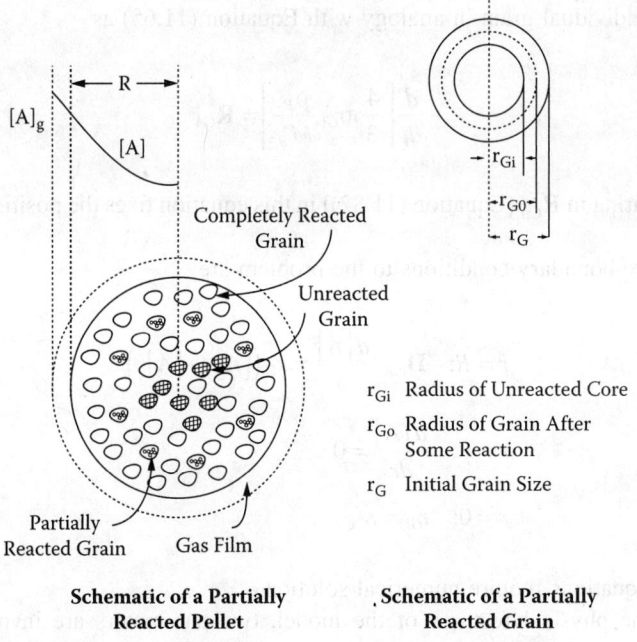

FIGURE 11.11 The particle-pellet model for gas-solid noncatalytic reactions.

The mathematical formulation of the model requires consideration of the rate processes within an individual grain, and the overall mass balance for the gaseous reactant in the pellet and its stoichiometric relationship with the extent of solid consumed. As far as the individual grain is concerned, the rates of diffusion through the reacted portion of the grain and of reaction at the interface can be obtained in analogy with Equations (11.60) and (11.61) as

$$\mathbf{R}_{GA} = \frac{4\pi \mathbf{D}_{eG} r_{Gi} r_{G0}}{r_{Gi} - r_{G0}} \left([A]_S - [A]_i\right) \tag{11.79}$$

$$\mathbf{R}_{GA} = 4\pi r_{Gi}^2 k_S [A]_i \tag{11.80}$$

Eliminating the unknown concentration $[A]_i$ at the interface, we obtain the overall rate per unit grain as

$$\mathbf{R}_{GA} = \frac{4\pi r_{Gi}^2 k_s [A]}{\left(1 + r_{Gi} k_s / \mathbf{D}_{eG}\right)\left(1 - r_{Gi} / r_{G0}\right)} \tag{11.81}$$

Once the rate of reaction for the individual grain is known, the overall pellet equation can be written:

$$\nabla^2 [A] = \mathbf{R}_{GA} \frac{3(1-\varepsilon_p)}{4\pi r_{G0}^3} \tag{11.82}$$

where the term in parentheses refers to the grains in the pellet volume. The term $\mathbf{R}_{GA}$ involves a knowledge of the interfacial position r_{Gi} within each grain, which is a function of both the time and position within the pellet. To evaluate this, a stoichiometric balance on the solid reactant B can be written for an individual grain in analogy with Equation (11.65) as

$$-\frac{d}{dt}\left[\frac{4}{3}\pi r_{Gi}^3 \frac{\rho_B}{M_B}\right] = \mathbf{R}_{GA} \tag{11.83}$$

The term $[A]$ appearing in $\mathbf{R}_{GA}$ [Equation (11.82)] in this equation fixes the position of the individual grain in the pellet.

The appropriate boundary conditions to the problem are

$$r = R: \quad \mathbf{D}_{e,As} \frac{d[A]}{dr} = k_G \left([A]_b - [A]_S\right)$$

$$r = 0: \quad \frac{d[A]}{dr} = 0 \tag{11.84}$$

$$r = 0: \quad r_{Gi} = r_{G0}$$

In general, these equations require numerical solution.

Considering the physical features of the model, two parameters are involved: τ_G, the time required for complete conversion of the grain in the $[A]$ environment; and τ_p, the time for complete conversion of the particle by diffusion if the grain conversion process is extremely fast. In the limiting case of grain diffusion controlling, the simple homogeneous model is recovered. The individual grains could follow the shrinking core model with ash diffusion or reaction controlling. Because the processes within the grain determine the system behavior, the conversion-time relationship is independent of the pellet dimensions. On the other extreme, when diffusion within the pellet controls, one observes shrinking core behavior with ash diffusion control, and typically the system behavior is dependent on the pellet dimensions ($t \propto R^2$). In the intermediate region where τ_G and τ_p are of the same order of magnitude, one expects the pellet behavior to lie within the limiting cases of shrinking core with reaction and ash diffusion control.

The grain models are useful in cases where pellets are formed by compaction of particles in very fine sizes. This is not so in some naturally occurring minerals, in which case fictitious grains are assumed. Also, the model, in its simplest form, does not explain S-shaped behavior and leveling off of conversion.

11.3.2.1.5 Other Models

Nucleation effects are often significant in systems such as reduction of metallic oxides. In these systems, the process proceeds with the generation of nuclei, which subsequently grow and finally overlap. When the nuclei generation rate is faster, the whole surface becomes covered with the metallic phase, and the reaction proceeds topochemically. On the other hand, for a slow generation rate, the metal-oxide interface is irregular, and different considerations prevail in estimating the conversion-time relationships. The following empirical model, proposed by Arvani (1941), is often valid:

$$X_B = 1 - \exp(-at^b) \qquad (11.85)$$

where a and b are constants. Modified forms have been suggested by Erofeev (1961), Ruckenstein and Vavanellos (1975), Rao (1979) and Bhatia and Perlmutter (1980), and experimental data of Neuberg (1970) and El-Rahaiby and Rao (1979) validate the model.

11.3.2.2 Extensions to the Basic Models

Some basic macroscopic models were described in the previous sections, including the slightly more rigorous particle-pellet model. These models ignored several complexities, mainly the effects of bulk flow, nonisothermicity, and variations in structure due to reaction. The first two can be included in the basic models and become extensions of them. The effects of structural changes can, however, be better defined in newer models that incorporate them at a more basic level.

11.3.2.2.1 Bulk-Flow or Volume-Change Effects

In addition to diffusion, bulk flow can occur within a reacting pellet (Beveridge and Goldie, 1968; Gower, 1971; Sohn and Sohn, 1980). This effect is considerably magnified for reactions with volume change such as

$$C(s) + CO_2(g) \rightarrow 2CO(g) \qquad \text{R11.12}$$

$$FeCl_2(s) + H_2(g) \rightarrow Fe(s) + 2HCl(g) \qquad \text{R11.13}$$

Note that reaction R11.12 is different from those in which the molar volume of the solid itself changes, leading to a structural change as the reaction progresses. Both effects can occur simultaneously, as in the second reaction shown above.

For reactions with a change in the gas volume, a continuity equation can be written with appropriate boundary conditions (analogous to those for catalytic reactions by Weekman and

Goring, 1965) and nondimensionalized to incorporate the effect through a dimensionless quantity for volume change. The final asymptotic solution (Sohn and Sohn, 1980) obtained is

$$\frac{\ln(1+\theta)}{\theta}\frac{\hat{t}}{\phi^2} = 1 - \frac{(\beta+1)(1-X_B)^{2/(\beta+1)} - 2(1-X_B)}{(\beta-1)} = f(X_B) \qquad (11.86)$$

where β is the shape factor and ϕ and θ are, respectively, the Thiele modulus and a volume change modulus defined as

$$\phi = \frac{sV_p}{A_p}\sqrt{\frac{k_V}{2\mathbf{D}_{e,As}}} \qquad (11.87)$$

$$\theta = \left(\frac{v_R}{v_A} - 1\right)X_A$$

Clearly, $\theta = 0$ for a reaction with no volume change.

These equations are, however, not applicable to nonisothermal reactions. SIM equations are based on the applicability of the PSS assumption, which is not applicable to nonisothermal reactions.

11.3.2.2.2 Effect of Nonisothermicity

Although SIM generally cannot be applied to analyze a nonisothermal reaction, it is well suited for certain decomposition reactions (Narsimhan, 1961; Hills, 1968; Campbell et al., 1970), such as

$$A(s) \rightarrow R(s) + S(g) \qquad \text{R11.14}$$

The phase rule suggests one degree of freedom. Each temperature, therefore, has a fixed value of partial pressure of the product gas S. Once this value is reached, the reaction starts, and the front moves inside. The process typically yields SIM behavior and is controlled either by heat or gas diffusion through the product layer. For heat transfer through the product layer controlling, the interface stays isothermal, and the equation for SIM with $\mathbf{D}_{e,As}$ replaced by the corresponding heat transfer parameters in the definition of represents the conversion-time behavior. Where gas diffusion is controlling, the variation of $\mathbf{D}_{e,As}$ with temperature should be accounted for. This variation usually takes the form $\mathbf{D}_{e,As} = T^{1.5-2.0}$ in the bulk diffusion regime with $\mathbf{D}_{e,As} = T^{0.5}$ in the Knudsen regime. Luss and Amundson (1969) have provided a more rigorous analysis that incorporates the transient heat accumulation term and gives the interface temperature as a function of the interfacial position r_i.

11.3.2.3 Models That Account for Structural Variations

The main structural changes that occur in a solid are those due to reaction and sintering.

11.3.2.3.1 Effect of Reaction

The reaction effect is mainly the result of the difference in molal volumes of the product and reactant solids, leading to voidage and therefore diffusivity changes as the reaction progresses. To incorporate these effects in any model, it is necessary to relate the overall solids conversion to voidage and diffusivity. An important feature of the structural effect is that when the porosity at the surface of the solid becomes zero (pore closure), the governing equations predict incomplete conversion, so often observed in gas-solid reactions (and not predicted by the basic models).

One way to account for structural changes is to allow for changes in the grain size in the particle-pellet model (Garza-Garza and Dudukovic, 1982a, b). A more useful way is to incorporate

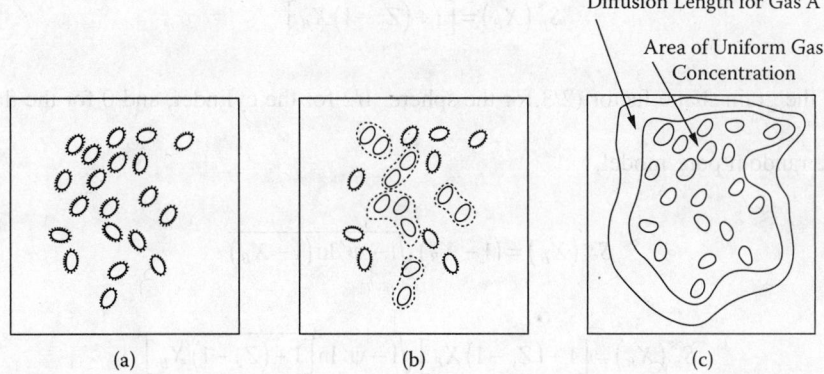

FIGURE 11.12 The random pore model of gas-solid noncatalytic reactions: stages in surface development.

the effect through changes in pore size distribution. The simplest of such models is the single pore model of Ramachandran and Smith (1977a) and Chrostowski and Georgakis (1978). In the Ramachandran–Smith approach, changes in a single pore are assumed to reflect changes in the pellet as a whole. The pore contracts, expands, or remains unchanged depending on whether there is an increase, decrease, or no change in the solid volume due to reaction. The model yields a simple conversion-time relationship based on the pore radius and length and the radius of the associated solid. Ulrichson and Mahoney (1980) have extended this model to incorporate the effects of bulk flow and reversibility of the reaction.

Perhaps the most realistic model is the random pore model of Bhatia and Perlmutter (1980; 1981a, b; 1983), which assumes that the actual reaction surface of the reacting solid B is the result of the random overlapping of a set of cylindrical pores. Surface development as envisaged in this model is illustrated in Figure 11.12. The first step in model development is therefore the calculation of the actual reaction surface, based on which the conversion-time relationship is established in terms of the intrinsic structural properties of the solid. In the absence of intraparticle and boundary layer resistances, the following relationship is obtained:

$$\hat{t} = \frac{k_S[A]_b^m S_0 t}{1-\varepsilon_0} = \frac{dX_{B1}}{S^*(X_{B1})} = \frac{\beta'}{2} \int_0^{X_B} \int_0^{X_{B1}} \left[\frac{1}{S^{*2}(X_B)} + \frac{Z_v - 1}{S_p^{*2}(X_B')} \right] dX_B' dX_{B1} \quad (11.88)$$

where X_{B1} and X_B' are dummy variables; S^* and S_p^* refer, respectively, to dimensionless reaction surface area and pore surface area; and

$$\beta' = \frac{2k_s \nu_A \rho_B (1-\varepsilon_0)}{\nu_B M_B D_{e,As} S_0} \quad (11.89)$$

This characterizes the diffusional resistance to the flow of gas (zero for kinetic control and infinity for product layer diffusion control). Expressions for S^* and S^*_p depend on the reaction model used. Thus:

For the grain model,

$$S^*(X_B) = (1-X_B)^g \quad (11.90)$$

$$S_p^*(X_B) = \left[1 + (Z_v - 1) X_B\right]^g \tag{11.91}$$

where g is the grain shape factor (2/3 for the sphere, 1/2 for the cylinder, and 0 for the flat plate).

For the random pore model,

$$S^*(X_B) = (1 - X_B)\sqrt{1 - \psi' \ln(1 - X_B)} \tag{11.92}$$

$$S_p^*(X_B) = \left[1 + (Z_v - 1) X_B\right]\sqrt{1 - \psi' \ln\left[1 + (Z_v - 1) X_B\right]} \tag{11.93}$$

where ψ' is a structural parameter defined by

$$\psi' = \frac{1}{\ln\left(\dfrac{1}{1 - \varepsilon_0}\right)} \tag{11.94}$$

for uniform pore radius. The equation becomes complicated for a nonuniform radius (see Bhatia and Perlmutter, 1983). Substituting the expressions for any of these models in Equation (11.88) and integrating leads to the desired conversion-time relationship. Although the random pore model appears more realistic, the predictions of the grain model are surprisingly close to those of this model. Several improvements, many marginal, have been suggested (see Bhatia and Gupta, 1993).

11.3.2.3.2 Effect of Sintering

The use of high temperatures in certain reactions, such as those in gas cleaning using lime-based adsorbents or exothermic reactions with generation of large amounts of heat, leads to sintering of the solid. It becomes more severe at higher temperatures (usually over 800 K), causing a decrease in the effective diffusivity of the solid or an increase in grain size leading to a lower specific area. Also, there could be a decrease in porosity and an increase in the tortuosity factor, both leading to a lowering of the effective diffusivity. Empirical models have been used for sintering, such as exponential decay for diffusivity and first-order decay for surface area (Ranade and Harrison, 1979, 1981). The combined effects of the two have been considered by Kim and Smith (1974), Chan and Smith (1976), and Ramachandran and Smith (1977b). The following is recommended:

$$\mathbf{D}_e = \frac{1}{F(f_p)} \left[1 - (1 - \varepsilon_0)\left(\frac{r_{Gi}}{r_{G0}}\right)^3\right]^2 (1 - f_p) \tag{11.95}$$

where f_p is the fraction of pores removed and is given by

$$\frac{df_p}{dt} = k_p (1 - f_p) \tag{11.96}$$

where k_p is the rate constant for pore removal.

TABLE 11.14
Functional Forms of $f(X_B)$ for Different Gas-Solid Noncatalytic Reaction Models

Number	Functional Form $f(X_B)$	Reaction Model
1	$(1 - X_B)^{-n}$	Volume reaction model
2	$-1 - \dfrac{(1-X_B)^{1/3} + (1-X_B)^{2/3}}{Sh}$	Grain model
3	$-1 - \dfrac{(1-X_B)^{1/3}}{Sh} + \dfrac{(1-X_B)^{2/3}}{Sh\left[Z_V + (1-Z_V)(1-X_B)^{1/3}\right]}$	Grain model with structural variations
4	$\dfrac{1}{n}\left(\dfrac{1}{1-X_B}\right)\ln\left(\dfrac{1}{1-X_B}\right)^{(1-n)/n}$	Nucleation model

11.3.2.4 A General Model That Can Be Reduced to Specific Ones

A general mathematical formulation that is applicable to most models described in the earlier sections begins with the volume reaction model described by Equations (11.69) through (11.72). Then, for a reaction first order in the gaseous component, we recast these equations as

$$\frac{1}{\xi^\beta}\frac{\partial}{\partial \xi}\left(\alpha \xi^\beta \frac{\partial \left[\hat{A}\right]}{\partial \xi}\right) = \phi^2 \frac{\partial X_B}{\partial \hat{t}} \tag{11.97}$$

$$\frac{dX_B}{d\hat{t}} = \frac{\left\lfloor \hat{A} \right\rfloor}{f[X_B]} \tag{11.98}$$

where α refers to the diffusivity ratio $\mathbf{D}_{e,As}/\mathbf{D}_{e,As0}$, X_B is the solid conversion, and ϕ is the Thiele modulus [Equation (11.87)]. Using the cumulative gas concentration defined by Equation (11.75), the following final equation can be developed:

$$f(X_B)\frac{d^2}{d\xi^2} + \beta \frac{f(X_B)}{\xi}\frac{dX_B}{d\xi} + f'(X_B)\left(\frac{dX_B}{d\xi}\right)^2 - k(X_B) = 0 \tag{11.99}$$

The solution of this equation gives conversion profiles. The equation is sufficiently general, since several different models can be incorporated in it. Typical forms of $f(X_B)$ for some of the more frequently used models are presented in Table 11.14. Prasannan et al. (1986) employed a collocation procedure to solve this equation for the models considered in the table.

11.3.3 FLUID-FLUID REACTIONS

Fluid-fluid (gas-liquid and liquid-liquid) reactions are of great industrial importance and include gas-liquid, liquid-liquid, gas-liquid-solid (noncatalytic and catalytic), and solid-liquid systems.

TABLE 11.15
Examples of Industrially Important Gas-Liquid Reactions

Regime 1: Very slow reaction
Air oxidation of a variety of aliphatic and alkyl aromatic compounds; air oxidation of p-nitrotoluene sulfuric acid; substitution chlorination of a variety of organic compounds; reaction between isobutylene and acetic acid; oxidation of ethylene to acetaldehyde (Wacker processes); hydrochlorination of olefins; absorption of phosphine in an aqueous solution of formaldehyde and hydrochloric acid; acetic acid from the carbonylation of methanol; oxidation of tri-alkyl phosphine; dimerization of olefins.

Regime 2: Slow reaction
Absorption of CO_2 in carbonate solution; absorption of O_2 in aqueous acid solutions of CuCl at concentrations less than 1×10^{-4} mol/cm^3; oxidation of organic compounds; preparation of the C-13 isotope.

Regime between 1 and 2
Absorption of CO_2 in carbonate buffer solutions in packed columns; oxidation of black liquor in the paper and pulp industry; absorption of CO_2 in carbonate solution; wet air oxidation of soluble compounds in waste water.

Regime 3: Fast reaction
Absorption of CO_2 and COS in aqueous solutions of amines and alkalies; absorption of oxygen in aqueous acidic and neutral solutions of cuprous chloride, and in cuprous and cobaltous amine complexes in aqueous and polar solutions; absorption of oxygen in aqueous alkaline solutions of sodium dithionite (hydrosulfite); absorption of oxygen in aqueous sodium sulfite; absorption of $NO/NO_2/N_2O_4$ in water and aqueous solutions containing reactive species; reaction between dissolved NO and O_2; absorption of oxygen and ozone in aldehydes; absorption of isobutylene, 2-butene, and 2-methyl-2-butene in aqueous solutions of sulfuric acid; absorption of isobutylene in aqueous solutions containing thallium (III) ions; absorption of lean phosphine in aqueous solutions of sodium hypochlorite and concentrated sulfuric acid; absorption of ozone in aqueous and non-aqueous solutions, with or without dissolved organic chemicals; oxidation of cyanide ions in aqueous alkaline solutions and organic media containing olefinic substances; waste water treatment; oxidation of organometallic compounds; hydrogenation of unsaturated compounds with homogeneous catalysts; absorption of Cl_2 in aqueous solutions containing phenolic substances and aromatic nitro and sulfuric compounds; absorption of Cl_2 in aqueous and non-aqueous solutions of ketones.

Regime 4: Instantaneous reaction
Absorption of CO_2 in aqueous solutions of MEA; absorption of H_2S and mercaptans in aqueous solutions of alkanolamines and caustic soda; absorption of carbon monoxide in aqueous cuprous ammonium chloride solutions; absorption of lower olefins in aqueous solutions of cuprous ammonium compounds; absorption of pure chlorine in aqueous solutions of sodium carbonate or sodium hydroxide; conversion of dithiocarbamates to thiuram disulfides; sulfonation of aromatic compounds with lean SO_3; recovery of bromine from lean aqueous solutions of bromides; reactions of importance in pyrometallurgy; absorption of CO_2 in aqueous solutions of caustic alkalies and amine; absorption of O_2 in aqueous solutions of sodium dithionite; absorption of O_2 in aqueous sodium sulfite solutions; absorption of O_2 in alkaline solutions containing the sodium salt of 1,4-napthaquinone- 2-sulfonic acid (NQSA); special case: role of diffusion in the absorption of gases in blood in the human body.

Representative lists for gas-liquid and liquid-liquid reactions are given in Tables 11.15 and 11.16, respectively. Doraiswamy and Sharma (1984) have covered in detail the theory of mass transfer with simple and complex reactions. A summary of simple irreversible reactions is given below.

11.3.3.1 Theory of Mass Transfer Accompanied by Irreversible Chemical Reaction

The gas phase contains A and the liquid phase contains reactive species B. A is sparingly soluble in the B phase, and the reaction occurs exclusively in the B phase. The overall reaction involves two steps.

TABLE 11.16
Examples of Industrially Important Liquid–Liquid Reactions

Regime 1: Very slow reaction
Alkaline hydrolysis of a variety of organic compounds; nitration of chlorobenzene; alkylation of benzene with straight-chain olefins; reduction of aromatic nitro compounds to corresponding aromatic amines with aqueous Na_2S or Na_2S_x.

Regime 2: Slow reaction
Pyrometallurgical operations: open-hearth steel furnace; reaction of dinitrochlorobenzene; removal of mercaptans from petroleum fractions; alkaline hydrolysis of esters; nitration of aromatic compounds; nitration of higher olefins; liquid-liquid reactions in phosphorous chemistry; special case: measurement of the solubility of A in the B phase.

Regime between 1 and 2
Alkaline hydrolysis of acetate ester; sulfonation of benzene, toluene, and p-xylene.

Regime 3: Fast reaction
Alkaline hydrolysis of formate and halo-substituted acetic acid ester; manufacture of cyclohexanone oxime; manufacture of dithiocarbamate;, alkylation of isobutane with butenes and of toluene and xylenes with acetaldehyde (or acetylene); nitration of toluene; extraction of metals.

Regime 4: Instantaneous reaction
Extraction of free fatty acids from naturally occurring glycerides; removal of HCl from chlorinated organic compounds; recovery of aliphatic acids; HF and HCl from aqueous solutions; nitration of phenol; solvent extraction in mineral processing; interfacial polycondensation and esterification; manufacture of organo-phosphate pesticides.

1. Mass transfer of A from the gas phase to the liquid phase; assuming that the resistance to mass transfer is confined to the B phase, the rate of mass transfer is then given by

$$R_A a = k_L a \left([A^*] - [A]_0 \right) \quad (11.100)$$

2. The irreversible chemical reaction between the dissolved A and the liquid-phase reactive species B:

$$A(1) + Z\, B(1) \rightarrow \text{products} \quad \text{R11.14}$$

The rate of chemical reaction is given by

$$R_A a = \epsilon_L\, k_{mn} [A]^{*m} [B]_0^n \quad (11.101)$$

where ϵ_L is the fractional liquid-phase holdup. Since the reaction occurs exclusively in the liquid phase and the rate is expressed on the basis of total reactor volume, the usual rate expression is multiplied by ϵ_L.

The combined effect of mass transfer and chemical reaction depends on the relative rates of these steps. The possibilities are conveniently classified into four regimes: regime 1, very slow reactions; regime 2, slow reactions; regime 3, fast reactions; and regime 4, instantaneous reactions. The concentration profiles in these regimes are shown in Figure 11.13. In very slow and slow reaction regimes, no reaction occurs in the liquid film, and the steps of mass transfer and chemical reaction occur in series. In regime 1, the rate of chemical reaction is the controlling step, whereas the reverse is true in regime 2. In regime 3, diffusion and reaction occur in parallel, and all the reaction is completed in the diffusion film. In other words, all of A is consumed in the film and no

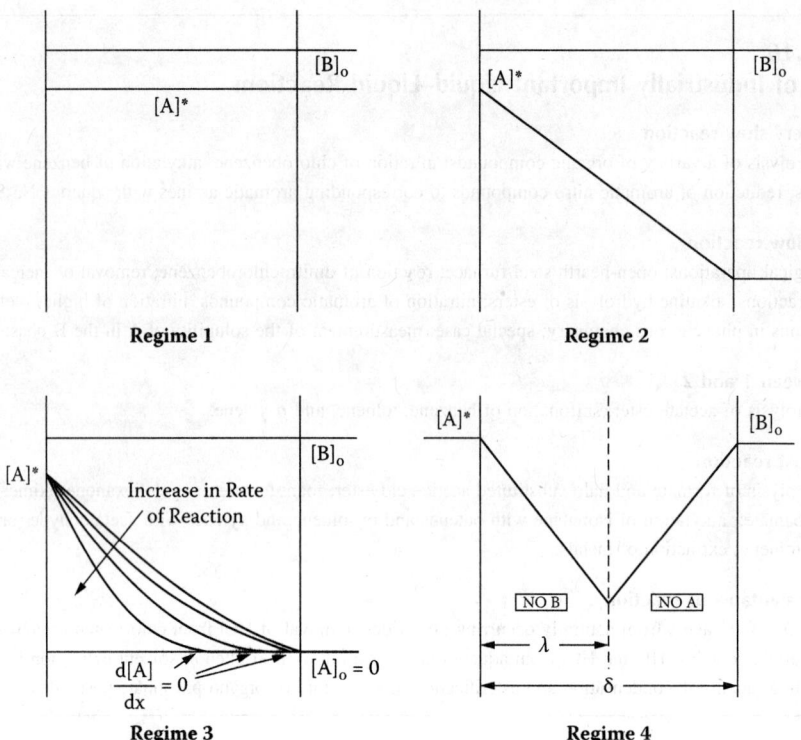

FIGURE 11.13 The different operating regimes in gas-liquid reactions.

A enters the bulk liquid. Furthermore, it can be seen from Figure 11.13 that the concentration profile of B is flat in the first three regimes. This means that the rate of diffusion of B is faster than the chemical reaction in the film. In regime 4, the reaction between A and B is instantaneous and occurs as soon as they meet at the reaction plane. The overall rate is controlled by the diffusion of A and B to the reaction plane.

Doraiswamy and Sharma (1984) have given detailed derivations for the overall rates of reaction and the conditions to be satisfied in each regime of operation. The final equations and conditions are summarized in Table 11.17. It is likely that a reaction falls in the intermediate range or more than one regime. For these cases, expressions for the overall rates are given in Table 11.18. Recently, Doraiswamy (2001) has given similar tables for some complex reactions.

11.3.3.2 Laboratory Reactors for Fluid-Fluid Reactions

For designing fluid-fluid reactors (described in Section 11.4), there are two basic requirements: (1) the flow behavior in reactors that governs the distribution of concentrations (including partial pressures) and temperature along the length and transverse directions and in all the phases, and (2) knowledge of overall rates at a given point in the reactor (with known concentrations and temperature). For the examples of industrial importance indicated in Tables 11.15 and 11.16, the equations for overall rates are given in Tables 11.17 and 11.18. For the estimation of overall rates of reaction, three types of information are needed: (1) physical properties of the system such as diffusivity, solubility, density, viscosity, and so on; (2) hydrodynamic parameters such as effective interfacial area, a, and true mass transfer coefficient $k_L a$; and (3) kinetic parameters such as rate constants, orders with respect to different reactants, and activation energy. The chapter by Harvey in this book suggests how to obtain the desired data. The work by Reid et al. (1977) is also helpful. The hydrodynamic parameters a and $k_L a$ depend on the type of reactor, flow rates of various phases,

TABLE 11.17
Theory of Mass Transfer Accomplished by an Irreversible Chemical Reaction: Overall Rates of Reaction and Conditions to Be Satisfied for the Four Regimes of Operation

No.	Regime	Conditions to be Satisfied	Overall Rate of Operation
1	Regime 1 (very slow reaction)	1. $k_L a[A]^* \gg \epsilon_L k_{mn}[A]^{*n}[B]_0^n$ 2. $\sqrt{M} \ll 1$	$R_A a = \epsilon_L k_{mn}[A]^{*n}[B]_0^n$
2	Regime 2 (slow reaction)	1. $k_A a[A]^* \ll \epsilon_L k_{mn}[A]^{*n}[B]_0^n$ 2. $\sqrt{M} \ll 1$	$R_A a = k_L a \{[A]^* - [A]_0\}$
3	Regime 3 (fast reaction)	1. $\sqrt{M} = \sqrt{\dfrac{2}{m+1} D_A k_{mn}[A]^{*n}[B]_0^n} > 3$ 2. $\sqrt{M} = \dfrac{[B]_0}{Z[A]_0}\sqrt{\dfrac{D_B}{D_A}} = \phi_a$	$R_A a = a[A]^* \sqrt{\dfrac{2}{m+1} D_A k_{mn}[A]^{*(n-1)}[B]_0^n}$
4	Regime 4 (instantaneous reaction)	$\sqrt{M} \gg \dfrac{[B]_0}{Z[A]^*}\sqrt{\dfrac{D_B}{D_A}}$	$R_A a = k_L a[A]^* \left[1 + \dfrac{[B]_0}{Z[A]^*}\sqrt{\dfrac{D_B}{D_A}}\right]$

and power consumption per unit volume. This subject is complex, and fundamental methods are not available in the published literature for the estimation of hydrodynamic parameters from basic principles. Often, empirical/semiempirical correlations have been reported in the literature. Several references, along with qualitative descriptions of fluid-fluid contactors, are given in Section 11.4.

The kinetic parameter can be estimated in laboratory reactors. For solid-fluid systems, this subject was described in Section 11.3.1.6. For fluid-fluid reactions, the commonly employed laboratory reactors include stirred cell, wetted wall column, rotating drum, laminar jet, stirred contractor, and others. These are schematically shown in Figure 11.14. In practically all of these reactors, the value of the fluid-fluid interfacial area is known. These reactors have been described by Treybal (1980) and Doraiswamy and Sharma (1984). As an illustration, the stirred cell will be described first, followed by a comparison with other laboratory reactors. The discussion of the stirred cell is restricted to gas-liquid systems, but it is also applicable (with minor variations) to liquid-liquid systems.

A schematic representation of a stirred cell is shown in Figure 11.14a. The cell consists of distinct regions of gas and liquid phases separated by a flat gas-liquid interface. Both the phases are stirred independently, usually from different drives. The impeller speeds are varied over a wide range but always ensuring that the interface remains flat so that the gas-liquid interfacial area can be taken equal to the cross-sectional area. There are several advantages to keeping the gas phase pure. First, for pure gases, the gas phase resistance is absent, and hence a gas-phase impeller need not be used. Second, the concentration of the gas phase can easily be varied by manipulating the total pressure. Third, for pure components, the extent of backmixing in the gas phase is not an issue. Finally, and most importantly, the rate of reaction can be easily followed by measuring the gas-phase pressure with respect to time. If there are restrictions in using a pure gas phase, its partial pressure is varied by adding an inert gas. The gas flow rate is adjusted such that the extent of conversion is small (low conversion to minimize the importance of gas-phase mixing). The gas-side resistance is eliminated by manipulating the

TABLE 11.18
Overall Rates of Reaction for Overlapping Regimes

1. Regime 1 and 2, $m = 1$, $n = 1$

$$R_A a = \frac{[A]^*}{\left[\dfrac{1}{k_L a}\right] + \dfrac{1}{\varepsilon_L k_{mn}[A]^{*m-1}[B]_0^n}}$$

2. Regime 2 and 3

$$R_A a = a[A]^* \sqrt{\dfrac{2}{m+1} \mathbf{D}_A k_{mn}[A]^{*m-1}[B]_0^n + k_L^2}$$

or

3. Regime 1, 2 and 3 ($m = 1$, $n = 1$)

$$R_A a = \dfrac{k_L a [A]^* \sqrt{M}}{\tanh \sqrt{M}}$$

$$R_A a = \dfrac{a[A]^*}{\dfrac{1}{\sqrt{\mathbf{D}_A k_2 [B]_0 + k_L^2}} + \dfrac{a}{\varepsilon_L k_2 [B]_0}}$$

4. Regime 3 and 4

$$R_A a = \Phi k_L a [A]^*$$

where

$$\Phi = \sqrt{M} \left[\dfrac{(\phi_a - \phi)}{(\phi_a - 1)}\right]^{0.5}$$

$$\Phi_a = 1 + \dfrac{[B]_0}{Z[A]^*} \sqrt{\dfrac{D_B}{D_A}}$$

5. Regime 1, 2, 3 and 4 together with gas side resistance ($m = 1$, $n = 1$)

$$R_A a = \Phi_L a [A]^*$$

where

$$\Phi_a = \sqrt{\dfrac{M^2}{4(\phi_a - 1)} + \dfrac{M \phi_a}{\phi_a - 1} + 1} - \dfrac{M}{2(\phi_a - 1)}$$

$$M = \dfrac{\mathbf{D}_A k_2 [B]_0}{k_L^2}$$

$$\phi_a = \_ + \sqrt{\beta^2 + \gamma}$$

$$\beta = 0.5(1 + f_1/f_3) + \sqrt{M}/f_3 - \sqrt{M}$$

$$\gamma = \sqrt{M}\,(1 + f_1/f_3)$$

$$f_1 = k_G / H_A k_L$$

$$f_2 = \dfrac{\mathbf{D}_B [B]_0}{Z \mathbf{D}_A [A]_0}$$

$$f_3 = f_1 / f_2$$

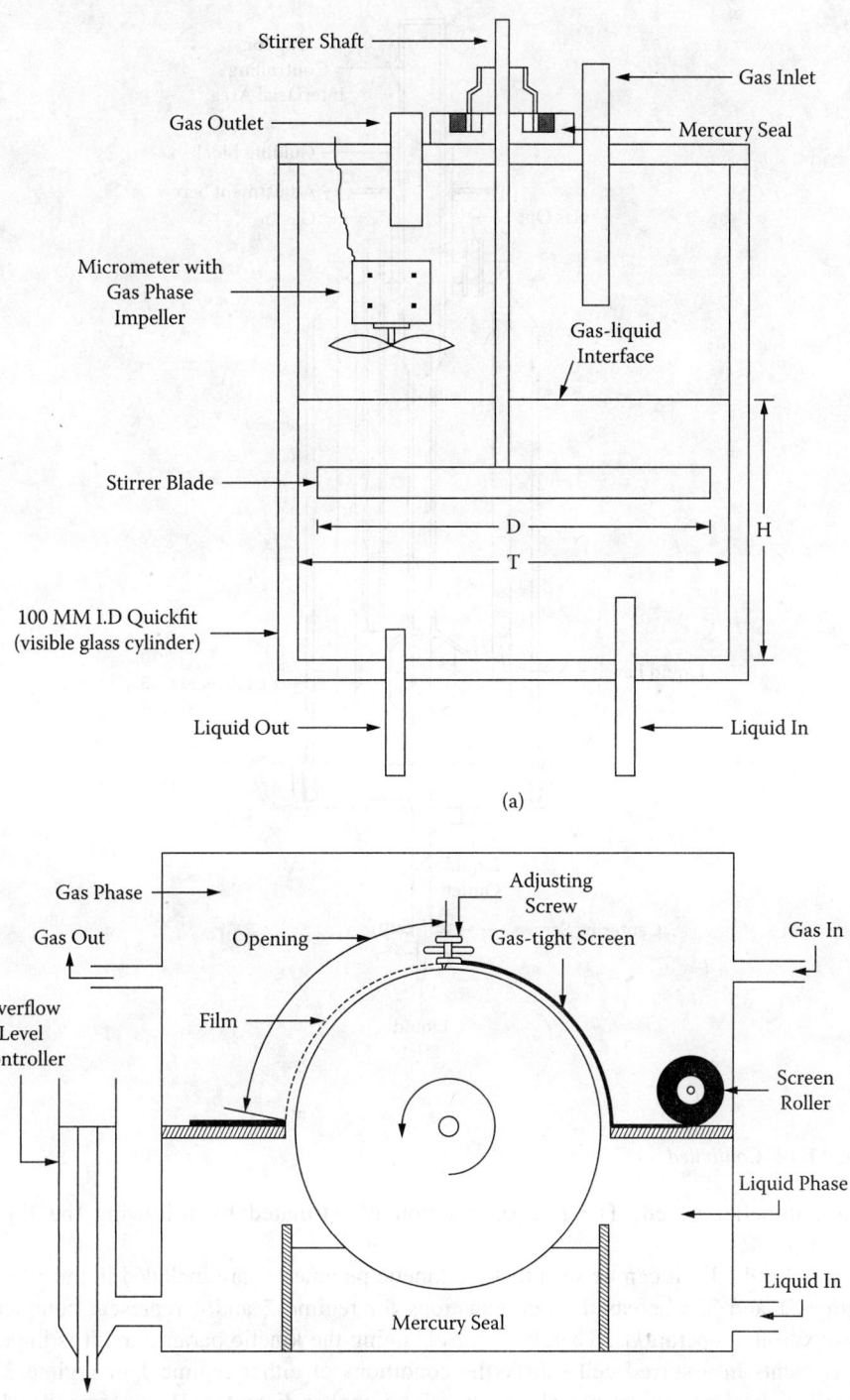

FIGURE 11.14 The stirred cell for obtaining precise kinetic data for fluid-fluid reactions: (a) stirred cell, (b) rotating drum, (c) cylindrical wetted wall, (d) wetted sphere, (e) laminar jet, and (f) stirred contactor. (From Danckwerts, P. V., *Gas-Liquid Reactions,* New York: McGraw-Hill, 1970.)

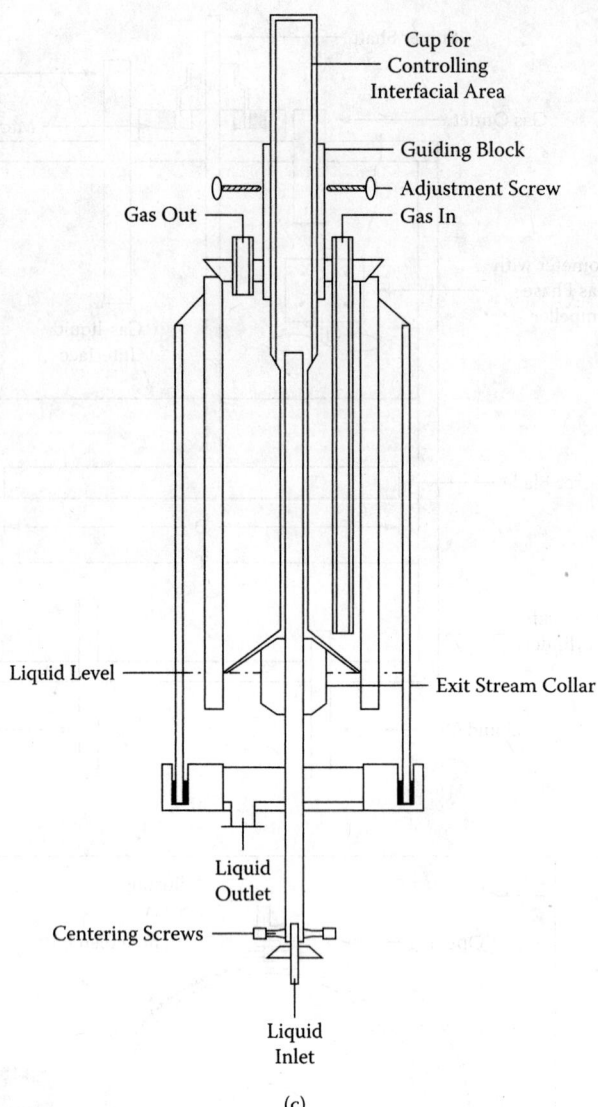

(c)

FIGURE 11.14 *Continued.*

gas-phase impeller speed. The rate of reaction is estimated by following the liquid-phase concentration.

From Table 11.17, it can be seen that the kinetic parameters are included in the rate equations for regimes 1 and 3, whereas the rate equations for regime 2 and 4 represent completely mass transfer-controlled operations. Therefore, for obtaining the kinetic parameters, it is important that the experiments in a stirred cell satisfy the conditions of either regime 1 or regime 3. A given stirred cell is characterized by vessel diameter (T), impeller diameter (D), and impeller design and location from the gas-liquid interface.

A typical stirred cell may have $T = 100$ mm, $D = 90$ mm, and a two-bladed paddle with $W/D = 0.1$ located 10 mm below the interface. The gas-phase impeller may be powered by a motor as shown in Figure 11.14a. A stepwise procedure for obtaining the kinetic parameters is outlined as follows:

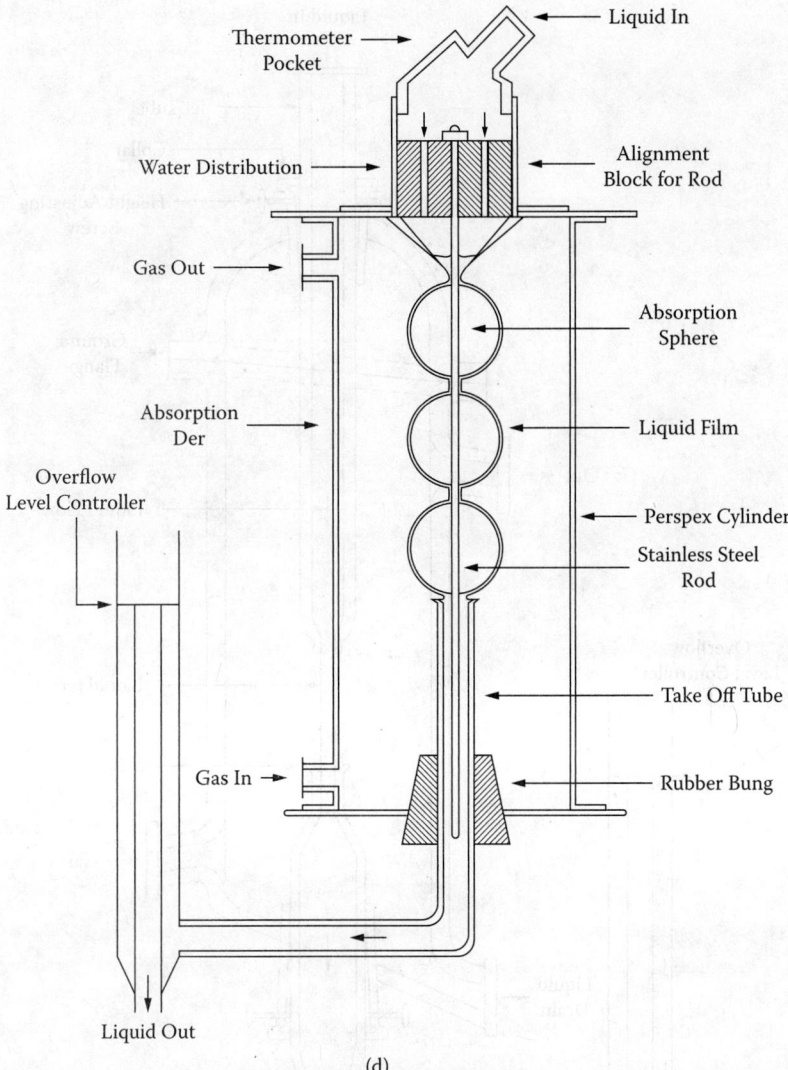

(d)

FIGURE 11.14 *Continued.*

1. For the gas-liquid system under consideration, the diffusivities of the gas-phase solute in both the phases and the liquid-phase reactant in the liquid phase are estimated. The solubility is also determined.
2. The values of the mass-transfer coefficient are measured at various impeller speeds using a model gas-liquid system such as CO_2–H_2O. Using these results of k_L vs. N, a similar relationship is obtained for the given gas-liquid system using the square root diffusivity relationship. Thus, at any impeller speed, the specific rate of absorption under mass transfer-controlled conditions (regime 2) can be obtained and is equal to $k_L[A]\,*$.
3. For a given gas-liquid system, the gas side should either be pure or used at a very high flow rate so that issues related to gas-phase back-mixing are eliminated. Furthermore, the gas-phase impeller speed is varied such that, beyond a particular impeller speed, the rate becomes independent of speed, ensuring that the gas-phase resistance is eliminated.
4. For a given gas-liquid system, the specific rate of absorption, R_A, is measured, and the enhancement factor is estimated using the following equation:

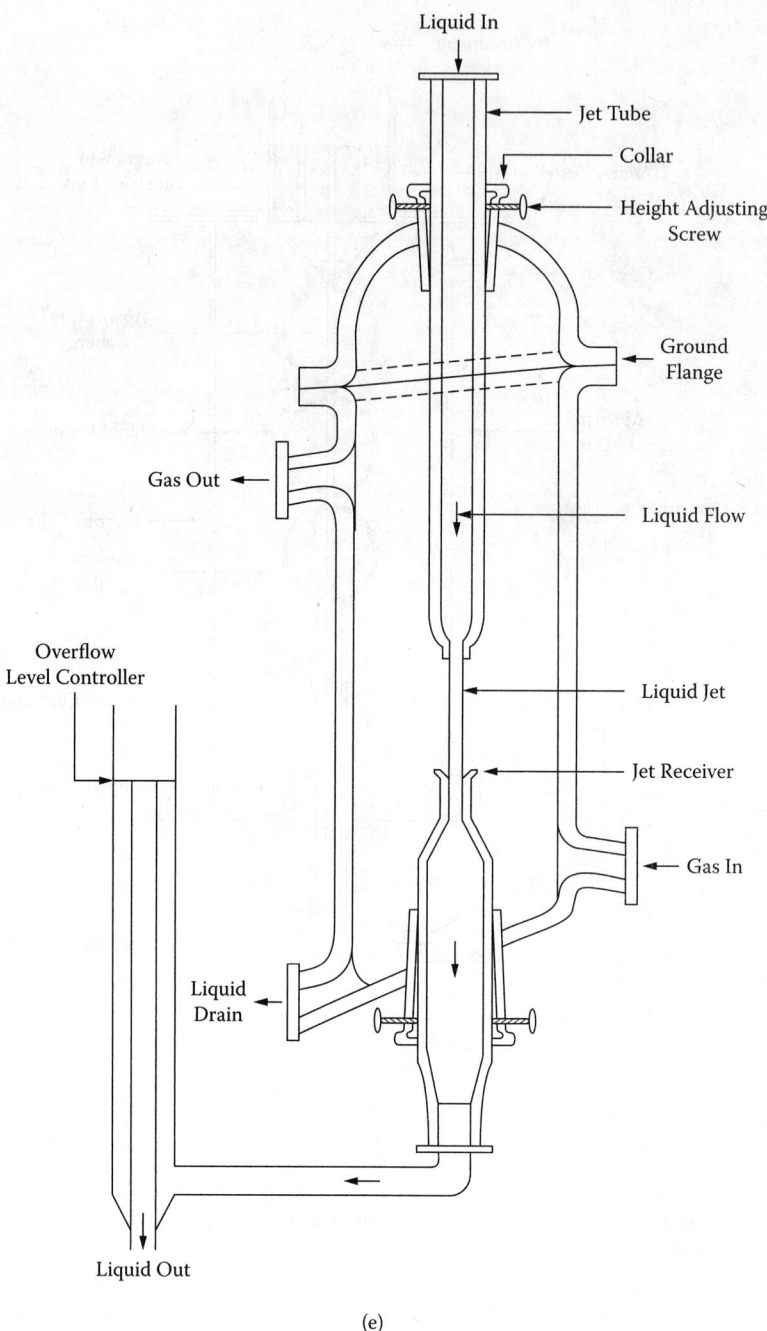

(e)

FIGURE 11.14 Continued.

$$\eta = \frac{R_A}{k_L[A]^*} \tag{11.102}$$

5. If $\eta < 1$, the absorption occurs between regimes 1 and 2. To achieve the conditions of regime 1, the impeller speed is increased such that R_A becomes independent of it, or ϕ becomes much less than 1. In regime 1, the overall rate of absorption (Table 11.17) is

Chemical Reaction Engineering

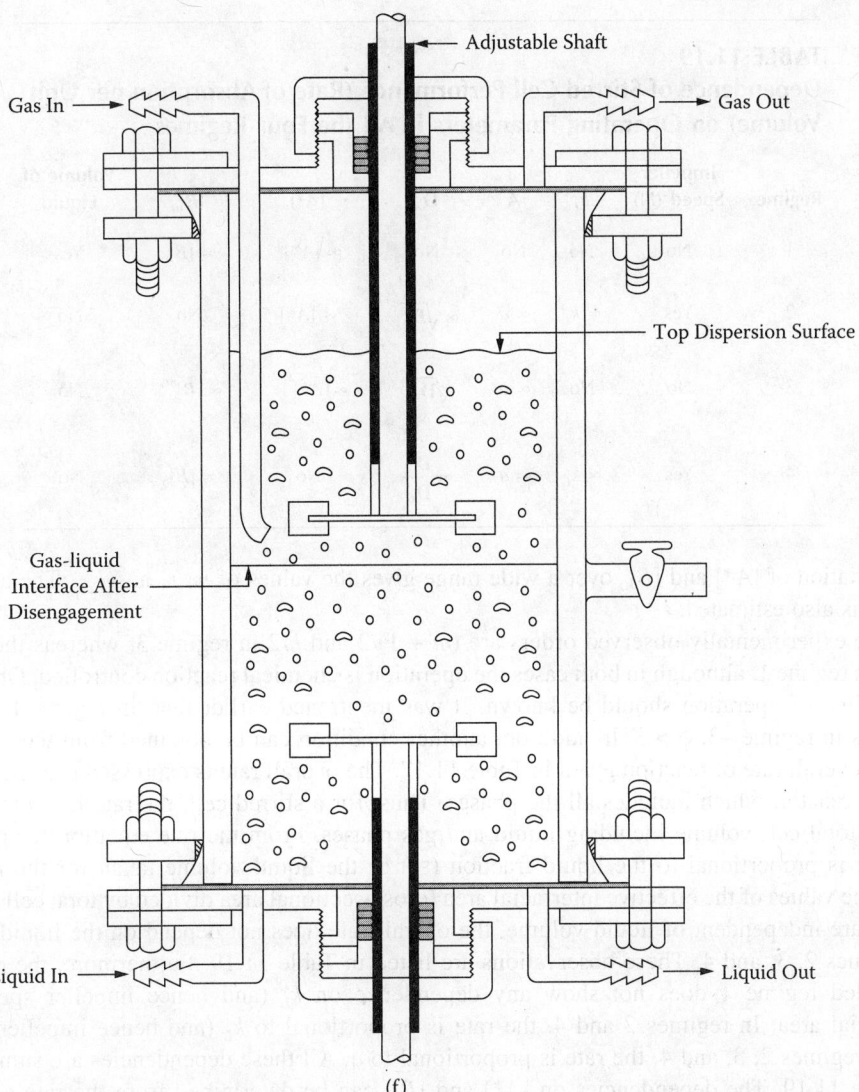

FIGURE 11.14 *Continued.*

given by Equation (11.101), in which ϵ_L is the ratio of liquid volume to total cell volume. The measurements of $R_A a$ over a wide range of gas-phase partial pressures (and hence $[A^*]$) and liquid-phase concentrations $([B]_0)$ give the values of the respective orders m and n and hence the rate constant k_{mn}. Experiments are also performed over a sufficient temperature range of practical interest to calculate the activation energy.

6. If $\phi \geq 1$, absorption occurs between regimes 2 and 3, and the specific rate is given in Table 11.18. Under these conditions, the impeller speed (and hence k_L) is reduced so that ϕ becomes greater than 3 and the absorption operation becomes fast, i.e., reaction-controlled (regime 3). The other condition $\sqrt{M} < 6 < \phi_a$ (Table 11.18), is also checked. The rate of absorption per unit liquid volume is given by

$$R_A a = a[A^*]\sqrt{\frac{2}{m+1}\mathbf{D}_A k_{mn}[A^*]^{m-1}[B]_o^n} \tag{11.103}$$

TABLE 11.19
Dependence of Stirred Cell Performance (Rate of Absorption per Unit Volume) on Operating Parameters in All the Four Regimes

Regime	Impeller Speed (N)	k_L	A	D_A	$[A^*]$	$[B]_0$	Volume of Liquid
1	No	No	No	No	$\propto [A^*]^m$	$\propto [B]_0$	Yes
2	Yes	$\propto k_L$	$\propto a$	$\propto \sqrt{D_A}$	$\propto [A^*]$	No	No
3	No	No	$\propto a$	$\sqrt{D_A}$	$\propto [A^*]^{\frac{m+1}{2}}$	$\propto [B]_0^{n/2}$	No
4	Yes	$\propto k_L$	$\propto a$	$\propto \dfrac{D_A}{D_B}$	No	$\propto [B]_0$	No

The variation of $[A^*]$ and $[B]_0$ over a wide range gives the values of m, n, and k_{mn}. The activation energy is also estimated.

The experimentally observed orders are $(m + 1)/2$ and $n/2$ in regime 3, whereas they are m and n in regime 1, although in both cases the operation is chemical reaction controlled. Obviously, the regime of operation should be known. It was mentioned earlier that, in regime 1, $\phi \ll 1$, whereas in regime –3, $\phi > 3$. In addition, another condition can be obtained from the equations for the overall rate of reaction given in Table 11.17. The overall rate is expressed per unit volume of total reactor, which includes all the phases. Thus, for a stirred cell, the rate is expressed per unit of total cell volume including liquid and gas phases. From the rate equation for regime 1, the rate is proportional to the liquid fraction (ε_L) or the liquid volume taken for the reaction. Since the values of the effective interfacial area (cross-sectional area divided by total cell volume) and k_L are independent of liquid volume, the overall rate does not depend on the liquid volume in regimes 2, 3, and 4. These observations are listed in Table 11.19. Furthermore, the reaction-controlled regime 1 does not show any dependence on k_L (and hence impeller speed) and interfacial area. In regimes 2 and 4, the rate is proportional to k_L (and hence impeller speed), and in regimes 2, 3, and 4, the rate is proportional to a. All these dependencies are summarized in Table 11.19. The dependencies on $[A^*]$ and $[B]_0$ can be determined from the rate equations given in Table 11.19.

In addition to the stirred cell, other laboratory reactors commonly used include rotating drum contactor, wetted wall column, wetted sphere column, laminar jet, and stirred contactor. These equipments are shown schematically in Figures 11.14b–f. All have several common features, the principal one being a well defined gas-liquid interfacial area and the ability to vary the area per unit reactor volume (a). In the stirred cell, it is achieved by varying the liquid height. As an alternative way, a solid circular baffle is placed at the gas-liquid interface. Holes are drilled on the baffle plate so that the hole opening area becomes the interfacial area. For varying a, baffle plates are made with different free (hole) areas.

In the rotating drum contactor (Figure 11.14b), the value of a is adjusted by sliding the gas-tight screen, thus permitting the opening over a wide range from fully closed (no interfacial area) to fully open where half the cylinder is exposed to the gas. In the wetted wall column (Figure 11.14c), a film is formed over a certain height of the rod, below which a tight collar removes the liquid and discharges it into the liquid pool at the bottom. The interfacial area is varied by changing the film height by positioning the collar. In the wetted sphere column (Figure 11.14d), a is varied by changing the number of spheres (Pigford and Pyle, 1951). In the four laboratory reactors just described, the range of k_L is practically the same. The value of k_L is varied by changing the liquid flow rate in the wetted wall column and wetted sphere column. In the stirred cell and rotating drum, the impeller and drum speeds are, respectively, manipulated for changing k_L.

As mentioned earlier, laboratory reactors are used for determining the intrinsic kinetic parameters. This objective is achieved by adjusting the parameters so the overall absorption operation falls in either regime 1 or regime 3, both being reaction controlled. It is preferable to obtain the regime 1 condition wherever possible. This is because, in regime 1, knowledge of diffusivity is not needed, and the reaction orders do not become modified (refer to Table 11.19). Although this may not be possible for all fluid-fluid reactions, in many cases regime 1 conditions can be obtained by lowering the value of $\sqrt{M}$. This objective is achieved mainly by selecting laboratory reactors that typically enhance k_L by a factor of 2 to 4. These include the laminar jet (Figure 11.14e) and stirred contactors. Typical ranges of k_L, a, contact time, and so forth of different laboratory reactors are given in Table 11.20.

In the laminar jet, a and k_L are varied by changing the jet length and liquid flow rate, respectively. The stirred contactor (Figure 11.14f) is similar to the stirred cell (Figure 11.14a), the chief differences being in impeller design and speed. In the stirred cell, the interfacial area is the cross-sectional area and the impeller speed is carefully adjusted to keep the interface visually flat. In the stirred contactor, turbines are used at high speed so that the contact area is in the form of bubbles. Very high values of $k_L a$ can be obtained in such a contactor so that the mass transfer resistance can be eliminated. However, it may be pointed out that the stirred contactor can be used only in regime 1, because the value of a cannot be estimated (this is the only laboratory reactor where *a priori* estimation of a is not yet possible).

As in the case of gas-solid catalytic reactors (Table 11.11), fluid-fluid reactors can also be rated in terms of ease of operation and cost. Such a rating is presented in Table 11.20. Further details about these reactors are given by Danckwerts (1970) and Doraiswamy and Sharma (1984).

11.3.4 Solid-Liquid Reactions

Several solid-liquid reactions are of commercial importance; sometimes the solid is slightly soluble or completely insoluble. Examples of the first category are dyeing with reactive dyes, sulfonation and nitration of aromatics at lower temperatures, recovery of paraffins by adductive crystallization with urea, and alkaline hydrolysis of solid esters such as dimethyl fumarate and nitrobenzoic acid esters. Those of the second category include manufacture of organometallic compounds (e.g., Grignard reagent), hydration of lime, generation of acetylene by reaction of water with CaC_2, and acidization of dolomite. Mass transfer associated with the liquid film surrounding the solid particles is then important. Doraiswamy and Sharma (1984) provide a detailed treatment of many aspects of these reactions.

11.3.5 Gas-Liquid-Solid Reactions

Such reactions are extremely important, particularly those where the solid acts as a catalyst and where the solid participates in the reaction. Examples are given in Tables 11.21 and 11.22, respectively. The basic principles of these reactions are explained in Case Studies 11.10 and 11.11, respectively.

Laboratory reactors for fluid-solid and fluid-fluid reactions were described in Sections 3.1.6 and 3.3.2, respectively. The discussion in these sections is also useful for gas-liquid-solid reactions. A combination of the Carberry reactor (Figure 11.7) and a stirred cell (Figure 11.14A) is useful for noncatalytic and catalytic reactions. Some discussion of these issues is presented in Case Studies CS8 and CS11 as well as by Joshi et al. (1985) and Joglekar et al. (1991).

TABLE 11.20
Laboratory Reactors for Fluid-Fluid Reactions: Their Principal Features and Ratings[a]

		Residence Time	Surface Renewal Time	Effective Interfacial Area m^2/m^3	Interfacial Area per Unit Liquid Volume m^2/m^3	Range of $k_L \times 10^5$ m/s	Ease of Operation	Accuracy of Liquid Phase Residence Time/Liquid Phase Mixing	Ease of Sampling and Accuracy of Measurement	Construction Difficulties and Cost
1.	Stirred cell	Wide range possible	0.05 to 4	4 to 25	10 to 40	1 to 20	G	F to P/G	G	G
2.	Stirred contactor	Wide range possible	0.005–0.1	40 to 300	50 to 400	20 to 50	F to G	F to P/G	G	F to G
3.	Cylindrical wetted wall	0.05 to 4	0.05 to 4	10 to 50	300 to 6000	1 to 20	F	G	F to G	F to G
4.	Spherical wetted wall	0.03 to 5	0.03 to 5	10 to 50	300 to 6000	1 to 20	F	G	F to G	F to G
5.	Rotating drum	0.002 to 0.5	0.002 to 0.5	10 to 50	300 to 3000	5 to 70	F	F to P	F to G	F
6.	Laminar Jet	0.001 to 0.1	0.001 to 0.1	40 to 500	1000 to 10,000	10 to 100	F to P	G	F to G	F

[a] F = fair, G = good, P = poor.

TABLE 11.21
Examples of Industrially Important Three-Phase (Gas-Liquid-Solid) Reactions Where the Solid Takes Part in the Reaction

Ammoniacal leaching of chalcopyrite; oxydesulfurization of coal; leaching of uranium ores; ammoniacal leaching of cobalt; acid leaching of bornite; ammoniacal leaching of nickel; leaching of copper sulphide; leaching of zinc sulphide; leaching of iron sulphide; leaching of lead sulphide; leaching of mixed metal ores; leaching of manganese oxides; acetylene-cuprous oxide reaction; manufacture of nickel hydroxide; thermal coal liquefaction; production of calcium acid sulfite; fluidized crystallization processes; production of acetylene; production of gas hydrates; melting of gas hydrates; reaction of phosphides with water; manufacture of calcium hypophosphite; carbonation of lime; wet oxidation of activated carbon; biological and photooxidation of organic solids; absorption of SO_2 in lime; absorption of SO_2 in $Mg(OH)_2$; manufacture of zinc dithionite; chorination of wood pulp; manufacture of sodium hydrosulphite; chlorination of polyethylene (and PVC) suspended in water; absorption of lean SO_2 in a slurry of $CaCO_3$ and MgO; absorption of H_2S, $COCl_2$, etc. in aqueous lime suspension; absorption of ethylene in naphthalene in presence of catalysts; absorption of SO_2 in MgO suspension; extraction of fatty acids in aqueous suspension of lime and $Ba(OH)_2$; absorption of CO_2 in a suspension of CaS; absorption of CO in a suspension of lime; leaching of copper, nickel, cobalt, etc.; carbonation of β-oxynaphthoic acid in kerosene; hydrogenation of styrene; air-oxidation of nylon particles in aqueous suspension; reaction between oxygen, silver and cyanide solutions; fermentation process; hydrogenation of sodium particles; absorption of NO_x in $Ca(OH)_2$ suspensions; chlorination of bauxite; delignification of pulp with O_2; making of pellets from concentrated solutions of molten materials.

TABLE 11.22
Examples of Industrially Important Three-Phase (Gas-Liquid-Solid) Reactions Where the Solid Acts as Catalyst

Hydrogenation of carbohydrates; hydrogenation of α-methyl styrene; hydrogenation of phenyl acetylene; hydrogenation of adiponitrile; hydrogenolysis of methyl esters of fatty acids; hydrogenation of olefinic compounds; hydrogenation of fatty oils; hydrogenation of nitro compounds to amines; reduction of nitrite/nitrate to hydroxylamine salt; Fischer-Tropsch synthesis; hydrogenation of crotonaldehyde; hydrogenation of allyl alcohol; oxidation of aqueous sodium sulfide; oxidation of alcohols in aqueous alkaline media; reduction of uranium by hydrogen; hydrocracking of α-cellulose; dehydrogenation of secondary alcohols to ketones; polymerization with Ziegler-Natta catalyst; simultaneous hydrogenation of ethylene and propylene; production of calcium acid sulfite; direct coal liquefaction; hydrodemetallization; hydrogenation of glucose; hydrogenation of butynediol; desulfurization of petroleum fractions; desulfurization of coal.

11.4 CLASSIFICATION OF REACTORS AND THEIR DESCRIPTION

11.4.1 Introduction

Reactors must often be designed to consider both chemical reactions and physical operations such as micromixing, macromixing, heat transfer, mass transfer, gas-liquid dispersing, liquid-liquid dispersing, and so forth. For all cases, energy is supplied in various ways to the equipment. The equipment can be classified on the mode of energy supply: (1) pressure energy, (2) kinetic energy, and (3) potential energy.

Class 1 equipment are also called column-type equipment. Under this category, there are the various multiphase contactors. Gas-liquid contactors include bubble columns, packed bubble columns, internal-loop and external-loop air-lift reactors, sectionalized bubble columns, plate columns, and others. Solid-fluid (liquid or gas) contactors include static mixers, fixed beds, expanded beds, fluidized beds, transport reactors or contactors, and so forth. For instance, fixed-bed geometry is used in unit operations such as ion exchange, adsorptive and chromatographic separations, and drying and in catalytic reactors. Liquid-liquid contactors include spray columns, packed extraction columns, plate extraction columns, static mixers, and so on. Similar equipments are widely used in gas-liquid-solid, gas-liquid-liquid, and gas-liquid-liquid-solid contactors.

In multiphase contactors, the dispersed phase is either bubbles, drops, particles, or combinations thereof. The continuous phase is either liquid or gas. The governing features of the dispersed phase are its size and velocity distributions, both of which have a major impact on the performance of these equipments. The other governing parameters include column diameter, column height, sparger design, and internal design. Furthermore, in class 1 equipment, the energy is supplied through the introduction of phases. For instance, in the case of bubble columns and gas-fluidized beds, the gas is supplied against the static pressure of the multiphase dispersion, and the energy input rate is given by the following equation:

$$E_i = Q_G \bar{\rho}_D H_D g \tag{11.104}$$

where $\bar{\rho}_D$ is the average density of the dispersion, and E_i is the pressure energy. Class 1 equipments, therefore, are classified under pressure energy.

Class 2 equipments include mainly stirred-tank reactors, in which impellers supply energy to the reactant(s). The fluid leaving the impeller has kinetic energy. The energy is used for a variety of objectives such as producing liquid-liquid, gas-liquid, solid-liquid, and higher-order dispersions. Stirred reactors, in which one or more impellers generate the desired flow and mixing, are among the most widely used reactors. Stirred reactors offer unmatched flexibility and control over the transport processes in the reactor. The performance of a stirred reactor can be optimized by appropriate adjustments of the reactor hardware and the operating parameters (reactor and impeller shapes; number, type, location and size of impellers; degree of baffling; control the performance of stirred reactors; and so on).

In class 3 equipments, the energy supplied is in the form of potential energy associated with the liquid. For instance, in the conventional packed column (used for either distillation or absorption), the liquid is pumped to the top of the column and distributed over the packing. If the gas is introduced at the bottom, the gas phase has the pressure energy; however, it is usually negligible as compared with the potential energy of the liquid. Even if the gas and liquid phases flow cocurrent downward, the major contribution to the energy is by the liquid phase. Since the liquid flows as films, these equipments (class 3) may also be termed *film contactors*. Other equipments in this category include trickle-bed reactors and falling-film reactors/evaporators. In packed columns, a variety of packing shapes (and of course sizes and materials of construction) are used in practice; these include Raschig rings (in old installations), Berl saddles, partition rings, intelox saddles, Pall rings, hipack rings, and structured packing. Equally important is the uniformity of liquid distribution. The more important ones (under each class) are described below. Table 11.23 presents a list of reactors under each category.

11.4.2 Pressure Energy

11.4.2.1 Gas-Liquid Reactors

11.4.2.1.1 Bubble Columns

The bubble column used as a reactor consists of a vertical cylindrical vessel with height-to-diameter ratio in the range of 1 to 20 (often 3 to 10). Gas is introduced at the bottom via a sparger. Spargers include sieve plate, ring, spider, radial sparger, ejector, injector, and so on. Schematic diagrams of a bubble column and spargers are shown in Figures 11.15 and 11.16, respectively. The column top is expanded to facilitate foam breakage and to reduce the entrainment of liquid in the exit gas. The bubble column is operated either in a semicontinuous mode (gas, continuous; liquid, batch) or in a continuous mode (gas and liquid both continuous). In the latter case, the liquid phase may flow either cocurrent or countercurrent to the gas.

TABLE 11.23
Classification of Reactors Belonging to Different Modes of Energy Input

Pressure Energy

Gas-Liquid Reactors
Bubble column, packed bubble column, sectionalized bubble column, plate column, external- and internal-loop air-lift reactors, static mixer, venturi scrubbers

Liquid-Liquid Reactors
Spray column, packed extraction column, liquid-liquid adaptations of loop reactors, plate extraction column, static mixers.

Solid-Liquid Reactors
Fluidized-bed reactor, fixed-bed reactor

Gas-Solid Catalytic Reactors
Fixed-bed reactors (multi-tubular and staged adiabatic), fluidized-bed reactors (bubbling bed, turbulent bed, fast, and transport or pneumatic), radial flow reactor, gauz reactor

Gas-Liquid-Solid Reactors
Fluidized-bed reactor, slurry reactor

Kinetic Energy
Single-stage and multistage stirred-tank reactors, self-inducing reactor, jet-loop reactor, plunging- jet reactor, surface aerator

Potential Energy
Packed column, trickle-bed reactor, film reactors (falling film, agitated film, scraped/wiped film), rotating disk (or rotating packed-bed) reactor

---

In bubble columns, since the gas bubbles are dispersed in the continuous liquid phase, fractional gas holdup (ε_G) is an important design parameter, affecting column performance. The most direct and obvious effect is on the column volume, since a significant fraction of the volume is occupied by the gas. The indirect influences are also important. For instance, the possible spatial variation of ε_G gives rise to pressure variation, which results in intense liquid phase motion. These secondary motions govern the rates of mixing plus heat and mass transfer.

The fractional gas holdup is defined as the volume fraction of gas in the gas-liquid dispersion and is measured in several ways (Joshi et al., 1990a). The measurements of average gas holdup $\bar{\varepsilon}_G$ and local gas holdup are both reasonably well developed. Despite the good agreement between different measuring methods, we have not yet reached a stage where holdup can be predicted for an unknown gas-liquid system, as next discussed.

The fractional gas holdup is principally governed by the following features of the bubbles: size, velocity, and number. The average bubble size depends on the sparger design, energy dissipation rate near the sparger, and the coalescing nature of the gas bubbles (which depends on liquid viscosity, surface tension, and other surface active properties). For a perforated plate and ring sparger, the energy dissipation rate is relatively small. The average bubble size decreases with a decrease in both the hole size and the surface tension. The coalescing nature has little influence on d_{bp} in the spargers. For sintered and ejector/injector type spargers, the coalescing nature of the liquid phase, however, has a substantial influence (Buchholtz et al., 1983; Heijnen and van't Riet, 1984; Zahradnik et al., 1997). The energy dissipation rate is also relatively high and, therefore, fine bubbles (a few micrometers to 1 mm) are generated in noncoalescing liquids. In the case of coalescing liquids, the bubbles grow a short distance from the sparger, and the bubble size is practically the same as that generated by perforated spargers.

Fractional gas holdup also depends strongly on superficial gas velocity and the nature of the gas-liquid system. Column diameter and height often have some influence on ε_G. The quantitative

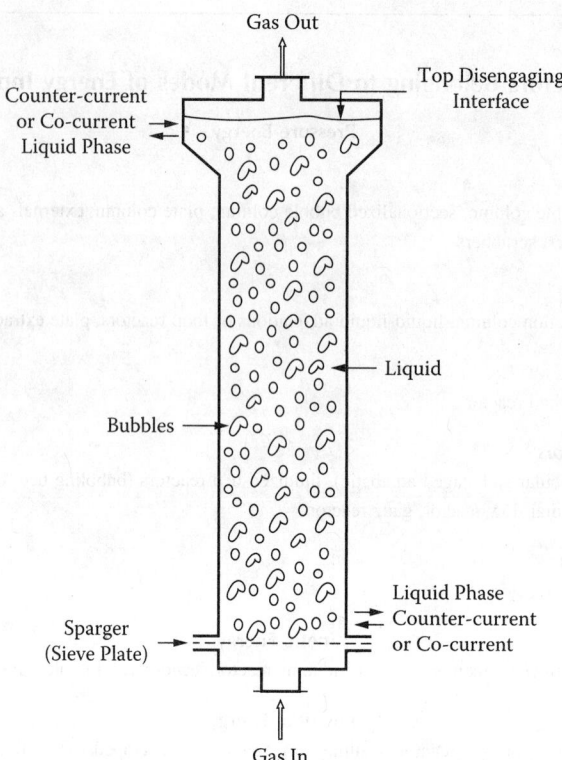

FIGURE 11.15 Schematic diagram of bubble column.

effect, however, depends on the regime of dispersion in bubble columns: homogeneous or heterogeneous. The homogeneous regime is characterized by almost uniform sized bubbles. Furthermore, the concentration of bubbles is also uniform, particularly in the transverse direction (Figure 11.17a). If the gas is sparged uniformly into the bottom of the column, it remains uniformly distributed across the column. All bubbles rise virtually vertically with only minor transverse and axial oscillations plus little coalescence or redispersion. Hence, the size of the bubbles in the homogeneous regime is governed mainly by the design of the sparger and the physical properties of the system.

In contrast, the heterogeneous regime is characterized by nonuniform bubble concentration (Figure 11.17b). The ε_G profile is usually parabolic and results in pressure profiles as shown in Figure 11.18a. These profiles cause upward liquid circulation in the central region and downward near the column wall (Figure 11.18b). In such a recirculatory flow, turbulence is much higher as compared to the homogeneous regime, where the recirculation is absent. In the heterogeneous regime, the bubbles retain their identity only over a small distance from the sparger, and this region is called the *sparger region*. In the bulk region, the bubble size is governed by the coalescence and redispersion phenomena. A wide bubble size distribution occurs in the bulk, and the average bubble size (which is called the *secondary bubble size*, d_{bs}) is decided by the balance between the breaking (viscous and turbulent shear stress), the retaining (surface tension) forces, and the coalescing nature of the liquid phase. The value of d_{bs} decreases with a decrease in surface tension and an increase in the power consumption per unit volume (**P/V**). Furthermore, for a given value of **P/V**, the average bubble size is smaller in noncoalescing liquids than in coalescing liquids. It is known that the coalescing nature increases with an increase in liquid viscosity and a decrease in gas density. Pure liquids usually show coalescing behavior. However, a mixture of two liquids gives some noncoalescing properties (Bach and Pilhofer, 1978). For example, water and aliphatic alcohol are individually of the coalescing type. The addition of small quantities of aliphatic alcohol to water makes

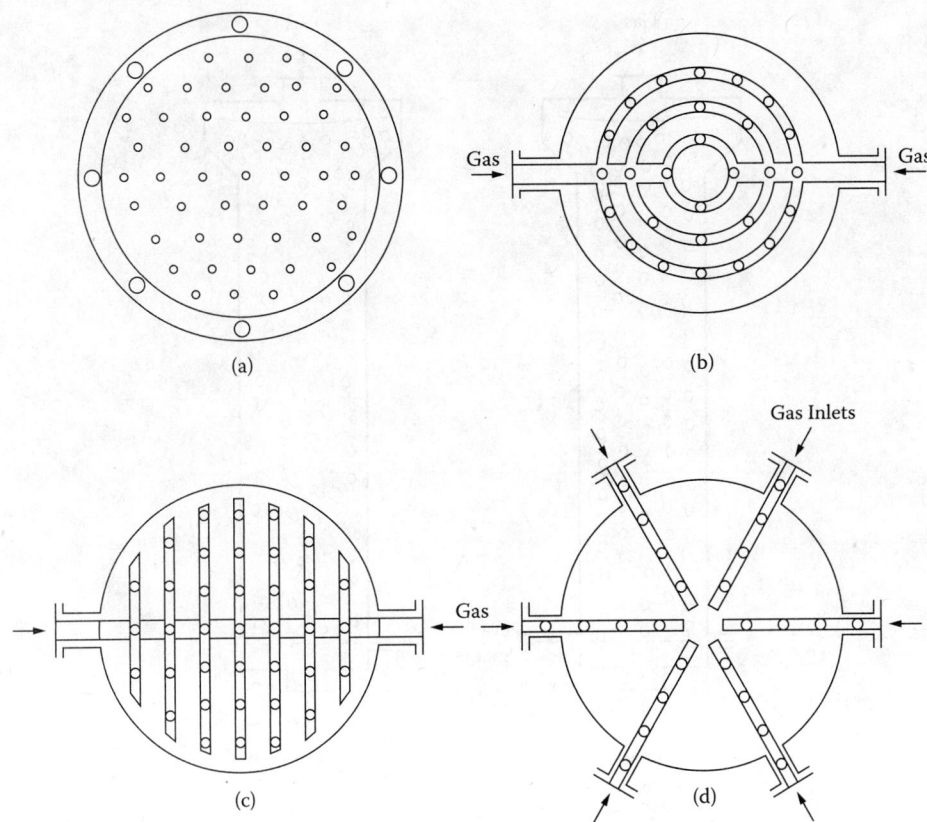

FIGURE 11.16 Schematic diagrams of spargers: (a) sieve-plate, (b) ring, (c) spider, and (d) radial.

the liquid mixture somewhat noncoalescing. The strength of the noncoalescing property increases with an increase in alcohol concentration (up to a limit) and an increase in the number of carbon atoms in the alcohol. An electrolyte in water imparts noncoalescing features, whose strength depends on its type and concentration. The presence of small amounts of surface active impurities may also impart noncoalescing features. For instance, tap water and distilled water often have quite different coalescing properties (Anderson and Quinn, 1970). The measurement and prediction of the strength of the noncoalescing nature is complex, which complicates the prediction of bubble size and fractional gas holdup.

The secondary bubble size, d_{bs}, generally is different from the primary bubble size, d_{bp}. In the near-sparger region, the primary size changes with distance from the sparger and finally attains the secondary size. The height of the sparger region depends on the difference between d_{bp} and d_{bs}. If special care is taken to bring d_{bp} close to d_{bs}, the height of the sparger region is small. When $d_{bp} < d_{bs}$, $\bar{\varepsilon}_G$ decreases with an increase in the column height. In contrast, when $d_{bp} > d_{bs}$ (for instance, in the case of single-point sparger), $\bar{\varepsilon}_G$ increases with an increase in the column height. For any difference between d_{bP} and d_{bS}, the height of the sparger region increases with an increase in the noncoalescing nature of the liquid phase. As a result, the height of the sparger region is in the range of $1 < H/D < 5$.

In the sparger region, the liquid flow pattern becomes developed. If the overall column height is much greater than the height of the sparger region, the holdup does not change with respect to the sparger design or the column height. However, care must be taken to account for the effect of the hydrostatic head on the superficial gas velocity. The value of u_G should be expressed at the average pressure between the bottom and the top.

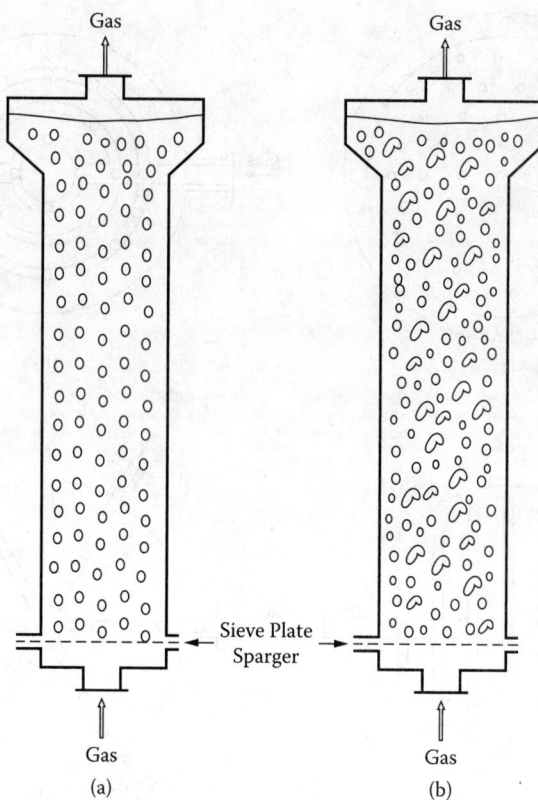

FIGURE 11.17 (a) Schematic representation of the homogeneous regime: uniform concentration and size distributions of bubbles in the transverse direction. (b) Schematic representation of the heterogeneous regime: nonuniform concentration and bubble size distributions in the transverse direction. (From Bhole, M. R. and Joshi, J. B. Stability analysis of bubble columns: Predictions for regime transition. *Chem. Engr. Sci.*, 4493–4507 (2005).)

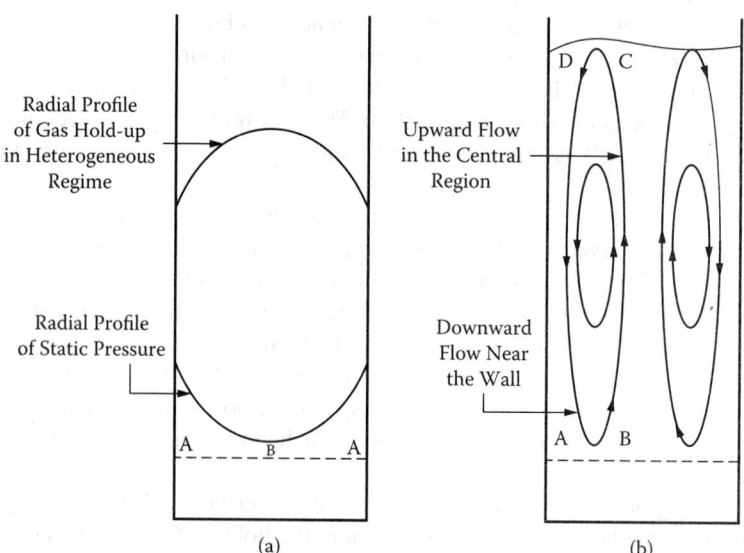

FIGURE 11.18 Some aspects of the heterogeneous regime: (a) radial profiles of gas holdup and static pressure, (b) scehematic of liquid circulation pattern.

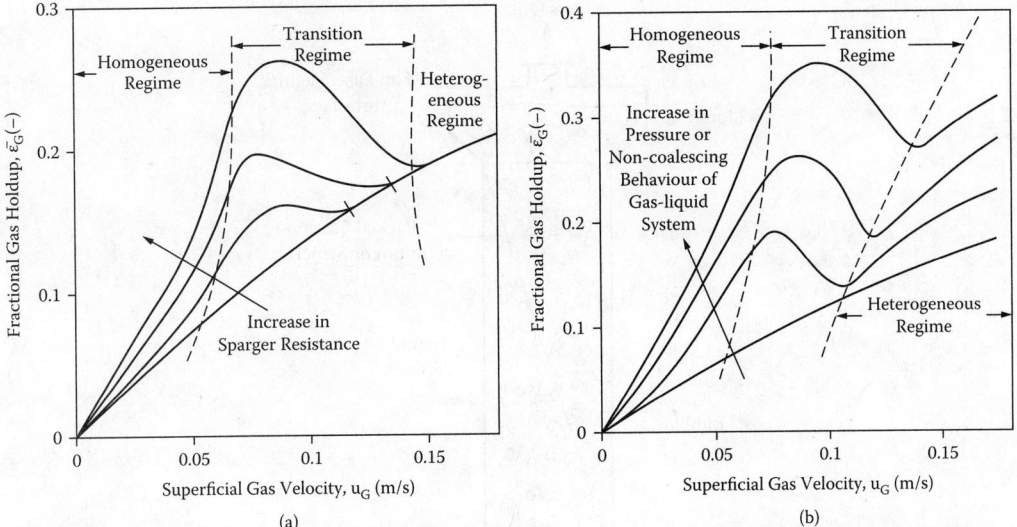

FIGURE 11.19 Relation between gas hold-up and superficial gas velocity for different regimes of operation: (a) effect of sparger resistance, (b) effect of total pressure and coalescing nature of the gas-liquid system.

The value of $\bar{\varepsilon}_G$ increases with an increase in the superficial gas velocity. However, the $\bar{\varepsilon}_G - u_G$ relationship depends on the regime of operation (Figures 11.19a and b). This in turn is governed by the design (sparger design, column diameter, column inclination) and operating (superficial gas and liquid velocities, μ_L, σ, ρ_G and the coalescing nature of the liquid phase) parameters. Figure 11.19a shows the effect of sparger design on regime transitions; the upper limit u_G (up to which the homogeneous or transition regime prevails) increases with an increase in sparger resistance. Once the heterogeneous regime is attained, the gas holdup is independent of sparger resistance. The effect of liquid-phase physical properties on regime transition is shown in Figure 11.19b. In this case, the upper limit of u_G increases with a decrease in bubble diameter, which happens with a decrease in liquid viscosity, increase in surface, increase in gas density, increase in the noncoalescing nature of the liquid phase, and so forth. Furthermore, in the heterogeneous regime and at the same u_G, the gas holdup increases with a decrease in bubble diameter. This observation is different from that shown for the sparger (Figure 11.19a). The $\bar{\varepsilon}_G - u_G$ relationship may be expressed in the form of a power law model

$$\bar{\varepsilon}_G \propto u_G^a \tag{11.105}$$

The value of a is greater than or equal to 1 in the homogeneous regime, whereas it is less than 1 (0.4 to 0.8) in the heterogeneous regime. In the transition regime, the value of x continuously decreases from a high homogeneous value to a low heterogeneous value. Other features of gas-liquid dispersions have been described by Joshi et al. (1998).

The fractional gas holdup is known to be independent of column diameter in both the homogeneous and heterogeneous regimes. However, for a given gas-liquid system, the value of $\bar{\varepsilon}_G$ is independent of the column diameter if the regime of operation is the same. The homogeneous regime prevails in small-diameter columns and the heterogeneous regime in large-diameter columns. Therefore, $\bar{\varepsilon}_G$ may depend on the column diameter. This problem is particularly noticeable at low values of u_G (< about 80 mm/s). At high values of u_G and for air-water type systems, the heterogeneous regime prevails, and the fractional gas holdup becomes independent of the column diameter when it exceeds about 150 mm. Such a critical value of column diameter increases with an increase

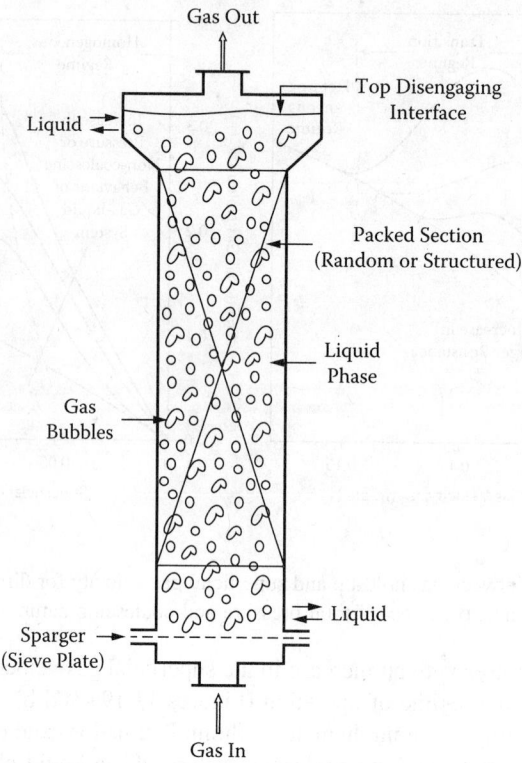

FIGURE 11.20 Packed bubble column.

in non-Newtonian behavior and an increase in viscosity beyond 100 mPa. Further details pertaining to bubble columns are reported by Deckwer (1992) and Joshi et al. (1998).

Bubble columns are simple in construction and operation, providing sufficient flexibility for the liquid-phase residence time. Also, they provide some flexibility even for the gas-phase residence time. However, the bubble columns have two limitations: back-mixing in the liquid phase and high pressure drop for the gas phase. To minimize these limitations, modified bubble columns are often used, as next reported.

11.4.2.1.2 Packed Bubble Columns

A bubble column filled with packings reduces liquid circulation (Figure 11.20). The extent of reduction depends on the resistance provided by the packings, which in turn depends on the voidage and packing factor. The packings may be random, wire mesh, or structured. Pandit and Joshi (1983) have reported that backmixing in packed bubble columns is three to five times lower than that in unpacked bubble columns. For a given duty, the volume of a packed bubble column is less than that of a bubble column, but the cost of the packed bubble column per unit volume is much higher. Therefore, the total equipment cost should be calculated before making any decision.

Packed bubble columns are also used as catalytic reactors where the packings are catalyst particles in the form of extrudates or tablets. These particles have a dual role: one as a catalyst and the second to reduce backmixing. In this case, there is the need to optimize column diameter, height, particle size, and shape.

11.4.2.1.3 Sectionalized Bubble Columns and Plate Columns

Liquid circulation is reduced by baffles. Figure 11.21 shows radial baffles with central openings. Figures 11.21a and b show flat and conical baffles, respectively (conical shape is preferable for a solid suspension). These baffles minimize liquid circulation. The resulting flow pattern is shown

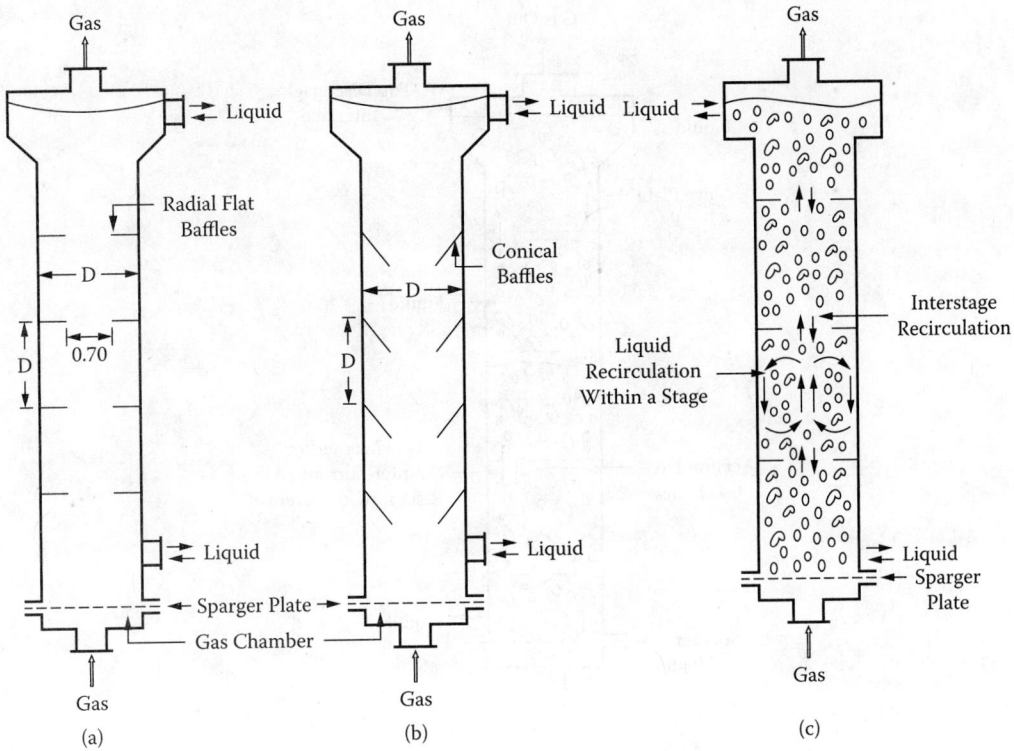

FIGURE 11.21 Sectionalized bubble columns: (a) flat baffles, (b) conical baffles, and (c) flow pattern.

schematically in Figure 11.21c. Joshi and Sharma (1979) analyzed the fluid mechanics of such a system and have recommended the optimum spacing to be $0.8\ D$ and the central hole diameter $0.7\ d_T$, where d_T is the column diameter. As indicated in Figure 11.21c, multiple stages equal to the number of baffles are formed. There may be interstage recirculation through the central opening, which reduces the effectiveness of multiple stages. The recirculation can be reduced by decreasing the hole diameter; however, it affects the gas distribution. Alternatively, a sieve plate may be selected instead of a baffle. Holes are provided for the flow of gas. The hole diameter and free area are selected so that sufficient resistance is provided for the flow of gas and, as a result, gas spaces are formed below each plate as shown in Figure 11.22. The net liquid upflow/downflow occurs through riser/downcomer in cocurrent/countercurrent cases, respectively. Each section is considered as one backmixed stage. Five such backmixed stages (Levenspiel, 1991) give flow behavior very close to plug-flow if interstage recirculation is completely absent. In reality, finite weeping and entrainment take place, and it is desirable to have up to ten stages to obtain plug-flow.

The extent of weeping and entrainment decreases with an increase in plate resistance resulting from smaller hole diameter. However, there is a limit, and micrometer-size sintered plates are not used because of clogging and maintenance difficulties. Furthermore, for a given plate, the extent of weeping decreases with a decrease in liquid submergence on a plate. Also, entrainment decreases with an increase in the height of gas space. Therefore, in a conventional plate column, the liquid submergence, empty space height, and plate resistance are optimized for the combined objective of maximizing plate efficiency and minimizing the pressure drop. The former height is in the range of 50 to 200 mm and the latter 200 to 600 mm. Plate efficiency is typically 40% to 75%. In plate columns, the trays are generally bubble cap, valve, or sieve plate. On each plate, the liquid and gas flows occur in a crossflow manner, whereas the interstage liquid flow occurs through the downcomers. Further details on trays and plate columns are found in Treybal (1980) and in the "Distillation" (Chapter 12) and "Absorption" (Chapter 13) of this book.

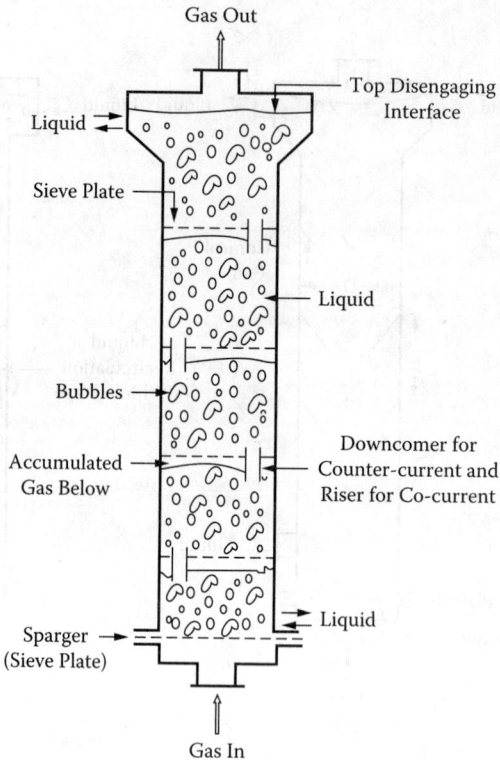

FIGURE 11.22 Sectionalized bubble column with sieve plate.

11.4.2.1.4 Air-Lift Reactors

The bubble column reactor suffers from two major limitations: (1) complete backmixing in the liquid phase and (2) high pressure drop. To overcome the first limitation, suitable modifications were described in previous sections. Now, let us consider the air-lift reactor, which offers the advantage of low pressure drop. Air-lift reactors are of two types: external-loop (Figures 11.23a and b) and internal-loop (Figure 11.24). An external-loop air-lift reactor (EL-ALR) consists of two columns that are connected at the bottom by a U-type connection as shown in Figure 11.23. At the top, a gas disengagement tank is provided. To start the operation of an air-lift reactor, gas is first introduced through the riser sparger. The gas phase ascends in the form of bubbles in the riser section and finally disengages at the top liquid surface in the disengagement tank. The gas bubbles are maintained on the riser side by a vertical baffle, which is located between the riser and downcomer sections. The downcomer section is kept bubble free during the startup. Since the gas-liquid dispersion is present only in the riser section, the average fluid density $(\varepsilon_L \rho_L + \varepsilon_G \rho_G)$ is less in the riser section. Therefore, the static pressure at the bottom of the riser (point C in Figure 11.23a) is low as compared to that at the corresponding location (point D) in the riser. The resulting pressure generates flow from D to C, which further ascends in the riser, overflows the top baffle, and flows downward in the downcomer. As a consequence, well-directed circulation is established. The average liquid circulation can be calculated, as a first approximation, by equating the pressure driving head with the velocity head:

$$P_C = (\varepsilon_{LR} \rho_L + \varepsilon_{GR} \rho_g) g H_d \qquad (11.106)$$

$$P_D = \rho_L g H_D \qquad (11.107)$$

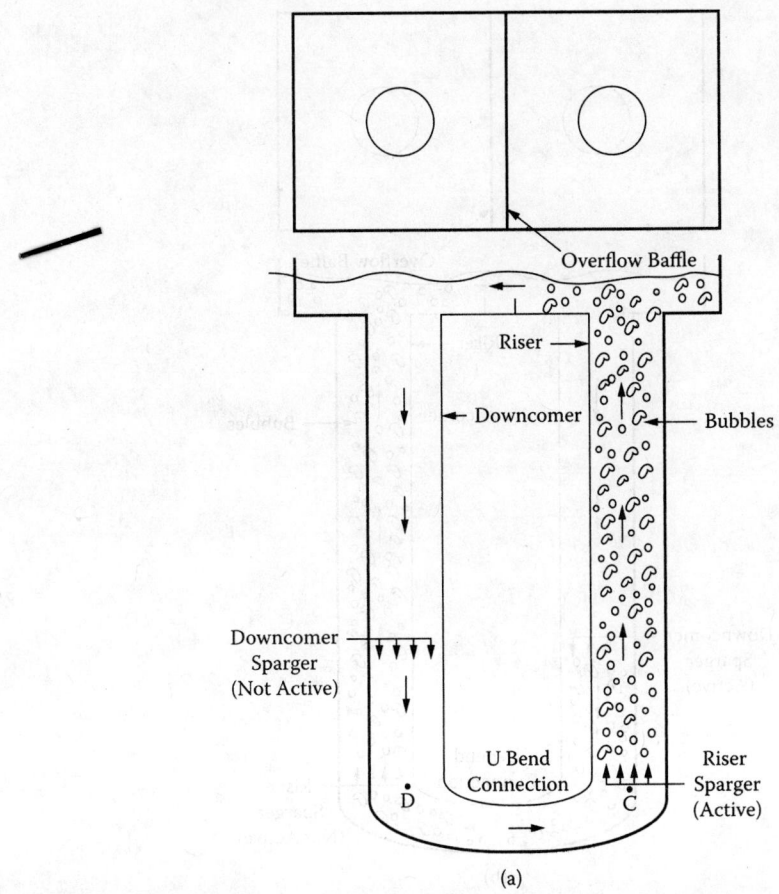

FIGURE 11.23 External-loop air-lift reactor: (a) riser sparging, (b) downcomer sparging. *Continued*

$$\rho_L \frac{u^2}{2} = P_D - P_C \tag{11.108}$$

Substituting Equations (11.106) and (11.107) in (11.108), assuming ρ_G to be negligible, and for simplicity assuming the same average liquid velocity in the downcomer and riser sections, we get

$$u = \sqrt{2\varepsilon_{GR} g H_D} \tag{11.109}$$

For instance, for a riser gas holdup, ε_{GR}, of 0.2 and a reactor height, H_D, of 10 m, the average liquid circulation velocity is 6.26 m/s. That is, liquid circulation velocity is very high. If a bubble is introduced in the downcomer, it is entrained downward if the liquid velocity is higher than the bubble rise velocity. It may be pointed out that the presence of gas on the downcomer side reduces the pressure driving force. However, the circulation continues even if part of the downcomer section is occupied by the gas phase. This is because (Figure 11.23b) part of the downcomer is bubble free. P_D is given by

$$P_D = [HSD(\varepsilon_{LD}\rho_L + \varepsilon_{GD}\rho_G) + (H - H_{SD})\rho_L]g \tag{11.110}$$

810

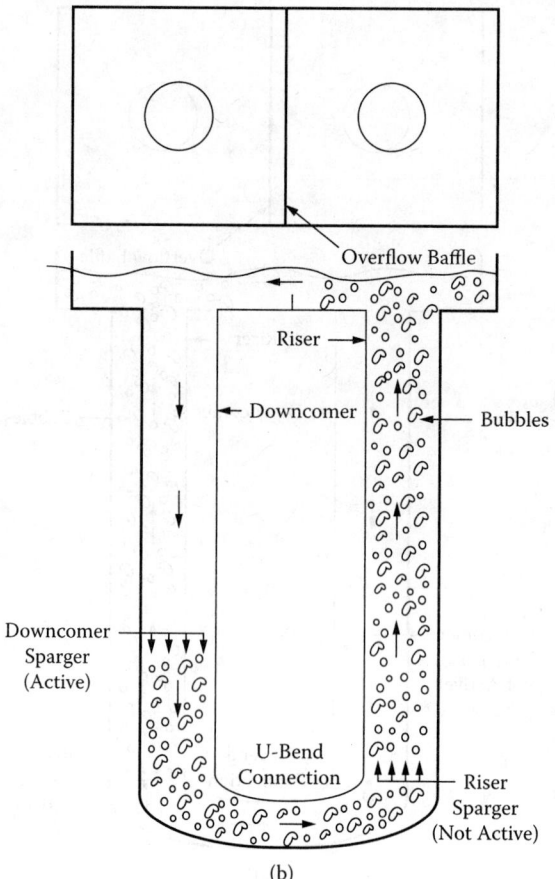

(b)

FIGURE 11.23 *Continued.*

where H_{SD} is the sparger location on the downcomer side and P_C is given by Equation (11.106). Now, the average liquid circulation velocity is given by

$$u = \sqrt{2(\varepsilon_{GR} g H_D - \varepsilon_{GD} g H_{SD})} \quad (11.111)$$

The liquid circulation continues until the value of u is higher than the bubble rise velocity. This is true even if the riser sparger is stopped (of course, after the circulation pattern is well developed). Furthermore, from Equation (11.111), it can be seen that the circulation velocity decreases with an increase in H_{SD}. At a critical location of $H_{SD}(H_C)$, u equals the bubble rise velocity. In reality, H_{SD} has to be lower than H_C. As a factor of safety, typically, H_{SD} is up to 40% of H.

The gas phase (Figure 11.23b) is introduced downward through the downcomer sparger, takes a U turn at the bottom, ascends in the riser, and leaves the column through the disengagement tank. Thus the gas phase need not be introduced at the bottom as in the conventional bubble column, and hence the pressure drop can be reduced up to 40% since sparger location is permissible at a higher level where the static pressure is obviously lower. Since the gas phase passes through the bottom section, high pressure in the bottom section results in a high saturation concentration of the solute gas.

In addition to low pressure drop, external-loop air-lift reactors have several advantages over bubble-column reactors. The former offer much flexibility in design in terms of height-to-diameter ratio, area ratio of downcomer to riser, sparger locations, and so on. Conditions can be manipulated

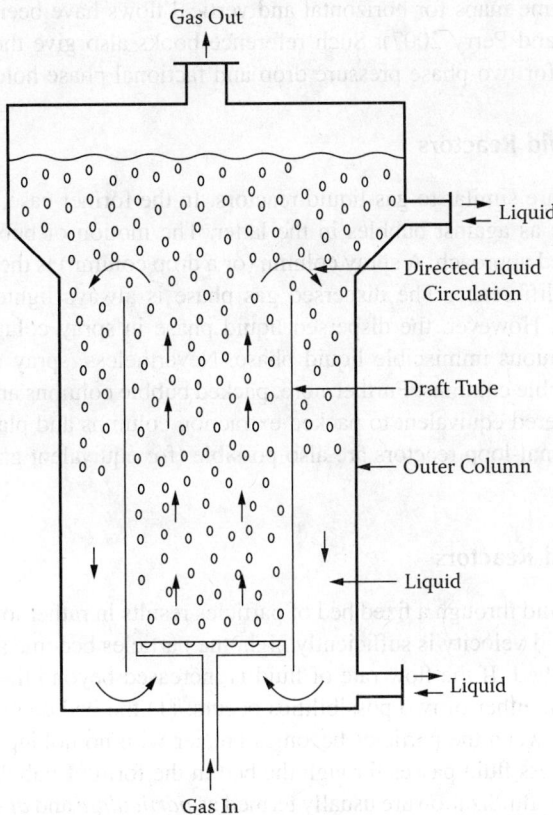

FIGURE 11.24 Internal-loop air-lift reactor.

for optimum performance to obtain the desired gas holdups, gas-liquid interfacial area, heat- and mass-transfer coefficients, and extent of mixing. If desired, the intensity of liquid circulation and turbulence can be kept low in an EL-ALR as compared to BCR. Therefore, the level of the shear stress (turbulent) is much smaller in EL-ALR as compared to BCR when compared at the same power consumption per unit volume. Low levels of shear in the loop reactor are generally desired for many biological reactions of shear-sensitive cells. Other design details of EL-ALR have been described by Joshi et al. (1990b) and Lele and Joshi (1993).

When the total reactor volume is small (< 5 to 10 m^3), an air-lift reactor can be used in the form of an internal loop as shown in Figure 11.24.

11.4.2.1.5 Two-Phase Pipe-Flow and Static Mixers

In previous sections, two-phase gas-liquid flow was considered in vertical bubble columns and modified bubble columns. In these reactors, the desired regime of gas-liquid flow is bubble flow, and therefore the column diameter is at least 150 mm. The superficial gas velocity usually varies from 10 to 500 mm/s and the superficial liquid velocity from 0.01 to 30 mm/s, with a liquid phase residence time from 10 min to a few days. In two-phase gas-liquid flows in horizontal pipes, the superficial gas and liquid velocities are in the range of 0.01 to 10 m/s and 0.1 to 5 m/s, respectively. For two-phase pipe-flow, the pipe diameter is usually up to 200 to 300 mm when used as a reactor.

In two-phase gas-liquid flows in horizontal pipes, various flow regimes occur, depending on the range of superficial gas and liquid velocities. These include (a) bubble or froth flow, (b) plug flow, (c) stratified flow, (d) wavy flow, (e) slug flow, and (f) spray or dispersed flow. In vertical pipes, the flow regimes are (a) bubbly flow (this regime was considered in the case of bubble column reactors); (b) piston, plug, or slug flow; (c) ripple or wavy flow; (d) annular or film flow;

and (e) mist flow. Regime maps for horizontal and vertical flows have been discussed in detail in various books (Green and Perry 2007). Such reference books also give the present status on the estimation procedures for two-phase pressure drop and factional phase holdup.

11.4.2.2 Liquid-Liquid Reactors

Liquid-liquid reactors are similar to gas-liquid reactors. In the former case, the dispersed phase is in the form of droplets as against bubbles in the latter. The motion of bubbles and drops can be described using a unified approach. A spray column (or a drop column) is the equivalent of a bubble column but with one difference. The dispersed gas phase is always lighter than the continuous liquid phase ($\rho_G < \rho_L$). However, the dispersed liquid phase in spray columns may be lighter or heavier than the continuous immiscible liquid phase. Nevertheless, spray columns can be easily described similar to bubble columns. Furthermore, packed bubble columns and sectionalized bubble columns can be considered equivalent to packed extraction columns and plate extraction columns. External-loop and internal-loop reactors are also possible (for equivalent gas-liquid reactors, refer to Section 11.4.2.1.4).

11.4.2.3 Solid-Liquid Reactors

The upward flow of a fluid through a fixed bed of particles results in rather low flow rates. However, when the superficial fluid velocity is sufficiently high, the particles become supported in the liquid, resulting in a fluidized bed. If the flow rate of fluid is increased beyond the minimum required to produce a fluidized bed, either of two possibilities occurs: (1) the bed continues to expand so that the average distance between the particles becomes greater with no holdup gradients in the radial direction, or (2) the excess fluid passes through the bed in the form of bubbles, resulting in holdup gradients. These types of fluidization are usually termed as *particulate* and *aggregative*, respectively. In general, particulate (or homogeneous) fluidization occurs with solid-liquid systems. For solid-gas systems, homogeneous fluidization occurs when the particle settling velocity is low (fine particles or/and high operating pressure) and is limited to a relatively small range of velocities. Heterogeneous fluidization occurs with all other solid-gas systems, and sometimes with solid-liquid systems as well when the particles are of high density or/and size. Particulate/aggregative fluidization has been discussed in detail by Joshi et al. (2001).

If a liquid is passed vertically upward through a bed of uniform particles, the pressure drop, ΔP, increases with an increase in the superficial liquid velocity, u_L. The relation between pressure drop and velocity is the same as for a fixed bed, as indicated by the following equation (Ergun, 1952):

$$\Delta P = \frac{150\mu_L u_L L}{d_p^2} \frac{\varepsilon_S^2}{\varepsilon_L^3} + 1.75 \frac{\rho_L u_L^2 L}{d_p} \frac{\varepsilon_S}{\varepsilon_L} \qquad (11.112)$$

For a given bed of particles, ρ_L and d_P, a log-log plot of ΔP vs. u_L (line LM in Figure 11.25) is linear. When the velocity reaches such a value that the frictional force on a particle equals the force due to the gravity minus the buoyancy, the particles become freely supported. At this stage (point M in Figure 11.25) the frictional pressure drop equals the buoyant weight of particles per unit bed cross-sectional area, and the bed is just fluidized and is said to be at the *point of incipient fluidization*. The superficial liquid velocity at this point is called the *minimum fluidization velocity*, u_{mf}. As the liquid velocity is further increased, the pressure drop across the bed remains constant (line MN in Figure 11.25) because the buoyant weight of solids remains constant. The wall pressure drop is usually negligible as compared with the frictional pressure drop at the particle-liquid interface.

The pressure drop in a fixed and fluidized bed can be given by the following single generalized equation (Pandit and Joshi, 1998):

Chemical Reaction Engineering

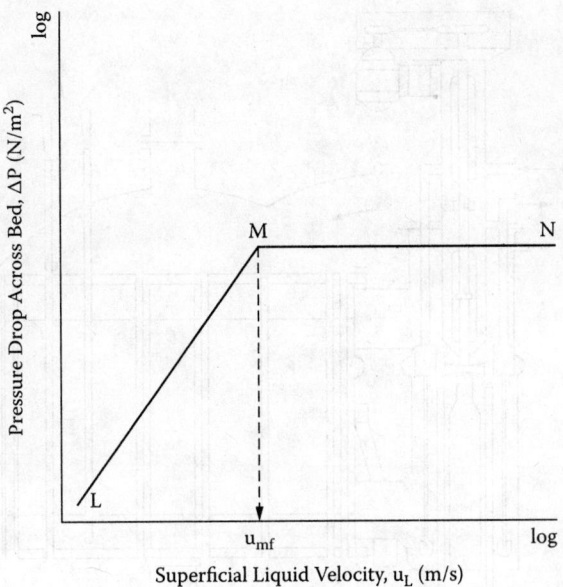

FIGURE 11.25 Log-log plot of pressure drop (ΔP) versus superficial liquid velocity for a liquid-liquid fluidized bed.

$$\frac{\Delta P}{L}\frac{d_p \varepsilon_L^{4.8}}{\varepsilon_s \rho_L u_L^2} = \frac{18}{\text{Re}_p} + 0.33 \tag{11.113}$$

In a solid-liquid fluidized bed, the bed expansion is determined by the Richardson-Zaki (1954) equation:

$$\frac{u_L}{u_{S\infty}} = \varepsilon_L^n \tag{11.114}$$

Solid-liquid fluidization details are reported by Joshi (1983) and Di Felice (1995).

11.4.2.4 Gas-Solid Catalytic Fixed-Bed Reactors

The fixed-bed reactor is the most widely used reactor for reactions catalyzed by solid catalysts. The most important of a number of designs are the nonisothermal nonadiabatic fixed-bed reactor (NINA-FBR) (also called the tube-and-shell or heat exchanger type reactor), the single or multistage adiabatic fixed-bed reactor (A-FBR), and the radial flow fixed-bed reactor (RF-FBR). A newer design is the spherical reactor (Hartig and Keil, 1993). It is important at the outset to note the difference between the approaches to the design of the NINA-FBR and the A-FBR. A typical NINA-FBR unit is shown in Figure 11.26. It consists of a tube bundle, usually around 3 cm in diameter, placed in a shell. Often, the catalyst is placed in the tubes, and a heat exchange fluid is circulated through the shell. A less common design is one in which the shell contains the catalyst, and the heat exchange fluid is circulated through the tube bundle. The height of the reactor usually is in the range of 3 to 10 m, resulting in a temperature profile for reactions with even a low heat effect (exothermic or endothermic). Exothermic reactions result in hot spots that frequently lead to catalyst deactivation and adverse changes in conversion/selectivity. The design objective is to find the height for a required conversion using tubes of fixed diameter. Then, depending on selectivity, stability,

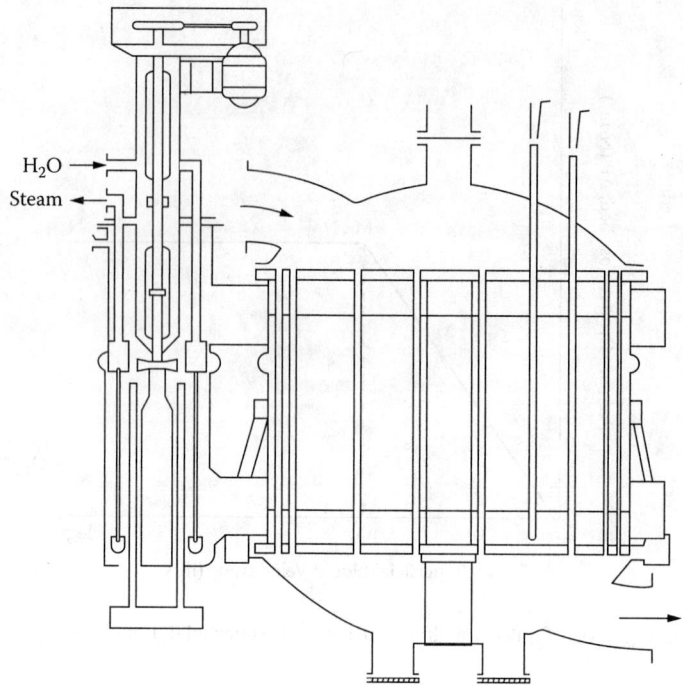

FIGURE 11.26 A typical NINA-FBR design: reactor for SO_2 oxidation (Suter, 1972).

radial and axial concentrations, and temperature profiles, the diameter and height are recalculated for optimum performance.

The approach to the design of an A-FBR is, however, quite different. There is no heat transfer in this reactor; radial transport of heat is absent, and that of mass can usually be neglected. Hence, the reactor diameter can be quite large, thus dispensing with the need for using small-diameter tubes as in the NINA-FBR. A single, large-diameter reactor should, in principle, be all that is needed. In practice, more than one stage may be needed to cool or heat the process fluid between stages. The design here calls for optimizing the inlet and outlet conditions (and therefore the stage height for a given diameter) for each stage to obtain the desired conversion/selectivity at the end of the final stage. The recommended method for optimizing is dynamic programming. A brief summary is presented below of the design procedures for both NINA-FBR and A-FBR units. An illustration of the use of the more important methods is provided in Case Study 11.4.

11.4.2.4.1 Fixed-Bed Reactors: Design of NINA-FBR

From the desired production rate and the allowable velocity in each tube, the number of tubes is obtained. Calculations are made to determine the height of the tube to obtain the required conversion. Since the tubes usually have ($L/D >> 50$), plug flow can usually be assumed. However, this assumption may not always be valid, because radial gradients often exist. Axial diffusion can also be present, thus modifying the conversion gradient that would normally exist due to flow as the reaction progresses from inlet to outlet. These gradients can cause severe deviations from the values corresponding to plug flow. To account for all these gradients and other possible nonidealities, several models have been proposed. Four are sketched in the Figure 11.27. Of these, the continuum model is the most common. Here, the solid-fluid system is considered as a single pseudohomogeneous phase with properties of its own. These properties (e.g., diffusivity, thermal conductivity, and heat transfer coefficient) depend on the properties of the gas and solid components of the pseudo phase. They are anisotropic; i.e., they have different values in the radial and axial directions.

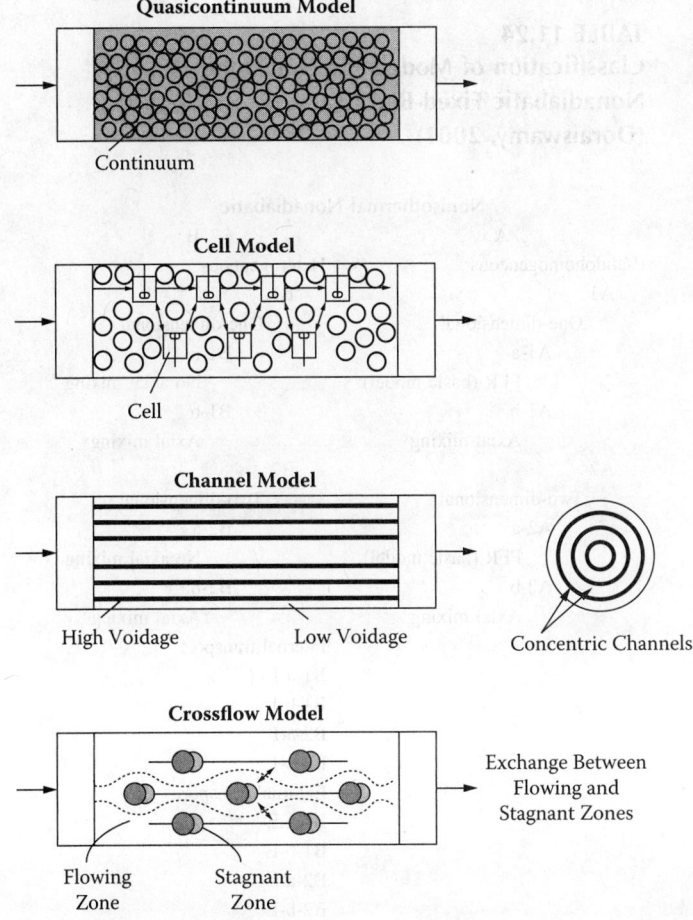

FIGURE 11.27 Models for NINA-FBR design.

Kulkarni and Doraiswamy (1980b) have compiled all the equations for predicting these effective properties. Both radial and axial gradients can be accounted for in this model; the system is really heterogeneous, involving transport effects both within the solid particles and between the particles and the flowing fluid. The simplest model is, of course, the homogeneous isothermal model with plug flow and no intra- or interphase effects. Varying degrees of complexity can be progressively added, as shown in Table 11.24.

Another important class of models are the cell models, which break up the reactor into many cells, with each cell (or microreactor) corresponding to a single pellet and its immediate neighborhood. By allowing for flow between cells in both the radial and axial directions, one- and two-dimensional models, as well as various degrees of mixing, can be simulated. The equations involved in these models are algebraic and not differential as in the continuum models. The continuum models are, however, most widely used, and hence we restrict our treatment to these models.

Taking model A1-a of Table 11.24, in which the assumption of isothermicity of the simplest model is relaxed, and applying it to the simple reaction

$$A \rightarrow R \qquad \text{R4.1}$$

the following mass and energy balance equations can be written:

TABLE 11.24
Classification of Models for Nonisothermal Nonadiabatic Fixed-Bed Reactors (NINA-FBR) (Doraiswamy, 2001)

Nonisothermal Nonadiabatic	
A	B
Pseudohomogeneous	Heterogeneous
A1	B1
One-dimensional	One-dimensional
A1-a	B1-a
PFR (basic model)	No axial mixing
A1-b	B1-b
Axial mixing	Axial mixing
A2	B2
Two-dimensional	Two-dimensional
A2-a	B2-a
PFR (basic model)	No axial mixing
A2-b	B2-b
Axial mixing	Axial mixing
	Internal transport
	B1-a-I
	B1-b-I
	B2-a-I
	B2-b-I
	External transport
	B1-a-E
	B1-b-E
	B2-a-E
	B2-b-E
	Internal and external transport
	B1-a-IE
	B1-b- IE
	B2-a-IE
	B2-b-IE

Source: Doraiswamy, L. K., 1991; *Organic Synthesis Engineering*, New York: Oxford, 2001.)

$$u\frac{d[A]}{d\ell} + (-r_A) = 0 \qquad (11.115)$$

$$u\rho\frac{dT}{d\ell} = \frac{4U}{d_T}(T - T_w) - (-\Delta H_r)(-r_A) = 0 \qquad (11.116)$$

with the initial conditions

$$\left.\begin{array}{l}[A] = [A]_0 \\ T = T_0\end{array}\right\} \text{at } \ell = 0 \qquad (11.117)$$

These equations can be recast in dimensionless form as

$$\frac{dc}{dZ} + \Re_M (-r_A) = 0 \tag{11.118}$$

$$\frac{d\tau'}{dZ} = \frac{4Ud_p}{urC_pT_0}(\tau' - \tau'_w) - \Re_H(-r_A) = 0 \tag{11.119}$$

and the initial conditions

$$\left.\begin{array}{c} c = 1 \\ \tau' = 1 \end{array}\right\} \text{at } Z = 0 \tag{11.120}$$

The parameters $\Re_M$ and $\Re_H$ are the mass and heat transfer groups defined in terms of common physical properties (see Nomenclature).

If the reaction is assumed to be isothermal and first order, the following simple equation for the exit concentration of A results:

$$\ln\frac{[A]_0}{[A]} = (1-\varepsilon_b)\frac{k_vL}{u} = \frac{kV_T}{F_T} \tag{11.121}$$

Equations for the more complex models are considered in Case Study 11.3.

An important aspect of NINA-FBRs is the role of hot spots in exothermic reactions. There is a certain region of parameter space (usually temperature and partial pressure) within which even a slight change in the value of one parameter (usually initial reactant partial pressure) results in a steep rise in the value of the other (usually temperature). In other words, in a well-defined initial pressure space, the temperature sensitivity of the reaction can be very high. Beyond a certain value, the reaction can "run away," sometimes leading to catastrophic consequences. This sensitivity varies for different controlling regimes. The reaction-controlled regime was considered by van Welsenaere and Froment (1970) and Rajadhyaksha et al. (1975) to avoid this runaway. Other criteria are also available (see, e.g., Hlavacek et al., 1969; Hlavacek, 1970; Oroskar and Stern, 1979), but the first two are perhaps the most useful of the safe limit ones.

Practical aspects of NINA-FBR design and operation include backmixing or axial dispersion, which depends to a large extent on the value of L/d_p. The effect of backmixing can be assumed to be absent if this ratio is greater than 50 (variously reported as 30 to 70) for gaseous reactions and 200 for liquid reactions (see Carberry and Wendel, 1963; Carberry, 1964, 1976). Since, in commercial reactors, this ratio is almost always much higher than the indicated values, backmixing is usually not a problem in their design. However, laboratory reactors are often shorter, so their data cannot be directly applied to large-scale reactors. A more rigorous analysis shows that the following criterion must be satisfied for the data to be free of backmixing effects (Mears, 1971b):

$$\frac{H}{d_p} > \frac{20}{Pe_{ma}} \ln\frac{[A]_0}{[A]_e} \tag{11.122}$$

where Pe_{ma} is the axial Peclet number, $d_pu/\mathbf{D}_{ea}$.

Another problem can arise if the catalyst is not uniformly distributed (leading to nonuniform pressure drops) among the thousands of tubes that are normally present in a NINA-FBR. If a particular tube has a lower pressure drop, more gas will flow through it, and some catalyst may be blown out until the tube becomes almost empty. Such a situation can be avoided by making incremental additions (say, in 10% lots) of catalyst and measuring the pressure drop in each tube after each addition. The tubes must be tapped or vibrated between additions to ensure uniform pressure drop at each height. Another strategy is to provide for a much larger pressure drop at the tube entrance than that created by the catalyst bed itself, so that any fluctuations in the latter will not affect the flow distribution. This can be accomplished by using a tube sheet with a single nozzle at the bottom of each tube, with a high pressure drop across it. This practice is, however, not generally followed, since tackling the problem at the source, as described earlier, is a much better alternative. The problem of fluid distribution is not critical except in the headers whose diameter corresponds to the shell diameter, which is very much larger than that of the tubes.

11.4.2.4.2 Fixed-Bed Reactors: Design of A-FBR

As already mentioned, the design of an A-FBR is quite different from that for a NINA-FBR. If it becomes necessary to limit the height of each stage in this reactor to avoid excessive temperature rise or fall, then a large number of stages with interstage cooling or heating are required to achieve the desired conversion (Figure 11.28). Two decisions must be taken for each stage: the inlet temperature and the outlet conversion to be achieved in it. Thus, if there are N beds, $2N$ decision variables much be simultaneously varied to optimize reactor operation to maximize profit. This almost always involves prohibitive calculations. A practical method of reducing the amount of computation is

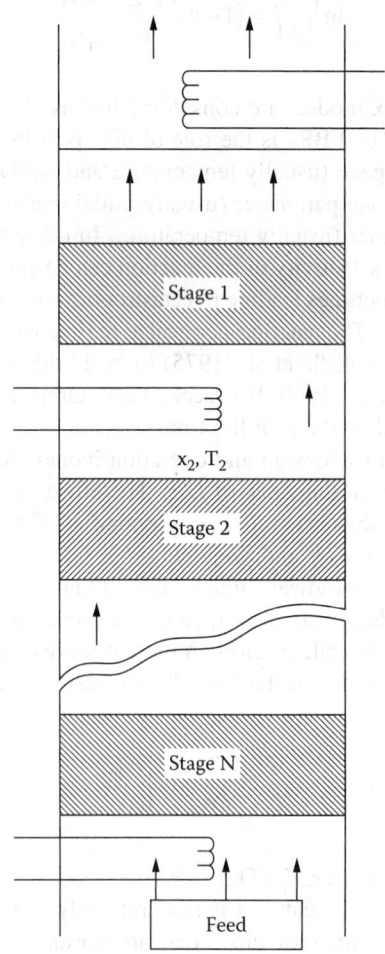

FIGURE 11.28 An interstage heat transfer scheme for A-FBR design.

dynamic programming (Bellman, 1957; Aris, 1961; Bellman and Dryfus, 1962; Roberts, 1964). Lee and Aris (1963) have applied a graphical procedure to optimize an SO_2 oxidation reactor.

Quite often, where the reaction is not highly exothermic or endothermic, a long, single bed is adequate. Because the reaction is adiabatic, there is a unique relationship between concentration and temperature for every reaction. In other words, two separate equations based on mass- and heat-balances are not required. For a reversible reaction of the type

$$A \leftrightarrow R \qquad \text{R4.2}$$

the following single relationship results:

$$X_A = \left(\frac{F_{t0}C_{Pm}}{F_{A0}(-\Delta H_r)}\right)(T - T_0) \qquad (11.123)$$

or, since $\Delta[A] = X_A[A]_0$,

$$\Delta[A] = J\Delta T \qquad (11.124)$$

where

$$J = \frac{C_{Pm}[A]_0}{(-\Delta H_r)}\left(\frac{F_{t0}}{F_{A0}}\right) \qquad (11.125)$$

This equation can compute conversion and temperature as functions of reactor length (for a given diameter) (Levenspiel, 1991; Doraiswamy, 2001). Procedures are also available (Levenspiel, 1999; see also Aris, 1965) for designing adiabatic reactors for various practically important situations, such as for the case of interstage cooling by cold feed (*cold shot cooling*) or by an inert external fluid.

Heat exchange in an overall adiabatic reactor is one of the most important aspects of its design. For example, heat is transferred from a hot to a cold stream within the insulated system; the exothermic heat is transferred (without removing it out of the system by external heat transfer media). The ammonia reactor is an excellent example of an adiabatic reactor for which many heat exchange schemes have been used. Four typical heat exchange schemes are sketched in Figure 11.29 (see Doraiswamy and Sharma, 1984; Aris, 1965, for a detailed analysis of adiabatic reactors).

11.4.2.4.3 Other Fixed-Bed Designs
In addition to the two most common reactors described above, the following have been used.

11.4.2.4.3.1 Radial-Flow Reactors
This configuration has been developed specifically to meet the high tonnage demands of basic chemicals such as ammonia (50 to 100 tons/h). In this reactor proposed by Haldor Topsøe, the catalyst is placed between coaxial cylinders, and the gas flows either from or to the center, as shown in Figure 11.30. The pressure drop is low, since only a short length of the catalyst bed is used. Based on several studies (Raskin et al., 1968a, b; Hlavacek and Kubicek, 1972; Hlavacek and Vortuba, 1977; Strauss and Buddle, 1978; Calo, 1978; Balakotaiah and Luss, 1981), some useful conclusions can be drawn:

1. The plug-flow assumption can be misleading; a simple and approximate criterion for plug flow is not generally satisfied:

$$Pe_{hr}(1 - \omega) > 50 \qquad (11.126)$$

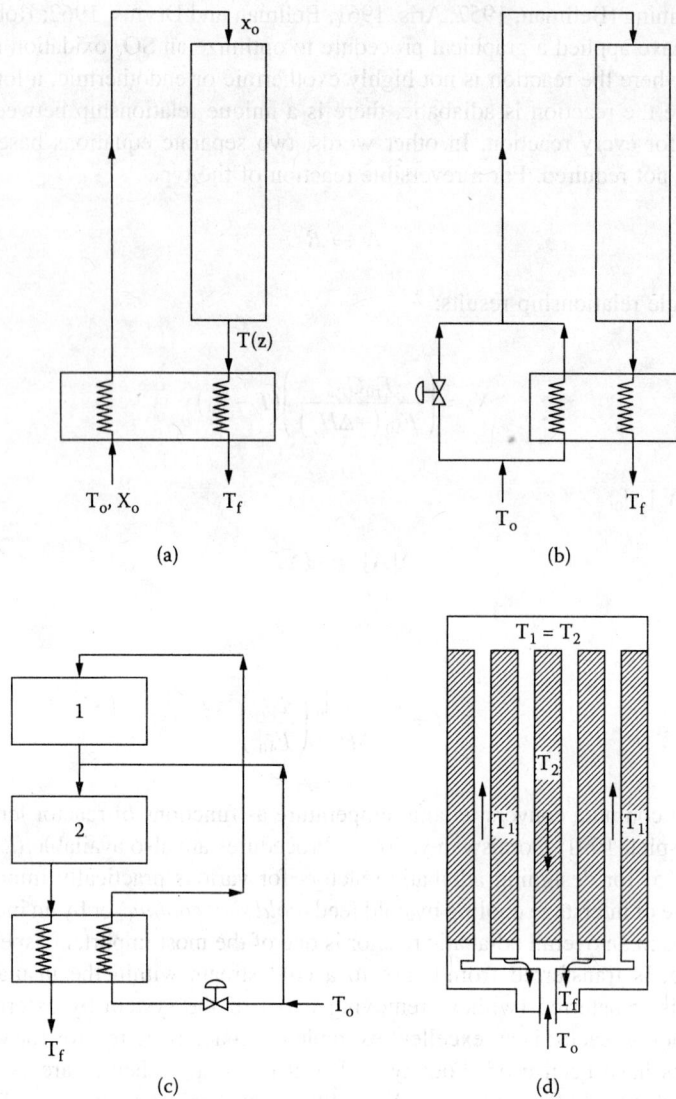

FIGURE 11.29 Typical heat exchange schemes for adiabatic operation.

2. For reactions with no volume change, outward flow of gas yields a higher conversion than inward flow for positive-order kinetics, whereas inward flow is superior for negative-order kinetics.
3. For a first-order reaction with increase in volume, outward flow gives higher conversions, whereas for a reaction with decrease in volume, inward flow is superior.
4. For ideal plug flow, the direction of flow is inconsequential. The difference in behavior between the two directions of flow is therefore due entirely to the dispersion effect. For example, for the ammonia reactor, which satisfies the plug-flow criterion given by Equation (11.126), the change of direction makes no difference.

11.4.2.4.3.2 Catalytic Wire-Gauze Reactors
In certain reactions involving precious metals like platinum, rhodium, or silver as catalyst, the catalyst is used in the form of wire gauze or filament. Examples of reactions that use wire-gauze

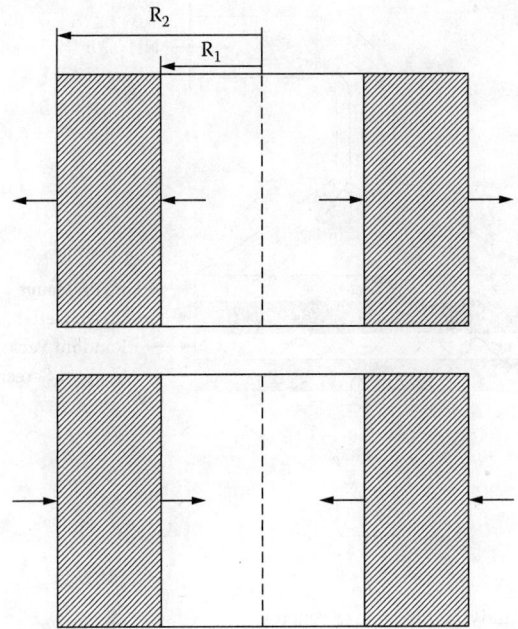

FIGURE 11.30 Radial-flow reactors.

catalysts are oxidation of ammonia to nitric oxide on Pt wire, oxidation of methanol to formaldehyde on Ag-Cu screen, oxidation of ethylene to ethylene oxide on Ag gauze, oxidation of some hydrocarbons on Pt screen, and the Andrussow process for manufacturing HCN on Pt gauze with NH_3, air, and CH_4. The most important consideration is prevention of the precious metal loss by vaporization. The L/D ratio for these reactors is usually very low, on the order of 1/200.

A condition known as *catalyst flicker* is mainly responsible for catalyst loss (e.g., Schmidt and Luss, 1971; Luss and Erwin, 1972; Edwards et al., 1973, 1974). Flicker is the result of nonuniform surface temperatures on the gauze caused by fluctuating transport coefficients. Changes in both wire temperature and surface concentration are important, but more work is needed before reliable predictions can be made. Studies on mass and heat transfer from/to wire gauzes have also been reported (e.g., Satterfield and Cortez, 1970; Shah and Roberts, 1974; Rader and Weller, 1974). The most important application of wire-gauze reactors is in the oxidation of ammonia. The catalyst employed consists of 20 to 60 layers of shining Pt wire screen along with other catalytic and support screens (Powell, 1969; Gillespie and Kenson, 1971). The principle, illustrated by Doraiswamy and Sharma (1984), is sketched in Figure 11.31.

11.4.2.5 Gas-Solid Catalytic Fluidized-Bed Reactors

Incipient fluidization, as outlined in Section 11.4.2.3 for liquid-fluidized beds, is equally valid for gas-fluidized beds. Following the onset of fluidization, however, bed behavior is different for the two cases. While fluidization by liquids results in an expansion of the bed with smooth internal movement of the individual particles of the suspension, fluidization by gas breaks up the flowing gas into bubbles (i.e., individual voids) after the onset of fluidization. As mentioned in Section 11.4.2.3, these two patterns of behavior are referred to as *particulate* or homogeneous and *aggregative* or heterogeneous fluidization, respectively. As the gas density increases, and/or the solid density decreases, the behavior approaches that of particulate fluidization (even if the fluid is not a liquid), and vice versa. The following criteria can be used to roughly distinguish between the two modes of fluidization (i.e., for defining the quality of fluidization):

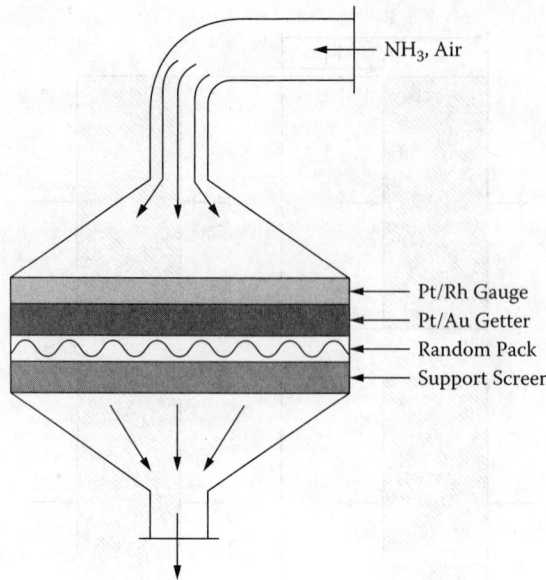

FIGURE 11.31 A typical catalytic wire-gauze reactor.

$$\left(Fr_{mf}\right)\left(Re'_{mf}\right)\left(\frac{\rho_s - \rho_G}{g}\right)\left(\frac{L_{mf}}{d_T}\right) < 100, \text{ particulate} \quad (11.127)$$

$$> 100, \text{ aggregative}$$

where $Fr_{mf} = u^2_{mf}/gd_p$ (Froude group at minimum fluidization), and $Re_{mf}' = d_p u_{mf} \rho_G / \mu$ (Reynolds number at minimum fluidization).

Although several studies over many years have added to the understanding of fluidization, six developments stand out as significant—providing the basic structure of fluid-bed reactor analysis and design:

1. Davidson's fluid dynamic approach to fluid-bed reactor design (see Davidson and Harrison, 1963)
2. Geldart's classification (1973) of solids in terms of their fluidization behavior
3. Grace's explicit recognition (1986) of different regimes of fluidization (drawing from other similar previous studies, e.g., Yerushelmi and Cankurt, 1978; Werther, 1980; Li and Kwauk, 1980; Squires et al., 1985; Horio et al., 1986), through a comprehensive map that demarcates the different regimes
4. Kunii and Levenspiel's modification (see their book, 1991, for original references) of the Davidson model and formulation of reactor design procedures for the different categories of Geldart's particles
5. The finding by Lewis and Gilliland (see Kunii and Levenspiel, 1991) that solids circulation between two fluidized beds (usually a reactor and a regenerator) and in the transport lines connecting them can occur stably
6. The finding that a fluidized-bed reactor can operate at more than one steady state (Elnashaie and Cresswell, 1973; Bukur and Amundson, 1975a, b; Furusaki et al., 1978; de Lasa et al., 1980), in particular the Kulkarni–Ramachandran–Doraiswamy criterion in 1980 for multiple solutions for a first-order reaction

In view of the importance of the Davidson, Kunii-Levenspiel and other models, we consider them at some length in Case Study 11.5. The groundwork for these models as well as the other important features of fluidization mentioned earlier are briefly outlined below.

11.4.2.5.1 Two-Phase Theory of Fluidization

Fluidization theory supported by experiment indicates that the gas-solid fluid bed can be divided into two phases:

1. The bubble phase formed by gas in excess of that required for the onset of fluidization. The bubble is usually surrounded by a *cloud* of gas-solid mixture and is characterized by an indentation caused by suction due to the upward movement of the bubble. The solids that fill up this region are called the *wake*. The bubbles are usually large and move faster than the surrounding emulsion gas flowing at u_{mf}, thus giving rise to the cloud. This behavior is usually characteristic of Geldart **A** and **B** particles (see Section 11.4.2.2).
2. The gas at incipient fluidization percolates through the solid particles, creating a liquid-like phase referred to as the *emulsion phase*. Although this so-called two-phase theory (Toomey and Johnstons, 1952) is not entirely accurate, it is generally valid (within acceptable error limits) and has served as the basis for several fluidized-bed reactor models. These models indicate that conversions increase due to mass transfer between the emulsion and bubble phases.

The application of the two-phase theory to fluidized-bed reactor design is illustrated in Case Study 11.5.

11.4.2.5.2 Geldart's Classification

Based on extensive studies involving a variety of solid particles and fluids with a wide range of properties, Geldart (1973) classified the fluidization behavior into four categories of particles **A**, **B**, **C**, and **D** (Figure 11.32 based on the original). **B** particles conform strictly to the two-phase theory of fluidization described earlier: the gas in excess of that needed for the onset of fluidization immediately breaks into bubbles. In the case of **A** particles, bubbling commences only after a certain amount of gas has passed through the bed beyond incipient fluidization. The other significant features of the particles are explained in the figure. Our concern in this section is essentially restricted to **A** and **B** particles. The smaller **C** particles are too cohesive and fluidize poorly, while the larger **D** particles are more relevant to reactions in which the solid also reacts, such as coal combustion.

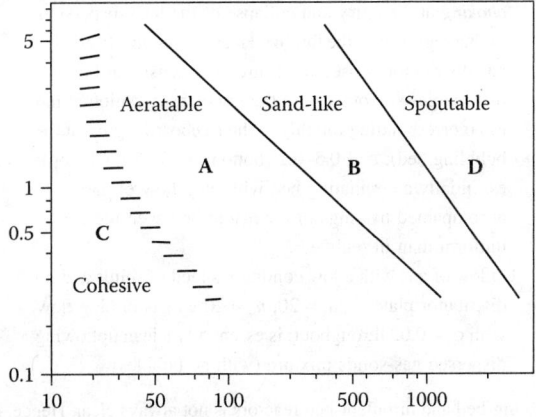

FIGURE 11.32 Geldart's classification of particles with respect to their fluidizability.

11.4.2.5.3 Classification of Fluidized-Bed Reactors

Several categories of fluidized-bed reactors are possible, depending on the mode of operation. The chief features of these reactors are summarized in Table 11.25 and sketched in Figure 11.33. As can be seen from this table, bed behavior (and category) is essentially determined by the fluidizing gas (reactant) velocity and particle size. This transition from fixed to pneumatic bed is usually depicted in terms of a *fluidization map* (e.g., Atipovi et al., 1978; van Deemter, 1980; Werther, 1980; Squires et al., 1985; Horio et al., 1986; Grace, 1986). The latest version, as consolidated by Kunii and Levenspiel (1991), is reproduced in Figure 11.34. Our main concern is with the bubbling bed, although the turbulent bed is only qualitatively different from it.

TABLE 11.25
Principal Features of Different Types (Regimes) of Fluidized-Bed Reactors

Type (Regime) of Fluidization	Main Features	Examples
1. Incipiently fluidized bed (stationary)	Upward flow of gas at about u_{mf} for **A** and **B** particles; no solids mixing; gas mostly in plug flow; solids mixing by a stirrer is sometimes useful; $u_o \leq 1.2\ u_{mf}$ with no bubbles; $\varepsilon_s \approx 0.5$–$0.6$ throughout bed	Methylchlorosilanes
2. Bubbling fluidized bed (stationary)	Upward flow of gas through a wide range of **A** and **B** particles; onset of bubbling depends on particle size, ranging roughly from $u_b = 40\ u_{mf}$ to $70\ u_t$ for small particles to a very narrow range ($u_{mf} < u_b \leq 2\ u_{mf}$) for large particles; $\varepsilon_s \approx 0.6$ (bottom)–0.4 (top)	Polymerization of ethylene to LD polyethylene, ethylene dichloride, vinyl acetate
3. Turbulent bed* (stationary)	Starts gradually at $u_o \gg u_t$ for small particles and $u_p \gg 0.5\ u_t$ for large particles, and merges smoothly into fast fluidized-bed region at higher velocities in each case; as u_o is not very high, internal cyclone is usually adequate; solids entrainment is usually high and, instead of bubbles, clusters of solids and voids of gas move through the bed; $\varepsilon_s \approx 0.3$–$0.4$ (bottom) to 0.2–0.3 (top); the void lifetime is short so that, overall, the bed looks more uniform than in regime 2.	Phthalic anhydride, *o*-cresol and 2–6 xylenol, acrylonitrile, chloromethanes
4. Fast fluidized bed* (circulating)	Continuous feed of both gas and solids; sufficiently high solids velocity—in excess of the upper limit for regime 5; the transition point (from the reverse direction) causes *choking* at the entry and collapse of the lean dispersion of that regime into the fluidized mass of regime 4; suitable gas distributor is used to ensure high density at bottom that merges smoothly with the low-density region at the top (corresponding roughly to the freeboard region of the bubbling bed); $\varepsilon_s \approx 0.5$–$0.2$ (bottom) to 0.05–0.01 (top); essentially a circulating bed with plug flow of gas accompanied by slugs of emulsion; bed even more uniform than in regime 3.	Fischer-Tropsch synthesis of hydrocarbons
5. Pneumatic or transport bed (circulating)	Upflow of gas with a low continuous feed of solids and no distributor plate; $u_o/u_s \approx 20$, $u_o \approx 20\ u_t$; gas in plug flow with $\sigma_s \approx 0.02$ throughout; is essentially a lean upflowing dispersed gas-solids mixture (with no bubbles)	FCC units for petroleum cracking

* The distinction between bubbling-bed and turbulent-bed reactors is not always clear. Hence, the classification of reactions under these categories is uncertain.

Chemical Reaction Engineering

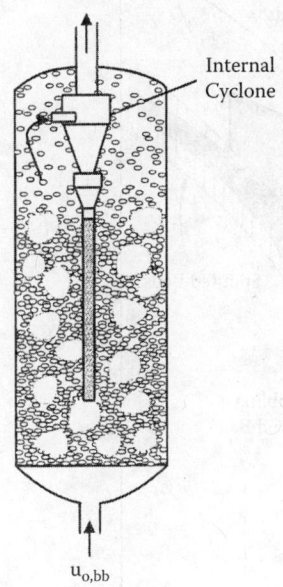

$u_{o,bb}$

A. Bubbling-bed Reactor

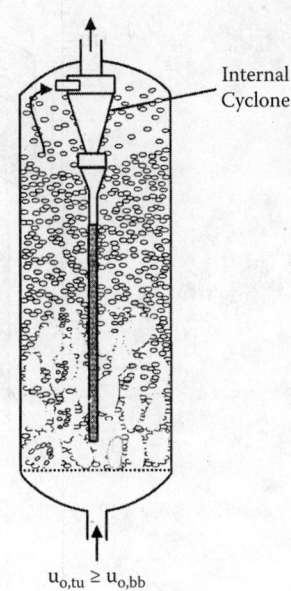

$u_{o,tu} \geq u_{o,bb}$

High Gas Velocity

B. Turbulent- (or fluid-) Bed Reactor

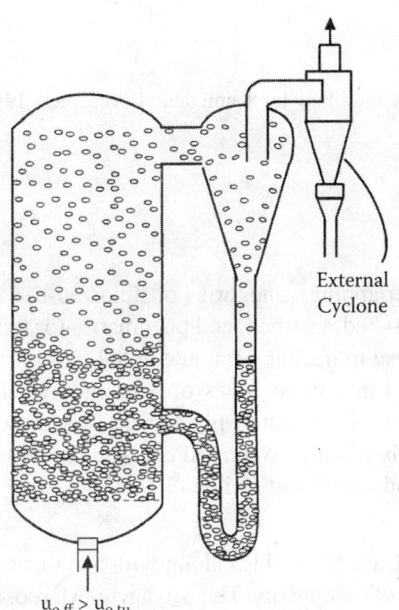

$u_{o,ff} > u_{o,tu}$

Very High Gas Velocity

C. Fast Fluidized-bed Reactor

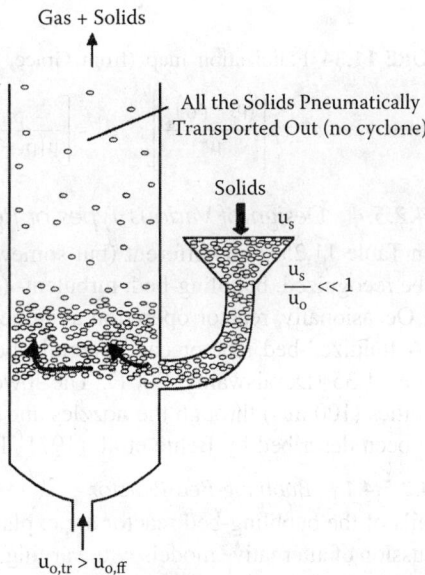

$u_{o,tr} > u_{o,ff}$

Extremely High Gas or Liquid Velocity

D. Transport Reactor

FIGURE 11.33 Classification of fluidized-bed reactors.

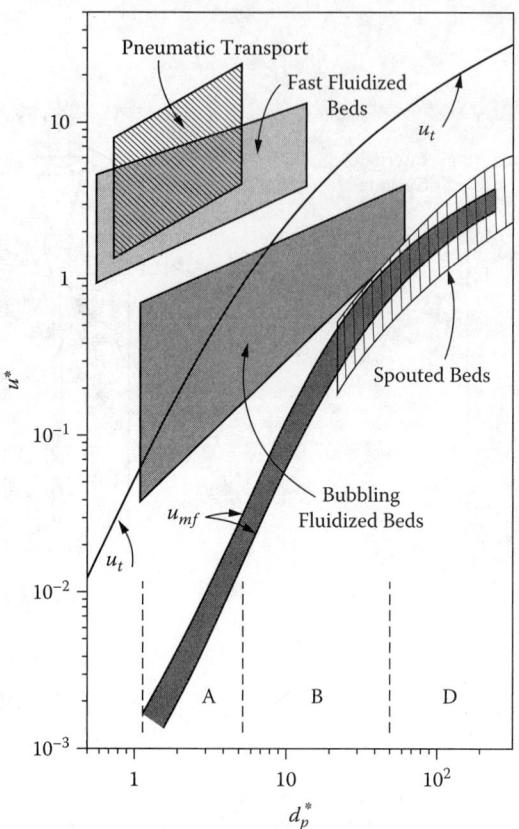

FIGURE 11.34 Fluidization map (from Grace, 1986, as modified by Kunii and Levenspiel, 1991). The parameters are $u^* = u\left[\dfrac{(\rho_s - \rho_G)g}{\mu^2}\right]^{1/3}$, $d_p^* = \left[\dfrac{\rho_G^2}{\mu(\rho_s - \rho_G)}\right]^{1/3}$.

11.4.2.5.4 Design of Various Types of Reactors

From Table 11.25, four different (but somewhat overlapping) categories of fluidized-bed reactors can be recognized: bubbling-bed, turbulent- (or fluid-) bed, fast-bed, and pneumatic- (or transport-) bed. Occasionally, reactor operation at velocities close to u_{mf} has been attempted.

A fluidized-bed reactor can generally be divided into three zones of operation, as shown in Figure 11.35 (Doraiswamy, 2001). The lower jet zone is relevant only for cases with very high velocities (100 m/s) through the nozzles and is usually not important. Salient features of its design have been described by Behie et al. (1971, 1976) and Behie and Kehoe (1973).

11.4.2.5.4.1 Bubbling-Bed Reactor

Details of the bubbling-bed reactor are explained in Case Study 11.5 along with a comparison and discussion of alternative models with varying degrees of complexity. The Miyanchi–Marooka model for the freebroad region (or the dilute phase) is among the models considered. Kunii and Levenspiel (1991) have also proposed a freeboard model in continuation of their bubbling-bed model.

11.4.2.5.4.2 Turbulent-Bed Reactor

Due to the higher gas velocities (which make the bubble-bed reactor more suitable for high-throughput reactions), there is greater turbulence at the interface, leading to more violent bursting of bubbles and splashing of emulsion clusters. Otherwise, it is not much different from the bubbling-bed reactor, despite the fact that the bubble and emulsion phases are not as clearly demarcated.

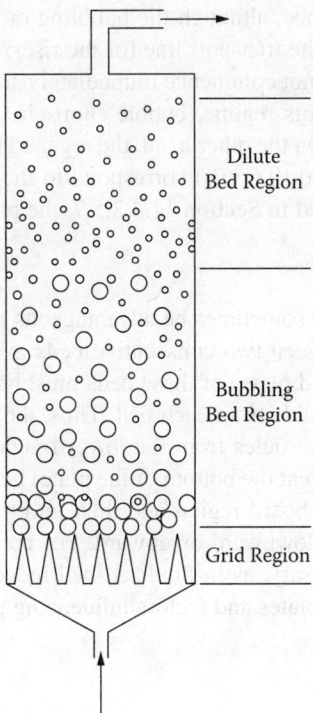

FIGURE 11.35 Operating zones of a fluidized-bed reactor.

Thus, the general procedure described for the bubbling bed including the dilute bed (Kunii and Levenspiel, 1991) or the Miyauchi–Marooka model described in Case Study 11.5 can be used. Then, the porosity distribution data will be different. In particular, the freeboard porosity will be higher. Strictly, new correlations for k_{bc} and k_{ce} are also required, but available correlations for the bubbling bed can be used as a first approximation.

11.4.2.5.4.3   Fast Fluidized-Bed Reactor

Here the gas velocity is even higher than in the turbulent bed. What we have is a bottom dense zone consisting of a mixture of bubbles with more solids in them than in the bubbling bed and clumps of emulsion rising through most of the bed's cross section, while some emulsion flows down at/near the wall. The overall coefficient K_f will be different from that defined by Equation (CS5.14) for Case Study 11.5. Kunii and Levenspiel suggest a relation for this; it requires data such as S_{core} and S_{wall}, which are not available. These observations are also true of the upper dilute region for which the parameter values would be quite different from those in Equation (CS5.19) for the freeboard region above the normal bubbling bed.

Obviously, more research is needed on the turbulent and fast fluidized-bed reactors. At present, tests on a series of pilot plant reactors ending with a semicommercial unit are unavoidable.

11.4.2.5.4.4   Transport (or Pneumatic) Reactor

Fluidized catalytic cracking (FCC) units are characterized by reaction in the transport lines between the reactor and the regenerator, in addition to that in the main reactor. Their design is complicated, involving more than one regime of fluidization. Innumerable studies have been reported on their modeling (e.g., de Croocq, 1984; Chuang et al., 1992). As already explained in Figure 11.33, the fluidization regime changes with increase in gas velocity (sometimes also expressed as the ratio of the superficial velocity to the minimum fluidization velocity). FCC units use very fine particles. In the more recent designs employing highly active zeolite catalysts, the main cracking occurs almost entirely in the riser, i.e., the section that transports the catalyst between the reactor and regenerator or back into the same reactor. Hence, although the bubbling or turbulent regime may be involved within the reactor or regenerator, the transport line (or the riser) operates in the pneumatic regime. In the main reactor, bubbling does not commence immediately after u_{mf} is reached, and the turbulent regime sets in far beyond u_t. In this regime, bubble short-circuiting is much less prevalent, and hence the conversions are higher. On the other hand, the regenerator operates in the bubbling regime even though the velocities involved (0.6 m/s) correspond to the turbulent regime. This is because of the absence of fines. As indicated in Section 11.4.2.5.7, the presence of fines plays an important role (see Yadav et al., 1994).

11.4.2.5.4.5 Staged Reactors

Vertical staging of the catalyst can sometimes be advantageous because the gas flow often approximates plug flow. The region between two consecutive beds is obviously the freeboard region of the lower bed. The holes in the grid plates of these beds must be carefully designed to balance the upward and downward flows of solids from each bed. Thus, the holes in the plate of a given stage should be large enough to allow particles from the lower freeboard region to flow into this stage (thus preventing their accumulation at the bottom of the plate) but small enough to prevent particles from the bed to leak into this freeboard region and then into the lower stage. There is, however, always a through flow of solids, downward or upward. For countercurrent contacting of gas and solid, downflow of solids is necessary, as in fluidized-bed reduction of metal ores. For details of particle interchange at perforated plates and factors influencing particle leakage in staged reactors, see Briens et al. (1978).

11.4.2.5.5 Circulation Systems

In Table 11.25, we saw that certain types of fluidization (fast and pneumatic) involve solids recirculation. Figure 11.36 indicates that the solids are circulating between two fluidized beds A and B. They are connected through the two curves of a U-tube in such a way that the difference in static pressures drives the solids from one bed to the other. The use of a second U-tube completes the circulation between the beds. As there is a frictional resistance associated with solids flow (increasing with increasing flow rate), the rate of circulation is controlled by a balance between the frictional resistance and the static pressure difference mentioned earlier. The frictional resistance can be controlled by varying the average densities of the flowing gas-solid mixtures in the various sections of the circulation loop.

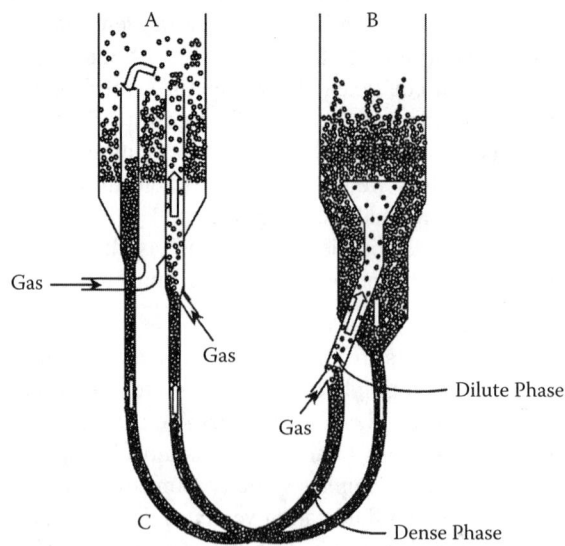

FIGURE 11.36 The main features of solids circulation in fluidized-bed reactors.

Chemical Reaction Engineering

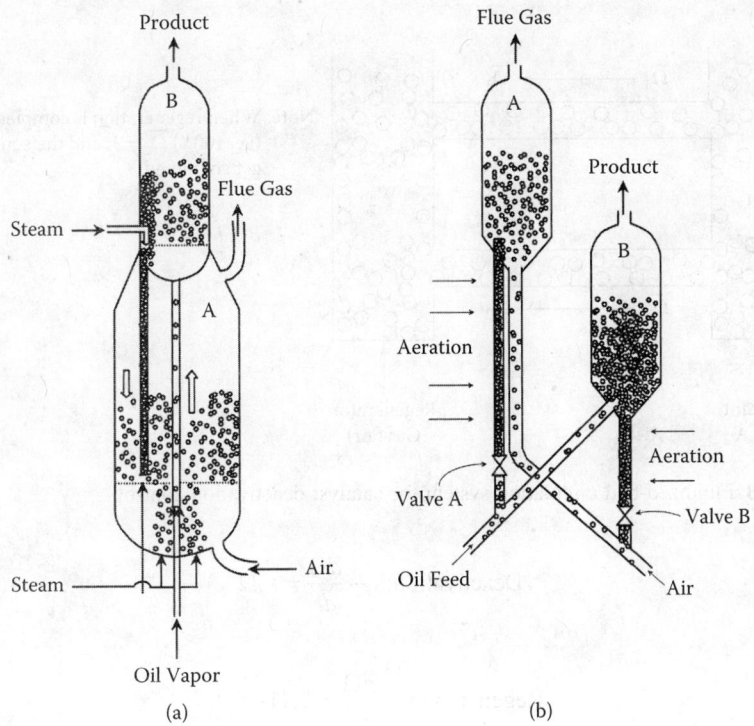

FIGURE 11.37 Two examples of circulation systems in fluidized-bed operation.

Two circulation systems are shown in Figure 11.37 (see Kunii and Levenspiel 1991). Solids circulation is most useful for a deactivating catalyst. Catalytic reaction including simultaneous deactivation by carbon deposition occurs in one location (the reactor), and catalyst regeneration by carbon burn-off with air occurs in a second location (regenerator). Although different fluidizing gases are used in the two reactors, the reactant (usually vaporized oil) in the case of the catalytic converter and air in the case of the regenerator, solids circulation can be restricted to a single loop for both the gases or can be accomplished in two loops, one for the oil and the other for the air.

The second important aspect of circulation is the heat balance between absorption in the endothermic reaction and release in the exothermic zone. The circulation system design depends on whether the overall scheme is deactivation controlled or heat transfer controlled. The final equations, with a brief reference to the principles, are provided for the two extreme cases.

11.4.2.5.5.1 Deactivation Control
To arrive at a balanced circulation rate of solids, let us first define the activity Ω of the catalyst as

$$\Omega = \frac{\text{rate of reaction on catalyst at a given condition}}{\text{rate of reaction on a fresh catalyst}} \quad (11.128)$$

We assume that the catalyst undergoes deactivation in the reactor but that the activity is not fully recovered in the regenerator. This general situation is depicted in Figure 11.38, which shows an average activity $\bar{\Omega}_1$ in the reactor and $\bar{\Omega}_2$ in the regenerator. Assuming first-order deactivation, we have

$$\text{Reaction:} \quad -\frac{d[A]}{dt} = k_v [A] \Omega \quad (11.129)$$

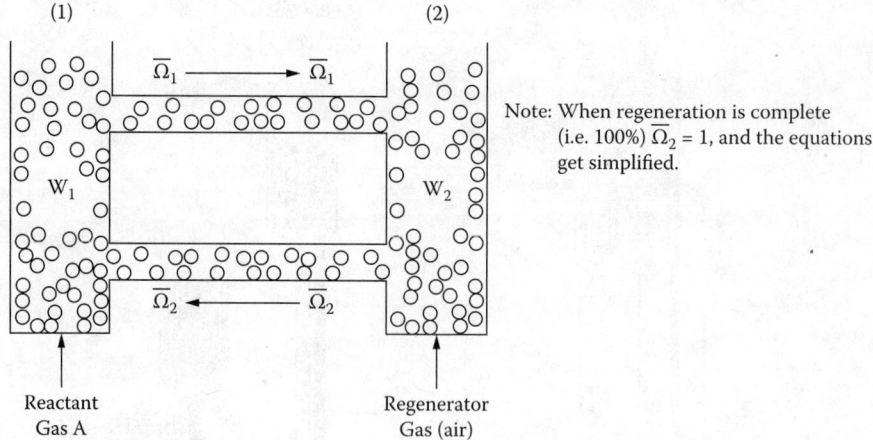

FIGURE 11.38 Fluidized-bed circulation system for catalyst deactivation control.

$$\text{Deactivation:} \quad -\frac{d\Omega}{dt} = k_d \Omega \qquad (11.130)$$

$$\text{Regeneration:} \quad \frac{d\Omega}{dt} = k_a(1-\Omega) \qquad (11.131)$$

It is reasonable to assume that the solids flow is fully backmixed with an exponential distribution of residence times. Based on this assumption and by writing an expression for the average activity of the leaving catalyst stream (which contains particles of all ages with their corresponding activities), the following equations are derived:

$$\text{Reactor:} \quad k_d \bar{t}_1 = \frac{\bar{\Omega}_2 - \bar{\Omega}_1}{\bar{\Omega}_1} \qquad (11.132)$$

$$\text{Regenerator:} \quad k_a \bar{t}_2 = \frac{\bar{\Omega}_2 - \bar{\Omega}_1}{1 - \bar{\Omega}_1} \qquad (11.133)$$

where

$$\bar{t}_1 = \frac{W_1}{F_S}, \quad \bar{t}_2 = \frac{W_2}{F_S}$$

Eliminating $\bar{\Omega}_2$ from Equations (11.132) and (11.133),

$$\bar{\Omega}_1 = \frac{k_a \bar{t}_2}{k_d \bar{t}_1 + k_d k_a \bar{t}_2 + k_a \bar{t}_2} \qquad (11.134)$$

This equation for average catalyst activity is then combined with the rate constant, k_v, to give $\bar{\Omega} k_v$ in the usual reactor equations that would normally use simply k_v. For the three gas flows, we obtain

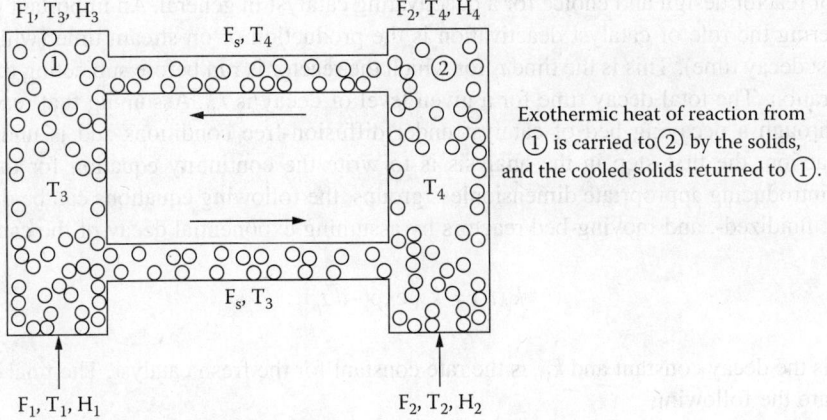

FIGURE 11.39 Fluidized-bed circulation system for heat transfer control.

$$\text{Fully mixed:} \quad \frac{[A]}{[A]_0} = \frac{1}{1 + k_v \bar{\Omega}_1 \bar{t}_1} \tag{11.135}$$

$$\text{Plug:} \quad \frac{[A]}{[A]_0} = \exp(k_v \bar{\Omega}_1 \bar{t}_1) \tag{11.136}$$

$$\text{Fluidized:} \quad \frac{[A]}{[A]_0} = \exp(K_f \bar{\Omega}_1 \bar{t}_1) \tag{11.137}$$

where K_f is the overall rate constant of the Kunii–Levenspiel model of the fluidized bed in the absence of deactivation [see Equation (CS5.14) of Case Study 11.5]. Solution of Equations (11.135) and (11.136) or (11.137) gives the circulation rate F_s and the bed weights for reactor (W_1) and regenerator (W_2).

11.4.2.5.5.2 Heat Transfer Controlled

An important aspect of solids circulation is that heat is transferred from its source (usually exothermic regeneration) to its destination (usually endothermic oil cracking). This can be the controlling feature of a reaction-regeneration system, and the necessary circulation rate for balancing the two depends on the enthalpies at various points in the scheme as shown in Figure 11.39 (see Kunii and Levenspiel 1991). The final equation derived by Kunii and Levenspiel (1991) assuming no heat loss is

$$F_s = \frac{F_1(-\Delta H_r + H_1 - H_3)}{C_{ps}(T_3 - T_4)} = \frac{F_2(H_4 - H_2)}{C_{ps}(T_3 - T_4)} \tag{11.138}$$

If the reaction is endothermic, the cooler is replaced by a heater, with no other change in the analysis.

11.4.2.5.6 Reactor Choice for a Deactivating Catalyst

Catalyst deactivation is always a problem in catalyst and catalytic reactor design. Empirical equations to represent deactivation rates in design calculations are reported by Weekman (1968), Sadana and Doraiswamy (1971), and Doraiswamy and Sharma (1984). Here, we briefly touch upon the problem of reactor design and choice for a deactivating catalyst in general. An important parameter in considering the role of catalyst deactivation is the production or on-stream time (which is also the catalyst decay time). This is the time t_p for which the reactor is run before subjecting the catalyst to regeneration. The total decay time for a given level of decay is t_{p1}. Assuming that reactant A is passing through a decaying bed of catalyst under diffusion-free conditions and is undergoing a simple reaction, the first step in the analysis is to write the continuity equation for the reactor. Then, by introducing appropriate dimensionless groups, the following equations can be derived for the fixed-, fluidized-, and moving-bed reactors by assuming exponential decay of the catalyst, i.e.,

$$k_v(t_p) = k_{v0} \exp(-d_c t_p) \tag{4.36}$$

where d_c is the decay constant and k_{V0} is the rate constant for the fresh catalyst. The final equations obtained are the following.

Fixed-bed reactor:

$$\frac{dy_A}{dZ} = -B' \exp\left(-\lambda \hat{t}\right) y_A^m \tag{11.140}$$

where

$$\left. \begin{array}{c} \hat{t} = \dfrac{t_p}{t_{p1}}, \quad \lambda = d_c t_{p1} \\ \\ B' = \dfrac{M}{\rho_A S_{v,A}} (1-\varepsilon_B) k_{v0} \\ \\ Z = \text{reduced length} \end{array} \right\} \tag{11.141}$$

When a decaying catalyst is used, the average conversion over a given period of time is determined from

$$\overline{X}_A = 1 - \overline{y}_A = 1 - \int_0^1 y_A d\hat{t} \tag{11.142}$$

Fluidized-bed reactor:
If the solids are fully mixed, we have

$$\frac{dy_A}{dZ} = \left(\frac{B'}{\lambda+1}\right) y_A^m \tag{11.143}$$

Since this is a steady-state operation, the conversion is simply

$$X_A = 1 - y_A \tag{11.144}$$

Chemical Reaction Engineering

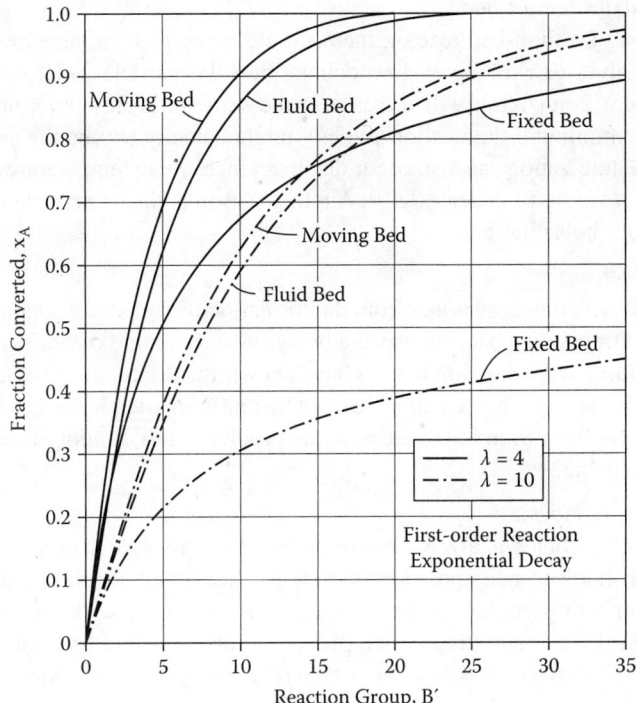

FIGURE 11.40 Comparison of performances of different reactor types for a reaction on a deactivating catalyst. (From Sadana, A. and Doraiswamy, L. K., 1971. *J. Catal.* 23: 147.)

Moving-bed reactor:
In this reactor, catalyst decay is a function of position; the decay time at any position z is given by zt_p. Hence, the reduced decay time is $zt_p/t_p = z$. The following equation is for the mole fraction of A:

$$\frac{dy_A}{dZ} = B' \exp(-\lambda Z) y_A^m \qquad (11.145)$$

where X_A is given by Equation (11.144).

The performances of the three reactors are compared for different parameters in Figure 11.40 for a first-order reaction. The importance of the decay parameter λ is clearly evident and provides a strong point (among many other issues to be considered) in reactor choice. Sadana and Doraiswamy (1971) and Prasad and Doraiswamy (1974) have extended the comparisons to non-first-order and complex reactions.

11.4.2.5.7 Some Practical Considerations

The fluidized-bed reactor is far more difficult to scale up than the fixed-bed reactor. Practical considerations involved in its operations are briefly outlined here.

11.4.2.5.7.1 Slugging

The bed was once assumed to slug when the ratio of the bubble to tube diameter is greater than 0.8 to 0.9 (Stewart and Davidson, 1967). Baeyens and Geldart (1974) give a more elaborate criterion for such a situation. Generally, the conversion in a narrow slugging bed is higher than that in the scaled-up version. Since slugging is not an "unacceptable" mode of operation, multitubular reactors operating under fluidizing conditions are a useful alternative.

11.4.2.5.7.2 Defluidization of Bed

During the operation of a fluid-bed reactor, there should be no sudden increase in pressure due to malfunctioning of valves or choking in downcomers by solid particles. If it does occur, the mass velocity for minimum fluidization will increase at the same total flow rate, and a stage may be reached at which the minimum fluidization velocity might actually exceed the gas velocity, leading to defluidization. Defluidization can also occur due to an increase in temperature or coke deposition of catalyst particles (leading to increased u_{mf}). Another reason is the switch from an inert fluidizing gas used at startup to the actual gas.

11.4.2.5.7.3 Gulf Streaming

The phenomenon of gulf streaming arises from the formation of violent circulating currents induced by bubbles. These currents may become especially significant in large commercial beds (especially shallow beds). Davidson and Harrison (1971) have shown that circulation velocities of 200 mm/s can exist in large beds. As a result, the bubble residence time is smaller, leading to lower conversions. No simple way has yet been found to overcome this problem; thus, a pilot plot of appropriate size seems almost unavoidable.

11.4.2.5.7.4 Effects of Fines

A size distribution of particles is always desired rather than a single size in a fluidized bed. The two-phase theory of fluidized-bed operation is suspect when a bed contains appreciable fines, and models based on uniform particles should be used with caution. The dense phase in such cases should really be regarded as consisting of two phases: emulsion and clusters of fines ($d_p < 40$ μm). Indeed, the results of Yadav et al. (1994) on commercial propylene ammoxidation catalyst clearly show that the fines agglomerate. A critical level of fines (30%) was found in terms of bed expansion, aeratability, and cluster size at which fluid-bed behavior is optimum. They proposed a model that takes the two dense phase components (emulsion and cluster) into account. Adding fines widens the limits of operable gas velocities and minimizes the segregation of particles.

11.4.2.5.7.5 Start-Up

The start-up of a fluidized bed requires an initial burst of pressure to lift the solids and initiates fluidization. This can severely damage the bed internals, and indeed the entire structural framework. It is therefore essential to begin the operation with an empty reactor (by removing the solids after each shutdown) and progressively increase the gas velocity and introduce the solids incrementally at the same time.

Other practical considerations are attrition of particles; caking of catalyst from malfunctioning of the reactor due to formation of tarry products (resulting sometimes in "cakes" as large as the reactor diameter); and the need to avoid premixing of reactants (particularly when they can form explosive mixtures) and fix their relative locations within the bed (e.g., in the chlorination of methane and ammoxidation of propylene). Refer to Doraiswamy and Sharma (1984) for further details.

11.4.2.5.8 Fluidized-Bed vs. Fixed-Bed Reactors

Following their first major success in 1942 in the refining/petrochemical industries, fluidized-bed reactors were hailed as the panacea for most reactor evils. This optimism was clearly too hasty in view of many spectacular failures. But more recently there has been a better understanding of fluid-bed behavior in general. The main advantages are a prevalence of near-isothermal conditions in the entire bed due to solids movement (and therefore the absence of hot spots, a key drawback of fixed-bed reactors), less danger of explosions and temperature excursions, better operational flexibility, and higher heat and mass transfer rates compared to other modes of fluid-solids contacting (often leading to smaller sized heat exchangers). The fixed-bed reactors are less frequently chosen, in spite of their many advantages, namely relatively easy scale-up, minimum catalyst loss due to attrition, and the theoretical possibility of imposing an optimum temperature profile (taking advantage of the inherent nonisothermicity of the bed).

Chemical Reaction Engineering

Pressurized fluidized-bed reactors are currently widely used in gas-solid noncatalytic systems such as coal gasification and coal combustion power plants (Dutta and Gualy, 1999). Higher pressures enhance productivity per unit reactor volume and often significantly increase equilibrium conversion and selectivity. An important example of reactor choice for gas-solid noncatalytic reactions, based on extensive studies on different alternatives, is the hot gas desulfurization process using zinc titanate as the regenerative adsorbent. ZnO, the active component of this adsorbent, reacts with H_2S to form ZnS, which is regenerated by air oxidation and recycled back to the adsorber. This process is part of the integrated gasification coal carbonization (IGCC) power plants. Experimental results on moving-bed, bubbling-bed, and circulating fluidized-bed reactors have clearly shown the circulating bed to be the preferred candidate (Dutta, 1994).

Note that gas-solid moving-bed catalytic reactors, once popular in petroleum cracking, are increasingly being replaced by circulating fluidized beds. They are still used in gas-solid noncatalytic reactors (see Section 11.4.2.6).

11.4.2.6 Gas-Solid Noncatalytic Reactors

For gas-solid catalytic reactors also, fixed-bed, fluidized-bed, and moving-bed reactors are commonly employed. However, since all gas-solid noncatalytic reactions are inherently time dependent, time becomes an unavoidable parameter in the analysis. We briefly outline the procedures for the three reactor types mentioned and also touch upon a few other types. In view of the special importance of fluidized-bed reactors for these systems, two case studies involving the use of this reactor are presented in Section 11.5 (Case Studies 11.6 and 11.12). Dutta and Gualy (1999) give a comparative evaluation of fixed- and fluidized-bed reactors.

11.4.2.6.1 Fixed-Bed Reactors

Fixed-bed reactors are employed for roasting and sintering of ores, incineration of solid wastes, reduction of metal oxides, and production of light-weight aggregates. They can be of the conventional types as shown in Figure 11.41a, in which a tube is packed with the reactant pellets and the gas passed through the bed. An alternative design, not used in catalytic reactors, is shown in Figure 11.41b. Here, the solids are continuously fed on a moving grate, and the gas is blown through the bed. The reaction front in the bed moves as the reaction progresses. The conventional design is usually restricted to batch reactions and hence is not applicable to large-scale production.

In an ideal fixed-bed reactor, plug flow of gas is assumed. This is, however, not a good assumption for reactive solids, because the bed properties vary with position, mainly due to changing pellet properties (and dimensions in most cases), and hence the use of nonideal models is often necessary. The dispersion model, with all its limitations, is still the most practical one. The equations involved are cumbersome, but their asymptotic solutions are simple, particularly for systems

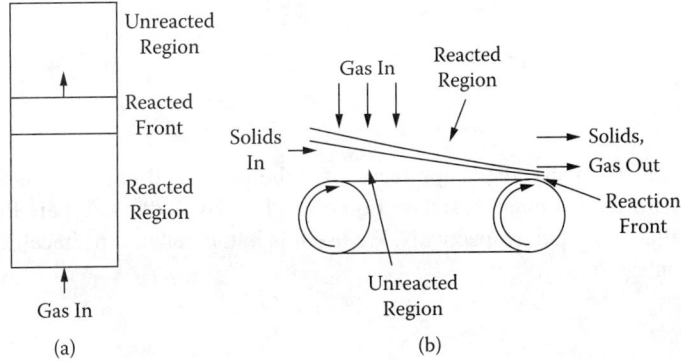

FIGURE 11.41 Reactors for gas-solid noncatalytic reactions.

conforming to pellet reaction control. Their use is especially appropriate for the following reasons. As brought out in Section 11.3.4, the controlling regime in a reactive pellet changes as the reaction progresses. For example, in the case of SIM, the reaction usually starts with reaction control but eventually becomes diffusion controlled, making it necessary to consider individual pellets in the reactor. Often, recourse to one- or two-dimensional heterogeneous models becomes necessary. In such cases, the choice of the gas-solid reaction model has an important bearing on reactor simulation (Sotirchos and Zarkanitis, 1989). Mutasher et al. (1989) have simulated a fixed-bed reactor using the zone model. The particle-pellet model used here has greater generality.

Consider a typical reaction,

$$A(g) + B(s) \rightarrow R(g) + S(s) \qquad \text{R4.3}$$

As pointed out in Section 11.3.4, it can be assumed that the pellet itself is isothermal, even for reactions involving large heat effects. One can also reasonably assume that axial diffusion effects are absent (particularly for long beds), and hence the one-dimensional model can be used. We shall also assume that the heat generated by reaction is lost through the sensible heat carried by the gas leaving the system and by convection at the outside surface of the reactor wall. The nonisothermicity of the reaction greatly affects the linear velocity and density of gas along the reactor, and these effects should be considered. However, in our relatively simple approach, we ignore them.

Derivation of the model equation is reported by Evans and Song (1974). The following equations are written: material and energy balances for the gas phase and the rate equations for the pellet assuming any one of the models described in Section 11.3.4. As mentioned earlier, we use the particle-pellet model. Appropriate initial and boundary conditions are also written. Assuming isothermal behavior, the material balance equation is

$$-G_A \frac{\partial c}{\partial t} = \frac{(1-\varepsilon_p)(1-\varepsilon_b)\rho_{Bm}}{\upsilon} \frac{dX^x}{dt} + (\varepsilon_p + \varepsilon_b)[\bar{A}]\frac{\partial c}{\partial t} \qquad (11.146)$$

where $[\bar{A}]$ is the external field concentration at some point in the bed, and c is the dimensionless external field concentration. The changing value of the concentration $[\bar{A}]$ with position allows for a more realistic simulation of the reactor.

Here the reactor equation is general and independent of the rate equation, which depends on the gas-solid reaction model used. It can also be adapted to other reactors with appropriate modifications. However, these involve unacceptable assumptions for the fluid-bed model.

The particle-pellet equation is written in terms of a generalized effectiveness factors defined as

$$\varepsilon = \frac{\int_0^1 \xi^{s-1} g \eta^{g-1} [\hat{A}]_p \, d\xi}{\int_0^1 \xi^{s-1} d\xi} \qquad (11.147)$$

where s and g are, respectively, the shape factors for the pellet and grain [3 for the sphere, 2 for the cylinder, and 1 for the flat plate, based on Equation (11.49)], and ξ and η are the dimensionless positions in the pellet and grain, respectively. We use this in the material balance Equation (11.146) recast in dimensionless form to give

$$-\frac{\partial c}{\partial Z^*} = \frac{dX^*}{dt^*} = c\varepsilon \qquad (11.148)$$

and then manipulate it into the form

$$\frac{dX^*}{dt^*} = \varepsilon \exp\left[-\int_0^{Z^*} \varepsilon d\lambda\right] \quad (11.149)$$

where λ is a dummy variable, and t^* and Z^* are dimensionless time and distance given by

$$t^* = \left(\frac{\upsilon k_s [\overline{A}]_0 S_g}{\rho_{Bm} g V_g}\right) t \quad (11.150)$$

$$Z^* = \frac{k_s S_g [\overline{A}]_0}{G_A g V_g}(1-\varepsilon_p)(1-\varepsilon b) L \quad (11.151)$$

and X^* is the extent of reaction of the pellet defined by

$$X^* = \frac{\int_0^1 \xi^{s-1}(1-\eta^g) d\xi}{\int_0^1 \xi^{s-1} d\xi} \quad (11.152)$$

The calculation procedure is straightforward. We compute ε for different values of X_B and a reaction modulus σ that is defined as

$$\sigma = \frac{s\, V_p}{A_p}\left[\frac{(1-\varepsilon_p) k_s}{\mathbf{D}_{e,As}} \frac{S_g}{V_g}\right]^{1/2} \frac{1}{(2sg)^{1/2}} \quad (11.153)$$

and may be considered to be a dimensionless pellet size. The precise table given by Evans and Song (1974) can be used to estimate σ, but the following equation for a sphere ($s = 3$) is usually adequate (Sohn and Szekely, 1972):

$$\varepsilon \cong \frac{1}{s}(1-X^*)^{1/g-1} + \sigma\left[2(1-X^*)^{-1/3} - 2\right]^{-1} \quad (11.154)$$

Knowing ε from Equation (11.147), Equation (11.148) can be integrated and X^* plotted as a function of t^* for a given value of σ.

Several complicating features can be added, such as a different bed porosity at the wall by using the correlation of Chandrasekhar and Vortmeyer (1979), allowing for nonisothermicity of the bed (Sampath et al., 1975), accounting for propagation of the reaction front (Bagajewiicz, 1992), and the variation of gas properties throughout the bed.

11.4.2.6.2 Moving-Bed Reactors

The same approach as for the fixed-bed reactor is employed here but by making allowance for the special features of the moving-bed reactor (Figure 11.42). The main difference is that the solid is also moving, and a mass balance equation for the solid phase is therefore needed—both for

FIGURE 11.42 A typical moving-bed reactor.

plug flow and complete mixing of the solids. The following are based on the more likely plug-flow behavior.

The equations for the gas and solid phases are

$$\text{Gas:} \quad \frac{dc}{dZ^*} = -c\varepsilon \tag{11.155}$$

$$\text{Solid:} \quad \frac{dX^*}{dZ^*} = \upsilon\left(\frac{G_A}{G_B}\right)c\varepsilon = R'c\varepsilon \tag{11.156}$$

The boundary conditions are

$$c = 1, Z^* = 0 \tag{11.157}$$

and

$$\text{for cocurrent flow,} \ X^* = 0, Z^* = 0$$

$$\text{for countercurrent flow,} \ X^* = 0, Z^* = L^* \tag{11.158}$$

where L^* is a dimensionless reactor length defined for concurrent flow, defined as

$$L^* = \left[\frac{k_s [\bar{A}]_0 S_g (1-\varepsilon_p)(1-\varepsilon b)}{G_A g V_g}\right] L \qquad (11.159)$$

From stoichiometry,

$$X^* - X_0^* = R'(c-1) \qquad (11.160)$$

Substituting Equation (11.160) in (11.156) and integrating, we get

$$\text{for cocurrent flow:} \quad Z^* = \int_0^{X^*} \frac{d\chi}{(R'-1)\varepsilon(x)} \qquad (11.161)$$

where χ is a dummy variable, and

$$\text{for counter-current flow:} \quad L^* = \int_0^{X_0^*} \frac{dX^*}{(R' - X^* + X_0^*)\varepsilon(X^*)} \qquad (11.162)$$

Using Equation (11.147) for ε, these equations can be solved to give c and X^* as functions of Z^* or L^* for given values of σ. Countercurrent operation normally gives significantly higher conversions.

11.4.1.6.3 Fluidized-Bed Reactors

As with the other two reactors considered, the prerequisite for the design of a fluidized-bed reactor is a reaction model for the single pellet. These models have been discussed in Section 11.3.4. The main features of the fluidized bed were outlined in Section 11.3.3 and are further explained and illustrated in Case Study 11.5. What is unique to the fluidized-bed reactor for gas-solid noncatalytic reactions is the particle size distribution, which changes with reaction mainly because the particle density changes with reaction. Thus, an important design feature is the prediction of particle size distribution in the product solids from a knowledge of the distribution in the reactant solids. Allowance must be made for elutriation of fines and their partial return to the reactor. Particle growth or shrinkage during reaction will also affect the particle size distribution and fluidization characteristics of the bed and add further complexity to the design. This complexity can also arise in fixed- and moving-bed reactors but is not as crucial. In the foregoing treatment of fixed- and moving-bed reactors, it was assumed that the particle size remains constant. Nondecaying catalytic systems can be operated in the batch mode with respect to the solids, but the consequences of reaction and solids consumption often require that noncatalytic gas-solid reactions be operated in the continuous mode.

The main features of the design of a noncatalytic fluidized-bed reactor are described in Case Study 11.12 using the manufacture of chlorosilanes as an illustrative example. Hence, no further description of these reactors will be attempted here.

11.4.3 KINETIC ENERGY (STIRRED TANK REACTORS)

11.4.3.1 Introduction

In chemical process industries, stirred tank reactors are frequently used. For hydrogenation/oxidation applications, the heats of reaction are large, and the overall operation may be heat transfer controlled. In reactions where selectivity is of importance (for example, oxidation, hydrogenation,

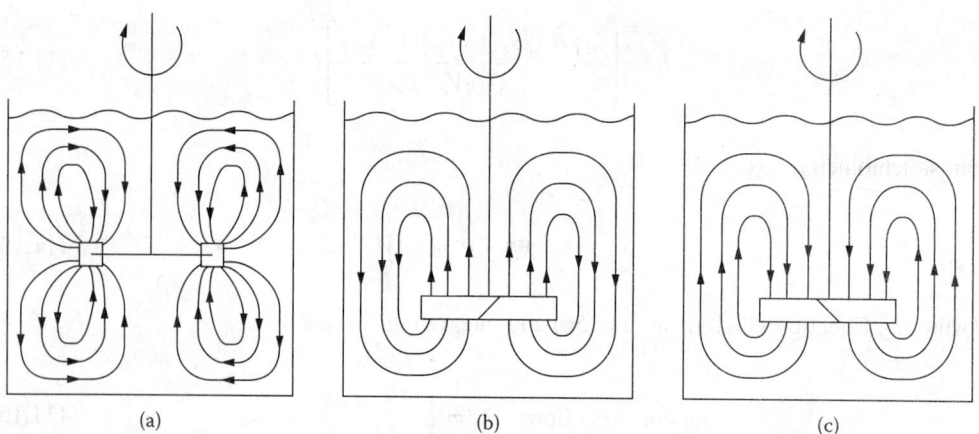

FIGURE 11.43 Flow patterns in a stirred tank reactor: (a), radial flow, (b) axial upflow, (c) axial downflow.

ethoxylation, alkylation), it may be necessary to maintain a uniform concentration in the reactor; in such cases, mixing is of importance. If the gas is sparingly soluble, its concentration in the liquid phase is fairly small, and the overall reaction may become mass transfer controlled. In many gas-liquid/gas-liquid-solid reacting systems, conversion of gas per pass is low. If it is undesirable (because of toxicity, explosion, or cost considerations) to vent this unreacted gas, then the unreacted gas has to be recycled back into the reactor. Hydrogenation, chlorination, alkylation, oxidation with pure oxygen, ozonolysis, hydrochlorination, and others are in this category. Recycling of the unreacted gas can be achieved with what is called a *dead-end reactor* in which the unreacted gas is not vented. These systems employ gas-inducing impellers, surface aerators, rotating disc contactors, jet-loop reactors, and plunging jet reactors to recontact the gas with the liquid phase.

All the above-mentioned processes require external energy input. In the stirred reactor, energy is supplied by the impeller. The fluid leaving the impeller receives mean and turbulent kinetic energy resulting from the impeller motion. The presence of a second or third phase may also introduce additional energy. The impeller designs are generally classified into two types: (1) radial flow impellers and (2) axial flow impellers. Rushton turbine, straight-blade turbines, curved-blade turbines, Brumagins, and so on belong to the radial flow category. The flow field generated by them is radially outward (Figure 11.43a), toward the wall. At the wall, this flow divides into two parts, one going above the impeller and the other going below. Both flows then return to the impeller near the vessel centerline. Axial flow impellers can be operated in either the "pumping-up" mode (Figure 11.43b) or "pumping-down" mode (Figure 11.43c). Marine propellers, pitched-blade turbines, hydrofoil impellers, and so forth fall under this category. In the pumping-down mode, the impeller generates a downward flow, which hits the vessel bottom, turns radially, and is ultimately converted to upflow before returning to the impeller. The mean flow generated causes bulk motion of the constituents, whereas the turbulence generated is responsible for eddy diffusion of momentum, heat, and mass. The size, shape, and number of impellers; vessel geometry such as number, location, and size of baffles; vessel diameter; liquid level; impeller clearance; and system properties all have a profound impact on the flow field, thus affecting the performance of the stirred vessel.

To design/select impellers on a rational basis, an understanding of two aspects is essential: (1) the relation between impeller design, vessel geometry, and the flow field produced; and (2) the relation between the flow field and design objectives such as mixing, heat transfer, solid suspension, gas dispersion, and others. From 1950 to 1970, impeller design was investigated mainly empirically. For this purpose, the investigators fabricated several different shapes and used them in laboratory-scale equipment. The performances of the various impellers in terms of a design objective (e.g., gas dispersion, solid suspension, blending, and so on) were studied, and the best design was chosen on the basis of such experiments (Kramers et al., 1953; Zwietering, 1958; van't Riet and Smith, 1973). Such experiments are costly and time consuming. Since the 1970s, the focus has shifted to modeling and understanding the phenomena at hand. Over the last 20 to 30 years, many models have been proposed for each of the above design objectives (Mann and Mavros, 1982; Warmoeskerken and Smith, 1985; Molerus and Latzel, 1987; Patwardhan and Joshi, 1999a). These models relate the flow generated by the impeller to the design objective. In the following paragraphs, an attempt is made to briefly discuss the important aspects of process design of stirred reactors for various objectives.

11.4.3.2 Blending

The blending process is characterized in terms of the blending time, commonly called the *mixing time*. This is considered to be the time between the introduction of a tracer at a certain point and the tracer concentration to reach a certain degree of homogeneity at some other point. The blending process occurs as a result of transport at three levels: molecular, eddy, and bulk (convection). Usually, bulk motion (or bulk diffusion) is superimposed on either molecular or eddy diffusion or both. In industrial practice, many blending operations are performed at turbulent conditions. In that case, molecular diffusion can be neglected in comparison to bulk diffusion and eddy diffusion. The blending time is commonly represented in terms of a dimensionless quantity (called the *homogenization number*, Nt_{mix}), which is the product of the impeller speed and mixing time. The value of Nt_{mix} depends on the impeller design, diameter, location, number, and size, and the location of baffles, presence of a draft tube, reactor dimensions, and so forth. Under laminar conditions, the value of Nt_{mix} is inversely proportional to impeller speed, but, at turbulent conditions, it is independent of impeller speed. Therefore, under turbulent conditions, the mixing time, t_{mix}, is proportional to N^{-1} or $P^{-1/3}$. Rewatkar and Joshi (1991a) investigated the effect of impeller design on mixing time. At a given power consumption, axial-flow impellers produce better mixing (lower t_{mix}) as compared to Rushton turbine impellers. Patwardhan and Joshi (1999a) investigated the blending characteristics of several axial flow impellers. Mixing is essentially controlled by the convective transport. They also correlated the mixing time data in terms of the secondary flow number of the impeller. Thus, the energy efficiency of the blending process depends on the impeller pumping capacity as well as the power consumed by the impeller. The authors give the values of the secondary flow number, N_{QS}, for a variety of impeller designs. The following correlation is recommended by Patwardhan and Joshi (1999a):

$$t_{mix} = \frac{12.66 N^{1/3} D^{2/3}}{N_{QS}(P/M)^{1/3}} \tag{11.163}$$

11.4.3.3 Gas Dispersion

In gas-liquid mechanically agitated contactors, the gas flow often alters energy input from the impeller. Typically, the quantity of gas fed to the reactor is characterized in terms of the superficial gas velocity, gas flow number, or volume of gas per unit volume of liquid per minute (VVM). For a given gas flow rate, an increase in impeller speed results in various modes of operation (called *regimes of operation*). The regimes of a pitched-blade downflow turbine with sparger below the impeller are shown schematically in Figure 11.44a. At very low speeds, the gas rises vertically with very little mixing, and the impeller does not disperse it. This is called *flooding* (LM in figure). As the impeller speed increases, the downward flow generated by the impeller becomes stronger and opposes the upward flow generated by the gas. As a result, the power consumption increases, and the P_G/P_O value can even be greater than 1 (point L in figure). At

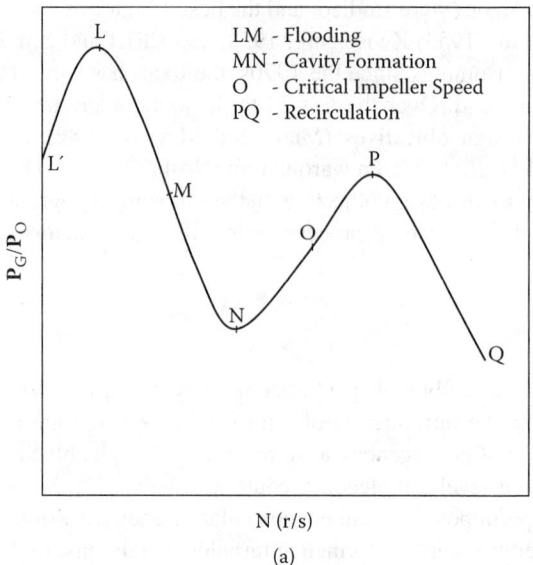
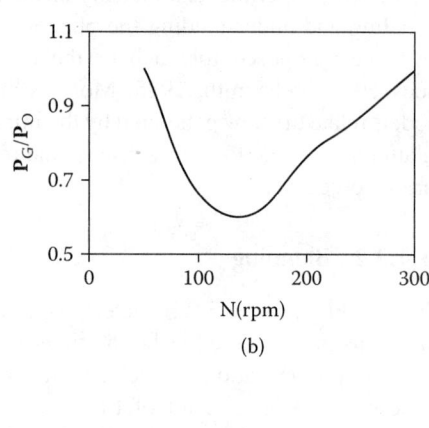

FIGURE 11.44 Regimes of operation in a stirred tank reactor: (a) with a pitched-blade downflow turbine with sparger below the impeller; (b) with upflow impeller.

this point, the gas becomes dispersed in the radial direction. As the impeller speed increases further, the gas is distributed radially from the impeller, i.e., loading occurs. Then the gas starts forming cavities behind the impeller blades (MN in figure). The size of the cavities increases with a higher impeller speed, and therefore P_G/P_O falls. At still higher speeds, the impeller-generated flows are dominant and overcome the upward flow generated by the bubbles. Under these conditions, the impeller disperses gas even below the impeller. This is referred to as complete dispersion. Beyond this point (point N in figure), the shear forces become strong enough to break the cavities, and the cavity size reduces. This leads to an increase in the P_G/P_O ratio (point O in the figure), and the so-called critical impeller speed is reached. At very high speeds, the impeller-dispersed gas phase is recirculated by the impeller-generated flow called recirculation (beyond point P in the figure). The gas holdup in the impeller region and therefore the average density in the impeller region start reducing; as a result, the P_G/P_O ratio again starts reducing. To obtain efficient gas-liquid dispersion, the impeller should be operated at a speed above the critical speed required for complete dispersion.

For the upflow impeller, the P_G/P_O vs. impeller speed curve is shown in Figure 11.44b. The flow generated by this impeller and the gas flow are complementary to each other, and thus formation of large cavities is prevented (Bujalski et al., 1988). Mhetras et al. (1994) have correlated the critical impeller speed for complete dispersion with the gas flow number as

$$Fr = aFl^b \qquad (11.164)$$

Values of the constants a and b have been reported for many impeller designs.

Ideally, an impeller should have a P_G/P_O ratio of unity. Efforts to reach this have resulted in the development of axial-flow impellers (up-pumping or down-pumping) with a high solidity ratio, as well as radial-flow impellers with convex-shaped blades. The power drawn by these impellers is practically constant, even at high gas flow rates. These are therefore called *gas-foil impellers*. When compared on the basis of equal power consumption per unit mass, gas-foil impellers show a larger gas holdup (that is, better gas dispersion) than conventional Rushton turbines (McFarlane and Nienow, 1996). Mhetras et al. (1994) found that axial-flow impellers give higher gas holdups than radial-flow impellers. They also investigated the effect of blade twist so as to convert an ordinary pitched-blade turbine to hydrofoil shape. Hydrofoil shape gives larger gas holdups at low gas velocities (<15 mm/s), and a conventional pitched-blade turbine gives higher gas holdup at larger gas velocities (>15 mm/s). The following correlation is recommended for estimation of gas holdup (Rewatkar et al., 1993):

$$\varepsilon_G = 3.54 \ (T/D)^{2.08} \ Fr^{0.51} \ Fl^{0.43} \tag{11.165}$$

This was developed for pitched-blade turbines for air and water as the working fluids. In industrial practice, the physical properties of the fluids are likely to be dramatically different. Therefore, the constant in the equation may vary, but its values can be obtained with a few experiments in the laboratory. The new correlation thus obtained can be used for scale-up. Holdup correlations for other impellers are given elsewhere (Mhetras et al., 1994; McFarlane and Nienow, 1996).

11.4.3.4 Solid Suspension

As the impeller speed increases in a reactor containing solid particles and a fluid, the rolling and sliding of particles (on the base of the vessel) results in the formation of either a dune at the center (observed with radial-flow impellers) or a ring at the periphery (observed with axial-downflow impellers) of the vessel. With increasing impeller speed, the dune or the ring gradually reduces in size until a speed is reached at which it completely disappears; this is called the *critical impeller speed* for solid suspension and is denoted as N_{CS}. Evidently, several transitions in particle motion are possible, and the numerical value of the impeller speed depends on the state of suspension. The problem of selecting a representative speed for suspension is complicated, since the transition from one state of motion to another is gradual and occurs over a range of impeller speeds. The critical impeller speed can be defined in various ways, namely, (1) all the particles moving on the tank base (no particle is stationary on the tank base for more than 1 to 2 s), N_{CM}; (2) complete off-bottom suspension (particles do not remain on the tank base for more than 1 to 2 s), N_{CS}; and (3) suspension height (impeller speed required to get a particular suspension height, for instance 50% or 70% of the height), N_{C50}, N_{C70}, etc. When a particle is at the base of the tank, several forces act on it. These are the gravity force, buoyancy force, drag and lift forces due to motion of the fluid over the particle, and frictional forces between the particles or between the particles and the base of the reactor. The gravity and buoyancy forces are solely related to the particle size, shape, and density, and the density of the liquid. However, the drag and the lift forces depend on the velocity field around the particle, which in turn is influenced by the particle motion, the presence of other particles, and so on. Thus, the solid suspension process is fairly complex. Zwietering (1958), in pioneering work, correlated the effect of various parameters such as impeller design, diameter, clearance, particle size, solid loading, particle density, physicochemical properties of the liquid, and so forth, on the critical impeller speed for solid suspension. Since solid suspension occurs from the base of the reactor, its shape is an important aspect influencing the value of N_{CS}. Chudacek (1982) performed systematic investigations by altering the shape of the reactor bottom. For a small particle size (116 μm), the power required at critical suspension conditions was highest for a flat-bottom tank. The power required for a cone and fillet tank (CFT) was smaller by about 40%, and that for a profiled bottom tank (PBT) was about 60% smaller. For a larger particle size (290 μm), the CFT and PBT required only about 40 and 35% of the power required for a flat-bottom tank. Raghav Rao et al. (1988) investigated the effect of impeller design in detail. The power required at critical suspension conditions for a pitched-blade downflow impeller was about 20% of that for a disc turbine and about 33% of that for a pitched-blade upflow impeller. The following correlation is recommended for the estimation of the critical impeller speed for solid suspension for pitched-blade downflow turbine (PBTD) impellers:

$$N_{CS} = 1.15 \frac{u_{S\infty}^{0.28} X^{0.1}}{D^{0.85}} (D/T)^{0.8} \qquad (11.166)$$

For other impellers, see Zwietering (1958) and Chapman et al. (1983).

11.4.3.5 Heat Transfer

To operate isothermally, it is generally necessary to remove heat from, or supply heat to, the reactor. This can be achieved by using jacketed reactors or helical coils inside the reactor. The heat transfer area provided by jackets per unit volume of the reactor is equal to $4/d_T$, where d_T is the reactor diameter. Thus, upon scale-up, the heat transfer surface area per unit volume of reactor decreases. Helical coils provide larger heat transfer area per unit volume and are generally preferred. Under certain conditions, especially for low-viscosity applications, it may be necessary to consider the heat transfer coefficient on the reactor side as well as the jacket/coil side and then determine the overall heat transfer coefficient. When only one impeller is to be used, it is desirable to keep the impeller centrally located to obtain good heat transfer coefficients over the entire length of the jacket/coil. The reactor-side heat transfer coefficients for a jacketed vessel have been correlated as follows for a wide range of impeller designs (Uhl and Gray, 1966); they also report values of the exponents:

$$\frac{hD}{k_T} = a \left(\frac{NT^2 \rho}{\mu} \right)^b \left(\frac{c_P \mu}{k_T} \right)^c \left(\frac{\mu_w}{\mu} \right)^d \qquad (11.167)$$

The values of b range from 0.66 to 0.75, those of c to from 0.25 to 0.50, and of d to from -0.14 to -0.25. The values of the constant a vary dramatically from 0.17 to 0.75. Since the heat transfer coefficient is typically proportional to $N^{0.67}$, it can be considered to be proportional to $(power)^{0.22}$ (Oldshue, 1983). Oldshue also provides a general correlation for the heat transfer coefficient for an immersed tube as

$$\frac{hT}{k_T} = 0.17 \left(\frac{ND^2 \rho}{\mu} \right)^{0.67} \left(\frac{c_P \mu}{k_T} \right)^{0.37} \left(\frac{D}{T} \right)^{0.1} \left(\frac{d}{D} \right)^{0.5} \left(\frac{\mu}{\mu_S} \right)^m \qquad (11.168)$$

where m is a function as defined by Oldshue. Fasano et al. (1991) compared the heat transfer coefficients for a pitched-blade turbine and HE-3 impeller. At the same T/D and C/D ratios and at equal impeller Reynolds numbers, the heat transfer coefficient of the Rushton turbine was marginally smaller than that for the four-bladed pitched-blade turbine. Since the power number of the Rushton turbine is three times that of the four-bladed pitched-blade turbine (used in their work), this would mean that at equal power, the pitched blade turbine would give a roughly 35% higher heat transfer coefficient. The six-bladed pitched-turbine resulted in an even larger heat transfer coefficient. Oldshue (1983) had also previously recommended axial-flow impellers for efficient heat transfer.

11.4.3.6 Blending in Gas-Liquid Systems

Gas in a stirred reactor alters the flow pattern dramatically, depending on the regime of operation (flooding, loading, complete dispersion, and so on). Rewatkar and Joshi (1991b) systematically investigated the mixing characteristics in various regimes of operation. The mixing time in the presence of gas is larger than that in its absence. However, the relative increase depends on the

Chemical Reaction Engineering

regime of operation. At very low speeds (flooding conditions), the mixing time decreases rapidly with increasing impeller speeds. As cavity formation starts and the cavity size increases (leading to a drop in P_G/P_O), the mixing time increases with an increase in impeller speed. At speeds higher than the critical impeller speed for complete dispersion of the gas, the cavity size reduces (leading to an increase in P_G/P_O ratio), and the mixing time again begins to reduce with an increase in impeller speed. Rewatkar and Joshi (1991b) correlated the liquid phase mixing time in gas-liquid systems as

$$Nt_{mix} = 38.58 D^2 T^{-1.94} u_G^{0.226} \tag{11.169}$$

leading to

$$\frac{(Nt_{mix})_{G-L}}{(Nt_{mix})_L} = 3D\, T^{-0.94} u_G^{0.226} \tag{11.170}$$

Pandit and Joshi (1983) correlated the mixing time in the presence of gas for the disc turbine and propeller.

11.4.3.7 Blending in Solid-Liquid Systems

Raghav Rao et al. (1988) investigated the effect of solid phase on blending characteristics over a wide range of particle sizes and solid loadings. The mixing time in the presence of a solid phase is generally larger than that in the absence of one. However, the extent of increase depends on particle size, solid loading, impeller speed, and other factors. For a given solid loading, higher impeller speed increases the fraction of suspended solids. Part of the energy supplied by the impeller is dissipated at the solid-liquid interface. Thus, the increase in the liquid circulation velocity with an increase in impeller speed is less pronounced compared to that in the absence of solid particles. Therefore, the mixing time of liquid increases with an increase in particle size and solid loading. This increase is most pronounced when the speed is close to the critical impeller speed for solid suspension. At speeds much higher than the critical impeller speed for solid suspension, the solid particles circulate along with the liquid flow, and hence the mixing time increases only to a small extent. The correlation of Raghav Rao et al. (1988),

$$Nt_{mix} = C X^{0.19} d_P^{0.11} D^{0.32} T^{-1.15} \tag{11.171}$$

is recommended for the estimation of the critical impeller speed for solid suspension. Comparing this correlation with that reported by Patwardhan and Joshi (1999a), we get

$$\frac{(Nt_{mix})_{S-L}}{(Nt_{mix})_L} = C' X^{0.19} d_P^{0.11} D^{-0.68} T^{-0.15} \tag{11.172}$$

11.4.3.8 Suspension in Gas-Liquid-Solid Systems

The introduction of the gas phase leads to the formation of cavities behind the impeller blades. As a result, the power number and the impeller pumping capacity are reduced. Hence, the impeller speed has to be increased to compensate for the loss of pumping capacity. Consequently, the critical impeller speed for solid suspension was always higher in the presence of the gas phase (N_{SG})

(Rewatkar et al., 1991). N_{SG} increased with an increase in the superficial gas velocity, particle size, and solid loading. Their correlation is recommended for design:

$$N_{SG} - N_{CS} = 132.7 u_{S\infty}^{0.5} T^{-1.67} Du_G \qquad (11.173)$$

Chapman et al. (1983) provide a similar correlation.

11.4.3.9 Heat Transfer in Gas-Liquid Systems

Heat transfer coefficients in gas-liquid systems are generally lower than those in the liquid alone (Karcz, 1999), a result of the reduction in power consumption in the presence of gas. The heat transfer coefficient is approximately proportional to *(power)*$^{0.22}$.

11.4.3.10 Dead End Systems

11.4.3.10.1 Gas-Inducing Impellers

A gas-inducing impeller generates a high velocity in the impeller region. The local pressure decreases sufficiently to overcome the hydrostatic head; gas is hence induced from the head space (Figure 11.45a). Many designs of these impellers have been reported in the published literature and are divided into three categories: hollow-pipe impellers, hollow pipes with dispersers, and stator-rotor systems. Saravanan and Joshi (1995, 1996) and Saravanan et al. (1996, 1997) investigated single- as well as multiple-impeller stator-rotor systems. When a single gas-inducing impeller is located close to the top liquid surface for maintaining a high rate of gas induction, the following drawbacks occur: poor dispersion of the induced gas, poor solid suspension ability, and so on. They arise from the fact that the liquid flow and turbulence characteristics are feeble in a major portion below the gas-inducing impeller. These drawbacks become more serious with a large-scale operation. These limitations have been overcome by using a second impeller below the gas-inducing impeller. The gas-inducing impeller can be positioned near the top liquid surface (about 100 to 200 mm below the liquid surface), irrespective of the scale of operation, to ensure a high gas induction rate. For performing gas dispersion (distribution of the induced gas throughout the equipment), solid suspension, heat transfer, mixing, and so forth, a second impeller is often used. The flow patterns for multiple-impeller combinations are depicted in Figures 11.45b and c. The design of the lower impeller was further optimized for different duties such as gas holdup, solid suspension

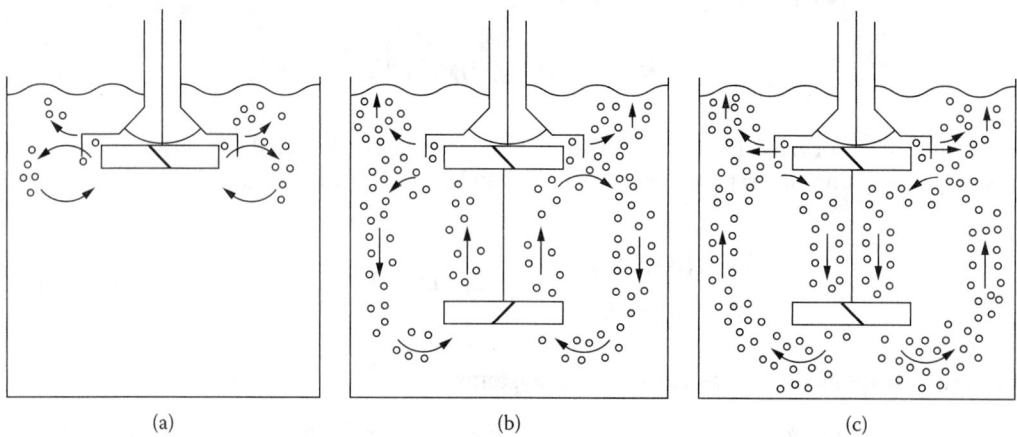

FIGURE 11.45 Gas-inducing reactors: (a) single inducing impeller, (b) inducing impeller together with upflow lower impeller, and (c) inducing impeller together with downflow lower impeller. (From Patwardhan, A. W. and Joshi, J. B., 1998.)

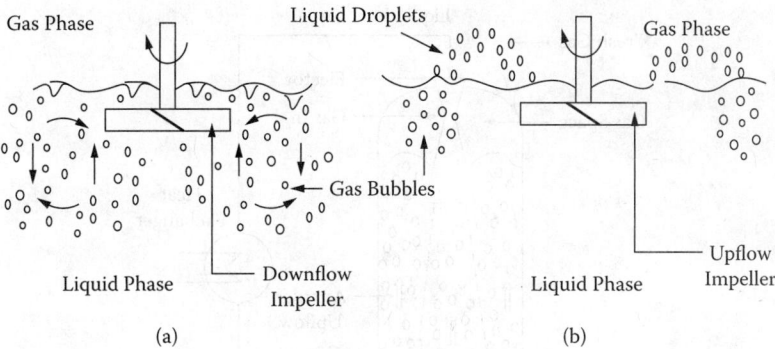

FIGURE 11.46 Surface aerators: (a) gas-in-liquid dispersion, (b) liquid-in-gas dispersion.

ability, and others. The optimum design of the lower impeller changes with the design objective. For example, to achieve a high fractional gas holdup, an upflow impeller should be at the bottom whereas, to achieve good solid suspension, a downflow impeller at the bottom is preferred. Recommended correlations are by Patwardhan and Joshi (1999b).

11.4.2.10.2 Surface Aerators

In baffled stirred reactors at sufficiently high speeds, entrainment of the gas bubbles into the liquid surface occurs. Alternatively, in unbaffled vessels, vortex formation occurs, which at high impeller speeds causes bubbles to entrain into the liquid phase. Under certain conditions, at high impeller speeds, a spray of liquid droplets in the gas phase also occurs. The three phenomena are called surface aeration, as depicted in Figures 11.46a and b. They depend on the impeller design, location, and speed of agitation. The important characteristics of surface aerators are the critical impeller speed for the onset of surface aeration, the intensity of surface aeration, power consumption, mass transfer coefficient, gas holdup, solid suspension, and so forth. Little information has been published on the heat transfer and solid suspension aspects of surface aerators. A characteristic feature of surface aeration is that the impeller has to be provided at the liquid surface to ensure a high intensity of surface aeration. With a single gas-inducing impeller, drawbacks often occur such as poor mixing in the rest of the reactor, poor dispersion of gas, and poor solid suspension/heat transfer characteristics. All difficulties can be eliminated by providing one or more impellers below the surface aerating impeller. The hydrodynamics and performance characteristics of such multiple-impeller surface aerators are similar to those of multiple-impeller gas-inducing systems. The recommended correlations are given by Patwardhan and Joshi (1998).

11.4.3.10.3 Jet-Loop Reactors

In a jet-loop reactor, liquid or gas and liquid are pumped into the reactor through a nozzle at either the top or the bottom. The jet of liquid or gas and liquid leaving the nozzle provides the energy for mixing and dispersion of the two phases. The unreacted gas can be recycled to the nozzle, forming a loop. In addition, the liquid may also be forced to circulate within the reactor (with the help of a draft tube) or externally (using a pump). Such reactors are referred to as jet-loop reactors, as shown in Figure 11.47. Jet-loop reactors can be further classified according to the construction of the nozzle as either injector or ejector. In injector nozzles or two-fluid nozzles, the gas is sparged along with the liquid through the nozzle. Intense gas-liquid mixing occurs in the nozzle, creating a fine dispersion. The gas-liquid dispersion exits the nozzle, and the gas is further dispersed in a draft tube. Unreacted gas escapes into the head space and can be either vented or recycled to the injector nozzle. Ejector nozzles, on the other hand, can suck gas from the atmosphere or from the reactor head space. The liquid (primary fluid) is pumped through a nozzle at high velocity. Based on Bernoulli's principle, a vacuum is created in the suction chamber. The gas (secondary fluid) therefore is sucked into the suction chamber. The liquid jet and the gas are mixed, and a gas-liquid

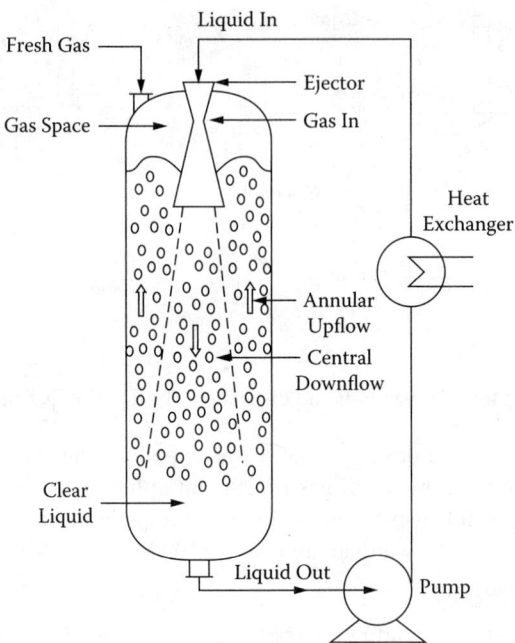

FIGURE 11.47 Jet-flow reactor.

dispersion is created in the mixing tube. The diffuser section at the end of the mixing tube helps in pressure recovery and compresses the gas-liquid mixture to the discharge pressure. The entire assembly (nozzle, mixing tube, and diffuser) is collectively called the *ejector*. The gas-liquid dispersion issuing from the ejector is sent to a separator, a reactor vessel, or a holding tank, which provides additional gas-liquid contact. The liquid jet performs two functions: it provides the momentum necessary for circulation of the vessel contents to ensure good mixing, and it disperses the entrained gas into small bubbles producing a large interfacial area. The unreacted gas escapes back into the head space, which can again be entrained into the ejector. Bhutada and Pangarkar (1987, 1989) investigated the effects of nozzle and ejector geometry, operating conditions, and physicochemical properties on the rate of gas entrainment and fractional gas holdup. These were found to be independent of the shape of the entry section of the nozzle. With a decrease in the mixing tube length, the rate of gas entrainment increases, whereas the fractional gas holdup is almost independent of the length. A pressure recovery diffuser section improves the rate of gas induction and the fractional gas holdup. The effects of geometrical parameters on the mass transfer characteristics need to be investigated systematically and related to the gas holdup and rate of gas entrainment.

11.4.3.10.4 Rotating Biological Contactors (RBC)

An RBC consists of a number of plastic disks mounted on a rotating horizontal shaft, and it is placed in a concrete tank containing the liquid to be treated. During rotation, the disk surface withdraws a thin film of liquid from the tank and transfers it through the gas space (i.e., air). This thin liquid film on the disk surface absorbs oxygen from the air, thereby providing gas-liquid contact. A pictorial representation of this reactor is shown in Figure 11.48. The RBC system provides high interfacial area per unit volume of contractor, resulting in efficient aeration. Vaidya and Pangarkar (1985) solved the simplified Navier–Stokes equation for a Newtonian fluid, assuming a half-immersed disk vertically withdrawn from a pool of liquid; various radial points on the disk had different linear speeds ($r\omega$) and hence different film thicknesses. Using computed profiles, the conservation equation for the gas phase was solved in addition to the mass transfer coefficient.

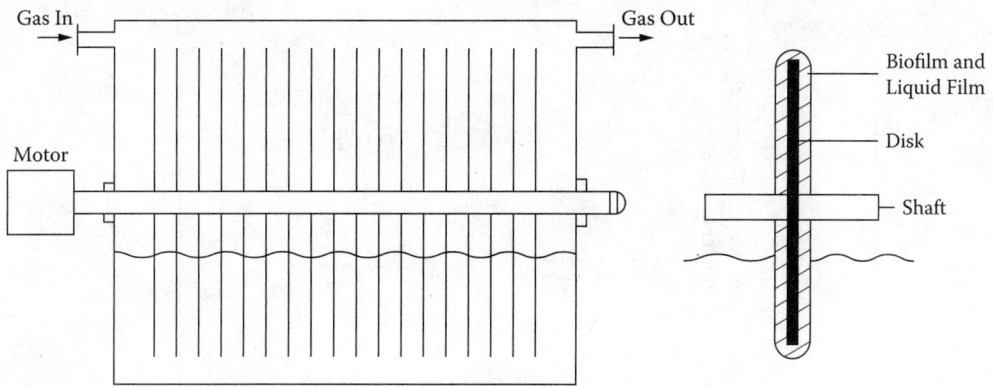

FIGURE 11.48 Rotating-disc reactor.

11.4.4 POTENTIAL ENERGY (FILM CONTACTORS)

As described in Section 11.4.1, most energy is supplied through the liquid phase as potential energy. Packed columns, trickle-bed reactors, falling-film reactors/evaporators, and wiped-film evaporators (and their variants, such as agitated-film and thin-film) fall into this category. In these cases, liquid flows in the form of a film and thus occupies only a very small fraction of the reactor/contactor volume (less than 10 to 20%) as compared to class 1 and class 2 equipment. Since the liquid phase occupies a small fraction of the equipment, only a low phase pressure drop results. However, it gives low values of interfacial area and mass transfer coefficient as compared to class 1 and class 2 equipment. In all film reactors (except wiped-film reactors to some extent), the design of the liquid distributor (and in some cases the liquid redistributor) is very crucial.

11.4.5 COMPARISON OF GAS-LIQUID REACTORS

It is instructive to compare gas-liquid reactors (from all the classes) on the basis of capacity, turn-down ratio (L/G), liquid-phase axial mixing, gas-side pressure drop, mass transfer coefficient, effective interfacial area, heat transfer coefficient, and the number of theoretical stages. Table 11.26 presents such a comparison using ratings 1 to 5 (poor to excellent). This table should be useful to design engineers.

11.5 CASE STUDIES

11.5.1 INTRODUCTION

We now consider 12 case studies that include simple homogeneous liquid-phase reactions, complex homogeneous gas-phase reactions, gas-solid catalytic and noncatalytic reactions, gas-liquid simple and complex reactions, gas-liquid-solid (noncatalytic) reactions, gas-liquid-solid (catalytic) reactions, and solid-solid reactions. The scope and coverage of each case study are summarized in Table 11.27. In the first, homogeneous reactions are considered. For these relatively simple reactions, the possibility of optimum design is discussed.

In the case of multiphase reactions, mass transfer and/or chemical reactions occur in series and/or parallel. Depending on the relative rates of these steps, the rate-controlling step is likely to be either one or a combination of two or more steps. The rate-controlling step may depend on the type of reactor. Furthermore, within a given reactor, it may change with location because of possible variations in the concentrations of different species, total pressure, temperature, and system properties. Hence, some case studies include a discussion of the rate-controlling step and estimation of the overall rate of reaction. Pertinent literature is cited in each case.

TABLE 11.26
Comparative Performances of Gas-Liquid Reactors

Equipment	Capacity	Turndown Ratio (L/G)	Liquid Phase Residence Time	Axial Mixing Liquid	Axial Mixing Gas	Gas-Side Pressure Drop	Mass Transfer Coefficient	G/L Interfacial Area	Heat Transfer Coefficient	No. of Theoretical Stages
Bubble Column	5	2,3,4	3,4,5	1,2	3,4,5	1	3,4	3,4,5	3,4	1
Packed Bubble Column	3	2,3,4	3,4,5	3	4,5	1	4	3,4,5	3,4	2
Sectionalized Bubble Column	5	2,3,4	3,4,5	4,5	5	1	4	3,4,5	3,4	4,5
Plate Column	3,4	1,2	2,3	5	5	2,3	4	4,5	3,4	5
External-Loop Airlift Reactor	5	2,3,4	4,5	2,3	4,5	2	4	4,5	4,5	2,3
Internal-Loop Airlift Reactor	5	2,3	3	1,2	3,4,5	1,2	4	3,4	3,4	1
Spray Column	5 (GP)	1,2	1	4	1,2	5	1,2	1,2	0	1
Pipeline Reactor	1,2	2,3	1	4,5	5	3,4	2	1,2	1,2	1
Static Mixer	1,2	2,3	1	4,5	5	3,4	5	2,3,4	3,4	1
Stirred Reactor Single Stage	2,3,4	4,5	3,4	1	1,2	2,3	4,5	3,4,5	4	1
Stirred Reactor Multistage	1,2,3	4,5		2,3	2,3,4	1,2	4,5	3,4,5	4	2,3
Packed Column	4	1,2	1,2	4,5	4,5	4,5	2,3	2,3	0	2,3,4
Falling Film Reactor	2,3	1,2	1	5	4,5	5	2	2	2	2
Rotating Disc Contactor	2,3	4,5	3,4	1,2	3,4	5	2	2,3,4	0	1,2
Jet-Loop Rector	2,3	4,5	2,3	1,2	3,4	5	5	3,4	5	1,2

1: poor, 5: excellent; GP: gas phase.

TABLE 11.27
List of Case Studies Covered with Brief Descriptions

No.	Case Study	Coverage
1	Homogeneous liquid-phase simple reaction: $A \rightarrow R$	The design of pipeline, coil, and heat exchanger (reaction on tube or shell side) reactors and their optimization are illustrated and discussed.
2	Homogeneous gas-phase complex reaction: Oxidation of NO using ozone	The reaction involves several intermediate, side and end products. For all these species, the concentration profiles in a pipeline reactor are obtained. The ozonation of NO is shown to be very fast and the importance of mixing of ozone is brought out.
3	Gas-solid (catalytic) reaction: Modeling of a complex reaction on a deactivating catalyst (a model scheme)	The chlorination of tetrachloroethane successively to the penta and hexa chlorides on a deactivating catalyst is used as a good complex model system for illustrating the application of rigorous statistical methods for model discrimination using a powerful discriminatory criterion.
4	Gas-solid (catalytic) reaction in fixed-bed NINA and adiabatic reactors: Reduction of nitrobenzene to aniline	The general principles of modeling a vapor-phase reaction on a solid catalyst in a fixed-bed reactor are illustrated by applying them to the specific case of nitrobenzene reduction to aniline. Several models of varying complexity are discussed.
5	Comparison of selected fluid-bed models for predicting the conversion in the catalytic reduction (by hydrogen) of nitrobenzene to aniline	The design of a fluidized-bed reactor for the same reaction as under (4) is discussed. Five important models are used to design the reactor and the results are compared. A brief discussion of a few other (more complex) models is also included.
6	Gas-solid (noncatalytic) reaction: Development of a solid sorbent for cleaning coal followed by modeling of the reaction involved and a conceptual reactor design	A problem of paramount environmental importance is the cleaning of coal gas for industrial use, such as in electrical power generation by the new IGCC process. The development of a calcium-based adsorbent for this purpose is described followed by the postulation of a possible reactor configuration.
7	Gas-liquid reaction: Air-oxidation of sodium sulfide (a simple reaction with typical problems of gas-liquid reactions)	For an industrially important system, oxidation of black liquor, six reactor types are designed. A comparison of all these six reactors is presented in terms of reactor dimensions (fixed cost) and power consumption (operating cost).
8	Gas-liquid reaction: Absorption of NO_x gases in the manufacture of nitric acid (a system with multiple complexities)	Absorption of NO_x gases in water is an important step in nitric acid manufacture. The absorption process involves multiple complexities. The design of a plate column, which incorporates many of these complexities, is discussed.
9	Gas-liquid-solid (noncatalytic) reaction: Oxydesulfurization of coal in a slurry reactor	A brief discussion of the rate-controlling step in coal oxydesulfurization is given on the basis of the shrinking core model. A three-phase slurry reactor is designed by including the nonideal behavior of the solid phase.
10	Gas-liquid-solid (noncatalytic) reaction with the particle size diminishing to less than the diffusion film thickness: Carbonation of lime	In this three-phase reaction, the solid phase undergoes a chemical reaction. The rate-controlling step is shown to vary with particle size. The case of particles smaller than the diffusion film thickness is considered, and a procedure is given for optimizing the production capacity of a given reactor.
11	Gas-liquid-solid (catalytic) reaction: Hydrogenation of an organic compound	A conventional stirred tank reactor and a self-inducing reactor are designed. Self-induction is shown to enhance the overall rate of reaction. A procedure is described for optimum design.
12	Solid-solid reaction followed by a gas-solid reaction: Manufacture of chlorosilanes	Silicon reacts with cuprous chloride autocatalytically to form a solid known as η-phase, which then catalyzes the reaction between silicon and methyl chloride to form methyl chlorosilanes. Mass transfer effects are neglected to illustrate the mechanism of this unique reaction and equations are developed for predicting the final product size distribution.

For sound process design, we need values of numerous design parameters such as fractional phase holdups, pressure drop, dispersion coefficients (the extent of axial mixing) of all the compounds, heat and mass transfer coefficients across a variety of fluid-fluid and fluid-solid interfaces depending on the type of multiphase system, type of reactor, and the rate-controlling steps. To clarify the scope of the case studies selected, their salient features are next listed.

Examples of several classes of industrially important multiphase reactions were given in Section 11.3. The case studies are chosen to represent different classes of reactions. Thus, we cover reactions ranging from simple homogeneous to quite complex multiphase reactions.

The rate-determining step is particularly important when a choice has to be made between competing mechanisms for a given reaction. One case study illustrates a rigorous statistical procedure.

In available texts, emphasis is often on the design of a single (often arbitrarily selected) reactor for a given reaction. In some case studies presented here, several reactor configurations are considered for a single reaction, and an optimum configuration (along with appropriate geometrical details) is recommended. Such a study can nontrivially alter reactor choice and performance.

Several models are sometimes available for the design of the reactor. Thus, in one case study, designs are made using several models, and a comparative assessment is presented.

Capital and operating costs are central to reactor selection. This aspect of design is illustrated in two case studies.

It is theoretically possible to include all considerations mentioned above in just two or three case studies. This has the disadvantage of severely restricting their breadth. Therefore, we decided to spread these considerations over 12 carefully selected case studies.

At the end of each case study, general remarks are given that bring out the lessons learned. Furthermore, at the end of the case studies we give our recommendations (morphology) for the selection and design of reactors.

CASE STUDY 11.1 HOMOGENEOUS LIQUID-PHASE SIMPLE REACTION: $A \to R$

Nomenclaure for Case Study 11.1

A	reactant
A_A	available area of heat transfer available (m²)
A_H	area of heat transfer (m²)
$[A]$	concentration of reactant A
$[A]_0$	inlet concentration of A
$[A]_e$	exit concentration of A (mol/m³)
A_S	shell area (m²)
a_S	cross-sectional area for shell-side flow (m²)
B	baffle spacing (m)
C	clearance (m)
C_{PC}	heat capacity of coolant (kcal/kg K)
D	diameter of tube (m)
D_C	diameter of coil helix (m)
D_O	outer diameter of tube (m)
D_S	shell diameter (m)
D	dispersion coefficient (m²/s)
d_E	equivalent diameter of the shell (m)
f	friction factor (–)
f_C	friction factor for a coil (–)
f_P	permissible stress for steel (N/m²)
f_S	friction factor on shell side (–)
G_S	shell-side mass velocity (kg/m²/s)
H	height (m)
ΔH	heat of reaction (kcal/kmol)

Chemical Reaction Engineering

h	height coordinate (m)
h_C	tube-side heat transfer coefficient of a coil (W/m² K)
h_S	shell-side heat transfer coefficient (W/m² K)
h_T	tube-side heat transfer coefficient (W/m² K)
k_A	first-order rate constant (1/s)
k_T	thermal conductivity (W/m K)
L_T	length of pipe (m)
M_C	mass flow rate of coolant (kg/s)
N_B	number of baffles
N_T	number of tubes (–)
Nu_L	Nusselt number for laminar flow (–)
Nu_T	Nusselt number for turbulent flow (–)
Pe	Peclet number (–) $Pe = DV\rho C_p/k$
P	pressure (N/m²)
Pr	Prandtl number (–)
Pr_C	Prandtl number of coolant (–)
Pt	tube pitch (m)
ΔP	pressure drop (N/m²)
ΔP_C	pressure drop of coil (N/m²)
ΔP_S	shell-side pressure drop (N/m²)
ΔP_T	tube-side pressure drop (N/m²)
Q	volumetric flow rate (m³/s)
Q_H	heat load (W, kcal/s)
R	product
Re	Reynolds number (–)
r_A	rate of reaction (mol/m³/s)
Re_S	shell-side Reynolds number (–)
Re_T	tube-side Reynolds number (–)
R_{ST}	ratio of shell area to tube area (–)
S	area of cross section (m²)
$(\Delta T)_{LM}$	logarithmic mean temperature difference (K)
T_{Ce}	coolant outlet temperature (K)
T_{C0}	coolant inlet temperature (K)
T_0	inlet temperature (K)
T_e	outlet temperature (K)
ΔT	temperature difference (K)
$\bar{t}$	residence time (s)
t_s	shell thickness (m)
t_T	tube thickness (m)
U	overall heat transfer coefficient (W/m² K)
u	average velocity (m/s)
V	reactor volume (m³)
V_C	volumetric flow rate of coolant (m³/s)
W_R	weight of pipe (kg)
W_S	shell weight (kg)
X	fractional conversion (–)

Greek

μ_C	viscosity of coolant (Pa·s)
μ_W	viscosity at the wall (Pa·s)
η	safety factor
ρ	density (kg/m³)
ρ_C	density of coolant (kg/m³)
ρ_S	density of steel (kg/m³)

Subscripts

C	coil/coolant
E	equivalent
e	outlet
o	inlet
S	shell side
T	tube side

The Problem

A continuous reactor is to be designed for a homogeneous liquid-phase reaction $A \to R$. The reaction is first order with respect to A, and the rate is given by

$$-r_A = k_1[A] \quad \text{(CS1.1)}$$

where $k_1 = 0.00167 \ s^{-1}$. The heat of exothermic reaction is -209 kJ/gmol. The flow rate of the reactant is 0.01 m³/s. The concentration of A at the inlet of the reactor is 1 kmol/m³, and a conversion of 99% is desired. Find the volume of the reactor if the same reaction is to be performed in a (1) backmixed reactor and (2) plug-flow reactor (PFR). Calculate the optimum length-to-diameter ratio for a plug-flow reactor. Also examine the alternative configurations, i.e., (3) a coil reactor and (4) a shell-and-tube heat exchanger type reactor.

Data

Density (ρ) = 1000 kg/m³; specific heat (C_P) = 4180 J/kg K; viscosity (μ) = 1×10^{-3} Pa·s; Prandtl number (Pr) = 5; number of working hours per year = 8000; cost of electric power = 0.03 \$/kWh; cost of mild steel = 0.4 \$/kg; density of steel = 8000 kg/m³; permissible stress for steel (f_P) = 8.5×10^7 N/m²; design pressure of the reactor = 0.5×10^6 N/m².

Solution

1. Completely Backmixed Reactor

Assumption: isothermal operation
Data: $[A]_0 = 1$ kmol/m³, $x = 0.99$, $F = 0.01$ m³/s, outlet concentration,

$$[A]_e = [A]_0(1-X)$$
$$= 0.01 \text{ kmol/m}^3 \quad \text{(CS1.2)}$$

For a backmixed reactor, the residence time is given by the following equation:

$$\bar{t} = V/Q = ([A]_0 - [A]_e)/k_1[A]e$$
$$= 59281 \text{ s} \quad \text{(CS1.3)}$$

Therefore, the reactor volume, $V = \bar{t} \times Q = 59400 \times 0.01 = 593$ m³

2. Plug-Flow Reactor: Pipeline Reactor

The operation is assumed to be isothermal. To start with, we assume plug flow, a condition that is validated later. For a plug-flow reactor, the residence time is given by

Chemical Reaction Engineering

$$\bar{t} = V/Q = -(1/k_1)\ln(1-X)$$

$$V = Q \times \bar{t} = 0.01 \times [-(1/0.00167) \times \ln(1-0.99)] = 27.63 \text{ m}^3 \tag{CS1.4}$$

Thus, the ratio of backmixed reactor (CSTR) volume to plug-flow reactor (PFR) volume is 594/27.63 = 21.5. The CSTR not only has 21.5 times the PFR volume, it also needs sufficient impeller power to achieve backmixing. Obviously, from fixed as well as operating costs points of view, PFR is a better option. However, let us now consider the length-to-diameter (L_T/D) ratio for PFR behavior.

For the case of turbulent flow in a pipeline, the dispersion coefficient (D) is given by (Taylor, 1953)

$$\boldsymbol{D} = 5.05\, u\, D\, (f/2)^{0.5} \tag{CS1.5}$$

The Peclet number is

$$Pe = uL_T/\boldsymbol{D} \tag{CS1.6}$$

PFR behavior is attained when Pe exceeds 10. However, considering the end effects, it is desirable to select a higher number (say, 50) for which the length-to-diameter ratio can be estimated from Equations (CS1.5) and (CS1.6) and that works out to be 12.5. Thus, the plug-flow condition is satisfied when $L_T/D > 12.5$ for a given volume of a PFR. For selection of optimum L_T/D, the following points need to be considered: (1) pressure drop, (2) heat transfer area of the wall, and (3) fixed cost. Let us estimate these for $L_T/D = 1000$:

$$V = (\pi/4)D^2 L_T$$

$$= (\pi/4)D^2(1000 D) = (\pi/4)D^3 \times 1000 \tag{CS1.7}$$

$$D = (4V/1000\pi)^{1/3} = 0.327 \text{ m}, \; L_T = 327 \text{ m}$$

Area of cross section,

$$S = (\pi/4)D^2$$

$$= 0.084 \text{ m}^2 \tag{CS1.8}$$

Average velocity,

$$u = Q/S = 0.119 \text{ m/s} \tag{CS1.9}$$

Reynolds number,

$$\text{Re} = Du\,\rho/\mu$$

$$= 0.327 \times 0.119 \times 1000/10^{-3} = 38913 \tag{CS1.10}$$

Friction factor,

$$f = 0.08\,\text{Re}^{-0.25} \tag{CS1.11}$$
$$= 0.0057$$

Dispersion coefficient $D = 5.05\,uD(f/2)^{0.5} = 0.0104$ m²/s
Peclet number $Pe = uL_T/D = 3742$
Peclet number is very high (>50), which confirms the assumption of plug flow.

Pressure drop,

$$\Delta P = 2f\,L_T\,u^2\,\rho/D \tag{CS1.12}$$
$$\Delta P = 161 \text{ N/m}^2$$

$$\text{Power} = Q \times \Delta P \tag{CS1.13}$$
$$= .01 \times 160 = 1.61 \text{ W}$$

Annual electric energy required = (1.6 W/1000) × (8000 h/year) = 12.8 kWh

Annual operating cost = 12.8 kWh × 0.03 $/kWh = $0.4

Estimating the fixed cost of the reactor:

Thickness,

$$t_T = PD/2f_P\eta$$
$$= 0.5 \times 10^6 \times 0.327 \left[2 \times 8.5 \times 10^7 \times 0.9 \right] \tag{CS1.14}$$

$t_T = 0.0011$ m ~ 1.1 mm, which is very low. In addition, we need to have some corrosion allowance:

Assuming a minimum thickness of 4 mm, $t_T = 0.004$ m

Weight of the material required,

$$W_R = \pi D L_T t_T \rho_s \tag{CS1.15}$$
$$= 3.142 \times 0.327 \times 327 \times 0.004 \times 8000 = 10750 \text{ kg}$$

Cost of reactor = W_R × unit cost of material = 10750 kg × 0.4 $/kg, i.e., Fixed cost = $4300

Considering 20% of the fixed cost as the annualized cost (depreciation cost),

Total annualized cost = operating cost + 20% of fixed cost = 0.4 + 0.2 × 4300 = $860/year

Chemical Reaction Engineering

Since the reaction is exothermic, heat transfer must be considered. The total heat load is

$$Q_H = Q([A]_0 - [A]_f)(-\Delta H) = 0.01 \text{ m}^3/\text{s} \times (1 - 0.01) \text{ kmol/m}^3 \times (209 \times 10^6) \text{ J/kmol} = 2.069 \times 10^6 \text{ W}$$

Heat transfer area, $A_H = \pi D L_T = \pi \times 0.327 \times 327 = 336 \text{ m}^2$

Estimation of the heat transfer coefficient:

$$Pr = Cp\ \mu/k_T = 5 \text{ (given)}$$

$$k_T = C_P\ \mu/5 = 0.836 \text{ W/m K}$$

$$h_T D/k_T = 0.023\ Re^{0.8}\ Pr^{0.33} = 0.023 \times (38913)^{08} \times (5)^{0.33} = 183.85$$

$$h_T = 183.85\ k_T/D = 183.85 \times 0.836/0.327 = 469.8 \text{ W/m}^2 \text{ K}$$

Assuming this heat transfer coefficient to be the controlling heat transfer resistance, the overall heat transfer coefficient $U = 469.8$ W/m² K. Since the reactor is a plug-flow reactor, the concentration of A is high along its entrance length, so the rate of reaction and hence the rate of heat generation are also high. Toward the end of the reactor, the rate of heat generation will be less. However, for simplicity, we calculate the driving force on the basis of log mean temperature difference. The rate of heat transfer is given by

$$Q_H = U A_H (\Delta T)_{LM} \tag{CS1.16}$$

Substitution of Q_H, U, and A_H for (L_T/D) of 1000, gives the log mean temperature difference:

$$(\Delta T)_{LM} = Q_H/U A_H = 2.069 \times 10^6/(469.8 \times 336) = 13.1°C$$

This is a reasonable driving force for fixing the temperature of the heat transfer fluid. It can be seen from Table CS1.1 that for a reasonable $(\Delta T)_{LM}$, the L_T/D ratio needs to be greater than 500 or the length greater than 200 m.

The above calculations are repeated for various L_T/D ratios with the help of a computer. The results are shown in Table CS1.1; for low L_T/D ratios, the fixed cost of the pipeline is the same. This is explained below:

Weight of empty pipe, $W_R = \pi D L_T t_T \rho_s = PD/2\ f_P \eta$

$$\begin{aligned}W_R &= \pi D L_T (PD/2\ f_P \eta) \rho s \\ &= (1/2)(\pi D^2 L_T)(P/f_P \eta)\rho_S = 2V(P/f_P\eta)\rho_S\end{aligned} \tag{CS1.17}$$

Thus, W_R is constant for a given reactor volume V.

However, at higher L_T/D ratios, the calculated thickness of small diameter pipelines is also small. At least a minimum thickness of 4 mm of the pipeline is required. Hence, at higher L_T/D ratios, the weight of the empty pipe is

TABLE CS1.1
Sensitivity Analysis of Length-to-Diameter Ratio for the Pipeline Reactor

L_T/D	D m	L_T m	A_H m²	u m/s	Re	f	ΔP N/m²	Power W	Annual Electric Energy Cost $	Final Tube Thickness m	Weight kg	Fixed Cost $	Annualized Cost $	h_T W/m² K	ΔT °C
1	3.28	3.28	33.68	0.001	3890	0.010	0.000	0.000	6.86E-08	0.011	2884	1154	231	7.44	6120.3
5	1.92	9.58	57.59	0.003	6651	0.009	0.001	0.000	2.56E-06	0.006	2884	1154	231	19.53	1362.7
10	1.52	15.20	72.56	0.006	8380	0.008	0.005	0.000	1.22E-05	0.005	2884	1154	231	29.60	713.6
50	0.89	44.45	124.08	0.016	14330	0.007	0.190	0.002	4.56E-04	0.004	3970	1588	318	77.75	158.9
100	0.71	70.56	156.33	0.026	18054	0.007	0.904	0.009	2.17E-03	0.004	5002	2001	400	117.84	83.2
500	0.41	206.32	267.32	0.075	30872	0.006	33.78	0.338	8.11E-02	0.004	8554	3422	684	309.52	18.5
1000	0.33	327.51	336.80	0.119	38896	0.006	160.70	1.607	3.86E-01	0.004	10778	4311	863	469.14	9.7
5000	0.19	957.63	575.92	0.347	66512	0.005	6007.67	60.08	1.44E+01	0.004	18429	7372	1489	1232.21	2.2
10000	0.15	1520.15	725.61	0.551	83800	0.005	28577.46	285.77	6.86E+01	0.004	23219	9288	1926	1867.68	1.1

$$W_R = \pi D L_T t_T \rho_S = \pi D L_T (0.004)\rho_S = 0.004\pi D^2 (L_T/D)\rho_S$$
$$= 0.004\pi[\{4V/\pi(L_T/D)\}^{1/3}]^2 (L_T/D)\rho_S \qquad \text{(CS1.18)}$$
$$= 0.004\pi (4V/\pi)^{2/3} \rho_S (L_T/D)^{1/3}$$

Thus, the fixed cost of the pipeline increases with increase in the L_T/D ratio. As L_T/D increases, the velocity and pressure drop increase. Hence, the operating cost increases. Thus, the total annualized cost increases with increase in the L_T/D ratio. However, for plug-flow behavior, L_T/D greater than 12.5 is needed. Furthermore, for the transfer of exothermic heat, sufficient heat transfer area is needed. The total annualized cost is $973 for an L_T/D of 500, which gives a reasonable driving force for heat transfer (25°C). At higher L_T/D ratios, the total annualized cost is higher, and at lower L_T/D ratios, the $(\Delta T)_{LM}$ is too high. Hence, we choose the case with L_T/D of 500 as the optimum for the pipeline reactor ($D = 0.413$ m, $L_T = 206$ m).

For simplicity, various assumptions were made, including the major assumption of isothermal behavior even though a temperature profile is expected due to the exothermic reaction. Also, in reality, one has to consider the cost of the reactor as well as the costs of the heat transfer system, fabrication of reactor, and insulation in the fixed cost. The cost of pressure drop should include a multiplying factor to account for pump efficiency. The operating cost should include the cost of pumping and recooling of the heat transfer medium and so on. Even for the simple case considered here, there is a need for optimization. The procedure can be extended to increasingly complex situations by progressively eliminating the above-mentioned assumptions.

At this stage, one very important point may be noted. The selection of the optimum L_T/D ratio depends on the available heat transfer area. Furthermore, the length of 327 m is rather large. Therefore, let us use a coil reactor for managing the reactor length and conserving the floor space and also a shell-and-tube exchanger (as reactor) for imparting flexibility in the selection of heat transfer area.

3. Helical Coil Reactor

Let us continue with the case of $L_T/D = 500$, the same as that of the pipeline reactor with a diameter of 0.413 m. Considering the helix diameter to be 10 times the pipe diameter, $D_C = 4.13$ m.

$$\text{Perimeter of the coil} = \pi D_C = 12.97 \text{ m} \qquad \text{(CS1.19)}$$

Number of turns of the coil required = total length/perimeter = 206/12.97 = 16

The pressure drop is higher due to helical flow (Srinivasan et al., 1968):

$$\text{Re} = D u \rho / \mu = 3.08 \times 10^4$$

$$\text{The coil friction factor, } f_C = 0.08 \text{Re}^{-0.25} + 0.01(D/D_c)^{0.5} \qquad \text{(CS1.20)}$$

$$= 9.2 \times 10^{-3}$$

Pressure drop, $\Delta P_C = 51.3 \text{N/m}^2$

Annual power consumption = 51.3 N/m² × 0.01 m³/s × 8000 h/yr/1000 = 4.21 kWh

Annual power cost = 4.21 kWh × 0.03 $/kWh = $0.13

From Table CS1.2, fixed cost = $3422

Total annualized cost = 0.13 + $3422 × 0.2 = $684.53

However, there is an enhancement in the heat transfer, too, compared to the pipeline reactor (Kern, 1950):

$$h_C / h_T = (1 + 3.5\, D / D_C)$$
$$h_C / h_T = 1.35 \qquad (CS1.21)$$

Therefore, the $(\Delta T)_{LM}$ requirement is reduced to $25.00 \times h_T/h_C = 18.52°C$ or a lower L_T/D ratio may be selected.

4. Multitubular Reactor

The same reaction can be performed in a multitubular reactor (with a shell-and-tube heat exchanger configuration). Two possibilities exist: the reactant is on the tube side or on the shell side. Both are considered below.

REACTANT ON THE TUBE SIDE

Tube-Side Calculations

Number of tubes:
Besides the diameter of the tube, the number of tubes, N_T, is also assumed and, correspondingly, the length required is

$$L_T = V / [(\pi / 4) D^2 N_T] \qquad (CS1.22)$$

If we consider 25 tubes with an inner diameter of 0.154 m and an outer diameter of 0.168 m, the length of the tube becomes $L_T = 27.63/[(\pi/4) \times 0.154^2 \times 25] = 59.3$ m.

Such a long tube needs many 1-1 shell-and-tube exchangers in series or a multipass exchanger with many passes on the tube side. A 1-1 type heat exchanger is considered here for simplicity.

Pressure Drop:

$$S = (\pi / 4) \times 0.154^2 = 0.0186 \text{ m}^2$$
$$u = Q / S N_T \qquad (CS1.23)$$

$u = [0.01 \text{ m}^3/\text{s}]/[0.0186 \text{ m}^2 \times 25] = 0.021$ m/s

Reynolds number $Re = Du\rho/\mu = 3306$

Friction factor $f = 0.08\, Re^{-0.25} = 0.011$

Dispersion coefficient $D = 5.05\, uD(f/2)^{0.5} = 0.00121$ m²/s

Peclet number $Pe = uL_T/D = 1028$ (>>50), which confirms our assumption of plug flow

TABLE CS1.2
Sensitivity Analysis of Length-to-Diameter Ratio for the Helical Coil Reactor

L_r/D	D m	L_r m	D_c m	No. of Turns	A_H m²	u m/s	Re	f	ΔP N/m²	Power W	Annual Electric Energy Cost $	Final Tube Thickness m	Weight kg	Fixed Cost $	Annualized Cost $	h_T W/m² K	T °C
100	0.71	70.56	7.06	3	156.33	0.026	18054	0.010	1.32	0.013	0.00	0.004	5002	2001	400	159.09	83.2
200	0.56	112.01	5.60	6	196.96	0.041	22747	0.010	6.39	0.064	0.02	0.004	6303	2521	504	241.13	43.6
300	0.49	146.77	4.89	9	225.46	0.053	26039	0.009	16.08	0.161	0.04	0.004	7215	2886	577	307.55	29.8
400	0.44	177.80	4.44	12	248.15	0.064	28659	0.009	30.97	0.310	0.07	0.004	7941	3176	635	365.49	22.8
500	0.41	206.32	4.13	15	267.32	0.075	30872	0.009	51.49	0.515	0.12	0.004	8554	3422	684	417.85	18.5
1000	0.33	327.51	3.28	31	336.80	0.119	38896	0.009	249.91	2.50	0.60	0.004	10778	4311	863	633.34	9.7
5000	0.19	957.63	1.92	159	575.92	0.347	66512	0.008	9821.33	98.21	23.57	0.004	18429	7372	1498	1663.48	2.2
10000	0.15	1520.15	1.52	318	725.61	0.551	83800	0.008	47797.07	477.97	114.71	0.004	23219	9288	1972	2521.37	1.1

$$\text{Pressure drop } \Delta P = 2 f L_T \, u^2 \, \rho/D = 3.7 \text{ N/m}^2$$

$$\text{Power} = Q \times \Delta P = 0.01 \times 3.7 = 0.037 \text{ W}$$

Heat Transfer Coefficient:
The tube-side Reynolds number is in the transition regime between laminar and turbulent flows $(2100 < Re < 4000)$: using the Sieder–Tate equation for laminar flow $(Re < 2100)$,

$$\left(\frac{h_T D}{k_T}\right) = 1.86 \left(Re \, Pr \, \frac{D}{L_T}\right)^{1/3} \left(\frac{\mu}{\mu_w}\right)^{0.14} \tag{CS1.24}$$

At $Re = 2100$, substituting the values of Re, Pr, D, and L_T in Equation (CS1.24), the Nusselt number for laminar flow is 5.6. At the other extreme, for turbulent flow, the Nusselt number is given by

$$\left(\frac{h_T D}{k_T}\right) = 0.023 \, Re^{0.8} \, Pr^{1/3} \left(\frac{\mu}{\mu_w}\right)^{0.14} \tag{CS1.25}$$

Assuming the value of $(\mu/\mu_w)^{0.14} \sim 1$, $Re = 4000$, and $Pr = 5$, the Nusselt number is 29.8. Using interpolation of the log-log relation of the Nusselt number and Reynolds number in the transition regime $(2100 < Re < 4000)$,

$$\frac{\ln(h_T D / k_T) - \ln(5.6)}{\ln(29.8) - \ln(5.6)} = \frac{\ln(Re) - \ln(2100)}{\ln(4000) - \ln(2100)} \tag{CS1.26}$$

At $Re = 3306$, $Nu = 18.18$. Substituting for D and k_T in the Nusselt number $(h_T D/k_T)$, the heat transfer coefficient, $h_T = 99$ W/m² K.

Shell-Side Calculations

Flow Rate of the Cooling Medium:
Assume the cooling medium to be chilled water with an inlet temperature $T_{C0} = 7°C$, density $(\rho_C) = 1000$ kg/m³, specific heat $(C_{PC}) = 4180$ J/kg K, viscosity $(\mu_C) = 1 \times 10^{-3}$ Pa·s, Prandtl number $(Pr_C) = 5$.
Assuming an outlet temperature (T_{Ce}) of 27°C,

$$Q_H = M_C C_{PC} (T_{Ce} - T_{C0}) \tag{CS1.27}$$

Mass flow rate of the chilled water required is $M_C = Q_H / \{C_{PC}(T_{Ce} - T_{C0})\} = 26.1$ kg/s

Chilled water volumetric flow rate $V_C = M_C/\rho_C = 0.026$ m³/s

Shell Diameter:
Assuming a triangular pitch of the tubes, with a pitch of $P_T = 1.33 \times D_o$

$$P_T = 1.33 \times 0.168 = 0.224 \text{ m}$$

Chemical Reaction Engineering

The ratio R_{ST} of shell area/tube area is given by

$$R_{ST} = \frac{\frac{1}{2}P_T \, 0.86 P_T}{\frac{1}{2}\frac{\pi D_O^2}{4}} \quad \text{(CS1.28)}$$

$$= 1.095 (P_T/D_O)^2 = 1.95$$

Shell area $A_S = R_{ST} \times$ tube area

$$A_S = R_{ST} \times N_T \times S \quad \text{(CS1.29)}$$

$$= 1.95 \times 25 \times 0.0186 = 0.91 \text{ m}^2$$

With this cross-sectional area, the shell diameter, D_S, is found to be 1.08 m. Providing an additional 0.3 m to the diameter of the shell, (so that the outermost tubes are away from the shell), $D_S = 1.08 + 0.3 = 1.38$ m.

Pressure Drop:
Let baffle spacing $B = $ (0.8 times of D_S) = 1.1 m:

$$\text{Number of compartments in the shell} = L_T/B \quad \text{(CS1.30)}$$

$$L_T/B = 59.3/1.1 = 54$$

Number of baffles $N_B = 54 - 1 = 53$:

$$B = L_T/(N_B + 1) \quad \text{(CS1.31)}$$
$$= 1.098 \text{ m}$$

Clearance,

$$C = P_T - D_O \quad \text{(CS1.32)}$$
$$= 0.056 \text{ m}$$

Cross-sectional area for shell side flow a_S is given by

$$a_S = \frac{D_S C}{P_T} B \quad \text{(CS1.33)}$$
$$= 0.40 \text{ m}^2$$

The mass velocity is given by

$$G_S = M_C/a_S \quad \text{(CS1.34)}$$

$$d_E = 4 \frac{\frac{1}{2}P_T 0.86 P_T - \frac{1}{2}\frac{\pi D_O^2}{4}}{\frac{1}{2}\pi D_O} \quad \text{(CS1.35)}$$

$G_S = 65.1$ kg/m²/s, and the equivalent diameter (d_E) of the shell is given by

$$d_E = 0.158 \text{ m}$$

Reynolds number,

$$\text{Re}_S = d_E G_S / \mu c$$
$$= 10263 \quad \text{(CS1.36)}$$

Friction factor, $f_S = 0.08 \, \text{Re}^{-0.25} = 0.008$:

$$\Delta P_S = \frac{f_S G_S^2 D_S \frac{L_T}{B}}{2\rho_c d_E} \quad \text{(CS1.37)}$$

Pressure drop on the shell side ΔP_S is given by

$$\Delta P_S = 8 \text{ N/m}^2$$

$$\text{Power} = Q_C \times \Delta P_S = 0.207 \text{ W}$$

$$\left(\frac{h_S D}{k_T}\right) = 0.36 \, \text{Re}^{0.55} \, \text{Pr}^{1/3} \left(\frac{\mu}{\mu_W}\right)^{0.14} \quad \text{(CS1.38)}$$

Shell-Side Heat Transfer Coefficient:
Assuming $(\mu/\mu_W)^{0.14} \sim 1$ and substituting the values of Re, Pr, D, and k_T in the above equation, the calculated value of the shell-side heat transfer coefficient $h_S = 524 \text{W/m}^2$ K.

Overall Heat Transfer Coefficient (U):
For simplicity, considering only the shell- and tube-side resistances (neglecting dirt factors), the overall heat transfer coefficient U is given by

$$\frac{1}{U} = \frac{1}{h_T} + \frac{1}{h_S D_O / D} \quad \text{(CS1.39)}$$

$$U = 88 \text{ W/m}^2 \text{ K}$$

Check for Adequacy of Heat Transfer Area:
Initially it was assumed that the chilled water outlet temperature is 27°C. The log mean temperature difference is given by

Chemical Reaction Engineering

$$\Delta T_{LM} = \frac{(T_e - T_{Ce}) - (T_0 - T_{Ce})}{\ln[(T_e - T_{co})/(T_0 - T_{Ce})]} \quad \text{(CS1.40)}$$

where T_0 and T_e are the inlet and outlet temperatures of the reaction mixture (process fluid), respectively. Assuming the shell side to be isothermal, $T_0 = T_e = 50°C$, $(\Delta T)_{LM} = 32°C$:

$$A_H = \frac{Q_H}{U \Delta T_{LM}} \quad \text{(CS1.41)}$$

The required heat transfer area is then calculated as

$$A_H = 734.7 \text{ m}^2$$

Actual available area (A_A) of the shell-and-tube exchanger = $A_A = N_T \pi D L_T = 717.4 \text{ m}^2$.

$A_A/A_H = 0.97$. Since the value of A_A/A_H is different from 1, the number of tubes needs to be changed and the whole procedure repeated to finally obtain a value of 1. The final solution is: number of tubes = 23, shell I.D. = 1.34 m, tube length = 63.7 m. The details of this reactor are given in Table CS1.3.

Cost Evaluation

Annual operating cost:
For case 1 in Table CS1.3,

$$\text{Power (shell side + tube side)} = 0.238 + 0.045 = 0.283 \text{ W}$$

$$\text{Annual electric energy required} = (0.283 \text{ W}/1000) \times (8000 \text{ hr/year}) = 2.3 \text{ kWh}$$

$$\text{Annual operating cost} = 2.3 \text{ kWh} \times 0.03 \text{ \$/kWh} = \$0.07$$

To estimate the fixed cost, let us calculate shell and tube weights.

Shell thickness:

$$t_S = PD_S / 2 f_P \eta$$
$$= (0.5 \times 10^6) \times 1.34 / [2 \times (8.5 \times 10^7) \times 0.9] = 0.0045 \text{ m} \sim 4.5 \text{ mm} \quad \text{(CS1.42)}$$

In addition, we need a corrosion allowance, say 3 mm. Thickness $t_S = 0.0075$ m. Weight of the material of construction required is

$$W_S = \pi D_S L_T t_S \rho_S$$
$$= 3.142 \times 1.34 \times 63.7 \times 0.0075 \times 8000 = 15{,}784 \text{ kg} \quad \text{(CS1.43)}$$

$$\text{Weight of tubes} = N_T \pi D L_T t_T \rho_S$$
$$= 40819 \text{ kg} \quad \text{(CS1.44)}$$

TABLE CS1.3
Sensitivity Analysis of Length-to-Diameter Ratio for the Multitubular Reactor

No.	Shell id	T_e °C	L_T/D	D m	D_o m	N_T	L_T m	M_c kg/s	t_s m	B m	Re (Tube Side)	Power (Tube Side) W	Power (Shell Side) W	Power W	Annual Electrical Energy Cost $	Weight kg	Fixed Cost $	Annual Cost $	U_T W/m²K	LMTD[a] °C
											Reactant on the Tube Side									
1	1.34	27	413	0.15	0.16	23	63.7	26.1	0.0074	1.08	3549	0.05	0.24	0.29	0.069	56603	22641	4528.06	91.4	31.5
2	1.38	20	383	0.15	0.16	25	59	41.3	0.0075	1.11	3287	0.04	0.7	0.74	0.17	56151	22461	4528.17	80.8	35.7
3	1.38	27	386	0.15	0.16	25	59.4	25.1	0.0075	1.11	3063	0.04	0.2	0.24	0.058	56172	22469	4494.05	84.7	32
4	1.41	15	364	0.15	0.16	26	56.1	70.7	0.0076	1.12	3127	0.03	2.83	2.86	0.69	55870	22348	4469.69	75.1	38.4
5	1.25	27	375	0.20	0.22	11	76.7	26.1	0.0071	1	5671	0.05	0.28	0.33	0.08	48920	19568	3913.08	121.1	31.6
6	1.37	20	293	0.20	0.22	14	59.9	41.3	0.0075	1.09	4433	0.03	0.56	0.58	0.14	47368	18947	3790.14	107.3	35.7
7	1.43	15	265	0.20	0.22	16	54.2	70.7	0.0077	1.15	4005	0.02	1.85	1.87	0.45	46799	18719	3744.40	99.5	38.4
											Reactant on the Shell Side									
8	1.07	25	216	0.20	0.22	7	44.1	29.1	0.0065	0.21	24818	1.18	0.88	2.06	0.49	19956	7982	1596.48	305.1	32.8
9	1.5	20	123	0.20	0.22	18	25.1	41.3	0.0079	0.3	14419	0.37	0.15	0.52	0.12	24328	9731	1946.12	203.1	35.7
10	2.15	15	200	0.20	0.22	42	13.5	70.7	0.01	0.44	10493	0.2	0.02	0.22	0.05	28835	11534	2307.05	147.7	38.4
11	1.7	25	135	0.15	0.17	43	20.9	29.1	0.0086	0.34	5662	0.1	0.11	0.21	0.05	31899	12759	2252.03	147.8	32.8
12	2.25	20	83	0.15	0.17	82	12.9	41.3	0.0104	0.44	4150	0.05	0.03	0.08	0.02	36621	14648	2930.02	113.5	35.7
13	2.97	15	51	0.15	0.17	154	7.9	70.7	0.0127	0.61	3802	0.05	0.01	0.05	0.01	40937	16375	3275.10	91.7	38.4

[a] LMTD = log mean temperature difference.

Chemical Reaction Engineering

For simplicity, we consider only the cost of the shell and tubes. In reality, however, one has to consider the tube sheets, head, nozzles, baffles, and cost of fabrication.

$$\text{Cost of reactor} = \text{Total weight} \times \text{unit cost of material}$$

$$= (15784 + 40819) \text{ kg} \times 0.4\$/\text{kg} = 56603 \times 0.4 = \$22641$$

$$\text{Fixed cost} = \$22641$$

Considering 20% of the fixed cost in the annualized cost,

$$\text{Total annualized cost} = \text{Operating cost} + 20\% \text{ of fixed cost}$$

$$= 0.07 + 0.2 \times 22641 = \$4773$$

In Table CS1.3, various options are given with respect to the coolant outlet temperature. For a lower outlet temperature of the coolant, the following occur: higher coolant flow rate and an increase in the operating, and hence the total, annualized cost. Also, the use of 0.204 m I.D. tubes (case 4, Table CS1.3) instead of 0.154 m I.D. tubes leads to lower fixed cost and hence lower total annualized cost (\$4199).

REACTANT ON THE SHELL SIDE

When the reactant is on the shell side, the volume of the reactor is approximately

$$V = (\text{Volume of empty shell}) - (\text{Volume of tubes}) \qquad (\text{CS1.45})$$

when we neglect the volume of baffles, spacer rods, and so forth. In this case, the design steps to be followed are as follows. (1) Assume diameter D, number of tubes N_T, and tube thickness. (2) Based on triangular (or square) pitch, calculate the shell diameter D_S. (3) From a knowledge of the shell cross-sectional area and tube cross-sectional area, calculate the length of the reactor for the desired volume. (4) Assume coolant inlet and outlet temperatures. (5) Calculate the coolant flow rate for the desired heat duty. (6) Calculate the tube-side pressure drop and heat transfer coefficient. (7) Assume baffle spacing and calculate the shell-side pressure drop and heat transfer coefficient. (8) Calculate the overall heat transfer coefficient and log mean temperature difference. (9) Calculate the heat transfer area required and compare it with the area available. (10) Vary N_T or D. Repeat steps 1 to 9 until $A_A/A_H = 1$. (11) Cost evaluation: since most steps are described in the previous case (reactant on the tube side), one case is presented briefly below (refer to case 10 in Table CS1.3).

1. Assume 43 tubes with I.D. = 0.154 m, O.D. = 0.168 m, $N_T = 43$, $D = 0.154$ m, $D_0 = 0.168$ m, pitch = $1.33 \times D_0 = 0.223$ m.
2. By following a procedure similar to that given in the previous case for reactant on the tube side, the shell diameter (D_S) works out to be 1.7 m.
3. Since the reactor volume is the difference between the empty shell volume and the volume of the tubes based on O.D., the reactor length is obtained as

$$L_T = \frac{V}{\frac{\pi}{4}\left(D_s^2 - N_T D_O^2\right)} \qquad (\text{CS1.46})$$

$$= 20.76 \text{ m}$$

4. Assume chilled water (coolant) inlet temperature $T_{C0} = 7°C$ and outlet temperature $T_{Ce} = 25°C$.
5. Mass flow rate of the chilled water required is $M_C = Q_H/\{C_{PC}(T_{Ce} - T_{C0})\} = 29.1$ kg/s.
6. Tube-side pressure drop and heat transfer coefficient:

   $Re_T = 5662$, pressure drop = 3.4 N/m², heat transfer coefficient $h_T = 213$ W/m² K

7. Shell-side pressure drop and heat transfer coefficient: Since the reactant is on the shell side, assume low baffle spacing so that the flow may be assumed to be plug.
   Assume baffle spacing of $0.2 \times D_S = 0.34$ m
   Shell-side pressure drop = 11 N/m²
   Heat transfer coefficient $h_S = 525$ W/m² K
8. Overall heat transfer coefficient and log mean temperature difference:
   For simplicity, considering only the shell- and tube-side resistances (neglecting the dirt factor),

   $$U = 148 \text{ W/m}^2 \text{ K}$$

   Log mean temperature difference is given by Equation (CS1.40), where T_0 and T_e are the inlet and outlet temperatures of the reaction mixture (process fluid), respectively. Assuming the tube side to be isothermal, $T_0 = T_e = 50°C$, $(\Delta T)_{LM} = 32.77°C$.

   $$A_H = \frac{Q_H}{U \Delta T_{LM}} = 427.27 \text{ m}^2$$

9. Required heat transfer area is calculated from the actual area available A_A of the shell-and-tube exchanger as $A_A = N_T \pi D L_T = 427$ m².
10. $A_A/A_H = 0.999 \sim 1$. This is a good match. If the value of A_A/A_H was found to be very different from 1, then the number and/or diameter of tubes would need to be varied and the whole procedure repeated to finally obtain a value of 1.
11. Annual operating cost:
    Shell-side power required = $Q \times \Delta P_S = 0.01 \times 11 = 0.11$ W
    Tube-side power required = $Q_C \times \Delta P_T = 0.029 \times 3.4 = 0.1$ W
    Power (shell side + tube side) = 0.21 W
    Annual electric energy required = (0.21 W/1000) × (8000 hr/year) = 1.68 kWh
    Annual operating cost = 1.68 kWh × 0.03 \$/kWh = \$0.05

Fixed cost of reactor:

Shell thickness, $t_S = PD_S/2f_P\eta = (0.5 \times 10^6) \times 1.7/[2 \times (8.5 \times 10^7) \times 0.9] = 0.0056$ m ~ 5.6 mm. In addition, since we need to have some corrosion allowance, say 3 mm, thickness $t_S = 0.0086$ m.
Weight of the material of construction required = $\pi D_S L_T t_S \rho_S = 7698$ kg.
Weight of tubes = $N_T \pi D L_T t_T \rho_S = 24409$ kg. For simplicity, we consider only the cost of the shell and tubes. In reality, however, one also has to consider the tube sheets, head, nozzles, baffles, and cost of fabrication.

Cost of reactor = Total weight × unit cost of material = (7698 + 24409) kg × 0.4\$/kg = \$12842.

$$\text{Fixed cost} = \$12842.$$

Chemical Reaction Engineering

Total annualized cost = operating cost + 20% of fixed cost = 0.05 + 0.2 × 12842 = $2568.6.

For the case of reactant on the shell side, it can be seen from a comparison of cases 7 to 12 of Table CS1.3 that use of tubes of 0.204 m I.D. and coolant outlet temperature of 20°C gives the lowest annualized cost ($2398). Hence, it is the optimum configuration. Also, Table CS1.3 shows that having the reactant on the shell side gives a lower annualized cost.

Lesson

The superiority of the plug-flow reactor over other reactors for many reactions is often taught in undergraduate CRE courses. The importance of this simple fact is quantitatively illustrated, but it is later shown that such a straightforward (known) result is not directly useful in practice. Although plug-flow behavior gives minimum volume, it is important to establish the critical length-to-diameter ratio to obtain plug-flow conditions. The selection of L/D ratio also depends on the operating cost (which is related to pressure drop) and the fixed cost. It is also shown that plug-flow behavior is advantageously reached by using a coil reactor, which also saves floor space. Since the reaction is generally accompanied by large heat effects, various geometrical alternatives and their cost implications must be considered.

CASE STUDY 11.2 HOMOGENEOUS GAS-PHASE COMPLEX REACTION: OXIDATION OF NO USING OZONE

The Problem

Flue gases from thermal power plants and industrial boilers are a major source of pollution. The polluting components are mainly nitric oxide, NO, and sulfur dioxide, SO_2. The solubility of NO in water is limited, but it can be oxidized to N_2O_5 using ozone. N_2O_5 readily reacts with water to form nitric acid. In the present problem, only the oxidation step is considered.

Fu et al. (2000) have reported rate constants for the following reaction scheme:

$$NO + O_3 \rightarrow NO_2 + O_2 \qquad \text{(CS2.R1)}$$

$$2\,NO_2 + O_3 \rightarrow N_2O_5 + O_2 \qquad \text{(CS2.R2)}$$

$$2\,O_3 \rightarrow 3\,O_2 \qquad \text{(CS2.R3)}$$

$$SO_2 + O_3 \rightarrow SO_3 + O_2 \qquad \text{(CS2.R4)}$$

$$CO + O_3 \rightarrow CO_2 + O_2 \qquad \text{(CS2.R5)}$$

$$NO + 1/2\,O_2 \rightarrow NO_2 \qquad \text{(CS2.R6)}$$

$$N_2O_5 + H_2O \rightarrow 2\,HNO_3 \qquad \text{(CS2.R7)}$$

$$SO_2 + N_2O_5 \rightarrow SO_3 + 2\,NO_2 \qquad \text{(CS2.R8)}$$

As a typical case, the mass flow rate of the gas from a power plant is 2,543,629 kg/h, inlet temperature is 50°C, and pressure is 1.1×10^5 N/m². A pipeline reactor is to be designed to give a conversion of total NO_x to N_2O_5 of 99%. The gas composition at the inlet to the reactor (flue gas

+ ozone) is as follows: NO- 389 ppm, NO_2- 39 ppm, HNO_2- 6 ppm, N_2O_5- 0 ppm, HNO_3- 1 ppm, O_3- 663 ppm, O_2- 6.5297 mol%, SO_2- 1944 ppm, CO- 0 ppm, CO_2- 11.9815 mol%, SO_3- 0 ppm, CO- 0 ppm, HNO_3- 1 ppm, O_3- 663 ppm, H_2O- 6.9974 mol%, N_2- rest. The theoretical requirement has been estimated on the basis of NO and NO_2 conversion.

Solution

The following assumptions are made: (1) Ozone and flue gas are thoroughly mixed in a short distance using a proper distributor. (2) Gas moves in plug flow. (3) Gas follows ideal gas law. (4) There are no radial gradients of temperature, concentration, and density. (5) The reactor is operating at steady state.

Component balances across a differential length can be established for O_3, O_2, NO, NO_2, N_2O_5, HNO_3, H_2O, SO_2, CO, and SO_3, and ten first-order ordinary differential equations are written. For the solution of this set of simultaneous differential equations, the Runge–Kutta fourth-order method was employed. The integration along the volume of the reactor was performed using the feed composition as the inlet boundary condition. The outlet boundary condition was specified to be 99% conversion of the total inlet NO_x to N_2O_5.

The level of ozone affects the reactor volume. The rate of ozonation of NO being fast, NO is consumed almost instantaneously, forming NO_2. NO_2 concentration decreases due to the formation of N_2O_5. The residence time required for 99% conversion is 7.4 s. The effect of percentage excess of ozone on reactor performance for 99% conversion of total inlet NO_X (NO and NO_2) can be easily calculated.

Lesson

The abatement of NO_x pollution is an important environmental problem that has alternative technologies. This case study focuses on technology that won the McGraw-Hill 2001 Kirkpatrick Award. The main reaction involves the ozonation of NO_X to give N_2O_5, followed by absorption in water/Ca(OH)$_2$. While Case Study 11.1 illustrates a detailed procedure for determining the geometric specifications for a simple reaction, the present study outlines a design methodology for a complex system and tells us that such a study is not only relevant but almost inescapable. Because this case study pertains to the treatment of exhaust gases, it demands very low pressure drops, even at high volumetric gas flow rates. Although not considered in this case study, the injection methodology of ozone in flue gases using a distributor should be such that the length required for mixing of ozone (in flue gases) is negligible compared to the length of the reactor.

CASE STUDY 11.3 GAS-SOLID (CATALYTIC) REACTION: MODELING OF A COMPLEX REACTION ON A DEACTIVATING CATALYST (A MODEL SCHEME)

Nomenclature for Case Study 11.3

(Some symbols are defined in the text; some are self-evident statistical terms.)

A_1, A_2, A_3	chemical species
f_1, f_2, f_3, f_4	functions
h	proportionality constant defined by Equation (CS3.13)
I_r	$(r \times r)$ identity matrix
K, K_1, K_3	vectors of rate constants, equilibrium constants, etc.
m	total number of models
n	total number of experiments at any stage
P	fouling complex
$P_{n+1,v}$	posterior probability of the vth model
$P_{n,v}$	prior probability of the vth model
p	total number of parameters

r	total number of responses
$\underline{r}, \underline{r}_1, \underline{r}_2$	reaction rates defined by Equations (CS3.1), (CS3.5), and (CS3.6)
S	multiresponse sum of squares function defined by Equation (CS3.7) a determinant
s	total number of independent variables
s^{11}, s^{12}, s^{22}	elements of the determinant S
t	on-stream line
U	utility function defined by Equation (CS3.17)
V	$(r \times r)$ precision matrix of responses
$\mathbf{X}_v^i$	$(n \times p)$ matrix of partial derivatives of the ith response in vth model
x	vector of independent variables
x_k	an $(s \times 1)$ vector of the kth experimental setting or kth design vector
y_k^i	observed value of the ith response in the kth experiment
y_k	vector of observation in the kth experiment
$Z_{n+1,v}$	$r \times p$ matrix of partial derivatives, where components are defined by Equation (CS3.1)
$z_{k,u,\ell}^i$	partial derivative of the ith response with respect to the ith parameter in the uth model using the kth experiment
z_{n+1}^i	partial derivative of $\eta_{n+1,v}^i$ with respect to th parameter evaluated after nth experiment

Greek

$\alpha^{i,j}$	(i, j)th component of V^{-1}
ε_k^i	error in the ith response in the kth experiment
ε_k	$\gamma \times 1$ vector of experimental errors for kth experiments
η_k^i	true value of the ith response using the kth experimental settings x_k
η_k	$r \times 1$ vector of true responses in kth experiment
$\hat{\eta}_{n+1,v}$	$r \times 1$ vector of calculated responses using vth model after n experiments using current best estimates of parameters
$\phi_{n+1,u}$	$(r \times r)$ precision matrix of the uth model after the $(n + 1)$th experiment
Ω	catalyst activity

As shown in Section 11.1.2.2, hyperbolic (LHHW) models are usually the most appropriate for reactions occurring on solid catalysts. These are nonlinear models whose parameters can be determined using statistical methods. Several examples are available.

An equally important system of reactions is one where the catalyst becomes deactivated either intrinsically or through deposition of carbonaceous products. The modeling of such systems becomes involved where deactivation has to be coupled with LHHW kinetics. Another complicated situation can arise where the reaction occurring on the deactivating catalyst is complex but each step can be represented by power law kinetics. In the present case study, we consider such a reaction to illustrate the application of rigorous statistical methods to complex reacting systems.*

The Problem

Tetrachloroethane chlorinates as follows using activated silica gel as catalyst (by-product HCl is not shown):

$$C_2H_2Cl_4 \xrightarrow[\text{cat}]{Cl_2} C_2HCl_5 \xrightarrow[\text{cat}]{Cl_2} C_2Cl_6 \quad \text{CS3.R1}$$
$$(A_1) \qquad (A_2) \qquad (A_3)$$

The catalyst deactivates with time. The rate of deactivation and level of activity depend on the variety of silica gel used. Basic kinetic data for this reaction were obtained by Prasad and

---

* Basic knowledge of statistics is assumed in describing this case study.

TABLE CS3.1
Experimental Rate Data

Sr. No.	Space Time g h/g mol	Decay Time h	Mol Fraction of A_1	Mol Fraction of A_2	Reaction Rate of A_1 mol/g h × 10^2	Reaction rate of A_2 mol/g h × 10^2
		t	y_1	y_2	r_1	r_2
1	15	0	0.871	0.120	−0.819	0.833
2	30	0	0.757	0.242	−0.697	0.707
3	45	0	0.661	0.337	−0.591	0.563
4	75	0	0.499	0.482	−0.511	0.439
5	105	0	0.363	0.599	−0.321	0.258
6	15	1	0.871	0.127	−0.776	0.728
7	30	1	0.766	0.229	−0.659	0.643
8	75	1	0.515	0.475	−0.495	0.370
9	90	1	0.443	0.520	−0.460	0.242
10	60	2	0.598	0.406	−0.531	0.483
11	30	3	0.784	0.216	−0.606	0.600
12	15	5	0.890	0.109	−0.715	0.668
13	90	5	0.478	0.496	−0.476	0,430
14	60	6	0.632	0.352	−0.518	0.450
15	15	9	0.903	0.095	−0.585	0.524
16	105	9	0.505	0.451	−0.303	0.320

Doraiswamy (1974) on a particular variety of silica gel in a fixed-bed reactor. The role of catalyst deactivation was studied by obtaining kinetic data at various times. Fortunately, by-products were negligible. This case study describes the reaction-deactivation kinetics and proposes a complete model using appropriate statistical methods. The emphasis will be on the use of a suitable algorithm for sequential updating of the models until one of them emerges as the "best."

Procedure

For model discrimination to be accomplished sequentially, leading to the selection of the "best" model, the following data/techniques are essential:

1. Experimental rate data in a suitable form.
2. A set of plausible models containing the true model.
3. An efficient method of parameter estimation.
4. A good sequential design strategy.
5. A powerful discriminatory criterion.

Experimental Data

For the reaction under consideration, values of the reaction rate as a function of concentration and catalyst decay time may be obtained by numerical differentiation of the experimental data obtained by Prasad and Doraiswamy (1974) at various space velocities and catalyst decay times in a fixed-bed reactor. The bed length was maintained small enough to give isothermal conditions to within 2°C. It was also ensured that the feed velocity was high enough, and the particle size small enough, to eliminate external and internal diffusion effects, respectively. The kinetic parameters, including decay time, will vary with the type of silica gel used. The data obtained for the silica gel used are summarized in Table CS3.1.

Models

Vapor-phase catalytic chlorination of A_1 on activated silica gel at 200°C is accompanied by slow deactivation of the catalyst. A_1 is successively chlorinated to A_2 and A_3. Trace quantities of by-products formed (trichloroethylene and tetrachloroethylene) may be neglected. The following assumptions are made:

1. Both the reaction steps in the above scheme are first order.
2. The catalyst activity decreases uniformly for both reaction steps.
3. The fouling product is the result of an unwanted side reaction involving reactants or products.
4. The catalyst decay is slower than the reaction rate by several orders of magnitude.

The last mentioned assumption allows the use of separable kinetics. Thus the reaction rate model in a deactivating system may be written as two simultaneous equations:

$$\underline{r} = f_1[\mathbf{x}, \mathbf{K}] f_2[\Omega] \tag{CS3.1}$$

$$\frac{d\Omega}{dt} = f_3[\mathbf{x}, \mathbf{K}] f_4[\Omega] \tag{CS3.2}$$

where $\underline{r}$ is the rate of reaction; $f_1, f_2, f_3,$ and f_4 are some functions depending on the system; $\mathbf{x}$ is the vector of controlled variables such as concentrations of species, temperature, and so on; $\mathbf{K}$ is the vector of rate constants, equilibrium constants, reaction orders, and so forth; t is the catalyst decay time; and Ω is the activity of the catalyst defined as

$$\Omega = \frac{\underline{r}|t=t}{\underline{r}|t=0} \tag{CS3.3}$$

If assumption 4 is valid, Equation (CS3.2) may be independently solved and substituted in Equation (CS3.1).

For the reaction under consideration, several combinations of series and parallel fouling schemes may be chosen, involving either one or more of A_1, A_2, A_3 or one or more of $A_1 + A_2, A_2 + A_3, A_3 + A_1,$ or $A_1 + A_2 + A_3$. The different forms of the function f_3 in Equation (CS3.2) resulting from these fouling reactions are summarized in Table CS3.2. Furthermore, f_2 and f_4 are assumed to be given by

$$f_2[\Omega] = f_4[\Omega] = \Omega \tag{CS3.4}$$

f_1 consists of two functions for the present system in which observations are made on two independent responses, namely, rate of disappearance of A_1 and rate of formation of A_2. Thus,

$$\underline{r}_1 = -\mathbf{K}_1 y_1 \Omega \tag{CS3.5}$$

$$\underline{r}_2 = [\mathbf{K}_1 y_1 - \mathbf{K}_2 y_2] \Omega \tag{CS3.6}$$

Using different forms for f_3, Equation (CS3.2) may be solved and substituted in Equations (CS3.5) and (CS3.6) to yield expressions for different models. These are included in Table CS3.2.

Method of Parameter Estimation

The method of parameter estimation depends on the models involved. All models listed in Table CS3.2 are nonlinear and algebraic. One might employ the Gauss–Newton method of nonlinear least

TABLE CS3.2
The Rival Models

Model No.	Fouling Reactions	Expressions for f_3 in Equation (CS3.2)	Model Equations for r_1	Model Equations for r_2
1.	$A_1 \to P$	$-K_3 y_1$	$-K_1 y_1 \exp[-K_3 y_1 t]$	$[K_1 y_1 - K_2 y_2] \exp[-K_3 y_1 t]$
2.	$A_2 \to P$	$-K_3 y_2$	$-K_1 y_1 \exp[-K_3 y_2 t]$	$[K_1 y_1 - K_2 y_2] \exp[-K_3 y_2 t]$
3.	$A_3 \to P$	$-K_3[1 - y_1 - y_2]$	$-K_1 y_1 \exp[-K_3 (1 - y_1 - y_2) t]$	$[K_1 y_1 - K_2 y_2] \exp[-K_3(1 - y_1 - y_2) t]$
4.	$A_1 + A_2 \to P$	$-K_3 y_1 y_2$	$-K_3 y_1 \exp[-K_3 y_1 y_2 t]$	$[K_1 y_1 - K_2 y_2] \exp[-K_3 y_1 y_2 t]$
5.	$A_1 + A_3 \to P$	$-K_3 y_1 [1 - y_1 - y_2]$	$-K_1 y_1 \exp[-K_3 y_1(1 - y_1 - y_2) t]$	$[K_1 y_1 - K_2 y_2] \exp[-K_3 y_1 (1 - y_1 - y_2) t]$
6.	$A_2 + A_3 \to P$	$-K_3 y_2 [1 - y_1 - y_2]$	$-K_1 y_1 \exp[-K_3 y_2(1 - y_1 - y_2) t]$	$[K_1 y_1 - K_2 y_2] \exp[-K_3 y_2(1 - y_1 - y_2) t]$
7.	$A_1 + A_2 + A_3 \to P$	$-K_3 y_1 y_2[1 - y_1 - y_2]$	$-K_1 y_1 \exp[-K_3 y_1 y_2(1 - y_1 - y_2) t]$	$[K_1 y_1 - K_2 y_2] \exp[-K_3 y_1 y_2(1 - y_1 - y_2) t]$
8.	$A_1 \to P$ $A_2 \to P$	$-K_3[y_1 + y_2]$	$-K_1 y_1 \exp[-K_3(y_1 + y_2) t]$	$[K_1 y_1 - K_2 y_2] \exp[-K_3(y_1 + y_2) t]$
9.	$A_1 \to P$ $A_3 \to P$	$-K_3[1 - y_2]$	$-K_1 y_1 \exp[-K_3(1 - y_2) t]$	$[K_1 y_1 - K_2 y_2] \exp[-K_3(1 - y_2) t]$
10.	$A_2 \to P$ $A_3 \to P$	$-K_3[1 - y_1]$	$-K_1 y_1 \exp[-K_3(1 - y_1) t]$	$[K_1 y_1 - K_2 y_2] \exp[-K_3(1 - y_1)t]$
11.	$A_1 \to P$ $A_2 \to P$ $A_3 \to P$	$-K_3[y_1 + y_2 + (1 - y_1 - y_2)]$ $= -K_3$	$-K_1 y_1 \exp[-K_3 t]$	$[K_1 y_1 - K_2 y_2] \exp[-K_3 t]$

Main reaction is $A_1 \to A_2 \to A_3$

squares but, for the present two-response system, the resulting equations are quite involved. Hence, the following least squares objective function may be minimized using the "complex" method of Box (1965):

$$S = \det \begin{bmatrix} s^{11} & s^{12} \\ s^{12} & s^{22} \end{bmatrix} \quad \text{(CS3.7)}$$

where s^{ij} is the sum of cross products of residuals of the ith and jth responses.

A Good Sequential Design Strategy

Basic Equations

Any experimental program involving r dependent and s independent variables and p parameters in the kth experiment may be described by algebraic equations,

$$y_k^i = \eta_k^i(\mathbf{x}_k, \mathbf{K}) + \varepsilon_k^i, \quad i = 1, 2, \ldots r \quad \text{(CS3.8)}$$

where $\mathbf{x}_k$ is an $(s \times 1)$ vector of independent variables for the kth experiment, is a $(p \times 1)$ vector of model parameters, η_k^i is the true value of the ith response for the kth experimental settings $\mathbf{X}_k$, y_k^i is the observed value of the ith response, and ε_k^i is the experimental error for the ith response, also for the kth experiment. The uth model for all the responses may be written in vector form as

$$\mathbf{y}_k = \eta_k(\mathbf{x}_k, \mathbf{K}) + \varepsilon_k^i \quad \text{(CS3.9)}$$

If the ith component of the $(r \times 1)$ vector of nonlinear functions η_k is expanded in Taylor's series about an estimate $\hat{\mathbf{K}}_u$ of the parameters and truncated after the first partial derivatives, Equation (CS3.8) may be written for the $(n + 1)$th experiment as

Chemical Reaction Engineering

$$\varepsilon_{n+1}^i = y_{n+1}^i - \eta_{n+1,u}^i(\mathbf{x}_{n+1}, \mathbf{K})$$

$$= y_{n+1}^i - \hat{\eta}_{n+1,u}^i(\mathbf{x}_{n+1}, \hat{\mathbf{K}}_u) - \sum_{\ell=1}^{p} z_{n+1,u,\ell}^i (K_\ell - \hat{K}_{\ell u}) \qquad \text{(CS3.10)}$$

$$i = 1\ldots r$$

where $\hat{K}_{\ell u}$ is the lth parameter in the ($p \times 1$) vector of parameters for the uth model estimated using the first n experiments, and $z_{n+1,u,\ell}^i$ is given by

$$z_{n+1,u,l}^i = \delta\eta_{n+1,u}^i(\mathbf{x}_{n+1}, \mathbf{K})/\delta K_l \big|_{\mathbf{K} = \hat{\mathbf{K}}_u}$$

$$i = 1,\ldots r \qquad \text{(CS3.11)}$$

$$u = 1,\ldots m$$

Bayesian Discrimination

Since competing models exist, Bayes' theorem provides a sound statistical method of discriminating between them. This theorem defines the posterior probability of the uth model after ($n + 1$) experiment as the product of the prior probability and the likelihood

$$P_{n+1,u} = P_{n,u} L_{n+1,u} (\mathbf{K}_u | \phi_{n+1,u}, \mathbf{y}_{n+1})/h \qquad \text{(CS3.12)}$$

where $P_{n+1,u}$ is the posterior probability of the uth model after y_{n+1} data have been collected; $P_{n,u}$ is the prior probability, i.e., the probability before performing the ($n + 1$)th experiment; $\phi_{n+1,u}$ is the ($r \times r$) precision matrix for the predictions made by the uth model after ($n + 1$) experiments; $L_{n+1,u}(\mathbf{K}_u | \phi_{n+1,u}, \mathbf{y}_{n+1})$ is the likelihood of the uth model after data $\mathbf{y}_{n+1}$ are obtained; and h is the proportionality constant given by

$$h = \sum_{v=1}^{m} P_{n,v} L_{n+1,v}(\mathbf{K}_V | \phi_{n+1,v}, \mathbf{y}_{n+1}) \qquad \text{(CS3.13)}$$

For a system of r responses, the likelihood is given by (Hill, 1966)

$$L_{n+1,v}(\mathbf{K}_V | \phi_{n+1,v}, \mathbf{y}_{n+1}) = \frac{|\phi_{n+1,v}|^{-1/2}}{(2\pi)^{r/2}} \times$$

$$\exp\left[-\frac{1}{2}(\mathbf{y}_{n+1} - \hat{\eta}_{n+1,v})^T \phi_{n+1,v}^{-1} (\mathbf{y}_{n+1} - \hat{\eta}_{n+1,v})\right] \qquad \text{(CS3.14)}$$

where $\hat{\eta}_{n+1,v}$ ($r \times 1$) is the vector of responses for the ($n + 1$)th experiment predicted using the vth model and the previous best estimates of parameters. The precision matrix $\phi_{n+1,v}$ is found from

$$\phi_{n+1,v} = \mathbf{V} + \mathbf{Z}_{n+1,v} \mathbf{M}_v^{-1} \mathbf{Z}_{n+1,v}^T, \quad v = 1,2,\ldots m \qquad \text{(CS3.15)}$$

where $\mathbf{Z}_{n+1,v}$ is the $(r \times p)$ matrix of partial derivatives whose (i,ℓ)th element is given by Equation (CS3.11), $\mathbf{M}_v^{-1}$ is the $(p \times p)$ precision matrix of parameters of the vth model, and $\mathbf{V}$ is the precision matrix. $\mathbf{M}_v$ is given by

$$\mathbf{M}_v = \sum_{i=1}^{r}\sum_{j=1}^{r} \alpha^{ij} \mathbf{X}_v^{i^T} \mathbf{X}_v^{j}, \quad v=1,2,\ldots m \qquad (CS3.16)$$

where α^{ij} is the (i,j)th component of $\mathbf{V}^{-1}$ and $\mathbf{X}_v^i$ is the $(n \times p)$ matrix of partial derivatives of the ith response for the vth model whose (k,ℓ)th element is obtained from Equation (CS3.11) by replacing subscript $(n+1)$ with k. Since Equation (CS3.12) is basic to the Bayesian procedure, the probabilities of all the models are sequentially updated by collecting additional data until one model emerges with the highest probability.

A Powerful Discriminatory Criterion

Often, no single model emerges as the most probable. In such a case, the next experiment should be performed at a setting of independent variables that maximizes the divergence between competing candidates. This is done by using a discriminatory design criterion, which is some suitably defined function of the vector of independent variables $\mathbf{x}_k$, called the *utility function* $U(\mathbf{x}_k)$. Several criteria have been proposed (Reilly and Blau, 1974). The Box–Hill (1967) criterion is perhaps the most commonly used and is given by

$$U(\mathbf{x}_k) = \frac{1}{2}\sum_{u=1}^{m}\sum_{v=u+1}^{m} P_{n,u}P_{n,v}\left[Tr\left(\boldsymbol{\phi}_u\boldsymbol{\phi}_v^{-1} + \boldsymbol{\phi}_v\boldsymbol{\phi}_u^{-1} - 2I_r\right)\right.$$
$$\left. + (\boldsymbol{\eta}_u - \boldsymbol{\eta}_v)^T\left(\boldsymbol{\phi}_u^{-1} + \boldsymbol{\phi}_v^{-1}\right)(\boldsymbol{\eta}_u - \boldsymbol{\eta}_v)\right] \qquad (CS3.17)$$

where $U(\mathbf{x}_k)$ is the utility function for the kth set of independent variables $\mathbf{x}_k$, $P_{n,u}$ is the posterior probability of the uth model after n experiments, $\boldsymbol{\phi}_u$ is the $(r \times r)$ precision matrix for the predictions made by the uth model after n experiments, and $\boldsymbol{\eta}_u$ is the $q \times 1$ vector of predicted responses for the uth model using the kth set of independent variables $\mathbf{x}_k$. This criterion weighs the divergences over the inverse of the precision matrix so that large divergences due to imprecisely known parameters do not contribute significantly to the utility function. The object is to find the $\max_k U(\mathbf{x}_k)$ that determines the next best experiment to be performed to maximize divergence between rival models. Hence, several new sets of $\mathbf{x}_k$ can be randomly generated to evaluate U_{max}.

A Sequential Discrimination Algorithm

A computational algorithm for sequential discrimination among rival models is presented in Figure CS3.1.

Choice of Model for the Present System (see Prasad and Rao, 1977)

The eleven models given in Table CS3.2 can be compared taking an initial probability of 1/11 or 0.0909 for each and with an initial set of six experiments taken at random from Table CS3.1. Ten sequential experiments (total of 16 experiments) were performed (Table CS3.3), at the end of which one of the models (model 9) attained a probability of 0.912, compared to 0.088 for the nearest rival (model 1). The probabilities of the other models are negligibly small, even after the first discriminatory experiment.

The computations show that the given system represents an example of series-parallel deactivation:

Chemical Reaction Engineering

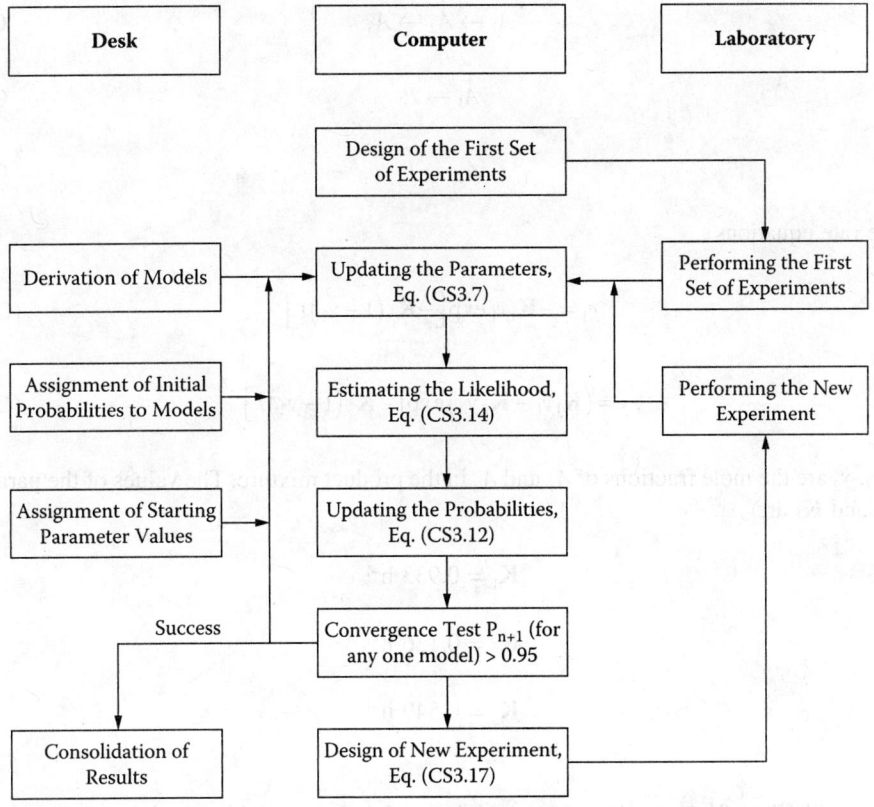

FIGURE CS3.1 A computational algorithm for sequential discrimination among rival models.

TABLE CS3.3
Posterior Probabilities of Models Using the Box-Hill Criterion

Discrimination Stage	Total Number of Experiments	Posterior Model Probability[a]	
		P_9	P_1
1	7	0.502	0.483
2	8	0.513	0.487
3	9	0.623	0.377
4	10	0.656	0.344
5	11	0.730	0.270
6	12	0.794	0.206
7	13	0.833	0.167
8	14	0.849	0.151
9	15	0.902	0.098
10	16	0.912	0.088

Computation stopped

[a] The posterior probabilities of the remaining models are negligibly small. Hence they are not reported here.

$$A_1 \to A_2 \to A_3 \qquad \text{CS3.R2}$$

$$A_1 \to P \qquad \text{CS3.R3}$$

$$A_3 \to P \qquad \text{CS3.R4}$$

with the rate equations

$$\underline{r}_1 = -K_1 y_1 \exp\left[-K_3(1-y_2)t\right] \qquad \text{(CS3.18)}$$

$$\underline{r}_2 = (K_1 y_1 - K_2 y_2)\exp\left[-K_3(1-y_2)t\right] \qquad \text{(CS3.19)}$$

where y_1, y_2 are the mole fractions of A_1 and A_2 in the product mixture. The values of the parameters K_1, K_2, and K_3 are

$$K_1 = 0.933 \text{ h}^{-1}$$

$$K_2 = 0.136 \text{ h}^{-1}$$

$$K_3 = 0.549 \text{ h}^{-1}$$

Lesson

This case study describes a statistical method for modeling a complex reaction. A model reaction occurring on a deactivating catalyst, in which the fouling reactions generate the complexity, is chosen for illustration, but the procedure is equally applicable to any complex reaction scheme with or without fouling, catalytic or noncatalytic. While a great reliance on statistics is not always needed, this study tells us that the statistical method can identify the controlling step. On the other hand, it does not consider the possibility that the true rate-determining mechanism may have been omitted from the list of candidates. Thus, any statistical procedure of this kind, so common in chemical engineering, is no more than one of convergence to the "best" and not a direct postulation of a model from first principles and validated by experiments.

When this procedure is applied to LHHW models, caution should be exercised in claiming a model to be the "best." The need for this increases with the number of unknown parameters, since several models may then show almost equal convergence. In general, it is advisable to confine the methods to models with no more than three or four unknown parameters. Furthermore, it should be ensured that all data points are of equal precision.

CASE STUDY 11.4 GAS-SOLID (CATALYTIC) REACTION IN FIXED-BED NINA AND ADIABATIC REACTORS: REDUCTION OF NITROBENZENE TO ANILINE

Nomenclature for Case Study 11.4

A	nitrobenzene
$[A]$	concentration of A (kmol/m^3)
B	hydrogen
$[B]$	concentration of B (kmol/m^3)
C_P	heat capacity [kcal/kg (reactant gases)°C]
C_{pm}	molar heat capacity (kcal/kg °C)

Chemical Reaction Engineering

$D_{A,er}$	effective radial diffusivity of A (m²/s)
d	diameter (m)
E	activation energy for reaction (kcal/mol)
E_{ad}	activation energy for adsorption (kcal/mol)
F	flow rate of A (kmol/s)
$-\Delta H_r$	heat of reaction (kcal/mol)
h	heat transfer coefficient (kcal/m² °C s)
K_B	adsorption constant of B (m³/mol)
K^o_B	Arrhenius constant for K_B (m³/mol)
$k_{T,er}$	effective radial thermal conductivity (kcal/m °C s)
k_w	reaction rate constant (kmol/g cat s)
k^o_w	Arrhenius constant (m³/g cat s)
R_g	gas law constant (kcal/mol K)
r	radial coordinate (m)
$-r_A$	rate of disappearance of A (kmol/m³ s)
$-r_{wA}$	rate of disappearance of A (kmol/kg cat s)
S	cross-sectional area (m²)
T	temperature (°C or K)
u	velocity (m/s)
W	weight of catalyst (kg)
x_A	conversion of A (–)
y_A	mole fraction (–)
z	length coordinate (m)

Greek

α	parameter defined by Equations (CS4.14) and (CS4.16)
β	parameter defined by Equations (CS4.15) and (CS4.17)
ε_b	bulk voidage (–)
ρ	fluid density (kg/m³)
ρ_b	bulk density (kg/m³)
ρ_s	solid (catalyst) density (kg/m³)

Subscripts

er	effective radial
S	solid (catalyst)
t	total
w	wall; weight based
0	initial (entrance)

NINA Reactor

Aniline is produced by the catalytic reduction (with hydrogen) of nitrobenzene in a fixed-bed reactor according to the reaction

$$C_6H_5NO_2(A) + 3H_2(B) \rightarrow C_6H_5NH_2 + 2H_2O \qquad \text{CS4.R1}$$

The rate equation is given by

$$-r_{wA} = \frac{k_w [A]}{\left(1+K_B[B]^2\right)}, \text{mol/g cat s} \qquad (CS4.1)$$

where

$$k_w = k_w^o \exp\left(-\frac{E}{R_g T}\right), \text{ cm}^3/\text{g cat s} \qquad \text{(CS4.2)}$$

$$K_B = K_B^o \exp\left(\frac{E_{ad}}{R_g T}\right), \text{ cm}^3/\text{mol} \qquad \text{(CS4.3)}$$

Assuming a pseudohomogeneous two-dimensional reactor model with plug flow of fluid and constant properties, calculate the axial concentration and temperature profiles at several radial positions along the axis of a single tube. Use the following property/parameter values (Doraiswamy, 2001):

$[A]_0 = 7.1 \times 10^{-4}$ mol/L, $(-\Delta H_r) = 180$ kcal/mol, $\rho_s = 2.18$ g/cm³, $\varepsilon_B = 0.312$, $C_p = 0.49$ cal/g (reactant gases) °C at 200°C, $k_{T,er} = 1.4$ cal/m °C s, $h_w = 9.5 \times 10^{-4}$ cal/cm² °C s, $\mathbf{D}_{A,er} = 4.74 \times 10^{-4}$ cm²/s, $\rho = 0.0944$ g/L, $u = 40$ cm/s, $k^o{}_w = 9.46 \times 10^{-3}$ L/g catalyst s, $E = 2631$ cal/mol, $K^o{}_B = 10.7$ cm³/mol, $E_{ad} = 8039$ cal/mol. $T_0 = 160°$C, $T_w = 100°$C.

Solution

Before proceeding with the solution, it is instructive to list the governing equations for a number of simple and complex situations. These situations have already been described in Section 11.4, and Table 12.2 of Doraiswamy and Sharma (1984) lists the more important equations. Case A2-a of this table is the so-called basic model that incorporates enough complexity to make the design sufficiently realistic and yet is relatively easy. The homogeneous nonisothermal nonadiabatic (NINA) model allows one to compute temperature and concentration profiles within the reactor tube. The equations for this model are

$$\mathbf{D}_{A,er}\left(\frac{\partial^2 [A]}{\partial r^2} + \frac{1}{r}\frac{\partial [A]}{\partial r}\right) - u\frac{\partial [A]}{\partial z} - (-r_A) = 0 \qquad \text{(CS4.4)}$$

$$-k_{T,er}\left(\frac{\partial^2 T}{\partial r^2} + \frac{1}{r}\frac{\partial T}{\partial r}\right) - u\rho C_p\frac{\partial T}{\partial z} + (-\Delta H_r)(-r_A) = 0 \qquad \text{(CS4.5)}$$

with boundary conditions

$$\frac{\partial [A]}{\partial r} = 0 \quad \text{at } r = 0,\ z > 0$$

$$[A] = [A]_0 \quad \text{at } r > 0,\ z = 0$$

$$\frac{\partial T}{\partial r} = 0 \quad \text{at } r = 0,\ z > 0 \qquad \text{(CS4.6)}$$

$$-k_{T,er}\frac{\partial T}{\partial r} = h_w(T - T_w) \quad \text{at } r = 0,\ z = 0$$

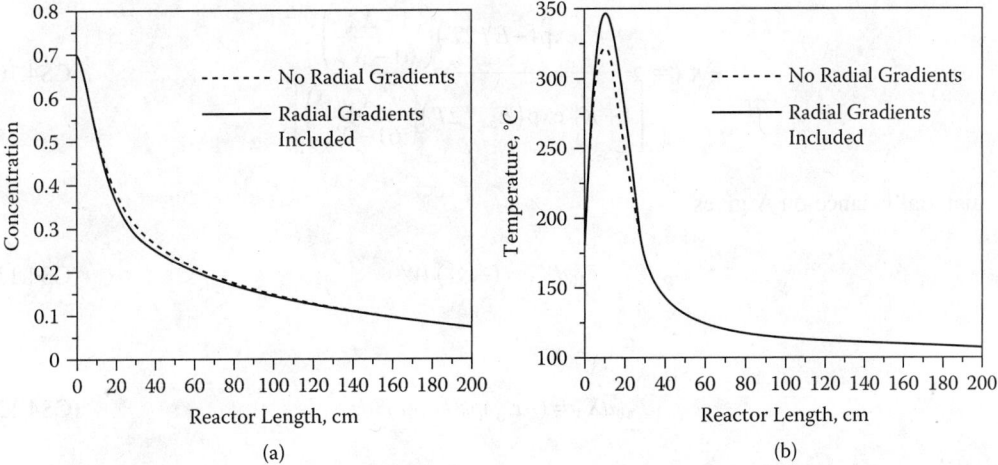

FIGURE CS4.1 Computed concentration and temperature profiles in the multitubular reactor for the catalytic reduction of nitrobenzene to aniline.

Using the data given above, the equations were solved (Doraiswamy, 2001) to obtain axial concentration and temperature profiles at different radial positions. From these profiles, an average value was calculated for the entire cross section for a given axial position. The resulting plots of the average temperature and concentration as functions of reactor length are shown as continuous lines in Figure CS4.1. For comparison, the results of computations of the model without radial gradients are described by

$$u\frac{d[A]}{dz}+(-r_A)=0 \tag{CS4.7}$$

$$u\rho C_P \frac{dT}{dz}+\frac{4U}{d_T}(T-T_w)-(-\Delta H_r)(-r_A)=0 \tag{CS4.8}$$

with boundary conditions

$$[A]=[A]_0, \; T=T_0 \quad \text{at } t=0 \tag{CS4.9}$$

are also included. The axial temperature maximum occurs close to the entrance. Also, there is very little difference between the one- and two-dimensional models. The desired conversion of 99.7% is achieved in a length of 8.3 m, whereas a length of 1.9 m provides a 90% conversion.

Extension to Adiabatic Operation

It is also possible to operate the aniline reactor adiabatically, as in the earliest reactors used in Germany. As mentioned in Section 11.4.2.4.2, adiabatic reactors are usually operated as multistage reactors with interstage cooling. Since the present reactor is reasonably exothermic, a single reactor can be used, provided a large excess of hydrogen is employed such that the hydrogen-nitrobenzene ratio is as high as 50:1 to 80:1. Without going into numerical details, we shall illustrate the procedure (see Doraiswamy, 2001, for details). Let us use the same rate equation as for the NINA reactor. However, we express the concentrations in terms of partial pressures; hence the numerical values of the rate and adsorption constants will be different.

Assuming a hydrogen-nitrobenzene ratio of 60:1, the rate equation can be recast as

$$-r_{wA} = \frac{k_w^o \exp(-E/2T)\left(\dfrac{1-X_A}{61-X_A}\right)}{\left[1 + K_B^o \exp(E_{ad}/2T)\left(\dfrac{X_A}{61-X_A}\right)\right]^2} \quad \text{(CS4.10)}$$

A material balance on A gives

$$F_{A0} dX_A = (-r_{wA}) dW \quad \text{(CS4.11)}$$

or

$$F_{A0} dX_A = (-r_{wA}) \rho_b (1-\varepsilon_b) S dz \quad \text{(CS4.12)}$$

and an energy balance gives

$$F_{A0}(-\Delta H_r) dX_A = F_{t0} C_{pm} dT \quad \text{(CS4.13)}$$

where F_{t0} is the total flow rate of the inlet gases. In view of the large excess of hydrogen, F_{t0} can be assumed constant.

From Equations (CS4.12) and (CS4.13), we can write

$$dX_A = \left[\frac{\rho_b (1-\varepsilon_b) S(-r_{wA})}{F_{A0}}\right] dz \quad \text{(CS4.14)}$$

$$dT = \left[\frac{(-\Delta H_r) \rho_b (1-\varepsilon_b) S(-r_{wA})}{F_{t0} C_{pm}}\right] dz \quad \text{(CS4.15)}$$

or

$$dX_A = \alpha dz \quad \text{(CS4.16)}$$

$$dT = \beta dz \quad \text{(CS4.17)}$$

Equations (CS4.16) and (CS4.17) are coupled ordinary differential equations in conversion and temperature. They can be solved for the initial conditions

$$z = 0, \quad X_A = 0, \quad T_0 = 473 \text{ K}$$

using the Runge–Kutta fourth-order method for different tube diameters and ratios [note that Equation (CS4.10) will be slightly different for different ratios].

The results (not given here) show that (1) conversion and temperature decrease with the reactor length with an increase in H_2–$C_6H_5NO_2$ ratio, and (2) for a given ratio, conversion and temperature rise throughout the reactor length with an increase in reactor diameter. Hence, in general, larger diameters and lower ratios are favored.

Chemical Reaction Engineering

Lesson

The present case study outlines a modeling procedure for a catalytic reaction but places it in context by including it as one of 12 case studies chosen to cover a variety of reactions. The main lesson here is that a relatively complex model often does little better than its simplest counterpart. While all models should be explored, the suitability of the simplest one should always be considered.

The choice between NINA and adiabatic reactors is not easy for moderately to highly exothermic reactions. The values of the adiabatic temperature change ATC and temperature sensitivity given by

$$ATC = \frac{(-\Delta H_r) y_{A0}}{C_{pm}}$$

$$TS = (-r_{wA}) \frac{E}{R_g T^2}$$

are useful indicators. ATC is usually in the range of 90 to 800 K, and TS is in the range of 2 to 20. Values in the lower half of the ranges would suggest a preference for adiabatic operation. Another important consideration is the cost of recycling the large excess of hydrogen used in adiabatic operation. The fixed cost depends on the number of stages (not considered here; see Froment and Bischoff, 1990; and Doraiswamy, 2001). Detailed calculations should be done on NINA and several adiabatic configurations. The final selection should be made from the two or three "best" alternatives that may emerge, based on convenience as the final test. Catalyst life is usually not a problem, since it is quite robust. If it lasts for a year or more, it is usually preferable to replace it rather than regenerate it. *A priori,* the present reaction is a candidate for more than one mode of operation.

CASE STUDY 11.5 COMPARISON OF SELECTED FLUID-BED MODELS FOR THE CATALYTIC REDUCTION (BY HYDROGEN) OF NITROBENZENE TO ANILINE

Nomenclature for Case Study CS11.5

A	nitrobenzene
A_{be}	bubble-emulsion interfacial area (m^2)
a_b	bubble area per unit volume (1/m)
d_b	bubble diameter (m)
D	gas diffusion coefficient (m^2/s)
g	acceleration due to gravity (m/s^2)
K', K_b, K_d, K_m	parameters defined in Figure CS5.3
K_0	kinetic group defined by Equation (CS5.2)
K_f	overall rate constant defined by Equation (CS5.14)
k_{bc}	bubble-cloud mass transfer coefficient (1/s)
k_{ce}	cloud-emulsion mass transfer coefficient (1/s)
k_G	mass transfer coefficient (m/s)
k_{ob}	overall mass transfer coefficient at the bubble-emulsion interface (m/s)
k_v	reaction rate constant (1/s)
L	height of fluidized bed (m)
L_0	initial fixed-bed height (m)
L_f	height of fluidized bed (m)
L_{mf}	height at minimum fluidization (m)
L_t	total bed height including the dilute bed region
Q_{be}	total volumetric flow rate between the bubble and emulsion phases (m^3/s)

q	flow rate between the bubble and emulsion phases induced by circulation (m^3/s)
r	radius (m)
s_{bb}	solids fraction in bubble, Equation (CS5.9)
s_{cb}	solids fraction in cloud, Equation (CS5.10)
s_{ce}	solids fraction in emulsion, Equation (CS5.11)
s_{wb}	fraction of solids in wake, Equation (CS5.12)
U	fluidization group defined by Equation (CS5.2)
u_{br}	slip velocity (m/s)
u_b	real bubble velocity
u_f	fluidization velocity (m/s)
u_{mf}	velocity at minimum fluidization (m/s)
V_b	bubble volume (m^3)
X_A	conversion of A (–)
Y	mass exchange group defined by Equation (CS5.2)
Z	dimensionless height of fluidized bed, L/L_f (–)
Z_t	total dimensionless height of fluidized bed including the dilute bed region, L_t/L_f

Greek

α	ratio u_b/u_f
δ	volume fraction of bubbles in bed
ε_b	static bed voidage
ε_{mf}	bed voidage at minimum fluidization
ρ_c	density of catalyst (kg/m^3)

Subscripts

b	bubble
c	cloud
e	emulsion
f	fluidization
mf	minimum fluidization

The Problem

Aniline is produced by the hydrogenation of nitrobenzene on a solid catalyst as was shown in CS4.R1 (Case Study 11.4). Kinetic data for this reaction using a copper-based catalyst were obtained during pilot plant studies at the National Chemical Laboratory. Although an LHHW model was proposed, it was also possible to fit a simpler, power law equation to the data, with a rate constant of 1.2 s^{-1}. The purpose of the present case study is to calculate the conversions obtained for this reaction in a fluid-bed reactor using several available models. Although more rigorous models have been proposed that account for bubble size distribution, volume change, and other complicating features of a fluidized bed, simpler models are usually quite adequate. In this study, we compare the predictions of the following five models: Davidson, Kunii–Levenspiel, Miyauchi, Fryer–Potter, and Jayaraman–Kulkarni–Doraiswamy. As the catalyst maintains its initial activity for over a year, a batch fluidized bed is used, and the catalyst is replaced after it begins to lose its activity.

To put the five models in perspective, and to understand the complications that may arise in more rigorous modeling, we conclude this case study with a brief discussion of a few complex models.

Data

The following data (calculated, assumed, or experimentally obtained) will be used:

Minimum fluidization velocity u_{mf} = 2 cm/s, feed velocity at inlet u_0 = 30 cm/s, bubble diameter d_b = 10 cm, initial (fixed-bed) height L_0 = 1.4 m, reaction rate constant k_v = 1.2 s^{-1}, bed voidage

Chemical Reaction Engineering

at minimum fluidization $\varepsilon_{mf} = 0.6$, temperature = 270°C, diffusion coefficient of gas $\mathbf{D} = 0.9$ cm²/s, density of catalyst $\rho_c = 2.2$ g/cm³, reactor diameter $d_T = 3.55$ m, solids fraction in the wake $s_{wb} = 0.33$.

Solution

The original references provide more details of the models. The solutions below illustrate only the use of the model equations. However, the main features are sketched in Figure CS5.1.

Davidson Model

The first hydrodynamic model proposed for fluid-bed reactor design (see Davidson and Harrison, 1963) is simple but is the basis of most models developed since. A sketch of the model appears in Figure CS5.1a. Three main groups are involved: U for fluidization, K_0 for reaction, and Y for mass transfer. Equations can be derived both for plug flow and mixed flow of emulsion gas. The simpler mixed-flow model is usually adequate (with predictions close to those of the plug-flow model) and is given by

$$[1-X_A] = Ue^{-Y} + \frac{1-Ue^{-Y}}{K_0 + \left[1-Ue^{-Y}\right]} \tag{CS5.1}$$

where

$$U = 1 - u_{mf}/u_0$$

$$K_0 = k_v L_{mf}/u = k_v L_f (1-\delta)/u_0 \tag{CS5.2}$$

$$Y = Q_{be} L_f / u_b V_b$$

The bubble velocity u_b is the real bubble velocity and is the sum of the slip velocity of bubbles u_{br} and the velocity at which the bed itself moves; it is given by

$$u_b = 0.711 \times (gd_b)^{1/2} + [u_0 - u_{mf}] = 98.4213 \text{ cm/s} \tag{CS5.3}$$

where $g = 981$ cm/s², and $V_b = \pi d_b^3 / 6 = 523.33$ cm³

The quantity Q_{be} appearing in the group Y is the sum of the volumetric flow between the bubble and emulsion phases induced by circulation (q) and diffusive flow ($k_g A_{be}$),

$$Q_{be} = q + k_G A_{be} \tag{CS5.4}$$

where

$$k_G = 0.975 \mathbf{D}^{1/2}[g/d_b]^{1/4} = 2.91 \text{ cm/s}$$

$$A_{be} = \pi d_b^2 = 3.14 \times 10^2 = 314 \text{ cm}^2 \tag{CS5.5}$$

$$q = 0.75 u_{mf} \pi d_b^2 = 471 \text{ cm}^3/\text{s}$$

Substituting for k_g and A_{be} in Equation (CS5.4),

$$Q_{be} = 471 + 9.14 \times 314 = 1384.74 \text{ cm}^3/\text{s}$$

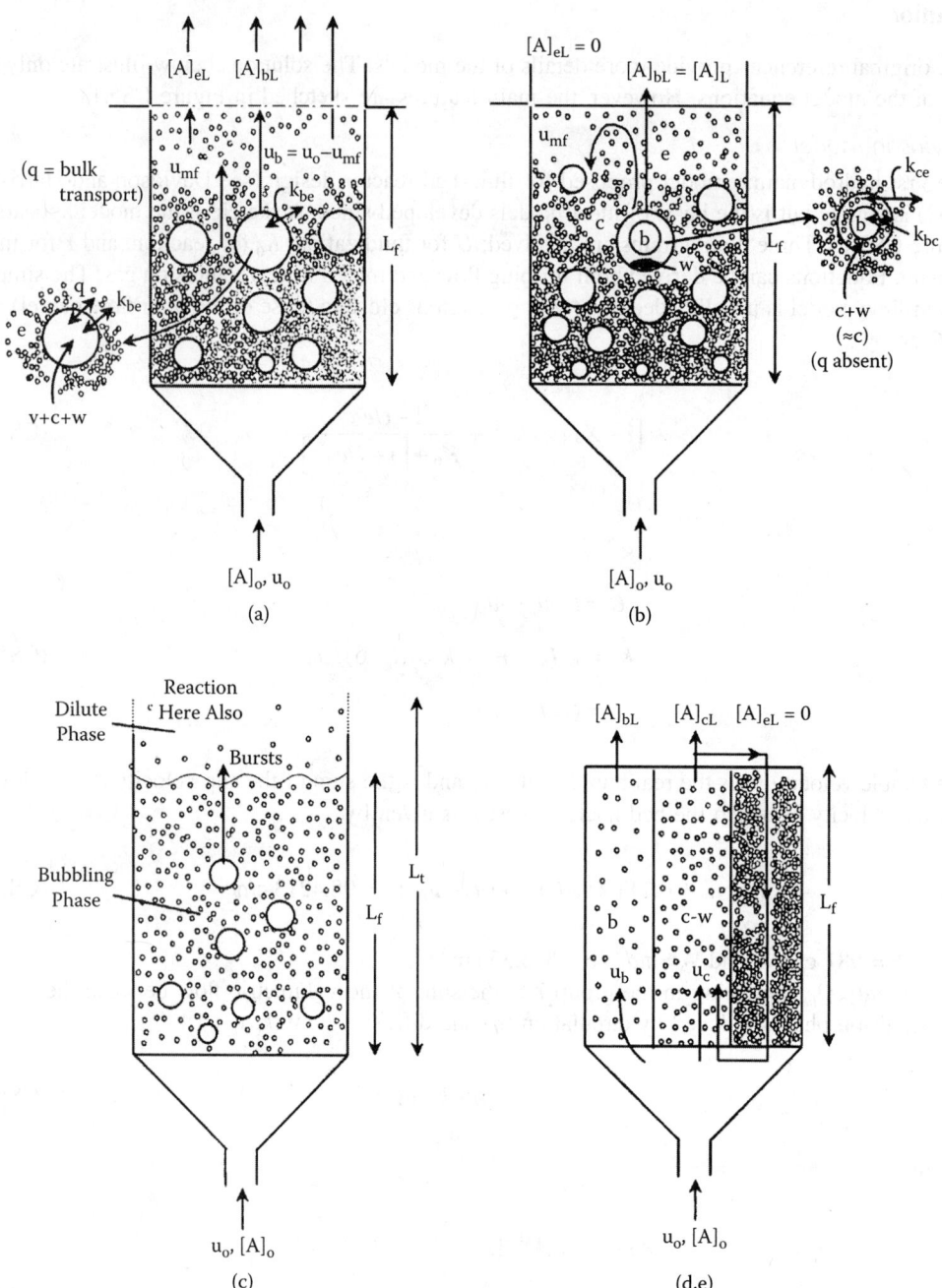

FIGURE CS5.1 Schematics of several fluidized-bed reactor models: (a) Davidson model, (b) Kunii-Levenspiel model, (c) Miyauchi model, (d), (e) Fryer-Potter and Jayaraman-Kulkarni-Doraiswamy models.

Chemical Reaction Engineering

L_f is related to L_0 by the following relationship (Kunii and Levenspiel, 1969):

$$L_f = \left(\frac{1-\varepsilon_B}{1-\varepsilon_{mf}}\right)\frac{L_0 u_b}{u_{br}} \quad \text{(CS5.6)}$$

where

$$u_{br} = 0.711(gd_b)^{1/2} = 70.4213 \text{ cm/s}$$

$$L_f = 290 \text{ cm} \quad \text{(CS5.7)}$$

$$\delta = (u_0 - u_{mf})/u_b = 0.2845$$

The quantities U, K_0, and Y are then easily calculated giving

$$U = 0.9333, \; K_0 = 8.3, \; Y = 7.79$$

Substituting the parameter values calculated above in Equation (CS5.1),

$$(1-X_A) = 0.9333\exp[-18.83] + \frac{1-0.9333\exp(7.79)}{8.2999+[1-0.933]\exp(7.79)} = 0.119$$

giving conversion $X_A = 88.1\%$.

Kunii–Levenspiel (K–L) Model

The main postulates of the K–L model (Kunii and Levenspiel, 1968a, b; 1991) are sketched in Figure CS5.1b. The bubble develops a "cloud" of particles around it as it moves upward at a velocity

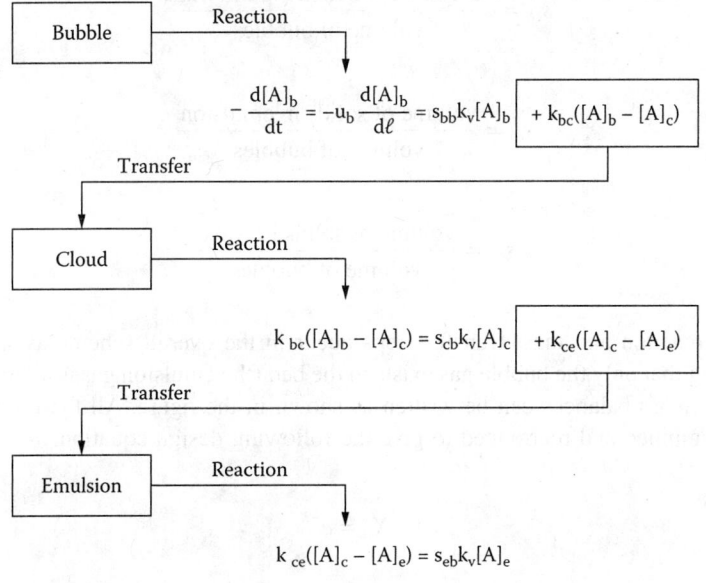

FIGURE CS5.2 Schematic representation of calculations for the Kuni–Levenspiel model. (From Doraiswarmy, L. K., 1991.)

greater than the emulsion gas velocity. The thickness of the cloud is a function of the bubble velocity, which in turn depends on the size of the bubbles and of the fluidizing particles. In a bed of small particles with large bubbles (with corresponding high velocities), a cloud of thickness calculated as

$$\frac{r_c - r_b}{r_b} \cong \frac{u_f}{u_{br}} = \frac{1}{\alpha}$$

$$\frac{r_c}{r_b} = \left(\frac{\alpha + 2}{\alpha - 1}\right)^{1/3} \text{ (more rigorous)}$$

(CS5.8)

develops around each bubble. The bubble gas does not circulate through the rest of the bed. The gas exchange is by diffusive transfer with no bulk circulation (unlike in the Davidson model, which has both). Furthermore, as already mentioned in Section 11.4.2.5.1, as the bubble moves upward, suction is created at the bottom of the bubble; solid particles are drawn into that region and lodged there. This is known as the *wake* and, for purposes of mass transfer, it is assumed to be fully mixed with the cloud-phase components. Because there are now two operative phases in the bubble-cloud-wake unit (bubble and cloud/wake), mass transfer between the emulsion and this unit occurs by two consecutive steps (unlike in the Davidson model, where only one step is involved): between emulsion and cloud, and between cloud and bubble/cloud/wake. For simplicity, the gas and solid particles are assumed to be fully mixed in each of the phases, cloud/wake and bubble. Reaction occurs in all three phases, and the extent of reaction in each phase depends on the solids fraction in that phase. These solid fractions are defined as

$$s_{bb} = \frac{\text{volume of solids in bubbles}}{\text{volume of bubbles}}$$

(CS5.9)

$$s_{cb} = \frac{\text{volume of solids in clouds/wake}}{\text{volume of bubbles}}$$

(CS5.10)

$$s_{eb} = \frac{\text{volume of solids in emulsion}}{\text{volume of bubbles}}$$

(CS5.11)

$$s_{wb} = \frac{\text{volume of solids in wake}}{\text{volume of bubbles}}$$

(CS5.12)

Hence, several reaction and mass transfer steps occur in the overall scheme as shown in Figure CS5.2. Assuming that only the bubble gas exists in the bed (the emulsion gas is assumed to circulate within the bed), mass balances can be written as shown in the figure. All these equations can be algebraically combined and rearranged to give the following design equation:

$$1 - X_A = e^{-K_f}$$

(CS5.13)

where K_f represents a composite constant given by

$$K_f = \frac{L_f k_v}{u_b} \left[s_{bb} + \cfrac{1}{\cfrac{k_v}{k_{bc}} + \cfrac{1}{s_{cb} + \cfrac{1}{\cfrac{k_v}{k_{ce}} + \cfrac{1}{s_{eb}}}}} \right]$$

$$s_{bb} \approx 0.05 \qquad \text{(CS5.14)}$$

$$k_v = 1.2 s^{-1}$$

Values of k_{be} and k_{ce} can be calculated using the Kunii–Levenspiel equations:

$$k_{bc} = 4.5 \left(\frac{u_{mf}}{d_b} \right) + \frac{5.85 g^{1/4} \mathbf{D}^{1/2}}{d_b^{5/4}} = 2.6466 \text{ s}^{-1} \qquad \text{(CS5.15)}$$

$$k_{ce} = \left(\frac{6.78 \varepsilon_{mf} \mathbf{D} u_b}{d_b^3} \right)^{1/2} = 0.60 \text{ s}^{-1} \qquad \text{(CS5.16)}$$

The other terms appearing in Equation (CS5.14) may be estimated as follows:

$$\delta = 0.2845 \text{ (already calculated)}$$

$$s_{cb} = (1 - \varepsilon_{mf}) \left[\frac{3 u_{mf} / \varepsilon_{mf}}{u_b - u_{mf} / \varepsilon_{mf}} + s_{wb} \right] = 0.1916 \qquad \text{(CS5.17)}$$

$$s_{eb} = \frac{(1 - \varepsilon_{mf})(1 - \delta)}{\delta} - s_{cb} - s_{bb} = 0.5678 \qquad \text{(CS5.18)}$$

Thus, from Equation (CS5.14),

$$K_f = 1.51$$

and from Equation (CS5.13),

$$1 - X_A = e^{-K_f} = e^{-3.3149} = 0.0363$$

$$X_A = 0.7790$$

or conversion = 77.90%

Miyauchi Model

As was pointed out earlier (Figure CS5.1c), it is often desirable to account for conversion in the end zones of the fluid bed. The model of Miyauchi (1974) accounts for conversion in the region

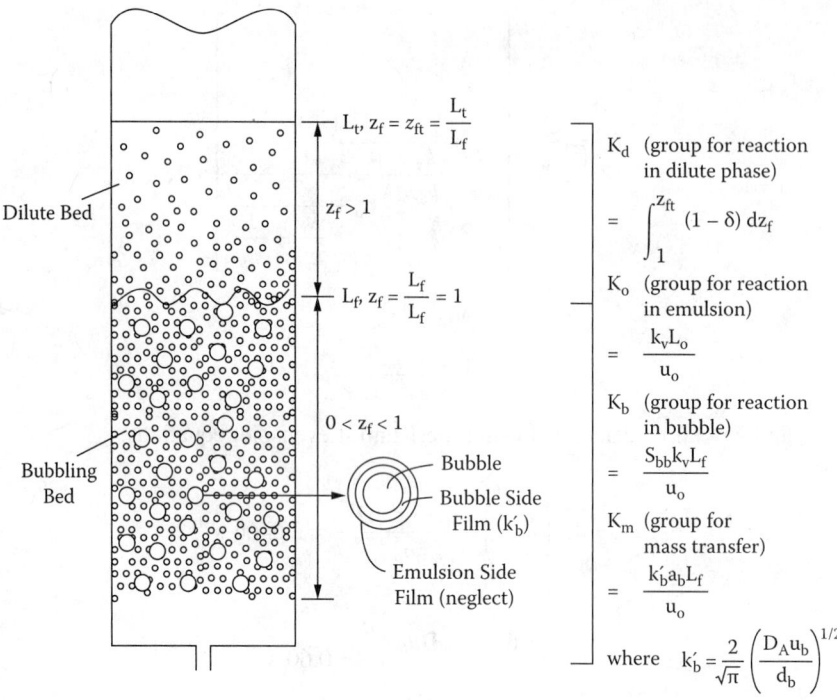

FIGURE CS5.3 Parameters of the Miyauchi model. (From Doraiswarmy, L. K., 1991.)

immediately above the bed. This region is characterized by higher voidage and by an increase of this voidage with height (Lewis et al., 1962; Fan et al., 1962). The final equation is developed in terms of four dimensionless groups: for reaction in the emulsion phase (K_0), for reaction in the bubble phase (K_b), for mass transfer (K_m), and for reaction in the dilute phase above the bubbling bed (K_d) based on experimental voidage distribution data. K_m and K_0 are combined into a single group K'. The various groups and their definitions are given in Figure CS5.3 (Doraiswamy, 2001), and the final equation may be written as

$$(1 - X_A) = \exp[-(K' + K_b + K_d)] \tag{CS5.19}$$

where

$$\frac{1}{K'} = \frac{1}{K_m} + \frac{1}{K_0} \tag{CS5.20}$$

Assigning numerical values,

$$K_m = k_{ob} a_b L_f / u_0 \tag{CS5.21}$$

The value of k_{ob} (the overall coefficient at the bubble-emulsion interface) can be calculated by an elaborate method described by Miyauchi and Marooka (1969) (see also Doraiswamy and Sharma, 1984), giving a value of 0.819 cm/s. Thus,

$$K_m = 4.7502$$

giving

$$K' = 3.0211$$

Now let us calculate K_b and K_d:

$$K_b = s_{bb}\delta k_v L_f / u_0 = 0.165 \qquad (CS5.22)$$

$$K_d = \frac{k_v L_f}{u_0} \int_1^{Z_t} (1-\delta) dZ \qquad (CS5.23)$$

where

$$Z_t = 3/2.9$$

To evaluate the integral in Equation (CS5.23), δ should be known as a function of height between $z = 1$ and $z = z_t$. Unfortunately, neither are available. However, as an approximate solution, we use the average experimental value of $= 0.310$ for this region. Thus,

$$K_d = \frac{1.2 \times 290}{30} \times (1 - 0.310) \times \frac{0.1}{2.9} = 0.276$$

thus

$$[1 - X_A] = \exp[-3.0211 + 0.165 + 0.276]$$

or

$$X_A = 0.923$$

Conversion = 92.3%

Fryer–Potter Model

In the models considered so far, no distinction was made between the velocities of the gas in the bubble, cloud/wake, and emulsion phases. In a model of Fryer and Potter (1972a, b), the following features are postulated:

1. The entire bubble phase is provided by part of the incoming gas.
2. The remaining incoming gas combines with the emulsion gas to provide the gas for the cloud/wake phase, part of which constitutes the exit gas along with the bubble-phase gas, and the rest constitutes the downflowing emulsion gas.

These two postulates, captured in Figure CS5.1d, e, provide the basis for the equations for the three velocities as well as for the boundary conditions for the mass balances for the three phases. The calculations are tedious but quite straightforward. The final conversion obtained is 92.35%.

Jayaraman–Kulkarni–Doraiswamy Model

The Fryer–Potter model assumes plug flow in all the phases, including the emulsion phase. Experimental evidence indicates that the emulsion phase gas is more nearly mixed than in plug flow. If it is fully mixed, the Fryer–Potter model converts from a boundary value problem to an initial value problem (Jayaraman et al., 1981). This leads to considerable simplification in the calculations. All equations other than the emulsion phase equation remain unchanged. All mass balance equations can be readily written and solved (see Jayaraman et al., 1981). The conversion obtained is 88.98%.

Summary of Results

Model	Conversion, %
1. Davidson	88.10
2. Kunii–Levenspiel	77.90
3. Miyauchi	92.35
4. Fryer–Potter	92.35
5. Jayaraman–Kulkarni–Doraiswamy	88.98
Mean	87.93

Other Models: A Brief Discussion

The five models used in the design just presented do not differ greatly in predicting the exit conversion. The average deviation is 2.41% and the maximum is 4.57%. These deviations are not greatly significant, considering the many simplifying assumptions made. The assumption of a single effective bubble size is the most limiting. The values of the wake fraction used and the assumption of complete mixing or plug flow of gas are other serious drawbacks. It is likely that the gas is only partially mixed, which is difficult to accommodate in any design. The assumption of isothermicity is questionable for highly exothermic or endothermic reactions and is inconsistent with the plug-flow assumption, notwithstanding the role of solids mixing in eliminating temperature variations.

Models have been proposed that eliminate the constant bubble size assumption and allow for variation in size. Features of the major models now available have been summarized by Doraiswamy and Sharma, Table 14.3 (1984).

The assumptions discussed above are obvious. Many others are less so and are inherent in almost all fluid-bed models. The most important is with respect to particle size range. Most models do not account for this explicitly, except the K–L model, which gives different equations for fine, intermediate, and coarse particles (Kunii and Levenspiel, 1991). Our calculations for all the five models are for mainly Geldart B class particles and should generally be valid for the so-called fine and intermediate size particles. With the inherent uncertainties of prediction in all the models now available, a finer distinction is not warranted between these two classes of particles.

Except the Miyauchi model, the models considered do not account for conversion in the dilute phase. For really fast reactions with high velocities through the grid plate nozzles, conversion in the jetting region immediately above the grid plate should also be considered (see Figure 11.35). It has to be calculated separately, and it becomes the conversion used as the inlet to the bubbling bed region of any of the models now available. On the other hand, the conversion in the dilute phase is part of the overall calculation in the Miyauchi model considered in our calculations. Kunii and Levenspiel (1991) also suggest a procedure by which this can be included in their model.

An Overall Strategy

Several factors have to be considered in the design of a fluidized-bed reactor. The designs just considered did not include all factors as required for a more rigorous design (see Doraiswamy and Sharma, 1984; Doraiswamy, 2001; and Kunii and Levenspiel, 1999).

Lesson

Fluidized-bed reactor modeling has attracted considerable attention over the years, but there also has been an increasing realization that so many "unpredictables" are involved in the modeling that a realistic design from first principles is difficult. This case study examines five important available models. The chief lesson from this study is that, for a complicated reactor system like the fluidized-bed reactor, none of the models completely accounts for real-life situations. Thus, a fairly simple one, the Kunii–Levenspiel model, is recommended for preliminary designs a detailed design of a commercial reactor should be based on data from a pilot plant reactor about 10% the size of the proposed commercial unit.

CASE STUDY 11.6 GAS-SOLID NONCATALYTIC REACTION: DEVELOPMENT OF A SOLID SORBENT FOR CLEANING COAL GAS FOLLOWED BY MODELING OF THE REACTION AND A CONCEPTUAL REACTOR DESIGN

Gasification of coal with steam and limited oxygen is a common method of removing the sulfur in coal. The reaction converts sulfur to hydrogen sulfide (H_2S) and some carbonyl sulfide (COS). The concentration of H_2S ranges from 0.1 to 3.0%, depending on the type of coal. To use this gas for electric power generation, preferably by the new *integrated gasification combined cycle (IGCC)* process, the H_2S has to be removed at the high temperatures and pressures existing in the gas turbines (unlike the less-efficient conventional process, in which the gasification is at atmospheric pressure). A solid adsorbent removes both H_2S and COS.

Calcium-based sorbents are highly effective (e.g., Squires et al., 1971; Westmoreland and Harrison, 1976; Heesink and van Swaaij, 1995; Fenouil and Lynn, 1995a, b; Yrjas et al., 1996; Zevenhoeven et al., 1996). The sulfurous gas is removed according to the reaction

$$CaO(s) + H_2S(g) = CaS(s) + H_2O(g) \tag{CS6.R1}$$

Other metal oxides have also been used, such as zinc oxide (see, e.g., Uysal et al., 1988; Khare et al., 1995).

The main requirements of a commercially viable sorbent are as follows: it should be inexpensive, be able to operate at the gasifier outlet conditions, be highly reactive with H_2S, be regenerable for at least 100 cycles involving sulfidation (to remove H_2S) and oxidation (to regenerate the original oxide), and possess high enough strength to withstand these repeated cycles. Wheelock and cowokers (Wheelock and Akiti, 2000; Jagtap and Wheelock, 1996; Akiti et al., 2001, 2002a, b) have developed sorbents that meet these requirements, including a cyclic oxidation-reduction scheme for generation. They have also proposed a model for the sulfidation step as described in this case study.

Sorbent Development

The pelletization of the calcium-based sorbent particles is often the cheapest and the most practical approach. However, to produce a strong pellet, it is necessary to mix a binding material with the active solid. The chief drawback is that the solid matrix formed tends to crack with repeated reaction and regeneration cycles, since there is a volume change when CaO is converted to CaS and other intermediates that may form (e.g., $CaSO_4$). Furthermore, the impurities present in the binding material may lead to sintering at temperatures lower than the reaction temperature, thus reducing the effectiveness of the sorbent.

To counter this problem, it is necessary to combine the strength of the binder with the reactive capacity of the sorbent in a configuration that will not diminish either. This requires physical separation of the two materials in the pellet and can be accomplished by coating the sorbent material with a protective layer of the binder that forms a tough shell. We call this configuration the *core-in-shell (CIS)* sorbent. Different variations of this basic configuration have been used by varying the core and shell materials and compositions. It has been found that combinations of the two in both the core and shell give pellets of optimum strength and reactivity. Clearly, these combinations should have different compositions.

Pellet formation essentially involves three steps: nucleation (consolidation of several grains into a nucleus), transition (joining of several nuclei to form larger entities), and ball growth (growth

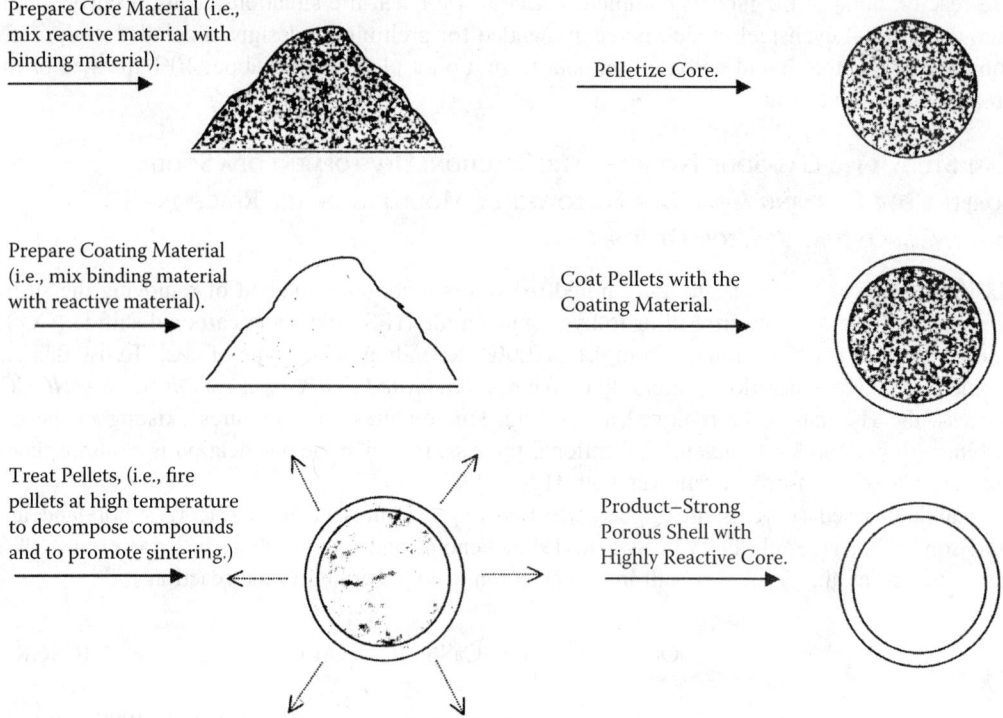

FIGURE CS6.1 A practical protocol for preparing core-in-shell pellets.

of larger entities into the required pellets at the expense of the smaller ones that get crushed). Based on these steps, a protocol for preparing CIS pellets is illustrated in Figure CS6.1.

As mentioned earlier, a viable procedure for pellet regeneration is an integral part of pellet development. Several procedures have been suggested (e.g., Yoo and Steinberg, 1983; Chou and Li, 1984a, b; Illerup et al., 1993; van der Ham et al., 1996; Jagtap and Wheelock, 1996; Wheelock, 1997, 2000). The recommended procedure (Jagtap and Wheelock, 1996; Wheelock, 1997, 2000, 2004) essentially involves a cyclic operation alternating between oxidizing and reducing atmospheres in the reactor. During the oxidation process, CaS is converted to CaO according to the reaction

$$CaS(s) + 3/2 O_2(g) = CaO(s) + SO_2(g) \quad\quad (CS6.R2)$$

However, under the conditions of high temperature (650 to 1000 K) and excess oxygen existing in the reactor, the following reaction also occurs:

$$CaS(s) + 2O_2(g) = CaSO_4(s) \quad\quad (CS6.R3)$$

which is unfortunately the favored reaction, both thermodynamically and kinetically. The larger volume of the $CaSO_4$ causes pore plugging, eventually leading to an impenetrable layer around the reactive solid (CaO). However, the $CaSO_4$ can be converted back to CaO by treatment with CO according to the reaction

$$CaSO_4(s) + CO(g) = CaO(s) + CO_2(g) + SO_2(g) \quad\quad (CS6.R4)$$

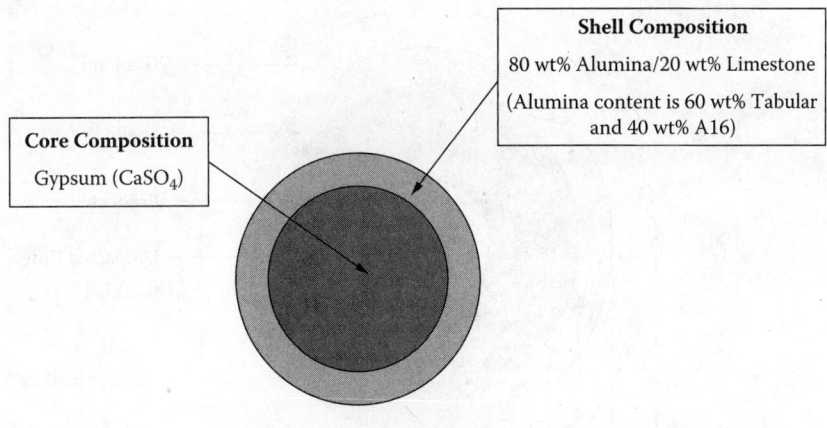

FIGURE CS6.2 Compositions of core and shell in the recommended CIS pellets.

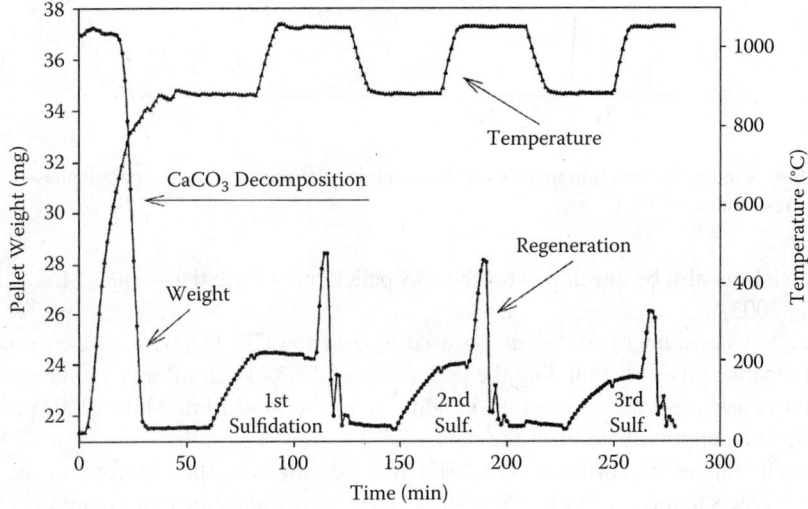

FIGURE CS6.3 Reaction-regeneration cycles involved in the overall reaction scheme.

Thus, regeneration is accomplished by alternating between oxidizing conditions [when reactions (CS6.R2) and (CS6.R3)] occur) and reducing conditions [when reaction (CS6.R4) occurs], leading to complete conversion of CaS to CaO.

Several shell materials and two core materials as well as different compositions of the shell and core were tested by Akiti (2001) and Akiti et al. (2001, 2002a,b). Although more research is still needed, the pellet described in Figure CS6.2 gives acceptable results. Using this pellet, the complete reaction-regeneration cycle involved in the overall reaction scheme is shown in Figure CS6.3.

Modeling of the Sulfidation Step (i.e., the Main Reaction)

The more important basic gas-solid reaction models and their extensions are described in Section 11.3.2. Heesink and van Swaaij (1995) proposed a particle-pellet model for reaction between calcined limestone and H_2S at temperatures ranging from 500 to 700°C using a pellet of conventional

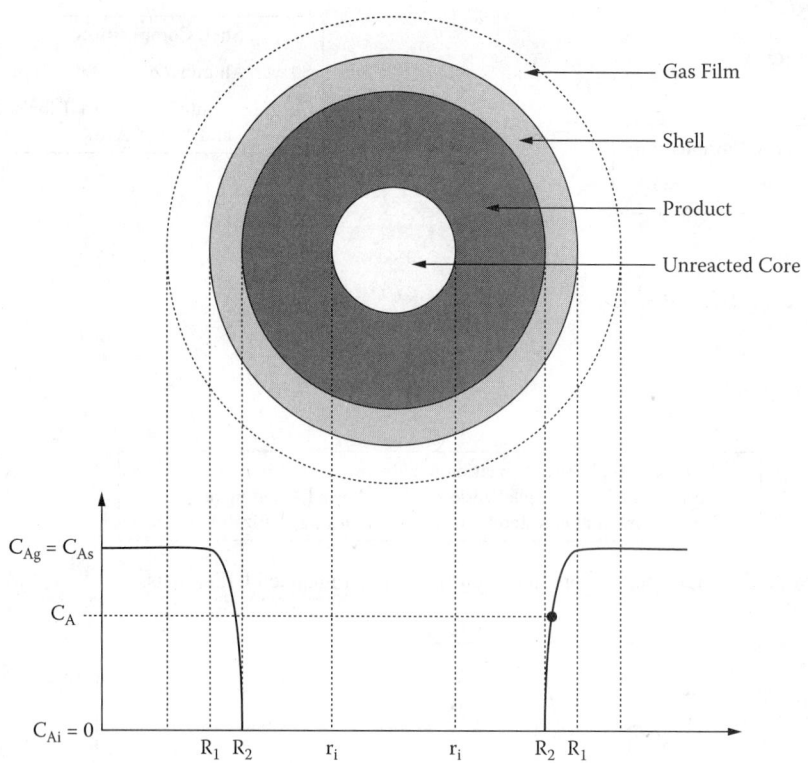

FIGURE CS6.4 A model for reaction in a CIS pellet assuming the shell acts as a separate phase offering pure mass transfer resistance.

design. A model can also be developed for the CIS pellet, but we use the simpler SIM (Akiti, 2001; Hasler et al., 2003).

Two simple extensions of SCM can be used to describe CIS behavior. The external shell is treated as a separate phase surrounding the pellet (Figure CS6.4) that offers one more step of pure mass transfer resistance with no reaction in addition to the fluid film. Akiti. (2001) used such a model to obtain an approximate fit of the data.

More recent studies by Hasler et al. (2003) showed, however, that the lime in the shell (see Figure CS6.2) reacts to the extent of 25% to 40%. The photograph shown in Figure CS6.5 suggests that one of the zone models described in Section 11.3.2.1.3 better describes the observed behavior. Since the zone thickness exhibits rapidly decreasing product density toward the core, one can, as a first approximation, assume SCM behavior with the ash and shell layers treated as separate layers as proposed by Akiti (2001). The outer shell acts as a "solid film." Although this shell contains some lime that undergoes reaction, the amount is so small relative to that in the core that it can be neglected. Diffusion through the two layers (ash and shell) is the controlling resistance, with a small contribution from the fluid film surrounding the pellet. Further studies are needed, however, to verify the zone model (or the particle-pellet model).

Reactor Design

In Section 11.4.2, we discussed the three main categories of reactors that can be used for gas-solid reactions. Since the present system uses pellets of reasonable size, fluid-bed reactors can be ruled out. We shall therefore consider fixed- and moving-bed reactors. Two features of this system should first be noted:

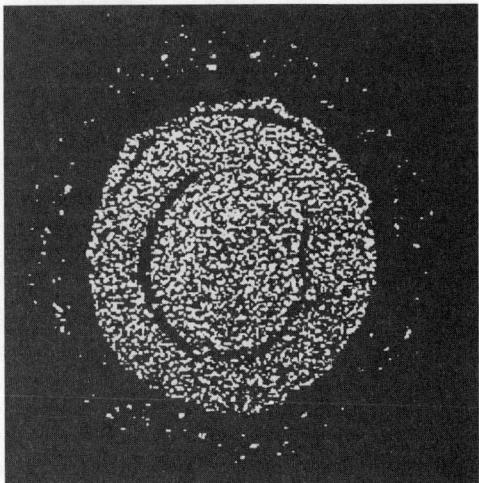

FIGURE CS6.5 Cross-sectional view of a reacted CIS pellet.

1. The reaction involves cyclic operation alternating between sulfidation, in which the absorbent (a metal oxide) removes the sulfur from the feed stream and oxidation, in which the sulfided adsorbent is oxidized to regenerate the original oxide of the adsorbent.
2. The dust present in the flue gas must be removed in the sulfidation unit. Dust can be removed only in fixed or moving beds of pellets by allowing it to accumulate in the interstitial space. This is not possible in a fluidized bed, providing one more reason why this bed is unacceptable.

Strictly, the total reaction scheme consists of three steps, one for sulfidation and two for regeneration. In other words, the total reaction scheme consists of a reaction-regeneration cycle into which a secondary regeneration cycle is embedded, consisting of alternating steps of oxidation and reduction (Figure CS6.5). Any conceptualization of a reactor scheme must take these reactions into account. Two alternatives are possible:

1. Use two reactors in which the main reaction (sulfidation) and regeneration (reduction and oxidation) are performed separately.
2. Use three reactors, one each for sulfidation, oxidation, and reduction.

Scheme 1 might appear to be the cheaper, but it is difficult to design a single reactor for regeneration involving periodic switching between oxidizing and reducing environments. Hence, a three-reactor system commends itself as the better alternative.

Each of the reactors in the three-reactor system can be a fixed- or moving-bed reactor. The fixed-bed reactor scheme for regeneration suffers from the need to subject each reactor to alternating environments, one for oxidation and the other for reduction. While one reactor is accomplishing oxidation, the other would be engaged in reduction. A very close control strategy is required to operate this scheme successfully. Even so, the residual environment of one reaction before the second reaction is allowed to occur in a given reactor can vitiate the results. These are powerful arguments against the fixed bed. Thus, purely from a qualitative discussion of the alternatives, the moving-bed reactor appears best.

A Practical Moving-Bed Design

Because dust removal is one of the main requirements of a good operating system, a special design proposed by Westinghouse (Yang et al., 1992), the *standleg moving granular bed filter,* appears to

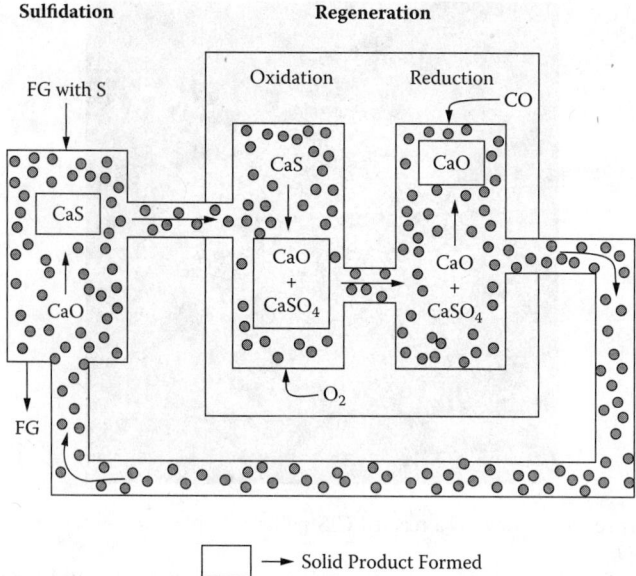

FIGURE CS6.6 A reaction-regeneration cycle into which a secondary regeneration cycle is embedded consisting of alternate cycles of oxidation and reduction.

be the best basis for designing the sulfidation reactor. A modification of the Westinghouse design was proposed by Colver et al. (2002), in which the gas and solids move countercurrently (as against the cocurrent flow in the original Westinghouse design); this modification is recommended here. This reactor differs from the conventional moving-bed reactor mainly with respect to the details of design, as brought out in Figures CS6.7a and b. The design equations given in Section 11.4.2.5.2

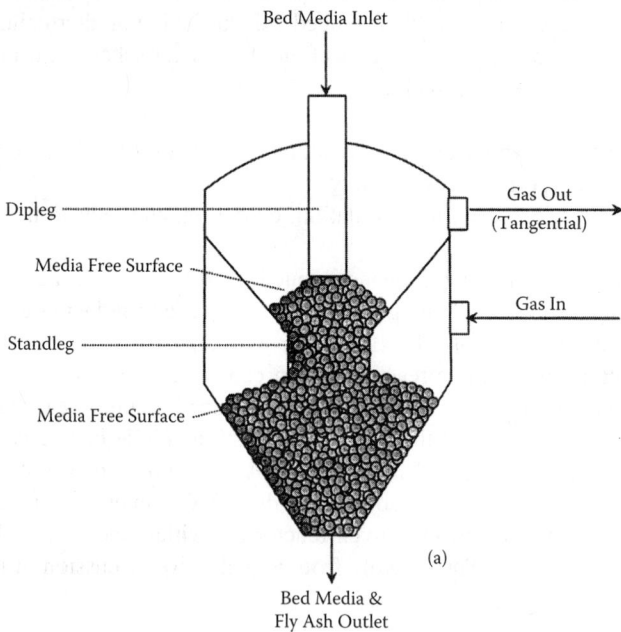

FIGURE CS6.7 Moving-bed reactor designs for the system at hand: (a) conventional design; and (b) a special design proposed by Colver et al. (2002). *Continued.*

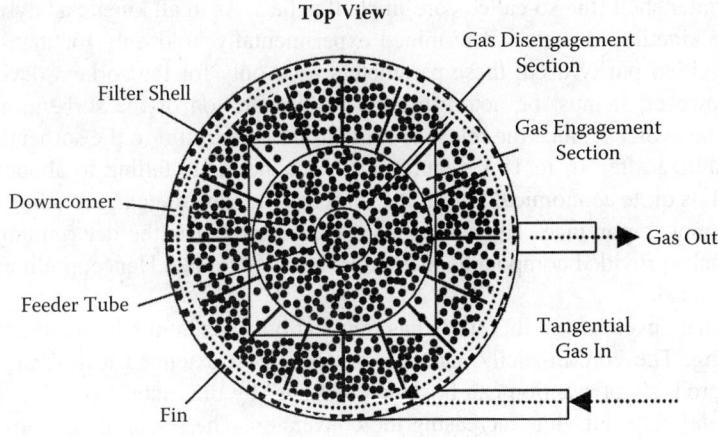

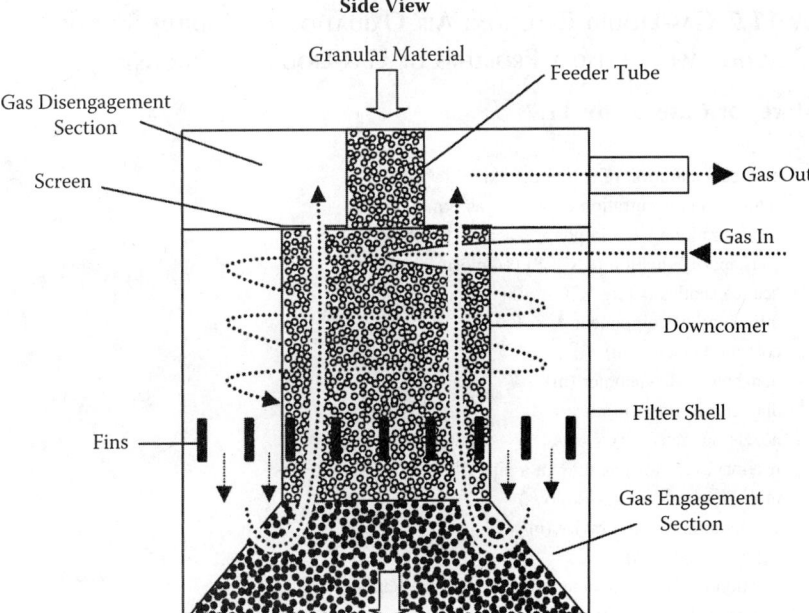

FIGURE CS6.7 *Continued.*

are equally applicable to this reactor, for both the cocurrent and countercurrent configurations. The reactors for the regeneration step can be of the conventional moving-bed type.

With insufficient data to design the reactor, we suggest this design be used only for the sulfidation step in the overall operating scheme shown in Figure CS6.6. It is necessary that a dust collector (not shown in the figure) be used to ensure dust-free feed to the regeneration system.

Lesson

The use of coal for producing electric energy has always been of importance. Since sulfur can be removed, there has been increased attention recently to developing solid sorbents to remove the sulfur as hydrogen sulfide and some sulfur carbonyls. Among the many solids tried, lime has been one of the most widely used. Because of its poor crushing strength, it has a tendency to crumble. The present case study describes the development of a pellet in which the softer calcium salt is encased in a hard outer shell (the so-called core-in-shell pellet). As in all kinetic analyses involving solid catalysts, the kinetic parameters determined experimentally hold only for that catalyst and for the feed of specified purity used; these parameters hold only for the sorbent developed for a particular coal. However, it must be noted that, in the preparation of the sorbent, any chloride compound should be avoided, since the released HCl would be harmful to the sorbent's life. CaO-based sorbents stabilize after 16 to 18 regenerations, with the value falling to about 60% of the original. Even so, it is quite economical to use this sorbent. The type of coal, including the maceral impurities, is of minor importance. The composition of the sorbent is the determining factor and works for many coals, provided compounds like chlorides are avoided. Hence, preliminary testing is always recommended.

The kind of pellet developed in this study has implications outside the limits of its present use in coal gas cleaning. The core-in-shell concept a priori can be extended to reversible reactions where one of the products of reaction can be "adsorbed out" by the inner core while the reaction continues in the catalytic shell, thus increasing the conversion. The lesson is that novel solutions can be found for challenging problems and, equally important, they can often find applications in a variety of other situations, if only one keeps looking for possibilities.

CASE STUDY 11.7 GAS-LIQUID REACTION: AIR OXIDATION OF SODIUM SULFIDE (A SIMPLE REACTION WITH TYPICAL PROBLEMS OF GAS-LIQUID REACTIONS)

Nomenclature for Case Study 11.7

$[A]$	bulk concentration of solute gas (kmol/m^3)
$[A^*]$	saturation concentration of solute gas (kmol/m^3)
$[B]$	concentration of liquid phase reactant (kmol/m^3)
C_1	constant in Equation (CS7.38) (1/kmol)
C_P, C_V	heat capacities (kJ/kg °C)
$\mathbf{D}_A$	diffusivity of component A (m^2/s)
D	column diameter (m)
D_D	diameter of downcomer (m)
D_R	diameter of riser (m)
d_N	nozzle diameter (m)
f_G	friction factor for gas flow in a pipe (–)
G	molar flow rate of gas (kmol/s)
g	acceleration due to gravity (m/s^2)
H	column height (m)
H_C	location of the downcomer sparger from the bottom (m)
H_D	height of dispersion (m)
H_W	Henry's coefficient for aqueous Na$_2$S solution (kmol/m^3 Pa)
h	height coordinate (m)
k_i	rate constant, $I = 1, 2, \ldots$ (conc^{1-n}/time)
k_L	true liquid-side mass transfer coefficient (m/s)
$k_L a$	volumetric liquid-side mass transfer coefficient (1/s)
L	volumetric liquid flow rate (m^3/s)
Lc	length of the contactor (m)
N_N	number of nozzles
P	power consumption (W)
p_{O_2}	oxygen partial pressure (Pa)
P	pressure (Pa)
$\bar{P}_T$	average pressure in the column (Pa)
R_A	specific rate of absorption of A (kmol/m^2 s)
$(Re)_G$	gas-phase Reynolds number (–)
$(Re)_L$	liquid-phase Reynolds number (–)
r_A	volumetric rate of absorption of A (kmol/m^3 s)

S	cross sectional area (m²)
T	temperature (°C)
u_G	gas velocity (m/s)
u_L	liquid velocity (m/s)
u_N	nozzle velocity (m/s)
u_S	slip velocity (m/s)
V_D	volume of dispersion (m³)
y	mole fraction of solute gas in gas phase (–)
Z	stoichiometric coefficient for the reaction between oxygen and sodium sulfide
z	length coordinate (m)
z_t	length of pipeline reactor (m)

Greek

α'	term defined in Equation (CS7.48) (m³/kmol)
β	term defined in Equation (CS7.9) (1/m)
β^*	term defined in Equation (CS7.57) (kmol$^{1/2}$/m$^{5/2}$)
δ	term defined in Equation (CS7.29) (m⁶/kmol)
δ'	term defined in Equation (CS7.40) (kmol/sec)
δ_1	term defined in Equation (CS7.58) (kmol$^{1/2}$/m$^{3/2}$)
η	term defined in Equation (CS7.58) (kmol$^{1/2}$/m$^{3/2}$)
ε_L	fractional liquid holdup (–)
ε_{GR}	fractional gas holdup in the riser (–)
ε_{GD}	fractional gas holdup in the downcomer (–)
ε_{LR}	fractional liquid holdup in the riser (–)
ε_{LD}	fractional liquid holdup in the downcomer (–)
Ψ	constant defined by Equation (CS7.7) (1/m)
Γ	term defined in Equation (CS7.16) (kmol/m³)
λ_1	term defined in Equation (CS7.19) (m³/kmol)
λ_2	term defined in Equation (CS7.20) (–)
μ_w	viscosity of water (Pa·s)
γ	term defined in Equation (CS7.49) (–)
ν	term given in Equation (CS7.12) (–)
ν^*	term defined in Equation (CS7.31) (–)
ν'	term defined in Equation (CS7.44) (kmol)
ν''	term defined in Equation (CS7.45) (kmol)

Subscript

b	bulk phase
B	bottom
e	exit value for continuous mode of operation
H	horizontal sparged reactor
n	stage number (Equation CS 7.15)
T	top
V	vertical sparged bubble column
0	inlet/initial value for continuous mode of operation

The Problem

Air oxidation of aqueous sodium sulfide solution is encountered in the pulp and paper industry where the black liquor, which contains sodium sulfide (Na_2S), is oxidized. In a typical process, 200 tons/day of an aqueous solution of sodium sulfide are oxidized with air at 75°C. The concentration of Na_2S is to be reduced from 12 to 0.5 kg/m³. The following contactors operated continuously are to be considered: (1) bubble column, (2) sectionalized bubble column, (3) external loop air-lift reactor, (4) horizontal sparged contactor, (5) pipeline contactor, and (6) packed column. The operating conditions for each contactor are given at the appropriate places. The black liquor contains additional dissolved solids, which consume O_2. Its properties are significantly different from those of an aqueous sodium sulfide solution. However, for illustration, we shall consider the oxidation of an aqueous sodium sulfide solution according to the following stoichiometric equation:

$$2O_2 + 2Na_2S + H_2O \rightarrow Na_2SO_3 + 2NaOH + S \qquad \text{(CS7.R1)}$$

Data

The reaction is first order in both oxygen and sodium sulfide, rate constant k_2 at 75°C = 4.64 m³/kmol s (there is some doubt about this constant, as it is dependent on several metal impurities; however, for the sake of illustration, this value is reasonably reliable), Henry's constant for O_2 in water at 75°C = 7.2×10^{-9} kmol/m³Pa, Henry's constant for O_2 in Na_2S solution $H_W = 6.42 \times 10^{-9}$ kmol/m³ Pa (at the average concentration), density of the solution $\rho_L = 1000$ kg/m³, vapor pressure of water at 75°C = 3.87×10^4 Pa (290 mm Hg), diffusivity of O_2 in Na_2S solution at 75°C = 6.8×10^{-9} m²/s.

1. Bubble Column (BC) (Countercurrent Operation)

Assumptions

The liquid phase is completely backmixed, and the gas phase moves in plug flow. The superficial velocity of the gas at the top of the column will be fixed at 5×10^{-2} m/s. It may be further assumed that the value of $k_L a$ in the column is 0.04 s⁻¹ and that of k_L is 2×10^{-4} m/s. Fractional liquid holdup in the bubble column = 0.8. The values of ε_L, k_L, and a actually depend on superficial gas velocity (u_G), column diameter (D), column height (H), sparger design, the nature of the gas-liquid system, and operating pressure and temperature. At relatively low pressures (<0.5 MPa), it is known (Deckwer, 1992; Joshi et al., 1998) that the values of ε_L, k_L, and a are independent of column diameter when D exceeds 0.15 (refer to Section 11.4.2.1.1). Furthermore, these are independent of column height and sparger design when H/D exceeds a certain critical value (3 for highly coalescing to 10 for highly noncoalescing systems). However, quantification of the nature of the gas-liquid system (and its variation with temperature and pressure) is not currently possible. Therefore, it is desirable to measure k_L and a in a small-scale bubble column ($D > 150$ mm, $10 > H/D > 3$) using the same system and operating conditions (T and P). Such measurements are preferably made at least at four levels of u_G and three of T and P in the vicinity of the proposed operating conditions. In the present case study, it is assumed that the experiments were performed as indicated above and that, at the operating condition of u_G, T, and P under consideration, the values of ε_L, k_L, and a are 0.8, 2×10^{-4} m/s, and 200 m²/m³, respectively.

Solution

The rate of oxygen absorption in aqueous Na_2S is assumed to be

$$R_A a = k_L a ([A]^* - [A]_b) \qquad \text{(CS7.1)}$$

The conditions to be satisfied are

$$k_L a \approx \varepsilon_L k_2 [B]_b \qquad \text{(CS7.2)}$$

$$\sqrt{\frac{D_A k_2 [B]_b}{k_L}} \ll 1 \qquad \text{(CS7.3)}$$

The material balance for A over a differential height dH_D of the column gives

$$G\, dy = R_A a\, S\, dH_D \tag{CS7.4}$$

Substitution of $r_A a$ from Equation (CS7.1) in Equation (CS7.4) gives

$$G\, dy = k_L a([A]^* - [A]_b) S\, dH_D \tag{CS7.5}$$

$[A^*]$ is given by the following equation:

$$[A]^* = H_W P_T (1 + \psi H_D) y \tag{CS7.6}$$

where

$$\psi = \frac{1}{P_T}\frac{dP}{dH_D} \tag{CS7.7}$$

Substitution of $[A^*]$ from Equation (CS7.6) in Equation (CS7.5) and subsequent rearrangement gives

$$\frac{dy}{dH_D} = \beta(1+\psi H_D)y - \left[\frac{k_L a [A]_b S}{G}\right] \tag{CS7.8}$$

where

$$\beta = \frac{k_L a H_W P_T S}{G} \tag{CS7.9}$$

$[A]_b$ can be obtained from the overall material balance

$$G(y_0 - y_e) = \varepsilon_L k_2 [A]_b [B]_e V_D \tag{CS7.10}$$

where V_D is the total volume of dispersion (reactor). Elimination of $[A]_b$ in Equations (CS7.5) and (CS7.10) gives

$$\frac{dy}{dH_D} = \beta(1+\psi H_D)y - \frac{\nu}{H_D} \tag{CS7.11}$$

where

$$\nu = \frac{k_L a(y_0 - y_e)}{\varepsilon_L k_2 [B]_e} \tag{CS7.12}$$

Equation (CS7.11) on integration gives

$$y_0 \exp\left[-\frac{\beta}{2\psi}(1+\psi H_D)^2\right] - y_e \exp\left[-\frac{\beta}{2\psi}\right]$$
$$+ \frac{v}{H_D}\sqrt{\frac{\pi}{2\beta\psi}}\left[erf\left(\sqrt{\frac{\beta}{2\psi}}(1+\psi H_D)\right) - erf\left(\sqrt{\frac{\beta}{2\psi}}\right)\right] = 0 \qquad \text{(CS7.13)}$$

Here we have $L = 2.32 \times 10^{-3}$ m³/sec, $[B]_0 = 0.154$ kmol/m³, and $[B]_e = 6.4 \times 10^{-3}$ kmol/m³.

The amount of oxygen provided is 200% in excess of the stoichiometric requirement. We also assume that G remains constant. Therefore, $y_e = 2/3 \times 0.21 = 0.14$.

The overall material balance can be written as

$$G(y_0 - y_e) = \frac{L}{Z}\left([B]_0 - [B]_e\right) \qquad \text{(CS7.14)}$$

Substituting the relevant values, we determine $G = 4.89 \times 10^{-3}$ kmol/s. From the values

$$\varepsilon_L k_2 [B]_e = 0.0238 \text{ s}^{-1},\ k_L a = 0.04 \text{ s}^{-1}, \text{ and } \sqrt{\frac{D_A k_2 [B]_e}{k_L}} = 0.071$$

we note that the condition given by Equations (CS7.2) and (CS7.3) is satisfied. The superficial gas velocity at the top of the column = 5×10^{-2} m/s.

The gas phase also contains water vapor. It is assumed that relatively small amounts of Na_2S do not alter the vapor pressure characteristics of water. Furthermore, the increase in the concentration of Na_2S due to the evaporation of water may be neglected, as the rate of evaporation of water is about 1.75% of the liquid flow rate.

Therefore, the overall volumetric gas flow rate at the outlet is given by

$$4.89 \times (760/470) \times (348/273) \times 22400 = 22.59 \times 10^4 \text{ cm}^3/\text{s} = 0.2259 \text{ m}^3/\text{s}$$

Hence, the cross-sectional area of the bubble column, $S = (0.2259/5 \times 10^{-2}) = 4.52$ m². Therefore, $D = 2.4$ m.

We thus have

$$P_T = \frac{760 - 290}{760} = 0.618 \text{ atm} = 6.26 \times 10^4 \text{ Pa},\ \beta = \frac{k_L a\, H_W P_T S}{G} = 1.49 \times 10^{-2}$$

$$\psi = \frac{\varepsilon_L \rho_L g}{P_T} = 0.125,\ v = \frac{k_L a\, (y_0 - y_e)}{\varepsilon_L k_2 [B]_e} = 0.118$$

After substituting the relevant values, Equation (CS7.13) was solved by trial and error to give $H_D = 29.1$ m.

Due to the foaming nature of the black liquor, the value of $k_L a$ is likely to be higher than 0.04 s⁻¹ at a u_G of 5×10^{-2} m/s. The problem is reworked for a value of $k_L a$ equal to 0.12 s⁻¹ at $u_G = 5 \times 10^2$ m/s and the value of H_D found to be 30.0 m. The height of the column changes little with such a large correction of $k_L a$. When $H_D = 29.1$ m, $\varepsilon_L = 0.8$, and $\rho_L = 1000$ kg/m³, the total pressure at the bottom (P_B) is 2.28×10^5 Pa. As a result, the oxygen partial pressure at the bottom $(p_{O_2})_B =$

TABLE CS7.1
Effect of $k_L a$ on the Height of Bubble Column (Under Conditions Such That No Desorption Occurs in Any Part of the Column)

$k_L a$ (at $u_c = 5$ cm/s) s^{-1}	Cross-Sectional Area cm^2	Diameter m	Height m
0.04	1.64×10^4	4.43	16.0
0.0616	2.08×10^5	5.15	12.35
0.127	3.28×10^5	6.46	7.24

TABLE CS7.2
Effect of Pressure on the Height of Bubble Column

Total Pressure at the Top atm	Height cm	Pressure at the Bottom atm
1	2860	3.2
2	1530	3.18
3	1100	3.85
4	872	4.68

$0.21 \, (P_B - 3.87 \times 10^4) = 3.98 \times 10^4$ Pa, whereas $(p_{O_2})_T = 8.76 \times 10^3$ Pa. It can be seen that the partial pressure of oxygen (and hence the saturation concentration $[A^*]$) varies by a large factor from bottom to top. In contrast, the concentration of dissolved oxygen $[A]_b$ is uniform throughout the reactor because of the complete backmixed behavior of the liquid phase. As a consequence, the driving force $([A^*] - [A]_b)$, which is positive at the bottom, may reverse its sign in the top region. Because of the large variation in p_{O_2}, absorption may occur in the bottom region and desorption in the top, as discussed by Parulekar et al. (1988).

Effect of Suspended Impurities
In practice, depending on the quality of wood, the black liquor contains dissolved and suspended impurities. These impurities in turn have a marked influence on the mass transfer coefficient. This problem was discussed earlier, and the methodology for the estimation of $k_L a$ has been given. To understand the parametric sensitivity of $k_L a$, the example has been reworked for two higher values of $k_L a$, and the results are given in Table CS7.1.

Operation of Bubble Column under Pressure
Because of desorption in the upper part of the column, it may be desirable to operate the column under pressure. For different total pressures (at the top), the problem has been reworked, and the results are reported in Table CS7.2.

Based on pressures at the bottom, it can be seen that an increase in the pressure at the top of the column from 1.0133×10^5 to 3.04×10^5 Pa gives nominal variation in the pressure at the bottom. Therefore, the operating cost would remain practically the same. However, there is a substantial reduction in the height of the column. An increase in pressure at the top from 1.0133×10^5 to 2.03×10^5 Pa increases the partial pressure of O_2 at the top from 1.32×10^4 to 3.44×10^4 Pa.

2. Sectionalized Bubble Column (SBC)
Since the extent of backmixing in the liquid phase is substantial, it is desirable to consider the sectionalization of the column (a packed bubble column has not been considered, since the black

liquor usually contains considerable suspended impurities). The sectionalized bubble column is shown schematically in Figure 11.21c.

For the nth section of the sectionalized bubble column, the relationship between y_n and y_{n+1} can be obtained from Equation (CS7.13) as

$$y_{n+1}\exp\left[-\frac{\beta}{2\psi}(1+\psi H_D)^2\right] - y_n\exp\left(-\frac{\beta}{2\psi}\right) + \Gamma\left(\frac{y_{n+1}-y_n}{[B]_{b_n}}\right) = 0 \quad \text{(CS7.15)}$$

where

$$\Gamma = \frac{k_L a}{\varepsilon_L k_2 H_D}\sqrt{\frac{\pi}{2\beta\psi}}\left[erf\left(\sqrt{\frac{\beta}{2\psi}}(1+\psi H_D)\right) - erf\left(\sqrt{\frac{\beta}{2\psi}}\right)\right] \quad \text{(CS7.16)}$$

Material balance for the nth section gives

$$y_{n+1} - y_n = \frac{L}{zG}\left([B]_{b_{n-1}} - [B]_{b_n}\right) \quad \text{(CS7.17)}$$

Substitution of y_{n+1} from Equation (CS7.17) into Equation (CS7.15) and subsequent rearrangement gives

$$\lambda_1[B]_{b_n}^2 - [B]_{b_n}\left(\lambda_1[B]_{b_{n-1}} + \lambda_2 y_n - 1\right) - [B]_{b_{n-1}} = 0 \quad \text{(CS7.18)}$$

where

$$\lambda_1 = \frac{1}{\Gamma}\exp\left(-\frac{\beta}{2\psi}(1+\psi H_D)^2\right) \quad \text{(CS7.19)}$$

$$\lambda_2 = \frac{ZG}{L\Gamma}\left(\exp\left[-\frac{\beta}{2\psi}(1+\psi H_D)^2\right] - \exp\left(-\frac{\beta}{2\psi}\right)\right) \quad \text{(CS7.20)}$$

Equation (CS7.18) can be solved for $[B]_{b_n}$ to give

$$[B]_{b_n} = \frac{\left(\lambda_1[B]_{b_{n-1}} + \lambda_2 y_n - 1\right) + \sqrt{(\lambda_1[B]_{b_{n-1}} + \lambda_2 y_n - 1)^2 + 4\lambda_1[B]_{b_{n-1}}}}{2\lambda_1} \quad \text{(CS7.21)}$$

Procedure for the Design of a Sectionalized Bubble Column

1. Calculate the area of cross section of the column from a consideration of superficial gas velocity.
2. Choose a suitable height for the section.

TABLE CS7.3
Comparison of Different Reactors

No.	Reactor	Diameter m	Height M	Volume m³	Pressure at Entrance MP	Power Consumption kW
1	Bubble column	2.31	30.35	127	3.35	20.5
2	Sectionalized bubble column	2.31	19.5	81.7	2.51	14.9
3	External-loop airlift reactor	$D_R = D_D = 1.88$	15.3	84.9	1.96	10.5
4	Horizontal sparged reactor	5.0	24.2 (length)	475	1.33	3.84
5	Pipeline reactor (pure O_2, 20% excess, operating pressure 3 atm.)	0.194	168	5.0	3.04	7.42
6	Packed column	1.69	407	913	Essentially atmospheric	—

3. Start from the top of the column. For section (1), $y_n = y_e$, $[B]_{b_{n-1}} = [B]_{b_o}$, $n = 1$, for $[B]_{b_0}$ from Equation (CS7.21).
4. Calculate y_2 from Equation (CS7.17).
5. Repeat the calculation for the next section.
6. Continue the calculation until $[B]_{b_n} = [B]_{b_e}$. If $[B]_{b_n} < [B]_{b_e}$, the height of last section can be calculated from Equation (CS7.13). If the superficial gas velocity at the top is maintained at 5×10^{-2} m/s, then, as shown earlier, $D = 2.4$ m.

Let the height of each section be 0.5 m. The above procedure gives the total number of sections as 34.64. We may provide 35 complete sections so that the total height is 17.5 m. The sectionalized bubble column is more attractive than a bubble column, because both the volume and the power consumption are approximately 64% of those in the bubble column.

3. External-Loop Air-Lift Reactor (EL-ALR)

As indicated in Section 11.4.2.1.4, the energy requirement of a bubble column can be reduced with an external-loop air-lift reactor (Figure 11.23a). As an illustration, consider case 2 in Table CS7.2. This case was shown to be optimum on the basis of bottom pressure or the pressure at which the gas phase enters the column. Furthermore, there is no complication of desorption.

Let us assume the following: height of reactor = 15.3 m, slip velocity = 0.15 m/s, and average superficial gas velocity at the column middle = 0.05 m/s.

The expressions for the slip velocity in the riser and downcomer region of the column are

$$\frac{u_G}{\varepsilon_{GR}} - \frac{u_L}{\varepsilon_{LR}} = u_S \quad \text{(CS7.22)}$$

$$-\frac{u_G}{\varepsilon_{GD}} - \frac{u_L}{\varepsilon_{LD}} = u_S \quad \text{(CS7.23)}$$

where ε_{LR}, ε_{LD}, ε_{GD}, and ε_{GR} are the fractional holdups of liquid (L) and gas (G) phases in the riser (R) and downcomer (D) of the loop reactor. The energy balance over the column can be used to obtain the average liquid circulation velocity (refer to Section 11.4.2.1.4):

$$u_L = [2(\varepsilon_{GR}H_D - \varepsilon_{GD}H_C)g]^{0.5} \qquad \text{(CS7.24)}$$

Solving the above equations simultaneously yields the values of fractional gas holdup for the riser as well as the downcomer, the liquid velocity, and the height of critical location of the sparger from the bottom. The downcomer sparger is located 5.4 m from the bottom. Since $H_D = 15.3$ m, the diameter of the downcomer and riser ($D_D = D_R$ assumed) was found in such a way that the total gas volume in EL-ALR equals that in the bubble column of case 2 in Table CS7.3. As a result, we get $D_R = D_D = 1.88$ m. It may be noted that the procedure has been considerably simplified. For rigorous calculations, see Lele and Joshi (1993).

4. Horizontal Sparged Contactor (HSC)

For all other contactors considered previously, we have used 200% excess oxygen. Hence, the gas utilization efficiency in these contactors is low. Let us consider a horizontal sparged contactor, which has the advantage of low pressure drops. The following assumptions are made: the liquid phase is completely backmixed, and the average pressures can be used (since the pressure drop is small).

The following dimensions may be considered: diameter of the contactor = 5 m, liquid submergence = 4 m, nozzle spacing = 0.5 m, distance of the nozzle tips from the bottom of the contactor = 0.2 m.

The material balance for oxygen can be written as

$$G dy = k_L a ([A]^* - [A]_b) dV_D \qquad \text{(CS7.25)}$$

$$= k_L a \left(H_W \overline{P}_T y - [A]_b \right) dV_D \qquad \text{(CS7.26)}$$

The direction of V_D has been taken as the direction of increasing y. Integration of Equation (CS7.26) gives

$$V_D = \frac{G}{k_L a \, H_W \overline{P}_T} \ln \left[\frac{H_W \overline{P}_T y_0 - [A]_b}{H_W \overline{P}_T y_e - [A]_b} \right] \qquad \text{(CS7.27)}$$

$[A]_b$ can be obtained from an overall material balance as

$$G(y_0 - y_e) = \frac{L}{Z}([B]_0 - [B]_e) = V_D \varepsilon_L k_2 [A]_b [B]_e \qquad \text{(CS7.28)}$$

Eliminating $[A]_b$ between Equations (CS7.27) and (CS7.28) gives

$$V_D = \frac{G}{k_L a \, H_W \overline{P}_T} \ln \left[\frac{y_o - \frac{\delta}{V_D}([B]_0 - [B]_e)}{y_e - \frac{\delta}{V_D}([B])_0 - [B]_e} \right] \qquad \text{(CS7.29)}$$

where

$$\delta = L / \left(\varepsilon_L Z k_2 H_W \overline{P}_T [B]_e \right)$$

In this example, we have selected the following parameter values:

$P_B = 1.0133 \times 10^5$ Pa (excluding vapor pressure of water)
$P_T = 6.27 \times 10^4$ Pa (excluding vapor pressure of water)
$\bar{P}_T = (P_B + P_T)/2 = 8.20 \times 10^4$ Pa
$k_L a = 0.02$ s^{-1}
$\varepsilon_L = 0.9$
$\delta = 165$

Substituting these values in the above equation for V_D and solving by trial and error gives $V_D = 333$ m³. This liquid volume corresponds to a submergence of 4 m. The cross-sectional area of the contactor corresponding to the liquid submergence of 4 m is equal to 16.83 m². Therefore, the length of the contactor

$$L_C = \frac{333}{16.83} = 19.8 \text{ m}$$

Number of nozzles, $N_N = 47$; nozzle gas velocity, $u_N = 70.0$ m/s
Nozzle diameter, $d_N = 0.00735$ m

(For a foaming system, the value of $k_L a$ of 0.02 s^{-1} appears to be reasonable for a horizontal sparged contactor.) The total volume of dispersion for the horizontal sparged contactor = 333 m³. Furthermore, from the earlier section of this case study under "Operation of Bubble Column under Pressure," it can be seen that the total volume of dispersion for the bubble column = 132 m³. A comparison between the horizontal sparged contactor and the bubble column can now be made on the basis of the total power consumption.

The power consumption for the horizontal sparged contactor can be calculated using the following equation:

$$\mathbf{P}_H = \frac{\pi}{4} d_N^2 u_N N_N \left[\rho_L \varepsilon_L g \frac{(H_D - 20)}{100} + \frac{\rho_G u_N^2}{2} \right] \quad \text{(CS7.30)}$$

ε_L = fractional liquid holdup = 0.9. Substitution of the relevant values in Equation (CS7.30) gives $\mathbf{P}_H = 5.12$ kW. For the bubble column, the power consumption can be calculated using the following equation:

$$\mathbf{P}_V = \frac{v^*}{v^* - 1} GR_g T \left[\left(\frac{P_B}{P_T} \right)^{\frac{v^*-1}{v^*}} - 1 \right] \quad \text{(CS7.31)}$$

where $v^* = C_p/C_v$, G is the molar flow rate of gas, P_T = pressure at the top of the bubble column, 62.66 kN/m², and P_B = pressure at the bottom of the bubble column, 228.38 kN/m². Substituting these values in Equation (CS7.31) gives $\mathbf{P}_V = 22.13$ kW. Hence, the horizontal sparged contactor can be advantageously used, as compared to the bubble column, when such an excess of air is used.

5. Pipeline Contactor (PC)

Data

Pure oxygen, 20% in excess over the theoretical requirement; operating pressure = 3.04×10^5 Pa, absolute; liquid-side Reynolds number = 40,000 (to keep the two-phase flow patterns as froth); liquid-side mass transfer coefficient at the entrance (u_G = 0.127 m/s, u_L = 0.077 m/s); $k_L a$ = 0.12 s^{-1}.

Assumptions

The two-phase flow pattern (in this case, froth) does not change along the length of the pipeline. The liquid-side mass transfer coefficient varies linearly with the superficial gas velocity at a constant liquid flow rate. As indicated later, the pressure drop can be neglected.

The fractional liquid holdup can be calculated by the procedure proposed by Lockhart and Martinelli (1949). The values of fractional liquid holdup are calculated to be 0.5 at the inlet (u_G = 0.127 m/s, u_L = 0.077 m/s) and 0.7 at the outlet (u_G = 0.0212 m/s, u_L = 0.077 m/s). The following correlation is assumed to hold for ε_L:

$$\varepsilon_L = -585 G + 0.74 \tag{CS7.32}$$

Overall material balance between the inlet and the location where oxygen flow rate is G gives

$$G_0 - G = \frac{L}{Z}\left([B]_0 - [B]\right) \tag{CS7.33}$$

or

$$[B] = [B]_0 - \frac{Z}{L}(G_0 - G) \tag{CS7.34}$$

The material balance for a differential length dz, z cm away from the entrance, gives

$$-dG = R_A a S dz \tag{CS7.35}$$

The values of $k_L a$ and $\varepsilon_L k_2 [B]$ are comparable at the entrance and at the exit, as shown below.

	$k_L a$ (s^{-1})	$\varepsilon_L k_2 [B]$ (s^{-1})
Entrance	0.12	0.357
Exit	0.02	0.0208

Under these conditions, the rate of absorption of oxygen is calculated by the following equation:

$$R_A a = [A] * \left[\frac{1}{\dfrac{1}{k_L a} + \dfrac{1}{\varepsilon_L k_2 [B]}}\right] \tag{CS7.36}$$

Substitution of Equation (CS7.36) in Equation (CS7.35) gives

$$[A]^* S dz = -dG\left[\frac{1}{k_L a} + \frac{1}{\varepsilon_L k_2 [B]}\right] \quad \text{(CS7.37)}$$

In the pipeline contactor, $k_L a$ is assumed to be proportional to G, i.e.,

$$k_L a = C_1 G \quad \text{(CS7.38)}$$

Substitution of $k_L a$ from Equation (CS7.38) and ε_L from Equation (CS7.32) into Equation (CS7.37) gives

$$dz = \frac{-dG}{S[A]^*} \left(\frac{1}{C_1 G} + \frac{1}{k_2(-585G + 0.74)\left[[B]_o - \frac{ZG_o}{L} + \frac{ZG}{L}\right]} \right) \quad \text{(CS7.39)}$$

Let

$$\delta' = L[B]_0 - ZG_0 \quad (Z=1) \quad \text{(CS7.40)}$$

Therefore,

$$dz = \frac{-dG}{S[A]^*}\left[\frac{1}{C_1 G} + \frac{L}{k_2(-585G + 0.74)(\delta' + G)}\right] \quad \text{(CS7.41)}$$

Integration of Equation (CS7.41) with the boundary conditions

$$\text{at} \quad z = 0 \quad G = G_0$$

$$z = z_T \quad G = G_e \quad \text{(CS7.42)}$$

$$z_T = \frac{1}{S[A]^*} \ln\left(\left(\frac{G_0}{G_e}\right)^{\frac{1}{C_1}} \left[\left(\frac{\delta' + G_0}{\delta' + G_e}\right)\left(\frac{-585G_e + 0.74}{-585G_0 + 0.74}\right)\right]^{v'}\right) \quad \text{(CS7.43)}$$

where

$$v' = \frac{L}{k_2(0.74 + 585\delta')} \quad \text{(CS7.44)}$$

In this problem, the theoretical oxygen requirement is

$$= \frac{L}{Z}\left([B]_0 - [B]_e\right)$$

$$= 2.32 \times 10^{-3}(0.154 - 0.0064) = 0.342 \times 10^{-3} \text{ kmol/s}$$

The oxygen supplied is 20% in excess of the theoretical requirement. Therefore,

$$G_o = 0.41 \times 10^{-3} \text{ kmol/s and } G_e = (0.41 - 0.342) \times 10^{-3} = 0.068 \times 10^{-3} \text{ kmol/s}$$

It is given that the Reynolds number of the liquid = 40,000 and the viscosity of the solution at 75°C = 0.38×10^{-3} Pa·s.

Therefore, diameter of pipeline = 0.197 m; cross-sectional area, S = 0.031 m². At the entrance, $k_L a = C_1 G_o = 0.12$ s^{-1}, $G_o = 0.41 \times 10^{-3}$; therefore $C_1 = 293.68$, and

$$\delta' = L\left([B]_0 - \frac{ZG_0}{L}\right)$$

$$= -5.27 \times 10^{-5}$$

$$[A]^* = (P - P_{H_2O}) H_W$$

$$= 1.703 \times 10^{-3} \text{ kmol/m}^3$$

$$v' = L / \left[k_2 (0.74 + 585\delta')\right]$$

$$= 0.705 \times 10^{-3}$$

Substitution of the relevant values in Equation (CS7.43) gives

$$z_t = 170 \text{ m}$$

In the above case of a pipeline contactor, pure oxygen is used instead of air. The excess of unreacted oxygen, which comes out of the pipeline contactor, can be compressed and reused with the fresh oxygen feed.

Let us consider the use of a large excess of oxygen to reduce the length of the contactor. The above problem is reworked for the case of 250% excess oxygen over the theoretical requirement:

$$G_0 = 1.197 \times 10^{-3} \text{ kmol/s}, \quad G_e = 0.855 \times 10^{-3} \text{ kmol/s}$$

The values of fractional liquid holdup, ε_L, at the inlet and exit are found to be 0.37 and 0.43, respectively. For this case, ε_L can be assumed to be constant and equal to the average value of 0.4.

Following the same procedure as per Equations (CS7.35) through (CS7.44), the length of the contactor can be obtained from the following equation:

$$z_t = \frac{1}{S[A]^*} \ln\left[\left(\frac{G_0}{G_e}\right)^{\frac{1}{C_1}} \left(\frac{\delta' + G_0}{\delta' + G_e}\right)^{v''}\right] \qquad \text{(CS7.45)}$$

where

$$v'' = L / (\varepsilon_L k_2) = 1.25 \times 10^{-3}, \quad C_1 = 292.68$$

$$\delta' = L\{[B]_0 - (ZG_0 / L)\} = -0.84 \times 10^{-3}$$

Substitution of these values in Equation (CS7.45) gives

$$z_t = 101 \text{ m}$$

Hence, the increase of oxygen from 20 to 250% excess gives a relatively small decrease in the contactor length from 170.0 m to 101.0 m.

Pressure drop in the pipeline contactor can be calculated from the procedure given by Lockhart and Martinelli (1949):

$$D = 0.194 \text{ m}, u_L = 0.0773 \text{ m/s}, u_G = 0.074 \text{ cm/s}, \rho_L = 1000 \text{ kg/m}^3, \rho_G = 3.4 \text{ kg/m}^3$$

$$(\text{Re})_L = 40{,}000, (\text{Re})_G = 2030, f_G = 0.00788$$

$$(\Delta P / \Delta L)_L = 3.384 \times 10^{-4} \left(\text{kN/m}^2 \right)/\text{m}$$

$$X_L = \left[\left(\frac{\Delta P}{\Delta L} \right)_L \bigg/ \left(\frac{\Delta P}{(\Delta L)} \right)_G \right]^{1/2} = 15.21$$

$$(\Delta P/\Delta L)_G = 1.462 \times 10^{-6} \left(\text{kN/m}^2 \right)/\text{m}$$

X_L is the X-parameter in the Lockhart–Martinelli correlation. For both the gas and the liquid phases under turbulent conditions, the ordinate values can be obtained from the Lockhart–Martinelli chart. Thus, $y_L = 2.5$, where y_L is the Y-parameter in the Lockhart–Martinelli correlation.

Therefore, the total two-phase pressure drop

$$= y_L \times (\Delta P/\Delta L)_L \times \text{total length}$$

$$= 143.82 \text{ N/m}^2$$

6. Packed Column (PC)

Data
Superficial gas velocity = 0.1 m/s, liquid-side mass transfer coefficient, $k_L a = 5 \times 10^{-3} \text{ s}^{-1}$, effective gas-liquid interfacial area, $a = 100 \text{ m}^2/\text{m}^3$. The material balance between any section of the column and the outlet gives

$$G(y - y_e) = \frac{L}{Z}([B]_0 - [B]) \quad \text{(CS7.46)}$$

Equation (CS7.46) on rearrangement gives

$$y = \gamma - \alpha'[B] \quad \text{(CS7.47)}$$

where

$$\alpha' = L/(ZG) \quad \text{(CS7.48)}$$

$$\gamma = y_e + \alpha'[B]_0 \qquad (CS7.49)$$

Material balance across a differential height dh of the column gives:

$$\frac{L}{Z}d[B] = R_A a S dH \qquad (CS7.50)$$

At the top of the column,

$$\varepsilon_L k_2 [B]_o = 0.0715 \text{ s}^{-1}; \quad k_L a = 0.005 \text{ s}^{-1} \quad \text{and} \quad \sqrt{M} = 1.39; \quad \varepsilon_L = 0.1$$

where

$$\sqrt{M} \text{ is } \sqrt{\mathbf{D}_A k_2 [B]_0} / k_L$$

From the above values, it can be seen that

$$k_L a \ll \varepsilon_L k_2 [B]_b \qquad (CS7.51)$$

and

$$\sqrt{M} = 1 \qquad (CS7.52)$$

At the bottom of the column,

$$\varepsilon_L k_2 [B]_e = 2.98 \times 10^{-3} \text{ s}^{-1}, \quad k_L a = 0.005 \text{ s}^{-1}, \quad \sqrt{M} = 0.284$$

Therefore,

$$k_L a = \varepsilon_L k_2 [B]_0, \quad \sqrt{M} \ll 1 \qquad (CS7.53)$$

Based on Equations (CS7.51) and (CS7.52) at the top of the column, the reaction falls in between regimes 1, 2, and 3. From Equation (CS7.53), at the bottom of the column, the reaction falls in between regimes 1 and 2. Thus, the following equation holds for $R_A a$ in the entire column:

$$R_A a = a[A]^* \left[\frac{1}{\frac{1}{\sqrt{\mathbf{D}_A k_2 [B] + k_L^2}} + \frac{a}{\varepsilon_L k_2 [B]}} \right] \qquad (CS7.54)$$

Substituting $R_A a$ from Equation (CS7.54) into Equation (CS7.50) and using Henry's law ($[A^*] = H_W P_T y$), we get

$$\frac{L}{Z}d[B] = \left[\frac{ayP_T H_W}{\dfrac{1}{\sqrt{\mathbf{D}_A k_2 [B] + k_L^2}} + \dfrac{a}{\varepsilon_L k_2 [B]}}\right] Sdh \qquad \text{(CS7.55)}$$

Substituting y from Equation (CS7.47) into Equation (CS7.55) and simplifying gives

$$d[B] = dH \left[\frac{\beta^*(\gamma - \alpha'[B])}{\dfrac{1}{\sqrt{[B] + \delta_1^2}} + \dfrac{\eta}{[B]}}\right] \qquad \text{(CS7.56)}$$

where

$$\beta^* = \frac{ZaSH_W P_T \sqrt{\mathbf{D}_A k_2}}{L} \qquad \text{(CS7.57)}$$

$$\delta_1 = \frac{k_L}{\sqrt{\mathbf{D}_A k_2}}, \quad \eta = \frac{a\sqrt{\mathbf{D}_A k_2}}{\varepsilon_L k_2} \qquad \text{(CS7.58)}$$

Integration of Equation (CS7.56) between $h = 0$, $[B]_b = [B]_e$, and $h = H([B]_b = [B]_0)$ gives

$$H = \frac{1}{\beta^* \alpha' \sqrt{(\delta_1^2) + (\gamma/\alpha')}} \ln\left[\frac{\sqrt{\delta_1^2 + [B]_0} + \sqrt{\delta_1^2 + (\gamma/\alpha')}}{\sqrt{\delta_1^2 + [B]_0} - \sqrt{\delta_1^2 + (\gamma/\alpha')}} \times \frac{\sqrt{\delta_1^2 + [B]_e} - \sqrt{\delta_1^2 + (\gamma/\alpha')}}{\sqrt{\delta_1^2 + [B]_e} + \sqrt{\delta_1^2 + (\gamma/\alpha')}}\right]$$
$$+ \frac{\eta}{\gamma\beta^*} \ln\left[\frac{[B]_0(\gamma/\alpha') - [B]_e}{[B]_e(\gamma/\alpha') - [B]_0}\right] \qquad \text{(CS7.59)}$$

$L = 2.32 \times 10^{-3}$ m³/s, $[B]_0 = 0.154$ kmol/m³, $[B]_e = 6.4 \times 10^{-3}$ kmol/m³

$$y_0 = \frac{0.21(760 - 290)}{760} = 0.13$$

The oxygen supplied is 200% in excess over that theoretically required. We also assume that G remains constant. Therefore, $y_e = (2/3) \times 0.13 = 0.0867$.

From overall material balance,

$$G(y_0 - y_e) = \frac{L}{Z}\left([B]_0 - [B]_e\right) \qquad \text{(CS7.60)}$$

Substitution of the relevant values in Equation (CS7.60) gives $G = 7.91 \times 10^{-3}$ kmole/s.
The superficial gas velocity, u_G, is equal to 0.1 m/s (assumed).

Hence, $S = 3.652$ m^2, and $D = 2.16$ m.

$$\alpha' = L/(ZG) = 0.293, \quad \gamma = y_e + \alpha'[B]_0 = 0.1319$$

$$\beta^* = \left[ZaSH_W P_T \sqrt{D_A k_2} \right]/L = 9.85 \times 10^{-3}$$

$$\delta_1 = k_L/\left(\sqrt{D_A k_2}\right) = 0.281$$

$$\delta = \left(a\sqrt{D_A k_2}\right)/\left(\varepsilon_L k_2\right) = 0.038$$

Substituting the relevant values in Equation (CS7.59), H can be calculated and is found to be 469 m, i.e., an abnormally large packed column. Such a column provides very low liquid holdup (0.03 to 0.1), and the present oxidation reaction occurs in the liquid phase. Therefore, a packed column should be rejected.

A comparison of all the equipment is presented in Table CS7.3. For all the cases, the reactor volume and power consumption are given. The former indicates the capital cost, whereas the latter indicates the operating cost. A preliminary comparison reveals that SBC and EL-ALR are comparable but superior to the conventional BC. The volume of HSR is approximately 5.4 to 5.8 times higher than SBC or EL-ALR. However, the power consumption is low (0.25 and 0.362 times lower, respectively). Although the power consumption of the PC is very low, the capital cost of the packed volume would be prohibitive.

Lesson

This case study on oxidation of sodium sulfide illustrates the design of a variety of gas-liquid reactors and compares their performances. Bubble column reactors are particularly attractive, as they offer advantages such as simplicity of construction and operation, but they suffer from such drawbacks as high pressure drop and backmixing in the liquid phase. To reduce the pressure drop, two modifications have been considered: an external-loop air-lift reactor and a horizontal sparger reactor. Both result in substantial energy savings (because of low ΔP) under similar conditions of capacity and conversions in the gas and liquid phases.

The other limitation of liquid-phase backmixing has been addressed through the modification of sectionalization. In this manner, near plug-flow behavior is achieved, and the column height is reduced to as much as 60%. The reduction in height reduces the equivalent pressure drop and hence the energy consumption for air compression.

The case study demonstrates yet another peculiar behavior of bubble columns: steep reduction in oxygen partial pressure from bottom to top, mainly a result of the change in total pressure due to hydrostatic head. Such a situation results in column underutilization. To overcome this problem, guidelines have been given, with detailed calculations indicating substantial savings.

The effects of a few other important parameters have also been considered, but the study should be extended to include investigation of all of them. Cost considerations in selecting parameters such as temperature, pressure, excess oxygen, height-to-diameter ratio, and so forth play an important role in final selection. Cost implications in the design and selection of downstream equipment also constitute a major factor. The philosophy and details of design outlined above for bubble columns should be extended to include other gas-liquid reactors. For instance, for packed columns, the type, size, and cost of different packings (offered by different vendors) should be examined by careful selection of liquid distributors, redistributors, and packing supports.

A practically useful table, presented in Section 11.4 (Table 11.26), provides valuable guidelines for the selection of gas-liquid reactors. Other reactor configurations should be considered and a table similar to Table 11.26 prepared.

Chemical Reaction Engineering

CASE STUDY 11.8 GAS-LIQUID REACTION: ABSORPTION OF NO_x GASES FOR THE MANUFACTURE OF NITRIC ACID (A SYSTEM WITH MULTIPLE COMPLEXITIES)

Nomenclature for Case Study 11.8

D	diffusivity (m²/s)
H_W	Henry's constant (kmol/m²(kN/m²))
$H_W \sqrt{kD}$	absorption factor for fast pseudo-first-order reaction (kmol/m²(kN/m²)s)

The Problem

Absorption of NO_x gases is an important step in the manufacture of nitric acid. It is one of the most complex of all absorption operations, because of the following. (1) The NO_x gas contains NO, NO_2, N_2O_3, N_2O_4, and so on, and the absorption of NO_x gases in water results in both nitric acid and nitrous acids. (2) Several reversible and irreversible reactions occur in both the gas and liquid phases. (3) Simultaneous absorption and desorption of several gases occurs, followed by chemical reaction; the absorption of NO_2, N_2O_3, and N_2O_4 is accompanied by chemical reactions, whereas the desorption of NO, NO_2, and HNO_2 is preceded by chemical reactions. (4) Equilibria prevail between various components in each phase. (5) Heterogeneous equilibria also prevail between the gas phase and liquid phase components. Sherwood et al. (1975) and Joshi et al. (1985) have reviewed NO_x absorption.

For the design of NO_x absorption towers, it is necessary to understand the several equilibria and the rates of mass transfer and chemical reaction. The mechanism of NO_x absorption in water has been given by Pradhan et al. (1997). Furthermore, substantial heat effects are associated with NO_x absorption resulting in temperature variations. These aspects of the reaction, along with a mathematical model, method of solution, and an optimum design strategy, have been described by Pradhan et al. (1997). *Note: the tables and figures referred to in this section are from the original reference.* Some results are given below.

Effect of Various Parameters

To get the number of stages needed for the desired absorption, operating variables such as pressure, temperature, and heights of submergence and empty sections were varied over a wide range. The capacity of the absorber was 750 tons/day (100% basis), and 30% excess oxygen was used. Values of the parameters are listed in Table 5 in the original reference.

The effects of total pressure, height of submergence, and height of empty section on the number of plates have been studied as a function of the outgoing NO_x concentration (up to 20 ppm level). Figures 3-A, 3-B, and 3-C show the effect of height of empty section at weir heights of 0.025, 0.05, and 0.1 m, respectively, and at 1.3 MPa pressure. At higher empty section heights, fewer plates are required. Although the number of plates decreases, this does not necessarily mean that the total height decreases. In many cases, the total height increases with an increase in the height of empty section. Tables 6, 7A, and 7B of the original reference indicate that the total height increases with an increase in the height of the empty section. The increase is nominal when the empty height is increased from 0.4 to 0.7 m. This result is useful because the cost associated with many stages can be reduced. However, a similar result is not obtained when the empty height is increased to 4 m and then to 10 m. The total height substantially increases in such a case.

The effect of liquid submergence has also been investigated. The results are summarized in Tables 6, 7A, and 7B for outlet NO_x concentrations of 700 and 20 ppm. Four levels of liquid submergence (weir height) were considered: 25, 50, 100, and 200 mm. It can be seen that the number of plates decreases with an increase in the height of liquid submergence. The decrease is small when the submergence is increased from 100 to 200 mm. Furthermore, from Tables 6, 7A, and 7B of the original reference, the effect of the empty section is more dominant than that of weir height.

The effect of temperature is shown in Table 7B of the reference. The number of plates increases with increased temperature. However, it is not advantageous to operate the absorption column at temperatures requiring significant refrigeration. The rate of heat removal due to vaporization of a liquid refrigerant (often ammonia) can be used advantageously in the top section, where the rate of oxidation is very low. Most heat is usually removed in the bottom section of the column by the ambient water. Hence, the operating temperature is decided by the ambient temperature, and the optimum heights for empty section and submergence section can be selected rather directly. However, the optimum temperature and pressure depend on the geographical location of the nitric acid plant and the cost of power. The costs will be considered in the optimum design strategy.

Optimum Design

The total cost ($/day) was estimated at two levels of outlet NO_x concentration, 700 and 150 ppm. The sensitivity of fixed cost was studied by assuming stainless steel prices at two levels: $3.52/kg and $7.04/kg. The results are in Tables 9A and 9B for different combinations of weir and empty section heights at five levels of total pressure, 0.4, 0.71, 1.01, 1.31, and 2.02 MPa. As indicated, pressure is a major parameter governing the fixed cost. To understand the overall effect of fixed and operating costs, the daily contribution was estimated on the basis of 25 and 45% of the fixed cost as the yearly (300 days) expenditure. For this case, 95% of the compressor energy was assumed to be recoverable, i.e., 5% of the cost of compression was included in the operating cost. The analysis was performed at a height of 0.025 m, with empty section height as parameter. The results are shown in Figures 6A and 6B for 700-ppm, and in Figures 6C and 6D for 150-ppm, outlet concentration. The following observations can be noted: (1) For empty heights of 0.4 and 0.7 m, the total cost decreases monotonically with increase in pressure. The reduction is sharp up to 1 MPa and thereafter minimal. (2) For empty heights of 1.5, 4, and 10 m, the total cost curves show minima. The curve for 1.5 m is shallow, whereas it is deep for 10 m and in between for 4 m. The optimum pressure decreases with an increase in empty height. However, it must be emphasized that the optimum costs (at minima) for 4- and 10-m empty heights are higher than those for 0.4-, 0.7-, and 1.5-m empty heights. (3) The total costs for three empty heights (0.4, 0.7, and 1.5 m) are comparable-being in the range 0.7 to 1.0 MPa. (4) The optimum plate spacing depends on the fixed cost components. For instance (from Figures 6E and 6F of the original reference), at 0.4 MPa pressure, the total cost is minimum for 4-m empty height. With an increase in pressure, the optimum height decreases. Thus, the optimum heights are 1.5, 0.7, and 0.4 m at pressures of 0.7, 1.0, and 1.3 MPa, respectively. Suchak and Joshi (1994) and Pradhan et al. (1997) have examined the effects of mass transfer characteristic and residence time on the designs of packed and plate column.

Lesson

Examples of simple reactions (such as considered in Case Study 11.7) are not hard to find in industry, but those of complex reactions are more widespread. One can envisage systems with increasing complexities such as the manufacture of nitric acid. The complexities include the following: (1) multiple (parallel) reactions by the same reactant, (2) reactions in series, (3) multiple reactions among multiple reactants, (4) reversible reactions, (5) reactions in multiple phases, and (6) simultaneous mass transfer with chemical reactions. These complex problems can be tackled if sufficient details of the following are provided: (1) reaction chemistry, (2) rate expressions, (3) rate constants, and (4) diffusivities and solubilities in the case of multiphase systems. The kinetics of each reaction can be determined in the laboratory by (1) ensuring conditions that make the desired step rate controlling and (2) minimizing/suppressing the other reactions. In some cases, lumped parameters such as $H\sqrt{kD}$ can be obtained and used directly.

Although one can progressively account for one additional complexity at a time, it is also possible to take a direct leap to a problem with a combination of complexities, as illustrated in this case study on NO_x absorption in the manufacture of nitric acid. The NO_x absorption system may contribute about 40% of the total equipment cost. Hence, optimal design of the NO_x absorption tower is important; major cost-determining factors should be identified and selectively addressed with greater rigor.

Chemical Reaction Engineering

Case Study 11.9 Gas-Liquid-Solid (Noncatalytic) Reaction: Oxydesulfurization of Coal in a Slurry Reactor

Nomenclature for Case Study 11.9

$[A]_s$	concentration of dissolved solute gas at the solid surface (kmol/m^3)
$[A]_e$	concentration at the exit (kmol/m^3)
b	stoichiometric coefficient
b_r	eigenvalues of Equation (CS9.10)
C	concentration of tracer (kmol/m^3)
c	dimensionless concentration of tracer
c_0	dimensionless concentration of tracer at the inlet
D	reactor diameter (m)
H	reactor height (m)
$\mathbf{D}_e$	effective diffusivity through ash layer (m^2/s)
$\mathbf{D}s$	dispersion coefficient (m^2/s)
$E(\theta)$	exit age distribution function
k_r	rate constant for surface reaction
L	length (m)
N_C	critical stirrer speed (rps)
P	pressure (MPa)
Pe	Peclet number
Q	amount of the pulse tracer
R	particle radius (m)
r	radius of shrinking core (m)
S_{wb}	ratio of wake to bubble volumes (–)
T	temperature (K)
T_r	eigenvalue Equation (CS9.10)
$\bar{t}$	average residence time (s)
t	time (s)
u_G	superficial gas velocity (m/s)
u_L	superficial gas velocity (m/s)
u_S	settling velocity (m/s)
u_{SL}	slurry velocity (m/s)
u_c	circulation velocity (m/S)
X	conversion (–)
$\bar{X}$	overall conversion (–)
z	distance
Z	dimensionless distance (z/L)

Greek

ε_S	solid holdup
ε_G	gas holdup (–)
δ	film thickness at the interface (m)
ρ_B	density of component B (kg/m^3)
τ	time required for complete conversion (s)
θ	dimensionless time (t/τ)

The Problem

Precombustion of sulfur from coal by selective oxidation is an important step in coal cleaning, without the need for a postcombustion cleanup step to remove sulfur oxides. This process, known as *oxydesulfurization of coal*, is a three-phase gas-liquid-solid system. As an example, a sparged reactor will be designed to oxydesulfurize 100 tons/day of coal in an aqueous slurry of coal with air.

Data

The reaction temperature and the water vapor pressure are 466 K and 1.36 MPa, respectively. The coal particle size is 100μm. At the operating temperature of 460 K and in the range of H_2SO_4 concentrations encountered in oxydesulfurization, the vapor pressure of sulfuric acid is below 1×10^{-2} Pa. The operating pressure is in the range of 2 to 7 MPa, and thus most of the sulfuric acid will remain in the liquid phase. For simplicity, all the sulfur is assumed to be in the form of FeS_2 (content = 2.7 wt% of coal). The bulk density of coal is 889 kg/m³ and of water for making the slurry is 1000kg/m³. The slurry concentration is 26 wt%.

Determination of the Rate-Controlling Step

The steps during the process of oxydesulfurization are as follows:

1. Transfer of oxygen through the gas-liquid film
2. Transfer of oxygen through the liquid-solid film
3. Diffusion of oxygen through the ash layer
4. Chemical reaction between the dissolved oxygen and FeS_2 at the solid surface

The mechanism of oxydesulfurization of coal has been extensively investigated by Vracar et al. (1970), Friedman and Warzinski (1977), Vetter (1967), and Joshi et al. (1982). The dissolution of sulfide minerals can be interpreted as an electrochemical surface reaction similar to the corrosion of metals. Oxygen is reduced at cathodic areas, and sulfides are dissolved liberating electrons to complete the couple.

$$FeS_2 + \frac{7}{2}O_2 + H_2O \rightarrow Fe^{2+} + 2SO_4^{2-} + 2H^+$$

$$2FeS_2 + \frac{15}{4}O_2 + H_2O \rightarrow 2Fe^{3+} + 4SO_4^{2-} + 2H^+ \quad \text{(CS9.R1)}$$

$$FeS_2 + 3.75O_2 + 2H_2O \rightarrow 0.5Fe_2O_3 + 2H_2SO_4$$

To determine the rate-controlling step, it is necessary to study all of the four possibilities by experiments over a wide range of particle sizes, temperatures, pressures, and impeller speeds.

The overall reaction may be represented by the shrinking core model (SCM). Laboratory-scale experiments were carried out in a 1-L vessel in the temperature range of 428 to 488 K and total air pressure range of 3.44 to 7.88 MPa. Coal particles of practically uniform size were used for the experiments, and coal loading was varied over a range of 0.025 to 0.22 kg/L of water. The experiments were carried out by changing the stirrer speed, and it was observed that, above a certain critical speed (N_c), the rate was independent of the speed. Therefore, the gas-liquid and liquid-solid mass transfer resistances can be considered to be negligible when the stirrer speed is higher than N_c. Thus, the process becomes either ash diffusion controlled or chemical reaction controlled. The experiments over the temperature range mentioned enabled the estimation of the energy of activation, which was found to be 46.5×10^{06} J/kmol. The value of activation energy indicates that the overall operation is reaction controlled. The reaction order with respect to oxygen was found to be 0.7 (for reasons reported by Joshi et al., 1981).

To get supporting evidence for the earlier conclusion of the reaction-controlled operation, the fractional conversion of FeS_2 (in coal) with time was studied, and the time required for complete conversion was also obtained. The following SCM relationships were used (see Section 11.3.2.1.1) to determine the controlling mechanism. For ash diffusion control,

$$\frac{t}{\tau} = 1 - 3(1-X)^{2/3} + 2(1-X) \qquad \text{(CS11.9.1)}$$

where

$$\tau = \frac{\rho_B R^2}{6Z\mathbf{D}_e[A]_s} \qquad \text{(CS9.2)}$$

is the time for complete conversion. For chemical reaction control,

$$\frac{t}{\tau} = 1 - (1-X)^{1/3} \qquad \text{(CS9.3)}$$

where

$$\tau = \frac{\rho_B R}{Zk_r [A]_s^{0.7}} \qquad \text{(CS9.4)}$$

The experimental conversion-time relationship was found to agree with Equation (CS9.3). Furthermore, the experimental value of t was found to be proportional to R. From these two observations, and also from the value of activation energy, it was concluded that the oxydesulfurization process of coal is chemical reaction controlled. For coal of 100 μm size, the time required for complete conversion was found to be 1160 s.

Design of Three-Phase Sparged Reactor

The following assumptions have been made:

1. The variation in total pressure due to static head of liquid is small as compared to the total pressure.
2. Conversion of solute in the gas phase is small. Thus, the extent of backmixing in the gas phase is unimportant. The interfacial concentration of the solute gas will be calculated at the log mean partial pressure of the inlet and the outlet.
3. As a consequence of assumptions 1 and 2, the superficial gas velocity, gas holdup, and liquid-solid mass transfer coefficient remain practically constant along the length of the reactor.
4. Solid loading is uniform in the reactor. This assumption is reasonable, since the particle size is small, and/or the density difference between the solid and liquid is small.
5. As discussed before, the rate-controlling step is the surface reaction between dissolved oxygen and FeS_2. It is assumed that, under the hydrodynamic and other operating conditions, the assumption of the rate-controlling step holds.
6. Since the particles retain their identity in the reactor and the rate equations are nonlinear, a residence time distribution model will be used. The particle slip velocity was found to be negligible as compared with the liquid-phase circulation velocities, which govern the dispersion coefficient. It will be assumed that the axial dispersion coefficients for the solid and liquid phases are practically the same. Thus, the exit age distribution for the solid particles can be found by following the procedure for the liquid phase.
7. The particles are spherical and of uniform size.
8. The reactor is operated under isothermal conditions at 460 K.

The residence time distribution is obtained by solving the following material balance equation for a tracer:

$$\frac{\partial c}{\partial \theta} = \frac{1}{Pe}\frac{\partial^2 c}{\partial Z^2} - \frac{\partial c}{\partial Z} \quad (CS9.5)$$

where c is the dimensionless concentration, and other dimensionless parameters are

$$Z = z/L, \quad e = \frac{L u_{SL}}{(1-\varepsilon_G)D_s}, \quad \theta = \frac{t u_{SL}}{L(1-\varepsilon_G)} \quad (CS9.6)$$

The boundary conditions are

$$c_0 = c\big|_{0^-} = c\big|_{0^+} - \frac{1}{Pe}\frac{\partial c}{\partial Z}\bigg|_{0^+} \text{ at } Z = 0 \quad \text{for all } \theta$$

$$\frac{\partial c}{\partial Z} = 0 \text{ at } Z = 1, \quad \text{for all } \theta \quad (CS9.7)$$

$$c = 0 \text{ at } \theta = 0, \quad \text{for all } Z$$

The parameters are

$$\int_0^\infty c\,dt = \int_0^\infty (C/Q)\,dt = 1 \quad (CS9.8)$$

and

$$Q = \int_0^\infty C\,dt \quad (CS9.9)$$

where C is the actual concentration at the exit, and Q is the amount of pulse tracer. When solved in dimensionless form, the exit age distribution is given as

$$E(\theta) = 2\exp(Pe/2)\sum (-1)^{r+1}\exp(T_r,\theta)\frac{b_r^2}{\frac{4}{Pe}+1+b_r^2} \quad (CS9.10)$$

where b_r is the root of

$$b_r = \tan\left(r\frac{\pi}{2} - \frac{Pe b_r}{4}\right) \quad \text{and} \quad T_r = \left(-\frac{Pe}{4}\right)(1+b_r^2)$$

For the case of continuous feed of particles to a bubble column slurry reactor, with an exit age distribution of particles $E(\theta)$, a mass balance gives

Chemical Reaction Engineering

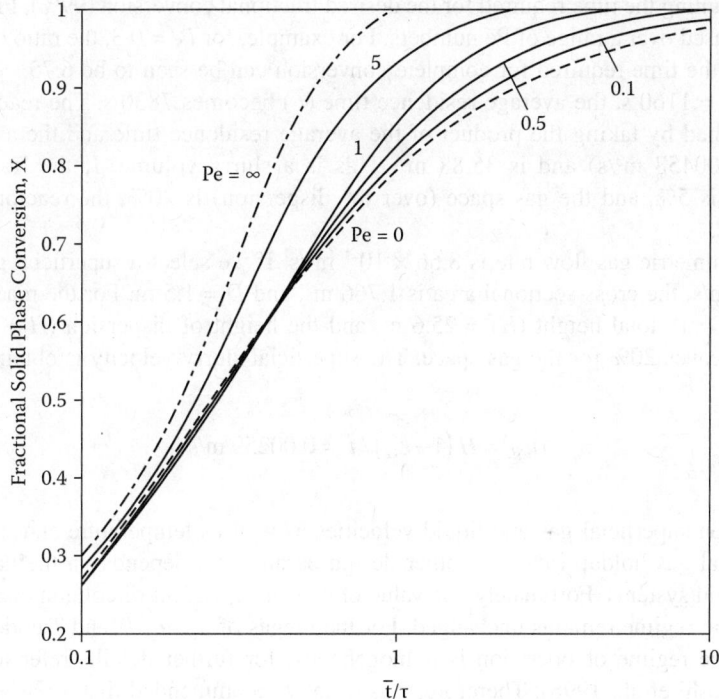

FIGURE CS9.1 Performance diagram of a three-phase reactor where the solid phase undergoes reaction: overall solid phase conversion with respect to dimensionless residence time, with Peclet number as parameter.

$$1 - \bar{X} = \int_0^{\tau/\bar{t}} (1-X) E(\theta) d\theta \tag{CS9.11}$$

where $E(\theta)$ is given by Equation (CS9.10), X is the conversion of particles with ages between θ and $(\theta + d\theta)$, and $\bar{X}$ is the overall average conversion. For the present reaction-controlled mechanism, $\bar{X}$ is given by Equation (CS9.3). The solution of Equation (CS9.11) is shown graphically in Figure CS9.1. The figure also includes the two extreme cases of plug-flow and backmixed conditions.

Design

Data

Coal to be treated = 100 tons/day, FeS_2 content in coal is 22.5 kmol/day. By stoichiometric balance given in reaction scheme CS9.R1, oxygen required = 84.37 kmol/day.

As a first case, let us consider 200% excess oxygen, which corresponds to 1205.35 kmoles of air per day. Let us assume an average total pressure of 7.6 MPa, which includes water vapor pressure of 1.36 MPa and air partial pressure of 6.24 MPa. At the operating pressure and temperature, the volumetric flow rate of gas = 31.86 m³/h or 8.66×10^{-3} m³/s.

Mass flow rate of slurry = 384.6 tons/day.

Total water required for slurry = 284.615 tons/day, and the volume of slurry = 395.93 m³/day.

Volumetric flow rate of slurry is 0.0046 m³/s, and the time required for complete conversion is 1160 s.

Typical Procedure

1. Assume a value for the Peclet number, say 0.3.

2. For obtaining the time required for the desired fractional conversion (99%), Figure CS9.1 can be used over a range of Pe numbers. For example, for *Pe* = 0.3, the ratio of residence time to the time required for complete conversion can be seen to be 6.75.
3. Since τ = 1160 s, the average residence time ($\bar{t}$) becomes 7830 s. The reactor volume is obtained by taking the product of the average residence time and the average flow rate (0.00458 m³/s) and is 35.88 m³. This is a slurry volume. If the fractional gas holdup is 5%, and the gas space (over the dispersion) is 20%, the reactor volume is 45.2 m³.
4. The volumetric gas flow rate is 8.66 × 10⁻³ m³/s. If we select a superficial gas velocity of 5 mm/s, the cross-sectional area is 1.766 m², and *D* = 1.5 m. For the reactor volume of 45.33 m³, total height (*H*) = 25.6 m, and the height of dispersion (*H_D*) = 21.33 m, which leaves 20% for the gas space. The superficial slurry velocity is obtained from

$$u_{SL} = H(1-\varepsilon_G)/\bar{t} = 0.00259 \text{ m/s} \qquad \text{(CS9.12)}$$

5. For given superficial gas and liquid velocities as well as temperature and pressure, the fractional gas holdup (and also other design parameters) depends on the nature of the gas-liquid systems. Fortunately, the value of ε_G is independent of column diameter if the operating regime remains unchanged. For the ranges of u_G, u_L, *P*, and *T* under consideration, the regime of operation is homogeneous (for further details, refer to Deckwer, 1992; Joshi et al., 1998). Therefore, it is usually recommended that a few experiments be performed in a small-scale column (at least 150 mm I.D.) in the ranges of interest of u_G, u_L, *P*, and *T* as used for large-scale columns. For the present values of these variables, let ε_G be 0.05 and the slip velocity 0.1 m/s.
6. Similar to ε_G, the dispersion coefficient also depends on the nature of the gas-liquid system. Therefore, it is recommended that the dispersion coefficient be estimated in a small-diameter column. Furthermore, the literature correlation is accepted except for the proportionality constant, which is estimated using the results from 150-mm I.D. column and for the system under consideration.

Let the resulting correlation be

$$\mathbf{D}_S = 326 \left[\frac{DS_{wb}\varepsilon_G u_S}{1-\varepsilon_G - S_{wb}\varepsilon_G} \right]^{1.7} \qquad \text{(CS9.13)}$$

where S_{wb} is the ratio of wake to bubble volumes. Kumar and Kuloor (1970) have recommended S_{wb} = 11/16. For ε_G = 0.05 and u_s = 0.1 m/s, $\mathbf{D}_s$ works out to be 0.0489 m²/s.
7. The estimated values of the dispersion coefficient can be used to calculate the Peclet number as

$$Pe = \frac{H_D u_{SL}}{(1-\varepsilon_G)\mathbf{D}_S} = \frac{21.39 \times 0.00259}{0.95 \times 0.0489} = 1.19 \qquad \text{(CS9.14)}$$

8. From Figure CS9.1, it can be seen that, for an overall average conversion of 99% at *Pe* = 1.19, $\bar{t}/\tau$ is 3.18. Therefore, $\bar{t}$ = 3683 s and slurry volume = 16.87 m³ and the reactor volume is 45.29 m³. Let us select, D = 1.5 m, H = 25.63 m, H_D = 21.36 m. Substitution of these values in Equation (CS9.14) gives *Pe* = 1.19 which is different from the starting

assumed value of 0.3. By following the steps 1 through 8 by trail and error, the dimensions work out to be D = 1.48 m, H = 16.77 m, H_D = 13.9 m, u_{SL} = mm/s and the value of Pe was found to be 0.801.

Lesson

Three-phase noncatalytic reactions are of great industrial importance. The present case study is concerned with such a reaction where the reactant goes through a number of steps in series, i.e., oxygen transfer through gas-liquid and liquid-solid films, diffusion through solid ash layer, and finally reaction with the solid. An important requirement in such cases is the use of laboratory information for discerning the rate-controlling mechanism from many plausible candidates prior to scaleup. Moreover, in cases where the solid particles take part in the reaction, the rate equations are nonlinear, adding mathematical complexity to the simulation. Simple dispersion models or even the tanks-in-series model, amply described in the literature, are not applicable in such cases. The present case study shows that models based on exit age distributions and extent of backmixing in the solid phase are useful. A new methodology is suggested for including the extent of backmixing of the solid phase in reactor design. The chief lesson of this study is that, while existing models and procedures may often be useful, the search for new, more directly relevant procedures can often be a rewarding exercise. In this case also, the possibility of optimization should be considered as in Case Study 11.7, and efforts should be made to produce a cost-effective column design.

CASE STUDY 11.10 GAS-LIQUID-SOLID (NONCATALYTIC) REACTION: CARBONATION OF LIME

Nomenclature for Case Study 11.10

$[A]$	concentration of component CO_2 in the liquid phase (kmol/m³)
$[A^*]$	saturation concentration of component A at the gas-liquid interface (kmol/m³)
a	interfacial area as defined by Equation (CS10.6) (m²/m³)
a_P	interfacial area of the particle (m²/m³)
a_{P0}	initial interfacial area of particles per unit volume of dispersion (m²/m³)
$[B]$	concentration of OH^- in the liquid phase (kmol/m³)
$[B]_b$	bulk concentration of component B (kmol/m³)
$[B]_S$	saturation solubility of component B (kmol/m³)
C_D	drag coefficient (–)
D	impeller diameter (m)
$\mathbf{D}_A$	diffusivity of component CO_2 (m²/s)
$\mathbf{D}_B$	diffusivity of component OH^- (m²/s)
d	characteristic length (m)
d_b	bubble diameter (m)
d_P	particle diameter (m)
d_{P0}	initial diameter of particle (m)
F_I	inert flow rate (kmol/s)
G_S	mass velocity (kg/m²/s)
g	gravitational constant (m/s²)
H	height of reactor (m)
H_x	Henry's constant for component x (kmol/m³)/(N/m²)
k_{CO2}	reaction rate constant for CO_2 hydrolysis (m³/kmol/s)
$k_L a$	gas-liquid mass transfer coefficient (1/s)
k_{SL}	solid-liquid mass transfer coefficient (m/s)
$\sqrt{M}$	term defined by Equation (CS10.13) (–)
M_{Wi}	molecular weight of component i (kg/kmol)
N	speed of rotation of impeller (rps)
N_{CD}	critical speed for complete dispersion (rps)

N_{JSG}	critical speed for solid suspension in presence of gas (rps)
N_{max}	maximum operating speed (rps)
n	moles of CO_2 required (kmoles)
P	total pressure (N/m²)
p_x	partial pressure of component x (N/m²)
PDN	production (kg)
$\mathbf{P}_O$	power number (–)
$\mathbf{P}_G$	power consumption in presence of gas (W)
$\mathbf{P}_L$	power consumption in absence of gas (W)
Q	volumetric flow rate of gas (m³/s)
R_A	specific rate of absorption of A (kmol/m² s)
Re_p	particle Reynolds number (–)
R_g	universal gas constant (J/(kgmol K))
r_A	rate of reaction (kmol/m³ s)
S	cross-section of reactor (m²)
Sc	Schmidt number, $\mu_L/\rho_L \mathbf{D}_B$ (–)
Sh	Sherwood number, $k_{SL}d_p/D_{AB}$ (–)
T	temperature (K)
$\mathbf{T}$	reactor diameter (m)
u	superficial velocity (m/s)
u_c	circulation velocity (m/s)
u_G	superficial gas velocity (cm/s)
u_{G0}	superficial gas velocity at the inlet (m/s)
$u_{S\infty}$	terminal settling velocity of particle (m/s)
V_D	volume of dispersion (m³)
V_S	volume of solids (m³)
W	blade width (m)
X_S	percentage solid loading (%)
x	mol CO_2/ mol inerts (–)
x_1	mole of CO_2/mole of inerts at the inlet (–)
Z	stoichiometric coefficient (–)

Greek

α	constant defined in Equation (CS10.4) (–)
α_1	constant defined in Equation (CS10.3) (–)
β	term defined in Equation (CS10.12) (–)
β_1	constant defined in Equation (CS10.3) (–)
ε_G	gas holdup (–)
ε_L	liquid holdup (–)
ε_S	solid holdup (–)
ε_{S0}	initial solids holdup (–)
γ_P	number of particles per unit volume (1/m³)
η	enhancement factor defined by Equation (CS10.11) (–)
μ_L	viscosity of liquid (kg/m s)
$\bar{\rho}$	average density of the medium (kg/m³)
ρ_S	density of solid (kg/m³)
ρ_L	density of liquid (kg/m³)

Subscripts

D	dispersion
f	final
G	gas

Chemical Reaction Engineering

L	liquid
P	particle
S	solid
S-L	solid-liquid
0	initial
1, 2	conditions 1, 2

The Problem

A company has a multipurpose batch plant to produce various chemicals on demand. The plant has a 10 m³ mechanically agitated sparged reactor (diameter = 2.19 m, height = 2.64 m) for pressures up to 2×10^5 N/m². Pure calcium carbonate ($CaCO_3$) is to be produced for toothpaste, cosmetics, and paints. Based on the logistics of the manufacture of $CaCO_3$ by the carbonation of lime, each batch run requires approximately 4 h to complete. Calculate the maximum production of $CaCO_3$ that can be achieved in the existing setup and find the corresponding process conditions using reasonable assumptions.

Data

Gas with 20 mol% CO_2 (rest 80% inerts, N_2) is available from a nearby plant; the reactor has a six-blade pitched (45°) downflow turbine agitator; maximum power delivered to the shaft by the motor = 60,000 W; density of solid $CaCO_3$ = 2700 kg/m³; density of solid $Ca(OH)_2(\rho_s)$ = 2080 kg/m³; impeller details: 45° pitch six-bladed downflow turbine, impeller diameter D = 0.9 m, impeller width W = 0.2 m, clearance = 0.7 m, power number P_o = 1.6; $Ca(OH)_2$ initial particle diameter = 0.1 mm; viscosity of water, μ_L = 10^{-3} Pa·s; diffusivity of CO_2 in water, $\mathbf{D}_A$ = 2.17×10^{-9} m²/s; diffusivity of OH in water, $\mathbf{D}_B$ = 3.472×10^{-9} m²/s; Henry's constant for CO_2 solubility in water, H_{CO_2} = 2.849×10^{-7} kmol/m³/(N/m²) at 30°C.

Reaction Mechanism

The production of $CaCO_3$ can be represented by the overall reaction

$$Ca(OH)_2(l) + CO_2(g) \rightarrow CaCO_3(s) + H_2O(l) \quad \text{(CS10.R1)}$$

The steps involved are

1. *Absorption of CO_2*, $CO_2(g) \rightarrow CO_2(l)$ (CS10.R2)
2. *Dissolution of* $Ca(OH)_2$, $Ca(OH)_2(s) \rightarrow Ca(OH)_2(l)$ (CS10.R3)
3. *Ionization of* $Ca(OH)_2$, $Ca(OH)_2(l) \rightarrow Ca^{2+}(l) + 2\,OH^-(l)$ (CS10.R4)
4. *Formation of bicarbonate ion,* $CO_2(l) + OH^-(l) \rightarrow HCO_3^-(l)$ (CS10.R5)
5. *Formation of carbonate ion,* $HCO_3^-(l) + OH^-(l) \rightarrow H_2O(l) + CO_3^{2-}(l)$ (CS10.R6)
6. *Formation of* $CaCO_3$, $Ca^{2+}(l) + CO_3^{2-}(l) \rightarrow CaCO_3(l)$ (CS10.R7)
7. *Precipitation of* $CaCO_3$, $CaCO_3(l) \rightarrow CaCO_3(s)$ (CS10.R8)

Juvekar (1976) has performed exhaustive experimental and modeling work on this system. According to this study, steps 3, 5, 6, and 7 are very fast. As a result, steps 1, 2, and 4 control the overall rate of reaction.

To calculate the maximum capacity of the reactor for the production of $CaCO_3$, we simulate the reaction for different levels of production per shift and check if the operation is feasible.

Solution

Initially assume a production of 5 tons $CaCO_3$ per batch. Assuming a handling loss of 2%, reaction product = $(102/100) \times 5 = 5.10$ tons = 5100 kg. Molecular weight of $CaCO_3$ = 100, $CaCO_3$ in reaction product = 5100/100 = 51 kmoles. Stoichiometric $Ca(OH)_2$ required = 51 kmoles, stoichiometric CO_2 required = 51 kmoles. An estimation of the operating parameters is given below.

Gas Flow Rate

CO_2 is available from another plant as a 20 mol% stream (remaining 80% inerts, N_2). Assuming 5% excess CO_2, the gas stream required = $(51/0.2) \times (105/100) = 267.75$ kmoles. Since the reactor can handle up to 2×10^5 N/m² maximum process pressure, choose a lower operating pressure of 1.8×10^5 N/m². Choose room-temperature operation, $T = 30°C = 303.16$ K. Initial gas volume = $nR_gT/P = (267.75 \times 8314 \times 303.16)/1.8 \times 10^5 = 3749$ m³. Since this volume is very high compared to the reactor volume of 10 m³, choose semibatch mode of operation with continuous supply of feed gas and removal of residual gas to maintain the reactor pressure. In view of this, a slurry of $Ca(OH)_2$ in water is prepared and CO_2 is bubbled through it for 4 h. The mother liquid phase will be initially saturated with $Ca(OH)_2$. As the reaction proceeds, $CaCO_3$ precipitates out. The mother liquor can be separated from the solid $CaCO_3$ precipitate by filtration. For ease of operation and control, assume constant gas flow rate.

Inlet gas flow rate = Q = 3749 m³/(4 h × 3600) = 0.26 m³/s
Reactor inner diameter T = 2.19 m, height H = 2.64 m
Area of cross section of the reactor $S = (\pi/4)T^2 = 3.767$ m²
Superficial gas velocity at the inlet, $u_{G0} = Q/S = 0.069$ m/s
F_1, kmol/s of inert = 267.75/(4 × 3600) × mol fraction of inerts
= {267.75/(4 × 3600)} × (1 − 0.2) = 0.0149 kmol/s
x at the outlet = 0.012

Initial solids holdup ε_{S0}:
Of the reactor volume of 10 m³, assume 20% head space.
Volume of the dispersion = 10 × [1 − (20/100)] = 8 m³.
Stoichiometric quantity of CO_2 required for the reaction = 51 kmol/s
Initial weight of $Ca(OH)_2$ = 51 × 74 = 3774 kg.
Density of solid $Ca(OH)_2$, ρ_S = 2080 kg/m³.
Volume of solids V_S = 3774/2080 = 1.814 m³.
Initial holdup of solids, ε_{S0}
$\varepsilon_{S0} = V_S/V_D = 0.2268$.

Gas holdup ε_G:
Assume an impeller speed (N) of 3.82 rps. The gas holdup is calculated using the following correlation (Rewatkar et al. 1993):

$$\varepsilon_G = 3.54 \left(\frac{D}{T}\right)^{2.08} \left(\frac{N^2D}{g}\right)^{0.51} \left(\frac{Q}{(N)D^3}\right)^{0.43} = 0.233 \quad \text{(CS10.1)}$$

Liquid holdup ε_L:

$$\varepsilon_L = 1 - \varepsilon_G - \varepsilon_{S0} = 0.54$$

Power consumption:
To calculate the power consumption in the absence of gas (P_L),

$$\text{Average density}, \bar{\rho} = \rho_L(1 - \varepsilon_{S0}) + \rho_S\varepsilon_{S0}$$

$$P_L = P_0 \rho N^3 D^5 = 65565 \text{ W} \tag{CS10.2}$$

In the presence of gas, the power consumption (P_G) is lower at the same speed. It is estimated using the following correlation of Rewatkar and Joshi (1991b, d):

$$\left(\frac{P_G}{P_L}\right) = 0.434 \left(\frac{Q}{ND^3}\right)^{-\alpha_1} \left(\frac{N^2 D}{g}\right)^{-\beta_1} \left(\frac{W}{D}\right)^{0.174} \left(\frac{D}{T}\right)^{-0.64} \tag{CS10.3}$$

where $\alpha_1 = 0.064 \, T^{-0.94}$, $\beta_1 = 0.148 \, T^{-1.42}$.

$P_G = 27{,}941$ W. This is less than the maximum available power of 60,000 W that the motor can supply to the impeller. Hence, 3.82 rps is permissible. It will be shown later that the available extra power can be used for extra production capacity.

Critical speeds of agitator: Critical impeller speed for solid suspension
The speed should be adequate to keep the particles suspended. The critical speed for suspension in the presence of gas, N_{JSG}, is calculated from the following correlation (Rewatkar and Joshi 1991d):

$$\frac{(u_{S\infty}^{0.5} - \alpha u_c)}{N_{JSG} D}\left(\frac{T}{D}\right) X_S^{0.1} u_G^{0.14} = 0.18 \tag{CS10.4}$$

where

$$\alpha = 0.9 u_{S\infty}^{1.27}$$

In this equation for N_{JSG}, $u_{S\infty}^{1.27}$ is the terminal settling velocity of the particle in an infinite medium and is calculated by the standard procedure (McCabe et al., 1993), giving $u_{S\infty}^{1.27} = 0.006$ m/s. X_S, the percent solids loading $= 100 \, \rho_S \varepsilon_S / (\rho_S \varepsilon_S + \rho_L \varepsilon_L + \rho_G \varepsilon_G) = 46.62$.

$$N_{JSG} = 1.15 \text{ rps}$$

Critical speed for complete dispersion N_{CD}:
The speed should be adequate for the gas to be completely dispersed. It is calculated from the following correlation (Harnby et al., 1992):

$$\left(\frac{Q}{N_{CD} D^3}\right) = 0.2 \left(\frac{T}{D}\right)^{0.5} \left(\frac{N_{CD}^2 D}{g}\right)^{0.5} \tag{CS10.5}$$

$N_{CD} = 1.94$ rps. The maximum operating speed $N = 3.82$ rps. This speed is much higher than N_{JSG} and N_{CD}; hence, the solids are suspended and the gas is completely dispersed.

Interfacial area for gas-liquid mass transfer (a):
The value of a was calculated from the following correlation (Lopes de Figueiredo and Calderbank, 1979):

$$a = 593 \left(\frac{P_G}{V_D \varepsilon_L}\right)^{0.25} u_G^{0.75} \tag{CS10.6}$$

$$= 716.9 \text{ m}^2/\text{m}^3$$

Bubble diameter d_b:

$$d_b = 6\varepsilon_G/a = 2 \times 10^{-3} \text{ m}$$

Interfacial area for solid liquid mass-transfer a_{P0}:
Initial particle surface area $a_{P0} = 6\varepsilon_{S0}/d_{P0} = 6 \times 0.2268/0.0001 = 13608.2 \text{ m}^2/\text{m}^3$. As the particle dissolves and reacts, the particle size diminishes until it vanishes.

Schmidt number Sc:

$$Sc = \left(\frac{\mu_L}{\rho_L D_B}\right) = 288.02, \text{ where } B \text{ denotes OH}^- \text{ ion}$$

As soon as $Ca(OH)_2$ dissolves into the aqueous phase, it ionizes almost completely, being a strong base.

Gas-liquid mass transfer coefficient $k_L a$:
The value of $k_L a$ is estimated using the following correlation (Chandrasekharan and Calderbank, 1981):

$$k_L a = \frac{0.0248}{D^4}\left(\frac{P_G}{V_D \varepsilon_L}\right)^{0.55} Q^{0.551/\sqrt{D}} = 2.28 \qquad (CS10.7)$$

In the presense of fine particles $k_L a$ decreases and has been assumed to be 0.49 sec^{-1}.

$$k_L = (k_L a)/a = 6.86 \times 10^{-4} \text{ m/s}$$

Solid-liquid mass transfer coefficient k_{SL}:
The value of k_{SL} is estimated using the following correlation (Rowe et al., 1965):

$$Sh = 2.0 + 0.72(Re_P)^{0.5} Sc^{0.333} \qquad (CS10.8)$$

where

$$Re_P = \frac{d_P u_{s\infty} \rho_L}{\mu_L} = 0.59, \quad Sh = \frac{k_{SL} d_P}{D_B} = 5.6$$

$$k_{SL} = 1.96 \times 10^{-4} \text{ m/s}$$

Estimation of batch time for the reaction:
It is assumed that the gas is completely backmixed. Gas-side resistance for mass transfer is neglected. The reaction is performed under isothermal conditions. Initially, a slurry of $Ca(OH)_2$ in water is prepared and the liquid is saturated with $Ca(OH)_2$. Then CO_2 gas is sparged to the reactor. It is assumed that the number of reacting $Ca(OH)_2$ particles per unit volume of the reactor is constant (no breakage or agglomeration) until the particles are completely consumed. The instantaneous rate of reaction is integrated with respect to time until all the $Ca(OH)_2$ is consumed. Thus, the batch time is obtained:

Chemical Reaction Engineering

$$CO_2 + 2\,OH^- \rightarrow CO_3^{-2} + H_2O \qquad \text{(CS10.R9)}$$

For this reaction, $Z = 2$ (stoichiometric coefficient, moles of OH^- required per mole of CO_2). At any instance t, the rate of reaction of the dissolved carbon dioxide with the OH^- is given by

$$R_A = -k_{CO_2}[A][B] \qquad \text{(CS10.9)}$$

where $[A]$ and $[B]$ denote the concentrations of CO_2 and OH^- in kmol/m³ and k-ion/m³ in the liquid phase, respectively. In the presence of mass transfer with fast chemical reaction in the liquid film, the rate of mass transfer of CO_2 is given by

$$R_A a = k_L a [A]^* \eta \qquad \text{(CS10.10)}$$

where the enhancement factor

$$\eta = \frac{\sqrt{M}}{\tanh\sqrt{M}} \sqrt{\frac{\beta - \eta}{\beta - 1.0}} \qquad \text{(CS10.11)}$$

with

$$\beta = 1 + \frac{[B]}{Z[A]^*}\sqrt{\left(\frac{D_B}{D_A}\right)} = 3.46 \qquad \text{(CS10.12)}$$

$$\sqrt{M} = \frac{\sqrt{D_A k_{CO_2}[B]_b}}{k_L} = 1.51 \qquad \text{(CS10.13)}$$

$$[A]^* = H_{CO_2} p_{CO_2}$$

where H_{CO_2} = Henry's law constant for CO_2
Partial pressure of CO_2 is calculated from

$$p_{CO_2} = P\frac{x}{(1+x)}$$

x being the mol CO_2 per mol inert in the gas.

The rate of formation of OH^- by dissolution of $Ca(OH)_2$ is given by $k_{SL}a_P([B]_s - [B]_0)$, where $[B]_S$ is the concentration of OH^- at the solid-liquid interface and is the saturation solubility of $Ca(OH)_2$, and $[B]_0$ = bulk concentration of OH^-.

In the initial stages of the reaction, when the particle size is large, the rate of solid dissolution is high as compared to the rate of CO_2 absorption with reaction. The bulk concentration of OH^- is equal to the saturation concentration $[B]_S$, i.e., $[B]_0 = [B]_S$:

$$\frac{k_{SL}a_P[B]_s}{Z} \gg k_L a[A]^*\eta \qquad \text{(CS10.14)}$$

Therefore

$$R_A a = k_L a \text{H}_{\text{CO}_2} P\left(\frac{x}{1+x}\right)\eta \tag{CS10.15}$$

where x = kmol CO_2/kmol inerts at the outlet and has the same value throughout the backmixed vessel. The above equation is solved for x at any time t and the rate of $CaCO_3$ production (kg/s) is calculated from

$$\frac{d}{dt}(PDN)_{\text{CaCO}_3} = R_A a V_D M_{W\text{CaCO}_3} \tag{CS10.16}$$

As the $Ca(OH)_2$ is consumed, its particle size decreases:

$$-\frac{d}{dt}\left[\gamma_P\left(\frac{\pi}{6}d_P^3\right)\rho_S V_D\right] = R_A a V_D M_{W\text{Ca(OH)}_2} \tag{CS10.17}$$

$$-\frac{d}{dt}(d_P) = \frac{2 R_A a V_D M_{W\text{Ca(OH)}_2}}{\pi \gamma_P \rho_S V_D d_P^2} \tag{CS10.18}$$

where γ_P = number of $Ca(OH)_2$ particles per unit volume of dispersion.

In the final stages of the reaction, when the particle size is very small, the rate of solid dissolution is very low as compared to the rate of CO_2 absorption with reaction. The bulk concentration of OH^-,

$$\frac{k_{SL} a_P [B]_S}{Z} \ll k_L a [A]^* \eta \tag{CS10.19}$$

$$[B]_0 = 0$$

Thus, the overall rate is governed by the supply of OH^- ions. In this phase, the production rate of $CaCO_3$ is given by

$$\frac{d}{dt}(PDN)_{\text{CaCO}_3} = \frac{1}{2} k_{SL}\left(\gamma_P \pi d_P^2\right)[B]_S V_D (M_{W\text{CaCO}_3}) \tag{CS10.20}$$

As the $Ca(OH)_2$ is consumed, there is a corresponding decrease in the particle size:

$$-\frac{d}{dt}\left[\gamma_P\left(\frac{\pi}{6}d_P^3\right)\rho_S V_D\right] = \frac{1}{2} k_{SL}\left(\gamma_P \pi d_P^2\right)[B]_S V_D M_{W\text{Ca(OH)}_2} \tag{CS10.21}$$

$$-\frac{d}{dt}(d_P) = \frac{k_{SL}[B]_S M_{W\text{Ca(OH)}_2}}{\rho_S} \tag{CS10.22}$$

The rates of production and decrease in particle diameter are integrated with time until the particle vanishes. In this way, the batch time is calculated and is 4.09 h, which is close to the desired batch time for the reaction.

TABLE CS10.1
Production Capacity of a Given Stirred Tank Reactor

Case	Batch Production (Tons)	ε_S	$k_L a$ (sec^{-1})	N (sec^{-1})	ε_G	N_{JSG} (sec^{-1})	N_{CD} (sec^{-1})	P_L (W)	P_G (W)
1	4	0.181	0.44	3.23	0.181	1.05	1.94	44151	19094
2	5	0.227	0.49	3.82	0.233	1.11	1.94	65565	27941
3	6	0.272	0.54	4.07	0.277	1.16	1.94	90229	38000
4	7	0.318	0.58	4.36	0.317	1.20	1.94	113945	47554
5	8	0.363	0.60	4.58	0.351	1.24	1.94	134737	55838
6	8.5	0.386	0.61	4.75	0.366	1.30	1.94	143716	59387

Note: In the above table it has been confirmed in all the cases that operating speed is more than critical impeller speed for solid suspension and gas dispersion.

To calculate the maximum attainable production, the above procedure is repeated for different assumed levels of production. The results are shown in Table CS10.1. However, for levels above 8.5 tons per batch, it can be seen that, even after utilizing the maximum motor power of 60 kW, the maximum feasible operating speed is ~30% higher than that required for suspension of the solids in the presence of gas, $N_{max} < N_{JSG}$. This maintains a safe margin to ensure complete suspension. If we increase the production further, then solid and gas holdups are very high, and N_{max} is marginally higher than N_{JSG}, which may not be a safe design.

Hence, the maximum production is 8.5 tons per batch, with a batch reaction time of ~4 h. Extra time of perhaps 0.75 to 1.5 h is of course required to unload, clean, and reload the reactor for the next batch run. The profile of particle diameter as percentage of initial diameter is plotted against time in Figure CS 10.1.

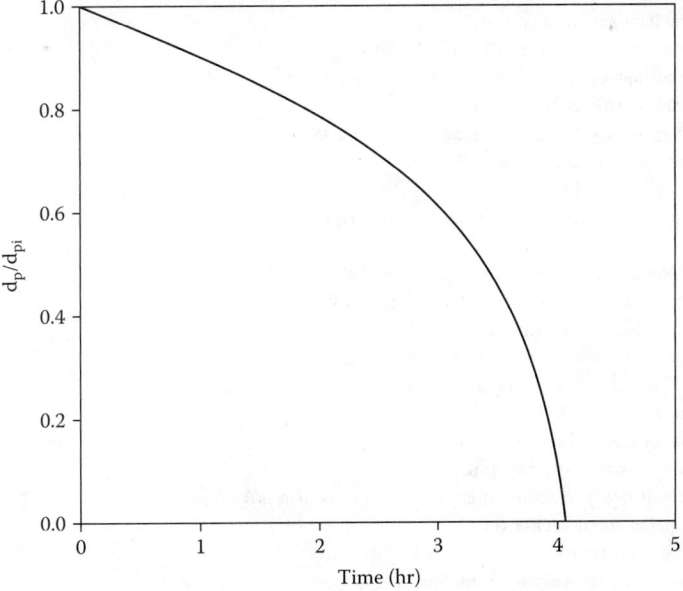

FIGURE CS10.1 The fate of a calcium hydroxide particle in the reactor.

Lesson

This case study is an example of how a common reaction can provide the basis for modeling a novel reaction system: a gas-liquid-solid reaction performed in the batch mode; the solid in this case is first dissolved followed by chemical reaction with a product of the reactive absorption of the solute gas. Unlike Case Study 11.9, where all steps were in series, here some steps occur in parallel. Moreover, the rate-controlling mechanisms often change with time and process conditions. These facets of the problem are dealt with to determine the maximum production capacity of a reactor, which can often be a cost-determining issue. The lesson here is that maximizing the use of an existing reactor is sometimes preferable to designing a new one.

CASE STUDY 11.11 GAS-LIQUID-SOLID (CATALYTIC) REACTION: HYDROGENATION OF AN ORGANIC COMPOUND

Nomenclature for Case Study 11.11

A	constant in Equation (CS11.7) (–)
$[A]_b$	bulk concentration of solute gas A (kmol/cm^3)
$[A^*]$	saturation concentration of gas A (kmol/cm^3)
$[A]_S$	surface concentration of gas A (kmol/cm^3)
a_1, b_1, c_1	constants in Equation (CS11.9) (–)
A	effective gas-liquid interfacial area per unit volume (m^2/m^3)
a_P	area of the catalyst particle per unit volume (m^2/m^3)
$[B]_b$	concentration of the liquid phase reactant (gmol/cm^3)
$[B]_S$	surface concentration of B (kmol/cm^3)
C_3	constant in Equation (CS11.6) (–)
D	impeller diameter (m)
$\mathbf{D}_0$	diffusivity (m^2/s)
d_b	bubble diameter (m)
d_P	diameter of catalyst particle (m)
d_P^*	critical particle size of diffusion control operation (m)
Fl	flow number, Q_G/ND^3 (–)
Fr	Froude number, u^2/gd_p (–)
g	gravitational constant (m/s^2)
k_i	rate constant, $i = 1,2,\ldots$ (conc^{1-n}/time)
k_{SL}	solid-liquid mass transfer coefficient (m/s)
k_L	mass transfer coefficient (m/s)
L	characteristic length of the catalyst particle (–)
$\sqrt{M}$	enhancement factor (–)
N	impeller speed (rps)
N_{CG}	critical impeller speed for gas induction (rps)
N_M	impeller speed at which mass transfer resistance is eliminated (rps)
$\mathbf{P}o$	power consumption in absence of gas (W)
$\mathbf{P}_G$	power consumption in presence of gas (W)
Q_G	gas induction rate (m^3/s)
Q_{GS}	rate of gas induction in presence of sparging (m^3/s)
Q_{UG}	net rate of escape of the unreacted gas into the head space (m^3/s)
R	impeller radius (m)
R_A	rate of reaction of A (kmol/m^2 s)
Re	Reynolds number, $d_p u_{G0}\rho/\mu$ (–)
r_A	overall rate of reaction of component A (kmol/(m^3 s))
S	impeller submergence (m)
T	tank diameter (m)
u_{G0}	superficial gas velocity at the inlet (m/s)
u_I	impeller tip speed in Equation (CS11.7) (m/s)
V_L	liquid volume (m^3)
W	solid loading (kg/m^3)

Chemical Reaction Engineering

Greek

α^*	constant in Equation (CS11.7) (–)
$\alpha_1, \alpha_2, \alpha_3, \alpha_4$	constants in Equation (CS11.6) (–)
$\beta_1, \beta_2, \beta_3$	constants in Equation (CS11.8) (–)
ε	effectiveness factor of the catalyst particles (–)
ε_S	solid holdup (–)
ε_G	gas holdup (–)
λ^*	constant in Equation (CS11.7) (m)
σ	surface tension (N/m)
ρ_L	density of liquid (kg/m³)
θ	constant in Equation (CS11.4) (–)
ψ	slip factor (–)

Subscript

G	gas
p	particle
S	catalyst surface
L	liquid

The procedure to determine the rate-controlling step is described below for a gas-liquid-solid reaction. The specific case of a gas-liquid reaction is only a part of this entire process. A reaction sequence between a gas and a liquid (catalyzed by solid particles) involves many steps. These are shown schematically in Figure CS11.1a. As a first step, the gas-phase reactant has to diffuse through a gas film to the gas-liquid interface. If the reaction involves a pure gas such as hydrogen, oxygen, and so on, then the diffusional resistance in the gas phase is generally absent. The gaseous reactant then dissolves in the liquid phase at the gas-liquid interface and is subsequently transported to the bulk liquid through the liquid film. The gaseous reactant and the reactant from the liquid phase then diffuse through the liquid film near the solid catalyst particles. Once they reach the solid surface, both reactants diffuse through the porous catalyst particles. These reactants are then adsorbed on the catalytically active sites. The reaction occurs on these active sites, and products are formed. The products of reaction then have to desorb from such active sites and diffuse through the pores of the catalyst and the liquid film into the bulk liquid.

Obviously, many steps are involved, and any step can be the rate-determining one. In addition, if the reaction is highly endothermic or exothermic (typical of oxidation, hydrogenation reactions), then heat has to be supplied or removed from the reactor. Sometimes the rate of heat transfer may control the overall rate of the reaction. In gas-liquid reactions catalyzed by solid particles, the suspension of catalyst particles can sometimes control the overall rate of reaction. As a first step in the process design portfolio, the rate-controlling step has to be determined, as described below.

The various resistances for mass transfer and the concentration driving force for the transport of the reactants can be seen in Figure CS11.1a. Based on these concentrations and the volumetric transport coefficients, the overall rate of reaction can be written as

$$R_A a = k_L a \left([A]^* - [A]_0\right)$$
$$= k_{SL} a_P \left([A]_0 - [A]_S\right) \quad \quad (CS11.1)$$
$$= \varepsilon k_2 [A]_S [B]_S$$

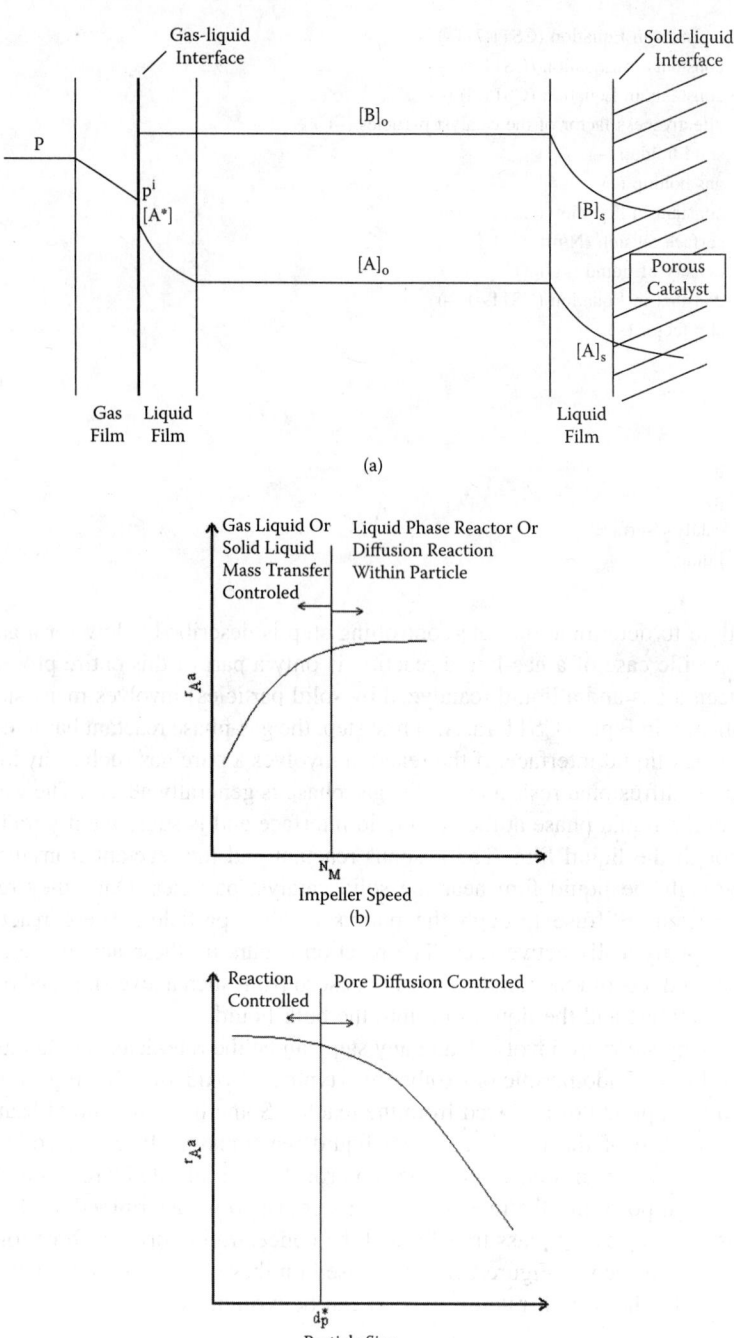

FIGURE CS11.1 The three-phase catalytic reactor: (a) concentration profiles of gas and liquid phase components; (b) effect of impeller speed on overall rate of reaction; and (c) effect of particle size on overall rate of reaction at stirrer speeds higher than N_M. (From Patwardham, A. W. and Joshi, J. B., 1998.)

Here, ε is the effectiveness factor of the catalyst particle (see Section 11.3). Eliminating $[A]_0$ and $[A]_s$ and rearranging,

$$\frac{[A]^*}{R_A a} = \frac{1}{k_L a} + \frac{1}{k_{SL} a_P} + \frac{1}{\varepsilon k_2 [B]_s} \tag{CS11.2}$$

The gas-liquid and the solid-liquid interfacial areas per unit volume can be written in terms of the holdup of the respective phases and the mean bubble/particle diameter. Assuming the bubbles and particles to be spherical, Equation (CS11.2) takes the following form:

$$\frac{[A]^*}{R_A a} = \frac{d_b}{6 k_L \varepsilon_G} + \frac{d_P}{6 k_{SL} \varepsilon_S} + \frac{1}{\varepsilon k_2 [B]_s} \tag{CS11.3}$$

This equation is very useful in determining the rate-controlling step.

The gas-liquid-solid reaction under investigation is performed in a laboratory-scale stirred contactor (500 mL to 2 L). The catalyst type, solid loading, particle size, and reaction conditions (temperature and pressure) are chosen. Under these conditions, the impeller speed is varied, and the overall rate of reaction is monitored. At low impeller speeds, the gas-liquid and solid-liquid mass transfer coefficients ($k_L a$ and $k_{SL} a_P$) are low. The overall reaction is therefore mass transfer controlled. As the impeller speed increases, the values of these mass transfer coefficients increase, and as a result the overall rate of reaction increases. This is shown in Figure CS11.1b. At high impeller speeds, the rates of gas-liquid and solid-liquid mass transfer are high, and the overall rate of reaction is limited by the reaction kinetics or diffusion through the porous catalyst particles. The speed above which the overall reaction is controlled by internal diffusion or chemical reaction is denoted as N_M.

At impeller speeds above N_M, the second or third term on the right-hand side of Equation (CS11.3) becomes controlling; i.e., the overall rate is controlled by intrinsic chemical reaction or pore diffusion. Under such conditions, if experiments are performed at different particle sizes, then the effects of internal diffusion and chemical reaction can be elucidated.

For a first-order chemical reaction occurring within the catalyst particle, the effectiveness factor of the catalyst particle is given as

$$\varepsilon = \frac{\tanh(\phi)}{\phi} \tag{CS11.4}$$

Here, ϕ is given by $\phi = s\sqrt{k_1/\mathbf{D}}$, where s is the generalized length defined by Equation (11.49). At speeds higher than N_M, the reaction occurs with different particle sizes, and a graph of the overall rate of reaction vs. particle size is plotted. Such a graph is shown in Figure CS11.1c. For small particle sizes, the pore diffusional limitations are essentially absent ($\varepsilon = 1$), and the overall rate of reaction is controlled by the chemical reaction. As the particle size is increased, the diffusional limitations become increasingly important, and above a certain particle size d_P^*, the overall rate of reaction is determined by the diffusion of the reactants into the catalyst pores. The evaluation of the kinetic parameters for the reaction should be performed at impeller speeds higher than N_M and particle sizes lower than d_P^*. The reaction taking place on the catalyst surface itself is composed of various steps, such as (1) adsorption of the reactants on the active sites, (2) chemical reaction at the active sites, and (3) desorption of the products from the active sites. The rate of reaction can be written in terms of these various steps (see Section 11.3).

At impeller speeds well below N_M, the reaction may be gas-liquid and/or solid-liquid mass transfer controlled. If the solid loading (ε_S) is varied under these conditions ($N \ll N_M$), then only

TABLE CS 11.1
Effect of Variable Parameters on the Rate of Reaction

Rate controlling step	Effect of Increase in Values of Variables on the Overall Rate of Reaction							
	$[A^*]$	$[A]_b$	$[A]_s$	$[B]_0$	$[B]_s$	ε_s	d_p	N
Gas-liquid mass transfer	$\propto [A^*]$	$\propto [A^*] - [A]_b$	$[A]_s = [A_b]$	No effect	No effect	No effect	No effect	increases
Solid-liquid mass transfer	$[A^*] = [A]_0$	$\propto [A]_b$	$\propto [A]_b - [A]_s$	No effect	No effect	$\propto \varepsilon_s$	$\propto (1/d_p)$	increases
Pore diffusion	$[A^*] = [A]_s$	$[A]_b = [A]_s$	$\propto [A]_s$	No effect	No effect	No effect	$\propto (1/d_p)$	No effect
Chemical reaction	$[A^*] = [A]_S$	$[A]_b = [A]_s$	$\propto [A]_s^m$	$[B]_b = [B]_S$	$\propto [B]_s^n$	$\propto \varepsilon_s$	No effect	No effect

the second term on the right-hand side of Equation (CS11.3) is affected. If the rate of reaction varies linearly with the solid loading, then it can be concluded that the overall reaction is solid-liquid mass transfer controlled. However, if the rate of reaction is unaffected by the solid loading, then gas-liquid mass transfer is the rate-controlling step.

Whenever heat effects are important, a proper energy balance has to be established. The rate at which heat is generated inside the reactor and the rate at which heat is transferred to and from the heat transfer media need to be quantified to determine whether the rate of heat transfer is the controlling step. Table CS11.1 summarizes the effect of various parameters such as $[A^*]$, $[A]_S$, $[B]_b$, d_P, ε_S, impeller speed, and so forth on the overall rate of reaction for different rate-controlling steps. For the geometry under consideration, the values of $k_L a$ and $k_{SL} a_p$ can be estimated from available correlations. Using these values and the estimated kinetic parameters, it is possible to determine the rate-controlling step. During the course of the reaction, one or more operating conditions often change. For example, reactant concentration, operating temperature, pressure, and other conditions may change. Then the rate-controlling step itself may change with time or location in the vessel. In such cases, the rate-controlling step has to be determined as a function of time/location in the reactor. Once the rate-controlling step is known, the overall rate of reaction can be written. This overall rate of reaction can then be integrated with respect to space or time to determine the throughput of the reactor.

Solved Example

Hydrogenation of an organic compound is a typical example of such reactions. The gas and liquid phase reactants are denoted by A and B, respectively. For the purpose of illustration, it will be assumed that the overall reaction is mass transfer controlled. It is desirable to perform such a reaction in a gas-including contractor along with a sparger. The lower impeller is a pitched blade up-flow turbine (PBTU), and a large ring sparger will be assumed. Let u_{G0} be the superficial velocity of the sparged gas. A part of this sparged gas reacts in the vessel, and the remainder escapes into the head space. Simultaneously, the gas is recycled into the liquid using a gas-inducing impeller. Part of the entrained/induced gas also reacts, and the remainder escapes into the head space. For an overall material balance, the maximum permissible superficial gas velocity (u_{Go}) should be such that the rate of gas sparging equals the rate of reaction of the gas. This ensures that the head space pressure remains unchanged with time. Alternatively, if an overall material balance is written for the head space, then it is clear that the net rate of escape of the unreacted gas into the head space (Q_{UG}) must be equal to the rate of gas induction.

Figure CS11.2a shows a comparison of fractional gas holdup under sparging conditions, with and without the use of a gas-inducing impeller. In the absence of a gas-inducing impeller, such a system behaves like a conventional mechanically agitated contactor (MAC). The comparison is made in terms of gas holdup as a function of power consumption per unit volume for different superficial velocities of sparged gas ($u_{Go} = 0$, 6, 18, and 29 mm/s). It can be seen that the fractional

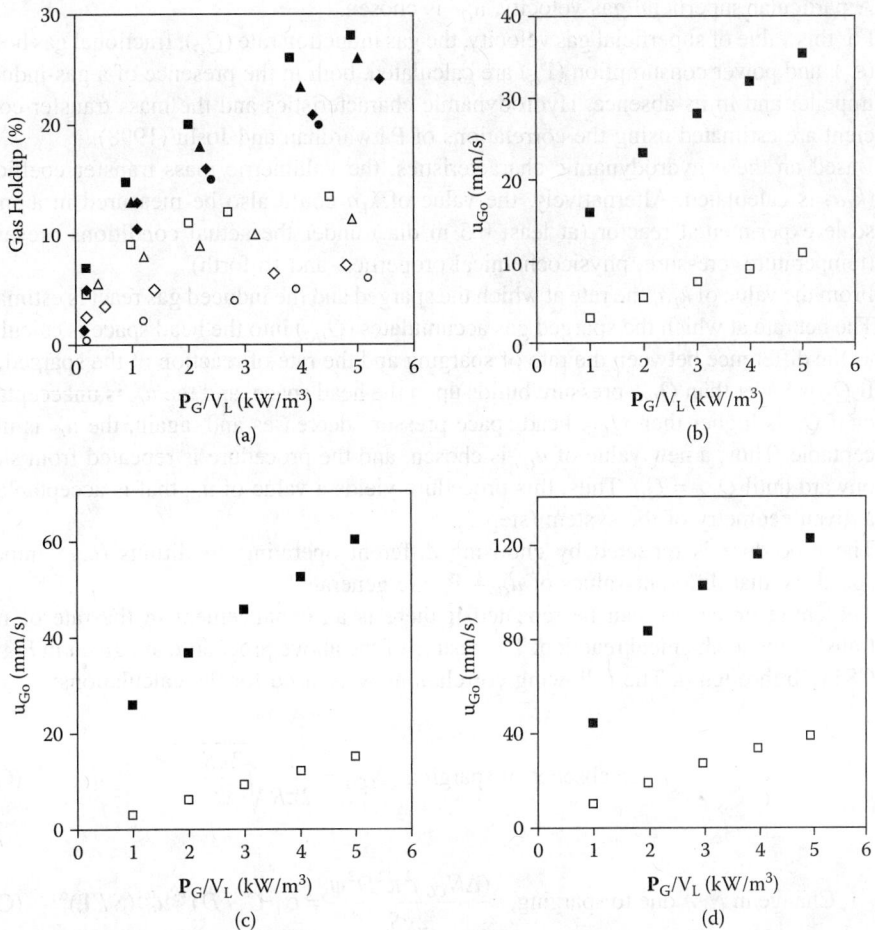

FIGURE CS11.2 Comparison of stirred tank reactor with and without a gas inducing impeller. (a) Comparison of fractional gas holdup under sparging conditions with and without a gas-inducing impeller. Keys: open symbols: without gas-inducing impeller, filled symbols: in presence of gas inducing impeller ○: 0 mm/s, ◊: 6 mm/s, △: 18 mm/s, o: 29 mm/s. (b) Permissible values of superficial gas velocity with respect to power consumption: Mass transfer controlled operations. Keys: open symbols: without gas-inducing impeller, filled symbols: in presence of gas-inducing impeller. (d) Permissible values of superficial gas velocity with respect to power consumption: Fast reaction controlled operations for $\sqrt{M}$ = 5. (d) Permissible values of superficial gas velocity with respect to power consumption: Fast reaction controlled operations for $\sqrt{M}$ = 50. (From Patwardnam, A. W. and Joshi, J. B., 1998.)

gas holdup in the presence of a gas-inducing impeller is substantially higher than for MAC. Similarly, u_{Go} = 0 corresponds to the case of gas induction alone (no sparging). It can also be seen that, as the rate of sparging increases, the gas holdup increases. The following stepwise procedure is recommended for design.

Stepwise Procedure

1. The stoichiometry of the reaction under consideration is known. Furthermore, the solubility of gas and the physicochemical properties of the gas and the liquid phases are known.
2. The geometry of the system is first chosen, e.g., tank diameter, impeller design, D/T ratio, clearance, submergence, sparger size, impeller speed, and others.

3. A particular superficial gas velocity, u_{Gi}, is chosen.
4. For this value of superficial gas velocity, the gas induction rate (Q_G), fractional gas holdup (ε_G), and power consumption ($\mathbf{P}_G$) are calculated, both in the presence of a gas-inducing impeller and in its absence. Hydrodynamic characteristics and the mass transfer coefficient are estimated using the correlations of Patwardhan and Joshi (1998).
5. Based on these hydrodynamic characteristics, the volumetric mass transfer coefficient ($k_L a$) is calculated. Alternatively, the value of $k_L a$ could also be measured in a small-scale experimental reactor (at least 0.5 m dia.) under the actual conditions prevailing (temperature, pressure, physicochemical properties, and so forth).
6. From the value of $k_L a$, the rate at which the sparged and the induced gas react is estimated.
7. The net rate at which the sparged gas accumulates (Q_{UG}) into the head space is calculated as the difference between the rate of sparging and the rate of reaction of the sparged gas.
8. If Q_G is lower than Q_{UG}, pressure builds up in the head space, and the u_{Gi} is unacceptable, or if Q_G is higher than Q_{UG}, head space pressure decreases and, again, the u_{Gi} is unacceptable. Thus, a new value of u_{Go} is chosen, and the procedure is repeated from step 3 onward until $Q_{UG} = Q_G$. Thus, this procedure yields a value of u_{Gi} that is acceptable for a given geometry of the system (step 2).
9. The procedure is repeated by choosing different operating conditions (e.g., impeller speed) so that different values of $u_{Go} - \mathbf{P}_G$ are generated.
10. The entire procedure can be repeated if there is an enhancement in the rate of mass transfer due to chemical reaction. The results of the above procedure are given in Figures CS11.2b through d. The following correlations were used for the calculations:

$$N_{CG} \text{ in absence of sparging, } N_{CG} = \frac{1}{2\pi R}\sqrt{\frac{2gS}{\psi}} \qquad \text{(CS11.5)}$$

$$\text{Change in } N_{CG} \text{ due to sparging, } \frac{(\Delta N_{CG})^2 \pi^2 D^2 \psi}{2gS} = \alpha_1 (C_3/D)^{\alpha_2} u_G^{\alpha_3} (S/T)^{\alpha_4} \qquad \text{(CS11.6)}$$

Gas induction rate in absence of sparging,

$$Q_G = \lambda^* N R^2 \left(1 - \frac{2gS}{\psi u_I^2}\right) + \alpha^* N R^3 \left(1 - A\left(\frac{2gS}{\psi u_I^2}\right)^{3/2}\right) \qquad \text{(CS11.7)}$$

$$\text{Change in gas induction rate due to sparging, } 1 - \frac{Q_{GS}}{Q_G} = \beta_1 Fl^{\beta_2}(C_3/D)^{\beta_3} \qquad \text{(CS11.8)}$$

$$\text{Fractional gas holdup, } \varepsilon_G = a_1(D/T)^{b_1} (\text{Re } Fr Fl)^{c_1} \qquad \text{(CS11.9)}$$

Power consumption,

$$\frac{\mathbf{P}_G}{\mathbf{P}_O} = 0.1 \left(\frac{NV_L}{Q_G}\right)^{1/4} \left(\frac{N^2 D^4}{W V_L^{2/3} g}\right)^{-1/5} \qquad \text{(CS11.10)}$$

$$\mathbf{P}_O = N_P \rho_L N^3 D^5$$

The enhancement in the rate of mass transfer due to chemical reaction in Figure CS11.2c is $\sqrt{M} = 5$, and in Figure CS11.2d is $\sqrt{M} = 50$.

For Figures CS11.2b, c, and d, it was assumed that the production rate was 50 tons/day of compound R by the following reaction:

$$3A(g) + B(l) \rightarrow R(l)$$

The solubility of A in the liquid $[A]$ was assumed to be 0.1 kmol/m^3. It was also assumed that the reaction takes place at 150°C and 10 atm pressure. The performance was estimated using the correlations given above. In addition, the following correlations were used:

$$\text{Mean bubble diameter } d_b = 4.15 \left(\frac{(P_G/V_L)^{0.4} \rho_L^{0.2}}{\sigma^{0.6}} \right)^{-1} \varepsilon_G^{0.5} + 0.09 \quad \text{(CS11.11)}$$

$$\text{Interfacial area per unit volume, } a = \frac{6\varepsilon_G}{d_b} \quad \text{(CS11.12)}$$

Volumetric mass transfer coefficient $k_L a = 0.0002\, a$.

Figure CS11.2b shows that, for a particular value of power consumption per unit volume, a higher value of u_{Gi} is permissible when a gas-inducing impeller is used along with a sparger. For example, for a power consumption of 2 kW/m^3, the permissible u_{Gi} is about 23 mm/s when a gas-inducing impeller is used as compared to about 6 mm/s when the gas-inducing impeller is not present. A higher permissible u_{Gi} in existing equipment implies that higher productivity can be realized when a gas-inducing impeller is used along with a sparger. If the productivity is to remain the same, then a higher permissible u_{Gi} implies that a smaller size of reactor can be used. If the reaction involves highly corrosive chemicals (e.g., hydrogenation of chloro compounds), then an exotic material of construction is needed (Hastelloy, titanium, tantalum, and so on). A reduction in size of such a reactor implies a considerable saving in equipment cost. Alternatively, if the same permissible u_{Gi} is to be maintained (corresponding to the desired rate of production), then much lower power consumption is required when the gas-inducing impeller is employed. That is, for a u_{Gi} of 10 mm/s, 5 kW/m^3 are required when a gas-inducing impeller is not used as compared to about 0.5 kW/m^3 when one is used. Reduced power consumption results in a significant reduction in operating costs and is particularly important when the cost of electricity is high. A comparison of Figures CS11.2b and d reveals that the difference in u_{Gi} or power consumption increases as the enhancement factor increases.

From the above discussion, it is clear that a given rate of production can be achieved by different combinations of reactor size and power consumption. The selection of the optimum reactor size is based on the annualized cost of the reactor. The annualized cost consists of the cost of capital (depreciation and interest on fixed cost) and the operating cost. The fixed cost consists mainly of equipment cost (that is, the material cost plus fabrication cost), and the operating cost consists mainly of electricity (power) cost. For the purpose of this illustration, five different materials of construction were selected having costs 0.3, 1.0, 3.0, 10.0, and 30.0 $/kg. This range of costs covers practically all of the materials commonly used in industry, such as mild steel, stainless steel, glass-lined vessels, Hastelloy, titanium-lined vessels, and so forth. Two levels of (depreciation + interest) were examined: 20% and 50% per annum. Three costs of electricity were used: 0.035, 0.10, and 0.30 $/kWh.

For a particular reactor volume and operating pressure, the thicknesses of shell and dished ends were calculated. Knowing the density of the material, the weight of the shell was estimated. An extra 50% was added for the nozzles, supports, shaft, impellers, agitator drive, and so on. In addition,

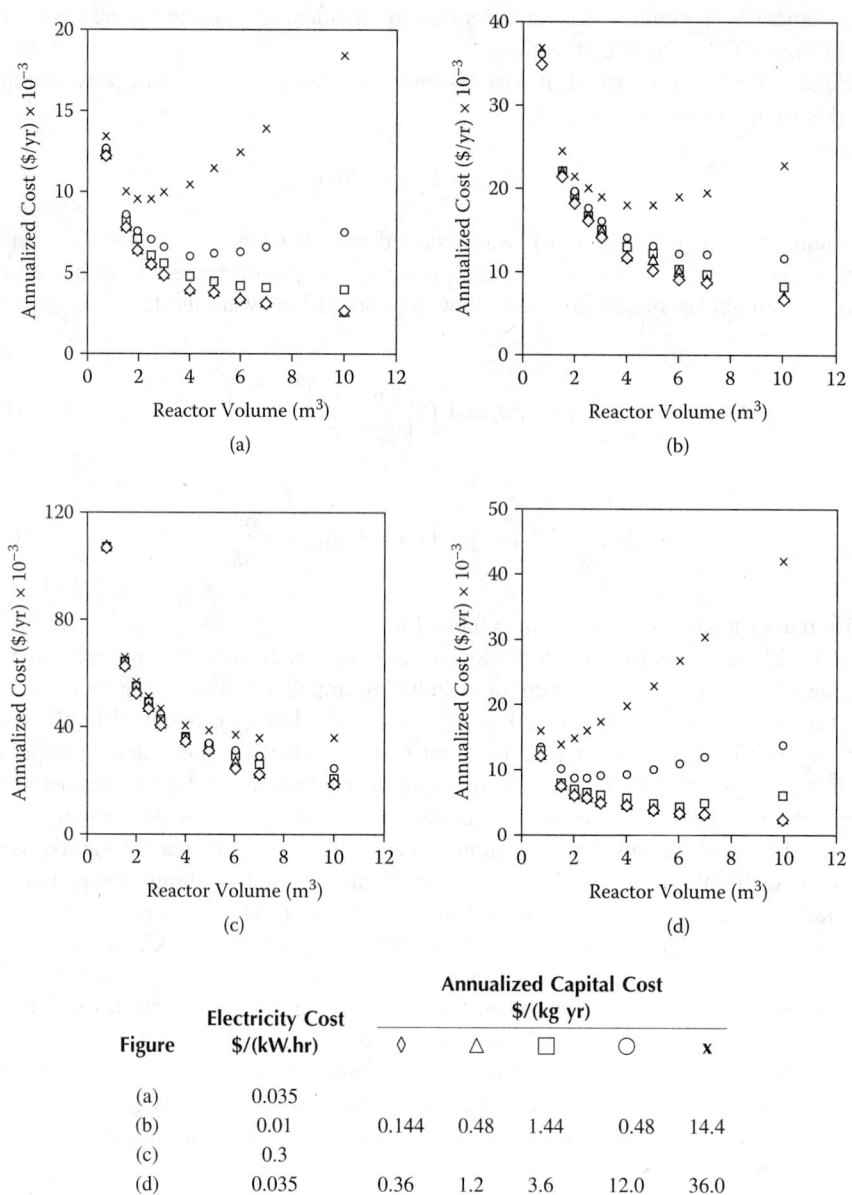

FIGURE CS11.3 Optimization of hydrogenation reactor on the basis of annualized cost. (From Patwardnam, A. W. and Joshi, J. B., 1998.)

the fabrication cost was considered as 60% of the material cost. The operating cost was calculated as the product of power consumption and the electricity cost. The optimum reactor size was determined based on the annualized cost (capital + operating).

Figure CS11.3 shows the annualized cost for various combinations of fixed and operating cost. The principal operating cost includes electricity, and this has been considered at four levels: 0.035, 0.01, 0.3, and 0.035 $/kWh in Figures CS11.3a through d. For every figure, the annualized capital cost consists of two parts:

1. Cost of material of construction
2. Level of depreciation, interest, maintenance, etc.

Chemical Reaction Engineering

For these two factors, five levels of annualized capital costs are considered in all four figures, and the values are shown by symbols. Several conclusions can be drawn. For low-cost materials like mild steel, the annualized cost decreases as the reactor volume is increased. This is because the fixed cost is low, and the operating cost contributes significantly to the annualized cost. Thus, choosing a larger reactor volume and consequently keeping a lower level of power consumption reduces the annualized cost. For medium-cost materials (10 $/kg), there is a certain reactor size at which the annualized cost is minimum. If the reactor size is larger than this value, the fixed cost increases more than the decrease in power consumption. On the other hand, if the reactor volume is lower than the optimum value, the operating cost increases more than the decrease in the fixed cost. This optimum reactor size shifts to lower volumes as the cost of the material increases. For example, if the material cost is 10 $/kg, the optimum reactor size is about 4 m^3, and if the material cost is 30 $/kg, the optimum reactor size is about 2 m^3. Thus, for any values of the material and electricity costs, the annualized cost can be calculated, and the optimum size of the reactor can be found.

Comparing Figures CS11.3a through c (electricity cost 0.035, 0.10, and 0.30 $/kWh), it can be seen that the optimum reactor size shifts toward higher volumes. Thus, as the cost of electricity rises, larger reactors and decreased power consumption reduce the operating costs. A comparison of Figures CS11.3a and d shows that, as the capital cost increases (from 20 to 50%), the optimum reactor size decreases. This shows the advantage of smaller reactor size to keep the fixed cost low, even though the power consumption (operating cost) increases.

Lesson

This case study is concerned with a three-phase gas-liquid-solid (catalytic) reaction. A systematic stepwise procedure has been described for determining the rate-controlling step, which depends on the catalyst type, particle size, operating pressure and temperature, mass transfer coefficient, and concentrations of reactants and products. As indicated, the rate-controlling step may change with location in a continuous reactor and with time in a batch reactor.

For catalytic hydrogenation, where complete gas utilization is of the utmost importance, the self-inducing type of reactor has been shown to be superior to the conventional gas-liquid stirred reactors. The selection of the reactor volume and level of power consumption per unit volume depends on the material of construction and power cost. By considering a wide range of material and power costs, the selection procedure has been elucidated. Such an optimization exercise should be extended to catalyst (type, particle size, loading, and resulting cost per ton of product), pressure, and temperatures of operation. Often, a significant cost reduction can be obtained by proper optimization of an operating plant. Suitable filtration and recycle of catalyst are crucial components of cost saving.

CASE STUDY 11.12 SOLID-SOLID REACTION FOLLOWED BY A GAS-SOLID REACTION: MANUFACTURE OF METHYLCHLOROSILANES

Nomenclature for Case Study 11.12

A	reactant (CuCl)
$[A]$	concentration of A (kmol/m^3)
$[A]_o$	concentration of A (kmol/m^3)
B	reactant (Ci or FeSi)
$[B]$	concentration of B (kmol/m^3)
$[\hat{B}]$	dimensionless concentration $[B]/[B]_0$
C	constant
$E^*(R)$	elutriation constant (1/s)
$F_{s,0}$	quantity of feed of uniform size R_0 (kg/s)
$F_{s,1}$	quantity of solid leaving the reactor (kg/s)

$F_{s,e}$	quantity of solids elutriated (kg/s)
f_{eff}	fraction defined by Equation (CS12.1)
G	methyl chloride
I	integral defined by Equation (CS12.42)
k	general notation for rate constant (appropriate units)
k_1	constant given by Equation (CS12.9)
k_2	constant given by Equation (CS12.9) (1/s)
k_{ac}	rate constant for the autocatalytic step (m³/mol s)
k_{hom}	rate constant for homogeneous reaction (m³/mol s)
$\tilde{k}$	rate of change of pore radius (constant) given by Equation (CS12.30) (m/s)
N	total number of particles in the feed (–)
N_B^o	total number of particles of A and B surrounding one central B particle
P	reaction product ($SiCl_4$)
Q	reaction product (Cu^*)
$[Q]$	concentration of Q (kmol/m³)
R	radius of solid or cylindrical coordinate (m)
R_0	initial radius (m)
R_A, R_B	radii of solids A and B (m)
R_M	smallest feed size for a growing particle or largest size for a shrinking particle (m)
$r(R)$	rate of change of particle size (m/s)
r_A	rate of reaction of A (mol/m³ s)
r_{hom}	rate of homogeneous reaction (mol/m³ s)
r_t	sum of rates ($r_{hom} + r_{ac}$) (mol/m³ s)
$r_{t,max}$	r_t at $X_{A,max}$
s	shape factor (–)
t	time (s)
$\bar{t}$	residence time (s)
t_i	induction period for the reaction (s)
W	weight of solids (kg)
W_A, W_B	weights of solids A and B (kg)
w	weight ratio, W_A/W_B
x_i	conversion of i (–)
$\bar{X}_A$	residence time averaged conversion (–)
$\bar{\bar{X}}_A$	residence time and particle size averaged conversion (–)

Greek

α	group defined by Equation (CS12.2)
δ	bubble fraction in a fluidized bed
$\tilde{\phi}$	size distributions function
$\tilde{\phi}_0, \tilde{\phi}_1, \tilde{\phi}_e$	size distributions in the feed, outflow and elutriation streams, respectively
η	η phase (Cu_3Si)
ρ_A, ρ_B	densities of solids A and B (kg/m³)
τ	time for complete conversion as defined in Table 11.13

Defining the Systems

Methylchlorosilanes are used in the manufacture of a variety of resins, elastomers, and silicone oils. They are produced as a mixture of chlorosilanes, mainly dimethyldichlorosilane, by the reaction between silicon and methyl chloride by a direct route discovered independently by Rochow (1945) and Muller (1950). In this route, metallic copper, with or without promoters, is used to accelerate the reactions. The form of copper is important and depends on its preparation and association with the silicon phase. The whole system of solids comprising silicon metal, copper

Chemical Reaction Engineering

catalyst, and other solid materials (present in impure silicon) is collectively called the *contact mass*. Numerous methods have been proposed (see Doraiswamy and Gokarn, 1993), and in this case study, the original method developed by Rochow as modified at the National Chemical Laboratory (NCL) will be used.

A mixture of silicon and cuprous chloride granules is heated in a slow stream of inert gas. Cuprous chloride is thus reduced to metallic copper and $SiCl_4$ gas. The copper produced is in an active state and forms an alloy, the so-called η-phase, with the silicon. This phase is far more active than the alloy prepared by the conventional method of reducing copper oxide by hydrogen and then alloying the copper with silicon. The NCL process uses ferrosilicon instead of pure silicon, because it is much cheaper. Also, the presence of iron in general is reported to add stability to the η-phase (Lobusevich et al., 1976). This phase catalyzes the reaction between silicon and methyl chloride. Thus we have here a solid-solid reaction to give the catalyst followed by a gas-solid reaction to give the final products. The entire scheme may be represented as

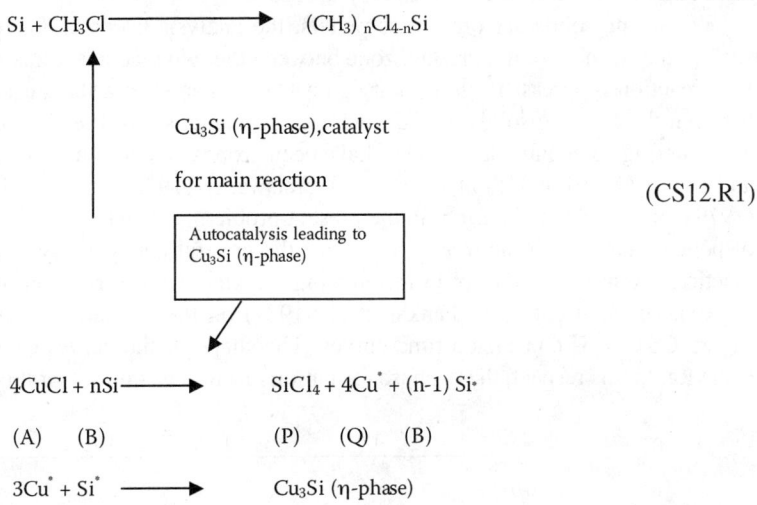

(CS12.R1)

The first step (starting at the bottom) is the formation of the η-phase, which then catalyzes the main reaction shown at the top (methylchlorosilane formation). The asterisks denote active states of the element. Then we model the formation of η-phase and develop an equation for predicting the size distribution of products from a given size distribution of the reactant solid.

Kinetics of η-phase Formation (a Solid-Solid Reaction)

As a class, solid-solid reactions are very difficult to analyze. Tamhankar and Doraiswamy (1979) have reviewed the important aspects of these reactions. When the two solids are present as mixed powders, the following situations may arise: product growth controlled by (1) diffusion of reactants through a continuous product layer, (2) nucleation and nuclei growth, and (3) phase-boundary reactions. In addition, kinetic equations based on the concept of an order of reaction can also be used. The last two are possible only when the effect of diffusion is negligible (or is embedded in the rate constant). Where diffusion is important, contact between the solids is of crucial importance. For this purpose, the concept of an *effective contact area* has been used with considerable success (see Doraiswamy and Sharma, 1984, for a brief discussion). In one study (Komatsu, 1965), a theory has been developed based on the number of contact points. For a system consisting of particles of A and B, the effective contact area of, say, B is given by the true surface area multiplied by a fraction f_{eff} defined as

$$f_{eff} = \frac{N(A/B)}{N_B^o} = \left(\frac{\alpha w}{1+\alpha w}\right)^m \quad \text{(CS12.1)}$$

Here, α is given by

$$\alpha = \frac{R_B^3 \rho_B}{R_A^3 \rho_A} \quad \text{(CS12.2)}$$

where R_A and R_B are the radii of A and B, respectively, ρ_A and ρ_B their densities, N_B^o is the total number of particles of A and B surrounding one central B particle, and w is the weight ratio of the components (W_A/W_B). The hypothetical particle with this surface area will be completely in contact with the other component (A). This analysis has the merit of bringing it in line with that for gas-solid reactions, and it is generally very useful.

Where the solids are present as pellets, the analysis becomes far more complicated. Factors such as the formation of a product zone between the two reactant solids, the change in contact area with reaction progress (from a point contact for spheres) to a flat surface, formation of a "neck" between the reacting solids, self-diffusion of reactants as well as their diffusion into the product, and sintering come into play. Models have been proposed to account for these; see, e.g., Arrowsmith and Smith (1966) and Tamhankar and Doraiswamy (1978) for the role of diffusion, and Ristic (1979) for the role of sintering. In the present problem, we shall not be concerned with the analysis of pellet behavior. Furthermore, in view of the general applicability of models based on order of reaction, we use this concept in formulating the kinetics of formation of the η-phase.

Experimental data (Tamhankar et al., 1981) on the formation of the η-phase are plotted in Figure CS12.1 as conversion-time curves. The shape of the curves clearly suggests autocatalytic behavior. In this respect, the η-phase formation reaction is unique, for it is perhaps the only instance

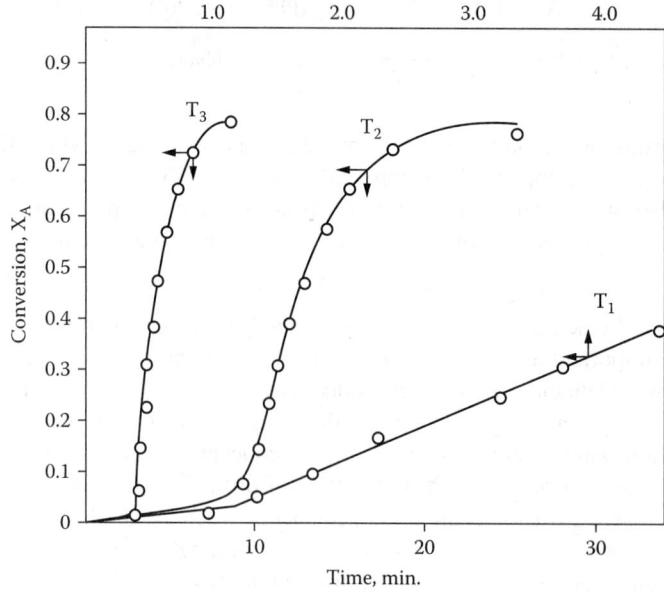

FIGURE CS12.1 Fractional removal of CuCl as a function of time (redrawn from the data of Tamhankar et al., 1981).

Chemical Reaction Engineering

of its kind in solid-solid reactions. However, in a complete analysis of this reaction, a simultaneous noncatalytic reaction all but overwhelmed by the autocatalytic step cannot be ignored. Thus, we begin the analysis by considering this noncatalytic step as the staring point. This reaction can be written as

$$A + B \rightarrow P + Q \qquad \text{CS12.R2}$$

giving

$$r_{\text{hom}} = -\frac{d[A]}{dt} = k[B][A] = k_{\text{hom}}[A] \qquad (CS12.3)$$

Since experimental curves show autocatalysis, we write

$$A + B + Q \rightarrow P + 2Q \qquad \text{CS12.R3}$$

$$r_{ac} = \frac{d[A]}{dt} = k_{ac}[A][Q] \qquad (CS12.4)$$

Total rate is given by

$$r_t = r_{\text{hom}} + r_{ac} = k_{\text{hom}}[A] + k_{ac}[A][Q] \qquad (CS12.5)$$

where $k_{\text{hom}} = k[B]$. If X_A is the conversion,

$$r_t = -\frac{d[A]}{dt} = A_0 \frac{dX_A}{dt} \qquad (CS12.6)$$

giving

$$\frac{dX_A}{dt} = \frac{r_t}{[A]_0} = \frac{k_{\text{hom}}[A]_0(1 - X_A) + k_{ac}[A]_0^2 X_A(1 - X_A)}{[A]_0} = R'$$

or

$$\frac{dX_A}{dt} = k_{\text{hom}}(1 - X_A) + k_{ac}[A]_0 X_A (1 - X_A) = \frac{r_t}{[A]_0} \qquad (CS12.7)$$

Thus, a plot of $v_t/(1 - X_A)[A]_0$ vs. X_A should give a straight line of slope $k_{ac}[A]_0$ and intercept k_{hom}. Another interesting method is to rewrite Equation (CS12.7) and integrate it to give

$$\ln \frac{X_A + k_1}{1 - X_A} = k_2(t - t_i) + \ln k_1 \qquad (CS12.8)$$

where

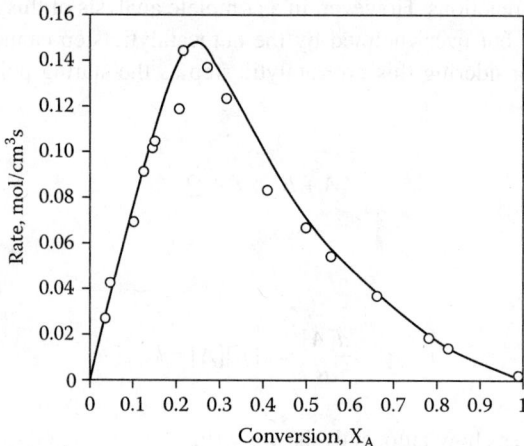

FIGURE CS12.2 Rate as a function of conversion for run T2 of Figure CS12.1 (redrawn from the data of Tamhankar et al., 1981).

$$k_1 = k_{\text{hom}} / [A]_0 k_{ac} \qquad \text{(CS12.9)}$$

$$k_2 = k_{\text{hom}} + [A]_0 k_{ac} \qquad \text{(CS12.10)}$$

t_i = induction period for the reaction

A plot of LHS vs. $(t - t_i)$ should give k_2 if k_1 is known (but it is not).

To get k_1, we differentiate the S-shaped X_A vs. t curves (Figure CS12.1). Because of the nature of autocatalysis, the resulting rate curves (see Figure CS12.2) show a maximum. Thus, at

$$X_A = X_{Am}$$

$$\frac{d(dX_A/dt)}{dX_A} = \frac{dr_t}{dt} = [A]_0^2 k_{ac} - k_{\text{hom}}[A]_0 - 2[A]_0^2 k_{ac} X_{Am} = 0 \qquad \text{(CS12.11)}$$

giving

$$k_{\text{hom}} / [A]_0 k_{ac} = k_1 = (1 - 2X_{Am}) \qquad \text{(CS12.12)}$$

where $X_{Am} = X_A$ at $r_{t,\max}$.

Using k_1 for getting the LHS in plotting Equation (CS12.8), k_2 can be obtained. Finally,

$$\left. \begin{array}{l} k_{\text{hom}} = \dfrac{k_1 k_2}{(k_1 + 1)} \\[2ex] k_{ac} = \dfrac{k_2}{[A]_0 (k_1 + 1)} \end{array} \right\} \qquad \text{(CS12.13)}$$

The actual values obtained are (Tamhankar et al., 1981):

$T_1 = 300°C \quad k_{hom} = 2.368 \; 10^{-4} \; s^{-1} \quad k_{ac} = 0.175 \; cm^3 \; mol^{-1} \; s^{-1}$
$T_2 = 330°C \quad k_{hom} = 1.333 \; 10^{-3} \; s^{-1} \quad k_{ac} = 1.778 \; cm^3 \; mol^{-1} \; s^{-1}$
$T_3 = 360°C \quad k_{hom} = 2.345 \; 10^{-3} \; s^{-1} \quad k_{ac} = 9.770 \; cm^3 \; mol^{-1} \; s^{-1}$

Determining the Size Distribution of Products from a Fluidized-Bed Reactor for Chlorosilanes

The proposed reaction will be performed in a fluidized-bed reactor. This operation consists of two stages. In the first, the solids CuCl (cuprous chloride) and FeS (ferrosilicon) or pure silicon are thoroughly mixed in the bed, which is continuously fluidized by an inert gas at a velocity close to u_{mf}. The following solid-solid reaction takes place (which is equivalent to the last step shown in reaction [CS12.R1]):

$$CuCl(s) + Si(s) \text{ or } FeSi(s) \rightarrow \eta\text{-phase } (Cu_3Si) \quad \quad CS12.R2$$

(15 parts)   (100 parts)

In the second step of the operation, the inert gas is replaced by methyl chloride (CH_3Cl), and the fluidizing conditions are maintained so that the gas-phase concentration is uniform. The following gas-solid reaction occurs:

$$CH_3Cl(g) + Cu_3Si(s) \rightarrow CH_3SiCl_3 + (CH_3)_2SiCl_2, (CH)_3SiCl + \dots \quad \quad CS12.R3$$

with the kinetics described by SCM. The products are a mixture of methyl chlorosilane, dimethyl dichlorosilane, and trimethyl monochlorosilane. A complete procedure for finding the exit size distribution of solids for an initial distribution $\tilde{\phi}_0(R)$ is given by Doraiswamy and Sharma (1984).*

Design of a Continuous Reactor: General Principles

The reaction characteristic of the present system are best performed in a semicontinuous reactor in which the solid is stationary, as described in the previous section. This easily permits the two steps. In general, however, continuous reactors in which both the gas and solid phases move continuously are more important. We therefore briefly consider in this section the mathematical basis for the design of such a reactor. The chief reactor and operating parameters are gas and solids feed rates, product size distribution, bed size, and so on, and procedures for determining them are described. With a size distribution $\tilde{\phi}_0(R)$, an elutriation stream $F_{s,e}$, and an arbitrary rate law for the changing particle size, a material balance on solids of size between R and $R + dR$ yields

$$\begin{pmatrix} \text{solids in} \\ \text{feed, kg/s} \end{pmatrix} - \begin{pmatrix} \text{solids in} \\ \text{outflow, kg/s} \end{pmatrix} - \begin{pmatrix} \text{solids leaving} \\ \text{in elutriation} \\ \text{stream, kg/s} \end{pmatrix}$$

$$- \begin{pmatrix} \text{growth of solids} \\ \text{into and out of} \\ \text{the interval, kg/s} \end{pmatrix} + \begin{pmatrix} \text{mass increase} \\ \text{of solids within} \\ \text{the interval, kg/s} \end{pmatrix} = 0 \quad \quad (CS12.14)$$

---

* There are some errors in the procedure which can easily be corrected. Also note: Although the chlorosilane gases are the desired products, we use this reaction as a model reaction for illustrating the procedure for predicting the solid product size distribution.

With complete backmixing of the solid, the size distribution of the outflow stream also represents that of the solids within the bed. This equation can be written as

$$F_{s,0}\breve{\phi}_0(R) - F_{s,1}\breve{\phi}_1(R) - WE^*(R)\breve{\phi}_1(R)$$
$$-W\frac{d}{dR}\left[r(R)\breve{\phi}_1(R)\right] \qquad \text{(CS12.15)}$$
$$+\frac{sW}{R}r(R)\breve{\phi}_1(R) = 0$$

The term $E^*(R)$ represents the elutriation constant defined by the equation

$$\begin{pmatrix}\text{rate of removal}\\ \text{of solids of size } R_j\end{pmatrix} = E^*\begin{pmatrix}\text{weight of that}\\ \text{size of solids}\\ \text{in the bed}\end{pmatrix} \qquad \text{(CS12.16)}$$

and s and $r(R)$ represent, respectively, the shape factor of the particle (3 for a sphere, 2 for a cylinder and 1 for a flat plate) and the rate of change of particle size. Equation (CS12.15), valid for a particular size range, can be supplemented by an overall balance equation over all sizes to give

$$F_{s,e} + F_{s,1} - F_{s,0} = sW\int_{\text{all }R} \frac{\breve{\phi}_1(R)r(R)}{R}dR \qquad \text{(CS12.17)}$$

This equation assumes positive or negative values depending on whether the particle grows or shrinks. For shrinking particles, Equations (CS12.15) and (CS12.17) can be rearranged to give the following expression for $W/F_{s,0}$:

$$\frac{W}{F_{s,0}} = \int_{R_t\to\infty}^{R_M}\frac{R^3}{r(R)}I\left[\int_{R_0=R}^{R_0=R_M}\frac{\breve{\phi}_0(R)dR_0}{R_0^3 I}\right]dR \qquad \text{(CS12.18)}$$

The corresponding outflow size distribution is given by

$$\breve{\phi}_1(R) = \frac{F_{s,0}R^3}{Wr(R)}I\left[\int_{R_0=R}^{R_0=R_M}\frac{\breve{\phi}_0(R)dR_0}{R_0^3 I}\right] \qquad \text{(CS12.19)}$$

For growing particles, the integration limits are reversed. In these equations, R_M represents the smallest feed size for a growing particle or the largest size for a shrinking particle, and I is an integral defined as

$$I = \exp\left[\int_R^{R_0}\frac{(F_s,1/W) + E^*(R)}{r(R)}dR\right] \qquad \text{(CS12.20)}$$

The set of Equations (CS12.18) through (CS12.20) generally cannot be solved analytically. However, they represent a total generalization with respect to feed size distribution, reaction kinetics, and presence of an elutriation stream. The use of these equations for calculating the various reactor and operating parameters such as solids feed rate, exit bed size rate, product size distribution, and so on is outlined below (see Kunii and Levenspiel, 1969, for details) for the simpler case of single-size feed.

1. Assuming that the solids weight W or the outflow rate of solids $F_{s,1}$ is known, guess a value of the other and calculate I from Equation (CS12.20) for a number of values of R.
2. Determine $F_{s,0}$ and $F_{s,1}$ or W from Equation (CS12.18). Calculation of $F_{s,0}$ is straightforward. But to calculate $F_{s,1}$ or W, guess values of the unknown quantity and solve Equation (CS12.18) until it is satisfied.
3. Using Equation (CS12.19), calculate the outlet solids size distribution $\phi_1(R)$.
4. Estimate the elutriation rate from Equation (CS12.17).

For specific situations, such as constant-size feed, linear kinetics, or no elutriation, these equations can be simplified to obtain analytical solutions.

Adapting the Procedure to the Chloromethane Reactor

This procedure has been adapted to the design of the gas-solid reactor (i.e., step 2) of the present reaction. The solid-solid reaction of step 1 is performed in the semicontinuous reactor mentioned earlier, but enough η-phase is produced and stored for the continuous second step. The outlet particle size distribution from this step becomes the inlet distribution for the second step. The procedure already outlined above is employed to develop the final equation for exit product size distribution.

General Conversion Equations

We have seen how problems of particle size distribution of reactant and solid products can be employed in the design of fluid-bed reactors. The conversion obtained at the reactor exit depends on these distributions plus various other factors. The equations presented so far were based on continuous solids feed. In calculating the conversions, it is easier to divide the solid reactants into discrete ranges (each with an average size) and express the conversion as the sum from all the ranges. Furthermore, size distribution is usually determined by screen analysis, which gives discrete measurements.

Table CS12.1 (which contains many equations from Kunii and Levenspiel 1969) summarizes the conversion equations based on discrete distributions for selected important situations. The residence time $\bar{t}$ and the time for complete conversion τ are both important. τ is defined in Table 11.13 for different shapes and the residence time (which should be specified to calculate conversion, or vice versa) can be readily computed from $\bar{t} = W/F_{s,0}$.

Obviously, the smaller particles would have completely reacted before the larger ones do. Thus, the lower limit of the summation indicates that particles smaller than $R\,(\bar{t} = \tau)$ are completely converted and should not be considered in calculating the unreacted fraction.

Models with Varying Gas-Phase Concentrations

The chief assumption made in the development presented previously was that the gas environment is known and essentially constant throughout the reactor. This assumption simplified the treatment considerably, since an analysis based on the solid phase alone could be used to describe the bed behavior. Many practical systems using large-particle beds fluidized with a large excess of gas conform to this situation. If vigorously fluidized beds of fine particles are involved, however, the composition of the gas seen by the solid would vary. The previous simple analysis would then be inapplicable. In these instances, the conservation equations for the gas-phase species

TABLE CS12.1
Conversion Equations for Gas-Solid Reactions at Specified Reactor Conditions

	State of Solids Feed	Flow Pattern of Solids	Gas Composition	Conversion Equations	Remarks
1	Single size particles	Plug flow	Uniform	Same as those given in Table 11.13 for all the three controlling steps	Equations give X_B directly; no averaging of any kind is involved since all particles spend equal times in the reactor in a PFR
2	Mixture of different sizes	Plug flow	Uniform	$1 - \bar{X}_B = \sum\limits_{R(\bar{t}=\tau)}^{R_M} \left[1 - X_B(R_j)\right] \dfrac{F_{s,0}(R_j)}{F_{s,0}}$	Averaging over size range required; for each size in the equation, the expression for the controlling step involved is used (from Table 3.9)
3	Single size particles	Mixed flow	Uniform	Film resistance control: $1 - \bar{X}_B = \dfrac{1}{2}\left(\dfrac{\tau}{\bar{t}}\right) - \dfrac{1}{3!}\left(\dfrac{\tau}{\bar{t}}\right)^2 + \dfrac{1}{4!}\left(\dfrac{\tau}{\bar{t}}\right)^3$ Ash diffusion control: $1 - \bar{X}_B = \dfrac{1}{5}\left(\dfrac{\tau}{\bar{t}}\right) - \dfrac{19}{4.20}\left(\dfrac{\tau}{\bar{t}}\right)^2 + \dfrac{41}{4620}\left(\dfrac{4620}{\bar{t}}\right)^3 - 0.000149\left(\dfrac{\tau}{\bar{t}}\right)^5 + \cdots$ Chemical reaction control: $X_B = 3\dfrac{\bar{t}}{\tau} - 6\left(\dfrac{\bar{t}}{\tau}\right)^2 + 6\left(\dfrac{\bar{t}}{\tau}\right)^3\left[1 - \exp(\tau/\bar{t})\right]$	Since the particles spend different times in the reactor, characterized by exponential distribution in a perfectly mixed reactor, averaging of conversions in particles of all residence times is involved.

4	Mixture of different sizes	Mixed flow	Uniform	Film resistance control: $$1-\bar{\bar{X}}_B = \sum^{R_M}\left\{1\left[\frac{\tau(R_j)}{t}\right] - \frac{1}{2!}\left[\frac{\tau(R_j)}{t}\right]^2 + \cdots\right\}\left[\frac{F_{s,0}(R_j)}{F_{s,0}}\right]$$ Ash diffusion control: $$1-\bar{\bar{X}}_B = \sum^{R_M}\left\{\frac{1}{5}\left[\frac{\tau(R_j)}{t}\right] - \frac{19}{420}\left[\frac{\tau(R_j)}{t}\right]^2 + \cdots\right\}\left[\frac{F_{s,0}(R_j)}{F_{s,0}}\right]$$ Chemical reaction control: $$1-\bar{\bar{X}}_B = \sum^{R_M}\left\{\frac{1}{4}\left[\frac{\tau(R_j)}{t}\right] - \frac{1}{20}\left[\frac{\tau(R_j)}{t}\right]^2 + \cdots\right\}\left[\frac{F_{s,0}(R_j)}{F_{s,0}}\right]$$	Here, two averagings are involved: for residence time distribution [as in (3)] and for particle size distribution [as in (2)]; this is denoted by a double bar over X_B
5	Single size or mixture of different sizes	Plug or mixed flow	Axial variation	See references under "Models with varying gas-phase concentrations"	Situation likely for vigorously fluidized beds of fine particles – encountered commonly in industry; no complete procedure yet available
6	Single size or mixture of different sizes	Plug flow with lateral variation of composition	Axial (and lateral) variation	See Chang et al. (1982)	Lateral variation of solids concentration is prominent in shallow beds

have to be incorporated into the model, and separate equations for the bubble and cloud-wake phases are necessary.

During the course of reaction, the solid reactant in the particles is increasingly displaced by the solid product formed. The volume and specific density of the product formed may not be the same as those of the reactant. For a realistic representation of the bed, it is therefore necessary to account for variations in both size and density of the particles.

Thus, the modeling of the gas-solid noncatalytic reactor is far more complex than that of its catalytic counterpart. Even so, simplified models have been used to get a qualitative (and to some extent, quantitative) feel for the performance of the reactor. Thus, the two-phase model of Davidson and Harrison (1963) has been used by Campbell and Davidson (1975) to analyze the data on the combustion of carbon particles for short periods of combustion in a batch reactor. The model has also been used and considerably extended by Amundson (see Bukur et al., 1977). Tigrel and Pyle (1971) have used this model for the not-too-different problem of catalyst deactivation. Kunii and Levenspiel (1969) and Kato and Wen (1969) have extended their models to gas-solid noncatalytic systems. A particularly useful model that takes account of some of the complexities in practical systems has been suggested by Chen and Saxena (1978).

Comments

In this case, no specific problem in quantitative terms has been posed and solved. For the first step, which is a solid-solid reaction, experimental details have been given because it is a unique solid-solid reaction, perhaps the only known autocatalytic one. Having clearly established this fact, general procedures are described for the two reactors involved, since no experimental data are available for attempting a specific design. The thrust here is the description of a procedure for calculating the outlet solid size distribution given the inlet size distribution. This is an important factor in solid-phase reactions.

Lesson

Silicon reacts with copper chloride catalytically to form a solid product known as η-phase, which then catalyzes the reaction between silicon and methyl chloride to form methyl chlorosilanes. The formation of the η-phase is a unique example of an autocatalytic solid-solid reaction (we are not aware of any other example of such a reaction). Procedures for determining the kinetics of this reaction and the size distribution of the final products of the second reaction are described in this case study. The most important feature of the study is the unexpected discovery of autocatalysis in a solid-solid reaction, which leads us to the important lesson that one should vigorously pursue any such observation that falls outside the "comfort zone" of research, for that is where true discoveries lie. This advice is particularly relevant to engineers, since they generally do not accept the unexpected and would prefer to tread on safer ground.

POSTSCRIPT

The case studies presented in this chapter illustrate reactor design procedures for a carefully selected set of reacting systems wherein the physical dimensions of the reactor (diameter, height) and fixed and operating parameters (catalyst loading, superficial velocity, impeller speed, and other) were calculated. As a postscript to these studies, we would like to consolidate and emphasize certain fundamental and practical considerations in reactor selection and design.

- For a given system, the necessary requirements for any proposed reactor design are that the desired capacity, selectivity, and purity be achieved. The exercise should also include pertinent aspects of environmental load and safety.
- Although the engineer may be required to design for a given capacity, it is advisable to provide for additional capacity (in case, this is inexpensive and not the cost-controlling

Chemical Reaction Engineering

bottleneck). At the same time, the reactor must be able to perform well at lower loads during slack market periods.
- In the case of small- and medium-scale plants, the reactor may need to have adequate flexibility to produce various similar/different compounds, so a broad-based design is necessary.
- The identification of the rate-controlling step(s) is crucial in the design of multiphase reactors. However, it may be emphasized that the rate-controlling step emerges from a complex combination of series and parallel steps. Often, it is difficult to find the rate-controlling step, since it is compounded by its dependence on location in the reactor. The importance of using a carefully selected laboratory reactor for obtaining precise rate data, uncontaminated by diffusional falsification, cannot be overemphasized. We have emphasized critical points by presenting tables that provide comparative assessments of laboratory reactors for different classes of reactions. The data should be collected over wide parameter ranges, covering possible ranges in industrial practice and extensions on either side. These data are not only useful in design but in subsequent diagnostic analyses, modifications, debottlenecking, replacement, and process identification.
- It is desirable to encourage the production staff to cooperate with the design team to generate plant scale data that can be used to modify/improve the design based on laboratory data. This is particularly true for transport coefficient estimation. The following simplified block diagram is useful for practicing engineers.

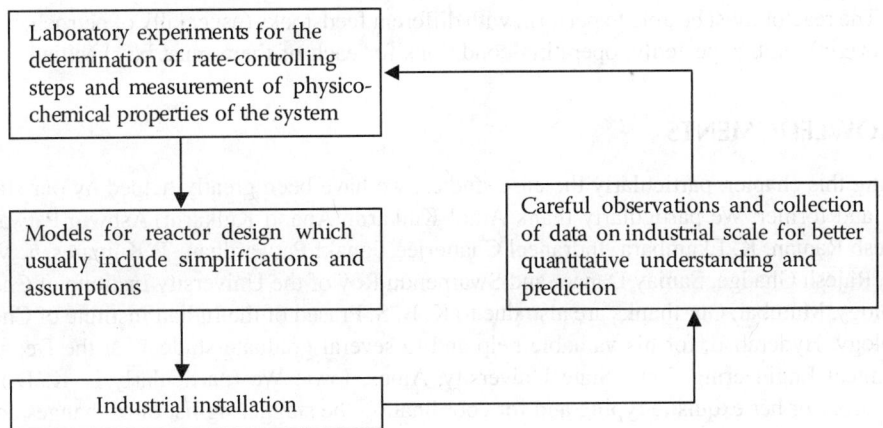

- Many reactors are available, and it is advisable not to reject any reactor configuration at the first glance.
- Provision of pressure relief valves, drains, and so on is mandatory.
- Induction of cost considerations at an early stage is strongly recommended.

In addition, an engineer must bear in mind several more specific aspects of design. An attempt to make an exhaustive list would be futile, but some illustrative considerations are mentioned here. The designer must envisage all such aspects that are pertinent to a specific case and incorporate them in the design portfolio.

- For oxidation of hydrocarbons, it is necessary to ensure that, at no time and at no point in the reactor, the flammability limits are approached.
- For catalytic reactions, there must be a provision to trap/recover the catalyst in the event that it is entrained.

- For highly exothermic reactions such as polymerization, it is advisable to include a means for quenching the reaction (e.g., by providing for addition of an inhibitor) in the event of a runaway situation.
- Not only should hazardous conditions be avoided during normal operation, but any such possibility during startup or shutdown should also be foreseen.
- In the case of reactions carried out in mechanically agitated reactors where heat transfer is crucial, provision must be made for adequate heat transfer. During power failure, agitation may stop. Emergency power must be provided to the crucial drives, and provision must be made to drain/flare the contents during an emergency.
- In the case of catalytic packed-bed reactors, provision of extra volume should be considered to accommodate additional catalyst in case of selectivity/activity loss over time.
- In some heat transfer-controlled reactions, it may be possible to partly evaporate the solvent itself, condense it in a reflux condenser, and send the solvent back to the reactor.
- The thermal sensitivities of all the species in the reactor must be known. Some compounds may decompose dangerously at certain conditions, and suitably lower temperatures must be chosen for operation.
- Some reactions may produce solid products that degrade within the reactor and deposit on the cooling coils, and so forth, and drastically affect the heat transfer and product quality. This possibility must be envisaged and accounted for in the design.
- Where handling of solid reactants is involved, the equipment must be designed to present clogging and choking.
- The reactor must be able to perform with different feedstocks (especially of petrochemical origin), and, expectedly, operating conditions for each of them must be specified.

ACKNOWLEDGMENTS

In writing this chapter, particularly the case studies, we have been greatly helped by our students, present and former. We particularly thank Amol Kulkarni, Anand Kulkarni, Ashwin Patwardhan, Dharmesh Ranjan, K. Ekambara, Indraneel Chatterjee, Janaki Patwardhan, T. Kumaresan, Mahesh Dhotre, Rajesh Ghadge, Sanjay Danao, and Swarnendu Roy of the University Institute of Chemical Technology, Mumbai. Our thanks are also due to K. B. S. Prasad of the Indian Institute of Chemical Technology, Hyderabad, for his valuable help and to several graduate students at the Department of Chemical Engineering, Iowa State University, Ames, Iowa. We (particularly L. K. D.) thank Linda Edson for her exquisite typing and for coordinating the staggering list of exchanges between the two authors, incorporating repeated additions, corrections, and revisions. L. K. D. also thanks C. E. Glatz, chair of the Department of Chemical Engineering, for his continued support, even in the author's "retirement."

Nomenclature

A	chemical species (–)
A_p	area of pellet (m^2)
$[A]_p$	concentration of A within a pellet (kmol/m^3)
$[A]_s$	concentration of A at catalyst or any solid surface (kmol/m^3)
$[\overline{A}]$	average concentration of A (kmol/m^3); external field concentration at some point in a bed (kmol/m^3)
$[A]_0$	entrance concentration of A (kmol/m^3)
$[A]_e$	exit or effluent concentration of A (kmol/m^3)
$[A]_f$	final concentration (in time) of A (kmol/m^3)
$[A]_i$	initial concentration of A (kmol/m^3)
$[\hat{A}]$	dimensionless concentration of A within a pellet, $[A]_p/[A]_s$ (–)
a	interfacial area (m^2/m^3); any exponent (–)
B	chemical species (–)

B'	group defined by Equation (11.141) (–)
$[\hat{B}]$	normalized concentration of B (–)
Bi_m	Biot number for mass transfer (same as Sh) (–)
B	any exponent (–)
C	chemical species (–)
C_p	heat capacity (kcal/kg °C)
C_{pm}	molar heat capacity (kcal/kmol °C)
C_{ps}	heat capacity of solid, (kcal/kg °X)
c	dimensionless concentration $[A]/[A]_0$ (–)
$\mathbf{c}$	vector ($N \times 1$ matrix) of component concentrations (–)
D	chemical species; impeller (or column) diameter (m)
$\mathbf{D}$	diffusivity in general (m^2/s)
$\mathbf{D}_A$	diffusivity of A (m^2/s)
$\mathbf{D}_b$	bulk diffusivity (m^2/s)
$\mathbf{D}_e$	general notation for effective diffusivity (m^2/s)
$\mathbf{D}_{eA}$	effective diffusivity of A (m^2/s)
$\mathbf{D}_{e,As}$	effective diffusivity of A in solid (m^2/s)
$\mathbf{D}_{ea}$	axial effective diffusivity (m^2/s)
$\mathbf{D}_{eG}$	effective diffusivity in a grain (m^2/s)
$\mathbf{D}_{er}$	radial effective diffusivity (m^2/s)
$\mathbf{D}_K$	Knudsen diffusivity (m^2/s)
Da	Damköhler number, $k_s R[A]_b^{m-1}/\mathbf{D}_{e(-)}$
d_{bp}	primary bubble size (m)
d_{bs}	secondary bubble size (m)
d_c	catalyst decay constant (1/s)
d_p	particle diameter (m)
d_T	tube diameter (m)
E	activation energy (kcal/kmol)
E_i	energy input rate (W)
F	function (–); feed rate (mol/s)
F_i	feed rate of i (kmol/s)
F_s	solids circulation rate (kg/s)
F_T	flow rate in a single tube (kmol/s)
F_t	total feed rate (kmol/s)
Fr	Froude group, $u^2/g\,d_p$ (–)
Fr_{mf}	Froude group at minimum fluidization, u_{mf}^2/gd_p (–)
F, f', f_1, f_2, f_3	functions (–)
f_p	fraction of pores removed (–)
G_A	specific gas (A) flow rate $u\rho_A/M_A$ (kmol/m^2/s)
G_B	specific solids (B) flow rate $u\rho_B/M_B$ (kmol/m^2/s)
g	shape constant for the grain: 3 for a sphere, 2 for a cylinder, and 1 for a flat plate (–); gravitation constant (m/s^2); functionality
H	enthalpy (kcal/kmol); height (m)
H_A	Henry's law constant of A (kmol/m^3 atm)
H_D	height of dispersion (m)
H_{SD}	sparger location on the downcomer side (m)
ΔH_r	heat of reaction (kcal/kmol)
J	group defined by Equation (11.125) (–)
K	reaction equilibrium constant (appropriate units)
K_i	adsorption equilibrium constant of species i (1/atm or m^3/kmol)
k	general notation for reaction rate constant (appropriate units)
$\tilde{k}$	term defined in Table 11.4
k_a	rate constant for catalyst activation (1/s)
k_d	rate constant for catalyst deactivation (1/s)
k_G	mass transfer coefficient (m/s)

K_m	rate constant for an mth-order reaction [(m^3/kmol)$^{m-1}$(1/s)]
k_P	rate constant for pore removal (1/s)
k_S	rate constant [(kmol/m^2 cat) (m^3/kmol)^m (1/s)]
k_V	rate constant [(m^3 reactor /kmol)$^{m-1}$ (1/s)]
k_v	rate constant [(m^3 cat/kmol)$^{m-1}$ (1/s)]
k_{vo}	rate constant for fresh catalyst (1/s)
k_{vs}	rate constant at surface conditions, conforming to any set of units of k_v
k_w	rate constant [(mol/kg cat) (m^3/mol)^m (1/s)]
k^o	Arrhenius pre-exponential factor (units of the corresponding rate constant)
L	reactor length (m)
L_{mf}	reactor length at minimum fluidization (m)
L^*	dimensionless reactor length defined by Equation (11.162) (–)
M	number of reactions; the "space factor" (appropriate units)
$\sqrt{M}$	ratio of reaction in film to that in bulk, $\sqrt{D_A k[B]_b / k_L}$
M_i	molecular weight of species i (kg/kmol)
m	reaction order; any exponent (–)
N	number of components; number of kmoles (–); speed (rps)
N_{cs}, N_{CM}	see text
N_{SG}	see text
N_{C50}	see text
N_{C70}	see text
N_i	number of kmoles of component i (–)
N_P	number of particles
N_{QS}	secondary flow number (–)
N_t	total number of kmoles (–)
Nu	Nusselt number (–)
n	reaction order (–)
P	pressure (atm or Pa)
P	power (W)
Pe_{hr}	radial Peclet number for heat transfer, $d_p u C_p \rho / k_{p\,er}$ (–)
Pe_{ma}	axial Peclet number for mass transfer, $d_p u / D_{ea}$ (–)
P_i	partial pressure of i (atm or Pa)
P_{nm}	parameter defined in Table 11.9 (–)
Q	volumetric flow rate (m^3/s)
Q_S	solids circulation rate (m^3/s)
R	chemical species (–); radius (m)
R'	group $\nu u_A / u_B$
R_A	specific rate of absorption (reaction) (kmol/m^2/s)
$\mathbf{R}_A$	reaction rate per pellet (kmol/s)
$\mathbf{R}_{GA}$	reaction rate of A in a grain (kmol/s)
R_g	ideal gas law constant (kcal/kmol K)
$\hat{R}$	dimensionless radius, r/R (–)
Re_p	particle Reynolds number ($d_p u_g \rho / \mu$ (–)
Re_{mf}'	Reynolds number at minimum fluidization, $u_{mf} d_p \rho / \mu$ (–)
r	general notation for reaction rate (appropriate units); radial coordinate (m)
r	vector ($M \times 1$ matrix) of reaction rates (–)
r_A, r_{VA}	reaction rate of A (kmol/m^3 reactor s)
R_{Go}	initial grain radius (m)
r_{Gi}	interface radius in a grain (m)
r_i	radial position at the core-shell interface in SIM (m)
r_j	reaction rate of the jth reaction (appropriate units)
r_{SA}, r_A'	reaction rate of A (kmol/m^2 cat s)
r_{vA}	reaction rate of A (kmol/m^3 cat s)
r_{WA}	reaction rate of A (kmol/kg cat s)
S	selectivity term defined variously as in Table 11.9

S	chemical species (–)
S_g	surface area of grain (m²)
S_0	initial surface area per unit volume (1/m)
S_p	surface area of pellet (m²)
S^*	dimensionless reaction surface area defined by Equations (11.90) and (11.92) (–)
S^*_p	dimensionless pore surface area defined by Equations (11.91) and (11.93) (–)
$S_{v,A}$	space velocity of A (1/s)
Sh	Sherwood number $k_g R/D_e$ (–)
s	selectivity k_{v1}/k_{V2}; generalized distance parameter or shape constant of a pellet, 3 for a sphere, 2 for a cylinder, and 1 for a flat plate (–)
s_a	observed (actual) selectivity $(k_{v1}/k_{V2})_a$ (–)
T	temperature (K)
$\hat{T}$	dimensionless temperature within a pellet. (T/T_s) (–)
$\mathbf{T}$	tank diameter (m)
t	time (s)
t_{mix}	mixing time (s)
t_{pl}	total decay time for a fixed final level of decay (s)
$\bar{t}$	residence time in a flow reactor (s)
$\bar{t}_1, \bar{t}_2$	average residence times in reactor (W_1/F_s) and regenerator (W_2/F_2), respectively (s)
$\hat{t}$	any normalized time, e.g., decay time t_f/t_{pl}; time given by Equations (11.72) and (11.88) (–)
T^*	dimensionless time defined by Equation (11.150) (–)
U	overall heat transfer coefficient (kal/m² K s)
u	velocity (m/s)
u_{mf}	minimum fluidization velocity (m)
$u_{s\infty}$	terminal settling velocity of particle (m/s)
V	volume (m³)
V^*	liquid volume in a variable volume reactor given by Equation (11.23)
V_g	pore volume of grain (m³)
V_p	volume of pellet (m³)
V_r	reactor volume where specifically mentioned (m³)
V_T	volume of a single tube (m³)
W	weight of catalyst (kg)
W_1, W_2	weights of catalyst in reactor and regenerator, respectively (kg)
X'_B, X_{Bl}	dummy variables (–)
X^*	extent of reaction of a pellet defined by Equation (11.152) (–)
X_i	conversion of i (–)
Y_i	yield of species i (–)
Y_{ia}	Y_i under actual (observed) conditions (–)
y_i	mole fraction of i (–)
Y_{ra}, Y_{rb}	yields of R defined as $[R]/[A]_0$, $[R]/[B]_0$ (–)
Z	dimensionless reactor length, ℓ/d_p (–)
Z^*	dimensionless distance defined by Equation (11.151) (–)
Z_v	ratio of solid product to solid reactant kmolar volumes (–)

Greek

α	Arrhenius parameter, $E/R_g T$; diffusivity ratio, $\mathbf{D}_{eAs}/\mathbf{D}_{e,Aso}$ (–)
α_s	Arrhenius parameter at surface conditions, $E/R_g T_s$ (–)
β	shape factor: 2 for sphere, 1 for cylinder, 0 for flat plate
β'	diffusional parameter given by Equation (11.89) (–)
β_m	heat generation term given by Equation (11.54) (–)
χ	dummy variable (–)
$\delta(t)$	residence time distribution function
δ_i	change in volume per unit volume of i upon reaction (see Table 11.1) (kmol)
ε	porosity (voidage) (–); holdup; catalyst effectiveness factor (–)

ε_b	bulk or bed porosity (voidage) (–)
$\bar{\varepsilon}_G$	average gas holdup (–)
ε_{gr}	riser gas holdup (–)
ε_{LD}	downcomer liquid holdup (–)
ε_{lr}	riser liquid holdup (–)
ε_i	change in total volume upon reaction per kmole of i reacting (m³/ kmol)
ϕ	generalized Thiele modulus independent of pellet shape and reaction order defined by Equation (11.50); also general notation for Thiele modulus; modulus defined by Equations (11.71) and (11.87) (–)
ϕ_a	actual or Weisz modulus defined by Equation (11.51) (–)
ϕ_c	a critical value of the Thiele modulus given by Equation (11.78) (–)
ϕ_n	shape generalized Thiele modulus for an n order reaction (–)
ϕ_{pn}	Thiele modulus for an nth-order reaction in a flat plate (–)
ϕ_{sn}	Thiele modulus for an nth-order reaction in a spherical pellet (–)
η	dimensionless position in grain (–); viscosity (Pa·s); enhancement factor (–)
λ	group defined by Equation (11.141)
ν	ratio of stoichiometric coefficients, ν_B/ν_A (–)
ν_i	stoichiometric coefficient of species i (–)
ν_{ij}	stoichiometric coefficient of A_i in the jth reaction (–)
$\mathbf{v}$	$(N \times M)$ matrix of stoichiometric coefficients (–)
θ	volume change modulus defined by Equation (11.87) (–)
θ_i	fraction of active sites occupied by species i (–)
θ_v	fraction of vacant active sites (–)
ρ	density (kg/m³)
ρ_B	density of B (kg/m³)
ρ_{Bm}	Molar density of B (mol/m³)
ρ_b	bulk density (kg/m³)
ρ_i	density of i (kg/m³)
$\bar{\rho}_D$	average density of dispersion (kg/m³)
σ	reaction modulus defined by Equation (11.153) (–); surface tension (dynes/m)
τ	total time given by Equation (11.66) (–)
τ'	dimensionless temperature in a reactor, T/T_0 (–)
τ_f, τ_a, τ_c	times for complete conversion of a solid if film transfer, ash diffusion, or reaction alone were controlling (s)
τ_G	time required for complete conversion of a grain (s)
τ_p	reaction or on stream time (s); time required for complete conversion of a pellet(s)
ω	dimensionless radial position in a reactor, r/R (–)
ξ	dimensionless position in a catalyst pellet or reacting solid (–)
ξ_i	dimensionless position of unreacted core, r_c/R (–)
ξ_j	moleculartiy of reaction j (i.e., number of kmoles reacting according to reaction j) (–)
ψ	ratio y_{i0}/y_{A0}, F_{i0}/F_{A0} or $[i]/[i]_0$; cumulative gas concentration given by Equation (11.75) (–)
ψ'	structural parameter (–)
ψ_B	initial concentrations ratio, $[B]_0/[A]_0$ (–)
Ω	catalyst activity (–)
$\bar{\Omega}$	average catalyst activity (–)
ℓ	length coordinate (m)
$\mathfrak{R}_m$	mass transfer group, $d_p/u[A]_0$ (m³s/ kmol)
$\mathfrak{R}_H$	heat transfer group, $(-_r)d_p/uC_pT_0$ (m³s/kmol)

Subscripts

b	bulk
bp	bubble generated at the sparger
bs	secondary bubble
CS	critical speed for solid suspension in the absence of gas
C	catalyst

D	dispersion
e	effective; exit (outlet) condition
er	effective radial
f	final condition (in time)
G	gas; grain
GD	gas in downcomer
GR	gas in riser
i	interface
L	liquid
LD	liquid in downcomer
LR	liquid in riser
mf	minimum fluidization
mix	mixing
O	initial (entry) condition
p	particle, pellet
r	reactor; radial
SG	critical speed in the presence of gas
s	solid; surface
	tube (reactor); tank
T	total
w	wall

REFERENCES

Aitken, R. A. and Kilênyi, S. N., *Asymmetric Synthesis*. Boca Raton, FL: CRC Press, 1992.

Akiti, T. T. Jr., Ames, Iowa: Ph.D. thesis, Iowa State University, 2001.

Akiti, T. T. Jr., Constant, K. P., Doraiswamy, L. K. and Wheelock, T. D. 2001. *Adv. Environ. Res.* 5: 31.

Akiti, T. T. Jr., Constant, K. P., Doraiswamy, L. K. and Wheelock, T. D. 2002a. *Adv. Environ. Res.* 6: 419.

____. 2002b. *Ind. Eng. Chem. Res.* 41: 587.

Anderson, J. L., and Quinn, J. A. 1970. *Chem. Eng. Sci.* 25: 373.

Arhancet, J. P., Davis, M. E. and Hanson, B. E. 1991. *J. Catal.* 129: 94.

Arhancet, J. P., Davis, M. E., Merola, J. S. and Hanson. B. E. 1990. *J. Catal.* 121: 327.

Aris, R. 1957. *Chem. Eng. Sci.* 6: 262;

____. *The Optimal Design of Chemical Reactors—A Study in Dynamic Programming*. New York: Academic Press, 1961.

____. *Introduction to the Analysis of Chemical Reactors*. Englewood, Cliffs, NJ: Prentice Hall, 1965.

____. *The Mathematical Theory of Diffusion and Reaction in Permeable Catalysts*. Oxford, UK: Clarendon Press, 1975.

____. Some problems in the dynamics of chemical reactions, in *Frontiers in Chemical Reaction Engineering*, Doraiswamy, L. K. and Mashelkar, R. A., eds. New Delhi, India: Wiley Eastern, 1984.

____. *Elementary Chemical Reactor Analysis*, New York: McGraw-Hill, 1969 (also Boston: Butterworth-Heinemann, 1989).

Arrowsmith, R. J. and Smith, J. M. 1966. *Ind. Eng. Chem. Fundam.* 5: 327.

Avrami, M. 1940. *J. Chem. Phys.* 8: 212.

Bach, H. F. and Pilhofer, T. 1978. *German Chem. Eng.* 1: 270.

Baeyens, J. and Geldart, D. 1974. *Chem. Eng. Sci.* 29: 255.

Bagajewicz, M. 1992. *Chem. Eng. Commun.* 112: 145.

Bailey, D. C. and Langer, S. H. 1981. *Chem. Rev.* 81: 109.

Balakotaiah, V. and Luss, D. 1981. *AIChE J.* 27: 442.

Balaraman, K. S., Mashelkar, R. A. and Doraiswamy, L. K. 1980. *AIChE J.* 26: 635.

Balogh, M. and Laszlo, P., *Organic Chemistry Using Clays*. New York: Springer-Verlag, 1993.

Barnard, J. A. and Mitchell, D. S. J. 1968. *J. Catal.* 12: 326.

Barrer, R. M. 1963. *Appl. Mat. Res.* 2: 129.

Behie, L. A. and Kehoe, P. 1973. *AIChE J.* 19: 1070.

Behie, L. A., Bergougnou, M. A. and Baker, C. G. J., in *Fluidization Technology*, vol. 1, Keairus, D. L., ed. Washington, DC: Hemisphere, 1976.
Behie, L. A., Bergougnou, M. A., Baker, C. G. J. and Base, T. E. 1971. *Can. J. Chem. Eng.* 49: 557.
Bellman, R., *Dynamic Programming*. Princeton, NJ: Princeton University Press, 1957.
Bellman, R. and Dryfus, S. E., *Applied Dynamic Programming*. Princeton, NJ: Princeton University Press, 1962.
Berty, J. M. 1973. Paper presented at AIChE 66th Annual Meeting, Philadelphia.
Beveridge, G. S. G. and Goldie, P. J. 1968. *Chem. Eng. Sci.* 23: 912.
Bhatia, S. and Gupta, J. S. 1993. *Rev. Chem. Eng.* 8: 177.
Bhatia, S. K. and Perlmutter, D. D. 1980. *AIChE J.* 26: 379; 1981a. *AIChE J.* 27: 226; 1981b. *AIChE J.*, 27: 247; 1983. *AIChE J.* 29: 287.
Bhutada, S. R. and Pangarkar, V. G. 1987. *Chem. Eng. Commun.* 61(1–6): 23; 1989. *Chem. Eng. Sci.* 44: 2384.
Bischoff, K. B. 1965. *AIChE J.* 11: 351; 1968. *Chem. Eng. Sci.* 23: 451.
Borman, P. C., Boss, A. N. R. and Westerterp, K. R. 1994. *AIChE J.* 40: 862.
Boudart, M. and Djeja Mariadasseau, G., *Kinetics of Heterogeneous Catalytic Reactions*. Princeton, NJ: Princeton University Press, 1984.
Bowen, J. H. and Cheng, C. K. 1969. *Chem. Eng. Sci.* 24: 1829.
Box, G. E. P. 1965. *Computer J.* 8: 42.
Box, G. E. P. and Hill, W. J. 1967. *Technometrics* 4: 30.
Briens, C. L., Bergougnou, M. A. and Baker, C. G. J. 1978. *2nd Fluid Proc. Eng. Found. Conf.*: 38.
Briggs, R. A. and Sacco, A., Jr. 1991. *J. Mater. Res.* 6: 574.
Buchholtz, R., Sepentoinides, T., Stienmann, J. and Onken, U. 1983. *Ger. Chem. Eng.* 6: 105.
Bujalski, W., Konno, M. and Nienow, A. W. 1988. *Proc. 6th Eur. Conf. Mixing* 24–26 May, BHRA, Leeuwenhorst, Pavia, Italy.
Bukur, B. D. and Amundson, N. R. 1975a. *Chem. Eng. Sci.* 30: 847; 1975b. *ibid.* 30: 1159.
Bukur, D. V., Caram, H. S. and Amundson, N. R., in *Chemical Reactor Theory—A Review*, Lapidus, L. and Amundson, N. R., eds., Englewood Cliffs, NJ: Prentice-Hall, 1977.
Butt, J. B., *Reaction Kinetics and Reactor Design*, 2nd ed. Englewood Cliffs, NJ: Prentice Hall, 1999.
Butt, J. B., Bliss, H. and Walker, C. A. 1962. *AIChE J.* 8: 42.
Calo, J. M. 1978. *ACS Symp. Ser.* 65: 550.
Campbell, E. K. and Davidson, J. F., in *Fluidization Technology*, vol. 2, Keairns, D. F., ed. Washington, DC: Hemisphere, 1975.
Campbell, R. R., Hills, A. W. D. and Paulin, A. 1970. *Chem. Eng. Sci.* 25: 929.
Carberry, J. J. 1962. *Chem. Eng. Sci.* 17: 675; 1964. *Ind. Eng. Chem.* 56: 39; *Chemical and Catalytic Reactor Engineering*. New York: McGraw-Hill, 1976.
Carberry, J. J. and Kulkarni, A. A. 1973. *J. Catal.* 31: 41.
Carberry, J. J. and Wendel, M. 1963. *AIChE J.* 9: 129.
Catipovic, N. and Levenspiel, O. 1979. *Ind. Eng. Chem. Process Des. Dev.* 18: 558.
Chakrabarti, A. and Sharma, M. M. 1993. *Reactive Polym.* 20: 1.
Chan, O. T. and Rinker, R. G. 1978. *Chem. Eng. Sci.* 33: 1201.
Chan, S. F. and Smith, J. M. 1976. *Indian Chem. Eng.* 18: 42.
Chandrasekharan, K. and Calderbank, P. H. 1981. *Chem. Eng. Sci.* 36: 819.
Chandrashekar, B. C. and Vortmeyer, D. 1979. *Warme-Stoff.* 12: 105.
Chang, C. C., Fan, L. T. and Rong, S. X. 1982. *Can. J. Chem, Eng.* 60: 272.
Chapman, C., Nienow, A. W., Cooke, M. and Middleton, J. C. 1983. *Chem. Eng. Res. Des.* 61: 82.
Chen, T. P. and Saxena, S. C. 1978. *AIChE Symp. Ser.*, no. 176, 74: 149.
Chou, C. L. and Li, K. 1984a. *Chem. Eng. Com.* 29: 153; 1984b. *ibid.*, 181.
Choudhary, V. R. and Doraiswamy, L. K. 1972. *Ind. Eng. Chem. Proc. Des. Dev.* 11: 420; 1975. *ibid* 14: 227.
Chrostowski, J. W. and Georgakis, C. 1978. *ACS Symp. Ser.* 65 *(Chem. React. Eng.)*, Houston, TX: 225.
Chuang, K. C., Young G. W. and Benslay R. M., *Advanced Fluid Catalytic Cracking Technology*. New York: AIChE, 1992.
Chudacek, M. W. 1982. *Proc. 4th Eur. Conf. on Mixing*, 27–29 April, BHRA, Leeuwenhorst, Netherlands.
Clark, J. H., Kybett, A. P. and Macquarrie, D. J., *Supported Reagents: Preparation, Analysis, and Applications*. Weinheim, Germany: VCH, 1992.

Colver, G. M., Brown, R. C., Shi, H., and Soo, D. 2002. *Symp. Gas Cleaning at High Temperatures*, Morgantown, WV.
Copelowitz, I. and Aris, R. 1970. *Chem. Eng. Sci.* 25: 885.
Corma, A., in *Zeolite Microporous Solids: Synthesis, Structure and Reactivity*, Derouane, E. G., Lemos, F., Naccache, C. and Ribeiro, F. R., eds. NATO ASI Series. Dordrecht, Germany: Kluwer Academic, 1991.
Danckwerts, P. V., *Gas-Liquid Reactions*. New York: McGraw-Hill, 1970.
Davidson, J. F. and Harrison, D., *Fluidized Particles*. New York: Cambridge University Press, 1963; *Fluidization*. London: Academic Press, 1971.
de Croocq, D., *Catalytic Cracking of Heavy Petroleum Fraction*. Paris: Inst. Francais du Petrole, Technip., 1984.
Deckwer, W. D., *Bubble Column Reactors*. New York: John Wiley, 1992.
DeLasa, H. I., Errazu, A., Barrero, E. and Solioz, S. 1980. *Can. J. Chem. Eng.*
del Borghi, Dunn, J. C. and Bischoff, K. B. 1976. *Chem. Eng. Sci.* 31: 1065.
Di Felice, R. 1995. *Chem. Eng. Sci.* 50: 1213.
Dodgson, I., in *Heterogeneous Catalysis and Fine Chemicals*, Guisnet, M., Barbier, J., Barrault, J., Bouchoule, C., Dupez, D., Montassier, C. and Pérot, G., eds. Amsterdam: Elsevier 1993.
Doraiswamy, L. K. 1991. *Prog. Sur. Sci.* 37: 1; *Organic Synthesis Engineering*, New York: Oxford, 2001.
Doraiswamy, L. K. and Gokarn, A. N., in *Catalyzed Direct Reactions of Silicon*, Lewis, K. M. and Retwisch, D. G., eds. New York: Elsevier, 1993.
Doraiswamy, L. K. and Kulkarni, B. D., in *Chemical Reaction and Reactor Analysis*, Carberry, J. J. and Varma, A., eds. New York: Marcel Dekker,1987.
Doraiswamy, L. K. and Sharma, M. M., *Heterogeneous Reactions: Analysis, Examples, and Reactor Design*, vol. 1, Gas-Solid and Solid-Solid Reactions, vol. 2, Fluid-Fluid and Fluid-Fluid-Solid Reactions. New York: John Wiley & Sons, 1984.
Doraiswamy, L. K. and Tajbl, D. D. 1974. *Cat. Rev. Sci. Eng.* 10: 177.
Dudukovic, M. P. and Lamba, H. S. 1978. *Chem. Eng. Sci.* 33: 303, 471.
Dutta, S. 1994. *AIChE Symp. Ser.* 301, 157.
Dutta, S. and Gualy, R. June 2000. *Chem. Eng.*: 72; July 1999. *Hydroc. Proc.*: 45.
Edwards, W. M., Worley, F. L. Jr. and Luss, D. 1973. *Chem Eng. Sci.* 28: 1479.
Edwards, W. M., Zuniga-Chaves, J. E., Worley, F. L. Jr. and Luss, D. 1974. *AIChE J.* 20: 571.
Elnashaie, S. S. E. H. and Cresswell, D. 1973. *Proc. Intl. Symp. Fluidization and Its Applications*, Tolouse, France.
El-Rahaiby, S. K. and Rao, Y. K. 1979. *Met. Trans.* 10B: 257.
Ergun, S. 1952. *Chem. Eng. Progr.* 48: 89.
Erofeev, B. V. 1961. *Proc. 7th Intl. Symp. Reactivity of Solids*.
Evans, J. W. and Song, S. 1974. *Ind. Eng. Chem. Process. Des. Dev.* 13: 146.
Fan, L. T., Lee, C. J. and Bailie, R. C. 1962. *AIChE J.* 8: 239.
Fasano, J. B., Brodkey, R. J. and Haam, S. J. 1991. *Proc. 7th Eur. Conf. Mixing*. 18–20 Sept., Brugge, Belgium.
Federsel, H. J. 1993. *Chemtech* 24 (24 December).
Fernouil, L. A. and Lynn, S. 1995a. *Ind. Eng. Chem. Res.* 34: 2324; 1995b.; *ibid.*, 2334.
Fogler, S. G., *Elements of Chemical Reaction Engineering*. Upper Saddle River, NJ: Prentice Hall, 1999.
Friedman, S., Warzinski, R. P. 1977. *Eng. Power* 99: 361.
Froment, G. F. and Bischoff, K. B., *Chemical Reactor Analysis and Design*. New York: John Wiley, 1990.
Fryer, C. and Potter, O. E. 1972a. *Ind. Eng. Chem. Fundam.* 11: 338; 1972b. *Powder Technol.* 6: 317.
Fu, Y., Diwekar U. M. and Suchak, N. J. 2000. *Adv. Environ. Res.* 3: 424.
Furusaki, S., Takahasi, M. and Miyauchi, T. 1976. *AIChE J.* 22: 354; 1978. *J. Chem. Eng. Jpn.* 11: 309.
Galiski, J. B. and Hightower, J. W. 1970. *Can. J. Chem. Eng.* 48: 151.
Garza-Garza, O. and Dudukovic, M. P. 1982a. *Chem. Eng. Sci.* 6: 131; 1982b. *Chem. Eng. J.* 24: 35.
Geldart D. 1973. *Powder Technology*. 7: 285; *Gas Fluidization Technology*. New York: John Wiley, 1986.
Gilbert, L. and Mercier, C., in *Heterogeneous Catalysis and Fine Chemicals*, Guisnet, M., Barbier, J., Barrault, J., Bouchoule, C., Dupez, D., Montassier, C. and Pérot, G., eds. Amsterdam: Elsevier, 1993.
Gillespie, G. R. and Kenson, R. E. 1971. *Chem. Tech.* 1: 627.
Gower, R. C. 1971. Ph.D. thesis, Lehigh University.
Grace, J. R., in *Recent Advances in Engineering Analysis of Chemical Reaction Systems*, Doraiswamy, L. K., ed. New Delhi: Wiley Eastern, 1984; *Gas Fluidization Technology*, Geldart, D. ed. New York: John Wiley, 1986.

Green, D. W. and Perry, R. H. *Perry's Chemical Engineers' Handbook,* 8th ed. New York: McGraw-Hill, 2007.
Guha, B. K. and Narasimhan, G. N. 1972. *Chem. Eng. Sci.* 27: 703.
Gullett, B. K., Bruce, K. R. and Machilek, R. M. 1990. *Rev. Sci. lustrum.* 61(2): 904.
Harnby, N., Edwards, M. F. and Nienow, A. W., *Mixing in the Process Industries*, 2nd ed. Oxford, UK: Butterworth-Heinemann, 1992.
Hartig, F. and Keil, F. J. 1993. *Ind. Eng. Chem. Res.* 32: 424.
Hasler, D. J. L., Doraiswamy, L. K. and Wheelock, T. D. *Ind. Eng. Chem. Res.* 42: 2003.
Heesink, A. B. M. and van Swaaij, W. P. M. 1995. *Chem. Eng. Sci.* 50: 2983.
Hegedus, L. L. and Petersen, E. E. 1973a. *Chem. Eng. Sci.* 28: 69; 1973b. 345; 1973c. *J. Catal.* 28: 150; 1974. *Catal. Rev., Sci. Eng.* 9: 245.
Heijnen, J. J. and van't Riet, K. 1984. *Chem. Eng. J.* 28: B21.
Hill, W. J. 1966. Thesis, University of Wisconsin, Madison, WI.
Hills, A. W. D. 1968. *Chem. Eng. Sci.* 23: 297.
Hlavacek, V. 1970. *Ind. Eng. Chem.* 62: 8.
Hlavacek, V. and Kubicek, M. A. 1972. *Chem. Eng. Sci.* 25: 1537.
Hlavacek, V. and Votruba, J., in *Chemical Reactor Theory—A Review,* Lapidus, L. and Amundson, N. R., eds. Englewood Cliffs, NJ: Prentice Hall, 1977.
Hlavacek, V., Marek, M. and John, T. M. 1969. *Coll. Czech. Chem. Commun.* 34: 3868.
Hoelderich, W. F. and van Bekkum, H. 1991. In *Introduction to Zeolite Science and Practice,* van Bekkum, H., Flanigen, E. M. and Jansen, J. C., eds. *Stud. Surf. Sci. Catal.* 58: 631.
Hofmann, H., in *Chemical Reactor Design and Technology: Overview of the New Development of Energy and Petrochemical Reactor Technologies*, de Lasa, H. I., ed. *Proceedings of the NATO Advanced Study Institute on Chemical Reactor Design and Technology*, London, Ontario, Canada, 2–12 June, 1985, Kluwer Academic, Dordrecht, Netherlands, 1986.
Horío, M., Nonaka, A., Hoshiba, M. L., Morisita, K., Kobukai, Y., Naito, J., Tashibana, O., Watanabe, K. and Yoshida, N., in *Circulating Fluidized Bed Technology, Basu, P.,* ed. New York: Pergamon, 1986.
Hughes, C. C. and Mann, R. 1978. *ACS Symp. Ser.* 65: 201.
Illerup, J. B., Dam-Johansen, K. and Johnsson, J. E. 1993. *Gas Clean High Temp. [Pap. Int. Symp.]* 2: 492.
Ishida, M. and Wen, C. Y. 1968. *AIChE J.* 14: 311.
Jagtap, S. B. and Wheelock, T. D. 1996. *Energy & Fuels* 10: 821.
Jayaraman, V. K., Kulkarni, B. D. and Doraiswamy, L. K. 1981. *Am. Chem. Soc. Symp. Ser.* 168: 19.
Joglekar, H. S.; Samant, S. D. and Joshi, J. B. 1991. *Water Res.* 25: 135.
Johnstone, R. A. W., Wilby, A. H. and Entwistle, I. D. 1985. *Chem. Rev.* 85: 129.
Jorgensen, K. A. 1989. *Chem. Rev.* 89: 431.
Joshi, J. B. 1983. *Trans. Inst. Chem. Eng.* 61: 143.
Joshi, J. B. and Sharma, M. M. 1979. *Trans, Inst. Chem. Eng.* 57: 244.
Joshi, J. B., Abichandani, J. S., Shah, Y. T., Ruether, J. A. and Ritz, H. J. 1981. *AIChE J.* 27: 937.
Joshi, J. B., Dinkar, M., Deshpande, N. S. and Phanikumar, D. V. 2001. *Adv. Chem. Eng.* 28: 1.
Joshi, J. B., Mahajani, V. V. and Juvekar, V. A. 1985. *Chem. Eng. Commun.* 33: 1.
Joshi, J. B., Parasu Veera, U., Prasad, Ch. V., Phanikumar, D. V., Deshpande, N. S., Thakre, S. S. and Thorat B. N. 1998. *PINSA-A.* 64: 441.
Joshi, J. B., Patil, T. A., Ranade, V. V. and Shah, Y. T. 1990a. *Rev. Chem. Eng.* 6: 73.
Joshi, J. B., Ranade, V. V., Gharat, S. D. and Lele, S. S. 1990b. *Can. J. Chem. Eng.* 68: 705.
Joshi, J. B., Shah, Y. T., Albal, R. S., Ritz, H. J. 1982. *Ind. Eng. Chem. Process Des. Dev.* 21: 594.
Juvekar, V. A. 1976. Ph.D. thesis, University of Mumbai, Mumbai, India.
Kammermeyer, K. A. and Rutz, L. O. 1959. *Chem. Eng. Prog. Symp. Ser.* 55 (24): 163.
Karcz, J. 1999. *Chem. Eng. J.* 72 (3): 217.
Kato, K. and Wen, C. Y. 1969. *Chem. Eng. Sci.* 24: 1351.
Kern, D. Q., *Process Heat Transfer.* Tokyo, Japan: McGraw-Hill Kogakusha, 1950.
Khan, A. R. 1999. *Chem. Eng. Res. Des.* 77: 11.
Khare, G. P., Delzer, G. A., Kubicek, D. H. and Greenwood, G. J. 1995. *Environ. Prog.* 14: 146.
Kijenski, S., Glinski, M. and Reinhercs, J., in *Heterogeneous Catalysis and Fine Chemicals,* Guisnet, M., Barbier, J., Barrault, J., Bouchoule, C., Dupez, D., Montassier, C., and Pérot, G. eds. Amsterdam: Elsevier, 1988.
Kim, K. K. and Smith, J. M. 1974. *AIChE J.* 20: 670.

Komatsu, W. 1965. *Proc. 5th Intl. Symp. Reactivity of Solids.*
Kramers, H. A., Baars, G. M. and Knoll, W. H. 1953. *Chem. Eng. Sci.* 2: 35.
Krasuk, J. H. and Smith, J. M. 1965. *Ind. Eng, Chem. Fundam.* 4: 102.
Kubota, H. and Yamanaka, Y. 1969. *J. Chem. Eng. Jpn.* 2: 238.
Kubota, H., Yamanaka, Y. and Lana, I. G. D. 1969. *J. Chem. Eng. Jpn.* 2: 71.
Kulkarni, B. D. and Doraiswamy, L. K. 1980a. *Chem. Eng. Sci.* 35: 817; 1980b. *Catal. Rev. Sci. Eng.* 22: 431; *Modelling of Noncatalytic Gas-Solid Reactions,* in *Handbook of Heat and Mass Transfer,* vol. 2, *Mass Transfer and Reactor Design,* Cheremisinoff, N. P., ed. Houston, TX: Gulf Publishing, 1986.
Kulkarni, B. D., Ramachandaran, P. A. and Doraiswamy, L. K., *Fluidization,* Grace, J R. and Matsen, J. M., eds. New York: Plenum Press, 1981.
Kumar, R. and Kuloor, N. R., in *Adv. in Chem. Eng.* vol. 8, Drew T. B., Cokelet, G. R., Hoops, J. W. and Vermeulen, T., eds. New York: Academic Press, 1970.
Kumbhar, P. S., Coq, B., Moreau, C., Moreau, P. and Figueras, F. 1994. *J. Phys. Chem.* 98: 10180.
Kunii, D. and Levenspiel, O. 1968a. *Ind. Eng. Chem. Fundam.* 7: 466; 1968b. *Ind. Eng. Chem. Proc. Des. Dev.* 7: 481; *Fluidization Engineering,* 1st ed. New York: Butterworth-Heinemann, 1969; 2nd ed., 1991.
Lee, K. U. and Aris, R. 1963. *Ind. Eng. Chem. Proc. Des. Dev.* 2: 300, 306.
Lele, S. S. and Joshi, J. B., *Encyclopedia of Fluid Mechanics,* suppl. 2, Cheremisinoff, N. P., ed. Houston, TX: Gulf Publishing, 1993.
Lele S. S., Joshi, J. B. and Parulekar, S. J. 1993. *Chem. Eng. Sci.* 48: 3631.
Levenspiel, O., *Chemical Reactor Omnibook.* Oregon State University Bookstore, Corvallis, OR, 1991; *Chemical Reaction Engineering,* 2nd ed., New York: John Wiley & Sons, 1999.
Lewis, W. K., Gilliland, E. R. and Girouard, H. 1962. *Chem. Eng. Prog. Symp. Ser.* 58: 38, 87.
Li, Y. and Kwauk, M., in *Fluidization,* Grace, J. R. and Matsen, J. M., eds. New York: Plenum Press, 1980.
Lobusevich, N. P., Golubstov, S. A., Malysheva, L. A., Lainer, L. I., and Trofimova, I. V. 1976. *J. Appl. Chem. USSR.* 41: 607.
Lockhart, R. W. and Martinelli, R. C. 1949. *Chem. Eng. Prog.* 45: 39.
Lopes de Figueiredo, M. M. and Calderbank, P. H. 1979. *Chem. Eng. Sci.* 34: 1333.
Luckner, C. R. and Wills, G. B. 1973. *J. Catal.* 28: 83.
Lunde, P. J. and Kester, F. L. 1974. *Ind. Eng. Chem. Proc. Des. Dev.* 13: 27.
Luss, D. and Amundson, N. R. 1967. *AIChE J.* 13: 759; 1969. *AIChE J.* 15: 194.
Luss, D. and Erwin, M. A. 1972. *Chem. Eng. Sci.* 27: 315.
Maki, T., Yokoyama, T. and Fuji, K. 1993. *Shokubai (Catalyst)* 35: 2.
Mandler, J. Lavie, R. and Sheintuck, M. 1983. *Chem. Eng. Sci.* 38: 979.
Mann, R. and Mavros, P. 1982. *Proc. 4th Eur. Conf. on Mixing.* April 27–29, BHRA, Leeuwenhorst, Netherlands.
Mantri, V. B., Gokarn, A. N. and Doraiswamy, L. K. 1976. *Chem. Eng. Sci.* 31: 779.
McCabe, W. L., Smith, J. C. and Harriot, P., *Unit Operations of Chemical Engineering,* 5th ed. New York: McGraw Hill, 1993.
McFarlane, C. M. and Nienow, A. W. 1996. *Biotech. Prog.* 12: 9.
McKillop, A. and Young, D. W. 1979. *Synthesis* 401: 481.
Mears, D. E. 1971a. *Ind. Eng. Chem. Proc. Des. Dev.* 10: 541; 1971b. *Chem. Eng. Sci.* 26: 1361; 1976. *Ind. Eng. Chem. Fund.* 15: 20.
Mhetras, M. B., Pandit, A. B. and Joshi, J. B. 1994. *Proc. 8th Eur. Conf. on Mixing, I. Chem. Symp. Ser.* no. 136.
Misono, M. 1987. *Catal. Rev. Sci Eng.* 29: 269; in *New Frontiers in Catalysis,* Guczi, L., Solymosi, F. and Tetenyi, P., eds. Amsterdam: Elsevier and Budapest: Akademiai Kiado, 1993.
Miyauchi, T. 1974. *J. Chem. Eng. Jpn.* 7: 201.
Miyauchi, T. and Marooka, S. 1969. *Int. Chem. Eng.* 9: 713; 1974. *J. Chem. Eng. Jpn.* 7: 201.
Molerus, O. and Latzel, W. 1987. *Chem. Eng. Sci.* 42: 1431.
Molnar, A., Smith, G. V. and Bartok, M. 1989. *Adv. Catal.* 36: 329.
Muller, R. 1950. *Chem. Tech.* (Berlin) 2: 41.
Musick, J. K., Thomas, F. S. and Johnson, J. E. 1972. *Ind. Eng. Chem. Process Des. Dev.* 11: 350.
Mutasher, E. I., Khan A. I. and Bowen, J. H. 1989. *Ind. Eng. Chem. Res.* 28: 1150.
Narasimhan, G. 1961. *Chem. Eng. Sci.* 16: 7.
Nauman, E. B., *Chemical Reactor Design,* New York: John Wiley & Sons, 1987.
Neuberg, H. J. 1970. *Ind. Eng. Chem. Proc. Des. Dev.* 9: 285.

Olah, G. A., Iyer, P. S. and Prakash, G. K. S. 1986. *Synthesis:* 513.
Oldshue, J. Y., *Fluid Mixing Technology.* New York: McGraw-Hill, 1983.
Oroskar, A. and Stem, S. A. 1979. *AIChE J.* 25: 903.
Otani, S., Wakao, N. and Smith, J. M. 1965. *AIChE J.* 11: 466.
Pandit, A. B. and Joshi, J. B. 1983. *Chem. Eng. Sci.* 38: 1189; 1984. *Rev. Chem. Eng.* 2: 1; 1998. *Rev. Chem. Eng.* 14: 321.
Pansing, W. F. and Malloy, J. B. 1962. *Chem. Eng. Prog.* 58: 12, 53.
Parulekar, S. J., Joshi, J. B. and Shertukde, P. V. 1988. *Chem. Eng. Sci.* 44: 543.
Patwardhan, A. W. and Joshi, J. B. 1998. *Can. J. Chem. Eng.* 76: 339; 1999a. *Ind. Eng. Chem. Res.* 38: 3131; 1999b. *ibid,* 49.
Pawalski, J. 1961. *Chem. Eng. Tech.* 33: 492.
Perkins, T. K. and Rase, H. F. 1958. *AIChE J.* 4: 351.
Pérot, G. and Guisnet, M. 1990. *J. Mol. Catal.* 61: 173.
Peters, M. S. and Skorpinski, E. J. 1965. *J. Chem. Educ.* 42: 329.
Pigford, R. L. and Pyle, C. 1951. *Ind. Eng. Chem.* 43: 1649.
Pilling, M. J. and Seakins, P. W., *Reaction Kinetics.* Oxford, UK: Oxford Science Publications, 1995.
Powell, R. 1969. Nitric Acid Technology, Recent Developments. *Chem. Proc. Rev.* 30.
Pradhan, M. P, Suchak, N. J., Walse, P. R. and Joshi, J. B. 1997. *Chem. Eng. Sci.* 52: 4569.
Prasad, K. B. S. and Doraiswamy, L. K. 1974. *J. Catal.* 32: 384.
Prasad, K. B. S. and Rao, M. S. 1977. *Chem. Eng. Sci.* 32: 1411.
Prasannan, P. C. and Doraiswamy, L. K. 1982. *Chem. Eng. Sci.* 37: 925.
Prasannan, P. C., Ramachandran, P. A. and Doraiswamy, L. K. 1986. *Chem. Eng. J.* 33: 19.
Pratt, K. C., in *Catalysis: Science and Technology,* vol. 8, Anderson, J. R. and Boudart, M., eds. Netherlands: Springer-Verlag, 1987.
Raab, C. G., Englisch, M., Marenelli, T. B. L. W. and Lercher, J. A. 1993. *Stud. Surf. Sci. Catal.* 78: 211.
Radar, C. G. and Weller, S. W. 1974. *AIChE J.* 20: 515.
Raghav Rao, K. S. M. S., Rewatkar, V. B. and Joshi, J.B. 1988. *AIChE J.* 34: 1332.
Rajadhyaksha, R. A. and Joshi, G. W., in *Heterogeneous Catalysis and Fine Chemicals,* Guisnet. M., Barbier, J., Barrault, J., Bouchoule, C., Dupez, D., Montassier, C. and Pérot, G., eds. Amsterdam: Elsevier, 1991.
Rajadhyksha, R. A., Vasudeva, K. and Doraiswamy, L. K. 1975. *Chem. Eng. Sci.* 20: 25; 1976. *J. Catal.* 41: 61.
Rajagopalan, K. and Luss, D. 1979. *Ind. Eng. Chem. Proc. Des. Dev.* 18: 459.
Ramachandaran, P. A. and Doraiswamy, L. K. 1982. *AIChE J.* 28: 881.
Ramachandaran, P. A. and Kulkarni, B. D. 1980. *Ind. Eng. Chem. Proc. Des. Dev.* 19: 717.
Ramachandaran, P. A. and Smith, J. M. 1977a. *Chem. Eng. J.* 4: 137; 1977b. *AIChE J.* 23: 353.
Ranade, P. V. and Harrison, D. P. 1979. *Chem. Eng. Sci.* 34: 427; 1981. *Chem. Eng. Sci.* 36: 1079.
Rao, Y. K. 1979. *Met. Trans. B.* 10: 243.
Raskin, A., Ja, et al. 1968a. *Theoret. Found. Chem. Tech.* 2: 220; 1968b. *Chim. Ind.* (Milan) 44: 199.
Reid, R. C., Prausnitz, J. M. and Sherwood, T. K., *The Properties of Gases and Liquids.* New York: McGraw-Hill, 1977.
Reilly, P. M. and Blau, G. E. 1974. *Can. J. Chem. Eng.* 52: 289.
Rewatkar, V. B. and Joshi, J. B. 1991a. *Chem. Eng. Commun.* 102: 1; 1991b. *Chem. Eng. Technol.* 14: 333; 1991c. *Chem. Eng. Technol.* 14: 386; 1991d. *Ind. Eng. Chem. Res.* 30: 1784.
Rewatkar, V. B., Deshpande, A. J., Pandit, A. B. and Joshi, J. B. 1993. *Can. J. Chem. Eng.* 71: 226.
Richardson, J. F. and Zaki, W. N. 1954. *Trans. Inst. Chem. Eng.* 32: 35.
Richardson, J. T. and Friedrich, H. 1975. *J. Catal.* 37: 8.
Ristic, M. M., *Sintering:* "New Developments" in *Proc. 4th Intl. Round Table Conf. on Sintering* (Durbronic, Poland). Amsterdam: Elsevier, 1979.
Roberts, G. W. 1972. *Chem. Eng. Sci.* 27: 1409.
Roberts, G. W. and Satterfield, C. N. 1965. *Ind. Eng. Chem. Fundam.* 4: 288; 1966. 5, 317.
Roberts, M. W. and McKee, C. S., *Chemistry of the Metal-Gas Interface.* New York: Oxford University Press, 1979.
Roberts, S. M., *Dynamic Programming in Chemical Engineering and Process Control.* New York: Academic Press, 1964.
Rochow, E. G. 1945. *J. Am. Chem. Soc.* 67: 963; 1945. U.S. Patents 2,380,995 and 2,380,996.

Rony, P. R. and Roth, J. F., in *Catalysis: Heterogeneous and Homogeneous,* Delmon, B. and Jannes, G., eds. Amsterdam: Elsevier, 1975.
Rowe, P. N., Claxton, K. T. and Lewis, J. B. 1965. *Trans. Inst. Chem. Eng.* 43: T14.
Ruckenstein, E. 1970. *AIChE J.* 16: 151.
Ruckenstein, E. and Vavanellos, T. 1975. *AIChE J.* 21: 756.
Ruthven, D. M. 1968. *Chem. Eng. Sci.* 23: 759.
Sadana, A. and Doraiswamy, L. K. 1971. *J. Catal.* 23: 147.
Sahimi, M., Gavalas, G. R. and Tsotsis, T. T. 1990. *Chem. Eng. Sci.* 45: 1443.
Sampath, B. S., Ramachandran, P. A. and Hughes, R. 1975. *Chem. Eng. Sci.* 30: 125.
Saravanan, K. and Joshi, J. B. 1995. *Ind. Eng. Chem. Res.* 34: 2499; 1996. *Can. J. Chem. Eng.* 74: 16.
Saravanan, K., Patwardhan, A. W. and Joshi, J. B. 1997. *Can. J. Chem. Eng.* 75: 664.
Saravanan, K., Patwardhan, A. W., Mundale, V. D. and Joshi, J. B. 1996. *Ind. Eng. Chem. Res.* 35: 1583.
Satterfield, C. N., *Mass Transfer in Heterogeneous Catalysis.* Cambridge, MA: MIT Press, 1970; *Heterogeneous Catalysis in Industrial Practice.* New York: McGraw-Hill, 1991.
Satterfield, C. N. and Cortez, D. H. 1970. *Ind. Eng. Chem. Fundam.* 9: 613.
Schmalzer, D. K. 1969. *Chem. Eng. Sci.* 24: 615.
Schmidt, L. D. and Luss, D. 1971. *J. Catal.* 22: 269.
Scott, D. S. 1962. *Can. J. Chem. Eng.* 40: 170.
Shah, M. A. and Roberts, D. 1974. *Adv. Chem. Ser.* 133: 259.
Sheldon, R. A., in *Aspects of Homogeneous Catalysis,* vol. 4, Ugo, R., ed. Dordrecht, Netherlands: Reidel. vol. 4, 1981; *Chirotechnology.* New York: Marcel Dekker, 1993; 1996. *J. Chem. Technol. Biotechnol.* 67: 1.
Sherrington, D. C. and Hodge, P., eds., *Syntheses and Separations Using Functional Polymers.* New York: John Wiley, 1988.
Sherwood, T. K., Pigford, R. L. and Wilke, C. R., *Mass Transfer.* New York: McGraw-Hill, 1975.
Shibata, M. and Masumoto, T., in *Preparation of Catalysts IV,* Delmon, B., Grange, P., Jacobs, P. A. and Poncelet, G., eds. Amsterdam: Elsevier, 1987.
Smith, K., *Solid Supports and Catalysts in Organic Synthesis.* New York: Ellis Harwood: PTR Prentice Hall, 1992.
Smith, T. G., Zahradnik, J. and Carberry, J. J. 1975. *Chem. Eng. Sci.* 30: 763.
Sofekun, O. A., Rollins, D. K. and Doraiswamy, L. K. 1994. *Chem. Eng. Sci.* 49: 2611.
Sohn, H. Y. and Sohn, H. J. 1980. *Ind. Eng. Chem. Process Des. Dev.* 19: 237.
Sohn, H. Y. and Szekeley, J. 1972. *Chem. Eng. Sci.* 27: 763.
Sohn, H. Y., Merrill, R. P. and Petersen, E. E. 1970. *Chem. Eng. Sci.* 25: 399.
Sotirchos, S. V. and Zarkanitis, S. 1989. *AIChE J.* 35 (7): 1137.
Squires, A. M., Graff, R. A. and Pell, M. 1971. *Chem. Eng. Prog. Symp. Ser.* 67: 23.
Squires, A. M., Kwauk, M. and Avidan, A. 1985. *Science.* 230: 1329.
Srinivasan, P. S., Nandapurkar, S. S., and Holland, F. A. 1968. *Chem. Eng.* (London), no. 218, CE113-CE119.
Stewart, P. S. B. and Davidson, J. M. 1967. *Powder Technol.* 1: 61.
Stewart, W. E. and Villadsen, J. 1969. *AIChE J.* 15: 28.
Strauss, A. and Buddle, K. 1978. *Chem. Tech.* (Berlin), 30: 73.
Suchak, N. J. and Joshi, J. B. 1994. *AJCh E J.* 40: 6.
Suter, H., *Phthalsaäureanhydrid.* Darmstardt, Germany: Steinkopf Verlag, 1972.
Syamala, M. S., Reddy, G. D., Rao, B. N. and Ramamurthy, V. 1986. *Curr. Sci.* 55: 875.
Szekely, J., Evans, J. W. and Sohn, H. Y., *Gas-Solid Reactions.* New York: Academic Press, 1976.
Tabushi, I. and Kuroda, Y. 1983. *Adv. Catal.* 32: 417.
Tajbl, D. G., Fledkerchner, H. L. and Lee, A. L. 1967. *Adv. Chem. Ser.* 69: 166.
Tajbl, D. G., Simons, J. B. and Carberry, J. J. 1966. *Ind. Eng. Chem. Fundam.* 5: 171.
Tamhankar, S. S. and Doraiswamy, L. K. 1978. *Ind. Eng. Chem. Fundam.* 17: 84; 1979. *AIChE J.* 25: 561.
Tamhankar, S. S., Gokarn, A. N. and Doraiswamy, L. K. 1981. *Chem. Eng. Sci.* 36: 1365.
Tanaka, Y., Imizu, Y., Hattori, H. and Tanabe, K. 1980. *Proc. 7th Intl. Cong. Catal.,* Tokyo, Japan.
Thomas, J. M. and Thomas, W. J., *Principles and Practice of Heterogeneous Catalysis.* New York: VCH, 1999.
Tigrel, A. Z and Pyle, D. L. 1971. *Chem. Eng. Sci.* 26: 133.
Tine, C. B. D. 1985. *Chem. Eng. Res. Des.* 63: 112.
Toomey, R. D. and Johnston, H. F. 1952. *Chem. Eng. Prog.* 48: 220.

Torok, B., Molnar, A., Borszeky, K., Toth-Kadar, E. and Bakonyi, I., in *Heterogeneous Catalysis and Fine Chemicals,* Guisnet, M., Barbier, J., Barrault, J., Bouchoule, C., Dupez, D., Montassier, C. and Pérot, G., eds. Amsterdam: Elsevier, 1993.
Treybal, R. E., *Mass-Transfer Operations.* 3rd ed. New York: McGraw-Hill, 1980.
Uhl, V. W. and Gray, J. B., *Mixing: Theory and Practice,* vol. 1, Orlando: Academic Press, 1966.
Ulrichson, D. L. and Mahoney, D. J. 1980. *Chem. Eng. Sci.* 35: 567.
Uner, D. O., Pruski, M., Gerstein, B. C. and King, T. S. 1994. *J. Catal.* 146: 530.
Usyal, B. Z., Aksahin, I. and Yusel, H. 1988. *Ind. Eng. Chem. Res.* 27: 434.
Vaidya, R. N. and Pangarkar, V. G. 1985. *Chem. Eng. Commun.* 39: 337.
van Deemter, J. J., in *Fluidization III,* Grace, J. R. and Matsen, J. M., eds. New York: Plenum, 1980.
van den Bleek, C. M. Van der Wiele, K. and Van den Berg, P. J. 1969. *Chem. Eng. Sci.* 24: 681.
van der Ham, A. G. J., Heesink, A. B. M., Prins, W. and van Swaaij, W. P. M. 1996. *Ind. Eng. Chem. Res.* 35: 1487.
van Welsenaere, R. J. and Froment, G. F. 1970. *Chem. Eng. Sci.* 25: 1503.
van't Riet, K. and Smith J. M. 1973. *Chem. Eng. Sci.* 28: 1031.
Varghese, P. and Wolfe, E. E. 1980. *AIChE J.* 26: 55.
Varghese, P., Varma, A. and Carberry, J. J. 1978. *Ind. Eng. Chem. Fundam.* 17: 195.
Vetter, K. J., *Electrochemical Kinetics, Theoretical and Experimental Aspects.* New York: Academic Press, 1967.
Villa, L. T. and Quiroga, O. D. 1989. *Chem. Eng. Res. Des.* 67: 76.
Vracar, R., Vucurovic, D., Rudarstvo, I. 1970. *Metalurgija.*
Wan, K. and Davis, M. E. 1993a. *J. Chem. Soc. Chem. Commun.* 1262; 1993b. *Tetrahedron: Asymmetry* 4(12), 2461.
Warmoeskerken, M. C. G. and Smith, J. M. 1985. *Chem. Eng. Sci.* 40: 2063.
Weekman, V. W., Jr. 1968. *Ind. Eng. Chem. Proc. Des. Dev.* 7: 90; 1974. *AIChE J.* 20: 833.
Weekman, W. W., Jr. and Gorring, R. L. 1965. *J. Catal.* 4: 260.
Weisz, P. B. 1957. *Physik. Chem.* (Frankfurt) 11: 1.
Weisz, P. B. and Hicks, J. S. 1962. *Chem. Eng. Sci.* 17: 265.
Weisz, P. B. and Prater, C. D. 1954. *Adv. Catal.* 6: 143.
Wen, C. Y. and Fan, L. T., *Models for Flow Systems and Chemical Reactors.* New York: Marcel Dekker, 1975.
Werther, J. 1980. *Intl. Chem. Eng.* 20: 529.
Westmoreland, P. R. and Harrison, D. P. 1976. *Environ. Sci. and Technol.* 10: 659.
Wheelock, T. D. 1997. U.S. patent 5,653,955; 2000. U.S. patent 6,083,862.
Wheelock, T. D. and Akiti, T. T., Jr. Filed January 27, 2000, U.S. patent pending.
Yadav, N. K., Kulkarni, B. D. and Doraiswamy, L. K. 1994. *Ind. Eng. Chem. Res.* 33: 2412.
Yang, K. H. and Hougen, O. A. 1950. *Chem. Eng. Prog.* 46: 146.
Yang, W. C., Newby, R. A., Lippert, T. E., Keairns, D. L., and Cicero, D. C. 1992. Fluidization VII. *Proc. 7th Eng. Found. Conf. on Fluidization,* Brisbane, Australia, May 2–8, Potter, O. E. and Nicklin, D. J., eds.
Yerushelmi, J. and Cankurt, N. T., in *Fluidization,* Davidson, J. F. and Keairns, D., eds. London and New York: Cambridge University Press, 1978.
Yoo, H. J. and Steinberg, M. 1983. Final Report DOE/CH00016–1494.
Yrjas, K. P., Cornelis, A. P. and Hupa, M. M. 1996. *Ind. Eng. Chem. Res.* 35: 176.
Yushchenko, V. V. Korneichirk, G. P., Usha-Kova, Stasevich. V. P. and Semenyuk, Yu. V. 1968. *Kinet. Katal.* 4: 154.
Zahradnik, J., Fialova, M., Linek, V., Sinkule, J., Reznickova, J. and Kastanek, F. 1997. *Chem. Eng. Sci.* 52: 4499.
Zevenhoven, C. A. P., Yrjas, K. P. and Hupa, M. M. 1996. *Ind. Eng. Chem. Res.* 35: 943.
Zwietering, T. N. 1958. *Chem. Eng. Sci.* 8: 244.

12 Distillation

James R. Fair

CONTENTS

- 12.1 Basic Models 972
  - 12.1.1 Phase Equilibrium 972
  - 12.1.2 The Contacting Stage 973
  - 12.1.3 Multiple Stages 975
- 12.2 Vapor-Liquid Equilibrium 975
  - 12.2.1 Estimation and Measurement of Activity Coefficients 980
  - 12.2.2 Summary—Vapor–Liquid Equilibrium 981
- 12.3 Stage Calculations 984
  - 12.3.1 Separation Specifications 984
  - 12.3.2 Single-Stage Process 984
  - 12.3.3 Multistage Processing 985
  - 12.3.4 Binary Systems 985
  - 12.3.5 Stages-Reflux Relationships 987
  - 12.3.6 Multicomponent Systems 987
  - 12.3.7 Optimum Reflux 990
  - 12.3.8 Design Procedure 990
  - 12.3.9 Computer Methods 991
- 12.4 Special Distillations 993
  - 12.4.1 Azeotropic Distillation 993
    - 12.4.1.1 General Principles 993
    - 12.4.1.2 The Azeotropic Distillation Process 997
    - 12.4.1.3 Prediction of Azeotropes 999
  - 12.4.2 Extractive Distillation 1000
  - 12.4.3 Complex System Distillation 1001
  - 12.4.4 Steam Distillation 1002
  - 12.4.5 Batch Distillation 1002
  - 12.4.6 Reactive Distillation 1005
- 12.5 Distillation Columns 1006
- 12.6 Tray Column Hydraulics 1009
  - 12.6.1 Crossflow Tray Columns–Sieve Trays 1009
    - 12.6.1.1 Performance Profile 1009
    - 12.6.1.2 Vapor-Handling Capacity 1009
    - 12.6.1.2 Other Crossflow Trays 1019
    - 12.6.1.3 Fixed Valve Trays 1021
  - 12.6.2 Counterflow Perforated Trays 1022
  - 12.6.3 Baffle Trays 1024
  - 12.6.4 Multiple Downcomer Trays 1026
  - 12.6.5 General Comments on Tray-Type Columns 1026

12.7 Packed Column Hydraulics 1029
    12.7.1 Packing Characteristics 1030
        12.7.1.1 Maximum Vapor–Liquid Capacity 1033
        12.7.1.2 Pressure Drop 1037
        12.7.1.3 Liquid Distribution 1038
        12.7.1.5 Liquid Distributors 1039
        12.7.1.6 Liquid Redistribution 1040
        12.7.1.7 Vapor Distribution 1041
        12.7.1.8 Turndown 1041
12.8 Mass Transfer Efficiency 1041
    12.8.1 General Mass Transfer Relationships 1042
    12.8.2 Tray Column Efficiency 1043
        12.8.2.1 Regimes on Trays 1043
        12.8.2.2 Tray Efficiency Definitions 1044
        12.8.2.3 Point Efficiency 1045
        12.8.2.4 Tray (Murphree) Efficiency 1046
        12.8.2.5 Overall Column Efficiency E_{oc} 1046
        12.8.2.6 Efficiency from Performance Data 1047
        12.8.2.7 Empirical Efficiency Methods 1048
        12.8.2.8 Efficiency from Laboratory Experiments 1049
        12.8.2.9 Efficiency from Mass Transfer Models 1050
        12.8.2.10 Conversion of Point Efficiency to Tray Efficiency 1052
        12.8.2.11 Entrainment Effects on Efficiency 1052
        12.8.2.12 Overall Column Efficiency 1054
        12.8.2.13 Multicomponent Systems 1054
    12.8.3 Packed Column Efficiency 1055
        12.8.3.1 Random Packings 1056
        12.8.3.2 Structured Packings 1057
        12.8.3.3 Mechanistic Model for Structured Packings 1059
        12.8.3.4 Scale-Up of Structured Packing Efficiency 1061
        12.8.3.5 Structured Packing Performance 1062
    12.8.4 Packings versus Crossflow Trays 1063
        12.8.4.1 Counterflow Trays 1065
12.9 Troubleshooting 1065
    12.9.1 Inadequate Vapor Capacity 1066
    12.9.2 Liquid Flow Capacity 1066
    12.9.3 Pressure Drop 1066
    12.9.4 Separation Efficiency 1066
    12.9.5 Stability 1066
12.10 Nomenclature 1066
References 1069

Distillation is a physical process for separating a liquid mixture into its constituents. When such a mixture is partially vaporized, the vapor normally has a composition different from that of the residual liquid. Implied in the method is the condensation of the vapor to form a product liquid, called the *distillate*. The residual liquid product is often called the *bottoms*.

    Distillation in crude form was practiced over 2000 years ago, usually for the concentration of alcoholic spirits. The first formalized documentation of distillation appears to be the treatise by Brunschwig in 1500.[1] Distillation has since emerged as the key method for separating liquid

Distillation

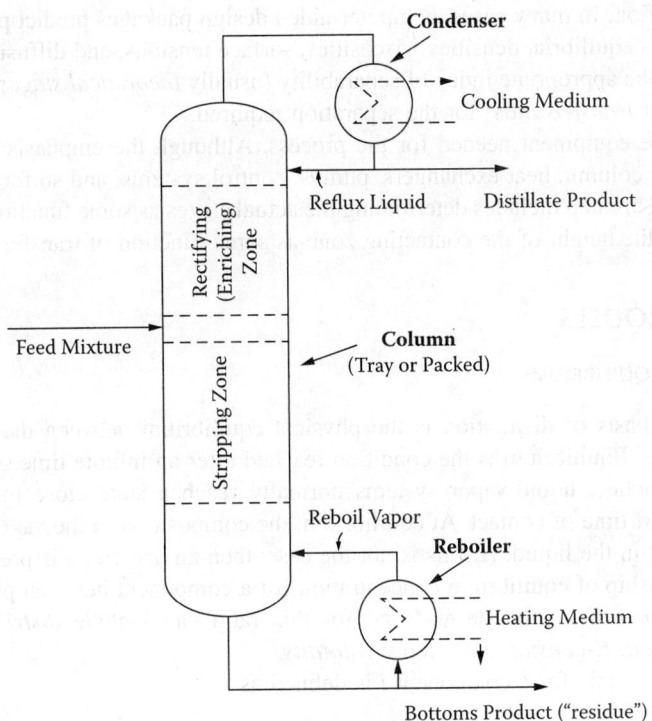

FIGURE 12.1 Distillation system operated in a continuous mode.

mixtures in chemical processing and related industries because of its versatility, simplicity, economy, and many years of experience.

Early distillations were of the batch, takeover type, sometimes called *simple distillation* or *differential distillation*. A charge of liquid mixture is vaporized from a *still*, or *stillpot*, by heat addition, and the product vapor is condensed into one or more fractions. Thus the term *fractional distillation*, or *fractionation*, has become associated with any distillation operation designed to obtain defined or specified constituent fractions.

Most distillations today are of the multistage rectification type, operated continuously or in the batch mode. They are characterized by vertical vessels (*distillation columns*) with internal contacting devices (usually trays or packings) that provide intimate contacting of vapor and liquid. When operated continuously (the usual preferred mode), the towers are normally fed with a liquid mixture near the center of the column, providing a *stripping zone* below the feed point and a *rectifying zone* (or "enriching zone") above the feed point. A diagram of a typical distillation column is shown in Figure 12.1. Heat is added at the base of the column by vapor from a *reboiler* and is removed at the top of the column in a *condenser* to provide the distillate product. Part of the distillate is returned to the column as reflux liquid.

The usual sequence of design steps for a distillation process as shown in Figure 12.1 is as follows:

1. Determine the feed composition including all minor components.
2. Establish the degree of separation to be achieved, including purities of key products as well as the fractions of certain feed materials to be recovered in one of the product streams.
3. Procure pertinent physical property data, especially including vapor-liquid equilibrium relationships ("equilibria") over the composition, temperature, and pressure ranges of

the separation. In many cases, computer-aided design packages predict physical properties such as equilibria, densities, viscosities, surface tensions, and diffusion coefficients.
4. Calculate the appropriate index of separability (usually *theoretical stages* but sometimes numbers of *transfer units*) for the separation required.
5. Specify the equipment needed for the process. Although the emphasis may be on the distillation column, heat exchangers, pumps, control systems, and so forth must also be included. This step includes determining the actual stages as some function of theoretical stages, or the height of the contacting zone as some function of transfer units.

12.1 BASIC MODELS

12.1.1 Phase Equilibrium

The fundamental basis of distillation is the physical equilibrium between the liquid and vapor phases of a system. Equilibrium is the condition reached after an infinite time of contact between the phases. In practice, liquid-vapor systems normally reach a state close to equilibrium in a comparatively short time of contact. At equilibrium, the composition in the vapor phase is usually different from that in the liquid. (If this is not the case, then an *azeotrope* is present, as discussed later.) The relationship of equilibrium concentrations of a component between phases is described by the *equilibrium ratio*. Other terms used for this ratio can include *distribution coefficient*, *equilibrium constant*, *K-constant*, or simply *volatility*.

The equilibrium ratio for a component i is defined as

$$K_i = y_i^* / x_i \quad \text{or} \quad y_i^* = K_i x_i \tag{12.1}$$

where

K_i = equilibrium ratio for component i
y_i^* = mole fraction of i in vapor at equilibrium
x_i = mole fraction of i in liquid

For example, if at equilibrium the liquid contains 0.1 mol fraction isobutane (iC_4), and if the equilibrium ratio of iC_4 under the system conditions is 2.5, then 0.25 mol fraction iC_4 is in the vapor. The equilibrium ratio is some function of temperature, pressure, and the composition of the liquid, which may be expressed as

$$K_i = f(T, P, x) \tag{12.2}$$

where

T = system temperature
P = system pressure
x = vector (set) of liquid molar concentrations for all components

The equilibrium ratios of two components in the system may be compared at the same conditions by means of the *relative volatility*, defined as

$$\alpha_{ij} = K_i / K_j \tag{12.3}$$

where α_{ij} = relative volatility of i with respect to j.

Distillation

Often the relative volatility is expressed as the ratio of the more volatile component to the less volatile component so that the numerical value of α_{ij} is greater than unity.

Relative volatility is a useful tool to judge the feasibility and ease of a distillation separation. In general, the larger the relative volatility between two key components, the easier and less costly will be the separation of those keys. If the relative volatility between the keys is unity, then their separation by ordinary distillation is impossible. Such a situation exists with azeotropes. The relationship between relative volatility and difficulty of separation may be illustrated by application of a simple relationship (the Fenske equation, to be discussed later):

$$N_m = \frac{\ln\left[\frac{(x_i/x_j)_D}{(x_i/x_j)_B}\right]}{\ln \alpha_{ij}} \quad (12.4)$$

where
 N_m = minimum number of equilibrium stages required at total reflux
 D = distillate product
 B = bottoms product

Since the optimum number of equilibrium stages for a column, based on economic considerations, is approximately 2.5 times the minimum value (to be discussed later), Equation (12.4) may be expressed in terms of design stages, while rearranging variables, as follows:

$$N_t = \frac{2.5 \ln\left(\frac{x_{iD}/x_{iB}}{x_{jD}/x_{jB}}\right)}{\ln \alpha_{ij}} \quad (12.5)$$

where N_t = number of equilibrium stages required.

For illustration, suppose a separation is to be made in which the light key component has a concentration in the distillate 100 times that in the bottoms, and the heavy key component has a concentration in the bottoms 100 times that in the distillate. Equation (12.5) may be applied with natural logarithms as follows:

$$N = \frac{2.5 \ln\left[\frac{100}{1/100}\right]}{\ln \alpha_{ij}} = \frac{2.5 \ln 10,000}{\ln \alpha_{ij}} \quad (12.6)$$

Application of Equation (12.6) over a range of relative volatilities gives results as shown in Table 12.1. Clearly, the phase equilibrium relationship is of paramount importance in distillation process economics.

12.1.2 The Contacting Stage

The basic building block for stagewise distillation processes is the contacting stage, illustrated in Figure 12.2. This stage receives liquid from the stage above and vapor from the stage below and passes vapor to the stage above and liquid to the stage below. It may also receive feed from the outside and/or produce a sidedraw product of liquid, vapor, or both. If a reaction occurs on the stage, as in reactive distillation (to be discussed later), the stage concentrations must reflect the

TABLE 12.1
Effect of Relative Volatility on Required Stages for a Typical Separation

Relative Volatility, α_{ij}	Required Equilibrium Stages, N
10	10
5	14
2	33
1.5	57
1.2	126
1.1	241
1.0	∞

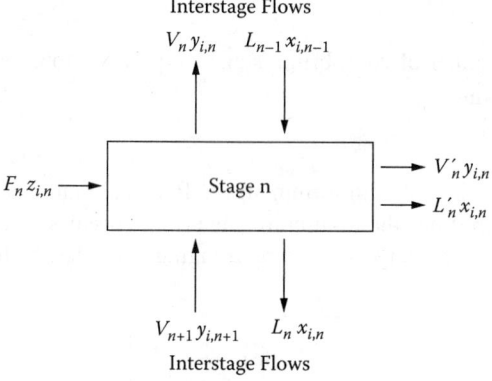

FIGURE 12.2 Component mass flows on a contacting stage.

conversion obtained on the stage at its temperature and pressure conditions. For the hypothetical *theoretical stage,* the exit streams of liquid and vapor must be in equilibrium with each other.

A mass balance may be written around stage n as follows:

$$F_n z_{i,n} + L_{n-1} x_{i,n-1} + V_{n+1} y_{i,n+1} = (V_n + V'_n) y_{i,n} + (L_n + L'_n) x_{i,n} \tag{12.7}$$

An energy balance can be written around the contacting stage using the enthalpies of all vapor and liquid streams (Figure 12.3):

$$F_n H_{F,n} + L_{n-1} h_{n-1} + V_{n+1} H + Q_n = (V_n + V'_n) H_n + (L_n + L'_n) h_n \tag{12.8}$$

where Q_n represents the heat added, removed, or generated within the stage either by a heat exchanger or an exothermic or endothermic reaction. In these balances, the stages are numbered downward from the top of the column.

The enthalpies of the vapor and liquid streams are functions of temperature, pressure, and composition:

$$H = f(T,P,y) \tag{12.9}$$

$$h = f(T,P,x) \tag{12.10}$$

Distillation

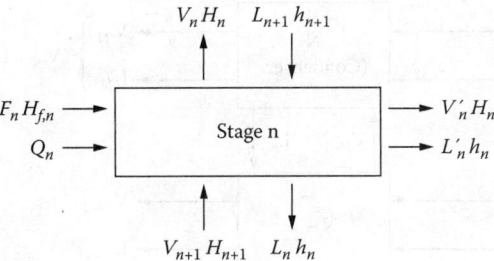

FIGURE 12.3 Stream energy flows on a contacting stage.

The approach to equilibrium for a contacting stage is called the *Murphree stage efficiency,* defined as the composition change in vapor across the stage divided by the composition change that would be achieved if true equilibrium existed:

$$E_{mv,i,n} = \frac{y_{i,n} - y_{i,n+1}}{K_{i,n} x_{i,n} - y_{i,n+1}} \qquad (12.11)$$

Equation (12.11) is stated in terms of vapor compositions; an alternate form, not often used, utilizes liquid compositions to give a liquid efficiency $E_{ML,i,n}$. The exit vapor equilibrium $y_{i,n}{}^* = K_{i,n} x_{i,n}$ is based on the exit liquid composition from the stage. For a real tray, the exit liquid may not have the same composition as the average liquid composition on the tray, i.e., there is often a composition gradient in the liquid. For this reason, Equation (12.11) sometimes gives efficiencies greater than 1.0. A more basic form of Equation (12.11) applies to any point within the stage, and this is the *Murphree point efficiency,* defined by Equation (12.11), but with point notations and given the symbol E_{og}. The point efficiency can never be greater than 1.0 and is a fundamental efficiency used in theoretical modeling.

12.1.3 MULTIPLE STAGES

The contacting stage shown in Figure 12.2 may be combined with other contacting stages to form a cascaded multiple-stage process as shown in Figure 12.4. This arrangement leads to the distillation column previously described. In more complex arrangements, the column can have multiple feeds and one or more sidedraw products and can be integrated with several columns in a *distillation train.*

12.2 VAPOR-LIQUID EQUILIBRIUM

The equilibrium ratio [Equation (12.1)] involves physical equilibrium between phases. The system involved may be binary or multicomponent and ideal or nonideal, according to the terminology used in solution thermodynamics. Methods for correlating or predicting equilibria are based on an application of thermodynamics to each phase and to the solutions of the components within each phase. The techniques are described in standard reference works such as Walas[2] and Reid et al.[3] For multicomponent systems, the usual approach is to model the equilibria for each of the binary pairs and then combine the binary models in special ways to obtain the multicomponent model.

The simplest situation occurs in a binary ideal system, such as benzene-toluene at low pressure (less than 2 atm). Here, Raoult's Law applies to the liquid phase:

$$p_i = p_i^{sat} x_i \qquad (12.12)$$

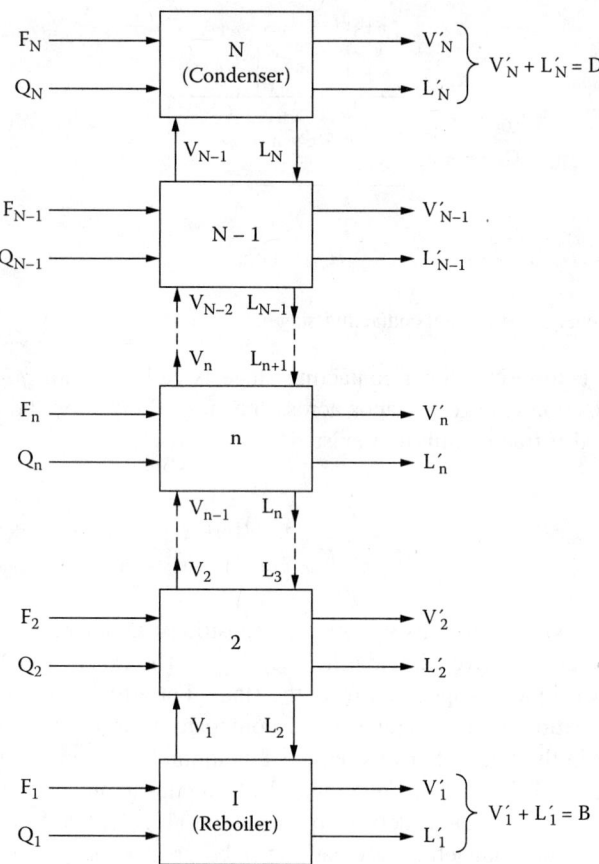

FIGURE 12.4 Cascaded stages in a distillation column.

where the vapor pressure p_i^{sat} is a function of temperature normally represented by the Antoine equation:

$$\ln p_i^{sat} = A + B/(T + C) \tag{12.13}$$

where constants A, B, and C have been evaluated for many elements and compounds.

If Dalton's law applies,

$$y_i^* = p_i/P \tag{12.14}$$

Combining Equations (12.1), (12.12), and (12.14) gives

$$K_i = \frac{y_i^*}{x_i} = \frac{p_i/P}{p_i/P_i^*} = \frac{P_i^*}{P} \tag{12.15}$$

The relative volatility between the two components i and j [Equation (12.3)] is

$$\alpha_{ij} = K_i/K_j \tag{12.16}$$

For the binary system, since $x_j = 1 - x_i$ and $y_j^* = 1 - y_i^*$,

Distillation

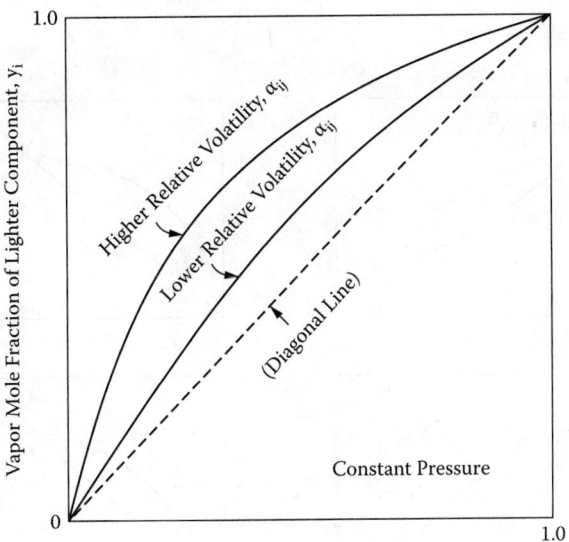

FIGURE 12.5 Equilibrium y-x diagram for ideal binary system i-j.

$$\alpha_{ij} = \frac{y_i^* / x_i}{(1-y_i^*)/(1-x_i)} \qquad (12.17)$$

or

$$y^* = \frac{\alpha x}{1+(\alpha-1)x} \qquad (12.18)$$

where x and y^* are the equilibrium concentrations of the more volatile component at equilibrium. These equilibrium concentrations may be computed from the relative volatility and, if the latter is constant, an equilibrium curve is obtained, as shown in Figure 12.5. For concentration ranges in which an average value of α is assumed, portions of the curve may be computed.

Most systems do not mix ideally in the liquid phase and, at lower pressures, their equilibria may be based on a modified form of Raoult's law:

$$K_i = y_i^*/x_i = (\gamma_i^L p_i^{sat})/P \qquad (12.19)$$

where γ_i^L is the liquid phase activity coefficient for species i (sometimes called a "Raoult's law correction factor") and is a function of temperature and composition. This coefficient can range from as low as 10^{-8} (strong acids) to as high as 10^5 (immiscible liquids), but it usually falls in the range of 0.1 to 20.

Representative phase diagrams for the most common types of binary systems are shown in Figure 12.6, along with example systems. For each type, phase diagrams are presented in two ways: T vs. x and y, and y vs. x. Typical isothermal curves of activity coefficients for binary systems are illustrated in Figure 12.7.

Isothermal or isobaric activity coefficient relationships are modeled from experimental data by a variety of equations, all with a thermodynamic basis. The more useful of these equations are Van

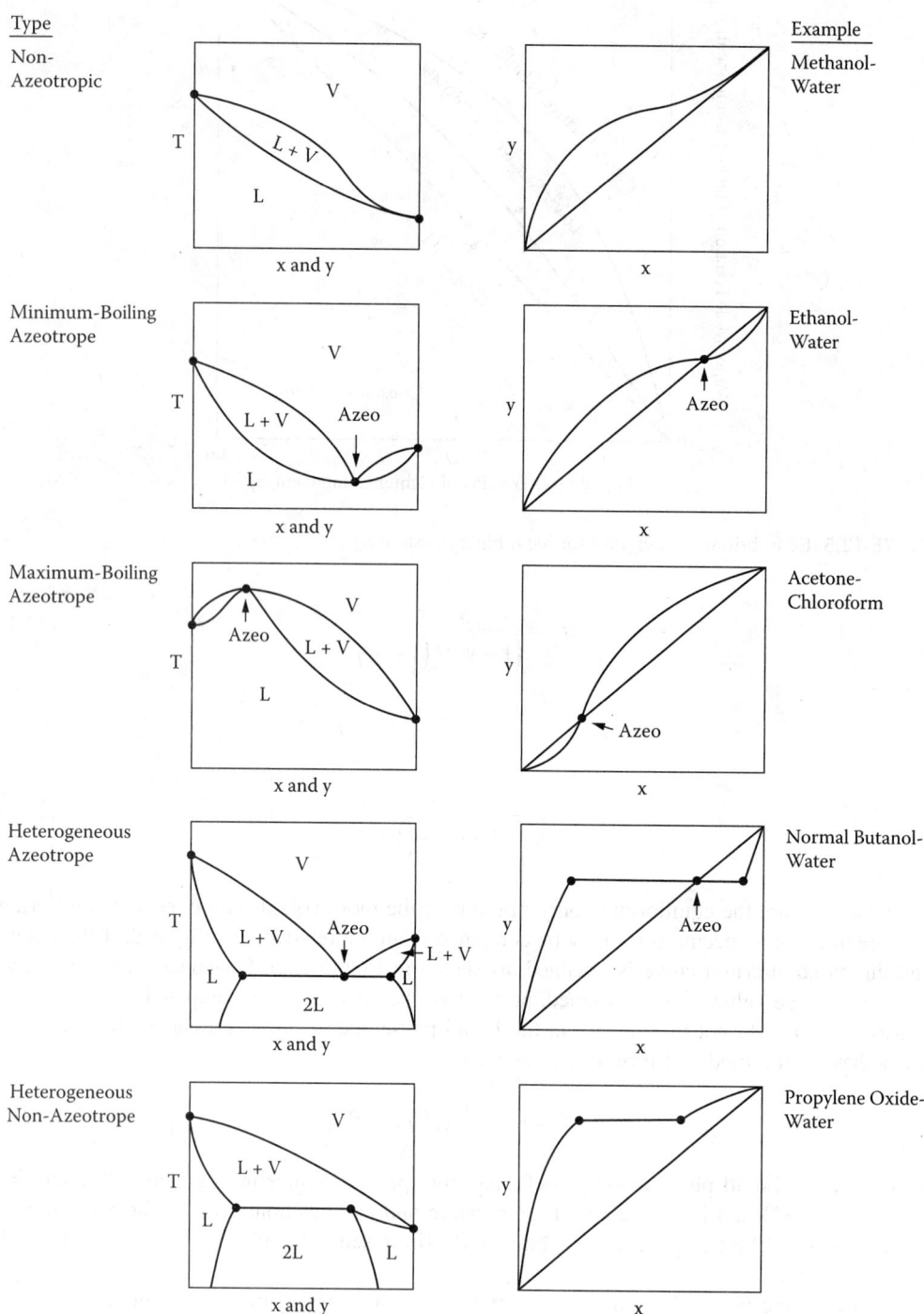

FIGURE 12.6 Common types of binary systems. (W. L. Bolles; J. R. Fair, 1982. *Encyclopedia of Chemical Processing and Design* 16: 51. Courtesy of Routledge/Taylor & Francis Group, © 1982.)

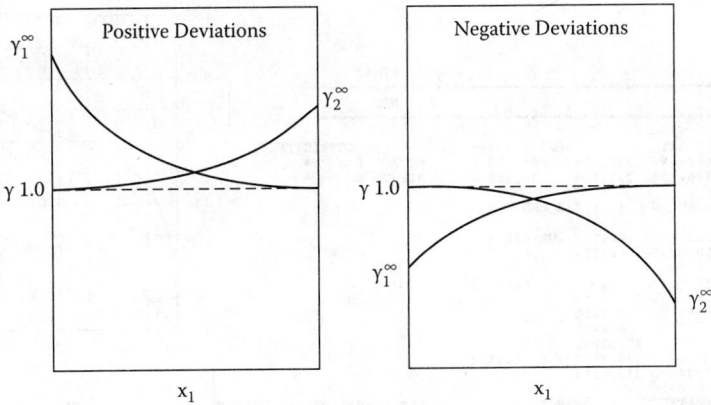

FIGURE 12.7 Typical isothermal curves of liquid-phase activity coefficients for binary systems.

Laar; Margules; Wilson; nonrandom, two liquid phases (NRTL), or Renon–Prausnitz; and Universal Quasi-Chemical Activity Coefficients (UNIQUAC). All of these equations have two constants except for the NRTL, which has three.

Perhaps the simplest, and surprisingly accurate, is the Van Laar.[4] Its use will be described here, but other equations can be handled in a similar fashion. References 2 and 3 provide details. For the binary system i-j, the Van Laar equation is

$$\ln \gamma_i = A_{ij} \left(\frac{A_{ji} x_j}{A_{ij} x_i + A_{ji} x_j} \right)^2 \tag{12.20}$$

$$\ln \gamma_j = A_{ji} \left(\frac{A_{ij} x_i}{A_{ji} x_j + A_{ij} x_i} \right)^2 \tag{12.20a}$$

where A_{ij} and A_{ji} are the constants determined from experimental y-x data or from azeotrope information. An important parameter in VLE studies is the nonideality at infinite dilution, where the activity coefficient normally is at a maximum. This terminal coefficient is called the *infinite dilution activity coefficient*, $\gamma_i^{L,\infty}$. From Equation (12.20), as $x_i \to 0$, $x_j \to 1.0$ and $\ln \gamma_i^{L,\infty} = A_{ij}$. Similarly, as $x_j \to 0$, $x_i \to 1.0$ and $\ln \gamma_j^{L,\infty} = A_{ji}$. There are special measuring devices that deal only with the infinite dilution case, and the total composition range is then handled with equations such as (12.20) and (12.20a). Evaluation of equilibria in the very dilute region can be critical when high-purity products are being obtained by distillation.

Activity coefficient modeling will be clarified later by Problem 12.1. The Van Laar equation, as well as the other equations listed above, can be used for multicomponent systems by first determining the constants for all possible binary pairs and then combining them according to certain rules. Values of the constants may be fitted from experimental data, and values for a large number of systems have been collected and checked by Gmehling et al.[5] One of their example pages is shown in Figure 12.8, covering one experimental study of the ethanol/water system at atmospheric pressure. Note that the activity coefficient constants for the above-listed equations are shown, together with their fit to the referenced experimental data (the fit is shown both for composition and equilibrium temperature).

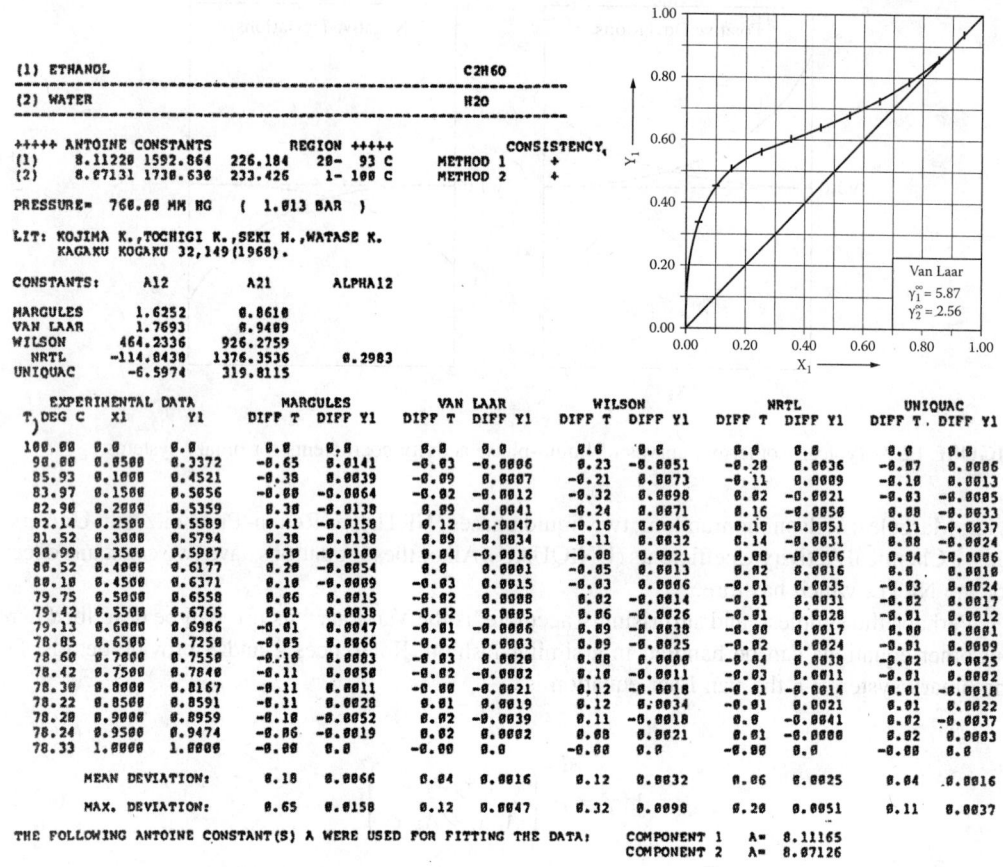

FIGURE 12.8 Example tabulation of equilibrium data from the Chemistry Data Series. [J. Gmehling et al., 1979. *Vapor-Liquid Equilibrium Collection*, Chemistry Data Series (continuing series), DECHEMA, Frankfurt, Germany.]

Equation (12.19) applies primarily to lower pressures (below 2 to 3 atmospheres), where the equilibrium vapor mixes ideally and also behaves ideally according to the perfect gas law. The rigorous thermodynamic expression for the K value, taking all conditions into account, is

$$K_i = \left(\frac{\gamma_i^L}{\gamma_i^V}\right)\left(\frac{P_i^*}{P}\right)\exp\left[\frac{1}{RT}\int_{P_i^*}^{P}\left(v_i^L dP\right)\right]\exp\left[\frac{1}{RT}\int_{P}^{P_i^*}\left(v_i^v - \frac{RT}{P}\right)dP\right] \quad (12.21)$$

The modeling of activity coefficients in multicomponent systems and the application of the general model [Equation (12.21)] are discussed in detail by Walas[2] and Reid et al.[3]

12.2.1 Estimation and Measurement of Activity Coefficients

When equilibrium data can be found, as for example in the Gmehling compilations,[5] they should, of course, be checked for thermodynamic consistency and viewed critically. They may not cover all possible temperature, pressure, and composition conditions, thus they should be incorporated into a model such as described above. Then they can be extrapolated with some measure of reliability. Many distillation simulators have VLE built into them.

Distillation

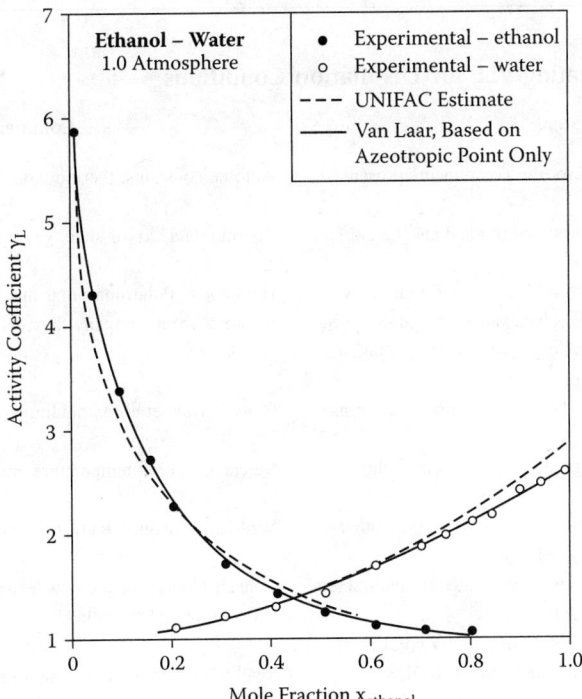

FIGURE 12.9 Representative activity coefficient–composition curves. Ethanol–water system, one atmosphere. Shown are the following: (i) experimental data [Figure 12.8], (ii) van Laar curves with constants computed from the aceotrope condition, and (iii) curves based on the UNIFAC model.

When the needed VLE data are simply not available, and for various reasons it is not feasible to measure them, then predictive methods might be considered. The most useful such method is called UNIFAC, based on Fredenslund et al.[6] and described by Walas[2] and Reid et al.[3] UNIFAC is a group contribution method, and one must have available the contribution parameters of the groups from which the molecules involved are derived. For example, the structure of the 2,2,4-trimethyl-pentane molecule comprises 5 CH_3 groups, 1 CH_2 group, 1 CH group, and 1 C group. Quantitative values of each group are available in the above references. Thus, one "builds" a total contribution of each molecule involved in the mixture, and the UNIFAC mixing rules are then applied. UNIFAC does not provide activity coefficients or K values of high accuracy, but they may be sufficient for exploratory purposes. The isobaric activity coefficient curves for the ethanol/water system, shown in Figure 12.9, show a comparison between the measured data (Figure 12.8) and the predictions by the Van Laar equation and the UNIFAC method.

When at least some experimental data are needed, a variety of equilibrium stills or other devices can be used. Perhaps the most complete descriptions of such devices are in the book by Hala et al.[7] Wiltech Research Co. in Provo, Utah is the best known place in the United States for custom VLE measurements. This company is prepared to make measurements on a variety of systems, including those that are corrosive or would foul the usual circulating equilibrium still. Laboratories in Europe for making such measurements include that of Professor Gmehling.

12.2.2 Summary—Vapor–Liquid Equilibrium

For mixtures that form ideal solutions in the liquid phase, handling of VLE relationships is relatively simple. One needs to know the vapor pressures of the components, and these are usually available.

TABLE 12.2
Procedure for Estimating VLE for Distillation Conditions

Steps	Comments
Obtain vapor pressure data for the components present in the mixture.	Antoine constants, thermo data, handbooks, etc.
Check vapor phase for departure from ideal gas, i.e., obtain fugacities.	Thermo data, eq. of state, generalized correlations.
View mixture for possible deviations from Raoult's law.	Homologs? Polar/non-polar mixtures? Azeotropes?
For hydrocarbon mixtures, K values may suffice at this point, with small departures from liquid phase idealities taken up by the convergence pressure.	K values from company files, API data book, GPSA data book, etc.
Find whether the mixture has been studied elsewhere (binary pairs).	Consult Hala et al., Gmehling et al., DIPPR data.
Find whether the mixture (total or by pairs) is in the Gmehling compilation.	Be careful about temperature and pressure variations.
If binary pair data are available from Ghehling or other sources, use appropriate model.	VanLaar, Wilson, NRTL, Uniquac, etc.
If still no data, check for azeotropes and calculate parameters for activity coefficient models	Consult Horsley or the new, extensive compilation by Gmehling and coworkers.
Failing the above, make estimates of activity coefficients:	
• Use UNIFAC structural contribution method.	Readily available on computer diskette (e.g., Sandler, S.I. *Chem. and Eng. Thermo.*, 2nd ed., Wiley, NY, 1989).
• As a variation for hydrocarbon mixtures, consider the Chao-Seader version of regular solution theory.	Occasionally the basic Scott-Hildebrand regular solution model might be useful for non-hydrocarbon mixtures.
• For critical cases where small errors could result in large costs, have experiments made to confirm estimated VLE.	Outside laboratories are available for this work if it is not feasible to do it in-house.
• Finally, recognize the importance of variations in composition, temperature, and pressure on the values of the K values used for design.	

Note: Process simulators can aid in some of the steps listed above.

If the pressure is high, corrections for departure from ideal gas behavior must be taken into account; equations of state or generalized compressibility charts are usually sufficient.

Mixtures with nonsimilar molecular structures, and in particular mixtures containing water, generally exhibit nonideal behavior. When this situation prevails, recourse must be made to published experimental data or correlations of liquid phase activity coefficients. With a minimum of experimental data, some very good models are available to provide activity coefficients over a broad range of conditions. If no experimental data are available, and there is no azeotrope, then approximate results can be obtained from the UNIFAC model. But it is better to make some experiments.

A procedure for approaching the problem of liquid phase nonideality through the use of activity coefficients is given in Table 12.2.

Problem 12.1

At 1.0 atm, ethanol and water form an azeotrope at 78.17°C and 0.9037 mole fraction ethanol. From this information, calculate the $\gamma^L - x$ relationships for ethanol (i) and water (j), and compare the results with the experimental data shown in Figure 12.8. Make a comparison also with curves generated by the UNIFAC method.

At 78.17°C and use of the Antoine constants in Figure 12.8,

Distillation

$$\text{Ethanol: } \log_{10} p^{sat} = 8.11220 - \frac{1592.864}{°C + 226.184}; p_i^{sat} = 756.27 \text{ mm Hg}$$

$$\text{Water: } \log_{10} p^{sat} = 8.07131 - \frac{1730.630}{°C + 233.426}; p_j^{sat} = 329.08 \text{ mm Hg}$$

[Note: This form of the Antoine equation is an alternate to that given in Equation (12.13) in that common logarithms are used. This form enables use of the Antoine constants given in Figure 12.8.]

Because

$$\frac{y_i}{x_i} = \frac{\gamma_i^L p_i^{sat}}{P} = 1.0 \text{ (at the azeotropic point)}$$

then

$$\gamma_i^L = \frac{760}{756.27} = 1.0049$$

Similarly,

$$\gamma_j^L = \frac{760}{329.08} = 2.309$$

Equations (12.20) and (12.20a) may be rearranged to become explicit in the constants:

$$A_{ij} = \ln \gamma_i^L \left[1 + \frac{x_j \ln \gamma_j}{x_i \ln \gamma_i}\right]^2 \quad A_{ji} = \ln \gamma_j^L \left[1 + \frac{x_i \ln \gamma_i}{x_j \ln \gamma_j}\right]^2$$

Substituting,

$$A_{ij} = 1.8100, A_{ji} = 0.9311$$

The infinite dilution activity coefficients may be obtained from $\ln \gamma^{L,\infty} = A$, giving

$$\gamma_i^{L,\infty} = 6.110 \quad \gamma_j^{L,\infty} = 2.537$$

These compare with the values of Gmehling (Figure 12.8) extrapolated from experimental data of

$$\gamma_i^{L,\infty} = 5.87 \quad \gamma_j^{L,\infty} = 2.56$$

Figure 12.9 shows the experimental data points (Figure 12.8), curves based solely on the azeotrope evaluation of activity coefficients, and the curves determined from the UNIFAC model. The agreement is remarkably good.

12.3 STAGE CALCULATIONS

As noted earlier, distillation columns may be designed as a series of contacting stages, with each stage having a finite approach to an equilibrium stage. (An alternative approach, to be discussed later, is needed when the column does not contain discrete stages or plates.) Calculations for multistage columns are based on the assumption that equilibrium is attained, and correction is made later to account for the failure to reach equilibrium. Stage calculations are made to evaluate the parameters required for the column to meet the separation specifications. These parameters are commonly the number of stages, reflux ratio, distillate:feed ratio, and feed location. More details on stage calculations may be found in references 9 and 10.

12.3.1 SEPARATION SPECIFICATIONS

These specifications define the desired degree of separation. For a simple distillation process (separation of a single feed into two products, distillate and bottoms), the specifications consist of one (and only one) variable related to the distillate quality and one (and only one) variable related to the bottoms quality. For example, consider a feed of 60 mole-% benzene and 40% toluene, benzene being the more volatile. One specification might be that the distillate contain no more than 1.0 mole-% toluene and the bottoms no more than 1.0 mole-% benzene. This would fix the overall material balance and show that over 99% of the entering benzene is recovered in the distillate. Alternatively, one could specify that 95% of the benzene in the feed be recovered in the distillate at 99.0% purity. This would fix the material balance.

12.3.2 SINGLE-STAGE PROCESS

In some cases, a single stage is sufficient to effect the desired degree of separation between feed components. The single-stage process may be either *continuous* (steady state) or *differential* (batch). The continuous process is frequently referred to as flash vaporization, because one way of carrying out this process is to superheat the liquid feed and "flash" it across a valve to a vessel operating at a lower pressure, where the single-stage equilibrium between vapor and liquid is achieved.

The principal design and operating variables in the single-stage process are pressure, temperature, and L/V (liquid/vapor) ratio. Application of Gibbs' phase rule shows that these three variables are not entirely independent: fixing two of them automatically fixes the third. Ordinarily, two of the variables are specified, and computations are employed to determine the value of the third variable as well as the compositions of the vapor and liquid products.

There are two limiting cases. One is the *dew point*, where $L/V = 0$ and the product is entirely vapor but at the threshold of condensation. The relationship between pressure and temperature at the dew point is obtained by a summation of mole fractions:

$$\sum x_i = 1 \tag{12.22}$$

where

$$x_i = z_{iF} / K_i \tag{12.23}$$

The other limit is the bubble point, where $L/V = \infty$, and the product is entirely liquid at the threshold of vaporization:

$$\sum y_i = 1.0 \tag{24}$$

Distillation

where

$$y_i = K_i z_{i,F} \tag{12.25}$$

In the region between bubble point and dew point conditions, application of a mass balance on the ith component results in

$$y_i V = \frac{z_i}{1 + L/VK_i} \tag{12.26}$$

Summing up for all components gives

$$V = \sum_i \frac{z_i}{1 + L/VK_i} \tag{12.27}$$

These equations can be combined to yield

$$y_i = \frac{z/(1 + L/VK_i)}{\sum_i z_i/(1 + L/VK_i)} \tag{12.28}$$

$$x_i = \frac{z_i - y_i V}{L} \tag{12.29}$$

In addition, it is required that

$$\sum y_i = 1 \quad \text{and} \quad \sum x_i = 1 \tag{12.30}$$

Equations (12.28) through (12.30) are solved by trial and error, by hand, or with the aid of a computer. For example, if P and T are specified, then the iteration must proceed on L/V. When a final temperature is specified, an *isothermal flash* results.

When no heat is added or withdrawn from the process, an *adiabatic flash* occurs. For this case, Equations (12.28) through (12.30) still apply, but in addition the enthalpy balance must be satisfied:

$$F h_F = L h_L + V H_v \tag{12.31}$$

The single-tray differential process will be discussed later in connection with batch distillation.

12.3.3 MULTISTAGE PROCESSING

If a single stage is insufficient to meet the desired separation, then multistage processing must be employed (see Figure 12.4).

12.3.4 BINARY SYSTEMS

The simplest case of multistage distillation is the continuous distillation of binary systems where the feed contains only two components. Here the most widely used method of analysis is that of

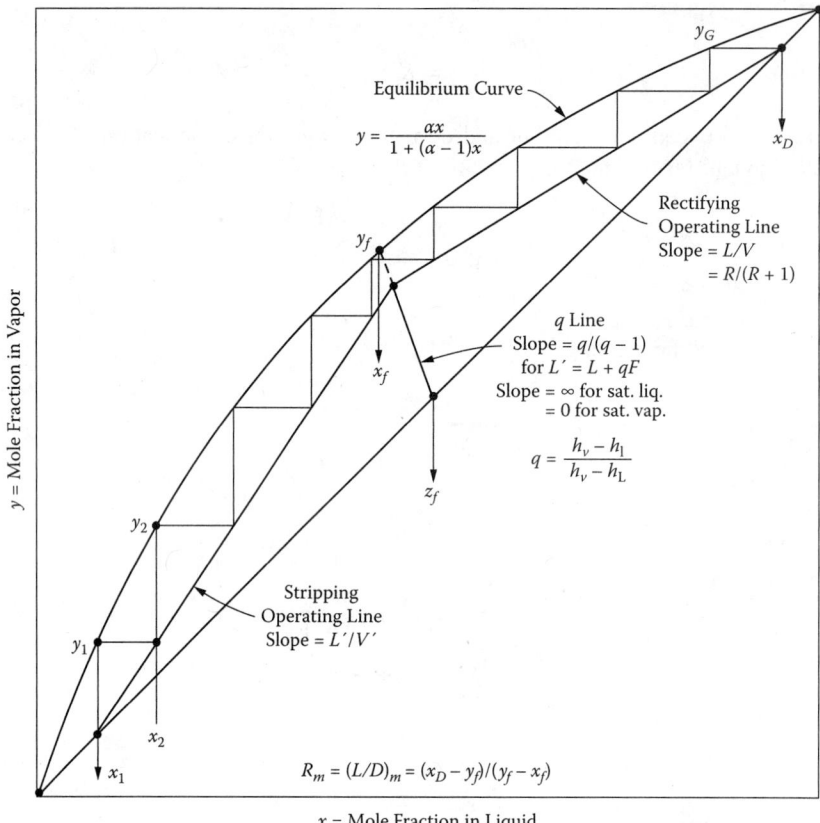

FIGURE 12.10 McCabe–Thiele diagram for a binary system.

McCabe and Thiele[8] as summarized in Figure 12.10. First, equilibrium mole fraction coordinates are plotted to obtain the y-x curve. Then the feed, distillate, and bottoms compositions are located on the diagonal. Next, the *q-line* is drawn from the feed point with a slope as determined by its enthalpy as noted on the graph. (For a saturated liquid feed, the line is vertical, and for a saturated vapor feed, the line is horizontal.) Then the *rectification operating line* is drawn from the distillate composition point with a slope determined from the *reflux ratio*. Next, the *stripping operating line* is drawn from the bottoms composition point up to the intersection of the q-line and the rectification operating line. (These lines will be straight if the molal latent heats of vaporization of the pure components are equal; if these heats are unequal, then the lines will be curved and can be located only by running intermediate heat balances.) Finally, the equilibrium stages are stepped off as indicated in Figure 12.10 to give the required number of equilibrium stages. Nine equilibrium stages are shown in Figure 12.10. Details on the construction and use of McCabe–Thiele diagrams (Figure 12.10) are found in standard texts.[4,9,10]

In the above procedure, the reflux ratio is specified and the required stages calculated. Alternatively, the number of stages is specified and the required reflux ratio calculated. This implies a fitting procedure on the chart, with the operating lines adjusted to give the correct number of stages. Such a procedure is useful when conditions are changed in an operating column; e.g., the feed composition changes and the proper reflux ratio adjustment can be predicted. The McCabe–Thiele approach can also be used to determine the minimum number of stages at total reflux and the minimum reflux ratio corresponding to an infinite number of stages.

The McCabe–Thiele approach is applicable to any binary system, even if nonideal, as will be evidenced by an unusual shape of the equilibrium curve. As normally used, the procedure requires

Distillation

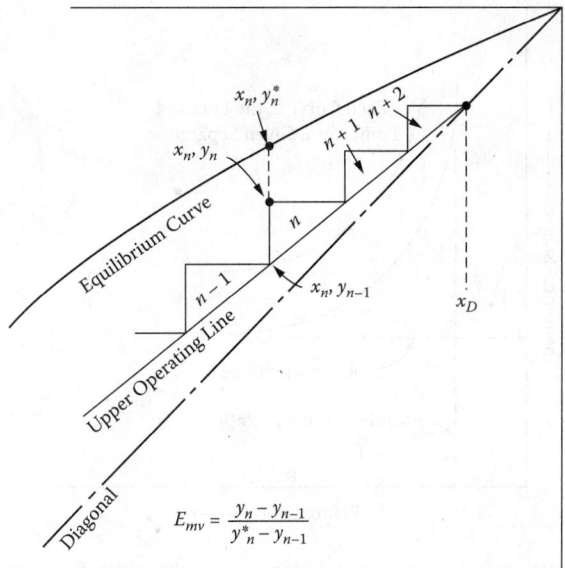

FIGURE 12.11 Modified McCabe-Thiele plot to show approach to equilibrium on real stages. The relative height of the real stage to the equilibrium stage height denotes Murphree stage efficiency.

the assumption of constant molal overflow in each of the sections (above and below the feed), leading to straight operating lines. Negligible heat losses to the surroundings is a usual assumption. The method can also be used for more than one feed and for side drawoffs of product.

In Figure 12.10, it is not required that the steps touch the equilibrium curve. They may take stage efficiency into account and go only part way to the curve, as shown in Figure 12.11. If for some reason the efficiencies of all stages are not the same, this can be taken into account easily. The fractional approach to the height of an equilibrium stage is the Murphree efficiency as defined in Equation (12.11) and discussed later.

Another graphical design method for binary systems is that of Ponchon[11] and Savarit.[12] This method includes an energy balance on each stage and is totally rigorous for binary systems. However, it requires mixture enthalpy data, often unavailable, and is cumbersome in application.

12.3.5 Stages–Reflux Relationships

For a specified separation of a feed stream, the required stages and the required reflux ratio are related as shown in Figure 12.12. While the curves are somewhat idealized, they show the limiting conditions of the design parameters. As the curve increases in the y direction, it approaches an asymptote value of reflux ratio. This is the *minimum reflux ratio* and corresponds to an infinite number of theoretical stages. Along the x-axis, the asymptote value represents the *minimum number of stages* equivalent to infinite (i.e., total) reflux. A practical design point lies somewhere in between these limiting conditions.

The case of total reflux also represents a practical condition for operating a distillation system during startup or for determining the efficiency of various contacting devices for a distillation column. Performance data for such devices, shown later herein, are likely obtained under total reflux conditions.

12.3.6 Multicomponent Systems

For systems with more than two components, there are two approaches to stage calculations: *approximate methods,* which can be done quickly by hand, and *rigorous methods,* which require a computer. The approximate methods will be discussed first.

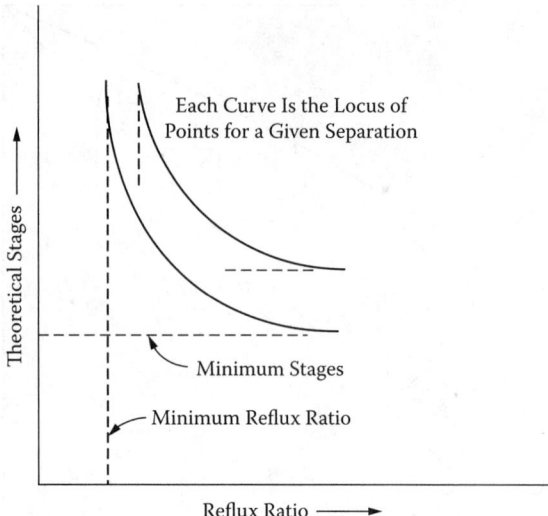

FIGURE 12.12 Representative plot of theoretical stages vs. reflux ratio for a given separation. Note the limiting conditions of minimum reflux and minimum stages.

For the limiting case of total reflux (Figure 12.12), the minimum number of stages may be determined by the Fenske[13] equation:

$$N_m = \frac{\ln\left[\frac{(x_i/x_j)_D}{(x_i/x_j)_B}\right]}{\ln \alpha_{ij}} = \frac{\ln\left[\frac{(x_D/x_B)_i}{(x_D/x_B)_j}\right]}{\ln \alpha_{ij}} \quad (12.32)$$

where i and j are usually the light and heavy key components. The Fenske method is rigorous if the relative volatility is constant throughout the section of the column being considered. It can be used for binary systems to obviate the need for the McCabe–Thiele graphics if the relative volatility is constant, i.e., the y-x curve is uniform.

The other limiting parameter, minimum reflux ratio at infinite stages, may be calculated by means of the Underwood[14] equations:

$$R_m = \sum_i \left[\frac{\alpha_i x_{iD}}{\alpha_i - \theta}\right] - 1 \quad (12.33)$$

where θ is the root of

$$\sum_i \frac{\alpha_i x_{iF}}{\alpha_i - \theta} = 1 - q \quad (12.34)$$

and q is computed as in the McCabe–Thiele method (see Figure 12.9).

The relationship between stages and reflux ratio was quantified empirically by Gilliland[15] and presented graphically in Figure 12.13a. Numerous authors have published mathematical fits to the Gilliland curve, a representative fit being that of Rusche:[16]

Distillation

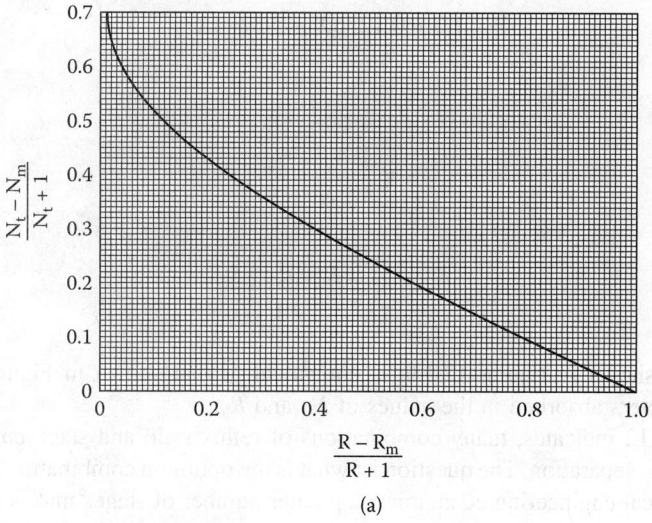

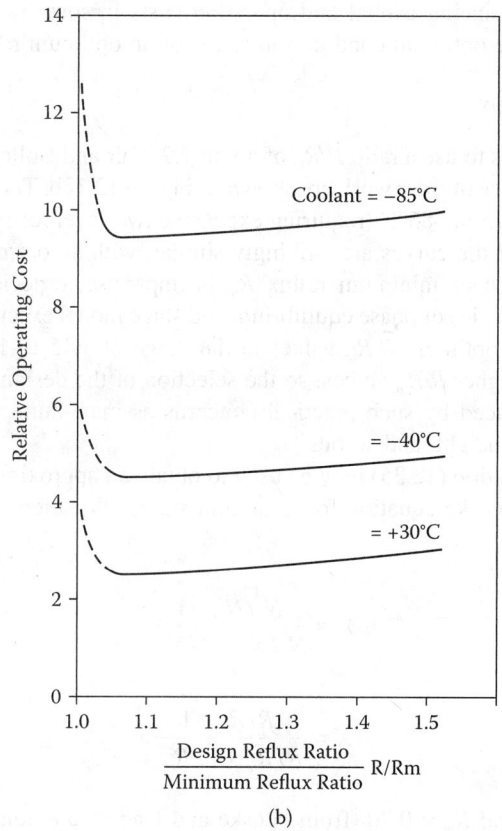

FIGURE 12.13 (a) Gilliland relationship between actual reflux ratio R, minimum reflux ratio R_m, theoretical stages N_t and minimum theoretical stages N_m. (b) Optimum reflux ratio as a function of operating temperature level. (E. R. Gilliland, 1940. *Ind. Eng. Chem.* 32: 1220.)

$$Y = 1.0 - 0.1256X - 0.8744X^{0.2910} \quad (12.35)$$

where

$$Y = \frac{N_t - N_m}{N_t + 1} \quad (12.36)$$

$$X = \frac{R - R_m}{R + 1} \quad (12.37)$$

The relationship shown in Figure 12.13a is equivalent to that shown in Figure 12.12, with the separation parameters absorbed in the values of N_m and R_m.

As Figure 12.12 indicates, many combinations of reflux ratio and stage number can be used to obtain a required separation. The question is: what is the optimum combination? This is a classical problem in chemical engineering economics: a greater number of stages makes the column higher and more expensive, and this is balanced against the savings of utilities at the lower reflux ratio. Thus it is a matter of balancing capital and operating costs in some rational fashion. The usual approach is to express the optimum condition in terms of an optimum ratio of R to R_m.

12.3.7 Optimum Reflux

A popular rule of thumb is to use a ratio R/R_m of about 1.2. Fair and Bolles[17] made a detailed study of this ratio, and three cases of their work are shown in Figure 12.13b. The cases represent different levels of coolant ranging from −85°C (requiring expensive two-level refrigeration) to +30°C (cooling water). The shapes of the curves are strikingly similar, with an optimum R/R_m of about 1.05. However, the computation of minimum reflux R_m is imprecise, especially for multicomponent systems, owing to inaccuracies of phase equilibrium and stage model execution. To avoid the danger of an impossible design, optimum R/R_m values in the range of 1.15 to 1.25 should be used, The curves are fairly flat at higher R/R_m values, so the selection of the design value has flexibility and may very well be influenced by such practical concerns as maximum allowable diameter, plant limitations on maximum height, and so on.

Figure 12.13a or Equation (12.35) may be used to obtain an approximate number of theoretical stages from the simple Fenske equation for minimum stages. Equations (12.36) and (12.37) may be modified to

$$Y = \frac{N_t/N_m - 1}{N_t/N_m + 1/N_m} \quad (12.36)$$

$$X = \frac{R/R_m - 1}{R/R_m + 1/R_m} \quad (12.37)$$

For example, if $N_m = 5$ and $R_m = 0.70$ (from Fenske and Underwood relationships), and optimum $R/R_m = 1.15$, $X = 0.058$ and, from Figure 12.13a, $Y = 0.57$. Solving, $Nt/N_m = 2.59$. While this ratio will vary with conditions, the heuristic of $N_t/N_m = 2.5$, given in Section 12.1.1, is often used.

12.3.8 Design Procedure

First, the minimum number of stages is computed by the Fenske method. Then, the minimum reflux ratio is computed by the Underwood method. Next, the design (operating) reflux ratio is chosen as some multiple of the minimum reflux ratio, e.g., $1.15 \times R_m$. (The optimum multiple is in the

Distillation

range of 1.15 to 1.25.) Next, the required number of equilibrium stages is determined by the Gilliland correlation. Finally, the required number of actual trays is determined using a stage efficiency [Equation (12.11)]; elaboration of this last step will follow in later sections of this work.

Problem 12.2

To demonstrate the application of approximate stage calculations, an example design problem will be solved. The elements of the problem are tabulated below, including the specified material balance in moles. The particular case is a six-component de-ethanizer required to separate between ethane (light key) and propylene (heavy key). Such a column is needed in some large ethylene plants. An elaboration of this problem may be found elsewhere.[17]

	Feed	Distillate	Bottoms
Methane	5.00	5.00	—
Ethane (LK)	35.00	31.89	3.11
Propylene (HK)	15.00	0.95	14.05
Propane	20.00	—	20.00
i-Butane	10.00	—	10.00
n-Butane	15.00	—	15.00
	100.00	37.84	62.16

Minimum reflux ratio (Underwood)	1.378
Minimum theoretical stages (Fenske)	6.8
Operating reflux ratio (1.25 × minimum)	1.722
Theoretical stages (Gilliland graph)	14.5
Theoretical stages (Rusche equation)	14.8

12.3.9 Computer Methods

Digital computer programs are widely available for making rigorous stage calculations. They can handle many components (25+) and many stages (100+), and their accuracy is limited only by the reliability of the input thermodynamic data. They can handle multiple feeds; multiple sidedraws; stage heat exchangers; and chemical reactions occurring in the liquid and/or vapor, for reactive distillation applications. They can also handle the approximate design approach discussed above.

Two approaches are available to solve the multicomponent separation problem: the design method and the rating method. These methods have different input-output specifications, as shown in Table 12.3.

The *design method* is best described by Lewis and Matheson.[18] It is called "design" because it starts with the design specifications in terms of key components in the terminal products. The concentrations of all other components in the terminal products must be assumed. The stage

TABLE 12.3
Design versus Rating Models for Rigorous Distillation Stage Calculations

Design Method		Rating Method	
Input	Output	Input	Output
Feed	No. of stages	No. of stages	Distillate composition
Distillate composition	Feed stage	Feed stage	Bottoms composition
Bottoms composition	Distillate rate	Distillate rate	
Reflux/min. reflux	Reflux ratio	Reflux ratio	
		Feed	

calculations proceed from the ends of the column toward the feed [or from one end to the other, taking into account the addition of the feed(s)]. In each stage, four criteria must be satisfied: material balance, equilibrium between phases, summarion of phase mole fractions to 1.00, and heat balance.

From the first letters, the familiar MESH equations are identified. These equations must be solved rigorously at each stage before moving to the next. When the traverse of the column is completed, there must be a match between the computed and assumed overall material balance. If not, another trial is necessary. Especially onerous are the points of addition of components not present in both distillate and bottoms streams. While the Lewis–Matheson method is straightforward (although extremely laborious) for hand calculations, it was the standard method before large computers became available. With large computers, it often presents serious programming problems with respect to the convergency algorithm—particularly the assumptions of splits of nonkey components.

The *rating method* for rigorous stage calculations was developed by Thiele and Geddes.[19] Their approach is a "rating" method in that it starts with a specified number of stages and reflux ratio and then rates the separation possible with that combination. The method is more amenable to computer implementation and now is the standard. The Thiele–Geddes derivation begins with a component mass balance on the general contacting stage, counting downward from the top of the column:

$$V y_{n+1,i} + L x_{n-1,i} - V y_{n,i} - L x_{n,i} + F z_{n,i} = 0 \tag{12.38}$$

Equation (12.38) may be rewritten in terms of the molar flow rates of a specific component:

$$v_{n+1,i} + l_{n-1,i} - v_{n,i} - l_{n,i} + f_{n,i} = 0 \tag{12.39}$$

The phase equilibrium relationship is defined by

$$y_{i,n} = K_{i,n} x_{i,n} \tag{12.40}$$

which may be rewritten as

$$\frac{v_{n,i}}{V} = K_{i,n} \frac{l_{n,i}}{L} \tag{12.41}$$

from which

$$l_{n,i} = \left(\frac{L}{K_{i,n} V}\right) v_{n,i} \tag{12.42}$$

Introducing the "absorption factor" A_n, defined as

$$A_{n,i} = L/K_{i,n} V \tag{12.43}$$

Equation (12.42) becomes

$$l_{n,i} = A_{n,i} v_{n,i} \tag{12.44}$$

Introducing Equation (12.44) into Equation (12.39) gives

$$v_{n+1,i} + A_{n-1,i}\ v_{n-1,i} - v_{n,i} - A_{n,i}\ v_{n,i} - A_{n,i}\ v_{n,i} + f_{n,i} = 0 \qquad (12.45)$$

which can be rearranged to

$$A_{n-1,i}\ v_{n-1,i} - (1 + A_{n,i})v_n + v_{n+1,i} = -f_{n,i} \qquad (12.46)$$

One should note that Equation (12.46) is a linear equation in three unknowns (for each component): v_{n-1}, v_n, and v_{n+1}. The equation may be written N times for every stage of an N-stage column. The result is a matrix of N linear equations in N unknowns for the ith component. This matrix is a *sparse matrix* of the *tridiagonal* type. The matrix is easily solved by a computer to yield the component flows on each stage of the column. This procedure is repeated for each component of the system.

The Thiele–Geddes method is simple and straightforward for the case of constant molal overflow (L = constant) and constant component equilibrium ratios K_i throughout the column, because this results in constant absorption factors. Actually, in the general case, the molal overflow is not constant but is governed by the stage energy balances, and the equilibrium ratios are not constant since they depend on changing temperature, pressure, and composition. Accordingly, some elaborate iterative schemes are necessary to make the method work

Problem 12.3

The design of Problem 12.2 was solved by means of a rigorous computer program based on the Thiele–Geddes model. Since the approach is the rating method, start with input values of the column configuration and operating conditions, and choose the results of the approximate calculations from Problem 12.2 (reflux ratio = 1.722, 14 equilibrium stages, excluding reboiler and condenser).

The resulting computer output is reproduced in Table 12.4. Now compare the predicted performance by the hand and computer methods:

	Approximate	Rigorous
Mole-% propylene in distillate	2.50	2.40
Mole-% ethane in bottoms	5.00	5.79

The approximate method gives an overall separation reasonably close to that computed by the rigorous method. This agreement is aided by the problem selected—distillation of a homologous series of hydrocarbons, a fairly ideal case. Even so, it is clear that some adjustment in the input conditions for the rigorous method is necessary if the original specifications are to be met.

12.4 SPECIAL DISTILLATIONS

12.4.1 Azeotropic Distillation

12.4.1.1 General Principles

An *azeotrope* is a mixture of two or more components that, when brought to boiling, issues a vapor with the same composition as the liquid. Hence, separation by simple distillation is not possible. Binary systems containing azeotropes have y-x equilibrium curves as shown in Figure 12.6. On either side of the azeotropic composition, separation by simple distillation is possible.

TABLE 12.4
Computer Solution for Problem 12.3 Six-Component Deethanizer

Part 1

Reflux ratio, mol/mol	1.722
Reflux rate, lb-mol/h	65.160
Column duties, Btu/h:	
Condenser	494,790
Reboiler	573,407

External stream data	Top liquid	Bottom liquid	Feed 1
Theoretical stage number	15	0	7
Flow rate, lb-mol/h	37.840	62.160	100.000
Mole ratio to Feed 1	0.378	0.622	1.000
Mean molecular weight	28.70	48.46	40.99
Flow rate, lb/h	1086.088	3012.558	4098.645
Temperature, °F	−10.92	174.84	90.38
Pressure, lb/in.^{2}	400.000	402.400	401.700
Vapor, mol%	0.0	0.0	0.0
Enthalpy, Btu/lb-mol	−960.167	3838.231	1236.349
Composition, mol%			
Methane	13.2129	0.0004	5.0000
Ethane	82.9823	5.7907	35.0000
Propylene	2.4010	22.6697	15.0000
Propane	1.3986	31.3236	20.0000
i-Butane	0.0043	16.0849	10.0000
n-Butane	0.0009	24.1307	15.0000
Composition, mass%			
Methane	7.3839	0.0001	1.9567
Ethane	86.9371	3.5928	25.6779
Propylene	3.5201	19.6832	15.4002
Propane	2.1485	28.4962	21.5144
i-Butane	0.0087	19.2894	14.1803
n-Butane	0.0018	28.9382	21.2704
Recovery, %			
1. Methane	99.9950	0.0050	
2. Ethane	89.7158	10.2842	
3. Propylene	6.0569	93.9341	
4. Propane	2.6462	97.3538	
5. i-Butane	0.0162	99.9838	
6. n-Butane	0.0022	99.9978	

Part 2

Stage	Temperature (°F)	Pressure (lb/in.^{2})	Liquid flow (lb-mol/h)	Vapor flow (lb-mol/h)	Molecular weight Liquid	Molecular weight Vapor	Murphree % Efficiency
15	−10.92	400.00	65.16	0.0	28.70	22.53	100.0
14	42.11	401.00	69.23	103.00	30.99	28.70	100.0
13	55.27	401.10	67.53	107.07	32.32	30.18	100.0
12	64.57	401.20	65.18	105.37	33.66	31.02	100.0
11	73.62	401.30	63.14	103.02	35.00	31.84	100.0

TABLE 12.4 (Continued)
Computer Solution for Problem 12.3 Six-Component Deethanizer

Stage	Temperature (°F)	Pressure (lb/in.^{2})	Liquid flow (lb-mol/h)	Vapor flow (lb-mol/h)	Molecular weight Liquid	Molecular weight Vapor	Murphree % Efficiency
10	82.08	401.40	61.44	100.98	36.27	32.64	100.0
9	89.92	401.50	59.71	99.28	37.54	33.39	100.0
8	98.03	401.60	57.58	97.55	39.02	34.11	100.0
7	107.74	401.70	163.09	95.42	40.91	34.93	100.0
6	115.92	401.80	165.07	100.93	41.50	36.25	100.0
5	122.89	401.90	166.00	102.91	42.17	37.29	100.0
4	130.37	402.00	166.78	103.84	49.95	38.40	100.0
3	138.80	402.10	167.63	104.62	43.86	39.67	100.0
2	148.40	402.20	168.35	105.47	44.97	41.15	100.0
1	159.83	402.30	168.48	106.19	46.41	42.92	100.0
0	174.84	402.40	62.16	106.32	48.46	45.21	100.0

Component No.	Name	Molecular weight	Antoine constants A	B	C
1	Methane	16.04	5.50960	−216.67	24.07
2	Ethane	30.07	6.04155	−480.81	−105.69
3	Propylene	42.08	6.36310	−729.61	−96.84
4	Propane	44.09	6.40701	−768.18	−95.11
5	i-Butane	58.12	6.65749	−1002.87	−83.50
6	n-Butane	58.12	6.75988	−1103.92	−78.18

Liquid enthalpy constants

H, liquid (cal/g-mol) = $A + BT + CT^2$ (°K)

Compound no.	Compound name	A	B	C
1	Methane	−3532.4	10.900	0.0
2	Ethane	−4687.5	16.850	0.0
3	Propylene	−6485.1	23.750	0.0
4	Propane	−6799.9	24.900	0.0
5	i-Butane	−8656.4	31.700	0.0
6	n-Butane	−8656.4	31.700	0.0

Vapor enthalpy constants

H, vapor (cal/g-mol) = $A + BT + CT^2$ (°K)

Compound no.	Compound name	A	B	C
1	Methane	−1877.8	9.620	0.0
2	Ethane	−2375.9	16.820	0.0
3	Propylene	−3152.1	21.850	0.0
4	Propane	−4273.7	25.350	0.0
5	i-Butane	−6555.5	33.600	0.0
6	n-Butane	−6581.0	33.700	0.0

Continued

TABLE 12.4 *(Continued)*
Computer Solution for Problem 12.3 Six-Component Deethanizer

Part 3
Stage Liquid Compositions, mol%

Stage	Methane	Ethane	Propylene	Propane	*i*-Butane	*n*-Butane
15	13.2129	82.9824	2.4010	1.3986	0.0043	0.0009
14	2.9845	86.6237	6.2324	4.1270	0.0255	0.0069
13	1.4660	79.6127	10.8481	7.9536	0.0892	0.0304
12	1.2486	70.0240	15.6894	12.6693	0.2572	0.1115
11	1.2266	60.5271	19.7462	17.5037	0.6421	0.3543
10	1.2311	52.5229	22.2506	21.5755	1.4232	0.9967
9	1.2376	46.3140	22.8969	24.1831	2.8463	2.5220
8	1.2450	41.4649	21.6470	24.7590	5.1421	5.7419
7	1.2552	37.3540	18.6561	22.9094	8.2599	11.5654
6	0.4215	34.8887	20.1297	24.4285	8.4259	11.7057
5	0.1395	30.9766	21.8767	26.3576	8.6912	11.9583
4	0.0457	26.0987	23.7688	28.6531	9.0898	12.3439
3	0.0148	20.6871	25.4766	31.0782	9.7342	13.0092
2	0.0047	15.2175	26.4363	33.0905	10.8664	14.3846
1	0.0014	10.1366	25.8269	33.6826	12.8729	17.4795
0	0.0004	5.7907	22.6697	31.3236	16.0849	24.1307

Stage Vapor Compositions, mol%

Stage	Methane	Ethane	Propylene	Propane	*i*-Butane	*n*-Butane
15	54.2878	45.0709	0.4272	0.2137	0.0003	0.0000
14	13.2122	82.9828	2.4011	1.3987	0.0043	0.0009
13	6.5986	85.3369	4.8787	3.1630	0.0180	0.0047
12	5.6835	80.8223	7.8155	5.6003	0.0587	0.0198
11	5.6424	74.7827	10.8093	8.5304	0.1643	0.0709
10	5.7174	68.9399	13.2477	11.4699	0.4031	0.2219
9	5.7973	64.1303	14.6862	13.8865	0.8825	0.6172
8	5.8823	60.5355	14.9477	15.3463	1.7440	1.5442
7	5.9905	57.9268	14.0158	15.4965	3.1049	3.4655
6	2.0280	56.7902	16.1847	17.7282	3.4413	3.8277
5	0.6758	52.4624	18.5958	20.2643	3.8003	4.2016
4	0.2228	46.0511	21.4021	23.3853	4.2659	4.6728
3	0.0725	38.1634	24.4218	27.0665	4.9342	5.3416
2	0.0232	29.4659	27.1308	30.9334	5.9916	6.4551
1	0.0072	20.7356	28.6410	34.1247	7.8117	8.6798
0	0.0020	12.6775	27.6728	35.0618	10.9949	13.5909

The occurrence of a minimum or maximum on the temperature vs. composition surface is caused by positive or negative deviations from Raoult's law [see Equation (12.19)]; a system is *positive* if the logarithm of the activity coefficient γ_i^L is greater than zero, and *negative* if the logarithm is less than zero. Deviation from Raoult's law is not sufficient to cause the occurrence of an azeotrope; the boiling points of the pure components must also be sufficiently close to permit a minimum or maximum temperature to occur. Close-boiling components with small deviations from ideality may form an azeotrope. On the other hand, compounds that form very nonideal liquid mixtures may not exhibit an azeotrope because of a wide difference in their boiling points. Azeotropes seldom occur between compounds whose boiling points differ by more than 30°C.

Distillation

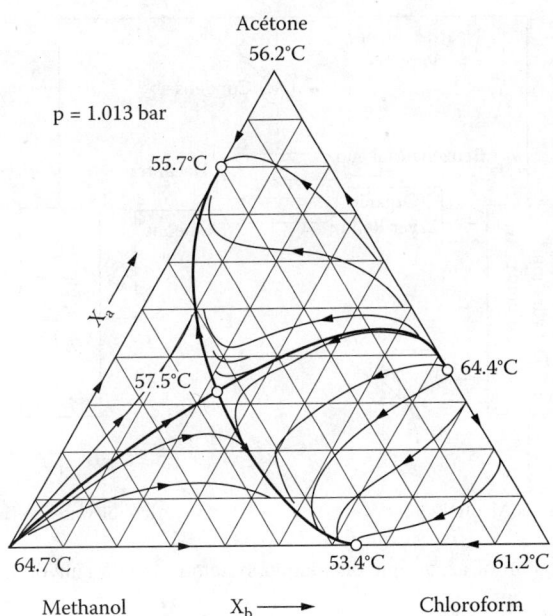

FIGURE 12.14 Ternary diagram of a system containing three binary azeotropes and one ternary azeotrope. Arrows denote direction the distillation will proceed in, based on starting mixtures. (J. Stichlmair, J. R. Fair, 1998. *Distillation—Principles and Practices*. New York: Wiley-VCH.)

Figure 12.6 shows five of the most common types of binary systems. Types B, C, and D are azeotropic, whereas Types A and E are not. Types B and C are called *homogeneous azeotropes* of a single liquid phase; in like fashion, a *heterogeneous azeotrope* derives from two immiscible liquid phases. Type B is a *minimum-boiling azeotrope* (boiling temperature lower than those of the pure components), whereas Type C is a *maximum-boiling azeotrope* (boiling temperature above the pure components). Minimum boiling azeotropes are the most prevalent. Azeotropes can occur in multicomponent systems; an example case is shown in Figure 12.14.[20] Here, a three-component system has three binary azeotropes and one ternary azeotrope. The curves in Figure 12.14 show likely distillation paths for a multistage system.

The only way to be certain that an azeotrope exists in a mixture is to make the equilibrium measurements. Measurements have been made and are reported by Horsley[21] and by Gmehling et al.[22] Most systems reported are binaries.

12.4.1.2 The Azeotropic Distillation Process

Azeotropic distillation involves either an *embedded azeotrope*, present in the feed mixture, or a *contrived azeotrope*, formed by the addition of an extraneous component called an *entrainer*. Benzene-water may be separated into high-purity benzene and the benzene–water azeotrope; this is frequently practiced to remove water from benzene when very dry benzene is needed for chemical processing. More commonly encountered are distillation separations that are enhanced through the addition of an entrainer to form an azeotrope. Perhaps the best known separation of this type is the production of anhydrous ethanol from the ethanol–water azeotrope. Here, benzene is added as the entrainer, with the result that a low-boiling ternary azeotrope is formed between benzene, ethanol, and water. This permits the higher-boiling ethanol to be taken from the bottom of the column. The distillate condenses to a heterogeneous mixture of benzene and alcohol–water phases.

In Problem 12.4, it may be recognized that the water layer can be further distilled to produce the ternary azeotrope (b.p. 64.86°C, 1 atm) from a bottoms material containing water and the water-

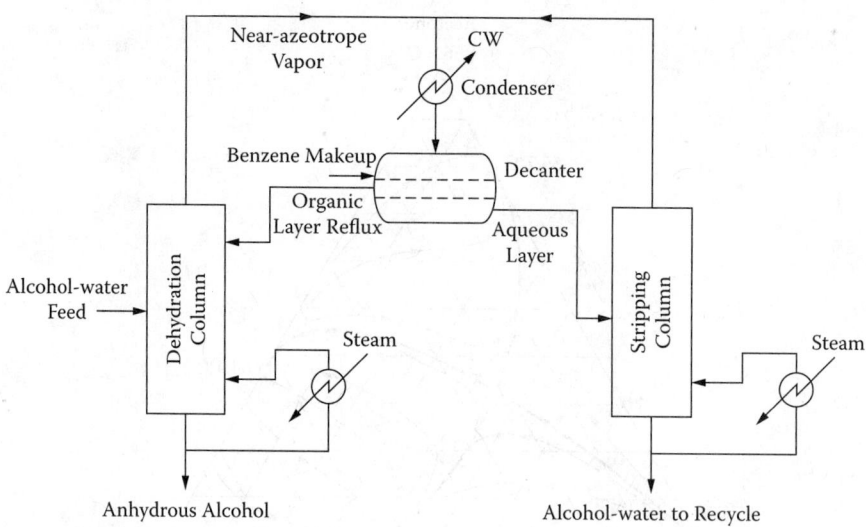

FIGURE 12.15 Flow diagram of azeotropic distillation system to obtain anhydrous ethanol.

alcohol azeotrope. Figure 12.15 shows a flow diagram of the entire process, starting with the water-ethanol azeotrope as feed. Thus, in the total process, the following could be involved:

	B.P., °C at 1 atm	Comp., wt-%
Benzene–alcohol–water azeotrope	64.86	H_2O 7.4, EtOH 74.1, C_6H_6 74.1
Benzene–alcohol azeotrope	68.3	EtOH 32.0
Benzene–water azeotrope	69.0	H_2O 8.8
Alcohol–water azeotrope	78.2	H_2O 4.0
Alcohol	78.3	
Benzene	80.1	
Water	100	

If alcohol is to be produced in high purity as a bottoms stream, then the water and benzene must be tied up in lower-boiling azeotropes and pure compounds and distilled overhead.

For approximate calculations, the azeotropes may be considered pseudocomponents with their individual vapor pressures, as indicated in Figure 12.16. The vapor pressure of the psuedocomponent (butanol-water azeotrope) roughly parallels the lines for n-butanol and water, and relative volatilities between the three components could be established.

Problem 12.4

Consider producing anhydrous ethanol from a feed stream containing 85 mole-% C_2H_5OH and the remainder water. Since this system exhibits a minimum-boiling azeotrope at 78.1°C containing 89.43 mole-% C_2H_5OH (at 1 atm), the separation is impossible by ordinary distillation.

However, the addition of benzene as an entrainer results in a minimum-boiling heterogeneous ternary azeotrope at 64.86°C (1 atm) of the following molar composition:

Ethanol	22.8%
Benzene	53.9%
Water	23.3%
	100.0%

Distillation

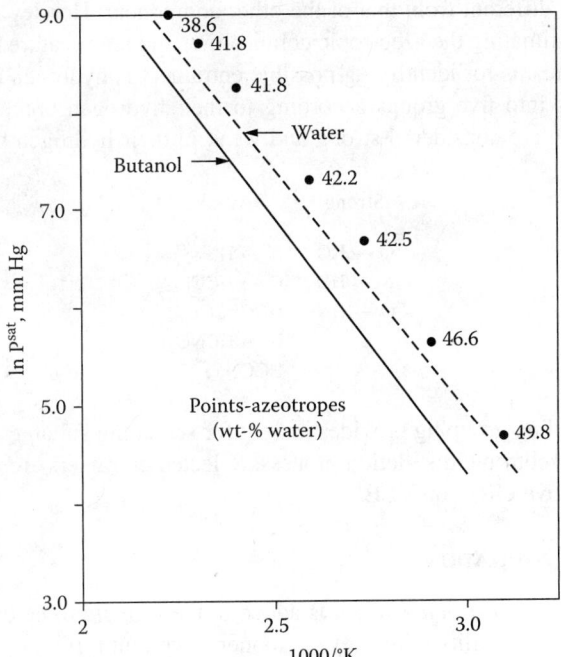

FIGURE 12.16 Vapor pressures of *n*-butanol, water, and *n*-butanol/water azeotrope. (Data from L. H. Horsley, 1973. *Azeotropic Data III*, Washington, DC. Am. Chem. Soc.)

On condensation, this azeotrope separates into two immiscible liquid phases of the following compositions (mole-%)

Component	Benzene layer	Water layer
Ethanol	23%	22%
Benzene	65%	3%
Water	12%	75%
	100%	100%

That benzene is an effective entrainer can be confirmed by comparing the ethanol/water molar ratio in the binary and ternary azeotropes:

Binary 89.43/10.57 = 8.46
Ternary 22.8/23.3 = 0.98

The fact that these two ratios are different indicates that separation of anhydrous alcohol is feasible.

Clearly, the azeotrope has vapor pressure characteristics of a single component. It is also clear that pressure has an effect on the location of the azeotrope, and this pressure influence is used commercially to make separations (see reference 23). In some cases, a change in pressure can cause the azeotrope to disappear; for ethanol-water, distillation at a pressure of 75 mm Hg or less results in an azeotrope-free operation.

12.4.1.3 Prediction of Azeotropes

For a system to be azeotropic, the components are not only close-boiling but of different structures. When an agent (entrainer) is added, its function is to interact so that the volatility of one component is affected to a degree different from that of the other component. Horsley[21] and Gmehling et al.[22] discuss methods for estimating the azeotropic composition and temperature for numerous systems.

One of the mechanisms for identifying possible entrainers is hydrogen bonding. Ewell et al.[23a] have classified liquids into five groups according to their hydrogen bonding characteristics. For example, the following are considered strong and weak in their hydrogen bonding:

Strong	Weak
O → HO	N → HN
N → HO	O → HCCl$_3$
O → HN	HCCl-CCl
	N → HCNO$_2$
	HCCN

The hydrogen bonding grouping provides insight for screening suitable entrainers in the development of a feasible azeotropic distillation process. Selected entrainers are then tested experimentally for their quantitative effect on VLE.

12.4.2 Extractive Distillation

In extractive distillation, an *extractive agent* is added to the mixture to be distilled for the purpose of modifying the relative volatility of the key components without forming an azeotrope. Extractive distillation is usually employed to improve the separability between close-boiling components for which ordinary distillation would not be economically feasible.

Like azeotropic distillation, extractive distillation involves highly nonideal phase equilibria as well as the addition of an agent, often called a *solvent*, that modifies and improves the phase equilibria among the system components. However, extractive distillation is different in that no azeotropes are involved, and the agent added is essentially nonvolatile and introduces no new azeotropes.

Extractive distillation is a simpler process than azeotropic distillation. Because of its low volatility, the solvent always leaves the column with the bottoms product, and thus an additional distillation step is required to separate the solvent for recycle. To maintain the solvent throughout the column, the solvent must be introduced with or above the fresh feed. It is customary to employ a rectification section with reflux, above the solvent feed point, so as to prevent loss of solvent with the distillate.

Methods of analysis based on simple distillation may often be used on a solvent-free basis. The relative volatility between key components, i and j, is based on the simple relationship

$$\alpha_{ij} = \frac{p_i^*}{p_j^*} \left(\frac{\gamma_i^L}{\gamma_j^L} \right)_{solvent} \tag{12.47}$$

When the vapor pressures are very close, searches for effective solvents involve simple measurement of the activity coefficient ratio. The searching process is easier than that for azeotropic distillation. The usual approach is to select a solvent that is similar in structure to the heavy key component. Since the light key component is of a different structure (as it must be, or simple distillation could be employed), the solvent serves to increase the volatility of the light key relative to the heavy key.

An example of an extractive distillation process is the separation of methylcyclohexane (MCH) from toluene using a phenol solvent, as shown in Figure 12.17. Since MCH boils at 101.0°C and toluene boils at 110.7°C (1 atm), their separation by ordinary distillation is very difficult even though they do not form an azeotrope. Phenol is an effective solvent, since it has a structure more similar to the aromatic than to MCH (a naphthene), and it is relatively nonvolatile. The rectification

Distillation

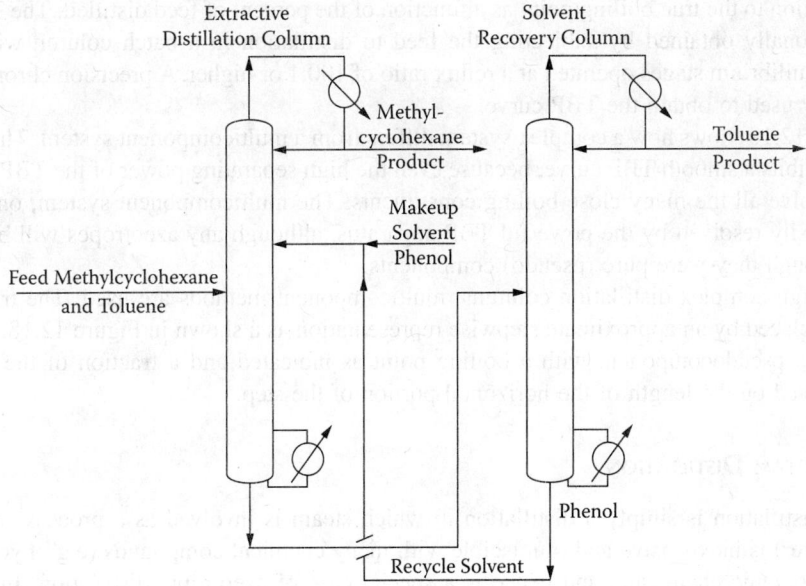

FIGURE 12.17 Extractive distillation process for separating methylcyclohexane and toluene.

section at the top of the extractive column prevents contamination of the MCH product with solvent. The second column employs ordinary distillation to recover the solvent from the toluene product. Since the toluene/solvent separation cannot be complete (excessive stages), and since minor amounts of solvent can be lost with the MCH, provision for a makeup solvent is required.

12.4.3 Complex System Distillation

A complex system is one containing so many components that they cannot be separated into discreet pure components by the distillation process. An example of such a system is naturally occurring petroleum, which contains hundreds of chemical constituents. Crude tall oil from paper pulping is another example of a complex system.

A complex system may be compared with a defined multicomponent system on the basis of a true boiling point curve (TBP) diagram as shown in Figure 12.18. The TBP curve is a close

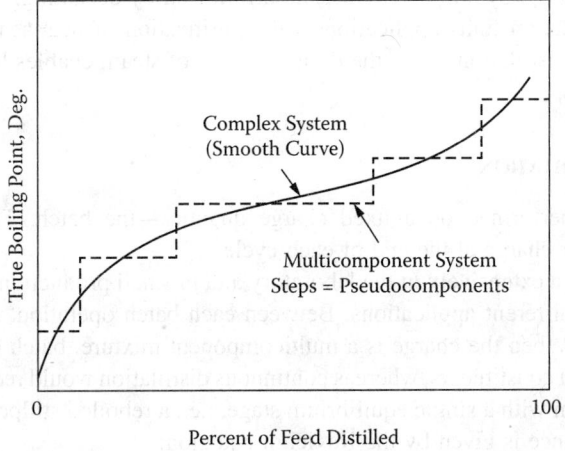

FIGURE 12.18 True boiling point curve diagram for a complex system and a defined multicomponent system.

approximation to the true boiling point as a function of the percent of feed distilled. The TBP curve was traditionally obtained by subjecting the feed to distillation in a batch column with several hundred equilibrium stages operated at a reflux ratio of 100:1 or higher. A precision chromatograph is currently used to obtain the TBP curve.

Figure 12.18 shows how a complex system differs from a multicomponent system. The complex system exhibits a smooth TBP curve, because even the high separating power of the TBP apparatus cannot resolve all the many close-boiling constituents. The multicomponent system, on the other hand, is easily resolved by the powerful TBP apparatus, although any azeotropes will be distilled over as though they were pure (pseudo) components.

To design complex distillation columns, multicomponent methods are used. The true boiling curve is replaced by an approximate stepwise representation as a shown in Figure 12.18. Each step represents a pseudocomponent with a boiling point as indicated and a fraction of the total feed mixture based on the length of the horizontal portion of the step.

12.4.4 Steam Distillation

A steam distillation is simply a distillation in which steam is involved as a process component. Steam (water) is inexpensive and immiscible with many chemical compounds (e.g., hydrocarbons and many organic chemicals) and hence is a special case of azeotropic distillation. In the usual application, very little of the steam condenses in the liquid phase, and thus problems of handling two liquid phases in the contacting equipment are avoided.

If an organic compound is essentially immiscible with water in the liquid phase, then vapor pressures are additive to make up the system total pressure:

$$p_{org}^{sat} + p_{water}^{sat} = P \qquad (12.48)$$

This simple equation is based on very little miscibility such that the activity coefficients for each phase are essentially unity.

Application of Equation (12.48) with the Antoine equation [Equation (12.13)] shows that steam distillation can be used to distill an organic compound at much lower temperatures than would otherwise be possible. Considering the system pressure P as constant, then the more steam introduced, the lower will be the partial and vapor pressures of the organic and thus the lower its boiling point. The other key element of the choice of water as the entrainer is that it is easily separated from the product (organic) by simple condensation followed by decanting.

Thus, steam distillation finds application in the purification of heat-sensitive materials as an alternative to vacuum distillation, since the dilution effect of steam enables lower effective boiling points of the materials.

12.4.5 Batch Distillation

Batch distillation is performed on a fixed charge quantity—the batch. The original charge is replenished by another charge at the end of each cycle.

This process is used extensively in the laboratory and in small production units where the same equipment can serve different applications. Between each batch operation, the equipment can be cleaned as necessary. When the charge is a multicomponent mixture, batch distillation in a single column can separate all constituents, whereas continuous distillation would require several columns.

For a binary system with a single equilibrium stage, i.e., a reboiled stillpot with no rectification column, the mass balance is given by the Rayleigh equation:

$$\ln\frac{L_o}{L_t} = \int_{x_t}^{x_o} \frac{dx}{y-x} \tag{12.49}$$

where the subscripts o and t refer to initial and final times, and L represents the amount of total liquid in the stillpot. The mole fraction of a given component in the stillpot at time t is denoted by x. By introducing relative volatility, Equation (12.49) can be integrated with the following result:

$$\ln\frac{L_o}{L_t} = \frac{1}{\alpha-1}\left(\ln\frac{x_o}{x_t} + \alpha \ln\frac{1-x_t}{1-x_o}\right) \tag{12.50}$$

Problem 12.5
One hundred moles of a 50-50 mixture of benzene and toluene are to be batch distilled in a simple takeover process. The relative volatility of benzene with respect to toluene is 2.5. The distillation is to proceed until the mole fraction of benzene in the stillpot is 0.20. How many moles remain in the stillpot at the end of the distillation?
By Equation (12.50),

$$\ln\frac{L_o}{L_t} = \frac{1}{2.5-1}\left(\ln\frac{0.5}{0.2} + 2.5\ln\frac{1-0.2}{1-0.5}\right) = 1.394$$

From which $L_t = 24.8$ moles are left in the stillpot.

For single-stage multicomponent batch distillation, the stage mass balance may be applied successively to any two components, j and k, as follows:

$$\ln\frac{L_{jt}}{L_{jo}} = \alpha \ln\frac{L_{kt}}{L_{ko}} \tag{12.51}$$

For the production of relatively pure products by batch distillation, the stillpot is usually augmented by a multistage rectification column with reflux. Shown in Figure 12.19 is a flow diagram for a typical batch distillation process. The feed in this example contains components A, B, and C. First, pure A is distilled into the first receiver. Then, an intermediate cut consisting of a mixture of A and B is collected in the second receiver. Following this, pure B is collected in the third receiver, followed by collection of an intermediate B/C mixture in the fourth receiver. At this point, relatively pure C remains in the stillpot. After removal of the products from their holding receivers, the intermediate cuts are drained back to the stillpot, to be joined by fresh feed for the next batch.

The plan for control of the reflux ratio in batch distillation is called the *operating policy*. One operating policy is to distill at constant reflux ratio throughout a given cut, collecting a composite mixture that meets specifications. The required reflux ratio is usually different for each cut. Another operating policy is to maintain a constant purity in the receiver by varying the reflux ratio. There are obviously other possible policies. Figure 12.20 shows a y-x diagram for operation at constant composition of the material being collected. For three theoretical stages, during the course of the distillation, the reflux ratio (indicated by the slope of the operating line) must be changed as shown. The diagram is representative only and applies to a binary mixture.

Stage calculations for multistage batch distillation are much more complicated than for continuous distillation, for two reasons. First, all conditions change with time. Second, the holdup in

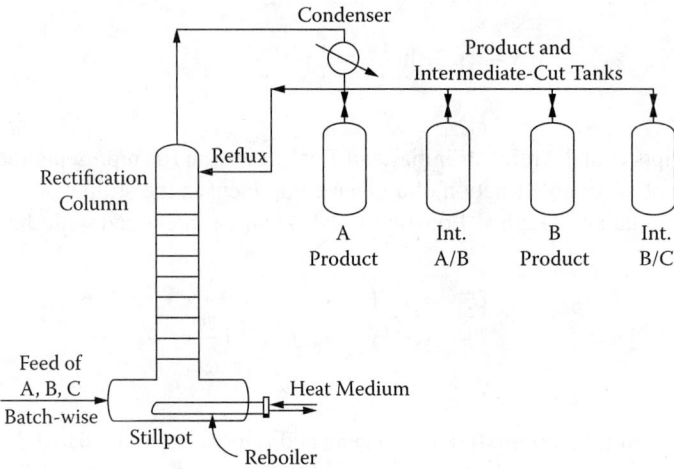

FIGURE 12.19 Typical batch distillation system.

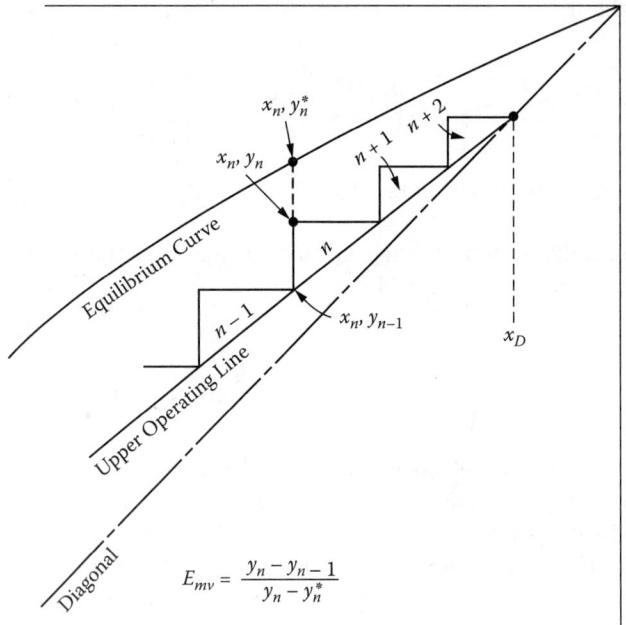

FIGURE 12.20 Upper quadrant of a McCabe–Thiele diagram for batch distillation at constant reflux. Four theoretical stages shown.

the column affects the concentration gradients in the column. The equation for the effect of holdup on the mass balance around a general batch contacting stage n is

$$J_n\left(\frac{dx_{i,n}}{dt}\right) = L_{n-1} x_{i,n-1} + V_{n+1} y_{i,n+1} - L_n x_{i,n} - V_n y_{i,n} \qquad (12.52)$$

where J_n is the stage molar holdup, and dx/dt is the rate of change of liquid concentration with respect to time.

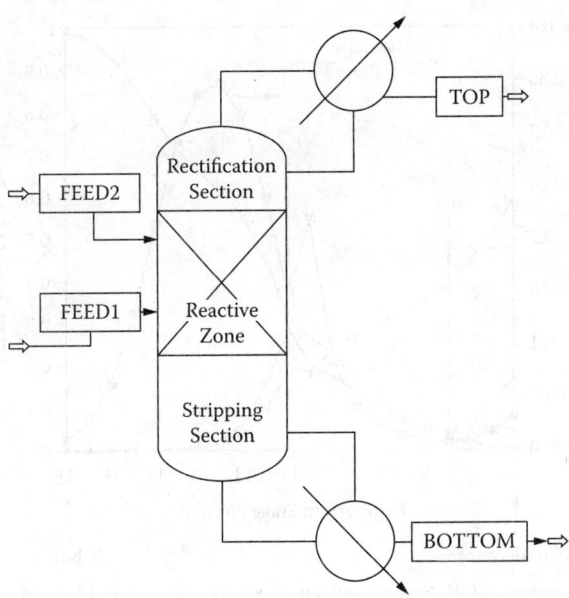

FIGURE 12.21 Typical reactive distillation configuration showing two feed streams and a reaction zone that can comprise multiple beds.

Equation (12.52) for batch distillation is the same as the mass balance equation for continuous distillation except for the term on the left side of the equation, which is normally zero for continuous distillation. Thus, it is theoretically possible to employ the same approach for batch distillation as previously presented for continuous distillation, provided an accumulation term is introduced. However, although apparently simple, it is actually very difficult in practice because of problems in solving the many simultaneous differential equations involved. In any event, it is erroneous to neglect tray and column holdup in stage computations for batch distillation.

The rigorous computation of multicomponent, multistage batch distillation is extremely complicated, and one should resort to available computer programs.[24–26] These programs can give a reasonable approximation for scaleup.

12.4.6 REACTIVE DISTILLATION

If a chemical reaction occurs inside a distillation column, with reactants and products subject to the usual requirements of the distillation process (phase equilibria, fractionation, and contacting device hydraulics), it is possible to shift the reaction equilibrium in a favorable direction. A soluble or insoluble catalyst is likely to be involved; thus, the operation is often known as *catalytic distillation*. Reactive distillation has been used successfully for etherification and esterification reactions and, to some extent, for alkylation, nitration, and amidation reactions. In most applications, the reaction has occurred in the liquid phase, and an example of this application, where methyl acetate is produced from methanol and acetic acid using a soluble catalyst, has been described in detail.[27] Flows for a generalized reactive column are shown in Figure 12.21.

Advantages of reactive distillation include the elimination of complicated product recovery, and separation and recycling of unconverted reactants, all of which lead to savings in equipment and energy costs. Compared to the conventional approach (reactor followed by distillation equipment), reactive distillation may also improve other factors such as selectivity and rates of mass transfer.

In deciding whether to consider reactive distillation, there are several key considerations:[28]

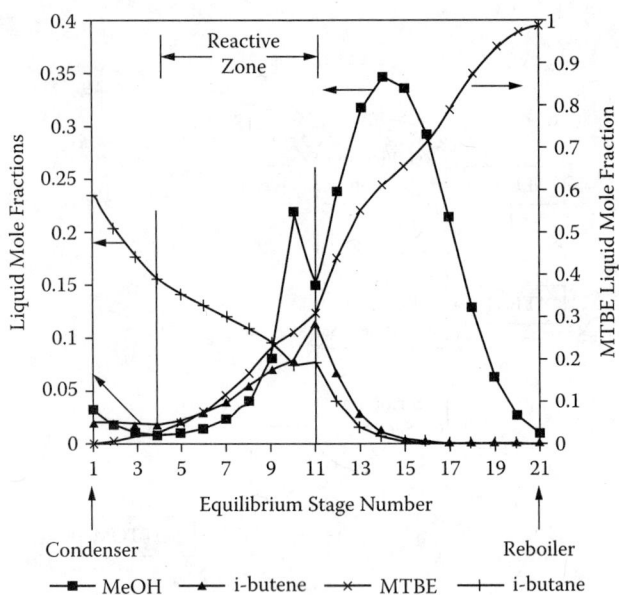

FIGURE 12.22 Computed composition profiles in an MTBE reactive column. Composition of the diluent, *n*-butane, may be obtained by difference. (H. E. Subawalla, 1997. Ph.D. dissertation, Univ. of Texas at Austin.)

- Is there an advantage in shifting the chemical equilibrium? A reaction with a highly favorable equilibrium is not likely to be a candidate.
- The reaction product must boil in an appropriate range and be separable by distillation. If the boiling point of the desired product instead falls within a range of other products, more distillation columns will be needed.
- The optimum temperature and pressure conditions for the reaction must be the same as those required for separation.
- Only one liquid phase should be present. A second liquid phase can introduce problems in the hydraulic design of the distillation column. In some cases, this problem can be avoided through the use of a separate cosolvent, which then must be separated for recycle.
- If the catalyst is a solid, it must create neither excessive pressure drop nor an excessively large column. An optimal arrangement of solid catalysts in the column is needed.
- A suitable model must be available for determining concentration profiles throughout the column and judging needs for reflux and boilup. This implies knowledge of the reaction kinetics, likely obtained separately in laboratory equipment.

A representative profile for a column in which methanol and *i*-butene are reacted to form methyl tert-butyl ether (MTBE) is shown in Figure 12.22. Note that an inert *n*-butane, is not shown. The reactants are fed at different stages, and the desired product, MTBE, is removed from the bottom of the column in high purity.

Many patents and publications relating to reactive distillation have been published. A paper on the heuristics of reactive distillation column design contains many references.[29]

12.5 DISTILLATION COLUMNS

As noted earlier, distillations are performed in vertical, cylindrical columns when more than a single contacting stage is required, as indicated in Figure 12.23.

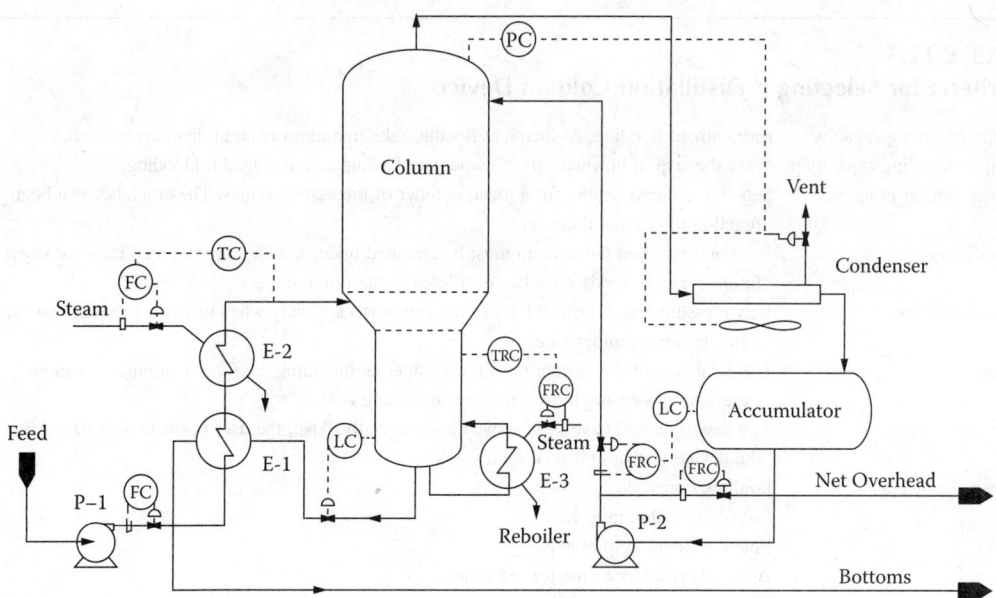

FIGURE 12.23 Flow diagram of a full distillation system. While the design of the column is emphasized in this chapter, auxiliaries such as heat exchangers, pumps, and controls must not be neglected.

After computations have been made to establish the number of required theoretical contacts, the equipment must be designed. Our central focus is the column, even though the system comprises heat exchangers, pumps, vessels, controls, and so forth. Supporting equipment design and characteristics are covered elsewhere in this handbook. The sequence of steps that applies to the design of a new distillation system, or to the analysis of an existing system, is as follows.

Define the system means identifying and quantifying each component in the mixture to be separated. Unexpected components such as intermediate boilers often play havoc with the intended separation. If the mixture is complex, as discussed earlier, proper steps to identify pseudocomponents are required.

Establish separation criteria means determining product purities, or feed component recoveries, or some combination of the two. A cogent comment here is that one should "look down the road" to anticipate future changes in product purity requirements, variations in feed composition, shifts in throughput rates, and so on.

Obtain physical data for the calculations. In some cases, the data might be stored in an easily retrieved format. In any case, however, the designer must verify data, especially for mixtures, and determine the possible effects of data errors on the outcome of the design work. This important step should be done with care. One of the most important areas of data gathering deals with vapor-liquid equilibria.

Determine stages or transfer units, which are parameters that characterize the difficulty of the separation. The model for obtaining the parameters might be rigorous or nonrigorous, depending on the needs.

Develop complete specifications for the distillation column: diameter, spacing of trays, type and size of packing, height of contacting zone, and so on, as reported in the present section.

The final requirement involves the development of the complete system design, for example, the typical system shown in Figure 12.23. The fact that more than just the column is involved was mentioned earlier.

The required theoretical stages or transfer units will have been determined. Also, some value of the stage efficiency will have been obtained for starting purposes and to be verified later, after

TABLE 12.5
Criteria for Selecting a Distillation Column Device

Vapor-handling capacity	Entrainment flooding. At incipient flooding, the minimum column diameter is fixed.
Liquid-handling capacity	Fixes the size of downcomers. Downcomer backup can also lead to flooding.
Mass transfer efficiency	Sets the required height, for a given number of theoretical stages. The efficiency can be a function of column diameter.
Flexibility	Is a concern when the column must be operated under a wide range of feed rates, or when future capacity needs must be considered in the initial design.
Pressure drop	Low pressure drop is critical for vacuum columns, especially when there are needs to maintain a low bottoms temperature.
Cost	The total cost of the system must be considered, including auxiliary equipment. A more expensive device might lead to lower operating costs.
Design limitations	The device should have been proven commercially. Also, the user needs to understand how the device was designed (if by a vendor)
Special concerns	Fouling, corrosion
	Ease of installation or removal
	Potential foaming problems
	Adequate residence time for reactions
	Special heat transfer needs

the effects of the geometry of the column have been taken into consideration. (This is the *rating approach,* common in chemical engineering practice, where a device geometry is contrived and then rated for capability to perform the needed service.) Thus, as a result of earlier studies, the following information will be at hand: actual stages (or packed height), vapor flow rate, liquid flow rate, vapor and liquid compositions, temperatures, and pressures.

Ideally, this information will be available for a number of points in the column. As a bare minimum, it must be available at the top, bottom, and center (above and below the feed point for a single column). Some approximations may be in order, e.g., assumed: pressure profile, equal molal downflow and upflow, and constant efficiency throughout the column. Normally, these assumptions will in turn be verified or modified.

The general objective of equipment process design is to establish specifications for the optimum fractionating system hardware. (Alternatively, an objective could be to gain an understanding of an existing system under analysis.) For the column only, criteria for selection of the internal contacting devices are given in Table 12.5. Each of these criteria should be considered carefully.

The function of the device is to bring vapor and liquid into intimate contact. Many devices have been developed through the years.[30] Three important tray-type devices are shown in Figure 12.24, and several packings are shown in Figure 12.25. All devices vary in their ability to provide intimate contacting, effect sharp separations, realize low pressure drops, and handle liquids that might foul or plug the system.

An enormous amount of study has been made to evaluate the performance of many contacting devices, using a variety of test mixtures. Only a portion of the results have been published in the open literature, on a random basis with unification through handbooks, texts, and review papers. Fractionation Research, Inc. (FRI), an industry-sponsored organization for conducting commercial-scale tests of devices under carefully controlled experimental distillation or absorption conditions, has been a leader, performing thousands of experiments.[31,32] An updated state-of-the-art review of equipment types and characteristics has been published by FRI and collaborators.[33]

Devices for effecting contact of vapor and liquid may be classified as shown in Table 12.6. Recently, there has been a trend away from the conventional crossflow devices, in the direction of counterflow devices, mostly packings. Still, an enormous number of older columns are operating with trays, and thus the technology of tray design and analysis is important.

Distillation

Sieve Tray
(a)

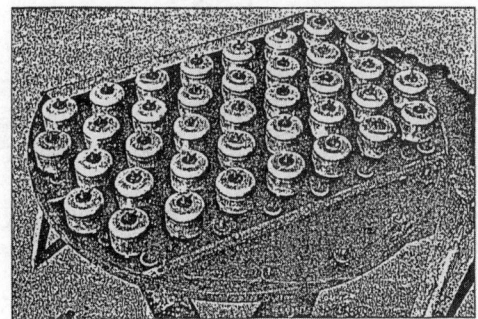

Bubble-cap Tray
(b)

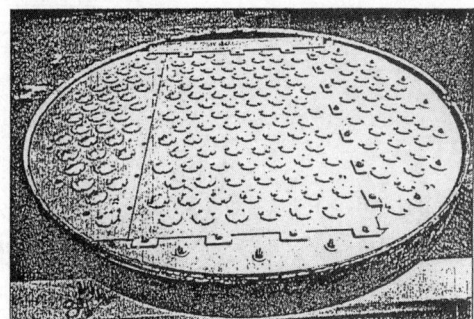

Valve Tray
(c)

FIGURE 12.24 Views of representative crossflow tray-type devices. (J. R. Fair, 1984. *AIChE Symp. Ser. No. 235*, 79: 1.)

12.6 TRAY COLUMN HYDRAULICS

12.6.1 CROSSFLOW TRAY COLUMNS–SIEVE TRAYS

12.6.1.1 Performance Profile

A representative performance profile for a commercial crossflow sieve tray is given in Figure 12.26.[34] The zone of "good operation" is variable in extent and rarely is covered by a constant value of efficiency. The profile is influenced by tray type and dimensions, system properties, and liquid-vapor throughput rates. A second profile, for pressure drop, will be discussed later.

12.6.1.2 Vapor-Handling Capacity

The first step in sizing a fractionator is usually to calculate its approximate diameter based on the required vapor capacity. In a simplified way, column cross-sectional area required is given by the following relationship:

$$A_t = A_a + 2A_d = A_n + A_d \tag{12.53}$$

where *net area* A_n is usually 85 to 95% of A_t, the total tower cross-sectional area. Thus, the total downcomer area is 5 to 15% of the total cross section. See Figure 12.27.

A schematic diagram of a perforated tray is shown in Figure 12.28. The zone marked "froth" (liquid-continuous) can sometimes invert to a "spray" (vapor-continuous), but quantitative procedures dealing with the spray regime have not yet been developed. Such a regime prevails at very

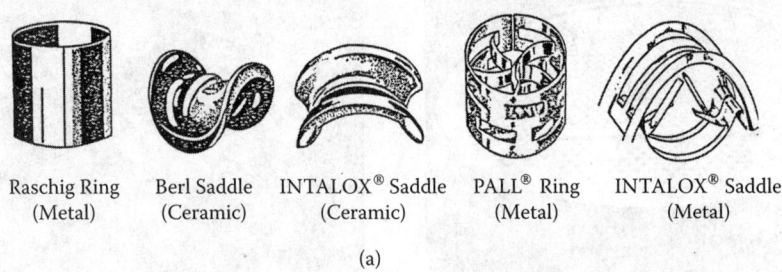

Raschig Ring (Metal)   Berl Saddle (Ceramic)   INTALOX® Saddle (Ceramic)   PALL® Ring (Metal)   INTALOX® Saddle (Metal)

(a)

(b)

FIGURE 12.25 Views of representative packing elements. (a) Random. (b) Structured.

TABLE 12.6
Classification of Contacting Devices

Crossflow trays	Bubble-cap	
	Sieve	
	Valve	
Counterflow Packings	Random	Rings
		Saddles
		Other
	Ordered	Structured
		Mesh
		Grid
Counterflow Trays	Dualflow	
	Multiple downcomer	
	Baffle trays (splash decks)	
Special devices	Sprays	
	CoFlo	
	Moving internals	

Distillation

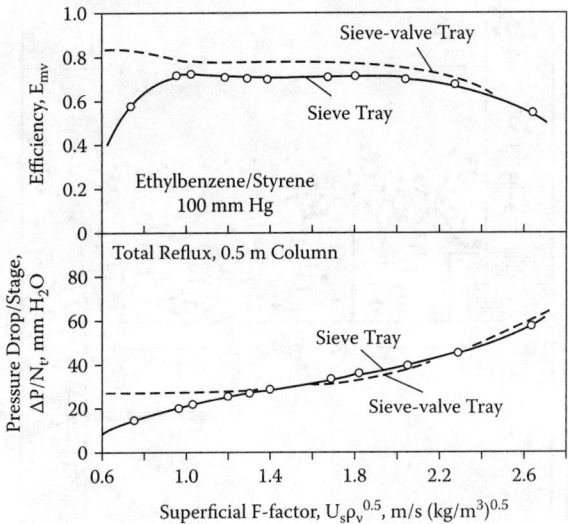

FIGURE 12.26 Performance profiles of representative crossflow trays. The upper plot is of efficiency vs. F-factor (= $U_s \rho_v^{0.5}$). The lower plot is of pressure drop per theoretical stage. Profiles for both sieve and sieve-valve hybrid devices are shown. (R. Billet et al., 1969. *IChemE Symp. Ser.* 32: 5, 111.)

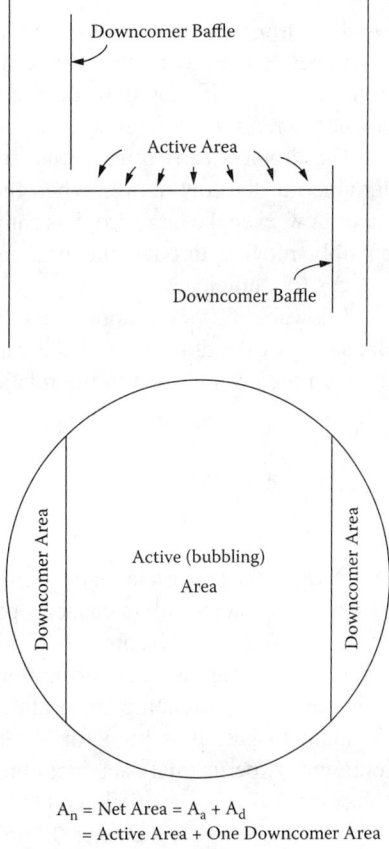

A_n = Net Area = $A_a + A_d$
  = Active Area + One Downcomer Area

A_t = Total Column Cross-sectional Area
  = $A_a + 2A_d$

FIGURE 12.27 Area designations for crossflow trays (sieve, valve, bubble-cap).

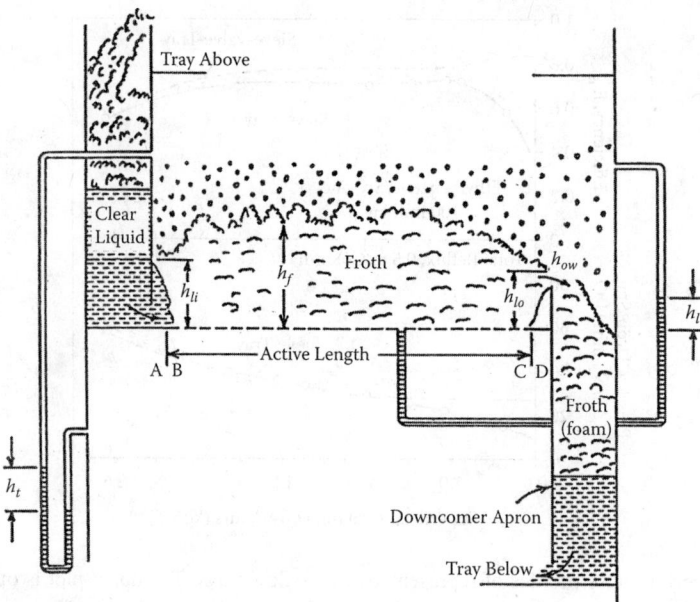

FIGURE 12.28 Schematic diagram of a crossflow sieve tray. Valve trays and bubble-cap trays have equivalent diagrams.

low liquid-to-vapor volumetric flow ratios such as occur in high-vacuum distillations. Liquid is distributed to the tray by flow under the downcomer apron; additional distribution can be obtained from an inlet weir (not shown). The active length BC normally includes the total distance from downcomer apron (or inlet weir) to outlet weir, i.e., A and D lengths are effectively zero.

The droplets carried above the froth may return to the froth or, for very small droplets, are entrained to the tray above. When the froth level approaches the tray above, as at high rates of vapor flow, even the large droplets cannot complete their trajectories and thus impact the tray above, possibly moving through the perforations as entrained liquid. Such a phenomenon drastically reduces tray efficiency.

Allowable flows of vapor through the froth are correlated on the basis of liquid entrainment. Balancing of the drag force of the vapor on a representative drop of liquid against the gravitation force on the drop has led to the relationship,

$$U_N = C_{sb} \left(\frac{\rho_L - \rho_v}{\rho_v} \right)^{0.5} \qquad (12.54)$$

where U_N is the superficial vapor velocity based on A_N (Figure 12.27) and C_{sb} is a correlating term called the Souders–Brown capacity parameter.

Values of C_{sb} may be obtained from Figure 12.29 and are based on tray spacing, surface tension, vapor velocity through the perforations, and liquid flow rate. The abscissa group, $L/G \sqrt{\rho_v / \rho_L}$, is an important correlating term called the *flow parameter* and represents a ratio of kinetic energies of liquid to gas. It is used for several correlations involving vapor-liquid ratios. Figure 12.29 represents experimental data, including recent FRI work, and should be used for design purposes.[35] It has the following restrictions: (1) low- to nonfoaming system, (2) weir height less than 15% of tray spacing, (3) hole diameter 0.5 in. (12.7 mm) or less (sieve trays), and (4) hole (or riser) area 10% or more of the active (bubbling) area. Smaller hole areas tend to produce jetting because of high hole velocities, and corrections are as follows:

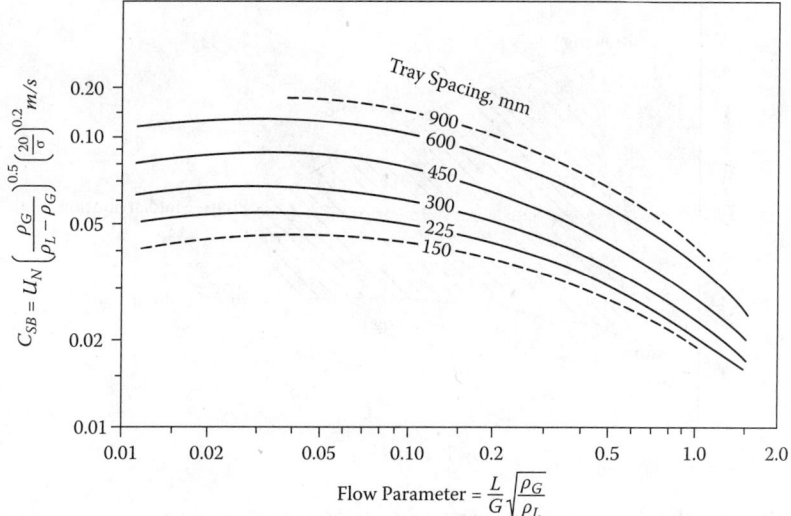

FIGURE 12.29 Design chart for estimating maximum vapor capacity of crossflow trays (both bubble cap and sieve trays). The lower values at the left indicate that liquid capacity prevails over vapor capacity as the limiting parameter.

A_h/A_a	U_{des}/U_{chart}
0.10	1.00
0.08	0.90
0.06	0.80

An alternative model for predicting entrainment flooding was published recently.[36] The concept of entrainment flooding is discussed by Silvey and Keller.[37]

The value of A_N has been modified by new tray designs in which the downcomer is suspended above the tray floor ("hanging downcomer"), enabling the net area to include the area under the downcomer. For this case, $A_N = A_t - A_d$.

Figure 12.30 shows qualitatively the region of "satisfactory operation." We have just looked at flooding; at low vapor rates, there are limits of weeping or dumping, i.e., a portion of the cross-flowing liquid flowing down through the perforations. At high liquid rates, the downcomers may not be able to pass the required amount of liquid. These limits are discussed below.

12.6.1.2.1 Entrainment

As the flood point is approached, liquid entrainment becomes a problem. Entrained liquid is recycled back to the tray above, negating the countercurrent effect and decreasing tray efficiency. The recirculation resulting from entrainment is indicated in Figure 12.31. The molar ratio of liquid entrainment rate to the "dry" liquid flow rate is $\psi = e/(L_{MD} + e)$. A chart for estimating ψ as a function of approach to flooding, $C_{sb}/C_{sb,flood}$, and the flow parameter, defined earlier, is given in Figure 12.32.[38] This graph has the same limitations as Figure 12.29. As discussed later, the term ψ corrects the "dry" tray efficiency:

$$\frac{E_w}{E_d} = \frac{1}{1 + E_d \psi / (1 - \psi)} \qquad (12.55)$$

where E_w and E_d are wet and dry efficiencies.

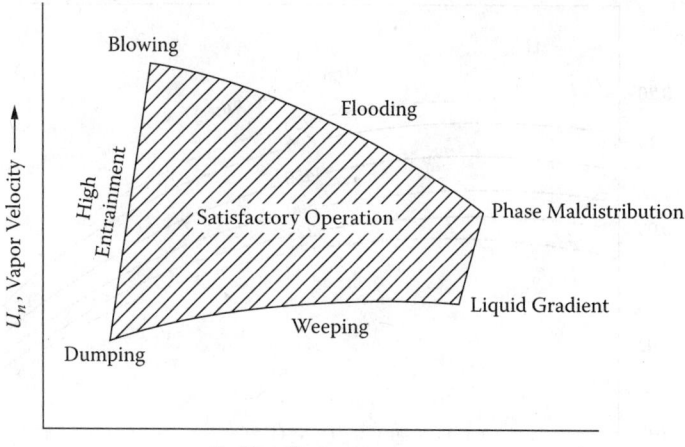

FIGURE 12.30 Generalized plot showing operating zones of a crossflow tray.

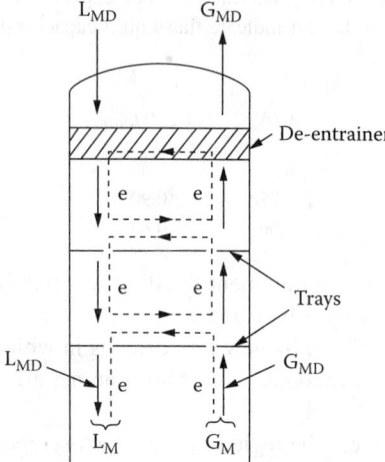

FIGURE 12.31 Diagram showing the recirculation effect of liquid entrainment, and the definition of terms used in design Equation (12.55).

12.6.1.2.2 Pressure Drop

The decrease in pressure, as the vapor flows upward through a tray, is the sum of the pressure drop through the perforations and the residual drop through the froth or spray:

$$h_t = h_d + h_L \tag{12.56}$$

where pressure drop is expressed in terms of head of clear liquid on the tray. For flow through the perforations, the simplified orifice equation is used (assuming uniform flow through all the holes):

$$h_d = \frac{50.8}{C_v^2} \frac{\rho_v}{\rho_L} U_h^2 \tag{12.57}$$

The discharge coefficient C_v may be obtained from Figure 12.33[39] or from the relationship

Distillation

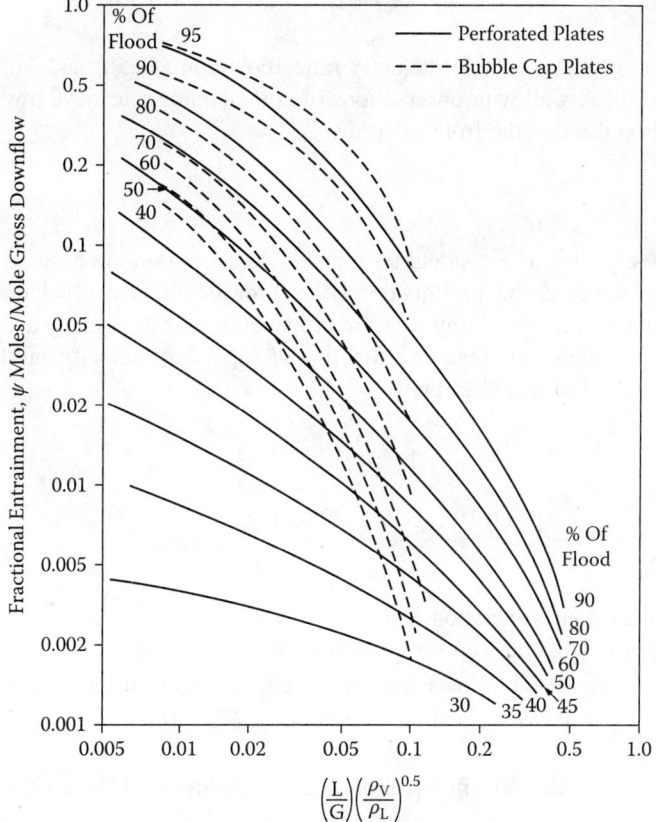

FIGURE 12.32 Chart for estimating entrainment parameter as a function of flow parameter and approach to entrainment flood. (J. R. Fair, 1997. *Perry's Chemical Engineers' Handbook*, 7th ed., R. H. Perry, D. Green, eds., New York: McGraw-Hill.)

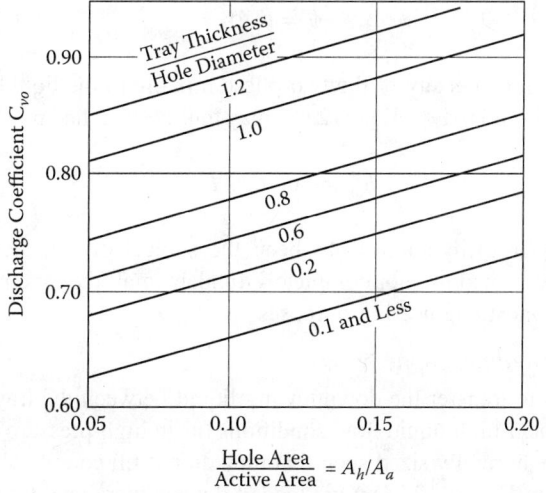

FIGURE 12.33 Chart for estimating the orifice coefficient for dry tray pressure drop.

$$C_v = 0.74(A_h/A_a) + \exp[0.29(t_t/d_h) - 0.56] \tag{12.58}$$

where t_t/d_h is the tray thickness/hole diameter ratio. Equation (12.58) and Figure 12.33 predict pressure drops that check well with observations of commercial scale sieve tray columns.

For pressure drop through the froth or spray,

$$h_L = h_t - h_d \tag{12.59}$$

where h_L is simply regarded as a residual term and is correlated on the basis of dry tray pressure drop measurements, standardizing on Equation (12.58) for the orifice coefficients when actual data are not available, along with overall tray pressure drop measurements (usually available in operating columns). Values of h_L are determined as a function of vapor hole velocity and liquid flow rate as shown in Equations (12.60) and (12.61):

$$\beta = h_L/(h_w + h_{ow}) \tag{12.60}$$

$$\beta = 0.19 \log_{10} L_w - 0.62 \log_{10} F_{vh} + 1.679 \tag{12.61}$$

where
- β = dimensionless aeration factor
- h_w and h_{ow} = weir height and weir crest (clear liquid basis), mm
- L_w = flow rate of liquid over the outlet weir, m³/s-m weir length
- F_{vh} = vapor "F-factor" based on hole velocity, $F_{vh} = U_h(\rho_v)^{0.5}$

The value of the weir crest h_{ow} in Equation (12.60) employs the classic Francis weir equation:

$$h_{ow} = 664 \, L_w^{0.667} \tag{12.62}$$

Even though froth actually flows over the weir (unless calming zones are used), h_{ow} is expressed on an equivalent clear liquid basis, assuming that the Francis relationship also represents froth flow.

Later, in connection with mass transfer in tray froths, we will discuss a relative froth density:

$$\phi_f = \rho_f/\rho_L \tag{12.63}$$

This is a ratio of the average density of the two-phase mixture to the liquid density (it approaches unity for lightly aerated liquids). Equation (12.62) may thus be modified by Equation (12.63) to give

$$h_{ow} = 664 \, \phi_f^{1/3} \, L_w^{2/3} \tag{12.64}$$

This area of sieve tray froth hydraulics has been the subject of considerable study. While the relationships thus far presented are not completely fundamental, they are simple to apply and are sufficiently reliable for most engineering purposes.

12.6.1.1.3 Liquid-Handling Capacity

The downcomer serves to transfer the downflowing liquid between the trays in a column and can become a bottleneck under high liquid flow conditions (as in high-pressure absorbers and fractionators). Downcomers are normally sized such that they do not fill completely with the combination of clear liquid and froth (Figure 12.28). Overloaded downcomers lead to a flood condition in the entire column. Thus, we distinguish between "entrainment flood" and "downcomer flood" but recognize that one often leads to the other.

Distillation

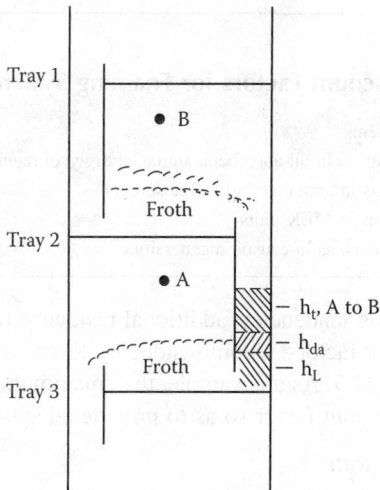

FIGURE 12.34 Pressure balance across two crossflow trays.

Downcomer backup h_{dc} is obtained from a pressure balance:

$$h_{dc} = h_t + h_L + h_{da} \tag{12.65}$$

which should be clear from Figure 12.34, with the datum being a level just above the floor of the lower tray. The head loss for liquid flow under the downcomer apron h_{da} is calculated from

$$h_{da} = 165 \, (q/A_{da})^2 \tag{12.66}$$

If an inlet weir is used, h_{da} obtained from Equation (12.66) should be increased by about 50%. If hanging downcomers are used, special and proprietary relationships may be needed, as discussed later.

The backup relationship [Equation (12.65)] is used with pressure head on a clear liquid basis. The actual height in the downcomer is

$$h'_{dc} = \frac{h_{dc}}{\phi_{dc}} \tag{12.67}$$

where ϕ_{dc} is the average value of relative froth density in the downcomer. The computed value of h_{dc}' should be less than the value of the tray spacing. A value of $\phi_{dc} = 0.5$ is often used, but its proper value depends on the disengaging tendency of vapor and liquid. (Note that a countercurrent phase flow occurs in the downcomer when vapor disengages from the froth and rises against the overflowing stream of froth.)

If the flow over the weir is a relatively clear liquid, then there is no disengagement problem, and $h_{dc}' \sim 1.0$. If the vapor bubbles are slow to rise against the flow of liquid, the superficial velocity of liquid in the downcomer (at its narrowest cross section) must not be greater than the free rise velocity of the vapor bubbles. A minimum velocity of 0.1 m/s (0.4 ft/s) is usually satisfactory. Conditions favoring rapid disengagement include low liquid viscosity and low vapor density. For high-pressure fractionators operating near the critical point, liquid and vapor densities tend to merge, and disengagement is slow. Hoek and Zuiderweg[40] analyzed FRI tests at high pressure and concluded that a considerable amount of vapor passes to the tray below as entrainment in the liquid, accompanied by a significant loss of tray efficiency.

TABLE 12.7
Capacity Discount Factors for Foaming Systems

Nonfoaming systems	1.00
Moderate foaming, as in oil absorbers, amine, and glycol regenerators	0.85
Heavy foaming, as in amine and glycol absorbers	0.73
Severe foaming, as in MEK units	0.60
Foam-stable systems, as in caustic regenerators	0.15

When the liquid has foaming tendencies, additional residence time must be allowed for phase disengagement. System discount factors, recommended by Koch–Glitsch, Inc. on the basis of field experience, are shown in Table 12.7. As an example, the "maximum allowable velocity" of 0.1 m/s would be multiplied by the discount factor so as to provide adequate downcomer volume.

12.5.1.1.4 Minimum Vapor Rate

As shown previously in Figure 12.25, the performance of tray devices drops off at low velocity because of weeping and/or poor vapor dispersion. Weeping is usually associated with sieve trays or fixed valve trays, which have no built-in protection against liquid flow through the openings at low vapor velocities. Movable valve trays can also weep, since the valve units are designed not to close completely at zero vapor rate. In theory, at steady state and with equal vapor velocity through all of the holes, there will be no weeping on a sieve tray when

$$(h_d + h_\sigma) > h_L \tag{12.68}$$

Then the pressure drop through the hole plus the head of liquid necessary to overcome surface tension can act against the liquid head above the hole to prevent liquid from entering the hole. The term h_σ in Equation (12.68) may be estimated from the following *dimensional* equation:

$$h_\sigma = 409 \left(\frac{\sigma}{\rho_L d_h} \right), \text{ mm liquid} \tag{12.69}$$

where

σ = surface tension, mN/m
d_h = hole diameter, mm
ρ_L = liquid density, kg/m^3

Equation (12.68) applies only to a steady-state condition, and this rarely happens. Some holes pass neither vapor nor liquid. There is considerable sloshing, froth pounding, oscillation, and so on. A modified Equation (12.68) is shown in Figure 12.35. The lines of the figure may be approximated by

$$h_d + h_\sigma = 0.036(h_w + h_{ow}) + 0.25 \text{ (20\% open)} \tag{12.70}$$

$$h_d + h_\sigma = 0.013(h_w + h_{ow}) + 0.25 \text{ (6--14\% open)} \tag{12.70a}$$

While weeping may occur over a broad range of operating rates, it does not necessarily detract from the mass transfer capability of the tray, so long as loadings are reasonably high. At some very low vapor rate, efficiency plunges, and the flow condition is often called the *dump point*. Figure

Distillation

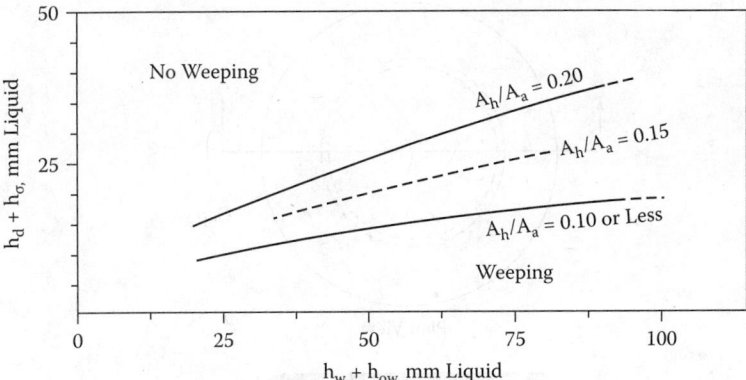

FIGURE 12.35 Chart for estimating the vapor load at the point of efficiency drop-off. The drop-off can be due to weeping or dumping. (B. D. Smith, 1963. *Design of Equilibrium Stage Processes*, New York: McGraw-Hill.)

12.35 should be used to develop the performance profile of a sieve tray, and no quantitative measures of weeping should be attempted.

Often, a question is asked about the *design turndown ratio* of a device. It is practical to express this as a ratio of the design loading to the loading at the "point of efficiency drop-off," whether represented by physical weeping or dumping. Figure 12.35 represents this lower limit.

12.6.1.2 Other Crossflow Trays

12.6.1.2.1 Bubble-Cap Trays

These devices are currently used only occasionally. They (see Figure 12.36) were the industry standard before 1950, and at very low vapor flows they do not weep, as do sieve trays and valve trays. Vapor flows upward through a central riser, reverses direction, flows downward through an annular space, and exits under the skirt (or through slots cut into the skirt). As the vapor issues from the cap, it is dispersed in the crossflowing liquid. Bubble caps are heavy and more expensive to fabricate than the simple holes of a sieve tray, and the dispersion of the vapor is not as effective as with a sieve tray.

The design of a bubble-cap tray is similar to that for a sieve tray. The vapor capacity is determined from Figure 12.29, which represents bubble-cap trays as well as sieve trays (see original reference). Entrainment may be estimated directly from Figure 12.32, which shows that the caps entrain more than the sieve holes. Pressure drop utilizes the same general equations as the sieve tray, except that the dry tray drop through the caps is much larger and utilizes different parameters. A detailed analysis of the pressure drop function is reported by Bolles.[41] Finally, computations for liquid capacity through the downcomer utilize exactly the same relationships as for sieve trays except for one correction term in Equation (12.65): on the right side, a term is used to represent hydraulic gradient on the tray. This gradient results from the resistance to crossflow provided by the caps themselves. Bolles deals with hydraulic gradient extensively since, if severe, it can cause vapor to maldistribute as it flows upward through each tray.

Today the main process application for bubble cap trays is in reactive distillation columns or in chemical absorption columns; in either case, it may be necessary to control very carefully the residence time of the liquid to complete a reaction step. For example, bubble-cap trays are used for the methyl acetate column described earlier and published by Agreda et al.[27] An abridged version of the Bolles treatment of bubble-cap tray design is given in the fifth edition of Perry's handbook.[42]

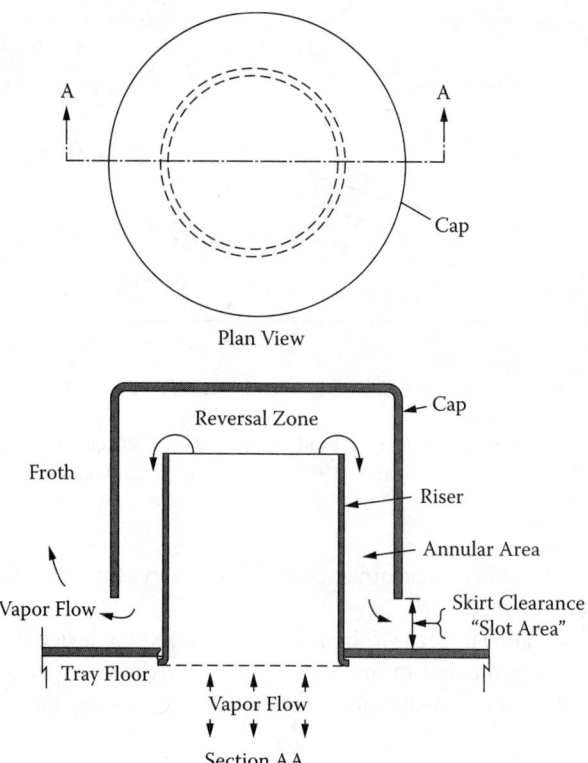

FIGURE 12.36 Diagram of a typical bubble cap.

12.6.1.2.2 Movable Valve Trays

These contacting devices can extend the operating range of a crossflow sieve tray, as indicated by Figure 12.26, which shows that the sieve tray under observation loses efficiency at an F-factor of about 1.0 m/s(kg/m^3)$^{0.5}$, whereas the addition of valves permits operation down to an F-factor of 0.6 or below. A movable valve is shown in Figure 12.37 alongside a fixed valve counterpart. Note that valves can have various shapes, often circular (see Figure 12.24).

Manufacturers of valve trays, such as Koch–Glitsch, Inc., of Wichita, Kansas (Ballast trays and Flexitrays), and Sulzer–Chemtech (formerly Nutter Engineering Co.), of Tulsa, Oklahoma (Float Valve Trays), have prepared proprietary design manuals. Hence, only limited discussion will be given here. As for bubble-cap trays, design methods follow those for sieve trays. The vapor capacity chart (Figure 12.29) covers valve trays, as does the alternate method of Kister and Haas.[36] Information on liquid entrainment is proprietary, but measurements have been made by Fractionation Research, Inc.[31] Because of the vapor flow reversal, one would not expect entrainment from valve trays to be greater than that from sieve trays. Liquid capacity considerations follow exactly those for sieve trays.

Pressure drop prediction follows the same approach as for sieve trays, with the primary difference being in the determination of dry tray pressure drop. Values of dry drop for fully open and fully closed valves may be estimated from relationships such as those of Klein[43]:

$$\text{Closed: } h_d = K_c \frac{\rho_v}{\rho_L} U_c^2 \qquad (12.71)$$

Distillation

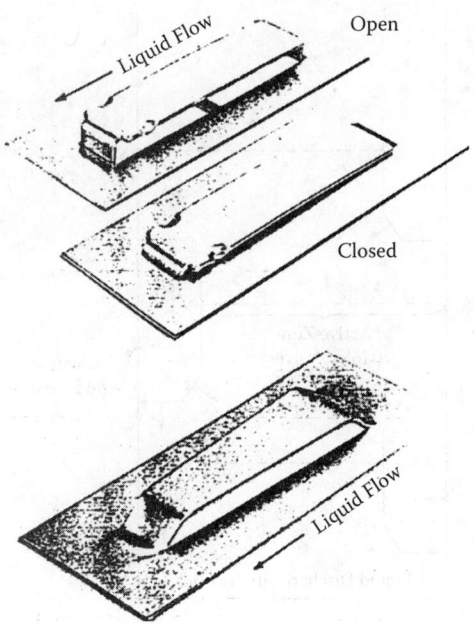

FIGURE 12.37 Views of a rectangular, movable valve and a rectangular fixed valve (the V-grid). (Courtesy of Sulzer Chemtech USA, Inc.)

$$\text{Open: } h_d = K_o \frac{\rho_v}{\rho_L} U_o^2 \qquad (12.72)$$

with Klein giving rather complicated equations for coefficients K_c and K_o. Klein also provides a relationship for predicting the aeration factor β so that the equation

$$h_t = h_d + \beta(h_w + h_{ow}) \qquad (12.60)$$

can be applied. Bolles[44] provides a good example of the hydraulic relationships for movable-valve trays.

12.6.1.3 Fixed Valve Trays

For this device, valves are fixed in place (extruded from the tray metal), and Figure 12.37 shows a comparison between fixed and movable valves of the same general design. The fixed valve, known as the "V-grid," was introduced in 1979 with a useful paper by Nutter.[45] More recently, a modern version of the V-grid was compared with a sieve tray with 12.7-mm holes by Nutter and Perry,[46] showing that the valve tray has about the same efficiency but slightly greater capacity when compared with a sieve tray. As might be expected, the design of the fixed valve tray follows that of the sieve tray, the basic difference between the trays being the simple hole versus an elevated opening that can be irregular in shape. Valve tray manufacturers offer many variations of the basic design.

A variation of the fixed valve tray design is the elevation of the downcomer floor and discharging downflow liquid through openings in the lower part of the downcomer. A sketch of such a device is given in Figure 12.38, which shows the "hanging downcomer." This modification frees up space for valves or perforations in the area normally used for the bottom of the downcomer. Depending on the downcomer area, this can promise 10 to 15% additional vapor flow capacity. This additional

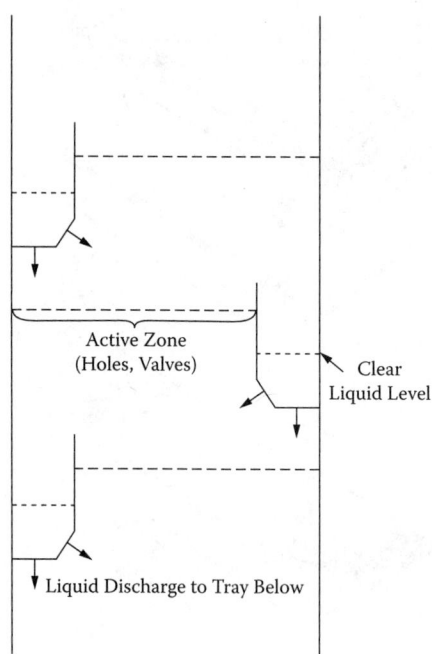

FIGURE 12.38 Hanging downcomer tray. Note that the part of the tray floor normally used as the bottom of the downcomer is converted to bubbling (active) area.

capacity has not always been achieved, however, and the design of the downcomer is critical, because too big of an opening at its bottom can allow vapor to bypass the contacting zone.

12.6.2 Counterflow Perforated Trays

These trays, often called *dualflow trays*, are perforated and occupy the entire column cross section. Similar devices are known as *turbogrid trays* and *ripple trays*. Their simplicity is evident from the view in Figure 12.39. Since there are no downcomers, liquid must compete with vapor for passage through the perforations, and this complicates the hydraulic analysis of the device. Observations show that a given perforation passes either vapor or liquid, but not both at the same time. Furthermore, the phase flowing through the perforation shifts with time in a random way. This alternative flow is believed to be the reason why trays of this type can often handle systems that foul conventional trays and packings.

Extensive commercial-scale performance data on counterflow devices were released by FRI in 2000. Detailed information on tray performance and modeling was reported.[47] Figure 12.40, from reference 48, shows the influence of open area on efficiency and also makes clear the sensitivity of efficiency to loading. The figure is quite revealing: One can obtain increased capacity at the expense of efficiency by going to larger open areas. The figure also shows that the counterflow device has a characteristically narrow operating range.

Geometric variables for design are few: perforation type and size, open area, and tray spacing. The perforations are usually round and in the range of 12.7- to 25.4-mm diameter. Open areas run from 15 to 25%. Vapor flow capacity (entrainment limitation) is correlated in a fashion similar to that for crossflow trays, as shown in Figure 12.41, from reference 47. The figure is based on FRI data taken in a 1.2-m diameter column.

Zuiderweg and coworkers[49] identified a segregation effect on turbogrid trays and found that towers in the range of 3- to 4-m diameter have 80% or less of the efficiency of a small (0.5 m) tower, because of phase segregation and channeling.

Distillation

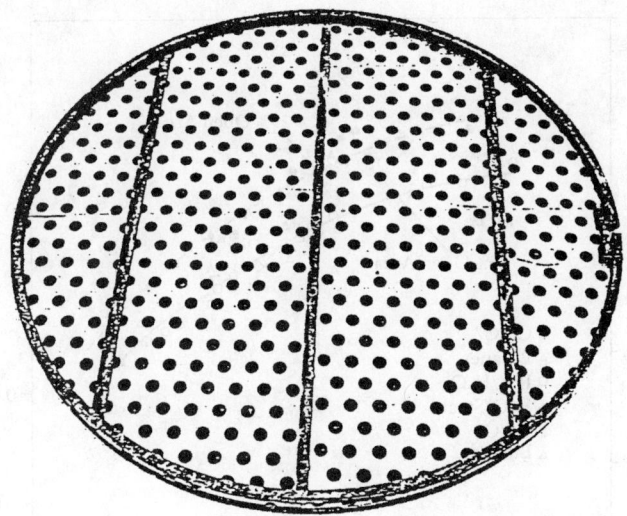

FIGURE 12.39 View of a 1.2-m dualflow tray. (Courtesy of Fractionation Research, Inc.)

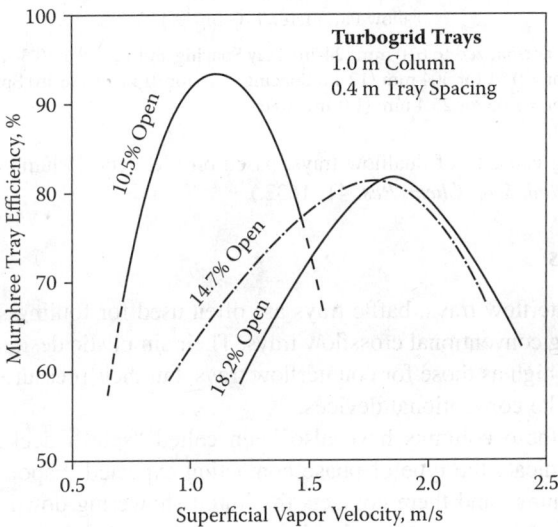

FIGURE 12.40 Efficiency of stamped turbogrid trays, methanol/water system. One-meter column, atmospheric pressure, total reflux. (Kastanek et al., 1969. *IChemE Symp. Ser. 32*, 5–100.)

Pressure drops of counterflow perforated trays are difficult to predict, because the open area used by the vapor varies with loading, and one must predict the fraction of opening for vapor at a given load point. Forms of the simple two-resistance equation are used:

$$h_t = h_d + h'_L \tag{12.61}$$

but with variable open area (vapor velocity through openings) and orifice coefficient.

The character of the two-phase mixture above the tray floor is different from that of a conventional crossflow tray because of the counterflow of the phases, and this makes it possible for a counterflow tray to achieve more than one theoretical stage. A paper by Garcia and Fair[47] provides approaches to modeling pressure drop and efficiency of dualflow trays, based on the FRI database.

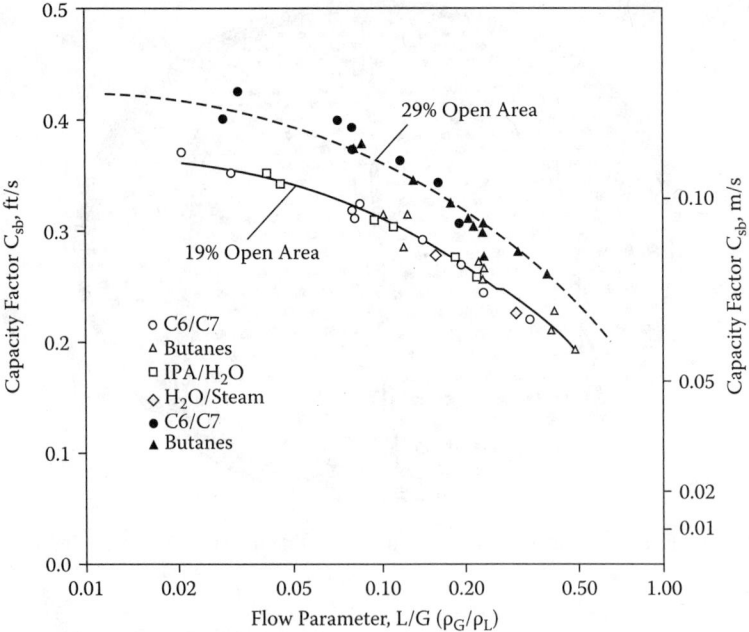

FIGURE 12.41 Flooding capacity of dualflow trays, based on FRI data. Column diameter = 1.2 m. (J. A. Garcia; J. R. Fair, 2002. *Ind. Eng. Chem. Res.* 41: 1632.)

12.6.3 Baffle Trays

As in the case of counterflow trays, baffle trays are often used for fouling services, e.g., handling slurries that would plug conventional crossflow trays. Their simplistic design does not lead to mass transfer efficiencies as high as those for counterflow trays, but their pressure drop and capacity can have advantages over the conventional devices.

The internals for these columns have also been called "splash decks" or "shower decks," descriptive terms to indicate the type of phase contacting expected. Vapor (or gas) flows upward through the baffle openings and there contacts the liquid showering down from one baffle to the next. Figure 12.42 shows a representative baffle tray column containing segmental baffles.

Geometric design variables for baffle tray columns are simple and few. The horizontal opening for gas flow is known as the *window area*. The vertical opening through which gas passes in a more horizontal direction is the *curtain area*. Much of the vapor-liquid contacting occurs within a zone bounded by the curtain area. Excessive constrictions in either area can cause higher pressure drop and flooding. Normally, it is desirable to make the two areas about equal in size.

A recent paper[50] provides a most complete analysis of design methodology for baffle columns. Figure 12.43 provides flooding prediction and uses the same parameters as the charts for crossflow trays (Figure 12.29) and counterflow perforated trays (Figure 12.41). The data points on Figure 12.43 include results from Fractionation Research, Inc. (FRI) for total reflux tests with the several systems indicated.[50a] These tests were run in a 1.2-m diameter column, and some variation in baffle geometry was employed. The term "cut" is the percentage of open area for phase flows.

Pressure drop through baffle columns is primarily a function of vapor flow rate, with about two velocity heads per opening being a good rule of thumb. For high liquid rates, this estimate must be increased. FRI tests indicate that the efficiency of a baffle column will be about half that for a

Distillation

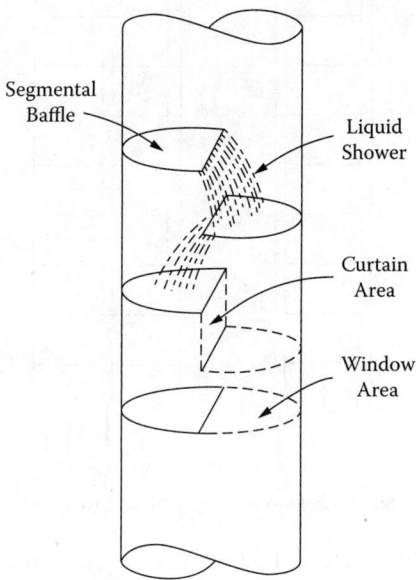

FIGURE 12.42 Section of a baffle tray column, containing side-to-side segmental baffles.

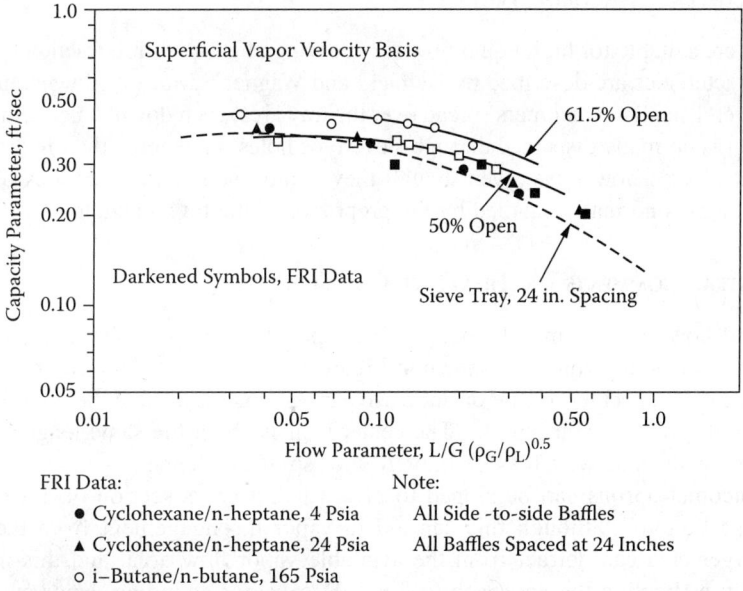

FRI Data:
- Cyclohexane/n-heptane, 4 Psia
- Cyclohexane/n-heptane, 24 Psia
- i–Butane/n-butane, 165 Psia

Note:
All Side-to-side Baffles
All Baffles Spaced at 24 Inches

FIGURE 12.43 Flooding capacity of baffle tray columns. [J. R. Fair, 1993. *Hydrocarbon Proc.* 72 (5): 75.]

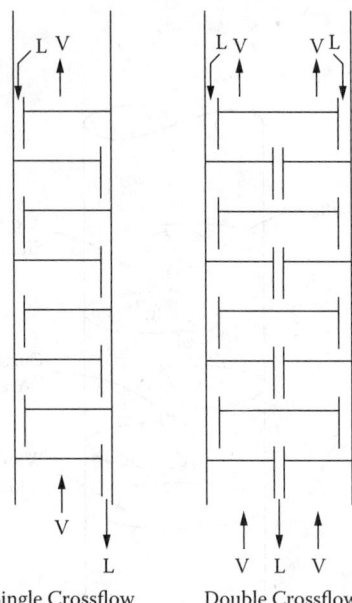

FIGURE 12.44 Multipass liquid flow arrangements for crossflow tray columns.

crossflow tray column under the same loading conditions. Hence, it may be better to estimate mass transfer rates from heat transfer rates using the mass transfer–heat transfer analogy. This is covered in some detail by Fair.[51]

12.6.4 Multiple Downcomer Trays

These devices are suitable for high liquid flow services (e.g., high-pressure fractionation), and their operating characteristics are described by Delnicki and Wagner.[52] Multiple downcomer trays (MD trays) have several small downcomers spread over the tray area, each downcomer being the hanging type. The vapor is normally dispersed through sieve-type holes. In general, their hydraulic behavior resembles that of crossflow trays, even though they exhibit some characteristics of counterflow trays. Their design is normally handled by the proprietor of the tray technology.

12.6.5 General Comments on Tray-Type Columns

For high liquid flow cases, it may be necessary to split the flow into two or more passes. This creates separate contacting zones, as shown in Figure 12.44. For the *two-pass tray* shown, liquid flows side to center and center to side on successive trays; this means that the effective weir crests differ because of differing weir lengths. The center weir is about the same length as the column diameter, whereas the side weir may be only 70% or so of the diameter.

The downcomer aprons can be sloped to give a larger cross section of downcomer at the top than at the bottom. Although this can aid in vapor disengagement from the downcomer liquid, the larger area can detract from the available vapor flow area, and this must be taken into account in estimating the approach to flood. Because of changing vapor and liquid flows throughout the column, hydraulic studies should include, as a minimum, conditions at top and bottom, and above and below the feed tray. This variation in flow is particularly pronounced in high-vacuum columns, where the vapor density can change twofold or more in a single column. With a feed stream that is partially vaporized, volume must be provided to allow adequate phase separation.

TABLE 12.8
Starting Dimensions for Crossflow Sieve Trays

	Vacuum	Atmospheric	Pressure
Tray spacing, in.	24	24	24
m	0.61	0.61	0.61
Downcomer area, % column	5	10	15
Active area, % column	90	80	70
Hole area, % active	12	10	8
Weir height, in	1	2	2
mm	25	50	50
Hole diameter, in	0.25	0.25	0.25
mm	6	6	6
Downcomer clearance, in	0.5	1.0	1.5
mm	12	25	38

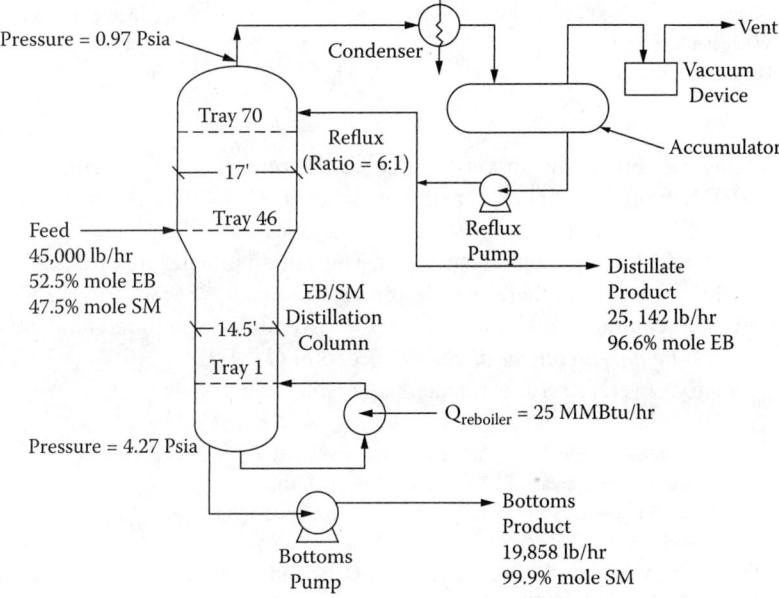

FIGURE 12.45 Flow diagram, ethylbenzene/styrene fractionation. For Problem 12.6. [J. C. Frank et al., 1969. *Chem. Eng. Prog.* 65(2): 79.]

As a starting point for crossflow tray design, the dimensions shown in Table 12.8 may be considered. While the table is designated for sieve trays, dimensions for valve trays are easily juxtaposed.

Problem 12.6

A large fractionator to separate ethylbenzene (EB) as overhead from styrene monomer (SM) plus small amounts of tar as bottoms is being designed along the lines of the flow diagram and material balance published by Frank et al.,[53] reproduced here as Figure 12.45. A point in the rectifying section is being considered, where the properties are as follows:

Temperature	172°F	78°C
Pressure	100 mm Hg abs.	13.4 kPa

Relative volatility, EB/SM = α = 1.40
Vapor density   0.030 lb/ft³   0.481 kg/m³
Liquid density   52.2 lb/ft³   837 kg/m³
Surface tension   25 dyn/cm   25 mN/m

The flow conditions are

Vapor   126,000 lb/hr = 1,200 lb-moles/hr = 1,167 ft³/s
           15.88 kg/s = 33.0 m³/s
Liquid   108,000 lb/hr = 1,028 lb-moles/hr = 0.57 ft³/s
           13.61 kg/s = 1.162 (10⁻²) m³/s

Develop dimensions of a crossflow sieve tray for these conditions, plus these specified parameters:

Tray spacing	24 in.	0.61 m
Weir height	1.0 in.	25.4 mm
Hole diameter	0.1875 in	4.8 mm
Tray metal thickness	0.078 in	2.0 mm
Downcomer area = 5% of total cross section	Hole area = 14% of active (bubbling) area	
Weir length	79.2 in.	2012 mm
Downcomer clearance	0.75 in.	19 mm

Solution:

Column Diameter: First, the abscissa value for Figure 12.29 is calculated: 13.61/15.88 $\sqrt{0.481/837}$ = 0.021. From that figure, for 0.61 m spacing, C_{sbf} = 0.37 ft/s = 0.11 m/s, based on net area, which is 95% of the total cross section. This gives a flooding vapor velocity (corrected for surface tension) = 4.91 m/s through the net area. Thus, minimum net area = 33.0/4.91 = 6.72 m².

Use 80% of flood for design; therefore, design vapor velocity = 0.8(4.91) = 3.93 m/s through net area. This gives a required net area = 6.72/0.80 = 8.40 m² and a required total cross section of 8.40/0.95 = 8.85 m². *Required column diameter = 3.36 m (11.0 ft).*

For this geometry, several useful parameters emerge:

Active area A_a = 7.98 m²   Active area vapor velocity U_a = 33.0/7.98 = 4.14 m/s
F-factor for active area F_{va} = 2.87 m/s(kg/m³)0.5
F-factor for hole area F_{vh} = 2.87/0.14 = 20.5 m/s(kg/m³)⁰·⁵

Pressure Drop: From Figure 12.33, for A_H/A_a = 0.14 and tray thickness/hole diameter of 2/4.8 = 0.42, the orifice coefficient is 0.76. Then, by Equation (12.5),

$$h_d = \frac{50.8}{(0.76)^2} \frac{0.481}{837} \left(\frac{33.0}{1.12}\right)^2 = 43.9 \text{ mm liquid}$$

By Equation (12.62), h_{ow} = 664 $L_w^{0.667}$, and L_w = 1.615(10⁻²)/(2.012)

$$\therefore h_{ow} = 26.6 \text{ mm}$$

From Equation (12.61), β = 0.47

$$\therefore h_t = h_d + \beta(h_w + h_{ow}) = 43.9 + 0.47(25.4 + 26.6) = 68.3 \text{ mm total drop/tray}$$

from which h_L = 24.4 mm liquid

Distillation

Downcomer Backup
By Equation (12.66), $h_{da} = 165\,(0.0162/0.0382)^2 = 29.5$ mm
By Equation (12.65), $h_{dc} = h_t + h_L + h_{da} = 68.3 + 14.6 + 29.5 = 122.2$ mm
(This is well below the tray spacing of 610 mm, $\therefore$ no backup problem.)

Lower operating limit
At 50% of design,
By Equation (12.69), $h_\sigma = 409\,25\,(837(4.76)) = 2.57$ mm
And $h_d + h_\sigma = 11.0 + 2.57$ mm. From Figure 12.35, this appears to be close to the lower limit. Accordingly, can operate down to 50% of design rate without serious weeping or dumping.

12.7 PACKED COLUMN HYDRAULICS

The packed column used for distillation separations is a vertical, cylindrical vessel containing a bed of packing elements, suitably supported and irrigated by liquid flowing through a special distributor. The height of the bed is determined by the mass transfer capability of the packing; limitations are governed by allowable pressure drop and overall column height. The overall arrangement of a column is shown in Figure 12.46.

Packings for distillation columns come in many types, shapes, and sizes. So-called *random packings* (or "dumped packings") are relatively small pieces, or elements, that are placed in the column (usually by pouring) in a random arrangement to produce a *packed bed*. The counterparts

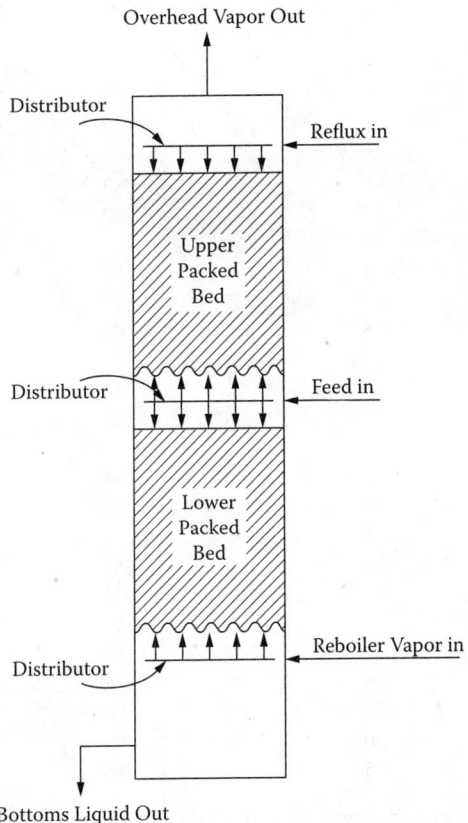

FIGURE 12.46 Overall arrangement of a simple packed distillation column.

of random packings are *ordered* or *structured packings*, which are relatively large, fabricated to close dimensional tolerances, and carefully positioned in the column. Random packings and ordered packings are usually competitive for a given problem, and the designer must know the characteristics of each, as discussed separately below.

12.7.1 Packing Characteristics

Representative random packings were shown in Figure 12.25; properties of important packings of the class are given in Table 12.9. The first three packings listed are the "through-flow" type in which the phases can flow freely through the elements with reduced form drag. The others listed are of the "around flow" type, with increased form drag and therefore higher pressure drop.

TABLE 12.9
Characteristics of Random Packings

Name	Material	Nominal Size, mm	Surface Area m²/m³	% Voids	Packing Factor, m⁻¹	Vendor
Intalox saddles*	M	(No. 25)		97	135	Koch-Glitsch, LP
"IMTP"		(No. 40)		97	24	Wichita, KN
		(No. 50)		98	59	
Pall rings*	M	16	—	92	266	(Generic)
		25	205	94	184	
		38	130	95	131	
		50	115	96	88	
		90	92	97	59	
Pall rings*	P	16	340	87	310	(Generic)
		25	205	90	180	
		50	130	91	131	
		90	100	92	85	
Berl saddles	C	13	465	62	790	(Generic)
		25	250	68	360	
		38	150	71	215	
Intalox saddles	C	13	625	78	660	Koch-Glitsch, LP
		25	255	77	197	Wichita, KN
		50	118	79	98	
		75	92	80	70	
Intalox saddles	P	25	206	91	131	Koch-Glitsch, LP
		50	108	93	92	Wichita, KN
		75	88	94	59	
Raschig rings	C	13	370	64	1900	(Generic)
		25	190	74	587	
		38	120	68	305	
		50	92	74	213	
		75	62	75	121	
Raschig rings	M	19	245	80	730	(Generic)
		25	185	86	470	
		38	130	90	270	
		50	95	92	187	
		75	66	95	105	

Note: Materials of construction: C = ceramic; M = metal; P = plastic (usually polypropylene).

* Through-flow packing.

TABLE 12.10
Characteristics of Structured Packings

Name	Material	Nominal Size, mm	Surface Area m²/m³	% Voids	Packing Factor, m⁻¹	Vendor
Flexipac	S	1	558	91	108	Koch-Glitsch
		2	223	93	72	Wichita, Kansas USA
		3	134	96	52	
Intalox	S	1T	315	95	66	Koch-Glitsch
		2T	213	97	56	Wichita, Kansas USA
Max-Pak	S		229	95	39	Jaeger Products
						Houston, Texas USA
Mellapak	S	125Y	125	97	33	Sulzer Chemtech
		125X*	125	97		Tulsa, Oklahoma USA;
		250Y	250	95	66	Winterthur, Switzerland
		250X*	250	95		
		350Y	350	93	75	
		500X*	500	91	25	
Sulzer	G	AX*	250	95		Sulzer Chemtech
		BX*	492	90	69	
		CY	700	85		
Montz-Pak	S	B1-125	125	97		Julius Montz
		B1-125.6*	125	97	72	Hilden, Germany
		B1-250	250	95		
		B1-250.6*	250	95		
		B1-350	350	93		
	G	A3-500	500	91		
	E	BSH-250	250	95		
		BSH-500	500	91		
Ralupak	S	250YC*	250	95		Raschig AG
						Ludwigshafen, Germany

Note: Material of construction: C = ceramic; E = expanded metal; G = metal gauze; S = sheet metal.

* 60-degree corrugation angle (with the horizontal); all others 45 degrees.

Source: Packing factors from Kister & Gill (*Chem. Eng. Prog.* 87 (2) 32 (1991), and Houston AIChE Meeting, March 19-23, 1995.

Random packings are fabricated from a wide variety of materials. Originally, most were made of ceramic and were popular for corrosive services despite their friability. Perhaps the most popular material of construction today is polypropylene, which gives a bed of light weight, reasonable resistance to corrosion by many fluids, and low cost. Plastic random packings are frequently used in air stripping columns where dissolved volatile organic compounds are removed from groundwaters.

A representative structured packing was shown in Figure 12.25. Packings of this type are designed specifically to do three things: minimize form drag (pressure drop), provide considerable active surface area for mass transfer, and maintain uniform liquid distribution so that most of the surface area is active for mass transfer.

Structured packings are generally more expensive than random packings on a volumetric basis but provide better efficiencies and significantly lower pressure drops. Table 12.10 shows typical characteristics of structured packings. Figure 12.47 indicates the key advantages of low pressure drop per theoretical stage for packed columns as compared to a sieve tray.

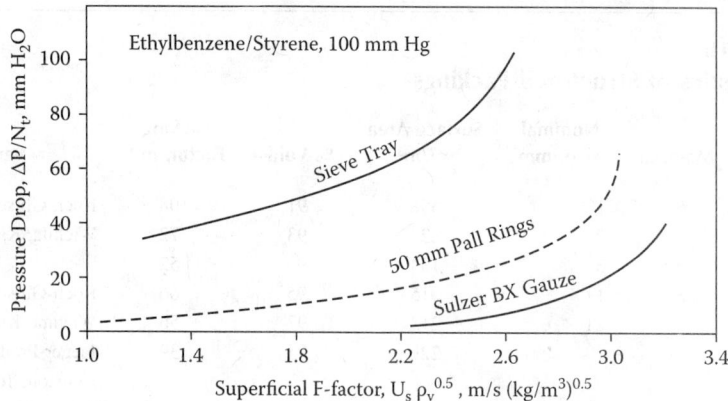

FIGURE 12.47 Comparison of devices for vacuum service, pressure drop/theoretical stage.

Most structured packings are of the corrugated sheet type. In these packings, crimped, corrugated sheets of metal or metal gauze are positioned to create flow channels for the gas and promote distribution. Some details of their fabrication and assembly are shown in Figure 12.48. (Less common types of ordered packings, involving meshes and grids, are not discussed here.) Structured packings can also be made from a variety of materials; the choice of material depends not only on corrosion resistance and cost but also on the wettability of the surface by the liquid being used. For example, water does not wet clean metals well but spreads almost completely on many ceramic surfaces. Aspects of packing wetting will be discussed below, in the "Mass Transfer Efficiency" section.

As in the case of trays, the important hydraulic considerations for packings are pressure drop and flooding. To a lesser extent, liquid holdup is also important. The hydraulics affect the performance profile for a given separation in a specific packing. A typical profile for efficiency and pressure drop is shown in Figure 12.49.[54] The equivalence to the tray profile is evident (see Figure 12.26). The ordinate is given in theoretical stages per unit of bed height, but a more common description of efficiency is the inverse of this, the *height equivalent to a theoretical plate (HETP)*. A generalized profile for a packed column is shown in Figure 12.50, which shows the importance of having uniform liquid distribution to the bed. (The distributor for the tests shown in Figure 12.49 was of exceptionally high quality.)

Some random packings correct a poor initial distribution of liquid, whereas others (especially the high-void, low-form-drag packings) must always be fed a uniformly distributed liquid at the top of the bed if they are to perform efficiently. Still other packings exhibit a deterioration of a good distribution with bed depth, making the use of *redistributors* mandatory at depths beyond around 3 to 6 m.

Liquid and vapor flow countercurrently through openings between and within packing elements. At low vapor rates, there is relatively little disturbance of liquid by the vapor, and mass transfer proceeds in a fashion similar to that in a wetted wall column. At higher rates, there is considerable interaction between the phases, with vapor flow causing increases in liquid turbulence and holdup. In the so-called "loading zone," there is an enhancement of mass transfer but, as rates are increased further, flooding occurs.

The loading/flooding phenomena are best understood through a study of pressure drop curves such as those shown schematically in Figure 12.51. For a dry packing (with no liquid irrigation), pressure drop follows a traditional friction factor, orifice-type relationship. When there is irrigation at a constant liquid rate, increasing the gas rate leads to a change in the slope of the pressure drop curve, indicating the onset of loading. Further gas rate increase leads to a rapid and continuing increase in pressure drop, indicating flooding. Flooding can also be reached by increasing the liquid rate at

Distillation

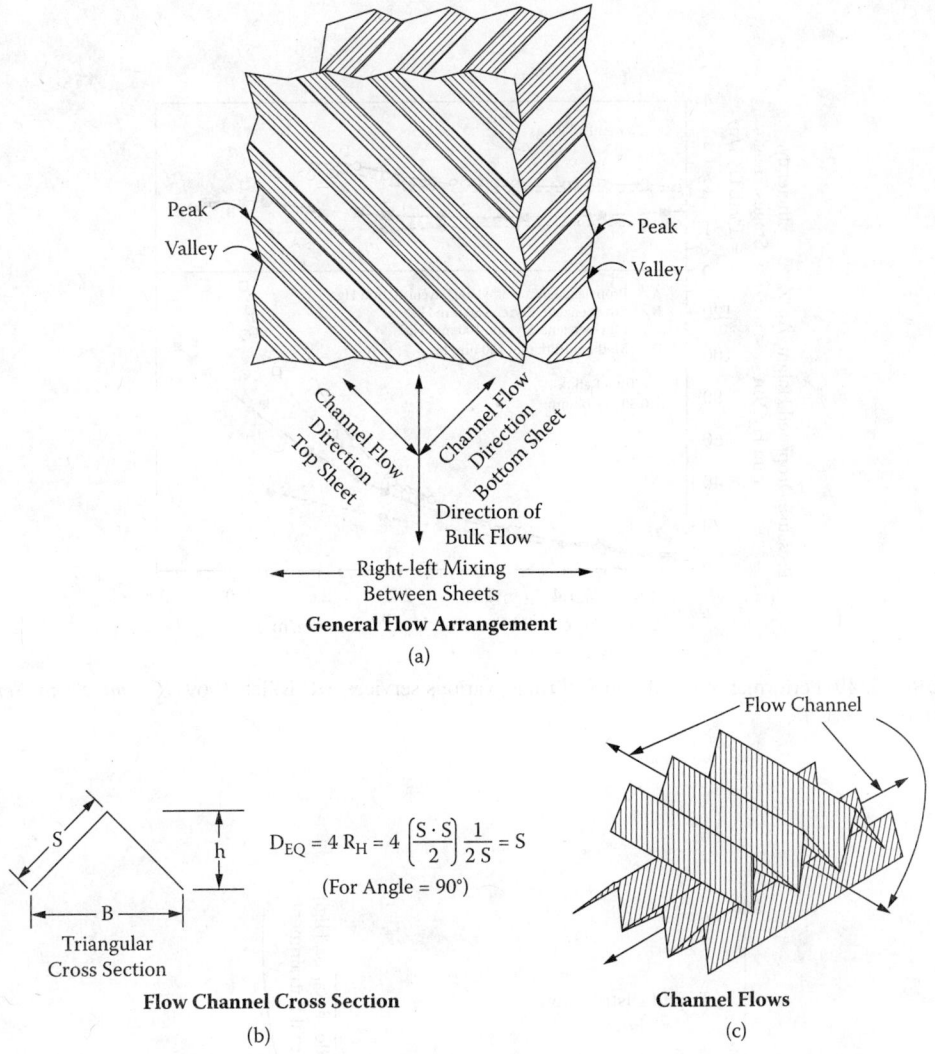

FIGURE 12.48 Dimension details for structured packing analysis.

constant gas rate. For distillation applications, both liquid and vapor rates are increased to approach the flood point. Vendors provide pressure drop curves for each of the packings listed in Tables 12.9 and 12.10, but the experiments conducted to develop such curves are usually done with the air-water system. For distillation at total reflux, curves such as those in Figure 12.49 are more realistic.

Regions below the loading point as well as the loading region itself are apparent from liquid holdup and interfacial area data as shown in Figure 12.52.[55] At low rates, there is little influence of vapor rate on holdup and interfacial area. This behavior is typical of random as well as structured packings. In the latter, the effects of the rates on interfacial area seem to be less pronounced than for random packings. The appearance of the flood point in the holdup and area plots will undoubtedly match the flood point in the pressure drop.

12.7.1.1 Maximum Vapor–Liquid Capacity

The hydraulic flood point, as illustrated in Figure 12.50, is reached at a combination of countercurrent gas and liquid rates such that the pressure drop exhibits a relatively large increase with a

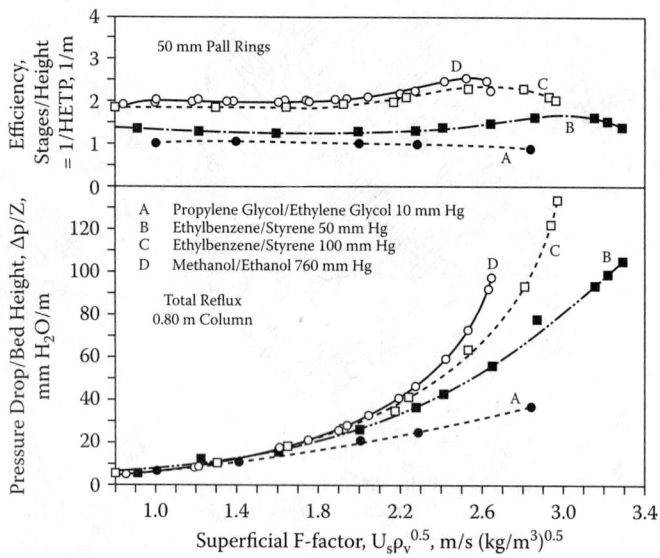

FIGURE 12.49 Performance of 50-mm Pall rings, various services. (R. Billet, 1969. *IChemE Symp. Ser. 32*, 5: 111.)

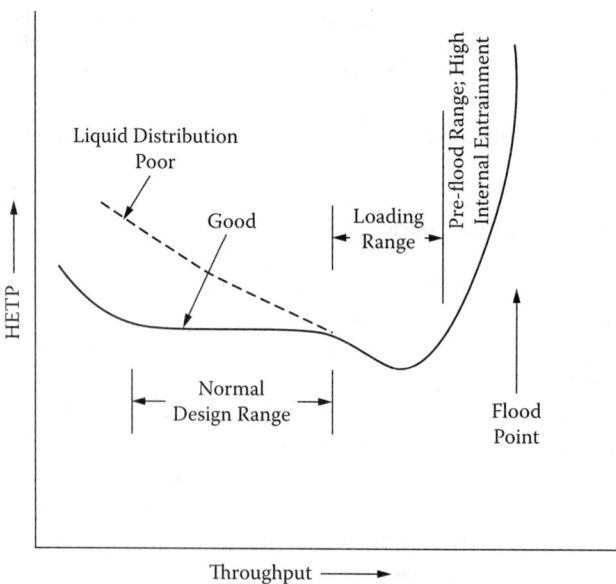

FIGURE 12.50 Generalized efficiency profile for a packed column.

Distillation

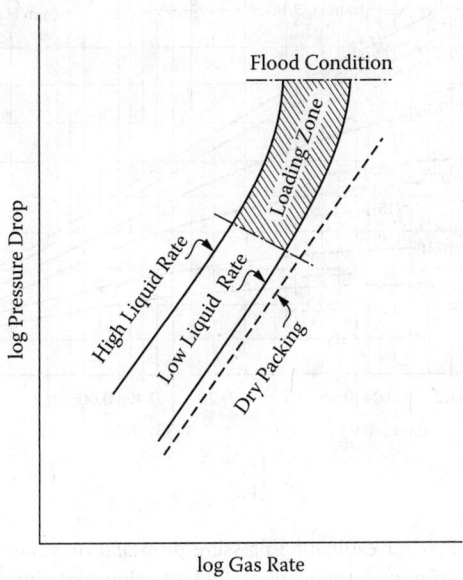

FIGURE 12.51 Pressure drop and loading characteristics of packed columns.

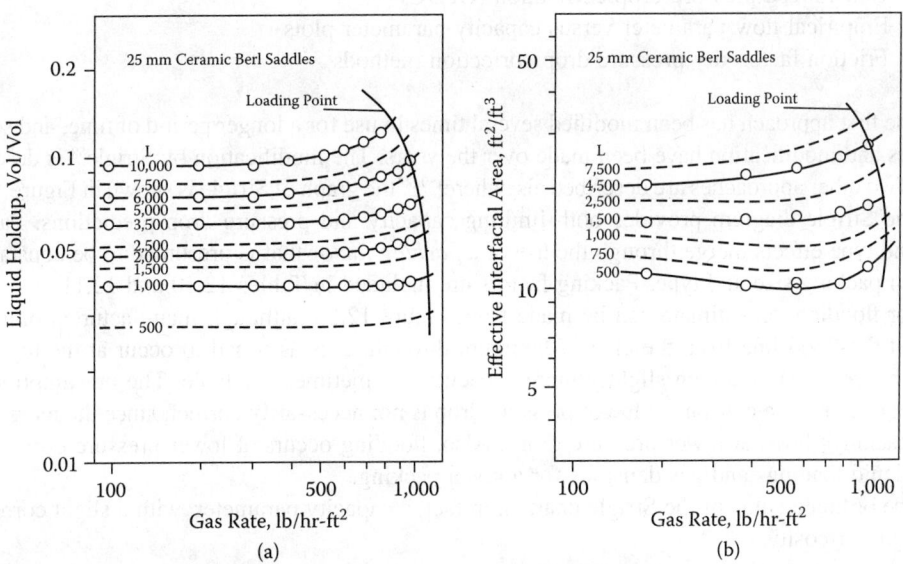

FIGURE 12.52 Liquid holdup and interfacial area for 25-mm Berl Saddles.

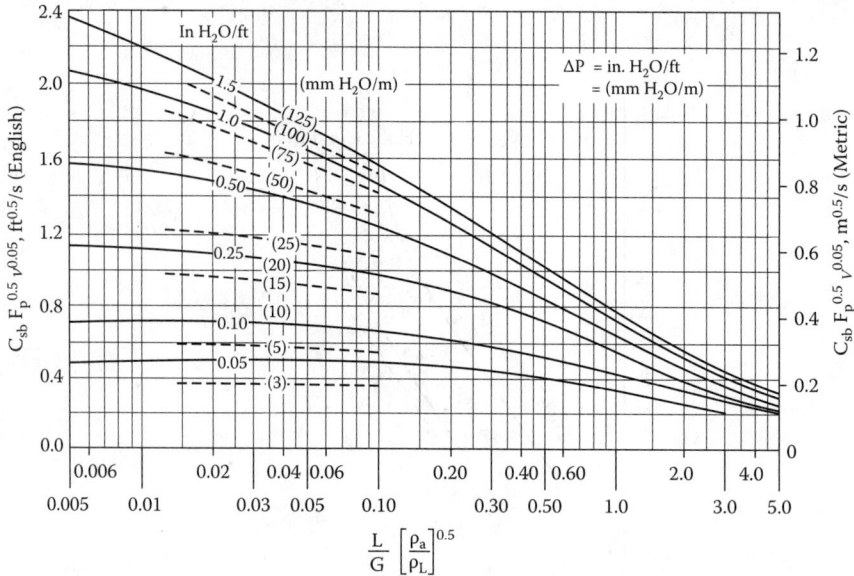

FIGURE 12.53 Generalized chart for estimating pressure drop and maximum capacity of packings. (R. F. Strigle, 1994. *Packed Tower Design and Applications*, 2nd ed., Houston: Gulf Publ. Co.)

small increase in flow rate of either phase. Methods for estimating maximum hydraulic capacity for a packing provide a basis for design (optimum approach to flood). Such methods employ one of three separate approaches:

- Generalized pressure drop correlation (GPDC)[56,57]
- Empirical flow parameter versus capacity parameter plots
- Friction factor–dry pressure drop correction methods

The first approach has been modified several times in use for a longer period of time, and several updates and modification have been made over the years. The modification by Strigle[58] is described here. Two other approaches are described elsewhere;[59,60] the graph of Strigle is shown in Figure 12.53.

The Strigle diagram provides both limiting capacity and pressure drop predictions. Packing size and type effects merge through the use of a *packing factor* that is presumed to be constant for a given packing size and type. Packing factors are included in Tables 12.10 and 12.11.

For flooding, an estimate can be made from Figure 12.53, although manufacturers often like to omit the flood line from the chart. Maximum throughput is assumed to occur at the top curve (125 mm H_2O/m), although slightly more capacity is sometimes available. The presumption that flooding occurs at a constant value of pressure drop is not necessarily correct, since the more open-type packings flood at lower pressure drops. Also, flooding occurs at lower pressure drops under high liquid loadings and gas densities for a given packing.

The ordinate value of the Strigle chart is, in fact, a capacity parameter, with a slight correction for liquid viscosity,

$$\text{ordinate} = C_s F_p^{0.5} \nu^{0.05} \text{ which has units of ft}^{0.5}/\text{sec or m}^{0.5}/\text{s}$$

where

$C_s = U_s [\rho_v/(\rho_L - \rho_V)]^{0.5}$, ft/sec or m/s
ν = liquid kinematic viscosity = centistokes (centipoises/density, g/ml)
F_p = packing factor, 1/ft or 1/m (Tables 12.9 and 12.10)

Distillation

For a given packing type and size (F_p = constant), and for the kinematic viscosity term close to unity (usually the case), the ordinate is obviously closely related to the Souders–Brown capacity parameter used for trays.

The abscissa value is the flow parameter, also used for trays:

$$FP = \frac{L}{G}\sqrt{\frac{\rho_v}{\rho_L}} \qquad (12.62)$$

The F-factor $F_s = U_s \rho_v^{0.5}$ ∴ $F_s \sim C_{sb}/\rho_L^{0.5}$. Thus, the ordinate is closely related to the capacity parameter used for trays, with a correction for packing type and size, and a minor adjustment for liquid viscosity. (Unlike trays, however, there is no correction for surface tension.)

12.7.1.2 Pressure Drop

Predictions of flooding and pressure drop are discussed by Kister,[60] who has developed charts similar to Figure 12.53 (Strigle chart) for individual packing types and sizes. Kister suggests that the most fundamental approach to estimating pressure drop and flooding has been provided by Stichlmair et al.[61] However, this approach requires two constants for each packing type and size, and sets of constants have been developed for relatively few packings.

For the usual design case, Figure 12.53 is recommended. When the estimate is critical and requires the best possible reliability [e.g., for high-vacuum distillation or high-throughput scrubbing (to minimize blower horsepower)], the method of Stichlmair et al.[61] should be used for random packings. A second choice would be the individual packing charts of Kister.[60] For structured packing, the method of Rocha et al.[62] is recommended, which takes into account surface wettability and texturing, liquid holdup, geometric details, and fluid properties. The model is complex and requires a computer program but is supported by considerable commercial test data, mostly under distillation conditions. The input data for a particular packing are straightforward. The Separations Research Program at the University of Texas at Austin can provide more information.

Problem 12.7

Performance data for 50-mm metal Pall rings in ethylbenzene/styrene (EB/SM) separation service are shown in Figure 12.45. For the conditions given in the figure, estimate (a) the flood point and (b) the pressure drop for a loading of F-factor = 2.4 m/s(kg/m³)$^{0.5}$. Compare the results with the values given in Figure 12.45.

Relevant properties at a selected point in the column are

$$\rho_L = 837 \text{ kg/m}^3 \qquad M_{EB} = 106$$
$$\rho_v = 0.480 \text{ kg/m}^3 \qquad M_{SM} = 104$$
$$\mu_L = 0.38 \text{ cP} \qquad \nu = 0.45 \text{ cSt}$$
$$\sigma = 25 \text{ dynes/cm}$$

At total reflux, the flow parameter is $FP = L/G[\rho_v/\rho_L]^{0.5} = 1.0[0.480/837]^{0.5} = 0.021$.

For 50m-mm metal Pall rings, $F_p = 88$ m^{-1} (Table 12.9).

From Figure 12.53, for FP = 0.021,

$$C_{sb}F_p\nu^{0.05} = 1.10 \text{ m}^{0.5}/\text{s} \text{ (highest } \Delta P \sim \text{flood limit)}$$

from which $C_{sb} = 0.122$ m/s, and $U_s = 16.7$ ft/s = 5.10 m/s flood superficial velocity.

Converting to F-factor, to make the comparison,

$$F_s = U_s \rho_v^{0.5} = 5.10\,(0.480)^{0.5} = 3.50$$

(a) Flood F-factor = 3.50 m/s (kg/m³)$^{0.5}$
Note that performance curve C of Figure 12.49 indicates flooding at F ~ 3.0 m/s(kg/m³)$^{0.5}$.
(b) For pressure drop at $F = 2.4$, convert to $C_{sb} = F_s/\rho_L^{0.5} = 2.4/(837)^{0.5} = 0.0830$.
$C_{sb} = 0.0830$ m/s. Then, ord. of Figure 12.53 = 0.0830 (88)$^{0.5}$(0.46)$^{0.05}$.
ord =0.748 m0.5/s for which, at $FP = 0.021$, $\Delta P = 40$ mm H$_2$O/m.
The experimental value was about 50 mm H$_2$O/m.

12.7.1.3 Liquid Distribution

Good liquid distribution is essential to the effective performance of packed columns. While there are no quantitative measures for "good distribution," experience leads to the following guidelines: for dumped packings, the ratio of column diameter to packing element diameter should be 8 or greater, and the liquid distributor should produce at least 10 inlet streams, evenly distributed, per square foot (100 per square meter) of cross section. The "around flow" bluff body packings (e.g., Raschig rings) may require fewer inlet liquid streams.

Problem 12.8
Problem 12.6 deals with the dimensioning of a large ethylbenzene/styrene tray-type fractionator. Because high pressure drop in this separation leads to increased tar production, a packing (instead of trays) is to be considered. The conditions are the same as for Problem 12.6. For 80% flood, determine the required column diameter and associated pressure drop for a 50-mm Pall ring random packing.
Solution. For both capacity and pressure drop, Figure 12.53 will be used. For the mixture, the absolute viscosity is 0.38 cP, giving a kinematic viscosity of 0.45 cSt. From Table 12.8, the packing factor F_p for 50-mm metal Pall rings is 88 m^{-1} or 26.8 ft^{-1}. The abscissa for the figure (the flow parameter) is the same as before: $FP = 0.021$. For Figure 12.53, the ordinate term is

$$C_{sbf}F_p^{0.5}v^{0.02} = C_{sbf}(5.18)(0.45^{0.02}) = 5.10\,C_{sbf}\text{ ft/s}$$

At the highest curve of Figure 12.53 (assumed equivalent to flooding), the ordinate value is 2.05, from which $C_{sbf} = 0.40$ ft/s. At 80%, the ordinate value is 2.05(0.80) = 1.64, and the pressure drop is 55 mm H$_2$O/m (0.66 in H$_2$O/ft) or 66 mm liquid/m.
The column diameter is based on $C_{sb} = 0.80$, $C_{sbf} = 0.32$ ft/s = 0.98 m/s. For a vapor flow of 33.0 m³/s, the required diameter is 3.21 m or 10.5 ft.
Comparison with Trays. The pressure drop for a tray with 0.61-m spacing was found to be 58.4 mm liquid. The comparative packing pressure drop is 66(0.61) = 40.3 mm liquid. The diameter for the packed column is 3.21 m vs. 3.36 m for the trays. If a lower approach to flood for the packing is used, to make the column diameter 3.36 m, the packing pressure drop becomes less than half that of the trays.

Once the liquid has been introduced to the top of the bed, the question arises as to whether the initial distribution is maintained throughout the bed. Silvey and Keller[63] used internal bed liquid samplers and total reflux, and found that bluff body packings such as ceramic Raschig rings correct an inadequate initial liquid distribution and also maintain a good distribution throughout the bed. But the "through-flow" random packings and the structured packings cannot correct an initial problem. Potthoff[64] elucidated in-bed distribution effects, and representative plots of his findings are shown in Figure 12.54. The difference between a single-point distributor (the worst possible case) and a multipoint distributor is dramatic. The effect of bed height on efficiency is discussed in the next section.

In any event, the optimum number of pour points per unit of tower cross section depends on the type and size of packing selected, and also on the practical limits of distributor design and

Distillation

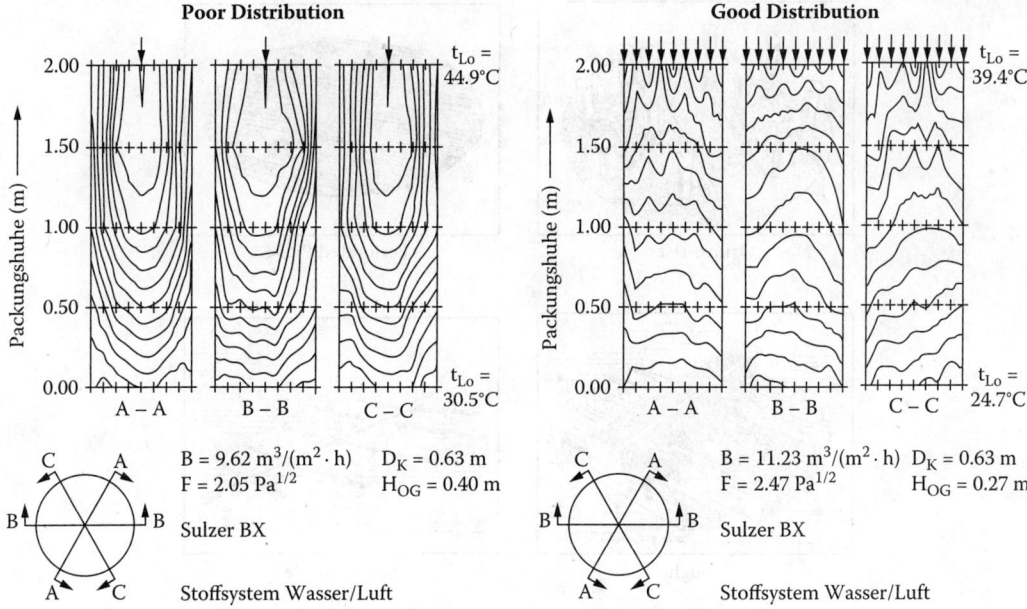

FIGURE 12.54 Constant composition profiles in a bed of structured packing. Left, single-point distributor; right, multipoint distributor. Packing: Ralu-Pak 250YC. Column diameter: 0.63 m. Air-water. B = liquid rate, m³/m²-hr H_{og} = height of an overall transfer unit, m, F = gas rate, m/s (kg/m³)$^{0.5}$. (R. Potthoff, 1992. Ph. D. dissertation, Univ. of Essen, Germany.)

operation. Several workers have found that it is not practical or even necessary to design for more than 10 streams per square foot (100 per square meter).

12.7.1.5 Liquid Distributors

For single- or multiple-bed columns, a distributor is needed for each bed. A center-fed packed distillation column, for example, requires at least two beds. In some cases, the cost of the distributor(s) can approach that of the packing. One can select from several different types of liquid distributors, including trough, orifice/riser, perforated pipe, and spray nozzle. These are illustrated in Figure 12.55 and described below.

The *trough distributor* provides good distribution under widely varying flow rates of gas and liquid. The liquid may flow through simple V-notch weirs or through tubes that extend from the troughs to near the upper level of the packing. Some deposition of solids can be accommodated. Some phase contacting occurs during liquid distribution, providing incremental mass transfer but also introducing possible entrainment problems. The trough is a general-purpose unit that can be adapted to special distribution requirements at relatively low cost.

The *orifice/riser distributor* provides numerous uniformly distributed pour points. The vapor risers must be designed for variations in flow rate, often with a minimum of pressure drop. Typically, risers occupy 20% of the column cross section; if the percentage is much lower, there is danger of liquid backing up into the risers. More exact allocation of riser area can be obtained from the simple relationship

$$h_L = h_{holes} + h_{risers} = 2.42\, u_{Lh} + 5.19\, (\rho_g/\rho_L)U_r^2 \qquad (12.63)$$

where

h_L = liquid height in distributor (around the risers), cm

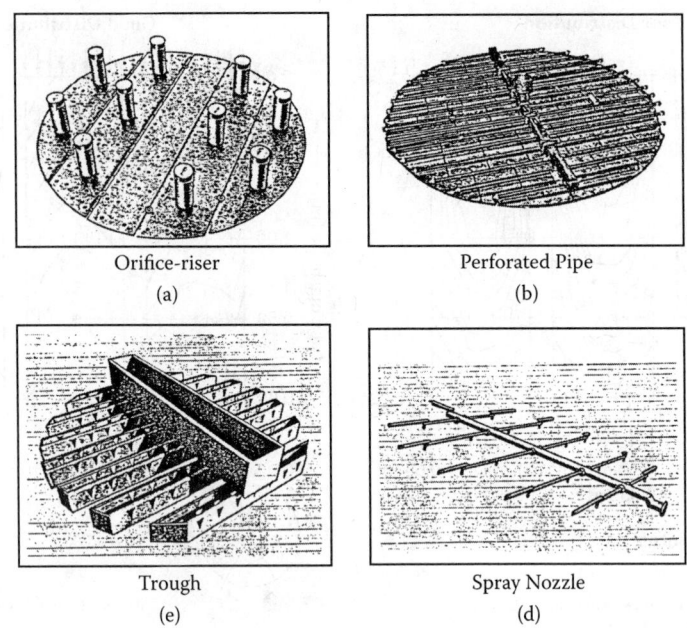

FIGURE 12.55 Examples of liquid distributors.

u_{Lh} = liquid velocity through the distributor holes, m/s
U_r = vapor velocity through the risers, m/s

Thus, liquid backup can be attributed to both orifice area and riser area.

For distribution-sensitive packings, some pour points are needed near the wall (to within 2 cm). For larger-diameter columns, and for low liquid rates, the distributor must be level within ±5 mm for a 3-m diameter column. In addition, the risers must be high enough to accommodate backup for high liquid and vapor flow rates.

The *drip tube/riser distributor* is a variation of the orifice/riser device, with tubes added to the orifices to ensure close contact with the packing (not shown in Figure 12.55). When it is necessary that all tubes discharge evenly, a double tubesheet affair can be used with liquid under pressure; pressure drop through the orifices and tubes must be estimated carefully for this arrangement.

The *perforated pipe distributor* has a central feed sump with pipes that branch out to provide the liquid discharge. The level in the sump varies with liquid total flow rate, and the size of the lateral pipes and their orifices must be dimensioned carefully to ensure that the ends of the pipes are not starved for liquid. The orifice sizes are typically 0.125 to 0.25 in. (3 to 6 mm) in diameter and can plug if foreign matter is present. The pipes must be leveled carefully, especially in large-diameter columns.

The *spray nozzle* is not widely used. If more than one nozzle is used, it is difficult to obtain a uniform spray pattern because of overlap and underlap of the patterns. Also, liquid entrainment from the sprays is a problem. Spray distributors are sometimes used in petroleum refinery vacuum columns.[65] The full cone nozzle is normally used, singly or in banks.

12.7.1.6 Liquid Redistribution

For bed heights greater than 6 to 8 m, the total bed is subdivided. This is because of natural deterioration of mass transfer efficiency with height. Since each bed must have good initial distribution, it is necessary to collect the liquid flowing from a bed (which may have become maldis-

Distillation

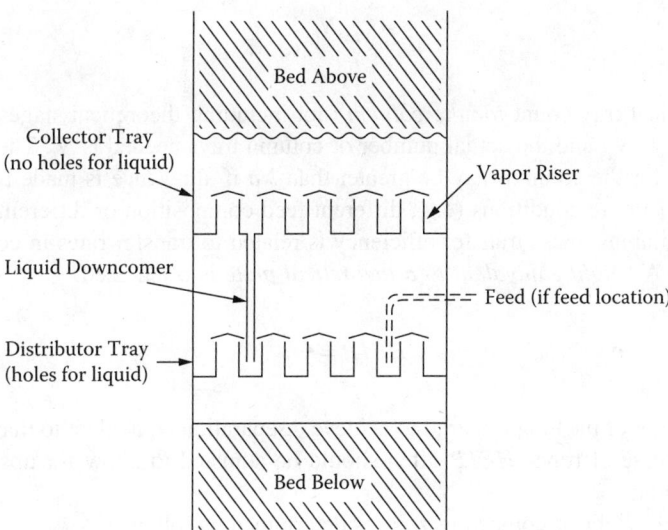

FIGURE 12.56 Collection/redistribution arrangement in a packed column.

tributed) and pass it through a liquid distributor for the bed below. The concept is shown in Figure 12.56. Packing vendors have different recommendations on maximum allowable bed height.

12.7.1.7 Vapor Distribution

While the emphasis is usually on promoting good liquid distribution to the top of a packed bed, poor vapor distribution also leads to poor liquid distribution. The assumption is often made that if there is not good initial vapor distribution, pressure drop in the bed will correct it. This assumption is good for higher pressure drop packings, but for the through-flow packings, an initially poor distribution often persists throughout the bed.

Reasonable care should be used to disperse a side-entering gas (e.g., vapor/liquid return from a reboiler) across the tower cross section. For extremely critical cases, one or two crossflow trays may be used to improve distribution—but they cause increased pressure drop.

The kinetic energy of the gas or vapor entering the tower and the position of the entrance are both critical. Packing vendors can provide guidelines; in some cases, multipoint gas introduction into a tower will be required. Also, the vertical distance between the point of injection and the bottom of the bed is important; for small columns, this distance should be at least one diameter, and at least 0.5 m for larger columns. Studies of vapor distribution in packed beds have been reported.[66]

12.7.1.8 Turndown

An attractive feature of packed beds for mass and heat transfer operations is that the bed itself offers better turndown characteristics than the equivalent tray column. The mass transfer efficiency of a packed column with a good distributor is about as good at both low and high loadings. This favorable comparison is evident from the data in Figures 12.26 and 12.49 (same column and test mixture). The liquid distribution is the key, as indicated in Figure 12.50.

12.8 MASS TRANSFER EFFICIENCY

Equipment design, as discussed in the preceding section, requires preliminary evaluation of column efficiency. There is a close coupling between geometry and efficiency. Several different efficiency terms will be used for tray columns, but the one ultimately needed for design is *overall column efficiency:*

$$E_{oc} = N_t / N_a \qquad (12.64)$$

where the theoretical tray count *for the column* (not including theoretical stages for reboiler and partial condenser) is N_a, and the actual number of column trays needed is N_t. The number of trays actually specified for the column may be greater than Na if allowance is made for column upsets or for alternate operating conditions (e.g., different feed composition or different separation).

For packed columns, mass transfer efficiency is related to transfer rates in counterflow vapor-liquid contacting. A *height equivalent to a theoretical plate* is often used:

$$HETP = Z / N_t \qquad (12.65)$$

where Z is the height of packing required to achieve a separation equivalent to that of N_t theoretical stages. As in the case of trays, *HETP* often should be adjusted to allow for upsets or alternative separating conditions.

Thus, the total height of contacting zone is determined as follows:

$$\text{Tray columns: } Z = \frac{N_t}{E_{oc}} TS \quad (TS = \textit{tray spacing}) \qquad (12.66)$$

$$\text{Packed columns: } Z = N_t\, HETP \text{ or } Z = N_{og} H_{og} \qquad (12.67)$$

12.8.1 General Mass Transfer Relationships

Mass transfer relationships in a distillation column are based on the basic interphase transfer model for a differential slice of the cross section, as shown in Figure 12.57. The slice is taken from the packed bed or from the tray froth. For component i,

$$L dx_i + x_i dL = V dy_i + y_i dV \qquad (12.68)$$

For equimolar counterdiffusion (approximately the case for distillation), $dV = dL$, and for net transfer of i from bulk vapor to bulk liquid,

$$V dy_i = K_g a_i (y_i - y_i^*) dZ \qquad (12.69)$$

where y_i is the mole fraction of i in the vapor, and y_i^* is the mole fraction of i in the vapor that would be in equilibrium with bulk liquid composition x_i. Note that y_i^* is not a real composition but rather a calculated term from the equilibrium expression $y_i^* = K_i x_i$.

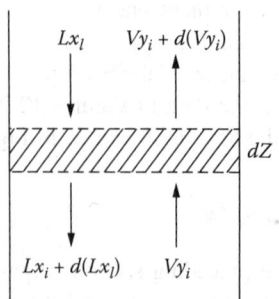

FIGURE 12.57 Flows in a differential section of a countercurrent contactor.

Distillation

Equation (12.69) may be rearranged and the *overall transfer unit* defined as follows:

$$\frac{dy_i}{y_i - y_i^*} = \frac{K_{og} a_i dZ}{V} = \frac{1}{H_{og}} dZ \qquad (12.70)$$

Thus, the overall transfer unit $H_{og} = V/(K_{og}a_i)$ is defined, where a_i is the effective interfacial area on a volumetric basis, e.g., $a_i = $ m^2/m^3 transfer volume. [The symbol H_{og} is often reported as $(HTU)_{og}$]. Integration of Equation (12.70) yields a convenient design relationship:

$$\int_1^2 \frac{K_{og} a_i dZ}{V} = \int_1^2 \frac{1}{H_{og}} = \int_1^2 \frac{1}{y_i - y_i^*} = N_{og} \qquad (12.71)$$

where N_{og} is the *number of overall transfer units* based on vapor concentrations. Equation (12.71) is normally used in connection with counterflow packed columns [Equation (12.67)] but is also the basis for mass transfer calculations for tray columns. N_{og} is related to theoretical stages, and H_{og} is an efficiency term based on the intensity of phase contacting, in turn related to the geometry of the contacting device, the flow rates of the phases, and the diffusional properties in the vapor-liquid dispersion. For more details see Hines and Maddox.[67]

12.8.2 Tray Column Efficiency

The mass transfer efficiency of a tray is a crucial parameter in the design or analysis of a distillation column (as well as absorption and stripping columns). The actual trays (plates) are determined from theoretical stages using overall column efficiency as indicated by Equation (12.64) or (12.66). A more basic parameter is the *point efficiency, E_{og}*. This is the efficiency at a single point on the tray, and for uniform tray concentrations (liquid and vapor well mixed) equals the overall tray efficiency E_{MV} (often called the "Murphree efficiency"). When all the trays in the column are taken into account, the average of the E_{MV} values becomes the overall column efficiency E_{oc}, which can sometimes equal overall column efficiency E_{oc}, based on mass transfer fundamentals, with conversion to E_{oc} being largely a matter of geometry and hydraulics. It is not uncommon for E_{oc} to be 20% to 40% greater than E_{og} in larger columns. For smaller columns, $E_{oc} \to E_{og}$.

12.8.2.1 Regimes on Trays

The manner by which liquid and vapor contact each other on a tray determines the vapor-liquid interfacial area for mass transfer. The character of the vapor-liquid dispersion is determined by the relative flow rates and properties of the phases. These dispersions are conveniently described in terms of two-phase flow regimes. The *froth regime* is most commonly found; it is a liquid-continuous mixture containing a variety of bubble sizes that provide the necessary interfacial area. The vapor bubbles circulate rapidly, undergo coalescence and breakup, and have a wide range of nonuniform shapes and sizes. For unusually high flow ratios of liquid to vapor, a free-bubbling or *emulsified regime* occurs, characterized as a simple extension of the froth regime, liquid-continuous bubbling with decreased mixing energy. For unusually high ratios of vapor to liquid, the dispersion is inverted to a vapor-continuous mixture, called the *spray regime*. Here, the liquid entering from the downcomer is atomized by the high energy of the vapor, with coalescence and recirculation feeding the atomization process across the tray.

These regimes appear to be an approximate function of the dimensionless *flow parameter*, introduced earlier in connection with tray and packed column hydraulics:

$$F_{lv} = \left(\frac{L}{G}\right)\left(\frac{\rho_G}{\rho_L}\right)^{1/2} \tag{12.72}$$

This parameter represents a ratio of liquid-to-vapor kinetic energies, and approximate ranges for the regimes are

$F_{lv} < 0.01$ spray (vapor continuous)

$0.01 < F_{lv} < 0.10$ froth (liquid continuous)

$F_{lv} > 0.10$ emulsion or free bubbling

The transition between spray and froth regimes is usually detected by light-scattering techniques. Mass transfer models discussed later are based on the froth regime, the one that normally prevails.

12.8.2.2 Tray Efficiency Definitions

The mass transfer efficiency of contacting trays is often expressed in several ways, but here only two efficiencies will be used: point efficiency and tray efficiency. The former deals with the approach to equilibrium at some point on the tray and cannot be greater than 1.0 (100%). Clearly, the equilibrium can vary across the tray as liquid composition varies; thus, there are a number of different values of point efficiency when the tray liquid is not completely mixed (the normal case).

The overall tray efficiency E_{MV} (often called "Murphree efficiency") takes into account concentration gradients on the tray. It has an arbitrary definition based on the average total vapor composition leaving the froth and the specific liquid composition *leaving* the tray [not the *average* composition on the tray (see Figure 12.58)]. Because of this arbitrariness, the overall tray efficiency value can exceed 1.0. When there are no composition gradients on the tray, $E_{og} = E_{MV}$.

The point efficiency is a fundamental criterion of mass transfer, whereas the overall tray efficiency is the value that can be measured, especially for larger distillation columns. Large-column

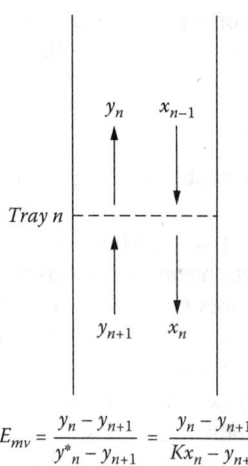

$$E_{mv} = \frac{y_n - y_{n+1}}{y^*_n - y_{n+1}} = \frac{y_n - y_{n+1}}{Kx_n - y_{n+1}}$$

Notes: y_n and y_{n+1} are Averaged Over the Cross Section

x_n Is Overflow Composition Only (which can differ from averaged x_n)

FIGURE 12.58 Compositions leading to the definition of overall tray efficiency, E_{MV}.

Distillation

efficiency must take both of these efficiencies into account, and the analysis can be divided into two parts:

- Understanding basic mechanisms associated with point efficiency
- Determining the concentration gradients across the tray

12.8.2.3 Point Efficiency

The point efficiency of tray n is expressed by

$$E_{OG} = \left(\frac{y_n - y_{n-1}}{y_n^* - y_{n-1}} \right)_{point} \tag{12.73}$$

The efficiency concept is based on two-resistance theory and is usually expressed in terms of the molar rate of diffusion:

$$N = k_G a_i (y_i - y) = k_L a_i (x - x_i) \tag{12.74}$$

Assuming phase equilibrium at the interface, the overall mass transfer coefficient can be obtained as the sum of mass transfer resistances:

$$\frac{1}{K_{OG}} = \frac{1}{k_G} + \frac{m}{k_L} \tag{12.75}$$

(overall resistance = vapor resistance + liquid resistance)

For distillations, the vapor resistance dominates, whereas for the stripping or absorption of slightly soluble materials, the liquid resistance dominates. In Equation (12.75), m is related to the VLE ratio K_i [Equation (12.19)], and for the binary distillation of an i-j mixture is defined as the slope of the equilibrium line on a y-x diagram:

$$m = \frac{dy_i}{dx_i} = \frac{\alpha_{ij}}{\left[1 + (\alpha_{ij} - 1) x_i \right]^2} \tag{12.76}$$

For low values of x_i and high values of α_{ij}, the slope is relatively high and the more important the second term on the right side of Equation (12.75) becomes, meaning that the liquid side resistance becomes more important.

Considering a mass balance across a differential element in the froth of a sieve tray (Figure 12.57), the expressions for the vapor and liquid phase mass transfer units are

$$N_G = k_G a_i' t_G \tag{12.77}$$

and

$$N_L = k_L a_i' t_L \tag{12.78}$$

Volumetric coefficients $k_G a_i'$ and $k_L a_i'$ are used when the interfacial area a_i cannot be determined, and residence times tG and tL depend on the nature and volume of the dispersion as well as the velocity of the gas. The number of overall transfer units [Equation (12.71)], based on gas concentrations, follows from the individual phase units:

$$\frac{1}{N_{OG}} = \frac{1}{N_G} + \frac{\lambda}{N_L} \qquad (12.79)$$

where λ is the ratio of slopes, equilibrium line to operating line:

$$\lambda = \frac{m}{L/V} = \frac{mV}{L} \qquad (12.80)$$

Thus, a continuous contacting approach (use of transfer units) is employed. The point efficiency is computed from the value of N_{OG}:

$$E_{OG} = 1 - \exp[-N_{OG}] \qquad (12.81)$$

12.8.2.4 Tray (Murphree) Efficiency

For complete mixing, $E_{MV} = E_{og}$. For the *plug flow* of liquid,

$$E_{MV} = \frac{1}{\lambda}\left(\exp[\lambda E_{og} - 1]\right) \qquad (12.82)$$

The point and overall efficiencies are the same when the phases in the froth are completely mixed. This is more likely for small columns and for very high gas-to-liquid ratios for larger columns. When the liquid is in plug flow, $E_{MV} > E_{og}$. To deduce point efficiency values from measured overall efficiency values, a model involving mixing tendencies must be available, as discussed later.

12.8.2.5 Overall Column Efficiency E_{oc}

This is the efficiency most used and understood [see Equation (12.66)]. At total reflux, E_{OC} and E_{MV} are about equal. Otherwise,

$$E_{oc} = \frac{\ln[1 + E_{MV}(\lambda - 1)]}{\ln \lambda} \qquad (12.83)$$

or

$$E_{MV} = \frac{(\lambda^{E_{oc}} - 1)}{(\lambda - 1)} \qquad (12.83a)$$

where λ is defined by Equation (12.80).

The relationships between these efficiencies are shown in Figure 12.59. For design, the sequence is

$$E_{og} \rightarrow E_{MV} \rightarrow E_{oc}$$

with the reverse applying to the analysis of existing columns.

Distillation

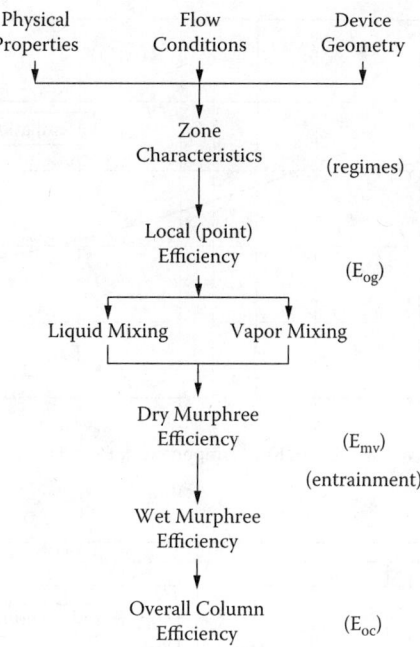

FIGURE 12.59 Relationships between the different efficiencies.

12.8.2.5 Predicting Tray Efficiency

The efficiency of a tray or packed column is a function of the effectiveness of mass transfer between the phases, i.e., *mass transfer efficiency*. This efficiency depends on three sets of design parameters:

- The system: composition and properties
- Flow conditions: rates of throughput
- Geometry/type of contacting device

The column designer has little control over the first set. The system may have inherently low mass transfer capability, e.g., low diffusion coefficients or high liquid viscosities (high Schmidt numbers). Conversely, the system may have such favorable mass transfer characteristics that almost any device chosen will be satisfactory. Clearly, for new designs, one may not know quite what to expect in terms of mass transfer favorability.

Four methods for predicting the efficiency of a commercial fractionator are

- Comparison with performance data for a *very similar* installation
- Use of an empirical or statistical efficiency prediction method
- Direct scaleup from carefully designed laboratory or pilot plant experiments
- Use of theoretical or semitheoretical mass transfer models

Often, more than one approach likely is used. For example, available data from a similar installation may be used for orientation, and a theoretical model then used to extrapolate to the new conditions. Each approach is covered in the material below, with less emphasis on the fourth (modeling) approach.

12.8.2.6 Efficiency from Performance Data

If an existing fractionation system is to be duplicated, careful measurements on the existing system can provide sufficient data for a confident design. For a new system, are the physical properties

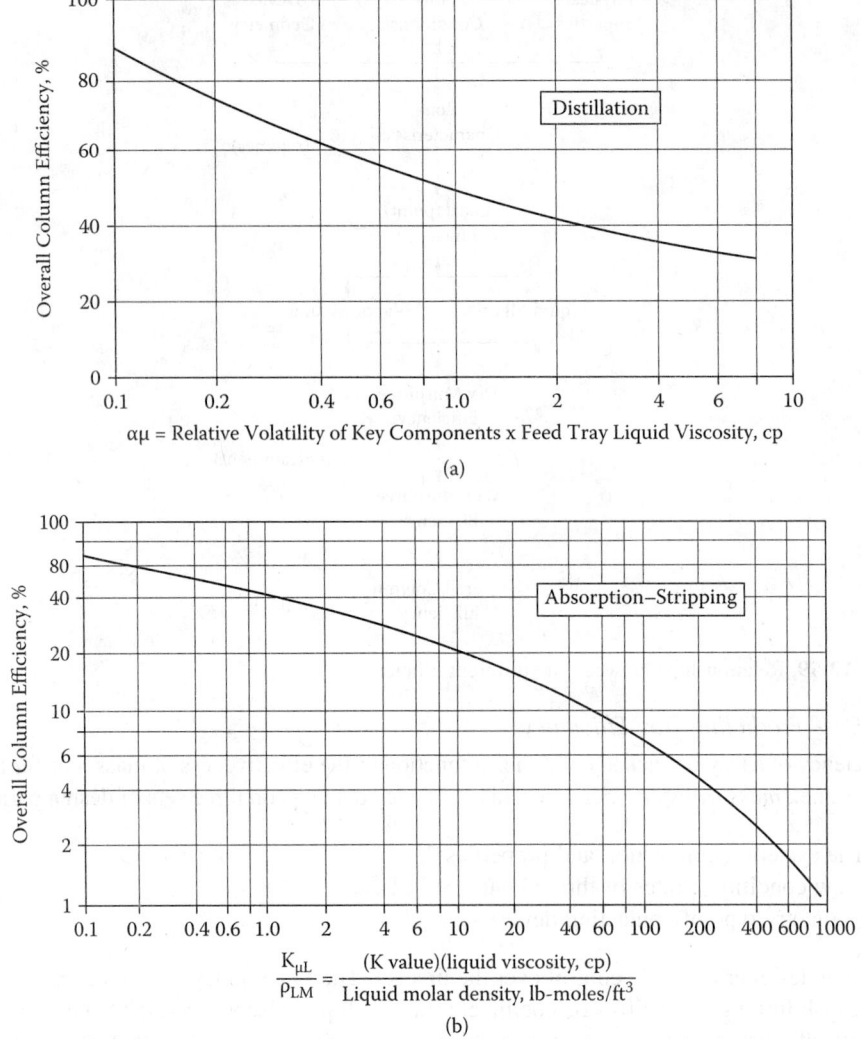

FIGURE 12.60 O'Connell method for estimating overall column efficiencies of distillation columns and absorbers. (H. E. O'Connell, 1946. *Trans AIChE* 42: 741.)

different? Is the loading about the same as for the existing system? Is the same contacting device to be used? Clearly, caution is needed, but it is the best basis to use. Those with access to Fractionation Research, Inc. (FRI) data can benefit from this approach if the system at hand is similar to a system studied by FRI. However, the same questions must be asked.

12.8.2.7 Empirical Efficiency Methods

The best known and most used empirical method is that of O'Connell[68] (Figure 12.60), for distillation columns and absorbers. The curves are based on plant data for several bubble-cap columns plus a few pilot-scale units. Efficiency is related to two properties of the feed mixture: liquid viscosity μ_L and relative volatility α. Higher values of the $\mu_L \alpha$ product indicate larger liquid-side mass transfer resistance and hence a lower efficiency. For a vapor feed or a mixed vapor-liquid feed, the correlating viscosity should be that of the feed tray liquid.

Distillation

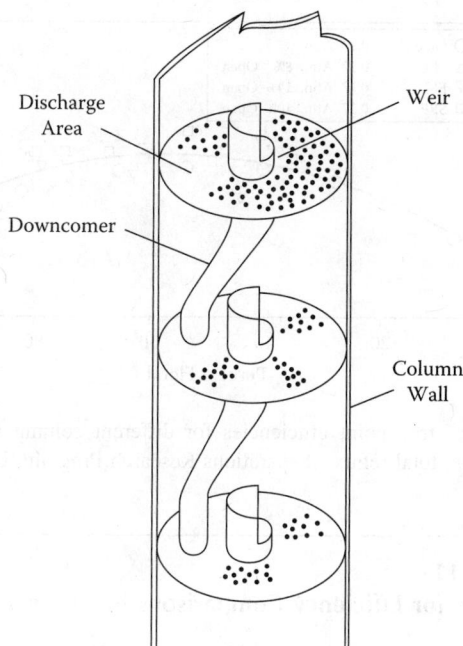

FIGURE 12.61 Diagram of a section of an Oldershaw column.

Such simplified correlations should be used with great discretion but may suffice for preliminary designs, e.g., to support technical feasibility and early economic studies of new or modified processes. In the related areas of absorption or stripping, the liquid phase resistances are higher than for distillation, and the separate correlation of Figure 12.60 for absorbers has been adapted from the O'Connell work. An example is the absorption or stripping of a sparingly soluble gas. The low solubility can result in a high value of the equilibrium ratio K (e.g., in the range of 2400 for toluene in water). The resulting abscissa value for Figure 12.60 gives a very low overall column efficiency.

12.8.2.8 Efficiency from Laboratory Experiments

Large-scale tray column efficiencies can be predicted from measurements in laboratory columns as small as 25 mm (1.0 in) diameter. Comparative studies (laboratory vs. commercial) have shown that the use of a special laboratory sieve tray column, the Oldershaw column, produces equivalent separations.[69] A glass Oldershaw column is shown in Figure 12.61. Each tray has a center downcomer that discharges to one side of the tray below. Comparisons between laboratory and commercial scale efficiencies are shown in Figure 12.62[70] and Table 12.11. The Oldershaw value of overall efficiency E_{oc} is equivalent to the point efficiency E_{og} of the larger column.

The procedure for using this scaleup approach is summarized as follows:

1. Arbitrarily select a number of Oldershaw trays and set up the equipment to operate either at total or at finite reflux, as the situation demands. Oldershaws are generic, off-the-shelf units with varying numbers of trays and can be assembled rapidly.
2. Determine the upper operating limit (flood limit) of the equipment by experiment or by general correlation. Then collect data at some selected approach to flood.
3. Obtain overhead and bottoms samples and determine the degree of separation.
4. If the separation is satisfactory, the commercial column will require the same or fewer trays than the Oldershaw trays used. Comparisons are made at the same approach to flood for both columns.

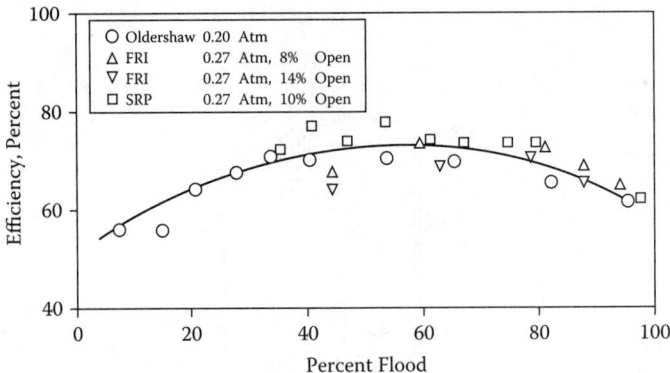

FIGURE 12.62 Comparison of tray point efficiencies for different column sizes. Cyclohexane/n-heptane system, 0.20–0.27 atm pressure, total reflux. (Separations Research Program, Univ. of Texas at Austin.)

TABLE 12.11
Conditions for Efficiency Comparisons in Figure 12.62

	Oldershaw[1]	SRP[2]	FRI-1[3]	FRI-2[4]
Head pressure, atm	0.20	0.27	0.27	0.27
Column diameter, m	0.025	0.43	1.2	1.2
Tray spacing, m	0.025	0.46	0.61	0.61
Tray open area, % of active area	10.0	8.7	8.0	14.0
Weir height, mm		51	51	51
Hole diameter, mm	0.89	4.8	12.7	12.7
Downcomer area, % of total area	8.2	7.2		

Notes: All runs with cyclohexane/n-heptane mixture at total reflux. The SRP column contained a splash baffle. FRI efficiencies are corrected from overall column to point.

Sources: [1]Fair, J. R.; Null, H. R.; Bolles, W.L. *Ind. Eng. Chem. Proc. Des. Devel.* 1983, 22, 53. [2]Garcia, J. A., Ph.D. Dissertation, The Univ. of Texas at Austin, 1999. [3]Sakata, M.; Yanagi, T. *IChemE Symp. Ser. No. 56*, 1979, 3.2/21. [4]Yanagi, T.; Sakata, M. *Ind. Eng. Chem. Proc. Des. Devel.* 1982, 21, 712.

12.8.2.9 Efficiency from Mass Transfer Models

The above relationships, coupled with experimental efficiency results, provide a semifundamental approach to modeling tray efficiency. Since the contacting process is not well understood, only an approximation of the efficiency can be expected. However, the application of mass transfer principles can aid the overall process of design by showing relative effects of variables.

The AIChE efficiency model[71] was the first to use a rational approach to mass transfer efficiency. Equations (12.74) through (12.83) are used to predict point efficiency. The values of $k_G a_i'$, $k_L a_i'$, t_G, and t_L are deduced from laboratory measurements with small test trays containing bubble caps. No attempt is made to evaluate interfacial area a_i'. The detailed procedure is in reference 71.

An adaptation of the AIChE model was published by Chan and Fair,[72] is specific to sieve trays, and is based on larger-scale sieve tray measurements. For the vapor phase,

Distillation

$$k_G a_i = \frac{316 D_G^{0.5} \left(1030 f - 867 f^2\right)}{h_L^{0.5}} \quad (12.84)$$

where

D_G = vapor phase diffusion coefficient, m²/s
h_L = liquid holdup, mm [obtain from Equation (12.67)]
f = fractional approach to flood, i.e., $C_{sb}/C_{sb,flood}$ (from Figure 12.30)
$k_G a_i$ = volumetric mass transfer coefficient for the vapor phase, sec⁻¹

The number of vapor phase transfer units is obtained from Equation (12.74), where the residence time t_G is based on the volume of gas flowing through the two-phase mixture on the tray.

The volume of the froth is based on the base (A_a) times the height (Z_f). Estimates of Z_f can be made from Figure 12.63, from which

$$t_G = \frac{\text{vol. of vapor space in froth}}{\text{volumetric flow of vapor}} = \frac{A_a \varepsilon Z_f}{Q_v} = \frac{\varepsilon Z_f}{U_a} \quad (12.85)$$

where

Z_f = height of froth
ε = void fraction of froth = $1 - h_L/Z_f$

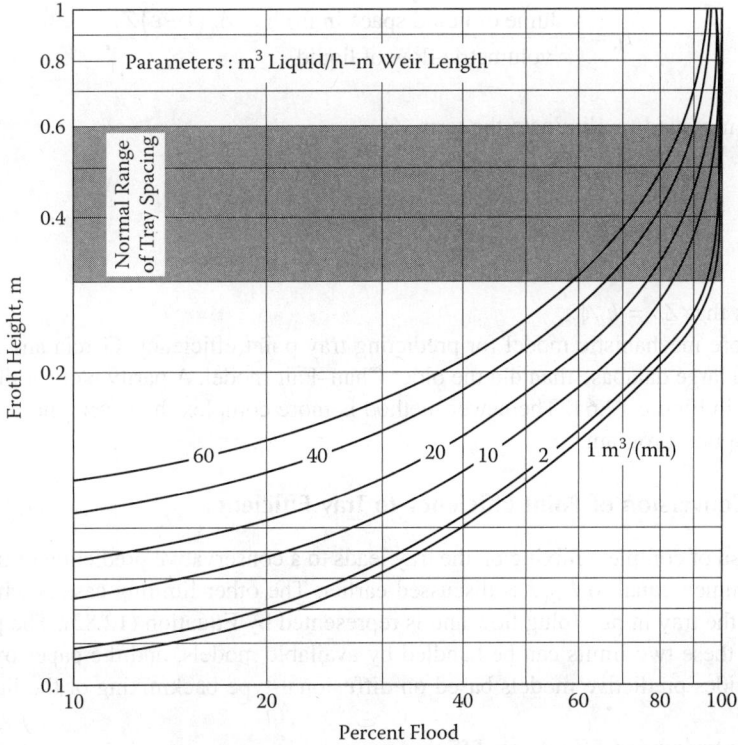

FIGURE 12.63 Estimation of froth height on a tray. (J. Stichlmair, J. R. Fair, 1998. *Distillation—Principles and Practices,* New York: Wiley-VCH.)

However, the volume $A_a Z_f$ is not all effective for mass transfer; its upper zone provides relatively little area for mass transfer. Accordingly, some effective value of Z_f applies. Because this value is not easily determined, an alternative approach to estimating vapor residence time is used:

$$t_G = \frac{(1-\phi_{ef})h_L}{\phi_{fe} U_a} \quad (h_L \text{ in m}) \tag{12.86}$$

where an *effective froth density* ϕ_{fe} [see Equation (12.63)] is used, based on the work of Bennett et al.,[73] who provided the basis for estimating ϕ_{fe}:

$$\phi_{fe} = \exp\left[-12.55(C'_{sb})^{0.91}\right] \tag{12.87}$$

with C_{sb}' based on the active area (rather than the net area), in m/s.

The number of liquid phase transfers is obtained from Equation (12.78), with the volumetric coefficient for *sieve trays*:

$$k_L a_i = (3.88 \cdot 10^8 \, D_L)(0.40 \, F_{va} + 0.17), \text{ sec}^{-1} \tag{12.88}$$

The residence time t_L is based on the volume of liquid flowing through the two-phase mixture on the tray:

$$t_L = \frac{\text{volume of liquid space in froth}}{\text{volumetric flow of liquid}} = \frac{A_a(1-\varepsilon)Z_f}{q_L} \tag{12.89}$$

which is also modified to eliminate the term Z_f:

$$t_L = \frac{h_L A_a}{q_L} \tag{12.90}$$

which implies that $Z_{f'} = h_L/\phi_{fe}$.

With a more mechanistic model for predicting tray point efficiency, Garcia and Fair[74] showed a better fit to a large database than did the older Chan–Fair model. A parity plot for the Garcia–Fair work is given in Figure 12.64. The newer method is more complex, however, and requires a fairly elaborate computer program.

12.8.2.10 Conversion of Point Efficiency to Tray Efficiency

The assumption of complete mixing on the tray leads to a conservative prediction of tray efficiency, i.e., E_{og} is assumed equal to E_{MV}, as discussed earlier. The other limiting case is where the liquid moves across the tray in pure plug flow and is represented by Equation (12.82). The partial mixing case between these two limits can be handled by available models, and the paper by Bennett and Grimm[75] provides predictive models based on diffusional-type backmixing of the liquid.

12.8.2.11 Entrainment Effects on Efficiency

The performance curve of a crossflow tray shows a lowering of efficiency as vapor velocity approaches a flooding value (Figure 12.27). This lowering results from liquid entrainment. Figure

Distillation

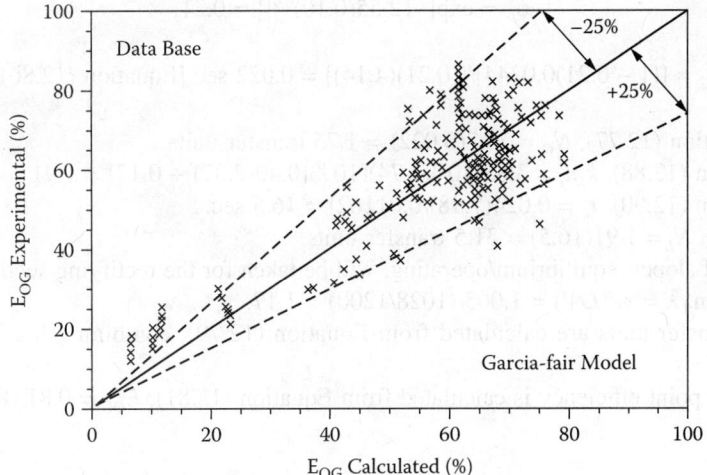

FIGURE 12.64 Parity plot for Garcia–Fair efficiency prediction method. (2002. *Ind. Eng. Chem. Res.* 41: 1632.)

12.33 is used to determine the entrainment ratio ψ. Then, an expression based on early work of Colburn[76] is used to correct the "dry" tray efficiency E_{MV} for entrainment:

$$\frac{E_{MV,wet}}{E_{MV,dry}} = \frac{1}{1 + E_{MV,dry}\left[\psi/(1-\psi)\right]} \quad (12.91)$$

Problem 12.9

Problem 12.6 deals with the dimensioning of a large ethylbenzene/styrene tray-type fractionator. The overall column efficiency for the sieve trays is to be estimated, based on the same conditions given for Problem 12.6. Two methods will be used: (a) the O'Connell (Figure 12.60) approximate method and (b) the Chan/Fair more basic method [Equation (12.84)], corrected for entrainment and liquid crossflow effects.

Solutions

(a) The O'Connell method requires values of relative volatility and liquid viscosity. These values are 1.40 and 0.38 cP, respectively. For the product $\alpha\mu = 1.40 (0.38) = 0.53$, the efficiency is read from Figure 12.60 as $E_{oc} = 0.56$ or 56%.

(b) The point efficiency model of Chan and Fair requires values of the phase diffusion coefficients. These are estimated to be:
$D_G = 2.09(10^{-5})$ m²/s = 0.209 cm²/s
$D_L = 3.74(10^{-9})$ m²/s = 3.74(10^{-5}) cm²/s

Vapor transfer units:
By Equation (12.84),

$$k_G a_i = \frac{316\left[2.09(10^{-5})\right]^{0.5}\left[1030(0.8) - 867(0.64)\right]}{(24.4)^{0.5}} = 78.7 \text{ sec}^{-1}$$

For $C_{sb}' = C_{sb}(A_n/A_a) = 0.098(8.40/7.98) = 0.10$ m/s, and by Equation (12.87),

$$\phi_{fe} = \exp[-12.55(0.10)^{0.91}] = 0.21$$

$$t_G = [(1 - 0.21)0.0244]/[(0.21)(4.14)] = 0.022 \text{ sec [Equation (12.86)]}$$

From Equation (12.77), $N_G = 78.7(0.022) = 1.75$ transfer units.
By Equation (12.88), $k_L a_i = 3.88(10^8)(3.74)(10^{-9})[0.40(2.87) + 0.17] = 1.91$ sec^{-1}.
By Equation (12.90), $t_L = 0.024(7.98)/0.01162 = 16.5$ sec.
From which $N_L = 1.91(16.5) = 31.5$ transfer units.

The ratio of slopes, equilibrium/operating, will be taken for the rectifying section where slope $m = 1.005$. Then, $\lambda = m/(L/V) = 1.005/(1028/1200) = 1.17$.

Overall transfer units are calculated from Equation (12.79), combining N_G, N_L, and λ; $N_{OG} = 1.64$.

Finally, the point efficiency is calculated from Equation (12.81), $E_{OG} = 0.81$ (81%).

Notes:

- The point efficiency may be enhanced by liquid crossflow effects and/or diminished by entrainment effects.
- The calculated efficiency is significantly greater than the estimate from the O'Connell method.

Commercial ethylbenzene/styrene fractionators show efficiencies in the range of 80 to 90%.

12.8.2.12 Overall Column Efficiency

As indicated by Figure 12.59, the final step is to convert E_{MV} (wet or dry) to the overall column efficiency E_{oc}. This is done using the Lewis[77] relationships:

$$E_{oc} = \frac{\ln[1 + E_{MV}(\lambda - 1)]}{\ln \lambda} \quad (12.92)$$

$$E_{MV} = \frac{\lambda^{E_{oc}} - 1}{\lambda - 1} \quad (12.93)$$

12.8.2.13 Multicomponent Systems

For binary systems, the efficiency for each component is the same and thus

$$(E_{og})_i = (E_{og})_j = (E_{og})_{i-j} \quad (12.94)$$

When more than two components are present, the efficiencies of each are not necessarily the same. The rigorous approach to handling multicomponent mixtures, outlined by Taylor and Krishna,[78] uses the Maxwell–Stefan diffusional equations. Chan and Fair[72] used the rigorous approach to compare multicomponent system separations with those predicted by the use of the equivalent pseudobinary systems. They found that if the dominant pair of components present in the mixture is used to determine efficiency for all of the components, the separation determined is quite close to that resulting from rigorous multicomponent procedures.

General rules proposed by Toor and Burchard[79] for multicomponent systems are summarized as follows:

Distillation

1. When all binary pair efficiencies are high, the multicomponent efficiencies are also high.
2. As the fraction of the total resistance in the liquid phase increases, differences between the component efficiencies diminish.
3. When vapor phase resistance dominates, differences between component efficiencies increase as differences between the binary pair diffusion coefficients increase. If all the vapor diffusion coefficients are equal, indicating minimum liquid-side interaction, all component efficiencies are equal.
4. In a ternary system where there are two similar and one dissimilar species, the dissimilar one will have an efficiency close to the binary value, and the efficiencies of the similar species will differ from each other and from the binary value.
5. The efficiencies of one or more minor components may differ from the binary value (assuming only two major components), but the efficiencies of the major components will approach each other (and the binary value).

12.8.3 Packed Column Efficiency

Packed columns operate in the counterflow mode, and thus it is not really appropriate to utilize stagewise concepts for their analysis and design. Despite this, the HETP approach [Equation (12.69)] is often used. With this approach, the height of packing needed to effect a given separation is simply,

$$Z = (HETP)(N_t) \tag{12.95}$$

Thus, if theoretical stages are computed by one of the methods described earlier, it is only necessary to multiply by some "characteristic value" of the packing, the HETP, to arrive at the required height. This approach is often used in practice, with the packing vendor supplying values of HETP. The problem is that the vendor may not take into account the basic requirements of determining efficiency, as stated earlier for trays:

- The system: composition and properties
- Flow conditions: degree of loading
- Geometry: type and size of packing

A more rigorous—and generally more useful—approach is to utilize transfer units as introduced earlier by Equations (12.69) through (12.71), in which the required height is a product of the height of a transfer unit and the number of transfer units:

$$Z = (H_v)(N_v) = (H_L)(N_L) = (H_{ov})(N_{ov}) \tag{12.96}$$

The transfer unit approach permits consideration of varying vapor and liquid flows, varying properties, and differences in packing geometry. However, computation of transfer units, especially for multicomponent systems, is not straightforward. Instead, a "hybrid" approach can be used:

$$N_{ov} = N_t \left[\frac{\ln \lambda}{(\lambda - 1)} \right] \tag{12.97}$$

where, as before,

$$\lambda = m / L/V$$

Equation (12.97) is strictly valid when the operating and equilibrium lines are straight. When the lines have equal slope, $N_{ov} = N_t$. The equation may be applied by dividing the column into sections where the equilibrium line can be considered straight.

For distillation,* the height of a transfer unit is defined as

$$H_{og} = \frac{V}{K_{og} a_i} \quad H_v = \frac{V}{k_g a_i} \quad H_L = \frac{L}{k_L a_i} \tag{12.98}$$

These relationships apply when the driving force for mass transfer is expressed as mole fractions. The overall mass transfer coefficient K_{og} is normally used in connection with

$$K_{og} = \frac{1}{\dfrac{1}{k_g} + \dfrac{m}{k_L}} \tag{12.99}$$

In these expressions, k and K have units of kg-moles/(s-m²-m.f.) or equivalent English units. Usually, the interfacial area a_i is not known and is not segregated from the mass transfer coefficient, giving volumetric coefficients $K_{og} a_i$, $k_g a_i$, and $k_L a_l$. These terms may be substituted in Equation (12.98).

The overall transfer unit height may be broken down into individual phase terms:

$$H_{oG} = \frac{V}{k_G a_i} + \lambda \frac{L}{k_L a_i} \tag{12.100}$$

or

$$H_{oG} = H_G + \lambda H_L \tag{12.100a}$$

from which

$$HETP = H_{OG} [\ln \lambda / (\lambda - 1)] \tag{12.100b}$$

12.8.3.1 Random Packings

Several methods have been proposed for predicting values of H_v and H_L as functions of system, flow conditions, and packing type.[80–83] The model of Wagner et al.[82] is directed specifically to the newer, open-style random packings. The Bolles–Fair[83] model has the broadest validation and will be used here and is summarized by the following equations:

$$H_G = \frac{\psi' Sc_G^{0.5} \left(Z_p / 3.05 \right)^{1/3} (3.28 D_c)^{n_1}}{3.28 (735 L' \Gamma)^{n_1}} \tag{12.101}$$

$$H_L = 0.305 \, \phi' \, C_F \left(Z_p / 3.05 \right)^{0.15} Sc_L^{0.5} \tag{12.102}$$

---

* For absorption and stripping, the equations are slightly different, in particular if concentrated solutions are being considered. See a mass transfer text, e.g., Hines and Maddox,[67] for a detailed discussion.

where

$$\Gamma = \left(\mu_L/\mu_W\right)^{0.16} \left(\rho_W/\rho_L\right)^{0.16} \left(\sigma_W/\sigma\right)^{0.8} \qquad (12.103)$$

and

$$Z_p = \text{bed height, m}$$

Values of exponent n_1 are 1.24 and 1.11 for ring-type and saddle-type packings, respectively. Values of exponent m_1 are 0.6 and 0.5 for ring-type and saddle-type packings. Parameter C_F accounts for enhancement of interfacial area near the flood point; suitable values are

50% flood......1.00
60% flood......0.90
80% flood......0.60

The D_c term (tower diameter) has an upper limit correction *if* there is very good liquid distribution. The upper limit is for $D_c = 0.61$ m (2.0 ft); for larger diameters, use the correction for 0.61 m. The height correction Z_p applies to each bed in a multibed column and is normalized to a bed height of 3.05 m (10 ft).

For values of parameters ψ' and ϕ', see Figure 12.65. Berl saddle parameters are applicable to ceramic Intalox saddles—but not to metal Intalox saddles, which have a quite different geometry and are of the "through-flow" type covered by Wagner et al.[82]

The smallest size packings in the Bolles–Fair correlation are nominally 13 mm. If the usually required minimum ratio of column diameter to packing diameter of 8 is to be retained (to avoid bypassing at the wall), scaleup studies require a minimum column diameter of about 100 mm. Research in progress may lead to the use of a special, small laboratory packing (e.g., Pro-Pak®, from Scientific Development Co.) for columns as small as 25 mm. *HETP* values of two sizes of Pro-Pak, for cyclohexane/*n*-heptane at 1.0 atm, are shown in Figure 12.66, from reference 84.

12.8.3.2 Structured Packings

Packings of this type have been described above. Vapor and liquid contact each other in channels that have the general characteristics of wetted wall columns. The design efficiency of these packings may be characterized by *HETP* values obtained experimentally. But the performance data should be based on the same or similar system at the same degree of loading. An approximate method for estimating *HETP* is that of Lockett[85] and derives from a parametric study of a more rigorous model of Bravo and coworkers, to be described later. At total reflux and $\lambda = m/(L/V) = 1.0$,

$$HETP = A_1 + B_1 (F_{vs,80}) \qquad (12.104)$$

where $F_{vs,80} = U_{vs}\rho_g$ = superficial F-factor at 80% of flood.

The constants for Equation (12.107) are

	A_1	B_1	a_p, m²/m³
Flexipac 1	0.018	0.106	443
Flexipac 2	0.023	0.156	223
Flexipac 3	0.061	0.235	111

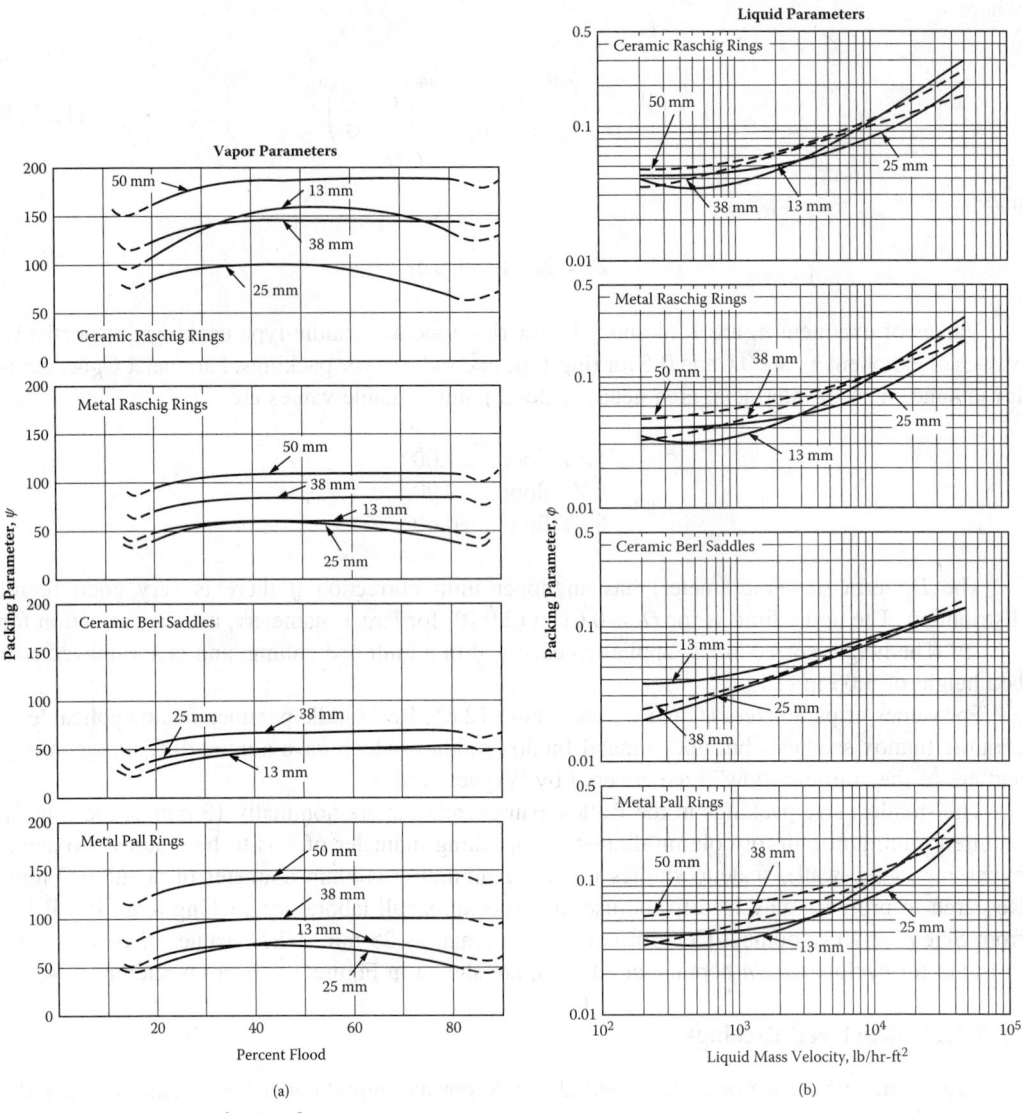

Note: to Convert from lb/hr-ft² to kg/s-m², multiply by 0.00136.

FIGURE 12.65 Packing parameters for random packing efficiency. (Bolles, W. L.; Fair, J. R. *Chem. Eng.* 1982, 89 (14) 109.

and *HETP* is in meters. The restriction of $\lambda = 1.0$ is not overly important, since many distillation columns have an average λ close to unity. Packings other than those listed may be used on an equivalent surface basis. Lockett suggests another useful approximation,

$$HETP = \frac{2450 F_{vs,80}}{a_p^{0.5}} \quad (12.105)$$

that allows the use of whatever packing surface is to be specified (Table 12.10). Equations (12.104) and (12.105) require a value of the flooding velocity, obtainable from the vendor or from the generalized methods given earlier.

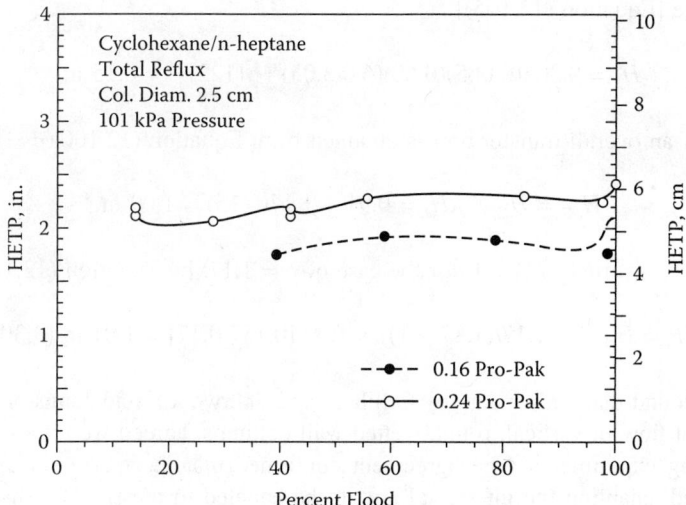

FIGURE 12.66 Efficiency of laboratory random packing, cyclohexane/*n*-heptane system. (Separation Research Program, Univ. of Texas at Austin.)

12.8.3.3 Mechanistic Model for Structured Packings

The first mechanistic and fundamental mass transfer model for structured packings was proposed by Bravo et al.[86] This model is based on a variety of test data obtained for Sulzer BX gauze-type packing. This packing has flow channels triangular in cross section and the total flow path is zigzag as individual elements are passed. Although somewhat oversimplified, because of cross-mixing opportunities, this model forms a reasonable basis for predicting rates of mass transfer between bulk vapor and liquid flowing over the metal surfaces.

Problem 12.10

To continue with the ethylbenzene/styrene separation of the immediately previous problems, now consider the use of a random packing for the fractionator. In Problem 12.8, for 50-mm metal Pall rings, a column diameter of 3.21 m (10.5 ft) would be required to operate at 80% of flood. Now estimate the efficiency of this same packing for the same service. The basic data are given in Problems 12.6 through 12.9. A packed height of 6.0 m will be assumed.

The model of Bolles and Fair [Equations (12.104) through (12.106)] is appropriate. Input data for the model are as follows:

Liquid Schmidt number = 122
Vapor Schmidt number = 0.80 (viscosity of the vapor = 0.008 cP)
Slope of equilibrium line, m = 1.005
Stripping factor $\lambda = m/(L/V)$ = 1.17

For 3.2-m column diameter and 13.61 kg/s liquid flow, L' = 1.69 kg/s-m².
The correction for departure from water properties Γ = (0.37/0.38)(61/52.2)(62/25) = 2.50.
Finally, the packing parameters for 50-mm Pall rings: ψ' = 140, ϕ' = 0.065 (Figure 12.65).
Substituting in Equation (12.104),

$$H_G = \frac{140(0.80)^{0.5}(6.0/3.05)^{1/3}(3.28(0.61))^{1.24}}{3.28\left[735(1.69)(2.50)\right]^{0.6}} = 0.91\,\text{m}$$

For the liquid side [Equation (12.105)],

$$H_L = 0.305(0.065)(0.60)(6.0/3.05)^{0.15} (122)^{0.5} = 0.15 \text{ m}$$

The height of an overall transfer unit is obtained from Equation (12.100a):

$$H_{oG} = H_G + \lambda H_L = 0.91 + 1.17(0.15) = 1.09 \text{ m}$$

This value can be converted to HETP for the case of $\lambda = 1.17$, by Equation (12.100b):

$$HETP = H_{oG} [\ln 1.17/(1.17 - 1)] = 1.09 [0.157/0.17] = 1.01 \text{ m } (3.30 \text{ ft})$$

Bravo et al. found that earlier data by Gilliland and Sherwood[87] and Johnstone and Pigford,[88] for countercurrent flow in vertical, round wetted-wall columns, agreed well with the results from structured packing experiments. The agreement confirmed that the gauze surface is essentially completely wetted, enabling the interfacial area to be equated to the specific surface area of the packing. The final model for the gas phase mass transfer coefficient is

$$Sh_v = 0.0338(Re_v)0.8(Sc_v)^{0.333} \tag{12.106}$$

where

Sh_v = Sherwood number = $(k_v d_{eq}/D_v)$ (12.107)
Re_g = Reynolds number = $(d_{eq}\rho_g/\mu_g)(U_{g,eff} + U_{L,eff})$ (12.108)
Sc_g = Schmidt number = $(\mu_g/\rho_v D_v)$ (12.109)

The effective gas and liquid rates, $U_{g,eff}$ and $U_{L,eff}$, take into account the flow angles, void fractions, and film thicknesses that derive from gross flow rates and packing geometry. The equivalent diameter basis is shown in Figure 12.48.

For the liquid phase, a penetration model is used, with the exposure time based on flow across the corrugation side of the packing. Thus,

$$k_L = 2\sqrt{\frac{D_L U_{L,eff}}{\pi S}} \tag{12.110}$$

where S is the length of a side of a corrugation.

Heights of transfer units for the individual phases are obtained from Equation (12.98), and Equation (12.100) provides the height of an overall transfer unit. For the gauze packings, with spreading aided by capillary action, interfacial area $a_i = a_p$ (specific surface area from Table 12.10). For sheet metal packings, which do not promote as much liquid spreading, an empirical correction factor was proposed by Fair and Bravo[89]:

$$a_i = \beta a_p \tag{12.111}$$

where the discount factor β is some function of liquid rate and surface wettability. For poorly wetted surfaces, where rivulet flow dominates, β is quite low—in the range of 0.1 to 0.3, even at high liquid rates. On the other hand, for well wetted surfaces, β can range up to 1.0 or higher, For very approximate estimates, the designer can use the relationship for well wetted metal surfaces:

$$\beta = 0.50 + 0.0058 \text{ (\% flood)} \tag{12.112}$$

for flooding percentages in the range of 0 to 85. Above 85% flood, $\beta = 1.0$. Once the value of H_{og} is determined, conversion to *HETP* for a given value of λ is

$$HETP = H_{og}[\ln \lambda/(\lambda - 1)] \tag{12.113}$$

with the same restrictions noted for Equation (12.95). When the operating and equilibrium lines are straight and parallel, $HETP = H_{og}$.

A more rigorous prediction of structured mass transfer efficiency is based on results from the University of Texas' Separations Research Program (SRP) and the Delft University of Technology, in the Netherlands. Models from these sources have been described and compared in a joint publication[90] and are essentially equivalent. The newer SRP model[91] follows that of Bravo et al.,[86] described above, and a brief summary of it follows.

The gas phase mass transfer coefficient is obtained through the Sherwood number, as in Equation (12.107):

$$Sh_v = \frac{k_v S}{D_v} = 0.54 \left[\frac{(U_{g,eff} + U_{L,eff})\rho_g S}{\mu_v} \right]^{0.8} Sc_v^{1/3} \tag{12.114}$$

The liquid phase mass transfer coefficient k_L is obtained from Equation (12.110). The effective interfacial area a_i is obtained from the rather complex relationship

$$\frac{a_i}{a_p} = F_{SE} \frac{29.12(We_L Fr_L)^{0.15} S^{0.359}}{Re_L^{0.2} \varepsilon^{0.6} (1 - 0.93\cos\gamma)(\sin\theta)^{0.3}} \tag{12.115}$$

where
- S = length of side of corrugation = equivalent diameter, m
- F_{SE} = factor to account for surface texturing
- θ = corrugation angle, measured from the horizontal, deg
- γ = contact angle, liquid/surface, deg, estimated from $\cos\gamma = 5.211 \cdot 10^{-16.835\sigma}$
- σ = surface tension, mN/m

The effective velocities of the phases are computed from

$$U_{g,eff} = \frac{U_{gs}}{\varepsilon(1-h_t)\sin\theta} \tag{12.116}$$

$$U_{L,eff} = \frac{U_{Ls}}{\varepsilon h_t \sin\theta} \tag{12.117}$$

where h_t is total liquid holdup, vol. liquid per vol. packed bed. Values of a_p, F_{SE}, ε, S, and θ for representative structured packings are given in Table 12.12.

12.8.3.4 Scale-Up of Structured Packing Efficiency

As noted earlier, test data from an Oldershaw column as small as 25 mm diameter can often be converted to commercial-scale designs. For random packings, a minimum test column diameter of 100 mm is recommended. For structured packings, a 25-mm column diameter can be used for

TABLE 12.12
Values of Parameters for Equations (12.114) through (12.117)

	a_p (m²/m³)	F_{SE} (–)	ε (–)	S (m)	θ (deg)
Flexipac 2	233	0.350	0.95	0.018	45
Gempak 2A	223	0.344	0.95		45
Intalox 1T	315	0.415	0.95	0.0152	45
Intalox 2T	213	0.415	0.95	0.0221	45
Jaeger Maxpak	229	0.364	0.95	0.0175	45
Mellapak 250Y	250	0.350	0.95		45
Mellapak 500Y	500	0.350	0.91		45
Sulzer BX	492	0.90		0.0090	60

scale-up studies. Scaled-down versions of some of the packings are available in sizes of about 25 and 50 mm diameter.

Hufton et al.[92] used a 23-mm diameter Sulzer BX laboratory packing and found that the original Bravo et al. model [Equation (12.106)] worked well at the small scale so long as the actual geometric dimensions of the laboratory packing were used.

12.8.3.5 Structured Packing Performance

The most common structured packing has a specific surface area in the range of 225 to 250 m²/m³, has a corrugation angle of 45°, and is fabricated from thin-gauge sheet metal; this is often called a "no. 2 packing." The metal may have a textured surface to promote liquid spreading and may have perforations to equalize pressure (and promote spreading). Performance tests are usually run at total reflux with several standard test mixtures. Figure 12.67 shows how the performance of a no. 2 packing compares with that of two random packings (50-mm Pall rings and no. 40 IMTP) and that of a structured packing with twice the surface and fabricated from metal gauze (Sulzer BX). The structured packings are more efficient (lower HETP values) than the random packings.

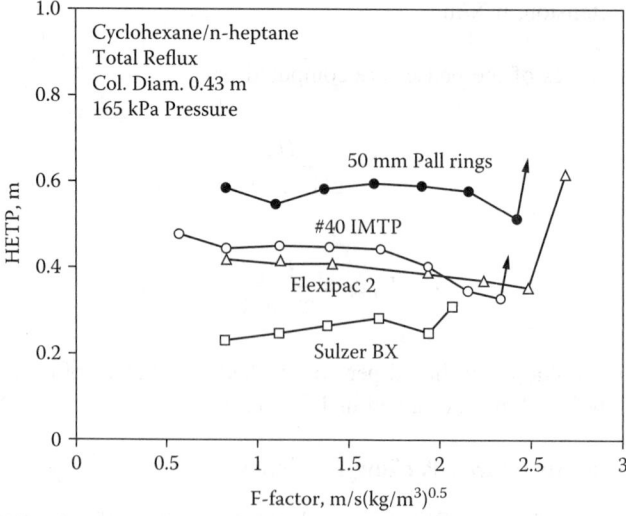

FIGURE 12.67 Efficiency comparison of several packings, cyclohexane/n-heptane test mixture. Total reflux, 0.43 m column, 165 kPa. (Z. Olujic et al., 2000. *Chem. Eng. Proc.* 39: 335.)

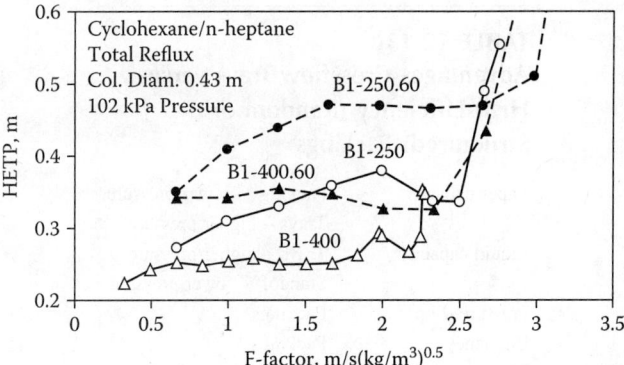

FIGURE 12.68 Efficiency comparison, two structured packings with 45° and 60° corrugation angles. Same conditions as for Figure 12.67. (Z. Olujic et al., 2000. *Chem. Eng. Proc.* 39: 335.)

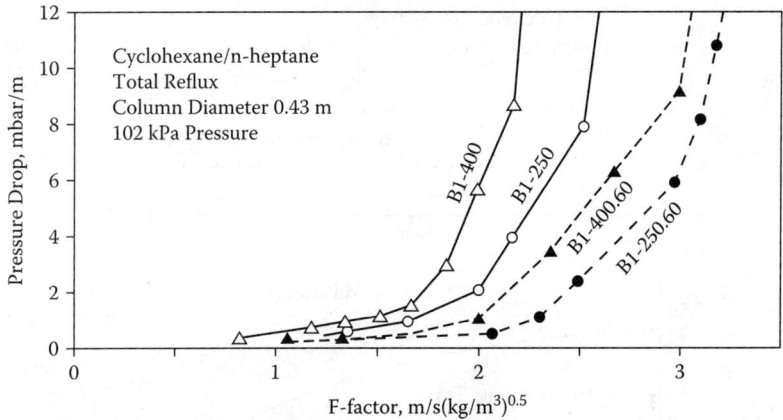

FIGURE 12.69 Pressure drop comparison, two structured packings. Same conditions as for Figure 12.67. (Z. Olujic et al., 2000. *Chem. Eng. Proc.* 39: 335.)

The capacities are about equal for all the packings. The pressure drops (not shown) of the structured packings are lower than those of the random packings. However, the costs of the structured packings are generally considerably greater on a volume basis than those of the random packings.

Figure 12.68 shows a comparison of two structured packings having the same surface texturing but different surface areas and corrugation angles. The packing with the higher area, Montzpak B1-400, has a better efficiency but a lower capacity. The higher-angle packing, 60° from the horizontal, has a lower efficiency but a higher capacity. Figure 12.69 gives a companion to Figure 12.67, showing comparative pressure drops. The higher-angle packing has a lower pressure drop and a higher capacity but lower efficiency. Obviously, several variables must be adjusted when optimizing the design of a bed of structured packing.

12.8.4 Packings versus Crossflow Trays

Choices between packing types and between packings and trays are listed in Table 12.13. In general, trays are cheaper, especially when the cost of expensive distributors and redistributors is added to the packing cost. But packings give higher efficiency and thus can permit lower reflux flows for the same separations. There are also capacity considerations, as indicated qualitatively in Figure 12.70. Packings have capacity advantages over trays at low flow parameter values, typically

TABLE 12.13
Advantages-Crossflow Trays versus High-Efficiency (Random or Structured) Packings

Vapor capacity	Packings – lower pressure
	Trays – higher pressure
Liquid capacity	Trays – higher pressure
	Standoff – lower pressure
Pressure drop	Packings
Efficiency	Packings
Efficiency modeling	Packings
Foaming	Packings
Fouling	Trays
Chemical reactions	Trays
Control	Trays
Ultra-pure products	Trays
Designer comfort	Trays – hydraulics
	Packings – efficiency
Scaleup	Trays
Cost	Depends on situation

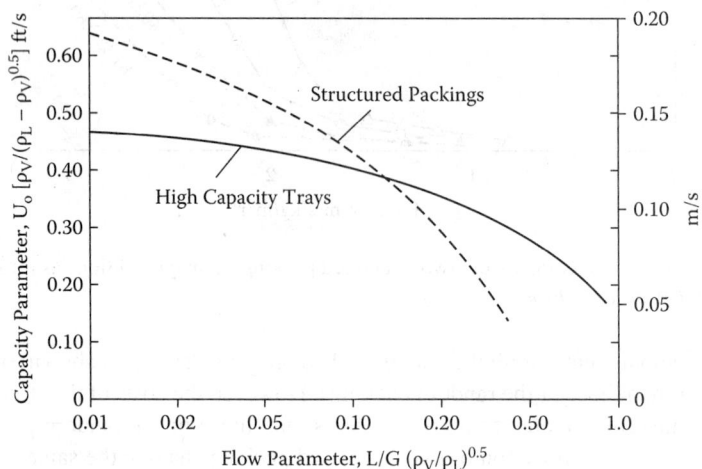

FIGURE 12.70 Generalized comparison of capacities, structured packings vs. high-capacity trays. [J. L. Bravo, 1998. *Chem. Eng.* 105 (2): 77.]

encountered in vacuum distillations; at high parameters, typical of high-pressure distillations, trays can handle the larger liquid flows.

Problem 12.11

Structured packing is now considered for the ethylbenzene/styrene separation. In Problem 12.10, 50-mm Pall rings were used in a 3.20-m column to give an HETP of 1.01 m. For the present case, a 3.05-m column will be used (i.e., rated) along with Mellapak 250Y (see Table 12.12). This is a sheet metal structured packing with about 250 m^2/m^3 surface and a corrugation angle of 45°. with the horizontal. For pressure drop and efficiency, the method of Rocha et al. [Equations (12.114)–(117)] is used.

Distillation

Results: From a computer program, the results are shown:

Flooding Calculations

Superficial vapor velocities	
At the flood point	6.113 m/s
At design conditions	4.519 m/s
Column diameter	3.050 m
Flooding factor (%)	73.9

Pressure Drop Calculations

Friction factor	0.190
Fractional liquid holdup	0.047
Dry pressure drop	0.0023 bar/m
Wet pressure drop	0.0036 bar/m
Flood pressure drop	0.0103 bar/m

Mass Transfer Calculations

Stripping factor	1.173
Effective interfacial area	148.0 1/m
Ae/ap	0.592
Mass transfer coefficient, liquid	1.372E−04 m/s
Mass transfer coefficient, vapor	7.443E−02 m/s
Height of a transfer unit, liquid	0.110 m
Height of a transfer unit, vapor	0.410 m
Overall height of a vapor transfer unit	0.539
Number of transfer units	11.135
HETP, plug flow	0.497

The HETP is 0.50 m, compared with 1.01 m for 50-mm Pall rings. A 3.05-m column diameter shows 74% flooding, a pressure drop of 0.0036 bar/m. The 6-m packed section can achieve 11.1 transfer units. At this loading, 59% of the total packing surface is wetted.

From the previous problems we can now show a comparison of devices:

	Sieve Tray	50 mm Pall Rings	Structured Packing
Col. Diam., m	3.36	3.21	3.05
HETP, m	1.33	1.01	0.50
ΔP, mm liq/m	72	66	44
Stages/P	0.012	0.015	0.045

12.8.4.1 Counterflow Trays

A model for predicting the efficiency of dualflow-type trays has been reported[47] and is mechanistic in character. A large efficiency database, provided by Fractionation Research, Inc. (FRI), was used in constructing the model. Efficiency data for counterflow baffle tray columns are sparse, and the only model available[51] is based largely on a heat transfer database, using the appropriate analogies.

12.9 TROUBLESHOOTING

Troubles with the performance of operating distillation columns generally fall into one of four categories: capacity for vapor or liquid flows falls short of design, pressure drop is higher than predicted, the separation is not as expected, or the column does not operate in a stable fashion.

Each of these will be discussed in further detail below. Other troubleshooting areas, not discussed, include mechanical failures, damages from fire and explosion, and problems with auxiliaries.

Kister[93], Lieberman,[94] and Saletan[95] provide details on troubleshooting.

12.9.1 Inadequate Vapor Capacity

The criterion here is usually a capacity correlation such as given in Figures 12.29, 12.41, 12.43, and 12.53, depending on the type of contacting device. If the capacity falls short of prediction, there are several possibilities: the correlation was not used properly, too much was expected of the correlation, there is unexpected foaming, or the column internals are not arranged according to specification or are disarranged as a result of an internal vapor surge.

A gamma-ray scan of the column can generally pinpoint problem areas or zones. The rays penetrate the column metal, and their absorption depends on the amount of fluids or solids in their paths. Several commercial organizations specialize in scanning, and they are capable of traversing the entire column, with rays emitted in several different directions.

12.9.2 Liquid Flow Capacity

For tray columns, liquid flow constrictions are normally found in the tray downcomers. Example causes are lack of adequate froth collapse time, excessive pressure drop for flow under the downcomer baffle, and foreign materials being left in the downcomer by construction people. For packed columns, clogged support plates, broken packings, or growth of fouling deposits can cause liquid flow restrictions.

12.9.3 Pressure Drop

The usual problem is excessive pressure drop. It can be caused by flow constrictions/plugging, a bad predictive model, faulty internals, flow rates higher than design, entrainment, and weeping.

12.9.4 Separation Efficiency

Possible explanations for this shortcoming are faulty equilibrium data (relative volatility different from design), use of an improper stage model, wrong material balance (e.g., trace impurities), an erroneous scaleup model, or column internals being less efficient than anticipated.

In the previous sections, the comment was made that the process engineer should make certain that all possible components of the feed are identified. The comment should be extended to a concern about whether "hidden" azeotropes might be present.

12.9.5 Stability

Unstable operation often results from control system problems, improper design of the reboiler, or very low flow rates of liquid (leading to pulsing). Some devices, such as dualflow trays (with no downcomers), tend toward oscillating flows. The most common cause of stability is the control system.

12.10 NOMENCLATURE

a_i = interfacial area for mass transfer, m²/m³
a_i' = effective interfacial area for mass transfer, m²/m³
a_p = total surface area of packing, m²/m³
A = areas of trays, m²
A_a = active area of tray
A_d = downcomer area
A_{da} = area under downcomer baffle
A_h = hole area
A_n = net area

Distillation

A_t	= total (superficial) area
A,B,C	= constants in Antoine equation [Equation (12.13)]
A_{ij}, A_{ji}	= constants in Van Laar relationship [Equation (12.20)]
A_1, B_1	= constants in Lockett equation [Equation (12.104)]
A_n	= absorption factor [Equation (12.43)]
C_F	= correction factor for approach to flood [Equation (12.102)]
C_{ov}	= discharge coefficient for sieve trays
C_s	= capacity factor for packings, m/s (Figure 12.53)
C_{sb}	= Souders–Brown capacity coefficient for trays, m/s
d_b	= bubble diameter, mm or m
d_h	= hole diameter, mm or m
D_c	= column diameter, m
D_e	= eddy diffusion coefficient, m²/s
D_G	= molecular diffusion coefficient for gas or vapor, m²/s.
D_L	= molecular diffusion coefficient for liquid, m²/s
E	= mass transfer efficiency, fractional
E_{oc}	= overall column efficiency
E_{og}	= point (local) efficiency, vapor basis
E_{mv}	= Murphree tray efficiency, vapor basis
f	= fractional approach to flooding condition [Equation (12.84)]
F	= feed rate, kg-moles/s
F_p	= packing factor (Figure 12.53)
FP, F_{lv}	= flow parameter (Figure 12.29)
Fr	= Froude number
F_{SE}	= surface texturing factor [Equation (12.115)]
F-factor	= vapor flow factor = $U\rho_v^{0.5}$
F_{va}	= active area F-factor for vapor, $U_a \rho_v^{0.5}$
F_{vs}	= superficial (total) area F-factor, $U_s \rho_v^{0.5}$
GM	= molar vapor rate, lb-moles/hr-ft2
h	= pressure head, mm or m of liquid
h_d	= head loss through holes
h_{ad}	= head loss under downcomer baffle
h_{dc}	= head of clear liquid in downcomer
h_{dc}'	= head of aerated liquid in downcomer
h_L	= head loss through liquid ~ liquid holdup
h_L'	= residual loss on dualflow trays
h_σ	= head loss in bubble formation
h_t	= total head loss across tray
h_{ow}	= clear liquid over weir
hw	= weir height, mm or m
h, H	= enthalpy of liquid, vapor, J/kg
H	= height of a transfer unit, m
H_L	= liquid
H_v	= vapor or gas
H_{og}	= overall, vapor or gas basis
HETP	= height equivalent to a theoretical plate, m
k	= mass transfer coefficient, kg-moles/(s-m²-mole fraction)
k_v	= vapor (gas)
k_L	= liquid
K_{ov}	= overall, vapor basis
K_c, K_o	= discharge coefficients for valve trays [Equations (12.71) and (12.72)]

K	= vapor-liquid equilibrium ratio ("K value")
L	= molar liquid rate, lb-moles/hr-ft^2
l_w	= weir length, in
L, L_M	= liquid molar flow rate, kg-moles/s
L'	= liquid flow rate, kg-/s-m^2
L/G	= liquid to vapor mass flow rate ratio
L_w	= liquid flow rate over the tray weir, ft^3/s-m
m	= slope of equilibrium curve
m_1	= exponent in Equation (12.101)
M_L	= molecular weight of liquid
n_1	= exponent in Equation (12.101)
N	= number of trays or stages
N_a	= actual stages or trays
N_m	= minimum stages
N_t	= total trays
N	= number of transfer units (also NTU)
N_L	= number of liquid phase transfer units
N_{ov}	= number of overall transfer units, liquid basis
N_v	= number of vapor phase transfer units
N_A	= molar flux, species A
p	= partial pressure, atm or bar
p^{sat}	= vapor pressure, atm or bar
P	= total pressure, atm or bar
q	= liquid flow rate, m^3/s
Q	= vapor flow rate, m^3/s
R	= reflux ratio
R_m	= minimum reflux ratio
Re	= Reynolds number, dimensionless [Equation (12.108)]
S	= length of a side of corrugation, structured packing, m
Sc	= Schmidt number, dimensionless [Equation (12.109)]
Sh	= Sherwood number, dimensionless [Equation (12.109)]
t	= residence time, s
T	= absolute temperature
u_a	= liquid velocity to active area of tray (q/A_a), m/s
u_{Lh}	= liquid velocity through distributor holes, m/s
U	= vapor or gas velocity, m/s
U_a	= vapor velocity through active area of tray (Qv/Aa), m/s
U_g	= actual vapor velocity in structured packing, m/s
U_L	= actual liquid velocity on structured packing surface (film flow), m/s
v	= vapor or gas flow for an individual component, m/s
V	= vapor or gas flow rate, kg-moles/s
V'	= vapor sidestream flow, kg-moles/s
W	= length of flow travel on tray, m
x	= liquid mole fraction x^* = equilibrium value
y	= vapor mole fraction y^* = equilibrium value
z	= mole fraction in feed
Z	= height of packing, m
Z_f	= height of froth on tray, m

Distillation

SYMBOLS

α	= relative volatility (α_{ij} for i-j binary)
β	= froth aeration factor [Equation (12.60)]
β	= fractional liquid coverage of packing surface [Equation (12.111)]
γ	= activity coefficient
γ	= liquid-solid contact angle, deg [Equation (12.115)]
Γ	= physical property correction factor [Equation (12.103)]
ε	= void fraction of packed bed
θ	= corrugation angle for structured packing (from horizontal), deg
λ	= ratio of slopes, equilibrium/operating lines
μ	= viscosity, Pa-s
ν	= kinematic viscosity, cSt
ρ	= density, kg/m³
σ	= surface tension, mN/m (dynes/cm)
ϕ'	= packing parameter [Equation (12.102)]
ϕ_f	= froth or spray relative density [Equation (12.63)]
ψ	= packing parameter, Eq 101)
ψ	= entrainment ratio, moles/mole gross downflow liquid

SUBSCRIPTS

A	= species A
B	= bottoms
D	= distillate
eff	= effective
F	= feed
G	= gas phase
i,j,k,m	= components i,j,k,m
L	= liquid phase
M	= molar
$n, n-1, n+1$	= stages n, $n-1$, $n+1$
org	= organics
V	= vapor
W	= water

SUPERSCRIPTS

L	= liquid phase
V	= vapor phase
F	= feed
∞	= infinite dilution

REFERENCES

1. Brunschwig, Hieronymus, *Liber de arte distillandi; Das buch der rechten kunst zu distillieren.* Strassburg, 1500.
2. Walas, S. M. *Phase Equilibria in Chemical Engineering.* Reading, MA: Butterworths, 1985.
3. Reid, R. C., Prausnitz, J. M., and Poling, B. E. *The Properties of Gases and Liquids,* 4th ed. New York: McGraw-Hill, 1987.

4. Van Laar Z. *Physik. Chem.* 1910, *72*, 723; 1913, *83*, 599.
5. Gmehling, J., Onken, U., and Arlt, W. *Vapor-Liquid Equilibrium Collection,* Chemistry Data Series (continuing series). Frankfurt, Germany: DECHEMA, 1979–.
6. Fredenslund, A., Gmehling, J., and Rasmussen, P. *Vapor-Liquid Equilibria Using UNIFAC.* Amsterdam: Elsevier, 1977.
7. Hala, E., Pick, J., Fried, V., and Vilim, O. *Vapor-Liquid Equilibrium,* 2nd ed., Oxford, U.K.: Pergammon, 1967.
8. McCabe, W. L. and Thiele, E. 1925. *Ind. Eng. Chem.* 17: 605.
9. Seader, J. D. and Henley, E. J. *Separation Process Principles.* New York: John Wiley & Sons, 1998.
10. Stichlmair, J. and Fair, J. R. *Distillation—Principles and Practices.* New York: Wiley-VCH, 1998.
11. Ponchon, M. 1921. *Tech. Mod.* 13 (20): 55.
12. Savarit, R. 1922. *Arts Metiers,* 65, 142, 178, 241, 266, 307.
13. Fenske, M. R. 1932. *Ind. Eng. Chem.* 24: 482.
14. Underwood, A. J. V. 1948. *Chem. Eng. Prog.* 44: 603.
15. Gilliland, E. R., 1940. *Ind. Eng. Chem.* 32: 1220.
16. Rusche, F. A. 1999. *Hydrocarb. Proc.* 78 (12): 41.
17. Fair, J. R. and Bolles, W. L. 1968. *Chem. Eng.* 75 (9): 156.
18. Lewis, W. K. and Matheson, G. L. 1932. *Ind. Eng. Chem.* 24: 494.
19. Thiele, E. and Geddes, R. L. 1933. *Ind. Eng. Chem.* 25: 289.
20. Stichlmair, J., Fair, J. R., and Bravo, J. L. 1989. *Chem. Eng. Prog.* 85 (1): 63.
21. Horsley, L. H. *Azeotropic Data—III.* Advances in Chemistry Series No. 116. Washington, DC: American Chemical Society, 1973.
22. Gmehling, J., Menke, J., Krafczyk, J., and Fischer, K. *Azeotropic Data.* New York/Weinheim: VCH Publishers, 1994.
23. Frank, T. 1997. *Chem. Eng. Prog.* 93: 52–63.
23a. Ewell, R. H., Harrison, J. M., and Berg, L. 1944. *Ind. Eng. Chem.* 36: 871.
24. Barton, P. and Roche, E. C. Batch Distillation, Section 1.3 in *Handbook of Separation Techniques,* P. A. Schweitzer, ed. New York: McGraw-Hill, 1997.
25. Diwekar, U. M. *Batch Distillation: Simulation, Optimal Design and Control.* Washington, DC: Taylor and Francis, 1995.
26. Aspen Technology, Cambridge, MA, Program *Batchfrak.*
27. Agreda, V., Partin, L., and Heise, W. 1990. *Chem. Eng. Prog.* 86 (2): 40.
28. Fair, J. R. 1998. *Chem. Eng.* 105 (11): 158.
29. Subawalla, H. and Fair, J. R. 1999. *Ind. Eng. Chem. Res.* 38: 3696.
30. Fair, J. R. 1984. Historical Development of Distillation Equipment, *AIChE Symposium Series No. 235, 79:* 1.
31. Fractionation Research, Inc., P. O. Drawer F, Bartlesville, OK 74005. Website www.fri.com.
32. FRI reports are available from Oklahoma State Archives, Stillwater, OK 74078.
33. Kunesh, J. G., Kister, H. Z., Lockett, M. J., and Fair, J. R. 1995. *Chem. Eng. Prog.* 91 (10): 43.
34. Billet, R., Conrad, S., and Grubb, C. M. 1969. *IChemE Symp. Ser. 32,* 5: 111.
35. Fair, J. R., 1997. *Perry's Chemical Engineers' Handbook,* 7th ed. R. H. Perry and D. Green, eds., New York: McGraw-Hill.
36. Kister, H. Z. and Haas, J. R., 1990. *Chem. Eng. Prog.* 86 (9): 69.
37. Silvey, F. C. and Keller, G. J. 1966. *Chem. Eng. Prog.* 67 (1): 69.
38. Fair, J. R. and Mathews, R. L. 1958. *Petrol. Refiner* 37 (4): 153–158; Fair, J. R. 1961. *Petro/Chem Eng.* 33 (10): 57–64.
39. Leibson, I., Kelley, R. E., and Bullington, L. A. 1957. *Pet. Refiner* 36 (2): 127.
40. Hoek, P. J. and Zuiderweg, F. J. 1982. *AIChE J.* 28: 535.
41. Bolles, W. L. 1963. *Design of Equilibrium Stage Processes,* ed. B. D. Smith, Chapter 14. New York: McGraw-Hill, 1963.
42. *Perry's Chemical Engineer's Handbook,* 5th ed., R. H. Perry and C. H. Chilton, eds. New York: McGraw-Hill, 1973, 18-10–18-12.
43. Klein, G. F. 1982. *Chem. Eng.* 89: 81 (May 3, 1982).
44. Bolles, W. L. 1976. *Chem. Eng. Prog.* 72 (9): 43.
45. Nutter, D. E. 1979. *IChemE Symp. Ser. 56,* 3.2: 47.

46. Nutter, D. E. and Perry, D. Sieve Tray Upgrade 2.0—The MVG® Tray, presented at AIChE Meeting, Houston, TX, March 1995.
47. Garcia, J. A. and Fair, J. R. 2002. *Ind. Eng. Chem.* 41: 1632.
48. Kastanek, F., Huml, M., and Braun, V. 1969. *Proc. Intl. Symp. Distillation (Brighton)* 5: 100.
49. Zuiderweg, F. J., De Groot, J. H., Meeboer, B., and van der Meer, D. 1969. *Proc. Intl. Symp. Distillation (Brighton)* 5: 78.
50. Fair, J. R. 1993. *Hydrocarbon Proc.* 72 (5): 75.
50a. Fractionation Research, Inc., Progress Report, Dec. 1951 [available from Oklahoma State Univ. Library].
51. Fair, J. R. 1990. *Trans. ASME/J. Solar Energy Eng.* 112: 216–222.
52. Delnicki, W. V. and Wagner, J. L. 1956. *Chem. Eng. Prog.* 52 (1): 28.
53. Frank, J. C., Geyer, G. R., and Kehde, H. 1969. *Chem. Eng. Prog.* 65 (2): 79.
54. Billet, R. 1969. *Proc. Intl. Symp. Distillation (Brighton)* 4: 42.
55. Shulman, H. L., Ullrich, C. F., and Wells, 1955. *N. AIChE J.*, 1: 247, 253.
56. Sherwood, T. K., Shipley, G. H., and Holloway, F. A. L. 1938. *Ind. Eng. Chem.* 30: 765.
57. Eckert, J. S. 1970. *Chem. Eng. Prog.* 66 (3): 39.
58. Strigle, R. F. *Packed Tower Design and Applications.* Houston: Gulf Publ. Co., 1994.
59. Fair, J. R. Chap. 5 In *Handbook of Separation Process Technology.* R. W. Rousseau, ed., New York: John Wiley & Sons, 1987, 313.
60. Kister, H. Z. 1992. *Distillation—Design.* New York: McGraw-Hill.
61. Stichlmair, J., Bravo, J. L., and Fair, J. R. 1989. *Gas Sepn. Purif.* 3: 19.
62. Rocha, J. A., Bravo, J. L., and Fair, J. R. 1993. *Ind. Eng. Chem. Res.* 32: 641.
63. Silvey, F. C. and Keller, G. J. 1969. *Proc. Intl. Symp. Distillation (Brighton)* 4: 18.
64. Potthoff, R. 1992. *Maldistribution in Füllkörperkolonnen*, Ph.D. dissertation, Univ. of Essen, Germany.
65. Trompiz, C. and Fair, J. R. 2000. *Ind. Eng. Chem. Res.* 39: 1809.
66. Yoeman, N. 1998. *Chem. Eng. Prog.* 94 (3): 17.
67. Hines, A. L. and Maddox, R. N. 1985. *Mass Transfer Fundamentals and Applications.* Englewood Cliffs, NJ: Prentice-Hall.
68. O'Connell, H. E. 1946. *Trans. AIChE* 42: 741.
69. Fair, J. R., Null, H. R., and Bolles, W. L. 1983. *Ind. Eng. Chem. Proc. Des. Dev.* 22: 53.
70. Garcia, J. A. 1999. *Fundamental Model for the Prediction of Distillation Sieve Tray Efficiency: Hydrocarbon and Aqueous Systems.* Ph.D. dissertation, Univ. of Texas at Austin.
71. *Bubble Tray Design Manual,* New York: American Institute of Chemical Engineers (AIChE) 1958.
72. Chan, H. and Fair, J. R. 1984. *Ind. Eng. Chem. Proc. Des. Dev.* 23: 814, 820.
73. Bennett, D. L., Agrawal, R., and Cook, P. J. 1983. *AIChE J.* 29: 434.
74. Garcia, J. A. and Fair, J. R. 2000. *Ind. Eng. Chem. Res.* 39: 1809, 1818.
75. Bennett, D. L. and Grimm, H. J. 1991. *AIChE J.* 37: 589.
76. Colburn, A. P. 1936. *Ind. Eng. Chem.* 28: 526.
77. Lewis, W. K. 1936. *Ind. Eng. Chem.* 28: 399.
78. Taylor, R. and Krishna, R. *Multicomponent Mass Transfer.* New York: John Wiley & Sons, 1993.
79. Toor, H. L. and Marchello, J. M. 1958. *AIChE J.* 4: 97.
80. Onda, K., Takeuchi, H., and Okumuto, Y. J. 1968. *Chem. Eng. Japan* 1: 56.
81. Bravo, J. L. and Fair, J. R. 1982. *Ind. Eng. Chem. Proc. Des. Devel.* 21: 162.
82. Wagner, I., Stichlmair, J., and Fair, J. R. 1997. *Ind. Eng. Chem. Res.* 36: 227.
83. Bolles, W. L. and Fair, J. R. 1982. *Chem. Eng.* 89 (14): 109.
84. Orts, P. 1995. *Performance and Scale-up of a Laboratory-Scale Distillation Column Containing Random Packing,* diploma thesis, Université de Liège, May.
85. Lockett, M. J. 1998. *Chem. Eng. Prog.* 94 (1): 60.
86. Bravo, J. L., Rocha, J. A., and Fair, J. R. 1985. *Hydrocarbon Proc.* 64 (9): 91.
87. Gilliland, E. R. and Sherwood, T. K. 1934. *Ind. Eng. Chem.* 26: 516.
88. Johnstone, H. F. and Pigford, R. L. 1942. *Trans. AIChE* 38: 25
89. Fair, J. R. and Bravo, J. L. 1990. *Chem. Eng. Prog.* 86 (1): 19.
90. Fair, J. R., Seibert, A. F., and Olujic, Z. 2000. *Ind. Eng. Chem. Res.* 39: 1788.
91. Rocha, J. A., Bravo, J. L., and Fair, J. R. 1996. *Ind. Eng. Chem. Res.* 35: 1660
92. Hufton, J. R., Bravo, J. L., and Fair, J. R. 1988. *Ind. Eng. Chem. Res.* 27: 2096.

93. Kister, H. Z. 2006. *Distillation-Troubleshooting*. New York: Wiley-Interscience.
94. Lieberman, N. P. 1983. *Process Design for Reliable Operations*. Houston: Gulf Publ. Co.
95. Saletan, D. 1994. *Creative Troubleshooting in the Chemical Process Industries*. New York/London: Chapman & Hall.

13 Absorption and Stripping

James R. Fair

CONTENTS

13.1 Introduction...1074
    13.1.1 Functions of Absorption and Stripping..1074
    13.1.2 Commercial Applications..1074
    13.1.3 Solubility and Phase Equilibria..1075
        13.1.3.1 Sources of Solubility Data..1077
        13.1.3.2 Multicomponent Solubilities...1078
        13.1.3.3 Graphical Representation of Solubility............................1079
    13.1.4 Solvent Selection...1079
    13.1.5 Theoretical Stages/Transfer Units...1080
13.2 Plate Columns...1080
    13.2.1 Absorption Stages..1081
        13.2.1.1 Multiple Solutes...1084
        13.2.1.2 Concentrated Feeds...1085
        13.2.1.3 Minimum Solvent Rate..1087
        13.2.1.4 Heat Effects..1087
    13.2.2 Stripping Stages...1087
        13.2.2.1 Minimum Stripping Gas Rate..1090
    Equipment Design...1090
        13.2.2.2 Capacity and Column Diameter..1092
        13.2.2.3 Pressure Drop..1093
    13.2.3 Stage Efficiency...1095
13.3 Packed Columns...1096
    13.3.1 Transfer Units..1096
    13.3.2 Number of Transfer Units...1097
    13.3.3 Heights of Transfer Units...1097
    13.3.4 Heights of Theoretical Stages...1098
    13.3.5 Equipment Design...1098
    13.3.6 Packing and Packed Bed Characteristics...1100
    13.3.7 Maximum Gas/Liquid Capacity..1100
    13.3.8 Pressure Drop..1101
    13.3.9 Liquid Distribution and Redistribution..1101
    13.3.10 Mass Transfer..1101
13.4 Other Devices...1104
    13.4.1 Spray Chamber...1106
    13.4.2 Crossflow Scrubber...1107
    13.4.3 Venturi Scrubber...1107
13.5 Steam Stripping..1107
13.6 Chemical Absorption..1111
    13.6.1 Computational Approach..1111

13.7 Summary 1114
13.8 Nomenclature 1114
  13.8.1 Greek Letters 1115
  13.8.2 Subscripts 1116
References 1116

13.1 INTRODUCTION

13.1.1 FUNCTIONS OF ABSORPTION AND STRIPPING

Absorption and stripping are counterpart processes used in the process industries for separating gas or liquid mixtures. Their functions are generally of two types: *recovering* one or more components from the mixture or *purifying* the mixture to meet specifications or standards. In absorption, a gas mixture is contacted with a suitable liquid *solvent* to remove preferentially one or more mixture components from the gas phase. Conversely, in stripping, a liquid mixture is contacted by a suitable *stripping gas* to remove preferentially one or more components from the liquid phase. In practice, absorption and stripping are often coupled processes, as shown later. Both processes involve gas-liquid contacting, generally in vertical, cylindrical towers called *absorbers* or *strippers*.

Processes of this type are distinguished between physical and chemical absorption or stripping. For physical processes, transfer of the solute between phases is by physical mechanisms without any chemical reactions. For chemical processes, the solute reacts with a component of the solvent, resulting in an increased capacity of the solvent for the solute. Conversely, stripping sometimes breaks the chemical bonds between solute and solvent.

13.1.2 COMMERCIAL APPLICATIONS

Absorption is often used to purify a gas stream; an alternate name of the operation is *scrubbing*. Contaminated air, for example, can be purified by scrubbing out the contaminants with a selective solvent. Another example is removing sulfur and carbon dioxide from natural gas using solvents such as alkanolamines. As the solutes are being removed from the gas, the solvent becomes richer in these materials. To purify the solvent for reuse, stripping is needed. The contaminants are removed from the liquid by the action of a stripping gas. The general flow diagram for this coupled operation is shown in Figure 13.1. Such a process is useful for reducing the level of gaseous pollutants. Representative commercial applications of physical and chemical absorption are shown in Table 13.1.

Not all absorptions and strippings are as straightforward as indicated in Figure 13.1. Absorption may be used to produce a product—for example, the absorption of nitrogen oxides in water to form nitric acid. Or absorption may be used to recover valuable products, such as propanes and butanes from natural gas. And stripping may be a once-through operation such as the removal of volatile organic compounds (VOCs) from groundwater using atmospheric air for the stripping medium.

Flow diagrams for representative absorption and stripping processes are shown in Figures 13.2 and 13.3. In Figure 13.2, acetone is scrubbed from air to meet an exit air specification. The acetone transfers to the water phase, and the rich solvent flows to a distillation column wherein the acetone is recovered in high purity and the water is recycled to the absorber. In Figure 13.3, trichloroethylene (TCE) is stripped from groundwater by air, with the discharge from the stripper flowing through a carbon bed in which the TCE is adsorbed.

Obviously, a solvent must be found that has an affinity for the constituent(s) to be absorbed, and the stripping medium must be such that the physical or chemical bond between solute and solvent can be easily broken. This affinity, or solubility, is a critical consideration in the design of absorption/stripping systems, as discussed next.

Absorption and Stripping

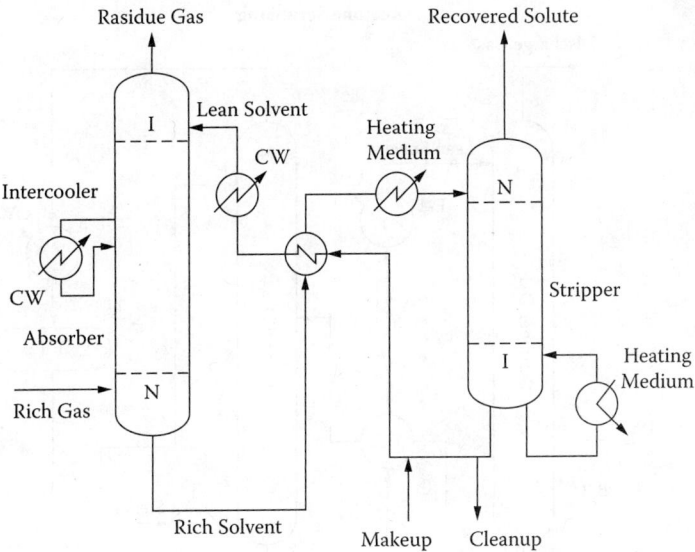

FIGURE 13.1 Typical flow diagram for a coupled absorption-stripping unit. Because the solvent flows in a closed circuit, allowance is made for losses ("makeup") as well as removal of accumulated impurities ("cleanup"). An intercooler is shown to suggest the need to remove heat of solution in the absorber.

TABLE 13.1
Representative Commercial Applications of Absorption

Solute	Solvent	Type of Absorption
Acetone	Water	Physical
Ammonia	Water	Physical
Formaldehyde	Water	Physical
Hydrochloric acid	Water	Physical
Sulfur trioxide	Water	Physical
Benzene and toluene	Hydrocarbon oil	Physical
Butanes and propane	Hydrocarbon oil	Physical
Carbon dioxide	Aq. NaOH	Irreversible chemical
Hydrochloric acid	Aq. NaOH	Irreversible chemical
Chlorine	Water	Reversible chemical
Carbon monoxide	Aq. cuprous ammonium salts	Reversible chemical
Carbon dioxide	Aq. ethanolamines	Reversible chemical
Hydrogen sulfide	Aq. ethanolamines	Reversible chemical
Nitrogen oxides	Water	Reversible chemical

13.1.3 Solubility and Phase Equilibria

If an absorber is to be designed for efficient and economical service, the solvent chosen must have a special affinity for the one or more components of the gas mixture that are to be selectively removed. In other words, a degree of *solubility* is needed. Solubility data, which relate the solute, the solvent, the concentrations, and the temperature-pressure conditions, may be found in various compendia and journal articles. In the past, it was common to report solubility as a *Bunsen solubility coefficient*:

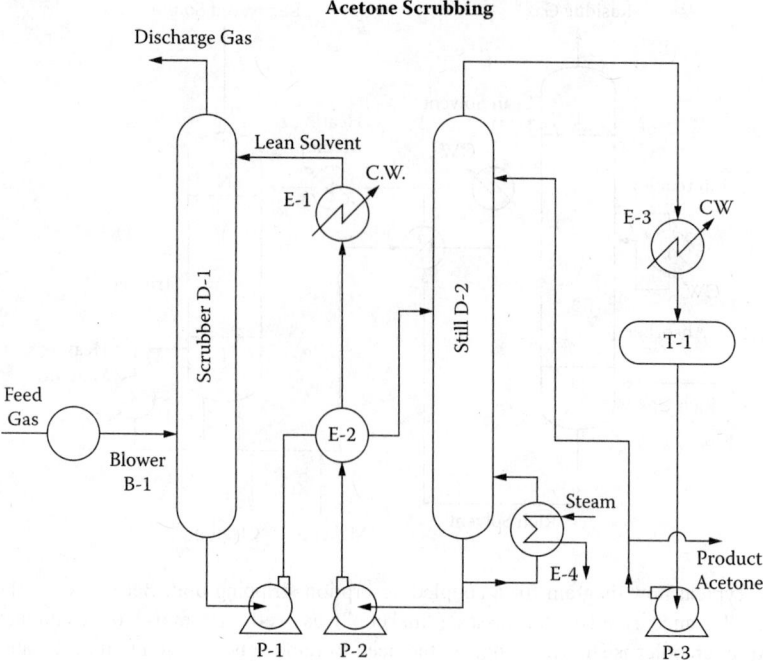

FIGURE 13.2 System for scrubbing acetone from air using a circulating solvent. Water or a non-volatile organic liquid can be used as the solvent.

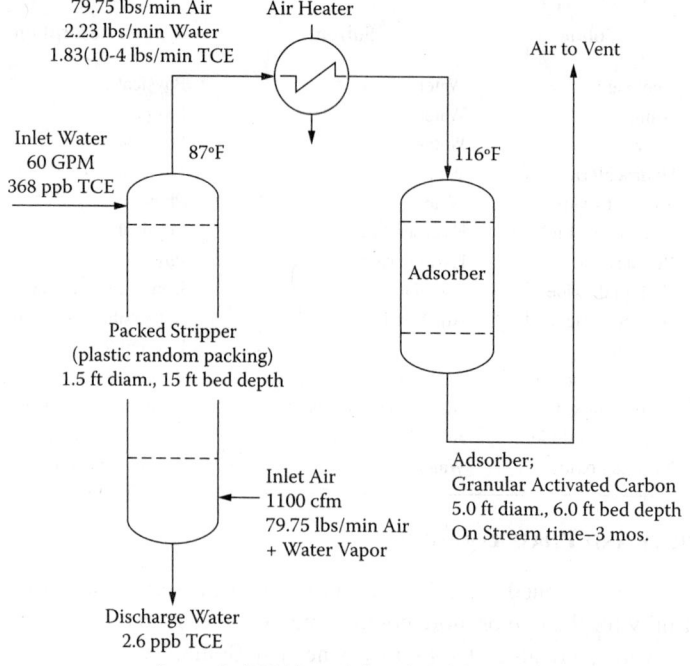

Removal of Trichloroethylene (TCE) from Groundwater

FIGURE 13.3 System for removing volatile organic compounds (VOCs) from groundwater, by stripping with atmospheric air. The exit gas contains the stripped VOCs and must be purified by flow through a bed of activated carbon.

Absorption and Stripping

$$\alpha'_i = \frac{vol\ i\ dissolved\left[0°, 1.0\ atm\right]}{(vol.solvent)(partial\ pressure\ of\ i\ in\ gas, atm)},\ atm^{-1} \quad (13.1)$$

(An older variation of this term is the *Ostwald coefficient*, which uses the gas volume at actual conditions instead of at standard conditions.) In general use today is the *Henry's law coefficient*, H_i:

$$H_i = \frac{partial\ pressure\ of\ i\ in\ gas, atm}{mole\ fraction\ of\ i\ in\ solvent} = \frac{p_i}{x_i} \quad (13.2)$$

A related term, widely used in distillation and absorption calculations, is the equilibrium ratio, K_i:

$$K_i = \frac{mole\ fraction\ of\ i\ in\ gas}{mole\ fraction\ i\ in\ liquid} = \frac{y_i^*}{x_i} \quad (13.3)$$

where the asterisk indicates an equilibrium condition. For certain absorption/stripping calculations, a modified K value will be used:

$$K'_i = \frac{moles\ i\ per\ mole\ solute\ free\ gas}{moles\ i\ per\ mole\ solute\ free\ solvent} = \frac{Y_i^*}{X_i} \quad (13.4)$$

The solute-free gas is sometimes called the "carrier gas."

For the nonideal solutions usually encountered in absorption/stripping problems, the K_i value can be expressed as follows:

$$K_i = \frac{\gamma_i p^{sat}}{P} \quad (13.5)$$

where γ_i is the liquid phase activity coefficient for i in the solvent, p^{sat} is the vapor pressure of i, and P is the total pressure. In many cases, especially those dealing with dilute solutions, the value of γ_i approaches the thermodynamically important infinite dilution coefficient γ_i^∞.

Most importantly, Equations (13.1) through (13.3) are based on thermodynamic equilibrium. A further discussion of the equilibrium ratio and its basis is found in Chapter 12, "Distillation," of this handbook. Note that the simple relationship

$$H_i = K_i P \quad (13.6)$$

where P is total pressure, can be useful in solubility studies because of the vast amount of work and reference data on the K value. One must be careful with units, particularly in utilizing data from various sources. In the following sections, the K value will be used, in some cases with molar ratios [Equation (13.4)].

13.1.3.1 Sources of Solubility Data

For solubility of various elements and compounds in water, there are many references, but only a few are mentioned here. Hwang et al.[1] have provided data for 404 common organic pollutants at 1.0 atm and temperatures of 25 and 100°C, the latter values being useful for steam stripping. Their data are for low-concentration conditions, the usual case for sparingly soluble organic compounds. For cases where experimental data do not exist, the predictive model UNIFAC,[2] discussed in Chapter 12 is recommended.

Problem 13.1

Acetone is to be scrubbed from dry air into water at 30°C and 1.0 atm. The acetone concentration is 2.60 vol-%. For the condition of thermodynamic equilibrium, calculate the following:

(a) The Henry's law coefficient
(b) The equilibrium ratio, K
(c) The liquid phase activity coefficient
(d) The Bunsen coefficient.

Data: molecular weights: acetone = 58, air = 29, water = 18
Vapor pressures at 30°C: acetone = 283 mm Hg, water = 31.8 mm Hg
According to the measurements of Othmer et al.,[17] for a partial pressure of acetone = 760(0.025) = 19.0 mm Hg, the equilibrium acetone composition in the liquid is 0.010 mole-%. Thus,

(a) H_i = 19/0.01 = 1900 mm Hg/m.f. = 2.5 atm/m.f.
(b) K value = 0.025/0.010 = 2.50 [$H_i = K_i P$ = 2.50(760) = 1900 mm/m.f.]
(c) Since $K_i = \gamma_i\ p^{sat}/P$, then γ_i = 2.5(760)/284 = 6.72
(d) Bunsen coefficient [Equation (13.1)]. Basis 1.0 mole equil. liquid. Moles acetone = 0.025. Vol. acetone at STP = 22,400(0.025) = 560 cm³. Vol. water = 18 (0.0975) = 1.755/1.00 = 1.755. Hence α_i = 560/(1.755 × 19/760) = 12,760 mm⁻¹ or 16.8 atm⁻¹

Yaws and coworkers[3] have presented many compilations and correlations of solubility data, for example, hydrocarbons in water. Methods are available for estimating the solubility of gases in organic liquids.[4-6] And special compendia include those in the Solubility Data Series,[7] Seidell and coworkers,[8-10] and Dack.[11] Extensive reviews have been provided by Markham and Kobe,[12] Long and McDevit,[13] Battino and Clever,[14] Wilhelm and Battino,[15] and Wilhelm et al.,[16] among others. Often data usually associated with vapor-liquid equilibrium in distillation can be used. Finally, various handbooks contain some useful solubility data.

13.1.3.2 Multicomponent Solubilities

The simplest case of absorption and stripping is the removal of a single component from the second component in a binary mixture. This means, however, that the solubility of the other component must also be considered. In Problem 13.1, the tacit assumption was made that air (the second component) is insoluble in water. For many situations, there is a mixture of varying solubilities to be removed in absorption, or varying effective vapor pressure to be stripped. Thus, *relative solubility* must be considered.

For dilute mixtures where the majority component is not being absorbed or stripped, the mixture components usually can be considered to behave as individual entities subject to their individual solubilities. Thus, for a mixture of 1000 ppm-vol trichloroethylene and 500 ppm-vol perchloroethylene, the solubility for each can be determined as if the other is absent. In general, for several components of varying solubility, the relative solubility is

$$\alpha'_{i...n} = K_i/K_{j...n} \qquad (13.7)$$

Absorption and Stripping

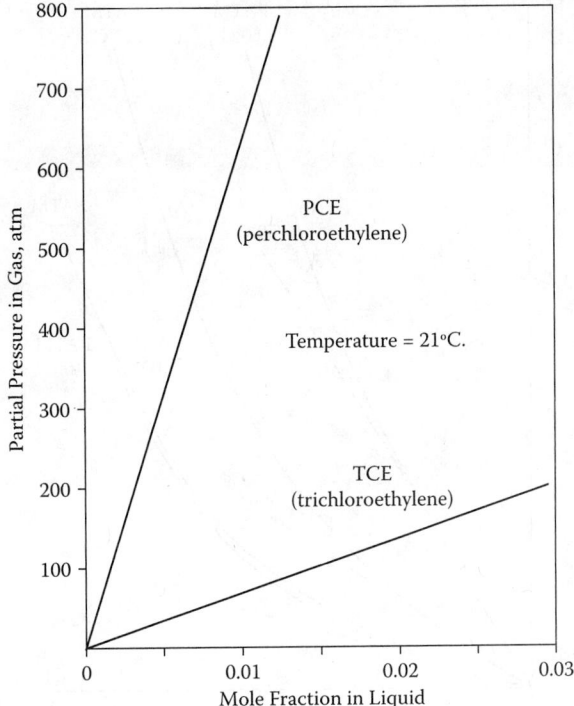

FIGURE 13.4 Solubility of trichloroethylene (TCE) and perchloroethylene (PCE) in water at 21°C. The PCE is much less soluble, requiring a relatively high partial pressure to achieve a given mole fraction liquid concentration.

13.1.3.3 Graphical Representation of Solubility

The equilibrium composition of a component in the gas is often plotted versus the equilibrium concentration of the same component in the liquid, at constant temperature. Figure 13.4 is a plot of solubilities of trichloroethylene (TCE) and perchloroethylene (PCE) in water at 21°C. The H_i values are $H_{TCE} = 6753$ and $H_{PCE} = 61{,}790$. The higher value means that PCE is less soluble than TCE. At 1.0 atm total pressure, the numerical values of K and H are the same.

Graphs, such as Figure 13.4, are useful for determining separation requirements in terms of stages or transfer units. An alternate form of plotting, useful for stage calculations, is the Y-X diagram, where Y represents the ratio of moles solute per mole of solute-free gas, and X represents the ratio of moles solute per mole of solute-free solvent. For mass transfer calculations, the simple mole fraction plot, y*-x, is convenient, as discussed later. For concentrated gas cases, particularly where heat effects can be significant, plots showing temperature effects (such as Figure 13.5) may be needed.

13.1.4 Solvent Selection

As discussed immediately above, an important criterion for solvent selection is solubility. The solvent must accommodate a reasonable amount of the species to be removed from the gas while still maintaining selectivity between the solute(s) and carrier gas. When the carrier is air or an inert gas, it is easy to designate solute vs. carrier; in other cases, it is a matter of relative solubility.

Other criteria for solvent selection include the following:

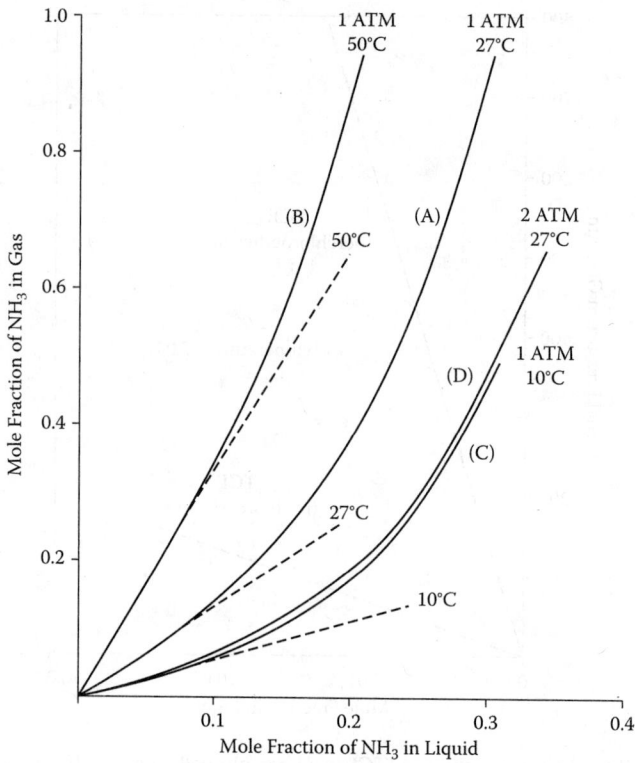

FIGURE 13.5 Solubility of ammonia in water from ammonia-air mixtures. The dashed lines indicate the slope of the lines near the origin. [Perry, R. H. (ed.), 1985. *Chemical Engineer's Handbook*, 5th ed., New York: McGraw-Hill.]

1. Availability and cost
2. Volatility (higher volatility can contaminate the absorber exit gas)
3. Stability, especially for recirculating systems
4. Corrosiveness
5. Ease of stripping
6. Convenience of disposal
7. Degree of hazard to humans and the environment

13.1.5 Theoretical Stages/Transfer Units

In designing an absorber or stripper, one must ascertain some index of "difficulty of separation." As in the case of distillation separations, this index is either the required *number of theoretical stages* or the required *number of transfer units*. These parameters are interchangeable, with stages often used for plate columns and transfer units for packed columns. Background on the parameters is provided in Chapter 12. For convenience in handling the associated hydraulics and mass transfer calculations, the remainder of this chapter is divided into three parts: "Plate Columns," "Packed Columns," and "Special Devices."

13.2 PLATE COLUMNS

Plate columns are vertical vessels, usually circular in cross section, that contain a multiplicity of plates that bring gas and liquid into intimate contact. These columns are the same as those used

Absorption and Stripping

for distillation. Most plates are connected by downcomers, where the downflowing liquid transfers from a plate to the next plate below. Typically, the plates have openings through which the upflowing gas passes and above which the gas and liquid make contact. If the gas leaving the plate is in thermodynamic equilibrium with the liquid leaving the plate, then a *theoretical stage* is provided.* To account for the failure to achieve equilibrium, *plate efficiency* is used. The computational approach is to determine the theoretical stages and then correct to actual stages (plates) by means of *plate efficiency*, in the same fashion as for distillation columns.

13.2.1 Absorption Stages

The number of stages required for a specified amount of absorption may be calculated from the Kremser–Brown equation[18] with stages numbered down from the top of the column:

$$E_{ai} = \frac{Y_{N+1} - Y_1}{Y_{N+1} - Y_o^*} = \frac{A_i^{N+1} - A_i}{A_i^{N+1} - 1} \tag{13.8}$$

where

E_{ai} = absorption efficiency of component i
Y_{N+1} = moles i per mole of solute-free gas, entering (bottom) condition
Y_1 = moles i per mole of solute-free gas, exit (top) condition
Y_o^* = moles i per mole of solute-free gas, determined as in equilibrium with i in the entering solvent
N = required theoretical stages
A_i = absorption factor for component $i = L'/(V'K_i)$
L' = moles of total liquid
V' = moles of total gas

For some studies, a rearrangement of Equation (13.8), to give explicit number of stages, is convenient:

$$N = \frac{\ln\left(\dfrac{E_{ai} - A_i}{E_{ai} - 1}\right)}{\ln A_i} - 1 \tag{13.9}$$

Different solutes require different numbers of stages; the "key solute," for which the separation is specified, prevails. Thus, the required stages for that solute are calculated, and then Equation (13.8) determines the degree of absorption for the other solutes (if any). There are other aspects of Equations (13.8) and (13.9) that should be noted. The value of the absorption factor can change with changes in the L/V ratio; for the very lean gas feeds, the value of L/V varies very little because the material balance changes very little. Also, the value of the absorption factor can change with changes in K. Again, this may not be important for the dilute case, which is essentially isothermal, but may present problems for the more concentrated cases. The net effect of these limitations is the problem of assigning some average value of A_i in Equations (13.8) and (13.9).

For varying L/V, an analytical approach utilizing the modified molar concentrations [see Equation (13.4)] may be used:

---

* The theoretical stage concept is discussed in detail in the section on distillation as well as in unit operations textbooks.

$$\text{Gas: } Y_i = \frac{y_i}{1-y_i} = \frac{\text{moles solute}}{\text{mole solute free gas}} = \frac{\text{moles solute}}{\text{mole "carrier gas"}} \quad (13.10)$$

$$\text{Liquid: } X_i = \frac{x_i}{1-x_i} = \frac{\text{moles solute}}{\text{mole solute free solvent}} = \frac{\text{moles solute}}{\text{mole solvent}} \quad (13.11)$$

Clearly, at very low concentrations, $y \sim Y$ and $x \sim X$, but as will be shown later, the modified relationships can be used for concentrated cases. When there is more than one solute, the denominators in Equations (13.10) and (13.11) do not change.

Consider a general-type absorber as diagrammed in Figure 13.6. Gas and liquid flow countercurrently through real or imagined stages. The usual specifications include (a) fraction of absorption of solute in the feed gas, (b) maximum allowable concentration of solute in the exit gas, and (c) optimal ratio of solvent to feed gas. Based on mass balances,

$$\text{Total: } G_{N+1} + L_o = G_1 + L_N \quad (13.12)$$

$$\text{Solute } i: G_{N+1}Y_{N+1} + L_oX_o + = G_1Y_1 + L_{N+1}X_{N+1} \quad (13.13)$$

By definition, $G_{N+1} = G_1 = G$, and $L_o = L_{N+1} = L$

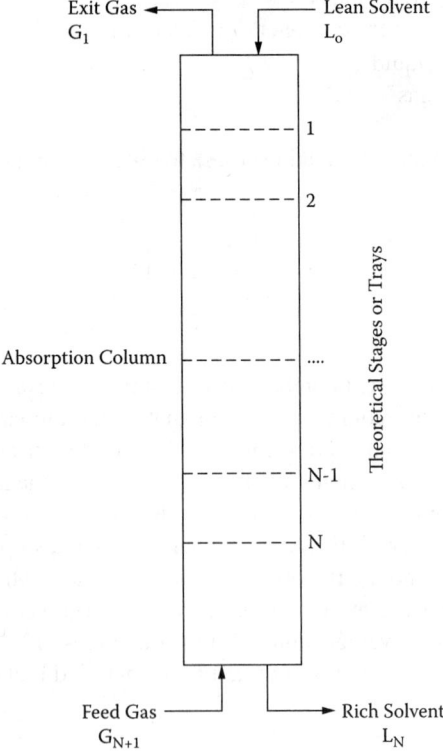

FIGURE 13.6 Arrangement and nomenclature for a plate-type absorber. Note that stages or trays are numbered from the top.

Absorption and Stripping

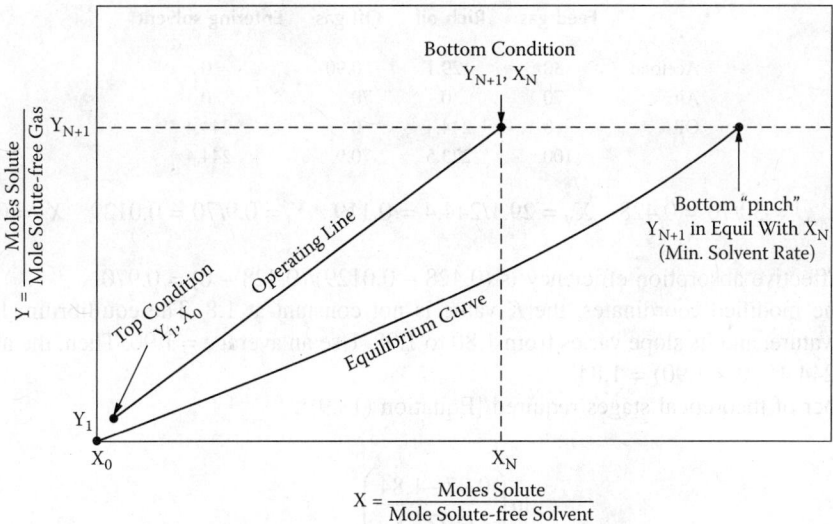

FIGURE 13.7 General layout for graphical determination of theoretical stages in an absorber. Note the bottom pinch point, which determines the minimum solvent rate.

$$\text{So that } Y_{N+1} = (L/G) X_N + Y_1 - L_o X_o/Y_{N+1} \quad (13.14)$$

Equation (13.14) is linear in a plot of Y vs. X. The slope is L/G, and the intercept is $Y_1 - L_o X_o/Y_{N+1}$; and for the often-encountered case of no solute in the entering solvent (once-through solvent or fully stripped solvent in a recirculating system), the intercept is simply Y_1. In Equations (13.8) and (13.9), the absorption factor is based on the slope as defined in Equation (13.14), i.e., on a solute-free basis. Figure 13.7 shows these concepts and can be generalized to apply to any stage n in the column. Note that while the operating line is always straight, the equilibrium curve may not be straight.

Problem 13.2

Acetone is to be absorbed from its mixture with air into a nonvolatile oil. The entering gas contains 30 mole-% acetone and 70% dry air. Removal of acetone from the air is to be 97%. Calculate the following:

(a) Minimum solvent rate, in terms of L_o/G_{N+1}
(b) Theoretical stages at a solvent rate 1.4 times the minimum
(c) Theoretical stages if the solvent contains 0.5 mole-% acetone (solvent incompletely stripped) and the exit gas concentration from part (b) is to be maintained

Data: The molar concentration of acetone in the liquid is linear with a molar concentration of acetone in the air: $y^* = 1.80\ x$.

Solution

(a) The minimum solvent ratio is based on a "pinch" at the bottom of the column. When the feed gas composition is in equilibrium with the rich solvent exit composition, an infinite number of stages would be required. On the basis of $y^* = 1.80\ x$, $x = 0.30/1.80 = 0.167$. Then $L_{o,min} = 29.1/0.167 = 173.6$ moles.

(b) If the solvent rate is 1.4 times the minimum, the bottoms composition is $0.167/1.4 = 0.119$. A molar balance across the column becomes (basis 100 total moles of feed gas)

	Feed gas	Rich oil	Off gas	Entering solvent
Acetone	30	29.1	0.90	0
Air	70	0	70	0
Oil	0	244.4	0	244.4
	100	273.5	70.9	244.4

$Y_{N+1} = 30/70 = 0.428 \quad X_N = 29.1/244.4 = 0.119 \quad Y_1 = 0.9/70 = 0.0129 \quad X_o = 0$

The effective absorption efficiency is $(0.428 - 0.0129)/(0.428 - 0) = 0.970$.

For the modified coordinates, the K value is not constant at 1.8. The equilibrium line has a slight curvature, and its slope varies from 1.80 to 1.98. Use an average = 1.90. Then, the absorption factor is $244.4/(70 \times 1.90) = 1.84$.

Number of theoretical stages required [Equation (13.9)]:

$$N_t = \frac{\ln\left(\frac{0.97 - 1.84}{0.97 - 1}\right)}{\ln 1.84} - 1 = 4.52$$

The material balance for this case is

	Feed gas	Rich oil	Off gas	Entering solvent
Acetone	30	30.3	0.9	1.2
Air	70	0	70	0
Oil	0	244.4	0	244.4
	100	274.7	70.9	245.6

$Y_{N+1} = 30/70 = 0.428 \quad X_N = 30.3/244.4 = 0.124 \quad Y_1 = 0.9/70 = 0.0129 \quad X_o = 1.2/244.4 = 0.005$

Now the effective absorption efficiency is $(0.428 - 0.0129)/(0.428 - 0.005) = 0.981$.

The number of theoretical stages by Equation (13.9) = 5.04. The acetone in the entering oil increases the stage requirement from 4.52 to 5.04. The graphical treatment of case (c) is shown in Figure 13.8.

13.2.1.1 Multiple Solutes

Equation (13.8) can be used for multiple solute cases, since it can accommodate absorption factors $A_i, A_j, A_k, \ldots$, each giving a different value of E_a. The liquid/gas ratio does not change, since it is based on overall flows. Thus, Equation (13.9) can give different values of N (theoretical stages). It is here that the key component enters into the picture. The required stages for that component are calculated, and then the same value of N is substituted back in Equation (13.8) to determine the relative absorption effectiveness of the other components.

The graphical counterpart of Equations (13.8) and (13.9) is shown in Figure 13.9 for the example of absorbing ethane, propane, butane, and pentane into absorption oil. Pentane is the key component, and its value of Y is reduced from 0.04 to essentially zero. The other components are restricted by the same operating line slope as required for pentane. Thus, butane is reduced from $Y = 0.08$ to $Y = 0.02$, propane is reduced from $Y = 0.08$ to 0.063, and ethane is reduced very little. The requirement for butane is about four stages. Ethane and propane fall into a "pinch" composition after one or two stages and then for the remaining stages have no change in composition. The absorption effectiveness values indicated in Figure 13.9 would also result from the use of Equations (13.8) and (13.9).

Absorption and Stripping

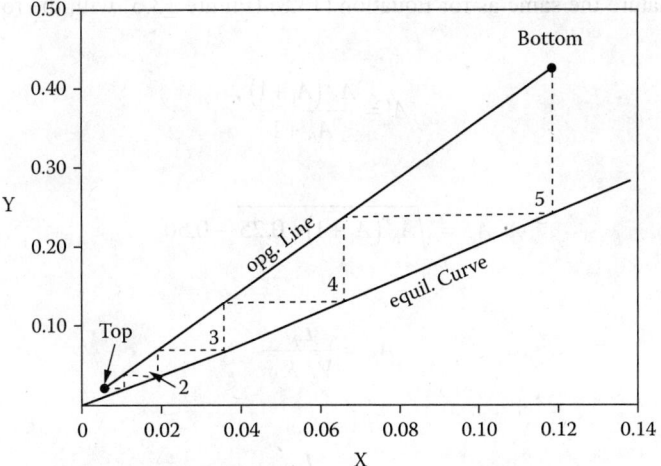

FIGURE 13.8 Graphical determination of equilibrium stages for Problem 13.2.

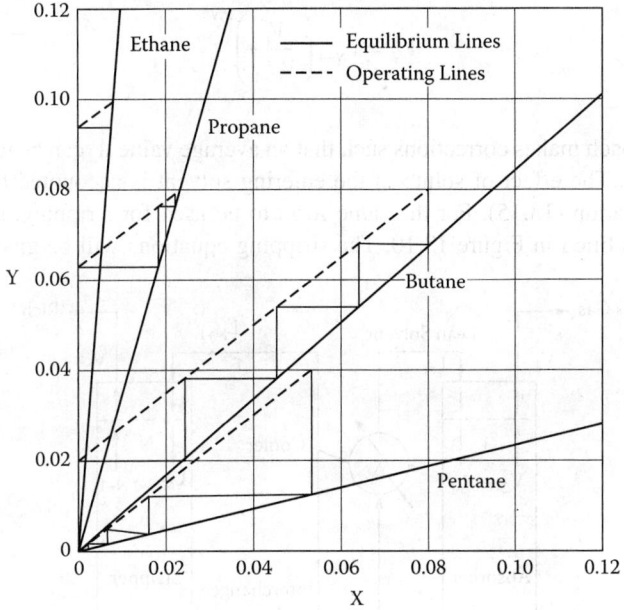

FIGURE 13.9 Equilibrium stage graph for a multiple solute system. Note that the operating lines are straight and parallel. If pentane is the key absorbing component, then the exit gas has significant amounts of butane, propane, and ethane.

13.2.1.2 Concentrated Feeds

Equations (13.8) and (13.9) apply to whatever concentration range can result in minor variations of the absorption factor, i.e., for essentially linear equilibrium lines. When this is not the case, a variant of Equation (13.8), due to Edmister,[19] may be used:

$$E_{ai} = \frac{Y_{N+1} - Y_1}{Y_{N+1}} = \left[1 - \frac{L_o X_o}{A' V_{N+1} Y_{N+1}}\right]\left[\frac{A_e^{N+1} - A_e}{A_e^{N+1} - 1}\right] \qquad (13.15)$$

with the nomenclature the same as for Equation (13.8) (Figure 13.6) with the following additions:

$$A' = \frac{A_N(A_1+1)}{A_N+1} \tag{13.16}$$

$$A_e = \sqrt{A_N(A_1+1)+0.25} - 0.50 \tag{13.17}$$

$$A_N = \frac{L_N}{V_N K_N} \tag{13.18}$$

$$A_1 = \frac{L_1}{V_1 K_1} \tag{13.19}$$

$$V_N = V_{N+1}\left(\frac{V_1}{V_{N-1}}\right)^{1/N} \tag{13.20}$$

The Edmister approach makes corrections such that an average value A_e can be used in the Kremser-Brown relationship. The effect of solute in the entering solvent is accounted for by the first term on the right in Equation (13.15). For the same form to be used for stripping, tray numbering and other terms are amplified in Figure 13.10. The stripping equations will be given later.

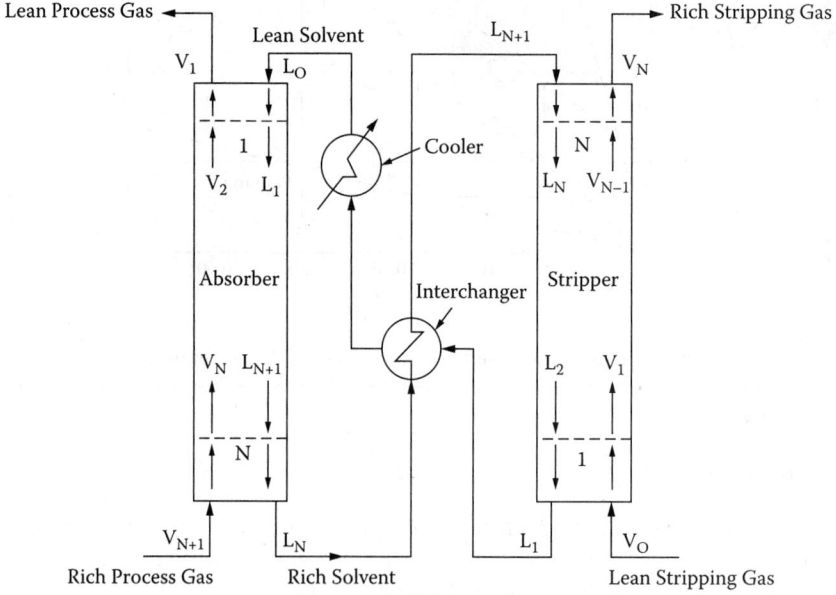

FIGURE 13.10 Nomenclature for use of the Edmister method for determining stages in absorbers and strippers. In order to maintain similarity of equations for absorption and stripping, the stages are numbered up from the bottom of the stripper.

13.2.1.3 Minimum Solvent Rate

As noted in the solution to Problem 13.2, when the solvent rate is lowered to a value at which the entering gas composition approaches equilibrium with the exit solvent, the required stages approach infinity. Recognize that this is not based on Equation (13.8), where it might appear that when $A = 1$, $N \to \infty$; in fact, when $A = 1$, $E_{ai} = N/(N + 1)$ by L'Hospital's rule. Rather, it is based on a "pinch" at the bottom of the column where the gas leaving the bottom stage is the same as the gas entering that stage (i.e., feed gas). In Figure 13.8, if an operating line is drawn from the bottom pinch condition to the desired top condition, an infinite number of steps would be required at the bottom condition.

The bottom pinch is the most common type but, under certain conditions, the pinch can occur at the top, wherein the off gas composition is in equilibrium with the entering solvent composition.

Many studies have indicated that an optimum solvent-to-feed ratio is about 40% greater than the minimum. The optimum represents a trade-off between additional capital cost (more stages at lower solvent/feed ratios) and increased operating costs owing to the circulation of more solvent. It is recommended that the minimum rate be determined first and, as a starting value, a design solvent rate be 1.4 times the minimum.

13.2.1.4 Heat Effects

With lean gas feeds, only relatively small amounts of feed components pass into the liquid. Hence, the heats of solution of these components are relatively small, and the absorption column operates approximately *isothermally*. For significant amounts of absorption, the exothermic heats of solution become important. In the latter case, the temperature of the liquid phase can increase measurably, resulting in lower solubility of solutes and more solvent required. For this adiabatic case, the exit solvent temperature can be calculated from heats of absorption and the heat capacity of the solvent. As the solvent temperature rises, two approaches can be used:

(1) Allow for decreased solubility in the design, and adjust the location of the equilibrium curve (or the value of K) accordingly. For example, Equation (13.8) can be used for segments of the column where the temperatures are different because of solution effects.
(2) Install intercooling capacity in the column to maintain a pseudo-isothermal condition. The cooling can be done with coils on each tray (as in nitric acid columns) or with external heat exchangers processing the total liquid downflow.

Figure 13.5 compares the slopes of the equilibrium curves at a low concentration of ammonia in the liquid and at a higher concentration. If the concentration of ammonia is less than about 0.1 mole fraction, the equilibrium line is fairly straight. However, heat effects associated with the absorption of ammonia can move the equilibrium curve to a higher slope. Problem 13.3[20] shows how even a low concentration case may require correction.

13.2.2 STRIPPING STAGES

Stripping is in many ways the reverse of absorption, and most of the same design equations apply. Figures 13.1 and 13.2 show examples of stripping coupled with absorption, while Figure 13.3 shows a once-through stripping operation. In this process, an inert gas is used to remove components from a liquid feed mixture, and the same solubility considerations presented earlier for absorption apply to stripping. To maintain the same form of the equations, stripping stages are numbered from the bottom to the top (Figure 13.10). The stripping gas is G_o (the counterpart of solvent L_o in absorption), the feed mixture is L_{N+1}, and so on.

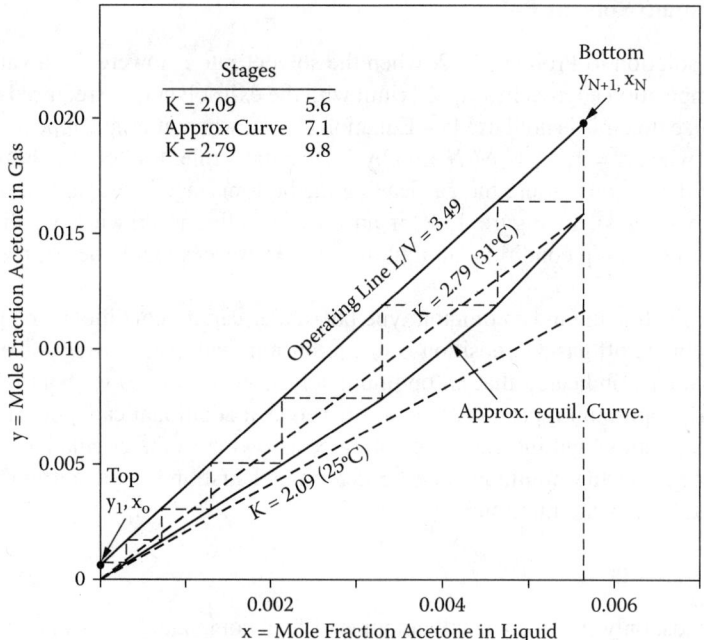

FIGURE 13.11 Diagram for Problem 13.3, showing effect of heat of absorption on required stages. Only the steps for the temperature-corrected equilibrium curve are shown.

Problem 13.3

Acetone is to be absorbed from wet air into pure water in a plate column. The laden air enters at 35°C, and the water solvent enters at 25°C. Pressure is atmospheric. The air contains 2.0 mole-% acetone and is 70% saturated with water vapor (4 vol-% water). The concentration of acetone in the air is to be reduced to 0.05 mole-% (E_{ai} = 0.975). The solvent rate is chosen such that the operating line has a slope of 3.49 on a mole fraction-plotting basis. How many stages are needed?

Solution

Based on an overall heat balance, with the exit air temperature closely approaching the inlet water temperature, the adiabatic temperature rise of the liquid is 6°C, giving an exit liquid temperature of 31°C. Isothermal K values (= y/x) are 2.09 at 25°C and 2.79 at 31°C. The exit water contains 0.57 mole-% acetone.

Figure 13.11 shows a y-x diagram with the isothermal equilibrium lines at 25 and 31°C. Also shown is the approximate actual equilibrium curve location, taking into account heat effects in the column. For the linear equilibrium line, stages are determined by Equation (13.9) or by stepping graphically. For the curved (approximate actual) line, stages are stepped off as shown in Figure 13.11:

1. For the straight line at 25°C: stages = 5.6
2. For the straight line at 31°C: stages = 9.8
3. For the actual curve: stages = 7.1

Clearly, the equilibrium line based on the outlet temperature (31°C) requires excessive stages—but neglecting heat effects is too optimistic. The design of 7.1 stages would be used.

The Kremser–Brown relationship [Equation (13.8)] can be modified for stripping:

Absorption and Stripping

$$E_{si} = \frac{X_{N+1} - X_1}{X_{N+1} - X_o^*} = \frac{S_i^{N+1} - S_i}{S_i^{N+1} - 1} \tag{13.21}$$

where
- E_{si} = stripping efficiency of component i
- X_{N+1} = moles i per mole of solute-free liquid, entering (top) condition
- X_1 = moles i per mole of solute-free liquid, exiting (bottom) condition
- X_o^* = moles i per mole of solute-free liquid, determined as in equilibrium with i in the entering stripping gas
- N = required theoretical stages
- S_i = stripping factor for component i = $(V K_i)/L'$ = $1/A_i$
- V/L' = vapor/liquid molar ratio

The gas and liquid concentrations in capitals are the same as those defined by Equations (13.10) and (13.11).

The stages needed to achieve a given degree of stripping are obtained from a counterpart of Equation (13.9):

$$N = \frac{\ln\left(\frac{E_{si} - S_i}{E_{si} - 1}\right)}{\ln S_i} - 1 \tag{13.22}$$

The most prevalent stripping gases are steam and air (or nitrogen). Thousands of air strippers are used to purify contaminated groundwaters, and many steam strippers are used to purify contaminated process aqueous wastes as well as groundwaters.

For the more concentrated feed liquids, such as those found in petroleum refinery stripping operations, the Edmister method [Equations (13.15) through (13.20)] are modified as follows (see Figure 13.10):

$$E_{si} = \frac{X_{N+1} - X_1}{X_{N+1}} = \left[1 - \frac{V_o Y_o}{S' L_{N+1} X_{N+1}}\right]\left[\frac{S_e^{N+1} - S_e}{S_e^{N+1} - 1}\right] \tag{13.23}$$

$$S' = \frac{S_N(S_1 + 1)}{S_N + 1} \tag{13.24}$$

$$S_e = \sqrt{S_N(S_1 + 1) + 0.25} - 0.50 \tag{13.25}$$

$$S_N = \frac{V_N K_N}{L_N} \tag{13.26}$$

$$S_1 = \frac{V_1 K_1}{L_1} \tag{13.27}$$

$$V_N = V_{N-1}\left(\frac{V_N}{V_o}\right)^{1/N} \qquad (13.28)$$

13.2.2.1 Minimum Stripping Gas Rate

An infinite number of stages will be required if the exit gas and the entering liquid approach equilibrium, i.e., $Y_N \sim K X_{N+1}$. It is recommended that a stripping gas rate of about twice the minimum be considered first. However, for once-through air strippers, where the stripping gas has no cost other than the energy needed to move it through the stripper, an optimum V/L ratio may depend on some trade-off of blower horsepower and height of contactor.

EQUIPMENT DESIGN

Plate columns for absorption and stripping are fairly standard in design, with the plates perforated to allow gas to pass upward and provisions for conveying liquid from a plate to the next plate below. Figure 13.12 shows an absorber of this type, which utilizes simple sieve trays for contacting. Only the rudiments of designing such a contactor will be given here; see Chapter 12 for more details. Predicting the performance of a plate absorber or stripper involves the following key elements:

(a) Capacity. This dictates the diameter of the needed column and is a function of the plate geometry, the vertical spacing of the plates, and the flow rates of liquid and gas.
(b) Pressure drop. This is a function of flow rates and plate geometry and is particularly important for large gas flows where blower horsepower may be an important element in the economics of system design.
(c) Efficiency. The foregoing discussions of absorption and stripping stages have led to a determination of theoretical stages or plates. For a practical design, a plate efficiency is needed to determine the number of *actual* plates needed in the column. The discussion on plate efficiency will follow.
(d) Liquid holdup. The amount of liquid present on the trays has implications for control system design; the more the holdup, the less responsive the column is to changes in control variables. It may be particularly important for chemical absorbers where a degree of residence time of the liquid is needed to complete the absorption/reaction. This is discussed further in a separate section on "Chemical Absorption/Stripping."

Problem 13.4
A contaminated groundwater stream contains 5 mg/L trichloroethylene (TCE) and 1.5 mg/L perchloroethylene (PCE) at 21°C. The stream is fed to the top of a tray column, where it is stripped with a countercurrent flow of air, also at 21°C.

(a) Calculate the minimum air rate needed (using infinite stages).
(b) For a molar feed-air ratio of 10.0, how many stages are needed to remove 99.8% of the TCE?
(c) What percentage of the PCE will be removed in the process?

Solubility data for PCE and PCE are given in Figure 13.4. Use $K = 6753$ for TCE and 61,790 for PCE.

Solution
The inlet composition of TCE may be converted to a mole fraction of $6.85(10^{-7})$. The equilibrium exit gas concentration is $6753(6.85)(10^{-7}) = 0.00463$ moles TCE per mole of air + TCE, or a

Absorption and Stripping

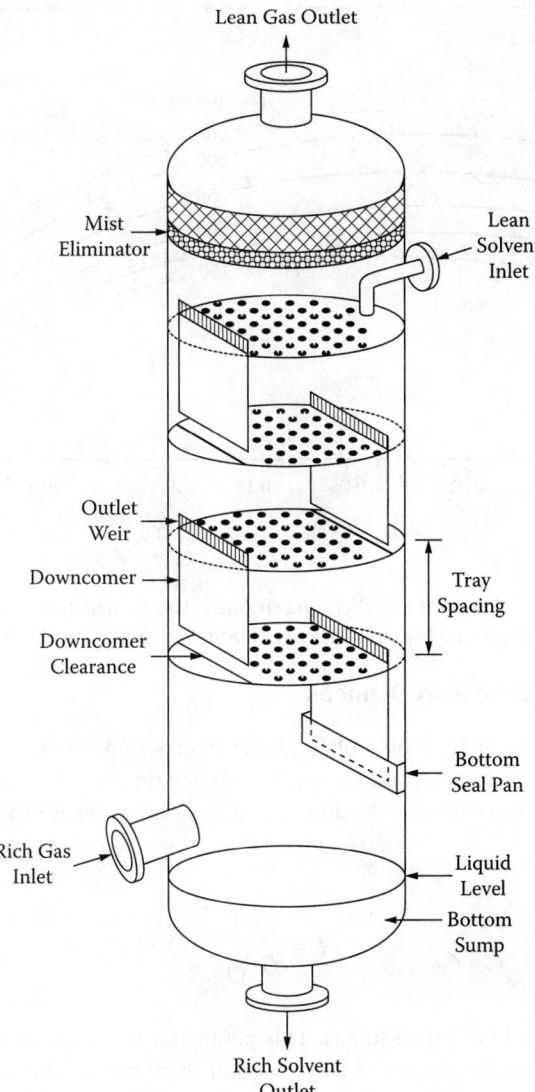

FIGURE 13.12 Plate-type absorber.

minimum air rate of 0.1465 moles/1000 moles feed. The design ratio of 100 is far greater than the minimum, and is based on field experience of optimum operation.

The stripping factor at the bottom is $VK/L = (6753)/100 = 67.53$. Because of the very dilute feed, the stripping factor changes very little across the column. By Equation (13.22),

$$N = \frac{\ln\left(\dfrac{0.998 - 67.53}{0.998 - 1.000}\right)}{\ln 67.53} - 1 = 1.47 \text{ stages}$$

For PCE, $S_{avg} = 67.53(61790/6753) = 618$. By Equation (13.22), 100% of the PCE is stripped.

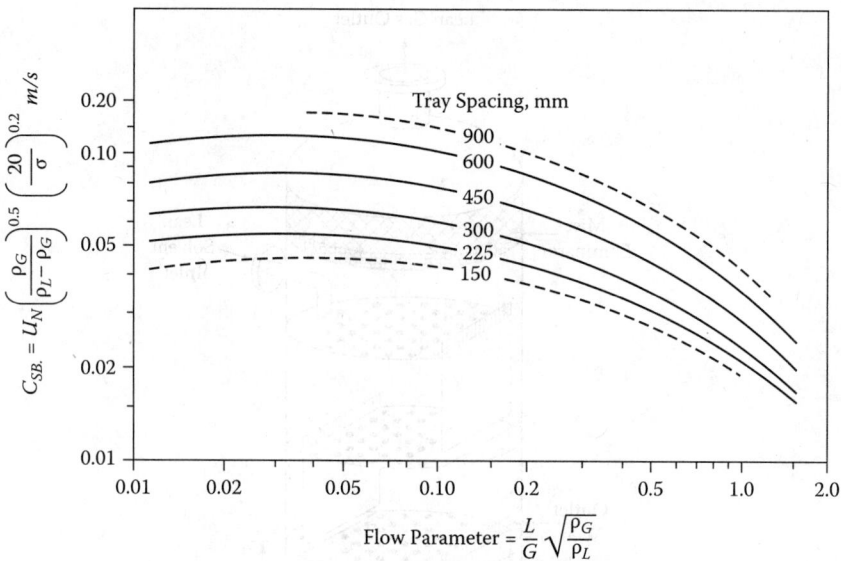

FIGURE 13.13 Capacity of plate-type absorbers and strippers. Curves denote the incipient flooding condition, and for design must be discounted. [Fair, J. R., 1961. *Petro/Chem. Engineer* 33 (9): 57.]

13.2.2.2 Capacity and Column Diameter

The maximum allowable capacity of a plate absorber or stripper is represented by an incipient flooding condition. The prudent operating capacity is based on about 85% of the predicted flooding condition. The first step is to calculate the dimensionless *flow parameter* for the point in the column being investigated:

$$FP = \frac{L}{G}\sqrt{\frac{\rho_G}{\rho_L}} \tag{13.29}$$

where L/G is the mass ratio of liquid to gas. This parameter is a high number for the typical high L/G ratios encountered in absorbers and strippers, but there are notable exceptions (e.g., the use of glycols to scrub water out of natural gas) where the L/G is extremely low. The allowable gas velocity for a given flow parameter is obtained from Figure 13.13,[21] making use of the geometrical relationships shown in Figure 13.14.

The value of FP is obtained directly from the overall material balance. For a selected tray spacing (usually 18 in (457 mm) of 24 in (610 mm) and FP as the abscissa, a value of the ordinate is obtained. This is the capacity parameter C_{sbf} at the maximum (flood) capacity. The ordinate value is

$$C_{sbf} = U_N \left[\frac{\rho_G}{\rho_L - \rho_G}\right]^{0.5} \left[\frac{20}{\sigma}\right]^{0.2} \tag{13.30}$$

where U_N = linear gas velocity through the net area of the tray, i.e., through the cross section open to flow, considering the space blocked off by the downcomer(s), and σ = the surface tension of liquid in mN/m.

Since $U_N = U_s (A_T/A_N)$, the superficial gas velocity is calculated as

Absorption and Stripping

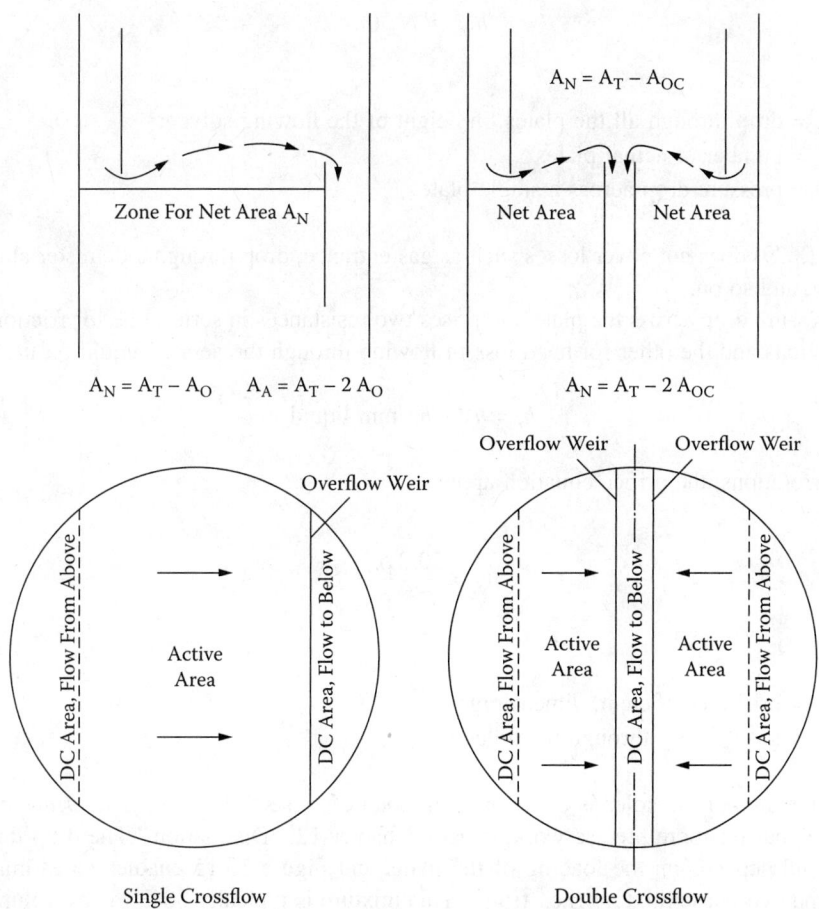

FIGURE 13.14 Geometrical relationships for plate-type absorbers and strippers. Liquid flows across the plates.

$$U_s = \frac{0.85\, C_{sbf}}{\dfrac{A_T}{A_N}\sqrt{\dfrac{\rho_G}{\rho_L - \rho_G}}\left[\dfrac{20}{\sigma}\right]^{0.2}} \tag{13.31}$$

so that

$$\text{col. diam.} = \sqrt{\frac{\text{vol. flow of gas}}{0.785\, U_s}} \tag{13.32}$$

Multiple pass trays are handled in the same fashion (see Figure 13.14); one should recognize that geometric variables are different for successive trays.

13.2.2.3 Pressure Drop

The total pressure drop through the column, in terms of liquid head, is

$$h_{total} = N_{act}(h_t) \tag{13.33}$$

where

h_{total} = drop through all the plates, in height of the flowing solvent
N_{act} = number of actual plates
h_t = pressure drop across a single plate

Equation (13.29) does not cover losses such as gas entrance, drop through a demister at the top of the column, and so on.

The pressure drop across the plate comprises two resistances in series, one for friction through the perforations and the other for head loss in flowing through the aerated liquid on the tray:

$$h_t = h_d + h_L, \text{ mm liquid} \tag{13.34}$$

For the perforations, the orifice equation applies:

$$h_d = \frac{50.8}{C_v^2} \frac{\rho_G}{\rho_L} U_h^2 \tag{13.35}$$

where

C_v = orifice coefficient, dimensionless
U_h = gas velocity through the holes

Typically, the area of the holes is 6 to 8% of the total cross-sectional area. The orifice coefficient is about 0.8, but for more precise work, consult Chapter 12, "Distillation." The drop through the aerated liquid depends on the loading of the plate, and Figure 13.15 enables an estimate of the height of the two-phase mixture (i.e., froth). This mixture is typically 80% gas by volume, so the froth height times 0.2 will give the equivalent height of clear liquid for use in Equation (13.30).

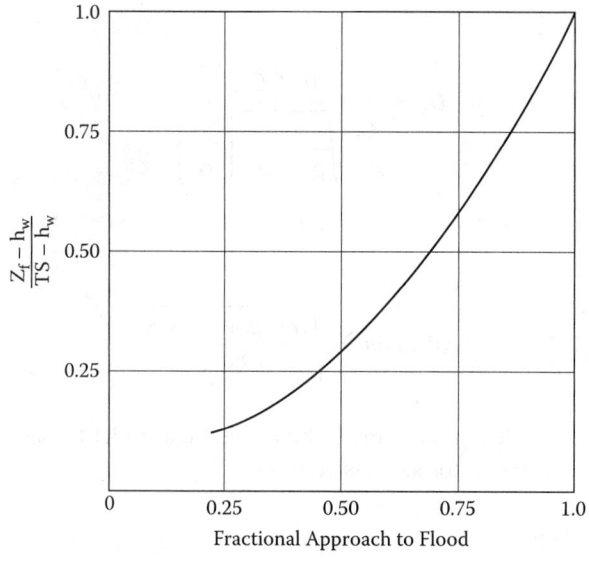

FIGURE 13.15 Chart for estimating the height of froth on a crossflow plate.

Absorption and Stripping

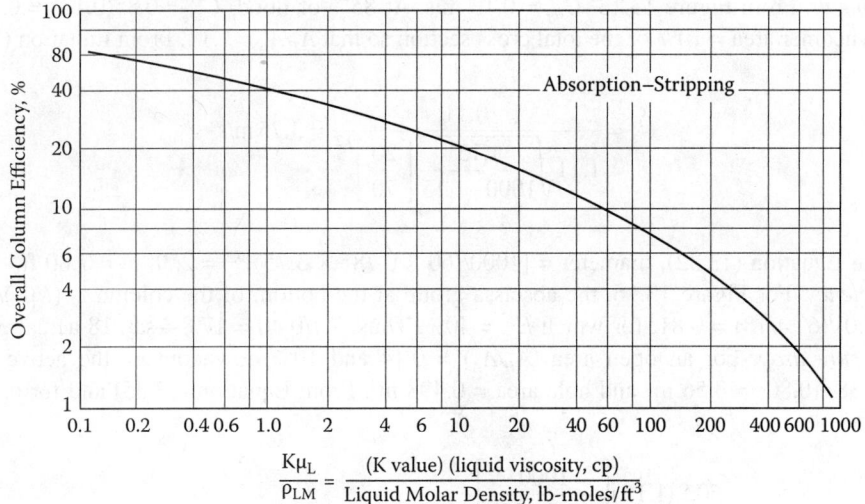

FIGURE 13.16 Chart for estimating efficiency of plate-type absorbers and strippers. [H. E. O'Connell, 1946. *Trans. AIChE* 42: 741.]

13.2.3 STAGE EFFICIENCY

In Sections 13.1 and 13.2, methods are given for the estimation of the number of theoretical stages required for a specified separation. The number of actual stages, or plates, is computed from an average plate efficiency for the overall column:

$$E_{oc} = \frac{\text{theoretical stages}}{\text{actual plates}} \times 100 \tag{13.36}$$

This efficiency term actually comprises a number of component efficiencies, and for meticulous work, it may be necessary to take them into account. Methods given in Chapter 12 can be used for this purpose, since the mass transfer mechanisms for absorption and stripping are the same as those for distillation.

Figure 13.16, from O'Connell,[22] may be used for preliminary designs—and often for final designs. It relates overall column efficiency to the equilibrium ratio K, the solvent phase viscosity, and the molar density of the solvent phase.

Problem 13.5

In Problem 13.3, 7.1 actual stages were required to remove a certain amount of acetone from wet air. What are the column diameter, pressure drop, and actual number of stages if the airflow rate is 35,300 actual ft³/min (1000 actual m³/min)? A tray spacing of 24 in (610 mm) will be used. Weir height = 50 mm.

Data: Liquid density = 1000 kg/m³; gas density = 1.15 kg/m³; liquid viscosity = 1.0 cp; surface tension = 70 mN/m.

Solution

Use bottom conditions, since in that zone both liquid and gas flows are greatest. The temperature is taken as an average of the entering gas and the exit liquid = 33°C. The operating line has a slope of 3.49 in molar units, or 3.49 (18/29) = 2.17 in mass units. The flow parameter FP = $2.17[1.15/1000]^{0.5}$ = 0.074.

Diameter. From Figure 13.13, $C_{sbf} = 0.13$ m/s. At 85% of flood, $C_{sb} = 0.85(0.13) = 0.11$ m/s. Use downcomer area = 10% of the total cross section so that $A_T/A_N = 1.11$. From Equation (13.31),

$$U_s = \frac{0.11}{1.11\sqrt{\frac{1.15}{1000-1.15}}\left[\frac{20}{70}\right]^{0.2}} = 3.75 \text{ m/s}$$

From Equation (13.32), diameter = $[1000/(60 \times 0.785 \times 3.75)]^{0.5} = 2.38$ m = 7.80 ft.

Efficiency. For Figure 13.16, the abscissa group at the bottom of the column is $(K\mu_L)/\rho_{LM}$, or $(2.8 \times 1.0)/(62.4/18) = 0.81$, for which $E_{oc} = 40\%$. Thus, 7.1/0.40 = 17.8—say 18 actual plates.

Pressure drop. For an open area $(A_H/A_A) = 0.14$ and 10% downcomers, the active area is $0.785(2.38)^2(0.80) = 3.56$ m^2 and hole area = 0.498 m^2. From Equation (13.35) and for one tray,

$$h_d = \frac{50.8(1.15)\left[\frac{1000}{60(0.498)}\right]^2}{(0.80)^2(1000)} = 102.2 \text{ mm liquid} = 4.02 \text{ in liquid}$$

For pressure drop through the liquid at 85% flood, from Figure 13.15,

$$\text{Froth height} = Z_f = 0.75(610 - 50) + 50 = 470 \text{ mm}$$

At 20% liquid in froth, liquid head = 0.2(470) = 94 mm.

Total drop/tray = 102.2 + 94 = 196 mm liquid. For 18 plates, 3528 mm = 3.528 m = 5.0 lb/in^2.

13.3 PACKED COLUMNS

As in distillation columns, a packed column contains small packing elements dumped into the column in a random fashion, and these elements can provide the needed intimate contacting of the descending liquid and the rising gas or vapor. Packings of an ordered, or structured, form are rarely if ever used for absorption or stripping.

13.3.1 TRANSFER UNITS

Instead of stages or plates, *transfer units* are introduced to represent the separation requirement of a continuous, countercurrent contactor such as a packed column. As will be shown later, theoretical stages can be converted to transfer units for design purposes. Thus, only a summary of transfer unit theory will be given here, since it is usual practice to determine the required stages, as discussed in the previous section, and then convert them to transfer units. The conversion is in order because the mass transfer characteristics of packings are usually correlated in terms of heights of a transfer unit. Thus,

$$\text{Packed height} = Z = N_{og}H_{og} \tag{13.37}$$

where N_{og} = number of overall (both phases considered) transfer units and H_{og} = height of an overall transfer unit.

Overall transfer units may be expressed in terms of individual phase transfer units:

$$H_{og} = H_g + \lambda H_L \tag{13.38}$$

Absorption and Stripping

$$\frac{1}{N_{og}} = \frac{1}{N_g} + \frac{\lambda}{N_L} \qquad (13.39)$$

where λ is the ratio of slopes, equilibrium line to operating line, for the transferring component i:

$$\lambda_i = \frac{dy_i^* / x_i}{(L/V)_{total,molar}} = \frac{m_i}{L/V} \qquad (13.40)$$

The λ term is equivalent to the stripping factor defined earlier. When the equilibrium line is essentially straight (e.g., as in Figure 13.9) and for dilute cases, λ is constant.

13.3.2 Number of Transfer Units

According to basic relationships, not included here,* the overall number of gas-phase transfer units is defined as

$$N_{og,i} = \int_{y_{i2}}^{y_{i1}} \frac{dy_i}{y_i - y_i^*} \qquad (13.41)$$

where y_i is the mole fraction of i in the gas and y_i^* is the mole fraction of i in the gas that would be in equilibrium with bulk liquid composition x_i. Note that y_i^* is not a real composition but rather a calculated term from the equilibrium expression $y_i^* = K_i x_i$. The subscripts 1 and 2 for the gas composition refer to the inlet and outlet conditions. If a stage model is being used to compute the separation, theoretical stages can be converted to transfer units by the relationship

$$N_{og} = N_{theo} \left(\frac{\ln \lambda}{(\lambda - 1)} \right) \qquad (13.42)$$

Note that when $\lambda = 1.0$, $N_{og} = N_{theo}$.

As in the case of absorption factors with stages, each component can have its individual number of transfer units.

13.3.3 Heights of Transfer Units

For systems with significant mass transfer resistance in both phases, Equation (13.38) is used. Values of H_g and H_L are usually obtained from generalized correlations supported by experimental work but can be expressed in terms of more fundamental relationships:

$$\text{Gas: } H_{gi} = \frac{G_m}{k_{gi} a_e P(1 - y_i)} \qquad (13.43)$$

---

* Detailed development of relationships for packed columns is included in mass transfer texts such as that of Hines and Maddox.[22a]

where

G_m = total gas flow, kg-moles/s–m²
k_{gi} = mass transfer coefficient, kg-moles/(s-m²-atm)
a_e = effective interfacial area, m²/m³
P = total pressure, atm

$$\text{Liquid: } H_{Li} = \frac{L_m}{k_{Li} a_e \rho_L (1-x_i)} \tag{13.44}$$

where

L_m = total liquid flow, kg-moles/s-m²
k_{Li} = mass transfer coefficient, kg-moles/(s-m²-kg/m³)
ρ_L = average liquid density, kg/m³

In Equations (13.43) and (13.44), the units shown are representative; other consistent sets may be used.

13.3.4 Heights of Theoretical Stages

There are instances when it is preferable to convert from heights of transfer units to theoretical stages. The term *height equivalent to a theoretical plate (HETP)* is used. This is the height of a packed bed in which one theoretical stage of separation is achieved. The conversion is related to Equation (13.42):

$$HETP = H_{og}\left(\frac{\ln \lambda}{(\lambda - 1)}\right) \tag{13.45}$$

Thus, when $\lambda = 1.0$, $HETP = H_{og}$.

13.3.5 Equipment Design

Packed columns for absorption and stripping are simple in design, if not in analysis. A vertical column is fitted with a packing support plate, filled with random packing, and equipped at the top with a suitable device to distribute the liquid to the bed. A screen or hold-down grid is usually placed on top of the bed to prevent movement of the packing, especially if the packing material has a low density (e.g., plastic) and the momentum of the gas is high. In a few cases, a special device at the bottom of the bed is used to ensure uniform gas distribution up through the bed. A representative packed absorber is illustrated in Figure 13.17; the liquid distributor is of the perforated pipe type, but other types, less subject to liquid fouling, can be used. See Chapter 12.

Problem 13.6

In Problem 13.3, theoretical stages were determined for a scrubber to remove acetone from air with a water solvent. For the same conditions, determine the required number of overall gas phase transfer units. The graphical work for the previous problem is shown in Figure 13.11.

Solution

Since this is a lean gas case, simplified expressions can be used. The number of gas phase transfer units can be obtained by integrating Equation (13.41), or from theoretical stages by Equation

Absorption and Stripping

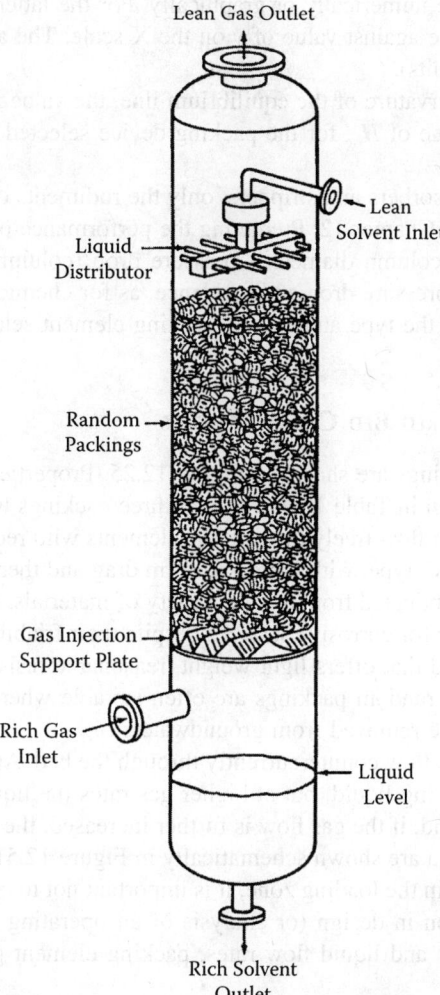

FIGURE 13.17 Packed-type absorber.

(13.42). Since the theoretical stage requirement is available from Problem 13.3, the stage conversion approach will be used first.

For the lower equilibrium line ($K = 2.09$), the value of λ is calculated as follows:

$$\lambda = \frac{m}{L/V} = \frac{2.09}{3.49} = 0.599$$

From Equation (13.42),

$$N_{og} = 5.6 \left(\frac{\ln 0.599}{0.599 - 1} \right) = 7.16 \text{ transfer units}$$

Similarly, for $K = 2.79$, $N_{og} = 11.4$ transfer units.

For the intermediate case, where there is significant curvature of the equilibrium line, Equation (13.41) can be integrated over the range of y values from 0.005 (top condition) to 0.020 (bottom condition). This can be done numerically or graphically. For the latter approach, values of $1/(y - y^*)$ are plotted on the Y scale against value of y on the X scale. The area under the plotted curve equals 8.2 (or 8.2 transfer units).

Since there is definite curvature of the equilibrium line, the value $N_{og} = 8.2$ should be used. It can be combined with a value of H_{og} for the packing device selected to provide the total packed height needed.

As in the case of tray absorbers and strippers, only the rudiments of design will be given here; more details are available in Chapter 12. Predicting the performance of a packed column involves four key criteria: capacity (column diameter), pressure drop (column diameter), efficiency (bed height), and liquid holdup (pressure drop and residence, as for chemical reactions). Each of these criteria depends strongly on the type and size of packing element selected. Thus, the first step is to choose a packing.

13.3.6 Packing and Packed Bed Characteristics

Representative random packings are shown in Figure 12.25. Properties of important packings for absorption/stripping are given in Table 13.2. The first three packings listed are the "through-flow" type, in which the phases can flow freely through the elements with reduced form drag. The others listed are of the "around flow" type, with increased form drag and therefore higher pressure drop.

Random packings are fabricated from a wide variety of materials. Originally, most were made of ceramic and were popular for corrosive services despite their friability. Today, polypropylene is widely used to provide a bed that offers light weight, reasonable resistance to corrosion by many fluids, and low cost. Plastic random packings are often suitable where dissolved volatile organic compounds (VOCs) are to be removed from groundwaters.

Normally, liquid and gas flow countercurrently through the bed. At low gas rates, there is little disturbance of the downflowing liquid, but at higher gas rates the liquid is dragged by the vapor into a "loading" situation, and, if the gas flow is further increased, the packed bed begins to flood. Loading/flooding phenomena are shown schematically in Figure 12.51. While it may be optimum to operate a packed column in the loading zone, it is important not to operate too close to flooding. Thus, as a first consideration in design (or analysis of an operating column), the flood point is located as a function of gas and liquid flow rates, packing element geometry, and properties of the phases.

13.3.7 Maximum Gas/Liquid Capacity

The hydraulic flood point, as illustrated in Figure 12.51, is reached at a combination of countercurrent gas and liquid rates such that the pressure drop exhibits a relatively large increase with a small increase in flow rate of either phase. Methods for estimating maximum hydraulic capacity for a packing provide a basis for design (optimum approach to flood). Although there are several approaches to predicting flooding, the most popular also provides a pressure drop estimating capability, most recently published by Strigle[23] as the diagram in Figure 12.53.

From Figure 12.53, both limiting capacity and pressure drop may be predicted. The abscissa term is the flow parameter [Equation (13.29)], used also for plate columns. The ordinate term is a capacity parameter CP, modified from the capacity parameter for plate columns [Equation (13.30)]:

$$CP = U_S \sqrt{\frac{\rho_G}{\rho_L - \rho_G}} F_p \, \upsilon^{0.05} = C_{sb} F_p \, \upsilon^{0.05} \qquad (13.46)$$

where U_S = superficial gas velocity, m/s; υ = kinematic viscosity, cs; and F_p = packing factor, characteristic of the packing type and size. Packing factors are given in Table 13.2.

For flooding, an estimate can be made from Figure 12.53, although manufacturers often like to omit the flood line from the chart. Maximum throughput is assumed to occur at the top curve (125 mm H_2O/m), although slightly more capacity is sometimes available. The presumption that flooding occurs at a constant value of pressure drop is not necessarily correct, since the more open-type packings flood at lower pressure drops. Also, flooding occurs at lower pressure drops under high liquid loadings and gas densities for a given packing.

13.3.8 Pressure Drop

Prediction of pressure drop is handled by Figure 12.53 for all but the most critical designs. A more fundamental, and more complicated, method has been developed by Stichlmair and coworkers.[24,25] Diagrams of the Strigle type are available for individual packings (thus eliminating the need for the packing factor) in Kister[26]. Finally, estimates of pressure drop can be obtained from graphs developed by the vendors of the packings. Figure 13.18 shows data for a particular packing, including liquid holdup and a mass transfer coefficient. The bases for information such as in Figure 13.18 should be understood. The basic tests were made with air-water in a column with a diameter of about 1 m. Holdup was measured by abruptly shutting off air and water flow and then collecting the water that drained from the bed. The mass transfer coefficient is on an overall (including gas and liquid resistances) basis, using carbon dioxide absorption in dilute aqueous sodium hydroxide. Since many absorption and stripping applications involve gases and liquids with properties not far different from those of air and water, the vendor charts can be quite useful.

13.3.9 Liquid Distribution and Redistribution

This topic is discussed in detail in Chapter 12, and only a brief summary is included here. Good initial liquid distribution to the packed bed is essential for good performance. This is done by introducing the liquid through a distributor device that provides a multiplicity of small, uniform streams to the bed. A minimum of 100 separate streams per sq. m. is needed. If a random packing without through-flow characteristics is needed, fewer streams can be accommodated, since the packing itself tends to provide uniform distribution.

Once the liquid has been introduced to the top of the bed, it is possible for the uniformity of the liquid to deteriorate—for example, to migrate toward the wall of the column. For beds greater than 10 m high, it may be necessary to collect the liquid and redistribute it for further flow down the bed. If redistribution is not used, then the loss of mass transfer efficiency must be taken into account. In some cases, the optimum design does not call for redistribution, even for bed heights much greater than 10 m.

13.3.10 Mass Transfer

Several methods have been proposed for predicting values of H_g and H_L as functions of system, flow conditions, and packing type. While the Bolles/Fair[27] random packing method has a broad validation, most of the supporting data are from distillation tests. It is described in Chapter 12. A better model for absorption is based on the work of Onda et al.,[28] as modified by Bravo and Fair,[29] and is detailed below. For the gas mass transfer coefficient,

$$k_g = \frac{5.23\, a_p D_g}{RT} \left(\frac{G}{a_p \mu_g} \right)^{0.7} \left(\frac{\mu}{\rho D} \right)_g^{1/3} \qquad (13.47)$$

This equation represents the usual functional relationship between the dimensionless groups (Sherwood, Reynolds, and Schmidt numbers) plus the assumption that the effective packing element

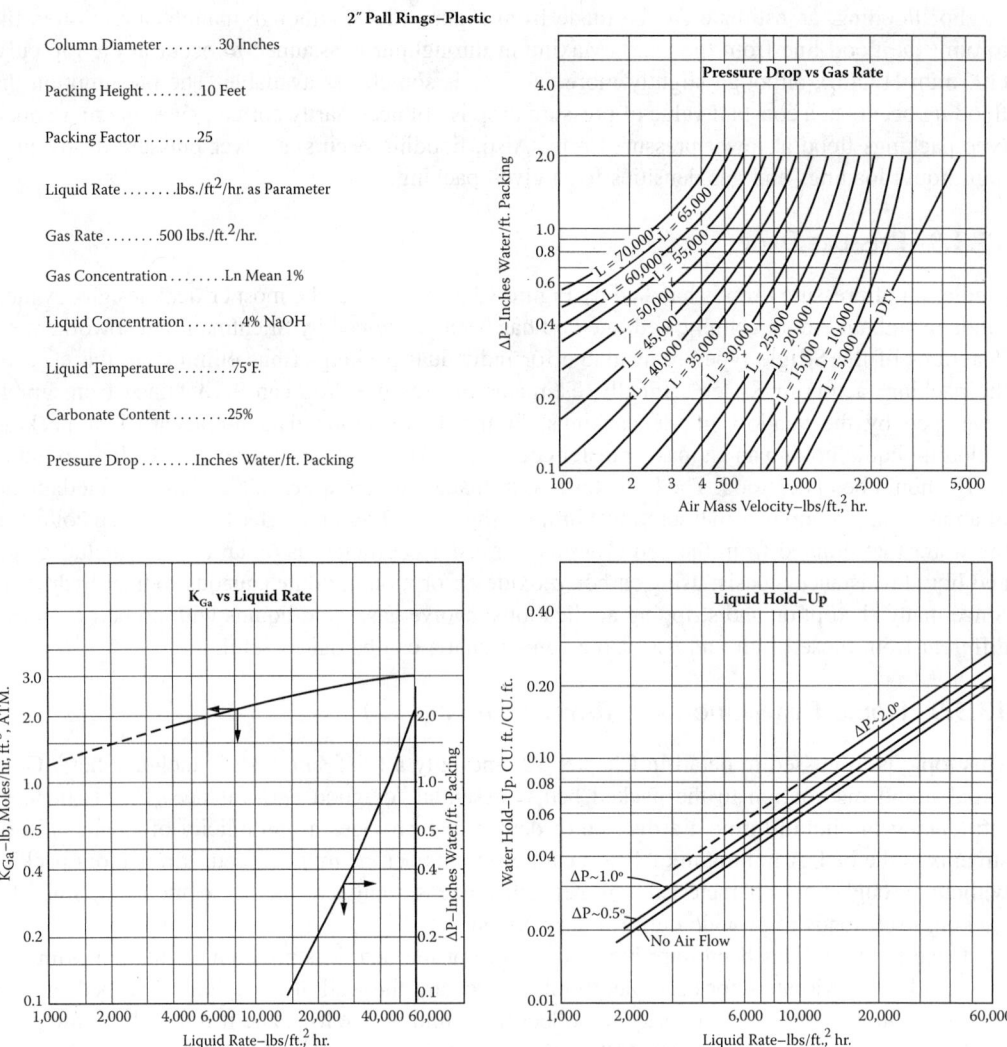

FIGURE 13.18 Representative vendor information on the characteristics of a random packing. Two-inch plastic Pall rings.

diameter can be expressed as the diameter of a sphere with the same surface area, i.e., $d_p = 6/a_p$. Values of packing surface area a_p may be obtained from Table 13.2 or equivalent listings.

For the liquid coefficient, Onda et al. departed from more usual formulations to obtain

$$k_L = 0.010 \left(\frac{g\mu_L}{\rho_L} \right)^{1/3} \left(\frac{L}{a_e \mu_L} \right)^{2/3} \left(\frac{\rho D}{\mu} \right)^{1/2}_L \quad (13.48)$$

where a_e is an effective interfacial area.

For absorption and stripping, Bravo and Fair[29,30] found the effective area to be

$$a_e = 9.79\, a_p \left(\frac{\sigma^{0.5}}{Z^{0.4}} \right) (Ca_L \operatorname{Re}_G)^{0.1603} \quad (13.49)$$

TABLE 13.2
Characteristics of Random Packings

Name	Material	Nominal size, mm	Surface area m²/m³	% Voids	Packing factor, m⁻¹	Vendor
Intalox saddles	M	(No. 25)		97	135	NorPro-Ste. Gobain
"IMTP"		(No. 40)		97	24	Akron, Ohio USA
		(No. 50)		98	59	
Pall rings*	M	16	—	92	266	(Generic)
		25	205	94	184	
		38	130	95	131	
		50	115	96	88	
		90	92	97	59	
Pall rings*	P	16	340	87	310	(Generic)
		25	205	90	180	
		50	130	91	131	
		90	100	92	85	
Intalox saddles	C	13	625	78	660	NorPro-Ste. Gobain
		25	255	77	197	Akron, Ohio USA
		50	118	79	98	
		75	92	80	70	
Intalox saddles	P	25	206	91	131	NorPro-Ste. Gobain
		50	108	93	92	Akron, Ohio USA
		75	88	94	59	
Raschig rings	C	13	370	64	1900	(Generic)
		25	190	74	587	
		38	120	68	305	
		50	92	74	213	
		75	62	75	121	
Raschig rings	M	19	245	80	730	(Generic)
		25	185	86	470	
		38	130	90	270	
		50	95	92	187	
		75	66	95	105	

Note: Materials of construction: C = ceramic; M = metal; P = plastic (usually polypropylene).

* Through-flow packing.

where

σ = surface tension, kg/s (0.001 • dynes/cm)

Z = bed height (or height between distributor and redistributor), m

Ca_L = dimensionless capillary number for liquid = $\dfrac{\mu_L L}{\rho_L \sigma}$

Re_G = dimensionless Reynolds number for gas = $6G/(a_p \mu_G)$

The units of the liquid mass transfer coefficient k_L are m/s; of the gas coefficient k_g, kg-moles/(s-m²-atm). Note that in Equation (13.47), the specific packing area a_p is used, whereas in Equation (13.48), the effective area a_e is used. The two terms are related by Equation (13.49).

The procedure for using the Onda–Bravo approach is as follows:

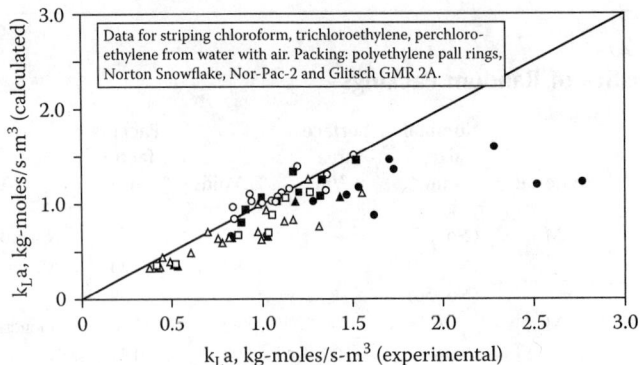

FIGURE 13.19 Parity plot for calculated vs. measured mass transfer coefficients, air-stripping of volatile organic compounds from water with air.

1. For the flow rates and properties, compute the effective area by Equation (13.49).
2. Calculate mass transfer coefficients k_G and k_L by Equations (13.47) and (13.48).
3. Compute heights of transfer units H_G and H_L by Equations (13.43) and (13.44), using volumetric mass transfer coefficients $k_g a_e$ and $k_L a_e$.
4. Compute overall H_{og} by Equation (13.38).
5. Multiply H_{og} by N_{og} to obtain required packed height, or
6. Convert H_{og} to HETP by Equation (13.45) and multiply by theoretical stages to obtain required packed height.

Problem 13.7 illustrates the procedure.

This model has been applied successfully to random packings of the ring and saddle type, in ceramic, steel, or plastic materials. A parity plot for tests of air-stripping of VOCs from water with air is shown in Figure 13.19, taken from reference 30. Several plastic packings were used. For details of the experiments, as well as use of the original Onda equations, see reference 31.

The Onda–Bravo approach has been found especially reliable for absorbers and strippers operating with lean mixtures, as is usually the case. For more concentrated cases (inlet gas > 5 vol-% solute for absorbers and inlet liquid > 5 vol-% solute for strippers), the Bolles/Fair[27] model is more appropriate.

13.4 OTHER DEVICES

While tray columns and packed columns are invariably the devices used for stripping, other devices are occasionally used for absorption or scrubbing. These are often proprietary, and the vendor does the design based on the specifications provided by the user and/or process designer. While the foregoing material is based on countercurrent contacting of gas and liquid, there may be some instances when cocurrent operation has advantages:

1. Throughput capacity is great, since flooding does not occur in the usual sense.
2. Pressure drop is lower than countercurrent.
3. Removal of solid particulates from the gas is favorable.

Problem 13.7

For Problem 13.4, a contaminated groundwater stream is used to illustrate the computation of theoretical stages. Now a stripping column is to be dimensioned based on the same removal of

Absorption and Stripping

TABLE 13.3
Properties and Data for Problem 13.7

Basis: 294 K, 1.0 atm.

	Liquid	Gas
Density, kg/m³	1000	1.234
Viscosity, kg/m-s	1.02E–03	1.75E–05
Surface tension, m²/s	0.070	—
Duffusivity, m²/s	8.8E–10	8.2E–05
Schmidt number = $\mu/(\rho D)$	1150	1.73
Reynolds number	45.3 L	2640 G
Capillary number	14.6E–06 L	—

Note: G and L units: kg/s-m².

trichloroethylene (TCE) and perchloroethylene (PCE) at ambient pressure (1.0 atm) and temperature (21°C). A higher gas/liquid ratio will be used to reduce the number of stages. The packing is to be 50-mm polypropylene Pall rings. Properties are given in Table 13.3. The material balance is to be as follows:

Flows in kg/s:

	Feed liquid	Feed air	Exit liquid	Exit air
Air	—	5.09	—	5.09
TCE	1.579E–04	—	1.579E–06	1.563E–04
PCE	4.737E–04	—	0	4.737E–04
Water	31.57	—	31.40	0.17

For 50-mm plastic Pall rings, $a_p = 130$ m²/m³ and $F_p = 131$ m⁻¹ (Table 13.3).

(a) Column diameter

$$\text{Flow parameter} = \frac{L}{G}\sqrt{\frac{\rho_g}{\rho_L}} = \frac{31.57}{5.09}\sqrt{\frac{1.234}{1000}} = 0.218$$

From Figure 12.53, for $\Delta P = 85$ mm H₂O/m (1.02 " H₂O/ft) and $FP = 0.218$, $C_{sb}F_p v^{0.05} = 0.65$ (metric)
From which, $C_{sb} = 0.65/[(131)^{0.5} (1.0)] = 0.057$
Superficial velocity of gas = $0.057[1000/1.234]^{0.5} = 1.63$ m/s
For the design air rate of 5.09 kg/s = 4.12 m³/s, the cross section is 2.40 m², and rounding,

Column diameter = 1.75 m, Column cross section = 2.54 m²

(b) Effective area
Assume a packed height of 10 ft (3.05 m)
For $A_t = 2.54$ m², $L = 31.57/2.54 = 12.43$ kg/s-m², $G = 5.09/2.54 = 2.00$ kg/s-m².
From Equation (13.49), $a_e/a_p = 9.79[0.070^{0.5}/3.05^{0.4}][(14.6E–06 \cdot 12.43)(2637 \cdot 2.00)]^{0.1603} = 1.65$.
Thus, $a_e = 1.65 (130) = 215$ m⁻¹.

(c) Transfer coefficients

[Equation (13.48)]

$$k_L = 0.0306 \left(\frac{9.81(1.02(10^{-3}))}{1000} \right)^{1/3} \left(\frac{31.57}{2.54(130)(1.02)(10^{-3})} \right)^{2/3} \left(\frac{1}{1150} \right)^{1/2} = 1.30(10^{-3}) \text{ m/s}$$

[Equation (13.47)]

$$k_G = \frac{0.145(130)(8.2)(10^{-6})}{(0.082)(294)} \left(\frac{2.00(10^5)}{(130)(1.75)} \right)^{0.7} (1.73)^{1/3} = 8.84(10^{-4}) \frac{\text{kg moles}}{s - m^2 - atm}$$

[Equation (13.43)]

$$H_g = \frac{2.00}{(29)(8.84E - 04)(215)(1.0)(1.0)} = 0.363 \text{ m}$$

[Equation (13.44)]

$$H_L = \frac{12.43}{18(1.30)(10^{-3})(215)(1000)(1.0)} = 2.47(10^{-3}) \text{ m} \quad \lambda = \frac{m}{L_m/V_m} = \frac{6753}{0.691/0.0690} = 675$$

$$H_{og} = H_g + \lambda H_L = 0.363 + 675(2.47(10^{-3})) = 0.363 + 1.667 = 2.03 \text{ m}$$

The stripping factor is 675 (same as λ). By Equation (13.22), 0.71 theoretical stages are needed.

By Equation (13.42), this translates to $N_{og} = 0.0066$, and $Z = N_{og}H_{og} = 0.013$ m. Thus, for the relatively high gas/liquid ratio used, and for, say, 3.05 m bed height, the removal of TCE is far greater than 99%. In fact, it is greater than 99.999%. It is not unusual for a feed of, say, 20 parts per million to be purified to less than one part per billion by air stripping.

Figure 13.20 shows how a cocurrent absorber may be analyzed. Note the equilibrium limitation, which limits the degree of recovery of solute. Basic models for estimating hydraulics and mass transfer are based on countercurrent flow, which is normally more efficient, and must be adjusted if the gas flow direction is reversed.

Other contacting devices are shown in Figure 13.21 and discussed below.

13.4.1 Spray Chamber

This is the ultimate low-pressure drop device, since the energy for dispersion and contacting is pumped into the entering liquid. Spray contactors suffer from departure from countercurrent flow, due to the high degree of backmixing or circulation in the gas. Drop coalescence and migration of drops to the vessel wall compromise the effective area for mass transfer. The mass transfer rate is highest in the close vicinity of the spray nozzles. As shown in the figure, it is economical to use more than one level of spray nozzles, and for large diameter units, each level comprises a bank of nozzles. Heights of transfer units for commercial-scale spray chambers have been reported by Pigford and Pyle.[32] More than two or three transfer units usually cannot be achieved in a spray contactor, but those units are achieved at very low pressure drop.

Spray chambers are frequently used for scrubbing sulfur dioxide and solid particulates from stack gases. The nozzles can be designed to handle slurries, e.g., lime slurry fed to the stack gas scrubber.

Absorption and Stripping

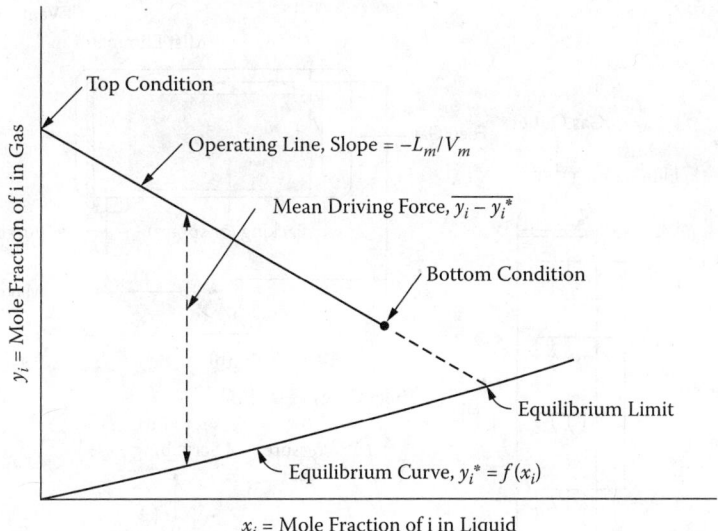

FIGURE 13.20 Calculation diagram for a cocurrent absorber.

13.4.2 CROSSFLOW SCRUBBER

The gas flows horizontally, contacting by downflowing liquid. The effective driving force for mass transfer is between that for counter- and cocurrent contactors. Crossflow scrubbers have low pressure drop and usually require a lower liquid/gas ratio than either counter- or cocurrent scrubbers. The time of contact between gas and liquid is relatively low, and crossflow units are not recommended for most chemical absorptions. Design procedures follow a finite-element approach; the scrubber volume is divided into cubes, each of which is assumed to reach equilibrium.

13.4.3 VENTURI SCRUBBER

This is a cocurrent scrubber with pressure drop advantages (venturi effect) and a high degree of turbulence at the point of contact. The equilibrium limitation as indicated in Figure 13.20 prevails. Venturi units have found applications for rapid reactions and for particulate removal. As shown in Figure 13.21, elaborate means for separating the effluent gas and liquid may be necessary.

13.5 STEAM STRIPPING

Steam stripping is used widely and differs from stripping with an inert gas (e.g., air) in that the vent gas can be condensed and the stripped liquids recovered by decantation or distillation. Thus, steam stripping can be used for removing VOCs and differs from air stripping in two important respects: steam is condensable, and the operating temperate is high, which reduces the solubility of the solute(s) in the feed liquid. Invariably, the feed is an aqueous solution, and the operating pressure can be varied over a wide range. An overall heat balance shows that there is a minimum rate of steam to prevent complete condensation (and thus no stripping) in the unit. A typical flow diagram for a steam stripper is shown in Figure 13.22. The steam can be extraneous (live steam) or generated in a reboiler as shown.

Steam stripping is often performed in plate columns since, in the past, few studies of mass transfer in packed columns have been made. A detailed study of stripping toluene from water in a 1.2-m diameter sieve tray column has been reported by Kunesh et al.[33] Earlier, Rush and Stirba[34] reported stripping methylisobutyl ketone (MIBK) from water in a 0.46-m sieve tray column.

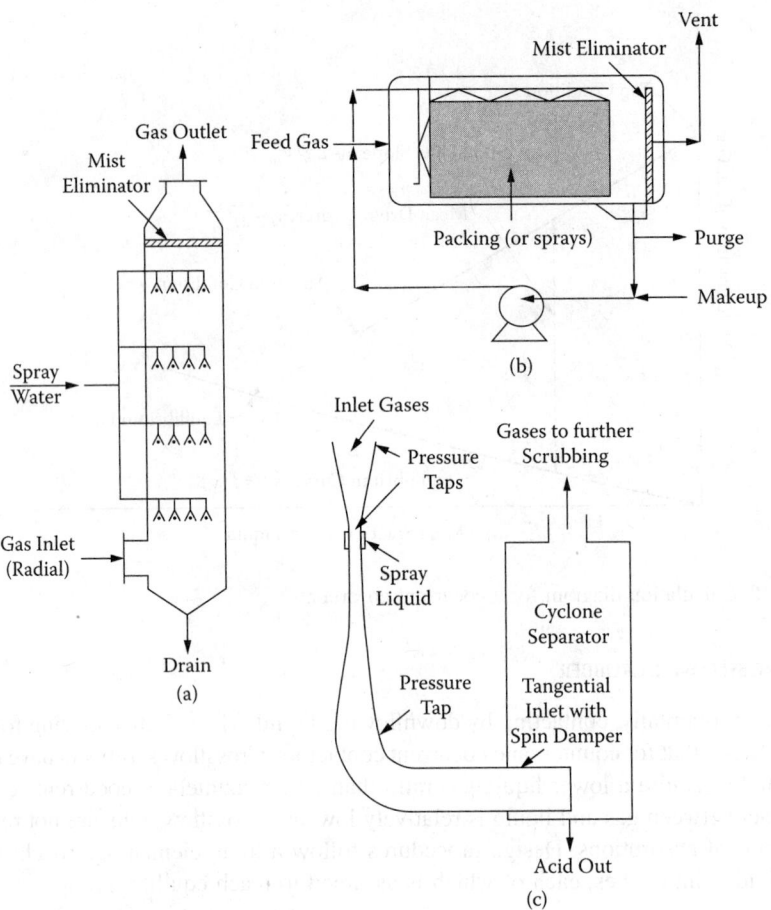

FIGURE 13.21 Representative vendor information on the characteristics of a random packing. Two-inch plastic Pall rings. [Courtesy of NorPro-Ste.Gobain Co., Akron, Ohio.]

Representative results from these studies are shown in Figure 13.23. The efficiencies tend to be low, and the liquid phase resistance dominates. Fair and Harvey[35] found that a slight modification of the Chan–Fair model[36] could be used for steam stripping (Figure 13.24).

While packed steam strippers are not widely used, there is no reason why they should not be. Performance data for such strippers are limited, however. Because of the temperature levels, metal pickings would normally be specified. Kolev et al.[37] studied chloroform stripping in a 0.40-m column using metal structured packing. Ortiz–Del Castillo et al.[38] studied toluene stripping in a 0.25-m column using metal structured and random packings. The latter group developed models for mass transfer based on air stripping experiences. The use of packing for commercial-scale steam stripping apparently has not been investigated.

The economics of vacuum steam stripping of volatile organic compounds have been reported by Rasquin et al.[39] The U.S. Environmental Protection Agency has favored steam stripping over air stripping, because the closed system prevents the stripped components from moving into the atmosphere.

Problem 13.8

A refinery aqueous stream flows at 600 gal/min and contains 15 lb/hr benzene. The benzene is to be removed at a stripping efficiency of 99.95%. A tray column is to be used. Consider two operating pressures, 2.0 psia and 14.7 psia. With no attempt at optimization, evaluate the parameters that should be considered for design. In keeping with refinery practice, use conventional English units.

Absorption and Stripping

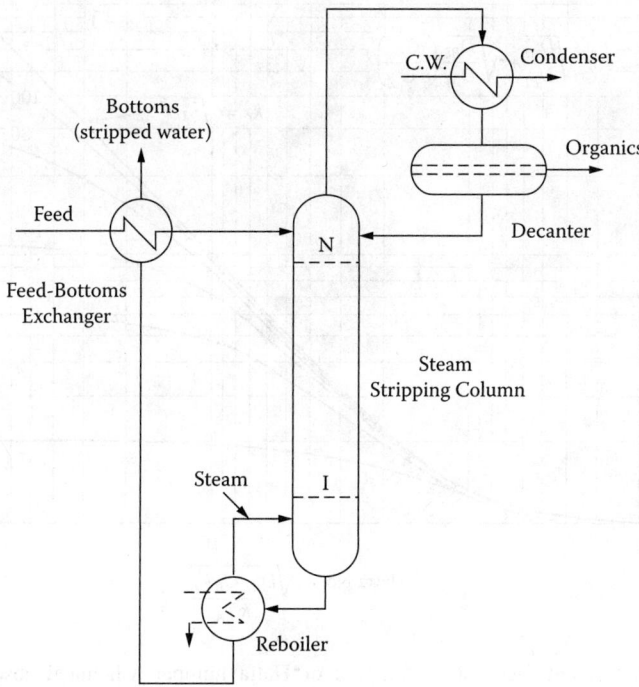

FIGURE 13.22 Typical flow diagram for a steam stripper.

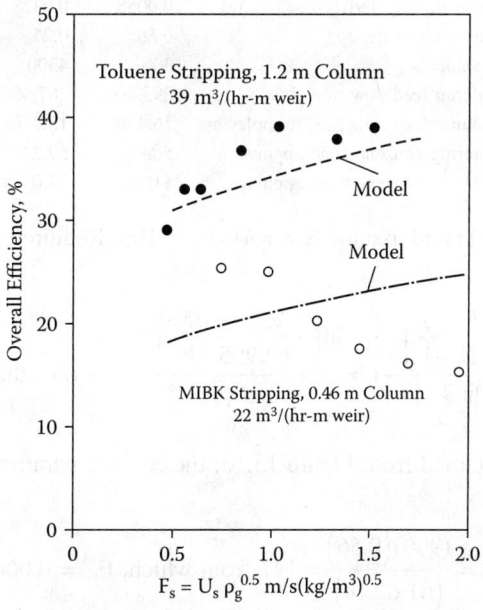

FIGURE 13.23 Overall tray efficiencies for steam stripping using sieve trays.

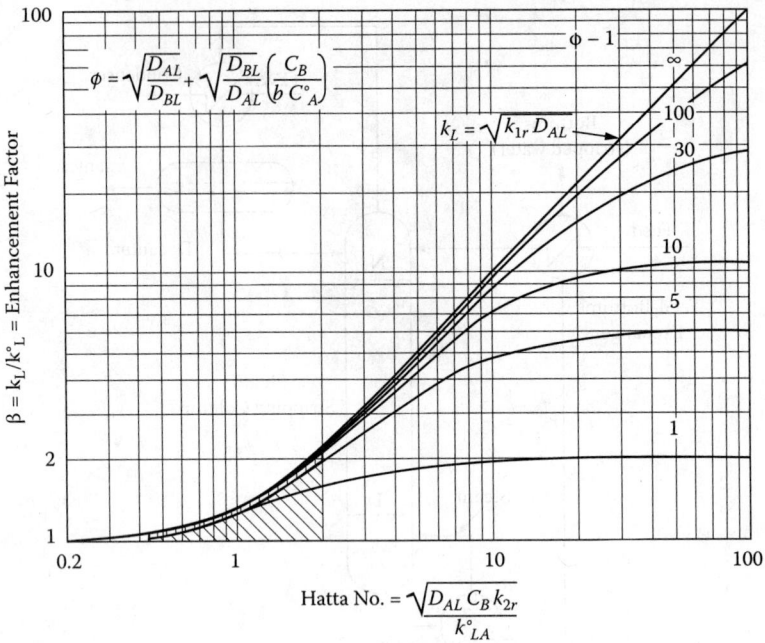

FIGURE 13.24 Enhancement factor as a function of Hatta number. Chemical absorption, irreversible reactions.

Properties:	2.0 psia	14.7 psia
Temperature, °F	126	212
Liquid density, lb/ft³	61.6	59.8
Vapor density, lb/ft³	0.0058	0.0373
Liquid viscosity, cp	0.56	0.27
K value	300	4300
Column feed flow rate, lb/hr	295,800	287,400
Column feed flow rate, lb-moles/hr	16,430	15,970
Entering benzene conc., ppmw	50.8	52.2
Entering benzene conc., ppmm	11.7	12.0

For the first try, use 2.0 psia and assume $S = K/(L/V) = 100$. Required stages [Equation (13.22)]:

$$N = \frac{\ln\left(\frac{E_s - S}{E_s - 1}\right)}{\ln S} - 1 = \frac{\ln\left(\frac{0.9995 - 100}{0.9995 - 1}\right)}{\ln 100} - 1 = 1.65 \quad \text{theo.stages}$$

The stage efficiency is obtained from Figure 13.16; the X-scale parameter is

$$K\mu_L/\rho_{LM} = \frac{(900)(0.56)}{(61.6/18)} = 147 \text{ from which, } E_{oc} = 0.060 = 6.0\%$$

Actual plates required in column = 1.65/0.060 = 28

Absorption and Stripping

A sample of other calculations follows:

P, psia	°F	S	L/V	N_{theo}	Eff, %	N_{act}	lb/hr steam	Vapor, acfs	Col. diam,ft*
2.0	126	75	12.0	1.76	6.0	30	24,600	1178	8.2
2.0	126	100	9.0	1.56	6.0	28	32,800	1570	9.1
14.7	212	75	57.3	1.76	3.2	55	5,200	39	4.6
14.7	212	100	43.0	1.65	3.2	52	6,900	51	4.5
14.7	212	200	21.5	1.76	3.2	45	13,700	102	5.1

* Using Figure 13.13 with 24-in tray spacing, 12% downcomer area, 80% approach to flood.

Interpretation. The examples used are far from optimum. An intermediate pressure may give the best combination of column height and column diameter. Refinery strippers usually have 20 to 30 plates, but such a heuristic depends on the stream being stripped, the cost of steam, and so on.

13.6 CHEMICAL ABSORPTION

Many commercial absorptions involve a chemical reaction between the solute (diffusing species) and the solvent. A common example is the removal of carbon dioxide from air or an inert gas by scrubbing with aqueous sodium hydroxide:

$$CO_2 + 2NaOH = Na_2CO_3 + H_2O \tag{13.50}$$

or for the removal of carbon dioxide by an amine solution:

$$2R-NH_2 + CO_2 = R-NHCOONH_3-R \tag{13.51}$$

where, for monoethanolamine, $R = HOCH_2CH_2$.

These are examples of commercially important chemical absorptions where simple physical mass transfer rate equations need modifications. Usually, the presence of a reaction speeds up mass transfer. For a rigorous design, information must be available on the rate of reaction as a function of concentrations, temperature, pressure, and degree of ionization in the liquid.

13.6.1 COMPUTATIONAL APPROACH

The following development applies to almost all chemical absorptions. Solute molecules in the gas diffuse to and across the interface, then diffuse in the liquid until meeting a reactant. If the reaction is very fast, the nonreactive mass transfer relations, previously discussed, apply—but very conservatively; the effective rate is higher. The flux equation for liquid phase transfer of component i is modified as follows:

$$N_i = \beta k_{Li}^o \left(C_i^* - C_i\right) \tag{13.52}$$

where N_i is the transport flux, kg moles/s-m² effective interfacial area, and k_{Li}^o is the physical mass transfer coefficient discussed previously. (The superscript is added to avoid confusion with the overall rate $k_{Li} = \beta k_{Li}^o$.) The coefficient β is called an "enhancement factor," which allows for chemical reaction. There are models available for estimating β from reaction rate data, but they are discussed only briefly here. For details, consult References 40 through 42.

Equation (13.52) may be made more general to include gas resistance and a reaction rate constant:

$$\frac{p_i}{N_i} = \frac{1}{k_{gi} a_e} + \frac{H_i}{k_{Li} a_e} + \frac{H_i}{k_r a_e} \qquad (13.53)$$

In other words, the total resistance to mass transfer is the sum of resistances of gas, liquid, and reaction:

$$R_t = R_g + R_L + R_r \qquad (13.54)$$

where H_i is the Henry's law constant $= p_i^*/C_i$, and k_r is the specific velocity constant for the reaction. When the reaction is very rapid and complete, k_r is very high, and R_r drops out. If there is negligible gas resistance, R_g drops out, and

$$\frac{p_i}{N_i} = \frac{H_i}{k_L a_e}$$

or the equivalent of Equation (13.52), with $C_i \sim 0$. Usually, a bimolecular reaction is of interest, involving species A from the gas and B from the liquid,

$$A + bB \rightarrow C$$

We can define a dimensionless Hatta number,

$$Ha = \frac{\sqrt{D_{AL} C_B k_{2r}}}{k_{LA}^o} = \frac{\text{effective rate}}{\text{rate for diffusional transport only}} \qquad (13.55)$$

where
- D_{AL} = liquid diffusivity for A, m²/s
- C_B = bulk liquid concentration of B, kg-moles/m³
- k_{2r} = reaction velocity constant (second-order), m³/(s-kg mole)
- k_{La}^o = physical liquid phase mass transfer coefficient, kg mole/(s-m²-kg mole/m³) = m/s

When the Hatta number (Ha) is evaluated, the value of the enhancement factor may be determined from Figure 13.24, based on reference 43.

The parameters in Figure 13.24 are $\phi - 1$, where

$$\phi = \sqrt{\frac{D_{AL}}{D_{BL}}} + \sqrt{\frac{D_{BL}}{D_{AL}}} \left(\frac{C_B}{b C_A^*} \right) \qquad (13.56)$$

The top curve in Figure 13.24 applies to first-order and pseudo-first-order irreversible reactions, and for Ha > 3,

$$k_L = \sqrt{k_{1r} D_{AL}} \qquad (13.57)$$

The above treatment has dealt only with irreversible bimolecular reactions and, as a special case, first-order irreversible reactions. Although these cases predominate, others are certainly possible. The references cited are sources for additional situations.

Absorption and Stripping

Problem 13.9

Eckert et al.[44] report test data for carbon dioxide absorption from air in a packed column. Conditions for a particular test are as follows:

Total pressure	1.0 atm	—
CO_2 partial pressure	0.01 atm	—
Temperature	24°C	75°F
Packing	25 mm ceramic Raschig rings	—
Gas rate	0.678 kg/s-m²	500 lb/hr-ft²
Liquid rate	6.78 kg/s-m²	5000 lb/hr-ft²
Column diameter	0.76 m	2.50 ft
Packed height	3.05 m	10.0 ft
Solvent	1.0 N NaOH, 25% conversion to carbonate	

Solution

For the degree of conversion, and taking into account the ionized liquid, the Henry's law coefficient is H_i = 2535 atm/m.f. = 45.63 atm/(kg moles/m³). Also, for the flow conditions given, the physical mass transfer coefficients, based on the method of Bolles and Fair,[27] are

$$(k_L a_e)° = 16.7E{-}03 \text{ sec}^{-1}$$

$$k_g a_e = 5.69E{-}02 \text{ kg-moles/(s-m}^3\text{-atm)}$$

$$\frac{1}{K_{og} a_e} = \frac{1}{k_g a_e} + \frac{H}{(k_L a_e)°} = \frac{1}{5.69E-02} + \frac{45.63}{16.7E-03} k_g a_e$$

For purely physical absorption, with component i = CO_2,

$$\frac{1}{K_{og} a_e} = \frac{1}{k_g a_e} + \frac{H_i}{(k_L a_e)°} = \frac{1}{5.69E-02} + \frac{45.63}{16.7E-03}$$

from which $K_{og} a_e$ = 3.64E–04 kg moles/(s-m³-atm).

Now, determine the effect of the chemical reaction between CO_2 (comp. A) and NaOH (comp. B).

- Bimolecular reaction velocity constant, k_{2r} = 11,000 m³/(kg mole-s) (reference 40)
- Conc. of NaOH = C_{OH^-} = 0.375 kg-ions/m³
- Diffusivity of CO_2 in ionized liquid = D_{AL} = 1.92E–09 m²/s
- Effective interfacial area: assume $a_e = a_p$ = 190 m²/m³ (Table 13.2)
- Physical mass transfer coefficient $k_L°$ = $(k_L a_e)°/a_e$ = 16.7E–03/190 = 8.79E–05 m/s

$$\phi = 1 + 1.0 \left(\frac{0.375}{1.0(2.191E-04)} \right) = 1.0 + 1711 = 1712$$

To evaluate the ϕ parameter for Figure 13.27, calculate the dimensionless Hatta number:

$$C_A^* = p_{CO2}/H_{CO2} = 0.01/45.63 = 2.191E{-}04 \text{ kg moles/m}^3 \text{ assume } D_{AL} \sim D_{BL} \text{ (reference 40)}$$

From Figure 13.24, for Hatta No. = 32.0 and ϕ = 1171, enhancement factor β = 32.

This converts to $K_{og}a_e$ = 2.18 lb moles/(hr-ft^3-atm), very close to the value of 2.20 measured by Eckert et al.[44] The absorption with reaction is about 32 times faster than for physical absorption alone.

13.7 SUMMARY

Absorption and stripping are counterpart processes, and the same general equations apply to both. Most applications of the processes deal with dilute mixtures, namely, absorbing small amounts of material from a large volume of gas or stripping small amounts of material from a large volume of liquid. When the dilute conditions prevail, the methodology for determining theoretical stages or transfer units is relatively simple. For concentrated mixtures, rigorous solutions normally are obtained from special computer programs. Approximate approaches to handling concentrated mixtures are given in this section.

Absorptions and strippings may be performed in conventional plate-type columns or countercurrent packed columns. In a few cases, specialized equipment is used. The traditional approaches to plate and packed column design, used in distillation, apply also to absorption and stripping. Thus, readers of this chapter should also refer to Chapter 12 particularly for dimensioning the contacting equipment.

A significant number of commercial absorption processes involve a chemical reaction in the liquid phase. The effect of the reaction is to speed up the rate of mass transfer and, to some extent, make the determination of transfer units or stages simpler.

As for all separation processes, basic thermodynamic and physical property data must be available. For absorption and stripping, key data relate to solubility, and fortunately there are large databases available to support predictive methods.

13.8 NOMENCLATURE

a_e = effective interfacial area, m^2/m^3
a_p = specific surface area of packing, m^2/m^3
A = area, m^2
A = absorption factor, dimensionless [Equation (13.8)]
A_T = total cross sectional area of column, m^2
A_N = net area cross section, m^2
A' = average absorption factor, dimensionless [Equation (13.16)]
C = concentration, kg moles/m^3
C_v = orifice coefficient, dimensionless
C_{sbf} = capacity parameter, m/s [Equation (13.30)]
Ca = dimensionless capillary number [Equation (13.49)]
CP = capacity parameter [Equation (13.46)]
D = diffusion coefficient (diffusivity), m^2/s
E_a = absorption efficiency, fractional
E_{oc} = overall column efficiency, fractional
E_s = stripping efficiency, fractional
F_p = pacing factor, 1/m (Table 13.1)
F_S = gas F-factor = $U_s \rho_G^{0.5}$, m/s(kg/m^3)$^{0.5}$
g = gravitational constant, m/s^2
G = gas flow, kg/s-m^2
G_m = gas flow, kg moles/s-m^2
h = pressure head, mm liquid
h_d = head loss for flow through holes, mm liquid

Absorption and Stripping

h_L = head loss for flow through froth on plate, mm liquid
h_w = weir height, mm
H_g = height of a gas phase transfer unit, m
H_i = Henry's law coefficient for component i, atm/mole fraction
H_L = height of a liquid phase transfer unit, m
H_{og} = height of a liquid phase transfer unit, m
Ha = dimensionless Hatta number [Equation (13.55)]
HETP = height of packing equivalent to theoretical plate, m
k_g = mass transfer coefficient for gas, kg moles/(s-m²-atm)
k_L = mass transfer coefficient for liquid, m/s [kg moles/(s-m²-kg/m³)]
$k°_L$ = mass transfer coefficient for liquid, physical basis only, m/s
k_r = reaction velocity constant (units dependent on reaction order)
K = gas-liquid equilibrium ratio = y^*/x
K_{og} = overall mass transfer coefficient, gas concentration basis, kg moles/(s-m²-atm)
L = liquid flow, kg/s-m²
L_m = liquid flow, kg moles/s
m = slope of equilibrium line
N_{theo} = number of theoretical stages
N_g = number of gas phase transfer units
N_i = molar flux of component i, kg moles/s-m²
N_L = number of liquid phase transfer units
N_{og} = number of overall transfer units, gas concentration basis
p = partial pressure, atm
p^{sat} = vapor pressure, atm
P = total pressure, atm
R = universal gas constant
Re = dimensionless Reynolds number
S = stripping factor, dimensionless [Equation (13.21)]
S' = average stripping factor, dimensionless [Equation (13.24)]
S_e = effective stripping factor, dimensionless [Equation (13.25)]
T = absolute temperature, Kelvin
TS = tray spacing, m or mm
U = gas velocity, m/s
U_h = gas velocity through holes, m/s
U_N = gas velocity through net area of tray, m/s
U_s = superficial velocity through column, m/s
V = vapor or gas flow, kg/s-m²
V_m = vapor or gas flow, kg moles/s
x = mole fraction in liquid
X = ratio, moles/mole solute free liquid
y = mole fraction in gas or vapor
y^* = mole fraction in gas or vapor, equilibrium condition
Y = ratio, moles/mole solute free gas
Z = height, m
Z_f = froth height, m or mm

13.8.1 Greek Letters

α_c = relative volatility
α_i' = Bunsen solubility coefficient, 1/atm [Equation (13.1)]
α = relative solubility

β	= enhancement factor for chemical absorption [Equation (13.52)]
γ	= activity coefficient
λ	= ratio of slopes, equilibrium line/operating line
μ	= absolute viscosity, mPa-s (cp)
ν	= kinematic viscosity, cs
ρ	= density, kg/m^3
ρ_{LM}	= liquid molar density, kg moles/m^3
σ	= surface tension, mN/m (dynes/cm)
ϕ	= parameter for Figure 13.24

13.8.2 Subscripts

1	= top plate in column for absorbers, bottom plate in column for strippers
A,B	= components A or B
g	= gas phase
i	= component i
L	= liquid phase
N	= bottom plate in column for absorbers, top plate in column for strippers

REFERENCES

1. Hwang, Y.-L.; Olson, J. D.; Keller, G. E. 1992. *Ind. Eng. Chem. Res.* 31: 1759.
2. Fredenslund, A.; Gmehling, J.; Rasmussen, P. 1977. *Vapor-Liquid Equilibria Using UNIFAC,* Amsterdam: Elsevier.
3. Yaws, C. L.; Pan, X.; Lin, X. 1993. *Chem. Eng.* 100 (2): 108.
4. Prausnitz, J. M.; Shair, F. H. 1961. *AIChE J.* 7: 682.
5. Yen, L. C.; McKetta, J. J. 1962. *AIChE J.* 8: 501.
6. Yaws, C. L.; Hopper, J. R.; Wang, X.; Rathinsamy, A. K. 1999, *Chem. Eng.* 106 (6): 102.
7. *Solubility Data Series* (continuing series). Elmsford, NY: Pergamon Press.
8. Seidell, A. 1952. *Solubilities of Inorganic and Organic Compounds.* New York: Van Nostrand.
9. Seidell, A. 1958. *Solubilities of Inorganic and Metal-Organic Compounds.* New York; Van Nostrand.
10. Linke, W. F.; Seidell, A., *Solubilities of Inorganic and Metal-Organic Compounds,* vol. 2. Washington, DC: American Chemical Soc., 1965.
11. Dack, M. J. R., ed. 1975. *Solutions and Solubilities.* New York: John Wiley & Sons.
12. Markham, A. E.; Kobe, K. A. 1941. *Chem. Rev.* 28: 519.
13. Long, F. A.; McDevit, W. F. 1952. *Chem. Rev.* 51: 119.
14. Battino, R.; Clever, H. L. 1966. *Chem. Rev.* 66: 395.
15. Wilhelm, E.; Battino, R. 1973. *Chem. Rev.,* 73: 1.
16. Wilhelm, E.; Battino, R.; Wilcock, R. J. 1977. *Chem. Rev.* 77: 219.
17. Othmer, D. F.; Kollman, R. C.; White, R. E. 1944. *Ind. Eng. Chem.* 36: 963.
18. Kremser, A. 1930. *Natl. Petroleum News.* 22 (21): 42; Souders, M.; Brown, G. G. 1932. *Ind. Eng. Chem. 24:* 519.
19. Edmister, W. C. 1943. *Ind. Eng. Chem.* 35: 837.
20. Fair, J. R., in *Perry's Chemical Engineers' Handbook,* 7th edition, R. H. Perry and D. W. Green, eds. New York: McGraw-Hill, 1997, 14–14, 14–15.
21. *Ibid.,* 14–27.
22. O'Connell, H. E. 1946. *Trans. AIChE* 42: 741.
22a. Hines, A. L.; Maddox, R. N. 1985. *Mass Transfer—Fundamentals and Applications.* Englewood Cliffs, NJ: Prentice Hall.
23. Strigle, R. F. 1994. *Packed Tower Design and Applications.* Houston: Gulf Publ. Co.
24. Stichlmair, J.; Bravo, J. L.; Fair, J. R. 1989. *Gas Sepn. Purif.* 3: 19.
25. Engel, V.; Stichlmair, J.; Geipel, W. A. "A New Correlation for Pressure Drop, Flooding and Holdup in Packed Columns." Presented at AIChE Meeting, Miami, FL, October 1998.

26. Kister, H. Z. 1992. *Distillation—Design*. New York: McGraw-Hill.
27. Bolles, W. L.; Fair, J. R. 1982. *Chem. Eng.* 89 (14): 109.
28. Onda, K.; Takeuchi, H.; Okumoto, Y. 1968. *J. Chem. Eng. Japan* 1 (1): 56.
29. Bravo, J. L.; Fair, J. R. 1985. *Ind. Eng. Chem. Proc. Des. Devel.* 21: 162.
30. Bravo, J. L.; Fair, J. R., "Mass Transfer Modeling of Packed Absorbers and Strippers," presented at AIChE Spring Meeting, Orlando, FL, March 1990. Copy available from Separations Research Program, University of Texas at Austin, Austin, TX 78712.
31. Dvorak, B. I.; Lawler, D. F.; Fair, J. R.; Handler, N. E. 1996. *Environ. Sci. Tech.*, 30: 945.
32. Pigford, R. L.; Pyle, C. 1951. *Ind. Eng. Chem.* 43: 1649.
33. Kunesh, J. G.; Ognisty, T. P.; Sakata, M.; Chen, G. X. 1996. *Ind. Eng. Chem. Res.* 35: 2660.
34. Rush, F. R.; Stirba, C. 1957. *AIChE J.* 3: 336.
35. Fair, J. T.; Harvey, R. L., "Modeling of Tray-Type Steam Stripping Columns." Presented at the Spring 1984 AIChE Meeting, Atlanta, April 1994.
36. Chan, H.; Fair, J. R. 1984. *Ind. Eng. Chem., Proc. Des. Devel.* 23: 814.
37. Kolev, N.; Darakchiev, R.; Semkov, K. 1997. *Ind. Eng. Chem. Res.* 36: 238.
38. Ortiz–Del Castillo, J. R.; Guerrero–Medina, G.; Lopez–Toledo, J.; Rocha, J. A. 2000. *Ind. Eng. Chem. Res.* 39: 731.
39. Rasquin, E. A.; Lynn, S.; Hanson, D. 1978. *Ind. Eng. Chem. Fundam.* 17: 170.
40. Sherwood, T. K.; Pigford, R. L. 1975. *Mass Transfer*. New York: McGraw-Hill.
41. Danckwerts, P. V. 1970. Gas-Liquid Reactions. New York: McGraw-Hill.
42. Fair, J. R., in *Perry's Chemical Engineers' Handbook*, 7th ed., R. H. Perry and D. W. Green, eds. New York: McGraw-Hill, 1997, 14-17—14-23.
43. Van Krevelen, D. W.; Hoftijzer, P. J. 1948. *Rec. Trav. Chim.* 67: 563.
44. Eckert, J. S.; Foote, E. H.; Rollinsen, L. R.; Walter, L. F. 1967. *Ind. Eng. Chem.* 59 (2): 41.

14 Adsorption

Kent S. Knaebel

CONTENTS

14.1 Introduction ... 1120
14.2 Adsorbents ... 1123
    14.2.1 Adsorbent Selection Criteria ... 1124
    14.2.2 Adsorbent Characteristics ... 1124
        14.2.2.1 Capacity, Selectivity, and Regenerability ... 1124
        14.2.2.2 Kinetics ... 1126
        14.2.2.3 Pore Structure and Surface Area ... 1127
        14.2.2.4 Particle Size Distribution ... 1129
        14.2.2.5 Durability and Cost ... 1129
    14.2.3 Common Adsorbents ... 1129
        14.2.3.1 Aluminas ... 1130
        14.2.3.2 Silicas ... 1130
        14.2.3.3 Zeolites ... 1130
        14.2.3.4 Carbons ... 1131
        14.2.3.5 Polymers ... 1132
        14.2.3.6 Biomass ... 1132
14.3 Adsorption Thermodynamics ... 1132
    14.3.1 Adsorption Equilibrium and Heats of Adsorption ... 1133
    14.3.2 Sources of Equilibrium Data ... 1134
    14.3.3 Isotherm Equations ... 1135
    14.3.4 Mixture Equilibria ... 1139
14.4 Dynamics of Adsorption ... 1140
    14.4.1 Overview of Rate Phenomena ... 1140
    14.4.2 Mass Conservation and Transport ... 1141
        14.4.2.1 Local Equilibrium Model ... 1145
        14.4.2.2 Modified Wheeler and Robell Method ... 1146
        14.4.2.3 Mass Transfer Resistances ... 1146
    14.4.3 Heat Transfer ... 1148
    14.4.4 Pressure Drop ... 1149
    14.4.5 Dispersion ... 1150
    14.4.6 Chemical Kinetics and Adsorption ... 1151
14.5 Design of Batch, Fixed-Bed Adsorbers ... 1152
    14.5.1 General Objectives ... 1152
    14.5.2 Empirical versus Theoretical Approaches ... 1152
    14.5.3 Shortcut Procedures ... 1153
        14.5.3.1 Length of Unused Bed Method ... 1153
        14.5.3.2 Direct Method Using Breakthrough Models ... 1155
    14.5.4 Examples ... 1155
14.6 Cyclic and Continuous Adsorption Processes ... 1159
    14.6.1 Inert-Purge Cycle ... 1161

14.6.2 Displacement-Purge Cycle 1162
　　14.6.3 TSA Cycle 1162
　　Example 14.5 TSA Solvent Recovery 1162
　　14.6.4 PSA Cycle 1163
Bibliography..1170

14.1 INTRODUCTION

Adsorption is the phenomenon that selectively segregates atoms or molecules between a fluid and a solid. In some ways, it is like *ab*sorption, except the liquid-phase *ab*sorbent is replaced with a solid-phase *ad*sorbent. Sometimes solids could be said to *ab*sorb or *ad*sorb fluids, i.e., the distinction is blurred.* In this chapter, the focus is on selective uptake (and release) by porous solids, especially for process applications. Another subtle distinction occurs when vapor is adsorbed from a gas; it is like condensation, except that the driving force is subtle molecular forces rather than a temperature gradient.

When applied as a unit operation, adsorption can be used to split mixtures containing significant percentages of adsorbable components or purify streams containing trace amounts of contaminants. These separations, whether for gases or liquids, are accomplished by allowing the fluid and solid phases to interact under controlled conditions. Molecules that are selectively taken up are called *adsorbates*, and the solid surface that attracts the adsorbate is called the *adsorbent*. Jargon used in connection with adsorbents and adsorption, as well as some prominent definitions, are listed in Table 14.1. Fundamental causes and effects of adsorption are discussed in Section 14.3.

This chapter presents different facets of adsorption. Unfortunately, there is not sufficient space to provide case studies for any of the hundreds of unique applications of adsorption. On the other hand, most of the topics discussed here are the subject of dozens of technical papers each year, many of which appear in specialized journals, including (alphabetically) Adsorption, Carbon, Langmuir, Microporous and Mesoporous Materials (formerly Zeolites), Reactive Polymers, and other more general titles. Some subjects are also treated in much greater detail in books, many of which are listed in the bibliography that follows.

It is an understatement to say that adsorption is a diverse field. It impacts separation processes, materials science, catalysis, soil science, pharmaceutical products, environmental applications, and other widely different fields. A brief overview of those subjects, mainly oriented toward applications, is presented here.

Adsorption can perform many separations that are impossible or impractical by conventional techniques, such as distillation, absorption (gas–liquid), and even membrane-based systems. Adsorption has found recent applications in solving environmental problems and meeting stringent quality requirements. Important applications of adsorption fall in the category of purification, as cited by Yang (1987), but they represent only a small fraction of the uses of adsorption. The largest single application of adsorption is for water treatment. It employs nearly 100 million pounds of activated carbon annually in the U.S. alone, to remove compounds that could be toxic or merely pose problems of taste or odor. It was first used for municipal water treatment in powdered form in the late 1930s, and as granules in the 1960s. In addition, activated carbon has been used to decolorize sugar since the 1920s. Another widespread purification application is the pressure-swing air dryer found on heavy trucks and buses for their air-brake systems. Those would probably go unnoticed were it not for the abrupt, audible "blowdown" of these adsorbers, which sounds like a tire blowing out or a hydraulic system failure. A third, very widespread application is in the seal between thermopane windows. They employ 3A zeolite (as a powder mixed with caulk or as

---

* When a solid behaves like a sponge (i.e., by capturing a fluid via surface tension or by nonselective penetration), it could be called *ab*sorbent (e.g., granules used to pick up oil that has spilled on a floor). When a solid exhibits selective uptake, it should be called an *ad*sorbent(e.g., granules contained in a column through which oil-laden water passed, selectively removing the oil).

TABLE 14.1
Definitions of Terms Used Commonly in Adsorption

Adsorbate	Component (usually dilute) taken up by the adsorbent; also called *adsorptive*. The remaining fluid (solvent or carrier) is viewed as nonadsorbing. Some prefer this be strictly the component in the adsorbed state.
Adsorbent	Porous solid with "active" sites that take up specific constituents such that the composition on the surface differs from that in the surrounding fluid. Some pores may act as a sieve. Examples: activated carbon, alumina, silica gel, and zeolite.
Adsorption	Enrichment (or depletion) of one or more constituents in an interfacial layer [IUPAC, 1972]. The net effect of chemical and/or physical forces exerted by the adsorbent or co-adsorbed species.
Adsorption potential	The free energy change associated with the desorption of a molecule (or mole). Employed by Polanyi and Dubinin and coworkers to fit isotherm data. See Equation (14.7).

Breakthrough	An observation in a fixed-bed adsorber as the effluent composition rises toward the feed composition. It can refer to a criterion, e.g., 5% of the feed concentration of a component, or to the shape of the composition vs. time profile. The profile can change sharply or gradually, depending on equilibrium properties and mass transfer rates.
	Classifications (see Figure 14.9 for illustrations):
	    Constant pattern or self-sharpening front—usually a sharp, symmetric transition observed during uptake under favorable (Type I) equilibrium. The physical limit (without mass transfer effects) is called a *shock wave*.
	    Proportional pattern or disperse front—usually a gradual and asymmetric transition during regeneration or uptake under unfavorable (Type II) equilibrium. The physical limit (without mass transfer effects) is a *simple wave*.

CTC index	Carbon tetrachloride index for activated carbon; weight percent of carbon tetrachloride adsorbed at a partial pressure of 0.1505 atm (114 torr) and a carbon temperature of 25°C; obsolescent.
Chemisorption	Chemical bonding via electron transfer: specific; slow (activation barrier); forms monolayer, $\Delta H_{ads} > 3\ \Delta H_{vap}$.
Co-adsorption	Uptake of two or more constituents, each potentially affecting the adsorbability of the other.
Constant pattern front	An axial composition profile for a component having a shape that does not depend on axial position. This occurs during uptake of a component that exhibits a favorable isotherm and no dispersion. Mathematically, this occurs when $C_A / C_{AF} = \bar{n}_A / n_A^*(C_{AF})$.
Favorable isotherm	Concave downward (also Type I; see Figure 14.4 and Section 14.3.2).
Heats of adsorption	1. Isosteric heat of adsorption: $\Delta H_{ADS_i} = q_{st_i} = -R \left. \dfrac{d \ln p_i}{d 1/T} \right|_{q_i} > 0$ Similar to Clapeyron equation; can be used to interpolate or extrapolate data to other temperatures; and $\Delta H_{ADS_i} = f(q_i^*)$ 2. Differential heat of adsorption: $\Delta H_{Diff_i} = q_{st_i} - RT$ 3. Equilibrium heat of adsorption: $$\Delta H_{Equil_i} = \left[\int_0^{q_i} \Delta H_{ADS_i} dq_i - \pi_i A \right] \Big/ q_i$$ where the spreading pressure, π_i, is (for ideal gases): $$\dfrac{\pi_i A}{RT} \doteq \int_0^{q_i} \dfrac{d \ln p_i}{d \ln q_i} dq_i$$ 4. Integral heat of adsorption: $\Delta E_{Int_i} = \dfrac{1}{q_i} \displaystyle\int_0^{q_i} \Delta H_{Diff_i} dq_i$

Continued

TABLE 14.1 (Continued)
Definitions of Terms Used Commonly in Adsorption

Hysteresis	Divergence of uptake and release isotherm data due to the difference in mechanisms of pore filling vs. emptying, or other irreversible phenomena of uptake and release.
Iodine number	Amount of iodine adsorbed from an iodine/potassium iodide solution (0.02 N) per mass of activated carbon (mg/g). A measure of capacity of activated carbon for small molecules.
Ion exchange	Sorption using "fixed" acidic or basic functional groups, such as $-SO_3^-$, $-NH^+$, etc. Rates of uptake or release may be affected by reaction and/or diffusion.
Isobar	Plot of equilibrium data as amount adsorbed vs. temperature, holding partial pressure constant.
Isotherm	Set of equilibrium data: amount adsorbed (n^*_i) vs. concentration (c_i) or partial pressure (p_i), at constant temperature. See Figure 14.4 for classifications. Constraints: (1) $\lim_{c_i \to 0} \frac{dn^*_i}{dc_i} < \infty$ (2) $\frac{dn^*_i}{dc_i} \geq 0$ Typical units: $n^*_i [=] \frac{\text{moles }(i)}{\text{mass (adsorbent)}}$ or $\frac{\text{mass }(i)}{\text{mass (adsorbent)}}$ or $\frac{\text{std. vol. }(i)}{\text{vol. (adsorbent)}}$
Macropores	$d_p > 500$Å. (pigments, bacteria)
Mesopores	$d_p \in (20, 500)$Å (colloidal silica, viruses)
Micropores	$d_p < 20$Å (common molecules)
Mass transfer zone	Region of the axial composition profile in which the composition varies from that of the feed (toward the feed end) to that of the initially present fluid (toward the product end). Despite its name, the shape of this zone can depend on mass transfer resistances, dispersion, or equilibrium effects. For uptake with a favorable isotherm, the width remains essentially constant beyond a certain axial position, leading to the term *constant pattern front*. In that case, the effects are exclusively due to intraparticle- and/or film-diffusion resistances.
Methylene blue factor	Volume (cm³) of 0.15% methylene blue solution decolorized by 100 g of adsorbent in 5 min.
Molasses number	Quantitative measure of decolorizing capacity based on the change in color of molasses; relates to adsorption of large molecules from a liquid.
Physisorption	Based on weak physical attraction: sensitive to temperature, nonspecific, rapid (no activation barrier), possibly multilayer $\Delta H_{ADS} < 3\Delta H_{VAP}$.
Proportional pattern	An axial composition profile (or breakthrough curve) having a shape that "relaxes" gradually as the front moves along the axis. This is characteristic of desorption for a system that has a favorable isotherm. Axial dispersion exhibits a similar effect but may affect uptake and release.
Sorption	Adsorption or desorption.
Structure	Described by densities, void fractions, porosity, pore-size distribution.
Unfavorable isotherm	Concave upward (also Type II; see Figure 14.4 and Section 14.3.2).
Working capacity	The effective change of loading of an adsorbent between the regenerated and exhausted states. May reflect either operating data or a theoretical limit based on isotherm data.

particles within an aluminum frame) to adsorb moisture in the air space, preventing it from fogging the internal glass surface.

Two types of *cyclic* adsorption technologies [pressure swing adsorption (PSA) and temperature swing adsorption (TSA)], developed since the mid-1960s, are widely used for industrial separations. Examples of PSA include air dryers (as mentioned previously), hydrogen from refinery off-gases (commercialized in the 1960s), oxygen from air (early 1970s), and nitrogen from air (late 1970s). Many TSA systems remove or recover VOCs from solvent-laden air, e.g., at polymer processing, painting, and pharmaceutical production sites. Those use steam, hot gases, or even electrical resistance heating for regeneration. In addition, simulated moving-bed (SMB) systems split xylene isomers, sugars, and pharmaceutical products.

Moving-bed systems are *continuous* and include vapor-phase recovery systems that employ staged fluidized beds in which the adsorbent circulates through a thermal regenerator. Generally, it is possible for the fluid phase and adsorbent to be contacted in a countercurrent flow pattern. An early version, called the hypersorption process, was patented in 1922 but was not fully commercialized until the late 1940s. It performed well but suffered from high attrition rates of the adsorbent. A later version, called the Purasiv^{SM} process, employed small, glassy activated carbon beads that resisted attrition. Other moving-bed systems have mostly been used for solvent recovery from contaminated air. Among them are the PolyAd^{SM} system of Chematur and the Blizzard^{SM} system of On-Demand Environmental Systems, which used polymeric adsorbents because of their inherent toughness.

SMB systems were created to exploit some of the countercurrent features of moving-bed systems, but employing fixed beds to avoid attrition. Liquid-phase SMB adsorption systems, such as UOP's Sorbex^{SM} processes, have been commercialized since the early 1960s. Among the Sorbex^{SM} family, the Molex^{SM} process separates normal paraffins from branched and cyclic isomers; the Olex^{SM} process splits olefins from paraffins; the Parex^{SM} process isolates *p*-xylens from *m*-, *o*-xylene, and ethyl benzene mixtures; and the Sarex^{SM} process splits fructose from corn syrup. These are discussed further in Section 14.6.

All of the various applications require special adsorbent characteristics. The broadest and generally the most significant are the inherent adsorption *capacity* and *selectivity*. In many cases, the adsorption and desorption rates or *kinetics* and pressure drop are also important; hence, particle size is important. In addition, nearly every different application has a different set of additional priorities. For example, the main prerequisite for municipal water purification and many other large-scale applications is *low cost*. Other adverse conditions can complicate adsorbent specifications. For example, properties such as density, color, fluid compatibility, and durability (e.g., attrition resistance, crush strength, and hardness) all may be important. Adsorbents that have the ideal combination of essential characteristics for a specific application may or may not exist. That implies that compromises are frequently necessary.

Going hand in hand with advances in process technology and adsorbent materials, engineers and scientists have developed a better understanding of the mechanisms of adsorption. The mechanisms are complicated, because adsorption is always transient vs. steady state (implying that partial differential equations are common), and various mass transfer resistances may be important (implying that a significant amount of data, as well as constitutive rate equations, are needed). For example, it is highly important to know the time, t, when the adsorbent at a particular location, z, achieves a certain level of saturation, $n^*(C,z,t)$; it is, hence, necessary to know the fluid phase concentration, $C(z,t)$. For most design purposes, four different technical aspects must be addressed: adsorbent properties, adsorbate characteristics, equipment performance, and operating conditions. These are intricately coupled, and they control feasibility, safety, and profitability. Of course, the ease and flexibility of numerical simulations in particular have been augmented by faster, more accessible computers.

14.2 ADSORBENTS

Specific classes of adsorbents are described herein, along with some typical applications, although no attempt has been made to be exhaustive. Many varieties (activated aluminas, silica gels, activated carbons, zeolites, and polymeric adsorbents) are practically commodities and are generally available off the shelf. New, customized adsorbents can also be synthesized to have different properties, which may translate into better performance. A new adsorbent may take months or years to perfect; hence, a rule of thumb is that there is never enough time to develop a new adsorbent for a specific application.

Following an Edisonian testing approach, most minerals and many synthetic inorganic materials have been tried as adsorbents. Some have proved desirable, despite being mediocre adsorbents, simply because they are inexpensive. Others, fortuitously, are effective adsorbents. Some inorganic materials act more as "absorbents" than adsorbents, since they are not porous solids with active surfaces. Those have applications from flue gas desulfurization to drying to recovery of polychlorinated biphenyls. Among them are metal chlorides ($CaCl_2$), oxides (CaO, MgO, and ZnO for life support in the space program), silicates ($MgSiO_3$), sulfates ($CaSO_4$, the familiar "Drierite"), kieselguhr (or diatomaceous earth), and even sodium bicarbonate and limestone (for flue gas treatment). Some are used in an anhydrous state, whereas others are hydrated. Other inorganic adsorbents have been developed recently (e.g., pillared clays, aluminophosphates, and mesoporous adsorbents) that have not yet been commercialized. Those covered below are commercial products and are frequently encountered.

14.2.1 Adsorbent Selection Criteria

Adsorbent selection criteria for any application generally include the following main attributes: capacity, selectivity, regenerability, kinetics, durability, and cost. These attributes represent combinations of properties that are strongly affected by the pertinent conditions. For example, the first four govern how much adsorbent is necessary for a particular application, and the last two affect the annual cost. Furthermore, the first three are tied to the equilibrium characteristics, about which much can be said (see Section 14.3). Likewise, kinetics is covered in more detail in Section 14.4. Both will be discussed briefly here, however. Finally, adsorbent cost obviously depends on both its price and lifetime, which can depend on its resistance to attrition, degradation, fouling, and so on.

Rarely will a single adsorbent be optimal in all of these respects. Frequently it will be possible to narrow the choice to one or two classes of adsorbents, leaving a vast array of possible particle sizes, shapes, pretreatment conditions, and so forth. Final decisions should always be based on data. To make budget estimates, however, a number of different approaches can be derived from rules of thumb to provide quick experimental feasibility tests. Potential sources of such information are adsorbent or equipment vendors, published or commercial databases, and in-house or external laboratories.

14.2.2 Adsorbent Characteristics

14.2.2.1 Capacity, Selectivity, and Regenerability

Adsorption capacity (or loading) is probably the most important characteristic of an adsorbent. More detail is provided in Section 14.3.2. The loading is the amount of adsorbate taken up by the adsorbent, per unit mass (or volume) of the adsorbent, and it depends on the fluid-phase concentration, the temperature, and other conditions (especially the initial condition of the adsorbent). Typically, adsorption capacity data are plotted as isotherms (loading of adsorbate on the adsorbent versus fluid-phase adsorbate concentration at constant temperature), isosteres, isobars, and others mentioned later. Examples are shown in Figures 14.1 through 14.3. Adsorption capacity is of paramount importance to the capital cost, because it sets the amount of adsorbent required and also the volume of the adsorber vessels; the costs of both are significant if not dominant. When comparing alternate adsorbents, it is fair to express their capacity on a per unit volume basis, since that fixes

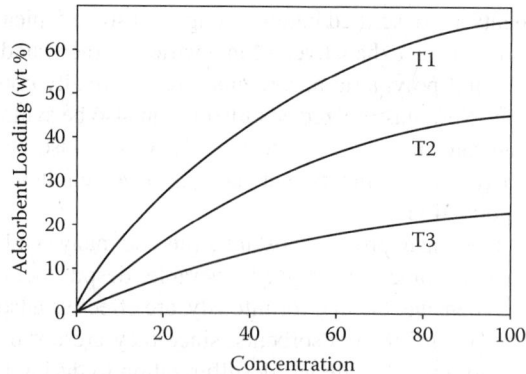

FIGURE 14.1 Typical Type I isotherms at increasing temperatures, $T_1 < T_2 < T_3$.

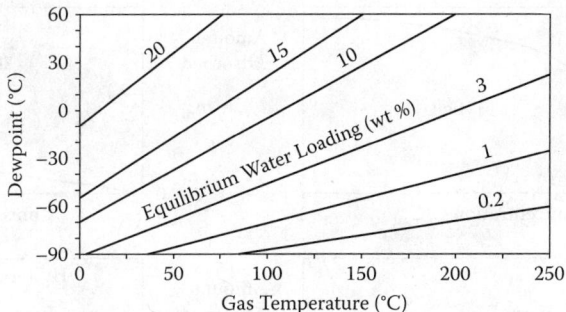

FIGURE 14.2 Isosteres: water vapor on 5A zeolite (activated at 350°C and 0.01 torr).

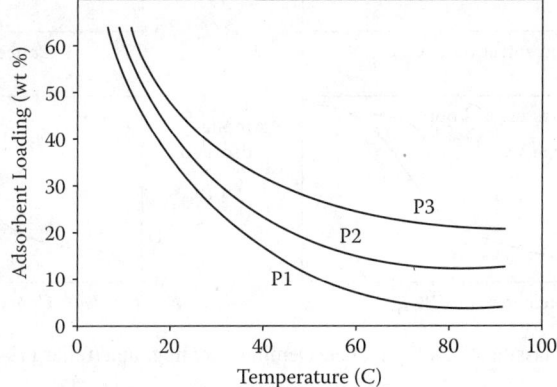

FIGURE 14.3 Isobars at various pressures, $P_1 < P_2 < P_3$.

the cost of the vessels, pipes, and other equipment. Conversely, the adsorbent is practically always sold on a per unit mass basis, so most isotherms are plotted on the basis of capacity per unit mass.

Six common isotherm shapes are shown in Figure 14.4. In fact, those are the classic isotherm types suggested by Brunauer et al. (1940). Each can be represented by numerous empirical equations, some of which are discussed later. The inherent shapes or types arise from the pore structure of the adsorbent, the nature of the forces between the adsorbent surface and adsorbate, and the dependence on concentration. Besides isotherms, other properties are related to adsorption capacity, especially surface area and pore size distribution. Some other properties are application oriented, such as CTC (carbon tetrachloride) index, iodine number, methylene blue factor, and molasses number, all defined in Table 14.1. They are frequently employed to describe activated carbons.

Selectivity is related to capacity, but there are several distinct definitions, as discussed later. The simplest is the ratio of the capacity of one component to adsorb on the adsorbent to that of another at a given fluid concentration. That ratio generally approaches a constant value as the concentration drops toward zero. Of course, the concentrations of interest may not be near zero, so the choice of definitions becomes subtle. The closest analogy is to relative volatility (e.g., in distillation) in that the closer it is to unity, the larger the required equipment. An ideal situation occurs when the major component is not adsorbed much (so it can be thought of as an inert "carrier"), which leads to a very large selectivity. Despite the attractiveness of an analogy to relative volatility (with a range from 0 to ∞), some people prefer a bounded selectivity (with a range from 0 to 1), and they employ the inverse of the ratio mentioned above. Thus, it is necessary to clarify the definition first, or to speak of "good" or "bad" instead of "large" or "small."

All cyclic adsorption applications rely on regenerability, which allows the adsorbent to operate in sequential cycles with uniform performance. This means that each adsorbable component

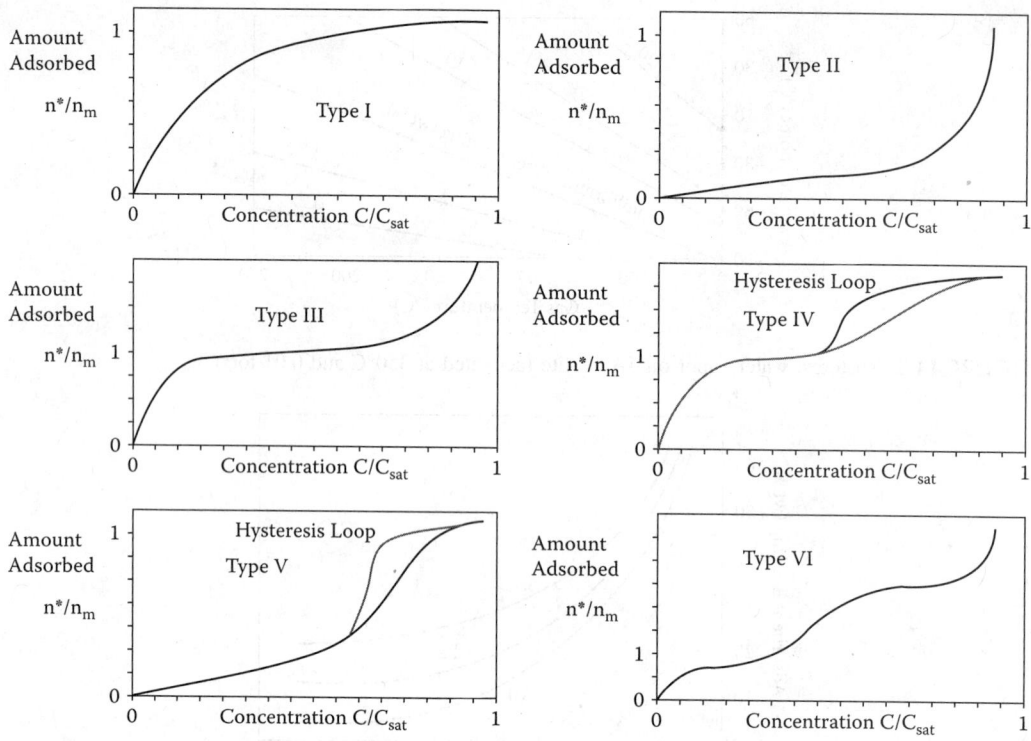

FIGURE 14.4 Isotherm classifications [Brunauer, Deming, Deming, and Teller (1940)].

(adsorptive or adsorbate) must be relatively weakly adsorbed (or physisorbed). The heat of adsorption provides a measure of the energy required for regeneration, and hence low values are desirable. Regeneration may be accomplished by a thermal swing, pressure swing, or chemical reaction (e.g., by displacement, elution, or supercritical extraction), or sometimes by a combination of those. Displacement would involve introducing a species that adsorbs more strongly than the adsorbate of interest, while elution would entail dissolving the adsorbed material by a solvent that is adsorbed weakly, if at all. Chemical methods require a separate separation operation that may be costly, plus the regenerated bed must be purged of the regenerant. In some cases, regeneration takes place by contacting the adsorbent with another fluid not used during loading. This requires draining or displacement, which might be time consuming and should be avoided whenever possible. The regenerability of an adsorbent affects the fraction of the original capacity that is retained for cyclic usage (sometimes called the *working capacity*) and the time, energy, and other requirements for regeneration.

14.2.2.2 Kinetics

Mass transfer kinetics is a catch-all term related to intraparticle mass transfer resistance. It is important because it controls the cycle time of a fixed-bed adsorption process. Fast kinetics provides a sharp breakthrough curve, while slow kinetics yields a distended breakthrough curve. The effect of a distended breakthrough curve can be overcome by increasing the amount of adsorbent and hence pressure drop in the bed. On the other hand, kinetics has been exploited as the basis of adsorptive separations. The most common example is the PSA process that splits nitrogen from air using a carbon molecular sieve, which relies on the relatively fast diffusion of oxygen compared with the much slower diffusion of nitrogen. In conventional adsorption processes, however, slow diffusion of any adsorbate is a disadvantage. To compensate for slow diffusion, it is also possible to use small particles, but there is a corresponding sacrifice resulting from the increased pressure

Adsorption

drop. The usual solution to that dilemma is to use relatively large particles and to employ an extra increment of adsorbent.

14.2.2.3 Pore Structure and Surface Area

Typical adsorbents have very large internal surface areas as a result of their porous structures, often varying from 5 to 3000 m²/g. Accordingly, pore structures are described by their size, shape, and connectivity. Three pore size (d_p) ranges are commonly cited: micropores ($d_p < 20$ Å), mesopores (20 Å $\leq d_p < 500$ Å), and macropores ($d_p \geq 500$ Å). Shapes range from slits to straight, cylindrical holes to narrow pore mouths that open to larger cavities. Connectivity can go from highly ordered and crystalline as in zeolites to more random and even amorphous, as in some silica gels and activated carbon. To envision how pore structure relates to surface area, it may help to consider an adsorbent particle as a partly filled bookshelf. The outer boundary of the shelf corresponds to the external surface of a particle. Gaps between the shelves correspond to macropores, the spaces between adjacent books correspond to mesopores, and the spaces between individual pages of the books correspond to micropores. Stretching that analogy, it is easy to see that, on a "unit volume" basis, the external area, as well as the area attributed to the macropores (shelves), is small. In contrast, the mesopores (book covers) contribute more area than the macropores, but the micropores (pages) can dwarf the others.

Surface area is a relative term, since the value depends on the way it is measured. As explained more fully below, the surface area is inferred from the apparent monolayer loading (plateau on the isotherm) of a standard adsorbate (e.g., nitrogen at its normal boiling point). The numerical value is obtained from the monolayer loading using the adsorbate's density and molecular dimensions. Values typically correlate with capacity and for various adsorbents are in the range of, say, 5 to 3000 m²/g. This means that if you held in your hand the same mass of adsorbent as that of a penny (2.5 g), the area available would range from 3.7 × 3.7 m² to nearly two football fields! Some specific ranges are listed for common adsorbents in the next section. Even though it can be measured repeatably and is widely accepted as a key criterion, surface area alone is not a proper basis for choosing an adsorbent.

The *BET surface area* measurement is attributed to Brunauer et al. (1941). It is usually conducted with nitrogen at its normal boiling point (−195.8°C) and measured up to atmospheric pressure ($p/p_{vap} = 1$). Essentially, the monolayer capacity (or loading) is estimated using the BET isotherm (see Table 14.3 and Section 14.3.2), and the numerical value is converted to a surface area using a factor that is believed to represent the coverage area per molecule for nitrogen. From the same data, it is possible to extract the cumulative pore volume.

The pore size distribution (PSD) indicates the fraction of the space within a particle occupied by micropores, mesopores, and macropores. An adsorbent's pore size distribution roughly indicates its potential uptake capacity, may reveal possible mass transfer constraints, and can show its potential for separating molecules by a sieving effect. In fact, sometimes pore size information is gleaned from diffusion rate data for molecules of known sizes, i.e., the extent to which sieving effects are observed.

Both measurement and modification of pore size distribution are rapidly growing fields. Pore size distributions are often reported in either differential or cumulative form. The former provides the percent or fraction of volume as a function of pore radius or diameter (i.e., dV/dr). That form makes it easy to discern the average pore diameter and skewness of the distribution. In contrast, the cumulative PSD clearly depicts total pore volume as a function of pore diameter and shows relative amounts of microporosity and macroporosity. Figures 14.5 and 14.6 show representative differential and cumulative PSDs, respectively, of some common adsorbents.

Figure 14.7 illustrates how the pore sizes of some common adsorbents relate to the sizes of several molecules and elements. The latter values are Lennard–Jones (6–12) apparent collision diameters, extracted from viscosity measurements. For many adsorbents, only a minimum pore size is shown

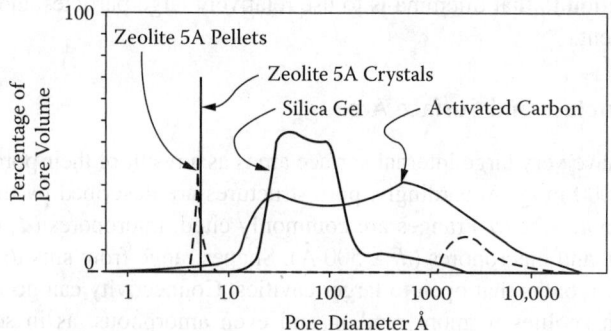

FIGURE 14.5 Differential pore size distribution for common adsorbents.

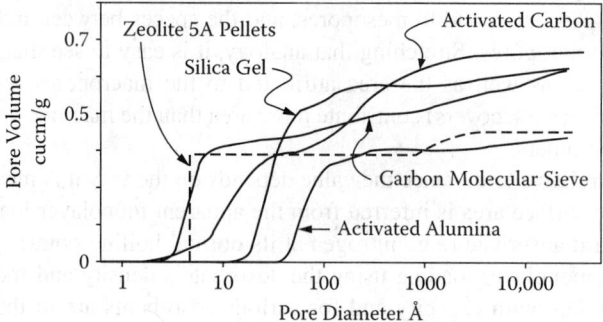

FIGURE 14.6 Cumulative pore size distribution for common adsorbents.

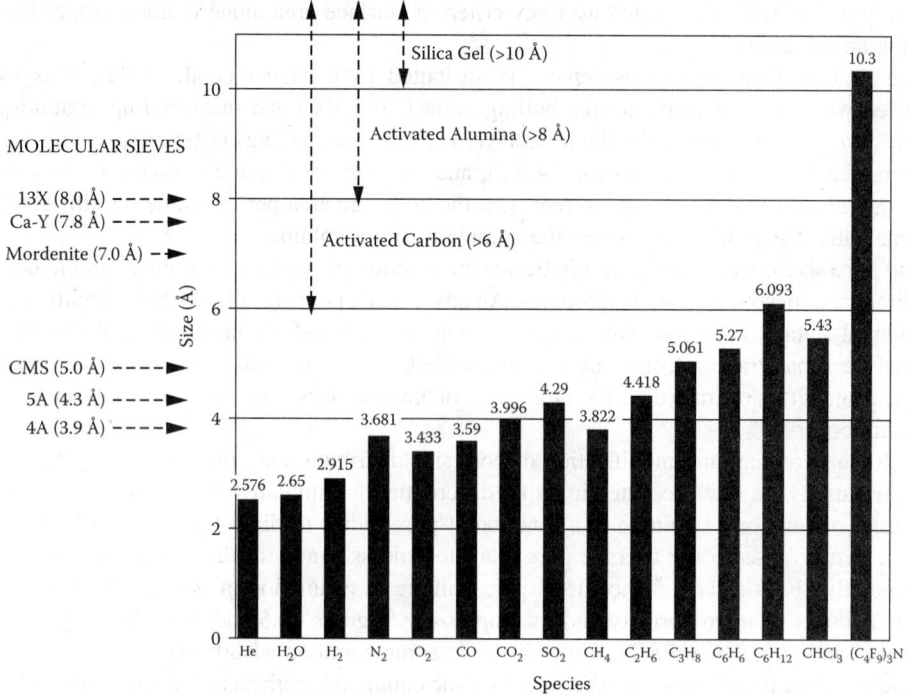

FIGURE 14.7 Molecular diameters and pore dimensions for common adsorbents.

(ignoring the average and range), while for molecular sieves the nominal micropore size is shown (ignoring the macropores). Perfluorotributylamine, as shown here, is larger than most micropores.

14.2.2.4 Particle Size Distribution

Adsorbent particles are available in diverse shapes and sizes. The most common forms are granules, flakes, pellets, powders, and beads. Granules can be nearly spherical or nearly flakes, depending on the starting material and the manufacturing methods. Pellets are generally cylindrical rods with length-to-diameter ratios of 1 to 10. Sometimes they are made in other shapes (e.g., hollow cylinders and trochoidal shapes), but they are more expensive and may not behave identically to ordinary pellets because of different stresses imposed during production. Beads generally offer superior attrition resistance and may promote better flow distribution than other shapes.

14.2.2.5 Durability and Cost

Durability relates to various possible modes of attack that could reduce the life expectancy of the adsorbent, such as chemical, physical, or biological fouling, degradation, and attrition. To combat those effects, the adsorbent, binder, and surface groups (depending on the type of adsorbent) should be physically strong and inert to the carrier or solvent, and should not irreversibly react with (or chemisorb) the adsorbate(s) or contaminants.

Attrition generally has three potential causes: crushing due to an imposed normal force, shearing or abrasion due to tangential stresses, and wear due to impingement of entrained particles at high velocity. Those causes of attrition are affected by operating conditions such as velocity and flow orientation, bed height, temperature, pressure, and vibration, which contribute to the stresses on the adsorbent particles. Normal forces frequently result from pressure gradients, especially in PSA systems, but they can also be associated with thermal expansion and contraction, mostly of the vessel wall (e.g., during TSA operation). Most adsorbents exhibit low coefficients of thermal expansion, so they do not induce much thermal stress. Resistance to normal forces can be quantified as an adsorbent's "hardness" or "crush strength." There are also standard tests for tangential stresses, i.e., to measure "abrasion resistance." The other forms of wear cannot be accurately quantified, but they are no less important.

Attrition can cause a host of problems, from increased pressure drop due to the presence of "fines" in the intersticies between whole particles to total failure of the adsorber when adsorbent dust is conveyed into the pipes and fittings downstream. In addition, a short-term loss of working capacity (mentioned earlier) commonly occurs during the first few cycles of operation of fresh adsorbent, followed by gradual decay, perhaps over hundreds of cycles, due to aging (partial collapse of the adsorbent's pore structure), chemisorption (poisoning), or other causes. It is that decay plus attrition losses that essentially govern the life of the adsorbent.

Adsorbent costs vary with time, supplier, and other factors, even for the same exact material. Prices often range from $0.50 per pound for some commodity adsorbents to $100 per pound or more for more exotic materials. Some that are impregnated with expensive metals are much higher priced. Price is also sensitive to quantity. For example, from a laboratory supply house, in 100-g quantities, activated alumina and silica gel cost as much as $70 to $140 per pound. But in tonnage quantities, they may sell for less than $1 to about $4 per pound.

14.2.3 Common Adsorbents

In this section, we focus on the properties, uses, and origins of a few generic types of commercial adsorbents. Neither the list nor the descriptions are comprehensive. The intent is to provide some background information regarding the most important aspects.

14.2.3.1 Aluminas

Activated alumina is produced from hydrated alumina, $Al_2O_3 \cdot n\, H_2O$, where $n = 1$ or 3, by dehydrating (calcining) under controlled conditions to get $n \approx 0.5$. This form, being hydroxylated, augments adsorption by hydrogen bonding. It is a white or tan opaque material that has a chalky appearance. Several grades are available from various manufacturers. The distinctions among products mostly are related to the specific crystal structures that alumina can exhibit. For example, stable crystalline forms are usually not thought of as adsorbents because of their low surface areas. Conversely, transitional forms, such as gamma and eta alumina, have defect spinel structures that lead to both higher concentrations of surface acid sites and effective surface areas from 200 to 400 m^2/g. Common forms are beads or balls 1 to 8 mm dia., granules, extrudates (pellets) 2 to 4 mm dia., and powder. Activated aluminas are widely used as catalysts (or catalyst supports) and as desiccants. Ancillary uses as an adsorbent are for removal of oxygenates and mercaptans from hydrocarbon feed streams, fluoride ions from water, HCl from hydrogen in catalytic reforming, and others. For most gas-phase applications, pretreatment requires heating to about 250°C.

14.2.3.2 Silicas

Silicas are also highly hydroxylated SiO_2, so adsorption results, to a considerable extent, from hydrogen bonding. Most silica gels are transparent or translucent. Some silica gels are alloyed with a few percent alumina, which yields an opaque white or tan appearance. Several forms are available that encompass diverse types of silica gels, porous borosilicate glass, and aerogels. The last is a relatively new, exceedingly porous material currently with few commercial applications. Its unique characteristics make it an interesting prospect for the future. Silica gel and porous glass are both nondusting and resistant to attrition.

There are more than a dozen varieties of SiO_2; most adsorbent forms are rigid (but not crystalline) assemblages of spherical microparticles made of colloidal silica. Aerogels are a form of open-celled porous glass. Effective surface areas range from 300 to 900 m^2/g, depending on the density, with more dense materials having finer pores and larger surface areas. Common forms are beads 1 to 3 mm dia., granules, extrudates (pellets) 2 to 4 mm dia., and powder. The largest uses of silica gel and porous glass are as a desiccant. Other uses as an adsorbent are for separation of hydrocarbons, dewpoint reduction for natural gas, and drying of liquid hydrocarbons. Pretreatment for gas-phase applications (especially as a desiccant) requires heating to about 200°C.

14.2.3.3 Zeolites

In the mid-1700s, a Swedish mineralogist named Cronstedt discovered a mineral that appeared to boil when heated. He named it *zeolite*, from the Greek *zeo* (to boil) and *lithos* (stone). The observed boiling effect is caused by the expulsion of adsorbed water from inside the zeolite. The term zeolite now refers to a class of aluminosilicates that are stoichiometric blends of silica and alumina. They are generally opaque and white or tan in appearance. There are about 40 natural and 150 synthetic types having the general formula $M_{x/n}[(AlO_2)_x(SiO_2)_y] \cdot w\, H_2O$, where M represents a cation of valence n, w is the number of water molecules per "cage," and x/y is the Al-to-Si ratio. The sum $x + y$ is the number of tetrahedra per unit cell. Those that contain roughly equal amounts of silica and alumina are hydrophilic, whereas those that contain predominately silica (e.g., with Al-to-Si ratios > 10) are hydrophobic. Frequently, there is water of hydration within the crystals, and to balance the charges, cations are associated with the alumina. The presence of water is largely associated with field-dipole interactions, which can be large for zeolites.

Internally, zeolites are inherently crystalline and exhibit micropores within those crystals that have uniform dimensions, as depicted in the pore size distribution shown in Figures 14.5 and 14.6. The micropores are so small and uniform that they commonly can distinguish among nearly identically sized molecules. As a result they are frequently called "molecular sieves." That term, however, is broad and encompasses materials that are made from aluminophosphates, titanosilicates, and so forth. As mentioned earlier, Figure 14.7 illustrates the sizes of a few atoms and some simple molecules along with the micropore diameters of some common zeolites.

Most zeolites sold as commercial adsorbents are composites of very fine crystals held together with a binder. Surprisingly, the binder may also exhibit substantial adsorption capacity. The adsorption capacity of zeolites is the result of micropores that are so small that it is impractical to express an effective surface area. To promote crystal uniformity, most commercial zeolites are synthesized in autoclaves under tightly controlled conditions. The resulting crystals exist in a metastable form, and more than 100 distinct forms have been produced. Typical zeolites include faujasite (also called zeolite X and zeolite Y), mordenite, silicalite, zeolite A, zeolite L, zeolite β, zeolite ρ, zeolite ω, Boggsite, chabazite (also called zeolite D), clinoptilolite, erionite, and ferrierite. Only a handful are commercially significant (e.g., A, X, Y, mordenite, ZSM-5, and silicalite), although the first four of those have a few different common "exchange" forms and are produced in different sizes and shapes. Some of the variety is summarized in Table 14.1. Particle selection includes 1- to 6-mm dia. extrudate, 0.5- to 3-mm beads, 20×40 to 6×12 mesh, and powders. Applications of zeolites include gas or liquid drying, separation of oxygen from air, normal paraffins from naphtha, and p-xylene from other isomers. Activation for gas-phase applications typically requires more stringent conditions than for silica or alumina, namely, 300°C under full vacuum or an inert purge gas.

14.2.3.4 Carbons

Activated carbons are produced from wood, coal, peat, coconut shells, saran, recycled tires, and other sources. The final adsorbents may look similar to the casual observer (i.e., black granules or pellets), but appearances can be deceptive.

Activation produces a wide range of internal pores and carbon surfaces (e.g., graphitic versus oxidized), which generally enhance the adsorptive capacity. By varying activation conditions, differences of the internal surfaces are produced, even for materials that appear to be identical. Another feature that varies, depending on the nature of the base material, is the ash content, which is of course inorganic; typical values are between 2% and 25%, but the average is about 7%. Alkali ash near or at the surface can be removed by acid washing, or other minerals may be deposited by impregnation. The microscopic structure (pore size distribution and surface area), surface qualities, and chemical composition affect adsorption characteristics and therefore affect the performance parameters (capacity, selectivity, regenerability, kinetics, compatibility, and cost). Specifications are maintained only through tight quality control.

Effective surface areas generally range from 300 to 1500 m^2/g, depending on the base material, activation method, density, and so on, although some made from petroleum coke exceed 3000 m^2/g. Larger surface areas are often assumed to be better, but surface area does not always correlate with capacity. Even when it does, kinetics and other aspects may outweigh the effect of capacity on overall cost and performance.

Common forms are beads 1 to 3 mm dia., granules, extrudates (pellets) 2 to 4 mm dia., and powder. Some typical applications are water and wastewater treatment to remove hazardous organic compounds or those that impart odor or taste, cleanup of off-gases containing volatile organic compounds (especially solvents which might be recovered and odoriferous chemicals that are merely trapped), upgrading methane from substandard natural gas wells, food decolorization, and pharmaceutical purification. Impregnated activated carbons are widely used in gas masks and to remove other specific contaminants in gas or water. Impregnants include sulfuric acid (for ammonia or mercury), iron oxide (for hydrogen sulfide or mercaptans), zinc oxide (for hydrogen cyanide), and a combination of heavy metal salts (for phosgene, arsine, and nerve gases). Pretreatment for gas-phase applications is often performed as the last step of manufacture.

Adding to the diversity are products called "carbon molecular sieves," which are analogous to zeolites and other crystalline molecular sieves mentioned previously. While micropores in zeolites tend to have rounded apertures, the carbon-based counterparts are more slit-like, as in the space between layers of graphite. To date, only one type of commercial separation employs this material: separation of nitrogen (at 97% to 99.99%) from air by PSA, which exploits the size difference between oxygen (3.43 Å) and nitrogen (3.68 Å).

14.2.3.5 Polymers

Polymeric adsorbents, including ion exchange resins, tend to be opaque spherical beads; most are white or tan, but some are brown, orange, or black. Some are inert, polystyrene-divinyl benzene, polymethacrylate, divinylbenzene/ethylvinylbenzene, or vinylpyridine beads, while others, which contain acidic or basic functional groups, are ion exchange resins. Internally, many of these polymer beads contain "microbeads" that are joined together at a few points each, creating a macropore structure. In addition, some polymeric adsorbents are activated via pyrolysis, in much the same way as carbon, yielding black materials.

Some polymeric adsorbents are sufficiently hydrophilic to be used as a desiccant, while others are quite hydrophobic. The effective surface area is usually smaller than for activated carbon, e.g., 5 to 800 m²/g. The corresponding pore diameters range from about 20 to 2000 Å, or from 3 to 2000 Å if activated. The available forms are generally limited to beads of 0.3 to 1 mm dia., usually in a relatively narrow range. Larger particles are not yet commercially available. A minor drawback of these materials is that they tend to shrink and swell with cyclic use. For gas-phase applications, they may require conditioning prior to use (e.g., washing with water and/or another solvent followed by drying).

The range of applications is somewhat restricted, since the cost of most polymeric adsorbents is typically about ten times more than that of other common adsorbents. In some instances, polymeric materials are the only choice. In other cases, they compensate for the cost differential by yielding much better performance, especially for high value-added uses. Current applications include recovery and purification of antibiotics and vitamins, decolorization, decaffeination, hemoperfusion, separation of halogenated light organics from water, and treatment of certain industrial wastes such as aqueous phenolics and VOC recovery from off-gases.

14.2.3.6 Biomass

Some adsorbents are organic materials that function as solid "absorbents" rather than adsorbents. Among these are cellulose (the most abundant biopolymer in nature), chitin (the second most abundant biopolymer in nature), collagen, wool, starch-polyacrylamide gels (which absorb many times their own weight of water at ambient temperature but release most of it by gentle heating), polysaccharides derived from corn, and miscellaneous forms of biomass (e.g., residue from crop harvests). The latter materials fall in the fuzzy gap between adsorbents and absorbents. Nevertheless, some of these may have niches, but none is a general-purpose adsorbent.

14.3 ADSORPTION THERMODYNAMICS

Adsorption relies mainly on the physisorption of molecules (and noble gases) and is analogous to condensation. Knaebel et al. (1999) indicated that two types of forces contribute to physisorption: *van der Waals* (or dispersion) forces and *electrostatic* forces. Van der Waals forces relate to polarizability, i.e., the ease with which an atom's or molecule's electron density can be distorted. That effect increases as atoms within a group in the periodic table become larger. The outcome is a favorable interaction energy that is significant for highly polarizable species. Similarly, electrostatic forces, which are relevant only if the surface is polar, are caused by polarization forces, field-dipole interactions, and field gradient-quadrupole interactions. In that vein, polarization forces are causative, while polarizability represents a latent tendency.

For sorbate, field-dipole interactions occur that exhibit a permanent dipole moment (e.g., H_2O), which interacts with the surface electric field to give an attractive energy. An example is hydrogen bonding, which is important for hydroxylated surfaces and sorbates such as water and alcohols. Basically, hydrogen atoms are often covalently bonded to highly electronegative atoms at the surface, and the latter draw the electron clouds of the former, leaving their nuclei exposed. The nuclei, in turn, form purely electrostatic bridges to other atoms of the adsorbate or surface. Field gradient-quadrupole interactions arise for nonpolar, electrically neutral molecules (such as CO_2 and N_2) that exhibit significant quadrupole moments. In the presence of a spatially varying electric field, the attractive energy is proportional to the product of the local field gradient and the quadrupole moment.

Thus, the magnitude of physisorption forces is most strongly affected by the size, polarity, polarizability, and quadrupolarity of sorbate atoms or molecules, as well as the electric field strength and the local field gradient of the solid surface.

Chemisorption differs from physisorption in that it involves the formation of chemical bonds and electron transfer between adsorbate and adsorbent. It is of primary concern with regard to catalysis, and it is only useful as an extreme measure for separation purposes. Removal of chemical warfare agents from breathing air is a prime example. One way to discern between physisorption and chemisorption is by heat effects and the effect of temperature. For example, physisorption is always exothermic, i.e., $\Delta H_{ads} < 0$. Furthermore, the adsorbed state is more ordered than the fluid state, so $\Delta S < 0$. To be thermodynamically feasible, $\Delta G_{ads} < 0$, so $\Delta H_{ads} < T\Delta S$. In contrast, endothermic adsorption is possible for dissociative chemisorption, even if the entropy of the adsorbate decreases, because the entropy of the adsorbent may increase to more than offset that, e.g., by expanding.

The loading due to physisorption decreases monotonically with temperature, while that due to chemisorption can increase with temperature, especially considering rate effects. The net contribution of chemisorption may be nil at low temperatures and rise to levels much higher, even substantially above the boiling point, where physisorption would sharply diminish. Physisorption is practically fully reversible, even when multilayered, whereas chemisorption may or may not be reversible. To clarify, it must be reversible to exhibit catalytic behavior, but it tends not to be when it acts as a poison. Physisorption is not very specific, except in molecular sieves, which discriminate based on size and steric effects. In contrast, chemisorption is much more specific, depending on conditions and the nature of the sorbate and adsorbent as well as its previous treatment.

14.3.1 Adsorption Equilibrium and Heats of Adsorption

The importance of adsorption equilibrium cannot be overstated, since it is often the constraint that has the greatest impact on adsorption applications. In fact, to address most adsorption applications (whether to design a process, to debottleneck an existing process, or even to solve problems in which adsorption is peripherally related), it is necessary to obtain equilibrium data, as described in Section 14.3.2, and then put it into a useful form—usually as a correlation, as described in Sections 14.3.3 and 14.3.4. This section first clarifies what is meant by equilibrium data and heats of adsorption.

The terms adsorption capacity and loading are used generically to express the amount of adsorbate taken up by the adsorbent. To solve problems, it is necessary to express loading quantitatively, e.g., as an explicit function of partial pressure or concentration and temperature:

$$n_i^* = f(p,T) \quad \text{or} \quad f(C,T) \tag{14.1}$$

In the next section, we will see specific functions that are commonly used to represent data. It should be noted that many additional factors affect the adsorption capacity, such as pretreatment, exposure history, presence of contaminants or competitively adsorbed substances, and so forth. Although those factors can strongly affect adsorption performance, they are difficult to predict or even correlate. From here on, we assume that the adsorbent has been properly pretreated and that the other adverse effects are negligible.

Most fundamental equations have direct analogies to vapor–liquid equilibrium and simple gas laws. For example, an adsorbed component is frequently viewed as a two-dimensional fluid that exhibits a spreading pressure on the adsorbent, in contrast to its partial pressure in the gas phase. There is a corresponding empirical relationship between the spreading pressure, area covered per mole adsorbed, and temperature, analogous to the ideal gas law, but it is called the "ideal surface gas" equation. When applied to adsorption of mixtures, it is called the "ideal adsorbed solution" theory, developed by Myers and Prausnitz (1965). Equilibrium calculations mainly involve the interactions of one molecule (or atom) of adsorbate with the adsorbent surface; i.e., they neglect adsorbate-adsorbate interactions. This viewpoint is valid only in the most idealized conditions. Normally, the surface is irregular or heterogeneous (both energetically and geometrically), which can be taken into account by many theories, and the adsorbed molecules (or atoms) interact with each other via mutual repulsion (or sometimes attraction).

As mentioned above, equilibrium data can be presented and used in a variety of forms: isotherms (loading vs. concentration at constant temperature), isosteres [partial pressure (or dewpoint, or some other form of concentration) vs. inverse absolute temperature at specific degrees of loading], and isobars [loading as a function of temperature for given partial pressures (or some other concentration)], listed in order of decreasing prevalence. The object of isosteres and isobars is to plot data on coordinates for which approximate linearity is expected, to make interpolation and extrapolation easier.

The heat of adsorption is a measure of the energy required for regeneration in gas- or vapor-phase applications, and low values are desirable. It also indicates the temperature rise that can be expected due to adsorption under adiabatic conditions. Again, there are several definitions: isosteric, differential, integral, and equilibrium, to name a few. The most relevant (because it applies to flow systems instead of batch systems) is the isosteric heat of adsorption, which is analogous to the heat of vaporization and is a weak function of temperature. The definition is

$$q_{st} = -R\left[\frac{\partial \ln p}{\partial (1/T)}\right]_{n^*} \tag{14.2}$$

where p and n^*, respectively, would be total pressure and loading for a pure gas or partial pressure and component loading for a mixture. Besides an indication of the energy required for regeneration, this term shows how the adsorbate interacts with the adsorbent. To illustrate, a plot of isosteric heat of adsorption vs. loading generally follows one of three trends (monotonically decreasing, increasing, or constant) as loading increases. The first case indicates that adsorption is strong at low concentrations, possibly due to a heterogeneous surface at which the "strong" sites are filled first. The net effect is that regeneration is likely to be difficult. The second case indicates the reverse, and regeneration is likely to be relatively easy, especially if the heat of adsorption is also low. The third case is neutral. Other, less utilitarian heats of adsorption are defined in Table 14.1.

14.3.2 Sources of Equilibrium Data

There are two types of sources of equilibrium data: literature and laboratory. In many ways, it is easier and cheaper to obtain data or correlations from the open literature. Unfortunately, there are not many compilations, so it may not be easy to find applicable data, even if they exist. A few that may prove useful are the *Adsorption Equilibrium Data Handbook* by Valenzuela and Myers (1989), *Carbon Adsorption Isotherms for Toxic Organics* by Dobbs and Cohen (1980), and *Adsorption-Capacity Data for 283 Organic Compounds* by Yaws, Bu, and Nijhawan (1995). Individual reports of equilibrium data are abundant and may be found in specialized journals such as *Adsorption, Adsorption Science and Technology, Journal of Colloid and Interface Science, Langmuir*, and *Separation Science and Technology*, as well as other more general journals such as *AIChE Journal*,

Chemical Engineering Science, *Industrial Engineering Chemistry Research*, and *Journal of Chemical and Engineering Data*.

Except for the most common applications, one is unlikely to find data or correlations that cover the proper range of conditions for a very wide range of adsorbents (so as to choose the best). In addition, accurate predictions are currently impossible, so it is *always* necessary to obtain data under relevant conditions. Consequently, most applications will require measurements in a laboratory. Isotherm measurements require great care. There are three basic types of equipment: volumetric, gravimetric, and chromatographic. The equipment and techniques are reviewed briefly here.

Volumetric measurements for gases generally involve a vessel containing an adsorbent that is subjected to a step change of pressure (measured via a pressure transducer). The final pressure can be used to calculate the amount adsorbed. For liquid–phase volumetric measurements, the dry adsorbent is contacted with a liquid of predetermined concentration, and the final concentration is measured. A variety of instruments can be used *in situ*, but it is also acceptable to extract small samples with a syringe for individual analysis. For liquids or gases, this method is generally the best in terms of flexibility, accuracy, and cost.

The gravimetric approach mainly applies to gas–phase adsorption. It involves measuring the amount taken up by the adsorbent by weight. These isotherm measurements are quick and accurate, and the interpretation is easy. Some types of equipment are elaborate, with a small adsorbent-bearing pan suspended from a quartz spring. With this approach, the main problem is cost, plus the fact that the equipment tends to be finicky (each seal is subject to leaks). Other problems sometimes overlooked are adsorption on the walls rather than on the adsorbent and buoyancy effects (which may cause a 10% error). Another version uses a column of adsorbent through which gas of a known concentration is passed. Periodically, the flow is stopped, and the column is sealed and weighed. The adsorption capacity can be determined once a steady state is reached. This is more tedious but is reliable and relatively inexpensive. Alternatively, employing thermogravimetric analysis (TGA), the adsorbent can be heated strongly, and the off-gases can be trapped and analyzed to infer the adsorbate composition.

Finally, chromatographic analysis is primarily a screening technique in which adsorbents are crushed and placed in a chromatographic column, then a pulse of the components of interest is injected into a nonadsorbing carrier fluid. In principle, the technique applies to both gases and liquids, but the former application is much more popular. The Henry's law coefficient can be determined readily from the retention of each peak. This general technique can also be applied in stepwise (rather than pulse) experiments using a material balance and integrating the breakthrough curve to ascertain the amount adsorbed.

14.3.3 Isotherm Equations

An isotherm is a set of equilibrium capacity (or loading) data over a range of fluid-phase concentrations at a fixed temperature. Many semitheoretical equations are used to fit isotherm data, and those having more parameters can account for more subtle effects. In most cases, simpler is better.

Brunauer, Deming, Deming, and Teller (1940) observed that isotherm data followed patterns, i.e., Types I–V shown in Figure 14.4. They also suggested a powerful but complex four-parameter equation (referred to as the B.D.D.T. equation) that fits all of the forms. Most other equations apply to Type I, II, or IV data, which represent "favorable" equilibrium (concave downward). A few others apply to Types III and V data, which represent "unfavorable" (concave upward) equilibrium.*
Frequently, if an isotherm exhibits hysteresis (shown for Types IV and V), there may be an adverse impact on kinetics and regenerability. Hysteresis occurs when desorption follows a different path

---

* Generally speaking, *favorable* isotherms are promising for uptake. In contrast, adsorbents that exhibit *unfavorable* isotherms would not perform well in most adsorption applications.

TABLE 14.2
Characteristics of Several Commercial Zeolites

Zeolite Type	Cationic Form	Nominal Pore Diameter (Å)	Si/Al
3A	K	3.0	1.0
4A	Na	3.9	1.0
5A	Ca	4.3	1.0
10X	Ca	7.8	1.2
13X	Na	8.0	1.2
Y	K	8.0	2.4
Mordenite	Na	7.0	5.0
ZSM-5	Na	6.0	31.0
Silicalite	—	6.0	∞

from adsorption, usually as a result of liquid filling pores in a certain way that is not the same as when they are emptied.

The following paragraphs describe the isotherms and explain the terms that are listed in Table 14.3. Most can accept any form of concentration, C, for the fluid phase (e.g., having units of mol/m^3, lb/ft^3, and so on) or a convenient variable (e.g., partial pressure, ppm, and so forth). Some, however, are restricted to relative saturation, i.e., expressed as a fraction. Generally, they can fit adsorbent loading, n^*, having units of mmol/g, lb/ft^3, g/100 g, and others. Finally, the parameters A, B, and so forth are determined empirically.

The simplest equilibrium concept is that the extent of adsorption is proportional to the fluid-phase concentration, i.e., Henry's law [a in Table 14.3]. A principle of thermodynamic consistency is that isotherms should reduce to Henry's law at the limit of zero loading. If not, the equation (or data) is thermodynamically inconsistent and therefore fundamentally flawed. Conversely, there is pragmatic appeal to the premise that if an isotherm equation fits, one may use it, even though some scientific principle may be violated. The hazard of that approach arises if the data are flawed, causing an erroneous fit, which could lead to a defective design or mistaken predicted performance.

The Langmuir isotherm [b in Table 14.3] accounts for surface coverage by balancing the relative rates of uptake and release, the former being proportional to the fraction of the surface that is open, and the latter proportional to the fraction that is covered. The equilibrium constant for those rates is K, which also is the Henry's law coefficient. When the fluid concentration is very high, a monolayer forms on the adsorbent surface, having a loading of n_M. Those two parameters help us understand the nature of adsorption (although for some isotherms, n_M does not strictly refer to a monolayer but instead may denote a maximum). Hence, they are used directly in the equations when appropriate. When possible, the equivalent terms are listed in Table 14.2.

Freundlich recognized that, when data do not fit well on linear coordinates, the next logical step is to try log-log coordinates, and that led to the isotherm [c in Table 14.14. 3] bearing his name. It is probably the most commonly used isotherm equation, despite being "thermodynamically inconsistent."

The Brunauer–Emmett–Teller (B.E.T.) and B.D.D.T. isotherms [d and e in Table 14.3] account for pore filling via multiple layers instead of just a monolayer, and they use C/C_{sat}, that tends toward unity as the pores are completely filled. The B.D.D.T. isotherm includes the number of layers explicitly (m), as well as a heat of adsorption term (q). The B.E.T. isotherm is mostly used to estimate surface areas, not for process calculations [see Equation (14.10)].

The dual-mode isotherm is merely a combination of Henry's law and the Langmuir isotherm. Other isotherms, such as the Redlich–Peterson, Langmuir–Freundlich, Sips, and Toth [g, h, i, and j, respectively, in Table 14.3] versions extend the Langmuir isotherm by accounting for subtle

TABLE 14.3
Pure Component Isotherm Equations

	Name	Equation Form	K[1]	n_M[2]
a.	Henry's law	$n^* = K n_M C$	K	∞
b.	Langmuir	$n^* = K n_M C/(1 + KC)$	K	n_M
c.	Freundlich	$n^* = AC^B$	∞	∞
d.	Brunauer–Emmett–Teller ($C_r = C/C_{sat}$)	$n^* = K n_M C_r/[(1 + (K-1)C_r)(1 - C_r)]$	K	n_M
e.	B.D.D.T.[3] ($C = C/C_{sat}$)	$n^* = \dfrac{K n_M C_r (1 + m((q-1)C_r^{m-1} + qC_r^{m+1} - (2q-1)C_r^m) - C_r^m)}{(1 - C_r)(1 + (K-1)C_r + K((q-1)C_r^m - qC_r^{m+1}))}$	K	n_M
f.	Dual-mode isotherm	$n^* = K_1 n_M C/(1 + KC) + KC$	$K_1 + K_2$	
g.	Redlich–Peterson	$n^* = K n_M C/(1 + KC^B)$	K	n_M
h.	Langmuir–Freundlich	$n^* = A n_M C^B/(1 + AC^B)$	∞	n_M
i.	Sips	$n^* = n_M [AC/(1 + AC)]^B$	∞	n_M
j.	Toth	$n^* = KC/(1 + C^B/A)^{1/B}$	K	$KA^{1/B}$
k.	UNILAN	$n^* = \dfrac{n_M}{2B} \ln \dfrac{D + C \exp(B)}{D + C \exp(-B)}$	$\dfrac{n_M \sinh(B)}{BD}$	n_M
l.	Dubinin–Radushkevich[4]	$n^* = n_M \exp(-((k_0 \varepsilon/\beta)^2))$	0	n_M
m.	Dubinin–Astakhov[4]	$n^* = n_M \exp(-((k_0 \varepsilon/\beta_0)^n))$	0	n_M
n.	Dubinin–Stoeckli[4]	$n^* = n_{M1} \exp(-((k_{01} \varepsilon/\beta_{01})^2)) + n_{M2} \exp(-((k_{02} \varepsilon/\beta_{01})^2))$	0	$n_{M1} + n_{M2}$

[1] The Henry's law constant, K, is commonly defined as the initial slope of "fractional coverage" (or $\theta = n^*/n_M$) vs. concentration or partial pressure, which leads to awkward units, and the maximum loading, n_M, must be known. Less commonly but more sensibly, the slope of n^* versus C or p can be used. That choice is unavoidable if saturation is not observed.

[2] n_M corresponds to the "monolayer" loading for Equations a through e, while for the rest it represents a sort of "maximum" loading. Details go beyond the scope of this chapter.

[3] B.D.D.T. = Brunauer–Deming–Deming–Teller; m is the number of monolayers, and q is a heat of adsorption.

[4] The adsorption capacity, n^*, and maximum loading, n_M, for this isotherm are expressed as the *volume* adsorbed per unit mass or volume of adsorbent. The function $n^* = f(\varepsilon)$ is called a characteristic curve and can fit data for one component over a range of temperature—and sometimes for a family of components. The *adsorption potential*, ε, is defined in Table 14.1. β is called an affinity coefficient that can force data for diverse components to fit a common characteristic curve.

nonlinearities and are "power-law" forms. The last of this type is the UNILAN isotherm (k). Its name comes from UNI for "*uni*form distribution" and LAN for "*L*angmuir local model." It has a strong theoretical basis but is difficult to grasp, since it employs the logarithm of a ratio containing exponential functions. Despite that, the UNILAN isotherm and the Toth isotherm, although slightly different from the one shown here, were used predominantly by Valenzuela and Myers (1989) in their thorough and precise compilation of isotherm data.

Polanyi recognized in 1914 that there was an analogy between adsorption and condensation. Through the use of free energy, he arrived at a term called the *adsorption potential*, ε_i, for component "i." The definition is

$$\varepsilon_i = RT \ln(f_i^* / f_i) \approx RT \ln(p_i^* / p_i) \approx RT \ln(C_i^* / C_i) \qquad (14.3)$$

where the *pure* fluid fugacity and vapor pressure (or solute solubility limit) are f_i^* and p_i^* (or C_i^*), and the *equilibrium* fugacity and partial pressure (or solute concentration) are f_i and p_i or (C_i), respectively, all at the temperature of interest, T. For vapors (neglecting fugacity coefficients), the adsorption potential is equivalent to the work required to compress the adsorbable component from its partial pressure to its vapor pressure. In addition, to apply this approach to liquids requires deducting a correction factor from ε_i to account for the displaced solvent.

Dubinin and coworkers showed that a specific adsorbent adsorbs nearly equal volumes of similar compounds when their adsorption potentials are equal. In this case, the adsorbed volume is estimated from the liquid density, usually at the same temperature. Their reasoning is that the adsorbed state closely resembles the liquid phase. They suggested a plot of volume adsorbed vs. adsorption potential would produce a "characteristic curve," applicable to that group of compounds for the specific adsorbent. Most people who use this type of isotherm equation have adopted W as the symbol for loading, volume (of liquid) adsorbed per unit mass, or volume of adsorbent, but we will retain n^* and recognize that the units are specialized, as noted in Table 14.3. From the measured data (moles or mass adsorbed), one calculates the volume adsorbed using V_m, the molar volume of the saturated liquid evaluated at the adsorption pressure, but sometimes evaluated at the normal boiling point or another condition. Regardless, it should be consistent and clearly stated. It is then easy to extrapolate to other temperatures and other similar adsorbates for a given adsorbent. The main drawback is that the characteristic curve does not reduce to Henry's law at low coverage.

The isotherms developed by Dubinin and coworkers employ a power to which the adsorption potential is raised that indicates the prevalent type of pores. The Dubinin–Radushkevich equation [1 in Table 14.3] was intended for microporous adsorbents, since the exponent is 2. The Dubinin–Astakhov equation (m) allows the exponent B to vary, but a reasonable lower limit is unity (for macroporous adsorbents). The Dubinin–Stoeckli equation (n) allows a distribution of pore sizes, which is a feature of many adsorbents.

For this type of isotherm, n_0 represents the maximum loading, which correlates with pore volume among different adsorbents. The other isotherm parameters, k_0 and β_0 [no relation to the terms in Equations (14.4) or (14.5)], represent the *characteristic parameter* of the adsorbent and an *affinity coefficient* of the compound of interest, respectively. The characteristic parameter, k_0, defines the shape of the n^* versus ε curve. The affinity coefficient, β_0, adapts the compound of interest to the characteristic curve. It is a "fudge factor" that has been correlated to the ratio of molar volumes, parachors, or polarizabilities (via the Lorentz–Lorenz equation) of the compound of interest to that of a reference component (e.g., benzene or n-heptane). These three methods are roughly equivalent in accuracy. The molar volume version is $\varepsilon_i = \varepsilon_{ref} V_i / V_{ref}$. The only controversy is whether to use the actual temperature to estimate volumes or some other temperature such as the normal boiling point.

For given a set of data, which isotherm equation (or equations) fits best? And what is the impact of the quality of fit on predicted performance? Unfortunately, neither question cam be answered fully. It is fair to say that the greater the number of parameters in an equation, the more likely it is to fit well; and the better it fits, the more valid will be subsequent process simulations. That should be balanced against the statistical significance of the parameters. Finally, the isotherm fit that best accommodates heat effects and multicomponent aspects, if any, will be superior. An example that illustrates different degrees of quality of fit of four equations to one set of data is provided in Section 14.5.4. Specialized programs are available that fit equations and plot the results.

The term *regenerability* was also referred to previously without giving an exact definition. To do so implies choosing the regeneration method. Regenerability would then revolve around the isotherm (or loading) under the process conditions vs. during regeneration. For a TSA cycle, this would mean looking at the appropriate isotherms for uptake and release and assessing the change of loading. Likewise, for chemical regeneration (e.g., by displacement or elution) or for a PSA cycle, it would mean looking at the loading under the relevant conditions. Of course, kinetics could affect the ability to attain those loadings, but that is covered later.

14.3.4 Mixture Equilibria

Selectivity is a means by which mixture equilibria can be grasped. Several different definitions exist, but all basically are ratios of "what is adsorbed" to "what remains in the fluid phase" at equilibrium. Selectivity, therefore, provides a simple description of the nature of multicomponent equilibria, although the values are seldom employed in mathematical models. Some common definitions are

$$\alpha_{ij} = \frac{y_i / y_j}{x_i / x_j} \tag{14.4}$$

$$\alpha'_{ij} = K'_i / K'_j \tag{14.5}$$

$$\beta_{ij} = \frac{1 + \frac{1-\varepsilon}{\varepsilon} K'_j}{1 + \frac{1-\varepsilon}{\varepsilon} K'_i} \tag{14.6}$$

$$\beta'_{ij} = K'_j / K'_i \tag{14.7}$$

where x_i and y_i are mole fractions in the fluid phase and adsorbed phase, respectively; $K'_i = K_i n_{Mi}$; and ε is the overall void fraction of the adsorbent. As mentioned in Tables 14.3 and 14.4, K_i is called the Henry's law constant, and n_{Mi} is the nominal monolayer loading. Following convention, the first component, i, is more strongly adsorbed than component j. In that case, α and α' vary between unity and infinity, as does the relative volatility, while β and β' vary between zero and unity. The ratio defined by Equation (14.4) may vary as the fluid composition varies. Conversely, Equation (14.5) is frequently used to compare adsorbents or operating conditions because it is

---

TABLE 14.4
Multicomponent Isotherm Equations

	Name	Equation Form
a.	Henry's Law	$n_i^* = K_i n_{Mi} C_i$
b.	Markham–Benton[1]	$n_i^* = K_i n_{Mi} C_i / (1 + \Sigma K_j C_j)$
c.	Schay[2]	$n_i^* = (K_i n_{Mi} C_i / \eta_i) / [1 + \Sigma (K_j n_{Mj} C_j / \eta_j)]$
d.	Yon–Turnock[2]	$n_i^* = n_{Mi} (A_i / \eta_i) C_i^{B_i} / [1 + \Sigma (A_j / \eta_j) C_j^{B_j}]$
e.	Sips–Yu–Neretnieks	$n_i^* = n_{Mi} K_i C_i (\Sigma K_j C_j)^{D_i - 1} / [1 + \Sigma (K_j C_j)]^{D_i}$
f.	Redlich–Peterson–Seidel ($C_{ref} = 1 \,[=]\, C_i$)	$n_i^* = A_i C_i / [1 + B_i C_{ref}^{D_{ii}} C_i^{E_i - D_{ii}} \Sigma (A_{ij} C_j)^{D_{ij}}]$

[1] Isotherm parameters can be estimated from pure component values.
[2] η_i is a lateral interaction parameter for component i. It is an additional empirical coefficient that can be determined from multicomponent data.

constant (at a given temperature), although it normally applies only at the limit of zero concentration. When discussing results, it is a good idea to clarify the definition of what is meant by selectivity or to speak of "good" or "bad" instead of "large" or "small."

There are five common means of dealing with mixtures rather than single adsorbable species. First and easiest, one may pretend that the mixture consists only of the major adsorbable component. That can be catastrophic. The second approach, treating the constituents independently, is useful and accurate when a nonadsorbing carrier contains very dilute contaminants. It is also very easy, since only pure component isotherms are required. Third, one may apply a method developed by Tien and coworkers (Jayaraj and Tien, 1984; Kage and Tien, 1987; Moon, Park, and Tien, 1991) called *species grouping*. The idea is to deal with a mixture of, say, ten components by identifying two or three (sometimes fictitious) components to represent the entire set. That reduces the complexity, saves time and money, and is fairly accurate if only an approximate answer is desired. It requires some pure component isotherm data to know how to group the species. The fourth method is to use one of several empirical isotherm equations that account for "competitive" adsorption of the relevant components. This method requires both pure component and mixture isotherm data. Depending on which equation is selected, the data analysis and fitting are more involved than for pure components, but not enormously so. In that case, the results are compact and relatively simple to use for design or simulation. Examples of the equations are listed in Table 14.4.

The fifth approach is more a field than a concise method, since it embodies so many theoretical concepts and associated methods. All are grouped together as "adsorbed mixture models." Basically, this involves treating the adsorbed mixture in the same manner that the liquid is treated when doing VLE calculations. The major distinction is that the adsorbed phase composition cannot be directly measured (i.e., it can only be inferred); hence, it is difficult to pursue experimentally. A *mixture model* is used to account for interactions, which may be as simple as Raoult's law or as involved as Wilson's equation. These correspond roughly to the Ideal Adsorbed Solution theory and Vacancy Solution model, respectively. Pure component and mixture equilibrium data are required. The unfortunate aspect is that they require iterative root-finding procedures and integration, which complicates adsorber simulation. They may be the only route to acceptably accurate answers, however. It would be nice if adsorbents could be selected to avoid both aspects, but adsorbate-adsorbate interactions may be nearly as important and as complicated as adsorbate-adsorbent interactions.

14.4 DYNAMICS OF ADSORPTION

14.4.1 Overview of Rate Phenomena

To perform adsorber simulations or even to select an adsorbent, one must appreciate the impact of processing conditions, geometry, and transport constraints on performance. This might mean solving the governing partial differential equations or just gaining an awareness of the variables and parameters that affect transport phenomena, especially those regarding the adsorbent. At least two constraints of importance include intraparticle diffusion and pressure drop in the packed bed (see Do, 1998; Kärger and Ruthven, 1992; and Tien, 1994).

Film diffusion resistance, axial dispersion, and axial and radial temperature gradients (implying significant resistances to intraparticle heat transfer, as well as axial and radial conduction and convection) may all be relevant. Furthermore, the adsorbent itself, and internal fittings and devices (e.g., for bed retention or flow distribution), may affect flow behavior and pressure drop. Accordingly, these may affect performance, but accounting for them is not generally practical. Even if the equations could be written, determining the applicable coefficients would be very time consuming.

Mass transport in an adsorbent particle can be viewed as a combination of several mechanisms, as shown in Figure 14.8. Macropore and micropore diffusion are shown and could be considered as examples of intraparticle diffusion. Bulk flow or conveyance through the particle (e.g., via connected pores) is shown in that figure but, to be significant, requires high porosity or a large

Adsorption

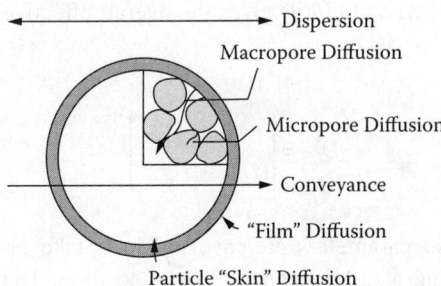

FIGURE 14.8 Illustration of mass transfer mechanisms for a single particle.

pressure gradient. Other facets not shown, but overlapping with those, are mesopore diffusion and surface diffusion. In the latter case, strongly adsorbed components tend to form two or more layers. When that occurs, the outer layer(s) may be mobile, exhibiting what is called *surface diffusion*. The effect is that the adsorbed molecules migrate quickly because of their high density relative to the fluid-phase counterparts.

14.4.2 Mass Conservation and Transport

This section is concerned with the resistances that govern sorption. It begins with the intraparticle aspects, extends to the fluid phase, and finally looks at overall adsorber behavior by considering performance and dynamics as affected by the effects of conditions and parameters on adsorber behavior.

The intraparticle diffusion mechanism is characterized by an effective diffusivity, D_{eff}. For macropore diffusion control alone,

$$D_{eff} = \frac{\varepsilon_p D_{AB}}{\tau} \tag{14.8}$$

and the dimensionless time is $D_{eff} t/\ell^2$, where D_{AB} is the solute diffusivity in the fluid; t is the elapsed time; ε_P is the particle void fraction; τ is its tortuousity; and ℓ is the characteristic length, which equates to the particle radius, r_p, for a bead or cylinder, microcrystal radius, r_c, or particle thickness, h_p, for granules or flakes.

More generally, adsorption is controlled by a combination of transport mechanisms in macropores or micropores, depending on the pore size distribution, the sorbate concentration, the isotherm, and other conditions. The combining bulk diffusion with surface diffusion gives the effective macropore diffusivity:

$$D_{eff} = \frac{\dfrac{\varepsilon_p D_{AB}}{\tau} + (1-\varepsilon_p)\dfrac{dq^*}{dc} D_s}{\varepsilon_p + (1-\varepsilon_p)\dfrac{dq^*}{dc}} \tag{14.9}$$

The effective Knudsen diffusivity (for rarefied gases) in micropores, in cm²/s, is

$$D_{K_{eff}} = 19,400 \frac{\varepsilon_p^2}{\tau a_{BET} \rho_p} \left(\frac{T}{M}\right)^{1/2} \tag{14.10}$$

The effective average pore diameter is $d_{pore} = \varepsilon_p/a_{BET}\rho_p$, and the BET surface area is a_{BET}.

Combining Equations (14.8) and (14.9) gives the overall effective diffusivity in the transition regime:

$$D_{\text{eff}}^* = \left[\frac{1}{D_{\text{eff}}} + \frac{1}{D_{K_{\text{eff}}}}\right]^{-1} \tag{14.11}$$

To understand how these parameters are involved in uptake or release rates, consider the fractional change, F, following a sudden change of composition. The expression is a solution of Fick's law. The *initial* response ($D_{\text{eff}} t/\ell^2 < 0.4$) and the *final* response ($D_{\text{eff}} t/\ell^2 > 0.4$) of a spherical particle, respectively, are approximated by

$$F \equiv \frac{C_t - C_0}{C_f - C_0} \approx \frac{6}{\sqrt{\pi}}\left(\frac{D_{\text{eff}} t}{\ell^2}\right)^{1/2} - 3\frac{D_{\text{eff}} t}{\ell^2}\bigg|_{\text{initial}} \text{ or}$$

$$\approx 1 - \frac{6}{\pi^2} e^{-\pi^2 D_{\text{eff}} t/\ell^2}\bigg|_{\text{final}} \tag{14.12}$$

where the C_0, C_f, and C_t represent the initial, final, and instantaneous values of concentration averaged over the particle. For a flow system, $t \approx \ell/v_i$, where $v_i = v_s/\varepsilon_B =$ the interstitial velocity, and $v_s = Q/A_{cs} =$ the superficial velocity, where Q is the volumetric flow rate and A_{cs} is the cross-sectional area of the bed. From these approximations, we can see that when $D_{\text{eff}} t/\ell^2$ is less than 0.001, the adsorbent is not effective. Conversely, when it exceeds unity, adsorption is largely complete. Thus, when searching for an effective (fast) adsorbent, it is usually a safe bet to choose one having a large diffusivity or small diameter. Other concerns may overrule the selection of small particles, as mentioned later.

Interstitial mass transfer in fixed beds is frequently a significant factor in adsorption dynamics. Sherwood et al. (1975) developed an equation for both gases and liquids that employs the Colburn–Chilton j-factor:

$$j_D = 1.17\, Re^{-0.415} \quad [\text{for } 10 < Re < 2500] \tag{14.13}$$

where $j_D = (k_f/v_s)Sc^{0.667}$, the Reynolds number is $Re = \rho v_s d_p/\mu$, and the Schmidt number is $Sc = \mu/\rho D_{AB}$. We are primarily interested in the fluid-to-particle mass transfer coefficient, k_f. It is mostly governed by the fluid properties (density, ρ; viscosity, μ; and diffusivity, D_{AB}).

Yoshida et al. (1962) suggested another correlation for estimating the mass transfer coefficient:

$$j_D = 0.91\, Re^{-0.51}\, \psi \quad [\text{for } Re' < 50] \tag{14.14}$$

where the modified Reynolds number is $Re' = \rho v_s/\mu a_p \psi$, in which $a_p = 6/d_p$ for beads; $\psi =$ shape factor is 1.0 for spherical beads, 0.95 for granular adsorbents, 0.91 for cylindrical pellets, and about 0.86 for flakes; and $D_{Abeff} = D_{AB}/\mathcal{T}$, where $\mathcal{T}$ is an apparent tortuosity factor that lumps the fluid phase resistance (film diffusion) with the intraparticle mass transfer resistance.

For practical purposes, these only depend on one adsorbent parameter, the particle diameter, d_p. As can be seen, $k_f \propto d_p^{-0.415}$ to $d_p^{-0.51}$, so for a given fluid and flow rate, a tenfold increase of the particle diameter would lead to a 60 to 70% decrease of the mass transfer coefficient. Conversely, a tenfold increase of the velocity often leads to a four- to threefold increase of the mass transfer coefficient. Generally, a large value of k_f is desired, although if achieved by employing high velocity, v_s, the associated pumping or compression costs will be high, and the time of exposure in Equation (14.12) will be short.

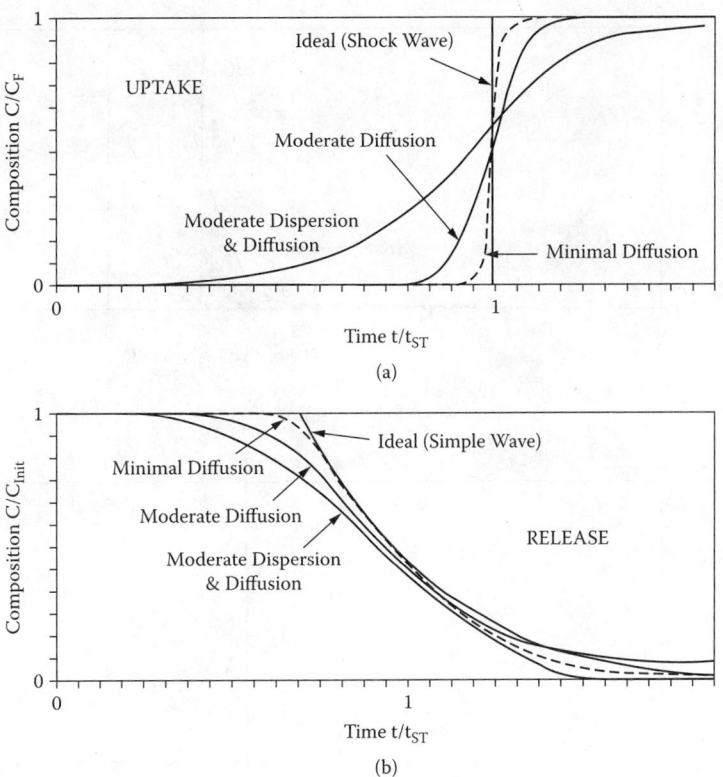

FIGURE 14.9 Realistic and ideal uptake and release breakthrough patterns.

The shape of a breakthrough curve is critical in predicting performance and depends on geometry, adsorbent properties, and operating conditions. An illustration of breakthrough behavior in fixed-bed adsorbers is shown in Figure 14.9. From that figure, we can classify breakthrough patterns for uptake as a *self-sharpening front* or *constant pattern front* or *shock wave*. In contrast, breakthrough patterns for regeneration are referred to as *gradual transitions* or *proportional patterns* or *simple waves*.

Figure 14.10 is a plot of single-component uptake breakthrough data in which a bed of length L_B initially contains fluid at C_0, and the feed concentration is C_F. The upper part of that figure shows the breakthrough curve, L_B, acquired at the end of the bed. That is, it shows the effluent concentration history, $C(t, L_B)$, expressed as $(C - C_0)/(C_F - C_0)$, vs. time. The lower part shows the corresponding loading change of the adsorbent, $(\bar{n}_A - n_0^*)/(n_F^* - n_0^*)$, versus position. This sketch is slightly more abstract than the fluid-phase concentration part for two reasons. First, data of this sort are seldom available. Second, any version would represent a "snapshot" of the state at a specific time. The one shown is the adsorbent bed at imminent breakthrough. The constant pattern assumption (mentioned previously) holds when $C_A/C_{AF} = \bar{n}_A/n_A^*(C_{AF})$. The data shown in Figure 14.10 require certain assumptions to be valid; for example, the flow rate and pressure are constant; the isotherm is "favorable," and the system is nearly isothermal; and the adsorbent is uniform along the axis and across the cross section.

Specifically, in the upper part of Figure 14.10, point 1 represents *imminent breakthrough*, where the effluent concentration first deviates from C_0 reaching C_{IB}, and the corresponding time is called t_{IB}. Likewise, we could consider point 3, where the concentration is practically that of the feed, to be *complete breakthrough*, represented by C_{IB} and t_{CB}. To be very specific, the values could be defined according to process or instrumentation limits, e.g., $C_{IB} = 0.01 \times C_F$ and $C_{CB} = 0.99 \times C_F$.

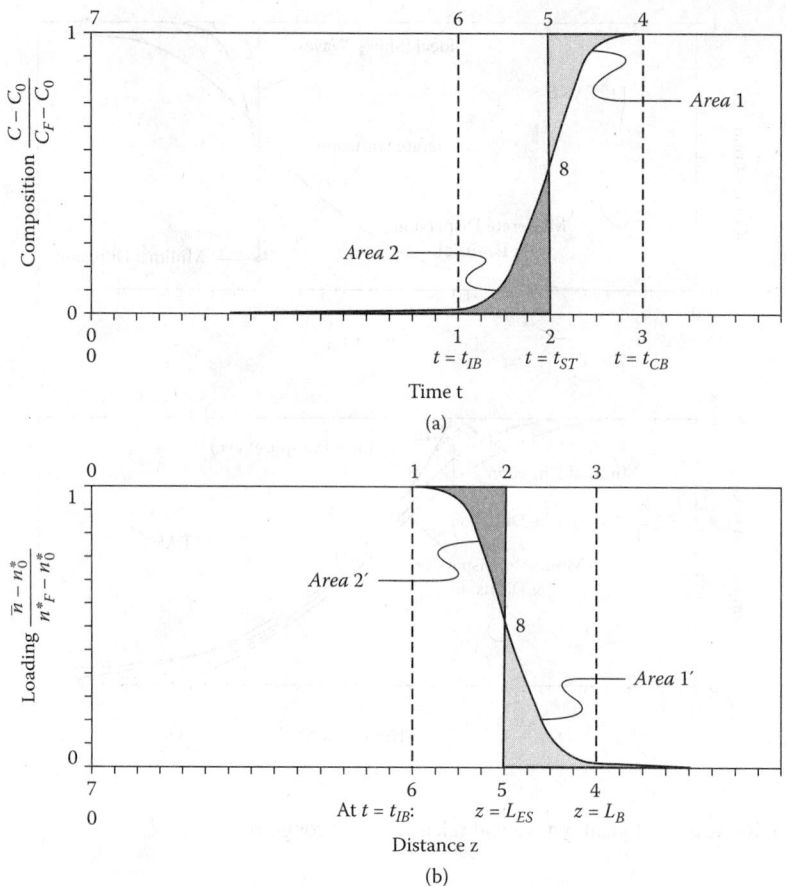

FIGURE 14.10 Breakthrough history (t) and profile (z). Times and positions are labeled to represent specific sets of conditions.

To determine the stoichiometric breakthrough time involves integration. Basically, the points 1 through 6 are chosen such that the area of the rectangle 1-2-5-6 equals that above (or below) the curve 1-8-4-6, and likewise the area of the rectangle 2-3-4-5 equals that below (or above) the curve 1-3-4-8. The result of those is that Area 1 = Area 2 and Area 1' = Area 2', and the time dividing them (at point 2) is by definition t_{ST}. If the curve is symmetric, point 8 will lie at the mid-point between the feed and initial concentrations, but that is only coincidental.

Before delving into specific mathematical models, it might be helpful to look at the basic mass conservation equation, because some terms that represent adsorbent properties appear in it. The material balance equation for solute A is

$$\varepsilon_B \frac{\partial C_A}{\partial t} + (1-\varepsilon)\rho_p \frac{\partial \bar{n}_A}{\partial t} + \varepsilon^v \frac{\partial C_A}{\partial z} = E_{ax} \frac{\partial^2 C_A}{\partial z^2} \qquad (14.15)$$

where ε = bed void fraction, ρ_p = particle density, ρ_B = bulk density = $(1-\varepsilon_B)\rho_p$, C_A = solute concentration in fluid, $\bar{n}_A$ = adsorbent loading (averaged), v = interstitial velocity (v_s = superficial velocity = ε^v), and E_{ax} = axial dispersion coefficient. If the column is well designed, the last term can be made negligible. Among these terms, the particle density and void fraction are inherent properties of the adsorbent. Of course, the adsorbent loading is, too, but it also depends on other operating conditions.

Adsorption

14.4.2.1 Local Equilibrium Model

The local equilibrium model assumes that all types of diffusion are instantaneous. It generally predicts shock waves and simple waves, which are ideal versions of the transitions that typically occur. For all that, it can produce useful results with extreme ease (compared with other, more sophisticated methods). Furthermore, it applies to any isotherm form and can be used for uptake or regeneration.

The local equilibrium model assumes that there are no concentration gradients within a particle or in the film surrounding a particle. Thus, the solid responds instantaneously: $\bar{n}_A \sim n_A^*(C_A,T)$. As we have seen in Equations (14.12) through (14.14), this implies that the particles are small or that the diffusivity is large. Also, neglecting dispersion (i.e., $E_{ax} = 0$) leads to the simplest possible set of equations. In virtually every case, the solution of those equations yields the best possible performance of an adsorber. One of the key results obtained by solving Equation (14.15) is the velocity of constant composition:

$$v_C = \left.\frac{dz}{dt}\right|_{C_A} = \frac{v_s}{\varepsilon_B + \rho_B \dfrac{\partial n_A^*}{\partial C_A}} \quad (14.16)$$

which can generally be used to evaluate breakthrough curves for regeneration.

In the local equilibrium model, the breakthrough curve for uptake is simply a step change from the initial concentration to that of the feed. The speed at which it moves is most important, because that relates the amount of adsorbent to the amount of material processed. This may be found by again assuming that $E_{ax} = 0$ and casting the material balance in difference form, analogous to Equation (14.15):

$$\varepsilon_B \left.\frac{\Delta C_A}{\Delta t}\right|_z + (1-\varepsilon_B)\rho_p \left.\frac{\Delta n_A^*}{\Delta t}\right|_z + \varepsilon^v \left.\frac{\Delta C_A}{\Delta z}\right|_t = 0 \quad (14.17)$$

Letting $v_{SH} = \Delta z/\Delta t$ (the velocity of the step change), $\Delta C_A = C_h - C_l$, and $\Delta n^* = n_h^* - n_l^*$, and rearranging, gives the *shock wave* velocity:

$$v_{SH} = \frac{v_s}{\varepsilon_B + \rho_B \dfrac{\Delta n_A^*}{\Delta C_A}} \quad (14.18)$$

To illustrate, the Langmuir isotherm [Equation b in Table 14.3] gives, for feed at C_h and the initial column contents at $C_l \sim 0$,

$$\frac{\Delta n^*}{\Delta C} = \frac{\dfrac{Kn_M C_h}{1+KC_h} - \dfrac{Kn_M C_l}{1+KC_l}}{C_h - C_l} = \frac{Kn_M}{(1+KC_h)(1+KC_l)} \quad (14.19)$$

The ideal breakthrough time is $t_{BT} = \Delta t_{SH}$, and it may be evaluated as follows:

$$\Delta t_{SH} = \left[\varepsilon_B + \rho_B \frac{\Delta n_A^*}{\Delta C_A}\right]\frac{L}{v_s} = \left[\varepsilon_B + \rho_B \frac{\Delta n_A^*}{\Delta C_A}\right]\frac{V_{ads}}{Q} \quad (14.20)$$

where Q is the volumetric flow rate and V_{ads} is the volume of the adsorbent bed ($= LA_{cs}$). This result applies to any favorable isotherm. It is restricted to isothermal plug flow with a constant fluid velocity. For uptake of a major gas component, allowance should be made for velocity variation.

14.4.2.2 Modified Wheeler and Robell Method

The Wheeler–Robell equation (1969) is an empirically adjusted local equilibrium model. It neglects axial dispersion but applies to any isotherm form:

$$\frac{C_A}{C_{AF}} = e^{k_w\left(\frac{C_{AF}}{\rho_B n^*_{A\,eff}} t - \frac{A_{CS}L}{Q_F}\right)} \tag{14.21}$$

where k_w and $n^*_{A\,eff}$ are adjustable. Determining $n^*_{A\,eff}$ empirically along with k_w leads to discrepancies compared with true equilibrium values. Equation (14.21) can be altered by determining $n^*_{A\,eff}$ independently; e.g., by measuring or predicting it. In addition, the breakthrough time can be predicted via Equation (14.20) (the local equilibrium model) using isotherm data. Solving Equation (14.21) with $C_A/C_{AF} = 0.5$ leads to the following relation for the effective capacity of the adsorbent, n^*_{AF}:

$$n^*_{AF} = \frac{C_{AF} t_{BT}/\rho_B}{\theta - 0.693/k_w} \tag{14.22}$$

where the superficial residence time is $\theta = A_{CS}L/Q_F$ in units consistent with k_w and t_{BT}.

Replacing t_{BT} with the local equilibrium model result $\approx \Delta t_{SH}$ gives a modified Wheeler equation:

$$\frac{C_A}{C_{AF}} = e^{\frac{k_w - 0.693/\theta}{\varepsilon_B + \rho_B n^*_{NF}/C_{AF}} t - k_w \theta} \tag{14.23}$$

The restrictions as written are to uptake in a clean bed of adsorbent, and only to $C_A/C_{AF} = 0.5$. They could be extended to arbitrary (but uniform) initial conditions, and they can be converted to symmetric breakthrough curve shapes.

14.4.2.3 Mass Transfer Resistances

Three main types of mass transfer resistances are recognized: *film diffusion* (which occurs at the external surface of the adsorbent), *intraparticle diffusion* (which occurs within the pores or amorphous structure of the adsorbent), and *adsorption/desorption kinetics* (which occurs at the internal surface of the adsorbent).

Film diffusion is expressed as a simple function, using the middle term of Equation (14.15):

$$(1 - \varepsilon_B)\rho_p \frac{\partial \bar{n}_A}{\partial t} = k_f a_p (C_A - C^*_A) \tag{14.24}$$

where k_f = film mass transfer coefficient and a_p = area (of surface) per unit volume.

Intraparticle diffusion is commonly expressed with Fick's law using terms of pore diffusion or surface diffusion (when adsorption loading is large):

$$\frac{\partial n_A}{\partial t} = \frac{D_{A\,eff}}{r^2}\frac{\partial}{\partial r}\left(r^2 \frac{\partial C_A}{\partial r}\right) \quad \text{or} \quad \frac{D_{A_s}}{r^2}\frac{\partial}{\partial r}\left(r^2 \frac{\partial n_A}{\partial r}\right) \tag{14.25}$$

where D_{Aeff} = effective diffusivity of A in pores, D_{AS} = diffusivity of A in adsorbed state, r = radial distance in adsorbent bead, and r_p is the nominal particle radius. Intraparticle diffusion may also be expressed with the "linear driving force" (LDF) approximation. This is equivalent to the film diffusion model when the isotherm is linear:

$$\frac{\partial \bar{n}_A}{\partial t} = \psi k_s a_p (n_A^* - \bar{n}_A) \tag{14.26}$$

where k_s = solid-phase mass transfer coefficient, a_p = area (of surface) per unit volume, and $\psi k_s a_p \approx 15 D_{Aeff}/r_p^2$ for most system, which was originally suggested by Glueckauf.

Adsorption/desorption kinetics is generally a simple rate equation based on Langmuir's isotherm derivation [see Equation (14b) in Table 14.3]:

$$\frac{\partial n_A}{\partial t} \cong k_r \left(C_A (n_{\max_A}^* - n_A) - \frac{1}{K} n_A \right) \tag{14.27}$$

where k_r = the reaction kinetic constant and K = the adsorption equilibrium constant.

Perhaps the simplest model that employs these principles was suggested by Hougen and Marshall (1947). Hines and Maddox (1985) generalized their treatment so that virtually any form can be reduced to a LDF form. Their basic equation assumes no axial dispersion, $D_z = 0$; Henry's law isotherm, $n_A^* = K_D C_A$ (which is equivalent to $n_A = K_D C_A^*$); and that film diffusion resistance predominates, as in Equation (14.24). The result is

$$\frac{C_A}{C_{AF}} = \bar{J}(\tau, \zeta) = 1 - \int_0^\zeta e^{-(\tau+\zeta)} I_0(i(4\zeta\tau)^{1/2}) d\zeta \tag{14.28}$$

where

$$\tau = \frac{k_f a_p}{(1-\varepsilon_B)\rho_p K_D}(t - z/v)$$

and

$$\zeta = \frac{k_f a_p z}{\varepsilon_B v}$$

where ζ is commonly called the "dimensionless time" and τ is commonly called the "number of transfer units." The mass transfer coefficient, k_f, can be adjusted to fit experimental breakthrough data by accounting for the dependence of flow rate, diffusivity, and so forth.

The integration is a bit complicated, so $\bar{J}(\tau, \zeta)$ is normally evaluated from a plot of $\bar{J}$ versus τ for various values of ζ. Frequently, the following approximations are useful.

For $\tau\zeta > 36$,

$$J(\zeta, \tau) \cong \frac{1}{2}[1 - \mathrm{erf}(\sqrt{\zeta} - \sqrt{\tau})] + \frac{e^{-(\sqrt{\zeta}-\sqrt{\tau})^2}}{\sqrt{\pi}[(\zeta\tau)^{1/4} + \sqrt{\tau}]} \tag{14.29}$$

For $\tau\zeta > 3600$,

$$\bar{J} = (\zeta, \tau) \cong \frac{1}{2}[1 + erf(\sqrt{\tau} - \sqrt{\zeta})] \quad (14.30)$$

Note that $erf(-x) = -erf(x)$, and that $erfc(x) = 1 - erf(x)$. An approximation for $erf(x)$ is

$$erf(x) \cong 1 - (A_1 t + A_2 t^2 + A_3 t^3)e^{-x^2} \quad (14.31)$$

where $t = (1 + A_4 x)^{-1}$, and the parameters are $A_1 = 0.34802$, $A_2 = -0.09588$, $A_3 = 0.7478556$, and $A_4 = 0.47047$. For example, $erf(1) = 0.842701$, while the equation above predicts 0.842718.

Extensions of the Hougen–Marshall model, developed by Rosen (1954) account for mass transfer resistances in both the fluid and solid phases. However, it is seldom practical to conduct all the experiments necessary to determine values of both coefficients and their dependence on conditions.

14.4.3 Heat Transfer

Adsorption from the gas or vapor phase is usually associated with significant heat release upon uptake, or cooling upon desorption. Three modes of operation are possible: isothermal, adiabatic, and intermediate (not quite either extreme). Isothermal operation can often be assumed for liquid-phase adsorption but generally not for gas- or vapor-phase systems. Temperature shifts affect adsorption capacity strongly but diffusivity to a lesser extent.

Following the form suggested by Equation (14.14), heat transfer to the fluid phase can be estimated by the Yoshida et al. (1962) correlations for particle to fluid heat transfer coefficients:

$$j_H = 0.91 Re'^{-0.51} \psi \quad [\text{for } Re' < 50] \quad (14.32)$$

$$j_H = 0.61 Re'^{-0.41} \psi \quad [\text{for } Re' > 50] \quad (14.33)$$

in which $j_H = (h/C_{pf}\rho v_s)(C_{pf}\mu/\lambda_f)^{2/3}$, in which the heat capacity, density, viscosity, and thermal conductivity of the fluid are C_p, ρ, μ, and λ, respectively, and the subscript f indicates "fluid phase and film temperature." As in Equation (14.14), the particle shape factor is ψ. Likewise, the modified Reynolds number, Re', is defined as before. An alternate for the previous equations is the Wakao and Funazkri correlation for the particle to the *gas* heat transfer coefficient:

$$Nu = 2.0 + 1.1 Pr^{1/3} Re^{0.6} \quad (14.34)$$

where $Nu = h_f d_p/\lambda_f$, in which the heat transfer coefficient is h_f and the gas phase thermal conductivity is λ_f.

Heat generation is frequently significant in adsorbers. Nevertheless, adsorbent particles are usually assumed to be at a uniform temperature. There are many reasons for this observation. Heats of adsorption (and desorption) are not sufficiently large to cause large temperature gradients within typical, small adsorbent particles, but the cumulative effects may generate swings of 10 to more than 50°C. The *thermicity* reveals the magnitude of temperature gradient expected in an adsorbent particle:

$$\beta = \frac{c_s(-\Delta H_{ads})D_{eff}}{\lambda_s T_{ext}} \approx \frac{\Delta T_{max}}{T_{ext}} \quad (14.35)$$

in which c_s is the intraparticle concentration, ΔH_{ads} is the isosteric heat of adsorption, λ_s is the solid thermal conductivity, and T_{ext} is the external fluid temperature. For gas-phase systems, an order-of-

Adsorption

magnitude estimate is $\lambda_s = 0.1$ to 0.3 W/mK, $\Delta H_{ads} = 1$ to 100 kJ/mol, $D_{eff} = 1.0\text{E}{-}06$ m²/s, $T_{ext} = 300$ K, and $c_s = P/RT = 400$ mol/m³ (at 10 bar): $\beta < 2$ K.

Heat transfer to column wall depends on convection. The relevant heat transfer coefficient at the internal surface of a column may be estimated from Leva's correlation:

$$Nu = \frac{h_w d_B}{\lambda_f} = 0.813 \text{Re}^{0.19} \exp\left(\frac{-6d_p}{d_B}\right) \tag{14.36}$$

14.4.4 Pressure Drop

The third major factor in dynamics is packed bed flow behavior. The pressure drop in the adsorbent media is caused by drag due to fluid flowing through the interstices. For slow flow, the pressure drop, ΔP, across a bed of length Δz and superficial velocity v_s can be represented by Darcy's equation:

$$\frac{-\Delta P}{\Delta z} = \frac{v_s \mu}{k} \tag{14.37}$$

where μ is the absolute viscosity, and k = permeability, which has units called a *darcy* = 1 (cm/s)cP/atm/cm ≈ 1.0E–8 cm². Kozeny modified Equation (14.36) using the bed voidage, ε, and the specific surface, S:

$$\frac{-\Delta P}{\Delta z} = \frac{v_s \mu S^2}{k'} \frac{(1-\varepsilon)^2}{\varepsilon^3} \tag{14.38}$$

This equation is valid when $\text{Re}' = v_s d_p/\mu(1-\varepsilon) < 10$ and $\varepsilon < 0.5$, and experimental evidence indicates that $S^2/k' = 150/d_p^2 g_c$. Ergun (1952) combined the Kozeny equation with the Burke–Plummer equation to cover both laminar and turbulent flow conditions:

$$\frac{\Delta P}{L} = \left(\frac{150}{\text{Re}'} + 1.75\right)\left(\frac{\rho_f v_s^2}{g_c d_p} \frac{1-\varepsilon}{\varepsilon^3}\right) \tag{14.39}$$

which is valid for the range of modified Reynolds numbers, $0.1 < \text{Re}' < 10{,}000$. An alternative, called Forchheimer's law, can be written after integration as

$$\frac{-\Delta P}{\Delta z} = \alpha v_s + \alpha' v_s^n \tag{14.40}$$

in which the first term predominates at low flow rates due to viscous forces, and the second term is more significant at high flow rates. The parameters α and α' must be determined empirically. The change from streamline or laminar flow to completely turbulent flow is gradual, because flow conditions are not the same in all interstices, because of their size and distribution. When high pressure drops develop in a bed, some convective transport occurs as described by suitable equations, but the effective particle size is reduced.

Most adsorbers are designed to operate with relatively small pressure drop; relatively large particles are used, and the velocity is typically low to allow equilibration of the fluid with the adsorbent. That typically means short, fat packed beds rather than long, thin ones. Generally, pipes, valves, and fittings contribute as much to the flow restriction as the pressure drop in the adsorbent bed.

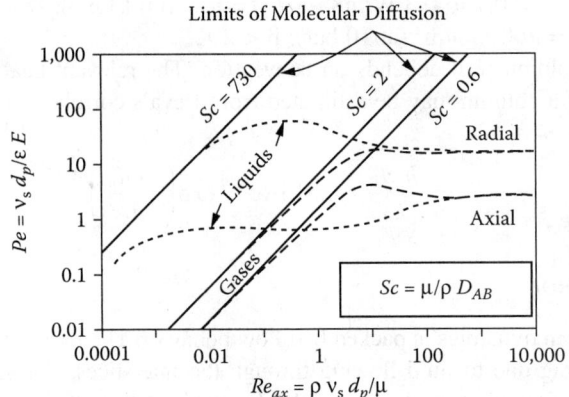

FIGURE 14.11 Axial and radial Peclet numbers for single-phase flow of a gas or liquid through a packed bed of spherical particles. The limits of molecular diffusion are shown by solid lines that represent $Pe = Re\, Sc\, \tau_B/\varepsilon_B$, where τ_B = bed tortuosity = 1.4 and ε_B = bed voidage = 0.4 (Sherwood et al., 1975).

14.4.5 Dispersion

The principal aspect that reflects flow distribution in fixed beds is dispersion. Flow through a packed bed is commonly represented by dispersed plug flow in which all mechanisms contributing to mixing are lumped together in effective dispersion coefficients, E_{ax} and E_r. A model analogous to Fick's law, as given by Equation (14.15), is applicable:

$$\frac{\partial c}{\partial t} + \frac{v_s}{\varepsilon}\frac{\partial c}{\partial z} = \frac{E_r}{r}\frac{\partial}{\partial r}\left[r\frac{\partial c}{\partial r}\right] + E_{ax}\frac{\partial^2 c}{\partial z^2} \qquad (14.41)$$

in which the terms for adsorption have been omitted, E_{ax} = the axial dispersion coefficient, and E_r = the radial dispersion coefficient. It is evident that there are two modes of dispersion: radial and axial. Axial dispersion always adversely affects adsorption performance. Despite that, axial dispersion can be offset by radial dispersion, meaning that radial dispersion promotes plug-flow behavior, which is always beneficial to adsorption performance. There are two main dispersive mechanisms: hindered molecular diffusion (corrected for tortuosity) and turbulent mixing arising from flow around the particles. Figure 14.11 indicates the limits of hindered molecular diffusion at low flow rates: $E_{ax} = E_r = D_{AB}/\tau_B$, where τ_B = bed tortuosity ≈ 1.4 (typical) for fixed beds.

The coefficient representing axial dispersion, E_{ax}, is measured using a tracer by pulse, sinusoidal, or step-change residence time distribution tests, or by measuring backflow. Sometimes the phenomenon is represented by a "cell" model, in which the number of well-mixed cells fits dispersion. The coefficient representing radial dispersion, E_r, is determined by measuring the radial spread of a tracer from the centerline toward the wall.

The Peclet number for dispersion is defined as $Pe = v_s d_p/\varepsilon_B E$, which is dimensionless, and $v_s/\varepsilon_B = v_i$ = the interstitial velocity. A convenient way to comprehend the data for fixed beds is via Figure 14.11, which shows the trends for gases and liquids in both axial and radial dispersions. A general correlating equation for gases is

$$\frac{1}{Pe_{ax}} = \frac{K_1}{Re\, Sc} + \frac{1}{Pe_{ax\infty}}\frac{1}{1 + K_2/(Re\, Sc)} \qquad (14.42)$$

where $Pe_{ax\infty}$ = 2.0, the Reynolds number is $Re = \rho v_s d_p/\mu$, and the Schmidt number is $Sc = \mu/\rho D_{AB}$. The product, $Re\, Sc = v_s d_p/D_{AB}$, is a Peclet number for ordinary diffusion. The coefficients K_1 and

K_2 depend on which study is consulted. In a pair of studies for relatively large particles ($d_p > 3$ mm), Wen and Fan (1975) found that $K_1 = 0.3$ and $K_2 = 3.8$ for $0.008 < Re < 400$ and $0.28 < Sc < 2.2$, while Edwards and Richardson (1968) found that $K_1 = 0.73\varepsilon_B$ and $K_2 = 9.49\varepsilon_B$, where $\varepsilon_B \approx 0.4$ to 0.6 so that the results are similar. In contrast, $Pe_{R\infty} = 11.0$. For liquids, $Pe_{Ax} \approx 0.45$ in the range $0.005 < Re < 20$.

14.4.6 Chemical Kinetics and Adsorption

Adsorption, and chemisorption in particular, is closely allied to heterogeneous catalytic reactions: both involve similar mass and heat transport constraints, in addition to bond formation at the solid surface. In fact, adsorption is viewed as a precursor to catalytic reaction, and desorption is viewed as the step subsequent to the reaction itself. Adsorption of the reactant(s) and product(s) must be strong enough to deflect the original bonds, but not so strong as to poison the catalyst. This phenomenon has been related to the adsorption potential suggested by Polanyi (see Section 14.3.2).

Regarding mass transport mechanisms, just as in adsorption, reactants transfer to the catalyst sites, and the products depart by intraparticle diffusion; either may be rate limiting. A schematic diagram showing the phenomena is given in Figures 14.12a and 14.12b. The former shows an ideal planar surface, while the latter shows crude pores. Both show counterdiffusion of the reactant, A, through the product, B, to the surface; transition from A to B at the surface; and counterdiffusion of B through A. In real systems, the mechanism is more complicated, since the surfaces are not ideal, plus diffusion through the pores is complex, even for the simplest stoichiometry.

Descriptive parameters for heterogeneous catalysts overlap with adsorbent properties, for example, the well-known B.E.T. surface area analysis and other properties mentioned in Section 14.2.2.3. The measured value may only be proportional to true catalysis, due to steric or energetic factors. Slightly different values are obtained when fluids other than nitrogen are used (e.g., CO_2 at $-78°C$ or n-butane at $0°C$).

Finally, another area of overlap is the potential presence of a false activation energy. This occurs when film diffusion is coupled with, say, a first-order reaction at the catalyst surface. The mass transfer rate $r = k\, a_s(c - c_s)$. In contrast, the inherent reaction rate is $r = k_r a_s c_s$. These two can be combined to give an expression for the surface concentration: $c_s = k_f c/(k_r + k_f)$, which is difficult to measure. This expression can be substituted in the original reaction rate expression to give $r = k_0 a_s c$, with an effective rate constant of $k_0 = 1/(1/k_r + 1/k_f)$. Thus, if the inherent rate obeys the Arrhenius dependence on temperature, $k_r = A\exp(-\Delta E_r/RT)$, and if k_f is constant, the observed Arrhenius activation energy, $\Delta E_0 = RT\, d\ln k_0/d(1/T)$, would be deceptively low.

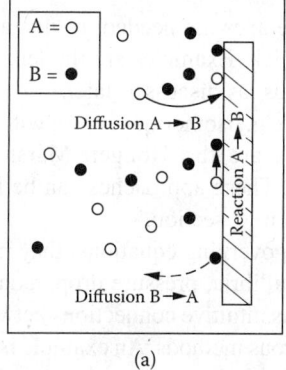
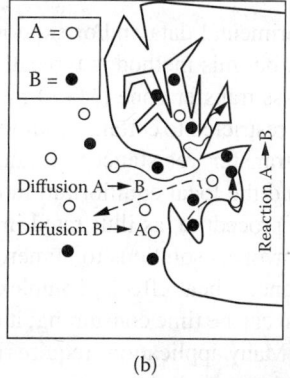

FIGURE 14.12 (a) Simplistic view of diffusion of the reactant, A, through the product, B, reaction at the surface, followed by diffusion of B through A. (b) View of interactions with intraparticle diffusion of the reactant, A, through the product, B, reaction at the surface, followed by diffusion of B through A.

14.5 DESIGN OF BATCH, FIXED-BED ADSORBERS

14.5.1 GENERAL OBJECTIVES

This section describes *batchwise* fixed-bed adsorbers in which the adsorbent is replaced with fresh material, or removed and regenerated after it is "exhausted," then reinstalled. Commercial examples include columns used for chemical feedstock purification, decolorizing solutions, and wastewater treatment. The goal is generally to employ material balance and rate equations to predict adsorber performance, possibly to analyze experimental data (e.g., breakthrough curves and temperature histories), to diagnose problems, or to assess properties or conditions. Unfortunately, various conditions often result in nearly identical behavior, so diagnosing causes may be difficult.

Often there is an economy of scale, meaning that the cost to process material drops substantially as the size of the unit increases. Unfortunately, there are rather small economic advantages for larger adsorbers, because the cost of the adsorbent typically depends on the amount purchased with an exponent of 0.8 to 1.0. Depending on the pressure rating required, the length-to-diameter ratio (L/D) of a typical adsorber may range from 1 to 10, with larger values favored for high pressures. Flow is oriented vertically as a rule, since the adsorbent tends to settle, which would cause void space and bypassing at the top of a horizontal bed. Control is generally based on fixed time cycles, although effluent composition is preferred for performance reasons.

Factors affecting performance are the component isotherm(s), geometry, operating conditions, mass transfer resistances (including film diffusion, intraparticle diffusion, and dispersion), and thermal effects, especially for gases. How these are taken into account varies depending on the application, the tools available, and the situation. For many applications, established methods for simulation and design are readily available. The simplest approach is to estimate the volume of the adsorber, knowing only the adsorbent's capacity at the relevant concentration and temperature, and add a large safety factor.

For more detailed designs, the equations describing breakthrough behavior in the previous section, plus a modest safety factor, may be sufficient. With complex heat and/or mass transfer issues, multiple components, or unusual equipment geometry, a standard model is likely not suitable. In such cases, experimental tests may be required to develop a suitable mathematical model; the required effort might be up to a man-year.

14.5.2 EMPIRICAL VERSUS THEORETICAL APPROACHES

Model classifications are as follows:

> *Empirical.* Experimental data and/or experience axes are needed; results are imprecise and prone to error, but this method is typically quick. Examples are the length of unused bed (LUB) and mass transfer zone (MTZ) methods, as discussed later.
> *Analytic.* This is restricted to certain conditions (e.g., isothermal uptake with clean adsorbent) *or* no mass transfer resistances. Examples are the Hougen–Marshall and Thomas approaches, and the local equilibrium model. These approaches can be incorporated in a simple design procedure, as illustrated in the next section.
> *Rigorous.* This involves solutions to numerous governing equations; they may include mass transfer resistances, heat effects, complex equilibria, pressure drop, axial dispersion, and others. The last can be time consuming; it lacks intuitive connections between performance and variables. Many applications require rigorous methods. An example is multicomponent adsorption of the type shown in Figure 14.13.

Beyond this, some methods and models do not fit these categories, because they have elements of two or all three, e.g., the modified Wheeler equation shown as Equation (14.23).

Adsorption

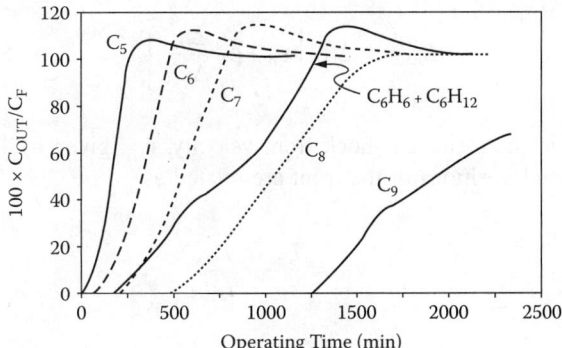

FIGURE 14.13 Breakthrough curves for hydrocarbons in silica gel at 30°C. The lower molecular weight species illustrate roll-up, which is a symptom of displacement by species of higher molecular weight.

14.5.3 Shortcut Procedures

14.5.3.1 Length of Unused Bed Method

A quick way to design an adsorber is by the *length of unused bed* (LUB) approach suggested by Collins (1967). It involves observations about the *mass transfer zone* (MTZ), so experimental data are required. A precursor was suggested by Michaels (1952) as a method of understanding ion exchange column performance. The concept assumes that the transition of concentration from that of the initial contents of the bed to that of the feed is due solely to mass transfer resistance. That is a misconception, but in many cases it suffices.

Figure 14.10 also illustrates the MTZ for an adsorbent bed of length L_B, which initially contains fluid at C_0, and the feed concentration is C_F. As before, the basic assumptions are that flow rate and pressure are constant, the system is isothermal, the adsorbent is uniform along the axis and across the cross section, and for this approach the shape of the breakthrough curve is fixed. t_{IB} is the time of incipient breakthrough, t_{ST} is the stoichiometric breakthrough time [e.g., as predicted by Equation (14.15)], and t_{CB} is the time at which breakthrough is complete. At t_{IB}, the MTZ is contained within the rectangle 1-3-4-6. Ideally, however, the length of adsorbent bed that would be at equilibrium with the feed is L_{ES}. The portion of the bed that has become saturated (to the equilibrium limit) is represented by the rectangle 0-1-6-7. Similarly, in an ideal system, the length of adsorbent bed that would be still at equilibrium with the fluid initially in the bed at t_{IB} is the "length of unused bed," $L_{UB} = L_B - L_{ES}$, represented by the rectangle 2-3-4-5. That is, the concentration in this portion of the bed would ideally be at the initial state, and that upstream would be that of the feed. In reality, of course, both segments of adsorbent are partially used.

At equilibrium, the net amount adsorbed (moles or mass) of component A, n, at any elapsed time prior to breakthrough, t, equals the amount admitted:

$$\dot{n}_A = \dot{m}(C_F - C_0)t = L_{ES}\rho_B(n_F^* - n_0^*) \tag{14.43}$$

where $\dot{m}$ is the flow rate (mass or molar, e.g., ρQ or $\rho Q/\bar{M}$, where $\bar{M}$ is the average fluid molecular weight). Thus, the length of the equilibrium section is

$$L_{ES} = \frac{\dot{m}t}{\rho_B} \frac{(C_F - C_0)}{(n_F^* - n_0^*)} \tag{14.44}$$

or, following Equation (14.20),

$$L_{ES} = v_s \Delta t_{SH} \left[\varepsilon_B + \rho_B \frac{\Delta n_A^*}{\Delta C_A} \right]^{-1} \tag{14.45}$$

If the front moves at the idealized shock front velocity, v_{SH}, given by Equation (14.18), then $t_{ST} = t_{SH}$ and the times and positions of the front are related as

$$v_{SH} = \frac{L_{ES}}{t_{IB}} = \frac{L_B}{t_{ST}} \tag{14.46}$$

This equation allows a concise expression of L_{UB}:

$$L_{UB} = L_B - L_{ES} = v_{SH}(t_{ST} - t_{IB}) \tag{14.47}$$

Combining the previous two equations yields

$$L_{UB} = L_B(1 - t_{IB}/t_{ST}) \tag{14.48}$$

Equations (14.44) through (14.48) apply to a single flow rate and cross-sectional area (not to mention temperature, pressure, feed concentration, and other conditions). By analyzing breakthrough data from one column, it is easy to determine the relationships between bed length and time. For example, given the *actual* effluent concentration history, $C(t,L_{Ba})$, and bed length, L_B, for one system, it is straightforward to identify the *actual* times, t_{IBa} and t_{STa}. From those, the *actual* length of unused bed, L_{UBa}, can be found from Equation (14.47). Likewise, the *actual* length of the equilibrium section, L_{ESa}, can be found from Equation (14.45) or (14.46). A design situation could require choosing a bed with a different, "desired" length of unused bed, L_{UBd}, based on the *desired* and *actual* stoichiometric breakthrough times, t_{STd} and t_{STa}. Under those constraints, the length of the *desired* bed can be found from Equation (14.45), $L_{Bd} = L_{Ba}t_{STd}/t_{STa}$. Subsequently, Equation (14.46) can be used to estimate L_{ESd} for the *desired* bed, $L_{ESd} = L_{Bd} - L_{UBd}$.

To scale up by this technique, i.e., to accommodate other flow rates, Q, or bed geometries, masses should be used. They are found by using the cross-sectional area of the bed and the bulk density of the adsorbent. For a cylindrical bed,

$$W_B = \rho_B \pi d_B^2 L_B / 4 \tag{14.49}$$

where W_B is the mass of the bed, and by analogy W_{UB} and W_{ES} are the mass of the unused bed and mass of the equilibrium section, calculated from their lengths, respectively:

$$W_{ES} = \rho_B Q t_{ES} \left[\varepsilon + \rho_B \frac{\Delta n_A^*}{\Delta C_A} \right]^{-1} \tag{14.50}$$

The relationship between W_B, W_{UB}, and W_{ES},

$$W_B = W_{ES} + W_{LUB} \tag{14.51}$$

The only problem occurs when $L_{UB} = f(Q)$, which must be determined empirically.

TABLE 14.5
Procedures for Design of a Conventional Adsorber

1. Choose adsorbent candidates.
    a. Measure basic adsorbent properties: densities, void fractions, particle size distribution, possibly the B.E.T. surface area and pore size distribution, as well as crush strength and attrition resistance.
    b. Evaluate isotherm(s) over the relevant range of concentrations, temperature(s), including multicomponent interference if applicable.
    c. Analyze intraparticle diffusivity at relevant concentrations, temperature(s).
    d. Ensure that the materials are chemically compatible (including minor contaminants).
    e. Consider pretreatment conditions or follow manufacturer's recommendations.
    f. Estimate life and/or replacement schedule (e.g., 20% per year).
2. Determine bed dimensions and layout.
    a. The "ideal" bed volume is determined by the feed flow rate, feed concentration, isotherm, and allowable time. The "real" bed volume is always larger and depends on the desired product concentration. In addition, most vessels will have dead volumes to accommodate bed support (or retention) and flow distribution devices.
    b. Use a model to calculate effluent concentration as a function of time for a given flow rate, feed concentration, etc. It is important to recognize that mass transfer parameters are not transferable among the models.
    c. If desired, the length-to-diameter ratio can be varied systematically for a fixed volume and compared to the required uptake times. It is critical to take into account the dependence of the mass transfer coefficient on Reynolds number, which will vary as L/D varies.
3. Assess other performance factors.
    a. Pressure drop, and associated power consumption
    b. Flow distribution (fixtures or devices)
    c. Energy requirement to maintain processing temperature range (and need for thermal insulation—internal or external)
    d. Devise control system and necessary instruments and elements.
4. Determine cost (net present value, payback time, etc.).
    a. Rank-order the candidate adsorbents at the extremes of the operating conditions.
    b. For the best adsorbent(s), optimize the operating conditions, vessels, and energy consumption.
    c. Select the best combination of adsorbent and operating conditions.
5. If performance and cost are acceptable, stop. If not, return to step 1 and repeat for other adsorbents. If no plausible adsorbents remain, consider an alternative version of adsorption (e.g., TSA or PSA).

14.5.3.2 Direct Method Using Breakthrough Models

A conventional adsorber can be designed using a breakthrough model relatively easily. Begin with the local equilibrium model, and Equations (14.18) and (14.20) in particular. The former would be useful for determining the flow rate for a given bed length and allowable breakthrough time. The latter would be useful for determining the breakthrough time for a given bed length and flow rate. That method provides no information regarding the shape of the breakthrough front, however. To take into consideration the shape requires the Wheeler–Robell, Hougen–Marshall, or Thomas model. In addition, there is a need for experimental data to assess the mass transfer coefficient associated with the selected model.

A typical design situation for a conventional adsorber begins with the feed flow rate and concentration, allowable time, temperature, pressure, and/or desired product concentration. The last quantity may be either an *instantaneous* or *average* cutoff value, depending on the situation. The sequence of steps necessary to convert that information into a design is given in Table 14.5.

14.5.4 Examples

For relatively simple applications of adsorption, the first goal is normally to find a suitable adsorbent and the quantity required, and then to settle a variety of ancillary issues regarding how the adsorbent will be used, discarded, or recycled. Typically, one would buy the adsorbent, or in some cases a modular, prefilled adsorber, which is replaced when the problem recurs or after a predetermined time.

Example 14.1 Isotherm Fitting
Consider the following adsorption isotherm data for water vapor on silica gel at 25°C. Fit the data to the Langmuir, Freundlich, Redlich–Peterson, and B.D.D.T. equations.

Relative Humidity	Loading Mol/cm^3
0.0000	0.00000
0.0116	0.00173
0.0198	0.00223
0.0378	0.00353
0.0600	0.00484
0.1330	0.00868
0.1770	0.01120
0.1800	0.01182
0.2180	0.01367
0.2380	0.01526
0.2410	0.01559
0.2560	0.01708
0.2780	0.01740
0.2790	0.01794
0.2920	0.02003
0.3070	0.02043
0.3500	0.02345
0.3690	0.02528
0.5260	0.03097
0.5390	0.03204
0.6490	0.03303
0.6970	0.03332
0.7580	0.03387
0.8250	0.03411
0.8250	0.03415

Results
The data and isotherm fits are shown in Figure 14.14. The corresponding parameters and statistics are provided in Table 14.6. The average percentage deviations for those fits are 8.8, 12.0, 7.9, and 2.7%, respectively. The average errors for the Langmuir and Redlich–Peterson equations, in particular, seem high due to relatively large percentage deviations for the first few points. These deviations are exaggerated at low loadings. That is, if an adsorbent has a maximum loading of 1 kg/kg, a discrepancy of 0.01 kg/kg would be equivalent to a 100% deviation if it occurred at 1% of the maximum loading, but only a 1% deviation if it occurred at saturation. When, however, a weighted scale is used, in which the weights are simply proportional to loading, a discrepancy of 0.01 kg/kg amounts to a 1% deviation. Applying that for this example, the weighted percentage deviations are 3.9, 4.8, 1.9, and 1%, respectively.

The B.D.D.T. isotherm essentially provides an ideal fit at low, intermediate, and high concentrations. The parameter m indicates that 4.4 layers are adsorbed on the surface at saturation. This isotherm form also discerns the "real" Henry's law coefficient, which is about four times larger than the value that the simpler equations estimate. The Redlich–Peterson equation fits the data well, although it incorrectly predicts a local maximum in loading. Hence, it is unreliable for extrapolations, and it underpredicts the loading at low concentrations (i.e., below 10% RH). The Langmuir isotherm predicts well at low concentrations but not near saturation. The Freundlich isotherm is least satisfactory in nearly every respect. It might be adequate for estimating overall life, but not for design purposes.

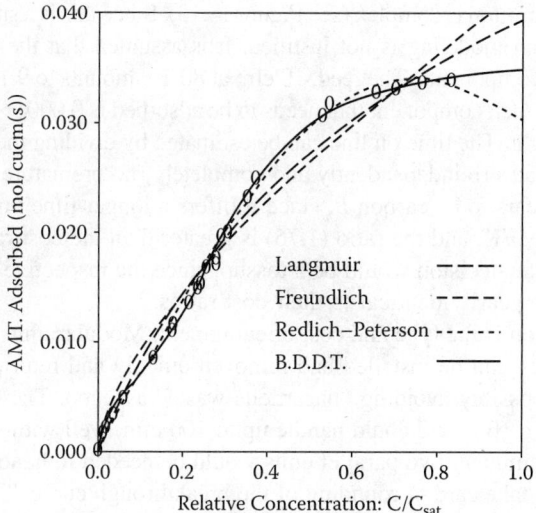

FIGURE 14.14 Adsorption data of water vapor on silica gel at 25°C. Isotherm fits: Brunauer–Deming–Deming–Teller Freundlich, Langmuir, and Redlich–Peterson equations.

TABLE 14.6
Isotherm Parameters and Statistics for Water Vapor on Silica Gel at 25°C (Units: n^*, n_M = mol/cm³ and $C_r = C/C_{sat}$ = relative humidity. See Table 14.3 for equation forms.)

	Langmuir	Freundlich	Redlich–Peterson	B.D.D.T.
	Kn_M = 0.09006	A = 0.04235	$K\,n_M$ = 0.06748	n_M = 0.00822
	K = 1.27700	B = 0.66980	K = 1.21870	m = 4.38560
	n_M^* = 0.07052		B = 3.25100	K = 20.33870
			n_M^* = 0.02076	q = 58.5475
Avg. % Abs. Dev.	8.767	12.045	7.940	2.695
Wtd.	0.9804	0.9658	0.9955	0.9984

* Denotes calculated result.

Example 14.2 Dilute Gaseous Emissions

Consider a process vent containing 100 ppm (vol.) each of MEK, n-hexane, and toluene, at a flow rate of 10 cfm and at 80°F, operating 24 hours per day.

Results

Because the contaminants are organic, activated carbon is a candidate adsorbent, with one reservation: safety. Ketones, in particular, sometimes cause fires in activated carbon. To avoid that complication, other adsorbents might be considered. We will assume that activated carbon is acceptable, in part being the least expensive on a single-use basis.

Table 14.6 lists loadings for two activated carbons: one produced from coal, A, and one produced from coconut shell, B. Carbon B costs 50% more than carbon A.

Multicomponent adsorption is complex (see Figure 14.13). Since this is a small application, delving into equilibria and column modeling is not justified. It is assumed that the lighter components are displaced by the heavier components. The feed, 10 cfm at 80°F, amounts to 9.1 scfm or 1.52 lb-mol/hr. Thus, the *loading rate* of each component that needs to be adsorbed is $0.0001 \times 1.52/MW$ (lb/h), where *MW* is its molecular weight. The time on line can be estimated by dividing the loading by the *loading rate*, assuming that they adsorb independently and completely (no premature breakthrough).

The best choice appears to be carbon B, since it offers a longer time on line than carbon A for the critical component, MEK, and the ratio (1.76) is greater than the cost ratio (1.5). If there were no MEK in the stream, the decision would be a tossup, since the respective carbons' loading ratios of those components are nearly identical to their cost ratios.

The final consideration is the type and cost of equipment. Modular units come packed with 150 to 500 lb of carbon. They can be installed and removed quickly and returned to the manufacturer on an exchange basis (possibly avoiding "hazardous waste" aspects). The smaller units would be packed with 4×6 mesh carbon and could handle up to 100 cfm, well within the range needed. For 50 lb per batch per contaminant, two parallel units would be needed to handle the MEK for 600 hr, allowing some excess to take care of rounding of the breakthrough curve. The cost would be in the range of $600 each. Even at this low cost ($17,520 per year, not including labor, shipping, and other costs), it might be possible to justify a small PSA unit that would recover concentrated vapor (for incineration) or a condensed product. Such systems are not yet available off the shelf, however.

Example 14.3 Dilute Liquid Solutions

Consider a feed of 0.3 mg/l (or 300 ppm) of diphenylamine in 100 gpm water, with a minimum on-line time of 500 hr.

Results

Dobbs and Cohen (mentioned earlier) reported that Filtrasorb-300 (Calgon Corp.) adsorbs 80 mg/g at this concentration. Thus, $\Delta n^*/\Delta C = 80$ (mg/g)/0.0003 (mg/cm³). $\rho_B = 31.2$ lb/ft³ (or 0.5 g/cm³), $\varepsilon = 0.70$. They also noted that a fine powder (200×400 mesh, or 0.05-mm diameter) required about 5 hr to equilibrate. Equation (14.12) suggests the time required to equilibrate a 0.5-mm diameter granule. Namely, the parameter $D_{eff} t/\ell^2$ would be constant, as would the effective diffusivity. Thus, $t_{3mm} = t_{0.05mm}$ (0.5 mm/0.05 mm)² or 500 hr! Granules will not exhibit a sharp breakthrough, so a large safety factor is needed. Accordingly, facilities often employ a slurry of powdered carbon with a downstream filter to remove the waste.

Equation (14.20) is used to estimate the bed volume, given the equilibrium breakthrough time. We find that the minimum volume of adsorbent is 112.5 gal = 15 ft³ or roughly 470 lb. To be conservative, a *safety factor* of 3 is recommended, but even that should be tested. Thus, the bed size would be about 45 ft³ (roughly 3 ft dia. $\times$ 6.5 ft long), containing about 1400 lb of carbon. The length-to-diameter ratio, L_B/d_B, could be optimized. To obtain a large mass transfer coefficient, that ratio should be large, but to minimize pressure drop, that ratio should be small. Generally, the optimum ratio lies in the 1:1 to 5:1 range.

Example 14.4 Gaseous Emissions

Predict the breakthrough patterns for uptake of CO_2 from air by activated carbon under the following data, and compare with the breakthrough data (Note: the symbol [=] denotes "has units of"):

$$\rho_B = 0.360 \text{ g/cm}^3 \quad \varepsilon_B = 0.345 \quad d_p = 0.5 \text{ mm}$$
$$d_B = 0.718 \text{ cm} \quad L_B = 72 \text{ cm} \quad p_{AF} = 0.1316 \text{ atm}$$
$$\dot{Q} = 1.85 \text{ std. cm}^3/\text{s} \quad T = 0°C \quad P = 1 \text{ atm}$$

$$n_A^* = \frac{0.0112 p_A}{1 + 7.258 p_A} [=] \frac{\text{mol (A)}}{\text{g (adsorbent)}}; \quad p_A [=] \text{ atm}$$

t (min)	10.73	11.567	11.983	12.4	12.817	13.233	14.067
C/C_F	0.000	0.117	0.250	0.517	0.700	0.808	0.933

Results

(1) *Local Equilibrium Model*:
Uptake time =

$$\Delta t_{SH} = \left[1 + \frac{\rho_b}{\varepsilon_b}\frac{\Delta n_A^*}{\Delta C_A}\right]\frac{L}{v}$$

where

$$v = \frac{\dot{Q}}{A_{CS}\varepsilon_b} = 13.24\ \frac{\text{cm}}{\text{s}}$$

and

$$\frac{\Delta n^*}{\Delta C_A} = RT\frac{\Delta n^*}{\Delta p} = RT\frac{0.0112}{(1+7.258\times 0.1316)} = 128.4\ \frac{\text{cm}^3}{\text{g}}$$

$$\therefore \Delta t_{SH} = 744.\ s = 12.4\ \text{min}$$

(2) *Modified Wheeler Equation*:

$$\frac{C_A}{C_{AF}} = e^{\frac{k_w - 0.693/\theta}{\varepsilon_b + \rho_b n_A^*/C_{AF}}t - k_w\theta}$$

where

$$\theta = V_{bed}/\dot{Q}_F = 29.56/1.85 = 15.98\ \text{s} = 0.266\ \text{min},$$

$$n_A^*/C_{AF} = 0.0007539/0.000005871 = 128.4\ \text{cm}^3/\text{g, and}$$

$$\varepsilon_B + \rho_B n_A^*/C_{AF} = 0.345 + 0.36 \times 128.4 = 46.57.$$

At $t = 11.98$ min, choosing $k_w = 1$ s^{-1} yields $C_A/C_{AF} = 0.2992$, while $k_w = 3$ s^{-1} yields $C_A/C_{AF} = 0.1022$, but the actual value is $C_A/C_{AF} = 0.250$. Using the two to interpolate, we find $k_w = 1.3$ s^{-1} yields $C_A/C_{AF} = 0.255$.

Breakthrough curves for values of k_w from 1.3 to 100 are shown in Figure 14.15.

14.6 CYCLIC AND CONTINUOUS ADSORPTION PROCESSES

Cyclic adsorbers are periodically regenerated in place after they are "exhausted." The four main categories are inert-purge, displacement-purge, TSA, and PSA. Generic characteristics of those cyclic systems are explained in substantially more detail later in this section.

Continuous processes employ circulating adsorbent or moving ports so that a portion of the adsorbent is always being regenerated. Design procedures for this type of process are complicated

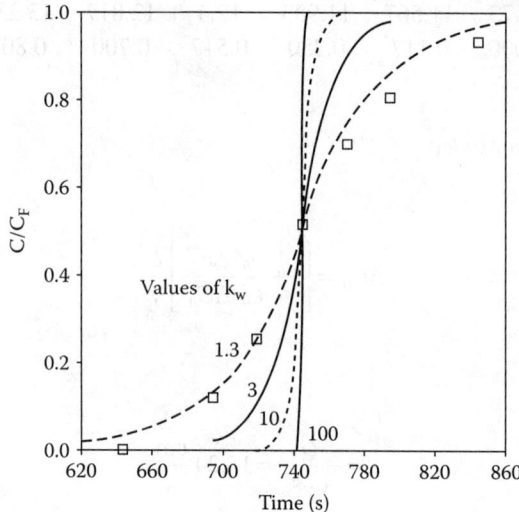

FIGURE 14.15 Comparison of breakthrough predictions for the Wheeler equation for k_w = 1.3, 3, 10, and 100 1/s. The data are shown as square symbols.

and are beyond the scope of this chapter. Nevertheless, it may be helpful to consider the features of typical continuous systems, since they may be transferred to other situations. Example flowsheets of three types of continuous process are given in Figures 14.16, 14.17, and 14.18. The first is a stirred system, which functions like a CSTR from the standpoint of the liquid. The adsorbent, however, undergoes a batchwise experience. The second flowsheet a continuous TSA system in which the adsorbent circulates through uptake and regeneration zones. The third flowsheet depicts a simulated moving-bed (SMB) system, in which the adsorbent is in a fixed bed and the feed, product, and regenerant entry and exit are shifted synchronously via valves such as in UOP's Sorbex^{SM} processes.

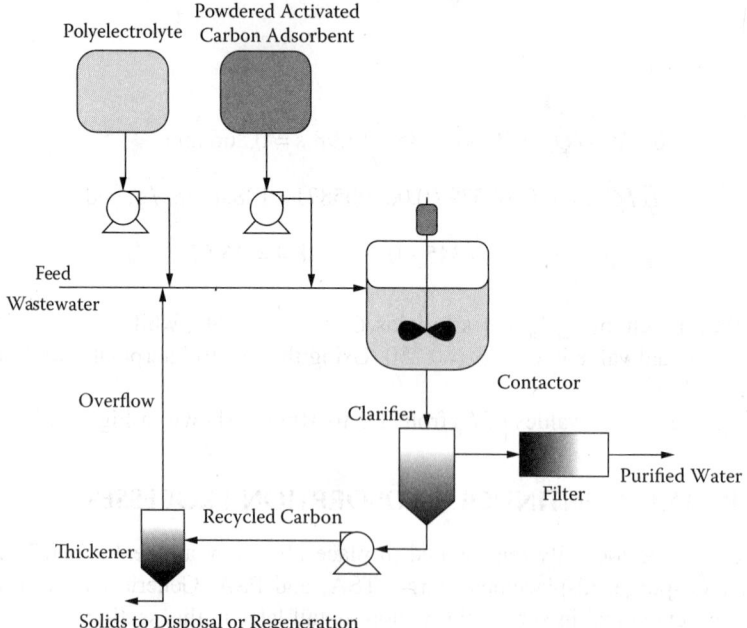

FIGURE 14.16 Powdered activated carbon system for cleanup of wastewater.

Adsorption

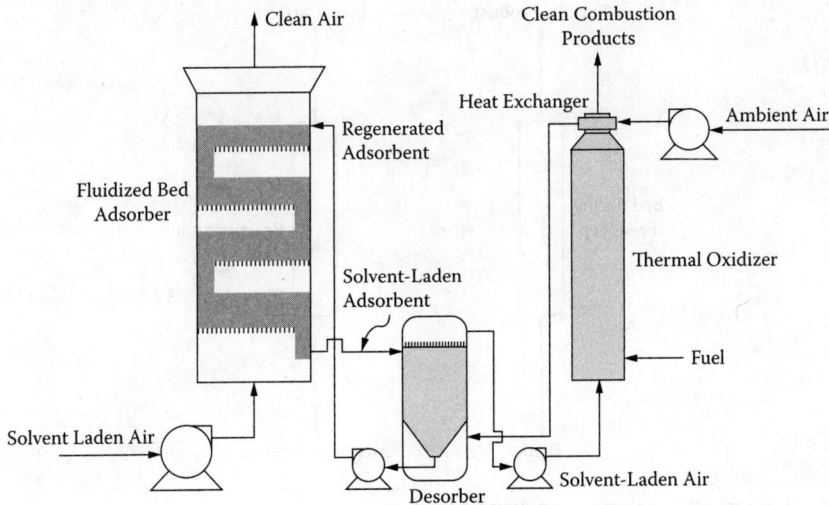

FIGURE 14.17 "Blizzard" adsorption system for cleanup of solvent-laden gas, by On-Demand Environmental Systems, Inc.

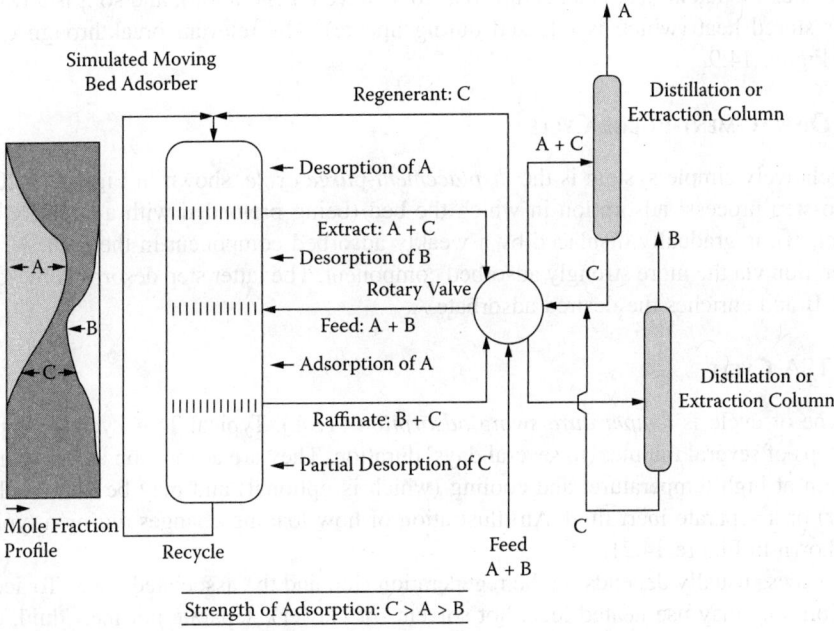

FIGURE 14.18 Simulated moving-bed adsorption system, after UOP Sorbex concept.

Generalizations about the relative merits of cyclic and continuous processes are nearly futile. Few applications are amenable to both. Rather, one or the other will usually be a clear choice. The criteria for making that choice boil down to cost and performance, and it is not possible to assert which version will prevail except under extreme conditions.

14.6.1 Inert-Purge Cycle

The simplest cycle is the so-called *inert-purge cycle*, shown in Figure 14.19. It is basically a two-step process of a few minutes' or hours' duration. First is adsorption (often with heat evolution),

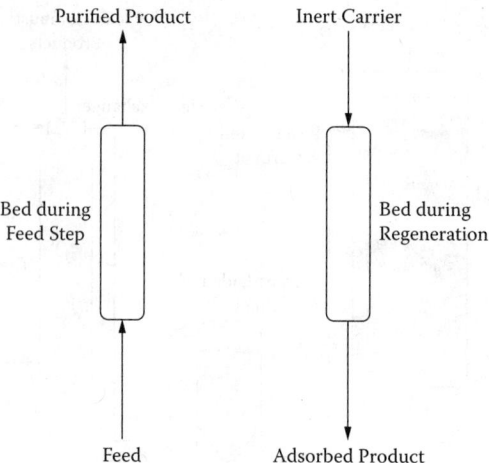

FIGURE 14.19 Simple two-column inert-purge adsorption system.

followed by regeneration with an inert fluid. It essentially transfers the adsorbed constituent from one stream to another, but in so doing it dilutes (instead of enriching) adsorbate. It also cools by desorption (i.e., the latent heat is taken up so as to achieve regeneration), and so it is advantageous to recover stored heat (which is released during uptake). The relevant breakthrough curves are shown in Figure 14.9.

14.6.2 Displacement-Purge Cycle

Another relatively simple system is the *displacement-purge cycle,* shown in Figure 14.20. This is also a two-step process: adsorption in which the bed (being preloaded with a strongly adsorbed component, B), is gradually displaced by a weakly adsorbed component in the feed, A, followed by regeneration via the more strongly adsorbed component. The latter step desorbs A by selectively adsorbing B and enriches the desired adsorbate A.

14.6.3 TSA Cycle

A third type of cycle is *temperature swing adsorption* (TSA). Typical TSA cycles comprise two or three steps of several minutes' to several days' duration. They are adsorption at low temperature, regeneration at high temperature, and cooling (which is optional) and may be done with solvent (or carrier) or a separate inert fluid. An illustration of how loading changes during a typical TSA cycle is shown in Figure 14.21.

TSA success usually depends on the regeneration step and the associated costs. To accomplish regeneration, one may use heated feed, hot solvent (or carrier), separate hot inert fluid, electrical current, microwaves, or steam. Steam, although widely used, usually accomplishes regeneration by conveyance rather than displacement, so the effluent composition profile is not sharp. Nevertheless, the adsorbate is enriched because the amount of fluid required for desorption is much less than that during uptake. The energy requirements of regeneration strongly affect overall cost. A typical cycle is shown in Figure 14.22.

Example 14.5 TSA Solvent Recovery

Consider *n*-heptane–laden air at 5000 ppm, 1000 metric tons of solvent/yr (= 125 kg solvent/hr), operated 8000 hr/yr. The TSA cycle duration is 6 hr (half adsorption, half desorption), and the equipment consists of two parallel beds operating sequentially.

Adsorption

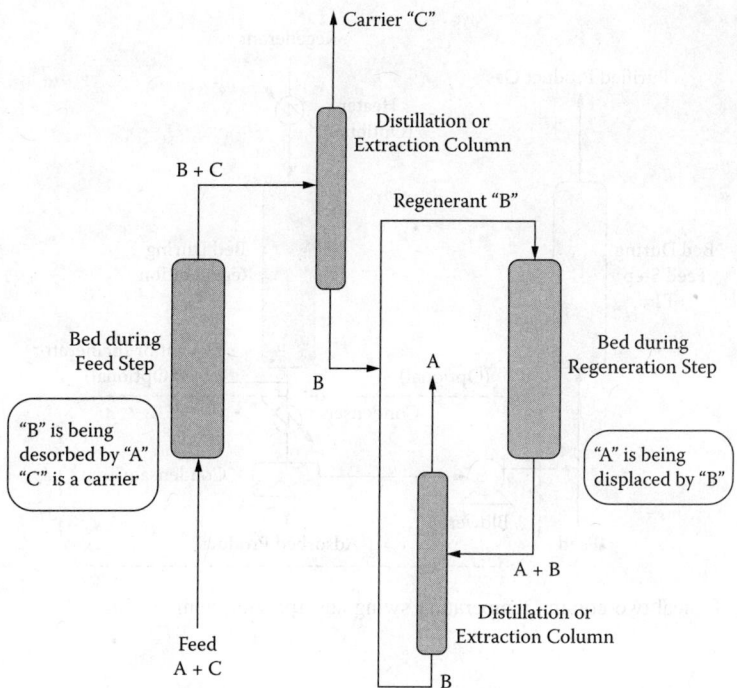

FIGURE 14.20 Two-column displacement-purge adsorption system.

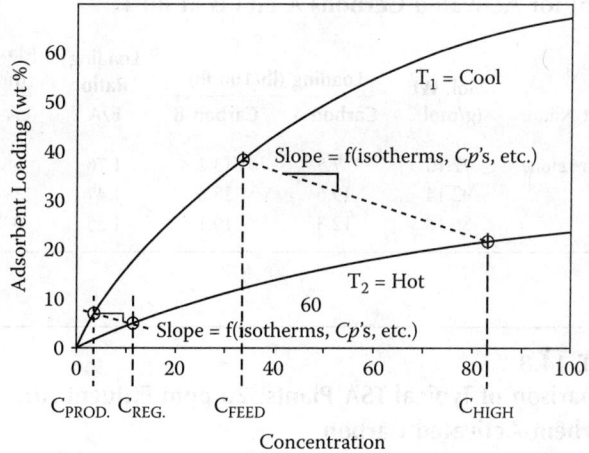

FIGURE 14.21 Isotherms at increasing temperatures, showing operating lines for a TSA cycle.

Data for typical systems are provided in Table 14.8. There we find, for n-heptane, the working capacity is 6 kg/100 kg loading, and the steam requirement is 4.3 kg steam/kg solvent. A mass balance yields an estimate for the required amount of adsorbent = 6250 kg/bed, and the required steam flow rate = 537.5 kg/hr.

14.6.4 PSA Cycle

The fourth and final type of system reviewed is *pressure swing adsorption* (PSA). A very simple cycle is shown in Figure 14.23. The factors governing PSA are basically no different from those

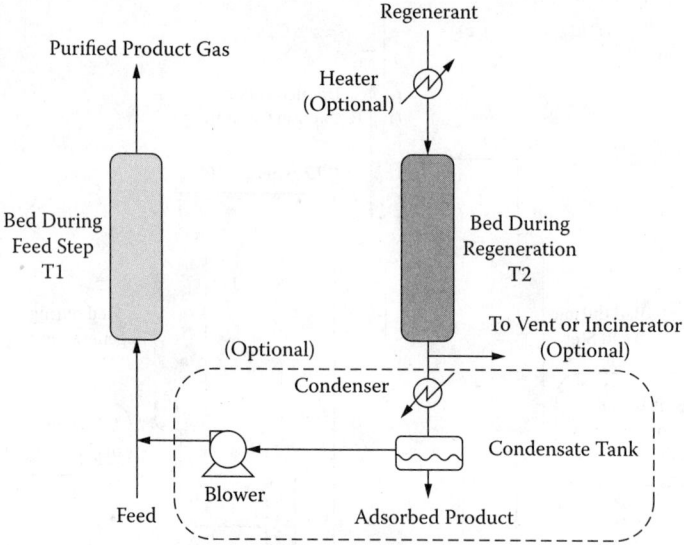

FIGURE 14.22 Typical two-column temperature swing adsorption system.

TABLE 14.7
Estimated Loadings and Service Times of Emitted Components (at 100 ppm each) for Activated Carbons A and B at 80°F

Component Name	Mol. Wt (g/mol)	Loading (lb/100 lb) Carbon A	Loading (lb/100 lb) Carbon B	Loading Ratio B/A	Max. On Line (hr/100 1b) A	Max. On Line (hr/100 1b) B
Methyl ethyl ketone	72.10	7.5	13.2	1.76	683	1203
Toluene	92.14	19.6	28.8	1.47	1398	2054
n-Hexane	86.18	12.3	19.1	1.55	938	1456

TABLE 14.8
Comparison of Typical TSA Plants: 20 ppm Effluent Air, Adsorbent-Activated Carbon

	Inlet Concentration (ppm)	Working Capacity (wt% of bed)	Steam (kg/kg)
Methylene dichloride	10,000	17	1.4
Trichloroethylene	5,000	20	1.8
Tetrahydrofuran	5,000	9	2.3
Ethyl acetate	5,000	13	2.1
Methyl isobutyl ketone	2,000	9	3.5
n-Hexane	5,000	8	3.5
n-Heptane	5,000	6	4.3
Toluene	4,000	9	3.5

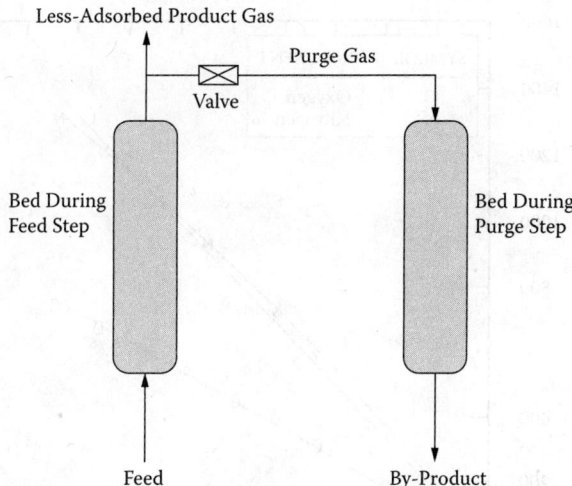

FIGURE 14.23 A simple two-column pressure swing adsorption system.

of ordinary adsorption. For example, the tendency of an adsorbent to take up one component selectively from a mixture is crucial to ordinary adsorption as well as to PSA. It is easy to envision the magnitude of the adsorbent *selectivity* by viewing the isotherms and/or by examining the effective diffusivities of each component. The isotherms indicate the *equilibrium* selectivity of the adsorbent, while the effective diffusivities provide a measure of the *kinetic* selectivity of the adsorbent. This section deals mainly with the former, although the latter is discussed briefly.

It is primarily the span of absolute pressures, coupled with differences in mole fractions (caused by pressure shifts and/or by admitting different streams to the column), that drive PSA separations. Equilibrium selectivity can cause composition shifts to occur simply by changing the pressure. To illustrate this point, consider an adsorbent bed of zeolite 5A filled with air (assumed to be only oxygen and nitrogen) at 3 atm. The mole fractions in the gas phase of oxygen and nitrogen are 0.21 and 0.79, and their initial concentrations are 24.1 and 90.8 mol/m^3, respectively. The isotherms on zeolite 5A at 45°C are plotted in Figure 14.24. Assuming that equilibrium exists, the respective loadings are 109 and 748 mol/m^3. Subsequently depressurizing the bed to 0.5 atm will shift the gas-phase composition approximately to the ratio of moles evolved. On that basis, a simple material balance shows that the new gas-phase mole fractions would be 0.137 and 0.863 for oxygen and nitrogen, respectively. More precise estimates, obtained by solving the governing PDEs, are 0.080 and 0.920 [see Equation (14.62)]. This shift, although substantial, is less than the shift that occurs on uptake, say, during the feed step. In that step, it is possible to shift the nitrogen mole fraction from less than 0.003 to 0.79. It is surprisingly easy and exciting to develop PSA systems in which such shifts occur, creating high-performance gas separations.

Skarstrom (1972), widely recognized as the inventor of PSA, cited four principles; (1) Employ short cycles and low throughput to conserve the heat of adsorption within the bed. (2) Regenerate at low pressure using some of the product for countercurrent purge. (3) Use an equal (or greater) volume for purge as fed during a cycle. (4) To obtain a pure product, the pressure ratio $\mathcal{P}\,(= P_H/P_L)$ should exceed the inverse of the light component feed mole fraction, $1/y_{BF}$. These rules of thumb continue to be useful but have been reinforced and supplemented by simple theoretical expressions that have been experimentally validated.

Various terms characterize PSA performance. Chief among them are product purity, product recovery, adsorbent productivity, and energy requirements. The purity and recovery normally refer to the light product or sometimes the heavy product. Purity is expressed in terms of "the number of nines," e.g., five nines would be 99.999% contaminant free. Frequently, only specific contaminants are considered in the remainder, leading to delusions of absolute purity. Recovery is expressed

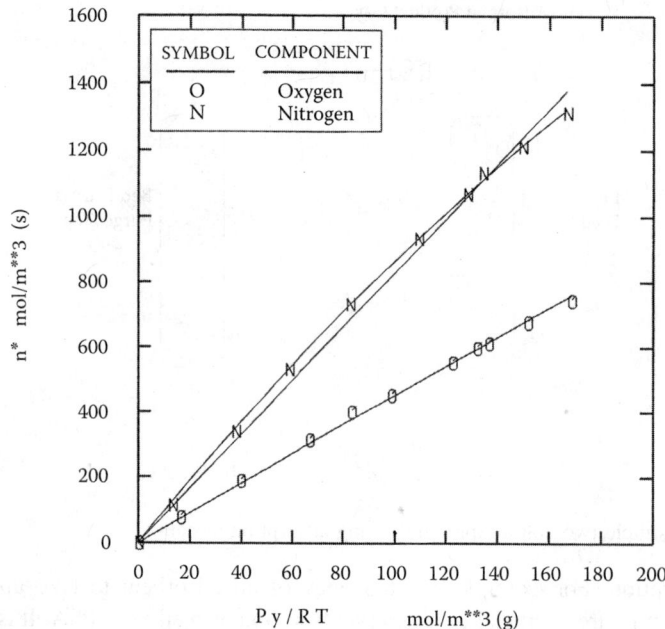

FIGURE 14.24 Isotherms of oxygen and nitrogen on zeolite 5A at 45°C.

as the fraction (or percentage) of the desired component in the product with respect to that in the feed. For example, the fractional recovery of the light component in the high-pressure product might be denoted $R_{HB} = y_{HB}\dot{Q}_{HB}/y_{FB}\dot{Q}_{FB}$, where the $\dot{Q}$'s represent molar flow rates. Adsorbent productivity is normally related to the space time and mass of adsorbent, as $P_{ads} = \dot{Q}_{HB}/M_{HB}m_{ads}$, where M_{HB} represents the mean molecular weight of the product, and m_{ads} is the mass of adsorbent. Occasionally, the units of productivity are expressed in terms of the volumetric ratio of flow to adsorbent rather than the mass ratio.

Common PSA cycles consist of 2 to 12 steps of a few seconds' to several minutes' duration. The most common steps are *pressurization* (with feed or purified light product), *feed* (at high pressure), *blowdown, and purge* (with purified product at low pressure). *Feed* involves a "wave" of the more adsorbable component moving toward the product end at high pressure. Flow stops when breakthrough is imminent and usually yields the less strongly adsorbed, or "light," product. Regeneration always occurs by *blowdown* or *depressurization*, which may be supplemented by *purging*, so it is ready for *pressurization*. Blowdown usually yields the more strongly adsorbed, or "heavy," by-product. Rinse (at high pressure) is optional, inserted after the feed step that permits adsorption of the "heavy" component, which allows the residual feed to be recovered. The heavy component can be subsequently recovered by depressurizing. Operating two beds in parallel, but 180° out of phase, generally permits continuous feeding and production.

Most cycles exploit *equilibrium selectivity* differences between species. Some exploit differences in *intraparticle diffusion rates*. These are becoming more prevalent with advances in adsorbent modification techniques. Efficient PSA cycles result from a good adsorbent, full exploitation of the input power, and careful control. The input power is exploited fully when a gas product or by-product is used to accomplish tasks as "reversibly" as possible. The shift in adsorbent loading may be 1 to 10 wt% for concentrated feeds. Generally, the energy requirements of regeneration most strongly affect overall cost.

The ratio of absolute pressures, $\mathcal{P}$, governs performance, since it governs performance and power consumption. Isotherm curvature, which becomes relevant as absolute pressure increases, leads to poorer performance at higher absolute pressures. Flow rates and step times are closely coupled. To maximize adsorbent productivity, it is common to maximize the flow rate (and minimize the step time). Conversely, to achieve high purity and sharp breakthrough patterns, it is common to minimize the flow rate (and maximize the step time). Obviously, there exists an optimum dictated by economics and product quality constraints. The details of these trade-offs are beyond the scope of this chapter, but it is usual to maintain relatively sharp profiles at the expense of reduced adsorbent productivity.

Besides the effects mentioned above, the flow rate dramatically affects pressure drop. To compensate for the high flow rate needed to achieve high productivity, the column length and diameter may be manipulated for a given volume of adsorbent, as discussed later. Likewise, particle size may be involved in the choice of L/D and flow rates, as discussed below. Of course, the product of flow rate and pressure drop contributes to power consumption. The other variables directly affecting pressure drop, besides flow rate, particle size, bed diameter, and bed length, are adsorbent void fraction and fluid viscosity and density. The feed composition dramatically affects performance, especially the product recovery. It is always the case that the maximum possible recovery of a component increases as its mole fraction in the feed increases. In some situations with adsorbates having relatively low volatility, caution is warranted. This is because performance may deteriorate near the dewpoint pressure due to adsorption hysteresis, thermal effects related to conserving the heat of adsorption, and (in extreme cases) the possibility of two-phase flow.

In view of the many properties and parameters that affect PSA performance, it is possible to characterize the adsorbates and adsorbent for a prospective system in either very simple or very complex terms. The simplest version requires only information regarding the adsorption isotherms and adsorbent void fraction. In the case of linear isotherms, the properties can be combined as in Equation (14.6),

$$\beta = \beta_A / \beta_B \tag{14.52}$$

where the terms β_A and β_B can be determined via

$$\beta_i = \frac{\varepsilon}{\varepsilon + (1-\varepsilon)K_i'} \tag{14.53}$$

The values of K_i' and ε' can be measured independently in separate batchwise experiments. In fact, the isotherm slopes for oxygen and nitrogen with zeolite 5A at 45°C, as shown in Figure 14.24, are 4.51 and 8.24, respectively. The void fraction was determined to be 0.478, and the bulk density was 810.0 kg/m³. Thus, the values for that system are $\beta_A = 0.100$, $\beta_B = 0.169$, $\beta = 0.593$. Alternatively, β may be determined for many bulk separations (e.g., oxygen from air) via a so-called breakthrough experiment. In such an experiment, a fixed bed of adsorbent having the same (or scaled-down) features of the full-scale bed is initially pressurized with the light component. Then the feed is admitted to the bed at the same pressure. By monitoring just the flow rates, the value of β may be found from

$$\beta = 1 - \frac{(1-\dot{Q}_P/\dot{Q}_F)}{y_{AF}} \tag{14.54}$$

where $y_{AP} = 0$, the bed pressure is kept constant with $\Delta P = 0$, and $\dot{Q}$'s are the constant volumetric flow rates.

Two distinct approaches can be taken to model a PSA system: develop a rigorous model that accounts for all significant effects, or settle for the simplest possible model and add terms if necessary. One must balance the fact that the former type requires more time to develop, a variety of empirical coefficients, and a powerful computer, while the latter may ignore effects that *could* be important, but it is generally simple and quick to implement. For equilibrium-based separations, the latter is in most ways adequate. For kinetics-based separations, the former is essential.

Fortunately, many applications can be predicted accurately via the simplest models. They ignore transport phenomena, mainly account for mass conservation, and are called equilibrium models. Though the concept is simple, such a model can account for a wide variety of effects, including flow, pressure, and composition variations, of course, as well as nonlinear isotherms and dispersive effects due to dead space, say, at the entrance or exit of the adsorbent bed.

The model considered here is the simplest equilibrium model that applies to many cases; it is very simple to use. The coefficients required for this model have already been introduced as β, β_A, and β_B. Furthermore, in tests conducted to date, the predictions have agreed with the detailed experimental observations. The theory usually reduces to simple performance equations that relate the operating conditions and design parameters. A fairly complete review of such models appears in the book by Ruthven et al. (1994). The model proposed by Knaebel and Hill (1985) is explained briefly here.

The inherent assumptions of this equilibrium theory are

1. Local equilibrium (i.e., at each axial location) is achieved instantaneously between the adsorbent and adsorbates.
2. The feed is a binary, ideal gas mixture.
3. Axial dispersion within and at the entrance and exit of the adsorbent bed is negligible.
4. Axial pressure gradients are negligible.
5. There are no radial velocity or composition gradients.
6. Temperature is constant.
7. One hundred percent of the adsorbent is utilized during the feed and purge steps.
8. Pressure is constant during the feed and purge steps.
9. The isotherms are linear and uncoupled.

The following individual component molar balance is similar to Equation (14.15). For a binary mixture, two equations are coupled in the gas-phase but are assumed independent in the adsorbed phase:

$$\varepsilon\left(\frac{\partial Py_i}{\partial t} + \frac{\partial vPy_i}{\partial z}\right) + RT(1-\varepsilon)\frac{\partial n_i}{\partial t} = 0 \qquad (14.55)$$

The general subscript i refers to component A or B. By convention, the more strongly adsorbed component is identified as component A. The linear isotherm equation gives the moles adsorbed per unit volume of adsorbent in terms of the gas-phase concentration in moles per unit open volume:

$$n_i = K'_i Py_i / RT \qquad (14.56)$$

Combining Equations (14.55) and (14.56) yields

$$\frac{\partial Py_i}{\partial t} + \beta_i \frac{\partial vPy_i}{\partial z} = 0 \qquad (14.57)$$

When pressure varies, and when one end is closed (e.g., during blowdown or pressurization), the overall material balance can be solved to find the interstitial velocity in the packed bed, as

$$v = \frac{-z}{\beta_B[1+(\beta-1)y_A]} \frac{1}{P}\frac{dP}{dt} \qquad (14.58)$$

Alternatively, when pressure is fixed, the following result relates the velocities and compositions at any two points, as long as the composition profile is continuous between them, i.e.,

$$\frac{v_2}{v_1} = \frac{1+(\beta-1)y_{A_1}}{1+(\beta-1)y_{A_2}} \qquad (14.59)$$

The simplicity of equilibrium-based theories results from the fact that the coupled first-order partial differential equations (material balances) can be recast as two ordinary differential equations (but these are still coupled). The mathematical technique employed is called the method of characteristics. The results are

$$\frac{dz}{dt} = \frac{\beta_A v}{1+(\beta-1)y_A} \qquad (14.60)$$

$$\frac{dy_A}{dP} = \frac{(\beta-1)(1-y_A)y_A}{[1+(\beta-1)y_A]P} \qquad (14.61)$$

Equation (14.60) defines *characteristic* trajectories in the z, t-plane. These are paths along which the composition varies, if at all, according to Equation (14.61). Thus, the latter expresses the dependence of composition on pressure along those characteristics. It should be clear from Equation (14.61) that, when pressure is constant, composition is fixed along each characteristic.

Integrating Equation (14.61) yields the composition shift due to a pressure shift as follows:

$$\frac{y_{A\,final}}{y_{A\,initial}} = \left[\frac{1-y_{A\,final}}{1-y_{A\,initial}}\right]^{\beta} \mathcal{P}^{\beta-1} \qquad (14.62)$$

The prediction mentioned earlier concerning the composition shift of air in a PSA system was based on $\beta = 0.593$, as explained earlier (see Equations (14.52) and (14.53)).

Commonly, the purge and blowdown steps are governed by Equations (14.60) and (14.61). To analyze other steps typically requires additional equations that reflect the feature that enables PSA (as well as other adsorption processes) to operate efficiently. The feed, rinse, and (possibly) pressurization steps involve uptake of the more strongly adsorbed component (while the blowdown and purge steps usually do not). This phenomenon results in a sharp concentration front, sometimes called a *constant pattern* profile or a *shock wave*. Usually, it is the velocity of this sharp front through the packed bed (of a specific length) that governs the duration of the particular step.

The necessary condition for such a front to exist is that the feed must be enriched in the heavy component, compared to the initial column contents. Likewise, the velocity is controlled by the molar (or volumetric) flow rate into and out of the system; hence, the material balance is also affected.

The velocity of the shock wave depends on the interstitial fluid velocities at the leading and trailing edges of the wave. These are related, since the shock wave velocities for both components A and B are equal, for cases in which pressure varies:

$$v_{SH} = \frac{-\beta_A z}{[1+(\beta-1)y_{A_1}][1+(\beta-1)y_{A_2}]} \frac{1}{P} \frac{dP}{dt} \qquad (14.63)$$

When pressure is fixed, the following expression is obtained:

$$v_{SH} = \beta_A \frac{u_2 y_{A_1} - u_1 y_{A_1}}{y_{A_2} - y_{A_1}} \tag{14.64}$$

Propagation of this composition front or shock wave dominates the feed step and pressurization with feed.

By combining the equations for the various steps, expressions for product recovery can be obtained. Some examples are given here. The result for the light product recovery of the conventional four-step cycle is

$$R_B = (1-\beta)\left[1 - \frac{1}{\rho y_{B_F}}\right] \tag{14.65}$$

The light and heavy product recoveries from the five-step cycle including rinse are

$$R_B = 1 - \frac{1}{\rho y_{B_F}} \tag{14.66}$$

$$R_A = 1 - \frac{1}{(1-\beta)\rho y_{A_F}} \tag{14.67}$$

Although many PSA applications exhibit some degradation of performance from that predicted by the equilibrium theory, only those that *rely* on differences in intraparticle diffusivities *require* detailed mathematical models. Carbon molecular sieves and special zeolitic molecular sieves have pore sizes tailored to admit compact molecules readily and to exclude marginally larger molecules. For example, splitting nitrogen from air by a carbon molecular sieve is reported to exhibit a ratio of effective diffusivities of oxygen to nitrogen of about 45. There happens to be essentially no equilibrium selectivity in that case. Similarly, for the same separation, a modified 4A zeolite exhibited an oxygen-to-nitrogen diffusion rate selectivity of about 50, but the equilibrium selectivity was about 2.0 in the reverse direction.

BIBLIOGRAPHY

Basmadjian, D., 1997. *The Little Adsorption Book*. Boca Raton, FL: CRC Press.
Brunauer, S., L. S. Deming, W. E. Deming, and E. Teller, 1940. On a Theory of the van der Waals Adsorption of Gases, *J. Am. Chem. Soc*. 62, 1723–1732.
Breck, D. W., 1974. *Zeolite Molecular Sieves*. New York: John Wiley & Sons.
Cheremisinoff, N. P. and P. N. Cheremisinoff, 1993. *Carbon Adsorption for Pollution Control*, Englewood Cliffs, NJ: Prentice Hall.
Clark, R. M. and B. W. Lykins, 1991. *Granular Activated Carbon*. Chelsea, MI: Lewis Publishers.
Collins, J. J., 1967. *Chem. Eng. Progress, Symp. Ser*. 63(74): 31.
Cooney, D. O., 1999. *Adsorption Design for Wastewater Treatment*. Boca Raton, FL: CRC Press.
Do, D. D., 1998. *Adsorption Analysis: Equilibria and Kinetics*. London: Imperial College Press.
Dobbs, R. A. and J. M. Cohen, 1980. *Carbon Adsorption Isotherms for Toxic Organics*, EPA-600/8-80-023.
Edwards, M. F. and J. F. Richardson, 1968. *Chem. Eng. Sci*. 23: 109.
Ergun, S., 1952. *Chem. Eng. Prog*. 48, 89–94.
Ganetsos, G. and P. E Barker, eds., 1993. *Preparative and Production Scale Chromatography*, Chromatographic Sci. Ser., vol. 61. New York: Marcel Dekker.
Gregg, S. J. and K. S. W. Sing, 1982. *Adsorption, Surface Area and Porosity*. London: Academic Press.
Hougen, O. A. and W. R. Marshall, 1947. *Chem. Eng. Progress* 43: 197.

Jayaraj, K. and C. Tien, 1984. *Proc. Env. Engng. Conf.* ASCE, New York, 394.
Kage, H. and C. Tien, 1987. *Ind. Eng. Chem. Res.* 284,
Kärger, J. and D. M. Ruthven, 1992. *Diffusion in Zeolites and other Microporous Solids.* New York: John Wiley & Sons.
Keller, G. E. II and R. T. Yang, 1989. *New Directions is Sorption Technology.* Stoneham, MA: Butterworths.
Knaebel, K. S., D. M. Ruthven, J. L. Humphrey, and R. W. Carr, 1999. Adsorption Technologies, in *Emerging Separation and Separative Reaction Technologies for Process Waste Reduction*, P.P. Radecki et al., eds. New York: American Inst. of Chem. Eng.
Knaebel, K. S. and F. B. Hill, 1985. Pressure Swing Adsorption: Development of an Equilibrium Theory, *Chem. Eng. Sci.* 40, 2351–2350.
Kohl, A. L. and F. C. Riesenfeld, 1985. *Gas Purification.* Houston: Gulf Publishing Co.
LeVan, M. D., G. Carta, and C. Yon, 1997. Adsorption and Ion Exchange, in *Perry's Chemical Engineers' Handbook*, 7th ed. New York: McGraw Hill.
Liapis, A. I, ed., 1987. *Fundamentals of Adsorption.* New York: Engineering Foundation.
Mersmann, A. B. and S. E. Scholl, eds., 1991. *Fundamentals of Adsorption.* New York: Engineering Foundation.
Michaels, A. S., 1952. *Ind. Eng. Chem.*, 44(8): 1922.
Moon, H., H. C. Park, and C. Tien, 1991. *Chem. Eng. Sci.* 46, 23.
Myers, A. L. and G. Belfort, eds., 1984. *Fundamentals of Adsorption.* New York: Engineering Foundation.
Myers, A. L. and J. M. Prausnitz, 1965. *AJChE J.* 11: 121.
Perrich, J. R., 1981. *Activated Carbon Adsorption for Wastewater Treatment*, Boca Raton, FL: CRC Press.
Reid, R. C., J. M. Prausnitz and B. E. Poling, 1987. *The Properties of Gases and Liquids.* New York: McGraw Hill.
Rosen, J. B., 1954. Kinetics of a Fixed Bed System for Solid Diffusion into Spherical Particle, *End. Eng. Chem.* 46, 1590.
Rouquerol, F., J. Rouquerol, and K. S. W. Sing, 1999. *Adsorption by Powders and Porous Solids.* London: Academic Press.
Rousseau, R. W., ed., 1987. *Handbook of Separation Process Technology.* New York: John Wiley & Sons.
Ruthven, D. M., 1984. *Principles of Adsorption & Adsorption Processes.* New York: John Wiley & Sons.
Ruthven, D. M., S. Farooq, and K. S. Knaebel, 1994. *Pressure Swing Adsorption.* New York: VCH Publishers.
Said, A. S., 1981. *Theory and Mathematics of Chromatography.* Heidelberg: Alfred Hüthig, Verlag.
Schweitzer, P. A., ed., 1988. *Handbook of Separation Techniques for Chemical Engineers.* New York: McGraw Hill.
Sengupta, A. K., ed., 1995. *Ion Exchange Technology.* Lancaster, PA: Technomic.
Sherwood, T. K., R. L. Pigford, and C. R. Wilke, 1975. *Mass Transfer.* New York: McGraw Hill.
Slejko, F. L., 1985. *Adsorption Technology.* New York: Marcel Dekker.
Smallwood, I., 1993. *Solvent Recovery Handbook.* London: Edward Arnold.
Suzuki, M., 1990. *Adsorption Engineering.* Amsterdam: Elsevier.
Suzuki, M., ed., 1993. *Fundamentals of Adsorption IV.* Tokyo: Kodansha.
Szostak, R., 1992. *Handbook of Molecular Sieves.* New York: Van Nostrand Reinhold.
Tien, C., 1994. *Adsorption Calculations and Modeling.* Newton, MA: Butterworth-Heinemann.
Valenzuela, D. B. and A. L. Myers, 1989. *Adsorption Equilibrium Data Handbook.* Englewood Cliffs, NY: Prentice Hall.
van Bekkum, H., E. M. Flanigen, and J.C. Jansen, eds., 1991. *Introduction to Zeolite Science and Practice.* Amsterdam: Elsevier.
Vansant, E. F., ed., 1994. *Separation Technology.* Amsterdam: Elsevier.
Vansant, E. F., 1990. *Pore Size Engineering in Zeolites.* Chichester, UK: John Wiley & Sons.
Vansant, E. F. and R. Dewolfs, eds., 1989. *Gas Separation Technology.* Amsterdam: Elsevier.
Wankat, P. C., 1986. *Large Scale Adsorption and Chromatography* (2 vols.). Boca Raton, FL: CRC Press.
Wankat, P. C., 1990. *Rate Controlled Separations.* London: Elsevier Applied Science.
Wen, C. Y. and L. T. Fan, 1975. *Models for Flow Systems and Chemical Reactors.* New York: Marcel Dekker.
Wheeler, A. and A. J. Robell, 1969. *J. Catal.* 13: 299–305.
Yang, R. T., 1987. *Gas Separation by Adsorption Processes.* Boston: Butterworths.
Yaws, C. L., L. Bu, and S. Nijhawan, 1995. *Environ. Engng. World* 1(3): 16–20.
Yoshida, F. D. Ramaswami, and O.A. Hougen, 1962. *AIChE J.* 8: 5.

15 Process Control

James B. Riggs, William J. Korchinski, and Arkan Kayihar

CONTENTS

- 15.1 Introduction...1175
  - 15.1.1 Introduction..1175
  - 15.1.2 General Dynamic Behavior 1177
  - 15.1.3 First-Order Plus Deadtime (FOPDT) Model 1179
  - 15.1.4 Closed-Loop Dynamic Behavior...1181
- 15.2 Control Loop Hardware and Troubleshooting 1182
  - 15.2.1 Introduction..1182
  - 15.2.2 Distributed Control System 1184
    - 15.2.2.1 Background..1184
    - 15.2.2.2 Structure of a DCS 1184
    - 15.2.2.3 Approach 1185
    - 15.2.2.4 Programmable Logic Controllers 1186
    - 15.2.2.5 Fieldbus Technology 1186
  - 15.2.3 Actuator Systems (Final Control Elements) 1187
    - 15.2.3.1 Control Valves 1187
    - 15.2.3.2 Valve Actuators 1190
    - 15.2.3.3 I/P Transmitters 1190
    - 15.2.3.4 Optional Equipment 1190
  - 15.2.4 Sensor Systems...1191
    - 15.2.4.1 Temperature Measurements 1192
    - 15.2.4.2 Thermowells 1192
    - 15.2.4.3 Repeatability, Accuracy, and Dynamic Response 1192
    - 15.2.4.4 Pressure Measurements 1193
    - 15.2.4.5 Flow Measurements 1193
    - 15.2.4.6 Level Measurements...1193
    - 15.2.4.7 Chemical Composition Analyzers..1193
    - 15.2.4.8 Sampling System 1193
    - 15.2.4.9 Transmitters 1194
  - 15.2.5 Troubleshooting Control Loops 1194
    - 15.2.5.1 Overall Approach 1195
    - 15.2.5.2 Final Control Element..1195
    - 15.2.5.3 Sensor Systems...1197
    - 15.2.5.4 Controller/DCS System..1197
    - 15.2.5.5 Process 1199
    - 15.2.5.6 Testing the Entire Control Loop 1199
- 15.3 PID Controllers..1201
  - 15.3.1 Introduction..1201
  - 15.3.2 PID Algorithms..1201
  - 15.3.3 Analysis of P, I, and D Action 1206

		15.3.3.1	Proportional Action	1206
		15.3.3.2	Integral Action	1206
		15.3.3.3	Derivative Action	1207
	15.3.4	Controller Design Issues		1208
		15.3.4.1	P-Only Control	1208
		15.3.4.2	PI Control	1208
		15.3.4.3	PID Control	1208
	15.3.5	Analysis of Typical Control Loops		1209
		15.3.5.1	Flow Control Loop	1209
		15.3.5.2	Level Control Loop	1210
		15.3.5.3	Pressure Control Loop	1211
		15.3.5.4	Temperature Control Loop	1211
		15.3.5.5	Composition Control Loop	1212
15.4	PID Tuning			1213
	15.4.1	Introduction		1213
	15.4.2	Tuning Criteria and Performance Assessment		1213
		15.4.2.1	Tuning Criteria	1213
	15.4.3	Effect of Tuning Parameters on Dynamic Behavior		1214
		15.4.3.1	PI Control	1214
		15.4.3.2	PID Control	1216
		15.4.3.3	Control Interval	1217
	15.4.4	Recommended Approach to Controller Tuning		1218
	15.4.5	Controller Reliability		1218
	15.4.6	Selection of Tuning Criterion		1219
	15.4.7	Tuning the Filter on Sensor Readings		1220
	15.4.8	Fast-Response Loops		1221
	15.4.9	Slow-Response Processes		1222
	15.4.10	PID Tuning		1225
	15.4.11	Level Controller Tuning		1225
15.5	Advanced PID Control			1227
	15.5.1	Introduction		1227
	15.5.2	Cascade Control		1227
	15.5.3	Ratio Control		1229
	15.5.4	Feedforward Control		1230
		15.5.4.1	Tuning	1231
		15.5.4.2	Overview	1232
	15.5.5	Inferential Control		1233
		15.5.5.1	Inferential Temperature Control for Distillation	1234
		15.5.5.2	Inferential Reaction Conversion Control	1234
		15.5.5.3	Inferential Measurement of the Molecular Weight of a Polymer	1235
		15.5.5.4	Soft Sensors Based on Neural Networks	1235
	15.5.6	Scheduling Controller Tuning		1236
		15.5.6.1	Nonstationary Behavior	1237
	15.5.7	Override/Select Control		1238
	15.5.8	Computed Manipulated Variable Control		1239
	15.5.9	Antiwindup Strategies		1240
	15.5.10	Bumpless Transfer		1241
	15.5.11	Split-Range Flow Control		1242
	15.5.12	MIMO Process Control		1242
		15.5.12.1	SISO Controllers and (c, y) Pairings	1242
		15.5.12.2	Tuning Decentralized Controllers	1245

15.6 Model Predictive Control 1246
    15.6.1 What Is MPC? 1246
    15.6.2 History of MPC 1248
    15.6.3 When Should MPC Be Used? 1248
    15.6.4 General Method for Designing a Multivariable Controller 1249
        15.6.4.1 Understand the Process 1249
        15.6.4.2 Design the Plant Test 1251
        15.6.4.3 Conduct the Plant Test and Collect the Data 1254
        15.6.4.4 Structure the Controller and Analyze the Data 1255
        15.6.4.5 Tune the Controller 1259
        15.6.4.6 Commission the Controller 1260
        15.6.4.7 Post Audit 1261
    15.6.5 Multivariable Controller Troubleshooting 1261
        15.6.5.1 Commonsense Approach 1262
        15.6.5.2 Detailed Troubleshooting—Use a Checklist 1262
        15.6.5.3 Is the Controller Still Making Money? 1264
References, Section 15.2 1264
References, Section 15.3 1264
References, Section 15.4 1264
Reference, Section 15.5 1264
References, Section 15.6 1264

15.1 INTRODUCTION

15.1.1 INTRODUCTION

The process control engineer is responsible for the safe operation of a plant while reliably and efficiently producing the desired product quality. The control engineer does this by transferring the variability that otherwise goes into important controlled variables to manipulated variables and less important controlled variables. In order to meet these objectives, process control engineers use a complete knowledge of the process to attain the most desirable performance.

The refining and chemical industries are intimately involved in the effort to obtain the most efficient operation of the plant. Minimizing product variability is often a key operational objective and is directly affected by the performance of the process control system. In fact, the performance of an overall process control system is often expressed in terms of the variability in the products produced by the process. Figure 15.1 shows the measurement of the impurity in a product for the original control system (case A). Case B represents the performance of a new control system that produces a more uniform product than for case A. In addition, since case B has lower variability, the average impurity level can be moved closer to the impurity specification (case C). Many times, operating closer to the limit allows greater production rates or lower utility usage, both of which result in increased profits. Other types of operational limits are encountered, resulting from environmental regulations, capacity limits on equipment, and safety limits. In a similar manner, operating close to these limits can be economically important.

An endothermic continuous stirred tank reactor (CSTR) is shown schematically in Figure 15.2. The symbols used in the control schematics are listed in Table 15.1 The controlled variable is the temperature of the product leaving the reactor, and the manipulated variable is the flow rate of steam to the heat exchanger, which adds heat to the recycle line. The final control element is the control valve and associated equipment on the steam line. The sensor is a temperature sensor/transmitter that measures the temperature of the product stream leaving the reactor. The controller compares the measured value of the product temperature with its desired temperature (setpoint) and makes changes to the control valve on the steam to the heat exchanger. The process is the

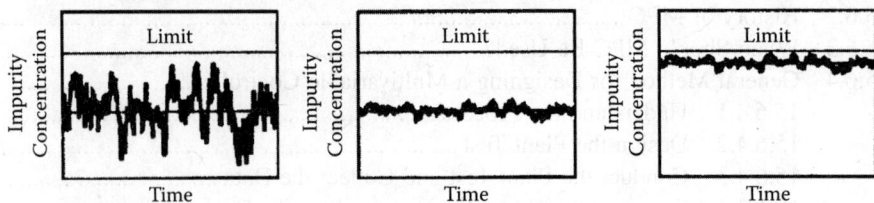

FIGURE 15.1 Comparison between impurity measurements and the upper limit on the impurity in a product for the original control system (case A), the improved control system with the original impurity setpoint (case B), and the improved control with new setpoint (case C).

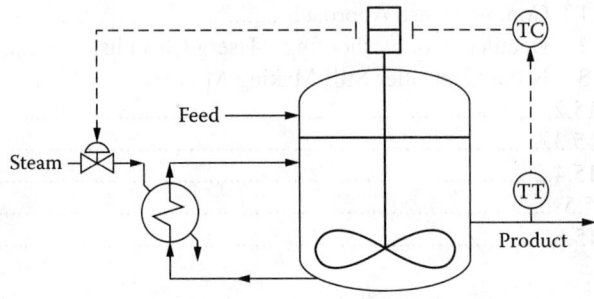

FIGURE 15.2 Schematic of an endothermic CSTR. TT = temperature sensor/transmitter, TC = temperature controller.

TABLE 15.1
Definition of Symbols for Control Diagrams

AC – analyzer controller (i.e., composition controller)
AT – analyzer transmitter (i.e., composition analyzer/transmitter)
DPC – differential pressure controller
DPT – differential pressure sensor/transmitter
HS – high select (this element selects the larger of two or more inputs)
LC – level controller
LS – low select (this element selects the lower of two or more inputs)
LT – level sensor/transmitter
PC – pressure controller
pHC – pH controller
pHT – pH sensor/transmitter
PT – pressure sensor/transmitter
RSP – remote setpoint (i.e., the setpoint calculated by another controller)
TC – temperature controller
TT – temperature sensor/transmitter
⊗ – summation block (minus signs for an input compartment denote subtraction)
+ – addition function (i.e., two inputs are added to yield the output)
× – multiplication function (i.e., two inputs are multiplied to yield the output)

Process Control

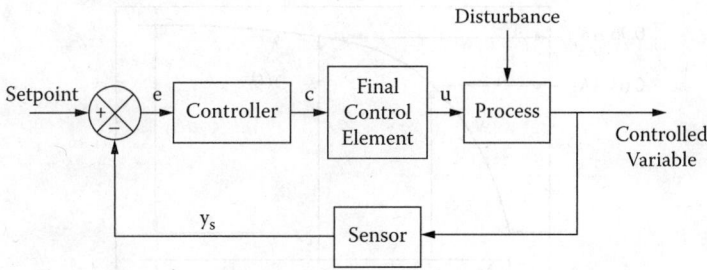

FIGURE 15.3 Block diagram of a generalized feedback system; e is the error from setpoint, c is the controller output, and u is the manipulated variable.

reaction mixture in the reactor and in the heat exchanger and the lines to and from the heat exchanger. Changes in the feed composition, feed rate, and quality of the steam are examples of disturbances to this process.

Figure 15.3 shows a block diagram of a generalized feedback control system for the system shown in Figure 15.2. That is, this example has a controller, a final control element, a process, and a sensor, in that order, along with feedback of the measured value of the controlled variable to the controller. In addition, the example process is affected by disturbances. Note that the sensor reading, y_s, is compared with the setpoint, and the controller chooses control action based on this difference. The final control element is responsible for implementing changes in the level of the manipulated variable. The "process" for a control loop is only the part of the system that determines the value of the controlled variable from the inputs. The overall process can be based on a number of processing units.

The symbol ⊗ in Figure 15.3 represents a summation block. The negative sign on the measurement of the controlled variable results in forming the difference between the setpoint and the measured value of the controlled variable, which is referred to as the error from setpoint. A block diagram of an open-loop process involves only the actuator, process, and sensor without the feedback of the measurement of the controlled variable to the controller.

15.1.2 General Dynamic Behavior

The simplest type of dynamic behavior results from a first-order process. Figure 15.4 shows the response of a first-order process (y) to a step change in u. The analytical solution of a first-order process subjected to a step input change of A is given by

$$y(t) = AK_p \left(1 - e^{-t/\tau_p}\right)$$

The process gain, K_p, and the size of the step change, A, determine the new steady-state value of y. The time constant, τ_p, determines the dynamic path the process takes as it approaches the new steady state, i.e., how long it takes to reach a new steady state. Note that 63.2% of the final change occurs in one time constant after the input change. Ninety-five percent occurs in three time constants, and 98% occurs in four time constants. The process gain is the steady-state change in y divided by the corresponding change in u, i.e.,

$$K_p = \frac{\Delta y}{\Delta u}$$

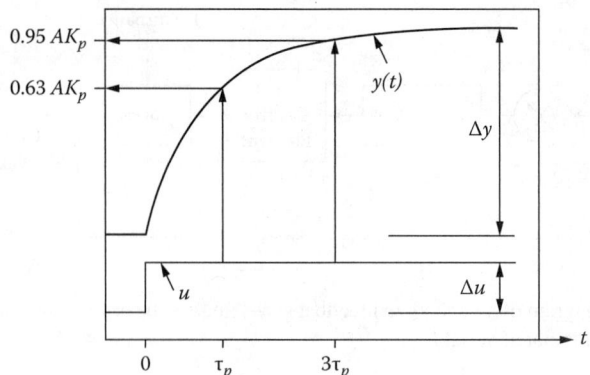

FIGURE 15.4 Dynamic response of a first-order process to a step input change.

The time constant for a well mixed vessel without chemical reaction is the volume of liquid in the vessel divided by the total volumetric feed rate, i.e., the residence time of the vessel.

Figure 15.5 shows the three general types of dynamic behavior of a second-order process, which can also be used to describe the dynamic behavior of feedback systems: overdamped, critically damped, and underdamped. Overdamped behavior is characterized by a monotonic approach to steady state. Underdamped behavior is characterized by an oscillatory approach to steady state. A critically damped response marks the boundary between overdamped and underdamped behavior.

Figure 15.6 shows the key features of an underdamped response. The rise time is the time required to first cross the final steady-state value. The response time or settling time is the time required for the response to settle to within ±5% of the steady-state change (±5% D). The decay ratio is the ratio C/B, which indicates how fast the oscillations damp out.

Distillation, absorption, and extraction columns have a number of combined stages, which yield high-order dynamic behavior. Figure 15.7 shows the dynamic response of 3, 5, and 15 combined stages. Note that the more stages that are combined, the slower the response becomes; i.e., the process exhibits a sluggish response. For the case corresponding to 15 stages, it requires a relatively large amount of time before a significant change in the output, y, can be observed. The time after an input change before a significant change in the output variable can be observed is an indication of the process deadtime.

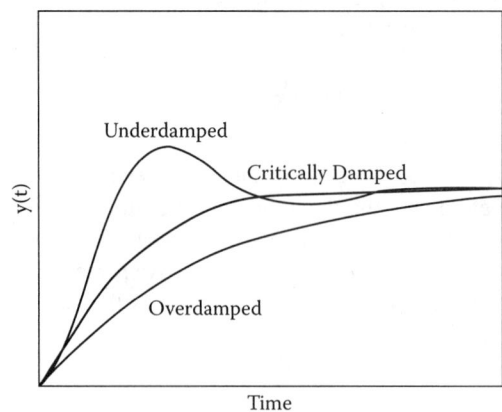

FIGURE 15.5 The three general types of dynamic behavior for a second-order process.

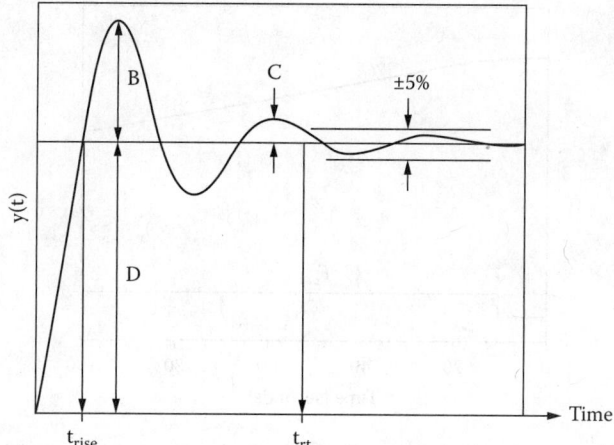

FIGURE 15.6 The general characteristics of an underdamped response.

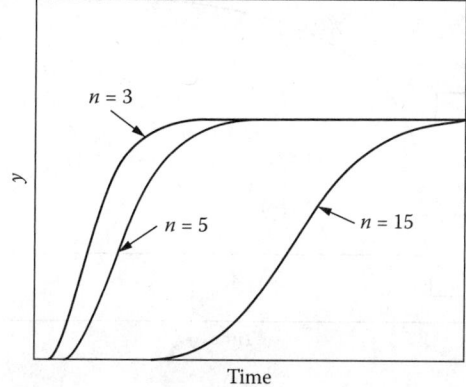

FIGURE 15.7 Dynamic response of several high-order systems.

If a process settles at a new steady state after an input change, the process is referred to as self-regulating. Levels in tanks, accumulators, and reboilers, and many pressure systems, behave as integrating processes. Consider the level in a tank for which both the flow in and the flow out are set independently. Initially, the flow out, F_{out}, is equal to the flow in, and the level is constant. Figure 15.8 shows the level as a function of time for a step change in the flow out at time equal to 10 s. Note that the level in the tank begins to decrease at a constant rate. This is an example of a non-self-regulating process, since the process does not move to a new steady state.

An inverse response can occur when opposing factors exist within a process for which there is a fast-acting, low-gain response combined with a slower-acting, high-gain response. An example of the dynamic response of an inverse-acting process is shown in Figure 15.9. Note that, initially, a decrease in $y(t)$ results, ultimately followed by an increase.

15.1.3 First-Order Plus Deadtime (FOPDT) Model

A FOPDT model is the combination of a first-order model with deadtime, which can be represented by the combination of the process gain (K_p), the process time constant (τ_p), and the process deadtime (θ_p). The process gain represents the steady-state change in the output of the process (y) for a change in the input (u). The time constant indicates how fast the process settles to the new steady

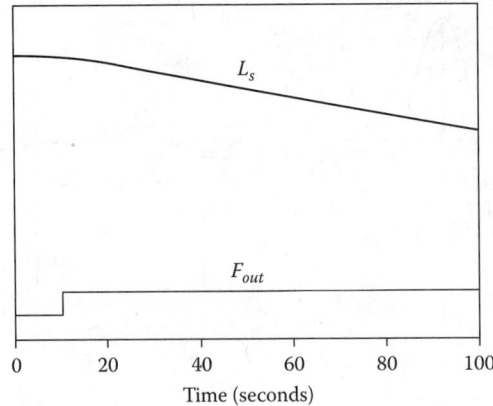

FIGURE 15.8 Dynamic response of a level in a tank to a change in the flow out of the tank.

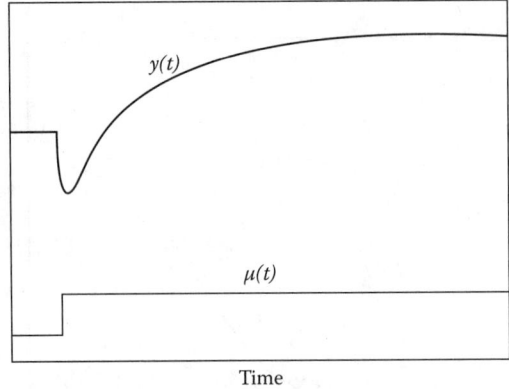

FIGURE 15.9 Dynamic response of an inverse-acting process.

state; i.e., the response time of the process is related to the time constant of the process. The deadtime indicates how soon after an input change has occurred that a noticeable change in y can be observed.

A step test (i.e., a step input change) in certain cases can be used to develop an FOPDT model. Figure 15.10 shows one such approach. First, identify the resulting change in y (i.e., Δy) and the step change in the input, Δu. Then, from the step response, identify the time required for one third of the total change in y to occur, $t_{1/3}$. Next, identify the time required for two thirds of the total change in y to occur, $t_{2/3}$. Then, the following estimates can be used:

$$\tau_p = \frac{t_{2/3} - t_{1/3}}{0.7}$$

$$\theta_p = t_{1/3} - 0.4 t_p$$

$$K_p = \frac{\Delta y}{\Delta u}$$

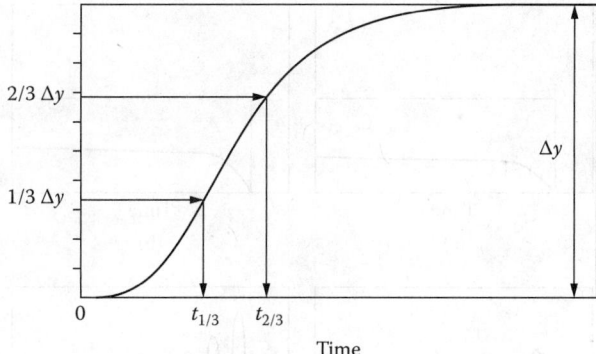

FIGURE 15.10 Graphical representation of an approach for determining the parameters of a FOPDT model.

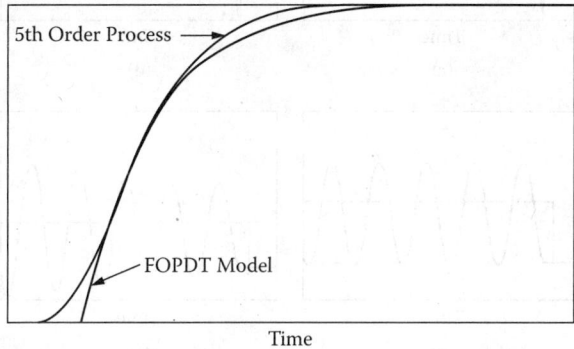

FIGURE 15.11 Comparison between a FOPDT model and an overdamped fifth-order process.

Note that $t_{1/3}$ and $t_{2/3}$ are based on the assumption that time is equal to zero when the step change in u is implemented. This modeling approach is particularly well suited for modeling high-order processes because the deadtime approximates the initial response of the high-order system before significant change has occurred. Figure 15.11 shows a fifth-order process and a FOPDT model that was selected to match this high-order process. Overall, the FOPDT model provides a good approximation for overdamped process behavior; therefore, since most industrial processes behave as overdamped systems, the FOPDT model is one of the best idealized models to represent industrial processes.

15.1.4 CLOSED-LOOP DYNAMIC BEHAVIOR

Figure 15.12 shows the general range of dynamic behavior of a feedback system. Figure 15.12a shows a sluggish response of a feedback process, which is also an overdamped response. Figure 15.12b demonstrates a critically damped response, which represents the transition between overdamped and oscillatory behavior. Figure 15.12c shows an oscillatory response, which is also underdamped behavior. Although Figure 15.12c shows a response that results in a rapid decay of the overshoot, Figure 15.12d shows the case in which damping occurs much more slowly, which is referred to as *ringing*. Figure 15.12e shows the case for sustained oscillations because the amplitude of the oscillations remains constant. Figure 15.12f demonstrates unstable behavior in which the amplitude of the oscillations grows with time.

Most feedback control loops exhibit the same stages of dynamic behavior as the aggressiveness of the controller is increased. Since most processes in the chemical industry are overdamped under open-loop conditions (no feedback control action applied), the process remains overdamped when

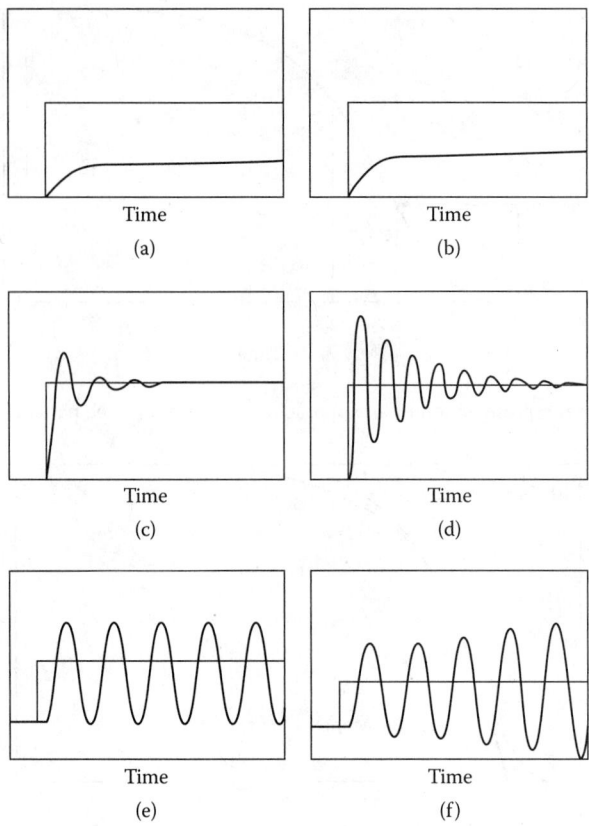

FIGURE 15.12 The general stages of dynamic behavior of a feedback system.

a small amount of feedback action is applied. As the aggressiveness of the controller is increased, the dynamic response becomes critically damped immediately before it becomes underdamped (exhibits oscillations). A further increase in the controller aggressiveness results in an increase in the rate of oscillations and a slower damping of the amplitude of the oscillations. As the controller aggressiveness is increased further, the amplitude of the oscillations eventually begins to increase, and unstable dynamic behavior results.

15.2 CONTROL LOOP HARDWARE AND TROUBLESHOOTING

15.2.1 INTRODUCTION

To apply control to a process, one measures the controlled variable and compares it to the setpoint and, based on this comparison, typically uses the actuator to make adjustments to the flow rate of the manipulated variable. The industrial practice of process control is highly dependent upon the performance of the actuator system (final control element) and the sensor system as well as the controller. If either the final control element or the sensor is not performing satisfactorily, it can drastically affect control performance regardless of controller action. Each of these systems (i.e., the actuator, sensor, and controller) is made up of several separate components; therefore, the improper design or application of these components, or an electrical or mechanical failure of one of them, can seriously affect the resulting performance of the entire control loop. The present description of these devices focuses on their control-relevant aspects. Later, troubleshooting approaches and control loop component failure modes are discussed.

Process Control

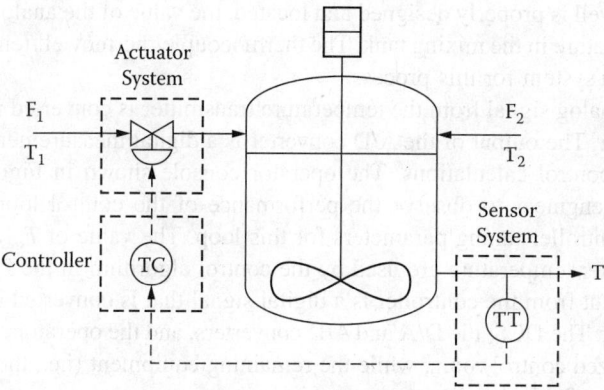

FIGURE 15.13 Schematic of the CST thermal mixer showing the actuator system, sensor system, controller, and the process.

Figure 15.13 is a schematic of a feedback control loop for a temperature controller on a mixer in which stream 1, with flow rate F_1 and temperature T_1, is mixed with stream 2, with flow rate F_2 and temperature T_2. This loop consists of a controller, a final control element, a process, and a sensor. Figure 15.14 is a schematic of the hardware that constitutes this feedback temperature control loop as well as the signals that are passed between the various hardware components. The sensor system shown in Figure 15.13 corresponds to the thermowell, thermocouple, and transmitter in Figure 15.14, while the actuator system in Figure 15.13 corresponds to the control valve, I/P converter, and instrument air system in Figure 15.14. Likewise, the controller in Figure 15.13 is made up of the analog-to-digital (A/D) and digital-to-analog (D/A) converters, the distributed control system (DCS), and the operator console in Figure 15.14. The abbreviations used in this paragraph are described in the next several sections.

A thermocouple is placed in a thermowell, which is in thermal contact with the process fluid leaving the mixing tank, providing a measurement of the temperature of the fluid inside the mixing tank. The temperature transmitter converts the millivolt signal generated by the thermocouple into a 4–20 mA analog electrical signal. When the thermocouple/transmitter system is calibrated prop-

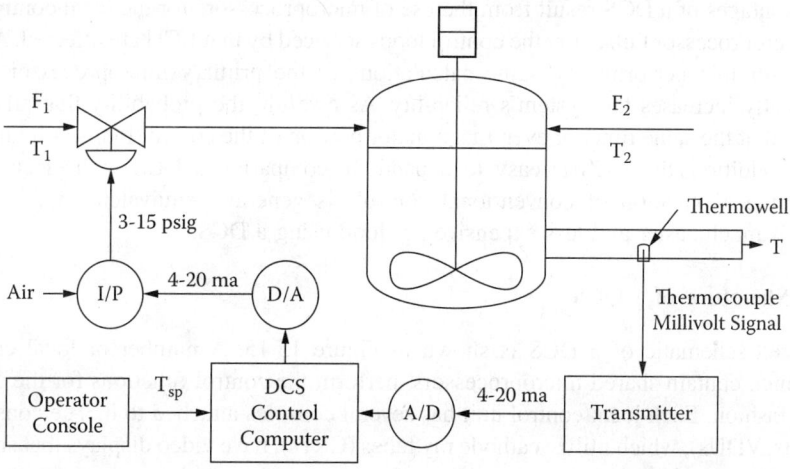

FIGURE 15.14 Schematic of the control system on the CST thermal mixer showing each component along with the various signals.

erly and the thermowell is properly designed and located, the value of the analog signal corresponds closely to the temperature in the mixing tank. The thermocouple/thermowell/temperature transmitter makes up the sensor system for this process.

The 4–20 mA analog signal from the temperature transmitter is converted into a digital reading by the A/D converter. The output of the A/D converter is a digital measurement of the temperature that is used in the control calculations. The operator console shown in Figure 15.14 allows the operator or control engineer to observe the performance of the control loop and to change the setpoint, T_{sp}, and controller tuning parameters for this loop. The value of T_{sp} and the digital value of the measured mixer temperature are used by the control algorithm in the DCS (i.e., the control computer). The output from the controller is a digital signal that is converted into an analog signal by the D/A converter. The DCS, the D/A and A/D converters, and the operator consoles are typically located in a centralized control room, while the remaining equipment (i.e., the actuator and sensor system) resides in the field near the process equipment.

The analog signal from the D/A converter goes to the current-to-pressure (I/P) converter. The I/P converter uses a source of instrument air to change the air pressure applied to the control valve (3 to 15 psig). Changes of instrument air pressure change the opening of the control valve, which result in changes in the flow rate to the process. These changes in the flow rate to the process cause changes in the temperature of the mixer, which are measured by the sensor. This completes the feedback control loop. The final control element consists of the I/P converter, the instrument air system, and the control valve.

15.2.2 Distributed Control System

15.2.2.1 Background

Based on technological breakthroughs in computers and associated systems in the 1970s, a new computer control architecture was developed and introduced by vendors in the latter part of the decade. It is based on using a number of local control units (LCUs), which have their own microprocessors and are connected together by shared communication lines (i.e., a data highway) as well as connected to operator/engineer consoles, a data acquisition system, and a general-purpose computer. This computer control architecture became known as a distributed control system (DCS) (Figure 15.15) since it involved a network with various control functions distributed for a variety of users.

The advantages of a DCS result from the use of microprocessors for the local control function. Even if a microprocessor fails, only the control loops serviced by that LCU are affected. A redundant microprocessor that performs the same calculations as the primary microprocessor (i.e., a hot backup) greatly increases the system's reliability. As a result, the probability that all the control loops will fail at the same time, or even that a major portion of the control loops will fail, is greatly reduced. In addition, the DCS is easy to expand. In comparing a DCS with electronic analog controllers, the application of conventional controls is generally equivalent, but implementing controllers is much easier and less expensive per loop using a DCS.

15.2.2.2 Structure of a DCS

A generalized schematic of a DCS is shown in Figure 15.15. A number of local control units (LCUs), which contain shared microprocessors, perform the control functions for the process in a distributed fashion. Each local control unit has several consoles attached to it. The consoles (video display units, VDUs), which utilize cathode ray tubes (CRTs), have video displays that show process schematics with current process measurements. Operators and control engineers use these displays to monitor the behavior of the process, set up control loops, and enter setpoints and tuning parameters. Normally, these consoles have touch screen capability so that, if operators want to make a change to a control loop, they touch the icon for the desired controller. Then a screen pops

Process Control

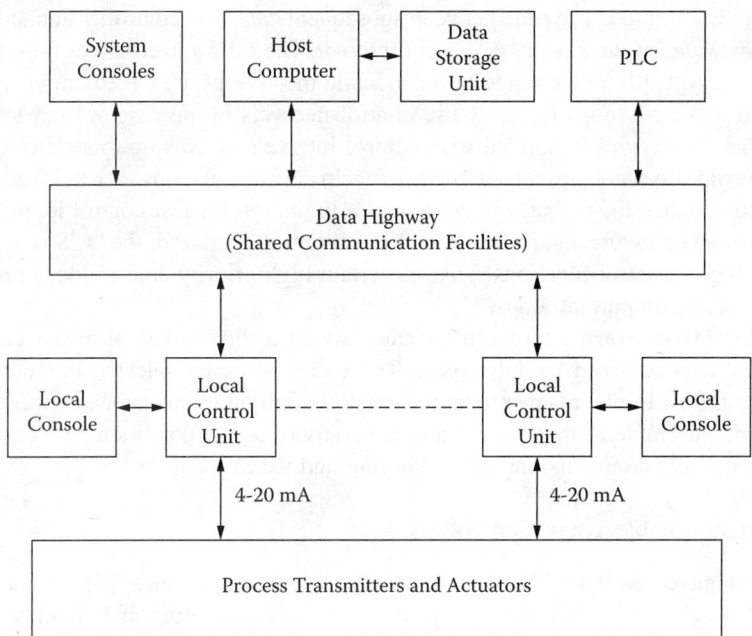

FIGURE 15.15 Generalized diagram of a DCS.

up that allows the operator to make the desired changes. On some DCSs, control loops can be conveniently set up by clicking and dragging on the tags for the desired sensor readings and the final control elements, and connecting these to the type of controller chosen. Since the local control unit is attached to the shared communications facility, a local display console can show schematics and current operating data for other parts of the plant but typically can make changes only to the control loops associated with its LCU. The local console can also be used to display historical trends of process measurements. To do this, the local console must access historical data in the data storage unit by using the data highway (i.e., the shared communication facilities).

Data acquisition is accomplished by transferring the process measurements through the LCUs onto the data highway and into the host computer, where the process data are passed on to the data storage unit. The archived process data can be accessed from one of the system consoles or one of the local consoles. In control rooms that used analog controllers, data storage for important control loops was typically accomplished using a strip chart recorder, which provided a record of previous measurements on a small roll of paper, using different colors of ink to record different process measurements.

The data highway holds the entire DCS together by allowing each modular element and each global element to share data and communicate with each other. The data highway is composed of one or more levels of communication hardware and the associated software.

System consoles attached to the data highway act as a local console for any of the local control units. In addition, system consoles can be used to change linking functions of the distributed elements.

The host computer is a mainframe that is used for data storage, process optimization calculations, and applying advanced process control approaches. Attached to the host computer is the data storage unit (usually a magnetic tape system) where archived data are stored.

15.2.2.3 Approach

The goal of a DCS is to apply the control calculations for each control loop so quickly that the control appears continuous. Since DCSs are based on sequential processors, each control loop is applied at a discrete point in time, and the control action is held constant at that level until the next time the control is executed. The time between subsequent calls to a controller applied by the DCS is called the *controller cycle time* or the *control interval*. The fastest cycle times for controller calls within a DCS are typically in the range of 0.2 s, while most loops are called only every 0.5 to 1.0 s. The regulatory control loops typically use control intervals in the range of 0.5 to 2.0 s, while supervisory control is typically applied with control intervals of 20 s up to several minutes. This controller cycle time does not present a limitation for slower control loops such as level, temperature, and composition control loops, but it does present a limitation for fast control loops such as flow controllers and some pressure controllers. A real-time control system for the DCS is used to enforce a priority ranking of control functions. That is, certain high-priority control loops are maintained at the expense of less-important loops.

Since DCSs are based on digital controller calculations, a wide variety of special control options are available in self-contained modular form. These can be easily selected by "click and drag" action on some DCSs. In this manner, complex control configurations can be conveniently assembled, interfaced, and implemented. In addition, a variety of signal-conditioning techniques can be applied to process measurements, including filtering and validity checks.

15.2.2.4 Programmable Logic Controllers

Programmable logic controllers (PLCs) are used primarily in the chemical process industries (CPI) for controlling batch processes and for sequencing of process startup and shutdown operations. PLCs are traditionally based on ladder logic, which allows the user to specify a series of discrete operations; e.g., start the flow of feed to the reactor until the level reaches a specified value, next start steam flow to the heat exchanger until the reactor temperature reaches a specified level, next start catalyst flow to the reactor, and so on. A small PLC can be responsible for monitoring 128 separate operations, while a large PLC can handle over 1000 operations. Today, the distinction between PLCs and DCSs has become less clear, as PLCs are being designed to implement conventional and advanced control algorithms, and DCSs are offered that provide control for sequenced operations. PLCs are typically attached to the data highway in a DCS (Figure 15.15) and provide sequenced control functions during startup, shutdown, and override of the normal controllers in the event of an unsafe operating condition.

15.2.2.5 Fieldbus Technology

The fieldbus approach to distributed control is shown in Figure 15.16. Control is distributed to intelligent field-mounted devices (i.e., sensors, valves, and controllers with onboard microproces-

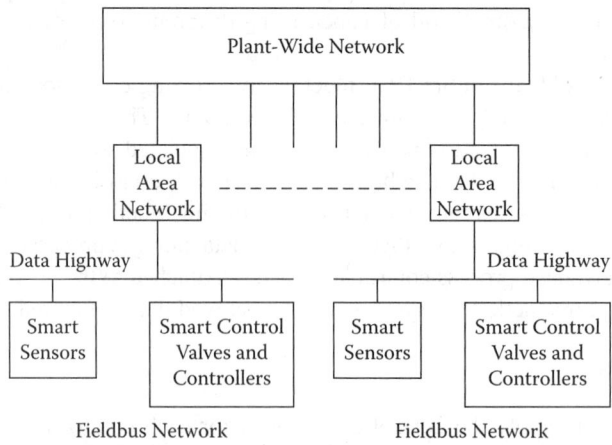

FIGURE 15.16 Schematic of the integration of the fieldbus with plant networks.

sors, which are used for complex operations and diagnostics) using a high-speed, digital, two-way communication system that connects the field-mounted devices with local area networks (LANs), process automation systems, and the plant-wide network. This communication system is similar to the data highway used by DCSs. Supervisory and advanced controls are implemented in the LANs, while the regulatory control functions are handled by the field-mounted devices on the fieldbus network. The advantage of the fieldbus design comes from the fact that many field-mounted devices can be attached to a single two-wire communication line instead of running electrical wires from each sensor/transmitter to the centralized control room and from the control room to each final control element. This results in a significant reduction in the time and cost associated with system installation. In addition, due to the standardization of fieldbus technology, all fieldbus-certified control components can be interchanged regardless of the manufacturer. Fieldbus technology is just beginning to be available commercially but is expected to remove regulatory controls from DCSs and move them into the field in the future.

15.2.3 Actuator Systems (Final Control Elements)

The actuator system for a process control system in the CPI is typically composed of the control valve, the valve actuator, the I/P transmitter, and the instrument air system. The actuator system is known industrially as the *final control element*. In addition, a variety of optional equipment is designed to enhance the performance of the actuator system, such as valve stem positioners and instrument air boosters.

15.2.3.1 Control Valves

The most common type of control valve in the CPI is the globe valve. Figure 15.17 shows a detailed cross section of a globe control valve with a plug in a cage-guided valve arrangement, along with notation indicating some of the key components of a control valve and valve actuator. The cage provides guidance for the plug as the plug is moved toward or away from the valve seat. The cage also provides part of the flow restriction produced by the control valve. The packing reduces the leakage of the process stream into the environment but provides resistance to movement of the valve stem. The travel indicator indicates the valve stem position.

Sizing of control valves is important because, if the valve is oversized or undersized, accurate control of the flow rate can be poor and a contrain for the flow rate can be contrained. A simplified valve flow equation for an incompressible fluid is given by

$$F_m = KC_v(x)\sqrt{(P_1 - P_2)/\rho}$$

where F_m is the mass flow rate through the valve, K is a constant that depends on the units used in this equation, $C_v(x)$ is the valve coefficient [which depends on the stem position (x) i.e., $C_v(x) = C_v^{max}$ when the valve is fully open and $C_v(x) = 0$ when the valve is closed], ρ is the density of the fluid, P_1 is the upstream pressure, and P_2 is the downstream pressure. In general, control valves should be designed so that the valve provides accurate metering of the flow over a wide operating range between almost fully open $(x \sim 0.9)$ and almost closed $(x \sim 0.1)$. The valve should be sized to provide a wide operating range.

Figure 15.18 shows how $f(x)$ varies with stem position for three types of valves: a quick opening valve, an equal percentage valve, and a linear valve, where

$$f(x) = \frac{C_v(x)}{C_v^{max}}$$

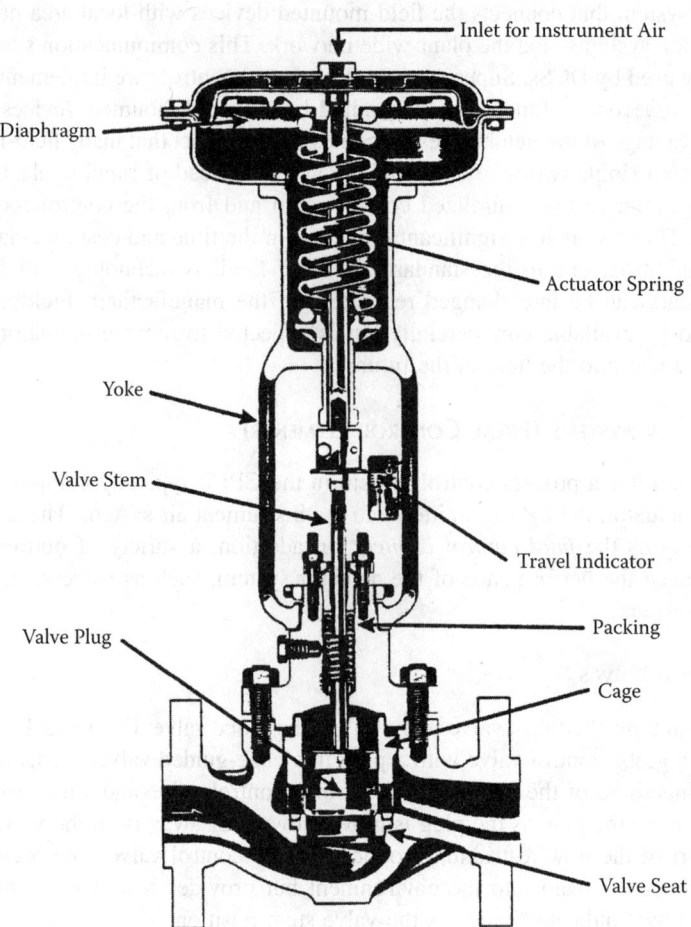

FIGURE 15.17 Cross section of a globe valve with an unbalanced plug. Courtesy of Fisher Controls.

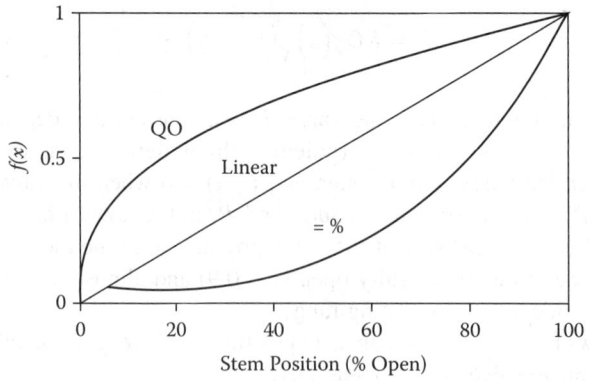

FIGURE 15.18 Inherent valve characteristics for a quick opening (QO), linear, and equal percentage valve (=%).

Process Control

This figure shows the inherent valve characteristics for these valve types, which indicate how the flow rate through the valve varies with stem position for a fixed pressure drop across the valve. The design of the plugs, valve seats, and cages (where applied) determine the particular flow vs. stem position. For a quick opening valve, as the valve opens, the cross-sectional area of the restriction of the valve increases much faster than the linear or equal percentage valves. From a process control standpoint, a control valve should exhibit a linear relationship between flow rate and stem position over a wide range for the installed valve. Normally, the pressure drop across a valve changes as the flow through the valve changes and does not remain constant (e.g., Figure 15.19). An equal percentage control valve generally provides a much more linear installed valve characteristic than linear or quick opening valves, as shown in Figure 15.20. For this reason, over 90% of the globe valves used in the CPI are equal percentage valves. For applications where the pressure drop across the control valve remains relatively constant (e.g., control of the flow rate of condensing steam to a heat exchanger), linear valves are preferred. The flow rate of steam to a heat exchanger should provide a fairly constant pressure drop across the valve, since the upstream steam supply pressure remains relatively constant, as does the pressure of the condensing steam within the heat exchanger. Quick opening valves are not usually used for feedback flow control applications but are used in cases where it is important to start a flow rate as quickly as possible (e.g., coolant flow through a bypass around a control valve for an exothermic reactor). In general, if the pressure drop across a control valve going from the maximum flow rate to the minimum flow rate changes by a factor of four or more, an equal percentage valve should be used. Otherwise, use a linear valve.

Figure 15.20 indicates that this equal percentage valve should be able to perform well for a stem position between 10 and 80% open due to its superior linearity. For economic considerations,

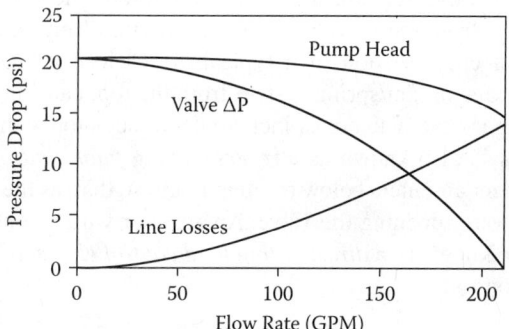

FIGURE 15.19 Pressure drop vs. flow rate for a typical flow system.

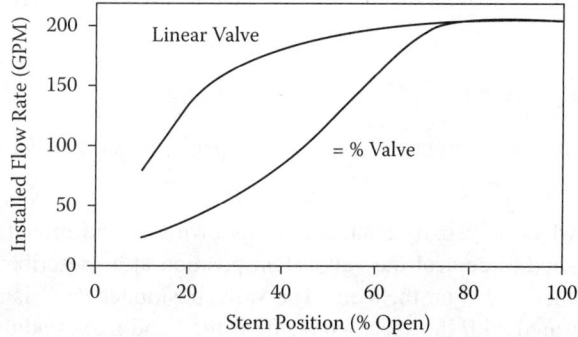

FIGURE 15.20 Installed valve characteristics for a typical equal percentage (=%) valve and linear valve.

butterfly valves, which are generally less expensive than globe valves, are also used for flow control applications for which the line diameter is larger than 6 in. Butterfly values have a significantly smaller range of valve position within which they perform well, compared with globe valves.

An important characteristic of a valve is the valve *deadband*, which is a measure of how precisely a control valve can control the flow rate. The deadband for a steering system on an automobile is the maximum positive and negative turn in the steering wheel that does not result in a noticeable change in direction of the automobile. For a control valve, deadband is the maximum positive or negative change, expressed in percent, in the signal to a control valve that does not produce a measurable change in the flow rate. Valve deadband is caused by the friction between the valve stem and valve packing and other forces on the valve stem. Typically, industrial control valves have a deadband of 10 to 25%. Generally, the larger and older the control valve is, the larger the deadband. A properly functioning valve with a valve positioner typically should have a deadband <0.5%.

Cavitation results when the liquid vaporizes and implodes inside the control valve. As a fluid flows through a control valve, the pressure drops sharply near the restriction between the valve plug and the valve seat because of high velocity in this region. As the fluid passes the valve restriction region and enters a region with a larger cross section, the pressure increases sharply (i.e., pressure recovery) as a result of the drop in the fluid velocity. Cavitation results in noise and vibration, reduced flow, and potentially rapid erosion of the body of the valve. It can be prevented by proper design using an increased downstream pressure.

15.2.3.2 Valve Actuators

The valve actuator provides the force necessary to move the valve stem position and alter the flow rate through the valve. The valve actuator must provide the force necessary to overcome pressure forces, flow forces, friction from the valve packing, and friction from the guide surfaces.

Figure 15.17 shows a cross section of a typical air-to-close actuator. The pressure of the instrument air acts on the diaphragm/spring system from the top, causing the valve to close as the air pressure supplied to the valve actuator is increased. An actuator with a control valve with an air-to-close valve actuator is also known as a *reverse-acting final control element*. For an air-to-open actuator, the instrument air enters below the diaphragm so that, as the air pressure is increased, the valve stem moves upward, opening the valve. An actuator with a control valve with an air-to-open valve actuator is also known as a *direct-acting final control element*. Valve actuators generally provide a fail-safe function.

15.2.3.3 I/P Transmitters

The I/P transmitter is an electromechanical device that converts the 4–20 mA signal from the controller to a 3–15 psig instrument air pressure to the valve actuator, which, in turn, affects the valve stem position.

15.2.3.4 Optional Equipment

Several devices are available for improving the overall performance of final control elements.

15.2.3.4.1 Valve Positioners

The valve positioner, which is usually contained in its own box and mounted on the side of the valve actuator, is designed to control the valve stem position at a prescribed position in spite of packing friction and other forces on the stem. The valve positioner itself is a feedback controller that compares the measured with the specified stem position and makes adjustments to the instrument air pressure to provide the proper stem position. In this case, the setpoint for the valve positioner can be a pneumatic signal coming from an I/P converter or the 4–20 mA analog signal coming directly from the controller. A valve with a deadband of 25% can provide flow rate precision to less than ±0.5% using a valve positioner. Valves with low levels of valve friction can control the average flow rate to a precision approaching ±0.1% using a valve positioner.

For flow control loops that are controlled by a DCS, a valve positioner is a necessity, because the control interval for a DCS (0.5 to 1.0 s) is not short enough for accurate flow control for most flow control loops. There are three general types of valve positioners: pneumatic, electronic, and digital. Pneumatic and electronic positioners receive a pneumatic or electronic signal from the I/P converter and send a pneumatic signal to the valve actuator. A more modem type of valve positioner is a digital positioner, which receives digital analog signal directly and adjusts the instrument air pressure sent to the actuator. Digital positioners have the advantage that they can be more easily calibrated, tuned, and tested remotely, and they can be equipped with self-tuning capabilities.

15.2.3.4.2 Booster Relays

Booster relays are designed to provide extra flow capacity for the instrument air system, which decreases the dynamic response time of the control valve (i.e., the time for most of a change to occur). Booster relays are used on valve actuators for large valves that require a large volume of instrument air to move the valve stem. Booster relays use the pneumatic signal as input and adjust the pressure of a high flow rate capacity instrument air system that provides pressure directly to the diaphragm of the valve actuator.

15.2.3.4.3 Adjustable-Speed Pumps

Adjustable-speed pumps can be used instead of the control valve systems just discussed. A centrifugal pump directly driven by a variable-speed electric motor is the most commonly used form of adjustable-speed pump. Another type is based on using a variable-speed electric motor combined with a positive displacement pump. Adjustable-speed pumps have the following advantages compared with control valve-based actuators: (1) They use less energy; (2) they provide fast, accurate flow metering without additional device requirements; and (3) they do not require an instrument air system. Their major disadvantage is capital cost, particularly for large flow rate applications. Another disadvantage of adjustable-speed pumps, based on safety considerations, is that they do not fail open or closed like a control valve with an air-to-close actuator or with an air-to-open actuator, respectively. As a result, the CPI almost exclusively uses control valve-based actuators except for low-flow applications, such as catalyst addition systems or base injection pumps for wastewater neutralization, which typically use adjustable-speed pumps.

15.2.4 Sensor Systems

Sensor systems are composed of the sensor, the transmitter, and the associated signal processing. The sensor measures certain quantities (e.g., voltage, current, or resistance) associated with devices in contact with the process such that the measured quantities correlate strongly with the actual controlled variable value. There are two general classifications for sensors: *continuous measurements* and *discrete measurements*. Continuous measurements are, as the term implies, generally continuously available, whereas discrete measurements update at discrete times. Pressure, temperature, level, and flow sensors typically yield continuous measurements, whereas certain composition analyzers (e.g., gas chromatographs) provide discrete measurements.

Several terms are used to characterize the performance of a sensor:

- *Span* is the difference between the highest and lowest measurement values made by the sensor/transmitter.
- *Zero* is the lowest reading available from the sensor/transmitter, i.e., the sensor reading corresponding to a transmitter output of 4 mA.
- *Range* is the maximum and minimum sensor readings. For example, the range of a pressure sensor can be expressed as a maximum of 150 psig and a minimum of 50 psig, corresponding to a span of 100 psig.

- *Accuracy* is the difference between the value of the measured variable indicated by the sensor and its true value. The true value is never known; therefore, accuracy is estimated by the difference between the sensor value and an accepted standard.
- *Repeatability* is related to the difference between sensor readings while the process conditions remain constant (see Figure 2.14).
- *Process measurement dynamics* indicate how quickly the sensor responds to changes in the value of the measured variable.
- *Calibration* involves the adjustment of the correlation between the sensor output and the predicted measurement so that the sensor reading agrees with a standard.

Smart sensors are available that have built-in microprocessor-based diagnostics. For example, smart pH sensors are available that can identify the buildup of coatings on the pH electrode surface and trigger a wash cycle to reduce the effect of the coatings. In general, smart sensors are moderately more expensive than conventional sensors, but, when they are properly selected and implemented, they can be an excellent investment in terms of greater sensor reliability and reduced maintenance. Best practice[2] for instrument selection, for instrument installation, and to reduce maintenance costs has been identified for the CPI.

A wide variety of sensors are available for measuring process variables.[3] Choosing the proper sensor for a particular application depends on the controlled variable that is to be sensed, the properties of the process, accuracy and repeatability requirements, and costs, both initial and maintenance. The following is a coverage of the most commonly used sensors in the CPI that are used for feedback control.

15.2.4.1 Temperature Measurements

The two primary temperature-sensing devices used in the CPI are thermocouples (TCs) and resistance thermometer detectors (RTDs). Thermocouples are less expensive and more rugged than RTDs but are an order of magnitude less precise than RTDs. Typically, RTDs should be used for important temperature control points, such as on reactors and distillation columns.

15.2.4.2 Thermowells

Thermowells typically are cylindrical metal tubes that are capped on one end and protrude into a process line or vessel to bring the TC or RTD into thermal contact with the process fluid. Thermowells provide a rugged, corrosion-resistant barrier between the process fluid and the sensor that allows for removal of the sensor while the process is still in operation. Thermowells that are coated with polymer or another adhering material can significantly increase the lag associated with the temperature measurement, i.e., significantly increase the response time of the sensor.

15.2.4.3 Repeatability, Accuracy, and Dynamic Response

TCs typically have a repeatability of ±1°C, whereas RTDs have a repeatability of ±0.1°C. Accuracy is a much more complex issue. Errors in the temperature reading can result from heat loss along the length of the thermowell, electronic error, sensor error, error from nonlinearity, calibration errors, and other sources.[4]

The dynamic response time of a TC or RTD sensor within a thermowell can vary over a wide range and is a function of the type of process fluid (i.e., gas or liquid), the fluid velocity past the thermowell, the separation between the sensor and inside wall of the thermowell, and material filling the thermowell (e.g., air or oil). Typical well-designed applications result in time constants of 6 to 20 s for measuring the temperature of most liquids.

15.2.4.4 Pressure Measurements

Pressure sensors respond very quickly and are commonly used to measure flow rates and levels. Repeatability for pressure measurement is generally less than ±0.1%.

15.2.4.5 Flow Measurements

The most commonly used flow meter is an orifice meter. An orifice meter uses the measured pressure drop across a fixed area flow restriction (an orifice) to predict the flow rate. The pressure drop across an orifice is usually measured using a differential pressure (DP) cell. A straight run of pipe preceding the orifice meter is required. If not, an error as large as 15% can result in the predicted flow rate. Since the orifice meter is based on a measured pressure drop, it is a very fast-responding measurement. Orifice meters typically provide repeatability in the range of ±0.3 to ±1%. Other types of flow meters are used for flow rate control in special situations, including vortex shedding flow meters and magnetic flow meters. Flow measurement devices, whichever type is chosen, are typically installed upstream of the control valve to provide the most accurate, lowest-noise measurement. Installing the flow sensor downstream subjects the sensor to flow fluctuations and even two-phase flow, and these reduce sensor accuracy and increase measurement noise.

15.2.4.6 Level Measurements

The most common type of level measurement is based on measuring the hydrostatic head in a vessel using a differential pressure measurement. This approach typically works well as long as there is a large difference between the density of the light and heavy phases. This approach usually has relatively fast measurement dynamics, since it is based on a pressure measurement. Level measurements typically have repeatability of approximately ±1%.

Differential pressure measurements can be used to determine the level in a vessel. This approach directly measures the hydrostatic head. Because of plugging and corrosion problems, it may be necessary to keep the process fluid from entering the differential pressure transmitter. In addition, it is important to keep vapor from condensing in the upper tap and collecting in the low-pressure side of the differential pressure transmitter. This usually can be accomplished by insulating the pressure tap and wrapping it with resistive heating tape. Other level-measuring approaches are based on a variety of physical phenomena and are used in special cases.

15.2.4.7 Chemical Composition Analyzers

The most commonly used on-line composition analyzer is the gas chromatograph (GC), but infrared, ultraviolet, and visible radiation analyzers have made inroads recently. On-line composition measurements are generally relatively expensive.

15.2.4.8 Sampling System

The sample system is responsible for collecting a representative sample of the process and delivering it to the analyzer for analysis. Obviously, the reliability of the sample system directly affects the reliability of the overall composition analysis system. The transport delay associated with the sample system contributes directly to the overall deadtime associated with an on-line composition measurement. This difference in sampling deadtime can have a drastic effect on the performance of a control loop. Table 15.2 summarizes the dynamic characteristics and repeatability of typical control valve systems and several different types of sensors.

TABLE 15.2
Summary of Control-Relevant Aspects of Actuators and Sensors

	Time Constant (sec)	Valve Deadband or Sensor Repeatability	Turndown Ratio, Rangeability or Range
Control valve*	3–15	10–25%	9:1
Control valve w/valve positioner*	0.5–2	0.1–0.5%	9:1
Flow control loop w/valve positioner*	0.5–2	0.1–0.5%	9:1
TC w/thermowell	6–20	±1.0°C	−200 to 1300°C
RTD w/thermowell	6–20	±0.1°C	−200 to 800°C
Magnetic flow meter	<1	±0.1%	20:1
Vortex shedding meter	<0.1	±0.2%	15:1
Orifice flow meter	<0.2	±0.3–±1%	3:1
Orifice meter w/smart transmitter	<0.2	±0.3–±1%	10:1
Differential pressure level indicator	<1	±1%	9:1
Pressure sensor	<0.2	±0.1%	9:1

* Based on globe valves.

15.2.4.9 Transmitters

The transmitter converts the output from the sensor (i.e., a millivolt signal, a differential pressure, a displacement, and so forth) into a 4–20 mA analog signal that represents the measured value of the controlled variable. Consider a transmitter that is applied to a temperature sensor. Assume that the maximum temperature the transmitter is expected to handle is 200°C and that the minimum temperature is 50°C; then the span of the transmitter is 150°C and the zero of the transmitter is 50°C. Transmitters are typically designed with two knobs that allow for independent adjustment of the span and the zero of the transmitter. Properly functioning and implemented transmitters are so fast that they do not normally contribute to the dynamic lag of the process measurement. Modern transmitters have features that, if not applied properly, can reduce the effectiveness of the control loop. For example, excessive filtering of the measurement signal by the transmitter can add extra lag to the feedback loop, thus degrading control loop performance.

15.2.5 TROUBLESHOOTING CONTROL LOOPS

Control engineers spend a major portion of their time troubleshooting control loops. An operator may point out that a particular loop has been behaving erratically and ask the control engineer to improve its performance. The control engineer may discover that an important loop is under manual operation (open-loop operation). A final product may have excessive variability in its impurity levels, and the control engineer's job is to reduce the variability to an acceptable level. In this latter example, several control loops may require scrutiny. When one or more loops is not performing properly, troubleshooting is required to return them to the expected performance levels, or at least to identify the source of the problem. To effectively troubleshoot control loops, the control engineer must understand the proper design and expected performance of the hardware that compose a control loop.

Troubleshooting control loops involves identifying the source of the problem from an overwhelming number of possible causes. The size of this problem requires a systematic approach when troubleshooting. Control loop troubleshooting is too often treated as an afterthought and performed haphazardly. This section presents a general troubleshooting procedure as well as a detailed analysis of fault detection for the final control element, the sensor system, the control computer or DCS, and the process.

15.2.5.1 Overall Approach

The key to effective troubleshooting is expressed in the old adage, "divide and conquer." It is important to locate the portion of the control loop hardware that is causing the poor performance: the final control element, the sensor system, the controller, or the process. The place to start is to test each system separately to determine whether that portion of the control loop is operating properly. The final control element can be evaluated by applying a series of input step tests. That is, the input to the final control element, which is normally set by the controller, can be manually adjusted. The test allows the determination of the dynamic response and deadband of the actuator system. If the performance in these two areas is satisfactory, there is no need to evaluate the actuator system further.

Sudden changes in control loop performance from satisfactory to unacceptable may be caused by recent changes in the process. Considering what changes have been made to the process can expedite the troubleshooting process. For example, the controller could have been retuned. A new analyzer could have been installed. A new instrument technician could be responsible for calibrating and maintaining an analyzer. The feed to the unit could have significantly changed. These examples and many more can be directly related to the source of poor control loop performance. When something significant has changed, it can provide a valuable clue that allows quicker determination of the source of the problem with a poorly performing control loop.

When troubleshooting a more complex control system, it is advisable to start by comparing the existing control loops with those from the piping and instrumentation diagram (P&ID). The current problem may be the result of inappropriate modifications to the original control configuration.

15.2.5.2 Final Control Element

The final control element consists of the instrument air system, the I/P converter, and the control valve (the valve and the valve actuator). The fastest way to identify gross problems with the final control element is to plot both the manipulated variable and the controller output. If the manipulated variable does not follow the controller output, there is probably a problem with the final control element.

Even if the manipulated variable seems to follow the controller output, there could be a problem with the actuator. Estimates of the actuator deadband and dynamic response are required to determine if the actuator system is performing properly (both of which can be determined by a block sine wave test). This test is shown in Figure 15.21. A block sine wave is a series of equally sized step changes that approximate a sine wave. For the test shown in the figure, the amplitude

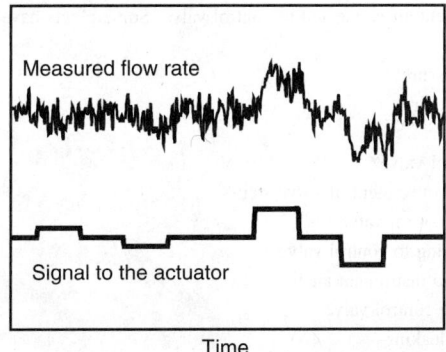

FIGURE 15.21 Results of a series of block sine wave tests to determine the deadband of the actuator. Note that the signal to the actuator and the measured value of the flow rate are plotted on different scales.

of the step change used in the block sine wave is initially small enough that consistent positive and negative changes in the measured value of the flow rate of the manipulated variable are not observed. The next block sine wave uses a larger-amplitude step change and, for this case, the measured manipulated variable can be seen to make both positive and negative changes corresponding to the positive and negative changes in the controller output. Therefore, the deadband of the actuator in this case is larger than the step size used in the first block sine wave and smaller than the one used in the second. The same block sine wave test can be used to estimate the time constant of the final control element. If the time between step changes used in the block sine wave is large enough, the settling time of the actuator can be estimated. Then, the time constant of the actuator is estimated as one quarter of the settling time.

Once the deadband and time constant of the actuator have been determined, the performance of the actuator can be assessed. The deadband for valves with positioners typically ranges from 0.1 to 0.5% of the flow rate for properly implemented systems and depends on the size of the valves, the pressure drop across the valve, the fluid properties, and other factors. The deadband for an industrial valve without a positioner typically ranges from 10 to 25%, and even higher for older valves that have not been maintained. The time constant of a properly functioning final control element is less than 2 s for a valve with a positioner or a control valve in a flow control loop. Otherwise, the actuator time constant is usually between 3 and 15 s.

If it has been determined that the actuator system is not functioning properly, one should first determine if the instrument air pressure system is operating properly. This can be done by observing that the instrument air pressure at the control valve after a step change in the signal to the final control element has been implemented in the DCS or in the control computer. If the instrument air pressure at the control valve increases sharply after the step test has been implemented, then the control valve is the source of the slow or erratic response. Another common problem is valve packing that is overtightened, which primarily increases the valve deadband. A valve that is operating below 10% or above 90% opening typically performs below standards. Another problem is an improperly tuned valve positioner. If the valve positioner is tuned too aggressively, oscillatory control performance results for the control valve, resulting in an increase in the actuator deadband. If the valve positioner is not tuned aggressively enough, the dynamic response of the actuator is slower than it should be. Table 15.3 lists a number of common problems with the components of the actuator system.

TABLE 15.3
Common Problems with the Final Control Element

- Excessive lag in the instrument air system
- Wrong type of instrument air connected to control valve. Some plants have high- and low-pressure instrument air
- Low instrument air pressure
- Wet or dirty instrument air
- Excessive deadband*
- Improperly sized control valve*
- Excessive resistance to movement of valve stem*
- Leak in diaphragm of control valve
- Debris is stuck in opening to control valve
- A plugged or obstructed instrument air line
- Plug/seat erosion in the control valve
- A bypass line open or leaking
- Flashing and cavitation
- Improperly tuned valve positioner*

* More frequently observed problems

15.2.5.3 Sensor Systems

The sensor system is composed of the sensor, the transmitter, and the sampling system that allows it to make measurements. The performance of a sensor can be assessed by determining its repeatability, time constant, and (sometimes) accuracy. Accuracy is important for a composition analyzer on a final product to ensure that the product meets specifications. The accuracy of a flow transmitter is usually not important, because the flow rate is adjusted incrementally by a supervisory controller so that its actual flow rate is unimportant. The dynamics of the sensor can affect the feedback control performance if it is too slow, and a large repeatability can increase the variability in the controlled variable.

The accuracy of the sensor can be checked by comparing the sensor reading to a standard or known condition. For example, a composition standard can be processed to verify the accuracy of the GC, or a thermocouple can be placed in boiling water to check its accuracy. The repeatability of a sensor can be estimated by observing sensor readings during a period of relatively steady-state operation. The repeatability is the variation in the sensor reading caused by noise. One assumes that, during steady operation, the process was not in fact changing.

Determining the time constant of the sensor is usually more difficult than estimating the repeatability. To determine the time constant of the sensor system, one needs to know the actual process measurement. Consider a temperature measurement. A measurement of the actual process temperature is required to estimate the time constant of a sensor. Instead, the thermal resistance, which causes the excessive thermal lag of the temperature sensor, can be evaluated. The location of the thermowell should be checked to ensure that it extends far enough into the line that the fluid velocity past the thermowell is sufficient; the possibility of buildup of insulating material on the outside of the thermowell should be assessed, and the thermal contact between the end of the temperature probe and the thermowell walls should be evaluated. In this manner, an indirect estimate of the responsiveness of the temperature sensor can be developed. The velocity of a sample in the line, which delivers a sample from a process line to a GC, can indicate the transport delay associated with the sample system. A low velocity in the sample line from the process stream to a GC can result in excessive transport delay, which can greatly reduce controller effectiveness.

One should be careful to determine if the sensor used is really measuring the controlled variable of interest. Differential pressure sensors are used for pressure, flow, and level measurements. They are particularly susceptible to plugging of the sensing lines that connect the differential pressure sensor to the process itself. Plugging of the sensing lines can result from the buildup of coatings or solids or from freezing of the fluid in the pressure taps. The calibration of a differential pressure sensor is quite sensitive to the conditions of the fluid in the sensing lines. Condensate buildup in lines that should be dry can lead to large calibration errors. Table 15.4 lists some commonly encountered sources of problems for sensor systems. For a complete analysis of the sensor system, an instrument engineer or other expert familiar with that particular sensor may be required. Table 15.2 lists the expected ranges for the repeatability and the time constants for several commonly used sensors in the CPI. Deviation of the apparent repeatability or time constant from these expected values identifies a poorly performing sensor.

15.2.5.4 Controller/DCS System

The control computer/DCS system consists of controllers, A/D and D/A converters, and the signal conditioning hardware and software, i.e., filtering and validation. Each of these components requires separate evaluation. Table 15.5 lists possible problems with the controller/DCS system. One way to initially check controller tuning is to place the control loop in manual (open the control loop) and observe whether the controlled variable lines out to a steady-state or near steady-state value. Comparing the open-loop and closed-loop performance indicates whether the controller is upsetting the process. If not, disturbances to the control loop in question are the primary source of the upsets.

TABLE 15.4
Commonly Encountered Problems with Components of a Sensor System

Sensor	Common Problems
Transmitter	Not calibrated correctly*
	Low resolution
	Excessive signal filtering*
	Slow sampling
Thermocouple/RTD	Off-calibration*
	Short in the electrical circuit/grounding problems
	Improperly located thermowell*
	Thermowell with excessive thermal resistance (e.g., stainless steel thermowells)
	Partially burned out thermocouple
	Interference from heat tracing
Pressure Indicators	Plugged line to pressure indicator*
	Confusion about absolute pressure readings, gauge pressure readings, and vacuum pressure readings
	Condensation in lines to pressure indicator*
Sampling System For GC	Excessive transport delay for an analyzer
	Sample drawn from wrong process point
	Plugged sample system*
	Sample system closed off
GC	Out of calibration
	Plugging in the GC column
	Failure of electrical components in GC
	Excessive noise on measurement
	Frozen signal
	Spiking due to inadequate flushing for GCs that run multiple samples
Flow Indicator	Square root compensation applied for non-differential pressure type flow indicator
	Square root compensation not applied for differential pressure type flow indicators
	Square root compensation applied twice, i.e., once in transmitter and once in DCS
	Orifice plate installed backwards
	Damaged orifice plate
	Plugged line to differential pressure sensor*
	Flashing of liquids as they flow through an orifice meter
Level Indicator	Plugged line from process to DP cell*
	Leak in line to DP cell or in DP cell itself
	Boiling of liquid in line to or from DP cell due to a steam leak in the steam tracing line
	Solidification of liquid in line to or from process to DP cell due to failure in steam tracing
	Formation of emulsions that can confound interface level measurements
	Leak in float type level indicators
	Formation of foams that can interfere with level measurements

* More frequently observed problems.

TABLE 15.5
Possible Problems with the Controller/DCS System

- Improperly tuned controller*
- Wrong scaling for A/D and D/A converter
- Improper or lack of pressure/temperature compensation for flow measurement
- Improper selection of reverse-acting or direct-acting controller
- Too much or not enough filtering of the measured controlled variable*
- Signal aliasing due to excessive control interval (see Appendix B)
- Poor resolution on A/D or D/A converters
- Derivative action based on error from the setpoint instead of the measurement.

* More frequently observed problems.

15.2.5.5 Process

The effect of the process on the closed-loop behavior can be examined directly by opening the control loop in question and observing the process behavior. Open-loop oscillatory behavior indicates a problem internal to the process. The noise level on the sensor reading can also be assessed under open-loop conditions.

Fluctuating disturbances and process gain changes due to nonlinearity are a natural part of process control. Extraordinary disturbances combined with nonlinearity can cause an otherwise properly tuned controller to oscillate or go unstable during upset periods. Severe process nonlinearity can be identified if a closed-loop process exhibits ringing and sluggish behavior during different periods with the same controller tuning. Scheduling the controller tuning (Section 15.5.6) is one way to compensate for process nonlinearity. It may be possible to reduce the magnitude of the disturbances to acceptable levels by modifying the upstream operations (e.g., tuning upstream controllers). Excessive disturbances, when not measurable, can be inferred by observing the range of the average manipulated variable levels. If large-magnitude disturbances are affecting the process, large changes in the average manipulated variable level are required to maintain the process near its desired operating point. Excessive fouling of heat exchangers or deactivation of the catalyst can result in process gain changes that produce sluggish or unstable behavior.

Process changes that require manipulated variable levels in excess of what is physically available can also occur. After feed rate increases to a distillation column, the reboiler might be unable to provide enough heat transfer to maintain the purity of the bottom product. When this occurs, it is a physical limitation of the process and not the fault of the controller. Loss of steam pressure can also cause a constraint that can affect control loop performance. Downstream pressure changes can cause a constraint on the maximum flow rate due to an inadequate pressure driving force. Constraint control techniques should be used when manipulated variables saturate. It should be clear that a thorough understanding of the process is a prerequisite for control loop troubleshooting.

15.2.5.6 Testing the Entire Control Loop

After each of the components has been evaluated and corrected wherever possible, the closed-loop system should be checked. From an overall point of view, there are three general factors that affect the closed-loop performance of a control loop: (1) the type and magnitude of disturbances, (2) the lag associated with the components that compose the control loop, and (3) the precision to which each component of the control loop performs. Actuator deadband affects the variability in the controlled variable. The addition of lag to a control loop (e.g., sensor filtering) results in slower disturbance rejection, which can increase the variability in the controlled variable. Disturbance magnitude directly affects variability.

The performance of a closed-loop system can be assessed by the settling time, closed-loop deadband, and the variability of the controlled variable evaluated over an extended period of time. The settling time and the closed-loop deadband can be determined using a closed-loop block sine wave test. For a closed-loop block sine wave test, the setpoint for the control loop is applied in the form of a block sine wave, and the amplitude of the block sine wave is varied until the deadband is determined. During these tests, the settling time of the controller can also be estimated. An accurate determination of the variability of a controlled variable generally requires an extended period of operation. An evaluation of the variability based on a short period of time may not be representative of true system performance.

Consider a control system with an excessive lag (e.g., buildup of scale on the exterior of a thermowell) added to a control loop. The controller can be tuned for any tuning criterion (e.g., a decay ratio of 1/6 to critically damped); therefore, tuning a control loop to the desired tuning criterion is not, in itself, an indication of the performance of the control system. The settling time and the variability provide a measure of the performance of the control system. A process with the additional lag exhibits a longer settling time than a process without it. The average variability in the controlled variable over an extended period also shows that the system without the additional lag exhibits superior control performance. The variability is usually a direct measure of controller performance that generally can be related to the overall objectives of the process, but it requires a significant operating period (e.g., a week) to accurately determine. On the other hand, the settling time can be determined much more quickly and easily, but it provides only a relative measure of control performance. One can determine the relative change in controller performance by comparing the settling time before control loop troubleshooting was undertaken to it afterward.

The closed-loop deadband is an indication of the variability in the controlled variable that results from the combined effects of actuator deadband, sensor noise, and resolution of the A/D and D/A converters. The closed-loop settling time is an indication of the combined lags of the control loop components. The closed-loop performance assessment is a means of determining whether all the major problems within a control loop have been rectified.

Example 15.1 Troubleshooting Example

The following is a step-by-step troubleshooting process along with intermediate results for a temperature controller that was observed to result in sluggish closed-loop performance.

> Step 1. Determine the deadband of the final control element using a series of block sine wave tests. Result: the deadband of the final control element was less than 0.4%, and the dynamic response time of the final control element was 2 s; therefore, the final control element was found to be functioning properly.
> Step 2. Retune the temperature controller. Result: the controller settings did not change significantly; therefore, the controller tuning does not appear to be the problem.
> Step 3. Evaluate the sensor. Check the repeatability of the sensor by observing the temperature measurements during a steady-state or near steady-state period. Result: the repeatability was less than 0.1°C, which is good for an RTD. An independent measurement of the temperature is made and compared with the sensor reading. Result: the sensor reading is observed to have excessive lag, i.e., a dynamic response time for the sensor was estimated to be about 5 min. It was determined on further examination that there was an excessive air space between the RTD element and the surface of the thermowell. The position of the RTD in the thermowell was changed, and the dynamic lag of the sensor was found to be in the proper range. The controller was retuned, and control performance was significantly improved.

Process Control

15.3 PID CONTROLLERS

15.3.1 INTRODUCTION

PID controllers are simple to implement and are extremely flexible and are used for processes ranging from refineries to spacecraft to electronic devices to power plants. The PID algorithm is quite computationally efficient, and much of the flexibility of a PID controller comes from the unique characteristics of proportional, integral, and derivative action (hence PID).

15.3.2 PID ALGORITHMS

The Instrument Society of America (ISA) standard for the PID algorithm in the position form is given as

$$c(t) = c_0 + K_c \left[e(t) + \frac{1}{\tau_I} \int_0^t e(t)dt + \tau_D \frac{de(t)}{dt} \right] \qquad (15.1)$$

where K_c, τ_I, and τ_D are the user-selected tuning parameters; $c(t)$ is the output from the controller; and $e(t)$ is $[y_{sp} - y_s(t)]$. Note that c_0 is the value of the controller output when the controller is turned on. K_c is the controller gain and should not be confused with the process gain, K_p. K_p has units of y_s/c when the actuator, process, and sensor are lumped together, and K_c has units corresponding to c/y_s. Equation (15.1) is written in a form corresponding to a reverse-acting controller. Equation (15.2) lists the position form for a direct-acting controller:

$$c(t) = c_0 - K_c \left[e(t) + \frac{1}{\tau_I} \int_0^t e(t)dt + \tau_D \frac{de(t)}{dt} \right] \qquad (15.2)$$

The decision between using a direct-acting controller [Equation (15.2)] or a reverse-acting controller [Equation (15.1)] depends on the sign of the process gain and whether a direct- or reverse-acting final control element is used. The usual convention is that Equation (15.2) is called a direct-acting controller; it is direct acting with regard to the process measurement, because to form $e(t)$ the process measurement is subtracted from the setpoint. Consider a heat exchanger in which the steam flow rate to the exchanger is manipulated to control the temperature of the process stream leaving it. Since an increase in steam flow to the heat exchanger results in an increase in the outlet temperature of the process stream, the process gain of this system is positive. Also, consider a direct-acting final control element on the steam that causes an increase in steam flow rate when the signal to the final control element is increased. Furthermore, consider the case in which the measured outlet temperature is below its setpoint [i.e., $e(t)$ is positive]. Since it is desired to move the controlled variable toward its setpoint, the steam flow rate to the heat exchanger should be increased; therefore, from an examination of Equations (15.1) and (15.2), it is clear that a reverse-acting controller [Equation (15.1)] should be used. On the other hand, if the direct-acting final control element is replaced with a reverse-acting final control element, a decrease in the signal to the final control element is required; therefore, a direct-acting controller [Equation (15.2)] is required. The choice between a reverse- and direct-acting final control element usually depends on whether the control element should fail open or closed when instrument air pressure is lost.

TABLE 15.6
Usual Convention for Selecting Direct- and Reverse-Acting Controllers

Process Gain	Direct-Acting Actuator	Reverse-Acting Actuator
Positive	Use reverse-acting PID	Use direct-acting PID
Negative	Use direct-acting PID	Use reverse-acting PID

Now consider a heat exchanger in which the cooling water flow rate to the exchanger is manipulated to control the temperature of the process stream leaving it. Since an increase in cooling water flow rate to the heat exchanger results in a decrease in the controlled variable for this process, the process gain is negative. Also, consider a direct-acting final control element. Similar to the previous example, consider the case in which the controlled variable is below its setpoint. Under these conditions, a decrease in cooling water flow rate is required; therefore, a direct-acting controller should be used. Finally, if a reverse-acting final control element was substituted for the direct-acting one, a reverse-acting controller should be used. Table 15.6 summarizes these results. Obviously, these different combinations of positive and negative process gains and reverse- and direct-acting final control elements can occur in the implementation of industrial process control. As a result, the process control engineer needs a way to conveniently choose a direct-acting or a reverse-acting controller. On a modern control computer (i.e., DCS), when a control loop is set up, there is typically a box to check to select a direct- or reverse-acting controller. For analog controllers, there is a switch on the back that allows the user to select the proper form.

Another way to represent the controller gain is the proportional band (PB), which is an approach that was in more common use 10 to 15 years ago. Proportional band can be expressed (as a percent) in terms of K_c when K_c is in dimensionless form. For example, the controller output and the error from setpoint can be scaled 0 to 100%, yielding a dimensionless K_c:

$$PB = \frac{100\%}{K_c}$$

The proportional band is small when the controller gain is large, and PB is large when K_c is small.

When a setpoint change is made using the forms given by Equations (15.1) and (15.2), a spike in the calculated value of $de(t)/dt$ will occur, causing a spike in $c(t)$. This behavior is called *derivative kick* and can be eliminated by replacing $de(t)/dt$ with $-dy_s(t)/dt$, yielding

$$c(t) = c_0 - K_c \left[e(t) + \frac{1}{\tau_I} \int_0^t e(t)dt - \tau_D \frac{dy_s(t)}{dt} \right] \qquad (15.3)$$

for a direct-acting controller. The derivative-on-measurement form of the PID algorithm is recommended because it is not susceptible to derivative kick.

Equation (15.3) is applied within a DCS by implementing it in a digital form using the following approximations:

$$\int_0^t e(t)dt \approx \sum_{i=1}^n e(i\Delta t)\Delta t$$

where n is equal to $t/\Delta t$, and

$$\frac{dy_s(t)}{dt} \approx \frac{y_s(t) - y_s(t - \Delta t)}{\Delta t}$$

where Δt is the time interval between applications of the PID algorithm, i.e., the control interval. These approximations result in the digital version of the position form of the PID controller:

$$c(t) = c_0 + K_c \left[e(t) + \frac{\Delta t}{\tau_I} \sum_{i=1}^{n} e(i \cdot \Delta t) - \tau_D \left[\frac{y_s(t) - y_s(t - \Delta t)}{\Delta t} \right] \right] \quad (15.4)$$

The PID algorithm can also be applied in velocity form. Applying Equation (15.4) at $t - \Delta t$ results in the following equation:

$$c(t - \Delta t) = c_0 + K_c \left[e(t - \Delta t) + \frac{\Delta t}{\tau_I} \sum_{i=1}^{n-1} e(i \cdot \Delta t) - \tau_D \left[\frac{y_s(t - \Delta t) - y_s(t - 2\Delta t)}{\Delta t} \right] \right] \quad (15.5)$$

Subtracting Equation (15.5) from Equation (15.4) results in the velocity form of the PID algorithm:

$$\Delta c(t) = K_c \left[e(t) - e(t - \Delta t) + \frac{\Delta t}{\tau_I} e(t) - \tau_D \left[\frac{y_s(t) - 2y_s(t - \Delta t) + y_s(t - 2\Delta t)}{\Delta t} \right] \right] \quad (15.6)$$

and then

$$c(t) = c(t - \Delta t) + \Delta c(t) \quad (15.7)$$

Note that Equation (15.7) is written as a reverse-acting controller. A reverse-acting controller subtracts $\Delta c(t)$ from $c(t - \Delta t)$. The velocity form for the derivative on the error from setpoint is given by

$$\Delta c(t) = K_c \left[e(t) - e(t - \Delta t) + \frac{\Delta t}{\tau_I} e(t) + \tau_D \left[\frac{e(t) - 2e(t - \Delta t) + e(t - 2\Delta t)}{\Delta t} \right] \right] \quad (15.8)$$

Another popular version of the velocity form of the PID can be developed by eliminating proportional action for setpoint changes. Noticing that the proportional part of Equation (15.6) is simply the difference between $y(t)$ and $y(t - \Delta t)$ when the setpoint remains unchanged leads one to replace the difference between errors from setpoint with the difference between measured values of the controlled variable in the velocity form of the PID controller, i.e.,

$$\Delta c(t) = K_c \left[y_s(t - \Delta t) - y_s(t) + \frac{\Delta t}{\tau_I} e(t) - \tau_D \left[\frac{y_s(t) - 2y_s(t - \Delta t) + y_s(t - 2\Delta t)}{\Delta t} \right] \right] \quad (15.9)$$

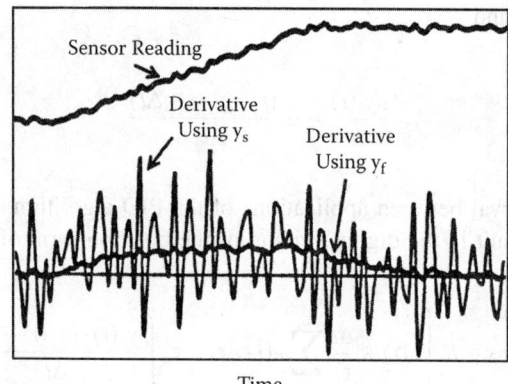

FIGURE 15.22 The instantaneous and filtered values for the derivative of a noisy sensor reading. Also shown is the value of the derivative calculated using filtered values of $y_s(t)$ with a filter factor of 0.08.

The advantage of this form of the PID controller is that it will not act as abruptly to setpoint changes as Equation (15.8). In fact, from Equation (15.8) it can be seen that only the integral action will move the process toward a new setpoint. This reduction in aggressive setpoint tracking has an effect that is similar to *bumpless transfer*, which is discussed later.

The position form of the PID algorithm calculates the absolute value of the output of the controller, whereas the velocity form calculates the change in the controller output that should be added to the current level of the controller output. The position and velocity modes are different forms of the same equation; therefore, they are generally equivalent. The velocity form is usually used industrially. In general, DCSs offer the velocity form of the PID controller in three versions: the velocity form in which P, I, and D are based on the error from setpoint [Equation (15.8)]; the form in which only P and I are based on the error from setpoint [Equation (15.6)]; and the form in which only integral action is based on the error from setpoint [Equation (15.9)].

When derivative action is applied to a process where there is significant noise on the sensor reading, erratic derivative action can result, since the difference between successive sensor readings can be dominated by the noise. Figure 15.22 shows a sensor reading with noise and the corresponding derivative value. A digital filter can be used to "smooth out" the noisy sensor reading:

$$y_f(t) = fy_s(t) + (1-f)y_f(t-\Delta t) \quad (15.10)$$

where $y_f(t)$ is the sensor reading after a digital filter has been applied, $y_s(t)$ is the current sensor reading, and f is the filter constant, which is normally between 0.01 and 0.5. The digital filter provides a running average and tends to absorb short-term variations caused by the noise. In this manner, filtered values of the controlled variable can be used where the derivative is calculated in the PID control equation, i.e.,

$$c(t) = c_0 + K_c\left(e(t) + \frac{1}{\tau_I}\sum_{i=1}^{m}e(i\Delta t)\Delta t + \tau_D\frac{y_f(t) - y_f(t-\Delta t)}{\Delta t}\right) \quad (15.11)$$

When the ratio of noise to measured value of the controlled variable is large, the measured value of the controlled variable used for proportional action may also require filtering [Equation (15.11)]. Figure 15.23 shows the results of a PI controller with and without filtering on the measurement of the controlled variable. For the case without filtering, the noise on the measured

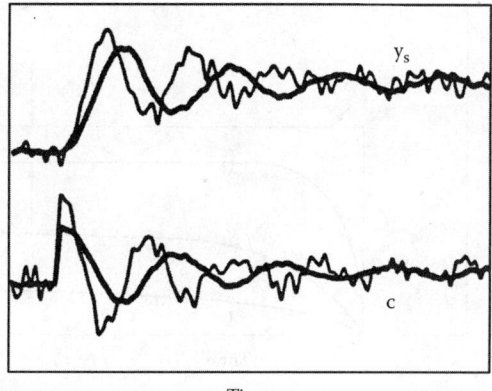

Time

FIGURE 15.23 Comparison of the results of a PI controller with and without filtering on the measured value of the controlled variable. Thick line = with filtering, thin line = without filtering.

value of the controlled variable is passed into the controller output by the proportional portion of the controller, which in turn affects the level of the manipulated variable and thus the process. Figure 15.23 also shows the filtered value of the controlled variable and the corresponding controller output. The filtering removes most of the noise on the controlled variable so the controller output is considerably smoother. The aggressiveness of the controller, when filtering was applied, was reduced to produce the same general dynamic response. That is, filtering of the controlled variable puts additional lag into the overall process; therefore, less-aggressive tuning must be used. The detuning of the controller and the delay caused by the filtering process itself cause the filtered case to respond more slowly than the unfiltered case. It should be clear from this example that filtering is necessary in certain cases, but the minimum required filtering should be used to limit the detrimental effect of filtering on control performance.

An older version of the PID algorithm that was originally applied using analog devices is called an interactive PID controller, which is also referred to as "rate before reset" controller, since the derivative action is in series with and precedes the integral action. A PI or a P-only interactive controller is no different from the earlier form presented [i.e., noninteractive PID; Equation (15.1)]. The only difference between an interactive and a noninteractive controller relates to the PID controller. Both controllers apply the PID algorithm, but there are differences in the tuning constants. That is, for the same amount of proportional, integral, and derivative action, the interactive and noninteractive controllers have different values of the tuning constants (i.e., K_c, τ_I, and τ_D). The following are formulas for converting from interactive tuning parameters (the tuning parameters with primes) to tuning parameters for the conventional noninteractive PID form:

$$K_c = K_c'\left(1 + \tau_D' / \tau_I'\right)$$

$$\tau_I = \tau_I'\left(1 + \tau_D' / \tau_I'\right)$$

$$\tau_D = \tau_D'\left[\frac{1}{1 + \tau_I' / \tau_I'}\right]$$

Even though interactive controllers are an option on most DCSs, they are not recommended, because of the possible confusion concerning the tuning parameters; they offer no advantage over the

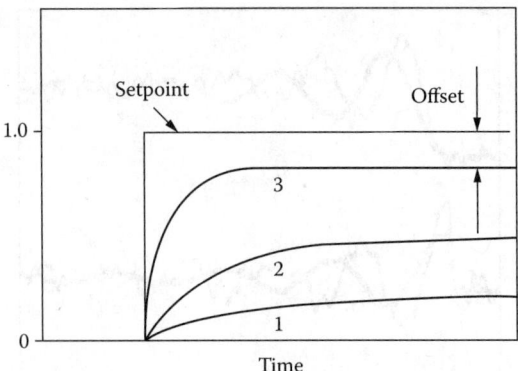

FIGURE 15.24 The effect of K_c on the response of a P-only controller for a first-order process to a setpoint change. Note that K_c is increased from 1 to 3.

noninteractive form of the PID controller. As a result, only the formulas for converting the settings from the interactive form to the noninteractive form are presented here.

15.3.3 Analysis of P, I, and D Action

The results of a theoretical analysis of proportional-only, integral-only, and derivative-only controllers[1] are summarized as follows.

15.3.3.1 Proportional Action

The key characteristics of proportional action are summarized as follows:

1. In general, proportional action does not change the order of the process.
2. The closed-loop time constant is smaller than the open-loop time constant. That is, proportional action makes the closed-loop process respond faster than the open-loop process.
3. The steady-state gain is not equal to unity. Figure 15.24 shows setpoint changes for three different values of $K_c K_p$. The steady-state value differs from the setpoint value, which indicates offset. Offset is the error between the new setpoint and the new steady-state controlled variable value. Also note that as K_c increases, the offset is reduced.

Figure 15.25 shows the portion of the controller signal resulting from proportional action for a PI controller that is applying a setpoint change. The proportional control action is positive when y is below y_{sp} and negative when y is above y_{sp}; its magnitude is directly proportional to the error from setpoint. Initially, the setpoint change causes a spike in proportional action, but as y moves toward the setpoint, the proportional action is reduced and eventually goes to zero as y settles at the setpoint.

15.3.3.2 Integral Action

The following are the fundamental characteristics of integral action:

1. All steady-state corrections for disturbances or setpoint changes must come from integral action.
2. There is no offset at steady state.

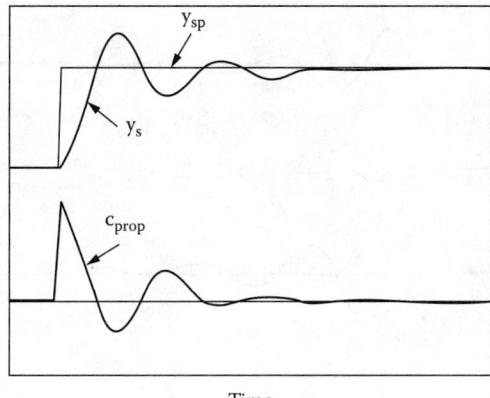

FIGURE 15.25 The portion of the controller output resulting from proportional control (c_{prop}) for a setpoint change applied by a PI controller.

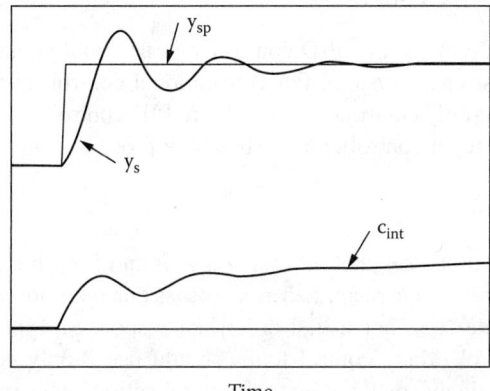

FIGURE 15.26 The portion of the manipulated variable level resulting from integral action (c_{int}) for a setpoint change applied by a PI controller.

3. Integral action increases the order of the process dynamics by 1.
4. The process becomes faster, but at the expense of larger overshoots and more sustained oscillations.

Figure 15.26 shows the portion of the controller output resulting from integral action for a PI controller for the same process shown in Figure 15.25. The peaks in c_{int} occur when y_s crosses y_{sp}. Also, as the process lines out at the setpoint, c_{int} lines out at a nonzero value.

15.3.3.3 Derivative Action

Derivative action (1) does not change the order of the process, (2) does not eliminate offset, and (3) tends to reduce the oscillatory nature of feedback control.

Figure 15.27 shows the portion of the manipulated variable level resulting from derivative action (c_{der}) for a setpoint change using a PID controller. Note that c_{der} is zero at the peaks and valleys of y, since it is directly related to the slope of y_s.

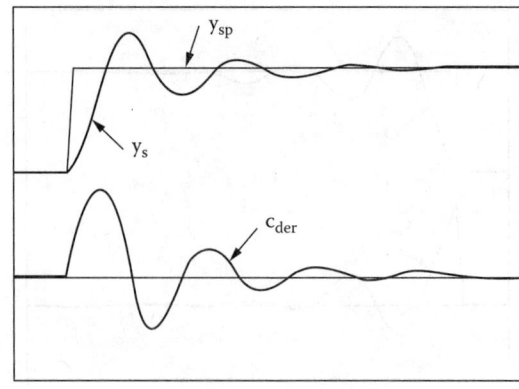

FIGURE 15.27 The portion of the manipulated variable level resulting from derivative action (c_{der}) for a setpoint change applied using a PID controller.

15.3.4 Controller Design Issues

When choosing between P-only, PI, or PID controllers, one should consider the dynamics of the combined actuator/process/sensor system. For conventional control loops in the CPI, about 93% are PI controllers, 2% are P-only controllers, and 5% are PID controllers.[2] The following guidelines can be used to choose the proper controller mode based on process dynamics and control objectives.

15.3.4.1 P-Only Control

P-only control is used for processes that are not sluggish and for which some degree of offset is acceptable. A sluggish process is characterized as a process that does not respond quickly to changes in the manipulated variable (i.e., not a first-order-like response). Typical applications are level control and pressure control. Many control loops should use P-only controllers but instead use typical PI or PI with a relatively small amount of integral action, since most operators do not want offset from setpoint.

15.3.4.2 PI Control

PI controllers are used for processes that are not sluggish and for which it is necessary to have offset-free operation. Typical applications are flow control, level control, pressure control, temperature control, and composition control.

15.3.4.3 PID Control

PID controllers are useful for certain sluggish processes. Typical applications are temperature control and composition control. A sluggish process often has a tendency to cycle under PI control due to inertia; therefore, derivative action tends to reduce the tendency to cycling and allows more proportional action to be used, both of which contribute to improved control performance. A key issue here is to determine whether a process is sluggish enough to warrant a PID controller. Assume that an FOPDT model has been fit to an open-loop step test. If the resulting deadtime, θ_p, and time constant, τ_p, are such that

$$\frac{\theta_p}{\tau_p} < \frac{1}{2}$$

the process is not sufficiently sluggish to warrant a PID controller. If

$$\frac{\theta_p}{\tau_p} > 1$$

the process is sufficiently sluggish that a PID controller should offer significant benefits over a PI controller. For

$$\frac{1}{2} < \frac{\theta_p}{\tau_p} < 1$$

either PI or PID control could be preferred. In the event that FOPDT models are not available, excessive oscillations of a PI controller or a sluggishly responding PI controller are indications that a PID controller may provide improved control performance. In addition, measurements of the controlled variable with significant noise levels can make the use of derivative action ineffective due to the sensitivity of the derivative to noise on the measurement. That is, since the measurements of the controlled variable have so much noise that, if filtering is used, the lag added by the filter can negate any benefit produced by the derivative action.

15.3.5 Analysis of Typical Control Loops

All feedback control processes involve a final control element, a process, and a sensor. That is, to change the manipulated variable level, a final control element is required. A sensor is also required to measure resulting changes in the controlled variable. The input to the final control element/process/sensor system is the controller output, c, and its output is the sensor reading. When evaluating the dynamic behavior of a process, the relative dynamics of the final control element, the process, and the sensor should be considered. For example, there are processes for which one or two of the dynamic components (final control element, process, or sensor) respond substantially faster than the slowest element and therefore can be neglected when analyzing the dynamic behavior of the overall process.

15.3.5.1 Flow Control Loop

Figure 15.28 shows a schematic of a flow control process. The dynamics of the process (i.e., flow rate changes for changes in the valve stem position) and the sensor (i.e., changes in the measured pressure drop for changes in the flow rate) are relatively fast compared with the dynamics of the control valve (i.e., changes in valve stem position for changes in the signal to the final control element). Since the overall process is relatively fast, and accurate control to setpoint is required, a PI controller is the proper choice for most flow control applications.

In spite of the fact that industrial control valves have a deadband of 10 to 25%, flow control loops are able to precisely meter deadband in the average flow typically to within 0.5% and down to 0.1% in certain cases. To understand how a flow control loop can very accurately control the flow rate using such an imprecise actuator, consider the measured flow rate and specified flow rate shown in Figure 15.29. The significant variation in the flow rate (i.e., sustained oscillations) is due to the deadband of the valve, but the average flow is precisely controlled via the high-frequency feedback control provided by the flow controller. Since the period of the flow variations is in the range of seconds, and most chemical processes have time constants several minutes or larger, the process is sensitive only to the average flow and not to flow fluctuations. Therefore, flow control loops can provide very precise metering of the average flow rate in spite of the fact that a very

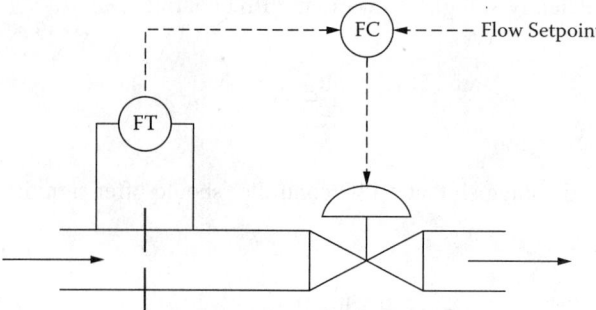

FIGURE 15.28 Schematic of a flow control loop.

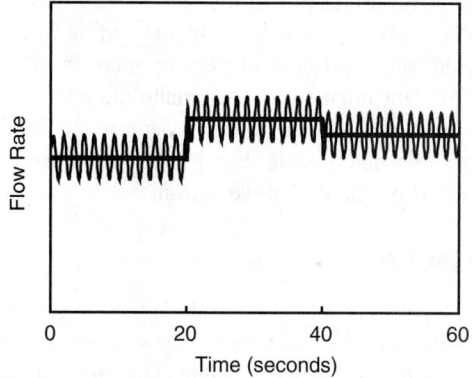

FIGURE 15.29 Measured flow rate and the specified average flow rate for a valve with a positioner. Thin line = measured flow rate, thick line = flow rate setpoint.

imprecise actuator is used. If a valve positioner is used, the valve positioner will provide the high-frequency feedback necessary to counteract the detrimental effects of the control valve deadband on the metering precision of the average flow rate. That is, the high-gain P-only controller applied by the valve positioner will open and close the valve in a manner similar to the results shown in Figure 15.29. A flow control loop applied to a control valve with a positioner will eliminate the offset that the positioner does not account for and absorb unmeasured disturbances such as changes in upstream and downstream pressures.

15.3.5.2 Level Control Loop

A schematic for a level control loop used to control the level in a tank is shown in Figure 15.30. A differential pressure sensor is used to measure the liquid level, and a flow control loop is used to control the flow rate from the tank. The output of the level controller is the setpoint for the flow controller on the line leaving the tank. Some level controllers send their outputs directly to the valve on the line, but most level controllers in the CPI are implemented as shown in Figure 15.30, using flow control loops. The objective of this loop is to maintain the level within a certain range—for example, from 30 to 40% of full level for changes in this feed rate to the tank and changes in operating conditions. On the other hand, many operators want levels controlled to specified setpoints and are not satisfied if, for example, the level is 32 or 38% when the setpoint is 35%.

The dynamics of the sensor are quite fast, and the dynamics of the flow control loop are usually fast compared with the dynamics of the process (i.e., percentage level change for change in flow

Process Control

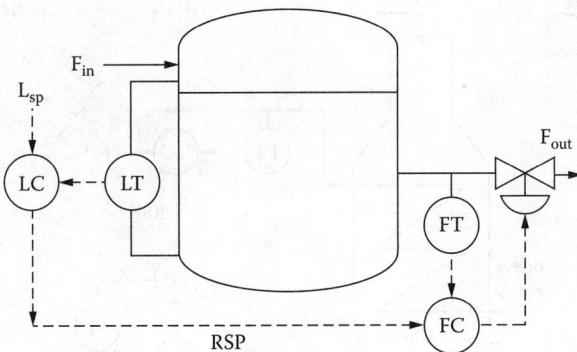

FIGURE 15.30 Schematic of a level controller applied to a tank.

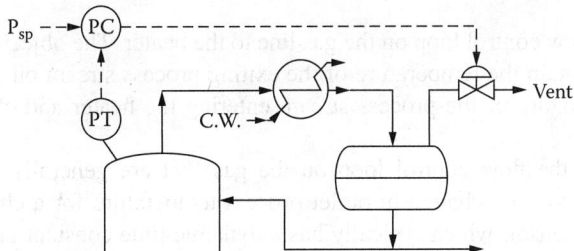

FIGURE 15.31 Schematic of a pressure controller for the overhead of a distillation column.

leaving the tank). For a typical system under open-loop conditions, a 5% level change can occur in about 1 min for about a 10% change in feed rate to the tank. Thus, the response of the final control element/process/sensor system is typically controlled by the process dynamics. Since the overall process is not generally sluggish, a P-only controller is the proper choice when offset elimination is not required. When offset elimination is required (e.g., level control for a reactor), a PI controller should be used.

15.3.5.3 Pressure Control Loop

Pressure control loops are used to maintain system pressure for distillation columns, reactors, and other process units. A pressure control loop for maintaining overhead pressure in a column is shown in Figure 15.31. The final control element is a control valve on the vent line, and the sensor is a pressure sensor mounted on the top of the column. The output from the pressure controller goes directly to the control valve on the vent line. The objective of this loop is to maintain the column overhead pressure at or near setpoint for changes in condenser duty and changes in vapor flow rate up the column.

The pressure sensor is quite fast, whereas the process (change in pressure for change in vent valve stem position) and the actuator are generally the slowest elements in the feedback system; therefore, this is also a relatively fast-responding process. The P-only controller can be used if offset elimination is not important, and a PI controller can be used when offset elimination is important.

15.3.5.4 Temperature Control Loop

Temperature control loops can be applied to control the temperature of a stream exiting a heat exchanger, of a tray in a distillation column, or of a CSTR. Figure 15.32 shows a schematic of a temperature controller applied to control the temperature of a process stream leaving a gas-fired heater. The sensor is an RTD element placed in a thermowell located in the line leaving the heater,

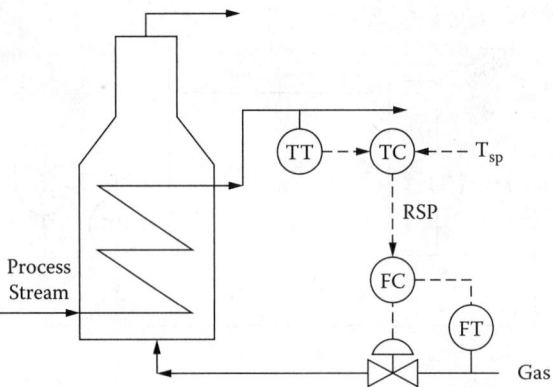

FIGURE 15.32 Schematic of a temperature controller for a gas-fired heater.

and the actuator is a flow control loop on the gas line to the heater. The objective of the temperature control loop is to maintain the temperature of the exiting process stream on setpoint in the face of changes in the temperature of the process stream entering the heater and changes in the heating value of the gas.

The dynamics of the flow control loop on the gas fuel are generally much faster than the dynamics of the process (i.e., change in outlet process temperature for a change in gas flow rate to the heater) and the sensor, which typically has a dynamic time constant between 6 and 20 s for a properly installed RTD. The process fluid entering the gas-fired heater flows by plug flow through the heat exchanger tubes that are exposed to high-temperature combusted gas. There is a thermal lag associated with changing the temperature of the metal of the heat exchanger tubes as well as transport delay caused by plug flow through the heater tubes. The transport delay and resulting overall process deadtime will increase as the feed rate of the process fluid is reduced. Since the heater is likely to behave as a sluggish process, a PID controller is generally the preferred choice in this example. Excessive sensor noise can make the use of derivative action ineffective. If this process were less sluggish, a PI controller would be preferable.

15.3.5.5 Composition Control Loop

Composition control loops are used to keep products produced by distillation columns on specification, to maintain constant conversion in a reactor, and to maintain oxygen levels in the flue gas of a boiler to eliminate carbon monoxide emissions. Figure 15.33 shows a schematic of a compo-

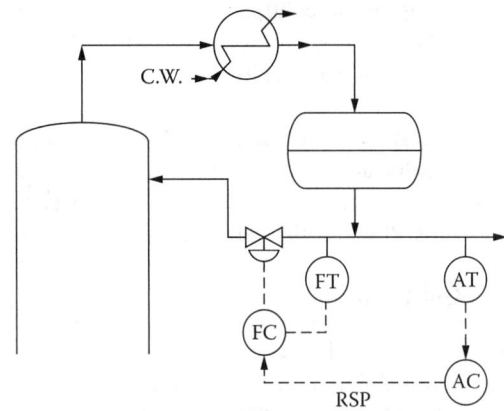

FIGURE 15.33 Schematic of a composition controller for the overhead of a distillation column.

sition loop that controls the impurity level in the overhead product of a distillation column. The sensor is a gas chromatograph that samples the distillation product, and the output of the controller for this loop is the setpoint for the reflux flow controller. The objective of the composition control loop is to keep the impurity level in the overhead product on setpoint during changes in the feed flow rate and feed composition.

The dynamics of the flow controller on the reflux are relatively fast, whereas the sensor typically can have 3 to 10 min of analyzer delay. The process (i.e., the change in impurity level in the overhead product for a change in the setpoint for the reflux flow controller) can be quite slow. If the process and analyzer delay result in a sluggish actuator/process/sensor system, a PID controller may be preferred.

15.4 PID TUNING

15.4.1 Introduction

Tuning PID loops is one of the major responsibilities of a process control engineer, and the resulting controller settings have a dominant effect on the performance of a PID control loop. Tuning a PID controller requires selecting values for K_c, τ_I, and τ_D that meet the operational objectives of the control loop, which usually requires making a proper compromise between performance (minimizing deviations from setpoint) and reliability (the controller's ability to remain in service while handling major disturbances).

15.4.2 Tuning Criteria and Performance Assessment

15.4.2.1 Tuning Criteria

To guide the tuning process, the following tuning objectives should be considered:

Deviations from setpoint should be minimized.
Good setpoint tracking performance should be attained.
Excessive variation of the manipulated variable levels should be avoided.
The controlled process should remain stable for major disturbance upsets.
Offset elimination may or may not be important.

Simultaneously satisfying each of these objectives is never possible; therefore, tuning is a compromise. For example, tuning for minimum deviation from setpoint for normal disturbances is contrary to tuning the controller to remain stable for major disturbances. That is, if the controller is tuned for normal disturbances, the closed-loop system may go unstable when a major disturbance enters the process. On the other hand, if the controller is tuned for the largest possible disturbance, control performance is likely to be excessively sluggish for normal disturbance levels.

Industrial control performance is often assessed by the variability in the final products produced. The standard deviation from setpoint (σ) can be used as a measure of variability in industrial products and thus a measure of control performance:

$$\sigma = \sqrt{\frac{\sum_{i=1}^{N}[y_s(t_i) - y_{sp}]^2}{N}}$$

where $y_s(t_i)$ is the sampled controlled variable value at time equal to t_i, and N is the number of samples. Note that the smaller the deviation from setpoint, the better the control performance.

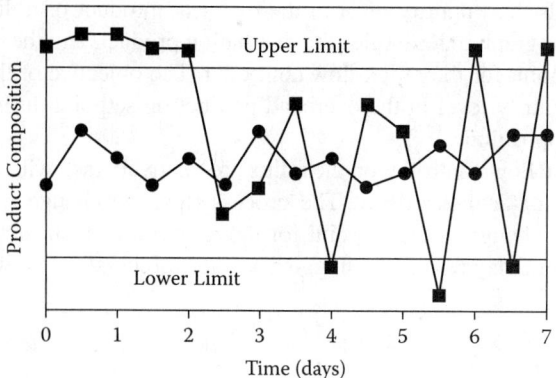

FIGURE 15.34 SPC chart based on a seven-day period of data for two different controllers on the same process.

Remember that the standard deviation is based on the error from the average value of a set of data, whereas this statistic is based on the error from setpoint.

Most companies keep statistical process control (SPC) charts that track the laboratory analysis of final products, which are typically sampled one to three times daily. Figure 15.34 is an example of an industrial SPC chart for two different controllers, for two different seven-day periods. It is easy to see which controller performed better.

15.4.3 Effect of Tuning Parameters on Dynamic Behavior

For an open-loop overdamped process, the closed-loop dynamic behavior will go through the same stages as the controller aggressiveness is increased: overdamped, critically damped, underdamped, ringing, and unstable (see Figure 15.12). For a PID controller, controller aggressiveness is increased as K_c is increased or as τ_I is decreased.

15.4.3.1 PI Control

Figure 15.35 shows the dynamic behavior of a process for a setpoint change with different amounts of proportional action. Figure 15.35b shows the results for a PI controller tuned for quarter-amplitude damping (QAD, a decay ratio of 1/4). In addition, the QAD tuning was modified by increasing K_c (Figure 15.35c) while keeping τ_I constant and by decreasing K_c (Figure 15.35a) while keeping τ_I constant. Note that the increase in K_c resulted in ringing, while the decrease in K_c resulted in sluggish behavior. In addition, larger controller gains result in longer settling times. Figure 15.36 shows similar results for the effect of variations of τ_I. Figure 15.36b shows the results for QAD tuning and is the same result as shown in Figure 15.35b. A decrease in τ_I from QAD settings results in ringing (Figure 15.36c) and an increase results in a slow removal of offset (Figure

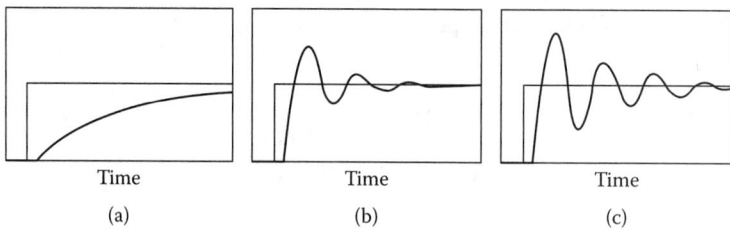

FIGURE 15.35 PI controller responses for a FOPDT process with varying amounts of proportional action. (a) K_c too low, (b) K_c tuned for QAD, (c) K_c too large.

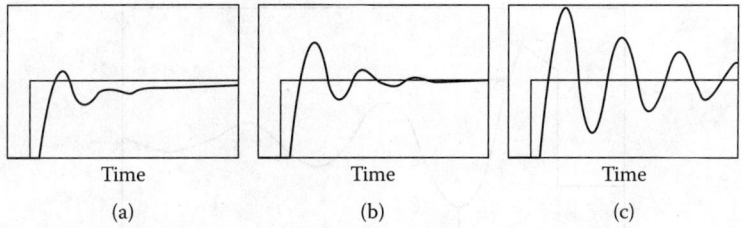

FIGURE 15.36 PI controller response for an FOPDT process with varying levels of integral action. (a) τ_I too large, (b) τ_I tuned for QAD, (c) τ_I too small.

15.36a). By comparing Figures 15.35a and 15.36a, it can be seen that when K_c is too low, long rise times and overdamped behavior result, and when integral action is too low (τ_I is too large), offset elimination is slow. Also note that ringing from too much proportional action (Figure 15.35c) and ringing from too much integral action (Figure 15.36c) are quite similar; therefore, when controller ringing results, it is difficult to tell whether there is excessive proportional action, excessive integral action, or both.

Figure 15.37 shows the manipulated and controlled variables for a QAD-tuned controller. The controller output "lags" behind the controlled variable for this case. Figure 15.38 shows the same system, except the K_c is increased by 25% and τ_I is increased by a factor of 2 (i.e., the ringing is caused by too much proportional action). The lag between the controlled variable and the controller output is significantly reduced; therefore, excessive gain reduces the lag between the controlled variable and the controller output. Figure 15.39 shows the case in which K_c and τ_I are reduced by a factor of 2 compared with the QAD settings. In this case, the lag increases significantly; therefore, excessive integral action results in an increase in the lag of the sytem. As a result, the lag between the controller output and the controlled variable can be used to determine if a controlled process is ringing from too much proportional action or too much integral action. This analysis can also be used to compare the ringing results shown in Figures 15.35 and 15.36. These results also show that the lag between the controlled variable and the controller output is larger when there is excessive integral action, but the differences are less distinct than those shown in Figures 15.38 and 15.39. With pure proportional action, the maximum c occurs at the maximum deviation from setpoint, which corresponds to zero lag (i.e., in phase). For pure integral action, the maximum c occurs when the error from setpoint changes sign, which corresponds to a large lag.

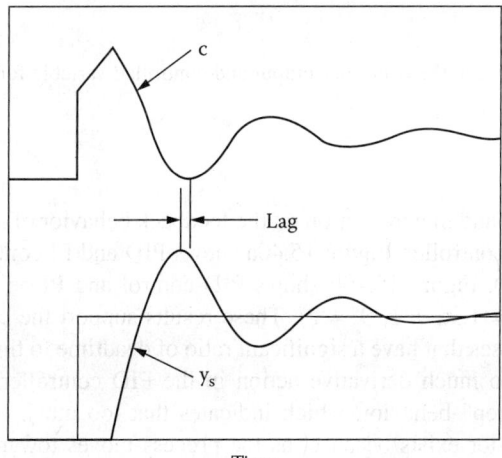

FIGURE 15.37 The lag between the controller output and controlled variable for a QAD-tuned PI controller.

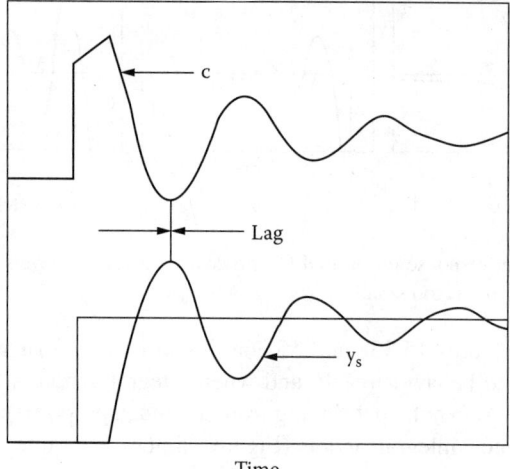

FIGURE 15.38 The lag between the controller output and controlled variables for a controller with too much proportional action.

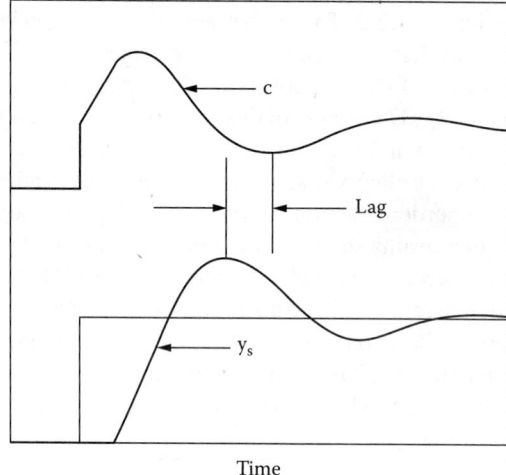

FIGURE 15.39 The lag between the controller output and controlled variable for a controller with too much integral action.

15.4.3.2 PID Control

The effect of proportional and integral action on the feedback behavior of a PID controller is similar to that observed for a PI controller. Figure 15.40a shows PID and PI control on a FOPDT process ($K_p = 1$, $\tau_p = 1$, $\theta_p = 0.1$). Figure 15.40b shows PID control and PI on another FOPDT process with more deadtime ($K_p = 1$, $\tau_p = 1$, $\theta_p = 2$). These results support the conclusion that derivative action is useful for processes that have a significant ratio of deadtime to time constant. Figure 15.41 shows a case that has too much derivative action in the PID controller. Note that the feedback response shows a "stairstep" behavior, which indicates that too much derivative action is being used. The stairstep behavior exists because, as the process moves toward the setpoint, excessive derivative action causes the process to stall or level out. When the process stalls, the proportional and integral actions act on the process to move the controlled variable toward the setpoint. When

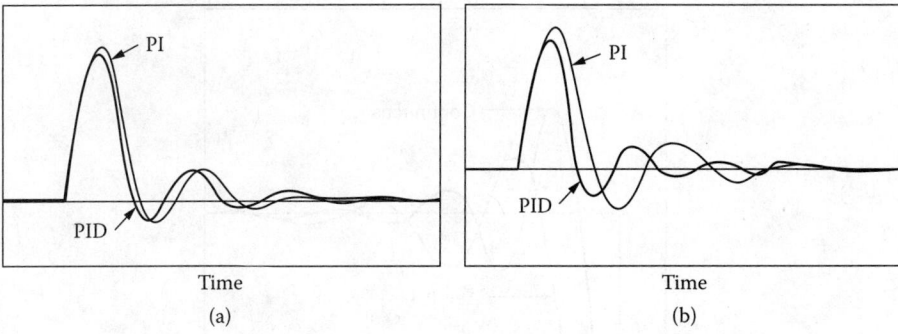

FIGURE 15.40 Comparison between PI and PID controllers for a process with (a) low deadtime and (b) larger deadtime.

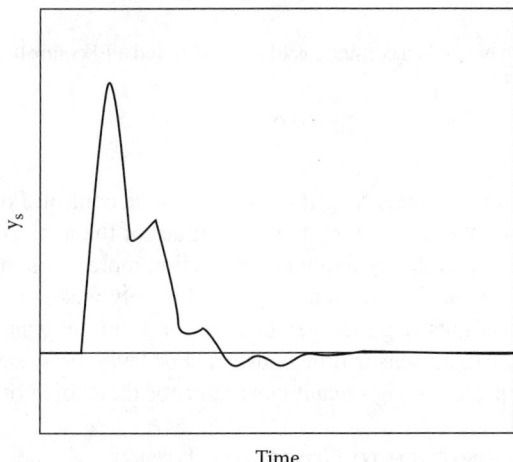

FIGURE 15.41 The control performance of a PID controller with too much derivative action.

this occurs, the derivative of the controlled variable will build up, and the derivative action will act against it, causing the stairstep effect.

15.4.3.3 Control Interval

The PID control results presented so far have been based on a continuous application of the controller. The digital application of feedback control is applied at discrete points in time. DCSs use sequential microprocessors that perform control calculations for many control loops. Typical control loops are executed every 0.2 to 0.5 s for regulatory loops and 30 to 120 s for supervisory loops. The time between control applications is the control interval, Δt. PID control is applied industrially on DCSs using digital formulas that are applied at discrete control intervals [Equations (15.6), (15.8), and (15.9)].

Consider the QAD timed continuous PI controller that was applied to the FOPDT process ($K_p = 1$, $\tau_p = 1$, $\theta_p = 5$). If these settings are applied using a control interval of 0.5, the control behavior is unstable. The PI controller was retimed using a control interval of 0.5, and the results are compared with continuous control for a setpoint change in Figure 15.42. The controller gain for discrete control ($\Delta t = 5$) was reduced by 60% compared to the continuous controller. Because of the detuning and the delayed response, the discrete controller resulted in a longer settling time.

As a general rule,[1] the control interval should be selected such that

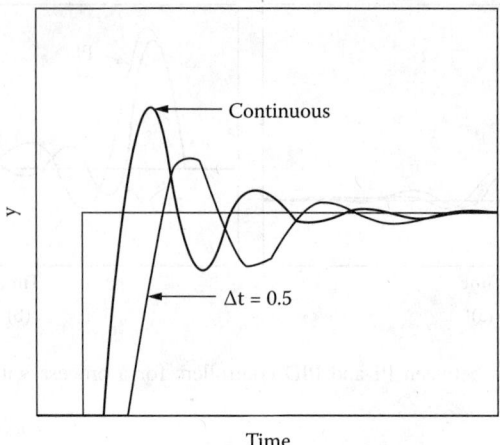

FIGURE 15.42 Comparison between a continuous PI controller and a PI controller applied each 0.5 time unit.

$$\Delta t \leq 0.05\left(\theta_p + \tau_p\right)$$

to obtain control performance approaching that of continuous control. For feedback control using an on-line GC, the control interval is set by the cycle time for the analyzer updates (typically 3 to 10 min). No advantage is gained by applying control action more frequently than the GC updates, since new information on the process response is available only when the GC updates. For sensors that provide continuous readings (e.g., temperature sensors), the maximum recommended control interval is typically equal to one sensor time constant. For level, pressure, and flow loops, sensor dynamics do not usually present a significant constraint for the choice of the control interval.

15.4.4 Recommended Approach to Controller Tuning

The following procedure is recommended for tuning PID control loops:

1. Select the tuning criterion for the control loop. The tuning criterion depends on how the control loops affect the overall process objectives and can involve applying a compromise between performance and reliability.
2. Apply filtering to the sensor reading. Sensor filtering reduces the effect of sensor noise on the variability in the controlled variable but introduces lag to the feedback system, which is detrimental to control performance. Therefore, filtering should be applied carefully.
3. Determine if the control loop is a fast- or slow-responding control loop. The distinction between fast- and slow-responding control loops is concerned with the closed-loop response time of the system. If a set of controller settings can be tested using setpoint changes in a reasonable period of time (e.g., less than 10 min), the process is a fast-responding control loop. If not, it is a slow-responding control loop.
4. For fast-responding control loops, apply field tuning.
5. For slow-responding control loops, apply the ATV-based tuning procedure.

15.4.5 Controller Reliability

Controller reliability has to do with whether a controller will stay in service during major upsets. Major upsets cause significant deviations from setpoint that result in variations in the process behavior for nonlinear processes, which can significantly affect the tuning performance. That is, when the inputs to a nonlinear process change, the effective gain, time constant, and deadtime can change. For example, if a disturbance to a process causes the process gain to increase, the time constant to decrease, and the deadtime to increase, the closed-loop behavior will move toward (if not into) unstable operation. On the other hand, if a disturbance to a process causes the gain to decrease, the time constant to increase, and the deadtime to decrease, the closed-loop behavior will become more sluggish. Therefore, a nonlinear process subjected to large disturbances should be tuned more conservatively than a more linear process with low-level disturbances to reduce the likelihood that the operation of the nonlinear process will become unstable.

The general behavior of a second-order process (Figures 15.5 and 15.6) can be used to characterize the dynamic behavior of a closed-loop process and used as the tuning criterion for a controller. For example, a controller tuned for critically damped behavior is more conservatively tuned than a controller tuned for QAD (decay ratio of 1/4). A more conservative tuning criterion should be chosen for more nonlinear processes that are subjected to large-magnitude disturbances, whereas a more aggressive tuning criterion should be used for more linear processes that are subject to low-level disturbances. More conservative settings are chosen for more difficult control problems (i.e., high nonlinearity and severe disturbances) so that larger variations in process characteristics (effective gain, time constant, and deadtime) can occur and yet the controller can remain stable. On the other hand, a more aggressive tuning should be selected for easier control problems so as not to unnecessarily sacrifice control performance (variability from setpoint).

15.4.6 Selection of Tuning Criterion

Selecting the tuning criterion for a control loop is equivalent to determining the aggressiveness of the feedback controller. If one were to choose a critically damped response as the tuning criterion, a controller with a low aggressiveness would result. On the other hand, QAD tuning criterion corresponds to a very aggressively tuned controller.

The first factor that should be considered when choosing the proper tuning criterion is how the control loop affects the overall objectives of the process. Consider the accumulator of a distillation column that feeds a plug flow reactor (Figure 15.43). Tight level control for the accumulator results in sharp changes in the feed to the reactor. Sharp, frequent changes to the feed rate to the plug flow reactor significantly upset the operation of the reactor, because the residence time of the reactor is directly related to the feed rate, and changes in the residence time affect the degree of conversion. On the other hand, if the level controller for the accumulator is tuned for sluggish performance, the changes in the feed rate to the reactor are much more gradual and smooth, allowing better operation of the reactor. From an overall process point of view, maintaining smooth operation of the reactor is much more economically important than maintaining tight level control in the accumulator; therefore, a critically damped or overdamped tuning criterion is the proper choice for the level controller on the accumulator in this example.

As another example of how the overall process objective can affect the selection of the tuning criterion, consider the CSTR and separation train shown in Figure 15.44. If the level controller for the CSTR is tuned loosely, the CSTR level can change significantly. Because the production rates of the various products are related to the residence time in the reactor, large variations in the CSTR level directly affect the product distribution. The resulting composition changes in the stream

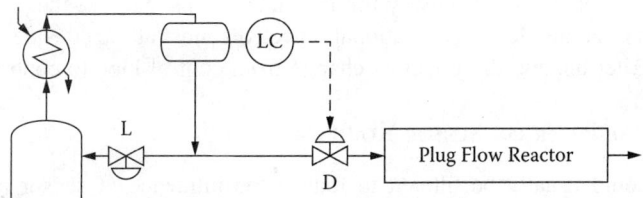

FIGURE 15.43 Schematic of a level controller that feeds a plug flow reactor.

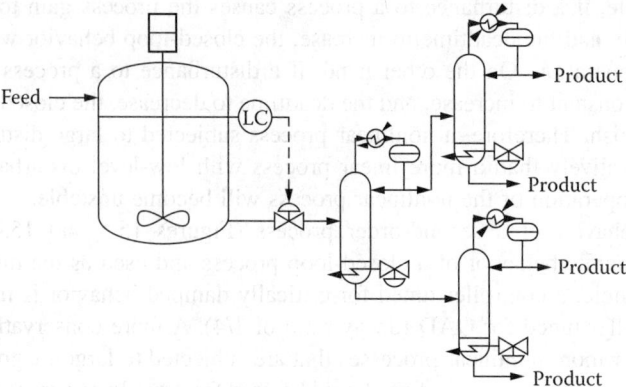

FIGURE 15.44 Schematic of a level controller that feeds a separation train.

leaving the CSTR represent major upsets for the composition controllers for the distillation columns in the separation train. Even though tight level control for the CSTR causes short-term variations in the feed to the first column, these upsets are much easier to handle than composition changes. Therefore, considering the overall process objectives, a tuning criterion corresponding to tight level control should be chosen for the level controller on the CSTR. These two examples indicate that it is important to consider how a control loop affects the overall process when selecting its tuning criterion.

If consideration of the overall process objectives does not place a limit on the aggressiveness of the tuning criterion, or if the overall process objectives dictate that minimum variability in the controlled variable of the loop in question is desirable, the selection of the tuning criterion should be based on a compromise between performance and reliability. The previous section indicates that process nonlinearity and disturbances determine the reliability of a controller. If a process is highly nonlinear and subject to large disturbances, controller reliability will likely be a problem, and a more conservative tuning criterion should be selected (e.g., a critically damped response). On the other hand, if the process is relatively linear and the disturbances are relatively mild, a more aggressive tuning criterion should be selected (e.g., a 1/6 decay ratio or 40% overshoot). Therefore, when control engineers choose a tuning criterion, they compromise between performance and reliability; they must use their knowledge of the process to evaluate the relative nonlinearity of the process and the relative degree of severity of the disturbances.

Even though QAD provides the best overall performance in terms of errors from setpoint, many companies are reluctant to have their control engineers tune even well-behaved control loops for QAD because of the 50% overshoot associated with QAD and because QAD is too close to the onset of instability. In addition, since QAD causes significant variation in the manipulated variable levels, QAD can result in unduly upsetting other parts of the process. For these reasons, it is probably better to tune well-behaved loops for decay ratios of 1/6 to 1/8. For a process that is more nonlinear with more severe disturbances, 1/10 amplitude damping or a critically damped response is more appropriate. In extreme cases, an overdamped tuning criterion may be the proper choice. No single tuning criterion works effectively for all control loops, because the process nonlinearity, disturbance type and magnitude, and operational objectives must all be considered when choosing the proper tuning criterion, and these factors change from control loop to control loop.

15.4.7 Tuning the Filter on Sensor Readings

Sensor readings should usually be filtered to reduce the influence of sensor noise on feedback control performance. Filtering, however, adds lag to the closed-loop response. In certain cases, tuning a filter on a sensor can involve balancing the benefits of reducing the noise against the

Process Control

detrimental effects of adding lag to the overall process. In other cases, if possible, the time constant of the filter should be significantly smaller than the other dominant time constants in the actuator/process/sensor system so that filtering does not slow down the response of the system.

It is usually more convenient to use the filter time (i.e., the time constant of the first-order filter), τ_f, to specify the amount of filtering applied to a sensor reading. In this manner, the time constant of the filter can be directly compared to the time constants of the actuator, process, and sensor to determine whether it affects the closed-loop dynamics. The filter factor, f, and the cycle time for applying the filter, Δt_f, can be used to calculate the filter time constant by the following equation:

$$\tau_f = \Delta t_f \left[\frac{1}{f} - 1 \right]$$

For most DCSs, sensor readings are updated five or six times per second (i.e., Δt_f is 0.16 to 0.2 s). For a filter time constant of 3 s, the filter factor is equal to 0.06; therefore, relatively extensive sensor filtering can result for most sensors using high-frequency updating by a DCS.

The amount of sensor filtering required depends on the amount of noise on the reading. For example, a reading from a thermocouple is expected to require more filtering than a reading from an RTD, which has an order of magnitude smaller repeatability (Table 15.1), for the same application. From an examination of Table 15.2, one can see that a properly functioning sensor usually has a relatively small amount of noise. As a result, most flow, level, pressure and temperature sensors can be filtered effectively using a filter time constant of 3 to 5 s, which does not normally affect the closed-loop dynamics. Composition analyzer readings from GCs are updated so infrequently that filtering is usually not used for them. On the other hand, if composition measurements are available at a sufficiently high frequency, filtering can be used effectively.

In certain cases, the filtering of a noisy sensor is required. Noisy sensors present a challenge. A nuclear-based level sensor, a pressure sensor located too close to a 90° elbow in a line, or an orifice flow meter located immediately downstream of a control valve are examples of noisy sensors. For these cases, tuning the filter is a compromise between removing the noise from the sensor reading and adding lag to the closed-loop response when one is forced to use a noisy sensor reading.

15.4.8 Fast-Response Loops

For fast-responding loops, such as flow control and pressure control loops, the simplest and quickest tuning method available is field tuning, which is based on a trial-and-error selection of tuning parameters. Some level and temperature loops also behave as fast-responding control loops. A fast-responding control loop is defined here as a control loop that has a closed-loop response time of 10 min or less. Since these processes respond quickly, trial-and-error tuning is effective. It is usually easier to field tune a fast-responding loop rather than using other tuning methods. Use initial tuning parameters from a chosen tuning method, and adjust the tuning to meet the selected tuning criterion. The recommended procedure for field tuning, assuming that the tuning criterion has been selected and the sensor reading filtered, follows:

1. Turn off the derivative action ($\tau_D = 0$) and the integral action ($\tau_I \to \infty$).
2. Use an initial estimate of K_c, e.g.,

$$K_c = \frac{1}{2K_p}$$

Estimate K_p from process knowledge.

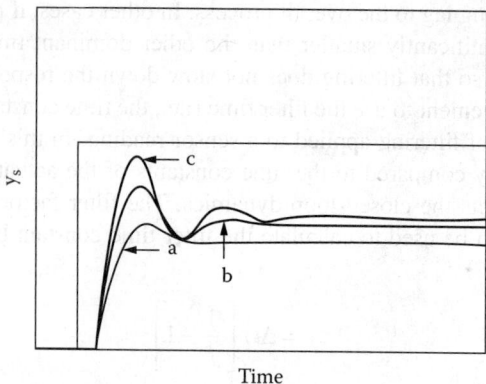

FIGURE 15.45 Selection of K_c during field tuning. (a) Results for initial value of K_c, (b) results for an increase in K_c, and (c) results for final value of K_c (1/6 decay ratio).

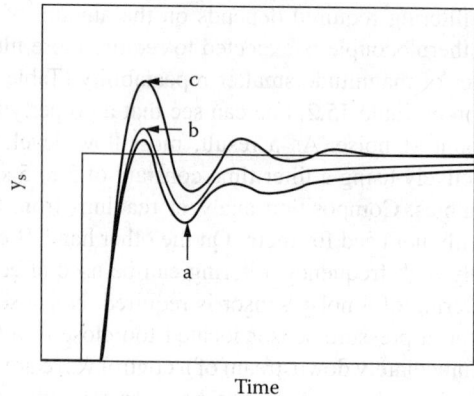

FIGURE 15.46 Selection of τ_I during field tuning. (a) Results for initial value of τ_I, (b) results for a decrease in τ_I, and (c) results for the final value of τ_I (1/6 decay ratio).

3. Using setpoint changes, increase K_c in small increments until the response meets the tuning criterion. (See Figure 15.45, which is based on a 1/6 decay ratio.) For tuning a P-only controller, the tuning procedure is completed.
4. Decrease K_c by 10%.
5. Use an initial value of τ_I, i.e., $\tau_I \cong 5\tau_p$. Estimate τ_p from process knowledge.
6. Decrease τ_I until offset is eliminated and the tuning criterion is met for setpoint changes. (See Figure 15.46, which is also based on a 1/6 decay ratio.)
7. Check to ensure that adequate levels of proportional and integral actions are being used.

15.4.9 Slow-Response Processes

For slow-response loops (e.g., certain temperature and composition control loops), field tuning can be a time-consuming procedure that leads to less than satisfactory results. Step test results can be used to generate FOPDT models, and tuning parameters can be calculated from a variety of techniques. This approach suffers from the fact that it takes approximately the open-loop response time of the process to implement a step test, and during that time, measured and unmeasured disturbances can affect the process, thus corrupting the results from the step test. In addition, it is unlikely that the selected tuning approach will result in the proper balance between reliability and

Process Control

TABLE 15.7
Typical Tuning Parameters for Common Loops in the CPI

Loop Type	PB	τ_I (s)	τ_D (s)
Flow controller	100–500%	0.2–2.0	0
Gas pressure controller	1–15%	5–100	0
Liquid pressure controller	100–500%	0.2–2.0	0
Level controller	5–50%	5–60	0
Temperature controller	10–50%	40–4000	30–2000*
Composition controller	100–1000%	100–5000	30–4000*

* τ_D should always be smaller than τ_I.

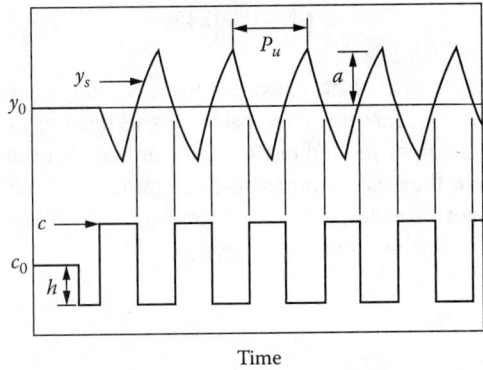

FIGURE 15.47 Graphical representation of an ATV test.

performance. Because of model mismatch and the likely selection of an inappropriate tuning criterion, significant adjustments to the tuning are still required.

The ATV² (autotune variation) method determines the ultimate gain and period in a manner similar to the ultimate method, but ATV tests can be implemented without unduly upsetting the process. Controller settings can be calculated, and the controller can then be tuned on-line to meet the selected dynamic performance.

Figure 15.47 graphically demonstrates the ATV method. The user must select h, the relay height used or the change in the manipulated variable that is applied. The value of h should be small enough that the process is not unnecessarily upset yet is large enough that the resulting amplitude, a, can be accurately measured.

To initiate an ATV test, the process should be at steady-state or near steady-state conditions, c_0 and y_0. Next, the controller output is set to $c_0 + h$ (or $c_0 - h$) until y deviates significantly from y_0. At that point, the controller output is set to $c_0 - h$ (or $c_0 + h$), which will turn the process back toward y_0. Then, each time y crosses y_0, the controller output is switched from $c_0 + h$ to $c_0 - h$ or from $c_0 - h$ to $c_0 + h$. The process is also referred to as a *relay feedback experiment*. A standing wave is established after 3 to 4 cycles; therefore, the values of a and the ultimate period, P_u, can be measured directly, and the ATV test is concluded. The ultimate gain, K_u, is calculated by

$$K_u = \frac{4h}{\pi a} \quad (15.12)$$

K_u and P_u can be used in one of several tuning schemes. One tuning approach is the Ziegler–Nichols (ZN) ultimate settings.[3] Consider the ZN settings for a PI controller:

$$K_c^{ZN} = 0.45 K_u$$

$$\tau_I^{ZN} = P_u / 1.2$$

ZN settings are fairly aggressive and can lead to ringing behavior for nonlinear processes due to the relatively small value of τ_I (i.e., large integral action).

Another tuning approach that was developed for processes that behave like an integrator plus deadtime system is the Tyreus and Luyben (TL) settings[4]:

$$K_c^{TL} = 0.31 K_u$$

$$\tau_I^{TL} = P_u / 0.45$$

The TL settings are less aggressive, with considerably less integral action than the ZN settings. The TL settings are recommended for more sluggish processes that are well represented as integrator plus deadtime for a good portion of its step test (e.g., a sluggish distillation column). After the ZN or TL settings are calculated, they may require on-line tuning, particularly for the ZN settings, to meet the desired dynamic performance (e.g., 1/6 decay ratio or critically damped). For example, the ZN settings are tuned on-line as follows:

$$K_c = K_u^{ZN} / F_T$$

$$\tau_I = \tau_I^{ZN} \times F_T \qquad (15.13)$$

by adjusting F_T on line. Note that as F_T is increased, K_c decreases while τ_I increases by the same proportion (detuning). The tuning factor, F_T, can be adjusted to meet the performance requirements for each individual application. Therefore, on-line tuning has been reduced to a one-dimensional search for the proper level of controller aggressiveness for a PI controller. If the controller is too aggressive, F_T is increased. If the controller is too sluggish, decrease F_T.

Note that the procedure based on ATV identification with on-line tuning is applicable for tuning PI controllers. It should be pointed out that, for certain cases, after this procedure has been applied, it will be evident that the proper balance between proportional and integral action has not been used, e.g., if offset elimination is slow. In these cases, adjustments in the relative amount of proportional or integral action may be required. For example, if the TL settings were used and not enough integral action resulted, the 0.45 factor in the TL settings for integral action (i.e., $P_u/0.45$) could be increased to speed up offset elimination. Figures 15.35 and 15.36 can be helpful in determining if insufficient proportional or integral action is being used.

As an example of an ATV test, consider its application to a dynamic simulator of a C_3 (propylene/propane) splitter. Figure 15.48 shows an ATV test and an open-loop test on the same time scale for the bottom product composition control loop. Note that the four cycles of the ATV test required 6 to 8 hr, while the open-loop test required in excess of 60 hr. The ATV results were used with TL settings, and the results for three different tuning factors are shown in Figure 15.49.

Summarizing, identifying the ultimate gain and ultimate period of a slow-response loop using the ATV method is relatively fast, providing a "snapshot" of the process without unduly upsetting the system. In addition, the on-line tuning procedure provides a systematic method of selecting the proper degree of controller aggressiveness. Therefore, the ATV test with on-line tuning represents

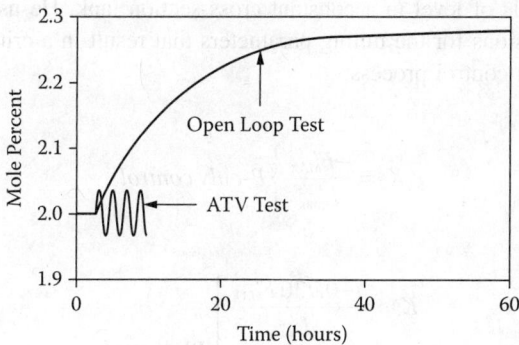

FIGURE 15.48 Comparison of an ATV and an open-loop test.

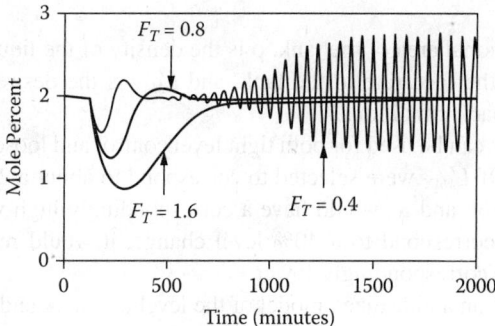

FIGURE 15.49 Effect of F_T on dynamic response.

an industrially relevant means of attaining high-quality controller tuning for loops with large response times.

15.4.10 PID Tuning

PID tuning of slow-response processes is less systematic than tuning PI controllers, since the on-line tuning procedure [Equation (4.2)] is not generally effective for PID controllers. That is, applying a tuning factor, F_T, to the derivative time and tuning a PID controller by adjusting only F_T does not, in general, lead to a well-tuned PID controller. The recommended procedure for tuning PID controllers is as follows:

1. Tune a PI controller using ATV identification with on-line tuning. Make sure that the proper balance between proportional and integral action is used. It may be necessary to reduce τ_I to produce symmetric oscillations about the new setpoint.
2. Add derivative action and tune τ_D for minimum response time. Initially set τ_D equal to $P_u/8$, where P_u comes from the ATV test.
3. Increase K_c and τ_D by the same factor until the desired dynamic response is obtained.
4. Check the response to ensure that the proper level of integral action is being used.

15.4.11 Level Controller Tuning

If a level control process is fast responding, then field tuning is effective. If the level control process is relatively slow responding, it can be helpful to use the following approach to select the initial settings for the level controller. Marlin[5] developed closed-form solutions for the dynamic behavior of PI and P-only controls of level in a constant cross-section tank. He used these expressions to derive analytical expressions for the tuning parameters that result in a critically damped response for the closed-loop level control process:

$$K_c = \frac{-F'_{MAX}}{L'_{max}} \quad \} \text{P-only control} \qquad (15.14)$$

$$\left. \begin{array}{l} K_c = \dfrac{-0.736\, F'_{MAX}}{L'_{max}} \\[1em] \tau_I = \dfrac{4 A_c \rho}{-K_c} \end{array} \right\} \text{PI control} \qquad (15.15)$$

where A_c is the cross-sectional area of the tank, ρ is the density of the liquid, F'_{MAX} is the maximum expected step change in the feed rate to the tank, and L'_{MAX} is the desired level change that F'_{MAX} should cause under feedback conditions.

These tuning relations can be used for both tight level control and loose level control, depending on the selection of L'_{MAX}. If L'_{MAX} were selected to correspond to about a 2% level change, it would represent tight level control, and K_c would have a correspondingly high value. On the other hand, if L'_{MAX} were selected to correspond to a 40% level change, it would represent quite loose level control, and K_c would be correspondingly lower.

This analysis is based on an idealized model of the level of a tank and does not consider sensor or actuator dynamics and does not consider that horizontal tanks do not have a constant cross section. For these reasons, it is recommended that Equations (15.14) and (15.15) be used as initial estimates of the tuning parameters and that an on-line tuning factor, F_T, be used to tune for the desired level control performance:

$$K'_c = K_c / F_T$$

$$\tau'_I = \tau_I \times F_T$$

Example 15.2 Calculation of Initial Tuning Parameters for a Level Controller

Problem statement. Consider level control in a horizontal cylinder tank that is 6 ft in diameter and 20 ft long. Normally, the feed rate to the tank is 10,000 lb/hr of a dilute aqueous solution. Feed rate step changes are normally within the range of ±10% of the normal feed rate. The setpoint for the level is usually set at 20%. The pressure taps for the level indicator are located at the top and bottom of the tank. Determine the tuning parameters for a PI controller that will keep the level within ±5% of setpoint based on Equation (15.15) for ±10% feed rate changes.

Solution. By geometric analysis, the width of the liquid level in the tank at 20% full is 4.8 ft; therefore, the cross-sectional area is 96 ft². Using the density of pure water,

$$F'_{MAX} = (0.1)(10{,}000\, \text{lbs/h}) \left(\frac{\text{h}}{60\, \text{min}} \right) = 16.67\, \text{lbs/min}$$

$$K_c = \frac{-0.736(1000\, \text{lbs/h})}{5\%} = -147.4 \frac{\text{lbs/h}}{\%}$$

$$\tau_I = \frac{(4)(96\,ft^2)(62.4\,lbs/ft^3)(6\,ft/100\%)}{147.4\dfrac{lbs/h}{\%}} = 585\,\min$$

15.5 ADVANCED PID CONTROL

15.5.1 Introduction

The performance of PID controllers suffers from several limitations: disturbances, analyzer deadtime, process nonlinearity, constraints, windup, abrupt startup of a loop, and flow control over a wide operating range. This section addresses approaches that have been developed to improve the performance of PID controllers with respect to this set of control problems.

15.5.2 Cascade Control

Cascade control significantly reduces the effect of certain types of disturbances by applying two control loops in tandem, i.e., the output of one controller is the setpoint for the other controller. The secondary or slave controller receives its setpoint from the primary or master controller and operates on a much faster cycle time than the primary. As a result, the secondary controller can eliminate certain disturbances before they are able to affect the primary control loop.

Figure 15.50a shows a schematic of a steam-heated heat exchanger without cascade control. Assume that the steam pressure increases and results in an increase in the temperature of the process stream leaving the heat exchanger. As the outlet temperature begins to rise, the PID controller on the temperature of the outlet stream begins to take corrective action by reducing the stem position of the valve on the steam line. By the time the PID controller starts to take corrective action, an excessive amount of heat has already been transferred from the steam to the process fluid in the heat exchanger.

Figure 15.50b shows the steam heated-heat exchanger with a cascade control configuration. The pressure control loop is the secondary or slave loop. The temperature control loop is the primary or master loop. Hence, the output of the temperature controller for the cascade control case is the setpoint for the pressure controller, while the output from the temperature controller for the case without cascade control goes directly to the control valve on the steam line. For the cascade control case, when the steam supply pressure increases, the pressure of steam inside the heat exchanger increases, but the pressure controller reacts quickly by closing the valve until the desired pressure of steam in the heat exchanger is reinstated. As a result, the steam pressure disturbance is almost completely absorbed by the slave loop before it can affect the master loop. In addition, the pressure control loop overcomes the detrimental effects of valve deadband by using high-frequency feedback action. Since the pressure control loop (slave loop) is much faster responding than the temperature control loop (master loop), the pressure control loop quickly compensates for specific disturbances that affect the steam pressure before they affect the temperature loop.

Figure 15.51 is a multiple-cascade configuration designed to maintain the impurity level in the bottoms product of a distillation column at its setpoint. The innermost control loop is a flow control loop on the steam to the reboiler. Cascade controllers that use flow controllers as the slave loop are the most common form of cascade control in the CPI. The flow controller provides fast response to steam pressure changes in spite of valve deadband. The setpoint for the flow control loop is set by the tray temperature controller. Tray temperature strongly correlates with product composition for a large class of industrial columns, which is an example of inferential control, as discussed in Section 15.5.5. The advantage of controlling tray temperatures on distillation columns comes from the fact that composition changes are measured much faster using tray temperatures than using on-line analyzers. Moreover, for fast-acting columns (i.e., when the reflux ratio is relatively low),

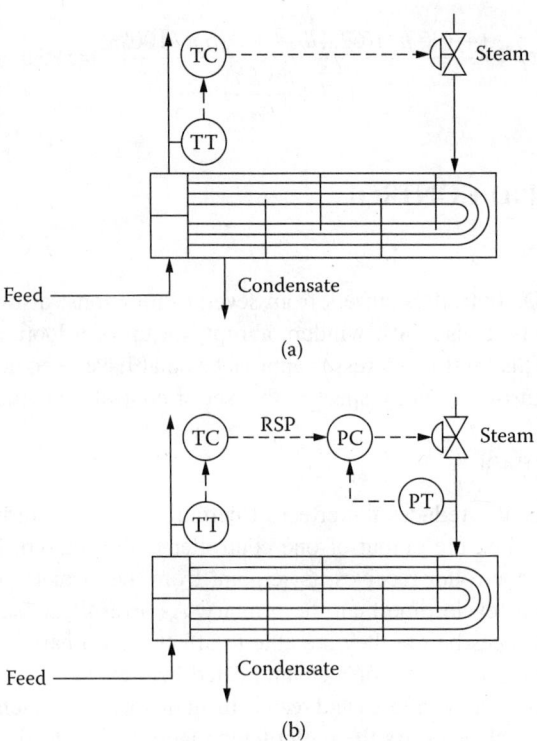

FIGURE 15.50 Schematic of a steam-heated heat exchanger with a temperature controller for controlling the temperature of the exiting process fluid. (a) Without cascade and (b) with cascade control.

feedback control using the GC can result in poor control performance, because the resulting deadtime to time constant ratio of the process is too large. For these cases, tray temperature control loops have a much smaller deadtime to time constant ratio because of the fast response of temperature sensors. Therefore, tray temperature control loops exhibit better control performance with shorter closed-loop response times than control directly off the GC. As the feed composition changes, the proper tray temperature setpoint changes. Therefore, adjustments to the setpoint for the tray temperature controllers are made by the composition control loop, which is the overall master loop for this cascade arrangement. This multiple-cascade arrangement works effectively, because the flow control loop is much faster than the temperature control loop, which is much faster than the composition control loop.

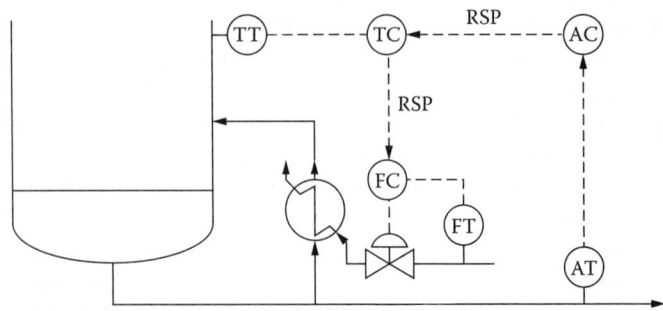

FIGURE 15.51 Schematic of a multiple-cascade configuration applied for bottoms composition control of a distillation column.

Process Control

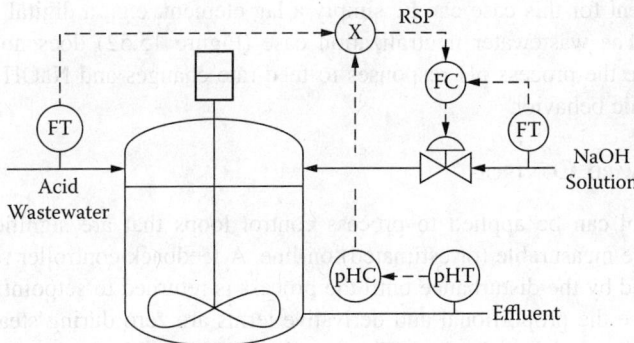

FIGURE 15.52 Schematic of ratio control applied for pH control of an acid wastewater neutralization process.

15.5.3 Ratio Control

Many processes scale directly with the feed rate to the process, e.g., distillation columns and wastewater neutralization. For distillation columns, all the liquid and vapor flow rates within the column are directly proportional to the column feed rate if the product purities are maintained and the tray efficiency is constant. For wastewater neutralization, the amount of reagent necessary to maintain a neutral pH for the effluent varies directly with the flow rate of the wastewater feed, as long as the titration curve of the wastewater remains constant. When the manipulated variable of a process is, in general, directly proportional to the feed rate, ratio control can significantly reduce the effect of feed rate disturbances on the process.

Figure 15.52 shows the application of ratio control to the effluent pH for a wastewater neutralization process applied in a mixing tank. This controller can effectively handle wastewater feed flow rate changes when the chemical makeup of the wastewater remains relatively constant. Small changes in the chemical makeup of the wastewater can usually be handled by the feedback controller, which adjusts the reagent-to-wastewater ratio to maintain the specified effluent pH.

A schematic of the stripping section of a distillation column with a ratio controller for the bottom products composition is shown in Figure 15.53. This application is similar to ratio control except that dynamic compensation is added to the measured column feed rate. If the steam flow to the reboiler were increased immediately for an increase in column feed rate, the corrective action would initially be an overcorrection. This results because, when a feed rate change occurs, it takes some time for the bottom product composition to respond. The purpose of the dynamic compensation (DC) element is to allow for the correct timing for the compensation for feed rate changes.

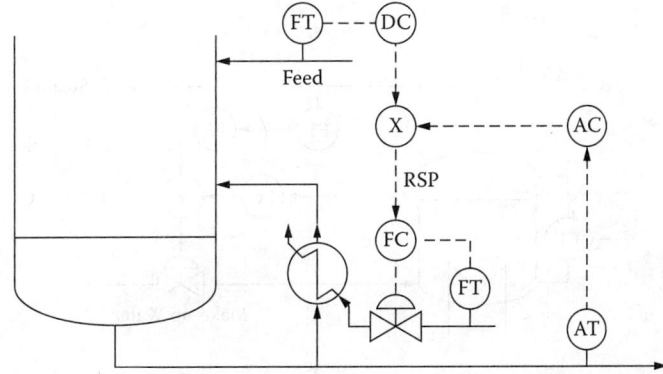

FIGURE 15.53 Schematic of ratio control for feed rate changes applied to the stripping section of a distillation column.

The dynamic element for this case can be simply a lag element, e.g., a digital filter described by Equation (15.10). The wastewater neutralization case (Figure 15.52) does not require dynamic compensation, since the process pH responses to feed rate changes and NaOH flow rate changes have similar dynamic behavior.

15.5.4 Feedforward Control

Feedforward control can be applied to process control loops that are significantly affected by disturbances that are measurable (or estimated) on-line. A feedback controller reacts to deviations from setpoint caused by the disturbance until the process is returned to setpoint. As pointed out in Section 15.3.3, since the proportional and derivative terms are zero during steady-state operation at the setpoint, the integral term in the PID controller is responsible for long-term compensation for disturbances. A feedforward controller anticipates the effects of a measured change in a disturbance (i.e., a load change) and takes corrective action before the disturbance affects the process. In effect, the feedforward controller applies corrective manipulated variable changes corresponding to the integral action that a feedback controller would generate; therefore, when a feedback controller and feedforward controller are used together, the feedback controller has much less "work" to do to compensate for a measured disturbance.

Example 15.3 Feedforward Example

Figure 15.54a shows a feedback controller applied for the level control of a boiler drum. The feedback controller compares the measured value of the level with the setpoint for the level and adjusts the flow rate of the feedwater to the drum. Therefore, when changes in the demand for steam occur, changes in the drum level result. If large swings in steam demand occur, a large gain is required for the feedback controller to maintain the level near its setpoint. But for large controller gains, the process is more susceptible to oscillatory behavior in the level and feedwater flow rate to the drum. Also, high-gain controllers are sensitive to noisy measurements of the controlled variable, and, in this case, level indicators can have significant noise levels.

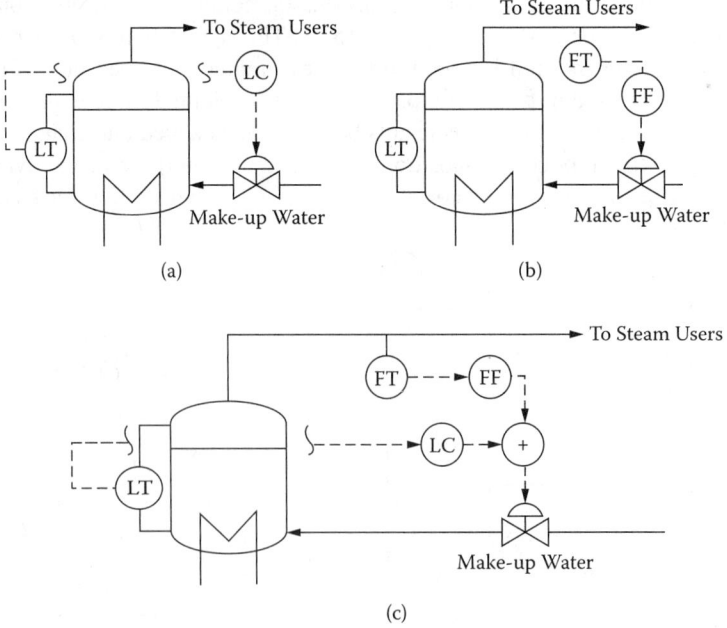

FIGURE 15.54 Boiler drum level control. (a) Feedback, (b) feedforward, and (c) feedback and feedforward combined.

Process Control

Figure 15.54b is a schematic of a feedforward controller applied for steam drum level control. If the flow rate of the makeup feedwater is equal to the steam usage, the drum level remains constant. One is tempted to conclude that the feedforward controller is all that is needed for this application. Unfortunately, the measurements of the steam usage and the feedwater flow rate are not perfectly accurate. Even small errors in measured flow rates add up over time, leading to one of two undesirable extremes. The drum can fill with water and put water into the steam system, or the liquid level can drop, exposing the boiler tubes, which can damage them. As a result, neither feedback nor feedforward are effective by themselves for this case. In general, feedforward-only controllers are susceptible to measurement errors and unmeasured disturbances, and, as a result, some type of feedback correction is typically required.

Figure 15.54c shows a combined feedforward and feedback controller for the control of the level in the steam drum. Note that the feedforward controller provides most of the required control action by responding to the measured steam usage. The feedback controller can be a relatively low-gain controller, since it needs to compensate only for measurement errors and unmeasured disturbances.

A lead-lag element is usually used to implement feedforward control. A lead-lag element has four tuning parameters: the gain (K_{ff}), the lead (τ_{ld}), the lag (τ_{lg}), and the deadtime (θ_{ff}). The gain indicates the change in feedforward control action for a change in the measured disturbance. The lag indicates how quickly the feedforward control action should approach its steady-state level. Whether the lead is greater or less than the lag determines the general shape of the feedforward action. The deadtime indicates how long it takes to make a correction after a measured change in the disturbance.

15.5.4.1 Tuning

Tuning a feedforward controller involves selecting the values of K_{ff}, τ_{ld}, τ_{lg}, and θ_{ff}. The following field tuning procedure is recommended:

1. Make initial estimates of K_{ff}, τ_{ld}, τ_{lg}, and θ_{ff} based on process knowledge.
2. Under open-loop conditions, adjust K_{ff} to minimize deviation from setpoint after a disturbance has had its steady-state effect on the process. Figure 15.55a shows the dynamic response of a feedforward controller for a step change in the disturbance based on initial feedforward controller settings and after K_{ff} has been adjusted to eliminate offset.
3. By analyzing the dynamic mismatch, adjust θ_{ff}. The direction of the deviation should indicate whether the feedforward correction is applied too soon or too late, causing dynamic mismatch. Figure 15.55b shows the feedforward control performance after θ_{ff} is tuned.
4. Finally, adjust ($\tau_{ld} - \tau_{lg}$) until approximately equal areas above and below the setpoint result. Figure 15.55c shows the results after ($\tau_{ld} - \tau_{lg}$) is adjusted. It is recommended to adjust the difference between τ_{ld} and τ_{lg}, since this difference (the relative dynamics of

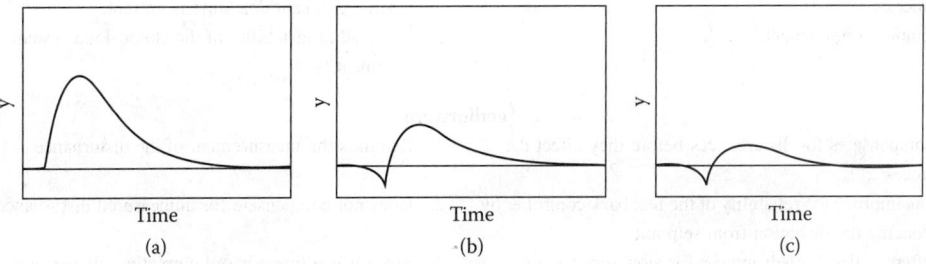

FIGURE 15.55 Tuning results for a feedforward controller for a step change in the disturbance. (a) Results for initial settings with correct feedforward gain, (b) results after deadtime tuned, and (c) final tuning results.

the process to manipulated variable and disturbance changes) has a profound effect on the shape of the response.

15.5.4.2 Overview

Table 15.8 summarizes the advantages and disadvantages of feedforward and feedback control. Note that feedforward and feedback control are complementary, i.e., each can overcome the disadvantages of the other so that together they are superior to either method alone. Feedforward control does not offer a significant advantage for fast-responding processes, because a feedback-only controller can usually absorb disturbances efficiently for these cases. But for slow-responding processes or processes with significant deadtime, by the time a feedback-only controller starts to respond to the effects of a disturbance, the process can already be severely upset. For these cases, the disturbance can cause the controlled variable to change significantly from its setpoint, resulting in relatively large process parameter changes (K_p, τ_p, and θ_p). In some cases, this can lead to closed-loop instability. When feedforward is added to a slow process or a process with significant deadtime, the deviation of the controlled variable from setpoint can be significantly reduced, resulting in smaller process parameter changes. Therefore, feedforward can provide significantly more reliable feedback control performance when the feedforward control compensates for a major disturbance to the process. In general, feedforward is useful when (1) feedback control by itself is not satisfactory, i.e., for slow-responding processes or processes with significant deadtime, and (2) the major disturbance to a process is measured on-line.

Feedforward control provides a linear correction and therefore can provide only partial compensation to a nonlinear process. Nevertheless, feedforward control can be effective when properly implemented, since it can reduce the amount of feedback correction required. When tuning a feedforward controller for a nonlinear process, care should be taken to ensure that the feedforward controller is tuned with consideration to both increases and decreases in the disturbance level.

Figure 15.56 shows the effect of the ratio of τ_{ld}/τ_{lg} on the dynamic response of a lead/lag element. When τ_{ld}/τ_{lg} is greater than 1, overcompensation is used. That is, when the process responds faster to the disturbance than to the controller output, larger than steady-state changes in the controller output are required to compensate for dynamic mismatch. On the other hand, when τ_{ld}/τ_{lg}

TABLE 15.8
Comparison of Feedback and Feedforward Control

Advantages	Disadvantages
Feedback	
1. Does not require a measurement of the disturbance.	1. Waits until the disturbance has affected the process before taking action.
2. Can effectively reject disturbances for responding process.	2. Susceptible to disturbances when the process is slow or when significant deadtime is present.
3. Simple to implement.	3. Can lead to instability of the closed-loop system due to nonlinearity.
Feedforward	
1. Compensates for disturbances before they affect the process.	1. Requires the measurement of the disturbance.
2. Can improve the reliability of the feedback controller by reducing the deviation from setpoint.	2. Does not compensate for unmeasured disturbances.
3. Offers noticeable advantages for slow processes or processes with significant deadtime.	3. Since it is a linear-based correction, its performance deteriorates with nonlinearity.

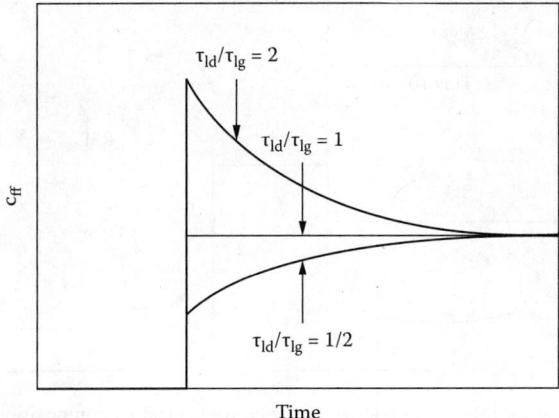

FIGURE 15.56 The effect of the ratio of τ_{ld} to τ_{lg} on the dynamic response of a lead/lag element; c_{ff} is the output from the feedforward controller.

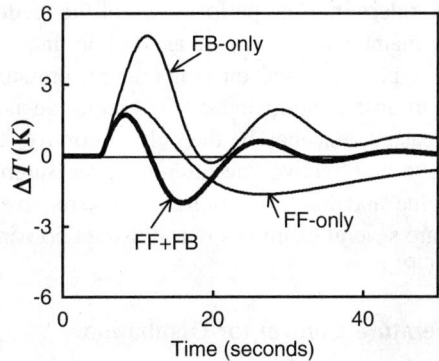

FIGURE 15.57 Comparison among feedforward (FF-only), feedback (FB-only), and combined feedforward and feedback (FF + FB) for a disturbance upset.

is less than 1, the application of the controller output is more gradual, eventually approaching its steady-state level.

Figure 15.57 shows a comparison between a feedback-only controller, a feedforward-only controller, and a combined feedforward and feedback controller. The feedforward-only controller significantly reduces the initial deviation from setpoint compared to the feedback-only controller but is sluggish in returning to the setpoint. The combined feedforward and feedback controller uses the feedforward action to reduce the initial deviation from setpoint and the feedback action to quickly settle at the setpoint.

15.5.5 Inferential Control

Up to this point, it has been assumed that the sensor in a control loop provides a direct measurement of the controlled variable. In fact, the output of the sensor only correlates with the value of the measured variable. A thermocouple exposed to a process stream at a specific temperature generates a millivolt signal that correlates strongly with the temperature of the process stream. In this section, it is shown that easily measured quantities, such as pressures, temperatures, and flow rates, can be effectively used to infer quantities that are more difficult to measure, such as compositions, molecular weight, and extent of reaction. Then, the inferred value of the controlled variable can

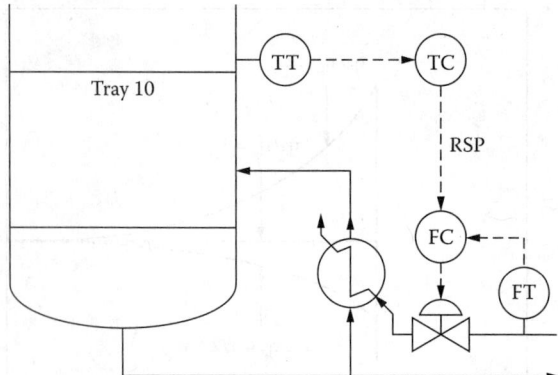

FIGURE 15.58 Schematic for inferential control of the bottoms product composition of a distillation column.

be used as the value of the controlled variable in a feedback control loop, greatly reducing the associated measurement delay.

There are three main reasons for using inferential measurement of a controlled variable: (1) Excessive analyzer deadtime undermines the performance of the feedback loop. (2) The total cost (i.e., the purchase price and maintenance cost) of an on-line analyzer can be excessive. Since inferential measurements are typically based on temperature, pressure, and flow measurements, they are much less expensive to install and maintain. (3) An on-line analyzer may not be available. In that case, an inferential measurement may be the only option for feedback control.

For an inferential control to be effective, the inferential measurement must correlate strongly with the controlled variable value, and this correlation should be relatively insensitive to unmeasured disturbances. The following are several examples that illustrate how inferential measurements can be effectively applied in the CPI.

15.5.5.1 Inferential Temperature Control for Distillation

Tray temperatures correlate very well with product compositions for many distillation columns; therefore, inferential control of distillation product composition is a widely used form of inferential control. Figure 15.58 shows the arrangement for inferential temperature control of the bottoms product composition for this column. Note that the tray temperature controller is cascaded to a flow controller.

15.5.5.2 Inferential Reaction Conversion Control

Consider an adiabatic fixed-bed reactor. For a single irreversible reaction, $A \to B$, the macroscopic energy balance assuming no phase change is given by

$$X_A C_{A_{in}} \left(-\Delta H_{rxn} \right) = \rho C_P \left(T_{out} - T_{in} \right)$$

where X_A is the fractional conversion of reactant A, $C_{A_{in}}$ is the inlet concentration of A to the reactor, ΔH_{rxn} is the heat of reaction, ρ is the average density of the process stream, C_P is the average heat capacity of the process stream, T_{out} is the temperature of the outlet stream from the reactor, and T_{in} is the temperature of the inlet stream to the reactor. Rearranging the previous equation,

$$X_A = \frac{\rho C_P}{C_{A_{in}} \left(-\Delta H_{rxn} \right)} \left(T_{out} - T_{in} \right)$$

Process Control

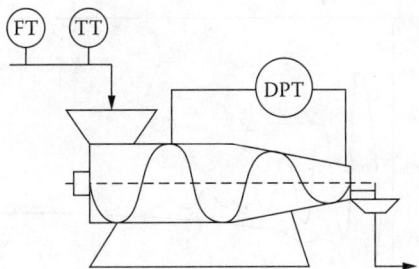

FIGURE 15.59 Schematic of a polymer extruder.

Note that this relationship is not affected by changes in the feed rate, although feed rate affects T_{out} and thus X_A. In an industrial reactor, there are heat losses, side reactions, and variations in the physical parameters; therefore, the assumed inferential relationship is

$$X_A = a\left(T_{out} - T_{in}\right) + b$$

A plot of the experimental data for a reactor (X_A, T_{out}, and T_{in}) can be used to determine a and b as well as check the validity of this functional form. For this approach to be effective, the temperature difference across the reactor needs to be large enough that temperature measurement noise does not significantly affect the results of the measured temperature. Once a and b are identified, the inlet temperature, T_{in}, can be adjusted to maintain a fixed reaction conversion, X_A. Periodically, composition measurements for the product leaving the reactor can be made and the results used to update the value of b in the previous equation, since a is less likely to change significantly compared with b.

15.5.5.3 Inferential Measurement of the Molecular Weight of a Polymer

Figure 15.59 shows a schematic for a polymer extruder. The procedure for estimating the molecular weight of a polymer is as follows. First, the flow rate, F, temperature, and pressure drop across the extruder, ΔP, are measured. Next, a fluid dynamic relationship is used to calculate the corresponding viscosity of the polymer melt at the prevailing temperature, $\mu(T)$. Then, the viscosity is corrected for temperature so that the viscosity of the melt is calculated at a standard temperature, T_o. Finally, a correlation between $\mu(T_o)$ and the molecular weight of the polymer melt is developed using laboratory measurements of the molecular weight (M_{wt}) for a range of $\mu(T_o)$ values. Therefore, from measurements of the flow rate of the polymer melt, the pressure drop, and the temperature, the molecular weight of the polymer can be estimated online. This value can be used by a feedback controller to make adjustments to the polymer reactor to control the molecular weight of the polymer product. Without this inferential estimator, samples of the extruded polymer have to be tested in the laboratory requiring in the range of 10 hr to perform each test. Since the residence time of the reactor/extruder process can be less than 1 hr, a 10-hr analysis deadtime makes feedback molecular weight control extremely difficult, if not impossible. The samples, which are taken one to three times per day, are used to make corrections to the correlation functions used to infer molecular weight.

15.5.5.4 Soft Sensors Based on Neural Networks

In electric power generating stations, restricting the NO_x (nitrogen oxide compounds) emissions from the flue gas to acceptable levels is important, because NO_x compounds contribute to air pollution. Typically, on-line analyzers are used to measure the NO_x in the flue gas from the boilers.

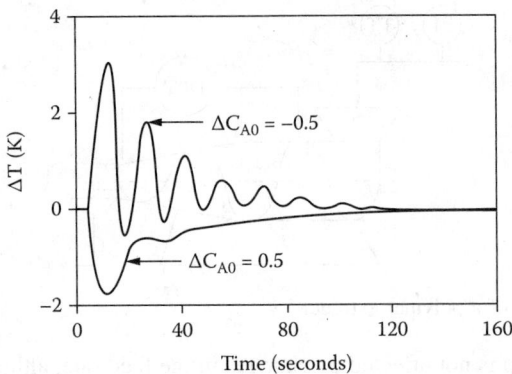

FIGURE 15.60 The effect of feed composition upsets on the PI feedback behavior for an endothermic CSTR.

Occasionally, the NO_x analyzers on a boiler fail. If the NO_x level is not measured, the power companies must pay a fine for emissions. Instead of installing additional on-line NO_x analyzers, which are quite expensive, some power companies have applied a type of inferential estimator to predict the NO_x level in their flue gas.

Instead of using one or two process measurements, all the measured process conditions (e.g., fuel feed rate, oxygen in the flue gas, heating value of the fuel, ambient air temperature, and so on) have been empirically correlated to predict the NO_x concentration in the flue gas. The empirical correlation is based on training an artificial neural network to predict the flue gas NO_x concentration from all the available data.

15.5.6 Scheduling Controller Tuning

In Section 15.4.5, it was demonstrated that nonlinear process behavior can result in a controller becoming unstable in certain situations; in others, it can become extremely sluggish (see Figure 15.60). Tuning PID controllers for the case with the largest process gain can eliminate unstable operation, but at the expense of largely sluggish performance. For some processes, certain measurements directly indicate whether the process parameters have increased or decreased and by how much; therefore, scheduling of the controller tuning based on process measurements can often compensate for process nonlinearity. The controlled variable and the feed rate are examples of such key process measurements that typically can be used to schedule the chemical controller tuning.

Consider a heat exchanger used to heat a process stream with steam as the heating medium. As the feed to the heat exchanger flows through the tube bundle, it is heated by steam condensing on the shell side. As the feed rate changes, the residence time of the feed in the tubes exposed to the steam changes. Figure 15.61 shows the open-loop responses for three different feed rates for a step change in the setpoint of the steam pressure controller. The feed rate is represented by the average fluid velocity (v) in the tubes. Both the gain and the dynamic response change as the feed rate is changed. Table 15.9 lists the FOPDT parameters for each flow rate. Note that the gain and the deadtime each change by a factor of about 2.5. Using these FOPDT parameters, the controller settings for each flow rate are also listed in Table 15.9. Note that the controller gain changes by a factor of 5, but the reset time changes are more gradual. It is clear from these results that it is not reasonable to expect one set of PI controller settings to work effectively for significant changes in the feed rate to this heat exchanger. For example, if the temperature controller for the outlet of the heat exchanger were tuned for $v = 7$ ft/s, when the feed rate is reduced to $v = 4$ ft/s, the controller becomes unstable. Conversely, if the controller were tuned for the low flow rate condition, it would perform sluggishly for the high flow rate conditions. Figure 15.62 shows results with and without scheduling of the controller tuning based on feed rate for a step decrease in the feed rate corre-

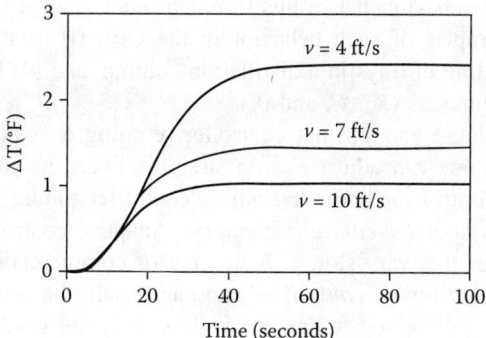

FIGURE 15.61 Open-loop response for a heat exchanger for different feed rates.

TABLE 15.9
FOPDT and PI Tuning Parameters for Heat Exchanger Case as a Function of Feed Rate

	v = 4 ft/s	v = 7 ft/s	v = 10 ft/s
K_p	0.25	0.15	0.11
τ_p	10.7	9.9	10.0
θ_p	10.2	5.8	4.0
K_c	4.2	10.8	21.3
τ_I	12.0	8.8	7.3

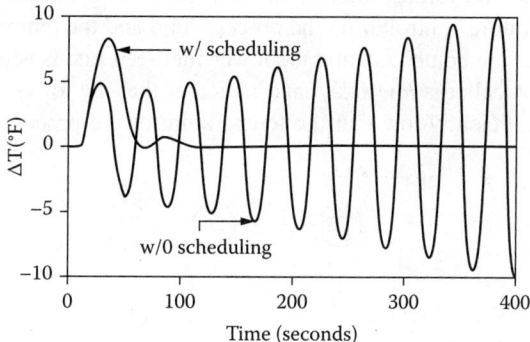

FIGURE 15.62 Closed-loop results for a step change in feed rate with and without scheduling of the controller tuning for the heat exchanger case.

sponding to a change in the velocity through the tubes from 7 to 4 ft/s. Note that the controller without scheduling was tuned for a feed rate corresponding to $v = 7$ ft/s.

15.5.6.1 Nonstationary Behavior

Consider a wastewater neutralization process. If the titration curve of the wastewater and the other process parameters remain fixed, the process is referred to as *stationary*. On the other hand, if the titration curve changes, the process is *nonstationary*. In the case of this pH control example, changes in the titration curve can have an overwhelming effect on the process gains. There are many more examples of nonstationary behavior that results in much more gradual process gain changes. The following are several examples of such behavior in the CPI: (1) catalyst deactivation, (2) heat exchanger fouling, (3) fouling of trays in a distillation column, and (4) feed composition changes that affect the process parameters (K_p, τ_p, and θ_p).

These effects can be large enough that controller retiming is required. If an overall tuning factor, F_T, has been used, one can adjust F_T in a straightforward manner to compensate for the nonstationary behavior. Control methods that adjust controller tuning to adapt to nonstationary behavior are referred to as *adaptive control techniques*. Adaptive control techniques can be effectively applied for processes that vary slowly. A number of commercially available adaptive controllers are referred to as *self-tuning controllers* and can usually be installed on a DCS. While a range of approaches are used for self-tuning controllers, they are generally limited to processes that vary in a gradual, consistent manner.

15.5.7 Override/Select Control

Constraints are a natural part of industrial process control. As processes are pushed to produce as much product as possible, process limits are inevitably encountered. When an upper or lower limit on a manipulated variable is encountered, or when an upper or lower value of a controlled or output variable from the process is reached, it can become necessary to apply different control loops from those previously used. That is, effective industrial controller implementation requires that safeguards be installed to prevent the process from violating safety, environmental, or economic constraints. These constraints can be met using override/select controls.

Consider the furnace-fired heater shown in Figure 15.63. Under normal operating conditions, the fuel flow rate is adjusted to control the exit temperature of the process fluid. As the feed rate of the process fluid is increased, the furnace tube temperature increases. At some point, the upper limit on furnace tube temperature (an operational constraint) is encountered. The fuel flow rate to the furnace must be adjusted to keep the furnace tube temperature from exceeding its upper limit, at which point damage to the furnace tubes results. Figure 15.63 shows that the output of both control loops (the temperature controller on the process fluid and the temperature controller on the furnace tube temperature) are combined, and the lower fuel feed rate is actually applied. The "LS" symbol in Figure 15.63 is called a *low select* and indicates that the lower fuel feed rate is chosen. When the feed rate is sufficiently low that the temperature of the process fluid can be controlled

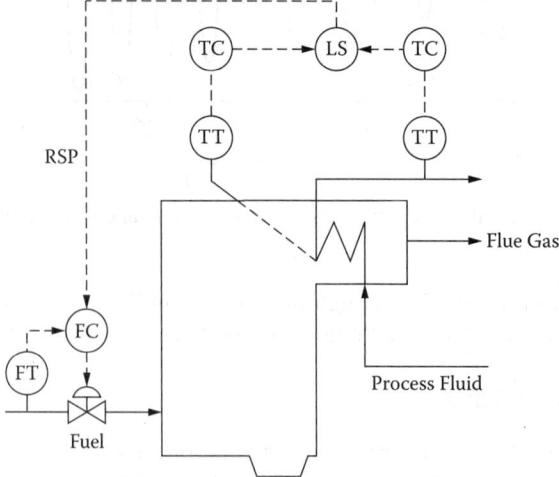

FIGURE 15.63 A schematic of a furnace-fired heater with low select firing controls.

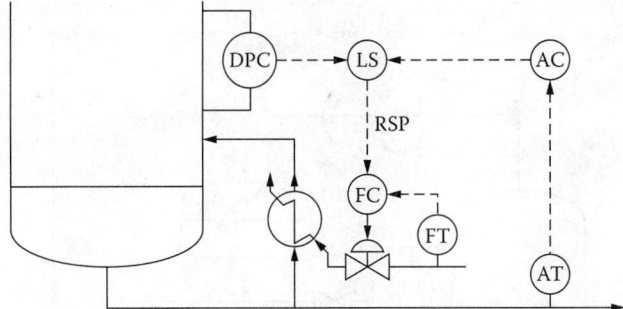

FIGURE 15.64 Schematic of the stripping section of a distillation column with override/select control to maintain bottom product purity when a maximum reboiler constraint is encountered.

to setpoint, the output of the process fluid temperature controller is selected, since it is lower than the output of the tube temperature controller. Likewise, when the tube temperature approaches its upper limit, the output of the tube temperature controller is selected.

Figure 15.64 shows a distillation column that reaches an upper limit on the reboiler duty. When the remote setpoint for the steam flow rate to the reboiler is consistently greater than the measured steam flow, an override controller switches to using the column feed rate as a manipulated variable to keep the bottom product purity on specification. When the column feed rate is adjusted back to its normal level and the control valve on the steam to the reboiler is no longer saturated (i.e., fully open), the control configuration is changed so that the reboiler duty is manipulated to control the bottom product purity.

15.5.8 Computed Manipulated Variable Control

In certain cases, the desired manipulated variable for a particular process cannot be directly adjusted. Distillation columns can be particularly sensitive to sharp changes in ambient conditions due to weather fronts or thundershowers, since both of these cases can cause significant increases in the reflux subcooling. Equating the heat lost by the condensing vapor to the heat required by the subcooled reflux results in the following equation:

$$C_p F_{ex}(T_{oh} - T_r) = \Delta F_{int} \Delta H_{vap}$$

where C_p is the heat capacity of the reflux, T_{oh} is the overhead temperature, T_r is the subcooled reflux temperature, F_{ex} is the external reflux flow (the setpoint for the flow controller on the reflux), ΔF_{int} is the change in the reflux caused by the condensing vapor, and ΔH_{vap} is the heat of vaporization of the vapor. ΔF_{int} combines with the external reflux to form the internal reflux. Then, the equation for the internal reflux flow rate (F_{int}) is given by

$$F_{int} = F_{ex}\left(1 + C_p[T_{oh} - T_r]/\Delta H_{vap}\right)$$

This equation can be rearranged to calculate the external reflux that maintains a specified internal reflux control (F_{int}^{spec}), i.e.,

$$F_{ex} = \frac{F_{int}^{spec}}{1 + C_p(T_{oh} - T_r)/\Delta H_{vap}}$$

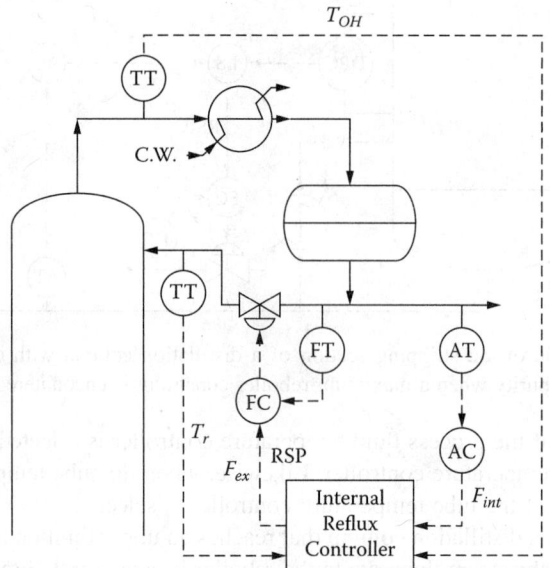

FIGURE 15.65 Schematic of an internal reflux controller applied for composition control of the overhead of a column.

This approach, called *internal reflux control,* is shown schematically in Figure 15.65. Note that the composition controller outputs the internal reflux flow rate, and the internal reflux controller calculates the external reflux flow rate, which is used as the setpoint for the flow controller on the reflux.

15.5.9 Antiwindup Strategies

Figure 5.17a shows the manipulated and controlled variables for a standard PI controller for which the manipulated variable reaches its upper limit, i.e., the control valve is fully open or fully closed, which is referred to as a *saturated control valve.* This can occur when a large disturbance enters the process. Since the manipulated variable cannot be increased further, the PI controller is unable to return the controlled variable to its setpoint. As long as there is an error between the controlled variable and its setpoint, the integral term in the PI controller [Equation (15.1)] continues to accumulate, which is referred to as *reset windup* or *integral windup.* After some time, the disturbance level returns to its original value. At this point, integral windup in the PI controller keeps the manipulated variable at its maximum level, even though the value of the controlled variable is now above its setpoint. In effect, before the process can return to steady state, an equal area above the setpoint must be generated to compensate for area "A" shown in Figure 5.66a.

This behavior occurs because the integral is allowed to continue accumulating after control of the process has been lost (i.e., the manipulated variable saturates). Figure 5.66b shows the same case as Figure 5.66a except that, when the manipulated variable saturates, the integral is not allowed to accumulate (windup). Note that when control returns to the process (i.e., when the manipulated variable is no longer saturated), the controlled variable moves directly back to its setpoint and does not exhibit prolonged deviations from setpoint as before. Because the integral action was turned off when the manipulated variable became saturated, the PI controller does not have to generate an area equivalent to area "A" above the setpoint.

Antireset windup can be implemented by simply not allowing the integral to accumulate when the manipulated variable is saturated. The manipulated variable is saturated when the control valve on the line supplying the manipulated variable is either closed or fully open. A saturated control valve can be identified when there is sustained offset between the manipulated variable level

Process Control

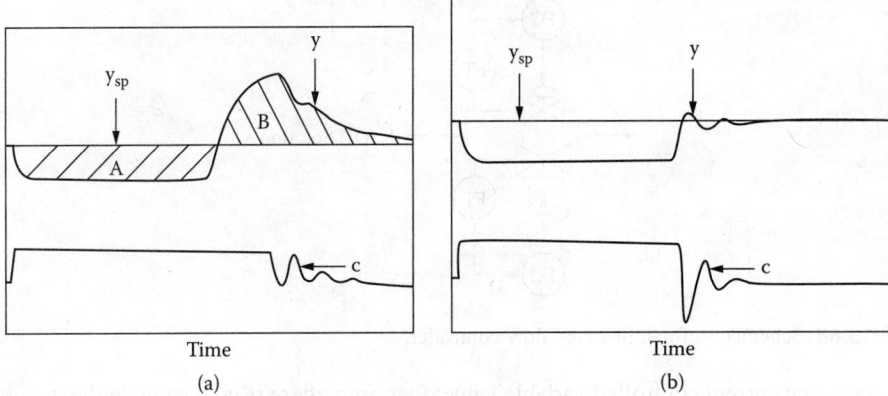

FIGURE 15.66 Response of a feedback system to a saturated manipulated variable. (a) Conventional PI controller and (b) PI controller with antireset windup.

requested by the flow controller and the actual flow rate of the manipulated variable. Modern DCSs have built-in provisions for preventing windup.

15.5.10 Bumpless Transfer

Figure 15.67 shows the process behavior with and without bumpless transfer. Without bumpless transfer, if the controller is turned on when the controlled variable is far removed from setpoint, the controller takes immediate action and drives the process to setpoint in an underdamped fashion. In certain cases, the controlled variable can be far enough away from setpoint, and the process can be sufficiently nonlinear that the control loop becomes unstable. Even if the control loop does not become unstable, the abrupt action of the feedback controller can significantly upset other control loops in the process. As a result, operators find that the behavior of a controller without bumpless transfer is generally unacceptable, particularly for key loops such as composition and temperature control loops.

For bumpless transfer, there are two types of setpoints: the true setpoint, which corresponds to the desired operating point, and the internal setpoint, which is used for bumpless transfer. When a control loop is turned on, the setpoint used by the controller is actually different from the true setpoint when applying bumpless transfer. When the controller is turned on, the internal setpoint

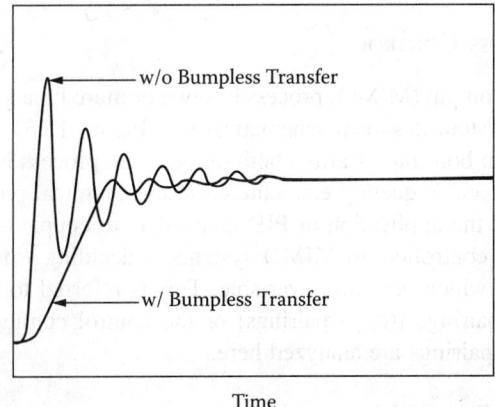

FIGURE 15.67 The startup response of a feedback system without bumpless transfer and with bumpless transfer.

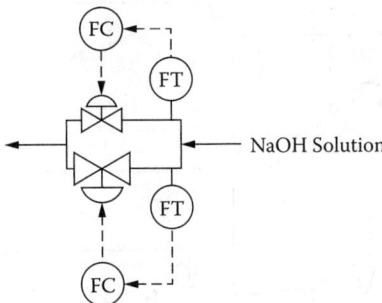

FIGURE 15.68 Schematic of a split-range flow controller.

is set equal to the current controlled variable value; therefore, there is no change in the manipulated variable level. After this, the internal setpoint is ramped toward the true setpoint, and the process gradually begins moving toward the true setpoint. After the internal setpoint reaches the true setpoint value, it remains constant. By selecting a proper setpoint ramping rate, smooth and consistent startups for control loops result.

15.5.11 SPLIT-RANGE FLOW CONTROL

Consider a wastewater neutralization process with a titration curve for the wastewater that exhibits a high gain at neutrality. To control the pH to ±1.0 pH units at a setpoint of pH 7, the base flow rate must be metered accurately to within ±0.5%. A single-flow control loop with a control valve with a positioner can meet this metering precision. But if the total flow rate of base were to range from 0.1 to 10 GPM, one flow control loop could not meter the base flow rate to within ±0.5% at both 0.1 and 10 GPM.

Two flow control loops that work together can meet this requirement, as shown in Figure 15.68. At low flow rates, the large control valve is closed, and the flow control loop with the smaller control valve can accurately meter the low-flow operation. As the total flow increases, the smaller control valve begins to approach saturation. Before this happens, the flow control loop with the larger control valve comes into service. At large flow rates (>1 GPM), the small control valve is completely open, and the flow control loop with the larger valve is accurately metering the base flow rate. This is an example of split-range flow control, which is used when accurate flow control is required over a wider operating range than one control valve can provide.

15.5.12 MIMO PROCESS CONTROL

A multiple-input/multiple-output (MIMO) process has two or more inputs and two or more outputs. A two-input/two-output system is shown schematically in Figure 15.69. Note that both c_1 and c_2 affect both y_1 and y_2. When both inputs affect both outputs, the process is referred to as a *coupled process*. MIMO processes are frequently encountered in the chemical processing industries.

This section considers the application of PID controllers to coupled MIMO processes. A key issue when applying PID controllers to MIMO systems is deciding which manipulated variable should be used to control which controlled variable. This is referred to as choosing the manipulated/controlled variable pairings [(c, y) pairings] or the control configuration. The factors that affect the choice of (c, y) pairings are analyzed here.

15.5.12.1 SISO Controllers and (c, y) Pairings

Figure 15.70 shows two single-loop PID controllers applied to a two-input/two-output process (2 × 2 system). Applying single-loop PID controllers to a MIMO process is called *decentralized*

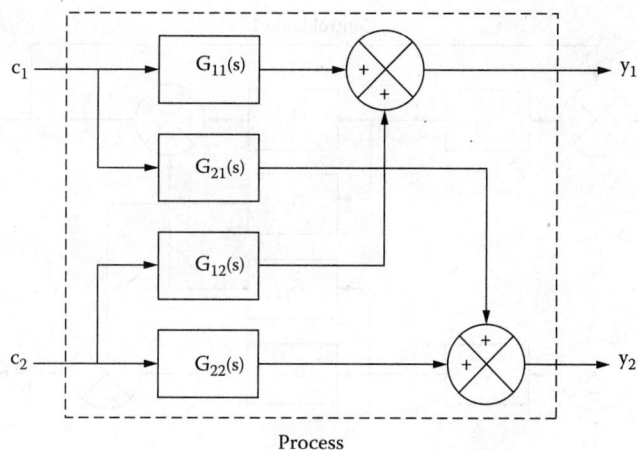

FIGURE 15.69 Block diagram of a two-input/two-output process. Note that G represents an input/output function.

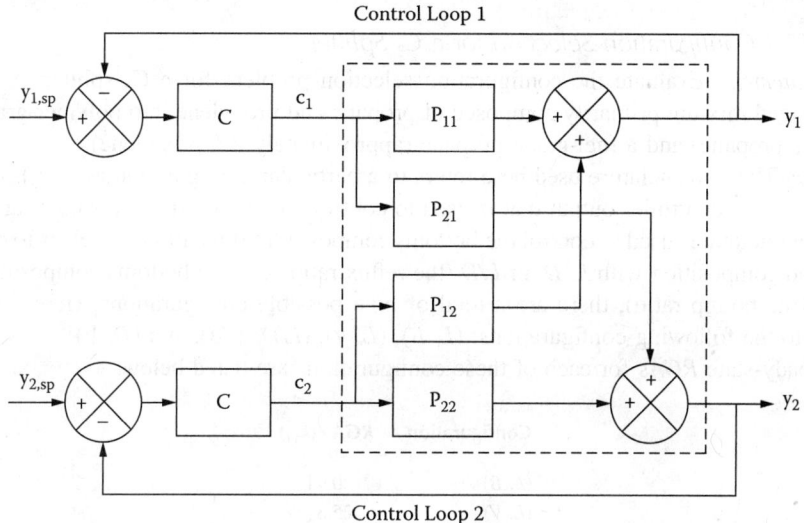

FIGURE 15.70 Block diagram of a 2 × 2 process with single-loop controllers applied (decentralized control). Note that C represents a controller and P represents an input/output function.

control. Note that the coupling in this 2 × 2 system causes the two control loops to interact. That is, while control loop 1 adjusts c_1 to keep y_1 at its setpoint, it upsets control loop 2. Likewise, the operation of control loop 2 can act as an upset for control loop 1. Figure 15.71 shows schematically the coupling effect of control loop 2 (indicted by heavy lines) as an additive disturbance to control loop 1. For this 2 × 2 example, when tuning control loop 1, the effects of control loop 2 must be taken into account, and vice versa. When tuning single-loop PID controllers applied to a MIMO process, one must take into account the effects of coupling.

The selection of pairings for a decentralized controller can have a dramatic effect on the resulting overall control performance. Three factors determine the best pairings for a MIMO process: coupling, dynamic response, and sensitivity to disturbances. Process control engineers typically rely on their understanding of the process and their experience when selecting a control configuration for a MIMO process.

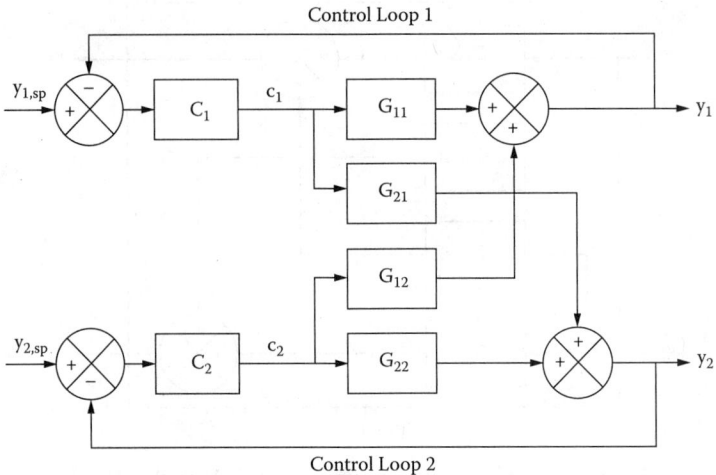

FIGURE 15.71 A block diagram of a 2 × 2 process with single-loop controllers showing the coupling effect of loop 2 on y_1 for changes in c_1. Note that C represents a controller and P represents an input/output function.

Example 15.4 Configuration Selection for a C_3 Splitter

Problem statement. Evaluate the configuration selection problem for a C_3 splitter. A C_3 splitter separates a feed mixture primarily composed of propane and propylene into polymer-grade propylene (<0.5% propane) and a fuel-grade propane (approximately 2% propylene).

Solution. The nomenclature used here refers to a particular configuration as (c_1c_2), where c_1 is assumed to be the controller output that is used to control the overhead composition, and c_2 is the controller output that is used to control the bottoms composition. If we limit ourselves to controlling the overhead composition with L, D, or L/D (the reflux ratio) and the bottoms composition with V, B, or V/B (the boilup ratio), there are a total of nine possible configurations. Here we limit the discussion to the following configurations: (L, B), (L, V), (L/D, V/B), and (D, V).

The steady-state *RGAs* for each of these configurations are listed below:

Configuration	RGA[1] (λ_{11})
(L, B)	0.94
(L, V)	25.3
(L/D, V/B)	1.70
(D, V)	0.06

Based on these results, the (L, B) and (L/D, V/B) configurations appear to be the most promising.

The dynamics of distillation columns can be understood by recognizing that product composition changes result from changes in the internal vapor/liquid traffic in the column. On the other hand, changes in B and D must depend on the level controllers to change the vapor/liquid traffic of the column; therefore, the dynamic response of the product compositions is significantly slower when B and D are changed, compared with changing L and V. The dynamic response to changes in L/D and V/B are intermediate between L and V on the fast side and B and D on the slow side. Based on this analysis, (L, B) is expected to perform better for the overhead composition control than for the bottoms, but there is no clear winner between the (L, B) and the (L/D, V/B) configurations with regard to the overall dynamic response.

Table 15.10 shows the relative changes in each manipulated variable for a change in feed composition. This table is based on steady-state results in which the product compositions are maintained at a constant level. A lower relative change for a manipulated variable indicates a reduced sensitivity to feed composition changes for that manipulated variable. Note that L, L/D,

TABLE 15.10
Relative Changes in the Manipulated Variables to Maintain the Product Purities for a 5-Mole % Increase in Feed Composition

Manipulated Variable	Percentage Change
L	4.2
D	7.4
L/D	−3.0
V	4.4
B	−16.8
V/B	25.5

TABLE 15.11
Control Performance (IAE) for a Step Change in Feed Composition

Configuration	IAE for Overhead	IAE for Bottoms
L,B	0.067	1.49
L,V	0.250	13.3
L/D,V/B	0.095	2.00
D,V	0.098	1.91

and V show the least sensitivity to feed composition changes. Table 15.11 lists the integral absolute error (IAE) for each configuration for each product for a feed composition upset. A lower IAE value indicates closer control to setpoint. Note that the (L, B) configuration provided the best overall control performance, especially for the overhead product. This is consistent with the observation that L is dynamically fast and relatively insensitive to feed composition changes coupled with the relatively moderate steady-state coupling as indicated by the RGA.

The (L, V) configuration has the advantages of fast overall dynamics and insensitivity to feed composition upsets. These advantages are negated by the extreme degree of steady-state coupling as indicated by its steady-state RGA value. The control performance of the (L, V) configuration is the poorest of the four configurations listed in Table 15.11. The $(L/D, V/B)$ configuration has a good steady-state RGA and dynamic characteristics, but is particularly sensitive to feed composition upsets for the bottom composition control loop. The steady-state RGA value of the (D, V) configuration indicates that this configuration does not function properly, but it is quite reasonable, i.e., the IAEs for the (D, V) configuration were only about 30% larger than those for the (L, B) configuration, which is not substantial in this case.

For complex configuration selection problems, such as distillation columns, the previous analysis is helpful but does not always guarantee that the best configuration will be identified. The performance differences between reasonable configuration choices and the best configuration can be substantial. Therefore, in these cases, the use of detailed dynamic simulations for the analysis of the control performance of feasible configurations is recommended wherever possible.

15.5.12.2 Tuning Decentralized Controllers

The recommended tuning procedure for a single PID loop can be extended to tuning the single-loop PID controllers applied for decentralized control of a MIMO process. The first step in tuning a decentralized controller is to apply ATV tests for each manipulated variable/controlled variable pair. While an ATV test is applied to one loop, the other loops should be maintained in an open-loop condition.

Next, determine if any of the loops are significantly faster responding than the other loops. This can be done by comparing the values of the ultimate periods, P_u, obtained in the ATV tests. If the smallest value of P_u is at least five smaller than the next larger P_u, that loop alone should be implemented first, before tuning the other loops. It can be tuned as a single PID loop as discussed in Chapter 3. Next, ATV tests on the remaining loops should be rerun with the tuned fast loop in service (closed-loop operation). Then, the remaining control loops can be tuned using the following procedure.

Assume that it is required to tune PI controllers on a 2×2 MIMO process. The ATV results are used to select the controller gain and reset time based on, for example, Zeigler–Nichols tuning. Then, a single tuning factor, F_T, is applied to the tuning parameters for both control loops:

$$\left. \begin{array}{l} K_c = K_c^{ZN} / F_T \\ \tau_I = \tau_c^{ZN} \times F_T \end{array} \right\} \text{First control loop}$$

$$\left. \begin{array}{l} K_c = K_c^{ZN} / F_T \\ \tau_I = t_I^{ZN} \times F_T \end{array} \right\} \text{Second control loop}$$

F_T is adjusted until the proper dynamic response is obtained. While tuning, if the closed-loop response is sluggish, decrease the value of F_T. Likewise, if the controller exhibits periods of ringing, increase the value of F_T.

After F_T has been adjusted to tune the set of decentralized PI controllers, fine tuning of the controller settings should be used. For example, if one observes that one of the control loops is slow to settle at setpoint in a manner similar to Figure 15.3a, an increase in integral action for that loop should be tested. If one of the loops exhibits ringing, derivative action should be tested to determine if it improves the feedback control performance of that loop. In the latter case, derivative action should be tuned in the manner that was described in Section 15.3.

15.6 MODEL PREDICTIVE CONTROL

This section covers model predictive control (MPC). It describes what it is, how to design it, how to install it, and how to make it work. This work is both fun and useful, but there is one rule: *Understand the process*. This section contains tips and clues about how to analyze and learn process behavior.

15.6.1 WHAT IS MPC?

There are many ways to control plant operations. Among these are the following.

Batch control is used to produce batches of material to specification. For example, certain polymer reactors fall into this category.
Programmable logic controllers (PLCs) are typically used to control very high-frequency equipment such as steel mill rolls or compressors.
PID (the old standby of continuous control) is largely dedicated to controlling a single variable, such as flow in a pipe.

Process Control

Neural networks are relatively new and are used in nonlinear multivariable situations where a rigorous process model is unavailable or unsuitable for traditional implementation. Neural networks are used where traditional identification techniques are not well suited for the problem due to nonlinearities.

Model predictive control (MPC), the subject of this section, is sometimes referred to as *multivariable control or MVC*.

A short definition of MPC is: a way to control many variables simultaneously, using a plant model to manipulate more than one variable to produce improved plant operations.

Some other useful definitions are as follows:

Controlled variable (CV). Controlled variables are those that you want to control.
Disturbance variable (DV). A DV affects the process in a known way but cannot be manipulated.
Feedforward (FF) variable. Same as DV.
Input. See MV.
Linear program (LP). A linear economic optimizer.
Model plant model. Normally dynamic and linear, often described in matrix notation as A.
Manipulated variable (MV). Manipulated variables are those moved by the controller.
Output. See CV.

Referring to the above definitions, a model predictive controller has at least one CV, more than one MV, and possibly DVs. It can be used to optimize the plant in an economic sense with an LP. With today's technology, the model, A, is linear and dynamic, and it describes the approximate relationships between the CVs and MVs (and DVs). In this section, we will assume that the process being controlled is continuous (as opposed to batch or discrete).

In the step response model, a change in MV of size u causes a corresponding change in the CV, y. In matrix notation, this is described as

$$y = A\Delta u \qquad (15.16)$$

where y and u are vectors, and A is a matrix. A prediction of the CV's future value, $\tilde{y}$, can be calculated using the plant model and previous changes in the MVs. By including future changes made in the MVs, a prediction of the future response of the CV with control $\tilde{y}_C$ at each discrete time interval, k, can be made:

$$\tilde{y}_C(k) = \tilde{y}(k-1) + A\Delta u(k-1) + \mathbf{d}(k) \qquad (15.17)$$

where $\mathbf{d}$ is a disturbance that may effect the plant. The objective of MPC is to have the CVs reach their targets, y_T, or setpoints, y_{sp}, in the most efficient manner by a certain time. The move sizes to be made are calculated by minimizing the error, e, between the CV target (y_T) and the predicted CV ($\tilde{y}_C$) while the MVs are being moved. This unconstrained solution is formed as

$$\mathbf{e}(k) = \mathbf{y}_T(k-1) - \tilde{\mathbf{y}}_C(k-1)$$
$$\min_{\Delta u} \mathbf{\Phi} = [\mathbf{e}(k)]^2 \qquad (15.18)$$

At each time step, the first step (k) of the calculated moves is then implemented, and the calculation proceeds again (see Figure 15.72).

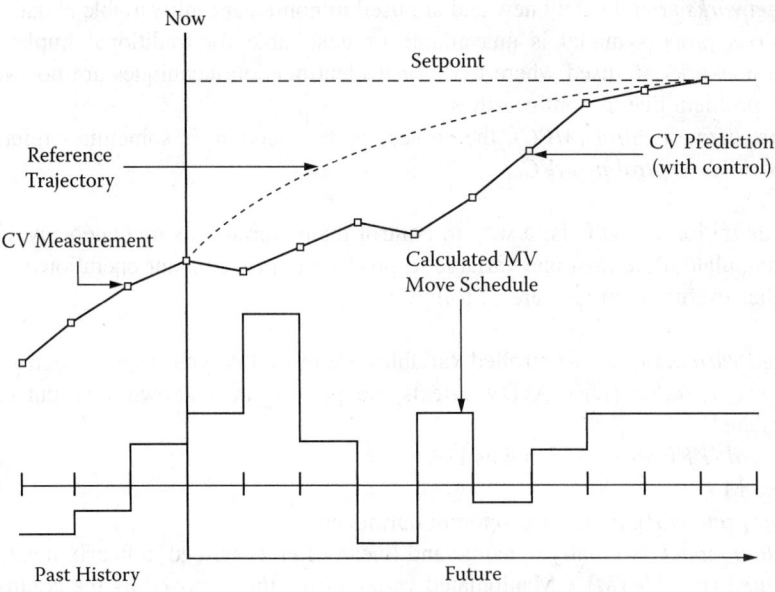

FIGURE 15.72 Representation of CV prediction and scheduled MV moves at each calculation step for MPC.

15.6.2 History of MPC

Industrial model predictive control (MPC) is based on algorithms that were developed many years ago. They share several common traits:

They are based on discrete linear dynamic models.
The algorithm for identifying linear dynamic models is derived from plant data.
The algorithm for calculating on-line control actions is based on plant models.

They have been refined and applied successfully over the past 40 years. Currently, DMC is the predominant industrial MPC.

15.6.3 When Should MPC Be Used?

Because MPC requires significant investments in software and time, it is best to observe certain considerations:

There is significant economic payout. A typical MPC takes about a man-year to build. Software costs are typically about one third of the investment in engineering time.
The process is interactive. This means that a single MV affects more than one CV.
There are significant process constraints.
The plant must have suitable infrastructure, particularly skilled engineers and operators, as well as modern and well-maintained instrumentation.
All of the necessary final control elements must exist and be able to work properly.

One should not to use MPC if

The problem is easily solved with one or two PID controllers or other standard control blocks. This happens when the process is naturally decoupled, i.e., either there are few

Process Control

process interactions or very fast-acting parts of the process are naturally separated from very slow-acting parts.

Special logic is required, for example, safety shutdown systems.

15.6.4 GENERAL METHOD FOR DESIGNING A MULTIVARIABLE CONTROLLER

There is a well-defined methodology for designing MPCs. It is quite general and can be applied to almost any situation where a MPC is required. In summary, the method is as follows:

Understand the process
Design the plant test
Conduct the plant test and collect data
Structure the controller and analyze the data
Tune the controller
Commission the controller
Post audit

15.6.4.1 Understand the Process

15.6.4.1.1 General

Process control is a blend of different disciplines, including mathematics, chemical engineering, computer science, economics, and process understanding. These skills are not equally weighted. Process understanding accounts for about 70 to 80% of the total knowledge required to install a successful MPC application; the other skills make up the remaining 20 to 30%. The lesson is clear: you must understand the process before proceeding. The following are things you can do to better understand the process:

Read and understand process flow drawings (PFDs).
Read and understand the piping and instrumentation diagrams (P&IDs).
Spend a few hours in the control room and talk to operators.
Interview plant engineers.
Interview plant economic planners.
Run a simple steady-state simulation of the process if one is readily available.
Read the plant operating manuals.
Cultivate a thorough understanding of such things as
　Why does this plant exist, i.e., what is the purpose of this facility?
　Where does the feed come from, and how much does it cost?
　Where do the products go, and how much are they sold for?
　How much flexibility is there in the feed supply and the product demand?
　Are there any seasonal operations?
　What are the top three or four constraints that operators worry about?
　Where is energy used and how much does it cost?
　What are the important plant specifications?
　What are typical plant daily operating orders?

The following is a good equation to keep in mind:

$$\text{Probability of MPC success (\%)} = \text{completeness of process understanding (\%)}$$

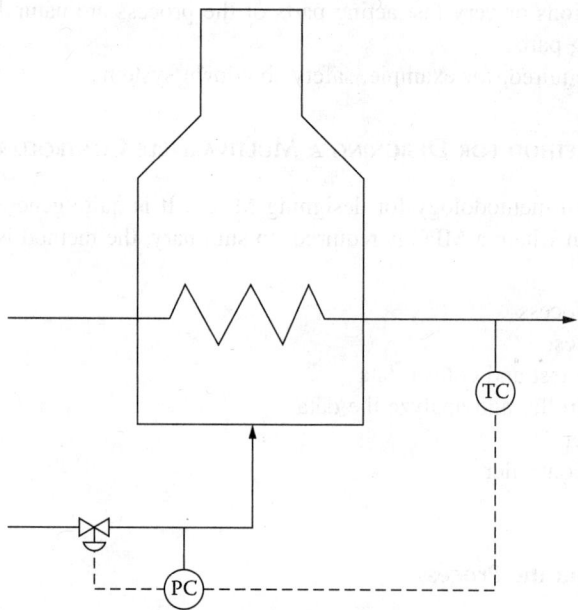

FIGURE 15.73 Fired heater outlet temperature control in the DCS.

15.6.4.1.2 Regulatory Control Strategy

Early in the project, learn which variables to keep in cascade, automatic, or manual modes. This is commonly referred to as the *state of the plant.* Normally, you want the PID loops to do as much work as possible, allowing the MPC to focus on higher-level functions.

The following are situations in which you probably want to leave the PID loops closed:

Fired heater outlet temperature control
Drum level control (see exceptions later)
Controls on rotating equipment
Properly functioning distillation tower temperature controls

The fired heater in Figure 15.73 shows the outlet temperature in the DCS. One reason for leaving this loop closed is that disturbances in these systems are usually very high in frequency and are rejected well by simple PID loops in the DCS.

Situations in which you may want to break the PID loops in the DCS, taking them out of cascade, or possibly putting them in manual, include the following:

Distillation columns with temperature controllers on both overhead and bottom
Process lines with two control valves in series (e.g., flow/level or flow/flow)
Loops with very long dynamics (e.g., composition loop with an hour or more of deadtime)

Consider a distillation column that has a temperature controller on the overhead and one on the bottom as well. Usually, this configuration leads to instability, variable saturation, or both. The best solution is to break one of the temperature loops (put in manual) in the DCS before beginning the plant test. Leave the one closed at the end of the column, where composition matters the most.

In general, think very carefully about the state of the plant before doing plant tests. If you choose incorrectly, you can always repeat the tests, but this will cost you many weeks of engineering time and a loss of credibility with the plant operators.

15.6.4.1.3 Inventories and Process Decoupling

Drums (tanks) are common in process plants. Every drum represents inventory, which in some cases can be usefully exploited to decouple the plant. This is especially useful in plants with significant heat integration.

For example, let us assume that the overhead stream from a main fractionator is condensed, enters the overhead accumulator (inventory), and then feeds a gas plant downstream. Assume also that the main fractionator has pumparounds lower down the tower that are used to supply reboiler heat to the same gas plant. This is an example of a heat-integrated plant.

In such a plant, the following can happen. Imagine that a big rainstorm causes increased condensing of the main fractionator overhead product. Assuming that the overhead accumulator is on level control, product flow to the gas plant increases, affecting the gas plant. The gas plant reboilers (being on temperature control) call for more heat, which results in increased main fractionator heat removal (via the pumparounds). This heat imbalance in the main fractionator again affects the overhead product flow, which again affects the gas plant, and so on. This whole chain of events can take a day or so to subside.

By breaking the main fractionator overhead level control loop, the vicious cycle described above can be avoided, because many of the heat integration imbalances simply show up as fluctuations in the overhead accumulator level. MPC will do a very good job of controlling overhead drum level so as to keep the gas plant stable.

Pay attention. Carefully determine which level loops should be broken before conducting plant tests.

15.6.4.2 Design the Plant Test

The plant test is the most important part of designing and installing MPC. Normally, a plant test consists of making changes to an MV and watching how the rest of the plant (CVs) responds. Usually, a single MV is moved while all others are held constant. In some situations, more than one MV can me moved simultaneously—for example, in different parts of a plant that are not significantly coupled. Although it is theoretically possible to move all MVs simultaneously in a carefully constructed plant test, such tests are nearly impossible to conduct in commercial plants and therefore are not recommended.

The following are requirements for a good test:

At least 5 moves per MV are required, and 15 moves would be better.
The signal (MV movement) must be as random as possible so as to minimize the chance that it correlates in time with some other variables in the plant.
The MV movement must be large enough to elicit a process response (1% of current MV value is a good minimum, and 10% of current MV value is a good maximum).
At least some of the moves in the MV must be held long enough that the plant comes to steady state.
Use a series of steps in the MVs. Other waveforms are possible (e.g., sinusoids) but are much more difficult to analyze visually.

Before you conduct the test, make a detailed list of which MVs you want to move and when you will move them. Discuss the move sizes with the plant operations people—they will let you know how big the moves can be. (Often, you can start with conservative move sizes and, as the test progresses, increase the move sizes to get better information.)

Above all, make sure you understand what the product specifications are and where important equipment and environmental constraints are located. You must accept responsibility for keeping the plant in a safe and economic operating condition during the plant tests. A law of nature is that

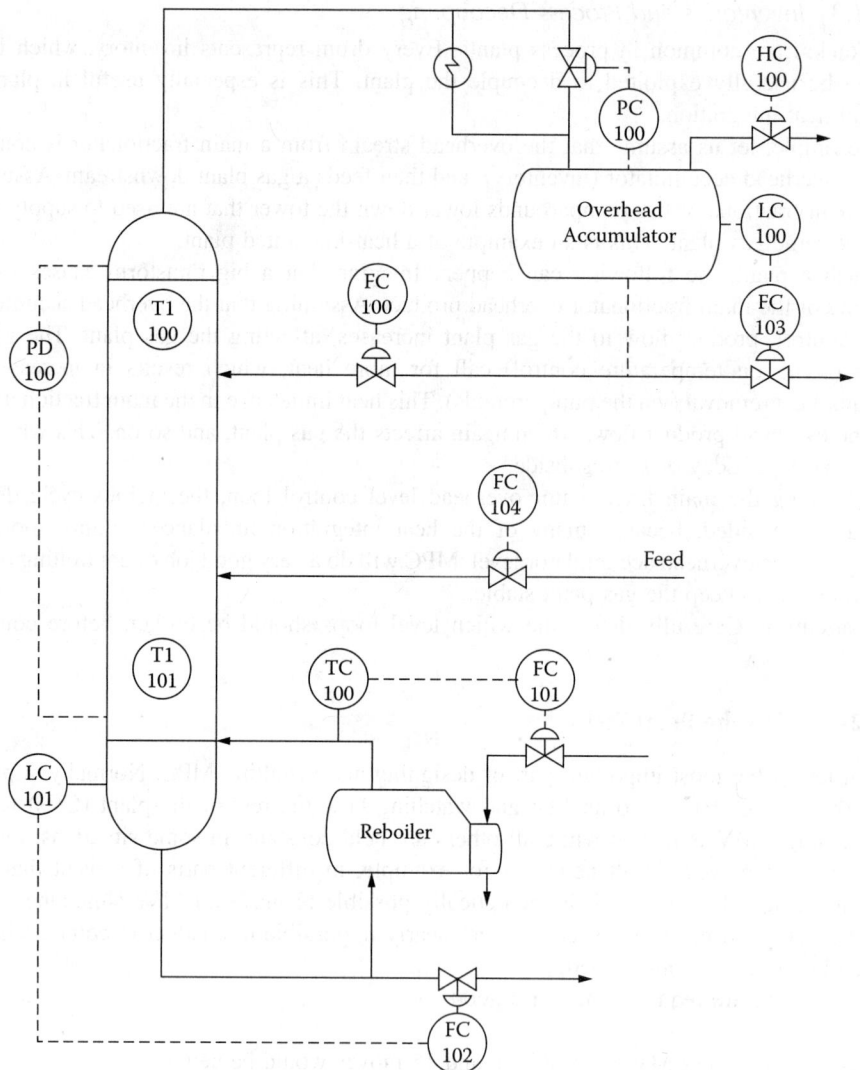

FIGURE 15.74 A distillation column with temperature control on the reboiler return.

as soon as a problem occurs in the plant, you will be blamed because you are doing something new. Be prepared, and don't let that happen.

For example, assume that you want to perform tests on the plant, represented by Figure 15.74. The plant is a simple distillation column with overhead accumulator pressure controlled by moving the hot vapor bypass, bottoms level maintained by bottoms product draw rate, and the overhead accumulator level controlled by adjusting the overhead product draw rate. Reflux is on flow control, and the reboiler is on temperature control. Typical move sizes for this plant are shown in Table 15.12.

15.6.4.2.1 Designing a Good Input Waveform

There are many ways to make moves to independent variables during a plant test. The most practical approach is as follows:

Use a series of step changes in a single variable.
Move only one independent variable at a time. If you move more, the resulting test data often become difficult to analyze visually.

Process Control

TABLE 15.12
Typical Move Sizes for the Plant Shown in Figure 15.73

Tag	Description	Nominal Value	Typical Move Size
FC100	Tower reflux	2000 BPH	20–200 BPH
TC100	Reboiler Return temp.	200°F	2–5°F
HC100	Vent to fuel	5%	0.1–0.5%
PC100	Overhead accum. pressure	200 PSIG	2–10 PSIG

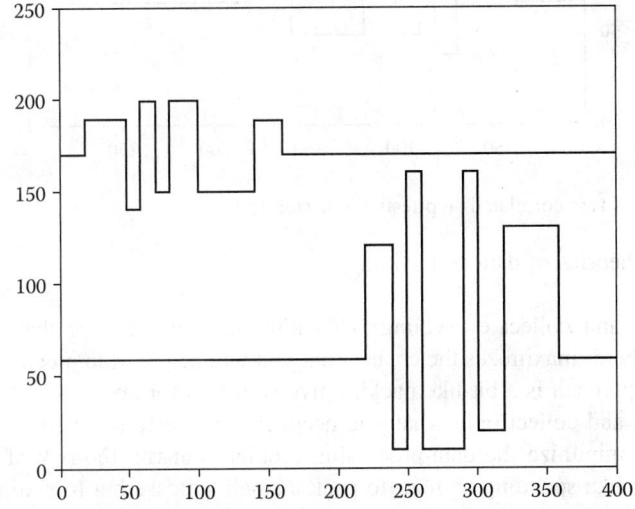

FIGURE 15.75 Plot of two unconnected step tests (good step test).

Make sure the input series is pseudorandom. This means that the behavior of the input variable being moved is not correlated with itself (no autocorrelation).

Figures 15.75 and 15.76 are examples of good and poor testes, respectively, being plots of input vs. time.

15.6.4.2.2 Set Up the Data Collection

15.6.4.2.2.1 Sample interval
The normal data collection in use today is one minute. If you are dealing with a really fast process, you may need to sample as often as every few seconds. If the plant is extremely slow, consider one sample every five minutes. There is nothing to prevent you from sampling either faster or slower, but you need to consider the trade-off between increased time resolution (good) and increased data set size (bad). Also, you should consider the dynamics of the system that you are dealing with—there is no point in sampling a system every ten seconds when the time constant is eight hours.

15.6.4.2.2.2 Collecting the Right Instrument Tags
In addition, you need to carefully define which data to collect. With modern digital instrumentation systems, it is tempting to collect all of the tags in a plant. Instrument tags contain the identification numbers assigned to process measurements. In a typical process unit, the number of tags is between 1000 and 3000, so keep in mind that if you collect them all, you will pay a price later in terms of the raw time it takes to handle the data.

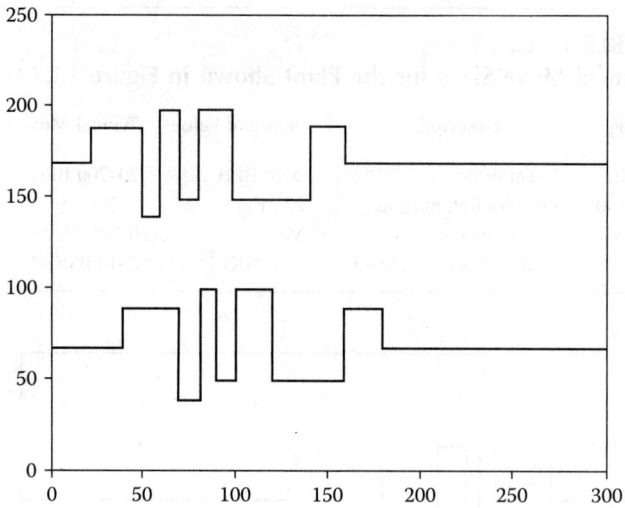

FIGURE 15.76 Plot of two correlated step tests (poor step test).

There are two theories of data collection:

1. Be cautious and collect everything. This minimizes the chance that you will forget something, but it maximizes the chance that you will acquire too much data to deal with later. This approach is a bit like packing five suitcases for an overnight trip.
2. Be practical and collect only what you need. If you carefully determine which data to collect, you minimize the data-processing problem but run the risk of missing useful data. This is like spending an hour to pack a small suitcase but forgetting your comb.

15.6.4.2.2.3 Laboratory Data

You need to identify which stream samples to collect during the plant test. Later, you will analyze the sample data to build inferred properties. Make sure you agree with the plant operations people about how much lab data you will need, for which streams, and how often it should be collected. Finally, remember that lab analyses are very expensive, so be careful to take just enough, but not too many, samples (30 to 60 samples per stream is about right). A good rule of thumb is to take at least ten data samples for each inferred properties coefficient being modeled. For a simple inferred property with 3 coefficients, at least 30 samples should be taken; e.g., $Y = a_1 x_1 + a_2 x_2 + a_3 x_3$, where a_i = the coefficients.

15.6.4.3 Conduct the Plant Test and Collect the Data

15.6.4.3.1 Preparation

The objective of doing a plant test is to get *good* data. The requirements of getting good data include

The instrumentation must work properly.
The PID controllers must be properly tuned.
Control valves must be properly sized and functioning correctly.
The process equipment must be in a normal state.

Before investing the time to conduct a plant test, you should spend a few days making sure that everything in the plant works properly. Move the setpoints to the main controllers to verify that the PID loops and final control elements work as they should. Look at the heat, mass, and

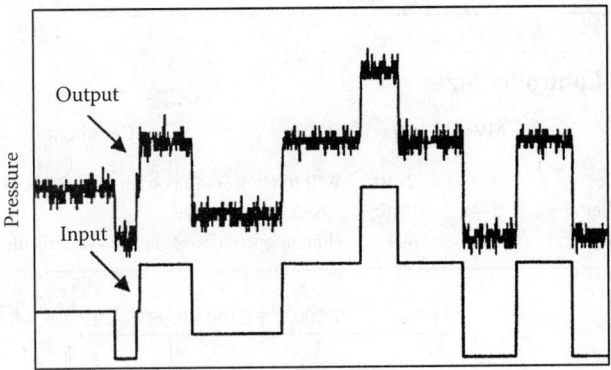

FIGURE 15.77 A good test.

composition balances on the unit to track down problems in process equipment. Check all of the major equipment and make sure that it is actually operating. For example, make sure that compressors, turbines, pumps, and furnaces are operational and running well.

In summary, first look at the big things, such as major actuators and major equipment, and make sure they are all in working order. You can do it first, or you can do it after investing three months of work and discovering that the big power generation expander, which makes most of the money in the plant, has been shut down for maintenance during the entire plant test.

15.6.4.3.2 Plant Test
Conducting a proper plant test is probably the most important step in a multivariable control project. A checklist for conducting a successful plant test follows:

Meet with plant operators at the start of each shift and explain what you are doing and why. The meeting should take five to ten minutes.
Make sure each MV move gets made on time, in the proper direction, and by the proper amount.
Once per shift, make sure that your data collection is still running.
Make real-time plots and try to explain all of the behavior you see. You should be doing this on a more or less continuous basis. Often, this will help you identify operating problems before the operators do, and will help to keep the plant stable during your tests.
Make notes during the plant test. Record any unusual events, along with the time they happened.
Talk to plant operators. Learn as much as you can about the process.
Buy plenty of pizzas and donuts for operators. A happy operator is a cooperative operator.

When you have finished the plant test, you will have an excellent understanding of how the unit operates. If you do not have this understanding, you will have failed to achieve one of your main objectives.

Figure 15.77 shows an example of a good test. There are nine input moves. The input series is not autocorrelated (i.e., the moves are of different lengths and do not show a pattern). The move sizes are large enough to get a good response in the output.

15.6.4.4 Structure the Controller and Analyze the Data

15.6.4.4.1 How to Pick MVs, CVs, and DVs
Make sure that you select the right number of MVs; too many, and the controller is cumbersome and hard to understand, too few, and you may miss important interactions and constraints (and

TABLE 15.13
Relative MPC Controller Sizes

	MVs	CVs	Comment
Very small controller	2–4	2–10	Will likely miss important constraints and interactions.
Reasonable controller size	5–40	10–40	About the right size.
Controller is too large	40+	40+	Human operators begin to have difficulty in understanding.

	AP100	PD100	PC100.VP	FC101.VP
FC100	–	+	+	+
TC100	+	+	–	+
HC100	0	0	+	0
PC100	–	–	+	–

FIGURE 15.78 Roughed-out gain matrix for the control design of the tower in Figure 15.73.

therefore benefits). Typically, a single controller spans a large part of a process unit. Table 15.13 illustrates the concept of proper controller size.

At this point in a control project, you should have a good idea of the controller structure—which variables are MVs, which are CVs, and which are DVs. Draw a matrix with the MV and DV tags as rows and the CV tags as columns (or vice versa). Using your newly gained knowledge of the process, write in a "+" or a "–" for each MV/CV pair (or DV/CV pair) where you expect to see a positive response and where you expect to see a negative response, respectively. Tape the matrix to your wall and look at it from time to time just to make sure you are on track as you proceed with more detailed analyses.

Be sure to look for patterns in your roughed-out matrix. For example, if two blocks have no common interactions, you may be able to divide it into two smaller controllers. Keep this in mind as you progress with the detailed analysis. See Figure 15.78 for an example.

15.6.4.4.2 Designing for Closed-Loop Optimization

Today's multivariable controllers are mainly linear algorithms. Depending on the process, there can be significant advantages to optimizing the process using a nonlinear model. Most nonlinear models today are steady-state, rigorous, heat and mass balance models and are built separate from the multivariable controller.

Even at this early stage in designing your controller, keep in mind that some of its targets may be set by a nonlinear optimizer. It is useful to decide which CVs or MVs will have their targets set by an optimizer. This becomes important later, particularly when tuning the controller.

15.6.4.4.3 Building the Power Model for the MPC Controller

Contemporary model building methodology is well established. Essentially, it is a matter of identifying which plant test data are usable; which variables should be CVs, MVs, or DVs; and what

Process Control

is the best controller frequency. Different software packages vary in which steps are involved, but when all is said and done, the designer has a few important decisions left:

Is the controller structure correct? Have the correct variables been chosen for MVs, CVs, and DVs?

Do the resulting models make sense? Are the process gains the correct sign? Should some models be zero? Are the gains roughly the correct magnitude?

Does the controller contain parallel CVs or MVs (singular matrix)? If yes, is this the best design, or can the controller be simplified to remove the parallel variables? (This is an important issue, because parallel variables usually result in some form of unacceptable controller behavior.)

Is the scope (size) of the controller correct? Are all of the right interactions and constraints present?

Will the plant operators understand the final controller, or is it too complex to understand?

Model building is a very important step. Get this part right, and the rest follows fairly easily. Get it wrong, and you will be in for a rough time later.

Extensive information is available on how to use modern MPC modeling software. Most of this takes the form of training courses and material supplied by the various technology vendors. This software does a regression between MVs and CVs and produces linear dynamic models as shown in Figure 15.79.

For those who are unfamiliar with these models, here is a brief explanation. The matrix represents the distillation tower shown earlier in Figure 15.74. The rows of the matrix are the MVs,

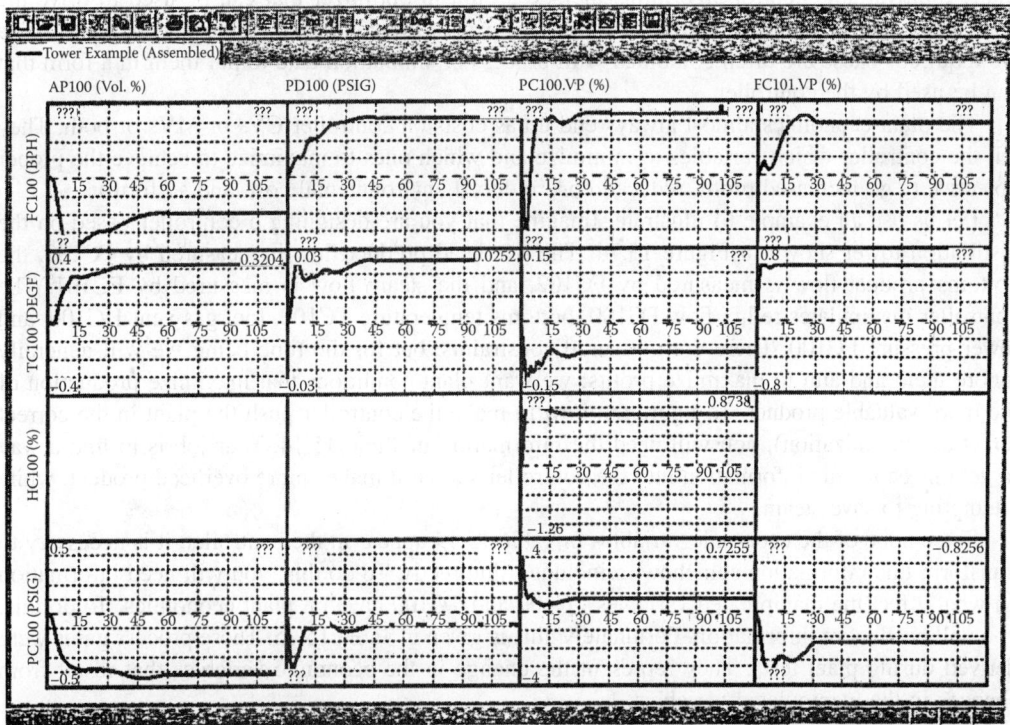

FIGURE 15.79 Calculated controller matrix for the control design of the tower in Figure 15.73. DMCplus® was used to create this matrix.

TABLE 15.14
Economic Data for Distillation Column in Figure 15.74

Tag	Description	Cost	Units
FCl01	Steam flow	0.003	$/lb
FC102	Bottoms product	12	$/BBL
FC103	Overhead product	20	$/BBL

TABLE 15.15
Process Gains for Distillation Column in Figure 15.73

	FC101 Steam (lb/hr)	FC102 Bottoms (BPD)	FC103 Overhead (BPD)
FC100 (reflux, BPD)	1090	0.2	−0.2
TCl00 (reboiler, Dee F)	2800	−10	10

and the columns are the CVs. At the intersection of each row and column is a single model (blanks indicate that no relationship exists between a given MV/CV pair).

15.6.4.4.4 Plant Economics

Many multivariable control algorithms include a built-in optimizer that can be used to drive the plant in a better economic direction. To tune this part of the controller, reasonable economic data are required. These are often combined with process or simulation data to put them in a form that can be used by the controller.

The plant economics almost always end up as costs on controller CVs or MVs or both. They tell the controller which variables to minimize and which ones to maximize to achieve the proper objective (e.g., increased profit, reduced cost, reduced environmental cost, and so forth).

Let us use an example to illustrate. Imagine that you are designing a controller based on the distillation tower shown in Figure 15.74. The overhead product flow is measured by FC103, the bottoms product flow is measured by FC102, and the steam flow is measured by FC101. The controller manipulates reflux flow FC100, bottoms temperature TC100, fuel gas vent HC100, and tower pressure PC100. There are measured constraints, but for the time being we will generally ignore them and aim to maximize profits; we want plant conditions that maximize production of the more valuable product. To figure out how to make the controller push the plant in the correct direction (optimization), you will need the information in Table 15.14. Your job is to find a way to get this economic information into the controller so that it makes more overhead product, while attempting to save steam.

Since none of the economic variables in Table 15.14 appear in the controller, it is necessary to transform this information somehow to include it indirectly. To do this, you will need information on what effect the manipulated variables FC100 and TC100 have on plant economics. To do this, you will need more information. Fortunately, this is given in Table 15.15. These process gains were derived during plant tests. They represent the change in the economic variables that result from changes in the manipulated variables.

So, combining the information in Tables 15.14 and 15.15, we can compute the economic costs to be included in the controller as in Table 15.16. Note that one cost is positive and the other negative. The positive cost for reflux will cause the controller to reduce reflux flow. The negative cost for reboiler temperature will cause the controller to increase reboiler temperature. In practice,

TABLE 15.16
Final Controller Economics

	Cost	Units
FC100 (reflux)	4.87	$/BPD
TC100 (reboiler)	−71.6	$/°F

the controller will reduce reflux and increase reboiler temperature until a constraint is encountered. As conditions change, the controller will continue to push the MVs in this direction, thus increasing profit, while always respecting constraints.

15.6.4.5 Tune the Controller

The next step after building controller models is to tune the controller. This involves using the dynamic model to simulate the plant and then tuning the controller against that. Even though every commercial algorithm is slightly different, you need to get three types of behavior right before installing the controller in the plant:

Economics
Constraints
Dynamics

15.6.4.5.1 Controller Tuning—Economics

Most MVC algorithms have a way to impose an economic solution on the controller. This approach will drive the controller to a most favorable operating point (when there are degrees of freedom available). Once you have derived appropriate plant economics (see Section 15.6.4.4.4), the next step is to run the plant/controller simulation many times with different combinations of MVs and CVs at constraints while making sure that the controller pushes the plant in a direction that makes sense.

15.6.4.5.2 Controller Tuning—Constraints

The next important aspect of tuning a multivariable controller is to determine what happens when different constraint sets become active. For example, assume that you have designed a distillation tower controller with constraints on overhead composition, bottom composition, and tower ΔP. The MVs are reboiler duty and reflux rate. The following are examples of the types of constraint trade-offs that you need to consider:

Which constraint is more important: overhead composition or tower ΔP?
What should the controller do when all three constraints are active (since the controller has only two MVs, it must give up on one of the CVs).
When the reboiler is at the maximum limit, which constraint should be dropped last?

The plant/controller simulation should be exercised many times, with many different combinations of active constraints. For each constraint combination, it is important to make sure that when constraint violations are unavoidable (more active constraints than MVs), the CV violations occur in the correct order, with the highest priority given to variables related to safety.

As expected, most algorithms have a way to deal with this situation. Typically, each CV can be given a weight, with lower-weighted CVs being violated first. The most important consideration in tuning for constraints is to ensure that a representative number of different constraint combinations are tested in an off-line environment, before the controller is run in the plant.

15.6.4.5.3 Controller Tuning—Dynamics

All multivariable control algorithms include a way to tune how quickly they respond to setpoint changes or disturbances. Two general classes of tuning parameters affect the speed of response.

The first is often referred to as *reference trajectory* and can be thought of as a filter on setpoint changes. The filtered setpoint in essence becomes a trajectory—the controller attempts to move the MVs so that the CV follows their individual trajectories. Increased filtering produces slower controller action.

The second class of dynamic tuning focuses on how much the MVs move for a given disturbance (either setpoint change or unmeasured disturbance). This type of tuning parameter is commonly called *move suppression* and provides a mechanism to slow down some MVs relative to others. Move suppression is important when a given MV affects a critical piece of equipment, for example, a large compressor with many automatic overrides and shutdown interlocks.

The general procedure for tuning the dynamics of a new controller is to run the controller/simulation combination under many different combinations of setpoint changes and unmeasured disturbances. Typically, plant operators will provide valuable feedback for how much the MVs should move in different circumstances and how fast the CVs should get to setpoint and away from constraints.

15.6.4.5.4 Operator Training

In many states in the U.S.A., it is a legal requirement that all operator training take place before a multivariable closed-loop controller is activated in an operating plant for the first time. Even in places where this is not a legal requirement, common sense dictates that this step be carried out first.

Typically, the plant operators each receive an hour or two of training while working on their normal shift. In some exceptional cases, the operators receive formal classroom training. Better training leads to increased project success.

15.6.4.5.5 Watchdog Timer

A multivariable controller is expected to run continuously. Sometimes, however, the control software or control hardware fails. This situation should be detected quickly and the operator warned, normally by an alarm in the DCS.

The most common way of detecting and alarming this problem is to use a small program called a *watchdog timer* running in the DCS. The watchdog timer communicates frequently with the multivariable controller program. If the watchdog timer fails to get a response after a predetermined period of time, the watchdog forces each MV into its fallback (shed) position and writes an alarm to the operator.

15.6.4.6 Commission the Controller

After a controller has been designed, built, and tuned in an off-line environment, the next big step is to install it in the plant computing hardware and commission it. Since, at this stage, it becomes possible for the controller to directly modify how the plant is running, one must proceed methodically and with caution. Use checklists. Be observant and, at least initially, monitor all of the variables related to the controller frequently and thoroughly. Below is a good suggested procedure.

15.6.4.6.1 Use a Checklist

Installing a multivariable controller is not difficult, but, if done carelessly, it can lead to serious problems such as an off-specification product or equipment damage. To maximize the chances of a smooth installation, use a checklist. Here is a simple example.

1. Make sure that operators are trained on how the controller works. Do not proceed until they are.

Process Control

2. Install software, database, and other real-time components.
   - ❏ Control algorithm installed in on-line computer.
   - ❏ Controller on-line database built.
   - ❏ Watchdog timer installed.
   - ❏ Controller definition information (models, tuning parameters).
3. Check controller data base.
   - ❏ Controller CVs, DVs, and MVs match raw DCS inputs *exactly*.
   - ❏ Controller models, tuning parameters, etc., match off-line versions *exactly*.
4. Turning controller on in prediction mode.
   - ❏ Make sure that each MV is configured correctly in the DCS.
   - ❏ Check the shed mode for each DCS.
   - ❏ All MVs out of cascade (MFC controller cannot change).
   - ❏ Controller turned on in prediction mode.
   - ❏ Watch calculated move sizes. Check that the sizes make sense.
   - ❏ Watch the CV predictions. Check that they make sense.
5. First-time MV check.
   - ❏ Narrow upper and lower limits of MVs (or rate limits).
   - ❏ Make sure the controller is running.
   - ❏ Put MVs into cascade *one at a time*. Ensure no bad behavior in DCS setpoint.
   - ❏ Make sure that the DCS setpoint changes by *exactly* (to within machine precision) the size of the calculated move.
6. Closing the loop.
   - ❏ Make sure controller is running.
   - ❏ Activate controller master ON switch (if there is one).
   - ❏ With MV limits still clamped, put one or two MVs into cascade.
   - ❏ Relax the MV bounds a little and check for good controller behavior.
   - ❏ Add more MVs gradually, making sure controller behavior is OK.

For a reasonably large controller, steps 2 through 5 should take a few days. Step 6 should take a day or so, assuming that there are no problems. From this point on, the main activity is to gradually relax the controller bounds until it is clear that the actual behavior in the plant matches the behavior that was simulated during design. Over the next few weeks, make sure that the standard deviations on the CVs decrease (normally only when they are at constraint limits), that the MVs are not moving around too much, and that the economics are driving the controller in the correct direction.

15.6.4.7 Post Audit

The last step in a properly run control project is to conduct a post audit. Collect postcommissioning data and compare them to precommissioning data. Make sure the new operating point is economically better than the old one. Check to ensure that the standard deviations of economically important CVs (often associated with product specifications) have decreased significantly. Evaluate whether the MV bounds are still set correctly. Can they be widened? Is the controller forcing the process to run at the best set of constraints most of the time?

15.6.5 MULTIVARIABLE CONTROLLER TROUBLESHOOTING

A newly commissioned multivariable controller works well. Over time, the performance can deteriorate to the point at which the controller is turned off by plant operators. This section can be used to diagnose common problems that can lead to disabling a multivariable controller.

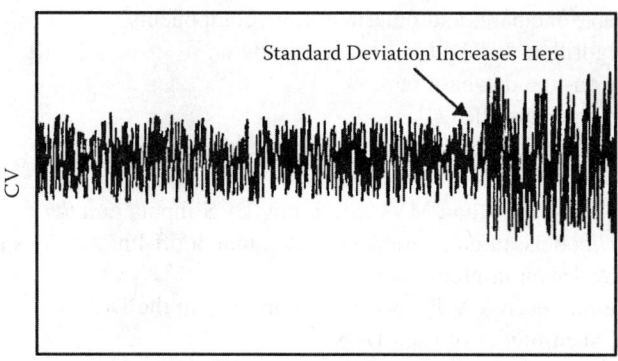

FIGURE 15.80 Standard deviation of a CV increases suddenly.

15.6.5.1 Commonsense Approach

Ninety percent of troubleshooting multivariable controllers is common sense. For example, below is a daily checklist that can be applied to diagnose the last 24 hr of operation for a MPC application.

15.6.5.1.1 Quick Screening

The first step in controller troubleshooting is to check overall controller behavior during the recent past. This quick screening is often a simple set of questions or a few simple reports. For example, here are some questions to ask each morning:

- What percent of time was the controller running?
- Which constraint set was most active?
- How many MVs are at bounds?
- How many CVs are at bounds?
- How many MVs are active (being manipulated by the controller)?
- What is the standard deviation of CVs that are at bounds?
- Is the controller pushing the plant in the right direction?
- How far apart are the MV bounds?

15.6.5.1.2 Deeper Analysis

If the quick screening turns up potential problems, deeper analysis can help narrow down the problem areas. One good way to do this is to examine time plots for all CVs and MVs for the last 24 hr. Abnormal behavior can be quickly traced to one or a few variables by this approach.

Figure 15.80 is an illustration of what happens when a multivariable controller develops a problem; in this case, a CV in the controller exhibits a sudden increase in standard deviation. Now, having narrowed the problem, an engineer can conduct even more detailed analyses.

15.6.5.2 Detailed Troubleshooting—Use a Checklist

Finally, once the problem behavior has been narrowed down, a good final step is to try to pinpoint the problem using a checklist. It is surprising how many problems can be found by running quickly through a checklist—for example,

- ❏ Controller is on, but MVs don't move
- ❏ Controller turned off by itself
- ❏ Operator turned controller off

Process Control

- ❑ Controller is on, but control is poor
- ❑ General controller error
- ❑ One or more inputs is bad
- ❑ MVs are moving erratically
- ❑ Controller is on, but MVs don't move:
  - MV upper and lower bounds are the same
  - MV step bounds are very small (or zero)
  - Most CVs are turned off (removed from controller)
  - Valves are wide open or closed
- ❑ Controller turned off by itself:
  - Look at error log, check for numerical problems
  - Critical CVs are not available or bad
  - Critical MVs are not available or bad
  - Bounds on CVs or MVs do not make sense (e.g., lower bound greater than upper bound)
  - There is a hardware failure (often DCS or computer)
  - Check watchdog timer alarm—often indicates hardware problem
- ❑ Operator turned controller off:
  - Ask operator why controller was turned off
  - Are operators adequately trained?
  - Process is very upset, wait for normal behavior and turn on again
- ❑ Controller is on, but control is poor:
  - Has the process operating point changed significantly?
  - Have the controller targets changed significantly?
  - Is there a new feed?
  - Is the weather very different (e.g., snow, rainstorm, etc.)?
  - Have valves been bypassed?
  - Has new process equipment been put in service (or existing equipment taken out of service)?
  - Has a controller model been recently changed?
  - Has the multivariable controller been retuned?
  - Have PID loops been retuned?
  - Have actuators broken or been replaced?
  - Have the controller economic parameters been changed?
- ❑ General controller failure:
  - Most algorithms have diagnostics based on error number. Check error number
- ❑ One or more input is bad:
  - AI needs to be reset
  - DP cell plugged
  - FI plugged
  - LI float stuck
  - TI burned out
  - Calculated variable has error (usually, a flow has gone to zero)
  - Check software interface between DCS and multivariable controller hardware
- ❑ MVs are moving erratically:
  - Has multivariable controller tuning changed?
  - Has a new control matrix been loaded into the controller?
  - Has DCS database changed?
  - Has an instrument been reranged in the DCS?

15.6.5.3 Is the Controller Still Making Money?

In the final analysis, only one thing really matters—is the controller still making money? You should get in the habit of periodically evaluating the controller behavior; every few months is probably often enough. Check to see whether

- The controller is pushing the plant in a direction that makes economic sense.
- Controller economics match the current process economics.
- Standard deviations of key variables are acceptable compared to a good base case.
- CV and MV constraints still make economic sense and are safe.
- The variable set is correct. (Should some be removed? Others added?)
- The scope of the controller is still OK. (Should it be expanded? Made smaller?)
- It is time to add a real-time nonlinear optimizer.
- The operators are still adequately trained in the use of the controller.

REFERENCES, SECTION 15.2

1. Lucas, M. P., 1986. *Distributed Control Systems*. New York: Van Nostrand Reinhold, 4.
2. McMillan, G. K., G. E. Mertz, and V. L. Trevathan, 1998. Troublefree Instrumentation, *Chem. Eng.*, Nov.: 80–88.
3. Liptak, B., 1995. *Instrument Engineers Handbook*. Philadelphia: Chilton.
4. McMillan, G. K., 1995. *Advanced Temperature Control*. Research Triangle Park, NC: Instrument Society of America, 133–155.

REFERENCES, SECTION 15.3

1. Riggs, James B., 1999. *Chemical Process Control*. Lubbock, TX: Ferret Publishing, 154–160.
2. Private communication, Jim Downs, Tennessee Eastman Company, Nov. 1998.

REFERENCES, SECTION 15.4

1. Marlin, T. E., 1995. *Process Control*. New York: McGraw-Hill, 393.
2. Astrom, K. J. and T. Hagglund, 1988. *Automatic Tuning of PID Controllers*. Research Triangle Park, NC: Instrument Society of America, 233.
3. Ziegler, J. G. and N. B. Nichols, 1942. Optimum Settings for Automatic Controllers, *Trans ASME* 64: 759.
4. Tyreus, B. D. and W. L. Luyben, 1992. Tuning PI Controllers for Integrator/Deadtime Processes. *Ind. Eng. Chem. Res.* 31: 2625.
5. Marlin, T. E., 1995. *Process Control*. New York: McGraw-Hill, 588–590.

REFERENCE, SECTION 15.5

1. Bristol, E. H., 1966. *IEEE Trans Auto. Con.*, AC-11: 133.

REFERENCES, SECTION 15.6

Cutler, C. R. Ramaker, B. L., 1979. Dynamic Matrix Control—A Computer Control Algorithm, AIChE National Meeting, Houston, TX.

Cutler, C. R. Ramaker, B. L., 1980. Dynamic Matrix Control—A Computer Control Algorithm, *Proceedings of the Joint Automatic Control Conference.*

Cutler, C. R., Yocum, F. H., 1991. Experience with the DMC Inverse for Identification, in Y. Arkun and W. H. Ray, eds., *Chemical Process Control—CPC IV*, Fourth International Conference on Chemical Process Control. Amsterdam: Elsevier, 297–317.

Froisy, J. B., 1994. Model Predictive Control: Past, Present and Future. *ISA Trans.* 33: 235–243.

Froisy, J. B. Matsko, T., 1990. IDCOM-M Application to the Shell Fundamental Control Problem, AIChE Annual Meeting.

Grosdidier, P., Froisy, B. Hammann, M., 1988. The IDCOM-M Controller, in T. J. McAvoy, Y. Arkun E. Zafiriou, eds., *Proceedings of the 1988 IFAC Workshop on Model Based Process Control.* Oxford, UK: Pergamon Press, 31–36.

Korchinski, W. J., Kaneko, R., Bunya T., et al., 1999. Economic Analysis of FCC Re-examined. Hydrocarbon Processing, April.

Prett, D. M. Gillette, R. D., 1980. Optimization and Constrained Multivariable Control of a Catalytic Cracking Unit, *Proceedings of the Joint Automatic Control Conference.*

Richalet, J., Rault, A., Testud, J. L. Papon, J., 1976. Algorithmic Control of Industrial Processes, *Proceedings of the 4th IFAC Symposium on Identification and System Parameter Estimation*, 1119–1167.

Richalet, J., Rault, A., Testud, J. L., Papon, J., 1978. Model Predictive Heuristic Control: Applications to Industrial Processes." *Automatica* 14: 413–428.

16 Conceptual Process Design, Process Improvement, and Troubleshooting

Donald R. Woods, Andrew N. Hrymak, and James R. Couper

CONTENTS

16.1	Problem-Solving Process		1273
	Additional Reading		1280
16.2	Rules of Thumb for Problem Solving		1280
	Recommended Reading		1283
16.3	Engineering Economics		1283
	16.3.1	The Role of Economics in Financial Reporting	1284
		16.3.1.1 Balance Sheet (Anon., 1992; Couper et al., 2000)	1284
		16.3.1.2 Income Statement (Anon., 1992; Couper et al., 2000)	1287
		16.3.1.3 Accumulated Retained Earnings	1288
		16.3.1.4 Changes in Financial Position	1288
		16.3.1.5 Other Financial Terms of Significance	1288
		16.3.1.6 10K Report	1289
		16.3.1.7 Concluding Comments	1289
		16.3.1.8 Financial Ratios (Couper et al., 2000)	1289
		Additional Reading	1292
	16.3.2	Financial Attractiveness	1292
		16.3.2.1 Quantitative Measures	1292
		16.3.2.2 Qualitative Measures	1292
		16.3.2.3 Cost of Capital, Minimum Acceptable Return, and Risk	1293
	16.3.3	Operating Expense Estimation	1295
		16.3.3.1 Manufacturing Expense Sheet	1295
		16.3.3.2 Operating Expense Items	1295
		Additional Reading	1300
	16.3.4	Estimation of Fixed and Total Capital Investment	1300
		16.3.4.1 Definitions and Terminology	1300
		16.3.4.2 Cost Contributions for One Type of MPI	1301
		16.3.4.3 Accounting for Time and Location	1305
		16.3.4.4 Estimating the $L + M$ Cost of Instrumentation	1306
		16.3.4.5 Accounting for Materials of Construction Other Than Carbon Steel	1308
		16.3.4.6 Different Methods for Different Levels of the Gating Process	1310
		Additional Reading	1312
16.4	Build a Reliable Data Base		1313
	16.4.1	Input Data from Scientists	1313

	16.4.2	Minimize Uncertainty in Input Data through Experimentation: Principles of Scale-Up	1313
		Additional Reading	1314
	16.4.3	Units of Measurement and Communication	1314
	16.4.4	Gather Input Data from Data Bases for Design and Process Improvement	1314
		16.4.4.1 Checklist for Reactivity	1317
		16.4.4.2 Checklist for Chemical Persistence	1317
		Additional Reading	1320
	16.4.5	Gather Data for Troubleshooting	1321
		Additional Reading	1321
16.5	Sustainability from the Start		1322
	16.5.1	Impact of Sustainability on Design	1323
	16.5.2	Impact of Sustainability on Process Improvement	1323
		Additional Reading	1324
16.6	Operability and Control Considered Throughout		1324
	16.6.1	Control and Design	1324
		16.6.1.1 Design: Impact of Control on the Initial Design Statement	1324
		16.6.1.2 Design: Impact of Control on the Evolving Flowsheet	1324
		16.6.1.3 Design: Impact of Control on the Type of Equipment Selected	1325
		16.6.1.4 Design: Impact of Control on the Sizing of Equipment	1325
		16.6.1.5 Design: Impact of Control on Issues beyond Conceptual Design	1325
	16.6.2	Control and Process Improvement	1326
	16.6.3	Control and Troubleshooting	1326
		Additional Reading	1326
16.7	Consider Safety, Waste Minimization, and Environmental Sensitivity Throughout		1327
	16.7.1	Step 1: Target Opportunities	1327
		16.7.1.1 Know the Legal and Economic Context	1327
		16.7.1.2 Identify the Opportunity	1327
	16.7.2	Steps 2 through 5: Move from "Eliminate" to "Isolate" during the Design Process	1328
		16.7.2.1 Step 2: Eliminate the Source	1328
		16.7.2.2 Step 3: Minimize the Source	1330
		16.7.2.3 Step 4. Minimize the Impact	1332
		16.7.2.4 Step 5: Isolate the Source	1333
		16.7.2.5 Step 6: Isolate the Impact	1333
	16.7.3	For Process Improvement	1333
	16.7.4	For Troubleshooting	1333
		Additional Reading	1333
16.8	Aids for Design, Process Improvement, and Troubleshooting		1333
	16.8.1	Steady-State Flowsheeting	1336
	16.8.2	Formulating a Well-Posed Problem	1336
	16.8.3	Flowsheet Architectures	1338
		16.8.3.1 Sequential Modular Architecture	1338
		16.8.3.2 Equation-Based Architecture	1341
	16.8.4	Importance of Thermodynamics Packages	1341
		16.8.4.1 Vapor-Liquid Equilibrium	1342
		16.8.4.2 At High Pressure or near the Critical Region	1342
		16.8.4.3 Equation of State, EOS	1342

Conceptual Process Design, Process Improvement, and Troubleshooting

    16.8.5 Tools for Flowsheet Development 1342
        16.8.5.1 Mathematically Based Programming 1343
        16.8.5.2 Hierarchical Use of Heuristics for Evolutionary Development 1343
    Additional Reading 1343
16.9 Process Optimization 1344
    16.9.1 Define the Problem 1344
    16.9.2 Select the Optimization Solution Scheme 1344
        16.9.2.1 Nonlinear Objective Function Problems 1345
        16.9.2.2 Linear Programming 1346
        16.9.2.3 Nonlinear Programming 1346
        16.9.2.4 Discrete Decision Variables 1346
    16.9.3 Linking Optimization with the Flowsheeting Tool 1346
    Additional Reading 1347
16.10 Rules of Thumb for Interpersonal Skills and Teamwork 1347
    16.10.1 Interpersonal Skills 1347
        16.10.1.1 Self-Awareness 1347
        16.10.1.2 Five Key Principles 1347
    16.10.2 Effective Teams 1348
    Additional Reading 1349
16.11 Rules of Thumb 1349
    16.11.1 Overall Process 1351
        16.11.1.1 Process Control 1351
        16.11.1.2 Properties of Fluids 1353
    16.11.2 Transportation 1353
        16.11.2.1 Gas Moving: Pressure Service 1353
        16.11.2.2 Gas Moving: Vacuum Service 1355
        16.11.2.3 Liquid 1356
        16.11.2.4 Gas-Liquid (Two-Phase Flow) 1357
        16.11.2.5 Pumping Slurries: Liquid-Solid Systems 1357
        16.11.2.6 Solids 1358
        16.11.2.7 Ducts and Pipes 1359
        16.11.2.8 Steam 1359
    16.11.3 Energy Exchange 1359
        16.11.3.1 Drives 1360
        16.11.3.2 Thermal Energy: Furnaces 1360
        16.11.3.3 Thermal Energy: Fluid Heat Exchangers, Condensers, and
                    Boilers 1361
        16.11.3.4 Thermal Energy: Fluidized Bed (Coils in Bed) 1363
        16.11.3.5 Thermal Energy: Motionless Mixers 1364
        16.11.3.6 Thermal Energy: Direct Contact L-L Immiscible Liquids 1364
        16.11.3.7 Thermal Energy: Direct Contact G-S Kilns 1364
        16.11.3.8 Thermal Energy: Direct Contact G-S Fluidized Beds 1365
        16.11.3.9 Thermal Energy: Direct Contact G-S Multiple Hearth
                    Furnaces 1365
        16.11.3.10 Thermal Energy: G-S Drying of Solids 1365
        16.11.3.11 Thermal Energy: Direct Contact G-L Cooling Towers 1366
        16.11.3.12 Thermal Energy: Direct Contact G-L Quenchers 1366
        16.11.3.13 Thermal Energy: Direct Contact G-L Condensers 1366
        16.11.3.14 Thermal Energy: G-G Thermal Wheels, Pebble Regenerators,
                    and Regenerators 1367

16.11.3.15 Thermal Energy: Solidify Liquids 1367
 16.11.3.16 Thermal Energy: Heat Loss to the Atmosphere 1367
 16.11.3.17 Thermal Energy: Refrigeration 1367
 16.11.3.18 Thermal Energy: Steam Generation 1368
16.11.4 Homogeneous Separation 1368
 Overall Guidelines 1368
 16.11.4.1 Evaporation 1368
 16.11.4.2 Distillation 1369
 16.11.4.3 Freeze Concentration 1372
 16.11.4.4 Melt Crystallization 1372
 16.11.4.5 Zone Refining 1373
 16.11.4.6 Solution Crystallization 1373
 16.11.4.7 Precipitation 1374
 16.11.4.8 Gas Absorption 1374
 16.11.4.9 Gas Desorption/Stripping 1376
 16.11.4.10 Solvent Extraction, SX 1376
 16.11.4.11 Adsorption: Gas 1378
 16.11.4.12 Adsorption: Liquid 1378
 16.11.4.13 Ion Exchange 1379
 16.11.4.14 Foam Fractionation 1380
 16.11.4.15 Membranes: Gas 1381
 16.11.4.16 Membranes: Dialysis 1381
 16.11.4.17 Membranes: Electrodialysis 1382
 16.11.4.18 Membrane Configurations 1382
 16.11.4.19 Membranes: Pervaporation 1382
 16.11.4.20 Membranes: Reverse Osmosis (RO) 1384
 16.11.4.21 Membranes: Nanofiltration 1385
 16.11.4.22 Membranes: Ultrafiltration (UF) 1385
 16.11.4.23 Membranes: Microfiltration 1386
16.11.5 Heterogeneous Separations 1387
 General Guidelines 1387
 16.11.5.1 Gas-Liquid 1387
 16.11.5.2 Gas-Solid 1388
 16.11.5.3 Liquid-Liquid 1390
 16.11.5.4 Liquid-Solid: General Selection 1391
 16.11.5.5 Dryer 1392
 16.11.5.6 Screens for "Dewatering" 1396
 16.11.5.7 Settlers 1396
 16.11.5.8 Hydrocyclones 1397
 16.11.5.9 Thickener 1397
 16.11.5.10 CCD: Countercurrent Decantation 1397
 16.11.5.11 Sedimentation Centrifuges 1398
 16.11.5.12 Filtering Centrifuge 1399
 16.11.5.13 Filter 1400
 16.11.5.14 Leacher 1404
 16.11.5.15 Liquid-Solid: Dissolved Air Flotation, DAF 1404
 16.11.5.16 Liquid-Solid: Expeller and Hydraulic Press 1405
 16.11.5.17 Solid-Solid: General Selection 1405
 16.11.5.18 Froth Flotation 1405
 16.11.5.19 Electrostatic 1406
 16.11.5.20 Magnetic 1406

Conceptual Process Design, Process Improvement, and Troubleshooting

  16.11.5.21 Hydrocyclones 1407
  16.11.5.22 Air Classifiers 1407
  16.11.5.23 Rake Classifiers 1408
  16.11.5.24 Spiral Classifiers 1408
  16.11.5.25 Jig Concentrators 1408
  16.11.5.26 Table Concentrators 1408
  16.11.5.27 Sluice Concentrators 1409
  16.11.5.28 Dense Media Concentrators (DMSs) 1409
  16.11.5.29 Screens 1409
 16.11.6 Reactors and Vessels 1410
  16.11.6.1 General Rules of Thumb 1410
  16.11.6.2 Reaction Type and Typical Reactor Configuration 1411
  16.11.6.3 PFTR: Empty Tube 1411
  16.11.6.4 PFTR: Fire Tube 1412
  16.11.6.5 PFTR: Fixed Bed Catalyst in Tube or Vessel: Adiabatic 1412
  16.11.6.6 PFTR: Multitube Fixed Bed Catalyst: Nonadiabatic 1414
  16.11.6.7 PFTR: Multibed Adiabatic with Interbed Quench 1414
  16.11.6.8 PFTR: Fixed Bed with Radial Flow 1414
  16.11.6.9 PFTR: Transported or Slurry, Transfer Line 1414
  16.11.6.10 PFTR: Motionless Mixer in Tube 1415
  16.11.6.11 PFTR: Bubble Reactor 1415
  16.11.6.12 PFTR: Spray Reactor and Jet Nozzle Reactor 1416
  16.11.6.13 PFTR: Trays 1417
  16.11.6.14 PFTR: Packing 1417
  16.11.6.15 PFTR: Trickle Bed 1418
  16.11.6.16 PFTR: Monolithic 1418
  16.11.6.17 PFTR: Thin Film 1419
  16.11.6.18 PFTR: Multiple Hearth 1419
  16.11.6.19 PFTR: Traveling Grate 1419
  16.11.6.20 PFTR: Rotary Kiln 1420
  16.11.6.21 PFTR: Shaft Furnace 1420
  16.11.6.22 PFTR: Melting Cyclone Burner 1420
  16.11.6.23 PFTR Via Multistage CSTR 1420
  16.11.6.24 STR: Batch (Backmix) 1421
  16.11.6.25 STR: Semibatch 1421
  16.11.6.26 CSTR: Mechanical Mixer (Backmix) 1421
  16.11.6.27 STR: Fluidized Bed (Backmix) 1422
  16.11.6.28 Tank Reactor (TR) 1423
  16.11.6.29 Mix of CSTR, PFTR with Recycle 1424
  16.11.6.30 STR: PFTR with Large Recycle 1425
  16.11.6.31 Reaction-Injection Molding 1425
  16.11.6.32 Reactive Distillation, Extraction, Crystallization 1425
  16.11.6.33 Membrane Reactors 1425
  16.11.6.34 Process Vessels 1426
  16.11.6.35 Storage Vessels and Bins 1426
 16.11.7 Mixing 1426
  16.11.7.1 Gas-Solid 1427
  16.11.7.2 Gas-Liquid 1427
  16.11.7.3 Liquid 1427
  16.11.7.4 Liquid-Liquid 1428
  16.11.7.5 Liquid-Solid 1428

   16.11.7.6 Dry Solids...1429
   16.11.7.7 Pastes, Polymers, Foodstuffs, Clay, and Fertilizers........................1429
  16.11.8 Size Reduction..1429
   16.11.8.1 Gas in Liquid (Bubbles in Liquid) 1429
   16.11.8.2 Liquid in Gas (Sprays)..1430
   16.11.8.3 Liquid-Liquid..1430
   16.11.8.4 Solids: Crushing and Grinding 1431
   16.11.8.5 Solids: Modify Size and Shape: Extruders, Pug Mills, and
        Molding Machines..1432
   16.11.8.6 Solids: Solidify Liquid to Solid: Flakers, Belts, and Prill
        Towers..1432
  16.11.9 Size Enlargement...1432
   16.11.9.1 Size Enlargement: Liquid-Gas: Demisters.....................................1432
   16.11.9.2 Size Enlargement: Liquid-Liquid: Coalescers 1433
   16.11.9.3 Size Enlargement: Solid in Liquid: Coagulation/Flocculation.........1433
   16.11.9.4 Size Enlargement: Solids: Spray Drying...1433
   16.11.9.5 Size Enlargement: Solids: Fluidization..1433
   16.11.9.6 Size Enlargement: Solids: Spherical Agglomeration.......................1434
   16.11.9.7 Size Enlargement: Solids: Disc Agglomeration 1434
   16.11.9.8 Size Enlargement: Solids: Drum Granulator 1434
   16.11.9.9 Size Enlargement: Solids: Briquetting..1434
   16.11.9.10 Size Enlargement: Solids: Tabletting..1434
   16.11.9.11 Size Enlargement: Solids: Pelleting..1435
   16.11.9.12 Size Enlargement: Solids: Sintering/Pelletizing.............................1435
   16.11.9.13 Size Enlargement: Solids: Crystallization 1435
   16.11.9.14 Solids: Modify Size and Shape: Extruders, Pug Mills, and
        Molding Machines..1435
   16.11.9.15 Solids: Solidify Liquid to Solid: Flakers, Belts, and Prill
        Towers..1436
Additional Reading 1436

Consult this chapter first if you need to accomplish any of the following:

- Design a piece of equipment
- Propose and design new equipment or a new process
- Evaluate a current piece of equipment or a process
- Suggest changes for equipment or a process
- Debottleneck by increasing the limiting capacity in a process
- Troubleshoot to correct faulty operation

We classify these problems as being design, process analysis and improvement, and troubleshooting.
  Engineering is based on fundamentals and the dynamic balance among economic, operability, sustainability, safety, environmental, legal, ethical, and time constraints. It is *dynamic,* because the development of new technology, catalysts, and materials of construction, and the shifting economical, social, and political scenes, mean that the processes we develop, improve, and operate in the future will be dramatically different from past practice. We say it is based on fundamentals, because mass, energy, momentum, and charge will still be conserved.
  The goal is a win-win balance (not a lose-lose compromise) among competing forces. Moderation and patient creative problem solving are the keys. Regrettably, plants have been built that have so many safety interlocks, so much heat integration, and such extremely minimized inventory that

Conceptual Process Design, Process Improvement, and Troubleshooting

they cannot be started up or operated safely. Likewise, process options with the highest economic return may not be the best. A win-win balance among apparently conflicting pressures is needed.

In this chapter, we remind you how to balance astutely the competing issues. Ideally, we would discuss this as an integrated whole. However, for quick reference, the issues and sections break down as follows:

Section 16.1: The Problem-Solving Process
Section 16.2: Problem-Solving Rules of Thumb
Section 16.3: Engineering Economics
Section 16.4: Building a Data Base
Section 16.5: Sustainability
Section 16.6: Integrating Process Control with Engineering
Section 16.7: Safety, the Environment, and Waste Reduction (see also Section 16.11 for pertinent rules of thumb)
Section 16.8: Software Options
Section 16.9: Optimization Methods
Section 16.10: Rules of Thumb for Teamwork
Section 16.11: Rules of Thumb for Selection and Sizing Equipment

16.1 PROBLEM-SOLVING PROCESS

The key to effective conceptual design, process improvement, and troubleshooting is the thinking process and how we use our basic knowledge. The process and procedure to solve any design, process improvement, or troubleshooting problems has six characteristics:

1. *The problem-solving process is systematic, organized, and cyclical.* It is systematic and organized to keep track of many details. The process is cyclical; we start by solving a very simple version of the big problem using the rules of thumb in Section 16.11. If this very simple concept seems to be satisfactory, then we repeat the cycle, spending more time and resources and using technical details given in other chapters in this handbook. This progression is illustrated in Figure 16.1. The problem-solving strategy is applied many times, as illustrated for the first phases of the hierarchical conceptual design sequence.

    We start at the top central spiral at point A and solve our initial simplified conception of the problem. Once the first cycle is completed, a person assigned the decision-making responsibility will assess the "deliverables" from this cycle. These deliverables usually include an estimate of the technical size, the total capital cost, the operating cost, the financial attractiveness (such as in terms of a measure called the *net present value*, described in Section 16.3), a summary of the legal constraints, the hazard rating, the sustainability implications, and an estimate of the control and operability. The decision being made is either to *stop* (we are no longer interested) or *go* (this idea is worth investigating in more detail). The stop-go decision is like a gate that stops the poor ideas and admits the good ideas. The process is sometimes called the "gating process," and the person making the decision is the gatekeeper. The problem is solved over and over again, in ever increasing detail and with ever increasing accuracy, provided that the idea is given a *go* at each gate (Doyle, 1980). Table 16.1 illustrates this gating process from a slightly different perspective. In column 1, the general business cycle for any commodity is shown: startup, market entry, market leader, and then phaseout. The typical marketing activities are illustrated in column 2. The name and "deliverables" for the gating process are listed in column 4, with data from market research in column 3

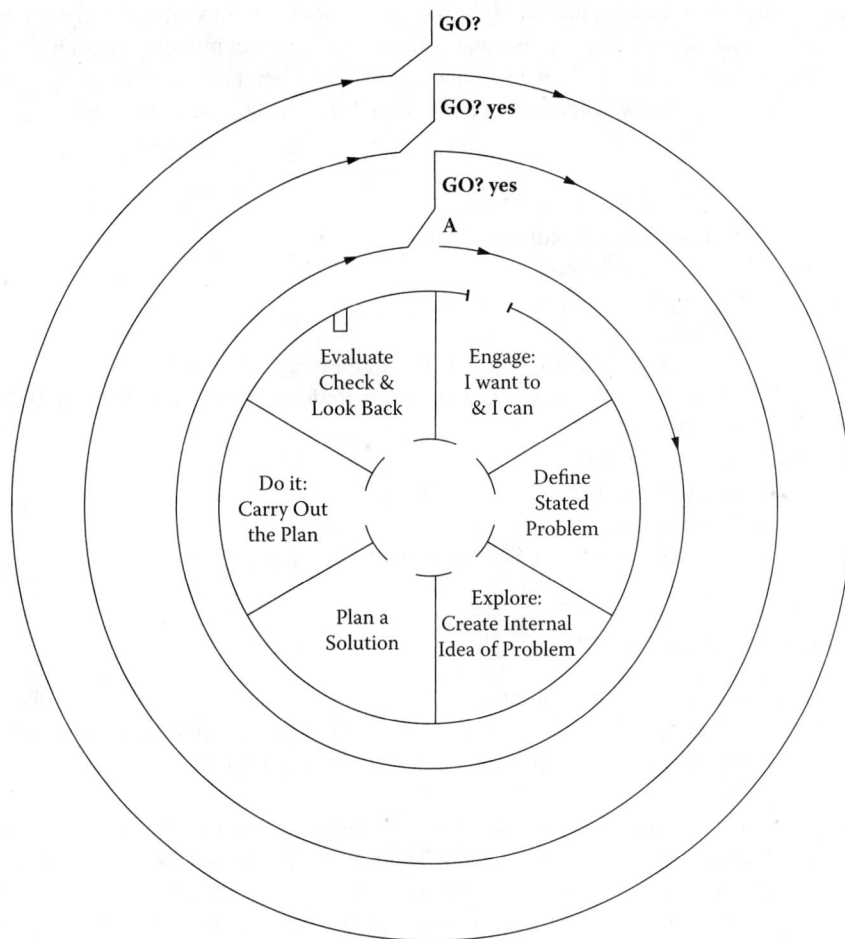

FIGURE 16.1 The problem solving strategy is applied many times in solving problems as illustrated for the first phases of the hierarchical conceptual design sequence.

suggesting the number of ideas or options considered at the different gates (Cooper, 1987). Listed in columns 5 and 6 are the different technical and economic dimensions of the overall task. Column 5 also shows when the three different types of problems (conceptual design, process improvement, and troubleshooting) are usually encountered. The "startup phase" includes conceptual design through construction.

2. The problem-solving process applies to design, process improvement, and troubleshooting. To manage the process, we usually consider the problem in a hierarchical series of cycles that reflect the systematic evolution of our ideas. For conceptual design, these stages in the cycle are listed across the top of Table 16.2 (based on Douglas, 1988). The constraints are usually (1) the production capacity, the purity of the products, and the anticipated disturbances; (2) economic (amount of capital available and the projected return); (3) time (when); (4) the budget, time, and resources for the design process; and (5) the potential for expansion.

For *process improvement,* the cyclical stages and criteria are as illustrated in Table 16.3. The constraints are usually (1) the existing plant, equipment, and layout, and (2) the existing feedstocks (although one might allow different trace contaminants) and economic constraints of budget and profitability. For process improvement and waste

Conceptual Process Design, Process Improvement, and Troubleshooting

TABLE 16.1
Relating the Business Cycle, the Gating Process, and Engineering and Marketing Activities

Business Cycle	Marketing	Illustrative Number of Options	Gating Process: Levels of Decision Making	Technical Engineering	Economic Dimension of Engineering and Expected Accuracy in Economic Values
Period 1: Startup	Preliminary market assessment	50	**0.** Idea **1.** Preliminary evaluation	**Order of magnitude** using rules of thumb	Use ratios ± 50% or scale-up of process cost correlations
	Concept identification & market study	20	**2.** Concept generation: produce *Business plan*	**Conceptual design,** flowsheet development and rules of thumb sizing of equipment. Use of the hierarchical stages illustrated in Table 16.2.	Equipment costs using flowrate correlations ±25 to 40%
	Develop market plan	10	**3.** Development: budget authorization	**Preliminary Engineering design** Flowsheeting; semi-detailed design of the major pieces of equipment.	No quotations; detailed factor method used ±15 to 30%
	Prototype	4	**4a.** Pilot plant, prototype	Confirm key data; revise detailed design as needed	Recalculate based on new data
	Finalization of market plan	2	**4b.** Finalized design	**Defined equipment design** details of equipment but estimated piping	Vender quotations, ±6%
		1	Select the contractor	**Detailed contractor bid design** detailed takeoffs done; bid documents	Detailed takeoffs and bid documents, ±3%
	Implement market plan	1	**4c.** Construct, produce, and launch	Manage construction, authorize modifications	Manage budgets; authorize changes
Period 2: Market entry	Search for new markets		Annual *Balance sheet and Income statement*	Startup **trouble shooting** Use of hierarchical stages illustrated in Table 16.4	
Period 3: Market leader	Strategies for price reduction and increased functionality		Gating process for improvements; Annual *Balance sheet and Income statement*	**Continual process improvement** Use of the hierarchical stages illustrated in Table 16.3. And trouble shooting	Economic justification
Period 4: Phase out	Identify new user needs		Annual *Balance sheet and Income statement*	Search for new opportunities	

TABLE 16.2
Full Problem-Solving Cycle Is Applied to Each of the Hierarchical Stages in Conceptual Design

Criteria	I. Batch versus Continuous	II. Gather the Data for Inputs and Target Outputs; Set Goals	III. Preliminary Scouting of the Reactor	IV. Explore Mass Recycle	V. Explore Separations	VI. Explore Energy Integration	V. Polish
			Hierarchical Stages in Conceptual Design				
Technical feasibility	Use when difficult to scale up; identify as multiproduct, general multipurpose, or campaign operation of multipurpose.	Define product amount, purity, and range of production rates	Reaction conditions: Excess reagent? Heat removal method? Shift equilibrium? Add diluent? Combine reactor-separator?	Buildup of trace materials? Overall mass balance? Changes in corrosion?	Prefer not to add agents	Pinch technology	
Economic feasibility	< 2,500 Mg/a <0.1 kg/s	Use cost correlation to estimate investment; estimate product cost using %. Define risk and expected return; Define tax and insurance impacts			Use cost correlation based on flowrate		Minimum cost may not give minimum waste
Sustainability		TNS: renewable resources? Bioassimilimlation of feedstock, product, byproducts, intermediates, and wastes					

Category							
Control and operability	Setpoint and random, seasonal and gradual disturbances should be specified	Heat removal, runaway reactions?	Control purge? Monitor and measure wastes, increased scaling of heat exchangers? Reduced reliability because of intimate interdependence	Sufficient surge volume?	Thermal disturbance because of recycle? Add trim heaters/coolers? Reduced reliability because of intimate interdependence; additional startup-shutdown equipment?	Increased fugitive emission for each valve added;	
Hazard: safety, environmental, waste minimization	Remove impurities in feed	Heat removal? Runaway reactions?	Recycle waste to eliminate reversible reaction production?			Fugitive emissions: sampling: monitor and measure wastes	
Legal and ethical	Know the law; know patents	Ethical: sustainability? Legal: hazard	Loss of water rights	Impact of adding agents			
People	Teams of key people; bring members together early to increase ownership; seek operator input	Few design errors occur because of technical incompetence. Most result from poor communication. Use e-mail. Use rules of thumb for effective teams, given in Section 16.10.					

TABLE 16.3
Full Problem-Solving Cycle Is Applied to Each of the Hierarchical Stages for Process Improvement

Criteria	Hierarchical Stages in Process Improvement							
	I. Better Inventory; Reduce Fugitives	II. Identify Realistic Needs for Process Units	III. Optimization of Reactor/ Separation	IV. Cycle Time Optimization; Improved Control	V. Process Modifications to Debottleneck: Simple Switch	VI. Process Modifications to Debottleneck; Redesign, Equipment Replacement	VII. Recycle Management; Heat and Mass Exchange Networks	VIII. Substitute Raw Materials, Reagents, Solvents, and Additives
Technical feasibility	Minimum risk with greatest economic potential	Realistic specifications for products, utilities;	Focus here because usually raw material cost predominates: consider issues from design		Relocate feed, replace internals, vertical condenser to horizontal condenser		Pinch technology	
Economic feasibility	Incremental analysis							
Sustainability			Purity of feedstocks: renewable? and bioassimilation?					
Control and operability	Minimum surge volume needed to damp out disturbances	Four levels	Better sensors, relocate sensors	Shift to statistical process control?	Impact on control	Impact on control, startup		
Hazard: safety, environmental, waste minimization	Main focus here!	Consider zero discharge; guidelines for acceptable risk	Minimize waste production		Minimize holdup	Increase in fouling? Increase in corrosion?		
Legal and ethical: social/political	Ethical concern for community; legal requirements							
People								

TABLE 16.4
Full Problem-Solving Cycle Applied to Each Hierarchical Stage When Applied in Trouble Shooting

	Hierarchical Stages in Trouble Shooting				
Criteria	I. Prioritize the Trouble	II. Problem Definition, Hypothesis Generation	III. Gather Data and Identify Cause	IV. Correct Cause	V. Prevent Reoccurrence of Cause
Technical feasibility		Focus on changes? or on fundamentals?			
Economic feasibility	Third priority		Use inexpensive detectors		
Sustainability					
Control and operability			Put on manual control		
Hazard: safety, environmental, waste minimization		Top priority: safety, environmental impact	Ensure system is safe		
Legal and ethical		Legal: accidental spill or discharge top priority	Follow internal procedures		
People			Use Jungian typology to identify preference on the P-J dimension; avoid fixation and pseudodiagnosticity		

management, our sequence of issues to consider is approximately the reverse of the order used for design: better inventory practices, scrutiny of operating practices and procedures, reduction of fugitive emissions, optimization of reactor and separation operating conditions and control of operations, recycle management plus energy and mass exchange networks, and substitution of raw materials.

For *troubleshooting*, the cycles and criteria are as listed in Table 16.4. The constraints are usually safety and time: A process often operates 25 to 28 days per month to cover costs. On the remaining days, it is making a profit. We need to keep the process operating. The tests, changes, and remedy should be safe, environmentally acceptable, and ethically and legally possible.

3. *The problem-solving process focuses on money*. The prime concern of the gatekeeper is "will this idea be profitable for the company?" Without profit, companies fold. The feasibility criterion is economic feasibility: The financial investment should provide a return commensurate with the risk.
4. *Throughout the problem-solving process, we consider a host of "feasibility" criteria other than money*. These include (1) market feasibility (the product provides customers with benefits they want and are willing to buy), (2) technical feasibility (the process must work, the desired amount and purity of the target product must be produced), (3) sustainability (the process should use our world's resources wisely so that we evolve toward sustainable development), (4) operability and controllability, (5) minimal hazards,

(6) satisfaction of all legal and ethical concerns, and (7) providing the company with a good corporate image. More is given under "Qualitative Measures" in Section 16.3 and in Sections 16.5 through 16.7.
5. *The problem-solving process uses all available current computer assists such as flowsheeting, the Web, data sourcing, and data-processing techniques.*
6. The problem-solving process is usually a team effort, with many people contributing.

Additional Reading

J. M. Douglas, *Conceptual Design of Chemical Processes*. New York: McGraw-Hill, 1988. This book and articles published by Douglas and his research group give more on the process of conceptual design.
R. G. Cooper, *Winning at New Products*, 3rd ed. New York: Perseus Books Group, 2001. Gives more on the business aspects of the design process.
D. J. Doyle, *Making Technology Happen*. Nepean, Ontario, Canada: Doyletech Corp., 1980. An excellent 120-page summary that integrates engineering, business, and entrepreneurship.

16.2 RULES OF THUMB FOR PROBLEM SOLVING*

Below are 18 rules of thumb about the process of problem solving.

1. Be aware. Be able to describe the thought processes being used and to identify where in the process you are.
2. Know the systematic stages for each cycle of the problem-solving process (as illustrated in Figure 16.1).
3. Focus on accuracy. Check and double check.
4. Actively write things down. Make charts and raw diagrams, write down goals, list measurable criteria, and record ideas in brainstorming.
5. Monitor and reflect. Mentally keep track of the problem-solving process and monitor about once per minute.
6. Be organized and systematic. Use an organized strategy such as those illustrated in Figure 16.1 and in Tables 16.2 to 16.4.
7. Define the "real" problem by creating a rich perspective of the problem. See it from many different points of view. Be willing to spend at least half the total available time defining the problem. Ask many "what-if" questions. Try to bound the problem space. "Swim with the data" to see how they respond. Identify the real problem by asking a series of "why" questions to generalize the situation. In Figure 16.1, this activity of identifying the real problem is called the *explore* stage and is the heart of the problem-solving process.
8. Be flexible.
9. Use your creativity effectively. Defer judgment; be succinct; list 50 ideas in 5 min; create a risk-free environment; encourage free and forced association of ideas; piggyback on previous ideas; use triggers, such as those listed in Table 16.5, to maintain the flow of ideas; don't be discouraged—in the last 2 min of a 5-min brainstorming session, over 85% of the ideas are impractical—but spend time to identify the treasures among the 15%. Use impractical and ridiculous ideas as steppingstones to innovative, practical options.
10. Critically assess the knowledge and data used. Too often, we hope that the data are applicable. A colleague, in designing a petrochemical plant, was unable to locate the physical properties of the organics. He decided to assume that they were the same as water and hoped that this would work out. Just a short time spent in critical assessment of this assumption would have saved six months of wasted work. Too often, we accept

---

* Reproduced with permission from Donald R. Woods, *Rules of Thumb in Engineering Practice*, pages 22 to 25, 2007 copyright Wiley-VCH Verlag GmbH & Co. KGaA, Weinheim, ISBN: 978-3-527-31220-7.

TABLE 16.5
Checklist of Triggers for Brainstorming

Name of Trigger to Change Point of View		Elaboration of How It Is to Be Used	Comments
To be used for objects "How to improve *ion exchange resin*"	Function	What is the function of this object? How else can we achieve this function? What function **can** it **not** do?	Objects for engineers are "products" and hardware. Usually engineers deal with situations. However, these triggers are needed to help apply triggers for situations.
	Physical uses	What are its physical properties and characteristics? What are they **not**? How else might we obtain or use these physical properties?	Probably the most useful perspective; use first. The negative view is often extremely illuminating.
	Chemical uses	What are its chemical properties? How can these be exploited? What are they **not**?	Easy for us to use; use second after physical uses.
	Personal uses	What personal uses can we make of the this object?	These three give a unique perspective that is often overlooked.
	Interpersonal uses	What interpersonal uses can be made?	
	Aesthetic uses	How can we create pictures? music? artistic creations with the object? sculpt? weave?	
	Mathematical or symbolic properties	What are the mathematical or symbolic uses? What can they not be used for? What are they often confused with?	Only applicable for certain objects or ideas.
Situations "I need to design an ion exchange unit." "The reformer is not functioning; fix it."	Checklist	Of the many checklists published, SCAMPER is probably the most effective: **S**: substitute who? What? Other processes? Other places? **C**: combine purposes? Ideas? Appeals? Uses? **A**: adapt what else is like this, new ways? **M**: modify, maximize, minimize? **P**: put to other uses, other locations **E**: eliminate **R**: reverse, rearrange	Effective for objects and some situations. Try each viewpoint. Easy to do.
	Wildest fantasy	Think of the craziest idea and use	Need to establish self-confidence and group confidence before really outlandish ideas are presented. Build confidence in the merit of this approach by later bridging to unique ideas. Brings laughter and tension relief.
	How nature does it	Identify the situation and then think of how nature fulfills this function. Bridge to engineering reality.	Great potential; depends on the situation. Successful in designing new bridges; new bog machines.
	What if? In the extremes	Extrapolate from the unfamiliar to a simplified version	Extension of problem-solving skill in defining simple problems. Relatively easy to apply.
	Boundary exploration	Identify the constraints and remove them	Much easier for some to do than for others.

Continued

TABLE 16.5 (Continued)
Checklist of Triggers for Brainstorming

Name of Trigger to Change Point of View	Elaboration of How It Is to Be Used	Comments
Functional analogy	How else is the function achieved?	
Appearance analogy	How else might we get the appearance of this situation?	
Morphology	Break the problem situation into a series of parts. For each part list 10+ options. Then, systematically combine one option from each part and ask why not?	More mechanical; easy to computerize; most of the work is in setting up the parts and the options; fun and surprising to see some of the results.
Symbolic replacement	Replace the original problem by an interesting idea generated in the session; refocus.	Occurs naturally in many brainstorming sessions. Add this trigger explicitly if needed.
Juxtaposition	Bring in three completely random words and bridge to current situation. Example: "refrigerator, light switch, clock."	Very effective. Don't fake it by using words "you think will help." Use random words.
Personal analogy	Imagine yourself as part of the situation. Describe your feelings and what you are experiencing. Be imaginative.	Tricky. Works for some people; not for others. Effective with fluid dynamic problems.
Reversal	Do the reverse.	More challenging than one thinks. Focus systematically on the reverse of different elements of the situation at a time.
Book title	Create a title for a "best selling" novel. The title should sum up the current situation. "Will exchanging bring happiness to Mary Lou?"	Interesting. Worth a try for 1 min.
Letter, word, sentence	Focus on three different levels of detail corresponding to the big view (a sentence), an intermediate view (a word in the sentence), and a letter (in a word). In ion exchange this might be "the ions, the resin, the packed bed, the separation."	Very effective for most of our problems. Consider a 5-min brainstorm at each level.

data from the published literature, yet about 8% of published data are mistakes. "The temperature into the hydrodealkylation reactor is >1150°C," states one reference. It should read >1150°F. A major handbook published an incorrect value of the heat of vaporization through several editions. Check the data coming from computer programs and simulations. Check the physical property package estimates.

11. See challenges and failures as opportunities for new perspectives.
12. Spend time where it benefits you the most. Use Pareto's principle (80% of the results can be found from 20% of the effort). Find the key 20%.
13. Be an effective decision maker. Express the goal as results to be achieved rather than as actions to be taken. Make decisions based on some criteria that are explicit and measurable. Distinguish between *must* criteria (the process must have an internal rate of return of 35%) and *want* criteria (the process might have the potential to be licensed). Reject options that do not meet the must criteria. Use a rating system to score the want criteria.

14. Be willing to take risks.
15. Manage stress well. Solving problems is stressful. When we initially encounter a problem, we experience distress because of the uncertainty. Such stress tends to immobilize us. When we successfully solve a problem, we experience the joy and exhilaration of stress (which distracts us from checking and double checking that our answer is the best). A certain level of stress motivates us. Excessive stress makes us make mistakes. Data suggest that operators with confidence and training working under high stress make one mistake in ten actions. Operators with confidence and training who receive feedback about their actions and are under low stress make 1 mistake in 1000 actions. Although these data refer to plant operators, the same trends can be extended to suggest how stress and a lack of reflection and feedback might interfere with engineering practice. High stress would be a rating of over 450 on the Holmes–Rahe scale (Holmes–Rahe, 1967).

    Ten suggested approaches to managing stress include the following: worry only about things over which you have control, include physical exercise as part of your routine, have hobbies and destimulating activities in which you can lose yourself, plan ahead, avoid negative self-talk, rename the events that are stressful to you, build a support system, be decisive, put the situation into perspective, and use role models of others who have succeeded.
16. Manage your time well. Covey (1990) offers excellent suggestions on time management. Identify problems and decisions according to their importance and urgency. Shift the important situations to being nonurgent. Learn to say "no."
17. Understand your strengths, limitations, and preferred style. See Section 16.10.1.
18. For problems involving people, use the 85/15 rule: 85% of the problems occur because of rules and regulations; 15% of the problems are because of people.

Recommended Reading

H. S. Fogler and S. LeBlanc, *Strategies for Creative Problem Solving*. Indianapolis, IN: Prentice Hall PTR, 1994. Provides interesting examples and convenient summaries of key approaches to problem solving, including the approach of Kepner–Tregoe.

Peter Checkland, *Systems Thinking, Systems Practice*. New York: John Wiley & Sons, 1981. A classic on how to define messy problems. Use their CATWOE criteria.

Min S. Basadur, *Simplex: A Flight to Creativity*. Buffalo, NY: Creative Education Press, 1994. Excellent for problem mapping and creativity, http://www.basadursimplex.com.

S. R. Covey, *The Seven Habits of Highly Effective People*. New York: Simon and Schuster, 1990. Provides the basics of self- and time management and building trust.

Peter Block, *The Empowered Manager*. Hoboken, NJ: Jossey-Bass, 1990. Book on leadership. Leads you through creating the vision of greatness, building support, balancing autonomy and dependence, and taking risks and being courageous.

D. R. Woods, *Problem Based Learning: How to Gain the Most from PBL*. Waterdown, Ontario, Canada: Woods, 1994. Provides the details for the citations in this section. Chapter 3 summarizes ideas about problem solving. Distributed by McMaster University Bookstore, http://titles.mcmaster.ca/.

16.3 ENGINEERING ECONOMICS

Engineering economics is the critical criterion to consider early in design, process improvement, and troublshooting. In this section, we review the role of economics in financial reporting through the balance sheet and the income statement, options for measuring financial attractiveness, operating expense estimation, and capital cost estimation.

16.3.1 Role of Economics in Financial Reporting

Some basic knowledge of economics and financial reporting is essential for a professional engineer moving up the corporate ladder from a technical professional to a corporate manager. The engineer must recognize the importance of financial management and learn to speak and understand the language used by corporate management. The information in this section is beyond what one would encounter in college courses in business school economics or in engineering economics courses. A company financial report, sometimes called an annual report, contains much information. It is designed to tell the reader how well the company performed in the previous year and how this performance is measured against various standards. Two important documents in the annual report are the *balance sheet* and the *income statement*. Two ancillary documents often included are the *accumulated retained earnings* and the *changes in working capital*. As each of these documents is discussed, the reader should have Tables 16.6 and 16.7 available while reading the following sections. These are the financial documents for a fictitious chemical company, ALLCHEM, for the current year 20X6 with data for the previous year 20X5.

16.3.1.1 Balance Sheet (Anon., 1992; Couper et al., 2000)

The balance sheet is the financial status of the company on a given date. The choice of date is arbitrary and depends on the company's operations. A *consolidated* report means that all the financial data for the parent company as well as for the subsidiary companies are presented in one document. A balance sheet contains some real figures, e.g., cash and marketable securities; some estimated amounts or allowances, e.g., inventories and accounts receivable; as well as some fictitious figures, e.g., intangibles for which numbers are difficult to determine.

Two main parts of the balance sheet are the *assets,* which is what the company owns, and the *liabilities and stockholders' equity,* which is what the company owes. The total assets equal the total liabilities plus the stockholders' equity.

16.3.1.1.1 Assets

The assets of a company are divided into three categories: *current assets, fixed assets,* and *intangibles.*

Current assets are those that may be converted to cash within a year from the date of the balance sheet. The current assets include cash and money on deposit in a bank as well as *marketable securities* (e.g., commercial paper and/or government bonds) that can be easily converted to cash. *Accounts receivable* are goods sold to customers on a 30-, 60-, or 90-day basis for which full payment has not been received. An allowance is made for uncollected bills, because some customers were unable to pay. *Inventories* consist of raw materials on hand, goods in process, supplies, and finished goods ready for shipment. Raw materials are charged at cost, goods in process at raw material cost plus one half the conversion cost, and finished goods are valued at selling price. Frequently, inventory costs are carried at values slightly less than these figures to account for deterioration, decline in prices, and other factors. *Prepaid expenses* include prepaid insurance premiums as well as leases for equipment, computers, and office equipment. These expenses are listed under current assets because, although a company has paid for them, full benefit has not been received, but the company expects to do so within the year. The *total current assets* are the sum of cash and marketable securities, accounts receivable, and prepaid expenses.

Fixed assets include land, buildings, manufacturing equipment, office equipment, and so forth, that the company owns. These items are carried on the books at cost less the accumulated depreciation, except for land, which is generally entered at the same value year to year. The sum of these items is the *net fixed assets.*

Intangibles are assets that have substantial value to the company, such as patents, licenses, trademarks, goodwill, and so forth. There is no consistent way to evaluate these assets, so the company often balances both sides of the balance sheet by making this value "close" the sheet.

TABLE 16.6
Consolidated Balance Sheet for ALLCHEM, Inc.

CONSOLIDATED BALANCE SHEET*
As of December 31

Assets	20X6	20X5
Current Assets		
Cash	$63,000	$51,000
Marketable Securities	41,000	39,000
Accounts Receivable**	135,000	126,000
Inventories	149,000	153,000
Prepaid Expenses	3,200	2,500
Total Current Assets	$391,200	$371,500
Fixed Assets		
Land	35,000	35,000
Buildings	101,000	97,500
Machinery	278,000	221,000
Office Equipment	24,000	19,000
Total Fixed Assets	$438,000	$372,500
Less Accumulated Depreciation	128,000	102,000
Net Fixed Assets	$310,000	$270,400
Intangibles	4,500	4,500
TOTAL ASSETS	$705,700	$646,500

Liabilities	20X6	20X5
Current Liabilities		
Accounts Payable	$ 92,300	$ 81,300
Notes Payable	67,500	59,500
Accrued Expenses Payable	23,200	26,300
Federal Income Taxes Payable	18,500	17,500
Total Current Liabilities	$201,500	$184,600
Long Term Liabilities		
Debenture Bonds, 10.3% due in 2015	110,000	110,000
Debenture Bonds, 11.5% due in 2007	125,000	125,000
Deferred Income Tax	11,600	10,000
TOTAL LIABILITIES	$448,100	$429,600
STOCKHOLDERS' EQUITY		
Preferred stock, 5% cum		
$5.00 par value - 200,000 shares	$ 10,000	$ 10,000
Common Stock, $1.00 par value		
20X5 28,000,000 shares		
20X6 32,000,000 shares	32,000	28,000
Capital Surplus	8,000	6,000
Accumulated Retained Earnings	207,600	172,900
TOTAL STOCKHOLDERS' EQUITY	$257,600	$216,900
TOTAL LIABILITIES AND STOCKHOLDERS' EQUITY	$705,700	$646,500

* All figures are in thousands of dollars.
** Includes an allowance for doubtful accounts.

TABLE 16.7
Consolidated Income Statement for ALLCHEM, Inc.

CONSOLIDATED INCOME STATEMENT*

	20X6	20X5
Net Sales (Revenue)	$932,000	$850,000
Cost of Sales and Operating Expenses		
Cost of Goods Sold	692,000	610,000
Depreciation and Amortization	40,000	36,000
Selling, General and Administrative Expenses	113,500	110,000
Operating Profit	$86,500	$94,000
Other Income (Expenses)		
Dividends and Interest Income	10,000	7,000
Interest Expense	(22,000)	(22,000)
Income Before Provision for Income Taxes	$74,500	$79,000
Provision for Federal Income Taxes	24,500	26,000
Net Profit for Year	$ 50,000	$ 53,000
ACCUMULATED RETAINED EARNINGS STATEMENT*		
Balance as of January 1	$172,900	$141,850
Net Profit for Year	50,000	53,000
Total for Year	$222,900	$194,850
Less Dividends Paid on		
Preferred Stock	700	700
Common Stock	14,600	21,250
Balance December 31	$207,600	$172,900

* All figures are in thousands of dollars.

Other assets, such as investments in affiliates or *deferred charges* for which full benefit has not been received, may be included before the total assets are summed.

Total assets are the sum of current assets, fixed assets, intangibles, and deferred charges.

16.3.1.1.2 Liabilities

The *total liabilities* are what a company owes and consist of the sum of *current* and *long-term liabilities*. *Current liabilities* are debts that must be paid within a year from the date of the balance sheet. The total current liabilities are the sum of the *accounts payable, notes payable, accrued expenses payable,* as well as *income taxes payable* from current assets. *Accounts payable* include such items as invoices for raw materials, supplies, and others, that a company has purchased from suppliers and for which payment is due on a 30-, 60-, or 90-day basis. *Notes payable* include monies owed to banks and other creditors as well as promissory notes. *Accrued expenses payable* include such entries as salaries, wages, interest on borrowed funds, insurance premiums, pensions, and so on. *Income taxes payable* are the debt owed to federal, state, and local governments. These taxes are usually paid on a quarterly basis and commonly isolated from other expenses.

Long-term liabilities are debts due after one year from the date of the financial report and include bonds, loans, and deferred income tax. *Bonds and loans* include *first mortgage bonds* (issued at a stated rate due in a stated year and backed by the company's property), *debenture bonds* (backed by the general credit of the company rather than by company property), and *long-term loans* from insurance companies and investment houses. *Deferred income taxes* are encouraged by the government as a tax incentive that will benefit the economy. Accelerated depreciation, which provides the rapid write-off in the early years of the investment, is an example. The net effect is to reduce what the company owes in the early years and will be paid in the future. To smooth out wide fluctuations in a company's earnings, this entry shows what the taxes would be without accelerated depreciation write-offs.

16.3.1.1.3 Stockholders' Equity

The *stockholders' equity* is the interest that stockholders have in the business and is the *net worth* of the company, namely, total assets minus total liabilities. The *total stockholders' equity* is the sum of the *preferred stock, common stock capital surplus,* and *accumulated retained earnings.*

Capital stock is classified into two broad categories: preferred and common stock. *Preferred stock* has preference over other shares regarding dividends and/or distribution of assets. Some preferred stock is termed *cumulative,* which means that, if in any given year the company does not pay dividends, the dividends accumulate. When dividends are paid, the preferred stockholders receive these dividends before common stockholders. Preferred stockholders, however, do not normally have a voice in company affairs or voting rights unless the company fails to pay them dividends. Preferred stock is carried on the books at the stated par value. For *common stock,* there are no requirements that dividends be paid. If the company's earnings are high, dividends may be paid, but, if the earnings are poor, dividends may not be paid. Common stock is valued at a stated par value.

Common stock capital surplus is the amount of money the stockholders paid for the stock over and above the par value of the stock.

The *accumulated retained earnings* or *earned surplus* are calculated by subtracting the dividends paid to stockholders from the net profit. If the profits in any year are not distributed, they are retained by the company and added to the next year's earnings. They may be used for research and development activities and/or the purchase of capital equipment.

16.3.1.1.4 Total Liabilities and Stockholders' Equity

The sum of the total liabilities and stockholders' equity is what the company owes. It must equal the total assets for the balance sheet to "balance."

16.3.1.2 Income Statement (Anon., 1992; Couper et al., 2000)

The *income statement*, sometimes called the *earnings statement*, displays the operating activities for the year and may be an indication of how well the company is doing. A typical statement will show the numbers for the current year and at least one year previous. Frequently, an annual report will have a five- or ten-year summary near the end of the report. Table 16.7 is an example of an income statement.

16.3.1.2.1 Net Sales

Net sales is the amount of money received for the goods sold less the amount for returned goods and allowances for reduction in prices, e.g., allowing freight on goods shipped. Some income statements use the term *net revenue,* which is the income from all sources such as net sales, income from patents and licensing technology, and others.

16.3.1.2.2 Cost of Goods Sold and Operating Expenses

This item includes all expenses in converting raw materials into finished product, including cost of goods sold; depreciation; and sales, administration, research, and engineering (SARE) expenses.

The *cost of goods sold* represents the cash operating expenses for raw materials, labor, utilities, supplies, maintenance, waste disposal, plant indirect expenses, and so forth.

The *depreciation, amortization,* and *depletion* are paper transactions. The federal government allows a company to charge off a portion of an asset due to wear and tear as well as obsolescence each year as an operating expense. This *depreciation* is not a cash item. *Amortization* is the decline in useful value of a tangible asset, such as a patent. *Depletion* is the diminution of a natural resource, such as a coal mine. All of these paper allowances appear as one item in most income statements.

The *SARE expenses* are expenses associated with the maintenance of sales offices, paying corporate officers and their staffs, and research and engineering expenses not attributable to a specific project.

16.3.1.2.3 Operating Profit (Operating Income)

This entry is the difference between net sales or net revenue and all operating expenses.

Other income may be derived from dividends or interest received by the company from other investments, income from patents, licenses, and so forth. When other income is subtracted from the operating profit, the result is *income before provision for federal income taxes*.

Every company has a basic federal income tax rate that it must pay. However, the actual *federal income taxes* paid may be less than the basic rate, due to tax incentives, tax credits, depreciation write offs, capital gains, and so on.

16.3.1.2.4 Net Profit for the tear after Income Taxes

This entry is obtained by subtracting the provision for federal income taxes from the income before provision for federal income taxes.

16.3.1.3 Accumulated Retained Earnings

Accumulated retained earnings is an important part of the financial report and is often found along with the income statement. It shows how much money the company has retained for growth and how much was paid out as dividends to stockholders. When the accumulated retained earnings increase, the company has more value. The balance of the accumulated retained earnings at the end of one year is the opening balance on January 1 of the next year. To that figure, the net profit after taxes for that year from the income statement is added, giving the total for the year. From the total for the year, the dividends paid to the preferred and common stockholders are subtracted, and the remainder is the balance of the accumulated retained earnings at the end of the year on December 31.

16.3.1.4 Changes in Financial Position

Scrutiny of the balance sheet and the income statement reveals how much money flowed through the company; how much profit was made; how the funds provided by the net profit, depreciation, sale of common stock, and other items were used; and how the cash generated affected the company operations. The changes in financial position show how the company managed its funds.

16.3.1.5 Other Financial Terms of Significance

Working capital, which allows a company to carry on day-to-day operations, is determined by subtracting current liabilities from current assets. (See also item 20 in Section 16.3.4.2.) *After-tax cash flow* is defined as the net profit after taxes plus depreciation. It is essential to have adequate cash flow to operate the company. *Earnings per share* interests common stockholders and stockbrokers. It is found by dividing the net profit after taxes by the number of shares of common stock.

Depreciation methods in use today are *straight-line* or *accelerated* depreciation. In the former method, the cost of the asset is divided by the asset's life. Accelerated methods were introduced in the 1950s to stimulate capital spending. In the intervening years, there have been many changes to the tax laws as well as how depreciation is handled. The net effect of accelerated depreciation was to provide a larger amount of depreciation in the early years of a venture, thereby increasing the cash flow when the plant is coming on stream and the market for a product is being established.

The current depreciation system is the Modified Accelerated Cost Recovery System (MACRS), and chemical industries are in the seven-year life category. The rates for years one through eight are 14.29, 24.49, 17.49, 12.29, 8.93, 8.92, 8.93, and 4.46, respectively. The depreciation is spread over an eight-year period for a seven-year asset. It is assumed that, during the first year of the life of the asset, full benefit will not be received from the asset. Therefore, a half-year convention is adopted, and the remaining recovery is made in the eighth year.

16.3.1.6 10K Report

The federal government requires a public company to submit annually a 10K report to the Securities and Exchange Commission. It contains many more financial details than the annual reports discussed above.

16.3.1.7 Concluding Comments

When reading an annual report, one should be cognizant of the following:

- The independent accountants certification, which states that the financial statements have been prepared in accordance with generally accepted practices, and the auditing steps meet the accounting world's approved standards of practice.
- In reading an annual report, one will find numerous notes referring to the balance sheet and/or the income statement. These notes indicate any litigation pending and generally how some numbers in the report were obtained. The notes often alert the reader to potential financial problems.

16.3.1.8 Financial Ratios (Couper et al., 2000)

Four classes of financial ratios are useful measures of a company's performance: liquidity, leverage, activity, and performance ratios. They are of particular interest to financial analysts, investment bankers, and corporate executives. As one reads this section, it is advisable to look at the financial ratios for ALLCHEM, Inc. in Table 16.8.

16.3.1.8.1 Liquidity Ratios
Liquidity ratios measure a company's ability to pay its short-term debts when due.

Two measures of a company's liquidity are the *current ratio* and the *cash (quick) ratio*. The current ratio is defined as the current assets divided by the current liabilities, which is a measure of the firm's ability to meet its current obligations from current assets. A "comfortable" level of 1.5 to 2.0 is considered adequate. (*Note*: Numbers presented here are typical as of the date of publication, but they will vary depending on a company's style of management.) The cash ratio is the ability of a company to cover bills from its assets in an emergency. It is the cash plus marketable securities divided by current liabilities. A typical figure is about 1.0.

16.3.1.8.2 Leverage Ratios
Leverage ratios measure the company's overall debt burden.

The *debt-to-assets ratio* is determined by dividing the total debt by the total assets and is usually expressed as a percentage. The industry average is about 35%. Another measure is the *debt-to-equity ratio*, also expressed as a percentage. Both of these ratios are measures of the amount of debt that a company maintains. The higher the ratios, the greater the financial risk. In an economic turndown, it might be difficult for a company with high ratios to meet creditor's demands. The *times-interest-earned ratio* is a measure of the extent to which profits could decline before a firm is unable to pay interest charges on debts. This ratio is calculated by dividing the earnings before interest and taxes (EBIT) by the interest charges. A comfortable figure is 7 to 8 times.

TABLE 16.8
Financial Ratios for ALLCHEM, Inc.

Liquidity

$$\text{Current Ratio} = \frac{\$391,200}{\$201,500} = 1.94$$

$$\text{Cash Ratio} = \frac{\$391,200 - 149,000}{\$201,500} = 1.20$$

Leverage

$$\text{Debt to Assets} = \frac{\$448,100 - 201,500}{\$705,700} \times 100 = 35\%$$

$$\text{Times Interest Earned} = \frac{\$74,500 - 22,000}{\$22,000} = 4.39$$

$$\text{Fixed Charge Coverage} = \frac{\$86,500}{\$22,000} = 3.93$$

Activity

$$\text{Inventory Turnover} = \frac{\$932,000}{\$149,000} = 6.25$$

$$\text{Average Collection Period} = \frac{\$135,000}{(\$932,00/365)} = 52.8 \text{ days}$$

$$\text{Fixed Assets Turnover} = \frac{\$932,000}{\$438,000} = 2.13$$

$$\text{Total Assets Turnover} = \frac{\$932,000}{\$705,700} = 1.32$$

Profitability

$$\text{Gross Profit Margin} = \frac{\$932,000 - 692,000}{\$932,000} \times 100 = 25.8\%$$

$$\text{Net Operating Margin} = \frac{\$74,500}{\$932,000} \times 100 = 7.99\%$$

$$\text{Profit Margin on Sales} = \frac{\$50,000}{\$932,000} \times 100 = 5.36\%$$

$$\text{Return on Net Worth (Return on Equity)} = \frac{\$50,000}{\$705,700 - 448,100} \times 100 = 19.4\%$$

$$\text{Return on Total Assets} = \frac{\$50,000}{\$705,700} \times 100 = 7.09\%$$

The *fixed charge coverage* is obtained by dividing the income available for meeting fixed charges by the fixed charges. Some firms enter into long-term lease agreements, and these are a part of the fixed costs in doing business. The numerator of this ratio is the operating profit before deducting the interest expense, lease costs, and income taxes divided by lease costs and interest expenses.

16.3.1.8.3 Activity Ratios

Activity ratios measure how effectively a company manages its assets.

Activity ratios are based on the assumption that there are proper relationships between a company's assets and the sales and income that they generate. Most analysts compile averages from balance sheet data, which are end-of-year information. This will be used here.

Two *inventory-turnover ratios* are in common use today. The *inventory-sales ratio* is determined by dividing the sales by the inventory. Another method is to divide the cost of sales by the inventory. In either case, the industry average is between 7 and 9. The *average collection period* is the number of days that customers' invoices are unpaid. This figure is found by dividing the annual sales by 365 days to obtain the average daily sales and then dividing that figure into the accounts receivable balance. The average period for the chemical process industries is about 45 days. The *fixed assets* and *total assets turnover* indicate how well the fixed and total assets are being used. They are determined by dividing annual sales by the fixed assets or total assets, respectively. A reasonable figure for the former is 2 to 3 times, and the latter is 1 to 2 times.

16.3.1.8.4 Profitability Ratios

Profitability ratios indicate a firm's management of both income and assets.

The *gross profit margin* is calculated by dividing the gross profits by the net sales, and the result is reported as a percentage. This ratio is an indication of the effectiveness of a company's pricing, purchasing, and production policies. The *net operating margin* is equal to the earnings before interest and taxes (EBIT) divided by the net sales expressed as a percentage. This is a measure of a firm's income performance before interest and taxes. These margins vary considerably. *Profit margin on sales* is calculated by dividing net profit after taxes by the net sales, expressed as a percentage; a reasonable value is 5 to 8%.

The *return-on-equity ratio* is the net income after taxes and interest divided by the stockholders' equity. It sometimes is called the *return on net worth*, and this ratio is probably the best measure of management's performance. Returns on equity vary depending on company performance, but a figure of 15% is not unreasonable. The *return-on-total assets ratio* is the net profit after taxes divided by the total assets expressed as a percentage. It reflects the overall return that a company has earned on its assets; a reasonable value is about 10%.

In conclusion, a company's performance should be compared with the performance of companies in the same line of business, i.e., a chemical company should be compared to a chemical company, and an oil company with another oil company. Companies in the same line of business frequently use the same style of management.

Example 16.1

Use financial ratios to assess ALLCHEM, Inc. for the year 20X6.

Solution

The financial ratios calculated for the year 20X6 are given in Table 16.8. The following conclusions can be drawn about the operation of this company:

- From the liquidity ratios, the company is able to pay its short-term debts when due.
- With respect to the leverage ratios, the "fixed charge coverage" and "times interest earned" are a bit low. The debt-to-asset ratio is about average, but the company needs to reduce fixed and/or interest charges or increase the profit or income.
- The activity ratios are average for a chemical company, and it appears the company manages its assets effectively. One area that bears watching is the average collection period, lest it increase.
- The profitability is about average, although the return on total assets is low; the rest of the operating performance appears satisfactory.

In conclusion, the company overall seems to be operated by management satisfactorily, except as noted above.

Additional Reading

Anon., *How to Read an Annual Report*. New York: Merrill Lynch, 1992.
J. R. Couper, *Process Engineering Economics*. New York: Marcel Dekker, 2003.
J. R. Couper, O. T. Beasley, and W. R. Penny, *The Chemical Process Industries Infrastructure*. New York: Marcel Dekker, 2000.

16.3.2 FINANCIAL ATTRACTIVENESS

In the free enterprise system, if profits are not maintained, the firm's growth might be stifled. The objective is to maximize profit on invested capital. The attractiveness of a venture is viewed by corporate officers and boards based on "quantitative" and "qualitative" measures of profitability. Although we use the term "profitability" loosely to measure a project's worthiness, it may not be a good measure. Drucker (1974) has said that "profitability is not a perfect measurement; no one has ever been able to define it, and yet it is a measurement despite all its imperfections."

16.3.2.1 Quantitative Measures

Although there are many quantitative measures of profitability, the four most commonly used are return on investment, payout period, net present value, and internal rate of return. The first two measures do not include the time value of money; however, the payout period can be modified to do so, as we shall see later.

The *return on investment* method is defined as the net profit after taxes divided by the total capital investment. Many variations on this measure have been used in the past. It does not include the time value of money but is simple enough that it can be used for simple screening purposes. The *payout period* is defined as the fixed capital investment divided by the after-tax cash flow. It indicates how long it will take to recover the fixed capital investment from the after-tax cash flow. Again, this is used for screening purposes. The payout period with interest takes into account the time value of money (Couper and Rader, 1986) but is a tedious calculation, and the results are not as informative as the net present value or the internal rate of return methods.

The *net present value* method is the most popular one in use today. An arbitrary time frame (i.e., time zero or the present time) is selected as the basis for the calculations. All investment expenditures made prior to time zero are compounded forward to time zero, and all income items are discounted back to time zero using an interest rate (or arbitrary "barrier return") set by management at a few percentage points above the *cost of capital*, depending on the project risk. The equation for determining the net present value is

Net present value (NPV) = net present value of all cash flows derived from income minus net present value of all investment items

If the NPV is positive, then the venture will earn more than the arbitrary barrier return. Conversely, if the NPV is negative, then the venture will earn less.

The *internal rate of return (IRR)* method is similar to the NPV method in principle. In the NPV method, the interest rate is stated, and the NPV is calculated. In the IRR method, an interest rate is sought by trial and error such that the NPVs sum to zero. That interest rate is the return that the venture will earn. There has been much discussion in the literature about which method is preferred (Couper et al., 2000).

16.3.2.2 Qualitative Measures

Before an investment decision is made, consideration must be given to certain qualitative factors and their impact on the decision-making process. In some cases, these may be the controlling factors. The qualitative measures or criteria are outlined in Tables 16.2 to 16.4.

The *technical feasibility* and the *control and operability* are usually accounted for in the selection of the project risk and the barrier return used in the quantitative NPV method described above.

The *environmental* issues—increasingly tighter restrictions on water, air, land, and noise pollution—have forced management to reconsider existing operations as well as future investments. It is a simple matter to determine the capital requirements to meet the current and projected future environmental standards, but the benefits are manifested in continued operation and community recognition of a company's good citizenship.

Indeed, the *company's good citizenship* or *corporate image* is a qualitative measure that considers most of the remaining criteria. How a company is perceived by the public and by its employees is an important factor in capital investment decisions. In terms of *safety and health* issues, a safe working place not only reduces the cost of operation through reduced injury and lost production but contributes markedly to employee morale. Regarding *legal* constraints, local and federal laws are intangibles that must be considered whenever an investment decision is made. A proposed venture that violates or infringes on a statute is doomed. It is essential, before locating a plant, to review the existing laws and seek advice on their influence on a site location. Most firms want to avoid litigation, as it can be costly and damage a firm's image. Regarding *product liability*, consumer advocates and the public demand a safe product. Management is wise to forego the installation of a facility until there is a high probability that the product will meet safety requirements. Management must take a responsible posture with regard to this item. Considering *plant maintenance*, a poorly maintained plant is an eyesore to the public and indicates an indifference about the locale. It may be an indication of management's sloppy attitude that may carry over in maintaining product quality and may affect employee morale. The *contributions to the community* and the *company's approach to sustainability* affect the corporate image. *Employee morale* is related directly to the efficiency of operation. If employees do not consider their jobs or working conditions to be good, surely this will affect not only the amount of material produced but also the quality of the product. Quality circles and seeking ISO designation have done much to improve workers' interest in their jobs and in their company. Employees who feel they "belong" to an organization will take more interest in it, be more content, and display high morale.

It is not possible to assign rankings to these qualitative factors, but they may be as important as the quantitative measures when making an investment decision. Attempts to present numerical criteria to rank these qualitative measures have been subjective and unsuccessful.

16.3.2.3 Cost of Capital, Minimum Acceptable Return, and Risk

16.3.2.3.1 Cost of Capital

The *cost of capital* is what it costs a company to borrow money from all sources. Three general sources of capital available are borrowed money, equity capital, and retained earnings. For borrowed capital (from investment houses, banks, insurance companies, and venture capitalists), the interest rate on loans is a few percentage points above the prevailing prime interest rate. The interest rate charged also depends on the length of the loan, size of the loan, and potential risk perceived by the lender. For equity capital, obtained from the sale of preferred and/or common stock, companies may float new stock issues or have shares of stock that may be released to secure capital funds. Retained earnings or reserves may be used to the extent of their availability.

A company must keep a balance between the cheaper "borrowed monies" and "equity capital." If too much money is borrowed, the leverage is affected: the debt/equity ratio increases, and the times-interest-earned ratio drops. If equity borrowing is relied on too heavily, the average cost of capital increases, the stock is diluted, and the stock price is reduced. Lending institutions as well as investors look at the debt/equity ratio and may become suspicious if it is too high. On the other hand, if it is too low, concern may develop that management may not be exercising good judgment.

An example of an average weighted cost of capital is presented in Table 16.9. It is based on a fictitious company. The dollar amount of the loans, bonds, and stock is multiplied by the after-tax

TABLE 16.9
Cost of Capital

Balance Sheet 12/31/20X6	Millions ($)	After-Tax Yield to Maturity (%)	After-Tax Weight Ave. Cost (%)
Long-Term Debt			
Revolving Account	5.0	4.5	0.02
4 3/8's debentures	12.0	4.0	0.05
6 1/2 DM	3.4	4.7	0.02
6 3/4 DM*	9.4	4.2	0.04
7 1/2 DM	74.5	4.2	0.30
9 3.8 DM	125.0	4.4	0.53
Other	23.2	4.4	0.10
	252.5		1.06
Deferred Taxes	67.7	0	0
Reserves	16.1	0	0
Preferred Stock	50.0	8.6	0.42
Shareholder's Equity	653.9	15.6	9.80
	1,040.2		11.28

* The after-tax weighted average cost of capital is 11.28%.

yield to obtain an after-tax weighted cost for each capital item. The after-tax weighted average costs are then summed to give the average weighted cost of capital. In the example, it is noted that the stockholders' equity is the predominate factor in the cost of capital calculation, while the bonds contribute about 10% to the cost of capital. Equity capital is an expensive way to obtain funds. The cost of capital should be reviewed periodically by the company's financial personnel because this figure is used as a basis for funding projects.

16.3.2.3.2 Minimum Acceptable Rate of Return

Every business enterprise has limitations on the amount of capital available for investment each year. A minimum rate of return or barrier rate is established by management for different projects, depending on the risk involved, the type of venture, the cost of capital, and the duration of the venture. Different styles of management often use different methods for establishing the minimum rate of return.

16.3.2.3.3 Risk

Companies release funds for capital expenditures on a quarterly, semiannual, or annual basis. Executive committees determine on a scheduled basis how the funds should be released, based on risk and measures of profitability. The riskier the project, the higher the barrier return. For the discussion here, let us assume a company has a venture classification system as follows: necessity projects, product improvement projects, process improvement projects, and new ventures.

Necessity projects are mandated by environmental, health, safety, legal, or image issues. The company must invest in these projects to stay in business and avoid fines and/or litigation. The calculation of an NPV is meaningless, as the company must comply. These projects have zero risk. *Product improvement projects* involve the addition of perhaps a processing step, requiring a low capital investment to improve quality so that more sales or increased sales price is possible. These projects are short term, and there is a low risk. The minimum rate of return for these projects is set at a few percentage points above the cost of capital.

Conceptual Process Design, Process Improvement, and Troubleshooting

Process improvement projects may involve some laboratory research, pilot plant, or plant studies to reduce operating expenses and to improve overall process efficiency. These projects are more expensive and of longer duration than product improvement projects. The assurance of generating a greater return is not present, so there is a higher (moderate) level of risk, and the minimum acceptable return is higher than in the previous case.

New ventures involve either the introduction of a new product to the market or an entirely new process to manufacture a product. In the domestic field, there is a risk of marketing a new product related to its potential acceptance. The number of options considered was illustrated in column 3 of Table 16.1. Research and development costs are high and require considerable technical effort. As a result, there is a chance of failure, so higher minimum barrier returns are required to satisfy management. New ventures become more risky when joint ventures are involved because of the different operating philosophies. The risk is high. Foreign new ventures are even riskier, especially joint ventures, as there are different cultures involved, time delays, market volatilities, delays in equipment deliveries, as well as potential economic crises in foreign countries. These projects are of the highest risk. The minimum acceptable rate of return for foreign projects is frequently double that of domestic ventures.

16.3.3 Operating Expense Estimation

Few methodologies have been published for the estimation of operating expense, probably because much of the information is proprietary.

Operating or *manufacturing expenses* are recurring and significantly affect profitability. Such expenses consist of the expense of manufacturing as well as of packaging and shipping, selling and distribution, and company overhead. Direct expenses are those directly associated with the manufacture of the product, exclusive of raw material expenses, which are often a separate item. Indirect expenses refer to those that are not affected by the production rate, such as depreciation, local taxes, insurance, and plant indirect expenses, such as plant roads, docks, plant security, and so on.

16.3.3.1 Manufacturing Expense Sheet

A typical manufacturing expense sheet is presented in Table 16.10. On this illustrative sheet, the bottom line is the expense to manufacture a product at the manufacturing department door and excludes packaging, shipping, and overhead expenses.

This manufacturing expense sheet has four distinct parts: the heading, raw material section, direct conversion expenses, and indirect conversion expenses. The *heading* identifies the product to be made, the amount produced, the fixed capital investment, and sometimes the location, yields, operating time, and other items. The *raw material* includes the raw materials used to produce the product as well as any by-products for which credit may be received. The *direct expenses* include utilities, labor, and maintenance. The sum of these two sections is the *direct conversion expense*. The *indirect expenses* include the depreciation and plant indirect expenses, such as plant security, fire protection, roads, docks, and so forth. These expenses occur continually regardless of whether a product is produced. The sum of these expenses is the *total indirect conversion expense*.

16.3.3.2 Operating Expense Items

1. *Raw materials*. A material balance is always needed to determine the amount of raw materials and their costs. The quality grade and the form of the raw material (powder, flake, crystal, or liquid) affect the cost. Quantity discounts on large contract amounts

TABLE 16.10
Manufacturing Cost Estimate

MFG. CAPITAL:	M&E	$8,000,000	Product	Plasticizer M
	Bldg.		Process	
	Total	$8,000,000	Annual Production	7,500,000 kg/annum

RAW MATERIAL

Material	Unit	Quantity	$/unit	$1,000/annum	$/Mg
A	___	___	___	2,000	___
B	___	___	___	1,300	___
___	___	___	___	___	___
___	___	___	___	___	___
___	___	___	___	___	___
		GROSS RAW MATERIAL COST		___	___
By Product (Credit)					
___	___	___	___	___	___
___	___	___	___	___	___
		TOTAL CREDIT		___	
		NEW RAW MATERIAL COST		3,300	___

DIRECT EXPENSE

Steam	Mg	30,000	$7.00	210	___
	Mg	___	___	___	___
Water-Plant Cooling Tower	m³	300,000	$0.01	3	___
Sanitary City	m³	300,000	$0.08	24	___
Electricity	MJ	15,000,000	$0.01	150	___
Fuel	___	___	___	___	___
___	___	___	___	___	___
Supervision				36	___
Labor				441	___
Payroll Charges				215	___
Repairs				480	___
Laboratory				36	___
Supplies				3	___
Clothing & Laundry				6	___
Miscellaneous				12	___
TOTAL DIRECT CONVERSION				1,616	___

INDIRECT EXPENSE

Depreciation	1,143	___
Factory Indirect Expense	240	___
TOTAL INDIRECT CONVERSION	1,383	___
TOTAL CONVERSION	2,999	___
TOTAL MANUFACTURING COST	6,299	___

have potential savings, but storage facilities, which increase the capital investment, may be needed. The cost is also affected by the way the material is delivered—drums, leverpaks, carboys, tank trucks, tank cars, barges, or pipeline. Some raw materials may be seasonally produced, which might present a storage problem.

2. *By-products.* By-products are reported the same way as raw materials but are entered separately. The pricing of by-products depends on the current market for the material. If there is no market, then credit cannot be claimed. If by-products have a market value, they may be credited at their net value.

3. *Utilities.* The utility requirements for a process are obtained from both a material and energy balance. The unit costs for each utility may be obtained from plant expense sheets, the accounting department, or a utility superintendent. These costs often increase continually, so they should be reviewed frequently. Utilities are usually steam (high-, medium-, and low-pressure) and their associated unit costs, electricity, natural gas, cooling tower water, and treated or city water. Sometimes instrument air, demineralized water, and refrigeration are considered utilities if they come from a central source and are not tied to a given process.

4. *Labor.* Operating labor is usually the second largest expense. The most reliable way of projecting labor requirements is to prepare a table of shift, weekend, and vacation coverage. If the estimator has not had previous experience, it is advisable to seek help from operating personnel. For continuous coverage of one laborer per shift, 4.2 laborers are required, assuming a 40-hr work week. The labor rates may be obtained from the union contract or from the plant labor-relations person. In the southwest United States, a rate for a qualified operator is between $38,000 and $49,000 per year (2007). Information on labor requirements in chemical plants in the literature is meager. The Wessel Equation (1952) tends to predict high labor requirements and does not take into account increases in labor productivity.

5. *Supervision.* For preliminary purposes, use 20% of the labor expense. Some estimators consider supervision as a fixed expense at 100% of capacity operation. If a supervisory position can be identified, then a typical salary may be obtained from a plant accountant.

6. *Payroll charges.* These include paid vacations, pensions, group insurance premiums, disability pay, social security premiums, and unemployment taxes. Estimate these as 30 to 35% of the labor plus supervision expense.

7. *Repairs (maintenance).* The best source of information about the materials and labor costs is a company's records. Little useful information is in the open literature. For preliminary estimates, the total of the materials and labor expense is 6 to 10% of the fixed capital investment per year. The higher percentages in this range should be used for processes that have a large amount of rotary equipment, and lower percentages for processes that operate at or near ambient conditions. If the repair expenses must be split into materials and labor, use 40% for materials and 60% for labor.

8. *Supplies.* This expense includes chart paper, computer paper, filter cloths, brooms, mops, and so on. Company records are the best source, but use 6% of the operating labor for preliminary estimates.

9. *Laboratory expenses.* The use of on-line analyzers and sophisticated analytical equipment often eliminates expensive and time-consuming tests. There are, however, certain tests that can be performed only in a laboratory. The charges for these services are escalating rapidly and may be between $130 and $300 per laboratory hour (2007). As an alternative, for complex processes, an estimate of 10 to 20% of the operating labor is reasonable.

10. *Clothing and laundry.* Manufacturing departments that produce a highly toxic material or a pharmaceutical or food product often provide clothing and laundry services for their operating personnel. For estimating purposes, use 6 to 10% of the operating labor expenses.

11. *Other expenses.* Technical services for troubleshooting process problems may employ a young engineer at a salary of about $50,000 to $60,000 per year (2007). Expenses for environmental control often include the costs of incineration, chemical treatment of the waste, landfill, and perhaps notification, documentation, reporting, manifesting, and labeling under RCRA legislation. The most reliable source for these figures is the environmental engineering staff in the specific plant, with some elaboration as given in Section 16.7.
12. *Total direct conversion expense.* The sum of items 3 through 11 is the *total direct conversion expense.*
13. *Indirect conversion expenses* include (a) depreciation and (b) plant indirect expenses. (a) Depreciation is the major component of indirect expenses. For estimation purposes, the annual depreciation is assumed to be straight-line and is calculated by dividing the fixed capital investment by the number of years over which the investment is to be depreciated. (Accelerated depreciation is not normally entered into the manufacturing expense sheet but is used in cash flow analyses after backing out straight-line depreciation from the manufacturing expenses.) (b) Plant indirect expenses include many items such as *ad valorem* and real estate taxes, insurance, maintenance of yards, docks, and roads, cafeterias, fire protection, and security. Plant accounting departments often compile factors to be applied for each plant site. To estimate this expense, use 2 to 4% of the fixed capital investment per year.
14. *Total indirect conversion expense* is the sum of depreciation and plant indirect expenses.
15. *Total manufacturing expense* is the sum of the net raw material expense, the total direct conversion expense, and the total indirect conversion expense.
16. *Packaging, loading, and shipping expenses.* The packaging expense includes the container and the labor to package the product. The loading expense may entail moving the product from the manufacturing department to a centralized storage facility before shipping to a customer. Under the shipping expense may be labor charges for loading on trucks or rail cars and any materials, such as dunnage, required to protect the product containers. Transportation charges vary widely depending on whether the product is shipped by pipeline, barges or tankers, railroad cars, or tank trucks. Companies maintain good expense control records on this item. Depending on the manner in which a product is packaged and shipped, this expense could vary between $0.013 and $0.034 per kilogram of product (2000).
17. *Total product expense* is the sum of the total manufacturing expense and the packaging, loading, and shipping expenses.
18. *General overhead expenses* refers to the expense in maintaining sales offices throughout the country, staff engineering departments, research laboratories, and administrative offices. All products are expected to share in these burdens, so an appropriate charge is made for each product. Company accounting departments frequently develop formulas or factors based on sales. The charge is highly dependent on the amount of customer service provided. A factor between 6 to 15% of the annual sales revenue may be assessed. The former figure is for established commodity products, and the latter for new specialty or fine chemicals.
19. *Total operating expense* is the sum of the total product expense and the general overhead expense.

Example 16.2

The Acme Specialty Company is considering manufacturing Plasticizer M. The engineering department has estimated that a fixed capital investment of $8 million will be required to produce 7.5 million kilograms of product per year. The product is expected to sell for $1.24 per kilogram.

Conceptual Process Design, Process Improvement, and Troubleshooting

The net raw material expense is $0.44 per kilogram. It is estimated that three operators per shift will be required. The process is semicontinuous and is assumed to operate at 90% stream time. Other costs and usages are

Utilities	Usage/kg Product	Cost
Steam	4.0 kg	$7/Mg
Electricity	2 MJ	1 cent/MJ
Cooling water	40 L	1 cent/1000 L
City water	40 L	8 cents/1000/L

Maintenance—6% per year of the fixed capital investment
Average operating labor expense—$35,000 per year
Supervision—$3,000 per month
Payroll charges—45% of labor plus supervision
Supplies—$250 per month
Clothing and laundry—$500 per month
Laboratory charges—40 hours per month at $75 per hour
Packaging, loading, and shipping charges—1 cent per kg of product
Other direct expenses—$1,000 per month
Depreciation—straight-line for seven years
Plant indirect expenses—3% of the fixed capital investment per year
General overhead expense—6% of the annual net sales

(a) Prepare an annual manufacturing expense sheet based on full production.
(b) Calculate the total annual operating expense.

Solution

Table 16.10 has been filled in with the data from this example. Below are the details of the calculations:

Basis: One year of operation
Annual sales: 7,500,000 kg/yr × $1.24/kg = $9,300,000
Manufacturing expense sheet:
    Net material expense: 7,500,000 kg/yr × $0.44/kg = $3,300,000
    Direct conversion expenses:
      Utilities:
        Steam: 7,500,000 kg product × 4 kg steam/kg product × $7/Mg steam = $210,000
        Electricity: 7,500,000 kg product × 2 MJ/kg product × $0.01/MJ = $150,000
        Cooling water: 7,500,000 kg product × 40 L/kg product × $0.01/1,000 L = $3,000
        City water: 7,500,000 kg product × 40 L/kg product × $0.08/1000 L = $24,000
        Total utilities: $387,000
      Labor: 4.2 operators × 3 operators per shift × $35,000 = $441,000
      Supervision: $3,000/mo × 12 mo/yr = $36,000
      Maintenance: 0.06/yr × $8,000,000 = $480,000
      Payroll charges: 0.45 × ($441,000 + $36,000) = $215,000
      Supplies: $250/mo × 12 mo/yr = $3,000
      Clothing and laundry: $500/mo × 12 mo/yr = $6,000
      Laboratory charges: 40 hr/mo × 12 mo/yr × $75/hr = $36,000
      Other direct expenses: $1,000/mo × 12 mo/yr = $12,000
      Total direct conversion expense: $1,616,000

Indirect conversion expenses:
   Depreciation: $8,000,000/7 yr = $1,143,000
   Plant indirect expenses: $0.03/yr × $8,000,000 = $240,000
   Total indirect conversion expenses: $1,383,000

Total manufacturing expenses: $6,299,000 (answer a)

   Packaging, loading, and shipping expense: 7,500,000 kg product × $0.01/kg = $75,000
   General overhead expenses: $9,300,000 × 0.06 = $10,000

Total operating expenses: $6,984,000 (answer b)

Actual operating expense reports, given in Table 16.10, form the basis for information needed for the income statement, Table 16.8, such as the cost of goods sold, depreciation, amortization and depletion, and general overhead expense. This information would be used in the quantitative measures of financial attractiveness.

Additional Reading

J. R. Couper, *Process Engineering Economics*. New York: Marcel Dekker, 2003.
J. R. Couper, O. T. Beasley, and W. R. Penny, *The Chemical Process Industries Infrastructure*. New York: Marcel Dekker, 2000.
J. R. Couper and W. H. Rader, *Applied Finance and Economic Analysis for Scientists and Engineers*. New York: Van Nostrand Reinhold, 1986.
P. Drucker, *Management: Tasks, Responsibilities and Practices*. New York: Harper & Row, 1974.
J. Hackney, *Control and Management of Capital Projects*. New York: McGraw Hill, 1992.
W. E. Wessel, 1952. *Chem. Eng.* 59 (7): 209–210.

16.3.4 Estimation of Fixed and Total Capital Investments

The fixed capital investment is the cost to build the manufacturing facility. Corresponding to the different levels in the gating process given in Table 16.1 are methods of cost estimation that have different levels of accuracy. Here we describe estimation methods for the conceptual design stages that use process cost correlations (which are usually accurate to ±40 to 50%) and the "bare module" factor method (±30%). Some estimation methods provide improved accuracy but require vendor quotations; detailed estimates of material costs of piping, valves, and insulation; and estimates of installation labor hours and the mix of labor rates. Such methods are beyond the scope of this chapter.

The methods described here work well if clear definitions, unambiguous terminology, and correct methods to account for alloys, inflation, and location are used, and if care is taken to include all of the cost contributions.

16.3.4.1 Definitions and Terminology

Battery limits cost (BL). This includes the cost of the process itself and excludes the costs of utilities, storage, and services unless these are required only for this process.

BL cost estimation. The simplest and least accurate method uses a correlation of the cost with the capacity of final product. When more accuracy is needed, the process equipment is costed, and costs are attributed to all the components needed to convert the equipment items into a complete working BL process. Many different methods have been developed, depending on the accuracy needed.

Outside battery limits or *offsite costs*. Factory-site (as opposed to process-specific) facilities such as (a) access roads, rail spur lines, paving, unloading and loading facilities; (b) utilities facilities

Conceptual Process Design, Process Improvement, and Troubleshooting

such as boiler houses, electrical transformer stations, water supplies, cooling towers, and oxygen or nitrogen supply systems; (c) water treatment facilities; (d) environmental impact facilities for wastewater, air, solids, odors, and noise; (e) storage, including bulk storage for feed, products, and by-products; and (f) support facilities such as administration buildings, laboratories, cafeterias, gatehouses, maintenance shops, and firefighting, communication, purchasing, and safety facilities.

Grass roots or green field cost. This accounts for the all of the facilities needed to convert a grass field into a complete working plant site: BL plus offsites.

Main plant items (MPI). These include all "major" process- and utility-related equipment. "Major" includes pumps, fans, in-line filters, and installed spare equipment but excludes valves, piping, fittings, pipe hangers, heat tracing, lighting, vents, hoists, interlocks, breakers, and starter panels. Skid-mounted packages are considered to be one MPI.

FOB cost. The cost of the crated, fabricated equipment loaded onto a carrier at the location of fabrication (or another specified location). The acronym FOB stands for "free-on-board," meaning the equipment has been placed on board the carrier without an additional loading charge, e.g., FOB Pittsburgh. Such a cost does not include tax, duties, freight, and delivery costs. The term "purchase cost" is ambiguous.

Bare module cost (BM). A BL process is divided into a collection of *modules* for each type of MPI: reactor module, heat exchanger module, and membrane module. The boundary for a module has been selected from experience so that the sum of the costs of all modules for all MPIs for the process equals the cost of the whole process. In general, the *module* boundary is about 2 m from the outside of the equipment. Thus, the cost of a "pump" module would include the FOB cost of the pump plus the material and labor costs for the yard work, sewers, piping, electrical, instruments, concrete, structural steel, a section of the pipe rack through the module, painting, and electrical plus the cost of inspection, hauling to site, attaching the baseplate to the plinth, aligning the drive and the pump, and hooking up the pump to the electrical and piping. The BM cost usually does not include offsites or buildings. The BM cost includes duties, freight and delivery charges, and taxes and expenses for the design; obtaining vendor quotes and ordering the equipment; and the field work to oversee the installation. However, it excludes contingency, contractors' fees, lawyers' fees, royalties, land, and expendables such as catalyst and solvent.

BM factor method of cost estimation. Component costs within the BM are estimated as percentages (or factors) of the FOB cost. The factors differ depending on the type of equipment in the module and the alloys. Two types of factors have been developed: (1) $L + M^*$ factors to account for the FOB cost plus the materials and labor for the supports, concrete, electrical, process and utility piping, vents and sewers, insulation, and painting, plus labor for inspection and installation but excluding instruments and instrumentation. (2) *Indirect cost factors* for home office and field expense.

16.3.4.2 Cost Contributions for One Type of MPI

For one type of MPI, 21 contributions accrue to be the total capital cost for that MPI. The sum of these for all MPIs represents the total capital cost for the process. The cost contributions are shown below.

1. *The FOB cost.* FOB costs can be obtained from correlations or vendor quotations. For most installations, the FOB cost is about one third to one fifth of the total fixed capital cost. "Delivered costs" should be separated into an FOB cost with the delivery, taxes, and duties expense to be accounted for in item 6. For larger equipment that must be fabricated in the field, the FOB for this MPI is replaced by the *field erected cost.*
2. *Direct module labor and material cost, $L + m^*$ cost, and the $L + M^*$ cost.*
   (a) $L + m^*$ *cost.* The costs for the materials, m^*, needed within the module are expressed as a factor of the FOB carbon steel equipment cost. Some illustrative values for

different takeoffs are supports (0 to 25%), concrete (5 to 10%), electrical (3 to 5%, with 10 to 15% for fans and compressors and 30 to 75% for pumps), process and utility piping, vents, and sewers, including the "local" pipe rack within about 2 m of the equipment (15 to 60%), insulation (3 to 25%), and painting (1 to 3%). For carbon steel equipment, the material cost, m^*, is usually 60 to 70% of the FOB cost for process equipment handling fluids and 20 to 30% of the FOB cost for process equipment handling primarily solids. Instrumentation costs are handled separately in item 3. The other "material" cost within the module is the FOB cost, item 1.

The costs for the labor, L, include those to receive, uncrate, inspect, haul to the site, and install the equipment, and the labor to install all materials needed to provide the local environment to operate the equipment. If the unit is a skid-mounted package, then the labor cost is 20 to 45% of the FOB cost, which represents the minimum labor cost.

The sum of $L + m^*$ accounts for materials and labor within the BM but excludes the FOB cost (item 1) and excludes freight, tax and duties, instrumentation, buildings, land and site preparation, and long runs of piping needed to connect equipment modules separated by more than the usual distance because of safety. These latter costs are site specific and are added separately.

(b) $L + M^*$. For ease in calculation, the FOB costs from item 1 are combined with $L + m^*$ costs and expressed as $L + M^* = L + m^* + FOB$. That is, $L + M^*$ cost $= FOB_{carbon\ steel} \times L + M^*$ factor. This approach would be used for the purposes of decisions made at Gate 3 in Table 16.1.

The * indicates that the instrumentation costs have been excluded from the factor, and instrumentation is estimated in item 3. Typically, $L + M^*$ factors are in the range 1.2 to 3 times, depending on the type of equipment and the number of such pieces being installed. Table 16.11 gives example values of the factors. Care is needed, because published data report $L + M$ factors but use them as $L + M^*$; others confuse $L + m$ with $L + M$. All $L + M^*$ factors are based on FOB carbon steel equipment; the $L + M^*$ factor varies dramatically with material of construction. Section 16.3.4.5 gives the details.

The $L + M^*$ can be separated into a labor cost and total materials M^* cost if the L/M ratio is known. The L/M ratio ranges from 0.15 to 0.65, with 0.4 as the usual value. Figure 16.2 illustrates how the factors interact.

TABLE 16.11
Illustrative Breakdown of Material and Labor Components

	Centrifuge			Rotary Drum Filter			Crystallizer		
	Total	Labor, L	Material, M	Total	Labor, L	Material, M	Total	Labor, L	Material, M
FOB	1.00			1.00			1.00		
Setting		0.101		1.104	0.104			0.076	
Piping		0.109	0.25		0.07	0.59		0.17	0.41
Insulation		0.024	0.04		—	—		0.024	0.05
Paint		0.01	0.01			0.05		0.005	0.01
Elect		0.122	0.27		0.01	0.10		0.036	0.05
Concrete		0.038	0.09		0.01	0.13		0.048	0.06
Steel		0.049	0.10		0.01	0.15		0.048	0.07
L+M*	2.21	0.453	0.76	2.224	0.204	1.02	2.06	0.407	0.65
L/M	0.26			0.10			0.24		

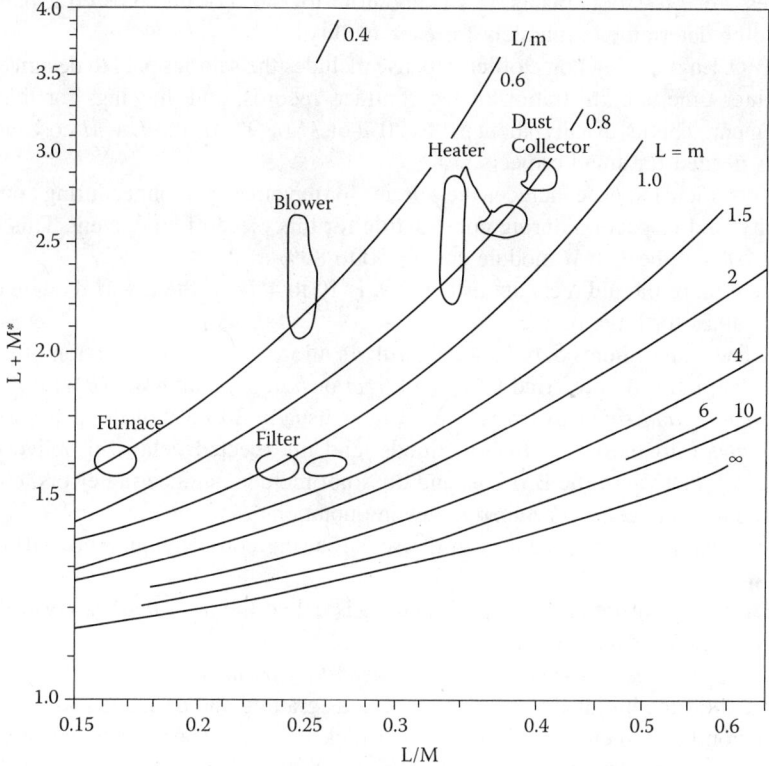

FIGURE 16.2 $L+M^*$, L/m, L/M.

3. *Direct $L + M$ instrumentation cost.* Costs for instrumentation are relatively independent of the FOB cost and of the size of the equipment. Hence, the factor method described in item 2 should not be used; instead, methods given in Section 16.3.4.4 are recommended.
4. *Within BL building and housing costs.* This includes the concrete and structural steel for the building, heating, lighting, and ventilation but excludes the foundations and supports for the MPI and process piping for vents and drains that have already been accounted for in the module cost, item 2. Process buildings can be estimated based on a cost per square meter, based on methods given by Richardson (annual), Navarrete (1995), or (as a last resort) 10 to 45% of the FOB cost for fluid processing equipment or 15 to 60% of the FOB for solids processing equipment.
5. *Subtotal.* The sum of the $L + M^*$ cost (item 2, which includes item 1, the FOB cost), the cost of instruments (item 3), and the cost of buildings (item 4) is called the $L + M$ *module cost.*
6. *Tax, duties, freight, and insurance costs.* These typically total 15 to 25% of the FOB cost. The freight is 2 to 5% of the FOB cost, with 3% being usual. For overseas, use 5 to 10% of the FOB cost.

    Some processes (e.g., water and wastewater treatment facilities) include concrete vessels plus traditional pumps, heat exchangers, and mixers. To cost such facilities, it is more realistic to report the sum of the $L + M$ module costs and the delivery, taxes, and duties. We refer to this cost as the *physical module (PM) cost.*
7. *Offsites.* These are usually costed separately (Woods et al., 1979) or, as an approximation, allow 1 to 5% of FOB for small modification of the offsites, 5 to 15% of FOB for restructuring the offsites, and 15 to 45% of the FOB cost for a major expansion of the

offsites. For grassroots plants, this is about 45 to 150% of the FOB cost, but the costs should be determined separately for each facility.

8. *Indirects*. Engineering home office expense includes the salaries paid to design engineers, computer time, administration of the contract, records, and drawings for this piece of equipment. This is usually about 10 to 20% of M or 9% of the $L + M$ cost, with larger values needed for small projects.
9. *Indirects*. Field expense includes the salaries for the onsite personnel during construction, security, and inspection during construction for this piece of equipment. This is usually about 20% of the $L + M$ module cost or 50 to 80% of L.

    The total of the indirects, items 8 and 9, is 10 to 45% of the $L + M$ module cost, with small values for large projects.
10. *Total*. The sum of items 5, 6, 8, and 9 (with an allowance for modification of the offsites as needed, item 7) is referred to as the *direct* or *bare module costs (BM)*.
11. *Contractors' and subcontractors' fees*. This is usually 3 to 5% of the BM cost.
12. *Contingency* for strikes, tornados, floods, and unexpected delays in delivery. This is usually 10 to 15% of the BM cost and does not include engineering errors, engineering uncertainties in design, or uncertainties in quotations.
13. *Design contingency* for changes in the scope during construction. Allow 10 to 30% of the BM cost.
14. *Total*. The sum of items 10, 11, 12, and 13 is called the *total module cost (TM)* or the *fixed capital investment cost*.
15. *Royalties, licenses, know-how fees, and regulatory permits*.
16. *Land* and (a) the initial site preparation such as grading and drainage and (b) the special foundation costs such as pile driving and rock blasting. These costs are important for many wastewater treatment facilities, which are often located on low-grade land. This is often about 1 to 2% of the TM cost. Site development is sometimes expressed as 10 to 20% of the FOB cost.
17. *Spare parts* (these are especially important for remote factory sites). Use 4 to 8% of the FOB equipment cost for the MPI or 1 to 2% of the TM cost.
18. *Interest* during the construction period and/or the cost of borrowing capital and paying it back, as with a municipal bond. This can be estimated to be 6% of the TM cost.
19. *Legal fees*. An estimate might be 1% of the TM cost.
20. *Working capital*. This accounts for the initial charge of solvent, catalyst, heat transfer media, and initial inventory of raw materials and of product awaiting sale. For commodity and pseudocommodity products produced at a uniform rate, use 15 to 20% of the TM cost. For seasonal products such as herbicides and fertilizers, use 25 to 40% of the TM.

    For specialty products like perfumes, cosmetics, and flavors with high raw material costs and relatively low TM costs, it may be preferred to express working capital as 15 to 40% of sales, with 30% being a reasonable value.
21. *Startup expenses*. These include a living allowance for the startup team, special tests and test equipment, and computer costs until the process is producing specification product. Estimate as 8 to 10% of the TM cost. Operator training is 0 to 4% of M. The total startup expense is 15 to 40% of M.

Although the 21 items usually represent the total capital investment, different items are excluded for different purposes. For example,

- Item 20. This cannot be included as the cost of fixed assets for government depreciation purposes.
- Item 18. For a municipal waste treatment facility, the project must be self-financed, because the municipality does not have a continual reserve. Hence, the cost of capital

Conceptual Process Design, Process Improvement, and Troubleshooting

and paying it back must be included. The cost is called the *bonded fixed capital cost* and is about 25% larger than the usual fixed capital investment, TM cost.
- Item 15. A waste recycling facility requires permits under legislation. Although the cost of the permit may be nominal, the cost of the public hearings, consultants, and expert witnesses needed as part of the permit application increases the cost to be about $1 million (2000). For more, see Section 16.7.
- Item 8. Home office expense may be paid for by the government on certain projects such as wastewater treatment. That is, a government agency may design the facility.

In summary, the total capital investment is the sum of several dozen cost components. The challenges in estimating the costs are to include all of the needed items and not to double count, be fully aware of local and federal regulations and legislation so that appropriate items are included and excluded, use clearly defined terms, include error estimates so that the accuracy of estimate is understood, and include the currency, location, and date.

16.3.4.3 Accounting for Time and Location

16.3.4.3.1 Time

Prices change with time. However, because of the development of inflation indices specific to the process industry, capital costs from the past can be scaled to current conditions by multiplying by the ratio of inflation index values:

$$Cost_{now} = cost_{then} \, [index \, value_{now}/index \, value_{then}]$$

More than 40 capital cost indices have been developed. These include

- *Marshall and Swift, MS* (formerly Marshall and Stevens, 1926 = 100); cost compiled quarterly for U.S. conditions, primarily for insurance companies interested in the changing value of the assets of companies. A separate MS index is compiled for the overall industry and for eight process and four related industries. Published in each issue of *Chemical Engineering*.
- *Chemical Engineering Index, CE* (1957–1959 = 100); cost developed for U.S. conditions primarily as a plant construction index. Separate CE indices are published for process equipment, heat exchangers and tanks, process machinery, and for each of the cost components and for the overall weighted total for a typical process. *CE instruments* is used in Section d. Published in each issue of *Chemical Engineering*.
- *Vatavuk Air Pollution Control Cost Index* (VAPCCI; 1994 = 100); FOB vendor quotes for U.S. conditions for nine types of air pollution control equipment. Published in each issue of *Chemical Engineering*.
- *EPA-STP Treatment Plant Index* (1957–1959 = 100); fixed capital investment costs for U.S. conditions of primary and secondary wastewater treatment plants averaged monthly over 20 U.S. cities.
- *Nelson Refinery Index* (1946 = 100); fixed capital investment costs for U.S. conditions for refineries (without consideration of changes in technology). The Nelson true-cost index accounts for changes in technology. Published in the *Oil and Gas Journal*.

In addition, process industry cost indices have been published for Canada, the United Kingdom, France, India, Italy, Japan, Germany, and The Netherlands. Table 16.12 gives values of the CE and MS indices that are general and reliable for North American conditions. All data reported in this chapter have been scaled into the future using CE = 1000 or CE instruments = 1000.

Costs obtained from the current value of CE can be inflated to future times based on an estimate of the annual inflation rate.

TABLE 16.12
Inflation Indices

	CE Index (1957–59 = 100)	CE Instruments (1957–59 = 100)	MS; Process Industry (1926 = 100)	M&S Index: All Industry (1926 = 100)
1980	261.1	249.5	676	659.6
1981	297.0	287.9	744.9	721.3
1982	314.0	297.6	774.4	745.6
1983	316.9	308.4	786	760.7
1984	322.7	319.1	806.5	780.4
1985	325.3	322.8	813.4	789.6
1986	318.4	324.5	816.9	797.7
1987	323.8	330.0	830.4	813.6
1988	342.5	341.9	870.1	852.0
1989	355.4	352.1	914.2	895.1
1990	357.6	353.6	934.5	915.1
1991	361.3	353.8	951.8	930.6
1992	358.2	356.6	960.5	943.1
1993	359.2	358.4	975.3	964.2
1994	368.1	365.4	1000	993.4
1995	381.1	378.3	1037.3	1027.5
1996	381.7	372.2	1051.3	1039.2
1997	386.5	372.4	1068.3	1056.8
1998	389.5	365.5	1077	1061.9
1999	390.6	363.5	1083.1	1068.3
2000	394.1	368.5	1102.7	1089
2001	394.3	363.3	1109.0	1093.9
2002	395.6	363.5	1121.1	1104.2
2003	402.0	365.8	1143.5	1123.6
2004	444.2	374.3	1201.8	1178.5
2005	468.2	381.8	1295.1	1244.5
2006	499.6	420.1	1364.8	1302.3

16.3.4.3.2 Location

Construction costs differ depending on the location, since labor costs and productivity differ. The labor component for North America can be extracted from $L + M^*$ data through the L/M ratio data and adjusted to reflect location costs and productivity. Illustrative location indices, given in Table 16.13, compare construction costs for complete processes in U.S. dollars to the cost in U.S. dollars in a different country. Specialists in your company should be consulted to obtain pertinent data.

16.3.4.4 Estimating the $L + M$ Cost of Instrumentation

Three options to estimate the cost are to use an average cost per MPI, cost the control elements per MPI, and cost the instrumentation for different types of equipment.

Option 1: Main Plant Items and an Average Cost for Instrument

For typical process plants, use $47,000 U.S. (CE instruments = 1000) per MPI; for highly automated batch processes, use $69,000 to $82,000 U.S. (CE instruments = 1000) per MPI. CE instruments is the *Chemical Engineering* inflation index for process instruments. Details are given in Section c and in Table 16.12. Scale the instrument cost to the correct time.

Option 2: Take-Offs for Different Types of Processes

Table 16.14, based on Navarrete (1995) and Page (1996), reports the number of different elements in the distributed control system (DCS) for the MPIs in different types of plants. The average number of field instruments decreases as the scope of the project increases.

TABLE 16.13
Trends in Location Factors: Capital Investment Relative to United States = 1.00

Country	1963	1970	1976	1978	1993
Australia		1.1		1.3	1.6
Austria				1	
Belgium		0.96		1	1.26
Canada, imported		1.05		1.0	1.32
indigenous				0.7	
Central Africa				2	
Central America		1.04		1	
China imported				1.1	
China indigenous				0.55	
Croatia					1.2
Denmark		0.86		1.0	1.46
Finland				1.2	
France	1.13	0.98	0.94	0.95	1.64
Germany			0.99	1	1.19
Greece				0.9	
India imported			0.9	1.8	
India indigenous				0.65	
Ireland		0.79		0.8	
Italy		0.96	0.94	0.9	2.15
Japan			0.64	0.9	0.95
Malaysia				0.8	
Mexico		1.08			
Middle east				1.1	
Netherlands				1.0	1.04
New Zealand				1.3	
North Africa, imported				1.1	
North Africa, indigenous				0.75	
Norway				1.1	
Portugal				0.75	
South Africa				1.15	
South America, north		0.92		1.35	
South America, south		1.06		2.25	
Spain, imported				1.2	2.32
Spain, indigenous		0.76		0.75	
Sweden				1.1	1.79
Switzerland				1.1	
Turkey				1.0	
UK		0.89	0.96	0.9	1.76
USA			1.0	1.0	1.0
Yugoslavia				0.9	

TABLE 16.14
Number of Elements per MPI (Based on Page, 1996, and Navarrete, 1996)

Elements	Different Types of Plants			
	Usual Overall	Batch Process	Continuous Process	Storage Area
DCS Points	2	3.6	1.7	1.0
DCS wiring units	3	5.4	2.5	1.5
Field instruments	7.5–2.5	12	7	6.5
Air hookups	1	1.8	0.9	0.5

Total $L + M$ cost = DCS FOB cost + 2.1 (FOB cost for field instrumentation), where

DCS FOB, hardware including computer = $5000 (CE = 1000)/point
FOB field instrumentation c/s equipment = $1800 (CE = 1000)/balloon for carbon steel.
The alloy cost factors are low alloy $\times$ 1.33, high alloy $\times$ 1.5.

Option 3: Costs for Instrumentation of Particular Types of Equipment

Table 16.15 lists "typical" DCS (control loops, C; alarms, A; and transmitters, T) and the types of field instruments (indicators, I, and relief valves, R) for different types of equipment and their FOB cost. For this list, we give an estimate of the installed instrument cost in Table 16.15. These results are reasonably consistent with the previous two methods.

16.3.4.5 Accounting for Materials of Construction Other Than Carbon Steel

In Section 16.3.4.2, the $L + m^*$ costs were expressed as a fraction of the cited FOB cost (which is usually carbon steel fabrication). If the equipment is made of an alloy, the FOB cost of the equipment is larger, say $70,000 instead of $30,000. However, the $L + m$ costs of concrete, painting, insulation, and other components in the BM remain about the same as they were for the carbon steel equipment. To account for this, we need to multiply the FOB alloy cost by an $L + M^*$ factor that is smaller than the factor reported for carbon steel. This is counterintuitive, because we intuitively may think "alloys" = "more expense for everything." The amount of reduction in the $L + M^*$ factor for carbon steel is determined from Figure 16.3.

Example 16.3

a. The FOB cost of a carbon steel pressure tank is $70,000, CE = 400. What is the $L + M$ cost if the $L + M^*$ factor is 2.5?
b. The vessel needs to be 316 stainless steel. If the FOB alloy cost is 1.4 times the carbon steel cost, what is the $L + M$ cost?

Solution to Part a

The $L + M^*$ cost is $L + M^* \times$ FOB = 2.5 $\times$ $70,000 = $175,000.
Comment: In practice, a date or inflation index value and error range should be given with every cost estimate.
For preliminary estimates using FOB cost correlations and $L + M^*$ factors, the error is usually $\pm 30\%$.
In addition, we need to add the $L + M$ cost of instrumentation. For a pressure tank, from Table 16.15, the $L + M$ cost of instruments is $8300 at CE = 1000. Therefore, the cost at

TABLE 16.15
Installed Instrumentation Costs

	Different Types of Equipment									
	Reactor: Gas Phase	Reactor: Liquid	Condenser	Fired Heater	Heat Exchanger/ Reboiler	Distillation	Evaporator	Storage Tank	Pressure Tank	Intermediate Process Tank	Pump, Stage of Compressor
Flow	C	C	C	C	C	C, I					
Level		C, A		C, A		2C, 2A	C, A	I	I	C	
Pressure	2C, 2A R	2C, 2A R	I, R	I		3I, T, A, R	I, R	R	I, R	R	I, R
Temperature	C,A,2I	C, A, 2I	C, A	C, A	C, 2I	6I, 4C	I	I	I	I	I
Cost, US $	63,000	70,500	40,000	40,000	27,000	150,00	25,000	7,000	8,300	17,400	7,000
Cost US $ method (i)						46,000					
method (ii)	57,400	69,500	29,800	37,700	16,240	117,300	23,800	11,340	15,120	16,240	11,340

Note: CE instruments = 1000, for different process equipment (excluding control valves) based on Cran, 1981.

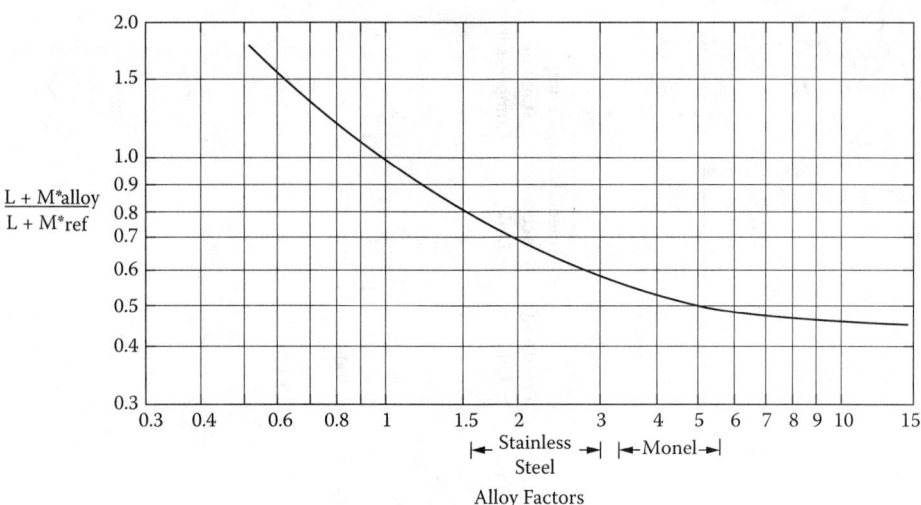

FIGURE 16.3 $L+M*$ and alloys (reproduced with permission from Donald R. Woods, *Rules of Thumb in Engineering Practice*, page 377, 2007, copyright Wiley-VCH Verlag GmbH & Co. KGaA, Weinheim, ISBN: 978-3-527-31220-7).

CE = 400 would be found by the ratio of the indices = \$8300 × 400/1000 = \$3320. The $L + M$ cost at CE = 400 is \$175,000 + 3,320 = \$178,320 or \$178,000 ± 30%.

Solution to Part b

The FOB cost of the stainless steel vessel is \$70,000 × 1.4 = \$98,000.

The $L + M*$ ratio for an alloy of 1.4 is 0.82, from Figure 16.3. The $L + M*$ factor for the alloy is therefore $L + M*$ carbon steel × correction factor = 2.5 × 0.82 = 2.05. Therefore, the $L + M*$ cost is \$98,000 × 2.05 = \$196,000 ±30% at CE = 400.

Comment: we need to add the cost of instruments = \$3320 from part a for an $L + M$ cost = \$199,320 or 199,000 ± 30% at CE = 400.

16.3.4.6 Different Methods for Different Levels of the Gating Process

For *preliminary evaluation* (Gate 1 in Table 16.1), correlations for complete processes scaled up for capacity and for time are commonly used. For a *conceptual design* (Gate 2), FOB equipment cost correlations based on flow rate are often used and scaled up to BM cost based on $L + M*$ factors. For a *preliminary engineering design* (Gate 3), the equipment is sized more accurately; based on simple rules of thumb, the cost is estimated from FOB cost correlations related to equipment size and scaled up to BM cost based on $L + M*$ factors. The process and FOB cost correlations are all of the form

$$\text{cost} = \text{cost}_{\text{reference}} \left(\frac{\text{size}}{\text{size}_{\text{reference}}} \right)^n$$

Tables of cost data are given for the reference cost corresponding to the reference size, for example, *flow rate, annual production, filter area*, the exponent n (e.g., usually $n = 0.6$ to 0.7), the range of the size variable over which the correlation is valid, and the error (e.g., ± 30%). Sometimes the costs for one piece of equipment will be correlated with a different exponent, n, for different size ranges. For example, for flow rates 10 to 100, $n = 0.4$. For flow rates 100 to 1,000, $n = 1.0$.

Conceptual Process Design, Process Improvement, and Troubleshooting 1311

That is, at flow rate 100, with $n = 1$, the cost doubles when the size doubles, and there no longer is economy of scale. Since the key size is where $n = 1$, polynomial curve-fit correlations of costs should be avoided, since the size corresponding to $n = 1$ is not obvious.

16.3.4.6.1 Process Cost Correlations

For this method, we need to know the names of the products, raw materials, process or the chemical reactions involved, and capacity. Treat with caution correlations developed from *construction boxscore* data because of ambiguity in process details.

An example correlation is BL ammonia (ex natural gas including reforming, CO converter, centrifugal compressors, methanation, low-temperature gas purification, and low-temperature/low-pressure synthesis); costs $290 million for 400,000 Mg/annum product; $n = 0.7$ for range of 300,000 to 750,000 Mg/a ± 40% at CE = 1000.

Example 16.4

Estimate the BL capital cost of a plant to produce 600,000 Mg/annum of ammonia from natural gas. The current CE = 390; we estimate inflation to be 4% per annum and determine the cost for one year from now.

Solution

The desired production rate is within the range of the correlation. For CE = 1000, the estimated BL cost would be

$$\text{cost} = \$290 \text{ million} \left(\frac{600{,}000 \text{ Mg/a}}{400{,}000 \text{ Mg/a}} \right)^{0.7} = \$385 \text{ million US} \pm 40\%$$

Correct to the current time by using the ratio of CE values:

$$= \$385 \text{ million} \times 390/1000 = \$150 \text{ million}$$

Adjusting for one year of inflation at 4%:

$$= \$150 \text{ million} \times 1.04$$

$$= \$156 \text{ million U.S.} \pm 40\% \text{ one year from now}$$

16.3.4.6.2 Equipment Cost Correlations Based on Flow Rate and BM Factor Method

We need to know the flowsheet, the mass and energy balances, and the names and materials of construction of the processing equipment selected.

An example correlation is as follows. The FOB cost of a rotating biological contactor, including rotating mechanism, motor, and drive but excluding the basin, is $900,000 at a design flow rate of 1000 m³/day with exponent $n = 0.56$ valid over the range 45 to 1100 m³/day ± 35% at CE = 1000. $L + M^* = 2$. $L/M = 0.35$.

Example 16.5

Estimate the cost at CE = 400 for an RBC for a design flow rate of 400 m³/day day installed in an existing concrete basin.

Solution

The correlation applies. The FOB cost for CE = 1000 is

$$\text{FOB cost} = \$900,000 \left(\frac{400}{1000}\right)^{0.56}$$

$$= \$539,000$$

$$\text{FOB cost at CE} = 400 = \$215,600$$

$$L + M^* \text{ cost} = \text{FOB} \times 2 = \$431,200$$

Accounting for instrumentation, from Table 16.15, estimate minimal controls needed as $8,000 at CE = 1000 or $8,000 × 400/1000 = $3,200. Hence, $L + M$ cost = $431,200 + 3,200 = $435,000 ± 35% at CE = 400.

16.3.4.6.3 FOB Equipment Cost Correlations Based on Design Size and BM Module Factor Method

For this method, the size, materials of construction, and conditions of operation for the process equipment must be known. The calculation is the same as that used in Example 16.3. The difference is an increase in accuracy for the estimate of the FOB cost because the correlation requires an estimation of *size* of the unit instead of the *flow to the unit*. For example, for a bag filter, we could use a cost correlation based on feed gas flow rate to the filter or, for increased accuracy in engineering design, use a correlation based on the bag filtration area. Since the gas flow rate/unit filter area depends on the specific application, the cost correlation based on filtration area is site specific and therefore more accurate. Cost correlations based on flow rate are usually less accurate (±25 to 40%) than those based on design size (±15 to 30%). Rules of thumb, given in Section 16.11, can be used to obtain the size.

Additional Reading

D. E. Garrett, *Chemical Engineering Economics,* New York: Van Nostrand Reinhold, 1989. Provides sound basic advice with data in appendices (general).

The American Association of Cost Engineers is an excellent organization that publishes data and holds annual conferences. http://www.aacei.org.

Richardson Engineering Services, *Process Plant Construction Estimating Standards,* vols. 1– 4. Annual publications give FOB and take-off costs for equipment, piping, concrete, mechanical and electrical. http://www.resi.net.

Woods, D. R. et al. have published a series of papers on estimating offsites:

1979. *Canadian Journal of Chemical Engineering* 57: 533–566 (utilities).

1982. *Canadian Journal of Chemical Engineering*, 60: 173–201 (offsite industrial gases).

1993, 1994. *Canadian Journal of Chemical Engineering*, 71: 575–590; 72: 342–351 (liquid disposal).

Page, J. S. *Conceptual Cost Estimating Manual*, 2nd ed. Houston: Gulf Publishing, 1996. Gives FOB cost and erection man-hours for about 200 types of process equipment plus take-offs for site preparation, concrete, structural steel, buildings, piping, electrical, instrumentation, insulation, painting, paving, overheads, and indirects.

Guthrie, K. M. *Process Estimating, Evaluation and Control.* Solana Beach, CA: Craftsman, 1974. The original classic that we still refer to often. A gold mine of data.

J. Cran. 1981. *Improved Factor Method gives Better Preliminary Cost Estimates. Chemical Engineering*, April, 65–79.

P. F. Navarrete. *Planning, Estimating and Control of Chemical Engineering Construction Projects.* New York: Marcel Dekker, 1995.

Conceptual Process Design, Process Improvement, and Troubleshooting 1313

16.4 BUILD A RELIABLE DATABASE

At the start of the project, and continuing throughout the project and for troubleshooting, we need a range of input data about the chemicals and reactions and about species used and produced in the process. Information may be supplied from chemists who have unraveled the mysteries of a synthesis or from a pilot plant or small-scale operations that provide key design data (as illustrated in step 4a in Table 16.1).

16.4.1 INPUT DATA FROM SCIENTISTS

Traditionally, chemists take an approach different from that of engineers, so information needed by engineers may not be collected. For example,

Chemists Tend to:	Engineers Tend to:
Ask "Why?"	Ask "How can I use it?"
Work on small scale.	Work on large scale.
Process the chemicals "once through."	Often use "recycle" with a buildup of trace materials.
Consider a linear, once-through sequence.	Consider system.
Operate batchwise.	Operate continuously.
Neglect cost considerations.	Consider costs critical.
Use glass as the material of construction.	Use carbon steel or 316 stainless.
Work at atmospheric conditions.	Use a range of conditions.
Neglect disposal problems.	Consider waste disposal as a major concern.
Work in fume hood.	Place major interest on environment impact.
Control the process by watching & adjusting.	Use computers and control schemes to automatically control.
Focus on the yield of target product.	Need to know system information for mass and energy balances

Without good communication, scientists may collect data that are incomplete for design purposes, or engineers may infer incorrect conditions.

Example 16.6
"I crystallized the product in the usual way." Translation: crystallization was from slightly supersaturated conditions in the metastable region to promote homogeneous nucleation to yield large, easy-to-filter crystals. Scratching the inside of the glass vessel was needed to start the crystallization. The crystals were removed by filtration. No record was made of the volume of filtrate. The crystals were dissolved in pure solvent and then recrystallized. This was done three times until the crystals appeared pale. Engineers might infer that crystallization occurred in the labile region by seeding and that high purity and high yield of crystals were obtained after one cycle. They might infer that the crystals obtained were pure, robust, and easy to filter.

16.4.2 MINIMIZE UNCERTAINTY IN INPUT DATA THROUGH EXPERIMENTATION: PRINCIPLES OF SCALE-UP

Some process information is unknown or uncertain. To minimize the uncertainty, in step 4a, Table 16.1, we can gather data on a smaller scale. For the information to be valuable and transferable, the small-scale experiment should be geometrically, mechanical, thermally, and chemically the same as the proposed large-scale operation.

Geometrical similarity. Two bodies are geometrically the same when to every point in one body there exists a corresponding point in the other. *For example, a laboratory Florence flask reactor should be changed to a reactor of the same shape as visualized in the plant.*

Mechanical similarity. Similar bodies deform geometrically similarly when subjected to constant stress or when fluid tracers exhibit geometrically similar patterns on both scales for static, kinematic, and dynamic conditions. To achieve this, the dimensionless numbers are kept the same for small- and large-scale operations. *For example, a mix in the lab should not be done with a magnetic stirrer, because this does match geometrically the large-scale operation from, say, a propeller agitator. For mechanical similarity, for some operations, operate at the same impeller Reynolds numbers for small- and large-scale equipment.*

For drops, bubbles, particles, or catalysts, we cannot scale down, because the area, volume, and surface/volume ratio change. For this situation, we often use the "element" strategy in which the small scale consists of a small slice of the real situation that contains the real size of the drops, bubbles, packing, or catalyst. Recurring sources of scale-up problems include (a) the inability to have similar equipment shape; (b) faulty accounting for the batch vs. continuous operation and the scale of operation; (c) changes in the surface/volume ratio that cannot be appropriately accounted for; (d) unexpected flow patterns; (e) flow instability and mixing characteristics, (f) wall, edge, and end effects; (g) materials of construction; (h) trace impurities; and (i) heat removal from gaseous, highly exothermic catalytic reactions.

Additional Reading

A. Bisio and R. L. Kabel, *Scaleup of Chemical Processes*. New York: John Wiley & Sons, 1985. Chapters from 17 experts. Six of the 18 chapters are on reactors.
H. F. Johnstone and M. Thring, *Pilot Plants, Models and Scale-Up Methods in Chemical Engineering*. New York: McGraw Hill, 1957. A classic on how and what to consider.
S. J. Kline, *Similitude and Approximation Theory*. New York: McGraw Hill, 1960. A fundamental view of creating dimensionless groups.

16.4.3 UNITS OF MEASUREMENT AND COMMUNICATION

My calculations for the first plant I designed were 10% too small. That is because I was working in *short* tons (2000 lb), whereas my supervisor in this UK-based company assumed I was using *long* tons (2240 lb). The definitions of such terms as ton, yard, inch, gallon, and barrel vary around the world. Fortunately, the SI (Le Systeme International d'Unites) eliminates most ambiguity, and the use of the time unit of seconds and length unit of meters allows simple determination of force and energy without additional factors. However, care is still needed. Some engineers and texts use metric units instead of SI and assume they are using SI. Examples include the archaic use of mm mercury, bar, torr, and kg/ cm^2 for pressure instead of Pa; the use of calorie instead of joule; and the use of minutes, hours, or days instead of the standard use of second.

Concerning powers, often the SI prefixes work well *provided* everyone knows the definitions. A common error is to use lowercase "m" or "MM" instead of capital "M" to represent 1000 kg. Care is needed to distinguish between gauge pressure and absolute pressure; it is recommended that all pressures (except for differences in pressure) be reported as kPa-g or kPa-a.

16.4.4 GATHER INPUT DATA FROM DATABASES FOR DESIGN AND PROCESS IMPROVEMENT

Data needed include expected variations in conditions; costs of materials, labor, equipment, and utilities; disposal limitations; sources; legal definitions and restrictions; environmental impact measures; and numerical values for the criteria. For all the species involved in the process, we need physical and thermodynamic data, and such reactivity and safety properties as flammability, corrosivity, abrasiveness, and propensity for dust explosions of solids, stability, environmental persistence and health indicators such as the LD_{50}, carcinogenicity, mutagenicity, and toxicity and those listed in Table 16.16.

TABLE 16.16
Definitions of Key Terms

ACGIH threshold limit value	American Congress of Governmental Industrial Hygienists threshold limit, see TLV.
Autoignition	Temperature above which a flammable mixture is capable of extracting enough energy from the environment to self-ignite. AIT. Benzene, 560°C; toluene, 480°C.
BLEVE physical explosion	Boiling liquid expanding vapor explosion is an explosion that occurs when a liquid stored above its boiling point is released into the atmosphere. Such a release immediately causes a shock wave; if the liquid is flammable, it can create a VCE. Example: A propane tank ruptures during a fire and causes a BLEVE followed by a VCE.
Carcinogenic	Risk of inducing cancer in humans or to increase its incidence.
Deflagration	An explosion caused by a very rapid chemical combustion in which the shock wave moves slower than the speed of sound. A deflagration proceeds by transport processes from the reaction front to the unreacted front. Contrast with *detonation*.
Detonation	An explosion caused by a very rapid chemical combustion in which the shock wave moves faster than the speed of sound. A detonation proceeds by a shock wave. Contrast with *deflagration*.
Dow rating	Hazard ratings are based primarily on flammability.
Dust explosion	Rapid combustion of fine solid particles. Upper and lower concentration limits of the dust explosion conditions are available.
Explosion	A rapid expansion of gas resulting in a rapidly moving pressure or shock wave. Three types of explosions are mechanical, thermal chemical, detonation, and deflagration. A mechanical explosion is caused by the pressure difference causing a vessel to rupture. A thermal chemical explosion has no reaction front separating unreacted from reacted materials: the rate of heat generated exceeds the rate of heat removal. In a chemical explosion, a reaction front separates reacted from unreacted materials. If the shock wave travels at less than the speed of sound, the explosion is referred to as a *deflagration*. If the shock wave speed is greater than the speed of sound, the explosion is called a *detonation*.
Fire point	Lowest temperature at which a vapor above a liquid will continue to burn once it is ignited; fire point > flash point.
Flammability limits	Upper and lower concentrations of combustible material in air that bound the region for combustion or explosion. Benzene in air at 298K and 101 kPa: 1.2 and 8.0% v/v respectively; toluene, 1.2 to 7.1% v/v, respectively. Increased temperature and/or pressure tends to lower the lower limit and raise the upper limit.
Flash point of a liquid	The temperature at which the vapor in equilibrium with the standard atmosphere above a pool of that liquid is at the lower flammability limit. Reported as open-cup or closed-cup. For materials that vaporize rapidly, the two values are about the same. For materials that vaporize slowly, the closed-cup temperatures are always less than open-cup data because of the conditions of the test. Benzene: −11.15°C; toluene: 4.0°C.
Hodge–Sterner toxicity classes	Rates 1 to 6 based on amount orally taken that causes death. Rating: 1, practically nontoxic, (15 g/kg) to 6, supertoxic (<5 mg/kg).
HON	Hazardous organic NESHAP (National Emissions Standards for Hazardous Air Pollutants). Rating of emissions and their impact on the environment. Subpart F denotes species as hazardous.
IDLH	Immediately dangerous to life and health concentrations are the maximum concentrations from which one could escape within 30 min without experiencing escape-impairing or irreversible health effects.
LD_{50} lethal dose	Administered orally or by skin absorption will cause death 50% of the time within 14 days. This can also be used as a crude index of biological activity or for an impact on the environment (Hushon et al., 1984, p. 95). Example: LD_{50} for 1-2 dichloroethane for rats for 30-min exposure is 12,000 ppm. For humans, the critical anesthesia limit is probably about 20,000 ppm.

Continued

TABLE 16.16 (Continued)
Definitions of Key Terms

LC_{50} lethal concentration	Administered via 4-hr inhalation will cause death 50% of the time within 14 days.
LEL	Lower explosive limit, the minimum concentration of combustible material that will support combustion.
Light liquid	One that contains a liquid compound to the extent >20% w/w that has a vapor pressure greater than 0.3 kPa at 20 C.
Light liquid petroleum	One that has >10% evaporate at 150°C.
MAK	Maximale Arbeitsplatzkonzentration or maximum workplace concentration.
Mond rating	Ratings are based primarily on flammability.
Mutagenic	Risk of hereditable genetic defects.
NFPA	National Fire Protection Association. System gives a qualitative rating for health, flammability, and spontaneity. These values range from 0 to 4, with 4 being the most hazardous. These can be used for an initial screen to sensitize engineers to the potential hazard. Quantitative values are published elsewhere for each separate issue: health (TLV, STEL, IDLH); flammability explosivity (upper and lower explosive limits).
OSHA	Rating of employee exposure and working conditions.
Overpressure	Pressure exerted on an objective from an impacting shock wave.
PDEP corrosion index	Process Design and Engineering Practice qualitative rating on the corrosivity of pure components contacting carbon steel, copper, and 316 stainless steel. These can be used for an initial screen to sensitize engineers to the potential hazard (Woods, 1994).
PEL	Permissible exposure limits; typical worker can be exposed to this time-weighted average concentration of the substance for 8 hr per day, five days per week for a working lifetime, without ill effects.
RCRA	Resource Conservation and Recovery Act (U.S.). Legislation dealing with licenses, fees, and accountability for the processing of hazardous wastes.
STEL	Short-term exposure limit, a 15-min time-weighted average concentration. Example: TLV-STEL for monochloroethane is 1250 ppm or 3250 mg/m^3.
Teratogenic	Risk of subsequent nonhereditable birth defects.
TWA	Time-weighted average.
TLV-STEL	Threshold limit values as measured by short-term exposure. A typical worker can be exposed to this time-weighted average concentration of the substance for 15 min without ill effects. (UK based on 10-min TWA), benzene, 50 ppm.
TLV-TWA	Threshold limit values as measured on a time-weighted average (sometimes referred to as the ACGIH permissible limit values). Typical worker can be exposed to this time-weighted average concentration of the substance for eight hours per day, five days per week for a working lifetime, without ill effects. Example: TLV-TWA for monochloroethane is 1000 ppm or 2600 mg/m^3; benzene, 10 ppm.
UFL UEL	Upper explosive limit; the maximum concentration of combustible material that will support combustion.
VCE	Vapor cloud explosion; occurs when a cloud of combustible material is within the flammable or explosive limits.
VOC	Volatile organic compound.

Data from each of the HON, OSHA, NFPA, Dow, and Mond hazard ratings should be gathered. Example 16.7 highlights the disadvantages of using only one type of hazard indicator and the importance of knowing the legal ratings.

Example 16.7

Acetophenone is on the HON and HON Section F list (defined in Table 13.16) as a hazardous chemical. Thus, legally we must monitor and minimize emissions from vents, storage vessels, and transfer racks, and discharges to wastewater, and minimize leaks from rotating shafts, sampling, and valves. On the other hand, the NFPA rating for *acetophenone* indicates only a modestly high flammability rating (Health 1, Fire 2, and Stability 0). Contrast this with *ethane*, which is not on the HON list and has a higher NFPA rating (defined in Table 13.16) of Health 1, Fire 4, and Stability 0.

16.4.4.1 Checklist for Reactivity

The key data to identify the reactivity of species include the following:

1. The *heat of decomposition*, ΔH_d, of the products in the absence of oxygen. The reaction is considered very hazardous, with potential for deflagration if $\Delta H_d > 0.2$ to 0.3 MJ/kg (AIChE, 1995), but the absolute value varies. King and CHETAH (King, 1990) cite $\Delta H_d > 1.25$ MJ/kg as being hazardous with $\Delta H_d > 2.9$ MJ/kg being likely to explode when subjected to mild heat or shock.
2. *Potentially hazardous reactants*. Spontaneous polymerizations with exothermic heat generation include styrene, substituted styrene, vinyl chloride, vinyl pyridine, acrylonitrile, butadience, isoprene cyclopentadience, and methyl isocyanate; reactions involving peroxides as illustrated in Table 16.17, azides, perchlorates, or nitro compounds; and decompositions, nitrations, oxidations, alkylations, aminations, combustions, condensations, diazotizations, halogenations, or hydrogenations.
3. The *exothermic heat of reaction* of the target and the by-product reactions gives an adiabatic temperature rise of >100 to 200°C (AIChE, 1995). Table 16.18 suggests some ranges for heats of reaction.
4. *Potential reactions between the process chemicals and the utilities*: water, air, coolants, lubricants, and oxidizers, for example. Table 16.19 lists materials that react strongly with water.
5. The *spontaneous stability* of the species: a high value of the NFPA value, i.e., those with an impact sensitivity of <60 J for solids and <10 J for liquids (AIChE, 1995).
6. The *explosive limits* for atmospheric and for reactor conditions. For example, for the oxidation of acetaldehyde to acetic acid, explosive conditions occur outside the operating window. However, for some processes during startup, the reactor passes through the explosive conditions to reach the operating window.
7. The *rate of temperature rise*, dT/dt, is a useful measured value for active control.
8. *Explosive conditions for dusts*, especially if the particles are <200 μm.

16.4.4.2 Checklist for Chemical Persistence

Table 16.20 gives example data for chemicals and their persistence in the environment. If such data are not available, such as from the database of Mackay and colleagues (1992 to 1997) or from Allen and Shonnard (2002), then we need to estimate the "ecological half-life" by estimating (1) the natural distribution in the environment and (2) the bioassimilation in the dominant compartment.

> Step 1. *Natural distribution of a species* among the compartments (air, water, soil, and biota) can be estimated from its solubility in water, vapor pressure, and octanol-water partition coefficient. Use Figure 16.4 for estimation purposes. Species with a high ratio of vapor pressure/solubility and a low ratio of K_{OW}/solubility are likely to reside in the air. Species with lower values of the vapor pressure/solubility and higher values of the ratio of K_{OW}/solubility are likely to be in the sediments and concentrate in the biota. In Figure 16.4, the numeral in brackets is the ppm concentration in biota; a high value implies that the species would move up the food chain. The octanol-water partition coefficient, K_{OW}, suggests the ease with which a species is absorbed across a cell membrane and thus is a measure of the potential uptake of the target chemical up the food chain.

TABLE 16.17
Bonds That Are Potentially Hazardous

Type		Examples	
Vinyl	–CH=CH2	Styrene, vinyl chloride, ethyl acrylate	Exothermic polymerizations
	>C=CH–CC–	Vinyl acetylene	Contains peroxidisable hydrogen with potential for explosion
Congugated double bonds with carbon, nitrogen, and oxygen atoms	–CH=CH–CH=CH–	Butadience, isoprene, chloroprene, cyclopentadience	Exothermic polymerizations
	–CH=CH–CH=O	Acrolein, crotonaldehyde	Exothermic polymerizations
	–CH=CH–C=N	Acrylonitrile	Exothermic polymerizations
Adjacent double bonds	–CH=C=O	Ketene	Exothermic polymerizations
	–N=C=O	Methyl isocyanate, toluene diisocyanate	Exothermic polymerizations
Three-membered rings	O / \\ –CH–CH2	Ethylene and propylene oxide, epichlorhydrin	Exothermic polymerizations
	NH / \\ –CH–CH2	Ethylene imine	Exothermic polymerizations
Aldehydes	–CH=O	Acetaldehyde, buyraldehyde	Exothermic polymerizations: contains peroxidisable hydrogen with potential for explosion
	>CH–O–	Acetals, ethers, oxygen heterocycles	Contains peroxidisable hydrogen with potential for explosion
Isopropyl compounds	–CH2 >CH– –CH2	Isopropyl compounds, decahydronaphthalenes	Contains peroxidisable hydrogen with potential for explosion
Allyl compounds	>C=C–CH2–	Allyl componds	Contains peroxidisable hydrogen with potential for explosion
Haloalkenes	>C=CH–X		Contains peroxidisable hydrogen with potential for explosion
Dienes	>C=CH–C=CH<		Contains peroxidisable hydrogen with potential for explosion

TABLE 16.18
Heats of Reaction

Classification	Heat per Mass of Total Reactants Solvents and Diluents Fed to the Reactor, MJ/kg [King, 1990, 174]	Heat per Mol of Target Reactant, MJ/kmol	Examples
Extremely exothermic	>3	>150	Direct oxidation of hydrocarbons with air, chlorination, polymerization of ethylene without diluent
Strongly exothermic	>1.2 but < 3	100 to 130	Nitrations, polymerizations of propylene, styrene and butadiene
Moderately exothermic	> 0.6 but < 1.2	30 to 90	Condensation or polymerization reactions of species of molar mass 60 to 200
Mildly exothermic	>0.2 but < 0.6		
Thermally neutral	> −0.2 but < 0.2		Esterifications
Endothermic	< −0.2		Cracking

TABLE 16.19
Materials That React Aggressively with Water (Based on King 1990)

Solid Species	Gas, Liquid, Solid	Reacts with Water	To Produce
Acetic anhydride (liquid)	L	Moderately	steam
Acetyl chloride	L	Vigorously	steam and acid fumes
Alumina, activated	S	Moderately	steam
Aluminum alkyls	L	Vigorously	alkanes
Aluminum alkyl halides	L	Vigorously	alkanes
Aluminum chloride	S	Vigorously	steam and acid fumes
Aluminum phosphide	S	Moderately	phosphine
Calcium	S	Moderately	hydrogen
Calcium carbide	S	Moderately	acetylene
Calcium hydride	S	Vigorously	hydrogen
Calcium oxide	S	Vigorously	steam
Calcium phosphide	S	Moderately	phosphine
Ethyl aluminum dichloride	L	Vigorously	alkanes
Fluorine	G	Vigorously	oxygen and ozone
Lithium	S	Moderately	hydrogen
Lithium hydride	S	Vigorously	hydrogen
Molecular sieves, activated	S	Moderately	steam
Phosphorus pentoxide	S	Vigorously	steam and acid fumes
Phosphorus trichloride	L	Vigorously	steam and acid fumes
Potassium	S	Explodes	hydrogen
Potassium hydroxide	S	Moderately	steam
Silica, activated	S	Moderately	steam
Silicon tetrachloride	L	Vigorously	steam and acid fumes
Sodium	S	Vigorously	hydrogen
Sodium hydroxide	S	Moderately	steam
Sodium peroxide	S	Moderately	oxygen
Sulfur trioxide	S	Vigorously	steam and acid fumes
Sulfuric acid	L	Vigorously	steam and acid fumes
Thionyl chloride	L	Vigorously	steam and acid fumes
Titanium tetrachloride	L	Vigorously	steam and acid fumes
Zinc alkyls	L	Vigorously	alkanes

TABLE 16.20
Fate of Chemical Species: Distribution among Compartments and Degradation Rate within Each Compartment (Mackay and Paterson, 1991)

Synthetic Chemical Species	Illustrative Amount in the Compartment or Distribution among Compartments	Rate of Assimilation within a Compartment		Net Overall Rate of Assimilation in the Ecosphere
		Degradation Rate Constant, h^{-1}	Comments	
Benzene	air [2500 ng/m^3]	8.6×10^{-4}	partitions mainly to the air	about 50 to 200 h depending on the atmospheric temperature
	soil [0 01 µg/kg]	0		
	water [290 ng/L]	0.0048		
	biota [0.002 mg/kg]	0		
	suspended solids sediment [0.5 ng/g]	0		
Benzopyrene	air [3.0 ng/m^3]	0		
	soil [0.04 mg/kg]	3.5×10^{-5}		
	water [11 ng/L]	3.5×10^{-5}		
	biota [0.26 mg/kg]	0		
	suspended solids sediment [190 ng/g]	0		
Hexachlorobiphenol	air [1.4 ng/m^3]	0.		> 8 years
	soil [0.04 mg/kg]	1.5×10^{-5}		
	water [5 ng/L]	1.5×10^{-5}		
	biota [0.2 mg/kg]	1.0×10^{-6}		
	suspended solids sediment [80 ng/g]	1.5×10^{-5}		
Hexachlorobenzene	air [0.2 ng/m^3]	0.0144		> 10 years
	soil [0.0009 mg/kg]	1.9×10^{-5}		
	water [1.3 ng/L]	0		
	biota [0.01 mg/kg]	0	high bioconcentration potential	
	suspended solids sediment [7 ng/g]	0		

Example 16.8

Mackay and Patterson (1981) estimated that 79% of DDT would end up in sediments, 98% of benzene in the air, and 61% of p-cresol in water. Hence, the biodegradation of most interest would be for DDT in sediments, for benzene in air, and for p-cresol in water.

Step 2. *Estimate* the *bioassimilation* of the species in the dominant compartment.

Additional Reading

D. T. Allen and D. R. Shonnard, *Green Engineering: Environmentally Conscious Design of Chemical Processes.* Englewood Cliffs, NJ: Prentice Hall, 2002. http://www.epa.gov/oppt/greenengineering.

D. R. Woods, *Data for Process Design and Engineering Practice.* Englewood Cliffs, NJ: Prentice Hall, 1994. Gives order of magnitude values of properties for more than 1200 compounds and a guide to sources for more accurate data. This includes the PDEP Corrosion Index referred to in Table 13.16.

R. C. Reid, J. M Prausnitz, and B. E. Poling, *The Properties of Gases and Liquids,* 4th ed. New York: McGraw Hill, 1987. Gives methods for estimating properties.

R. King, *Safety in the Process Industries.* Oxford, UK: Butterworth-Heinemann, 1990.

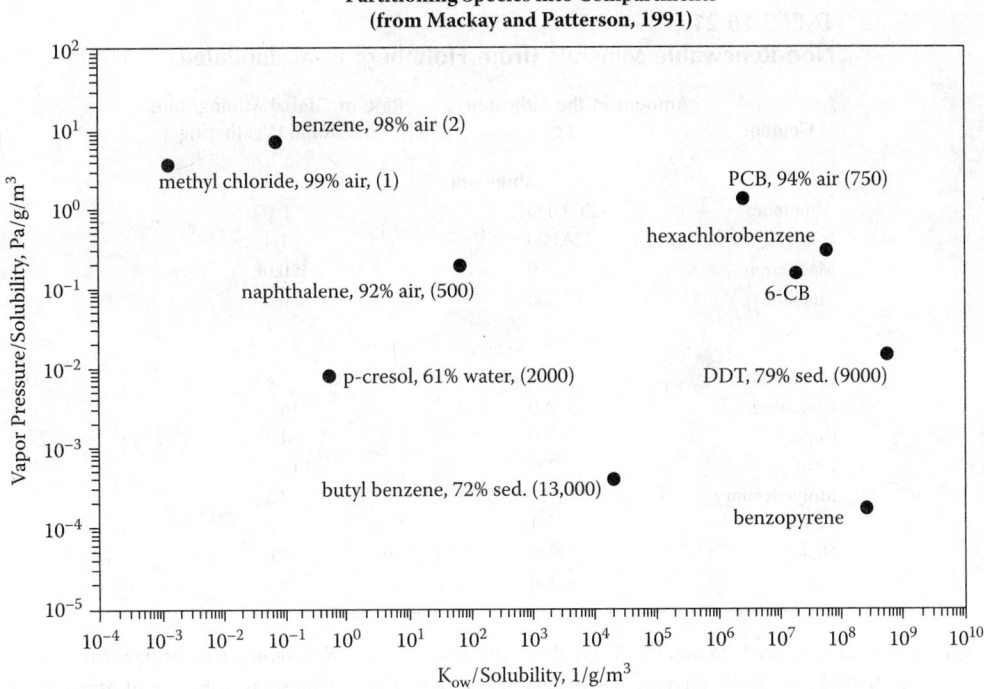

FIGURE 16.4 Partitioning species into compartments. (From Mackay and Patterson, 1991.)

AIChE/CCPS, *Guidelines for Chemical Reactivity Evaluation and Application to Design.* New York: Center for Chemical Process Safety, AIChE, 1995.

D. Mackay, W. Y. Shiu, and K-C. Ma, *Illustrated Handbook of Physical, Chemical Properties and Environmental Fate for Organic Chemicals.* Boca Raton, FL: CRC Press, 1992–1997.

16.4.5 GATHER DATA FOR TROUBLESHOOTING

The general principles are as follows:

1. Believe the facts and be cautious about the inferences. For example, the pressure gauge reads 230 kPa-g is a fact; an inference is that the pressure is 230 kPa + 110 = 340 kPa.
2. Be sensitive to your own style in collecting data and interpreting data. Those dominant J personalities in Jungian typology (MBTI) often prefer to act rather than collect data to determine a cause, whereas dominant P personalities may prefer to gather extensive data before deciding on a corrective action.
3. Beware of "pseudo diagnosticity," a fixation on one cause and a refusal to change even when data prove the cause to be faulty.
4. Prioritize: consider safety first, then short-term choices, and finally long-term options. An option might be to live with the "trouble."
5. If pertinent, focus on gathering data about *changes* that occurred.
6. Test both positive and negative parts of the hypothesis.
7. Use the general problem-solving guidelines from Section 16.2.

Additional Reading

C. H. Kepner and B. B. Tregoe, *The New Rational Manager.* Princeton, NJ: Kepner Tregoe Institute, 1981. Offers strategies for troubleshooting when a *change* has caused the trouble.

TABLE 16.21
Non-Renewable Minerals (from Holmberg et al., undated)

Element	Amount in the Lithosphere, Eg	Rate of Global Mining/Rate of Global Weathering
Abundant		
Aluminum	1,200,000	0.02
Iron	720,000	1.4
Magnesium	15,000	< 0.01
Titanium	82,000	0.06
Scarce		
Cadmium	8.3	3.8
Chromium	1,200	16
Copper	760	24
Lead	200	12
Molybdenum	17	7.5
Mercury	1.2	3.8
Nickel	920	3.0
Zinc	1,200	8.1

A. S. Elstein et al., *Medical Problem Solving: An Analysis of Clinical Reasoning.* Cambridge, MA: Harvard University Press, 1978. Gives a superb description of how to troubleshoot when we must *hypothesize* about the cause. Written in a medical context, but most of the ideas apply directly.

16.5 SUSTAINABILITY FROM THE START

Robèrt's (1992) program, The Natural Step (TNS), is preferred to such options as life cycle, ecological footprinting, and green engineering. The four principles of TNS are as follows:

1. Substances extracted from the core, mantle, and crust of the Earth (such as oil and fossil fuels, metals, and other minerals) must not systematically accumulate in the ecosphere. The rate of mining from the Earth's crust must not be at a pace faster than the extracted species can be redeposited and reintegrated into the crust. The rate of extraction can be reduced by recycling or by substitution. Table 16.21 lists minerals and compounds present in the lithosphere that are deemed to be economically extractable based on today's best practice. Also shown is the ratio of annual rate of mining to the rate of natural weathering. All abundant minerals (except iron) have values < 1. However, all scarce metals have values > 1. For the sources of energy given in Table 16.22, shift toward renewable resources.
2. Substances produced by society must not systematically increase in the ecosphere. Yet some synthetic substances are produced at rates faster than they can be broken down and integrated into the cycles of nature.
3. The physical conditions for productivity and assimilation within the ecosphere cannot be systematically diminished. Forests, wetlands, prime agricultural land, and natural plants and animals cannot be systematically diminished. The assimilative capacity of the ecosystem must exceed any rates of pollution so that nature can regenerate itself. The rates of use of renewable resources do not exceed their rates of regeneration. Table 16.23 illustrates the net rate of natural regeneration.
4. Since we have limited resources and a human population with needs, the basic human needs must be met with the most resource-efficient methods possible. The satisfaction

TABLE 16.22
Energy Sources (Harte, 1988)

Source of Energy	Estimated Reserves, EJ	Energy Usage, EJ/annum
Renewable		
Solar	—	
Wind	—	
Wood and dung combustion	—	30
Water	—	6.1
Nonrenewable		
Petroleum [43 GJ/kg]	10,000	135
Natural gas [50 MJ/kg]	10,000	60
Coal [29 GJ/kg]	250,000	90
Tar sands	>2,000	0
Oil shale	2,000,000	0
Uranium [83 TJ/kg]	20,000	6.3
Thorium & uranium	10,000,000	0
Deuterium and lithium in seawater [250 TJ/kg]	10^{13}	0

TABLE 16.23
Some Example Biosystems and Their Net Rates of Productivity (Harte, 1988)

	Area, 10^{12} m^2	Mean Plant Biomass, kg (Carbon)/m^2	Net Primary Productivity, kg (Carbon)/m^2/annum
Forests			
Tropical	24.5	18.8	0.83
Temperate	12	14.6	0.56
Boreal	12	9.0	0.36
Woodland and scrubland	8.0	2.7	0.27
Savanna	15.0	1.8	0.32
Grassland	9.0	0.7	0.23
Swamp and marsh	2.0	6.8	1.13
Cultivated land	14.0	0.5	0.29

of the basic needs must take precedence over the production of luxuries. Furthermore, the focus should be on increasing efficiency rather than on increasing quantity.

16.5.1 Impact of Sustainability on Design

TNS principle 1 applies to Stage II in Table 16.2 and suggests that using renewable resources as the feedstocks (e.g., fuel derived from the fermentation of corn, *gasahol*) shifts the raw material from nonrenewable to a renewable resource. TNS principles 1, 2, and 4 are satisfied if the hydrocarbon raw materials are obtained from fermentation instead of from fossil fuels. TNS principles 1 and 4 suggest that aluminum be used instead of copper in electrical applications. TNS principle 4 suggests the use of electric motors instead of internal combustion engines.

16.5.2 Impact of Sustainability on Process Improvement

As engineers debottleneck, upgrade, and improve processes, The Natural Step criteria provide a convenient, and sometimes unsettling, set of guidelines.

Additional Reading

K.-H. Robèrt, *The Necessary Step.* Stockholm: Ekerlids forlag. This is the original publication. However, sites on the Web provide more accessible data and examples. Search for "The Natural Step." Some example sites are http://www.globalideasbank.org/BOV/BV-273.html, http://www.naturalstep.org, and http://iisdl.iisd.ca/business/naturalorg.htm.

D. T. Allen and K. S. Rosselot, *Pollution Prevention for Chemical Processes.* New York: John Wiley & Sons, 1997. The authors do not use The Natural Step, but they provide excellent insight into sustainability.

D. T. Allen and D. R. Shonnard, *Green Engineering: Environmentally-Conscious Design of Chemical Processes.* Englewood Cliffs, NJ: Prentice Hall, 2001.

16.6 OPERABILITY AND CONTROL CONSIDERED THROUGHOUT

Chemical processes must be controlled to make them operable, profitable, safe, and environmentally friendly. Issues of control affect process design, process improvement, and troubleshooting.

16.6.1 CONTROL AND DESIGN

Control is the heart of process design; how to control the plant effectively should be in the designer's mind right from the very beginning of the design process.

16.6.1.1 Design: Impact of Control on the Initial Design Statement

The initial design statement usually identifies the target product, amount, and purity. However, the initial design goal should also include the base case conditions, the expected setpoint changes, and the likely process disturbances.

16.6.1.1.1 Range of Setpoints

The range in production modes is determined by *setpoint*:

- Production rate (processes typically operate at 80% design capacity).
- Product specifications and the distribution among products on multiproduct processes.
- The range in product purity. The product purity has a major impact on both capital investment costs and operating costs. For example, for distillation, increasing the purity specifications by 1% may increase the energy requirement by about 10%. For sulfur removal from waste gases, for the same volume of gas treated, the fixed capital investment triples when the sulfur removal is changed from 90 to 99%.
- The different grades of a product customers want and may want in the future, for example, the grades of acid.

16.6.1.1.2 Range of Anticipated Process Disturbances

Anticipated disturbances include the range in feedstock composition, the range in expected utility properties (summer and winter temperatures of the cooling water and of the air), the slow drift in reaction kinetics, the gradual deactivation of the catalyst, and the gradual fouling of heat exchangers. Downs and Vogel (1993) provide a good example of a well-defined problem with the setpoint specifications in Table 7 and the disturbances in Table 8 in their paper.

16.6.1.2 Design: Impact of Control on the Evolving Flowsheet

Major issues include surge volume, recycle, and the control system:

> *Surge volume.* As the flowsheet design progresses, sufficient (but not excessive) surge volume must be provided. The trade-off is between a steady-state and a safety viewpoint, in which the surge volumes are eliminated to minimize the capital cost and the volumes of hazardous

Conceptual Process Design, Process Improvement, and Troubleshooting

material stored in intermediate surge vessels, and the control viewpoint, in which surge volumes provide sufficient liquid inventory to reduce the effects of changes in inlet stream flow rate and properties.

Recycle. Self-regulation must be provided for components and energy in recycle loops. Recycling of material can recycle disturbances if the control is, for example, based on level control. Disturbances in the heat balance can be recycled if extensive heat integration is used. Add trim heaters and coolers to damp out thermal disturbances.

Control system. For subsequent selection and sizing of pumps and compressors, we need to map out the number and location of the control valves. Since the number of control valves is related to the number *of control degrees of freedom*, identify the *control degrees of freedom*. For example, a typical hydrodealkyllation process with a reactor, furnace, vapor-liquid separator, recycle compressor, two heat exchangers, and three distillation columns has 23 *control degrees of freedom* (Luyben et al., 1997). This requires 23 control valves whose location affects the rest of the design and the safety and hazards (see Section 16.7).

Provide fast feedback dynamics for control: include an adjustable bypass around slow processes, inject cooling material to quench fluid streams, adjust flows that more directly affect the key controlled variable rather than flows that affect the process environment slowly. Select fast manipulated variables, even when costly, because they can provide a "trim" to slower, less costly manipulated variables.

16.6.1.3 Design: Impact of Control on the Type of Equipment Selected

Identify the process operating equipment constraints, conditions for reliable operation, and shutdown limits. For example, reactor operating temperatures should be 25°C less than the maximum temperature of the catalyst.

16.6.1.4 Design: Impact of Control on the Sizing of Equipment

For transportation equipment: Controllability affects the sizing of pumps and compressors. In determining the system resistance, the appropriate anticipated variation in flow and pressure (from both setpoints and disturbances) and the location and pressure drop across the control valves must be included. Ensure enough pressure drop across the control valves to provide good control. Although the rules of thumb vary, the general guideline is to allow 50% of the system frictional resistance or 140 kPa. The available net positive suction head (NPSH) used in the design affects the selection of the pumps for liquids.

For energy exchange equipment: Supply sufficient excess of heat transfer area in reboilers, condensers, cooling jackets, and heat removal systems for reactors to be able to handle the anticipated upsets and dynamic changes. Sometimes extra area is needed in overhead condensers to subcool the condensate to prevent flashing in the downstream control valves. Too frequently, overzealous engineers size the "optimum heat exchangers" based on an economic minimum based on steady-state conditions and produce uncontrollable systems.

For distillation columns: Add sufficient trays to account for disturbances and anticipated expansion of production. The expected turndown ratios affect the choice of internals in a distillation column with a large turndown ratio, suggesting the use of bubble caps; a low turndown ratio may point to structured packings. At the same time, safety and hazard analysis indicates that we want a minimum of liquid holdup in the system.

16.6.1.5 Design: Impact of Control on Issues beyond Conceptual Design

As the process design moves through the gating process of Table 16.1, attention to the details of piping layout is crucial for reliable control. Minimize process delays because of piping and large time constants, minimize sensor dynamics (especially sample systems for analyzers), and provide the required upstream and downstream lengths of straight pipe for flow meter and sensor reliability. Prevent maldistribution of flow in parallel systems. The selection and location of steam traps and air vents are important.

Consider using statistical process control, reexamine the types and locations of sensors, and use the four basic levels of control: (1) basic process control system, (2) alarm system, (3) safety interlock system, and (4) relief system. Consider active/passive controls. Active controls include sensors to detect hazardous or unsafe conditions; for example, a flame sensor that is used to shut down the fuel supply if no pilot flame is present and sensors for trace impurities that trigger deflagration. Passive controls (or self-regulating control systems) include fail-open and fail-shut valves, safety relief valves, and bursting disks to release pressure; large solvent or heat transfer fluid heat sinks to absorb heats of reaction should the cooling system fail; and the addition of a cold shot or reaction inhibitor. Prevent overheating by limiting the steam pressure; use tempered water instead of steam to ensure that the heating medium cannot overheat sensitive reactants. Include process conditions that have a negative temperature coefficient on reaction conditions. Self-limiting reactor configurations could be used. Marlin (2000) provides details.

16.6.2 Control and Process Improvement

To improve processes and remove bottlenecks, one of the first areas of focus is control. For batch processes, consider installing surge volumes to smooth out variations and make overall batch operations more efficient. Add the four levels of control.

The location and type of sensor can greatly affect controllability, safety, and on-time reliability. Controllability can also be improved when a different type of sensor is used. For example, a temperature sensor might be used to infer composition. Such a sensor is easy, robust, and accurate in measuring the temperature. However, the new specifications for the purity have shifted the composition to a range where temperature measurements are no longer sensitive to the composition. Refractive index measurements should be used.

Control can often be improved by using statistical process control. Montgomery and Runder (1994, Chapter 14) and Kourti and MacGregor (1996) describe this approach.

16.6.3 Control and Troubleshooting

When trouble occurs, one of the first steps is to put the system on manual control and see if the trouble disappears. Other suggestions are given in Section 16.11.1.1 on rules of thumb for process control.

Additional Reading

T. Marlin, *Process Control*, 2nd ed, New York: McGraw Hill, 2000. http://chemeng.mcmaster.ca.
J. J. Downs and E. F. Vogel. ,1993. A Plant-Wide Industrial Process Control Problem. *Computer Chem. Eng.* 17, 3: 245–255.
Luyben, M. L. et al., 1997. Plantwide Control Design Procedure. *AIChE Journal* 43, 3161–3174.
Luyben, W. L., *Process Modeling, Simulation and Control for Chemical Engineers*. New York: McGraw Hill, 1989.
Montgomery, D. C. and G. C. Runger, Statistical Process Control, chap. 14 in *Applied Statistics and Probability for Engineers*. New York: John Wiley & Sons, 1994.
Kourti, T., and J. F. MacGregor, 1996. Multivariate Statistical Process Control methods for Monitoring, Diagnosing Process and Product Performance. *J. Qual. Tech.* 28: 409–428.

16.7 CONSIDER SAFETY, WASTE MINIMIZATION, AND ENVIRONMENTAL SENSITIVITY THROUGHOUT

Safety, waste minimization, environmental concerns, and health hazards should be considered together because of their commonality in impact and in the database, and because if we solve one problem, say in waste reduction, we often solve problems in the related areas such as safety and environment. Minimizing waste, environmental impact, and unsafe conditions are not always consistent with the minimum cost. The criteria given in Tables 16.1, 16.2, and 16.4 do not always agree.

16.7.1 STEP 1: TARGET OPPORTUNITIES

Know the legal and economic contexts and identify the opportunity.

16.7.1.1 Know the Legal and Economic Contexts

Consider each in turn.

16.7.1.1.1 The Legal Context

Engineering decisions must comply with laws related to hazards, the environment, wastes, recycling, and occupational health and safety. Federal, state, and community agencies regulate source reduction, waste reduction, pollution prevention, and hazardous materials. The legal definitions may be counterintuitive to traditional engineering ideas. To illustrate, the legal definition of "in-process recycling" may be *recycling unreacted feedstock to the reactor* whereas, from an engineering point of view, we might incorrectly think it to be immaterial whether *unreacted feedstock is recycled to a reactor in this process or in another*. The legal definition affects whether reporting is required.

In 1996, for example, a Pollution Prevent (P2) plan was required if >1 Mg/month of hazardous waste was generated. P2 plans were also needed for storage volumes that exceed legally defined volumes if regulated substances were used in quantities exceeding the thresholds, and if toxic pollutant discharges required NPDES permits.

16.7.1.1.2 The Economic Context

The economic context is that the use of designated species will usually add operating and capital expenses. Typically, annual overhead costs for processes using "hazardous species" can range from $400 to $200,000 per annum (2000) for notification, documentation, reporting, manifesting, and labeling required by law under the Resource Conservation and Recovery Act (RCRA). Annual general liability and insurance cost for wastes could be $600 to $1000 (2000) per Mg of waste. For some facilities, an RCRA permit may be needed, adding an additional cost to be included in item 15 for the capital cost estimation (Section 16.3.4). The initial permit fee could range up to $3 million (2000), with most of the cost resulting from public meetings, consultants, and expert witnesses.

16.7.1.2 Identify the Opportunity

Identify hazardous chemicals from databases as discussed in Section 16.4.4 and from local legal regulations. Identify the potential for hazards from the checklist for reactivity (in Section 16.4.4) and from HAZOP-type studies or fault-tree analysis (Crowl and Louvar, 1990). In HAZOP studies, use checklist key words as triggers to systematically analyze the impact of changes to *flow rate, temperature, pressure, composition, level, viscosity, heat transfer, reaction,* and *conditions* and the potential for *barrier failure* and *startup* and *shutdown* to cause hazards.

Identify the amount of wastes from mass and energy balances.

Use checklists, such as given in Table 16.24, to identify potential waste reduction and safety opportunities. Identify opportunities by challenging the nominal specifications.

16.7.2 Steps 2 through 5: Move from "Eliminate" to "Isolate" during the Design Process

Guidelines for inherent safety, environmental protection, or waste minimization can be summarized in the following five steps that move from elimination to isolation.

16.7.2.1 Step 2: Eliminate the Source

Eliminate the source at the reactor and elsewhere in the process.

16.7.2.1.1 At the Reactor: Eliminate, Substitute, and Recycle

Eliminate
Eliminate reactants that pose problems with sustainability, the environment, and the hazard.

Substitute
Modify the reaction pathway, the operating conditions, the heat management system, and the reactor configuration.

Change the reaction pathway. Select reaction pathways between the raw material and the desired product to eliminate the use of hazardous intermediaries.

Change the operating conditions. Alter the reaction conditions to eliminate the hazards or production of toxic waste. Waste by-products from the reactor can be eliminated if the reaction to produce the by-products is reversible (as for biphenyl production in the hydrodealkyllation process).

Change heat management systems. Substitute sources of thermal energy and provide active and passive control. For example, for the catalytic liquid-phase oxidation of a substituted acetophenone to produce carboxylic acid operating at 80°C, the reaction is very exothermic (–920 MJ/kmol acetophenone) and poses great potential for temperature runaway. Temperature runaway can be prevented by the use of a catalyst that deactivates at 100°C.

Change the reactor configuration to minimize hazard.

Recycle
Recycle by-products until the rate of production of the by-product equals its rate of consumption.

16.7.2.1.2 Elsewhere in the Process: Eliminate, Substitute, and Recycle.

Eliminate
Additives. Eliminate catalysts, solvents, and adsorbents. For example, use thermal hydrodealkylation instead of catalyzed hydrodealkylation.

Trace contamination. Eliminate by locating air and water intakes in pollution-free locations. For example, instrument air compressors for a plant site were incorrectly located beside ammonia refrigeration units. Throughout the plant, the ammonia reacted with the mercury to produce azides in the instruments.

VOCs. Eliminate by eliminating rotating shafts that must be sealed or by selecting a sealing mechanism that produces no emissions. Use welded pipe instead of flanges. Table 16.25 provides data for options.

The source of electrical hazards. Eliminate electric sparks and charge buildup; replace electrical with pneumatic instrumentation, replace electric-driven with steam-driven devices. Eliminate charge buildup by increasing conductance, e.g., increasing the conductivity of the atmosphere (via ionization) and increasing the conductivity of nonconductors (by humidification or additives).

TABLE 16.24
Potential Sources of Wastes and Safety Hazards

Wastes associated with the reactor (approximately 5 to 50 kg/Mg product)
Unreacted feedstock
Impurities in the feedstocks
Undesirable by-products
Startup and shutdown losses
Gas purges
Reactor washings
Catalyst usage and losses
Off-spec material from upsets; power failures; freezing

Safety hazards associated with the reactor (main source of hazard)
Uncontrolled temperature excursion because of insufficient heat removal and/or loss of agitation (insufficient area, fouled surface, loss of solvent)
Catalyst hot spot because of flow maldistribution or instabilities
Accumulation of reactants
Hazardous material
Induced unexpected reactions because of contamination, heat, friction, or electrostatic
Plugging and temperature/pressure buildup
Explosive limits
Incorrect sequencing of feedstocks
Mischarging
Corroded relief valves
Corrosion and vessel failure

Wastes because of type of reactor operation
Cleanout of lines between batches

Safety hazards because of type of reactor operation
Incorrect switching sequence
Hazardous residual left
Single reactor for multistage reactions
No tests for impurities
No lockouts or interlocks

Wastes caused by operating conditions or because of equipment selected
Tar bottoms from columns
Filtrate washing; filter aids
Interstitial purging of any packed system between cycles
Coagulants
Adsorbents, IX resins

Safety hazards because of conditions or equipment
Hazardous residual left from previous cycle in batchwise adsorption, IX
Flammable materials at high pressure and high temperature
Flammable liquids above their atmospheric boiling temperature because of vacuum/pressure or refrigeration
Operation near explosive limits (for dust or fluid systems)
Processing spontaneously polymerizable materials or unstable materials.
Large inventory (above critical value for each material)
Unrecognized byproducts that interfere with operation: silica carry over from reactor to foul downstream heat exchanger; elemental sulfur buildup in crude oil distillation

Safety hazards because of solvents
Flammability and toxicity
Trace cross contamination with water

Waste spent solvents (approximately 20 to 150 kg/Mg product)
Solvent losses and purges of recycled
Cross contamination with water

Trace surfactant buildup causing stable emulsion and decanter flooding/spillover
Static charge buildup (especially with hydrocarbon/water flow)

Continued

TABLE 16.24 *(Continued)*
Potential Sources of Wastes and Safety Hazards

Waste from general operation and maintenance	Safety hazards from general operation and maintenance
Oils and lubricants	Poor housekeeping,
Equipment cleaning	Electrical lines and water
Sampling and analytical losses	Maintenance done on wrong equipment
Purges of internal environments at startup and shutdown	Inadequate isolation of equipment; venting of atmosphere; removal of toxics
	Insufficient room around equipment
Safety hazards from packaging and storage	**Wastes from packaging and storage**
Storage mass larger than safe maximum	Packing, use of non-reusable drums
	Failure to use disposable liners
	Damage during storage
	Storage vessel residue
	Fugitive via vents
Safety hazards because of fugitive emissions	**Wastes from fugitive emissions**
	Seals around rotating shafts: valves, pumps, compressors, mixers
	Breathing from storage vessels
Safety hazard because of accidental spills	**Wastes from accidents**
	Incidental spills
	Wastes from metal sludges (approximately 0 to 150 kg/Mg product)
	Wastes from slags and ashes (approximately 10 to 50 kg/Mg product)
	Non-chemical industrial wastes (approximately 10 to 80 kg/Mg product)

Substitute

Mechanical gauges for mercury gauges. Nonhazardous additives: solvents, catalysts, filter aids, adsorbents, absorbents. Nonhazardous solvents can be identified by matching the Hildebrand hydrogen bond and polar solubilities parameters and the molar volumes. Example substitutions are water > nonpolar organic > aliphatics with flashpoint < 60°C > aliphatics with flashpoint > 60°C > aromatics > halohydrocarbons > polar organics > alcohols > organic acids > other oxygenates > nitrogen-bearing compounds > halogenated organics > chlorinated organics.

Recycle

Water. Caution: account for the buildup of trace amounts of other species; salts may precipitate if the concentration of salts becomes too high. Increased corrosion can occur because the recycle causes the buildup of unexpected species that enhance corrosion. Trace surfactants can build up and stabilize foams, causing foaming in distillation and stable emulsions in decanters. Although the buildup of trace species can be controlled by purging or by separation, the location of the purge stream can affect the overall performance. Purging minimizes but does not eliminate waste.

16.7.2.2 Step 3: Minimize the Source

If one cannot eliminate, one can at least minimize.

TABLE 16.25
VOC Release from Various Equipment Options

Source:	Equipment Option	Rate, mg/s [King, 1990]	Rate, mg/s [Allen and Rosselot, 1977]
Pump shaft seals	Regular packing with external lube sealant	140	Industry average for light liquid: 5–30. Industry average for heavy liquid: 2–6
	Regular packing with lantern ring oil-injection	14	
	Grafoil packing	14	
	Single mechanical seal	1.7	
	Double mechanical seal	0.006	
	Bellows seal, diaphragm pump, canned	Nil	
Valve stems excluding pressure relief	Rising stem regular packing rating < 4 MPa	1.7	Industry average for gas: 1–7. Industry average for liquid: 0.05–3
	Rating > 4 MPa	0.03	
	Non-rising stem	0.005	
Pressure relief valves	Without upsteam protection by bursting disk	2.8	Industry average for gas: 29–50. Industry average for liquid: 2
Compressors	Reciprocating: single rod packing	45	Industry average: 60–175
	Reciprocating: double rod packing	3.6	
	Rotary: labyrinth seal	45	
	Rotary: mechanical	Same as pumps	
	Rotary: liquid film seal	0.006	
Agitators	[pump seal data x shaft speed [m/s]]/1.91		Industry average: 5–30
Piping and flange connections	Open ended pipe	0.63	Industry average: 0.5–6
	Flange with asbestos gasket: 2–4 MPa	0.056	Industry average: 0.07–0.3
	Flange with asbestos gasket: > 4 MPa	0.056	
	Threaded connection	0.056	
	Welded connection	Nil	
Sampling connections	General		Industry average: 2.7

Intensify

Use lower volumes in all process vessels. This has probably the most dramatic effect on both safety and cost. For example, *minimize the inventory.* Use smaller storage, eliminate surge drums, use packing instead of plate trays, reduce the interim storage, use thermosyphon reboilers instead of kettle reboilers, use plate exchangers instead of shell-and-tube models, keep transfer lines to a minimum length, transport as a gas instead of a liquid, and consider pumping at higher velocities to keep holdup volume in the line less. *Eliminate storage* by making the hazardous materials *in situ.* An example is the *in situ* production of hydrogen selenide for microelectronics production; this innovation eliminates the need to store this toxic material under high pressures. *Combine or segregate equipment.* Combine equipment: use reactive distillation, membrane reactors. Segregating the reaction into a series of separate reactors for each reaction in a multiple reaction sequence creates a safer environment. *Segregate and concentrate waste streams* to minimize the volume of waste, recycle, concentrate the wastes to minimize the volume, and manage containers; use the smallest equipment. *Eliminate the accumulation of reactants.* In a liquid-liquid mass transfer controlled reaction system (for example, nitration), an increase in agitation will increase the mass transfer, and reduce buildup of reactants and decrease the potential of a runaway reaction. *Minimize the amount produced* by using closed-loop sampling instead of once-through sampling.

TABLE 16.26
Dust Emissions from Various Equipment Options

Source	Description of Equipment Option	Emission Rate, mg/s [King, 1990]
Vibratory screens	Open top	5.5 times top surface area in m^2
	Closed top with open port access: 15 cm diam. port	0.11
	20 cm diam. port	0.21
	30 cm diam. port	0.44
	Closed cover, no ports	Nil
Bag dumping	Manual slitting and dumping	3
	Fully automatic with negative pressure	Nil
Bagging machines	No ventilation	1.5
	Local ventilation	0.01
	Totally enclosed with negative pressure	Nil

Substitute

Employ less-hazardous species. *Use less-hazardous additives* (solvents, catalysts, filter aids, adsorbents, and absorbents) if nonhazardous ones are not available.

Attenuate or Minimize

Use equipment, seals, and valve stems that leak less. Tables 16.25 and 16.26 give estimates of VOC leak rates and dust generation from different types of equipment.

Minimize waste by changing the operating conditions: substitute continuous processing for batch processing to minimize cleanout cycles, and clean vessels with a dry ice spray instead of sandblasting. *Change the operating conditions.* Use hazardous materials under less hazardous conditions. Use hot water instead of steam and water instead of hot oil; use chemicals with higher flash temperatures. Use a vacuum to reduce the boiling temperature, reduce process temperatures and pressures, dissolve hazardous species in safe solvents, add diluents, and operate at conditions far from explosive limits and from reactor runaway conditions. Add inert gas (such as steam, nitrogen, or carbon dioxide) to combustible mixtures to reduce the concentration of oxygen to about 4% below the minimum. *Minimize the hazard from storage tanks* by providing double-walled tanks with water between the walls (provided that water does not react with the stored material). *Reduce the intensity of electric sparks;* ensure that the voltage used is so low that low-energy sparks do not ignite a flammable mixture. For mixtures of flammable vapors in air, these voltages are <0.01 to 1 mJ; for sensitive explosives, <0.001 mJ; for dust explosions, <0.1 mJ.

Recycle

Minimize the source through recycle.

16.7.2.3 Step 4. Minimize the Impact

If hazardous materials must be used and unsafe environments exist, *limit the potential for leakage*; use finned tube heat exchangers for vaporization on the shell side. *Develop special procedures during the startup phase,* when conditions go through the explosive limits. *Measure key variables to trigger safety measures*; for example, for reactors, include instrumentation to detect temperature runaways. For example, AIChE/CCPS (1995) recommends $d^2T/dt^2 > 0$ and $d(T_{reactor} - T_{coolant})/dt > 0$. Monitor the various streams for impurities. *Install explosion suppressors* in potentially explosive spaces for dusts, such as the exit ducts from grinders. *Dump cold liquid or inert solids* into systems on the verge of temperature runaway. *Add an inhibitor to neutralize* the catalyst for a potential temperature runaway (for example, add gaseous ammonia to neutralize the Lewis acid catalyst BF_3

Conceptual Process Design, Process Improvement, and Troubleshooting

complex). *Add inhibitor to short-stop* a polymerization reaction. (For example, add hydoquinone to stop vinyl polymerizations.) *Use fireproofing.*

16.7.2.4 Step 5: Isolate the Source

Sometimes the hazard and unsafe condition cannot be eliminated or minimized. One option is to let it exist in isolation. Solutions include barricades, separate pump rooms, and establishing a hot end of the plant vs. spark- and flame-free zones. The hazard can be enclosed using pneumatic conveying instead of belt conveying and by enclosing screens and grinders. Place the equipment under negative pressure, shield high-temperature surfaces, and seal sewers. Provide ventilation and use a vacuum, pressure, or sweep-through purging. Fires can be isolated via flame arresters.

For electrical safety, allow sparking to happen but in a safe environment. *For electrical sparks*, isolate/segregate, use flameproof enclosures, pressurize, and purge. *For static charge*, let the charge build up but supply a safe route to discharge by grounding or bonding the source to the ground.

Create alliances. Instead of a company producing a raw material and shipping it to a client, the company can form an alliance with the client and produce the material on the client's site.

16.7.2.5 Step 6: Isolate the Impact

If a fire, explosion, excessive pressure, or unsafe system occurs, we can isolate it from people and equipment by dumping the liquid and gaseous materials or by venting.

16.7.3 FOR PROCESS IMPROVEMENT

The six-step approach described above can be applied in the sequence given in Table 16.3. Use mass and thermal pinch technology, given in Section 16.8, to guide the direction.

16.7.4 FOR TROUBLESHOOTING

For troubleshooting, we use the safety, health, and pollution impacts as the major criteria to prioritize actions.

Additional Reading

AIChE/CCPS, *Guidelines for Chemical Reactivity Evaluation and Application to Process Design*. New York: Center for Chemical Process Safety, AIChE, 1995.
Allen, D. T., and K. S. Rosselot, *Pollution Prevention for Chemical Processes*. New York: John Wiley & Sons, 1997. The authors elaborate on the issues and give excellent examples.
Crowl, D. A., and J. F Louvar, *Chemical Process Safety*, Englewood Cliffs, NJ: Prentice Hall, 1990.
King, R., *Safety in the Process Industries*. Oxford, UK: Butterworth-Heinemann, 1990.
Smith, R., *Chemical Process Design*. New York: McGraw Hill, 1995.
Wells, G. L., and L. M. Rose, *The Art of Chemical Process Design*. Amsterdam: Elsevier, 1986, 174. Measures for continuous exothermic reactors; 448 ff.
Woods, D. R., *Data for Process Design and Engineering Practice*. Englewood Cliffs, NJ: Prentice Hall, 1995.
Woods, D. R., *Process Design and Engineering Practice*. Englewood Cliffs, NJ: Prentice Hall, 1995.

16.8 AIDS FOR DESIGN, PROCESS IMPROVEMENT, AND TROUBLESHOOTING

A broad range of software is available to help engineers. Table 16.27 lists some of the software that aids in design and process improvement. Software for troubleshooting is developed for specific types of process equipment. Most applications support *flowsheeting*, or the use of computer aids to predict the steady-state mass and energy balances, the size of the equipment, and the cost and engineering economics. In this section, the focus is on flowsheeting.

TABLE 16.27
Software Aids

Vendor	Process Synthesis and Analysis	Physical Properties	Steady-State and Dynamic Simulation, Optimization	Equipment Sizing and Rating, Costing
Aspen Technology www.aspentech.com	**Aspen Pinch**™: minimize energy usage **Aspen Split**™: conceptual design of azeotropic systems **Aspen Water**™: minimize water usage **Aspen PEP Process Library**™: comparison of processes development, route synthesis, and implementation **DISTIL**: synthesis of column sequences for multicomponent azeotropic systems **HX-NET**: conceptual design tool for heat exchanger networks; uses pinch technology **HYSYS.Concept:data** regression and thermodynamic database access for conceptual design of separation systems	**Aspen Properties**™: Databank, property and estimation methods **Aspen OLI**™: aqueous mixtures **DETHERM**™: compilation of pure physical properties	**Aspen Plus**™: steady-state simulation **Batchfrac**™: batch distillation **Ratefrac**™: distillation and absorber column distillation **Aspen Plus Optimizer**™: equation oriented optimization **Aspen Custom Modeler**™: new model builder **Aspen OnLine**™: online connection with knowledge capture **Aspen ADSIM**™: adsorption processes **Aspen Chromatography**™: chromatographic processes **Aspen FCC**™: fluidized catalytic cracker simulation **Batch Plus**™: batch process recipe oriented modeler	**Aspen Aerotran**™ **Aspen Hetran**™ **Aspen Teams**™: various heat exchanger design tools **CFX**: computational fluid dynamics tool for unit design **FLARENET**: design, rating, and debottlenecking of flare systems **HTFS**: thermal design and simulation of heat exchangers **PIPESYS**: pipeline flow modeling **STX/ACX**: rate and design shell-and-tube and air-cooled heat exchangers

Simulation Sciences Inc.
www.simsci.com

HEXTRAN: heat transfer simulation and pinch analysis

Polymers Plus™: steady-state and dynamic polymer process modeller and units

HYSYS.Refinery: modeling of complete refining processes including economics

HYSYS.RTO+: real-time, online multivariable optimization

POLYSIM: steady-state and dynamic polymer process simulation

PRO/II: general-purpose flowsheeting and optimization

INPLANT™: multiphase fluid flow simulation for piping networks

VISUALFLOW™: design and modeling of safety systems and pressure relief networks

Chemstations Inc.
www.norpar.com

ChemCAD: steady-state process simularo, including equipment sizing and costing

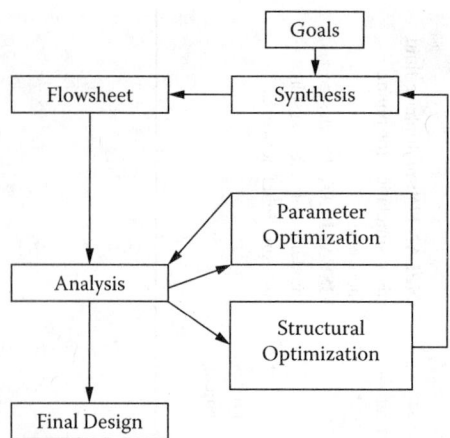

FIGURE 16.5 Synthesis, analysis, and optimization activities.

16.8.1 Steady-State Flowsheeting

Software flowsheeting aids are usually used for the preliminary engineering design (step 3 in Table 16.1) or perhaps in the conceptual design (step 2). The flowsheeting activity is divided into three basic steps:

Step 1. **Synthesis** or selecting the structure of the flowsheet: identification of the equipment, interconnecting and specifying the initial design values. In this step, a crude approximate flowsheet is created to consider recycles, purges, and possible separation schemes; make a simple, linear model to assess the effects of major parameters and structural variations; do necessary laboratory work; get more details for physical properties, thermodynamics, utilities, and process units; write unit models, if necessary; and fix the flowsheet layout.

Step 2. **Analysis** or solving the heat and material balances; sizing and costing the equipment; evaluating the economic worth of the project; assessing the safety and environmental impact; and analyzing the operability.

Step 3. **Optimization** of the structure and the parameters. For the structure, optimize the choice of equipment and the interconnection of units. For parameters, select values for the state variables (such as temperature, pressure, and level) within a given process configuration.

In an additional step 4, move from the steady-state condition to study the dynamics and controller effects and the operability.

The six key components in a process flowsheet simulator are the component database, the thermodynamic model solution package, the flowsheet graphic tool, the models of the unit operations, the data output generation, and the flowsheet solver strategy.

16.8.2 Formulating a Well-Posed Problem

To obtain a solution to a process model, first formulate a *well-posed problem* (Himmelblau, 1996). This requires that we (1) identify the stream variables entering and leaving a unit and the variables that describe the unit, (2) determine the number of independent equations for each unit, (3) calculate the degrees of freedom or number of decision variables for each unit, and (4) specify the values of variables equal to the number of degrees of freedom.

The degrees of freedom (DOFs) are the variables in a set of independent equations that must have their values assigned. DOF = no. of variables − no. of equations and constraints, or

Conceptual Process Design, Process Improvement, and Troubleshooting

$$N_d = N_v - N_c$$

For the DOF, important process *variables* include temperature, pressure, mass (mole) component flow rates, concentration and total flow rates, specific enthalpies, heat flow, work, and flow ratios (e.g., recycle, feed/product, reflux). The number of process variables to describe a stream containing N_{sp} species is given by $N_v = N_{sp} + 2$.

For the DOF, *equations and constraints* include the independent material balances for each species or a total flow balance and ($N_{sp} - 1$) species balances, the energy balance, the phase equilibrium relationships that link the compositions between phases, and the chemical equilibrium relationships. The unit and composition constraints may be explicit (e.g., a given stream fraction is condensing) or implicit (e.g., a species concentration is zero). Himmelblau (1996) provides the full derivation for the degrees of freedom for the following Examples 16.11–16.14.

Example 16.11
Determine the number of process variables, equations, and constraints and the overall DOF for an adiabatic stream splitter.

N_v:	Stream variables	$3(N_{sp} + 2)$
N_c:	Material balance	1
	Composition specification	$2(N_{sp} - 1)$
	Temperature specification	2
	Pressure specification	2
Overall DOF:	$N_d = \{3(N_{sp} + 2)\} - \{2N_{sp} + 3\} = N_{sp} + 3$	

Example 16.12
Determine the number of process variables, equations, and constraints and the overall DOF for a nonadiabatic mixer in which two streams are mixed to produce a single exit stream.

N_v:	Stream variables	$3(N_{sp} + 2)$
	Heat gain/loss (Q)	1
N_c:	Component balance	N_{sp}
	Energy balance	1
Overall DOF:	$N_d = \{3(N_{sp} + 2) + 1\} - \{N_{sp} + 3\} = 2N_{sp} + 6$	

Example 16.13
Determine the number of process variables, equations, and constraints and the overall DOF for a flash tank with three input liquid streams, a single vapor stream, and Q heat input. Assume that the component flows are known.

N_v:	Stream variables	$4(N_{sp} + 2)$
	Heat, Q	1
N_c:	Component balance	N_{sp}
	Energy balance	1
	Phase equilibrium	N_{sp}
	Temperature (exit)	1
	Pressure (exit)	1
Overall DOF:	$N_d = \{4(N_{sp} + 2) + 1\} - \{2N_{sp} + 3\} = 2N_{sp} + 6$	

Example 16.14

Determine the number of process variables, equations, and constraints and the overall DOF for a combination of a mixer and flash separator.

Mixer:
N_v: $3(N_{sp} + 2) + 1$
N_c: $N_{sp} + 1$
N_d: $2N_{sp} + 6$

Flash separator:
N_v: $3(N_{sp} + 2) + 1$
N_c: $2N_{sp} + 5$
N_d: $N_{sp} + 2$

Sum of DOF of individual units:
$N_d = 3N_{sp} + 8$

Less redundant variables and constraints:
Stream 3 = stream 4 $\qquad\qquad N_{sp} + 2$
Need one energy balance $\qquad\qquad$ 1
Total DOF $\quad N_d = \{3(N_{sp} + 8)\} - \{N_{sp} + 3\}$

16.8.3 Flowsheet Architectures

The two basic flowsheet software architectures are sequential modular and equation-based. In *sequential modular,* we write each unit model so that it calculates output(s), given feed(s), and unit parameters. This is the most commonly used flowsheeting architecture at present, and examples include Aspen+ plus Hysys (AspenTech), ChemCAD, and PROII (SimSci). In *equation-based* (or open-system) architectures, all equations are written describing material and energy balances as algebraic equations in the form $f(x) = 0$. This is the preferred architecture for new simulators and optimization, and examples include Speedup (AspenTech) and gPROMS (PSE plc). Each is discussed in turn.

16.8.3.1 Sequential Modular Architecture

Unit models are written to *calculate* output stream values, *given* input stream values and unit parameters. The recycle stream values are then calculated and checked against the estimated values for that iteration. If they agree within a tolerance, then the flowsheet has converged. This procedure is called *tearing* a recycle stream. The important questions for this approach are

1. How is a process flowsheet analyzed to determine whether there are recycles and which stream to tear (estimate)?
2. How do we handle many recycles, which may be nested?
3. How are calculated values for the recycle stream used for subsequent iterations?
4. What solution procedure should be used if not all the inputs to the process are known, but some outputs are specified?

16.8.3.1.1 Tearing

There are a number of approaches to develop better tearing strategies, but they require analysis of the number of components and number of recycles. Biegler et al. (1997) describe algorithms to minimize the number of tears and the number of recycle variables in the tears.

Conceptual Process Design, Process Improvement, and Troubleshooting

16.8.3.1.2 Convergence Block Schemes: Successive Substitution

Let X^k be the estimate for the tear stream variables at the kth iteration, Let $F(X^k)$ be the calculated result for that tear stream's values. The simplest iteration scheme is successive substitution, or fixed-point iteration:

$$X^{k+1} = F(X^x)$$

The next estimate on the tear stream is the newly calculated variable values. A more general, weighted form is

$$X^{k+1} = a F(X^k) + (1-a) X^k$$

that is useful if the iterations start to oscillate, e.g., $a = 0.5$.

Given the feed/process streams to a process unit, the sequential modular solution scheme is

1. Estimate the value of the recycle stream to that unit (tear).
2. Calculate all the units between the first unit and the recycle stream.
3. Check if new recycle values are approximately equal to guessed values:
   a. If not, go to step 1 and use new values as a next guess.
   b. If OK, then it is converged.

16.8.3.1.3 Wegstein Acceleration

Let X^k be the guess for the tear stream variables at the kth iteration. Let $F(X^k)$ be the calculated result for that tear stream's values. Treat each variable with a secant method. Do two or more successive substitution iterations to generate X^k, $F(X^k) = X^{k+1}$, $F(X^{k+1})$. Then accelerate:

$$X^{k+2} = X^{k+1} - \underline{\{(X^{k+1} - X^k)/(F(X^{k+1}) - F(X^k))\}\, F(X^k)}$$

The rate of change of the stream estimates (underlined term) is usually bounded between upper and lower bounds to prevent divergence (usually between –5 and 25).

16.8.3.1.4 Control Blocks for Design Situations where Some Outputs Are Specified

Information flow in a standard *simulation* problem calculates unit outputs (stream values) given input streams and unit parameters (simulation problem). *Design* requires specification of an output variable and then calculating an input value or equipment parameter (design problem).

Example 16.15

Compare simulation with design for the adiabatic flash shown in Figure 16.6.

	Simulation case:	Design case:
Given:	F, z_i, T, P	F, z_i, T, V
Calculate:	V, y_i, L, x_i	P, y_i, L, x_i

In Example 16.15, for the design case, assume that a vapor overflow V is to be calculated. The unit parameter, pressure (P) in this case, is to be adjusted to attain the vapor flow rate, V_{sp}. Key constraints include the dew and bubble pressures (to maintain a two-phase region in the flash) and

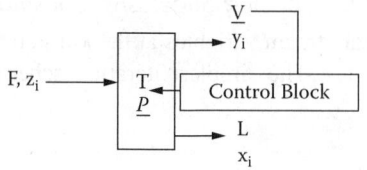

FIGURE 16.6 Symbols and configuration for the adiabatic flash in Example 16.15.

safety and operational constraints on the temperature and pressure in the vessel. A control block solves the equation:

$$Error = set\ point - value < tolerance$$

$$E = V_{sp} - V < error$$

The calculation procedure may proceed as follows:
For iteration k in the flash example,

1. If $k = 1, 2$, estimate P^k (make sure flash is in the two-phase region).
2. Calculate V^k.
3. Calculate $E^k = V_{sp} - V^k$.
4. If, $|E| <$ error, exit

   else, $|E| >$ error, use secant method to calculate P^{k-1}:

$$P^{k+1} = P^k - \{(P^k - P^{k-1})/(E^k - E^{k-1})\}E^k$$

In many problems, there are recycle loops and design specifications. Recycle loops require convergence blocks. *Explicit* iterations occur to satisfy material, temperature, and pressure agreement in a given recycle stream. Design specifications require control blocks. *Implicit* iterations occur to satisfy a specification by adjusting an input stream variable or equipment parameter.

For a design problem, converge the entire flowsheet (close all the recycles) for every intermediate value of the adjust variable. This is very expensive computationally and is a major drawback to the sequential modular approach. Alternative and sometimes faster approaches include

- Partially converge the flowsheet, a less stringent recycle tolerance, for the intermediate adjusted variable values.
- For each iteration of the control block, converge only the units between the specified and adjusted variables; solve the entire flowsheet after the adjust variable is converged.

These alternatives are not guaranteed to converge, and partially converged flowsheets may provide poor estimates for the monitoring of the specified variable.

16.8.3.1.5 Partitioning and Precedence Ordering

Partitioning locates groups of units that must be solved together. *Precedence ordering* solves these groups of units within a process in the proper computational order. A partition is unique, but the precedence ordering is not. Partitions may exist where the order of the solution may have groups of units calculated in serial or parallel order. An effective algorithm developed by Sargent and Westerberg is

Conceptual Process Design, Process Improvement, and Troubleshooting

1. Trace from an arbitrary starting unit to the next unit, through a connecting stream, to form a "string" of units.
2. Continue, until
   a. A unit in the string reappears. Then all the units between the repeated unit occurrences, including the repeat unit, become a "group." The group is treated as a single unit in subsequent steps.
   b. A unit has no more outputs. Then the unit, or group, that has no output is placed at the top of a list of groups and is deleted from the problem.

The list will contain the groups in correct precedence order.

16.8.3.2 Equation-Based Architecture

Model equations are written in the open form, $F(x,u) = 0$, where x represents the stream variables and u the unit parameters. All the nonlinear equations are assembled and solved simultaneously. This creates a huge nonlinear equation set, typically between 10^3 and 10^5 equations. The solution methodology is complicated, because most equations include only a few variables in each equation. Typically, each equation has less than 5% (95% sparsity) of the total number of variables in the problem. The larger the problem, the higher the sparsity.

The advantage of the approach is that any parameter or stream variable can be specified as long as the problem has an equal number of equations and variables, and it is nonsingular with linearly independent equations. The two basic solution approaches for the equation-based flowsheeting systems are tearing and linearization. In *tearing*, we find the minimum number of variables that need to be guessed and iterate to find the rest of the variables. The equations are grouped to allow a few variables to be used to solve a group of equations. This requires preanalysis and does not generalize. In *linearization*, the set of nonlinear equations is solved by linearization and solution, such as Newton–Raphson.

The challenges with an equation-based strategy are the following:

1. It is difficult to use equations from existing sequential modular packages.
2. Equation management is complex.
3. Large computer storage is required.
4. There are numerical problems in obtaining derivatives and with singularity.
5. Starting values are needed.
6. The interface to thermodynamics packages must provide derivative information.
7. The ordering of equations affects the equation solution time.
8. We need input data verification to prevent singular problems (structural and numerical singularity).
9. Backup plans are needed if the solution fails to converge.

The advantages of equation-based strategies are that they remove the need for convergence and control blocks and allow for more flexible problem formulation. As such, they are the basis for all new simulators being currently developed.

16.8.4 IMPORTANCE OF THERMODYNAMICS PACKAGES

In using a flowsheet simulator, one of the most important decisions is the choice of thermodynamics package. The choice of thermodynamics options affects the accuracy of the material and energy balances. Appropriate choices depend on the compounds in the system, temperature, pressure, and the availability of parameters. Equations of state and activity models are used to calculate stream properties: number of phases, phase composition, PVT relationships, enthalpy, and entropy.

TABLE 16.28
Select the Model for the System

System	Margules	VanLaar	Wilson	NRTL	UNIQUAC
Multi	?	?	OK	OK	OK
Azeotropic	OK	OK	OK	OK	OK
Liquid-liquid-vapor	OK	OK	NO	OK	OK
Self-associating	NO	NO	OK	OK	OK
Polymers	NO	NO	NO	NO	OK

16.8.4.1 Vapor-Liquid Equilibrium

Liquid activity models must be used in vapor-liquid equilibria calculations, with the appropriate model tested against available data. Models often used include Margules, Van Laar, Wilson, nonrandom two-liquid (NRTL), and universal quasi-chemical (UNIQUAC). For mixtures, mixing rules are used to combine pure component parameters. Table 16.28 suggests regions of applicability for different models.

16.8.4.2 At High Pressure or near the Critical Region

At high pressure or near the critical region, the thermodynamic functions for the liquid phase cannot be assumed to be independent of pressure. Simple equations of state are not adequate for vapor-phase predictions. Chao–Seader method includes the calculation of liquid-phase fugacity coefficient by Pitzer's correlation. Chao–Seader uses the Redlich–Kwong equation of state with a temperature range of –20 to 250°C and pressure <10,000 kPa. The Grayson–Streed extension of the model of Chao–Seader is used for systems rich in hydrogen and methane at a temperature range of –20 to 450°C and pressure <20,000 kPa. It is advisable to check the simulator manual.

16.8.4.3 Equation of State, EOS

Commonly used EOS models include the ideal, virial, PengRobinson, Soave–Redlich Kwong, and Lee–Kesler. The reduced form of the EOS is particularly significant. Substances with the same reduced properties are in *corresponding states*. Van der Waal's EOS is a poor predictor of state properties, but the experimental data do correlate well with reduced conditions. Many of the cubic EOS models are based on the van der Waal equation.

EOS models are used to determine liquid and vapor densities, vapor pressures, fugacities, thermal property deviations from ideal, and enthalpies. EOS parameters can be adjusted to fit some data listed above, but usually one EOS set of parameters cannot fit all of the pure component data well. Flowsheet simulators allow the user to mix EOSs to get the best match between EOS parameters and desired state variables. In general, EOSs are very good predictors of the vapor-phase properties and of the liquid-phase density and composition. Cubic EOSs are relatively simpler to use. The Lee–Kesler model (based on the theory of corresponding states, and using reduced temperature and pressures) can cover a wider range of pressures and temperatures. It is recommended practice not to extrapolate the EOS results outside of conditions for which the parameters were fitted. Check the simulator manual for temperature pressure and composition restrictions. Generally, the higher the pressure (>10 bar), the poorer the prediction. If polar molecules are present (alcohols, acids), then EOS should not be used. Use an activity model for the liquid phase or a specially fitted EOS for those systems.

16.8.5 Tools for Flowsheet Development

Flowsheet development involves the selection of the equipment and the interconnects that integrate the equipment into a working system. Two approaches are used for the development of the flowsheet: mathematically based programming and the use of heuristics and evolutionary development.

16.8.5.1 Mathematically Based Programming

Mathematical programming approaches tend to use a superstructure approach with many alternative solutions embedded within a complex superstructure containing many alternate units and feed and recycle configurations. The mathematical program then solves the corresponding model to determine the flows and units for the optimal process configuration contained within the superstructure that meets the overall process design objectives. Engineering insights are used to develop the superstructure, identify candidate units, and weight the mathematical model toward a practical process configuration. Biegler et al. (1997) review modeling techniques.

16.8.5.2 Hierarchical Use of Heuristics for Evolutionary Development

The approach to conceptual design outlined in Table 16.2 is hierarchical and based on heuristics. In such evolutionary approaches, key insights set targets for attainable performance. Flowsheet changes act to improve performance. Software programs have been developed to assist in some of the tasks. For example, the task of energy integration (and optimization) can be assisted by the use of heat exchanger network analysis based on thermodynamics, called *pinch technology* or *methodology*. It is the most developed of the various process synthesis and integration methods. The method starts with the identification of streams as sources or sinks of heat (hot and cold streams). Streams have initial and final temperatures (supply and target temperatures) to be met. Typically, they are plotted as a function of temperature vs. heat load in the stream (Q), where Q = *mass flow rate* * *heat capacity* * ($T_{target} - T_{supply}$). The streams are dealt with as composite streams where, within a given temperature range, all the hot streams are treated as one composite hot curve, and all the cold streams are treated as a composite cold curve. The plot of the two composite curves is separated by the *minimum approach temperature (MAT)*, which is the closest temperature approach permitted in the heat exchanger design (the *pinch* point). As the MAT decreases, the required heat exchanger area increases, but the utilities decrease. An economic optimum exists as the optimal trade-off between capital and operating costs. The composite curves are plotted, and the overlaps denote the required utility loads. The overlap of the hot curve over that of the cold composite curve at the low temperature end is the required cold utility (where heat is rejected) target. The overlap of the cold composite curve to the hot composite curve at the high temperature is the required hot utility target. The key insight is that, at temperatures above the pinch temperature, the process is a net energy sink and that, below the pinch temperature, the process is a net energy source. This observation effectively partitions the process synthesis problems into simpler subproblems with a defined set of utilities that will be used in each part of the subproblem. Matching streams starting at the pinch temperature and moving outward to higher and lower temperatures, respectively, determine heat exchanger networks.

Most process simulators can provide plots of the composite curves and determine the pinch temperature, target hot and cold utility loads, and temperatures. Specialized tools are available to generate the heat exchanger network configuration; size; and cost of the heat exchangers, e.g., Aspen Pinch (Aspen Technology), SuperTarget (Linnhoff–March), Hextran (Simulation Sciences), and HX-NET (AEA Hyprotech). Similar techniques have been developed for water use minimization, hydrogen use in refineries, among other applications (Halale, 2001).

As suggested in Table 16.3, *pinch technology* can be used for process improvement as well as for flowsheeting.

Additional Reading

Biegler, L. T., I. E. Grossmann, A. W. Westerberg, *Systematic Methods of Chemical Process Design*. Englewood Cliffs, NJ: Prentice Hall, 1997.
Douglas, J. M., *Conceptual Design of Chemical Processes*. New York: McGraw-Hill, 1988.
Halale, N. 2001. Burning Bright: Trends in Process Integration. *CEP* 7: 30–41.

Himmelblau, D. M., *Basic Principles and Calculations in Chemical Engineering,* 6th ed. Englewood Cliffs, NJ: Prentice Hall, 1996.
King, C. J., *Separation Processes,* 2nd ed. New York: McGraw-Hill, 1980.
Seider, W. D., J. D. Seader, and D. R. Lewin, *Process Design Principles: Synthesis, Analysis and Evolution.* New York: John Wiley & Sons, 1999.
Smith, R., *Chemical Process Design.* New York: McGraw-Hill, 1995.
Turton, R., R. C. Bailie, W. B. Whiting. J. A. Shaeiwitz, *Analysis, Synthesis, and Design of Chemical Processes.* Englewood Cliffs, NJ: Prentice Hall, 1998.

16.9 PROCESS OPTIMIZATION

In Step 3, optimization of the structure and the parameters (Section 16.8.1), the two types of process optimizations are parameter and structural optimization. *Parameter optimization* is the process of determining the best value of a process unit parameter or stream quantity in terms of improving performance within a given set of constraints. Parameter optimization is usually a nonlinear continuous variable (over a range of variable values defined by upper and lower bounds). *Structural optimization* involves the determination of the best set of units and their interconnections such that the process configuration provides the best performance within a given set of constraints. Structural optimization requires discrete decisions. Pinch technology, described in Section 16.8.5, is a form of structural optimization.

16.9.1 DEFINE THE PROBLEM

The first key steps are to determine the *objective function*, the *decision variables* (or design variables or degrees of freedom), and the *constraints*. The *objective function* is the performance measure to be used to determine the degree of improvement in the process. Examples include maximizing profit, minimizing cost, maximizing production, and minimizing a waste stream. *Decision variables* are those variables that can be changed, within a given process design, to improve the objective function. Examples of continuous decision variables include the pressure or temperature of a stream, and the number of stages in an absorber column would be a discrete decision variable. *Constraints* are of three types: variable bounds, equality constraints, and inequality constraints. Variable bounds are the upper and lower limits on a variable due to operational, physical, or economic considerations. Equality constraints are functions of process variables that define a relationship that must be enforced in a feasible solution, e.g., material balance on a unit. Equality constraints reduce the dimensionality of a problem, since each equation reduces the number of independent decision variables. Inequality constraints are usually one-sided feasible regions (less-than or greater-than relationships) defined by a function of process variables; e.g., distillation columns must be operated within flooding constraints.

16.9.2 SELECT THE OPTIMIZATION SOLUTION SCHEME

Combinations of these objectives, variables, and constraints are so common that the optimization solution schemes are tailored for the function properties. A key step in solving an optimization problem is to match the appropriate solution algorithm with the problem type as outlined in Table 16.29.

Optimization algorithms all require the definitions of the objective function and decision variables, and constraints if applicable. In addition, most solution algorithms require a reasonable initial set of values for the decision variables, preferably a feasible starting point. Optimization algorithms generally follow a pattern:

1. Provide a starting set of decision variables.
2. Determine a *search direction* for the decision variable to improve the objective function.

TABLE 16.29
Problem Types, Objective Functions, Constraints, and Decision Variables

Problem Type	Objective	Constraints	Variables
Single variable	Nonlinear	None	Continuous
Linear programming	Linear	Linear	Continuous
Quadratic programming	Quadratic	Linear	Continuous
Nonlinear programming	Nonlinear	Nonlinear	Continuous
Mixed integer linear programming	Linear	Linear	Continuous-binary (0, 1)

3. Determine a *step length* (change in the decision variables) that will be taken on any one step.
4. Set the criteria for *termination* of the algorithm at a feasible solution.

A preliminary *exploratory search* of the problem should be done by changing a few of the decision variables, within the constraint space (or *feasible region*), to ensure that the objective function is affected by changes in the decision variables. This is also known as preparing a *response surface* or *sensitivity analysis*. If the objective function is insensitive to changes in the decision variables, within the feasible region that is probed, then there is no optimization problem, because the problem will have a multitude of solutions with the same objective function value.

Solution algorithms, for continuous variable problems, generally increase in complexity with increasing nonlinearity of the constraints and increasing number of discrete variables (binary or integer variable values).

Edgar et al. (2001) survey optimization algorithms as applied to chemical engineering problems. A variety of simple algorithms for relatively small problems are available in Press et al. (1992).

16.9.2.1 Nonlinear Objective Function Problems

For *single*, nonlinear, continuous decision variable problems, common techniques are either nonderivative or derivative.

Nonderivative methods compare the value of the function with changes in the decision variable. Methods include polynomial approximation, Fibonnaci search, and golden section.

Derivative methods use the necessary condition for optimality where the derivative of the function with respect to the decision variables is zero at the optimum. Methods include Newton and quasi-Newton methods.

For *multivariable*, nonlinear, continuous decision variable problems, the choice of nonderivative or derivative methods (to determine the search direction) depends on the availability of derivatives of the objective function with respect to the decision variables.

Nonderivative methods include random search, grid search, simplex search, and conjugate directions (or Powell's method). The nonderivative methods use various patterns for generating new test points for decision variables, and then a comparison of the new objective function value against previous values. A subsequent test point is then generated, either based on the immediate comparison or using the previous history of test points.

Derivative methods and derivative approximation methods use steepest ascent/descent (or Cauchy's method), conjugate gradients, Newton's, or quasi-Newton methods.

Generally, for multivariable methods, a *line search algorithm* is used to determine the step length that the decision variables will be allowed to change along a search direction. If the steps are too small, then convergence is ensured, but it will take much iteration to reach the optimum. If the steps are too large, the procedure may become unstable.

16.9.2.2 Linear Programming

Linear programming is probably the most commonly used optimization method. Edgar et al. (2001) and Press et al. (1992) describe many available sophisticated algorithms.

16.9.2.3 Nonlinear Programming

Nonlinear programming problems are generally formulated as

$$
\begin{aligned}
&\text{Min. or max.} \quad f(x) \\
&\text{Subject to} \quad h(x) = 0 \\
&\quad\quad\quad\quad\quad\; g(x) \leq 0 \\
&\quad\quad\quad\quad\quad\; 1 \leq x \leq u
\end{aligned}
$$

where x is the set of decision variables bounded between 1 and u, $f(x)$ is the objective function, $h(x)$ is the set of equality constraints, and $g(x)$ is the set of inequality constraints. Nonderivative methods are generally slow and unreliable for problems with more than a few decision variables. Some variations on the simplex method have been made to incorporate constraints (Press et al., 1992). Most algorithms use derivative information, or approximations to the derivatives, for $f(x)$, $h(x)$, and $g(x)$. Example algorithms include successive linear programming, reduced gradient methods, and successive quadratic programming. Many programs are available to solve these classes of problems. Most optimization algorithms require continuity of the objectives and constraints within the exploratory region, which can be difficult for real process design problems. While some advances have been made in partitioning problems to seek the global optimum, this is still a relatively difficult area of optimization and not routinely used.

16.9.2.4 Discrete Decision Variables

Design problems in particular usually have discrete decisions. Few large-scale optimization algorithms are available that can handle nonlinear constraints with a mix of continuous and discrete variables.

16.9.3 Linking Optimization with the Flowsheeting Tool

The most important part of process optimization is linking the process flowsheeting tool to the optimization algorithm. With an *equation-based architecture*, the unit equations (material and energy balances, operating constraints, and specifications) are constraints in a general nonlinear programming formulation. The main problems are

- Problem size, as most process optimization problems have on the order of 10^4 to 10^6 constraints and at least 10 to 100 decision variables. Generally, this is done with sparse matrix algorithms or by partitioning the problem to take advantage of the equation structure due to the unit equations and recycles.
- Starting point for the decision variables, which generally requires a complete simulation of the process to be completed.
- Derivative information is required, which is especially problematic for thermodynamics packages and the avoidance of discontinuities in the search space.

With a *sequential modular architecture*, optimization is generally done on much smaller problems (usually less than 20 decision variables) using either of the following:

- Nonderivative search methods for multivariable problems, with simple bounds and testing of points to ensure that they are feasible.
- Derivative-based optimization methods, where the optimization algorithm is used to simultaneously converge the recycles and determine the decision variables that improve the objective function (Biegler et al., 1999; Edgar et al., 2001).

Additional Reading

Biegler, L. T., I. E. Grossmann, A. W. Westerberg, *Systematic Methods of Chemical Process Design.* Englewood Cliffs, NJ: Prentice Hall, 1997.

Edgar, T. F., D. M. Himmelblau, L. S. Lasdon, *Optimization of Chemical Processes,* 2nd ed. New York: McGraw-Hill, 2001.

Press, W. H., et al., *Numerical Recipes in Fortran 77: The Art of Scientific Computing,* 2nd ed. New York: Cambridge University Press, 1992. Also available for other computer languages.

16.10 RULES OF THUMB FOR INTERPERSONAL SKILLS AND TEAMWORK

We need interpersonal skills and skill at teamwork.

16.10.1 Interpersonal Skills

The basis for effective teamwork in design, process improvement, and troubleshooting is good interpersonal skills that are developed through self-awareness and the application of five key principles.

16.10.1.1 Self-Awareness

Become aware of your own uniqueness and personal style and how you might differ from the style of others, through such validated inventories and readings as Kirton's KAI inventory, which illustrates how you wish to apply your creativity (Kirton, 1976); Schutz's FIRO-B inventory, which explores how you address the three different phases of group evolution of form, storm, and norm (Whetton and Cameron, 1984, p. 80); Jungian typology inventory, which suggests how you prefer to seek validation of your ideas, how collecting data contrasts with taking action, the relative weight placed on facts vs. feelings, and the relative importance of the big picture vs. details (Keirsey and Bates, 1984); self-awareness of the difference between men and women's approach to conversation (Tannen, 1990); and Johnson's approaches to resolving conflict (Johnson, 1986).

We move from self-awareness to self-acceptance, to acceptance of others with different styles, and to self-confidence.

16.10.1.2 Five Key Principles

We usually accept the best of people when we first meet and begin to work together. We trust; we expect them to behave honorably toward us. Our relationships can improve or deteriorate depending on how we handle the following five key ideas about effective interpersonal relationships:

> *Key 1.* Honor the seven fundamental rights of individuals: R, to be Respected; I, Inform or to have an opinion and express it; G, have Goals and needs; H, have feelings and express them; T, trouble and make mistakes and be forgiven, S, select your response to others' expectations, and claim these rights and honor these in others.
> *Key 2.* Avoid the four behaviors that destroy relationships: contempt, criticism, defensiveness, and withdrawal/stonewalling.

Key 3. Build trust. Trust glues relationships together. We build trust by *keeping* commitments to ourself and others; *clarifying* expectations that we have of ourself and of others; *showing* personal integrity, honesty, and loyalty to others, especially when they are not present; *apologizing* promptly and sincerely when we know we are wrong; *honoring* the fundamental *rights* listed above and avoiding the killers; *listening* and *understanding* another's perspective; *being truthful*; and *accepting others,* "warts and all." We destroy trust by the reverse of the builders of trust listed above and by selectively listening, reading, and using material out of context; not accepting experience of others as being valid; making changes that affect others without consultation; blind-siding by playing the broken record until we are eventually worn others out or subtly make changes in the context/issues/wording gradually so that they are unaware of what is happening until it is too late.

Key 4. The 12:1 rule applies to rebuild relationships. Twelve positive experiences are needed to overcome one negative experience.

Key 5. To improve and grow, we need feedback about performance. Give feedback to others to encourage and help them; it not for you to get your kicks by putting them down.

16.10.2 Effective Teams

A team is not just a collection of individuals. In a collection of individuals, each has a personal goal, trusts only himself, and rarely reveals personal skills. Decisions are usually not made, and conflict is ignored. In a team, all unanimously accept common goals, each is clear about his role, trust and involvement levels are high, personal unique skills are used effectively, decisions are made by unanimity, and the team has methods for handling conflict. Our meetings and teamwork improve when we strive for the characteristics of teams. Some target behaviors of teams follow:

- Have a purpose for each team and each meeting. Set and follow agendas to get the task done.
- No agenda, no meeting! If a meeting must be held without a circulated agenda, then spent the first five minutes creating the agenda.
- The team must have the correct membership and resources to achieve the goal.
- The team should be empowered and accountable to achieve the goal.
- Both *task* (getting the job done) and *morale* (feeling good about the group work and about how you have interacted with the other group members) are important.
- Have a chairperson whose role is to facilitate the team process: this person thinks through the tasks to be done, decides on the need for a team meeting, identifies the time and place for the meeting, sets and circulates an agenda, facilitates the meeting, and starts and stops the meeting on time.
- Being the chairperson and being a leader are different; different people may become leaders at different times.
- Group evolution tends to follow a pattern described by such descriptors as *"forming, storming, norming, and performing."*
- Establish norms. Agree on terminology and procedures for problem solving, for brainstorming, for decision making. Agree on the role of the chair in decision making (e.g., *vote or no vote);* roles, minutes, and records of decisions (e.g., *format, details, who prepares them, whether they are circulated, their use for subsequent meetings*); how to handle conflict and the level of intervention; combating "groupthink"; how to handle emergencies and criteria; and procedures for asking a member of the group to resign.
- Each has a clear idea of roles and of group norms.
- When groups are functioning effectively, about 70% of the time is spent on the task, 15% on morale-building activities, and 15% on task process activities.
- The products from groups or teams are improved when members have different "styles," as outlined in Section 16.10.1.1.

Conceptual Process Design, Process Improvement, and Troubleshooting

- The quality of decisions, product, and task is improved if group members offer different perspectives, disagree, and seem to introduce conflict into the process. The trick is to manage the apparent conflict well.
- Use the 20-min rule. After 20 min, either make a decision or identify the key information that is missing and arrange to obtain that information for the next meeting.
- Document decisions and record actions with "what," "who," and "when." *For example,* "Decide to compare two optional control strategies." Action: JBM by March 30.
- Systematically assess the quality of the teamwork and set goals for improvement. Table 16.30 is an example form for such an assessment.

Additional Reading

Francis, D., and D. Young, *Improving Work Groups: a Practical Manual for Team Building.* San Diego, CA: University Associates, 1979.

Woods, D. R., Group Skills, chap. 5 in *Problem-Based Leaning: How to Gain the Most from PBL.* Waterdown, Ontario, Canada: Woods Publisher, distributed by McMaster University Bookstore, Hamilton, Ontario, 1994.

Fisher, K., et al., *Tips for Teams.* New York: McGraw-Hill, 1995.

Kirton, M. J., 1976. Adaptors and Innovators: a Description and Measure. *J. Applied Psychology* 61: 622–629.

Keirsey, D., and M. Bates, *Please Understand Me: Character and Temperament Types.* Del Mar, CA: Gnosology Books, 1984. http://www.keirsey.com.

Schutz, W. C., *FIRO: A Three-Dimensional Theory of Interpersonal Behaviour.* New York: Holt, Rinehart and Winston, 1958, with the instrument and scoring available from Whetton, D. A., and K. S. Cameron, *Developing Management Skills.* Glenview, IL: Scott Foresman, 1984, 80.

Tannen, D., *You Just Don't Understand: Women and Men in Conversation.* New York: Ballantine, 1990.

Johnson, D. W., *Reaching Out.* Englewood Cliffs, NJ: Prentice Hall, 1986.

16.11 RULES OF THUMB

Rules of thumb are suggested values that are reasonable to assume based on experience. Rules of thumb are based on the application of fundamentals. They do not replace fundamentals, but, rather, they enrich the correct use of fundamentals to solve problems. Rules of thumb

- Help us judge the reasonableness of answers
- Allow us to assess quickly which assumptions apply
- Are used to guide our better understanding of complex systems and situations
- Allow us to supply rapid order-of-magnitude estimates

These rules of thumb are organized by prime function, and within each function are listed the usual types of equipment. Some equipment is multifunctional. For example, a fluidized bed is primarily a device for mixing a gas-solid mixture. However, it is also used as a dryer, a heat exchanger, a reactor, and an agglomerator.

For each piece of equipment, the following are listed:

- *Area of application.* How to select the equipment, when you would use it, and the usual available size range (describing how big or small a single unit is usually built).
- *Guidelines.* How to size the equipment: rules of thumb for size estimation. In general, these usually work within a factor of ten but a factor of four is often preferable.

The following information is reproduced from the book, *Rules of Thumb for Process Design and Engineering Practice,* by D. R. Woods (Wiley-VCH, 2007).

TABLE 16.30
Rating Form for Teams

                    Assessment of your team & team meeting   Name: _____

                                                                                                   Date: _____

Purpose of team: _____   unclear: ☐
Purpose of this meeting: _____   unclear: ☐
Agenda for this meeting: detailed, clear, and circulated ahead of time ☐
                bare minimum circulated ahead of time ☐
                none ☐

Three minute team task to seek consensus about the rating of the Task and Morale:

- **Teamwork: Task** all members clear about and committed to goals; all assume roles willingly; all influence the decisions; know when to disband for individual activity; all provide their unique skills; share information openly; the team is open in seeking input; frank; reflection and building on each other's information; team believes it can do the impossible; all are seen as pulling their fair share of the load.

The degree to which these descriptors describe your team's performance (as substantiated by evidence: meetings, engineering journal, interim report, presentations).

None of these behaviors	Few of these behaviors but major omissions			Most features demonstrated		All of these behaviors
☐	☐	☐	☐	☐	☐	☐
1	2	3	4	5	6	7

- **Teamwork: Morale:** Trust high, written communication about any individual difficulties in meeting commitments; cohesive group; pride in membership; high *esprit de corps*; team welcomes conflict and uses methodology to resolve conflicts and disagreements; able to flexibly relieve tension; sense of pride; "we" attitude; mutual respect for the seven fundamental rights of all team members; absence of contempt, criticism, defensiveness, and withdrawal.

The degree to which these descriptors describe your team's performance (as substantiated by evidence: meetings, engineering journal, interim report, presentations).

None of these behaviors	Few of these behaviors but major omissions			Most features demonstrated		All of these behaviors
☐	☐	☐	☐	☐	☐	☐
1	2	3	4	5	6	7

Each, in turn, gives a 30-second summary of his/her perception of his/her contribution. This is presented *without discussion*.
Individual, 30-second reporting of his/her contribution to this meeting:

_____
_____
_____

Four-minute team task to reach consensus about the five strengths and the two areas for growth.

Strength of your team   Areas to work on for growth

_____   _____
_____   _____
_____

                                                                                          D. R. Woods (2001)

This form should be completed after each meeting and copies used as evidence of growth.

Conceptual Process Design, Process Improvement, and Troubleshooting

16.11.1 Overall Process

In this section, we consider rules of thumb for process control and the properties of fluids.

16.11.1.1 Process Control*

Area of Application

For all processes, provide the four levels of control with (1) the basic control system, (2) an alarm system, (3) a safety interlock system (SIS), and (4) a relief system. Related topics are discussed in Section 16.6 and the chapters on *safety* and *process control*.

Guidelines

Sensors. What are we to measure? Variables are measured by sensors to achieve the following objectives, in hierarchical sequence:

Safety, environmental protection, and equipment protection (see Section 16.7; this could include redundant temperature sensing and alarms on reactors and reboilers handling corrosive chemicals such as HF), smooth operation, product quality, profit, monitoring, diagnosis, and troubleshooting. Identify the objective and select a pertinent variable. Direct measurement of the variable is preferred. If direct measurement is impractical, select an inferential or calculated variable. *For example, temperature can infer conversion and composition.*

Variables must be measured that might quickly deviate from acceptable range, such as (1) non-self-regulating variables (e.g., *level*), (2) unstable variables (e.g., some temperatures in reactors), and (3) sensitive variables that vary quickly in response to small disturbances (e.g., pressure in a closed vessel).

How are we to measure? Select sensor to balance accuracy and reproducibility, to cover the range of normal and typical disturbed operations, and to provide minimum interference with the process operation and costs. *For example, prefer a low-pressure loss flow sensor when compression costs are high.* Use a second sensor for extremely large ranges due to startup, large disturbances, or different product specifications. The sensors should be consistent with the process environment. For example, for flow measurements, the instrument should be located downstream of sufficient straight pipe to stabilize the flow patterns reaching the instrument—at least ten diameters upstream and five diameters downstream of straight pipe. The sensors should be located to assist operators in performing their tasks and engineers in monitoring and diagnosing performance.

Sensors for control should compensate for known nonlinearities before the measurement is used for monitoring or control. Prefer sensors that do not need calibration.

Specifics

Temperature. Prefer resistance temperature detectors, RTD. Prefer narrow span transmitters instead of thermocouples.

Differential pressure. Prefer precision-filled diaphragm seals or remote heads for Δp, because signal depends in part on the density of the fluid in the sensing lines.

Flow. Prefer coriolis, vortex, or magnetic flow meters over orifice or venturi. Keep fluid velocity >0.3 m/s.

Composition

Level. Prefer tuning fork, radar, or nuclear, or consider radio frequency admittance if the composition changes.

Manipulated variables and final elements. The manipulated variable, usually flow rate, has a causal effect on a key controlled variable, can be manipulated by an automated final element, provides fast feedback dynamics, has the capacity to compensate for expected disturbances, and can be adjusted without unduly upsetting other parts of the plant. The final element, usually a valve,

---

* Based on communication from T. E. Marlin, McMaster University, 2001.

has a causal effect on the controlled variable. The number of *final elements* is greater than or equal to the number of measured variables to be controlled, and we must provide an independent means for controlling every variable.

Final elements must provide the desired capacity with the required precision of flow throttling over the desired range, usually 10% to 95% of maximum flow. The valve characteristic should provide a linear closed-loop gain, *except* choose linear or quick-opening characteristics for valves that are normally closed but must open quickly. Select the valve failure position for safety. The valve body should satisfy such requirements as required flow at 0% stem position, plugging, pressure drop, or flashing. The nonideal final element behavior, such as friction and deadband, should be small, as required by each application. Control valves should have manual bypass and block valves to allow temporary valve maintenance when short process interruptions are not acceptable. However, the bypass should never compromise safety interlock systems.

Specifics

Include the same length of upstream straight-run piping before control valves as for orifices. This is particularly important for rotary valves.

Globe valves. Permissible stroke range: 10% to 90%; sliding stem gives the highest sensitivity, and the actuator stem feedback position more closely represents the final element position (but not for fouling or solids).

Rotary ball valves. Permissible stroke range: 20% to 80%. Δp across the valve is small, but accounts for the pipe reducers needed for installation. Be sensitive to the need for upstream straight pipes.

Rotary butterfly valves. Permissible stroke range: 25% to 65%.

Signal transmission. Use sensor-matched transmitters. All measurements used for control should be transmitted using high-level 4–20 mA transmission.

Feedback controller. Match the type to the process requirements as follows:

Manual. When close regulation of the variable near its desired value is not required, and when knowledge is required that is not available in the control computer.

On/off. When the system responsed slowly to disturbances, and close regulation of the variable is not required.

Regulator. A self-contained P-only regulator offers low-cost and reliable control of noncritical variables that can be permitted to deviate from their setpoints for long periods.

PID. The proportional-integral-derivative algorithm is used for most single-loop applications.

PID control. Always determine the form of the PID algorithm being used. Select the PID modes: *P*, always; *I*, when the controlled variable should return to its set point; *D*, for processes that are undamped, unstable, or have a very large ratio of dead time/time constant.

Tuning. Typical values are $K_c = 0.8/K_p$, $T_I = 0.75 (\theta + \tau)$, and $T_d = 0.0$. The proportional band ($100/K_c$) and the reset time ($1/T_I$) can be calculated from these. When fine tuning, observe the behavior of both the controlled and manipulated variables. Always use an implementation that includes antireset windup protection for use when the manipulated variable encounters a constraint. Use an implementation that includes initialization that starts automatic control bumplessly from the last manual condition. The digital execution period should be fast with respect to the feedback dynamics, with $\Delta t \leq 0.05 (\theta + \tau)$ where possible.

Filtering. For control variables, when filtering is needed, use a first-order filter to reduce the effects of high-frequency noise. Do not excessively filter measurements unless absolutely necessary. The filter time constant, τ_f, should be much less than feedback dynamics.

For monitoring variables. Use filters to reduce the noise at frequencies higher than the effects being observed. Recall that "averaging" is a filter that is often performed by the DCS historian features.

Loop pairing. Where options exist, pair the most important variables with manipulated variables that have fast feedback dynamics and large capacity. Select pairings that give good integrity. For example, select direct loop designs in which controllers will function when other controllers are not functioning. Usually avoid nested loops (a condition when some controllers will not function while other controllers are in manual or have their outputs saturated) that do not have individual causal relationships. Select loop pairings that require little adjustment to controller tuning when the controllers switch from manual to automatic. Tune the most important loops aggressively, and the less important loosely (when interaction is unfavorable).

DCS structure. All safety and basic regulatory controls should be highly reliable and executed in the lowest-level digital processes, with direct input and output wiring contact signals within the process. Should the LAN fail, the control system should provide means for personnel to operate the process, possibly through displays and adjustable parameters at the digital processor. The system should provide redundancy (with automatic switching) for key elements such as processors, power supplies, and LANs. The operator consoles should provide sufficient access to displays and adjustments for off-normal operations such as startup and disturbances.

Implementation hierarchy. Implement process control in a hierarchy based on frequency of decision making. The first level is protection (safety, environment, and equipment), and the second is smooth operation and stability (through control of flows, temperature, pressures, and levels and through alarms). The third level is product quality. The fourth level is profitability. The final level is monitoring and diagnosis.

16.11.1.2 Properties of Fluids

1. Vapor pressure doubles every 20°C.
2. Latent heat of vaporization of steam is five times that of most organics.
3. If two liquids are immiscible, the infinite dilution activity coefficient is >8.
4. Ten percent salt in water doubles the activity coefficient of a dissolved organic.
5. Infinite dilution is essentially <1000 ppm of dissolved organic.
6. Freezing temperature may be suppressed 1°C for every 1.5 mol% impurity present.
7. A ratio of impurity concentration between a solid/liquid phase greater than 0.2 is probably due to solid solution.
8. Dissolving 2% to 20% organic solute usually reduces the interfacial tension.

16.11.2 Transportation*

In this section, we consider gas moving equipment for pressure service and then for vacuum liquid moving equipment. Then, two-phase flow is considered in Section 16.11.2.4. Hydraulic conveying of solids is discussed in Section 16.11.2.5. Conveying solids is considered in Section 16.11.2.6, while Sections 16.11.2.7 and 16.11.2.8 address piping and steam.

16.11.2.1 Gas Moving: Pressure Service

Area of Application

Fans: 0.1 to 30 kPa; 1 to 10^5 dm^3/s.
Blowers: 10 to 300 kPa; 1 to 10^3 dm^3/s.
Rotary screw: 20 to 2500 kPa; 10 to 10^4 dm^3/s.
Centrifugal compressors: 20 to 30,000 kPa; 200 to 3×10^5 dm^3/s.

---

* For more details, see D. R. Woods. *Rules of Thumb in Engineering Practice.* Wiley-NCH, 2007.

Reciprocating piston compressors: 30 to 400,000 kPa; 5 to 10^4 dm^3/s. Usually economical for >6 MPa and <150 dm^3/s or any discharge pressure and flows <100 adm^3/s.

Axial compressors: 20 to 2,000 kPa; 4×10^3 to 10^6 dm^3/s.

Guidelines

Fans: Power: up to 7.5 kW/m^3/s.
Blowers: Power: up to 125 kW/m^3/s.
For compressors,
    Adiabatic pV^k = constant
    Isothermal pV = constant
    Polytropic pV^n = constant: for uncooled internally, $n > k$; for internally cooled, $1 < n < k$, with $n \approx k$. $k = c_p/c_v$ = 1.04, increases to 1.67 as the molar mass decreases with air = 1.4 and such gases as ethylene, carbon dioxide, steam, sulfur dioxide, methane, and ammonia = 1.2 to 1.3.

Temperature rise between feed 1 and exit 2:

$$T_2/T_1 = (p_2/p_1)^{[(n-1)/n]}$$

$$(n-1)/n = (k-1)\eta/k_p$$

η_p = polytropic efficiency

For each stage, keep the exit temperature $(T_2 - 298) < 120 - 150°C$. For diatomic gases, $k = 1.4$, which limits the compression ratio (p_2/p_1) to 4; for triatomic gases, 6.

$\eta_p > \eta_{adiabatic}$. For uncooled compressors, polytropic, hydraulic, and temperature rise efficiencies are the same and range from 0.7 to 0.8 with the usual value of 0.72.

Rotary screw: Power: 100 to 750 kW/m^3/s.

Centrifugal compressors: These deliver actual volumetric flows (cubic decimeters per second, and performance should not be expressed as mass, moles, or standard volumetric flow). Assume compression ratios to be equal in all stages. The maximum number of stages that can be on one shaft or fit in the "frame" = eight minus one stage for each side nozzle. The compression ratio is 2.5 to 4. The pressure coefficient = 0.5 to 0.65; assume 0.55. The pressure differential increases with increase in suction gas density (increased molar mass or suction pressure, or decrease inlet temperature, decrease in k). Power: Up to 7.5 kW/m^3/s. Efficiency of large centrifugal compressors: 76 to 78%.

Centrifugal compressors operate between low volumetric flow rate "surge" conditions and high volumetric flow rate limited by the sonic velocity at the eye of the impeller. At "surge" conditions, the gas flow back through the compressor causes damage to the thrust bearings. The surge point is usually 0.33 to 0.5 of the normal operating capacity of the compressor. During startup, the machine goes through the surge region. The point of surge is a minimum for a single impeller. The range of stable operation decreases 5% with the addition of each impeller. High molar mass decreases the range of operation. Surge may be caused by a system disturbance (especially changes in the molar mass of the feed gas) and insufficient flow.

Surge is related to power used:

- When the molar mass of the inlet gas increases, the motor amps increase.
- If the molar mass increases by 20% and we control the suction drum pressure by recycling exit gas to the inlet (spill back control), the motor amps increase by 20%; if control is by throttle of the suction line, the motor amps increase by 10%. For every

Conceptual Process Design, Process Improvement, and Troubleshooting

10% decrease in the total number of moles compressed, the amp load on the motor drive decreases by 5%.

Reciprocating piston compressors. Compression ratios are 1.2 to 6; select to keep outlet temperature <150°C. Efficiencies for reciprocating compressors are 65% for compression ratio 1.5; 75% compression ratio 2, and 80% to 85% for compression ratios 3 to 6. Power: 70 to 1200 kW/m^3/s.

Axial compressors. Compression ratios of 1.2 to 1.5 per stage and 5 to 6.5 per machine. Efficiency: 70% except for liquid ring 50%. Power: 35 to 950 kW/m^3/s.

In general,

Velocity: pump gas 30 to 60 m/s
pump oxygen/chlorine: 20 m/s
pump steam 60 m/s

Pressure drop through pipes: 1 velocity head per 45 to 50 length/pipe diameter
Through shell and tube exchangers: 5 kPa/pass
Through wet sieve tray: 0.3 to 0.65 kPa/theoretical stage
Through packing tower: 0.2 to 0.75 kPa/m packing
    moderate- to high-pressure distillation   0.3 to 0.6 kPa/m
    vacuum distillation   0.08 to 0.16 kPa/m
    absorbers   0.15 to 0.5 kPa/m
Through porous bed Δp (particle dia.)^3 depth: 2 to 50 kPa (mm^3/m)
Through cyclone: 0.5 to 1.6 kPa
Through venturi scrubber: 0.5 to 6 kPa

16.11.2.2 Gas Moving: Vacuum Service

Area of Application

Liquid-piston pump: down to 12 kPa absolute; 0.01 to 1,000 kg/h air
Rotary sliding vane: down to 4 kPa abs.; 0.01 to 300 kg/h
Wet reciprocating: down 3 kPa abs.; 0.01 to 100 kg/h
Dry reciprocating: down to 0.1 kPa abs.; 0.01 to 300 kg/h
Mechanical vacuum: down to 0.01 kPa abs.; 0.01 to 40 kg/h
Steam ejector:
    1 stage: down to 5 to 6.7 kPa abs.; 0.01 to 100 kg/h air exhausted
    2 stage: down to 0.5 to 1.4 kPa abs.; 0.01 to 100 kg/h
    3 stage: down to 0.1 to 0.2 kPa abs.; 0.01 to 50 kg/h
    4 stage: down to 0.01 to 0.25 kPa; 0.01 to 10 kg/h
    5 stage: down to 0.001 to 0.0025 kPa abs.; 0.01 to 8 kg/h
    6 stage: 0.0004 kPa abs.

The practical limit is 0.0004 kPa abs.

Guidelines

Air leakage into unit: 50 kg/h
Liquid-piston pump: 100 to 200 kW/m^3/s exhausted air
Rotary sliding vane: 130 to 250 kW/m^3/s air exhausted
Wet reciprocating: 2 to 50 kW/m^3/s air exhausted
Dry reciprocating: 2 to 50 kW/m^3/s air exhausted

Mechanical vacuum: 2 to 50 kW/m³/s air exhausted
Steam ejector:
    1 stage: 0.002 to 10 kg steam/kg air exhausted/kPa abs
    2 stage: 100 kg steam/kg air exhausted/kPa abs
    3 stage: 1 Mg steam/kg air exhausted/kPa abs
    4 stage: 2 Mg steam/kg air exhausted/kPa abs
    5 stage: 40 Mg steam/kg air exhausted/kPa abs
Steam ejector, general:
    down to 10 kPa abs., 1 to 200 kg/h air; 1.3 kg steam/kg air exhausted/kPa abs.
    down to 30 kPa abs., 1 to 20 kg/h air exhausted; 0.1 kg steam/kg air exhausted

The compression ratio of the first-stage ejector is set primarily by the intercondenser cooling water temperature.

Assume discharge pressure to atmosphere after the last stage = average barometric pressure + 7 kPa.

Assume motive steam = minimum steam pressure in header less 5 to 10%. Pressure drop on shell side of surface condenser usually <5% of absolute design operating pressure.

A related topic is covered in Section 16.11.3.13 for interstage direct contact G-L condensers, although current practice is to use surface shell-and-tube condensers.

16.11.2.3 Liquid

Area of Application

    Centrifugal pumps: head = fluid viscosity <300 mPa·s
        end suction, single stage: head = 0.2 to 100 m; 0.05 to 4000 L/s; 0.05 to 0.7 kW/L/s; efficiency 40 to 60%
        end suction, multistage: head = 50 to 800 m; 10 to 400 L/s; 0.2 to 10 kW/L/s; efficiency 40 to 60%
    Peripheral: 10 to 300 m; 0.1 to 2 L/s; 4 kW/L/s
    Centrifugal axial: fluid viscosity <300 mPa·s: 0.3 to 10 m; >150 L/s; 0.1 kW/L/s
    Reciprocating pumps: fluid viscosity <5,000 mPa·s, diaphragm or piston: 1 to 1,000 m; <50 L/s; 0.1 to 3 kW/L/s
    Rotary screw: fluid viscosity, usually above 10 Pa·s; 0.2 to 300 m; <120 L/s

Guidelines

    Optimum exit pipe size selection:
        pump liquids at >1.5 m/s
        hydrocarbons with low conductivity <0.3 m/s
    Suction pipe of larger diameter to prevent cavitation
    NPSH requirement [m] = $\{rpm/(5400 \text{ volumetric flow rate } [L/s])\}^{1.33}$
    Centrifugal pumps operate on their operating head-capacity curve. Head-capacity curves are independent of the fluid, although the curve drops slightly at higher capacities for higher-viscosity fluids. For centrifugal pumps, the drive power required and pressure at the exit flange depend on the fluid density. Reciprocating pumps are constant-volume devices, producing essentially constant "pressure."
    Pressure drop through pipes: 1 velocity head per 45 to 50 length/diameter; for water, 23 kPa/100 m pipe
    Pressure allowance across a control valve for good operability: 20 to 50% of the dynamic head loss or 70 to 140 kPa

Pressure drop:
   shell and tube exchanger: 70 kPa
   plate exchanger: 50 kPa
   dialysis: 50 to 10 MPa
   RO: 0.1 to 4 MPa
   Filter press: 70 kPa
   Porous bed: 0.3 to 7 kPa/m depth of bed/L/m²·s usual: 5 to 10 bed volumes per h; 1 to 10 L/m²·s, usually limited to <80 kPa

Similitude for centrifugal devices: volumetric flow rate ratios = (diameter of impeller ratio) times (impeller rpm ratio); head ratios = (diameter ratio)² times (the impeller rpm ratio)²; power ratio = (diameter ratio)³ times (impeller speed ratio)³.

16.11.2.4 Gas-Liquid (Two-Phase Flow)

Table 16.31 illustrates different flow regimes for vertical cocurrent upflow, vertical countercurrent, and horizontal flows. This also pertains to boilers (Section 16.11.3.3) and to evaporators (Section 16.11.4.1).

Two-phase flow over a packing is given in Section 16.11.6.15 for trickle bed reactors.

16.11.2.5 Pumping Slurries: Liquid-Solid Systems

Area of Application

Solid concentration: 25% to 65% w/w; solid particle size: 20 to 300 μm. For finer particle size, watch for non-Newtonian behavior. For higher concentrations, consider short pumping runs only; for small concentrations, beware of settling out. For larger particles, watch for settling out.

Guidelines

Pumping velocity: 1 to 4 m/s; 0.06 to 0.7 kW/Mg/h per km distance. Loading 0.3 to 1.7 kg solids/kg water. Economic trade-off between solids loading and pumping costs. Try to use the heterogeneous flow regime and consider adjusting pH to alter stability of dispersion.

TABLE 16.31
Superficial Velocities of Liquids and Gases, m/s

	Horizontal		Vertical Cocurrent Up		Vertical Cocurrent Down		Vertical Countercurrent Up	
	Liquid	Gas	Liquid	Gas	Liquid	Gas	Liquid	Gas
Dispersed: nearly all liquid entrained	Same as gas	>60	>1	<15				
Annular: liquid ring; central gas	<0.15	>6	<8	15–80	0.01–0.15	<0.5–10	0.01–0.2	<15
Stratified: liquid along bottom	<0.15	0.15–3						
Slug: periodic liquid slug	4.5 but < vapor	0.9–15	<1	1–15	0.15–0.8	<20		
Plug: alternating slugs of liquid & gas	<0.6	<1.2	<0.5	0.1–1.5				
Bubbles	1.5–4.5	0.15–0.6	<8	<0.1–1	>0.8	<8		
Choked/flooded			>10	>80	>1–100	>80	0.01–>0.2	0.1–>15
Part dry			<0.05	>20	<0.01	<80	<0.01	

16.11.2.6 Solids

Choice depends on particle characteristics (size, flowability, corrosiveness, abrasiveness, handling characteristics, safety-hazard [static electrification, fumes, flammability]), and vertical versus horizontal distance. A related topic is covered in Section 16.11.2.5, and bins for storage (Section 16.11.6.35).

Area of Application

> *Belt conveyors:* <10° incline; 10 to 100 m horizontal distance; capacity 3 to 270 kg/s. OK for most diameters of particles but not for particles that cake or are light and fluffy.
> *Bucket/belt elevators:* usually for >25 m vertical; 15 to 150 Mg/h; usually not for particles <150 µm dia. and not for particles that cake or are light and fluffy.
> *Screw conveyors:* 2 to 75 m horizontal distance; 0.3 to 275 kg/s. Not for particles that cake or are light and fluffy. Can be used for inclines up to 20°.
> *Vibratory feeder:* <20 m; 1 to 400 Mg/h; not for light and fluffy materials or particles <150 µm dia.
> *Apron feeder:* <18° incline; 2 to 12 m horizontal; 10 to 150 Mg/h; not for particles that cake or are light and fluffy or are fine with <50 µm dia.
> *Pneumatic transportation:* limited by solids loading that plugs pipe.
> > *Dilute phase:* pressure: continuous: particle dia. 60 µm to 0.3 cm; pressure drop <100 kPa. Distance <600 m. One-point collection and several-point delivery.
> > *Dilute phase:* vacuum: continuous: particle dia. 60 µm to 0.3 cm; Δp pressure drop <50 kPa. Distance <50 m. One-point delivery and several-point delivery.
> > *Dense phase:* pressure: batch: particle dia. 60 µm to 0.3 cm; pressure drop 550 to 700 kPa. High capacity (<10 kg/s) over long distances <2300 m. For materials that degrade easily or are highly abrasive.

Guidelines

Caution for all: dust explosions: dust explosion potential for particle dia. <200 µm. Minimum ignition temperature >300°C; minimum ignition energy 10 to 30 mJ.

> *Belt conveyors:* keep speeds <1 m/s for fines; otherwise speeds in the range 2.7 to 4 m/s; burden thickness (cm) = 0.17 (volumetric capacity, dm^3/s)/{(belt speed, m/s)(belt width, m)}. Belt width 0.5 to 0.8 m.
> *Speed:* 0.8 to 2 m/s.
> *Power:* 0.02 to 0.4 kW/Mg/h per km horizontal distance.
> *Bucket conveyor:* vertical: velocity <1.5 m/s; for material of density 0.4 Mg/m^3 capacity <16 kg/s; for 2 Mg/m^3 capacity <85 kg/s or <40 m^3/s. Power 0.15 to 0.35 kJ/kg or 0.013 to 0.023 kJ/kg/m of height.
> *Screw conveyors:* 10 to 120 rpm and trough loading 15 to 95%, depending on the particle size, flowability, abrasiveness. Dia. 0.3 to 0.4 m. Power 10 to 20 kW/Mg/h per km horizontal distance.
> *Rotary, star valve feeder:* used especially as solids feeders for dilute-phase pneumatic conveying to provide an air lock and to feed solids. Seal/wear depends on Δp and abrasiveness of powder. For pressure systems keep Δp <80 kPa; for vacuum systems Δp <40 kPa. Provide an air vent to take the air loss away from the gravity flow of the solids and to control the filling of the star.
> *Pneumatic conveying:* dilute phase: pressure: continuous: nominal gas velocity 5 to 35 m/s with usual 11 to 25 m/s. Solids loading 3.5 to 15 kg solid/kg air with usual 6 to 15 kg solids/kg air or 1 to 7 m^3 solids/m^3 air. Power 7 to 11 kJ/kg. Problems: about 30% air leakage out of the system. Rotary/star valve problem/bridging: overcome with bin

Conceptual Process Design, Process Improvement, and Troubleshooting

agitation, astute bin design, and vent star valve to prevent the backflow of gas through the valve into the bin. Minimize bends.

Pneumatic conveying: dilute phase: vacuum: continuous: nominal gas velocity 20 to 35 m/s. Solids loading 2.8 to 11 kg solids/kg, gas with usual 3 to 4 kg solids/kg/air. Power 11 to 18 kJ/kg. Rate and distance sensitive.

Pneumatic conveying: dense phase: pressure: pressure, batch; cycle time 1 to 4 min; charge 10 s; convey 0.5 to 3 min; 20 to 50 batches/h, nominal gas velocity: horizontal 4.5 to 35 m/s with usual 6 to 10 m/s; vertical 1.5 to 27 with usual 3 m/s. Solids loading: horizontal 12 to 130 kg/kg; vertical 10 to 800 with usual 250 to 400 kg solids/kg air. Blow tank <15 m^3. Line dia. 3 to 10 cm. Power 3 to 5.5 kJ/kg. Watch for humid air and line plugging. The longer the distance, the lower the solids loading or the more gas needed to convey.

Enclosed conveyor: 0.09 to 9 dm^3/s; conveyor size 7.5 to 25 cm; travel speed 0.02 to 0.3 m/s.

16.11.2.7 Ducts and Pipes

Guidelines

Pressure drops: see gas, Section 16.11.2.1; and liquid, Section 16.11.2.3.

Velocities, see gas, Section 16.11.2.1; and liquid, Section 16.11.2.3. Keep the velocity of compressible gases <0.6 sonic velocity.

Sonic velocity, m/s = 1.97 $[(c_p/c_v)$ pressure, kPa, $\times$ density, kg/m$^3]^{0.5}$.

Sewer pipes have maximum flow rate when liquid level is 93% of the diameter; flow rate when full = flow rate when liquid level is 80% of the diameter.

Schedule 40 = usual specification; schedule 80 is heavy duty.

16.11.2.8 Steam

Area of Application

Steam for power via turbine drives (Section 16.11.3.1) and via vacuum ejectors (Section 16.11.2.2). Thermal heating (Section 16.11.3.3), separations via evaporators (Section 16.11.4.1), distillation (Section 16.11.4.2), and dryers (Section 16.11.5.5).

Guidelines

For delivery, gas velocity 60 m/s. Typical steam gauge pressures: 100 to 200 kPa, 120 to 130°C. *Comment*: tends to be sluggish in response to changes in demand. 1 MPa, 185°C; 1.7 MPa, 210°C; 2.75 MPa, 230°C; 4.1 MPa, 255°C.

Steam leakage: 40 g/s at 0.7 kPa through a 6-mm dia. hole. Leak rate proportional to (hole diameter)^2 × pressure.

See also steam traps to separate condensate from steam (Section 16.11.5.1) and steam generation, furnaces (Section 16.11.3.2).

Steam flow through a control valve (called *wiredrawing*) occurs at constant enthalpy with a resulting increase in temperature or superheat.

16.11.3 Energy Exchange*

In this section, we consider drives and then a wide range of thermal energy exchangers: furnaces, heat exchangers, indirect fluidized beds, motionless mixers, direct contact liquid-liquid, kilns, fluidized beds, and multiple-hearth furnaces. Drying is considered in Section 16.11.3.10. Various direct-contact gas-liquid exchangers are considered in Sections 16.11.3.11 through 3.13. Thermal

---

* For details, see chapter on heat transfer. Reproduced from *Rules of Thumb for Process Design and Engineering Practice* with permission from Donald R. Woods, ©2001.

wheels are discussed in Section 16.11.3.14. Sections 16.11.3.15 through 16.11.18 discuss solidification of solids, heat loss to the atmosphere, refrigeration, and steam generation.

16.11.3.1 Drives

Area of Application

Gasoline/diesel engines: 200 to 800 rpm; >80 kW; efficiency: 28 to 38%.
Electric motors (synchronous): <500 rpm; 35 to 500 kW; use 480 V for motors up to 115 kW; 4160 V for >115 kW.
Electric motors (induction): >500 rpm; 10 to 15000 kW.
Steam turbine: single-stage, single-valve: 1000 rpm to 12,000 rpm; 50 to 1500 kW
  Single-valve, multistage: 1500 to 2800 kW
  Multivalve, multistage: 2800 to 30,000 kW
Gas combustion turbine: cogeneration

Guidelines

For less than 75 kW, select motor or turbine.

Gasoline-diesel engines usual application: 200 to 400 rpm; 500 to 1200 kW.
Electric motors: select synchronous for low speed; usual application for either synchronous or induction: 500 to 2100 rpm; 150 to 500 kW. Usually use total enclosed Fan cooled (TEFC) enclosure, efficiency: 84 to 95%.
Induction: available for large power requirements, relatively low efficiency, power factor is low if rpm <500 and at starting and fractional loads.
Synchronous: high efficiency at any speed, suitable for direct coupling for <1000 rpm. Power factor >1; constant speed without slip. Power consumption, kW = amps (0.001 volt $\times$ PF $\times 0.95 \times \sqrt{3}$).

Power, kW	Volts, V
0.1 to 1	110
1 to 75	220 to 440 3 phase
50 to 200	440 to 2,300
175 to 2000	2,300 to 4,160
>2000	11,000 to 13,200

Steam turbines: competitive above 75 kW; condensing: 2 kg/h steam/kW with 1.8 m^2 condenser surface area per kg/h steam. Multivalve, multistage efficiency: 42 to 78%.

16.11.3.2 Thermal Energy: Furnaces

Multiuse including heating, boiling, reactions. Related topics are distillation (Section 16.11.4.2) and reactors (Section 16.11.6.4).

Area of Application
250 to 1300°C; <30 MPa; thermal efficiency: 70 to 75%.

Guidelines
Heat flux in radiant section: 10 to 60 kW/m^2 based on outside tube area with fluid velocity inside tubes 0.1 to 3 m/s. Use 1.5 m/s. In the convection section: 12 kW/m^2.
  Equate heat duties in the radiant and convection sections; 80% efficiency based on net heating values.
  Size radiant section to absorb 50% of the radiant energy with 1.22 m^3 chamber per m^2 tube area.

Field fabricated for sizes above 5 MW total heat load.
Gas catalytic endothermic reaction inside tube, heat transfer coefficient $U = 0.045$ kW/m^2·K.

16.11.3.3 Thermal Energy: Fluid Heat Exchangers, Condensers, and Boilers

Exothermic processes should supply all the heat requirements for the process.

Area of Application

Shell-and-tube: −200 to 600°C; <30 MPa; 0.15 to 4 THTU/pass and 5 to 150 kPa/THTU; fluid viscosities <100 mPa·s. Area: 2 to 2000 m^2.
   Fixed tube sheet limited to low thermal expansion or if $\Delta T \leq 30°C$.
   Usually need U-tube or floating head.

Spiral: −100 to 400°C; <1.8 MPa; use with sludges, slurries, high-viscosity materials (especially 4×10^4 to 4×10^5 mPa·s); >90% heat recovery. High heat transfer coefficients; low pressure drop. Ratio of flow rates being handled <3.5. Not when $\Delta T_1 \gg \Delta T_2$. Area: 4 to 100 m^2.

Lamella: −200 to 500°C; <3 MPa. Ratio of flow rates 1 to 1.8. Area: 100 to 10,000 m^2.

Gasket/plate: −30 to 180°C; <2.5 MPa, depending on the gasket material; areas <2,000 m^2; 0.2 to 3 THTU/pass and 15 to 200 kPa/THTU; fluid viscosities $< 4 \times 10^4$ mPa·s. Holdup low, 1.5 L/m^2. Ratio of flow rates 0.7 to 1.3. Not when $\Delta T_1 \gg \Delta T_2$. Area: 10 to 600 m^2.

Double-pipe: −200 to 600°C; <30 MPa; usually <20 m^2. Use for high pressure. Area: 0.3 to 200 m^2.

Air-cooled: OK when air can be used as coolant. Low maintenance, area: 20 to 2000 bare tube m^2. Finned area 16:1.

Cubic/monolithic: corrosive liquids, acids, bases, or used as catalyst/heat exchanger for reactors. Usually made of graphite or carbon that has high thermal conductivity, area: 1 to 20 m^2. Ceramic monoliths are used as solid catalyst for highly exothermic gas-catalyst mass transfer-controlled reactions.

Agitated falling film: usually to concentrate slurry; see evaporation, Section 16.11.4.1.

Agitated horizontal film (votator): usually to condition foodstuffs, crystallize, and react. Especially for viscous feed. OK for foaming, fouling, crystal formation, and suspended solids. Viscosities >2000 mPa·s. Relative to agitated film retention time of 1:1 and volume 1:1. Overall heat transfer coefficient 2 kW/m^2·°C, decreasing with increasing viscosity; 3 to 12 kW/m^2. See also evaporation, Section 16.11.4.1.

Coil in tank: area: 1 to 30 m^2.

Jacketed: usually to exchange heat in a reactor or storage tank.

Cooling finger: added to tanks to increase the exchange area.

Guidelines

Media for heating and cooling:
   Water: 18°C to maximum outlet 50°C with velocity >1.2 m/s
   Air: 18°C to maximum outlet 50°C
   Steam: 1.7 MPa: 203 to 220°C
            43 MPa 260°C
For pressures 0.1 to 0.2 MPa:
   −100°C, consider ethane
   −50°C, consider propane, propylene
   −30°C, consider ammonia
   0°C, consider butane

175°C, consider Dowtherm J
200 to 400°C, consider molten salt
275°C, consider Dowtherm A
310°C, consider Dowtherm G

Shell-and-tube:
Use shell-and-tube exchangers for gas-gas and for low-viscosity liquid-liquid systems (<200 mPa·s).

Use floating head if the temperature difference between shell-and-tube fluids exceeds 30°C (to minimize impact of thermal expansion).
Surface compactness: for ordinary tubes: 70 to 500 m^2/m^3; for finned tubes 65 to 3300 m^2/m^3.

Number of Shell Passes Affected by Temperature Ratio $\dfrac{(t_{hot/in} - t_{hot/out}) + (t_{cold/out} - t_{cold/in})}{(t_{hot/in} - t_{cold/in})}$	No. of Shell Passes
0–0.8	1
0.8–1.1	1 or 2
1.1–1.3	2
1.3–1.4	2 or 3
1.4–1.5	3

Tube velocity >1 m/s; for crude preheat exchangers, tube velocity is 1 to 2 m/s; for overhead water condensers, tube velocity is 1 to 3.5 m/s. To minimize fouling, keep tubeside velocities 3 to 4.5 m/s, reduce the wall temperature, and use single-flow channels.

Nozzle velocity <2 m/s and Δp, kPa = 0.75 to 0.9 (nozzle velocity, m/s)^2.

Baffle cut <30%; baffle pitch: minimum should be the maximum of 5 cm or 1/5 of ID shell diameter. The maximum should be the smaller of 0.75 m or shell ID diameter.

For shell-and-tube heat exchange: numerous related topics including evaporation (Section 16.11.4.1), distillation (Section 16.11.4.2), crystallization (Section 16.11.4.6), freeze concentration (Section 16.11.4.3), melt crystallization (Section 16.11.4.4), and PFTR reactors (Section 16.11.6.5 and .6). Approach temperature 5 to 8°C; use 0.4 THTU/pass; design so that the total pressure drop on the liquid side is about 70 kPa. Allow four velocity heads pressure drop for each pass in a multipass system. Put inside the tubes the more corrosive, higher-pressure, dirtier, hotter, and more viscous fluids. Recommended liquid velocities 1 to 1.5 m/s with maximum velocity increasing as more exotic alloys are used. Use triangular pitch for all fixed tube sheets and for steam condensing on the shell side. Try $U = 0.5$ kW/m^2·°C for water/liquid; $U = 0.3$ kW/m^2·°C for hydrocarbon/hydrocarbon; $U = 0.03$ kW/m^2·°C for gas/liquid, and 0.03 kW/m^2·°C for gas/gas. Finned tubes: see air-cooled.

For shell-and-tube condensation: related topics include evaporation (Section 16.11.4.1) and distillation (Section 16.11.4.2). Prefer condensation outside horizontal tubes; use vertical tubes when condensing immiscible liquids to subcool the condensate. Assume pressure drop of 0.5 of the pressure drop calculated for the vapor at the inlet conditions. Baffle spacing is 0.2 to 1 times the shell diameter with the baffle window about 25%. Limit pressure drop for steam to 7 kPa on the shell side. $U = 0.5$ to 0.85 kW/m^2·°C.

For shell-and-tube boiling: approach temperature <25°C to ensure nucleate boiling. Related topics include evaporation (Section 16.11.4.1), distillation (Section 16.11.4.2), solution crystallization (Section 16.11.4.6), and reactors, PFTR nonadiabatic (Section 16.11.6.6).

Kettle: use for clean, relatively low-pressure, nucleate pool boiling; not for foams; 75 to 100% vaporized. $U = 1.1$ kW/m^2·°C.

Vertical/horizontal thermo syphon: use for clean, relatively low-pressure, nucleate pool boiling; 5 to 25% vaporized. Use 45 to 56 kW/m² heat flux for hydrocarbon and petrochemicals, 62 to 75 for aqueous, and 37 to 44 for vacuum. Rarely used for vacuum or very high-pressure service. $U = 1$ kW/m² K.

Forced circulation: use for vacuum or viscous/fouling fluids. Sensible heat only. Tube velocity 3 to 4.5 m/s.

For various configurations of evaporators, see evaporation, Section 16.11.4.1.
For steam usage, see steam traps, Section 16.11.5.1.
For shell-and-tube gas reactor with fixed catalyst inside tubes: related to reactor, Sections 16.11.6.6 and 16.11.6.7.

Endothermic reaction in catalytic fired tube: $U = 0.045$ kW/m²·°C.
Exothermic reaction in catalytic multitube, nonadiabatic, with boiling water or molten heat transfer salt or liquid: 0.05 to 0.12 kW/m²·°C.

Spiral: for large flows and alloys construction select spiral over shell-and-tube. Heat transfer coefficients about double that of shell and tube, liquid-liquid $U = 2.8$ to 4 kW/m²·°C. Surface compactness: up to 185 m²/m³.

Lamella/plate (Raman): approach temperature 2°C; for liquid-liquid $U = 1.7$ to 2.5 kW/m²·°C.

Surface compactness: 150 to 5900 m²/m³.

Gasket/plate: approach temperature 1°C; heat transfer coefficients about 3 to 5 times higher than shell-and-tube, for liquid-liquid $U = 4$ to 4.5 kW/m²·°C. Surface compactness: 120 to 1000 m²/m³.

Double-pipe: for heat exchange: use for <20 m²; attractive for the condensation of reacting gases because it provides least area per unit volume.

Air-cooled: two fins/cm of tube length; use finned tubes on gas side to give less weight, less pressure drop, but not for dirty gases on the fin side because of difficulty in cleaning. Use when the potential finned side coefficient is small (0.05 to 0.1 kW/m²·°C) and when the ratio of tube side to gas side coefficient is 75 to 300, but not when it is 1 to 25. Use $h_o A_o$, not area only, for optimization (because heat transfer coefficient decreases as the fin area increases). Finned tubes with 15 to 20 m² total area/m² bare tube area, 0.4 to 0.6 kW/m²·°C of bare tube; approach temperature >26°C; fan power is 5 to 13 kW/kW thermal energy transferred or 0.15 kW/m² of bare tube.

Cubic/monolithic: corrosive acids and bases: liquid-liquid $U = 0.5$ to 1.1 kW/m²·°C. For gas-catalyst heat exchange, $U = 3.5$ to 7.5 kW/m²·°C. Heat transfer areas up to 2200 m².

Agitated falling film: see evaporation, Section 16.11.4.1.

Agitated horizontal film (votator): thin liquids $U = 2.2$ to 4 kW/m²·°C; viscous films $U = 0.8$ to 2.3 kW/m²·°C; crystallization $U = 0.15$ to 3.7 kW/m²·°C; sulfonation $U = 0.8$ to 2.2 kW/m²·°C; polymerization $U = 1$ to 2.8 kW/m²·°C. See also evaporation, Section 16.11.4.1.

Coil in tank: heating $U = 0.6$ to 2.5 kW/m²·°C; cooling $U = 0.2$ to 1.2 kW/m²·°C.

Jacketed: heating $U = 0.2$ to 1.7 kW/m²·°C; cooling $U = 0.1$ to 0.6 kW/m²·°C.

Cooling finger: cooling $U = 0.5$ to 1.2 kW/m²·°C.

16.11.3.4 Thermal Energy: Fluidized Bed (Coils in Bed)

Related topics include direct heat transfer (Section 16.11.3.8), reactors (Section 16.11.6.27), mixing (Section 16.11.7.1), dryers (Section 16.11.5.5), and size enlargement (Section 16.11.9.5).

Area of Application

Use for highly exothermic reactions or for heating/cooling solids. Mixing provides uniform temperature in the bed and solids increase the heat transfer coefficients 5 to 25 times that of the gas alone.

Guidelines

Coils in fluidized bed: $U = 0.2$ to 0.4 kW/m²·°C; bed to wall $U = 0.45$ to 1.1 kW/m²·°C; solids to gas $U = 0.017$ to 0.055 kW/m²·°C; gas-solid thermal conductivity 0.17 to 42.5 kW/m·K.

16.11.3.5 Thermal Energy: Motionless Mixers

Related topics include reactors (Section 16.11.6.10), mixing (Section 16.11.7.3), size reduction, gas-liquid (Section 16.11.8.1), and liquid-liquid (Section 16.11.8.3).

Area of Application

Use for viscosity liquids <50 mPa·s and liquid reactions (Section 16.11.6.9).

Guidelines

Cooling melts $U = 0.15$ to 0.25 kW/m²·°C. Polymerization: 10 kW/m²·°C. Internal heat transfer coefficient $h_{inside} = 3 \times$ value for fluid flowing in empty pipe of same diameter and length.

16.11.3.6 Thermal Energy: Direct Contact Liquid-Liquid Immiscible Liquids

A related topic is size reduction (Section 16.11.8.3).

Area of Application

Exchange heat across a liquid-liquid interface. Droplet diameter and area per unit volume estimated from size reduction (Section 16.11.8.3).

Guidelines

> *Tray, sieve:* $U_v = 160$ kW/m³ °C
> *Wetted wall:* $U_v = 20$ kW/m³ °C
> *Baffles:* $U_v = 20$ kW/m³ °C;
> *Spray tower:* $U_v = 2$ to 100 kW/m³ °C
> *RTD:* $U_v = 260$ kW/m³ °C
> *Pipeline:* $U_v = 200$ to 1000 kW/m³ °C

16.11.3.7 Thermal Energy: Direct Contact Gas-Solid Kilns

A related topic is reactors (Section 16.11.6.17).

Area of Application

Temperature 520 to 1700°C; atmospheric pressure; particle diameter 7 µm to 20 mm; solid residence time 2000 to 35,000 s.

Used for drying (Section 16.11.5.5), incineration, and gas-solid reactions (Sections 16.11.6.17 through 16.11.20).

Guidelines

> *Rotary cascading kiln dryer:* heat transfer coefficient $U_m = 0.1$ W/kg solids °C for coarse materials and 0.4 W/kg °C for fine materials.

Incineration: volumetric loading 5% solids; temperatures 820 to 1600°C; heat release from combustion 260 to 415 kW/m³; gas velocity 4.5 to 6 m/s; solids residence time 1 to 2 h; gas residence time 2 to 4 s. L/D 3.4 to 4/1.

Reactor: volumetric solids loading 3 to 12%; heat usage: 25 to 60 kW/m³ of kiln volume; solids capacity for cement 0.4 to 1.1 Mg/d·m³ kiln volume; solids capacity pigments, dolomite 0.1 to 2 Mg/d·m³ kiln volume.

16.11.3.8 Thermal Energy: Direct Contact Gas-Solid Fluidized Beds

Related topics are indirect heat transfer (Section 16.11.3.4), reactors (Section 16.11.6.24), mixing (Section 16.11.7.2), dryers (Section 16.11.5.5), and size enlargement (Section 16.11.9.5).

Area of Application

Particle diameter 40 to 100 µm.; surface area 20,000 to 100,000 m²/m³ bed. Used for drying (Section 16.11.), incineration, gas-solid, and gas catalyst reactions (Section 16.11.6.24).

Guidelines

Gas and particles leave the bed at the same temperature. Particle-gas heat transfer coefficient $U = 0.01$ to 0.06 kW/m²·K.

Incineration: volumetric loading 5% solids; temperatures 650 to 980°C; heat release from combustion 200 to 340 kW/m³; superficial gas velocity 60 to 75 g/s·m² of cross-sectional area for wet solids; 15 to 20 g/s·m² of cross-sectional area for dry solids combustion.

16.11.3.9 Thermal Energy: Direct Contact Gas-Solid Multiple-Hearth Furnaces

Related topics include dryers (Section 16.11.5.5) and reactors (Section 16.11.6.15).

Area of Application

Used for drying (Section 16.11.5.5), reactions (Section 16.11.6.15), and incineration.

Guidelines

Incineration: temperatures 790 to 980°C; heat release from combustion 150 to 250 kW/m³.
Reactors: solids residence time 4 to 5 h.

16.11.3.10 Thermal Energy: Gas-Solid Drying of Solids

Topics in Sections 16.11.3.7 through 16.11.3.9 can be used to dry solids. Other dryers are described in Section 16.11.5.5.

Area of Application
See Section 16.11.5.5.

Guidelines

Solutions of salts: overall heat transfer coefficient: $U = 0.12$ to 0.35 kW/m²·°C.
Slurries, powders, and granules: overall heat transfer coefficient: $U = 0.03$ to 0.23 kW/m²·°C.
Drum dryer (conduction): overall heat transfer coefficient: $U = 0.001$ to 0.002 kW/m²·°C.
Jacketed screw (conduction): overall heat transfer coefficient: $U = 0.015$ to 0.06 kW/m²·°C.
Agitated pan (conduction): overall heat transfer coefficient: $U = 0.01$ to 0.05 kW/m²·°C.
Tray/pan shelf (conduction; vacuum): overall heat transfer coefficient: $U = 0.02$ to 0.5 kW/m²·°C.

Pneumatic or flash dryers: wall to gas/particle systems: $U = 0.1$ kW/m²·°C.
Spray dryer: volumetric heat transfer coefficients: $U_v = 0.13$ to 0.18 kW/m³·°C.
Rotary steam tube dryer: heat transfer coefficient: $U = 0.03$ to 0.09 kW/m²·°C.
Rotary cascading dryer: heat transfer coefficient: $U_m = 0.1$ W/kg solids °C for coarse materials and 0.4 W/kg·°C for fine materials.
Continuous freeze dryer: overall heat transfer coefficient: $U = 0.001$ to 0.01 kW/m²·°C, depending on the vacuum and residence time, with smaller values for lower total pressure and longer residence time.

16.11.3.11 Thermal Energy: Direct Contact Gas-Liquid Cooling Towers

Area of Application

Use of the latent heat of evaporation of 9% of hot water to cool the water by removing sensible heat from hot water; size range: 60 to 1500 L/s.

Guidelines

Volumetric heat transfer coefficients $U_v = 0.5$ to 15 kW/m³ °C; gas superficial velocity 1 m/s; liquid loading 1 L/s·m² cross-sectional area.

16.11.3.12 Thermal Energy: Direct Contact Gas-Liquid Quenchers

Area of Application

The quenching media can be any liquid. Gas flow rate 4 to 100 m³/s.

Guidelines

Tray: heat transfer coefficient, $U = 7$ to 20 kW/m²·°C. of tray area; volumetric heat transfer coefficient, $U_v = 3$ kW/m³ °C; superficial gas velocity =1 to 1.6 m/s; mass loading liquid/gas = 10 to 20/1.
Packing: volumetric heat transfer coefficient, $U_v = 3$ kW/m³·°C; superficial gas velocity = 1 m/s; mass loading liquid/gas = 1.5 to 3/1.
Spray tower: volumetric heat transfer coefficient, $U_v = 1.8$ to 5 kW/m³·°C; superficial gas velocity = 1 m/s; mass loading liquid to gas = 1 to 50/1.
Baffles: volumetric heat transfer coefficient, $U_v = 0.5$ kW/m³ °C; superficial gas velocity = 0.7 to 1 m/s; mass loading liquid to gas = 15 to 60/1.

16.11.3.13 Thermal Energy: Direct Contact Gas-Liquid Condensers

Area of Application

The condensing medium is water to condense steam or to cool gas. Use in steam ejector systems, although these are being replaced by indirect condensers. A related topic is vacuum service (Section 16.11.2.2). Body dia., 0.1 to 2.5 m; liquid flow rate, 3 to 700 L/s.

Guidelines

Liquid loading 43 to 85 L/s·m² cross-sectional area. Height/diameter = 5/1. Keep temperature increase in the cooling water limited to 8.3°C for operating pressures <6 kPa, limited to 17°C for operating pressures 6 to 25 kPa, limited to 25°C for operating pressures >25 kPa; and keep overall temperature <50°C to minimize erosion and corrosion.

Gas cooling: $\Delta T = 16$ to 18°C.
Steam condensation: $\Delta T = 2.75$°C; approach temperature at both ends of the condenser = 2.75°C.

Volumetric heat transfer coefficient U_v = 100 to 150 kW/m^3 °C, but value is very sensitive to sensible heat load; usually about 30 kg water/kg steam condensed.

16.11.3.14 Thermal Energy: Gas-Gas Thermal Wheels, Pebble Regenerators, and Regenerators

Area of Application

Hot gas temperature: 180 to 950°C or cryogenic conditions; 80 to 180 kPa absolute. Usually used to recover "waste" heat. For pebble beds, temperatures up to 1700°C. Nominal gas flow rate to one side, 0.6 to 20 m^3/s.

Guidelines

For air to solids: U = 0.02 kW/m^2·°C; for hydrogen to solids = 0.1 kW/m^2·°C.

> *Pebble bed:* heat transfer coefficient, U_v = 280 kW/m^3 K volume of pebbles; contact time, fractions of seconds. Surface compactness: up to 15,000 m^2/m^3.
> *Thermal wheel:* contact time fraction of seconds; area 500 to 650 m^2/m^3 per wheel; gas face velocity 1 to 5 m/s, use 3.5 m/s; rpm 10 to 20. Surface compactness: up to 6600 m^2/m^3.

16.11.3.15 Thermal Energy: Solidify Liquids

A related topic, size enlargement, is discussed in Section 16.11.9.15.

Guidelines

> *Flaker:* heat transfer coefficient, U = 0.35 kW/m^2·°C.
> *Chilled belt:* heat transfer coefficient, U = 0.35 kW/m^2·°C.
> *Prilling tower:* volumetric heat transfer coefficient, U_v = 0.005 kW/m^3·°C.

16.11.3.16 Thermal Energy: Heat Loss to the Atmosphere

Area of Application

Depends on the type and thickness of insulation and the external weather conditions.

Guidelines

Heat transfer coefficient, uninsulated into still air, U = 0.005 to 0.01 kW/m^2·°C; insulated into still air, U = 0.003 kW/m^2·°C. Values increase as wind velocity increases: for 10 km/h, multiply U by 2; for 20 km/h, multiply U by 3.

Walls to wet earth, U = 0.0015 kW/m^2·°C; walls to dry earth, U = 0.001 kW/m^2·°C; floor to ground, U = 0.0007 kW/m^2·°C.

16.11.3.17 Thermal Energy: Refrigeration

Area of Application

Mechanical vapor recompression for temperatures <20°C, ranging from 10 to –62°C. Refrigeration capacity 20 to 5000 kW.

Guidelines

About 0.2 kW compressor power/kW refrigeration.

16.11.3.18 Thermal Energy: Steam Generation

See thermal energy furnaces/boilers (Section 16.11.3.2). See Section 16.11.2.8 for steam distribution.

16.11.4 HOMOGENEOUS SEPARATION*
Overall Guidelines

1. Keep it simple.
2. Exploit differences in properties between the key target species.
3. Consider mixing/blending of streams before considering separation.
4. Remove lightest component, most abundant, least expensive first; or remove the most difficult last.
5. Leave high specific recoveries until last.
6. Try to avoid adding an agent to achieve the separation.
7. Try to avoid extremes in operating conditions.
8. Consider shifting from homogeneous phase separation to heterogeneous phase separation.
9. Consider reaction to shift the species to another form.

16.11.4.1 Evaporation

This is related to crystallization (Section 16.11.4.6).

Area of Application

α_{vp} = 20 to 5000. Liquid feed concentration of target solute 2 to 50%; 1 to 40 kg/s feed rate. One product. Usually used when no solids, nonfoaming and doesn't thermally degrade, although special designs can accommodate these. Can be operated as multistage with up to 3 to 10 stages.

Guidelines

Dissolved solids elevate the boiling temperature between the solution and saturated vapor by 1.5 to 5.5°C.

Temperature-sensitive materials: use the Hickman decomposition hazard index, *HDHI*, which is expressed as the amount of time the material is kept under temperature (as measured by pressure, Pa) in units of Pa·s.

> *External short tube, vertical exchanger, natural circulation:* OK for foaming but not for fouling, crystal formation, or suspended solids. Viscosities <20 mPa·s. Relative to agitated film retention time of 16:1 and volume 10:1. Overall heat transfer coefficient 1 to 1.5 kW/m² ·°C. HDHI = 10^8 to 10^9 Pa·s. Area 3 to 1000 m².
>
> *External short tube, vertical exchanger, forced circulation:* OK for foaming, fouling, crystal formation, and some suspended solids. Viscosities <150 mPa·s. Relative to agitated film retention time of 42:1 and volume 13:1. Overall heat transfer coefficient 0.75 to 3 kW/m²·°C, decreasing with increasing viscosity. Power 0.13 to 0.5 kW/m². HDHI = 10^7 to 10^8 Pa·s. Area 3 to 600 m².
>
> *External kettle reboiler, natural circulation:* viscosities <200 mPa·s.
>
> *Internal calandria, vertical short tube:* preferred for fouling and crystal forming systems. OK for foaming, fouling, crystal formation, and some suspended solids. Viscosities <1000 mPa·s. Relative to agitated film retention time of 168:1 and volume 3:1.

---

* Reproduced with permission from Donald R. Woods, *Rules of Thumb in Engineering Practice,* pages 87 to 135, 2007, copyright Wiley-VCH Verlag GmbH & Co. KGaA, Weinheim, ISBN: 978-3-527-31220-7. See this book for more details, especially about Good Practice, Trouble shooting and Capital Costs.

Long tube, rising film: especially for clear and relatively dilute feed, OK for foaming, but not for fouling, crystal formation, or suspended solids. Viscosities 150 to 1000 mPa·s. ΔT > 14°C. Select ΔT to create annular or mist-annular flow. Overall heat transfer coefficient 2 to 5 kW/m^2·°C, decreasing with increasing viscosity. HDHI = 10^6 to 10^7 Pa·s. Area 3 to 300 m^2.

Long tube falling film: especially for heat-sensitive and moderately viscous feeds. Not for foaming, fouling, crystal formation. or suspended solids. Viscosities 150 to 1000 mPa·s. ΔT > 3°C. Select ΔT to create annular or mist-annular flow. Outlet temperature 5 to 6.5°C < steam temperature. Overall heat transfer coefficient 2.5 to 5.6 kW/m^2·°C, decreasing with increasing viscosity. Residence time 15 to 30 s per stage. HDHI = 10^3 to 10^9 Pa·s. Consider multistaging with optimum. Area 3 to 1000 m^2. For condenser, use U = 1.1 to 1.25 kW/m^2·K.

Multistage rising or falling film evaporators: when the boiling temperature rise is large, use four to six multiple-effect stages with forward feed. When the boiling temperature rise is small, use eight to ten multiple effects. For multistaging, feed forward is the most common, but feed backward is used for highly viscous liquids. In feed backward, the more concentrated solution is heated with the highest temperature steam, and pumping is required between stages. The steam economy for n stages is $0.8n$ kg evaporated/kg of outside steam. Boost the interstage steam pressure with steam jet compressors (efficiency 20 to 30%) or with mechanical compressors (efficiency 70 to 75%).

Agitated falling film: especially for viscous feeds. OK for foaming, fouling, crystal formation, and suspended solids. Viscosities >1000 mPa·s. Relative to agitated film retention time of 1:1 and volume 1:1. Retention time about 1 to 30 s. Keep Δt high at 27 to 50°C to keep high heat transfer. Overall heat transfer coefficient 1 to 2 kW/m^2·°C, decreasing with increasing viscosity, although heat transfer coefficients as high as 150 kW/m^2·°C have been reported. Power 1.8 to 4 kW/m^2. HDHI = 10^4 to 10^6 Pa·s. A related topic is thin film reactors (see Sections 16.11.6.17 and 16.11.3.3).

Agitated horizontal film (votator): especially for viscous feed. OK for foaming, fouling, crystal formation, and suspended solids. Viscosities >2000 mPa·s. Relative to agitated film retention time of 1:1 and volume 1:1. Overall heat transfer coefficient 2 kW/m^2·°C, decreasing with increasing viscosity. Three to 12 kW/m^2.

Rising and falling film: especially for heat-sensitive. OK for foaming but not for fouling, crystal formation, or suspended solids. Viscosities <1500 mPa·s. Relative to agitated film retention time of 0.5:1 and volume 0.8:1.

16.11.4.2 Distillation

Area of Application

Liquid feed concentration 15 to 80% w/w; α_{vp} > 1.2; with materials not temperature-sensitive, negligible solids, nonfoaming, 99%+ purity possible for both product streams. Maximum column diameter 12 m or feedrate = 300 kg/s; maximum height 35 m.

Guidelines

This is usually the first choice for homogeneous separation.

Configurations include ordinary, vacuum, molecular (small capacity and very high vacuum <3.5 kPa, usually for material with molar mass 250 to 1200) cryogenic (operating at temperatures <–100°C), steam stripping (steam provides direct heating or "inert" steam is added to provide very high "vacuum" when used with organics that are immiscible with water), extractive (solvent is added to the top of the column), azeotropic distillation (solvent is added to the feed to the column), HIGEE (unique spinning tray design), and reactive (reaction and distillation are combined, e.g., for selective catalytic hydrogenation of C4 to C6). The operation can be batch or continuous.

Prefer to remove the most valuable as distillate.
Prefer to remove components one by one as overhead.
Prefer separations that give equimolar splits.
Set the column pressure to try to use water as the coolant and steam as the energy source.
Overhead concentrations 95 to 99% mol.
Internals are trays, packing, or high-performance packing.
If trays are selected, then calculate the theoretical stages required.
Tray options: valve: usually cheaper than sieve, use for large-diameter columns at moderate to high pressures, moderate turndown ratio. Usually about 120 to 140 values per active m^2; sieve: use for large-diameter columns at moderate to high pressures, moderate turndown ratio; bubble cap: use for columns with small liquid flow rate, high turndown ratio, heat transfer needed internally.
Typical Δp/HETS = 0.4 kPa.
Downcomer: unaerated liquid velocity 0.5 m/s; head loss via underflow clearance <0.3 kPa. Allow 3 s liquid residence time and extend to 6 s for foaming systems. Weir overflow velocity = 5 to 20 L/s·m of outlet weir.
If packings are selected, then calculate the number of transfer units, NTU, required or number of theoretical stages (NTS).
  Conventional random dumped packings: pall rings, Tellerettes, raschig rings, beryl or intalox saddles, lessing rings: use for columns <1 m dia., vacuum service, small pressure drop, high liquid flow rates, viscous liquids, minimum liquid holdup. Prefer Pall rings and Tellerettes; if ceramic, use saddles.
  High-performance structured packing: Flexipac, gauze, Glitsch grid, Leva film: not for viscosities >2 mPa·s, pressures >200 kPa;
Δp/HETS 0.001 to 0.5 kPa; very small superficial gas velocity.

To estimate the NTS,

1. Estimate minimum reflux ratio from the Underwood equation, or as an approximation if the distillate is almost pure:

$$R_{min} = \frac{1}{(a-1)(X_{LK,Feed})}$$

   Select operating reflux ratio that is 1.2 to 1.5 times minimum. For vacuum distillation, the reflux ratio usually is >10:1, especially for packed columns.

2. Estimate the minimum number of theoretical stages using the Fenske equation, or as approximations:

   Douglas': T = absolute temperature for overhead distillate, D, and for the bottoms, W:

$$N_m = \frac{T_D + T_W}{3(T_W - T_D)}$$

   Latour's: x = mol fraction:

$$N_m = 0.11 \frac{T_D + T_W}{(T_W - T_D)} \log_{10} \left[\frac{x_D}{x_W} \frac{(1-x_W)}{(1-x_D)} \right]$$

   The number of theoretical trays = twice the minimum number, N_m.

Conceptual Process Design, Process Improvement, and Troubleshooting

3. *For trays:* estimate the actual number of trays using the following tray efficiencies:
   Sieve, valve, or plate trays, tray efficiency 60%; pressure drop 0.7 to 1.4 kPa/tray or 0.3 to 0.65 kPa/theoretical stage (see Section 16.11.2.1)
   Sieve, valve, or plate trays with high viscosity feed, tray efficiency 25%
4. *For trays:* estimate the column height by the number of actual trays × 0.6 m tray spacing plus 1.2 m at the top and 1.8 m at the bottom.
5. *For trays:* estimate the column diameter with trays, assume acceptable superficial vapor velocity.

   0.3 m/s for high-pressure operation
   0.9 m/s for atmospheric operation
   2.5 m/s for vacuum <13 kPa
   or superficial density-weighted vapor flow rate, F factor, of 1.2 to 1.4 m/s $(kg/m^3)^{0.5}$ or boilup rate = 1.35 kg/s m²

   L/D <30; maximum height 55 m limited by windload
3. *For packing:* estimate the height of packing from the NTU × HTU (or NTS × HETS). For distillation, HTU ≈ HETS, because mass transfer resistance is in the gas phase, or $k_L/k_G \gg 1$. For conventional dumped packing: column diameter/packing diameter >15; use 2.5-cm packing for gas flow rates about 250 dm³/s; use 5-cm dia. packing for gas flow rates >1000 dm³/s. In general, HETS 0.3 to 1.8 m; 2.5-cm Pall rings: HETS 0.4 to 0.8 m; 5-cm Pall rings, HETS 0.7 to 0.9 m.

   Split total height into self-supported sections:

Plastic packings: 3 to 4.5 m max. height to prevent weight from collapsing the packing
Metal packings: 6 to 7.5 m height max.

Use liquid distributors every five to ten tower diameters of height with at least one every 6 m. Distributors have 30 to 50 discharge nozzles/m² for diameters <1 m.

5. *For packing:* estimate the column diameter:
   For conventional dumped packing: acceptable superficial vapor velocity, expressed as a percentage of gas flooding velocity, is

   For intalox saddles, 80% of flooding; surface areas 50 to 700 m²/m³, with increasing values as the diameter decreases
   For raschig rings, 60% of flooding
   For Pall rings, 90% of flooding
   or 1.4 to 2.2 m/s or superficial gas density-weighted velocity, F factor, of 0.6 to 3.5 m/s $(kg/m^3)^{0.5}$

Superficial mass ratio weighted velocity for flooding = 0.35 m/s.
The liquid loading must be sufficient to wet the packing, which is 0.015 to 4 kg/s·m of wetted periphery of the packing (0.015 to 4 L/s·m) or superficial liquid loading of 3 to 35 L/m²·s. Must be at least 0.5 L/m²·s. If there is insufficient liquid, then increase the size of packing to reduce the area per unit volume.
Flooding at the density ratio-weighted superficial vapor velocity, k factor, of 0.1 m/s.
Δp for moderate to high pressure, 0.3 to 0.6 kPa/m packing height:
   for vacuum 0.08 to 0.16 kPa/m and see Section 16.11.2.1.
   HETP

Packing size, cm	HETP, m
2.5 cm	0.46 m (vacuum operation, 0.67 m)
3.8 cm	0.67 m (vacuum operation, 0.82 m)
5 cm	0.91 m (vacuum operation, 1.06 m)

If structured packing is selected: design for 70 to 80% of flooding; for foaming, use 40% of flooding. Usual liquid flow rate of 0.007 to 5 L/s·m². Usual gas density-weighted superficial velocity, F factor, of 0.5 to 2.5 m/s (kg/m³)$^{0.5}$. Δp/HETS = 0.01 to 0.05 kPa. If the Δp > 1.2 to 1.6 kPa/m, then the packing is flooded. HETS 0.12 to 0.45 m, with this increasing to 1.8 m for high liquid capacity Glitsch grid.

For *extractive distillation*, the solvent should have a boiling temperature 50 to 100°C higher than those of the products, with solubility parameters and molar volumes similar to those of one product but different from the other.

For molecular distillation, 0.2 to 0.5 g/s·m² with about 4.5 m²/unit; 80 to 90% efficient.

For *steam distillation*:

For direct heating, 25 to 120 kg/m³. For inert steam added, flow rate calculated.

For *HIGEE*:

Residence time 0.1 to 1 s; voidage 90 to 95%; 2 m dia. max., area per volume 2000 to 5000 m²/m³. 1000 rpm; film thickness 100 μm. Overall heat transfer coefficient U = 10 kW/m²·K. See also reactive distillation, Section 16.11.6.32.

For *batch distillation*: size based on cycle time: fill, distill, discharge bottoms, clean.

for α_{vp} = 2, NTS = 10; reflux ratio 20:1
α_{vp} = 4, NTS = 6; reflux ratio 12:1
α_{vp} = 10, NTS = 3; reflux ratio 7.5:1

To convert TS to height of packing:

1 TS = 1 m of >5 cm dumped Pall rings or saddles
1 TS = 0.9 m of 4 cm dumped Pall rings or saddles
1 TS = 0.6 m of <2.5 cm dumped Pall rings or saddles
1 TS = 0.15 m of structured mesh packing in column of diameter <0.3 m
1 TS = 0.3 m of structured mesh packing in column of diameter >0.3 m

For *batch distillation*: As the more volatile species is removed, the separation becomes progressively more difficult. Too low a reflux gives can unattainable product, regardless of the number of trays. Gives high-purity recovery from a small concentration of a low-boiling species in a single operation. Keep the reflux holdup <10% to 15% of the initial batch charge.

16.11.4.3 Freeze Concentration

Area of Application

Liquid feed 1 to 30% w/w solute; operates below the freezing temperature of the *solvent*.

Guidelines

Don't exceed the K_{sp} of the solute. For juices, the solute concentration increases from about 10 to 50% when the freezing temperature is −10°C.

16.11.4.4 Melt Crystallization

Area of Application

Liquid concentration 60 to 90% w/w. Operates below the freezing temperature to solidify the target *solute*. Capacity usually <3 kg/s. Use for temperature-sensitive materials, when α_{vp} < 1.4, when azeotropes form, or for solid product. A freeze test shows >50% reduction in impurities.

Guidelines

Design based on heat transfer. Limited by eutectic formation. Multistaging possible but usually <10 stages.

Suspension crystallization: crystals and melt same temperature; design on degree of supersaturation; separation of crystals from melt depends on density difference in countercurrent operation. Scraped wall crystallizer (Section 16.11.4.6).

Progressive freezing: crystals colder than melt; design on temperature gradient; separation of crystals from melt by gravity draining. Static crystallization at crystal growth rates of 0.2 cm/h; falling film crystallization at crystal growth rates of 2 cm/h. Reflux ratio 1.3:1.

16.11.4.5 Zone Refining

Area of Application

Operates below the freezing temperature. Solid feed with input concentration of 20 to 70%, but usual is 95 to 99.9% w/w purity; can obtain 99.999999% purity. Temperatures 480 to 600°C. Impurity must lower the freezing temperature for method to be effective. Particulate impurities must not be encapsulated by the solidification front.

Guidelines

Design based on heat transfer with ΔT = 50 to 100°C from the freezing temperature and stop before the eutectic temperature. Ultrapurity of solids, low throughput at 10^{-7} to 10^{-6} kg/s. Multiple cycles might be needed to yield improved purity.

Design (velocity of freezing front)(liquid boundary layer thickness that depends on mixing)/(diffusivity of solute impurity in the liquid) = 1.

Typical velocities:
Semiconductors and pharmaceuticals: 10 to 70 μm/s
Organics: 0.5 to 10 μm/s

16.11.4.6 Solution Crystallization

Related topics include size enlargement (Section 16.11.9.13) and evaporation (Section 16.11.4.1).

Area of Application

Liquid feed 20 to 60% w/w; target solute becomes insoluble by producing supersaturation. Operates about the freezing temperature of the solute. Use for heat-sensitive materials, when α_{vp} < 1.4 or when a solid product is desired. Usual operation temperature 50 to 120°C. Very high-purity product with low energy usage. Never possible to obtain complete separation of the target species, because other species simultaneously crystallize with the target. Use for separation or to obtain a solid product.

Guidelines

Limited by K_{sp} and hydrate formation. Usually design with ΔT = 2 to 3°C from the solubility curve. Washing crystals is critical: too little, and product is contaminated; too much, and crystals redissolve and loss of yield occurs.

Solution concentration/saturation concentration = 1.02 to 1.05. Five options include: evaporate, use vacuum to both cool and evaporate, cool, use chemical reaction, or add an antisolvent.

If the solute solubility is relatively temperature-independent, add heat to remove the solvent or add an antisolvent to "drown out" crystals. For evaporation, 14 to 20 g vapor evaporated/s·m² exchanger area. For the exchangers, use 2.5-cm dia. tubes with fluid velocities 1.5 to 3 m/s to minimize plugging. Caution: if the vapor pressure rise >3.4 kPa/°C, then potential problems with control.

If the solute solubility is strongly temperature-dependent, use cooling (e.g., scraped surface: U = 0.15 to 0.3 kW/m²·°C) or might use vacuum cooling without external heat; that is, preheated feed followed by crystallizer.

If the solute solubility shows some temperature dependence, use vacuum cooling plus external heat.

If the reactants are highly soluble but the product is relatively insoluble, generate crystals by reaction.

If fouling is a problem, consider flash growth type Pachuca, draft tube MSMPR units, or classified product removal CPR type: Oslo, krystall type.

May be operated batch or continuous.

Seed the process because the nucleation sets the crystal habit and influences crystal size distribution (CSD). Importance of "contact nucleation" where crystals strike pump and mixer impellers. Crystal growth kinetics increase with temperature increase. Crystal growth rate = 0.1 to 0.8 mm/h with growth approximately the same in all directions

Batch: size based on cycle time: fill, crystallize, dump, clean. Use for production capacity of <0.15 kg/s and where target is relatively uniform and narrow CSD; although reproducibility may be difficult between batches. Batch gives narrower CSD than continuous.

Continuous CPR with classified fines and classified product removal gives increased crystal size and narrow CSD. Multistaging in series gives larger crystals and narrow CSD.

16.11.4.7 Precipitation

Area of Application

Liquid feed; operates at temperatures above the freezing temperature of the solute, usually in the range 18 to 35°C. Solute formed by reaction is insoluble, key parameter is K_{sp} of product. Provides a sharp first cut removal of solute.

Guidelines

Design based on mass transfer. Reactions are usually very rapid and design is based on mixing to distribute the reactant. For most precipitation reagents, allow 5-min residence time. If secondary reagents are needed to change the oxidation state of the target species before precipitation, then example residence times are

Arsenic: 30 min
Hexavalent chromium or iron: 20 min

Either pre- or post-pH change, allow about 0.180 kg acid or base/m^3 water.

16.11.4.8 Gas Absorption

See also heterogeneous gas-liquid separations, turbulent bed contactor (TCA, TVA) (Section 16.11.5.2), distillation (Section 16.11.4.2), reactors (Sections 16.11.6.11 through 16.11.6.17), and direct-contact heat exchange (Sections 16.11.3.11 through 16.11.13), with gas-liquid contacting characteristics described in Sections 16.11.8.1 and 16.11.8.2.

Area of Application

α = 2000 to 100,000, and gas feed concentration of target solute is 0.1 to 20%; 98% purity possible. Target species is soluble.

For high concentration and solubility plus particulates, use

Jets, venturis: very soluble gas only with target species Henry's law constant <10^3 kPa/mol fraction; feed gas concentration >1 vol%. Efficiencies 50% to 85%.
Turbulent bed contactor: both gas absorption and removal of heavy, sticky participates. Related to Section 16.11.5.2.

For high concentration and solubility, use

Spray column: target species Henry's law constant 10^3 to 10^4 kPa/mol fraction; feed gas concentration 0.3 to 4 vol%. Efficiencies 85 to 95%.

Spray chamber: target species Henry's law constant $<10^4$ kPa/mol fraction; feed gas concentration >1 vol%. Efficiencies 50 to 85%.

For low concentration, low range of solubility, and no particulates, use

Countercurrent packed column: target species Henry's law constant $<10^7$ kPa/mol fraction; feed gas concentration <1 vol%. Efficiencies 95+%. Vulnerable to plugging. OK for foaming and corrosive.

Countercurrent tray column: target species Henry's law constant $<10^7$ kPa/mol fraction; feed gas concentration <1 vol%. Efficiencies 95+%. Not for foaming, corrosive, or particulates.

For adsorption plus chemical reaction, consider cocurrent falling film multitube system. See also long tube falling film evaporator, Section 16.11.4.1.

Guidelines

To minimize the cost of recovering and recycling the solvent, contact the minimum amount of liquid with the maximum amount of gas. For packings, this means the liquid loadings about 3 L/m²·s with high gas flow rates of superficial density-weighted velocity F-factor of 3.5 m/s (kg/m³)$^{0.5}$.

Selection of the solvent: select lowest molar mass liquid whose solubility parameters are similar to those of target species in the gas. Choose a solvent in which the target species are highly soluble. >700 mg/g solvent; Henry's law constant: $<10^3$ kPa/mol fraction; or $(HTU_G/HTU_L)(F_L/F_G)$(pressure, kPa/Henry's constant) >20: then gas phase controls mass transfer. This is most likely for many absorptions.

If target species have medium solubility: 100 to 700 mg/g solvent; Henry's law constant 10^3 and to 10^5 kPa/mol fraction: both liquid and gas phases control the mass transfer. These systems are usually processed in packed and tray columns.

If target species have low solubility: <50 mg/g solvent; 1/Henry's law constant: $<8 \times 10^{-5}$ mol fraction/kPa: liquid phase controls mass transfer.

Regeneration of the absorbent by distillation or pressure-reduction desorption.

Jets, venturis: economic NTU = 3.5; critical energy-consuming phase is the gas at about 20 kJ/m³ with liquid-to-gas ratio about 1.3 to 1.6 L/m³; design on gas phase controlling. Power usage 1 to 8 kW·s/m³. Δp gas = 1.2 to 6 kPa.

Peabody absorber/impingement baffle: economic NTU usually <5; liquid-to-gas ratio 0.2 to 0.7 L/m³; superficial gas velocity 1 to 1.6 m/s. Power usage 0.5 to 2 kW·s/m³. Δp gas = 1.5 to 3.7 kPa.

Spray column: economic NTU = 2.5; critical energy-consuming phase is the liquid; gas energy 8 kJ/m³; with liquid-to-gas ratio high; design on gas phase controlling, superficial gas velocity 1 m/s. Power usage 0.03 to 0.5 kW·s/m³. Δp gas = 0.6 to 1.2 kPa.

Spray chamber: economic NTU = 4; gas energy consumption 2.5 kJ/m³ with liquid-to-gas ratio about 1.5 L/m³; design on gas phase controlling, superficial gas velocity 0.9 to 1.2 m/s. Power usage 0.03 to 0.5 kW·s/m³. Δp gas = 0.5 kPa.

NTU needed:

NTU + 2 = 6 $\log_{10} (y_{in}/y_{out})$, or

Use the Kremser equation with a molar stripping factor or molar absorption factor of F_L/F_G m = 1.4. A molar stripping or adsorption factor = 5 is approximately the same as ∞.

Countercurrent packed column: Economic NTU usually <5; critical energy-consuming phase is the gas at about 3 kJ/m^3 with liquid-to-gas ratio about 0.7 to 1.5 L/m^3. Liquid loading on usual packings 3 to 35 L/m^2·s; (0.007 to 5 on structured packings) molar ratio liquid to gas 1.4 to 80; superficial gas velocity 1.4 to 2.2 m/s; design on mass transfer in both gas and liquid. Power usage 0.15 to 0.6 kW·s/m^3. Δp gas = 0.07 to 0.4 kPa/theoretical stage; HETS = 0.5 to 1 m. For low liquid loadings and high gas flow rates, $HTU_G/HTU_L \approx 2$; for high liquid loading and low gas flow rates, 0.2.

Assume 99% absorption; set molar flow rates F_L/F_G = 1.2 to 2 times the minimum. Use the Kremser equation for the design of dilute units with molar absorption factor F_L/F_G m = 1.2 to 2; use 1.4. NTU$_{OG}$ = 20 for 99% recovery.

Assume HTU$_{OG}$ = 0.6 m with maximum packing height 12 m. Superficial gas velocity 1 m/s and/or superficial density-weighted velocity F-factor of 0.6 to 3.5 m/s (kg/m^3)$^{0.5}$. Use 2.5-mm Pall rings or Tellerettes of metal or plastic.

Countercurrent tray column: economic NTU usually <5; liquid-to-gas ratio 1.5 to 18 L/m^3; superficial gas velocity 1 to 1.6 m/s and/or superficial density-weighted velocity F-factor of 1.2 to 1.8 m/s (kg/m^3)$^{0.5}$. Power usage 0.5 to 2 kW·s/m^3; Δp gas = 0.7 to 1.5/tray. HETS = 0.7 m. Tray efficiencies 10 to 20%.

Assume 99% absorption; use the Kremser equation for the design of dilute units with molar absorption factor for economic recovery of the solvent F_L/F_G m = 1.4 with general range 1.2 to 2. Tray spacing = 0.6 m with valve or sieve trays. See also Section 16.11.4.2 for such details as downcomer sizing.

For *multitube cocurrent falling film*: U = 0.6 kW/m^2·K.

16.11.4.9 Gas Desorption/Stripping

Area of Application

α = 2000 to 100,000 and liquid feed concentration of target solute is 0.1 to 5%; 98% purity possible.

Guidelines

Because the target solute usually has low solubility in the liquid, usually the desorption is liquid-phase controlled.

For packings the goal is high liquid loadings, about 30 L/m^2·s, and minimum gas flow rates. Superficial density-weighted velocity F-factor of 0.6 m/s (kg/m^3)$^{0.5}$.

Use Kremser equation for the design of dilute units with molar L/absorption factor for economic stripping of $F_G m/F_L$ = 1.2 to 2 with starting value of 1.4; m = Henry's constant/total pressure. For packing, HETS = 1.83 m.

Edible oil deodorizing: high vacuum. For <0.6 kg/s, irregular production, use batch. Requires processing time = 4 h/batch, low heat recovery. For >0.6 kg/s, continuous with processing time = 1 h. Keep liquid films thin to promote mass transfer of volatiles and use astute distribution of sparge steam.

16.11.4.10 Solvent Extraction, SX

A related topic is size reduction (Section 16.11.8.3).

Area of Application

General: feed concentration 0.03 to 95% w/w; for minerals, typically 0.01 to 2% w/w; separation factor α = partition coefficient ratio with values 2 to 500 and should be >5.

Spray and packed columns: gravity flow (spray, plate, packed column): superficial liquid velocity, 0.001 to 0.02 m/s; area per unit volume 7 to 75 m^2/m^3. Product of the density difference with the interfacial tension (Mg/m^3, mN/m) >1 and number of theoretical stages needed <3; interfacial tension <10 mN/m.

Gravity flow (raining bucket, RTL): product of the density difference with the interfacial tension (Mg/m^3, mN/m) >1 and number of theoretical stages needed <3; handles dirty liquids and ones that tend to emulsify.

Stirred tanks; mixer settler (including Lurgi): superficial liquid velocity, 0.00015 to 0.004 m/s; area per unit volume 400 to 10,000 m^2/m^3. Product of the density difference with the interfacial tension (Mg/m^3, mN/m) >4 and number of theoretical stages needed >3. Usually about one theoretical stage per unit. Rarely build more than 5 stages; can handle high phase ratios.

Stirred or pulsed columns: superficial velocity, 0.002 to 0.02 m/s; area per unit volume 75 to 3,000 m^2/m^3.

Reciprocating plate: product of the density difference with the interfacial tension (Mg/m^3, mN/m) between 1 and 4 and number of theoretical stages needed >2. Can handle dirty liquids.

Pulsed plate or packed: product of the density difference with the interfacial tension (Mg/m^3, mN/m) between 1 and 4 and number of theoretical stages needed >2. sensitive to contamination. Difficult to pulse large columns.

Rotating disk contactor, RDC, ARD contactor; Mixco, Scheibel, Treybal, Oldshue–Rushton, Kuehni: product of the density difference with the interfacial tension (Mg/m^3, mN/m) between 1 and 4 and number of theoretical stages needed >2. Low HETS; can handle dirty liquids, large throughputs. Needs flow ratios 1:1. Difficulty handling low interfacial tension systems that tend to emulsify.

Centrifugal extractor: product of the density difference with the interfacial tension (Mg/m^3, mN/m) <1 and number of theoretical stages needed <6. Cannot handle dirty systems or high phase ratios. For continuous differential type machines (density differences >0.05 kg/L; drop diameter >200 μm): six to eight stages per machine. For discrete, disc type machines (density differences >0.02 kg/L and handle drops <200 μm) 2 to 3 stages per machine.

Guidelines

Design flow rate about 50% to 90% of flooding with the superficial velocity selected varies directly with the density difference and interfacial tension. Solvent/feed = 0.5 to 1.5/1; with usual 1/1.

Gravity:

Spray: HETS increases exponentially with diameter; 10 to 20 m at 1 m dia.; superficial velocity about 5.5 L/s·m^2.

Packed: HETS increases exponentially with diameter; 2.5 m at 1 m dia.; prefer diameter <0.6 m; superficial velocity about 5.5 L/s·m^2, 2.5 cm Pall rings. Redistribute the dispersed phase every 1.5 to 2 m.

Sieve tray: HETS increases exponentially with diameter; 1 m at 1 m dia.; usually diameter >0.6 m; superficial velocity about 5.5 L/s·m^2. Efficiency inversely proportional to the interfacial tension: 40% at 5 mN/m.

RTL: HETS increases exponentially with diameter; 0.5 m at 1 m dia.; superficial velocity about 0.3 to 2 L/s·m^2.

Mixer settler: one theoretical stage per unit; input energy 1 kW/m^3; superficial velocity for the settler 0.5 to 7 L/s·m^2 with 1.4 L/s·m^2 being typical. Low values for small density differences. Height increases with increasing total feed flow rate with 1 m height for 10 L/s. Typical drop diameter is 150 μm; extraction efficiencies about 80%.

Lurgi contactor: HETS increases exponentially with diameter; 0.7 m at 1 m dia.; diameters <8 m; superficial velocity about 5.5 L/s·m².

Pulsed packed column: HETS increases exponentially with diameter; 0.7 m at 1 m dia.; max. diameter 2.5 m; superficial velocity about 5.5 L/s·m².

Pulsed sieve plate column: HETS increases exponentially with diameter; 0.4 m at 1 m dia.; max. diameter 3 m; superficial velocity about 5.5 L/s·m² sieve holes 3 to 8 mm; velocities through the holes <0.2 m/s to minimize the formation of small drops. Tray efficiencies about 20 to 30%.

Reciprocating plate: HETS increases exponentially with diameter; 0.35 m at 1 m dia.; max. diameter 1.5 m; superficial velocity about 11 L/s·m².

Rotating disk contactor, RDC, ARD contactor; Mixco, Scheibel, Treybal, Oldshue–Rushton, Kuehni: HETS is sensitive to rotor speed. HETS increases slightly with diameter; 0.5 m at 1 m dia.; superficial velocity about 5.5 L/s·m² with Kuehni 9.7 L/s·m². Mixco, Scheibel, Treybal, Oldshue–Rushton, diameter <2.5 m; RDC diameter <9 m.

Total flow through a column = 10 L/s·m² with a density difference of 0.2 Mg/m³.

Centrifugal: two to six units per machine, depending on the machine.

16.11.4.11 Adsorption: Gas

Area of Application

Use when feed concentration of the more volatile species is small, 0.15 to 10%, and when α_{ads} > 2; when the target species is difficult to condense.

Guidelines

Select adsorbent based on pore size related to the target species.

Alumina: surface area: 210 to 350 m²/g; pore volume 0.21 cm³/g, temperature <320°C; superficial gas velocity 125 to 500 dm³/m²·s; usually adsorb 800 kg/m³ or 0.14 to 0.22 kg organics/kg dry solid; 0.15 kg water/kg dry solid. Lifetime: 150 cycles.

Silica: surface area: 750 to 830 m²/g; pore volume 0.4 to 0.45 cm³/g, temperature <230°C; superficial gas velocity 125 to 500 dm³/m²·s: usually adsorb 720 kg/m³; 0.3 to 0.6. kg organics/kg dry solid; 0.4 kg water/kg dry solid.

Four-Å molecular sieve: surface area: 640 to 80 m²/g; pore volume 0.27 cm³/g, temperature <300°C: superficial gas velocity 150 to 250 dm³/m²·s; usually adsorb 480 to 720 kg/m³; 0.05 kg nitrogen/kg dry solid; 0.22 to 0.36 kg water/kg dry solid. Lifetime: 400 cycles.

Activated carbon: surface area: 1000 to 1500 m²/g; pore volume 0.6 to 0.8 cm³/g, temperature <540°C: superficial gas velocity 100 to 600 dm³/m²·s; capacity depends on organic; range 0.06 to 0.2 kg organics/kg dry solid adsorbent.

For fixed bed: batch: size on cycle time: load, swing out of service, regenerate, swing back into service. Loading time: 100 to 3000 bed volumes (BV)/h with time based on the ratio of the adsorption isotherm to the feed concentration of the target species (usual range 50 to 300 corresponding to load times of 0.2 to 2 h). Regenerate with steam (at 3 to 5 kg steam per kg organic removed), solvent, reduced pressure, combustion, or via vacuum/pressure shift. Use superficial gas velocity of 60 to 600 dm³/dm²·s or recommended value for the adsorbent to determine *cross-sectional* area. Residence time 0.03 to 0.8 Bed volumes/second: (F/V) with depth >0.33 dia.

16.11.4.12 Adsorption: Liquid

A related topic is ion exchange (Section 16.11.4.13).

Conceptual Process Design, Process Improvement, and Troubleshooting

Area of Application

Prime option for dilute concentrations with a need for greater selectivity than solvent extraction can provide.

Guidelines

Select adsorbent:

> *Activated carbon:* surface area: 1000 to 1500 m^2/g; pore volume 0.6 to 0.8 cm^3/g, temperature <540°C; loading very dependent on molar mass of target solute, solubility in the carrier liquid, and pH. Example loading 0.01 kg organic molar mass 100/kg dry solid. The value varies with the (molar mass)$^{3.5}$.
> *Acid-treated clay:* surface area: 225 to 300 m^2/g.
> *Fuller's earth:* surface area: 130 to 250 m^2/g.

Batch:

Fixed bed: batch: size on cycle time: load, backwash/clean, regenerate, swing on-stream. Load: typical flow rate 2 to 3.5 BV/h with the load time based on the ratio of the adsorption isotherm to the feed concentration of the target species (usual range varies with the application: 18 to 100 min, while for water treatment: 70 to 100 days). Backwash with velocity to fluidize the bed (see liquid fluidization, Section 16.11.7.5); velocity 0.8 BV/h. Time such that <5% feedrate used in backwash. Usually, carbon is removed and regenerated about four times per annum. Try to match the loading cycle to the regeneration cycle.

Use fixed bed if <20 L/s and carbon usage >180 kg/day. Superficial velocity 1 to 15 L/m^2·s; use 5 L/m^2·s to estimate cross-sectional area. Too low a superficial velocity (<3 L/m^2·s) gives poor feed distribution. Use 2 to 3.5 bed volumes/h to give the required residence time of 18 to 100 min. Bed depths in the range 3 to 10 m, but keep the pressure drop <75 kPa. Height/diameter 1:1 to 4:1. Maximum size is based on carbon usage <9 Mg dry carbon per day.

Continuous:

Moving bed: use if >20 L/s. A related topic is transfer line reactor (Section 16.11.6.9).
Fluidized bed: use if slimes or fine particles in feed. Use superficial velocities of 8 to 14 L/m^2·s. Feed contacts bed for 30 min or 0.8 bed volumes/h. Related topics include reactors (Section 16.11.6.27), heat transfer (Section 16.11.3.8), drying (Section 16.11.5.5), size enlargement (Section 16.11.9.5), and mixing liquid-solid (Section 16.11.7.5).
Slurry approach: use if the carbon usage is <180 kg/day. Mix and suspend powdered adsorbent and then filter exit line. Often uses up to three stages of countercurrent contacting. Used for continuous bleaching of edible oils. Batch process is simple, flexible, and easy to change feedstocks. Continuous operation offers better protection against oxidation, provides shorter holdup, and has the potential of heat recovery. Bleach time 25 min. A related topic is transfer line reactor (Section 16.11.6.9).

Loading times and elution-regeneration times should be approximately equal.

16.11.4.13 Ion Exchange

Related topics include the use of ion exchange resins as catalysts in reactors (Section 16.11.6.32), liquid-solid fixed bed reactors (Sections 16.11.6.14 and 16.11.6.15), and adsorption-liquid (Section 16.11.4.12).

Area of Application

High-valence ionic species in liquid phase with $\alpha_{IX} = \Gamma^+(1 - c^+)/c^+(1 - \Gamma^+) = 1.01$ to 1.04 and feed concentration 0.02 to 2% w/w. Γ^+ = surface concentration of cations.

Guidelines

Select the ion exchange resin based on the pH of the environment and the valence of the target ion. For high efficiency, try to use weak electrolyte resin.

Weakly acidic cationic (WAC) exchange resin: carboxylic; pH >4; $T < 100°C$; hydrogen or sodium form, depending on regeneration preference; loading 3.5 equivalent/L of resin; good selectivity; greatest affinity for alkaline earth metals in the presence of alkalinity.

Strongly acidic cationic (SAC) exchange resins: sulfonic; full range of pH, $T < 120°C$; hydrogen or sodium forms; loading 1.9 equivalents/L resin; high capacity, high activity. Ionic sequence Ba > Pb > Sr > Ca > Ni > Mn > Be > Cd^{2+} > Cu > Co > Zn > Mg.

Weakly basic anionic (WBA) resin: aromatic polyamine: pH<7; temperature <50°C; chloride (more thermally robust) or OH form; loading 1.6 eq/L resin.

Strongly basic anionic (SBA) exchange resins: type I trimethyl amine: full range of pH; chloride (more thermally stable) or OH forms; loading 0.46 to 1 eq/L resin.

Ions being exchanged must have a higher valence than ions in bed from regeneration.

Batch:

Fixed bed: batch: size on cycle time: load, backwash/rinse, elute, standby. Try to match loading with off-line time. Load: loading feedrate 5 to 10 BV/h with loading time depends on the ratio of exchange capacity of the resin to the feed concentration of the target species. Usual ratio 10 to 100 corresponding with a loading time of 1.5 to 15 h. Backwash with a velocity to fluidize the bed for 1 to 10 BV (corresponding to about 1.5 h). Eluent feedrate 1 to 3 BV/h for typically 8 to 20 h. Standby is typically 20% of the backwash time. When the eluate is valuable, try to match the loading time with the backwash, eluate recovery, and standby.

Feed flow rate: superficial velocity 1 to 15 $L/m^2 \cdot s$; use 5 $L/m^2 \cdot s$. Superficial velocity <3 $L/m^2 \cdot s$ gives poor feed distribution; 5 to 10 bed volumes/hour; too deep a bed leads to excessive pressure drop; keep below 350 kPa. Add 2.5 m to height to allow for bed support and head room for backwashing. Typical bed depths 1 to 5 m; usual industrial bed cross section 5 m^2.

Use fixed bed if <20 L/s. Pressure, gravity, cocurrent, countercurrent, series, single bed, or mixed bed. Gravity is economical where the bed volume > 80 m^3.

Loading times and elution-regeneration times should be approximately equal.

For WAC and WBA systems, the backwash, regeneration, and rinse cycles are 30 to 60 min.

Continuous:

Moving bed: use if >20 L/s.

Fluidized bed: use if slimes or fine particles in feed, elution time >> IX time, eluant is expensive. Use 16 to 20 mesh resin with superficial velocities of 40 $L/m^2 \cdot s$. Try to operate at 50 to 200% bed expansion. Related topics include adsorption-liquid (Section 16.11.4.12).

16.11.4.14 Foam Fractionation

A related topic is size reduction, gas in liquid (Section 16.11.8.1).

Area of Application

Solute feed concentration usually 1 ppm to 0.1% with some applications up to 10% (10^{-3} to 10^{-9} mol).

Guidelines

Surface concentration 1 to 5 µmol/m^2 bubble area. Height to diameter 5:1 to 15:1. Usual liquid feed superficial velocity 0.1 to 1 L/s·m^2. Bubble diameter = 0.8 to 1 mm. Volumetric feedrate ratio of inlet gas to liquid flow rates: 4 to 7.5 dm^3 gas/L liquid. Volumetric ratio of gas in foam to overhead liquid in draining foam: 100 to 250:1. Liquid in the foam/liquid in the feed = 6 to 12% v/v; draining foam density 0.003 Mg/m^3. Superficial gas velocity for foam drainage section 3 to 250 dm^3/s·m^2. Scaleup based on constant superficial gas velocity.

16.11.4.15 Membranes: Gas

Area of Application

In general, there are two applications: remove organics (VOC) from air and separate permanent gases (such as nitrogen, hydrogen, carbon dioxide).

To remove organics from air, VOC: feed concentration is >0.1% v/v and the air flow rate is <480 dm^3/s. For lower concentrations, use adsorption (Section 16.11.4.11); for higher flow rates, use adsorption and condensation.

To separate permanent gases: feed concentration is 10 to 85% and the flow rates range from 0.05 to 65 Nm^3/s.

α = 1.4 to 4; gas feed concentration 5 to 75% w/w; 75 to 90% gas purity possible. Diameter of the target species: 0.1 to 1 nm.

Guidelines

> *Membrane type:* asymmetric: homogeneous or microporous; composite of a homogeneous polymer film on a microporous substructure or symmetrical homogeneous or porous polymer film. Driving force for the rate of separation: hydrostatic pressure concentration.
> *Membrane*: nonporous membrane elastomer or glassy.
> *Pressure*: >0.7 MPa for separation of permanent gases.
> *Temperature*: 0 to 60°C.

For VOC removal from air, use elastomeric membranes (rubbery membranes). The membrane has a 0.1 to 1 µm skin layer on a 100- to 300-µm substrate. These membranes tend to have lower selectivities and higher permeabilities, and are relatively impermeable to nitrogen. The permeability of organics tends to be 4 to 100 times higher than that of inert gases. For these membranes, the selectivity increases with increasing boiling temperature of the species. The permeability increases 10% per degree Celsius increase in temperature. Typical permeabilities for the organics are >200 Ndm^3·µm/s·m^2·MPa.

For the separation of permanent gases, use glassy membranes with higher selectivity and lower permeability.

Typical permeances are

for hydrogen methane separation: 0.22 to 3.75 Ndm^3/s·m^2 ·MPa.
for carbon dioxide-methane separation: 0.075 to 1.5 Ndm^3/s·m^2·MPa.
for air separation: 0.003 to 1.9 Ndm^3/s·m^2 ·MPa.

The temperature of the feed should be >20 to 40°C above the feed dew point.

16.11.4.16 Membranes: Dialysis

Area of Application

$\alpha = c_{21}\mathbf{D}_2/c_{11}\mathbf{D}_1$ = 6 to 25%; liquid feed concentration 2 to 25% w/w; 99% purity possible; c_2 = concentration of solute in the feed; $\mathbf{D}$ = diffusivity. Diameter of the target species: 0.5 to 5 nm.

Guidelines

Driving force for the rate of separation: concentration of target species.

Membrane: symmetric microporous with 0.1 to 10 mm pore diameter. Hydraulic permeability: 10^{-3} to 8 g/s·m²·MPa. Membrane-solute permeability 0.05 to 9 µm/s, depending on the solute and the membrane. Dialysis transfer coefficient: 1 to 10 µm/s.

16.11.4.17 Membranes: Electrodialysis

Area of Application

Diameter of the target species: 0.2 to 0.8 nm. Feed concentration <20% ionic. Feed range 0.005 to 5%.

Guidelines

> *Driving force for the rate of separation:* electropotential
> *Membrane:* ion exchange, homogeneous or microporous polymer with positively or negatively charged fixed ions
> *Hydraulic permeability:* 10^{-3} to 8 g/s·m²·MPa
> Pretreat feed until ferric < 0.3 mg/L
>   Mn < 0.1 mg/L
> H_2S < 0.3 mg/L
> *Temperature:* >10 and <43°C
> *Optimum feed concentration:* 1000 to 5000 mg/L
> *Energy:* about 5.5 to 9 MJ/m³ product
> *Pressure drop:* horizontal stack 0.2 to 0.4 MPa; vertical stack 0.02 to 0.07 MPa

16.11.4.18 Membrane Configurations

Can be operated *dead-end* or *crossflow:* "batch" with 100% recycle of retentate; "continuous" with recycle ratios 15/1 to 30/1 and purge and "multistage," where the purge from one stage becomes the feed of the next.

Batch process: stop feed because the membrane performance has deteriorated or because the target concentrations or volumes of permeate or retentate have been achieved. Deteriorated membrane performance can be (1) because of a buildup of particulates or biological deposits (this can be corrected by backflushing with permeate or cleaning with detergents, acids, or bases) or (2) because the membrane needs to be replaced (the life of a membrane is one or two to ten years, depending on the operating conditions and the type of membrane). See Table 16.32.

Membranes:

> *Hydrophilic:* cellulosic (temperature <30°C), polyacrylic, nylon 66, ceramic. Life: polymers: 1 year; ceramic: <10 years.
> *Hydrophobic:* polysulfones, polyolefins, carbon. Life: polymers: 1 to 2 years; fluoropolymers <4 years.
> Neither hydrophilic nor phobic: sintered.

16.11.4.19 Membranes: Pervaporation

Area of Application

$\alpha = (c_{i3}/c_{j3})/(c_{i1}/c_{j1})$, where i is the target permeate species, j is the reject; 3 = permeate, and 1 = feed, separates dissolved organics = 1.3 to 41. Particle diameter, 0.2 nm. Target concentration 5 to 20% w/w, but 0.1 to 10% w/w for economical. Temperature <120°C.

TABLE 16.32
Characteristics of Various Membrane Configurations

	Tubes: Fine Hollow Fibers to Tubules				Channel				Tubular Monolithic Elements	Tubes
	RO Type	UF/Microfilter			Flat Sheet: Plate/Frame	Flat Sheet: Plate/Frame	Spiral Wound	Pleated Sheet		
Diameter or spacing, mm	0.01–0.03	0.02–0.1	0.5	1.1	0.2–0.4	1.2–2.5	0.75 or 1.1		2.5–6	12–25
Length, cm	< 100				0.5–1	45				1800
Reciprocal volume, l/cm^3	300–7500	0.9–6								
Holdup, L/m^2	0.33	0.120	0.3		0.8–30					
Flow conditions, Reynolds no.		Laminar, 500–3000				Turbulent				Turbulent, Re > 10,000
Velocity, m/s	0.01–2.5	0.6–1	0.6	0.6	0.06–1.6		0.1–2			2–6
Feed rate/ surface area, L/s·m^2	0.002	0.002			0.25–0.5		0.25–0.5			1–5
Shear stress/ length, l/cm·s	10–100				80–300		80–300			500–4000
Pressure drop along the membrane, kPa	30–130				70–1000		100–140			14–20
Area per unit volume, m^2/m^3	6000	600–1200	500–1100	320–505	400–1000		650–1000			25–70
Max size/unit, m^2	1	0.05–5			0.02–0.07				8–22	0.4

Guidelines

Driving force for the rate of separation: concentration gradient in the vapor pressure.

Membrane: asymmetric: homogeneous, or microporous (cellulose acetate, polyamide, polysulfone, polyacrylonitrile); composite of a homogeneous polymer film on microporous substructure (cellulose acetate, polyamide, polysulfone, polyimide, polyvinyl alcohol). Usual pore 0.1 to 0.2 μm.

0.1 μm PVA: water >> methanol > ethanol > other organics.

Silicone rubber: methanol > ethanol > aldehydes > ketones >> water; paraffins > olefins.

Cellulose esters: aromatics > paraffins; olefins > paraffins; dienes > olefins; n-paraffins > branch; low molar mass paraffins > high molar mass paraffins.

Capacity/unit: feed <1.5 kg/s.

Feed pressure: atmospheric.

Feed temperature: close to the normal boiling temperature; usually 50 to 100°C.

Permeate conditions: pressure 0.5 to 2 kPa absolute with condenser temperatures: −20 to +30°C.

For dehydration, use a difference in partial vapor pressure at least one order of magnitude; the permeate flux doubles for an increase in temperature of 10°C.

Permeate flux: depends on membrane but in the range 0.008 to 5 g/s·m².

Configuration: see Section 16.11.4.18.

Spiral wound, transverse hollow fiber.

Use *crossflow* with recycle ratio 15 to 30/1. Criterion: 10% of feed volume is permeate or purity of the permeate or retentate.

Membrane life: 2 to 4 years.

Cycle time: usually limited by life of membrane.

16.11.4.20 Membranes: Reverse Osmosis (RO)

Area of Application

α = 6 to 25; feed concentration 0.05 to 20% w/w; with suggested economic feed concentration <0.5%; 99% purity possible. Diameter of the target species: 0.2 to 0.8 nm. Must have a difference in osmotic pressure. The osmotic pressure coefficient in mass ratio units for different solutes = 20 to 80 MPa·kg/kg at 25°C. The higher the valence, the better the rejection.

Guidelines

Driving force for the rate of separation: hydrostatic pressure.

Membrane: asymmetric: homogeneous or microporous; active dense 20- to 50-μm layer of cellulose acetate with total thickness 100 μm; composite of a homogeneous polymer film on microporous substructure.

Pressure: 1.4 to 10 MPa (1.4 to 4.2 for brackish water; 5.6 to 10 for seawater). Inlet pressure > twice the inlet osmotic pressure.

Temperature: <45°C.

Capacity/unit: <7 L/s.

For cellulose acetate membranes: $1/\Theta^+ = Ap/B\rho$ = 1 to 500 with usual value 300 (dimensionless).

For aromatic polyamide membranes: $1/\Theta^+$ = 0.7 to 20.

A = permeate hydraulic permeability, g/s·m²·MPa.

p = total operating pressure, MPa (1.4 to 10 MPa).

B = target solute transport coefficient, μm/s (10^{-6} to 10 μm/s).

ρ = mass density of the feed stream.

Hydraulic permeability, A: 0.0005 to 8 g/s·m²·MPa (0.1 to 10 for cellulosic).
Permeate flux: 0.001 to 0.1 L/s·m²; for cellulose acetate: 0.006 to 0.0075 L/s·m²; for hollow fiber: 0.001 to 0.002 L/s·m²; for thin film composite: 0.007 to 0.009 L/s·m². Permeate flux increases about 3% for every 1°C increase. Permeate flux decreases by 10 to 50%, depending on the concentration polarization. Permeate flux is reduced because of particulates and bacterial adhesion so that flux for tubular < spiral wound < hollow fiber.
Configuration: see Section 16.11.4.18.
Spiral wound, hollow fiber and tubular (used for low-volume, high-value commodities), pleated sheet, tubular monolithic elements, or plate and frame.
Use *crossflow* batch with 100% recycle; continuous with recycle ratio 15 to 30/1 or multistage (often three stages). Criteria to back wash and clean: operate until a given concentration or volume reduction is reached in the retentate or a given purity or volume is achieved in the permeate.
Membrane life: 2 to 4 years.
Cycle time: pretreat to prevent scaling or buildup or operate a short cycle, 2 to 12 h; cleaning with dilute nonionic detergent. Degree of pretreatment: hollow fiber > spiral > tubular.

16.11.4.21 Membranes: Nanofiltration

Area of Application

Molar mass cutoff = 200. Can handle fluid with significant osmotic pressure, sugars, dissociated acids, and divalent salts.

Guidelines

> *Driving force* for the rate of separation: pressure
> *Membrane:* asymmetric thin film, noncellulose membrane; membrane usually negatively charged to reject anions.
> *Pressure:* (between UF and RO) = 0.3 to 1.4 MPa.

16.11.4.22 Membranes: Ultrafiltration (UF)

A related topic is filters (Sections 16.11.5.12 and 16.11.5.13).

Area of Application

$\alpha = \kappa_2 D_2 / \kappa_1 D_1$ = 6 to 60; liquid feed concentration 0.04 to 20% w/w; 99.9% purity possible, κ = partition coefficient; D = diffusivity. Diameter of target species 0.8 to 200 nm; removes soluble macromolecules, colloids, salts, and sugars but cannot separate dissolved salts, species with molar mass <1000, or species exhibiting a significant osmotic pressure. Feed concentration <50% dissolved organics.

Guidelines

> *Driving force* for the rate of separation: hydrostatic pressure.
> *Membrane:* most UF membranes are polysulfone. Asymmetric microporous with thin skin 0.1 to 1 μm supported on a porous layer 50 to 250 μm. Pore size 0.001 to 0.2 μm. This is too porous for RO. Pore size prevents concentration polarization (limiting RO), but performance is limited by gel polarization with x_{gel} = 0.2 to 0.4. x_{gel} = 0.25 to 0.35 for macromolecules, 0.75 for colloids. Need to have membrane life >1 year.
> *Pressure:* 0.1 to 0.7 MPa. Hydraulic permeability, A: 0.8 to 800 g/s·m²·MPa. Feed concentration: 0.05 to 15% w/w.
> *Temperature:* <90°C and pH 0.5 to 13 for polysulfone. Capacity/unit: 0.1 to 25 L/s.

Select diameter or channel spacing so that diameter of the target species is 0.1 of the diameter or channel spacing; except for spiral wound where, for 0.75-mm spacing, the particles must be <5 to 25 μm; or 0.006 to 0.034 diameter of spacing; for 1-mm spacing, particles should be <25 to 50 μm or 0.025 to 0.05 of the diameter or channel spacing.

Permeate flux: depends on the membrane and configuration: hollow fibers/polysulfone: 0.005 to 0.016 L/s·m²; spiral wound/polysulfone: 0.08 to 0.14 L/s·m²; tubes/polysulfone: 0.06 to 0.2 L/s·m².

Liquid permeability increases 25% for every 10°C increase in temperature.

Power depends on target species and configuration: water treatment 1.8 kJ/L permeate; food application: 32 kJ/L permeate; electropaint: 60 kJ/L permeate. Configuration: hollow fiber 6 kJ/L; plate and frame 9 kJ/L; spiral 3 to 6 kJ/L; tube 15 kJ/L; or hollow fibers: 100 to 280 W/m²; plate and frame: 180 to 280 W/m²; spiral wound 25 to 120 W/m².

Configuration: see Section 16.11.4.18.

Spiral wound, hollow fibers, plate and frame, and tubular (use for small flow, high value, and severe fouling applications).

For laminar flow operation of hollow fiber, plate and frame, and spiral wound, keep the operating pressure 0.1 to 0.2 MPa; for turbulent flow operation of plate and frame, spiral wound and tubes, operate at 0.5 to 0.7 MPa.

Use *dead-end* for low concentrations of particles >0.1 μm.

Use *crossflow* batch with 100% retentate recycle, continuous bleed with recycle ratio 15 to 30/1 and multistage: when a concentrated retentate is desired or when particle diameter <0.1 μm. Criterion to backwash/or clean: when 90% reduction in retentate volume is achieved; given quality or volume of permeate; the permeate flux <15% of initial flux; or when the viscosity of the retentate is 100 to 300 mPa·s. This corresponds to concentrations for pigments of 30 to 70%; for microorganisms of 1 to 10%.

Membrane life: 2 to 4 years.

Cycle time: clean by backwashing with permeate every 6, 24, 170, 360, 1000 h (depending on the criteria, the membrane, operating conditions, and the amount of pretreatment); example, 8 h on and 1 to 2 h off to clean; or short pulses every 150 to 300 s so that steady-state flux operation is never achieved.

Clean by steam, detergents, solvents, acids, or bases. Another option to lengthen cycle time is to add solids to the feed to mechanically wear away the gel layer. The choice affects membrane life. Steam cleaning gives 50 to 150 cycles before membrane replacement; nonsteam cleaning gives 200 to 500 cycles.

16.11.4.23 Membranes: Microfiltration

A related topic is filters (Section 16.11.5.13).

Area of Application

Particulate diameter 0.05 to 800 μm; feed solids concentration <75% w/w; <50% v/v. Remove solid or gelatinous particulates by pore size in the membrane. Pore size: 0.2 to 1 μm with the membrane cutoff sizes in the range 0.05 to 10 μm.

Guidelines

Driving force for the rate of separation: pressure.

Membrane: symmetric or asymmetric microporous. Ceramic, sintered metals, or polymers with pores 0.2 to 1 μm. Symmetric polymers have a porosity of 60 to 85%; asymmetric ceramic membranes, porosity 30 to 40%, are used for high pressure and higher temperature <200°C.

Conceptual Process Design, Process Improvement, and Troubleshooting

Hydraulic permeability, A: 70 to 10,000 g/s·m². MPa. Pressure: 0.3 to 0.5 MPa for ceramic. Capacity/unit: 0.001 to 1 L/s. Liquid permeate flux: 0.001 to 0.2 L/s·m² with the permeate flux through ceramic membranes 2 to 3 times higher than through symmetric polymeric or sintered metal membranes and 5 to 10 times higher than through asymmetric polymeric membranes because ceramic operates at higher pressure.

Configuration: see Section 16.11.4.18.

Use tubular for feed concentrations of 10 to 80% w/w; spiral wound or thin channels for low concentrations with particulates <100 μm. For more see UF, Section 16.11.4.22.

16.11.5 Heterogeneous Separations*

General Guidelines

1. Consider shifting from heterogeneous phase separation to homogeneous phase separation.
2. If possible, separate gas phase first, then the liquid, and then solid-solid.
3. Consider using dense media separation to preconcentrate before grinding to final liberation size.
4. Use feed assay and liberation size as criteria to guide selection of options.
5. Try froth flotation as the first option. Suggested heuristics are given by Woods (1994, pp. 5–71).

For solid-solid separation, see Section 16.11.5.17.

16.11.5.1 Gas-Liquid

Area of Application

Knockout pot: drop diameter >100 μm; feed concentration >1% liquid v/v.

Zigzag baffled chamber: >100 μm; >0.01% v/v, but keep superficial gas velocity <1 m/s to prevent reentrainment.

Wet cyclone: 10 to 400 μm; 4 to 60% liquid v/v.

Spray chamber: 10 to 100 μm; 0.1 to 8% v/v, with collection efficiency decreasing from 90% to 50% as mist diameter decreases.

Venturi: <100 μm; <0.1% liquid v/v, with collection efficiency decreasing from 95% as mist diameter decreases.

Mesh demister: 10 μm; 0.001 to 0.01% liquid v/v; see size enlargement, Section 16.11.9.1.

Crossflow packed tower: <10 μm; 0.001 to 0.1% liquid v/v.

Afterburner: <0.8 μm; <0.1% liquid v/v.

Steam traps: separate condensate from steam.

Guidelines

Knockout pot (drums and accumulators for high ratios of liquid/gas, as in distillation column overheads): use horizontal cylinder: size vapor space to provide the residence time for drops to settle out. Vapor volume between 20 and 50% with a minimum of 0.3 m. Design vapor-phase cross-sectional area to allow drops to settle in assigned length of the drum. Assume drops 0.1 to 200 μm. The Souders–Brown "separation" maximum superficial gas velocity is

---

* Reproduced with permission from Donald R. Woods, *Rules of Thumb in Engineering Practice*, pages 137 to 184, 2007, copyright Wiley-VCH Verlag GmbH & Co. KGaA, Weinheim, ISBN: 978-3-527-31220-7. See this book for more details, especially about Good Practice, Trouble shooting and Capital Costs.

TABLE 16.33
Selection Guide within Section 16.11

		Major Feed Component		
		Solid	Immiscible Liquid	Gas/Vapor
Minor feed component	Solid	General **5.17** Zone refine **4.5** Screen **5.29** Classifier **5.21–24** Separator **5.25–28** Flotation **5.18** Electrostatic **5.19** Magnetic **5.20**	Settler **5.7** Thickener **5.9** Screen, **5.6** Hydrocyclone, **5.8** DAF **5.15** Filter **5.13** Centrifuges, **5.11, 5.12** Ultrafiltration, **4.22** Microfiltration, **4.23**	General **5.2**
	Liquid	General options **5.4** Dryer **5.5** Screen **5.6** Leach **5.14** Expeller **5.16**	General options **5.3**	**5.1** Knockout pots Zigzag baffled chambers; wet cyclones; spray chambers, venturis, cross flow; wetted packing
	Gas/vapor		Stripping **4.9**	Homogeneous: membranes **4.15**

$v_{o\,max} = k((\rho_L - \rho_G/\rho_G))^{0.5}$, where k is usually 0.13. $k_{horiz} = 1.25\, k_{vertical}$

and use superficial design value of 0.5 to 0.85% of v_{max}. (This separation superficial gas velocity is used to design/size many types of equipment: distillation columns, demisters.)

Design liquid volume usually 80% with sufficient volume for 300 s residence time to satisfy process control requirements. Typical length-to-diameter ratio of 3:1 to 5:1.

Knockout pot (for low ratios of liquid/gas, as in demisters): use vertical cylinder. Size the same as horizontal with height of the vapor space 1.5 times dia. with 15 cm minimum above the top of the inlet nozzle and use lower value for k in the above equation. For the liquid phase: maximum liquid level at least 18 cm below bottom of inlet nozzle; liquid residence time about 300 to 600 s.

Steam traps: ball float, open bucket, inverted bucket, liquid expansion, and thermodynamic. Float: continuous discharge, operating principle of buoyancy, OK for low loads but not high pressure. Inverted bucket: intermittent discharge, operating principle = weight of the bucket, robust, okay for high pressure and corrosive condensate, use check valve before trap. Balanced pressure, thermostatic: operating principle = vapor pressure of fluid inside bellows. Thermodynamic/kinetic energy: intermittent, operating principle is Bernoulli's principle/impulse, poor air handling, larger sizes more susceptible to back pressure. Usually for steam pressure <1.2 MPa, affected by ambient temperature.

Select inverted bucket traps based on condensate flow rate, pressure differential, and "safety factor."

The other units are sized using approaches used for gas-solid separations (Section 16.11.5.2).

16.11.5.2 Gas-Solid

Area of Application

"Fumes" are particles <1 μm. In general: use cyclones and settling basins for solids loading >20 g/m³. Then, select bag filters unless fumes are also present, low temperatures (<100°C

Conceptual Process Design, Process Improvement, and Troubleshooting

for natural fibers; <300°C for synthetic glass), noncorrosive conditions and not close to the dew point. Select scrubbers if fumes are present. For high temperatures, consider electrostatic precipitators. Design can be "high efficiency" or "standard," with mass collection efficiency decreasing as size of target particle decreases.

Dry cyclone: dust diameter 10 to 1000 μm; feed concentration 5 to 75 g/m^3; temperature <400°C; gas phase Δp = 0.2 to 1.6 kPa; mass collection efficiency 50%; power usage 0.8 to 20 kW/m^3/s.

Settling basin: dust diameter >100 μm; feed concentration 2 to 75 g/m^3; temperature <400°C; gas phase Δp = 3 to 7 kPa; power usage 0.04 to 0.7 kW/m^3/s.

Bag filter: very efficient removal of small-diameter and small-particle loadings; dust diameter 0.5 to 70 μm; feed concentration 0.01 to 100 g/m^3; temperature <100°C for natural fibers and <300°C; gas phase Δp = 0.5 to 1 kPa: power usage 0.8 to 30 kW/m^3/s.

Wet cyclone: dust diameter 0.75 to 10 μm; feed concentration 5 to 75 g/m^3; temperature <100°C; gas phase Δp = 0.5 to 1 kPa, power usage 0.8 to 20 kW/m^3/s.

Crossflow scrubber: dust diameter >3 μm; feed concentration <10 g/m^3; temperature <100°C; gas phase Δp = 0.2 to 1.6 kPa.

Wet scrubbers: countercurrent wet packing: dust diameter 0.2 to 3 μm; feed concentration <0.1 g/m^3; temperature <100°C; gas phase Δp = 1.25 to 6 kPa.

Turbulent bed contactor (see also Section 16.11.4.8, absorber): temperature <100°C; OK for heavy, sticky particles; allows high gas and liquid flow rates with high mass transfer efficiencies for gas absorption; 1 to 2 μm; 2.5 to 20 kW·s/m^3.

Venturi scrubbers: dust diameter 0.02 to 3 μm; feed concentration 0.1 to 20 g/m^3; temperature <100°C; gas phase Δp = 1.25 to 6 kPa; mass collection efficiency 99%; power usage 3 to 40 kW/m^3/s.

Low-voltage electrostatic precipitator: dust diameter 1 to 100 μm and conducting particles; feed concentration <30 g/m^3; temperature <800°C; gas phase Δp = 0.02 to 0.1 kPa; power usage 0.04 to 0.7 kW/m^3/s. Mass collection efficiency 5%, increasing to 90% as the particle size increases from 0.01 to 5 μm.

High-voltage electrostatic precipitator: dust diameter 0.01 to 5 μm and conducting particles; feed concentration <25 g/m^3; with pressures <7 MPa and temperature <800 °C; gas phase Δp = 0.02 to 0.1 kPa; mass collection efficiency 99.5%; power usage 0.04 to 0.7 kW/m^3/s. Gas velocity 0.3 to 5 m/s across the face of the collecting surface. One high-voltage system/2500 m^2 collecting surface, 30 to 100 kV DC. High initial capital investment.

Afterburners: dust diameter <0.1 μm; combustible; feed concentration <0.1 g/m^3.

SO$_2$ scrubbers (double alkali, Catox, Tyco): power usage 30 to 40 kW/m^3/s.

Guidelines

Dry cyclone: size based on an inlet gas velocity based on the particle loading: for particle loadings of <7 g/m^3, use inlet gas velocity of 11 to 23 m/s to size inlet nozzle and then scale configuration from this dimension. For loadings of 10 g/m^3, use 20 m/s; loadings of 100 g/m^3, use 10 m/s; loadings of 1000 g/m^3, use 5 m/s; loadings of 10,00 g/m^3, use 1.8 m/s.

Gravity settler: 5 to 7 m/s at inlet baffle; but the gas superficial velocity should be <3 m/s to avoid reentrainment.

Bag filter: batch: load cycle and clean: intermittent shaking, reverse pulse, reverse blow ring or sonic cleaning. Load filter until the gas pressure drop across the filter is >1.5 kPa, then clean. Choice of fabric is critical: static charge on fabric, operating temperature, potential for fumes to absorb with moisture to deteriorate bag, and need to select dust removal option to keep the Δp across the bag of 0.5 to 1.5 kPa. Felted material gives higher gas flow rate per unit area than woven, costs 3 to 4 times more, and cannot be cleaned by shaking. Gas-to-cloth ratio of 25 to 150 dm^3/s·m^2, but usually design on <75 dm^3/s·m^2. Usual range for woven fabric 7.5 to 50 dm^3/s·m^2; for felted fabric, use 7.5 to 100 dm^3/s·m^2; for microporous tubes, use 9 to 20 dm^3/s·m^2. Gas loading depends on density of particles, size, inlet dust concentration, type of fabric. Batchwise loading with dust removal by reverse jet or blowring is about 0.04 kW/m^2 of bag area. Bag length: diameter <33:1.

Wet scrubbers: limited to lower temperatures <100°C.

Wet cyclone: size on internal superficial gas velocity of 1 m/s; height to diameter of 3:1 and water usage of 0.4 L/m^3; water flow rate 1.3 to 2.5 L/m^2·s.

Crossflow scrubber: Δp = 0.3 kPa/m of width; 1 to 1.5 m width usual; water flow rate 2.7 L/m^2·s of horizontal cross-sectional packing; size on actual inlet gas flow rate to the packing face of 1 m^3/m^2·s.

Countercurrent wet packing: pressure drop 0.3 to 0.5 kPa/m of packing. Liquid loading about 0.4 L/m^3 gas or 0.6 to 1 L/s·m^2; superficial gas velocity 0.5 to 1 m/s; mass loading liquid/gas = 0.7 to 1.5.

Turbulent bed contactor: Liquid loadings 20 L/m^2·s; superficial gas velocity based on actual inlet gas flow rate 2 to 11 m^3/s·m^2 horizontal cross section; mass loading liquid/gas = 4 to 8. Related topics include fluidized bed, drying (Section 16.11.5.5); heat transfer (Section 16.11.3.8); reactors (Section 16.11.6.27); mixing (Section 16.11.7.1), and size enlargement (Section 16.11.9.5).

Venturi scrubbers: size on throat velocity of 15 to 150 m/s selected based on particle size to be removed with 40 m/s (and 25 kPa pressure drop) for 1 μm and 120 m/s (and 25 kPa) for 0.1 μm. Water usage is in the range 0.5 to 5 L/m^3 gas, with pressure drop increasing as throat velocity increases; mass loading liquid/gas = 1.3 to 1.6.

Wet impingement baffle (Peabody) scrubbers: height/diameter 1.3:1 to 4.6:1; liquid loading 1 to 2 L/m^2·s of horizontal cross-sectional area; mass loading liquid/gas = 0.2 to 0.7.

Electrostatic precipitator: Batch process: load electrodes, then clean: via wet spray or mechanical rapping. For particle conductivity between 10^{-8} and 0.01 reciprocal ohm·m. Prefer negatively charged configuration. May need to adjust conditions to get particle conductivity into acceptable range. For particles >10 μm, use 38-m^2 plate area per m^3/s gas flow; 1 μm, use 100-m^2 plate area per m^3/s gas flow; 0.4 μm, use 120-m^2 plate area per m^3/s gas flow.

16.11.5.3 Liquid-Liquid

Usual drop size 200 μm; interfacial tension 30 mN/m.

Area of Application

Decanter: drop diameter >100 μm:, feed concentration >2% v/v.

Hydrocyclone: drop diameter >20 μm; feed concentration 6 to 60% v/v. Interfacial tension must be >10 mN/m to prevent drop breakup.

Sedimentation centrifuge: disc type: drop diameter >20 and <200 μm; feed concentration 6 to 60% v/v; suited for low surface tension, density differences >0.02 Mg/m^3. Solids contamination <0.1% v/v. Use differential type for drop diameter >200 μm; feed concentration 6 to 60% v/v; suited for low surface tension, density differences >0.05 Mg/m^3. Solids contamination <0.1% v/v.

Electrodecanter: drop diameter 9 to 500 μm; feed concentration 0.8 to 8% v/v.

Fibrous bed coalescer: drop diameter 3 to 75 μm; feed concentration <3% v/v. See size enlargement, Section 16.11.9.2.

API separator: drop diameter >75 μm; feed concentration 0.015 to 3% v/v.

Dissolved air flotation: drop diameter >8 μm with feed concentration 0.005 to 0.015% v/v and drop diameter <8 μm; feed concentration 0.0075 to 0.1% v/v. See Section 16.11.5.15.

Coagulation/flocculation: drop diameter >8 μm with feed concentration 0.005 to 0.015% v/v and drop diameter <8 μm; feed concentration 0.0001 to 20% v/v. See size enlargement, Section 16.11.9.3.

Deep bed filtration: drop diameter >8 μm with feed concentration 0.0002 to 0.005% v/v, and drop diameter <8 μm; feed concentration 0.002 to 0.05% v/v. See filters, Section 16.11.5.13.

Solvent extraction: drop diameter 0.1 to 1 μm; feed concentration 0.001 to 10% v/v. See Section 16.11.4.10.

Guidelines

16.11.5.3.1 Decanter

Ill-behaved dispersions usually drift with time, are sensitive to incoming drop-diameter distribution and to upstream energy input. Examples include most systems with kerosene-based immiscible systems. First approximation: allow 20-min residence time or total overflow velocity of 0.35 L/s·m².

Feed concentration <10% v/v, size as sedimentation-controlled provided surfactants and contamination negligible and mixture is not "ill-behaved." Use overflow total flow rate velocity of 0.5 to 3 L/s·m² based on horizontal cross-sectional area with a usual value of 1.4 L/s·m². This is for a horizontal cylinder with length-to-diameter ratios of 3.5. Allow both phases to have >20% of the diameter and no less than 0.2 m to ensure that the exit phases do not become cross contaminated. For process control, the minimum distance between the high and low levels of the interface should be 0.36 m or at least 2 min residence time.

Feed >10% v/v, or contamination present, or ill-behaved, size as coalescence controlled. For vertical decanters, allow a total residence time that depends on density difference and interfacial surface tension. For a typical 0.5-m height of coalescent band (or a decanter of 0.7-m height), use an overflow total flow rate velocity of $1.5 \, (\Delta\rho/0.1)^{0.5}$ L/s·m², where the density difference is in units of Mg/m³ to determine the horizontal cross-sectional area. For horizontal configurations, use half of the vertical overflow velocity. Can add parallel plates or high- and low-energy combination coalescer promoters (see size increase, liquid-liquid, Section 16.11.9.2).

16.11.5.3.2 Hydrocyclone

Design using the same principles as liquid-solid hydrocyclone, Section 16.11.5.8. For flooded underflow, the pressure drop is about two to seven times greater than for air core operation.

16.11.5.3.3 Sedimentation Centrifuge

Disc type: (Westfalia, Alfa–Laval, Robatel) continuous: centrifugal field about 10^4 g and 100 rps with residence times of 1 to 10 s. Power 3 to 10 kW·s/L of feed.

Differential type: (Podbielniak, Quadronic) continuous: centrifugal field about 500 g and 25 rps with about 10 to 75 s residence time. Power 1 kW·s/L.

16.11.5.4 Liquid-Solid: General Selection

Effect of Particle Diameter and Solid Concentration on the Choice

For particle diameter greater than 1000 μm and solid concentration >3%, use screens, Section 16.11.5.6. For particle diameter <2 cm and >5 μm and solid concentration 1 to 50%, consider settlers, filters, or centrifuges. For particle diameter <300 μm and solid feed concentration 0.01 to 20%, consider thickeners, Section 16.11.5.9. For particle diameter >20 μm and solid feed concentration greater than 50%, consider dryers, Section 16.11.5.5. For particle diameter 0.01 to 150 μm, consider deep bed filter, Section 16.11.5.13, or dissolved air flotation, Section 16.11.5.15. For particle diameter 0.6 to 40 μm and solids concentration <0.1%, consider homogeneous separation via ultrafiltration, Section 16.11.4.22. For particle diameters from 0.8 to 20 μm, consider using a filter aid to precoat on the filter medium. For example, use diatomous earth or perlite. A *fine* filter aid is 8- to 20-µm diameter to give a precoat bed of permeability 0.05 to 0.5 µm^2; a *medium* filter aid is 30- to 60-µm diameter to give a precoat bed of permeability 1 to 2 µm^2; a *coarse* filter aid is 70- to 100-µm dia. to give a precoat bed of permeability 4 to 5 µm^2. For particle diameter less 1 µm, consider size increase via coagulation/flocculation, Section 16.11.9.3. An example coagulant is starch.

Effect of Recovery on the Choice

To recover liquid: in the order of preference of filters, Section 16.11.5.13, and filtering centrifuges, Section 16.11.5.12: deep bed, horizontal vacuum, pressure leaf, gravity flat table; cartridge, precoat drum, and plate and frame, or vertical basket filtering centrifuge. For high fluid viscosity, use plate and frame or horizontal filtering cone centrifuge.

To recover liquid: in order of preference of settlers, Section 16.11.5.7, thickeners, Section 16.11.5.9, and sedimentation centrifuges, Section 16.11.5.11: clarifier, settler, washing tray thickener, reactor-clarifier, hydrocyclone, batch tubular bowl centrifuge, batch automatic (horizontal or vertical bowl, disc with intermittent nozzle discharge); continuous disc bowl centrifuge with nozzle discharge.

To recover liquid and solids: in order of preference of filters: expellers and presses, Section 16.11.5.16.

To recover liquid and solids: in order of preference of settlers, Section 16.11.5.7, CCD, Section 16.11.5.10, and sedimentation centrifuges, Section 16.11.5.11: continuous countercurrent decanter circuit CCD, horizontal solid bowl centrifuge with scroll discharge.

To recover solids: in order of preference of filters: requiring good washing: pressure, vacuum, gravity table/pan, horizontal pressure or vacuum, horizontal belt, vacuum drum, cylindrical screen scroll discharge filtering centrifuge, and plate and frame. Requiring good washing and the crystals break easily: gravity, vacuum table/pan; vacuum, pressure, gravity drum, and plate and frame. If the cake is compressible, use low pressure, <200 kPa, or vacuum rotary drum. For dry solids: consider the following filtering centrifuges, Section 16.11.5.12: basket, basket automatic constant speed vertical, basket automatic variable speed horizontal; continuous conical with scroll conveyor, oscillating conical screen; cylindrical screen with pusher conveyor, or horizontal solid-screen scroll conveyor. To recover solids: in order of preference of thickeners, Section 16.11.5.9, and sedimentation centrifuges, Section 16.11.5.11: thickeners, deep thickener, rake thickener, and tray thickener or hydrocyclone, batch automatic solid bowl centrifuge, continuous conical bowl centrifuge, continuous contour bowl vertical or horizontal centrifuge.

16.11.5.5 Dryer

Use when the goal is solid recovery. Related topics are screens (Section 16.11.5.6), centrifugal filters (Section 16.11.5.12), and dewatering expellers (Section 16.11.5.16).

Area of Application

Particle diameter 20 µm to 1 cm; feed solid concentration >50% solids. Liquid contamination in exit solids 0 to 20% v/v liquid. Use batch for <40 g/s; use continuous for >280 g/s. Select initially on temperature sensitivity. Prefer "adiabatic" over conduction.

Indirect conduction options: (nonadiabatic) for temperature-sensitive solids ($T < 40°C$).

Batch:

Pan/tray/shelf, jacketed, atmosphere, or vacuum (conduction): batch, feed: thin or thick liquids, soft or stiff pastes, moist crumb, grains (>150 µm) and grits (<150 µm). Product: solid cake. *Pan, agitated,* atmosphere or vacuum (conduction): batch, feed: thin or thick liquids, soft or stiff pastes, moist crumb, grains (>150 µm), and grits (<150 µm). Product:

solid cake. *Rotary indirect* (steam or hot fluid) atmosphere or vacuum (conduction): batch feed: moist crumb, grains (>150 μm), and grits (<150 μm). Dries in <60 min. Product: crumb/powder. *Cone/double cone,* jacketed, atmosphere or vacuum (conduction): batch, feed: sticky, fine, moist crumb, grains (>150 μm), and grits (<150 μm). Product: crumb. *Freeze,* batch, atmosphere or usually vacuum (conduction): batch, feed: thin or thick liquids, soft or stiff pastes, moist crumb, grains (>150 μm), coherent sheets, discontinuous sheets. *Dielectric:* feed: thin or thick liquids, soft or stiff pastes, moist crumb, grains (>150 μm), and grits (<150 μm).

Continuous:

Drum dryer, atmosphere or vacuum (conduction): feed: thin or thick liquids, soft and stiff pastes, wet paper. Dries within 2 to 30 s. Product: flakes, dry sheet. Screw, jacketed, atmosphere or vacuum (conduction): feed: sticky, fine, moist crumb, grains (>150 μm), and grits (<150 μm). Product: crumb. *Conveyor,* jacketed, atmosphere or vacuum (conduction): feed: sticky, fine, moist crumb, grains (>150 μm), and grits (<150 μm). Product: crumb.

Indirect convection options: for moderately temperature sensitive, sensitive and very sensitive solids ($5 < T < 40°C$). Indirect heating means heat source temperature <200°C.

Batch:

Pan/tray/shelf, crossflow (indirect convection): batch, feed: soft paste, preform, hardpaste, granular, fragile particles, fibrous, discontinuous sheets, and shaped pieces. *Pan/tray/shelf, through flow* (indirect convection): batch feed: preform, granular, fibrous. *Open sand bed:* (natural convection): batch feed of wastewater sludge.

Continuous:

Spray (indirect convection) feed: thin liquids and slurries. Dries within 1 to 10 s. Product: 1- to 300-μm powder. *Flash/transported* (indirect convection) feed: preformed paste, granular, fibrous solids. Dries within 0.5 to 3 s. Product: powder. Particle diameter <2 mm. *Fluidized bed* (indirect convection): feed: soft paste, sludge, preformed paste, granular, fibrous solids. Product: powder. Particle size 0.05 to 15 mm. *Pan/tray/shelf* continuous (Turbo) (indirect convection). Feed: soft paste, preform, granular, fibrous. Product: solid cake. *Desolventizer:* feed: leached vegetable seeds contaminated with solvent (such as hexane). Product: solvent-free cake. *Tunnel/truck* continuous (indirect convection): feed: grains (>150 μm), discontinuous sheets and shaped pieces, soft paste, preform, granular, fibrous. Product: solid cake. *Belt, through flow,* continuous (indirect convection): grains (>150 μm), discontinuous sheets, and shaped pieces. *Rotary, continuous* (including steam tube) (indirect convection): feed: hard, granular, fibrous. Dries within <60 min, <50 kg/s/unit.

Direct convection options (adiabatic) for temperature-sensitive, moderately sensitive, and insensitive solids ($10 < T < 150°C$).

Options similar to *indirect convection,* but hotter gases are used.

Guidelines

Indirect conduction at reduced pressure: evaporative capacity 42 to 5 g water/s·m^2 as moisture goes from 0.5 to 0.015 kg/kg dry solids.

Convection with hot gas through the bed: evaporative capacity 1.2 to 0.25 g water/s·m² as moisture goes from 0.9 to 0.025 kg/kg dry solids.

Convection plus radiation: evaporative capacity 0.8 to 0.15 g water/s·m² as moisture goes from 0.9 to 0.025 kg/kg dry solids.

Indirect conduction options:

Batch: size on cycle time: load, dry, discharge, clean.

Pan/tray/shelf, jacketed, batch (conduction): evaporative capacity 0.13 to 0.27 g water/s·m² tray area for crystals, 0.07 to 0.14 g water/s·m² tray area for finely divided solids; loading 10 to 35 kg wet paste/m² tray area; pressure 7 to 27 kPa. *Pan, agitated, batch, atmosphere* (conduction): area 1.5 to 15 m²; evaporative capacity 1 to 12 g water evaporated/s·m²; power 4 kW/m³. Heat transfer coefficient 10 to 50 W/m² °C; solids capacity 2.8 to 4.2 g dry solids/s·m³. *Pan, agitated, batch, vacuum (conduction):* evaporative capacity 1.35 to 6.8 g/s·m²; time 7.5 to 35 h. *Tray/pan/shelf* (conduction, vacuum): area 1 to 20 m²; evaporative capacity 0.27 to 7.7 g/s·m²; time 4 to 48 h. Heat transfer coefficient U = 0.002 to 0.5 kW/m²·K. *Rotary indirect,* batch, vacuum (conduction): area 2 to 35 m²; evaporative capacity 0.3 to 5 g water evaporated/s·m² with values increasing as initial feed moisture content increases; 0.7 to 5 g organic evaporated/s·m². Power 0.5 kW/m² area. *Cone/double cone,* jacketed, batch, vacuum (conduction): area 1 to 10 m²; evaporative capacity 2.7 to 5.4 g water evaporated/s·m² of actual surface area. Power 0.5 kW/m² area. Freeze, batch, vacuum (conduction) drying temperature serum, –9 to –12°C; plasma, –20 to –25°C; penicillin –28 to –32°C; size vacuum pump to remove the water vapor; avoid air leaks. Pressure 10 to 200 Pa or 1/4 to 1/2 the vapor pressure at the temperature; 7 to 8 h drying cycle.

Continuous:

Drum dryer, atmosphere (conduction): area 2 to 50 m²; evaporative capacity 7 to 11 g water/s·m²; residence time: 6 to 15 s; solids capacity 5 to 50 kg/m²; heat transfer coefficient 0.001 to 0.002 kW/m²·°C; 1 to 10 rpm; 1 to 2.5 kW/m². 1 to 10 rpm; drive power 1 to 2.5 kW/m² drum surface area. Related topic is flakers, Section 16.11.9.15. *Fourdrinier machine* for paper: 37 cylinders, 1.5-m dia.; velocity of paper = 4 to 5 m/s; residence time 31 s; heat transfer coefficient U = 0.34 kW/m²·°C. Water/solid = 0.4; 2.8 g water evaporated/s·m². *Screw, jacketed,* atmosphere (conduction): 2 to 30 rpm; area 4 to 60 m², volume 0.1 to 3 m³, 0.03 to 0.5 kg water evaporated/s. Heat transfer coefficient 4 to 60 W/m² °C with 4 to 10 for hollow screw and 5 to 35 for hollow paddles; power 140 MJ/Mg. *Continuous band, vacuum;* (conduction): 2.2 kg steam/kg water evaporated; solid capacity 2 to 5 g dry solids/s·m² belt area; belt size <8 m². *Continuous band, atmospheric* (conduction): evaporative capacity 0.8 g water/s·m²; dry solids output 2 g solid/s·m²; area 56 m². *Freeze,* continuous tray: 50-mm thick bed; vacuum 13 Pa absolute: heat transfer coefficient, U = 0.01 to 0.02 kW/m²·°C; relatively independent of stirring speed but dependent on residence time. Residence time 10 to 200 s. *Dielectric:* 315°C.

Indirect convection options: for moderately temperature sensitive, sensitive, and very sensitive solids ($5 < T < 40°C$).

Batch: size on cycle time: load, dry, discharge, clean.

Tray/shelf (indirect convection) crossflow: 0.05 to 0.2 g water evaporated/s·m² tray area; residence time 4 to 48 h; 1.5 to 5 m/s gas flow; steam 1.8 to 2 kg steam/kg water evaporated; power 8 to 15 kJ/kg; air temp. 50 to 110°C. solids capacity <6 g/s. *Open sand bed:* (natural convection) 1000 to 100,000 m² including piping, sand/gravel beds, and underground collection.

Continuous:

Spray (indirect convection): residence time 3 to 30 s; gas velocity 0.2 m/s; thermal efficiency 50%; adiabatic efficiency 100%; solid temperature = adiabatic saturation temperature; volumetric heat transfer coefficient 0.13 to 0.18 kW/m^3 K; 1.8 to 2.7 kg steam/kg water evaporated. Δp = 1.5 to 5 kPa. See size reduction sprays, Section 16.11.8.2; spray reactor, Section 16.11.6.12; heat exchange, Section 16.11.3.12; and size enlargement, Section 16.11.9.4. *Flash/transported* (indirect convection): 175 to 630°C, gas velocity 3 to 30 m/s or 2.5 to 3 times the terminal velocity of the particles; gas requirement 1 to 5 Nm^3/kg solid or 1 to 10 kg air/kg solid; exit air temperature 20°C greater than exit dry solid temperature; 4000 to 10,000 kJ/kg water evaporated. See transported slurry, transfer line reactors, Section 16.11.6.9. Heat transfer coefficient for gas drying: h = 0.2 kW/m^2·K. Fluidized bed (indirect convection): residence time 30 to 60 s for surface fluid vaporization; 15 to 30 min for internal diffusion; 3500 to 4500 kJ/kg water evaporated. *See fluidized bed reactors, Section 16.11.6.27; heat transfer, Sections 16.11.3.4 and 16.11.3.8; size enlargement, Section 16.11.9.5; and mixing, Section 16.11.7.1. Tray/gas flow through the bed:* 0.24 to 3.3 g water evaporated/s·m^2 tray area. Residence time 2 to 8.5 h; superficial air velocity 0.2 to 1 m/s; steam 2 to 6.8 kg steam/kg water evaporated. Fan power 1.6 to 2.5 kJ/g solids throughput, area 4 to 15 m^2; air temperature 9 to 100°C. Solids capacity <6 g/s. *Continuous tray, turbo* (indirect convection), 1 to 2 rpm; 10 to 50 trays/unit; residence time 0.25 to 2 h; diameter 1.25 to 11 m; height 1.5 to 18 m. Area 10 to 1000 m^2; evaporative capacity 0.05 to 15 g water evaporated/s·m^2. Gas velocity 0.6 to 2.4 m/s; fan power 8 to 15 kJ/kg solids handled; 0.08 to 15 kW/m^2, 0.1 kg steam/kg dry solid; capacity 0.003 to 2.3 kg/s. 5 to 8 kW/Mg dried solids; 5 to 7 kW/Mg water evaporated. Energy 50 to 100 kJ/m^2 drying area for dry solids; 70 to 200 kJ/m^2 drying area for wet solids. *Desolventizer:* combination of live and indirect steam; inlet solvent concentration about 30%; capacity 2 to 50 kg/s. Flakes 0.25- to 0.3-mm thick. *Tunnel/truck* (indirect convection) area 10 to 100 m^2; evaporative capacity 0.15 to 0.77 g water evaporated/s·m^2. *Rotary* (indirect convection) temperature 300°C; area 10 to 1000 m^2; evaporative capacity 9 g water evaporated/s·m^3 or 3.5 g water/s·m^2 of peripheral area; 5 to 10% solids, residence time 0.1 V/volumetric feed rate; gas velocity 1 to 1.5 m/s; peripheral velocity 0.1 to 0.5 m/s. Steam heated gas temp. 120 to 175°C; size on volumetric heat transfer coefficient. Power 0.15 to 0.25 kW/m^2 nominal circumferential area. L/D 5.5/1. *Rotary steam tube* (indirect convection): temperature 150 to 180°C; area 10 to 1000 m^2; evaporative capacity 0.77 g water/s·m^2; gas velocity 0.3 m/s; heat transfer coefficient U = 0.03 to 0.09 kW/m^2·°C. *Continuous metal band:* (heated by forced convection air, IR, direct steam, or direct hot water): heat transfer coefficient from impinging hot air: U = 0.06 to 0.09 kW/m^2·K. Air velocity 15 to 25 m/s. 5 to 50 kg water evaporated/m^2 drying surface.

Direct convection options: for temperature-sensitive, moderately sensitive, and insensitive solids [10 < T < 150°C]. *Hot gas temperature* 550 to 800°C

Continuous band: crossflow: <3-mm thick bed area: 20 to 100 m^2; evaporative capacity 2 to 13 g water evaporated/s·m^2; residence time 0.2 to 1.3 h; steam 1.7 to 1.9 kg/kg water evaporated; fan power 35 to 1300 kJ/kg; belt drive power 1 kW/m^2 belt. *Continuous band: gas flow through the bed:* 1 to 4 g water evaporated/s·m^2; steam 2 to 5 kg steam/kg water evaporated; 1.25 m/s gas velocity through the bed; area 5 to 25 m^2. *Rotary cascading dryer/kiln roto-louvre* area 10 to 1000 m^2; evaporative capacity 18 g water evaporated/s·m^3 or 7 g water/s·m^2 of peripheral area; 5 to 10% solids, residence time 0.1 V/volumetric feedrate; gas velocity 1 to 1.5 m/s or 1/2 terminal velocity of particles; peripheral velocity 0.1 to 0.5 m/s, 4 to 5 rpm; rpm times diameter = 3 to 12 rpm·m; L/D = 4 to 15/1; average temperature of evaporation throughout = 3°C above the inlet wet-bulb temperature. Countercurrent exit air temperature about 100°C; for cocurrent exit air, temperature = 10 to

20°C higher than the exit solid temperature; heat transfer coefficient 0.1 W/kg solids °C for coarse materials and 0.4 W/kg °C for fine. NTU = 0.5 for vegetables. NHTU (for air water) = 1.0 to 3.5 = $\ln(T_{hot\,gas\,in} - T_{wet\,bulb})/(T_{hot\,gas\,out} - T_{wet\,bulb})$. Drive power 0.075 to 0.16 kW/m² nominal circumferential area; area in the range 200 to 2500 m². Related topics are kilns (Section 16.11.3.7) and reactors (Section 16.11.6.20). *Open bed:* residence time: minutes; gas through bed 0.5 to 1 m/s.

16.11.5.6 Screens for "Dewatering"

See also Section 16.11.5.29, screens for solid-solid separation. Related topics are filters (Section 16.11.5.13), centrifugal filters (Section 16.11.5.12), and expellers (Section 16.11.5.16).

Area of Application

In general, particle diameter >1000 μm and solid concentration >3%. Special types include microscreens for >20 μm.

Batch:
Deep bed or granular: batch, for 0.01 to 50 μm (see filter, Section 16.11.5.13). *Fixed bar screen;* batch, grizzly (the filter cloth is made of rods and bars) removal of very coarse material of diameter >2 to 5 cm; low concentration of solids <15 mg/L; bars at 30 to 60° to the horizontal that can be cleaned manually or automatically. *Microscreen* (rotary drum or disk). Batch, removal of particles of diameter >20 μm; 20 mg/L solids feed concentration.

Continuous:
Fixed, inclined wedge wire screen (sieve bends, DSM): removal of particles >0.15 cm. Variable inclination from 65 to 45° to the horizontal; feed concentration 200 mg/L. For dewatering minerals, dewatering particles of diameter >40 μm. *Vibrating screens:* typical exit liquid concentration for <8-mm particle size, 20 to 45 vol% liquid; for >40-mm particle size, 2 to 10% v/v liquid. *Belt, gravity:* see Section 16.11.5.13.

Guidelines

> *Batch:* consider cycle time: load, clean.
> *Bar screen:* batch: fluid velocity through the screen 0.6 to 1.2 m/s; head loss 15 to max. 75 cm. Mechanical clean. *Rotating microscreen:* batch, fluid loading 3 to 6 L/s·m² of submerged area; usually 66% area submerged; solids loading 0.05 to 0.1 g/s·m²; head loss 7 to 14 cm to max. of 45 cm. Clean by backwash at 2 to 5% volumetric throughput capacity.

Continuous:

> *Fixed, inclined wedge wire screen,* sieve bend: fluid loading 6.7 to 20 L/s·m²; or 10 to 40 L/s·m of width; solids loading 1.4 to 4.2 g/s·m²; exit solids concentration 12 to 15% w/w solids. For dewatering minerals, fluid capacity of 0.0015 to 0.03 L/s·m² with larger fluid capacity for larger size particles (1 to 2 mm) and smaller than the included angle of the bend. The bend in the screen is such that the oversized are continually sluiced off. *Vibrating screens:* see Section 16.11.5.29.

16.11.5.7 Settlers

Area of Application

Particle diameter <2 cm; solids concentration 0.2 to 50%.

Conceptual Process Design, Process Improvement, and Troubleshooting

Guidelines

Upward liquid overflow rate 0.25 to 0.6 L/s·m², with the value increasing as the feed concentration, particle diameter, and density increase.

For solids concentration 0.2 to 2, use 0.4 to 1 L/s·m²; for 1 to 5, use 0.5 to 1.4 L/s·m². Usual depth is 4 to 5.2 m with drive for rake 0.5 to 10 kW.

16.11.5.8 Hydrocyclones

See also Section 16.11.5.21 for solid-solid separations.

Area of Application

Particle diameter 4 to 400 μm; feed solids concentration 4 to 30% solids v/v, but usual application is separating particles >50 μm and concentrations <10% v/v with pressure loss of 20 to 100 kPa. Regular hydrocyclones: particle "settling velocity" 2 μm/s to 5 mm/s and clarified liquid product of 0.1 to 70 L/s. Miniature hydrocyclones: particle "settling velocity" 0.05 μm/s to 0.04 mm/s and clarified liquid product of 0.02 to 30 L/s.

Guidelines

Determine diameter of hydrocyclone from D (cm) = 6×10^5 [Feed flow rate (L/s)/target diameter² (μm²) Δp (kPa)] [ρ_L, density liquid (Mg/m³) × liquid viscosity (mPa·s)/$\Delta \rho$, density difference, (Mg/m³)].

Typical target diameter is 5 to 100 μm and is the diameter that 50% reports to the overflow and 50% reports to the underflow. Typically three times target diameter is the diameter below which all particles in distribution are removed. The standard hydrocyclone has an inlet diameter of 0.28 D; the overflow exit diameter = 0.34 D; the vortex finder length is 0.4 D; cylindrical body of height of 0.4 D, vertical length of cone = 5 D or cone angle about 10°. Underflow diameter adjustable to adjust the volume split between the overflow and underflow.

16.11.5.9 Thickener

Area of Application

Particle size 0.1 to 300 μm; try to avoid using a thickener for particle diameters >200 μm, especially if their density is >2 Mg/m³. Feed solids concentration 0.01 to 20% v/v; exit liquid contamination in the solids exit 80 to 90% v/v liquid. Deep cone thickener: exit liquid contamination in the solids exit 30 to 40% v/v liquid.

Guidelines

Downward solid flux of 0.2 to 500 g/s·m² with low values of 0.2 to 2 for wastewater treatment, pickle liquor, and higher values of 40 to 70 for mineral processing. Higher flux rates by factors of 3 to 10 up to fluxes of 800 g/s·m² can be obtained with the addition of flocculants. Fluid residence times 2 to 4 h; solids residence times 4 to 24 h. Torque for rake (Nm) = K(thickener diameter, m)², where K = 15 to 30 for solid flux loadings of <2 g/s·m²; = 70 to 130 for solid flux loadings of 2 to 7 g/s·m²; = 150 to 300 for solid flux loadings of 7 to 23 g/s·m², and = >300 for solid flux loadings of >23 g/s·m². Operate the rake at 1 to 20% of design torque at the drive head. Rake tip speed 15 to 25 cm/s, but for appreciable amounts of particles of diameter >200 μm, increase tip speed to 25 to 40 cm/s. Pump the underflow at 1 to 2.4 m/s. See Section 16.11.2.5.

16.11.5.10 CCD: Countercurrent Decantation

Area of Application

When dilute liquid overflow is acceptable because high wash water ratios are used; for high-temperature operation when anticipate changes in feed materials and if the filtration rate is <140 g/s·m².

Guidelines

Wash ratio 2 to 2.2 water to 1 solid. Wash ratio and the number of stages are selected for expected recovery.

16.11.5.11 Sedimentation Centrifuges

Primarily to recover liquid, although it can be used to recover solids. Incomplete separation.

Area of Application

Liquid contamination in the exit solids: 10 to 15% v/v liquid.

Batch:

> *Tubular bowl:* batch, very fine particles with settling velocities 5×10^{-8} to 5×10^{-7} m/s; particle diameter 0.1 to 500 µm, feed concentrations 0.1 to 20% v/v; <5% w/w. Clarify low concentrations. *Vertical solid basket:* batch, particle diameter >30 µm; feed solids concentration 3 to 5% w/w to prevent frequent cleaning. *Multichamber vertical solid basket, manual batch discharge:* particle settling velocity 2×10^{-6} to 7×10^{-5} m/s; particle diameter 1 to 50 µm; feed solids concentration <4 to 5% v/v. Very dry product and higher capacity. *High-speed disc vertical solids retaining:* batch, particle settling velocity 8×10^{-8} to 2×10^{-7} m/s; particle diameter <1 µm; feed concentration solids <1% w/w; <20% v/v.

Continuous:

> *High-speed disc, intermittent solids ejecting:* particle diameter 1 to 500 µm; feed concentration solids 2 to 6% v/v and <5% w/w. *High-speed disc, continuous nozzle discharge:* particle diameter 0.1 to 500 µm; feed concentration solids 5 to 30% v/v; <10% w/w. Usually select when have a large amount of fines. *Horizontal scroll discharge:* (solid bowl decanter) 10^{-6} to 5×10^{-6} m/s; particle diameter 2 to 5,000 µm; feed solids concentration 0.5 to 50% v/v. Usually >10% v/v. Only option when feed concentration >40% v/v solids.

Guidelines

> Scaled up based on liquid handling capacity.
> *Batch:* size on cycle time: separate, remove solids, clean.
> *Tubular bowl:* batch, up to 50,000 rpm; 3×10^3 to 6×10^4 g; L/D 4 to 8; liquid flow rates 0.03 L/s at high rpm, 0.1 to 10 L/s at lower rpm. Clarify low concentrations. Vertical solid basket: batch, 450 to 3500 rpm, 900 to 1100 g; L/D = 0.8; 1.6 to 2.8 L/s; 70- to 120-cm diameter bowl; 200 to 500 g/s·m²; area 0.5 to 3 m²; 3 to 17 kW. Multichamber vertical solid basket, batch discharge: 4500 to 8500 rpm; 3×10^3 to 3×10^4 g; 0.7 to 2.8 L/s; 30- to 60-cm dia. bowl. *High-speed disc vertical solid basket:* <12,000 rpm; 3×10^3 to 3×10^4 g; L/D = 1; <28 L/s; 15- to 100-cm bowl diameter; cone 35 to 50°. Solids retaining: 5 to 20 L/s;

Σ, m²	Bowl dia., cm
1000	10
10,000	20 to 30
100,000	60 to 90

Continuous:

> *Horizontal scroll discharge:* 1600 to 6000 rpm; 1500 to 5000 g; L/D 1.5 to 3.5; 0.1 to 16 L/s;

Σ, m²	Bowl dia., cm
200	12 to 15 cm
1000	25 to 35
5000	50 to 70

16.11.5.12 Filtering Centrifuge

Table 16.34 summarizes the characteristics of different rates of filtration.

Area of Application

Liquid contamination in the exit solids: 4 to 13% v/v liquid.

Batch: vertical basket: batch, particle diameter >2 µm; feed solids concentration 2 to 50% w/w. Minimize damage to crystals. Low capacity: 0.003 to 5 kg/s. Product dryness 70% solids. Possible for sterilization between batches. *Vertical basket with bottom plough discharge:* batch, particle diameter 5 to 500 µm; feed solids concentration 9 to 90% w/w; capacity 0.1 to 1.25 kg/s. Product dryness 25 to 60% solids. Use with free draining, medium capacity, and where need multiple rinse. Horizontal basket, plough discharge: particle diameter 200 to 20,000 µm; feed solids concentration 10 to 70% w/w. Free slow draining, multiple rinses, low to medium capacity: 0.5 to 5.5 kg/s. Product dryness 87 to 96%. Keep feed consistent of high solids concentration.

Semicontinuous: horizontal basket, pusher discharge: particle diameter 50 to 5000 µm; feed solids concentration 15 to 70% w/w. Capacity 0.3 to 7 kg/s. Product dryness 85 to 98%. Free draining. Main advantage is to rinse the solids.

Continuous: horizontal combo solid-screen, scroll discharge: particle diameter >150 µm; feed solids concentration 10 to 75% w/w. Solvent-based reactions. *Vertical or horizontal cone: slip discharge:* continuous, particle diameter >200 µm; feed solids concentration 10 to 80% w/w, capacity 0.25 kg/s. Dewater crystals and free draining fibers. *Vertical cone, scroll discharge:* continuous, particle diameter 200 to 5,000 µm; feed solids concentration 5 to 60% w/w. Capacity 0.02 to 8 kg/s. *Horizontal cone oscillating/torsional vibration*

TABLE 16.34
Characteristics of Different Rates of Filtration

	Fast	Medium	Slow	Very Slow	Very, Very Slow
Intrinsic permeability, m⁴/N·s	$> 20 \times 10^{-10}$	$1\text{--}20 \times 10^{-10}$	$0.2\text{--}1 \times 10^{-10}$	$0.02\text{--}0.2 \times 10^{-10}$	$< 0.02 \times 10^{-10}$
for viscosity 1 mPa·s	> 50 µm	> 15 µm	> 4 µm	> 1.5 µm	< 1.5 µm
for viscosity 60 mPa·s	> 350 µm	> 100 µm	> 50 µm	> 17 µm	< 17 µm
Cake buildup rate, g/s·m²	> 700	70–700	7–70	0.02–7	<0.02
Possible option	Continuous pusher	Peeler centrifuge	Vertical basket centrifuge	Multichamber vertical solid bowl sedimentation centrifuge, **16.11.5.11**	Sedimentation centrifuge, **16.11.5.11**

discharge: particle diameter <6 mm; feed solids concentration 40 to 80% w/w. capacity 7 to 40 kg/s.

Guidelines

Batch: size on cycle times: accelerate to load speed, load and cake formation, accelerate to wash, wash, accelerate to full speed, spin dry, decelerate to unload speed, unload. Cake volume in the centrifuge is key. During load and cake formation, the particles migrate to the periphery to form a cake, the supernatant liquid moves through the bed to the cake surface, and then the liquid drains out of the interstices in the wet cake. The latter two times usually control. *Vertical basket:* batch with cycle: cake formation, wash, discharge, clean. Total cycle >10 min, e.g., 180 min. Cake buildup rate 200 to 1000 g/s·m^2; area per unit: 2.8 m^2; cake volume 0.009 to 0.5 m^3; cake thickness 25 to 150 mm with smaller thickness in smaller diameter centrifuges. 600 to 2100 rpm; 300 to 800 G. Basket diameter 0.3 to 1.5 m; L/D = 0.5 to 0.6/1. Power 3 to 6 kW/m^2 filter area. Temperature <175°C. *Vertical basket, bottom discharge:* batch with cycles: accelerate to load speed, load and cake formation, accelerate to wash, wash, accelerate to full speed, spin dry, decelerate to unload speed, unload: 3 to 120 s. Cycle time for free draining 2 to 6 min; slower draining materials cycle time 20 to 30 min with some >60 min, area per unit 0.4 to 4.5 m^2; cake volume 0.018 to 0.53 m^3; about 67% of feed volume; cake thickness 50 to 150 mm with smaller thickness in smaller diameter centrifuges. 900 to 1800 rpm. Basket diameter 1 to 1.2 m; L/D = 0.5. *Horizontal basket, plough discharge:* batch with cycles: load 7 to 25 s; wash 5 to 25 s, spin dry 12 to 30s, unload cake 3 to 12 s; total cycle time 20 s to 15 min but usually <3 min. Cake buildup rate 500 to 3000 g/s·m^2; area per unit 0.1 to 3.8 m^2; cake volume 0.03 to 0.17 m^3. Bowl dia. 0.3 to 1.2 m. 1000 to 2500 rpm; <1250 G. Power 25 to 40 kW/m^2 filter area.

Semibatch: particles move through the cycle continually. *Horizontal basket, pusher discharge:* 20 to 120 cycles/min: cake formation, wash, discharge. Cake buildup rate 1000 to 10,000 g/s·m^2; area per unit: 1.4 m^2; cake thickness <75 mm. Basket diameter 0.2 to 1.2 m; 300 to 600 G. Power 40 to 60 kW/m^2 filter area or 7 to 15 kJ/kg "salt" (about 3000 g/s·m^2). Keep feed solids concentration >40% w/w for optimum. Example data: adipic acid, urea 1.4 kg/s·m^2; crude sodium bicarbonate 1.45 kg/s·m^2; phosphate rock 2.2 kg/s·m^2; ammonium sulfate 2.36 kg/s·m^2; and sodium chloride 3.1 kg/s·m^2;

Continuous: horizontal combo solid-screen, scroll discharge: area per unit, 1.2 m^2; power 12 to 15 kW/m^2 filter area. *Vertical or horizontal cone:* slip discharge: cake buildup rate 3000 to 20,000 g/s·m^2; area per unit 4 m^2; cone angle 20 to 35° and greater than angle of repose of the solids. 2500 G. rinsing efficiency: fair. *Vertical cone, scroll discharge:* diameter 0.5 to 1 m. 2500 to 3800 rpm; 1800 to 2500 G. Product solids 24%. *Horizontal cone oscillating/torsional vibration discharge:* cone angle 13 to 18°; cycles >20 cycles/min. 300 to 500 rpm; <500 G. Diameter 0.5 to 1 m; L/D = 0.5 to 1/1. Product 92% solids. Limited liquid handling capacity. Power 0.2 kW/Mg solids.

16.11.5.13 Filter

Table 16.35 relates filtration rate to process operation.

For *incompressible* cakes, the filtration rate is directly proportional to the specific cake resistance and the pressure/vacuum, and inversely proportional to the viscosity and cake thickness. The filtration rate is inversely proportional to the ratio of solids to filtrate, while the rate of cake formation is directly related to this ratio. For *compressible* cakes, the filtration rate is relatively independent of pressure. The more flocculated the solids, the more compressible will be the filter cake.

TABLE 16.35
How Process Operation Relates to Filtration Rates

	Fast	Medium	Slow	Very Slow	Very Very Slow; Clarification
Cake formation rate, mm/s	>1	0.12–1	0.02–0.12	0.001–0.02	<0.001
Cake formation rate, g/s·m²	700	70–>700	7–70	0.02–7	<0.02
Filtrate rate, L/s·m²	>3	0.3–3	0.03–0.3	0.003–0.03 rarely filter for flux < 0.01	< 0.003
Typical cake resistance, m/kg	$<8 \times 10^8$	$8 \times 10^8 – 10^{10}$	$10^{10} – 5 \times 10^{10}$	$5 \times 10^{10} – 3 \times 10^{11}$	$>3 \times 10^{11}$
Comments	Usual prefer continuous Vacuum belt, screens, belts, top feed drum	Tilting pan, vacuum drums or disks	Continuous filters	Usually prefer batch Vacuum filter, rotary vacuum, pressure filters	Batch cartridge, precoat drums, deep bed

Area of Application

Prefer *batch* when cake formation rate is <0.01 mm/s (low concentrations of small diameter particles). OK for higher liquid viscosities and higher temperatures. Batch can be vacuum or pressure operation. Prefer vacuum operation for particle diameter >30 μm, if there is a small mass of fines (<50% with diameter <5 μm) and bed has permeability between 0.1 to 1000 μm². Prefer pressure filters for particle diameter 1 to 70 μm, if there is a large mass of fines (with >50% w/w with diameter <10 μm) and bed has permeability between 0.001 to 0.1 μm².

Prefer *continuous* for cake formation >0.01 mm/s (high concentrations of larger diameter particles); viscosity <50 mPa·s. Continuous is usually gravity or vacuum operation. For gravity operation particle diameter >1 mm and bed has permeability >1000 μm². For vacuum, usually particle diameter >30 μm, and bed has permeability between 0.1 to 1000 μm².

For *compressible* cakes, prefer mechanical compression via diaphragm plates or belt presses. Compressed volume = 0.6 to 0.75 volume before compression. Compression cycle 0.33 to 0.4 h for diaphragm plate and frame press. For *highly compressible*: latex, highly flocculated materials > silica, talc, attapulgite > kaolin >> barite, diatomaceous earth >> incompressible = polystyrene, carbonyl iron.

Batch: *leaf, pressure vertical:* particle diameter 1 to 120 μm; feed solid concentration 0.08 to 0.5% w/w. Recover liquid. *Leaf, vacuum:* particle diameter 1 to 500 μm; feed solid concentration 0.07 to 2% w/w; solids capacity <5.5 kg/s. *Leaf, horizontal pressure:* particle diameter 1 to 100 μm; feed solid concentration 0.003 to 0.05% w/w. Recover wet solid with good washing (rank 2). *Plate, horizontal vacuum:* recover liquid (rank 1). Recover wet solid with good washing (rank 2); liquid contamination in exit solids: 15 to 50% v/v liquid. Dry solids capacity <0.0027 kg/s. *Plate and frame:* particle diameter 1 to 100 μm; feed solid concentration 0.003 to 25% w/w. Recover liquid (rank 5); use for liquids with viscosity >50 mPa·s. Recover wet solid with good washing (rank 6); with good washing and fragile crystals (rank 3). Relatively incompressible solid bed. Liquid contamination in exit solids: 30 to 70% v/v liquid. Solid capacity <55 kg/s. *Diaphragm plate and frame:* compressible cake. *Deep bed:* particle diameter 0.01 to 50 μm; feed solid concentration 0.002 to 0.02% w/w. Recover liquid (rank 1); 50 to 90% removal efficiency. 10- to 300-mg/L feed solids concentration but usually <0.5 mg/L. Dry solids capacity <0.5 mg/s. *Cartridge:* particle diameter 0.8 to 50 μm; feed solid concentration 0.003 to 0.02% w/w. Recover liquid (rank 3). Dry solids capacity 0.0014 kg/s.

TABLE 16.36
Particle Diameter and Cake Permeability and Resistance

Particle Diameter, μm	Permeability (μm)^2	Intrinsic Permeability, m^4/N·s	Cake Resistance, m/kg
1	10^{-3}–0.1 depending on $\in$	10^{-13}–10^{-10} depending on $\in$ and liquid viscosity	10^{12}
10	0.1–10 depending on $\in$	10^{-12}–10^{-8} depending on $\in$ and liquid viscosity	10^{10}
100	10–1000 depending on $\in$	10^{-10}–10^{-6} depending on $\in$ and liquid viscosity	10^{8}
1 mm	1000–10^5 depending on $\in$	2×10^{-8}–10^{-4} depending on $\in$ and liquid viscosity	10^6
10 mm	10^5–10^7 depending on $\in$		<10^6

Continuous: drum, gravity: particle diameter 40 to 5,000 μm; feed solid concentration 0.08 to 0.8% w/w. Recover wet solid with good washing and fragile crystals (rank 2). *Drum, pressure:* particle diameter 5 to 200 μm; feed solid concentration 0.75 to 5% w/w. Recover wet solid with good washing and fragile crystals (rank 2). *Drum, rotary vacuum:* particle diameter 1 to 700 μm; feed solid concentration 5 to 60% w/w. Recover wet solid with good washing (rank 4). Recover wet solid with good washing and relatively compressible cake (rank 1). Recover wet solid with good washing and fragile crystals (rank 2). Dry solids capacity <50 kg/s. *Drum, precoat rotary vacuum:* particle diameter 0.5 to 80 μm; feed solid concentration 0.02 to 0.1% w/w. Recover liquid (rank 4). *Disk, rotary vacuum:* particle diameter 15 to 500 μm; feed fraction solids 0.02 to 0.70. Free flowing. *Table/pan gravity:* recover liquid (rank 1). Recover wet solid with good washing (rank 1). Recover wet solid with good washing and fragile crystals (rank 1). Particle diameter 40 to 50 μm; feed fraction solids 0.02 to 0.20. Dry solids capacity 280 kg/s. *Table/pan vacuum:* particle diameter 40 to 50 μm; feed fraction solids 0.02 to 0.20. dry solids capacity: 280 kg/s. *Belt, gravity:* recover wet solid with good washing (rank 3). Liquid contamination in exit solids 20 to 30% v/v liquid. *Belt, vacuum:* particle diameter 20 to 70 μm; feed fraction solids 0.05 to 0.6. Dry solids capacity <30 kg/s. *Belt press:* dewater and handle flocculated solids. Rotary press: dewater and handle flocculated solids. Particle size 1 to 50 μm. Feed solids 0.1 to 25% w/w. Related to extruders, Section 16.11.7.7. *Ultrafiltration:* see Section 16.11.4.22. Particle diameter 0.4 to 200 nm; feed solids concentration <20% w/w. *Microfiltration:* see Section 16.11.4.23. Particle diameter 0.05 to 800 μm; feed solids concentration <75% w/w. *Dissolved air flotation, DAF:* see Section 16.11.5.15. Particle diameter 0.1 to 50 μm with typical target diameter 2 μm; feed solid concentration 0.002 to 0.08% w/w; >80% removal efficiency.

Guidelines

General porosity of cake over the usual range of pressure difference for filtration:
0.9: silica, polystyrene; 0.8 to 0.85: talc, zinc sulfide; 0.7: calcium carbonate; 0.6 titanium dioxide, ignition plug; 0.5 kaolin; 0.42 iron carbonyl.

Table 16.36 relates particle diameter to parameters important in filtration.

Batch:
Usually size unit on cycle and based on volume of cake removed. Select required cake volume in batch unit. Two thirds of the cake forms in 1/3 of filtration time: consolidation of the last 1/3 of the cake during the last 2/3 of the filtration time. Typically filtration cycle stops when the filtrate flux is <0.01 L/s·m^2. Cycle: filter, drain, 3 min, fill with wash, 2 min; wash to with volume = five times the cake void volume; air blow; unload, 6 min. *Leaf, pressure vertical:* cycle (for wet cake): filter 2 to 80 h; open, dump, close 0.4 to 4 h; cake volume/unit 0.1 to 2 m^3 corresponding to 5 to 90 m^2 filter area; leaves on 75 mm spacing. Δp 250 to 400 kPa. Cake buildup flux: 0.001 kg/s·m^2.

Precoat: 0.68 kg/m² filter. Solid flux for precoat 0.06 to 0.18 kg/s·m². Filtrate flux through precoat = 0.1 to 1.1 L/s·m². *Leaf, vacuum:* cycle: precoat; filter 0.6 h to 20 h; wash-clean. Cake formation rate 1.3 mm/s; solids flux 0.009 to 0.02 kg/s·m² filter area. Cake thickness <10 cm. Area per unit 1.2 to 180 m²; Δp <80 kPa. Cake volume per unit: 0.1 to 4.5 m³. Precoat: liquid filtrate flux for pharmaceuticals: 0.0003 to 0.0017 L/s·m². *Leaf, horizontal pressure:* cycle: precoat; filter 8 h; wash-clean. Availability per cycle for filtration 65 to 85%. Cake thickness <10 cm; solids flux 0.001 to 0.04 kg/s·m² of filter area; filtrate flux: 0.34 to 1 L/s·m². Area per unit <280 m²; Δp 100 to 700 kPa. Cake volume per unit: 0.2 to 4.5 m³. *Plate, horizontal vacuum:* cycle: precoat, filter, 70 h; wash-clean: 2 to 4 h. Solids flux: 0.003 kg/s·m² of filter area; area per unit <6 m²; Δp <80 kPa. Cake volume per unit: 0.06 m³. *Plate and frame:* cycle time: feed, wash, unload, clean; filtration: 10 min to 24 h, usually 2 to 8 h; wash 10 to 25 min. Cake formation rate <0.7 mm/s; solids flux 0.02 to 0.07 kg/s·m² of filter area; filtrate flux 0.004 to 0.8 L/s·m² of filter area. Area per unit <500 m²; Δp 400 to 1600 kPa. Cake thickness 2.5 to 5 mm. Cake volume per unit: 0.012 to 2 m³ corresponding to 1 to 160 m² filter area for 25-mm plates; 0.024 to 4 m³ for 50-mm plates. *Diaphragm plate and frame:* cycle: fill, stop when about 80% of the plate has been filled: 0.2 to 0.5 h; squeeze, 0.25 to 0.4 h; discharge, 0.33 to 0.4 h. Postsqueeze volume/presqueeze volume = 0.63 to 0.75. Cake volume 0.003 to 7.8 m³. Fill at 700 kPa; squeeze at 0.85 to 1.5 mPa. *Deep bed:* cycle: load 8 h; backwash 15 min; 93% availability. Cake formation rate 1×10^{-5} mm/s; solids flux 0.02 to 0.5 g/s·m² of filter area; filtrate flux 2.7 to 10 L/s·m² of filter area with fluid loading decreasing as solids feed concentration increasing. Area per unit <25 m²; maximum depth 4.5 m. *Gravity upflow or downflow with single or multiple media;* pressure upflow or downflow. Use pressure units for small to medium fluid capacities where high terminal head loss is expected. Backwash at 6 L/s·m²; see Section 16.11.7.5. *Cartridge:* cycle: days to months. Cake formation rate <0.000067 mm/s; area per unit 03. to 0.7 and 1.5 to 15 m².

Continuous: drum, gravity: cake formation rate >3 mm/s; solids flux 0.15 to 0.47 kg/s·m² of effective filter area. Actual filter area 0.7 to 15 m². *Drum, rotary vacuum:* cake formation rate 0.01 to 17 mm/s; solids flux 0.008 to 0.16 kg/s·m² of submerged filter area; filtrate flux 0.08 to 1 L/s·m² of submerged filter area. Total drum area per unit 0.59 m²; Δp <80 kPa. rpm 0.4 to 1. Variety of options to remove cake: *belt:* submerged 35%. Air required 13 to 22 dm³/s·m². Thickness of cake: >3 to 5 mm. *Roll:* submerged 35%. Air required 13 to 22 dm³/s·m². Thickness of cake: >1 mm. *Standard scraper:* submerged 35%. Air required 13 to 22 dm³/s·m². Thickness of cake: >6 mm. *Coil:* submerged 35%. Air required 13 to 22 dm³/s·m². Thickness of cake: >3 to 5 mm. *String*: submerged 35%. Air required 13 to 22 dm³/s·m². Thickness of cake: >6 mm. Wash ratio 1.5 water/solid. Agitate the feed pan to keep the feed solids in suspension but not interfere with cake formation, solid flux for fine diameter minerals 0.09 kg/s·m² of total drum area at pressure drop across the cake = 60 kPa; coarse diameter minerals 0.35 kg/s·m² of total area at pressure drop across the cake = 10 to 20 kPa. *Drum, precoat rotary vacuum:* submerged 35, 55, 85%. Usually 85%; air required 28 to 42 dm³/s·m². Thickness of cake: <3 mm. Need low-viscosity liquid. Filtrate flux: 0.004 to 0.07 L/s·m². Procedure to form the precoat: use maximum drum speed, gradually increase the submergence from 5 to 85%. *Disk, rotary vacuum:* cake formation rate 0.01 to 1.6 mm/s; solids flux 0.01 to 0.55 kg/s·m² of submerged filter area. Total disk area per unit 0.4 to 300 m²; Δp <80 kPa. Submergence 35%; cake thickness >13 mm. Air/vacuum: 8 to 25 dm³/s·m². *Table/pan gravity:* cake formation rate 0.5 to 16 mm/s; solids flux 0.02 to 0.25 kg/s·m². Total table area per unit 8 to 200 m²; Δp <80 kPa. Cake thickness >20 to 25 mm. Air vacuum needed: 10 to 40 dm³/s·m². *Belt, gravity:* dilute sludges with 0.5 to 8% solids, typically dewater in 20 to 80 s. Length 2.4 m. (Often combined with downstream belt press.) *Belt, vacuum:* solids flux 0.08 to 2.2 kg/s·m²; filtrate flux 0.038 to 1 L/s·m² area: for particles <0.3 mm, 0.08 to 2 L/s·m²; for particles >0.3 mm, 2 to 12 L/s·m². Area per unit 1 to 120 m²; Δp <80 kPa. Cake thickness 5 to 100 mm, usually 12 mm. Size on the filtrate flux. Vacuum air 25 dm³/s·m². Velocity <0.75 m/s. Wash ratio 1.1 to 1.2 water/solids. Width <3 m; 22 h continuous operation plus 2 h routine maintenance; 1 L/s of sludge/m width. *Belt press:* belt width 0.5 to 2.6 m with areas from 6.9 to 35.7 m²; belt speed 0.04 to 0.15 m/s. For feed concentrations <5%, usually liquid dewatering controls: liquid load 2.5 to 3.5 L/s·m of belt width. For feed concentrations >5% solids throughput limiting: 0.15 to 0.275 kg/s·m of belt width. Δp <80 kPa. *Rotary press:* <3 rpm; dewatering area: 1.25 to 5 m^2. Solids flux 0.003 to 0.127 kg/s·m^2, depending on the feed solids concentration in range 0.1 to 6% solids. Product 20 to 75% solids. See also Section 16.11.7.7. *Ultrafiltration:* see Section 16.11.4.22. *Microfiltration:* see Section 16.11.4. *Dissolved air flotation (DAF):* see Section 16.11.5.15.

16.11.5.14 Leacher

Area of Application

Percolation leach: particle diameter >700 μm; liquid concentration 0.8 to 20%; relatively fragile solid (e.g., seeds). *Immersion leach:* particle diameter <700 μm; liquid concentration <20%; relatively robust solids (e.g., minerals). *Combo leach:* high feed concentration of solute, relatively robust solid. *Supercritical solvent:* usually CO_2 for small capacity of high value products, especially for temperature sensitive foods, cosmetics, and pharmaceuticals.

Guidelines

> *Percolation leach:* high solute leach rate; solvent percolation through the bed >3 mm/s; 3 to 10 L/s·m^2; bed permeability >200 μm^2; low feed concentration of solute; 0.5 to 0.7 kg liquid solvent carryover from stage to stage/kg inert solid; solute diffusivities of essential oils 10^{-7} to 10^{-14} cm^2/s; of sugar in sugar beets 10^{-5} cm^2/s. Tend to use series of counter-current contactors with time for diffusion and separation in each stage. *Rotating cells/baskets: capacity:* 30 kg/s flaked soybeans in 10-m diameter unit; prepressed cottonseed × 0.66; unpressed cottonseed × 0.33; sugar cane × 0.33; canola × 0.33. *Buckets:* 10 kg/s flaked soybeans with 40 buckets. *Belt:* 10 kg/s flaked soybeans with belt 20 m long; 2 to 5 kg solvent makeup/Mg oil seed feed. *Drag chain:* 10 kg/s flaked soybeans with belt 20 m long.
> *Immersion leach:* low solute leach rate. Product of residence time × concentration of leachant for acid leach = constant for a given particles diameter. Tends to have a separate leacher followed by system to separate and wash solids. *Pachuca:* plus CCD, settler; particle diameter <70 μm; solids 30 to 60% w/w; agitator 0.07 to 0.2 kW/m^3; for CCD 1.5 to 2 kg liquid solvent carryover/kg inert solid. *Autoclaves:* plus CCD: particle diameter <70 μm; solids 30 to 60% w/w; 0.7 to 1.3 kW/m^3.
> *Combo leach:* tend to use countercurrent contactors with time for diffusion and separation in each stage. *Sloped diffuser with rotating screw:* 10 kg/s sugar beets in 2 m diameter. *Tower with rotating screw:* 10 kg/s sugar beets in 0.3-m diameter unit. *Trough-scroll:* 10 kg/s sugar beets in three troughs. *Rotary diffuser:* 10 kg/s sugar beets in 5-m dia. unit.
> *Supercritical solvents:* batch process operating at $T > 31°C$ and pressures >7.3 MPa. Solvency is intermediate between nonpolar and weakly polar solvents. Contact times 0.1 × usual leach times.

16.11.5.15 Liquid-Solid: Dissolved Air Flotation, DAF

Area of Application

Particle diameter 0.1 to 50 μm with typical target diameter 2 μm; feed solid concentration 0.002 to 0.08% w/w; >80% removal efficiency.

Guidelines

Bubble size 70 to 90 μm; air: solids 0.005 to 0.1 kg/kg; liquid loading 0.5 to 2 L/s·m^2. Liquid residence time 20 to 200 min with a minimum depth of 1.8 m. If feed concentration is <400 mg/L, use about 50% recycle.

Conceptual Process Design, Process Improvement, and Troubleshooting

16.11.5.16 Liquid-Solid: Expeller and Hydraulic Press

Area of Application

Particle diameter >1 cm; liquid concentration 10 to 60%.

Guidelines

Batch: hydraulic press: 30 to 60 MPa. Cycle: load, press, discharge.

Continuous: expeller: *expelling essential oils:* prepress capacity 0.02 to 5 kg/s; prepress capacity (which will be followed by leaching, Section 16.11.5.14) = 2 × full press capacity (where only expelling is used to remove the essential oils); prepressing 2.2 Mg/d cottonseed/rpm; relative capacities depend on seed being processed: cottonseed and soybean × 1; copra, wet corn germ, canola × 0.75 to 0.82; flax seed, safflower × 0.62; power 100 to 200 kJ/kg cottonseed prepress with the value increasing with feed capacity.

Dewatering polymers: capacity 1 to 1.5 kg/s; power 200 increasing to 450 kJ/kg as capacity increases. Related topics are dryer (Section 16.11.5.5), screens (Section 16.11.5.6), and centrifugal filters (Section 16.11.5.12).

16.11.5.17 Solid-Solid: General Selection

To separate solids having about the same density and particle size, use flotation, electrostatic and magnetic separators. Particle diameter must be >20 µm. To separate solids having about the same density but with a range of particle size, separate based on cut diameter and use air or liquid *classifiers* such as cyclones, hydrocyclones, or spiral classifiers. Particle size 25 to 2000 µm; feed solids concentration 5 to 40%. Cut diameter is the particle diameter that has equal chance to report to either the overflow or the underflow streams. To separate solids having about the same particle size but a different density, separate based on cut density and use *concentrators* such as jigs, tables, sluices, or dense media separators (DMS). Particle diameter must be >40 µm. Concentrate before flotation if the feed assay of mineral is <0.3%. To separate solids having different densities and particles sizes, use combinations, such as a screen, to provide narrow size range followed by concentrators. For minerals, the liberation size is 0.01 of the diameter of the mineral crystal.

16.11.5.18 Froth Flotation

Area of Application

For systems with a narrow range of both density differences and particle sizes. Density ratio 1 to 1.3; particle diameter usually 20 to 50 µm, although occasionally the diameter could be as large as 200 µm. Feed concentration of target solid >0.5%.

Guidelines

Condition the solids to alter the wettability of the mineral and the gangue. The fundamental surface wettability for sulfide ores is different from oxides, silicates, and salt-type minerals. pH is a critical variable. Typical conditioning chemical additions include collector about 0.01 to 0.1 kg/Mg solids; frother about 0.01 to 0.5 kg/Mg solids; activator about 1 to 4 kg/Mg solids; depressant about 0.02 to 2 kg/Mg solids. Allow 6-min contact for conditioning. Bubble size about 1000 µm. Flotation rate constant is 0.2 to 1 min^{-1}; sink rate constant is 0.005 min^{-1}. Flotation cells: *mechanical cell:* for fast float, sequential separation, and relatively coarse particle diameter; 1.6 to 2.4 kW/m^3 cell volume. *Pneumatic cell:* for relatively dilute feed concentrations and smaller particle diameters. Air blower 0.5 kW/m^3 cell volume. Typical solids throughput 0.4 to 0.8 kg/s·m^3; feed concentration 10 to 40% w/w. Air escape velocity 0.02 m/s. Float times 6 to 20 min. Feed concentration to rougher or scavenger: 30% w/w; to cleaner: 10% w/w.

16.11.5.19 Electrostatic

Area of Application

For systems with dry particles, no slimes or organic coatings, a narrow range of both density differences and particle sizes, and a difference in conductivity. Particle diameter 40 to 3,000 µm: usually 80 to 1000 µm; feed concentration of target species 5 to 75%. Good conductors have relative permittivity >11; poor or nonconductors have relative permittivity <10; threshold voltage to make species conducting is 1 to 10 kV/cm.

Guidelines

Select voltage where one species conducts and the other does not. *Corona-active electrode rotary drum:* handles wide range of particle diameters from 75 to 1000 µm; high capacity, <0.75 kg/s·m of drum width; high efficiency 95%; separates good from poor conductors. Zero to 40 kV DC, 0.5 to 1 mA/electrode; insensitive to humidity and temperature; often can recycle the middlings with recycling 10 to 30% OK. *Active electrode rotary drum:* trouble handling fines; capacity <0.5 kg/s·m of drum width, moderate efficiency; separates good from poor conductors or two semiconductors. Zero to 30 kV DC, 0.04 mA/electrode. The voltage gradient must be sufficient to charge target particles, operates best in a controlled environment.

For roughers and scavengers, set the active polarity so that the valuable minerals become conductors. For cleaners and recleaners, set the active polarity so that the gangue and middlings become the conductors. Collect middlings if the valuable target is trapped in the gangue; if the density of the nonconducting is heavier than the conducting and if the feed has a wide range of particle diameters.

16.11.5.20 Magnetic

Area of Application

For systems with a narrow range of both density differences and particle sizes; particle diameter >50 µm; feed concentration of target species 0.4 to 40%. Magnetization = product of the mass magnetic susceptibility and the magnetic field, T m^3/kg. Ferromagnetic species: magnetization >10^{-4} T m^3 kg. Paramagnetic species: magnetization 10^{-8} < value < 10^{-5} T m^3/kg.

Batch: cycle: load, clean. *Plate:* batch: magnetization >10^{-4} T m^3/kg; particle diameter >6 mm; feed concentration of magnetic <0.01% w/w. Primarily to remove tramp ferrous metal. *Grate:* batch: magnetization >10^{-4} T m^3/kg; particle diameter <1.8 mm; feed concentration of magnetic <0.01% w/w. Primarily to remove tramp ferrous metal. *WHGMS (wire, Kolm–Marston):* batch: magnetization >2 × 10^{-9} T m^3/kg; particle diameter 1 to 30 µm; feed concentration of magnetic <0.05% w/w. Solids concentration in water 20%. *WHGMS (grooves, Jones):* batch: magnetization >2 × 10^{-9} T m^3/kg; particle diameter 200 to 1000 µm; feed concentration of magnetic <50% w/w; solids concentration in water <50%.

Continuous: pulley: magnetization >10^{-4} T m^3/kg; particle diameter >6 mm; feed concentration of magnetic <0.01% w/w. Primarily to remove tramp ferrous metal. *Belts/cross or inline:* magnetization >10^{-4} T m^3/kg; particle diameter >6 mm; feed concentration of magnetic 0.01 to 10% w/w. Primarily to remove tramp ferrous metal. *Belts/cross HGMS;* magnetization >10^{-4} T m^3/kg; particle diameter 150 to 1500 µm; feed concentration of magnetic 0.5 to 2% w/w. Primarily to remove tramp ferrous metal. *Wet belt:* magnetization >10^{-4} T m^3/kg; particle diameter 100 to 2500 µm; feed concentration of magnetic 0.3 to 2% w/w. *Dry drum/LGMS:* magnetization >2 × 10^{-5} T m^3/kg; particle diameter 0.1 to 100 mm; feed concentration of magnetic 1.5 to 75% w/w. *Dry drum high-speed/MGMS:* magnetization >2 × 10^{-5} T m^3/kg; particle diameter 100 µm to 30 mm; feed concentration of magnetic <10% w/w. *Dry rotor HGMS (induction):* magnetization >4 × 10^{-8} T m^3/kg; particle diameter 70 to 2000 µm; feed concentration of magnetic <0.05 w/w. *Wet drum/LGMS:* magnetization >10^{-4} T m^3/kg; particle diameter 70 µm to 6 mm; feed concentration of magnetic

>25% w/w. *WHGMS (carousel, BoxMag, Frantz):* magnetization >2 × 10^{-9} T m^3/kg; particle diameter 200 to 1000 μm; feed concentration of magnetic <50% w/w; solids concentration in water <50%.

Guidelines

Consider wet processing for particle diameter <6 mm to minimize dusting and electrostatics. For particle diameter >150 μm, prefer wet belt; <150 μm, prefer wet drum. Match magnetic gradient to the diameters of the particles. Match the machine to the liberation size (liberation size is 0.01 of the diameter of the mineral crystal). For wet machines, pump 4 Mg water/Mg solids, although 90% of the water can be recirculated.

Batch: plate: batch, remove tramp metal from solids moving in ducts or on conveying belt. Usually keep burden depth <20 cm; magnetic filed 0.3 T; gradient 0.04 T/cm. *Grate:* batch, magnetic field 0.05 T; gradient 0.02 to 0.5 T/cm. *WHGMS (wires, Kolm–Marsden):* batch; 100-μm wires; 10 to 80 kg/s per pole; 10 to 160 L/s·m^2 gap *cross-sectional* area; solids concentration in water 20 to 25% w/w; induced magnetic field 1 to 2 T; gradient 1000 T/cm; 35 to 42 kW/pole; 1 to 18 kJ/kg solids for magnet and 4.5 kJ/kg pumping; solids loading before unload 0.1 to 0.5 Mg/m^3 of matrix. *WHGMS (grooves):* batch, 0.03 to 0.04 kg/s; 8 to 12 kg/s per pole; solids in water <40% w/w; induced magnetic field 1 to 2 T; gradient 0.1 to 1000 T/cm, usually 10 to 200 T/cm; 16 kW/pole; 1.8 kJ/kg solids collected for magnet; feed <50%w/w solids in water.

Continuous: pulley: remove tramp metal from solids exiting from a conveyor belt. Usually keep burden depth <20 cm; magnetic field 0.1 to 0.3; gradient 0.04 T/cm. *Belts/cross or inline:* <1 kg/s·m width; gradient 0.1 to 1 T/cm; belt speed 1 to 2 m/s. *Belt with rotating disk:* 0.2 kg/s·m width; magnetic field 0.04 to 0.1 T; gradient 0.1 to 1000 T/cm. *Wet belt:* 3 kg/s·m width; magnetic field 0.3 T; gradient 0.1 to 1 T/cm. *Dry drum/LGMS:* 80 to 200 kg/s·m width of drum, 60 to 90 dm^3/s·m width for drum diameter <1.2 m; 25 to 35 rpm, peripheral speed 1 to 1.5 m/s; magnetic field 0.04 to 0.1 T; gradient 0.001 to 0.5 T/cm; drive power 2 to 5 kW/m width. *Dry drum high-speed/MGMS:* 0.5 to 20 kg/s·m width, 0.8 to 20 dm^3/s·m width with capacity decreasing as magnetization decreasing; 2.5 to 7 m/s with peripheral speed increasing as the number of poles increases from 6 to 44; 50 to 200 rpm; as particle diameter decreases, the number of poles and the rpm increase; for highest recovery, operate at lower rpm; for higher incoming concentration of mags and more efficient recovery of middlings, operate at higher speeds; magnetic field 0.04 to 0.1 T; gradient 0.001 to 0.5 T/cm; drive power 10 kW/m width. *Dry rotor HGMS (induction):* 0.8 to 1 kg/s·m width; induced magnetic field 1 to 2 T; gradient 0.001 to 1000 T/cm, usually 10 T/cm. *Wet drum/LGMS:* 1 to 5 kg/s·m width; 4 to 25 L/s·m width; water velocity 0.5 m/s; drum peripheral speed 2 m/s; 20 rpm; diameter 0.75 m; magnetic field 0.04 to 0.1 T; gradient 0.001 to 0.5 T/cm; constraint is the amount of magnetic material discharged from the drum; feed <25% w/w solids in water; drive power 1 to 2 kW/m width. *WHGMS (carousel, BoxMag, Frantz):* continuous 1 to 5 kg/s pole, 10 to 50 L/s·m^2 gap cross-sectional area; usual canister cross-sectional area 0.5 to 3 m^2; induced magnetic field 0.1 to 2 T; 1 to 6 MJ/Mg solids collected; feed solids concentration in water 30 to 50%.

16.11.5.21 Hydrocyclones

See Sections 16.11.5.3 and 16.11.5.8, separation of liquid-solid, for guidelines.

Area of Application

Cut diameter 5 to 1000 μm; capacities up to 50 kg/s per unit.

16.11.5.22 Air Classifiers

Area of Application

Separate solids with similar densities based on cut diameter. Conveying fluid is a gas such as air. Contrast with hydrocylcones and spiral classifiers, where the conveying fluid is a liquid such as water. Particle cut diameter 30 to 1000 μm. Particle diameter >1.5 μm. Feed concentration of target solid 4 to 60% w/w.

Guidelines

Solids-to-gas ratio 2 to 8 w/w. Solids loading 2 to 4 kg/s·m². *Zigzag:* cut diameter 100 to 10,000 μm; capacity 0.01 to 0.08 kg/s. *Gas centrifugal separator:* cut diameter >20 μm; capacity <15 kg/s. *Gas cyclone:* cut diameter 10 to 50 μm; capacity <15 kg/s. *Gas gravitational inertial classifier (GIC):* cut diameter 50 to 200 μm; capacity <300 kg/s. *Gas Mikroplex spiral:* cut diameter 2 to 20 μm; capacity 0.01 to 1 kg/s. *Gas Nauta–Kosakawa:* cut diameter 3 to 300 μm; capacity 0.01 to 1 kg/s. *Gas classifiers for MSW:* cut diameter >100 μm with capacities up to 20 kg/s.

16.11.5.23 Rake Classifiers

Superceded by hydrocyclones (Sections 16.11.5.21 and 16.11.5.8).

16.11.5.24 Spiral Classifiers

Area of Application

Bird number (mass of solids with a ±0.1 density variation from the cut density) <15; particle cut diameter >50 μm but usually 1000 to 20,000 μm.

Guidelines

Two types: cut diameter >200 μm, use *high weir configuration;* cut diameter <200 μm, use *submerged spiral.* Size based on the target exit overflow concentration and "limiting" particle diameter. Diameter of the limiting particle should have a settling velocity double that of the cut diameter (cut diameter is about 70% of the limiting diameter). Separation efficiency is 50%; that is, use double the cross-section area obtained from the settling velocity of the "limiting" particle diameter. Capacity 2 to 250 kg/s. Underflow solids capacity is a function of the spiral diameter and solids density; independent of the separation. Angle of inclination 16°.

High weir: 1 m² pool area produces about 5 kg/s dry solids in the overflow with 400 μm limiting diameter or about 280 μm cut diameter and 30% concentration. *Submerged weir:* 1 m² pool area produces about 0.5 kg/s dry solids in the overflow with 150 μm limiting diameter or about 100 μm cut diameter and 16% concentration. For both configurations, increase the area by a factor of 10, increase the amount of overflow solids by 10 with the same cut and limiting diameters and overflow concentrations. Rotational speed increases with the cut diameter.

16.11.5.25 Jig Concentrators

Area of Application

Bird no. (mass of solids with a ±0.1 density variation from the cut density) <15; relative density ratio 2 to 2.5 and particle diameter 2 to 10 mm. Feed concentration 1.5 to 30% w/w.

Guidelines

Loading 1 to 8 kg/s·m² with loading increasing with increase in particle diameter. Power 1 kW/m² and water usage 13 L/s·m².

16.11.5.26 Table Concentrators

Area of Application

Bird number (mass of solids with a ±0.1 density variation from the cut density) <15; relative density ratio >2 to 2.5; usual particle diameter 70 to 2000 μm; only one valuable mineral. *Tilting frames:*

Conceptual Process Design, Process Improvement, and Troubleshooting

density ratio 2.5 and particle diameter >50 μm ranging to density ratio of 1.25 with particle diameter >5 mm. *Holman slimes table* for <70 μm.

Guidelines

Seven to 12 m² per unit; 6° incline; usual capacity 0.03 to 0.06 kg/s·m². The higher the Bird no., the lower the capacity. Power 0.15 kW/m²; wash water 0.1 to 0.3 L/s·m².

16.11.5.27 Sluice Concentrators

Area of Application

Bird number (mass of solids with a ±0.1 density variation from the cut density) <15; relative density ratio >1.8; usual particle diameter 400 to 3000 μm; only one valuable mineral.

Guidelines

Humphreys and Reichert spiral: capacity/unit: 0.25 to 1.5 kg/s to handle particle diameters 75 μm to 2 mm.

16.11.5.28 Dense Media Concentrators (DMS)

Area of Application

This is for systems with a narrow range of particle sizes but with a range in densities. Bird number (mass of solids with a ±0.1 density variation from the cut density) $15 > Bi > 25$; relative density ratio >1.25 to 1.5; usual particle diameter >10 mm.

Guidelines

Choose media with density between desired cuts. For media with density 1.2 to 2.2, use magnetite; for 2.9 to 3.4, use ferrosilicon. Use mixtures for intermediate densities; 2.5 to 15 kg/s·m² of pool area.

16.11.5.29 Screens

See also Section 16.11.5.6 for liquid-solid separation.

Screens can be used to wash or dewater (see Section 16.11.5.6); *scalp:* remove 5% of oversize and have >50% half size; *screen:* remove fines <425 μm; *size or separate:* choice depends on particle size: coarse >4.75 mm; intermediate size between 425 and 4750 μm, and fines between 45 and 425 μm. See filters, Section 16.11.5.13; centrifugal filters, Section 16.11.5.12; and expellers, Section 16.11.5.16.

Area of Application

Grizzly: particle diameter 15 to 30 cm and >30 cm; Use: primarily to scalp, screen fines. *Rod grizzly:* particle diameter 8 to 30 cm. Use: primarily to scalp. Rod deck screen: particle diameter 0.75 to 8 cm. Use: scalp, dewater, and separate. *Sieve bend:* particle diameter: 45 to 2000 μm. Use: dewater, separate intermediate and fines: 45 to 4750 μm. *Revolving screen, Trommel:* particle diameter 3 mm to 50 cm; feed solids concentration 5 to 25%. Use: separation. *Revolving screen, centrifugal:* particle diameter: 400 to 1200 μm, Use: dewater, separate intermediate diameters, and use high speed for fines. *Revolving screen, probability:* particle diameter <6 mm. Use: separate intermediate and fines 45 to 4750 μm. *High-speed vibrating horizontal screen* (600 to 3000 rpm with low amplitude <2.5 cm): 3 to 100 mm. Use for dense granular materials >0.3 Mg/m³. Use to wash, dewater, scalp, screen fines, separate coarse. *High-speed vibrating inclined screen* (600 to 7000 rpm with low amplitude <2.5 cm): 200 to 100,000 μm. Use for dense granular >0.3 Mg/m³. Use to wash, scalp, and separate wide range of particle diameters. *Low-speed oscillating screen* (25 to 500 rpm with 15 to 30 mm amplitude): 74 to 12,000 μm, especially for lower density solids, although some suppliers recommend this for only >12,000 μm. Use to separate. *Gyratory in the plane of the screen* (500 to 600 rpm; 5° inclination): fine separations particle diameter 50 to 4000 μm.

Guidelines

Screen: efficiency 85 to 95%; length/width of screen 2:1 to 1.5:1; rate of travel of solids along the screen face 0.3 to 0.5 m/s. The flux of solids passing through the screen is about: for coarse particle diameters, 2 to 5 kg/s·m²; for intermediate particle diameters, 0.4 to 4 kg/s·m²; for fine particle diameters, 0.08 to 0.4 kg/s·m², with fluxes decreasing as the density decreases.

Trommels: rotational speed should be slow enough that the particles free fall (about 45% of the transition rpm). L/D 2:1 to 5:1; residence time 30 to 60 s with flux rates of 0.03 to 0.1 kg/s·m².

16.11.6 Reactors and Vessels*

Reactors are considered in Sections 16.11.6.1 through 16.11.6.33; vessels are discussed in Sections 16.11.6.34 and 16.11.6.35. For most process equipment, equipment is sized based on the optimization criterion of cost. For reactors, the optimization criterion may be selectivity, yield, flexibility, ability to control, or cost. Section 16.11.6.1 lists the general rules of thumb. Section 16.11.6.2 gives examples of the reactor type based on the type of reaction. Sections 16.11.6.3 onward give rules of thumb organized by reactor type.

Symbols: adiabatic factor = adiabatic temperature change accompanying complete consumption of the reactant without regard to equilibrium considerations; $Bd:$ = Bodenstein number; D = diffusivity; D = tube diameter; D_p = particle or catalyst diameter; Eo = Eotvos number; GL = gas-liquid; GS = gas-solid; $GLcS$ = gas-liquid reaction with solid catalyst; GrS = gas reacting with a solid; H = height; Ha = Hatta number; heat generation potential = reaction activation energy/RT_{inlet}^2; L = length of tubes; Nu = Nusselt number; OD = outside diameter; Pe = Peclet number; R = ideal gas law constant. RTD = residence time distribution; Re = Reynolds number; T = temperature; and β = volume of liquid in the reactor/volume of mass transfer liquid "film" in the reactor.

16.11.6.1 General Rules of Thumb

1. The rate of reaction doubles for a temperature increase of 10°C.
2. Prefer continuous to semicontinuous to batch.
3. Prefer PFTR to CSTR.
4. Prefer minimum inventory in reactor by providing better mixing, increased temperature or pressure, or changing the catalyst.
5. Prefer internally heated/cooled to externally heated/cooled.
6. Consider light and ultrasound as optional forms of energy.
7. For gas reactions, usually use plug flow tubular reactors (PFTRs), static mixers, or fluidized beds.
8. For gas fixed-bed catalytic reactions, use $L/D_p > 100$ and $D_p/D < 0.10$ to ensure good gas distribution and negligible back mixing.
9. For liquid reactions, usually use continuous stirred tank reactors (CSTRs) (unless the pressure is very high), static mixers.
10. Prefer operating temperatures below the boiling temperature or dilute the liquid with a safe solvent.
11. For gas-liquid reactions, usually use CSTR or bubble column.
12. For reactions using solid catalysts: although many of the reactors described below are classed as "continuous," whenever a solid catalyst is used, the performance of the catalyst decays over time. The catalyst needs to be regenerated or replaced periodically. Time

---

* Reproduced with permission from Donald R. Woods, *Rules of Thumb in Engineering Practice*, pages 196, 225 to 278, 2007, copyright Wiley-VCH Verlag GmbH & Co. KGaA, Weinheim, ISBN: 978-3-527-31220-7. See this book for more details, especially about Good Practice, Trouble shooting and Capital Costs.

for reaction between regeneration is usually years. If the catalyst decays rapidly, consider fluidized- or moving-bed operation. Regeneration options include combustion, alternate oxidation and reduction, high-temperature operation. *Catalyst life:* some illustrative life: iron, nickel, palladium promoted catalysts for the production of ammonia or sulfuric acid, the life years are 5 to 10, with the cause of decay being slow sintering. Copper on zinc and aluminum oxides catalysts for methanol synthesis, the life years are 2 to 4, with the cause of decay being slow sintering. Nickel on alumina catalysts for steam reforming, the life years are 2 to 4, with the cause of decay being sintering and carbon formation. Silver on alumina catalyst for ethylene production, life years are 1 to 4, with decay caused by sintering. Silver granule catalyst for the partial oxidation of methanol, the life years are 0.3 to 1, with decay caused by poisoning. Platinum catalyst for catalytic reforming, the life years are 0.01 to 0.5, with coking causing the decay. Zeolite catalyst for catalytic cracking, the life years are 10^{-5}, with decay caused by coking.

13. Prefer monolithic configuration for intensification with surface area/volume usually 1.5 to 4 times greater than traditional pellets. Excellent for mass transfer controlled reactions.
14. For gas reacting with solid: usually heat transfer controlled, because these are highly exothermic or endothermic reactions. Particle size and size distribution are critical. These reactions may follow different patterns:
    a. For multigranule reactions, the time for reaction is independent of particle diameter.
    b. For shrinking core without ash reactions, the time for reaction = particle diameter (for reaction controlled); = $\mathbf{D}^{1.5}$ (for fluid diffusion controlled with fluidized, fixed, or moving beds); = $\mathbf{D}^2$ (for fluid diffusion controlled with transported bed).
    c. For shrinking core plus ash, the time for reaction = D_p (for reaction controlled); = D_p (for fluid diffusion controlled); = D_p^2 (for fluid diffusion through the ash controlled).
15. For biological treatment of wastewater, $BOD_u \approx 1.43 \times BOD_5$; approximately 1.4 to 1.5 kg O_2/kg BOD_5. Microorganisms use the organic substrate, characterized by BOD_5 or COD, for growth and for endogenous respiration. Illustrative reaction rate terms include, for COD at 20°C: substrate concentration at which the specific growth rate is 0.5 maximum, K_s = 25 to 100 mg COD/L. Maximum specific utilization rate of the substrate, k = 6 to 8 kg COD/day·kg VSS. Biomass lost to endogenous respiration per unit time per unit biomass, K_d = 0.05 to 0.1 1/day. Yield of biomass produced per unit of substrate removed, Y_T = 0.35 to 0.45 mg VSS/mg COD.

16.11.6.2 Reaction Types and Typical Reactor Configurations

Below are listed typical reactions and the usual reactor configurations used. *Alkylation* C-C: LL series CSTR; *animation* by ammonolysis: GL trickle bed; *combustion (incineration): burning in presence of oxygen of liquids and solids:* fluidized bed backmix. *Cracking, steam:* PFTR; *esterification:* carboxylic acid plus alcohol (cat: acid catalyst, usually sulfuric acid): usually CSTR. *Fermentation:* batch. *Halogenation:* addition of halogen: multiphase usually CSTR. *Hydration:* trickle bed. *Hydroformylation:* addition of hydrogen and –CHO to the double bond of an olefin with subsequent reaction of CO + H2 in the presence of cobalt catalyst (OXO synthesis): usually CSTR. *Hydrogenation:* addition of hydrogen: for multiphase usually semibatch STR or CSTR (series of CSTR). *Hydrolysis* (hydration): usually CSTR. *Nitration:* usually CSTR. *Oxidation:* GL trickle. *Polymerization:* batch, semibatch, CSTR, and multistage CSTR. *Sulfonation* of aromatic by oleum, batch.

16.11.6.3 PFTR: Empty Tube

Bodenstein number = ∞. Peclet number > 100. Usually controlled by reaction kinetics, heat transfer, or both.

Area of Application

Phases: gas, liquid, gas-liquid, liquid-liquid. Use if the order of the reaction is positive and >95% conversion is the target, and for consecutive reactions with an intermediate as the target product. *Gas* (or gas with homogeneous catalyst): residence time: 0.4 to 2000 s; heat of reaction: endothermic; reaction rate, fast; capacity: 0.001 to 200 L/s; good selectivity for: consecutive reactions and irreversible first order; volume of reactor 1 to 10,000 L; OK for high pressures. For temperatures <500°C. *Liquid* (or liquid with homogeneous catalyst): residence time: 0.4 to 2000 s; heat of reaction: endothermic; reaction rate, fast or slow; capacity: 0.001 to 200 L/s; good selectivity for consecutive reactions; volume of reactor 1 to 10,000 L; OK for high pressures. *Gas-liquid:* residence time: short; heat of reaction: primarily for endothermic reactions. Beware of highly exothermic reactions because of inability to control temperature; good selectivity for consecutive reactions in which the product formed can react further. See transfer line, Section 16.11.6.9, or bubble reactors, Section 16.11.6.11. *Liquid-liquid:* see transfer line, Section 16.11.6.9, or bubble reactors, Section 16.11.6.11.

Guidelines

Gas: residence time 0.5 to 1.3 s; gas velocity 3 to 10 m/s; Re > 10^4, L/D > 100. To eliminate backmixing, Pe > 100. *Liquid:* residence time 1 to 6 s; liquid velocity 1 to 2 m/s; Re > 10^4, L/D > 100. PFTR is smaller and less expensive than CSTR. PFTR is more efficient/volume than CSTR if the reaction order is positive with simple kinetics. For fast reactions, use small-diameter empty tube in turbulent flow. For slow reactions, use large-diameter empty tubes in laminar flow. If reaction is complex and a spread in RTD is harmful, consider adding motionless mixer (Section 16.11.6.10). *Examples: hydrolysis of corn starch to dextrose; polymerization of styrene; hydrolysis of chlorobenzene to phenol; esterification of lactic acid. Gas-liquid:* see transfer line, Section 16.11.6.9, or bubble reactors, Section 16.11.6.11. *Liquid-liquid:* see transfer line, Section 16.11.6.9, or bubble reactors, Section 16.11.6.11.

16.11.6.4 PFTR: Fire Tube

For temperatures >500°C, put tubes in furnace. Same area of application, guidelines, and good practice as empty tube, Section 16.11.6.3. Radiant heat flux in furnaces 30 to 80 kW/m^2. Reformers:

Gas oil: heat flux: 40 to 50 kW/m^2; fluid velocity inside tubes 1.5 to 2.5 m/s
Light oil: heat flux: 25 to 40 kW/m^2; fluid velocity inside tubes 1.4 to 2.3 m/s
Heavy oil: heat flux: 25 to 35 k.W/m^2; fluid velocity inside tubes 1.7 to 2.1 m/s

Cracking: ethylene as product from

Ethane: heat flux: 23 to 28 kW/m^2 at exit conditions with double values at inlet
Propane: heat flux: 14 to 17 kW/m^2 at exit conditions with double values at inlet
Butane or naphtha: heat flux: 11 to 15 kW/m^2 at exit conditions; at inlet the values are twice as large.

See thermal energy: furnaces, Section 16.11.3.2.

16.11.6.5 PFTR: Fixed-Bed Catalyst in Tube or Vessel: Adiabatic

Area of Application

Phases: gas, liquid, gas-liquid. Use if the order of the reaction is positive and >95% conversion is the target, and for consecutive reactions with an intermediate as the target product. *Gas or Liquid with fixed bed of solid catalyst:* residence time: 0.4 to 2000 s; heat of reaction: endothermic;

Conceptual Process Design, Process Improvement, and Troubleshooting

reaction rate, fast; capacity: 0.001 to 200 L/s; good selectivity and activity for: consecutive reactions and for irreversible first-order reactions; volume of reactor 1 to 10,000 L; OK for high pressures. High conversion efficiency, simple, flexible, gives high ratio of catalyst to reactants. *Gas-liquid downflow over fixed catalyst:* see trickle reactors, Section 16.11.6.15, or bubble reactors, Section 16.11.6.11.

Guidelines

Catalyst diameter 1 to 5 mm. *Gas with fixed catalyst bed:* residence time <1 s; favored if the life of the catalyst is >3 months; select fluidized (Section 16.11.6.27) or slurry reactors (Section 16.11.6.9) if the catalyst deactivates rapidly. Catalyst must have an axial crush strength >50 to 80 kg/cm². To ensure good gas distribution and negligible backmixing, $Pe > 2$; length/catalyst particle diameter $L/D_p > 100$ and $D_p/D < 0.10$. Usually, bed volume porosity 0.42, which decreases to 0.38 as the bed ages. PFTR lesser volume than slurry or fluidized-bed reactors. If the main reactant undergoes 90% conversion within a reactor length of bed height/catalyst particle diameter = 100, then the reaction is not mass transfer–controlled. Select a bed height such that the height of the bed/mass flow velocity (kg/s·m²) is >0.5 m³·s/kg fluid. Δp is <1 to 10% of total pressure; if too high, then use larger catalyst or change catalyst. For adiabatic operation with exothermic reactions, limit the height of the bed to keep temperature increase <50°C. Tube diameter <50 mm to minimize extremes in radial temperature gradient. For fast reactions, catalyst pore diffusion mass transfer may control if the catalyst diameter is >1.5 mm. Temperature gradients within the catalyst and in the external bulk phase:

1. Within a catalyst pellet the internal temperature gradient is rarely >1 to 2°C between the surface and the center. Assume temperature at the center of the catalyst = surface temperature.
2. The external temperature gradient is usually high, with the catalyst surface temperature possibly being 10 to 30°C hotter than the temperature in the gas phase.

Concentration gradients within the catalyst and in the bulk phase:

3. Within the catalyst pellet, the internal concentration gradient is often very high (with 0 concentration at the center).
4. External concentration gradient is usually small except for very fast reactions.

A related topic is gas adsorption (Section 16.11.4.11).

Liquids with fixed catalyst bed: to minimize backmixing, $Pe > 1$; use $H/D_p > 200$ and $D_p/D < 0.10$. Temperature gradient within the catalyst and in the external bulk phase:

1. Within the catalyst, the internal temperature gradient is low. Assume the temperature at the center of the catalyst = surface temperature.
2. The external temperature gradient from the catalyst surface temperature to the bulk is low.

Concentration gradient within the catalyst and in the external bulk liquid phase:

3. Internal concentration gradient depends on the reaction.
4. External concentration gradient may be high because of slow mass transfer or fast reaction.

For IX resin, height/vesel = 0.5. If the main reactant undergoes 90% conversion with the bed height/catalyst particle diameter = 10,000, then the reaction is not mass transfer–controlled. Since this height/diameter ratio is usually achieved, check if mass transfer controls.
A related topic liquid is adsorption (Section 16.11.4.12).

Gas-liquid downflow over fixed catalyst: see trickle reactors, Section 16.11.6.15.

16.11.6.6 PFTR: Multitube Fixed-Bed Catalyst: Nonadiabatic

Area of Application

Phase: gas, liquid. Use if the order of the reaction is positive and >95% conversion is the target, and for consecutive reactions with an intermediate as the target product. Exchange heat generated if the product of the adiabatic factor × heat generation potential >10.

Guidelines

Shell-and-tube exchanger with reactants and catalyst inside the tubes, 250 to 400 m^2/m^3. Tube diameter <50 mm. *Gas* with fixed bed of catalyst: use high mass gas velocity to improve heat transfer kg/s·m^2 > 1.35. To ensure good gas distribution and negligible backmixing, $Pe > 2$; height/catalyst particle diameter $H/D_p > 100$ and $D_p/D < 0.10$. Gas velocity 3 to 10 m/s; residence time 0.6 to 2 s. Heat transfer coefficient $U = 0.05$ kW/m^2·K. For fast reactions, catalyst pore diffusion mass transfer may control if catalyst diameter is >1.5 mm. *Liquids* with fixed bed of catalyst: to minimize backmixing, $Pe > 1$; use $L/D_p > 200$ and $D_p/D < 0.10$. Liquid velocity 1 to 2 m/s; residence time 2 to 6 s. Heat transfer coefficient $U = 0.5$ kW/m^2·K.

16.11.6.7 PFTR: Multibed Adiabatic with Interbed Quench

Area of Application

Phase: gas. For fast reactions that are strongly exothermic or endothermic. Use if the order of the reaction is positive and >95% conversion is the target, and for consecutive reactions with an intermediate as the target product.

Guidelines

Limit the height of the bed to keep temperature increase <50°C to minimize effects of radial temperature gradients. Bed can be shallow and wide. Quench can include injection of cold reactants, internal or external heat exchangers.

16.11.6.8 PFTR: Fixed Bed with Radial Flow

For fast reactions, strongly exothermic or endothermic reactions. Phases: gas with solid catalyst. Use to minimize pressure drop limitations.

16.11.6.9 PFTR: Transported or Slurry, Transfer Line

A related option is fluidized reactor (Section 16.11.6.27).

Area of Application

Phases: gas-solid, gas-solid reactant, gas-liquid, liquid-solid, gas-liquid-solid. *Gas plus catalytic solid:* gas residence time in milliseconds. Reaction rates very fast and very rapid deactivation of catalyst. Solid particle diameter 0.007 to 1.5 mm. *Gas plus solid reactant:* solid residence time, 0.8 to 300 s: gas residence time, <1 s; solid particle diameter 0.007 to 1.5 mm. *Gas-liquid:* see size reduction, Section 16.11.8.1. *Gas liquid plus solid catalyst:* for fast hydrogenation reactions. Compared with trickle bed, Section 16.11.6.15, or PFTR fixed bed with upflow, Section 16.11.6.5. (1) Catalyst particles are small, so less chance of diffusional resistance to mass transfer. (2) Better control of temperature (because of better heat transfer efficiency and high heat capacity of slurries), attractive for exothermic reactions. (3) No need to shut down for catalyst replacement or reactivation. (4) Partial wetting and need to maintain a coating film of liquid (as needed in the trickle bed) are not issues. (5) The space time yield is usually better in slurry reactors (under comparable conditions).

Guidelines

Gas plus catalytic solid and *Gas plus solid reactant:* see pneumatic conveying, Section 16.11.2.6. *Gas-liquid:* see size reduction, Section 16.11.8.1, and two-phase flow, Section 16.11.2.4. Surface area 50 to 400 m^2/m^3; power 1 to 80 kW/m^3 total volume. *Gas-liquid plus solid catalyst:* usually operate in the churn-slug and piston-slug flow regimes with gas velocities >0.05 m/s for water-like liquids. Flow regimes are given in Table 16.31, Section 16.11.2.4. Gas holdup is proportional to the (superficial gas velocity)^n, where $n = 0.7$ to 1.2 in the bubbling regime and $n = 0.4$ to 0.7 in the churn turbulent regime. Gas holdup is independent of diameter but very sensitive to trace contaminants and foaming. Jet loop recycle reactor k_L about 0.04 to 0.05 cm/s; $k_L a = 7$ 1/s; power 2 kW/m^3; air lift $k_L a = 0.27$ 1/s.

16.11.6.10 PFTR: Motionless Mixer in Tube (Static Mixer)

Area of Application

Phases: gas with mixer as catalyst, gas-liquid, liquid, liquid-liquid. Fast competing parallel or consecutive reactions that are highly exothermic. For gas-liquid: fast reactions in the liquid phase.

Guidelines

See heat transfer, Section 16.11.3.5, and mixing, Section 16.11.7.3. For *gas reactions:* example oxidation of ammonia to nitric acid, production of maleic anhydride, xylene, styrene, vinyl chloride monomer, ethylene dichloride. L/D for mass transfer mixers 6:1 to 20:1. Gas velocity for turbulent flow. For gas-liquid reactions: cocurrent mass transfer in bubble flow: gas superficial velocity 0.6 to 2 m/s; liquid superficial velocity 0.3 to 3 m/s; spray flow: gas superficial velocity 3 to 25 m/s; see size reduction, Section 16.11.8.1. For *liquid-liquid reactions:* dispersed phase drops diameter 100 to 2000 μm, with diameter decreasing as the velocity increases, the surface tension decreases, and the hydraulic radius of the mixing element decreases; surface area 100 to 20,000 m^2/m^3, depending on the drop diameter and the concentration of dispersed phase. Turbulent flow. Example reactions as a PFTR; polymerizations of polystyrene, nylon, urethane; sulfonation reactions and caustic washing; see size reduction, Section 16.11.8.3.

16.11.6.11 PFTR: Bubble Reactor

Area of Application

Phases: gas-liquid, gas-liquid catalytic solid, and liquid-liquid. Use if the order of the reaction is positive and >95% conversion is the target, and for consecutive reactions with an intermediate as the target product. *Gas-liquid:* For large liquid holdup, slow reactions that are kinetically controlled reactions that require long residence times, and low-viscosity liquids. Preferred if large gas volumes needed or if liquid vol. is >40 m^3; OK for high pressure. Cocurrent: surface area 50 to 400 m^2/m^3; downflow: surface area 20 to 1000 m^2/m^3; see size reduction, Section 16.11.8.1. *Gas-liquid catalytic solid:* surface area 50 to 350 m^2/m^3. *Liquid-liquid:* surface area 7 to 75 m^2/m^3. Rotating disk contactor (RDC) can handle dirty fluids and large throughputs. Need flow ratios 1:1; difficulty in handling systems with low interfacial tensions that tend to emulsify. Related topics are solvent, extraction, Section 16.11.4.10, and size reduction, Section 16.11.8.3.

Guidelines

Gas-liquid: can operate countercurrently or cocurrently. Holdup: volumetric liquid holdup per total reactor volume: >0.7 and usually 0.95. Gas holdup = 0.05 to 0.4, increasing with increase in gas velocity. Superficial gas velocity 1 to 30 cm/s, although it has been as high as 50 cm/s. Mass transfer coefficients: typical liquid mass transfer coefficient = $k_L a = 0.005$ to 0.01 1/s; $k_L = 0.6$ to $0.7 \cdot 10^{-4}$ m/s. For gas phase $k_G a = 1$ to 3 1/s. Surface area gas-liquid per volume of reactor: 20 to 1000 m^2/m^3 volume reactor depending on flow conditions. Surface area gas-liquid per volume of liquid phase: 120 to 700 m²/m³ liquid phase. Volumetric ratio liquid to mass transfer liquid film, $\beta > 100$. Fast reaction systems, consider mass transfer: film diffusion. Power: cocurrent: 0.03 to 2 kW/m³; countercurrent: 0.04 to 1 kW/m³. See size reduction, Section 16.11.8.1. *Gas-liquid catalytic solid:* catalyst diameter, <0.1 mm. Operate semibatch. Holdup: volume fraction liquid 0.8 to 0.9; volume fraction catalyst 0.01; volume fraction gas 0.1 to 0.2. Gas holdup slightly less than for GL systems. Backmixing: solids backmixing $Pe = 2$ to 5 for superficial gas velocities of 2 to 7 cm/s; Liquid-phase backmixing about the same as gas-liquid systems; gas-phase backmixing, about the same as gas-liquid systems. Surface area: surface area solid 500 m²/m³; surface area gas-liquid 100 to 400 m²/m³; power 0.1 to 2 kW/m³ sufficient to keep catalyst in suspension. Heat transfer solid wall to mixture >1 kW/m²·K; presence of solids increases the heat transfer coefficient. Catalyst activity: variable but often able to avoid diffusion limitations because of small-diameter catalyst. Catalyst selectivity: OK. Catalyst stability: change catalyst between batches, heat exchange OK. Consider complications because of catalyst deposition and erosion. *Liquid-liquid:* superficial dispersed drop velocity 0.001 to 0.02 m/s with usual values 5.5 L/s·m². Smallest reactor volume. *For example, for esterification: reactor volume divided by the daily production, $m^3 \cdot day/kg$.*

PFTR	0.7
3 CSTR in series	0.85
Batch STR	1.04
CSTR	1.22

RDC: sum of the superficial velocities for both phases is 1 to 2.5 cm/s; dia. <9 m for SX but usually <2.5 m for reactions. A related topic is SX (Section 16.11.4.10).

16.11.6.12 PFTR: Spray Reactor and Jet Nozzle Reactor

Area of Application

Phases: gas-liquid, liquid-liquid. *Gas-liquid:* residence time, very short. Reaction rates, very fast; need rapid absorption. Very high gas capacity. Used for neutralization reactions with one of the reactants in the gas phase. For surface area see size reduction, Section 16.11.8.2. *Gravity spray:* surface area 30 to 70 m²/m³; target species Henry's law constant 10^3 to 10^4 kPa/mol fraction; feed gas concentration 0.3 to 4 vol%. *Venturi jet nozzle:* surface area: 200 to 2500 m²/m³; very soluble gas only with target species Henry's law constant <10^3 kPa/mol fraction; feed gas concentration >1 vol%. *Liquid-liquid:* surface area: 7 to 75 m²/m³. Related topics are solvent extraction (Section 16.11.4.10) and size reduction (Section 16.11.8.3).

Guidelines

Gravity spray towers: superficial velocity about 5.5 L/s·m²; mass transfer coefficients liquid phase: $1.5 \times 10^{-4} < k_L < 3 \times 10^{-4}$ m/s: for the gas phase $0.4 < k_G < 1$ mol/m²·atmos·s, or in other units, $0.01 < k_G RT < 0.25$ m/s; critical energy-consuming phase is the liquid; gas energy 8 kJ/m³; with liquid-to-gas ratio high; design on gas-phase controlling. Superficial gas velocity 1 m/s. Power usage 0.03 to 0.5 kW·s/m³. Δp gas = 0.6 to 1.2 kPa. Related topics are gas-solid separation (Section 16.11.5.2) and size reduction (Section 16.11.8.2). *Gas-liquid venturi:* Δp gas = 5 kPa, velocity in the throat 30 to 100 m/s and usually 100 m/s; mass transfer coefficient for liquid phase $k_L = 7 \times 10^{-4}$ m/s; for gas phase, $k_G = 10^{-2}$ to 3×10^{-2} m/s; gas-liquid surface area 150 to 300 m²/m³. For fluids with surface tension 40 to 70 mN/m and viscosity of liquid = 1 mPa·s; critical energy-consuming phase is the gas at about 20 kJ/m³ with liquid-to-gas ratio about 1.3 to 1.6 L/m³; design on gas-phase controlling. Power usage 0.04 to 8 kW·s/m³. Δp gas = 1.2 to 6 kPa. Related topics are gas-solid separation (Section 16.11.5.2) and size reduction (Section 16.11.8.2). *Liquid-liquid:* superficial velocity = 50% flooding.

Conceptual Process Design, Process Improvement, and Troubleshooting

16.11.6.13 PFTR: Trays

Related topics are gas absorption (Section 16.11.4.8), distillation (Section 16.11.4.2), and reactive distillation (Section 16.11.6.32).

Area of Application

Phases: gas-liquid, liquid-liquid. *Gas liquid:* residence time, short. Use for very fast reactions; all reaction is in the liquid film and is mass transfer–controlled. Gas-liquid surface area, max. observed 800 m²/m³ with the usual range 200 to 500 m²/m³ or slightly higher than a bubble column. Target species Henry's law constant $<10^7$ kPa/mol fraction; feed gas concentration <1 vol%. Not for foaming, corrosive, or particulates. *Liquid-liquid:* can operate as gravity flow or with fluid pulsed operation. For pulsed operation: surface area 75 to 3000 m²/m³. Related topics are solvent extraction (Section 16.11.4.10).

Guidelines

Gas-liquid: liquid holdup = 0.15; typical liquid mass transfer coefficient = $k_L a$ = 0.01 to 0.05 1/s; k_L = 1.5 × 10⁻⁴ to 4.5 × 10⁻⁴ m/s; for gas phase k_G = 0.02 to 0.2 m/s sieve tray; power 0.03 to 2 kW/m³. *Liquid-liquid:* sieve tray holes 3 to 8 mm with a smaller diameter as the surface tension increases; spacing between the holes three to four times the diameter of the holes to prevent coalescence soon after the drops are formed. For pulsed operation: superficial velocities 2 × 10⁻³ to 0.02 m/s.

16.11.6.14 PFTR: Packing

Related topics are gas absorption (Section 16.11.4.8), distillation (Section 16.11.4.2), and reactive distillation (Section 16.11.6.32). Porosity 0.6 to 0.95, depending on the packing. (Contrast with trickle bed with solid catalyst "packing," Section 16.11.6.15.)

Area of Application

Phases: gas-liquid, liquid-liquid, gas-liquid bio solids. *Gas-liquid:* use for very fast reactions; all reaction is in the liquid film and is mass transfer–controlled. Gas resistance important with very soluble gases. Surface area gas-liquid per volume of reactor: 50 to 250 m²/m³ volume reactor; surface area gas-liquid per volume of liquid phase: 1000 to 1600 m²/m³ liquid phase. Cocurrent over packed: surface area 400 to 3000 m²/m³. Target species Henry's law constant $<10^7$ kPa/mol fraction: feed gas concentration <1 vol%. Vulnerable to plugging. OK for foaming and corrosive. *Liquid-liquid:* surface area 7 to 75 m²/m³. Sensitive to contamination. Related topics are solvent extraction (Section 16.11.4.10) and size reduction (Section 16.11.8.3). *Gas-liquid bio solids:* gravity and rotating: *gravity:* trickling filter reactor (carbon removal): surface area 45 to 115 m²/m³. Standard-rate loading: 1.3 to 4.2 kg BOD_5/s·m³. High-rate loading: 4.2 to 21 kg BOD_5/s·m³; (carbon oxidation/nitrification combo): plastic media: loading <4 kg BOD_5/s·m³; (nitrification): biofilter: (carbon removal) organic loading: 20 to 50 kg BOD_5/s·m³. *Rotating:* rotating biological contactor (RBC): (carbon oxidation): loading 0.250 g BOD_5/L; capacity 20 to 5000 L/s; (nitrification): loading 10 to 20 mg NH_3 – N/L.

Guidelines

Gas-liquid: liquid holdup per total reactor volume: volume fraction liquid 0.05 to 0.15; superficial velocities: select for loading on packings that are 0.5 to 0.7 times flooding conditions; backmix: for short column heights = 0.2 to 0.3 m, significant backmixing can occur with Peclet for the liquid = 0.4 to 2; Peclet for the gas = 4 to 20. For short columns, double design height to account for backmixing. Typical liquid mass transfer coefficient = $k_L a$ =0.005 to 0.02 1/s; volumetric ratio liquid to mass transfer liquid film. β = 10 to 100. *Packing:* plastic packings have effective surface area = 1/2 corresponding value for ceramic if fluids are polar. OK for nonpolar; metal stainless steel less wetted than ceramic but better than plastic. Preferred packings include unglazed ceramic, Intalox 2 to 3.8 cm, but >5 cm, the interfacial area is too small, and <2 cm, the capacity is reduced because of flooding. Other recommended packings include Pall rings, mini-rings, Sulzer, Multinit, and tellerettes. *Liquid-liquid:* Superficial velocity of continuous phase = 30 to 50% of flooding. Backmixing less than in spray column or tray columns. The walls and packing must be preferentially wetted by the continuous phase. Packing size = 0.5 to 1 cm. Superficial velocities 0.001 to 0.02 m/s. Prefer diameter <0.6 m; superficial velocity about 5.5 L/s·m²; 2.5 cm Pall rings. Redistribute the dispersed phase every 1.5 to 2 m. *Gas-liquid bio solid:* gravity: trickling filter reactor: gas holdup 0.46 to 0.94; (carbon oxidation): *standard rate:* liquid superficial velocity 0.01 to 0.04 L/s·m², depth 1.8 to 3 m; recycle ratio 0; *high rate:* liquid superficial velocity 0.1 to 0.4 L/s·m², depth 0.9 to 2.4 m; recycle ratio 1/1 to 4/1; (carbon oxidation/nitrification combo): liquid superficial velocity = 0.095 to 0.18 L/s·m². Six-meter depth, recirculation ratio 1:1; (nitrification): liquid superficial velocity = 0.3 to 1.3 L/s·m². Biofilter reactor: media depth 1.5 to 6.5 m, usually 4 m; liquid recycle ratio 0.4; liquid loading: 2.3 L/s·m².

Rotating: RBC: (carbon oxidation) liquid residence time <1 h; 3 to 3.6 m dia.; 40% submerged; 1 to 2 rpm; bio layer 2 to 4 mm thick. Liquid loading 0.0005 to 0.004 L/s·m²; temperature >13°C; Module 10,000 m²; power = 3.5 kW; (nitrification): liquid loading 0.0004 to 0.0025 L/s·m². Temperature >13°C.

16.11.6.15 PFTR: Trickle Bed

Gas and liquid flow cocurrently down through a packed bed of catalyst. Porosity 0.38 to 0.42. (Contrast with packing described in Section 16.11.6.14.)

Area of Application

Phases: gas-liquid catalytic solid. Gas-liquid catalytic solid: use for very fast reactions; all reaction is in the liquid film and is mass transfer–controlled.

Guidelines

Gas-liquid with solid catalyst: catalyst particle diameter 1 to 5 mm. Operate close to the boundary between two-phase (trickle) and pulse flow. Superficial liquid velocity $0.005 < v_{Lo} < 9$ mm/s (although some operate in the range 0.8 to 25 L/s·m² or mm/s). The velocity should be reduced if liquid tends to foam. Superficial gas velocity is >0.010 m/s. Holdup volume fraction liquid 0.05 to 0.25. Static liquid holdup is constant for low Eotvos number, $Eo < 4$ (Eo = density liquid × gravitational constant × particle diameter squared/liquid-gas surface tension). Liquid holdup increases with Eo for $Eo > 4$. Dynamic liquid holdup increases with liquid flow rate but is independent of gas flow rate. The liquid axial backmixing is negligible if the height/particle diameter is >150. Holdup volume fraction catalyst 0.6 to 0.7. Holdup volume fraction gas 0.2 to 0.35. Surface area solid 1000 to 2000 m²/m³. Surface area gas-liquid 100 to 3500 m²/m³. Power input 1 to 100 kW/m³. Catalyst activity: variable but often reduced because of mass transfer limitation. Plug flow is favorable. Catalyst selectivity: often reduced because of mass transfer limitation; plug flow is favorable. Catalyst stability: should have stability because of difficulty in replacing. Heat exchange: challenging, so usually work adiabatically.

16.11.6.16 PFTR: Monolithic

Related topics are motionless mixer in tube (Section 16.11.6.10) and trickle bed (Section 16.11.6.15).

Area of Application

Phases: gas with solid catalyst; liquid with solid catalyst; gas-liquid with solid catalyst.

Guidelines

Prefer because of intensification with 1.4 to 4 times the surface/volume. Possible to install in pipelines. Use when mass transfer affects selectivity or reactivity. Perhaps not for highly exothermic reactions because of limited radial heat transfer unless crossflow is used. See Section 16.11.3.3, cubic/monolithic.

16.11.6.17 PFTR: Thin Film

Related topics are evaporation (Section 16.11.4.1) for gravity and agitated falling films, absorbers (Section 16.11.4.8), and shell-and-tube heat exchangers (Section 16.11.3.3).

Area of Application

Phases: gas-liquid, liquid-liquid. *Gas-liquid:* residence time 3 to 600 s; reaction rate: mass transfer controlled, fast adsorption and highly exothermic reactions; highly endothermic reactions producing a volatile compound whose desorption is desirable to shift the equilibrium or prevent side reactions; volume 1 to 80 L; capacity 0.02 to 5 kg/s. Gravity falling film or agitated falling film for viscous fluids. *For gravity film:* liquid residence time: 5 to 100 s; surface area gas-liquid per volume of reactor: 3 to 100 m^2/m^3 volume reactor; surface area gas-liquid per volume of liquid phase: 300 to 600 m^2/m^3 liquid phase; film surface area per unit 0.1 to 100 m^2/unit; viscosity <1500 mPa·s. *For agitated film:* liquid residence time: 5 to 600 s; film surface area gas-liquid per total reactor 0.1 to 25 m^2/unit reactor; viscosity: <2,000,000 mPa·s. *Liquid-liquid:* surface area 5 to 120 m^2/m^3. Related topics are solvent extraction (Section 16.11.4.10) and size reduction (Section 16.11.8.3).

Guidelines

Gas-liquid: gravity falling film: holdup: liquid holdup per total reactor volume: volume fraction liquid 0.01 to 0.15; liquid holdup per total reactor volume: 0.0002 to 0.5 m^3; liquid loading: 0.06 to 1.1 L/s·m of length. Volumetric ratio liquid to mass transfer liquid film, β = 10 to 200; energy for a falling film = energy as packed tower. *Gas-liquid: agitated thin film:* holdup: liquid holdup per total reactor volume: 0.0002 to 0.2 m^3; liquid loading: 0.06 to 1.25 L/s·m of length; backmixing: liquid plug flow or as a series of at least five backmix stages; energy needed = 1 kW/m^2 film surface. *Liquid-liquid:* gravity wetted wall; superficial velocities 2×10^{-5} to 8×10^{-4} m/s.

16.11.6.18 PFTR: Multiple Hearth

Phases: gas plus reactant solid. A related topic is energy exchange (Section 16.11.3.9).

Area of Application

Solid residence time 5,000 to 30,000 s; solid particle diameter 0.2 to 20 mm.

Guidelines

Solids loading 1.25 to 2 g/s·m^2 of single hearth or 2 to 20 g/s·m^2 of total hearth. Hearth area is 42 to 62% of nominal read based on overall OD. For fast reactions, bulk phase film diffusion may control, and pore diffusion may control if the solid diameter is >1.5 mm.

16.11.6.19 PFTR: Traveling Grate

Phases: gas plus reactant solid.

Area of Application

Solid residence time 2,500 to 20,000 s; gas residence time <1 s; solid particle diameter 8 to 30 mm. Product diameter 80 to 150 mm; capacity 15 to 300 kg/s; product crush strength >10^4 kPa.

Guidelines

Sintering: capacity 0.015 to 0.04 kg/s·m^2. For fast reactions, bulk phase film diffusion may control, and pore diffusion may control if the solid diameter is >1.5 mm. *Incineration:* temperature 1000°C; solid residence 1200 to 2700 s. See heat transfer, Section 16.11.3.7.

16.11.6.20 PFTR: Rotary Kiln

Phases: gas plus reactant solid. Related topics are incineration (Section 16.11.3.7) and drying (Section 16.11.5.5).

Area of Application

Solid residence time 2,500 to 20,000 s; gas residence time <1 s; solid particle diameter 7 μm to 20 mm. Sinter product: product diameter 80 to 150 mm; capacity 15 to 300 kg/s; product crush strength >10^4 kPa.

Guidelines

See heat transfer (Section 16.11.3.7) and related topic (drying, Section 16.11.5.5). Sintering: 0.012 kg/s·m^3 volume. For fast reactions, bulk phase film diffusion may control, and pore diffusion may control if the solid diameter is >1.5 mm. Incineration: see heat transfer, Section 16.11.3.7.

16.11.6.21 PFTR: Shaft Furnace

Phases: Gas plus reacting solid.

Area of Application

Solid residence time 20,000 to 200,000 s; gas residence time <1 s; solid particle diameter 8 to 300 mm.

Guidelines

For fast reactions, bulk phase film diffusion may control, and pore diffusion may control if the solid diameter is >1.5 mm.

16.11.6.22 PFTR: Melting Cyclone Burner

Phases: Gas plus reacting solid.

Area of Application

Solid residence time 0.004 to 2 s; gas residence time <1 s; solid particle diameter, 0.002 to 0.4 mm.

Guidelines

For fast reactions, bulk phase film diffusion may control.

16.11.6.23 PFTR Via Multistage CSTR

$Bd = 2n$, where n = number of CSTRs in series. About 100 CSTR in series to provide negligible backmixing.

Area of Application

Phases: liquid, gas-liquid bio solids. *Liquid:* use for residence time >4 h; for systems where PFTR conditions are desired. Heat of reaction: highly exothermic reaction rate, slow. Capacity 0.2 to 100 kg/s. *Gas-liquid bio solids:* variety of reactor configurations to remove soluble BOD_5 from wastewater: aerated lagoon. *Pure-oxygen backmix-activated sludge,* CSTR in series: 85 to 95% removal, compact unit for use where space is limited.

Guidelines

Gas-liquid bio solids: aerated lagoon: three CSTRs in series: residence time: 1 day for each CSTR, removal 80 to 95%; GL bubble reactor: retention time less than a facultative lagoon, biomass suspended, plus subsequent settling = no recycle activated sludge reactor. Oxygen requirements control if residence time is <1 d; mixing controls if residence time is >1 d; length/width <1.5; depth 2.4 to 5.4 m; SS concentration = 1 to 5 g/L; power 0.011 to 0.023 kW/m^3; 80 to 95% removal; food/microorganism ratio = 22 mg BOD_5/s·kg MLSS; depth 2.4 to 5.4 m; 0.01 to 0.25 kW/m^3; SS

1 to 5 g/L. *Pure-oxygen backmix-activated sludge,* CSTR in series with recycle: mean cell residence time = 8 to 20 days; food/microorganism ratio = 3 to 12 mg BOD_5/s·kg MLVSS; volumetric loading = 1.6 to 4 kg BOD_5/m^3; MLSS = 6 to 8 g/L; residence time =1 to 3 h; recycle ratio = 0.25 to 0.5.

16.11.6.24 STR: Batch (Backmix)

$Bd = 0$; $Pe < 1$. A related topic is mixing (Section 16.11.7.3).

Area of Application

Batch, semibatch, and continuous stirred tank reactors: residence time 600 to 15,000 s (10 min to 4 h); heat of reaction: primarily exothermic; reaction rate slow to moderate. High-pressure autoclaves <100 L. *Unique to batch STR:* phases: L, LL, LcS, L bioS. Low capacity, <3 L/s, but usually <0.3 L/s; use if the concentrations of all reactants are high and need good selectivity for consecutive reactions. For low and variable production rates, use a variety of similar products or several products sequentially; low fixed capital cost, but usually the product cost is high because of the cycles. Details of other applications and guidelines are given in Section 16.11.6.26, CSTR. Liquid plus biosolid: anaerobic digesters: (first stage): STR batch: batch microbiological treatment of municipal sludge: high rate.

Guidelines

Batch operation: size on operating cycle: load, temperature and pressure adjustment, react, return to usual conditions, discharge and, perhaps, clean. Loading and discharge times are proportional to the volume of the reactor (0.75 to 1 h each); reactant heating and product cooling depend on the reaction temperature (about 1 to 3 h each); cleaning (for polymer reactors 0.5 to 1 h). Liquid plus bio solid: *Anaerobic digesters:* (first stage): residence time 10 to 15 d; 2 to 10 × washout retention time; operating temperature 18 to 38°C, usually 35; organic VS loadings =18 to 25 mg/s·m³ digester volume; pH 7 to 7.1; volatile acid concentration 0.2 to 0.8 g/L; alkalinity concentration 2 to 3.5 g/L; Circular, diameter 6 to 35 m; depth 6 to 14 m; power 0.006 to 0.03 kW/m^3.

16.11.6.25 STR: Semibatch

$Bd = 0$; $Pe < 1$. A related topic is mixing (Section 16.11.7.3).

Area of Application

Batch, semibatch, and continuous stirred tank reactors: residence time 600 to 15,000 s (10 min to 4 h); heat of reaction: primarily exothermic; reaction rate slow to moderate. High-pressure autoclaves <100 L. *Unique to semibatch:* phases: liquid, gas-liquid, liquid-liquid, gas-liquid catalytic solid. Use where a batch operation is appropriate (Section 16.11.6.24), but one reactant (e.g., gas) needs to be added continuously or if the initial reaction rate is very high. Selectivity is best for parallel reactions. For more details, see CSTR, Section 16.11.6.26.

Guidelines

Batch operation: size on operating cycle: load, temperature and pressure adjustment, react, return to usual conditions, discharge and, perhaps, clean. Loading and discharge times are proportional to the volume of the reactor (0.75 to 1 h each); reactant heating and product cooling depend on the reaction temperature (about 1 to 3 h each); cleaning (for polymer reactors) 0.5 to 1 h.

16.11.6.26 CSTR: Mechanical Mixer (Backmix)

$Bd = 0$; $Pe <1$. Related topics mixing, 16.11.7.3, and solvent extraction, 16.11.4.10.

Area of Application

Batch, semibatch, and continuous stirred tank reactors: residence time 600 to 15,000 s (10 min to 4 h); heat of reaction: primarily exothermic; reaction rate slow to moderate. High-pressure autoclaves <100 L. *Unique for CSTR:* phases: liquid, gas-liquid, liquid-liquid, liquid catalytic solid, gas-liquid catalytic solid, gas-liquid bio solid. Capacity 0.0001 to 100 L/s and usually >0.4 L/s; volumes 1 to 1,000,000 L. Autothermal reactions. Usually used if the concentration of reactants is low, and need low concentration of reactants for selectivity. CSTR is larger and more expensive than PFTR. For multiphase, STRs are characterized by high liquid holdups; holdup of the reactive phase is important if the reaction is slow, $Ha < 1$; phase ratio is easy to control. *Liquid:* <300,000 mPa·s; volume <75 m^3. Use for kinetically controlled reactions that require long residence times. *Gas liquid:* surface area 60 to 500 m^2/m^3; surface area gas-liquid per volume of reactor: 200 to 2000 m^2/m^3 volume reactor; surface area gas-liquid per volume of liquid phase: 220 to 2500 m^2/m^3; liquid phase can handle viscous liquids and suspensions, only appropriate for smaller size reactors <10 to 20 m^3. *Liquid-liquid:* surface area 400 to 3500 m^2/m^3 with area increasing with decreasing surface tension and increasing velocity. Drop diameter 4 to 5000 µm; for viscosities <10^4 mPa·s. Phase ratio is easy to control. *Liquid with catalytic solid:* catalyst diameter <0.1 mm; surface area solid 500 m^2/m^3. *Gas-liquid with catalyst solid:* catalyst diameter <0.1 mm; surface area 50 to 1200 m^2/m^3; surface area solid 500 m^2/m^3; surface area gas-liquid 100 to 1500 m^2/m^3. *Gas-liquid bio solid:* aerobic sludge digesters: reduce the volume of and render biologically stable the sludge from a variety of sources: conventional activated sludge and primary clarifier.

Guidelines

Reactor size 8 to 32 m^3, 1.5 MPa, with jacketed heat transfer surface 1.5 to 2.5 m^2/m^3 volume. Heat transfer coefficient $U = 0.06$ to 0.35 kW/m^2·K for jacket to inside reactor contents; coil 0.7 to 0.8 kW/m^2·K. Power = 0.2 to 2 kW/m^3, see mixing, Section 16.11.7.3, and heat transfer, Section 16.11.3.3. Liquid: power input to promote heat and mass transfer: 1 to 6 kW/m^3 reactor volume. Gas-liquid: holdup: liquid holdup >0.7, gas holdup <0.1; bubble diameter = 2.5 mm regardless of the agitation and has a mean upward velocity of about 0.27 m/s; superficial gas velocity, 0.05 to 1 m/s; backmix; complete; typical liquid mass transfer coefficient = $k_L a$ = 0.02 to 0.2 1/s; volumetric ratio liquid to mass transfer liquid film, $\beta > 100$; power input 0.1 to 4 kW/m^3. Height of liquid = tank diameter or use multiple impellers if height of liquid/tank diameter >2. Impeller diameter 0.3 to 0.5 of tank diameter. *Liquid-liquid:* holdup: volume fraction dispersed liquid 0.01 to 0.5; typical drop diameter is 150 µm; power input 0.2 to 3 kW/m^3. A related topic is solvent extraction (Section 16.11.4.10). *Liquid with catalytic solid:* holdup: volume fraction catalyst 0.01; volume fraction liquid 0.99; power input to facilitate heat and mass transfer, suspend solids, and promote mass transfer: 1 to 4 kW/m^3 reactor volume. *Gas-liquid with catalyst solid:* holdup: volume fraction catalyst 0.01; volume fraction liquid 0.8 to 0.9; volume fraction gas 0.1 to 0.2. Power input 0.05 to 2 kW/m^3. Catalyst activity: variable but often able to avoid diffusion limitations because of small-diameter catalyst. Catalyst selectivity: OK. Catalyst stability: change between batches. Heat exchange: OK. *Gas liquid bio solids:* aerobic sludge digesters: CSTR designed on the basis of VSS reduction. Mixing to keep the solids suspended plus oxygenation. Cell residence time for cells 12 to 22 days, depending on the source of the sludge. Typical organic load 4 to 26 µg VSS/s·m^3; dissolved oxygen concentration 1 to 2 mg/L; air requirement for activated sludge = 0.25 to 0.33 dm^3/s·m^3; mixture of primary plus activated sludge = 0.4 to 0.5 dm^3/s·m^3; 1.42 kg O$_2$/kg biosolids digested. Oxygen usage 1.4 to 11 mg O$_2$/s·kg VSS, depending on the source of the sludge. For diffused air 0.33 to 1 dm^3/s·m^3, depending on the sludge; surface aeration = 0.025 to 0.033 kW/m^3. Power = 0.015 to 0.02 kW/m^3.

16.11.6.27 STR: Fluidized Bed (Backmix)

Related topics are heat transfer (Section 16.11.3.4 and Section 16.11.3.8), mixing (Section 16.11.7.1), dryers (Section 16.11.5.5), and size enlargement (Section 16.11.9.5).

Area of Application

Phases: gas with solid catalyst, gas reacting with a solid, liquid reacting with solid catalyst, gas-liquid with solid catalyst: residence time: for gas = seconds; for solids = minutes to hours. Primarily for highly exothermic, very fast reactions where the need is for uniform, closely controlled temperature; need fast reaction that occurs at the bottom of the bed. Not good if have consecutive reactions to produce the product (because of backmixing). Relatively inflexible. Preferred over PFTR for strongly exothermic or endothermic reactions. Easier than fixed bed for catalyst regeneration, excellent heat transfer for high exothermic reactions; lower pore diffusional resistance because of smaller-diameter catalyst particles. Select if catalyst life is <3 months. Catalyst must withstand attrition. *Gas with catalytic solid:* solid residence time, 300 to 15,000 s; gas residence time <1 s; solid particle diameter 0.005 to 7 mm. *Gas with reacting solid:* solid residence time, 300 to 15,000 s; gas residence time <1 s; solid particle diameter 0.005 to 7 mm. *For combustion, gasification, incineration.* Advantages: ease in solids handling, uniform temperature, and thermal stability, even for highly exothermic and endothermic reactions. Cannot be used for solids with partial fusion or softening of particles. *Liquid with catalytic solid:* catalyst diameter 0.1 to 5 mm; surface area solid 500 to 1000 m^2/m^3. *Gas-liquid with catalytic solid:* catalyst diameter 0.1 to 5 mm; surface area solid 500 to 1000 m^2/m^3; surface area gas-liquid 100 to 1000 m^2/m^3.

Guidelines

Gas-solid (catalytic or reacting): operate 10 to 100 × minimum fluidizing velocity; operate in the bubbling regime. Heat transfer coefficient 0.2 $kW/m^2 \cdot K$. For fast reactions, gas film diffusion may control, and catalyst pore diffusion mass transfer may control if catalyst diameter is >1.5 mm. The fluidization is characterized as "bubbling" (aggregate fluidization). Bubbling consists of two phases: (1) gas bubbles: assume move through in plug flow; superficial gas velocity (0.1 m/s) >> minimum superficial gas velocity to cause fluidization (0.01 m/s); mass transfer coefficient between the bubble and the emulsion phase = 0.01 m/s; the fraction of the volume of the fluidized bed occupied by the bubbles is 0.04, and (2) homogeneous emulsion phase: emulsion phase; reaction occurs in the emulsion phase. Area between the bubble phase and the emulsion phase = 10 m^2/m^3. Diffusivity of the reactant in the emulsion phase = diffusivity in the bubble phase = 10^{-5} to 10^{-4} m^2/s. The six options include (1) single bubbling bed (BFB), (2) BFB/BFB combo with catalyst as the oxygen carrier, (3) circulating fluidized bed (CFB), (4) CFB/BFB combo with catalyst as the oxygen carrier, (5) multistage BFB, and (6) multistage BFB with split air flow and temperature programming. Heat transfer: heat transfer coefficient wall to fluidized bed is 20 to 40 × gas wall at the same superficial velocity, h = 0.15 to 0.3 $kW/m^2 \cdot K$. Nu = 0.5 to 2. Fluidized bed usually expands 10 to 25%. Superficial fluid velocity is usually 5 to 10 × the minimum fluidization velocity; backmix type reactor, which increases the volume of the reactor and usually gives a loss in selectivity. Usually characterized as backmix operation or, more realistically, as a series of CSTRs if the height/diameter is >2. Usually 1 CSTR for each H/D = 1. *Liquid with catalytic solid:* particulate fluidization: homogeneous with no bubbles. For highly exothermic reactions, consider external recirculation through a heat exchanger. *Gas-liquid with catalytic solid:* holdup volume fraction catalyst 0.1 to 0.5; volume fraction liquid 0.2 to 0.8; volume fraction gas 0.05 to 0.02. Catalyst activity: variable but often reduced because of mass transfer limitation, backmixing is unfavorable. Catalyst selectivity: often reduced because of mass transfer limitation; backmixing is unfavorable. Catalyst stability: must withstand attrition; can be removed for regeneration. Heat exchange; good heat transfer. Isothermal efficiency may be 1, but with larger diameter particles this can decrease. May have lower isothermal efficiency because of diffusion into the pellet and the larger size pellet.

16.11.6.28 Tank Reactor (TR)

Minimal to no internal mixing.

Area of Application

Phases: gas-liquid bio solid; liquid plus bio solid. Liquid plus bio solid: anaerobic digesters: (conventional first stage): batch microbiological treatment of municipal sludge; no mixing: use for <50 m^3/s. Other options include anaerobic ponds, facultative lagoons and ponds, and aerobic ponds.

Guidelines

Liquid plus bio solid: *anaerobic digesters:* (conventional first stage): Residence time 30 to 60 days (35°C); organic loading = 4 to 9 mg VS/s·m^3; circular, diameter 6 to 35 m; depth = 6 to 14 m. *Anaerobic pond:* residence time: 5 days; surface loading 600 to 3500 μg BOD$_5$/s·m^2 with 150 μg BOD$_5$/s·m^2 for winter conditions; volumetric loading 3 to 100 kg BOD$_5$/s·m^3. pH 6.7 to 7.1; depth 1 to 2.5 m. *Facultative lagoon:* residence time = 7 to 20 days, usually 4 to 8 days, longer in cold temperatures; loading 40 to 130 μg BOD$_5$/s·m^2; depth 2.4 to 4.8 m; surface aeration power to oxygenate. *Facultative pond:* residence time: 7 to 50 days; surface loading 26 to 65 μg BOD$_5$/s·m^2; depth 1 to 2.5 m; length/width = 3/1. No surface aeration; photosynthesis is source of oxygen; recycle ratio = 0.2 to 8, usually 4 to 8. *Aerobic pond:* residence time: 2 to 6 days; surface loading 130 to 260 μg BOD$_u$/s·m^2; depth 0.15 to 0.45 m; recirculation ratio = 0.2 to 2. A related topic is aerobic lagoon (Section 16.11.6.23).

16.11.6.29 Mix of CSTR, PFTR with Recycle

Area of Application

Phases: gas-liquid and bio solid. Bio solid removes the soluble organics, COD or BOD$_5$, from wastewater. Variety of reactor configurations. Related topics are trickling filter (Section 16.11.6.14), CSTR (Section 16.11.6.26), and CSTR in series (Section 16.11.6.23). *Conventional PFTR-activated sludge:* low-strength domestic wastewater, susceptible to shock loads; 85 to 95% removal. *Conventional backmix-activated sludge:* usual-strength domestic wastewater, resistant to shock loads, 85 to 95% removal. *Step aeration, modified aeration PFTR-activated sludge:* higher-strength domestic wastewater, 85 to 95% removal. *Contact stabilization:* PFTR: OK for domestic wastewater; unsuitable for most industrial wastewater, flexible, 80 to 90% removal. *Extended aeration backmix-activated sludge* (oxidation ditch): low organic loadings, small capacity: <40 L/s; flexible, 75 to 95% removal. *High rate aeration backmix-activated sludge:* high-strength domestic wastewater, 75 to 90% removal.

Guidelines

Conventional PFTR-activated sludge: mean cell residence time = 5 to 15 days; food/microorganism ratio = 2.3 to 4.6 mg BOD$_5$/s·kg MLVSS; volumetric loading = 0.3 to 0.6 kg BOD$_5$/m^3; MLSS = 1.5 to 3 g/L; residence time = 4 to 8 h; recycle ratio = 0.25 to 0.5. Air requirements = 100 m^3/kg of input BOD$_5$. Conventional backmix-activated sludge: mean cell residence time = 5 to 15 days; food/microorganism ratio = 2.3 to 7 mg BOD$_5$/s·kg MLVSS; volumetric loading = 0.8 to 2 kg BOD$_5$/m^3; MLSS = 3 to 6 g/L; residence time = 3 to 5 h; recycle ratio = 0.25 to 1. Air requirements = 100 m^3/kg of input BOD$_5$. Step aeration, modified aeration PFTR-activated sludge: mean cell residence time = 5 to 15 days; food/microorganism ratio = 2.3 to 4.6 mg BOD$_5$/s·kg MLVSS; volumetric loading = 0.6 to 1 kg BOD$_5$/m^3; MLSS = 2 to 3.5 g/L; residence time = 3 to 5 h; recycle ratio = 0.25 to 0.75. Air requirements = 100 m^3/kg of input BOD$_5$. Contact stabilization: PFTR with recycle: mean cell residence time = 5 to 15 days; food/microorganism ratio = 2.3 to 7 mg BOD$_5$/s·kg MLVSS; volumetric loading = 1 to 1.2 kg BOD$_5$/m^3; MLSS = 4 to 10 g/L; residence time = 3 to 6 h; recycle ratio = 0.25 to 1. Air requirements = 100 m^3/kg of input BOD$_5$. Extended aeration-activated sludge backmix with recycle: mean cell residence time: = 20 to 30 days; food/microorganism ratio = 0.6 to 1.8 mg BOD$_5$/s·kg MLVSS; volumetric loading = 0.16 to 0.4 kg BOD$_5$/m^3; MLSS = 3 to 6 g/L; residence time = 18 to 36 h; recycle ratio = 0.75 to 1.5. Air requirements = 125 m^3/kg of input BOD$_5$. High rate aeration-activated sludge, backmix with recycle:

Conceptual Process Design, Process Improvement, and Troubleshooting

mean cell residence time = 5 to 10 days; food/microorganism ratio = 4.5 to 17 mg BOD_5/s·kg MLVSS; volumetric loading = 1.6 to 16 kg BOD_5/m^3; MLSS = 4 to 10 g/L; residence time = 0.5 to 2 h; recycle ratio =1 to 5. Air requirements = 25 to 100 m^3/kg of input BOD_5.

16.11.6.30 STR: PFTR with Large Recycle

Area of Application

Phases: any. Recycle ratio = 20/1 gives backmix. Only exceed this recycle ratio if this is required for heat transfer.

16.11.6.31 Reaction-Injection Molding

Area of Application

Phases: liquid, liquid-liquid. Viscosities <1000 mPa·s; time for reaction to gel under adiabatic conditions >0.1 s; small capacity: 0.150 to 2k g/s.

Guidelines

Viscosity between 10 to 100 mPa·s to prevent bubbles; $Re > 300$; fill time >1 s and less than the reaction time to gel under adiabatic conditions. Mold temperature <100°C (or <200°C for high-temperature operation). Mold temperature plus the adiabatic reaction exotherm must not exceed the degradation temperature. Reaction should be 95% complete in <3 min.

16.11.6.32 Reactive Distillation, Extraction, Crystallization

Gas-liquid reaction with catalytic solid. Include the catalyst with structured packing in a distillation column. A related topic is distillation (Section 16.11.4.2).

Area of Application

Phases: gas-liquid, gas-liquid catalytic solid, gas-liquid plus catalytic solid: minimizes catalyst poisoning, lower pressure than fixed bed. Used for hydrogenation reactions and MTBE and acrylamide production. For example, 90% conversion via reactive distillation contrasted with 70% conversion in fixed-bed option. Liquid with homogeneous catalyst: etherification, esterification. Liquid-liquid: HIGEE for fast, very fast, and highly exothermic liquid-liquid reactions such as nitrations, sulfonations, and polymerizations. Equilibrium conversion <90%. Use a separate pre-reactor when the reaction rate at 80% conversion is >0.5 initial rate. The products should boil in a convenient temperature range. The pressure and temperature for distillation and reaction should be compatible.

Guidelines

Use concentration profiles developed from either equilibrium or nonequilibrium reaction-separation to identify the reactive zone. The reflux ratio for reactive distillation is greater than for distillation. Use 1.2 to 1.4 × minimum. For catalytic structured packing, use liquid loadings up to 14 L/s·m^2 and vapor "capacities," F factor, of 2.5 m/s $(kg/m^3)^{1/2}$ (based on velocity and the root of vapor density).

16.11.6.33 Membrane Reactors

Area of Application

Phases: gas-liquid, gas-liquid catalytic solid. Equilibrium reactions where selective removal via membrane will shift equilibrium or use of membrane as catalyst.

Guidelines

Try to match permeation rate and reaction rate.

16.11.6.34 Process Vessels

Area of Application

Used for a wide range of temperatures, pressures, and uses. Can be decanters, reflux drums, intermediate storage, towers, KO pots, reactors.

Guidelines

Operating pressure and temperature constrain practical size of vessel. Design codes for pressure vessels vary slightly with the country. In general, for operating pressures >10 MPa, vessel volume usually <1 m^3. Pressure decreases as temperatures exceed 250°C. For temperatures above 350°C, consider carbon/molybdenum, and for temperatures >500°C, consider austenitic steels. *Corrosion allowance* 1.5 mm for corrosion rates 0.08 mm/a; 3 mm for rates 0.09 to 0.3 mm/a; 4.5 mm for 0.31 to 0.4 mm/a; 6 mm for >0.4 mm/a. If pressure <400 kPa, use *L/D* of 2 to 3:1. For pressures >400 kPa, use *L/D* of 4 to 5:1. *For surge,* allow 2-min liquid residence time; for draw-off, use 15 min; for reflux, use 5 min, provided this allows sufficient time for controllers to function. Total volume = 1.3 × holdup if the holdup volume is >3 m^3.

16.11.6.35 Storage Vessels and Bins

Guidelines

For liquids, consider floating-head vertical cylinders for pressures <100 kPa; spheres, vertical cylinders with dome ends for pressures <250 kPa; small spheres and horizontal cylindrical tanks for pressures <800 kPa. For gases, consider pressure cylinders and small horizontal cylindrical tanks for pressures >800 kPa. For solids, bins, and hoppers: promote mass flow with cone angle to the vertical related to the wall friction angle of the solid.

Wall Friction Angle	Cone Angle with the Vertical	Wedge Angle with the Vertical
	<	<
30°	10°	20°
20°	25°	38°
10°	35°	48°
0°	45°	60°

Both bottom angle and surface smoothness are important. For a circular cone, the outlet should be at least six to eight times the diameter of the largest particle. For wedge hoppers, the discharge opening should be at least three to four times the diameter of the largest particle. The length of the slot opening should be at least three times the width. The desired mass flow requires the entire discharge opening to be active by the use of: tapered interface belt conveyor, a tapered shaft screw conveyor, a screw conveyor with an increasing pitch (providing the length:screw diameter is <3:1), or a combination of screw conveyor with a tapered shaft plus increased pitch, provided the length:screw diameter is <6:1.

16.11.7 Mixing*

The major situations are gas-solid, gas-liquid, liquid, and solids. Solids blending can include dry solids blenders; extruders for foodstuffs and polymers; and pug mills for clays, thick pastes, and fertilizers.

---

* Reproduced with permission from Donald R. Woods, *Rules of Thumb in Engineering Practice,* pages 282 to 291, 2007, copyright Wiley-VCH Verlag GmbH & Co. KGaA, Weinheim, ISBN: 978-3-527-31220-7. See this book for more details, especially about Good Practice, Trouble shooting and Capital Costs.

Conceptual Process Design, Process Improvement, and Troubleshooting

16.11.7.1 Gas–Solid

Gas fluidization is used for reactions (Section 16.11.6.26), to facilitate heat transfer (Section 16.11.3.8), dryers (Section 16.11.5.5), and size enlargement (Section 16.11.9.5).

Area of Application

Solid particle diameter 60 to 80 μm, solid particles with ratio of maximum to minimum diameter about 11 to 25.

Guidelines

Minimum gas fluidization velocity of 0.5 mm/s to 15 mm/s or about a factor of 0.01 to 0.1% of the pneumatic conveying velocity. Ratio of height to diameter ≥1. Bed depth usually 0.3 to 15 m. When used for heat transfer: particles and gas tend to leave the bed at the same temperature. Heat transfer from the bed to the walls: $U = 0.45$ to 1.1 kW/m$^2\cdot$°C; from bed to immersed tubes: $U = 0.2$ to 0.4 kW/m$^2\cdot$°C; from solids to gas in the bed: $U = 0.017$ to 0.055 kW/m$^2\cdot$°C.

16.11.7.2 Gas-Liquid

The flow characteristics are given in Section 16.11.2.4. The area and other contacting characteristics are summarized in Sections 16.11.8.1 and 16.11.8.2. A related topic is bubble reactors (Section 16.11.6.11).

Area of Application

Mechanical agitator: viscosity <50,000 mPa·s; volumes <75 m^3. *Air agitation:* viscosity <1000 mPa·s; volumes >750 m^3. See aeration, size reduction, Section 16.11.8.1. Bubble columns: see size reduction, Section 16.11.8.1; a related topic is bubble reactors (Section 16.11.6.11).

Guidelines

Mechanical agitators: see size reduction (Section 16.11.8.1). Air agitation: see aeration, size reduction (Section 16.11.8.1). Bubble columns: see size reduction (Section 16.11.8.1).

16.11.7.3 Liquid

Related to stirred tank reactors (Sections 16.11.6.23 through 16.11.6.25).

Area of Application

Fluid viscosity and volume to be mixed are the most significant factors. *Propellers:* viscosity <3000 mPa·s; volumes <750 m^3. *Turbines and paddles:* viscosity <50,000 mPa·s; volumes <75 m^3. *Liquid jets:* viscosity <1000 mPa·s; volumes >750 m^3. *Air agitation:* viscosity <1000 mPa·s; volumes >750 m^3. *Anchors:* viscosity <100,000 mPa·s; Re <10,000; volumes <30 m^3. *Kneaders:* viscosity 4,000 to 1.5×10^6 mPa·s; volumes 3 to 75 m^3. *Roll mills:* viscosity 10^3 to 200,000 mPa·s; volumes 60 to 450 m^3. For viscosity >10^6 consider extruders, Banbury mixers, and kneaders. *Paddle reel/stator-rotor:* gentle mechanical mixing for coagulation, viscosity <20 mPa·s; volumes large. *Motionless mixers:* viscosity ratio <100,000:1; continuous and constant flow rates; residence times <30 min and flow rate ratio of <100:1. Other related sections are size reduction (Sections 16.11.8.1 and 16.11.8.3), reactors (Section 16.11.6.10), and heat transfer (Section 16.11.3.5).

Guidelines

Turbines, propellers, and paddles: power = 0.2 to 1.5 kW/m^3 for mixing liquids with impeller discharge rate >20 × liquid flow rate into tank. Heat transfer, 0.4 to 2 kW/m^3, but don't neglect heat input from the mixer; mass transfer 2 to 4 kW/m^3. *Turbines, propellers, and paddles:* power = 1 to 4 kW/m^3 for mass transfer. *Air agitation:* diffused air: 0.3 to 0.5 Ndm^3/s·m^3; 1.5 to 6 dm^3/s·m of linear distance along the basin. Diffusers 15 to 30 dm^3 air/s·m^2 diffuser area. For 45-min detention time, 0.6 to 1 dm^3/L. See related topics, flocculation (Section 16.11.9.3) and size reduction, gas in liquid (Section 16.11.8.1). *Anchor:* power = 4 to 9 kW/m^3. *Gentle mechanical mixing, such as paddle reel or stator-rotor for flocculation* (see related topic, Section 16.11.9.3): 0.035 to 0.04 tapering to 0.001 to 0.009 kW/m^3. *Motionless mixers:* for viscosity ratio <100:1 and Reynolds no., *Re* >10,000, use turbulent vortex mixer; for viscosity ratio >100:1 and *Re* <10,000, use helical element. For pipe diameter <0.3 m, element is 1.5 × pipe diameter; for pipe diameter >0.3 m; element = pipe diameter. For *Re* <10, use 18 elements with the number of elements reducing to 2 as Reynolds no. increases to >5000. *Annular sparger:* annular sparger blends liquids of equal viscosity and density in 50 pipe diameters; central injection blends liquids of equal viscosity and density in 80 pipe diameters; in mixing tee after 10 pipe diameters. For viscosity differences <10:1, inject the viscous liquid into the thin liquid. For *blending and heat transfer*, identify viscosity when turbulent mixing occurs. Usually, impeller *Re* of 200.

16.11.7.4 Liquid-Liquid

Related sections are solvent extraction (Section 16.11.4.10), size reduction (Section 16.11.8.3), and reactors (Sections 16.11.6.9 through 16.11.6.16, 16.11.6.23 through 16.11.25, and 16.11.6.30).

Area of Application
Fluid viscosity and volume to be mixed are the most significant factors. Propellers: viscosity <3000 mPa·s; volumes <750 m^3. Turbines and paddles: viscosity <300,000 mPa·s; volumes <75 m^3.

Guidelines
Propellers and turbines: power: 0.2 to 1.5 kW/m^3 for mixing immiscible liquids, with values decreasing as the interfacial tension decreases, and for heat transfer. Power: 1 to 4 kW/m^3 for emulsification and mass transfer.

16.11.7.5 Liquid-Solid

Area of Application
Stirred tank: crystallizers, see Section 16.11.4.6; and reactors, Sections 16.11.6.23 through 16.11.6.25. Liquid fluidized bed: reactors, Section 16.11.6.26; liquid adsorption, Section 16.11.4.12; ion exchange, Section 16.11.4.13; backwash fixed-bed operations such as deep-bed filters, Section 16.11.5.13; liquid adsorbers, Section 16.11.4.12; and ion exchangers, Section 16.11.4.13.

Guidelines
Stirred tank: paddles: power input: suspend solids, 0.2 to 1.6 kW/m^3; *L/D* = 0.7 to 1.05/1. Baffle, four @ 90°; baffle width = 0.08 × tank diameter; off-the-wall distance = 0.015 × tank diameter. Minimum level of liquid = 0.15 × tank diameter for impeller:tank diameter 0.28:1 and minimum level = 0.25 × tank diameter for impeller:tank diameter = 0.4:1. Use a foot bearing plus a single, main axial hydrofoil impeller diameter = 0.28 × tank diameter located 0.2 × tank diameter from the bottom plus a pitched blade impeller diameter = 0.19 × tank diameter located 0.5 × tank diameter from the bottom. *Liquid fluidized bed:* in general, particle diameter 0.5 to 5 mm with density and diameter of the particle dependent on the application. The superficial liquid velocity to fluidize the bed depends on both the diameter and the density difference between the liquid and the particle. Usually, the operation is particulate fluidization. Particle diameter 0.2 to 1 mm reactors; superficial liquid velocity 2 to 200 mm/s. *Fluidized adsorption:* bed expands 20 to 30%; superficial liquid velocity for usual carbon adsorbent = 8 to 14 mm/s. *Fluidized ion exchange:* bed expands 50 to 200%; superficial liquid velocity for usual ion exchange resin = 40 mm/s. *Backwash operations:* fixed-bed adsorption: superficial liquid velocity = 8 to 14 mm/s; fixed-bed ion exchange: superficial backwash velocity = 3 mm/s.

Conceptual Process Design, Process Improvement, and Troubleshooting

16.11.7.6 Dry Solids

Area of Application

For free-flowing particles that do not segregate, use mixers where outside shell moves (cone, double cone, zigzag). For pastes, plastics, and powders that tend to segregate, use mixers with outside shell fixed (ribbon, edge mill, double arm, Banbury, extruder).

Guidelines

Working capacity is 50 to 60% of total internal volume. Mixing residence times are 3 to 10 min. Mixing time increases with (particle diameter)$^{0.5}$. Rpm 20 to 100. Speed of rotation (rpm) × mixing time (minutes) = 300. *Double cone:* 0.1 to 10 m^3 working capacity; 10 kW/m^3 reducing to 2 kW/m^3 as the working capacity increases. *Ribbon:* 0.02 to 15 m^3 working capacity; 50 kW/m^3 reducing to 4 kW/m^3 as the working capacity increases. *Edge mill:* 1 to 3 m^3 working capacity; 40 to 50 kW/m^3 and relatively independent of working volume. *Double arm:* 0.05 to 3 m^3 working capacity; 100 kW/m^3 and relatively independent of working capacity. *Banbury:* 0.1 to 0.3 m^3 working capacity; 10,000 kW/m^3 and relatively independent of working capacity.

16.11.7.7 Pastes, Polymers, Foodstuffs, Clay, and Fertilizers

See size enlargement, Section 16.11.9.14. Related topics are size reduction (Section 16.11.8.5) and dewatering rotary press (Section 16.11.5.13).

Area of Application

Use mixers with outer shell fixed.

16.11.8 Size Reduction*

16.11.8.1 Gas in Liquid (Bubbles in Liquid)

Gas-liquid contacting for reactions, separations, mixing. See reactions (Section 16.11.6.11), separations (Sections 16.11.4.8 and 16.11.4.9), and mixing (Section 16.11.7.2) for separate rules of thumb. Flow characteristics are summarized in Section 16.11.2.4.

Area of Application

Aeration: gentle mixing and oxygenation of bio reactors. *Bubble column:* area per unit volume 5 to 70 m^2/m^3; can handle solids, high-pressure drop. *Spray column:* area per unit volume 10 to 100 m^2/m^3; can handle foaming and solid-laden gases, low pressure drop. Reactions: good for reaction with highly soluble gases. *Packed column:* area per unit volume 75 to 200 m^2/m^3; low pressure drop, cannot handle solids, can handle foaming by operating in the cocurrent upflow bubble region. Little flexibility in varying the gas/liquid volumetric flow rates because of flooding. *Tray column:* area per unit volume 75 to 400 m^2/m^3; can handle solids, cannot handle foaming. Some flexibility in varying the gas/liquid volumetric flow rates. *Agitated tank:* area per unit volume 20 to 200 m^2/m^3; can suspend solids, maximum pressure limited by the seal around the rotating mixing shaft. *Motionless mixer:* area per unit volume 1,000 to 7,000 m^2/m^3; cannot handle foaming; see also liquid-liquid, Section 16.11.8.3. *Pipe/tubes, transfer line:* area per unit volume 50 to 2.000 m^2/m^3; high pressure drop, Reactions: use with irreversible reactions and pure gas feed.

Guidelines

Aeration: 0.3 to 0.5 dm^3/s·m^3. Fifteen to 30 dm^3/s·m^3 diffuser area; or 1.5 to 6 dm^3/s·m of linear length of diffuser. Need 75 to 110 m^3 air/kg BOD removed; 0.018 to 0.04 g oxygen absorbed/dm^3

---

* Reproduced with permission from Donald R. Woods, *Rules of Thumb in Engineering Practice*, pages 291 to 297, 2007, copyright Wiley-VCH Verlag GmbH & Co. KGaA, Weinheim, ISBN: 978-3-527-31220-7. See this book for more details, especially about Good Practice, Trouble shooting and Capital Costs.

air sparged into the liquid. $k_La = 0.0008$ 1/s. A related section is mixing (Section 16.11.7.3). *Bubble column:* superficial gas velocity, 0.03 to 0.04 m/s holdup <0.2, $k_La = 0.005$ to 0.01 1/s. k_La is independent of the diameter if column diameter >0.15 m; k_La is not affected by the type of gas sparger if the gas velocity exiting the orifice is >0.03 m/s. If <0.03 m/s, then use a sintered plate. Height: 3 < height < 12 m; allow 0.75 of diameter or 1 m at the top for foam disengagement. Energy 0.01 to 1 kW/m³. *Spray column:* superficial gas velocity, 0.75 to 2 m/s; holdup <0.8, energy: high for atomization; $k_La = 0.0007$ to 0.015 1/s. *Packed column:* superficial gas velocity, 0.75 to 1.5 m/s for atmospheric pressure; 0.2 m/s for pressure operation and 2.5 m/s for high vacuum; holdup <0.95; energy 0.01 to 0.2 kW/m³; $k_La = 0.005$ to 0.02 1/s; packing >6 mm; catalyst >3 mm. Column diameter/packing diameter >8 and prefer >30 to prevent liquid channeling. Redistributors every 3 to 4.5 m. Higher mass transfer coefficient, $k_La = 0.15$ 1/s for cocurrent upflow in the bubble regime. *Tray column:* superficial gas velocity, 0.75 to 1.5 m/s for atmospheric pressure; 0.2 m/s for pressure operation and 2.5 m/s for high vacuum; holdup <0.8; energy 0.01 to 0.2 kW/m³; k_La = 0.01 to 0.05 1/s. *Agitated tank:* superficial gas velocity, 0.05 to 1 m/s; holdup <0.1, energy 0.1 to 4 kW/m³; $k_La = 0.02$ to 0.2 1/s. Height of liquid = tank diameter or use multiple impellers if height of liquid/tank diameter is >2. Impeller diameter 0.3 to 0.5 of tank diameter. *Motionless mixer:* superficial gas velocity, 1 to 2 m/s; holdup 0.5, energy 10 to 700 kW/m³; $k_La = 0.1$ to 3 1/s. Volumetric flow rate of gas to liquid = 1 at the nozzle. *Pipe/tubes, transfer line:* superficial gas velocity, 0.01 to 0.4 m/s; holdup 0.05 to 0.95; energy 0.1 to 100 kW/m³. For <10-cm dia. tubes, $k_La = 0.01$ to 0.7 1/s.

16.11.8.2 Liquid in Gas (Sprays)

Area of Application

Pressure nozzle: spray diameter 70 to 10,00 μm; capacity 0.03 to 0.3 L/s; low viscosity and clean fluids. *Spinning disc:* spray diameter 50 to 250 μm; capacity 0.0015 to 0.4 L/s; for usual fluids and for viscous fluid or fluid containing solids. *Twin fluid:* spray diameter 2 to 80 μm; capacity 0.03 L/s; increasing the ratio of atomizing fluid to liquid from 1 to 10, will decrease the spray diameter by a factor of 10. *Rayleigh breakup* to produce uniform drops of diameter 1.8 × diameter of orifice. A related topic is prilling (Section 16.11.9.14). *Surface aerators:* for activated sludge oxidation (instead of diffused air aeration, Section 16.11.8.1). *Brush aerators:* for oxidation ditches. *Motionless mixers:* spray flow.

Guidelines

Pressure nozzle: pressure 0.45 to 14 MPa; increasing the pressure increases the capacity. *Spinning disc:* increasing the capacity increases the drop size. *Twin fluid:* high energy input. *Surface aeration:* 0.01 to 0.025 kW/m³ or 0.3 to 1.2 kg O_2/MJ. *Brush aeration:* 0.015 to 0.018 kW/m³ 0.6 to 0.8 kg O_2/MJ. *Motionless mixers:* spray flow: gas superficial velocity 3 to 25 m/s; liquid superficial velocity 0 to 0.6 m/s. Turbulent flow.

16.11.8.3 Liquid-Liquid

Liquid-liquid contacting for reactions, separations, mixing. See reactions, Section 16.11.6.26; separations, Section 16.11.4.10; and mixing, Section 16.11.7.4, for separate rules of thumb.

Area of Application

Mixer in tank: drop diameter 4 to 5000 μm; capacity >0.05 L/s; for viscosities <10^4 mPa·s. *Colloid mill:* drop diameter 1 to 8 μm; capacity 0.01 to 3 L/s; for viscosities <10^4 mPa·s but usually >1000 mPa·s. *Homogenizer:* drop diameter 0.1 to 2 μm; capacity 0.03 to 30 L/s; for viscosities <10^3 mPa·s but usually <200 mPa·s; decrease the drop diameter by increasing the exit pressure. *High shear disperser:* for viscosities 10^3 to 5×10^6 mPa·s. *Roller mills:* for viscosities >10^3 mPa·s. *Motionless mixer:* drop diameter 100 to 1000 μm (about 0.15 times drop diameter for fluid velocity in a pipe without the mixer); capacity 0.3 to 5 L/s. The densities and flow rates of the two phases should be about equal; viscosities <50 mPa·s. See also gas in liquid, Section 16.11.8.1. *Ultrasonic:* drop diameter 1 to 2 μm; capacity 1 L/s.

Guidelines

Mixer in tank: surface area 400 to 10,000 m^2/m^3, with area increasing with decreasing surface tension and increasing velocity. Power 3 kW/m^3. *Colloid mill:* surface area 10,000 to 2,000,000 m^2/m^3, with areas increasing with decreasing drop diameter and increasing volume fraction of dispersed phase. Power 40 to 200 kW·s/L. *Homogenizer:* surface area 20,000 to 10,000,000 m^2/m^3, with areas increasing with decreasing drop diameter and increasing volume fraction of dispersed phase, power 25 to 120 kW·s/L; power increases as exit pressure increases from 3.5 to 55 MPa. *Motionless mixer:* surface area 100 to 20,000 m^2/m^3, depending on the drop diameter and the concentration of dispersed phase. Velocity 0.25 to 2.5 m/s. Turbulent flow. Δp is 100 × Δp in pipe without mixer; L/D about 33; power 0.001 to 0.015 kW·s/L. *Ultrasonics:* 18 to 30 kHz for 1 to 2 μm dia. drops.

16.11.8.4 Solids: Crushing and Grinding

Capacities are expressed for open-circuit operation. For closed-circuit with the same size reduction ratio, power, reduce the capacity by a factor of 2. Rod mill product is usually larger in diameter than from a ball mill. Selection depends on the size of feed, the reduction ratio, the target diameter, the capacity, hardness, toughness, fibrosity, and stickiness, and whether wet grinding is OK. Ball and rod mills can be used for most hardnesses including fibrous, friability, and stickiness. Use jaw crushers for hard materials but shift to cone crushers when Mohs hardness is <8, but not for sticky materials. Size reduction is about 1 to 5% efficient; most of the energy generates heat. Very ductile materials are difficult to break mechanically; use cold temperature to make brittle. Impact mills give products with less area per unit mass than ball or rod mills.

Area of Application

Jaw crusher: feed diameter 0.1 to 1.5 m; reduction ratio 5:1 to 10:1; capacity 1 to 300 kg/s; Mohs hardness <9 (reduction by compression). *Gyratory crusher:* feed diameter 0.75 to 1.5 m; reduction ratio 5:1 to 10:1, usually 8:1; capacity 140 to 1,000 kg/s; Mohs hardness <9. More suitable for slabby feeds than jaw crusher (reduction by compression). *Roll crusher:* feed diameter 1 cm; reduction ratio 5:1 to 10:1; capacity 0.3 to 20 kg/s; Mohs hardness <7.5. Suitable for softer, friable, and nonabrasive materials. OK for wet and sticky materials. *Cone crusher and short head cone:* feed diameter <25 cm; reduction ratio 5:1 to 10:1, usually 7:1; capacity 5 to 300 kg/s; Mohs hardness <8. Usually secondary or tertiary crusher. *Impact crusher:* pulverizers, shredders, or smooth roll: feed diameter 1 cm with a reduction ratio of 7:1 to 10:1; capacity 0.3 to 50 kg/s. *Mills, hammer,* feed diameter 10 mm, reduction ratio 10:1 to 50:1, capacity 0.01 to 5 kg/s; Mohs hardness <4.5. Maximum fines; feed not hard or abrasive. *Mills, ball, and rod:* feed diameter 0.5 mm with a reduction ratio of 10:1 to 50:1; Mohs hardness <9. *Mills, autogenous, semiautogenous;* feed diameter 200 mm, reduction ratio 10:1 to 50:1; capacity 0.1 to 100 kg/s; Mohs hardness <6. *Mills, fluid energy:* feed diameter 50 μm; reduction ratio 10:1 to 50:1; capacity <2 kg/s; Mohs hardness <4.5.

Guidelines

Two-stage grinding has lower capital cost but higher operating cost than single-stage primary grinding. Ball mills have lower capital costs but higher operating costs than pebble mills. Semiautogenous usually have lower capital and operating costs than fully autogenous. With breakage by compression, the power needed increases with an increase in hardness of the solid being processed. With breakage by tumbling, the power increases with an increase in the reduction ratio and relatively independent of the hardness. *Jaw crushers:* power 0.5 to 5 MJ/Mg; rpm 300 to 100; maximum capacity occurs under choke feed; minimum of fines (breakup by compression). Product diameter determined by the adjustment clearance between compressing plates. *Gyratory crusher:* power 3 to 10 MJ/Mg; rpm 450 to 110; minimum of fines (breakup by compression), product diameter determined by the adjustment clearance between compressing plates. *Cone crusher:* selected as secondary and tertiary reducers; power 0.9 to 5 MJ/Mg; rpm 290 to 220; (breakup by compression) product diameter determined by the adjustment clearance between compressing plates. *Short head cone crusher:* power 3 to 12 MJ/Mg; (breakup by compression) product diameter determined by the adjustment clearance between compressing plates. Often, choke fed as tertiary crusher. *Roll crusher:* power 3 to 15 MJ/Mg; (breakup by compression) product diameter determined by the space between the rolls. Speed determines the capacity. *Roller mill:* 50 to 500 kPa, (breakup by compression) product diameter determined by the space between the rolls. Speed determines the capacity. *Shredders:* power 25 to 250 MJ/Mg. *Hammer mill:* power 2 to 80 MJ/Mg, (breakup by impact against a plate traveling at 20 to 60 m/s) product diameter determined by exit screen size. *Cage mill:* (breakup by impact against a plate traveling at 20 to 30 m/s) product diameter determined by exit screen size. Handles amorphous materials. *Pin-disc mill:* (breakup by impact against a plate traveling at 200 m/s) product diameter determined by feed flow rate and speed of the pins. Ideal for soft material. *Impact mill:* (breakup by impact against a plate traveling at 50 to 110 m/s) product diameter determined by exit screen size. *Autogenous mill:* length:diameter 0.2 to 0.5 with 0.33 usual (breakup by impact among particles). *Rod mill:* power 5 to 80 MJ/Mg; length:diameter 1.4:1 to 1.6:1 with length <6.8 m; 35 to 40% v/v rod charge to give total charge of 45% v/v. (Breakup by variety of mechanisms with revolving media.) *Ball mill:* power 30 to 10,000 MJ/Mg; length:diameter 1:1 to 2:1; 50% v/v charge of balls (breakup by variety of mechanisms with revolving media); large balls give coarse particles, small balls give fine particles. *Fluid energy mill:* about 6 to 9 kg air/kg of solid or 1 to 4 kg steam/kg solid; (breakup by impact with other particles traveling at 100 to 300 m/s) product size determined by the feedrate. Power 700 to 1000 MJ/Mg.

16.11.8.5 Solids: Modify Size and Shape: Extruders, Pug Mills, and Molding Machines

See size enlargement, Section 16.11.9.14.

16.11.8.6 Solids: Solidify Liquid to Solid: Flakers, Belts, and Prill Towers

See Section 16.11.9.15.

16.11.9 SIZE ENLARGEMENT*

16.11.9.1 Size Enlargement: Liquid-Gas: Demisters

Area of Application

In general, liquids should not bind to the inserts or walls; they should flow as rivulets or drops along the surface of the insert. *Vane separators:* droplet diameter >20 μm; droplet concentration >0.1 mg/m^3. *Mesh pads:* droplet diameter 3 to 20 μm; droplet concentration 0.01 to 0.1 mg/m^3. *Fiber beds designed for impaction* (usually cylindrical or "candles"): droplet diameter 0.2 to 3 μm; droplet concentration 0.001 to 0.01 mg/m^3. *Fiber beds designed for Brownian motion* (usually cylindrical or "candles"): <0.1 μm; droplet concentration <10^{-3} mg/m^3.

---

* Reproduced with permission from Donald R. Woods, *Rules of Thumb in Engineering Practice*, pages 298 to 309, 323 to 324, 2007, copyright Wiley-VCH Verlag GmbH & Co. KGaA, Weinheim, ISBN: 978-3-527-31220-7. See this book for more details, especially about Good Practice, Trouble shooting and Capital Costs.

Guidelines

For all devices, inlet gas flow rates 80 to 120% of design. *Vane separators:* collecting fiber >300 μm; gas velocity 2.5 to 5 m/s; Δp 0.03 to 0.25 kPa. *Mesh pads:* collecting fiber 100 to 300 μm; gas velocity 2 to 4 m/s with smaller values used as liquid loading increases; Δp 0.1 to 0.75 kPa. *Fiber beds/impact:* collecting fiber 10 to 40 μm; gas velocity 1.25 to 2.5 m/s; Δp 1 to 2.5 kPa. *Fiber beds/Brownian motion:* collecting fiber 8 to 10 μm; gas velocity 0.05 to 0.25 m/s; Δp 1 to 4.5 kPa.

16.11.9.2 Size Enlargement: Liquid-Liquid: Coalescers

Area of Application

Stacked trays: droplet diameter >10 μm; concentration <15% v/v. *Packed bed:* droplet diameter 1 to 10 mm. *Mesh:* droplet diameter 50 to 500 μm. *Fibrous bed:* droplet diameter <20 μm; usually 1 to 9 μm; concentration <1% v/v. *Deep bed:* droplet diameter <1 μm. See Section 16.11.5.13. Ultrafiltration: see Section 16.11.4.21.

Guidelines

Stacked trays: 1 to 25 L/s per pack of trays on 2-cm spacing with 75 m² area per pack. Diameter of captured drop increases as the flow rate increases and the density difference and interfacial tension decrease. *Packed beds:* at lower capacities, the exit drop diameter is proportional to the void diameter in the packing. *Mesh:* use mix of high- and low-energy materials in mesh about 100 to 500 μm; flood velocities for mixed high- and low-energy fiber mesh <1.2 m/s; pressure drop 6 to 140 kPa. *Fibrous bed:* select fiber diameter that is about the diameter of droplets, fibers about 10 to 40 μm; typical exit drop diameter two to four times the inlet diameter; flood velocities 1 cm/s; usually design for 0.5 cm/s or 5 L/s·m² with velocity decreasing as surface tension decreases. Try for surface tension >20 mN/m.

16.11.9.3 Size Enlargement: Solid in Liquid: Coagulation/Flocculation

Area of Application

Particle diameter <1 μm; solids concentration <0.1%. Related to flocculants for thickening (Section 16.11.5.9).

Guidelines

Add coagulant: usually alum in the sweep floc concentration of 20 to 50 mg/L; adjust pH to 6 to 9. *Rapid mix:* 45 s residence time with 1.5 kW/m³ turbine agitation. *Basin:* velocity gradient, G, at the inlet 150 to 200 s⁻¹ reduced to 50 s⁻¹ later in the basin. Allow residence time for $Gt = 10^5$. For choice of mixer, see Section 16.11.7.3.

16.11.9.4 Size Enlargement: Solids: Spray Drying

A related topic is dryer (Section 16.11.5.5). For size of spray, see Section 16.11.8.2.

Area of Application
Product diameter: 0.15 to 2 mm.

16.11.9.5 Size Enlargement: Solids: Fluidization

Related topics are gas-solid mixing (Section 16.11.7.1) and reactors (Section 16.11.6.27).

Area of Application

Product diameter: 0.6 to 2.5 mm; batch process. For spouted bed with feed diameter >1 mm, gives product diameter of 3 to 3.5 mm.

Guidelines

Fluidized bed: 30 to 50 min for batch of 200 to 700 kg. Shallow bed, 0.3 to 0.6 m deep; gas velocity 0.1 to 2.5 m/s or 3 to 10 × minimum fluidization velocity. Evaporation rates 0.005 to 1 kg/s·m² cross-sectional area.

16.11.9.6 Size Enlargement: Solids: Spherical Agglomeration

Area of Application

Product diameter: 2 to 3 mm; batch; tensile strength of agglomerate: 10 to 100 kPa.

Guidelines

Power 10 to 40 kW/m³. Mixing time 30 to 300 s.

16.11.9.7 Size Enlargement: Solids: Disc Agglomeration

Area of Application

Product diameter: 10 mm, fertilizer 1.5 to 3.5 mm; ore 10 to 25 mm. Capacity <25 kg/s. Tensile strength of agglomerate 10 to 200 kPa, depending on the binder. Produces more nearly uniform granules than drum (Section 16.11.9.8).

Guidelines

Rotational speed about 50% of critical speed or 30 rpm decreasing to 6 rpm as diameter increases. Disc area = 0.1 to 200 m². L:D = 0.1 to 0.22; angle of inclination with the horizontal 40 to 70°. Power = 7 to 9 kW·s/kg or MJ/Mg.

16.11.9.8 Size Enlargement: Solids: Drum Granulator

Area of Application

Product diameter 2 to 5 mm; capacity 0.001 to 30 kg/s. Tensile strength of the agglomerate: 10 to 200 kPa, depending on the binder.

Guidelines

Drum volume: 5 to 100 m³. L:D = 2 to 4; angle of inclination with the horizontal, 10°. For fertilizer: 5 to 7 kW·s/kg or MJ/Mg. For iron ore: 2 kW·s/kg. 10 to 20 rpm. Product diameter controlled by speed, residence time, and binder.

16.11.9.9 Size Enlargement: Solids: Briquetting

Use of pressure to create agglomerate.

Area of Application

Product diameter 15 to 80 mm; capacity 1 to 30 kg/s. Crushing strength of agglomerate 1 to 10 MPa.

Guidelines

Constant product volume; operating pressure <50 MPa. Power 2 to 50 MJ/Mg.

16.11.9.10 Size Enlargement: Solids: Tabletting

Use of pressure to create agglomerate.

Area of Application

Product diameter 15 to 100 mm; capacity 0.011 to 1.5 kg/s. Crushing strength of agglomerate 1.5 to 10 MPa.

Conceptual Process Design, Process Improvement, and Troubleshooting 1435

Guidelines
Operate either as *constant volume machine:* mechanical tabletter with operating pressure <50 MPa, or *constant mechanical tensile strength (constant pressure) machine:* hydraulic tabletter with operating pressure 150 to 250 MPa. Power: 50–60 MJ/Mg.

16.11.9.11 Size Enlargement: Solids: Pelleting

Use of pressure to create agglomerate.

Area of Application
Product diameter 2 to 30 mm; capacity <6 kg/s. Cylindrical-shaped product; used primarily for foodstuffs. Feed is usually a viscous paste.

Guidelines
Power 18 to 70 MJ/Mg.

16.11.9.12 Size Enlargement: Solids: Sintering/Pelletizing

Use of high temperature to create strong agglomerates. Related sections are reactors (Sections 16.11.6.19 and 16.11.6.20), and kilns (Section 16.11.3.7).

Area of Application
Product diameter 80 to 150 mm; capacity 15 to 300 kg/s; product crush strength >10^4 kPa.

Guidelines
Capacity 0.015 to 0.04 kg/s·m^2 traveling grate; 0.012 kg/s·m^3 volume of rotating kiln.

16.11.9.13 Size Enlargement: Solids: Crystallization

See Section 16.11.4.6.

16.11.9.14 Solids: Modify Size and Shape: Extruders, Pug Mills, and Molding Machines

See related topic dewatering press (Section 16.11.5.13).

Area of Application
Batch: injection molding machine: thermoplastics: commodity resins, polyolefins (LDPE, HDPE, PP), styrenics (PS, PMMA. polycarbonates, ABS, PET), and engineered resins for higher impact strength. *Continuous: extruder:* thermoplastics. *Casting:* PP: fine film: 10 to 50 µm; *cast film:* 100 to 400 µm; *thermoformable sheet:* 200 to 2500 µm. *Film blow:* LDPE, HDPE, PP. *Pug mill:* clay materials for bricks, tiles, and ceramics.

Guidelines
Batch: cycle: fill, cool, unload. *Injection molding machine:* cycle: injection fill, 1/4 cycle or about 3 s; cooling time 3/4 cycle cool such that a release in pressure does not cause distortion, 17 to 30 s; machine open close, 7 s. Feed temperature = heat distortion temperature + 55°C; example temperature 200°C; mold temperature for commodity resins = 25°C. Viscosity 0.1 to 100 kPa·s; injection pressure 100 MPa. Clamping force 38 MN per m^2 of projected area surface part for polyolefins; 25 to 30 MN/m^2 for styrenics. Cooling time: for polyolefins, 1 s/0.1 mm wall thickness; for styrenics, × 1.3 to 1.8 longer. Cooling time also dependent on type of machine; toggle takes longer than hydraulic, which takes longer than electric. *Continuous: extruder:* for processing polymers: 0.4 to 1 kW·s/g or 400 to 1800 kJ/kg. *Casting:* polyproylene: sheet 0.6 m × 50 µm; extrude at 200°C; drum diameter 0.45 m. Film velocity 1 m/s. *Film blow:* LDPE, PP, HDPE: draw

velocity 0.35 m/s; draw ratio 4:1; extrude at 180°C; mass flow 0.2 g/s; thickness 34 µm. *Extruders for dewatering:* 1 to 4 kW·s/g or 1000 to 4000 kJ/kg. Related to rotary press (Section 16.11.5.13). *Extruders for foodstuffs:* rpm 30 to 40; shear 5 to 10 1/s; power 200 kJ/kg or 0.1 to 0.4 kW·s/g with values increasing with shear. *Extruders and cookers:* for foodstuffs: rpm 60 to 500; shear 20 to 180 1/s; power 75 to 500 kJ/kg. *Pug mills* for clays, thick pastes, and fertilizers: 0.004 kW·s/g or 3 to 12 kJ/kg.

16.11.9.15 Solids: Solidify Liquid to Solid: Flakers, Belts, and Prill Towers

Related topics are size decrease (Section 16.11.8.6) and dryers (Section 16.11.5.5), but use refrigeration (instead of steam) to solidify liquid.

Area of Application

Flaker: liquid feed: product: flakes about 1 cm × 1 cm × 1 mm thick; capacity <10 kg/s. *Chilled belt:* liquid feed: product: pastilles, flakes, pellets; capacity usually <10 kg/s per unit. *Prilling tower:* liquid feed, product diameter: spheres 1 to 3 mm. Capacity <5 kg/s per unit.

Guidelines

Flaker: 20 to 300 g/s·m^2. Heat transfer coefficient 350 W/m^2 C; power: 1 to 50 MJ/Mg, depending on the material; lower values for ammonium nitrate, benzoic acid, tetrachlorobenzene, sodium hydroxide; higher power usages for waxes and resins. *Chilled belt:* feeder: heated overflow weir, viscosities <1000 mPa·s, produces flakes 1 to 3 mm thick; overhead double roll, viscosities <10^8 mPa·s, produces flakes; rotoformer to produce pastilles 1 to 10 mm diameter; heated strip former (for brittle products). Heat transfer coefficient 350 W/m^2 C; 20 to 300 g/s·m^2. Power 1 to 50 MJ/Mg, depending on the material. *Prilling tower:* gas velocities less than the terminal velocity of the prill, <1 to 2 m/s; gas-to-solids ratio 10 kg air/kg solids. Assume solid surface temperature = solidification temperature, volumetric heat transfer coefficient 5 W/m^3 C. Height <60 m.

ADDITIONAL READING

For rules of thumb:

Woods, D. R. 2007. *Rules of Thumb in Engineering Practice,* Weinheim: Wiley-VCH.

For short cut methods:

Branan, C. R. 2005. *Rules of Thumb for Chemical Engineers,* 4th ed., Amsterdam: Elsevier.
Couper, J. R. et al. 2005. *Chemical Process Equipment,* 2nd ed., Amsterdam: Elsevier.
Coker, A. K. 2007. *Ludwig's Applied Process Design for Chemical and Petrochemical Plants,* 4th ed., Vol. 1, Amsterdam: Elsevier.
Ludwig, E. E. 1997. *Applied Process Design for Chemical and Petrochemical Plants,* 3rd ed., Vol. 2, Amsterdam: Elsevier.
Ludwig, E. E. 2001. *Applied Process Design for Chemical and Petrochemical Plants,* 3rd ed., Vol. 3, Amsterdam: Elsevier.

For trouble shooting:

Woods, D. R. 2007. *Successful Trouble Shooting,* Weinheim: Wiley-VCH.

17 Chemical Process Safety

Richard W. Prugh

CONTENTS

17.1 Introduction 1438
17.2 Process Safety Management 1438
17.3 Elements of the Process Safety Management Standards 1439
    17.3.1 Management System 1439
    17.3.2 EPA Off-Site Hazard Assessment 1440
        17.3.2.1 Release Source Terms (Gas or Vapor Flow Rates) 1441
        17.3.2.2 Release Source Terms (Liquid Flow Rates) 1442
        17.3.2.3 Release Source Terms (Two-Phase Flow) 1442
        17.3.2.4 Vapor Formation as a Result of Adiabatic Flashing 1443
        17.3.2.5 Pool Formation from Liquid Spills 1444
        17.3.2.6 Evaluation of Toxicity Hazards 1444
        17.3.2.7 Explosion-Hazards Evaluation 1447
        17.3.2.8 Flash-Fire Hazards Evaluation 1449
        17.3.2.9 Pool-Fire Hazards Evaluation 1451
        17.3.2.10 Pressure-Vessel-Burst Hazards Evaluation 1452
        17.3.2.11 BLEVE Hazards Evaluation 1455
        17.3.2.12 Jet-Fire Hazards Evaluation 1456
        17.3.2.13 "Condensed-Phase" Detonation Hazards Evaluation 1456
    17.3.3 Incident Identification/History 1457
    17.3.4 Employee Participation 1458
    17.3.5 Process Safety Information 1458
    17.3.6 Process Hazards Analysis 1458
    17.3.7 Operating Procedures 1460
    17.3.8 Training 1460
    17.3.9 Contractor Safety 1461
    17.3.10 Pre-Startup Safety Review 1461
    17.3.11 Mechanical Integrity (Equipment Maintenance) 1461
    17.3.12 Safe Work Practices; Safety Procedures for "Non-Routine" Work 1461
        17.3.12.1 Specified Safety Procedures (Hot Work) 1461
        17.3.12.2 Opening Process Equipment or Piping 1462
        17.3.12.3 Excavation 1463
        17.3.12.4 Entrance Control 1463
        17.3.12.5 Hot-Tapping 1464
    17.3.13 Management of Change 1464
    17.3.14 Incident Investigation 1465
    17.3.15 Emergency Planning and Response 1465
    17.3.16 Compliance Audits 1465
    17.3.17 Trade Secrets 1466

17.4 Other Aspects of Process Safety Management 1467
Table of Symbols 1468
Conversion Factors 1469
Appendix A 1470
Appendix B 1478
Appendix C 1479
Appendix D 1481

17.1 INTRODUCTION

Chemical engineers are faced with a variety of safety concerns, from both federal and state authorities as well as management at the corporate and local levels. At the federal level, the "Process Safety Management" standard of the Occupational Safety and Health Administration (OSHA) [1] and the "Risk Management Program" rule of the U.S. Environmental Protection Agency (EPA) [2] present comprehensive requirements for the control of processes to prevent releases of hazardous materials. Similarly, several states (notably, New Jersey, Delaware, and California) require chemical plants and other facilities that handle chemicals to protect the public and employees. Further, many companies and insurers have even stricter requirements for process control and for preventing and reporting of process incidents, including those that do not cause injury or property loss.

This chapter presents methods for meeting the requirements of the federal regulations concerning public and employee safety. Additional information is provided for some aspects of process safety that are not included in this legislation but can affect the safety of operations in facilities handling hazardous materials. There are other resources for analysis and control of general industrial hazards [3, 4].

Calculation procedures presented here help to evaluate the consequences of leaks or spills of hazardous material. However, the methodology and equations can be used for other calculations that are frequently encountered in chemical engineering studies.

17.2 PROCESS SAFETY MANAGEMENT

Table 17.1 shows the aspects of process safety for which actions are required by OSHA in Title 29 of the Code of Federal Regulations, Part 1910, Section 119 (29 CFR 1910.119) [1] and by the EPA in Title 40 of the Code of Federal Regulations, Part 68 (40 CFR 68) [2]. This "Chemical Process Safety" section concentrates on the engineering aspects of "Process Safety Information"—on the "consequences of failure of engineering and administrative controls" and the "qualitative evaluation of a range of the possible safety and health effects of failure of controls" requirements of the OSHA and EPA "Process Hazards Analysis" and the "Off-Site Hazard Assessment."

Appendices B and C of the OSHA Process Safety Management standard provide some guidance concerning the elements of the OSHA standard and, therefore, of similar elements of the EPA standard. To assist in compliance with the Offsite Consequence Analysis element of the EPA standard, dispersion-analysis Reference Tables 1 through 22 and calculation-methods Appendices A through E are provided in the accompanying rule [9].

The type and extent of the EPA requirements depend on the placement of a "stationary source" in Programs 1, 2, or 3. Program 1 eligibility is limited to sites having good histories and presenting no off-site hazards. Program 3 applies to sites subject to OSHA requirements and sites that are identified with the following North American Industrial Classification System [NAICS] codes: 32211, 32411, 32511, 325181, 325188, 325192, 325199, 325211, 325311, and 32532 [2]. Thus, program 3 applies to most large chemical-handling plants. Program 2 applies to other sources.

Additional guidance concerning management systems and calculation methods are provided in the following sections. Pertinent references are presented in Appendices C and D and are indicated by brackets [ ] in the text.

Chemical Process Safety 1439

TABLE 17.1
Resources for Evaluation of and Compliance with the Other OSHA and EPA "Administrative" Requirements

Subject	OSHA 1910 [1]	EPA 68 [2]	References [Reference Numbers in Brackets]	
			CCPS Ref. [6] & Page	Lees [7] Page
Management system	N/a	.15	G25: 129, 145, 147, 206	
Off-site hazard assessment	N/a	.20-.33	G 6 (book); G 9 (book)	
Incident identification/history	(e)	.42	G19: 2; 55–64	26/1 27/3, 24; A24/9
Employee participation	(c)	.83	G5: 32; G19: 80; G27: 20	
Process safety information	(d)	.48; .65	G27: 45–65	
Process hazards analysis	(e)	.50; .67	G18 (book); G20: 54; G27: 73	8/54
Operating procedures	(f)	.52; .69	G27: 191	5/15; 20/3
Training	(g)	.54; .71	G8/G10: 105, 203, 288; G20: 113; G22: 285; G27: 203; G29: 89, 178	14/37; 19/6 25/34; 28/7
Contractor safety	(h)	.85	G22: 230; G27, 283; G29: 239	6/15; 21/5;21
Pre-startup safety review	(i)	.77	G10: 324; G29: 103	8/57; 8/85; 19/12
Mechanical integrity	(j)	.65; .73	G10: 149; G20: 73; G22: 187; G27: 123 G29: 95, 203	7/48; 7/64; 19/14; 21/2; 21/42
Hot work (and other) permits	(f); (k) .146; .147	.69; .85	G10: 166; G22: 222, 315 G27: 293; G29: 226	21/7
Management of change	(l)	.75	G 8: 73; G10: 105; G20: 63; G27: 177 G29: 214	21/45
Incident investigation	(m)	.60; .81	G 8: 113; G10: 235; G19: 91 (book); G20: 93; G27: 253; G29: 184	2/2; 26/2; 27/5
Emergency planning and response	(n) .38; .120	.90; .95	G20: 121; G27: 217; G29: 102	20/7; 24/5
Compliance audits	(o)	.58; .79	G10: 301; G20: 6 (book); G27: 245	
Trade secrets	(p)	N/A		

17.3 ELEMENTS OF THE PROCESS SAFETY MANAGEMENT STANDARDS

17.3.1 MANAGEMENT SYSTEM

The OSHA standard indicates the need for a Process Safety Management (PSM) system by the following statements:

1. Employers need to develop the necessary expertise, experiences, judgement, and proactive initiative within their workforce to properly implement and maintain an effective process safety management program [1, Appendix C].
2. Employers may wish to form a safety and health committee of employees and management representatives to help the employer meet the obligations specified by this standard [1, Appendix C].
3. Employers shall consult with employees and their representatives on the conduct and development of the elements of process safety management in this standard [1, (c2)].

The EPA standard is more specific in its requirement for a Process Safety Management system [2, 68.15]:

1. The owner or operator of a stationary source shall develop a management system to oversee the implementation of the risk management program elements.
2. The owner or operator shall assign a qualified person or position that has the overall responsibility for the development, implementation, and integration of the risk management program elements. When responsibilities for implementing individual requirements of this part are assigned to other persons, the names or positions of these people shall be documented, and the lines of authority shall be defined through an organization chart or similar document.

A PSM system has been defined as the "comprehensive sets of policies, procedures, and practices designed to ensure that barriers to episodic incidents are in place, in use, and effective" [6(G20)], and "the set of formal and informal procedures and activities used by a facility to control and direct process safety" [6(G20)].

The PSM system involves planning, organizing, implementing, and controlling activities at the corporate and site-management ("strategic") levels, at the middle ("managerial") level and at the employee ("task") level [6(G8)]. It depends on the inherent and demonstrated hazards of the processes, their complexity, the numbers and potential exposures of employees, and many other factors, and guidance for PSM system design and implementation is available [6(G25); 6(G8); 6(G10)] (Table 17.2).

Each of the above aspects should normally be assigned to one or more persons in the organization, with periodic (e.g., monthly) reports concerning the success in implementation.

17.3.2 EPA Off-Site Hazard Assessment

Important aspects of process safety are minimizing the frequency and size of releases of hazardous materials and minimizing the exposure of persons to the toxicity, fire, and explosion consequences of hazardous-materials releases. To evaluate the potential hazards of such releases, it is necessary to estimate the rates and durations (or quantities) of possible releases, and then evaluate the consequences of such releases. The following sections provide methods for estimating the rates of releases from process containers and piping.

The EPA standard requires that—for the "worst case" evaluations—the total quantity of toxic or flammable material is released from the vessel that contains the greatest amount, over a 10-min period, and a liquified gas is assumed to be released as a gas (that is, 100% "flashing" to gas or vapor). If the largest container contains liquid or a gas maintained as a liquid by refrigeration, it is assumed that a pool is formed instantaneously (1 cm deep), with evaporation of the hazardous

TABLE 17.2
Factors That Should Be Addressed in the PSM System

Elements of the PSM System	Some Components of the PSM System
New-project design review	Process design, siting, and hazards reviews
Legislation, standards/codes, and internal	Design, construction, operation, and maintenance guidance
Process knowledge	Process design, operating conditions, and protective systems
Training	Orientation, initial, and refresher
Human factors	Operator interactions with controls and equipment
Process and equipment integrity	Materials of construction, and maintenance procedures
Process risk management	Hazards identification and reduction of risk
Management of change	Procedures and authorizations
Incident investigation	Injury, loss, and near-miss, and communication
Audits and corrective actions	Self-examination and third-party studies of practices
Continuous improvement	Updating of control and safety system and practices
Accountability at all levels	Continuity of operations and protection against injury/loss

Chemical Process Safety

vapor from the surface of the pool. For evaluation of "alternative" scenarios, the release rates may be calculated as shown in the following sections.

17.3.2.1 Release Source Terms (Gas or Vapor Flow Rates)

The rate of gas or vapor flow through a "hole" in process equipment is a function of (in descending order of importance) the diameter of the hole, the pressure within the equipment, the molecular weight of the gas or vapor, the temperature of the gas or vapor within the equipment, and the specific-heat ratio.

17.3.2.1.1 Hole in Piping or Equipment

The following simplified equation estimates the rate of gas or liquid flow through a hole for "critical-flow" (sonic) conditions (for pressures above about 13 pounds per square inch, gauge [psig]):

$$w_{vapor/gas} = 0.035\ D_{hole}^2\ P_o\ [M/T_o]^{0.5}\ \text{pounds/second} \tag{17.1}$$

where P_o and T_o are the absolute pressure and absolute temperature upstream of the hole (in pounds per square inch, absolute [psia] and °K, respectively), D_{hole} is the diameter of the hole (in inches), and M is the molecular weight of the gas or vapor. If the pressure is less than 13 psig (thus, subsonic flow), the pressure P_o can be replaced by a "virtual" pressure P_o' (to account for zero flow when the container pressure equals the ambient pressure), as given by

$$P_o' = 14.7\ \{[(P_g + 14.7)/14.7]^{2.5} - 1\}^{0.4}\ \text{psia} \tag{17.2}$$

where P_g is the gauge pressure, in psia.

17.3.2.1.2 Guillotine Failure of Piping

The following simplified equation can be used to estimate the rate of gas or vapor flow from the open end of a pipe, for "unchoked" (sub-sonic) conditions [10]:

$$W_{vapor/gas} = 0.15\ D_{hole}^2\ P_o\ [M/T_o]^{0.5}\ [D/L]^{0.5}\ \text{pounds/second} \tag{17.3}$$

where D and L are the pipe diameter and pipe length (in inches and feet, respectively). Because the above equation would give an "infinite" flow for zero pipe length, it should not be used if the pipe length is less than approximately 40 pipe diameters, and an orifice-flow equation should then be used to estimate the flow.

17.3.2.1.3 Time to Empty a Gas-Filled or Vapor-Filled Container

The quantity of gas or vapor in a container is a function of (in descending order of importance) the container volume, including the connected piping and other nonisolated equipment, the pressure, the molecular weight of the gas or vapor, the temperature, and the "compressibility factor" for the gas or vapor. The following equation can be used with sufficient accuracy for hazards evaluations:

$$W_{vapor/gas} = P_o V_T M / Z_v R T_o\ \text{pounds} \tag{17.4}$$

$$T_f = T_o\ (P_f / P_o)^{(k-1)}\ °K \tag{17.5}$$

$$W_{lost} = [V_T M P_o / 19.32 T_o][1/Z_o) - (14.7^{(1/k)}/[Z_f P_o^{(1/k)}])]\ \text{pounds for ideal gases} \tag{17.6}$$

where V_T is the volume of the container, in cubic feet; Z_o and Z_f are the initial and atmospheric-pressure compressibilities of the vapor or gas (typically near 1.0); R is the ideal-gas constant (19.32 psia-cu.ft./lbmol °K); P_f is the final pressure (14.7 psia, for discharge to the atmosphere); and k is the specific-heat ratio (about 1.4 for air)

$$W_{lost} = [V_T M P_o \, 19.32 \, T_o][(14.7/[Z_o P_o])] \text{ pounds for real gases} \tag{17.7}$$

$$t_c = (\ln [P_o t_c/14.7])/(0.035 \, D_H^2 \, [19.32/V_T][T_O/M]^{0.5}) \text{ seconds for ideal gases} \tag{17.8}$$

$$t_c = (\ln [P_O t_c/14.7])/(0.23 \, D_P^{\,2} \, [19.32/V_T][T_O/M]^{0.5} \, [D_P/L]^{0.5}) \text{ for real gases} \tag{17.9}$$

The preceding equations assume that the temperature of the gas or vapor remains constant (isothermal conditions), and this would apply only if the initial temperature was near-ambient and the duration of leakage was very long.

The EPA standard states that the total quantity of gas or vapor is to be divided by 10 min, to obtain a rate of flow that could then serve as input to the dispersion tables.

17.3.2.2 Release Source Terms (Liquid Flow Rates)

The rate of liquid flow through a "hole" in process equipment is a function of (in descending order of importance) the diameter of the hole, the pressure within the equipment (including the "head" of liquid above the hole), the density of the liquid, and the viscosity of the liquid within the equipment.

17.3.2.2.1 Hole in Piping or Equipment

The following simplified equation can be used to estimate the rate of liquid flow through a hole [8, 10]:

$$w_{liquid} = 0.31 \, D_{hole}^{\,2} \, [dP_g]^{0.5} \text{ pounds/second} \tag{17.10}$$

where P_g is the pressure upstream of the hole, and d is the density of the liquid (in psig and pounds/feet^3, respectively).

17.3.2.2.2 Guillotine Failure of Piping

The following equation can be used to estimate the rate of liquid flow from the open end of a pipe:

$$w_{liquid} = 0.31 \, D_{hole}^{\,2} \, [dP_g/(1 + [0.08 \, L/D])]^{0.5} \text{ pounds per second} \tag{17.11}$$

where L and D are the pipe length and diameter (in feet and inches, respectively).

17.3.2.2.3 Time to Empty a Container of Liquid

The time required to empty a tank of liquid is given by [10]

$$t_{empty} = 1.0 \, [D_{tank}/D_{hole}]^2 \, H_{liquid}^{\,0.5} \text{ minutes} \tag{17.12}$$

where D_{tank} and D_{hole} are the tank and hole diameters (in feet and inches, respectively), and H_{liquid} is the initial height of liquid above the hole. The above equation applies only if the pressure in the tank is atmospheric. If other conditions apply, a more complicated procedure should be used [11].

17.3.2.3 Release Source Terms (Two-Phase Flow)

A sudden release of pressure in a container of liquefied gas or a liquid that is at a temperature above the atmospheric-pressure boiling point can result in a rapid conversion of some of the liquid to vapor, with a consequent swelling of the now two-phase fluid. Since this fluid is not totally gas or liquid, the equations that are used to calculate flow rates for these phases do not apply. Similarly, venting of a liquefied gas or a liquid at a temperature above the atmospheric-pressure boiling point

Chemical Process Safety

can result in a conversion of some of the liquid to vapor, with a consequent reduction in flow through a restriction, such as a relief valve, a rupture disk, or piping.

For flow through an orifice, the following equation can be used [10]:

$$W_{two\text{-}phase,hole} = 0.3 D_{hole}^2 \, [d_{liquid}(P_o - P_{vapor})]^{0.5} \text{ pounds/second} \quad (17.13)$$

where d_{liquid} is the liquid density (in pounds/feet^3), P_o is the pressure upstream of the hole, and P_{vapor} is the saturated vapor pressure (both in psia).

For flow through a pipe that is of moderate length (but more than about 4 in. long), the following equation can be used [10, 12]:

$$W_{two\text{-}phase,hole} = 0.08 (D_{hole}^2 P_o M H_{vaporization}) [T_o^{1.5} C_{liquid}^{0.5}] \text{ pounds per second} \quad (17.14)$$

where M is the molecular weight of the liquid and vapor, $H_{vaporization}$ is the heat of vaporization (in BTU/pound), T_o is the upstream temperature (in °K), and C_{liquid} is the heat capacity of the liquid (in BTU/pound/°F).

17.3.2.4 Vapor Formation as a Result of Adiabatic Flashing

If a liquid is released while its temperature is above the atmospheric-pressure boiling point, some of the liquid will adiabatically and instantaneously "flash" to vapor; that is, the energy that is released as the liquid temperature drops to the boiling point is utilized in vaporizing some of the liquid. The fraction of liquid that is flash-vaporized often can be obtained from a thermodynamic diagram (such as a pressure-enthalpy plot [8, 13]) for the material of interest, where values of "x" (if shown) indicate the weight-fractions of vapor ("quality") in the two-phase mixture.

If tabulated thermodynamic data are available, the isenthalpic (irreversible adiabatic) "flashing fraction" can be calculated from [10]

$$f = (H_{liquid,Po} - H_{liquid,14.7})/(H_{vapor,14.7} - H_{liquid,14.7}) \quad (17.15)$$

where H is the enthalpy of the liquid or vapor at the initial pressure P_o or at 14.7 psia.

A similar isentropic (reversible adiabatic) calculation yields a smaller flashing fraction.

If thermodynamic data are not available, the "flashing fraction" can be approximated by [14]

$$f = 1 - e^{-2.63(C_{liquid}/H_{vaporization})(T_c - T_b)(1 - [(T_c - T)/(T_c - T_b)]0.38)} \quad (17.16)$$

where T_o, T_c, and T_b are the initial temperature; the critical temperature and the atmospheric-pressure boiling point, respectively, are in the same dimensions as the liquid heat capacity C_{liquid} and the heat of vaporization $H_{vaporization}$.

The preceding equation is based on an energy balance, with "Watson"-type corrections [8] for (1) the approach to zero of the latent heat of vaporization at the critical temperature and (2) the approach to infinity of the heat capacity at the critical temperature [15].

Some of the escaping liquid is carried with the "flashing" vapor as mist or aerosol, which then vaporizes as it mixes with warmer air and enters the cloud as vapor. It is usually assumed that, if the pressure in the container is the vapor pressure of the liquid, the amount of mist or aerosol is equal to that of the amount of vapor that results from "flashing," except that if the flashing fraction exceeds 33%, the amount of mist or aerosol would be half of the remaining liquid [16], with "rain-out" of the rest of the cold liquid. If, however, the pressure in the vapor space above the liquid is substantially above the vapor pressure, discharge of the contents may result in entrainment of all of the liquid, with no "rain-out" or deposition of liquid [16].

17.3.2.5 Pool Formation from Liquid Spills

The hazards of a spill of a volatile toxic or flammable liquid are primarily dependent on the size of the spill, in terms of its diameter or area, since the rate of evaporation depends primarily on the surface area. If the spill is into a dike or curbed area, the area of the spill that is exposed to evaporation is the area of the confinement. If the spill is onto a surface without confinement, the spill area is related to the viscosity and is limited by the rate of evaporation at the outside edge of the spill.

The area of a confined pool would be governed by the dike area (or curbed area, if the spill can be contained by the curbing). The area of an unconfined spill on an essentially flat surface depends on the average depth of the pool. The following equation gives the approximate pool depth:

$$h_{pool} = 0.3 W_{pool}^{0.5}/d_{liquid} = 0.1\ G_{pool}^{0.5}/d_{liquid}^{0.5} \text{ inches} \tag{17.17}$$

where W_{pool} is the weight of liquid in the pool (in pounds), and G_{pool} is the volume of liquid in the pool (in gallons).

For the purposes of the EPA "worst case" evaluation, the pool is assumed to be 0.39 in. (1 cm) deep [2]. The above equation assumes that no vaporization occurs as the pool is developing.

17.3.2.6 Evaluation of Toxicity Hazards

17.3.2.6.1 Toxicity Hazards Analysis

Of the many routes of toxic-material entry into humans, inhalation is the most likely to occur and is the route specified in current legislation. Toxicity data have been developed for most of the important industrial chemicals [19] and form the basis for "threshold" or "allowable" concentrations to which persons cam be exposed with no significant physiological effect. Evaluation of toxicity hazards is then based on the duration of exposures to concentrations above these "threshold" levels. The concentrations of toxic materials in air are primarily functions of vapor pressure and temperature (for toxic liquids); the rate of gas, vapor, mist, or dust release or generation; and distance from the source.

After the rate of toxic material entry into a cloud has been determined (using the equations in the preceding section), the effects of atmospheric conditions on the dispersion (dilution) of the material—particularly gas or vapor—can be evaluated.

Computer programs are available for calculating (1) gas and vapor concentrations; (2) concentration × time products (doses); (3) cloud lengths, widths, and heights; and (4) travel times, as functions of (in decreasing order of importance) wind speed, atmospheric stability, surface roughness, and averaging time. Evaluation and listing of these programs are outside the scope of this publication, but resources are available [6(G40)].

The EPA has selected conservative models for (1) neutrally buoyant gases and vapors and (2) dense gases and vapors. For compliance with the EPA standard [2], the values of the variables that are shown in Table 17.3 were specified by the EPA and should be used [9].

For evaluation of flash-fire hazards involving dispersion of flammable gases and vapors, an averaging time of 0.1 min was used in the EPA tables. Only "passive" mitigation (such as a dike or a confining building) is to be considered for the "worst-case" evaluation, but "active" mitigation (such as valves, transfers, and operator responses) can be considered for the "alternative" evaluations.

Each of the EPA tables can be represented by one or two "exponential" equations, having the form

$$X = K w_{vapor}^{a}/[MC v_{wind}]^{b} \text{ feet} \tag{17.18}$$

where w_{vapor} is the rate of vapor release to the atmosphere (in pounds per minute); M is the molecular weight of the vapor; C is the concentration of the vapor in air (in mg/L); v_{wind} is the wind speed

Chemical Process Safety

TABLE 17.3
EPA Values for Gases and Vapors

Property	Condition	Value
Toxic endpoint	No irreversible effects	Emergency Response Planning Guide–2 or the EPA Level of Concern: 0.1 IDLH (which is about 0.1 LCLo, or 0.01 LC50), or the TWA/TLV
Gas/vapor density	Buoyant or neutral	M < 28 or (vp)M < 500 mm Hg
	Dense	M > 30 or (vp)M > 500 mm Hg
Wind speed	Reference height	10 me
Stability	"F" stable atmosphere	1.5 me/s (3.4 MPH)
	"D" neutral atmosphere	3.0 m/s (6.7 MPH)
Surface roughness	Urban environment	Urban: 1 m; Rural: 3 cm
Averaging time	Short-duration releases	Short: up to 10 minutes; Long: 30 minutes or more
Environment	Ambient	25°C (77°F); 50% relative humidity

(in meters/second), and K, a, and b are functions of the vapor density, atmospheric stability, "surface roughness" (urban, or rural), and the duration of the release. The EPA guidance document provides tables for 10-min releases and for 60-min releases (for release durations that exceed 10 min).

Each of the EPA tables show a "break" in the exponential relationships, and it is necessary to solve each of the pair of equations to determine the distance that could be attained by "toxic endpoint" concentrations. If the distance exceeds the "break point," the "long-distance" equation would apply.

Table 17.4 provides values of K, a, and b for the equations for neutral and dense gases and vapors, rural and urban environments, short-duration releases (10 min), long-duration releases (30 min), and short distances and great distances. Values of EPA toxic endpoints for the EPA listed materials are presented in Appendix A. Also shown are values of toxic endpoints for OSHA-listed materials, based on (in order) ERPG-2 values, 1/10 of the IDLH, or 1/100 of the 4-hour LC50.

TABLE 17.4
EPA Values of K, a, and b for Gases and Vapors

Gas/Vapor Density	Atmos. Stability	Release Duration (min.)	Ref. Table	Short Distances			Long Distances			Break (feet)
				K	a	b	K	A	b	
Neutral	"F" stable; rural	10	1	455,000	1.0	0.48	800,000	1.0	0.60	55,000
		60	2	500,000	1.0	0.50	4,500,000	1.0	0.76	7500
	"F" stable; urban	10	3	200,000	1.0	0.46	220,000	1.0	0.49	40,000
		60	4	255,000	1.0	0.50	700,000	1.0	0.63	4500
Dense	"F" stable; rural	10	5	670,000	0.50	0.53	370,000	0.43	0.45	$6800/w_v^{0.5}$
		60	6	3,600,000	0.63	0.66	930,000	0.48	0.49	$22,000/w_v^{0.2}$
	"F" stable; urban	10	7	720,000	0.50	0.58	540,000	0.47	0.54	$35,000/wv$
		60	8	3,000,000	0.64	0.69	1,000,000	0.51	0.55	$30,000/w_v^{0.4}$
Neutral	"D" neutral; rural	10	14	340,000	1.0	0.55	210,000	1.0	0.46	15,000
		60	15	185,000	1.0	0.49	770,000	1.0	0.69	6000
	"D" neutral; urban	10	16	97,000	1.0	0.46	97,000	1.0	0.46	None
		60	17	70,000	1.0	0.44	235,000	1.0	0.65	5000
Dense	"D" neutral; rural	10	18	350,000	1.0	0.55	300,000	0.53	0.52	$20,000/w_v^{0.4}$
		60	19	500,000	0.61	0.59	500,000	0.61	0.59	None
	"D" neutral; urban	10	20	460,000	1.0	0.60	240,000	1.0	0.53	$2600/w_v^{0.1}$
		60	21	510,000	0.61	0.62	510,000	0.61	0.62	None

The equations in Table 17.4 thus can be used in "spreadsheet" calculations, with "logic" statements to determine which of the short-distance or long-distance equations applies, to interpolate between values in the EPA Reference Tables [9].

In the above equations, w_v is the rate of vapor release (as gas or vapor, or as vapor from the surface of a pool) in pounds per second, M is the molecular weight of the gas or vapor, C is the concentration of interest in parts per million by volume, and v_w is the wind speed in feet/second.

17.3.2.6.2 Consequence Analysis for Toxicity Hazards

The EPA standard requires that the consequences of a toxic-material release be described in terms of

1. The population within a circle with its center at the point of the release and a radius determined by the distance to the endpoint as determined from the dispersion analysis (release rate, gas or vapor properties, toxic endpoint, and atmospheric conditions). The population is to be estimated to two significant digits. U.S. Census data are available on CD-ROM disks individually, by section of the country, or as a set of 11 from the U.S. Bureau of the Census [301-457-4100].
2. The presence of institutions (such as hospitals, nursing homes, retirement centers, schools, day-care centers, and prisons); parks and recreational areas (such as stadiums and swimming pools); and major commercial, office, and industrial buildings (such as shopping malls and industrial parks) within the circle. Some schools and hospitals are shown on U.S. Geological Survey maps, and the locations of other concentrations of population generally may be found on local street maps. The EPA standard does not require a determination of the population at these locations.
3. The presence of environmental receptors (such as national or state parks, forests, or monuments, wildlife sanctuaries, preserves, refuges, or areas, and Federal wilderness areas) within the circle and as identified on U.S. Geological Survey maps.

The current EPA standard does not require an estimation of the number of persons that might be injured, the types or severity of injuries, the areas of environmental receptors that might be damaged, or the types or severity of damage that might result from a release of hazardous material.

17.3.2.6.3 Risk-Mitigation Methods for Toxic Materials

Mitigation efforts can take several forms [6(G4), 6(G24), 20]. However, the EPA standard allows consideration only of "passive" mitigation systems (that function without human, mechanical, or other energy input) when evaluating "worst case" scenarios. For "alternative" scenarios, the proper functioning of both "passive" and "active" mitigation devices or equipment can be assumed. Typical "post-release" mitigation methods for the protection of personal are as follows, in an approximate decreasing order of effectiveness in each category:

1. Lower "source strength" (reduction in toxic vapor/gas concentration)
    a. "Passive" methods for liquids [2]
        (1) Reducing both the evaporation rate and out-of-doors concentration (by design, so that a spill would be within a confining building)
        (2) Minimizing the surface area of a spill pool (by confining the spill in a diked area)
    b. "Passive" methods for gases or vapors
        (1) Reducing the out-of-doors concentration (by design, so that a release would be within a confirming building)
    c. "Active" methods for liquids [9]
        (1) Dilution of spilled liquid, if miscible
        (2) Neutralization/detoxification of spilled liquid
        (3) Spill-area-limitation devices
    d. "Active" methods for gases or vapors [9]

(1) Water or chemical sprays
(2) Dispersion with air curtains, fans, etc.
2. Lower "source duration" (reduction in duration of release)
   a. "Passive" methods for gases, vapors, or liquids
      (1) Excess-flow valves, to limit the duration of a high-rate leak
      (2) Check valves, for protection against back-flow toward a leak
   b. "Active" methods for gases, vapors, or liquids
      (1) Leak-sensor-actuated block or shutoff valves
      (2) Leak-sensor-actuated de-inventory valves
      (3) Operator-controlled remote-operated shutoff valves
      (4) Operator-controlled de-inventory to a scrubber, absorber, adsorber, or condenser
      (5) Applying spill-covering materials to liquid spills: foam; sheeting; floating granular materials; water (if less dense); adsorbent or reactive
      (6) De-inventory to a stack
      (7) Leak-stopping: patching, plugging, freezing
      (8) Deliberate ignition of flammable vapors or gases
3. Reduced off-site effects of a toxic-material release
   a. "Passive" methods
      (1) Walls or nonporous fences (for dense or cold gases and vapors)
   b. "Active" methods
      (1) Off-site alarms (audible, radio, television, local authorities) to "shelter in place" by entering structures, shutting-off air inlets and exhaust systems, closing doors and windows, etc.
      (2) Evacuation of persons where long-duration exposure would be likely
      (3) Providing self-contained breathing protection to potentially exposed persons

Toxic materials ordinarily would not be expected to cause significant structural damage or property loss, unless the toxic material was also flammable (as addressed in the following section) or corrosive (such as HCl or SO_3).

Prerelease methods for minimizing the size of a hazardous-material release (rate and duration, or quantity) and/or the frequency of any such release involve recognition of potential hazards early in the design of a facility and then incorporating appropriate mitigation features in the design [6(G4), 7, 20, 21].

17.3.2.7 Explosion-Hazards Evaluation

The EPA standard lists 46 flammable materials that are gases at 25°C (77°F) and atmospheric pressure, and 17 "Class IA" flammable liquids (that have flash points below 22.8°C/73°F and have boiling points below 37.8°C/100°F). This standard requires an evaluation of at least one "worst case" explosion involving a listed flammable gas, vapor, or liquid and one or more "alternative" explosions. It may be necessary to evaluate several possible explosions to determine which of them would give the greatest distance to the "endpoint" of 1.0 psig overpressure, and this scenario would then be the "worst case" incident.

17.3.2.7.1 Explosion Hazards Analysis (Unconfined Vapor Cloud Explosion)

The distances attained by a "side-on" blast overpressure of 1.0 psig are given in Reference Table 9 of the Risk Management Program Guidance [9]. A 1.0-psig overpressure was chosen as an explosion-hazards endpoint because this pressure can cause window breakage and damage to houses.

The relationships shown in Table 9 can be described by the following equation:

$$X_{1psig} = 3.4[eW_{flammable}H_{combustion}]^{1/3} \text{ feet} \qquad (17.19)$$

where e is the vapor cloud explosion efficiency (0.1, for 10%), $W_{flammable}$ is the weight of flammable vapor (in pounds) in the vapor/air cloud, and $H_{combustion}$ is the heat of combustion (in BTU/pound).

Heats of combustion for the materials that are listed in the EPA guidance are given in Appendix A. Also included in Appendix A are the heats of combustion for several of the materials that are listed in the OSHA standard.

For the "worst case" evaluation, the weight of flammable material in a vapor cloud is assumed to be the total quantity of the substance that could be released from a vessel or pipeline. For liquids, this assumption infers that the liquid is above its atmospheric-pressure boiling point and that 100% flashing to vapor occurs. Also, the entire quantity of vapor is assumed to have concentrations between the lower and upper flammability limits and that the entire quantity explodes, with an energy-conversion efficiency e of 0.10. For mixtures, the weight-average heat of combustion would be used in the above equation.

For "alternative" evaluations, the weight of flammable material in a vapor cloud can be calculated as the gas-release or vapor-release rate multiplied by the estimated time required to stop the release. For liquid releases, the vapor-release rate would be calculated from the liquid-release rate multiplied by twice the flashing fraction (to account for aerosol vaporization) or, for liquids released below the boiling point, as the rate of vaporization from a pool multiplied by the estimated time required to cover or dilute the pool. Also, a lower energy-conversion efficiency (such as 0.03) can be used in the calculation.

The EPA guideline for evaluation of the "far-field" (distant) explosion hazards is based on a "TNT" [TriNitro Toluene] model. Evaluation of the "near-field" (within the vapor cloud, or near the explosion center) requires departure from the blast-pressure and blast-impulse curves for TNT or modification of the distance/quantity relationship [22].

17.3.2.7.2 Consequence Analysis for Explosion Hazards

The EPA standard requires that the consequences of a flammable material release be described in terms similar to those listed for toxicity hazards in Section 17.3.2.6.2 above.

17.3.2.7.3 Risk-Mitigation Methods for Flammable and Explosive Materials

Mitigation of the effects of releases of flammable gases or vapors, or liquids above the boiling point, at high rates is very limited. Further, the EPA standard allows consideration only of "passive" mitigation systems (that function without human, mechanical, or other energy input) when evaluating "worst case" scenarios. For "alternative" scenarios, the proper functioning of both "passive" and "active" mitigation devices or equipment can be assumed.

Typical "post-release" mitigation methods are as follows, in an approximate decreasing order of effectiveness in each category:

1. Lower "source strength" (reduction in vapor concentration)
    a. "Passive" methods
        [None would apply to releases of flammable gases, vapors, or liquids]
    b. "Active" methods [6(G4)]
        (1) Dispersion with existing fans or air curtains, etc.
        (2) Water or chemical sprays
2. Lower "source duration" (reduction in vapor-cloud size)
    a. "Passive" methods
        (1) Excess-flow valves, to limit the duration of a high-rate leak
        (2) Check valves, for protection against back-flow toward a leak
    b. "Active" methods
        (1) Leak-sensor-actuated block or shutoff valves
        (2) Leak-sensor-actuated de-inventory valves
        (3) Operator-controlled remote-operated shutoff valves
        (4) Operator-controlled de-inventory to a scrubber, absorber, adsorber, or condenser
        (5) De-inventory to a stack, for high-elevation release

3. Reduced off-site effects of release
   a. "Passive" methods
      [None would apply to releases of flammable gases, vapors, or liquids]
   b. "Active" methods
      (1) Minimize the probability of ignition by shutting off all ignition sources: welding; vehicles; sparking electrical tools and other "non-explosionproof" electrical equipment; flares; incinerators and boilers (unless protected with flame arresters in the air inlets), actuated by a site-wide alarm
      (2) Off-site alarms (audible; radio; television; local authorities) to "shelter in place" by taking refuge in interior rooms or basements, closing drapes, curtains, and shades over windows, to minimize injury from flying glass [12]

Pre-release methods for minimizing the size of a hazardous-material release (rate and duration, or quantity) and/or the frequency of any such release involve recognition of potential hazards early in the design of a facility and then incorporating appropriate mitigation features in the design [6(G4), 7, 20, 21].

17.3.2.8 Flash-Fire Hazards Evaluation

The hazards of a flash fire, which involves ignition of a cloud containing a mixture of flammable gas or vapor and air, usually are limited to locations where the concentration of gas or vapor exceeds the lower flammable limit (LFL). Because of the relatively short duration of the "flash," persons or combustible materials outside the LFL boundary of the cloud are subjected to a low "dose" of thermal radiation. Somewhat in contrast, persons within the cloud and wearing combustible clothing may be at great risk of serious injury, because the clothing might be ignited by the flash fire [23].

OSHA has recently extended the Personal Protective Equipment standard [24] and its hazard-assessment requirement [25] to include an assessment of flash-fire and jet-fire hazards, with a requirement to consider the wearing of noncombustible clothing where a flash fire (or a jet fire) might occur.

17.3.2.8.1 Flash-Fire Hazards Analysis

The distances attained by LFL concentrations define the hazardous region around a flammable liquid spill or a flammable gas or vapor release. The distances are given in Reference Tables 18, 19, 20, and 21 of the Risk Management Program Guidance [9].

The relationships shown in these tables can be described by the following equations given in Table 17.5.

The dispersion analyses that yielded the EPA tables were based on an averaging time of 6 s as compared with the 10 min and 60 min used for the toxic gas and vapor dispersion analyses. As with toxic materials, the neutral density equations would be used for listed gases and vapors that have a molecular weight <28 or where the product of vapor pressure and molecular weight is <500 millimeters of mercury (mm Hg). In contrast, the dense vapor equations would be used for listed

TABLE 17.5
EPA Equations for Gas or Vapor Lower Flammable Limits

Gas/Vapor Density	Stability; Environment	CAA Table	Distance Equation (feet)
Neutral	"D" neutral; rural	18	$R_{LFL} = 3,200[w_v/M(LFL)v_w]^{0.58}$
Neutral	"D" neutral; urban	19	$R_{LFL} = 1,800[w_v/M(LFL)v_w]^{0.53}$
Dense	"D" neutral; rural	20	$R_{LFL} = 5,500[w_v^{0.54}/M(LFL)v_w^{0.71}]$
Dense	"D" neutral; urban	21	$R_{LFL} = 19,000[w_v^{0.53}/M(LFL)v_w]$

vapors that have a molecular weight >30 or where the product of vapor pressure and molecular weight >500 mm Hg.

17.3.2.8.2 Consequence Analysis for Flash-Fire Hazards

The EPA standard does not require an offsite consequence analysis for flash fires unless the flammability endpoint (LFL) is outside the property boundary, or unless there are locations within the property boundary to which the public has routine and unrestricted access during or outside business hours.

If either of these situations could exist, the EPA standard recommends that the consequences of a flash fire be evaluated as an "alternative" scenario for a flammable-liquid release and be described in terms similar to those listed under Toxicity Hazards.

17.3.2.8.3 Risk-Mitigation Methods for Flash Fires

Mitigation efforts can take several forms [6(G4), 6(G24), 20]. For "alternative" scenarios, the proper functioning of both "passive" and "active" mitigation devices or equipment can be assumed.

Typical "post-release" mitigation methods are as follows, in an approximate decreasing order of effectiveness in each category:

1. Lower "source strength" (reduction in concentration)
   a. "Passive" methods for liquids [2]
      (1) Minimizing the surface area of a spill pool (by confining the spill in a diked area)
      (2) Designing so that spills would occur in a building, which would have "explosion-proof" electrical equipment and no other ignition sources
   b. "Passive" methods for gases or vapors
      [None would apply to releases of flammable gases or vapors]
   c. "Active" methods for liquids [9]
      (1) Dilution of spilled liquid, if miscible
      (2) Spill-area-limitation devices
      (3) Ventilating a building in which a spill had occurred [The concentration above a spill would decrease approximately as the 0.22 power ($v^{0.78}/v$) of the air velocity over the spill]
   d. "Active" methods for gases or vapors
      (1) Water or chemical sprays
      (2) Dispersion with air curtains, fans, etc.
2. Lower "source duration" (reduction in duration of release)
   a. "Passive" methods for gases, vapors, or liquids
      (1) Excess-flow valves, to limit the duration of a high-rate leak
      (2) Check valves, for protection against back-flow toward a leak
   b. "Active" methods for gases, vapors, or liquids
      (1) Leak-sensor-actuated block or shutoff valves
      (2) Leak-sensor-actuated de-inventory valves
      (3) Operator-controlled remote-operated shutoff valves
      (4) Operator-controlled de-inventory to a scrubber, absorber, adsorber, or condenser
      (5) Applying spill-covering materials to liquid spills: foam; sheeting; floating granular materials; water (if less dense); adsorbent or reactive
      (6) De-inventory to a stack for elevation above ignition sources
      (7) Leak-stopping: patching, plugging, freezing
3. Reduced off-site effects of a flammable-material release
   a. "Passive" methods
      [None would apply to releases of flammable gases, vapors, or liquids]
   b. "Active" methods

Chemical Process Safety

(1) Minimize the probability of ignition, by shutting off all ignition sources: welding; vehicles; sparking electrical tools and other "non-explosionproof" electrical equipment; flares; incinerators and boilers (unless protected with flame arresters in the air inlets), actuated by a site-wide alarm
(2) Off-site alarms (audible; radio; television; local authorities) to "shelter in place" by entering structures, shutting off air inlets and exhaust systems, closing doors and windows, taking refuge in interior rooms or basements [12]

Pre-release methods for minimizing the size of a hazardous material release (rate and duration, or quantity) and/or the frequency of any such release involve recognition of potential hazards early in the design of a facility and then incorporating appropriate mitigation features in the design [6(G4), 7, 20, 21].

17.3.2.9 Pool-Fire Hazards Evaluation

The hazards of a fire above a pool of flammable liquid can be described in terms of thermal radiation intensity and duration of exposure. The EPA standard specifies a radiant-heat endpoint of 5 kilowatts per square meter (or 0.12 calorie per second per square centimeter) and a duration of 40 seconds. The thermal "dose" thus corresponds to 20 joules/cm^2 or 4.8 calories/cm^2 and could cause second-degree burns on unexposed skin, but is unlikely to ignite clothing [26, 27].

17.3.2.9.1 Pool-Fire Hazards Analysis

The EPA standard recommends use of the following equation to determine the hazard distance from pool fires [9], converted to English units:

$$X_{pool} = 0.004 H_{combustion} [\varepsilon\tau\, A_{pool}/q_{hazard}/H_{vaporization} + C_{liquid}[T_{boiling} - T_{ambient}])]^{0.5} \text{ feet} \quad (17.20)$$

where $H_{combustion}$ and $H_{vaporization}$ are the heats of combustion and vaporization, respectively (in BTU per pound), q_{hazard} is the hazardous thermal-radiation intensity, A_{pool} is the pool area (in square feet), C_{liquid} is the heat capacity of the liquid in the pool (in BTU/lb/°F), and $T_{boiling}$ and $T_{ambient}$ are the boiling points and ambient temperature, respectively (in °F).

For the EPA evaluations, it is to be assumed that the hazardous intensity of thermal radiation q would be 5 kilowatts per square meter (0.44 BTU/sec/ft^2) and that the duration of exposure would be 40 seconds. Thus, if the exposed person(s) retreated from the pool fire or took refuge in the "shadow" of a structure, the hazard of thermal-radiation exposure could be reduced significantly. It was also assumed that the fraction of heat ε that would be radiated from a pool fire would be 0.4 and that the atmospheric transmittivity τ would be 1.0. Values for ε, τ, and q are available in the literature [26, 27, 28].

17.3.2.9.2 Consequence Analysis for Pool-Fire Hazards

The EPA standard does not require an off-site consequence analysis for pool fires unless the thermal-radiation endpoint (5 kilowatts/m^2) is outside the property boundary, or unless there are locations within the property boundary to which the public has routine and unrestricted access during or outside business hours. If either of these situations could exist, the EPA standard recommends that the consequences of a pool fire be evaluated as an "alternative" scenario for a flammable-liquid release and be described in terms similar to those listed under Toxicity Hazards.

17.3.2.9.3 Risk-Mitigation Methods for Pool Fires

Mitigation efforts can take several forms [6(G4), 6(G24), 20]. For "alternative" scenarios, the proper functioning of both "passive" and "active" mitigation devices or equipment can be assumed.

Typical "post-release" mitigation methods are as follows, in an approximate decreasing order of effectiveness in each category:

1. Lower "source strength" (reduction in concentration)
   a. "Passive" methods for liquids or liquified gases [2]
      (1) Minimizing the surface area of a spill pool (by confining the spill in a diked area)
      (2) Designing so that spills would occur in a building, which would have "explosion-proof electrical equipment and no other ignition sources
   c. "Active" methods for liquids and liquified gases [9]
      (1) Dilution of spilled liquid, if miscible
      (2) Spill-area-limitation devices (personnel wearing flash suits over Level A protective clothing)
      (3) Ventilating a building in which a spill had occurred [The concentration above a spill would decrease approximately as the 0.22 power ($v^{0.78}/v$) of the air velocity over the spill]
      (4) Water or chemical sprays
      (5) Dispersion with air curtains, fans, etc.
2. Lower "source duration" (reduction in duration of release)
   a. "Passive" methods for liquids or liquified gases
      (1) Excess-flow valves, to limit the duration of a high-rate leak
      (2) Check valves, for protection against back-flow toward a leak
   b. "Active" methods for liquids or liquified gases
      (1) Leak-sensor-actuated block or shutoff valves
      (2) Leak-sensor-actuated de-inventory valves
      (3) Operator-controlled remote-operated shutoff valves
      (4) Operator-controlled de-inventory to a scrubber, absorber, adsorber, or condenser
      (5) Applying spill-covering materials to liquid spills: foam; sheeting; floating granular materials; water (if less dense); adsorbent or reactive
      (6) De-inventory to a stack for elevation above ignition sources
      (7) Leak-stopping: patching, plugging, freezing
3. Reduced off-site effects of a flammable-material release
   a. "Passive" methods for pool fires involving flammable liquids or liquefied gases [None would apply to releases of flammable liquids or liquified gases]
   b. "Active" methods for pool fires involving flammable liquids or liquified gases
      (1) Minimize the probability of ignition, by shutting off all ignition sources: welding; vehicles; sparking electrical tools and other "non-explosionproof" electrical equipment; flares; incinerators and boilers (unless protected with flame arresters in the air inlets), actuated by a site-wide alarm
      (2) Off-site alarms (audible; radio; television; local authorities) to "shelter in place" by taking refuge in structures and avoiding exposure to the thermal radiation.

Pre-release methods for minimizing the size of a hazardous-material release (rate and duration, or quantity) and/or the frequency of any such release involve recognition of potential hazards early in the design of a facility and then incorporating appropriate mitigation features in the design [6(G4), 7, 20, 21].

17.3.2.10 Pressure-Vessel-Burst Hazards Evaluation

The explosive bursting of a pressure vessel is not at present identified by the EPA as a likely cause of off-site injury or property damage. However, if a pressure vessel is large and the pressure within the vessel is very high, explosive rupture of the vessel can create blast pressures and impulses at considerable distances from the vessel. The following evaluation is for pressure vessels that contain gas or vapor, or where most of the volume is occupied by gas or vapor; evaluation of the explosive-rupture hazards of pressure vessels containing superheated liquid (resulting in a "BLEVE") is presented in a later section.

17.3.2.10.1 Pressure-Vessel-Burst Hazards Analysis

17.3.2.10.1.1 Non-Flammable Contents

If the gas or vapor that is confined in a pressure vessel is not flammable, the chief hazards of vessel burst would be blast effects and missiles or shrapnel. The TNT [TriNitro Toluene] equivalent of a pressure-vessel burst is a function of the burst pressure of the container, its volume, and the specific-heat ratio of the vapor.

An approximate TNT equivalent for the bursting of a gas-filled or vapor-filled container can be obtained from an isentropic-expansion equation [10]:

$$W_{TNT} = [0.0001 P_{burst} V_{tank}/(k-1)][1 - (14.7/P_{burst})^{(k-1)/k}] \text{ pounds of TNT} \quad (17.21)$$

where V_{tank} and P_{burst} are the volume and bursting pressure (typically, four times the design pressure) of the container or tank, respectively (in cubic feet and psia), and k is the specific-heat ratio.

If a possible cause of vessel bursting is explosion of a mixture of combustible material and an oxidant (such as flammable vapor and air, or combustible dust and air), the maximum TNT equivalent (for a "stoichiometric" mixture) of the explosion can be calculated from

$$W_{TNT} = 0.0005 W_{combustible} H_{combustion} \text{ pounds of TNT} \quad (17.22)$$

The approximate pressure that results from an explosion of a near-stoichiometric mixture of flammable gas or vapor and air can be calculated from

$$P_{max} = 8 P_{initial} + 103 \text{ psig} \quad (17.23)$$

where $P_{initial}$ is the initial pressure of the fuel/air mixture (in psig).

As a general rule, a container will not burst unless the energy equivalent of an internal explosion (expressed as a TNT equivalent) exceeds the energy equivalent of the container bursting (also expressed as a TNT equivalent), and it will not burst unless the explosion pressure (based on the initial pressure) exceeds the burst pressure of the container.

The EPA hazard distance, for a "side-on" or "incident" overpressure of 1 psig, can then be obtained from

$$X_{1psig} = 43 W_{TNT}^{1/3} \text{ feet} \quad (17.24)$$

Approximate values for the "far-field" surface-burst reflected blast pressure and blast impulse [29] can be obtained as functions of TNT equivalent and distance (from pressures of ~20 psig down to zero) from

$$P_{reflected} = [60/Z][1 + (30/Z)] \text{ psig} \quad (17.25)$$

where Z is the "scaled distance" and is equal to $X/W_{TNT}^{1/3}$ (feet/pound of TNT)$^{1/3}$ and

$$I_{reflected} = [125/Z][W_{TNT}^{1/3}][1 + (1.6/Z^{0.5})] \text{ psig/milliseconds} \quad (17.26)$$

The consequences of exposure to blast pressures and impulses are discussed in the next section.

TABLE 17.6
Probability Relationships for Hazard Distances

	Probability of Injury		
Type of Injury	1%	50%	99%
First-degree (sunburn)	[Not determined]	$X_{fireball} = 5.6W_f^{0.49}$	[Not determined]
Second-degree (blisters)	[Not determined]	$X_{fireball} = 5.3W_f^{0.48}$	[Not determined]
Third-degree (fatality)	$X_{fireball} = 5.0W_f^{0.46}$	$X_{fireball} = 3.6W_f^{0.46}$	$X_{fireball} = 2.5W_f^{0.46}$

17.3.2.10.1.2   Flammable Contents

If the gas or vapor that is confined in a pressure vessel is flammable, there may be a fireball hazard in addition to the explosion hazard, particularly if the cause of vessel rupture is fire exposure or impact.

The EPA equation for the second-degree burn hazard distance at which the combination of thermal-radiation intensity and duration of exposure would correspond to 5000 watts/m² and 40 seconds, respectively, is (approximately)

$$X_{fireball} = 0.10 H_{combustion}^{0.5} W_{flammable}^{0.43} \text{ feet} \qquad (17.27)$$

where $H_{combustion}$ and $W_{flammable}$ are the heat of combustion and weight of flammable material, respectively, in BTU per pound and pounds.

The preceding equation was developed for fireballs in which some of the flammable material was in the form of aerosol and in which the mixing of flammable material with air might be significantly slower (thus, "rich" conditions within the fireball) as compared with that of a flammable gas or vapor. The energy effects would be similar, although the cloud burning time might be shorter, and the maximum diameter of the cloud might occur at a lower elevation (as a result of faster mixing of the gas or vapor with air).

Relationships for hazard distances as a function of probability of injury are listed in Table 17.6 [26].

The equations in Table 17.6 were developed for propane (which has a heat of combustion near 12,000 calories/gram). For other materials, the weight of flammable material would be multiplied by the ratio of the heat of combustion (in calories/gram) to 12,000.

17.3.2.10.2   Consequence Analysis for Pressure-Vessel-Burst Hazards

The extent of injury and property damage from an explosion depends on both the blast overpressure and the blast impulse at the point of interest [30]. Table 17.7 presents examples of overpressure and impulse combinations as a function of distance for an explosion having an energy equivalent of 10,000 pounds of TNT, together with the approximate limits of various types of injury and property damage.

The EPA standard requires that the consequences of a flammable material release be described in terms similar to those listed under Toxicity Hazards.

17.3.2.10.3   Risk-Mitigation Methods for Pressure-Vessel-Burst Hazards

There are no "passive" or "active" methods for mitigating the blast or missile effects of an explosion after an explosion has taken place, other than the self-preservation actions of persons turning and running away if there is any visual (speed of light) warning. For this reason, all practical measures should be taken to prevent explosions in the design and operation of high-pressure and exothermic processes; thus, "pre-release" mitigation. This could include barricading, both to reduce the pressure effects outside the barricade and to limit the number, size, and range of missiles [6(G4), 7, 20, 21].

TABLE 17.7
Examples of Overpressure and Impulse Combinations

Distance (feet)	Value of "Z" (feet/$lb_{TNT}^{1/3}$)	Reflected Properties		Personal Injury	Property Damage
		Pressure (psig)	Impulse (psig-ms)		
115	5.3	76	4,000	100% Fatal	Reinf. conc. demolished
135	6.3	27 (s.o.)	265 (s.o.)	99% Fatal	Reinf. conc. unusable
210	9.7	10 (s.o.)	175 (s.o.)	1% Fatal	Panel on steel demolished
310	14.5	13	975	50% Eardrum rupture	Residences demolished; reinf. masonry usable
520	24	5.6	475	10% Eardrum rupture	Residences unusable; tall columns toppled
750	35	1.2 (s.o.)	55 (s.o.)	1% Eardrum rupture	Masonry block demolished; tall columns undamaged
930	43	2.1 1.0 (s.o.)	190 50 (s.o.)	Flying glass	Masonry block repairable; panel on steel usable
2000	94	0.8	75	Flying glass	Residences usable
3000	142	0.5	45	Flying glass	50% Windows broken
3400	155	0.4	40	None	1% Windows broken

s.o. indicates "side-on" blast pressures and impulses as contrasted with "reflected" values.

17.3.2.11 BLEVE Hazards Evaluation

A boiling-liquid expanding-vapor explosion (BLEVE) can occur if a container of liquid or liquified gas at temperatures above their atmospheric-pressure boiling points were to rupture. Sudden loss of containment and reduction in pressure can result in explosive vaporization of some of the liquid, and the sudden increase in volume can propel parts of the container to great distances and create blast effects (pressure and impulse). Further, if the superheated liquid or liquified gas is flammable, a fireball involving the ejected vapor, aerosol, and liquid can result, particularly if the cause of the overpressure and rupture is exposure to fire.

17.3.2.11.1 BLEVE Hazards Analysis

17.3.2.11.1.1 BLEVE Involving Nonflammable Contents
If the superheated material is not flammable, the hazards of a BLEVE would be blast effects and missiles or shrapnel. The TNT equivalent of a BLEVE is a function of the burst pressure of the container, its volume, and the specific-heat ratio of the vapor.

An approximate TNT equivalent for a liquid-full container can be obtained from [14]

$$W_{TNT} = \{0.0019 f V_{tank} d_{liquid} T_{burst}/[M(k-1)]\}\{1 - (14.7/P_{burst})^{(k-1)/k}\} \text{ pounds TNT} \quad (17.28)$$

where d_{liquid}, M, T_{burst}, and f are the density of the liquid (in pounds/feet³), the molecular weight of the material, the temperature of the liquid (in °K) at the burst pressure, and the flashing fraction, respectively. The resulting blast pressures, impulses, and effects can then be determined as described in previous sections.

17.3.2.11.1.2 BLEVE Involving Flammable Contents (Fireball)
If the liquid or liquefied gas that is confined in a pressure vessel is flammable, there may be a fireball hazard in addition to the explosion hazard. Ignition of the ejected contents is almost a certainty if the cause of the BLEVE is fire exposure, and ignition is likely if the BLEVE cause is

impact or a runaway reaction. Ignition simultaneous with release of flammable contents is less likely if the cause of container rupture is a liquid-full condition and solar or ambient heating.

The consequences of exposure to a fireball resulting from a BLEVE can be determined as described in the previous section.

17.3.2.11.2 Risk-Mitigation Methods for BLEVEs

There are no "passive" or "active" methods for mitigating the blast or missile effects of a BLEVE-type explosion after the BLEVE has taken place, other than the self-preservation actions of persons turning and running away if there is any visual (speed of light) warning. However, the primary cause of BLEVE is fire exposure, and several precautions can be taken during fire-fighting to minimize the risk of injury from BLEVE:

1. Provide water-spray cooling of fire-exposed containers either by remote-operated "monitors" or "water cannons" or by mobile "water cannons" placed near the fire-exposed containers "early" in the fire, with fire-lighters then retreating to a safe distance. The time available to set up and aim mobile "water cannons" would depend on the size of the container, its design pressure, the size of relieving devices, and the volatility of the contents.
2. Ensure that there are no fire-fighters or spectators in the "axial" directions for horizontal cylindrical containers, and evacuate buildings in the axial directions, since a BLEVE frequently projects parts to great distances (up to a quarter mile) in an axial direction.
3. Ensure that there are no fire-fighters or spectators within a radius of about 1000 feet (for a 55-ton tank car) to minimize exposure to a BLEVE fireball.

Because there are no reasonable methods for mitigating risk following the occurrence of a BLEVE, all practical measures should be taken to prevent the causes of a BLEVE. This could include fixed water-spray systems over process vessels containing flammable liquids or which could be exposed to flammable-liquid fire, ensuring that relief devices do not direct vented flammable vapors onto the container, systems for diverting spilled flammable liquid to locations that would not expose process vessels, and public-address systems to warn persons to evacuate locations where loss of process control could lead to a BLEVE [14].

17.3.2.12 Jet-Fire Hazards Evaluation

It is unlikely that a release of flammable fluid could result in a jet fire that would cause an off-site thermal-radiation hazard, and the EPA guidelines [9] do not provide guidance concerning jet-fire hazards assessment. However, the length of a jet flame can be calculated from [7]

$$L_{jet} = 0.37[W_{flammable}H_{combustion}]^{0.46} \text{ feet} \tag{17.29}$$

where $W_{flammable}$ is the rate of release of the flammable material (in pounds/second).

The hazard radius is usually assumed to be twice the jet-flame length [12].

There are no "passive" methods for mitigating the risk of exposure to a jet fire other than the self-preservation action of turning and running away from the fire or taking refuge in the "shadow" of intervening equipment or a structure. "Active" methods would include stopping the flow of flammable fluid to the leaking equipment and de-inventorying the equipment to another container or to a stack or flare.

17.3.2.13 "Condensed-Phase" Detonation Hazards Evaluation

Detonation of a self-reactive material or mixture of materials results in blast effects that can be related to an energy-equivalent weight of TNT, for which there are good blast-effects data. The heat of detonation (and, thus, the TNT equivalent) is related to the oxygen balance, being a maximum when the oxygen balance is near zero (except for strongly "endothermic" compounds, such as acetylene, ethylene, and lead azide).

Oxygen balance (OB) can be calculated from the molecular structure (or the relationships between combustible materials and oxidants, in a mixture) by [31]

$$OB = -100\{8(4[C] + [H] - 2[O])/M\} \text{ percentage} \quad (17.30)$$

An approximate relationship between oxygen balance and heat of detonation is

$$H_{detonation} = 1550\{1 - [1 - e^{K(OB^2)}]^{0.5}\} \quad (17.31)$$

where K is the slope of the line on probability graph paper, and is ~0.00003 for negative OB and ~0.0007 for positive OB.

Thus, an approximate value for the TNT equivalent of a detonation-type explosion can be obtained from chemical structure (or, in many cases, from [29, 31]) for assessing blast effects. Similarly, the TNT equivalent for explosion of a mixture of a combustible material in air (vapor-cloud explosion) can be obtained from the heat of combustion (when multiplied by an "explosion-efficiency" factor, which may be of the order of 10%).

The OSHA standard includes the manufacture, storage, transportation, and use of explosives, blasting agents, and pyrotechnics, with no threshold or minimum quantity. The EPA standard includes all Division 1.1 explosives, as listed and defined in the U.S. DOT regulations [36]. The threshold quantity is set at 5000 pounds, based on the potential to detonate and yield a blast wave overpressure of 3 psi (gauge) at a distance of 100 meters (~328 feet) [5]. It should be noted that many explosives are listed in 49 CFR 172.101 as "forbidden" rather than "1.1," and these materials include "Type A" peroxides and particularly ketone peroxides [36].

Fragments frequently are projected to great distances by explosions. The hazards to personnel and property depend on the number of fragments or missiles, and their weight and velocity [8]. There are no "passive" or "active" methods for "post-release" mitigation of the blast or missile effects of an explosion (after an explosion has taken place), other than the self-preservation actions of persons turning and running away if there is any visual (speed of light) warning. For this reason, all practical measures should be taken to prevent explosion in the design and operation of explosives-handling operations, particularly to avoid friction, impact, flame, spark, or other thermal or mechanical sources of initiation. "Pre-release" mitigation measures also would include barricading, both to reduce the pressure effects outside the barricade and to limit the number, size, and range of missiles [6(G4), 7, 20, 21].

17.3.3 INCIDENT IDENTIFICATION/HISTORY

The OSHA standard requires a study of previous incidents involving listed chemicals, to determine which had "a likely potential for catastrophic consequences in the workplace" [1(e)]. The EPA standard further requires a listing of incidents that had occurred during the previous 5 years, with details concerning the incident and, particularly, the on-site and off-site impacts (if any) of the incidents [2(42)]. This is particularly important in establishing "Program 1" eligibility, which requires that the process not have had an incident involving a regulated substance that led to off-site injury or restoration of the environment. Also, the EPA standard requires a statement concerning any operational or process changes that resulted from investigation of the incident.

During analysis of a process for hazards, it may be helpful to review the descriptions of incidents that have occurred at other locations worldwide and in other industries. There are several compilations of incidents that may be useful [7, 37–39].

17.3.4 Employee Participation

The OSHA regulations require employee participation in several aspects of process safety [l(c)]:

1. Specifically, the conduct and development of Process Hazard Analyses
2. Generally, the development of the other elements of the Process Safety Management Standard, which would include
   a. Compilation of Process Safety Information, particularly concerning the evaluation of the consequences of deviations from safe upper and lower limits of operating conditions
   b. Development and review of Operating Procedures
   c. Development of the Training program and, specifically, the frequency of refresher training
   d. Development of the Contractor Safety program, particularly concerning the evaluation of the performance of contractors in training employees to safely perform their jobs, in following the safety rules of the facility, and unique hazards presented by contract employer's work
   e. Participation in Pre-Startup Safety Reviews
   f. Development of Mechanical Integrity (maintenance) procedures, including inspections and tests and the frequencies of such inspections and tests
   g. Development of safe work practices, including lockout/tagout, confined space entry, opening process equipment, facility-entrance procedures, and hot-work procedures
   h. Development of the Management of Change procedure
   i. Participation in Incident Investigations
   j. Development of Emergency Plans
   k. Participation in Compliance Audits

A written plan of action concerning Employee Participation is required. Also, all of the information and the results of analyses, investigations, and audits are to be made available to employees and their representatives.

17.3.5 Process Safety Information

The OSHA and EPA require a compilation of Process Safety Information concerning the regulated chemicals and the processes in which they are employed [1(d), 2(65)]. Although Material Safety Data Sheets (MSDS) may serve as resources for this information, it should be recognized that some of the quantitative data that may be important to hazards analysis are not at present required by the pertinent standard [40]. In particular, quantitative thermal-stability (self-reactivity), incompatibility, and corrosivity data are frequently not found in the MSDS publications shipped with chemicals; further, some materials that are considered to be "stable" by manufacturers and distributors are not stable (can decompose, polymerize, form peroxides, auto-oxidize, or auto-ignite) at the temperatures that are frequently employed or accidentally occur in chemical processes [33]. Under the "general duty" clause [41], the employer would be responsible for determining the degree of stability or regions of instability of chemicals used in processes.

17.3.6 Process Hazards Analysis

A Process Hazards Analysis (PHA) is described as [1(C)]:

1. "An organized and systematic: effort to identify and analyze the significance of potential hazards associated with the processing or handling of highly hazardous chemicals."

2. "Directed toward analyzing potential causes and consequences of fires, explosions, releases of toxic or flammable chemicals, and major spills of hazardous chemicals."
3. "Focused on equipment, instrumentation, utilities, human actions (routine and nonroutine), and external factors that might impact the process."
4. "Provides information which will assist employers and employees in making decisions for improving safety and reducing the consequences of unwanted or unplanned releases of hazardous chemicals."

Several types of PHAs are suggested by the OSHA and EPA standards (Table 17.8). They include the following (in approximate order of increasing complexity), with pertinent references indicated for details concerning their application [5, 42].

TABLE 17.8
OSHA and EPA Standards for Process Hazards Analysis

Method	Description	References
Checklist	The PHA team is guided by a checklist previously prepared by an experienced, knowledgeable person to identify potential causes of incidents and to determine the adequacy of existing incident-prevention systems.	CCPS [6(G18)]: 48, 54, 77, 99, 204, 318; Lees [7]: 8/11
"What if..?" checklist	The PHA team formulates "What if.. ?" questions, as guided by a checklist of process-equipment components and possible causes of component failure, such as inadequate design, construction, or maintenance, or unsafe operation, to determine the adequacy of existing incident-prevention systems.	CCPS [6(G18)]: 62, 77, 122, 204, 383
"What if..?"	The PHA team formulates "What if.. ?" questions, as guided by their cumulative experience with a process and the existing design, construction, maintenance, and operating practices at the site, and external factors, to determine the adequacy of existing incident-prevention systems.	CCPS [6(G18)]: 48, 60, 77, 117, 204, 253, 257 Lees [7]: 8/57
Failure modes and effects	Each component in a process is listed, together with possible modes of failure and the effects of such failures on the process, to determine the adequacy of existing incident-prevention systems.	CCPS [6(G18)]: 66, 77, 151, 204, 377 Lees [7]: 8/77
Hazard and operability (HAZOP)	At each "node" in a process (vessel or pipe section), possible deviations in process variables (such as temperature) from the design intent are formed by combinations of variables and "guide words" (such as "more" or "high"), to determine the adequacy of existing systems to prevent hazardous deviations.	CCPS [6(G18)]: 48, 64, 77, 131, 204, 289, 330, 351 Lees [7]: 8/59 to 8/77; 8/80
Fault tree	The PHA team constructs a "tree" that has as its starting point an unwanted hazardous event. Causes of this "top event" are identified, together with precursor causes, to develop the "tree" downward to basic causes, for which occurrence frequencies are available. Calculation procedures yield a "top event" frequency that, if not "tolerable," can be reduced by adding safety systems to minimize the effects of component or human failures.	CCPS [6(G18)]: 67, 77, 160, 205, 304 CCPS [6(G6)]: 22, 192 to 211

The employer is obliged to determine a priority order for the analysis of processes involving regulated materials and to use an appropriate method [1(e), 2(67)]. The prioritizing should consider (1) the potential severity of an incident, (2) the number of potentially affected employees, (3) the operating history of the process (such as the frequency of incidents), and (4) the age of the process. The method selection should consider (1) the amount of existing knowledge about the process, (2) type of operation (batch, semi-batch, or continuous), (3) size and complexity of the process, and (4) previous operating experience with the process (length of time, and frequency of hazardous or "near-miss" incidents).

In addition to the Process Hazards Analysis, the following are to be addressed in the PHA report:

1. Previous incidents with potential for catastrophic consequences.
2. Applicable engineering and administrative controls.
3. Setting of the facility.
4. Human factors.
5. A qualitative evaluation of safety and health effects of failure of controls.

Updating and revalidation of Process Hazards Analyses is required at least every 5 years after May 26, 1997, for OSHA and after June 21, 1999, for the EPA.

17.3.7 Operating Procedures

The OSHA and EPA require that written operating procedures be developed and implemented for processes in which regulated chemicals are used. These procedures are to provide guidance for operators for each step of operation, including [1(f), 2(69)]:

1. Initial startup of a process, following construction.
2. Startup following a normal shutdown, a shutdown for maintenance, or an emergency shutdown.
3. Normal operations.
4. Temporary operations, with modified materials, equipment, or conditions.
5. Emergency operations, to regain control of a process.
6. Emergency shutdown, if control of a process is not recoverable.
7. Normal shutdown.

These procedures are to describe the steps to be performed, how the steps are to be performed, the desired values of the operating variables at each step, the limits for process variables, the reasons why the limits are not to be exceeded, steps to be taken to correct deviations from the desired operating conditions, the data to be recorded, the samples to be collected, and the safety and health precautions to be taken. Of particular importance are the conditions under which emergency shutdown is required, and the assignment of shutdown responsibility to qualified operators to ensure that emergency shutdown is executed in a safe and timely manner.

The operating procedures also should include precautions to ensure quality control and inventory control, and descriptions of safety systems and their functions. The operating procedures should be reviewed and corrected to reflect current operating practices or validated at least annually.

17.3.8 Training

Initial and periodic refresher re-training (at least every 3 years) are required by the OSHA and EPA standards. This training is to ensure that each employee involved in operating a process (1) understands and adheres to the current operating procedures of the process (including emergency shutdown); (2) understands the specific safety and health hazards; (3) can perform the safe work practices applicable to the employee's job tasks; and (4) has the required knowledge, skills, and abilities to safely carry out the duties and responsibilities as specified in the operating procedures [1(g), 2(71), 5, 42]. All training should be documented.

17.3.9 Contractor Safety

The safety of contractors who may be exposed to process hazards has become a great concern since the occurrence of very serious incidents at Pasadena, Texas (October 23, 1989), and Channelview, Texas (July 5, 1990). Employers are now required to (1) evaluate the safety performance of contractors; (2) inform contract employers concerning fire, explosion, and toxicity hazards; (3) ascertain that contractor employees understand the site's emergency plan; (4) control the presence of contractors in process areas; and (5) maintain an injury and illness log related to a contractor's work in process areas [1(h), 2(87), 5, 42].

17.3.10 Pre-Startup Safety Review

Prior to startup of a new or significantly modified facility, a pre-startup review is required by the OSHA and EPA standards. This review is to ascertain that (1) construction and equipment are in accordance with design specifications; (2) safety, operating, maintenance, and emergency procedures are in place and are adequate; (3) a process hazards analysis has been performed on new processes, and recommendations have been resolved, or a management-of-change procedure has been performed for modified processes; and (4) training of each employee involved in operating a process has been completed [1(i), 2(77), 5, 42].

17.3.11 Mechanical Integrity (Equipment Maintenance)

Written procedures to ensure the integrity of process equipment (such as storage tanks, piping systems, valves, pumps, pressure vessels, process controls, emergency shutdown systems, and relief and vent systems) are required by OSHA and EPA. This includes inspections and tests following recognized and generally accepted good engineering practices at frequencies consistent with manufacturers' recommendations and good engineering practices. The results of such inspections and tests are to be documented with names, dates, test descriptions, and serial numbers. Also, equipment deficiencies are to be corrected either (1) before further use, or (2) "in a safe and timely manner when necessary means are taken to assure safe operation" in the interim.

Newly installed equipment is to be checked to assure that it is suitable for the intended process application and that the equipment is installed properly. Spare parts and equipment are to be checked to assure that they also are suitable for the intended processes [1(j), 2(73), 5, 42].

17.3.12 Safe Work Practices; Safety Procedures for "Non-Routine" Work

The development and implementation of safety procedures for "non-routine" work are required by OSHA and EPA. These specifically include "hot work" (such as welding), lockout or tagout, line-breaking, and confined-space entry. Other examples of non-routine work for which procedures should be developed would be excavation, electrical hot work (on energized conductors), hot-tapping (on pressurized piping), and, in some instances, personal protective equipment.

17.3.12.1 Specified Safety Procedures

References to pertinent regulations and examples are shown in the Table 17.9.

The following sections provide some guidance concerning the other "non-routine" tasks, for which "open-literature" guidance may not be available.

TABLE 17.9
"Non-Routine" Work Regulations

Nonroutine Work	References	Examples
Hot work (welding, etc.)	[25] 29 CFR 1910.252	Lees [7], p. 21/28
Lockout/tagout	[25] 29 CFR 1910.147	"Hazards in the Workplace" [41], p. 39
Confined-space entry	[25] 29 CFR 1910.146	"Hazards in the Workplace" [41], p. 17
Hot work (electrical)	[25] 29CFR 1910.269(j; 1); 333(a)(2); (c)(ii); 335	"Hazards in the Workplace" [41], p. 97
Personal protective equipment	[25] 29 CFR 1910.132; 120	"Personal Protective Equipment" [41], p. 5

17.3.12.2 Opening Process Equipment or Piping

It is often necessary to open process equipment or piping for cleaning, maintenance, or modification. The "work-permit" procedure should address the following:

1. Establishing the levels and areas of authorizations (operations, maintenance; landlord, supervisor, site manager) required for the work to proceed.
2. Evaluating the hazards of the material(s) that might be in the equipment or piping.
3. Positively identifying the equipment or piping to be opened.
4. Determining whether the equipment or piping (a) can be and has been drained and depressurized, or (b) cannot be or has not been drained or depressurized, including a study of piping to identify undrained low points, unvented high points, and potential syphons.
5. Determining whether the equipment or piping has been flushed, purged, or evacuated.
6. Locking shut and tagging valves, using blind-flanges, or disconnecting sources of hazardous material associated with the equipment or piping to be opened, locking open vent valves, and locking off energy sources such as pumps, compressors, and electrical tracing.
7. Safeguarding against external hazards, such as ignition sources if the piping or equipment could contain flammable material, and sealing off drains or other possible sources of noxious liquids, vapors, or gases.
8. Preparing a safe working position, through proper use of platforms, scaffolds, ladders, safety harness, ventilation, etc.
9. Isolating the area, by barricades or shields, to exclude inadequately protected persons, particularly at lower elevations.
10. Preparing or maintaining exits from the work area, in event of a release of hazardous material.
11. Preparing for accidental releases or spills from the equipment or piping to be opened, by establishing spill control, having fire extinguishers ready, etc. ("contingency planning").
12. Ensuring the presence and testing of safety showers, eyewash fountains, or other washing or flushing facilities.
13. Establishing the types and degree of protection that should be worn for "drained" and "undrained" equipment or piping, considering the flammability, corrosivity, toxicity, thermal, and pressure hazards and quantity of material that might remain in the equipment or piping.
14. Providing for one or more standby "buddy" persons, who is equally protected, at a safe distance from the work area but maintaining visual contact with the worker.
15. Developing and implementing the procedure for gradually loosening bolts or connections, with periodic tests for "residual" pressure, to avoid exposure from unexpected sprays, pressure impacts, or other discharges.

Chemical Process Safety

16. Determining when protective clothing can be removed, after the equipment or piping has been opened.
17. Establishing a "turnover" procedure, which documents flow isolation and electrical lockout locations and protective equipment requirements, if the equipment or piping is to remain open for more than one shift.

17.3.12.3 Excavation

Removal of earth, concrete, asphalt, or other ground cover is often necessary prior to, or as part of, construction activities. The "work-permit" procedure should address the following [43]:

1. The depth that can be excavated without completing a permit, such as 2 inches in concrete and 6 inches in earth.
2. The types of underground hazards or facilities (such as electrical, gas, and telephone utilities; process or steam piping; drains and sewer lines; spilled or buried hazardous materials; etc.) that might be encountered during excavation. Identifying such hazards with the assistance of site maps and plans, utility companies, and knowledgeable employees.
3. De-energizing electrical services in the area to be excavated, particularly those over 50 volts AC or 100 volts DC.
4. De-pressurizing and/or emptying all piping in the area to be excavated.
5. Barricading or isolating the work site, to prevent close approach by unauthorized persons.
6. Placement locations for earth or other materials that are removed from the excavation.
7. Protection provided against cave-in or collapse of trench or excavation walls, using shoring and/or sloping.
8. Protecting the excavation from surface water, and ensuring that water or other liquids do not accumulate in the excavation.
9. Periodically testing the atmosphere in the excavation, to ensure the proper concentration of oxygen and absence of hazardous contaminants, such as hydrogen sulfide, methane, and carbon dioxide, and providing forced fresh-air ventilation.
10. Procedures for working above, near, and around obstructions, wiring, conduit, or piping that is expected in the area to be excavated, including manual/hand excavation (with high-voltage gloves and nonconductive paddles, if wiring is encountered) instead of using powered digging equipment.
11. Procedures for protecting exposed equipment, piping, conduit, and wiring that is exposed during excavation, to avoid damage during subsequent work activities.
12. Procedure for correcting maps or plans if unexpected lines or piping are discovered or are not found at the expected locations, and for adding equipment or piping that is being installed in the excavation.
13. Marking or physically protecting (with creosoted boards, for example) electrical cables and process piping that is laid within an excavation, for protection against future excavations.

17.3.12.4 Entrance Control

Unauthorized or unexpected entry of employees, contractors, employees of service organizations (such as postal and utilities), or other visitors, and the activities of these persons, could cause process upsets, could aggravate the consequences of hazardous incidents, and could interfere with emergency or rescue activities. A procedure should be established to control the entry and document the exit of persons from areas in which hazardous materials are handled.

The entrance-control procedure should address the following:

1. Sign-in and sign-out of visitors, with dates, times, printed name, and signature.
2. A warning statement concerning the hazardous materials that are used in the area to be visited.
3. Protective-equipment requirements (such as eye and head protection).
4. The layout of the area (using, for example, a map of the area).
5. Whether or not an authorized employee is required to escort the visitor at all times in the area.
6. An alarm provided to initiate evacuation, and the desired response to such an alarm.

17.3.12.5 Hot-Tapping

Hot-tapping is the drilling of a hole into, and fitting a branch on, a pipe while it is still on line, by welding or "strapping" a sleeve on the pipe. The "work-permit" procedure for hot-tapping should address the following [7, 44]:

1. Establishing the levels and areas of authorizations (operations, maintenance; landlord, supervisor, site manager) required for the work to proceed.
2. Establishing the pipeline fluid conditions: liquid at temperatures below the atmospheric-pressure boiling point; liquified gas at temperatures above the atmospheric-pressure boiling point; or gas, temperature, and pressure.
3. Establishing the pipe material, the wall thickness, and the condition of the pipe (particularly internal or external corrosion); and the effect of welding on the pressure-resistance strength of the pipe, if welding is to be performed.
4. Establishing the type of fitting to be placed on the piping and the attaching method(s), and whether or not reinforcing pads are required.
5. Positively identifying the piping to be hot-tapped.
6. Preparing a safe working position, through proper use of platforms, scaffolds, ladders, safety harness, ventilation, etc.
7. Isolating the area, by barricades or shields, to exclude inadequately protected persons.
8. Eliminating external hazards, such as ignition sources if the piping contains flammable material, and sealing off drains or other possible sources of noxious liquids, vapors, or gases.
9. Preparing and maintaining access to and exits from the work area, in the event of a release of hazardous material.
10. Preparing for accidental releases or spills from the piping to be hot-tapped, by establishing spill control, having fire extinguishers ready, etc. ("contingency planning").
11. Ensuring the presence of, and testing of, safety showers, eyewash fountains, or other washing or flushing facilities.
12. Establishing the types and degree of protection (eye, face, head, hands, feet, and body) that should be worn for the work, considering for flammability, corrosivity, toxicity, thermal, and pressure hazards.
13. Providing for one or more standby "buddy" persons, who is equally protected, at a safe distance from the work area but maintaining visual contact with the worker.

17.3.13 MANAGEMENT OF CHANGE

The development and implementation of a procedure for the control of changes in process chemicals, technology, equipment, and procedures are required by OSHA and EPA [1(1), 2(75)]. Replacement of a defective component "in kind" (same manufacturer, model, and size, and other characteristics) would not require following the procedure. Replacement of a defective, obsolete, or obsolescent device with a "similar" device would require a "minimal" procedure, with limited authorizations.

Changes in chemicals, operating conditions, process equipment, instrumentation, repairs to damage, and other facets of process technology require the following prior to the change:

1. The technical basis for the proposed change, such as laboratory study, purchased technology, or desired change in production rate.
2. An evaluation of the impact of the proposed change on safety and health of employees and the population in the area surrounding the site, and the environment.
3. Revision of the operating procedures, to correspond to the proposed change in chemicals, operating conditions, process equipment, or technology.
4. The date on which the proposed changes would become effective and, if a temporary change, the date on which the modified process would revert to the original process. If the changes are to be permanent, the Pre-Startup Review procedure should be implemented. Many companies require that the modified process be analyzed for hazards, prior to startup, using a Process Hazards Analysis method that would be consistent with the inherent or demonstrated hazards.

Temporary changes should be authorized by knowledgeable members of the site staff (such as the operations and maintenance managers, the research director or process engineer, and the safety, health, and environmental staff members). Operating, maintenance, and contractor employees who will be affected by the proposed change should be informed of, and trained in, the changed process prior to startup. The operating procedures and safety information should be revised to correspond to the proposed change.

17.3.14 Incident Investigation

Investigation of incidents that resulted in, or could reasonably have resulted in, a catastrophe (a major uncontrolled emission of a highly hazardous chemical, a fire, or an explosion) is required by OSHA and EPA [1(m), 2(81)]. The investigation should be initiated within 48 hours, and the investigation team should include (1) at least one person knowledgeable in the process involved, (2) a contract employee if contractors were involved, and (3) other persons having appropriate experience.

A written report is required. The contents of the report are to be reviewed with all affected personnel, and the report is to be retained for 5 years. The report is to address the following: (1) the date and time of the incident, (2) the date on which the investigation began, (3) a description of the incident, (4) the factors that contributed to the incident, and (5) recommendations resulting from the incident.

The employer is required to establish a system to promptly address and resolve the recommendations and to document the resolutions and corrective actions.

17.3.15 Emergency Planning and Response

The EPA and OSHA require development of an emergency plan and training in the implementation of the plan [1(n), 2(95), 25]. Because the requirements differ, a site's plan should include the requirements of both standards, as shown in Table 17.10.

17.3.16 Compliance Audits

The EPA and OSHA require self-audits every 3 years by at least one person who is knowledgeable in the process(es) that involve regulated chemicals, beginning June 21, 1999, and May 26, 1995, respectively, to determine conformance. The employer shall promptly determine and document appropriate responses to the findings of the audit and document that the deficiencies have been corrected [1(o), 2(58)].

TABLE 17.10
OSHA and EPA Emergency Requirements

Requirement	OSHA Standard	EPA Standard
Procedures and measures for emergency response	Emergency escape procedures (types of evacuation) and routes	Provide to the local emergency planning committee
Procedures for employees who remain to operate critical operation before they evacuate	Required	Not required
Procedures to account for all employees after emergency evacuation	Required	Not required
Rescue and medical duties for those employees who are to perform them	Required	Not required
First-aid and emergency medical treatment of persons exposed to regulated substances	Required [29 CFR 1910.119(f)(1)(iii)(C)]	Documentation required (for off-site treatment)
Preferred means of reporting fires and other emergencies	Required (to other employees)	Required (to public and local emergency response agencies)
Names or titles of persons who can be contacted for further details	Required	Required [40 CFR 68.15(b)]
Establish an alarm system, with distinctive signals for brigade members and other purposes	Required [29 CFR 1910.165]	Not required
Training for evacuation aides	Those employees who are to assist in safe and orderly emergency evacuation	Not required
Training for all employees	Initially and following any changes	Required
Procedures for handling small releases	Required	Not required
Hazardous waste operations and on-site emergency response requirements	Required unless releases are "incidental" and can be absorbed, neutralized, or otherwise controlled at the time of release by employees, and where there is no "reasonable possibility" for employee exposure to safety or health hazards.	Not required
Emergency response equipment	Required: Annual testing of alarm (or every two months if not a "supervised" system)	Required: Inspection, testing, and maintenance of emergency response equipment
Procedures to review and update the emergency response plan	Not specifically required	Required
Ensure that employees are informed of changes to the plan	Required	Required

The EPA requires, in addition, periodic audits by the implementing agency, such as the state or local air permitting agency. The frequency of such audits would be based on (1) accident history, (2) quantities of regulated substances, (3) proximity to public receptors, and (4) hazards identified in the Risk Management Program [2(220)].

17.3.17 TRADE SECRETS

The OSHA standard requires that all pertinent information be provided to employees involved in (1) compiling Process Safety Information, (2) performing Process Hazards Analysis, (3) developing

Chemical Process Safety 1467

Operating Procedures, (4) performing Incident Investigations, (5) preparing the Emergency Response Plan, and (6) performing the Compliance Audit, without regard to possible trade secret status [1(p)]. However, employers can require such employees to enter into confidentiality agreements. Further, employees and their designated representatives shall have access to trade secret information contained within the Process Hazard Analysis and other documents required to be developed under the OSHA standard. There are no similar requirements or stipulations in the EPA standard.

17.4 OTHER ASPECTS OF PROCESS SAFETY MANAGEMENT

Several other pertinent aspects of Process Safety Management are not directly addressed by the OSHA or EPA standards. They are listed in the Table 17.11, with pertinent references.

TABLE 17.11
Other aspects of Process Safety Management

Subject	OSHA Ref. [23]	CCPS [6] Ref. & Page	Lees [7] Page	Others
Carcinogens	1910.1000		18/22	
Decommissioning		G 8:97; G20:79; G22:80;G29:271	21/35	
Dispersion analysis		G 5 (book)	15/70	
Emergency relief systems		G42 (book)	12/60	
Explosion hazards		G9 (book); G26 (book)	17/5	
Fire protection			22/20	
Flame arresters			17/53	
Flammability hazards	1910.106	G 9 (book)	16/13	
Hazard communication	1910.1200			
Hazards of inert gases			16/285	[49]
Hazards of vacuum				[49]
Human error and human factors		G 8: 99; G10: 199; G15 (book); G20: 103; G27: 95; 153; G29: 16	14/4	[50]
Industrial hygiene		G22: 327	18/17; 25/6	
Inherently safer design		G19: 173; G41 (book)	11/16	[45]; [46]
Institutional memory				[48]; [51]
Laboratory safety	1910.1450			[47]
Leadership and accountability		G8: 17; G10: 15; G20: 33; G27: 15; G29: 12	28/3	
Plant siting and layout			10/2	[52]; [53]; [54]
Power-operated hand tools	1910.302			
Protective measures and mitigation		G 4 (book); G13: 159; G24 (book)	9/101	
Public notification				[42]
Quantitative analysis		G18: 22; 67; 160	9/13	
Quantitative consequence analysis		G 8: 59;G17: 5, 23, 28, 34, 38, 49, 62, 69, 79; G27: 107	9/7	
Respiratory protection	1910.134			
Safety manuals		G22: 322; G27: 20		
Scrubbers				
Software security		G27: 319		
Source term analysis (flow)			15/2	
Static electricity			16/85	
Storage of materials		G 3 (book); G30 (book)	22/7; 22/57	
Transportation safety		G28 (book)	23/2	

TABLE OF SYMBOLS

English units, with °K.

A_p	Area of the pool (the area of a dike, or $[\pi/4]\ D_p^2$ for an unconfined pool, in feet^2).
C	Concentration of interest, in parts/million, by volume.
$[C]$	Number of carbon atoms in the molecule.
C_d	Coefficient of discharge, which is a function of Reynold's number (~0.62); dimensionless.
C_L	Heat capacity of the liquid or liquified gas or vapor, in BTU/pound/°F.
C_p	Specific heat of a fluid at constant pressure, in BTU/pound/°F.
D_h	Diameter of the hole, in inches.
D_P	Diameter of the pool, in feet.
D_p	Inside diameter of the pipe, in inches.
D_T	Diameter of the vertical cylindrical tank, in feet.
f	Fanning friction factor; dimensionless.
g_c	Conversion constant (32.2 ft/lb$_m$/lb$_f$/sec^2).
G_P	Volume of the liquid in the pool, in gallons.
$[H]$	Number of hydrogen atoms in the molecule.
H_c	Heat of combustion of the flammable material, in BTU/pound.
$H_{L,Pb}$	Enthalpy of the liquid at the container-burst pressure, in BTU/pound.
$H_{L,14.7}$	Enthalpy of the liquid at atmospheric pressure, in BTU/pound.
$H_{V,14.7}$	Enthalpy of the vapor at atmospheric pressure, in BTU/pound.
H_v	Latent heat of vaporization, in BTU/pound.
K	Slope of a line on probability graph paper.
k	Specific-heat ratio; dimensionless.
L	Length of the pipe from the source container to the open end, in feet.
LFL	Lower flammable limit, in volume-percentage.
M	Molecular weight of the gas or vapor.
N_M	Mach number for gas or vapor flow; dimensionless.
N_{Re}	Reynold's number; dimensionless.
$[O]$	Number of oxygen atoms in the molecule.
OB	Negative or positive oxygen balance.
P_a	Pressure outside the equipment (usually 14.7 psia).
P_b	Bursting pressure of the container, in psia.
P_f	Final pressure (usually 14.7 psia).
P_g	Gauge pressure in the vapor space above the liquid, in psig.
P_o	Initial absolute pressure within the equipment, in psia.
P_o	"Virtual" absolute pressure, for subsonic flow, in psia.
P_s	Saturation vapor pressure of the flashing liquified gas at the temperature of interest, in psia.
q	Thermal-radiation intensity, in BTU/second/feet^2.
R	"Ideal gas law" constant, which is equal to 19.32 psia/feet^3/lb-mole/°K.
r_c	Critical pressure ratio, given by $(2/[k+1])^{(k/[k-1])}$.
RH	Relative humidity, in %.
S	Entropy value (similar to values of enthalpy H), in BTU/lb/°F.
$[S]$	Number of sulfur atoms per molecule.
$[Si]$	Number of silicon atoms per molecule.
t	Time required to empty a container of fluid, in seconds.
T_a	Ambient temperature, in °K (typically about 298°K [25°C/77°F]).
T_b	Atmospheric-pressure boiling point, in °K.
T_c	Critical temperature, in °K.
T_f	Final temperature, in °K.
T_P	Temperature of the liquid in the pool, in °K.
T_o	Initial temperature, in °K.
v	Velocity of liquid or gas through a pipe, in feet/second.
vp	Vapor pressure, in mm Hg.

$V_{L,Pb}$	Specific volume of the liquid at the container-burst pressure, in feet³/pound.
$V_{L,14.7}$	Specific volume of the liquid at atmospheric pressure, in feet³/pound.
$V_{V,14.7}$	Specific volume of the saturated vapor at atmospheric pressure, in feet³/pound.
V_T	Volume of a container, in feet³.
vw	Wind speed over the pool, in feet/second.
W_c	Burning rate, in pounds/second.
W_f	Weight of flammable material in the vapor cloud or vessel, in pounds.
W_L	Liquid flow rate, in pounds/second.
W_P	Weight of the liquid in the pool, in pounds.
W_{TNT}	TNT equivalent of the gas-filled or vapor-filled vessel burst, in pounds.
W_V	Weight of gas or vapor in a container, in pounds.
W_v	Rate of vapor release (as gas or vapor, or as vapor from the surface of a pool), in pounds/second.
W_2	Rate of two-phase release, in pounds/second.
$[X]$	Number of halogen (chlorine, bromine, fluorine, etc.) atoms per molecule.
X	Hazard distance from the center of the source to the receptor, in feet.
Z	"Scaled distance" $R/W_{TNT}^{1/3}$, in feet/pound of $TNT^{1/3}$.
Z_f	Compressibility of the vapor phase at the final temperature and pressure conditions; dimensionless.
Z_L	Compressibility of the liquid phase ($PM/RT\rho_L$), dimensionless.
Z_o	Compressibility of the vapor phase at the initial temperature/pressure conditions; dimensionless.
Z_S	Elevation of the liquid surface above the hole, in feet.
Z_V	Compressibility of the gas or vapor, dimensionless.
ρ_G	Density of the gas or vapor, in pounds/feet³.
ρ_L	Density of the liquid or liquified gas, in pounds/feet³.
ε_p	Roughness of the internal surface of piping, in inches.
ε_B	Fractional energy-conversion efficiency of the flammable-vapor explosion, as compared to TNT.
ε_T	Fractional conversion of energy to thermal radiation.
μ	Viscosity of the gas or vapor, in centipoises.

CONVERSION FACTORS

Symbol	Value	Metric	per	English
A	0.093	Square meters	per	Square foot
C_L; S	1.00	Calories per gram-°C	per	BTU per pound-°F
D; L; X	0.305	Meters	per	Foot
D	2.54	Centimeters	per	Inch
G	0.0038	Cubic meters	per	Gallon
g_c	30.5	Centimeters per second²	per	Foot-lb_{mass}/lb_{force}-second²
H	0.555	Calories per gram	per	BTU per pound
P	6,895	Pascals	per	pound per square inch
q	11.3	Kilowatts per square meter	per	BTU per second per square foot
R	0.848	Kilogram-meters per g-mol-°K	per	psia-cuft per lb-mol-°K
R	0.00424	Liter-atma per gmol-°K.	per	psia-cu.ft. per lb-mol-°K
T	0.55	C° [$T_C = (T_F - 32)/1.8$]	per	F° [$T_F = (1.8 T_C) + 32$]
V	0.0283	Cubic meters	per	Cubic foot
v	0.305	Meters per second	per	Foot per second
vp	133	Pascals	per	Millimeter of mercury [mm Hg]
W; (w)	0.453	Kilograms (per second)	per	Pound (per second)
ρ	16.0	Kilograms per cubic meter	per	Pound per cubic foot
μ	0.001	Pascal-seconds	per	Centipoise (0.000672 lb_m/ft-sec per cp)

APPENDIX A
OSHA AND EPA LISTED CHEMICALS

The EPA and OSHA regulations [A-l, A-2, and A-3] apply to facilities from which a release of hazardous material could occur, at or above the quantities specified, as shown in **bold** in Table 17.A.1. Also shown are the Immediately Dangerous to Life and Health [IDLH] concentrations (for 30-minute exposure) [A-4]; the Lower Flammable Limits and Heats of Combustion for combustible materials; and the EPA toxic, thermal-radiation, and overpressure "endpoints" for public-exposure evaluations. "N/L" indicates that the chemical is not listed in the pertinent document. Estimated values are shown with a superscript (^{e}). Where no inhalation-toxicity data were available, the oral dose that caused 50% fatalities is shown only to indicate qualitatively the systemic toxicity (for example, an oral LD50 of 1000 mg/kg would be considered relatively nontoxic). The data are for pure chemicals, except where otherwise indicated, that is, without added diluents. Additional hazardous-properties information can be obtained via the "Sax No." [A-6].

REFERENCES

A-1. U.S. Department of Labor, Occupational Safety and Health Administration. "Process Safety Management of Highly Hazardous Chemicals." Title 29, Subtitle B, Chapter XVII, Part 1910, Subpart H, Paragraph 119, Code of Federal Regulations [29 CFR 1910.119], vol. 57, no. 36, p. 6403; Federal Register, February 24, 1992.

A-2. U.S. Environmental Protection Agency. "Risk Management Program Rule," Title 40, Part 68, Subpart F, Code of Federal Regulations [40 CFR 68], vol. 61, p. 31717; Federal Register, June 20, 1996.

A-3. U.S. Environmental Protection Agency. "List of Regulated Substances and Threshold Quantities for Accidental Release Prevention." Title 40, Part 68, Subpart C, Paragraph 130, Tables 1–4, Code of Federal Regulations [40 CFR 68], vol. 59, no. 20, p. 4495; Federal Register, January 31,1994; rev. October 1997.

A-4. U.S. Department of Health and Human Services, National Institute for Occupational Safety and Health. "Pocket Guide to Chemical Hazards," June 1994.

A-5. American Conference of Governmental Industrial Hygienists. "Threshold Limit Values for Chemical Substances," 1995.

A-6. Lewis, R.J. *Sax's Dangerous Properties of Industrial Materials*. 9th ed., New York: Van Nostrand Reinhold, 1996.

A-7. U.S. Department of Labor, Occupational Safety and Health Administration. "Hazard Communication." Title 29, Subtitle B, Chapter XVII, Part 1910, Subpart Z, Paragraph 1200(c), Code of Federal Regulations [29 CFR 1910.1200]; Federal Register, July 1, 1995.

A-8. National Fire Protection Association. *Fire Protection Guide to Hazardous Materials*. 11th ed. Quincy, MA: NFPA, 1994.

A-9. U.S. Environmental Protection Agency. "Technical Guidance for Hazards Analysis – Emergency Planning for Extremely Hazardous Substances", December 1987: 2–13, Appendix C, .

A-10. American Industrial Hygiene Association. *Emergency Response Planning Guidelines*. Fairfax, VA: AIHA, 1995.

Chemical Process Safety

TABLE 17.A.1
OSHA and USEPA Listed Chemicals

OSHA Highly Hazardous Chemical or EPA Regulated Substance	"Sax" No.	Mole Wgt.	OSHA Threshold Quantity (pounds)	IDLH Conc. (ppm)	EPA Threshold Quantity (pounds)	Toxic Endpoint (mg/l)	Toxic Endpoint (ppm)	Lower[c] Flammable Limit (v%)	Heat of Comb. (cal/g)
Acetaldehyde	AAG250	44.1	2500	2000 Ca	**10,000** (Flamm.)	0.36	200[h]	4.0	6015
Acetylene	ACI750	26.0	10,000[a]	2500[d]	**10,000** (Flamm.)	0.27	250[h]	2.5	11,575
Acrolein [acrylaldehyde]	ADR000	56.1	150	2	5000	0.0011	0.5	2.8	6940
Acrylonitrile	ADX500	53.1	10,000[b]	85 Ca	20,000	0.076	35	3.0	
Acrylyl chloride	ADZ000	90.5	250	N/L	5000	0.0009	0.25	4.6[e]	
Alkylaluminums	AHD500	—	5000	N/L	N/L	0.002	0.5[j]	Various	Various
Allyl alcohol	AFV500	58.1	10,000[b]	2500[d]	15,000	0.036	14	2.5	7610
Allylamine	AFW000	57.1	1000	N/L	10,000	0.0032	1.4	2.2	
Allyl chloride [3-chloro,1-propene]	AGB250	76.5	1000	250 Ca	N/L	0.12	40[g]	2.9	5500
Ammonia, anhydrous	AMY500	17.0	10,000	300	10,000	0.14	200	15	5400
Ammonia solutions [aqua ammonia]	ANK250	35.1	15,000, >44 wt%	300	20,000 >20 wt%	0.14	200	15	
Ammonium perchlorate	ANP250	117.5	7500	N/L	N/L	Solid (No Data)		Oxid.; expl.	
Ammonium permanganate	PCI750	137.0	7500	N/L	N/L	0.0005	0.08[j]	Oxid.; expl.	
Arsenous trichloride	ARF500	181.3	N/L	1.6 Ca	15,000	0.010	1.3	Not combustible	
Arsine	ARK250	78.0	100	3 Ca	1000	0.0019	0.6	5.1	
Boron trichloride	BMG500	117.2	2500	N/L	5000	0.010	2	Not combustible	
Boron trifluoride	BMG700	67.8	250	25	5000	0.028	10	Not combustible	
Boron trifluoride/methylether	BMH000	113.9	<10,000[b]	N/L	15,000	0.023	4		
Bromine	BMP000	159.8	1500	3	10,000	0.0065	1	Not combust; oxidizer	
Bromine chloride	N/L	115.4	1500	N/L	N/L	No Data		Not combust; oxidizer	
Bromine pentafluoride	BMQ000	174.9	2500	N/L	N/L	0.0007	0.1[j]	Not combust; oxidizer	
Bromine trifluoride	BMQ325	136.9	15,000	N/L	N/L	No Data		Not combust; oxidizer	
Bromotrifluoroethylene	BOJ000	160.9	10,000[a]	N/L	**10,000** (Flamm.)	No Data		Pyro.	470
Butadiene	BOP500	54.1	10,000[a]	2000[d]	**10,000** (Flamm.)	0.11	50[g]	2.0	10,690
Butane	BOR500	58.1	10,000[a]	1900[d]	**10,000** (Flamm.)	0.45	190[h]	1.5	10,970
Butene	BOW250	56.1	10,000[a]	1600[d]	**10,000** (Flamm.)	0.37	160[h]	1.7	10,860
Tertbutyl hydroperoxide	BRM250	90.1	5000	N/L	N/L	0.013	3.5[j]	2.0[e]; expl.	
Tertbutyl perbenzoate	BSC500	194.2	7500	N/L	N/L	Oral: 914 mg/kg		0.9[e]; expl.	
Carbon disulfide	CBV500	76.1	<10,000[b]	500	20,000	0.16	50	1.3	3240
Carbon oxysulfide	CCC000	60.1	10,000[a]	12,000[d]	**10,000** (Flamm.)	2.9	1200[h]	12	2190

Continued

TABLE 17.A.1 (Continued) OSHA and USEPA Listed Chemicals

OSHA Highly Hazardous Chemical or EPA Regulated Substance	"Sax" No.	Mole Wgt.	OSHA Threshold Quantity (pounds)	IDLH Conc. (ppm)	EPA Threshold Quantity (pounds)	Toxic Endpoint (mg/l)	Toxic Endpoint (ppm)	Lower[c] Flammable Limit (v%)	Heat of Comb. (cal/g)
Carbonyl fluoride	CCA500	66.0	2500	N/L	N/L	0.0054	2[j]		
Cellulose nitrate >12.6% n	CCU250	504.3	2500	N/L	N/L	Oral:5000 mg/kg		Solid	
Chlorine	CDV750	70.9	1500	10	2500	0.0087	3	Not combust.; oxidizer	
Chlorine dioxide	CDW450	67.5	1000	5	1000	0.0028	1	10	
Chlorine monoxide	N/L	86.9	10,000[a]	23,500[d]	10,000 (Flamm.)	8.4	2350[h]	23.5	240
Chlorine pentafluoride	CDX250	130.4	1000	N/L	N/L	0.3	57[i]	Not combust.; oxidizer	
Chlorine trifluoride	CDX750	92.4	1000	20	N/L	0.0038	1[g]	Not combust.; oxidizer	
Chloro diethylaluminum	DHI885	110.6	5000	N/L	N/L	0.11	24[j]	Pyro.	
Chloro dinitrobenzene	CGM000	202.6	5000	N/L	N/L	Oral: 780 mg/kg		2.0	
Chloroform	CHJ500	119.4	N/L	500 Ca	20,000	0.49	100	Not combust.	750
Bischloro methylether	BIK000	115.0	100	Ca	1000	0.00025	0.05	5.7[e]	
Chloromethyl methylether	CIO250	80.5	500	Ca	5000	0.0018	0.5	4.6[e]	
Chloropicrin	CKN500	164.4	500	2	N/L	0.0013	0.2[g]	Not combustible	
Chloropicrin/methyl bromide	N/L	Mix	1500	N/L	N/L	Vendor's MSDS		Not combustible	
Chloropicrin/methyl chloride	N/L	Mix	1500	N/L	N/L	Vendor's MSDS			
2-chloro,1-propylene	CKS000	76.5	10,000[a]	4500[d]	10,000 (Flamm.)	1.4	450[h]	4.5	5520
1-chloro,1-propylene	PMR750	76.5	10,000[a]	4500[d]	10,000 (Flamm.)	1.4	450[h]	4.5	5520
Crotonaldehyde	COB250	70.1	10,000[b]	50	20,000	0.029	10	2.1	7730
Cumene hydroperoxide	IOB000	152.2	5000	N/L	N/L	0.012	2[i]	1.0[e]; expl.	
Cyanogen	COO000	52.0	2500	N/L	10,000 (Flamm.)	0.021	10[j]	6.0	5055
Cyanogen chloride	COO750	61.5	500	N/L	10,000	0.03	12	Not combustible	
Cyanuric fluoride	TKK000	135.1	100	N/L	N/L	0.0002	0.03[j]	5.1[e]	
Cyclohexylamine	CPF500	99.2	10,000[b]	N/L	15,000	0.16	40	1.5	
Cyclopropane	CQD750	42.1	10,000[a]	2400[d]	10,000 (Flamm.)	0.41	240[h]	2.4	11,175
Diacetyl peroxide >70%	ACV500	118.0	5000	N/L	N/L	Oral: 283 mg/kg		Oxid.; expl.	
Diazomethane	DCP800	42.1	500	2	N/L	0.00034	0.2[h]	6.7[e]	
Dibenzoyl peroxide	BDS000	242.2	7500	N/L	N/L	Oral: 5700 mg/kg		Oxid.; expl.	
Diborane	DDI450	27.7	100	15	2500	0.0011	1	0.8	
Tertdibutyl peroxide	BSC750	146.3	5000	N/L	N/L	Oral: 10,200 mg/kg		1.0[e]; oxid.	

Chemical	Code								
Dichloro acetylene	DEN600	94.9	250	Ca	N/L	0.0007	0.2^i	7.4^e	1975
Dichlorosilane	DGK300	101.0	2500	4000^d	10,000 (Flamm.)	1.6	400^h	4.0	
Diethyl zinc	DKE600	123.5	10,000	N/L	N/L	No data		Pyro.	2755
Difluoroethane	ELN500	66.1	10,000^a	3700^d	10,000 (Flamm.)	1.0	370^h	3.7	
Diisopropyl peroxydicarbonate	DNR400	206.2	7500	N/L	N/L	Oral:2140 mg/kg		Oxid.; expl.	
Dilauroyl peroxide	LBR000	398.7	7500	N/L	N/L	No data		Oxid.; expl.	
Dimethylamine, anhydrous	DOQ800	45.1	2500	500	10,000 (Flamm.)	0.18	100^g	2.8	8595
Dimethyl dichlorosilane	DFE259	129.1	1000	N/L	5000	0.026	5	3.4	
Dimethyl hydrazine	DSF400	60.1	1000	15 Ca	15,000	0.12	5	2.0	
Dimethyl propane [neopentane]	NCH000	72.2	10,000^a	1400^d	10,000 (Flamm.)	0.41	140^h	1.4	10,810
Dinitroaniline	DUP600	183.1	5000	N/L	N/L	Oral: 285 mg/kg		Solid; expl.	
Epichlorohydrin	EAZ500	92.5	<10,000^a	75 Ca	20,000	0.076	20	3.8	11,400
Ethane	EDZ000	30.1	10,000^a	3000^d	10,000 (Flamm.)	0.37	300^h	2.9	10,935
Ethyl acetylene [butyne]	EFS500	54.1	10,000^a	1400^d	10,000 (Flamm.)	0.31	140^h	2.0	8450
Ethylamine	EFU400	45.1	7500	600	10,000 (Flamm.)	0.11	60^h	3.5	4780
Ethyl chloride	EHH000	64.5	10,000^a	3800^d	10,000 (Flamm.)	1.0	380^h	3.8	11,315
Ethylene	EIO000	28.1	10,000^a	2700^d	10,000 (Flamm.)	0.31	270^h	2.7	
Ethylene fluorohydrin [fluoroethanol]	FIE000	64.1	100	N/L	N/L	0.002	0.8^i	4.6^e	
Ethylene diamine	EEA500	60.1	<10,000^b	1000	20,000	0.49	200	2.5	7530
Ethyleneimine	EJM900	43.1	1000	100 Ca	10,000	0.018	10	3.3	
Ethylene oxide	EJN500	44.1	5000	800 Ca	10,000	0.090	50	3.0	6850
Ethyl ether	EJU000	74.1	10,000^a	1900^d	10,000 (Flamm.)	0.58	190^h	1.9	8110
Ethyl mercaptan	EMB100	62.1	10,000^a	2800^d	10,000 (Flamm.)	0.71	280^h	2.8	6710
Ethyl methyl ketone peroxide >60%	MKA500	176.2	5000	N/L	N/L	0.0015	0.2^i	1.1^e; expl.	
Ethyl nitrite	ENN000	75.1	5000	4000^d	10,000 (Flamm.)	1.2	400^h	4.0	4320
Explosives [29 CFR 1910.109k]	ERF000 ERF500	N/A	Classes A, B, C, & (any) pyrotechnics	N/A	5000	1-psig "side-on" blast overpressure		N/A	Various
Flammable gas or liquid			10,000	10% LFL	N/L	5 kw/m^2; 40 sec.		Various	Various
Fluorine	FEZ000	38.0	1000	25	1000	0.0039	2.5	Not combust.;oxidizer	
Formaldehyde	FMV000	30.0	1000	20 Ca	15,000 (Sol.)	0.012	10	7.0	4470
Furan	FPK000	68.1	500	N/L	5000	0.0012	0.45	2.3	
Hexafluoro acetone	HCZ000	166.0	5000	N/L	N/L	0.0068	1^g	10^e	
Hydrazine	HGS000	32.1	<10,000^b	50 Ca	15,000	0.011	8.5	2.9	4990
Hydrochloric acid >37% (10/97)	HHL000	36.5	N/L	50 HCl	15,000	0.030	20	Not combustible	
Hydrofluoric acid >50%	HHU500	20.0	N/L	30 HF	1000	0.016	20	Not combustible	
Hydrogen	HHW500	2.02	10,000^a	4000^d	10,000 (Flamm.)	0.033	400^h	4.0	28,790

Continued

TABLE 17.A.1 (Continued)
OSHA and USEPA Listed Chemicals

OSHA Highly Hazardous Chemical or EPA Regulated Substance	"Sax" No.	Mole Wgt.	OSHA Threshold Quantity (pounds)	IDLH Conc. (ppm)	EPA Threshold Quantity (pounds)	Toxic Endpoint (mg/l)	Toxic Endpoint (ppm)	Lower[c] Flammable Limit (v%)	Heat of Comb. (cal/g)
Hydrogen bromide	HHJ000	80.9	5000	30	N/L	0.0099	3[h]	Not combustible	
Hydrogen chloride	HHX000	36.5	5000	50	5000	0.030	20	Not combustible	
Hydrogen cyanide	HHS000	27.0	1000	50	2500	0.011	10	5.6	5970
Hydrogen fluoride	HHU500	20.0	1000	30	N/L	0.016	20	Not combustible	
Hydrogen peroxide >52 wt%	HIB000	34.0	7500	75	N/L	0.010	7.5[h]	Not combust.; oxidizer	
Hydrogen selenide	HIC000	81.0	150	1	500	0.00066	0.2	6.7[e]	
Hydrogen sulfide	HIC500	34.1	1500	100	10,000	0.042	30	4.0	
Hydroxylamine	HLM500	33.0	2500	N/L	N/L	Oral: 29 mg/kg		Solid; expl.	
Iron pentacarbonyl	IHG500	195.9	250	N/L	2500	0.00044	0.05	Not combustible	
Isobutane	MOR750	58.1	10,000[a]	1800[d]	10,000 (Flamm.)	0.43	180[h]	1.8	10,940
Isobutyronitrile	IJX000	69.1	<10,000[b]	N/L	20,000	0.14	50	1.9[e]	
Isopentane	EIK000	72.2	10,000 [a]	1400[d]	10,000 (Flamm.)	0.41	140[h]	1.4	10,780
Isoprene	IMS000	68.1	10,000[a]	1500[d]	10,000 (Flamm.)	0.42	150[h]	1.5	10,510
Isopropylamine	INK000	59.1	5,000	750	10,000 (Flamm.)	0.18	75[h]	2.0	8760
Isopropyl chloride	CKQ000	78.5	10,000[a]	2800[d]	10,000 (Flamm.)	0.90	280[h]	2.8	5690
Isopropyl chloroformate	IOL000	122.6	<10,000[b]	N/L	15,000	0.1	20	2.7[e]	
Ketene	KEU000	42.0	100	5	N/L	0.00086	0.5[h]	5.2[e]	5860
Methacrylaldehyde	MGA250	70.1	1000	N/L	N/L	0.036	12.5[i]	2.2[e]	
Methacryloyl chloride	MDN899	104.5	150	N/L	N/L	0.0006	0.14[i]	2.7[e]	
Methacryloyl oxyethyl isocyanate	IKG700	155.2	100	N/L	N/L	0.00025	0.04[i]	1.5[e]	
Methane	MDQ750	16.0	10,000[a]	5000[d]	10,000 (Flamm.)	0.33	500[h]	5.0	12,010
Methyl acrylonitrile	MGA750	67.1	250	N/L	10,000	0.0027	1	2.1[e]	
Methylamine, anhydrous	MGC250	31.1	1000	100	10,000 (Flamm.)	0.13	100[g]	4.9	7540
Methyl bromide	MHR200	95.0	2500	250 Ca	N/L	0.78[f]	200	Not combustible	
Methyl butene	MHT000	70.1	10,000[a]	1500 [d]	10,000 (Flamm.)	0.43	150[h]	1.5	10,680
Methyl chloride	MIF765	50.5	15000	2000 Ca	10,000	0.82	400	8.1	3250
Methyl chloroformate	MIG000	94.5	500	N/L	5000	0.0019	0.5	7.4[e]	
Methyl ether	MJW500	46.1	10,000[a]	3400[d]	10,000 (Flamm.)	0.64	340[h]	3.3	6920
Methyl fluoroacetate	MKDOOO	92.1	100	N/L	N/L	0.00025	0.07[i]	3.9[e]	
Methyl fluorosulfate	MKG250	114.1	100	N/L	N/L	0.00023	0.05[i]	8.6[e]	

Chemical	Code								
Methyl formate	MKG750	60.1	10,000ᵃ	4500ᵈ	10,000 (Flamm.)	1.1	450ʰ	4.5	3680
Methyl hydrazine	MKN000	46.1	100	20 Ca	15,000	0.0094	5	2.5	1370
Methyl iodide	MKW200	141.9	7500	100 Ca	N/L	0.29	50ᵍ	8.1ᵉ	
Methyl isocyanate	MKX250	57.1	250	3	10,000	0.0012	0.5	5.3	4720
Methyl mercaptan	MLE650	48.1	5000	150	10,000	0.049	25	3.9	
Methyl propene	IIC000	56.1	10,000ᵃ	1800ᵈ	10,000 (Flamm.)	0.41	180ʰ	1.8	10,800
Methyl thiocyanate	MPT000	73.1	<10,000ᵇ	N/L	20,000	0.085	3	2.9ᵉ	
Methyl trichlorosilane	MQC500	149.5	500	N/L	5000	0.018	3	7.6	
Methyl vinyl ketone	MQM100	70.1	100	N/L	N/L	0.0001ᶠ	0.02	2.1	
Nickel tetracarbonyl	NCZ000	170.7	150	2 Ca	1000	0.00067	0.1	2.0	
Nitric acid	NED500	63.0	500, >94.5 wt%	25	15,000 >80 wt%	0.026	10	Not combust.; oxidizer	
Nitric oxide	NEG100	30.0	250	100	10,000	0.031	25	Not combust.; oxidizer	
Nitro aniline	NEO500	138.1	5000	53	N/L	0.030	5.3ʰ	Solid	5510
Nitrogen dioxide	NGR500	46.0	250	20	N/L	0.0094ᶠ	5	Not combust.; oxidizer	
Nitrogen tetroxide	NGU500	92.0	250	N/L	N/L	0.012	3.1ⁱ	Not combust.; oxidizer	
Nitrogen trifluoride	NGW000	71.0	5000	1000	N/L	0.29	100ʰ	Not combust.; oxidizer	
Nitrogen trioxide	N/L	76.0	250	N/L	N/L	No data		Not combust.; oxidizer	
Nitromethane	NHM500	61.1	2500	750	N/L	0.19	75ʰ	7.3	2770
Oleum 65 wt%–80 wt%	SOJ520	178.1	1000	(See SO₃)	10,000	0.010	3	Not combustible	
Osmium tetroxide	OKK000	254.2	100	0.13	N/L	0.0001ᶠ	0.01	Not combustible	
Oxygen difluoride	ORA000	54.0	100	0.5	N/L	0.00011	0.05ʰ	Not combust.;oxidizer	
Ozone	ORW000	48.0	100	5	N/L	0.002ᶠ	1	Not combust.;oxidizer	
Pentaborane	PAT750	63.1	100	1	N/L	0.00026	0.1ʰ	0.42	
Pentadiene	PBA250	68.1	10,000ᵃ	2000ᵈ	10,000 (Flamm.)	0.56	200ʰ	2.0	10,440
Pentane	PBK250	72.2	10,000ᵃ	1500ᵈ	10,000 (Flamm.)	0.44	150ʰ	1.5	10,730
Pentene	PBQ000	70.1	10,000ᵃ	1500ᵈ	10,000 (Flamm.)	0.43	150ʰ	1.5	10,680
Perchloric acid >60 wt%	PCD250	100.5	5000	N/L	N/L	Oral: 400 mg/kg		Not combust.;oxidizer	
Perchloromethyl mercaptan	PCF300	185.9	150	10	10,000	0.0076	1	10ᵉ	
Perchloryl fluoride	PCF750	102.5	5000	100	N/L	0.042	10ʰ	Not combust.;oxidizer	
Peroxyacetic acid >60%	PCL500	76.1	1000	N/L	10,000	0.0045	1.5	6.7ᵉ;Oxid.	
Phosgene [carbonyl chloride]	PGX000	98.9	100	2	500	0.00081	0.2	Not combustible	
Phosphine	PGY000	34.0	100	50	5000	0.0035	2.5	2.1	
Phosphorus oxychloride	PHQ800	153.3	1000	N/L	5000	0.0030	0.5	Not combustible	
Phosphorus trichloride	PHT275	137.4	1000	25	15,000	0.028	5	Not combustible	
Piperidine [hexahydropyridine]	PIL500	85.2	<10,000ᵇ	N/L	15,000	0.022	6	1.4ᵉ	
Propadiene [allene]	AFR000	40.1	10,000ᵃ	2100ᵈ	10,000 (Flamm.)	0.34	210ʰ	2.1	11,120

Continued

TABLE 17.A.1 (Continued)
OSHA and USEPA Listed Chemicals

OSHA Highly Hazardous Chemical or EPA Regulated Substance	"Sax" No.	Mole Wgt.	OSHA Threshold Quantity (pounds)	IDLH Conc. (ppm)	EPA Threshold Quantity (pounds)	Toxic Endpoint (mg/l)	Toxic Endpoint (ppm)	Lower[c] Flammable Limit (v%)	Heat of Comb. (cal/g)
Propane	PMJ750	44.1	10,000[a]	2100[d]	10,000 (Flamm.)	0.38	210[h]	2.0	11,120
Propargyl bromide	PMN500	119.0	100	N/L	N/L	0.0000[f]	0.01	3.4[e]; Expl.	
Propionitrile	PMV750	55.1	<10,000[b]	N/L	10,000	0.0037	1.5	2.6[e]	
Propyl chloroformate	PNH000	122.6	<10,000[b]	N/L	15,000	0.010	2	2.7[e]	
Propylene	PMO500	42.1	10,000[a]	2000[d]	10,000 (Flamm.)	0.34	200[h]	2.0	10,980
Propyleneimine	PNL400	57.1	<10,000[b]	100 Ca	10,000	0.12	50	2.3[e]	
Propylene oxide [epoxypropane]	PNL600	58.1	<10,000[b]	400 Ca	10,000	0.59	250	2.3	
Propyl nitrate	PNQ500	105.1	2500	500	N/L	0.21	50[h]	2.0; Expl.	
Propyne [methyl acetylene]	MFX590	40.1	10,000[a]	1700[d]	10,000 (Flamm.)	0.28	170[h]	1.7	11,080
Sarin ["gb" nerve gas]	IPX000	140.1	100	N/L	N/L	0.0001[f]	0.01	1.9[e]	
Selenium hexafluoride	SBS000	193.0	1000	2	N/L	0.0016	0.2[h]	Not combustible	
Silane	SDH575	32.1	10,000[a]	1400[d]	10,000 (Flamm.)	0.18	140[h]	Pyro.	10,630
Stibine [antimony hydride]	SLQ000	124.8	500	5	N/L	00026	0.5[h]	6.8[e]	
Sulfur dioxide, liquid	SOH500	64.1	1000	100	5,000	0.0078	3	Not combustible	
Sulfur pentafluoride	SOQ450	254.1	250	1	N/L	0.0010	0.1[h]	Not combustible	
Sulfur tetrafluoride	SOR000	108.1	250	N/L	2500	0.0092	2	Not combustible	
Sulfur trioxide [as sulfuric acid]	SOR500	80.1	1000	0.3	10,000	0.010	3	Not combustible	
Tellurium hexafluoride	TAK250	241.6	250	1	N/L	0.001	0.1	Not Combustible	
Tetrafluoroethylene	TCH500	100.0	5000	10,000[d]	10,000 (Flamm.)	4.1	1000[h]	11.0	310
Tetrafluorohydrazine	TCI000	104.0	5000	N/L	N/L	0.002	0.5[i]	Explosive	
Tetramethyl lead	TDR500	267.3	1000	4.7	10,000	0.0040	0.4	1.4[e]	
Tetramethyl silane	TDV500	88.2	10,000[a]	1300[d,c]	10,000 (Flamm.)	0.047	130[h]	1.3[e]	10,010
Tetranitromethane	TDY250	196.0	<10,000[b]	4	10,000	0.0040	0.5	Oxid.;Expl.	
Thionyl chloride	TFL000	119.0	250	N/L	N/L	0.024	5[i]	Not Combust.;Oxidizer	
Titanium tetrachloride	TGH350	189.7	N/L	N/L	2500	0.020	2.5	Not Combustible	
Toluene diisocyanate	TGM740	174.2	<10,000[b]	2.5 Ca	10,000	0.0070	1	0.9	
Trichloro chloromethyl silane	CIY325	183.9	100	N/L	N/L	0.0003[f]	0.04	8.8[e]	
Trichloro dichlorophenyl silane	DGF200	280.4	2500	N/L	N/L	0.008	0.7	2.2[e]	
Trichloro silane	TJD500	135.5	5000	7000[d]	10,000 (Flamm.)	3.9	700[h]	7.0	900
Trifluoro chloroethylene	CLQ750	116.5	10,000	8400[d]	10,000 (Flamm.)	4.0	840[h]	8.4	440

Trimethoxy silane	TLB750	122.2	1500	N/L	N/L	0.006	1.25[i]	2.6[e]	
Trimethyl amine	TLD500	59.1	10,000[a]	2000[d]	10,000 (Flamm.)	0.24	100[g]	2.0	9110
Trimethylchlorosilane	TLN250	108.6	<10,000[b]	N/L	10,000	0.050	11	2.4[e]	
Vinyl acetate	VLU250	86.1	10,000[a]	N/L	15,000	0.26	75	2.6	10,890
Vinyl acetylene	BPF:109	52.1	10,000[a]	2100[d]	10,000 (Flamm.)	0.45	210[h]	2.2	4520
Vinyl chloride	VNP000	62.5	<10,000[b]	Ca	10,000 (Flamm.)	Oral: 500 mg/kg		3.6	7900
Vinyl ethyl ether	EQF500	72.1	10,000[a]	1700[d]	10,000 (Flamm.)	0.50	170[h]	1.7	530
Vinyl fluoride	VPA000	46.0	10,000[a]	2600[d]	10,000 (Flamm.)	0.49	260[h]	2.6	2480
Vinylidene chloride	VPK000	96.9	10,000[a]	6500[d]	10,000 (Flamm.)	2.6	650[h]	6.5	2590
Vinylidene fluoride	VPP000	64.0	10,000[a]	5500[d]	10,000 (Flamm.)	1.4	550[h]	5.5	7330
Vinyl methyl ether	MQL750	58.1	10,000[a]	2600[d]	10,000 (Flamm.)	0.62	260[h]	2.6	

Ca: The compounds for which "Ca" [carcinogenic] is shown in the above table have been characterized by the National Institute for Occupational Safety and Health [NIOSH] as "potential occupational carcinogens" [A-4].

[a] Flammable liquids or gases and meet the definition of the OSHA "Hazard Communication" standard [A-7] as a liquid having a flash point <100°F or a gas having a lower flammable limit of 13% by volume or having a flammable range that is wider than 12% by volume. The OSHA Threshold Quantity for flammable liquids and gases is 10,000 pounds.

[b] Not specifically listed in the OSHA standard and are flammable liquids or gases, but a 10,000-pound Threshold Quantity probably would be considered excessively hazardous because of the toxicity hazard, as indicated by low values for the EPA "Toxic Endpoint" [A-9] or for the IDLH.

[c] Estimated values that were obtained by first calculating the "stoichiometric" concentration in air (for combustion to carbon dioxide, water, silicon dioxide, sulfur dioxide, and halogen acids) and then multiplying by 0.55 for nonhalogen materials and by 0.60 for halogen-containing materials. The "stoichiometric" concentration can be calculated from:

$$C_{st} = 83.8/(4[C] + 4[Si] + 4[S] + [H] - [X] - 2[O] + 0.84)$$

where lower flammable limits data are not given in the EPA Guidance, the data were obtained from NFPA 325 or NFPA 49 [A-8].

[d] The IDLH concentrations for flammable (and relatively nontoxic) materials are 10% of the lower flammable limits [A-8].

[e] Estimated values, based on the "stoichiometric" concentrations, for flammable and combustible materials.

[f] Calculated, from (MW × PPM)/(1000 × 24.4).

[g] Emergency Response Planning Guide [ERPG-2] values obtained from the current listing of the American Industrial Hygiene Association [A-10].

[h] One-tenth of the Immediately Dangerous to Life and Health [IDLH] value [A-4].

[i] One-hundredth of the concentration that was lethal to 50% of the exposed population [LC50], or one-tenth of the lowest observed lethal concentration [LCLo], for exposure durations of 4 hours.

[j] Threshold limit values or time-weighted average values, for 8-hour exposures.

APPENDIX B
FIRE-HAZARD PROPERTIES OF OSHA AND EPA LISTED HAZARDOUS MATERIALS

Flammable Liquid	Liquid Density (g/ml)	Boiling Point (°C)	Heat of Combust. (cal/g)	Heat of Vapor (cal/g)	Specific Heat (cal/g/°C)	$H_c/H_v+C_L(T_b-T_a))$ for $T_a=25°C$ (dimensionless)	Burning Rate (kg/sec-m²)	Rate/Ratio (kg/sec-m²)
Ammonia	0.82	33.3	5400	328	1.10	16.5	Not Determined	—
Butane	0.57	0.5	10,970	92	0.55	119	0.078	0.00065
Carbon Disulfide	1.26	46	3240	85	0.24	36	Not Determined	—
Cyclohexane	0.78	80.7	11,160	86	0.55	89	0.090	0.00101
DiMethyl Amine	0.68	6.8	8595	140	0.73	61	Not Determined	—
Ethane	0.54	88.6	11,400	117	0.57	97	0.122	0.00125
Ethyl Chloride	0.89	12.3	4780	92	0.37	52	Not Determined	—
Ethylene Oxide	0.88	10.6	6850	139	0.48	49	Not Determined	—
Ethyl Ether	0.71	34.5	8110	90	0.52	90	0.085	0.00094
Hydrogen	0.07	252.9	28,790	107	2.32	269	0.017	0.00006
Hydrogen Cyanide	0.69	26	5970	272	0.62	22	Not Determined	—
Methane	0.42	161.5	12,010	122	0.83	98	0.078	0.00079
Pentane	0.63	36	10,730	92	0.53	116	0.076	0.00065
Propane	0.49	42.1	11,120	102	0.58	109	0.099	0.00091
TetraFluoro-Ethylene	1.52	75.9	310	40	0.28	7.8	Not Determined	—
Vinyl Chloride	0.91	13.3	2480	100	0.30	25	Not Determined	—

References: SFPE "Handbook" [25], 2–50; NFPA *Handbook* [26], 21–37.

APPENDIX C
BIBLIOGRAPHY

1. U.S. Department of Labor, Occupational Safety and Health Administration. "Process Safety Management of Highly Hazardous Chemicals." 29CFR1910.11.9, February 24, 1992.
2. U.S. Environmental Protection Agency. "Risk Management Program Rule." 40 CFR 68, June 20, 1996. "List of Regulated Substances and Threshold Quantities for Accidental Release Prevention." 40 CFR 68, January 31, 1994; rev. October 1997.
3. National Safety Council. Accident Prevention Manual – Engineering and Technology. 9th ed. Itasca, IL: NSC, 1988.
4. Prugh, R.W. "Plant Safety", In *Kirk-Othmer Encyclopedia of Chemical Technology*, 4th ed, vol. 19, eds. J. I. Kroschwitz et al. New York: John Wiley & Sons, 1996.
5. W. H. Stewart et al. Thompson Publishing Group. *Chemical Process Safety Report*. Paragraphs 220, 370, 440, 510, 518, 548, 578, 603, 610, and 641 (updated monthly).
6. Center for Chemical Process Safety, American Institute of Chemical Engineers [refer to Appendix B for the titles of the "G" publications].
7. Lees, F.P. *Loss Prevention in the Process Industries,* 2nd ed. New York: Butterworth-Heinemann, 1996: 16/226, 20/10, 21/29, A7/25, Appendices 1–6, 16, 19, 21, 22.
8. Perry, J. H., Perry, R. H., and D. W. Green, eds. *Perry's Chemical Engineers' Handbook.* 4th ed. New York: McGraw-Hill, 1963: 3–158 (carbon dioxide), 3–221; 6th ed., 1984: 3–190 (ethylene), 3–219 (propylene), 5–8, 5–15; 7th ed., 1997: 2–215 (ammonia), 2–269 (R-11), 6–23, 26–18.
9. U.S. Environmental Protection Agency. "Risk Management Program Offsite Consequence Analysis Guidance." May 24, 1996. Paragraphs 3.2, 3.2.2B.2., 8.1.2, 8.2.2, 10.0, D.4.1, D.7.1, D.9; Reference Tables 1–22; also [5, Appendix 2].
10. Woodward, J.L. *Discharge Rates Through Holes in Process Vessels and Piping.* New York: Van Nostrand Reinhold, 1993: 105, 153 in "Prevention and Control of Accidental Releases of Hazardous Gases."
11. Crowl, D.A., and J.F. Louvar. *Chemical Process Safety: Fundamentals with Applications.* New York: Prentice Hall, 1990: 65, 86, 88, 89,113, 114, 115, 188.
12. Federal Emergency Management Agency. *Handbook of Chemical Hazard Analysis Procedures.* Washington, DC: USGPO, 1989: B-7, B-10, B-36, C-8.
13. Sandler, S.I. *Chemical and Engineering Thermodynamics.* 2nd ed. New York: John Wiley & Sons, 1989: 46 (steam).
14. Prugh, R.W. "Quantitative Evaluation of BLEVE Hazards." *Chemical Engineering Progress* 87 (February 1991): 66; *Journal of Fire Protection Engineers* 3 (January 1991): 9.
15. Reid, R.C, et al. *The Properties of Gases and Liquids.* 4th ed. New York: McGraw-Hill, 1987: 137
16. Koopman, R.P., et al. "A Review of Recent Work in Atmospheric Dispersion of Large Spills." In *1988 Hazardous Material Spills Conference.* American Institute of Chemical Engineers and National Response Team, 1988: 457, 464.
17. Klein, P.F. (chairman). "Methodology for Chemical Hazard Prediction." U.S. Dept. of Defense Explosives Safety Board, Technical Paper No. 10 (AD/A 008159), Annex C, March 1975.
18. Opschoor, G. "Evaporation." In *Methods for the Calculation of the Physical Effects of the Escape of Dangerous Material* [the "Yellow Book"]. Netherlands Division of Technology for Society [TNO], March 1980: Chapter 5.
19. Prugh, R.W. "Quantitative Evaluation of Inhalation-Toxicity Hazards" (paper 5c). In the *Proceedings of the 29th Annual Loss Prevention Symposium,* August 2, 1995.
20. Prugh, R.W. "Hazardous Fluid Releases: Prevention and Protection by Design and Operation." *Journal of Loss Prevention in the Process Industries* 5(2) (1992): 67.
21. U.K. Health and Safety Executive. *Canvey: An Investigation of Potential Hazards from Operations in the Canvey Island/Thurrock Area.* London: HSE, 1978: Appendix 3.
22. The Netherlands Organization of Applied Scientific Research [TNO]. *Methods for the Determination of Possible Damage to People and Objects Resulting from Releases of Hazardous Materials* [the "Green Book"]. Chapter 1, para. 6.4, Figure 2.3.
23. Thompson Publishing Group. "Chemical Process Safety Report." 1(11) (September 1991): 4; 1(12) (October 1991): 6; 4(7) (May 1994): 8.

24. U.S. Department of Labor, Occupational Safety and Health Administration. "Occupational Safety and Health Standards." Title 29, Subtitle B, Chapter XVII, Part 1910, Subpart I, of the Code of Federal Regulations, 29 CFR 1910.38(a) and 29 CFR 1910.132; July 1995.
25. Prugh, R. W. "Quantitative Evaluation of Fireball Hazards", *Process Safety Progress* 13(2) (April 1994): Table II.
26. Society of Fire Protection Engineers. *Handbook of Fire Protection Engineering*. Bethesda, MD: SFPE, 1988: 2–54, 2–61, 2–63, 2–83.
27. National Fire Protection Association. *Fire Protection Handbook*. 16th ed. Quincy, MA: NFPA, 1986: 21–37.
28. U.S. Army. "Structures to Resist the Effects of Accidental Explosions." Technical Manual TM 5–1300. Hyattsville, MD: USACE, 1990: 2–11, 2–13; Figures 2-7, 2-15; Table 2-1.
29. Prugh, R.W. "The Effects of Explosive Blast on Structures and Personnel" (paper 4c). In the *Proceedings of the 32nd Loss Prevention Symposium,* March 10, 1998.
30. Prugh, R. W. "Evaluation of Unconfined Vapor Cloud Explosion Hazards." In *International Conference on Vapor Cloud Modeling,* November 4, 1987: 720.
31. U.S. Army. *Engineering Design Handbook – Explosives Series – Properties of Explosives of Military Interest.* AMCP 706-177 (AD-764-340). Hyattsville, MD: USACE, 1971: 4; and individual compounds.
32. Meyer, R. *Explosives.* 3rd ed. Berlin: Wiley-VCH, 1987: 255; and individual compounds.
33. Urben, P.G. *Bretherick's Handbook of Reactive Chemical Hazards.* 5th ed., vols. 1, 2. New York: Butterworth-Heinemann, 1995: 74.
34. American Society of Testing and Materials. *Chemical Thermodynamic and Energy Release Evaluation [CHETAH].* Version 7.0. West Conshohocken, PA: ASTM, 1994.
35. U.S. Army. *Engineering Design Handbook – Principles of Explosive Behavior.* AMCP 706-180. Hyattsville, MD: USACE, 1972: 3–2.
36. U.S. Department of Transportation. "Hazardous Materials Regulations." Title 49, Subtitle B, Chapter 1, Subchapter C, Code of Federal Regulations, 49 CFR 172.101, 49 CFR 173.50, and 49 CFR 173.21(j); October 1995.
37. Manufacturing Chemists Association (now Chemical Manufacturers Association). *Case Histories of Accidents in the Chemical Industry.* Vols. 1–4. Washington, DC: MCA, 1962, 1966, 1969, 1975.
38. Marsh & McLennan Protection Consultants. *Large Property Damage Losses in the Hydrocarbon – Chemical Industries* [published annually].
39. Lenoir, E.M., and J.A. Davenport. "A Survey of Vapor Cloud Explosions – Second Update", *Process Safety Progress* 12(1) (January 1993): 12.
40. U.S. Department of Labor, Occupational Safety and Health Administration. "Hazard Communication", Title 29, Subtitle B, Chapter XVII, Part 1910, Subpart Z, Paragraph 1200, Code of Federal Regulations [29 CFR 1910.1200(g)]; July 1995.
41. J. J. Keller & Associates. "OSHA Compliance Manual – 29 CFR 1910 Plant Safety." April 1997: OSHA-19.
42. Thompson Publishing Group. *Risk Management Program Handbook.* Paragraphs 414, 418, 420, 424, 434, 458, 460, 710 (updated monthly).
43. U.S. Department of Labor, Occupational Safety and Health Administration. "Occupational Safety and Health Standards", Title 29, Subtitle B, Chapter XVII, Part 1926, Subpart P, Paragraph 650, Code of Federal Regulations [29 CFR 1926.650]; July 1989.
44. American Institute of Chemical Engineers. *Loss Prevention.* Vol. 9. New York: AIChE, 1975: several pertinent articles on "hot-tapping".
45. Kletz, T.A. *Plant Design for Safety.* Newport, Australia: Hemisphere Publishing, 1991: 11, 132, 134.
46. Crowl, D.A., ed. *Inherently Safer Chemical Processes.* Center for Chemical Process Safety. New York: AIChE, 1996.
47. Furr, A.K. *CRC Handbook of Laboratory Safety.* 4th ed. Boca Raton, FL: CRC Press, 1995.
48. Kletz, T.A. *Learning from Accidents in Industry.* New York: Butterworth-Heinemann, 1988: 149.
49. Kletz, T.A. *What Went Wrong?* Houston, TX: Gulf Publishing, 1985: 73, 143.
50. Kletz, T.A. *Computer Control and Human Error.* Houston, TX: Gulf Publishing, 1995.
51. Kletz, T.A. *Lessons from Disaster – How Organizations Have No Memory and Accidents Recur.* Rugby, UK: Institution of Chemical Engineers, 1993.
52. American Petroleum Institute. "Management of Hazards Associated with Location of Process Plant Buildings." API Recommended Practice 752. Washington, DC: API, 1995.
53. Mecklenburgh, J.C. *Process Plant Layout.* New York: Halsted Press, 1985.
54. Wells, G.L. *Safety in Process Plant Design.* Lodnon: George Godwin Ltd., 1980: 145, 255.

APPENDIX D
PUBLICATIONS OF THE CENTER FOR CHEMICAL PROCESS SAFETY (AICHE)

1. "Guidelines for Hazard Evaluations Procedures," G-1, 1985; superseded by G-18, 2nd Ed. (1992).
2. "Guidelines for Use of Vapor Cloud Dispersion Models," 1st ed., G-2, 1987; superseded by G-40 (1996).
3. "Guidelines for Safe Storage and Handling of High Toxic Hazard Materials," G-3, 1988.
4. "Guidelines for Vapor Release Mitigation," G-4, 1988.
5. "Workbook of Test Cases for Vapor Cloud Source Dispersion Models," G-5, 1989.
6. "Guidelines for Chemical Process Quantitative Risk Analysis," G-6, 1989; superseded by G-42, 2nd ed., 2000.
7. "Guidelines for Process Equipment Reliability Data," G-7, 1989.
8. "Guidelines for Technical Management of Chemical Process Safety," G-8, 1989.
9. "Guidelines for Evaluating the Characteristics of Vapor Cloud Explosions, Flash Fires, and BLEVEs," G-9, 1994.
10. "Plant Guidelines for Technical Management of Chemical Process Safety", G-10 (1992).
11. "Guidelines for Pressure Relief and Effluent Handling Systems," G-11, 1998.
12. "Guidelines for Safe Automation of Chemical Processes," G-12, 1993.
13. "Guidelines for Chemical Reactivity Evaluation and Application to Process Design," G-13, 1995.
14. "Directory of Chemical Process Safety Services," G-14, 1991.
15. "Guidelines for Preventing Human Error in Process Safety," G-15, 1994.
16. "Safety, Health, and Loss Prevention in Chemical Processes—Instructor's Guide for Undergraduate Engineering Curricula," G-16, 1990.
17. "Safety, Health, and Loss Prevention in Chemical Processes—Student Problems," G-17, 1990.
18. "Guidelines for Hazard Evaluation Procedures," 2nd ed., G-18, 1992.
19. "Guidelines for Investigating Chemical Process Incidents," G-19, 1992; superseded by G-82, 2nd ed., 2003.
20. "Guidelines for Auditing Process Safety Management Systems," G-20, 1993.
21. "Tools for Making Acute Risk Decisions," G-21, 1995.
22. "Guidelines for Process Safety Fundamentals in General Plant Operations," G-22, 1995.
23. "Guidelines for Engineering Design for Process Safety," G-23, 1993.
24. "Guidelines for Postrelease Mitigation Technology in the Chemical Process Industry," G-24, 1997.
25. "Guidelines for Implementing Process Safety Management Systems," G-25, 1994.
26. "Guidelines for Evaluating Process Plant Buildings for External Explosions and Fires," G-26, (996.
27. "Guidelines for Process Safety Documentation," G-27, 1995.
28. "Guidelines for Chemical Transportation Risk Analysis," G-28, 1995.
29. "Guidelines for Safe Process Operations and Maintenance," G-29, 1995.
30. "Guidelines for Safe Storage and Handling of Reactive Materials," G-30, 1995.
31. "Guidelines for Technical Planning for On-Site Emergencies," G-31, 1996.
32. "Guidelines for Writing Effective Operating and Maintenance Procedures," G-32 1996.
33. "Guidelines for Safe Warehousing," G-33, 1997.
34. "Contractor and Client Relations to Assure Process Safety," G-34, 1996.
35. "Concentration Fluctuations and Averaging Time in Vapor Clouds," G-35, 1995.
36. "Expert Systems in Process Safety," G-36, 1995.
37. "Understanding Atmospheric Dispersion of Accidental Releases," G-37, 1995.
38. "Guidelines for Integrating Process Safety Management, Environment, Safety, Health, and Quality," G-38, 1997.
39. "Guidelines for Design Solutions for Process Equipment Failures," G-39, 1998.
40. "Guidelines for Use of Vapor Cloud Dispersion Models," 2nd ed., G-40, 1996.
41. "Inherently Safer Chemical Processes," G-41, 1996.
42. "Guidelines for Chemical Process Quantitative Risk Analysis," 2nd ed., G-42, 2000.
43. "Loss Prevention in the Process Industries," 2nd ed., G-43, 1996.
44. [Title Not Yet Assigned], G-44 (not issued).
45. [Title Not Yet Assigned], G-45 (not issued).

46. "Chemical Process Safety: Case Histories," G-46, 1993.
47. [Title Not Yet Assigned], G-47 (not issued).
48. [Title Not Yet Assigned], G-48 (not issued).
49. [Title Not Yet Assigned], G-49 (not issued).
50. "RELEASE: A Model with Data to Predict Aerosol Rainout in Accidental Releases," G-50, 1998.
51. "Evaluating Process Safety in the Chemical Industry: A User's Guide to Quantitative Risk Analysis," G-51, 2000.
52. [Title Not Yet Assigned], G-52 (not issued).
53. "Practical Compliance with the EPA Risk Management Program," G-53, 1999.
54. "Local Emergency Planning Committee Guidebook: Understanding the EPA Risk Management Program Rules," G-54, 1999.
55. [Title Not Yet Assigned], G-55 (not issued).
56. "Guidelines for Improving Plant Reliability," G-56, 1998.
57. [Title Not Yet Assigned], G-57 (not issued).
58. "Electrostatic Ignitions of Fires and Explosions," G-58, 1997.
59. [Title Not Yet Assigned], G-59 (not issued).
60. "Estimating the Flammable Mass of a Vapor Cloud," G-60, 1999.
61. "Understanding Explosions," G-61, 2003.
62. "Guidelines for Process Safety in Batch Reaction Systems," G-62, 1999.
63. "Guidelines for Consequence Analysis of Chemical Releases," G-63, 1999.
64. "Deflagration and Detonation Flame Arrestors," G-64, 2002.
65. [Title Not Yet Assigned], G-65 (not issued).
66. "Layers of Protection Analysis," G-66, 2001.
67. "Avoiding Static Ignition Hazards in Chemical Operations," G-67, 1999.
68. "Guidelines for Process Safety in Outsourced Manufacturing Operations," G-68 2000.
69. [Title Not Yet Assigned], G-69 (not issued).
70. [Title Not Yet Assigned], G-70 (not issued).
71. "Revalidating Process Hazards Analyses," G-71, 2000.
72. [Title Not Yet Assigned], G-72 (not issued).
73. [Title Not Yet Assigned], G-73 (not issued).
74. [Title Not Yet Assigned], G-74 (not issued).
75. "Wind Flow and Vapor Cloud Dispersion at Industrial and Urban Sites and Surface Roughness," G-75, 2002.
76. "The Design and Evaluation of Physical Protection Systems," G-76, 2002.
77. [Title Not Yet Assigned], G-77 (not issued).
78. [Title Not Yet Assigned], G-78 (not issued).
79. "Guidelines for Analyzing the Security Vulnerabilities of Fixed Chemical Sites," G-79, 2002.
80. [Title Not Yet Assigned], G-80 (not issued).
81. "Essential Practices of Managing Chemical Reactivity Hazards," G-81, 2003.
82. "Guidelines for Investigating Chemical Process Incidents," 2nd ed., G-82, 2003.
83. "Guidelines for Fire Protection in Chemical, Petrochemical, and Hydrocarbon Processing Facilities," G-83, 2003.
84. "Guidelines for Facility Siting and Layout," G-84, 2003.
85. "Guidelines for Safe Handling of Powders and Bulk Solids," G-85, 2005.
86. [Title Not Yet Assigned], G-86 (not issued).
87. [Title Not Yet Assigned], G-87 (not issued).
88. [Title Not Yet Assigned], G-88 (not issued).
89. [Title Not Yet Assigned], G-89 (not issued).
90. [Title Not Yet Assigned], G-90 (not issued).
91. [Title Not Yet Assigned], G-91 (not issued).
92. [Title Not Yet Assigned], G-92 (not issued).
93. "Dust Explosions", 3rd Ed., G-93 (2003).
94. [Title Not Yet Assigned], G-94 (not issued).
95. "Guidelines for Safe Handling of Powders and Bulk Solids," G-95, 2004.
96. "Guidelines for Mechanical Integrity Systems," G-96, 2006.

97. "Guidelines for Performing Effective Pre-Startup Safety Reviews," G-97, 2007.
98. [Title Not Yet Assigned], G-98 (not issued).
99. [Title Not Yet Assigned], G-99 (not issued).
100. [Title Not Yet Assigned], G-100 (not issued).
101. [Title Not Yet Assigned], G-101 (not issued).
102. [Title Not Yet Assigned], G-102 (not issued).
103. [Title Not Yet Assigned], G-103 (not issued).
104. [Title Not Yet Assigned], G-104 (not issued).
105. "Safe Design and Operation of Process Vents and Emission Control Systems," G-105, 2006.
106. "Making EHS an Integral Part of Process Design," C-19, 2001.
107. "Emergency Relief System Design Using DIERS Technology," X-123, 1992. 108. "Human Factors Methods for Improving Performance in the Process Industries," 2007.
108. "Safety and Chemical Engineering Education [SACHE], "Process Safety Curricula," "Y" and "N" series of publications.

18 Environmental Engineering: A Review of Issues, Regulations, and Resources

Bradly P. Carpenter, Douglas E. Watson, and Brooks C. Carpenter

CONTENTS

18.1 Background 1486
18.2 The Clean Air Act (http://epa.gov/air/caa/) 1486
    18.2.1 Background and Pollutant Types 1486
        18.2.1.1 Criteria Pollutants 1487
        18.2.1.2 Criteria Pollutants and Smog 1487
        18.2.1.3 Hazardous Air Pollutants 1487
        18.2.1.4 Maximum Achievable Control Technology Standards 1488
    18.2.2 Compliance Requirements 1488
    18.2.3 Source Delineations under Title V 1488
    18.2.4 Methodologies for Determining Air Emissions 1489
    18.2.5 New Source Reviews and Modifications 1489
    18.2.6 New Regulations 1490
    18.2.7 Title IV (Acid Rain) and Title VI (Stratospheric Ozone) Considerations 1490
        18.2.7.1 The Title IV (Acid Rain) Permit (http://epa.gov/air/caa/title4.html) 1490
        18.2.7.2 Permitting Requirements 1490
        18.2.7.3 Emissions Trading 1491
    18.2.8 The Title VI (Stratospheric Ozone) Permit (http://epa.gov/air/caa/caa602.txt) 1491
        18.2.8.1 Permitting Requirements 1491
    18.2.9 Control Technologies: Hardware Options 1491
    18.2.10 Consultation 1492
18.3 The Clean Water Act (http://www.epa.gov/r5water/cwa.htm) 1492
    18.3.1 Existing Use 1492
    18.3.2 Total Daily Maximum Load Levels 1493
    18.3.3 National Permit Discharge Elimination System 1493
    18.3.4 Publicly Owned Treatment Works 1494
    18.3.5 "Wet Weather Flow" Conditions 1494
    18.3.6 CWA Section 319: Nonpoint Source Pollution 1495
    18.3.7 CWA Section 404: Wetland Protection 1495
    18.3.8 Final CWA Notes: Sections 303, 305, and Section 401 Certification 1495
18.4 Nuclear Power Generation 1496
    18.4.1 Hazardous Waste Disposal 1496

18.4.2 Lessons Learned: 3-Mile Island and Chernobyl..1496
18.4.3 Next Generation Designs: VHTR Reactors 1497
18.5 OSHA 1497
18.6 The "Superfund" Law 1498
18.6.1 Amendment to the Superfund Law: SARA 1498
18.7 EPCRA (Emergency Planning and Community Right to Know Act) 1498
18.8 Toxic Substances Control Act (TSCA)...1498
18.8.1 TSCA General Sections References 1499
18.9 Resource Conservation and Recovery Act (RCRA)..1499
18.10 Spill Prevention Control and Countermeasures (SPCC) Act 1499
18.11 Greenhouse Gases and (Global Warming) Climactic Change 1499
18.11 Summary..1500
Acknowledgments..1500

18.1 BACKGROUND

Increased attention to both local and global environmental issues over the past few decades has resulted in heightened focus on these issues. This new attention has increased the workload, knowledge base, and responsibilities of environmental professionals. Issues ranging from the impact of pollution on human health, global climactic change, national security risks, and nuclear fuel processing pose serious concerns.

As with any issue within our society, consideration to health and environment is made with attention to economics and commerce. Known cases exist where deleterious financial impact resulted from negative consumer perceptions of a corporation's environmental or human health–related policies. "Green marketing" wears a friendly neighbor label and instills a barrier-to-entry against competitive products.

The costs associated with environmental and personal safety policies (the "compliance burden") pose real challenges to corporations. However, these costs serve as prudently invested dollars when weighed against the ramifications of environmental policy inaction, whether as regulatory fines or market share loss. Adherence to the various Code of Federal Regulations (CFR) that apply to a large number of U.S. industries can prevent public embarrassment and help maintain customer loyalty.

This chapter will focus on statutes and mandates of the U.S. Environmental Protection Agency (EPA) enforceable by various federal regulations, as well as discuss the state of engineering and management tools available for managing corporate compliance, and provide numerous online resources suitable for in-depth investigation of each topic.

As a quick reference, government agencies that have important bearing on the environment can be contacted via the following sites:

- The U.S. Environmental Protection Agency (EPA)   www.EPA.gov
- The U.S. Nuclear Regulatory Commission (NRC)   www.nrc.gov
- The U.S. Department of Energy (DOE)   www.doe.gov
- Occupational Safety & Health Administration (OSHA)   www.osha.gov
- U.S. Department of Health and Human Services (HHS)   www.hhs.gov

18.2 THE CLEAN AIR ACT (http://epa.gov/air/caa/)

18.2.1 BACKGROUND AND POLLUTANT TYPES

The Clean Air Act (CAA) establishes monitoring and reporting mechanisms, air pollution inventories and reporting thresholds, and economic incentives for controlling and/or reducing pollutant emissions from manufacturing sectors of all types. The CAA monitors the National Ambient Air Quality Standards (NAAQS) via State Implementation Plans (SIP) to ensure consistent air quality for all U.S. residents. Each SIP adheres to the requirements of the CAA by assigning applicable air pollutants to one of two categories:

- Criteria Pollutants: Pollutants for which a primary (personal safety threshold) or secondary (environmental threshold) limit exists. These limits reflect the efforts of experimental science, with criteria pollutants including common pollutants and those that are harmful based on long-term cumulative exposure, or that typically occur in very low amounts.
- Hazardous Air Pollutants (HAPs): Pollutants possessing known carcinogenics or exhibiting other extreme pathological symptoms in humans and/or in the environment.

18.2.1.1 Criteria Pollutants

U.S. regions that exhibit criteria pollutant concentrations at or below defined threshold levels receive "attainment zone" classification. Roughly 90 million Americans currently live in nonattainment zones.

18.2.1.2 Criteria Pollutants and Smog

The adverse effects on human health due to smog include shortness of breath in those with compromised pulmonary function, and may include increased asthma frequencies in young children. Long-term exposure often permanently impairs human respiratory and immune system function, and can lead to death; historical records indicate that 4000 Londoners died in December 1952 due to exposure to unsafe smog levels.

Smog contains high concentrations of ground level ozone (O_3). Ozone naturally exists at high (stratospheric) altitudes, but volatile organic compounds (VOCs) and other criteria pollutants found in gasoline, organic solvents, and paints promote increases at ground level.

Criteria pollutants include combustion products, participates (of various sizes), and volatile organics. Of significance, the EPA includes some slow evaporating, low-vapor pressure compounds such as butyl cellosolve as volatile organic materials. A full list of such pollutants is at the EPA site, http://epa.gov/oar/oaqps/peg_caa/p/egcaa11.html.

18.2.1.3 Hazardous Air Pollutants

Many chemicals pose serious risk to humans and to the ecosystem if not controlled. Many VOCs are within this group, and are regulated within the National Emissions Standards for Hazardous Air Pollutants (NESHAP). They often are carcinogens at low exposure levels, impact reproduction efficiency and fetal development, or yield serious injury or death upon even slight exposure. A current list of such pollutants is located under Section (b)(1) at the EPA site, http://epa.gov/air/caa/caa112.txt.

Since HAP chemical lists often change, this site and its related links need to be revisited periodically. Delisting of certain chemicals (e.g., acetone) can have significant regulatory and fiscal ramifications.

Listed HAP materials may appear as pure compounds, as components of a chemical mixture, or as a by-product of a chemical reaction. Regardless of a facility's pollutant type, the pollutant must be categorized, its levels quantified and then compared to regulatory listings and *de minimis* reporting levels. If a facility emits criteria or other regulated pollutants in excess of specified levels, it is classified as a major source (see description below). The major source threshold varies by location and pollutant.

Material Safety Data Sheets (MSDS) are needed for all materials utilized and/or produced within a facility, whether as a raw material, an intermediate, or a by-product, and must be available to assist in determining a chemical's standing relative to its level of hazard within several categories. When available, even more detailed information is preferred.

18.2.1.4 Maximum Achievable Control Technology Standards

To limit the risk of hazardous air pollutant exposure over time, the EPA mandates that all major sources of HAP emissions operate under Maximum Achievable Control Technology (MACT) standards. The EPA intends to augment and add to the current list of MACT standards; additionally, there are "catch-all" regulations for all major sources of HAPs. One such example is the "MACT Hammer," which allows states to implement MACT standards for industries not currently addressed by EPA definitions. A list of such MACT standards, including the Miscellaneous Organic NESHAP (MON), can be referenced at http://www.dep.state.pa.us/dep/deputate/airwaste/aq/permits/neshaps/hammer_table.pdf.

18.2.2 Compliance Requirements

To implement the CAA, the EPA administers regulations via six discrete "Titles." These titles are defined as follows (courtesy of the EPA):

Title I – Air Pollution Prevention and Control (http://epa.gov/air/caa/title1.html)

- Part A – Air Quality and Emission Limitations
- Part B – Ozone Protection (replaced by Title VI)
- Part C – Prevention of Significant Deterioration of Air Quality
- Part D – Plan Requirements for Nonattainment Areas

Title II – Emission Standards for Moving Sources (http://epa.gov/air/caa/title2.html)

- Part A – Motor Vehicle Emission and Fuel Standards
- Part B – Aircraft Emission Standards
- Part C – Clean Fuel Vehicles

Title III – General (http://epa.gov/air/caa/title3.html)
Title IV – Acid Deposition Control (http://epa.gov/air/caa/title4.html)
Title V – Permits (http://epa.gov/air/caa/title5.html)
Title VI – Stratospheric Ozone Protection (http://epa.gov/air/caa/title6.html)

Requirements associated with Title V are of vital importance to industry, state, and local agencies. A facility's air permit defines the air pollutant threshold limits available to it. The permit declares essential operational parameters of the holder's facility with regard to product and raw material throughputs, and establishes the basic relationships between those throughputs and regulated pollutant emissions amounts.

18.2.3 Source Delineations under Title V

The EPA applies NESHAP, as well as other classifications, in defining an emitter's standing under the Title V program. Where new sources of emissions are created, or where existing sources are modified, other regulations (including both new source review (NSR) and new source performance standards [NSPS]) must be considered.

Agencies grant permits to manufacturers or distributors generally based on one of three regulated emitter definitions, as follows:

- Major Source
  - A stationary source having the potential to emit 25 tons of total HAPs to the air per annum
  - A stationary source having the potential to emit 10 tons of a single HAP to the air per annum
  - A stationary source having the potential to emit 100 tons of any criteria pollutant to the air per annum
  - A stationary source having the potential to emit HAPs in greater quantities than regionally established thresholds. This addresses nonattainment zones, such as the "extreme nonattainment" zone in the Los Angeles basin (10 TPY)
- Synthetic Minor Source
  - A stationary source that accepts federally enforceable limitations on their operations (production) to maintain their emissions below major source thresholds
- Minor Source
  - A stationary source with potential to emit, for all pollutants, less than major source threshold limits

Local and state regulations may further restrict emissions. Details concerning the requirements involved with applying for Title V permits are explained at the EPA site, http://epa.gov/air/caa/title5.html.

18.2.4 Methodologies for Determining Air Emissions

The EPA allows both simple conservative estimation as well as more precise and advanced approaches to emissions estimation. For example, facilities involved in the manufacture of limited products via a continuous process may obtain emission factors from experimental observation (e.g., stack testing) or via continuous emissions monitoring systems (CEMS).

When experimental data are sought, the EPA regulates testing methods. A list of approved methods and their descriptions is available at the Technology Transfer Network Emission Measurement Center (http://www.epa.gov/ttn/emc/tmethods.html).

Facilities unable to implement CEMS or to derive suitable experimental emission factors for their operations can utilize the factors and methods available at the EPA's Clearinghouse for Inventories & Emissions Factors, commonly referred to as CHIEF. A listing of the emission factors, free software, and other useful information is located at http://www.epa.gov/ttn/chief.

In all cases, the emission factors and methods outlined in AP-42 (a compendium of industry-specific EPA emission factors available at the CHIEF site) provide conservative estimates, thereby yielding an auditable "worst case" inventory of air pollutants nationally and regionally. Manufacturers adhering to AP-42 likely overestimate (and over report) air emissions.

The added expense attributable to an increased compliance burden resulting from inaccurate estimation of emissions, whether through administrative headcount or capital equipment, likely far exceeds the costs of implementing advanced methods for estimating air emissions. AP-42 can usually be excelled, and the variance between estimated emissions and actual emissions reduced considerably.

Available environmental management information systems (EMIS) may or may not provide advanced methods for emissions estimation, and should be reviewed versus their use of AP-42 or better.

18.2.5 New Source Reviews and Modifications

The Clean Air Act requires a revised permit for any change to a major source facility's "emissions fingerprint" resulting from the addition of physical emission points or the change of a facility's mode of operation (NSPS). All new and modified major sources of HAPs must comply with MACT standards.

These modifications may be positive or negative with regard to Title V permitted emission levels, but such a change requires an update to the air pollutant inventory under SIP rules. In many cases, a net zero change in emissions will be sought by combining parallel activities—one positively divergent, one negatively divergent. Substituting non-HAP for a HAP classified solvent in cleaning operations is a simple example whereby reduction of use in one area may allow for increased use, or possibly retooling, in another.

18.2.6 New Regulations

The Miscellaneous Organic NESHAP (MON) is structured to reduce the floor levels of major sources and to reassess the definitions of and methods by which MACT standards and levels are achieved (http://www.epa.gov/ttn/atw/mon/monpg.html)

Certain sites currently classified as minor emissions sources will find themselves classified as major sources under MON, and adherence to the requirements of the MACT guidelines posed by the major source classification will likely be costly. It is recommended that manufacturers of VOC-related products acquaint themselves with the MON and make necessary preparations. Costs associated with major source classification within MON often range upwards of $5–10 million per facility.

18.2.7 Title IV (Acid Rain) and Title VI (Stratospheric Ozone) Considerations

18.2.7.1 The Title IV (Acid Rain) Permit (http://epa.gov/air/caa/title4.html)

The environmental, health, and cultural impacts of acid rain were well documented in the late 20th century. Acid rain results from combustion of the sulfur and nitrogen contents of fossil fuels. Sulfur and nitrogen oxides combine with atmospheric water to form acids. The prevalence of mobile (e.g., automotive and aircraft) sources globally hints at the requirement for control of Title IV compounds and precursors.

Automotive catalytic converters (http://www.all-catalytic-converters.com/techtip1.html) alleviate this issue (as well as those associated with smog formation) within the mobile source category; however, they are not the perfect solution to emissions problems.

Within the stationary source category, acid rain forming compounds primarily result from combustion-based power plant facilities.

18.2.7.2 Permitting Requirements

Title IV addresses "affected sources," which comprise "affected units." These affected units are subject to emissions reductions or limitations under Title IV, and are granted allowances per annum, with each allowance equal to 1 ton of SO_2 per year.

Under Title IV, facilities must document and report annually regarding the aggregate amount of acid-rain precursors in the form of SO_2 and NO_X. Determination of these amounts are traditionally handled by stack testing, with the emissions factor subsequently calculated in units of pound of species emissions per pound of fuel combusted, per hour of operation, or other. Title IV thresholds are subject to penalty when exceeded. Additionally, in the case of violation, the EPA can enforce a mandatory emissions reduction in the subsequent year to offset the excess emission value.

Power generation facilities are evaluated within the Title IV requirements versus baselines established within the National Acid Precipitation Assessment Program (NAPAP) Emissions Inventory. Descriptions of source and unit types applicable under Title IV can be found at http://epa.gov/air/caa/caa402.txt.

18.2.7.3 Emissions Trading

Emission allowances under Title IV are eligible for trading and/or carry forward contingent upon adherence to EPA administrative and regulative rules. A summary discussion of such activity is provided at http://www.epa.gov/airmarkets/progsregs/arp/docs/clearingtheair.pdf.

18.2.8 THE TITLE VI (STRATOSPHERIC OZONE) PERMIT (http://epa.gov/air/caa/caa602.txt)

Decreasing levels of stratospheric ozone (O_3) over the polar caps have lead to the speculation that a reduction in high-atmospheric (stratospheric) ozone levels will lead to increased skin cancers rates (caused by excessive, nonhistorical UV-band radiation exposure).

Chlorine or other halogen bearing hydrocarbons (typically chlorofluorocarbons) defined as either class 1 or class 2 substances comprise the major stratospheric ozone depleting substances. EPA has established a midterm objective of an atmospheric chlorine content <2 parts per billion (ppb).

Glacial ice core samples indicate pre-industrial-age levels of atmospheric chlorine at <1 ppb. By the late 20th century, the concentration had reached 3.6 ppb. Continuation of current efforts may return the Earth's atmosphere to <1 ppb levels shortly after year 2100. The following link provides a summary analysis of CFC/HCFC atmospheric conditions compiled by the Canada Chemical Producers Association: http://en.wikipedia.org/wiki/ozone_layer.

18.2.8.1 Permitting Requirements

Operators must report all class 1 and class 2 compounds whether imported, exported, or manufactured. Additionally, users of nonessential (as determined by administrator review) class 1 or class 2 substances are prohibited from sale or transfer of such substances.

18.2.9 CONTROL TECHNOLOGIES: HARDWARE OPTIONS

See examples of pollution control devices at http://www.epa.gov/ebtpages/treapollutioncontrol.html or http://cn.wikipedia.org/wiki/air_polution/control_devices.

Numerous control technology hardwares are available for the reducing airborne emissions. Several types are reviewed below with general comments applied to each.

1. Oxidizers. Thermal oxidizers rely on high temperature combustion of VOC/HAP species to reduce emissions levels. Although often expensive, they provide high destruction efficiencies, often >99.8%. Hence, they often prove a useful asset with high measurable return.
2. Condensers. Condensers condense volatile species present in the vapor stream resulting in sidewall condensation and reflux to the liquid phase. Vapor pressure data are used to predict control efficiencies as functions of condenser temperature.
3. Scrubbers and absorbers. Scrubbers capture particles, gases, and/or acidic vapors. In treating particulate matter, scrubbers remove particulates in mists of liquid (typically water) and transport the particles away from the vapor stream. In these cases, high flow rates of liquid mist droplets increase the removal efficiency of the solids. In the treatment of gases, it is essential to select a liquid in which the gas is soluble and to use liquid and vapor flow rates that match the targeted capture percentage based on the system's solubility information. In both cases, the liquid to vapor ratio is an important operating variable.
4. In acid gas removal, the scrubber employs a solid alkaline "sorbent" material mixed with water. The gas phase acid adsorbs onto the sorbent surface, and then forms a salt.

Available sorbent surface area, vapor-liquid stream mixing, and reactivity will dictate the effectiveness of such "dry sorbent injector" scrubbers.
5. Biofilters. Although relatively new, biofiltration has been used in some instances for reduction of emissions and odors. Biofilters consist of microbes grown on porous media, and in small-scale implementations, they provide >90% reduction in emissions levels.
6. Cogeneration. Several new technologies combine destructive oxidation technologies with cogenerative site-based power facilities. The energy utilized in emissions oxidation can sometimes be reclaimed to yield economic benefits.

18.2.10 CONSULTATION

In general, numerous professional consulting service providers can help prepare Title V permits, annual emissions inventory reports, and other state-specific documents. An Internet search can identify potential regional candidates.

The EPA expects that EMIS systems be implemented at all major categorization facilities, and site implementation of such systems typically constitutes the first requirement of an EPA corrective action plan. As stated above, more accurate systems of emissions estimation can lead to reduced compliance burden and potentially eliminate the need for source classification adjustment.

In all cases, adherence to the full range of Federally Applicable Requirements, the crux of which are provided above, combined with the goal of maintaining operational flexibility, should be the objective of any permit-focused environmental engineer.

18.3 THE CLEAN WATER ACT
(http://www.epa.gov/r5water/cwa.htm)

The Clean Water Act (CWA) provides water quality standards (WQS) for surface waters in the United States. The CWA gives the EPA the authority to implement pollution control and regulatory strategies.

Water quality criteria (WQC) provide chemical-specific numerical values of amount (magnitude), interval of test, and frequency of occurrence that must be met in tandem to meet the requirements of the WQS. For chemicals with WQC documented at EPA, states and Indian tribes must also assign WQC (not necessarily of identical values per the EPA metrics).

The EPA typically allows exemptions from the WQS for locations downstream of point source discharges ("stream mixing") as well as for relatively rare conditions where a significant increase in flow diminishes water quality (for example, storm sewer flooding). Waivers based on mixing are regulated and require that mixing zones do not extend throughout a body of water or from bank to bank of a river or stream. Additionally, waivers will not allow deleterious impact to designated use (DU) zones. A contiguous region meeting WQS must be maintained.

18.3.1 EXISTING USE

Designated use nomenclatures indicate the level of quality to which a community aspires; the "existing use" establishes a benchmark. The intent of the existing use delineation assures that no commercial, public, or private activity reduces water quality to subexisting use levels, and regulating bodies will not authorize subexisting use conditions ("Tier I rule").

The EPA protects zones that significantly exceed current WQS from negative impacts, thus making it difficult to impact a water body while ensuring the quality remains above its minimum. Prior to initiation of such an adverse activity, several criteria must be met as defined under the "Tier II rule."

Certain waterways are designated as outstanding national resource waters, and no significant degradation in water quality is permitted for any significant period of time ("Tier III rule"). These

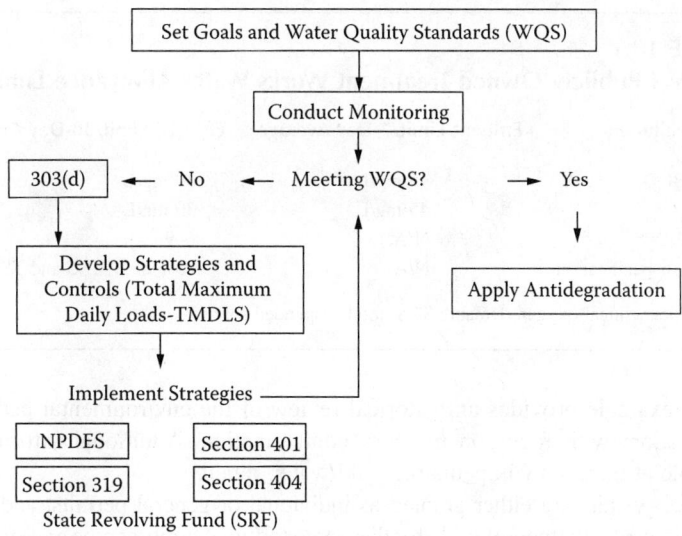

FIGURE 18.1 The EPA chart "Introduction to the Clean Water Act."

three "anti-degradation rules" provide high-level management metrics to regulating and enforcement bodies within the state, national, or regional level.

18.3.2 TOTAL DAILY MAXIMUM LOAD LEVELS

Infrequent measurements of water quality result in large statistical errors in estimation of performance versus WQS. When the available evidence indicates achieving WQS standards unlikely or impossible via application of source control technologies, the EPA imposes total maximum daily load (TDML) levels of individual pollutant discharge from point sources.

Unfortunately, application of TMDL rules does not typically result in achievement of WQS due to the large impact of nonpoint sources (e.g., soil runoff). Nonetheless, TMDLs for such defined pollutants as clean sediments, nutrients (nitrogen and phosphorus), pathogens, acids/bases, heat, metals, cyanide, and synthetic organic chemicals are enforced.

Caps are established, with margins of safety, and completed facility TDML plans and strategies are provided to the EPA for approval or disapproval. The time defined within a TDML plan does not necessarily have to equal 1 day, and a reserve for future polluting activities is typically comprehended. Approved agencies or parties will regularly audit site performance against a facility's TDML plan.

The EPA chart "Introduction to the Clean Water Act" (Figure 18.1) highlights the key focus areas of the CWA (http://www.epa.gov/watertrain/cwa/cwa1.htm).

18.3.3 NATIONAL PERMIT DISCHARGE ELIMINATION SYSTEM

Although the CWA states the illegality of pollution discharge from point sources, it does allow for controlled discharges via the *National Permit Discharge Elimination System* (NPDES). NPDES permits limit the levels of point source pollutant discharge allowable over defined periods (some sectors of industry are exempt; see http://www.epa.gov/watertrain/cwa/cwa37a.htm). Indirect discharges, i.e., those treated by a publicly owned treatment works (POTW) facility prior to release, are not covered under the NPDES permit program (see below).

The EPA defines waste load allocations (WLA) to point source pollution sources (generally or specifically), which are then regulated against the defined TMDL (if applicable) established in the NPDES permit site. A useful example of the administration of Waste Load Allocations within the State of California can be found at http://www.swrcb.ca.gov/rwqcb2/Agenda/03-19-03/03-19-

TABLE 18.1
General Publicly Owned Treatment Works Water Allowance Limits

Pollutant	Effluent Limit/7-Day Average	Effluent Limit/30-Day Average
5-day BOD	45 mg/L	30 mg/L
TSS	45 mg/L	30 mg/L
pH	N/A	6–9
Removal specification	N/A	85% of BOD 5 and TSS

BOD, biochemical oxygen demand; TSS, total suspended solids.

5bmou.doc. This example provides an historical review of the environmental performance of the waterway in question, with focus on mercury contamination. Additional information on waste quality is available at http://en.wikipedia.org/wiki/water_quality.

Often, NPDES permits are either granted as individual or general permits, and contain specific mass amounts of defined pollutants eligible for discharge (effluent limits at pipe end) from the permitted operation. To ensure compliance, internal auditing plans and quality control policies are defined. Further requirements for using best management practices, monitoring, and reporting are included.

Detailed information on NPDES permits is available at http://www.epa.gov/watertrain/cwa/cwa40.htm.

As with the CAA, the CWA establishes performance standards (the maximum effluent concentrations allowable) and control technology guidelines applicable to various industries and sectors. These guidelines reflect data compiled from numerous industry sources, and define expected discharge levels as a function of best available technology economically achievable (BAT or BATEA). More information on effluent streams is at http://www.epa.gov/waterscience/guide/.

For cases where application of BAT to effluent stream discharges will not meet the requirements of a water source's DU, and for which TMDL limits are not established, the permitting agency will apply water quality–based effluent limits (WQBEL). WQBEL includes an analysis of the level of effluent stream dilution required to retain DU levels. Where WQBEL is employed, a high risk has been assigned to the water source, and economics will often be secondary to technology.

18.3.4 Publicly Owned Treatment Works

Both primary and secondary levels of sequential treatment specifications exist for POTW plants. The former involves the settling and filtering of solids (activated sludge). Table 18.1 provides typical limits imposed on secondary treatment criteria.

The EPA offers supplemental information regarding activated sludges involving biosolids, including rules and standards (see http://www.epa.gov/watertrain/cwa/cwa42.htm).

18.3.5 "Wet Weather Flow" Conditions

Due to heavy rainfall, wastewater overflows from sewers or industrial facilities may result in underperformance versus WQS. Nevertheless, affected sewer systems, typically combined sewer overflow (CSO) or municipal separate storm sewer systems (MS4), remain subject to NPDES regulations. CSO systems, found in older residential or municipal areas, allow transportation of both rainwater and raw sewage within a common stream; MS4 systems separate the two flows.

Attention to the public health challenges posed by CSO sewers focus on control of pathogen dispersion. From NPDES perspectives, CSO systems are restricted from discharging untreated sewage during periods of dry weather.

NPDES regulates MS4 sewers that discharge to surface waters because of the potential for the water line to contain various levels of contaminants such as metals, oils, and pesticides. Permits for MS4s tend to focus on response planning and "design against failure" rather than on actual effluent stream concentrations. More information is available at http://www.epa.gov/water-train/cwa/cwa46.htm.

Industrial sites may or may not require individual NPDES permits for storm water runoff pertaining to overflow drainage into surface water or MS4 sewage systems. Frequently, these conditions will be covered under NPDES documentation via the site's storm water pollution prevention plan (SWPPP) (see http://cfpub.epa.gov/npdes/stormwater/indust.cfm).

SWPPP phase I applies to industrial facilities and to construction sites affecting more than 5 land acres, while SWPPP phase II regulations affect those construction sites affecting between 1 and 5 acres. SWPPPs are expected to provide plans for minimization of soil erosion, protect vegetation and wetlands, and define runoff source points to standing water.

Additionally, a spill prevention control and countermeasure (SPCC) plan may be required at facilities where oil materials are stored. Importantly, edible oils are often included in this designation. SPCC is discussed in more detail in Section 18.9.

18.3.6 CWA Section 319: Nonpoint Source Pollution

Nonpoint source runoff is the major source of water pollution in the United States. Topsoil runoff from farms and pastures represents the major source of this pollution.

Contrasting the federal position on point sources, the EPA provides grants ("319 funds") and other modes of assistance and incentive to such polluting states. The EPA's objective is to curtail pollution from such erosion sources and to define auditable levels of watershed TMDL. For more on nonpoint pollution, see the EPA CWA Web site, http://www.epa.gov/watertrain/cwa/cwa52.htm.

18.3.7 CWA Section 404: Wetland Protection

The "wetland protection program" (administered by the Army Corps of Engineers) addresses the displacement of dredged or fill-material into defined wet areas such as bottomland hardwood swamps, intermittent streams, or oceans (http://www.epa.gov/owow/wetlands/facts/fact10.html). The Army Corps of Engineers has ultimate authority for defining a Section 404 wetland, and for assuring compliance within Section 404 boundaries.

The wetland protection program requires that any commercial activity requiring the destruction of existing wetlands be essentially dependent on the presence of water. If not, the activity will be relocated to an unprotected area. Prior to construction on a protected wetland area, demonstration of destruction mitigation plans along with a wetland replacement strategy and/or a plan for improving an existing wetland must be presented.

18.3.8 Final CWA Notes: Sections 303, 305, and Section 401 Certification

Special attention is given waters that are directly threatened or do not meet WQS. These receive Section 303(d) or Section 305 (threatened or impaired waters) classification. Section 303(d) includes those surface water sources impacted by pollutants, whereas "nonpollutants" impair Section 305 waters. Where the source of aquatic life impairment is unclear, the source receives Section 303(d) classification. Waters classified as threatened or impaired require biennial progress reporting, with prioritization of TMDL levels and WQC established accordingly.

Section 401 certification ensures that all states indicate compliance status under the CWA. By requiring such a certification, states downstream of the discharge can become involved in the certification and permitting processes (http://www.epa.gov/watertrain/cwa/cwa58.htm).

The hydropower dam presents an interesting case study, as dams by definition cause great variance between existing and DU classifications and the final state of the dammed water.

18.4 NUCLEAR POWER GENERATION

Currently, nuclear energy serves as the second largest source (behind coal) of electrical power in the United States. Nuclear sources are inherently clean, but significant risks and therefore governmental regulations apply to the waste disposal methods.

The Nuclear Regulatory Commission (http://www.nrc.gov/) oversees all permitting, guidelines, and definition of operational requirements via the Office of Nuclear Reactor Regulation.

18.4.1 Hazardous Waste Disposal

Nuclear reactor facilities undergo hazardous waste management procedures every 12 to 24 months. Older fuel rods are removed, generating approximately 2000 metric tons of used radioactive material annually (the National Energy Institute presents the analogy of one football field covered to the depth of 5 yards during the past 40 years).

Based on laws against the proliferation of nuclear weapons, nuclear waste cannot be recycled, and is assigned a once-through single use designation. Department of Energy policies as administered by the U.S. Department of Energy Office of Health, Safety, and Security can be consulted at http://www.hss.energy.gov/NuclearSafety/techstds/standard/standard.html.

The Office of Environmental Management maintains the DOE's waste management policies. Two discrete levels of nuclear wastes are considered:

1. Low-Level Waste
   a. Radioactive by-products of NRC-licensed or DOE-permitted activities that are not high-level waste.
   b. Radioactivity, containing mostly beta and gamma emissions, persists for 5 to 50 years; contains elements with atomic numbers less than that of uranium (92), and is eligible for disposal via authorized independent or government-approved facilities, as determined by the Low-Level Waste Disposal Facility Federal Review Group (LFRG).
2. High-Level Waste
   a. Irradiated nuclear fuel ("spent fuel"), plus both liquid products of reprocessing and the solids into which such liquids have been incorporated.
   b. Contains numerous radioactive materials of high levels of radiation and with long half-lives. Such materials are generated either through nuclear power generation or Department of Defense activities. In general they are assigned for disposal at the U.S. Government Yucca Mountain National Depository.

For the centralized disposal of high-level nuclear wastes, Yucca Mountain, Nevada, was selected in large part because of national security concerns.

18.4.2 Lessons Learned: 3-Mile Island and Chernobyl

The incident at Unit 2 of the 3-Mile Island (Pennsylvania) reactor during March 28 to April 1, 1979, led to the formation of the Institute of Nuclear Power Operations (INPO). The INPO sets objectives, guidelines, and criteria for all U.S.-based nuclear power generating facilities. In 1985, this group established the National Academy for Nuclear Training, which accredits all plant operators and supervisors.

In April 1986, fundamental flaws in Soviet reactor designs, combined with procedural human error, resulted in a steam explosion and the expulsion of 5% of the Chernobyl-4 (Ukraine, USSR) reactor core into the atmosphere. This disaster sparked a review of (especially Russian) reactor designs and further emphasized the need for strictly enforced operational standards, guidelines, and policies. Differences between current "East" and "West" reactor designs can be reviewed at the site http://www.insc.anl.gov/neisb/neisb4/NEISB_1.1.html.

18.4.3 Next Generation Designs: VHTR Reactors

Very high temperature reactors (VHTR), currently being developed, not only address energy generation but also provide a relatively cheap source of molecular hydrogen. Hydrogen is currently proposed as an alternate fuel source for vehicular power, but current methods for obtaining it pose severe environmental problems.

In a VHTR cogeneration scheme, a fraction of the energy generated acts to produce and store hydrogen. When combined with an adequate logistics system, hydrogen can serve as a clean, renewable combustion energy source for personal or industrial use. Health and safety concerns, combined with economics considerations, will dictate the future of VHTR reactors. For further review of this and other nuclear power generation roadmap technologies, consult the site http://energy.inel.gov/gen-iv/default.shtml.

18.5 OSHA

The Occupational Safety and Health Act has the objective of protecting workers' lives and health by the enforcement of policies, regulations, and guidelines defining working and workplace conditions. OSHA lists regulations and compliance at the following sites:

http://www.osha.gov/comp-links.html
http://www.osha.gov/html/comp-guides.html

Employers must operate within these guidelines except for rare exceptions (variances). Variance types include:

1. Temporary
    a. Issued when an employer-operator cannot meet the requirements of an OSHA regulation due to a lack of available staff, equipment, or materials. Employers governed by a state-run worker protection program must contact their local plan representative for such a variance, but operators that span geographic areas or states that include OSHA-regulated regions must apply directly through OSHA.
2. Permanent
    a. Issued to an employer that demonstrates pre-existing methods superior to OSHA requirements. In attaining a permanent variance, the employer must inform the employee workforce of the intention to seek a variance, and allow for an employee hearing if requested.
    b. At any time within 6 months of the award of a permanent variance, employees and employers maintain the right to petition for the rescinding of the variance. OSHA may also follow this course at its own volition.
3. Experimental and Other Variances
    c. Granted by the Secretary of Health and Human Services to ascertain the impact of operations changes that require variances. These experimental variances are intended to improve existing OSHA methods.
    d. Because of National Security reasons, the Department of Defense may hold other variances.

Variances are not retroactive. During the period of operation prior to receipt of a variance, if issued, OSHA may provide an interim order, allowing continued activity based on current methods. Details of such an interim order are subject to site publication and employee-employer review.

18.6 THE "SUPERFUND" LAW

The "Superfund" law derives from citizen concerns over the health impacts of decades of accumulated hazardous wastes within or near communities. More accurately labeled the Comprehensive Environmental Response, Compensation, and Liability Act (CERCLA), the Superfund governs the removal and destruction of toxics and also establishes disposal procedures through the Office of Superfund Remediation Technology Innovation (OSRTI). Prioritization activities are within the jurisdiction of the OSTRI.

18.6.1 Amendment to the Superfund Law: SARA

The Superfund Amendments and Reauthorization Act (SARA) provides guidelines to improve Superfund enforcement methods and criteria. SARA also refines the Hazardous Rankings System, and by OSTRI prioritizes remediation activities. Rules and regulations under the Superfund and its SARA amendment are accessible at http://www.epa.gov/superfund/action/index.htm.

18.7 EPCRA (EMERGENCY PLANNING AND COMMUNITY RIGHT TO KNOW ACT)

EPCRA, also known as "Title III of SARA," protects local communities from the hazards of toxic pollutant discharge (http://www.epa.gov/region5/defs/html/epcra.htm). EPCRA mandates regulations surrounding the maintenance and site availability of such forms as MSDS, toxic chemical release forms, and emergency and hazardous chemical inventory forms (see http://www4.law.cornell.edu/uscode/html/uscode42/usc_sup_01_42_10_116_20_II.html

Under EPCRA, state emergency planning commissions and local emergency planning councils must be formed. EPCRA ensures the documentation of quantity and location of hazardous materials present in excess of Toxic Substance Controls Act (TSCA) thresholds. SARA Title III information must be submitted to state and local (including fire) agencies under SARA Sections 312 and 313.

EPCRA, in combination with the Pollution Prevention Act (PPA), administers the toxic release inventories (TRI) program. Industry and federal facilities emitting toxic materials must report on annually released quantities under SARA Section 313 (Form R). Although the EPA does not host a site containing all TRI materials, versions of this document can be obtained via the Government Printing Office (http://www.access.gpo.gov/). An executive summary, including requirements for submittal to the TRI via Form R submittals, is available at http://www.epa.gov/tri/tri-data/tri01/press/executivesummarystandalone.pdf.

18.8 TOXIC SUBSTANCES CONTROL ACT (TSCA)

More than 75,000 chemicals enter, are manufactured within, and/or are consumed within the United States each year. The TSCA (http://www.epa.gov/region5/defs/html/tsca.htm) provides methods for categorizing newly created and existing chemicals, and supplements the Clean Air Act and the Toxic Release Inventory portion of EPCRA. The TSCA assists in quantifying types and amounts of chemicals both produced and released to the environment annually.

Commercial activities that result in the creation of new chemicals must receive authorization from the EPA prior to production or distribution. To receive such authorization, a pre-manufacture notice (PMN) must be filed with the EPA's Office of Pollution Prevention and Toxic Substances.

Environmental Engineering: A Review of Issues, Regulations, and Resources 1499

For a listing of toxic substance categories as defined within the Chemicals on Reporting Rules database, access https://www.unh.edu/ehs/pdf/TSCA-CORR-By-CAS.pdf.

18.8.1 TSCA GENERAL SECTIONS REFERENCES

Asbestos:

http://www4.law.cornell.edu/uscode/html/uscode15/usc_sup_01_15_10_53_20_II.html

Indoor radon abatement:

http://www4.law.cornell.edu/uscode/html/uscode15/usc_sup_01_15_10_53_20_III.html

Lead exposure reduction:

http://www4.law.cornell.edu/uscode/html/uscode15/usc_sup_01_15_10_53_20_IV.html

18.9 RESOURCE CONSERVATION AND RECOVERY ACT (RCRA)

Administered by the Office of Solid Wastes, the RCRA establishes regulations regarding the effective handling of spilled, leaked, or otherwise improperly disposed wastes (http://www.epa.gov/rcraonline/). The RCRA also manages hazardous wastes from their origin to disposal, and oversees garbage and industrial waste handling. Example wastes not covered under RCRA are nuclear (see above), medicinal, and animal.

The site http://www.epa.gov/epaoswer/osw/topics.htm provides an information source addressing those wastes covered under the RCRA, including regulations surrounding underground storage tanks.

18.10 SPILL PREVENTION CONTROL AND COUNTERMEASURES (SPCC) ACT

The SPCC, ultimately a component of the Clean Water Act, focuses on prevention of oil spills into U.S. waters or shorelines. To minimize the impact of spills, the SPCC requires filing of facility response plans (FRP) under its oil pollution regulations.

The SPCC addresses owners or operators that drill, produce, gather, store, use, process, refine, transfer, distribute, or consume oil and oil products. With such a broad classification, de minimis standards are established that set maximum vessel or throughput quantities. The SPCC defines threshold EPA reportable event levels as:

- Two spills in excess of 42 gallons each within any given 12-month period
- One spill in excess of 1000 gallons

Highlights, de minimis standards, and additional information including FRP guidelines, are available at http://www.epa.gov/oilspill/spccrule.htm.

18.11 GREENHOUSE GASES AND (GLOBAL WARMING) CLIMACTIC CHANGE

No regulations currently apply to greenhouse gas (GHG) emissions. The United States and other nations are conducting research on global warming while establishing the current air inventory of GHG. (For a review, see http://yosemite.epa.gov/oar/globalwarming.nsf/content/emissions.html.)

Although few dispute that current industrial processes, automobiles, power production, etc., impact GHG levels, there is still much debate regarding human-induced and naturally occurring variations. Recently excavated ice core samples indicate that CO_2 levels are rising at a much faster rate than at any time in the past 400,000 years. Such cores show current atmospheric CO_2 levels of 380 ppmv, exceeding previous cyclical highs by roughly 18%. The following links to graphical data can assist in drawing conclusions:

- http://www.grida.no/climate/vital/02.htm (400,000-year CO_2 atmospheric measurements, Vostok ice sheet, Antarctica)
- http://www.grida.no/climate/vital/06.htm (recent data from Mauna Loa test site, altitude 4000 meters)
- http://www.grida.no/climate/vital/07.htm (CO_2 levels, 1870 to 2000)

Although the scientific community generally believes that decreasing polar ice cap size and commensurate rising ocean levels correlate with increased GHG levels and global warning, the full ramifications of the evidence remain contested.

18.11 SUMMARY

The daily tasks of compliance management in the face of potentially changing government regulations are facilitated by a strong industry complement of consulting expertise. Additionally, software packages of varied types are available to assist in automating difficult compliance tasks.

Environmental engineering issues span a wide gamut of technologies and requirements both mundane and essential. Ensuring compliance with government regulations, though never a glorified chore, nonetheless provides a necessary function that ensures the protection of interests ranging from the economic to the vital.

The roles of environmental engineers in ensuring and monitoring the healthy progress of developing nations, assisting in defining the directions of new and necessary technologies, and in protecting our children and ecosystems cannot be underestimated.

The economic impact of a dearth of environmental stewardship is too great to gamble.

ACKNOWLEDGMENTS

The authors thank the EPA (http://www.epa.gov/) for providing many of the links cited above for public use. "Exit links" provided by the EPA have been referenced in many instances. The EPA or the authors do not verify the accuracy and content of these links.

Additionally, the authors wish to thank their colleagues at Greenfield Environmental, Inc., for their general assistance and numerous contributions.

19 Biochemical Engineering

James M. Lee

CONTENTS

19.1 Introduction 1502
    19.1.1 Biotechnology 1502
    19.1.2 Biochemical Engineering 1502
19.2 Cells and Enzymes 1504
    19.2.1 Microbial Cells 1504
        19.2.1.1 Microbial Nomenclature 1505
        19.2.1.2 Bacteria 1505
        19.2.1.3 Fungi 1505
        19.2.1.4 Culture Media 1506
    19.2.2 Animal Cells 1506
    19.2.3 Plant Cells 1507
    19.2.4 Cell Growth Measurement 1508
    19.2.5 Enzymes 1508
        19.2.5.1 Nomenclature of Enzymes 1509
        19.2.5.2 Commercial Applications of Enzymes 1509
19.3 Kinetics 1509
    19.3.1 Cell Kinetics 1509
        19.3.1.1 Growth Cycle for Batch Cultivation 1509
        19.3.1.2 Unstructured Cell Kinetic Models 1510
        19.3.1.3 Structured Models 1512
    19.3.2 Simple Enzyme Kinetics 1513
        19.3.2.1 Michaelis–Menten Approach 1513
        19.3.2.2 Briggs–Haldane Approach 1514
        19.3.2.3 Numerical Solution 1515
        19.3.2.4 Evaluation of Michaelis-Menten Parameters 1516
        19.3.2.5 Inhibition of Enzyme Reactions 1516
19.4 Bioreactor Design 1518
    19.4.1 Bioreactors 1518
        19.4.1.1 Stirred-Tank Bioreactor (or Fermenter) 1518
        19.4.1.2 Alternative Fermenters 1519
    19.4.2 Mode of Bioreactor Operations 1520
        19.4.2.1 Batch or Plug-Flow Bioreactors 1520
        19.4.2.2 Continuous Stirred-Tank Bioreactor 1522
19.5 Agitation and Aeration 1525
    19.5.1 Introduction 1525
    19.5.2 Determination of Oxygen-Absorption Rate 1526
        19.5.2.1 Estimation of Equilibrium Oxygen Concentration C_L^* ...1526
        19.5.2.2 Sodium Sulfite Oxidation Method 1526
        19.5.2.3 Dynamic Technique 1527

    19.5.3 Correlation for $k_L a$ 1528
        19.5.3.1 Bubble Column 1528
        19.5.3.2 Mechanically Agitated Vessel 1528
    19.5.4 Power Consumption 1528
    19.5.5 Scale-Up 1529
References 1529

19.1 INTRODUCTION

Biochemical engineering involves the scale-up of biological processes. One of the earliest examples of the successful scale-up of a biological process was the production of penicillin during World War II. Industrial microbiologists used stirred-tank fermenters to culture molds on a large scale. During the 1960s and 1970s, biochemical engineering grew rapidly with the increased needs for the development of biological processes in the food, beverage, and pharmaceutical industries. The activities in bioprocessing grew even further as Arab oil embargos prompted the use of renewable energy sources, such as biomass. For example, cellulose, a major component of wood, can be broken down enzymatically to glucose, hence producing ethanol via yeast fermentation.

The research for the utilization of renewable resources dwindled in the early 1980s because of the reduction in energy prices. However, the area found another opportunity to grow as biologists discovered a way to manipulate genes of living organisms. The manipulation of genes is known as genetic engineering or biotechnology.

19.1.1 BIOTECHNOLOGY

Biotechnology is broadly defined as "commercial techniques that use living organisms, or substances from those organisms, to make or modify a product, including techniques used for the improvement of the characteristics of economically important plants and animals and for the development of microorganisms to act on the environment" [1]. This very broad definition includes not only the newly developed recombinant DNA technology, but also the traditional areas involving agriculture, animals, human heath, and industrial microbiology. Since ancient days, people utilized microorganisms to produce fermented beverages (alcohol, sweet refreshments) and food (cheese, soy sauce, fermented vegetables, sweets, and bread), though they did not understand those biological changes. People crossbred plants and animals for better yields.

However, in recent years, the term biotechnology is being used to refer to novel techniques such as recombinant DNA. This technique is also known as genetic engineering, a misleading description of scientific endeavors. The applications are numerous, as listed in Table 19.1. Previously expensive and rare pharmaceuticals (such as insulin for diabetics, a human growth hormone to treat children with dwarfism, interferon to fight infection, vaccines to prevent diseases, and monoclonal antibody for diagnostics) are now produced from genetically modified microbial cells or hybridoma cells inexpensively and in large quantities. Disease-free seed stocks and healthier, higher-yielding food animals have been developed. Important crop species can be modified to have traits that resist stress, herbicides, and pests. Furthermore, recombinant DNA technology can be applied to develop genetically modified microorganisms so that they will produce various chemical compounds with higher yields than available from unmodified microorganisms.

19.1.2 BIOCHEMICAL ENGINEERING

In order to cultivate genetically modified cells in large quantities, we need to develop a large-scale process that is technologically efficient and economically viable. A typical biological process (bioprocess) involving microbial cells can be illustrated as in Figure 19.1. Raw materials are treated and mixed with other ingredients that are required for cells to grow well. The liquid medium,

TABLE 19.1
Applications of Biotechnology

Area	Products or Applications
Pharmaceuticals	Antibiotics, antigens, neurotransmitters, gamma globulin, growth hormone, serum albumin, immune regulators, insulin, interferon, interleukins, lymphokines, monoclonal antibodies, neuroactive peptides, tissue plasminogen activators, vaccines
Animal agriculture	Development of disease-free seed stocks; healthier, higher-yielding food animals
Plant agriculture	Transfer of stress-, herbicide-, or pest-resistance traits to crop species; development of biological insecticides
Specialty chemicals	Amino acids, enzymes, vitamins, lipids, hydroxylated aromatics, biopolymers
Environmental applications	Mineral leaching, metal concentration, pollution control, toxic waste degradation, enhanced oil recovery
Commodity chemicals	Acetic acid, acetone, butanol, ethanol
Bioelectronics	Biosensors, biochips

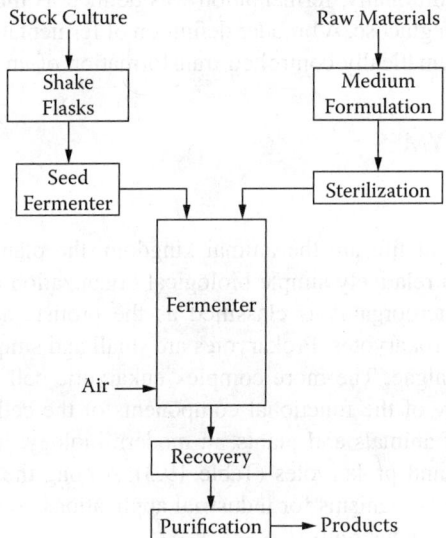

FIGURE 19.1 Typical biological process.

sterilized and introduced to a bioreactor or fermenter, is typically equipped with agitators, baffles, air spargers, and sensing devices for the control of the operating conditions. A pure strain of microorganisms is introduced into the vessel. The number of cells multiplies exponentially after a certain period of lag time and reaches a maximum cell concentration as the medium is depleted. The fermentation is then stopped and the contents are pumped out for the product recovery and purification. This process is operated either by batch or continuous mode.

For large-scale operations, biochemical engineers work with biological scientists to: (1) obtain the best biological catalyst (microorganism, animal cell, plant cell, or enzyme) for a desired process, (2) design and operate the bioreactor in the most efficient way, and (3) separate the desired products from the reaction mixture in the most economical way. Biological processes have advantages and disadvantages as compared with traditional chemical processes (Table 19.2). Bioprocessing is most promising for high-price, low-volume products such as pharmaceuticals, medical enzymes, and analytic reagents (Table 19.1). There are, however, several products produced through bioprocessing that are low-price, high-volume products. Some example of these are high-fructose corn syrup (and

TABLE 19.2
Characteristics of Bioprocessing

Advantages	Disadvantages
The reaction conditions are mild: typically atmospheric conditions.	Complex product mixture and low productivity make the product separation expensive.
Enzyme catalysts are highly specific. A great variety of enzymes exist that can catalyze a very wide range of reactions.	Cells tend to lose desirable characteristics. Enzymes are sensitive and unstable in harsh conditions.
The major raw material for bioprocesses is biomass, which is renewable.	Fermenter and medium have to be sterilized and kept in sterile conditions.

enzymes involved on the process such amylase and glucose isomerase) and laundry additives (protease enzyme).

Lastly, the word *fermentation* needs to be discussed due to the confusion caused by different definitions of the word. Traditionally, fermentation was defined as the process for the production of alcohol or lactic acid from glucose. A broader definition of fermentation, which has been adopted in this handbook, is an enzymatically controlled transformation of an organic compound.

19.2 CELLS AND ENZYMES

19.2.1 Microbial Cells

The three major categories of life are the animal kingdom, the plant kingdom, and the protists (Table 19.3). Protists have a relatively simple biological organization and do not differentiate into separate cell types. The microorganisms classified as the protists are further divided into two categories: eukaryotes and prokaryotes. Prokaryotes are small and simple. It is the unit of structure of bacteria and blue-green algae. The more complex eukaryotic cell has internal unit membrane systems that segregate many of the functional components of the cell. Since eukaryotic cells are also the unit of structure of animals and plants, in modern biology, life is broadly classified into two categories: eukaryotes and prokaryotes (Table 19.3). Among the protists, bacteria and fungi are two most important microorganisms for industrial applications. Table 19.4 gives several examples of microorganisms used industrially.

TABLE 19.3
Major Categories of Life

Type	Haeckel	Modern
Multicellular	Animals	Eukaryotes
	Plants	
Unicellular	Protists	
	Algae	
	Protozoa	
	Fungi	
	Molds	
	Yeasts	
	Bacteria	Prokaryotes
	Blue-green algae	

TABLE 19.4
Industrial Applications of Microorganisms

Microorganism	Industrial Applications
Bacteria	• Production of acetone, butanol, enzymes (such as amylase, protease, invertase, and rennin), lactic acid, sorbose, dextran, and hormones • Bioremediation of hazardous chemicals • Production of mammalian proteins (biopharmaceuticals) from recombinant bacteria
Yeasts	• Production of alcoholic beverages and industrial-grade alcohol • Production of baking goods and food supplements • Production of mammalian proteins (biopharmaceuticals) from genetically modified yeasts
Molds	• Production of penicillin, hormones, and other antibiotics • Production of citric acid, enzymes (such as amylases, invertase, proteases, glucose oxidase, and pectinase), fumaric acid, gluconic acid, and lactic acid

19.2.1.1 Microbial Nomenclature

Microbiologists use the binomial system, in which each organism has two names, for example, *Bacillus subtilis*. Proper names of organisms are always italicized. The first word is the name of the genus and is capitalized. The genus name is usually Latin or Greek; examples are *Bacillus* (a small rod), *Lactobacillus* (a small milk rod), *Micrococcus* (a small grain), and *Clostridium* (a small spindle). The second word in the name of a microorganism is the species name and is not capitalized. There may be several species with the same genus name, for example, *Bacillus subtilis, B. albus*, and *B. coagulans*. When the same genus name is repeated several times, it is often abbreviated.

19.2.1.2 Bacteria

Bacteria are unicellular microscopic organisms; about 1500 known species occur in practically all natural environments. The typical diameter of the cell ranges from 0.5 to 1 μm; the lengths vary greatly. Bacteria occur in a variety of shapes such as *cocci* (spherical or ovoid), *bacilli* (cylindrical or rod shaped), and *spirilla* (helically coiled). Bacteria reproduce predominantly by binary fission: a single cell divides into two after the development of a transverse cell wall to separate the intracellular content.

The nutritional medium to cultivate bacteria provides energy, carbon, nitrogen, sulfur, phosphorus, metallic elements, and vitamins. The physical conditions important for the optimum growth include temperature, the gaseous environment, and pH. The principal gases in the cultivation of bacteria are oxygen and carbon dioxide. Aerobic bacteria grow in the presence of free oxygen. Anaerobic bacteria grow in the absence of free oxygen. Generally, the optimum pH for growth lies between 6.5 and 7.5, while the minimum and maximum limits vary from 4 to 9.

Some bacteria form spores when growth ceases due to starvation or other causes. Spores are more resistant than normal cells to heat, drying, radiation, and chemicals. Spores can remain alive for many years, and can convert back to normal cells at proper conditions. Spore-forming bacteria are found most commonly in the soil.

19.2.1.3 Fungi

Fungi are plants devoid of chlorophyll and are therefore unable to synthesize their own food. They range in size and shape from single-celled yeasts to multicellular mushrooms.

Yeasts are found in fruits, grains, and other foods containing sugar. They are also in or on the soil, the air, the skin, and the intestines of animals. Yeasts are generally unicellular organisms and their shape is spherical to ovoid. Their size is 1 to 5 μm in width and from 5 to 30 μm in length.

TABLE 19.5
Typical Growth Medium for Yeasts

Glucose	100 g
Yeast extract	8.5 g
NH$_4$Cl	1.32 g
MgSO$_4$	0.11 g
CaCl$_2$	0.06 g
Anti-foam	0.2 mL
Water to make	1 L

The most common growth pattern for yeasts is budding, which is an asexual process. A small bud (or daughter cell) is formed on the surface of a mature cell.

Molds are filamentous fungi. A single reproductive cell or spore is germinated to form a long thread, or hyphae, which branches repeatedly as it elongates to form a vegetative structure called a mycelium. The most important classes of molds industrially are *Aspergillus* and *Penicillium*. Molds can produce antibiotics, industrial chemicals, enzymes, and food additives. The production of penicillin was scaled up between 1941 and 1945 from virtually nothing to more than 650 billion units per month; meanwhile, its cost dropped from $20 to 60¢ per 100,000 units [2]. The dramatic development was due to the improvements in composition of the medium, the development of the submerged culture technique, the production of mutant strains of *Penicillium chrysognum*, and the refinements of downstream separation techniques [3].

19.2.1.4 Culture Media

There are two main types of culture media: natural and synthetic. They vary widely in form and composition, depending on the species of organism to be cultivated and the purpose of the cultivation.

Natural media (or complex media) usually contain peptones, beef extract, or yeast extract. When a solid medium is desired, a solidifying agent such as gelatin or agar may be incorporated into the medium.

Synthetic media (or chemically defined media) consist of dilute solutions of chemically pure compounds. They may be simple such as inorganic ammonium salt plus minerals and a sugar, or complex such as purified casein with added vitamins, minerals, and a sugar. They can be produced with constant compositions year after year. Table 19.5 shows a typical growth medium for the cultivation of yeasts.

The medium must be sterilized to eliminate any living organisms in the vessel, often by moist heat (steam under pressure) in an autoclave. Generally, the autoclave is operated at approximately 15 psig at 121°C. The time of sterilization depends on the nature of the material, the type of container, and the volume; test tubes of liquid media can be sterilized in 15 to 20 minutes at 121°C.

Inoculation is the seeding of a culture vessel with the microbial material (inoculum). The inoculum is introduced with a metal wire or loop that is rapidly sterilized just before its use by heating it in a flame. Transfers of liquid culture are often made by using a sterilized pipette.

19.2.2 ANIMAL CELLS

Animal cells are classified as eukaryotic cells, and they can be cultivated in nutritional medium outside the donor's body. Such cultured cells grow in number and size. Tissue culture methodologies are employed to study cancer cells and malignant tumors, to determine tissue compatibility in transplantation, and to study specific cells and their interactions.

Biochemical Engineering

The mammalian cell culture technique can be employed to produce human growth hormones, interferon, plasminogen activators, viral vaccines, interleukins, and monoclonal antibodies. Traditionally, these biochemicals had been produced using living animals. However, the quantity obtained from these methods is limited for clinical usage.

Blood and lymph are rather atypical connective tissues with liquid matrices. Cells from blood or lymph fluids are suspension cells, or non-anchorage dependent when grown in culture. Most normal mammalian cells are anchorage dependent, that is, they require a surface for attachment and growth. The most widely used anchorage-dependent cell types are epithelial or fibroblast (broadly classified as connective) cells. Anchorage-dependent cells require a wettable surface such as glass or plastic Petri dishes and roller bottles. Bottles are laid on a slowly rotating roller in an incubator. A 1-liter bottle typically contains 100 mL of medium to facilitate cells both to grow on the wall and to be exposed to medium and gas. However, roller bottles are suitable only for small-scale laboratory use.

The nutritional requirements of mammalian cells are more stringent than those of microorganisms since they do not metabolize inorganic nitrogen. Therefore, many amino acids and vitamins should be provided. Typical medium contains amino acids, vitamins, hormones, growth factors, mineral salts, and glucose. Furthermore, the medium needs to be supplemented with 2% to 20% (by volume) of mammalian blood serum. The serum provides components that have not yet been identified but are necessary for culture viability.

The serum in the medium is not only expensive but also can be the source of virus or mycoplasma contamination. Since the chemical nature of serum is not well defined, its contents may vary batch after batch, which can affect cell growth. Many different proteins in serum often complicate the downstream separation processes. For these reasons, serum-free media have been formulated, which contain purified hormones and growth factors that can substitute for serum supplements [4].

19.2.3 Plant Cells

Plants are a valuable source of chemical compounds such as pharmaceuticals, flavors, pigments, fragrances, and agrochemicals. These products, known as secondary metabolites, are usually produced in trace quantities in plants and have no obvious metabolic function. They serve as a chemical interface between the producing plants and their surrounding environment, such as adaptations to environmental stresses and chemical defenses against microorganisms. Despite substantial advances in synthetic organic chemistry, many secondary metabolic compounds are either too difficult or costly to synthesize (e.g., rose oil). Most plants that produce commercially useful substances are grown in tropical and subtropical regions of the world. As a result, the availability and costs of these materials depend on the political and economic circumstances of the countries involved.

To produce secondary metabolic products from plants, exogenous plant tissue instead of a whole plant may be cultivated as a suspension culture in an aseptic condition. Such cultivation provides several advantages: (1) Plant cells can be cultivated where they are needed regardless of weather and geographical conditions. (2) The product quality and yields can be well controlled by eliminating problems encountered in the processing of botanicals. (3) Some metabolic products can be produced from suspension cultures in higher quantities than those observed in whole plants.

It is a challenge to cultivate plant cells on a large scale and to maximize the production and accumulation of secondary metabolites. This can be accomplished by selecting proper genotypes and high-yielding cell clones, formulating suitable media to cultivate the cells, and designing and operating effective cell culture systems. Plant cells and microbial organisms are, however, so different that the culture conditions and reactor configuration generally need to be modified.

Since plant cells are 10 to 100 times larger than bacterial and fungal cells, the metabolism of plant cells is slower than that of microbial cells by about one order of magnitude. Hence, only high value-added, low market-volume products are economically feasible to be produced from plant cell culture techniques. Potential products are

1. Food products: color, flavors, oils, sweeteners, and spices
2. Pharmaceuticals: alkaloids, steroids, shikonin, rosmarinic acid
3. Agricultural chemicals

Plant tissue cultures can be divided into two major types: unorganized growth cultures and organized growth cultures.

Unorganized growth cultures lack any recognizable structure of the original plant. Callus cultures are amorphous cell aggregates arising from the unorganized growth of explants on an aseptic solid nutrient medium. *Suspension (or cell) cultures* consist of cells and cell aggregates, growing dispersed in liquid medium. They are usually initiated by placing pieces of a friable callus culture in moving liquid medium. Suspension culture is generally preferred for mass propagation of plant cells because it can be maintained and manipulated similar to submerged microbial fermentation.

Organized growth cultures maintain their original organ structure such as root cultures and embryo cultures. *Root cultures* can be established from root tips taken from many plants. The "hairy root" clones that are produced can be cultivated to produce metabolites. *Embryo cultures* may be established from embryos removed from sterilized seeds, ovules, or fruits. Embryo cultures can be employed for the rapid production of seedlings from seeds that have a protracted dormancy period.

19.2.4 Cell Growth Measurement

Microscopic counts: The number of cells in a culture can be counted under a microscope using a hemocytometer, which has counting chambers etched on the surface of a glass slide. A sample of cell suspension is allowed to flow under the cover slip and to fill the counting chambers. Dense suspensions need to be diluted.

Viable plate count: An aliquot of cell suspension is spread over the agar surface. The plate is then incubated until colonies appear, and the number of colonies is counted. To obtain the appropriate number of cells per unit volume, the sample has to be diluted. For the larger dilution, it is common to use the serial dilution technique. For example, to make $1/10^6$ dilution, three successive 1/100 dilutions or six successive 1/10 dilutions can be made.

Cell dry weight: An aliquot of cell suspension is centrifuged to discard the supernatant. The cells are thoroughly washed with distilled water and recentrifuged to eliminate all soluble matter. The washed cells are dried in an oven and weighed. This is the most direct and probably the most reliable method. However, such determinations are time consuming and relatively insensitive to small changes of cell mass.

Turbidity: The cell mass can be estimated optically by determining the amount of light scattered by suspended cells. A calibration curve is obtained by measuring the absorbency of a sample with a known cell concentration. The measurements are usually made at a wavelength between 600 and 700 nm.

19.2.5 Enzymes

Enzymes are biological catalysts and proteins, which are produced by living organisms. Almost every reaction in a cell requires the presence of a specific enzyme. Their major function is to catalyze the making and breaking of chemical bonds. Therefore, like any other catalysts, they increase the rate of reaction without themselves undergoing permanent chemical changes. An enzyme is highly specific and catalyzes only one or a small number of chemical reactions. The rate of an enzyme-catalyzed reaction is usually much faster than that of the same reaction when

Biochemical Engineering

directed by non-biological catalysts. Reaction conditions (such as temperature, pressure, and pH) are mild.

19.2.5.1 Nomenclature of Enzymes

Originally enzymes were given nondescriptive names such as *pepsin* (hydrolyzes proteins) and *lysozymes* (capable of destroying bacterial cell walls). The nomenclature was later improved by adding the suffix *-ase* to the name of the substrate with which the enzyme functions, or to the reaction that is catalyzed. For example

1. Name of substrate
   lactase: splits lactose into glucose and galactose
   lipase: hydrolyzes lipids into glycerol and fatty acids
   maltase: hydrolyzes maltose to glucose
2. Reaction that is catalyzed
   dehydrogenase: removes hydrogen from a substrate
   isomerase: converts one isomer to another
   oxidase: catalyzes oxidation

A systematic scheme of nomenclature for enzymes was proposed by the International Enzyme Commission in 1964. They are now categorized into six major classes depending on the type of chemical reaction that they catalyze.

19.2.5.2 Commercial Applications of Enzymes

Enzymes have been used since early human history without knowledge of what they were or how they worked. They were used for such things as making sweets from starch, clotting milk to make cheese, and brewing alcohol. Enzymes have been utilized commercially since the 1890s, when fungal cell extracts were first used to break down starch into sugars. Fungal amylase takadiastase was employed as a digestive aid by 1894.

Enzymes are usually made by microorganisms grown in a pure culture or obtained directly from plants and animals. The enzymes produced commercially can be classified into three major categories: industrial enzymes, analytical enzymes, and medical enzymes. Industrial enzymes are those that are produced and used in large quantities such as starch conversion enzymes (α-amylase, glucoamylase, and glucose isomerase), which are used to convert starch into high-fructose corn syrup, largely used in soft drinks. Alkaline protease is another example of an industrial enzyme, which is added to laundry detergents as a cleaning aid. Analytical and medical enzymes are produced in the range of milligrams to grams and usually required to be highly purified; hence, their production costs are high.

19.3 KINETICS

19.3.1 Cell Kinetics

Understanding the growth kinetics of microbial, animal, or plant cells is important for the design and operation of the required fermentation systems.

19.3.1.1 Growth Cycle for Batch Cultivation

Four phases are observed during a batch culture: lag, exponential, stationary, and death (Figure 19.2).

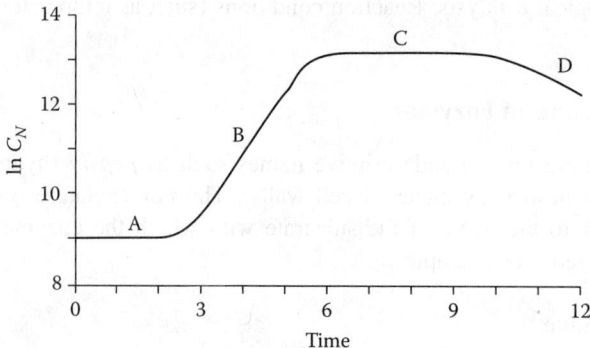

FIGURE 19.2 Typical growth curve of unicellular organisms: (A) lag phase; (B) exponential growth phase; (C) stationary phase; (D) death phase.

The lag phase is the initial period when the change of cell number is negligible; the cells may grow in size, but not in number. The length of the lag period depends on the type and age of the microorganisms, the size of the inoculum, and other culture conditions. The lag usually occurs as the cells adjust to the new medium before growth begins, and as the cells produce the enzymes necessary for the metabolization of the available nutrients.

At the end of the lag phase, growth begins and cell concentration increases exponentially. As the available nutrients are exhausted in the medium, the rate of growth declines and growth eventually stops. The stationary phase is usually followed by a death phase, in which the organisms in the population die due to the depletion of the cellular reserves of energy or due to the accumulation of toxic products.

19.3.1.2 Unstructured Cell Kinetic Models

Cell growth involves numerous complicated networks of biochemical and chemical reactions. A simple unstructured kinetic model was formulated based on the assumptions that

> Cells can be represented by a single component, such as cell mass or cell number.
> The cell suspension can be regarded as a homogeneous solution.

The rate of the cell growth r_X is proportional to cell concentration C_X as

$$r_X = \frac{dC_X}{dt} = \mu C_X \tag{19.1}$$

where μ is the specific growth rate [hr^{-1}]. Equation (19.1) can be integrated from time t_0 to t to give

$$C_X = C_{X_0} \exp[\mu(t - t_0)] \tag{19.2}$$

where C_{X_0} is the cell concentration at t_0 when the exponential growth starts. Equation (19.2) shows the increase of cells exponentially with respect to time. The cell concentration C_X can be replaced by cell number density C_N.

Actually, M does not remain constant during the exponential period, because it will be affected by the culture condition, such as substrate and production concentrations, which are constantly

Biochemical Engineering

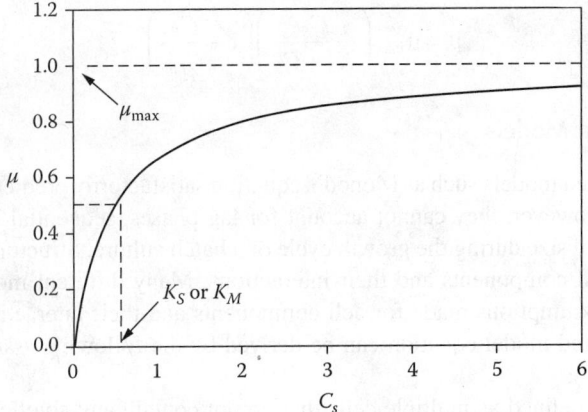

FIGURE 19.3 The effect of substrate concentration on the reaction rate.

TABLE 19.6
Typical Values of Monod's Kinetic Parameters (μ_{max} and K_s) with Glucose as the Limiting Substrate [5]

Organism	Temperature (°C)	μ_{max} (hr^{-1})	K_s (mg/L)
Escherichia coli	37	0.8–1.4	2–4
Saccharomyces cerevisiae	30	0.5–0.6	25
Candida tropicalis	30	0.5	25–75

changing during the period. One of the most widely employed expressions for the effect of substrate concentration on μ is the Monod equation [18]:

$$\mu = \frac{\mu_{max} C_S}{K_S + C_S} \tag{19.3}$$

where C_S is the concentration of the limiting substrate in the medium and K_S is a system coefficient. The value of K_S is equal to the concentration of nutrient when the specific growth rate is half of its maximum value μ_{max}, as shown in Figure 19.3.

While the Monod equation is an oversimplification of the complicated mechanism of cell growth, it often adequately describes fermentation kinetics. The Monod kinetic parameters can be determined by making a series of ideal continuous stirred-tank bioreactors, which will be discussed later. Table 19.6 shows the typical values of the Monod's kinetic parameters when glucose is a limiting substrate.

As cells grow, they produce metabolic by-products that accumulate in the medium. The growth of microorganisms is usually inhibited by these products (C_P), whose effect can be added to the Monod equation as follows:

$$\mu = \mu_{max} \left(\frac{C_S}{K_S + C_S} \right) \left(\frac{K_P}{K_P + C_P} \right) \tag{19.4}$$

or

$$\mu = \mu_{\max}\left(\frac{C_S}{K_S + C_S}\right)\left(1 - \frac{C_P}{C_{Pm}}\right)^n \tag{19.5}$$

19.3.1.3 Structured Models

Unstructured distributed models such as Monod's equation satisfactorily predict the growth behavior in many situations. However, they cannot account for lag phases, sequential uptake of substrates, or changes in mean cell size during the growth cycle of a batch culture. Structured models recognize the multiplicity of cell components and their interactions. Many different models have been proposed based on the assumptions made for cell components and their interactions.

A general structured model equation can be derived by the following assumptions:

1. The system is defined as multiple cells that do not contain any abiotic phase of culture. The mass of the system is m on a dry basis and the specific volume is $\hat{v}$.
2. There are c components in the cell and the mass of the jth component per unit volume of the system is $\hat{C}_{X_j}$.
3. There exist kinetic rate expressions for p reactions occurring in the system and the rate of the jth component formed from the ith reaction per unit volume of the system is $\hat{r}_{X_{i,j}}$.

Then, during batch cultivation, the change of the jth component in the system with respect to time can be expressed as [6]

$$\frac{d(m\hat{v}\hat{C}_{X_j})}{dt} = m\hat{v}\sum_{i=1}^{p}\hat{r}_{X_{i,j}} \tag{19.6}$$

If the specific volume $\hat{v}$ is assumed constant with time, the above equation can be rearranged to

$$\frac{d\hat{C}_{X_j}}{dt} = \sum_{i=1}^{p}\hat{r}_{X_{i,j}} - \frac{\hat{C}_{X_j}}{m}\frac{dm}{dt} \tag{19.7}$$

where the last term represents the dilution of intracellular components by the growth of biomaterial. All variables denoted by circumflexes in the preceding equations are intracellular properties. Since the structural models recognize the multiplicity of cell components and their interactions in the cell, it makes more sense to express the model with intrinsic variables.

The concentration terms in the preceding equations can be expressed as mass per unit culture volume V instead of that per biotic system volume $m\hat{v}$. The two different definitions of concentration are related as

$$\hat{C}_{X_j} = \frac{V}{m\hat{v}}C_{X_j} \tag{19.8}$$

Substituting the above equation into Equation (19.7) and simplifying for the constant V yields

$$\frac{dC_{X_j}}{dt} = \frac{m\hat{v}}{V}\sum_{i=1}^{p}\hat{r}_{X_{i,j}} \tag{19.9}$$

Biochemical Engineering

Even though concentration terms are expressed based on the total culture volume, kinetic parameters still remain on a biotic phase basis in the formulation.

One of the simplest structured models is the two-compartment model proposed by Williams [7]. It is based on the assumption that the cell comprises two basic components: a synthetic portion and a structural portion. The synthetic portion is fed by uptake from a substrate and the structural portion is fed from the synthetic portion. The detailed derivation of the model was described by Lee [8].

19.3.2 Simple Enzyme Kinetics

Assume that a substrate (S) is converted to a product (P) by the addition of an enzyme (E) as

$$S \xrightarrow{E} P \qquad (19.10)$$

Brown [9] postulated that the enzyme and the substrate form an intermediate complex that enables the substrate to be converted to a product as

$$S + E \underset{k_2}{\overset{k_1}{\rightleftarrows}} ES \qquad (19.11)$$

$$ES \xrightarrow{k_3} P + E \qquad (19.12)$$

The reaction rate equation can be derived from the preceding mechanism based on the following assumptions:

1. The total enzyme concentration remains constant during the reaction, that is, $C_{E_0} = C_E + C_{ES}$.
2. The amount of enzyme is very small compared to the amount of substrate. Therefore, the enzyme-substrate complex does not deplete the substrate.
3. The product concentration is so low that product inhibition is negligible.

In addition to the preceding assumptions, there are three different approaches to derive the rate equation: Michaelis-Menten approach [10], Briggs-Haldane approach [11], and numerical solution.

19.3.2.1 Michaelis–Menten Approach

In this approach, the product-releasing step, Equation (19.12), is assumed to be much slower than the reversible reaction, Equation (19.11). Since the slower reaction, Equation 19.12, determines the overall rate of reaction, the rate of product formation and substrate consumption is proportional to the concentration of the enzyme-substrate complex C_{ES} as

$$r = \frac{dC_P}{dt} = -\frac{dC_S}{dt} = k_3 C_{ES} \qquad (19.13)$$

The concentration is expressed as a molar unit, such as kmol/m^3 or mol/L. The concentration of the enzyme-substrate complex C_{ES} in Equation (19.13), is related to the substrate concentration C_s and the free-enzyme concentration C_E from the assumption that the first reversible reaction Equation (19.11) is in equilibrium as

$$k_1 C_S C_E = k_2 C_{ES} \tag{19.14}$$

Since we assume that the total enzyme contents are conserved, the free-enzyme concentration C_E can be related to the initial enzyme concentration C_{E0}:

$$C_{E0} = C_E + C_{ES} \tag{19.15}$$

From the previous three equations, C_E and C_{ES} can be eliminated by substituting Equation (19.14) into Equation (19.15) for C_E and rearranging for C_{ES} as

$$C_{ES} = \frac{C_{E0} C_S}{\dfrac{k_2}{k_1} + C_s} \tag{19.16}$$

Substitution of Equation (19.16) into Equation (19.13) results in the Michaelis-Menten equation.

$$r = \frac{dC_P}{dt} = -\frac{dC_S}{dt} = \frac{k_3 C_{E0} C_S}{\dfrac{k_2}{k_1} + C_S} = \frac{r_{\max} C_S}{K_M + C_S} \tag{19.17}$$

The Michaelis constant K_M is equal to the dissociation constant K_1 or the reciprocal of equilibrium constant K_{eq} as

$$K_M = \frac{k_2}{k_1} = K_1 = \frac{C_S C_E}{C_{ES}} = \frac{1}{K_{eq}} \tag{19.18}$$

When K_M equals C_S, r is equal to one-half of $r_{\max}$, according to Equation (19.17). The value of K_M equals the substrate concentration when the reaction rate is half of the maximum rate $r_{\max}$, as shown in Figure 19.3. K_M is of importance since it characterizes the interaction of an enzyme with a given substrate.

Another kinetic parameter in Equation (19.17) is the maximum reaction rate $r_{\max}$, which is proportional to the initial enzyme concentration. To express the enzyme concentration in molar units, one needs to know the molecular weight of the enzyme and the exact amount of pure enzyme added.

Enzyme concentration may be expressed in mass units instead of molar units. However, the amount of enzyme is not well quantified in mass units because the actual contents of an enzyme can differ widely depending on its purity. Therefore, enzyme concentration is expressed as an arbitrarily defined unit based on its catalytic ability. For example, one unit of an enzyme, cellobiase, is defined as the amount of enzyme required to hydrolyze cellobiose to produce 1 μmol of glucose per minute. Whatever unit is adopted for C_{E0}, the unit for $k_3 C_{E0}$ should be the same as that of r, kmole/m³/s. Care should be taken for the consistency of units when enzyme concentration is not expressed in molar units.

19.3.2.2 Briggs–Haldane Approach

In this approach, the change of the intermediate concentration is assumed to be negligible. From the mechanism described by Equations (19.11) and (19.12), the rates of product formation and of substrate consumption are

Biochemical Engineering

$$\frac{dC_P}{dt} = k_3 C_{ES} \tag{19.19}$$

$$-\frac{dC_S}{dt} = -k_1 C_S C_E - k_2 C_{ES} \tag{19.20}$$

The change of C_{ES} with time, dC_{ES}/dt, is assumed to be negligible as

$$\frac{dC_{ES}}{dt} = k_1 C_S C_E - k_2 C_{ES} - k_3 C_{ES} \cong 0 \tag{19.21}$$

Substituting Equation (19.15) into Equation (19.21) for C_E and rearranging for C_{ES} gives

$$C_{ES} = \frac{C_{E_0} C_S}{\frac{k_2 + k_3}{k_1} + C_S} \tag{19.22}$$

Substitution of Equation (19.22) into Equation (19.19) results in

$$r = \frac{dC_P}{dt} = -\frac{dC_S}{dt} = \frac{k_3 C_{E_0} C_S}{\frac{k_2 + k_3}{k_1} + C_S} = \frac{r_{\max} C_S}{K_M + C_S} \tag{19.23}$$

which is similar to the Michaelis-Menten equation, Equation (19.17). In the Michaelis-Menten approach, K_M equals the dissociation constant k_2/k_1, while in the Briggs-Haldane approach, it equals $(k_2 + k_3)/k_1$. Equation (19.23) can be simplified to Equation (19.17) if $k_2 \gg k_3$, which means that the product-releasing step is much slower than the enzyme-substrate complex dissociation step. Since the formation of the complex involves only weak interactions, the rate of dissociation of the complex is likely rapid.

19.3.2.3 Numerical Solution

From the mechanism described by Equations (19.11) and (19.12), two rate equations are written for C_{ES} and C_S as

$$\frac{dC_{ES}}{dt} = k_1 C_S C_E - k_2 C_{ES} - k_3 C_{ES} \tag{19.24}$$

$$\frac{dC_S}{dt} = k_1 C_S C_E + k_2 C_{ES} \tag{19.25}$$

The rate equation for C_P [Equation (19.19)] and Equations (19.24) and (19.25) with Equation (19.15) can be solved simultaneously. Since the analytical solution of the preceding simultaneous differential equations are not possible, they can be solved numerically using software packages such as Advanced Continuous Simulation Language (ACSL), MathCad (MathSoft, Inc., Cambridge, MA), and Mathematica (Wolfram Research, Inc., Champaign, IL).

This solution procedure requires the knowledge of elementary rate constants, k_1, k_2, and k_3. The elementary rate constants can be measured by experimental techniques, such as pre–steady-state kinetics and relaxation methods [12], which are more complicated compared to the methods to determine K_M and r_{max}. A numerical solution with the elementary rate constants often provides a more precise picture of what is occurring during the enzyme reactions.

19.3.2.4 Evaluation of Michaelis-Menten Parameters

The kinetic parameters can be estimated by making a series of batch runs with different levels of substrate concentration. Then the initial reaction rate is calculated as a function of initial substrate concentration. The results can be plotted graphically so that the validity of the kinetic model can be tested and the values of the kinetic parameters estimated. The most straightforward way is to plot r against C_S as shown in Figure 19.3. The asymptote for r will be r_{max} and K_M is equal to C_S when $r = 0.5\, r_{max}$. However, this is an unsatisfactory plot in estimating r_{max} and K_M because it is difficult to estimate asymptotes accurately and also to test the validity of the kinetic model.

Therefore, the Michaelis–Menten equation (Equation (19.17)) is usually rearranged so that the results can be plotted as a straight line. Some of the better known methods are

Langmuir plot [13]:
$$\frac{C_S}{r} = \frac{K_M}{r_{max}} + \frac{C_S}{r_{max}} \qquad (19.26)$$

Lineweaver-Burk plot [14]:
$$\frac{1}{r} = \frac{1}{r_{max}} + \frac{K_M}{r_{max}} \frac{1}{C_S} \qquad (19.27)$$

Eadie-Hofstee plot [15,16]:
$$r = r_{max} - K_M \frac{r}{C_S} \qquad (19.28)$$

If the Michaelis–Menten equation is applicable, the Langmuir plot of C_S/r versus C_S will result in a straight line with slope $1/r_{max}$ and intercept K_M/r_{max}. Similarly, the Lineweaver-Burk plot of $1/r$ versus $1/C_S$ results in a straight line with slope K_M/r_{max} and intercept $1/r_{max}$. The Eadie-Hofstee plot of r versus r/C_S results in a straight line with slope $-K_M$ and intercept r_{max}.

The Lineweaver-Burk plot is the most popular among the three because it shows the relationship between the independent variable C_S and the dependent variable r. However, $1/r$ approaches infinity as C_S decreases, which gives undue weight to inaccurate measurements made at low substrate concentrations. On the other hand, the Eadie-Hofstee plot gives slightly better weighting of the data than the Lineweaver-Burk plot. A disadvantage of this plot is that the rate of reaction r appears in both coordinates while it is usually regarded as a dependent variable. Therefore, the Langmuir plot (C_S/r versus C_S) is the most satisfactory of the three, since the points are equally spaced.

19.3.2.5 Inhibition of Enzyme Reactions

An inhibition (I) decreases enzymes activity, causing the rate of reaction to decrease competitively or noncompetitively (I). A competitive inhibitor has a strong structural resemblance to the substrate. Therefore, both the inhibitor and substrate compete for the active site of an enzyme as

Biochemical Engineering

$$E + S \underset{k_2}{\overset{k_1}{\rightleftarrows}} ES$$

$$E + I \underset{k_4}{\overset{k_3}{\rightleftarrows}} EI \qquad (19.29)$$

$$ES \xrightarrow{k_5} E + P$$

The formation of an enzyme-inhibitor complex reduces the amount of enzyme available for interaction with the substrate and, hence, the rate of reaction decreases. Based on the above mechanism, the rate of product formation can be derived by using the Michaelis-Menten approach to give

$$r_P = \frac{r_{max} C_S}{C_S - K_{M_I}} \qquad (19.30)$$

where

$$K_{M_I} = \frac{k_2}{k_1}\left(1 + \frac{C_I}{k_4/k_3}\right) = K_M\left(1 + \frac{C_I}{K_I}\right) \qquad (19.31)$$

Previous equations show that the maximum reaction rate is not affected by the presence of a competitive inhibitor while the K_{M_I} value is increased.

Noncompetitive inhibitors interact with enzymes in many different ways. They bind to the enzymes reversibly or irreversibly at the active site or at some other region, so the resultant complex is inactive. The mechanism of noncompetitive inhibition is expressed as follows:

$$E + S \underset{k_2}{\overset{k_1}{\rightleftarrows}} ES$$

$$E + I \underset{k_4}{\overset{k_3}{\rightleftarrows}} EI$$

$$EI + S \underset{k_6}{\overset{k_5}{\rightleftarrows}} EIS \qquad (19.32)$$

$$ES + I \underset{k_8}{\overset{k_7}{\rightleftarrows}} ESI$$

$$ES \xrightarrow{k_9} E + P$$

Similarly, the rate equation can be derived by employing the Michaelis-Menten approach as follows:

$$r_P = \frac{r_{I\,max} C_S}{C_S - K_M} \qquad (19.33)$$

where

$$K_M = \frac{k_2}{k_1} = \frac{k_6}{k_5}$$

$$r_{I,\max} = \frac{r_{\max}}{1 + \dfrac{C_I}{K_I}} \quad (19.34)$$

$$K_I = \frac{k_4}{k_3} = \frac{k_8}{k_7}$$

Therefore, the maximum reaction rate will be decreased by the presence of a noncompetitive inhibitor, while the Michaelis constant K_S will not be affected by the inhibitor.

19.4 BIOREACTOR DESIGN

A bioreactor is a reactor in which enzymes or living cells catalyze the biochemical transformations. It is frequently called a fermenter whether the transformation is carried out by living cells or in vivo cellular components (enzymes). Fermentation originally referred to the metabolism of an organic compound under anaerobic conditions. However, modern industrial fermentation includes both aerobic and anaerobic cultures of organisms. Currently, bioreactor and fermenter can be regarded as synonyms.

19.4.1 BIOREACTORS

19.4.1.1 Stirred-Tank Bioreactor (or Fermenter)

In laboratories, cells are usually cultivated in Erlenmeyer flasks on a shaker. The gentle shaking effectively suspends the cells, enhances the oxygenation through the liquid surface, and aids the transfer of nutrients without damaging the structure of the cells.

For a large-scale operation, the stirred-tank fermenters (STF) are widely used (Figure 19.4). They are employed for both aerobic and anaerobic fermentation of a wide range of cells including microbial, animal, and plant cells. The mixing intensity can be varied widely by choosing suitable impellers and by varying agitating speeds. The mechanical agitation and aeration are effective for the suspension of cells, oxygenation, mixing of the medium, and heat transfer. The STF was one of the first large-scale fermenters developed in the pharmaceutical industries for the production of

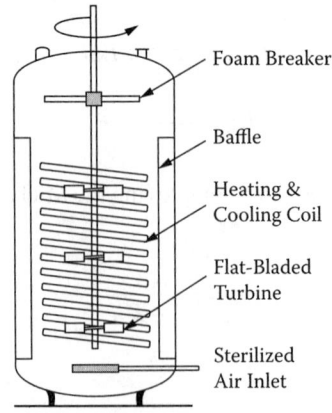

FIGURE 19.4 Schematic diagram of a stirred-tank fermenter.

Biochemical Engineering

penicillin. Its performance and characteristics have been extensively studied. Large industrial STFs are usually built with stainless steel, but laboratory-scale fermenters are often made of glass with a stainless steel top plate. Although the agitator is effective in mixing the fermenter content, it can damage a shear-sensitive cell system such as mammalian or plant cells.

The physical configuration of a typical fermenter is as follows: The height of the vessel is two to three times the vessel diameter and a fermenter is usually agitated with two or three turbine impellers. The impeller diameter to tank diameter ratio is generally 0.3 to 0.4. For multiple impeller systems, the distance between impellers is usually 1 to 1.5 impeller diameters. Four equally spaced baffles are usually installed to prevent a vortex formation that reduces the mixing efficiency. For aerobic fermentation, a single orifice sparger or ring sparger is used to aerate the fermenter. The pH in a fermenter can be maintained by employing either a buffer solution or a pH controller. The temperature is often controlled by heat transfer through a heating coil.

19.4.1.2 Alternative Fermenters

Many alternative fermenters have been proposed in order to provide better aeration, heat transfer, and cell retention. They are usually classified based on their vessel type such as tank, column, or loop fermenters [17]. Another way to classify fermenters is based on how the fermenter contents are mixed: by compressed air, by mechanical agitators, or by external liquid pumping. Representative fermenters in each category are listed in Table 19.7, and the advantages and disadvantages of three basic fermenter types are listed in Table 19.8.

The most simple fermenter is the bubble column fermenter (or tower fermenter), which is usually a long cylindrical vessel with a sparging device at the bottom as shown in Figures 19.5a and b. The fermenter contents are mixed by rising air bubbles that also provide the oxygen needs

TABLE 19.7
Classifications of Fermenters

Vessel Type	Primary Source of Mixing		
	Compressed Air	Internal Moving Parts	External Pumping
Tank	—	Stirred tank	—
Column	Bubble column, tapered column	Multistage (or cascade)	Sieve tray packed bed
Loop	Air-lift pressure cycle	Propeller loop	Jet loop

TABLE 19.8
Advantages and Disadvantages of Three Basic Fermenter Configurations

Type	Advantages	Disadvantages
Stirred tank	• Flexible and adaptable • Wide range of mixing intensity • Can handle high viscosity media	• High power consumption • Damage shear sensitive cells • High equipment costs
Bubble column	• Simple, no moving parts • Low equipment costs • High cell concentration	• Poor mixing • Excessive foaming • Limited to low viscosity system
Air-lift	• Simple, no moving parts • High gas absorption efficiency • Good heat transfer	• Poor mixing • Excessive foaming • Limited to low viscosity system

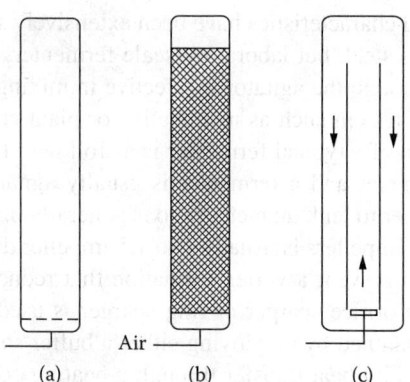

FIGURE 19.5 Alternative fermenters: (a) bubble column; (b) packed column; (c) air lift.

of the cells. Since the fermenter does not have any moving parts, a bubble-column fermenter is energy efficient with respect to the amount of oxygen transfer per unit energy input. As the cells settle, high cell concentrations occur in the lower portion of the column without any separation device. However, bubble-column fermenters are limited to aerobic fermentation. Furthermore, the rising bubbles may not provide adequate mixing for optimal growth. As the cell concentration increases in a fermenter, high airflow rates are required to maintain the cell suspension and mixing. The increased airflow rate can cause excessive foaming and high retention of air bubbles, which decreases the productivity of the fermenter. As bubbles rise in the column, they often coalesce rapidly, leading to a decrease in the oxygen-transfer rate. Therefore, column fermenters tend to be inflexible and limited to a relatively narrow range of operating conditions.

To overcome the weaknesses of the column fermenter, alternatives have been proposed. A tapered column fermenter can maintain a high airflow rate at the lower section of the fermenter where the cell concentration is high. Several sieve plates can be installed in the column for the effective gas–liquid contact and the breakup of the coalesced bubbles. The cylindrical column can be divided into multiple stages with stirrers. This configuration is analogous to stirred-tank fermenters connected in series. To enhance the mixing, the fermentation broth can be recirculated by using a pump.

A loop fermenter is a tank or column fermenter with a liquid circulation loop. Depending on how the liquid circulation is induced, it is often classified into three types: air-lift, stirred loop, and jet loop. The liquid circulation of the air-lift fermenter, Figure 19.5c, is induced by sparged air, which creates a density difference between the bubble-rich liquid in the riser (inner column) and the bubble-depleted liquid in the downcomer (space between the inner and the outer tubes). The liquid circulation and mixing are enhanced by installing a propeller or by circulating liquid externally using a pump. However, the addition of a propeller or pump diminishes the advantages of an air-lift fermenter as being simple and energy efficient.

19.4.2 Mode of Bioreactor Operations

19.4.2.1 Batch or Plug-Flow Bioreactors

An ideal stirred bioreactor is assumed to be well mixed so that the contents are uniform in composition at all times. The plug-flow bioreactor (PFB) is an ideal tubular-flow bioreactor without radial concentration variations. The nutrient concentration of an ideal batch bioreactor after time t will be the same as that of a steady-state PFB at the longitudinal location of the residence time. Therefore, the following analysis applies for both the ideal batch bioreactor and the steady-state PFB.

Biochemical Engineering

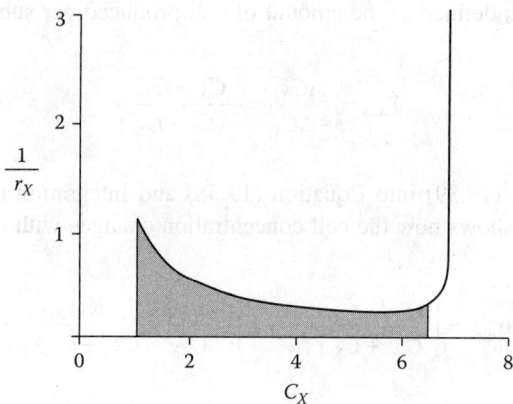

FIGURE 19.6 A graphical representation of the batch growth time $t - t_1$ (*shaded area*). The *solid line* represents the Monod model with $\mu_{max} = 0.935$ hr^{-1}, $K_s = 0.71$ g/L, $Y_{X/S} = 0.6$, $C_{X_0} = 1$ g/L, and $C_{S_0} = 10$ g/L.

19.4.2.1.1 Cell Cultivation

If a batch culture is initiated by the inoculation, cells grow exponentially after the lag phase. The change of the cell concentration is expressed as

$$r_x = \frac{dC_X}{dt} = \mu C_X \tag{19.35}$$

Equation (19.35) can be integrated from time t_0 to t as

$$\int_{C_{X_0}}^{C_X} \frac{dC_X}{r_X} = \int_{C_{X_0}}^{C_X} \frac{dC_X}{\mu C_X} = \int_{t_0}^{t} dt = t - t_0 \tag{19.36}$$

Equation (19.36) only applies when r_X is larger than zero. Hence, t_0 is the time that the cells start to grow.

According to Equation (19.36), the batch growth time $t - t_0$ is the area under the $1/r_X$ versus C_X curve between C_{X_0} and C_X, as shown in Figure 19.6. Though the batch growth time is rarely estimated by this method, the graphical representation is useful in comparing the performances of various fermenter configurations. The curve is U-shaped, which is characteristic of autocatalytic reactions. The rate is slow at the start because the cell concentration is low. It increases as cells multiply and reaches a maximum rate. As the substrate is depleted and the toxic products accumulate, the rate decreases to a low value.

If Monod kinetics [18] represents the growth rate during the exponential period

$$r_X = \frac{\mu_{max} C_S C_X}{K_S + C_S} \tag{19.37}$$

the substitution of Equation (19.37) to Equation (19.36) results in

$$\int_{C_{X_0}}^{C_X} \frac{(K_S + C_S)dC_X}{\mu_{max} C_S C_X} = \int_{t_0}^{t} dt \tag{19.38}$$

The growth yield ($Y_{X/S}$) is defined as the amount of cell produced per substrate consumed as

$$Y_{X/S} = \frac{\Delta C_X}{-\Delta C_S} = \frac{C_X - C_{X_0}}{-(C_S - C_{S_0})} \tag{19.39}$$

Substitution of Equation (19.39) into Equation (19.38) and integration of the resultant equation gives a relationship that shows how the cell concentration changes with respect to time:

$$(t - t_0)\mu_{max} = \left(\frac{K_S Y_{X/S}}{C_{X_0} + C_{S_0} Y_{X/S}} + 1\right) \ln \frac{C_X}{C_{X_0}} + \frac{K_S Y_{X/S}}{C_{X_0} + C_{S_0} Y_{X/S}} \ln \frac{C_{S_0}}{C_S} \tag{19.40}$$

19.4.2.1.2 Enzyme Reactions

Assume that an enzyme reaction is initiated at $t = 0$ by adding enzyme and the reaction mechanism can be represented by the Michaelis-Menten equation

$$-\frac{dC_S}{dt} = \frac{r_{max} C_S}{K_M + C_S} \tag{19.41}$$

Equation (19.41) is similar in form to the Monod equation, but is quite different when you compare it with Equation (19.37), which has an extra term, C_X.

An equation expressing the change of the substrate concentration C_S with respect to time can be obtained by integrating Equation (19.41), as follows:

$$\int_{C_{S_0}}^{C_S} -\left(\frac{K_M + C_S}{C_S}\right) dC_S = \int_0^t r_{max} dt \tag{19.42}$$

and

$$K_M \ln \frac{C_{S_0}}{C_S} + (C_{S_0} + C_S) = r_{max} t \tag{19.43}$$

With known values of r_{max} and K_M, the change of C_S with time in a batch reactor can be predicted from this equation.

19.4.2.2 Continuous Stirred-Tank Bioreactor

For an ideal continuous stirred-tank bioreactor (CSTB), the concentrations of the various components of the outlet stream are assumed to be the same as the concentrations in the bioreactor. Continuous operation of a bioreactor can increase the productivity of the reactor significantly by eliminating the downtime and the ease of automation.

19.4.2.2.1 Cell Cultivation

Microbial populations can be maintained in a state of exponential growth for extended time by using a system of continuous culture, operated either as chemostat or as a turbidostat. In a chemostat, the flow rate is set at a particular value and the rate of growth of the culture adjusts to this flow rate. In a turbidostat, the turbidity is set at a constant level by adjusting the flow rate. It is easier to operate a chemostat than a turbidostat, because the former is at a constant flow rate, whereas

Biochemical Engineering

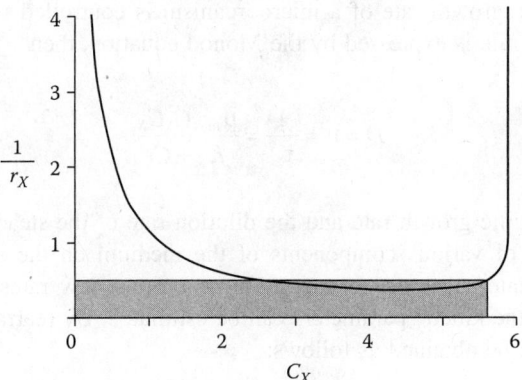

FIGURE 19.7 A graphical representation of the estimation of residence time for the CSTB. The *line* represents the Monod model with $\mu_{max} = 0.935$ hr^{-1}, $K_S = 0.71$ g/L, $Y_{X/S} = 0.6$, $C_{Si} = 10$ g/L, and $C_{X_i} = 0$.

the latter requires an optical sensing device and a controller. However, the turbidostat is recommended when continuous fermentation is performed at high dilution rates near the washout point. Turbidostats prevent washout by regulating the flow rate when cell loss in the output stream exceeds the cell growth in the fermenter.

When a CSTB (volume V) is operated with a constant inlet and an outlet flow rate (F) with the cell concentration of C_{X_i} and substrate concentration C_{S_i}, the material balance for the microorganisms in a CSTB can be written as

$$FC_{X_i} - FC_X + Vr_X = V\frac{dC_X}{dt} \tag{19.44}$$

where r_X is the rate of cell growth in the CSBR and dC_X/dt represents the change of cell concentration in the bioreactor with time.

For the steady-state operation of a CSTB, the change of cell concentration with time is equal to zero ($dC_X/dt = 0$); the microorganisms in the vessel grow just fast enough to replace those lost through the outlet stream. Therefore, Equation (19.44) becomes

$$\tau_m = \frac{V}{F} = \frac{C_X - C_{X_i}}{r_X} \tag{19.45}$$

which indicates that the required residence time (τ_m) is equal to $(C_X - C_{X_i})$ times $1/r_X$, which is equal to the area of the rectangle of width $C_X - C_{X_i}$ and height $1/r_X$ on the $1/r_X$ versus C_X curve as shown in Figure 19.7 (the shaded rectangular area). This graphical illustration of the residence time aids in comparing the effectiveness of fermenter systems.

If the input stream is sterile ($C_{X_i} = 0$), and the cells in a CSTB are growing exponentially ($r_X = \mu C_X$), Equation (19.45) becomes

$$\tau_m = \frac{1}{\mu} = \frac{1}{D} \tag{19.46}$$

where D is known as the dilution rate and is equal to the reciprocal of the residence time (τ_m). Therefore, for the steady-state CSTB with sterile feed, the specific growth rate equals the dilution rate. Hence, the specific growth rate of a microorganism is controlled by changing the medium flow rate. If the growth rate is expressed by the Monod equation, then

$$D = \mu = \frac{1}{\tau_m} = \frac{\mu_{max} C_S C_X}{K_S + C_S} \tag{19.47}$$

The equality of the specific growth rate and the dilution rate of the steady-state CSTB is helpful in studying the effects of various components of the medium on the specific growth rate. By measuring the steady-state substrate concentration at various flow rates, kinetic models can be tested and the value of the kinetic parameters can be estimated. By rearranging Equation (19.47), a linear relationship can be obtained as follows:

$$\frac{C_S}{\mu} = \frac{K_S}{\mu_{max}} + \frac{C_S}{\mu_{max}} \tag{19.48}$$

$$\frac{1}{\mu} = \frac{K_S}{\mu_{max}} \frac{1}{C_S} + \frac{1}{\mu_{max}} \tag{19.49}$$

$$\mu = \mu_{max} - K_S \frac{\mu}{C_S} \tag{19.50}$$

These are similar to the equations for the Langmuir, the Lineweaver-Burks, and the Eadie-Hofstee plots that were discussed earlier with the Michaelis-Menten kinetics.

From Equation (19.47), C_S can be calculated with a known residence time and the Monod kinetic parameters as

$$C_S = \frac{K_S}{\tau_m \mu_{max} - 1} \tag{19.51}$$

However, the preceding expression is only valid when $\tau_m \mu_{max} > 1$. If $\tau_m \mu_{max} < 1$, the growth rate of the cells is less than the rate of cells leaving with the outlet stream. Consequently, all of the cells in the fermenter will be washed out.

If the growth yield ($Y_{X/S}$) is constant, then

$$C_X = Y_{X/S}(C_{S_i} - C_S) \tag{19.52}$$

Substituting Equation (19.51) into Equation (19.52) yields the correlation for C_X as

$$C_X = Y_{X/S}\left(C_{S_i} - \frac{K_S}{\tau_m \mu_{max} - 1}\right) \tag{19.53}$$

Similarly, one can derive a relationship for the product concentration C_P as

$$C_P = C_{P_i} + Y_{X/S}\left(C_{S_i} - \frac{K_S}{\tau_m \mu_{max} - 1}\right) \tag{19.54}$$

19.4.2.2.2 Enzyme Reaction

The substrate balance of a CSTB can be set up as follows:

$$FC_{S_i} - FC_S + r_S V = V \frac{dC_S}{dt} \tag{19.55}$$

where r_S is the rate of substrate consumption for the enzymatic reaction, while dC_S/dt is the change of the substrate concentration in the bioreactor. As shown in Equation (19.55), r_S is equal to dC_S/dt only when F is zero, which is the case in batch operation.

For the steady-state CSTB, the substrate concentration of the reactor should be constant. Therefore, dC_S/dt is equal to zero. If the Michaelis-Menten equation can be used for the rate of substrate consumption (r_S), Equation (19.55) can be rearranged as

$$\frac{F}{V} = D = \frac{1}{\tau_m} = \frac{r_{max} C_S}{(C_{S_i} - C_S)(K_M + C_S)} \tag{19.56}$$

Equation (19.56) can be rearranged to give the linear relationship

$$C_S = -K_M + \frac{r_{max} C_S \tau_m}{C_{S_i} - C_S} \tag{19.57}$$

Michaelis-Menten kinetic parameters can also be estimated with a series of steady-state CSTB runs with various flow rates and plotting C_S versus $(C_S \tau_m)/(C_{S_i} - C_S)$. However, the initial rate approach in a batch mode already discussed is a better way to estimate the kinetic parameters than this method because steady-state CSTB runs are more difficult to make than batch runs.

19.5 AGITATION AND AERATION

19.5.1 INTRODUCTION

A key factor in designing a fermenter is to provide adequate mixing of its contents. Mixing in fermentation disperses the air bubbles, suspends the microorganisms (or animal and plant cells), and enhances both heat and mass transfer in the medium. Because most nutrients are highly soluble in water, little mixing is required during fermentation to transfer nutrients to the surfaces of living cells from the bulk solution. However, dissolved oxygen in the medium is an exception because its solubility of oxygen is very low, while its demand for the growth of aerobic microorganisms is high.

For example, the typical maximum concentration of oxygen in an aqueous solution in equilibrium with air is often 6 to 8 mg/L. The oxygen requirement is on the order of 10 mg/L/min. Therefore, the saturated dissolved oxygen will be consumed in less than 1 minute if not replenished continuously. Adequate continuous oxygen supply to cells is often critical in aerobic fermentation.

Mixing by a laboratory shaker is adequate to cultivate microorganisms in flasks or test tubes. Rotary or reciprocating action of a shaker is effective to provide gentle mixing and surface aeration. For bench-, pilot-, and production-scale fermenters, the mixing is usually provided by mechanical agitation.

Correlations are next reported for gas-liquid mass transfer, interfacial area, bubble size, gas hold-up, agitation power consumption, and volumetric mass-transfer coefficients, which are vital tools for the design and operation of fermenter systems.

TABLE 19.9
Solubility of Oxygen in Water at 1 atm[a]

Temperature (°C)	Solubility	
	mmol O_2/L	mg O_2/L
0	2.18	69.8
10	1.70	54.5
15	1.54	49.3
20	1.38	44.2
25	1.26	40.3
30	1.16	37.1
35	1.09	34.9
40	1.03	33.0

[a] Data from *International Critical Tables*. 1928. Vol. III, 271. New York: McGraw-Hill Book Co.

19.5.2 Determination of Oxygen-Absorption Rate

The oxygen absorption rate per unit volume q_a/v can be estimated by

$$\frac{q_a}{v} = K_L a (C_L^* - C_L) = k_L a (C_L^* - C_L) \tag{19.58}$$

where a is the interfacial area per unit volume and C_L^* is the liquid-side oxygen concentration in equilibrium with the gas-phase oxygen. Because the oxygen is a sparingly soluble gas, the overall mass-transfer coefficient K_L is approximately equal to the individual mass-transfer coefficient k_L. An objective in fermenter design is to minimize power consumption and airflow rate. To increase the oxygen absorption rate, larger k_L, a, and, $C_L^* - C_L$ are needed. Since the concentration difference is limited because the value of C_L^* is low, the main parameters of interest are the mass-transfer coefficient and the interfacial area.

19.5.2.1 Estimation of Equilibrium Oxygen Concentration C_L^*

Table 19.9 lists the dissolved concentration of oxygen in water that is in equilibrium with pure oxygen at atmosphere pressure at various temperatures. Since air supplies the oxygen to a bioreactor, the maximum concentration of oxygen is about one-fifth of the solubility listed, according to Henry's law:

$$C_L^* = \frac{P_{O_2}}{H_{O_2}(T)} \tag{19.59}$$

where P_{O_2} is the partial pressure of oxygen and $H_{O_2}(T)$ is Henry's law constant of oxygen at a temperature, T. The value of Henry's law constant can be obtained from the solubility listed in Table 19.9.

19.5.2.2 Sodium Sulfite Oxidation Method

One of the simplest techniques to determine $k_L a$ is the sodium sulfite oxidation method [19], which is based on the oxidation of sodium sulfite to sodium sulfate in the presence of a catalyst (Cu^{++} or Co^{++}) as

Biochemical Engineering

$$Na_2SO_3 + \frac{1}{2}O_2 \xrightarrow{Cu^{++} \text{ or } Co^{++}} Na_2SO_4 \tag{19.60}$$

The rate of this reaction is independent of the concentration of sodium sulfite (within the range of 0.04 to 1 N) and is also much faster than the oxygen transfer rate. Therefore, the rate of oxidation is controlled by the rate of mass transfer alone. The sodium sulfite solution, however, has quite different physical and chemical properties. This technique is helpful though in comparing the performance of bioreactors and studying the effect of scale-up.

19.5.2.3 Dynamic Technique

By using the dynamic technique [20], the k_La value for the oxygen transfer during a fermentation process can be estimated. This technique is based on the oxygen material balance in an aerated batch bioreactor while microorganisms are actively growing as

$$\frac{dC_L}{dt} = k_La(C_L^* - C_L) - r_{O_2}C_X \tag{19.61}$$

where r_{O_2} is the cell respiration rate [g O_2/g cell/h].

While the dissolved oxygen level of the bioreactor is steady, if the air supply is suddenly stopped, the oxygen concentration will be decreased (Figure 19.8) as follows:

$$\frac{dC_L}{dt} = r_{O_2}C_X \tag{19.62}$$

Therefore, by measuring the slope of the C_L versus t curve, we can estimate $r_{O_2}C_X$. If the air flow is restarted, the dissolved oxygen concentration will increase according to Equation (19.61), which can be rearranged to result in a linear relationship as

$$C_L = C_L^* - \frac{1}{k_La}\left(\frac{dC_L}{dt} + r_{O_2}C_X\right) \tag{19.63}$$

The plot of C_L versus $dC_L/dt + r_{O_2}C_X$ results in a straight line that has a slope of $-1/(k_La)$ and a intercept of C_L^*.

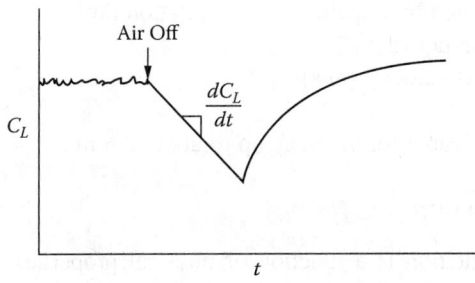

FIGURE 19.8 Dynamic technique for the determination of k_La.

19.5.3 CORRELATION FOR $k_L a$

19.5.3.1 Bubble Column

Akita and Yoshida [21] correlated the volumetric mass-transfer coefficient $k_L a$ [s^{-1}] for the absorption of oxygen in various aqueous solutions in bubble columns, as follows:

$$k_L a = 0.6 D_{AB}^{0.5} v_c^{-0.12} \left(\frac{\sigma}{\rho_c}\right)^{-0.62} D_T^{0.17} g^{0.93} H^{1.1} \tag{19.64}$$

where

D_{AB} = diffusivity of A through B [m^2/s]
v_c = kinematic viscosity [m^2/s]
σ = interfacial tension [kg/s^2]
ρ_c = the density of liquid [kg/m^3]
D_T = tank diameter [m]
g = gravity constant [m/s^2]
H = the height of the liquid level in a tank [m]

19.5.3.2 Mechanically Agitated Vessel

For aerated mixing vessels in an aqueous solution, the mass-transfer coefficient is proportional to the power consumption. Van't Riet [22] correlated the data of several investigators as follows:

1. For "coalescing" air-water dispersion,

$$k_L a = 0.026 \left(\frac{P}{v}\right)^{0.4} V_S^{0.5} \tag{19.65}$$

2. For "noncoalescing" air-electrolyte solution dispersions,

$$k_L a = 0.002 \left(\frac{P}{v}\right)^{0.7} V_S^{0.2} \tag{19.66}$$

where

P = power dissipated by impeller during agitation [W]
v = volume of the liquid [m^3]
V_S = superficial gas velocity [m/s]

These correlations are applicable for volumes up to about 2.6 m^3.

19.5.4 POWER CONSUMPTION

Power consumption by agitation is a function of physical properties, operating conditions, and vessel and impeller geometry. For fully baffled agitated vessels,

$$\frac{P}{\rho N^3 D_I^5} = \alpha \left(\frac{\rho N D_I^2}{\mu} \right)^\beta \quad (19.67)$$

The dimensionless group in the left-hand side of Equation (19.67) is the power number N_P and that in the right-hand side is the impeller Reynolds number N_{Re_i}. D_I and N are the diameter and the speed of an impeller, respectively.

The power number decreases with an increase of the Reynolds number and reaches a constant value (6 for a flat six-bladed turbine) when the Reynolds number is greater than 10,000. At this point, the power number is independent of the Reynolds number. For the normal operating conditions of gas-liquid contact, the Reynolds number is usually greater than 10,000. For example, for a 3-inch impeller with an agitation speed of 150 rpm, the impeller Reynolds number is 16,225 when the liquid is water. Therefore, for a six flat-bladed turbine, the power number is

$$\frac{P}{\rho N^3 D_I^5} = 6 \quad \text{when } N_{Re_i} > 10,000 \quad (19.68)$$

19.5.5 Scale-Up

For the optimum design of a production-scale fermentation system (prototype), we must translate the data on a small scale (model) to the large scale. The fundamental requirement for scale-up is that the model and prototype should be similar. To ensure similarity, two conditions must be satisfied: (1) geometric similarity of the physical boundaries and (2) dynamic similarity of the flow fields. The first requirement is obvious and easy to accomplish. The second is achieved when the values of the nondimensional parameters (such as Reynolds number and power number) are the same. However, it is difficult, if not impossible, to satisfy the dynamic similarity when more than one dimensionless group is involved in a system, which creates the need for scale-up criteria.

No one scale-up rule applies to different mixing operations. However, some principles for the scale-up are as follows [23]:

1. The system can be scaled up so that one of the important properties (such as pumping capacity, oxygen transfer coefficient, and shear rate) can be maintained.
2. The big tank has a longer blend time, a higher maximum impeller shear rate, and a lower average impeller shear rate.
3. The power per volume can be used as a scale-up criterion.

REFERENCES

1. Congress of the United States. 1984. *Commercial Biotechnology: An International Analysis*. Washington, DC: Office of Technology Assessment; 589.
2. J.L. Sturchio. 1988. *Today's Chemist*, February: 20–22.
3. M.J. Pelczar, R.D. Reid. 1972. *Microbiology*, 820–823. New York: McGraw-Hill.
4. M. Butler. 1987. *Animal Cell Technology: Principles and Products*, 3–11. Milton Keynes, UK: Open University Press.
5. H.W. Blanch, D.S. Clark. 1996. *Biochemical Engineering*, 187. New York: Marcel Dekker.
6. A.G. Fredrickson. 1976. *Biotechnology and Bioengineering* 18: 1481–1486.
7. F.M. Williams. 1967. *Journal of Theoretical Biology* 15: 190–207.
8. J.M. Lee. 1992. In *Biochemical Engineering*, ed., 176–178. Englewood Cliffs, NJ: Prentice Hall (ebook version is available at http://jmlee.net).

9. A.J. Brown. 1902. *Journal of the Chemical Society* 81: 373–388.
10. L. Michaelis, M.L. Menten. 1913. *Bio-chem Zeitschr* 49: 333–369.
11. G.E. Briggs, J.B.S. Haldane. 1925. *Biochemical Journal* 19: 338–339.
12. J.E. Bailey, D.F. Ollis. 1986. *Biochemical Engineering Fundamentals*, 111–113. New York: McGraw-Hill.
13. J.J. Carberry. 1976. *Chemical and Catalytic Reaction Engineering*, 364–366. New York: McGraw-Hill.
14. H. Lineweaver, D. Burk. 1934. *Journal of the American Chemical Society* 56: 658–666.
15. G.S. Eadie. 1942. *Journal of Biological Chemistry* 146: 85–93.
16. B.H.J. Hofstee. 1952. *Journal of Biological Chemistry* 199: 357–364.
17. K. Schüugerl. 1982. *International Chemical Engineering* 22: 591–610.
18. J. Monod. 1949. *Annual Review of Microbiology* 3: 371–394.
19. C.M. Cooper, G.A. Fernstrom, S.A. Miller. 1944. *Industrial & Engineering Chemistry* 36: 504–509.
20. H. Taguchi, A.E. Humphrey. 1966. *Journal of Fermentation Technology Japan* 44: 881–889.
21. K. Akita, R. Yoshida. 1973. *Industrial & Engineering Chemistry Process Design and Development* 12: 76–80.
22. K. Van't Riet. 1979. *Industrial & Engineering Chemistry Process Design and Development* 18: 357–364.
23. J.Y. Oldshue. 1985. In *Mixing of Liquids by Mechanical Agitation*, ed. J.J. Ulbrecht, G.K. Patterson, 309–342. New York: Gordon and Breach Science Publishers.

20 Measuring Physical Properties

Lyle F. Albright

CONTENTS

20.1 Introduction 1531
20.2 Temperature-Measuring Devices 1531
      20.2.1 Thermcouples 1532
      20.2.2 Resistance Thermometers and Thermistors 1532
      20.2.3 Radiation and Infrared Pyrometers 1533
      20.2.4 Filled-Bulb and Glass-Stem Thermometers 1534
      20.2.5 Bi-Metallic Thermometers 1535
      20.2.6 Pyrometric Cones 1535
20.3 Flowmeters 1535
      20.3.1 Several Popular Types of Flowmeters 1535
20.4 Pressure Measurements 1536
20.5 Level Measurements 1537
20.6 Other Measurements 1537
References 1537

20.1 INTRODUCTION

To measure physical properties reliably, the following must occur: The measuring instruments must be carefully selected and the instrument must be properly installed and operated. Factors to be considered in the selection process include the range of property to be measured, accuracy and reproducibility of the readings desired, ease of installation and servicing, and both capital and expected maintenance costs.

    Advice is offered here on selecting instruments or techniques to measure temperature, pressure, flow rates, etc. More details are in references [1, 2]. In general, a relatively large number of instruments are candidates for any specific application.

20.2 TEMPERATURE-MEASURING DEVICES

Numerous types of sensors are employed in industrial plants, pilot plants, and laboratory units. Often, several are considered before the final selection is made for a specific application. The method of installing and operating the sensor is also most important. Attempts are often needed to minimize the temperature difference between the sensor and the process stream (or unit) to be measured, as discussed later.

    The most commonly used thermometers (or temperature-measuring devices) in chemical or related processes are as follows: thermocouples employed in the range of about −260°C to 2500°C; resistant measuring instruments from −250°C to 650°C; radiation and infra-red pyrometers from 0°C to 4000°C, plus an assortment of bi-metallic and liquid-filled thermometers, color indicators, pneumatic cones, etc. These thermometers are often calibrated using the following materials: liquid hydrogen at its triple point of −259.34°C, gold at its freezing point of 1064.33°C, plus freezing and boiling points, respectively, of water at 0°C and 100°C (depending on barometric pressure). The current discussion will be limited to the most popular thermometers. More advice and information on various sensors relative to their installation and operation are in the literature.

20.2.1 THERMCOUPLES

Two dissimilar wires are connected in two places to form a loop. The first connection is positioned where the temperature is to be measured; the second connection is positioned where the temperature is known. This temperature difference causes a potential (or voltage) difference in the wires. Tables are available for predicting the temperature changes of several combinations of two dissimilar wires [3]. Temperature ranges for common thermocouples are as follows:

Chromel/alumel	–180°C to 1370°C
Iron/constantan	–180°C to 760°C
Copper/constantan	–180°C to 400°C
Platinum 10% rhodium/platinum	0°C to 1760°C
Tungsten/tungsten 26% rhenium	0°C to 2850°C
Tungsten 5% rhenium/tungsten 26% rhenium	0°C to 2850°C

Thermocouples unfortunately develop only relatively small voltage (electromotive force, or emf) changes for a temperature difference of 1°C. Hence errors of predicted temperatures are often relatively large. The first three thermocouples often have errors of 2°C to 3°C or 0.5% to 0.75%. The latter three (and more expensive) thermocouples typically have smaller errors. With specially constructed thermocouples having purer metals, better connections, etc., errors are reduced by perhaps 50% as compared with the values reported above.

Thermocouples tend to be relatively inexpensive, but the instruments to measure voltages and to record the temperatures are generally more expensive. To obtain reliable temperature measurements, the following factors are important relative to the installation of the thermocouple plus other types of thermometers:

1. The tip of the thermocouple (or other thermometer) often cannot be inserted directly in the liquid, gas, slurry, etc., due to chemical attack or erosion of the thermocouple. Obtaining a leak-proof system is frequently a problem.
2. A thermowell is often employed for the temperature probe in order to protect it against direct contact with the material being investigated. A gas such as air is often in the thermowell in direct contact with the thermocouple. As a result, the thermocouple may become oxidized or corroded, causing the calibration to change. Thermocouples for ethylene furnaces operated at about 800°C to 1000°C often need to be replaced after several months of operation.
3. Sometimes appreciable heat transfer may occur in both the thermocouple and the thermowell, e.g., in ethylene furnaces. In these furnaces, there are large temperature gradients along the length of the thermocouple.
4. Frequently, thermocouples are provided with protective sheaths, and their purchasing price is considerably higher.

A well-designed system, including thermocouple, thermowell, and often protective sheath, generally provides more reliable measurements.

20.2.2 RESISTANCE THERMOMETERS AND THERMISTORS

For these thermometers, the electrical resistance of a metal or metal oxide changes with temperature. This resistance change can be measured by three-wire or four-wire null bridges (or Wheatstone bridges) in order to predict temperature. The following metals, and their range of temperature, are all used commercially:

Platinum (wound and thin film)	−259°C to 482°C
Nickel	−195°C to 426°C
70% nickel/30% iron (Balco)	−195°C to 260°C

Platinum resistance thermometers that are carefully constructed and also carefully operated have been used as international standards from −259°C up to 1095°C. In general, resistance thermometers predict considerably more accurate temperatures than thermocouples. Resistance thermometers are often accurate to <0.01°C; they tend to retain their accuracy after extended use. In general, they are more expensive, more fragile, and larger than thermocouples.

For platinum resistance thermometers, the platinum is often wrapped around a ceramic or glass core; the above is often encapsulated in a ceramic or glass capsule. An alternative design is a deposit of platinum film on a substrate; next, the film is encapsulated. Sometimes highly inert Teflon is used as the outer encapsulating layer. With Teflon, the thermometer can often be immersed in highly corrosive liquids. On occasion, the resistance thermometer is attached (or "cemented") by using a suitable adhesive to solid surfaces to measure temperatures from about −200°C to 260°C. Thermowells are also frequently employed. Resistance thermometers are available in a wide variety of sizes (diameters, lengths, etc.), designs, and protective or encapsulating materials. Numerous companies sell the instruments to measure the electrical resistance and to record the predicted temperature [1].

Another type of resistance thermometer uses metal oxides, instead of metals; it is frequently referred to as a thermistor. Electrical resistance of these metal oxides changes rapidly with even rather small temperature changes. Hence, thermistors are often emplyed to measure small temperature changes such as 1°C to 5 °C. The thermistor proper tends to have low purchase prices. Metal oxides, which are semiconductors, include mixtures of the following oxides: nickel, manganese, copper, cobalt, tin, germanium, etc. [1].

20.2.3 Radiation and Infrared Pyrometers

Pyrometers predict temperatures for the following:

- Interior of high-temperature furnaces; pyrolysis operations; production of steels, alloys, etc.
- Moving materials on conveyor belts, rollers, etc.
- Highly corrosive materials or materials that are easily contaminated
- Distant surfaces in a plant or those not easily accessible

Prediction of the temperature based on radiation considerations is based on the Stefan-Boltzmann law, which states the amount of radiation depends on the absolute temperature raised to the fourth power. To obtain accurate temperature predictions of a solid, liquid, or gas requires consideration of several factors:

1. Absorbance and emissivity of the material being investigated; they are defined as the ratio of that property as compared with that of a blackbody. Hence, both have values between 0.0 and 1.0
2. Reflectivity of the material; once again, it is defined as the ratio compared with that of a blackbody that absorbs all the radiation to which it is exposed, i.e., there is no reflection with a blackbody.
3. Geometry of the system.

4. Gas or suspended solids or liquids in the space between the material being measured and the radiation detector measuring the radiation.
5. Wavelengths of the radiation being recorded by the measuring device. Wavelengths being emitted from the object being investigated vary significantly with its temperature often in range of 0.3 to 100 microns. At ambient temperatures, wavelengths in the visible range predominate, but at about 1000°C, infrared wavelengths of about 2 to 20 microns predominate. At still higher temperatures, even shorter wavelengths result, i.e., near infrared, ultraviolet, etc. Modern pyrometers often record the emissions of two or more wavelengths.
6. Most desired instrument for recording the radiation of the object being investigated.
7. Method of collecting radiation from the object being investigated. The size of the aperture or opening to view the object plus area being sited is important.

Several temperature readings sometimes need to be recorded. In an ethylene furnace, for example, a hydrocarbon feedstock is pyrolyzed to produce an ethylene-rich stream. At the inlet to the coil reactor, much lower temperatures occur as compared with the outlet. In addition, the ceramic wall of the furnace is often relatively close to that of the hot combustion gases. These gas temperatures are much higher than those of the metal walls. Normally, the measuring detector collects information at different locations inside the furnace through a small aperture on the wall of the furnace. De Witt and Albright [4] report the results of a test of a furnace for distilling crude oil in a refinery about 20 years ago. In this test, flat targets were employed in which calibrated thermocouples were embedded. One target was essentially at the temperature of the combustion gas (and hence was considerably hotter) while another target was cooled to a much lower temperature. The pyrometers of those tests resulted in temperature measurements that were too low for the low-temperature target. Newer instruments and improved methods of predicting the temperatures have since significantly improved the accuracy of measurements.

Emissivities of different metals vary greatly. Liptak et al. [1] report values for common metals including 0.03 for gold and 0.94 for dull oxidized wrought iron. Calibrations based on emissivities correct to a high degree predicted values. Part of the radiant energy is adsorbed by intervening smoke, carbon dioxide, etc. Narrow-band pyrometers use only a narrow band of the spectrum, often in the red zone. By proper selecting of the band, the desired temperature of the solid, gas, or glass can be predicted. They recommended the wavelength in microns for vacuum depositions (0.2 to 0.3); furnaces, refractory metals, etc. (0.653); metal processing (0.6 to1.0); textiles, plastics, food products (2.0 to 2.6); thin film polymers (3.43 ± 0.14); and carbon dioxide gas (4.8 to 5.6).

20.2.4 Filled-Bulb and Glass-Stem Thermometers

These thermometers contain either mercury, alcohol, etc., as liquids. The thermal expansion of these liquids is greater than the glass, so the height of liquid in the glass capillary rises as the temperature increases. A major problem is that the glass can be easily broken. Furthermore, mercury causes toxicity problems if the thermometer breaks. Visual observations are usually required to read the thermometers. Often these instruments are restricted to temperatures from about –40°C to 400°C. Their advantages are low costs, long life if properly protected, and reasonable accuracy. They still are widely used in experimental setups and for various home uses.

Some filled-bulb thermometers take account of the pressure increase in the capillary stream. This increase is due to the vapor-pressure rise of the liquid phase plus the rise due to the compression of the gas. A relatively high pressure may hence occur. Bending the capillary tube is sometimes used, and the tube straightens to a degree as the pressure rises. The amount of straightening can be calibrated with a pointer on a dial to measure the temperature.

20.2.5 Bi-Metallic Thermometers

Bi-metallic thermometers are based on the principle that different metals expand at different rates as the temperature increases. Two strips of the dissimilar metals are connected in the form of a spiral or helix. A pointer connected to these assemblies indicates the temperature on a dial. Such thermometers are widely used to measure temperatures of homes or buildings. They are often used over a temperature span of about 80°C.

20.2.6 Pyrometric Cones

These cones change shape over a rather wide range of temperatures. They have found use in the ceramic industries and in applications to determine the approximate final temperatures in materials such as films, throw-away items, and onetime applications.

20.3 FLOWMETERS

Numerous flowmeters are often available for a specific application. Factors of importance in making a choice include the following:

1. Character of fluid (gas, liquid, slurry or dispersion, etc.). What are the physical properties of the fluid including viscosity, density, surface tension, etc.?
2. Number of phases flowing. If multiple phases, what are the degree of dispersion, ratio of phases, continuous phase, etc.?
3. Cleanness and/or impurities in the fluid.
4. Corrosiveness of fluid must be known when choosing the materials of construction of the flowmeter.
5. Flowrates expected for fluids and Reynolds number of fluids. Hence, the expected diameter of the tubing to and from the flowmeter must be chosen. Are flows steady or pulsating?
6. Pressures range to be monitored. Pressures may vary greatly. A pressure drop is often experienced in the flowmeter, but pressure increases occur if metering pumps are employed.
7. Expected maintenance costs. Corrosion, deposition of solids, etc., add significantly to costs, changes in calibration of flowmeter, etc.
8. Reproducibility, accuracy, and calibration. Some applications require high accuracy.
9. Both the initial capital costs and expected maintenance costs are obviously important factors.
10. Devices to transmit the signals from the flowmeter to the recorders are often required. The costs of such devices are often relatively large.
11. Possible problems in startup and shutdown of the units need to be considered.

20.3.1 Several Popular Types of Flowmeters

Flowmeters can often be divided into the following categories:

1. Meters that measure differential pressures over the flowmeter and such pressure changes that can be interpreted as flowrates. Such flowmeters with a large number of designs include orifices, venturi tubes, pitot tubes, elbow taps, etc. Fluids that result in changes of the cross-sectional area due to erosion, corrosion, or deposition of solids obviously change the calibrations. These meters tend to be relatively cheap but are often not very accurate.
2. Several flowmeters are based on electromagnetic properties of the fluid. Often electrical conducive fluids are required. Coriolis mass flowmeters are widely used; an angular

momentum is imparted to flowing fluid by providing harmonic vibrations. Sometimes a U-bend is provided, and in other designs, two tubes are operated in parallel. Costs of these units tend to be high, but the electromagnetic detectors often provide highly reliable flowrate predictions. Reigner [6], however, indicates they are not reliable for two-phase flows such as aerated liquids. Omega Engineering [7] also provides important recommendations.

3. Displacement flowmeters for both liquids and gases. In some cases, positive displacement pumps are employed for liquids.
4. Rotameters can often be used for liquids and gases, but safety issues such as loss of toxic materials if the tube breaks are concerns.
5. Weirs and flumes are sometimes employed.
6. Flowmeters are available for granular solids.

20.4 PRESSURE MEASUREMENTS

Measuring pressure is obviously of great importance in chemical processes, and numerous instruments and techniques are available. Key considerations in selecting the best instrument for a specific application include the following:

1. Range of pressures to be measured may vary from high vacuums to several thousand atmospheres.
2. Short-time pulsations of pressure. For example, some pumps cause rapid pulsations.
3. Desired accuracy and precision of pressure reading.
4. Temperature range of materials being measured.
5. Character or nature of materials including corrosiveness, freezing point, viscosity, etc. Sometimes special precautions are needed in transferring pressure from process stream to pressure measuring device. For common Bourdon tubes, fluids such as silicone oils, light hydrocarbon oils, glycerine, etc., are employed for such transfers.
6. The required or preferred method of reading or recording the pressure. Pressures, for example, can be read usually from a gauge or a recorder.

Methods and instruments for measuring pressure are available for $100 or less up to several thousand dollars. It should be emphasized that many pressure devices read pressure differences between the process stream and atmospheric pressure.

1. Pressure gauges containing Bourdon tubes are widely used for either pressures from atmospheric to high pressures, such as over 700 bars, or from atmospheric to high vacuums. With pressure changes, the tube changes shape causing a needle to change direction so the pressure can be read. Gauges calibrated by the manufacturer often have inaccuracies up to ±2.0% over at least a portion of their range. Better accuracy can be obtained with more carefully manufactured and hence more expensive gauges.
2. Pressure gauges involving spring and bellows, bell-cased mounting, diaphragms, and other devices have also been developed.
3. Strain-gauge transducers are also employed for pressure measurements; mechanical strain caused by pressure changes causes the electrical resistance of the transducer to change. Semiconductors such as geranium or silicon are often employed. Thermal errors are minor at near-ambient temperatures. A wide range of prices occurs for such transducers.
4. Manometers are employed for relatively small pressure changes. Often costs are low.
5. Dead-weight piston gauges are employed to calibrate pressures up to large pressures. Well-designed gauges of this type are relatively expensive.

20.5 LEVEL MEASUREMENTS

A variety of methods are available to measure liquid or solid particle levels in storage tanks, reactors, etc. Methods vary from cheap, such as measuring pressure changes as gas is bubbled upward through liquid, to considerably more expensive using capacitane and radio frequency, conductivity, lasers, microwaves, radar, etc. In storage tanks with relatively large diameters and cross-sectional areas, a large amount of material is stored in a volume only 1 to 2 mm thick, so relatively accurate level control is sometimes needed.

Variables of importance in selecting instruments for level control include:

- Liquid, slurry, or solid
- Temperature range
- Pressure range
- Viscosity and physical properties of liquids
- Character of solid particles
- Capital and expected maintenance costs
- Expected precision and accuracy of readings
- Possible foam, boiling, and agitation

20.6 OTHER MEASUREMENTS

Other instruments are needed in many plants, including electrical ones, viscometers, densities, etc. Often more than one instrument can be chosen for a specific application, so the engineer must choose with consideration to operating, economics, and safety concerns.

REFERENCES

1. Liptak, B.G. 1995. *Process Measurement and Analysis*. 3rd ed. Randor, PA: Chilton Book Co.
2. Omega Engineering, Inc., Stamford, CT
   Temperature Handbook
   Pressure Strain and Force Handbook
   Flow and Level Handbook
   Electric Heaters Handbook
   Other handbooks available; various encyclopedias have been published since 2000 (www.omega.com).
3. Liptak 1995: 501–508.
4. DeWitt, D.P., and L.F. Albright. *Measurement of High Temperatures in Furnaces and Processes*. AIChE Symposium Series. Vol. 82. New York: AIChE, 1986: 13–27.
5. Liptak 1995: 459.
6. Reizner, J.R. "Exposing Coriolis Mass Flowmeters Dirty Little Secret." *Chemical Engineering Progress* March (2004): 24–30.
7. Omega Engineering, Inc. 2005. *Flow and Level Measurements Transactions*. Vol. 4.

21 Selecting Materials of Construction (Steels and Other Metals)

David A. Hansen

CONTENTS

21.1 An Overview of the Materials Selection Process 1540
    21.1.1 Materials Selection Criteria 1541
    21.1.2 Mandatory Requirements 1541
    21.1.3 Governing Criteria 1541
    21.1.4 Special Requirements 1542
    21.1.5 Materials Selection Template 1542
21.2 Basic Metallurgy 1542
    21.2.1 Heat Treatments 1542
        21.2.1.1 Annealing 1542
        21.2.1.2 Normalizing 1543
        21.2.1.3 Preheat 1543
        21.2.1.4 Stress Relief/Postweld Heat Treatment 1544
        21.2.1.5 Quench and Temper 1545
    21.2.2 Microstructural Terms 1546
        21.2.2.1 Austenite 1546
        21.2.2.2 Ferrite 1546
        21.2.2.3 Martensite 1546
        21.2.2.4 Pearlite 1547
    21.2.3 Metallurgical Terms 1547
    21.2.4 Alloy Designations 1549
    21.2.5 Manufacturing of Metals and Alloys 1549
    21.2.6 Metals and Alloys 1550
        21.2.6.1 Cast Iron 1550
        21.2.6.2 Carbon Steel 1552
        21.2.6.3 Alloy Steel 1553
        21.2.6.4 Stainless Steel 1554
        21.2.6.5 Common Alloys and Metals 1558
21.3 Corrosion 1561
    21.3.1 Corrosion Basics 1561
        21.3.1.1 Cathodes 1562
        21.3.1.2 Anodes 1562
    21.3.2 Corrosion Control 1562
        21.3.2.1 Barrier Coatings 1562
    21.3.3 Stress Corrosion Cracking 1564

	21.3.4	Microbiologically Influenced Corrosion (MIC)..1566
		21.3.4.1 Effect on Materials of Construction 1567
		21.3.4.2 Mitigation Methods...1567
21.4	Failure Modes...1567	
	21.4.1	Embrittlement Phenomena 1567
	21.4.2	Carbon and Low-Alloy Steel 1569
		21.4.2.1 Stainless Steel...1571
	21.4.3	High Alloys..1572
	21.4.4	Hydriding..1572
	21.4.5	High-Temperature Effects 1572
		21.4.5.1 Mechanical Effects..1572
		21.4.5.2 Metallurgical Effects 1573
	21.4.6	Low-Carbon Stainless Steel 1574
		21.4.6.1 Chemically Stabilized Stainless Steel..1574
	21.4.7	Hydrogen Gas...1577
	21.4.8	Nitriding..1578
	21.4.9	Oxidation 1579
	21.4.10	Sulfidation and Sulfidic Corrosion...1579
21.5	The Process of Materials Selection..1581	
	21.5.1	Designing a Template 1581
	21.5.2	Materials Selection Criteria..1588
		21.5.2.1 Product Contamination...1588
		21.5.2.2 Reliability 1588
	21.5.3	Organizing the Materials Selection Procedure 1589
	21.5.4	Materials Selection Procedure: Exceptions...1589
		21.5.4.1 Piping..1589
		21.5.4.2 Pumps 1589
		21.5.4.3 Fabricated Equipment 1589
	21.5.5	Grouping Process Regions 1590
	21.5.6	Materials Selection Procedure..1590
		21.5.6.1 Low Temperature 1590
		21.5.6.2 High Temperature..1590
		21.5.6.3 Corrosion 1591
		21.5.6.4 Upset Conditions: Review..1592
	21.5.7	Materials Selection Diagram..1593
21.6	Conclusions..1594	
References 1595		

21.1 AN OVERVIEW OF THE MATERIALS SELECTION PROCESS

In selecting metals and alloys as materials of construction, one must have knowledge of how materials fail, for example is, how they corrode, become brittle with low-temperature operation, or degrade as a result of operating at high temperatures. Corrosion, embrittlement, and other degradation mechanisms such as creep will be described in terms of their threshold values. Transient or upset operating conditions are common causes of failure. Examples include start-ups and shutdowns, loss of coolant, the formation of dew point water, and hot spots due to the formation of scale deposits on heat transfer surfaces. Identification and documentation of all anticipated upset and transient conditions are required.

Corrosion debris is a related problem. In some applications, materials with nominally low corrosion rates produce considerable corrosion debris. Such problems are always associated with units that contain large surface areas exposed to the corroding fluid. Examples include heat

Selecting Materials of Construction (Steels and Other Metals)

exchanger surfaces and packed beds. The materials selection process must document and organize such information.

This section will discuss problems that can complicate the selection procedure. Such problems include the variability of materials selection criteria, conflicting project objectives such as minimizing capital cost, and mandatory criteria such as those in governing engineering codes. Other topics discussed include organization of the information needed for the materials selection process, a procedure for materials selection, and the use of a materials selection diagram.

21.1.1 Materials Selection Criteria

Many engineers associate materials selection only with the design and construction of new facilities, plant additions, or plant renovations. However, it is also part of a plant's routine maintenance activities, being a subject of discussion between operations, planning, and maintenance personnel. Such discussions usually illustrate the need for both short-term and long-term solutions.

For simple jobs such as replacements in-kind or for jobs with which the responsible engineer has prior experience, materials selection is usually a straightforward task. However, some jobs involve complex combinations of requirements, which may include (1) demanding mechanical requirements such as thermal cycling, (2) conservative design requirements such as "leak-before-break", (3) stringent process-related requirements such as avoiding process contamination, (4) special fabrication requirements such as postweld heat treatment (PWHT), (5) aggressive corrodents or crack-inducing agents, (6) upset or transient operating conditions (i.e., nonstandard operating conditions), (7) limited capital budgets, and (8) aggressive project schedules.

21.1.2 Mandatory Requirements

Most localities mandate compliance with national engineering codes such as the ASME (American Society of Mechanical Engineers) Boiler & Pressure Vessel Code [1]. These codes govern mechanical design and provide the maximum allowable stresses and required low-temperature toughness, as a function of temperature, for approved materials. The codes also define requirements for fabrication procedures such as PWHT. All codes contain rules that ensure the selected material of construction will not be susceptible to brittle fracture at the minimum design temperature. Users of the codes must develop a familiarity with these rules. For example, code rules help avoid specifying materials at temperatures for which they are not permitted.

The codes do not normally include guidelines on materials selection. However, they often provide information about degradation mechanisms. Most designers use this information as if it were mandatory. For the same reason, nonmandatory recommended practices, such as NACE (National Association of Corrosion Engineers) MR0175 [2] and API (American Petroleum Institute) Publication No. 941 [3], are often customarily used as mandatory documents. Because of safety concerns and potential liabilities related to process guarantees such as yield or purity, materials selection guides provided by process licensors are usually regarded as mandatory. The materials recommended by process licensors are conservative. All engineering codes used for mechanical design contain tables defining allowable stresses for each material, usually as a function of temperature. Most, if not all, codes restrict materials selection to the listed materials only. Thus the design engineer must know the materials permitted.

21.1.3 Governing Criteria

All design data such as pressures and temperatures are regarded as mandatory conditions that must be satisfied. Upset and transient operating conditions are sometimes governing. For example, the maximum design pressure and coincident temperature will determine the wall thickness of a carbon steel process vessel containing dry H_2S. However, even traces of liquid water in the presence of H_2S can initiate sulfide stress corrosion cracking in carbon steel. This should add the requirement of PWHT.

Operating Temperature (Minimum/Maximum): _____

Operating Pressure (Minimum/Maximum): _____

Commodity: _____ **Phases:** _____ **Liquid Water** (Y/N): _____

Corrodents: _____

Crack-Inducing Agents: _____

Metallurgy: _____

PWHT(Y/N): _____ **Valve Trim*:** _____ **Corrosion Allowance:** _____

Notes: _____

* For most services, valve trim will be 13Cr stainless steel. Use full hard facing for temperatures >600°F (315°C).

FIGURE 21.1 A simplified materials selection template. (Copyright Marcel Dekker, Inc.; reprinted with permission [5].)

21.1.4 SPECIAL REQUIREMENTS

The required design life will generally affect materials selection and/or the determination of the recommended corrosion allowance. Normally, the user or process licensor will define design life requirements. It is helpful to define the design life requirements in the "Notes" addendum of the template.

Materials selection for some projects is affected by special objectives such as minimal capital cost, minimal maintenance, short project schedule, extended design life, concerns about product purity, or the consequences of a leak or rupture. If such objectives are governing, they should be included in the materials selection design basis in the "Notes" addendum of the template. Occasionally, objectives may be in conflict, such as minimal capital cost vs. short schedule. When this occurs, compromises are made. Or a superior material may not be selected if its delivery would delay start-up. Safety and environmental concerns must always be considered.

21.1.5 MATERIALS SELECTION TEMPLATE

A materials selection template, containing critical design and operating conditions for each piece of equipment and piping run, is used as part of a structured method to select suitable materials of construction. It is usually prepared by a plant or project process engineer, and is customized for the plant or project (see Figure 21.1 for an example of a simple template).

The template serves two major purposes: (1) it organizes all the technical information needed for materials selection; and (2) on a large job or project, the template is a convenient means to transmit process and materials selection information to design engineers.

21.2 BASIC METALLURGY

Metallurgical descriptions often contain jargon and arcane words, as explained below.

21.2.1 HEAT TREATMENTS

21.2.1.1 Annealing

For carbon and low-alloy steels, full annealing requires heating to a temperature of 1350°F to 1750°F (730°C to 955°C). The steel is held at this temperature long enough to ensure through-

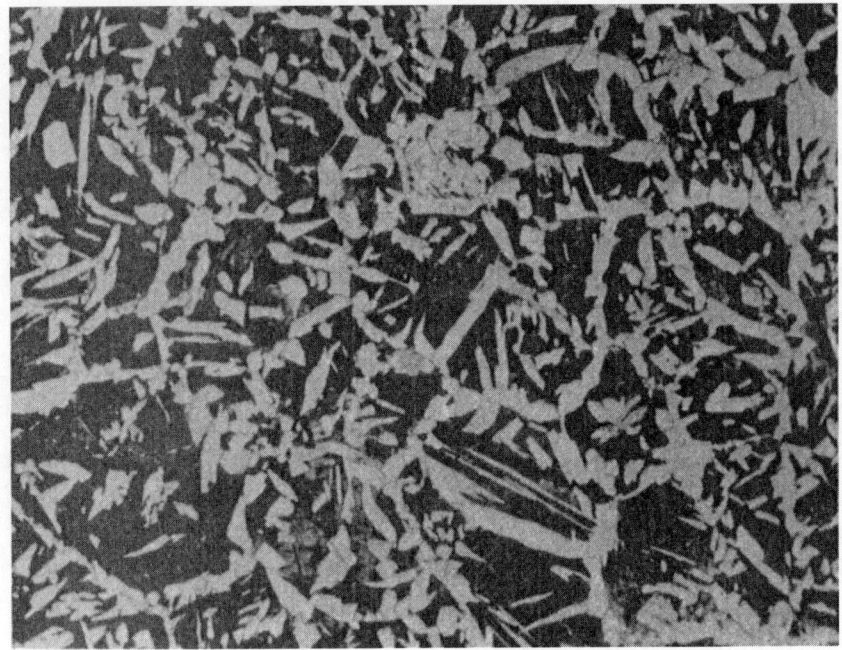

FIGURE 21.2 Carbon steel that has been furnace annealed. © Copyright by Marcel Dekker, Inc.; reprinted with permission [5]. (Courtesy of Dr. E.V. Bravenec, Anderson & Assoc.)

thickness heating; then the material is furnace cooled, producing a "dead soft" carbon steel (Figure 21.2). This condition is usually unacceptable, because carbon and low-alloy steels can be very brittle in the fully annealed condition. Such steels are usually subsequently normalized. Figure 21.3 shows the microstructural effects due to normalizing.

For austenitic stainless steels, annealing is usually at about 2000°F (1095°C), followed by either a water quench or rapid cooling in air. This procedure is more properly called a "solution anneal," since the objective is to redissolve any chromium carbides formed during prior processing. Rapid cooling minimizes formation of carbides during the cooling period. The minimum temperature for solution annealing depends on the composition of the alloy.

Chemically stabilized alloys such as Type 321 SS are available in a heat-treated condition known as "stabilization annealed." In this condition, such alloys have maximum resistance to sensitization and its effects. Sensitization is discussed in Section 21.4.

21.2.1.2 Normalizing

In this process, a carbon or low-alloy steel is heated to about 1650°F (900°C) and is then air-cooled. The process partially relieves stresses from prior processing and "refines" the material. "Refining" reduces the grain size and the grain structure becomes more homogeneous, thereby producing a tougher, more ductile product. Figures 21.2 and 21.3 illustrate the microstructural benefits of normalizing.

21.2.1.3 Preheat

Many steels are susceptible to cracking during or after welding. Examples include high-strength carbon steels and low-alloy steels such as the air-hardenable Cr-Mo steels. Preheating the base metal or substrate reduces the risks. Engineering codes [4] usually contain preheat temperatures and procedures.

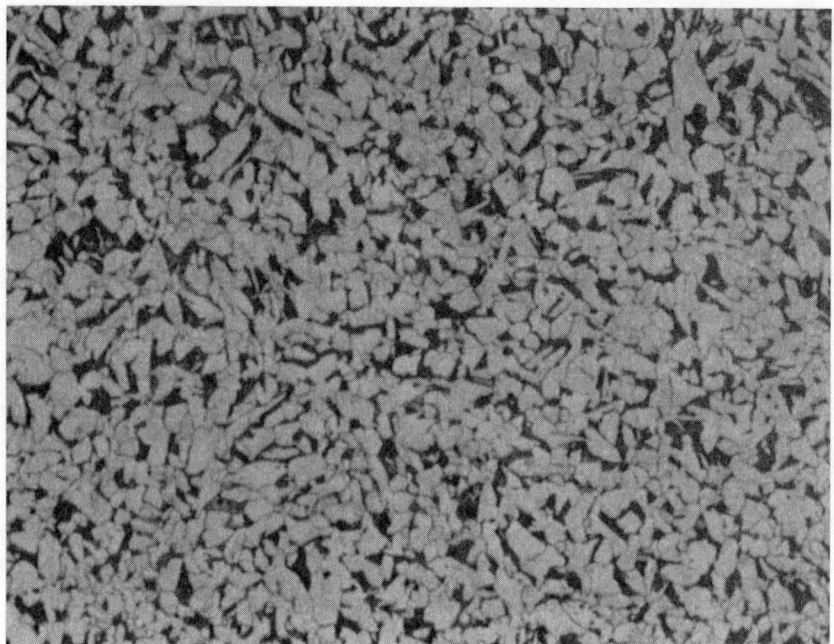

FIGURE 21.3 Carbon steel that has been normalized. © Copyright by Marcel Dekker, Inc.; reprinted with permission [5].

21.2.1.4 Stress Relief/Postweld Heat Treatment

Residual stresses can be introduced into a metal by fabrication processes such as forging or rolling, by uneven heating or cooling, or by welding. The magnitude of such stresses is usually on the order of the yield strength, but sometimes approaches the tensile strength.

To stress relieve or postweld heat treat carbon and low-alloy steels, they are typically heated to 1100°F to 1350°F (595°C to 730°C) for extended time, followed by air cooling. The minimum time is specified by the relevant engineering code, and the temperature must be less than the lower transformation temperature of the steel, which is the lowest temperature at which austenite starts to form, for example, 1333°F (720°C) for plain carbon steels. In order to avoid degrading the required mechanical properties of a heat treated alloy, subsequent fabrication heat treatment temperatures, such as those for stress relief and PWHT, must not exceed the tempering temperature (discussed in the next section).

The yield strengths of metals and alloys decrease with increasing operating temperature. Residual stresses in excess of the reduced yield strength are eliminated via plastic deformation. Upon cooling, the maximum residual stress possible is the yield strength at the holding temperature. For carbon steels, heat treatment will reduce residual stress by about two-thirds. When done for the purpose of removing residual stresses caused by cold work, the process is termed stress relief.

As the liquid metal in a weld solidifies, it becomes at least partially constrained by the surrounding parent metal. When the weld metal cools to ambient temperature, the resulting residual stress in the weld is approximately equal to the ambient temperature yield strength of the parent metal. Stress relief heat treatment for welds is called postweld heat treatment. An additional benefit of such treatment is a reduction of the hardnesses of the weld metal and heat-affected zone (HAZ), thus reducing the risk of stress corrosion cracking.

Stress corrosion cracking in heat treated weldments indicates that postweld heat treatments are sometimes ineffective. Sometimes this result can be attributed to microalloying. (Microalloying is discussed later.) In some cases, subsequent stress corrosion cracking can be traced to an improperly executed PWHT, for example, where the PWHT temperature was too low. In other cases, it has been speculated that even the lowered level of residual stress was sufficient for the initiation of stress corrosion cracking. When experience indicates that PWHT may be ineffective, a stress corrosion cracking mitigation measure (such as a cracking-resistant weld overlay) or a change in the process or the material of construction is indicated.

Some engineering codes allow low-temperature PWHTs if the weldment is held at the lower temperature for an extended time. Such PWHTs should not be permitted to avoid or minimize the risk of stress corrosion cracking.

Austenitic stainless steels are not usually stress relieved or postweld heat treated. When they are heat treated, they are held at 1600°F to 1650°F (870°C to 900°C), followed by rapid cooling. Such cooling avoids sensitization. (Sensitization involves thermally-induced precipitation of carbides in microstructural grain boundaries). Stress relief or postweld heat treatments of ordinary austenitic stainless steels at less than 1600°F (870°C) may sensitize the alloy. For this reason, local stress relief of unstabilized austenitic stainless steel is usually impractical, since the "runout" areas immediately adjacent to the region being heat treated will be grossly sensitized. An exception is stress relief of low-carbon grades of thin-section products such as tubing. These can be stress relieved fast enough to avoid gross sensitization. Stabilized grades of austenitic stainless steels such as Type 321 SS are much less susceptible to sensitization if they have been stabilization annealed (see Section 21.4 for a discussion of sensitization and stabilization annealing). Stabilized grades are usually selected if local stress relief or PWHT is required.

21.2.1.5 Quench and Temper

This procedure produces materials with improved strength and/or toughness. It is usually restricted to alloys whose microstructures are transformed upon cooling. In this procedure, a ferritic steel is first heated to a temperature of about 1650°F (900°C) or higher, then quickly cooled (i.e., quenched) in air, water, water spray, oil, or salt bath. The required conditions depend on alloy chemistry, section thickness, and desired mechanical properties.

Quenching produces a hard, strong—but brittle—phase called martensite. Tempering is usually performed at 1100°F to 1300°F (595°C to 705°C). The tempering procedure is very similar to that used for stress relief or PWHT. Tempering promotes the diffusion of some carbon from the martensite, thereby greatly improving the ductility and toughness of the quenched steel.

Occasionally, satisfying the simultaneous requirements for an ASTM product form heat treatment and for its construction code postweld heat treatment can be a problem. For example, the minimum tempering temperature for ASTM A335 1-1/4Cr-1/2Mo pipe is 1200°F (650°C). The ASME B31.3 piping code postweld heat treatment minimum temperature for this material is 1300°F (705°C). Thus, it can happen that 1-1/4Cr-1/2Mo pipe ordered to ASTM A335 will be subsequently postweld heat treated at a temperature exceeding the tempering temperature. If this happens, the minimum mechanical properties indicated by the ASTM specification may deteriorate, thus threatening the integrity of the design. This situation can be avoided by modifying the purchase order in either of two ways: (1) Establish a minimum ASTM tempering temperature that exceeds the subsequent PWHT temperature. To ensure that this strategy is followed, it will be necessary to instruct the fabricator regarding the maximum permitted PWHT temperature. In this manner, the minimum requirements of both the ASTM specification and the construction code are satisfied. (2) Require that coupons, exposed to the same heat treatments to be used for actual fabrication, be tested to demonstrate that the proposed heat treatments do not degrade the minimum required mechanical properties.

Thick sections of many ferritic steels cannot be cooled quickly enough in air to obtain the desired normalized microstructure. In such cases, quenching is often used to hasten the cooling rate. The objective is to produce the same microstructure as obtained from normalizing a thinner section of the same material. In thick sections of such materials, even quenching does not ordinarily generate the cooling rates necessary to develop martensite. Hence, tempering is primarily for stress relief rather than for softening martensite. Many codes, such as the ASME Boiler & Pressure Vessel Code, Section VIII [1] and materials specifications such as ASTM A516, permit such thick sections, properly quenched and tempered, to be equivalent to normalized material. Tempering is also sometimes done in conjunction with other heat treatments such as normalizing in order to soften and/or toughen the steel. In some cases, stress relief may be a secondary or even a primary objective.

Fabricators occasionally propose multiple heat treatments or heat treatments having unusually long holding times or unusually high holding temperatures. The user should be wary of such proposals. Some multiple heat treatments may cause degradation of carbon and low-alloy steels with loss of strength and/or loss of toughness. The fabricator should be required to demonstrate, by testing, that the proposed procedure will not result in material degradation.

21.2.2 Microstructural Terms

21.2.2.1 Austenite

Austenite is a high-temperature form of carbon steel, having a face-centered cubic crystal structure. The lowest temperature at which ordinary carbon steel can be fully austenitic is 1333°F (720°C). During normalizing heat treatments, the holding temperature and time are specified so that the alloy becomes fully austenitic. For the common carbon steels, the austenitizing temperature is typically specified as 1650°F (900°C).

Austenite has a much higher solubility for carbon than other forms of steel. Heating the steel to an austenitizing temperature causes any carbides present to dissolve. Alloys capable of forming austenite at high temperatures, but that transform to other crystal structures at lower temperatures, are said to be hardenable by heat treatment. Martensitic steels are an example. Most carbon and low-alloy steels are hardenable by heat treatment.

When alloying elements such as nickel or manganese are added to carbon steel, the austenitic microstructure becomes stable at low temperatures. For example, most austenitic stainless steel and high-nickel alloys exhibit stable austenitic microstructures at temperatures approaching absolute zero. These alloys have excellent low-temperature fracture toughness and are resistant to hydrogen embrittlement from causes other than cathodic charging. Most austenitic alloys are not hardenable by heat treatment, the major exception being a few precipitation hardenable types.

21.2.2.2 Ferrite

Ferrite is essentially pure iron with a body-centered cubic crystal structure. Ferrite forms from austenite at about 1675°F (915°C), as the austenite cools from a normalizing heat treatment. Because ferrite does not contain enough carbon to permit the formation of martensite, it is not hardenable by heat treatment. The most common truly ferritic steel is Type 405 SS, a ferritic stainless steel. The generic term "ferritic steel" often refers to carbon or low-alloy steels that contain other phases in addition to ferrite. Such steels are usually hardenable by heat treatment. Ferritic steels become brittle at low temperatures. This phenomenon is reversible, that is, the steels regain their former toughness after being warmed up. Ferritic steels are also susceptible to hydrogen embrittlement.

21.2.2.3 Martensite

Martensite is formed from high-temperature austenite, in heat-treatable alloys, by cooling the austenite fast enough to prevent the formation of ferrite. For some heat-treatable alloys, quenching in water or some other liquid such as oil or a molten salt is often required. Some steels have sufficient alloying additions that quenching is not necessary. Air cooling produces the martensitic microstructure in Type 410 SS and other martensitic stainless steels. Since martensite is usually brittle, it is normally subsequently tempered. The tempering temperature is less than the austenite transformation temperature. Tempering permits some carbon to diffuse from the martensite. Tempered martensite is significantly stronger and tougher than the parent ferritic alloy. It should never be stress relieved or postweld heat treated at a temperature exceeding the final tempering temperature, since its mechanical properties can be seriously degraded.

Although heat-treated martensitic steels have superior fracture toughness, they do tend to be brittle at low temperatures. Most martensitic steels are sensitive to hydrogen embrittlement. Martensitic alloys are widely used as high-strength bolting such as ASTM A193 Type B7, high-strength quenched and tempered plates such as ASTM A543, and martensitic stainless steels such as Type 410 SS.

21.2.2.4 Pearlite

Pearlite can form in most carbon and low-alloy steels with enough carbon to be hardenable by heat treatment. However, carbon steels usually are not intended to be hardened by heat treatment. Instead, they are normally produced with a more ductile, lower strength microstructure that forms during slow cooling from austenitic temperatures. This microstructure is a mixture of ferrite and pearlite. During cooling, ferrite starts to form from austenite. The ferrite contains essentially no carbon. As the ferrite forms, it leaves behind an increasing concentration of carbon in the remaining austenite. The excess carbon is eventually ejected from austenite. Normally, the excess carbon combines with iron to form iron carbide (Fe_3C), called "cementite." If the austenite cools relatively slowly, as in air cooling, pearlite forms. Pearlite is laminar, consisting of very fine alternating layers of ferrite and cementite. Such a microstructure is often found in normal carbon steels, as shown in Figure 21.3.

21.2.3 Metallurgical Terms

The terms defined in this section are frequently used in purchasing specifications. Other, less frequently used terms are defined as necessary in the text. For more complete definitions, see reference [6].

Base metal is used interchangeably with the term parent metal; it refers to the material of the components in a weldment, to differentiate such material from the weld metal and the HAZ of the base metal.

Carbon equivalent (CE) is used in evaluating the weldability of a carbon steel. Many different formulas are used to calculate the *CE* value, the most common being

$$CE = C + Mn/6 + (Cr + Mo + V)/5 + (Ni + Cu)/15 \qquad (21.1)$$

where the concentration of each element is expressed in weight percent (wt%).

The *CE* value is used to evaluate the risk of developing hard HAZs and the susceptibility of the weldment to delayed hydrogen cracking. Table 21.1 shows typical limits for carbon steels. When the maximum allowed *CE* value is exceeded, additional fabrication measures such as preheat, PWHT, and/or inspection for the effects of delayed hydrogen cracking are usually necessary.

Cold working is when a metal is stressed above its yield strength and plastic deformation occurs. Such deformation raises the internal energy of the material. The mechanism of energy storage creates distortions in the crystal structure of the material, i.e., an increase in unit entropy. Cold work involves plastic deformation at temperatures up to several hundred degrees. Deformation is irreversible, unless the material is subsequently heat treated, such as normalizing or solution annealing. Cold-worked materials sometimes have increased yield strengths and reduced ductilities; such materials are sometimes said to have been strain hardened.

Plastic deformation is caused by a stress that exceeds the yield strength of the material. Deformation at stresses less than the yield strength is elastic deformation. It is reversible, i.e., the

TABLE 21.1
Typical Carbon Equivalent (CE) Limits for Carbon Steels

Wall Thickness	1/2" (12.7 mm)	5/8" (15.9 mm)	3/4" (19 mm)	7/8" (22 mm)	1.0" (35.4 mm)
Max. CE Value	0.4	0.39	0.37	0.35	0.34

deformation disappears with the removal of the stress. Plastic deformation (sometimes called permanent strain) is permanent, i.e., plastic deformation remains after removal of the stress that caused it.

Galling is related to adhesive wear. When two metals of similar chemistry and hardness are in moving contact and under pressure, in the absence of lubrication, surface asperities (high points) tend to momentarily weld together. Continued movement ruptures these very local welds, resulting in metallic particles being torn from one or both surfaces. These particles result in rapidly increasing friction. Galling is an extreme form of adhesive wear, causing gross surface damage. Austenitic stainless steels are the most common susceptible materials. Examples include threaded fasteners (galling occurs in the threaded region) and valve closures.

Hardenable describes alloys that can be hardened and strengthened by cold working, for example, strain-hardened bolting. Hardenability describes the ability of an alloy to be hardened and strengthened, usually by heat treatments such as quenching and tempering.

Heat-affected zone (HAZ) is the volume of parent metal in which the mechanical properties and/or the microstructure have been changed by the heat of welding or thermal cutting. For most welds in carbon and low-alloy steels, the HAZ is a band, usually about 1/8 in. (3 mm) wide, adjacent to the fusion line of the weld. In austenitic stainless steels, a narrow, secondary HAZ may be generated some distance from the fusion line as illustrated in Figure 21.4.

Hot working causes plastic deformation to occur at a temperature high enough to prevent the material from becoming strain hardened. Instead, it spontaneously "recovers" plastic deformation. Hot-worked materials therefore do not have the internal energy characteristic of cold-worked materials. In practice, hot-work threshold temperatures are dictated by factors such as tool life. These temperatures range from as low as 350°F (175°C) for aluminum alloys to as high as 2300°F (1260°C) for steels and nickel alloys.

Oxide stabilized refers to materials, such as aluminum and the stainless steels, whose corrosion resistance depends on the formation and stability of a very thin surface oxide layer that is inert, easily "healed" if damaged, and tenacious. When the oxide layer has been disrupted and not healed, the material usually has little corrosion resistance. Both active and passive states sometimes exist adjacent to each other on the surface, resulting in rapid local corrosion. Crevice corrosion in stainless

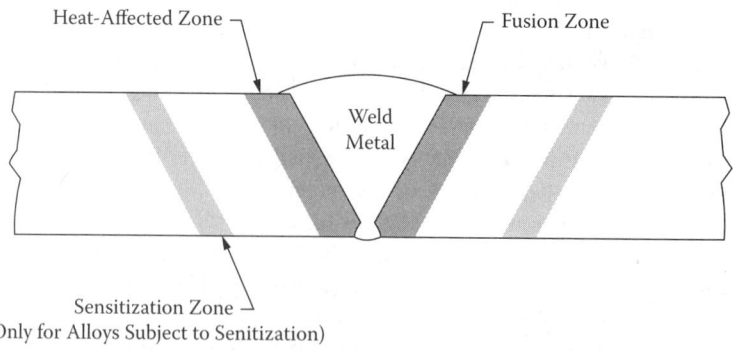

FIGURE 21.4 The typical features of a weld. © Copyright by Marcel Dekker, Inc.; reprinted with permission [5].

Selecting Materials of Construction (Steels and Other Metals)   1549

steel socket welds is a good example of rapid local corrosion. The potential difference between the anodic and the cathodic states drives the corrosion cells (this is an example of galvanic corrosion). Corrosion due to adjacent active-passive sites can be particularly rapid if the corrosion cell has an unfavorable anode/cathode area ratio.

Product forms include plate, strip, sheet, wire, pipe, tubing, bolting, bars, forgings, extrusions, and castings.

Toughness describes the ability of a material to deform plastically and absorb energy before fracturing. Fracture toughness is the energy per unit area necessary to create a fracture surface. Brittle materials fracture easily, requiring relatively little energy to crack. Fracture-tough materials, such as the ordinary austenitic stainless steels, require much more energy. Toughness is usually measured by the energy absorbed during an impact test. The most common example of such a test is the Charpy V-notch impact test (see ASTM A370).

Weldment is an assembly whose parts are joined by welding.

21.2.4 ALLOY DESIGNATIONS

Alloys are usually designated by the Unified Numbering System (UNS) [7, 8], which incorporates earlier identification systems developed for particular alloy families such as aluminum and copper alloys. The UNS system is particularly useful when designating proprietary alloys such as the nickel-based alloys. Ordinary alloy designations that are nonproprietary, such as those of the 300-series stainless steels, are commonly used instead of the UNS number.

21.2.5 MANUFACTURING OF METALS AND ALLOYS

Metals and alloys are available as *wrought* and *cast*. Products created by other methods such as powder metallurgy are uncommon and not discussed here. Wrought products are formed from solid metal, usually hot. Wrought processes generally employ compressive forces, which may be either continuous or cyclic, with or without dyes. Examples of wrought processes include rolling, forging, extrusion, and drawing, to produce plate, pipe, tubing, sheet, wire, forgings, extrusions, and bars. Wrought products are sometimes heat treated during the manufacturing process. The terms "hot finished" and "hot rolled" are usually not regarded as substitutes for heat treatment. If the materials selection process indicates a heat treatment requirement, check the purchasing specification to see if the product is in the heat-treated condition or if heat treatment is still needed.

Castings are products formed in a mold by solidification of a liquid. (Welds are a type of casting.) Virtually all wrought products begin as castings, such as ingots or continuously cast strands. Castings dominate pump cases, where geometry favors the simplicity of castings. Almost all castings are heat treated (the primary exceptions being some high alloys). Occasionally, a casting requires repair welding, as part of either a fabrication or maintenance procedure. Postweld heat treatment (or solution annealing in the case of austenitic stainless steels) may be indicated. Heat treatments can, however, warp previously machined surfaces. Special welding procedures, hardness controls, shot peening, etc., can avoid heat treatment. To avoid or minimize warping, special fabrication measures use strong-backs or bracing.

The chemical compositions of cast materials are usually different from those of their wrought equivalents. Castings usually contain more silicon than the wrought equivalent. (Silicon improves the "pourability" or fluidity of the liquid alloy.) Some alloys are available only in cast form, as they are too unstable or brittle to be formed by wrought methods. A few alloys are provided only in wrought form. Typically, an alloy in cast form usually has a different name than its wrought counterpart. For example, Grade CF-8 is the cast version of Type 304 SS. See reference [9] for a compilation of specifications, compositions, names, and typical mechanical properties of many casting materials.

Wrought products are usually preferred to castings. The hot-forming procedures employed break up and weld shut defects in the ingot or strand, while such defects remain present in castings. Wrought products tend to have a uniform, fine, partially isotropic grain structure. However, they are normally more expensive, reflecting the fabrication costs of hot working, machining, welding, etc.

Castings typically have lower strength, lower toughness, higher defect concentrations, and coarse anisotropic grain structures. Depending on the casting process, the surface may be less corrosion resistant than the bulk material. Because of this, corrosion tests should represent the expected surface condition of the casting. The advantages of castings include relatively low cost, ease of obtaining complex shapes, and minimal machining. In some alloys, the silicon addition and/or cast grain structure produces exceptional corrosion resistance. For example, austenitic stainless steel castings are more resistant to chloride stress corrosion cracking than are their wrought equivalents. In addition, castings are often less weldable, usually because of their greater silicon and/or carbon contents. Thus, repairability may be an issue and may be enhanced by specifying tighter compositions. A common example is the use of Grade CA-6NM (12Cr-4Ni-Mo) (UNS J91540) instead of Grade CA-15 (13Cr) (UNS J91150) for 12Cr castings. (The CA-6NM alloy has a much lower carbon allowable than the CA-15 alloy.) When choosing between wrought and cast components, the lower flaw density and/or better repairability of the wrought product or the lower cost or a unique property of the cast product should be considered.

Welding results in a metallurgical discontinuity* with different microstructures than the parent metal. Mechanical properties of the weld metal usually differ. Even with PWHT, weldments retain a residual stress field. Heat affected zones often contain a coarsened grain structure and/or "hard spots." Corrosion testing of the welds may be a critical part of the testing program.

As discussed earlier, cold work can change the mechanical properties of a material, including adversely affecting the fracture toughness. For applications such as bolting that will not be exposed to crack-inducing agents, cold working is deliberately used to increase yield strengths. However, cold-worked areas are typical sites for the development of strain aging (see Section 21.4). In addition, cold-worked areas can be susceptible to corrosion pitting, stress corrosion cracking, hydrogen embrittlement, and other damaging phenomena. For applications in which excessive cold working may be harmful, it is normal practice to stress relieve materials that have received more than 5% permanent outer fiber strain.

21.2.6 Metals and Alloys

21.2.6.1 Cast Iron

Cast irons typically contain at least 2 wt% carbon; the cast carbon steels used for plant construction rarely contain more than 0.35%. The high carbon content of cast iron makes it difficult to weld. Two types of cast iron are commonly used:

1. *Gray cast iron* such as ASTM A48 material is plain cast iron, composed of ferrite containing graphite stringers. Figure 21.5 shows the microstructure. It is relatively brittle and is used for applications in which toughness is not a concern, including utility services, as pump cases, and valve bodies.
2. *Ductile cast iron* (also known as *nodular* or *spheroidal iron*) contains a small amount of magnesium, which greatly improves ductility and toughness; an example is ASTM A536. Figure 21.6 shows the microstructure. Nodular cast iron is occasionally used in valve bodies, in pumps in various utility services, and in large reciprocating compressors. Malleable cast iron such as ASTM A47 material is a related alloy that is relatively expensive. Mildly acidic water can graphitize both gray and ductile cast iron as the iron

---

* A metallurgical discontinuity is an interruption in the normal physical structure of a metal. Examples include welds, cracks, laps, seams, bolted closures, inclusions, and porosity. A metallurgical discontinuity may or may not be a defect, that is, a discontinuity that is regarded as harmful.

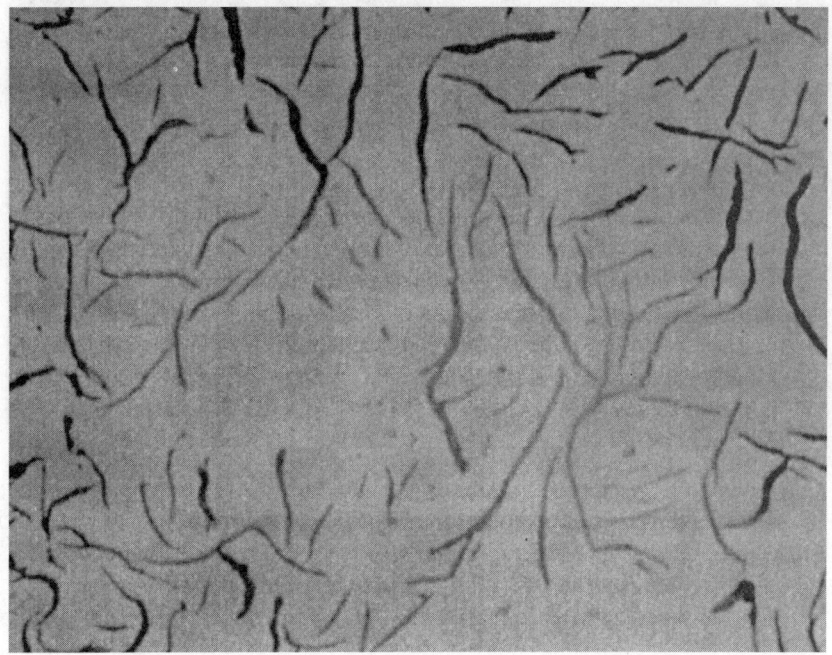

FIGURE 21.5 Gray cast iron. © Copyright by Marcel Dekker, Inc.; reprinted with permission [5].

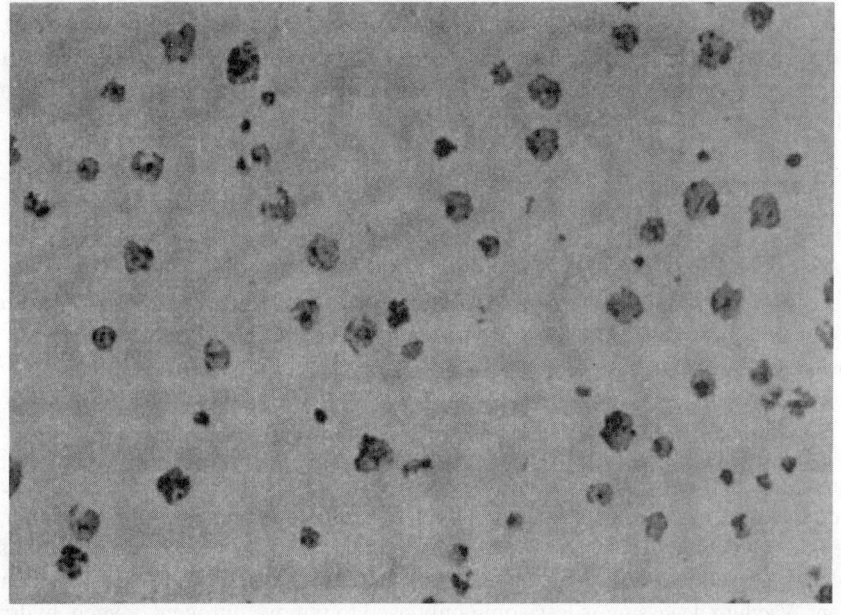

FIGURE 21.6 Ductile cast iron. © Copyright by Marcel Dekker, Inc.; reprinted with permission [5].

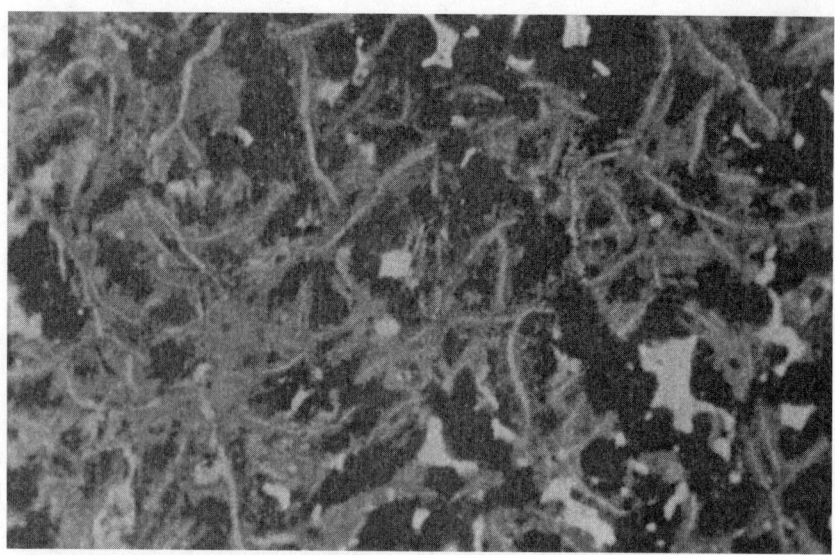

FIGURE 21.7 "Graphitization" of a gray cast iron pipe, caused by long-term service in slightly acidic water. © Copyright by Marcel Dekker, Inc.; reprinted with permission [5].

is slowly leached from the casting forming a network of graphite (Figure 21.7), resulting in low mechanical strength.

Three specialty cast irons are occasionally employed. Corrosion and erosion resistant *silicon cast irons* such as those of ASTM A518 find use in acid and abrasive services. They are relatively brittle and difficult to machine, limiting their usefulness. *White cast irons* such as ASTM A532, containing up to 25% chromium, are used in highly abrasive services such as pumping abrasive slurries; they are also relatively brittle. Nickel-rich cast irons, known as *Ni-resist cast irons* (such as those of ASTM A436), find use in both low- and high-temperature applications that require resistance to wear, and as seawater alloys. Because of its brittleness, cast iron is normally not permitted for hydrocarbon streams.

21.2.6.2 Carbon Steel

Carbon steels are the most widely used materials of construction. Unalloyed carbon steels typically contain nominal amounts of manganese, silicon, phosphorus, and sulfur. They are normally supplied with a pearlitic-ferritic microstructure (see Figure 21.3) produced by air cooling a hot-formed product (e.g., hot-rolled plate) or by a normalizing heat treatment. They are available as either killed carbon steel or plain carbon steel.

21.2.6.2.1 Killed Carbon Steel

Raw liquid steel is saturated with gaseous oxygen. By adding an oxygen scavenger such as silicon to the liquid steel before it is poured, the oxygen is removed as slag. The resulting "killed" steels are cleaner and contain fewer defects than "unkilled" (sometimes called "wild") steels or partially deoxidized steels. ASTM A106 pipe, A105 forgings, and A516 plate are examples. Cast carbon steel products are also killed, even though typical ASTM specifications do not mention this requirement. Vacuum degassing is less commonly used to remove nitrogen, oxygen, hydrogen, and carbon dioxide.

Steels killed with silicon, such as ASTM A515 plates, tend to have a coarse grain structure usually with a silicon content of 0.15 to 0.30 wt%. They characteristically have relatively high brittle-ductile transition temperatures, making them unsuitable for applications requiring low-temperature toughness. However, these coarse grained steels are more resistant to creep, graphitization, and some forms of corrosion. Steels killed with both silicon and aluminum or aluminum alone have a fine austenitic grain size. They have good low-temperature toughness; ASTM A516 (plate) is an example. Such steels are usually described in ASTM specifications as being made to "fine grain practice."

21.2.6.2.2 Plain Carbon Steel

The terms "semikilled," "rimmed," and "capped" refer to many carbon steels that have been partially deoxidized or not deoxidized at all. Examples include ASTM A53 and API 5L [10] for pipe and ASTM A36, a structural steel specification. (While these specifications do no require a killing practice, some, such as API 5L, are normally supplied in the killed condition). Although plain carbon steels are often permitted in benign services such as chemically treated utility water or air lines, killed carbon steels are generally used for at least three reasons: (1) There is virtually no cost difference between the two steels. The killed carbon steel is usually preferred because of its lower defect density and higher maximum code-allowable stress at higher service temperatures. (2) Common specifications such as ASTM A53 permit killed steel. (3) Because silicon killed and unkilled steels have different fracture toughness and also different design strengths, unintentional mix ups could cause fabrication, commissioning and/or operating problems.

21.2.6.3 Alloy Steel

Microalloyed steels (sometimes called high-strength, low-alloy steels, or HSLA steels) are intermediate between carbon steels and low-alloy steels. They are killed steels with small amounts of elements such as vanadium, titanium, and niobium; the combined total is usually about 0.1 wt% or less. These elements modify the microstructure, forming a small and uniform grain size that improves toughness and strength (typical specified minimum yield strengths are 60 ksi (410 MPa), or higher). These steels are usually used when section thickness or gross weight (and cost) is a concern, for example, large-diameter long pipelines. Such steels are also sometimes selected for improved toughness for piping and vessels. Microalloyed steels require care in selecting weld joint geometries and welding procedures.

Microalloyed steels can produce hard HAZs, increasing their risk of being susceptible to hydrogen stress cracking. Double-sided welds such as those preferred for vessels are much more likely to produce hard HAZs than single-sided welds normally used in piping and pipelines.

The common carbon steels are permitted by ASTM specifications to contain unreported microalloying elements capable of producing excessively hard HAZs. Occasionally, a weldment of a conventional carbon steel contains small regions in the HAZs having excessive hardness ("hard spots"). Current hardness testing is incapable of detecting hard spots in the HAZs.

Some users either prohibit the use of microalloyed carbon steels or place limits on both the carbon equivalent and the microalloying content. The microalloying limits vary from user to user, but usually fall in the range of 0.03 to 0.10 wt% for the sum of Ti, V, and Cb.*

Low-alloy steels are iron-based alloys containing less than a total of 12 wt% intentional alloying elements. They are all killed. Alloying enhances mechanical properties and/or improves corrosion resistance.

Alloying for improved mechanical properties can improve strength, toughness, and fatigue resistance. Such steels are normally heat treated to enhance their properties. Welding these alloys can degrade them.

Low-alloy Cr-Mo steels such as 1Cr-1/2Mo and 1-1/4Cr-1/2Mo steels are often used at temperatures above 800°F (425°C). Carbon steels become susceptible to creep at temperatures above about 750°F (400°C). In addition, carbon steels are weakened by carbide spheroidization and/or graphitization if exposed to sustained temperatures above 850°F (455°C).

---

* The element niobium is often called columbium (Cb) in engineering applications.

The mechanical properties of several Cr-Mo low-alloy steels are improved by adding vanadium. Examples include vanadium-enhanced 1Cr-1Mo for turbine rotors and vanadium-enhanced 1Cr-1/2Mo bolts, widely used for temperatures up to 1100°F (595°C). Vanadium-enhanced versions of the 2-1/4Cr-1Mo and 3Cr-1Mo plate and forging alloys are employed in heavy-wall vessel construction. These alloys can permit substantially reduced wall thicknesses. Alloys such as AISI 4140 steel (1Cr-0.2Mo, with carbon between 0.37 and 0.49%) (UNS G41400) are used as rotating equipment shafts, bolts, high-strength forgings, etc.

Ni-Mn (with either 3-1/2 or 9Ni) alloys are used for moderately low-temperature services, in the range of –50°F to –320°F (–46°C to –195°C). They are commonly used in liquified petroleum gas and liquified natural gas plants. (See ASTM A203, A333, A334, A350, A352, and A420 for various product forms of these materials.) Type 304L SS, while more costly, may result in a lower fabricated cost by avoiding welding problems.

Various enhanced strength plate steels are used in pressure vessels for high-pressure applications. Conventional carbon steels would require excessive wall thickness. Most enhanced steels also have increased low-temperature toughness. Postweld heat treatment is usually mandatory for these materials. Plate materials of this class include A302 (Mn-Mo and Mn-Mo-Ni steels), A517 (high-strength, low-alloy steels, quenched and tempered), A537 (Mn-Si steels; depending on section thickness, some grades can be qualified by impact testing at temperatures of –90°F (–68°C) or colder), A542 (micro-alloyed Cr-Mo steels, quenched and tempered), and A543 (Cr-Ni-Mo steels, quenched and tempered).

Alloying steels for improved corrosion resistance in chemical and hydrocarbon plants is based on chromium and molybdenum additions. The lowest of these alloys, 1Cr-1/2Mo and 1-1/4Cr-1/2Mo, are often used above 800°F (425°C). Low-alloy Cr-Mo steels (with 5% or greater Cr) are resistant to high-temperature sulfidic corrosion. However, the Cr-Mo alloys find their most critical use in high-temperature, high-pressure hydrogen service. The most commonly used alloys are the 1-1/4Cr-1/2Mo, 2-1/4Cr-1Mo, and 3Cr-1Mo steels. 9Cr-1Mo is available as piping, but is not commonly used in pressure vessel construction. A vanadium-enhanced version of 9Cr-1Mo is used in heavy wall vessels intended for high-pressure, high-temperature hydrogen service. All Cr-Mo alloys are air-hardenable and usually require PWHT. (See ASTM A182, A199, A217, A335, A336, A387, A541, and A739).

Weathering steels, many often classified as HSLA steels, are a common class of low-alloy steels employed for corrosion resistance. They commonly have small chromium and copper additions to form a stable patina-type rust in mildly corrosive atmospheres. These steels are used primarily in structural applications. (See ASTM A242, A588, and A618 for various product forms.)

21.2.6.4 Stainless Steel

21.2.6.4.1 Straight Chromium Stainless Steels

The 12Cr stainless steels are the least expensive and most commonly used alloys in this class. They are available in both ferritic and martensitic microstructures. Type 405 SS and Type 410S (ferritic) and Type 410 SS (martensitic) are corrosion resistance, particularly in wet CO_2 and in hot (T > 500°F [260°C]) services containing organic sulfur compounds or hydrogen sulfide. Type 410 SS, because it is martensitic, is used only when welding is not required. For welding, either Type 405 SS or Type 410S SS is specified. All 400-series stainless steels often grain coarsen in welds, causing relatively high brittle-ductile transition temperatures. Martensitic grades, being air-hardenable, often produce very brittle HAZs. Consequently, straight chromium stainless steels are usually not recommended for pressure containment utilizing welded construction. Their major use is in heat exchanger tubing, valve and pump internals, vessel internals, and as clad or weld overlayed linings in pressure vessels and heat exchangers.

Four hundred-series alloys are resistant to chloride stress corrosion cracking, but not chloride pitting. Accordingly, they are rarely used in aqueous chloride services. However, "superferritic" stainless steels, containing up to 29% chromium and 4% molybdenum, are now available. Some contain up to about 4% nickel without affecting their ferritic microstructure. One example is 25Cr-4Ni-4Mo (UNS S44635). They have satisfactory resistance to chloride pitting and chloride stress corrosion cracking except for the most severe services. Utilization of the superferritic alloys is restricted because they form embrittling phases while being cooled after annealing. This tendency limits useful thickness to about 0.2 in. (5 mm). Stabilized superferritic alloys, resistant to sensitization, are also available, for example, 26Cr-3Ni-3Mo, stabilized with niobium and titanium (UNS S44660).

The higher chromium grades such as Type 430 SS are susceptible to "885°F (475°C) embrittlement" at temperatures above ~750°F (400°C). Such embrittlement is usually mild in the straight 12Cr grades but may be severe in grades with 15% or more chromium. Straight chromium stainless steels are usually avoided for pressure containment above 650°F (345°C). The higher chromium grades should not be used above 650°F (345°C) unless subsequent embrittlement is of no concern. In addition, all ferritic and martensitic stainless steels are susceptible to hydrogen stress cracking phenomena, hydrogen embrittlement, and low-temperature embrittlement. If sensitized, these steels are susceptible to intergranular corrosion.

21.2.6.4.2 Austenitic Stainless Steel

Austenitic stainless steels including the 200-series austenitic Cr-Mn-Ni stainless steels (exemplified by Types 201, 202, and 216) are generally as corrosion resistant and stronger than their 300-series Cr-Ni cousins. These alloys are uncommon in chemical or hydrocarbon plants. As a consequence, the term "austenitic stainless steel" has come to mean the 300-series Cr-Ni alloys such as Type 304 SS. Sometimes called the "18-8s" (representing a nominal 18Cr-8Ni composition), the 300-series austenitic stainless steels are the workhorses for corrosion resistance in industry. Figure 21.8 shows a typical austenitic stainless steel microstructure. The 300-series finds extensive use as internals, cladding, and overlays in vessels exposed to corrosive services. They are also widely used in pumps, valves, and piping. Most are available in both wrought and cast forms.

FIGURE 21.8 Solution annealed austenitic stainless steel. © Copyright by Marcel Dekker, Inc.; reprinted with permission [5].

The austenitic stainless steels do not require PWHT as a hardness control measure. They are sometimes stress relieved or postweld heat treated to reduce residual stresses, thereby improving their resistance to stress corrosion cracking. The austenitic stainless steels are sometimes chosen in preference to the Cr-Mo low-alloy steels because PWHT can be avoided (the fusion zones of welds of the Cr-Mo low-alloy steels may be air hardenable).

The workhorse alloy of the 300-series is Type 304 SS, while the workhorse casting alloy is CF-8M. The low-carbon grades (Type 304L for the wrought alloy, CF-3M for the cast alloy) are preferred for welded construction. The low-carbon grades (as well as the stabilized grades such as Types 321 SS and 347 SS) are also preferred if sensitization is a problem. Type 316 SS (its small molybdenum addition differentiates it from Type 304 SS) is specified when chloride pitting or crevice corrosion is a problem. Type 316 SS also has a higher maximum allowable stress. The high-carbon H grades such as Type 304H SS are specified for high-temperature use (T > 1000°F [540°C]), since they have an allowable stress advantage over the conventional grades. The H grades should be used with caution if service-induced carburization may be a problem.

Higher chromium-nickel austenitic alloys are used extensively in high-temperature applications such as heaters, in both cast and wrought form. Examples include Type 310 stainless steel (25Cr-20Ni), a cast analogue called HK-40 (25Cr-20Ni) (UNS J94204), the Alloy 800-series (20Cr-32Ni, with Ti and Al) (UNS N08800, N08810, and N08811), and proprietary alloys such as the "HP-Mod" materials. These alloys can have problems such as weldment cracking, embrittlement, carburization, nitriding, oxidation, and metal dusting. (These phenomena are discussed in Section 21.4.) Industry experience or consultation with alloy specialists is recommended before material selection.

Galling is sometimes a problem with austenitic stainless steels; lubricants such as molybdenum bisulfide and graphite, and coatings are sometimes employed. The problem is also avoided by using two mating surfaces with a hardness difference of at least 50 BHN. With components such as valve closures, the hardness differential is usually obtained by using a hard face weld overlay or electroless nickel plating on one of the two components. In threaded connectors, the differential is usually obtained from cold working one of the components. Sometimes the hardness differential is obtained by specifying two materials having appropriately different hardnesses. Some users have employed electroplated or vapor-diffused aluminum on fasteners (preferably on the male component). Galling may also be mitigated by specifying one component to be a free machining grade such as Type 303 SS. Free machining materials are, however, sensitive to stress cracking problems. For example, NACE MR0175 [2] does not allow free machining grades in wet sour service.

Wrought austenitic stainless steels are more susceptible to chloride stress corrosion cracking than are the cast austenitic alloys. Accordingly, cast austenitic stainless steel valve bodies and pump casings are often useful in services in which higher alloys are necessary for the wrought components (pipe, tubing, fittings, plate, etc.). Ferritic stainless steels such as Type 410 SS are also subject to chloride pitting. Superferritic grades are an alternative to the plain ferritic grades if the component can be fabricated in thin sections, such as for heat exchanger tubing; for neutral or near-neutral pH applications, high-nickel stainless steels such as Alloy 20 Cb-3 (20Cr-35Ni-2.5Mo-Cb) (UNS N08020) have elevated chloride stress corrosion cracking threshold temperatures.

Alloys resistant to chloride stress corrosion cracking include: (1) "superaustenitic" alloys with high chromium and nickel content, as well as 2% to 6% molybdenum; Alloy AL-6XN (21Cr-25Ni-6.5Mo-N) (UNS N08367) is an example of a superaustenitic stainless steel; (2) nickel-chromium alloys such as Alloy 825 (22Cr-42Ni-3Mo, Ti stabilized) (UNS N08825), with nickel contents of approximately 40% or more; (3) duplex austenitic-ferritic alloys such as Alloy 2205 (22Cr-5Ni-3Mo-N) (UNS S31803); and (4) Ni-Cu alloys such as Alloy 400 (67Ni-30Cu) (UNS N04400).

Unless heavily cold worked, the austenitic stainless steels are resistant to hydrogen stress cracking such as that caused by hydrogen sulfide. They are also resistant to hydrogen embrittlement caused by phenomena other than cathodic charging. If sensitized, austenitic stainless steels can also be susceptible to intergranular corrosion.

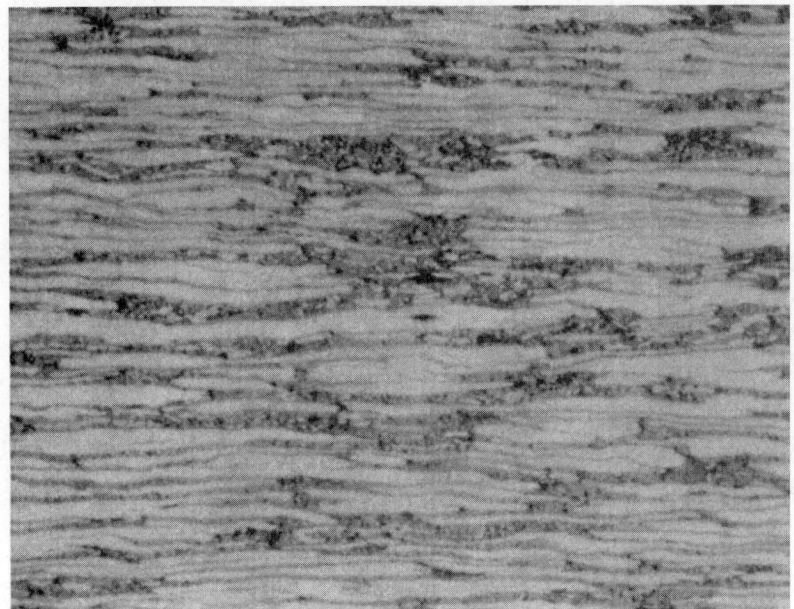

FIGURE 21.9 Duplex stainless steel (plate). © Copyright by Marcel Dekker, Inc.; reprinted with permission [5].

Cast austenitic alloys are usually more resistant to high-temperature corrosion than are their wrought equivalents. The cast alloys usually have a small, but significant, ferrite content, and they differ from their wrought equivalents as follows: (1) they are more susceptible to corrosion in acidic services; (2) they are more susceptible to sigma phase embrittlement; and (3) they are usually slightly magnetic, while the wrought alloys are nonmagnetic unless severely cold worked.

21.2.6.4.3 Duplex Stainless Steel

Duplex stainless steels contain both ferrite and austenite in approximately equal amounts; Alloy 2205 is an example. Figure 21.9 illustrates the microstructure of a duplex stainless steel microstructure in plate material. Typically, the duplex stainless steels contain 17 wt% or more chromium and <7% nickel. The more corrosion-resistant types contain at least 2% molybdenum. They are much stronger than the austenitic stainless steels, permitting a thinner section thickness. Thus, while they may cost more per pound, they may cost less per piece.

With the desired microstructure, these alloys are resistant to hydrogen stress cracking and much more resistant to chloride stress corrosion cracking than are the austenitic stainless steels. (The threshold temperature for chloride stress corrosion cracking of duplex alloys in neutral pH aqueous chlorides is about 300°F [150°C].) The chloride stress corrosion cracking resistance of the duplex alloys is similar to that of superaustenitic alloys such as Alloy AL-6XN. Because they contain about 50% ferrite, the duplex stainless steels are more susceptible to hydrogen embrittlement.

Welds of duplex stainless steels can become susceptible to chloride stress corrosion cracking and/or to hydrogen stress cracking. Welding and manufacturing of duplex stainless steel are usually more costly than those with conventional austenitic stainless steels. Because the duplex stainless steels can be embrittled by high temperature service, they are not recommended for operating temperatures greater than 650°F (345°C). The national engineering codes do not provide for design temperatures exceeding 600°F (315°C).

21.2.6.4.4 Precipitation Hardening Stainless Steel

Precipitation hardening stainless steels are those alloys ending with the suffix "PH" (i.e., precipitation hardening), for example, 17-4 PH (17Cr-4Ni-4Cu) (UNS S17400). They are hardenable by heat treatment. They are most often used for springs, valve stems, the internals of rotating equipment, and other applications where both high strength and superior corrosion resistance are desirable. Their corrosion resistance is superior to the 12Cr stainless steels but is somewhat inferior to Type 304 SS. They can be susceptible to both chloride stress corrosion cracking and hydrogen stress cracking.

Further information on stainless steels is available from the trade organization:

Specialty Steel Industry of North America
3050 K. Street, N.W.
Washington, DC 20007
Tel. (202) 342-8630
www.ssina.com

21.2.6.5 Common Alloys and Metals

21.2.6.5.1 Nickel Alloys

Nickel alloys plus nickel are widely used for services, including acids, caustics, corrosive waters, and for numerous corrosive process applications and low- and high-temperature applications. For further information contact:

Nickel Institute
55 University Ave., Suite 1801
Toronto, Ontario, Canada M5J 2H7
Tel. (416) 591-7999
www.nickelinstitute.org

21.2.6.5.2 Copper Alloys

Copper alloys, including brasses and bronzes, find extensive use in heat transfer systems exposed to corrosive waters (primarily brackish or saline waters). Naval brasses such as UNS C46400, usually as a cladding on carbon steel, are used for tubesheets and plate components. Inhibited admiralty alloys such as UNS C44300 and the 70/30 (UNS C71500) and 90/10 (UNS C70600) Cu/Ni alloys are often used for piping and heat exchanger tubes. The Cu/Ni alloys are usually preferred, as they have better impingement resistance and can tolerate higher velocities. Aluminum bronzes such as UNS C60800 are relatively high-strength alloys used for pump and valve components. The upper temperature permitted by most engineering codes for copper alloys 400°F (205°C). A few alloys with higher allowable temperatures include aluminum bronzes and Cu/Ni alloys.

Most copper alloys are unsuitable with ammonia, sometimes in even trace amounts and for wet sour services. Most are corroded by caustics. Some brass alloys contain zinc in excess of 15%. Unless properly "inhibited" by arsenic, antimony, or phosphorus, such alloys can "dezincify" in brackish or saline waters. To evaluate copper alloys, contact an alloy specialist or the copper alloy trade organization:

Copper Development Association
260 Madison Ave., 16th Floor
New York, NY 10016
Tel. (212) 251-7200
www.copper.org

21.2.6.5.3 Cobalt Alloys

Cobalt alloys are used for hardfacing to improve the resistance to abrasion, galling, and/or impact; Stellite 6* (60Co-29Cr-5W) (UNS R30006) is an example. Most alloys are in closure

---

* Registered trademark of Deloro Stellite Group, Goshen, Indiana.

applications such as valve seats, where both galling resistance and leak tightness are required, and in abrasive services such as mixers and nozzles. Grinding, which requires wear resistance, is also a common use. Cobalt hardface alloys are typically as corrosion resistant as the 300-series stainless steels and contain 30 to 60 wt% cobalt, typically with additions of carbon, nickel, chromium, tungsten, and/or molybdenum. They are applied, usually in one or two layers, by welding or thermal spray processes. The harder, more wear resistant alloys are difficult to apply and often develop cracks (called "crazing" or "check" cracks) in the overlay. Applied hardnesses are typically in the range of 20 to 50 HRC. Some alloys can be further hardened by cold work. Hardface alloys can be applied to almost any metallic substrate, with thicknesses of 1/16 to 1/4 in. (1.5 to 6.4 mm).

Cobalt-based alloys, in both wrought and cast forms, have also been developed for high-temperature applications such as gas turbine components, furnaces, and kilns. Alloy 25 (55Co-20Cr-10Ni-15W) (UNS R30605) is an example. Manufacturers are usually consulted in selecting these alloys.

21.2.6.5.4 Reactive and Refractory Metals

Reactive and refractory metals plus their derivative alloys are oxide-stabilized and can become reactive if the oxide layer is disrupted. These metals and their alloys often display both active and passive behavior. The definition of a refractory metal is somewhat arbitrary, and refractory metals are defined here as having melting points greater than that of iron (2795°F [1535°C]). Those of most interest (titanium, zirconium, and tantalum) are all sensitive to hydriding. Galvanic cells that promote hydriding can be particularly damaging. Zirconium and tantalum find applications in severe services such as hot concentrated inorganic acid processes. Most of these metals are relatively resistant to corrosion attack in aqueous oxidizing environments. However, each is subject to attack by specific corrodents and/or crack-inducing agents. All are, however, subject to catastrophic oxidation in certain environments. Selection should be on the basis of successful prior experience or a testing program.

21.2.6.5.5 Aluminum

Aluminum is a reactive (but not a refractory) metal having a melting point of 1221°F (660°C). Aluminum alloys are available in many variations, emphasizing properties such as strength, fatigue resistance, toughness, and enhanced corrosion resistance. Some aluminum alloys are hardenable (i.e., strengthened) by heat treatment. The mechanical properties and, to a large extent, the corrosion resistance are determined by their heat treatment. Aluminum and most of its alloys have excellent low-temperature toughness, permitting cryogenic applications such as liquified natural gas and liquified air. However, they lose strength rapidly with increasing temperature. The upper temperature limit is 400°F (205°C).

Many low- to moderate-strength aluminum alloys (primarily in the 1000-, 3000-, 5000- and 6000-series) have useful corrosion resistance. They are used in mildly corrosive atmospheres (such as those offshore or at seacoasts and for cable trays and fins on air cooler tubes). Several 5000-series alloys are moderately resistant to chloride pitting and find applications in clean seawater and other aqueous chloride services. They are very resistant to corrosion by wet CO_2 and are compatible with organics such as acetic acid.

Aluminum and alloys are not suitable for (1) alkalis, (2) acids at pH 4.5, and (3) mercury, which can be a significant risk in some liquified natural gas operations. The heat treatable, high-strength aluminum alloys of the 2000- and 7000-series are rarely used because of environmental cracking susceptibility. Aluminum and its alloys are susceptible to chloride pitting and to concentration cell problems such as crevice corrosion and under-deposit corrosion.

Due to its galvanic activity (aluminum is anodic to most common metals), aluminum is often used as a sacrificial anode and hence it corrodes preferentially to protect less active metals such as carbon steels. For evaluating aluminum alloys, contact:

The Aluminum Association, Inc.
1525 Wilson Boulevard, Suite 600
Arlington, VA 22209
Tel. (703) 358-2961
www.aluminum.org

21.2.6.5.6 Chromium

Chromium is a refractory metal having a melting point of 3375°F (1857°C). Neither chromium metal nor chromium-based alloys are widely in the hydrocarbon or chemical industries. Chromium plating is useful for aesthetic purposes, and "hard" chromium plating finds some use in hardface applications. It is extensively used as an alloy addition to low-alloy steels (usually for the purpose of stabilizing carbides) and in cast irons (to produce wear-resistant products) and nickel alloys (for increased corrosion resistance). Chromium is the main alloying addition in the 400-series stainless steels and is used extensively in the 200- and 300-series stainless steels.

21.2.6.5.7 Titanium

Titanium is a reactive refractory metal with a melting point of 3034°F (1668°C). It and its alloys find heat transfer applications in hydrocarbon and chemical process industries. It is resistant to both organic and inorganic corrodents; it is employed in heat exchanger tubing for corrosive processes on one side and corrosive cooling water such as seawater on the other side; it finds use for wet chlorine and for concentrated hot caustic solutions. It can be useful in mildly reducing applications, such as wet alkaline sour overhead condensing systems. However, titanium (and the other reactive and refractory metals) are generally unstable in strongly reducing environments. It can also become unstable in the presence of powerful oxidizers including dry chlorine, red fuming nitric acid, and liquid oxygen. Titanium is embrittled by the formation of hydrides.

Only three titanium alloys are commonly used in the chemical industries, grades 2, 7, and 12. Grade 2 is pure titanium, while grade 7 has a small Pd (palladium) addition. Grade 12 has small molybdenum and nickel additions and has a corrosion resistance between that of grades 2 and 7. Grade 7 alloy is intended for most severe services, and is especially useful in hot, low pH, chlorine-saturated brines. Grade 12 is useful to about 500°F (260°C) for pH <4, and grade 7 is resistant to such corrosion at temperatures of at least 500°F (260°C). Titanium alloys lose strength rapidly with increasing temperature. The upper temperature limit permitted by the common engineering codes 600°F (315°C). Titanium alloys are available in most, if not all, product forms.

21.2.6.5.8 Zirconium

Zirconium, having a melting point of 3365°F (1852°C), is a reactive refractory metal, but it and its alloys are relatively difficult to fabricate. Welded construction should be heat treated at 1425°F (775°C) and cooled rapidly to achieve the best corrosion resistance. These materials, although expensive, can be very useful in severe applications, such as hot concentrated alkalies and inorganic acids. They are favored in high-temperature processes for the production of urea, hydrogen peroxide, many organic acids, including formic and acetic acids, etc., and for high-temperature organic processes utilizing hydrogen chloride. It is one of the better metals for hydrochloric acid and sulfuric acid, being resistant up to 70% at the boiling point and up to 75% at 265°F (130°C). Stress corrosion cracking sometimes occurs in 64% to 69% sulfuric acid at elevated temperatures. An advantage of zirconium over nickel alloys is that it can handle these acids when oxygen or other oxidants are present. It is not suitable, however, for hydrofluoric acid. Corrosion of zirconium sometimes produces compounds that are pyrophoric; the corrosive products can ignite when equipment is taken out of service. Zirconium is resistant to oxidizing acids such as nitric acid. Its corrosion rate is less than 5 mpy* (0.1 mm/yr) in zero to 70% acid at temperatures up to 500°F (260°C). However, it is susceptible to stress corrosion cracking in concentrations exceeding 70%.

---

* mpy = mils per year, where a mil is 0.001.

Zirconium has greater resistance to caustics than tantalum. Because of its expense, zirconium is used only if no other metals or alloys are suitable. Utilization of zirconium and its alloys is generally confined to clad plate and to thin-wall applications, such as heat exchanger tubing or sheets used for "strip lining" or "wall papering." Zirconium alloys lose strength rapidly with increasing temperature. The upper temperature limit is 700°F (370°C).

21.2.6.5.9 Tantulum

Tantalum is the most expensive refractory metal used for corrosion resistance; it has a melting point of 5425°F (2996°C). Its resistance is often compared to that of glass, except that it tolerates higher temperatures. It is however attacked by hydrofluoric acid, caustics, oleum, sulfur trioxide, and chlorosulfonic acid. Tantalum is susceptible to catastrophic oxidation by air, dry oxygen, or chlorine gases. Because it is expensive, tantalum is used mainly in thin-section applications such as bayonet heaters, heating coils, plate heaters, or sheets used for strip lining or wall papering. Typical applications include heaters and condensers for organic acid recovery, ammonium chloride concentrators, hydrochloric acid absorbers, bromine condensers, and ferric chloride heaters.

21.3 CORROSION

Corrosion causes many plant problems. An understanding of electro-chemical principles is necessary in order to choose the best mitigation measures such as barrier coatings, inhibitors, and cathodic protection. Reference [11] provides further information.

21.3.1 CORROSION BASICS

The three simultaneously necessary conditions for electrolytic corrosion to occur are (1) an electrolyte, which is usually water containing dissolved salts; (2) a corrosion cell is composed of an *anode*, the area being corroded (i.e., oxidized), and a *cathode*, the area where electrons enter the electrolyte; and (3) the anode and cathode are connected by an electronically conductive (i.e., metallic) path. Figure 21.10 illustrates corrosion of iron exposed to wet CO_2 (carbonic acid).

Electrons provided by the iron at the anode flow through the metal to the cathode. Simultaneously, protons (hydrogen ions, H^+) generated at the anode diffuse through the electrolyte to the cathode. At the cathode, the protons combine with electrons flowing from the anode to generate hydrogen gas. The anode is an *electron donor* while the cathode is an *electron acceptor*. The two sites differ (e.g., grain orientation, degree of cold work, state of stress, composition, etc.). The differences can be transient, leading to situations in which a site alternates between being an anode and a cathode. (Cyclic stresses can cause this behavior.)

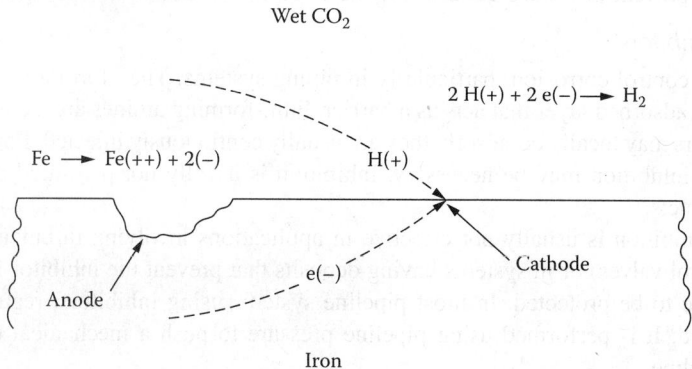

FIGURE 21.10 The essential features of an electrolytic corrosion cell. © Copyright by Marcel Dekker, Inc.; reprinted with permission [5].

Thermodynamically, an anode has a higher Gibbs free energy. For example, both cold work and residual tensile stresses increase the local free energy in metals and alloys. For two dissimilar metals, the different inherent corrosion resistances of the two metals translate into one having a greater Gibbs free energy. Because of the induced current and in accordance with Ohm's law,* corrosion rates become a function of current flow and electrical resistance.

21.3.1.1 Cathodes

In hot-formed carbon steel, the cathode is often the area covered by mill scale (Fe_3O_4), while the anode sites are cracks in the mill scale.

The cathode area is sometimes the noncold-worked area in a component that has been partially cold worked, for example, the straight-run tubing in a U-bend heat exchanger bundle.

In heat exchangers, pumps, and vessels, the internals are often the more corrosion-resistant alloys as compared with the material of the pressure retaining component. Thus, the internals are cathodic with respect to the pressure retaining component and the pressure retaining component is in effect a galvanic anode.

21.3.1.2 Anodes

The anode area may be the metal under a deposit on an otherwise clean surface.

For crevices such as in those in socket welds, the metal in the crevice is likely to be anodic. Crevice corrosion and under-deposit corrosion can be serious problems in oxide-stabilized materials such as aluminum and the stainless steels. Crevices and deposits can also accelerate corrosion in metals (such as carbon steel) that do not exhibit both active and passive states. However, the rate of corrosion is much slower in such materials because they lack the galvanic driving force of the active-passive states characteristic of the oxide-stabilized metals and alloys. The anode areas in crevices and under deposits are typically smaller than the cathode areas. This difference accelerates the corrosion rate.

The welding-induced sensitized area in a stainless steel is usually anodic.

21.3.2 CORROSION CONTROL

The rate of corrosion at the anode is directly proportional to the anode current density (expressed as amps per unit area). This knowledge aids in developing methods to control or eliminate corrosion.

21.3.2.1 Barrier Coatings

Barrier coatings prevent or reduce current flow between the corrodent and the metal surface.

21.3.2.1.1 Inhibitors

Inhibitors act to control corrosion, particularly in piping systems. They form a very thin (perhaps monomolecular) adsorbed layer that acts as a barrier. Film-forming amines are a common example. Because inhibitors may locally de-adsorb, they are usually continuously injected. For large-diameter pipelines, batch inhibition may be necessary. Inhibition is usually not permitted in processes for high-purity products.

Chemical inhibition is usually not effective in applications involving turbulent flow (such as pumps and control valves) or in systems having deposits that prevent the inhibitor from contacting the metal surface to be protected. In most pipeline systems using inhibitors, regular cleaning by "pigging" is used. It is performed using pipeline pressure to push a mechanical cleaning device through the pipeline.

---

* $V = IR$, where V is the electrical potential (in volts), I is the current (in amperes), and R is the resistance (in ohms).

21.3.2.1.2 Coatings and Linings

Coatings and linings that are dielectric are electrical insulators. In immersion services, large anode current densities can occur if the coating has "holidays."* Holidays act as very small anode areas. With thick linings, pinholing is not regarded as a problem since most are usually quickly plugged with corrosion products. The subsequent slow corrosion rate is controlled by diffusion and polarization. However, thin-film coatings, such as most paints, do have a fairly high risk of either initial or age-induced holidays resulting in locally high corrosion rates. Thin-film coatings should not be used in most immersion services without a cathodic protection system. In such situations, coating the anode without controlling the cathode can lead to unfavorable anode/cathode area ratios, e.g., coating the carbon steel channel/channel cover in a seawater heat exchanger having a more noble aluminum bronze tubesheet. In such a case, any holiday in the anode coating could result in an enormous anode current density. Two good mitigation responses for this example are as follows: (1) Coat the tubesheet as well as the channel/channel cover, using sacrificial anodes to handle holiday problems. (2) Coat only the tubesheet (cathode), without requiring the use of supplemental cathodic protection.

Galvanically noble metal coatings such as electroless nickel or chromium plating are often used as barrier coats on carbon steel. The coatings are cathodic, making any pinhole an anode, with a very large cathode/anode area ratio. The current densities at such anodes can be enormous: rapid pitting results. Such coatings, being galvanically noble, generate a large electrical potential between the anode and the cathode. For high-conductivity fluids such as seawater, resistivity is small. Ohm's law indicates why such couples have increased current densities.

The galvanic series in seawater is often used to estimate the risk of galvanic corrosion in other media, for which the series may not be available. The risk of galvanic corrosion depends as much on the corrosivity and conductivity of the medium as on the separation of the two metals in the galvanic series. Fresh waters generally have neither the corrosivity nor the conductivity to support galvanic activity.

In rare cases, a relatively small area near the weld will be an anode to the relatively large cathodic surface area of the parent metal. In moderately corrosive media, this zone may corrode much faster than either the weld metal or the parent metal. Postweld heat treatment is usually helpful. In some instances, normalizing (or even solution annealing in the case of an austenitic stainless steel) the weldment is necessary, a measure that can cause significant distortion problems. In most cases, the weld metal, HAZ, and parent metal do not have significant galvanic differences.

21.3.2.1.3 Cathodic Protection

Cathodic protection converts the surface to be protected into a cathode in one of the following two ways: (1) It is electronically and electrolytically connected to an inert material such as graphite or silicon iron. A power supply imposes a voltage that makes the inert material an anode, i.e., an electron donor. Such an *impressed current* cathodic protection system is frequently used for buried pipelines and submerged structures, and in plants to provide external cathodic protection to tank bottoms, buried piping runs, etc. (2) It is electronically and electrolytically connected to a more reactive material. For example, iron can be protected by connecting it to a less noble material (for example, zinc or aluminum) that acts as a *sacrificial anode*. Galvanized carbon steel is a common example of this application. Sacrificial anodes provide cathodic protection to offshore structures and pipelines. In plants, they are used for short, buried piping runs and for internal cathodic protection of tanks and vessels.

21.3.2.1.4 Anodic Protection

Anodic protection uses an impressed current to protect oxide stabilized metals and alloys, such as stainless steel or titanium, that can exist in both active and passive states. A power supply, an inert

---

* In the paints and coatings industry, "holidays" is the name sometimes used for pinholes.

impressed current electrode, and a potentiostat provide a potential that keeps the material in the passive state. The most common application is for stainless steel tanks in strong mineral acids and for coolers in sulfuric acid plants. The technique should only be used with expert advice. A similar application, without an impressed current system, involves spreading the cathode current over a large anode area. This minimizes the cathode area and thus minimizes the total cathode current available for corrosion on a carbon steel tank. It has been shown that turning the entire tank bottom (uncoated) into an anode, by abrasive blasting, reduces local pitting rates.

21.3.2.1.5 Passivation

Passivation consists of exposing a clean metal surface to an oxidizing environment to form an oxide film. This surface is much more corrosion resistant than it would be in an unpassivated state. In materials such as carbon steel, which form weak oxides, passivation can be destroyed rather easily. In oxide-stabilized alloys such as the stainless steels, this corrosion resistance is not easily destroyed.

Passivation is usually part of a chemical cleaning process using a sodium nitrite solution. (Chromates, once used extensively, are now considered to be too toxic.) Austenitic stainless steels are usually passivated in air after pickling and neutralization. *Pickling* is a chemical process to descale or clean new stainless steel. (See ASTM A380 for recommended procedures.) For heavily oxidized materials, the pickling process removes the chromium-depleted surface beneath the layer of scale. The acid solutions used contain sufficient nitric acid (a good oxidizer) so that a subsequent passivation step is unnecessary.

21.3.2.1.6 Polarization

Polarization occurs because of ion concentration buildup near the anode and/or cathode. Once the ion concentration reaches saturation, corrosion essentially stops. Polarization can occur when: (1) Hydrogen ions concentrate at an active cathode in the absence of a cathodic depolarizer. Dissolved oxygen acts as a cathodic depolarizer. (2) Metal ions saturate the electrolyte around an anode. (Soluble Fe^{++} may saturate the anode, perhaps as the result of the precipitation of an insoluble iron salt, inhibiting the diffusion of Fe^{++}. For example, insoluble surface compounds such as carbonate scales in a fresh water often occur on carbon steel.)

Examples of polarization are: (1) In deaerated, but otherwise corrosive, water, hydrogen polarization all but shuts down the corrosion mechanism. For example, seawater deaerated to less than about 10 ppbw is noncorrosive to carbon steel. (2) Many waters form insoluble dense scales on the corroded surface. The scale acts as a barrier to the diffusion of new corrodent and dissolved oxygen to the substrate surface. Scaling tendencies and corrosivities of waters can be estimated by analysis [12]. Polarization is prevented by: (1) Ions such as sulfides and cyanides that are "cathode poisons." Instead of the hydrogen ions forming hydrogen gas, which acts to polarize the cathode, cathode/poisons promote the formation of nascent hydrogen atoms. This hydrogen then diffuses into the substrate material. In ferritic steels, the following may result: hydrogen embrittlement, hydrogen stress cracking, and various forms of hydrogen induced cracking, including blistering. (2) By fluid flow phenomena such as turbulent flow or particulate erosion. (3) Cathodic depolarizers; the most common is dissolved oxygen.

The selection of materials must also consider oxidation/reduction processes that occur in the absence of an aqueous electrolyte. Examples include sulfidation, destructive oxidation of alloys in air or steam at high temperatures, carburization, nitriding, fuel ash corrosion, and high-temperature hydrogen attack.

21.3.3 Stress Corrosion Cracking

Failures due to stress corrosion cracking, while not common, may occur without warning, with catastrophic consequences. This cracking occurs only if both of the following conditions occur simultaneously: (1) a susceptible material and (2) an appropriate combination of stress (*tensile stress is required*), temperature, crack-inducing agent, pH, aeration, etc. For example, chloride stress corrosion cracking of austenitic stainless steels can be dependent on the presence of dissolved oxygen. Sometimes, stress corrosion cracking can be avoided by applying practical knowledge of the mechanism, for instance, keeping the temperature <140°F (60°C) for saline water in contact with an austenitic stainless steel.

There are at least three types of stress corrosion cracking: (1) hydrogen stress cracking (e.g., sulfide stress cracking); (2) anodic stress-cracking phenomena, such as alkaline stress cracking of carbon steels and chloride stress cracking of austenitic stainless steels; and (3) intergranular stress cracking (e.g., polythionic acid stress corrosion cracking). Cracking is usually associated with the residual stress fields of welds, although cracking can occur in cold-worked parent metal or metal subject to applied tensile stresses. Stress corrosion cracking depends on an electrolyte, usually liquid water containing the crack-inducing agent. Special metallurgical and fabrication requirements are customarily specified to prevent such cracking.

Hardness controls reduce the risk of *crack initiation*. *Crack propagation* risks are reduced by either increasing the fracture toughness of the material (usually by heat treatment) or by reducing stresses. The objective is to eliminate crack propagation or to ensure that it is slow and stable in order to obtain "leak-before-break" (i.e., failure is not catastrophic).

The following measures are recommended for hardness controls: (1) NACE RP0472 ("Methods and Controls to Prevent In-Service Cracking of Carbon Steel Welds in P-1 Materials in Corrosive Petroleum Refining Environments" [13]) limits the hardness of carbon steel weld metal to 200 BHN. Weld testing ensures that HAZ hardnesses do not exceed 248 VHN. Industry practice limits the hardness of weld metals for air-hardening Cr-Mo low-alloy steels as follows: for $1 < Cr < 3$, 225 BHN and for $3 < Cr < 9$, 241 BHN. Potential problems in welds may involve dissimilar metals and unusually different thicknesses at the weld joint. Tube-to-tubesheet welds and thin fillet welds used to attach tray support rings are examples. (2) Materials subject to hydrogen stress cracking mechanisms should also be required to meet the limitations of NACE MR0175 ("Sulfide Stress Cracking Resistant Metallic Materials for Oilfield Equipment" [2]).

Normalizing, sometimes in conjunction with tempering, increases the fracture toughness of carbon and some low-alloy steels. Other heat treatment methods such as quench and tempering may be used, e.g., for some Cr-Mo steels. NACE MR0175 [2] requires that most carbon and low-alloy steels in wet H_2S service have a toughness-enhancing heat treatment. Hot rolling is acceptable for carbon steel in wet H_2S service. Most carbon steel pipe is exempt from this requirement.

Postweld heat treatment (PWHT) is a stress reduction technique. It is widely used to avoid alkaline stress corrosion cracking in carbon steels and for enhancing fracture toughness in carbon and low-alloy steels. PWHT can also soften hard welds and HAZs, making them more resistant to crack initiation in services subject to hydrogen stress cracking. PWHT is often required based on some domestic recommended practices and foreign engineering codes for services known to be subject to stress corrosions.

Stress relief also reduces stress cracking in materials that have been cold worked. Common applications are for spun or pressed heads and for the U-bends in heat exchanger bundles.

Hardness control, normalizing, and PWHT and stress relief may be unnecessary if the combined stress in tension (applied) is less than the "10% rule." Examples of such applications are small tanks, atmospheric and low-pressure vessels, equipment and piping, and drains. Under such conditions, the residual stresses and/or untempered hardnesses of welds and/or HAZs may indeed initiate cracking and may even eventually produce through-thickness cracks. However, the combined stresses are too low to generate catastrophic cracking. Through-thickness cracks usually occur only after many years; risk-limiting measures are often not cost effective for such low-stress situations. If minimum capital cost is a project objective, the 10% rule may justify avoiding expenses such as those of hardness controls, PWHT and stress relief, and normalizing.

Four exceptions where risk-limiting measures may be justified include: (1) For new equipment subject to local laws that mandate conformance to an engineering recommended practice. (2) When through-thickness cracks, even though stable, would release a lethal, flammable, or corrosive substance. (3) For thin-ligament pressure-containing components where leaks could develop quickly and would require unscheduled maintenance (e.g., tube-to-tubesheet welds). (4) In new equipment or piping where subsequent inspection and/or repairs after long-term service would be excessively expensive. For example, sites where congestion may make it difficult to mobilize the necessary repair equipment (e.g., cranes).

Shot peening, in which hard shot is impinged on a surface, produces surfaces that are in residual compression, thus making them resistant to stress corrosion cracking.

Cladding or *overlays* are sometimes applied to carbon and low-alloy steel to avoid stress corrosion cracking problems. For example, carbon steel clad with Type 304L SS is often used in severe amine services.

Paint coatings are occasionally used to minimize stress corrosion cracking. They are often provided for external stainless steel surfaces exposed to wet chlorides, marine atmospheres, or in services under wet insulation. For under-insulation surfaces, some users coat only the welded areas where chloride stress corrosion cracking usually occurs. Paint coatings are also sometimes used to minimize stress corrosion cracking in old equipment that may be exposed to a crack-inducing agent. Internal coatings in vessels and tanks are occasionally employed.

Crack-inducing agents generally require the presence of an electrolyte in order to be active. For example, H_2S cracks carbon steel only in the presence of water or other electrolytes. The most common electrolytic crack-inducing agents for carbon and low-alloy steels are aqueous solutions of amines, caustics such as caustic soda (NaOH), hydrogen sulfide, dilute sulfuric acid, hydrofluoric acid, and mixtures of carbon monoxide and carbon dioxide. The austenitic stainless steels are sensitive to chloride stress corrosion cracking, caustic stress corrosion cracking, and, at high temperatures, to intergranular stress cracking by liquid zinc. When sensitized, these materials often experience intergranular stress corrosion cracking by weak oxidizing acids such as polythionic acid. High-strength aluminum alloys can be susceptible to chloride stress corrosion cracking. Most copper alloys are rapidly cracked by liquid mercury.

A few crack-inducing agents do not require an electrolyte, e.g., the attack of copper alloys by ammonia. Accordingly, the template should include the information necessary to determine if a crack-inducing agent is either active or can become active.

21.3.4 Microbiologically Influenced Corrosion (MIC)

This corrosion is defined as the corrosion of materials caused by microorganisms. The combination of unexpected attack and rapid failure is sometimes a concern. Generally it occurs in stagnant water systems or the water legs of mixed-phase, quiescent process streams at ambient temperatures (but may occur at temperatures up to 200°F [93°C]). Most materials of construction are susceptible. A typical MIC problem is the pitting corrosion in the bottom of a pipeline, before commissioning, due to microbial activity in residual hydrotest water. Systems subject to stagnant operation, shutdowns, or velocities less than 2 ft/sec (0.6 m/sec) are at risk. In some media, such as seawater, colonization by microorganisms may occur at velocities up to about 5 ft/sec (1.5 m/sec). For stagnant water or quiescent water-wet systems, exposures exceeding 30 days often lead to severe MIC.

The microorganisms responsible for MIC are primarily bacteria and fungi. They may be anaerobes (which do not function in the presence of oxygen), or aerobes (which require oxygen), or a mixture of several microorganisms living in colonies. While strictly anaerobic environments are not common in nature, anaerobes can develop in even highly aerated systems. For example, anaerobic conditions can be established under the slime film formed by some aerobic microbes. Common organisms include sulfate-reducing bacteria, sulfur/sulfide-oxidizing bacteria, iron/manganese-oxidizing bacteria, aerobic slime formers, methane producers, and acid-producing bacteria and fungi. They usually form discrete biodeposits, either nodules or flat shiny deposits. An exception is when they exist in high concentrations in anaerobic soil environments, where they do not need to form either a film or a deposit to become active. Under such conditions, these organisms often cause corrosion.

Selecting Materials of Construction (Steels and Other Metals)

There are four prerequisites for MIC: (1) The material of construction must be susceptible to MIC. (2) Microbes must be present in the environment. (3) The service temperature range must support microbial metabolism. It is often in a narrow temperature range (10°F to 20°F [5°C to 11°C]). However, different organisms can be active from below freezing to above 200°F (93°C). (4) The environment must supply the appropriate nutrients and provide the aerobic or anaerobic conditions required.

21.3.4.1 Effect on Materials of Construction

Carbon and low-alloy steels and *stainless steels* are the most commonly affected materials. The morphologies created by MIC attack seem to be related to the organisms involved. Pinhole openings under nodules, accompanied by extensive tunneling, may be typical of Gallionella bacteria on stainless steels, while shallow surface attack beneath nodules or open pits may be typical of sulfate-reducing bacteria on stainless and carbon steels.

Aluminum alloys can also be attacked by microorganisms. For example, there have been MIC problems with aluminum fuel tanks and transfer lines. In this case, microorganisms grow in the water layer under the fuel to produce volcano-shaped tubercles, frequently evolving gas. Pitting occurs under the tubercles. MIC attack on *copper alloys* is usually insignificant. However, corrosion of copper condenser tubes by microbially produced ammonia has been reported. In addition, sulfuric acid has been produced by microbial activity by corrosion of underground copper pipes.

21.3.4.2 Mitigation Methods

To prevent microbiological activity, avoid the development of stagnant or quiescent water legs. Use of barrier materials, biocide treatments, or using materials resistant to damage by MIC is also often effective. Once a MIC problem develops, cleaning and chemical treatment are necessary. Cleaning disrupts the colonies protected by their biofilms or by occlusion due to corrosion products such as tubercles. Mechanical cleaning employs brushes, pigs, sponge balls, water, or gas jets. Backwashing or flow jogging is also used. Acid cleaning, sometimes incorporating surfactants and/or dispersants, is usually an effective cleaning method (and can be an effective biocide treatment as well). Other chemical treatment options include biocides and biostatic inhibitors.

21.4 FAILURE MODES

Materials degradation, including electrolytic corrosion, high-temperature corrosion, and stress corrosion cracking, must be considered when selecting materials of construction in order to achieve the desired design. In addition, embrittlement by high- and low-temperature phenomena, as well as by various chemicals, must be properly addressed. Engineering codes often provide little guidance. In this section, the most common degradation phenomena and methods of mitigation are discussed.

21.4.1 EMBRITTLEMENT PHENOMENA

The following terms are employed relative to embrittlement.

Brittleness is the opposite of toughness. A "brittle" crack propagates with little or no macroscopic plastic deformation. Crack propagation requires very little energy and is rapid, usually resulting in rupture. Engineering codes contain rules to avoid brittle fractures.

Ductile crack propagation is accompanied by gross plastic deformation. Such propagation absorbs significant energy and is typically very slow. Fast ductile fractures do occur, in a ductile material, when the remaining ligament is too small to support the gross load; they usually exhibit gross tearing. Brittle fractures can be virtually eliminated by materials that are qualified in accordance with construction code requirements and by avoiding or accommodating service-induced embrittlement. In practice, most fractures involve ductile crack propagation due to plastic instability occurring in pressure-containing components that have been thinned by corrosion or erosion.

Risk of fracture differs from brittleness. A brittle material will not fracture as long as the applied stress is below the crack propagation threshold. In practice, brittle materials such as gray cast iron and glass can be used safely if the applied stresses are kept small so that inherent flaws in the material do not propagate into unstable cracks (i.e., brittle fracture). In ductile materials, the risk of fracture usually depends on the thickness in a locally thinned area. Evaluating the risk of fracture involves employing *fitness-for-service* or *remaining life assessment* techniques. In some localities, such analysis is a regulatory requirement. A conservative rule-of-thumb is to require further evaluation only if the corrosion allowance has been consumed. A less conservative criterion is to evaluate if the remaining minimum thickness is insufficient to limit the pressure hoop stress* to the specified minimum yield strength of the pressure-containing material. For simple cylindrical geometries such as vessel shells and piping, the pressure hoop stress can be adequately estimated by the "Barlow formula":

$$\text{Stress (sigma)} = \frac{(P)(r)}{t} \qquad (21.2.)$$

where P is pressure, r is the internal radius, and t is the remaining thickness.

Fracture-safe design is when there is little likelihood of a serious accident even if a through-thickness crack occurs. For each metal and alloy, there is a combination of critical crack size, stress intensity, and thickness above which cracks will no longer propagate in a ductile manner. Brittle fractures of this type are rarely encountered in low- to medium-strength carbon and low-alloy steels; they typically have tensile strengths of 70 ksi (480 MPa). As a result, fracture mechanics is not commonly used in most plant designs, except in the chemical and hydrocarbon industries for high-pressure processes in which relatively high-strength materials of construction are employed. The ASME Boiler and Pressure Vessel Code, Section VIII, Div. 3 [1], is used in such cases.

Leak-before-break is a common application of fracture-safe design, used to ensure that a pressure-retaining system will leak before breaking. This design procedure utilizes fracture mechanics to ensure that the applied stress is insufficient to cause a through-thickness crack to propagate in a brittle manner. Propagation of a through-thickness crack will be ductile. The crack will start leaking before there is a break. The 10% rule is one application of leak-before-break; it involves piping and equipment in relatively low-pressure services. For carbon and low- to medium-strength low-alloy steels, brittle fracture will not occur if the combined stress in tension is less than about 10% of the material's tensile strength. (This will hereafter be referred to as the "10% rule"). For carbon and low-alloy steel equipment, in compliance with the 10% rule, leak-before-break will prevail. This principle has been incorporated into many engineering codes. While low-stress conditions do not permit brittle fracture, stable crack propagation by other mechanisms such as stress corrosion cracking can still be active.

Embrittlement refers to a loss of ductility and fracture toughness. Embrittlement causes crack growth to change from ductile to brittle. In most cases, brittle or embrittled materials have a threshold temperature range above which they respond to crack propagation stresses in a ductile manner. Cracking below the threshold temperature is at least partially brittle. Such cracking is often catastrophic. Cracking above the threshold temperature is by a ductile mechanism. Embrittlement examples include: (1) The term "embrittlement" is applied to the ambient temperature brittleness of an alloy that has become embrittled by high-temperature service. (2) Process fluids can embrittle an otherwise ductile material. Hydrogen embrittlement is an example of this effect. (3) In carbon

---

* Pressure hoop stress is the circumferential tangential (tensile) stress on a cylinder as a result of internal pressure.

Selecting Materials of Construction (Steels and Other Metals)   1569

and low-alloy steels and in ferritic and martensitic stainless steels, the term sometimes refers to the brittleness that develops at temperatures below the brittle-ductile transition range.

21.4.2 CARBON AND LOW-ALLOY STEEL

Temper embrittlement occurs in 2-1/4Cr-1Mo and 3Cr-1Mo steels, often used for the pressure shells of heavy-walled reactors such as hydrotreaters. Temper embrittlement occurs at 700°F to 1100°F (370°C to 595°C); it is reversible by exposure to higher temperatures. In the 700°F to 1100°F (370°C to 595°C) range, elements such as tin, phosphorus, and arsenic segregate at the grain boundaries. Silicon and manganese also contribute to grain boundary embrittlement. While these (tramp) elements cannot be completely eliminated from the alloy, modern steel mills can control them, so that temper embrittlement of these alloys should not be a concern. Temper embrittlement does not affect high-temperature ductility at >250°F (120°C). Accordingly, components suspected to be temper embrittled can be safety operated if they are not pressurized at temperatures <250°F (120°C).

Creep embrittlement is a misnomer. It causes a loss of strength due mostly to a decrease in load-bearing capacity as gross cracks form. It can occur in 1Cr-1/2Mo and 1-1/4Cr-1/2Mo steels exposed to temperatures exceeding 850°F (455°C), usually within 8 to 15 years. The mechanism is unpredictable, since cracking occurs mainly in the HAZs of welds, usually at nozzles having sharp changes in cross section. This embrittlement is irreversible, and there is no consensus on how to prevent or control it. In low-stress applications, leak-before-break will govern. For higher stress applications, there are two choices: (1) Select the less expensive 1Cr-1/2Mo or 1-1/4Cr-1/2Mo material, with the risk of replacement or repair; the anticipated life is probably 15 to 20 years minimum. (2) Select 2-1/4Cr-1Mo, a more expensive alloy; however, this alloy is susceptible to temper embrittlement, as discussed above.

Strain aging occurs in many carbon and low-alloy steels. A cold-worked material ages at ambient or relatively low temperatures, developing anomalously high strength and hardness, but with reduced ductility. The rate of aging depends on the temperature. At ambient temperature, the effect may not be apparent for months, but when exceeding 400°F (205°C), aging may occur in minutes. Aluminum killed steels are quite resistant to strain aging. Such embrittlement is relatively rare, since materials that are deliberately cold-worked are usually stress relieved before being placed in service. Also, most modern carbon and low-alloy steels are aluminum killed. Typically, a limit of 5% cold work (defined in terms of outer fiber strain) is permitted without subsequent heat treatment. Because cold work is sometimes used to straighten or repair dented or bent structurals or equipment, there is a risk of such embrittlement.

Strain aging can also occur in susceptible steels by welding near a crack, causing embrittlement (called *dynamic strain aging*). For this and other reasons, many users require that plate material intended for pressure-containing components be scanned by ultrasonic inspection equipment for cracks and flaws near edges to be welded.

Hydrogen embrittlement can occur in carbon and low-alloy steels, in ferritic and martensitic stainless steels, and in duplex stainless steels. It is normally not a problem in either the austenitic stainless steels or nickel-based high alloys. Hydrogen can dissolve in a steel as a result of a number of phenomena: (1) Corrosion creates nascent hydrogen, usually in the presence of a cathodic poison. (2) High-temperature, high-pressure hydrogen. (3) Excessive cathodic protection or cathodic charging saturates a steel with nascent hydrogen. The most common example is due to an electroplating procedure in bolting. (4) Welding with moist consumables may also cause embrittlement in carbon and low-alloy steels.

At less than about 250°F (120°C), dissolved hydrogen sometimes results in a loss of ductility. Hydrogen embrittlement is reversible; dissolved hydrogen is driven from the material by a "bakeout" at high temperature, e.g., 600°F (315°C) for 4 hours. Bakeout is unnecessary if the component is allowed to "outgas" for a minimum of 7 days at ambient temperature. It is also temperature dependent, occurring from subambient to ~250°F (120°C). The maximum effect is in the range of

0 to 100°F (–18°C to 38°C). The risk rapidly diminishes >175°F (80°C). Hence, low-alloy steels such as the Cr-Mo steels are usually not pressurized, if in gaseous hydrogen service, until the operating temperature is brought up to 250°F (120°C). Ordinary low- to medium-strength steels (with tensile strengths of up to 70 ksi [480 MPa]) are moderately susceptible to this embrittlement. Between 70 and 90 ksi (480 and 620 MPa) tensile strength, steels are susceptible to it. Higher-strength steels (with tensile strengths in excess of about 90 ksi [620 MPa]) can be severely embrittled. Bolting is the most common product form in which hydrogen embrittlement is observed in plants, since they are often electroplated without a subsequent hydrogen bakeout (the electroplated material does not allow trapped dissolved hydrogen to escape by diffusing to a free surface).

For carbon and low-alloy steels, the relationship between susceptibility to hydrogen embrittlement and strength is very similar to that between susceptibility to hydrogen stress cracking and strength. This embrittlement often occurs in carbon steels, low-alloy steels, and ferritic or martensitic stainless steels subject to hydrogen stress cracking environments. The measures to minimize embrittlement are essentially the same as those to minimize the risk of stress corrosion cracking (hardness controls, control of microalloying additions, PWHT, etc.)

Hydrogen embrittlement is normally not a problem in most chemical plants, because the material strengths are too low to propagate cracks. Four situations should, however, be given special attention: (1) The rate and extent of hydrogen embrittlement are affected by the amount of residual cold work; hence, stress relieve components that have been cold worked. Five percent cold work is often used as the threshold. Designs should avoid sharp notches, as these can subsequently become cold worked. (2) Metals for high-temperature, high-pressure hydrogen service are often at risk, especially upon cooldown. Such components (usually Cr-Mo low-alloy steels) may be subject to brittle fracture if exposed to tensile or bending stresses during maintenance, revamp fabrication, etc. (3) Hard HAZs may be susceptible to both hydrogen embrittlement and hydrogen stress cracking. Welding and joint configurations are generally not susceptible to these phenomena. However, if a carbon steel parent metal has excessive microalloying, or if the weld cools too rapidly, excessive HAZ hardness can be created when a thin section is welded to a thick section, as in tube-to-tubesheet welds. Heat-affected zone hardnesses of 200 BHN or less are not likely to be affected by dissolved hydrogen. Similarly, hard spots susceptible to cracking can form at the fusion zone in dissimilar metal welds involving relatively high-carbon plain or low-alloy steels to stainless steels or Cr-Ni alloys. (4) *Delayed hydrogen cracking* (also called *underbead cracking* or *cold cracking*) is sometimes associated with hydrogen embrittlement; it is a form of hydrogen stress cracking. The problem occurs in freshly made welds, usually because of hydrogen generated during welding if a moist welding flux is used. However, such cracking can occur in repair welds because of hydrogen dissolved in the steel while in service. Cracking in repair welds is prevented by a bakeout. The delayed hydrogen cracking mechanism requires an incubation period before cracking occurs. Such cracking may not be visible if the weld is inspected immediately after it is completed. Inspection should be delayed 2 or 3 days after welding. The primary mitigation measures are (a) bakeout, if necessary; (b) preheat; and (c) control of welding flux to avoid moisture absorption.

Caustic embrittlement is a misnomer. The loss of ductility causes a reduction in load-carrying capability because of the formation of cracks due to caustic stress corrosion cracking.

Low-temperature embrittlement occurs in carbon and low-alloy steels at temperatures below their brittle-ductile transition temperature range. The effect is reversible: when the alloy is heated above the transition range, ductility is restored. This embrittlement is avoided by following the Charpy impact test requirements of the relevant engineering codes. The need to test depends primarily on the material, its thickness and the minimum design temperature.

Aluminum, copper, and other face-centered cubic metals and alloys (such as the austenitic stainless steels and nickel-base alloys) do not become brittle at low temperatures, except if heavily cold worked. Most such alloys are exempt from impact testing for design temperatures down to –320°F (–195°C). Some types, such as Type 304, are exempt down to –425°F (–255°C). The exemption temperatures for weld metals and HAZs are usually higher than those for the parent metal.

21.4.2.1 Stainless Steel

Ferritic stainless steels, such as straight chromium stainless steels containing 12% chromium, may be embrittled at 750°F to 975°F (425°C to 525°C). Embrittlement is not evident at high temperatures, but lack of ductility can occur at ambient temperatures; it is called *885°F (475°C) embrittlement* (or "blue embrittlement"). This embrittlement is reversible at higher temperatures. Alloys that embrittle, if used in pressure-containing services, should not be exposed to temperatures >650°F (345°C). However, some common domestic codes allow design stresses up to 1200°F (650°C). In most cases, these codes caution that these materials may embrittle.

Ferritic stainless steels are sometimes used for their corrosion resistance above the 885°F embrittlement threshold. Such applications usually involve vessel trays and internal cladding. Embrittlement 885°F (475°C) is not normally a problem in the 12Cr alloys such as Type 410S stainless steel. Type 405 stainless steel, with its slightly higher chromium content and small aluminum addition, and Type 409, containing a titanium addition, are more susceptible. For ferritic stainless steels containing 15% or more chromium, embrittlement can become severe. One should not accept "upgrades" of straight chromium stainless steels without first checking on their thermal history and the intended operating temperatures. Table UHA-109 in ASME Section VIII, Div. 1 [1] provides guidance on the risk of 885°F embrittlement of high-chromium stainless steels.

The higher chromium grades of the ferritic stainless steels such as Type 446 become susceptible to embrittlement, caused by the formation of intermetallic phases such as sigma or chi phases, at temperatures ~1050°F (565°C). Most straight chromium grades are also susceptible to sensitization-induced corrosion problems. These alloys should be avoided for high pressures when the operating temperature exceeds 650°F (345°C). Such alloys should be confined to nonpressure components such as valve trim, vessel internals, and pressure boundary liners such as weld overlays. These applications occur for high-temperature sulfur-containing streams such as refinery coke drums and crude units processing sour crudes. Ferritic stainless steels are also susceptible to low-temperature embrittlement. The engineering codes typically require such steels to be impact tested.

Martensitic stainless steels are susceptible to low-temperature embrittlement and engineering codes typically require they be qualified for low-temperature service by impact testing. As discussed earlier, the martensitic stainless steels can grain coarsen in the HAZs of welds. Grain coarsening causes relatively high brittle-ductile transition temperatures. Martensitic grades, being air hardenable, can also produce brittle HAZs. Consequently, martensitic stainless steels are not usually recommended for pressure vessels containing welds. However, some codes provide allowable design stresses for a few martensitic stainlesses up to 1200°F (650°C).

Austenitic stainless steels are often susceptible to *sigma-phase embrittlement* after extended exposure at 1050°F to 1700°F (565°C to 925°C). Occurrence depends primarily on the temperature and the presence of ferrite. While Type 304 SS can develop sigma-phase embrittlement, it is more common in austenitic products that contain small amounts of ferrite. Examples include austenitic weld metal and castings.

The upper temperature limit for sigma-phase formation varies from about 1600°F to 1800°F (870°C to 980°C), depending primarily on alloy composition. Embrittlement occurs slowly. Type 304 SS will usually show only 2% to 3% sigma phase in its microstructure after 10 years at 1200°F (650°C). At temperatures near the upper limit of the embrittlement range, embrittlement may develop in a few weeks. The rate of embrittlement is increased by cold work prior to exposure to embrittling temperatures. The nonstabilized alloys such as Type 304 SS embrittle more rapidly than do the stabilized alloys, typically represented by Types 321 and 347 SS. This embrittlement is reversible by solution annealing. The effect of sigma-phase embrittlement on toughness depends somewhat on both temperature and alloy composition. For nonstabilized austenitic stainless steels, a sigmatized alloy can be brittle up to 1400°F (760°C). Above this temperature, sigma phase has little effect. The behavior of the stabilized grades is less clear, but they recover some ductility at increasing temperature.

Duplex stainless steels are susceptible to 885°F (475°C) embrittlement and to sigma-phase formation, and they are usually not selected for temperatures above 650°F (345°C). Because of their ferrite content, they are susceptible to low-temperature embrittlement. However, the duplex stainless steels tend to have relatively low brittle-ductile transition temperatures. The engineering codes typically require the duplex stainless steels to be qualified for low-temperature service by impact testing. They can be susceptible to hydrogen embrittlement, but are less susceptible than are the ferritic and martensitic stainless steels.

21.4.3 High Alloys

High alloys with little exception suffer some embrittlement if exposed to sustained high-temperature service due to the formation of intermetallic compounds. Conditions and rates of embrittlement vary with the alloy. Check with alloy manufacturers for specific information. High alloys containing enough nickel to ensure an austenitic microstructure are, like austenitic stainless steels, unaffected by low-temperature embrittlement.

21.4.4 Hydriding

Hydriding is a common problem for all refractory metals, including titanium, zirconium, columbium, and tantalum. Galvanic cells that promote hydriding can be particularly damaging. Instances of iron sacrificial anodes causing hydriding in titanium heat exchanger components have been reported. Hydriding is related to hydrogen embrittlement. Hydrides are brittle, thermodynamically stable compounds. Re-refining the metal is required to destroy the hydrides.

Titanium is the most common material of construction that can be hydrided; it has a threshold temperature of ~175°F (80°C). Hydriding can be caused by: (1) Titanium can absorb hydrogen directly from anhydrous process streams containing hydrogen gas. Because only a small amount of moisture is necessary to inhibit such embrittlement, it is relatively uncommon. (2) Nascent hydrogen generated by a corrosion reaction involving cathodic poisons (such as sulfides, cyanides, and arsenic or antimony compounds) diffuses into titanium to form hydrides. (3) Cathodic protection of titanium charges the metal with hydrogen (most often the cathodic protection is inadvertent, being caused by a materials mismatch, such as titanium tubes rolled into a nontitanium tubesheet).

21.4.5 High-Temperature Effects

21.4.5.1 Mechanical Effects

More expensive materials of construction are often needed at high temperatures. Whenever possible, determine if the maximum design temperature can be reduced.

Creep frequently occurs. Most metals and alloys exhibit a temperature above which the grain boundaries become weaker than the grains themselves. Fabricated equipment such as furnaces, heaters, and combustion gas turbines often experience creep. Creep begins for carbon steel at ~750°F (400°C), for Cr-Mo steels at ~900°F (480°C) and higher, and for conventional austenitic stainless steels at 1050°F to 1100°F (565°C to 595°C). A safe estimate for the creep threshold temperature of a material is the upper temperature limit permitted by ASME Section VIII, Div. 2 [1]. Engineering codes such as ASME B31.3 [4] for piping and ASME Section VIII, Div. 1 [1] for vessels contain provisions for creep design. If creep is a concern, coarse grained materials are favored. Carbon steels killed with silicon are usually recommended when creep is a concern, e.g., ASTM A106 for pipe and ASTM A515 for plate.

Stress rupture can occur as a result of sustained operation in the creep range. Stress rupture design keeps allowable stresses much lower than for noncreep designs. The stress rupture design life is the expected period of *secondary* creep, during which cracking damage slowly accumulates, often over several years. During this period, component dimensions such as heater tube diameter

Selecting Materials of Construction (Steels and Other Metals)

and length increase. In wrought materials such as Alloy 800, secondary creep strains of 10% to 20% are common before failure. In high temperature cast materials such as HK-40 (25Cr-20Ni) (UNS J94204), total secondary creep strains, at failure, are usually 1% to 3%.

When operating in the creep range, temperature excursions are far more damaging as compared with pressure excursions. A 50°F (28°C) excursion may reduce the equipment life by 50% or more. If operating temperatures are high enough to permit creep, thermal fatigue can produce fractures that are virtually identical with those of creep failures. (Thermal fatigue is caused by thermal stresses, generated by thermal cycling. For heaters and furnaces, the most severe cause of thermal fatigue damage is due to operating "trips.") Some codes permit high-temperature maximum allowable stresses high enough to generate thermal fatigue. Accordingly, if thermal cycles are a feature of equipment design (such as for coke drums), thermal fatigue analysis is usually required.

21.4.5.2 Metallurgical Effects

Sensitization of conventional stainless steels, both the austenitic 300-series alloys and the straight chromium grades such as Types 405 and 410 SS, are subject to intergranular corrosion or cracking as a result of *sensitization* in both wrought and cast alloys. Chromium carbides precipitate in the grain boundaries of the alloy at 800°F to 1600°F (425° to 870°C). As the chromium carbides develop, the nearby metal is depleted in chromium, creating a local zone of corrosion-susceptible iron-nickel alloy. For straight chromium grades, this composition may approach that of plain carbon steel. A chromium-depleted zone can also interact galvanically with nearby nondepleted grains to accelerate intergranular corrosion rates.

The rate of sensitization depends strongly on temperature. For conventional austenitic stainless steels, sensitization at about >850°F (455°C) requires extremely long exposure times. The high-carbon H grades have both lower threshold temperatures and faster sensitization rates than do the regular and low-carbon grades. Cold-worked austenitic stainless steels sensitize as low as 700°F (370°C). Sensitization occurs most rapidly above 1500°F (815°C). For example, welding can sensitize the HAZs in stainless steels that have conventional carbon content. Sensitization is often caused by cooling too slowly following solution annealing or stress relief treatment.

Weld rusting occurs when mildly acidic liquids cause the locally chromium-depleted iron-nickel alloy to slowly rust. An example is HAZ rusting of stainless steel by dew (containing CO_2 dissolved from the air) condensing on the outside of a pipe. While normally this is only an aesthetic problem, in some contamination-sensitive processes, such rusting is unacceptable. If chlorides are involved, such corrosion may become aggressive as ferric chloride forms.

Intergranular corrosion results when some fluids, such as oxidizing acids, cause intergranular corrosion in the chromium-depleted grain boundaries. It is sometimes called "weld decay." In some cases, intergranular attack is stress-related and is more properly referred to as *intergranular stress corrosion cracking*. Such corrosion in hydrocarbon plants is usually caused by *polythionic acid*,* in the process-exposed side of a pipe, vessel shell, exchanger bundle, heater tube, etc. The phenomenon usually starts when the steel surface forms a thin iron sulfide film, after exposure to small amounts of sulfur, usually from hydrogen sulfide. During a shut-down, air and liquid water (often dew point water) convert the sulfides to polythionic acid. The polythionic acid then corrodes the chromium-depleted grain boundaries of the sensitized alloy. Because stainless steels are usually supplied to fabricators in the solution annealed condition, sensitization is usually confined to the HAZs of welds. Upon subsequent start-up, leaks may develop after two or more shut-downs. In some cases, the problem becomes obvious during a shut-down while repair welding. A massive growth of a crack network sometimes occurs.

Process controls often protect against such polythionic attack: (1) *Prevent air ingress* by monitoring a positive pressure to ensure that any leaks that do occur are from the inside to the

---

* $H_2S_nO_6$, where n may vary from 2 to 5.

outside. (2) *Prevent the formation or ingress of water.* Purge with a dry inert gas such as nitrogen and then maintain a slightly positive pressure. (3) *Use a neutralizing wash.* Both ammonia and soda ash solutions are used to maintain a basic pH. Refer to NACE RP0170: "Protection of Austenitic Stainless Steel from Polythionic Acid Stress Corrosion Cracking During Shutdown of Refinery Equipment" [14] for recommended practices.

External polythionic acid attack may occur in plants having atmospheric sulfide pollution including gas applications with water condensation. However, the problem does not usually occur on external surfaces in fired equipment because excess air causes sulfates to form, instead of sulfides. Wet sour systems containing carbon steel heat transfer surfaces may generate substantial iron sulfide. Such systems should not be flushed or drained into stainless steel piping or equipment prior to a shut-down unless appropriate precautions are taken. Several metallurgical methods address intergranular corrosion caused by sensitization.

21.4.6 Low-Carbon Stainless Steel

The "L" grades such as Type 304L SS with a low carbon content, limit the level of sensitization. However, they have lower allowable stresses than either the conventional grades or the stabilized grades. The low-carbon grades are typically chosen for services in which welded fabrication is required but the operating temperatures will be less than the sensitization threshold, about 800°F (425°C). With a low-carbon content, the rate of sensitization is low, and the postweld cooling rate in the HAZ is fast enough to avoid significant sensitization. 29Cr-4Mo (UNS S44700) is an example of a low-carbon ferritic stainless steel.

21.4.6.1 Chemically Stabilized Stainless Steel

Types 321, 347, and 348 SS and their H grades, and Types 316Ti and 316Cb SS are resistant to sensitization. They contain carbon scavengers (titanium in Type 321 and Type 316Ti and niobium in Type 347, Type 348, and Type 316Cb) that inhibit chromium depletion. These materials are referred to as *stabilized alloys.*

ASTM specifications for austenitic stainless steels, including the stabilized grades, are usually provided in the solution annealed condition, in which these alloys are not resistant to sensitization caused by long-term, high-temperature service. ASTM specifications warn the user that solution annealing the stabilized grades may result in inferior resistance to intergranular corrosion. With these specifications, the mill solution annealing heat treatment should be followed by a *stabilization anneal.* A widely used practice is to hold the alloy at about 1650°F (900°C) for 2 to 4 hours followed by air cooling to promote stable carbides, formed from either titanium for Types 321 and 316Ti SS or from niobium for Types 347, 348 SS, and 316Cb, without chromium depletion. The purchaser should specify stabilization annealing.

Special alloy composition requirements may be necessary to assure proper stabilization annealing [15]. These requirements limit the ratios of Ti/C in Type 321 stainless steel and of Cb/C in Type 347. The protection provided by a stabilization anneal can be partially destroyed by welding, since some chromium carbides may form. For full protection, welds made after the stabilization anneal should be restabilized. Type 321 SS is more susceptible to this welding effect than is Type 347.

Several stabilized nickel-based high alloys (such as Alloy 825 (22Cr-42Ni-3Mo, Ti stabilized) (UNS N08825) are resistant to sensitization. They are usually furnished in the stabilization annealed condition but may be made susceptible to sensitization by subsequent PWHTs. The alloy manufacturer should be consulted before undertaking stress relief or PWHTs.

Hot bending may cause cracking in stabilized stainless steel pipe, and should be avoided. Virtually all nonstabilized Cr-Ni and nickel-based high alloys are susceptible to sensitization and intergranular corrosion. If fabrication involves aggressive pickling, or if the alloy is to be exposed to polythionic acid or other oxidizing acids, sensitization and resistance to intergranular corrosion should be considered. Check the technical literature or consult with the alloy manufacturer.

Sensitization requires the following considerations: (1) Alloys susceptible to sensitization are acceptable if not welded and if the maximum design temperature does not exceed 800°F (425°C). For cold working of these alloys, the maximum design temperature should be <700°F (370°C). (2) The low-carbon "L" grades are acceptable in weldments for which the maximum design temperature does not exceed 800°F (425°C); in a cold-worked condition, <700°F (370°C). (3) Stabilized alloys are selected for up to 800°F (425°C). Types 316Ti and 321 SS are acceptable in weldments and in processes for temperatures <900°F (480°C). Types 347, 348, and 316Cb SS are acceptable in weldments and selected for processes in which the maximum design temperature exceeds 900°F (480°C). In processes for maximum design temperatures <975°F (525°C), these materials are resistant to sensitization. Type 347 SS is more resistant to sensitization than is Type 321 SS. (4) Stabilized ferritic stainless steels may be substituted for higher grades when superior corrosion resistance is not needed. However, they are susceptible to 885°F (475°C) embrittlement. (5) Stabilized Cr-Ni and nickel-based high alloys are available for processes in which the corrosion resistance of the 300-series is inadequate.

Solution annealing dissolves carbides formed by sensitization. It is required by ASTM specifications for austenitic stainless steel product forms, including castings. However, solution annealing is usually avoided in welds because of distortion problems above 800°F to 850°F (425°C to 455°C).

Spheroidization and *graphitization* can occur in carbon and C-1/2Mo steels >850°F (455°C). Chromium contents of ~0.7 wt% or more prevent these effects. Since most users avoid the selection of carbon steels for use >800°F (425°C), the effects of these two mechanisms are often not considered. However, many plants use carbon steel lined with refractory for such high-temperature services. Refractory failures may occur, resulting in metal temperatures substantially in excess of 800°F (425°C).

Iron carbide (cemenite, Fe_3C) is thermodynamically unstable. Decomposition of cementite in carbon and C-1/2Mo steels is negligible at <850°F (455°C), but accelerates with temperature exponentially. At 900°F (480°C), 50% conversion to graphite occurs in about 10,000 hours, but at 1100°F (595°C) in about 1000 hours.

Cementite decomposition to iron and graphite is called *graphitization*. Decomposition is accompanied by a moderate reduction in the strength, which may accelerate creep damage if the applied stress is large enough to cause creep. Decomposition of the iron carbide can "embrittle" the steel if the graphite that develops forms a continuous (or closely spaced discontinuous) band. Such ruptures are most frequent in C-1/2Mo steel, and this steel is no longer recommended for high-temperature services unless operating stresses are low. Aluminum-killed steels are more susceptible to graphitization than are silicon-killed steels, which are hence preferred for high-temperature services.

Spheroidization refers to the formation of iron carbide spheroids from the pearlite present in most carbon steels, at temperatures >900°F (480°C); the rates are exponential with temperature. However, for sustained high temperatures, graphitization becomes the dominant mechanism. A small amount of spheroidization may be beneficial, since the impact toughness of carbon steel increases with only minor loss of strength. Too much spheroidization causes unacceptable loss of strength. Above 850°F (455°C), with prolonged exposure, carbon steels may be susceptible to stress rupture. Above 1000°F (540°C), oxidation and spalling can result in early failure. Short-term excursions to 1000°F (540°C) may be tolerable if the applied stress is less than the maximum code-allowable stress.

Welding often causes several problems: (1) *Microstructural problems.* Grain coarsening that may reduce toughness; stainless steels may be sensitized. Postweld heat treatments, such as normalizing for ferritic steels and stabilization annealing or solution annealing for austenitic stainless steels, mitigate these problems. (2) *Residual stresses* contribute to cracking. Stress relief by PWHT is usually required for processes sensitive to stress corrosion cracking. (3) *Thermal gradients* cause mechanical distortion. Such problems are usually addressed by a combination of mechanical stabilization, such as bracing, slow heating and cooling, and cold straightening. Cold straightening should be avoided for steels susceptible to strain aging or dynamic strain aging.

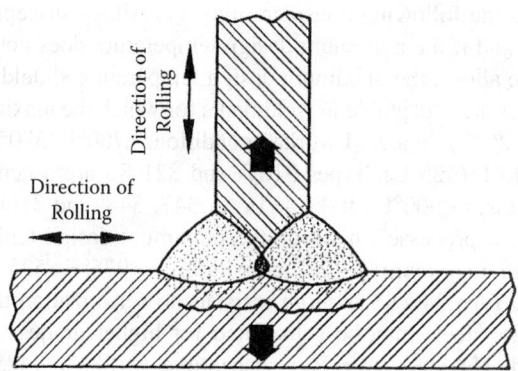

FIGURE 21.11 Lamellar tearing in a T-joint. © Copyright by Marcel Dekker, Inc.; reprinted with permission [5].

Welds of dissimilar metals result in: (1) *Warping, buckling, and/or excessive residual stresses* caused by different thermal expansion coefficients. (2) *"Hard spots"* in heat-affected zones and fusion zones are caused by formation of hard intermetallic compounds such as carbides. The heating/cooling cycle can harden HAZs and can cause a sensitized HAZ in nonstabilized alloys. (3) *Galvanic effects* between a sensitized HAZ and its adjacent parent metal can also occur.

Lamellar tearing is an unusual type of failure in plate and plate products, as the weldment cools (Figure 21.11); restraint causes the region just below the weld to be in tension and tearing can occur. (Poor through-thickness toughness is usually caused by the presence of a relatively high density of nonmetallic inclusions, that is, a "dirty" steel.) Lamellar tearing is normally not a problem in plate <1 in. (25 mm) thick.

Methods to avoid lamellar tearing include (1) *Materials control* utilizes plate with a low concentration of nonmetallic inclusions, using purchasing specifications developed for plate resistant to hydrogen-induced cracking (HIC). ASTM A770, "Standard Specification for Through-Thickness Tension Testing of Steel Plates for Special Applications," can be used to ensure plate with proven through-thickness toughness. ASTM A435, "Standard Specification for Straight-Beam Ultrasonic Examination of Steel Plates," ensures that the plate manufacturer or vendor has inspected the plate for injurious nonmetallic inclusions. (2) A *welding technique* option is to use a "butter" layer of low-strength weld metal on the surface of the plate for which lamellar tearing is to be avoided. This technique may require adjusting the weld joint efficiency if the minimum specified tensile strength of the filler metal is less than that of the base metal. Another option is to modify the joint or weld to minimize restraint. Using fillet welds rather than full penetration welds is sometimes useful.

Postweld inspection for lamellar tears is difficult unless the tears penetrate to the surface (usually at the toe of the weld). Dye penetrant inspections are reliable for tears that penetrate the surface. For embedded tears, ultrasonic inspection is sometimes used, but tears are difficult to differentiate from noninjurious inclusions.

Carburization is the development of a carbide-rich layer at high temperatures on a metallic surface exposed to a reducing hydrocarbon environment. Mild carburization of ordinary 300-series austenitic stainless steels is sometimes observed, e.g., in the plenum of a fluid catalytic cracking unit. Special alloys, such as the Alloy 800 series (20Cr-32Ni with Ti and Al) (UNS N08800/8810/8811) or HP-Mod, are generally used in the cracking tubes of ethylene furnaces. Carburization can cause premature failures such as cracking, due to the large difference in the coefficients of thermal expansion between the parent alloy and the carburized layer. Thermal cycling is the normal cause of such failures. Metal loss can also occur by the mechanism known as *metal dusting*. It is observed in the "syngas" process to convert natural gas to hydrogen. This very limited mechanism usually involving reducing hydrocarbon process streams is worst for Fe-Cr or Fe-Cr-Ni alloys having chromium contents of 25% [16] and is very sensitive to temperature. The proba-

Selecting Materials of Construction (Steels and Other Metals)

bility of dusting increases rapidly at the temperature increases ~850°F (450°C). Process licensors and alloy manufacturers usually provide excellent guidance.

Fuel ash corrosion can be a problem with fuels containing elements that promote corrosion. Major culprits are vanadium and sodium. Above ~1200°F (650°C), vanadium oxide vapor and sodium sulfate react to form sodium vanadate, which reacts with metal oxides on the surfaces of heater tubes, hangers, tubesheets, etc. The resulting slag can become a low-melting eutectic mixture, acting as a flux. The slag dissolves protective metal oxides and prevents their reformation. Sulfur in the fuel contributes to sulfidation and lowers the melting point of the vanadium oxide flux. Such corrosion depends on both the level of metal content of the fuel plus excess air used for combustion. Concentrations of <5 ppmw vanadium appear to have little effect. Concentrations up to ~20 ppmw are regarded as safe at <1550°F (845°C). Alloys rich in nickel and chromium (50Cr-50Ni) offer good protection. Reducing the excess air to <5% can control fuel ash corrosion. Vapor deposition coatings of aluminum and chromium have been tried as tube coatings with some success; various coatings have also been used for turbine blades.

21.4.7 Hydrogen Gas

When selecting a metal, hydrogen service is defined as any service in which the partial pressure of the hydrogen gas exceeds 100 psia (0.7 MPa). Hydrogen gas can cause two problems: *hydrogen embrittlement* (already discussed) and *hydrogen attack*. Hydrogen attack causes internal decarburization and fissuring in carbon and low-alloy steels above ~430°F (220°C). Dissolved hydrogen attacks iron carbide (Fe_3C-cementite), generating methane gas (CH_4), which is trapped in the grain boundaries because the methane molecule is too large for diffusion. As the concentration of methane gas increases, increasing pressure tears the grain boundary apart, first causing fissures, then cracks, then networks of cracks. Simultaneously, the loss of carbides lowers the strength of the material, substantially reducing the expected life of the metal. Both chromium and molybdenum form more stable carbides, and Cr-Mo alloys are preferred in hydrogen service. (Vanadium-enhanced versions of the Cr-Mo often have even greater resistance to hydrogen attack, but are not yet included in the Nelson curves, discussed below.) Unless severely cold worked, austenitic stainless steels are unaffected by hydrogen attack.

Hydrogen attack starts at an exposed surface and proceeds inward via diffusion; detection is relatively easy. The most common inspection technique involves *replicas*. Replication uses metallurgical polishing and etching to prepare the surface for study. A wetted plastic film is then pressed against the surface, producing a negative relief of the surface microstructure. The replica is then examined with a conventional microscope to determine the degree, if any, of decarburization. See ASTM E1351 for guidance on suitable procedures for replication.

The Nelson curves are used to select materials that resist hydrogen attack. Refer to API Publication No. 941, "Steels for Hydrogen Service at Elevated Temperatures and Pressures in Petroleum Refineries and Petrochemical Plants" [3]. Because of the "scatter" in the Nelson curve data, it is common to use the maximum operating temperature plus a 25°F (14°C) margin. Some, however, use a 50°F (28°C) margin. If the maximum design temperature is used for materials selection, the use of an additional operating temperature margin should be unnecessary. For the vertical portion of the curves, it is customary to use a 25 to 50 psia (170 to 345 kPa) margin for the maximum operating hydrogen partial pressure. Compliance with the Nelson curves usually precludes the possibility of hydrogen attack on carbon steel. However, carbon steels damaged by hydrogen blistering or related phenomena can suffer rapid and severe hydrogen attack if stress relieved or postweld heat treated. Hence such treatments should be avoided.

Unintentional materials substitutions have sometimes caused hydrogen attack that resulted in dangerous ruptures. Unfortunately, the Cr-Mo low-alloy steels most often used for high-temperature, high-pressure hydrogen service are not easily distinguishable from ordinary carbon steel. Laboratory testing or on-site testing using portable fluorescent x-ray analyzers distinguishs the Cr-Mo alloys.

As a result, most users employ a *positive materials identification* program, which includes welds as well as parent metals. Positive materials identification testing should be done on installed materials. When this is not practicable, alternatives such as testing immediately before installation are permitted.

Hydrogen attack is accelerated by inclusions and slag-type defects. Therefore, killed steels and inclusion-free welds are often specified. To further protect materials exposed to hydrogen, PWHT is recommended for all Cr-Mo alloy steels. Components cold worked more than 5%, such as cold-formed heads, should be stress relieved. Pockets, such as those created by strip linings, should be avoided. When the metal is exposed to infrequent and short-term transient combinations of high temperature and moderate hydrogen partial pressure, a significant incubation time often occurs before the effects of such attack become detectable. Investigation of incubation times can often justify the choice of a lower cost material of construction. API Publication No. 941 [3] provides details on incubation times. Such analyses are particularly valuable in applications such as bypass loops in which gaseous hydrogen may be blocked in or is stagnant and subsequently heated during start-ups or shut-downs.

To summarize, when selecting materials for gaseous hydrogen service

1. The Nelson curves, utilizing the maximum operating temperature plus 25°F (14°C), should be used.
2. Carbon steels should be killed or otherwise fully deoxidized.
3. Low-alloy steels such as the Cr-Mo steels should be postweld heat treated.
4. Cold-worked materials should be stress relieved.
5. Seamless tubing and pipe are preferred, as they avoid potential problems associated with longitudinal welds.

Hardness controls should be specified as follows:

1. NACE RP0472 [13]. The maximum weld metal hardness permitted for carbon steel is 200 BHN. Weld procedure qualification testing should be done to ensure that HAZ hardnesses do not exceed 248 BHN.
2. NACE MR0175 [2]. This practice, which limits the hardnesses of parent metals and HAZs, should be required. Recommended hardness limits for the weld metal of Cr-Mo low-alloy steels is 225 BHN for Cr < 3 and 241 BHN for 3 < Cr < 9.

Designs for gaseous hydrogen service should include the following features: (1) Plate materials, including clad plate, should be ultrasonically examined for laminations (see ASTM A578). Weld repair of laminations should not be permitted. (2) Pressure-containing and internal attachment welds should be full penetration. (3) Pocketed weldments such as reinforcement pads should be vented. Internal attachment welds that cannot be full penetration, such as tray support welds, should also be vented.

Since hot high-pressure H_2 can crack the Cr-Mo steels favored for this service, the following should be kept in mind: (1) Hydrogen embrittlement can occur at operating temperatures at about <250°F (120°C). This concern is usually addressed by not pressurizing at operating temperatures <250°F (120°C). (2) Delayed hydrogen cracking that can occur in weld repairs subsequent to service. This concern is addressed by an appropriate bakeout prior to repair welding.

21.4.8 Nitriding

Stainless steels and many higher alloys such as Alloy 800 slowly develop a brittle nitride layer if exposed to a nitriding atmosphere at temperatures exceeding about 750°F (400°C). By far the most common nitriding atmosphere is ammonia or a mixture rich in ammonia. Nitriding of stainless

TABLE 21.2
Oxidation/Scaling Threshold Temperatures for Commonly Used Metallic Materials

Material	Scaling Temperature
Carbon Steel	1000°F (540°C)
1-1/4Cr-1/2Mo	1050°F (565°C)
2-1/4Cr-1Mo	1075°F (580°C)
3Cr-1Mo	1100°F (595°C)
5Cr-1/2Mo	1150°F (620°C)
9Cr-1Mo	1200°F (650°C)
12Cr	1500°F (815°C)
3-1/2Ni	1000°F (540°C)
9Ni	1000°F (540°C)
18Cr-8Ni	1650°F (900°C)
Types 309 & 310 SS	2000°F (1095°C)

steels has also been reported with nitrogen-bearing organic compounds such as urea, but not nitrogen. Nitriding usually is much slower than carburization. Special alloys and/or aluminum including aluminum vapor-deposited coatings resist nitriding. Materials selection often accommodates nitriding by providing a nominal *nitriding allowance* such as 1/16 in. (1.5 mm).

21.4.9 OXIDATION

All metals and alloys have threshold temperatures above which they become susceptible to rapid scale formation and spalling when heated in air or steam. Table 21.2 shows the oxidation/scaling threshold temperatures for common metallic materials. Materials in applications subject to thickness losses due to oxidation are usually provided with a nominal *oxidation allowance*; 1/16 to 1/8 in. (1.5 to 3 mm) is typical.

Hot lines and equipment are usually thermally insulated to conserve energy and to ensure safe service at temperatures above the oxidation limits of the materials. Care must be taken to ensure that the process stream chemistry is either nonoxidizing or that the process-side surface is protected by an insulating or refractory liner. In such cases, the limiting factor will be the availability of a code maximum allowable stress.

Alloys containing molybdenum can potentially experience *catastrophic oxidation*. The superaustenitic stainless steels such as Alloy AL-6XN, a 21Cr-25Ni-6.5Mo-N alloy (UNS N08367), are an example. A heavy molybdenum oxide scale forms, usually as a result of an improper heat treatment or a severe thermal excursion. Removal of such scales prior to service (or return to service), usually by pickling, prevents this problem.

Heat tinting, such as the blue tinge often seen on welds, is a common condition associated with welds and thermally cut surfaces. For carbon and low-alloy steels, such tinting is usually ignored. However, the subsurface areas of heat-tinted stainless steels may be significantly depleted of chromium. For demanding environments, heat tinting is usually removed by mechanical methods such as grinding or by chemical cleaning or both.

21.4.10 SULFIDATION AND SULFIDIC CORROSION

Sulfur and sulfur compounds attack carbon and low-alloy steels at temperatures exceeding 500°F (260°C) and nickel-base alloys such as Alloy 600 (15Cr-72Ni-8Fe) (UNS N06600) above about

600°F (315°C). The term *sulfidic corrosion* refers to corrosion of carbon and low-alloy steels by sulfur compounds while the term *sulfidation* is normally used for higher alloys. Sulfidic corrosion is usually associated with sulfur in crude oil (as organic sulfides and/or as H_2S). Severe pitting and general wastage can occur in carbon and alloy steels above 500°F (260°C). Corrosion rates are estimated using the McConomy and Couper-Gorman curves.

The original McConomy curves [17] have been found by experience to usually be conservative [18]. They are used for services containing no hydrogen gas. The total sulfur content (in wt%) is not a precise indicator of the corrosivity of a crude oil since not all organic sulfur compounds are corrosive to carbon and alloy steels, even at elevated temperatures. The curves apply to sour crude oils and sour crude fractions, operating at temperatures of >500°F (260°C). They are used for processes containing gases having hydrogen partial pressures no greater than 50 psia (0.34 MPa). They provide *average* corrosion rates and are used to predict the time-to-first leak or to estimate corrosion allowances. (For low-pressure applications, some users utilize the entire wall thickness for making time-to-first leak estimates.) Because of uncertainties, McConomy curve estimates often do not agree with previous plant experience. In such cases, one obviously relies on plant experience with similar operating conditions.

The Couper-Gorman curves [19] are used for sour services with hydrogen gas having a partial pressure of at least 50 psia (345 kPa). Materials selected in compliance with the Couper-Gorman curves must also satisfy the restrictions involving gaseous hydrogen service. When the McConomy or Couper-Gorman curves indicate that carbon steel is appropriate, silicon-killed steels are generally preferred, as they are more resistant to sulfidic corrosion than are aluminum-killed steels. However, coarse-grained silicon-killed steels may be precluded by a requirement for the low-temperature toughness provided by the fine grain practice steels killed with aluminum. When these two curves indicate that the corrosion rates of both carbon or low-alloy steels are sufficiently low, consider the effect of corrosion on downstream equipment. For some processes, equipment such as reactors contain catalyst beds that may be plugged by sulfides. Hence corrosion-resistant steels are needed.

For temperatures at which sulfidic corrosion rates are excessive, two alternatives may be needed: (1) Refractory linings in both vessels and piping. A low iron-containing refractory is required, because spalling of the refractory has been associated with iron oxide contaminants in the refractory. (2) Aluminum diffusion coatings on steel. These coatings are applied by proprietary processes for vapor-diffusing aluminum on and into the steel. For heater tubing, the useful temperatures are extended to 800°F (425°C). Such coatings on low-alloy steels protect heater tubes from external sulfidic corrosion, corrosion-resistant alloys used for close-tolerance components such as internal bolting, and very thin components such as screens and catalyst baskets. Imperfections of the coatings can lead to early failure.

For applications requiring corrosion-resistant alloys, either low-carbon or stabilized stainless steels such as Type 321 SS are normally selected. Sensitization-induced polythionic acid corrosion is a concern in such applications. High-nickel alloys and copper-based alloys often corrode rapidly in the presence of high-temperature sulfur compounds.

Pure liquid sulfur is stored in pits made of Type V concrete. Carbon steel piping is usually specified for the liquid. Nitrogen blanketing or alloys such as Alloy 20 Cb-3 or Type 310 SS are employed since liquid sulfur and air are highly corrosive to carbon steel.

Sulfur and sulfur compounds may cause *sulfidation* of nickel-based alloys >600°F (315°C). The threshold temperature depends on both the process chemistry and the alloy composition. At least one user regards the threshold to be as low as 300°F (150°C) for Alloy 400 (67Ni-30Cu) (UNS N04400) in the presence of hydrogen sulfide. Copper-based alloys corrode quickly and should generally be avoided. Sulfidation can cause pitting, intergranular attack, or fluxing of molten sulfides. Ordinary austenitic stainless steels are subject to fluxing by molten Fe-Ni sulfides. Alloy manufacturer literature and assistance should be sought for differentiating among alloys.

21.5 THE PROCESS OF MATERIALS SELECTION

A four-step materials selection process is described here. In the first step, identify special requirements that will affect materials selection, including the specification of unusual design life and concerns about product contamination. Include such requirements in the "Notes" addendum of the template. Indicate whether operating conditions or design conditions are the basis for materials selection. Second, information about basic metallurgy, corrosion, and degradation phenomena is collected and used to design templates and "Notes" addenda tailored to the specific needs of a plant. Third, materials of construction are selected. Fourth, the materials selection diagram (MSD) is used to check for consistency and to document any special measures needed for corrosion or degradation control.

21.5.1 Designing a Template

The template serves two major purposes: (1) It organizes all the technical data needed for materials selection and (2) on a large job or project, it is a convenient way to transmit process and materials selection information to design engineers.

A template should be as simple as possible. An overly elaborate template is costly and slows the process of materials selection. Decide what information is necessary, then format the template to highlight the required information. After the template has been developed, it is good practice to ask that all requested information be provided, even if the answer may be "trace amount," "not applicable," "per Code," etc. To decide what information should be provided, review the discussions in Section 21.1 regarding template information. The threshold information for the "Notes" addendum of the template is discussed in Sections 21.2 through 21.4; also consider previous experience, plant or pilot plant testing, process licensors, and published literature or material manufacturers.

For small or uncomplicated jobs, a simple template may be preferred. A customized template is not usually needed for replacing or revamping a couple of vessels or a small piping system, or for a unit involving only a few, if any, corrodents. A rubber stamp template, with a process flow diagram, may be used to quickly create a materials selection diagram, as illustrated in Example 21.1.

For jobs involving complex combinations of corrodents, crack-inducing agents, upset conditions, processes, and/or design conditions, a detailed template is usually required. Templates suitable for a refinery are shown in Examples 21.2 (a small job) and 21.3 (a large job).

Many jobs will benefit from a job-specific customized template. The customized template should request information only about corrodents, crack-inducing agents, and upset conditions known to be characteristic to the job. Example 21.4 is a template customized for an ammonia plant. Example 21.5 is a template that could be customized for a chemical plant operated batchwise.

Example 21.1. Simplified Template

Operating Temperature (Minimum/Maximum): _____

Operating Pressure (Minimum/Maximum): _____

Commodity: _____ **Phases:** _____ **Liquid Water** (Y/N): _____

Corrodents: _____

Crack-Inducing Agents: _____

Metallurgy: _____

PWHT (Y/N): _____ **Valve Trim:** _____ **Corrosion Allowance:** _____

Notes: _____

Example 21.2. General Template for a Refinery

STREAM OR EQUIPMENT NUMBER: _____

Design Temperature (Minimum/Maximum): _____

Design Pressure (Minimum/Maximum): _____

Commodity: _____ **Phases:** _____ **Liquid Water** (Y/N): _____

Corrodents (1): _____

Crack-Inducing Agents (2): _____

Upset Conditions: _____

    Liquid Water (Y/N): _____

    Corrodents (1): _____

    Crack-Inducing Agents (2): _____

    Autorefrigeration (Y/N): _____

        If Yes, indicate minimum temperature: _____

    Other: _____

Wet Sour Service (Y/N): _____

    If Yes, indicate severity: _____

Metallurgy: _____

PWHT (Y/N): _____ **Valve Trim:** _____ **Corrosion Allowance:** _____

Notes: _____

The template should (1) Accommodate all of the processes and services to be analyzed plus process chemistries. (2) Contain all design information such as temperatures and pressures. Any anticipated operating conditions that affect mechanical design should be indicated. Examples include pressure and/or thermal cycling. This information will be used to prepare the piping and equipment data sheets and to alert designers to special conditions such as cyclic operation. (3) Contain the maximum operating temperatures and pressures. For most mature plant technologies, materials selection is based on operating conditions rather than design conditions. The major exception is for the minimum temperature, for which engineering codes require materials resistant to brittle fracture. (4) Define any upset (nonstandard) and transient design conditions. (5) Contain "Notes" to indicate supplementary materials selection information such as referring to NACE or API documents. The "Notes" section helps describe upset conditions, or a materials testing requirement such as positive materials identification. (6) Include, as an attachment, an addendum (referred to as the "Notes" addendum). It defines temperature, pressure, and concentration threshold values for applicable corrosion and other materials degradation phenomena, as obtained from data supplied by process licensors, or from testing in plants, pilot plants, laboratories, or the open literature. Addendum information helps evaluate risks of early failure. Examples include: (a) a widely used threshold definition for "sour" water is a dissolved H_2S concentration of at least 50 ppmw; (b) austenitic stainless steels are generally resistant to chloride stress corrosion cracking in neutral saline water at less than 140°F (60°C); (c) sensitization-induced intergranular corrosion of low-carbon austenitic stainless steel welded construction is normally not a problem at less than 800°F (425°C).

Notes for Example 21.2.

(1) Corrodents. Indicate:
   (a) Concentrations of acidic components (only if liquid water or other electrolyte is present).
      (a1) Inorganic or organic acids; indicate in wt% (indicate Total Acid Number if naphthenic acid is present).
      (a2) Acid gases such as H_2S, NO_2, SO_2, and CO_2 (indicate in mole% if in vapor and wt% if dissolved in water).
      (a3) Acid salts such as ammonium bisulfide or ammonium chloride (indicate in wt%).
      (a4) Anticipated pH.
   (b) Oxidants such as oxygen and chlorine (in wt%).
   (c) Other corrodents suspected of being significant, such as dissolved oxygen content in water or microbiological agents.
   (d) Total sulfur:
      (d1) As wt% sulfur, only if T >500°F (260°C) and the hydrogen partial pressure is <50 psia (0.34 MPa).
      (d2) As mole% H_2S, only if T > 500°F (260°C) and the hydrogen partial pressure 50 psia (0.34 MPa).
(2) Crack-inducing agents (list only if liquid water or other electrolyte is present [except for H_2]; indicate concentration of the following agents, if present above their threshold concentrations):
   (a) Amines, if present at >2 wt% (indicate wt%). Indicate type of amine.
   (b) Carbonates and bicarbonates, when the concentration of either or both (combined) exceeds 1 wt% (indicate wt%).
   (c) Hydrogen, when partial pressure is 100 psia (0.69 MPa) (indicate either mole% or partial pressure in psia [MPa or kPa]).
   (d) Chlorides, in any concentration (in ppmw).
   (e) Cyanides, if present at >20 ppmw (indicate ppmw).
   (f) NaOH, in any concentration, if T > 115°F (46°C) (indicate wt%).
   (g) Hydrogen sulfide:
      (g1) Gas phase: the partial pressure of H_2S exceeds 0.05 psia (0.34 kPa) (indicate either mole% or partial pressure in psia [MPa or kPa]).
      (g2) Sour water: H_2S is dissolved in water at a concentration of at least 50 ppmw (indicate ppmw).
   (h) Other known crack-inducing agents (e.g., HF).

Example 21.3. General Template for a Refinery

Stream Description	Units	References	Stream No. 1	2	Etc.
Commodity					
Phase (Liq/Vap/Solids)	L/V/S				
Design Temperature: Min/Max	°F				
Design Pressure: Min/Max	psig				
Operating Temperature: Normal/Min/Max	°F				
Operating Pressure: Normal/Min/Max	psig				
Fluid Velocity	ft/sec				
Free Water	Yes/No	Note 5			
Crack-Inducing Agents		Note 4			
Chloride	ppmw	Note 4			
Cyanide	ppmw	Note 4			
Hydrogen (partial pressure)	psia	Note 4			
Hydrogen sulfide	Note 4	Note 4			
Amine	wt%	Note 4			
NaOH or other caustics	wt%	Note 4			
Other		Note 4			
Corrodents		Note 3			
Sulfur	Notes 1&2	Notes 1&2			
Acids	wt%				
Acid gases	mole%				
Acid salts	wt%				
pH					
Other		Note 3			
Upset Conditions		Note 6			
Wet Sour Service	yes/no				
If yes, simple or severe					
Metallurgy		Note 7			
Corrosion Allowance	inch				
Valve Trim					
Notes					

Notes for Example 21.3.

1. Indicate in wt%, only if T > 500°F (260°C) and the hydrogen partial pressure is <50 psia (0.34 MPa).
2. Indicate in mole%, only if T > 500°F (260°C) and the hydrogen partial pressure is at least 50 psia (0.34 MPa).
3. Corrodents. Indicate:
   (a) Mole% of acidic corrodents such as inorganic or organic acids and acid gases such as H_2S, NO_2, SO_2, and CO_2. Indicate Total Acid Number if naphthenic acid is present.
   (b) Wt% of acid salts such as ammonium bisulfide or ammonium chloride.
   (c) Wt% of oxidants such as oxygen and chlorine.
   (d) Other corrodents suspected of being significant such as dissolved oxygen content in water or microbiological agents.
4. Crack-inducing agents (list only if liquid water or other electrolyte is present [except for H_2]; indicate concentration of the following agents, if present above their threshold concentrations):
   (a) Hydrogen, if partial pressure exceeds 100 psia (indicate either mole% or partial pressure in psia).
   (b) Amines, if present at >2 wt% (indicate wt%). Indicate type of amine.
   (c) Carbonates and bicarbonates, either or both (combined) when the concentration exceeds 1 wt% (indicate wt%).
   (d) Chlorides, in any concentration (indicate ppmw).
   (e) Cyanides, if present at >20 ppmw (indicate ppmw).
   (f) NaOH, in any concentration, if T > 115°F (46°C) (indicate wt%).
   (g) Hydrogen sulfide:
      (g1) Gas phase, if the partial pressure of H_2S exceeds 0.05 psia (in either mole% or partial pressure in psia).
      (g2) Sour water, if the concentration of H_2S dissolved in water is at least 50 ppmw (indicate ppmw).
   (h) Other known crack-inducing agents (e.g., HF).
5. Include indication of liquid water, for normal operation.
6. For upset conditions, indicate autorefrigeration, liquid water, wet sour service, carry-over of crack-inducing agents or corrodents, etc. Consider start-ups, shut-downs, regeneration, presulfiding, loss of flow, etc.
7. Provide metallurgy in generic form (CS, 18Cr-8Ni SS, 3Cr-1Mo, etc.).

Example 21.4. Ammonia Plant Template

STREAM OR EQUIPMENT NUMBER: _____

Design Temperature (Minimum/Maximum): _____

Design Pressure (Minimum/Maximum): _____

Commodity: _____ **Phases:** _____ **Liquid Water** (Y/N): _____

Corrodents (1): _____

Crack-Inducing Agents (2): _____

Upset Conditions: _____

    Liquid Water (Y/N): _____

    Corrodents (1): _____

    Crack-Inducing Agents (2): _____

    Autorefrigeration (Y/N): _____

        If Yes, indicate minimum temperature: _____

    Other: _____

Metallurgy: _____

PWHT (Y/N): _____ **Valve Trim:** _____ **Corrosion Allowance:** _____

Notes: _____

Notes for Example 21.4.

(1) Corrodents (only if liquid water or other electrolyte is present).
    (a) Indicate wt% for any acids.
    (b) Indicate partial pressure for wet CO_2.
    (c) Be alert to the danger of metal dusting in hot, high-alloy streams with CO/CO_2 ratios >0.5.

(2) Crack-inducing agents (list only if liquid water, or other electrolyte, is present [except for NH_3 and H_2]; indicate concentration of the following agents, if present above their threshold concentrations):
    (a) Anhydrous ammonia: water content in wt%. 0.1 wt% water is required to inhibit stress corrosion cracking in carbon steel. Note that inhibition will be ineffective in vapor spaces.
    (b) Hydrogen, if the partial pressure is 100 psia.
    (c) In lines and equipment in the CO_2 recovery unit (in wt% caustic or wt% amines).
    (d) Chlorides, any concentration (in ppmw).

Selecting Materials of Construction (Steels and Other Metals)

Example 21.5. Template for Batch Processes

STREAM OR EQUIPMENT NUMBER: _____

Step (1): _____

Mechanical Design Conditions

	Operating		Design	
	Low	High	Low	High
Temperature:	_____	_____	_____	_____
Pressure:	_____	_____	_____	_____

Process Chemistry

Chemicals Present (2): _____

Phases Present: _____

Corrodents Present (3): _____

Crack-Inducing Agents Present (4): _____

Upset Conditions (5): _____

Material of Construction: _____

PWHT (Y/N): _____ **Corr Allowance:** _____ **Value Trim:** _____

Notes: _____

Notes for Example 21.5.

(1) For batch processes, a template must be prepared for each step in the process. Indicate the step for which the template is intended.
(2) Chemicals present:
  (a) List all chemicals present, including contaminants and impurities as well as the major constituents.
  (b) Indicate if the process fluid is an electrolyte. If not, are other electrolytes present?
(3) Corrodents: Indicate which chemicals are known corrodents, including chemicals known to damage nonmetallic materials by mechanisms such as permeation and swelling.
(4) Crack-inducing agents: Indicate which chemicals are known crack-inducing agents.
(5) Upset conditions: Describe the nature and duration of possible anticipated upset conditions. Consider whether the upset conditions are for start-of-run, end-of-run, both, or will occur during the run. For each upset condition, indicate the presence of corrodents, crack-inducing agents, or electrolytes introduced because of the upset. For autorefrigeration, indicate the anticipated minimum temperature.

21.5.2 Materials Selection Criteria

While the normal criteria for materials selection address design life, sometimes other criteria govern. Examples include using design conditions rather than operating conditions as the basis for materials selection, requirements for unusual reliability or unusually long life, and corrosion debris or product contamination concerns. Such information is sometimes referred to as the *materials selection design basis*. It is helpful to indicate this information in the "Notes" addendum of the template.

21.5.2.1 Product Contamination

Food, drug, polymer, and fine chemical processes are generally sensitive to contamination, including flaking of scales. While carbon steels may have acceptably low corrosion rates, they often result in iron contamination. Damage by *corrosion debris* is a closely related consideration. Examples include processes in which downstream catalysts are damaged by corrosion debris or where debris can affect flow or heat transfer efficiencies or may cause mechanical problems. The amount of corrosion debris depends primarily on the amount of exposed surface area. Packed beds and heat transfer surfaces of heat exchangers are usually the primary sources of corrosion products. Even though the corrosion rate of the exposed material may be acceptable from the design life, the relatively large exposed surface area may contaminate the product.

21.5.2.2 Reliability

Some systems demand operating reliability. For example, fire water systems must have fast response and operation under severe conditions, including hydraulic surges and fire exposure. The latter is often cited as the reason for not using plastic or fiberglass piping in corrosive fire water systems. In other cases, reliability may not be very important, permitting the choice of less expensive materials of construction, for example, drain valves in low-pressure benign services. Some users have design standards and materials selection limitations that override materials selected. Galvanized steel is sometimes prohibited where a plant fire could cause liquid zinc to drip onto austenitic stainless steel piping, vessels, or equipment. Process licensors sometimes override selections made in accordance with a template. Normally, such recommendations are more conservative than those of the template. Nevertheless, process licensor recommendations are always subject to review.

In selecting a material, always remember the "common sense" of materials selection: (1) Ease of maintenance, replacement, and/or repairability of the component being evaluated. For example, for a design that calls for 100% spares (e.g., one pump running, one on standby), ease of maintenance or replacement may permit selection of less expensive, or even nonrepairable, materials such as cast iron pump internals. In some cases, repairability may influence selection, such as using cast steel, which is repairable by welding, instead of cast iron. (2) Plant experience is particularly useful for processes having a broad base of experience and for selecting materials for water services. In some cases, plant experience may indicate that a lower grade of material is adequate even if the available nomographs and corrosion charts indicate otherwise, for example, some high-temperature sulfur services in chemical and hydrocarbon plants. (3) Once the design, operating and upset conditions have been defined for a piping run or equipment item, it is normal to assume that all components are subject to the same exposure factors. Some components may, however, be exposed to less than the full set of template conditions. Blinds in a piping run are a good example. Such blinds are usually used only to isolate the piping run from equipment or other runs, for the purpose of hydrostatic testing or as part of a maintenance procedure. Most blinds will never experience normal operating or upset conditions. In applications in which the piping run requires special materials, the blinds often can be made of carbon steel. Care taken to differentiate among such service-related applications is often rewarded with significant cost savings.

21.5.3 ORGANIZING THE MATERIALS SELECTION PROCEDURE

For selecting a material of construction, a two-stage process is recommended. In the first stage, a template is used to select a material of construction. The first step is to use the minimum design temperature to choose a material of adequate toughness. The maximum design temperature is then used to modify the selection, if necessary, to obtain satisfactory resistance to corrosion or to thermal degradation.

If an upgrade is necessary, one must iterate to ensure that the upgrade material has adequate toughness at the minimum design temperature. In some cases, the upgrade candidates are one or more families of materials such as high alloys. In such cases, alternatives are evaluated before proceeding. The evaluation criteria depend to a large extent on the project objectives and constraints. Minimal cost, minimal maintenance, short schedule, extended design life, and consequences of a leak or rupture are typical objectives or constraints. A checklist is then used to determine if special requirements such as PWHT, hardness controls, external coating, etc., are necessary. The final step is to ensure that the template is properly filled out and that the template contains all necessary notes.

In the second stage, the materials selection information on the various templates is entered on a simplified process flow diagram (PFD), creating a MSD. The MSD is then reviewed for consistency. A checklist determines if factors such as excessive pressure drop must be addressed.

21.5.4 MATERIALS SELECTION PROCEDURE: EXCEPTIONS

The procedure used for materials selection is, for the most part, independent of the component. The primary exceptions are piping, pumps and fabricated equipment.

21.5.4.1 Piping

Materials selected for piping in mild to moderately corrosive services are sometimes less conservative than those for vessels, heat exchangers, tanks, and pumps in the same services. In this case, piping materials may be chosen on the basis of a shorter design life. This is usually justified because piping is easier to inspect, both online and offline. Also, piping is usually easier to replace and does not have the long lead-time problem often associated with fabricated equipment, vessels, etc.

21.5.4.2 Pumps

Guides used to select pumps include API Standard 610, "Centrifugal Pumps for General Refinery Service" [20]; ASME B73.1M, "Specification for Horizontal End Suction Centrifugal Pumps for Chemical Processes" [21]; and ASME B73.2M, "Specification for Vertical In-Line Centrifugal Pumps for Chemical Processes" [22]. API 610 provides considerable guidance on materials selection, but neither ASME B731.3 nor B73.2M do. Pump manufacturers usually provide guidelines for materials selection for specific process applications.

When selecting materials for pumps, first determine if the pump will be operating continuously or intermittently and whether or not it will be spared. Less expensive materials can sometimes be specified for pumps that will be spared because they can be withdrawn from service for repairs. Pumps expected to operate after being stagnant for long periods of time may require either upgraded materials or special lay-up procedures that control corrosion.

21.5.4.3 Fabricated Equipment

Fabricated equipment such as blowers, turbines, lube oil skids, etc. is usually ordered as "Manufacturer's Standard" for the intended service. Other "standard equipment," such as valves, is also usually supplied with off-the-shelf materials of construction. Most often, any deviations in materials proposed by the fabricator or supplier will exceed the minimum material of construction selected in accordance with the template. However, the proposed materials should be reviewed for compliance with template requirements, including any special fabrication specifications such as PWHT for process reasons, NACE MR0175 [2], and safety and design life requirements.

Equipment to be subjected to two or more sets of process conditions may require two or more templates. Examples include: (1) For shell and tube heat exchangers, use one template for the shellside and another for the tubeside. (2) For separators and distillation towers, use one template for the overhead section and another for the bottoms section. Towers with multiple feeds and/or draws usually require multiple templates.

21.5.5 GROUPING PROCESS REGIONS

When different sections of the process are exposed to essentially the same environment, they can be grouped together for materials selection. The following example is for a refinery; only four types of commodities are needed. (1) *Hydrogen and hydrogen mixtures.* Either pure hydrogen gas or commodities that are mixtures of hydrogen gas with other components, such as hydrocarbons. (2) *Hydrocarbons.* Commodities that are hydrocarbons or mixtures of hydrocarbons with other materials such as water, hydrogen, steam, etc. (3) *Noncorrosive gases.* Commodities such as nitrogen and dry plant air at ambient temperature. (4) *Other services.* Commodities such as amines, cooling water, fire water, and chemicals such as sulfur, caustics, acids, oxidants, etc.

21.5.6 MATERIALS SELECTION PROCEDURE

The first step is to consider the effect of the design temperatures.

21.5.6.1 Low Temperature

Ensure that the upset conditions listed in the template have been considered in establishing the minimum design temperature. A candidate material of construction must comply with minimum code requirements for low-temperature toughness. As the template information is subsequently reviewed for the effects of corrosion, crack-inducing agents, embrittlement, etc., make sure that the low-temperature toughness requirements are retained after making any changes in materials.

21.5.6.2 High Temperature

The preliminary material of construction should be checked for the risk of thermally induced degradation at the maximum design temperature: (1) Is it susceptible to thermally induced embrittlement or thermal degradation that could case failure during high-temperature service? For example, creep embrittlement and spheroidization or graphitization. (2) Will sustained operation at the maximum design temperature cause the material to be brittle at lower operating temperatures? Sigma-phase embrittlement of stainless steels is an example. (3) Will sustained operation at the maximum design temperature cause the material to be susceptible to corrosion at lower temperatures? Polythionic acid attack of stainless steels is an example.

The material is upgraded as necessary, always ensuring that it has adequate toughness at the minimum design temperature. The suitability of the upgraded material for the anticipated corrosion/degradation environment will be evaluated later. If sustained operation at the maximum design will cause the material to be brittle at lower operating temperatures, the design engineer should prepare appropriate operating instructions. Next, the candidate material is evaluated for the anticipated external environment, e.g., external chloride stress corrosion cracking, induced by a marine atmosphere. Next, the following considerations of corrosion, upset and transient conditions, and final review are organized to use the grouped process regions mentioned above.

21.5.6.3 Corrosion

The maximum design temperature should be used for assuring corrosion resistance, as follows:

Hydrogen and hydrogen mixtures: The Nelson curves [3] are used for all hydrogen services, including any mixtures of hydrogen with other commodities. Establish the minimum acceptable material, using the maximum design temperature or the maximum operating pressure plus a design margin (usually 25°F or 50°F [14°C or 28°C]). The material selected will be the minimum acceptable material for pressure containment.

Hydrocarbons: The McConomy [17] or Couper-Gorman [19] curves are used for streams containing either sulfur and/or H_2S for temperatures exceeding 500°F (260°C). When using the Couper-Gorman curves, choose the curve (either naphtha or gas oil diluent) most similar to the process stream hydrocarbon. If an 18Cr-8Ni SS is indicated, select a stabilized grade (e.g., Type 321 SS) if the design temperature exceeds 800°F (425°C). If naphthenic acid attack is probable, Type 316 SS or Type 317 SS ("L" grade if it is to be welded) or Types 316Ti or 316Cb SS (for plate and plate products) should be selected. In the "Notes" space on the template, indicate that the molybdenum content of the Type 316 grades shall not be <2.5 wt%. The following notes should be included for stainless steels: (1) for Types 316 and 316L, air and/or liquid water must be excluded during shut-down; or (2) the operating manual should include instructions regarding a neutralizing wash during the front end of a shut-down. Refer to NACE RP0170, "Protection of Austenitic Stainless Steel from Polythionic Acid Stress Corrosion Cracking During Shutdown of Refinery Equipment" [14].

If hydrogen is involved in the process, the material selected for resistance to sulfur must also meet the minimum requirements, based on the Nelson curves. In many cases, combined hydrogen-sulfur or hydrogen-hydrogen sulfide service require cladding or overlays, to provide resistance to the effects of high-temperature sulfidic corrosion and hydrogen attack, at an affordable cost.

Carbon steel will be selected for most hydrocarbon services, the major exception being for temperatures exceeding 800°F (425°C). Carbon steel is a viable candidate for higher temperature services under some circumstances: (1) For low-pressure applications such as decoking, carbon steel is sometimes used up to 1000°F (540°C) despite its tendency to graphitize above 800°F (425°C). (2) In some low-pressure services with intermittent excursions exceeding 1000°F (540°C), carbon steel is chosen, usually with the expectation of early replacement. (3) For higher pressure applications, the minimum acceptable material will be 1-1/4Cr-1/2Mo. Above 1050°F (565°C), a higher chromium alloy is required. (4) Internal refractory lining can extend the usefulness of carbon and low-steel alloys at temperatures above their oxidation threshold temperatures.

Noncorrosive gases: Carbon steel is the normal material of construction.

Other services: For chemicals, refer to Section 21.4 for guidance on materials selection for common chemicals. For other chemicals, refer to the available literature. For proprietary technologies, follow the guidance of the process licensor. Such guidance is always subject to review for compliance with safety and design life requirements. Plant experience and pilot plant testing, in some cases, will indicate suitable materials.

Extensive literature is available on various water services. Whenever possible, start with plant history, utilizing the same or similar water chemistries. Up to about 200°F (93°C), paint coatings may control corrosion in immersion service.

If corrosion concerns require a material upgrade, the upgraded material should have adequate toughness at the minimum design temperature. If the upgrade involves evaluating one or more families of materials, make sure that the job or project objectives and constraints, such as minimal cost or consequences of a leak or rupture, are considered. Complete the evaluation and choose a candidate material before proceeding.

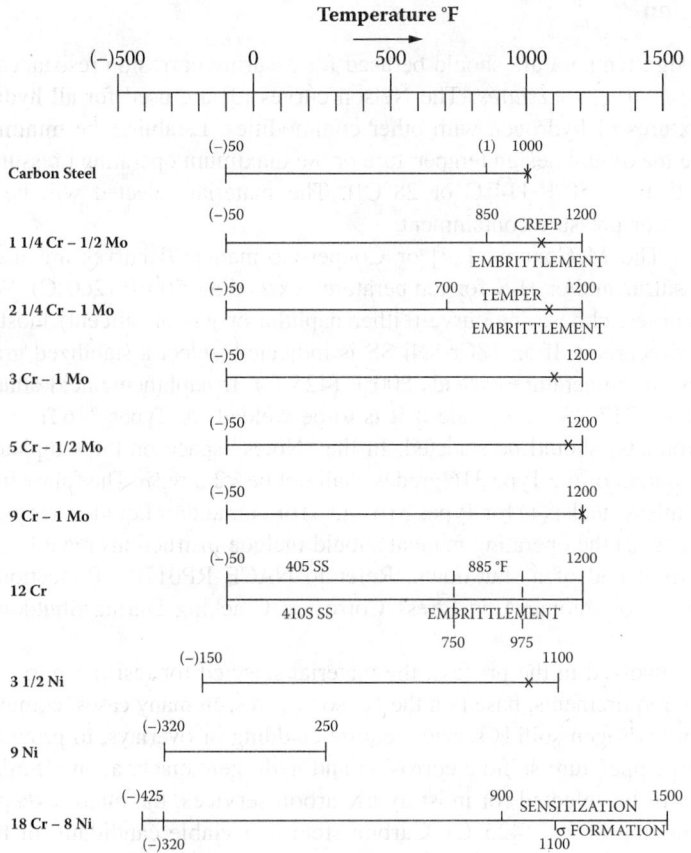

X Forms excessive scale above this temperature if exposed to air or steam.
(1) Weakens by graphitization of carbides at prolonged temperature above 850 °F.

FIGURE 21.12 Materials selection as a function of temperature. (The upper value of the temperature range represents the limit for which code-allowable stresses are available.) © Copyright by Marcel Dekker, Inc.; reprinted with permission [5].

21.5.6.4 Upset Conditions: Review

Several items on the template should be reviewed, as follows:

1. All material selections for conditions above 1000°F (540°C), to avoid oxidation, scaling, or spalling problems. Figure 21.12 is useful for evaluating such problems, including thermally induced embrittlement >700°F (370°C).
2. For carbon steel in hydrogen stress cracking services, determine if heavy section/sharp thickness gradients such as thick nozzles will be required. Such gradients may indicate a need for HIC-resistant plate and PWHT to prevent stress-oriented hydrogen-induced cracking. Determine where HIC-resistant plate is to be used for vessels, heat exchangers, and/or piping. Determine if normalizing and PWHT should be specified.
3. Review all carbon steel templates for which PWHT is indicated. Consider the maximum operating pressure. If it is <65 psia (0.45 MPa) or if the combined stress in tension is less than the 10% rule, PWHT may be unnecessary.
4. Corrosion allowance is probably unnecessary for pressures <65 psia (0.45 MPa), particularly for piping.

Selecting Materials of Construction (Steels and Other Metals)

5. Consider recommending "L" grades if nonstabilized austenitic stainless steels have been recommended and welding is part of the fabrication process. This will minimize potential problems caused by sensitization, and the "L" grades are easier to weld.
6. For fired heaters subject to creep problems, make sure that the tube metal temperature was considered in materials selection. In the absence of better information, assume the fireside temperature is 100°F (38°C) higher than the process temperature. (If tube-side fouling is anticipated [e.g., coke formation], assume the tube metal temperature is 150°F [85°C] higher than the process temperature.) If necessary, make a note on the template to ensure that creep is accommodated during design of heater tubes, in accordance with API 530 [23].
7. For all heat exchangers, evaluate the effect of leaks that permit mixing of the nonprocess side with the process side. In some processes, leaks require an immediate shut-down. In such cases, an upgrade in materials may be justified. In addition, evaluate the potential effects of a loss of flow on both the nonprocess and process sides. If such events are regarded as likely, an upgrade in materials may be required. For shell-and-tube heat exchangers, make sure that the front- and backside metallurgies and corrosion allowances of the tubesheet(s) are consistent with the channel and shellside processes, respectively.
8. Indicate in the "Notes" space of the template: (a) Where galvanic couples or alloy crevices are exposed to electrolytic corrosion, cathodic protection may be needed (for example, at tube-to-tubesheet joints in heat exchangers). (b) For austenitic stainless steels exposed to chloride-bearing external environments, a note requiring external coatings to prevent stress corrosion cracking or HAZ pitting may be needed. (c) Where an alternate process may provide better or more economical service, indicate the alternative. For example, steam tracing to keep water from condensing in a carbon steel line containing CO_2, versus using austenitic stainless steel as the material of construction. Such notes should be discussed with process engineers or designers during subsequent review(s) of the MSD. (d) For processes in which hydrogen stress cracking may occur, identify weld joints having significantly mismatched component thicknesses. Joints such as tray attachment welds and tube-to-tubesheet welds could have excessive weld metal and HAZ hardnesses and may need PWHT.

Indicate special corrosion-based requirements, such as

- Hardness limits in accordance with NACE MR0175 [2] and NACE RP0472 [13] for hydrogen stress cracking services.
- Paint/coating requirements for insulated piping and equipment or buried piping.
- Paint/coating and cathodic protection for submerged or buried structures or piping.
- Coatings/cathodic protection for the internal surfaces of tanks or vessels in corrosive service.

Complete the final steps in the materials selection process:

1. For each template, make sure that the metallurgy, valve trim, PWHT, and corrosion allowance requirements are filled in.
2. Use the "Notes" space on the template to indicate special requirements such as inspection categories, positive material identification, special flange face machined finishes, etc.
3. Specify PWHT in the template only if required by the process, for example, carbon steel in amine service. If PWHT is required, but not for process reasons, indicate "per code."

21.5.7 Materials Selection Diagram

Obtain a simplified process flow diagram; it need not contain detailed process data. All piping and equipment for which templates have been generated should be indicated, using the same stream numbers and equipment designations utilized on the respective templates. This PFD should show design temperatures and pressures for all piping and equipment. There should be sufficient room to create a "Notes" section, which includes any special materials selection requirements such as concerns about product contamination or operating reliability. Templates should be included as attachments.

The preliminary material selections are entered on the simplified PFD, along with corrosion allowances and any special features such as hardness controls or PWHT. Use arrows with legends, color codes, dotted or dashed lines, or some other method to identify the material/corrosion allowance/fabrication requirements for each pipe run and piece of equipment. In most cases, the "Notes" section simplifies the identification process, since many of the process services will refer to common notes. Examples include NACE hardness controls, PWHT, requirements for internal and/or external paint coatings, anodic or cathodic protection, and design-related measures such as a requirement for self-draining. This generates the materials selection diagram. The MSD is useful for several activities:

1. Compare the metallurgies, corrosion allowances, and fabrication requirements of the incoming and outgoing lines for each piece of equipment versus those of the equipment. This is a consistency check. Highlight any inconsistencies for later resolution.
2. If materials selection depends on corrosion control by process-related measures (such as chemical treatment), these should be indicated on the MSD. Indicate the intended injection points and the type of chemical to be injected. Examples include corrosion inhibitors, scale inhibitors, biocides, pH control chemicals, wash water, etc. Also indicate the location of proposed corrosion monitoring and sampling sites. If anodic or cathodic protection is to be part of the corrosion control design, the MSD or its "Notes" section should indicate the piping and/or equipment to be protected.
3. If degradation processes such as high-temperature embrittlement or autorefrigeration will affect operating procedures such as pressurization during start-up, indicate such limitations as general notes to the MSD.
4. Check for large pressure drops such as those that can occur at control valves. Determine if pressure drops will induce corrosive flashing. Flash spools or splash plates may be required downstream of the affected control valve.
5. Check for potential "mixing point" corrosion problems. A typical example of this type of problem occurs where a nominally noncorrosive electrolyte such as water is mixed with a dry acid gas or acid salt. The turbulence associated with mixing usually requires a "mixing" spool of corrosion-resistant material for at least 10 pipe diameters downstream of the mixing point. Neutralization may also be required.
6. Indicate convenient specification breaks. Specification breaks are points where the materials of construction change from one type to another.
7. Indicate the need for check valves, to protect upstream piping and equipment from damage by corrosive reverse flows. In most cases, an upgrade of the upstream piping is less expensive and more reliable than a corrosion-resistant alloy check valve.

If review of the MSD causes any changes in the templates, make sure that the changes are documented.

21.6 CONCLUSIONS

The procedure we have discussed is not a "cook book" process. Prior plant experience should always be an important part of the materials selection process. Common sense plays a key role in evaluating candidate materials. Time should be taken to determine the leeway provided by ease of maintenance, repairability, and sparing. In some cases, these considerations require upgrading of the materials or fabrication procedures (e.g., PWHT). In other cases, it may make sense to recommend changes in the design conditions or in the process itself, in order to specify a lower cost or more reliable material. Finally, unusual project conditions, such as an anticipated short design life or concern about product contamination, will often lead to nonconventional material selections.

Sometimes a user pursues a more conservative course and demands a more expensive material or material processing that exceeds the minimum requirements, for example, PWHT when it is not otherwise required by the construction code and is not justified by the process. Once the user understands the reasons for the recommended material or material processing, the issue becomes a management decision. Occasionally, a user will demand the use of a material that will not meet the minimum requirements. In this situation, safety and design life requirements as well as potential consequences should be reviewed and the results documented to the user.

Finally, new materials and materials technologies are continually coming into the market. In some cases, conditions will encourage their use on a prototype basis. However, experience with prototype technologies strongly suggests that it is best to let someone else be the first to try them. Correcting unanticipated problems can be difficult and costly.

REFERENCES

1. American Society of Mechanical Engineers. Current edition. *ASME Boiler and Pressure Vessel Code*. New York: ASME.
2. NACE International. *Sulfide Stress Cracking Resistant Metallic Materials for Oilfield Equipment*. NACE MR0175. Houston, TX: NACE International.
3. American Petroleum Institute. Current edition. *Steels for Hydrogen Service at Elevated Temperatures and Pressures in Petroleum Refineries and Chemical Plants*. API Publication No. 941. Washington, DC: API.
4. American Society of Mechanical Engineers. Current edition. *Process Piping*. ASME B31.3. New York: ASME.
5. Hansen, David A., and Robert P. Puyear. 1996. *Materials Selection for Hydrocarbon and Chemical Plants*. New York: Marcel Dekker.
6. American Society for Metals. 1983. *ASM Metals Reference Handbook*. 2nd ed. Metals Park, OH: ASM, 1–80.
7. Society of Automotive Engineers and American Society for Testing and Materials. Current edition. *Metals and Alloys in the Unified Numbering System*. Warrendale, PA: ASTM International.
8. American Society for Testing and Materials. Current edition. *Standard Practice for Numbering Metals and Alloys (UNS)*. ASTM E 527. Warrendale, PA: ASTM International.
9. Steel Founders Society of America. Current edition. *Steel Casting Handbook*. Supplement 2. Des Plaines, IL: SFSA.
10. American Petroleum Institute. Current edition. S*pecification for Line Pipe*. API Specification 5L. Washington, DC: API.
11. Schweitzer, Philip A., ed. 1996. *Corrosion Engineering Handbook*. New York: Marcel Dekker.
12. Caplan, F. 1975. "Is Your Water Scaling or Corrosive?" *Chemical Engineering* September 1: 29.
13. NACE International. Current edition. *Methods and Controls to Prevent In-Service Environmental Cracking of Carbon Steel Weldments in Corrosive Petroleum Refinery Environments*. NACE RP0472. Houston, TX: NACE International.
14. NACE International. Current edition. *Protection of Austenitic Stainless Steels and Other Austenitic Alloys from Polythionic Stress Corrosion Cracking During Shutdown of Refinery Equipment*. NACE RP0170. Houston, TX: NACE International.
15. Beggs, D.V., and R.W. Howe. 1993. *Effects of Welding and Thermal Stabilization on the Sensitization and Polythionic Acid Stress Corrosion Cracking of Heat and Corrosion-Resistant Alloys*. CORROSION/93, Paper No. 541. Houston, TX: NACE International.
16. Schillmoller, C.M. "Solving High-Temperature Problems in Oil Refineries and Petrochemical Plants." *Chemical Engineering* January 6, 1986: 83–87.
17. McConomy, H.F. 1963. *High Temperature Sulfidic Corrosion in Hydrogen Free Environment*. API Subcommittee on Corrosion, May 12.

18. Gutzeit, J. 1986. In *High Temperature Sulfidic Corrosion of Steels, Process Industries Corrosion-Theory and Practice*, eds. B. J. Moniz and W. I. Pollock, 367–372. Houston, TX: NACE International.
19. Couper, A.S., and J.W. Gorman. "Computer Correlations to Estimate High Temperature H_2S Corrosion in Refinery Streams." *Materials Protection and Performance* 10(1) (1971): 31–37.
20. American Petroleum Institute. Current edition. *Centrifugal Pumps for General Refinery Service*. API Standard 610. Washington, DC: API.
21. American Society of Mechanical Engineers. Current edition. *Specification for Horizontal End Suction Centrifugal Pumps for Chemical Processes*. ASME B73.1M. New York: ASME.
22. American Society of Mechanical Engineers. Current edition. *Specification for Vertical In-Line Centrifugal Pumps for Chemical Processes*. ASME B73.2M. New York: ASME.
23. American Petroleum Institute. Current edition. *Recommended Practice for Calculation of Heater Tube Thickness in Petroleum Refineries*. API Recommended Practice 530. Washington, DC: API.

22 Solid/Liquid Separation

Frank M. Tiller, Wenping Li, and Wu Chen

CONTENTS

22.1 Overview of Solid/Liquid Separation 1599
    22.1.1 Introduction 1599
    22.1.2 Particles 1599
    22.1.3 Classification of SLS Operations by Particle Size 1599
    22.1.4 Solid/Liquid Separation Operations 1599
        22.1.4.1 Sedimentation 1599
        22.1.4.2 Straining 1600
        22.1.4.3 Cake Filtration 1600
        22.1.4.4 Cross-Flow Filtration 1600
        22.1.4.5 Deep-Bed Filtration 1601
        22.1.4.6 Centrifugation 1601
        22.1.4.7 Washing 1601
        22.1.4.8 Deliquoring 1601
        22.1.4.9 Membrane Filtration 1601
        22.1.4.10 Hydrocyclone Separation 1601
        22.1.4.11 Flotation 1601
    22.1.5 Stages of Solid/Liquid Separation 1602
22.2 Pretreatment—Coagulation and Flocculation 1603
    22.2.1 Interparticle Forces and Zeta Potential 1604
    22.2.2 Coagulation and Flocculation 1604
    22.2.3 Laboratory Tests 1607
22.3 Pretreatment—Filter Aids 1608
    22.3.1 Diatomaceous Earth 1611
    22.3.2 Perlite 1612
    22.3.3 Other Filter Aids 1613
22.4 Filtration Fundamentals 1613
    22.4.1 Theory 1613
    22.4.2 Material Balance 1614
    22.4.3 Volume vs. Time 1616
    22.4.4 Data Analysis 1616
    22.4.5 Deviations From Parabolic Theory 1620
    22.4.6 Filtration of Super-Compactible Materials 1623
        22.4.6.1 Effective Pressure 1623
        22.4.6.2 Darcy's Law 1625
        22.4.6.3 Empirical Constitutive Equations 1626
        22.4.6.4 Integration of the Darcy Equation 1626

22.4.7 Strange Behavior of Super-Compactible Materials 1627
22.4.8 Critical Pressure Drop 1628
22.5 Solid/Liquid Separation Equipment and Operation 1629
    22.5.1 Batch Pressure Filters 1630
        22.5.1.1 Pressure Nutsches 1630
        22.5.1.2 Leaf Filter 1631
        22.5.1.3 Candle Filter 1632
        22.5.1.4 Bag Filters 1633
        22.5.1.5 Cartridge Filters 1634
        22.5.1.6 Filter Presses 1634
    22.5.2 Continuous Filters 1636
        22.5.2.1 Rotary Drum and Disc Filters 1636
        22.5.2.2 Horizontal Belt Filter 1639
        22.5.2.3 Indexing Belt Filter 1639
    22.5.3 Deep-Bed Filter 1640
    22.5.4 Cross-Flow Filters 1641
    22.5.5 Membrane Filters 1641
    22.5.6 Thickening and Clarification 1643
    22.5.7 Centrifugation 1645
    22.5.8 Hydrocyclone 1646
    22.5.9 Expression Equipment 1647
22.6 Cake Washing 1648
    22.6.1 Displaced Washing and Repulping 1648
    22.6.2 Terms and Definitions for Displacement Washing 1648
    22.6.3 Washing Curves 1649
22.7 Laboratory SLS Test 1650
    22.7.1 Laboratory SLS Test 1650
    22.7.2 Problems in Laboratory Tests 1653
        22.7.2.1 Samples 1653
        22.7.2.2 Adverse Effect of Sedimentation on Filtration Experiments 1653
        22.7.2.3 Wall Friction in C-P Cell and Small-Scale Cells 1653
        22.7.2.4 Test Numbers 1653
22.8 SLS System Design 1654
    22.8.1 Selection of SLS Equipment 1654
        22.8.1.1 Specifications of a SLS System 1654
        22.8.1.2 Selection of Equipment and Process 1654
        22.8.1.3 Comparison of SLS Equipment 1656
    22.8.2 Decision Network for Design for Cake Filtration Systems 1656
    22.8.3 Selection of Filter Media 1658
        22.8.3.1 Introduction 1658
        22.8.3.2 Woven Filter Media 1659
        22.8.3.3 Selection of Filter Media 1659
    22.8.4 Selection of Pumps in Filtration Operation 1660
    22.8.5 Cycle Analysis in Selection of Filter Aids 1661
    22.8.6 Scale Up 1663
Table/Figure Source Lines 1664
References 1665

22.1 OVERVIEW OF SOLID/LIQUID SEPARATION

22.1.1 Introduction

Solid/liquid separation (SLS) is encountered in many processes ranging from raw material purification, products separation, to waste management. SLS operations include straining with screens; cake, deep-bed, and membrane filtration; expression; gravitational sedimentation; sedimenting or filtering centrifugation; hydrocycloning; and flotation. Due to the complex nature of fluid/particle systems and a general lack of fundamental training, selection of proper SLS equipment and optimum operating conditions are always challenges for engineers.

In designing SLS processes, it is essential to have information on filtration and settling properties such as settling velocities, cake permeabilities, cake resistances, etc., and other properties including particle size, particle size distribution, slurry concentration, etc. In general, knowledge and experience reside in equipment suppliers, and users frequently furnish vendors with slurry samples for testing and recommendations. Potential problems include aging of slurries and changes in characteristics while being transported and handled by laboratory technicians. In addition, plant process variation and improper sampling (Svarovsky 1990) may lead to unrepresentative samples. Users should carefully evaluate all steps involved in sampling, shipping, testing, and analysis. Another challenge users face is appropriate decision making for equipment selection and system design among the bewildering variety of SLS equipment and techniques provided by manufacturers.

22.1.2 Particles

In the area of SLS, liquid can be water, oil, solvent, or solutions; and solids are particles with sizes starting in the molecular range and continuing up to millimeters. In general, particles have odd shapes. "Size" or "dimensions" cannot be answered with a single arithmetic number. Nevertheless, it is useful to have some idea of the size of apertures in a screen or membrane that permit passage of a particle. In the size range of importance in SLS, a micrometer (μm; 1 μm = 0.001 mm), which is often called a micron, is frequently employed. The limit of visibility of particles is frequently said to be 40 to 50 microns. With an increased rate of development of particle technology, emphasis has shifted to smaller particles; and the nanometer (1 nm = 0.001 μm; 1 angstrom = 0.1 nm) has come into prominence. The nanometer is used extensively in electronic circuits, nanotubes, thin films, foils, etc.

22.1.3 Classification of SLS Operations by Particle Size

Separation operations are frequently classified according to the size of particles. No precise size definitions exist for terms such as ultra-, micro-, nano-, colloidal, and fine particle filtration. Different authors use different nomenclature in classifying separations in terms of "size." A simplified classification is shown in Table 22.1.

22.1.4 Solid/Liquid Separation Operations

22.1.4.1 Sedimentation

Separation involving sedimentation is dependent upon settling velocity, which requires a difference in density between solid particles and the suspending liquid. Gravitational sedimentation operations are divided into clarification and thickening. Clarification involves dilute suspensions and frequently has the objective of liquid recovery. Thickening refers to solid recovery by forming more concentrated slurries. Particle size, liquid and particle densities, and liquid viscosity are important factors in sedimentation processes.

TABLE 22.1
Terminology in Relation to Size Range

Size (μm)	Terminology	Examples	Solid/Liquid Separation Method
0–0.001	Ionic	Aqueous salts	Nanofiltration, reverse osmosis, chromatography
0.001–0.1	Macro-molecular	Virus, colloids	Ultrafiltraton
0.1–10	Fine particle	Pigments, clay, bacteria	Microfiltration, cake filtration, deep-bed filtration, centrifugation
10–100	Medium size	Bacteria, yeast, fibers, fine sand	Cake filtration, gravity sedimentation, floatation, cycloning
100–1000	Large	Coarse sand	Screens, shakers

22.1.4.2 Straining

Screens with openings smaller than the particles are used to remove the particles (Figure 22.1a). Screens, strainers, and bag filters belong to this category. Bar screens and trash rakes are employed for removal of large debris. Traveling, passive, vibrating, and rotating screens are employed. Sieves have openings ranging from 25 μm (400 mesh) to 8 mm (2 1/2 mesh). Larger sizes are specified by the actual size of the aperture. Shale shakers are screens used for separating cuttings from drilling fluids in the production of petroleum. Bag filters represent the simplest and most commonly used type.

22.1.4.3 Cake Filtration

In cake filtration (Figure 22.1b), solids and liquid are separated by filter medium, which retains the solids as a cake and permits the liquid to pass through under pressure, vacuum, or centrifugal forces. At the start of filtration, some particles may pass through the medium, and the filtrate may be turbid. Once the cake is formed, it becomes the primary filter medium and particles finer than the openings of the medium can be separated.

22.1.4.4 Cross-Flow Filtration

Cross-flow filtration (Figure 22.1c) is principally used on difficult to filter materials. When a cake of low permeability forms, filtration rates can drop to unacceptable values. The cake thickness can be reduced by introducing shear forces at the cake surface. Low-shear cross-flow operation is accomplished by pumping slurries through channels with porous walls or through porous tubes. Rotating turbines or mixers are used in high-shear devices. Cross-flow filtration is essentially a filter-thickening operation in which the suspension thickens as it flows through the unit. Ultimately, the concentration reaches a point where pumping or mixing is no longer feasible. Due to the fine

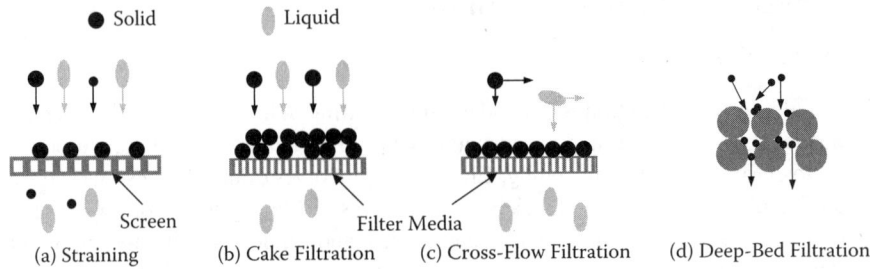

FIGURE 22.1 Schematic mechanisms of three types of filtration.

particles involved, the cross-flow filtration mechanism is almost exclusively used in membrane filtration applications.

22.1.4.5 Deep-Bed Filtration

In deep-bed filtration (Figure 22.1d), particles are caught inside rather than on the surface of the filter medium. Examples of deep-bed filters are granular beds and some cartridge filters. These filters are used for dilute suspensions (<100 ppm) which contain fine particles that are not easy to be removed by sedimentation or cake filtration. They form highly resistant cakes and are preferably removed in the pores of granular beds and cartridge filters.

22.1.4.6 Centrifugation

Centrifugation increases the separating driving force by developing centrifugal forces on particles. A difference in density between the liquid and the particles is also a prerequisite. Centrifuges are employed in filtering and sedimenting modes. The terms *perforated bowl* and *solid bowl* differentiate the two types of operations.

22.1.4.7 Washing

Following cake or sediment formation, removal of soluble materials by washing is a common operation. Washing can be accomplished by reslurrying or washing involving displacement and diffusion. A clean liquid is used in this step.

22.1.4.8 Deliquoring

Deliquoring is accomplished by expression or blowing with gas. Expression operations include squeezing with stationary diaphragms, moving belts, rollers, or screws.

22.1.4.9 Membranes Filtration

Membranes are used for a wide variety of separations. A membrane serves as a barrier to some particles while allowing others to selectively pass through. The pore size, shape, and electrostatic surface charge are fundamental to particle removal. Synthetic polymers (cellulose acetate, polyamides, etc.) and inorganic materials (ceramics, metals) are generally the principal materials of construction. Membranes may be formed with symmetric or asymmetric pores, or formed as composites of ultra thin layers attached to coarser support material. Reverse osmosis, nanofiltration, ultrafiltration, and microfiltration relate to separation of ions, macromolecules, and particles in the 0.001 to 10 µm range (Rushton et al. 1996).

22.1.4.10 Hydrocyclone Separation

Hydrocyclones (see Figure 22.55) are closely related to centrifuges in that centrifugal forces effect the separation of particles. Rotational motion is effected by bringing the slurry radially into the upper periphery of the cyclone at high velocity. Solids are thrown out to the wall, flow down the inclined walls, and exit at the bottom. In general, hydrocyclones operate as classifiers with large particles in the underflow and small particles in the overflow.

22.1.4.11 Flotation

Flotation is commonly employed for beneficiation of minerals. The valuable components are separated from the waste or gangue by preferential floating of one of the components to the top while the other sinks to the bottom. Organic collectors preferentially wet one of the materials and

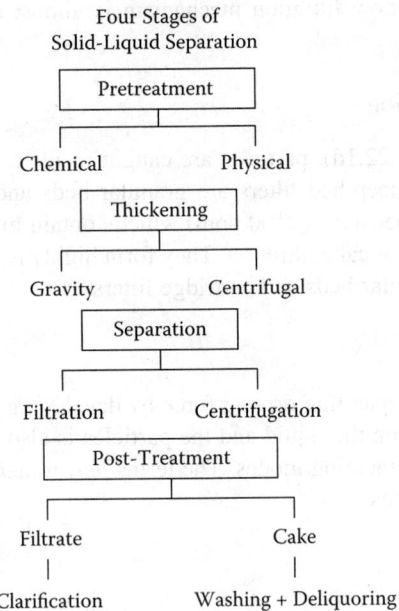

FIGURE 22.2 Stages of solid/liquid separation.

render it hydrophobic. Frothers are added to the vigorously agitated mixture. Air bubbles attach to the collector-mineral particles, which then rise and are removed along with the froth or foam.

22.1.5 Stages of Solid/Liquid Separation

Separation of solids from liquids usually consists of four stages including pretreatment, thickening, separation, and post-treatment, as shown in Figure 22.2.

An overall perspective is essential to solution of SLS problems which are basically complex. Engineers must consider the four steps and their interrelationships in approaching system design or improving existing operations.

Frequently, slurries settle slowly, and are difficult to filter or centrifuge. Pretreatment techniques such as coagulation and flocculation or filter aid addition can be employed to improve separation characteristics. When a suspension consists of fine particles that are difficult to separate, increasing the effective size through aggregation or flocculation leads to improved separability. Aggregated particles are called flocs. A floc is illustrated in Figure 22.3. Degradation of flocs by shear forces is a problem in coagulation and flocculation. It is necessary to study the effects of pumping and

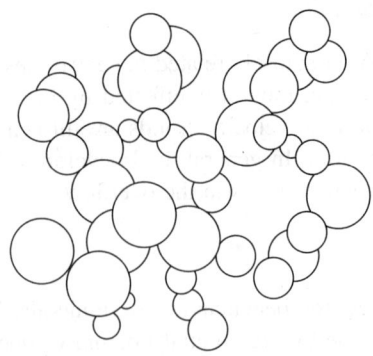

FIGURE 22.3 Arrangement of flocculated particles.

mixing on the possible breakup of the aggregated particles. Filter aids are inert material that can be used to increase the cake porosity and decrease the cake resistance so as to improve the filterbility of the process material.

Thickening frequently follows pretreatment and can be achieved by gravity, centrifugal, or cross-flow filtration. For large-scale operations, removal of liquid is generally accomplished more economically in gravity thickeners as compared to filters and centrifuges. The core separation step can be accomplished with filters or centrifuges, which are necessary for producing cakes with a high percentage of solids. Following separation, both cake solids and filtrate may require further processing in post-treatment operations. Various expression operations are used to reduce the liquid content. Filtrates from filters, centrates from centrifuges, and the overflow from thickeners generally contain suspended solids. Further clarification using cartridges, membranes, or deep beds provides a means for meeting specifications on the product liquid. To reduce the mother liquor in the cake, washing can be applied. Reslurrying wash is more effective, but more wash liquid and additional separation steps are required. Cake washing by displacement followed by diffusion, although less effective, is more convenient.

The following summarizes the elements of the four stages:

1. Pretreatment
   a. Aggregation of particles through use of
      (1) Bi- and tri-valent metallic ions (coagulation)
      (2) Polyelectrolytes (flocculation)
      (3) pH control
   b. Physical processes
      (1) Control of crystal size
      (2) Adjusting temperature
   c. Addition of filter aids to increase cake permeability
2. Thickening and clarification
   a. Gravity sedimentation
   b. Cross-flow filtration (to minimize cake thickness)
      (1) Low-shear operation with flow through porous tubes and parallel to cake surface
      (2) High-shear operation with rotating turbines
3. Separation (production of cakes or sediments)
   a. Pressure and vacuum filtration
   b. Centrifugal filtration
   c. Centrifugal sedimentation (decanters)
4. Post-treatment
   a. Washing by displacement and dilution reslurrying
   b. Deliquoring (reducing liquid content of cakes)
      (1) Expression (cake squeezing)
      (2) Utilizing of high pump pressure
      (3) Drainage under vacuum
      (4) Blowing with gas
      (5) Drainage in filtering centrifuge under high centrifugal forces
      (6) Reducing differential between scroll and centrifuge bowl rotational rates to increase detention time of cake
   c. Clarification of filtrate, centrale, or thickener overflow with flow-through membranes, cartridges, or granular beds

22.2 PRETREATMENT—COAGULATION AND FLOCCULATION

Suspensions generally contain a proportion of particles in the colloidal range. Stable colloidal systems with small particles do not form aggregates when there are large, mutually repulsive

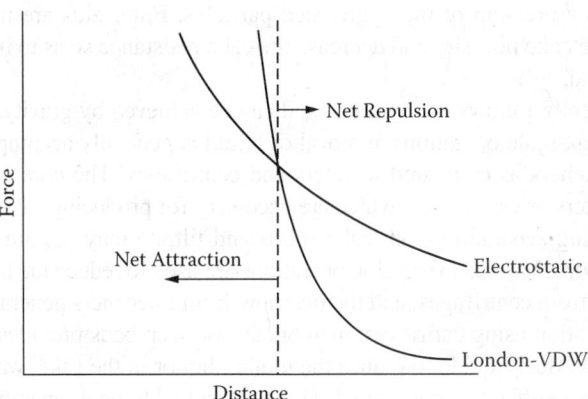

FIGURE 22.4 Interparticle forces.

electrostatic charges. Pretreatment operations aim at neutralization of the repulsive force and stimulation of particle aggregation.

22.2.1 Interparticle Forces and Zeta Potential

Most particles suspended in water and wastewater, e.g., clays, silica, hydrated metal oxides, paper fibers, biological cells, etc., possess negative surface charges in the neutral pH range. The repulsive electrostatic forces due to the electrostatic surface charge and the attractive London–van der Waals (LVDW) force represent the two principal forces among particles. In Figure 22.4, the electrostatic and LVDW forces are shown as a function of the distance r between particles. Repulsive electrostatic forces vary with the inverse square (r^{-2}) of the distance between particles. Attractive LVDW forces vary inversely with the sixth power (r^{-6}) of the distance. In order for particles to be coagulated, the LVDW forces must be greater than the electrostatic forces. In Figure 22.4, as the distance r decreases, the LVDW force increases very rapidly; and at the point, where the two forces intersect, the net force is zero. At a shorter distance, the net force is attractive; and the particles form aggregates that increase the ease of separation. The magnitude of electrostatic force of colloidal particles is evaluated by the term "zeta potential" explained as follows:

A particle in a suspension with negative electrostatic surface charge is shown in Figure 22.5. Positive ions in the solution are attracted to the negatively charged surface where they may be strongly adsorbed. The adsorbed layer remains rigidly attached and forms what is known as the Stern layer with a thickness equal to δ. Outside the Stern layer is a diffuse layer in which negative ions outnumber positive ions and balance the excess in the Stern layer. The electrostatic potential at the particle surface A decreases through the Stern layer along AB and along BC in the diffuse layer and reaches zero at point C, where the bulk of the solution is encountered. The Stern layer and diffuse layer are called double layers. They are important in determining forces between colloidal particles. Zeta potential BC is the difference between the charge at the edge of the Stern layer and the bulk of the suspending liquid, as shown in Figure 22.5a,b. High zeta potential leads to stability of colloidal systems. The relationship of colloid stability to values of zeta potential is shown in Table 22.2.

Zeta potential can be measured by tracking the rate of a negatively or positively charged particle moving across an electric field. The most common commercial instrument is the zeta meter. It can be expensive and requires some operational experiences to get reliable testing results.

22.2.2 Coagulation and Flocculation

Coagulation generally refers to aggregation by charge neutralization with inorganic salts, and flocculation is brought about by large molecular weight polyelectrolytes. Coagulation and floccu-

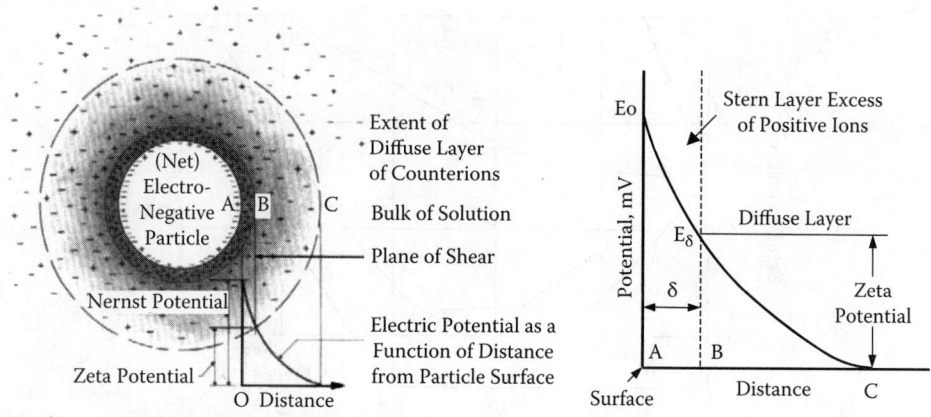

FIGURE 22.5 a, b: Double layer.

TABLE 22.2
Stability of Suspensions with Relation to Zeta Potential

Stability Characteristics	Avg. Zeta Potential (mV)
Maximum agglomeration and precipitation	0 to +3
Range of strong agglomeration and precipitation	+5 to −5
Threshold of agglomeration	−10 to −15
Threshold of delicate dispersion	−16 to −30
Moderate stability	−31 to −40
Fairly good stability	−41 to −60
Very good stability	−61 to −80
Extremely good stability	−81 to −100
From Riddick (1968)	

lation processes can be evaluated and controlled by zeta potential in relation to the stability of suspensions, as shown in Table 22.2.

According to Table 22.2, zeta potential values less negative than −15 mV usually represent the onset of agglomeration. A plateau region marking the threshold of either coagulation or dispersion exists from about −14 mV to −30 mV. Values more electronegative than −30 mV generally represent sufficient mutual repulsion to result in stability (i.e., no agglomeration). When coagulation is the desired end, the zeta potential must be rendered less electronegative in order to reduce the forces of mutual repulsion. Gentle agitation will then create impingements of particles that enter the region where the short range LVDW forces can furnish the attraction required to cause agglomeration. Maximum coagulation usually occurs in the zeta potential range of 0 to +3 mV. One example of the effect of type and dosages of metal coagulants on zeta potential according to Riddick (1968) is given in Figure 22.6.

Various salts of aluminum, calcium, and iron, and a few acids and bases used as coagulants are listed in Table 22.3.

As can be seen in Figure 22.6, there is a distinct difference in the effect of the coagulants depending on the valence of the ions in accord with the Schultz-Hardy rule, which states that the salt concentration required to neutralize an electrostatic charge is inversely proportional to the sixth

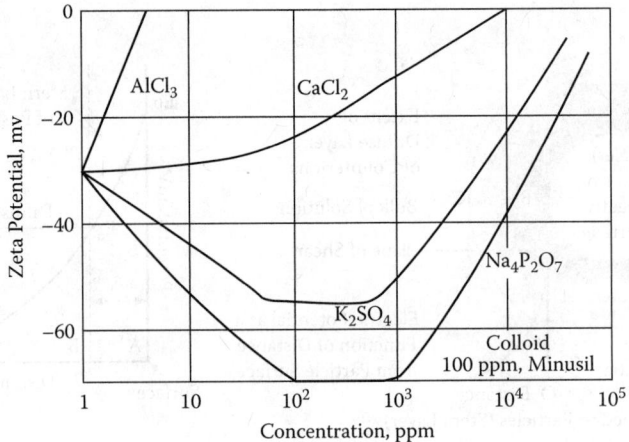

FIGURE 22.6 An example of the impact of type and dosage of electrolyte on zeta potential.

TABLE 22.3
Inorganic Coagulants

Chemical Name	Common Name	Formula
Acids		
Hydrochloric acid		HCl
Sulfuric acid		H_2SO_4
Bases		
Calcium hydroxide	Slaked lime	$Ca(OH)_2$
Sodium hydroxide		NaOH
Salts		
Aluminum chloride		$AlCl_3$
Aluminum chlorohydrate		$Al(OH)_2Cl$
Aluminum sulfate		$Al_2(SO_4)_3$
Calcium chloride		$CaCl_2$
Calcium oxide	Quick lime	CaO
Ferrous chloride		$FeCl_2$
Ferric chloride		$FeCl_3$
Ferrous sulfate	Copperas	$FeSO_4$
Ferric sulfate		$Fe_2(SO_4)_3$
Sodium aluminate	Soda alum	$NaAlO_2$

power of the valence. Although not precisely correct, the rule correctly predicts that the quantity of $AlCl_3$ to achieve zero potential is much less than that of $CaCl_2$. In general, Al^{+3} and Fe^{+3} have proven to be more effective in the coagulation of negatively charged particles. On the other hand, the pyrophospate ion $P_2O_7^{-4}$ is highly effective in increasing the negative nature of the particles and leads to a highly stable colloidal suspension. This property makes phosphate ions useful components of detergents.

In addition to coagulation with electrolytes or polyelectrolyte by controlling the zeta potential, agglomeration is usually accomplished by employing a long-chain polymer to provide mechanical bridging, which is called flocculation. Coagulation and flocculation can occur simultaneously or with some degree of overlapping.

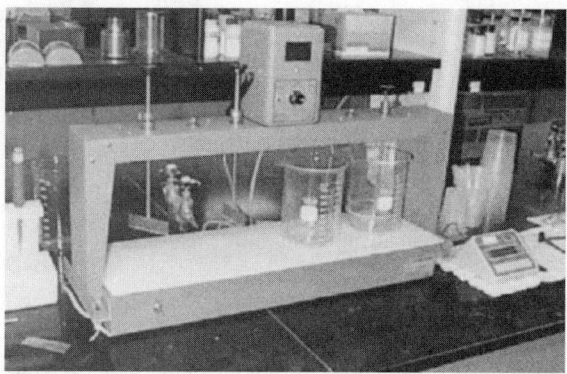

FIGURE 22.7 Jar test apparatus.

There are natural flocculants like Guar gums or starches, but synthetic polyacrylamide derivatives are the predominant flocculants used today. They are commonly referred to as "polymers" in the industry. Polymer manufacturers apply their proprietary recipe in attaching different functional groups to the polyacrylamide backbone to make cationic, anionic, or neutral flocculants. Type of charge, charge intensity, and the molecular weight are used to categorize flocculants. Since charge neutralization is not the primary mechanism for flocculation, neutral or even anionic flocculants may work well for negatively charged particles.

The commercial polymers are in the form of powder, emulsion, or water solution. They need to be diluted before being used to floc the particles. Emulsion polymers require vigorous mixing to "invert" the mixture from water emulsion in polymer to polymer solution in water. Some polymers will be wasted if mixing is not sufficient. For the powder form polymers, medium to longer agitation time is required to completely dissolve the polymer. If the polymer is in a water solution, light mixing is sufficient. Excess mixing can cause the breakage of the polymer chain and loss of polymer activity. Therefore, it is important to consult with the polymer suppliers about the proper mixing methods. After dilution, the polymer solution should be allowed to sit for 1 to 2 hours. This is for the polymer chain to "stretch out" so the full bridging effect of polymers can be utilized.

For laboratory usage, normally a 0.5% stock solution is prepared and should be discarded after one week due to the reduction in polymer activities. Further dilution to 0.05% is necessary before adding to the slurry. The shelf-life for the 0.05% solution is about 1 to 2 days, so fresh samples need to be prepared daily. For plant applications, a polymer solution of 0.5% should be prepared in the storage tank and added to the slurry in a mixing tank or by in-line injection with sufficient mixing.

22.2.3 Laboratory Tests

The efficiency of the coagulation-flocculation process depends on the type and dosage of the coagulant or flocculant, suspension pH, sequence of chemical addition, intensity and duration of mixing, retention time, etc. (Bratby 1980). Due to the complexity of the SLS process, lab testing is usually necessary to determine an efficient and economic coagulation and flocculation operation. A procedure for lab scale testing of polymer selection for wastewater and sludge treatments was developed by the Water Environment Research Foundation (Dentel 1993). A standard jar test apparatus is frequently used for assessing the performance of coagulation and flocculation processes (Figure 22.7). It consists of a rack of stirrers, driven by one motor, under which 600-mL, or preferably 1-L, glass beakers are arranged (Bratby 1980). Before the jar test, criteria for evaluation of coagulation or flocculation must be determined. The most common criteria to evaluate the performance of coagulation or flocculation process are listed as follows:

1. Settling velocity, supernatant turbidity, and volume fraction of solids of the sediment (average solidosity)
2. Filtration rate, specific cake resistance, CST (capillary suction time; slurry is poured into a small cylinder placed on a filter paper [see Figure 22.62a]. A cake is formed and filtrate is sucked out radially. The time required to pass between two fixed radii is termed CST), filtrate turbidity, and cake solidosity
3. Determining the time from coagulant addition to the first appearance of a visible floc
4. Visually recording and comparing flocs as they are formed
5. Floc density measurements
6. Zeta potential

Criteria (1) can be evaluated by sedimentation experiments and (2) can be tested by vacuum, pressure filtration, or CST experiments. Filtration and CST experiments will be discussed in Section 22.7.

A jar test procedure based on the bench stirrer method is basically as follows:

1. Choose the desired coagulant and/or flocculant and the desired dosage.
2. Prepare the coagulant and flocculant samples according to the guidelines in the previous section.
3. Measure and pour slurry samples into each beaker (Figure 22.7).
4. Mix the coagulant or flocculant at a chosen dosage with slurry: place the beakers with slurry under the bench stirrer with paddles centered in the beakers, approximately 1 in. from the surface of the sludge sample. Use graduated syringes or graduated cylinders to add various dosages of chemicals to each of the beakers. Mix the chemicals and slurry for 30 to 60 sec, with a paddle speed of 200 rpm for thorough chemical/particles mixing. Then reduce the speed to 30 rpm for sufficient time (1 to 15 min) to allow the flocs to grow.
5. Results evaluation based on specific criteria.

A jar test evaluation based on sedimentation rate as a function of coagulant or flocculant concentration is shown in Figure 22.8. Intervening time between coagulant and flocculant additions is a third variable. Results of CST (Dentel 1993) as a function of polymer dosage from CST test are shown in Figure 22.9. Minimum dosage of coagulants or flocculants can be determined from the two plots.

In wastewater and sludge treatment plants, the lab-scale jar test is used to predict full-scale performance of additives. Testing in the full-scale plant under controlled conditions is ideal but is seldom done. Tiller undertook full-scale tests in urban wastewater treatment plants using belt presses or centrifuges (Tiller and Li 2001b, 2002). Normally, the optimization of polymer usage is determined by the dryness of the dewatered sludge and overall economics are often neglected. In Tiller's test, an economic analysis based on the cost of the polymer and the cost of gas to dry solids is shown in Figure 22.10. Although increased polymer dosage led to increased cake solids as shown in Figure 22.11, the cost of additional polymer exceeded the cost of drying the sludge. Therefore, the total cost increases as more polymer is used.

22.3 PRETREATMENT—FILTER AIDS

"Filter aids" are inert materials that can be used in filtration pretreatment. The principle characteristics of filter aids are (1) increasing cake permeability, (2) increasing cake porosity, and (3) increasing cake rigidity (decreasing cake compactibility).

There are two objectives related to the addition of filter aids. One is to form a layer of second medium that protects the basic medium (filter cloth) of the system. This is commonly referred to

Solid/Liquid Separation

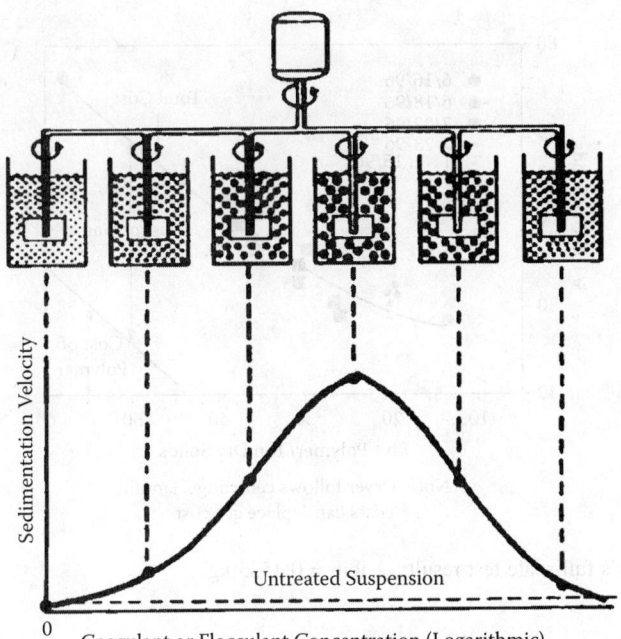

FIGURE 22.8 Jar test evaluation based on sedimentation.

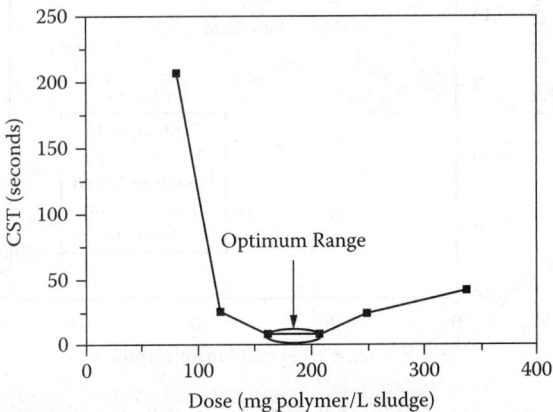

FIGURE 22.9 Optimum polymer dose determination by CST experiment.

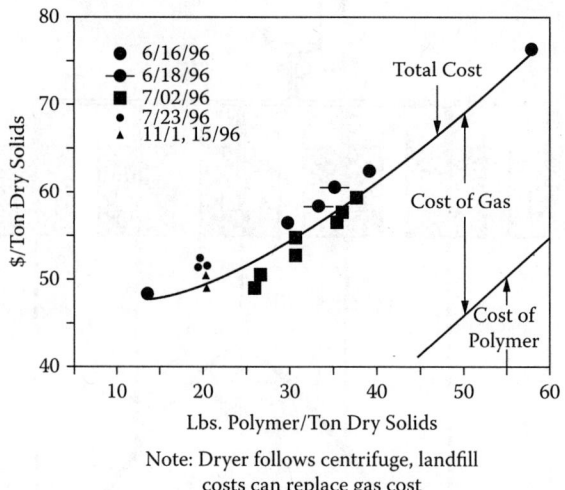

FIGURE 22.10 Tiller's full-scale test results, 1 lbm = 0.4535 kg.

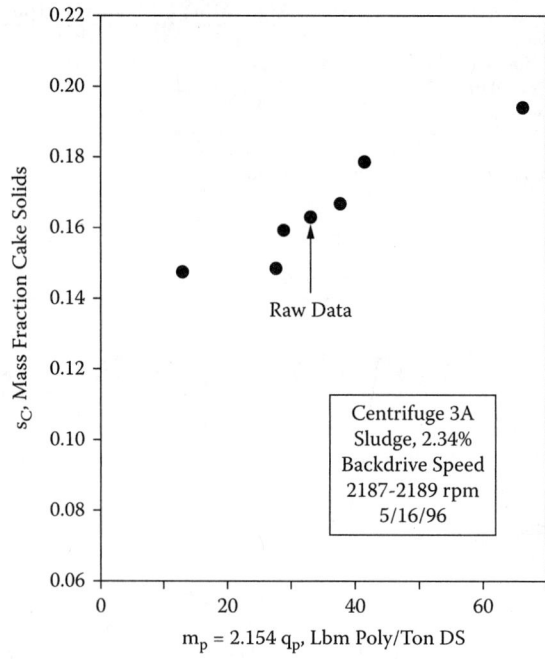

FIGURE 22.11 Cake water content against polymer dosages in Tiller's full-scale test, 1 lbm = 0.4535 kg.

Solid/Liquid Separation

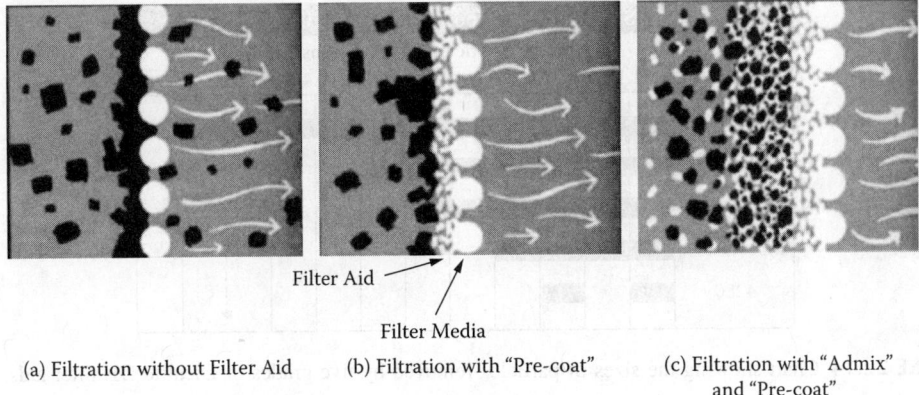

(a) Filtration without Filter Aid   (b) Filtration with "Pre-coat"   (c) Filtration with "Admix" and "Pre-coat"

FIGURE 22.12 Mechanism of filtration with filter aids.

as a "precoat." When the slurry is very dilute, a thick precoat may function as the medium in a deep-bed filtration. The second objective of filter aids is to improve the flow rate. In such cases, the filter aid is termed as "admix" or "body feed." Filtration without precoat, with precoat, and with precoat and admix is shown in Figure 22.12 (Eagle-Pitcher Minerals, 1970).

Commonly used filter aids include diatomaceous earth, perlite, asbestos, cellulose, agriculture fibers, etc.

22.3.1 Diatomaceous Earth

Diatomaceous earths (DE) are the skeleton of ancient diatoms (Figure 22.13a) mined from ancient seabeds, processed, and classified to make different grade of filter aids. Diatomaceous earth is the most commonly used filter aid today. However, the crystalline type DE is a suspicious carcinogen and inhalation needs to be avoided during handling.

Used as a precoat, a finer grade DE may be employed to increase the clarity of filtrate. In Figure 22.14 (Tiller 1978), an example of the size of solid particles removed by different grades of DE is shown based on laboratory tests. The smaller the filter aid particle size, the smaller the process particles can be removed. However, the filtration rate is lower. There is always a balance between initial filtrate clarity and filtration rate. The particle size capture by various filter aids may also vary because of liquid viscosity, surface charge, etc.

When DE is used as an admix to produce a less resistant cake, selection of the type and amounts of filter aids depends on laboratory tests with a constant pressure filter cell. A graph illustrating the effect of filter aid addition on the average flow rate is shown in Figure 22.15 (Tiller 1978). Overdose of filter aid resulted in a decrease of filtrate rate.

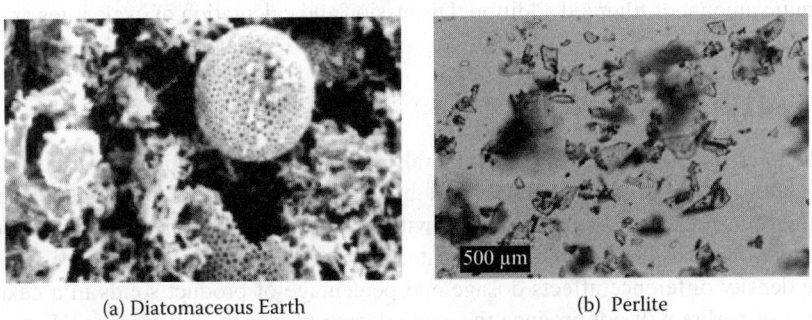

(a) Diatomaceous Earth   (b) Perlite

FIGURE 22.13 Typical filter aids.

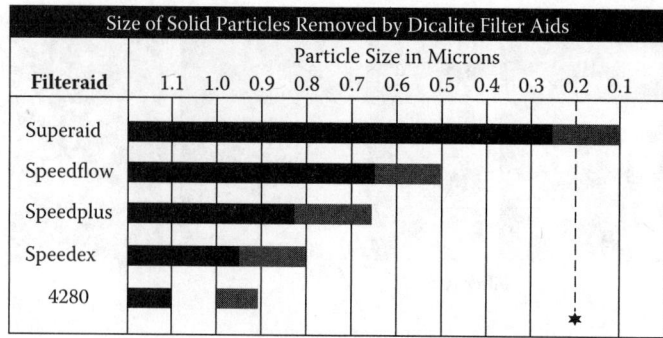

FIGURE 22.14 Chart showing the sizes of particles removed by five grades of Dicalite DE filter aids.

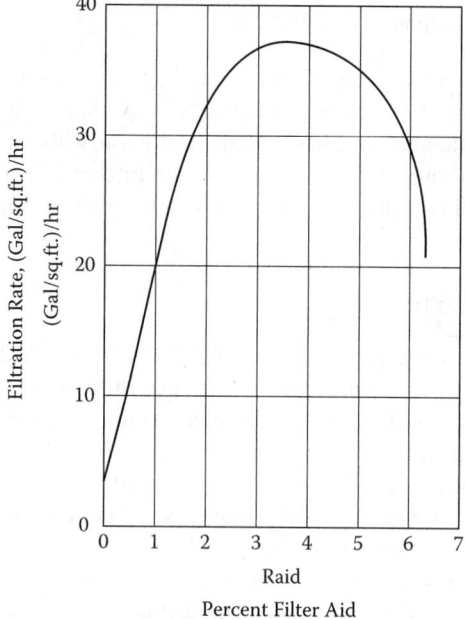

FIGURE 22.15 Optimum amount of filter aids for maximum flow.

Usually, as a filter aid is added to a system, the average flow rate is increased. However, the added solids not only decrease the percentage of product solids in the cake, but also fill the cake space sooner in batch pressure filters. Therefore, cycle analysis must be applied for determination of the optimum amount of filter aid addition. Discussion and calculation of cycle rates are presented in Section 22.8.5.

22.3.2 Perlite

Perlite is another important silica based mineral filter aid. It is a particular variety of naturally occurring glassy volcanic rock, characterized by onionlike, splintery breakage planes (Figure 22.13b). After crushing, this rock will expand with heat to about 10 times its original volume.

There are differences between DE and perlite. First, the bulk density of perlite is less than that of DE. The density difference affects dosage and percentage of product solids in a cake. Second, a packed bed of perlite will not produce the same degree of clarity as diatomite. However, due to the potential industrial hygiene issues by inhalation, perlite has replace DE in some applications.

Solid/Liquid Separation

22.3.3 OTHER FILTER AIDS

Other materials used as filter aids include asbestos, cellulose, rice hull ash, paper fibers, etc.

Cellulose is widely used for filtration of products that cannot tolerate silica. The filterability of cellulose is not as good as DE or perlite, but cellulose can be incinerated. Calcined rice hull ash and fibers from used newspapers are newly available filter aids. They are used for wastewater sludge dewatering as well as other applications.

22.4 FILTRATION FUNDAMENTALS

22.4.1 THEORY

Solid/liquid separation involves the relative motion of liquid and solids. The liquid may move faster in operations like filtration or the particles may have greater velocity as in thickening and sedimentation. Several classes of problems arise depending on the size, shape, state of aggregation, and concentration of particles in slurries. Particles involved in conventional separation operations range from the millimeter size to macromolecules measured in nanometers. The properties and separation methods vary enormously in accord with size and degree of flocculation. Flocs arising in pretreatment processes are deposited at a cake or sediment surface under a null stress in thickeners, filters, and centrifuges. As increasing deposits cover the surfaces, developing stresses continually compact the particulate bed primarily by particle movement into open pores. Principal sources of stress are (1) unbuoyed weight in gravity thickeners, (2) centrifugal forces in filtering and sedimenting centrifuges, (3) pump pressure converted into friction drag, and (4) surface forces developed by pistons or pressure actuated impermeable diaphragms.

Basic elements required for developing the theory of flow through compactible, porous cakes include:

1. Static force balance over cake yielding the compressive stress in the matrix of particles. The sum of the accumulative frictional drag and body (gravitational or centrifugal) forces per unit of cross-sectional area is called "effective pressure," a term first used in soil mechanics.
2. Relationship of superficial liquid flow rate q to liquid pressure gradient, Darcy's law.
3. Empirical constitutive equations relating permeability K, specific resistance α, and solidosity (vol fraction of solids) ε_s, to the effective pressure.
4. Integration of the Darcy equation to obtain flow rate as a function of the pressure drop across the cake Δp_c, and the thickness L, or the volume of solids per unit area ω_c.
5. Overall material balances involving slurry, filtrate, and cake and determination of pressure and filtrate volume as a function of time.

Conventional filtration theory is based on a two-resistance model in the form

$$q = \frac{dv}{dt} = \frac{p}{\mu(R_c + R_m)} = \frac{p}{\mu R} \tag{22.1}$$

where R_c and R_m are the cake and supporting medium resistances, R is the total resistance, p is applied pressure, v is filtrate volume/unit area of filtration, and t is time. As flow is laminar, the viscosity μ is separated from the resistances. The cake resistance is given by the product of the average specific resistance α_{av} and the volume of inert cake solids/unit area, ω_c, leading to

$$q = \frac{dv}{dt} = \frac{p}{\mu(\alpha_{av}\omega_c + R_m)} \quad (22.2)$$

Although the usual definition encountered in the literature employs w_c, mass of solids/unit area, there are distinct advantages accruing to the use of volumetric units in simplicity of formulas, correlations, and predictive power. The dimensions of α depend on whether ω_c (m³/m²) or w_c(kgm/m²) is used. The two quantities are related by $w_c = \rho_s\omega_c$, where ρ_s is density of solids. Resistance takes the form of $R = p/\mu q$ with dimensions $1/L$. Cake resistance is given by

$$R_c = \alpha_{av}\omega_c = \alpha^*_{av} w_c \quad (22.3)$$

where the superscript is used to differentiate the two types of specific resistance. The units of α_{av} and α^*_{av} are, respectively, $1/L^2$ and L/M. A closely packed bed of 10 µm spheres would be expected to have a specific resistance α of about 3.0E(12)m⁻². The resistance of the bed to flow would depend entirely on the size and shape of the pores and would be independent of the mass inside the particles. Using α^* requires that 3.0E(12)m⁻² be divided by the particle density, thereby leading to different values for every material for geometrically identical beds. Correct comparison of the resistances of different materials can only be made with the specific resistance as used in Equation (22.2) with dimensions of L^{-2}.

Integration of Equation (22.2) to obtain filtrate volume as a function of time t requires:

1. Specification of pump characteristics, that is, p as a function of q with constant pressure and constant rate being important special cases.
2. Relation between v and ω_c or w_c as obtained from a volumetric or mass balance over slurry, cake, and filtrate.
3. Equations relating the average values of the specific resistance α_{av}, permeability K_{av}, and volume fraction of solids in the cake ε_{sav} to the pressure drop Δp_c across the cake.

Before using Darcy's law to theoretically develop Equation (22.2), simplified procedures used by many authors (Rushton et al. 1996; Svarovsky 1990) will be employed to derive volume vs. time equations. Although widely used and valuable in interpreting data, highly significant information important to full-scale operation is missed when Equations (22.1) and (22.2) are the sole basis for developing the theory of cake filtration.

Although Equation (22.2) is often considered to be a "fundamental" formula governing filtration, it must be used with caution. It is very useful in analysis when viewed as an empirical approximation. It provides an instantaneous picture of the relationship among the variables, which include the instantaneous flow rate q, applied pump pressure p, and volume ω_c or mass w_c of solids/unit area.

22.4.2 Material Balance

Volumetric and mass balances are necessary in developing filtrate volume/unit area v as a function of t. A schematic diagram in Figure 22.16 shows how the slurry relates to the cake and filtrate. Quantities necessary for balances include:

	Total Solids	Concentration of Solids	
		Slurry	Cake
Volume basis	ω_c, vol./unit area	ϕ_s, vol. fraction	ε_{sav}, av. vol. fraction
Mass basis	w_c, mass/unit area	s, mass fraction	s_c, av. mass fraction

Solid/Liquid Separation

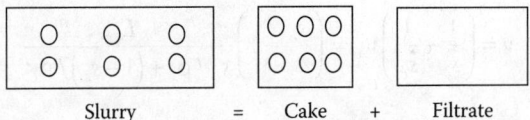

FIGURE 22.16 The volume or mass of slurry equals the volume or mass of cake and filtrate.

Other quantities needed are cake thickness L and liquid density ρ_L. A volumetric balance leads to slurry volume = cake volume + filtrate volume:

$$\frac{\omega_c}{\phi_s} = \frac{\omega_c}{\varepsilon_{sav}} + v = L + v \tag{22.4}$$

where L is cake volume/unit area and ω_c is $\varepsilon_{sav} L$. Equation (22.4) can be rearranged into the convenient form

$$v = \left(\frac{\varepsilon_{sav}}{\phi_s} - 1\right) L = \left(\frac{1}{\phi_s} - \frac{1}{\varepsilon_{sav}}\right) \omega_c \tag{22.5}$$

A mass balance leads to slurry mass = cake mass + filtrate mass:

$$\frac{w_c}{s} = \frac{w_c}{s_c} + \rho_L v \tag{22.6}$$

Solving for v yields

$$v = \left(\frac{1}{s} - \frac{1}{s_c}\right) w_c \tag{22.7}$$

Introducing cake thickness into the mass relations requires that the average mass fraction of cake solids s_c be converted to the average volumetric fraction ε_{sav} as follows:

$$\varepsilon_{sav} = \frac{s_c / \rho_s}{s_c / \rho_s + (1 - s_c) / \rho_L} \tag{22.8}$$

where ρ_s is density of the solids. A similar equation can be used for the slurry with s replacing s_c and ϕ_s replacing ε_{sav}. The vol/unit area of solids is given by

$$\omega_c = \varepsilon_{sav} L = \frac{w_c}{\rho_s} = \frac{s_c / \rho_s}{s_c / \rho_s + (1 - s_c) / \rho_L} L \tag{22.9}$$

A relation similar to Equation (22.5) takes the form

$$v = \left(\frac{1}{s} - \frac{1}{s_c}\right) w_c = \left(\frac{s_c}{s} - 1\right) \frac{L}{s_c/\rho_s + (1-s_c)/\rho_L} \tag{22.10}$$

It is clear that the volumetric balance leads to easier to handle expressions.

22.4.3 Volume vs. Time

Solving for ω_c in Equation (22.5) yields

$$\omega_c = \frac{\phi_s}{1 - \phi_s/\varepsilon_{sav}} v = cv \tag{22.11}$$

Substituting for ω_c in Equation (22.2) and rearranging the terms leads to

$$\frac{p}{\mu} dt = \left(\alpha_{sv} cv + R_m\right) dv \tag{22.12}$$

Historically, this equation has been integrated assuming α_{av}, ε_{sav}, and c to be constant even though they vary. When the variation is not large, integration from $t = 0$, $v = 0$ to (t, v) yields a parabolic relation in the form

$$\frac{pt}{\mu} = \alpha_{av} c \frac{v^2}{2} + R_m v \tag{22.13}$$

This is the well-known parabola that has been used extensively to calculate α_{av} and R_m based on v vs. t data for constant pressure filtrations performed in the laboratory. It must be used with caution.

22.4.4 Data Analysis

A simplified but useful approach to analysis of constant pressure data can be developed with the use of Equations (22.12) and (22.13) placed in resistance form. Dividing Equation (22.12) by dv and Equation (22.13) by v produces

$$\frac{p}{\mu} \frac{dt}{dv} = \frac{p}{\mu q} = \alpha_{av} cv + R_m \tag{22.14}$$

$$\frac{p}{\mu} \frac{t}{v} = \frac{p}{\mu q_{av}} = \frac{1}{2} \alpha_{av} cv + R_m \tag{22.15}$$

where $q_{av} = v/t$ is the average rate and $q = dv/dt$ is the instantaneous rate. Substituting R_c for $\alpha_{av} cv$ leads to

$$R = p/\mu q = R_c + R_m \tag{22.16}$$

$$p/\mu q_{av} = 0.5R_c + R_m \qquad (22.17)$$

These relationships between the resistances and pressure drops are illustrated in Figure 22.17. Plots of data in accord with Equations (22.14) and (22.15) are shown in Figure 22.18. Line AB corresponds to Equation (22.14) and line AC to Equation (22.15). Both lines have the same intercept, point A, which is the medium resistance. The slopes are in a ratio of 2:1 and yield α_{av} and $\alpha_{av}/2$. Although it is easier to plot $(p/\mu)(t/v)$ vs. v and avoid the difficulty of obtaining differential dt/dv required by Equation (22.14), we strongly recommend plotting both. Nevertheless, most investigators have simply plotted t/v vs. and have not used the resistance forms that we advocate. As filtration data are seldom precise and a number of important factors have been neglected in deriving Equations (22.14) and (22.15), plotting both equations provides a basis for detecting errors and deviations from the simplified parabolic theory.

As the volume vs. time data are parabolic, the construction in Figure 22.19 can be employed to obtain the slope, dv/dt, of the v versus t data. If two points $A(t_1, v_1)$ and $B(t_2, v_2)$ are chosen, the slope of AB is precisely equal to the tangent taken at M where $= (v_1 + v_2)/2$ but not at $(t_1 + t_2)/2$. This procedure provides a convenient method for obtaining slopes directly from experimental data. Frequently it is necessary to smooth the data before starting the analysis.

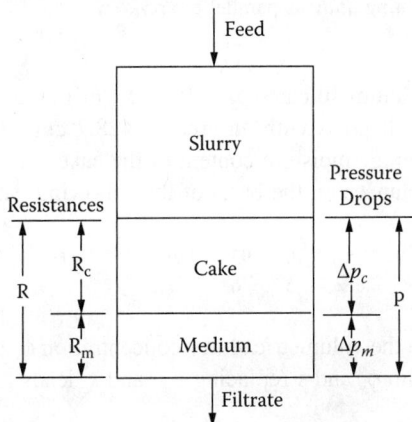

FIGURE 22.17 Diagram illustrating the relationship between pressure drops and resistances.

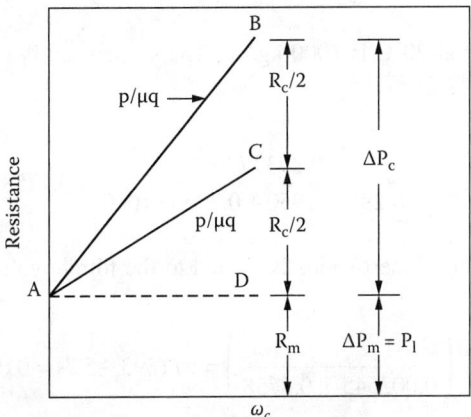

FIGURE 22.18 Plots used to obtain resistances and the average specific resistance.

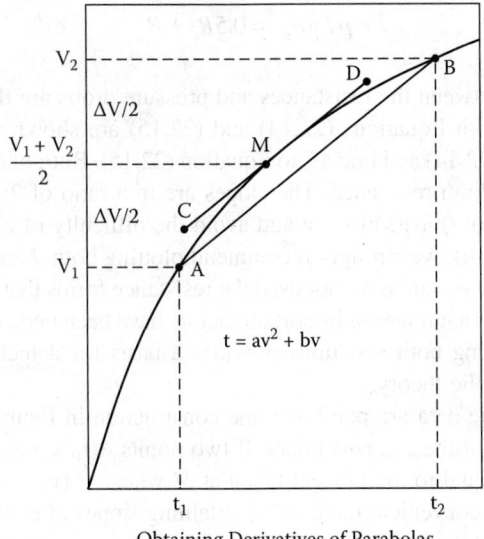

FIGURE 22.19 Tangent representing dt/dv as parallel to arc.

Example 22.1
An aqueous suspension of calcium silicate (ρ_s = 1950 kg/m³) at 20°C was filtered at a constant pressure of 68.9 kPa in a small press with an area of 428.7 cm². The slurry concentration was 0.495% by weight and the average moisture content of the cake was 70.63% by weight. Calculate the specific and medium resistances on the basis of the following data (Hosseini 1977):

Time, sec.	70	93	120	152	187	227	270
Vol. of filtrate, liters		5	6	7	8	9	10	11

It is necessary to calculate the volumetric slurry concentration ϕ_s and the average cake porosity ε_{sav}. Using Equation (22.8) with ϕ_s and s replacing ε_{sav} and s_c leads to

$$\varphi_s = \frac{0.00495/1950}{0.00495/1950 + (1-0.00495)/1000} = 0.002545 \tag{22.18}$$

where the density of water at 20°C is 1000 kg/m³. The suspension is very dilute. Calculating ε_{sav} based on s_c = 1.0 0.7063 = 0.2937,

$$\varepsilon_{sav} = \frac{0.2937/1950}{0.2937/1950 + 0.7063/1000} = 0.1758 \tag{22.19}$$

The volume/unit area ω_c of inert solids is related to the filtrate volume/unit area by

$$\omega_c = v \bigg/ \left(\frac{1}{0.002545} - \frac{1}{0.1758}\right) = v/(393-5.7) = 0.00258v \tag{22.20}$$

The ratio p/μ = 68,900/0.001 = 6.89E(7) is required for calculating the resistance terms

TABLE 22.4
Calculations for Example 22.1

Time t, sec	V, m³	Filtrate V, m³/m²	$(v_1 + v_2)/2$	Cake $\omega_c = .00258 v$ v/0.04287	dv/dt $(v_2 - v_1)/(t_2 - t_1)$	$P/\mu q$	$P/\mu q_v$
70	.005	0.117		3.02 E(4)			4.14 E(10)
			0.129	3.32	10.0 E(4)	6.89 E(10)	
93	.006	0.140		3.36			4.58
			0.152	3.92	8.52	8.09	
120	.007	0.163		4.20			5.07
			0.175	4.52	7.50	9.19	
152	.008	0.187		4.83			5.60
			0.199	5.14	6.57	10.5	
187	.009	0.210		5.43			6.14
			0.222	5.74	5.75	12.0	
227	.010	0.233		6.02			6.71
			0.245	6.33	5.58	12.3	
270	.011	0.257		6.64			7.24

$$p/\mu q_{av} = 6.89 E(7) t / v \qquad (22.21)$$

$$p/\mu q = 6.89 E(7)(t_2 - t_1)/(v_2 - v_1) \qquad (22.22)$$

Table 22.4 provides a systematic procedure for obtaining the terms required to obtain the average specific resistance α_{av}. The first two columns provide the raw volume (m³) vs. time (sec). The third column represents the total volume divided by the area (= 0.04287 m²). The fourth column corresponds to the average volume in accord with the construction shown in Figure 22.4. Equation (22.11) is used to obtain ω_c in the fifth column. The rate $q = dv/dt$ corresponding to the average volume is shown in the sixth column. Calculations for columns 4 and 6 follow for the first two points:

$$v = (v_1 + v_2)/2 = (0.117 + 0.140)/2 = 0.129 m \qquad (22.23)$$

$$dv/dt = (v_2 - v_1)/(t_2 - t_1) = (0.140 - 0.117)/(93 - 70) = 10.0 E(10)\ 1/m \qquad (22.24)$$

The two resistance terms are calculated and placed in the last two columns, thus

$$p/\mu q = 68,900/(0.001 \cdot 10 E(-4)) = 6.89 E(10)\ 1/m \qquad (22.25)$$

$$p/\mu q_{av} = [68,900/0.001] \cdot (70/0.117) = 4.14 E(10)\ 1/m \qquad (22.26)$$

Two lines are drawn through the two sets of points in Figures 22.5 such that they both have the same intercept and the slopes are in the ratio of 2 to 1.

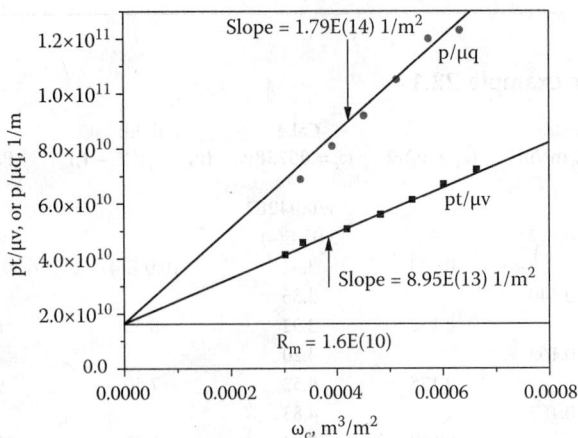

FIGURE 22.20 Resistance plot used to obtain medium and specific resistances.

The lines shown in Figure 22.20 yield $R_m = 1.6E(10)$ 1/m, and $\alpha = 1.69(14)$ 1/m². The specific resistance on a mass basis can be obtained from Equation (22.3) as

$$\alpha^*_{av} = \alpha_{av}/\rho_s = 1.69E(14)/1950 = 8.67E(10) \ m/kg \qquad (22.27)$$

The use of units m/kgm predominates in the literature. If regression techniques had been employed with the points in Figure 22.20, two lines with different intercepts and slopes not in the ratio of 2/1 would have resulted. The construction and methodology illustrated in example 22.1 represent a compromise, and there is generally no precise answer.

22.4.5 Deviations From Parabolic Theory

The equations employed in Example 22.1 require that the t versus v relation be a perfect parabola in constant pressure filtration. Errors are both theoretical and experimental. Introducing a suspension into a filter and suddenly increasing the pressure to the operating value results in a transient state in which a finite time is required to reach equilibrium conditions. If high pressures are involved, it is undesirable to raise the pressure too suddenly. Most slurries undergo some sort of pretreatment and mixing to improve filtration characteristics. If insufficient time is allowed for equilibrium to be reached, the cake permeability and porosity of successive layers will change continuously. Aging is an almost universal problem making it unwise to use other than fresh slurries. Biological solids are particularly susceptible to change.

Sedimentation in the filter vessel is a difficult problem that is frequently neglected. If vertical surfaces are involved, the cake will be thicker at the bottom compared with the top. In filters with horizontal surfaces, dense particles settle and increase the slurry concentration at the cake surface. Destruction of aggregates may occur during pumping, pouring, and mixing. Tiller, Hsyung, and Cong (1995) analyzed the sedimentation problem and developed approximate techniques to obtain the average specific resistance.

Slurries frequently involve a wide range of particle sizes that include submicron particles in the Brownian diffusional region. When cakes are deposited, the finest particles may diffuse into the filtrate and continuously clog the supporting medium, leading to increasing values of R_m. Tiller and Leu (1984) showed that clogging was a major problem in the removal of ash and asphaltenes from liquefied coal.

Processing of liquefied coal in a pilot plant at Wilsonville, Alabama, provides a good example of filtration operations that deviate from the simple constant pressure equations. After liquefying

Solid/Liquid Separation

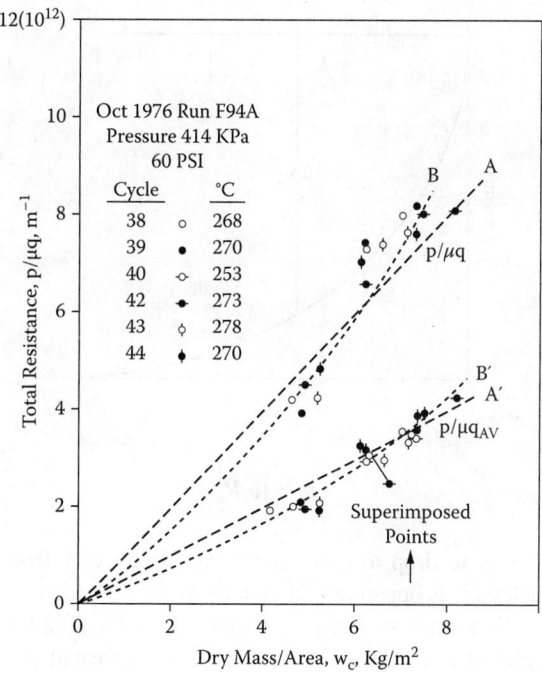

FIGURE 22.21 Clogging phenomena.

the coal, submicron particles arising from mineral residue and colloidal asphaltenes were removed in a horizontal leaf filter with centrifugal cake removal. A resistance plot for a series of six separate runs is shown in Figure 22.21. A straight line A was drawn through the $p/\mu q_{av}$ points. A second line A with twice the slope was drawn through the $p/\mu q$ data. It is clear that it is impossible to draw two lines independently through the two sets of points that have the same R_m and α_{av}. The location of the points along BC that lies above OA is an indication that total resistance is increasing more rapidly than predicted. As solid densities were not known, w_c was used rather than ω_c.

If only the $p/\mu q_{av}$ points had been plotted, it would have been assumed that $R_m = 0$ and that α_{av}^* was constant and equal to 5.0E(11) m/kg. As q_{av} represents an average rate over the entire cycle, it does not reflect instantaneous conditions. On the other hand, $q = dv/dt$ as it appears in Equation (22.2) corresponds to the instantaneous state of the cake prior to the averaging process resulting from integration. The specific resistance α_{av} in Equation (22.2) represents the value at a specific time. In the integrated form of Equation (22.13), α_{av} corresponds to the average over the entire cycle. Examination of Figure 22.21 indicates that the $p/\mu q$ plot is responding more rapidly than the $p/\mu q_{av}$ plot to the appearance of increased resistance. Tiller, Hsyung, and Cong (1995) showed that similar behavior accompanies the presence of sedimentation on horizontal cakes formed during constant pressure filtration. In general, whenever the $p/\mu q$ plots yield tangents with negative intercepts as in Figure 22.21 unaccounted for, resistances have appeared that were not involved in Equations (22.13) and (22.14). In the case of the filtration of the liquefied coal, transport and diffusion of sub-micron asphaltenes and mineral residue in the pores of the cake led to increased medium and cake resistances.

Theoretical problems with the traditional theory of the parabolic equation relate to the assumption of constant α_{av} and ε_{sav}, both of which are functions of the pressure drop across the cake. The pressure drop Δp_c varies when centrifugal and constant rate pumps are employed. Even in constant pressure filtration, Δp_c varies with time. At $t = 0$, there is no cake, and all of the pressure is absorbed by the medium, as shown in Figure 22.22. As the cake builds up, the flow rate q drops; and the pressure drop across the medium, p, decreases in accord with Equation (22.28):

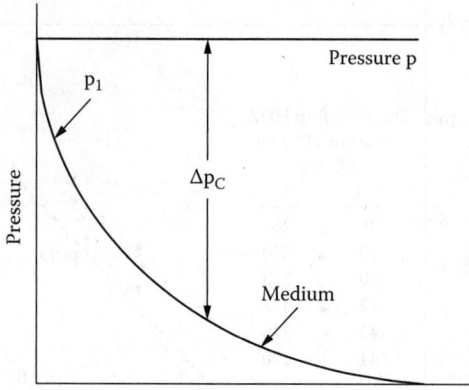

FIGURE 22.22 Variation of Δp_c.

$$p_1 = \mu q R_m \qquad (22.28)$$

The time required for p_1 to drop to a negligible value may vary from seconds to a relatively lengthy period. Nevertheless, it is apparent that as long as $R_m > 0$, there cannot be a filtration with constant Δp_c, and the simplified analysis shown in Figures 22.18 and 22.20 must be used with caution.

In Figure 22.23, a plot of $p/\mu q$ vs. ω_c is illustrated for a constant pressure filtration in which R_m is not negligible. As Δp_c increases with time, the slope of the curve and α_{av} also increase. The total resistance is given by the curve ABD, and the cake resistance R_c is represented by BC. In accord with Equation (22.3), α_{av} is obtained by dividing R_c by $\omega_c(AC)$. As point moves up the curve, the angle of the triangle ABC increases and ultimately approaches the slope given by BD. The pressure drop across the cake can be calculated from

$$\Delta p_c = \left(\frac{R_c}{R_c + R_m} \right) p \qquad (22.29)$$

Equation (22.29) can be used when pressure is constant or a function of time.

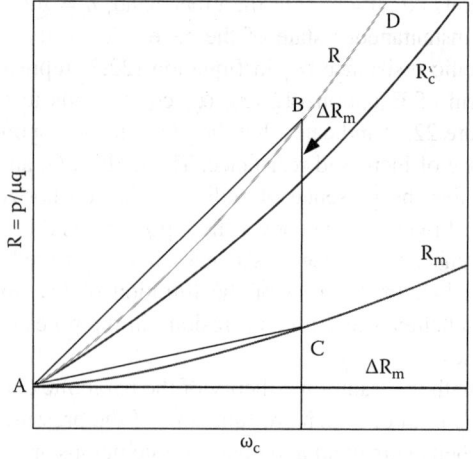

FIGURE 22.23 Medium and cake clogging.

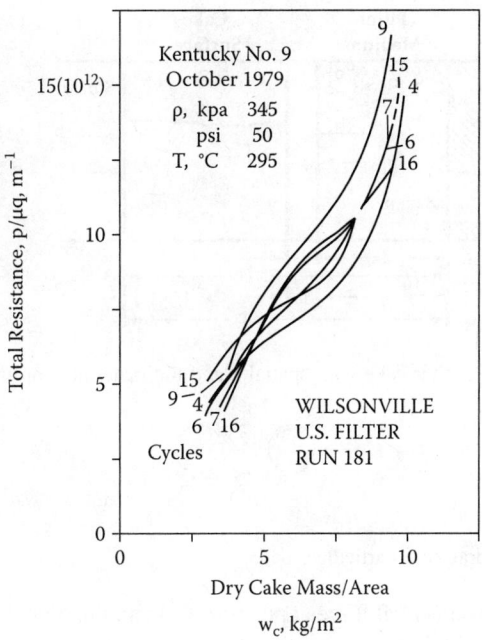

FIGURE 22.24 Clogging Phenomena in liquefied coal industry (10).

In Figure 22.24, resistance plots for $p/\mu q$ versus w_c are shown for a series of runs involving Kentucky No. 9 liquefied coal filter in the U.S. Filter Co. vertical leaf filter. They follow the general trends of filtrations subject to both medium and cake clogging as illustrated in Figure 22.23.

22.4.6 Filtration of Super-Compactible Materials (Tiller and Li 1999, 2000, 2001a)

The simplified theory used to develop Equation (22.13) giving filtrate volume as a function of time is based on having constant values of the average specific resistance and the average cake solidosity. In order to derive more basic equations, it is necessary to integrate local values based on Darcy's law to obtain equations relating average values of solidosity, specific resistance, and permeability as functions of the pressure drop across the cake.

22.4.6.1 Effective Pressure

It is generally assumed that the structure of a sediment or cake is a unique function of the local stress or effective pressure (termed compressive pressure by W.K. Lewis). A filter cake is shown in Figure 22.25 along with the spatial coordinate x and the material coordinate ω that represents the volume/unit area of solids in the distance x. The cake deposited on the filter medium grows as solids are deposited on the surface where the spatial coordinate $x = L$, and the pump or applied pressure is p. As the liquid flows through the cake, the hydrostatic pressure p_L drops and finally exits at the medium with a value equal to p_1. As the liquid flows through the pores, a drag force dF is exerted on each particle (Figure 22.26). The sum of all the forces on the particles equals the total drag F_s. The term F_s is the cumulative drag on particles, communicated from particle to particle and increasing in the direction from $x = L$ to $x = 0$. Assuming the accelerations of the liquid and solid are negligible and neglecting inertial terms, a stress balance is written over the portion of the cake lying over x and L. The total force Ap at the cake surface given by the product of p and the

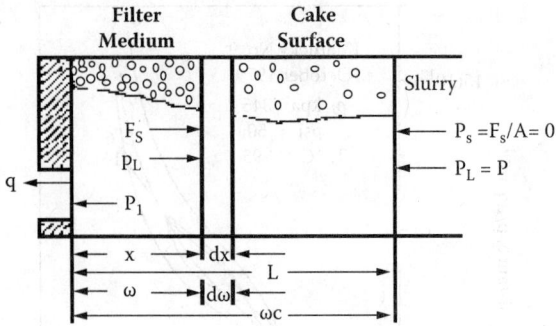

FIGURE 22.25 Diagram of a filter cake with spatial x and the material coordinate.

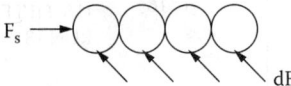

FIGURE 22.26 Fractional drag in a particulate bed.

area A must be balanced by an equal force at position x. Assuming that there is point contact among particles and that the local liquid pressure p_L is effective over the entire cross-sectional area leads to

$$Ap_L + F_s = Ap \tag{22.30}$$

Dividing Equation (22.30) by A and then defining the compressive drag or effective pressure by $p_s = F_s/A$, we have

$$p_L + p_s = p \tag{22.31}$$

The drag on the particle is an accumulation of form drag force resulting from friction along the surface of the particles and is transmitted through the point contacts over which the stress acts. The term p_s is a fictitious pressure used for convenience. It is called "effective pressure" in soil mechanics. In actual cakes, of course, there is a small area of contact, and the pressure at the contact points is much larger than the average given by p_s.

In thickening and centrifugation when gravitational and centrifugal stresses are involved, Equation (22.31) is replaced by more complex expressions Tiller and Hsyung (1993).

The quantities p_L and p_s vary with position and time and are functinos of (x,t). The applied pressure is either constant or a functin of time alone. If we take differentials with respect to x in the interior of the cake at constant t, we have

$$dp_s + dp_L = 0 \tag{22.32}$$

The effective pressure increases and hydraulic pressure decreases as the liquid flows from the upstream face of the cake toward the filter medium.

Variations of the hydraulic and solid or effective pressures in a typical filter cake are illustrated in Figure 22.27. The liquid pressure drops from the pump pressure at the cake surface to a value p_1 at the medium. At the surface of the cake where $p_L = p$, the effective pressure is zero as no drag on the particles has developed. Consequently, the surface layer of the cake is unstressed, and the solidosity has its minimum value at that point. The solidosity reaches a maximum at the medium

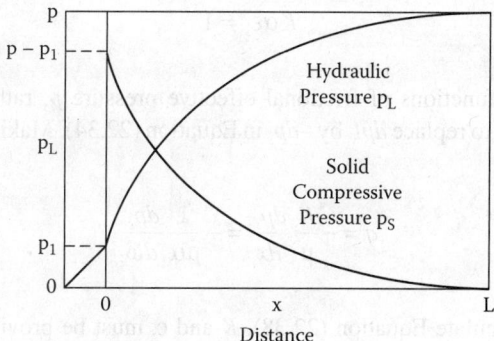

FIGURE 22.27 Variation of hydraulic and effective pressures in a compressible filter cake.

where the effective pressure has risen to its maximum value and equals the pressure drop across the cake. Whereas the solidosity at the cake surface theoretically remains the same regardless of the pressure drop, the solidosity at the medium increases as the pressure drops becomes larger.

22.4.6.2 Darcy's Law

The Darcy equation was first developed to describe the linear flow of liquid through incompressible sand beds in the form

$$q = \frac{K_{av}\Delta p}{\mu L} \tag{22.33}$$

in which K_{av} the average permeability was a constant for the sand bed; and Δp was the overall pressure drop across the sand bed of thickness L.

Differential forms involving both spatial and material coordinates illustrate the relationship of local values of permeability and specific resistance α to the gradients of p_L with respect to x and ω:

$$q = \frac{K}{\mu}\frac{dp_L}{dx} = \frac{1}{\mu\alpha}\frac{dp_L}{d\omega} \tag{22.34}$$

The spatial coordinate x is the distance from the medium, and the material coordinate is the volume of solids/unit area in distance x as shown in Figure 22.25. The volume of solids $d\omega$ in dx is related to local solidosity by

$$d\omega = \varepsilon_s dx \tag{22.35}$$

Integrating Equation (22.35) from $x = 0$ to L and $\omega = 0$ to ω_c produces

$$\int_0^{\omega_c} d\omega = \omega_c = \int_0^L \varepsilon_s dx = \varepsilon_{sav} L \tag{22.36}$$

where ε_{sav} is the average solidosity of the cake. Substituting Equation (22.35) into Equation (22.34) yields the following relationship involving the characteristic parameters, α, and ε_s:

$$K\alpha\varepsilon_s = 1 \tag{22.37}$$

As K, α, and ε_s are functions of frictional effective pressure p_s rather than hydraulic liquid pressure p_L, it is necessary to replace dp_L by $-dp_s$ in Equation (22.34). Making this substitution yields

$$q = -\frac{K}{\mu}\frac{dp_s}{dx} = -\frac{1}{\mu\alpha}\frac{dp_s}{d\omega} \tag{22.38}$$

To integrate the particulate Equation (22.38), K and α must be provided as functions of p_s.

22.4.6.3 Empirical Constitutive Equations

Empirical models of local solidosity, permeability, and specific resistance as functions of p_s can be in many forms. Tiller and Leu (1980) presented the following equations:

$$\left(\frac{\varepsilon_s}{\varepsilon_{so}}\right)^{1/\beta} = \left(\frac{K}{K_o}\right)^{-1/\delta} = \left(\frac{\alpha}{\alpha_o}\right)^{1/n} = 1 + \frac{p_s}{p_a} \tag{22.39}$$

in which p_a is an empirical parameter; and β, δ, n are cake compactibility coefficients. They provide a measure of the rate of change of ε_s, α, K with p_s. As a rough approximation, the exponents can be related by

$$\delta/5 = n/4 = \beta \tag{22.40}$$

22.4.6.4 Integration of the Darcy Equation

Combining the boundary conditions at the cake surface ($x = L$ or $\omega = \omega_c$, $p_s = 0$) and at the medium ($x = 0$ or $\omega = 0$, $p_s = \Delta p_c = p - p_1$) with the empirical constitutive equations ([Equation (22.39)]; assuming q is independent of x and is only a function of t, the particulate structure Equations (22.38) can be integrated to yield

$$\int_0^L \mu q\, dx = \mu q L = K_{av}\Delta p_c = \int_0^{\Delta p_c} K_o\left(1 + p_s/p_a\right)^{-\delta} dp_s = \frac{K_o p_a}{1-\delta}\left[\left(1 + \frac{\Delta p_c}{p_a}\right)^{1-\delta} - 1\right] \tag{22.41}$$

and

$$\int_0^{\omega_c} \mu q\, d\omega = \mu q \omega_c = \Delta p_c/\alpha_{av} = \int_0^{\Delta p_c} \frac{dp_s}{\alpha_o\left(1 + p_s/p_a\right)^n} = \frac{p_a}{\alpha_o(1-n)}\left[\left(1 + \frac{\Delta p_c}{p_a}\right)^{1-n} - 1\right] \tag{22.42}$$

These two equations give an instantaneous view of the cake and provide relationships among q, ω_c, L, K_{av}, α_{av}, and p_c. Equations (22.41) and (22.42) can be easily solved for q as a function of L or ω_c and Δp_c. Dividing Equation (22.42) by (22.41) leads to

Solid/Liquid Separation

$$\frac{\omega_c}{L} = \varepsilon_{sav} = \varepsilon_{so} \left(\frac{1-\delta}{1-n}\right) \frac{\left(1+\Delta p_c / p_a\right)^{1-n} - 1}{\left(1+\Delta p_c / p_a\right)^{1-\delta} - 1} \qquad (22.43)$$

where $1/\alpha_o K_o$ has been replaced by ε_{so} in accord with Equation (22.37). It is clear from the above equations that the flow rate and the volume fraction of cake solids are dependent on the null-stress cake structure parameters ε_{so}, K_o, α_o, and cake compressibility coefficients n, δ, β.

22.4.7 Strange Behavior of Super-Compactible Materials

The relative compactibility of cakes can be crudely classified in accord with the magnitudes of n and δ as shown in Table 22.5.

The cake compactibility parameters for an incompressible material carbonyl iron, a moderately compactible kaolin flat D, and a super-compactible activated sludge are given in Table 22.6.

For super-compactible materials when $n > 1$ and $\delta > 1$, $(1 - n)$ and $(1 - \delta)$ are negative; and Equations (22.42) and (22.43) are best rearranged as follows:

$$\mu q \omega_c = \Delta p_c / \alpha_{av} = \frac{p_a}{\alpha_o (n-1)} \left[1 - \frac{1}{\left(1+\Delta p_c / p_a\right)^{n-1}}\right] \qquad (22.44)$$

$$\varepsilon_{sav} = \varepsilon_{so} \left(\frac{\delta-1}{n-1}\right) \frac{1 - 1/\left(1+\Delta p_c / p_a\right)^{n-1}}{1 - 1/\left(1+\Delta p_c / p_a\right)^{\delta-1}} \qquad (22.45)$$

As Δp_c increases indefinitely in Equation (22.44), (22.45), the term with Δp_c in the denominator approaches zero. Plots of calculated values of flow rate and average solidosity as functions of pressure drop across cakes with $\omega_c = 0.02$ m³/m² are shown in Figures 22.28 and 22.29 for the three materials listed in Table 22.6. Whereas the flow rate q increases linearly with Δp_c for

TABLE 22.5
Classification of Cake Compactibility

Incompressible	$n = 0$	$\delta = 0$
Moderately compactible	$N \approx 0.4$–0.7	$\delta \approx 0.5$–0.9
Highly compactible	$N \approx 0.7$–0.8	$\delta \approx 0.9$–1.0
Super-compactible	$n > 1$	$\delta > 1.0$

TABLE 22.6
Cake Compactibility Parameters for Three Materials

Materials	ε_{so}	α_0, m⁻²	β	n	δ	p_a, Pa
Carbonyl Iron, grade (Grace 1953)	0.575	2.34(E13)	0.001	0.005	0.006	—
Kaolin flat D (Chen 1986)	0.14	2.98(E13)	0.12	0.4	0.52	11
Activated sludge (Kwon 1995)	0.05	3.62(E14)	0.26	1.4	1.66	190

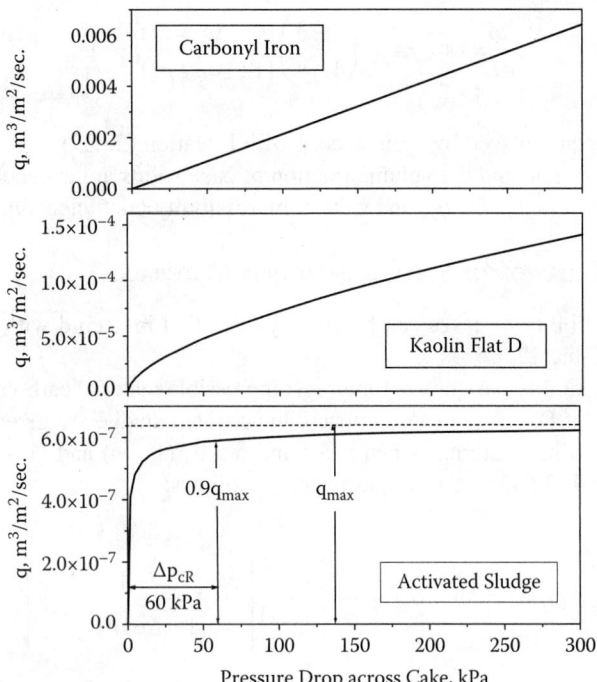

FIGURE 22.28 Variation of q against Δp_c.

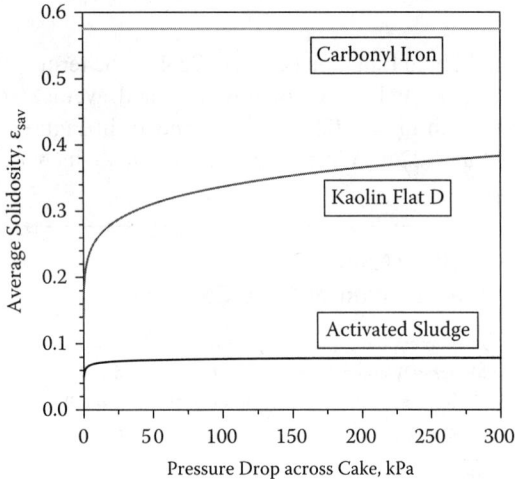

FIGURE 22.29 Variation of ε_{sav} against Δp_c.

incompressible materials like carbonyl iron, it increases with a power of Δp_c, which is less than unity for compactible materials like kaolin. Unfortunately, for super-compactible materials with $n > 1$ and $\delta > 1$, increasing the pressure drop beyond some low value has negligible effect on the flow rate and average solidosity, and both q and ε_{sav} approach maximum values.

22.4.8 Critical Pressure Drop

As Δp_c approaches infinity, the limiting values of the flow rate and average solidosity become

Solid/Liquid Separation

$$q_{max} = \frac{p_a}{\alpha_o (n-1)\mu\omega_c} \tag{22.46}$$

$$\varepsilon_{sav,max} = \varepsilon_{so}\left(\frac{\delta-1}{n-1}\right) \tag{22.47}$$

Dividing Equation 22.34 by Equation (22.46) and dividing Equation (22.45) by Equation (22.47) provides the ratios of the flow rate and the average solidosity to the maximum values:

$$q/q_{max} = \gamma = 1 - \frac{1}{(1+\Delta p_c/p_a)^{n-1}} \tag{22.48}$$

$$\varepsilon_{sav}/\varepsilon_{sav\,max} = \lambda = \frac{1 - 1/(1+\Delta p_c/p_a)^{n-1}}{1 - 1/(1+\Delta p_c/p_a)^{\delta-1}} \tag{22.49}$$

Substituting Equation (22.48) into Equation (22.49) yields a relationship between γ and λ:

$$\lambda = \frac{\gamma}{1-(1-\gamma)^{(\delta-1)/(n-1)}} \tag{22.50}$$

For activated sludge, when $\gamma = 0.90$, and $\lambda = 0.92$, the pressure reaches a critical point where a further increase of pressure drop will have little effect on flow rate or average solidosity as shown in Figures 22.28 and 22.29. A critical pressure drop can be defined as that value at which q reaches 90% of its ultimate rate given by Equation (22.48).

$$\Delta p_{cR} = p_a\left[\left(\frac{1}{1-\gamma}\right)^{1/(n-1)} - 1\right] = p_a\left[\left(\frac{1}{1-0.9}\right)^{1/(n-1)} - 1\right] = p_a\left[10^{1/(n-1)} - 1\right] \tag{22.51}$$

Equation (22.51) can be applied to calculate the operational pressure beyond which a further increase of pressure will have little effect on either the flow rate or the average cake solidosity. For the activated sludge in Table 22.5, if $\gamma = 0.90$, the critical pressure drop across the cake calculated from Equation (22.51) is 60kPa (8.7 psi), which corresponds to the results shown in Figures 22.28 and 22.29.

22.5 SOLID/LIQUID SEPARATION EQUIPMENT AND OPERATION

An overview of some of the major types of equipment in relation to the four stages of SLS is provided in Figure 22.30. After inorganic salts and polymers have been used in the pretreatment, the resulting slurry is fed to a gravity thickener. The overflow from the thickener passes through a deep-bed for removal of fine particles. The underflow goes to the solids separation operations, which are classified according to the driving force, e.g., gravity, vacuum, pressure, or centrifugal.

Gravity separation involving large particles is usually accomplished with screens that may be stationary, vibrating, or rotating cylinders. Large screens have slot openings greater than 0.5 in.

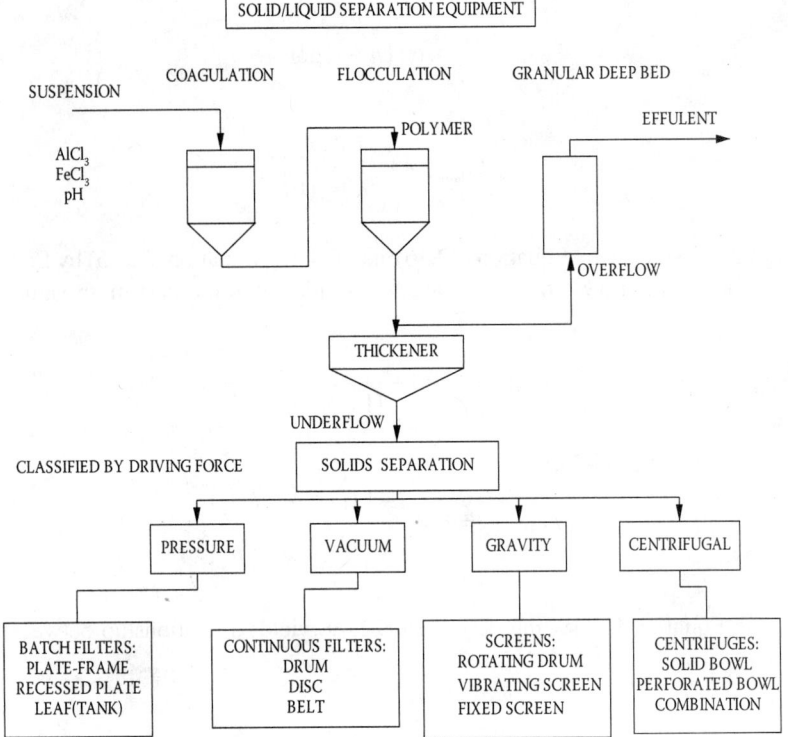

FIGURE 22.30 Equipment used for the four stages of solid/liquid separation.

and small screens have openings of less than 0.5 in. Slot openings in rotary screens run from 0.01 to 0.1 in. (254–2540 µm). Micro-screens frequently fall in the 15 to 60 micron range.

In vacuum filtration, the driving force (~20 in Hg = 10 psi, 1 psi = 4895 Pa) is slightly higher than the gravity. The vacuum operation frequently is tied in with continuous equipment such as drum, disc, or belt filters, in which cake is removed continuously. As the permeability of cakes diminishes, pressure becomes an important element in producing a satisfactory flow rate. Pressure filters are operated in the range of 30 to 60 psi and sometimes up to 100 psi. In contrast to vacuum filters, pressure filters normally operate in batch mode. In addition to vacuum and pressure, centrifugal forces are also used to increase the driving force in separation of particles from liquids.

Solid/liquid separation equipment can be also classified according to the principle of each unit operation, which is presented in Section 22.1.4. A list of equipment based on operating principle is given in the next section.

22.5.1 Batch Pressure Filters

Pressure filters are usually operated batch-wise. The batch pressure filters can be classified as tank (pressure vessel) filters or presses. Tank filters have filter elements of different types mounted in pressure vessels. Tank filters are divided into pressure nutsches, leaf filters, candle or tubular filters, bag filters, and cartridge filters. Presses (see Figure 22.38) consist of a series of filter surfaces (plates). The elements are mounted on a frame and are pressed together mechanically.

22.5.1.1 Pressure Nutsches

Nutsche filters contain a single horizontal filtering surface in a pressure vessel. Gas is used to provide pressure for filtration and avoid the use of a high-pressure pump. Due to the limited filtration

Solid/Liquid Separation

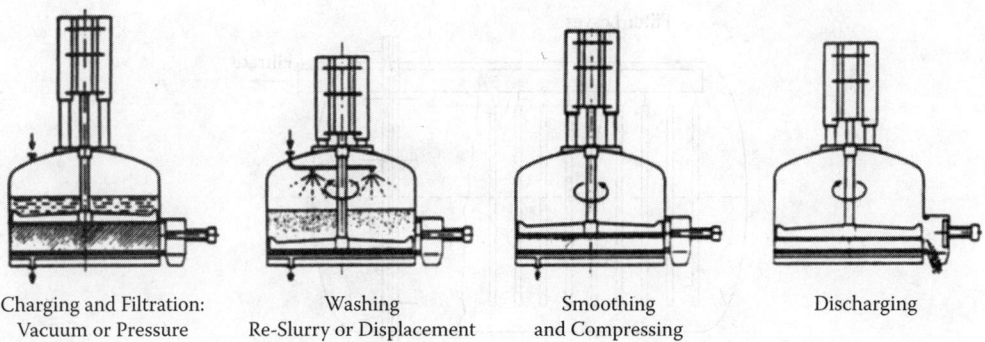

Charging and Filtration: Vacuum or Pressure

Washing Re-Slurry or Displacement

Smoothing and Compressing

Discharging

FIGURE 22.31 Operation of and automatic Nutsche filter.

area, they are often operated with thick cakes and are suitable for small batches. Automatic nutsches are available to perform reaction, crystallization, filtration, reslurry washing, drying, and cake discharging in the same vessel. GMP (Good Manufacturing Practices of FDA) design is also available for processing different pharmaceutical materials. Operation of an automatic nutsche filter is shown in Figure 22.31.

22.5.1.2 Leaf Filter

In comparison with nutsche filters, leaf filters provide more filtration area in the same volume of a pressure vessel. They are more suitable for handling larger quantities of slurry. Leaf filters can be subdivided into four classes in accord with the vertical or horizontal position of the tanks and leaves. In Figures 22.32 and 22.33, vertical leaves are shown in vertical and horizontal tanks. For vertical leaf filters, to prevent cake dropping, the cake thickness is normally restricted to 3.5 to 4.0

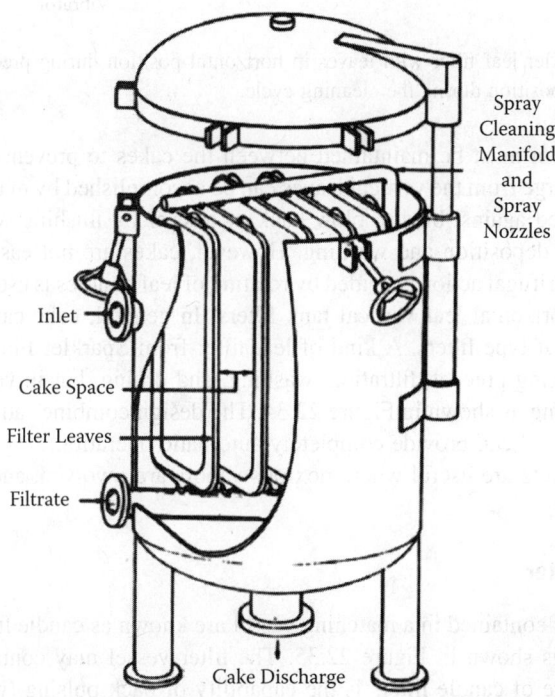

FIGURE 22.32 Vertical tank vertical leaf filter.

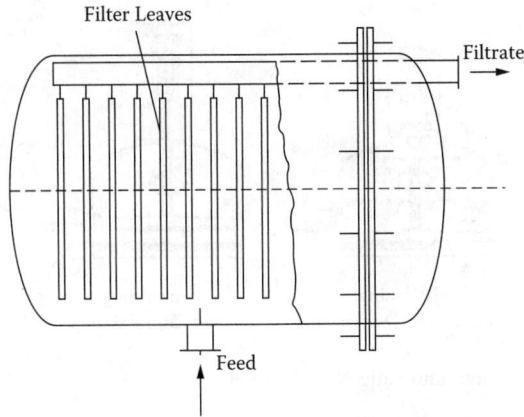

FIGURE 22.33 Horizontal-vessel, vertical-leaf filter.

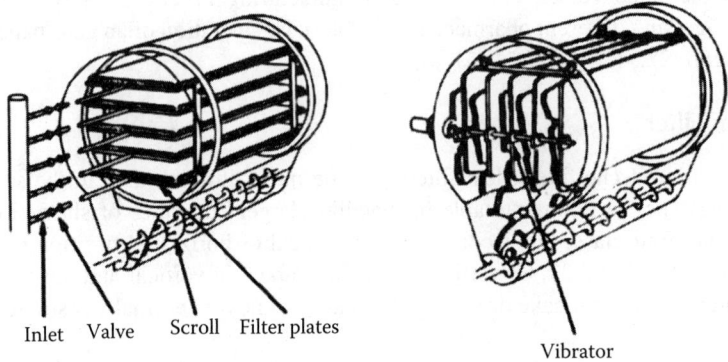

FIGURE 22.34 A Sparkler leaf filter with leaves in horizontal position during precoat, filtration, washing, and drying and vertical position during the cleaning cycle.

cm. A space of ~2.0 cm must be maintained between the cakes to prevent arching and facilitate discharge. Cake discharge from the vertical leaves can be accomplished by manual means, vibration, rotation at a slow speed against blades, blow back of gas, or by flushing with liquid. Horizontal leaves assure uniform deposition and washing. However, cakes are not easy to remove from the horizontal leaves. Centrifugal action provided by rotating of leaf bundles is used for cake discharging in some designs of horizontal leaf vertical tank filters. In general, cake can be discharged more easily from vertical leaf-type filters. A kind of leaf filter from Sparkler Filter, Inc. uses leaves in horizontal position during precoat, filtration, washing, and drying, but in vertical position during discharging and cleaning as shown in Figure 22.34. The design combines advantages of horizontal leaf and vertical leaf, and can provide completely automatic operation.

Pressure vessel filters are useful where noxious vapors are involved, and a completely closed system is desirable.

22.5.1.3 Candle Filter

Tubular filter elements contained in a matching vessel are known as candle filters. A typical candle filter (Purchas 1981) is shown in Figure 22.35. The filter vessel may contain one or more filter candles. The advantage of candle filters is the capability of back pulsing (with dry gas) or backwashing (with liquid) to automatically discharge the cake as shown in Figure 22.36. Some designs

Solid/Liquid Separation

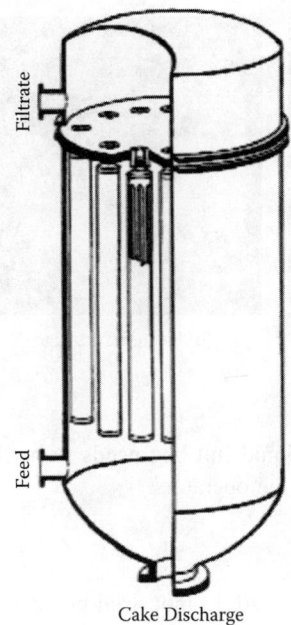

FIGURE 22.35 A candle filter.

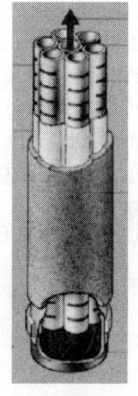 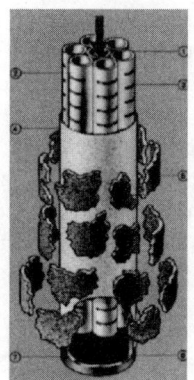

(a) Filtration   (b) Back Pulsing

FIGURE 22.36 a, b: Filtration and back pulsing.

use registers with easy release devices to permit easy isolation of separate candle bundles in case of candle failures, as well as elimination of heavy head plate.

The candle filter element can be textile cloth over a supporting tube, sintered metal, ceramic or spaced rings. It can also be used for air or gas filtration. Metallic candles are particularly suited for high temperature applications.

22.5.1.4 Bag Filters

Disposable bags made of synthetic fibers are fitted into a simple pressure vessel to remove particles from liquid. Such filters are least expensive and very commonly used. Due to the limited surface area, the solid loading should be very low in a bag filter unless the solids can form a permeable cake. It is important to replace the bag when pressure reaches 1 kg/cm² (15 psi) before it is

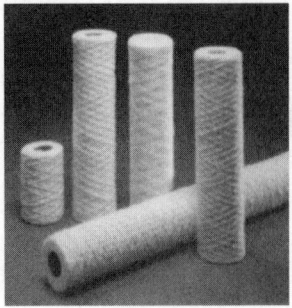  

(a) Wound Cartridge   (b) Pleated Cartridge   (c) Melt Blown Cartridge

FIGURE 22.37 a–c: Cartridge filters.

completely plugged. Otherwise, a liquid full bag needs to be removed, resulting in product loss and industrial hygiene concerns for the operators.

22.5.1.5 Cartridge Filters

Cartridge filters are very commonly used. In industrial processes, usage of bag filters or cartridge filters is greater than other SLS equipment combined. Three types of cartridge filters are shown in Figure 22.37.

In string wound cartridge, fiber yarn is wound around a mandrel to form a depth medium. Filtration is primarily done by deep-bed filtration and the separation efficiency is poorer than the pleated type. Because string wound cartridges are less expensive, they are commonly used as the prefilter in front of the more expensive pleated cartridges. Pleated cartridges are fabricated by pleating fabric sheet to provide additional surface area and strength. The filtration mechanism is surface straining so the filtration efficiency curve is steeper. They are used when high collection efficiencies for specific particles are required. The mechanisms of melt blown or resin bonded cartridges are deep-bed filtration, that will be discussed later.

A term called the β ratio is commonly used to describe the filtration efficiency of cartridge filters. The β ratio is calculated by

$$\beta \text{ ratio} = C_{feed} / C_{filtrate} \tag{22.52}$$

where C_{feed} and $C_{filtrate}$ represent the particle count per unit volume in the feed and filtrate.

22.5.1.6 Filter Presses

Filter presses play an important role in SLS since the nineteenth century. They are used when enclosed operation is not required. An excellent treatment of practical problems encountered with presses was provided by Alliot (1920). There are basically three types of filter presses: plate and frame, recessed, and membrane (diaphragm).

A plate and frame filter press is illustrated in Figure 22.38. A filter medium (usually cloth or paper) placed over a grooved plate serves as the support for the cake, which is deposited in an adjoining frame. Plates and frames are alternated as shown in Figure 22.38. The media serves as the gasket when mechanical closure is effected. Where washing is desired, every other plate is constructed so that liquid can be passed from one plate through the cake to the opposite plate. Thus the wash liquid passes through two cakes. Feed to a plate and frame press is generally through openings at the bottom corner. Large, dense particles tend to settle out, producing a nonuniform

Solid/Liquid Separation

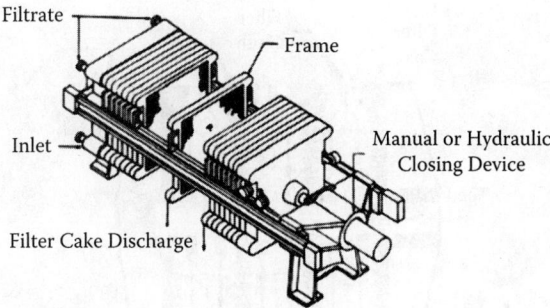

FIGURE 22.38 Plates and frames are alternated on a widebar mechanism.

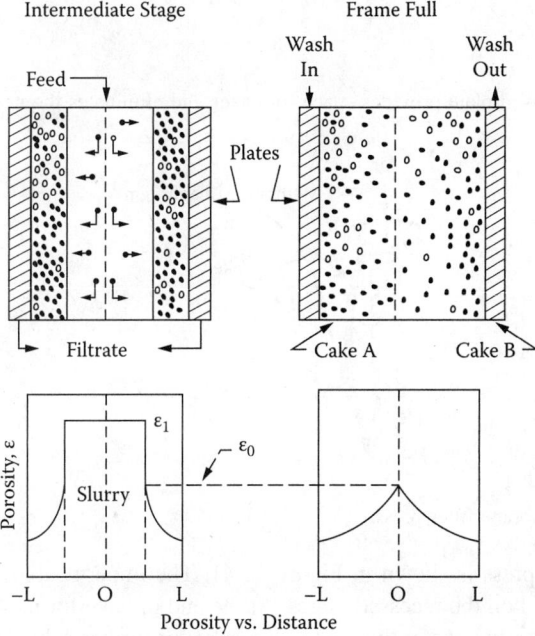

FIGURE 22.39 Ideal cake deposition in a press showing two cakes being laid down simultaneously.

cake with a greater permeability at the bottom. Upward feed has the advantage of reducing sedimentation tendencies.

In Figure 22.39, idealized deposition of cakes in a press is shown. Slurry enters the frame and deposits cakes on plates on either side as illustrated. The cakes, A and B, build out from the plates and meet at the center when the press is full. The cake surfaces meet at the dotted centerline, and a porosity vs. distance curve is developed as shown. Porosity varies from a maximum ε_0 at the unconsolidated cake surface to its minimum value ε_1 at the medium. In the washing mode for presses, liquid flows in one plate, passes through both cakes, and exits from the opposing plate. It traverses cake A in reverse flow with respect to the filtrate. The reversal in flow direction brings about a marked change in the frictional stresses and causes the cake to compact.

The recessed plate press as shown in Figure 22.40 does not require a frame. The edges of the plate are extended outward, leaving a space for a cake. Recessed plate presses are somewhat simpler than plate and frame presses. A center feed as shown in Figure 22.40 is common with recessed plates.

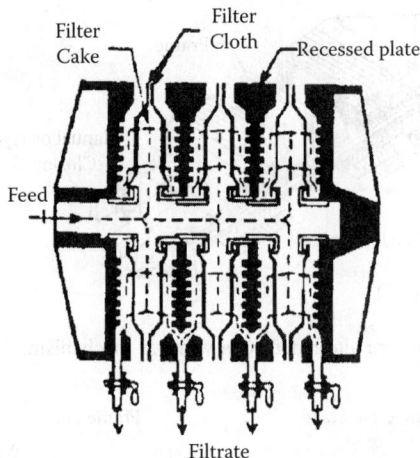

FIGURE 22.40 The recessed plate provides space for cakes and eliminates the need for frames.

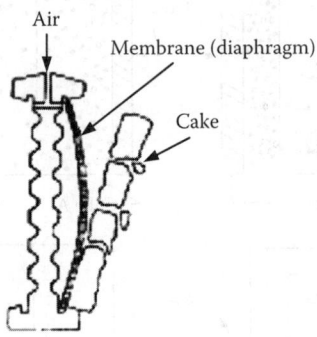

FIGURE 22.41 The membrane filter press.

A membrane filter press is shown in Figure 22.41. The impermeable flexible membranes, or diaphragms, are attached on the recessed plates. At the end of cake formation, the membranes can be inflated by air or pressured water to squeeze the cake for further cake deliquoring. This type of filter provides drier cakes compared with traditional plate and frame of recessed plate filter presses.

22.5.2 Continuous Filters

Continuous filters can be classified into rotary drum (Figure 22.42), disc (Figures 22.42, 22.43, 22.44), horizontal belt (Figure 22.45), table, and tilting pan. Drum filters are further divided according to the method used for cake removal. Most continuous filters use vacuum, and are best suited for materials that permit a reasonably fast rate of cake buildup. Continuous filters work best on medium-sized particles in the range of 5 to 50 μm. Larger particles generally exert minor capillary forces; therefore, cake drying or "drainage" can be accomplished by sucking air through the cakes under vacuum.

22.5.2.1 Rotary Drum and Disc Filters

The salient features of rotary drum and disc filters are illustrated in Figure 22.42. The rotary drum consists of a cylindrical drum having a permeable surface revolving partially submerged in a slurry. Disc filters consist of a series of thin discs revolving on a common shaft and partially submerged in a slurry, which is also shown in Figure 22.43. Continuous filters are normally used for materials

Solid/Liquid Separation

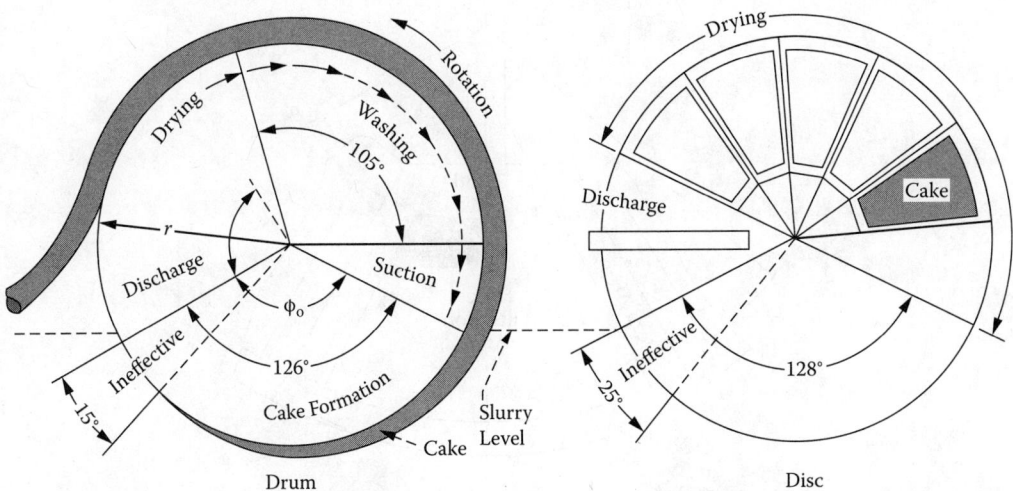

FIGURE 22.42 Cycles for disc and drum filters show cake formation, suction, washing, drainage, and discharge sections.

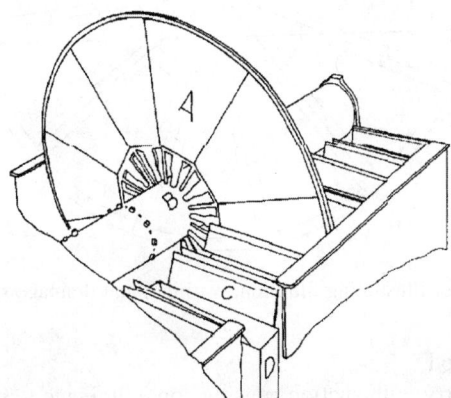

FIGURE 22.43 Isometric view of disc filter.

that are relatively concentrated and easy to filter. Rates of cake buildup are usually measured in cm/min. Submergence normally runs from 25% to 75% of the radius (40% being quite common) with rotation speeds from 0.1 to 3 rpm. With those conditions, filtration times usually range from 5 sec to 7.5 min. Drum diameters typically run from 1.8 to 3.6 m (6 to 12 ft), although larger values may be encountered.

A pressure differential is usually maintained between the outer and inner surfaces by means of a vacuum pump. However, the discs or drum can be enclosed and operated under pressure. In addition to the vacuum or pressure, each point on the periphery of the drum or disc is subjected to a hydrostatic head of slurry. With 40% submergence and a 12 ft (1 ft = 0.3048 m) diameter, the hydrostatic head ranges up to 5.7 ft, which is a significant fraction of the driving force in a vacuum filtration. As the slurry may have a density greater than water, the effective head may be as high as 6 to 7 ft.

The cycle shown in Figure 22.42 for a rotary drum filter consists of

1. Period during which the vacuum source is blocked to prevent air from being drawn through the medium. No filtration takes place.

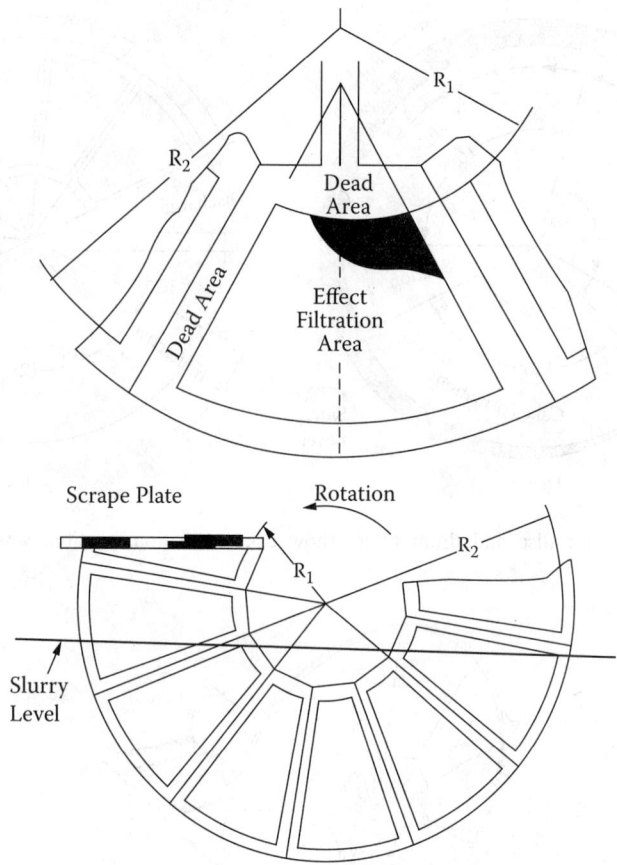

FIGURE 22.44 Horizontal filter illustrating filtration, washing, and drainage stages.

2. Cake formation period.
3. Emergence from slurry with suction causing some drainage to occur.
4. Application of a spray for displacement washing.
5. Drainage period that is frequently called *drying,* although no evaporation takes place. Steam heating reduces viscosity and surface tension and permits improved drainage. Displacement by influx of air may occur.
6. Cake discharge.

Although processes requiring the three steps of cake formation, washing, and drainage (drying) can be more easily done on a drum or horizontal belt filter, drainage can also be accomplished with discs. As illustrated in Figure 22.43, the filter consists of a series of discs *A* that rotate on a common axis in a slurry. The cake is scraped off at *C* and dropped into discharge chutes *D*. Each chute receives solids from facing sides of adjacent discs. The discs are divided into sections that have individual connections to the vacuum and air for the filtering, drainage, and blow zones. Piping is enclosed in the central revolving barrel connected to a wear plate that contacts a control block. Filtration takes place on both sides of the disc. Vacuum is applied inside, and the filter cake is deposited on the disc surface as it rotates through the slurry with only a part of the disc being submerged. At the end of the cycle, the vacuum is broken and the cake discharged. Cake removal from a rotary drum filter is carried out by a number of methods that are related to the properties of the cake and how it sticks to the medium. In Table 22.7, the minimum cake thickness for discharge is shown for different types of drums.

TABLE 22.7
Minimum Cake Thickness Design for Discharge

Filter Type	Minimum Cake Thickness (in.)
Drum	
Standard scraper	1/4
Belt	1/8 to 3/16
Roll discharge	1/32
Coil	1/8 to 3/16
String discharge	1/4
Horizontal belt	1/8 to 3/16
Disc	3/8 to 1/2

Comparison of disc and drum filters indicates that the former may be expected to give two to three times the flow rate of the latter for equal floor space and the same radii. For the disc filter, there is an optimum value of the inner radius (R_1 of Figure 22.44) that yields a maximum flow rate. The optimum ratio of inner to outer radii varies from about 0.6 to 0.75 and is independent of the properties of the slurry (Tiller 1974).

The cake formation time in drum and disc filters is strictly limited to about 40% of the cycle. The times required for adequate washing and drainage are a direct function of solid properties and cake thickness. The form time only occupies about 12% of the circumference. The various stages on the horizontal filter are completely adjustable, and the entire filter surface can be utilized in contrast to the limitations of the disc or drum types.

22.5.2.2 Horizontal Belt Filter

A horizontal filter is shown in Figure 22.45. It occupies more space than drum and disc filters for the same filter area. Normally drum or disc filters are more economical. When settling is acute, belt filters must be used in preference to the drum with the downward facing surface and the disc with the vertical filter surface.

22.5.2.3 Indexing Belt Filter

Like horizontal belt filters, the indexing belt filters are operated semicontinuously under vacuum or pressure. The belt is stationary during filtration and indexes forward for cake discharge. Dispos-

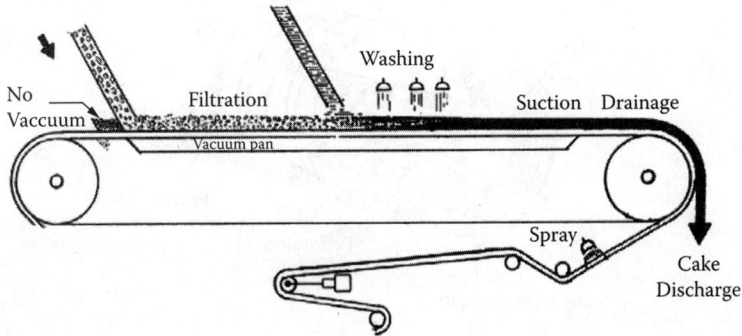

FIGURE 22.45 A continuous horizontal filter.

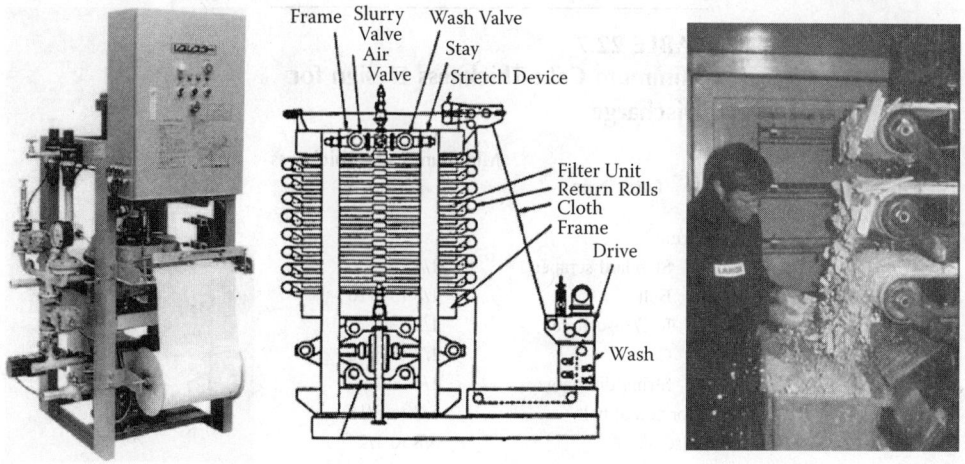

(a) With Disposable Media   (b) Tower Press

FIGURE 22.46 Indexing belt figures.

able media are employed when solids tend to stick to or blind the media (Figure 22.46a). They are especially useful for small waste stream clarifications. A few of these filters can be stacked to increase the filtration area and are called tower presses (Figure 22.46b). In tower presses, only permanent belts are used.

22.5.3 Deep-Bed Filter

Deep-bed filters are employed for slurries with very dilute concentration (<100 ppm = mg/L). The deep beds are in the form of granular media (sand, crushed anthracite coal, garnet) or cartridges (Figure 22.37), which are cylinders containing a variety of materials for trapping the particles. An example of the widely application of cartridge filters is removing particles from lubricating oils in automobiles and trucks.

A typical depth-type cartridge is shown in Figure 22.47. They are disposable. The illustration shows an outer region with large pores and high permeability and an inner region with smaller pores. This gradient design provides filtration efficiency without sacrificing on-line life.

Sand and anthracite are the two most commonly used media in granular bed filters. Single, dual, or multiple media can be used. Finer granular provides higher filtration efficiency but has a

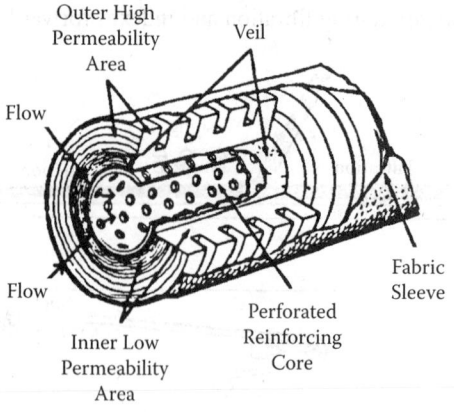

FIGURE 22.47 Johns-Manville cartridge.

FIGURE 22.48 Shriver continuous cross-flow filter thickener.

smaller bed porosity to accommodate the filtered particles. Coarser granular offers more solid loading capacity but with lower filtration efficiency. Dual or multiple media offer the advantage of graded porosity in the bed to enhance solids loading capacity while maintaining the filtration efficiency. In the filtration process, as the deposit builds up, the permeability ultimately drops to a point where the bed must be regenerated. Backwash is generally used to remove the deposit. Air scours can be employed to improve the cleaning efficiency. The backwash water flow rate is roughly twice the filtration rate. Insufficient backwash rate is the common reason for sand filter failure in the field. Even with good backwashes, the granular media need to be replaced every 4 to 5 years.

There are three basic types of granular bed filters: gravity, pressure, and continuous backwash. Higher filtration rate means a higher filter capacity. However, the higher the feed rate, the lower the filtration efficiency. The proper filtration rate for a particular application is best determined by experimenting with the actual filter. The normal operating range is 1 to 5 gpm/ft^2 (gpm: gallon per minute; 1 gallon = 0.003785 m^3; 1 ft = 0.3048 m) for gravity filters, 7 to 10 gpm/ft^2 for pressure filters and 3 to 6 gpm/ft^2 for continuous filters (Chen 2002). The filtration efficiency also depends on the granular size and bed height and is difficult to predict by theories. Any attempts to deviate from the manufacturer's specifications depend on test data.

22.5.4 Cross-Flow Filters

In cross-flow filtration (Figure 22.1c) or delayed cake filtration, the slurry flows parallel to the cake surface with sufficient velocity to prevent partially or entirely the deposition of cake. It is used successfully to increase flow rate in membrane filtration. It is also employed for concentrating and recovering very fine particles in dilute suspensions when deep-bed or cake filtration would not be applicable. In Figure 22.2, cross-flow filtration is employed for both slurry concentrating or thickening (stage 2) and separation (stage 3).

There are basically two types of cross-flow filters. One is without rotating element in which slurry is usually pumped into the filter in the direction parallel to the filter media to produce a cross flow. A modified plate-and-frame filter press provides channels through which the slurry flows (Figure 22.48). There are also units equipped with rotating elements. Either the agitator or the turbines are attached to a rotating shaft.

22.5.5 Membrane Filters

The major application today for cross-flow filtration is in the membrane filtration for bioprocessing or fine particle separations. Based on the size of the particles separated, membrane filtrations are categorized as microfiltration (MF), ultrafiltration (UF), nanofiltration (NF). and reverse osmosis (RO). The ratings of MF membranes are by micron ratings, just like other fabric filter media. In

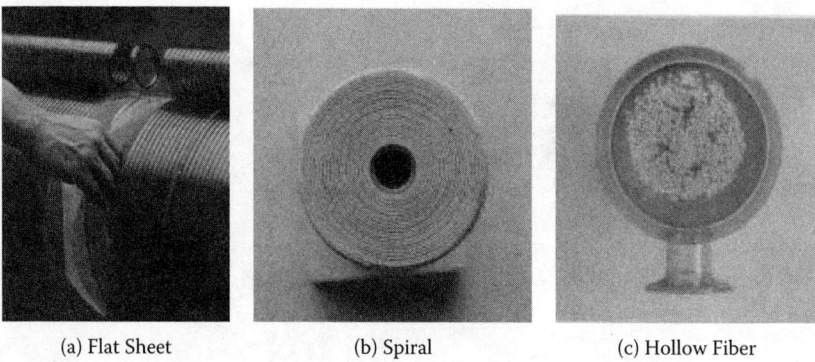

(a) Flat Sheet (b) Spiral (c) Hollow Fiber

FIGURE 22.49 Typical continuous thickener.

UF and NF, a term called molecular weight cut off (MWCO) is used. MWCO means molecules with molecular weight greater than the MWCO will be rejected by this membrane.

Membrane filters come in three configurations: flat sheet, spiral, and hollow fiber (Figure 22.49). Flat sheet membranes can be attached to the plates and by stacking many plates to achieve a large filtration area. The suspension flows across the membrane surface while the permeate flow through the membrane to the collection channels at the edge of the plates. This configuration is most suitable when concentrated suspension is encountered (a few wt%). The manufacturers tend to make special cassettes or plates that are not interchangeable with other suppliers' products.

Spiral elements are the most popular form of membrane filter today due to their relatively low cost. The membrane in a spiral element is wrapped with spacer sheets into a cylinder (Figure 22.49b). The suspension is flowing in parallel to the axis of the cylinder and the permeate flows spirally into the center collecting chamber. Although the design favors low solid concentrations, it is possible to run suspensions with high solid concentrations in a spiral membrane filter.

In hollow fiber systems (Figure 22.49c), a bundle of hollow fibers are contained within a shell and the suspension flows inside the hollow fiber and the permeate is collected in the shell side. Hollow fibers are best suitable for applications without solid particles. In addition to the above three types, there are also rotating discs, annual gap, and vibrating disc systems for membrane cross-flow filtration.

There are two modes that can be used to operate a membrane filter, concentration or diafiltration. In concentration operations, the permeate passes through the membrane and is collected as a filtrate while the retentate (the unfiltered part) is recycled back to the feed vessel. As more and more permeate is removed, the solid concentration in the retentate increases until the desired concentration is reached or the cross flow becomes inefficient due to the high viscosity. The retentate concentration can be calculate as (Cheryan 1998)

$$C_{retentate} = C_{initial} \left(\frac{V_{initial}}{V} \right)^R \quad (22.53)$$

where $C_{retentate}$ and $C_{initial}$ are the solid concentration in the retentate and the initial suspension, and $V_{initial}$ and V are the initial volume and current volume of the retentate in the feed tank. R is the rejection ratio of the species of interest. $R = 1 - C_{permeate}/C_{retentate}$.

In the diafiltration mode, fresh liquid is added to the retentate to maintain a constant retentate volume. This method is useful if some dissolved material in the retentate needs to be extracted to the permeated side. The concentration of retentate can be calculated as (Cheryan 1998)

Solid/Liquid Separation

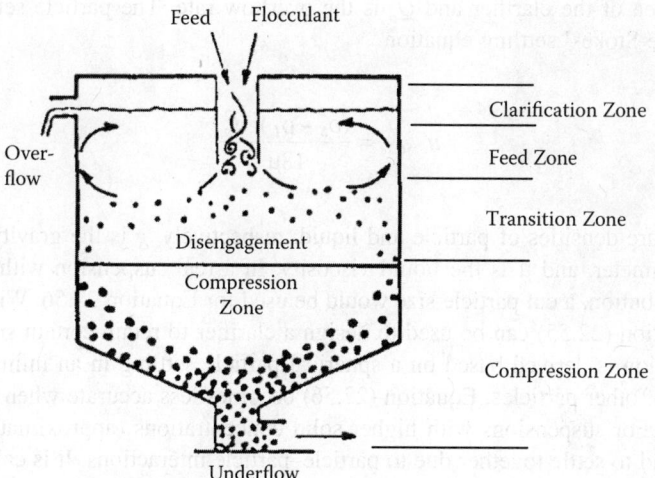

FIGURE 22.50 Typical continuous thickener.

$$C_{retentate} = C_{initial} e^{\frac{V_{permeate}}{V_{initial}}(1-R)} \qquad (22.54)$$

where $V_{permeate}$ is the total volume of permeate collected at the time.

22.5.6 Thickening and Clarification

Clarification and thickening are generally related to dilute suspensions. Clarifiers are employed to remove small quantities of solids where the liquids are the main product. Thickeners are used to concentrate dilute suspensions in preparation for further separation in filters and centrifuges. Thickeners frequently lead to the removal of a large fraction of the liquid in a slurry. Gravitational settlers are usually employed for clarification or thickening. A schematic continuous gravity thickener is shown in Figure 22.50. In the figure, a clear liquid overflows the top while a thickened sediment flows out of the bottom as underflow. Thickeners are widely used in fields such as water treatment, aluminum, coal, pulp and paper, cement, and sugar manufacturing.

Usually, there are four zones in a thickener: the clarification zone, the feed zone, the transition zone, and the compression zone. Not all of these zones will be present for all types of slurries. Liquid flows out of the feed zone, up through the clarification zone, and toward the overflow. Below the feed is the transition zone, where the particles start a downward motion. The concentration just below the supernatant-suspension (or sediment) interface assumes a value dictated by the dominance of either free settling or the presence of a compacting sediment. The region containing the particles offers resistance to liquid flow and accepts only a portion of the liquid. The remainder reverses direction and exits as overflow. Under steady-state conditions, both solids and liquid fluxes are constant and independent of depth in the suspension and/or sediment zone. No liquid is squeezed out and upward in the sediment unless channels exist.

A clarifier is designed by keeping the particles from flowing out of the clarifier. In order to do so, the liquid rising rate Q_o/A must be less than the particle settling rate u_s ($Q_o/A \leq u_s$). Therefore, the area of a clarifier can be calculated by

$$A = \frac{Q_o}{u_s} \qquad (22.55)$$

where A is the area of the clarifier and Q_o is the overflow rate. The particle settling rate can be estimated with the Stokes' settling equation

$$u_{S,Stoke} = \frac{(\rho_S - \rho_L)g d_p^2}{18\mu} \quad (22.56)$$

where ρ_S and ρ_L are densities of particle and liquid, respectively, g is the gravity acceleration, d_p is the particle diameter, and μ is the liquid viscosity. In a real suspension with a large range of particle size distribution, a cut particle size would be used for Equation 22.56. With known settling velocity u_s, Equation (22.55) can be used to design a clarifier to retain certain sized particles.

Stokes' equation is derived based on a spherical particle settling in an infinite liquid without the interference of other particles. Equation (22.56) becomes less accurate when the solid concentration increases. For suspensions with higher solid concentrations (approximately >1 wt%), the solid particles tend to settle together due to particle–particle interactions. It is called zone settling. For this type of settling, Stokes' law no longer applies and the settling rates are no longer dependent on the particle size but more dependent on the solid concentrations. Although there are correlations to calculate the zone settling velocity, it is better to run a settling test to obtain the settling velocity.

Almost all the design methods for continuous gravity thickener in the zone settling region are based on the following equation (Coe and Clevenger 1916):

$$G_T = \frac{u_S}{\dfrac{1}{C} - \dfrac{1}{C_u}} \quad (22.57)$$

where G_T is the total solid flux in the thickener, C and C_u are the solid concentration in the thickener and in the underflow, and u_s is the zone settling velocity corresponding to C that can be obtained by a batch settling test as shown in Figure 22.51. With known $u_s \sim C$ values, solid flux G_T as a function of C at a desired C_u can be calculated based on Equation (22.57), and is shown in Figure 22.52. The smallest G_T at point A represents the bottleneck for solid flux in the thickener. The area (A) of the thickener is designed based on this limiting flux ($G_{T,min}$):

$$A = \frac{Q_{feed} \cdot C_{feed}}{G_{T,min}} \quad (22.58)$$

For even higher solid concentrations where the particles are in contact with each other, "free settling" phenomena is replaced by "compression" or "consolidation." The Coe and Clevenger–type design methods are not sufficient. Sediment height and consolidation effect need to be included in the thickener design (Tiller and Tarng 1995).

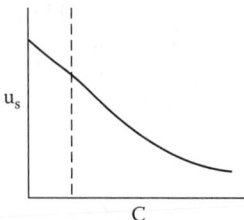

FIGURE 22.51 Settling test for thickener.

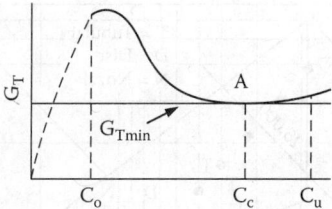

FIGURE 22.52 Designing settler.

22.5.7 Centrifugation

In centrifugation, high-speed rotating equipment is used to generate centrifugal forces as the driving force for separation. Centrifuges can be operated as classifiers, settlers, and filters. The centrifugal acceleration is calculated by $r\Omega^2$, where r = radius and Ω = angular velocity in radians/sec.

Stacked disc centrifuges (Figure 22.53a) are composed of a stack of closely spaced (a few millimeters) discs rotating at high speed (~6000 to 8000 g-force). This type of centrifuge is useful to separate fine particles. The narrow gaps between discs greatly reduce the distance for the particles to travel before they settle to the discs. For very dilute suspensions, a solid wall bowl is used. The machine is manually opened to clean the solid in the centrifuge. The bowl can also be automatically opened during the operation to discharge the collected solids. For even higher solid applications, a series of nozzles are installed along the periphery of the bowl to continuously discharge the solids.

Solid bowl decanters operate at lower g-forces (~2000 g-force) and normally are used for suspensions with high solid concentrations. The slurry enters the decanter as shown in Figure 22.53b. The solids settle out and are conveyed by a scroll mechanism, turning at a lower or faster speed than the bowl, toward the discharge. The clear liquid moves to the cylindrical (right) end and flows over weirs and is removed as *centrate*, corresponding to the filtrate. The region to the right of the slurry inlet is called the *pond*. The solids are conveyed to the conical (left) end along with some liquid, moving up an inclined portion of the bowl (called *beach*) and discharged as the cake.

In filtering centrifuges, the solids settle out as in the decanter. However, the liquid flows in the same direction and out through the perforated bowl. As the liquid flows in the same direction as the solids, it increases the rate of deposition. In addition, as the liquid flows through the cake, the frictional effect results in sharply increased stresses on the cake. At low rotation speeds, the stress due to the frictional flow may be larger than the centrifugal body forces (Tiller and Horng 1983).

Filtering and sedimenting centrifuges operate in sizes from 6 to 60 in. and 500 to 10,000 rpm. The rpm and bowl diameter are limited by the peripheral speed Ωr_b (maximum about 350 ft/sec) (Ambler 1997). A speed of 275 ft/sec corresponds to the heavy dots in Figure 22.49.

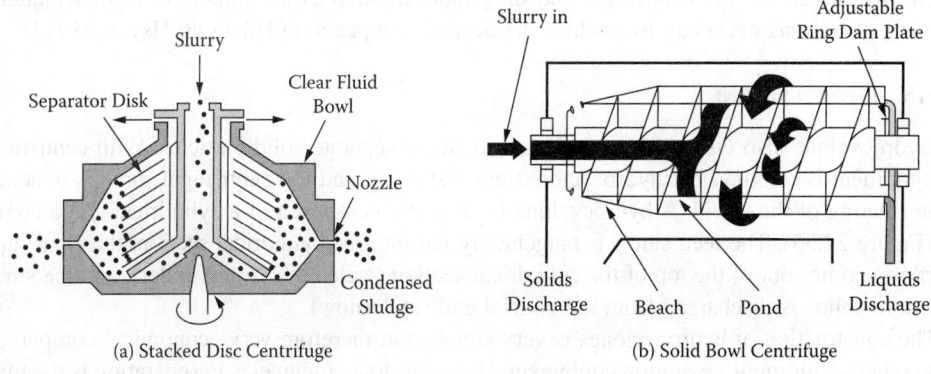

FIGURE 22.53 Typical centrifuges.

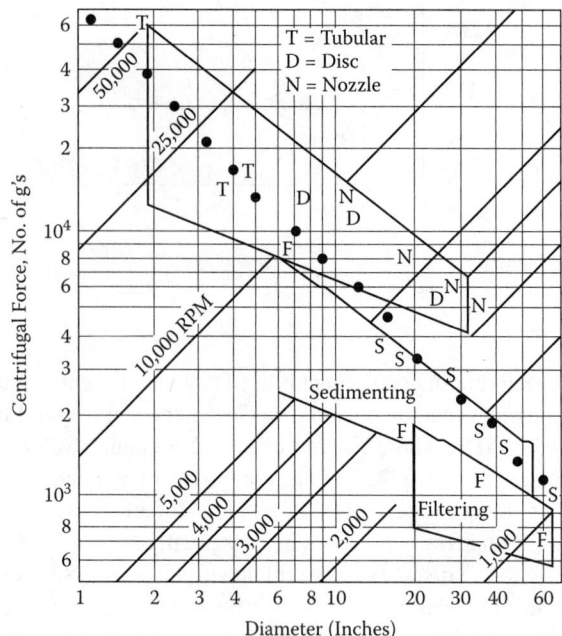

FIGURE 22.54 Classification of centrifuges.

In general, sedimenting centrifuges have smaller diameters and operate at higher speeds in comparison with filtering centrifuges. As a rough guide, centrifuges are designed to operate below the heavy dotted lines in Figure 22.54. The precise location of the dotted line depends up the elasticity and strength of the materials of construction.

In Figure 22.54, centrifuges are classified according to bowl diameter and the number of g-forces, which is defined by $r_b\Omega^2/g$, where r_b is the bowl radius. For a dilute suspension with particles in the free-settling region, the higher acceleration for small diameter units leads to improved rates of sedimentation. Tubular and disc types are used primarily for clarification where the throughput rate of solids is small. Solid bowl decanters can be used for processing large solid flow rates.

Compaction forces developed in filtering centrifuges differ markedly from those encountered in sedimenting centrifuges. In sedimenting centrifuges, the solids flow radially toward the bowl and displace liquid, which must then flow radially inward and resulting drag on the particle delays sedimentation and compaction. In a filtering centrifuge, liquid and solids flow in the same direction. Frictional drag due to the liquid flow through the cake adds to the centrifugal body force. The relative magnitudes of the centrifugal and drag induced stresses are important to understanding operating conditions necessary to produce the desired compaction (Tiller and Hsyung 1993).

22.5.8 Hydrocyclone

The hydrocyclone also utilizes the centrifugal forces to separate solid particles. With centrifuges, the equipment is rotating, but hydrocyclones are stationary and the centrifugal force is generated by the rotating of the liquid. A hydrocyclone is normally composed of a cylindrical and a conical part (Figure 22.55). The feed slurry is tangentially fed into the cylindrical section, the clear liquid (overflow) comes out of the top of the cylindrical section (called the vortex finder), and the stream with more solids is discharged from the conical end (underflow).

The construction of hydrocyclones is very simple and therefore very economical compared to all the other solid/liquid separation equipment. However, low efficiencies in separation is the major disadvantage.

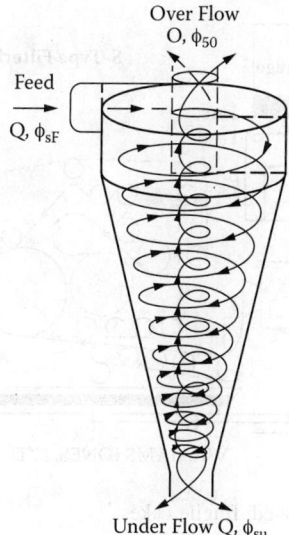

FIGURE 22.55 A hydrocyclone.

There are numerous models to predict the separation efficiencies in hydrocyclones. One example is (Besendorfer 1996):

$$X_{50} = 10 \cdot D^{0.66} \left(\frac{0.53 - \phi_{sf}}{0.53} \right)^{-1.43} \cdot \sqrt{\frac{1.65}{\rho_S - \rho_L}} \cdot \Delta P^{-0.28} \cdot \mu^{0.5} \qquad (22.59)$$

where X_{50} is the *cut size*, which means the particle size (in μm) with 50% separation efficiency in this hydrocyclone. D is the hydrocyclone diameter (in inches) at the cylindrical section. ϕ_{sf} is the feed slurry concentration expressed in volume fraction. ρ_s and ρ_L are densities of particle and liquid in g/mL. ΔP is the pressure drop (in psi) between feed and overflow. It needs to be noted that all the models [including Equation 22.59)] require validation with the actual hydrocyclone system under study. Severe discrepancy between operating data and model prediction values are possible (Chen 2004).

22.5.9 Expression Equipment

The extent of flocculation in the pretreatment step can have profound effects on the behavior of cakes and sediments in the succeeding stages. As flocs grow in size, the floc encompasses a growing proportion of liquid. Sedimentation velocities increase, and it appears that separation will be facilitated. However, the highly porous floc produces a cake with a high porosity (may be on the order of 95%), which may have unfavorable filtration characteristics. Because of the high porosity, the cake frequently must be squeezed in an expression operation to reduce the liquid content. Decreasing the liquid content of cakes has become a major objective of many SLS processes. Although the term *dewatering* is widely used, *deliquoring* is more general and covers both aqueous and nonaqueous liquids. Deliquoring operations consist of expression (squeezing), blowing, sucking with a vacuum, and gravitational and centrifugal drainage.

Figure 22.56 shows the mechanically deliquoring of flocculated, fragile and high porosity cakes by a belt filter press. In the process, large flocs settle out rapidly on the porous belt, and the liquid drains through quickly, producing a very soupy cake. Expression (the fourth stage of Figure 22.2) of the liquid from the cake is essential. It is accomplished by the squeezing action of the two

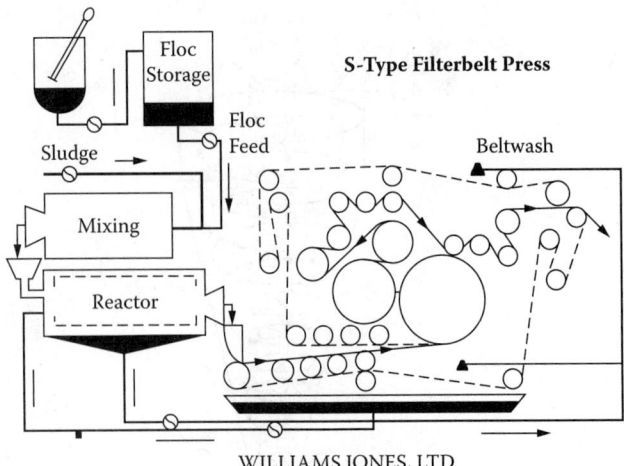

FIGURE 22.56 Deliquoring of flocculated, fragile cakes.

moving belts. A belt filter press is a common type of equipment used for deliquoring activated sludge at many wastewater treatment plants.

A laboratory instrument called a crown press is the best bench device to simulate belt filter press performance.

There are also compression filters which are operated at high pressures. These units are operated in batch mode. The filter press discussed earlier can be equipped with a diaphragm to squeeze the cake after filtration. They are commonly called membrane presses although they have nothing to do with the microfiltration membranes. One variation of the membrane press is the horizontal diaphragm press which also has a diaphragm to compress the cake, but the filter surface is horizontal instead of vertical as in membrane presses. For very high pressure expression, tubular presses can provide squeezing pressures as high as 2000 psi.

22.6 CAKE WASHING

22.6.1 Displaced Washing and Repulping

Cake washing can be classified by (1) displacement washing: using clear liquid to displace mother liquor in the cake; and (2) repulping: diluting with liquid, and filtering again. Initial liquid leaving the cake in displacement washing as shown in Figure 22.57 has a concentration equal to that of the mother liquor. When approximately half (a large variation in this value can be encountered with non-Newtonian liquids, nonuniform cakes, or where there is cake cracking) of the original liquid has been displaced, fresh wash breaks through, and the exit concentration begins to drop rapidly. Displacement becomes inefficient, and solute removal is controlled by diffusion from relatively stagnant areas. Where high degrees of solute removal are essential, it may be useful to follow displacement washing by reslurrying and replacing the displacement process. Only displaced washing will be discussed in this section.

22.6.2 Terms and Definitions for Displacement Washing

Wash ratio j: a ratio equaling the total volume of wash liquid exiting a cake to the volume of void of the cake.

Mother liquid concentration, C_o: concentration of liquid remaining in the cake before washing.

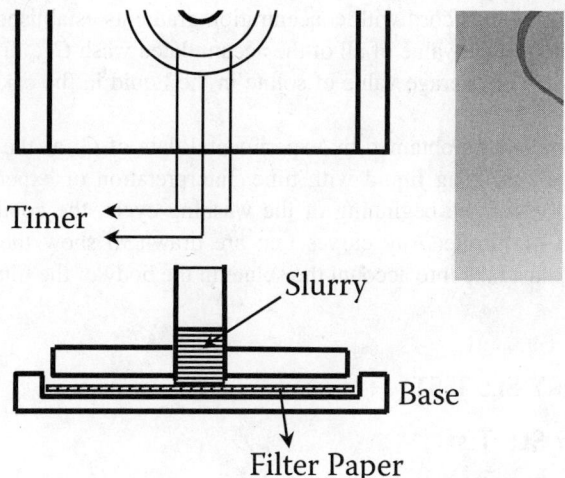

FIGURE 22.57 Displacement washing.

Instantaneous wash liquid concentration, C_w: instantaneous concentration of wash liquid when it is exiting the cake.

Average wash liquid concentration, C_{wav}: average concentration of total wash liquid coming out of the cake.

Average concentration of liquid in the cake, C_{av}: average solute concentration of total liquid remaining in the cake.

22.6.3 Washing Curves

Washing curves are used to describe washing efficiency. Frequently washing curves are represented by plots of C_w/C_o as functions of wash ratio j as shown in Figure 22.58. The ratio of C_w in mass to C_o indicating wash efficiency is constant for a portion of the wash cycle and drops rapidly after the wash ratio j increases beyond 0.6. The drop starts to slow down when the wash ratio j reaches about 1.5 when diffusion washing becomes the dominate factor. Solute is not uniformly distributed

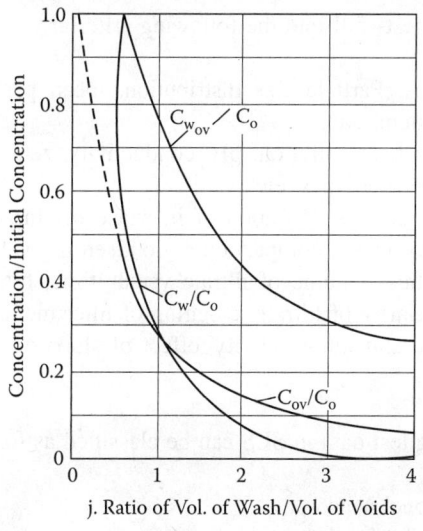

FIGURE 22.58 Washing curves.

in the cake, and it diffuses in accord with concentration gradients established in the interstices of the particulate bed. The average value of all of the accumulated wash C_{wav} divided by C_o lies above the instantaneous curve. The average value of solute in the liquid in the cake C_{av} is also shown in Figure 22.58.

The washing curves can be obtained by experimental data of C_w as the function of time, and instantaneous volume of washing liquid with time. Interpretation of experimental data must be approached with caution. At the beginning of the washing cycle, the conduits leading from the filter unit will be full of filtrate. Any curves that are drawn to show the percentage of solute remaining in the cake must take into account the solute in the body of the filter as the well as solute in the cake.

22.7 LABORATORY SLS TEST

22.7.1 Laboratory SLS Test

Separations can be classified as dealing with homogeneous or heterogeneous systems, e.g., fluid/particle systems. Homogeneous separations such as distillation, absorption, and extraction represent mature technologies where the fundamentals have been translated into well-developed design procedures with competing software programs being widely available. On the other hand, heterogeneous separations or fluid/particle separations involve more complex, less-well-understood phenomena. Parameters such as settling velocity in sedimentation and porosity and permeability of filter beds in filtration are very much depended on experimental results. Laboratory testing necessarily plays an important role in control and design of fluid/particle separation systems.

Objectives of laboratory SLS tests include

- Cake filtration characteristics such as porosity, permeability, specific resistance, and cake compactibility parameters n, β, δ, $\alpha_o \varepsilon_{so}$, or p_a in Equation (22.29)
- Selection of filter media
- Selection of filter aids
- Selection of coagulations and flocculants
- Cake washing efficiency
- Settling velocity for thickener design and scale-up

Parameters involved in SLS tests fall into the following criteria:

- Particle characteristics: Particle size distribution, mean particle size, particle shape, density, chemical content, etc.
- Slurry characteristics: Concentration, pH, conductivity, zeta potential, particle settling velocity, rheology characteristics, etc.
- Sedimentation characteristics: Variation of interface of slurry and clear liquid against time, zone-settling region and compaction region, settling velocity
- Filtration characteristics: Volume of filtrate versus time, filtrate rate, volume of filter cake, cake porosity, clarity of filtrate, clogging of filter media, effect of slurry concentration on filtrate rate and filtrate clarity, effect of slurry viscosity on filtrate rate and media clogging, etc.

Major methods for obtaining test data in SLS can be classified as follows:

1. Flow under gravitational head
    a. Filtration with falling head
    b. Filtration with constant head

Solid/Liquid Separation

2. Constant pressure filtration
   a. Vacuum leaf (Figure 22.59a)
   b. Buchner funnel (Figure 22.59b)
   c. Plate frame filter press (Figure 22.60)

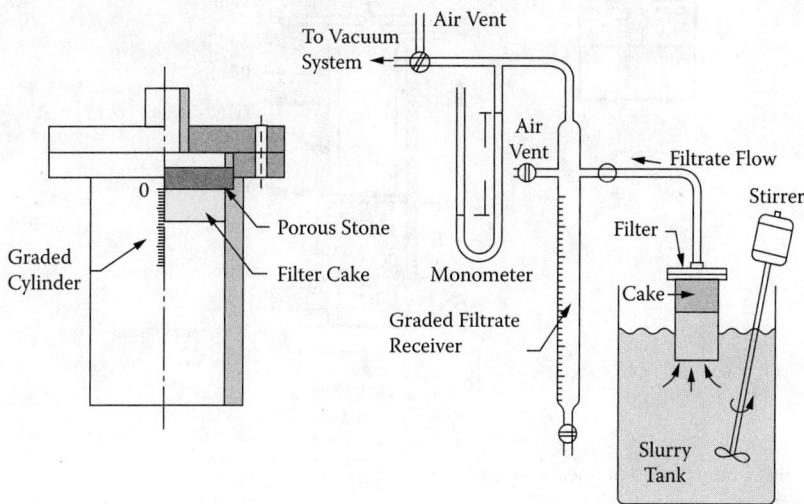

FIGURE 22.59a Vacuum horizontal leaf test filter.

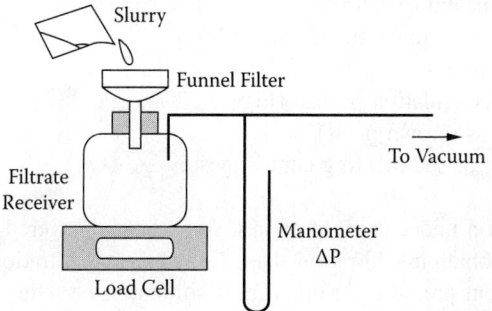

FIGURE 22.59b Vacuum Bucher funnel.

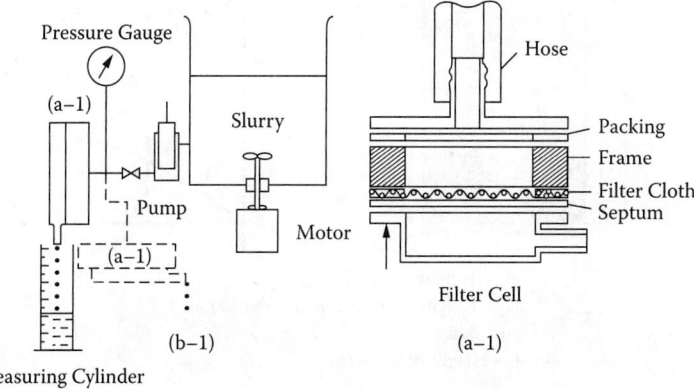

FIGURE 22.60 Plate-and-frame filter press.

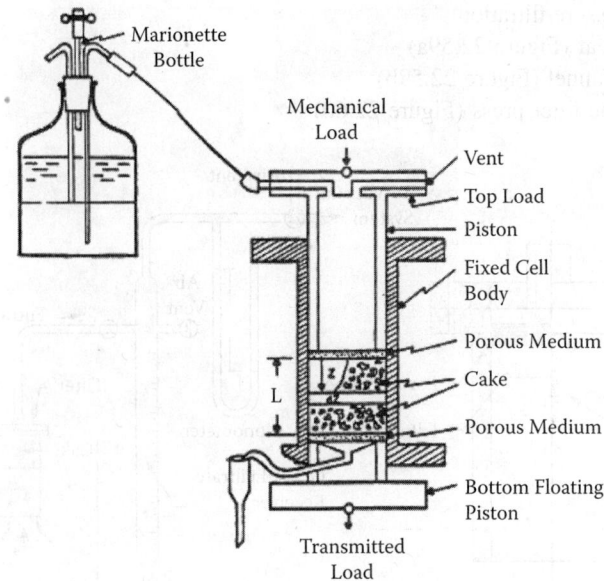

FIGURE 22.61 Compression permeability cell.

3. Constant rate filtration
4. Variable pressure and variable rate filtration
5. Compression-permeability cell (Figure 22.61)
6. Capillary suction time apparatus (Figure 22.62)
7. Batch sedimentation test
8. Flocculation and coagulation jar test (Figure 22.7)
9. Lab scale belt press (crown press)
10. Lab scale filtering or sedimenting centrifuges.

Constant pressure filtration under vacuum or in a small bomb powered by gas pressure has been the favorite method for obtaining filtration data. Constant rate filtration offers the advantage of providing data for different pressures in one run. It simulates the industrial filter operation where

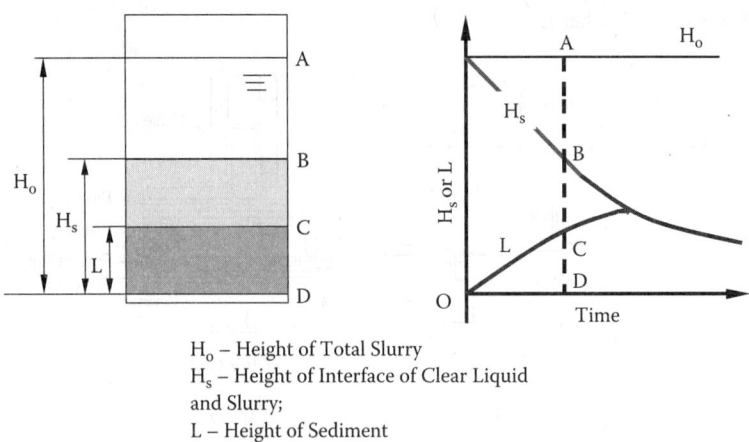

H_o – Height of Total Slurry
H_s – Height of Interface of Clear Liquid and Slurry;
L – Height of Sediment

(a) Batch Sedimentation Cylinders   (b) Batch Sedimentation Results

FIGURE 22.62 Capillary suction time apparatus.

the rate is constant but the pressure drop gradually increases. Variable-pressure, variable-rate filtration is similar to constant-rate operation but has the advantage of not requiring an absolutely constant rate. As the calculated flow resistance varies with the square of the rate, small errors in constant rate testing are magnified in determining resistance.

The capillary suction apparatus is particularly useful for rapid screening involving small samples. A tube containing several milliters of slurry is placed on filter paper. The liquid is sucked out of the tube and flows in a radial or rectangular manner through the paper. The time to flow through a specified distance from r_1 to r_2 is called the capillary suction time. Simplified theory for calculating CST is complicated by variable permeability parallel and perpendicular to the grain of the filter paper. Ellipses rather than circles result. In addition, sedimentation affects the rate of cake formation and the CST (Tiller and Li 2001b). For industrial applications, a commercially available capillary suction time apparatus is useful (Figure 22.62a).

22.7.2 Problems in Laboratory Tests

22.7.2.1 Samples

Representatives of samples and aging of samples are two big problems in lab tests. Special care should be paid to the sampling procedure. Due to aging of the sample slurry, and changing characteristics of slurry during shipping and storing, in-plant testing on fresh samples is strongly recommended.

22.7.2.2 Adverse Effect of Sedimentation on Filtration Experiments

All filtrations suffer from the difficulty of simultaneous sedimentation (Tiller et al. 1995). The relative rates of sedimentation and filtration or centrifugation underlie an assessment of the effect of sedimentation. Gravity filtration invariably requires agitation or recirculation. Sedimentation effects on a horizontal surface facing upward can be minimized by decreasing the slurry holding volume above the cake and providing adequate mixing in the reservoir. Particles with a wide size range in dilute slurries are almost impossible to filter satisfactorily in gravity filtration because of differential sedimentation. Large or dense particulates settle rapidly and form a cake that initially has an abnormally low resistance.

22.7.2.3 Wall Friction in C-P Cell and Small-Scale Cells

Percolation through compacted beds in compression-permeability (C-P) cells can be severely affected by sidewall friction which produces substantial cake nonuniformity (Tiller et al. 1972). It has been found that a large fraction of the applied load is absorbed by the wall. Risbud (1974) showed that thin cakes do not necessarily lead to uniform pressure distribution. For very thin cakes, the pressure just below the piston at the cake interface was distinctly nonuniform. The pressure at the centerline was higher than the applied pressure by as much as 10% to 15%. Radial movement of the solids involving friction on the piston as well as the walls accounts for this apparently strange behavior. For cakes of 20 to 75 mm in thickness in steel cells with diameters of 50 to 100 mm, as much as 50% of the load at the surface may be absorbed by the walls. The authors do not believe it is possible to produce a uniform cake in a compression-permeability cell or consolidometer. While the equipment can yield valuable information, it is complex, requires much time, and represents a relatively large capital investment if appropriately instrumented. It does not lend itself to routine testing.

22.7.2.4 Test Numbers

Determination of the effect of pressure drop and g-forces on flow rate and liquid content is the chief objective of tests involving compactible cakes. Accurate calculation of parameters in constitutive equations [Equation (22.29)] generally requires that a relatively wide range of pressures be utilized. Ideally, a very low pressure in the neighborhood of 300 mm of water (0.5 psi) should be used. Capillary suction apparatus employing Whatman #17 has a suction pressure of approximately 14.85 kPa (2.15 psi). Determination of compactibility necessitates the use of gravity, vacuum, or pressure. Due to the limited data obtained, errors from wall friction, or sedimentation during tests, most compressibility coefficients reported in the literature are of questionable quality.

22.8 SLS SYSTEM DESIGN

For a particulate application involving SLS, how to choose the best equipment, to use batch or continuous operation, which filter medium to select, and what is the optimum operating conditions are concerns of engineers. Although theories are available for some SLS operations, solutions for equipment selection, process design, and optimization are still very much dependent on test and experience, and are frequently qualitative or semi-qualitative. In this section, strategy and decision networks for selection of SLS equipment, introduction of filter media, centrifugal pumps for filtration operation, and selection of filter aids by cycle analysis will be discussed.

22.8.1 Selection of SLS Equipment

22.8.1.1 Specifications of a SLS System

In approaching the initial selection of equipment, questions include what are the product (liquid or solid), the feed and production rates of the system, the required filtrate clarity, and the liquid and solute contents in the cake. The product may be solids, liquid, or both. For example, in the liquefied coal manufacturing process, the dissolved carbon in the liquid represents the main product. At the same time, minerals in the original coal end up as cake along with unconverted carbon, which is also a valuable component and can be used for the production of hydrogen. As far as the SLS operation is concerned, both cake solids and filtrate are of value in liquefied coal processes.

Purchas (1977) provided a conceptual model for initial specifications of a SLS system as illustrated in Figure 22.63.

22.8.1.2 Selection of Equipment and Process

The selection depends to a great extent on the rate of growth of the cake. This rate has a major effect on the following choices: batch versus continuous process; pressure or vacuum; gravity filters; and separation technique (filters, centrifuges, etc.). Table 22.8 (features of slurry), Table 22.9 (selection of filters), and Table 22.10 (comparison of separation processes) aid in this effort.

The basic strategy for equipment selection relating to cake buildup rate in Table 22.8 can be summarized as follows:

1. Continuous filtration is suitable for material of rapid cake buildup.
2. Medium-speed filtering materials are suited to vacuum equipment and filtering centrifuges like peeler or vertical basket centrifuges.
3. Slow filtering materials require pressure equipment or sedimenting centrifuges.
4. Dilute materials that result in high-resistance cakes are generally best handled in batch pressure filters and filter aids can be considered. If the solid concentration is less than 100 ppm, deep-bed filters become options.
5. Rapid settling slurries lend themselves to gravity separation or filtering centrifuges with wedge wire screen (like pusher or screen bowl centrifuges).

Flood, Porter, and Rennie (1977) developed Table 22.9 as an expanded version 5 to Table 22.8.

Solid/Liquid Separation

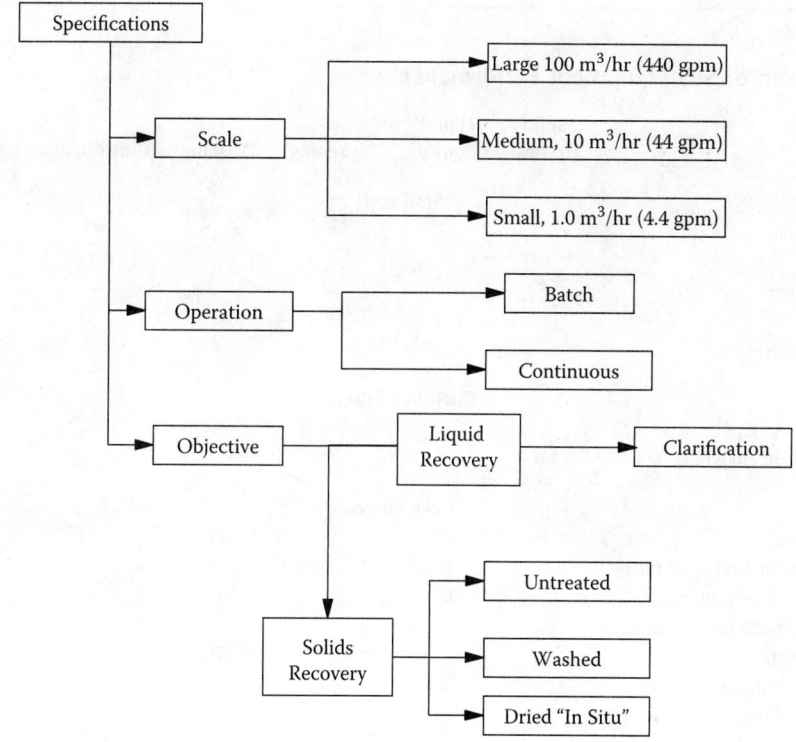

FIGURE 22.63 Specification of scale, type of operation, and objective.

TABLE 22.8
Slurry Characteristics

Type	Buildup	Characteristics
Rapid filtering	cm/sec	High concentration, fast settling, hard to suspend
Medium filtering	cm/min	Medium concentration forms, cake easily can be suspended
Slow filtering	cm/hr	Low concentration, high resistance, may blind media
Clarification	Little cake	Less than 1000 ppm, very high resistance

TABLE 22.9
Guide to Filter Selection

Slurry Settling Characteristics	Fast	Medium	Slow	Dilute	Very Dilute
Cake rate	in./sec	in./min	0.05–0.25 in./min	<0.05 in./min	No cake
Normal concentration	>20%	10%–20%	1%–10%	<5%	<0.1%
Settling rate	Rapid	Fast	Slow	Slow	—
Leaf test rate (lb/hr/ft^2)	>500	50–500	5–50	<5	—
Filtrate rate (gal/min/ft^2)	>5	0.2–5	0.01–0.02	0.01–2	0.01–2

Source: From Fitch (1974).

TABLE 22.10
Comparison of SLS Equipment Performance

	Solid Dryness	Liquid Clarity	Thickening	Washing	Classification	Particle Breakage
Strainers						
Screen strainers		G			G	G
Bag filters	F	—			—	G
Cartridge filters	F	—			—	G
Screens	F	F–P	G	G		G
Membrane filters					—	G
Deep-Bed Filters						
Granular (batch)			F	—	—	—
Granular (continuous)			F			—
Cake Filters						
Gravity filters	F	G		G	—	
Batch, semi-batch vacuum filters	G	G			—	
Continuous vacuum filters	G	G		F	—	
Batch, semi-batch pressure filters					—	G
Continuous pressure filters				F	—	G
Compression filters		F	—	F	—	
Filtering centrifuges	G				—	F–P
Precoat filters	—	—	—	—	—	—
Settlers						
Gravity, clarifier		G	—		—	
Gravity, thickener		G			—	
Tubular centrifuge			—	—	—	F
Disc centrifuge, solid bowl			—	—	—	G
Disc centrifuge, desludger					—	F
Disc centrifuge, nozzle		G	G	F		F
Solid bowl centrifuge	F–G	E–G				
Hydrocyclones			G	F		

Note: E, excellent; G, good; F, fair; P, poor; —, should not be considered.

Source: From Chen (2002).

22.8.1.3 Comparison of SLS Equipment

Fitch (1974) discussed the general problem of matching process specifications to SLS equipment and the necessity of considering trade-offs involving such items as filtrate clarity, cake dryness, reliability, maintenance, versatility, and cost. In Table 22.10, a comparative profile of all major types of SLS equipment is given. It should be emphasized that a wide range of performance occurs because of the endless variation in slurries, and the descriptive terms would undoubtedly vary some if another author were to prepare the same table.

22.8.2 Decision Network for Design for Cake Filtration Systems

A series of decisions must be made concerning the following factors in designing new or improving existing cake filtration systems:

Solid/Liquid Separation

1. Rate of cake buildup
2. Filtrate clarity
3. Solubles in cake
4. Liquid content of cake

In Figure 22.64, a framework for making decisions at different junctures is provided. At the first step, a decision must be made as to whether or not pretreatment may increase the rate of filtration and what type of filtration to use. Once a particular filter has been chosen, attention then focuses on the filtrate and wet cake. If the clarity is not satisfactory, a second-stage filtration (cartridge or granular bed) may be utilized. Alternatively, a tighter medium or a change in filter aid quality or quantity might provide a solution.

To remove mother liquor (or solubles in the mother liquor) in the cake, displacement or reslurrying washing, pressing, vacuum drainage, and blowing are employed in filters. Drainage or blowing can be employed separately or in combination with washing. At the start of displacement washing, pure mother liquor flows from the cake; and the washing efficiency is 100%. However, after approximately half of the liquid has been displaced, pure wash liquor breaks through the cake; and the displacement mechanism rapidly loses its effectiveness. Solubles may be trapped in "blind pores" where the wash does not effectively penetrate, and diffusion becomes the predominate mechanism of solute removal. Washing is inefficient in the diffusional stage, and reslurrying and refiltering may be the best path to follow.

Removal of residue liquid in the cake can be accomplished hydraulically or mechanically. The simplest procedure consists of increasing pressure at the end of filtration. If a centrifugal pump is employed, it may not be feasible to increase pressure, whereas a positive displacement pump provides a simple means for adjusting the final pressure. Highly compactible cakes do not generally lend themselves to reduction of average porosity by simple pressure increases. As demonstrated in Section 22.4.6, cakes that are highly compactible act differently under mechan-

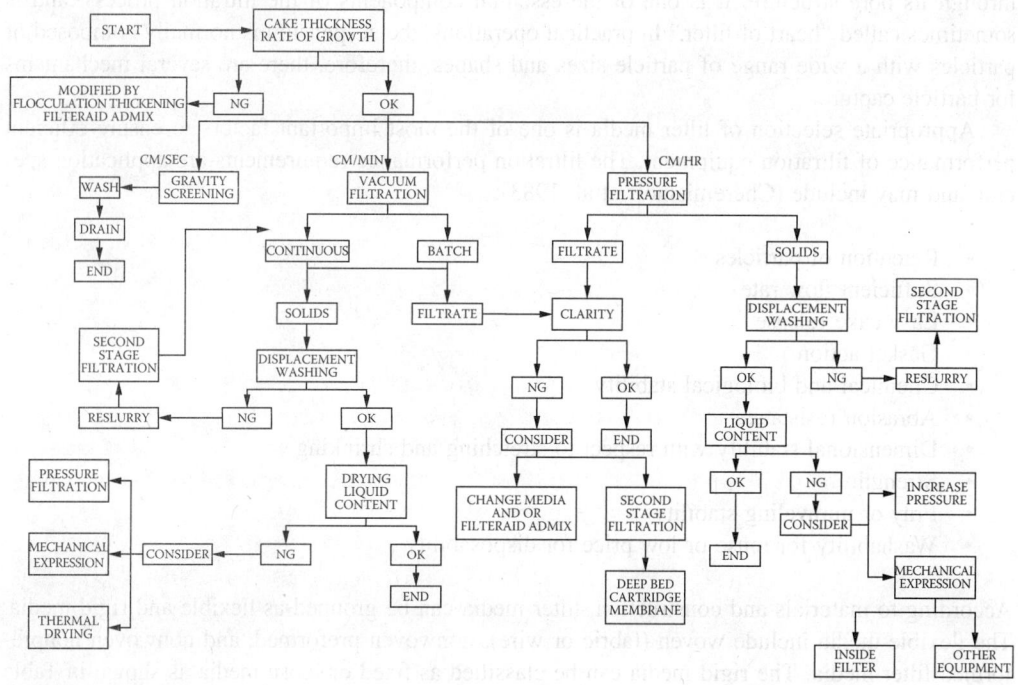

FIGURE 22.64 Decision network for design of filtration system.

TABLE 22.11
Classification of Filter Media

Type	Construction	Example
Flexible media	Woven fabric or woven wire	Including the wide variety of metallic, natural, and synthetic fabrics
	Nonwoven preformed	Including felt, melt blown, and similar materials
	Nonwoven, non-preformed	Cellulose pulp, asbestos
	Membranes	Organic or inorganic
Rigid media	Fixed	Including sintered metal, porous carbon, porous ceramic, formed plastics
	Loose	In packed beds of sand, carbon, and other loose materials

ical and hydraulic compression. A highly compactible material has very little reduction in porosity except in a layer close to the medium. There is little frictional drag in the first part of the cake, which results in a lack of consolidation except in the skin close to the medium. Comparatively, mechanical pressure directly applied on the cake can be quite effective. However, as the porosity decreases, cake permeability also decreases; and the rate of expression may drop substantially.

22.8.3 Selection of Filter Media

22.8.3.1 Introduction

The filter medium is the part of a filter that retains the solids while allowing passage of the fluid through its pore structure. It is one of the essential components of the filtration process, and is sometimes called "heart of filter." In practical operations, the suspension is normally composed of particles with a wide range of particle sizes and shapes, therefore, there are several mechanisms for particle capture.

Appropriate selection of filter media is one of the most important factors to ensure efficient performance of filtration equipment. The filtration performance requirements are application specific and may include (Cheremisinoff et al. 1983):

- Retention of particles
- Sufficient flow rate
- Easy cake release
- Gasket action
- Chemical and biological stability
- Abrasion resistance
- Dimensional stability with respect to stretching and shrinking
- Strength
- Fray or unraveling stability
- Washability for reuse or low price for disposability

According to materials and construction, filter media can be grouped as flexible and rigid media. The flexible media include woven (fabric or wire), nonwoven preformed, and nonwoven nonpreformed filter media. The rigid media can be classified as fixed or loose media as shown in Table 22.11. Among the various types of filter media, the woven flexible media is most commonly used for cake filters.

Solid/Liquid Separation

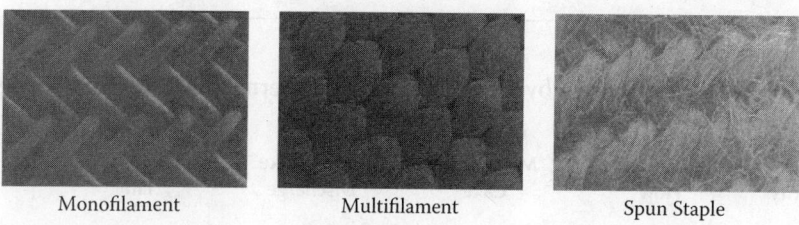

FIGURE 22.65 Three fibers widely used for filter media.

TABLE 22.12
Performances of Media of Three Types of Yarns in Descending Order

Maximum Filtrate Clarity	Minimum Resistance to Flow	Minimum Moisture in Cake	Easier Cake Discharge	Maximum Cloth Life	Least Tendency to Blind
Staple	Monofilament	Monofilament	Monofilament	Staple	Monofilament
multifilament	multifilament	multifilament	multifilament	multifilament	multifilament
monofilament	staple	staple	staple	monofilament	staple

Source: From Purchas (1967).

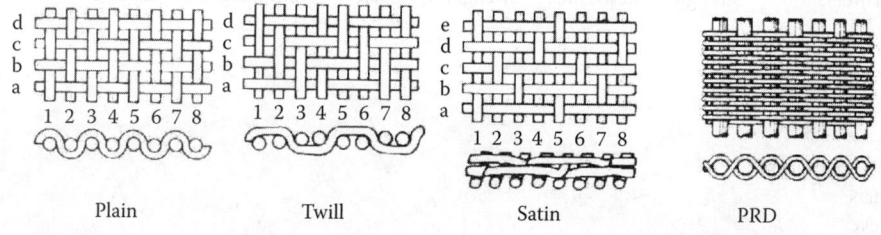

FIGURE 22.66 Four basic types of woven fabrics.

22.8.3.2 Woven Filter Media

Woven media are flexible and comprise metallic or nonmetallic materials formed by weaving. Metallic woven media are called woven wire, and nonmetallic woven media are called woven cloth or woven fabric. Woven wires are generally made with stainless steel, but can also be monel, nickel, inconel, or nichrome. They are stronger, better wear resistant, and more durable, but are more expensive compared with woven fabric. Selection of woven wire is sometimes solely on the basis of severe operating conditions.

Woven fabrics are usually made with natural or synthetic fibers from three different types of yarns: monofilament, multifilament, and spun staple, as shown in Figure 22.65. Performances of these three types of yarns are shown in Table 22.12.

There are many types of weaving patterns for woven fabrics and the four basic types are plain weave, twill weave, satin weave, and plain reverse Dutch weave (Svarovsky 1990) as shown in Figure 22.66. The filtration performances of the four types of weaves are shown in Table 22.13.

22.8.3.3 Selection of Filter Media

Selection of filter media is based mainly on testing results on the material performance and filtration performance. In addition to retention of variable particles and appropriate flow rate, special require-

TABLE 22.13
Performance of Woven Fabrics by Different Weave Patterns in Descending Order

Maximum Filtrate Clarity	Minimum Resistance to Flow	Minimum Moisture in Cake	Easier Cake Discharge	Maximum Cloth Life	Least Tendency to Blind
Satin	Plain	Plain	Satin	Satin	Plain
Twill	PRD	PRD	Twill	Twill	PRD
PRD	Twill	Twill	Plain	PRD	Twill
Plain	Satin	Satin	PRD	Plain	Satin

Note: PRD, plain reverse Dutch weave.

Source: From Purchas (1967).

TABLE 22.14
Desired Physical and Filtration Performance of Filter Media for Different Filters

	Requirements					
Filters	Strength	Abrasion Resistance	Tensile Strength	Gasket Action	Stability to Stretching and Shrinking	Easy Cake Release	Washability
Plate and frame	√	√		√		√	√
Rotary drum	√	√				√	√
Leaf filter					√		√
Disc	√	√			√	√	
Filtering centrifuges	√	√		√			√
Belt filters	√		√		√	√	√
Belt press	√		√		√	√	√

ments on physical and filtration performance of filter media for different filters are summarized in Table 22.14.

22.8.4 SELECTION OF PUMPS IN FILTRATION OPERATION

Pumps employed for slurries can be classified as (1) centrifugal, (2) progressive cavity or screw, (3) diaphragm, (4) piston pump, and (5) compressed gas either alone or combined with a pump. Pressure versus flow rate for various types of pumping mechanisms is illustrated in Figure 22.67. Arrows point in the direction of increasing time. Although constant pressure operation has dominated the literature and laboratory practice, generally diaphragm and single-screw rotary pumps are the best choices. Centrifugal pumps are widely employed in filtration in spite of their tendency to degrade aggregates because of large shear forces resulting from high-velocity impellers. Progressive cavity or screw pumps are capable of pumping high viscosity and high solid content slurries without particle breakup. The cost is high and maintenance may be a problem if the stator wears or seals leak. Piston pumps operate at constant rates that can be easily adjusted. There is a minimum breakup of particles. Abrasion can cause wear if particles invade the clearance between piston and cylinder. Severe pulsation is possible unless adequate damping devices are incorporated in the piping. Diaphragm pumps have some of the characteristics of piston pumps. However, with air driven pumps, the rate drops as pressure builds up. Progressive cavity, piston, and diaphragm pumps are also called positive displacement pumps. Gas driven filtration is only applicable in a

Solid/Liquid Separation

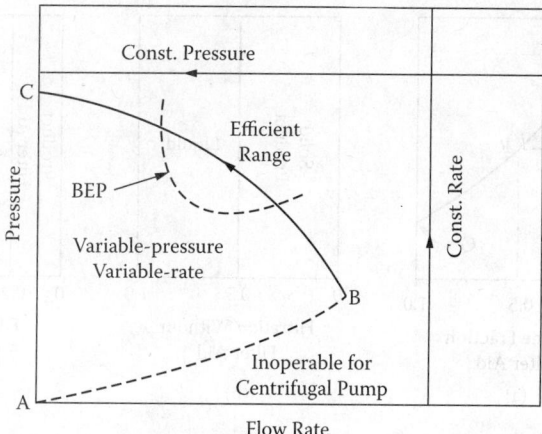

FIGURE 22.67 Characteristic curves for constant pressure and variable pressure operations.

few cases. Pumping of slurries frequently leads to erosion. Settling in feed tanks and piping sometimes is a problem.

With centrifugal pumps, pressure should be close to the best efficiency point (BEP), i.e., at intermediate pressures. Sometimes at low pressures, part of the liquid may vaporize, forming bubbles of gas. Such an occurrence may cause damage to the pump and surges and vibration coupled with low-energy efficiency. At low-flow rates, particles are retained for longer periods in the casing with increased possibility of particle degradation. Low pump speeds are best for fragile particles like flocs resulting from pretreatment operations. In general, it is advisable to test slurry characteristics before and after passing through a centrifugal pump to determine the possible degradation of aggregates resulting in increased specific resistance of cakes and decreased settling velocities in clarifiers.

A throttle valve is sometimes needed with a centrifugal pump or filtration is started at a pressure considerably above that at point B. The curve *AB* generally is essentially a parabola. The pressure drop across the cake is essentially the vertical difference between curves *CB* and *AB*. If the centrifugal pump is too large, overheating, particle breakup, and considerable maintenance problems are possible at startup.

22.8.5 Cycle Analysis in Selection of Filter Aids (Tiller and Li 2003)

Filter aids (diatomaceous earth, perlite, cellulose, etc.) are used to increase the porosity and decrease the flow resistance of cakes. Although the fundamentals are the same, design of a new filter and determination of the effect of adding filter aids to an existing filter (where the available cake thickness is fixed) lead to different conclusions concerning the optimum quantity. Determination of the optimum quantity of filter aid has generally rested on finding the amount that yields the maximum filtrate in a given time interval. Sometimes, this procedure can lead to gross errors if the final cake is saturated with liquid (such as no air blowing after filtration). It is essential to determine the average cake solidosity as a function of the fraction of filter aid and the cycle rate for the product solids. In a laboratory test, adding filter aid can substantially increase cake thickness. By contrast, cake thickness is fixed in an existing full-scale filter; and addition of filter aid replaces a portion of the product solid.

A typical relationship of the average solidosity of a cake as a function of the fraction of filter aid is shown in Figure 22.68a. The active solid product without filter aid has an average solidosity $\varepsilon_{sav} = 0.5$ as indicated by point *P*. With 50% filter aid, the mixture has a value of $\varepsilon_{sav} = 0.25$, point *R*. We first consider a filtration in which no filter aid is present. In Figure 22.68b, a cake with 50% solid product and 50% liquid is illustrated corresponding with point P in Figure 22.68a. In Figure 22.58c, the filtration is repeated with the addition of a volume of filter aid equal to the volume of

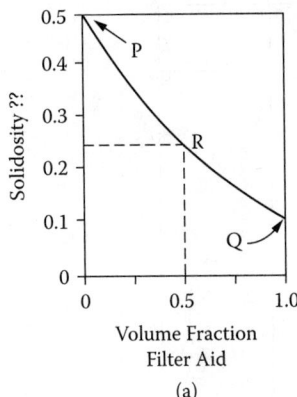 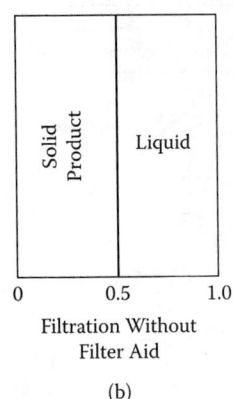 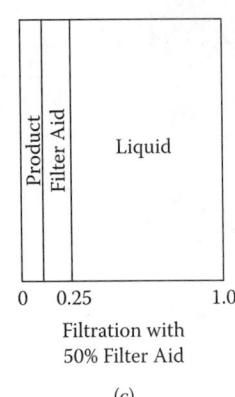

FIGURE 22.68

product solids corresponding with point R in Figure 22.68a. With the increase of the filter aid, the cake has 75% liquid, 25% solids, and the amount of product is reduced to one fourth of the quantity present in Figure 22.68b. As the active solid component in the filter has been reduced to 25% of the original amount in Figure 22.68b, four cycles will be necessary to yield the same quantity of active solids. It can easily be shown that after the addition of filter aids, overall cycle rates will be reduced and more cycles are required to produce the same amount of solids.

An example involving the addition of a perlite filter aid to a clay suspension with a volumetric concentration of $\varphi_s = 0.058$ will be discussed. Constant rate filtration (Camacho 1975) with different fractions of perlite was used to obtain cake resistance α_{av} and cake solidosity ε_{sav} as functions of Δp_c. The specific problem to be analyzed is a constant pressure filtration at 200 kPa (29 psi) of this clay suspension. For constant pressure and constant rate filtration with minimal medium resistance, maximum cycle flux is reached when the filtration time equals the time required to dump the cake, clean the filter, and start the next cycle (Rushton et al. 1996). In this example, it is assumed that the dead time between cycles is 15 minutes. With t (filtration) = 15 minutes, a cake thickness of $L = 0.014$ m results when the cycle rate is maximized for this highly resistant, moderately compactible clay.

Calculations involved in the analysis include (1) concentration of slurry after addition of perlite φ_{sF}; (2) time required to produce a cake of 0.014 m t; (3) number of cycles with filter aid addition f (volume fraction of filter aids in the total cake solids) to get the same amount of product solids without filter aid addition when $f = 0$; (4) volume/unit area of filtrate produced in 15 min; (5) the cycle rate based on liquid; and (6) cycle rate based on product clay solids. The calculated results are shown in Table 22.15.

TABLE 22.15
Cycle Rate Calculation Involving Filter Aids

1	2	3	4	5	6	7	8	9
				t, sec for	v, m³/m²	Number of	Cycle Rates (m³m²/hr)	
f	α_{av}(m⁻²)	ε_{sav}	φ_{sF}	L = 0.014 m	t = 15 min	Cycles	Liquid	Solid
0	2.21E(15)	0.47	0.058	3615	0.0496	1	0.0496	0.00328
0.1	2.22E(15)	0.45	0.064	2951	0.0466	1.04	0.0464	0.00312
0.2	7.27E(13)	0.33	0.071	43	0.233	1.42	0.0501	0.00365
0.333	4.09E(13)	0.26	0.085	11	0.265	1.81	0.029	0.00242
0.667	2.32E(13)	0.18	0.156	0.314	0.115	2.61	0.0021	0.00084

Starting with the fraction of perlite f in the first column, columns 2 and 3 contain the experimental values of α_{av} and ε_{sav} for a pressure filtration of 200 kPa. In column 4, values of the suspension concentration are shown as they reflect the addition of perlite. The time required to produce a cake with $L = 0.014$ m is tabulated in column 5. The short time of <1 sec for $f = 0.667$ results from the high slurry concentration of $\varphi_{sF} = 0.156$ compared with an average cake solidosity of $\varepsilon_{sav} = 0.18$. A more detailed analysis of the cake would provide the variation of ε_s with the distance x in the cake. An experiment with $f = 0.333$ indicated that ε_s at the cake surface had an approximate value of $\varepsilon_{so} = 0.1$; and at the medium, it reached a value in the range of 0.3 to 0.4. With $f = 0.333$, the concentration of $\varphi_{sF} = 0.085$ indicates the suspension is close to $\varepsilon_{so} = 0.10$ and when $f = 0.667$, the value of $\varphi_{sF} = 0.156$ is probably larger than ε_{so} indicates that the suspension is actually a cake. Calculations in the table for $f = 0.333$ and 0.667 cannot be taken seriously as the equations are only valid for true slurries. However, they indicate that concentrations in excess of $f = 0.333$ are impractical.

Column 6 contains the filtrate volume for time of 15 min. These values are frequently used to obtain the "optimum" amount of filter aid. They indicate that the maximum flow rate results from adding perlite when $f = 0.33$. However, the cycle rates in columns 8 and 9 show that the optimum dose of filter aid is $f = 0.2$. The example clearly shows that the choice of a filter aid and the value of f should be based on a cycle analysis and not on v versus t data.

22.8.6 Scale Up

SLS system design and scale up are frequently dependent on settling velocity of suspended particles, flow rate, specific cake resistance, cake % solids, medium resistance, clarity of filtrate, or overflow, which are obtained from tests. In-plant testing is suggested. If replacement equipment is to be considered for processes in operation, in-plant testing should be performed; and if a new product is involved, pilot-plant and bench-scale tests of separability should be conducted so that the maximum design information is available.

Truly satisfactory methods for scale up are not available. Purchas (1977) indicates that scale up has two meanings. While the obvious meaning refers to the prediction of the size of process equipment from small-scale tests, choice of suitable equipment is an equally important part of scale up. Purchas (1977) states "none of the test or scale up procedures presented merits being called a 'standard procedure'. The virtual total absence either of existing standards or of serious attempts to involve or develop standards in an area of such major importance can only be regarded as remarkable in the current era."

Included in *Solid/liquid Separation Equipment Scale Up* (1977) are methods advocated by different equipment manufacturers in the areas as shown in Table 22.16. Much useful information

TABLE 22.16
Scale Up Method in Different Areas by Different Authors

Title	Author	Company
Sedimenting centrifuges	F.A. Records	Pennwalt Ltd.
Filtering centrifuges	G. Hultsch, H. Wilkesmann	Krauss-Maffei AG
Hydrocyclones	H. Trawinski	Amberger Kaolinwerke
Deep-bed filters	K.J. Ives	University College London
Cartridge filters	G. W. Howard, N. Nickolaus	Pall Corp.
Pressure vessel filters	R. Bosley	Stockdale Engineering Ltd.
Filter presses	C.M. Thomas	Johnson-Progress Ltd.
Continuous filters	D.A. Dahlstrom, C.E. Silverblatt	Envirotech Corp.
Gravity separation equipment	E. Bryant Fitch	Formerly of Dorr-Oliver

Source: From Purchas 1977.

can be found in the treatment of separation operations by these authors. An excellent discussion of equipment is provided by Purchas (1981).

TABLE/FIGURE SOURCE LINES

FIGURE 22.5(A) Thomas Riddick, Control of Colloid Stability through Zeta Potential, Zeta Meter Inc, 1968.
FIGURE 22.6 Thomas Riddick, Control of Colloid Stability through Zeta Potential, Zeta Meter Inc, 1968.
TABLE 22.3 Thomas Riddick, Control of Colloid Stability through Zeta Potential, Zeta Meter Inc, 1968.
FIGURE 22.9 Frank M. Tiller et al., "Minimizing Total Cost Involving Use of Polyelectrolytes with Wastewater Centrifugation," Advances in Filtration and Separation Technology, vol. 16, American Filtration & Separations Society, 2002.
FIGURE 22.10 Frank M. Tiller et al., "Minimizing Total Cost Involving Use of Polyelectrolytes with Wastewater Centrifugation," Advances in Filtration and Separation Technology, vol. 16, American Filtration & Separations Society, 2002.
FIGURE 22.11 Frank M. Tiller et al., "Minimizing Total Cost Involving Use of Polyelectrolytes with Wastewater Centrifugation," Advances in Filtration and Separation Technology, vol. 16, American Filtration & Separations Society, 2002.
FIGURE 22.19 Frank M. Tiller, "Tutorial: Interpretation of Filtration Data, I," *Fluid/Particle Separation Journal*, vol. 3, no. 2, American Filtration & Separations Society, p. 91, 1990.
FIGURE 22.21 Frank M. Tiller, et al., "Filtering Coal Liquids: Clogging Phenomena in the Filtration of Liquefied Coal," Chemical Engineering Progress, December 1981, AIChE, p. 65, Figure 10, 1981.
FIGURE 22.24 Frank M. Tiller, "Filtering Coal Liquids: Clogging Phenomena in the Filtration of Liquefied Coal", Chemical Engineering Progress, December 1981, AIChE, Page 66, Figure 12, 1981
FIGURE 22.25 Frank M. Tiller, et al. Role of Porosity in Filtration: XIII. Behavior of Highly Compactible Cakes," *AIChE Journal*, vol. 44, *AIChE*, p. 2159, vol. 44, 1999.
FIGURE 22.26 Frank M. Tiller, et al. Role of Porosity in Filtration: XIII. Behavior of Highly Compactible Cakes", *AIChE Journal*, vol. 44, *AIChE*, p. 2159, vol. 44, 1999.
FIGURE 22.27 Frank M. Tiller, et al. Role of Porosity in Filtration: XIII. Behavior of Highly Compactible Cakes," *AIChE Journal*, vVol.44, *AIChE*, p. 2159, vol. 44, 1999.
TABLE 22.5 Source: Frank M. Tiller, et al, "Explaining Strange Behavior of Highly Compactible Materials," *Chemical Processing*, Sep. 2000, *AIChE*, September, 2000.
TABLE 22.6 Frank M. Tiller, Wenping Li, "Determination of the Critical Pressure Drop for Filtration of Super-Compactible Cakes,", Water Science and Technology, Vol. 44, No. 10, 2001, IWA, Table 3, Figure 3, Figure 4, no. 10, 2001.
FIGURE 22.28 Frank M. Tiller, Wenping Li, "Determination of the Critical Pressure Drop for Filtration of Super-Compactible Cakes," Water Science and Technology, vol. 44, No. 10, 2001, IWA, Table 3, Figure 3, Figure 4, No. 10, 2001.
FIGURE 22.29 Frank M. Tiller, Wenping Li, "Determination of the Critical Pressure Drop for Filtration of Super-Compactible Cakes", Water Science and Technology, vol. 44, no. 10, 2001, IWA, Table 3, Figure 3, Figure 4, no. 10, 2001.
FIGURE 22.35 Frank M. Tiller, et al., "The Role of Porosity in Filtration Part X: Deposition of Compressible Cakes on External Radial Surfaces," *AIChE Journal*, vol. 31, no. 8, *AIChE*, p. 1242, Figure 1, August 1985.
FIGURE 22.38 Frank M. Tiller, et al., "Hydraulic Deliquoring of Compressible Filter Cakes, Part I: Reverse Flow in Filter Presses," *AIChE Journal*, vol. 29, no. 2, *AIChE*, p. 298 Figure 1, Figure 2; p. 299, Figure 3, March 1983.
FIGURE 22.39 Frank M. Tiller, et al., "Hydraulic Deliquoring of Compressible Filter Cakes, Part I: Reverse Flow in Filter Presses," *AIChE Journal*, vol. 29, no. 2, *AIChE*, p. 298 Figure 1, Figure 2; p. 299, Figure 3, March 1983.
FIGURE 22.40 Frank M. Tiller, et al., "Hydraulic Deliquoring of Compressible Filter Cakes, Part I: Reverse Flow in Filter Presses," *AIChE Journal*, vol. 29, no. 2, *AIChE*, p. 298 Figure 1, Figure 2; p. 299, Figure 3, March 1983.
FIGURE 22.41 Frank M. Tiller, et al., "Hydraulic Deliquoring of Compressible Filter Cakes, Part I: Reverse Flow in Filter Presses," *AIChE Journal*, vol. 29, no. 2, *AIChE*, p. 298 Figure 1, Figure 2; p. 299, Figure 3, March 1983.

FIGURE 22.42 C. E. Silverblatt, Hemant Risbud and Frank M. Tiller, "Batch, Continuous Processes for Cake Filtration," *Chemical Engineering*, April 29, 1974; p. 129, Figure 2.
FIGURE 22.43 Frank M. Tiller, et al. "Analytical Formulas for Disk Filters," *AIChE Journal*, vol. 20, no. 1, *AIChE*, p. 36 Figure 1; p. 38 Figure 3 and Figure 4, January 1974.
FIGURE 22.44 Frank M. Tiller, et al. "Analytical Formulas for Disk Filters," *AIChE Journal*, vol. 20, no. 1, *AIChE*, p. 36 Figure 1; p. 38 Figure 3 and Figure 4, January 1974.
FIGURE 22.47 Alex Bagdasarian, KS Cheng and Frank M. Tiller, "Bench Scale Filter for Studying Thin-Cake Filtration," Filtration & Separation, January/February 1983 *Filtration & Separation*, p. 32, Figure 1, January/February 1983.
FIGURE 22.54 Masao Sambuichi, Hideo Nakakura, Kunihisa Osasa, F. M. Tiller, "Theory of Batchwise Centrifugal Filtration," AIChE Journal, vol. 33, no. 1, 1987, pp. 109-120, *AIChE*, p. 110, Figure 2, vol. 33, no.1, 1987.
FIGURE 22.57 Frank M. Tiller et al. "Solid-Liquid Separation: An Overview," CEP October 1977, *AIChE*, page 75, Figure 14, and p. 74, Figure 13, October 1977.
FIGURE 22.67 Frank M. Tiller et al. "Solid-Liquid Separation: An Overview," CEP October 1977, *AIChE*, p. 75, Figure 14, and p. 74, Figure 13, October 1977.
FIGURE 22.59-A Wei-Ming Lu, Frank M. Tiller, et al, "A New Method to Determine Local Porosity and Filtration Resistance of Filter Cakes," Journal of Chinese Institute of Chemical Engineers, vol. 1, 1970, *Journal of Chinese Institute of Chemical Engineers*, p. 50, Figure 8, vol. 1, 1970.
FIGURE 22.59-B Wei-Ming Lu, et al. Solid-Liquid Separation. GaoLi Publication Ltd., Taiwan, 2004, p. 326.
FIGURE 22.60 Wei-Ming Lu, et al. Solid-Liquid Separation. GaoLi Publication Ltd., Taiwan, 2004, p. 331.
FIGURE 22.61 Frank M. Tiller et al, "The Role of Porosity in Filtration VII: Effect of Side-Wall Friction in Compression-Permeability Cells," *AIChE Journal*, vol. 18, no. 1, AIChE, p. 14, Figure 1, vol. 18 no. 1.
FIGURE 22.63 Purchas, D.B., Solid-Liquid Separation Equipment Scale Up, Uplands Press Ltd., 1977.
TABLE 22.8 Fitch, B., Choosing a Separation Technique, *Chem. Eng. Prog.* vol. 70, 1974.
TABLE 22.9 Fitch, B., Choosing a Separation Technique, *Chem. Eng. Prog.* vol. 70, 1974.

REFERENCES

Alliot, E.A. 1920. Recessed plates, plate and frame, filter press: their construction and use. *Journal of the Society of Chemical Industry* 39: 261–285.
Ambler, C.M. 1997. Centrifuges. In *Chemical engineers handbook*. 7th ed. Eds. R.H. Perry and C.H. Chilton. New York: McGraw-Hill, 89–103.
Besendorfer, C. 1996. Exert the force of hydrocyclones. *Chemical EngIneering* September: XXX.
Bratby, J. 1980. *Coagulation and flocculation*. Croydon, UK: Uplands Press Ltd.
Camacho, Carlos. 1975. Studies in filtration of particulate mixtures. MS thesis, University of Houston, Texas.
Chen, W. 1986. Sedimentation and thickening. PhD diss., University of Houston, Texas.
Chen, W. 2002. *Solid/liquid separation fundamentals and practice*. AIChE Today Series. New York: AIChE.
Chen, W. 2004. The Use of Hydrocyclone Models in Practical Design. 9th World Filtration Congress, New Orleans, LA, April 19–22.
Cheremisinoff, N.P., and D.S. Azbel. 1983. *Liquid filtration*. Kent, UK: Butterworths Ltd.
Cheryan, M. 1998. *Ultrafiltration and microfiltration handbook*. Lancaster, PA: Technomic Publishing.
Coe, H.S. and G.H. Clevenger. 1916. Methods for determining the capacities of slime-settling tanks. *Trans. AIME* 55: 356–384.
Dentel, S.K., M.M. Abu-Orf, N. Griskowitz. 1993. Guidance Manual for Polymer Selection in Wastewater Treatment Plants. Water Environment Research Foundation.
Eagle-Picher Minerals. 1970. *Celaton diatomite filter aids catalog*. Inkster, MI: Eagle-Picher Minerals, Inc.
Fair, J.R. 1989. Commercially attractive bioseparation technology. *Chemical Engineering Progress* 85: 38.
Fitch, B. 1974. Choosing a separation technique. *Chemical Engineering Progress* 70: 33–37.
Flood, J.E., H.F. Porter, and F.W. Rennie. 1977. Filtration practice today. *Chemical Engineering* 73: 607–609.
Grace, H.P. 1953. Resistance and compressibility of filter cakes, part I. *Chemical Engineering Progress* 49: 6, 303.
Hosseini, M. 1977. Velocity and concentration effect in filtration. Masters thesis, University of Manchester, UK.
Kwon, J.H. 1995. Effects of compressibility and cake clogging on sludge dewatering characteristics. PhD diss., Seoul National University, Korea

Purchas, D.B. 1967. *Industrial filtration of liquids.* Cleveland, OH: CRC Press.
Purchas, D.B. 1977. *Solid/liquid separation equipment scale-up.* Croydon, UK: Uplands Press Ltd.
Purchas, D.B. 1981. *Solid/liquid separation technology.* Croydon, UK: Uplands Press Ltd.
Riddick, Thomas M. 1968. *Control of colloid stability through zeta potential.* New York: Zeta Meter Inc.
Risbud,... 1974. Mechanical expression, stresses at cake boundaries and new compression-permeability cell. PhD diss., University of Houston, Texas.
Rushton, A., A.S. Ward, and R.G. Holditch. 1996. *Solid-liquid filtration and separation technology.* New York: VCH Publishers, 411.
Svarovsky, Ladislav. 1990. *Solid-liquid separation.* 3rd ed. London, Boston: Butterworth & Co., 29–30.
Tiller, F.M. 1978. Characteristics of staged, delayed-cake filters. *Filtration and Separation* May/June: XXX.
Tiller, F.M. 1974. Continuous processes for cake filtration. *Chemical Engineering* April: 29.
Tiller, F.M., and T.M. Garrett. 1997. Developing methodology for improving dewatering characteristics of wastewater sludge with emphasis on optimal use of polyelectrolytes to minimize costs. Report to Greater Houston Wastewater Program, University of Houston, Texas.
Tiller, F.M., S. Haynes, and W.M. Lu. 1972. The role of porosity in filtration vii: effect of side-wall friction in compression-permeability cells. *AIChE Journal* 18: 13–19.
Tiller, F.M., and L.L. Horng. 1983. Hydraulic deliquoring of compressible filter cakes, part I: reverse flow in filter presses. *AIChE Journal* 29: 297–305.
Tiller, F.M., and N.B. Hsyung. 1993. Unifying the theory of thickening, filtration, and centrifugation. *Water Science Technology* 28: 1.
Tiller, F.M., N.B. Hsyung, and D.Z. Cong. 1995. Role of porosity in filtration: II. Filtration with sedimentation. *AIChE Journal* 41: 1153.
Tiller, F.M., and J.H. Kwon. 1999. Role of porosity in filtration: XIII. Behavior of highly compactible cakes. *AIChE Journal* 44: 2159.
Tiller, F.M., and W.F. Leu. 1980. Basic data fitting in filtration. *Journal of the Chinese Institute of Chemical Engineers.* 11: 61.
Tiller, F.M., and W. Leu. 1984. Solid-liquid separation for liquefied coal industries. Final Report for Project 1411-1, July.
Tiller, F.M., and P.J. Lloyd. 1978. *Theory and practice of solid/liquid separation.* Houston, TX: University of Houston.
Tiller, F.M., and D. Tarng. 1995. Try deep thickeners and clarifiers *Chemical Engineering Progress* March: 75–80.
Tiller, F.M., and Wenping Li. 1999. Comparing % cake solids in filtration, thickening, sedimenting centrifugation and expression. *Fluid/Particle Separation Journal* 12: 173.
Tiller, F.M., and Wenping Li. 2000. Explaining strange behavior of highly compactible materials. *Chemical Processing* September.
Tiller, F.M., and Wenping Li. 2001. Optimizing candle filters for super-compactible materials. *Advances in Filtration and Separation Technology* 15.
Tiller, F.M., and Wenping Li. 2001. Cost minimization study of a full scale urban wastewater treatment plant. Proceedings of the 6th European Biosolids and Organic Residuals Conference; West Yorkshire, UK; November, 11–14.
Tiller, F.M., and Wenping Li. 2002. Minimizing total cost involving use of polyelectrolytes with wastewater centrifugation. *Advances in Filtration and Separation Technology* 16.
Tiller, F.M., and Wenping Li. 2003. Dangers of lab-plant scaleup for solid/liquid separation systems. *Chemical Engineering Communications.*

23 Drying: Principles and Practice

Arun S. Mujumdar

CONTENTS

- 23.1 Introduction 1668
- 23.2 Basic Principles and Terminology 1669
  - 23.2.1 Thermodynamic Properties of Air-Water Mixtures and Moist Solids 1669
    - 23.2.1.1 Psychrometry 1669
    - 23.2.1.2 Equilibrium Moisture Content 1671
    - 23.2.1.3 Water Activity 1674
  - 23.2.2 Drying Kinetics 1676
- 23.3 Classification and Selection of Dryers 1683
  - 23.3.1 Classification of Dryers 1683
    - 23.3.1.1 Batch Dryers: Classification (Baker 1997) 1684
    - 23.3.1.2 Continuous Dryers: Classification 1684
    - 23.3.1.3 Direct Dryers 1685
    - 23.3.1.4 Indirect Dryers 1685
  - 23.3.2 Selection of Dryers 1686
- 23.4 Drying Equipment 1690
  - 23.4.1 Dryers for Particulates and Granular Solids 1690
    - 23.4.1.1 Tray Dryers 1690
    - 23.4.1.2 Rotary Dryers 1691
    - 23.4.1.3 Fluidized Bed Dryers 1693
    - 23.4.1.4 Freeze Dryers 1694
    - 23.4.1.5 Vacuum Dryers 1695
  - 23.4.2 Dryers for Slurries and Suspensions 1697
    - 23.4.2.1 Spray Dryers 1697
    - 23.4.2.2 Drum Dryers 1702
  - 23.4.3 Selected Dryers and Drying Systems 1703
    - 23.4.3.1 Two-Stage Dryers 1704
    - 23.4.3.2 Flash Dryers 1705
    - 23.4.3.3 Spin-Flash Dryers 1707
    - 23.4.3.4 Roto-Louvre Dryer 1707
    - 23.4.3.5 Tunnel Dryers 1708
    - 23.4.3.6 Band Dryers 1708
    - 23.4.3.7 Infrared Dryers 1709
    - 23.4.3.8 Microwave (MW) and Radio Frequency (RF) Drying 1709
    - 23.4.3.9 Superheated Steam Drying 1711
    - 23.4.3.10 Drying of Boards and Sheets 1712
- 23.5 Closing Remarks 1713
- Nomenclature 1714
- References 1715

23.1 INTRODUCTION

The solid/liquid separation operation of drying converts a solid, semi-solid, or liquid feedstock into a solid product by evaporation of the liquid via application of heat. In the special case of freeze drying, which takes place below the triple point of the liquid being removed, drying occurs by sublimation of the solid phase directly into the vapor phase. This definition thus excludes conversion of a liquid phase into a concentrated liquid phase (evaporation); mechanical dewatering operations such as filtration, centrifugation, sedimentation, etc.; supercritical extraction of water from gels to produce extremely high porosity aerogels (extraction); or so-called drying of liquids and gases by use of molecular sieves (adsorption), etc. Phase change and production of a solid phase as an end product are essential features of the drying process. Drying is an essential operation in the chemical, agricultural, biotechnology, food, polymer, ceramics, pharmaceutical, pulp and paper, mineral processing, and wood processing industries.

Drying is perhaps the oldest, most common, and most diverse of chemical engineering unit operations. Over 500 types of dryers have been reported in the literature, while over 100 distinct types are commonly available. It competes with distillation as the most energy-intensive unit operation due to the high latent heat of vaporization and the inherent inefficiency of using hot air as the (most common) drying medium. Various studies report national energy consumption for industrial thermal drying operations ranging from 10% to 15% for the United States, Canada, France, the U.K., etc., to 20% to 25% for Denmark and Germany. The latter figures have been obtained recently based on mandatory energy audit data supplied by industry and hence are more reliable.

Energy consumption in drying ranges from a low value of under 5% for the chemical manufacturing industries to 35% for papermaking operations. Capital expenditures for dryers are estimated to be in the order of only $800 million per annum for the United States. Thus, the major costs for dryers are in their operation rather than in their initial investment costs.

Drying of various feedstocks is needed for one or several of the following reasons: need for easy-to-handle free-flowing solids, preservation and storage, reduction in cost of transportation, achieving desired quality of product, etc. In many processes, improper drying may lead to irreversible damage to product quality and hence a nonsalable product.

Unique features of drying make it fascinating and challenging:

- Product size may range from microns to tens of centimeters (in thickness or depth)
- Product porosity may range from zero to 99.9%
- Drying times range from 0.25 sec (drying of tissue paper) to 5 months (for certain hardwood species)
- Production capacities may range from 0.10 kg/hr to 100 t/hr
- Product speeds range from zero (stationary) to 2000 m/sec (tissue paper)
- Drying temperatures range from below the triple point to above the critical point of the liquid
- Operating pressure may range from a fraction of a millibar to 25 atmospheres
- Heat may be transferred continuously or intermittently by convection, conduction, radiation, or electromagnetic fields

Clearly, no single design procedure applies to all or even several of the dryer variants. It is therefore essential to revert to the fundamentals of heat, mass, and momentum transfer coupled with knowledge of the material properties (quality) when attempting a design of a dryer or analysis of an existing dryer. Mathematically speaking, all processes involved, even in the simplest dryer, are highly nonlinear and hence scale-up of dryers is generally very difficult. Experimentation at laboratory and pilot scales coupled with field experience and know-how is essential to the development of a new dryer application. Dryer vendors are necessarily specialized and normally offer only a narrow range of drying equipment. The buyer must therefore be reasonably conversant with the basic knowledge of the wide assortment of dryers and be able to come up with an informal preliminary selection before going to the vendors. In general, several different dryers may be able to handle a given application.

23.2 BASIC PRINCIPLES AND TERMINOLOGY

Drying is a complex operation involving transient heat and mass transfer along with several rate processes, such as physical or chemical transformations, which may cause changes in product quality as well as the mechanisms of heat and mass transfer. Physical changes that may occur include: shrinkage, puffing, crystallization, glass transitions, etc. In some cases, desirable or undesirable chemical or biochemical reactions may occur leading to changes in color, texture, odor, or other properties of the solid product. In the manufacture of catalysts, for example, drying conditions can yield significant differences in the activity of the catalyst by changing the internal surface area.

Thermal drying caused by the vaporization of the liquid results as heat is supplied to the wet feedstock. As noted earlier, heat may be supplied by convection (direct dryers), conduction (contact or indirect), radiation, or volumetrically by placing the wet material in a microwave or radio frequency (RF) electromagnetic field. Over 85% of industrial dryers are of the convection type, with hot air or direct combustion gases as the drying medium. Over 99% of the application involves removal of water. All modes except the dielectric (microwave and radio frequency) supply heat at the boundaries of the drying object so that the heat must then diffuse into the solid primarily by conduction. The liquid must travel to the boundary of the material before it is transported away by the carrier gas (or by application of vacuum for nonconvective dryers).

Internal transport of moisture within the solid may occur by any one or more of the following mechanisms of mass transfer:

- Liquid diffusion, if the wet solid is at a temperature below the boiling point of the liquid
- Vapor diffusion, if the liquid vaporizes within material
- Knudsen diffusion, at very low temperatures and pressures, e.g., in freeze drying
- Surface diffusion (possible although not proven)
- Hydrostatic pressure differences created by internal vaporization rates exceeding the rate of vapor transport through the solid to the surroundings
- Combinations of the above mechanisms

Since the physical structure of the drying solid is subject to change during drying, the mechanisms of moisture transfer may also change with elapsed time of drying.

23.2.1 THERMODYNAMIC PROPERTIES OF AIR-WATER MIXTURES AND MOIST SOLIDS

23.2.1.1 Psychrometry

As noted earlier, a majority of dryers are of the direct (or convective) type. In other words, hot air is used both to supply the heat for evaporation and to carry away the evaporated moisture from the product. Notable exceptions are freeze and vacuum dryers, which are used almost exclusively for drying heat-sensitive products, tend to be significantly more expensive than dryers operated near atmospheric pressure. Another exception is the emerging technology of superheated steam drying (Mujumdar 1995a). In certain cases, such as the drum drying of pasty foods, some or all of the heat is supplied indirectly by conduction.

Drying with heated air implies humidification and cooling of the air in a well-insulated (adiabatic) dryer. Thus, hygrothermal properties of humid air are required for the design calculations of such dryers. Pakowski et al. (1991) have presented a comprehensive summary of the engineering

TABLE 23.1
Thermodynamic and Transport Properties of Air-Water System

Property	Expression
P_v	$P_v = 100 \exp[27.0214 - (6887/T_{abs}) - 5.321 \ln(T_{abs}/273.16)]$
Y	$Y = 0.622 RHP_v/(RHP_v)$
c_{pg}	$c_{pg} = 100926 \times 10^3 - 4.0403 \times 10^{-2} + 6.1759 \times 10^{-4}T^2 - 4.097 \times 10^{-7}T^3$
k_g	$k_g = 2.425 \times 10^{-2} - 7.889 \times 10^{-5}T - 1.790 \times 10^{-8}T^2 - 8.570 \times 10^{-12}T^3$
ρ_g	$\rho_g = PM_g/(RT_{abs})$
μ_g	$\mu_g = 1.691 \times 10^{-5} + 4.984 \times 10^{-8}T - 3.187 \times 10^{-11}T^2 + 1.319 \times 10^{-14}T^3$
c_{pv}	$c_{pv} = 1.883 - 1.6737 \times 10^{-4}T + 8.4386 \times 10^{-7}T^2 - 2.6966 \times 10^{-10}T^3$
c_{pw}	$c_{pw} = 2.8223 + 1.1828 \times 10^{-2}T - 3.5043 \times 10^{-5}T^2 + 30601 \times 10^{-8}T^3$

Source: Mujumdar (1995b) and Pakowski et al. (1991).

properties of humid air. Table 23.1 summarizes the essential thermodynamic and transport properties of the air-water system. In Table 23.2, a listing of brief definitions of various terms encountered in drying and psychrometry is given. It also includes several terms not explicitly discussed in the text.

Figure 23.1 is a psychrometric chart for the air-water system. It shows the relationship between the temperature (abscissa) and absolute humidity (ordinale, in g water per kg dry air) of humid air from 0°C to 130°C at one atmosphere absolute pressure. Line as representing percent humidity and adiabatic saturation are drawn according to the thermodynamic definitions of these terms. Equations for the adiabatic saturation and wet-bulb temperature lines on the chart are as follows (Geankoplis 1983):

$$\frac{Y - Y_{as}}{T - T_{as}} = -\frac{c_s}{\lambda_{as}} = -\frac{1.005 + 1.88Y}{\lambda_{as}} \qquad (23.1)$$

and

$$\frac{Y - Y_{wb}}{T - T_{wb}} = -\frac{h/M_{air}k_y}{\lambda_{wb}} \qquad (23.2)$$

The ratio ($h/M_{air}k_y$), termed the psychrometric ratio, lies between 0.96 and 1.005 for air-water vapor mixtures; thus it is nearly equal to the value of humid heat c_s. If the effect of humidity is neglected, the adiabatic saturation and wet-bulb temperatures (T_{as} and T_{wb}, respectively) are almost equal for the air-water system. Note, however, that T_{as} and T_w are conceptually quite different. The adiabatic saturation temperature is a gas temperature and a thermodynamic entity while the wet-bulb temperature is a heat and mass transfer rate-based entity and refers to the temperature of the liquid phase. Under constant drying conditions, the surface of the drying material attains the wet-bulb temperature if the heat transfer is by pure convection. The wet-bulb temperature is independent of surface geometry as a result of the analogy between heat and mass transfer.

Most handbooks of engineering provide more detailed psychrometric charts including additional information and extended temperature ranges. Mujumdar (1995aORb) includes numerous psychrometric charts for several gas-organic vapor systems as well.

TABLE 23.2
Definitions of Commonly Encountered Terms in Psychrometry and Drying

Term (Symbol)	Meaning
Adiabatic saturation temperature (T_{as})	Equilibrium gas temperature reached by unsaturated gas and vaporizing liquid under adiabatic conditions. (Note: For air-water system only, it is equal to the wet bulb temperature T_{wb}.)
Bound moisture	Liquid physically and/or chemically bound to a solid matrix so as to exert a vapor pressure lower than that of pure liquid at the same temperature.
Constant rate drying period	Under constant drying conditions, the drying period when the evaporation rate per unit drying area is constant (when surface moisture is removed).
Dew point	Temperature at which a given unsaturated air-vapor mixture becomes saturated.
Dry bulb temperature	Temperature measured by a (dry) thermometer immersed in vapor-gas mixture.
Equilibrium moisture content (X^*)	At a given temperature and pressure, the moisture content of a moist solid in equilibrium with the gas-vapor mixture (zero for nonhygroscopic solids).
Critical moisture content (X_c)	Moisture content at which the constant drying rate first begins to drop (under constant drying conditions).
Falling rate period	Drying period (under constant drying conditions) during which the rate falls continuously in time.
Free moisture (X_f) ($X_f = X = X^*$)	Moisture content in excess of the equilibrium moisture content (hence free to be removed) at given air humidity and temperature.
Humid heat	Heat required to raise the temperature of unit mass of dry air and its associated vapor through one degree (kJ kg^{-1} K^{-1} or Btu lb^{-10} F^{-1}).
Humidity, absolute	Mass of water vapor per unit mass of dry gas (kg kg^{-1} or lb lb^{-1}).
Humidity, relative	Ratio of partial pressure of water vapor in a gas-vapor mixture to equilibrium vapor pressure at the same temperature.
Unbound moisture	Moisture in solid that exerts vapor pressure equal to that of pure liquid at the same temperature.
Water activity (a_w)	Ratio of vapor pressure exerted by water in solid to that of pure water at the same temperature.
Wet-bulb temperature (T_{wb})	Liquid temperature attained when large amounts of air-vapor mixture are contacted with the surface. In purely convective drying, drying surface reaches T_{wb} during the constant rate period.

23.2.1.2 Equilibrium Moisture Content

The moisture content of a wet solid in equilibrium with air of given humidity and temperature is termed the equilibrium moisture content (EMC). A plot of the EMC at a given temperature versus dry-basis moisture content of the solid is termed the sorption isotherm. An isotherm obtained by exposing the wet solid to air of increasing humidity gives the adsorption isotherm. That obtained by exposing the solid to air of decreasing humidity is known as the desorption isotherm. Clearly, the latter is of interest in drying as the moisture content of the solids progressively decreases. Most drying materials display "hysteresis" in that the two isotherms are not identical.

Figure 23.2 shows the general shape of the typical sorption isotherms. They are characterized by three distinct zones, A, B, and C, which are indicative of different water binding mechanisms at individual sites on the solid matrix. In region A, water is tightly bound to the sites and is unavailable for reaction. In this region, there is essentially monolayer adsorption of water vapor and no distinction exists between the adsorption and desorption isotherms. In region B, the water is more loosely bound. The vapor pressure depression below the equilibrium vapor pressure of water at the same temperature is due to its confinement in the smaller capillaries. Water in region C is even more loosely held in larger capillaries. It is available for reactions and as a solvent.

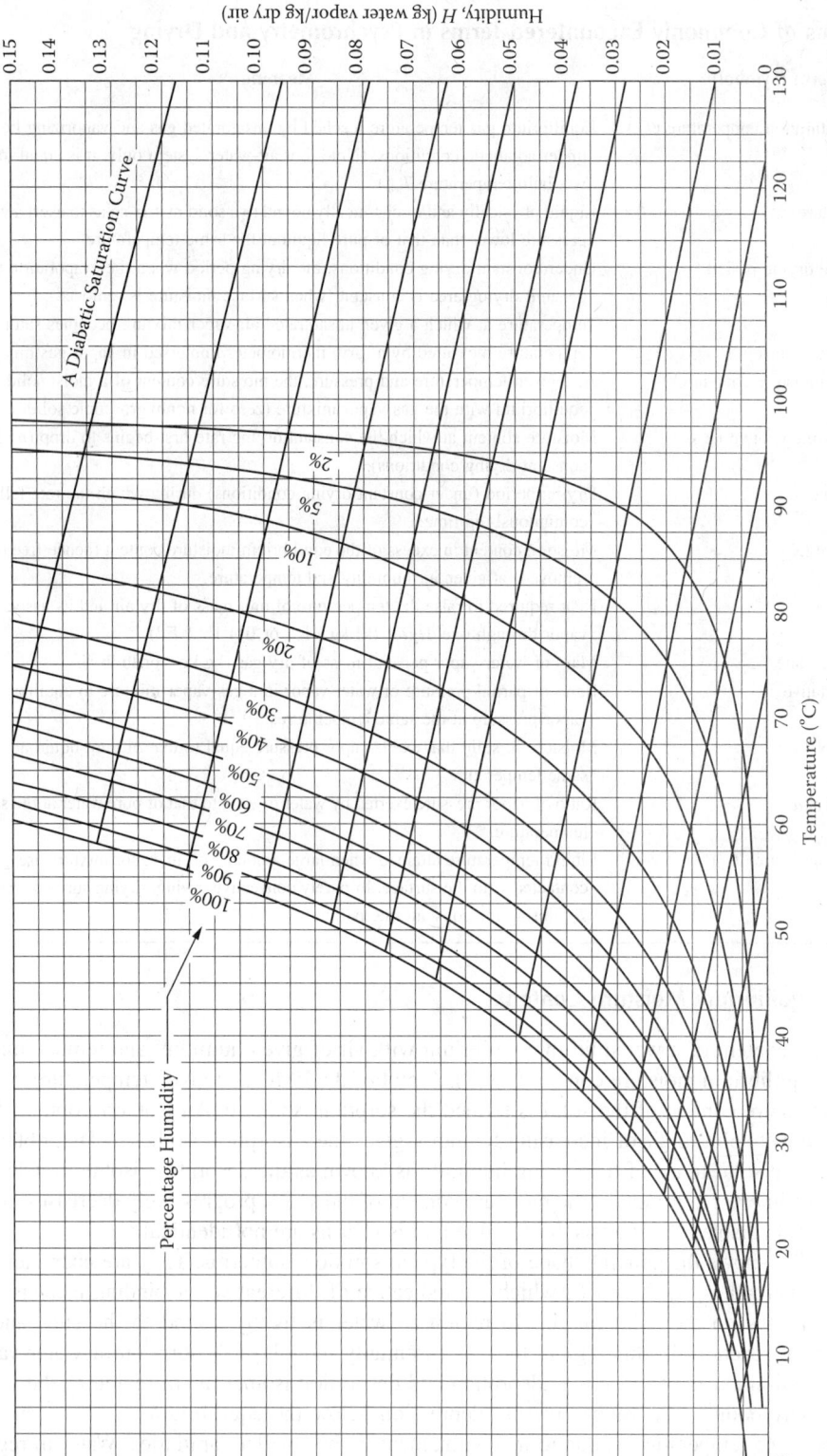

FIGURE 23.1 Humidity chart for mixtures of air and water (pressure of 101.325 kPa).

Drying: Principles and Practice

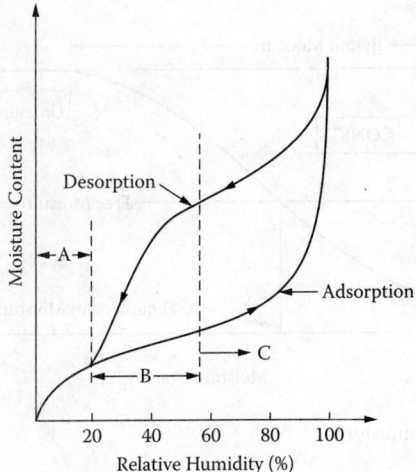

FIGURE 23.2 Typical sorption isotherms showing hysteresis.

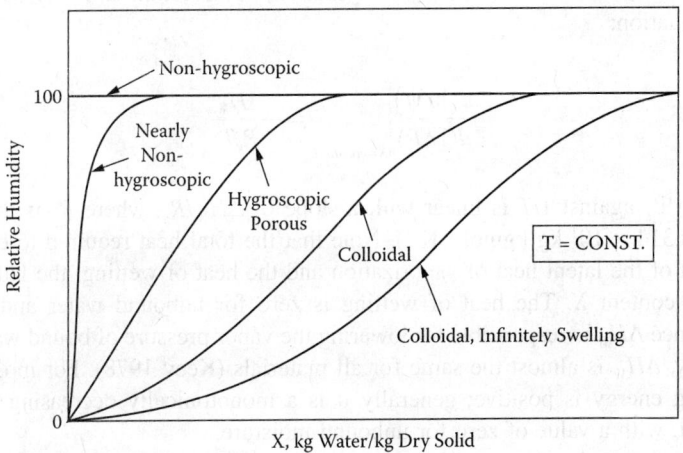

FIGURE 23.3 Equilibrium moisture curves for different types of solids.

Numerous hypotheses have been proposed to explain the hysteresis. Bruin and Luyben (1980), Fortes and Okos (1980), and Bruin (1988) provide more information on the topic.

Figure 23.3 shows schematically the shapes of the equilibrium moisture curves for various types of solids. Figure 23.4 shows the various types of moisture defined in Table 23.2. The desorption isotherms are also dependent on external pressure. However, in all practical cases of interest, this effect may be neglected.

According to Keey (1978), the temperature dependence of the equilibrium moisture content on temperature can be correlated by

$$\left[\frac{\Delta X^*}{\Delta T}\right]_{\Psi=constant} = -\alpha X^* \tag{23.3}$$

where X^* is the dry-basis equilibrium moisture content T is the temperature, and Ψ is the relative humidity of air. The parameter α ranges from 0.005 to 0.01 K^{-1}. This correlation may be used to estimate the temperature dependence of X^* if no data are available.

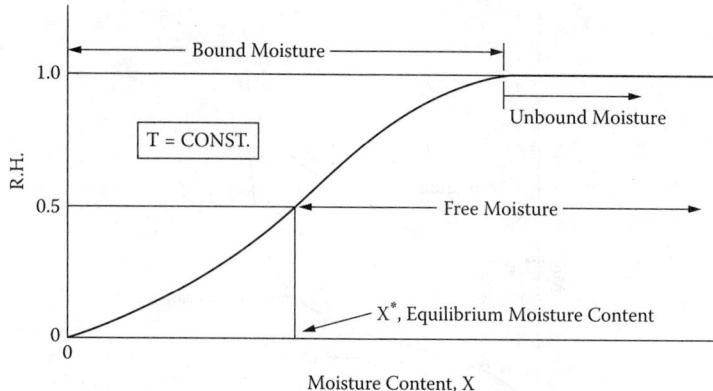

FIGURE 23.4 Various types of moisture.

For hygroscopic solids, the enthalpy of the attached moisture is less than that of pure liquid by an amount equal to this binding energy, which is also termed the enthalpy of wetting, ΔH_w (Keey 1978). It includes the heat of sorption, hydration, and solution and may be estimated from the following equation:

$$\left.\frac{d(\ln \psi)}{d(1/T)}\right|_{X=constant} = -\frac{\Delta H_w}{R_g T} \tag{23.4}$$

A plot of $\ln(\Psi)$ against $1/T$ is linear with a slope of $\Delta H_W/R_g$, where R_g is the universal gas constant ($R_g = 8.314 \times 10^3$ kg kgmol^{-1} K^{-1}). Note that the total heat required to evaporate bound water is the sum of the latent heat of vaporization and the heat of wetting; the latter is a function of the moisture content X. The heat of wetting is zero for unbound water and increases with decreasing X. Since ΔH_W is responsible for lowering the vapor pressure of bound water, at the same relative humidity, ΔH_W is almost the same for all materials (Keey 1978). For most materials, the moisture binding energy is positive; generally it is a monotonically decreasing function of the moisture content, with a value of zero for unbound moisture.

In general, water sorption data must be determined experimentally. Some 80 correlations, ranging from those based on theory to those that are purely empirical, have appeared in the literature. Two of the most extensive compilations are due to Wolf et al. (1985) and Iglesias and Chirife (1982). Aside from temperature, water sorption is also affected by the physical structure as well as the composition of the material. The pore structure and size, as well as the physical and/or chemical transformations during processing can cause significant variations in the moisture binding ability of the solid.

23.2.1.3 Water Activity

In drying of some materials, which requires careful hygienic attention, e.g., food, availability of water for growth of microorganisms, germination of spores, and participation in several types of chemical reactions becomes an important issue. This availability, which depends on relative pressure or water activity a_w, is defined as the ratio of the partial pressure p of water over the wet solid system to the equilibrium vapor pressure p_w of water at the same temperature. Thus, a_w, which is also equal to the relative humidity of the surrounding humid air, is defined as

$$a_w = \frac{p}{p_w} \tag{23.5}$$

Drying: Principles and Practice

TABLE 23.3
Minimum Water Activity for Microbial Growth and Spore Germination

Microorganism	Water Activity (a_w)
Organisms producing slime on meat	0.98
Pseudomonas, Bacillus cereus spores	0.97
B. subtilis, C. botulinum spores	0.95
C. botulinum, Salmonella	0.93
Most bacteria	0.91
Most yeast	0.88
Aspergillus niger	0.85
Most molds	0.80
Halophilic bacteria	0.75
Xerophilic fungi	0.65
Osmophilic yeast	0.62

Different shapes for the X versus a_w curves are observed, depending on the type of material (e.g., high, medium, or low hygroscopicity solids).

Table 23.3 lists the measured minimum a_w values for microbial growth or spore germination. If a_w is reduced below this value by dehydration or by adding water-binding agents like sugars, glycerol, or salt, microbial growth is inhibited. Such additives should not affect the flavor, taste, or other quality criteria, however. Since the amounts of soluble additives needed to depress a_w even by 0.1 are quite large, dehydration becomes particularly attractive for high moisture foods as a way to reduce a_w. Figure 23.5 shows schematically the water activity versus moisture content curve for different types of food. Rockland and Benchat (1987) provide an extensive compilation of results on water activity and its applications.

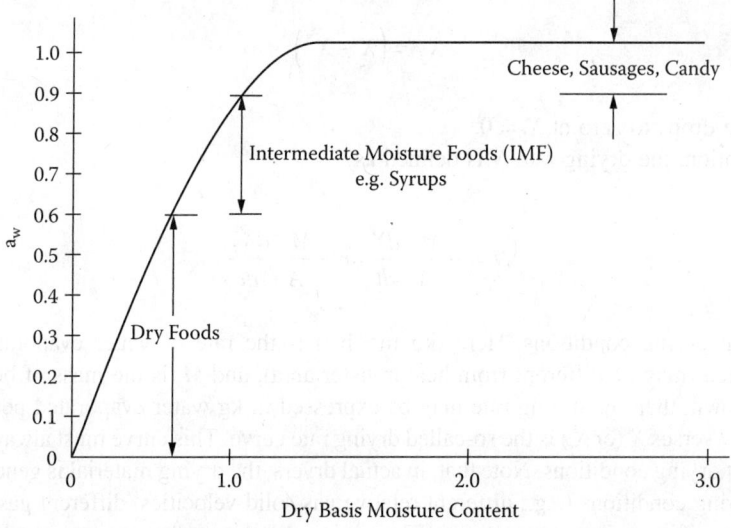

FIGURE 23.5 Water activity vs. moisture content curve.

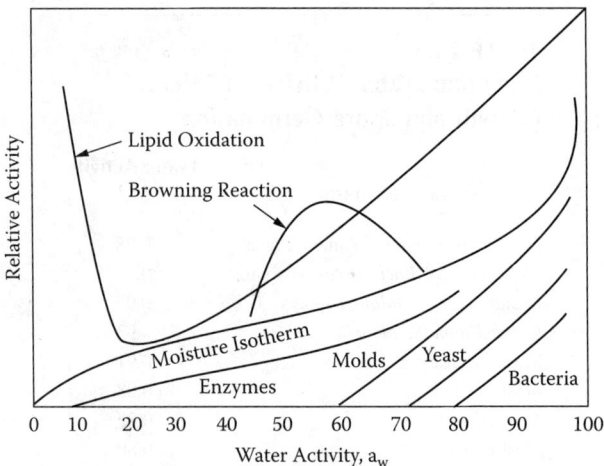

FIGURE 23.6 Deterioration rates as a function of water activity for food systems.

Figure 23.6 shows the general nature of the deterioration reaction rates as a function of a_w for food systems. Aside from microbial damage, which typically occurs for $a_w > 0.70$, oxidation, nonenzymatic browning (Maillard reactions), and enzymatic reactions can occur even at very low a_w levels during drying. Laboratory or pilot testing is essential to ascertain that no damage occurs in the selected drying process since this cannot in general be predicted.

23.2.2 Drying Kinetics

Consider the drying of a wet solid under fixed drying conditions. In the most general cases, after an initial period of adjustment, the dry-basis moisture content X decreases linearly with time t following the start of the evaporation. This is followed by a nonlinear decrease in X with t until, after a very long time, the solid reaches its equilibrium moisture content X^* and drying stops. In terms of free moisture content, defined as

$$X_f = \left(X - X^*\right) \tag{23.6}$$

the drying rate drops to zero at $X_f = 0$.

By convention, the drying rate N is defined as

$$N = -\frac{M_s}{A}\frac{dX}{dt} \; or \; -\frac{M_s}{A}\frac{dX_f}{dt} \tag{23.7}$$

under constant drying conditions. Here (kg m^{-2} h^{-1}) is the rate of water evaporation, A is the evaporation area (may be different from heat transfer area), and M_s is the mass of bone dry solid. If A is not known, then the drying rate may be expressed in kg water evaporated per hour.

A plot of N versus X (or X_f) is the so-called drying rate curve. This curve must always be obtained under constant drying conditions. Note that, in actual dryers, the drying material is generally exposed to varying drying conditions (e.g., different relative gas-solid velocities, different gas temperatures and humidities, different flow orientations). Thus, it is necessary to develop a methodology in order to interpolate or extrapolate limited drying rate data over a range of operating conditions.

Figure 23.7 shows a typical "textbook" drying rate curve displaying an initial constant rate period, where $N = N_c$ = constant. The constant rate period is governed fully by the rates of external

Drying: Principles and Practice

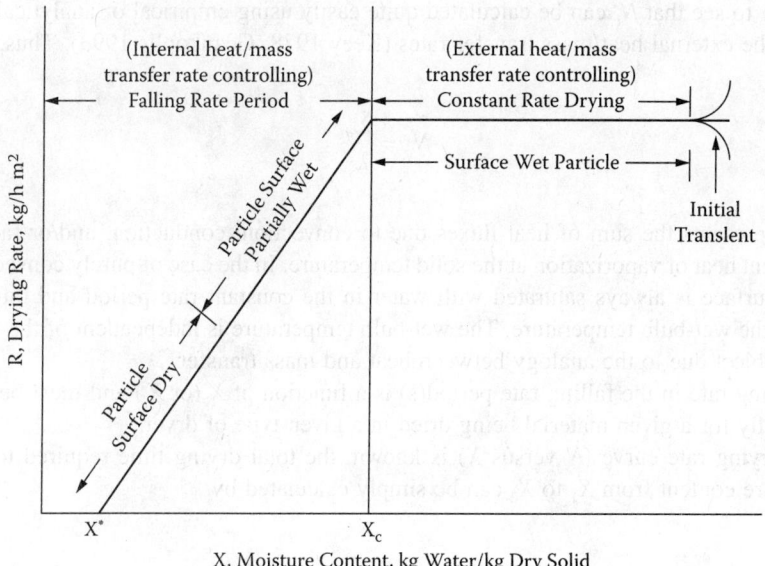

FIGURE 23.7 Typical textbook batch drying rate curve under constant drying conditions.

heat and mass transfer since a film of free water is always available at the evaporating surface. This drying period is nearly independent of the material being dried. Many foods and agricultural products, however, do not display the constant rate period at all since internal heat and mass transfer rates determine the rate at which water becomes available at the exposed evaporating surface.

At the so-called critical moisture content X_c, N begins to fall with further decreases in X since water cannot migrate at the rate N_c to the surface due to internal transport limitations. The mechanism underlying this phenomenon depends on both the material and drying conditions. The drying surface becomes first partially unsaturated and then fully unsaturated until it reaches the equilibrium moisture content X^*. Detailed discussions of drying rate curves are given by Keey (1992) and Mujumdar and Menon (1995). Approximate critical moisture content values for some selected materials are given in Table 23.4.

Note that a material may display more than one critical moisture content value at which the drying rate curve shows a sharp change of shape. This is generally associated with changes in the underlying mechanisms of drying due to structural or chemical changes. It is also important to note that X_c is not a material property. It depends on the drying rate under otherwise similar conditions. It must be determined experimentally.

TABLE 23.4
Approximate Critical Moisture Contents for Various Materials

Material	Critical Moisture Content (kg water/kg dry solid)
Salt crystals, rock salt, sand, wool	0.05–0.10
Brick clay, kaolin, crushed sand	0.10–0.20
Pigments, paper, soil, worsted wool fabric	0.20–0.40
Several foods, copper carbonate, sludges	0.40–0.80
Chrome leather, vegetables, fruits, gelatin, gels	>0.80

It is easy to see that N_c can be calculated quite easily using empirical or analytical techniques to estimate the external heat/mass transfer rates (Keey 1978; Geankoplis 1993). Thus,

$$N_c = \frac{\Sigma_q}{\lambda_s} \tag{23.8}$$

where Σ_q represents the sum of heat fluxes due to convection, conduction, and/or radiation, and λ_s, is the latent heat of vaporization at the solid temperature. In the case of purely convective drying, the drying surface is always saturated with water in the constant rate period and thus the liquid film attains the wet-bulb temperature. The wet-bulb temperature is independent of the geometry of the drying object due to the analogy between heat and mass transfer.

The drying rate in the falling rate period(s) is a function of X (or X_f) and must be determined experimentally for a given material being dried in a given type of dryer.

If the drying rate curve (N versus X) is known, the total drying time required to reduce the solid moisture content from X_1 to X_2 can be simply calculated by

$$t_d = -\int_{x_1}^{x_2} \frac{M_s}{A} \frac{dX}{N} \tag{23.9}$$

Table 23.5 lists expressions for the drying times for constant rate, linear falling rates, and a falling rate controlled by liquid diffusion of water in a thin slab. The subscripts c and f refer to the constant and falling rate periods, respectively. The total drying time is, of course, a sum of drying times in two succeeding periods. Different analytical expressions are obtained for the drying times t_f, depending on the functional form of N or the model used to describe the falling rate, e.g., liquid diffusion, capillarity, evaporation-condensation. For some solids, a receding front model (wherein the evaporating surface recedes into the drying solid) yields a good agreement with experimental observations. The principal goal of all falling rate drying models is to allow reliable extrapolation of drying kinetic data over various operating conditions and product geometries.

The expression for t_f in Table 23.5 using the liquid diffusion model (Fick's second law of diffusion form applied to diffusion in solids with no real fundamental basis) is obtained by solving analytically the following partial differential equation:

$$\frac{\partial X_f}{\partial t} = D_L \frac{\partial^2 X_f}{\partial x^2} \tag{23.10}$$

subject to the following initial and boundary conditions:

$$X_f = X_i, \text{ everywhere in the slab at } t = 0$$

$$X_f = 0, \text{ at } x = a \text{ (top, evaporating surface), and} \tag{23.11}$$

$$\frac{\partial X_f}{\partial x} = 0 \text{ at } x = 0 \text{ (bottom, non-evaporating surface)}$$

The model assumes one-dimensional liquid diffusion with constant effective diffusivity D_L, and no heat effects. X_2 is the average free moisture content at $t = t_f$ obtained by integrating the analytical

TABLE 23.5
Drying Times for Various Drying Rate Models

Model	Drying Time
Kinetic model $N = -\dfrac{M_S}{A}\dfrac{dX}{dt}$	t_d = Drying time to reach final moisture content X_2 from initial moisture content X_1.
$N = N(X)$ (General)	$t_d = \dfrac{M_s}{A}\displaystyle\int_{X_2}^{X_1} \dfrac{dX}{N}$
$N = N_c$ (Constant rate period)	$t_c = -\dfrac{M_s}{A}\dfrac{(X_2 - X_1)}{N_c}$
$N = aX + b$ (Falling rate period)	$t_f = \dfrac{M_s}{A}\dfrac{(X_1 - X_2)}{(N_1 - N_2)}\ln\dfrac{N_1}{N_2}$
$N = aX$, $X^* \le X_2 \le X_c$	$t_f = \dfrac{M_s X_c}{A N_c}\ln\dfrac{X_c}{X_2}$
Liquid diffusion model, D_L = constant, $X_2 = X_c$, Slab; one-dimensional diffusion, evaporating surface at X^*	$t_f = \dfrac{4a^2}{\pi D_L}\ln\dfrac{8X_1}{\pi^2 X_2}$ X, average free moisture content a, half-thickness of slab

Source: From Mujumdar (1997).

solution $X_f(x, t_f)$ over the thickness of the slab a. The expression in Table 23.5 is applicable only for long drying times since it is obtained by retaining only the first term in the infinite series solution of the partial differential equation.

The diffusivity of moisture in solids is a function of both temperature and moisture content. For strongly shrinking materials, the mathematical model used to define D_L must account for the changes in diffusion path as well. The temperature dependence of diffusivity is adequately described by the Arrhenius equation as follows:

$$D_L = D_{L0}\exp\left[-E_a/R_g T\right] \qquad (23.12)$$

where D_L is the diffusivity, E_a is the activation energy, and T is the absolute temperature. Okos et al. (1992) have given an extensive compilation of D_L and E_a values for various food materials. Zogzas et al. (1996) provide methods of moisture diffusivity measurement and an extensive bibliography on the topic. Approximate ranges of effective moisture diffusivity for some selected materials are given in Table 23.6.

It should be noted that D_L is not a true material property, and care should be taken in applying effective diffusivity correlations obtained with simple geometric shapes (e.g., slab, cylinder, or sphere) to the more complex shapes actually encountered in practice as this may lead to incorrect calculated results (Gong et al. 1997).

In addition to being dependent on geometric shapes, diffusivity depends as well on the drying conditions. Figure 23.8 compares the diffusivity values for air-dried, freeze-dried, and puff-dried potatoes as a function of water activity. At very high activity levels, no differences might be

TABLE 23.6
Approximate Ranges of Effective Moisture Diffusivity in Some Materials

Material	Moisture Content (kg/kg, DB)	Temperature (°C)	Diffusivity (m²/sec)
Alfalfa stems	3.70	26	2.6×10^{-10}–2.6×10^{-9}
Animal feed	0.01–0.15	25	1.8×10^{-11}–2.8×10^{-9}
Apple	0.10–1.50	30–70	1.0×10^{-11}–3.3×10^{-9}
Asbestos cement	0.10–0.60	20	2.0×10^{-9}–5.0×10^{-9}
Banana	0.01–3.50	20–40	3.0×10^{-13}–2.1×10^{-10}
Biscuit	0.10–0.60	20–100	8.6×10^{-10}–9.4×10^{-8}
Carrot	0.01–5.00	30–70	1.2×10^{-9}–5.9×10^{-9}
Clay brick	0.20	25	1.3×10^{-8}–1.4×10^{-8}
Egg liquid	—	85–105	1.0×10^{-11}–1.5×10^{-11}
Fish muscles	0.05–0.30	30	8.1×10^{-11}–3.4×10^{-10}
Glass wool	0.10–1.80	20	2.0×10^{-9}–1.5×10^{-8}
Glucose	0.08–1.50	30–70	4.5×10^{-12}–6.5×10^{-10}
Kaolin clay	<0.50	45	1.5×10^{-8}–1.5×10^{-7}
Muffin	0.10–0.95	20–100	8.5×10^{-10}–1.6×10^{-7}
Paper, thickness direction	~0.50	20	5×10^{-11}
Paper, in-plane direction	~0.50	20	1×10^{-6}
Pepperoni	0.16	12	4.7×10^{-11}–5.7×10^{-11}
Raisins	0.15–2.40	60	5.0×10^{-11}–2.5×10^{-11}
Rice	0.10–0.25	30–50	3.8×10^{-8}–2.5×10^{-6}
Sea sand	0.07–0.13	60	2.5×10^{-8}–2.5×10^{-6}
Soybeans	0.07	30	7.5×10^{-13}–5.4×10^{-12}
Silica gel	—	25	3.0×10^{-6}–5.6×10^{-6}
Starch gel	0.20–3.00	30–50	1.0×10^{-10}–1.2×10^{-9}
Tobacco leaf	—	30–50	3.2×10^{-11}–8.1×10^{-11}
Wheat	0.12–0.30	21–80	6.9×10^{-12}–2.8×10^{-10}
Wood, soft	—	40–90	5.0×10^{-10}–2.5×10^{-9}
Wood, yellow poplar	1.00	100–150	1.0×10^{-8}–2.5×10^{-8}

Source: Zogzas et al. (1996), Marinos-Kouris and Marouris (1995), and other sources.

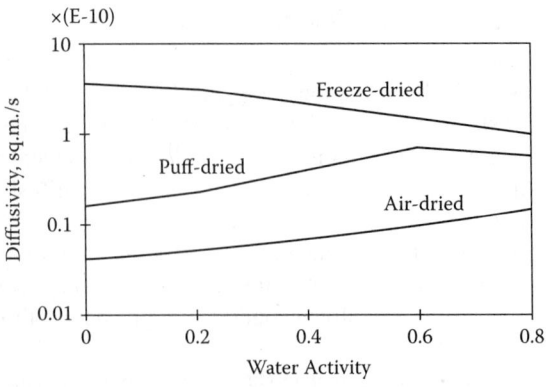

FIGURE 23.8 Moisture diffusivity in dehydrated potatoes.

TABLE 23.7
Solution to Fick's Second Law for Some Simple Geometries

Geometry	Boundary Conditions	Dimensionless Average Free Moisture Content
Flat plate of thickness $2b$	$t = 0$; $-b < z < b$; $X = X_0$ $t > 0$; $z = \pm b$; $X = X^*$	$X = \dfrac{8}{\pi^2} \sum_{n=1}^{\infty} \dfrac{1}{(2n-1)} \exp\left[-(2n-1)^2 \dfrac{\pi^2}{4b}\left(\dfrac{D_L t}{b}\right)\right]$
Infinitely long cylinder of radius R	$t = 0$; $0 < r < R$; $X = X_0$ $t > 0$; $r = R$; $X = X^*$	$X = 4 \sum_{n=1}^{\infty} \dfrac{1}{R^2 \alpha_n^2} \exp(-D_L \alpha_n^2 t)$ where α_n are positive roots of the equation $J_0(R\alpha_n) = 0$
Sphere of radius R	$t = 0$; $0 < r < R$; $X = X_0 t > 0$; $r = R$; $X = X^*$	$X = \dfrac{6}{\pi^2} \sum_{n=1}^{\infty} \dfrac{1}{n^2} \exp\left[\dfrac{-n^2 \pi^2}{R}\left(\dfrac{D_L t}{R}\right)\right]$

Source: From Pakowski and Mujumdar (1995).

observed, but at lower activity levels, the diffusivities may differ by an order-of-magnitude due to the inherently different physical structure of the dried product. Thus, the effective diffusivity is regarded as a lumped property that does not really distinguish between the transport of water by liquid or vapor diffusion, capillary or hydrodynamic flow due to pressure gradients set up in the material during drying. Further, the diffusivity values will show marked variations if the material undergoes glass transition during the drying process.

Keey (1978) and Geankopolis (1993), among others, have provided analytical expressions for liquid diffusion and capillarity models of falling rate drying. Table 23.7 gives solutions of the one-dimensional transient partial differential equations for cartesian, cylindrical, and spherical coordinate systems. These results can be utilized to estimate the diffusivity from the falling rate drying data or to estimate the drying rate and drying time if the diffusivity value is known.

The diffusivity D_L is a strong function of X_f as well as temperature and must be determined experimentally. Thus, the liquid diffusion model should be regarded purely as an empirical representation of the falling rate drying. More advanced models are, of course, available, but their widespread use in design of dryers is hampered by the need for extensive empirical information required to solve the governing equations. Turner and Mujumdar (1997) provide a wide assortment of mathematical models of drying and dryers, and also discuss the application of various techniques for the numerical solution of the complex governing equations.

One simple approach to interpolating a given falling rate curve over a relatively narrow range of operating conditions is that first proposed by van Meel (1958). He found that the plot of the normalized drying rate $v = N/N_c$ versus the normalized free moisture content $\eta = (X - X^*)/(X_c - X^*)$ was nearly independent of the drying conditions. This plot, called the characteristic drying rate curve, is illustrated in Figure 23.9. Thus, if the constant drying rate N_c can be estimated and the equilibrium moisture content data are available, then the falling rate curve can be estimated using this highly simplified approach. Extrapolation over wide ranges is not recommended, however.

Waananen et al. (1993) have provided an extensive bibliography of over 200 references dealing with models of drying for porous solids. Basically, such models are useful to describe

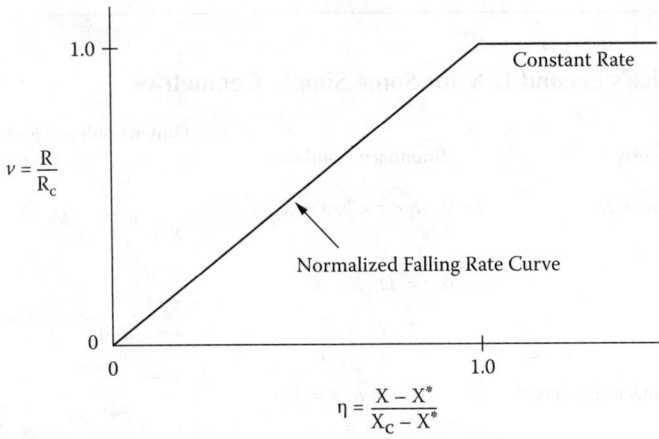

FIGURE 23.9 Characteristic drying rate curve.

drying processes for the purposes of engineering design, analysis, and optimization. A mathematical description of the process is based on the physical mechanisms of internal heat and mass transfer that control the process resistances, as well as the structural and thermodynamic assumptions made to formulate the model. In the constant rate period, the overall drying rate is determined solely by the heat and mass transfer conditions external to the material being dried, such as the temperature, gas velocity, total pressure, and partial pressure of the vapor. In the falling rate period, the rates of internal heat and mass transfer determine the drying rate. Modeling of drying is complicated by the fact that more than one mechanism may contribute to the total mass transfer rate and the contributions from different mechanisms may change during the drying process.

Diffusional mass transfer in the liquid phase, as discussed earlier, is the most commonly assumed mechanism of moisture transfer used in modeling drying that takes place at temperatures below the boiling point of the liquid under locally applied pressure. At higher temperatures, the pore pressure may rise substantially and cause a hydrodynamically driven flow of vapor, which, in turn, may cause a pressure driven flow of liquid in the porous material.

For solids with continuous pores, a surface tension driven flow (capillary flow) may occur as a result of capillary forces caused by the interfacial tension between the water and the solid particles. In the simplest model, a modified form of the Poiseuille flow can be used in conjunction with the capillary forces equation to estimate the rate of drying. Geankoplis (1993) has shown that such a model predicts the drying rate in the falling rate period to be proportional to the free moisture content in the solid. At low solid moisture contents, however, the diffusion model may be more appropriate.

The moisture flux due to capillarity can be expressed in terms of the product of a liquid conductivity parameter and moisture gradient. In this case, the governing equation has, in fact, the same form as the diffusion equation.

For certain materials and under conditions such as those encountered in freeze drying, a "receding-front" model involving a moving boundary between "dry" and "wet" zones often describes the mechanism of drying much more realistically than does the simple liquid diffusion or capillarity model. Examination of the freeze drying of a thin slab of frozen material indicates that the rate of drying is dependent on the rate of heat transfer to the "dry-wet" interface, and the mass transfer resistance offered by the porous dry layer to permeation of the vapor that sublimes from the interface. Because of the low pressures encountered in freeze drying, Knudsen diffusion effects may be significant. Liapis and Marchello (1984) have discussed models of freeze drying involving both unbound and bound moisture.

23.3 CLASSIFICATION AND SELECTION OF DRYERS

In this section, we will examine the key classification criteria for industrial dryers and then proceed to selection criteria with the explicit understanding that the latter is a complex process and is not entirely scientific, but also involves subjective judgment as well as considerable empiricism. It should also be noted that the pre-drying as well as post-drying stages have important bearing on the selection of appropriate dryer types for a given application. Indeed, for optimal selection of solid/liquid separation processes, one must examine the overall flowsheet as well as the "drying system." This section will be confined, however, only to the classification and selection of dryers.

Another important point to note is that several dryer types (or drying systems) may be equally suited (technically and economically) for a given application. A careful evaluation of as many of the possible factors affecting the selection will help reduce the number of options. For a new application (new product or new process), it is important to follow a careful procedure leading to the choice of the dryers. Characteristics of different dryer types should be recognized when selecting dryers. Changes in operating conditions of the same dryer can affect the quality of the dried product. So, aside from the dryer type, it is also important to choose the right operating conditions for optimal quality and cost of dehydration.

Baker (1997) has presented a "structural approach" for dryer selection that is iterative. It includes the following steps:

1. List all key process specifications.
2. Conduct preliminary selection.
3. Perform bench-scale tests, including quality test.
4. Make economic evaluation of alternates.
5. Conduct pilot-scale trials.
6. Select most appropriate dryer types.

Often, for same materials, a specific dryer type is indicated from the outset. If selection is based exclusively on past experience, it has three limitations:

1. If the original selection is not optimal (although it works satisfactorily), the new choice will also be less-than-optimal.
2. No new drying technologies are considered by default.
3. It is implicitly assumed the "old" choice was arrived at logically, which is often not the case.

23.3.1 Classification of Dryers

There are numerous schemes used to classify dryers (Mujumdar 1995b; van't Land 1991). Table 23.8 lists the criteria and typical dryer types. Types marked with an asterisk (*) are among the most common in practice. Such classification is rather coarse. Just the fluidized bed dryer can be subclassified into over 30 types.

Each type of dryer has specific characteristics, which make it suitable or unsuitable for specific applications. Details can be found in Mujumdar (1995). Certain types are inherently expensive (e.g., freeze dryers) while others are inherently more efficient (e.g., indirect or conduction dryers). Thus, it is necessary to be aware of the wide variety of dryers available in the market and their special advantages and limitations. Furthermore, note that the aforementioned classification does not include most of the novel drying technologies, which are applicable for very specific applications. The reader is referred to Kudra and Mujumdar (1995) for details on novel drying technologies.

Following is a general scheme proposed by Baker (1997) for classification of batch and continuous dryers. Note that there is a more limited choice of batch dryers—only a few types can be operated in both batch and continuous modes.

TABLE 23.8
Classification of Dryers

Criterion	Types
Mode of operation	Batch
	Continuous*
Heat input-type	Convection,* conduction, radiation, electromagnetic fields, combination of heat transfer modes
	Intermittent or continuous*
	Adiabatic or nonadiabatic
State of material in dryer	Stationary, moving, agitated, dispersed, etc.
Operating pressure	Vacuum*
	Atmospheric
Drying medium (convection)	Air*
	Superheated steam
	Flue gases
Drying temperature	Below boiling temperature*
	Above boiling temperature
	Below freezing point
Relative motion between drying medium and drying solids	Concurrent
	Countercurrent
	Mixed flow
Number of stages	Single*
	Multi-stage
Residence time	Short (<1 min)
	Medium (1–60 min)
	Long (>60 min)

* Most common dryer type in industry.

23.3.1.1 Batch Dryers: Classification (Baker 1997)

Particulate Solids: Major Classes—Layer (Packed Bed); Dispersion Type

1. Layer type
   a. Contact (conduction or indirect type), e.g., vacuum tray, agitated bed, rotary batch
   b. Convection (atmospheric tray)
   c. Special types (e.g., microwave, freeze, solar)
2. Dispersion type
   a. Fluidized bed/spouted bed
   b. Vibrated bed dryer

23.3.1.2 Continuous Dryers: Classification

Major Classes—Layer; Dispersion Type

1. Layer type
   a. Contact (conduction or indirect type), e.g. drum, plate, vacuum based, agitated bed; indirect rotary, etc.
   b. Convection, e.g., tunnel, spin-flush, throughflow, conveyor, etc.
   c. Special, e.g., microwave, radio frequency, freeze, solar, etc.

2. Dispersion type
   a. Fluid bed, vibrated bed, direct rotary, ring, spray, jet-zone

Classification of dryers on the basis of the mode of thermal energy input is perhaps the most useful since it allows one to identify some key features of each class of dryers.

23.3.1.3 Direct Dryers

Direct dryers, also known as convective dryers, are by far the most common. About 85% of industrial dryers are estimated to be of this type despite their relatively low thermal efficiency caused by the difficulty in recovering the latent heat of vaporization contained in the dryer exhaust in a cost-effective manner. Hot air produced by indirect heating or direct firing is the most common drying medium, although for some special applications superheated steam has recently been shown to yield higher efficiency and often a higher quality product. Superheated steam at near atmospheric pressure can be used only for materials that are stable at higher temperatures. For heat-sensitive materials it is possible to operate a superheated steam dryer at lower pressures where the saturation temperature is lower. Flue gases may be used when the product is not heat sensitive or affected by the presence of products of combustion. In direct dryers, the drying medium contacts the material to be dried directly and supplies the heat required for drying by convection; the evaporated moisture is carried away by the same drying medium.

Drying gas temperatures may range from 50°C to 400°C, depending on the material. Dehumidified air may be needed when drying highly heat-sensitive materials. An inert gas such as nitrogen may be needed when the product is susceptible to oxidation, when drying explosive or flammable solids, or when an organic solvent is to be removed. Solvents must be recovered from the exhaust by condensation so that the inert (with some solvent vapor) can be reheated and returned to the dryer.

Because of the need to handle large volumes of gas, gas cleaning and product recovery (for particulate solids) becomes a major part of the drying plant. Higher gas temperatures yield better thermal efficiencies subject to product quality constraints.

23.3.1.4 Indirect Dryers

Indirect dryers involve a supply of heat to the drying material without direct contact with the heat transfer medium, i.e., heat is transferred from the heat transfer medium (steam, hot gas, thermal fluids, etc.) to the wet solid by conduction. Since no gas flow is presented on the wet solid side, it is necessary to either apply vacuum or use gentle gas flow to remove the evaporated moisture so that the dryer chamber is not saturated with vapor. Heat transfer surfaces may range in temperature from –40°C (as in freeze drying) to ~300°C in the case of indirect dryers heated by direct combustion products such as waste sludges. In vacuum operation, there is no danger of fire or explosion. Vacuum operation also eases the recovery of solvents by direct condensation, thus alleviating serious pollution problems. Dust recovery is obviously simpler so that such dryers are especially suited for drying of toxic, dusty products that must not be entrained in gases. Furthermore, vacuum operation lowers the boiling point of the liquid being removed; this allows drying of heat-sensitive solids at relatively fast rates.

Heat may be supplied by radiation (using electric or natural gas-fired radiators) or volumetrically by placing the wet solid in dielectric fields in the microwave or radio frequency range. Since radiant heat flux can be adjusted locally over a wide range, it is possible to obtain high drying rates for surface-wet materials. Convection (gas flow) or vacuum operation is needed to remove the evaporated moisture. Radiant dryers have found important applications in some niche markets, e.g., drying of coated papers or printed sheets. However, the most popular applications involve the use of combined convection and radiation. It is often useful to boost the drying capacity of an existing convection dryer for sheets such as paper.

Microwave dryers are expensive in terms of both the capital and operating (energy) costs. They have found limited applications to date. However, they do have special advantages in terms of product quality when handling heat-sensitive materials. They are worth considering as devices to speed up drying in the tail end of the falling rate period. Similarly, RF dryers have limited industrial applicability. They have found some niche markets, e.g., drying of thick lumber and coated papers. Both microwave and RF dryers must be used in conjunction with convection or under vacuum to remove the evaporated moisture. Standalone dielectric dryers are unlikely to be cost-effective except for high-value products. See Schiffmann (1995) for a detailed discussion of dielectric dryers.

It is possible, indeed desirable in some cases, to use combined heat transfer modes, e.g., convection and conduction, convection and radiation, convection and dielectric fields, etc., to reduce the need for increased gas flow that results in lower thermal efficiencies. Use of such combinations increases the capital costs, but these may be offset by reduced energy costs and enhanced product quality. No generalization can be made a priori without tests and economic evaluation. Finally, the heat input may be steady (continuous) or time-varying; also, different heat transfer modes may be deployed simultaneously or consecutively depending on the individual application. In view of the significant increase in the number of design and operational parameters resulting from such complex operations, it is desirable to select the optimal conditions via a mathematical model.

23.3.2 Selection of Dryers

In view of the enormous choices of dryer types one could possibly deploy for most products, selection of the best type is a challenging task that should not be taken lightly, nor should it be left entirely to dryer vendors who typically specialize in only a few types. The user must take a proactive role and employ vendors' experiences and bench-scale or pilot-scale facilities to obtain data that can be assessed for a comparative evaluation of several options. A wrong dryer for a given application is still a poor dryer, regardless of how well it is designed. Note that minor changes in composition or physical properties of a given product can influence its drying characteristics, handling properties, etc., leading to a different dried product and, in some cases, severe blockages in the dryer itself. Thus, tests should be carried out with the "real" feed material and not a "simulated" one when feasible.

Although here we will focus only on the selection of the dryer alone, it is very important to note that in practice one must select and specify a drying system that includes pre-drying stages (e.g., mechanical dewatering, evaporation, pre-conditioning of feed by solids backmixing, dilution, or pelletization and feeding) as well as the post-drying stages of exhaust gas cleaning, product collection, partial recirculation of exhaust gas, cooling of product, coating of product, agglomeration, etc. The optimal cost-effective choice of a dryer will depend, in some cases, significantly on these stages. For example, a hard pasty feedstock can be diluted to a pumpable slurry, atomized, and dried in a spray dryer to produce a powder, or it may be pelletized and dried in a fluid bed or in a through circulation dryer, or dried as is in a rotary or fluid bed unit. Also, in some cases, it may be necessary to examine the entire flowsheet to see if the drying problem can be simplified or even eliminated. Typically, nonthermal dewatering is an order-of-magnitude less expensive than evaporation, which, in turn, is many-fold more energy efficient than thermal drying. Demands on product quality may not always permit one to select the least expensive option based solely on heat and mass transfer considerations. Often, product quality requirements have overriding influence on the selection process.

As a minimum, the following quantitative information is necessary to arrive at a suitable dryer:

- Dryer throughput; mode of feedstock production (batch/continuous)
- Physical and chemical properties of the wet feed as well as the product desired (specifications); expected variability in feed characteristics
- Upstream and downstream processing operations

- Moisture content of feed and product
- Drying kinetics data; moist-solid sorption isotherms
- Quality parameters
- Safety aspects, e.g., combustion and dust explosion, fire hazard, toxicity, etc.
- Value of the product
- Need for control
- Toxicological properties of the product
- Potential for fire or explosion hazards
- Turndown ratio, flexibility in capacity requirements
- Type and cost of fuel, cost of electricity
- Environmental regulations

For high-value products like pharmaceuticals, certain foods, and advanced materials, quality considerations override other considerations since the cost of drying is unimportant. Throughputs of such products are also relatively low.

In some cases, the feed may be conditioned (e.g., size reduction, flaking, pelletizing, extrusion, backmixing with dry product) prior to drying, which affects the choice of dryers.

As a rule, in the interest of energy savings and reduction of dryer size, it is desirable to reduce the feed liquid content by less expensive operations such as filtration, centrifugation, evaporation, etc. It is also desirable to avoid over-drying, which increases the energy consumption as well as the drying time, i.e., reduced throughput from a given dryer.

Drying of food and biotechnological products requires adherence to GMP (Good Manufacturing Practice of the Food and Drug Administration) and hygienic equipment design and operation. Such materials are subject to thermal as well as microbiological degradation during drying as well as in storage.

If the feed rate is low (<100 kg/hr), a batch-type dryer may be suitable. If the feed is produced in much larger quantities in batch mode (e.g., 2000 kg batches), one may consider a batch dryer. Note that there is a limited choice of dryers that can operate in the batch mode.

In <1% of cases of industrial drying, the liquid to be removed in dryers is a nonaqueous (organic) solvent or a mixture of water with a solvent. This is not uncommon in drying of pharmaceutical products. Special care is needed to recover the solvent and to avoid potential danger of fire and explosion.

Table 23.9 presents a typical checklist most vendors of dryers use to select and quote an industrial dryers.

Drying kinetics plays a significant role in the selection of dryers, aside from simply deciding the residence time required; it limits the types of suitable dryers. Location of the moisture (whether near the surface or distributed in the material), nature of moisture (free or strongly bound to solid), mechanisms of moisture transfer (rate limiting step), physical size of product, conditions of drying medium (e.g., temperature, humidity, flow rate of hot air for a convective dryer), pressure in dryer (low for heat-sensitive products), etc., have a bearing on the type of suitable dryer as well as the operating conditions. Most often, not more than one dryer type will likely meet the specified selection criteria.

Several attempts have been reported in the literature at development of automated expert systems that guide the selection process. This scheme has met with limited success because there are too many criteria and too many exceptions to rules encountered in practice. Slightly different rules result in different selection solutions. Some of the recommended dryers may not be available from local vendors or the delivery times may be unacceptably long. In some cases, none of the dryers may meet the specified criteria. For example, in drying a highly sticky liquid, none of the dryers may be appropriate unless the feed is backmixed, pelletized, dried in a through circulation tunnel dryer, and then milled to a desired specification. Local environmental regulations may dictate the choice of an indirectly heated rotating shelf vacuum dryer (plate dryer) if the material to be dried

TABLE 23.9
Typical Checklist for Selection of Industrial Dryers

Physical form of feed	Granular/particulate/sludge/crystalline/ liquid/pasty/suspension/solution/ continuous sheets, planks, odd-shapes (small large)
	Sticky, lumpy, other
Average throughput	kg/hr (dry/wet); continuous
	kg per batch (dry/wet)
Expected variation in throughput (turndown ratio)	
Fuel choice	Oil
	Gas
	Electricity
Pre- and post-drying operations (if any)	
For paniculate feed products	Mean particle size
	Size distribution
	Particle density
	Bulk density
	Rehydration properties
Inlet/outlet moisture content	Dry basis
	Wet basis
Chemical/biochemical/microbiological activity	
Heat sensitivity	Melting point
	Glass transition temperature
Sorption isotherms (equilibrium moisture content)	
Drying time	Drying curves
	Effect of process variables
Special requirements	Material of construction
	Corrosion
	Toxicity
	Nonaqueous solvents
	Flammability limits
	Fire hazard
	Color/texture/aroma requirements (if any)

is fragile and toxic at the same time. If it were not toxic, a through circulated rotating shelf dryer (e.g., Turbo dryer) may be more economic.

In Table 23.10, Kemp (1998) gives results that were obtained from the proprietary dryer selection algorithm developed by Separation Processes Service (SPS) of AEA Technology, Harwell, United Kingdom. Although only one choice is reported here, it should be noted that, in most cases, alternate dryers can also be recommended with nearly equal performance. If local cost of equipment and energy are factored in along with the value of the dried product itself, the results may be different as well.

A further caution to be exercised when selecting dryers is not to be biased by the way the product is made in the laboratory. Drying as well as filtration, at extremely small scales, are very different from that at a scale that is several orders-of-magnitude larger. Some materials may start to form hard lumps under their own weight—this may not show in pilot tests if the product depth is under the critical depth needed to produce lumps. A double-cone vacuum dryer, for example, has a tendency to form snowballs on full-scale but not necessarily on pilot-scale equipment (Kemp 1998).

Table 23.11 lists some key recommendations on dryer selection based on specific properties of the material. It is not all-inclusive, nor does it cover all special physical and/or chemical

TABLE 23.10
Selection Results for a Variety of Industrial Case Studies

Material	Solids Flow Rate (kg/hr)	Highest-Ranked Dryer	Remarks
Inorganic powder	360	Cascading rotary	Wide range of choices
Salt	500	Plug flow fluid bed	Attrition prevention
Organic peroxide	40	Batch fluid bed	Explosible; inerted
Organo-tin comp	100	Conical vacuum	Toxic, form cuts
Crystals	1000	Plate dryer	Fragile agglomerates
Pharmaceutical	30	Batch freeze dryer	Volatile retention
Mineral industry	500	Spray dryer	Attrition of nozzle
Sewage sludge	1500	Well-mixed fluid bed	Many alternatives
Corn cereal	2000	Cascading rotary	Min. residence time
Polymer	5000	Plug flow fluid bed	Also two-stage options
Inorganic paste	1000	Perforated band	Feed modified to pellets

Source: From Kemp (1998).

TABLE 23.11
Key Recommendations on Dryer Selection Based on Specific Properties of the Materials

Quality Parameter	Suggested Dryer/Remarks
Fragile, weak granules, agglomerates	Avoid agitation of material conveyor, tunnel, tray dryers
Flammable	Inert drying medium vacuum dryer, superheated steam dryer
Fire/explosion hazards	As above; lower oxygen concentration in drying; indirect drying
Very sticky feedstock	Pre-condition; solids backmixing, pelletization
High-heat sensitivity	Vacuum or freeze dryers
High-volatile content	Freeze dryers (to retain aroma)
Organic liquid present	Closed-system dryers, spray dryers, fluidized bed dryers
Gentle handling required	Batch or continuous tray, vibrated bed dryers
Excessive drying times	Use dielectric heating (microwave or radio frequency) over part of drying
No contamination permissible; filter cake	Dewater and dry in same unit, e.g., Nutsch filter/dryer

properties. Exceptions do exist, of course. Some other dryers can be modified to meet the specific requirements as well.

We do not focus on novel or special drying techniques here for lack of space. However, it is worth noting that many of the new techniques use superheated steam as the drying medium or are simply intelligent combinations of traditional drying techniques, e.g., combination of heat transfer modes, multistaging of different dryer types, etc. Superheated steam as the convective drying medium offers several advantages, e.g., higher drying rates under certain conditions, better quality for certain products, lower net energy consumption as the excess steam produced in the dryer is used elsewhere in the process, elimination of fire and explosion hazard, etc. Vacuum steam drying of timber, for example, can reduce drying times by a factor of up to four while enhancing wood quality and reducing net fuel and electricity consumption by up to 70%. The overall economics are also highly favorable. Some conventional and more recent drying techniques are listed in Table 23.11.

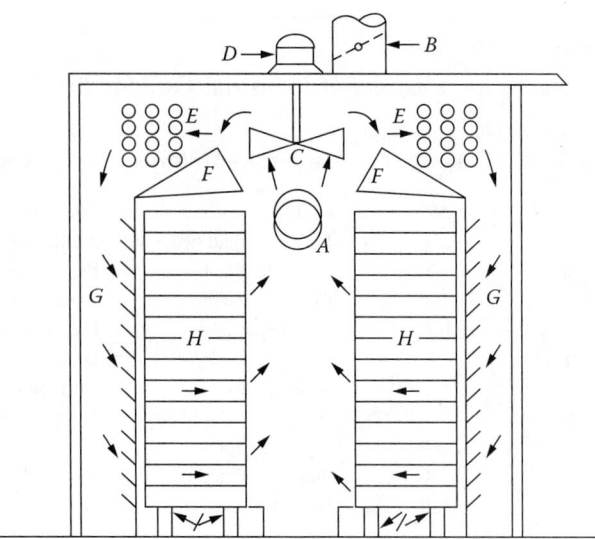

FIGURE 23.10 Schematic of a batch tray dryer.

23.4 DRYING EQUIPMENT

In this section, a number of dryer types commonly used in practice will be discussed based on their different capabilities of handling different types of feedstock, e.g., particulate/granular solids and slurries and suspensions. Some other selected dryers (e.g., infrared, microwave, radio-frequency dryers) will also be mentioned.

23.4.1 Dryers for Particulates and Granular Solids

23.4.1.1 Tray Dryers

By far the most common dryer for small tonnage products, a batch tray dryer (Figure 23.10) consists of a stack of trays or several stacks of trays placed in a large insulated chamber in which hot air is circulated with appropriately designed fans and guide vanes. Often, a part of the exhausted air is recirculated with a fan located within or outside the drying chamber. These dryers require a large amount of labor to load and unload the product. Typically, the drying times are long (10 to 60 hours). The key to successful operation is the uniform air flow distribution over the trays as the slowest drying tray decides the residence time required and hence dryer capacity. Warpage of trays can also cause poor distribution of drying air and hence poor dryer performance.

The batch tray dryer can often be converted into a continuous unit. Figure 23.11 shows the so-called Turbo dryer, which consists of a stack of coaxial circular trays mounted on a single vertical shaft. The product layer fed into the first shelf is leveled by a set of stationary blades that scratch a series of grooves into the surface layer of the solid particles. The blades are staggered to ensure mixing of the material. After one rotation, the material is wiped off the shelf by the last blade and falls onto the next lower shelf. Up to 30 trays or more can be accommodated.

Hot air is supplied to the drying chamber by turbine fans. In the design shown, the air is heated by internal heaters. The wet granular material is fed at the top and it falls under gravity to the next tray through radial slots in each circular shelf. A rotating rake mixes the solids and thus improves the drying performance. Such dryers can be operated under vacuum for heat-sensitive materials or when solvents must be recovered from the vapor. In a modified design, it is possible to heat the trays by conduction and apply vacuum to remove the moisture evaporated.

Drying: Principles and Practice

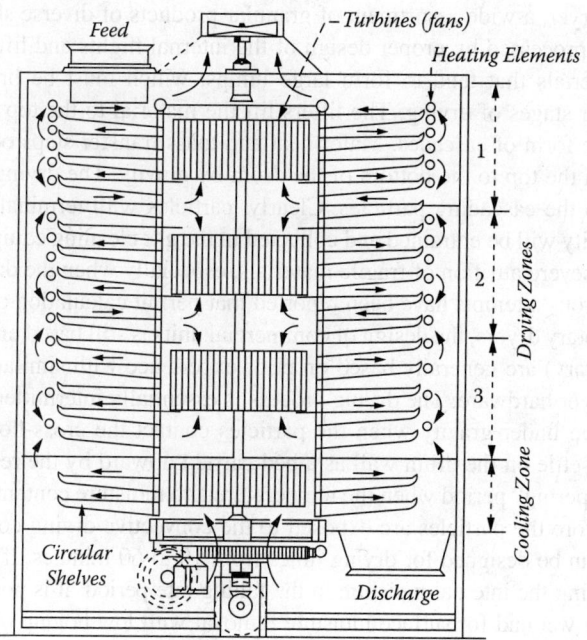

FIGURE 23.11 Schematic of a Turbo dryer.

23.4.1.2 Rotary Dryers

The cascading rotary dryer is a continuously operated direct contact dryer consisting of a slowly revolving cylindrical shell that is typically inclined to the horizontal a few degrees to aid in the transportation of the wet feedstock, which is introduced into the drum at the upper end and the dried product withdrawn at the lower end (Figure 23.12). To increase the retention time of very fine and light particles (e.g., cheese granules) in the dryer, in rare cases it may be advantageous to incline the cylinder with the product end at a higher elevation. The drying medium (hot air, combustion gases, flue gases, etc.) flows axially through the drum either concurrently with the feedstock or counter-currently. The latter mode is preferred when the material is not heat-sensitive and needs to be dried to very low moisture content levels. The concurrent mode is preferred for heat-sensitive materials and for higher drying rates in general.

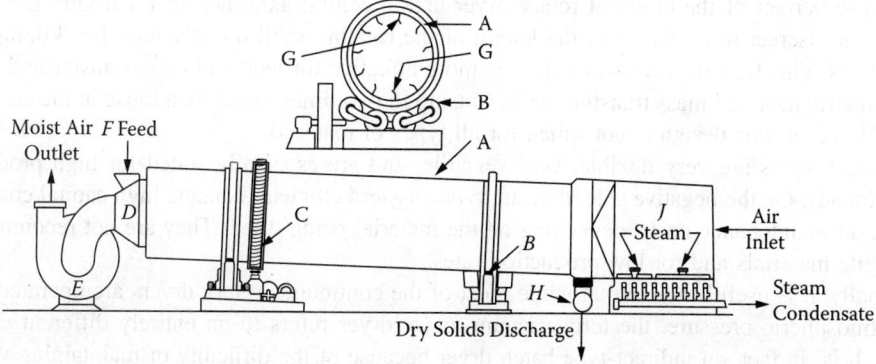

FIGURE 23.12 Countercurrent air-heated rotary dryer: (A) dryer shell; (B) shell-supporting rolls; (C) drive gear; (D) air-discharge hood; (E) discharge fan; (F) feed chute; (G) lifting flights; (H) product discharge; (J) air heater.

In this type of dryer, a wide assortment of granular products of diverse shapes, sizes, and size distributions can be processed by proper design of the internal flights and lifters. Special internals are needed for materials that tend to form large lumps, which must be broken to avoid major problems in the later stages of drying. The lifters lift the material to the top of the drum where it showers down in the form of cascades. Major heat and mass transfer steps occur during the flight of the particles from the top to the bottom of the drum by gravity. The drying medium is in cross-flow with respect to the cascading particles. Clearly, particles with terminal velocities below the cross-flow gas velocity will be entrained and collected in the gas cleaning equipment. The cascading action often causes severe attrition of fragile materials, especially when the drum diameter is large.

Although numerous attempts have been reported that permit calculation of the average particle residence times in rotary dryers, the design of commercial units is still based on pilot tests. Empirical rules (often proprietary) are generally based on prior experience with similar material and similar designs of rotary dryer hardware. The drying process is essentially intermittent. It is intense during the cascading motion under gravity when the particles contact the cross-flowing hot gas stream. When the particles settle on the drum wall as a bed carried upward by the revolving shell, there is a "soaking" or "tempering" period when the temperature and moisture content fields in the particles tend to equalize before the particles are exposed to the convective drying condition again.

Rotary dryers can be designed for drying times from 10 to 60 minutes. If a large retention time is needed for removing the internal moisture in the falling rate period, it is possible to use a smaller shell diameter at the wet end for surface moisture removal with low holdup of material in the drum and then increase the shell diameter at the dry end to allow a longer retention time with a larger holdup. In some designs, it is possible to use a pneumatic conveyor to carry the product out of the dryer.

Thermal efficiencies of rotary dryers vary widely in the range of 30% to 60%. For good efficiency, the product holdup (typically 10% to 15% of volume) should be such as to cover the flights or lifters fully. The lifters should be carefully designed to ensure good cascading action, avoiding large clusters of material falling from the flights. Length-to-diameter ratios of 4 to 10 are common in industrial practice.

Rotary dryers can be operated at very high temperatures to accomplish various reactions in addition to or instead of simple drying; these units are referred to as kilns. It is necessary to line the shell of rotary kilns with suitable refractory materials.

In order to enhance the drying rates in the rotary dryer without raising the gas temperature or gas flowrate excessively, it is possible to introduce steam heated tubes or coils within the shell. Aside from providing additional energy for drying, such internals can also help with redistribution or delumping of the material. Of course, it is possible to use internal heaters only if the material does not stick to the walls of the internals.

A new variant of the classical rotary dryer uses a central axial header for drying gas that is injected at discreet intervals along the length of the rotating shell directly into the "kilning" bed of particles. This type of flow distribution is more effective for heat and mass transfer and results in volumetric heat and mass transfer coefficients up to two times larger than those in the cascading dryer. However, this design is not suited for all types of materials.

Rotary dryers are very flexible, very versatile, and are especially suited for high production rate demands. On the negative side, they are typically less efficient, demand high capital costs, and significant maintenance costs, depending on the material being dried. They are not recommended for fragile materials and for low production rates.

Finally, it is useful to note that while most of the continuous rotary dryers are operated under near atmospheric pressure, the term vacuum-rotary dryer refers to an entirely different class of dryers. It is, in fact, an indirect-type batch dryer because of the difficulty of maintaining vacuum under continuous feeding and discharge conditions. Here, the horizontal cylindrical shell is stationary while a set of variously designed agitator blades revolves on a central shaft to agitate the material contained in the dryer shell. Heat is supplied by heating the shell jacket using condensing steam or a thermal fluid. In larger units, the central agitator shaft and the blades may also be heated.

Drying: Principles and Practice

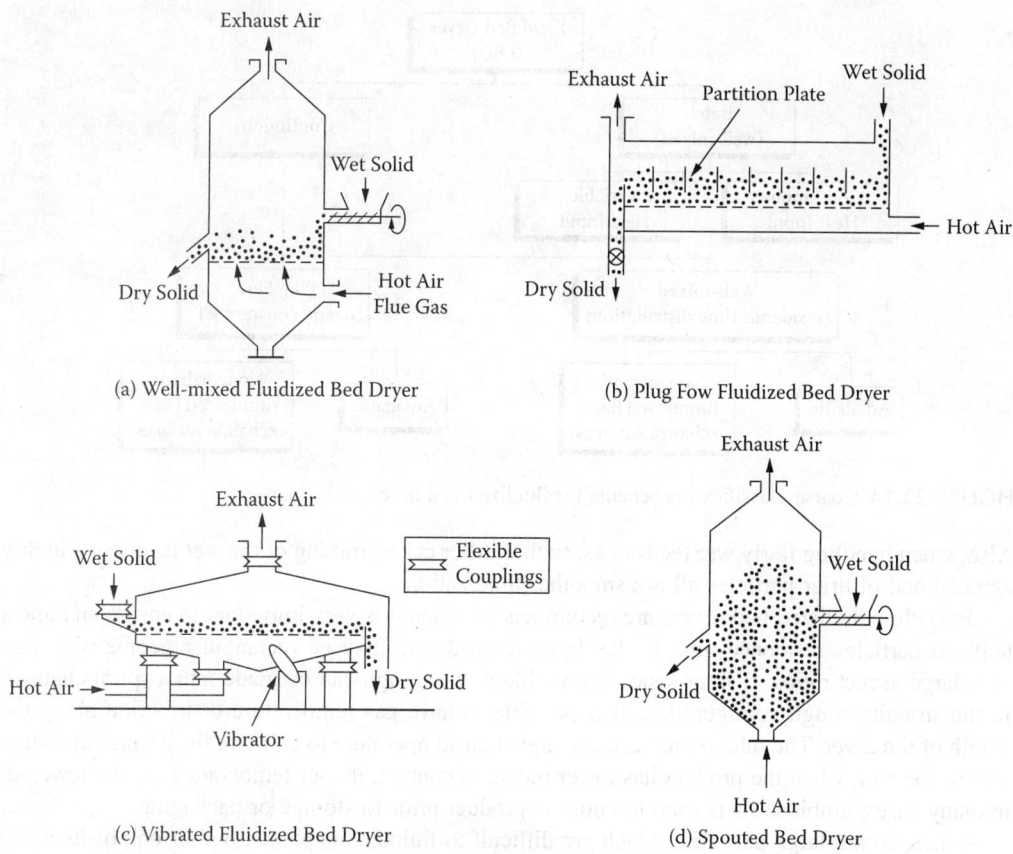

FIGURE 23.13 Fluidized and modified fluidized bed dryers.

The agitator may be a single- or double-spiral. The outer blades are set close to the wall and may have a scraper attached to keep the material from building up on the walls and deteriorating the thermal performance of the unit. This type of dryer is useful for handling heat-sensitive materials, which dry at lower temperatures because of the vacuum conditions.

23.4.1.3 Fluidized Bed Dryers

Fluidized bed dryers offer the advantages of high drying rates, high thermal efficiency, low floor area, ease of control, etc., for drying of particulate solids, sludges, and even slurries when inert particles are used as the bed media. Their limitations include high power consumption, attrition, and lower flexibility than competing dryers, e.g., rotary, conveyor dryers, etc. Mechanical agitation of vibration may be necessary to handle sticky or highly polydisperse particles.

Figure 23.13 shows the four common types of fluidized and modified fluidized bed dryers that may be operated in either the batch (lower capacities) or continuous mode. Numerous variants are possible and in use.

Figure 23.14 outlines a classification of fluidized bed dryers. For heat-sensitive materials, e.g., polymer pellets, use of immersed heat exchangers allows use of lower fluidizing air temperature without increasing drying time. Such exchangers can be used, however, only if the solids do not foul their surfaces.

Well-mixed continuous fluidized bed dryers are most commonly used if the final product moisture content distribution is permitted to vary somewhat due to the inherent residence time distribution in the dryer. Equilibration occurs during storage, resulting in uniform product moisture.

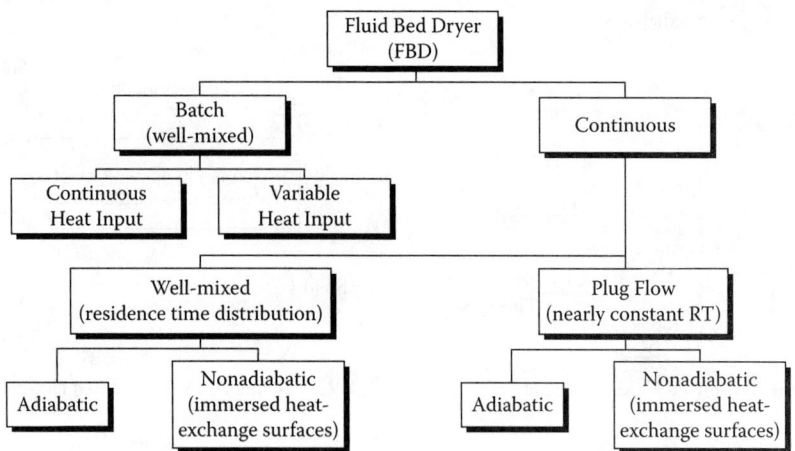

FIGURE 23.14 Coarse classification scheme for fluidized bed dryer.

Also, when handling fairly wet feedstocks, such as filter cakes, mixing of the wet feed into a highly agitated bed of drier particles allows smoother fluidization.

Plug-flow fluidized bed dryers are recommended when it is very important to ensure an almost uniform particles residence time in the dryer. Such dryers may be rectangular geometry trough with large aspect ratios. If floor space is a problem, the trough can be made into a spiral channel. In the straight trough configuration, it is possible to have gas temperature distribution along the length of the dryer. The inlet region can use higher air temperature to enhance the drying rate while toward the end, when the product has lower moisture content, the air temperature can be lowered; in many cases, ambient air is used to cool the product prior to storage or packaging.

When drying large particles, which are difficult to fluidize, a spouted bed or one of its many variants may be used instead of the conventional fluidized bed in which the air is distributed uniformly through a low open area–supporting grid. In a spouted bed, the air enters as a jet and causes a regular recirculatory motion of the particles within the vessel. Due to several limitations, the conventional spouted bed has not found significant commercial application as a dryer, however.

When handling difficult-to-fluidize granular products (because of large size, wide size distribution, or surface stickiness), mechanical assist may be needed to cause the particle bed to be dispersed. Mechanical agitation of the bed or application of vertical vibration at frequencies in the range of 5 to 20 Hz at half-amplitudes of 3 to 5 mm eases separation of the particles. In fact, very low velocities of the drying air are needed to cause pseudo-fluidization by vibration; the resulting drying rates are comparable to those obtained at higher velocities (and hence higher power consumption and particle attrition) in a conventional fluidized bed. The power required for vibration is rather small compared to that used for air handling in a fluidized bed. For fragile solids, e.g., crystals, weak agglomerates, tea leaves, etc., a vibrated fluidized bed may be preferable to conventional fluidized beds. It is used also for large, complex-shaped particles, e.g., chopped or cut fruits and vegetables.

23.4.1.4 Freeze Dryers

Highly heat-sensitive solids, such as some certain biotechnological materials, pharmaceuticals, foods with high flavor content, etc., may be freeze dried at a cost that is at least one order-of-magnitude higher than that of spray drying—itself not an inexpensive drying operation. Here, drying occurs well below the triple point of the liquid (often at temperatures in the vicinity of –40°C) by sublimation of the frozen moisture into vapor, which is then removed from the drying chamber by mechanical vacuum pumps or steam jet ejectors. Generally, freeze drying yields the highest quality

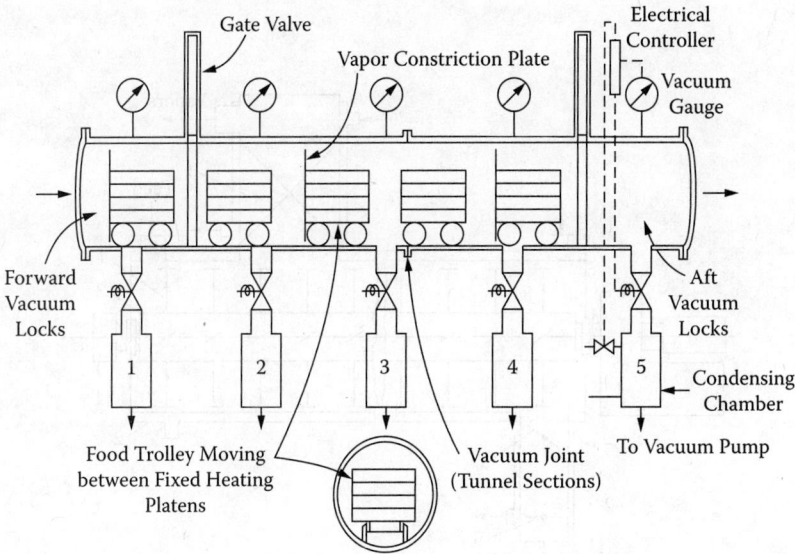

FIGURE 23.15 Schematic diagram of a typical tunnel freeze dryer.

product of any dehydration techniques. A porous, non-shrunken structure of the product allows rapid rehydration. Flavor retention is also high due to the low temperature operation. Living cells, e.g., bacteria, yeasts, and viruses, can be freeze dried and the viability on reconstitution can still be high. Mammalian cells, however, cannot be preserved by freeze drying. Because of its inherently high-cost nature, freeze drying is not common in the chemical industry.

Most freeze dryers are batch type with rather low capacities, although some continuous freeze drying units are in operation. Industrial freeze dryers can be of several types; simple tray freeze dryers are by far the most common. Heat for sublimation is supplied by conduction through the tray bottom. Vacuum pressure is typically under 25 Pa and the condenser operates at around –40°C. The heaters start at a higher temperature (say, 120°C) but the temperature drops with time, according to schedules determined empirically, to lower values, say –40°C over 8 to 10 hour runs. To minimize drying times, freeze dryers are program controlled. Multibatch freeze dryers are used to permit nearly equal loads on all systems throughout the drying cycle. A number of batch cabinets are programmed to operate with staggered, overlapping drying cycles.

A tunnel freeze dryer (Figure 23.15) is basically a large vacuum cabinet into which tray-carrying trolleys are loaded at intervals through a vacuum lock at one end of the tunnel. Figure 23.15 shows a unit with vacuum locks at each end, one for loading and the others for discharging. Low-pressure steam is used to heat the plates on which the trays sit. Liapis and Bruttini (1995) have provided a detailed analysis of the drying characteristics, costs, and details on various freeze-dried products. Recent developments in freeze drying (also called lyophilization) of pharmaceutical and biological products are discussed in Rey and May (1999).

23.4.1.5 Vacuum Dryers

For drying of granular solids or slurries, vacuum dryers of various mechanical designs are available commercially. They are more expensive than atmospheric pressure dryers but are suited for heat-sensitive materials or when solvent recovery is required or if there are risks of fire and/or explosion. Single-cone and double-cone mixers can be adapted to drying by heating the vessel jackets and applying vacuum to remove moisture. Figures 23.16 and 23.17 show two vacuum dryers available commercially. The paddle dryer is suited for sludge-like materials while the vacuum band dryer is

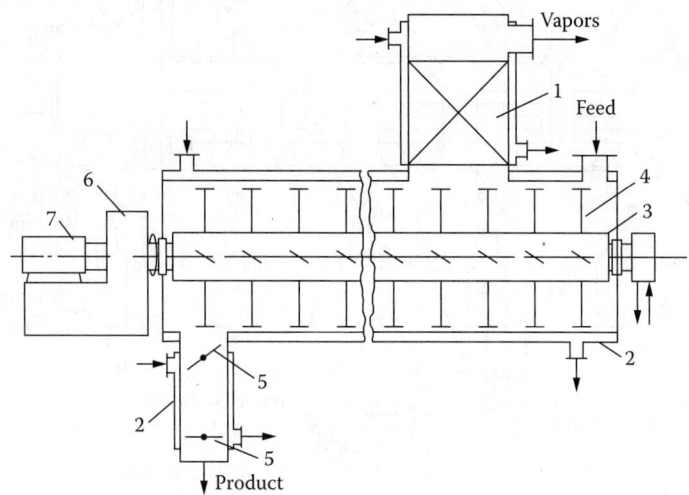

FIGURE 23.16 Continuous vacuum paddle dryer: (1) vapor filter; (2) steam jacket; (3) shaft with paddles; (4) paddles; (5) valves; (6) shaft drive; (7) shaft oscillator.

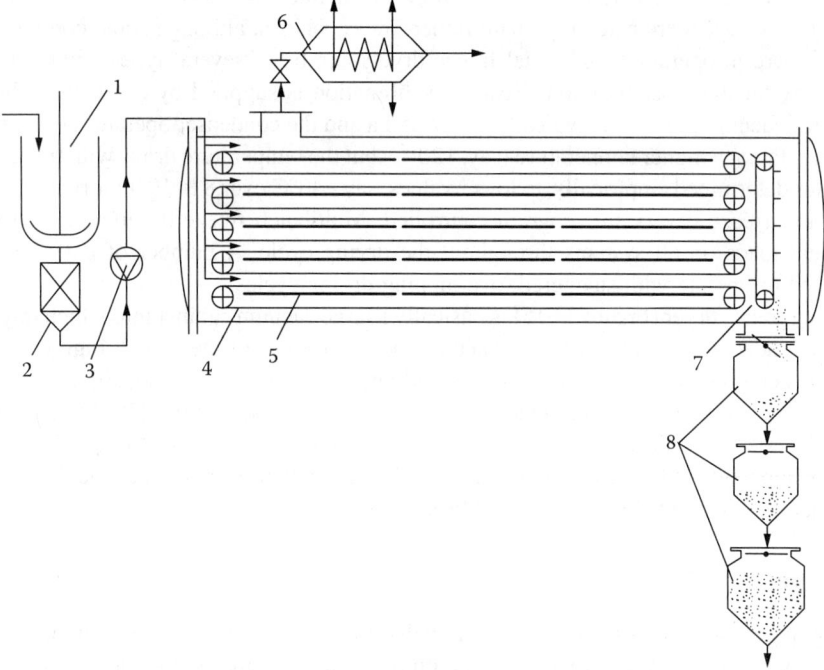

FIGURE 23.17 Band vacuum dryer installation: (1) feed mixer; (2) filter; (3) feed pump; (4) band; (5) heating panels; (6) vapor condenser; (7) scraper; (8) product collector system.

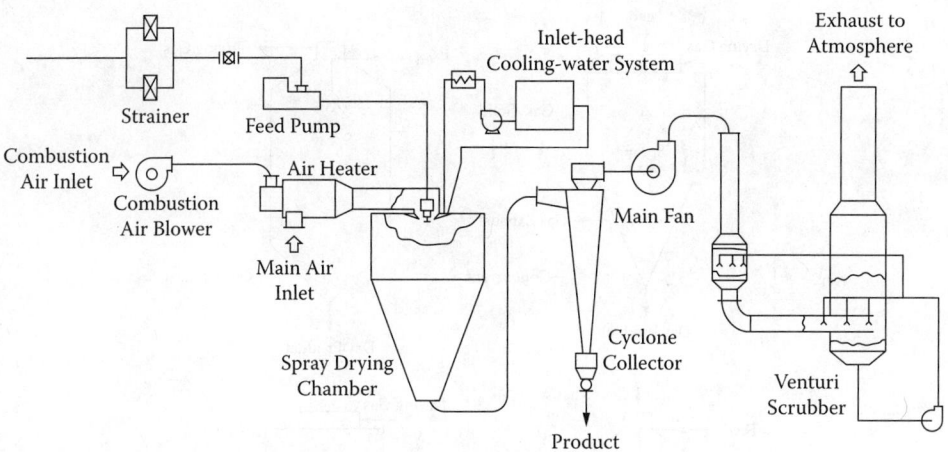

FIGURE 23.18 Spray drying process and plant.

good for thin pastes or slurries. The material forms a film over the heated band; it may boil and form a highly foamy, porous structure of very low-bulk density.

23.4.2 Dryers for Slurries and Suspensions

23.4.2.1 Spray Dryers

Over 20,000 spray dryers are presently in use commercially to dry products from agro-chemicals, biotechnological products, fine and heavy chemicals, dairy products, dyestuffs, and mineral concentrates to pharmaceuticals in capacities ranging from a few to 50 tons/hr evaporation capacity. Liquid feedstocks, such as solutions, suspensions, or emulsions, can be converted into powder, granular, or agglomerate form in a one-step operation in spray dryer. Figure 23.18 gives a process schematic for a spray dryer plant. Atomized feedstock in the form of a spray is contacted with hot gas in a suitably designed drying chamber. Proper selection and design of the atomizer is vital to the operation of the spray dryer as it is affected by the type of feed (viscosity), abrasive property of the feed, feed rate, desired particle size, and size distribution as well as the design of the spray dryer chamber geometry and mode of flow, e.g., concurrent, countercurrent, or mixed flow (see Figure 23.19).

Masters (1991) and Filkova and Mujumdar (1995) provide further information on spray drying in a tabular form. The design of spray dryers depends heavily on pilot-scale testing. It is impossible to scale-up quality criteria for spray dryers. Fortunately, in most cases, the larger scale dryer provides a better quality product than the one obtained in smaller scale-pilot tests. Aside from the drying rate and quality tests, it is also important to check for potential deposits in the drying chamber as they may lead to fire and explosion hazards.

Essentially, three major types of atomizers are used in practice: (1) rotary wheel (or disk) atomizers, (2) pressure nozzle, and (3) two-fluid nozzle. Figure 23.20 shows some typical atomizer designs. Ultrasonic and electrostatic atomizers can also be used for special applications to produce monodisperse sprays, but they are very expensive and low capacity. Most spray dryers operate at slight negative pressure. New designs with low-pressure chambers enhance drying rates at lower temperatures to dry heat-sensitive products.

The design of the spray drying chamber depends on the needed residence time (Table 23.12) as well as the type of atomizers used (Table 23.13). The mode of flow, e.g., concurrent, countercurrent, mixed flow, etc., depends on the desired characteristics of the product, as summarized in Table 23.14. Finally, Table 23.15 gives suggested spray dryer system layouts depending on the feedstock characteristics, e.g., presence of organic solvents, danger of fire or explosion, etc.

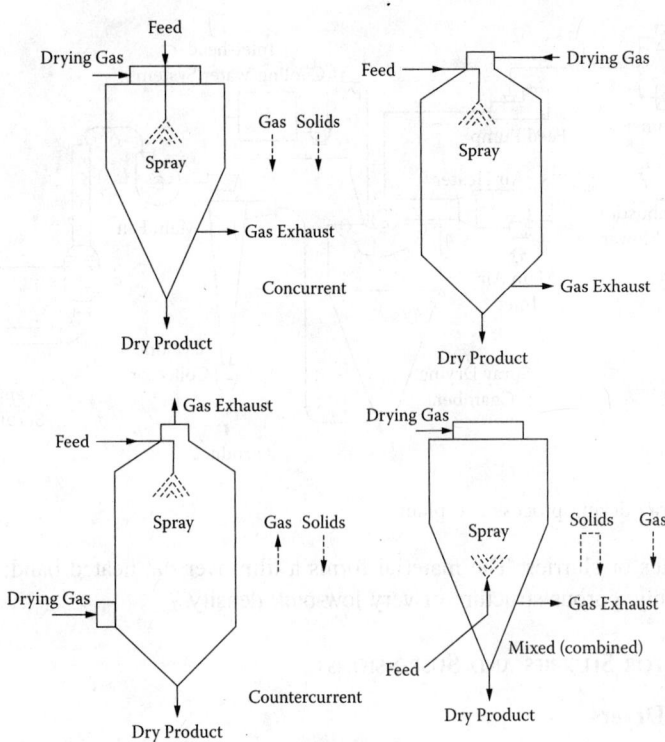

FIGURE 23.19 Schematic diagrams of co-current, countercurrent, and missed flow spray dryer configurations.

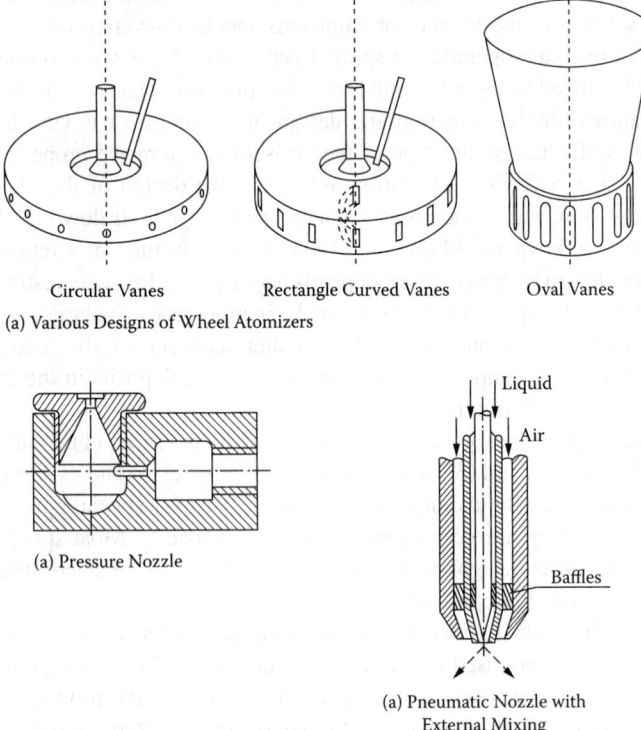

FIGURE 23.20 Atomizers.

TABLE 23.12
Residence Time Requirements for Spray Drying of Various Products

Residence Time in Chamber	Recommended for
Short (10–20 sec)	Fine, non-heat-sensitive products; surface moisture removal, nonhygroscopic
Medium (20–35 sec)	Fine-to-coarse sprays (d_{mean} = 180 μm)
	Drying to low final moisture
Long (>35 sec)	Large powder (200–300 μm)
	Low final moisture, low-temperature operation for heat-sensitive products

TABLE 23.13
Atomizer Selection Criteria

	Rotary Wheel	Pressure Nozzle	Two-Fluid Nozzle
Type of Chamber			
Concurrent	X	X	X
Countercurrent	—	X	—
Mixed (fountain)	—	X	X
Feed Type			
Solution/slurries	—	—	X
Low viscosity	X	X	X
High viscosity	X	—	X
Slurries			
Nonabrasive	X	X	X
Slightly abrasive	X	—	—
Highly abrasive	X	—	—
Feed Rate			
<3 m³/hr	X	X	X
> 3 m³/hr	X	X*	X*
Droplet Size			
30–120 μm	X	—	—
120–150 μm	—	X	—

Note: X, applicable; —, not applicable; X*, multinozzle assembly.

TABLE 23.14
Selection of Mode of Flow in Spray Drying Chamber Based on Desired Powder Characteristics

Dryer Design—Flow Type	Characteristics
Concurrent	Low product temperature
Mixed flow with integrated fluidized beds	To produce agglomerated powder
Mixed flow (fountain type)	For coarse sprays in small chambers; product not heat-sensitive
Countercurrent flow	Products that withstand high temperatures; coarse particles; high-bulk density powders

TABLE 23.15
Spray Dryer System Layout

System Layout	Characteristics
Open cycle	General; all aqueous feeds
Closed cycle	Recovery of solvents; prevention of vapor emissions; elimination of explosion or fire hazards
Semi-closed Self-inertizing	Prevent powder explosion (keep O_2 content low), yet use higher inlet temperature

TABLE 23.16
Selection of Dry Powder Collection System

Requirement	Recommended System
Low cost, efficient, easy to clean	Cyclones
Medium cost, very efficient, high running cost	Bag filter
Large air volumes	Electro-static precipitator
Product recovery; fines	Cyclone + wet scrubber

Collection of the dried powder from the spray dryer is also an important issue. Table 23.16 lists general recommendations for the selection of the dry powder collection system.

Since the choice of the atomizer is very crucial, it is important to note the key advantages and limitations of the wheel and pressure nozzles most common in practice. Although both types may be used for same feedstocks, the product properties (bulk density, porosity, size, etc.) will be different.

Rotary wheel (or disk) atomizers
    Advantages
        Handle large feed rates with a single wheel
        Suited for abrasive feeds with proper design
        Negligible clogging tendency
        Change of rpm controls particle size
        More flexible capacity
    Limitations
        Higher energy consumption compared to pressure nozzles
        More expensive
        Broad radial spray requires large drying chamber (cylindrical-conical type)

Pressure nozzles
    Advantages
        Simple, compact, cheap
        No moving parts
        Low-energy consumption
    Limitations
        Low capacity (flow rates)
        High tendency to clog
        Erosion can change spray characteristics

Figure 23.21 shows schematics of two spray dryers, one fitted with a wheel atomizer (cylindrical-conical) and the other with a nozzle atomizer (single or two-fluid), which is a cylindrical vessel.

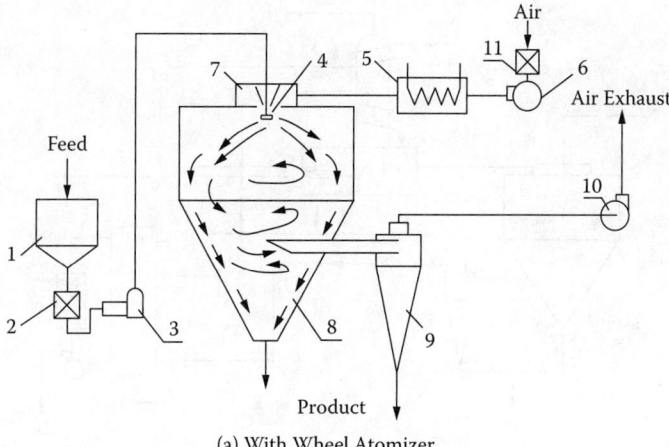

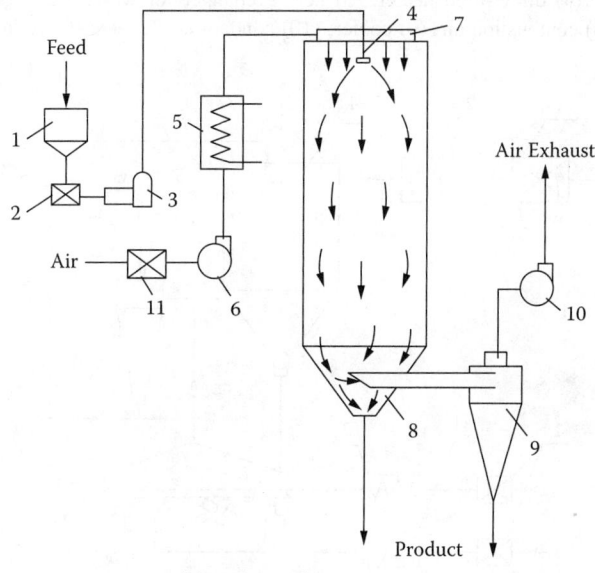

FIGURE 23.21 Spray dryer layout: (1) feed tank; (2) filter; (3) pump; (4) atomizer; (5) air heater; (6) fan; (7) air dispenser; (8) drying chamber; (9) cyclone; (10) exhaust fan; (11) filter.

These figures also show other components of the system (feed tank, filter, pump, air heater, fan cyclone, exhaust fan, etc.).

Figure 23.22 shows the layout of a spray dryer system, which is self-inertizing because the partial recycle from the combustion chamber decreases the oxygen level to a low enough value such that it does not sustain combustion of the product. Hence, it is used to handle materials with high risk of fire and explosion. Here, excess air entering the system passes through the burner flame and is used as combustion air, thus inactivating it.

When the product coming out of the spray dryer is too fine it does not wet readily and so is harder to reconstitute. To make the product more readily soluble, it is agglomerated in a small fluidized or vibrated fluidized bed, as shown in Figure 23.23. This two-stage arrangement is used in the production of instant coffee, milk powder, cocoa, etc. An extension of this basic concept is the so-called spray-fluidizer, which dries the material in two stages. In the first stage (spray dryer),

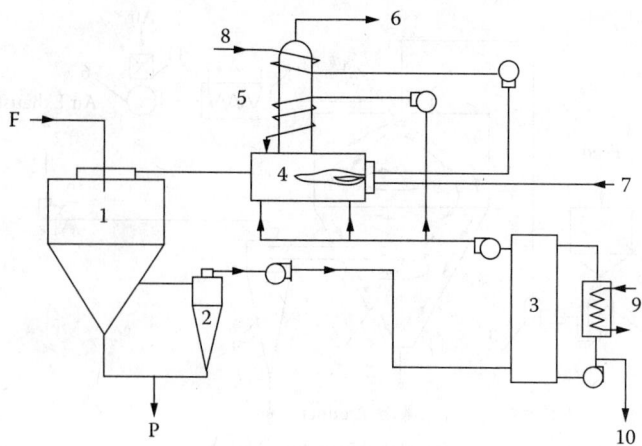

FIGURE 23.22 Self-inertizing spray drying system: (1) drying chamber with rotary atomizer; (2) cyclone; (3) scrubber condenser; (4) direct-fired heater; (5) heat exchanger for waste-heat recovery; (6) exhaust to atmosphere; (7) fuel; (8) combustion air; (9) cooler; (10) condensate; F, feed; P; product.

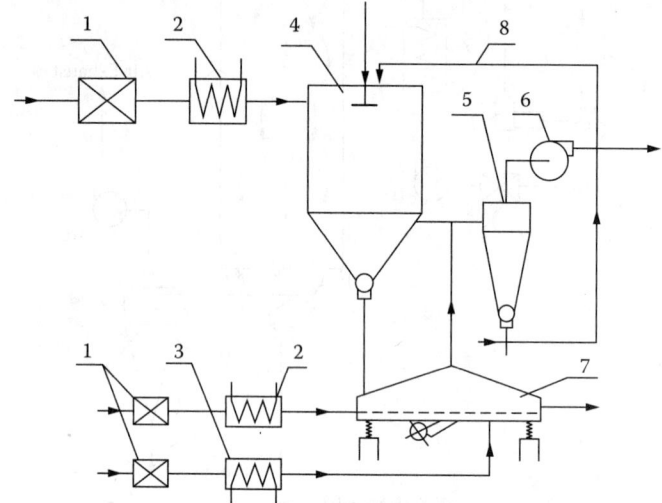

FIGURE 23.23 A two-stage spray drying process for milk: (1) air filter; (2) heater; (3) cooler; (4) spray dryer; (5) cyclone; (6) exhaust fan; (7) fluidized bed dryer; (8) return line of fine powder.

the surface moisture from droplets is removed fully, along with some internal moisture. The final moisture content is achieved in a fluidized bed located at the bottom of the spray chamber as an integral part of it. This two-stage drying process is very efficient and economic. The fluidized bed drying unit can be replaced with a through circulation band dryer at the bottom of the chamber; this concept is the basis of the so-called filtermat dryer used for sticky and sugar-rich materials, which are hard to dry. The bottom section of the spray chamber in this case is much wider, unlike the spray-fluidizer.

23.4.2.2 Drum Dryers

In drum dryers, slurries or pasty feedstocks are dried on the surface of a slowly rotating steam-heated drum. A thin film of the paste is applied on the surface in various ways. The dried film is removed (doctored off) once it is dry and is collected as flakes (rather than powder). Figure 23.24

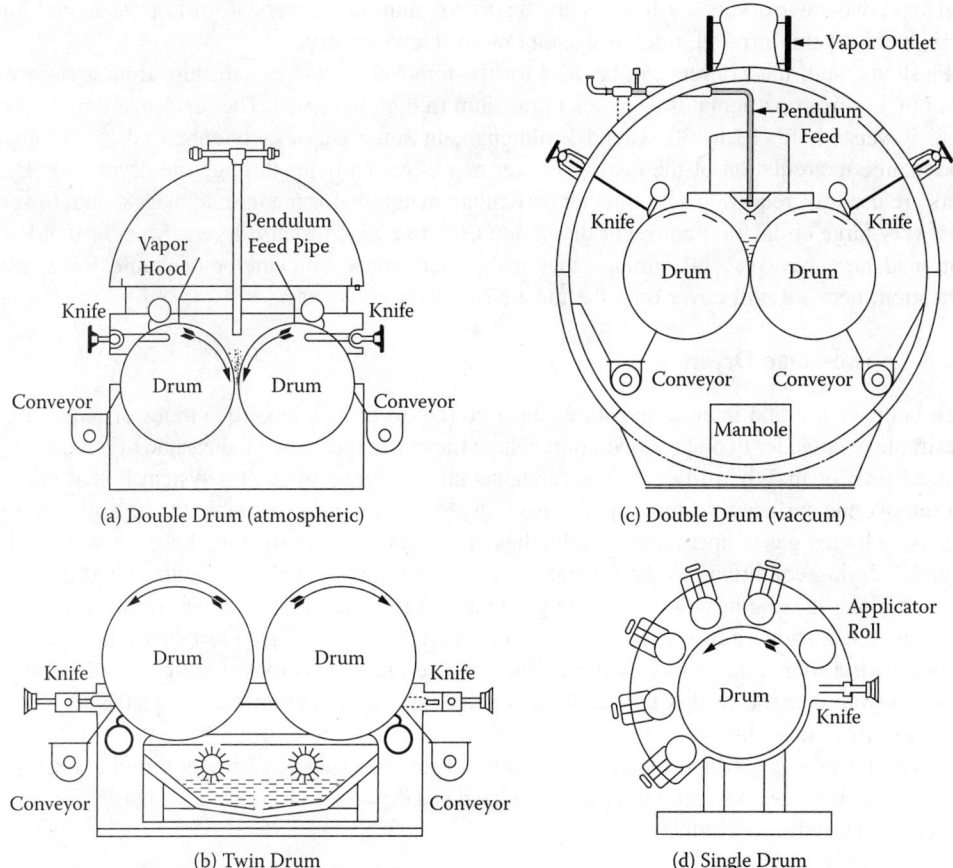

FIGURE 23.24 Types of drum dryers: (a) double drum (atmospheric); (b) twin drum; (c) double drum (vacuum); and (d) single drum.

shows four types of commonly used drum dryer arrangements, which are self-explanatory. The design of applicator rolls is important since the drying performance depends on the thickness and evenness of the film applied. The paste must stick to the surface of the drum for such a dryer to be applicable.

Five key variables influence the drum dryer performance. They are (1) steam pressure or heating medium temperature; (2) speed of rotation; (3) thickness of film; (4) feed properties (e.g., solids concentration, rheology, temperature, etc.); and (5) method of removing dust flakes by scraper or knife. Because they allow control of the temperature, drum dryers may be used to produce a precise hydrate of a chemical compound rather than a mixture of hydrates.

Vacuum operation of both single- and double-drum dryers enhances drying rates for heat-sensitive materials, such as pharmaceutical antibiotics. They are also used when a porous structure of product is desired. When recovery of solvents is an issue, vacuum operation is often recommended. When recovering high boiling point solvents such as ethylene glycol, lowering the pressure depresses the boiling point. For a detailed description and discussion of various drum dryers, the reader is referred to Moore (1995).

23.4.3 Selected Dryers and Drying Systems

Key features and applications of some specialized dryer types that are used in the chemical and ancillary industries but perhaps less commonly than the spray, rotary and fluidized bed types are listed here: two-stage dryers, flash or pneumatic dryers, spin-flash dryers, Roto-Louvre dryer, tunnel dryers, band dryers, infrared, microwave, and radio frequency dryers.

Flash and spin-flash dryers can be used for the removal of surface moisture from a variety of feeds ranging from particulates to pastes in medium to high tonnages. The residence time in these dryers is very short (10 to 30 seconds), although, in some cases, it is enhanced by automatic aerodynamic recirculation of the heavier (larger or wetter) particles through the dryer tube. Band dryers are used for relatively free-flowing particulate materials for medium tonnages, and, in some cases, very large circulation conveyor dryers are used to replace rotary dryers. Most of the dryers mentioned here have several variants that make them more efficient or desirable for a given application; here we will cover only the most basic dryer concepts.

23.4.3.1 Two-Stage Dryers

When both surface and internal moistures must be removed from large quantities of feedstock, it is desirable to consider two-stage operation, where the two stages may be the same dryer type (e.g., fluidized bed) or may be different. The fundamental advantage of such a system is that one can then remove the surface moisture rapidly using dryers or conditions suitable for its rapid removal (e.g., using higher gas temperatures or velocities), and using a dryer allowing longer residence time or gentler drying conditions as the second stage. A plug-flow continuous fluidized bed dryer can be zoned along its length by lowering the gas temperature from inlet to outlet, for example.

Figure 23.25 shows a two-stage arrangement, where the top first stage is a well-mixed fluidized bed dryer for a filter cake, which is difficult to fluidize unless it is mixed with a fluidized bed of lower moisture content. In this figure, the first stage also uses internal heating panels to increase the drying rate since this stage receives drying air, which is the exit air from the lower second stage. The lower stage, which receives the output of the first stage by gravity through a centrally located discharge tube, is a spiral plug-flow fluidized bed dryer, which controls the particle residence time to yield a uniform product.

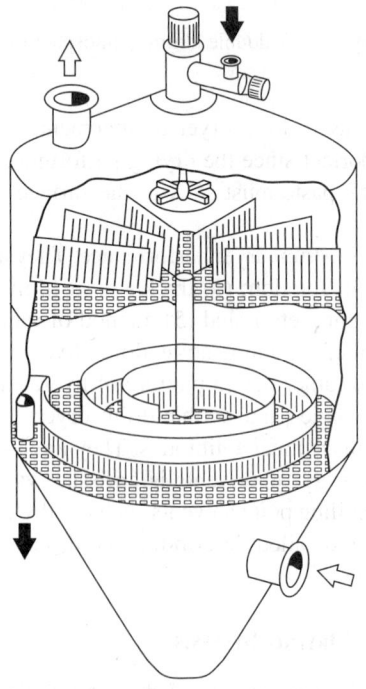

FIGURE 23.25 Schematic of a two-stage fluid bed.

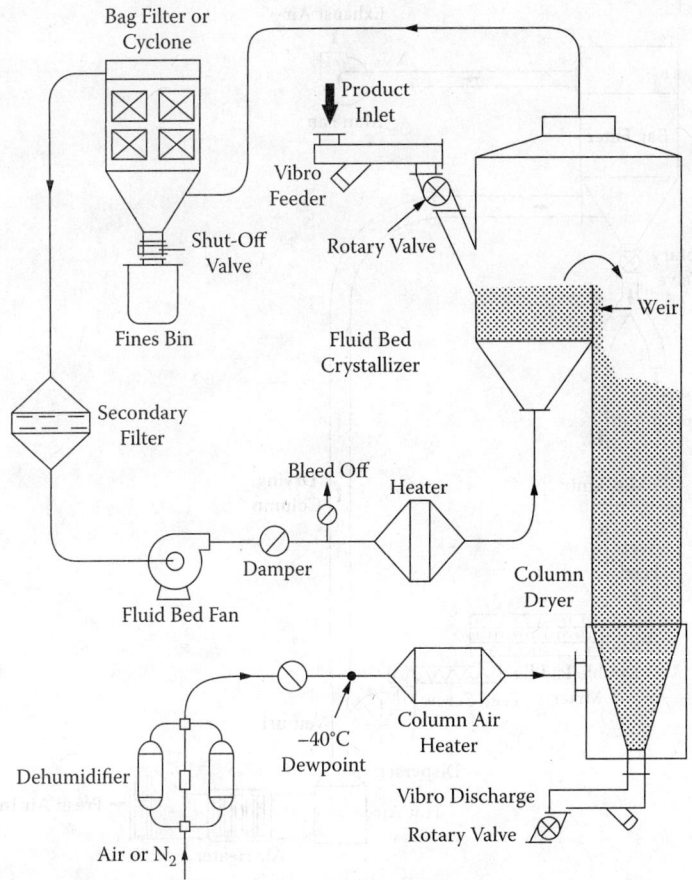

FIGURE 23.26 Continuous fluid bed crystallizer/column dryer for polyester.

Figure 23.26 is another example of a commercial two-stage dryer for crystallization/drying of polyester chips. A small fluidized bed, as the first stage, removes the readily removable liquid while the tall "column" dryer allows a very long residence time during which the material crystallizes and dries slowly.

Numerous examples of two-stage dryers are used commercially. They are viable options and perhaps need to be considered more often than they are today to reduce drying costs and even enhance product quality. Many examples can be found in Mujumdar (1995b).

23.4.3.2 Flash Dryers

Figure 23.27 shows a schematic of the simple flash (pneumatic) dryer system. Here, the wet feed is dispersed mechanically into a hot gas stream (commonly air or combustion gases) and conveyed for a long enough time to allow drying of the particulates in the size range 10 to 500 μm (microns) during their transport. Clearly, only the surface moisture of small particles can be removed economically in such a system of reasonable length of the insulated conveying tube. Most dryers are thus adiabatic and use a flash tube of circular and uniform cross-section. In some cases, the tube may diverge and converge, or have sudden expansions and contractions. The tube may be heated through the wall to maintain the temperature driving force, as the gas supplies heat to the particles as heat of vaporization and sensible heat. Noncircular cross-sections (e.g., rectangular with rounded corners) and tubes of nonrectilinear configurations (e.g., in the form of a ring) are also employed for special applications. For details, see Mujumdar (1995b).

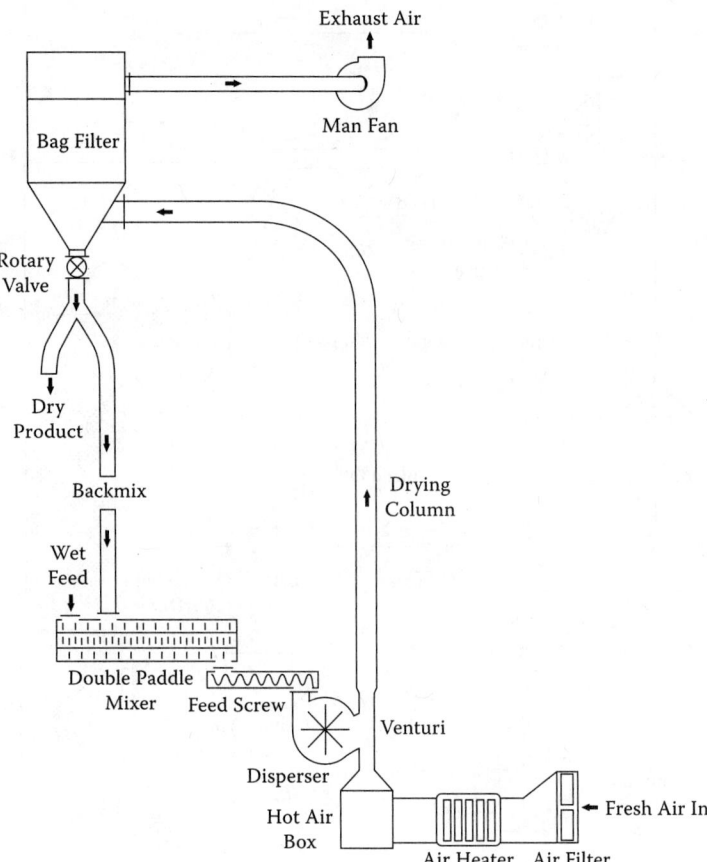

FIGURE 23.27 Flash dryer fitted with back mixer and bag filter.

Flash dryers may be used to dry heat-sensitive solids in view of the short exposure time to the drying medium. They have low capital cost, although, in some cases, the ancillary equipment (disperser, blender [if solids backmixing is needed prior to dispersion], heat exchangers, product collection devices, etc.) may cost much more than the basic flash dryer tube itself. Because of the risk of fire and explosion, care must be taken to avoid flammability limits in the dryer. The dryer must be designed with suitable rupture disks to minimize damage in the event of an explosion. The dryer has small "foot print" (e.g., small floor area) since the flash tube generally rises vertically so the flow of particulates against gravity increases the residence time in the tube of a given length. When it is feasible, a flash dryer should be considered. It does cause attrition, however. It can be used as the first stage of a two-stage dryer system to remove only the surface moisture rapidly and cheaply while a higher residence time dryer (e.g., fluidized bed) may be deployed as the second stage. Removal of the surface moisture also helps fluidize the material well, aside from reducing the size of the fluidized bed unit.

Design of the feed system is crucial for a flash dryer. For free-flowing powdery solids, a screw feeder or a rotary valve is effective. Pasty or sticky materials need to be pre-conditioned by blending them with dried product using a single- or twin-shaft paddle blender and then dispersing them mechanically using a kicker mill or one of several other designs of rotating disperser. The product may be collected in cyclones or baghouses and the very fine material removed prior to exhaust in wet scrubbers.

Flash dryers utilizing superheated steam as the drying medium often have unique quality and energy advantages over air drying systems. More recently, flash dryers consisting of inert media

Drying: Principles and Practice

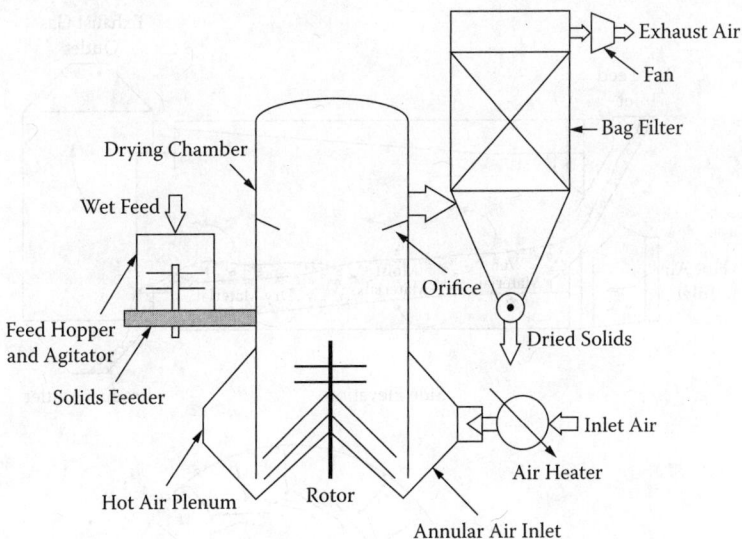

FIGURE 23.28 Spin-flash dryer.

have been employed at pilot scale to dry slurries and suspensions, which are sprayed on to them. The particles are coated thinly by the slurry and dried rapidly as a thin film. Attrition due to interparticle collisions and shrinkage-induced breakage of the dried film allow entrainment of the dry powder into the drying gas for collection in a cyclone or baghouse. This process is yet to be commercialized.

23.4.3.3 Spin-Flash Dryers

This dryer is basically a mechanically agitated fluidized bed designed for very short residence times, so it is suited for the removal of only the surface moisture. Sludges, pulps, pastes, filter cakes, high-viscosity liquids, etc., can be dried without the use of an atomizer. As shown in Figure 23.28, a rotor placed at the bottom of the dryer chamber serves to disperse the feed, which falls by gravity onto it; hot drying air enters the chamber tangentially and spirals upward, carrying and drying the dispersed particles. The exhaust containing the dried powder is cleaned and the powder recovered. Heavier wet particles remain within the chamber for a longer time and are broken up by the rotor—only the dried fine powder escapes to the gas cleaning system. This type of dryer can be a replacement for the more expensive spray dryer (which needs more thermal energy because the feed is wetter due to the pumpability requirements and also expensive because of the need for an atomizer). To date, numerous materials have been dried successfully in such units at capacities up to 10 tons/hr. They are more expensive than the conventional flash or fluidized bed dryers. Care must be taken to ensure in pilot tests that there is no danger of product sticking and accumulations on the walls.

23.4.3.4 Roto-Louvre Dryer

This dryer is a modification of the conventional rotary dryer, in which the drying gas contacts the wet particles rather inefficiently as the particles shower down from the flights and get exposed to the axial cross-flow of the gas. In a roto-louvre design (Figure 23.29), the slowly rotating horizontal drum (2 to 3 rpm) is fitted with longitudinal louvres, which make a tapered drum within the external drum. Diameters up to 3.5 m and lengths up to 12 m have been built commercially. The particles form a gently rolling fluidized bed at the bottom of the inner drum as the drying gas is introduced. The resulting heat and mass transfer rates are much greater than those achieved in a conventional

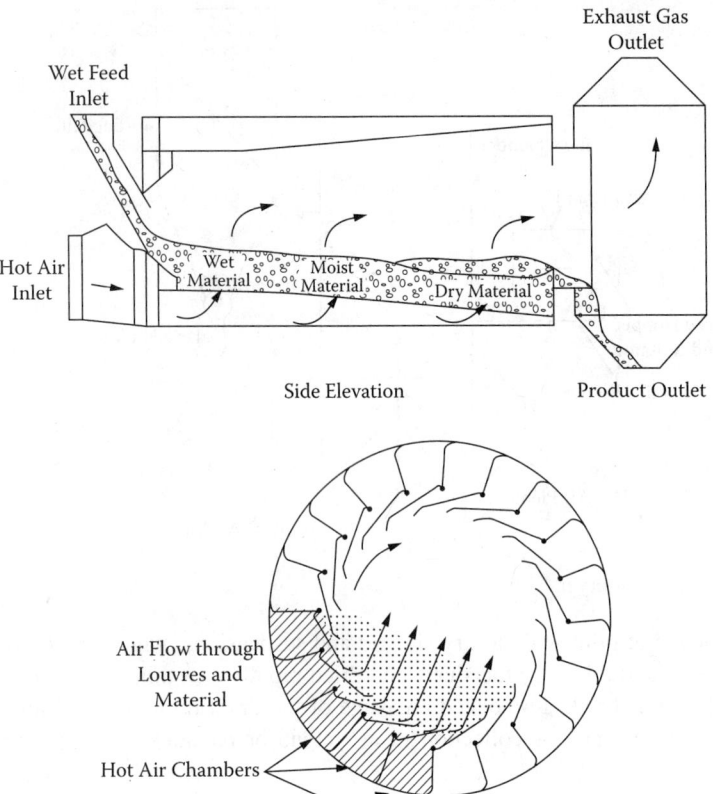

FIGURE 23.29 Roto-louvre dryer.

rotary dryer. Hence, the size of the dryer may be reduced by up to 50%. However, the additional complexity of the equipment increases the initial cost. Product handling is gentler and results in less attrition.

23.4.3.5 Tunnel Dryers

In this simple dryer concept, cabinets, trucks, or trolleys containing the material to be dried are transported at an appropriate velocity through a long insulated chamber (or tunnel). The hot drying gas flows in concurrent, countercurrent, cross-flow, or mixed-flow fashion (Figure 23.30). In the concurrent mode, the hottest and driest air meets the wetted product and hence results in high initial drying rates but with relatively low product temperature (wet-bulb temperature of surface moisture is present). Higher gas temperatures can be used in concurrent arrangements while in countercurrent dryers, the inlet drying gas must be at a lower temperature if the product is heat sensitive. If the material to be dried is not heat sensitive and low residual moisture content is a requirement, then one may employ higher gas temperatures in the countercurrent arrangement as well. Combination flow or cross-flow arrangements are used less commonly. The latter offer high drying rates, but the tunnels must be designed to fit the trolleys snugly so the drying gas flows through the material much like a through-circulation packed bed dryer. Total drying times often range from 30 minutes to 6 hours.

23.4.3.6 Band Dryers

For relatively free-flowing granules and extrudates, which may undergo mechanical damage if they are dispersed, band dryers (Figure 23.30) are a good option. It is essentially a conveyor dryer in which a perforated band supports the bed of drying solids. Drying air at rather low velocities flows

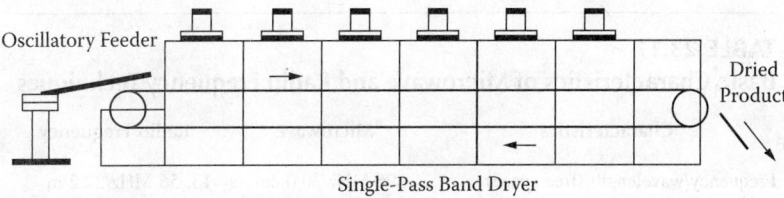

FIGURE 23.30 Band dryer.

upward through the band to accomplish drying. Clearly, this type of dryer is not a good choice for very wet or very fine solids. If the bed depth is large (>10 to 15 cm), a significant moisture profile may occur in the bed with the solids resting on the band overdried and overheated. One option to alleviate this problem is to reverse the gas flow direction alternately over the length of the dryer. This evens out the moisture profile while increasing the drying rate as well. Another option is to cause mixing of the bed at appropriate intervals of space. In some commercial designs, such as so-called multipass dryers, several bands are stacked one above the other and the material is made to drop under gravity from the higher to the next lower band, which causes some random mixing of the material before it undergoes further through-circulation drying. It is possible to use a temperature profile along the length of the conveyor so that the drier product can be exposed to lower gas temperatures if that is desired. Also, the final section may be a simple cooler so the product is ready for packaging or storage. Residence times from 10 to 60 minutes are economically feasible. These dryers are quite versatile and can handle relatively large and arbitrary-shaped particles, which may be heat sensitive and fragile at the same time. Gas cleaning requirements are minimal as low gas velocities are used. Also, power requirements for air handling are low due to the low pressure drops needed. In commercial designs of very large band dryers, it is important to ensure uniform distribution of the product on the band and also uniform distribution of the air flow within the chamber of the dryer to ensure uniform product moisture content.

23.4.3.7 Infrared Dryers

Infrared (IR) dryers may be gas-fired ceramic radiators or electrically heated panels. The IR wavelength range is from 0.1 to 100 μm, which generates heat in the exposed physical body. The wavelength ranges 0.75 to 3.0 μm, 3.0 to 25 μm, and 25 to 100 μm are referred to as near IR, middle IR, and far IR ranges, respectively. Industrial radiators are of two types: (1) light radiators (near IR), e.g., quartz glass with peak radiation intensity at 1.2 μm; and (2) dark radiators, e.g., ceramic (3.1 μm) or metal radiators (2.7 to 4.3 μm).

While convection can yield heat fluxes of the order of 1 to 2 kW m^{-2}, radiation can yield much higher levels of heat flux, e.g., 4 to 12 kW m^{-2} (light radiators) and 4 to 25 kW m^{-2} (dark radiators).

In many drying operations, the evaporation rates feasible are not high enough to require IR radiators, however. There are some niche applications for IR dryers in industry, e.g., drying of coated paper, booster drying of paper in paper machines, etc. They offer the advantages of compactness, simplicity, ease of local control, low equipment costs, etc. Also, in combination with convection, IR dryers offer the potential for significant energy savings and enhancement in drying rates with better product quality. On the negative side, the high heat flux may scorch the product and enhance fire and explosion hazards. Clearly, IR must be used in conjunction with convection or vacuum. Good control is essential for safe operation, i.e., the IR power source must be cut off if there is upset in the process, which may lead to overheating of the product.

23.4.3.8 Microwave (MW) and Radio Frequency (RF) Drying

Unlike conduction, convection, or radiation, dielectric heating heats a material containing a polar compound volumetrically, i.e., thermal energy supplied at the surface does not have to be conducted

TABLE 23.17
Basic Characteristics of Microwave and Radio Frequency Techniques

Characteristics	Microwave	Radio Frequency
Frequency/wavelength (free space)	896 MHz/30.0 cm	13.56 MHz/22.2 m
	915 MHz/32.8 cm	27.12 MHz/11.1 m
	2450 MHz/12.2 cm	40.68 MHz/7.4 m
Power source	Magnetron (klystron)	Class C triode
Typical efficiency	~50%	~60%
Maximum power of a single unit	60 kW at 915 MHz	600 kW
	6 kW at 2450 MHz	
Principal mechanism of heat generation	Dipole oscillation	Ionic conduction
Loss factor	May be low	Must be relatively high
Penetration depth	Medium or low	High
Product shape	Any	More or less regular
Product size	Small and medium	May be large
Generator service life	2000–5000 hr	5000–10,000 hr
Generator-to-applicator capital cost ratio	0.5	2.3

into the interior, as limited by Fourier's law of heat conduction. This type of heating provides the following advantages:

- Enhanced diffusion of heat and mass
- Development of internal pressure gradients that enhance drying rates
- Increased drying rates without increasing surface temperatures
- Better product quality

When an alternating electromagnetic field is applied to a "lossy" dielectric material, heat is generated due to friction of the excited molecules with asymmetric charges, e.g., water. This is a result of ionic conduction or dipole oscillations (Kudra and Strumillo 1998). The radio frequency range extends from 1 to 300 MHz while the microwave range is from 300 to 3000 MHz. However, only specific frequency ranges are permitted for industrial heating applications, i.e., in RF, ranges 13.56, 27.12, and 40 MHz and for MW, 915 (896 in Europe) and 2450 MHz. Bound and free water have different loss factors. Because loss factors increase with temperature, there is a danger of runaway, i.e., an accelerated heating rate causing a thermal damage to the material.

Table 23.17 summarizes the basic characteristics of MW and RF techniques. The main limitation of MW and RF drying is that the technique is highly capital intensive. It also consumes high-grade energy, i.e., electricity, and the conversion efficiency to the dielectric field is only in the order of 50%. Thus, these techniques are suited only for special applications involving very high-value products, extremely long drying time to remove traces of moisture, or to obtain products of special characteristics not obtained otherwise. It is therefore not surprising that MW/RF drying is used only in special niche applications. Further, these techniques are used mainly to boost drying capacity (to remove free water rapidly without generation of large thermal gradients in the material) or to remove the last few percent of water that comes out very slowly. Generally, dielectric heating is combined with convection or vacuum to reduce the energy consumption. Microwave vacuum drying and microwave freeze drying have found to date only limited commercial applications. Microwave freeze drying is typically carried out at well below the triple point of water. Typical conditions are pressure in the range of 500 Pa and temperature of –40°C. Use of excessive power as well as maldistribution of power due to nonhomogeneities in the frozen solids can cause problems in MW drying. The main hurdle to commercialization of MW freeze drying is the high cost.

Numerous laboratory- and pilot-scale studies have been reported on MW drying at atmospheric as well as vacuum conditions. It is also possible to "pipe" microwave energy in various dryer configurations, e.g., fluidized bed, spouted bed, vibrated bed, or tray dryers, to enhance convective drying rates. Unfortunately, while all these techniques do provide significant enhancement of the drying time required, the initial and operating costs are such that the enhancement obtained does not offset the added cost. Drying of treated grapes in combined microwave and convection dryers has been shown to be very rapid and energy efficient. However, the costs are prohibitively high. Recent research has shown that intermittent application of a microwave field results in the production of better quality dried product.

23.4.3.9 Superheated Steam Drying

Figure 23.31 classifies superheated steam dryers based on their operating pressures. The product temperature necessarily exceeds the saturation temperature of steam corresponding to the operating pressure. Thus, for products that may undergo undesirable physical transformations (e.g., melting) or chemical transformations (e.g., hydrolysis) at elevated temperatures with steam or with the small quantities of oxygen that necessarily leaks into any system, a low-pressure operation is desirable. The vapor evolving from the product may be withdrawn from the chamber, condensed, and the latent heat recovered. Alternatively, the vapor is reheated within the chamber by tubular or plate heat exchangers and recirculated as a convective drying medium to enhance the drying rate. Such a system is used commercially to dry timber with very attractive results; the net energy consumption is reduced several fold; the product quality is enhanced; and the environmental problem of emission of the volatile organic compound produced during drying is eliminated. The organic volatile components (boiling point ranging from room temperature to >200°C) are condensed out with the steam; since the two are immiscible, simple decantation allows recovery of the condensibles, which may be sold separately (e.g., terpenes, essential oils).

High-pressure superheated steam dryers need to be pressure vessels and hence are capital intensive. They are recommended only for special applications. For example, a pressurized, fluid-

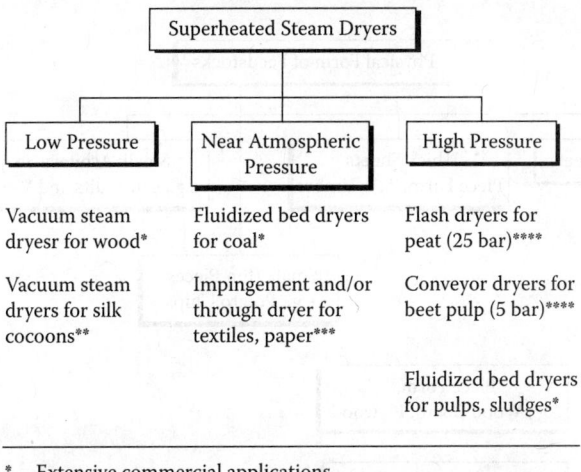

* Extensive commercial applications
** Laboratory scale testing
*** Pilot scale testing
**** At least one major installation

FIGURE 23.31 Classification of superheated steam dryers based on their operating pressures.

ized, bed superheated steam dryer is marketed commercially for drying of beat pulp. It yields a higher quality fiber that can be used for human consumption while the same product dried in hot air gives a lower quality fiber that is used for animal feedstocks. Higher pressures are typically not cost-effective, although the specific volume of superheated steam is reduced and its enthalpy per unit volume increased at higher pressures.

In general, superheated steam drying should be considered if one or more of the following conditions apply:

1. Energy cost is very high; product value is low or negligible (e.g., commodities like coal, peat, newsprint, tissue paper, and waste sludges that must be dried to meet regulatory requirements).
2. Product quality is much superior (e.g., newsprint, which yields superior strength properties in steam and permits use of lower chemical pulp content to attain same strength and runnability).
3. Risk of fire, explosion, or other oxidative damage is very high (e.g., coal, peat, pulps, etc.). Lower insurance premiums may partially offset added investment cost of a steam dryer.
4. Quantity of water to be removed as well as production capacity required are high. This affords economy of scale.

Such dryers are superior only for continuous operation because of the inherent problems associated with start-up and shut-down when water condensation on the product as well as presence of noncondensables (air) cause problems. See Mujumdar (1995a) for detailed information on superheated steam drying.

23.4.3.10 Drying of Boards and Sheets

Drying of sheet-form materials or materials in the form of large and small pieces often occurs for paper, textile, coated or impregnated fabrics, wood processing, food processing, polymer and photographic film production as well as in the graphic arts industries. Figure 23.32 outlines several possible dryers when the feedstock is other than particulate or liquid-like. The main dryer types suited for such materials will now be reviewed very briefly.

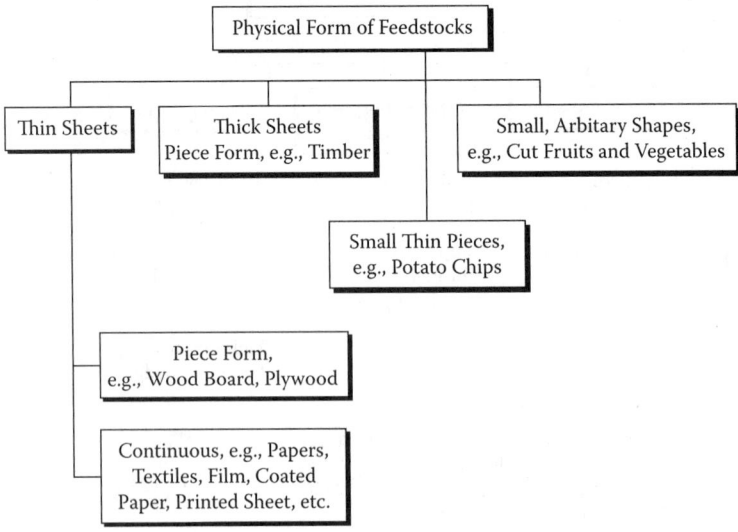

FIGURE 23.32 Classification of feedstock by physical form.

23.4.3.10.1 Dryers for Continuous Sheets

Convection, conduction, as well as infrared dryers can be used for such materials, although combined mode dryers often are more efficient. Paper, coated webs, or textiles are dried on steam-heated cylinders (conduction heating) or jets of hot air may be impinged on the sheet for concurrent convective heating as well. In some cases, it may be desirable to use infrared heating to augment the drying rate if the material is not very heat-sensitive. For drying of thin permeable sheets, drying air through the sheet highly enhances drying rates. Combined through and impingement drying is a particularly attractive option for drying of tissue or newsprint. For thin sheets, the total drying time may be in the order of seconds (e.g., tissue paper) to several minutes (e.g., textiles).

23.4.3.10.2 Dryers for Piece-Form Sheets

Materials like plywood or chip board require long residence times to dry. Impinging jets may be used initially to remove surface moisture; internal moisture removal can be achieved in a tunnel dryer with modest parallel flow of drying medium.

23.4.3.10.3 Dryers for Very Thick Sheets (or Odd Shapes)

The drying times may range from days to months. Wood, for example, is dried in hot air kilns from weeks to months, depending on the size of the pieces to be dried and the type of wood species. Superheated steam drying under vacuum conditions enhances drying rates as well as product quality. Only batch dryers are suited for these long drying time requirements.

23.4.3.10.4 Drying of Materials in the Form of Thin Wafers

Continuous conveyor, through-circulation, or impinging jet-fluidized bed dryers (sometimes called jet-zone dryers) are often used in the wood processing (e.g., wafer board) and food processing (e.g., potato chips, etc.) industries. In the jet-zone process, hot air jets impinge on a thin layer of wet chips, which are conveyed mechanically; the high-velocity jets "pseudo-fluidize" the material to accomplish uniform drying. Wood chips, for example, may be dried in a rotary or a conveyor dryer.

23.4.3.10.5 Drying of Arbitrary-Shaped, Non-Fluidizable Pieces

Vibrated beds, impinging jet-fluidized beds, mechanically fluidized beds, rotary, or conveyor dryers may be used. Each type of dryer has its own advantages and limitations, so a careful evaluation of all the key criteria noted earlier in this chapter must be made before the final selection.

23.5 CLOSING REMARKS

A general introduction has been provided to the basic principles and terminology of drying and drying equipment commonly used in practice as well as classification and selection criteria for industrial dryers for various forms of materials. For more detailed and current information, the interested reader may visit the Web site http://www.geocities.com/drying_guru. This site provides hot links to numerous other sites for literature, conferences, and other resources pertaining to all aspects of drying on a global scale. The recently published book edited by Devahastin (2000) is recommended for a concise yet in-depth discussion on industrial drying technologies. For information on ancillary equipment required for a drying system, Mujumdar (1995b) covers control of dryers, explosion and fire hazards, control of emissions, as well as new drying technologies not covered in this chapter. Also, research and development personnel may find the latest developments in drying technology by referring to *Drying Technology—An International Journal* as well as the proceedings of the biennial International Drying Symposium series and regional conferences devoted to drying.

NOMENCLATURE

A	Evaporation area, m²
a_w	Water activity
c_P	Specific heat, J kg⁻¹ K⁻¹
c_s	Humid heat, J kg⁻¹ K⁻¹
D_L	Effective diffusivity, m² s⁻¹
D_{L0}	Effective diffusivity at reference temperature, m² s⁻¹
E_a	Activation energy, J
ΔH_W	Enthalpy of wetting, J kg⁻¹
h	Convective heat transfer coefficient, W m⁻² K⁻¹
k_g	Thermal conductivity, W m⁻¹ K⁻¹
k_y	Convective mass transfer coefficient, kgmol s⁻¹ m⁻² mol frac⁻¹
M_{air}	Molar mass of air, kgmol⁻¹
M_s	Mass of bone dry solid, kg
N	Drying rate, kg m⁻² h⁻¹
P_v	Vapor pressure of pure water, Pa
p	Partial pressure, Pa
p_w	Equilibrium vapor pressure of water, Pa
R_g	Universal gas constant, 8.314 J mol⁻¹ K⁻¹
T	Temperature, °C
T_{abs}	Absolute temperature, K
T_{wb}	Wet-bulb temperature, °C
t	Time, sec (or hr)
X	Total moisture content, kg water/kg dry solid
X_c	Critical moisture content, kg water/kg dry solid
X_f	Free moisture content, kg water/kg dry solid
X^*	Equilibrium moisture content, kg water/kg dry solid
Y	Absolute air humidity, kg water vapor/kg dry air

GREEK LETTERS

η	Normalized moisture content
λ	Latent heat of vaporization, J kg⁻¹
μ_g	Dynamic viscosity, kg m⁻¹ s⁻¹
ν	Normalized drying rate
ρ_g	Density, kg m⁻³
Ψ	Relative humidity of air

SUBSCRIPTS

a^s	Adiabatic saturation
c	Constant rate period
f	Falling rate period
g	Gas
s	Solid
v	Vapor
w	Water
w^b	Wet-bulb

REFERENCES

Baker, C.G.J. 1997. Dryer selection. In *Industrial Drying of Foods*, ed. C.G.J. Baker, 242–271. London: Blackie Academic & Professional.
Bruin, S., Luyben, K.Ch.A.M. 1980. Drying of food materials: a review of recent developments. In *Advances in drying*. Vol. 1. Ed. A.S. Mujumdar, 155–216. Washington, DC: Hemisphere.
Bruin, S. 1988. *Preconcentration and drying of food materials*. Thijssen Memorial Symposium. Proceedings of the International Symposium on Preconcentration and Drying of Foods; Eindhoven, The Netherlands.
Devahastin, S., ed. 2000. *Mujumdar's guide to industrial drying: principles, equipment and new developments*. Montreal: Exergex.
Filkova, I., and Mujumdar, A.S. 1995. Industrial spray drying systems. In *Handbook of industrial drying*. 2nd ed. Ed. A.S. Mujumdar, 263–307. New York: Marcel Dekker.
Fortes, M., and Okos, M.R. 1980. Drying theories: their bases and limitations as applied to foods and grains. In *Advances in drying*. Vol. 1. Ed. A.S. Mujumdar, 119–154. Washington, DC: Hemisphere.
Geankoplis, C.J. 1993. *Transport processes and unit operations*. 3rd ed. Englewood Cliffs, NJ: Prentice Hall.
Gong, Z.-X., Devahastin, S., and Mujumdar, A.S. 1997. A two-dimensional finite element model for wheat drying in a novel rotating jet spouted bed. *Drying Technology—An International Journal* 15: 575–592.
Iglesias, H.A., Chirife, J. 1982. *Handbook of food isotherms: water sorption parameters for food and food components*, New York: Academic Press.
Keey, R.B. 1978. *Introduction to industrial drying operations*. Oxford, UK: Pergamon Press.
Keey, R.B. 1992. *Drying of loose and particulate materials*. Washington, DC: Hemisphere.
Kemp, I.C. 1998. Progress in dryer selection techniques. In *Drying '98*. Proceedings of the 11th International Drying Symposium; Halkidiki, Greece, eds. C. Akritidis, D. Marinos-Kouris, G. Saravacos, and A.S. Mujumdar, 668–675.
Kudra, T., and Mujumdar, A.S. 1995. Special drying techniques and novel dryers. In *Handbook of industrial drying*. 2nd ed. Ed. A.S. Mujumdar, 1087–1149. New York: Marcel Dekker.
Kudra, T., and Strumillo, C. 1998. *Thermal processing of bio-materials*. New York: Gordon & Breach.
Liapis, A.I., and Bruttini, R. 1995. Freeze drying. In *Handbook of industrial drying*. 2nd ed. Ed. A.S. Mujumdar, 309–343. New York: Marcel Dekker.
Liapis, A., and Marchello, J.M. 1984. Advances in modeling and control of freeze drying. In *Advances in drying*. Vol. 3. Ed. A.S. Mujumdar, 217–244. Washington, DC: Hemisphere.
Masters, K. 1991. *Spray drying handbook*. Essex, UK: Longman Scientific & Technical.
Moore, J.G. 1995. Drum dryers. In *Handbook of industrial drying*. 2nd ed. Ed. A.S. Mujumdar, 249–262. New York: Marcel Dekker.
Mujumdar, A.S. 1995a. Superheated steam drying. *Handbook of industrial drying*. 2nd ed. Ed. A.S. Mujumdar, 1071–1086. New York: Marcel Dekker.
Mujumdar, A.S., ed. 1995b. *Handbook of industrial drying*. 2nd ed. New York: Marcel Dekker.
Mujumdar, A.S. 1997. Drying Fundamentals. In *Industrial drying of foods,* ed. C.G.J. Baker, 7–30. London: Blackie Academic & Professional.
Mujumdar, A.S., and Menon, A.S. 1995. Drying of solids. In *Handbook of industrial drying*. 2nd ed. Ed. A.S. Mujumdar, 1–46. New York: Marcel Dekker.
Okos, M.R., Narsimhan, G., Singh, R.K., and Weitnauer, A.C. 1992. Food dehydration. In *Handbook of food engineering,* Eds. D.R. Heldman and D.B. Lund, 437–562. New York: Marcel Dekker.
Pakowski, Z., and Mujumdar, A.S. 1995. Basic Process calculations in drying. In *Handbook of industrial drying*. 2nd ed. Ed. A.S. Mujumdar, 71–112. New York: Marcel Dekker.
Pakowski, Z., Bartczak, Z., Strumillo, C., and Strenstrom, S. 1991. Evaluation of equations approximating thermodynamic and transport properties of water, steam and air for use in CAD of drying processes. *Drying Technology—An International Journal* 9: 753–773.
Rey, L., and May J.C., eds. 1999. *Freeze drying-lyophilization of pharmaceutical and biological products*. New York: Marcel Dekker, 477.
Rockland, L.B., and Beuchat, L.R. 1987. *Water activity: theory and applications to food*. New York: Marcel Dekker.
Schiffmann, R.F. 1995. Microwave and dielectric drying. In *Handbook of industrial drying*. 2nd ed. Ed. A.S. Mujumdar, 345–372. New York: Marcel Dekker.

Turner, I., and Mujumdar, A.S., eds. 1997. *Mathematical modeling and numerical techniques in drying technology.* New York: Marcel Dekker.

van Meel, D.A. 1958. Adiabatic convection batch drying with recirculation of air. *Chemical Engineering Science* 9: 36–44.

van't Land, C.M. 1991. *Industrial drying equipment: selection and application.* New York: Marcel Dekker.

Waananen, K.M., Litchfield, J.B., and Okos, M.R. 1993. Classification of drying models for porous solids. *Drying Technology—An International Journal* 11: 1–40.

Wolf, W., Spiess, W.E.L., and Jung, G. 1985. *Sorption isotherms and water activity of food materials.* Amsterdam: Elsevier.

Zogzas, N.P., Maroulis, Z.B., and Marinos-Kouris, D. 1996. Moisture diffusivity data compilation in foodstuffs. *Drying Technology—An International Journal* 14: 2225–2253.

24 Dry Screening of Granular and Powder Materials

A.J. DeCenso and Nash McCauley

CONTENTS

24.1 Fundamentals of Screening 1717
24.2 Factors Affecting Screening Performance 1717
24.3 Screen Motion and Configuration 1722
24.4 Specifying Screener Performance 1724
24.5 Screening Equipment Selection Criteria 1725
24.6 Screening Machine Operations and Maintenance 1727

24.1 FUNDAMENTALS OF SCREENING

Screening is a process that uses a media, such as mesh or a perforated plate, to separate a material based on particle size. Other techniques are used to separate particles based on characteristics such as density or shape. While screening is often the most accurate, economical, and common separation technique, size separation can also be achieved using other processes such as air classification or flotation. These latter techniques exploit differences in aerodynamic or hydrodynamic characteristics of different particle sizes to effect a separation. They are good choices for separating very fine materials (<40 microns) that present problems (e.g., screen blinding, high screen area requirements) for conventional mechanical screening. They cannot, however, provide a sharp separation, so serious compromises on product quality and separation efficiency must be made. Hence conventional dry screening is often the preferred method of separation.

In general, screening is used for separations in the particle size range of 40 microns (μm) to 50 mm. Table 24.1 lists the most common industrial applications for which screening is an important part of the process. There are three basic types of screening operations: scalping, fines removal, and grading. In scalping, a small percentage (<5%) of oversize material, considerably larger than the average particle size, is removed. An example is the removal of large lumps from fertilizer to improve its spreading characteristics. In fines removal, the opposite of scalping, a small percentage (<10%) of fine material, which is considerably smaller than the average particle size, is removed. An example is the removal of objectionable dust from a consumer product such as laundry detergent. In grading, material is separated into one or more grades or products, each having distinct particle size distributions and size limits. An example is the separation of activated carbon filter media into distinct granule size ranges.

24.2 FACTORS AFFECTING SCREENING PERFORMANCE

Screening performance is affected by many factors, the most important of which are material characteristics. Because screening is a separation based on particle size, the most important material characteristic is particle size distribution. This distribution is determined by taking a small repre-

TABLE 24.1
Common Screening Applications

Material	Scalping	Fines Removal	Grading
Carbon black, activated carbon	X		X
Ceramics, refractories	X		X
Detergents	X	X	
Electrolytics			X
Explosives, propellants	X		X
Fertilizers	X	X	X
Food additives	X	X	X
Herbicides, pesticides			X
Inorganics	X	X	X
Metal powders	X	X	X
Petroleum coke			X
Pharmaceuticals	X		
Pigments, toners	X		
Plastic beads			X
Plastic pellets	X	X	
Plastic resins	X		
Salts	X	X	X
Sugar, starch	X	X	X

sentative sample and determining the weight of particles within size ranges. This is most commonly accomplished by sieve analysis. In this technique, standardized test sieves with precise apertures defined by standards organizations such as ASTM, ISO, TYLER, DIN, and others, are used (Table 24.2). Sieves used in North America are defined by the ASTM E11 specification.

These test sieves are arranged in a stack from coarsest to finest with a collection pan below the finest sieve. The sample of material to be analyzed is introduced to the top sieve, and the stack is exposed to a combination of motions that cause the material to stratify through the sieves. After a set period of time, the stack is disassembled, and the weight of material retained on each sieve is measured and recorded as a percent of the total sample (Table 24.3).

While mechanical sieve analysis is the most common method for characterizing particle size distribution, other measurement techniques such as laser diffraction and video imaging are often utilized. Laser diffraction has the advantage of being able to measure small particles that cannot be effectively measured by sieves. It is particularly useful for fine powders. Video imaging systems have the advantage of being able to characterize not only particle size, but also particle shape. While their use may expand in the future, to date, neither laser diffraction nor video analysis has been embraced by many processors of granular materials, who continue to rely on mechanical sieve analysis techniques. Regardless of the measurement technique used, particle size distribution generally resembles a bell-shaped curve (Figure 24.1). Figure 24.1 also indicates the portion of the solids that might be separated in fines removal or scalping.

Performance, specifically screening machine capacity, is greatly influenced by the location, the desired separation point on the distribution curve. A typical capacity curve is shown in Figure 24.1. For scalping operations at the coarse end of the particle distribution (right side of curve), very high screening capacities are achievable. This is because most of the particles are much smaller than the screen opening and pass through quite easily, with multiple particles passing simultaneously. The capacities are also quite high for fines removal (left side of curve) because, again, the fine particles that should pass through the screen are much smaller than the screen opening. The capacities, in this case, are not quite as high as with scalping because the fine particles must first work their way through the bed of material that is traveling across the screen surface.

TABLE 24.2
Testing Sieve Openings

U.S. Sieve Number	Tyler Sieve Number	Sieve Opening	
		Inches	mm or μm
3 1/2	3 1/2	0.2230	5.6 mm
4	4	0.1870	4.75 mm
5	5	0.1570	4.00 mm
6	6	0.1320	3.35 mm
7	7	0.1100	2.80 mm
8	8	0.0937	2.36 mm
10	9	0.0787	2.00 mm
12	10	0.0661	1.70 mm
14	12	0.0555	1.40 mm
16	14	0.0469	1.18 mm
18	16	0.0394	1.00 mm
20	20	0.0331	850 μm
25	24	0.0280	710 μm
30	28	0.0236	600 μm
35	32	0.0197	500 μm
40	35	0.0167	425 μm
45	42	0.0140	355 μm
50	48	0.0118	300 μm
60	60	0.0098	250 μm
70	65	0.0083	212 μm
80	80	0.0071	180 μm
100	100	0.0059	150 μm
120	115	0.0049	125 μm
140	150	0.0042	106 μm
170	170	0.0035	90 μm
200	200	0.0030	75 μm
230	250	0.0025	63 μm
270	270	0.0021	53 μm
325	325	0.0017	45 μm
400	400	0.0015	38 μm

TABLE 24.3
Example of Sieve Analysis

U.S. Sieve No.	Opening		% Retained
6	3.35 mm	0.132 in.	4.8%
16	1.18 mm	0.0469 in.	18.8%
30	600 μm	0.0236 in.	26.3%
60	250 μm	0.0098 in.	28.1%
100	150 μm	0.0059 in.	11.3%
Pan			10.7%
			100.0%

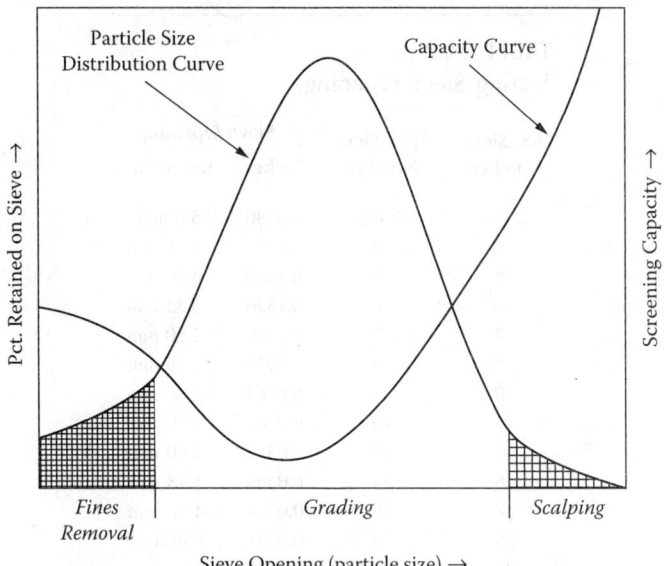

FIGURE 24.1 Particle size distribution curve.

Capacities are lowest when grading. This is because the required separations use screen openings that are near the peak of the particle size distribution curve, where there are numerous particles approximately the same size as the screen opening. These near-size particles that are smaller than the screen opening and should pass through the screen often require multiple exposures to the opening to pass through. In addition, the near-sizes that are slightly larger than the screen opening temporarily block openings, thus limiting the number of screen openings available for the smaller particles. Accurate gradations require relatively low feed rates and highly effective screening motions.

Particle shape affects screening performance. Particles can be various shapes: granular, spherical, cylindrical, etc. Regular shapes such as spherical often allow for relatively sharp separations. Screening of irregular shapes (e.g., elongated, sliver-like, plate-like) generally produce inaccurate separations since the particles enter the screen openings at many different angles. Particle shape can also cause screen blinding, in which particles slightly larger than the screen openings plug the openings. Crystalline particles and spherical particles are particularly prone to blinding.

Another material characteristic that affects screening performance is bulk density. In general, the higher the bulk density, the higher the screen capacity. There are two reasons for this. One is that the force that causes a particle to pass through an opening is proportional to the particle's mass. Consequently, heavy materials like metal powder screen quite readily. However, lightweight materials like sawdust are generally screened at very low-mass flow rates. The second reason is that screening is essentially a volumetric process, as volume defines the depth of material on the screen surface. So, for a given volumetric flow through a screener, the material with the higher bulk density will result in the higher screening rate for a given bed depth, i.e., mass flow.

Flowability of a material, indicated by its angle of repose, affects screening performance since materials that do not flow well do not spread out on the screen surface and properly present themselves to the screen openings. Materials with poor flow characteristics do not convey well along the screen surface. This leads to deeper bed depths and lower screening efficiency. While a material's angle of repose gives some indication of flowability, it is best to evaluate flowability by measuring the material's conveying rate under actual screening conditions.

Surface moisture generally has a significant negative impact on screening performance. Moisture level is measured directly using specialized laboratory equipment in which the solid granular sample is weighed before and after drying. Moisture can cause particles to agglomerate, thus

Dry Screening of Granular and Powder Materials

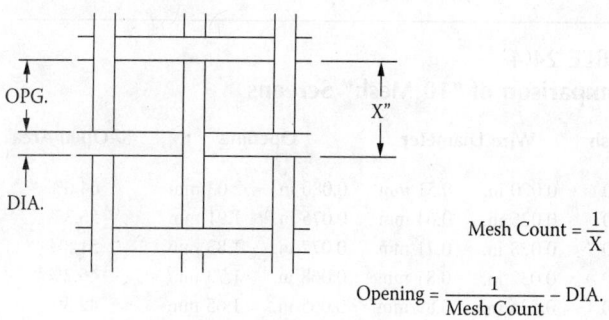

$$\text{Mesh Count} = \frac{1}{X}$$

$$\text{Opening} = \frac{1}{\text{Mesh Count}} - \text{DIA.}$$

$$\text{Percent Open Area} = (\text{Opening} \times \text{Mesh Count})^2 \times 100$$

FIGURE 24.2 Wire cloth openings.

preventing them from passing through the screen opening. Moisture can also cause particles to bridge across a screen opening, effectively blinding the screen media. Hence, many materials must be dried before they can be screened. While the allowable moisture level varies depending on the material, a good rule of thumb is that if the material can be agglomerated by hand (i.e., making a "snowball"), it is probably too moist to be screened.

The bridging of an opening can also be caused by static charge, which is most likely to occur in materials that have been excessively dried or have very fine particle sizes. For such particles, the force of static attraction (which is proportional to the particle's surface area) overcomes the gravitational screening force (which is proportional to the particle's mass). In addition, a static charge can result in a dust or solid particle explosion. While numerous accessories have been proposed for reducing static charges in screening equipment, none are completely effective. The use of additives such as aluminum oxide or the introduction of limited moisture in the form of steam is often quite successful.

The screen media selected has a significant impact on screening performance. Most production screens are specified by identification of at least two of the following four parameters: mesh counts, wire diameter, opening, and percentage open area. The mesh count is the number of wires (or openings) per linear inch. Subtracting the wire diameter from the inverse of the mesh count gives the actual opening in inches. Percentage open area is the ratio of the area of the openings to the total surface area of the screen multiplied by 100. The relationship of these wire cloth specifications is shown in Figure 24.2.

Wire cloth is produced to varying degrees of accuracy and in several commercial grades. It is important to avoid confusion between the sieve numbers used to identify test sieves and the mesh counts used to specify production screen clothing. Typically, multiple wire diameters are available for a given mesh count. This will result in several different available openings, and therefore screening characteristics. As shown in Table 24.4, there are five commercial grades of 10 mesh screen clothing available, yet none has the same opening as a No. 10 test sieve. This is not the problem it may appear to be, because in production screening one typically uses a screen opening slightly larger than the desired nominal separation point. Because of the confusion that can arise when screen media are described by "mesh," it is always best to define the screen opening in "inches" or "microns."

Selection of a screen goes beyond simply choosing a screen opening and wires material. Often, several commercial grades of screens are available with the same opening. In general, screens with a relatively heavy wire for a given opening will be the most durable; however, they are more likely to blind and will have lower capacities due to a lower open area percentage. Screens woven with a finer wire will be more flexible and therefore less prone to blinding, but may have a low service life. While square mesh screens are by far the most common, screens can be woven with a rectangular opening. This can be particularly effective in increasing screen capacity or making

TABLE 24.4
Comparison of "10 Mesh" Screens

Mesh	Wire Diameter		Opening		% Open Area
10	0.020 in.	0.51 mm	0.080 in.	2.03 mm	64.0%
10	0.025 in.	0.64 mm	0.075 in.	1.91 mm	56.3%
10	0.028 in.	0.71 mm	0.072 in.	1.83 mm	51.8%
10	0.032 in.	0.81 mm	0.068 in.	1.73 mm	46.2%
10	0.035 in.	0.89 mm	0.065 in.	1.65 mm	42.3%

U.S. No. 10 Test Sieve: 2.00 mm (0.078 in.) opening.

Tyler No. 10 Test Sieve: 1.70 mm (0.065 in.) opening.

length separations. Screen clothing for chemical applications is commonly constructed of Type 304 stainless steel.

Screens can be woven of synthetic material such as nylon and polyester. These alternatives offer performance advantages in certain screening applications. Another alternative is perforated steel plate. Weld profile and finish should also be consistent with the materials being screened. These plates are also effective and are quite durable. However, they have a low percentage of open area and therefore have lower capacity for a given screen area.

24.3 SCREEN MOTION AND CONFIGURATION

Screening machines are available in various shapes, sizes, and types of motion (Figure 24.3).

The two most common screen motions used in screening machines are vibratory and gyratory. Vibratory is characterized by a short stroke, high-frequency motion commonly used for coarse separations and for screening applications where a high degree of separation accuracy is not required. For finer separations, or when separation accuracy is a priority, gyratory motion is preferred due to its longer stroke and lower speed. Other types of screen motion such as drum screens or centrifugal screens are used for specialized applications.

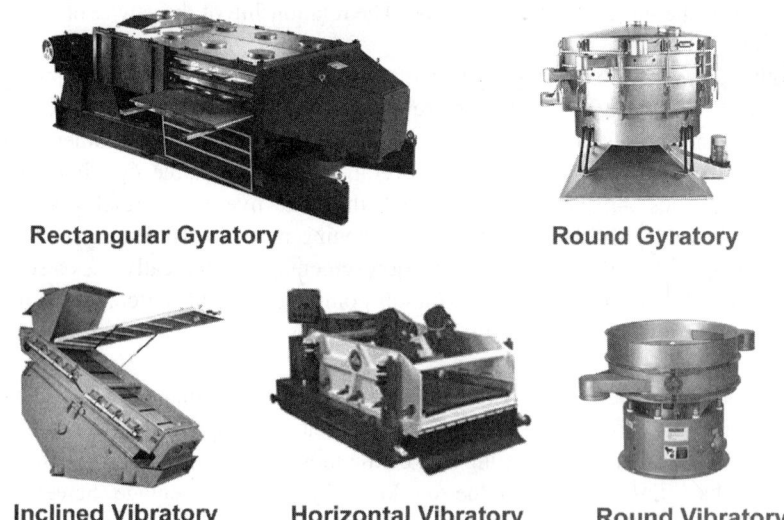

FIGURE 24.3 Screener types.

The performance of a screening machine is determined by its combination of stroke, frequency, plane of action, and screen shape. The stroke is the peak-to-peak distance traveled by the screen in one cycle. For vibrating screeners, the stroke is typically 3 to 6 mm, while for gyratory machines it is 50 to 90 mm. Frequency of a screening machine is the number of cycles of motion completed by the screen in a given time period, typically one minute. Vibrating screeners generally operate at 1000 to 3600 rpm, while gyratory screeners operate at 200 to 300 rpm.

The plane of action or slope of the screen deck affects performance. Increasing the slope of the screen surface often increases the conveying rate of the material and therefore increases the amount of material that can pass through the screener. However, this has a negative impact on screening accuracy. Horizontal screen decks typically provide better separation accuracy. Because the deck is horizontal, gravity has very little effect on the conveying of the material. To achieve reasonable capacities, the screening motion must be capable of effectively conveying the material across the horizontal screen deck.

Screen shape also affects performance. Relatively small screening machines that handle lower capacities use round horizontal screens. These machines are typically fed at the center of the screen, and the material moves toward the edges. The resulting short effective screen length limits their application when accurate grading is required. Rectangular screen decks are used for higher capacity applications and when greater separation accuracy is needed. These rectangular decks can be inclined to improve conveying and increase capacity. By feeding the material to one end of a rectangular screen, accurate separations can be achieved because the material must travel the full length of the screen, thereby maximizing exposure to the screen openings.

Because of their low equipment cost, vibrating screening machines are perhaps the most commonly used, and can be quite effective, particularly in scalping applications. For grading and fines removal, gyratory screening machines have significant advantages. Vibrating screeners often do not have sufficient stroke to properly spread and stratify the material as it is fed onto the screen deck. This prevents the finer particles from achieving maximum exposure to the screen openings. The long stroke of gyratory machines spreads the material across the full width of the machine and stratifies the bed of material, causing the finer particles to settle down to the screen surface. Because vibrating screeners have a large vertical component of motion, the material becomes suspended above the screen surface, reducing the probability that a particle will pass through a screen opening. Gyratory machines use a sifting motion to keep the material in contact with the screen surface, maximizing opportunities for fine particles to pass.

The performance of a screening machine can be significantly affected by the rate material fed to it. A screener has an ultimate volumetric capacity, which is a function of internal construction and material handling capabilities. However, because capacity is defined by acceptable separation performance, the operating capacity will be lower than the maximum volumetric capacity. This balance between volumetric capacity and screening performance must be evaluated when selecting screening equipment.

Blinding causes reduction in available screen area and therefore reduces screening accuracy and screener capacity. Blinding generally occurs when near-size particles, slightly larger than the screen opening, lodge in the opening. Blinding can also be caused by bridging, which occurs when groups of particles, all much smaller than the screen opening, bind together and block the screen openings. Surface moisture or electrostatic charge can also cause blinding. Fibrous materials can cause blinding when particles wrap themselves around screen wires. In screening applications where blinding is a concern, control measures must be employed to prevent the cumulative effects of this blinding. The most common method is to trap bouncing balls in pockets below the screen clothing. The motion of the screening machine itself causes the balls to bounce randomly within their pockets and cleans the screen clothing. Other measures such as vibrating rings, compressed air, rotating brushes, and ultrasonic vibrations have also been used.

24.4 SPECIFYING SCREENER PERFORMANCE

Screening machines are evaluated on their ability to maximize both product quality and screening efficiency. One or more of the fractions, or flows, produced is considered to be "product." This product has imposed on it a quality specification defining the acceptable particle size distribution. Often, this specification consists simply of particle size limits. For example, in a scalping application, the product quality specification may be stated as "0% + 10 U.S. allowable," i.e., a sample of material that passes through the screen should contain no material coarser than a No. 10 U.S. test sieve. Similar specifications can also be applied to define acceptable results of a fines removal operation. For example, a fines specification of 3% maximum allowable –60 U.S. means that the analysis of a sample taken from the material that passed over the screen would contain no more than 3% by weight passing the No. 60 U.S. test sieve.

Limits and tolerances can also be applied to define product quality in grading applications. Certain industries have arrived at other particle size quality measurements to be applied to their products. The fertilizer industry commonly uses SGN (size guide number) to define the desired mean particle size and UI (uniformity index) to define the allowable variation from that mean particle size. Industrial sands are characterized by the GFN (grain fineness number) that approximates the theoretical sieve opening that would equate to the mean grain size. Abrasives, such as shot and grit, have exacting specifications that set limits and tolerances on the product. Filter media, such as activated carbon, are specified in terms of ES (effective size) and UC (uniformity coefficient), which are analogous to the SGN and UI used in the fertilizer industry. Regardless of the product specification, the key factor affecting product quality is the screen opening selected. So, if a screener is not producing adequate product quality, typically a screen with a different opening must be used.

Although product quality is important, it is not the sole criteria for evaluating screening performance. The other key measure of screening performance is screening efficiency. Screening efficiency is generally measured in one of two ways. The most common of these is referred to as "yield." Yield is the amount of material separated as product as a percentage of that product fraction which is available in the feed. Thus yield measures the efficiency by which the screener produces the desired, on-specification product.

A more exacting measure of screening efficiency is referred to as product recovery efficiency. It is expressed as a percentage based on the amount of on-size product in the product fraction separated by the screener, divided by the amount of on-size material available in the feed to the screener. Product recovery is calculated as follows:

$$\text{Efficiency (Product Recovery)} = \frac{\text{percent on-spec material in product} \times \text{percent product production rate}}{\text{percent on-size material available in feed}}$$

The product production rate is defined as the percent of material separated by the screener as product times the feed rate to the screener. Production rates are meaningful only if the quality of the product meets the product specifications.

The example shown in Figure 24.4 is a typical screening application in which a screener is used to extract on-size product from a process flow. The example illustrates the relationship between product quality, product yield, and product recovery efficiency.

Product quality is always a factor in system design, but it is screening efficiency that determines production rates and overall system efficiencies. For systems that include recycle loops, screening efficiency can have a compound effect on overall system performance. Clearly, the more efficiently a screener extracts on-size product, the higher the production rate. Often, the reject streams from the screener are recycled and reintroduced to the screener. In such cases, on-size material that was mistakenly rejected by the screener will be needlessly reprocessed, resulting in higher energy costs

Dry Screening of Granular and Powder Materials

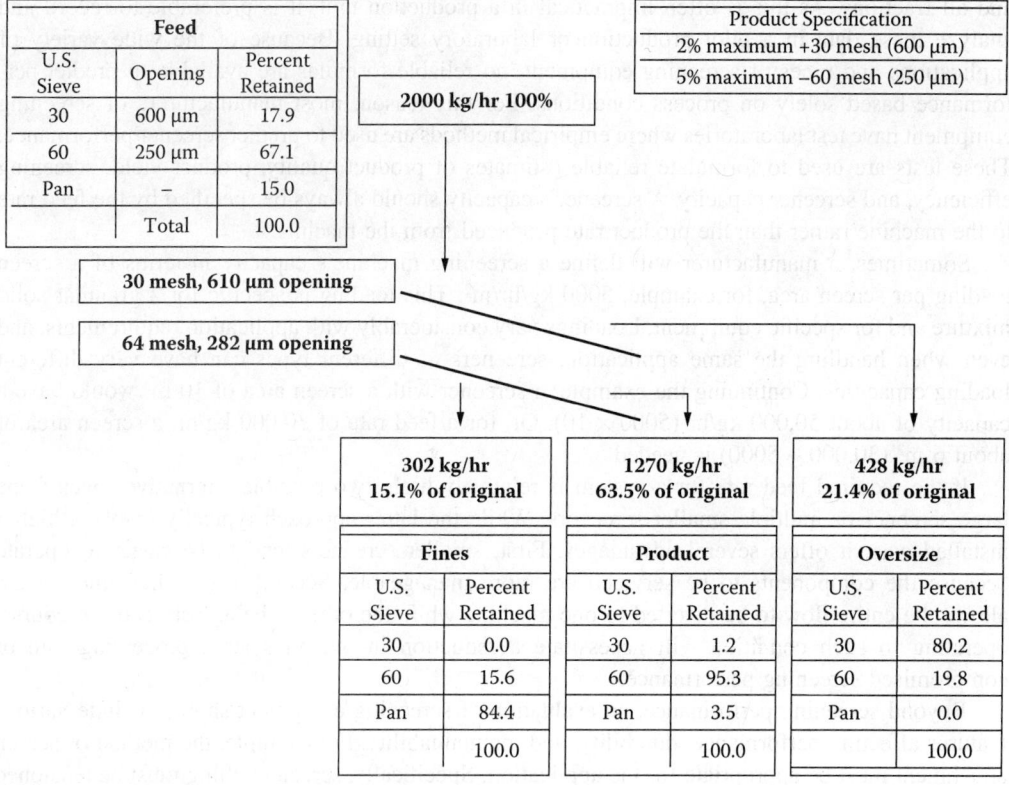

Oversize Rate: 2000 kg/hr × 21.4% = 428 kg/hr
Product Yield: 2000 kg/hr × 63.5% = 1270 kg/hr
Fine Rate: 2000 kg/hr × 15.1% = 302 kg/hr

Product Recovery Efficiency: $\dfrac{95.3\% \text{ on-specification} \times 63.5\% \text{ product yield}}{67.1\% \text{ available product}} = 90.2\%$

FIGURE 24.4 Typical screening application.

and lower overall system efficiency. This is particularly important in granulation systems and compaction circuits that typically produce material with a broad particle size range, in which only a relatively small percentage of the material produced is actually on-size product. In such cases, it is critical that the screening machine be as efficient as possible at extracting this on-size material.

While screen-opening selection has the greatest effect on product quality, it is the screening machine motion that has the greatest effect on screening efficiency. Certainly, changing the screen opening can increase screening efficiency. For example, in an oversize removal application, using a screen opening larger than the desired separation point can increase screening efficiency. However, this will have a negative impact on product quality, as the larger opening will allow oversize particles to pass through the screen and contaminate the product. By using an efficient screening motion, appropriate screen openings can be selected to ensure product quality without the penalty of inefficiently rejecting on-size material.

24.5 SCREENING EQUIPMENT SELECTION CRITERIA

To properly evaluate the performance of a screening machine, it is necessary to define the feed rates, and mass flow rates for each separated fraction, and particle size analyses for the feed material and all fractions. As this is often impractical in a production unit, it is preferable to record and analyze these data in a pilot production or laboratory setting. Because of the wide variety of applications and types of screening equipment, no reliable formulas are available to predict performance based solely on process conditions. For this reason, most manufacturers of screening equipment have test laboratories where empirical methods are used to predict screener performance. These tests are used to formulate reliable estimates of product quality, product yield, screening efficiency, and screener capacity. A screener's capacity should always be specified by the feed rate to the machine rather than the product rate produced from the machine.

Sometimes, a manufacturer will define a screening machine's capacity in terms of a screen loading per screen area, for example, 5000 kg/hr/m^2. This loading is specific for a granular solid mixture and for specific equipment. Loadings vary considerably with application requirements, and even when handling the same application, screeners of different types can have very different loading capacities. Continuing the example, a screener with a screen area of 10 m^2 would have a capacity of about 50,000 kg/hr (5000 × 10). Or, for a feed rate of 30,000 kg/hr, a screen area of about 6 m^2 (30,000 ÷ 5000) is needed.

If the required feed rate for a system is relatively high, two possible alternatives occur: one large screener or multiple smaller screeners. While the latter approach typically involves higher installed cost, it offers several advantages. First, smaller screeners tend to be easier to operate because the components to be serviced are more manageable. Second, a dual-machine system allows the entire flow to be diverted to one machine while the other is being serviced. Of course, operating in such condition will necessitate a reduction in overall system processing rate or compromised screening performance.

Beyond screening performance, an evaluation of screening equipment should include various features affecting performance, durability, and maintainability. For example, the method of screen attachment must be appropriate for the application. Specifically, screen clothing must be tensioned at the proper force level to ensure the screen deck remains taut under the material load. While coarse screens (>10 mm) can be tensioned by hand and fixed with mechanical fasteners, finer screen clothing should be tensioned by spring loaded devices or factory set pre-tensioning and attachment. If several screen decks are required to make the necessary separations, it is often desirable to use a single-screening machine with multiple decks installed. This is typically the most cost-effective approach if the amount of screen area required on each deck is approximately the same. If the number of decks required exceed three, it is often more cost-effective to use two separate screening machines arranged in series. This eliminates the complexity involved with a four- or five-deck screener and also allows for tailoring the screener sizes to the amount of screen area required for each separation.

A key element of any screening machine is the drive system. Two types of drive systems are generally used. Drives used on screeners (typically of the vibrating type) in which the screen box are mounted on springs or cables allow the box to move freely in a cyclical motion. The drive consists of a rotating eccentric weight and is attached to the screen box. This intentionally out-of-balance system produces a motion in the screen box that is synchronous with the drive speed and proportional to the ratio of the drive weight to the sprung weight of the screen box. Screen box stroke can therefore be increased or decreased by adjusting the drive weight. One drawback is that screener motion is dampened when a large amount of material is introduced to the screen box, resulting in a significant loss of screening performance.

The other type of screening machine drive system is the positive displacement or crank drive. In this system (typically used for gyratory screen motion), an eccentric crank produces the motion. The advantage is that screen motion is fixed at the designed stroke regardless of material load. The positive displacement drives must employ rotating weights, for counterbalancing the moving assembly. Gyratory screeners are more expensive than vibratory screeners.

When installing a screening machine, the surrounding structure should be properly designed to withstand the forces generated by the screening equipment. In addition to the static load of the

TABLE 24.5
Screening Machine Selection Guide

Screen Motion Screen Shape Screen Slope	Vibratory Round Horizontal	Vibratory Rectangular Inclined	Vibratory Rectangular Horizontal	Gyratory Round Horizontal	Gyratory Rectangular Horizontal
Applicability					
Coarse (>5 mm)					
Scalping	Fair	Good	Good	Good	Good
Grading	Poor	Fair	Good	Fair	Good
Fines removal	Poor	Fair	Fair	Fair	Good
Medium (0.5–5 mm)					
Scalping	Good	Fair	Good	Good	Good
Grading	Fair	Poor	Fair	Fair	Good
Fines removal	Poor	Poor	Fair	Good	Good
Fine (<0.5 mm)					
Scalping	Fair	Poor	Fair	Good	Good
Grading	Poor	Poor	Poor	Fair	Good
Fines removal	Poor	Poor	Poor	Fair	Good
Screening accuracy	Low-medium	Low	Medium	Medium-high	High
Screening capacity	Low	High	Medium	Medium	High
Relative cost	Low	Medium	Medium	High	High

screener and the material in it, the screening machine will generate dynamic loads, which may be produced at a frequency close to the resonant frequency of the surrounding structure. A thorough structural analysis is required when considering any screener installation. Also, consideration must be given to the accessibility of screener components requiring service or maintenance. For example, removal of screen decks for screen clothing replacement can necessitate an area of floor space equal to that of the machine itself. Accessibility must also be provided for servicing of the screener drive and material flow connections.

The proper selection of screening equipment must include a complete analysis of both installed (capital) and operating costs. For example, the cost advantage of an inclined screener over a horizontal screener could be outweighed by the additional cost of the building necessary to accommodate the taller inclined machine. Similarly, a rectangular vibrating screener typically costs less than a rectangular gyratory machine. However, because the vibrating screener may not spread the incoming material sufficiently, a separate, full width-vibrating feeder may be required. In such a case, the total cost for the vibrating feeder and screener could exceed that of the gyratory screener. When evaluating operating costs, the single most important factor is often screening efficiency. A more efficient screener will produce more product from a given feed rate, resulting in increased overall system efficiency and lower overall system operating costs.

The selection criteria described in this section are summarized in Table 24.5.

24.6 SCREENING MACHINE OPERATIONS AND MAINTENANCE

Most screening machines are designed to be configured for a specific set of application conditions and require no further adjustment during operation. However, it is necessary to monitor performance of the screener to ensure that it is operating properly. The frequency of this monitoring is driven by the cost impact of screened product that is out of specification. The cost impact may be a rejected shipment or an overloaded recycle loop. The best method of monitoring screener operation is to perform particle size analyses of samples of the input and output streams. Unfortunately, provisions for this kind of sampling are often not taken into account in the original design of the system. The results can identify various problems that may have developed since the last sampling. For example, if an analysis of material passing through a screen reveals the presence of particles larger than the installed screen opening, then one can infer that the screen has a tear in it or one of the seals in the machine has failed. Conversely, if an analysis of material passing over a screen reveals an unusually high percentage of fine material, then one can infer that the screen is blinded or the machine is overloaded, either of which would lead to poor fines removal performance.

Screener maintenance involves both preventive and corrective maintenance. Preventive maintenance includes lubrication, cleaning, and routine inspection of all critical components per the manufacturer's instructions. Corrective maintenance most commonly involves repair or replacement of the screen media. Screen media life is a function of many process variables. Some factors that lead to short screen media life are operating with high screen loadings, using very fine screen wire, and screening of abrasive materials. For example, a relatively coarse 5 mm opening screen with a large diameter wire might last several years, while a fine screen with a 100 micron opening might last only a few days or weeks. When designing processes that are expected to have low screen media life, consideration must be given to frequent servicing of the screen media. If the effects of a torn screen are intolerable, it may be necessary to establish a mean time to failure value for the screen media and make preemptive replacements before failures are expected.

Because most screening machines have relatively massive moving or rotating components, it is imperative that appropriate safety precautions be taken when working on or around the equipment. The manufacturers' manuals and labeling detail the safety considerations. Extra precaution must be taken when processing materials that present explosion hazards. Manufacturers of screening machines will often include special provisions for grounding the various subassemblies and components of the machine to ensure that an electrostatic charge differential does not develop between such components and result in a spark that could trigger an explosion. Machine maintenance can necessitate the removal of these grounding provisions, and care must be taken that they be properly replaced and secured before the machine is returned to operation.

Most screening machines are designed and fabricated such that their useful service lives are measured in decades rather than years in most applications. However, some applications can dramatically shorten service life. Abrasive materials can cause premature wear, particularly when process rates are high. Some materials attack the screener's metal components. Salts, such as sodium chloride, calcium chloride, and ammonium sulfate, are particularly aggressive. In such cases, selection of the materials of construction is critical to maximizing screener service life. In this matter, it is best to work closely with the manufacturer, sharing all pertinent process parameters so recommendations can be made based on their experience with similar systems.

25 Conveying of Bulk Solids

Fred Thomson

CONTENTS

25.1 Conveying of Bulk Solids 1729
25.2 Characterizing the Materials to Be Handled 1729
25.3 Belt Conveyors 1730
     25.3.1 Carrying Idlers 1732
     25.3.2 Belts 1732
          25.3.2.1 Take-Up 1733
          25.3.2.2 Backstops 1733
          25.3.2.3 Belt Cleaning 1733
          25.3.2.4 Power Requirements 1733
          25.3.2.5 Alternate Belt Design 1733
25.4 Screw Conveyors 1733
25.5 Bucket Elevators 1734
25.6 Vibrating Conveyors 1734
25.7 En-Masse Conveyors 1734
25.8 Air-Activated Gravity Conveyor 1735
25.9 Pneumatic Conveyors 1735
25.10 Summary 1735

25.1 CONVEYING OF BULK SOLIDS

Conveyors *transport* solid material streams. With appropriate geometry and speed control, they regulate the volumetric flow of the material and are called *volumetric feeders*. With the further mass sensing and control modules, they regulate the mass flow of material and are called *gravimetric feeders*.

25.2 CHARACTERIZING THE MATERIALS TO BE HANDLED

Before selecting a conveyor for a specific application, the properties and handling characteristics of the bulk material must be determined. Conveyor manufacturers often use empirical bench-top tests and observations of actual systems to measure bulk density, particle size, stickiness, abrasiveness, etc. Helpful information is often published by manufacturers' engineering associations in the various countries. Typical are the publications by the Conveyor Equipment Manufacturers Association (CEMA) in the United States. Used primarily for mechanical conveying, the CEMA provides terminology, definitions, and test procedures for describing 37 bulk materials. Each measured characteristic is assigned an alpha-numeric classification code designation. It is widely used and appears in many equipment catalogs as an aid in selecting a conveyor. The CEMA publication also includes an extensive list of handling characteristics for common bulk materials.

Although the characteristics of a solid are useful information, they are not a substitute for experience because only the user knows plant process conditions and potential process changes or upsets. Changes in moisture, particle size and size distribution, particle attrition, and even changes in raw material supplier, for example, affect the performance of mechanical and pneumatic conveyors. Actual operation must be considered before making a final choice of a conveyor. Often, material samples are tested or are sent to conveyor manufacturers for tests. Testing "similar" materials often helps.

Flow theories plus laboratory tests often permit the design of storage silos and hoppers. The following can be determined: power consumption for withdrawing solids and granules from hoppers and bins, prediction of frictional wear of conveyor and silo surfaces during handling of bulk materials, and attrition of particles during storage and handling. The American Society for Testing Materials (ASTM), Subcommittee D18.24, is currently developing improved Standard Testing Methods for Bulk Solids.

25.3 BELT CONVEYORS

A belt conveyor is made up of an endless belt that traverses between two or more pulleys and is supported at intermediate points by idler rolls. These conveyors can handle a wide range of materials, from fine powders to large lumpy stone or coal, at rates varying from several tons per hour to over 5000 ton/hr. Belt speeds vary from 35 to 300 m/min with belts extending over considerable distances. Such belts are widely used. Figure 25.1 shows a typical belt. It can be arranged horizontally and with inclined or declined sections combined with convex and concave curves in the belt. The desired path of travel is limited only by the strength of the belt and the permissible angle of incline or decline for the particular situation.

Bulk materials are sometimes conveyed on flat belts supported on horizontal idlers on the carrying and return runs. However, in most industrial systems, in order to increase the handling capacity, the belt is formed into a trough shape after it has been loaded with material and it is supported along the carrying run by troughing idlers. The most common troughing idler consists of three rolls. The belt is realigned to a flat position over the head pulley to discharge the material and returns supported on horizontal rolls. The viscoelastic properties of the conveyor belt and the distribution of stresses during start-up and shut-down must be considered when designing long conveyor belts and belts with horizontal curves.

Solids being transferred vary in size of particles; dusty material sometimes results in dust explosions. The materials can also vary in abrasiveness, be free-flowing, or cohesive, or friable. Wet or sticky materials, however, are often a problem if they are not continuously cleaned from the belt surface.

Material characteristics are typically used to determine the required belt width, the carrying capacity of a particular belt, and the maximum inclination at which the belt can be operated.

The angle of repose and angle of surcharge: The angle of repose is the angle a freely formed heap of bulk material makes with the horizontal. The measured value of this angle, when taken together with observations of particle shape and size, moisture, and flowability, determines the angle of surcharge. This is defined as the angle (to the horizontal) that a material assumes while the material is on a moving conveyor belt. For most materials, the angle of surcharge is 5° to 15° less than the angle of repose, but with some very free flowing materials it may be 20° less to avoid spillage from the edge of the belt. Because the cross-section area of the heap is defined by the belt trough geometry and the angle of surcharge, the belt speed required for a desired transport rate of a material with a known density can be easily determined.

The size and proportion of lumps: The larger the lump, and the greater the number, the wider the belt must be to prevent lumps from spilling from a horizontal conveyor; they are more likely to fall off of inclined conveyors.

Fluidizing or air retention properties: Materials that are easily aerated and have long air retention time limit belt inclination and speed.

Conveying of Bulk Solids

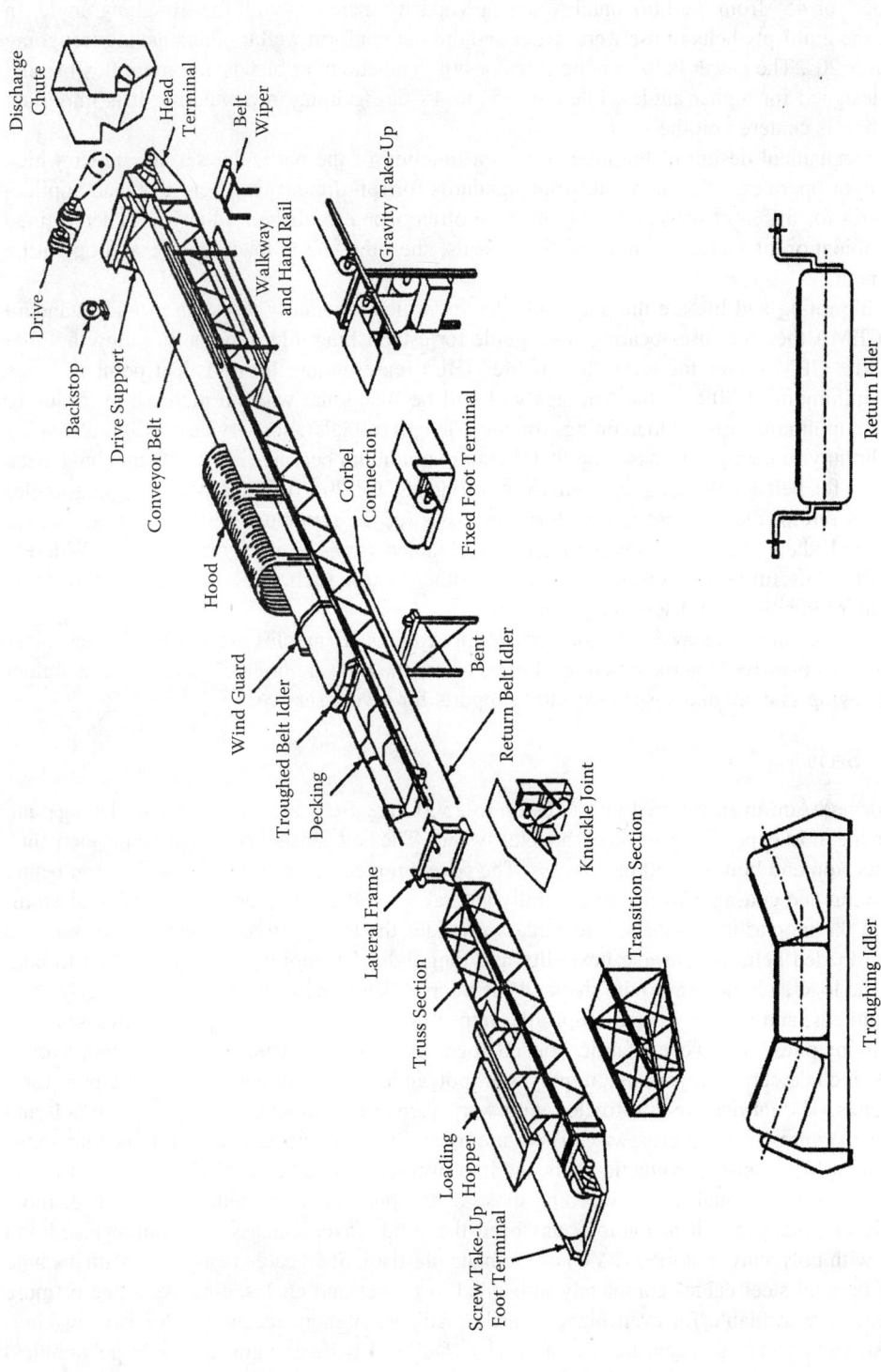

FIGURE 25.1 Troughing belt and support assembly.

25.3.1 Carrying Idlers

The most commonly used troughed carrying idler consist of three rolls with two outer rolls inclined at 20°, 35°, or 45° from the horizontal. Carrying capacity increases with the troughing angle. In the past, the multi-ply belts in use were stiffer and did not conform well to idlers having troughing angles over 20°. The newer belts, having carcasses of synthetic fiber blends, are more flexible and can be designed for higher angles. Idlers of 35° to 45° are gaining in popularity. It is important that the belt is centered on the idlers.

The mechanical design of the idler rolls is a function of the particular service under which the conveyor operates. Minimum industrial standards for roll dimensions, bearings, and application criteria for different service conditions have often been established. Idler life is determined by a combination of factors such as bearings, seals, shell thickness, load density, and operating environment.

Bearing rating and life are the only variables for which laboratory tests can provide standard values. CEMA therefore uses bearings as a guide for establishing idler ratings. In rating the idler bearing life, CEMA uses the term "useful life" (BU) representing the statistical point in hours where a minimum of 90% of the bearings will still be functional with no increase in torque or noise. The minimum required load ratings for equal-length roll idlers for service conditions ranging from light duty to heavy duty based on 90,000 hours minimum bearing life at 500 rpm have been determined for belt widths ranging from 18 to 96 inches, for 20°, 35°, and 45° troughing angles and for flat return idlers. These ratings form the basis for the minimum design requirements for CEMA-rated idlers. Actual figures for idlers are obtained from the idler manufacturer. Whereas bearing life is useful as an indicator of idler life, other factors such as bearing seal effectiveness may be more important in determining idler life.

Specialized idlers are available for certain applications. Examples are plastic disc catenary idlers for wet corrosive materials; two-roll idlers, where the rolls are oriented in a "V" for a lighter duty conveying system; and suspended idler supports for severe service.

25.3.2 Belts

Conveyor belts are manufactured in widths up to 2500 mm. Belts are susceptible to damage and are often the most expensive portion of the total system. The belt consists of two principal elements: the covers (top and bottom) and the carcass. The covers protect the carcass against environmental effects, wear, and cutting. Covers are generally natural or synthetic rubber, thermosetting elastomers, and thermoplastic materials. The carcass provides the tensile strength required to start and move the loaded belt, the traverse flexibility and longitudinal flexibility to allow the belt to both support the load and conform to the shape of the idlers when running empty and to properly wrap around pulleys, and the strength to absorb the impact forces at the loading point. Carcasses have one or more plies of a woven fabric bonded together with an elastomeric compound. Woven materials include cotton, rayon, nylon, polyester, polyamides (nylon), and glass, in the pure form or in blends. The fabrics are constructed with warp yarns that run lengthwise along the belt and filling (weft) yarns that run crosswise. There are a variety of weaves available for specific applications. New high-tenacity synthetic fibers and improvements in belt construction have led to belts with load capacities equal to or exceeding those of the previous belts with fewer plies. A more recent development in belt technology has been the solid-woven carcass belts impregnated and covered with poly-vinyl-chloride (PVC) or urethane plastisol. Steel cable belts made with a single layer of parallel steel cables completely imbedded in rubber and enclosed between one or more fabric plies are available for even higher tensile loadings. A more recent development in high-tensile strength belts has been the use of Kevlar (DuPont) high-strength (aromatic polyamides) fibers in woven, cord, solid woven, and cable construction. Careful cost analyses are needed in selecting the type of belt to be used.

25.3.2.1 Take-Up

A take-up is required on a belt conveyor to ensure the proper belt tension at the drive pulley and along the conveyor and to ensure the proper troughing contour between idlers. Belts stretch during start-up and as they age.

25.3.2.2 Backstops

A backstop is a device that permits rotation of the pulley in the forward direction but automatically prevents rotation in the opposite direction. A backstop should be mounted at the head shaft of an inclined belt to prevent the belt from moving in reverse in case of a power or mechanical failure.

25.3.2.3 Belt Cleaning

Idlers and snub pulleys on the return run support the belt on the material carrying surfaces. Cleaning these pulleys and the belt of materials that adhere is often necessary. A variety of scrappers or cleaners is available.

25.3.2.4 Power Requirements

Electric motors are commonly employed to operate the belts and to overcome the resistances. Power requirements contribute significantly to the cost of operation.

25.3.2.5 Alternate Belt Design

Modified belts currently available include the following:

Flexible sidewall belts: They consist of three elements: two flat sections for first loading and second unloading. An accordion-pleated flexible section is available with cleats or dividers for sections to raise or lower the solid material. In this latter section, sidewalls range from 40 to 400 mm, angles up to 90°, lift heights up to 80 m, and capacities up to 1000 metric ton/hr.

Air-cushioned belts: Air films are formed in U-shaped troughs beneath the belt. This air film reduces power requirements. Such belts tend to be limited to low levels of feeds.

Sandwich belts: Two belts sandwich the solid granules in order to transport them upward at angles up to 90°. Rates of transfer can be large, even more than 400 metric tons/hr.

Tubular-type conveyors: In these conveyors, the belt, after being loaded, is bent to form a closed tube. Such an arrangement minimizes the loss of solids, dust releases, cleaning, and weather-related problems.

Folded-belt and enclosed belt conveyors: Both types of conveyors can be used for moving solids at angles up to vertical.

25.4 SCREW CONVEYORS

A rotating screw is employed to push (or pump) solid powder through a tube or U-shaped trough. Screw conveyors are generally of relatively simple design and rather low cost, but power requirements are high. Sometimes cooling is required as considerable power is converted to heat. The abrasiveness and average size of the granular powder are important. For cohesive materials that pack or adhere, problems occur. Solid particles that easily break into smaller particles are much easier to convey via screw conveyors. Solids with hard and rather large lumps often cause problems in these conveyors. Screw conveyors that provide up to 30-m lifts have been built. Screw rpm often ranges from 200 to 400.

25.5 BUCKET ELEVATORS

A series of metal or plastic buckets are used to lift solid particles. These buckets are attached to an endless belt or chain. The buckets are first filled with the solid particles, and the buckets are unloaded after the moving belt has lifted the buckets. Continuous elevators can handle a wide range of materials, from light to heavy, free-flowing granular and pulverized materials with lumps up to 100 mm. This elevator handles material rather gently. Bucket speeds range from 30 to 46 m/min. The lower speeds are used in order to properly fill the buckets when fluffy, low-density materials are handled. Capacities range up to 75 m^3/hr on standard elevators.

Super-capacity elevators are continuous-type elevators in which buckets are mounted between two strands of chains. This arrangement makes it possible to handle a larger volume because the bucket can be extended in back of the chain. Elevators of this type are capable of handling up to 375 m^3/hr.

25.6 VIBRATING CONVEYORS

A vibrating conveyor consists of a trough supported by tuned springs and/or hinged links, having a drive system. The drive system is arranged to oscillate the trough, causing solid particles to be moved along the trough. Such conveyors are sometimes called oscillating conveyors. There are two types: reciprocating and vibrating.

On a reciprocating conveyor, material is carried forward in a horizontal direction by frictional contact with the trough. Inertia causes the material to be left in that position as the trough is quickly returned to its initial position. These conveyors are useful for granular free-flowing materials with a minimum of attrition. They provide gentle handling of food products, such as friable flakes and pellets, pharmaceuticals, powdered and granular chemicals, and minerals. Vibrating conveyors are uniquely suited for handling metal parts, abrasive, hot, and dusty materials, and can be designed to withstand heavy impact loads from materials such as rocks, iron and steel casting, and metal and wood scrap.

Tubular troughs are useful where dust-tight operation is required. Inclined flat bottom troughs with a sawtooth-shaped carrying surface can be used for conveying granular materials up to 10° to 20° slopes. Troughs can be provided with multiple discharge openings and can be designed to handle materials at temperatures up to 900°C (1652°F). Tubular or rectangular troughs wound in a vertical spiral can be used for vertical conveying of granular materials. This type of conveyor also operates at sub-resonance and is shop or field fine-tuned by weight adjustment on the trough before being put in service. This design requires fewer moving parts than the direct drive machine, but requires more attention to design and tuning. The exciters usually operate at frequencies of 15 to 20 Hz.

Excitation can also be supplied by an electromagnetic exciter that uses a rectified, pulsed AC power supply, or AC supply to an apposed electromagnet/permanent magnet drive. These units operate at very short strokes and frequencies of 50 to 60 Hz.

Compressed air or hydraulically driven reciprocating piston or rotary exciters are sometimes used in short conveyors. They are particularly useful where explosion hazards limit the use of electrical drives.

25.7 EN-MASSE CONVEYORS

An en-masse conveyor consists of an endless chain or cable pulling a series of spaced skeleton or solid plug flights through an enclosed casing or housing. Material is introduced through an opening in the casing, where it is captured by the flights and drawn through the casing until it reaches an opening in the housing where it discharges by gravity.

En-masse conveyors are compact and totally enclose the bulk material; they handle many materials with little particle attrition; they can have a L-shaped or a Z-shaped path, therefore eliminating transfer points that are required by conventional straight line conveyors; they can combine feeding, conveying, and elevating in one machine; and they can have multiple inlet and discharge openings.

These conveyors are best suited for nonabrasive, free-flowing, granular or powder materials. Sticky or smearing materials can build up between flight and casing, causing a mechanical overload.

25.8 AIR-ACTIVATED GRAVITY CONVEYOR

The air-activated gravity conveyor is a simple, low-cost, essentially maintenance-free, and low power-using device for conveying fine, dry, easily fluidized powders. There are two types of air activated conveyors: closed and open.

The closed type used for transporting powders consists of a downward-sloped rectangular trough, bisected by a porous membrane that defines a lower and upper channel. Air supplied to the lower channel permeates through the membrane and aerates powders in the upper channel, causing the fluidized powder to flow like a liquid on the surface of the inclined membrane. Powder flow occurs even if the membrane is only partially covered with material: the pressure drop through the membrane is such that the air flow is uniformly distributed across the membrane surface independent of the depth of material above. Simple gates or valves are used to control inlet and discharge at the end or at intermediate points. The air that permeates through the membrane exits with the powder to the receiving vessel, and this air must be vented through appropriate filters. On long runs, it is often necessary to vent the air at intermediate points to avoid a build-up within the powder flow channel. Conveying capacity is dependent on the material air retention properties, trough cross-section, and slope angle. Trough widths range from 100 to 850 mm; capacities range up to 1500 m^3/hr with slopes from 3° to 8°. Air pressures beneath the membrane range from about 0.07 to 0.2 atm. Air flows range from about 0.9 to 2.5 m^3/m^2 of the membrane surface.

In the open-type air-activated conveyors, the cover over the powder flow channel is omitted. They are installed on sloped walls of silo hoppers or on sloped supports mounted on flat silo bottoms, with the membrane exposed to the stored material. This arrangement promotes powder flow within the silo and causes the material to flow toward the outlet. Air pressure below the open type membrane usually ranges from 0.2 to 0.33 atm.

Manufacturers usually determine size, slope, and air requirements for a particular situation by scale-up from tests on simple air-activated laboratory test rigs.

25.9 PNEUMATIC CONVEYORS

In a pneumatic conveyor, bulk solids particles are transported through a closed duct in a gas stream. A wide range of particle sizes, from powders to large lumps, can be handled through these systems. Restrictions are that the conveying pipe should be able to accommodate the largest particle. Such conveyors are classified as either dilute phase conveying or dense phase conveying. In dilute phase systems, relatively high-air velocities lift and drag particles through the conveying pipe in a dilute suspension. The ratio of mass of solids to the mass of conveying air ranges from 0.25 to 15, with the higher values achieved only in short distances. In dense phase systems, air velocities are lower, pressures are higher, and the particles move a much denser stream, with a faster moving layer superimposed over a denser moving layer at the bottom of the pipe. Both horizontal and vertical conveying is possible.

25.10 SUMMARY

Numerous conveyors are available for transporting solid powders, granules, etc. Generally, considerable information needs to be assembled relative to the character and amount of solid to be transported, distances and heights of transport, etc. Manufacturers of conveyors can generally aid in the selection of a preferred conveyor. Both operating and capital costs need be considered in selecting the preferred conveyor.

26 Principles and Applications of Electrochemical Engineering

Peter N. Pintauro

CONTENTS

26.1 Introduction...1738
26.2 Fundamentals...1738
    26.2.1 Thermodynamics 1742
        26.2.1.1 Thermodynamics of Electrochemical Cells....................................1742
        26.2.1.2 Activity Coefficients and the Concept of an Electrochemical
                 Potential...1745
    26.2.2 Kinetics of Electrode Reactions..1749
    26.2.3 Mass Transport 1753
        26.2.3.1 Transport Equations 1754
        26.2.3.2 Transport of Electro-Active Species in the Mass Transfer
                 Boundary Layer...1757
        26.2.3.3 Infinite Dilution Transport Parameters 1763
    26.2.4 Driven vs. Self-Driven Cells and the Concept of Overpotential....................1763
26.3 Electrochemical Reactors and Reactor Design..1766
    26.3.1 Characterizing the Performance of an Electrochemical Reactor 1766
        26.3.1.1 Initial Design Experiments 1768
    26.3.2 Choosing an Electrochemical Reactor 1769
26.4 Inorganic and Organic Electrochemical Synthesis 1774
    26.4.1 Inorganic Processes 1774
        26.4.1.1 Aluminum Electrorefining 1774
        26.4.1.2 Chlor-Alkali Process 1775
        26.4.1.3 Water Electrolysis 1777
    26.4.2 Organic Electrochemical Syntheses 1778
        26.4.2.1 Essential Components of an Electro-Organic Reactor System.............1781
    26.4.3 Monsanto's Electrohydrodimerization Process...1784
    26.4.4 Electrocatalytic Reduction (Hydrogenation) Reactions in a Solid Polymer
           Electrolyte (Proton Exchange Membrane) Reactor..1784
26.5 Electroplating and Electroetching..1787
    26.5.1 Basic Principles...1787
    26.5.2 High-Speed and Laser-Assisted Plating..1791
    26.5.3 Electroless Plating 1792
    26.5.4 Alloy Plating..1794
    26.5.5 Electrochemical Metal Removal Processes 1794
26.6 Environmental Clean-Up Processes...1796
    26.6.1 Metal Ion Removal..1797
    26.6.2 Electrochemical Destruction of Organics 1800
    26.6.3 Electrodialysis..1801

26.7 Corrosion Processes..1805
    26.7.1 Thermodynamics of Corrosion Processes..1806
    26.7.2 Kinetic Aspects of Corrosion Reactions..1808
    26.7.3 Mass Transfer Effects..1810
    26.7.4 Metal Passivation...1811
    26.7.5 Types of Corrosion and Methods of Corrosion Prevention 1812
        26.7.5.1 Uniform Attack 1812
        26.7.5.2 Galvanic Corrosion 1812
        26.7.5.3 Crevice Corrosion 1813
        26.7.5.4 Pitting 1814
        26.7.5.5 Stress Corrosion Cracking 1815
        26.7.5.6 Hydrogen Embrittlement..1815
26.8 Batteries and Fuel Cells...1816
    26.8.1 Batteries..1816
        26.8.1.1 Battery Performance Characteristics 1816
    26.8.2 Fuel Cells..1821
        26.8.2.1 Fuel Cell Types 1822
Nomenclature 1824
References..1826

26.1 INTRODUCTION

Electrochemical methods of producing energy (primary and secondary batteries and fuel cells), fabricating materials (e.g., electroplating, electrowinning, and electromachining operations), purifying chemical compounds (electrodialysis membrane separations), and synthesizing chemicals (e.g., the industrial electrosynthesis of chlorine, caustic soda, and organic chemicals) are attractive due to their inherent energy efficiency, the moderate temperature of operation, and the ease of measuring and controlling current and voltage. Electrochemical routes are often less polluting than chemical schemes because electrons perform the oxidation and reduction reactions rather than chemical reagents. The application of electrochemistry to real-world systems requires the same understanding of chemical engineering fundamentals (thermodynamics, reaction kinetics, and mass transport) as needed for traditional chemical or petrochemical processes, with two exceptions. There is the need to consider the ionic species in an electrolyte that may react heterogeneously at an electrode surface and to include electrical potential effects, which drive both electrochemical reactions and the mass transfer of ionic species in solution.

    The merging of electrochemistry and chemical engineering into electrochemical engineering formally began in the 1950s. Since its inception, this subfield of chemical engineering has grown considerably. This chapter provides an overview of electrochemical principles and applications. First, thermodynamics, kinetics, and mass transfer, as applied to electrochemical systems, are reviewed. Next, specific examples of electrochemical processes/applications are presented, including electroplating and etching, environmental clean-up technologies, corrosion of metals, electrochemical reactor design, organic and inorganic electrochemical syntheses, and batteries and fuel cells.

26.2 FUNDAMENTALS

All electrochemical cells consist of at least two electrodes, an anode where oxidation reactions occur and a cathode for reduction reactions, with a conductive electrolytic solution between the anode and cathode. To maintain an overall charge balance, the electrons produced at the anode are consumed at the cathode; an external wire connecting the electrodes provides the pathway for electron flow. The electrical circuit is completed by current flow through the electrolyte solution,

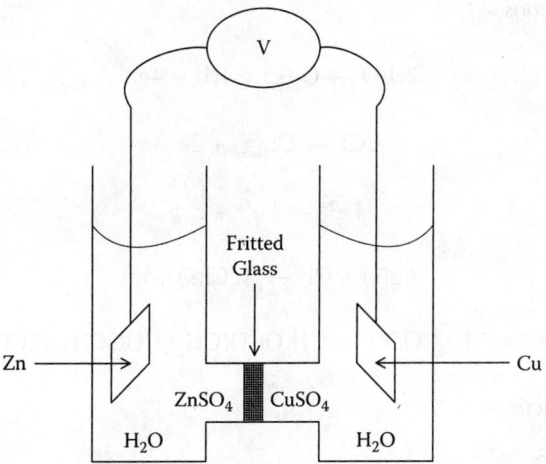

FIGURE 26.1 Electrochemical cell with zinc and copper electrodes.

where positively charged cations migrate to the cathode and negatively charged anions move to the anode. Figure 26.1 shows an electrochemical cell consisting of zinc and copper electrodes immersed in an aqueous electrolytic solution containing $ZnSO_4$ (which dissociates into Zn^{2+} and SO_4^{2-} ions) and $CuSO_4$ (Cu^{2+} and SO_4^{2-}). A porous glass frit or semipermeable membrane is often used to separate the two electrolytes; this separator minimizes the mixing of the two salt solutions and still permits ion movement. The electrochemical reactions at each electrode, which can be deduced from the composition of the electrodes and electrolyte solutions, are

$$Cu^{2+} + 2e^- = Cu(s) \qquad (26.1)$$

$$Zn^{2+} + 2e^- = Zn(s) \qquad (26.2)$$

As is standard practice, the above reactions are written as equilibrium reduction reactions and the total chemical change in an electrochemical cell is found by adding the individual anode and cathode reactions (the so-called half-cell reactions). Thus, the overall reaction for the Cu/Zn cell in Figure 26.1 is

$$Zn + Cu^{2+} = Cu + Zn^{2+} \qquad (26.3)$$

Equation (26.3) is written as an equilibrium reaction because we do not know a priori from the half-cell reaction chemistry if the overall reaction will proceed spontaneously in the forward or backward direction. Thermodynamic free energy arguments used below answer this question for any set of half-cell reactions. As written, the Gibbs free energy of Equation (26.3) is negative and the overall reaction will proceed to the right when the two electrodes are connected by an external wire, with spontaneous Cu^{2+} reduction to copper metal (at the cathode) and Zn oxidation to Zn^{2+} (at the anode). As the cathode solution chamber is depleted of Cu^{2+} and the Zn^{2+}, concentration increases in the anode compartment, SO_4^{2-} anions will move from the cathode to the anode compartment, and Zn^{2+} cations will migrate from the anode to the cathode chamber so that the electroneutrality of the electrolytic solutions will be maintained. To reverse the direction of the cell reaction (i.e., for Cu oxidation and Zn^{2+} reduction), electrical energy must be supplied to the cell by means of an external direct current power source.

In general, individual electrode reactions can involve a variety of differently charged or neutral inorganic or organic chemical compounds in either the solid, liquid, or gaseous state. Examples of electrochemical oxidation and reduction reactions are

Anode Oxidation Reactions

$$2H_2O \rightarrow O_2(g) + 4H^+ + 4e^- \tag{26.4}$$

$$2Cl^- \rightarrow Cl_2(g) + 2e^- \tag{26.5}$$

$$Fe^{2+} \rightarrow Fe^{3+} + e^- \tag{26.6}$$

$$Ag(s) + Cl^- \rightarrow AgCl(s) + e^- \tag{26.7}$$

$$2[CH_3OCO(CH_2)_4COO^-] \rightarrow CH_3OCO(CH_2)_8COOCH_3 + 2CO_2 + 2e^- \tag{26.8}$$

Cathode Reduction Reactions

$$2H_2O + 2e^- \rightarrow H_2(g) + 2OH^- \tag{26.9}$$

$$2H^+ + 2e^- \rightarrow H_2(g) \tag{26.10}$$

$$PbO_2(s) + SO_4^{2-} + 4H^+ + 2e^- \rightarrow PbSO_4 + 2H_2O \tag{26.11}$$

$$C_6H_{12}O_6 + 2H^+ + 2e^- \rightarrow C_6H_{14}O_6 \tag{26.12}$$

Examples of half-cell and overall reactions for industrially important electrochemical processes and commercial electrochemical devices are listed in Table 26.1.

Knowing that a given combination of anode and cathode half-cell reactions will proceed spontaneously does not ensure that the electrode reaction rates will be sufficiently high for practical applications. Reaction kinetics at the anode and cathode and mass transfer of reactants/products to/from the electrodes may play important roles in an electrochemical cell and may influence the choice of cell design and operating conditions. These important points will be addressed later in this chapter.

Since electrode reactions involve electron transfers, the measurement of current with units of amperes (where 1 ampere = 1 coulomb/sec) quantifies the rate of an oxidation or reduction reaction at an electrode. For the following generalized electrode reaction

$$\chi(Ox) + ne^- = \varepsilon\,(Red) \tag{26.13}$$

we see that χ/n moles of oxidized species (denoted as Ox) or ε/n moles of reduced species (abbreviated *Red*) react for each mole of electrons that is transferred into or out of the electrode.

Since the charge of an electron is 1.6×10^{-19} coulombs, one mole of electrons contains a total charge of 96,487 coulombs. This quantity, known as Faraday's constant, relates the current to the molar rate of product formation or reactant consumption in an electrochemical reaction. Molar rate expressions for the forward and backward reactions of Equation (26.13) are integrated to give Faraday's Law:

$$M_{oxid} = \frac{\int_0^{t^*} I(t)\,dt}{\left[\dfrac{n}{\chi}\right]F} \quad \text{and} \quad M_{red} = \frac{\int_0^{t^*} I(t)\,dt}{\left[\dfrac{n}{\varepsilon}\right]F} \tag{26.14}$$

TABLE 26.1
Half-Cell and Overall Cell Reactions for Electrochemical Processes

Process	Half-Cell Reactions	Overall Cell Reaction
Chlor-alkali membrane cell process (for the production of chlorine gas and concentrated NaOH)	Anode: $2Cl^- \rightarrow Cl_2(g) + 2e^-$ Cathode: $2H_2O + 2e^- \rightarrow H_2(g) + 2OH^-$	$2H_2O + 2Cl^- \rightarrow Cl_2 + H_2 + 2OH^-$
Electrolysis of water for the production of hydrogen and oxygen gases	Anode: $H_2O \rightarrow 1/2 O_2(g) + 2H^+ + 2e^-$ Cathode: $2H_2O + 2e^- \rightarrow H_2(g) + 2OH^-$	$2H_2O \rightarrow O_2(g) + 2H_2(g)$
Magnesium dioxide synthesis	Anode: $Mn^{2+} + 2H_2O \rightarrow MnO_2 + 4H^+ + 2e^-$ Cathode: $2H^+ + 2e^- \rightarrow H_2$	$Mn^{2+} + 2H_2O \rightarrow MnO_2 + 2H^+ + H_2$
Production of sodium chlorate	Anode: $6Cl^- \rightarrow 3Cl_2(aq) + 6e^-$ With subsequent chemical reactions: $3Cl_2(g) + 3H_2O \rightarrow 3HCl + 3HOCl$ $2HOCl + OCl^- \rightarrow ClO_3^- + 2HCl$ Cathode: $6H_2O + 6e^- \rightarrow 3H_2(g) + 6OH^-$	$NaCl + 3H_2O \rightarrow NaClO_3 + 3H_2(g)$
Copper electroplating	Anode: $Cu \rightarrow Cu^{2+} + 2e^-$ Cathode: $Cu^{2+} + 2e^- \rightarrow Cu$	$Cu(anode) \rightarrow Cu(cathode)$
Monsanto's acrylonitrile electrohydrodimerization	Anode: $H_2O \rightarrow 1/2 O_2(g) + 2H^+ + 2e^-$ Cathode: $2CH_2=CHCN + 2H_2O + 2e^- \rightarrow (CH_2CH_2CN)_2 + 2OH^-$	$2CH_2=CHCN + H_2O \rightarrow (CH_2CH_2CN)_2 + 1/2 O_2(g)$

where M denotes moles, I is the applied (measured) current in amperes, and F is Faraday's constant. The integral is simply the total charge passed during the electrolysis. For a constant current process, the numerator in Equation (26.14) simplifies to the product of I and t^*, hence,

$$\frac{M_{red}}{\varepsilon} = \frac{M_{ox}}{\chi} = \frac{It^*}{nF} \qquad (26.15)$$

When multiple oxidations or reductions occur simultaneously on an anode or cathode, the total moles of each product formed for a given time period is obtained by combining Faraday's Law with the total current and the current efficiency for each electrochemical reaction, where the current efficiency of the rth electrochemical reaction is defined as the fraction of the total current that is consumed by that reaction:

$$\text{Current efficiency for the rth reaction} = \frac{\int_0^{t^*} I_r(t)dt}{\int_0^{t^*} I_{total}(t)dt} \qquad (26.16)$$

where I_{total} is the total current, which is the summation of the partial currents associated with each electron transfer reaction at the anode or cathode.

In the next three sections of this chapter, fundamental theories and principles of thermodynamics, reaction kinetics, and mass transfer processes, as applied to electrochemical systems, are reviewed.

26.2.1 Thermodynamics

Thermodynamics is used in the analysis of electrochemical cells: (1) to predict which electrode reactions occur spontaneously in the anodic and cathodic directions if the two electrodes are in equilibrium with their respective adjacent solutions and are connected to one another via an external wire, and (2) to quantify chemical potentials and activity coefficients in nonideal electrolytic solutions.

26.2.1.1 Thermodynamics of Electrochemical Cells

In chemical systems, the state variable used to determine the spontaneity of a given reaction in the forward or reverse direction is the Gibbs free energy. For the electrochemical cell shown in Figure 26.1, we stipulate that the two electrodes are connected through a high-resistance voltmeter to essentially stop any current flow, in which case the overall cell reaction [given by Equation (26.3)] is at thermodynamic equilibrium. The Gibbs energy for this or any overall cell reaction at standard state conditions (where the temperature is 25°C, the pressure is one atmosphere, and the activities of reactants and products are unity) is given by

$$\Delta G^\circ = \Sigma \Delta G^\circ_{products} - \Sigma \Delta G^\circ_{reactants} \qquad (26.17)$$

where ΔG° denotes the standard state Gibbs energy. When the cell is not at standard state conditions, we can further write

$$\Delta G = \Delta G^o + RT \ln K = \Delta G^o + RT \ln \frac{\prod_i (a_{i,P})^{x_{i,P}}}{\prod_i (a_{i,R})^{y_{i,R}}} \qquad (26.18)$$

where K is the equilibrium constant of the reaction, R and T are the gas constant and absolute temperature, the $a_{i,P}$ and $a_{i,R}$ terms are the activities of the product and reactant species, respectively, $x_{i,P}$ and $y_{i,R}$ are the stoichiometric coefficients of the products and reactants in the reaction (e.g., ε and χ in Equation 26.13), and Π denotes the arithmetric product. When ΔG° or ΔG is < 0, the forward reaction as written is spontaneous, whereas the reverse reaction is spontaneous when the Gibbs energy change is >0. Electrochemists use the electric potential (E) rather than the Gibbs energy to assess the spontaneity of a given set of anode/cathode reactions. At constant temperature and pressure, the Gibbs energy and emf (electric potential) are related by

$$\Delta G^o = -nFE^o \quad \text{and} \quad \Delta G = -nFE \qquad (26.19)$$

where n is the number of electrons participating in the anode or cathode reactions ($n = 2$ for the Zn/Cu example above). Since the overall reaction in an electrochemical cell is given by the combination of anode and cathode "half-cell" reactions, the cell emf can also be decomposed into two individual electric potentials. An abbreviated listing of such standard half-cell reactions and their associate reduction potentials is in Table 26.2. Extensive listings of potentials are in electrochemical texts and chemical handbooks [see, for example, references (1–4)].

TABLE 26.2
Standard Half-Cell Reduction Potentials at 25°C

Reaction	$E°$ (V)
$Au = Au^{3+} + 3e^-$	+1.498
$O_2 + 4H^+ + 4e^- = H_2O$	+1.229
$Pt^{2+} + 2e^- = Pt$	+1.2
$Ag^+ + e^- = Ag$	+0.799
$Fe^{3+} + e^- = Fe^{2+}$	+0.771
$O_2 + 2H_2O + 4e^- = 4OH^-$	+0.401
$Cu^{2+} + 2e^- = Cu$	+0.337
$2H^+ + 2e^- = H_2$	0.000
$Pb^{2+} + 2e^- = Pb$	−0.126
$Ni^{2+} + 2e^- = Ni$	−0.250
$Cd^{2+} + 2e^- = Cd$	−0.403
$Fe^{2+} + 2e^- = Fe$	−0.440
$Zn^{2+} + 2e^- = Zn$	−0.763
$Al^{3+} + 3e^- = Al$	−1.662
$Mg^{2+} + 2e^- = Mg$	−2.363
$Na^+ + e^- = Na$	−2.714
$K^+ + e^- = K$	−2.925

Since only the electric potential *difference* between two electrodes can be measured, one half-cell reaction is chosen to fix the zero (reference) potential condition at all temperatures. Electrochemists have chosen the standard hydrogen electrode (SHE) as this reference, where the electrode reaction is

$$2H^+ + 2e^- = H_2 \qquad (26.20)$$

and where the proton and H_2 activities are unity and the temperature is 25°C. Half-cell potentials are measured in electrochemical cells against the standard hydrogen electrode. By convention, the emf of a cell is the difference between the right and left electrode half-cell reduction potentials,

$$E^o_{cell} = E^o(right) - E^o(left) \qquad (26.21)$$

which requires that the hydrogen electrode also be the left-side electrode in the cell. Similarly, for half-cell reactions where there is no listed standard reduction potential, one can combine the half-cell reaction of interest with the hydrogen evolution reaction [Equation (26.20)], compute the free energy difference between reactant and products, and then use Equation (26.19) to convert ΔG^o to E^o. For example, the equilibrium potential for the methanol/formic acid reaction is computed by writing the relevant electrode reaction as a reduction reaction:

$$HCOOH + 4H^+ + 4e^- = CH_3OH + H_2O \qquad (26.22)$$

This reaction is combined with the hydrogen evolution reaction, $H_2 = 2H^+ + 2e^-$, to give

$$HCOOH + 2H_2 = CH_3OH + H_2O \qquad (26.23)$$

The standard Gibbs energy (at 25°C) for this reaction is

$$\Delta G^o = \Sigma \Delta G^o(\text{products}) - \Sigma \Delta G^o(\text{reactants})$$

$$= [\Delta G^o(\text{CH}_3\text{OH}) + \Delta G^o(\text{H}_2\text{O})] - [\Delta G^o(\text{HCOOH}) + 2\Delta G^o(\text{H}_2)] \quad (26.24)$$

$$= [-39.9 \text{kcal/mol} - 56.7 \text{kcal/mol}] - [86.4 \text{ kcal/mol} + 0]$$

$$\Delta G^o = -10.2 \text{ kcal/mol} = -42.6 \text{ kJ/mol}$$

From Equation (26.19) (with $n = 2$), E^o is computed from this value of ΔG^o

$$E^o = \frac{-\Delta G^o}{nF} = \frac{42.6 \times 10^3}{2F} = +0.22 \text{ V vs. } SHE \quad (26.25)$$

Once the two half-cell reactions in an electrochemical cell have been assigned/identified and the standard half-cell potentials have been either ascertained from standard reduction potential tables or computed using Gibbs energy data, one can determine which reaction proceeds spontaneously in the anodic and cathodic directions. Using Equation (26.19) and our definition of zero potential, the half-cell reaction with the more negative standard reduction potential is the anode in an electrochemical cell and the reaction with the more positive half-cell potential proceeds in the cathodic direction. For the Cu/Zn cell given by Equation (26.3), we see from Table 26.2 that $(E^o)_{\text{Cu/Cu2+}} = +0.337$ V vs. SHE (an abbreviation for the standard hydrogen electrode) and $(E^o)_{\text{Zn/Zn2+}} = -0.763$ V vs. SHE, hence Zn will spontaneously oxidize and cupric ions will be reduced in the cell shown in Figure [26.1]). The standard reduction potentials in Table 26.2 are invariant in sign and independent of the position of the half-cell electrode reactions in an electrochemical cell. The cell potential (as defined by Equation 26.21), on the other hand, depends on the relative positions of the two electrodes (i.e., which electrode is on the left). For example, if the Zn electrode is the left-hand electrode in the cell, E_{cell} will be 0.34 V − (−0.76 V) = +1.1 V, whereas the cell potential will be −0.76 V − (0.34 V) = −1.1 V if the copper electrode were positioned on the left.

To determine the potential of a half-cell reaction when the reactants and products are not at unit activity, the Nernst equation is used. Consider again the general half-cell reaction given by Equation (26.13). When this reaction is combined with the standard hydrogen electrode reaction (Equation (26.20)), the overall cell reaction is

$$(n/2)\text{H}_2 + \chi \text{ (Oxid)} \rightarrow \varepsilon \text{ (Red)} + n\text{H}^+ \quad (26.26)$$

Combining Equations (26.18) and (26.19) for this reaction results in the following formula:

$$E^e = E^o + \frac{RT}{nF} \ln \frac{(a_{\text{Red}})^\varepsilon (a_{H+})^n}{(a_{\text{Oxid}})^\chi (a_{H2})^{n/2}} \quad (26.27)$$

where E^e is defined as the equilibrium half-cell potential with respect to the standard hydrogen electrode. For simplicity, the activities in Equation (26.27) are replaced with concentrations. Also, by definition of the standard hydrogen electrode, $a_{H+} = a_{H2} = 1$, in which case we arrive at the Nernst equation:

$$E^e = E^o + \frac{RT}{nF} \ln \frac{(C_{\text{Red}})^\varepsilon}{(C_{\text{Oxid}})^\chi} \quad (26.28)$$

Equations (26.20) and (26.27) are combined when two hydrogen electrodes are connected in an electrochemical cell. The left electrode is the standard hydrogen electrode with a half-cell potential of 0.0 V. The right electrode is a hydrogen electrode immersed in a solution at a particular pH (where $a_{H^+} \neq 1$). The resulting equilibrium cell potential, in terms of the right electrode compartment pH, is

$$E^e_{H_2/H^+} = -\frac{2.303RT}{F}\text{pH} \tag{26.29}$$

This equation is the basis for commonly used pH electrodes, where a pH variation in solution produces a measurable voltage signal.

26.2.1.2 Activity Coefficients and the Concept of an Electrochemical Potential

The electrochemical potential is used for charged ionic solutions in the same way as the chemical potential in uncharged systems. At constant temperature and pressure, any change in the system proceeds from a state of high to low Gibbs energy. If a system contains n_i components in a given phase, then we cannot specify its state without some elucidation of the composition of that phase. It follows that we can write

$$(dG)_{T,P} = \sum_i \left[\frac{\partial G}{\partial n_i}\right]_{T,P,n_j} dn_i \tag{26.30}$$

The term $[\partial G/\partial n_i]_{T,P,n_j}$ is the chemical potential (μ_i). The electrochemical potential, often denoted as $\bar{\mu}_i$, is the corresponding quantity for a system containing ionic charges, where the energy state of one ion depends on both chemical and electrical interactions with solvent molecules and other ions in the system. Guggenheim [5] separated the electrochemical potential of ion species i in phase α into chemical and electrical components:

$$(\bar{\mu}_i)^\alpha = (\mu_i)^\alpha + z_i F \Phi^\alpha = \mu^o_i + RT \ln a_i + z_i F \Phi^\alpha \tag{26.31}$$

where Φ^α is the electric potential in phase α, μ_i is the standard state chemical potential, z_i is the charge number of species i ($z_i < 0$ for anions and $z_i > 0$ for cations), and a_i is the activity of species i. Cation and anion activities in the above equation are usually defined in terms of molal (or molar) concentrations (denoted as m_i or C_i, respectively) and their corresponding activity coefficient (denotes as γ_i or f_i, respectively):

$$a_i = \gamma_i m_i = f_i C_i \tag{26.32}$$

The use of Equation (26.31) requires definitions of the standard state of a solid (a pure material at 25°C), an electrolyte solution (an infinitely dilute solution at 25°C and 1 atm pressure), and a gas (the ideal gas state at 1 atm pressure and 25°C). The combination of Equations (26.31) and (26.32) implies that the activity is independent of the electrical state of the system since there are separate a_i and Φ terms in the mathematical relationship for μ_i.

Unfortunately, the electrical potential in Equation (26.31) is not well defined and the individual ion activities in Equation (26.32) are not measurable because one cannot create an electrolytic solution containing only anions or cations (i.e., only neutral salt molecules, containing both cations and anions, can be dissolved in a solvent during an activity measurement experiment). To circumvent this problem, electrochemists have defined the electrochemical potential of neutral salt molecules in terms of mean activities and mean activity coefficients. For a salt $M_{\upsilon_+}X_{\upsilon_-}$ that dissociates completely into one type of cation species and one anion species when dissolved in a solvent, one can write the following equation:

$$M_{\upsilon_+}X_{\upsilon_-} \to (\upsilon+)M^+ + (\upsilon-)X \tag{26.33}$$

Relationships for the electrochemical potential of a neutral salt electrolyte ($\bar{\mu}_e$) are written as,

$$\begin{aligned}\bar{\mu}_e &= (\bar{\mu}_e)^o + \nu RT \ln a_\pm \\ &= (\bar{\mu}_e)^o + \nu RT \ln (\gamma_\pm m_e) \\ &= (\bar{\mu}_e)^o + \nu RT \ln (f_\pm C_e)\end{aligned} \tag{26.34}$$

where m_e and C_e are the molal and molar concentrations of the salt, respectively, $a_\pm$ is the mean activity of the salt, $\upsilon = (\upsilon+) + (\upsilon-)$, $\bar{\mu}_e = (\nu+)\bar{\mu}_+ + (\nu-)\bar{\mu}_-$, and $\gamma_\pm$ and $f_\pm$ are the mean molal and mean molar activity coefficients, which are defined in terms of the individual ion activity coefficients

$$(a_\pm)^\nu = (a_+)^{\nu+}(a_-)^{\nu-} \tag{26.35}$$

$$(\gamma_\pm)^\nu = (\gamma_+)^{\nu+}(\gamma_-)^{\nu-} \;;\; (f_\pm)^\nu = (f_+)^{\nu+}(f_-)^{\nu-} \tag{26.36}$$

In Equation (26.34), $\bar{\mu}_e$ is dependent on the electrical state of the system through the mean activity coefficient terms, which in turn are related to the individual ion activities.

Values of electrolyte activities, as measured by osmotic pressures, freezing point depression, and other experimental methods are in the literature (References 5 and 6, for example) or one can calculate activity coefficients based on models of molecular-level interactions between ions in electrolyte solutions. For illustrative purposes, mean molal activity coefficients for various salts at different aqueous molal (m_e) concentrations at 25°C are listed in Table 26.3 [7].

Numerous models predict the activity coefficient of individual ions in solution. The one by Debye and Hückel [8] considers only electrostatic (columbic) interactions between cations and anions in a dilute solution of a single, completely dissociated salt. It is assumed that ion–ion interactions (as opposed to other phenomena such as ion–solvent interactions, ion solvation effects, and variations in the solvent dielectric constant with salt concentration) cause the ion activity coefficients to deviate from 1.0. From a practical point, only the Debye-Hückel activity coefficient relationship is needed, along with some knowledge of the theory's shortcomings, which restrict its application. For a dilute electrolytic solution containing a binary salt (i.e., a salt with one type each of cation and anion species), the ion activity coefficient from Debye-Hückel theory is given by

$$\ln \gamma_i = \ln f_i = \frac{z_i^2 A \sqrt{I_S}}{1 + B\alpha \sqrt{I_S}} \tag{26.37}$$

where α is the mean diameter of the ion (nm) and

TABLE 26.3
Mean Molal Activity Coefficients ($\gamma_\pm$) of Aqueous Salt Solutions at 25°C

Molality (m_e)	CuSO$_4$	HCl	KCl	NaCl	NaOH
0.1	0.150	0.796	0.770	0.778	0.764
0.2	0.104	0.767	0.718	0.735	0.725
0.3	0.083	0.756	0.688	0.710	0.706
0.4	0.070	0.755	0.666	0.693	0.695
0.5	0.062	0.757	0.649	0.681	0.688
0.8	0.048	0.783	0.618	0.662	0.677
1.0	0.042	0.809	0.604	0.657	0.677
1.6		0.916	0.580	0.657	0.690
2.0		1.01	0.573	0.668	0.707
3.0		1.316	0.569	0.714	0.782
4.0		1.762	0.577	0.783	0.901
5.0		2.38		0.874	1.074
7.0		4.37			1.60
8.0		5.90			2.00

$$B = \frac{F\sqrt{\rho_o}}{\sqrt{\varepsilon RT/2}} \quad \text{and} \quad A = \frac{\sqrt{2\rho_o}\, F^2 e}{8\pi(\varepsilon RT)^{3/2}} \tag{26.38}$$

In Equations (26.37) and (26.38), α is the mean hard sphere diameter of the ion, ρ_o is the pure solvent density, ε is the permittivity of the electrolyte solution (the permittivity is equal to the product of the dielectric constant of the electrolyte solution and the permittivity of a vacuum, 8.8542 × 10^{-14} C/V-cm), e is the charge of an electron (1.60210 × 10^{-19} C), and I_S is the molal ionic strength of the electrolyte solution, defined as

$$I_S = \frac{1}{2}\sum_i z_i^2 m_i = \frac{\frac{1}{2}\sum_i z_i^2 C_i}{C_o M_o} \tag{26.39}$$

where C_o and M_o are the concentration and molecular weight of the solvent, respectively. The mean molal or mean molar activity coefficient is obtained by combining Equations (26.36) and (26.37):

$$\ln \gamma_\pm = \ln f_\pm = \frac{z_+ z_- A\sqrt{I_S}}{1 + B\alpha\sqrt{I_S}} \tag{26.40}$$

Values of the parameters A and B for aqueous solutions of any binary salt at four different temperatures are given in Table 26.4 [4].

For the asymptotic limiting condition where the ionic strength approaches zero, the denominators in Equations (26.37) and (26.40) reduce to unity and the following activity coefficient equations are obtained:

$$\ln \gamma_\pm = z_+ z_- A\sqrt{I_S} \tag{26.41}$$

TABLE 26.4
Debye-Hückel Activity Coefficient Parameters for Aqueous Solutions of Any Binary Salt

Temperature (°C)	0	25	50	75
A $(kg/mol)^{1/2}$	1.1324	1.1762	1.2300	1.2949
B $(kg/mol)^{1/2}/nm$	3.248	3.287	3.326	3.368

Data from [4].

$$\ln \gamma_i = -z_i^2 A \sqrt{I_S} \tag{26.42}$$

Equation (26.41) predicts to within approximately 10% mean molal activity coefficients for salt concentrations up to 0.1 molal. The more accurate form of the activity coefficient equation [Equation (26.40)] allows the model to be extended to salt concentrations up to 0.5 molal. To expand the applicability of the Debye-Hückel theory to higher concentrations, additional terms are added to Equation (26.40), such as [4]

$$\ln \gamma_\pm = \frac{z_+ z_- A\sqrt{I_S}}{1+B\alpha\sqrt{I_S}} + \frac{4(v_+)(v_-)\Theta}{(v_+)+(v_-)} m \tag{26.43}$$

where values of the parameter Θ for various aqueous electrolytes at 25°C [9] are listed in Table 26.5. For additional activity correlations and for a review of activity coefficients in multicomponent solutions, the reader is directed to [10, 11].

TABLE 26.5
Θ Parameter for Activity Coefficients in Aqueous Electrolytes at 25°C

HCl	0.27
$HClO_4$	0.30
LiCl	0.22
NaOH	0.06
NaCl	0.15
$NaNO_3$	0.04
NaF	0.07
KOH	0.13
KCL	0.10
KBr	0.11
$KClO_3$	−0.04
$AgNO_3$	−0.14
RbCl	0.06
$RbNO_3$	−0.14
CsOH	0.35
CsCl	0.00
CsI	−0.01
$CsNO_3$	−0.15

Data from [9].

26.2.2 Kinetics of Electrode Reactions

Electrochemical reactions at an electrode surface differ from normal heterogeneous chemical reactions in that they involve the participation of one or more electrons that are either added to (reduction) or removed from (oxidation) the reactant species. The explicit inclusion of electrons as reactants or products means that the reaction rate depends on the electric potential. Electron transfer processes occur within a small portion of the double layer immediately adjacent to the electrode surface (10 to 50 nm in thickness) where solution-phase electroneutrality does not hold and where very strong electric fields (on the order of 10^6 V/cm) exist during a charge transfer reaction. We begin the analysis of electrochemical kinetics by defining a generic electrode reaction:

$$Red = Ox^{n+} + ne^- \tag{26.44}$$

Several molecular-level processes occur within the double layer when the above reaction proceeds in either the forward or reverse direction. They include the transfer of the *Red* species through the double layer, the adsorption of *Red* onto the electrode, the removal of an electron from *Red*, the desorption of the *Ox* product from the electrode surface, and the reorientation of solvent molecules around *Ox* since its sign has changed from that of *Red*. These processes require an additional input of energy so that the anode and cathode reactions can proceed at an acceptable rate. A single electric voltage-dependent driving force (often called the surface or activation overpotential) is assigned to the sum total of all energy-related mechanistic events in the double layer. This voltage-dependent kinetic driving force distinguishes electrochemical reactions from heterogeneous chemical reactions. Thus, the electrode reaction rate can be altered by varying the magnitude and sign of the surface overpotential, and the analysis of reaction kinetics at an electrode requires the inclusion of reaction rate constants that are dependent on the electric potential.

For many, but not all, electrochemical reactions, the Butler-Volmer rate expression describes the interrelationship between the reaction rate (given in terms of the current density i, with units of A/cm^2) and the electrode potential [12]:

$$i = i_o \left[e^{(1-\beta)nF\eta_s/RT} e^{-\beta nF\eta_s/RT} \right] \tag{26.45}$$

where η_s is the surface or activation overpotential, defined as the deviation of the electrode potential away from its equilibrium potential, i.e., $\eta_s = \Phi - \Phi^e$. The first term on the right side of the Butler-Volmer equation denotes the anodic reaction rate, and the second term gives the rate of the cathodic reaction, i.e., anodic currents (forward reaction dominating) are positive, whereas cathodic currents are negative. The net reaction rate is thus defined as the difference between the anodic (forward) and cathodic (backward) rates, as is done in the kinetic analysis of strictly chemical reactions. We also see from Equation (26.45) that a positive value of η_s accelerates the reaction in the forward (oxidation) direction and a negative value of η_s promotes the reduction reaction.

There are two kinetic parameters in Equation (26.45): the symmetry factor (β) that is defined as the fraction of the potential change in the double layer that accelerates the backward (cathodic) reaction and the exchange current density (i_o). Normally, β is set equal to 0.5. Physically, the exchange current density is a measure of the inherent reactivity of the electrode for a given electrochemical reaction. At equilibrium, the net current density at the electrode is zero, the electrode potential is equal to the equilibrium potential for the electrode reaction as defined by the Nernst equation Equation (26.28), and the forward (anodic) and backward (cathodic) reaction rates are equal. The magnitude of this equilibrium reaction rate is the exchange current density. Thus, a large value of the exchange current density is characteristic of an electrode reaction that is said to be reversible and where a small overpotential (a small departure in voltage away from the equilibrium potential) produces a large current density. Electrode reactions with a low-exchange current density, on the other hand, are slow and are often referred to as irreversible because large overpotentials are required for appreciable current densities. Mathematically we can write an equation for i_o in terms of the chemical kinetic rate constants for the reaction in the anodic and cathodic directions (k_a and k_c, respectively) and the concentrations of the reduced and oxidized species at the electrode surface (C_{Red} and C_{Ox}, respectively) [4, 12]:

$$i_o = n F k_c^{(1-\beta)} k_a^{\beta} C_{Ox}^{(1-\beta)} C_{Red}^{\beta} \qquad (26.46)$$

Figures 26.2 and 26.3 show typical current density-overpotential plots where i varies exponentially with η_S, in accordance with the Butler-Volmer equation. In Figure 26.2, the effect of varying β on i-η_S curves is shown (as β decreases, i increases at a given value of η_S). The increase in current density at a given η_S for increasing values of i_o is shown in Figure 26.3. From these two figures it can be concluded that electrochemical reactions that follow Butler-Volmer kinetics will be facile when the kinetic parameter β is small and the value of i_o is large.

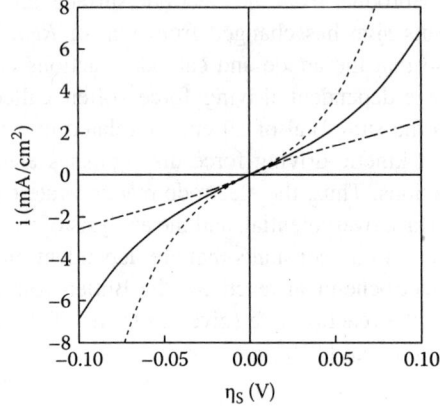

FIGURE 26.2 Current density-surface overpotential plots showing the effect of the transfer coefficient for a redox reaction (Equation (26.45)) with $n = 1$, $T = 298K$, and $i_o = 1.0$ mA/cm². - - - - $\beta = 0.74$; ——— $\beta = 0.5$; — — — — $\beta = 0.25$.

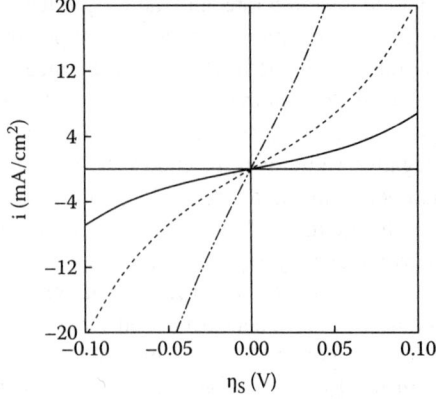

FIGURE 26.3 Current density-surface overpotential plots showing the effect of the exchange current density for a redox reaction [Equation (26.45) with $n = 1$, $T = 298K$, and $\beta = 0.5$. ——— $i_o = 1.0$ mA/cm²; - - - - $i_o = 3.0$ mA/cm²; ——— $i_o = 5.0$ mA/cm²].

The Butler-Volmer rate expression can be simplified for small and large surface overpotentials (i.e., for small or large deviations in the electrode potential away from the equilibrium potential). For large positive or large negative surface overpotentials (where $|\eta_s|$ is >50 to 100 mV), one of the exponential terms in the Butler-Volmer equation will dominate over the other. The resulting relationship can be rewritten in the general form

$$\eta_s = a + b \ln i \tag{26.47}$$

Equation (26.47) is known as the Tafel equation and b is defined as the Tafel slope. When $\eta_s \gg 0$ (large anodic surface overpotentials), the Tafel equation takes the form

$$\eta_s = \frac{2.303RT}{(1-\beta)nF} \log i_o + \frac{2.303RT}{(1-\beta)nF} \log i \tag{26.48}$$

and the anodic Tafel slope is given by

$$b_a = \frac{2.303RT}{(1-\beta)nF} \tag{26.49}$$

Similarly, the form of the cathodic Tafel equation and the expression for the cathodic Tafel slope are

$$\eta_s = \frac{2.303RT}{\beta nf} \log i_o + b_c \log i \tag{26.50}$$

$$b_c = \frac{-2.303RT}{\beta nF} \tag{26.51}$$

According to Equations (26.48) and (26.50), the extrapolated linear Tafel line will intercept the x-axis of a η_s vs. $\log i$ plot at $i = i_o$ (i.e., when $\eta_s = 0$ in Equations (26.48) and (26.50), $i = i_o$).

The Tafel relationship is a powerful tool for determining Butler-Volmer kinetic parameters. The experimental apparatus needed to collect current density-overpotential data for Tafel plots is diagramed in Figure 26.4. A three-electrode apparatus is used, with a working electrode (the anode or cathode electrode at which the measurements are made), a counter electrode (to complete the cell circuit), and a reference electrode (an electrode of known and near-constant potential). The reference electrode measures overpotential changes at the working electrode (since the overall cell potential includes the overpotential at the working and counter electrodes). Current voltage data can be collected either galvanostatically (by setting the current using a DC power supply and measuring the working electrode voltage relative to the voltage of the reference electrode) or potentiostatically (by setting the working electrode potential relative to that of the reference electrode and measuring the resulting current). A representative η_s versus $\log i$ plot is shown in Figure 26.5. At high anodic or cathodic overpotentials the plots are linear, in accordance with the Tafel approximation, and the intercept of the anodic and cathodic Tafel lines on the $\eta_s = 0$ axis gives the value of $\log i_o$. Additional details of the experimental setup and procedures for measuring kinetic parameters are found in many electrochemistry texts [2, 3, 12, 13].

Exchange current densities and cathodic Tafel slopes for the hydrogen evolution reaction ($2H^+ + 2e^- = H_2$) at different electrodes and with different electrolytes are listed in Table 26.6 [3, 12–14].

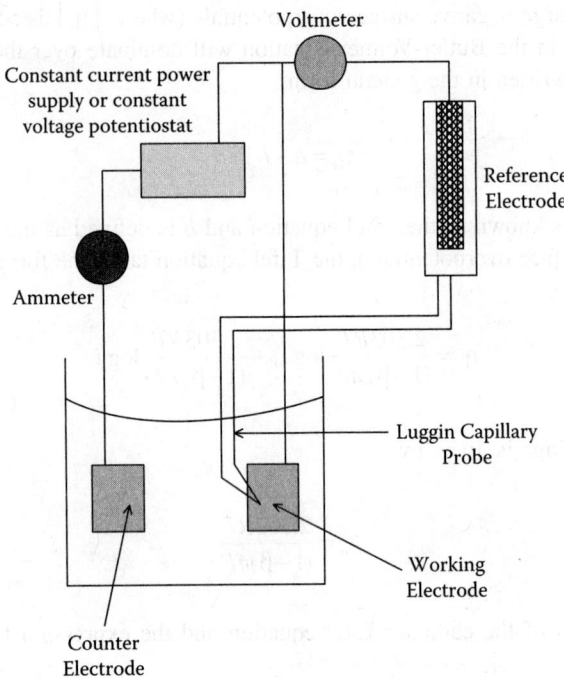

FIGURE 26.4 Three-electrode electrochemical cell for the collection of current-voltage data at the working electrode.

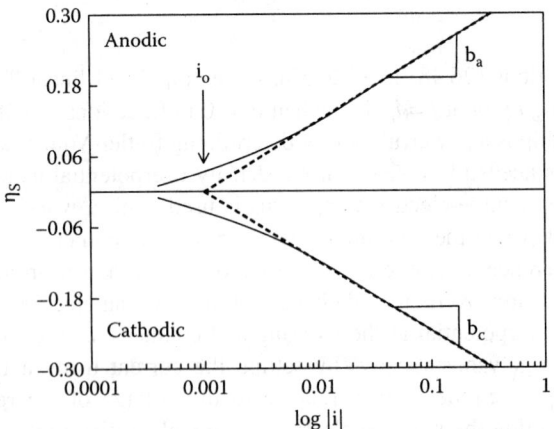

FIGURE 26.5 Surface overpotential vs. current density plots for a kinetically controlled redox reaction. Overpotentials less than zero refer to the reaction proceeding in the cathodic direction. The anodic and cathodic Tafel lines are drawn for high overpotentials.

TABLE 26.6
Exchange Current Densities and Cathodic Tafel Slopes for the Hydrogen Evolution Reaction at 25°C

Metal	Electrolyte	i_o (A/cm^2)	b_c (V/decade)
Pt	1N H$_2$SO$_4$	10^{-3}	0.03
Ni	1N H$_2$SO$_4$	10^{-5}	0.11
Ta	1N HCl	10^{-5}	0.10
Ag	0.1N HCl	$\approx 10^{-6}$	0.09–0.12
Cu	0.1N HCl	$\approx 10^{-7}$	0.115
Fe	0.5N HCl	$\approx 10^{-7}$	0.13
	0.1N NaOH	10^{-6}	0.12
Pb	0.1N HCl	$\approx 10^{-13}$	0.12
Hg	0.1N HCl	$\approx 10^{-12}$	0.115
	0.1N KOH	$\approx 10^{-15}$	0.10
Au	0.1N NaOH	$\approx 10^{-6}$–10^{-8}	0.119
W	0.5N HCl	$\approx 10^{-6}$–10^{-7}	0.070–0.112
Mo	0.1N HCl	$\approx 10^{-6}$–10^{-7}	0.078–0.104
	0.1N NaOH	10^{-3}	0.087–0.116

Whereas the Tafel slope (and β) is nearly constant, i_o differs by many orders of magnitude, depending on the concentration of H$^+$ and the catalytic activity of the electrode.

The Butler-Volmer equation can also be simplified for small values of η_s ($|\eta_s| \leq 10$ mV) since terms of the form e^x can be replaced by the function $(1 + x)$ for x small. The resulting linear current density equation is

$$i = \frac{i_o n F}{RT} \eta_s \tag{26.52}$$

From Equation (26.52), we see that the exchange current density can be obtained from the slope of a linear plot of η_s vs. i (for η_s small).

The Butler-Volmer equation is an empirical kinetic rate expression. It can be used for modeling and analysis purposes when the applied current density varies exponentially with the surface overpotential and when the two kinetic parameters, the exchange current density and the symmetry factor, are known. Readers interested in learning more about electrode kinetics, the Butler-Volmer equation, and mechanisms of electrode reactions should consult [2–4, 12]. Many electrode reactions of technical importance in electrochemistry, however, do not follow Butler-Volmer kinetics; chief among these is the anodic dissolution of some metals where a protective metal oxide film forms on the electrode surface at high values of η resulting in a drop in the current density. The formation of such passive films is an important phenomena during metal corrosion and electromachining (both of these topics will be discussed later).

26.2.3 Mass Transport

In Section 26.2, the rate of an electrochemical reaction was shown to be dependent on the electrode potential, the intrinsic rate constants for the forward and backward reactions at the electrode, and the concentrations of oxidized and reduced species at the electrode surface. When the transport of reactants and/or products to and/or from the electrode surface is the rate-controlling step, C_{Ox} and C_{Red} in Equation (26.46) will differ from those in the bulk solution. For uncharged species in solution, transport is governed by the usual processes of diffusion (where concentration or activity gradients are the driving force for transport) and convection, which is dependent on bulk fluid motion. Because of the electric field between the anode and cathode, there is an additional mode of transport for charged species in solution, called migration, where the driving force is the gradient in the electric potential.

26.2.3.1 Transport Equations

The analysis of mass transfer in electrochemical cells requires the use of equations that describe the condition of electroneutrality (which applies for the entire electrolyte outside the double layer at an electrode), species fluxes, mass conservation, current density, and fluid hydrodynamics. Often, mass transport events are rate limiting, as compared to kinetics processes at the electrode surface, in which case the overall electrode reaction rate is solely dependent on species mass transfer (e.g., during high-rate electroplating of some metals and for those electrochemical reactions where the concentration of reactant in solution is low).

Isothermal mass transport in electrochemical systems is described by the following relationships [4], which are strictly applicable for infinitely dilute electrolyte solutions:

1. The equation for electroneutrality in the bulk electrolytic solution,

$$\sum_i z_i C_i = 0 \qquad (26.53)$$

   where z_i is the charge number of species i ($z < 0$ for anions and $z > 0$ for cations) and C is concentration with units of mol/cm^3.

2. The molar flux (N_i) equation for charged ions in solution, with diffusion, migration, and convection terms,

$$N_i = -D_i \nabla C_i - z_i u_i F C_i \nabla \Phi + C_i v \qquad (26.54)$$

   where D_i is the diffusivity, u_i is the electric mobility of ion species i in solution and v is the fluid velocity.

3. A material balance equation for species i in the bulk electrolyte,

$$\frac{\partial C_i}{\partial t} = -\nabla \cdot N_i + R_i \qquad (26.55)$$

   where R_i represents the rate of homogeneous chemical reaction in solution that produces or consumes species i.

4. The current density equation,

$$i = F \sum_i z_i N_i \qquad (26.56)$$

To apply the above equations to electrochemical mass transfer, the solution velocity [v in Equation (26.54)] must be known, which requires the solution of a separate set of fluid mechanics equations. Also, the transport parameters, D_i and u_i, must be specified, as will be discussed in some detail below.

Manipulation of Equations (26.53)–(26.56) leads to a number of important mass transfer relationships. When the molar flux, current density, and electroneutrality equations are combined and simplified for the case where there are no concentration gradients in solution (i.e., when $\nabla C = 0$), the following Ohm's law expression for an electrolyte solution is obtained [4]:

$$i = -\kappa \nabla \Phi \tag{26.57}$$

where κ is the ionic conductivity of the electrolytic solution:

$$\kappa = F^2 \sum_i z_i^2 u_i C_i \tag{26.58}$$

Assuming no concentration gradients nor any homogeneous reactions in solution, the material balance equation (with $R_i = 0$) can be multiplied by $z_i F$, summed over all ionic species in solution, and simplified by substitution of the electroneutrality equation. The resulting expression describes charge conservation in solution:

$$\nabla \cdot i = 0 \tag{26.59}$$

Substitution of Ohm's law into Equation (26.59) (with κ constant since there are no concentration gradients) yields Laplace's equation, which is the governing relationship for the electric potential distribution in an electrolyte solution of uniform composition:

$$\nabla^2 \Phi = 0 \tag{26.60}$$

For the case of a binary electrolyte (i.e., a single salt that dissociates in solution into one cation and one anion species), we can rewrite the molar flux equations for positive and negative ions in terms of a salt concentration gradient diffusion term, a migration term explicit in the current density (as opposed to the $\nabla \Phi$ driving force term in Equation (26.54)), and a bulk convection term:

$$N_+ = -\nu_+ D \nabla C + \frac{t_+ i}{z_+ F} + \nu_+ C_\nu \tag{26.61}$$

$$N_- = -\nu_- D \nabla C + \frac{t_- i}{z_- F} + \nu_- C_\nu \tag{26.62}$$

where C is the concentration of the electrolyte salt, which is related to the individual ion concentrations and the number of cations and anions produced by the complete dissociation of one salt molecule:

$$C = \frac{C_+}{\nu_+} = \frac{C_-}{\nu_-} \tag{26.63}$$

where ν_+ and ν_- are the number of cations and anions, respectively, in a neutral salt molecule. In Equations (26.61) and (26.62), there are two transport parameters, the transference number t_i and the binary electrolyte diffusivity D, which are defined mathematically in terms of u_+, u_-, D_+, and D_-,

$$t_+ = 1 - t_- = \frac{z_+ u_+}{z_+ u_+ - z_- u_-} \tag{26.64}$$

$$D = \frac{z_+ u_+ D_+ - z_- u_- D_-}{z_+ u_+ - z_- u_-} \tag{26.65}$$

The convective diffusion equation for a binary electrolyte dissolved in an incompressible fluid is derived by combining the anion and cation flux relationships (using Equation [26.54]) with the corresponding material balance equations [Equation (26.55)], with $R_i = 0$ and $\nabla \cdot v = 0$) to give

$$\frac{\partial C}{\partial t} + v \cdot \nabla C = D \nabla^2 C \tag{26.66}$$

where D is the average ion diffusivity, as defined by Equation (26.65). Curiously, this equation for the movement of charged species in solution has no term explicit in the electric potential, although there is an electric potential gradient driving force acting on the ionic species when current is flowing between the anode and cathode. Thus, for the binary electrolyte case, one does not need a full description of the electric potential distribution between the anode and cathode to determine the concentration distribution of ion species in solution. This result greatly simplifies the mathematical analysis of transport in electrochemical systems. Also, the absence of an electric potential term in Equation (26.66) means that the convective transport equation for a binary electrolyte is of a similar form as that used in heat transfer and nonelectrolytic mass transfer systems. Consequently, many solutions to nonelectrochemical convective transport problems in the literature can be utilized by electrochemical engineers, as shown below.

To increase the electrolyte conductivity, an additional ionic component that does not participate in the electrochemical reactions is often added to a solution. This nonreactive component is called a supporting electrolyte or indifferent electrolyte. In the presence of a supporting electrolyte, there is a lowering of the electric field in solution, due to the electrolyte's high conductivity. Transport of the minor ionic species in solution is due primarily to diffusion and convection, in accordance with Equation (26.54) with $\nabla \Phi = 0$. Also, in the presence of a supporting electrolyte, the convective diffusion equation for a minor component in solution is written as

$$\frac{\partial C_i}{\partial t} + v \cdot \nabla C_i = D_i \nabla^2 C_i \tag{26.67}$$

This equation is similar in form to that obtained for the case of no supporting electrolyte (Equation (26.66)) except for the use of an individual tion concentration and diffusion coefficient.

To illustrate the use of the transport equations, the following problem is posed. An electrochemical cell containing vertical flat sheets of copper as the anode and cathode is operated with an aqueous $CuSO_4$ electrolyte. The copper plates are connected to a DC power supply so that oxidation and reduction reactions proceed at the anode and cathode ($Cu^{2+} + 2e^- \rightarrow Cu$ at the cathode; $Cu \rightarrow Cu^{2+} + 2e^-$ at the anode). For the case when there is no forced or natural convection during current flow, we derive a simple expression between the constant applied current density and the steady-state cupric ion concentration profile. The cation flux and current density equations for the flat plate electrode/no convection cell are

$$N_+ = -v_+ D \frac{dC}{dy} + \frac{t_+ i}{z_+ F} \tag{26.68a}$$

$$i = z_+ F N_+ \tag{26.68b}$$

where y is the perpendicular distance from the electrode, as measured from one electrode surface. Since only cupric ions react at the two electrodes, the steady-state flux of SO_4^- is zero and there is only one molar flux term in the current density equation (Equation (26.68b)). Equating N_+ in the above two equations and solving for current density yields

$$i = \frac{-z_+ \nu_+ DF \frac{dC}{dy}}{(1-t_+)} \tag{26.69}$$

We see from the above equation that the concentration profile will be a linear function of y (i.e., dC/dy = constant) for the entire electrolyte region between the anode and cathode when the current density is constant, fluid convection is nonexistent (diffusion and migration are the only modes of transport), and the diffusion coefficient and transference number are constant and independent of concentration.

If the flat plate cell were to contain an aqueous solution of $CuSO_4$ with an excess of H_2SO_4 supporting electrolyte (e.g., 0.05 M $CuSO_4$ and 1.0 M H_2SO_4), the current density/concentration gradient relationship analogous to Equation (26.69) would be derived by combining the following equations:

$$N_+ = -D_+ \frac{dC_+}{dy} \tag{26.70}$$

$$i = z_+ F N_+ \tag{26.71}$$

to give

$$i = -D_+ z_+ F \frac{dC_+}{dy} = -D_+ z_+ \nu_+ F \frac{dC}{dy} \tag{26.72}$$

Equation (26.72) differs from Equation (26.69) in two ways: (1) the absence of a transference number term (i.e., in the presence of a supporting electrolyte, the transference number of cupric cations is nearly zero and the $1 - t_+$ term in Equation (26.69) is equal to unity), and (2) the difference in the diffusion coefficients (with a supporting electrolyte, cupric ions diffuse independently of sulfate anions and the cation diffusivity, D_+, is used in Equation (26.70)).

26.2.3.2 Transport of Electro-Active Species in the Mass Transfer Boundary Layer

We saw above that the concentration gradient at an electrode will be linear with respect to the spatial coordinate perpendicular to the electrode surface if the anode/cathode cell were operated at a constant current density and if the fluid velocity were zero. In actuality, there will always be some bulk liquid electrolyte stirring during current flow, either an imposed forced convection velocity or a natural convection fluid motion due to changes in the reacting species concentration and fluid density near the electrode surface. In electrochemical systems with fluid flow, the mass transfer and hydrodynamic fluid flow equations are coupled and the solution of the relevant differential equations is often a formidable task, involving complex mathematical and/or numerical solution techniques. The concept of a stagnant diffusion layer or Nernst layer parallel and adjacent to the electrode surface is often used to simplify the analysis of convective mass transfer in

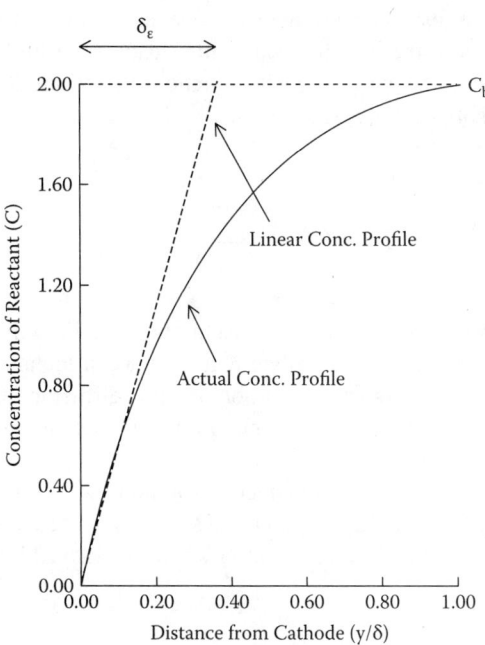

FIGURE 26.6 Concentration profile of reacting reacting species near an electrode surface under mass transfer controlled conditions.

electrochemical systems. One assumes that there is a small region very close to the electrode surface where all fluid velocities are zero and transport to and from the electrode surface occurs by diffusion (and migration) only, hence the terminology stagnant diffusion layer. Within this motionless fluid, the concentration profiles of the electro-active species are linear, as was the case in the example given above. The actual concentration profile in the mass transfer boundary layer is nonlinear, due to the presence of finite fluid velocities near the electrode (the velocity is only zero at the electrode surface). A typical nonlinear concentration profile and the region over which the concentration differs from that in the bulk fluid, which defines the actual mass transfer boundary-layer thickness (denoted as δ), are shown in Figure 26.6. The effective mass transfer boundary layer is also shown in this figure, where the linear concentration profile is tangent to the actual concentration variation at the electrode surface.

Equation (26.69) (or Equation (26.70) for the case of a supporting electrolyte) was originally derived under the assumption of no convective velocities. These same relationships can be utilized within the effective mass transfer boundary layer when fluid stirring exists. A current density/concentration equation with fluid stirring can now be generated starting with Equation (26.69) or Equation (26.70). The dC/dy term in Equation (26.69) is replaced by $\Delta C/\delta_e$ (where δ_e is the effective mass transfer boundary-layer thickness):

$$i = \frac{-z_+ \nu_+ DF \left.\dfrac{dC}{dy}\right|_{y=0}}{(1-t_+)} = \frac{-z_+ \nu_+ DF \left(\dfrac{C_\infty - C_o}{\delta_e}\right)}{(1-t_+)} \tag{26.73}$$

where $y = 0$ is the location of the electrode surface, and C_∞ and C_o are the concentrations of the reacting species in the bulk and at the electrode surface, respectively. Similar arguments can be used to generate a linear concentration profile current density equation when an excess supporting electrolyte is present (starting with Equation (26.70)).

Changes in the bulk electrolyte velocity far from the electrode and/or variations in the electrolyte flow pattern (as might occur if the cell geometry were altered) can be accounted for in Equation (26.73) by a change in the value of δ_e, as is well known from boundary-layer theory in nonelectrochemical systems. For given values of C_∞ and C_o, an increase in bulk electrolyte velocity will decrease δ_e, resulting in an increase in the current density. Frequently, δ_e is related to the mass transfer coefficient (k_d), a common mass transfer parameter,

$$k_d = \frac{D}{\delta_e} \tag{26.74}$$

so that Equation (26.73) can be rewritten as

$$i = \frac{-z_+ \nu_+ F k_d (C_\infty - C_o)}{(1 - t_+)} \tag{26.75}$$

As the current density is increased during a metal deposition process (or any electrochemical reaction where the electroactive species in the bulk electrolyte is consumed at the electrode surface) with a constant value of δ_e (i.e., for a given forced convection velocity), the surface concentration (C_o) decreases since C_∞ and δ_e are constant and independent of the electrode reaction rate. The linear concentration gradient in Equation (26.73) or (26.75) has a maximum value when $C_o = 0$. The current density corresponding to this surface concentration condition is known as the limiting current density (i_L):

$$i_L = \frac{-z_+ F D C_\infty}{\delta_e (1 - t_+)} = \frac{-z_+ F k_d C_\infty}{(1 - t_+)} \tag{26.76}$$

When there is a supporting electrolyte present, $t_+ \approx 0$ and D is replaced by D_+ in the above equation. If there are multiple ions in the electrolyte that can react at an electrode, then each species will have its own limiting current density, which will be dependent on its charge number (z_i) and bulk concentration. Also, once i_L is known, the surface concentration of reacting species at any current density less than i_L is given by

$$C_o = C_\infty \left(1 - \frac{i}{i_L}\right) \tag{26.77}$$

The limiting current density is an important parameter for the analysis of mass transfer controlled electrochemical processes and represents the maximum possible reaction rate for a given bulk reactant concentration and fluid flow pattern. During anodic metal dissolution, a mass transfer limiting current does not exist because the surface concentration of the dissolving ion (e.g., Cu^{2+} when the anode is composed of copper metal) increases with increasing current density, eventually leading to salt precipitation that blocks the electrode surface.

A typical current-voltage curve showing the limiting current density condition is shown in Figure 26.7. These data were collected at a vertical flat sheet copper cathode during Cu^{2+} reduction with assisting (upward direction) natural and forced convection velocities [15]. The limiting current density is identified by the horizontal plateau in this figure. When the surface concentration of reactant reaches zero, the electrode potential shifts (increases) until there is sufficient electrode energy for a secondary reaction to begin. During copper deposition from a $CuSO_4/H_2SO_4$, for example, the second reaction at high-cathodic overpotentials is hydrogen gas evolution (i.e., $2H^+$

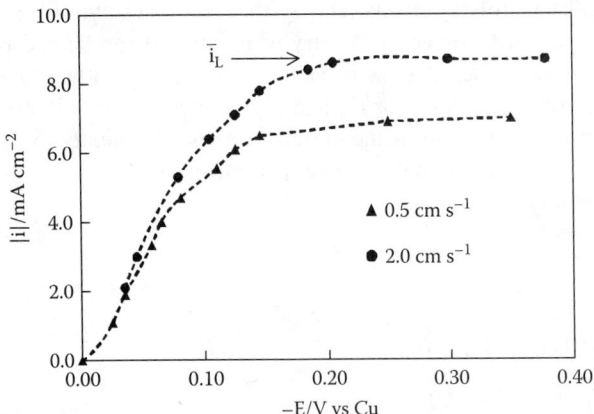

FIGURE 26.7 Current density (i) – voltage (E) plots showing the limiting current density during copper plating at a vertical flat plate cathode with assisting (upward) natural and force convection stirring (the forced convection fluid velocity is either 0.5 or 2.0 cm/sec). Data from [15]. Copyright Elsevier (1993).

+ 2e$^-$ → H$_2$). As one would expect, the data in Figure 26.7 show that i_L increases (δ_e decreases) as the forced convection velocity is increased from 0.5 to 2.0 cm/sec.

There may be situations where there is a significant amount of hydrogen gas evolution during metal deposition. In this case, i_L may not be easily observed in a current density-voltage plot, as is the case in Figure 26.8 [16] for copper deposition at a rotating disk electrode from a CuSO$_4$ solution containing H$_2$SO$_4$ as the supporting electrolyte and Gleam-PC, a polyethylene glycol–based copper plating additive manufactured by LeaRonal, Inc. In order to determine the limiting current density, separate current density-voltage measurements must be made in the absence of CuSO$_4$. Such data gives the hydrogen evolution rate on the cathode as a function of cathode potential. Subtraction of the hydrogen current densities for the i-V curve representing hydrogen evolution and copper reduction then gives the copper deposition limiting current density, as shown in Figure 26.8.

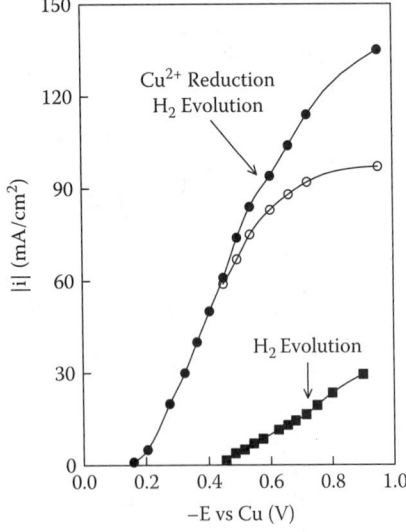

FIGURE 26.8 Current density (i) – voltage (E) plots for simultaneous cupric ion reduction + hydrogen gas generation and only hydrogen gas generation. The open circle curve (obtained by subtracting the two *solid line curves*) shows the Cu2+ limiting current. Data from [16]. Reproduced by permission of ECS—The Electrochemical Society.

The limiting current density for a given electrode reaction, cell geometry, and fluid flow condition is most easily determined experimentally, from a polarization curve such as that shown in Figures 26.7 and 26.8. The experimental apparatus to perform such measurements is shown in Figure 26.4 and consists of a working electrode (where the limiting current density is determined), a counter electrode, and a reference electrode. The use of a reference electrode permits the determination of the potential at the working electrode without interference from the counter electrode.

Rather than measure i_L experimentally, limiting current relationships can be derived theoretically by solving the relevant fluid flow and convective mass transport equations for a given electrode geometry and fluid stirring system. The solution of such equations gives k_d or δ_e in the limiting current equation. Alternatively, mass transfer coefficient correlations in the literature for nonelectrochemical systems have been used to derive i_L correlations in terms of the Sherwood (Sh) number, a dimensionless mass transfer coefficient, the Reynolds (Re) number for forced convection fluid flow, the Grashof (Gr) number for natural convection stirring, and the Schmidt (Sc) number. Such an approach is permissible because the convective diffusion equation for a binary salt solution with or without a supporting electrolyte is identical to that for an uncharged chemically reacting system. Here, k_d is determined from a Sh number correlation of the form

$$Sh_F = f(\mathrm{Re}, Sc) \tag{26.78a}$$

$$Sh_N = f(Gr, Sc) \tag{26.78b}$$

where

$$Sh = \frac{k_d L}{D} \tag{26.79a}$$

$$\mathrm{Re} = \frac{L v \rho}{\mu} \tag{26.79b}$$

$$Sc = \frac{\mu}{\rho D} \tag{26.79c}$$

$$Gr = \frac{\rho_b g L^3 \Delta \rho}{\mu^2} \tag{26.79d}$$

In Equations (26.78a,b), the subscripts F and N refer to forced and natural convection stirring. In Equations (26.79a–d), L is a characteristic electrode dimension, D is the diffusion coefficient, v is velocity, g is the gravity acceleration constant, μ is viscosity, ρ_b is bulk solution density, and $\Delta \rho$ is the difference in density between the bulk solution and the solution at the electrode surface. Once a Sh or k_d expression has been found/generated, it is combined with Equation (26.70) (or the analogous equation for a supporting electrolyte system) to obtain an i_L relationship. Examples of mass transfer correlations follow.

1. Mass transport to a rotating disk electrode with laminar flow over the disk surface:

$$Sh = 0.62(\mathrm{Re})^{1/2}(Sc)^{1/3} \tag{26.80}$$

or

$$Sh = 0.62\left[\frac{r^2\omega}{v}\right]^{1/2}\left[\frac{v}{D}\right]^{1/2} \quad (26.81)$$

where ω is the disk rotation speed (with units of s^{-1}), r is the disk radius, v is the kinematic viscosity ($v = \mu/\rho$), and the characteristic length for the Sh number is the disk radius (r). For this system, the fluid velocity can be varied by changing the disk rotation speed. The above correlation has been generated theoretically by Levich and verified in numerous experimental studies [17–19]. The corresponding limiting current correlation is

$$|i_L| = 0.62 z_+ F C_\infty D^{2/3} v^{-1/6} \omega^{1/2} \quad (26.82)$$

2. Laminar flow parallel to a flat plate [20]:

$$Sh = 0.677(Re)^{1/2}(Sc)^{1/3} \quad (26.83)$$

where both Sh and Re are defined in terms of the flat plate length, measured in the direction of fluid flow, hence the value of k_d in this correlation represents an average mass transfer coefficient over the entire plate length.

3. Laminar flow in a tube [21]:

$$Sh = 1.07(Re)^{1/3}(Sc)^{1/3}\left(\frac{d}{L}\right)^{1/3} \quad (26.84)$$

In the above equation, L is the tube length, the characteristic length in Sh is the tube diameter d, and Re is defined in terms of the tube diameter and the average velocity in the tube. Equation 26.84 does not take into account possible disruptions in the laminar flow condition at the entrance of the tube, nor can it be used at distances far down the tube (it has been shown that the correlation is accurate for $L < 5000\,d$) [4].

4. Natural convection at a vertical flat plate [22, 23]:

$$Sh = 0.66(ScGr)^{1/4} \quad (26.85)$$

where Gr and Sh are defined in terms of the plate length in the direction of the natural convection fluid flow.

5. Combined forced and free convection at a vertical flat plate, where the forced convection velocity is in the same direction as the natural convection flow (the so-called assisting mixed convection case). Here, researchers have combined Sherwood numbers for the pure forced and natural convection cases in the following way [15, 24–26]:

$$Sh_{mixed}^{3.2} = Sh_N^3 + Sh_F^3 \quad (26.86)$$

where Sh_N for natural convection is given by Equation (26.85) and Sh_F (forced convection) is defined by Equation (26.83). Additional limiting current correlations can be found in [27].

26.2.3.3 Infinite Dilution Transport Parameters

Transport parameters, which appear in the various forms of the infinite dilute transport equations, are the electrolyte conductivity, the ion mobility, the ion diffusion coefficient, and the ion transference number. All of these parameters can be determined from ionic equivalent conductances (λ_i, with units of (S-cm^2)/equiv) of cations and anions in solution. The ion mobility u_i, which appears in Equation (26.54), is related to λ_i by

$$u_i = \frac{\lambda_i}{|z_i|F^2} \tag{26.87}$$

The electric conductivity of an electrolyte solution (κ with units of S/cm) containing a single salt of concentration C is

$$\kappa = z_+ \nu_+ C(\lambda_+ + \lambda_-) \tag{26.88}$$

The ion diffusion coefficient and mobility are related by the Nernst-Einstein equation:

$$D_i = RTu_i \tag{26.89}$$

where the units for R are 8.314 J/mol K. By combining Equations (26.87) and (26.89), the ion diffusion coefficient is written in terms of the equivalent ionic conductance:

$$D_i = \frac{RT\lambda_i}{|z_i|F^2} \tag{26.90}$$

Finally, the transference number of an ion in a binary salt solution is

$$t_+ = 1 - t_- = \frac{\lambda_+}{\lambda_+ + \lambda_-} \tag{26.91}$$

The temperature and viscosity dependence of the ionic diffusion coefficient can be estimated from

$$\frac{D_i \mu}{T} = \text{constant} \tag{26.92}$$

where μ is the solution viscosity. Representative values of equivalent conductances of cations and anions in 25°C water at infinite dilution are listed in Table 26.7.

26.2.4 Driven vs. Self-Driven Cells and the Concept of Overpotential

We saw earlier that one can predict the thermodynamic open circuit (zero current) potential of an electrochemical cell by combining two half-cell reactions, one for the anode and the second for the cathode. The half-cell reaction with the lower (i.e., more negative) equilibrium potential will proceed spontaneously in the anodic direction, where the electrode acts as an electron sink for the anodic de-electronation (oxidation) reaction and the higher equilibrium potential reaction will occur spontaneously at the cathode, where the electrode acts as an electron source for the electronation (reduction) reaction. A cell with spontaneous reactions at the anode and cathode is called a self-

TABLE 26.7
Equivalent Conductances of Cations and Anions in 25°C Water at Infinite Dilution

Ion	z_i	λ_i (S-cm²)/equiv	Ion	z_i	λ_i (S-cm²)/equiv
H⁺	+1	349.8	OH⁻	−1	197.6
Li⁺	+1	38.69	Cl⁻	−1	76.34
Na⁺	+1	50.11	Br⁻	−1	78.3
K⁺	+1	73.52	I⁻	−1	76.8
NH₄⁺	+1	73.4	NO₃⁻	−1	71.44
Ag⁺	+1	61.92	HCO₂⁻	−1	54.6
Mg²⁺	+2	53.06	SO₄⁻	−1	80
Ca²⁺	+2	59.50	IO₄⁻	−1	54.38
Sr²⁺	+2	59.46	ClO₄⁻	−1	67.32
Cu²⁺	+2	54	BrO₃⁻	−1	55.78
Zn²⁺3	+2	53	HSO₄⁻	−1	50
La³⁺	+3	69.5	CH₃COO⁻	−1	40.9

Data from [4].

driven cell. One can reverse the direction of spontaneous anode and cathode reactions in an electrochemical cell by introducing a power supply in the external circuit between the electrodes. In such a driven cell, the negative terminal of the power supply is connected to the cathode and the positive lead is connected to the anode.

We have already discussed equilibrium electrode potentials and how they can be used to determine the tendencies for anodic de-electronation and cathodic electronation reactions at the electrode–solution interfaces in an electrochemical cell. For a self-driven cell, the current density will set up a number of current-produced potentials, known as overpotentials, that decrease the overall cell potential from its equilibrium value. In the development of energy-producing devices such as batteries and fuel cells, the variation of the cell potential with current is often more important than its open circuit, equilibrium potential. By proper engineering designs, one seeks to lower these overpotentials in order to obtain as much energy as possible from the cell. Similarly, in a driven cell, the open circuit, equilibrium potential represents the minimum thermodynamic potential driving force needed to carry out the anodic and cathodic reactions. Once these reactions begin and current flows, overpotentials are generated that increase the cell potential above that at open circuit. These overpotentials need to be minimized to ensure operation of the cell with the lowest possible energy input.

The overpotentials (denoted as η) that must be taken into account when designing and operating both driven and self-driven electrochemical cells describe energy consuming processes that occur during current passage. Due to the electrical conductivity of the electrolyte that separates the anode and cathode, the IR drop (sometimes called the resistance or ohmic overpotential η_{ohm}) depends on the magnitude of the current density, the distance between the anode and cathode, and the conductivity of the electrolyte solution. Normally, the concentration of ionic species in the bulk electrolyte is constant during current flow and variations in concentration are relegated to Nernst diffusion layers adjacent to the electrode surfaces, as described above. For this situation, the composition and conductivity of the bulk electrolyte are constant and one can write for η_{ohm}

$$|\eta_{ohm}| = \frac{|i|}{\kappa}\Delta y \qquad (26.93)$$

Overpotentials associated with concentration variations in the electrolyte are called concentration overpotentials (η_c). When the concentration near an electrode surface differs from that in the bulk, a concentration cell is established to equalize the two concentrations. Also, when there is diffusion of one ionic species (anion or cation) in the electrolyte due to a spatial variation in its concentration, a diffusion potential will be established to slow down the movement of the diffusing ion and speed up the movement of the oppositely charged ion so that electroneutrality in the solution is maintained. For the case where the electrolyte conductivity is assumed to be a weak function of concentration, the concentration overpotential can be expressed as the sum of a concentration cell potential and diffusion potential in the following way [4]:

$$\eta_c = \frac{RT}{nF} \ln \frac{C_{i0}}{C_{i\infty}} - \frac{F}{\kappa} \int_0^{\delta} \sum_i z_i D_i \frac{(C_{i\infty} - C_{i0})}{\delta_e} dy \qquad (26.94)$$

where C_{i0} is the electrode surface concentration of species i (which is reacting at the electrode surface), $C_{i\infty}$ is the bulk solution concentration of species i, δ_e is the effective boundary layer thickness, and y is the distance perpendicular from the electrode surface. Of course, κ, the electrolyte conductivity, would have to included within the integral if it were not assumed constant. When there is an excess of supporting electrolyte present, the diffusion potential term in the above equation is small and we can write

$$\eta_c = \frac{RT}{nF} \ln \frac{C_{i0}}{C_{i\infty}} \qquad (26.95)$$

We have already seen by way of Equation (26.77) that the electrode surface concentration of a reacting species is related to its bulk solution concentration, the applied current density, and the diffusion limiting current density. Substitution of Equation (26.77) (which is applicable to an electro-active species that is consumed at the electrode) into Equation (26.95) gives a more useful form of the concentration overpotential, since it does not contain the surface concentration, which is often difficult to measure:

$$\eta_c = \frac{RT}{nF} \ln \left[1 - \frac{i}{i_L} \right] \qquad (26.96)$$

One can now substitute various mass transfer/limiting current density correlations for i_L (such as Equation [26.82]) in order to relate η_c with the electrolyte flow conditions.

The third category of overpotential, known as the surface or activation overpotential, deals with those changes in the cell potential that are needed to create a sufficient driving force for an appreciable reaction rate at the electrode surface. From a thermodynamic viewpoint, once the electrode potential is negative to the equilibrium (open circuit) potential for a given electrochemical reaction, that reaction will proceed in the cathodic direction. Similarly, the reverse (anodic) reaction will occur if the electrode potential were more positive than the equilibrium potential. Thus, once the electrode potential deviates even a small amount from the equilibrium potential, there should, according to thermodynamics, be significant reaction rates and large anodic or cathodic currents. This does not usually occur because of finite reaction kinetics, in which case an additional potential (the surface or activation overpotential) is needed to drive the reaction at a given rate. For many, but not all, electrochemical reactions, the surface overpotential (η_s) is related to the current density and the surface concentration of reactant and product species by the Butler-Volmer equation. This equation as discussed earlier is as follows:

$$i = i_o[e^{(1-\beta)nF\eta_S/RT} - e^{-\beta nF\eta_S/RT}] \qquad (26.97)$$

where i_o is a function of the electrode surface concentrations of reactants and products. By definition, $\eta_s < 0$ favors reduction reactions, whereas $\eta_s > 0$ favors anodic reactions.

We now estimate the anode/cathode potential difference during current flow for driven and self-driven cells. First, we define the cell potential (V) as the difference in the anode and cathode potentials. For a self-driven cell, the cell voltage at a given current density will be less than the difference in equilibrium electrode potentials (ΔE_e) due to the presence of various overpotentials,

$$V = \Delta E_e - |\eta_S(anode)| - |\eta_c(anode)|$$
$$\quad - |\eta_S(cathode)| - |\eta_c(cathode)| - |\eta_{ohm}| \qquad (26.98)$$

where $\Delta E_e = E_e(\text{cathode}) - E_e(\text{anode})$. Since the thermodynamic quantity ΔE_e is always greater then zero for a self-driven system, the overpotentials in Equation (26.98) decrease the magnitude of the available anode/cathode electric potential difference. For a driven electrochemical cell, Equation (26.98) also holds, but $E_e(\text{cathode}) - E_e(\text{anode}) < 0$ and the various overpotentials add to ΔE_e. Thus, the anode/cathode voltage drop will be greater than that predicted from thermodynamic arguments alone.

In the remaining sections of this chapter, overviews of various electrochemical processes/systems are presented. Where appropriate, these overviews refer back to the fundamental principles and theories presented above.

26.3 ELECTROCHEMICAL REACTORS AND REACTOR DESIGN

An important task of electrochemical engineers is to design and operate electrochemical reactors (cells) for the manufacture of chemical products, including inorganic compounds (e.g., Cl_2 and concentrated NaOH), organic chemicals (e.g., adiponitrile), and purified metals (e.g., aluminum electrorefining). This aspect of electrochemical engineering deals with fundamental electrochemistry, heat and mass transfer, fluid flow, reactor configuration, scale-up, ancillary equipment for product purification, capital and operating costs analyses, and process optimization.

The configuration of an electrochemical reactor in an industrial process is geared to certain performance criteria. For example, in an organic electrochemical synthesis, reactors are designed to maximize product yield and current efficiency. A number of practical and general rules should be considered when deciding upon the design of an electrochemical reactor. The most general rules include (1) *simplicity of cell design and operation*, to lower capital and operating costs, minimize reactor down-time, and facilitate plant expansion; (2) *operational stability and reliability*, to ensure product purity and to enable automated operation; (3) *low cell cost and long lifetime*, to improve process economics; and (4) *safety of operation*, to minimize the dangers associated with electrical sparking and the possible production of explosive side-products during reactor operation.

26.3.1 CHARACTERIZING THE PERFORMANCE OF AN ELECTROCHEMICAL REACTOR

There are numerous ways of quantifying the energy efficiency and product selectivity of an electrochemical reactor, for both scale-up calculations and capital/operating cost analyses. Although products are formed at both the anode and cathode in such reactors, the cell performance is normally characterized in terms of the electrode where the desired product is generated.

1. Current efficiency: The current efficiency (*CE*) of an electrochemical reaction is defined as the ratio of the electric charge used in forming the product of interest to the total

charge Q passed through the reactor during an electrolysis. The CE can be viewed as the product yield based on the electrical charge and is given by

$$CE = \frac{nF(Vol)}{Q} \Delta C_P \qquad (26.99)$$

where *Vol* denotes the volume of the reaction medium, n is the number of electrons needed to synthesize one product molecule (in accordance with the electrode reaction), F is Faraday's constant, and ΔC_P is the concentration change of the product over the course of the electrolysis. The overall current efficiency is found by setting Q equal to the total charge passed during an electrolysis, whereas the differential current efficiency is determined by setting Q equal to the charge passed during a finite time interval during an electrolysis. In both cases, the charge is found by the time-integral of the current (for constant current operation, Q is the product of the current and reaction time).

2. Product selectivity: The product selectivity (S_p) is the ratio of the moles of the desired product to the total products formed from the starting material:

$$S_p = \frac{\text{number of moles of desired product}}{\Sigma \text{ moles of all products}} \qquad (26.100)$$

3. Electrical energy consumption: The electrical energy required to synthesize a given product is dependent on the CE of the electrode reaction and the voltage difference between the anode and cathode (E_{cell}) during reactor operation. On a mass basis, we can write the specific energy consumption as follows:

$$\text{Specific Energy Consumption} = \frac{nFE_{cell}}{CE \times MW} \qquad (26.101)$$

where MW is the molecular weight of the desired product. On a volume basis, the energy consumption is

$$\text{Volumetric Energy Consumption} = \frac{nFE_{cell}}{CE \times V_m} \qquad (26.102)$$

where V_m is the partial molar volume of the desired product. For constant current I operation for time period t, the specific energy consumption can be written as

$$\text{Specific Energy Consumption} = \frac{E_{cell} It}{\Delta C_P(vol) MW} \qquad (26.103)$$

4. Space-time-yield: The space-time-yield (τ) is the mass of product per unit time that can be obtained in a given electrochemical reactor of volume:

$$\tau = \frac{A_S \times I \times CE \times MW}{nF} \qquad (26.104)$$

where A_S is the active electrode area per unit reactor volume.

26.3.1.1 Initial Design Experiments

The first step in designing an electrochemical reactor is to identify the electrode and electrolyte components and to determine baseline operating conditions. Data are collected using bench-scale batch or semicontinuous reactor experiments, with a single anode and cathode, to determine (1) the choice of electrode materials, (2) the nominal operating current density and voltage of the reactor, (3) the cell temperature and pressure, (4) the proper reactant concentration, (5) the type of solvent and the type and concentration of supporting electrolyte, and (6) the method and extent of electrolyte agitation. The following general rules assist in designing an electrochemical reactor.

Choice of Electrode Materials

The anode and cathode should be stable in the electrolysis medium, allow the desired oxidation/reduction reactions at the highest possible rates with minimal by-product formation, and be of reasonable cost. In actuality, the electrodes may corrode or undergo physical wear during reactor operation, which may limit their lifetime. Often, if an expensive electrode material is needed for a given reaction, it can be plated or physically coated on a less costly, inert, and electronically conducting substrate. Common anode and cathode materials are listed in Table 26.8.

The form and shape of the electrodes are tailored for the specific reactor configuration. Typical shapes include flat metal sheets, perforated or expanded metal grids, metal foams and meshes, and three-dimensional packed bed electrodes formed by stacking metal meshes, pressing catalyst powders, or by use of microporous carbon felts and cloths (three-dimensional electrodes are particularly attractive for low-current density reactions because the electrode surface per unit reactor volume can be made very large) [28].

Electrolysis Medium

The reaction medium normally consists of a solvent into which the electroactive species is dissolved. Often, a supporting electrolyte salt is added to increase the conductivity of the solution. The concentration of electroactive species should be made as high as possible in order to pass the maximum feasible current through the reactor. When choosing a solvent, one must consider such factors as proton activity (the required pH for the electrode reactions), dielectric constant (which affects ion conductivity), solubility of electrolyte salts and other inorganic/organic substrates, accessible temperature range, vapor pressure, viscosity, toxicity, and price. Water is the solvent of choice, based on its low cost, high-salt solubility, and lack of toxicity. The solubility of organic compounds in an aqueous electrolytic solution, however, may be limited, necessitating the use of other protic solvents (e.g., methanol or acetic acid). In some cases an aprotic solvent such as

TABLE 26.8
Common Electrode Materials

Cathodes	Anodes
High hydrogen overpotential metals such as Hg, Pb, Sn, Zn, and metal amalgams (Zn-Hg)	Precious metals on an inert substrate (e.g., Ti) including Pt, Ir, and Pt-Ir
Graphite and other forms of carbon (including porous carbon and carbon powders bound in an inert polymer matrix)	Graphite or other forms of carbon
Steels and stainless steels	PbO_2 on Ti, Nb, or carbon
Coatings of low hydrogen overpotential metals (e.g., Ni, Ni/Zn, Ni/Al) on steel	Ni (for alkaline media)
Low hydrogen overpotential Raney metal powders pressed into a bed or bound in a polymer matrix	Dimensionally stable anodes such as mixed Ru-Ti oxide or Ti for Cl_2 production or IrO_2 on Ti for O_2 generation
Precious metal powders (Pt, Pd, and Pt on carbon)	Magnetite ($Fe_{3-x}O_4$)
Hastelloys (such as Bi-Mo-Fe and Ni-Mo-Cr alloys)	Conducting ceramics, such as Ti_4O_7
TiO_x	

TABLE 26.9
Accessible Potential Range of Some Solvents

Solvent	Anode/Ref. Electrode	Supporting Electrolyte	Anodic Limit (V)	Cathode/Ref. Electrode	Supporting Electrolyte	Cathodic Limit (V)
Water	Pt/SCE	$HClO_4$	1.5	Hg/SCE	TBAP	−2.7
Methanol				Hg/Hg pool	TEAB	−2.2
Acetic Acid	Pt/SCE	NaOAc	2.0	Hg/SCE	TEAP	−1.7
CH_3CN	Pt/Ag-Ag$^+$	$LiClO_4$	2.4	Pt/Ag-Ag	$LiClO_4$	−3.5
DMF	Pt/Hg pool	$LiClO_4$	1.5	$^+$Hg/Hg-pool	TEAP	−3.5
DMSO	Pt/SCE	$NaClO_4$	0.7	Hg/SCE	TEAP	−2.8

Note: DMF, dimethylformamide; DMSO, dimethyl sulfoxide; TEAP, tetraethylammonium perchlorate; TEAB, tetraethylammonium bromide; TBAP, tetrabutylammonium perchlorate; SCE, saturated calomel electrode.

From [29].

acetonitrile, dimethylformamide, or propylene carbonate may work best. On the other hand, it may be desirable to employ a molten salt electrolyte with no separate solvent species, as is done in aluminum electrorefining. The primary criteria in choosing a solvent is its useful potential range. In an electrochemical reactor, the solvent must be less reactive than the electroactive substrate(s) so that it is not oxidized or reduced at the electrodes. Approximate values for the anodic and cathodic limits of some commonly used solvent are given in Table 26.9 [29].

Operating Current Density and Voltage
While the combination of the applied current and current efficiency in an electrochemical reactor is a measure of the overall rate of product output, it is the product of the current and cell voltage that will determine the reactor's electrical power consumption, as indicated by Equation (26.103). The overall voltage in an electrochemical reactor is composed of the following components: (1) thermodynamic cell potential, (2) anode kinetic and mass transfer overpotentials, (3) anolyte IR drop, (4) diaphragm/membrane IR drop, (5) catholyte IR drop, and (6) cathode kinetic and mass transfer overpotentials. For more information on each of these terms, the reader should refer to Section 26.1.

Temperature and Pressure
Electrolyses at an elevated or reduced pressure are generally avoided because of the added complexity in cell design, unless high pressures are needed to avoid evaporation of the solvent. Electrochemical reactors are often operated at elevated temperatures in order to enhance electrode reaction rates, improve mass transfer of electroactive species to the electrode surface, and to increase the conductivity of the electrolyte. Joule (IR) heating during current flow through the reaction medium in conjunction with exothermic electrode reactions may require extensive cooling of the electrochemical reactor.

Mass Transport Effects
Mass transfer of reactants/products to/from the electrode surface may limit electrochemical reaction rates. To increase the concentration of reactants at the electrode surface, high Reynolds number flows are favored. High flow rates, however, increase pumping costs and could lead to undesirable mixing of the anolyte and catholyte.

26.3.2 Choosing an Electrochemical Reactor

Factors to consider when selecting an electrochemical reactor are listed in Table 26.10 [30]. Batch reactors are chosen for the synthesis of high value-added products in small quantities, whereas continuous reactors are more amenable to large-scale production. When the anolyte and catholyte

TABLE 26.10
Factors to Consider during Preliminary Reactor Design

Batch vs. continuous reactor operation
Single vs. multiple anodes and cathodes
Two-dimensional vs. three-dimensional electrode geometry
Monopolar vs. bipolar electrical connections
Divided vs. undivided
Open vs. closed cells
Static vs. moving electrodes
Finite interelectrode gap vs. zero gap or capillary cells

are to remain unmixed (to prevent unwanted side reactions at the electrodes, to isolate electrode products, and/or to prevent the formation of explosive or toxic mixtures), an ion-exchange membrane or some other kind of ionically conducting separator is placed between the anode and cathode chambers (the divided cell configuration). For simplicity of design and operation, most electrochemical reactors utilize flat plate electrodes.

Electrochemical reactors operate on DC current and, thus, high-voltage alternating current must be rectified and transformed before being sent to the cells. When the reactor contains multiple electrodes, electrical connections can be made using either a monopolar or bipolar design (see Figure 26.9). In a monopolar reactor, there are alternating anodes and cathodes, and both sides of each flat sheet electrode have the same polarity. External electrical contact is made to each electrode and the cell voltage is applied between each anode and cathode. In a bipolar design, electrical connections are made only to the two end electrodes, and the opposite faces of each electrode will have opposite polarities. Monopolar reactors require a low-voltage, high-current power supply, whereas the current is low and voltage high for a similar reactor geometry with bipolar connections. Although the electrical connection scheme is much simpler in a bipolar reactor, there may be leakage currents between cells (also known as bypass or shunt currents).

To minimize the cell voltage in an electrochemical reactor, the anode and cathode electrodes are placed as close as possible to minimize the resistance (IR) overpotential. Such voltage minimization is achieved in zero-gap and capillary cells [31, 32], by placing the electrodes directly adjacent to the membrane/diaphragm that separates the anode and cathode compartments. Figure 26.10 shows a bipolar capillary gap cell that was used in electro-organic processing [29], where

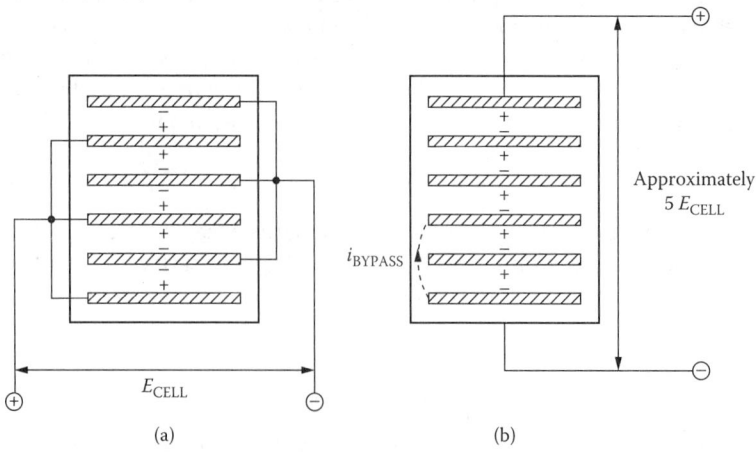

FIGURE 26.9 Electrical connections in multi-electrode cells. (a) Monopolar connection, (b) Bipolar connection (showing bypass current). From [30] (with kind permission from Springer Science and Business Media).

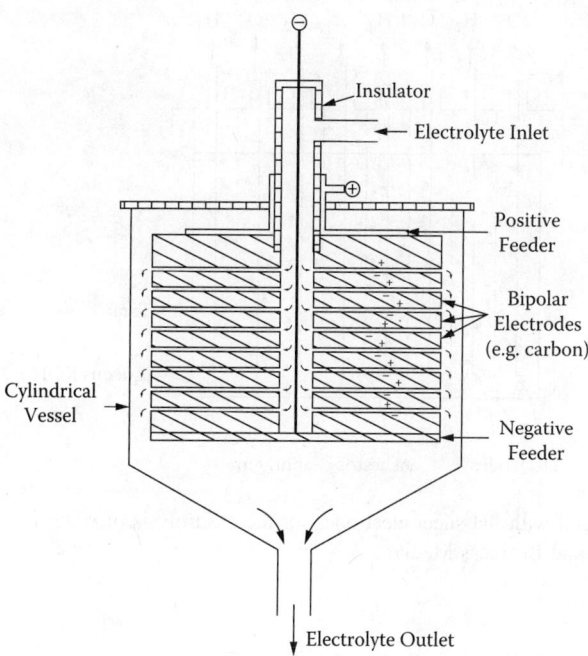

FIGURE 26.10 A capillary gap electrochemical reactor with bipolar connections. The electrodes are a stack of closely spaced disks and the electrolyte flows outward through the inter-electrode gap in the radial direction [31]. Pletcher and Walsh, Figure 2.33(a).

the narrow anode-to-cathode spacing (125 to 200 μm) is advantageous for low-conductivity electrolytes. The cell consists of a series of cylindrical bipolar graphite plates, precisely spaced by means of thin, radially positioned polyester strips. Electrolyte enters through a central channel and flows radially to the periphery of the cell, where it is collected and withdrawn.

Different reactors have been employed for various bench-scale and industrial-scale electrochemical processes. A partial listing of typical cell designs follows. More information on electrochemical reactors can be found in [28, 30, 33].

1. Tank cells: The tank cell shown in Figure 26.11 is a simple electrochemical reactor design where a series of alternating anode and cathode plates or expanded metal grids are placed in a single tank, either open or closed, with the electrodes extending the full width and depth of the tank. Batch tank reactors are used in metal electroplating, while semi-batch cells are employed for aluminum extraction, fluorine generation, and water electrolysis (i.e., the generation of H_2 and O_2 from water). The electrodes are usually placed vertically in the tank so that electrochemically generated gases can be easily removed from the inter-electrode gap. Electrical connections to the electrodes can be either monopolar or bipolar. The anode/cathode gap is made as small as possible to reduce the overall cell voltage, and the anolyte and catholyte chambers can be separated by means of a series of ion-exchange membranes or diaphragms (e.g., in water electrolysis cells, diaphragms are used to prevent mixing of hydrogen and oxygen gases). Normally, there is no externally applied stirring of the electrolyte in tank cells, although rising gas bubbles evolving at an electrode can create fluid agitation.
2. Parallel-plate flow cells: Most electrochemical flow cells are based on a parallel-plate electrode design with either horizontal or, more commonly, vertical electrodes in a monopolar or bipolar configuration (see Figure 26.12). With vertical electrodes, the cell is usually constructed in a plate-and-frame arrangement and mounted on a filter press.

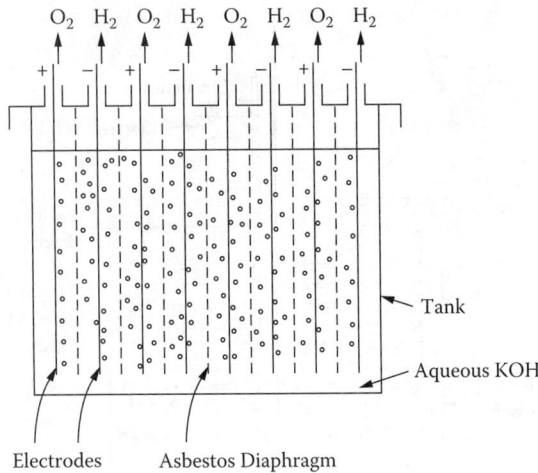

FIGURE 26.11 Tank cell with flat sheet electrodes for the electrolysis of water [30] (with kind permission from Springer Science and Business Media).

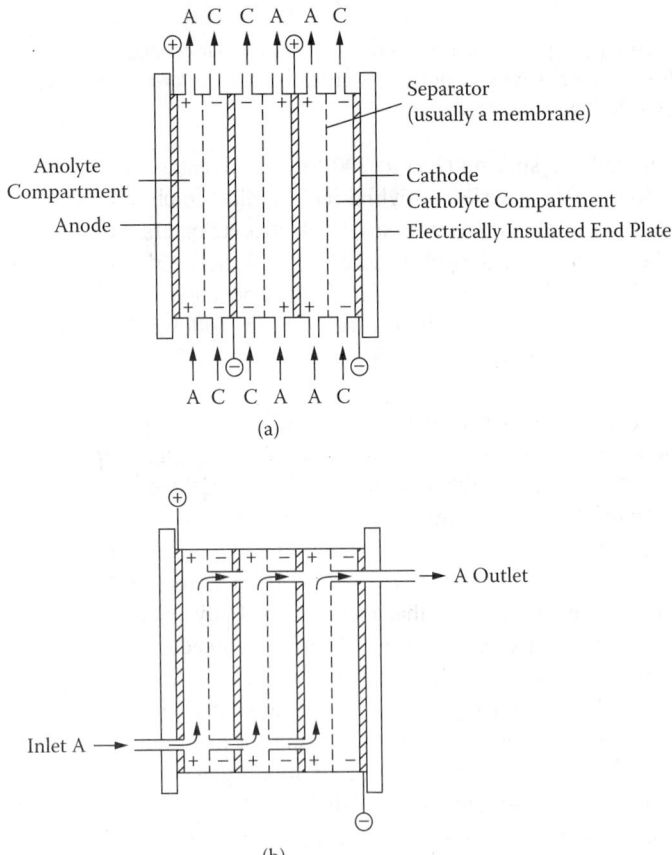

FIGURE 26.12 Plate and frame (filter-press) electrochemical flow cells. (a) Monopolar connection with external fluid flow manifolding. (b) Bipolar connections with internal manifolding [30] (with kind permission from Springer Science and Business Media).

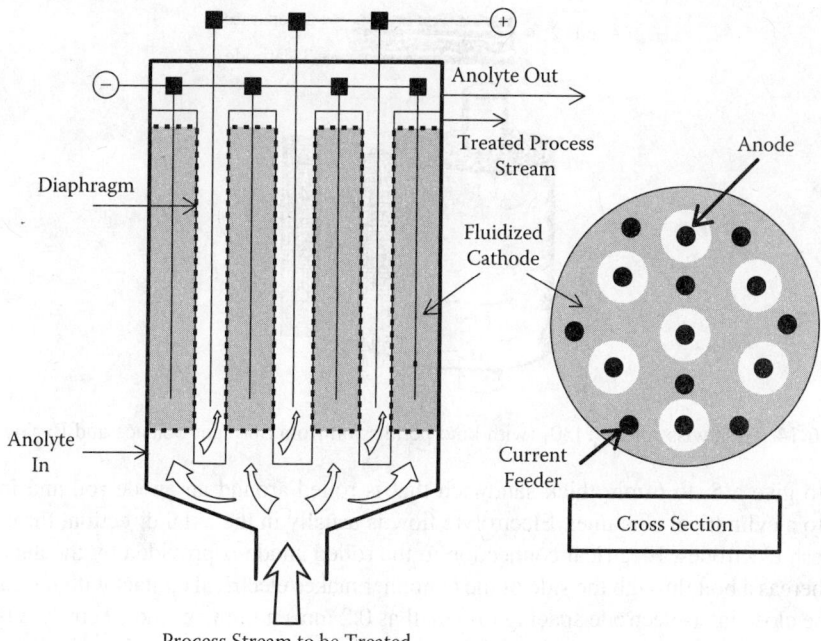

FIGURE 26.13 A schematic diagram of the Akzo fluidized bed electrochemical reactor (which resembles a shell-and-tube heat exchanger) [30].

Individual anode/cathode cells (as many as 100 cells in one reactor) can be divided (i.e., where a membrane or diaphragm separates the anolyte and catholyte solutions) or undivided. The interelectrode gap is normally between 0.5 and 5.0 cm and electrolyte is pumped at high flow rates over the electrodes The parallel-plate design is popular for many electrochemical processes e.g., water electrolysis, mercury cells for the production of Cl_2 and NaOH, and the electrohydrodimerization of adiponitrile. The advantages of this reactor type are (a) wide availability of electrode materials and separators, (b) simplicity of construction, (c) the ease of enhancing mass transfer by use of turbulence promoters and high fluid velocities, (d) the filter-press construction is similar to hardware commonly used by chemical engineers, and (e) scale-up simplicity, by either increasing the electrode size and/or increasing the number of electrodes.

3. Fluidized bed cells: In a fluidized bed reactor [34–36], one or both electrodes are composed of electrically conductive particles (e.g., metal-coated glass of polystyrene beads) that are fluidized by the upward flow of electrolyte, resulting in a bed expansion of 10% to 20% (Figure 26.13). The flow of current to the particle bed(s) can be provided by metal gauze conductors. The high electrode surface area to volume ratio (often >100/1) in this type of reactor is advantageous when the operating current density is small. The reactor can be operated in either a divided or undivided mode (i.e., with or without a separator to minimize anolyte and catholyte mixing). Although the space-time-yield may be high, the potential and current distribution may be highly nonuniform, leading to a low-product current efficiency and poor reaction selectivity. Additionally, the scale-up of these reactors is less predictable than flat plate electrode cells.

4. Swiss-roll cell: This cell, shown in Figure 26.14, has been proposed for both electro-organic syntheses and wastewater treatment [29, 37, 38]. It is fabricated from a multilayer stack of the following sheet materials: (a) a polymeric cathode spacer cloth or mesh, (b) a cathode mesh, (c) an ion-exchange membrane, (d) a polymeric anode spacer cloth or mesh, and (e) an anode mesh. These sheets are assembled in the sequence a-b-a-c-de-d-

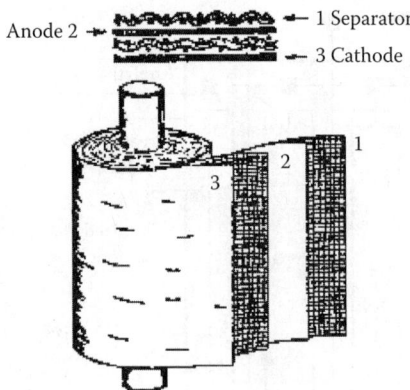

FIGURE 26.14 The Swiss-roll cell [30] (with kind permission from Springer Science and Business Media).

c to give a 5- to 6-mm thick sandwich that is rolled around an anode rod and inserted into a cylindrical container. Electrolyte flow is usually in the axial direction, through the mesh electrodes. Electrical connection to the rolled anode is provided by the anode rod, whereas a bolt through the side of the container makes electrical contact with the cathode. The close inter-electrode spacing (as small as 0.2 mm) minimizes the electrolyte IR drop and provides a high space-time-yield (a high electrode area per unit reactor volume), while the mesh spacers and electrodes promote turbulence and enhance mass transfer.

26.4 INORGANIC AND ORGANIC ELECTROCHEMICAL SYNTHESIS

26.4.1 Inorganic Processes

26.4.1.1 Aluminum Electrorefining

In our natural environment, metals are most stable in an oxidized state, e.g., Fe_2O_3 or Al_2O_3. As a consequence, one step of metal ore refining is the reduction of the metal oxide to its zero oxidation state. An electrochemical reduction process, where the electrolysis medium is a molten salt, is preferred for very electropositive metals (e.g., aluminum, sodium, lithium, and magnesium) and for metal refining where the chemical route suffers from environmental problems.

The electrolytic production of aluminum is carried out in Hall-Heroult cells that have changed little in nearly 100 years [39]. The Hall-Heroult process operates at a high temperature (about 1250 K) and utilizes a molten salt electrolyte of alumina (Al_2O_3) and cryolite (Na_3AlO_2), with additives such as calcium fluoride and aluminum trifluoride. The cathode reaction is the reduction of Al^{3+}, with a consumable carbon anode. The overall reaction in the Hall-Heroult cell (shown schematically in Figure 26.15) is

$$2Al_2O_3 + 3C \rightarrow 4Al + 3CO_2 \qquad (26.105)$$

The cells are strong steel boxes, lined with alumina (to act as a refractory), a thermal insulator, and carbon. The cathode is a liquid pool of aluminum that lies at the base of the cell, above a current collector consisting of a number of carbon blocks inlaid with steel bars. A frozen crust of electrolyte protects the cell housing from erosion. The cell has ports for the periodic addition of alumina through the crust, for the removal of Al metal, and an extractor to vent anode gases (mainly CO_2). As the carbon anode is consumed, it is lowered to maintain a constant anode/cathode gap (about 5 cm). In a typical plant for the production of 70,000 tons of Al per year, 200 Hall-Heroult cells, each 3 m × 8 m in size with 15 m² of anode area, are arranged in series. The operating current density is

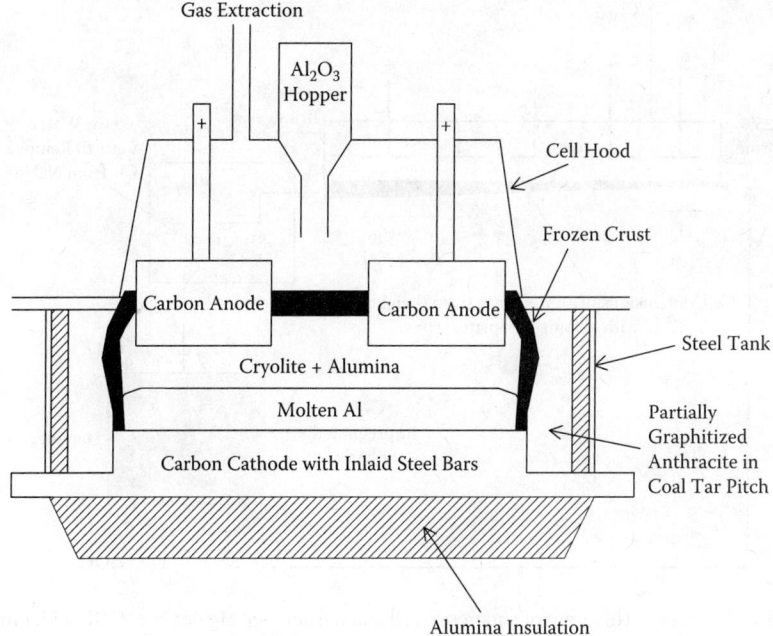

FIGURE 26.15 Hall-Heroult cell for aluminum extraction process [30] (with kind permission from Springer Science and Business Media).

about 1.0 A/cm² with an anode/cathode cell voltage of 4.0-4.5 V and a current efficiency of 85% to 90%. The energy requirement of the process is 14,000 to 16,000 kWh per ton of aluminum.

26.4.1.2 Chlor-Alkali Process

In the chlor-alkali process, an aqueous solution of NaCl is electrolyzed to give Cl_2, H_2, and NaOH, according to the following anode and cathode reactions:

$$\text{Anode: } 2Cl^- \rightarrow Cl_2 + 2e^- \tag{26.106}$$

$$\text{Cathode: } 2H_2O + 2e^- \rightarrow H_2 + 2OH^- \tag{26.107}$$

This process is the single largest of the electrolytic industries; U.S. chlorine demand in 2002 was 14,000,000 mt [40].

Several cell designs have evolved over the years for the electrochemical production of Cl_2 and NaOH [30]. The oldest commercial process utilized a mercury cell, where the anode and cathode reactions are

$$\text{Anode: } 2Cl^- \rightarrow Cl_2 + 2e^- \tag{26.108}$$

$$\text{Cathode: } 2Na^+ + 2Hg + 2e^- \rightarrow 2NaHg \tag{26.109}$$

A subsequent hydrolysis step outside the electrochemical cells (and in the presence of a catalyst) was used to convert the sodium amalgam to OH and hydrogen gas:

$$2NaHg + 2H_2O \rightarrow H_2 + 2NaOH + 2Hg \tag{26.110}$$

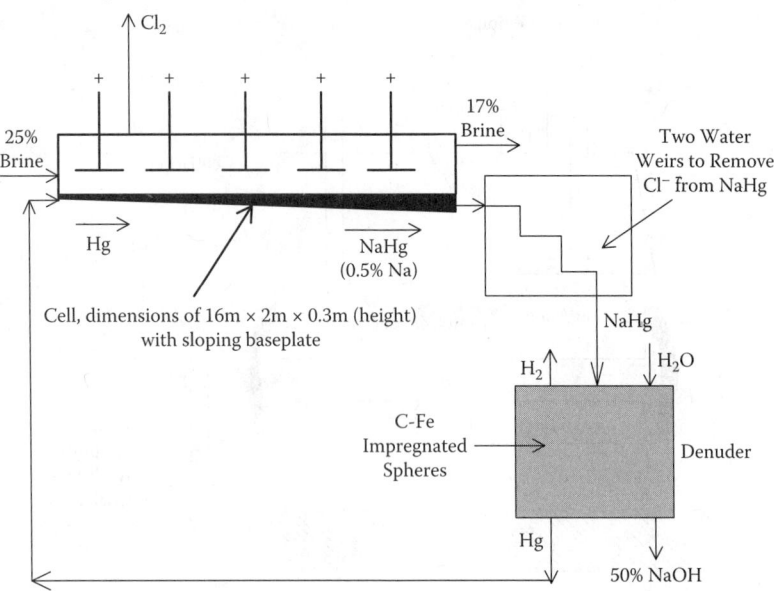

FIGURE 26.16 A schematic diagram of a mercury cell and the coupled Hg denuder [30] (with kind permission from Springer Science and Business Media).

A schematic diagram of a mercury cell is shown in Figure 26.16. The cell consists of a large shallow trough with a sloping base. Mercury flows along the bottom of the cell and acts as the cathode, while graphite or RuO_2-coated titanium (dimensionally stable anode) is used as the anode, with an anode/cathode gap of about 1 cm. Brine at an NaCl concentration of about 25% and a temperature of 60°C enters the cell and is electrolyzed, exiting at a concentration of about 17%. Cl_2 exits the top of the cell and the sodium amalgam leaves the base, where it is first washed with water to remove NaCl and then enters a denuder where a concentrated (50%) NaOH solution is generated from the amalgam and the proper amount of water. Hydrogen gas is vented from the top of the denuder and the Hg is recycled to the cells. Mercury cells typically operate at a current density of 1.0 A/cm^2, at an anode/cathode cell voltage of –4.4 V. Energy consumption is approximately 3150 kWh per ton of NaOH. Due to environmental concerns over the use of Hg, this cell design is being phased out in the U.S. and Japan and replaced by diaphragm and membrane cells.

In a diaphragm cell (shown schematically in Figure 26.17), an asbestos separator is deposited directly onto a steel cathode (or a cathode composed of steel with a nickel coating to lower the overpotential for hydrogen evolution) to physically inhibit mixing of the NaOH cathode product and brine feed (typically about 30%). The anode (graphite or dimensionally stable anode) is placed close to the cathode. A small fraction of the brine feed is allowed to diffuse through the asbestos diaphragm and H_2 and NaOH are formed on the opposite (cathode) side of the separator. The concentration of NaOH at the cathode is restricted to below 12%, otherwise OH ions will diffuse across the diaphragm and react with Cl_2 to form hypochlorite. A high pH brine also increases the rate of oxygen evolution (from water) at the anode, which lowers the current efficiency of the process and contaminates the Cl_2 gas. Also, the NaOH solution leaving the cathode contains some NaCl impurity. To synthesize a higher quality/higher concentration caustic product, water can be evaporated from the catholyte to give a 50% caustic solution. Removal of water also results in the crystallization of NaCl, but it is difficult to lower the salt concentration to <1%. Of course this added purification step is energy intensive and adds to the overall cost of the process. Diaphragm cells operate at an elevated temperature of 60°C to 80°C, a current density of about 0.2 A/cm^2, and a cell voltage of –3.45 V. When the water evaporation step to produce 50% caustic is added to the electrical energy for the electrolysis, a total of 3260 kWh is required per ton of NaOH.

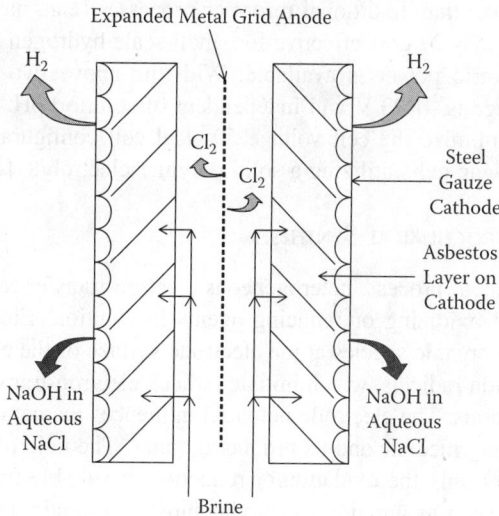

FIGURE 26.17 Principles of operation of a diaphragm cell for Cl_2 production.

In the 1970s, cation-exchange membranes composed of chemically resistant, perfluorinated polymers (e.g., DuPont's Nafion® and Flemion® from Asahi Glass) were developed and incorporated into chlorine-caustic cells. The resulting membrane cells utilized the same electrode reactions as the diaphragm cells, but the cation-selective permeability of the membrane (i.e., the membrane's ability to restrict the passage of OH^- and Cl^-) allowed for the direct synthesis of 30% to 40% sodium hydroxide solutions, essentially free of chloride ions. The very low leakage of OH^- across the membrane also lowered the oxygen contamination of the Cl_2 product and minimized current losses due to electrochemical oxygen evolution. Most membrane cells are based on a filterpress design, with vertical electrodes (louvered or expanded metal grids) and a "zero-gap" configuration where the anode and cathode are in contact with the opposite sides of the membrane. A coating on the membrane surface ensures proper gas bubble release. The anolyte feed is 25% NaCl in water and the catholyte is dilute caustic. A typical filterpress cell, for the production of 10,000 tons of Cl_2 per year, consists of 50 to 100 anode/cathode pairs, with each electrode 1 m × 1 m. The cell operates at a temperature of about 80°C, a current density of about 0.4 A/cm², and an anode/cathode cell voltage of −2.95 V. Electrical energy consumption by the cells is about 2520 kWh per ton of NaOH. Extensive brine feed purification is normally required for membrane cells in order to remove divalent cations (in particular Mg^{2+} and Ca^{2+}) that absorb into the membrane and lower the membrane's ability to reject Cl^- and OH^-.

26.4.1.3 Water Electrolysis

The electrolysis of water molecules in an electrochemical cell is carried out to generate high purity oxygen and hydrogen gases according to the following anode and cathode reactions:

$$\text{Anode: } 2H_2O \rightarrow O_2 + 4H^+ + 4e^- \tag{26.111}$$

$$\text{Cathode: } H_2O + 2e^- \rightarrow H_2 + 2OH^- \tag{26.112}$$

The most important product of water electrolysis is hydrogen gas, which is used as in the chemical catalytic hydrogenation of organic compounds (e.g., liquid oils that are hydrogenated to synthesize solid or semisolid fats), the refining of high-purity metals, the synthesis of ammonia, and the manufacture of semiconductors. The water-splitting route to H_2 gas, however, is much more energy-intensive (expensive) than traditional reformer processes (e.g., natural gas reforming), but the electrochemical route may be cost-effective for small-scale hydrogen production plants and/or where low-cost hydroelectric power is available. With the above two electrode reactions, the thermodynamic cell voltage is −1.23 V and independent of solution pH. Industrial water electrolyzers are designed to minimize the cell voltage. Typical cell configurations include tank cells, filterpress cells, and zero-gap cells utilizing a solid polymer electrolyte [28].

26.4.2 Organic Electrochemical Syntheses

In an organic electrochemical process, heterogeneous electron transfer reactions at the anode and cathode replace chemical oxidizing or reducing agents in solution. Electrons may be added or removed directly from an organic species at the electrode surface or the electron transfer reactions can generate anion or cation radicals, which initiate radical, electron-transfer, nucleophilic/electrophilic and acid/base reactions. The electrode potential influences the nature of the organic product or the electro-generated intermediate and its production rates. The field of organic electrochemical synthesis encompasses not only the oxidation or reduction of suitable functional groups, but also more intricate reactions, such as anodic or cathodic substitution, addition, coupling, cleavage of bonds, and polymerization [29, 41–44]. Thousands of organic compounds have been synthesized in the laboratory by means of electrochemical methods. In over 100 cases, the electrochemical process was sufficiently promising to build either pilot- or commercial-scale plants (see for example, Table 6.2 in [30] and Chapter 30, Table 2, in [29]). Several types of chemical transformations that can be performed electrochemically are presented in Table 26.11.

Electrochemical methods for the production of organic compounds are attractive for a number of reasons:

1. High yields and high selectivity are attained under appropriate electrolysis conditions; the desired compounds may be produced at a lower cost, as compared with traditional methods.
2. Electrical energy is used instead of chemical reagents (e.g., $LiAlH_4$ for organic reductions); accordingly, electrochemical processes may be less expensive with less pollution. Also, it may be possible to regenerate electrochemically spent redox agents in situ, so that large quantities of product can be produced from a small inventory of reagents.
3. Reactions occur at moderate temperature and pressure.
4. Oxidation and reduction reaction rates can be controlled by the current or electric potential.

Possible complications of an electrochemical route are listed below.

1. The rate of electron transfer at the anode or cathode may be slow for a complex reorganization of the substrate. This problem may be circumvented by: (a) three-dimensional packed, stacked, or fluidized bed electrodes with a high ratio of surface area to volume; (b) mediated reactions (a fast redox reaction at the electrode followed by a bulk solution homogeneous reaction with the organic substrate which consumes the electro-generated reagent); and (c) phase-transfer catalysts, where the active species is electro-generated, usually in an aqueous phase and then transferred by the catalyst into a nonaqueous phase where it reacts with a water-insoluble organic.
2. The high charge requirement of 96,487 coulombs (1 Faraday) per mole of electrons is a detriment for multiple electron transfer steps. Consequently, ideal candidates are (a) the synthesis of high value-added products, (b) compounds that cannot be made by traditional catalytic methods, and (c) organic compounds with a high molecular weight

TABLE 26.11
Examples of Organic Electrochemical Reactions

Difficult Oxidations

$$HOCH_2C\equiv CCH_2OH \xrightarrow{-8e^-} HO_2CC\equiv CCO_2H$$

Anodic Decarboxylation

[Structure: dihydrouracil with CO_2^- group $\xrightarrow{-2e^-, -CO_2}$ uracil]

Anodic Substitution

[Morpholine-N-CHO $\xrightarrow{-e^-, -H^+}$ iminium intermediate $\xrightarrow{CH_3OH}$ N-CHO morpholine with OCH_3]

Dehydrodimerization

$$2CH_2(COOC_2H_5)_2 \xrightarrow[-2H^+]{-2e^-} (C_2H_5OCO)_2CHCH(CooC_2H_5)_2$$

Fluorination

$$C_nH_{2n+1}COF \xrightarrow[HF]{\text{oxidation}} C_nF_{2n+1}COF$$

Oxidative Coupling

$$O_2S\begin{matrix}\diagup \overline{NR} \\ \diagdown \overline{NR}\end{matrix} \xrightarrow{-2e^-} RN\!\!=\!\!NR + SO_2$$

Selective Reduction

[4-nitrobenzyl cyanide $\xrightarrow{6e^-}$ 4-aminobenzyl cyanide]

Difficult Reductions

[Anthranilic acid (2-aminobenzoic acid) $\xrightarrow{4e^-}$ 2-aminobenzyl alcohol]

[Benzene $\xrightarrow[\text{aq }(C_4H_9)_4NOH]{2e^-}$ 1,4-cyclohexadiene]

Continued

TABLE 26.11 (Continued)
Examples of Organic Electrochemical Reactions

Carboxylation

$$\underset{R^2}{\overset{R^1}{\diagdown}}C=NR^3 + CO_2 \xrightarrow[2H^+]{2e^-} R^2-\underset{\underset{COOH}{|}}{\overset{\overset{R_1}{|}}{C}}-NHR^3$$

Selective Dehalogenation

$$\underset{Br\ \ S\ \ Br}{\overset{Br}{\diagup\!\!\diagdown}} \xrightarrow[2H_2O]{4e^-} \underset{S}{\overset{Br}{\diagup\!\!\diagdown}}$$

Desulfurization

$$ArCH_2S(CH_3)_2 \xrightarrow[H^+]{2e^-} ArCH_3 + (CH_3)_2S$$

In Situ Electro-Generated Base (EGB)

$$\text{Ph}-N=N-\text{Ph} \xrightarrow{2e^-} \text{Ph}-\overline{N}=\overline{N}-\text{Ph}$$
$$\text{EGB}$$

$$RCH_2\overset{+}{P}(C_6H_5)_3 \xrightarrow[R'CHO]{EGB} RCH=CHR'$$

In Situ Electrogeneration of Oxidant and Base

$$2Br^- \xrightarrow{-2e^-} Br_2 \underset{}{\overset{H_2O}{\rightleftharpoons}} HOBR + HBr$$

$$2H_2O \xrightarrow{2e^-} H_2 + 2OH^-$$

$$\underset{CH_3}{\overset{H_3C}{\diagdown}}\!\!=\!\!\diagup \xrightarrow[OH^-]{HOBr\ or\ Br_2} H_3C\overset{O}{\underset{\diagdown}{\diagup\!\!\!\diagdown}}CH_3$$

Oxidative Coupling by Decarboxylation (Kolbe Reaction)

$$2RCOO^- \xrightarrow{-2e^-} RR + 2CO_2$$

Electrohydrodimerization (Reductive Coupling)

$$2CH_2=CHCN \xrightarrow[2H_2O]{2e^-} NC(CH_2)_4CN + 2OH^-$$

(where the number of Faradays per unit mass of product is low even though the required number of Faraday's per mole is large).

3. Solubilizing an inorganic supporting electrolyte salt in the organic starting material or dissolving the organic substrate in an aqueous electrolyte solution is often a problem. To address this problem, the following have been used: (a) nonreactive aprotic solvents; (b) a single or mixed polar organic solvent (such as an alcohol or alcohol/water mixture, which can solubilize the organic reactant); (c) special supporting electrolyte salts, such

as quaternary ammonium and hydrotropic salts (e.g., tetramethylammonium acetate or tetraethylammonium p-toluenesulfonate [45]); (d) an emulsified reaction medium (where the organic reactant is dispersed in a water/supporting electrolyte solution [46]); (e) close placement of the anode and cathode to compensate for the low-salt concentration and low conductivity of the reaction medium; and (f) the use of a solid polymer electrolyte reactor (described below).

4. A counter electrode reaction is needed in the organic electrochemical reactor. The reactant and product(s) of this reaction must not interfere with the primary organic electrode reaction. Hence, a membrane separator is often used to divide the anode and cathode compartments of the reactor; the membrane adds to the cost of the reactor and usually increases its operating voltage.

5. Usually, only one product is synthesized in a reactor, with the counter electrode reaction forming a product that interferes minimally with the primary organic reaction. Thus, 50% of the reactor's energy consumption is wasted. This problem has been addressed by employing "paired" electro-organic syntheses, where the reactions at both the anode and cathode generate one or more useful organic product(s) [47].

26.4.2.1 Essential Components of an Electro-Organic Reactor System

When formulating an electro-organic synthesis on a laboratory scale, the proper choice of reactor, electrolyte composition, and electrode materials must be made.

26.4.2.1.1 Reactor Design

Batch Cells

Normally, a batch reactor is employed for small-scale preparative syntheses. Two-dimensional (flat sheet) electrodes provide proper operating conditions (e.g., electrode potential, ohmic resistance, and mass transfer) because high space-time-yields are not required. Cells can be either undivided (no separation of the anolyte and catholyte) or divided. In its simplest configuration, a laboratory beaker is used as an undivided organic electrochemical reactor, with working and counter electrodes, a conducting electrolytic solution, and means for stirring the reaction medium. To avoid unwanted side reactions and mixing of the anolyte and catholyte (which could complicate organic product extraction from the electrolytic solution downstream from the reactor), the anode and cathode chambers can be separated by a microporous barrier such as a fritted glass disk or ion-exchange membrane. The most widely used divided batch reactor for organic syntheses is the H-cell (shown schematically in Figure 26.18 [29, 48] for a cell open to the atmosphere with a Hg cathode and graphite anode). In more elaborate experiments, it is often desirable to use a closed cell [49] with accessories for working under an inert atmosphere or in vacuum [50], at elevated temperatures and pressures, with reflux, or to allow for the identification of gaseous products. For operation at a constant electrode potential, a reference electrode (calomel, Ag/AgCl, hydrogen, etc.) is inserted into the reaction medium, close to the working electrode.

Flow Cells

Electro-organic flow reactors can also be either divided or undivided. One well-studied flow reactor employs two-dimensional parallel plate electrodes in a plate-and-frame filter press design [28, 30, 51] (see Figure 26.12). Other reactors use three-dimensional packed bed electrodes; these cells have a variety of configurations, including trickle bed [52], fluidized bed [35, 36], flow-by [53], flow-through [54], and radial-flow [55] designs. Recently used solid-polymer electrolyte reactors have been employed for organic electrosyntheses, where the liquid electrolyte solution between the anode and cathode is replaced by a wetted ion-exchange membrane [56]. Thus, the need for a solvent/supporting electrolyte is eliminated, which is a significant improvement. This type of reactor will be discussed later.

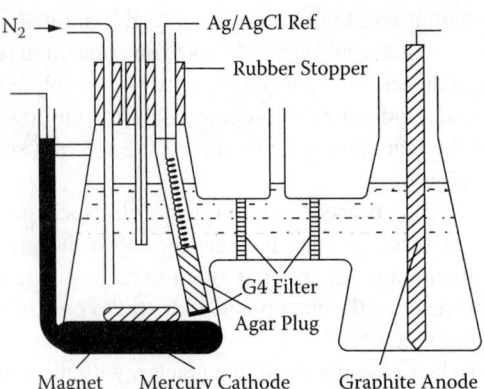

FIGURE 26.18 H-cell for semi-macroscale electro-organic syntheses [48]. (Used with permission from the *Journal of Chemical Education,* Vol. 48, No. 2, 1971, 136–137; © 1971, Division of Chemical Education, Inc.)

26.4.2.1.2 Separator Materials

The separator for a divided electrochemical reactor should ideally be chemically inert and an electric insulator, with a porosity that allows for the facile movement of supporting electrolyte cations and/or anions but prevents the passage and intermixing of organic compounds. Inert porous materials, such as sintered glass, ceramic refractories, filter paper, nylon cloth, and porous plastic provide a barrier for inter-diffusion. Ion-exchange membrane separators (such as DuPont's Nafion® perfluorosulfonic acid cation-exchange membranes) generally provide effective barriers (see Section 26.6.3 and Table 26.15 for more information regarding ion-exchange membranes).

26.4.2.1.3 Electrode Materials

The proper choice of electrode materials is important when designing a cell [57]. A general discussion of electrode materials was presented in Section 26.3.1.1. Factors that must be considered include: the catalytic properties of the electrode material, substrate absorption, the presence of solution-phase impurities (often due to electrode corrosion/degradation when using water as the electrolyte solvent), the physical state of the electrode, and the electrode history [58]. Popular cathode materials for organic electrochemical reactions include:

Mercury—The very high hydrogen overvoltage of Hg is useful for organic reduction reactions in protic solvents, where competing H_2 gas electrogeneration must be suppressed. The liquid state of mercury places certain restrictions on the design of such electrochemical cells (see Figure 26.18). Environmental concerns over the toxicity of mercury also limit its use.

Lead and Cadmium—The hydrogen overvoltage on Pb and Cd is high, and they are easy to work mechanically. For many reactions, a lead or cadmium cathode yields the same chemical products as mercury.

Tin—Electrodes composed of Sn are used mostly for the reduction of nitro compounds.

Nickel and Platinum—These two metals (in the form of Raney nickel and Pt-black) are used for electrocatalytic organic hydrogenation reactions (i.e., the electrochemical generation of hydrogen on the catalytically active, high surface area cathode followed by the chemical reaction of adsorbed hydrogen with the organic substrate).

Aluminum and Iron (Steel)—These metals are widely used in large-scale electro-organic processes, due primarily to their low cost.

The choice of anode material for electro-organic reactions is limited, due to anodic dissolution of most metals. The most common materials are

Platinum—The high standard reduction potential for platinum makes it an ideal anode material (although Pt corrosion does occur during some oxidations reactions, such as the Kolbe oxidative coupling reaction [29, 58, 59]). For anode potentials greater than about 0.50 V (on the hydrogen scale), an oxide film covers the Pt surface, so the electrode material is often platinum oxide. Large anodes are often titanium coated with platinum in order to reduce costs.

Carbon—Graphite electrodes are often used, but they are less stable toward corrosion/physical degradation than platinum. Graphite is available in many forms, including woven cloths, reticulated foam, and glassy (vitreous) carbon rods and plates. Organic products at different types of graphite anodes may differ considerably [60].

Lead—Alloying lead with silver, tin, or cobalt often improves the corrosion resistance of lead anodes. In many cases, the surface of a lead anode is actually lead dioxide [58]. PbO_2 electrodes are stable in sulfate media at low pH and the oxygen overpotential is high, but the material has poor mechanical properties and corrodes in HCl.

Dimensionally Stable Anodes—These anodes are composed of a base metal such as titanium, coated with a precious metal oxide (e.g., ruthenium dioxide). Such anodes can be used instead of Pt or carbon for oxygen evolution counter electrodes in an organic electrosynthesis. They have also found some applications for organic oxidation reactions [61].

26.4.2.1.4 Solvents

Important properties of the solvent in an organic electrosynthesis are (a) electrochemical stability (nonreactivity) at high anodic and cathodic potentials, (b) control of proton activity (which often influences cathodic reduction reactions), (c) the ability of the solvent to dissolve both supporting electrolyte salts and organic substances, (d) temperature stability, (e) low vapor pressure, (f) low/moderate viscosity, (g) minimal toxicity, and (h) reasonable cost. Solvents can be generally categorized as protic (proton containing) and aprotic. Aprotic solvents are preferred when aqueous or proton-containing solvents have undesirable properties, as is the case for electro-generated radical anions, which are more stable in the absence of protons. A general discussion of electrolysis media and solvents for electrochemical reactions was presented in Section 26.3.1.1 (see Table 26.9). Commonly used protic and aprotic solvents for organic electrochemical reactions are

Protic solvents—(1) Acids (the most common are sulfuric, fluorosulfonic, hydrogen fluoride, trifluoroacetic acid, and acetic acid), (2) neutral solvents (water, methanol, and other alcohols), and (3) basic solvents (ammonia, methylamine, and ethylenediamine).

Aprotic solvents—Include acetonitrile, dimethylformamide, N-methylpyrrolidone, hexamethylphosphoramide, pyridine, dimethyl sulfoxide, sulfolane, nitromethane, propylene carbonate, and methylene chloride.

26.4.2.1.5 Supporting Electrolyte Salts

During an electrosynthesis, the dissociated supporting electrolyte salt in the solvent/organic reaction medium provides ions for solution-phase current flow. It is desirable to employ a salt with a high solubility, complete dissociation, and high ionic mobility in the solvent medium. Additionally, the salt should be resistant to oxidation and reduction reactions at the anode and cathode. The choice of anions is of importance for anodic reactions (during current flow, anions migrate toward the anode), whereas one must carefully choose the cation component of the supporting electrolyte for cathodic reactions. Anions difficult to oxidize include perchlorate, tetrafluoborate, hexafluorophosphate, and nitrate. In practice, only alkali and alkaline earth cations (e.g., Li^+, Na^+, and Ca^{2+}) and tetraalkylammonium ions are used for organic electrochemical reduction reactions. In aqueous solutions, quaternary ammonium cations, e.g., tetraethylammonium and tetrabutylammonium ions, are particularly effective in minimizing unwanted hydrogen evolution during an organic electroreduction reaction by forming a low proton activity organic layer on the cathode surface [62].

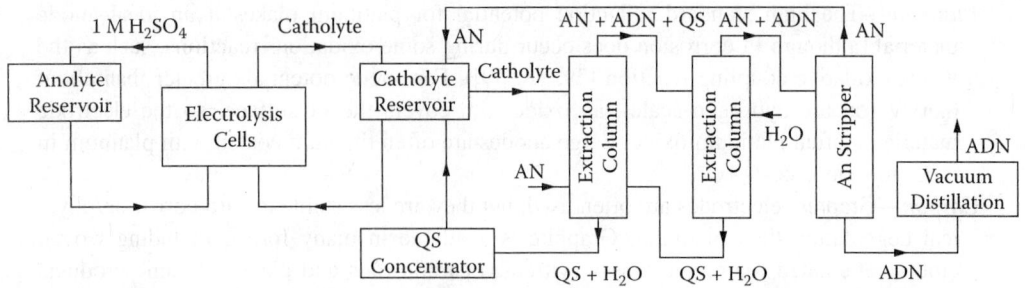

FIGURE 26.19 Schematic diagram of the adiponitrile process flowsheet. AN = acrylonitrile, ADN = adiponitrile, QS = quaternary ammonium salt [30] (with kind permission from Springer Science and Business Media).

26.4.3 Monsanto's Electrohydrodimerization Process

In 1965, the Monsanto Company started production of adiponitrile from acrylonitrile via an electrochemical route [Equation (26.113)]. Adiponitrile was then converted to either adipic acid or hexamethylenediamine by chemical means; these two compounds are reactants in the production of nylon 66.

$$2CH_2=CHCN + 2e^- + 2H_2O \rightarrow CN(CH_2)_4CN \qquad (26.113)$$

The key to this process [63, 64] was the addition of a quaternary ammonium salt (tetraethylammonium p-toluenesulfonate or tetraethylammonium ethylsulphate) to the reaction medium, which increased the solubility of acrylonitrile in the aqueous electrolyte and generated an aprotic organic layer on the cathode surface that inhibited the synthesis of propionitrile by-products (formed by the simple reduction of acrylonitrile).

Initially, divided electrochemical cells were employed (24 cells, each with 16 anode/cathode pairs in a bipolar filterpress design), with a lead cathode, a lead-silver anode, and a cation-exchange membrane separator. Due to the low conversion rate per pass, the cells were operated in a batch recycle mode with an external catholyte reservoir. A fraction of the reservoir was bled continuously into an extraction plant and fresh acrylonitrile was added to the reservoir. In the downstream extraction plant, quaternary ammonium salt supporting electrolyte and unreacted acrylonitrile were recovered and recycled, and adiponitrile was isolated and purified. Adiponitrile and acrylonitrile were extracted from the aqueous electrolyte in a sieve tray column by addition of acrylonitrile starting material. The organic phase from this unit was washed with water to remove the last traces of supporting electrolyte salt, and the adiponitrile and acrylonitrile were then separated by distillation. The final stage of the process was a vacuum distillation purification of adiponitrile. A schematic flowsheet is shown in Figure 26.19 [30].

During the 1970s, both a larger plant and a second plant were built. These expansions employed a second-generation, undivided cell design, with an emulsified reaction medium composed of 7% acrylonitrile in an aqueous solution of 15% disodium hydrogen phosphate with a low concentration (0.4%) of a quaternary ammonium salt. More information on the Monsanto adiponitrile process are found in [28–30, 42 (Part III), 65, 66].

26.4.4 Electrocatalytic Reduction (Hydrogenation) Reactions in a Solid Polymer Electrolyte (Proton Exchange Membrane) Reactor

Many electro-organic reduction reactions are performed using an aqueous reaction medium (the H atoms in water supply the necessary hydrogen for the reaction) and a high hydrogen overpotential cathode such as lead or cadmium to minimize unwanted H_2 gas evolution (from the electro-reduction of water molecules). An alternative method for organic hydrogenation is a low-temperature electrocatalytic route, where an electrically conducting catalyst (e.g., Raney nickel or platinum black) is used as the cathode. Atomic hydrogen is generated on the catalyst surface by the reduction of protons or water molecules from the electrolytic medium. The overall hydrogenation sequence is as follows:

$$2H^+ + 2e^- = 2H_{ads} \tag{26.114}$$

$$2H_{ads} + Organic \rightarrow Organic\text{-}H_2 \tag{26.115}$$

An unwanted side reaction that consumes current but does not affect the product yield is the formation of H_2 gas, either by the combination of two adsorbed hydrogen atoms or the electrochemical reduction of adsorbed hydrogen atoms

$$2H_{ads} \rightarrow H_2(gas) \tag{26.116}$$

or

$$H_{ads} + H_2O + e^- \rightarrow H_2(gas) + OH^- \tag{26.117}$$

Since the concentration of H_{ads} on the cathode surface is dependent on the applied current density and the rate of the chemical reaction of hydrogen with the organic substrate, Equations (26.116) and (26.117) are not equilibrium relationships. Unlike a chemical catalytic organic hydrogenation process, the predominant reaction for supplying hydrogen to the catalyst surface is not the equilibrium form of Equation (26.116), but rather Equation 26.114, which is not at equilibrium when electrons are being pumped into the cathode via the flow of current.

For a water-based electrolytic solution, the anode reaction can be the production of O_2, which can easily be removed from the reaction medium

$$H_2O \rightarrow 1/2 O_2 + 2H^+ + 2e^- \tag{26.118}$$

Numerous studies have shown that low hydrogen overpotential electrically conducting catalysts (e.g., Raney nickel, platinum and palladium on carbon powder, and Devarda copper) can be used to electrocatalytically hydrogenate a variety of organic compounds including benzene and multiring aromatic compounds, phenol, ketones, nitro compounds, dinitriles, and glucose [45, 46, 54, 55, 67–71]. These reactions have been carried out in both batch and semicontinuous flow reactors; in most cases, the reaction products were similar to those obtained from a traditional chemical catalytic scheme at elevated temperatures and pressures.

To circumvent the need for a supporting electrolyte, organic-electrohydrogenation reactions have recently been performed using a solid polymer electrolyte (denoted as SPE) reactor. In such a reactor, the anode and cathode chambers are separated by a thin, hydrated cation-exchange membrane. Thin film electrodes composed of precious metal catalyst powders (one anode and one cathode) are fixed to opposing faces of the membrane, forming a "membrane-electrode-assembly" (MEA), similar to that employed in proton-exchange membrane (PEM) hydrogen/oxygen fuel cells. Water or humidified hydrogen gas is circulated past the backside of the anode, where either water molecules are oxidized to O_2 gas and protons, according to Equation (26.118) or H_2 gas is oxidized to two protons and two electrons. The protons then migrate through the ion-exchange membranes under the influence of the applied electric field to the cathode catalyst component of the MEA, where they are reduced to atomic and molecular hydrogen [Equations (26.114) and (26.116)]. This electro-generated hydrogen can then react with the organic substrate that is circulated through the cathode chamber and flows past the backside of the cathode. Ion (proton) conductivity between the anode and cathode occurs through the wetted cation-exchange membrane, so that pure organic and distilled water (or H_2 gas) can be circulated in the cathode and anode chambers, respectively. In this regard, the SPE reactor represents a significant advancement in electrochemical reactor design since inorganic supporting electrolyte salts are not needed (such salts would have to be removed from the reaction medium downstream from the reactor, often with considerable difficulty). The close proximity of the anode and cathode on an MEA (the electrodes are separated by the ion-exchange membrane, which is at most 200 μm thick) and the high ion-exchange capacity of the cation-exchange membrane ensures facile H^+ transport between the anode and cathode.

Solid polymer electrolyte reactors have been examined previously for oxidation and organic electrochemical reduction of organic compounds [56, 72–78]. Ogumi and co-workers studied the electrochemical hydrogenation of nitrobenzene to aniline using a Cu-Pt cathode/Nafion membrane/Pt anode MEA [76] and the reduction of cyclo-octene, diethyl maleate, ethyl crotonate, and n-butyl methacrylate (dissolved in either ethanol, diethyl ether, or n-hexane) with MEAs composed of Pt, Au, or Au-Pt cathode layers deposited onto the surface of a Nafion membrane [72]. In addition, electrochemical reduction of edible oils (such as soybean or cottonseed) results in a quite different product [79].

The high temperature of a chemical catalytic hydrogenation process promotes the undesirable *cis*-to-*trans* isomerization of fatty acid double bonds; recent studies have shown that the injection of *trans* fatty acids in edible oils increases cholesterol blood levels and contributes to coronary heart disease. The low-temperature electrochemical hydrogenation scheme reduces such isomerization.

A schematic diagram of the SPE oil hydrogenation reactor and the principle reactions associated with electrochemical H_2 generation and oil hydrogenation are shown in Figure 26.20. The key component of the reactor is an MEA, composed of either RuO_2 or Pt/C powder (for the anode) and either Pt-black or Pd-black powder (for the cathode) that are hot-pressed as thin films onto the opposing surfaces of a cation-exchange membrane. During reactor operation at a constant applied current, water of H_2 gas is back-fed to the RuO_2 anode, where it is oxidized electrochemically to produce H^+. Protons migrate through the membrane under the influence of the applied electric field and contact the Pt or Pd cathode where they are reduced to atomic and molecular hydrogen. Oil is circulated past the backside of the cathode and unsaturated triglycerides react with the electro-generated hydrogen species. The SPE reactor was operated successfully with a low anode/cathode voltage drop for a variety of oil feeds at a constant temperature between 50°C and 80°C and an applied current density between 0.10 and 0.490 A/cm². Partially hydrogenated oil products had a lower percentage of total *trans* isomers (6% to 10%) and a somewhat high saturated fat (stearic acid) content, as compared to the products from a traditional chemical catalytic reaction scheme [80].

Various factors that might affect the oil hydrogenation current efficiency were investigated [81, 82], including the type of cathode catalyst, catalyst loading, the cathode catalyst binder loading, current density, and reactant flow rate. The current efficiency ordering of different cathode catalysts powders was found to be Pd > Pt > Rh > Ru > Ir (this is the same order as for chemical reactions). Oil hydrogenation current efficiencies with a Pd-black cathode decreased with increasing current density and ranged from about 70% at 0.050 A/cm² to 25% at 0.490 A/cm² (current efficiencies less than 100% were attributed solely to H_2 gas evolution at the catalytic cathode). The optimum cathode catalyst loading for both Pd and Pt was 2.0 mg/cm². When the oil feed flow rate was increased from 80 to 300 mL/min, the oil hydrogenation current efficiency at 0.10 A/cm² increased from 60% to 70%. A high (70%) current efficiency was achieved at 80 mL/min by inserting a nickel screen turbulence promoter into the oil stream. The concentration of *trans* isomers ranged from about 2% (with a Pt cathode) to about 9% with Pd-black. When a second metal (Ni, Cd, Zn, Pb, Cr, Fe, Ag, Cu, or Co) was electrodeposited onto a Pd-black powder cathode, substantial increases in the fatty acid selectivities were observed (due presumably to changes in oil adsorption/desorption on the catalyst surface) [82].

Principles and Applications of Electrochemical Engineering

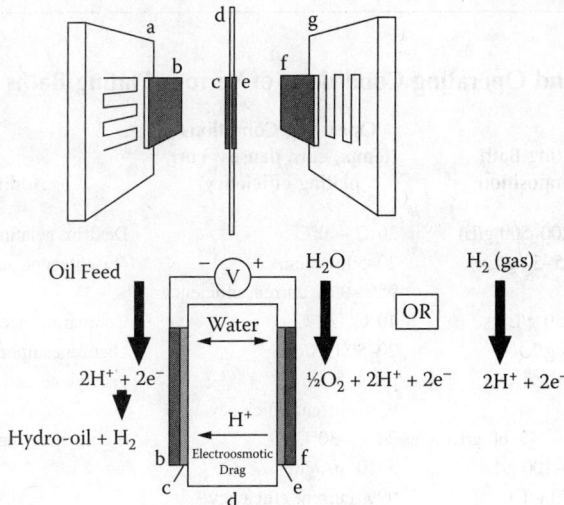

(a and g) Graphite (stainless steel) blocks with cross-patterned flow field.
(b) Gas/oil permeable backing, cathode side.
(c) Cathode thin film catalyst layer.
(d) Nafion cation-exchange membrane.
(e) Anode thin film catalyst layer.
(f) Gas/water permeable bcking, anode side.

FIGURE 26.20 Schematic diagram of the SPE reactor for edible oil hydrogenation using either water of H_2 gas as the source of H.

26.5 ELECTROPLATING AND ELECTROETCHING

26.5.1 Basic Principles

Electroplating is the most widely used electrochemical surface treatment process, involving the electrochemical deposition of a layer of metal or alloy onto a substrate by means of one or more cathodic metal ion reduction reactions of the form

$$M^{n+} + ne^- \rightarrow M \quad (26.119)$$

This technology is used to fabricate functional coatings (e.g., hard, nonporous, corrosion-resistant, and/or wear-resistant coatings) and decorative surface finishes on many objects used in everyday life such as auto components, screws, and kitchen utensils, as well as electronic components (e.g., printed circuit boards). The goal of an electroplating process is to create reproducibly a deposit of uniform thickness (ranging from a fraction of a micrometer to tens of micrometers) and composition that adheres well to the substrate and exhibits the necessary mechanical and physical properties. Most electroplating processes are directed toward the deposition of: (1) single metals (e.g., Cu, Ni, Sn, Cr, Zn, Cd, Pb, Ag, Au, and Pt), (2) metal alloys (e.g., Cu-Zn, Ni-Fe, Cu-Sn, Pb-Sn, Sn-Ni, Ni-Co, and Ni-Cr), or (3) metal/nonmetal composites (e.g., dispersions of Al_2O_3, SiC, WC, polytetrafluoroethylene, or graphite in Ni or some other metal matrix) [83]. The purpose of an electrodeposit is to impart a specific surface property on a material, such as corrosion and wear resistance, hardness, decorative appeal, optical or thermal reflectivity, thermal or electrical conductivity, or solderability.

The main components of an electroplating apparatus are (1) an electroplating bath containing a soluble species of the metal to be deposited; (2) an inert supporting electrolyte to increase the

TABLE 26.12
Components and Operating Conditions of Various Plating Baths

Metal	Plating Bath Composition	Operating Conditions (temp., curr. density, curr. plating efficiency)	Additives	Anode
Cu	$CuSO_4$ (200–500 g/L) H_2SO_4 (25–50 g/L)	20°C –40°C 20–50 mA/cm^2 95%–99% current efficiency	Dextrin, gelatin, S-containing brighteners, sulphonic acids	P-containing rolled Cu
Ni	$NiSO_4$ (250 g/L) $NiCl_2$ (45 g/L) H_3BO_3 (30 g/L)	40°C –70°C 20–50 mA/cm^2 pH 4–5 95% current efficiency	Coumarin, saccharin, benzenesulphonamide, acetylene derivatives	Ni
Ag	$KAg(CN)_2$ (40–60 g/L) KCN (80–100 g/L) K_2CO_3 (10 g/L)	20°C –30°C 3–10 mA/cm^2 99% current efficiency	S-containing brighteners	Ag
Sn	$SnSO_4$ (40–60 g/L) H_2SO_4 (100–200 g/L)	20°C –30°C 10–30 mA/cm^2 90%–95% current efficiency	Phenol, cresol	Sn
Zn	NH_4ZnCl_3 (200–300 g/L) NH_4Cl (60–120 g/L)	25°C 10–40 mA/cm^2 pH 5 98% current efficiency	Dextrin, organic additives	Zn
Cr	CrO_3 (450 g/L) H_2SO_4 (4 g/L)	45°C –60°C 100–200 mA/cm^2 8%–12% current efficiency	Complex fluorides to increase current efficiency	Pb-Sb or Pb-Sn coated with PbO_2

electric conductivity of the plating solution; (3) complexing agents (such as cyanide, hydroxide, and sulphamate ions) to suppress unwanted reactions at the plating surface; and (4) a variety of organic additives, including brighteners (e.g., aromatic sulphones or sulphonates, thiourea, and coumarin for nickel plating), levelers, structure modifiers, and wetting agents that improve the morphology and uniformity of the electroplate thickness, as well as the physical appearance, and physical properties of the deposit.

Most electroplating processes use a DC power supply and cells of simple geometry. Where possible, the anode in an electroplating cell is made from the same metal as that being plated at the cathode so that the concentration of the metal ion in solution remains nearly constant during the plating processes (so long as the anode and cathode current efficiencies are matched). Electrolyte stirring is accomplished by either moving the cathode through the plating bath or by pumping the electrolyte over the cathode surface. Representative examples of plating conditions are listed in Table 26.12 [30].

Electroplating is rarely used for coatings of thickness >75 μm. The upper limit on deposit thickness is governed by cost and time factors, the generation of unacceptably high internal stresses in the deposit, and/or irregular deposit growth. Most plating processes operate at a current density between 10 and 70 mA/cm^2 and are completed in a few seconds to several hours.

The key factors that control the rate of electrodeposition and the structure, physical properties, uniformity, and composition of electrodeposited metals and alloys are (1) thermodynamics (where the electric potential is based on the standard electromotive series); (2) electrode kinetics (which may vary with the structure of the electrodeposit); and (3) mass transport (which is important at high current densities, where the delivery of reactant to the cathode surface affects the local deposition rate and the structure of the deposit).

The normal goal in plating is to achieve a uniform distribution of the metal or alloy deposit, regardless of the geometry of the system. For the deposition of a single metal, the thickness of the electrodeposit depends on the local rate of metal deposition, which, in turn, is determined by (1) the local current efficiency for the metal reduction reaction (metal reduction current efficiency may be lowered by the presence of a secondary reaction, such as H_2 evolution) and by the local electric potential on the electrode (the potential should be identical over the entire cathode surface). Studies that have analyzed theoretically current and metal distribution on a cathode surface fall into four categories:

1. *Primary current and potential distribution* [84]. This analysis is applicable at low-current densities where concentration variations near the electrode surface are neglected and the electrode reaction is reversible. The potential (Φ) distribution in solution is governed by Laplace's equation,

$$\nabla^2 \Phi = 0 \tag{26.120}$$

with a constant potential boundary condition at the electrode surface and zero potential on adjacent insulating surfaces. Once the potential distribution is computed, the current distribution on the electrode surface is found using Ohm's law (i = $-\kappa\nabla\Phi$, where $\nabla\Phi$ is the potential gradient in the solution at the electrode). Alternatively, one can solve for the potential and current distribution using Ohm's law directly as the boundary condition for Equation (26.120) at the electrode and by using $\partial\Phi/\partial y = 0$ on insulating surfaces (where y is the perpendicular distance from the electrode surface). For either solution method, the resulting current distribution depends only on the geometry of the electrochemical cell. For example, the primary current distribution for a circular disk electrode of radius r_o embedded in an infinite insulating plane with a remote counter electrode is given by [4, 84]

$$\frac{i}{i_{avg}} = \frac{0.5}{\sqrt{1 - r^2/r_o^2}} \tag{26.121}$$

This equation indicates that the current density becomes infinite at the edge of the electrode, as shown in Figure 26.21.

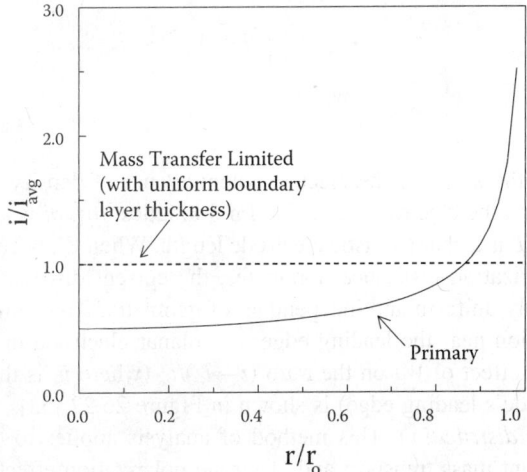

FIGURE 26.21 Primary, secondary, and mass transfer limited current distributions on a rotating disk electrode of radius r_o.

2. *Secondary current distribution* [85, 86]. Here, mass transfer effects are not controlling, but reaction kinetics are considered because of a non-negligible electrode polarization (i.e., electrode reactions that require an appreciable surface overpotential to sustain a high reaction rate). Once again, Laplace's Equation (Equation [26.120]) is solved for the potential distribution, but the boundary condition for Φ on the electrode surface ($y = 0$) is given by

$$-\kappa \nabla \Phi = f(\eta_s) \quad \text{at} \quad y = 0 \tag{26.122}$$

where $f(\eta_s)$ denotes a current density/voltage electrode kinetic expression (such as the Butler-Volmer electrochemical rate equation given by Equation (26.45) in Section 26.2.2 or simplified versions thereof) and η_s (the surface overpotential) is defined as $V - \Phi$ (where V is the imposed potential of the electrode and Φ is the electric potential in solution adjacent to the electrode). For example, when the surface overpotential is sufficiently small, the Butler-Volmer kinetic equation reduces to a linear form and Equation (26.122) is written

$$-\kappa \frac{\partial \Phi}{\partial y} = (\alpha_c + \alpha_a) \frac{i_o F}{RT} \eta_s \quad \text{at} \quad y = 0 \tag{26.123}$$

where α_a and α_c are the anodic and cathodic transfer coefficients, with $\alpha_a = (1 - \beta)n$ and $\alpha_c = \beta n$. A boundary condition expression also can be generated using the linear Tafel equation to relate i and η_s (when η_s is large) The secondary current distribution depends on the same electrode geometry conditions as the primary current distribution, but it also depends on electrode kinetic parameters (such as the transfer coefficient, α, and the exchange current density i_o in Equation (26.123)). The secondary current distribution on an electrode is finite at the electrode edges and is more uniform than the primary current distribution, as can be seen in Figure 26.21 for a rotating disk.

The uniformity of the secondary current distribution has been related to the value of the Wagner (*Wa*) number [87], which expresses the ratio of the polarization resistance at the electrode–electrolyte interface to the ohmic resistance of the electrolyte:

$$Wa = \frac{\kappa \left(\frac{\partial \eta_s}{\partial i}\right)_{i_{av}}}{L} \tag{26.124}$$

where $d\eta_s/di$ is the slope of the electric potential/current density curve at the average current density on the electrode (e.g., for Tafel kinetics, $\partial \eta_s/\partial i$ is given by the cathodic Tafel slope) and L is a characteristic electrode length. When *Wa* becomes large, i.e., when the kinetic polarization resistance dominates, the current distribution on the electrode becomes perfectly uniform and independent of geometry. An example of the secondary current distribution near the leading edge of a planar electrode in a parallel plate flow channel and the effect of *Wa* on the ratio $(i - i_\infty)/i_\infty$ (where i_∞ is the current density far from the electrode's leading edge) is shown in Figure 26.22 [33].

3. *Tertiary current distribution.* This method of analysis applies to those systems where there is significant mass transport and electrode polarization effects. Electrode kinetics is considered, with electrode surface concentrations of reactant and/or products that are no longer equal to those in the bulk electrolyte due to finite mass transfer resistance. The analysis of tertiary current distributions is complex, involving the solution of coupled

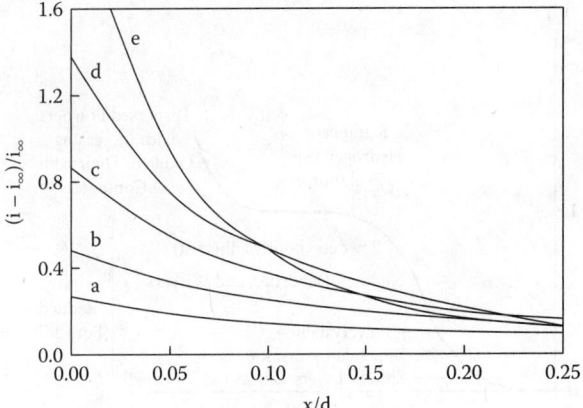

FIGURE 26.22 Secondary current distribution near the edge of a flat sheet electrode in a channel, where d (the inter-electrode gap) is $<<L$ (L = electrode length). X is the distance along the electrode. Each curve corresponds to a different Wagner number: (a) Wa = 0.80; (b) Wa = 0.40; (c) Wa = 0.20; (d) Wa = 0.10; (e) Wa = 0. Figure from [33] (with kind permission from Springer Science and Business Media).

algebraic and differential equations for electrode kinetics, the electric potential distribution, mass transfer, and even fluid flow. For further information on this topic, see [88–90].
4. *Limiting current distribution* [91]. When the deposition rate is controlled solely by mass transfer processes and is independent of the electrode potential distribution, the current density is proportional to the concentration gradient at the electrode surface in accordance with the limiting current density formula presented previously (see Equation [26.76]). The key parameter for this analysis is the effective (Nernst) diffusion layer thickness (δ_e), which depends on the physical properties of the electrolyte, the electrode geometry, and the electrolyte stirring rate, as discussed in Section 26.2.3.2. If δ_e were everywhere the same on the electrode surface, the current density distribution would be uniform, as shown in Figure 26.21 for a rotating disk electrode.

Identifying the limiting current density (i_L) during electrodeposition is important since it represents the maximum rate of metal plating for a given bulk reactant concentration and hydrodynamic flow pattern. Also, the structure of the electrodeposit varies with current density; see Figure 26.23 [92].

The effect of current density on the structure of electrodeposits has been investigated in a special trapezoidal cell with a purposely nonuniform current distribution known as the Hull cell [93]. The shape and typical dimensions of this cell are shown in Figure 26.24 [94]. Due to the nonuniform primary current distribution on the inclined cathode (with a local current density that decreases with increasing anode-cathode separation distance), a single electroplating experiment is used to study the effect of current density on deposit morphology (and deposit composition, for the case of alloy plating).

26.5.2 HIGH-SPEED AND LASER-ASSISTED PLATING

Decreasing the time required for plating a metal, without a loss in the electrodeposit's physical/mechanical properties, tends to decrease manufacturing costs. High-speed plating has been studied repeatedly, in which the mass transfer limiting current density has been increased significantly by increasing fluid convection across the cathode surface (thereby decreasing the mass transfer boundary-layer thickness). Submerged and unsubmerged impinging jets have been widely studied for the plating of electronic and machine parts (see [95, 96]). A plating current density as high as 4 A/cm^2 can be achieved with an impinging jet. Additional studies have focused on the

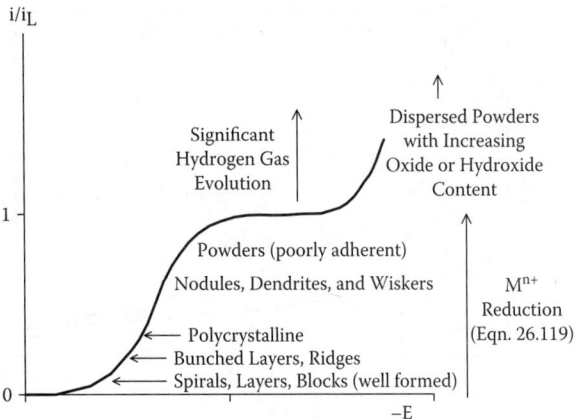

FIGURE 26.23 Variation in metal electrodeposit structure as a function of the applied current density and limiting current density. Figure from [92] (copyright Elsevier 1994).

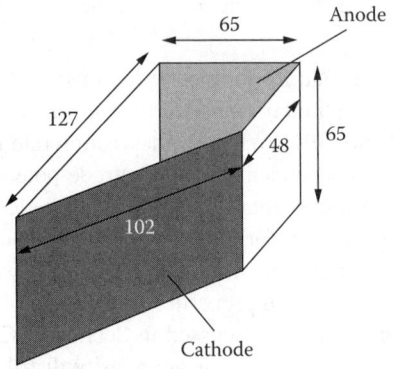

FIGURE 26.24 The classical Hull cell for electroplating. All dimensions are in mm. Figure from [94] (copyright Elsevier 1994).

current distribution in an impinging jet cell [97] and the effects of current density, electrolyte composition, electrolyte velocity, and nozzle height on the metal-thickness distribution [98].

The plating rate and the quality of the electrodeposit can also be increased by using a laser beam during the plating process. Such a technique has enhanced plating rates by factors of 1000 for nickel, copper, and gold [99, 100]. The elevated rates have been attributed to localized heating of the substrate by the laser beam [101]; higher temperatures increase the electrodeposition reaction rates when mass transfer is not controlling. Figure 26.23 shows that better quality electrodeposits are formed at current densities well below the mass transfer limiting current density. With a laser-enhanced electroplating technique, finely detailed electrodeposits can be created, with plated spots as small as 4 µm in diameter [102] (such plating precision is important when fabricating micro-circuitry and electronic devices). More recently, laser-enhanced plating, combined with an impinging jet has achieved localized high-speed plating of metal spots and fine lines of high quality [103]. In addition to increasing convective mass transport, the electrolyte jet limits the region of plating and also acts as a light pipe for the laser beam.

26.5.3 Electroless Plating

During a normal electrolytic metal deposition process, the electrons required for the reduction of the metal ions are supplied by an external current source. In an electroless deposition process, the catalytic or electrocatalytic oxidation of a soluble reducing agent supplies the electrons required for reduction and drives the cathodic metal deposition reaction. The oxidation, reduction, and overall reactions during this plating are

$$\text{Oxidation: } R - ne^- \to O \tag{26.125}$$

$$\text{Reduction: } M^{n+} + ne^- \to M \tag{26.126}$$

$$\text{Overall reaction: } M^{n+} + R \to M + O \tag{26.127}$$

where R denotes the soluble reducing agent. A variety of reducing agents have been investigated, including formaldehyde and hypophosphite [104]. For example, in an electroless nickel plating bath, Ni^{2+} reduction occurs simultaneously with hypophosphite oxidation on a single substrate surface. The process is represented by the following reactions:

Reduction: $\quad Ni^{2+} + 2e^- \to Ni \quad E° = -0.25$ V vs. SHE

Oxidation: $\quad H_2PO_2^- + H_2O \to H_2PO_3^- + 2H^+ + 2e^- \quad E° = +0.50$ V vs. SHE

The overall reaction has a negative ΔG and thus is spontaneous:

Overall: $\quad Ni^{2+} + H_2PO_2^- + H_2O \to Ni + H_2PO_3^- + 2H^+$

An electroless plating bath typically includes: (1) a soluble source of the metal to be plated; (2) a soluble reducing agent; (3) a metal-ion complexant (e.g., EDTA); (4) anion additives (also known as exaltants) such as succinate or fluoride ions that increase the plating rate by catalyzing the oxidation reaction; (5) stabilizers, such as thiourea, which adsorb onto nucleation sites in the plating bath (e.g., metallic particles or dust); and (6) buffers to control the plating bath pH, since the oxidation reaction often produces protons, as is the case for the $H_2PO_2^-$ oxidation above.

During electroless plating, the metal ion and reducing agent are consumed and the products of the oxidation reaction increase in concentration. The composition of the plating bath must therefore be monitored and the bath constituents replenished in order to maintain the consistency of the plating rates and deposit quality. The plating rate is also strongly dependent on temperature. For example, during acidic electroless nickel deposition, no plating will occur below 70°C. Above 70°C, the nickel deposition rate increases markedly, but above ~90°C, the electrolyte is unstable (i.e., at high temperatures metal deposition can occur spontaneously on the walls of the plating tank or there may be homogeneous metal ion reduction in solution). Examples of electroless plating baths, along with the plating temperature and pH conditions, are listed in Table 26.13 [30].

The cost of electroless metal coating is generally higher than conventional applied current electrodeposition processes because (1) bath chemicals are more expensive, (2) the plating bath must be heated, and (3) the plating rate tends to be slow. The use of electroless methods, therefore, is restricted to plating applications where: (1) metal is deposited on a nonconducting (e.g., plastic) substrate; (2) a uniform coating is needed; and (3) hard deposits of low porosity and good wear resistance are required. Thus, electroless plating is particularly attractive to produce small scale features on circuit boards and other electronic components since the process is not governed by the usual primary and secondary current distributions (the anode and cathode reactions occur on the same deposit substrate) and for plating metals on plastics parts in the electronics and automotive industries.

TABLE 26.13
Compositions and Operating Conditions of Electroless Plating Baths

Plating Bath	Bath Composition
Acid nickel	Nickel chloride: 20 g/dm^3
	Sodium hypophosphite: 20 g/dm^3
	Sodium acetate: 10 g/dm^3
	Sodium succinate: 15 g/dm^3
	pH: 4.5
	Temperature: 93°C
Alkaline nickel	Nickel chloride: 30 g/dm^3
	Sodium hypophosphite: 10 g/dm^3
	Ammonium citrate: 65 g/dm^3
	Ammonium chloride: 50 g/dm^3
	pH: 8–10
	Temperature: 90°C
Copper	Copper sulphate: 12 g/dm^3
	Formaldehyde: 8 g/dm^3
	Sodium hydroxide: 15 g/dm^3
	Rochelle salt: 14 g/dm^3
	EDTA: 20 g/dm^3
	pH: 11
	Temperature: 25°C

26.5.4 Alloy Plating

During alloy plating, more than one metal is deposited on the electrode. The relative importance of current distribution and mass transfer may differ for different metals and current densities; a nonuniform current density may lead to an electrodeposit of varying thickness during metal deposition and nonuniform partial currents lead to variations in alloy composition [94,105]. Thus, electroplating of specialty alloys requires tight control of the plating operating conditions, including the electrolyte bath composition (the ratio of metal ion concentrations in solution), the plating temperature, the current density, and the electrolyte flow conditions. For example, when nickel and iron are simultaneously plated onto a rotating disk cathode from a $NiCl_2/FeCl_2/H_3BO_3/NaCl$ bath at pH 3 and 25°C, the alloy composition and current efficiency for electrodeposition are dependent on the plating current density i_p and the electrolyte agitation rate, as shown in Figures 26.25 and 26.26 [106]. In this system, iron deposition is mass transfer controlled and Ni^{2+} reduction is kinetically controlled. The loss in current efficiency (<100%) is due to H_2 evolution, which is mass transfer controlled and independent of the electrode potential in the voltage regime where Ni and Fe co-deposit.

Some examples of alloy plating baths and plating conditions are listed in Table 26.14 [30]. For additional information on alloy as well as pure metal electrodeposition, see [92, 94, 105, 107–112].

26.5.5 Electrochemical Metal Removal Processes

Electrochemical metal removal processes have been developed and utilized by industry for a variety of different applications. Electropolishing as an industrial finishing operation was first demonstrated by Jaquet in 1930 [113]. Metals such as aluminum, steel, brass, cooper, and silver/nickel alloys are anodized to produce a highly reflective mirror finish. The surface to be polished is the anode with a current density in the range of 0.100 to 0.800 A/cm^2. The electrolyte is typically phosphoric

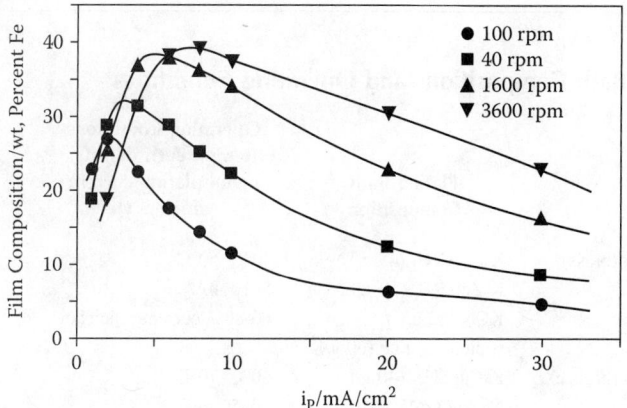

FIGURE 26.25 Composition of a NiFe film plated on a rotating disk electrode as a function of the total plating current i_p for different disk rotation speeds at 25°C. Solution composition was 0.2 M $NiCl_2$, 0.4 M H_3BO_3, and 0.5 M NaCl (pH = 3). Data from [106]. (Reproduced by permission of ECS—The Electrochemical Society.)

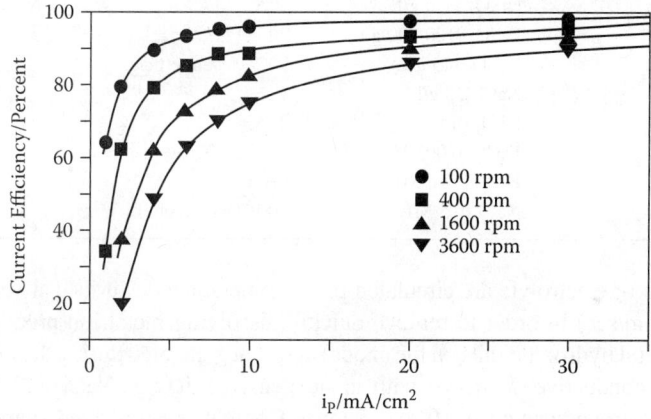

FIGURE 26.26 Plating current efficiency for NiFe film deposition on a rotating disk electrode vs. the total plating current density, for different disk rotation speeds. Plating bath composition was the same as that in Figure 26.25. Data from [106]. (Reproduced by permission of ECS—The Electrochemical Society.)

acid at a temperature of 60°C. Electropolishing involves selective anodic metal electro-oxidation (the potential distribution on the anode surface favors metal dissolution at surface peaks rather than troughs) as well as oxide film formation [30].

Electrochemical machining operations utilize very high current densities (5 to 400 A/cm²) to dissolve and bore through hard and tough metals and alloys. The technique was developed in the late 1950s and early 1960s in the aerospace and power generation industries for shaping and metal finishing operations [114, 115]. The anodic metal dissolution reaction is

$$M \rightarrow M^{n+} + ne^- \tag{26.128}$$

and the cathodic reaction is hydrogen evolution from water,

$$2H_2O + 2e^- \rightarrow H_2 + 2OH^- \tag{26.129}$$

TABLE 26.14
Alloy Plating Bath Compositions and Operating Conditions

Alloy	Plating Bath Composition	Operating Conditions (temp., curr. density, metal plating current efficiency)	Anode
70% Cu—30% Zn (brass)	$K_2Cu(CN)_3$ (45 g/L)	40°C–50°C	Brass
	$K_2Zn(CN)_4$ (50 g/L)	5–10 mA/cm^2	
	KCN (12 g/L)	60-80% current efficiency	
	sodium tartrate (60 g/L)		
40% Sn—60% Cu (bronze)	$K_2Cu(CN)_3$ (40 g/L)	60°C–70°C	Bronze or mixture of
	Na_2SnO_3 (45 g/L)	20–50 mA/cm^2	Sn and Cu
	NaOH (12 g/L)	70-90% current efficiency	
	KCN (14 g/L)		
65% Sn—35% Ni	$NiCl_2$ (250 g/L)	60°C–70°C	Separate Ni and Sn
	$SnCl_2$ (50 g/L)	10–30 mA/cm^2	plates
	$NH_4F \cdot HF$ (40 g/L)	97% current efficiency	
	NH_4OH (30 g/L)		
80% Ni—20% Fe	$NiSO_4$ (300 g/L)	50°C–70°C	Ni + Fe
	$FeSO_4$ (100-200 g/L)	20–50 mA/cm^2	
	H_3BO_3 (45 g/L)	90% current efficiency	
	NaCl (30 g/L)		
	$NiCl_2$ (0.2 M)	25°C	
	$FeCl_2$ (0.005 M)	5–30 mA/cm^2	
	H_3BO_3 (0.4 M)	pH 3	
	NaCl (0.5 M)	70-97% current efficiency	

Large volumes of electrolyte are circulated past the metal anode surface at very high velocities (typically 9 to 60 m/sec) in order to remove quickly dissolving metal ion product (so there is no precipitation of metal hydroxide salts on the anode) and heat generated during the machining process. Normally a highly conductive electrolyte with an inexpensive salt (e.g., NaCl or $NaNO_3$) is employed with small interelectrode spacing (0.10 to 1.5 mm). Complex features and shapes, including fine holes, angled holes, engine casting, and turbine components, can be produced with great precision and reproducibility. In contrast to conventional mechanical metalworking techniques, electromachining leaves the metal without scratches, unwanted sharp edges, and burrs. In addition, the machined surface is almost free of all induced stresses. More recently, electromachining has been used in the electronics industry to drill fine holes (0.25 to 0.4 mm in diameter) in tough cast alloys and to fabricate groove patterns and holes in thin metal foils. A review of such micromachining operations in the electronic industry has been compiled by Datta and Romankiw [116].

26.6 ENVIRONMENTAL CLEAN-UP PROCESSES

Effluent and waste water treatments are of growing importance due to legislation that limits toxic discharge, plus the need to recycle and reuse raw materials. Electrochemical processes are inherently less polluting because oxidation and reduction reactions at the electrodes involve "clean" electrons, as opposed to chemical oxidizing and reducing agents where the spent reagents themselves may be toxic. Several useful electrochemical methodologies have been developed to treat streams containing dissolved metal ions and/or organics, including the removal and recovery of trace metal ions in solution, the destruction of organics via electrochemically generated oxidants, and water purification by electrodialysis membrane processes.

26.6.1 Metal Ion Removal

Electrochemical cells for the removal (i.e., cathodic reduction) and recovery of metal ions from dilute aqueous waste streams have been well studied and several companies market such reactors. Electrolytic processes for metal ion clean-up offer a cost-effective alternative to traditional chemical remediation strategies, such as precipitation with sodium hydroxide, ion-exchange, and solvent extraction, because the metal is removed in its pure and valuable form. Whereas electroplating, electrorefining, and electrowinning processes utilize electrolytes containing relatively high concentrations of metal ions, the waste treatment processes described in this section deal with the removal of ppm concentrations of dissolved metals. The feeds to the electrochemical cell are often from sources associated with: (1) primary ore leaching, (2) spent and contaminated electroplating baths, (3) etching solutions and rinse waters, (4) metal-cleaning solutions, (5) catalyst liquors, (6) photographic processing solutions, and (7) reprocessing of spent batteries.

Continuous and semicontinuous electrochemical reactors are normally employed for effluent metal ion remediation, where the anode reaction is usually oxygen evolution from water [compare with Equation (26.4)]. After the metal contaminant is captured on the cathode, the cathode can be discarded, the collected metal can be resold, or the deposited metal can be chemically or electrochemically etched into a small volume of a suitable leaching liquor (e.g., water) so as to increase its concentration substantially.

In many metal ion removal processes, the waste stream contains few conducting ions and has a low electrical conductivity, thus the electrochemical cell potential will be high. The maximum electrochemical driving force is achieved by operating the reactor at or near the cathodic limiting current density for metal ion reduction. As the metals content drops during the clean-up process, mass transfer and electrolyte conductivity limitations become more severe, resulting in very low-current efficiencies (where cathodic current losses are usually due to excessive H_2 gas evolution) and high-energy costs. For a limiting current density reduction reaction that consumes n moles of electrons per mole of metal ions, and operates at the limiting current (I_L), we can write

$$I_L = k_d A n F C_\infty \qquad (26.130)$$

where k_d is the mass transfer coefficient, A is the cathode area, and C_∞ is the metal ion concentration in the bulk electrolyte. To decrease the reactor size for a given effluent throughput (i.e., for a given electrolyte flow rate and ion concentration change), the $k_d A$ product in the above equation must be maximized. This can be achieved by increasing the solution velocity so as to increase k_d and/or enlarging the surface area of the cathode. One simple reactor design is the plate-in-tank cell [30, 51, 117–119], where numerous vertical flat plate anodes and cathodes are either contained in a single tank or are mounted in a filterpress apparatus with suitable gasket materials between components to seal the assembly. A schematic diagram of a plate-and-frame cell is shown in Figure 26.12. Due to mass transport limitations, such a reactor cannot lower metal concentrations much below 200 ppm. Lower effluent concentrations can be achieved by using high surface area reticulated foam cathodes, often composed of vitreous carbon [30]. Another reactor, for small-scale removal of precious metals (e.g., silver or gold), utilizes rotating cylinder cathodes with a high ratio of cathode surface area to catholyte volume and high fluid velocities (produced by high cylinder rotation speeds) to ensure an acceptable metal ion removal rate Periodically, during reactor operation, the electrodeposit is scraped or peeled from the cylinder cathode surface. This reactor is used in the Eco-Cell process [120].

In recent years, more attention has been drawn to high surface area, three-dimensional particle bed or graphite felt cathodes, where the bed is either stationary or fluidized [52, 121, 122]. Such reactors (e.g., the AKZO/Billiton fluidized-bed reactor or the Enviro-Cell [30, 123, 124]) provide a very high surface area for metal deposition, but the inter-relationship between the fluid flow rate, local metal ion concentration, and the current density distribution in the bed is complex. Fixed,

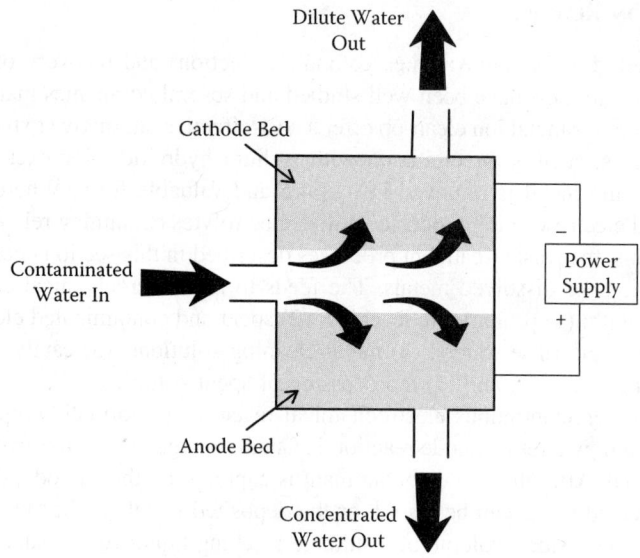

FIGURE 26.27 Schematic diagram of a flow-through packed bed reactor for the removal of dissolved metal ions.

flow-through particle bed electrode reactors have been mathematically modeled in order to better understand and improve the metal ion removal rate (see 4, Chapter 22). A schematic diagram of a flow-through particle bed electrode reactor is shown in Figure 26.27.

In this reactor, the feed solution enters via a central channel between the anode and cathode beds and then flows in the upward and downward vertical directions (where the majority of the solution passes through the porous cathode). When the cathode bed is filled to capacity with deposited metal, the polarity of the electrode beds is reversed and the metal is electrochemically etched into a small liquid volume to create a concentrated solution. The longer the contact time of the metal-laden solution in the porous cathode, the greater the extent of metal removal (where the contact time is inversely proportional to the catholyte flow rate and directly proportional to the cathode bed thickness). To maximize the energy efficiency for metal removal, the entire bed should operate at or near the metal reduction limiting current density, but this is difficult to achieve because of unwanted hydrogen gas evolution. The relevant differential equations are solved to obtain the metal ion concentration, electric potential, and current density distributions in the cathode bed are [125]

1. A material-balance relationship:

$$\frac{dC}{dy} = -\frac{ak_d C}{v} \qquad (26.131)$$

2. A charge-balance equation:

$$\frac{di}{dy} = -nFak_d C = nvF\frac{dC}{dy} \qquad (26.132)$$

3. Ohm's law in solution:

$$i = -\kappa \frac{d\Phi}{dy} \tag{26.133}$$

where a is the area per unit volume of the particle bed cathode, v is the average superficial catholyte velocity in the bed, n is the number of electrons involved in the metal deposition reaction, y is the distance into the cathode as measured from the front of the bed (i.e., that part of the bed closest to the anode), κ is the effective electrical conductivity of the solution in the bed, and Φ is the electric potential in solution.

The above equations are solved with the following boundary conditions: at $y = 0$, $C = C°$ (the initial concentration of metal ion); at $y = L$ (the bed thickness), $i = 0$; and at $y = 0$, $\Phi = 0$. The resulting bed profiles for $C(y)$, $i(y)$, and $\Phi(y)$ [125] are

$$\Phi(y) = \beta[\exp(-\alpha y) - 1 + \alpha y C(L)/C°] \tag{26.134}$$

$$i(y) = nFvC°[\exp(-\alpha y) - C(L)/C°] \tag{26.135}$$

$$C(y) = C° \exp(-\alpha y) \tag{26.136}$$

Here $\alpha = ak_d/v$, $\beta = nFv^2C°/ak_d\kappa$, and $C(L)$ is the metal concentration in the exiting catholyte for a cathode of thickness L. The bed thickness and catholyte flow rate are dictated by the maximum potential drop within the porous cathode. According to Equation (26.134), the cathode electric potential driving force for metal deposition decreases nonlinearly as one moves into the bed, away from the anode (i.e., as y increases); the greatest driving force occurring at the front face of the cathode bed, closest to the anode. Ideally, the potential should be sufficiently negative everywhere in the bed to ensure metal ion reduction, but not so negative that there is appreciable hydrogen evolution or some other unwanted side reaction. Normally, this potential drop is no more than 0.1 or 0.2 V, which then fixes the catholyte velocity according to the following equation:

$$v = \left(\frac{|\Delta\Phi|\kappa ak_d}{nFC°}\right)^{1/2} \tag{26.137}$$

The thickness of the cathode bed is now computed from $|\Delta\Phi|$ and the required/desired change in metal ion concentration within the bed:

$$L = \frac{v}{ak_d} \ln \frac{C°}{C(L)} \tag{26.138}$$

According to the above equations, we can estimate the performance of a particle bed cathode that treats a 667 ppm (667 mg/l) Cu^{2+} aqueous waste stream and generates a product effluent containing 1 ppm Cu^{2+}, as was done in [125]. The cathode bed particles in the reactor are assumed to have a surface area/volume ratio of 25 cm^{-1} with a bed porosity of 0.3. The reactor operates with a superficial solution velocity of 0.0036 cm/sec and a 0.2 V potential variation in the cathode bed, i.e., the front face potential of the cathode is 0.2 V more negative (more cathodic) than the back of the bed. For these conditions, the cathode bed thickness should be 5.0 cm. The length and width of the bed are determined by the volumetric feed flow rate into the reactor, hence, the flow rate divided by the cross-sectional bed area available for flow (width × length × porosity) must be equal to the superficial fluid velocity.

26.6.2 Electrochemical Destruction of Organics

Anodic oxidation processes can effectively destroy toxic organic compounds in aqueous waste streams. The oxidation reaction may occur directly at the anode or indirectly in solution via an oxidizing agent that is electrochemically generated, reacts homogeneously in the bulk solution with the organic, and then is regenerated at the anode [124]. For many organic-laden waste streams, the electrolyte conductivity is low and a supporting electrolyte salt must be added (this salt may have to be removed eventually from the waste stream). It may also be necessary to add a solvent (e.g., water or alcohol) to the stream to increase the solubility of the electrolyte salt.

The electro-oxidation of various organic compounds directly at an anode (e.g., the oxidation of phenol to CO_2 in water) has employed either a Pt or SnO_2 anode [126–128]. SnO_2 is often preferred to minimize organic intermediates and to reduce the total organic carbon content in the water solution. Farmer and co-workers [129], studied the indirect oxidation of organic compounds using Ag^{2+} or Co^{3+} as the oxidizing agent. With HNO_3, the electrochemical oxidation of Ag^+ to Ag^{2+} forms an $AgNO_3^+$ complex. This complex then reacts with an organic pollutant (such as ethylene glycol, benzene, or a chlorinated hydrocarbon) to produce CO_2. For benzene destruction with $AgNO_3^+$, the following oxidation reaction occurred:

$$30AgNO^+ + C_6H_6 + 12H_2O \rightarrow 6CO_2 + 30Ag^+ + 30HNO_3 \qquad (26.139)$$

By re-oxidizing spent Ag^+ at the anode, the organic oxidizing agent is regenerated continuously in the electrochemical reactor, thus minimizing the amount of Ag^{2+} required to destroy the organic compound. Unfortunately, for most organic electrochemical degradation reactions (including Equation (26.139)), the total charge required to oxidize the pollutant to CO_2 is very large (30 Faraday's per mole of benzene for the above reaction, where one Faraday is the charge equivalent of one mole of electrons). Thus, a reactor for organic destruction to a low-molecular-weight innocuous compound such as to CO_2 must operate at a very high current density (which may generate secondary reactions at the anode and lower the current efficiency for organic oxidation) and the electrical cost per pound of degraded organic contaminant is usually high. Whereas Ag^{2+} cannot be used to degrade chlorinated hydrocarbons (due to the formation of insoluble AgCl), Co^{3+} is an effective oxidizing agent for many chlorinated hydrocarbons such as 1,2-dichloro-2-propanol [130]. The relevant homogeneous oxidation and heterogeneous electrochemical reactions for this organic pollutant are

$$C_2H_5(OH)Cl_2 + 5H_2O + 14Co^{3+} \rightarrow 3CO_2 + 16H^+ + 14Co^{2+} + 2Cl^- \text{ (in solution)} \qquad (26.140)$$

$$Co^{2+} \rightarrow Co^{3+} + e^- \text{ (at the electrode)} \qquad (26.141)$$

Kaba et al. [131] employed a Ce^{3+}/Ce^{4+} redox couple and ultrasonic solution agitation to degrade human biomass waste using either a Pt or PbO_2 anode and a 12M H_2SO_4 electrolyte. The total organic carbon content of the electrolyte decreased by a factor of about 3 and the total nitrogen dropped by a factor of 2 after a 70-hour electrolysis.

Since electrochemical oxidation of organics does not yield a useful product, this technology has been driven more by governmental regulation rather than a favorable return on investment. Hence, the development of electrochemical processes for organics destruction has not progressed beyond laboratory-size or pilot-plant reactors.

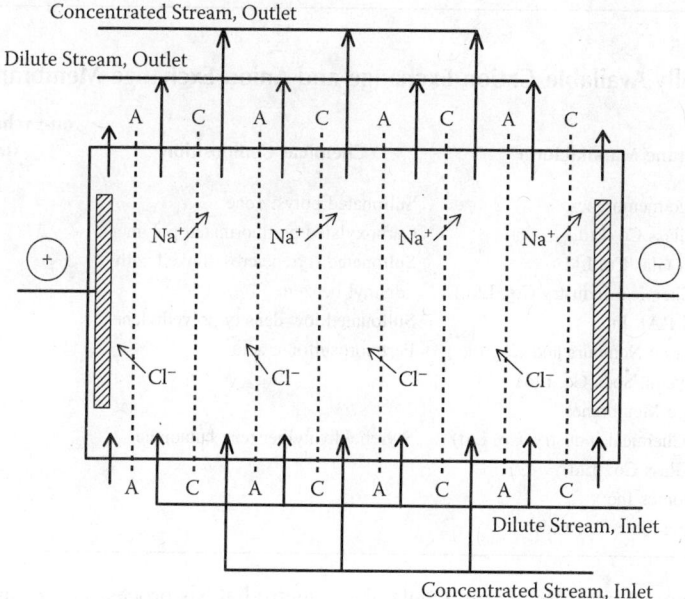

FIGURE 26.28 Diagram of an electrodialysis salt separation process. A denotes an anion-exchange membrane and C denotes a cation-exchange membrane.

26.6.3 Electrodialysis

Electrodialysis is a membrane-based process that utilizes electric fields to separate, remove, or concentrate ionic species in aqueous solutions. A schematic diagram of a typical electrodialysis cell is shown in Figure 26.28, where a series of alternating cation and anion exchange membranes are positioned between the anode and cathode. When an ion-containing solution is pumped through the feed compartments and an electric potential is established between the two electrodes, positively charged cations migrate to the cathode and negatively charged anions move to the anode. Anions can easily move through the anion exchange membranes (i.e., those membranes with fixed positive charges attached to the polymer) but cannot pass through the cation-exchange membranes (these membranes have negative fixed charges which repel mobile anions in solution). Similarly, cations pass through the cation exchange membranes but are blocked by the positively charged anion-exchange membranes. These events increase the salt concentration (i.e., both cations and anions) in alternating compartments, while the remaining compartments are depleted of salt. For an electrodialysis process to be effective, the ion-exchange membranes must repel coions (i.e., those mobile ions in solution with the same electric charge as the membrane's fixed charges) and the transmembrane flux of counterions (i.e., those mobile ions with a charge that is opposite to that of the membrane) must be high.

Electrodialysis was first developed for the desalination of brackish water in order to produce drinking water, and this application continues to be an important industrial application [132]. Another important application is the clean-up of rinse waters from electroplating processes. Process waters from a nickel plating process, for example, may contain up to 1 g/L of nickel sulfate. This solution can be concentrated to 60 g/L by electrodialysis and recyled directly to the plating bath. Similarly, electrodialysis can be used to concentrate Cr^{6+}-containing wastewaters from chromic acid plating baths [133]. Other uses of electodialysis include salt removal from effluent waters prior to reuse in industrial processes [134], the production of ultrapure water [135,136], and salt splitting, where acid and alkali are generated from a neutral salt (e.g., the electrochemical generation of NaOH and H_2SO_4 from Na_2SO_4) [137].

TABLE 26.15
Commercially Available Cation-Exchange and Anion-Exchange Membranes

Membrane Manufacturer	Chemical Composition	Ion-Exchange Capacity (mmol/g)
Cation-exchange membranes	Sulfonated polystyrene	2.4
CMV (Asahi Glass Co. Ltd.)	Carboxylated perfluorinated polymer	0.9
Flemion (Asahi Glas Co. Ltd.)	Sulfonated styrene/cross-linked with	1.4
K101 (Asahi Chemical Industry Co., Ltd.)	divinyl benzene	1.5
R-5010-L (Pall RAI, Inc.)	Sulfonated low density polyethylene	0.91
Nafion (DuPont de Nemours and Co., Inc.)	Perfluorosulfonic acid	2.0
CL-25T (Tokuyama Soda Co. Ltd)		
Anion-Exchange Membranes		
A 111 (Asahi Chemical Industry, Co. Ltd)	Styrene/divinylbenzene butadiene	1.2
AMV (Asahi Glass Co., Ltd)		1.9
103QZL386 (Ionics Inc.)		2.1
ACM (Tokuyama Soda Co. Ltd)		1.5

Clearly, the most important components of an electrodialysis process are the anion- and cation-exchange membranes. Ion-exchange membranes should possess (1) high permselectivity, where the membrane is highly permeable to counterions and impermeable to co-ions; (2) low electrical resistance, so that the transmembrane flux of ions will be high for a given electric potential driving force; (3) good mechanical strength and dimensional stability (i.e., low swelling/shrinkage); and (4) high chemical stability (since the membranes are routinely exposed to acid and alkali solutions as well as various oxidizing agents).

For cation-exchange membranes, one of the following charged moieties is normally attached to the polymer: $-SO_3^-$, $-COO^-$, $-HPO_2^-$, $-AsO_3^{2-}$, or $-SeO_3^-$, whereas an anion-exchange membrane will contain one of the following: $-NH_3^+$, $-RNH_2^+$, $-R_3N^+$, $=R_2N^+$, $-R_3P^+$, or $-R_2S^+$. These ionic groups have significant effects on the selectivity and ionic conductivity of the ion-exchange membranes. For example, sulfonic acid groups (SO_3^-) are completely dissociated over the entire pH range, but carboxylic acid groups (COO^-) are undissociated in acid solutions (at a pH <3). Similarly, quaternary ammonium groups ($-R_3N^+$) in anion-exchange membranes are completely dissociated over the entire pH range, whereas primary ammonium moieties ($-NH_3^+$) are only weakly dissociated. Table 26.15 lists some commercially available cation- and anion-exchange membranes and their associated ion-exchange capacities (with units of mmol per gram of dry membrane).

The permselectivity of an ion-exchange membrane is governed by the solubility of counterions and coions at the membrane–solution interfaces and by transport processes in the membrane. One approach to describe mathematically counterion and co-ion solubility in an ion-exchange membrane is Donnan equilibrium theory [138]. For a cation-exchange membrane (with negatively charged fixed ions of concentration X) immersed in a single 1:1 salt solution (at a concentration C^b), the counterion concentration inside the membrane (C_+) is given by

$$C_+ = \frac{X}{2} + \left[\left(\frac{X}{2}\right)^2 + \left(\frac{f_+^b}{f_+^m} C^b\right)^2 \right]^{1/2} \qquad (26.142)$$

where $f_\pm^b$ is the mean molar activity coefficient of the salt in the external bulk solution and $f_\pm^m$ is the activity of salt inside the membrane. A similar relationship can be derived for the concentration of anions in an anion-exchange membrane. Once the counterion concentration in the membrane is known, the coion concentration is computed from the macroscopic electroneutrality equation (i.e., $C_- = C_+ - X$ for a cation-exchange membrane or $C_+ = C_- - X$ for an anion-exchange membrane). The drawback in using the Donnan equation lies in the inability to predict accurately $f_\pm^m$. When activity coefficients are neglected, the Donnan theory works well for $C^b \ll X$ and $C^b \gg X$, but at salt concentrations comparable to or somewhat greater than the membrane fixed-charge concentration (an important concentration range for many problems), this model underestimates the counterion concentration in a variety of ion-exchange membranes (see, for example, [139, 140]). The application of Donnan theory to a multicomponent external salt solution results in the following relationship for the membrane-phase concentrations of cation species i and j [141]:

$$\left[\frac{f_i^m C_i^m}{f_i^b C_i^b}\right]^{z_i} = \left[\frac{f_j^m C_j^m}{f_j^b C_j^b}\right]^{z_j} \tag{26.143}$$

When activity coefficients were neglected, Equation (26.143) predicted incorrectly that the membrane-phase concentration ratio of any two similarly charged cations will be the same as that in the external solution [142].

Verbrugge and Hill [143] and more recently Bontha and Pintauro [144] and Pintauro et al. [145] extended the double layer theory of Gur et al. [146] to compute coion and counterion partition coefficients in a DuPont Nafion® cation-exchange membrane. This model has been successfully tested for the partition of aqueous H_2SO_4 solutions, aqueous salt mixtures containing alkali metal chlorides, alkali metal/divalent cations, and quaternary ammonium salts in Nafion. In the theoretical analysis, a cylindrical pore structure for the membrane was assumed and ion fixed-charge site electrostatic interactions and alignment of membrane-phase solvent molecules (by the strong electric field generated by fixed charges on the pore wall) were taken into account. Solvent orientation and the presence of an ion solvation interaction energy term in the model allowed one to compute different equilibrium partition coefficients for ions of like charge and different size immersed in the same solvent [143].

Cation, anion, and water transport in ion-exchange membranes have been described by several phenomenological solution-diffusion models and electrokinetic pore-flow theories. Phenomenological models based on irreversible thermodynamics have been applied to cation-exchange membranes, including DuPont's Nafion perfluorosulfonic acid membranes [147, 148]. These models view the membrane as a "black box" and membrane properties such as ionic fluxes, water transport, and electric potential are related to one another without specifying the membrane structure and molecular-level mechanism for ion and solvent permeation. For a four-component system (one mobile cation, one mobile anion, water, and membrane fixed-charge sites), there are three independent flux equations (for cations, anions, and solvent species) of the form

$$N_i = -L_1 \nabla C_i - L_2 \nabla P - L_3 \nabla \Phi \tag{26.144}$$

where Φ is the electric potential, P is pressure, and the L_j parameters are complicated functions of concentration, pressure, temperature, and binary interaction parameters (for an n component system, there are a total of $n(n-1)/2$ such parameters). The form of the flux equations does vary with membrane type, whereas the L_j terms contain information regarding molecular-level interactions and processes that distinguishes one membrane from another. Unfortunately, there have been very few studies linking any of the phenomenological parameters to membrane structure [149], and the large number of parameters for multicomponent systems make this modeling approach cumbersome and difficult to use. A review of irreversible thermodynamic transport models for ion-exchange membranes can be found in [150].

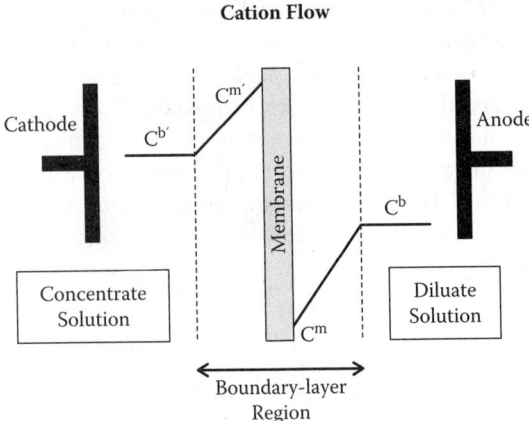

FIGURE 26.29 Schematic diagram of cation concentration profiles on the diluate and concentrate sides of a cation-exchange membrane during current flow.

Another structure/function transport model, often referred to as the "capillary" or "electrokinetic" model, predefines the microlevel structure of an ion-exchange membrane as an array of pores of known dimensions with a specified distribution of ion-exchange sites on the pore walls. Equations describing solute and solvent transport and theories for molecularlevel ion/solvent and ion–membrane interactions are then generated, based on this pore structure [151]. The fundamental transport equation for the molar flux of ionic species is the Nernst-Planck equation

$$N_i = -\frac{z_i F D_i C_i}{RT} \nabla \Phi - D_i \nabla C_i + v C_i \qquad (26.144)$$

When the radius of the membrane pores is of the same order of magnitude as the Debye screening length, electroneutrality within the pores is no longer valid. A non-zero space charge produces coupling between electrical forces, mass transfer, and fluid flow within the pore. This coupling manifests itself in two ways: (1) the interaction of fluid flow and the electric field generates a body force term in the momentum balance equation, and (2) the electrostatic potential in the space charge region of the pore causes counterion enrichment and coion exclusion. Also, the interaction between the space-charge region and the driving forces for transport directed parallel to the pore wall results in such electrokinetic phenomena as streaming potentials and electroosmosis [152]. For examples of electrokinetic transport models as applied to ion-exchange membranes, see [151, 153, 154].

Two important design parameters for an electrodialysis stack are the limiting current density and current utilization. The transport of charged species to the anode or cathode through a set of ion-exchange membranes leads to a concentration decrease of counterions in the boundary layer at the membrane surface that faces the dilute compartment and an increase at the surface facing the concentrated solution (Figure 26.29). The decrease in salt concentration at the membrane surface of the diluate limits the ion flux (or current density) through the membrane. At the absolute limiting current density, the ion concentration at the surface of the cation- and/or anion-exchange membrane drops to zero. The limiting current density (i_L) in this situation is given by [132, 155]

$$i_L = \frac{C^b z_+ F k_d}{t_+ - t_+^*} \qquad (26.146)$$

where C^b is the bulk solution concentration in the cell with the depleted solution, k_d is the mass transport coefficient in the diluate cell (the cell losing ion species), t_+ is the cation transference number in the diluate solution, and t^*_+ is the cation transference number in the membrane (which, for an ideal cation-exchange membrane, will be 1.0). According to the above equation, the limiting current density is proportional to the ion concentration and the mass transfer coefficient in the dilute cell (k_d is dependent on the solution flow velocity). If the limiting current density is exceeded in an electrodialysis stack, the process efficiency of the separation will be dramatically lowered, due to the increase in electrical resistance of the diluate solution (since there are no ions to carry the current in the solution in contact with the membrane) and due to a secondary reaction, such as the water splitting reaction (for aqueous solutions), which produces H^+ and OH^- (i.e., if there are no cations to carry the current through the membranes when the limiting current is exceeded, energy will be expended to generate H^+ and OH^- ions from water). As a consequence of the problems associated with operating the electrodialysis stack at the limiting current, the cells are often operated at approximately 80% of i_L.

The current utilization refers to the fraction of the total current that passes through an electrodialysis membrane stack that actually is used to transfer anions or cations from a feed solution. The current utilization is always less than 100% due to: (1) co-ion intrusion into the ion-exchange membrane (i.e., no ion-exchange membrane will completely exclude co-ions), (2) osmotic and ion-bound water transport (water will flow into the concentrate compartment due to osmosis and the electro-osmotic drag of water molecules with the transporting ions), and (3) shunt currents that skirt around the membranes and pass through the stack manifold.

The energy necessary to remove salts from a solution by electrodialysis is directly proportional to the total current flowing through the membrane stack and the total voltage drop between the two electrodes in the stack. The energy consumption is described mathematically by [132]

$$\text{Energy} = \frac{In R_e tz F Q_f \Delta C}{\xi} \tag{26.147}$$

where Q_f is the volumetric flow rate of the feed solution, ΔC is the concentration difference between the feed solution and the diluate, R_e is the resistance of an anion-exchange-membrane/cation-exchange-membrane cell pair (including the solution resistance between the membranes), n is the number of cell pairs in a stack, t is time, and ξ is the current utilization.

Capital costs for an electrodialysis stack include the stack housing, pumps, membranes, and electric rectifiers. The cost of an electrodialysis plant is strongly dependent on the total membrane area (denoted as A) needed to produce a given concentration change. The required membrane area, in turn, is directly proportional to the solution flow rate and specified concentration change and is inversely proportional to the applied current density and current utilization [132, 156]:

$$\text{Area} = \frac{z F Q_f n \Delta C}{i \xi} \tag{26.148}$$

where i is the applied current density.

26.7 CORROSION PROCESSES

Most metals occur naturally in their oxide or sulfide forms. The process of metal refining converts these ores into pure metals. Thermodynamically, a metal will return spontaneously to its original oxide form. Metal oxidation can occur at high temperatures, by direct reaction with O_2, or at a moderate temperature by reaction with water, O_2, and/or H^+. The latter oxidation, commonly referred to as wet corrosion, has as its basis the combination of electrochemical cathodic reduction and anodic metal oxidation reactions into a "corrosion cell." Thus, many corrosion processes are basically rooted in electrochemistry. The corrosion cell may be viewed as a short-circuited electrochemical cell, where a voltage difference between the anode and cathode provides the driving force for spontaneous oxidation and reduction reactions. The anode and cathode may be physically separated from one another or, more commonly, may occur on a single heterogeneous metal surface, where discrete microcells arise due to differences in metal composition and/or spatial variations in the composition of the solution in contact with the metal surface.

The primary oxidation reaction involves the anodic electrochemical dissolution of metal species M, according to the following reaction:

$$M \rightarrow M^{n+} + ne^- \tag{26.149}$$

This reaction is driven to the right and controlled by one or more reduction reactions, typically,

Proton reduction (in acid media): $\quad 2H^+ + 2e^- \rightarrow H_2 \quad$ (26.150)

Oxygen redution (in aerated solutions): $\quad O_2 + 4H^+ + 4e^- \rightarrow 2H_2O \quad$ (26.151)

Water reduction (in alkaline or pH-neutral solutions): $2H_2O + 2e^- \rightarrow H_2 + 2OH^- \quad$ (26.152)

Oxygen reduction in neutral/alkaline solution: $\quad 1/2 O_2 + H_2O \rightarrow 2OH + 2e^- \quad$ (26.153)

The overall reaction in a corrosion cell is the sum of an anodic and cathodic reaction. For the case of metal dissolution and O_2 reduction, the overall reaction is

$$M + n/4 O_2 + n/2 H_2O \rightarrow M^{n+} + nOH^- \tag{26.154}$$

Dissolved metal ions can either pass into the bulk solution or react immediately with OH^- to form insoluble metal oxides/hydroxides that coat the metal surface. For iron corrosion, iron hydroxide species react further with water and oxygen to form rust ($Fe(OH)_3$) according to the following reaction sequence:

$$2Fe^{2+} + 4OH^- \rightarrow 2Fe(OH)_2 \text{ (ppt)} \tag{26.155}$$

$$2Fe(OH) + 2H_2O + O_2 \rightarrow 2Fe(OH)_3 \tag{26.156}$$

The rate of a corrosion reaction is affected by pH (via H^+ reduction and hydroxide formation), the partial pressure of O_2 (the solubility/concentration of oxygen in solution), fluid agitation, and electrolyte conductivity. Corrosion processes are analyzed using the thermodynamics of electrode reactions, mass transfer of the cathode reactants O_2 and/or H^+, and the kinetics of metal dissolution reactions [157, 158].

26.7.1 Thermodynamics of Corrosion Processes

We have seen in Section 26.2.1 that thermodynamics (i.e., equilibrium half-cell potentials) can be used to determine which of two half-cell reactions proceeds spontaneously in the anodic or cathodic direction when the two reactions occur on the same piece of metal or on two metal samples that are in electrical contact with one another. The half-cell reaction with the higher equilibrium potential will always be at the cathode. Thus, under standard conditions any metal dissolution (corrosion) reaction with an $E°$ less than 0.0 V vs. SHE will be driven by proton reduction while metal dissolution reactions with an $E°$ less than +1.23 V vs. SHE will be driven by dissolved oxygen reduction. Noble metals such as Pd and Au have very high standard reduction potentials and, therefore, are highly resistant to corrosion. The list of standard reduction potentials (as reported in Table 26.2) indicates that the presence of O_2 in solution increases significantly the number of metals that corrode under normal environmental conditions. Thus, a simple and effective step in lowering metal corrosion is to deaerate aqueous solutions in contact with metal surfaces. For example, the copper oxidation (dissolution) reaction ($E° = +0.337$ V vs. SHE) cannot be driven by the hydrogen evolution reaction, but it can occur spontaneously if it is coupled to the oxygen reduction reaction.

When the activity of ions in solution is not unity, the standard reduction potential must be corrected by use of the Nernst equation (Equation [26.28]) to give the equilibrium half-cell potential (as explained in Section 26.2.1). Additionally, corrosion reactions may produce or consume H^+, in which case the spontaneity of a corrosion reaction will be pH-dependent. In order to anticipate the effects of pH on electrochemical reactions and chemical reactions at an electrode during metal oxidation (such as metal hydroxide formation), electrochemists employ potential vs. pH plots for a given metal, also known as Pourbaix diagrams [159]. Such a diagram is a plot of redox potential (the ordinate) as a function of pH for a given metal in water under standard thermodynamic conditions. A Pourbaix diagram for iron in water at 25°C is shown in Figure 26.30. The chemical and electrochemical reactions that define the numbered lines are listed in the figure caption. When the applied electrode potential is greater than the equilibrium potential for a given reaction, that reaction is driven spontaneously in the anodic direction. Similarly, a reduction reaction is favored thermodynamically when the electrode potential is less than the equilibrium potential for that reaction.

Although Pourbaix diagrams provide important information regarding the stability of various metal species in different oxidation states, they are of limited use because: (1) they contain only thermodynamic information and thus provide no indication of the rate (speed) of the relevant reactions, (2) they usually refer to reactions at 25°C and one atmosphere pressure, and (3) they consider only pure metals.

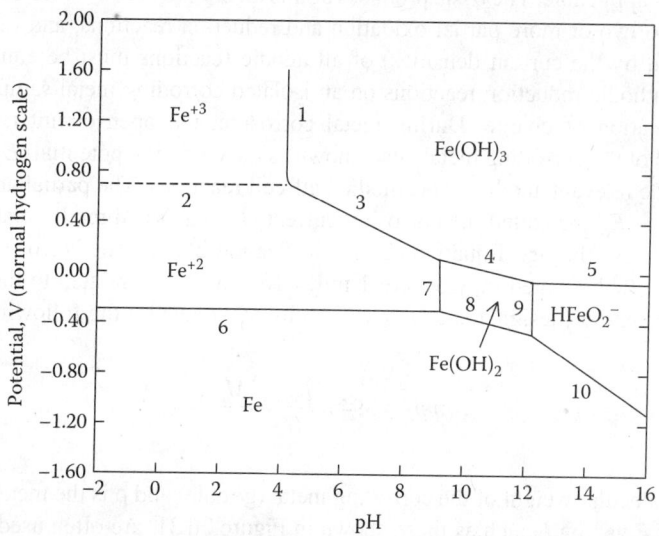

FIGURE 26.30 Simplified Pourbaix diagram for the iron-water system at 25°C. The numbers associated with each line correspond to the following reactions: (1) $2Fe^{3+} + 6H_2O = 2Fe(OH)_3 + 6H^+$, (2) $Fe^{2+} = Fe^{3+} + e^-$, (3) $2Fe^{2+} + 6H_2O = 2Fe(OH)_3 + 6H^+ + 2e^-$, (4) $2Fe(OH)_2 + 2H_2O = 2Fe(OH)_3 + 2e^-$, (5) $2HFeO_2^- + 2H_2O = 2Fe(OH)_3 + 2e^-$, (6) $Fe = Fe^{2+} + 2e^-$, (7) $Fe^{2+} + 2H_2O = Fe(OH)_2 + 2H^+$, (8) $Fe + 2H_2O = Fe(OH)_2 + 2H^+ + 2e^-$, (9) $Fe(OH)_2 = HFeO_2^- + H^+$, and (10) $Fe + 2H_2O = HFeO_2^- + 3H^+ + 2e^-$.

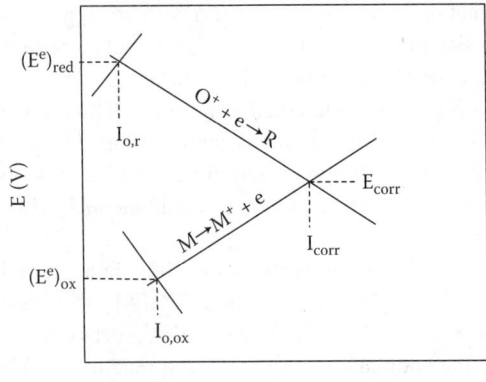

FIGURE 26.31 Current-voltage plots for metal corrosion, showing Tafel lines for metal (*M*) exidation and oxidant (*O*) reduction. The intersection of the Tafel lines gives the corrosion potential and corrosion current.

26.7.2 Kinetic Aspects of Corrosion Reactions

Almost all metals corrode, but many metals corrode very slowly under normal environmental conditions, due in part to kinetic limitations of the metal dissolution reaction. Thus, the rate of metal corrosion can be anticipated and controlled by developing kinetic rate expressions for metal oxidation reactions. There is a major difference, however, between "classical" electrochemical metal dissolution kinetics and metal dissolution in a corrosion system, that difference being the occurrence of one or more oxidation and reduction reactions on the same metal.

As stated above, wet-metal corrosion involves a metal dissolution (oxidation) reaction coupled to a reduction reaction. Electrons produced at anodic sites pass through the metal, and are consumed by reduction reactions so that there is no net charge build-up in the metal. For this situation, "mixed potential theory" is applicable. The basic premises of this theory are (1) any electrochemical reaction can be divided into two or more partial oxidation and reduction reactions, and (2) the sum of the rates (as quantified by the current densities) of all anodic reactions must be equal to the current densities for all cathodic reduction reactions on an isolated corroding metal sample because there is no net accumulation of charge. During metal corrosion, the open circuit (zero net current) electrode potential of the corroding metal, also known as the corrosion potential (E_{corr}), lies between the potentials of the relevant anodic and cathodic half-cell reactions. The partial anodic or cathodic current that flows at E_{corr} is called the corrosion current (I_{corr}) and is directly related to the rate of metal corrosion (g/sec) through Equation (26.14) in Section 26.2.2. The corrosion rate in mils of metal lost per year (abbreviated mpy, where 1 mil = 10^{-3} inches) is related to the current density at the corrosion potential (designated as i_{corr} with units of A/cm^2) by the following equation:

$$mpy = 394 \times \frac{i_{corr}}{nF} \times \frac{M}{\rho} \quad (26.157)$$

where M is the molecular weight of the corroding metal (g/mole) and ρ is the metal density (g/cm^3).

Tafel plots of E vs. log I, such as those shown in Figure 26.31, are often used to determine the rate of a corrosion process. For a corroding metal (anode) that is driven by a single kinetically controlled reduction reaction (such as hydrogen evolution from an acid-containing solution), one can write the following Tafel equations for cathodic proton reduction and anodic metal dissolution:

$$E - E_{H2}^e = -\beta_{H2}(\log I - \log I_{o,H2}) \quad (26.158)$$

$$E - E_M^e = \beta_M (\log I - \log I_{o,M}) \qquad (26.159)$$

where $1/\beta_{H2}$ and $1/\beta_M$ are the cathodic and anodic Tafel slopes (b_c and b_a, respectively, as defined by Equations (26.51) and (26.49) in Section 26.2.2), E^e is the equilibrium half cell potential, as defined by the Nernst equation [Equation (26.28)], and I_o is the exchange current. These equations are shown graphically as straight lines in Figure 26.31, where E is plotted vs. $\log |I|$. When the anodic and cathodic reactions are coupled to one another during a metal corrosion process, the electric potential driving force for both reactions is the same and equal to E_{corr} and the anodic and cathodic reaction rates are the same and equal to I_{corr}. Substituting E_{corr} and I_{corr} into Equations (26.158) and (26.159) for E and I and solving the resulting expressions simultaneously for I_{corr} and E_{corr} yields

$$I_{corr} = (I_{o,M})^{\beta_1} (I_{o,H2})^{\beta_2} \exp\left[\frac{2.3(E_{H2}^e - E_M^e)}{\beta_M + \beta_{H2}}\right] \qquad (26.160)$$

and

$$E_{corr} = \frac{\beta_M E_{H2}^e + \beta_{H2} E_M^e}{\beta_{H2} + \beta_M} + \frac{\beta_{H2} \beta_M}{2.3(\beta_{H2} + \beta_M)} \log\left[\frac{I_{o,H2}}{I_{o,M}}\right] \qquad (26.161)$$

where $\beta_1 = \beta_M/(\beta_M + \beta_{H2})$ and $\beta_2 = \beta_{H2}/(\beta_M + \beta_{H2})$. As can be seen from the above equations, the corrosion current and corrosion potential depend on both the thermodynamic equilibrium potential and the kinetic parameters I_o and β. Graphically, the anodic and cathodic Tafel lines intersect at the point I_{corr}, E_{corr}, as shown in Figure 26.31. Similarly, for two or more metal dissolution and/or reduction reactions, one first writes expressions similar to Equations (26.158) and (26.159) for each reaction and then substitutes these formulas into the charge-balance expression $\Sigma I_{anodic} = \Sigma I_{cathodic}$. Again, one can perform this analysis graphically using anodic and cathodic Tafel plots, by first summing all cathodic and anodic Tafel lines (where the currents from each anodic or cathodic Tafel line are summed at each potential) and then identifying the intersection of the anode and cathode summation lines, where the intersection is at I_{corr}, E_{corr} (Figure 26.32).

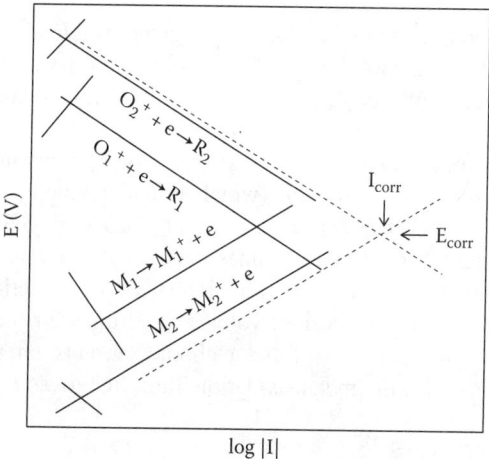

FIGURE 26.32 Current-voltage Tafel plots for metal corrosion showing multiple metal dissolution and oxidant oxidation reactions. The dotted line are the sum of all oxidation reactions and all reduction reactions. The intersection of the dotted lines gives the corrosion potential and corrosion current.

Combining the anodic and cathodic Tafel equations results in the following modified Butler-Volmer rate expression:

$$I = I_{corr}\left(\exp\left[\frac{2.3(E - E_{corr})}{b_a}\right] - \exp\left[\frac{2.3(E_{corr} - E)}{|b_c|}\right]\right) \quad (26.162)$$

Using the above equation, the deviation in the corroding metal potential away from its open circuit potential (E_{corr}) generates a net anodic or cathodic current (in the same way that a deviation of the electrode potential away from E^e for a single electrode reaction produces an oxidation or reduction current). The magnitude of the current, as given by Equation (26.162), depends on the level of electrode polarization ($E - E_{corr}$), the cathode and anodic Tafel slopes (b_c and b_a, respectively), and I_{corr} (which is analogous to the exchange current in the Butler-Volmer equation).

For a small electrode polarization (denoted as ΔE, which can be no more than ±20 mV) that produces a current of I, the exponential terms in Equation (26.162) can be linearized, resulting in the following current-voltage expression:

$$I = 2.3 \cdot I_{corr}\left[\frac{\Delta E}{b_a} + \frac{\Delta E}{|b_c|}\right] \quad (26.163)$$

When a metal corrodes, there is no external flow of current (electrons) to indicate the metal dissolution rate. One can, however, place the corroding metal in a three-electrode electrochemical cell (working electrode, counter electrode, and reference electrode, as shown in Figure 26.4), polarize the working electrode metal either anodically or cathodically away from its open circuit (corrosion) potential by no more than ±20 mV (this potential difference is measured between the working and reference electrodes), and then measure the resulting current. For this case, Equation (26.163) can be rearranged to give the following equation for the corrosion current

$$I_{corr} = \frac{b_a \cdot |b_c|}{2.3(b_a + |b_c|)} \cdot R_P \quad (26.164)$$

where R_p is known as the polarization resistance and is given by $\Delta I/\Delta E$ (i.e., the slope of the linear portion of an experimental I vs. E plot, for small polarizations of the working electrode). Once I_{corr} is known for a metal specimen of area A, then i_{corr} can be determined and the metal corrosion rate in mpy can be calculated using Equation (26.157).

The use of Tafel plots for the analysis of metal corrosion systems indicates how dissolved O_2 in solution and the subsequent O_2 reduction (which is under kinetic control) accelerates metal corrosion. Due to the high value of E^e for O_2 reduction (+1.23 V vs. SHE), the intersection of the oxygen reduction and metal dissolution Tafel lines occurs at high values of E_{corr} and I_{corr}. When the reduction of both H^+ and O_2 drives metal corrosion (with the reduction reactions under kinetic control), one simply adds together the current-voltage Tafel lines for the two reduction reactions. A new line is then drawn for the sum of the cathodic currents on the corroding metal. The intersection of this new line with the metal oxidation Tafel line gives E_{corr} and I_{corr}.

26.7.3 Mass Transfer Effects

In some situations, the oxygen or proton reduction reaction during metal corrosion will be mass transfer (diffusion) controlled, due to poor fluid agitation and/or a low concentration of H^+ or dissolved oxygen in solution (this is especially true for O_2, which has a low solubility in water at

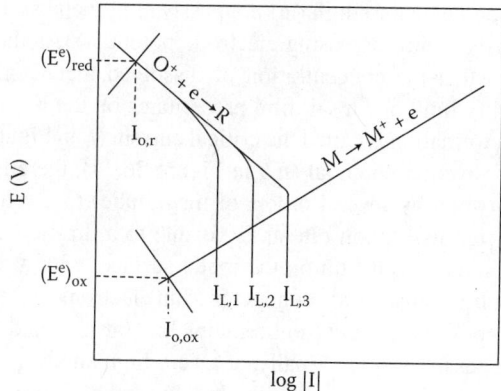

FIGURE 26.33 Current-voltage Tafel plot for a mass transfer controlled metal corrosion process, where the corrosion current is equal to the limiting current for oxidant reduction.

typical environmental conditions, i.e., atmospheric pressure and a temperature of about 25°C). Now the reduction reaction that drives metal dissolution is at or near its limiting current, as shown in Figure 26.33 (a vertical line on a E versus log $|I|$ Tafel plot is characteristic of an electrode reaction where the rate is independent of the electrode potential). In accordance with mixed potential theory, the intersection of the anodic and cathodic current-voltage curves gives I_{corr} and E_{corr}. As can be seen in Figure 26.33, I_{corr} is equal to the mass transfer limiting current for the reduction reaction, and, as such, the metal corrosion rate will increase with increasing I_L (i.e., with increasing stirring and/or increasing concentration of the cathode reactant in solution).

26.7.4 METAL PASSIVATION

When some metals or alloys are immersed in an oxidizing solution, corrosion does not occur, although it is favored thermodynamically. For example, iron will dissolve in dilute, but not in concentrated nitric acid. Corrosion is prevented because of the formation of a protective oxide film on the metal surface. This loss in chemical reactivity of a metal is known as passivation. Metals which readily undergo passivation in damp air include Fe, Al, Cr, Ni, Ti, and Pt.

Passivation can be induced and observed in electrochemical experiments. Figure 26.34 shows a typical E versus log $|I|$ plot for a metal that exhibits active-passive behavior [157]. The metal

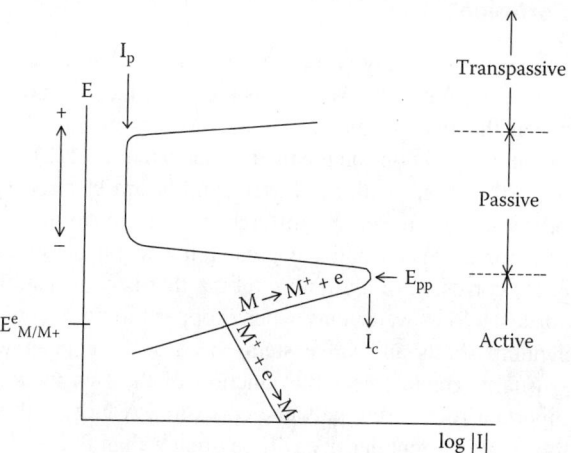

FIGURE 26.34 Current-voltage curve showing active-passive metal dissolution behavior.

initially displays dissolution behavior similar to nonpassivating metals (i.e., there is an increase in the anodic dissolution current with increasing electrode potential). As the potential increases, the dissolution rate is so high that the concentration of dissolved metal ions adjacent to the anode surface exceeds its solubility limit and a salt film precipitates on the metal. The maximum current achievable before salt film formation is called the critical current (I_c in Figure 26.34). At a designated potential known as the passivation potential (E_{pp} in Figure 26.34), there is a sharp decrease in the metal dissolution current (often by several orders of magnitude) to a value defined as the passive current I_p. The drop in metal dissolution rate at E_{pp} is due to a drastic change in the composition and properties of the precipitated salt film on the metal surface, with a reduction in film porosity and the conversion of the film from an ionic conductor to an electronic conductor. For an appreciable electrode potential range above E_{pp}, the current remains low (at I_{pass}) and independent of potential. This regime is termed the passive region. Finally, at a very high anodic potential, metal dissolution commences again, as the protective passive film breaks down. In this transpassive region, the dissolution current once again increases with increasing electrode potential. As will be discussed below, the presence of a passive film on certain metals has been exploited as a convenient way of protecting the underlying metal from further corrosion.

26.7.5 Types of Corrosion and Methods of Corrosion Prevention

26.7.5.1 Uniform Attack

Uniform attack or general overall corrosion is the most common form of corrosion, causing most metal destruction on an overall tonnage basis. During this form of corrosion, the metal becomes thinner and eventually fails. Although widespread, the form of corrosion is not of too great a concern because the corrosion rate can be accurately predicted using simple tests (immersing the metal specimen in the liquid of interest for an extended period of time and measuring the weight loss or performing electrochemical polarization experiments to determine the corrosion current) so that the metal lifetime can be estimated. As an example, this type of corrosion would occur when a steel or zinc plate is immersed in a dilute sulfuric acid solution, with rusting (corrosion) over the entire metal surface.

Prevention—Uniform corrosion can be prevented or reduced by (1) the proper choice of metal for a given environment; (2) the use of metal coatings, such as a paint coating or nonoxidizing organic inhibitors (e.g., benzotriazole or organic amines [160]) in solution that adsorb onto and protect the metal surface; or (3) cathodic protection (discussed below).

26.7.5.2 Galvanic Corrosion

A potential difference exists between any two dissimilar metals that are exposed to a corrosive or electrically conductive solution. When the two metals are in contact, the potential difference causes electrons to flow between the two metals. To a first approximation, the determination of the equilibrium potential of the two reactions on the metals (see Section 26.2.1.1) can be used to predict which metal will corrode (that metal with the lower equilibrium electrode potential will become the anode and will corrode). The greater the difference of the potentials of the two metals, the greater the driving force for corrosion and the greater the metal dissolution rate. Examples of galvanic corrosion are: (1) corrosion of aluminum tubing that is connected to brass (Cu-Zn alloy) fittings, (2) corrosion of a steel hot water tank where copper tubing is connected to the tank, and (3) corrosion of steel pump shafts and valve stems because of contact with graphite packing. Galvanic corrosion is usually greatest near the junction of the two metals. Anode/cathode area effects also play an important role during galvanic corrosion. A large cathode and small anode is unfavorable because the anode current density will be greater than that at the cathode even though both the anodic and cathodic currents are equal.

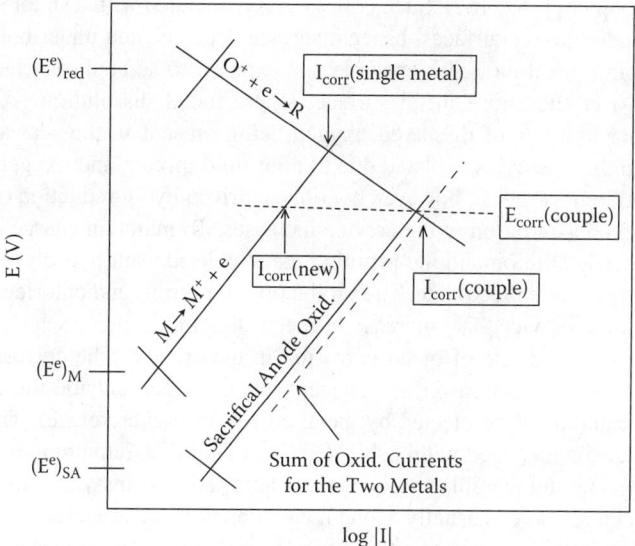

FIGURE 26.35 Current-voltage Tafel plot showing cathodic protection by use of a sacrificial anode. E^e_{red} = equilibrium potential of the reduction reaction; E^e_M = equilibrium potential of the primary metal dissolution reaction; E^e_{SA} = equilibrium potential of the sacrificial anode oxidation reaction.

Prevention—The following procedures combat galvanic corrosion: (1) dissimilar metals in contact should be as close together as possible in the galvanic series, i.e., in the list of standard reduction potentials (compare with Table 26.2 in Section 26.2.1); (2) avoid the unfavorable effect of a small anode and large cathode by fabricating small metal parts from highly corrosion resistant materials; (3) insulate dissimilar pieces of metal, wherever practical; (4) design equipment with readily replaceable anodic parts or make them thicker for longer life; (5) use an impressed DC electric current to pump electrons into the metal to be protected and thus force the mixed potential of the couple lower; or (6) connect a third metal with an equilibrium potential that is anodic to both metals in the galvanic couple (the third metal is known as a sacrificial anode and it will corrode preferentially and pump electrons into the metal to be protected).

The process of protecting either a metal undergoing uniform corrosion or one metal of a galvanic couple by methods (5) and (6) in the preceding paragraph is called cathodic protection [157, 161]. Here, electrons are pumped into the metal to be protected, thus converting it from a dissolving anode into a noncorroding cathode. Figure 26.35 shows how Tafel plots can be used to explain how cathodic protection with a sacrificial anode lowers the dissolution rate of a metal. When the metal to be protected is coupled to a second metal with a lower equilibrium potential, the E_{corr} of the two-metal couple is lowered. The overall corrosion rate of the couple increases (I_{corr} for the couple is greater than I_{corr} for the single metal, as shown in Figure 26.35), but this increase is at the expense of the sacrificial anode. The corrosion rate of the metal to be protected actually decreases (the horizontal line at $E = E_{corr}$ for the couple intersect the M → M⁺ oxidation line at a value of I_{corr} that is smaller than the I_{corr} for the single metal system). A practical example of cathodic protection by use of a sacrificial anode is galvanized steel, where a surface coating of zinc ($E°_{Zn/Zn+}$ = –0.763 V vs. SHE) becomes the anode and electrons are pumped into the underlying steel ($E°_{Fe/Fe2+}$ = –0.440 V vs. SHE), thus protecting it from corrosion.

26.7.5.3 Crevice Corrosion

Crevice corrosion is the intensive localized corrosion of a metal that occurs within crevices, overlapping metal parts, and shielded metal surfaces, where part of the metal surface is exposed to a corrosive environment [157, 162]. Such corrosion is associated with a small volume of stagnant solution in holes, under gasket surfaces, beneath surface deposits, and under bolts and rivet heads. When two overlapping metal parts, for example, are exposed to aerated seawater, there is initially uniform corrosion over the entire metal surface, where metal dissolution is driven by oxygen reduction (low concentrations of dissolved oxygen being present in the seawater). After a short time, oxygen within the crevice is depleted due to poor fluid mixing and oxygen reduction ceases. Dissolution of metal in the crevice, however, continues (driven by O_2 reduction outside the crevice) and the metal cation concentration in the crevice increases. To maintain charge neutrality, Cl^- ions migrate into the crevice solution and the resulting metal chloride salt hydrolyzes into H^+, Cl^-, and an insoluble metal hydroxide precipitate. The production of protons and chloride anions accelerates metal dissolution in the crevice. The increase in metal dissolution increases Cl^- migration, which in turn further heightens the rate of metal corrosion in the crevice. The corrosion process within the crevice drive oxygen reduction on the adjacent metal surface outside the crevice (this metal surface is actually cathodically protected by metal corrosion in the crevice). Thus, during crevice corrosion, metal attack is localized within shielded areas while the remaining metal surface suffers little or no damage. The unfavorable cathode/anode area ratio means very rapid disintegration of metal within the crevice. There is usually a long incubation period associated with crevice corrosion (when dissolved O_2 within the crevice is depleted). Once crevice corrosion starts, it often proceeds at an accelerating rate. Metals or alloys that depend on oxide films or passive layers for corrosion resistance are particularly susceptible to crevice corrosion (the films are destroyed by H^+ and/or Cl^-).

Prevention—Crevice corrosion is prevented or minimized by: (1) closing crevices and overlapping metal parts by continuous welding, caulking, or soldering; (2) removing solid deposits from metal surfaces; and (3) increasing fluid stirring to minimize the difference in dissolved O_2 concentration within and outside of a crevice [163].

26.7.5.4 Pitting

Pitting is a destructive and insidious form of corrosion, where localized dissolution occurs in small holes or pits on the metal surface [157, 161]. Although metal weight loss is small, equipment failure occurs due to perforation of the metal. Pitting occurs in passivating metals, where certain aggressive species in solution, e.g., chlorides, bromides, and hypochlorites, break down the passive oxide film. An autocatalytic and accelerating metal dissolution process then begins at film rupture. As is the case for crevice corrosion, rapid metal dissolution creates an excess of metal cations within the pit and the electromigration of Cl^- ions into the pit to maintain electroneutrality. The subsequent hydrolysis of metal chloride species produces H^+ and metal hydroxides (which precipitates at the pit entrance). The solubility of oxygen drops to virtually zero as the solution in the pit acidifies. Cathodic reduction of oxygen on the large metal surfaces adjacent to the pit drives the dissolution reactions, in a manner similar to crevice corrosion.

Prevention—The methods outlined above to combat crevice corrosion also apply to pitting. Metals that are resistant to pitting should be used as alloying agents; their passive films are more protective and more stable to halogen attack. For example, the addition of 2% molybdenum to 18-8 (type 304) stainless steel to produce 316 stainless steel significantly increases pitting resistance.

Anodic Protection—Anodic protection procedures make use of a passive film on a metal surface to minimize corrosion. Normally, the application of an anodic current to a metal structure would increase its electrochemical dissolution rate. With metals that exhibit active-passive behavior (e.g., Ti, Ni, Fe, Cr, and their alloys), application of an anodic current or an anodic potential results in passive film formation and a dramatic reduction in the metal corrosion rate. There are essentially two methods of anodically protecting a metal: (1) the imposition of an anodic potential greater than the passivating potential (E_{pp} in Figure 26.34) or (2) the use of oxidizing inhibitors (e.g., NO_3^-, CrO_4^{2-}, SO_4^{2-}, CH_3COO^-, or OH^-) [164] at a sufficiently high concentration so that the mixed

Principles and Applications of Electrochemical Engineering

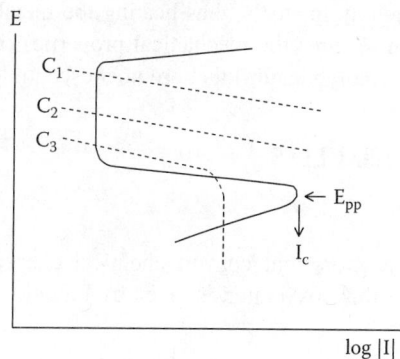

FIGURE 26.36 Tafel plot showing the anodic protection of a passivating metal by use of an oxidizing inhibitor, where the inhibitor concentrations are $C_1 > C_2 > C_3$ (C_3 is not high enough for metal passivation).

potential for metal dissolution coupled to inhibitor oxidation is raised into the metal's passive region (Figure 26.36) [157, 162].

26.7.5.5 Stress Corrosion Cracking

Stress corrosion cracking is a form of localized corrosion, where the simultaneous presence of tensile stresses and a specific corrosive environment produces metal cracks [157, 168]. Stress corrosion cracking generally occurs only in alloys (e.g., Cu-Zn, Cu-Al, Cu-Si, austenitic stainless steels, titanium alloys, and zirconium alloys) and only when the alloy is exposed to a specific environment (e.g., brass in ammonia or a titanium alloy in chloride solutions). Removal of either the stress on the metal (which must have a surface tensile component) or the corrosive environment will prevent crack initiation or cause the arrest of cracks that have already propagated. Stress corrosion cracking often occurs where the protective passive film breaks down. The continual plastic deformation of the metal at the tip of the crack prevents repassivation of the metal surface and allows for continued localized metal corrosion.

Prevention—Stress corrosion cracking can be prevented by increasing the tensile strength of the metal or alloy (to prevent formation of nucleating cracks) and/or reducing the stress on the metal (by redesigning the part or by stress-relieving heat treatments). Corrosion inhibitors and coatings (e.g., electrodeposited zinc, chromium, nickel, copper, and nitride coatings) are also effective in eliminating this type of corrosion.

26.7.5.6 Hydrogen Embrittlement

Hydrogen embrittlement is caused by penetration of hydrogen into a metal, which results in a loss of ductility and tensile strength [157, 165]. High-strength metals and alloys are particularly susceptible to this kind of corrosion. Its exact mechanism is not well understood. For titanium and other strong hydride-forming metals, dissolved hydrogen reacts within the metal to form brittle hydride compounds. Another mechanism is based on the accumulation of hydrogen near dislocation sites or microvoids in the metal. Hydrogen embrittlement is distinguished from stress corrosion cracking by the polarity of the metal when corrosion occurs. While stress corrosion cracking is associated with highly anodic metal surfaces, hydrogen embrittlement often occurs due to the presence of cathodic surface metal sites, where H_2 is produced during normal (uniform) corrosion.

Prevention—Hydrogen embrittlement is prevented or minimized by (1) reducing the uniform corrosion rate on a metal surface (by coating/painting the metal surface, for example), to decrease the rate of hydrogen evolution on the metal surface: (2) baking the metal (hydrogen evolution is an almost reversible process, especially in steels, thus heating the metal to 200°F to 300°F liberates trapped hydrogen), which often restores the mechanical properties of the metal; and/or (3) using alloying agents (e.g., nickel and molybdenum) that are not susceptible to hydrogen embrittlement.

26.8 BATTERIES AND FUEL CELLS

26.8.1 Batteries

Batteries are electrochemical reactors that convert chemical energy into electrical energy. The thermodynamic relationship for this conversion is given by Equation (26.19):

$$E = -\frac{\Delta G}{nF}$$

where E is the electromotive force (voltage) of the battery system, ΔG is the free energy change for the overall battery reaction (the sum of the primary anode and cathode reactions), and n is the number of electrons involved in the battery reactions. The available thermodynamic voltage depends on the chemical reactants. In a battery, reactants for the oxidation and reduction reactions do not meet directly, but are consumed at different sites, the anode and cathode, which causes electrons to flow through an external circuit between the battery terminals. The energy transfer is manifested as a voltage (a measure of the energy difference of electrons at the anode and cathode) and current (the flow of electrons through the external circuit connecting the anode and cathode). All batteries have four common features: (1) the anode and cathode electrodes must be electronically conducting; (2) the electrolyte between the anode and cathode must be ionically conducting; (3) there must be physical separation of the electrodes and the fuel and oxidant reactants; and (4) the available fuel and oxidant materials for the electrode reactions must be contained within the battery itself. Primary batteries [166–170] are designed only for a single discharge, while a secondary battery is expected to be capable of repetitive charge/discharge cycles. The original chemistry of a primary battery cannot be restored by electrical means internal to a cell due to (1) irreversibility in phase changes and chemical interconversion of the electrode active material; (2) isolation, electrically or electronically, of active material; and/or (3) local poor conductivity of electrode materials in the discharged state [169]. In contrast to primary batteries, secondary batteries [168, 171, 172] can be recharged electrically to their original condition many times by passing current through them in the opposite direction to that of the discharge current. Thus, the anode and cathode battery reactions in a secondary battery must be reversible.

26.8.1.1 Battery Performance Characteristics

Battery performance is related to the available voltage and current while the battery is discharging. The terminal voltage of a battery depends on the thermodynamic free-energy change of the anode and cathode reactions, according to Equation (26.19) and by the various activation (kinetic), concentration, and resistance overpotentials that lower the battery's voltage (as discussed in Section 26.3.4):

$$E_{cell} \equiv E^C - E^A = (E^e)^C - (E^e)^A - \left|\Sigma\eta^A\right| - \left|\Sigma\eta^C\right| - \left|\eta_{ohm}\right| \qquad (26.165)$$

where the superscripts A and C refer to the anode and cathode, respectively. According to the above equation, one should choose the anode and cathode reactions with a large $(E_e^C - E_e^A)$. At the same time, the various overpotentials (primarily the kinetic and resistance overpotentials) must be minimized, by ensuring fast electrode reactions and by designing the battery with a low electrical resistance (e.g., a high-conductivity electrolyte, a small interelectrode gap, and a low-resistance

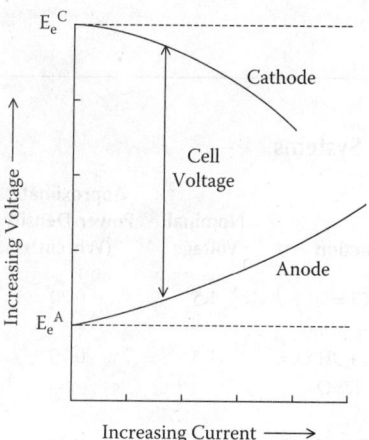

FIGURE 26.37 Cell polarization as a function of current density for the anode and cathode of a battery. The displacement of the electrode potentials from their equilibrium (thermodynamic) potentials is due to the presence of overpotentials.

anode/cathode separator). Unfortunately, these overpotentials increase with increasing current, causing the battery voltage to decrease, as shown schematically in Figure 26.37.

The key components of a primary or secondary battery are: (1) the container, which must possess the necessary mechanical strength and resistance to internal and external corrosion; in some batteries, such as the Lechlanche cell (Table 26.16), the zinc container acts as one of the electrodes; (2) separators, to minimize intermixing of the reactants and products of the anode and cathode reactions (e.g., microporous polyethylene for lead-acid batteries); (3) metal grid or sheet current collectors, which provide mechanical support for and improve the electrical conductivity of the electrodes (the electrodes are often composed of a thick, porous layer of active material with limited mechanical strength and electrical conductivity); (4) the electrolyte, which is determined by the electrode reactions, along with the need for substantial conductivity; and (5) active electrode materials, which are in solid form and in intimate contact with the electrolyte (often the active material structure is a porous electrode paste that is spread on a current collector).

Current is a measure of the rate at which a battery is discharging. As can be seen in Figure 26.37, a battery's current and voltage behaviors are linked. Ideally, one would like the battery to deliver a high current without excessive voltage losses, which means that there must be rapid electron transfer reactions at the two electrodes. Additionally, the electrodes should ideally be designed to ensure an adequate supply of electro-active material near sites where electron transfer reactions are occurring. Battery properties that are dependent on current/voltage behavior are [168]:

1. Capacity. The capacity of a battery depends on how much of the active materials at each electrode are consumed during battery discharge. The theoretical capacity of each electrode, with units of ampere hours, is calculated from Faraday's law, with the weight (w) and molecular weight (MW) of active (reactant) material and the number of electrons involved in the electrode reaction (n):

$$\text{Capacity} = \frac{w \times n \times F}{MW} \quad (26.166)$$

The capacity of the battery is determined by the electrode of lowest capacity. The actual capacity depends on the discharge current, commonly measured by the decline in the battery voltage with time at a fixed discharge current (see Figure 26.38).

TABLE 26.16
Commercial Primary Battery Systems

Common Name	Overall Cell Reaction	Nominal Voltage	Approximate Power Density (Wh/cm³)	Comments
Leclanche	$Zn + 2MnO_2 + 2NH_4Cl = Zn(NH_3)_2Cl_2 + H_2O + Mn_2O_3$	1.5	0.20	Low cost; general purpose; wide range of sizes
Zinc chloride	$4Zn + 8MnO_2 + ZnCl_2 + 9H_2O = 8MnOOH + ZnCl_2 \cdot 4ZnO \cdot 5H_2O$	1.5	0.20	Intermediate cost and performance
Alkaline	$2Zn + 3MnO_2 = 2ZnO + Mn_3O_4$	1.5	0.25	Sets standard for cylindrical cells
Silver	$Zn + AgO_2 = 2Ag + ZnO$	1.6	0.50	Good pulse power, higher voltage than Zn/HgO or Zn/Air
Mercury	$Zn + HgO = Hg + ZnO$	1.35	0.50	Sets the standard for button-size cells
Zinc-air	$2Zn + O_2 \text{ (air)} = 2ZnO$	1.4	1.0	Twice the capacity of a mercury battery
Li/CuO	$2Li + CuO = Li_2O + Cu$	1.5	0.60	Potential replacement for leclanche and zinc chloride batteries
Li/FeS	$2Li + FeS = Li_2S + Fe$	1.6	0.40	Replacement for Zn/HgO and Zn/Ag$_2$O
Li/SO$_2$	$2Li + 2SO_2 = Li_2S_2O_4$	2.8	0.4	Military battery; low temperature; excellent storage
Li/SOCl$_2$	$Li + SOCl_2 = LiCl + SO_2$	3.6	1.0	High voltage; high-energy density
Li/CF$_x$	$xLi + CF_x = xLiF + C$	2.7	0.50	High voltage; long shelf-life; wide operating temperature
Li/MnO$_2$	$Li + MnO_2 = LiMnO_2$	2.8	0.50	High voltage; long shelf-life; wide operating temperature
Li/I$_2$	$2Li + I_2 = 2LiI$	2.8	0.60	Battery for heart pacemaker

Data from [167–170].

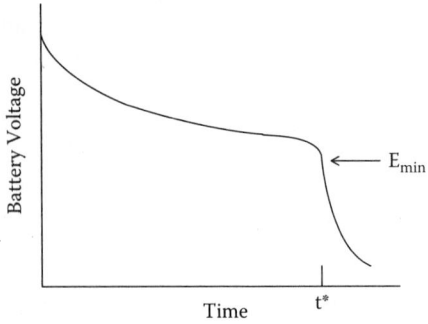

FIGURE 26.38 Voltage vs. time curve during battery discharge at a constant current.

From such data, the capacity is calculated from the product of I (the discharge current) and t^* (the time required for the battery voltage to decrease to a value of E_{min}, below which the battery is no longer useful). The flatness of the discharge curve and the ability of the battery to deliver its expected capacity at increasing currents (i.e., discharge rates) are important battery properties. The real capacity of a battery is usually well below its theoretical value.

2. Theoretical specific energy density. The theoretical specific energy density (with units of kW-hr/kg) is defined as the maximum energy available from a battery ΔG, divided by the anode and cathode reactant masses M_i:

$$\text{Theoretical Energy Density} = \frac{\Delta G}{\Sigma M_i} \qquad (26.167)$$

3. Actual specific energy density. The actual specific energy density is determined from the weight of active material w, the discharge current I, the discharge time t^*, and the average battery voltage during discharge E_{av}:

$$\text{Actual Energy Density} = \frac{I \times t^* \times E_{av}}{w} \qquad (26.168)$$

4. Storage density. The storage density of a battery is a measure of the charge per unit weight stored in a battery and is given by the capacity per unit weight of the battery. Low-molecular-weight electrode reactants translate into high storage densities, such as lithium, where the electrode reaction Li → Li$^+$ + e produces 1 F (96,487 coulombs) of charge during the consumption of only 7 g of metal.

5. Power density. The power density of a battery is given by the product of the current and battery voltage divided by the battery weight:

$$\text{Power Density} = \frac{I \times E_{cell}}{\text{weight}} \qquad (26.169)$$

As the battery voltage decreases during discharge at a given current, the power density will also decrease.

6. Cycle life. The cycle life of a secondary battery is the number of cycles that are possible before failure occurs. The charging period of a secondary battery's charge/discharge cycle should return the electrodes' active material to a suitable state for further discharge. The active material must have the proper chemical composition, morphology, and distribution in the battery after charging. Since the electrode reactions are not completely reversible, secondary batteries have a finite cycle life (with either a slow decay in performance or a sudden and catastrophic failure). The most common secondary battery failures are: (1) corrosion of the current collector or contacts, (2) shorting due to dendrites of electrically conductive material between electrodes, (3) changes in electrode morphology (so-called shape changes), and (4) shedding of active material from the electrodes.

The anode, cathode, and overall cell reactions during discharge of a lead-acid battery are

Anode (negative electrode): $\quad Pb + SO_4^{2-} \rightarrow PbSO_4 + 2e^-$ $\qquad$ (26.170)

Cathode (positive electrode): $\quad PbO_2 + 4H^+ + SO_4^{2-} + 2e^- \rightarrow 2H_2O + PbSO_4$ $\qquad$ (26.171)

Overall cell reaction: $\quad PbO_2 + Pb + 4H^+ + 2SO_4^{2-} \rightarrow 2PbSO_4 + 2H_2O$ $\qquad$ (26.172)

In the lead-acid battery, the actual capacity is well below its theoretical value due to the formation during discharge of poorly conducting $PbSO_4$, which forms a barrier between the electroactive PbO_2 and the current collector. Examples of various primary and secondary batteries are listed in Tables 26.16, 26.17, and 26.18.

TABLE 26.17
Commercial Secondary (Rechargeable) Battery Systems

Common Name	Overall Cell Reaction	Nominal Voltage	Approximate Power Density Wh/l	Approximate Power Density Wh/kg	Comments
Lead-Acid	$Pb + PbO_2 + 2H_2SO_4 = 2PbSO_4$	2.10	80	35	Lowest cost; largest sales
Nickel-Cadmium	$Cd + NiOOH + 2H_2O = Cd(OH)_2$	1.30	80	38	High rate; available sealed
Nickel-Iron	$Fe + 2NiOOH + 2H_2O = Fe(OH)_2 + 2Ni(OH)_2$	1.37	90	30	Very long cycle life; almost indestructible; old technology
Nickel-Hydrogen	$H_2 + 2NiOOH = 2Ni(OH)_2$	1.35	90	45	Special space battery; very long cycle life; high self-discharge
Silver-Zinc	$Zn + AgO + H_2O = Zn(OH)_2 + Ag_2OZn + Ag_2O + H_2O = Zn(OH)_2 + Ag$	1.86 / 1.60	200	100	Two-step discharge; limited cycle life; high-energy density
Lithium-MoS$_2$	$Li + MoS_2 = LiMoS_2$	2.3	150	80	Small sealed cell

Data from [168, 169, 171, 172].

TABLE 26.18
Battery Systems Formerly/Currently in Various Stages of Development

Common Name	Overall Cell Reaction	Nominal Voltage	Approximate Power Density Wh/l	Approximate Power Density Wh/kg	Comments
Nickel-Zinc	$Zn + 2NiOOH + H_2O = Zn(OH)_2 + Ni(OH)_2$	1.65	95	60	Limited cycle life; high rate capability
Zinc-Bromine	$Zn + Br_2 = ZnBr_2$	1.85	75	65	Bromine complexed by quaternary ammonium salt; circulating electrolyte
Lithium-FeS	$Li + FeS = LiFeS$	1.33	90	95	Low cost; high temperature fused salt
Sodium-Sulfur	$2Na + 3S = Na_2S_3$	2.1	120	160	Solid -Al_2O_3 separator; high temperature operation
Aluminum-Air	$4Al + 3O_2 + 2H_2O = 2Al_2O_3 \cdot H_2O$	1.6		300	Circulating electrolyte; low cost; low-energy efficiency
Lithium-TiS$_2$	$Li + TiS_2 = LiTiS_2$ (intercalate)	2.1		150	Room temperature; organic electrolyte
Solid-State Lithium	$C_6 + LiMO_2 = Li_xC_6 + Li_{1-x}MO_2$	3.4–3.7		90–120	

Data from [169, 173–176].

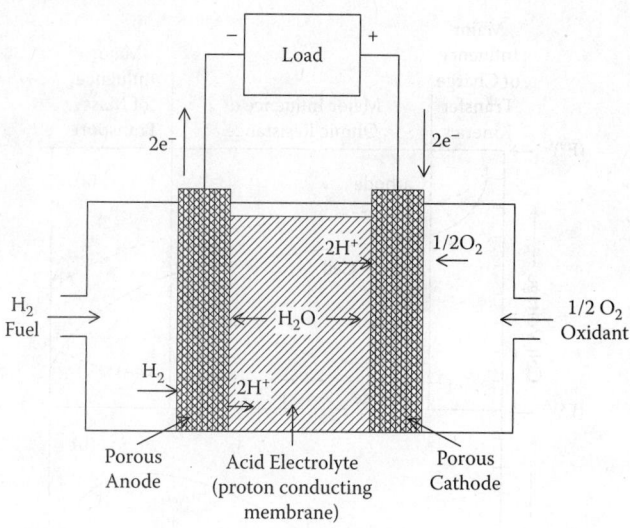

FIGURE 26.39 Principle components in a hydrogen/oxygen fuel cell.

26.8.2 Fuel Cells

A fuel cell is an electrochemical device that converts the chemical energy of a fuel and oxidant directly into low-voltage, direct current electricity. Unlike batteries, the fuel and oxidant are stored externally and are fed to the electrodes as needed. For most terrestrial fuel cells, the oxidant is atmospheric oxygen, whereas the fuel can be H_2 or a low-molecular-weight hydrogen-containing compound such as methanol. For H_2 and O_2 reactants, operation and components of a fuel cell are illustrated in Figure 26.39.

H_2 gas is fed to the anode, where it spontaneously oxidizes to form two protons (H^+) and two electrons. Protons migrate through an acid electrolyte to the anode (under the influence of the electric field), where they react with oxygen and electrons to form water. The electrons that are created at the cathode flow through an external metallic circuit and are consumed at the anode. The electrons provide electrical energy to an external load (e.g., an electric motor) in the circuit. The anode, cathode, and overall reactions for this fuel cell are

Anode: $\quad\quad\quad\quad\quad H_2 \rightarrow 2H^+ + 2e^-$ \quad\quad\quad\quad (26.173)

Cathode: $\quad\quad\quad\quad 1/2 O_2 + 2H^+ + 2e^- \rightarrow H_2O$ \quad\quad (26.174)

Overall fuel cell reaction: $\quad\quad H_2 + 1/2 O_2 \rightarrow H_2O$ \quad\quad\quad (26.175)

The overall reaction is basically the spontaneous cold combustion of hydrogen. In contrast to a combustion process, however, the fuel and oxidant in a fuel cell are kept separate and are never mixed. In the fuel cell, little heat is liberated and instead the free energy is released directly as electrical energy. Hence, the maximum energy efficiency is written as

$$\text{Max. Fuel Cell Eff.} = \frac{\Delta G_T}{\Delta H_o} \quad\quad (26.176)$$

where ΔG_T is the change in free energy for the overall anode/cathode reaction at the cell operating temperature and ΔH_o is the corresponding enthalpy change. ΔG_T is related to the open circuit

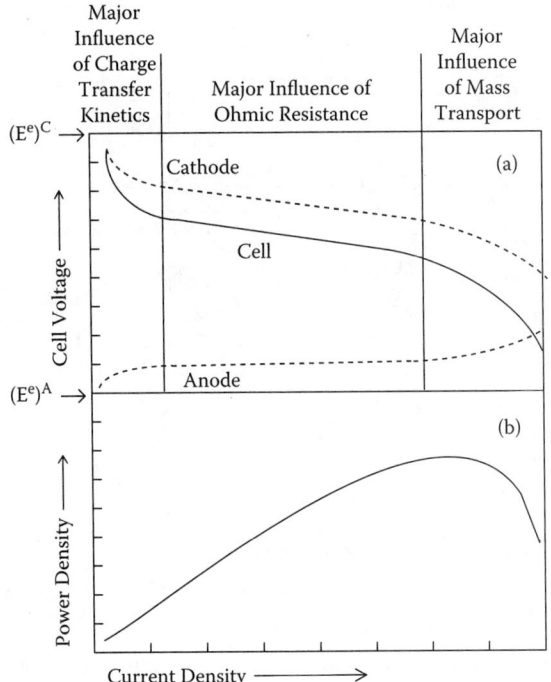

FIGURE 26.40 Typical current density/voltage (a) and current density/power density (b) plots for a fuel cell.

theoretical anode/cathode cell potential, which is about 90% to 95% that of ΔH_o at moderate temperatures. Consequently, the intrinsic maximum energy conversion efficiency of a fuel cell is much greater than that of a heat engine (typically 30% to 40%, as determined by the Carnot efficiency) [177–179]. The actual operational efficiency of a fuel cell is in the range of 30% to 60%) due to activation (kinetic), concentration (mass transfer), and resistance overpotentials, which lower the cell voltage, according to Equation (26.166). The current density regimes where these overpotentials dominate in a typical fuel cell are shown in Figure 26.40a, and the resulting power density vs. current density curve is shown in Figure 26.40b.

26.8.2.1 Fuel Cell Types

A summary of the electrolyte, anode, and cathode gases, operating temperature, and efficiency of different fuel cells is presented in Table 26.19.

Considerable research is currently in progress to find applications for fuel cells. Fuel cells may be useful for large-scale power generation during intermittent peak electricity demand and for load leveling (where electricity is stored in a fuel cell during off-peak hours) [179, 192, 193]. Automobiles equipped with fuel cells are being tested [194–196], but the cost of an automotive fuel cell power plant is too high. Military and space applications of fuel cells are also being explored [177, 179, 192] and portable power applications (replacement for lithium batteries) look promising. Research is in progress to develop cheaper sources of hydrogen gas (the fuel used in most fuel cells) and to find improved methods to store and transport hydrogen.

Molten Carbonate Fuel Cells—The anode fuel is hydrogen, with the following oxidation reaction:

$$H_2 + CO_3^{2-} - 2e^- \rightarrow CO_2 + H_2O \qquad (26.177)$$

TABLE 26.19
Types of Fuel Cells

Fuel Cell Type	Electrolyte	Anode Gas	Cathode Gas	Temperature	Efficiency	Refs.
Molten carbonate	Alkali-carbonates	Hydrogen	Atmospheric oxygen	650°C (1200°F)	40%–55%	177, 179, 180
Phosphoric acid	Phosphorous	Hydrogen	Atmospheric oxygen	210°C (400°F)	35%–50%	177, 178, 181, 182
H_2/O_2(air) proton exchange membrane	Solid polymer membrane	Hydrogen	Pure or atmospheric oxygen	75°C (180°F)	35%–60%	178, 183–187
Direct methanol	Solid polymer membrane	Methanol solution in water	Atmospheric oxygen	75°C (180°F)	35%–40%	188–190
Alkaline	Potassium hydroxide	Hydrogen	Pure oxygen	<80°C	50%–70%	177, 181
Solid oxide	Ceramic oxide	Hydrogen, reformed methane	Atmospheric oxygen	800°C–1000°C (1500°F–1800°F)	45%–60%	177, 181, 191

At the cathode, oxygen is reduced in the molten carbonate medium:

$$O_2 + 2CO_2 + 4e^- \rightarrow 2CO_3^{2-} \qquad (26.178)$$

Carbon dioxide transfers between the anode, where it is produced, and the cathode, where it is consumed. The anode electrode material is porous nickel or a Ni-Cr alloy, while the cathode is porous NiO. The electrolyte is a mixture of $LiAlO_2$, K_2CO_3, and Li_2CO_3, which is absorbed into a porous inert matrix. These cells operate at about 650°C, where a cell voltage of +0.9 V is obtained at a current density of about 150 mA/cm². Systems of 10 kW to 2 MW have been tested for electric utility applications.

Phosphoric Acid Fuel Cells

Hydrogen gas fuel and air (O_2) are fed to anode and cathode Pt catalyst powder layers, respectively. The Pt catalysts is Teflon-bonded to porous carbon sheets to form gas-diffusion electrodes, with a catalyst loading of about 1.0 mg/cm². The Pt anode and cathode are separated by a thin inert porous matrix that is filled with concentrated phosphoric acid. The cell operates at 200°C (to improve the electrode kinetics), with a cell voltage of about 0.67 V at a current density of 0.150 A/cm². Most voltage losses occur at the air cathode. The hydrogen gas must be pure because sulfur and carbon monoxide poison the Pt anode catalyst. This type of fuel cell is commercially available today, with more than 200 systems installed all over the world in hospitals, hotels, office buildings, and utility power plants.

Proton-Exchange Membrane (PEM) Hydrogen/Oxygen Fuel Cells

This type of fuel cell has received much attention in recent years due to its moderate operating temperature and high-power density. The anode and cathode are Pt-loaded carbon powders. Air is normally used as the source of oxygen for the cathode reaction. Of major importance is the removal/management of product water from the fuel cell. A hydrated cation-exchange membrane separates the anode and cathode.

Direct Methanol PEM Fuel Cells

Liquid methanol has been investigated as the anode fuel in PEM fuel cells, due to ease of handling methanol, its low cost, and widespread availability. The fuel oxidation reaction here is

$$CH_3OH + H_2O \rightarrow 6H^+ + CO_2 + 6e^- \qquad (26.179)$$

and the cell operates normally at a temperature of about 80°C. High-temperature operation (120°C to 140°C) with vaporized methanol has also been studied, to improve the oxidation kinetics and minimize anode catalyst poisoning by CO (a by-product of the methanol oxidation reaction). The problem of methanol crossover (leakage) through the solid polymer electrolyte cation-exchange membrane is yet to be solved completely. An alternative approach is to reform methanol and then pump the resulting H_2 gas product to a PEM H_2/O_2 fuel cell.

Alkaline Fuel Cells

This low-temperature fuel cell uses H_2 and O_2 reactants and a highly alkaline aqueous KOH electrolyte. The advantages of this fuel cell are the faster oxygen reduction reaction in the alkaline electrolyte and the possibility of using low-cost, nonprecious metal electrode catalysts, such as Ag-loaded carbon powder. The greatest problem with alkaline fuel cells is that the electrolyte reacts with traces of CO_2 to produce insoluble carbonates.

Solid Oxide Fuel Cells

Such fuel cells operate at high temperatures (~1000°C) and utilize a ceramic oxide material to separate the anode and cathode. At elevated temperatures, the ceramic electrolyte becomes conductive to oxide anions. In these fuel cells, either hydrogen or hydrogen/carbon monoxide mixtures (produced by reforming natural gas) are consumed at the anode and air or oxygen is fed to the cathode. This is a promising fuel cell for large, high-power applications, including industrial and large-scale central electricity generating stations.

NOMENCLATURE

a_i	Activity of species i
a	Electrode area per unit volume (cm^{-1})
A	Area (cm^2)
b_a, b_c	Anodic and cathodic Tafel slopes (V/decade)
C	Concentration (mol/cm^3)
C_∞, c^b	Bulk solution concentration (mol/cm^3)
CE	Current efficiency of an electrode reaction (%)
d	Tube diameter in Equation (26.80) (cm)
D	Diffusion coefficient of the electroyte (cm^2/sec)
D_i	Diffusion coefficient of species i (cm^2/sec)
e	Charge of an electron (1.6021 × 10^{-19} C)
E	Electromotive force (EMF) (V)
E^e	Equilibrium reduction potential (V)
E°	Standard reduction potential (V)
f_i	molar activity coefficient of species i
$f_\pm$	Mean molar activity coefficient
F	Faradays constant (96487 C/equiv.)
g	Acceleration of gravity (cm/s^2)
G	Gibbs energy (J)
Gr	Grashof number (defined by Equation [26.79d])
H	Enthalpy (J)
i	Current density (A/cm^2)
i_L	Mass-transfer limited current density (A/cm^2)
i_o	Exchange current density in the Butler-Volmer equation (A/cm^2)
I	Current (amps)

I_s	Ionic strength (mol/kg)
k_a, k_c	Chemical kinetic rate constants for the anodic and cathodic directions of an electrode reaction
k_d	Mass transfer coefficient (cm/sec)
L	Characteristic electrode dimension (cm)
m	Molality (mol/kg)
MW	Molecular weight (g/mol)
n	Number of electrons involved in the electrode reaction
N_i	Molar flux of species i (mol/cm^2-s)
P	Pressure (N/cm^2)
q	Total charge passed during an electrolysis (C)
r	Reaction rate in Equation (26.14) (mol/s)
r	Disk radius in Equation (26.81) (cm)
R	Universal gas constant (8.3142 J/mol-K)
R_i	Rate of homogeneous production of species i (mol/cm^3-s)
Re	Reynolds number
Sc	Schmidt number [defined by Equation (26.79c)]
Sh	Sherwood number [defined by Equation (26.79a)]
t	Time
t_i	Transference number of species i
T	Temperature (K)
u_i	Electric mobility of species i (cm^2-mol/J-s)
v	Fluid velocity (cm/s)
V	Volume (cm^3)
V_m	Partial molar volume (cm^3/mol)
w	Weight (g)
Wa	Wagner number, as defined by Equation (26.124)
x	Distance along the electrode surface (cm)
X	Membrane ion-exchange capacity (mol/cm^3)
y	Distance perpendicular from the electrode surface (cm)
z_i	Charge number of species i

SYMBOLS

β	Symmetry factor in the Butler-Volmer equation
γ_i	Molal activity coefficient of species i
$\gamma_\pm$	Mean molal activity coefficient
δ	Actual mass transfer boundary-layer thickness (cm)
δ_e	Effective mass transfer boundary-layer thickness (cm)
ε	Permittivity of the electrolyte solution
η_s	Activation overpotential (V)
η_{ohm}	Ohmic overpotential (V)
η_c	Concentration overpotential (V)
κ	Conductivity (ohm^{-1}-cm^{-1})
λ	Ionic equivalent conductance (cm^2-ohm^{-1}-equiv.$^{-1}$)
μ_i	Chemical potential of species i (J/mol)
$\bar{\mu}_i$	Electrochemical potential of species i (J/mol)
μ	Viscosity (g/cm-sec)
v_+, v_-	Numbers of cations and anions into which a molecule of electrolyte dissociates
v	Kinematic viscosity (cm^2/sec)

ρ	Density (g/cm^3)
Φ	Electric potential (V)
ω	Rotation speed of disk (compare Equation (261.81)) (rad/sec)

REFERENCES

1. Johnthan Dean, ed. *Handbook of Lange's Chemistry*. 13th ed. New York: McGraw-Hill, 1985: 2-6, 6-19.
2. A.J. Bard and L.R. Faulkner. *Electrochemical Methods: Fundamentals and Applications*. New York: John Wiley & Sons, 1980.
3. J.O'M Bockris and A.K.N. Reddy. *Modern Electrochemistry*. New York: Plenum Press, 1973.
4. J.S. Newman. *Electrochemical Systems*. 2nd ed. Englewood Cliffs, NJ: Prentice-Hall, 1991.
5. E.A. Guggenheim. *Thermodynamics*. Amsterdam: North-Holland Publishing, 1959.
6. G.N. Lewis and M. Randall. *Thermodynamics* 2nd ed. New York: McGraw-Hill, 1961.
7. R.A. Robinson and R.H. Stokes. *Electrolyte Solutions*. New York: Academic Press, 1959.
8. P. Debye and E. Hückel. *Physik Zeit* 24 (1923): 185.
9. E.A. Guggenheim and J.C. Turgeon. *Transactions of the Faraday Society* 51 (1955): 747.
10. G. Prentice. *Electrochemical Engineering Principles*. Englewood Cliffs, NJ: Prentice-Hall, 1991.
11. H.S. Harned and R.A. Robinson. *Multicomponent Electrolyte Solutions*. Oxford, UK: Pergamon Press Ltd., 1968.
12. Klaus J. Vetter. *Electrochemical Kinetics*. New York: Academic Press, 1967.
13. Paul Delahay. *Double Layer and Electrode Kinetics*. New York: Interscience Publishers, 1965.
14. N. Tanaka and R. Tamamushi. *Electrochimica Acta* 9 (1964): 963.
15. S. Roy and P.N. Pintauro. *Electrochimica Acta* 38 (1993): 1461.
16. S. Roy and P.N. Pintauro. *Journal of the Electrochemical Society* 140 (1993): 3167.
17. B. Levich. *Acta Physicochimica URSS* 17 (1947): 837.
18. V.G. Levich. *Physicochemcial Hydrodynamics*. Englewood Cliffs, NJ: Prentice-Hall, 1962.
19. A.C. Riddiford. "The Rotating Disk System." *Advances in Electrochemistry and Electrochemical Engineering* 4 (1966): 47.
20. C.W. Tobias and R.G. Hickman. *Zeit Physik Chemie* (1965) 229: 145.
21. T.K. Ross and A.A. Wragg. *Electrochimica Acta* (1965)10:1093.
22. N. Ibl and U. Braun. *Chimia* 21 (1967): 395.
23. C.R. Wilke, M. Eisenberg, and C. W. Tobias. *Chemical Engineering Progress* 49 (1953): 674.
24. A.A. Wragg and T.K. Ross. *Electrochimica Acta* 12 (1967): 1421.
25. E. Ruckenstein. *Advances in Chemical Engineering*. Vol. 13. New York: Academic Press, 1987: 11.
26. J. Jorne and T. Cheng. *Chemical Engineering Science* 42 (1987): 1635.
27. J.R. Selman and C.W. Tobias. In *Advances in Chemical Engineering*. Vol 10. T. Drew, ed. New York: Academic Press, 1978: 211.
28. H. Wendt and G. Kreysa. *Electrochemical Engineering: Science and Technology in Chemical and Other Industries*. Berlin: Springer, 1999.
29. M.M. Baizer and H. Lund, eds. *Organic Electrochemistry*. 2nd ed. New York: Marcel Dekker, 1983.
30. D. Pletcher and F.C. Walsh. *Industrial Electrochemistry*. London: Chapman and Hall, 1993.
31. F. Beck and H. Guthke. *Chem Ing Tech* 41 (1969): 943.
32. L. Eberson, K. Nyberg, and H. Sternerup. *Chemica Scripta* 3 (1973):12.
33. F. Goodridge and K. Scott. *Electrochemical Process Engineering: a Guide to the Design of Electrolytic Plant*. New York: Plenum Press, 1995.
34. D. Degner. "Scale-Up of Electroorganic Processes: Some Examples for a Comparison of Electrochemical Syntheses with Conventional Syntheses." In *Techniques of Electroorganic Synthesis, III*. N.L. Weinberg and B.V. Tilak, eds. New York: John Wiley & Sons, 1982.
35. J.R. Backhurst, J.M. Coleman, F. Goodridge, R.E. Plimley, and M. Fleischmann, *Journal of the Electrochemical Society* 116 (1969): 1600.
36. F. Goodridge and A.R. Wright. "Porous Flow-Through and Fluidised Bed Electrodes." In *Comprehensive Treaties of Electrochemistry*. Vol. 6. E. Yeager, J. O'M. Bockris, and S. Sarangapany, eds. New York: Plenum Press, 1983.

37. P.M. Robertson, P. Berg, H. Reimann, K. Schleich, and P. Seiler. *Journal of the Electrochemical Society* 130 (1983): 591.
38. J.F. Patzer, S.J. Yao, and S.K. Wolfson. *Industrial & Engineering Chemistry Research* 35 (1996): 1316.
39. K. Grjotheim, C. Krohn, M. Malinovsky, K. Matiasorvsky, and J. Thonstadt. *Aluminum Electrolysis*. 2nd ed. Dusseldorf: Verlag, 1982.
40. V. Srinivasan and L. Lipp. *Journal of the Electrochemical Society* 150 (2003): K15.
41. R. Jannson. *Chemistry & Engineering News* 62 (1984): 43.
42. N.L. Weinberg, ed. *Techniques of Electroorganic Synthesis*. Parts I, II, and III. Chichester, UK: John Wiley, 1974, 1982.
43. A.P. Tomilov, S.G. Mairanovskii, M.Y.A. Fioshin, and V.A. Smirnov. *Electrochemistry of Organic Compounds*. New York: Halsted Press, 1972.
44. S. Torii, ed. *Recent Advances in Electroorganic Synthesis*. Amsterdam: Kodansha-Elsevier, 1987.
45. P.N. Pintauro and J. Bontha. *Journal of Applied Electrochemistry* 21 (1991): 799.
46. G.J. Yusem and P.N. Pintauro. *Journal of the American Oil Chemists' Society* 69 (1992): 399.
47. P.N. Pintauro, D.K. Johnson, K. Park, M.M. Baizer, and K. Nobe. *Journal of Applied Electrochemistry* 14 (1984): 209.
48. P.E. Iversen. *Journal of Chemical Education* 48 (1971): 136.
49. S.D. Ross, M. Finkelstein, and R.C. Petersen. *Journal of the American Chemical Society* 98 (1967): 4088.
50. R. Lines, B.S. Jensen, and V.D. Parker. *Acta Chemica Scandinavica B*32 (1978): 510.
51. F. Walsh and D. Robinson. *Interface* 7 (1998): 40.
52. M. Fleischmann and R.E.W. Jansson. *Electrochimica Acta* 27 (1982): 1023.
53. V. Tricoli, N. Vatistas, and P.F. Marconi. *Journal of Applied Electrochemistry* 23 (1993): 390.
54. K. Park, P.N. Pintauro, M.M. Baizer, and K. Nobe. *Journal of the Electrochemical Society* 132 (1985): 1850.
55. G. Yusem, P.N. Pintauro, P.-C. Cheng, and W. An. *Journal of Applied Electrochemistry* 26 (1996): 989.
56. J. Jörissen. *Electrochimica Acta* 41 (1996): 553.
57. N. Sato, T. Sekine, and K. Sugino. *Journal of the Electrochemical Society* 115 (1968): 242.
58. A.M. Couper, D. Pletcher, and F.C. Walsh. *Chemical Reviews* 90 (1990): 837.
59. J.K. Hammond. *Proceedings of the Second International Forum on Electrolysis in the Chemical Industry*; Deerfield Beach, FL; 1988.
60. M.P.J. Brennan and R. Brettle. *Journal of the Chemical Society* [Perkin Trans.] 1 (1973): 257.
61. D. Vasudevan, S.S. Vaghela, G. Ramachandraiah. *Journal of Applied Electrochemistry* 30 (2000): 1299.
62. F. Beck, Angew. *Chemistry Engineering Education* 11 (1972): 760.
63. M.M. Baizer. *Journal of the Electrochemical Society* 111 (1964): 215.
64. M.M. Baizer and J.P. Petrovich. *Progress in Physical Organic Chemistry* 7 (1970): 189.
65. D.E. Danly. *Hydrocarbon Processes* 60 (1981): 161.
66. D.E. Danly. *Journal of the Electrochemical Society* 131 (1984): 435C.
67. A.P. Tomilov, S.L. Kirilyus, and I.P. Andriyanova. *Soviet Electrochemistry* 8 (1972): 1050.
68. T. Chiba, M. Okimoto, H. Nagai, and Y. Takata. *Bulletin of the Chemical Society of Japan* 56 (1983): 719.
69. G. Belot, S. Dejardins, and J. Lessard. *Tetrahedron Letters* 25 (1984): 5347.
70. A. Cyr, P. Hout, G. Belot, and J. Lessard. *Electrochimica Acta* 35 (1990): 147.
71. Y. Song and P.N. Pintauro. *Journal of Applied Electrochemistry* 21 (1990): 21.
72. Z. Ogumi, K. Nishio, and S. Yoshizawa. *Electrochimica Acta* 26 (1981): 1779.
73. J. Sarrazin and A. Tallec. *Journal of Electroanal Chemical and Interfacial Electrochemistry* 137 (1982): 183.
74. V.A. Grinberg, V.N. Zhuravleva, Y.B. Vasil'ev, and V.E. Kazarinov. *Electrokhimiya* 19 (1983): 1447.
75. Z. Ogumi, H. Yamashita, K. Nishio, Z. Takehara, and S. Yoshizawa. *Electrochimica Acta* 28 (1983): 1687.
76. Z. Ogumi, M. Inaba, S. Ohashi, M. Uchida, and Z. Takehara. *Electrochimica Acta* 33 (1988): 365.
77. R. Liu and P. Fedkiw. *J. Electrochem. Soc.*, 139 (1992): 3514.
78. Y.-L. Chenand T.-C. Chou. *Journal of the Electrochemical Society* 360 (1993): 247.
79. K. Warner, W. Neff, G.R. List, and P.N. Pintauro. *Journal of the American Oil Chemists' Society* 77 (2000): 1113.

80. W. An, J.K. Hong, P.N. Pintauro, K. Warner, and W. Neff. *Journal of the American Oil Chemists' Society* 75 (1998): 917.
81. P.N. Pintauro, M.P. Gil, K. Warner, G. List, and W. Neff, *Industrial & Engineering Chemistry Research* 44 (2005): 6188.
82. W. An, J.K. Hong, P.N. Pintauro, K. Warner, and W. Neff. *Journal of the American Oil Chemists' Society* 76 (1999): 215.
83. J.W. Dini. *Electrodeposition: The Materials Science of Coatings and Substrates.* Park Ridge, NJ: Noyes Publications, 1993.
84. J. Newman. *Journal of the Electrochemical Society* 113 (1966): 501.
85. J. Newman. *Journal of the Electrochemical Society* 113 (1966): 1235.
86. W.R. Parrish and J. Newman. *Journal of the Electrochemical Society* 116 (1969): 169.
87. C. Wagner. *Journal of the Electrochemical Society* 98 (1951): 116.
88. N. Ibl. *Comprehensive Treatise of Electrochemistry.* Vol. 6. E. Yeager, ed. New York: Plenum Press, 1983: 239.
89. W.R. Parrish and J. Newman. *Journal of the Electrochemical Society* 117 (1970): 43.
90. K. Viswanathan and D.-T. Chin. *Journal of the Electrochemical Society* 124 (1977): 709.
91. J. Newman. *Advances in Electrochemistry and Electrochemical Engineering.* C.W. Tobias, ed. 5 (1967): 87.
92. R. Winand. *Electrochimica Acta* 39 (1994): 1091.
93. M. Matlosz, C. Creton, C. Clerc, and D. Landolt *Journal of the Electrochemical Society* 134 (1987): 3015.
94. D. Landolt. *Electrochimica Acta* 39 (1994): 1075.
95. D.-T. Chin and K. L. Hsueh. *Electrochimica Acta* 31 (1986): 561.
96. A. Bensmaili and F. Coeuret, *Journal of the Electrochemical Society* 137 (1990): 3086.
97. R. C.alkire and J.B. Lu, J. Electrochem. Soc., 134 (1987): 294.
98. C. Karakusand C.-R. Chin. *Journal of the Electrochemical Society* 141 (1994): 691.
99. R. J. von Gutfeld and J. Cl. Puippe. *Oberfläche-Surface* 11 (1981): 294.
100. L. Kulynych, L. Romankiw, and R. J. von Gutfeld. *IBM Technical Disclosure Bulletin* 23 (1980): 1262.
101. J.Cl. Puippe, R.E. Acosta, and R.J. von Gutfeld. *Journal of the Electrochemical Society* 128 (1981): 2539.
102. R.J. von Gutfeld, E.E. Tynan, R.L. Melcher, and S.E. Blum. *Applied Physics Letters* 35 (1979): 651.
103. M.H. Gelchinski, L.T. Romankiw, D.R. Vigliotti, and R.J. von Gutfeld. *Journal of the Electrochemical Society* 132 (1985): 2575.
104. P. Bindra and J. Roldan. *Journal of the Electrochemical Society* 132 (1985): 2581.
105. A. Brenner. *Electrodeposition of Alloys: Principles and Practice.* Vols. 1, 2. New York: Academic Press, 1963.
106. P. C. Andricacos, C. Arana, J. Tabib, J. Dukovic, and L.T. Romankiw. *Journal of the Electrochemical Society* 136 (1989): 1336.
107. A. T. Kuhn, ed. *Techniques in Electrochemistry, Corrosion, and Metal Finishing – a Handbook.* London: Chichester, 1987.
108. J. Edwards. *Electroplating: a Guide for Designers and Engineers.* Middlesex, UK: Finishing Publications, 1983.
109. F.A. Lowenheim, ed. *Modern Electroplating.* 3rd ed. New York: Wiley, 1974.
110. W. Canning and Co., Ltd. *The Canning Hanbook: Surface Finishing Technology.* 23rd ed. Birmingham, UK: W. Canning in association with Spon, 1982.
111. J.W. Dini. *Electrodeposition: the Materials Science of Coatings and Substrates*, Park Ridge, NJ: Noyes Publications, 1993.
112. C.W. Ammen. *The Electroplater's Handbook.* Blue Ridge Summit, PA: Tab Books, 1986.
113. P.A. Jaquet. *Nature* 135 (1935): 1076.
114. J.F. Wilson. *Practice and Theory of Electrochemical Machining.* New York: Wiley Interscience, 1971.
115. J.A. McGeough. *Principles of Electrochemical Machining.* London: Chapman and Hall, 1974.
116. M. Datta and L.T. Romankiw. *Journal of the Electrochemical Society* 136 (1989): 185C.
117. W.N. Brooks. *AIChE Symposium Series* 98 (1986): 1.
118. L. Carlsson, B. Sandegren, D. Simonsson, and M. Rihovsky. *Journal of the Electrochemical Society* 130 (1982): 342.

119. P. Millington and I.M. Dalrymple. European Patent Appl. 64417, 1983. *Chemistry Abstract* 98 (1983): 62142.
120. D.R. Gabe and F.C. Walsh, *Journal of Applied Electrochemistry* 25 (1985): 807.
121. S. Langlois and F. Coeuret. *Journal of Applied Electrochemistry* 19 (1989): 51.
122. J. Gonzalez-Garcia, V. Montiel, A. Aldaz, J.A. Conesa, J.R. Perez, and G. Codina. *Industrial & Engineering Chemistry Research* 37 (1998): 4501.
123. G. van der Heiden, C.M.S. Roats, and H.F. Boon. *Chemistry and Industry* (London) (1978): 465.
124. K. Juettner, U. Galla, and H. Schmieder. *Electrochimica Acta* 45 (2000): 2575.
125. D.N. Bennion and J. Newman. *Journal of Applied Electrochemistry* 2 (1972): 113.
126. M. De Francesco and P. Costamagna. *Journal of Cleaner Production* 12 (2004): 159.
127. F. Walsh and G. Mills. *Chemistry and Industry (London)* 15 (1993): 576.
128. C. Comminellis and C. Pulgarin. *Journal of Applied Electrochemistry* 23 (1993): 108.
129. J.C. Farmer, F.T. Wang, R.A. Hawley-Fedder, and P.R. Lewis. *Journal of the Electrochemical Society* 139 (1992): 654.
130. J.C. Farmer, F.T. Wang, P.R. Lewis, and J. Summers. *Journal of the Electrochemical Society* 139 (1992): 3025.
131. L. Kaba, G.D. Hitchens, and J.O'M. Bockris. *Journal of the Electrochemical Society* 137 (1990): 1341.
132. H. Strathmann. In *Membrane Handbook*. W. Ho and K. Sirkar, eds. New York: Van Nostrand Reinhold, 1992: 219-262.
133. N.R. Khalili, R. Kizilel, J.R. Selman, and V.S. Donepudi. *Plating and Surface Finishing* 89 (2002): 62.
134. H. Strathmann. *Separation and Purification* 14 (1985): 41.
135. O. Kedem and Y. Maoz. *Desalination* 19 (1983): 465.
136. *Millipore Product Bulletin*. Lit. No. CI001 (1987).
137. H.P. Gregor and C.D. Gregor. *Scientific American* 239 (1978):88.
138. F. Helfferich. *Ion Exchange*. New York: McGraw-Hill, 1962: 339-417.
139. D.G. Petropolous, D.G. Tsimboukis, and K. Kouzeli. *Journal of Membrane Science* 26 (1983): 379.
140. P.N. Pintauro and D.N. Bennion. *Industrial and Engineering Chemistry Fundamentals* 23 (1984): 234.
141. M.A. Lake and S.S. Meisheimer. *AIChE Journal* 24 (1978): 130.
142. M.J. Manning and S.S. Meisheimer. *Industrial and Engineering Chemistry Fundamentals* 22 (1983): 311.
143. M.W. Verbruggeand and R.F. Hill. *Journal of Physical Chemistry* 92 (1988): 6778.
144. J.R. Bontha and P.N. Pintauro. *Chemical Engineering Science* 49 (1994): 3835.
145. P.N. Pintauro, R. Tandon, L. Chao, W. Xu, and R. Evilia. *Journal of Physical Chemistry* 99 (1995): 12915.
146. Y. Gur. I. Ravina, and A.J. Babchin. *Journal of Colloid and Interface Science* 64 (1978): 333.
147. P.N. Pintauro and D.N. Bennion. *Industrial and Engineering Chemistry Fundamentals* 23 (1984): 230.
148. A. Narebska, S. Koter, and W. Kujawski. *Journal of Membrane Science* 25 (1986): 153.
149. S. Capeci, P. N. Pintauro, and D. N. Bennion. Journal of the Electrochemical Society 136 (1989): 2876.
150. M. W. Verbrugge and P. N. Pintauro. In *Modern Aspects of Electrochemistry*. Vol. 19. B. E. Conway, J O'M. Bockris, and R. E. White, eds. New York: Plenum, 1989: 1-67.
151. A. Guzman-Garcia, P.N. Pintauro, M.W. Verbrugge, and R.F. Hill. *AIChE Journal* 36 (1990): 1061.
152. J.C. Fairand J. F. Osterle. *Journal of Chemical Physics* 54 (1971): 3307.
153. G.B. Westerman-Clark and J.L. Anderson. *Journal of the Electrochemical Society* 130 (1983): 839.
154. Y. Yang and P.N. Pintauro. *AIChE Journal* 46 (2000): 1177.
155. T. Davis. In *Handbook of Industrial Membrane Technology* M. C. Porter, ed. Park Ridge, NJ: Noyes Publications, 1990: 482–510.
156. R. E. Lacey. *Electrosorption and Desorption Process for Demineralization*. Report 398. U.S. Office of Saline Water Research and Development Program, 1969.
157. M.G. Fontana. *Corrosion Engineering*. New York: McGraw-Hill, 1986.
158. N.D. Tomashov. *Theory of Corrosion and Protection of Metals*. London: Macmillan, 1967.
159. M.J.N. Pourbaix. *Atlas of Electrochemical Equilibria in Aqueous Solutions*. New York: Pergamon Press, 1966.
160. S. Ramesh and S. Rajeswari. *Electrochimica Acta* 49 (2004): 811.
161. V. Ashworth and C.J.L. Booker, eds. *Cathodic Protection: Theory and Practice*. Chichester, UK: Ellis Horwood, 1986.

162. H.H. Ulig. *Corrosion and Corrosion Control: an Introduction to Corrosion Science and Engineering.* New York: Wiley, 1971.
163. L.L. Shreir. *Corrosion and Corrosion Control.* 2nd ed. Vols. 1, 2. London: Butterworth, 1976.
164. H. Bohni and H. Uhlig. *Journal of the Electrochemical Society* 116 (1969): 906.
165. J.C. Scully. *The Fundamentals of Corrosion.* 2nd ed. Oxford, UK: Pergamon Press, 1975.
166. G.W. Heise and N.C. Cahoon, eds. *The Primary Battery.* Vols. I, II. New York: Wiley, 1971, 1976.
167. P. Bro. *Interface* 4 (1995): 42.
168. D. Linden, ed. *Handbook of Batteries.* 2nd ed. New York: McGraw-Hill, 1995.
169. A.J. Salkind and R. J. Brodd. *Electrochemical Society Proceedings* 87–3 (1987): 127–145.
170. P. Bro and S.C. Levy. *Quality and Reliability Methods of Primary Batteries.* New York: Wiley, 1990.
171. *Rechargable Batteries Application Handbook.* London: Butterworth-Heinemann, 1992.
172. P.D. Bennet, K.R. Bullock, and M.E. Fiorino. *Interface* 4 (1995): 26.
173. R. Ishikawa, T. Hazama, M. Miyabashi, and H. Andoh. In *Proceedings of the International Workshop on Advanced Batteries.* Osaka, Japan; February 22–24, 1995.
174. S. Megahed and B. Scrosati. *Interface* 4 (1995): 34.
175. G. Pistoia. *Lithium Batteries.* New York: Elsevier, 1994.
176. D.S. Clive Tuck. *Modern Battery Technology.* Chichester, UK: Ellis Horwood, Ltd., 1991.
177. J.H. Hirschenhofer, ed. *Fuel Cell Handbook.* 4th ed. Orinda, CA: Business/Technology Books, 2000.
178. S.A. Angrist, ed. *Direct Energy Conversion.* 3rd ed. Boston: Allyn and Bacon, 1976.
179. K. Kordesh and G. Simader. *Fuel Cells and Their applications.* New York: VCH Publishers, 1996.
180. W. Vielstich, A. Lamm, and H.A. Gasteiger, eds. *Handbook of Fuel Cells: Fundamentals, Technology, and Applications.* New York: Wiley, 2003.
181. T. Fuller. *Interface* 3 (1997): 26.
182. W. Vielstich, A. Lamm, and H.A. Gasteiger. *Handbook of Fuel Cells: Fundamentals Technology and Applications.* vol. 4, Fuel Cell Technology and Applications. Pt. 2. Chichester, England: Wiley, 2003.
183. C. Zawodzinski, M.S., Wilson, and S. Gottesfeld. In *Proton Conducting Membrane Fuel Cells I.* Electrochemical Society Symposium Series 65-23. Pennington, NJ: The Electrochemical Society, Inc., 1995: 57.
184. R. Lemons. *Journal of Power Sources* 29 (1990): 251.
185. M. Wakizoe, O.A. Velev, and S. Srinivasan. *Electrochimica Acta* 40 (1995): 335.
186. G. Hoogers, ed. *Fuel Cell Technology Handbook.* Boca Raton, FL: CRC Press, 2003.
187. T.A. Zawodzinski, J. Davey, J. Valerio, and S. Gottesfeld. *Electrochimica Acta* 40 (1995): 297.
188. S.T. Narayanan, A. Kindler, B. Jeffries-Nakamura, et al. In *Proton Conducting Membrane Fuel Cells I.* Electrochemical Society Symposium Series 65-23. Pennington, NJ: The Electrochemical Society, Inc., 1995: 261.
189. S. Surampudi, S.R. Narayanan, E. Vamos, H. Frank, and G. Halpert. *Journal of Power Sources* 47 (1994): 377.
190. J.S. Wainright, J-T. Wang, D. Weng, R.F. Savinell, and M. Litt. *Journal of the Electrochemical Society* 142 (1995): L121.
191. S.C. Singhal and K. Kendall, eds, *High Temperature Solid Oxide Fuel Cells: Fundamentals, Design, and Applications.* Amsterdam: Elsevier, 2003.
192. J. Larminie and A. Dicks. *Fuel Cell Systems Explained.* New York: Wiley, 2000.
193. E.A. Gillis. *Chemical Engineering Progress* October (1980): 88.
194. M.S. Vreeke, D.T. Mah, and C.M. Doyle. *Journal of the Electrochemical Society* 145 (1998): 3668.
195. F. Barbir, ed., *PEM Fuel Cells: Theory and Practice.* New York: Elsevier Academic Press, 2005.
196. J. Motavalli. *Forward Drive: The Race to Build "Clean" Cars for the Future.* San Francisco: Sierra Club Books, 2001.

27 Patents and Intellectual Property

M. Henry Heines

CONTENTS

27.1 Introduction 1831
27.2 United States Patents and Related Rights and Documents 1832
    27.2.1 Utility Patent 1832
    27.2.2 Design Patent 1832
    27.2.3 Provisional Patent Application 1832
    27.2.4 Trademarks 1833
    27.2.5 Copyrights 1833
27.3 Patents Outside the United States 1834
27.4 Requirements for Patentability 1834
    27.4.1 Novelty 1834
    27.4.2 Utility 1835
    27.4.3 Nonobviousness 1835
27.5 Preliminary Steps for the Inventor 1835
    27.5.1 Representation before the USPTO 1835
    27.5.2 Prior Art Searching 1835
    27.5.3 Patents as Prior Art 1836
    27.5.4 Record Keeping 1836
    27.5.5 Minimizing Chances for Invalidation 1837
27.6 Trade Secrets 1837
27.7 Intellectual Property between the Employer and Employee 1838
    27.7.1 Patents—Ownership and Other Rights and Obligations 1838
    27.7.2 Trade Secrets—Obligations to Present and Past Employers 1838
    27.7.3 Rights and Obligations Created or Eliminated by Express Agreement 1839

27.1 INTRODUCTION

Proprietary engineering technology can be preserved as a corporate or personal asset by patents and trade secrets. In many segments of the chemical process industry, these assets contribute significantly to corporate valuation and the ability to compete. The purpose of this chapter is to set forth the basic elements of patents, trade secrets, and related issues of "intellectual property," as a guide to the chemical engineer. To obtain, maintain, and enforce patent or trade secret rights or to apply these guidelines to specific situations, the engineer is advised to consult with an attorney.

27.2 UNITED STATES PATENTS AND RELATED RIGHTS AND DOCUMENTS

27.2.1 UTILITY PATENT

The most common and well-known type of patent is the "utility patent." Utility patents cover functional features of an invention with utility as a key requirement. Inventions that qualify for utility patents include articles of manufacture, compounds or compositions of matter, manufacturing processes, and methods of using articles, compounds, or compositions.

The term of enforceability of a utility patent extends from its issue date to its expiration date. This is in contrast to certain other countries, where a patent once issued may qualify for enforcement retroactively to a publication date that precedes the issuance of the patent. When originally created by federal law, U.S. patents expired 17 years from the issue date. This was changed in 1995 by federal legislation implementing the GATT (General Agreement on Tariffs and Trade) Treaty for international harmonization of patent laws. Under the 1995 legislation, a utility patent applied for after June 8, 1995, will expire 20 years from the filing date of the earliest United States application that the patent derives benefit from. Thus, a utility patent issuing on an original application filed in the United States Patent and Trademark Office (USPTO) after June 8, 1995, expires 20 years from the filing date of the application; and a utility patent issuing on an application filed either before or after June 8, 1995, as a continuation, continuation-in-part, or division of an application filed before June 8, 1995, expires 20 years from the filing date of the previously filed application regardless of the status of that application. A patent that was either (a) issued prior to June 8, 1995, and had not yet expired by that date, or (b) issued after June 8, 1995, on an application filed before that date expires either 17 years from the issue date or 20 years from the filing date, whichever results in a longer period of time between issuance and expiration. The term of a patent is generally not extendable beyond these expiration dates, except for those patents that were issued as the result of an appeal before the USPTO Board of Patent Appeals and Interferences, and for patents on inventions requiring federal regulatory approval prior to commercialization.

A utility patent bestows on its owner the right to exclude others from making, using, offering to sell, or selling the invention as defined by the scope of the patent claims, within the geographical boundaries of the United States and its territories and possessions. Thus, patents to articles of manufacture, compounds, or compositions of matter are infringed by the unauthorized manufacture, use, offer of sale, or sale of the article, compound, or composition within these boundaries. Patents to methods of use are infringed by the unauthorized practice of the patented method within these boundaries. Patents to manufacturing processes are likewise infringed by the unauthorized practice of the patented process within these boundaries, but are also infringed by the importation into the United States of a product that was produced outside the United States by the patented process, or by the offer of sale, the sale, or the use of such a product in the United States. No patent insulates its owner from potential liability as an infringer of another patent.

27.2.2 DESIGN PATENT

Design patents are issued only for articles and cover only the ornamental features of the article rather than functional features. The term of enforceability of a design patent extends from its issue date to an expiration date 14 years from the issue date, and remains unaffected by GATT implementation legislation. Like a utility patent, a design patent bestows on its owner the right to exclude others from making, using, or selling a product that embodies the patented design, within the geographical boundaries of the United States and its territories and possessions.

27.2.3 PROVISIONAL PATENT APPLICATION

A provisional patent application is a document filed in the USPTO as a preliminary step toward applying for a utility patent. The primary advantage of a provisional patent application is the official recognition by the USPTO of the date of receipt of the document at the USPTO, and the option of transferring that date to a subsequently filed (nonprovisional) application for a utility patent to serve as an effective filing date for the utility application. Further advantages include a lower filing fee than an application for a utility patent and fewer filing requirements in terms of the content and form of the application. Provisional applications find their greatest use when an early filing date is desired with insufficient time to assemble the information needed to accompany an application for a utility patent or to prepare an application that will meet all requirements of a utility patent.

Provisional patent applications are regarded by the USPTO as abandoned 1 year from the application date. Hence, the filing date benefit of a provisional application can only be transferred to an application for a utility patent that is filed before the 1-year anniversary of the provisional application. For any U.S. patent application, whether utility, design, or provisional, the date of mailing to the USPTO with appropriate certification can be recognized as the official filing date. Patent applications can be filed electronically through the USPTO's Electronic Filing System (EFS), details of which are available on the USPTO Web site www.uspto.org.

A provisional application does not confer upon an inventor any ability to exclude others from practicing the invention, nor does it affect the term of any patent issued on a nonprovisional application subsequently filed.

27.2.4 TRADEMARKS

Trademarks confer property rights on names, symbols, or devices that are associated with goods or services. Trademarks reflect such features as the value of name recognition, quality assurance, and good will, and the association of these qualities with the source or origin of the good or service. While the source need not be known to the purchaser, the value of a trademark is a recognition by the purchaser that all goods bearing the trademark originate from a common source and share the value associated with that source. A trademark is infringed when the name, symbol, or device is used on a product or service from a different supplier in a manner that is likely to confuse a purchaser into believing that the product or service is from the same source as products or services sold by the true owner of the trademark.

A trademark becomes a property right simply upon use of the trademark in commerce, and as such is immediately enforceable against infringers. Trademarks can be federally registered with the USPTO, and when so registered are permitted to bear the symbol ®. Federal registration serves as official evidence of the validity of the trademark, the registrant's exclusive right to use the trademark, and the registrant's ownership of the trademark, all of which are considerable advantages to the trademark owner seeking to enforce the trademark against an infringer. Registrations used or renewed prior to Nov. 16, 1989 remain in force for 20 years from the date of issue or renewal, while registrations issued or renewed on or after Nov. 16, 1989 remain in force for 10 years from the date of issue or renewal. In all cases, registrations can be further renewed indefinitely for additional periods of 10 years by supplying evidence toward the end of the 10-year period that the trademark is still in use.

27.2.5 COPYRIGHTS

Copyrights confer property rights on original works of authorship that are fixed in a tangible medium of expression. Examples of qualifying works of authorship are literary works, musical works, pictorial and graphic works, sound recordings, audiovisual works, and computer programs. Copyrights do not confer proprietary rights in the underlying ideas, concepts, procedures, principles, or similar component parts, even though these components may be central to the work; instead, the copyright resides only in the form of expression.

The duration of a copyright for a work created after January 1, 1978, is the life of the author (or, for joint works, the last surviving author) plus 50 years. When the author is unknown or is a business entity rather than an individual, the duration is either 75 years from the year of first publication or 100 years from the year of creation, whichever expires first. Copyrights for works created before January 1, 1978, and still in force on that date have a duration of 28 years from the securement of the copyright, subject to renewal for an additional 47 years.

Copyrights need not be federally registered to be effective, but federal registration is required before suing an infringer, and early registration offers advantages to the copyright owner when enforcing the copyright, particularly in terms of the monetary damages and costs that the successful owner can collect. Copyrights are registered with the United States Copyright Office.

27.3 PATENTS OUTSIDE THE UNITED STATES

As in the United States, patents in countries throughout the world are generally enforceable only in the countries where the patents are issued. There is no global patent; patents must be applied for in each country where patent rights are desired.

Patent treaties between certain groups of countries serve to simplify the process of applying for patents in multiple countries within the groups and to promote consistency in standards for patentability. Included among the multi-country patent treaties are the European Patent Convention (31 countries), the Eurasian Patent Convention (9 countries), the African Regional Industrial Property Organization (ARIPO) (16 countries in western Africa), the African Intellectual Property Organization (OAPI) (16 countries in eastern Africa), the Gulf Cooperation Council (6 countries in the Persian Gulf), and the Patent Cooperation Treaty (PCT) (encompassing the latter 4 conventions plus approximately 60 additional countries including the United States). Each of these treaties offers the establishment of a filing date for a patent application in each of its member countries by filing the application at a single location. In addition, the European Patent Convention, the ARIPO, the OAPI, and the Gulf Cooperation Council each provide a single examination applicable to all of their own member countries.

The oldest and most widely used treaty is the Paris Convention, which permits an application initially filed in a single country to be subsequently filed in other countries with recognition of its original filing date, provided that the subsequent filings are made within a year of the original filing. The Paris Convention has approximately 170 signatory countries worldwide (including the United States). The Paris Convention can be used for foreign filings through any of the other treaties listed above.

The purpose of a patent in any country is to confer a state-recognized property right to an invention. The laws of individual countries differ in such matters as the types of inventions that can be patented, the degree to which an invention must advance the state of the art, the types of prior documents and activities that will defeat patentability, and whether or not the patentee can be required to issue licenses to others and under what conditions, and what constitutes infringement. U.S. patent law, for example, is notable for its difference from the patent laws of certain other countries in the patentability of pharmaceutical compositions and medical procedures (both patentable in the United States), and in cases where competing inventors apply for conflicting patents on the same invention (the United States awards the patent to the first to invent, while most other countries award the patent to the first to file, although, as of the date of publication of this chapter, legislation is pending in the United States to change the standard to first-to-file).

27.4 REQUIREMENTS FOR PATENTABILITY

To qualify for a patent in the United States, an invention must meet the fundamental requirements of novelty, utility, and nonobviousness.

27.4.1 NOVELTY

Novelty is met by the presence of any element or feature of the invention distinguishing it from the "prior art," the legal term used to designate certain categories of information that are considered by the patent statute to place the information in the possession of the public. The statute sets forth seven categories of prior art, each with qualifying factors that vary depending on such questions as whether or not the information is reduced to written or printed form or is represented by a use, sale, or offer of sale; if printed, whether or not it was published and to whom it was distributed; if published, the type of publication and when publication or distribution occurred; if a use rather than a publication, the context, degree, and purpose of the use; if a sale or offer of sale, the status or condition of the product being sold or offered. The novelty requirement is met by any distinction, however slight, including the identification of critical parameters not previously recognized in subject matter that is already known but only in a broader context.

27.4.2 Utility

Utility is met by stating that a function is served by the invention and that it has a value, even if the value is only potential. Commercial viability is not a requirement of utility. Nor is the inclusion of laboratory test data, except in inventions where such data are normally considered necessary by "those skilled in the art" (the scientific community) to establish that the invention will serve its stated function.

27.4.3 Nonobviousness

This requirement addresses the type and degree of the difference(s) distinguishing the invention from the prior art. Although originally derived from the conventional meaning of the word "obvious," the term "nonobviousness" has developed into a complex set of legal parameters that vary with both the type of invention and the type of distinction. Establishing nonobviousness in some cases can be achieved by a simple description of the invention, while others require comparative test data, and still others require expert testimony. Secondary factors such as long-felt need and evidence of commercial success are also relevant in many cases.

27.5 PRELIMINARY STEPS FOR THE INVENTOR

27.5.1 Representation before the USPTO

While inventors are legally entitled to represent themselves before the USPTO, it is advisable to use a qualified patent attorney for all communications with the USPTO and for consultation in all preliminary steps. Qualified patent attorneys are registered by the USPTO to represent inventors, and the appropriate patent attorney for any invention is one having a university degree in a technical field in which the invention resides or in a field that enables the attorney to understand the technical nature of the invention. Many patent attorneys handling inventions in chemistry or chemical engineering have advanced degrees in these fields, as well as industry experience as practicing chemists or engineers prior to entering law.

27.5.2 Prior Art Searching

A search of the prior art is often conducted by or for the inventor before applying for a patent. There is no legal requirement that a search be performed, but a search can be of use to an inventor seeking to decide whether or not to apply for a patent. The search can also help the inventor focus the invention or assign the invention an appropriate scope before submitting a patent application to the USPTO. Searches can be performed through computer databases as well as by professional searchers who have direct access to the USPTO records and the assistance of USPTO examiners. Prior art can take the form of published technical papers, oral presentations at technical symposia, news reports, previously issued patents, published patent applications, applications for federal regulatory approval, and in general any form of written or oral presentation throughout the world that is available to those in the technical field of the invention, or any commercial activity in the United States. The type of search and the manner in which it is conducted will determine what type of prior art is found.

27.5.3 Patents as Prior Art

A form of prior art commonly encountered is previously issued patents. The discovery of a relevant patent cannot only affect the patentability of a subsequent invention, but can also raise the possibility of infringement of the patent by the subsequent inventor. These considerations are often independent and must be addressed separately. The freedom to *patent* a new invention over a prior patent requires scrutiny of everything disclosed in the prior patent. The most relevant portion of the prior patent for this purpose is its specification (the text preceding the patent claims), since the specification often contains description or disclosure that does not appear or is not reflected in the claims, or is peripheral to the invention appearing in the claims. The freedom to *practice* a new invention requires scrutiny of the claims of the prior patent, and is achieved when the new invention avoids fully meeting the recitations in each of the claims.

> *Free to patent but not to practice.* An invention of this type is one that falls within the scope of one of the claims of a prior patent that is still in force, while still offering a patentable distinction over what is disclosed in the patent. This includes such inventions as those representing an improvement in a process or product whose initial discovery is the subject of the prior patent that claims the process or product in a generic manner, and inventions whose use requires the concurrent use of the invention claimed in a prior patent. An example is an invention residing in a new reaction to achieve a particular product, followed by an invention residing in a new catalyst or new reaction conditions to enhance the rate, product yield, or selectivity of the reaction without departing from the reaction itself. For inventions in this category, anyone seeking to practice the second invention will require permission from the owners of both patents. The owner of any one of the two patents will likewise require a license from the owner of the other patent to practice the second invention.
>
> *Free to practice but not to patent.* This describes an invention that is fully disclosed in a prior patent (that is still in force), although outside the scope of the claims of that patent. This category also includes an invention that is fully disclosed (and possibly claimed) in a prior patent that is either expired, dedicated to the public, or declared invalid.
>
> *Free to neither patent nor practice.* Inventions of this type are those that are fully disclosed in the specification of a prior patent and fall within the scope of the claims of that patent. The discovery of two patents, two patent applications, or a patent and an application claiming the same invention can lead to an "interference," a procedure that determines which of the two competing parties has priority over the other, usually in terms of the date of invention, and awards the patent to the prevailing party.
>
> *Free to patent and to practice.* Inventions of this type are those that are distinct from anything disclosed in the prior patent and fall outside the scope of the patent claims.

27.5.4 Record Keeping

The date of conception of an invention can often be used to remove a prior patent or publication from consideration as prior art. Laboratory notebooks or other written records are particularly useful for this purpose, provided they bear a fully contained description of the idea conceived or the work performed, and each page is signed and dated by the inventor. The value of the record as evidence is strengthened when the inventor's signature is supplemented by the signature and date of a witness, and further strengthened when the signatures are notarized.

27.5.5 Minimizing Chances for Invalidation

Those seeking to apply for a patent or to preserve the option to do so can minimize the risk of losing potential patent rights by being aware of actions that defeat patentability or that place a time limit on applying for a patent. Such actions include:

- Publications or public disclosures of the invention
- Sales of the invention or offers to sell the invention (i.e., selling or offering to sell a product or performing on a commercial basis or offering to perform a process, as opposed to selling or offering title or license rights to the invention)
- Uses of the invention in a commercial or public setting

When one or more of these acts is performed by the inventor, the inventor has 1 year from the earliest such act in which to apply for a patent. When the act is performed by anyone other than the inventor, with or without the inventor's knowledge or consent, the inventor has the right to establish conception of the invention prior to the act, but only if the patent was applied for within 1 year from the earliest such act.

Exceptions to these acts are as follows: For publications, neither the date of submission of a manuscript for review prior to publication nor the date of acceptance of the manuscript initiates the 1-year time limit, only the actual publication date. Disclosures of the invention to potential licensees, customers, or users can avoid initiating the time limit if the number of recipients is limited and the disclosure is made on a confidential basis. Confidentiality generally requires a written agreement including provisions requiring that the disclosed information be used only for evaluation of the invention and that neither party disclose the information to outsiders without the prior consent of the other party. An offer for sale of the invention can avoid initiating the time limit if the offer is made before the invention is reduced to a fully workable form. A use of the invention, even in a commercial setting, can avoid initiating the time limit if the use is experimental (i.e., expressly for purposes of evaluation, with direction and monitoring of performance by the inventor or owner).

Patentability can also be jeopardized during the application process by failure to disclose critical items in connection with the invention. One is the "best mode" of the invention, which is the favored embodiment or means of implementing the invention, particularly the one example that functions the best or enables the invention to perform to its utmost advantage, as known to the inventor at the date of applying for the patent. The "best mode" must be included in the text of the patent application, and cannot be reserved as a trade secret. A second disclosure requirement is the disclosure of relevant "prior art" that the inventor is or becomes aware of. This is a continuing requirement that lasts until the issuance of the patent, and encompasses any publications, disclosures, uses, sales or offers, and other acts or documents that qualify as prior art.

When the "best mode" requirement is not violated by maintaining a trade secret, a trade secret can coexist with a patent in a single area of technology. In general, however, a trade secret is an alternative to patent protection.

27.6 TRADE SECRETS

Any information considered proprietary can be a trade secret, regardless of whether the information would also qualify for a patent. Trade secrets are established by the manner in which information is treated, used or disseminated; no application or registration with a government authority is required.

Trade secrets are created and maintained by enforcing secrecy, either by employment agreements or policies or by confidential disclosure agreements. Trade secrets are extinguished by disclosures that are inconsistent with maintaining the secrecy of the information. Violation and extinguishment of a trade secret most often occur when one who is under an obligation to the owner to maintain the secret makes an unauthorized disclosure of the secret. The information in a patent application can have trade secret status until the application becomes a patent or is published prior to issuance, as often occurs in countries outside the United States. One can thus apply for a patent with the option of maintaining the invention as a trade secret if a patent is not granted.

Unlike patents, there is no limitation on the duration of a trade secret that is not otherwise extinguished by non-secret disclosure.

27.7 INTELLECTUAL PROPERTY BETWEEN THE EMPLOYER AND EMPLOYEE

The obligations of an inventor to the inventor's employer extend to both patents and trade secrets, and to both present and past employers. Issues regarding these obligations often arise when the inventor changes jobs.

27.7.1 PATENTS—OWNERSHIP AND OTHER RIGHTS AND OBLIGATIONS

While patents are by law issued in the name of the inventor, the patent of an employed inventor is most often owned by the employer, the transfer having been made through an assignment executed by the inventor as a requirement of employment. When no assignment has been executed, the employer may still claim ownership through a contract obligating the employee to assign. The contract is often an express contract included in a more general employment contract. If no employment contract has been signed, the contract to assign patent rights can be inferred from statements of policy in such documents as employee handbooks or company memoranda issued by the employer. Whether express or implied, the employer's claim to ownership of patent rights can be challenged on such grounds as a failure of the employer to clearly explain the obligation, an open disagreement and lack of acquiescence by the employee in the obligation, or the claim that an invention did not involve any materials, facilities, or trade secret information of the employer and bore no connection to work performed for the employer. The latter is often supported by state law.

When neither a patent assignment clause in an employment contract or in a policy statement in an internal company document exists, employer ownership of patent rights can be implied for employees who are hired to invent, i.e., those hired either for a specific project or problem or for their expertise in a specific area. If the inventor is an officer or director of a corporation or other business entity, a special obligation to assign may exist as part of the officer's or director's fiduciary duty toward the entity.

When an inventor refuses to sign the documents required to apply for a patent, a present or former employer claiming ownership of the invention can apply for a patent in the inventor's name, provided that an appropriate showing of the respective rights and obligations can be made. The inventor will be informed by the USPTO of the existence of the application, and the USPTO will independently invite the inventor to participate in the application process.

Employers that do not claim ownership in patented inventions may claim a "shop right," which is a nonexclusive, royalty-free (although nontransferable) right to use the invention. Shop rights arise when an inventor has used the employer's resources to conceive the invention, reduce it to practice, perform scale-up, or develop the invention for commercial use. Shop rights can be voided by actions of either the inventor or the employer that are inconsistent with a shop right, or by express agreement or the payment of compensation by the employer to the inventor, the compensation voiding the royalty-free aspect of the shop right.

27.7.2 TRADE SECRETS—OBLIGATIONS TO PRESENT AND PAST EMPLOYERS

Trade secrets are some of the most valuable assets of employers. Present employees are obligated to respect the trade secrets of their employers by maintaining confidentiality of the secrets. Former employees have a greater obligation because they are prohibited from using or disclosing the trade secrets of their past employers in any new employment situation. This prohibition does not extend to the use of one's general knowledge and skills, but only to information that is specific to the past employer's business and that is truly proprietary to that employer. The risk of violating a trade secret is highest when the new and previous employers are direct competitors and the new employer hires the employee to supervise, design, or work in connection with the same product or service in which the employee had intimate involvement under the previous employer. The risk is lowest when the project assignments under the previous and new employer are unrelated, and when both the employee and the new employer agree to avoid any involvement of the employee in any project that would compete with the previous employer or that would benefit from specific technical knowledge gained by the employee in the previous employment.

27.7.3 RIGHTS AND OBLIGATIONS CREATED OR ELIMINATED BY EXPRESS AGREEMENT

All rights and obligations that are implied by the employment situation can be negated or modified by written agreement. Additional rights and obligations that are not otherwise implied can also be created by written agreement. Some of the most common among these are obligations not to compete with a former employer, obligations to cooperate with the former employer, and the rights of a former employer to patents in inventions subsequently developed. Noncompetition provisions are usually limited in scope, either in terms of geographical area or time, and obligations to cooperate are usually directed to assisting in the securement of patent rights when such assistance is needed. In general, pre-employment agreements, employment contracts, and statements of company policy should be studied with care by the employee to identify the rights and obligations of the employee.

28 Communication

F. S. Oreovicz

CONTENTS

28.1 Introduction 1842
    28.1.1 Some Preliminaries 1842
28.2 Principles of Good Editing 1843
    28.2.1 Paramedic Method 1844
    28.2.2 Style: Dependent Clauses and Sentence Combining 1845
    28.2.3 Avoidance of First Person? 1846
    28.2.4 Verb Tense 1847
    28.2.5 A Final Word 1847
28.3 Grammatical Problems Common to Engineering Writing 1848
    28.3.1 Agreement 1848
    28.3.2 Agreement and Ambiguity 1848
    28.3.3 Conciseness—or, Wordiness from Insecurity 1848
    28.3.4 Dangling Modifiers—"Using" 1849
    28.3.5 Smothered Verb 1850
    28.3.6 Weak (or Strange) Predication 1851
28.4 Formats 1851
    28.4.1 Related Format Matters: Equations, Tables, Figures, References ...1852
28.5 Basic Principles of Oral Reporting in the Workplace 1854
    28.5.1 Considerations of Delivery 1855
    28.5.2 Coping with Anxiety 1857
    28.5.3 Six Signals All Audiences Want to Hear 1857
28.6 Tools and Resources for Effective Writing 1857
    28.6.1 Online Guides 1857
    28.6.2 Style Guides: Professional Societies 1857
    28.6.3 Usage Guides 1858
    28.6.4 Manuals and Guides 1858
    28.6.5 Graphics 1858
    28.6.6 Oral Presentations 1858

First, controlling language. What matters is insignificant detail. Spelling mistakes, typos, mistakes in idiom, unfashionable usages, all these characterize you as a writer controlled by language rather than controlling it … It is not a question of being clear. These revelations of self don't usually obscure ideas: they obscure you. Worse, they reveal you. They reveal that you have not paid attention to your own writing and invite the reader to respond in kind. They model the history of your mind.

Richard Lanham (*Revising Prose*. New York: Scribner, 1979, 81)

28.1 INTRODUCTION

At some point in your work—after you have used all the resources at your disposal, such as this handbook—you will have to write up your results and share them with the world, possibly giving a verbal presentation as well. This moment of communication is the point at which you begin to exist as an engineer or scientist. It is these nontechnical skills, the so-called "soft skills," that set you apart from others in your field. If readers or listeners of your work are constantly distracted or, worse, amused, by how you present yourself in writing or speaking, they may begin to question what you are trying to say. As Professor Lanham notes, these errors reveal you—and in a way that you don't want to be revealed.

Effective writing and editing involves many considerations, and a brief checklist shows how detailed the process can become. In no particular order, each exercise at writing involves the following:

- Coherence and unity
- Transitions
- Subject-verb agreement
- Clarity and logic
- Pacing
- Conciseness
- Sentence patterns
- Sentence variety
- Awkwardness
- Pronoun reference
- Correct modification
- Word choice

Anyone using this guidebook will likely have already been exposed to a large body of written material and probably be an experienced writer. Many of these steps will be handled in the course of writing, and rarely does one person at this level have trouble with all of the above. What's more likely is that he or she may have a blind spot, say, dangling modifiers and conciseness. For that person, it becomes a matter of identifying the problems and working on them. This is where paying attention becomes important.

Tip: Keep a record of the comments made on your writing and look for patterns.

Welcome the information, for it will help you improve. Technical writing is a skill like any other: the more you practice, the better you get. But remember: it is also possible to practice bad habits. Anyone who has ever played golf will know the truth of this. Practice doesn't make perfect—it ingrains. Perfect practice makes perfect.

A complete writing/speaking treatise is beyond the scope of this chapter, but there are some basic areas and problems that, once corrected, can lead to rapid improvement. Certain errors tend to occur often in technical writing and speaking, and these are dealt with first. Some suggestions follow on refinements of style, along with examples.

28.1.1 SOME PRELIMINARIES

Audience Analysis. This staple of all communication goes beyond merely thinking about the occupations or titles of your readers/listeners. At some point, the following considerations will come into play as you plan your communication. The more of these you can match, the likelier your message will be received successfully. Clearly, it is unrealistic to expect to match more than, say, the cognitive needs and style, but be aware that your readers will likely be reading you through a lens colored by these variables.

Variables affecting a receiver's (reader/listener) behavior:

- Self-esteem
- Anxiety
- Authoritarianism
- Open/close-mindedness
- Cognitive needs
- Cognitive style
- Commitment

Pay Attention. Getting in the habit of seeing what is actually written on the paper, instead of hastily scanning what you thought you had written, will get you off in the right direction. For example, look at the following sentences (all from students' final reports) with comments in brackets:

Viscosity is determined by combing Equations 6 and 13.

[With a fine-toothed comb?]

The water flew through heat exchanger 1 to 2 when the inlet valve was opened.

[On wings of....?]

It was necessary to calibrate the flow meters in conjunction with the gasses interested in running the mass spec.

[Gases were interested in running a mass spec?]

It takes little knowledge of grammar, mechanics, or style to figure out that something is amiss here.

Give the Draft Time to Cool Down. Errors somehow manage to slip in between the brain and the keyboard, between intent and pen. In revising or proofreading, we tend to review the script that unscrolled in our brains as we composed. Often we don't "see" what is actually on the page, reading only what we expect to be there. The less time we wait after writing, the clearer that script is, and the less likely we are to see the words on paper as simply that—words on paper. That's why it is so easy to spot errors in someone else's writing: we have no script to follow, so we only look at what's written. Putting a draft aside for a while lets us objectify what we've written, remove it from that familiar script in our heads, and get a fresh perspective. Letting someone else read the draft can accomplish the same, but then you're at the mercy of that person's "error-spotting" ability. If time is unavailable, then proceed to proactive editing below.

Read the Draft Aloud. Hear the rhythms. Hear the omissions. Get a feel for the lengths of sentences.

Be a Proactive Editor (Using the paramedic method described below, for example.). Look for particular problems (as well as strengths—it's okay to appreciate your own work and to pat yourself on the back for a well-turned phrase). Too often in looking over a draft we actually overlook it—hoping something will jump out at us. When nothing does, we are later surprised that someone else, usually a supervisor or higher, has returned the draft awash in red ink. Most writers tend to make the same mistakes in each document, or have similar problems in each writing. Ideally, you'll find out what your areas for improvement are—by working with an editor, for example. Failing that, you can work off the list provided below of common problems often arising in technical writing.

28.2 PRINCIPLES OF GOOD EDITING

Richard Lanham's "paramedic method" (*Revising Prose*, 34) is a direct and simple approach that focuses on aspects of writing that often occur in technical prose. This section also deals with an important aspect of style—the use of sentence combining—and concludes with some thoughts about the prevailing practice of avoiding the use of the first person.

28.2.1 Paramedic Method

Although there are five steps to the method, they are not sequential. For example, number 4 may be considered a first step; numbers 1 to 3 can be done at the same time; and number 5 acts as a final check. According to this method,

1. Circle the prepositions.
2. Circle the "is" forms (be, am, is, are, was, were).
3. Put the action where it belongs—in a simple active verb.
4. Start fast—no mindless "zero-content" openings.
5. Read your draft aloud and listen for sound patterns.

Circle the Prepositions. Prepositional phrases act as adjectives or adverbs, adding qualifying detail. Used properly, they do their little jobs quite effectively, but in technical writing they often march along in a parade of details—an ineffective catalog because all details are treated the same way without regard to levels of importance. Consider the following example (from T.P. Johnson. *Effective Communication for Engineers*, 6), which contains eight prepositional phrases.

Before

The unlikelihood **of** meeting orders **from** the majority **of** its new customers is **of** concern **to** the company, due **to** tardiness **in** the installation **of** its new manufacturing line.

Written in this way, all the details are equally weighted—nothing is more important than anything else. Read aloud, the sentence also lacks any life. The two- or three-word phrases all tramp along in the same rhythm. This is a poetic device (of two unstressed syllables and a stressed) that poets use to slow down the movement of the verse. Rarely is that a goal of technical writing. After a few such sentences, it is not surprising that readers find themselves nodding off.

The sentence gains new energy with the use of dependent clauses that subordinate details with respect to their importance.

After

The company does not know whether it can meet most new-customer orders, because its new manufacturing line is not yet installed.

This revision has zero prepositional phrases and two dependent clauses. It also exemplifies steps 2 and 3 in the paramedic method. See Section 28.2.2 for more about dependent clauses. Prepositional phrases are, of course, an integral part of writing and are not to be avoided. It is only their excessive use and without thought to what is being said that creates problems. The lack of such phrases in this sentence is only coincidental.

Circle the "Is" Forms (be, am, is, are, was, were). The "to be" verbs are important, but by their nature they make existence statements rather than action statements. When overworked, they rob a sentence of energy. For example,

Before

The calibration of the thermocouple **was** accomplished by the technician.

The intended action of this statement is the act of calibrating, but it is buried in the subject of the sentence, not in the action word, the verb, where it belongs, as in the "After" example below.

Put the Action Where It Belongs—in a Simple Active Verb. The section on smothered verbs (Section 28.3, Common Grammar Problems) shows how pervasive this problem can be. The sentence that expressed no action now becomes an active statement.

After

The technician **calibrated** the thermocouple.

Start Fast—No Mindless "Zero-Content" Openings. If a section heading is "Results," why make the first sentence: "This section contains the results of the experiment."

Read Your Draft Aloud and Listen for Sound Patterns. The rhythm of your writing will become apparent when you read aloud. Reread example 1 above and listen to the effect of the preposition parade. For example, you may also discover that all your sentences are basically "This is the house that Jack built" sentences—all beginning with "the" and all verbs lacking any force. Reading aloud will make this painfully clear. It will also tell you if your sentences are too long or too short.

28.2.2 STYLE: DEPENDENT CLAUSES AND SENTENCE COMBINING

(See Also Section 28.3.3.)
Becoming a good stylist in one's writing isn't always a matter of genes or nurture, but simply of paying attention, which the writer of the following two sentences clearly was not doing:

The attic tank is used as the water supply for the system. This tank is located in the attic.

The strings of prepositional phrases, mentioned above, are also often a signal that we haven't been paying attention. The solution is to emphasize the important details, such as by using dependent clauses to subordinate details with respect to their importance.

Recognizing a Dependent Clause.

- It is not a complete statement.
- It must have an introductory conjunction (if, when, which, that, because, since, etc.) or relative pronoun.
- It must have a subject and a verb, though not be independent.

Dependent clauses also act as adjectives and adverbs, and as nouns.

Noun: I saw **what you did**. (Object of the verb "saw")

Adjective: The car **(that) he owns** is very fast. (Modifies the noun "car")

Adverb: She sings **when she takes a shower**. (Modifies the verb "sings")

Dependent clauses make sentences more informative.

- They analyze detail, highlighting what is important (and tell me "when," "if," "because," "who," and so forth that help the reader interpret what is being described).
- They force the writer to break up the equating of detail (preposition parade).
- They make writing sound more natural.

Sentence Combining. Look for short sentences, especially if related by information, and combine them. Common sense has to prevail here. The goal is not just to write long sentences. Balance is important, but usually a series of short statements indicates undeveloped ideas or ideas lacking proper emphasis. The use of dependent clauses is at the heart of improving style by sentence combining. The following sentences illustrate what is involved.

Before

Before starting experiments each day, the spectrophotometer was calibrated. This was accomplished by determining the absorbance of a standard solution of dye.

After

We first calibrated the spectrophotometer by determining the absorbance of a standard solution of dye.

Before (The following example contains undifferentiated statements.)

The temperature difference is the driving force in a heat exchanger. The temperature difference can be the difference between the temperatures of the steam and the water.

After

The temperature difference, which is the driving force in a heat exchanger, is the difference, for example, between the temperatures of the steam and the water.

or, depending on the emphasis

The temperature difference, which is the difference between the temperatures of the steam and the water, is the driving force in a heat exchanger.

Before (This example contains an awkward shift from active to passive.)

To begin the experiment, we adjusted the water flow rate to the desired value. The exit stream valve was then opened and the other steam exit valves were closed.

After

To begin the experiment, we adjusted the water flow rate to the desired value, opened the exit stream valve, and closed the other steam exit valves.

28.2.3 Avoidance of First Person?

One of the ironies of writing in the engineering profession is the disappearance of the author. At every stage of the work, the engineer is the chief actor in the drama. From the birth of the idea to the presentation of the results to the world, he or she is the central figure. Yet, when it comes time to tell the world, the writer has to disappear. By being forced to drop the use of the word "I," the writer adopts the third person, which leads to more pronouns and passive verbs, and this in turn leads to all the problems with dangling and misplaced modifiers.

A workable compromise might be the use of the first person in moderation. For example, in certain sections of a report it makes sense to let the author's voice come through. A description of apparatus, for example, rightly focuses on the devices themselves, unless the writer wants to emphasize his or her use of the equipment. A description of procedure, however, is a record of that person's work; as such, the use of first person not only improves the style but also makes the writing more accurate, especially if legal questions arise later. A discussion of results also warrants a stronger authorial presence, whereas a treatment of method or theory reads better with an emphasis on the ideas themselves. And because teamwork is usually involved in a great deal of engineering work, the first person "we" doesn't carry as much of a stigma as "I."

28.2.4 Verb Tense

The choice of verb tense is usually a matter of common sense: if the report deals with a completed project, the appropriate choice is past tense. However, there are some confusing gray areas, such as the use of the "historical present" or a universal truth. For example, there is a large difference in implication between the following two statements:

Newton's law **states** that.....

Newton's law **stated** that....

The implication in the second is that the law has possibly been superseded. Even though it's several hundred years old, we use the present tense with Newton's law because the law still holds true. However, of the phlogiston theory we would probably write: "The phlogiston theory maintained that...."

Use past tense for stating the problem and describing the procedure. For example (with the words emphasized in boldface):

The objective of this work **was** to....

The unit **was** brought to a steady state....

Present tense is appropriate for theory and for discussing results, since the theory holds true and the results still stand. For example:

The theory **is** based on the conservation of....

Table 1 **shows** that....

The values for Run 1 **are** consistent with....

Use past tense to refer to what happened in the actual work:

The significant difference in output shows that the pump **was** not working....

28.2.5 A Final Word

Crafting a cabinet, refining a golf swing, or writing clear prose—all require effort. The rewards in all three begin with satisfaction and end with a product that speaks well of your effort. Begin by paying attention to what you have put on the page. Look at the words as words and the relationships between them. How do they connect? Which direction do they point? Who is doing what to whom? Does the verb do for the noun what it should be doing? Or is it pointing to some other word later in the sentence? And so on. Much can be inferred even if grammatical categories escape you. Read what is on the paper, not what you think you have written.

28.3 GRAMMATICAL PROBLEMS COMMON TO ENGINEERING WRITING

Handbooks, textbooks, usage manuals, and any number of general works are filled with the problems common to technical writing. Improvement in these typical areas will immediately improve your writing overall.

28.3.1 AGREEMENT

Problems in subject-verb and noun-pronoun agreement in technical writing often arise because the verb is separated from the subject by one or more clauses that often contain a plural element. In the following examples, the writers allowed themselves to be misled by the plurals in the intervening words:

> The **kerosene hold-up** in the column at agitator speeds of 200, 250, and 300 rpm **are** shown in Figure 5.

> **Interstage eddy flow** at the two agitator speeds also **confirm** the above prediction.

28.3.2 AGREEMENT AND AMBIGUITY

The following sentence illustrates a number of problems:

> The exit temperature of the water and the water flow rate was observed to give an approximation of how the constants functioned.

In this example, the writer creates several problems. First, it appears that the water flow rate also has a temperature! The order of the first two statements implies that the preposition "of" applies to both items. Writing "water flow rate" first would remove that problem and make it clear that the verb should be "were." Were observed to give what? That is, a dangling modifier (infinitive, see below) is next. Here's what must have been the original meaning:

> To give an approximation of how the constants functioned, we observed the water flow rate and the exit temperature of the water.

28.3.3 CONCISENESS—OR, WORDINESS FROM INSECURITY

First, conciseness is not the same as brevity. Clarity is the goal. Sometimes, a sentence with 40 words may be more concise than one with half that many. Wordiness is almost always the sign of an early draft. In an early draft, one's main concern is to get the ideas down; revision is for later.

An effective way of achieving conciseness is through subordination of sentences (see also Section 28.2.2). Another step is to write confidently. The following are examples of degrees of authorial insecurity: "I feel that," "It should be noted," and "The reason is that." If something should be noted, then do so. The note itself is sufficient. To preface it by saying "It should be noted that" or "It is worthy to note that" is to show your insecurity in your own words.

> I feel that the approach will not work under these conditions.

If you trust yourself and your work, be bold:

> The approach will not work under these conditions.

> It should be noted that the average value does fall within the range of literature values for a fouled system.

The average value falls within the range of literature values for a fouled system.

The reason a high uncertainty is associated with the Wilson plots is most likely due to random error coupled with the inapplicability of the Wilson equation.

The high uncertainty associated with the Wilson plots is most likely due to random error coupled with the inapplicability of the Wilson equation.

In each of these examples the revision reads with more conviction.

28.3.4 Dangling Modifiers—"Using"

Words point to words when we write, but the connections between and among them often become confused. Consider the following statement:

> Using this equation, we calculated the area.

Who is doing the using? Clearly, it is the subject of the sentence: "we." However, engineers are often told not to use first person. Their first instinct is to convert the sentence to passive voice:

> When this equation was used, the area was calculated.

But this revision sounds awkward, so they then convert only half the sentence to passive voice:

> Using this equation, the area was calculated.

That appears to sound better, but is it? "Using" (a verb used as an adjective, called a participle) must modify the subject of the main clause, which was originally "we," but instead it is left dangling without a sensible subject. Another way to look at it is to consider that an action is being indicated at the beginning of the sentence. Does a logical agent of the action follow? In the original sentence it makes sense that "we" were using; in the most recent revision, the area is now using the equation! The restriction on first person has pushed the writer into this corner of writing something that "sounds good" but makes no sense. "Ah," you might say, "but I know what the writer means." Yes, but why waste your time doing what was the writer's responsibility? Take the time you spend reading all the literature you are required to read—not including all that's out there in the professional literature that you'd always hoped to keep up with—and then add to that the time you spend rewriting what you are reading. This kind of writing wastes your time.

The same problem occurs if the clause comes later in the sentence:

> The area was calculated using this equation.

This is a very common expression in technical writing; in fact, writers tend to make "using" into an all-purpose preposition. For example,

> The permeability of a pure gas can be found using the pressure in both the shell and the permeate and the molar flow rate of the permeate.

> Models were developed using linear regression.

> Permeabilities were calculated using the numerical integration technique.

> Four runs were made using the co-current flow pattern at two conditions.

The error and the humor can be seen by considering the following sentences:

> The rabbits were seen using binoculars.

> The skeleton was discovered while digging for gold.

Just as the rabbits are holding binoculars and Mr. Bones is digging here, in the previous sentences the permeability is using pressure, models are using linear regression, and runs are using flow patterns! Sometimes the absurdity is immediately evident:

> After drying in the sun for a few days, the workers finished the roadway.

And if attention is not paid, a dangling modifier can strain common sense:

> After exiting the first exchanger, the water temperature was measured again by another thermocouple.

> The inoculum was prepared a period of time before arriving in the laboratory.

28.3.5 Smothered Verb

A smothered verb is also referred to as nominalization. A sure-fire way to make writing sound impressive, or so think ineffective writers, is to convert a verb into a noun and then add another weak verb to the sentence. This is guaranteed to add more words to your text. Consider the following sentence:

> The experimental engineer calibrated the thermocouple.

Simple and straightforward but not very "impressive." What you will often find in student writing is the following version:

> The calibration of the thermocouple was accomplished by the technician.

The main action of the sentence is contained in the act of calibrating. The verb has been *nominalized*—turned into a noun; two prepositional phrases have been added ("of the thermocouple" and "by the experimental engineer"); and another verb created ("was accomplished").
Another example:

> We compared the permeabilities during co- and countercurrent flow by graphing the permeability versus run number.

With the onus on the use of first person, this sentence quite easily becomes

> The comparison of the permeabilities during co- and countercurrent flow was done by graphing the permeability versus run number.

But it could just as easily have been written

> The permeabilities during co- and countercurrent flow were compared by graphing the permeability versus run number.

With no gain in clarity or meaning, the writers of these sentences have taken a perfectly good concise sentence and larded it with fat! Whenever possible, keep the main action of the sentence in the main verb.

28.3.6 WEAK (OR STRANGE) PREDICATION

Predication relates to the part of the sentence that expresses what is said of the subject; it consists of the verb plus any modifiers or complements. Weak predication arises when the main action of the sentence is not in the verb or when a weak verb is used. The following examples usually have a number of problems as a result.

Sentences 1a and 1b take on different meanings when considered in conjunction with this example:

1. Karen was waiting for the system to come to equilibrium.

1a. The final step was plugging in the vacuum pump.

1b. The hazard in the laboratory was climbing a ladder.

Of course it is possible that the final step and the hazard did the plugging and climbing on their own initiative, but what the writer intended was more likely the following:

1c. The final step involved plugging in the vacuum pump.

1d. Climbing the ladder was hazardous.

Other problems arise with "is when" and "is where" sentences. Often used in defining terms, they fail to include a category and are indirect. They equate ("is") a noun (the subject of the sentence) with an adverb of time ("when") or place ("where").

2. This limit is when the gas flow holds up the liquid flow.

2a. This limit is the point at which the gas flow holds up the liquid flow.

2b. This limit occurs when the gas flow holds up the liquid flow.

28.4 FORMATS

Formats are either organization or discipline specific. In general, though, they will contain some or all of the following sections.

Abstract/Executive Summary. The abstract is a mini-report independent of the main body. Placed before each report, the abstract distills the essence of the report. It abbreviates a longer text without changing any aspect of it and includes nothing that is not in the original. It is specific, includes key numbers, and clearly states the main conclusions but generally does not have tables and figures, unless their production was the main goal of the work. An executive summary is traditionally longer and more detailed than an abstract, often written for individuals who will read nothing else of the report.

The abstract must

- Reduce principal matters to the fewest words and eliminate details.
- Give a general view of the overall situation.
- Separate cause from effect.
- Retain a sense of substance and quality.

The main types of abstracts are the descriptive and the informative. A *descriptive* abstract tells what the report discusses but does not reveal its conclusions. It says more about the report than about the work. An *informative* abstract, however, contains information from the report, including the facts on which the conclusions are based, the conclusions themselves, and any recommendations made. Basically, it is a miniature report or distillation of an entire report.

Table of Contents. Keep the reader in mind and include everything that will make the work accessible to the reader. Remember to include figures and tables in a separate listing.

Introduction. The importance and relevance of the project and the objectives of the work are presented here. The level of detail depends on the type of report or the intended audience. A proposal or memorandum will require a different opening than a formal report, laboratory report, or technical manual. In any case, you should state your subject and purpose for writing and include some idea of the scope of the work. Again, the level of detail and size of the report/project will determine how much detail is needed in the introduction. For a technical manual, most of the detail should be left to the body of the report, with the introduction doing just that: introducing.

Method/Theoretical Information. Ideas, principles, theories, and techniques as well as relevant equations that apply to the project are included here. Derivations, calibrations, and other ancillary material are usually put in an appendix. Be careful not to turn this section into a list of equations. Provide enough information, both as introductory material and as transitions, to demystify the equations. In short, keep readers' needs in mind as you write.

Apparatus and Procedure. Describe the main features of the apparatus including schematics as well as major pieces of equipment. Use the past tense if reporting on work already completed. Use present tense when describing an apparatus or procedure in general.

Results and Discussion. In general, use plots (figures) rather than tables to discuss your results. Describe what is shown in the figures and tables and tell what it means. Do not make the reader figure it out. For tables and figures in the main body, describe each in a caption and refer to each in the text. Do not put all your figures and tables in the Results and Discussion section; show examples of some plots and discuss the general results. Present results of your statistical work, including regression, propagation of error, and analysis of variance. It is acceptable to refer to the appendix for some details.

Conclusions and Recommendations. Results and Discussion have led to concrete conclusions. Discuss those here as short sentences in a numbered list that must be supported by the data and the theory. Relate your conclusions to the objectives presented in the introduction and add the recommendations supported by the results of your project.

Notation. List symbols used in alphabetical order, English first and Greek in a second list. (Be sure to also introduce them in the report the first time they are used, in an indented table.)

References. List books and articles referred to in the report (see Section 28.4.5).

Appendices. Appendices contain details needed to support your results and to be used by people who may continue your work. Make it easy for them by organizing the material under subtitles or sub-appendices listed in the table of contents. Use enough words in appendices to define units and manipulations, but do not worry about smooth text with introductions and transitions. Include data sheets, sample calculations including error propagation, tables of intermediate variables in the calculations, working plots, calibration curves, and all the figures and tables that you did not include in the main body.

28.4.1 Related Format Matters: Equations, Tables, Figures, References

Equations. State the nature or origin of each equation that you use (definition, balance, kinetic expression, or empirical correlation) to take the mystery out of your development. For example,
   The rate of heat transfer in a heat exchanger is frequently reported as follows:

$$q = UA\Delta T \tag{44.6}$$

where
    q = heat transfer between hot and cold fluids in exchanger
    A = area over which heat is transferred
    ΔT = average temperature difference between hot and cold fluids in the exchanger

Equations should appear on separate lines from the text. Definitions of symbols should appear in an indented section also separate from the text. All of this material must be at hand for those interested in following the details, but all of it must be carefully separated so that a busy reader can get the sense quickly from the text. Even interested engineers are likely to read all the text first to find out where you are going and then go back and read the equations.

All definitions of symbols can be included in a nomenclature section at the end of the report or article, or right after the equation, as in the example above.

Tables and Figures. Tables are organized collections of numbers or words. Figures include plots, drawings, and photographs. Tables and figures are numbered consecutively in the order of their appearance in the report.

Refer in the text to each table and figure that appears in the main body, giving the table or figure number, and state the nature of its contents. Call attention in the text to trends or other features shown in the table or figure important to your development; never assume it is obvious to the reader. For example:

> In Figure 3 on page 8, the Nusselt number is plotted against the Reynolds number for each of the four exchangers. The prediction of the Dittus-Boelter equation is also shown. For each exchanger, the Nusselt number increases with the Reynolds number at a rate not significantly different from that of the prediction, but the 95% confidence intervals show that the Nusselt numbers for all exchangers are significantly different from the predicted values.

Labeling Tables and Figures. In general, table labels appear above the table and consist of a number and title, and frequently a line under the title giving more specification. (When any footnotes are used, they appear below the table.) For example

> Table 1. Heat-Transfer Coefficients
>
> *3 in. ID Exchanger*

Figure labels usually appear below the figure, and consist of a number, title, and usually a caption, which is needed to specify the content. For example

> Figure 2. Heat-transfer coefficients as a function of Reynolds number. Runs made after replacing the pressure gage., 0.5-in. tube; —, 1-in. tube;, 3-in. tube.

References to the Literature. Direct quotations and equations taken from the literature must be acknowledged. Special methods may also be acknowledged, but methods in general use need not be. As in most format matters, convention often rules, whether it be industrial or professional society prescribed. The guiding principle in whatever form you are involved with is to keep the reader in mind. And as a reader, what kind of information would you like to have as you read? Few things are as frustrating as trying to locate a reference that is poorly made.

In the main text show numbers in parentheses that are the numbers in your list of references. Show a page number also, if it is pertinent. For example

These results agree with those in Jones and McLaughlin (4, 77).

The derivation below follows that in Bird, Stewart, and Lightfoot (2, 255).

Hardcastle et al. (6, 215) showed that this method is incorrect.

The style of reference lists varies widely from one discipline to another, from one journal to another, and from one corporation to another. Professional societies often have a style that is used in all their journals. See Section 28.6 for lists of professional and general style guides.

Several types of references are needed. The following examples cover general usage:

Authored Text

Bird, R.B., W.E. Stewart, and E.N. Lightfoot, *Transport Phenomena,* New York: Wiley, 1960, chapter 12.

Greenkorn, R., and D. Kessler, *Transfer Operations*, New York: McGraw-Hill, 1972, 78–82.

Edited Text

CRC Handbook of Chemistry and Physics, 87th ed., R.C. Weast, ed., Cleveland, OH: CRC Press Inc., 2006.

Journal Article

Kramers, H., and P.J. Kreyger, *Chem. Eng. Sci.*, 6 (42) (1956).

Online Sources. In addition to the information provided in the previous examples (author, title, publication source, page numbers), these citations require the medium (online), the computer network (e.g., Internet), date, and electronic address (the URL).

28.5 BASIC PRINCIPLES OF ORAL REPORTING IN THE WORKPLACE

An oral report is simply a conversation—with attention to some constraints imposed by the situation. If, as you are walking down the hall, the boss asks, "How's the project coming?", your response is basically an oral report. Rather than have, say, a dozen such strolls—to talk with everyone who might have an interest or a stake in your work, it is more practical to talk to all of them at once. So why is this situation so much more fraught with terror than the former? It need not be. What is different about it? For one, with more people in the room you have to make eye contact with more than one person. Because they are further away from you, you have to speak louder. And since sound waves dampen rather quickly, you need to enunciate clearly so that even the last row will understand you. You also may not know the expectations or biases of all of your listeners, so it is best to err on the side of caution and to dress formally and maybe speak a little more formally than in everyday conversation. And the same holds true for the level of formality and quality of visuals or other materials. But these are all practical accommodations dictated by the situation.

For some reason we develop this fear of "The Formal Presentation." Not confident in ourselves or our speaking style, we adopt a "false formal" style that quickly sounds like the worst nineteenth-century legalese. See Section 28.5.2 for some tips on dealing with anxiety.

Oral presentations come in all types and lengths—from the stroll down the hall to a professional conference or major policy address. Though dramatically different, they all share some common features.

Target the audience: What is obvious to you may not be so to them. Consider their:

- Background
- Familiarity with the subject
- Information needs

Connect with the audience personally:

- Look at them
- Smile
- Talk to, not at, the audience
- Use movement to relax yourself and the audience

Use graphics to explain and connect. Make them

- Relevant
- Easy to read
- Interesting (color, clip art, etc.)

28.5.1 Considerations of Delivery

Delivery Method. The following are the four most common delivery methods for oral reporting:

1. **"Off the cuff"**: Too risky for a technical report where accuracy is important.
2. **Reading** (or manuscript speech): Acceptable as a last resort, but it is easy to lose touch with the audience.
3. **Memorization**: Very hard to do well, as memory lapse may unsettle the speech. Speaker thinks in words, not thoughts. Voice and body actions become stylized; lack spontaneity.
4. **Extemporaneous** (best choice): The speech is not written out, only outlined with any vital facts and figures that must be presented accurately (to guard against lapses of memory). It is practiced several times, but no attempt is made to memorize. The advantages are

- Better eye contact (only occasional glances at notes needed)
- Greater flexibility (you work from blocks of thought, not of words)
- Greater spontaneity (because you are not bound to words)

Ideally, you will know the material so well that all you will need are your visual aids.

Posture. Stand naturally and try to convey the impression of calm control. The audience seldom sees your knees shaking.

Quirks. Beware of jingling coins, waving a pointer, swaying, playing with your hair, and so on. These actions will so fascinate your listeners that they will be more interested in watching the show you are presenting than listening to what you have to say.

"uh...um...er...like...you know..."

Speakers often say "um" before they launch into their subject, and these "prefatory" um's are not too distracting. Pause-filling noises can be very annoying, though, and should be avoided. Don't be afraid to be silent. Your listeners will welcome the pause as an opportunity to reflect on your wisdom.

Eye Contact. Look at the audience as much as possible. Move your eyes slowly around the room, giving the back rows equal time. Learn to glance quickly at notes and to take in a block of information (which will be easy if you have practiced the talk). Also, practice "dropping" the eyes instead of bending down to look at notes. This motion is not as large as a movement of the head and is not as noticeable. Look back at the screen only for as long as it takes you to orient yourself.

Voice. Why should the audience show any enthusiasm if you do not? Enunciate clearly—sloppiness here may signal a lack of preparation. Worse yet, the audience may see it as a sign of disrespect, which is a sure way to lose their sympathy. You will also have to speak up if the room is fairly large. Have a friend signal to you if you are not making yourself heard. If talking too fast is your habit, focus on enunciating clearly. Consonants take time to form and doing so will give you the slowdown you need. There's no excuse, in a formal presentation, for "wanna," "gonna," and so on. Mispronouncing technical terms is especially bad practice. Like a performer in a one-person show on Broadway, you are in control of your material.

> **Tip:** Always give yourself positive cues. A basketball player who thinks, "Don't miss this freethrow!" is in trouble already. He or she has given the muscles no positive action to perform. Focusing on what is in one's control is important. That player should be focusing on technique, not outcome. In the same way, focusing on enunciating clearly will accomplish more than telling yourself, "Don't talk too fast" or "Slow down"—both of which detract from your concentration.

Accent. Don't worry about any accent you may have or think you have. Accents vary in all countries from region to region, but there's no excuse for sloppy pronunciation. Look up the pronunciation of words that give you trouble.

Movement. Don't be afraid to gesture when speaking, if it is natural for you to do so. Purposeful movement (gestures also) can help in several ways:

- Uses nervous energy and puts it to work
- Attracts attention; emphasizes points
- Adds confidence; keeps speaker and audience awake
- Refreshes the visual scene for the audience

Speakers on the radio gesture emphatically even though they are out of sight of the audience. Use normal descriptive gestures that everyone uses in conversation to indicate length, height, speed, roundness, and so forth.

Notes. They should be unobtrusive. Avoid holding an 8.5 x 11 sheet of paper. Make notes very brief, using key words or phrases.

Timing. Clearly, the most important element for your professional audience will be your explanation of your results. In planning your talk, do this part first; then, with the time available, fill in the other sections.

Visuals. Time limitations and volume of material to cover make it necessary that you use an overhead projector, slide projector, or some computer-aided variation. Quality is very important here. If your lettering is too small or sloppy, your audience will not only have trouble reading what you have written but will likely form a negative opinion about the quality of your work or effort. In general:

- Print, using large lettering
- Use brief action statements, not complete sentences
- Put only one equation on a slide (unless they are very brief)
- Draw a neat schematic of the apparatus, highlighting the main features
- Label axes of graphs; include data points and confidence intervals

28.5.2 Coping with Anxiety

1. **Expect it.** Anxiety is a fact of life. World-class athletes and performers experience it. You might as well anticipate it and prepare for it, realizing that for a few minutes, most likely at the beginning, you are going to be a little shaky. On the other hand, this same nervous energy can be a creative force, helping you to be alert and focused. Focused nervousness equals creative tension. So, literally follow step 2.
2. **Plan for it.** Ignoring or suppressing it is not going to make it go away. As you outline your preparations, just add "will be nervous" to the list, thereby reducing it to just another aspect of the talk needing a little attention, not the sole attention given it otherwise.
3. **Focus on the audience's needs,** not on your own success or failure. Ask yourself continually: "What can I do to help them understand....?" Your main goal is to communicate a message to the audience, not to perform. If you are caught up in your own success or failure, then any little mistake can loom large. If you are focused on the audience, then you will do what it takes to help them understand.

It may help you to remember that the audience is on your side. Have you ever noticed your reaction to a speaker who is struggling? Generally you sympathize with the person. Your listeners will feel the same sympathy, and even empathy, for you.

28.5.3 Six Signals All Audiences Want to Hear

One author suggests that you make sure the audience gets the following signals (Hoff, 1992; see also Section 28.6.6):

1. I will not waste your time.
2. I know who you are.
3. I am well organized.
4. I know my subject.
5. Here is my most important point.
6. I am finished.

28.6 TOOLS AND RESOURCES FOR EFFECTIVE WRITING

28.6.1 Online Guides

http://owl.english.purdue.edu (excellent resource for all kinds of writing help)
http://www.bucknell.edu/x3393.xmL (resources for writers)
http://www.bartleby.com/141/ (a cyber version of Strunk and White's *The Elements of Style*)
http://www.refdesk.com (a collection of reference links)
http://www.gpoaccess.gov/stylemanual/index.html (the latest *United States Government Printing Office Style Manual* available in a searchable database)

28.6.2 Style Guides: Professional Societies

American Chemical Society. *ACS Style Guide*: *Effective Communication of Scientific Information*. 3rd ed. Coghill, A.M., and L.R. Garson, eds. New York: Oxford University Press, 2006.

American Institute of Chemical Engineers. (See the journals for general format requirements.)

American Institute of Physics. *AIP Style Manual.* 4th ed. Woodbury, NY: American Institute of Physics, 1990.

28.6.3 Usage Guides

The American Heritage Guide to Contemporary Usage and Style. New York: Houghton Mifflin, 1996.

Fowler's Modern English Usage. 3rd rev. ed. R. W. Burchfield, ed. New York: Oxford University Press, 2004.

28.6.4 Manuals and Guides

Alred, G.T., C.T. Brusaw, and W.E. Oliu. *Handbook of Technical Writing.* 8th. rev. ed., New York: St. Martin's Press, 2006.

The Chicago Manual of Style. 15th ed. Chicago: University of Chicago Press, 2003.

National Information Standards Organization. *Scientific and Technical Reports—Preparation, Presentation and Preservation.* ANSI/NISO Z39.18 – 2005. Baltimore, MD: NISO, 2005.

Turabian, K.L., et al. *A Manual for Writers of Term Papers, Theses, and Dissertations.* 7th ed. Chicago: University of Chicago Press, 2007.

Strunk, W., and E.B. White. *The Elements of Style.* 4th ed. New York: Allyn & Bacon, 1999.

Principles of English Usage in the Digital Age. San Francisco: Hardwired, 1996.

28.6.5 Graphics

Tufte, E.R. *The Visual Display of Quantitative Information.* 2nd ed. Cheshire, CT: Graphics Press, 2001.

28.6.6 Oral Presentations

Dayne, M.B. *The Performers Voice,* New York: Norton, 2005.

Hoff, R. *I Can See You Naked.* Kansas City, MO: Andrews and McMeel, 1992. Out of print but if found, a delightful and perceptive account for giving all kinds of professional presentations.

29 Ethical Concerns of Engineers

Lyle F. Albright

CONTENTS

29.1 Introduction 1859
29.2 Case Studies 1860
    29.2.1 Case of the Eager Applicant 1860
    29.2.2 Case of the Hospitality Suite 1861
    29.2.3 Case of the Super Super Heat Exchanger 1861
    29.2.4 Spy in Sky 1862
    29.2.5 Case of Reactor That May Run Away 1862
    29.2.6 Competing Researchers 1863
    29.2.7 Questionable Company Ethics 1863

29.1 INTRODUCTION

Engineers will almost certainly be tempted to do something that is either unethical, if not illegal, on more than one occasion. Selecting a course of action that is ethical is not always easy and often requires careful planning.

In general, an engineer signs a legal contract with his employer to protect propriety information for at least several years. The engineer also needs to consider how to maintain good relations with his associates and the public. As an analogy, ethics in many ways is like a stool with three or four legs. The rights of the employer (or client), the public, industrial associates, plus the engineer, all need to be considered and balanced.

First, ethics and legal matters are not equivalent. A course of action may be on occasion legal, but not necessarily ethical. How is ethics defined? As defined here, and by many, ethics is the code adapted by a group, profession, and a country. For example, offering money to a political official is generally considered in the United States as a bribe. Hence, it would often be considered not only unethical but also illegal. Yet in certain countries, providing a gift to a key government official might be considered as a way of life and a method of taxation. The old saying was, "When in Rome, do as the Romans do." Since industry is rapidly becoming more internationalized, many engineers will have experiences with individuals from other cultures. Hence, another complication is added to maintaining high ethical standards.

The American Institute of Chemical Engineers plus other professional societies have adapted, as guidelines, codes of ethics. The engineers should serve with fidelity the employer and the public. At all times, the engineer uses his skills to advance human efforts, to increase his technical competence, and to increase the prestige of the engineering profession. Based on personal experience, nothing, and let me repeat nothing, can tarnish one's reputation in the eyes of his colleagues more than to be thought to act unethically.

Several areas in which disagreements can occur between an employer and an employee are as follows:

1. Confidential information: What is confidential information? Incidences have occurred in which an employer claims that specific information is confidential. Yet the company has permitted the information to be published in the open technical literature. Often the engineer is not certain just what is and what is not confidential.
2. What information can an employee take from one company to a second? Often details on the manufacture of a new catalyst are considered proprietary; yet if the employee developed a new method of designing a heat exchanger, the information might not be considered proprietary. An engineer supposedly improves and becomes more of an expert while he is with the first employer.
3. Assuming an engineer becomes convinced that a process or certain equipment has serious safety problems, but the employer insists that the current set-up is safe. What course of action should the engineer follow? With current laws, an engineer can be held liable for suppressing information on potential hazards. At the very least, it is recommended that the engineer address written correspondence on this matter to the top management of his company.
4. Stolen information: Situations have occurred in which an individual leaving one company fills his briefcase with confidential reports. When an employee at either the first or second company learns of this, what action should he take?
5. Technical reports have been published in which incorrect or misleading information has been published on a competitor's process.

Many industries and individuals find that they need to consult to an increased extent with attorneys in planning their actions and programs.

29.2 CASE STUDIES

The following case examples have been used by the author in numerous talks to engineers, students, and others throughout the United States. Several examples were published in the magazine *Chemical Engineering* (September 2, 1963, 87–90; December 9, 1963, 177–184; September 22, 1980, 177–184; September 2, 1987, 40–43; and September 28, 1987, 108–120). The author has, on occasion, modified the original. In some cases, he has used personal experiences. The questions raised in these talks are first reported. The answers received during these presentations are summarized.

29.2.1 CASE OF THE EAGER APPLICANT

Mr. A of ABC Chemical Co. interviews engineer X of XYZ Chemical Co. for an opening in ABC. ABC and XYZ are competitors. As engineer X explains his capabilities, he discloses confidential information about XYZ. Such information is of value to ABC.

Engineer X is obviously a top-notch engineer and he would be able to do the required engineering for ABC.

Questions

(a) Is it ethical for Mr. A to pass on the confidential information to the engineers of ABC? The information was volunteered freely by engineer X.
(b) If the information was passed on, is it ethical for engineers in ABC to use the information if they recognize how it was obtained?
(c) Should ABC hire engineer X?
(d) Should ABC inform XYZ that engineer X talks too much?

Ethical Concerns of Engineers

Comments

About 50% of engineers vote yes to both (a) and (b). Older engineers sometimes vote affirmably more than younger engineers. Of interest, the ethics of engineer A can be questioned. Did he in any way pump engineer X? Most engineers agree that engineer X acted unethically. Most engineers indicate that ABC Chemical should not hire engineer X since he is too talkative. I personally am aware of a real-life experience similar to this.

29.2.2 Case of the Hospitality Suite

Engineer B is working on a rush project for Supreme Chemical Co. Company X has already commercialized the specific product. Company X has a hospitality suite at an AIChE convention. It was advertised in the lobby.

Engineer B goes to the hospitality suite and starts talking with a tipsy sales representative; he is not asked his company affiliation and he does not volunteer it. He guides the conversation into an area relative to his rush project. The sales representative reports on sales projections for the product, which will be of value to Supreme.

Questions

(a) Was it ethical for engineer B to go to the hospitality suite of the competitor?
(b) Although not requested, should engineer B have made his affiliation known?
(c) Did engineer B act ethically in getting information from the company X sales representative?
(d) Assuming the sales representative was not really tipsy and he recognized engineer B, is it ethical for him to feed engineer B "misinformation"?

Comments

Most engineers agree that they can go to hospitality suites of competitors to get free refreshments. Most believe that engineer B should have identified himself, but also that company X representatives should have asked for identification. Also, engineer B should not have have tried to pump the company X representative. But then company X representatives were apparently not truly vigilant. So both engineer B and company X representatives seem to be at fault. One can also question the ethics of the company X representative in spreading "misinformation." That, too, is not really ethical!

29.2.3 Case of the Super Super Heat Exchanger

Ken initially works for Superbouncy Rubber Co. He has signed a secrecy agreement not to divulge information that Superbouncy considers proprietary. While there, he makes an important improvement on a heat exchanger for a highly viscous elastomeric solution. The company decides to keep the new equipment as a trade secret.

Ken leaves Superbouncy and goes to work for Yum Yum Candy Co. The improved heat exchanger would result in a substantial savings in operating expenses for fudge stream. Superbouncy and Yum Yum are not direct competitors. Ken should do which of the following?

(a) Not report the improved heat exchanger to Yum Yum.
(b) Suggest a development program to Yum Yum that will lead to a heat exchanger similar to the one used by Superbouncy.
(c) Recommend an improved heat exchanger that, based on his calculations, would be essentially identical to the one that he developed for Superbouncy.
(d) Other.

Comments

Most engineers consider this as a grey area. I agree. Is this new heat exchanger truly proprietary? Ken has had courses on heat exchangers in college. Is this new exchanger based on a truly new and unexpected concept?

Ken needs to report to Yum Yum that an improved exchange is possible but also to report possible legal problems. If Ken believes that the new exchanger is really his idea and that it is a logical extension of his scientific growth, then he should receive a favorable legal report first. If there is any question on the ethics, Yum Yum should inform Superbouncy, who would likely license at a rather low cost.

29.2.4 SPY IN SKY

Engineer Z of Super Plastics, Inc. is to estimate production costs of a new type of polyethylene produced by a competitor. She reviews the literature and her analysis indicates that the design and size of reactor are of key importance. She needs more information on the size of the reactor. Driving by the plant gives her a good idea on auxiliary equipment but not of the reactor. She is convinced that aerial photography would provide important information.

Questions

(a) Are photos taken from the road ethical?
(b) Are photos taken from an airplane ethical?
(c) Are photos taken from a helicopter ethical?

Comments

Most engineers and students agree by a large majority to a yes answer for question (a). Less than 50% vote affirmatively to question (b) and almost none to question (c). Yet, if the competitor really has information that should be hid, that company could either hide it by placing siding around the reactor or by building dummy reactors.

Engineers do not like the idea of snooping on a competitor by aerial photography. Yet a U.S. government agency is known to have used it on at least one occasion. In another case, I have been told that one company aerially spied on another.

29.2.5 CASE OF REACTOR THAT MAY RUN AWAY

Engineer L works for KO Acid Co. and has just been assigned a new supervisor (engineer M). Engineer L is asked to make laboratory batch runs of a process described in a memo from Alpine Chemical Co. Rumor is that this memo arrived with the new supervisor, who had a job with Alpine Chemical 30 days ago. If batch laboratory runs are satisfactory, plant runs will be started shortly with engineer L in charge. Batch runs are conducted. Three runs result in major boilover of strong acid, including two on the day before the first plant run is to be made.

Questions

(a) Should engineer L ask about the ethics of the situation from the new supervisor, who seems unfriendly?
(b) Should engineer L perform the batch runs as requested if circumstantial evidence indicates that the memo was taken without permission from Alpine?
(c) Should engineer L perform the plant runs as requested? Take any action relative to ethics before making the plant runs?
(d) If engineer L resigns from KO Acid, does he have any responsibilities to other employees who may be endangered if there is a "runway"?

Note: This case is based to a considerable extent on an incident in which more than one person was killed when the plant reactor blew up.

Comments

Engineer L needs to be assured that safety, ethical, and legal matters are above board. Both the EPA and OSHA potentially can become involved later. Possibly at the very least, engineer L should put his concerns in writing with a letter addressed to the top management of his company.

29.2.6 COMPETING RESEARCHERS

Twenty-four months ago Prof. Q of the University of Some Excellence (USE) submitted a research proposal to Supreme, Inc. Shortly thereafter Supreme sent a check to support the first year of research and indicated they expected to support a second and third year of research as well. No formal contract was signed, but the possibility of a patent was discussed between them.

Two Supreme personnel visited Prof. Q about 20 months ago, and the proposed research was discussed in considerable detail. Prof. Q also telephoned or wrote Supreme personnel on several occasions to discuss the proposed research. Supreme decided after reading Prof. Q's proposal to start their own in-house research, but did not at the time inform Prof. Q. When Prof. Q started, after several months, his research, he was told that the Supreme had started research and already knew some of the results to be expected from the first-year program. After research was started at USE, it was "suggested" that Prof. Q modify his first-year program and was requested not to publish research results already obtained until at least Supreme's patent application on the subject was acted upon.

(a) Should Supreme have informed Prof. Q or USE that they planned to do research in this area in their own laboratories?
(b) Should Supreme have tried to hire Prof. Q as a consultant for this project?
(c) Prof. Q and his research associate obtained much information in their first several months of research, which was sent as a progress report to Supreme. They asked Prof. Q not to publish, especially until a patent was granted; they did not want to inform their competition. Should Prof. Q agree to this request?
(d) Prof. Q had proposed industrial applications, and he believes that he should be listed as an inventor. Supreme disagrees but fails to provide any information pertaining to their specific claims on the patent application. Based on ethical considerations only, what action, if any, would you recommend to Prof. Q?

Note: Based on an actual example of several years ago.

29.2.7 QUESTIONABLE COMPANY ETHICS

Company A: Company A placed advertisements in local newspapers in their search for new employees. Company A was clearly seeking individuals who were employees of a competitor and who were privy to highly confidential information. This incident was repeated for employees of two different competitors of company A.

Company B: Several companies who licensed a process of company B became concerned about the safety of the process, which used a liquid catalyst known to be highly hazardous.

(a) Several users of company B's process were sufficiently concerned that they decided to test the safety of the process. Company B declined to become involved (and declined to contribute to the costs). *The test proved the process and catalyst of company B were even more dangerous* in at least one respect than expected.
(b) Company B, 2 years later, distributed to their potential customers a report suggesting a competing process *using a very different catalyst* was equally as dangerous as the catalyst

used by company B; company B reported no specific experimental evidence on the dangers of the other catalyst. *They merely postulated on the dangers.* Many people did not agree that this other catalyst was that dangerous.
(c) As a result of the report distributed by company B, several companies felt obligated to test the hypothesis publicized by company B. These other companies asked company B to be a participant and pay their share of the costs of the testing. Company B declined.
(d) Tests indicated that company B's postulate was incorrect.

Comments

If you were an engineer of either company A or company B, what action if any should you take when you learned of your company's actions?

Appendix: Conversion Factors

The following table gives conversion factors from various units of measure to SI units. It is reproduced from NIST Special Publication 811, *Guide for the Use of the International System of Units (SI)*. The table gives the factor by which a quantity expressed in a non-SI unit should be multiplied in order to calculate its value in the SI. The SI values are expressed in terms of the base, supplementary, and derived units of SI in order to provide a coherent presentation of the conversion factors and facilitate computations (see the table "International System of Units" in this section). If desired, powers of ten can be avoided by using SI prefixes and shifting the decimal point if necessary.

Conversion from a non-SI unit to a different non-SI unit may be carried out by using this table in two stages, e.g.,

$$1 \text{ cal}_{th} = 4.184 \text{ J}$$

$$1 \text{ Btu}_{IT} = 1.055056 \text{ E}+03 \text{ J}$$

Thus,

$$1 \text{ Btu}_{IT} = (1.055056 \text{ E}+03 \div 4.184) \text{ cal}_{th} = 252.164 \text{ cal}_{th}$$

Conversion factors are presented for ready adaptation to computer readout and electronic data transmission. The factors are written as a number equal to or greater than one and less than ten with six or fewer decimal places. This number is followed by the letter E (for exponent), a plus or a minus sign, and two digits that indicate the power of 10 by which the number must be multiplied to obtain the correct value. For example:

$$3.523\ 907 \text{ E}-02 \text{ is } 3.523\ 907 \times 10^{-2}$$

or

$$0.035\ 239\ 07$$

Similarly:

$$3.386\ 389 \text{ E}+03 \text{ is } 3.386\ 389 \times 10^{3}$$

or

$$3\ 386.389$$

A factor in boldface is exact; i.e., all subsequent digits are zero. All other conversion factors have been rounded to the figures given in accordance with accepted practice. Where less than six digits after the decimal point are shown, more precision is not warranted.

It is often desirable to round a number obtained from a conversion of units in order to retain information on the precision of the value. The following rounding rules may be followed:

1. If the digits to be discarded begin with a digit less than 5, the digit preceding the first discarded digit is not changed.

Example: 6.974 951 5 rounded to 3 digits is 6.97

2. If the digits to be discarded begin with a digit greater than 5, the digit preceding the first discarded digit is increased by one.

Example: 6.974 951 5 rounded to 4 digits is 6.975

3. If the digits to be discarded begin with a 5 and at least one of the following digits is greater than 0, the digit preceding the 5 is increased by 1.

Example: 6.974 851 rounded to 5 digits is 6.974 9

4. If the digits to be discarded begin with a 5 and all of the following digits are 0, the digit preceding the 5 is unchanged if it is even and increased by one if it is odd. (Note that this means that the final digit is always even.)

Examples:
6.974 951 5 rounded to 7 digits is 6.974 952
6.974 950 5 rounded to 7 digits is 6.974 950

REFERENCE

Taylor, B. N., *Guide for the Use of the International System of Units (SI)*, NIST Special Publication 811, 1995 Edition, Superintendent of Documents, U.S. Government Printing Office, Washington, DC 20402, 1995.

Factors in **boldface** are exact

To convert from	to	Multiply by	
abampere	ampere (A)	**1.0**	**E+01**
abcoulomb	coulomb (C)	**1.0**	**E+01**
abfarad	farad (F)	**1.0**	**E+09**
abhenry	henry (H)	**1.0**	**E−09**
abmho	siemens (S)	**1.0**	**E+09**
abohm	ohm (Ω)	**1.0**	**E−09**
abvolt	volt (V)	**1.0**	**E−08**
acceleration of free fall, standard (g_n)	meter per second squared (m/s^2)	**9.806 65**	**E+00**
acre (based on U.S. survey foot)[9]	square meter (m^2)	4.046 873	E+03
acre foot (based on U.S. survey foot)[9]	cubic meter (m^3)	1.233 489	E+03
ampere hour (A · h)	coulomb (C)	**3.6**	**E+03**
ångström (Å)	meter (m)	**1.0**	**E−10**
ångström (Å)	nanometer (nm)	**1.0**	**E−01**
apostilb (asb)	candela per meter squared (cd/m^2)	3.183 098	E−01
are (a)	square meter (m^2)	**1.0**	**E+02**

Appendix: Conversion Factors

To convert from	to	Multiply by	
astronomical unit (ua or AU)	meter (m)	1.495 979	E+11
atmosphere, standard (atm)	pascal (Pa)	1.013 25	E+05
atmosphere, standard (atm)	kilopascal (kPa)	1.013 25	E+02
atmosphere, technical (at)[10]	pascal (Pa)	9.806 65	E+04
atmosphere, technical (at)[10]	kilopascal (kPa)	9.806 65	E+01
bar (bar)	pascal (Pa)	1.0	E+05
bar (bar)	kilopascal (kPa)	1.0	E+02
barn (b)	square meter (m^2)	1.0	E−28
barrel [for petroleum, 42 gallons (U.S.)](bbl)	cubic meter (m^3)	1.589 873	E−01
barrel [for petroleum, 42 gallons (U.S.)](bbl)	liter (L)	1.589 873	E+02
biot (Bi)	ampere (A)	1.0	E+01
British thermal unit$_{IT}$ (Btu$_{IT}$)[11]	joule (J)	1.055 056	E+03
British thermal unit$_{th}$ (Btu$_{th}$)[11]	joule (J)	1.054 350	E+03
British thermal unit (mean) (Btu)	joule (J)	1.055 87	E+03
British thermal unit (39°F) (Btu)	joule (J)	1.059 67	E+03
British thermal unit (59°F) (Btu)	joule (J)	1.054 80	E+03
British thermal unit (60°F) (Btu)	joule (J)	1.054 68	E+03
British thermal unit$_{IT}$ foot per hour square foot degree Fahrenheit [Btu$_{IT}$ · ft/(h · ft^2 · °F)]	watt per meter kelvin [W/(m · K)]	1.730 735	E+00
British thermal unit$_{th}$ foot per hour square foot degree Fahrenheit [Btu$_{th}$ · ft/(h · ft^2 · °F)]	watt per meter kelvin [W/(m · K)]	1.729 577	E+00
British thermal unit$_{IT}$ inch per hour square foot degree Fahrenheit [Btu$_{IT}$ · in/(h · ft^2 · °F)]	watt per meter kelvin [W/(m · K)]	1.442 279	E−01
British thermal unit$_{th}$ inch per hour square foot degree Fahrenheit [Btu$_{th}$ · in/(h · ft^2 · °F)]	watt per meter kelvin [W/(m · K)]	1.441 314	E−01
British thermal unit$_{IT}$ inch per second square foot degree Fahrenheit [Btu$_{IT}$ · in/(s · ft^2 · °F)]	watt per meter kelvin [W/(m · K)]	5.192 204	E+02
British thermal unit$_{th}$ inch per second square foot degree Fahrenheit [Btu$_{th}$ · in/(s · ft^2 · °F)]	watt per meter kelvin [W/(m · K)]	5.188 732	E+02
British thermal unit$_{IT}$ per cubic foot (Btu$_{IT}$/ft^3)	joule per cubic meter (J/m^3)	3.725 895	E+04
British thermal unit$_{th}$ per cubic foot (Btu$_{th}$/ft^3)	joule per cubic meter (J/m^3)	3.723 403	E+04
British thermal unit$_{IT}$ per degree Fahrenheit (Btu$_{IT}$/°F)	joule per kelvin (J/k)	1.899 101	E+03
British thermal unit$_{th}$ per degree Fahrenheit (Btu$_{th}$/°F)	joule per kelvin (J/k)	1.897 830	E+03
British thermal unit$_{IT}$ per degree Rankine (Btu$_{IT}$/°R)	joule per kelvin (J/k)	1.899 101	E+03
British thermal unit$_{th}$ per degree Rankine (Btu$_{th}$/°R)	joule per kelvin (J/k)	1.897 830	E+03

To convert from	to	Multiply by	
British thermal unit$_{IT}$ per hour (Btu$_{IT}$/h)	watt (W)	2.930 711	E−01
British thermal unit$_{th}$ per hour (Btu$_{th}$/h)	watt (W)	2.928 751	E−01
British thermal unit$_{IT}$ per hour square foot degree Fahrenheit [Btu$_{IT}$/(h · ft² · °F)]	watt per square meter kelvin [W/(m² · K)]	5.678 263	E+00
British thermal unit$_{th}$ per hour square foot degree Fahrenheit [Btu$_{th}$/(h · ft² · °F)]	watt per square meter kelvin [W/(m² · K)]	5.674 466	E+00
British thermal unit$_{th}$ per minute (Btu$_{th}$/min)	watt (W)	1.757 250	E+01
British thermal unit$_{IT}$ per pound (Btu$_{IT}$/lb)	joule per kilogram (J/kg)	**2.326**	**E+03**
British thermal unit$_{th}$ per pound (Btu$_{th}$/lb)	joule per kilogram (J/kg)	2.324 444	E+03
British thermal unit$_{IT}$ per pound degree Fahrenheit [Btu$_{IT}$/(lb · °F)]	joule per kilogram kelvin [J/(kg · K)]	**4.1868**	**E+03**
British thermal unit$_{th}$ per pound degree Fahrenheit [Btu$_{th}$/(lb · °F)]	joule per kilogram kelvin [J/(kg · K)]	**4.184**	**E+03**
British thermal unit$_{IT}$ per pound degree Rankine [Btu$_{IT}$/(lb · °R)]	joule per kilogram kelvin [J/(kg · K)]	**4.1868**	**E+03**
British thermal unit$_{th}$ per pound degree Rankine [Btu$_{th}$/(lb · °R)]	joule per kilogram kelvin [J/(kg · K)]	**4.184**	**E+03**
British thermal unit$_{IT}$ per second (Btu$_{IT}$/s)	watt (W)	1.055 056	E+03
British thermal unit$_{th}$ per second (Btu$_{th}$/s)	watt (W)	1.054 350	E+03
British thermal unit$_{IT}$ per second square foot degree Fahrenheit [Btu$_{IT}$/(s · ft² · °F)]	watt per square meter kelvin [W/(m² · K)]	2.044 175	E+04
British thermal unit$_{th}$ per second square foot degree Fahrenheit [Btu$_{th}$/(s · ft² · °F)]	watt per square meter kelvin [W/(m² · K)]	2.042 808	E+04
British thermal unit$_{IT}$ per square foot (Btu$_{IT}$/ft²)	joule per square meter (J/m²)	1.135 653	E+04
British thermal unit$_{th}$ per square foot (Btu$_{th}$/ft²)	joule per square meter (J/m²)	1.134 893	E+04
British thermal unit$_{IT}$ per square foot hour [(Btu$_{IT}$/(ft² · h)]	watt per square meter (W/m²)	3.154 591	E+00
British thermal unit$_{th}$ per square foot hour [Btu$_{th}$/(ft² · h)]	watt per square meter (W/m²)	3.152 481	E+00
British thermal unit$_{th}$ per square foot minute [Btu$_{th}$/(ft² · min)]	watt per square meter (W/m²)	1.891 489	E+02
British thermal unit$_{IT}$ per square foot second [(Btu$_{IT}$/(ft² · s)]	watt per square meter (W/m²)	1.135 653	E+04
British thermal unit$_{th}$ per square foot second [Btu$_{th}$/(ft² · s)]	watt per square meter (W/m²)	1.134 893	E+04
British thermal unit$_{th}$ per square inch second [Btu$_{th}$/(in² · s)]	watt per square meter (W/m²)	1.634 246	E+06
bushel (U.S.) (bu)	cubic meter (m³)	3.523 907	E−02
bushel (U.S.) (bu)	liter (L)	3.523 907	E+01
calorie$_{IT}$ (cal$_{IT}$)[11]	joule (J)	**4.1868**	**E+00**
calorie$_{th}$ (cal$_{th}$)[11]	joule (J)	**4.184**	**E+00**
calorie (cal) (mean)	joule (J)	4.190 02	E+00
calorie (15 °C) (cal$_{15}$)	joule (J)	4.185 80	E+00

Appendix: Conversion Factors

To convert from	to	Multiply by	
calorie (20 °C) (cal$_{20}$)	joule (J)	4.181 90	E+00
calorie$_{IT}$, kilogram (nutrition)[12]	joule (J)	4.1868	E+03
calorie$_{th}$, kilogram (nutrition)[12]	joule (J)	4.184	E+03
calorie (mean), kilogram (nutrition)[12]	joule (J)	4.190 02	E+03
calorie$_{th}$ per centimeter second degree Celsius [cal$_{th}$/(cm · s · °C)]	watt per meter kelvin [W/(m · K)]	4.184	E+02
calorie$_{IT}$ per gram (cal$_{IT}$/g)	joule per kilogram (J/kg)	4.1868	E+03
calorie$_{th}$ per gram (cal$_{th}$/g)	joule per kilogram (J/kg)	4.184	E+03
calorie$_{IT}$ per gram degree Celsius [cal$_{IT}$/(g · °C)]	joule per kilogram kelvin [J/(kg · K)]	4.1868	E+03
calorie$_{th}$ per gram degree Celsius [cal$_{th}$/(g · °C)]	joule per kilogram kelvin [J/(kg · K)]	4.184	E+03
calorie$_{IT}$ per gram kelvin [cal$_{IT}$/(g · K)]	joule per kilogram kelvin [J/(kg · K)]	4.1868	E+03
calorie$_{th}$ per gram kelvin [cal$_{th}$/(g · K)]	joule per kilogram kelvin [J/(kg · K)]	4.184	E+03
calorie$_{th}$ per minute (cal$_{th}$/min)	watt (W)	6.973 333	E−02
calorie$_{th}$ per second (cal$_{th}$/s)	watt (W)	4.184	E+00
calorie$_{th}$ per square centimeter (cal$_{th}$/cm^2)	joule per square meter (J/m^2)	4.184	E+04
calorie$_{th}$ per square centimeter minute [cal$_{th}$/(cm^2 · min)]	watt per square meter (W/m^2)	6.973 333	E+02
calorie$_{th}$ per square centimeter second [cal$_{th}$/(cm^2 · s)]	watt per square meter (W/m^2)	4.184	E+04
candela per square inch (cd/in^2)	candela per square meter (cd/m^2)	1.550 003	E+03
carat, metric	kilogram (kg)	2.0	E−04
carat, metric	gram (g)	2.0	E−01
centimeter of mercury (0 °C)[13]	pascal (Pa)	1.333 22	E+03
centimeter of mercury (0 °C)[13]	kilopascal (kPa)	1.333 22	E+00
centimeter of mercury, conventional (cmHg)[13] pascal (Pa)	1.333 224	E+03	
centimeter of mercury, conventional (cmHg)[13] kilopascal (kPa)	1.333 224	E+00	
centimeter of water (4 °C)[13]	pascal (Pa)	9.806 38	E+01
centimeter of water, conventional (cmH$_2$O)[13]	pascal (Pa)	9.806 65	E+01
centipoise (cP)	pascal second (Pa · s)	1.0	E−03
centistokes (cSt)	meter squared per second (m^2/s)	1.0	E−06
chain (based on U.S. survey foot) (ch)[9]	meter (m)	2.011 684	E+01
circular mil	square meter (m^2)	5.067 075	E−10
circular mil	square millimeter (mm^2)	5.067 075	E−04
clo	square meter kelvin per watt (m^2 · K/W)	1.55	E−01
cord (128 ft^3)	cubic meter (m^3)	3.624 556	E+00
cubic foot (ft^3)	cubic meter (m^3)	2.831 685	E−02
cubic foot per minute (ft^3/min)	cubic meter per second (m^3/s)	4.719 474	E−04
cubic foot per minute (ft^3/min)	liter per second (L/s)	4.719 474	E−01
cubic foot per second (ft^3/s)	cubic meter per second (m^3/s)	2.831 685	E−02

To convert from	to	Multiply by	
cubic inch (in³)¹⁴	cubic meter (m³)	1.638 706	E–05
cubic inch per minute (in³/min)	cubic meter per second (m³/s)	2.731 177	E–07
cubic mile (mi³)	cubic meter (m³)	4.168 182	E+09
cubic yard (yd³)	cubic meter (m³)	7.645 549	E–01
cubic yard per minute (yd³/min)	cubic meter per second (m³/s)	1.274 258	E–02
cup (U.S.)	cubic meter (m³)	2.365 882	E–04
cup (U.S.)	liter (L)	2.365 882	E–01
cup (U.S.)	milliliter (mL)	2.365 882	E+02
curie (Ci)	becquerel (Bq)	**3.7**	**E+10**
darcy¹⁵	meter squared (m²)	9.869 233	E–13
day (d)	second (s)	**8.64**	**E+04**
day (sidereal)	second (s)	8.616 409	E+04
debye (D)	coulomb meter (C · m)	3.335 641	E–30
degree (angle) (°)	radian (rad)	1.745 329	E–02
degree Celsius (temperature) (°C)	kelvin (K)	$T/K = t/°C + 273.15$	
degree Celsius (temperature interval) (°C)	kelvin (K)	**1.0**	**E+00**
degree centigrade (temperature)¹⁶	degree Celsius (°C)	$t/°C \approx t/\text{deg.cent.}$	
degree centigrade (temperature interval)¹⁶	degree Celsius (°C)	1.0	E+00
degree Fahrenheit (temperature) (°F)	degree Celsius (°C)	$t/°C = (t/°F - 32)/1.8$	
degree Fahrenheit (temperature) (°F)	kelvin (K)	$T/K = (t/°F + 459.67)/1.8$	
degree Fahrenheit (temperature interval) (°F)	degree Celsius (°C)	5.555 556	E–01
degree Fahrenheit (temperature interval) (°F)	kelvin (K)	5.555 556	E–01
degree Fahrenheit hour per British thermal unit$_{IT}$ (°F · h/Btu$_{IT}$)	kelvin per watt (K/W)	1.895 634	E+00
degree Fahrenheit hour per British thermal unit$_{th}$ (°F · h/Btu$_{th}$)	kelvin per watt (K/W)	1.896 903	E+00
degree Fahrenheit hour square foot per British thermal unit$_{IT}$ (°F · h · ft²/Btu$_{IT}$)	square meter kelvin per watt (m² · K/W)	1.761 102	E–01
degree Fahrenheit hour square foot per British thermal unit$_{th}$ (°F · h · ft²/Btu$_{th}$)	square meter kelvin per watt (m² · K/W)	1.762 280	E–01
degree Fahrenheit hour square foot per British thermal unit$_{IT}$ inch [°F · h · ft²/(Btu$_{IT}$ · in)]	meter kelvin per watt (m · K/W)	6.933 472	E+00
degree Fahrenheit hour square foot per British thermal unit$_{th}$ inch [°F · h · ft²/(Btu$_{th}$ · in)]	meter kelvin per watt (m · K/W)	6.938 112	E+00
degree Fahrenheit second per British thermal unit$_{IT}$ (°F · s/Btu$_{IT}$)	kelvin per watt (K/W)	5.265 651	E–04
degree Fahrenheit second per British thermal unit$_{th}$ (°F · s/Btu$_{th}$)	kelvin per watt (K/W)	5.269 175	E–04
degree Rankine (°R)	kelvin (K)	$T/K = (T/°R)/1.8$	
degree Rankine (temperature interval) (°R)	kelvin (K)	5.555 556	E–01
denier	kilogram per meter (kg/m)	1.111 111	E–07
denier	gram per meter (g/m)	1.111 111	E–04

Appendix: Conversion Factors

To convert from	to	Multiply by	
dyne (dyn)	newton (N)	1.0	E−05
dyne centimeter (dyn · cm)	newton meter (N · m)	1.0	E−07
dyne per square centimeter (dyn/cm^2)	pascal (Pa)	1.0	E−01
electronvolt (eV)	joule (J)	1.602 177	E−19
EMU of capacitance (abfarad)	farad (F)	1.0	E+09
EMU of current (abampere)	ampere (A)	1.0	E+01
EMU of electric potential (abvolt)	volt (V)	1.0	E−08
EMU of inductance (abhenry)	henry (H)	1.0	E−09
EMU of resistance (abohm)	ohm (Ω)	1.0	E−09
erg (erg)	joule (J)	1.0	E−07
erg per second (erg/s)	watt (W)	1.0	E−07
erg per square centimeter second [1obrkt1ru|/(cm^2 · s)]	watt per square meter (W/m^2)	1.0	E−03
ESU of capacitance (statfarad)	farad (F)	1.112 650	E−12
ESU of current (statampere)	ampere (A)	3.335 641	E−10
ESU of electric potential (statvolt)	volt (V)	2.997 925	E+02
ESU of inductance (stathenry)	henry (H)	8.987 552	E+11
ESU of resistance (statohm)	ohm (Ω)	8.987 552	E+11
faraday (based on carbon 12)	coulomb (C)	9.648 531	E+04
fathom (based on U.S survey foot)[9]	meter (m)	1.828 804	E+00
fermi	meter (m)	1.0	E−15
fermi	femtometer (fm)	1.0	E+00
fluid ounce (U.S.) (fl oz)	cubic meter (m^3)	2.957 353	E−05
fluid ounce (U.S.) (fl oz)	milliliter (mL)	2.957 353	E+01
foot (ft)	meter (m)	3.048	E−01
foot (U.S. survey ft)[9]	meter (m)	3.048 006	E−01
footcandle	lux (lx)	1.076 391	E+01
footlambert	candela per square meter (cd/m^2)	3.426 259	E+00
foot of mercury, conventional (ftHg)[13]	pascal (Pa)	4.063 666	E+04
foot of mercury, conventional (ftHg)[13]	kilopascal (kPa)	4.063 666	E+01
foot of water (39.2 °F)[13]	pascal (Pa)	2.988 98	E+03
foot of water (39.2 °F)[13]	kilopascal (kPa)	2.988 98	E+00
foot of water, conventional (ftH$_2$O)[13]	pascal (Pa)	2.989 067	E+03
foot of water, conventional (ftH$_2$O)[13]	kilopascal (kPa)	2.989 067	E+00
foot per hour (ft/h)	meter per second (m/s)	8.466 667	E−05
foot per minute (ft/min)	meter per second (m/s)	5.08	E−03
foot per second (ft/s)	meter per second (m/s)	3.048	E−01
foot per second squared (ft/s^2)	meter per second squared (m/s^2)	3.048	E−01
foot poundal	joule (J)	4.214 011	E−02
foot pound-force (ft · lbf)	joule (J)	1.355 818	E+00
foot pound-force per hour (ft · lbf/h)	watt (W)	3.766 161	E−04
foot pound-force per minute (ft · lbf/min)	watt (W)	2.259 697	E−02

To convert from	to	Multiply by	
foot pound-force per second (ft · lbf/s)	watt (W)	1.355 818	E+00
foot to the fourth power (ft⁴)[17]	meter to the fourth power (m⁴)	8.630 975	E–03
franklin (Fr)	coulomb (C)	3.335 641	E–10
gal (Gal)	meter per second squared (m/s²)	1.0	E–02
gallon [Canadian and U.K. (Imperial)] (gal)	cubic meter (m³)	**4.546 09**	**E–03**
gallon [Canadian and U.K. (Imperial)] (gal)	liter (L)	**4.546 09**	**E+00**
gallon (U.S.) (gal)	cubic meter (m³)	3.785 412	E–03
gallon (U.S.) (gal)	liter (L)	3.785 412	E+00
gallon (U.S.) per day (gal/d)	cubic meter per second (m³/s)	4.381 264	E–08
gallon (U.S.) per day (gal/d)	liter per second (L/s)	4.381 264	E–05
gallon (U.S.) per horsepower hour [gal/(hp · h)]	cubic meter per joule (m³/J)	1.410 089	E–09
gallon (U.S.) per horsepower hour [gal/(hp · h)]	liter per joule (L/J)	1.410 089	E–06
gallon (U.S.) per minute (gpm)(gal/min)	cubic meter per second (m³/s)	6.309 020	E–05
gallon (U.S.) per minute (gpm)(gal/min)	liter per second (L/s)	6.309 020	E–02
gamma (γ)	tesla (T)	1.0	**E–09**
gauss (Gs, G)	tesla (T)	1.0	**E–04**
gilbert (Gi)	ampere (A)	7.957 747	E–01
gill [Canadian and U.K. (Imperial)] (gi)	cubic meter (m³)	1.420 653	E–04
gill [Canadian and U.K. (Imperial)] (gi)	liter (L)	1.420 653	E–01
gill (U.S.) (gi)	cubic meter (m³)	1.182 941	E–04
gill (U.S.) (gi)	liter (L)	1.182 941	E–01
gon (also called grade) (gon)	radian (rad)	1.570 796	E–02
gon (also called grade) (gon)	degree (angle) (°)	**9.0**	**E–01**
grain (gr)	kilogram (kg)	**6.479 891**	**E–05**
grain (gr)	milligram (mg)	**6.479 891**	**E+01**
grain per gallon (U.S.) (gr/gal)	kilogram per cubic meter (kg/m³)	1.711 806	E–02
grain per gallon (U.S.) (gr/gal)	milligram per liter (mg/L)	1.711 806	E+01
gram-force per square centimeter (gf/cm²)	pascal (Pa)	**9.806 65**	**E+01**
gram per cubic centimeter (g/cm³)	kilogram per cubic meter (kg/m³)	1.0	E+03
hectare (ha)	square meter (m²)	1.0	E+04
horsepower (550 ft · lbf/s) (hp)	watt (W)	7.456 999	E+02
horsepower (boiler)	watt (W)	9.809 50	E+03
horsepower (electric)	watt (W)	**7.46**	**E+02**
horsepower (metric)	watt (W)	7.354 988	E+02
horsepower (U.K.)	watt (W)	7.4570	E+02
horsepower (water)	watt (W)	7.460 43	E+02
hour (h)	second (s)	**3.6**	**E+03**
hour (sidereal)	second (s)	3.590 170	E+03
hundredweight (long, 112 lb)	kilogram (kg)	5.080 235	E+01
hundredweight (short, 100 lb)	kilogram (kg)	4.535 924	E+01
inch (in)	meter (m)	**2.54**	**E–02**

Appendix: Conversion Factors

To convert from	to	Multiply by	
inch (in)	centimeter (cm)	**2.54**	**E+00**
inch of mercury (32 °F)[13]	pascal (Pa)	3.386 38	E+03
inch of mercury (32 °F)[13]	kilopascal (kPa)	3.386 38	E+00
inch of mercury (60 °F)[13]	pascal (Pa)	3.376 85	E+03
inch of mercury (60 °F)[13]	kilopascal (kPa)	3.376 85	E+00
inch of mercury, conventional (inHg)[13]	pascal (Pa)	3.386 389	E+03
inch of mercury, conventional (inHg)[13]	kilopascal (kPa)	3.386 389	E+00
inch of water (39.2 °F)[13]	pascal (Pa)	2.490 82	E+02
inch of water (60 °F)[13]	pascal (Pa)	2.4884	E+02
inch of water, conventional (inH$_2$O)[13]	pascal (Pa)	2.490 889	E+02
inch per second (in/s)	meter per second (m/s)	**2.54**	**E−02**
inch per second squared (in/s^2)	meter per second squared (m/s^2)	**2.54**	**E−02**
inch to the fourth power (in^4)[17]	meter to the fourth power (m^4)	4.162 314	E−07
kayser (K)	reciprocal meter (m^{-1})	**1.0**	**E+02**
kelvin (K)	degree Celsius (°C)	.t/°C = T/K − **273.15**	
kilocalorie$_{IT}$ (kcal$_{IT}$)	joule (J)	**4.1868**	**E+03**
kilocalorie$_{th}$ (kcal$_{th}$)	joule (J)	**4.184**	**E+03**
kilocalorie (mean) (kcal)	joule (J)	4.190 02	E+03
kilocalorie$_{th}$ per minute (kcal$_{th}$/min)	watt (W)	6.973 333	E+01
kilocalorie$_{th}$ per second (kcal$_{th}$/s)	watt (W)	**4.184**	**E+03**
kilogram-force (kgf)	newton (N)	**9.806 65**	**E+00**
kilogram-force meter (kgf · m)	newton meter (N · m)	**9.806 65**	**E+00**
kilogram-force per square centimeter (kgf/cm^2)	pascal (Pa)	**9.806 65**	**E+04**
kilogram-force per square centimeter (kgf/cm^2)	kilopascal (kPa)	**9.806 65**	**E+01**
kilogram-force per square meter (kgf/m^2)	pascal (Pa)	**9.806 65**	**E+00**
kilogram-force per square millimeter (kgf/mm^2)	pascal (Pa)	**9.806 65**	**E+06**
kilogram-force per square millimeter (kgf/mm^2)	megapascal (MPa)	**9.806 65**	**E+00**
kilogram-force second squared per meter (kgf · s^2/m)	kilogram (kg)	**9.806 65**	**E+00**
kilometer per hour (km/h)	meter per second (m/s)	2.777 778	E−01
kilopond (kilogram-force) (kp)	newton (N)	**9.806 65**	**E+00**
kilowatt hour (kW · h)	joule (J)	**3.6**	**E+06**
kilowatt hour (kW · h)	megajoule (MJ)	**3.6**	**E+00**
kip (1 kip=1000 lbf)	newton (N)	4.448 222	E+03
kip (1 kip=1000 lbf)	kilonewton (kN)	4.448 222	E+00
kip per square inch (ksi) (kip/in^2)	pascal (Pa)	6.894 757	E+06
kip per square inch (ksi) (kip/in^2)	kilopascal (kPa)	6.894 757	E+03
knot (nautical mile per hour)	meter per second (m/s)	5.144 444	E−01

To convert from	to	Multiply by	
lambert[18]	candela per square meter (cd/m^2)	3.183 099	E+03
langley (cal$_{th}$/cm^2)	joule per square meter (J/m^2)	4.184	E+04
light year (l.y.)[19]	meter (m)	9.460 73	E+15
liter (L)[20]	cubic meter (m^3)	1.0	E−03
lumen per square foot (lm/ft^2)	lux (lx)	1.076 391	E+01
maxwell (Mx)	weber (Wb)	1.0	E−08
mho	siemens (S)	1.0	E+00
microinch	meter (m)	2.54	E−08
microinch	micrometer (μm)	2.54	E−02
micron (μ)	meter (m)	1.0	E−06
micron (μ)	micrometer (μm)	1.0	E+00
mil (0.001 in)	meter (m)	2.54	E−05
mil (0.001 in)	millimeter (mm)	2.54	E−02
mil (angle)	radian (rad)	9.817 477	E−04
mil (angle)	degree (°)	5.625	E−02
mile (mi)	meter (m)	1.609 344	E+03
mile (mi)	kilometer (km)	1.609 344	E+00
mile (based on U.S. survey foot) (mi)[9]	meter (m)	1.609 347	E+03
mile (based on U.S. survey foot) (mi)[9]	kilometer (km)	1.609 347	E+00
mile, nautical [21]	meter (m)	1.852	E+03
mile per gallon (U.S.) (mpg) (mi/gal)	meter per cubic meter (m/m^3)	4.251 437	E+05
mile per gallon (U.S.) (mpg) (mi/gal)	kilometer per liter (km/L)	4.251 437	E−01
mile per gallon (U.S.) (mpg) (mi/gal)[22]	liter per 100 kilometer (L/100 km)	divide 235.215 by number of miles per gallon	
mile per hour (mi/h)	meter per second (m/s)	4.4704	E−01
mile per hour (mi/h)	kilometer per hour (km/h)	1.609 344	E+00
mile per minute (mi/min)	meter per second (m/s)	2.682 24	E+01
mile per second (mi/s)	meter per second (m/s)	1.609 344	E+03
millibar (mbar)	pascal (Pa)	1.0	E+02
millibar (mbar)	kilopascal (kPa)	1.0	E−01
millimeter of mercury, conventional (mmHg)[13]	pascal (Pa)	1.333 224	E+02
millimeter of water, conventional (mmH$_2$O)[13] pascal (Pa)	9.806 65	E+00	
minute (angle) (')	radian (rad)	2.908 882	E−04
minute (min)	second (s)	6.0	E+01
minute (sidereal)	second (s)	5.983 617	E+01
nit	candela per meter squared (cd/m^2)	1.0	E+00
nox	lux (lx)	1.0	E−03
oersted (Oe)	ampere per meter (A/m)	7.957 747	E+01
ohm centimeter (Ω · cm)	ohm meter (Ω · m)	1.0	E−02
ohm circular-mil per foot	ohm meter (Ω · m)	1.662 426	E−09

Appendix: Conversion Factors

To convert from	to	Multiply by	
ohm circular-mil per foot	ohm square millimeter per meter ($\Omega \cdot mm^2/m$)	.1.662 426	E–03
ounce (avoirdupois) (oz)	kilogram (kg)	.2.834 952	E–02
ounce (avoirdupois) (oz)	gram (g)	.2.834 952	E+01
ounce (troy or apothecary) (oz)	kilogram (kg)	.3.110 348	E–02
ounce (troy or apothecary) (oz)	gram (g)	.3.110 348	E+01
ounce [Canadian and U.K. fluid (Imperial)] (fl oz)	cubic meter (m^3)	.2.841 306	E–05
ounce [Canadian and U.K. fluid (Imperial)] (fl oz)	milliliter (mL)	.2.841 306	E+01
ounce (U.S. fluid) (fl oz)	cubic meter (m^3)	.2.957 353	E–05
ounce (U.S. fluid) (fl oz)	milliliter (mL)	.2.957 353	E+01
ounce (avoirdupois)-force (ozf)	newton (N)	.2.780 139	E–01
ounce (avoirdupois)-force inch (ozf · in)	newton meter (N · m)	.7.061 552	E–03
ounce (avoirdupois)-force inch (ozf · in)	millinewton meter (mN · m)	.7.061 552	E+00
ounce (avoirdupois) per cubic inch (oz/in³)	kilogram per cubic meter (kg/m³)	.1.729 994	E+03
ounce (avoirdupois) per gallon [Canadian and U.K. (Imperial)] (oz/gal)	kilogram per cubic meter (kg/m³)	.6.236 023	E+00
ounce (avoirdupois) per gallon [Canadian and U.K. (Imperial)] (oz/gal)	gram per liter (g/L)	.6.236 023	E+00
ounce (avoirdupois) per gallon (U.S.)(oz/gal)	kilogram per cubic meter (kg/m³)	.7.489 152	E+00
ounce (avoirdupois) per gallon (U.S.)(oz/gal)	gram per liter (g/L)	.7.489 152	E+00
ounce (avoirdupois) per square foot (oz/ft²)	kilogram per square meter (kg/m²)	.3.051 517	E–01
ounce (avoirdupois) per square inch (oz/in²)	kilogram per square meter (kg/m²)	.4.394 185	E+01
ounce (avoirdupois) per square yard (oz/yd²)	kilogram per square meter (kg/m²)	.3.390 575	E–02
parsec (pc)	meter (m)	.3.085 678	E+16
peck (U.S.) (pk)	cubic meter (m^3)	.8.809 768	E–03
peck (U.S.) (pk)	liter (L)	.8.809 768	E+00
pennyweight (dwt)	kilogram (kg)	.1.555 174	E–03
pennyweight (dwt)	gram (g)	.1.555 174	E+00
perm (0 °C)	kilogram per pascal second square meter [kg/(Pa · s · m²)]	.5.721 35	E–11
perm (23 °C)	kilogram per pascal second square meter [kg/(Pa · s · m²)]	.5.745 25	E–11
perm inch (0 °C)	kilogram per pascal second meter [kg/(Pa · s · m)]	.1.453 22	E–12
perm inch (23 °C)	kilogram per pascal second meter [kg/(Pa · s · m)]	1.459 29	E–12
phot (ph)	lux (lx)	**1.0**	**E+04**
pica (computer) (1/6 in)	meter (m)	4.233 333	E–03
pica (computer) (1/6 in)	millimeter (mm)	4.233 333	E+00
pica (printer's)	meter (m)	4.217 518	E–03

To convert from	to	Multiply by	
pica (printer's)	millimeter (mm)	4.217 518	E+00
pint (U.S. dry) (dry pt)	cubic meter (m^3)	5.506 105	E−04
pint (U.S. dry) (dry pt)	liter (L)	5.506 105	E−01
pint (U.S. liquid) (liq pt)	cubic meter (m^3)	4.731 765	E−04
pint (U.S. liquid) (liq pt)	liter (L)	4.731 765	E−01
point (computer) (1/72 in)	meter (m)	3.527 778	E−04
point (computer) (1/72 in)	millimeter (mm)	3.527 778	E−01
point (printer's)	meter (m)	3.514 598	E−04
point (printer's)	millimeter (mm)	3.514 598	E−01
poise (P)	pascal second (Pa · s)	**1.0**	**E−01**
pound (avoirdupois) (lb)[23]	kilogram (kg)	4.535 924	E−01
pound (troy or apothecary) (lb)	kilogram (kg)	3.732 417	E−01
poundal	newton (N)	1.382 550	E−01
poundal per square foot	pascal (Pa)	1.488 164	E+00
poundal second per square foot	pascal second (Pa · s)	1.488 164	E+00
pound foot squared (lb · ft^2)	kilogram meter squared (kg · m^2)	4.214 011	E−02
pound-force (lbf)[24]	newton (N)	4.448 222	E+00
pound-force foot (lbf · ft)	newton meter (N · m)	1.355 818	E+00
pound-force foot per inch (lbf · ft/in)	newton meter per meter (N · m/m)	5.337 866	E+01
pound-force inch (lbf · in)	newton meter (N · m)	1.129 848	E−01
pound-force inch per inch (lbf · in/in)	newton meter per meter (N · m/m)	4.448 222	E+00
pound-force per foot (lbf/ft)	newton per meter (N/m)	1.459 390	E+01
pound-force per inch (lbf/in)	newton per meter (N/m)	1.751 268	E+02
pound-force per pound (lbf/lb) (thrust to mass ratio)	newton per kilogram (N/kg)	**9.806 65**	**E+00**
pound-force per square foot (lbf/ft^2)	pascal (Pa)	4.788 026	E+01
pound-force per square inch (psi) (lbf/in^2)	pascal (Pa)	6.894 757	E+03
pound-force per square inch (psi) (lbf/in^2)	kilopascal (kPa)	6.894 757	E+00
pound-force second per square foot (lbf · s/ft^2)	pascal second (Pa · s)	4.788 026	E+01
pound-force second per square inch (lbf · s/in^2)	pascal second (Pa · s)	6.894 757	E+03
pound inch squared (lb · in^2)	kilogram meter squared (kg · m^2)	2.926 397	E−04
pound per cubic foot (lb/ft^3)	kilogram per cubic meter (kg/m^3)	1.601 846	E+01
pound per cubic inch (lb/in^3)	kilogram per cubic meter (kg/m^3)	2.767 990	E+04
pound per cubic yard (lb/yd^3)	kilogram per cubic meter (kg/m^3)	5.932 764	E−01
pound per foot (lb/ft)	kilogram per meter (kg/m)	1.488 164	E+00
pound per foot hour [lb/(ft · h)]	pascal second (Pa · s)	4.133 789	E−04
pound per foot second [lb/(ft · s)]	pascal second (Pa · s)	1.488 164	E+00
pound per gallon [Canadian and U.K. (Imperial)] (lb/gal)	kilogram per cubic meter (kg/m^3)	9.977 637	E+01
pound per gallon [Canadian and U.K. (Imperial)] (lb/gal)	kilogram per liter (kg/L)	9.977 637	E−02

Appendix: Conversion Factors

To convert from	to	Multiply by	
pound per gallon (U.S.) (lb/gal)	kilogram per cubic meter (kg/m^3)	1.198 264	E+02
pound per gallon (U.S.) (lb/gal)	kilogram per liter (kg/L)	1.198 264	E−01
pound per horsepower hour [lb/(hp · h)]	kilogram per joule (kg/J)	1.689 659	E−07
pound per hour (lb/h)	kilogram per second (kg/s)	1.259 979	E−04
pound per inch (lb/in)	kilogram per meter (kg/m)	1.785 797	E+01
pound per minute (lb/min)	kilogram per second (kg/s)	7.559 873	E−03
pound per second (lb/s)	kilogram per second (kg/s)	4.535 924	E−01
pound per square foot (lb/ft^2)	kilogram per square meter (kg/m^2)	4.882 428	E+00
pound per square inch (*not* pound-force) (lb/in^2)	kilogram per square meter (kg/m^2)	7.030 696	E+02
pound per yard (lb/yd)	kilogram per meter (kg/m)	4.960 546	E−01
psi (pound-force per square inch) (lbf/in^2)	pascal (Pa)	6.894 757	E+03
psi (pound-force per square inch) (lbf/in^2)	kilopascal (kPa)	6.894 757	E+00
quad (10^{15} Btu$_{IT}$)[11]	joule (J)	1.055 056	E+18
quart (U.S. dry) (dry qt)	cubic meter (m^3)	1.101 221	E−03
quart (U.S. dry) (dry qt)	liter (L)	1.101 221	E+00
quart (U.S. liquid) (liq qt)	cubic meter (m^3)	9.463 529	E−04
quart (U.S. liquid) (liq qt)	liter (L)	9.463 529	E−01
rad (absorbed dose) (rad)	gray (Gy)	**1.0**	**E−02**
rem (rem)	sievert (Sv)	**1.0**	**E−02**
revolution (r)	radian (rad)	6.283 185	E+00
revolution per minute (rpm) (r/min)	radian per second (rad/s)	1.047 198	E−01
rhe	reciprocal pascal second [(Pa · s)$^{-1}$]	**1.0**	**E+01**
rod (based on U.S. survey foot) (rd)[9]	meter (m)	5.029 210	E+00
roentgen (R)	coulomb per kilogram (C/kg)	**2.58**	**E−04**
rpm (revolution per minute) (r/min)	radian per second (rad/s)	1.047 198	E−01
second (angle) (")	radian (rad)	4.848 137	E−06
second (sidereal)	second (s)	9.972 696	E−01
shake	second (s)	**1.0**	**E−08**
shake	nanosecond (ns)	**1.0**	**E+01**
skot	candela per meter squared (cd/m^2)	3.183 098	E−04
slug (slug)	kilogram (kg)	1.459 390	E+01
slug per cubic foot (slug/ft^3)	kilogram per cubic meter (kg/m^3)	5.153 788	E+02
slug per foot second [slug/(ft · s)]	pascal second (Pa · s)	4.788 026	E+01
square foot (ft^2)	square meter (m^2)	**9.290 304**	**E−02**
square foot per hour (ft^2/h)	square meter per second (m^2/s)	**2.580 64**	**E−05**
square foot per second (ft^2/s)	square meter per second (m^2/s)	**9.290 304**	**E−02**
square inch (in^2)	square meter (m^2)	**6.4516**	**E−04**
square inch (in^2)	square centimeter (cm^2)	**6.4516**	**E+00**
square mile (mi^2)	square meter (m^2)	2.589 988	E+06
square mile (mi^2)	square kilometer (km^2)	2.589 988	E+00

To convert from	to	Multiply by	
square mile (based on U.S. survey foot) (mi²)⁹	square meter (m²)	2.589 998	E+06
square mile (based on U.S. survey foot) (mi²)⁹	square kilometer (km²)	2.589 998	E+00
square yard (yd²)	square meter (m²)	8.361 274	E−01
statampere	ampere (A)	3.335 641	E−10
statcoulomb	coulomb (C)	3.335 641	E−10
statfarad	farad (F)	1.112 650	E−12
stathenry	henry (H)	8.987 552	E+11
statmho	siemens (S)	1.112 650	E−12
statohm	ohm (Ω)	8.987 552	E+11
statvolt	volt (V)	2.997 925	E+02
stere (st)	cubic meter (m³)	**1.0**	**E+00**
stilb (sb)	candela per square meter (cd/m²)	**1.0**	**E+04**
stokes (St)	meter squared per second (m²/s)	**1.0**	**E−04**
tablespoon	cubic meter (m³)	1.478 676	E−05
tablespoon	milliliter (mL)	1.478 676	E+01
teaspoon	cubic meter (m³)	4.928 922	E−06
teaspoon	milliliter (mL)	4.928 922	E+00
tex	kilogram per meter (kg/m)	**1.0**	**E−06**
therm (EC)²⁵	joule (J)	**1.055 06**	**E+08**
therm (U.S.)²⁵	joule (J)	**1.054 804**	**E+08**
ton, assay (AT)	kilogram (kg)	2.916 667	E−02
ton, assay (AT)	gram (g)	2.916 667	E+01
ton-force (2000 lbf)	newton (N)	8.896 443	E+03
ton-force (2000 lbf)	kilonewton (kN)	8.896 443	E+00
ton, long (2240 lb)	kilogram (kg)	1.016 047	E+03
ton, long, per cubic yard	kilogram per cubic meter (kg/m³)	1.328 939	E+03
ton, metric (t)	kilogram (kg)	**1.0**	**E+03**
tonne (called "metric ton" in U.S.) (t)	kilogram (kg)	**1.0**	**E+03**
ton of refrigeration (12 000 Btu$_{IT}$/h)	watt (W)	3.516 853	E+03
ton of TNT (energy equivalent)²⁶	joule (J)	**4.184**	**E+09**
ton, register	cubic meter (m³)	2.831 685	E+00
ton, short (2000 lb)	kilogram (kg)	9.071 847	E+02
ton, short, per cubic yard	kilogram per cubic meter (kg/m³)	1.186 553	E+03
ton, short, per hour	kilogram per second (kg/s)	2.519 958	E−01
torr (Torr)	pascal (Pa)	1.333 224	E+02
unit pole	weber (Wb)	1.256 637	E−07
watt hour (W · h)	joule (J)	**3.6**	**E+03**
watt per square centimeter (W/cm²)	watt per square meter (W/m²)	**1.0**	**E+04**
watt per square inch (W/in²)	watt per square meter (W/m²)	1.550 003	E+03

Appendix: Conversion Factors

To convert from	to	Multiply by	
watt second (W · s)	joule (J)	1.0	E+00
yard (yd)	meter (m)	9.144	E–01
year (365 days)	second (s)	3.1536	E+07
year (sidereal)	second (s)	3.155 815	E+07
year (tropical)	second (s)	3.155 693	E+07

[9] The U.S. survey foot equals (1200/3937) m. 1 international foot = 0.999998 survey foot.

[10] One technical atmosphere equals one kilogram-force per square centimeter (1 at = 1 kgf/cm^2).

[11] The Fifth International Conference on the Properties of Steam (London, uly 1956) defined the International Table calorie as 4.1868 J. Therefore the exact conversion factor for the International Table Btu is 1.055 055 852 62 kJ. Note that the notation for the International Table used in this listing is subscript "IT." Similarly, the notation for thermochemical is subscript "th." Further, the thermochemical Btu, Btu_{th}, is based on the thermochemical calorie, cal_{th}, where cal_{th} = 4.184 J exactly.

[12] The kilogram calorie or "large calorie" is an obsolete term used for the kilocalorie, which is the calorie used to express the energy content of foods. However, in practice, the prefix "kilo" is usually omitted.

[13] Conversion factors for mercury manometer pressure units are calculated using the standard value for the acceleration of gravity and the density of mercury at the stated temperature. Additional digits are not justified because the definitions of the units do not take into account the compressibility of mercury or the change in density caused by the revised practical temperature scale, ITS-90. Similar comments also apply to water manometer pressure units. Conversion factors for conventional mercury and water manometer pressure units are based on ISO 31-3.

[14] The exact conversion factor is 1.638 706 4 E–05.

[15] The darcy is a unit for expressing the permeability of porous solids, not area.

[16] The centigrade temperature scale is obsolete; the degree centigrade is only approximately equal to the degree Celsius.

[17] This is a unit for the quantity second moment of area, which is sometimes called the "moment of section" or "area moment of inertia" of a plane section about a specified axis.

[18] The exact conversion factor is $10^4/\pi$.

[19] This conversion factor is based on 1 d = 86 400 s; and 1 Julian century = 36 525 d. (See *The Astronomical Almanac for the Year 1995*, page K6, U.S. Government Printing Office, Washington, DC, 1994).

[20] In 1964 the General Conference on Weights and Measures reestablished the name "liter" as a special name for the cubic decimeter. Between 1901 and 1964 the liter was slightly larger (1.000 028 dm^3); when one uses high-accuracy volume data of that time, this fact must be kept in mind.

[21] The value of this unit, 1 nautical mole = 1852 m, was adopted by the First International Extraordinary Hydrographic Conference, Monaco, 1929, under the name "International nautical mile."

[22] For converting fuel economy, as used in the U.S., to fuel consumption.

[23] The exact conversion factor is 4.535 923 E–01. All units that contain the pound refer to the avoirdupois pound unless otherwise specified.

[24] If the local value of the acceleration of free fall is taken as g_n = 9.806 65 m/s^2 (the standard value), the exact conversion factor if 4.448 221 615 260 5 E+00.

[25] The term (EC) is legally defined in the Council Directive of 20 December 1979, Council of the European Communities (now the European Union, EU). The therm (U.S.) is legally defined in the Federal Register of July 27, 1968. Although the therm (EC), which is based on the International Table Btu, is frequently used by engineers in the United States, the therm (U.S.) is the legal unit used by the U.S. natural gas industry.

[26] Defined (not measured) value.

Index

A

AAD, *see* Average absolute deviation
Absolute temperature, 267
Absorption and stripping, 1073–1117
    chemical absorption, 1111–1114
    commercial applications, 1074
    functions, 1074
    Kremser–Brown relationship, 1088
    nomenclature, 1114–1116
    other devices, 1104–1107
        crossflow scrubber, 1107
        spray chamber, 1106
        Venturi scrubber, 1107
    packed columns, 1096–1104
        equipment design, 1098–1100
        heights of theoretical stages, 1098
        heights of transfer units, 1097–1098
        liquid distribution and redistribution, 1101
        mass transfer, 1101–1104
        maximum gas/liquid capacity, 1100–1101
        number of transfer units, 1097
        packing and packed bed characteristics, 1100
        pressure drop, 1101
        transfer units, 1096–1097
    plate columns, 1080–1096
        absorption stages, 1081–1087
        stage efficiency, 1095–1096
        stripping stages, 1087–1094
    plate efficiency, 1081
    solubility and phase equilibria, 1075–1079
        graphical representation of solubility, 1079
        multicomponent solubilities, 1078
        sources of solubility data, 1077–1078
    solvent selection, 1079–1080
    steam stripping, 1107–1111
    stripping gas, 1074
    summary, 1114
    theoretical stages/transfer units, 1080
Activity-coefficient models, 329–343
    complete local-composition equation, 339–341
    Flory-Huggins equation, 334–336
    nonrandom two-liquids equation, 338–339
    Redlich–Kister equation, 330
    regular solutions, 332–334
    UNIQUAC equation, 341–343
    van Laar equation, 330–332
    Wilson's local-composition equation, 336–338
Actuator systems (final control elements), 1187–1191
    control valves, 1187–1190
    I/P transmitters, 1190
    optional equipment, 1190–1191
    valve actuators, 1190
Adiabatic fixed-bed reactor (A-FBR), 813
Adiabatic reactors, gas-solid reaction in, 878–883
Adsorbents, 1123–1132
    adsorption/desorption kinetics, 1146
    characteristics, 1124–1129
        capacity, selectivity, and regenerability, 1124–1126
        durability and cost, 1129
        kinetics, 1126–1127
        particle size distribution, 1129–1130
        pore structure and surface area, 1127–1129
    Colburn–Chilton j-factor, 1142
    common adsorbents, 1129–1131
        aluminas, 1130
        biomass, 1132
        carbons, 1131–1132
        polymers, 1132
        silicas, 1130
        zeolites, 1130–1131
    film diffusion, 1146–1147
    Ideal Adsorbed Solution theory, 1140
    isosteres, 1124–1125, 1134
    isotherm classifications, 1126
    multicomponent isotherm equations, 1139
    procedures for design of conventional adsorber, 1155
    pure component isotherm equations, 1139
    selection criteria, 1124
    shock wave velocity, 1145
    terms, 1121–1122
    Vacancy Solution model, 1140
    Wheeler–Robell equation, 1146
Adsorption, 1119–1172
    adsorbents, 1123–1132
        characteristics, 1124–1129
        common adsorbents, 1129–1132
        selection criteria, 1124
    cyclic and continuous adsorption processes, 1159–1171
        displacement-purge cycle, 1162–1163
        example TSA solvent recovery, 1165
        inert-purge cycle, 1159
        PSA cycle, 1163–1171
        TSA cycle, 1162–1163
    design of batch, fixed-bed adsorbers, 1152–1159
        direct method using breakthrough models, 1155
        empirical versus theoretical approaches, 1152–1153
        examples, 1155–1159
        length of unused bed method, 1153–1155
        objectives, 1152
        shortcut procedures, 1153–1155
    dynamics, 1140–1152
        chemical kinetics and adsorption, 1150–1152
        dispersion, 1150–1151
        heat transfer, 1148–1149

local equilibrium model, 1145–1146
mass conservation and transport, 1141–1144
mass transfer resistances, 1146–1148
modified Wheeler and Robell method, 1146, 1155
pressure drop, 1149–1150
rate phenomena, 1140–1141
thermodynamics, 1132–1140
adsorption equilibrium and heats of adsorption, 1133–1134
isotherm equations, 1135–1139
mixture equilibria, 1139–1140
sources of equilibrium data, 1134–1135
A-FBR, see Adiabatic fixed-bed reactor
AIChE, see American Institute of Chemical Engineers
AIChE efficiency model, 1050
Air-lift reactors, 808, 811
Algebraic equations, 40, 81–92
nonlinear equations, 85–92
system of linear equations, 81–85
Aluminas, 1130
American Institute of Chemical Engineers (AIChE), 28
Analysis of variance (ANOVA), 243
Analytical solution of groups (ASOG) method, 345
Aniline
comparison of fluid-bed models for catalytic reduction (by hydrogen) of nitrobenzene to, 883–893
reduction of nitrobenzene to, 878–883
ANOVA, see Analysis of variance
Applied reaction kinetics, 739
Argument principle, 151
Arrhenius equations, chemical reaction engineering, 743
ASOG method, see Analytical solution of groups method
Austenite, 1546
Average absolute deviation (AAD), 365, 366
Azeotrope(s)
composition, 292
distillation, 993–1000
embedded, 997
prediction of, 999

B

BACL eos, see Boublik–Alder–Chen–Kreglewski eos
Batch distillation, 1002–1005
Batteries, 1816–1820
BC, see Bubble column
Bernoulli equation, 103, 419
Berty reactor, 770
Bidirectional reflectance distribution function (BRDF), 574
Bingham plastic, 399, 426, 639
Bingham plastic fluids, 426–428
all flow regimes, 428
laminar flow, 427–428
Biochemical engineering, 1501–1530
agitation and aeration, 1525–1529
correlation for $k_L a$, 1528
oxygen-absorption rate, 1526–1527
power consumption, 1528–1529
scale-up, 1529
biochemical engineering, 1502–1504

bioreactor design, 1518–1525
alternative fermenters, 1519–1520
batch or plug-flow bioreactors, 1520–1522
continuous stirred-tank bioreactor, 1522–1525
stirred-tank bioreactor (or fermenter), 1518–1519
biotechnology, 1502
cells and enzymes, 1504–1509
animal cells, 1506–1507
cell growth measurement, 1508
enzymes, 1508–1509
microbial cells, 1504–1506
plant cells, 1507–1508
kinetics, 1509–1518
cell kinetics, 1509–1513
enzyme kinetics, 1513–1518
Biomass, 1133
Bioreactor design, 1518–1525
alternative fermenters, 1519–1520
bioreactor operations, 1520–1525
batch or plug-flow bioreactors, 1520–1522
continuous stirred-tank bioreactor, 1522–1525
stirred-tank bioreactor (or fermenter), 1518–1519
Boiling heat-transfer correlations, 532–535
critical heat flux in pool boiling, 533
natural and forced convection vaporization, 533–535
nucleate boiling, 532–533
Bolles–Fair correlation, 1057
Bolles–Fair model, 1056
Bonded fixed capital cost, 1305
Boublik–Alder–Chen–Kreglewski (BACK) eos, 301
Box–Hill criterion, 876, 877
Boyko-Kruzhilin equation, 527
Brainstorming, triggers for, 1281–1282
BRDF, see Bidirectional reflectance distribution function
Bubble column (BC), 902
Bubbling-bed reactor, 826
Buckingham's theorem, 78, 80
Buckingham–Reiner equation, 427, 428
Bulk solids, conveying of, 1729–1736
air-activated gravity conveyor, 1735
belt conveyors, 1730–1733
belts, 1732–1733
carrying idlers, 1732
bucket elevators, 1734
en-masse conveyors, 1734–1735
materials characterization, 1729–1730
pneumatic conveyors, 1735
screw conveyors, 1733
vibrating conveyors, 1734

C

Calorimetry, 22
Capital investment (conceptual process design), estimation of fixed and total, 1300–1312
accounting for materials of construction other than carbon steel, 1308–1310
accounting for time and location, 1305–1306
cost contributions for one type of MPI, 1301–1305
definitions and terminology, 1300–1301

Index

different methods for different levels of gating process, 1310–1312
    estimating $L + M$ cost of instrumentation, 1306–1308
Carbons, 1131–1132
Carnahan-Starling (CS) equation of state, 300
Carnot cycle, 265
Carreau-Yashuda model, 402
Catalyst flicker, 821
Catalytic distillation, 1005
Cathodic protection, 1813
Cauchy integral formula, 148
Cauchy–Riemann equations, 146
Cauchy–Schwartz inequality, 58
Cell kinetics, 1509–1513
    growth cycle for batch cultivation, 1509–1510
    structured models, 1512–1513
    unstructured cell kinetic models, 1510–1511
Cell models, 815
Centrifugal pump characteristics, 444–449
    cavitation and NPSH, 447–449
    composite curves, 446
    required head, 446
    specific speed, 449
CFD, *see* Computational fluid dynamics
CFM, *see* Computational fluid mixing
CFT, *see* Cone and fillet tank
Chain-of-rotators (COR) equation, 302
Chaotic problems, 40
Chapman-Kolmogorov equation, 167
Chemical process industries (CPI), 1186
Chemical process safety, 1437–1483
    Center for Chemical Process safety, publications of, 1481–1483
    compliance audits, 1465–1466
    contractor safety, 1461
    conversion factors, 1469
    emergency planning and response, 1465
    employee participation, 1458
    EPA off-site hazard assessment, 1440–1457
        BLEVE hazards evaluation, 1455–1456
        "condensed-phase" detonation hazards evaluation, 1456–1457
        evaluation of toxicity hazards, 1444–1447
        explosion-hazards evaluation, 1447–1449
        flash-fire hazards evaluation, 1449–1451
        jet-fire hazards evaluation, 1456
        pool-fire hazards evaluation, 1451–1452
        pool formation from liquid spills, 1444
        pressure-vessel-burst hazards evaluation, 1452–1454
        release source terms (gas or vapor flow rates), 1441–1442
        release source terms (liquid flow rates), 1442
        release source terms (two-phase flow), 1442–1443
        vapor formation as result of adiabatic flashing, 1443
    incident identification/history, 1457
    incident investigation, 1465
    management of change, 1464–1465
    management system, 1439–1440
    mechanical integrity (equipment maintenance), 1461
    operating procedures, 1460
    OSHA and EPA listed chemicals, 1470, 1471–1477
    OSHA and EPA listed hazardous materials, fire-hazard properties of, 1478
    other aspects of process safety management, 1467
    pre-startup safety review, 1461
    process hazards analysis, 1458–1460
    process safety information, 1458
    process safety management, 1438, 1439
    safety procedures for "non-routine" work, 1461–1464
        entrance control, 1463–1464
        excavation, 1463
        hot-tapping, 1464
        opening process equipment or piping, 1462–1463
        specified safety procedures, 1461
    software security, 1467
    table of symbols, 1468–1469
    trade secrets, 1466–1467
    training, 1460–1461
Chemical properties, *see* Physical and chemical properties
Chemical reaction engineering (CRE), 737–968
    air-lift reactors, 808, 811
    applied reaction kinetics, 739
    Arrhenius equations, 743
    Bayesian discrimination, 875
    Box–Hill criterion, 876, 877
    bubbling-bed reactor, 826
    case studies, 849–954
        comparison of fluid-bed models for catalytic reduction (by hydrogen) of nitrobenzene to aniline, 883–893
        gas-liquid reaction, absorption of NO_x gases for manufacture of nitric acid (system with multiple complexities), 917–918
        gas-liquid reaction, air oxidation of sodium sulfide (simple reaction with typical problems of gas-liquid reactions), 900–916
        gas-liquid-solid (catalytic) reaction, hydrogenation of organic compound, 934–943
        gas-liquid-solid (noncatalytic) reaction, carbonation of lime, 925–934
        gas-liquid-solid (noncatalytic) reaction, oxydesulfurization of coal in slurry reactor, 919–925
        gas-solid (catalytic) reaction, modeling of complex reaction on deactivating catalyst, 870–878
        gas-solid (catalytic) reaction in fixed-bed NINA and adiabatic reactors, 878–883
        gas-solid noncatalytic reaction, development of solid sorbent for cleaning coal gas followed by modeling of reaction and conceptual reactor design, 893–900
        homogeneous gas-phase complex reaction: oxidation of NO using ozone, 869–870
        homogeneous liquid phase simple reaction, 852–869
        solid reaction followed by gas-solid reaction: manufacture of methylchlorosilanes, 943–954
    catalyst flicker, 821
    catalytic wire-gauze reactors, 820
    cell models, 815
    Choudhary–Doraiswamy reactor, 770

classification of reactors and their description, 799–849
    comparison of gas-liquid reactors, 849
    kinetic energy (stirred tank reactors), 839–848
    potential energy (film contactors), 849
    pressure energy, 800–839
column-type equipment, 799
complete dispersion, 842
computational fluid dynamics, 740
contact mass, 944
core-in-shell pellet, 900
critical impeller speed, 842, 843
Davidson model, 885, 888
dead-end reactor, 840
Denbigh reaction, 749
design of continuous reactor, 949
disguises, 764
effectiveness factor, 759, 760
ejector, 848
Eley–Rideal models, 758
emulsion phase, 823
film contactors, 800
flooding, 841
fluidized catalytic cracking units, 827
fractional gas holdup, 940
Fryer–Potter model, 891
gas dispersion, 841
gas-foil impellers, 842
gas-solid reactions, 764
Geldart's classification, 823
gradientless reactors, 739
gulf streaming, 834
heat transfer coefficient, 857, 862
helical coil reactor, 859
heterogeneous fluidization, 821
heterogeneous reactions, 753
homogeneous model, 776
homogenization number, 841
ideal reactors, 741
internal diffusional effect, 768
isothermal effectiveness factor, 759
isotherms, 756
Jayaraman–Kulkarni–Doraiswamy model, 891
jet-loop reactors, 847
Kunii-Levenspiel model, 887
Langmuir–Hinshelwood–Hougen–Watson model, 758
LHHW kinetics, 763
liquid-phase axial mixing, 849
liquid-solid reactions, 764
Lockhart-Martinelli correlation, 913
macroscopic models, 773
methylchlorosilanes, 944
microenvironmental aspects of solid catalysts, 756
minimum fluidization velocity, 812
mixed-flow model, 885
mixing time, 841
Miyauchi model, 889, 890
multiphase reactors, 741
multiple reaction, 742
NINA reactor, 879
nomenclature, 741, 956–961
nonisothermal effectiveness factors, 762
nonisothermicity, 782
operating at the edge, 763
overlapping regimes, 790
oxydesulfurization of coal, 919
particulate fluidization, 821
pellet regeneration, 894
plate efficiency, 807
point of incipient fluidization, 812
pseudo-steady-state assumption, 774
radial-flow reactors, 819
random pore models, 773, 783
rate constant, 742
reaction analysis, 739
reaction-controlled regime, 817
reaction coordinates, 749
reaction order, 742
reaction rate, 743
reaction rate definitions, 745
reaction and reactor fundamentals, 741–753
    complex (multiple and multistep) reactions, 742
    extension to complex (multiple) reactions, 745–750
    ideal reactors, 750–753
    reaction rates, 743–745
    scope, 741
    simple reactions, 741–742
    stoichiometry, 742–743
reactions with interface, 753–799
    fluid-fluid reactions, 785–797
    gas-liquid-solid reactions, 797
    gas-solid catalytic reactions, 753–770
    gas-solid noncatalytic reactions, 770–785
    solid-liquid reactions, 797
reaction space, 743
reactor analysis, 739
regimes of operation, 841
residence time, 743
rotating biological contactors, 848
Runge–Kutta fourth-order method for different tube diameters, 882
secondary bubble size, 802
segregated flow, 753
sequential discrimination algorithm, 876
shape factor, 760
shape selectivity, 754
sharp interface model, 770
Sherwood number, 764
shrinking core model, 770, 774
simulated moving-bed reactors, 740
single-pellet reactor, 770
sintering, 784
slugging, 833
solid catalysts in organic synthesis, 755
solid-liquid mass transfer coefficient, 930
sparger region, 802
stages reactors, 828
standleg moving granular bed filter, 897
stirred cell, 789, 791, 796
stoichiometric reactions, 742
supported aqueous phase catalyst, 754
supported organic phase catalyst, 756
surface aeration, 847

Index

tetrachloroethane chlorinates, 871
Thiele modulus, 759, 760
three-phase catalytic reactor, 936
three-phase sparged reactor, 921
turbulent-bed reactor, 826
two-zone model, 778
utility function, 876
volume reaction model, 776
wake, 888
Weisz modulus, 761
zone models, 778
Chernobyl, 1496–1497
Choudhary–Doraiswamy reactor, 770
Chromatography, dynamics of, 49
CIS sorbent, *see* Core-in-shell sorbent
Clapeyron equation, 352
CLC equation, *see* Complete local-composition equation, 341
Clean Air Act, 1486–1492
   background and pollutant types, 1486–1488
      criteria pollutants and smog, 1487
      hazardous air pollutants, 1487–1488
      maximum achievable control technology standards, 1488
   compliance requirements, 1488
   consultation, 1492
   control technologies (hardware options), 1491–1492
   methodologies for determining air emissions, 1489
   new regulations, 1490
   new source reviews and modifications, 1489–1490
   source delineations under Title V, 1488–1489
   Title IV (acid rain) and Title VI (stratospheric ozone) considerations, 1490–1491
      emissions trading, 1491
      permitting requirements, 1490
      Title IV (acid rain) permit, 1490
      Title VI (stratospheric ozone) permit, 1491
Clean Water Act, 1492–1496
   existing use, 1492–1493
   National Permit Discharge Elimination System, 1493–1494
   publicly owned treatment works, 1494
   Section 319 (nonpoint source pollution), 1495
   Section 404 (wetland protection), 1495
   Sections 303, 305, and Section 401 certification, 1495–1496
   total daily maximum load levels, 1493
   "wet weather flow" conditions, 1494–1495
Coal
   electric energy produced from, 899
   gas, development of solid sorbent for cleaning, 893–900
   oxydesulfurization of, 919
Colburn j-factor, 506
Colebrook equation, 437
Communication, 1841–1858
   formats, 1851–1854
   grammatical problems common to engineering writing, 1848–1851
      agreement and ambiguity, 1848
      conciseness, 1848–1849
      dangling modifiers, 1849–1850
      smothered verb, 1850
      weak (or strange) predication, 1851
   oral reporting in workplace, basic principles of, 1854–1857
      considerations of delivery, 1855–1856
      coping with anxiety, 1857
      signals all audiences want to hear, 1857
   preliminaries, 1842–1843
   principles of good editing, 1843–1847
      avoidance of first person, 1846–1847
      dependent clauses and sentence combining, 1845–1846
      final word, 1847
      paramedic method, 1844–1845
      verb tense, 1847
   tools and resources for effective writing, 1857–1858
      graphics, 1858
      manuals and guides, 1858
      online guides, 1857
      oral presentations, 1858
      style guides, 1857–1858
      usage guides, 1858
Complete local-composition (CLC) equation, 339–341
Complex system distillation, 1001–1002
Complex variables, 143–155
   analytic functions, 144–149
      Cauchy integral formula, 148
      elementary functions, 147
      Laurent series, 149
      logarithmic functions, 148
      multivalued functions, 147
   argument principle and Rouché theorem, 151–152
   conformal mapping, 152–155
      Dirichlet boundary-value problem, 152–154
      Neumann boundary-value problem, 154–155
   properties of complex numbers, 143–144
   residue theorem, 149–151
      application of complex integration, 150–151
      calculus of residues, 149–150
Compressors, 449–453
   isothermal, isentropic, and polytropic operations, 451–452
   staged operation, 452–453
Computational fluid dynamics (CFD), 131, 672, 740
Computational fluid mixing (CFM), 618
Conceptual process design, process improvement, and troubleshooting, 1267–1436
   ACGIH permissible limit values, 1316
   aids for design, process improvement, and troubleshooting, 1333–1344
      flowsheet architectures, 1338–1341
      formulation of well-posed problem, 1336–1338
      importance of thermodynamics packages, 1341–1342
      software aids, 1334–1335
      steady-state flowsheeting, 1336
      tools for flowsheet development, 1342–1343
   annual report, 1284
   balance sheet, 1284, 1285
   bare module costs, 1304
   bonded fixed capital cost, 1305

brainstorming, triggers for, 1281–1282
building of reliable database, 1313–1321
    gathering of data for troubleshooting, 1321
    gathering of input data from databases for design and process improvement, 1314–1320
    input data from scientists, 1313
    minimizing uncertainty in input data through experimentation, 1313–1314
    units of measurement and communication, 1314
business cycle, 1275
capital cost indices, 1305
capital stock, 1287
common stock, 1287
company good citizenship, 1293
cost of capital, 1293, 1294
debt-to-assets ratio, 1289
deflagration, 1315
depreciation, 1288
detonation, 1315
dust emissions from various equipment options, 1332
earnings statement, 1287
employee morale, 1293
energy exchange (rules of thumb), 1359–1368
    direct contact gas-liquid condensers, 1366–1367
    direct contact gas-liquid cooling towers, 1366
    direct contact gas-liquid quenchers, 1366
    direct contact gas-solid fluidized beds, 1365
    direct contact gas-solid kilns, 1364–1365
    direct contact gas-solid multiple hearth furnaces, 1365
    direct contact liquid-liquid immiscible liquids, 1364
    drives, 1360
    fluid heat exchangers, condensers, and boilers, 1361–1363
    fluidized bed (coils in bed), 1363–1364
    furnaces, 1360–1361
    gas-gas thermal wheels, pebble regenerators, and regenerators, 1367
    gas-solid drying of solids, 1365–1366
    heat loss to atmosphere, 1367
    motionless mixers, 1364
    refrigeration, 1367
    solidify liquids, 1367
    steam generation, 1368
energy sources, 1323
engineering economics, 1283–1312
    estimation of fixed and total capital investment, 1300–1312
    financial attractiveness, 1292–1295
    operating expense estimation, 1295–1300
    role of economics in financial reporting, 1284–1292
explore stage of problem solving, 1280
financial ratios, 1289, 1290
flowsheeting, 1333, 1336
gating process, 1273, 1275, 1310
HAZOP studies, 1327
heats of reaction, 1319
heterogeneous separations (rules of thumb), 1387–1410
    air classifiers, 1407–1408
    countercurrent decantation, 1397–1398
    dense media concentrators, 1409
    dryer, 1392–1396
    electrostatic, 1406
    filter, 1400–1404
    filtering centrifuge, 1399–1400
    froth flotation, 1405
    gas-liquid, 1387–1388
    gas-solid, 1388–1390
    general guidelines, 1387
    hydrocyclones, 1397, 1407
    jig concentrators, 1408
    leacher, 1404
    liquid-liquid, 1390–1391
    liquid-solid (dissolved air flotation), 1404
    liquid-solid (expeller and hydraulic press), 1405
    liquid solid (general selection), 1391–1392
    magnetic, 1406–1407
    rake classifiers, 1408
    screens, 1409–1410
    screens for "dewatering," 1396
    sedimentation centrifuges, 1398–1399
    settlers, 1396–1397
    sluice concentrators, 1409
    solid-solid (general selection), 1405
    spiral classifiers, 1408
    table concentrators, 1408–1409
    thickener, 1397
Holmes–Rahe scale, 1283
homogeneous separation (rules of thumb), 1368–1387
    adsorption, gas, 1378
    adsorption, liquid, 1378–1379
    distillation, 1369–1372
    evaporation, 1368–1369
    foam fractionation, 1380–1381
    freeze concentration, 1372
    gas absorption, 1374–1376
    gas desorption/stripping, 1376
    ion exchange, 1379–1380
    melt crystallization, 1372–1373
    overall guidelines, 1368
    precipitation, 1374
    solution crystallization, 1373–1374
    solvent extraction, 1376–1378
    zone refining, 1373
income statement, 1284, 1286, 1287
interpersonal skills and teamwork, 1347–1349
    effective teams, 1348–1349
    interpersonal skills, 1347–1348
    key principles, 1347–1348
    self-awareness, 1347
inventory-sales ratio, 1291
inventory-turnover ratios, 1291
Lee–Kesler model, 1342
liabilities, 1286
marketing activities, 1275
materials reacting aggressively with water, 1319
membranes
    configurations, 1382
    dialysis, 1381–1382
    electrodialysis, 1382
    gas, 1381
    microfiltration, 1386–1387

Index

nanofiltration, 1385
pervaporation, 1382–1384
reverse osmosis, 1384–1385
ultrafiltration, 1385–1386
minimum approach temperature, 1343
mixing (rules of thumb), 1426–1429
    dry solids, 1429
    gas-liquid, 1427
    gas-solid, 1427
    liquid, 1427–1428
    liquid-liquid, 1428
    liquid-solid, 1428
    pastes, polymers, foodstuffs, clay, and fertilizers, 1429
net operating margin, 1291
net present value, 1273
operability and control considered throughout, 1324–1326
    control and design, 1324–1326
    control and process improvement, 1326
    control and troubleshooting, 1326
    evolving flowsheet, 1324–1325
    initial design statement, 1324
    issues beyond conceptual design, 1325
    sizing of equipment, 1325
    type of equipment selected, 1325
overall process (rules of thumb), 1351–1353
    process control, 1351–1353
    properties of fluids, 1353
pinch technology, 1343
plant maintenance, 1293
potentially hazardous bonds, 1318
potential sources of wastes and safety hazards, 1329–1330
preferred stock, 1287
problem-solving cycle in hierarchical stages, 1276–1277, 1278, 1279
problem-solving process, 1273–1280
process improvement, 1274
process optimization, 1344–1347
    definition of problem, 1344
    discrete decision variables, 1346
    linear programming, 1346
    linking optimization with flowsheeting tool, 1346–1347
    nonlinear objective function problems, 1345
    nonlinear programming, 1346
    optimization solution scheme, 1344–1346
product liability, 1293
rating form for teams, 1350
reactors and vessels (rules of thumb), 1410–1426
    CSTR, mechanical mixer (backmix), 1421–1422
    general rules of thumb, 1410–1411
    membrane reactors, 1425
    mix of CSTR, PFTR with recycle, 1424–1425
    process vessels, 1426
    reaction-injection molding, 1425
    reaction type and typical reactor configuration, 1411
    reactive distillation, extraction, crystallization, 1425
    storage vessels and bins, 1426
    STR, batch (backmix), 1421
    STR, fluidized bed (backmix), 1422–1423
    STR, PFTR with large recycle, 1425
    STR, semibatch, 1421
    tank reactor, 1423–1424
reactors and vessels (rules of thumb), PFTR
    bubble reactor, 1415–1416
    empty tube, 1411–1412
    fire tube, 1412
    fixed bed catalyst in tube or vessel, 1412–1413
    fixed bed with radial flow, 1414
    melting cyclone burner, 1420
    monolithic, 1418–1418
    motionless mixer in tube, 1415
    multibed adiabatic with interbed quench, 1414
    multiple hearth, 1419
    multitube fixed bed catalyst, 1414
    packing, 1417–1418
    rotary kiln, 1420
    shaft furnace, 1420
    spray reactor and jet nozzle reactor, 1416
    thin film, 1419
    transported or slurry, transfer line, 1414–1415
    traveling grate, 1419
    trays, 1417
    trickle bed, 1418
    via multistage CSTR, 1420–1421
Redlich–Kwong equation of state, 1342
resistance temperature detectors, 1351
Resource Conservation and Recovery Act, 1327
return-on-total assets ratio, 1291
rules of thumb for problem solving, 1280–1283
safety interlock system, 1351
safety, waste minimization, and environmental sensitivity, 1327–1333
    economic context, 1327
    elimination of source, 1328–1330
    identification of opportunity, 1327–1328
    isolation of impact, 1333
    isolation of source, 1333
    legal context, 1327
    minimization of impact, 1332–1333
    minimization of source, 1330–1332
    move from "eliminate" to "isolate" during design process, 1328–1333
    process improvement, 1333
    targeting of opportunities, 1327–1328
    troubleshooting, 1333
sensitivity analysis, 1345
size enlargement (rules of thumb), 1432–1436
    liquid-gas, demisters, 1432–1433
    liquid-liquid, coalescers, 1433
    solid in liquid, coagulation/flocculation, 1433
size enlargement (rules of thumb), solids
    briquetting, 1434
    crystallization, 1435
    disc agglomeration, 1434
    drum granulator, 1434
    fluidization, 1433–1434
    modify size and shape, 1435–1436
    pelleting, 1435
    sintering/pelletizing, 1435

solidify liquid to solid, 1436
spherical agglomeration, 1434
spray drying, 1433
tabletting, 1434–1435
size reduction (rules of thumb), 1429–1432
    gas in liquid (bubbles in liquid), 1429–1430
    liquid in gas (sprays), 1430
    liquid-liquid, 1430–1431
    solids, crushing and grinding, 1431–1432
    solids, modify size and shape, 1432
    solids, solidify liquid to solid, 1432
stockholders' equity, 1287
structural optimization, 1344
sustainability from the start, 1322–1324
    impact of sustainability on design, 1323
    impact of sustainability on process improvement, 1323–1324
    The Natural Step, 1322, 1323
total module cost, 1304
transportation (rules of thumb), 1353–1359
    ducts and pipes, 1359
    gas-liquid (two-phase flow), 1357
    gas moving (pressure service), 1353–1355
    gas moving (vacuum service), 1355–1356
    liquid, 1356–1357
    pumping slurries (liquid-solid systems), 1357
    solids, 1358–1359
    steam, 1359
VOC release from various equipment options, 1311
Wegstein acceleration, 1339
wiredrawing, 1359
Condensation, special cases in, 530–531
    condensation of multicomponent vapor, 530
    condensation in presence of noncondensable gas, 530
    condensation of superheated vapor, 530–531
    enhanced surfaces, 530
Cone and fillet tank (CFT), 843
Construction materials, see Materials of construction (steels and other metals), selection of
Continuous stirred tank reactor (CSTR), 751
    endothermic, 1175, 1176
    mathematics, 41
    nonlinear equations, 45
    ordinary differential equations, 47
    singularity theory, 177
    steady-state multiplicity, 173
Controlled variable (CV), 1247
Core-in-shell (CIS) sorbent, 893
COR equation, see Chain-of-rotators equation
Corporate image, 1293
Corrosion
    current, 1808
    potential, 1808
    processes, 1805–1816
        crevice corrosion, 1813–1814
        galvanic corrosion, 1812–1813
        hydrogen embrittlement, 1815–1816
        kinetic aspects, 1808–1810
        mass transfer effects, 1810–1811
        metal passivation, 1811–1812
        pitting, 1814–1815
        stress corrosion cracking, 1815
        thermodynamics, 1806–1807
        uniform attack, 1812
COSMO, 12
CPI, see Chemical process industries
CRE, see Chemical reaction engineering
Critical impeller speed, 842
CS equation of state, see Carnahan-Starling equation of state
CSTR, see Continuous stirred tank reactor
CV, see Controlled variable

D

DAF, see Dissolved air flotation
Darcy equation, 1626–1627
Database, building of reliable, 1313–1321
    gathering of data for troubleshooting, 1321
    gathering of input data from databases for design and process improvement, 1314–1320
    input data from scientists, 1313
    key terms, 1315–1316
    minimizing uncertainty in input data through experimentation, 1313–1314
    units of measurement and communication, 1314
Davidson model, 885, 888
DCS, see Distributed control system
DDB, see Dortmund Data Bank
Deactivating catalyst, modeling of complex reaction on, 870–878
Dead-end reactor, 840
Debt-to-assets ratio, 1289
Debye-Hückel theory, 18
Degrees of freedom (DOFs), 1336
Denbigh reaction, 749
Design Institute for Physical Properties (DIPPR), 28
Dew point, 984
Difference equations, processes governed by, 42–44
    scalar difference equations, 42–43
    vector difference equations, 43–44
Differential Chapman-Kolmogorov equation, 167
Differential ebulliometry, 24
Differential pressure (DP) cell 1193
Differential scanning calorimetry (DSC), 23
Differential thermal analysis (DTA), 23
Dimensionless numbers, 504–507
    Colburn j-factor, 506
Graetz number, 506
    Grashof number, 507
    Nusselt number, 505
    Peclet number, 506–507
    Prandtl number, 506
    Reynolds number, 504–505
    Sieder-tate term, 507
    Stanton number, 506
DIPPR, see Design Institute for Physical Properties
Dirac delta function, 184
Dirichlet problem, 118
Dissolved air flotation (DAF), 1402, 1404
Distillation, 969–1072
    AIChE efficiency model, 1050

Index

basic models, 972–975
    contacting stage, 973–975
    multiple stages, 975
    phase equilibrium, 972–973
binary systems, 978, 985–987
Bolles–Fair correlation, 1057
Bolles–Fair model, 1056
bubble-cap trays, 1019
catalytic distillation, 1005
contacting devices, classification of, 1010
curtain area, 1024
dew point, 984
differential, 971
distillation columns, 1006–1008
downcomers, 1016, 1026
drip tube/riser distributor, 104
dry tray efficiency, 1013
dualflow trays, 1022
effective froth density, 1052
embedded azeotrope, 997
equilibrium ratio, 972
flash vaporization, 984
flow parameter, 1012
foaming systems, capacity discount factors for, 1018
froth regime, 1043
Garcia–Fair model, 1053
height equivalent to theoretical plate, 1042
infinite dilution activity coefficient, 979
isothermal flash, 985
mass transfer efficiency, 1041–1065
    conversion of point efficiency to tray efficiency, 1052
    counterflow trays, 1065
    efficiency from laboratory experiments, 1049
    efficiency from mass transfer models, 1050–1052
    efficiency from performance data, 1047–1048
    empirical efficiency methods, 1048–1049
    entrainment effects on efficiency, 1052–1054
    general mass transfer relationships, 1042–1043
    mechanistic model for structured packings, 1059–1061
    multicomponent systems, 1054–1055
    overall column efficiency, 1046–1047, 1054
    packed column efficiency, 1055
    packings versus crossflow trays, 1063–1065
    point efficiency, 1045–1046
    random packings, 1056–1057
    regimes on trays, 1043–1044
    scale-up of structured packing efficiency, 1061–1062
    structured packing performance, 1062–1063
    structured packings, 1057–1058
    tray column efficiency, 1043
    tray efficiency, 1046
    tray efficiency definitions, 1044–1045
Maxwell–Stefan diffusional equations, 1054
McCabe–Thiele diagrams, 986
movable valve trays, 1020
Murphree point efficiency, 975
Murphree stage efficiency, 975
nomenclature, 1066–1069

operating policy, 1003
orifice/riser distributor, 1039
packed column hydraulics, 1029–1041
    liquid distribution, 1038–1039
    liquid distributors, 1039–1040
    liquid redistribution, 1040–1041
    maximum vapor–liquid capacity, 1033–1037
    pressure drop, 1037–1038
    turndown, 1041
    vapor distribution, 1041
perforated pipe distributor, 1040
plug flow of liquid, 1046
point efficiency, 1043
random packings, 1029
Raoult's Law, 975, 977
rectification operating line, 985
relative volatility, 972
ripple trays, 1022
Souders–Brown capacity parameter, 1012
special distillations, 993–1006
    azeotropic distillation, 993–1000
    batch distillation, 1002–1005
    complex system distillation, 1001–1002
    extractive distillation, 1000–1001
    reactive distillation, 1005–1006
    steam distillation, 1002
splash decks, 1024
spray nozzle, 1040
spray regime, 1043
stage calculations, 984–993
    binary systems, 985–987
    computer methods, 991–993
    design procedure, 990–991
    multicomponent systems, 988–990
    multistage processing, 985
    optimum reflux, 990
    separation specifications, 984
    single-stage process, 984–985
    stages–reflux relationships, 987–988
stripping operating line, 985
stripping zone, 971
Sulzer BX gauze-type packing, 1059
Thiele–Geddes model, 993
tray column hydraulics, 1009–1029
    baffle trays, 1024–1026
    counterflow perforated trays, 1022–1023
    crossflow tray columns (sieve trays), 1009–1022
    general comments on tray-type columns, 1026–1029
    multiple downcomer trays, 1026
tray efficiency, predicting, 1047
troubleshooting, 1065–1066
    inadequate vapor capacity, 1066
    liquid flow capacity, 1066
    pressure drop, 1066
    separation efficiency, 1066
    stability, 1066
trough distributor, 1039
turbogrid trays, 1022
UNIFAC model, 981, 983
vapor-liquid equilibrium, 975–983

estimation and measurement of activity coefficients, 980–981
summary, 981–983
V-grid, 1021
VLE relationships, 981
window area, 1024
Distributed control system (DCS), 1184–1187
approach, 1185–1186
background, 1184
fieldbus technology, 1186–1187
programmable logic controllers, 1186
structure, 1184–1185
Disturbance variable (DV), 1247
DOFs, see Degrees of freedom
Dortmund Data Bank (DDB), 14
Downcomers, 1016, 1026
DP cell, see Differential pressure cell
Drying, principles and practice, 1667–1716
basic principles and terminology, 1669–1682
drying kinetics, 1676–1682
equilibrium moisture content, 1671–1674
psychrometry, 1669–1670
thermodynamic properties of air-water mixtures and moist solids, 1669–1676
water activity, 1674–1676
classification and selection of dryers, 1683–1689
batch dryers, 1684
classification of dryers, 168
continuous dryers, 1684–1685
direct dryers, 1685
indirect dryers, 1685–1686
selection of dryers, 1686–1689
drying equipment, 1690–1713
band dryers, 1708–1709
drum dryers, 1702–1703
dryers for particulates and granular solids, 1690–1697
dryers for slurries and suspensions, 1697–1703
drying of boards and sheets, 1712–1713
flash dryers, 1705–1707
fluidized bed dryers, 1693–1694
freeze dryers, 1694–1695
infrared dryers, 1709
microwave and radio frequency drying, 1709–1711
rotary dryers, 1691–1693
roto-louvre dryer, 1707–1708
selected dryers and drying systems, 1703–1713
spin-flash dryers, 1707
spray dryers, 1697–1702
superheated steam drying, 1711–1712
tray dryers, 1690
tunnel dryers, 1708
two-stage dryers, 1704–1705
vacuum dryers, 1695–1697
nomenclature, 1714
DSC, see Differential scanning calorimetry
DTA, see Differential thermal analysis
Duaflow trays, 1022
Duhamel's principle, 120, 121
DV, see Disturbance variable

E

Earnings before interest and taxes (EBIT), 1291
EBIT, see Earnings before interest and taxes
Ebulliometry, 23
Economics, see Engineering economics
EDD model, see Engulfment, deformation, and diffusion model
EL-ALR, see External-loop air-lift reactor
Electrochemical engineering, principles and applications, 1737–1830
activation overpotential, 1765
anodic metal dissolution reaction, 1795
aqueous electrolytes, activity coefficients in, 1748
assisting mixed convection case, 1762
batteries and fuel cells, 1816–1824
batteries, 1816–1820
fuel cells, 1821–1824
Butler-Volmer rate expression, 1751
capillary model, 1804
cathodic protection, 1813
concentration overpotentials, 1765
corrosion current, 1808
corrosion potential, 1808
corrosion processes, 1805–1816
crevice corrosion, 1813–1814
galvanic corrosion, 1812–1813
hydrogen embrittlement, 1815–1816
kinetic aspects, 1808–1810
mass transfer effects, 1810–1811
metal passivation, 1811–1812
pitting, 1814–1815
stress corrosion cracking, 1815
thermodynamics, 1806–1807
uniform attack, 1812
critical current, 1812
electrocatalytic reduction (hydrogenation) reactions in solid polymer electrolyte (proton exchange membrane) reactor, 1784–1786
electrochemical reactors and reactor design, 1766–1774
characterizing of electrochemical reactor performance, 1766–1769
choosing of electrochemical reactor, 1769–1774
electroplating and electroetching, 1787–1796
alloy plating, 1794
basic principles, 1787–1791
electrochemical metal removal processes, 1794–1796
electroless plating, 1792–1794
high-speed and laser-assisted plating, 1791–1792
environmental clean-up processes, 1796–1805
electrochemical destruction of organics, 1800
electrodialysis, 1801–1805
metal ion removal, 1797–1800
Faraday's constant, 1740
Faraday's Law, 1740
fundamentals, 1738–1766
driven vs. self-driven cells and concept of overpotential, 1763–1766
kinetics of electrode reactions, 1749–1753

mass transport, 1753–1763
    thermodynamics, 1742–1748
Gibbs free energy, 1739
Grashof number, 1761
half-cell reactions, 1739, 1742
Hull cell, 1791
infinite dilution, 1763, 1764
inorganic electrochemical synthesis, 1774–1778
    aluminum electrorefining, 1774–1775
    chlor-alkali process, 1775–1777
    water electrolysis, 1777–1778
limiting current density, 1759
migration, 1754
Monsanto's electrohydrodimerization process, 1784
nomenclature, 1824–1826
Ohm's law, 1755, 1798
organic electrochemical synthesis, 1778–1783
overpotentials, 1764
passivation, 1811
passive region, 1812
polarization resistance, 1810
Pourbaix diagrams, 1807
Schmidt number, 1761
self-driven cell, 1763–1764
SPE oil hydrogenation reactor, 1786, 1787
supporting electrolyte, 1756
Tafel equation, 1751
wet corrosion, 1805
Eley–Rideal models, 758
Ellis model, 401
Embedded azeotrope, 997
Emergency Planning and Community Right to Know Act (EPCRA), 1498
Engineering economics, 1283–1312
    accounting for materials of construction, 1308
    accumulated retained earnings, 1288
    assets, 1284
    balance sheet, 1284, 1285
    capital cost indices, 1305
    cost of capital, 1293, 1294
    depreciation, 1288
    employee morale, 1293
    estimation of fixed and total capital investment, 1300–1312
    financial attractiveness, 1292–1295
    financial ratios, 1289, 1290
    income statement, 1284, 1286, 1287
    inflation indices, 1306
    installed instrumentation costs, 1309
    inventory-turnover ratios, 1291
    liabilities, 1286
    major plant items, 1301
    manufacturing expense sheet, 1295
    net sales, 1287
    operating expense estimation, 1295–1300
    operating expenses, 1295–1298
    prepaid expenses, 1284
    role of economics in financial reporting, 1284–1292
    stockholders' equity, 1287
    total capital investment, 1305
    trends in location factors, 1307

Engineering statistics, 199–254
    appendix, 253–254
    confidence intervals, 213, 225, 226
    confidence intervals of population parameters, 212–230
        difference of two means, 226–228
        mean (population variance known), 213–225
        mean (population variance unknown), 225–226
        ratio of two variances, 229–230
        summary, 230
        variance, 228–229
    cumulative probability distribution functions, 202
    data types, 200
    design of experiments, 247–252
        blocks and new duplicates, 252
        computer-aided experimental design, 252
        design matrices, 247–248
        design for one-factor experiment, 248
        design for several factors, 248–251
        factorial design with more levels, 251
    design matrix, 247
    least squares regression, 233–245
        analysis of variance, 243–245
        generalized multiple linear regression, 239–243
        nonlinear regression, 245
        simple linear least squares regression, 234–239
    Poisson distribution, 205
    population statistical parameters, 204
    probability distributions, 201–202, 203–212
        binomial distribution (discrete variable), 204
        characteristic parameters of, 202–203
        chi-square distribution for sample variance (continuous variable), 210–211
        continuous probability density distributions, 201–202
        discrete probability distributions, 201
        F distribution for ratio of sample variances (continuous variable), 211
        geometric distribution (discrete variable), 205
        normal or Gaussian distribution (continuous variable), 206–207
        Poisson distribution (discrete variable), 205–206
        t distribution of sample means (continuous variable), 207–210
    propagation of error, 245
    random variables, 200
    statistical analysis of error propagation, 245–247
    statistical hypothesis testing, 231–233
    statistics of small sets of data, 203
    uncertainty, estimation of, 241
Engulfment, deformation, and diffusion (EDD) model, 647
Enthalpy balance, 261
Environmental engineering, 1485–1500
    background, 1486
    Clean Air Act, 1486–1492
        background and pollutant types, 1486–1488
        compliance requirements, 1488
        consultation, 1492
        control technologies (hardware options), 1491–1492
        methodologies for determining air emissions, 1489
        new regulations, 1490

new source reviews and modifications, 1489–1490
source delineations under Title V, 1488–1489
Title IV (acid rain) and Title VI (stratospheric ozone) considerations, 1490–1491
Title VI (stratospheric ozone) permit, 1491
Clean Water Act, 1492–1496
existing use, 1492–1493
National Permit Discharge Elimination System, 1493–1494
publicly owned treatment works, 1494
Section 319 (nonpoint source pollution), 1495
Section 404 (wetland protection), 1495
Sections 303, 305, and Section 401 certification, 1495–1496
total daily maximum load levels, 1493
"wet weather flow" conditions, 1494–1495
Emergency Planning and Community Right to Know Act, 1498
greenhouse gases and (global warming) climactic change, 1499–1500
nuclear power generation, 1496–1497
Chernobyl, 1496–1497
hazardous waste disposal, 1496
Three-Mile Island, 1496–1497
VHTR reactors, 1497
OSHA, 1497–1498
Resource Conservation and Recovery Act, 1499
SARA, 1498
Spill Prevention Control and Countermeasures Act, 1499
"Superfund" law, 1498
Toxic Substances Control Act, 1498–1499
Enzyme(s), 1508–1509
commercial applications, 1509
kinetics, 1513–1518
Briggs–Haldane approach, 1514–1515
evaluation of Michaelis–Menten parameters, 1516
inhibition of enzyme reactions, 1516–1518
Michaelis–Menten approach, 1513–1514
numerical solution, 1515–1516
nomenclature, 1509
EPA listed chemicals, 1470, 1471–1477
EPA listed hazardous materials, fire-hazard properties of, 1478
EPA off-site hazard assessment, 1440–1457
BLEVE hazards evaluation, 1455–1456
"condensed-phase" detonation hazards evaluation, 1456–1457
evaluation of toxicity hazards, 1444–1447
explosion-hazards evaluation, 1447–1449
flash-fire hazards evaluation, 1449–1451
jet-fire hazards evaluation, 1456
pool-fire hazards evaluation, 1451–1452
pool formation from liquid spills, 1444
pressure-vessel-burst hazards evaluation, 1452–1454
release source terms (gas or vapor flow rates), 1441–1442
release source terms (liquid flow rates), 1442
release source terms (two-phase flow), 1442–1443
vapor formation as result of adiabatic flashing, 1443

EPCRA, *see* Emergency Planning and Community Right to Know Act
Equation(s)
Arrhenius, 743
Bernoulli, 103, 419
Boyko-Kruzhilin, 527
Buckingham-Reiner, 427, 428
chain-of-rotators, 302
Chapman-Kolmogorov, 167
Clapeyron, 352
Colebrook, 437
complete local-composition, 341
Darcy, 1626–1627
elliptic, 118
Euler-Lagrange, 163, 164
filmwise condensation, 524–530
Flory-Huggins, 334–336
fluid statics, 408–410
Fokker-Planck, 167, 168
Fredholm, 42, 136, 140
Galerkin finite element, 114
Gibbs-Duhem, 281, 283
Hagen-Poiseuille, 419, 420, 435
ideal-gas, 258
integral, 42, 50
integrodifferential, 42
Îto stochastic, 53
Kremser, 725–726
Lagrange, 102
Laplace, 128
Maxwell–Stefan diffusional, 1054
multicomponent isotherm, 1140
Navier-Stokes, 618, 848
nonlinear, 45
nonrandom two-liquids equation, 338
ordinary differential, 47
parabolic, 119
partial differential, 48
Peng–Robinson, 11, 299
perturbation, 299
Poisson, 128, 130
polymer chain-of-rotator, 306
population balance, 52
pure component isotherm, 1138
radiative transfer, 583–585
Redlich–Kister, 330
Redlich–Kwong, 298, 1342
Ricatti, 105
Soave–Redlich–Kwong, 11
stochastic differential, 166
Tafel, 1751
UNIQUAC, 343, 369
van der Waals, 295, 330
van Laar, 330–332
vector difference, 43
Vogel–Tammann–Fulcher, 15
Volterra, 42, 132, 136
Wagner, 6
Weymouth, 440
Wheeler–Robell, 1146
Wilson's local-composition, 336–338

Index

Equations of state, 11, 295–313
  extended virial equations, 312–313
  perturbation equations, 299–309
  residual functions and energy functions from, 317–321
    enthalpy, 320
    entropy, 319
    general comments, 320–321
    Gibbs energy, 318–319
    Helmholtz energy, 317–318
    internal energy, 319–320
  van der Waals-type equations, 295–299
  virial and extended virial equations, 309–312
Equilibrium energy functions, 268–270
  diffusional instability, 275
  enthalpy as measure of heat effect, 272
  mechanical instability, 274
  work of expansion of steam, 272
Ethics, case studies, 1859–1864
  competing researchers, 1863
  eager applicant, 1860–1861
  hospitality suite, 1861
  questionable company ethics, 1863–1864
  reactor that may run away, 1862–1863
  spy in sky, 1862
  super super heat exchanger, 1861–1862
Euler-Lagrange differential equation, 163–164
Euler-Lagrange equations, 163, 164
  functional involving n-order derivative, 165
  functional of n-dependent variables, 164–165
  functional of two independent variables, 165
Evolution equations, 118
External-loop air-lift reactor (EL-ALR), 907
Extractive distillation, 1000–1001

F

Faraday's constant, 1740
Faraday's Law, 1740
FBT, see Flat-blade turbines
FCC units, see Fluidized catalytic cracking units
Feedforward (FF) variable, 1247
Ferrite, 1546
FF variable, see Feedforward variable
Fick's first law, 591
Film contactors, 800
Film diffusion, 1147
Filmwise condensation, design equations for, 524–530
  condensation inside horizontal tube, 528
  condensation outside horizontal tubes and tube banks, 529–530
  condensation on vertical plane and tubular surfaces, 524–528
Financial reporting (conceptual process design), role of economics in, 1284–1292
  accumulated retained earnings, 1288
  balance sheet, 1284–1287
  changes in financial position, 1288
  financial ratios, 1289–1291
  income statement, 1287–1288
  other financial terms of significance, 1288–1289
  10K report, 1289

First-order plus deadtime (FOPDT) model, 1179
Flash vaporization, 984
Flat-blade turbines (FBT), 623
Flory-Huggins equation, 334–336
Flow calorimetry, 23
Flowsheeting, 1333, 1336
  equation-based architecture, 1346
  steady-state, 1336
Fluid(s), see also Fluid flow
  compressible flows, 438–443
    adiabatic flow, 441
    choked flow, 441–442
    isentropic nozzle flow, 442
    isothermal pipe flow, 440–441
    supersonic flow, 442–443
  mixing during reaction, 640
  properties of, conceptual process design and, 1353
  statics, basic equation, 408–410
    constant density, 408–409
    ideal gas, 409
    standard atmosphere, 410
Fluid-bed models, comparison of for catalytic reduction (by hydrogen) of nitrobenzene to aniline, 883–893
  Davidson model, 885–887
  Fryer–Potter model, 891
  Jayaraman–Kulkarni–Doraiswamy model, 891–892
  Kunii-Levenspiel model, 887–889
  lesson, 892–893
  Miyauchi model, 889–891
  nomenclature, 883–884
  problem, 884–885
  solution, 885
Fluid flow, 393–478
  Bernoulli equation, 419
  Bingham plastic, 399, 426
  Buckingham-Reiner equation, 427, 428
  Carreau-Yashuda model, 402
  Colebrook equation, 437
  conservation principles, 404–408
    conservation of energy, 405–406
    conservation of mass, 404
    conservation of momentum, 407–408
  constitutive equation, 395
  control valves, 462–477
    compressible flows, 470–475
    incompressible flows, 464–470
    valve characteristics, 464
    viscosity correction, 475–477
  Ellis model, 401
  expansion factor, 460
  flow index, 400
  flow measurement and control, 453–462
    orifice meter, 455–462
    pitot tube, 453–454
    venturi and nozzle, 454–455
  fluid properties, 394–404
    classification of material/fluid properties, 395–397
    measurement of viscosity, 398–399
    viscous fluid models, 399–404
  fluid statics, 408–418
    basic equation, 408–410

moving systems, 410–411
    static forces on solid boundaries, 411–418
Hagen-Poiseuille equation, 419, 420, 435
hydraulic diameter, 435
inviscid fluid, 396
minimum required NPSH, 447
Newtonian fluid, 399
obstruction meters, 454
pipe flow, 419–443
    analysis, 431–433
    Bingham plastic fluids, 426–428
    compressible flows, 438–443
    economical diameter, 433–435
    fitting losses, 428–431
    flow regimes, 419
    Newtonian fluids, 419–422
    noncircular conduits, 435–437
    power law fluids, 422–426
    pressure-flow relations, 419
    turbulent drag reduction, 437–438
Poiseuille flow, 398
power law model, 400, 402, 422
properties of gases, 440
pumps and compressors, 443–453
    centrifugal pump characteristics, 444–449
    compressors, 449–453
    pump types, 443–444
rheological properties, 395
shear limiting viscosity, 401
shear rate, 396
shear thinning, 400
Sisko model, 401
structural viscosity, 401
vapor lock, 447
Weymouth equation, 440
Fluid-fluid reactions, 785–797
    laboratory reactors for, 788–797
    theory of mass transfer accompanied by irreversible chemical reaction, 786–788
Fluidized catalytic cracking (FCC) units, 827
Foaming systems, capacity discount factors for, 1018
Fokker-Planck equation, 167, 168
FOPDT model, see First-order plus deadtime model
Forced convection, correlations for common geometries in, 507–520
    flow across circular cylinder, 512–513
    flow across tube banks, 513–519
    heat transfer in packed and fluidized beds, 519–520
    inside annular channels, 510–511
    inside round tubes, 507–510
    internally enhanced tubes, 510
Fourier transform, 122, 157–160
    application, 158–160
    application of Fourier cosine transform, 160
    application of Fourier sine transform, 159–160
    application of Fourier transform, 158
    convolution property, 157
Fractionation Research, Inc. (FRI), 1065
Fredholm equations, 42, 140
Fredholm equations, methods of solution for, 136–142
    degenerate kernels, 136–137
    method of Fredholm resolvent kernels, 137–138
    method of iterated kernels, 138–139
    method of regularization, 142
    numerical solution of nonhomogeneous Fredholm equation of second kind, 139–140
    solution of ill-posed Fredholm equations of first kind, 140–142
    symmetric kernels, 139
Fredholm integral equations, 136
Fredholm resolvent kernels, 137
FRI, see Fractionation Research, Inc.
Froth regime, 1043
Fryer–Potter model, 891
Fuel cells, 1816–1824
Fugacity, definition of, 321

G

Galerkin finite element equation, 114
Garcia–Fair model, 1053
Gas(es)
    -foil impellers, 842
    holdup expression, 667
    -inducing impellers, 846
    properties of, 440
    superficial velocities of, 1357
Gas chromatograph (GC), 1193
Gas-liquid critical state, 285
Gas-liquid mass transfer, interfacial area for, 929
Gas-liquid mixing, 660–671
    equipment and its function, 662–63
    gas residence time, 668–669
    impeller characteristics, 663–665
    mass transfer, 661–662
    mass transfer and gas holdup, 666–668
    scale-up, 669–671
    scope, 660–661
Gas-liquid reaction(s)
    absorption of NO_x gases for manufacture of nitric acid (system with multiple complexities), 917–918
        effect of various parameters, 917–918
        lesson, 918
        nomenclature, 917
        optimum design, 918
        problem, 917
    air oxidation of sodium sulfide (simple reaction with typical problems of gas-liquid reactions), 900–916
        bubble column, 902
        external-loop air-lift reactor, 907–908
        horizontal sparged contactor, 908–909
        lesson, 916
        nomenclature, 900–901
        packed column, 913–916
        pipeline contactor, 910–913
        problem, 901–902
        sectionalized bubble column, 905–906
        solution, 902
    industrially important, 786

Index

Gas-liquid-solid (catalytic) reaction, hydrogenation of organic compound, 934–943
    nomenclature, 934–935
    solved example, 938–939
    stepwise procedure, 939–943
Gas-liquid-solid (noncatalytic) reaction
    carbonation of lime, 925–934
        data, 927
        gas flow rate, 928–934
        nomenclature, 925–927
        problem, 927
        reaction mechanism, 927
        solution, 927–928
    oxydesulfurization of coal in slurry reactor, 919–925
        data, 920
        design, 923
        design of three-phase sparged reactor, 921–923
        determination of rate-controlling step, 920–921
        lesson, 925
        nomenclature, 919
        problem, 919
        typical procedure, 923–925
Gas-solid (catalytic) reaction(s), 753–770, 870–878
    Bayesian discrimination, 875—876
    catalysis by solids, 753–756
    discriminatory criterion, 876
    effects of various factors on catalyst effectiveness, 764–765
    experimental rate data, 872
    fixed-bed NINA and adiabatic reactors, 878–883
        extension to adiabatic operation, 881–882
        lesson, 883
        nomenclature, 878–879
        solution, 880–883
    kinetics of reactions on solid surfaces, 756–759
    laboratory reactors for accurate kinetic data, 765–770
    lesson, 878
    method of parameter estimation, 873–874
    models, 872–873
    nomenclature, 870–871
    problem, 871–878
    relative roles of internal and external diffusion, 764
    role of diffusion within pellet (internal diffusion), 759–763
    role of external diffusion, 763–764
    sequential design strategy, 874—875
    sequential discrimination algorithm, 876–878
Gas-solid noncatalytic reaction(s), 770–785
    development of solid sorbent for cleaning coal gas followed by modeling of reaction and conceptual reactor design, 893–900
        lesson, 899–900
        modeling of sulfidation step, 895–896
        practical moving-bed design, 897–899
        reactor design, 896–897
        sorbent development, 893–895
    extensions to basic models, 781–782
    general model reduced to specific ones, 785
    modeling of gas-solid reactions, 770–781
    models accounting for structural variations, 782–784

Gas-solid reaction, solid reaction followed by, manufacture of methylchlorosilanes, 943–954
    chloromethane reactor, 951
    continuous reactor, 949–951
    fluidized-bed reactor, 949
    general conversion equations, 951
    kinetics of η-phase formation, 945–949
    lesson, 954
    models with varying gas-phase concentrations, 951–954
    nomenclature, 943–944
    systems definition, 944–945
GC, *see* Gas chromatograph
Gibbs-Duhem equation, 281, 283
Gibbs-Duhem relation, 280–284
Gibbs energy, 12, 318
Gibbs free energy, 1739
Gibbs phase rule, 385
Gradientless reactors, 739
Graetz number, 506
Granular and powder materials, dry screening of, 1717–1728
    equipment selection criteria, 1725–1727
    factors affecting screening performance, 1717–1722
    fundamentals, 1717
    operations and maintenance, 1727–1728
    screen motion and configuration, 1722–1723
    specifying screener performance, 1724–1725
Grashof number, 507, 1761
Green's function, 108, 110
Gulf streaming, 834

H

Hagen-Poiseuille equation, 419, 420, 435
Half-cell reactions, 1739, 1742
Hankel transform, 162–163
    application, 162–163
    property, 162
Heat exchangers, 536–563
    design principles, 550–560
        basic design integral, 552
        heat transfer between two fluids separated by wall, 550–552
        mean temperature difference concept, 552–560
    fouling, 562–563
    logic of design process, 560–562
    types and selection, 537–550
        air-cooled heat exchangers, 547–549
        double-pipe heat exchangers, 537–538
        gasketed-plate heat exchanger and related partially welded variants, 545–547
        mechanically aided heat exchangers, 549–550
        multitube ("hairpin") heat exchangers, 545
        plate-fin (matrix) heat exchangers, 547
        selection criteria, 537
        shell-and-tube heat exchangers, 538–545
Heat transfer, 479–566
    Boyko-Kruzhilin equation, 527
    condensation and vaporization heat transfer, 523–536

boiling heat-transfer correlations, 532–535
design equations for filmwise condensation, 524–530
mechanisms of condensation, 523–524
mechanisms of vaporization, 531–532
special cases in condensation, 530–531
special cases in vaporization, 535–536
conduction heat transfer, 481–503
extended surfaces, 487–493
mechanisms of conduction and basic equation, 481–482
numerical methods, 503
one-dimensional steady-state conduction, 482–486
thermal contact resistance, 486–487
transient conduction in simple solids, 497–503
two- and three-dimensional steady-state conduction, 493–496
friction factor, 519
Graetz-Nusselt problem, 508
heat exchangers, 536–563
design principles, 550–560
fouling, 562–563
logic of design process, 560–562
types and selection, 537–550
single-phase convection heat transfer, 503–523
correlations for common geometries in forced convection, 507–520
dimensionless numbers, 504–507
film coefficient of heat transfer, 504
mechanisms of convection, 503–504
single-phase heat transfer in natural convection, 520–523
Stanton number, 506
TEMA standards, 544
transition boiling regime, 532
Helgeson–Kirkham–Flowers (HKF) correlation, 19
Helical coil reactor, 859
Helmholtz energy, 12, 273, 317
Higbie model, 602
HKF correlation, see Helgeson–Kirkham–Flowers correlation
Homogeneous gas-phase complex reaction, 869–870
Homogeneous liquid phase simple reaction, 852–869
cost evaluation, 865–867
nomenclature, 852–854
problem, 854
reactant on shell side, 867–869
shell-side calculations, 862–865
solution, 854–860
tube-side calculations, 860–862
Horizontal sparged contactor (HSC), 908
HSC, see Horizontal sparged contactor
Hydrogen embrittlement, 1815

I

IAPWS, see International Association for the Properties of Water and Steam
Ideal Adsorbed Solution theory, 1141

Ideal gas
definition, 313
equation, 258
Gibbs energy, 316
Helmholtz energy, 316
law, fluid flow and, 409
properties, 8
Ideal reactors, 741, 750–753
batch reactor, 750
continuous flow reactors, 751–752
Ideal-solution law, 325
IEM model, see Interaction by exchange with the mean model
Immiscible liquid-liquid mixing, 671–682
characterization, 671–672
coalescence of suspended drops, 677–678
creation of dispersion (maintaining drop suspension), 678–679
dispersion of drops, laminar flow, and low viscosity, 673–674
dispersion of higher-viscosity drops turbulent flow, 676–677
dispersion of low-viscosity drops turbulent flow, 674–676
drop sizes, 672–673
equipment used for liquid-liquid mixing, 681
population-balance methods, 678
processes, 681–682
simultaneous suspension, dispersion, and coalescence, 680
Industrial mixing technology, 615–707
Bingham plastic, 639
blending, 630–639
flow patterns, 633
mixing time (laminar flow), 639
mixing time (turbulent and transitional flows), 635–638
nature of turbulent flow, 632–633
scope, 630
shear rates, 633–635
turbulent, transitional, and laminar flow blending, 630–632
Dämköhler number, 646
degree of mixing, 630
dimpled jacket, 701
engulfment, deformation, and diffusion model, 647
equilibrium, 675
gas holdup expression, 667
gas-liquid mixing, 660–671
equipment and its function, 662–63
gas residence time, 668–669
impeller characteristics, 663–665
mass transfer, 661–662
mass transfer and gas holdup, 666–668
scale-up, 669–671
scope, 660–661
half-pipe jacket, 700
heat transfer in mixing equipment, 697–705
external auxiliary devices, 701–702
heat transfer in agitated vessels, 699
heat-transfer surfaces and effective area, 700

Index

important considerations, 698–699
internal pipe coils, 701
jackets and other applied devices, 700–701
other internal devices, 701
process-side heat-transfer correlations, 702–703
service-side heat-transfer correlations, 703–705
hindered settling, 655
immiscible liquid-liquid mixing, 671–682
    characterization, 671–672
    coalescence of suspended drops, 677–678
    creation of dispersion (maintaining drop suspension), 678–679
    dispersion of drops, laminar flow, and low viscosity, 673–674
    dispersion of higher-viscosity drops turbulent flow, 676–677
    dispersion of low-viscosity drops turbulent flow, 674–676
    drop sizes, 672–673
    equipment used for liquid-liquid mixing, 681
    population-balance methods, 678
    processes, 681–682
    simultaneous suspension, dispersion, and coalescence, 680
    time for dispersion, 677
impellers
    characteristics of, 621–622
    dispersion, 680
    energy contours, 634
    Rushton, 675
inertia subrange, 657
intensity of segregation, 644
interaction by exchange with the mean model, 647
jet mixing, 694–697
    correlations, 695–697
    equipment, 694–695
    principles, 694
motionless mixer, 682
Newtonian fluid behavior, 638
noncoalescing suspension polymerization experiments, 674
overview, 617–618
    equipment and design, 618
    process effects, 617–618
    turbulent, transitional, and laminar mixing, 617
parallel reactions, 641
pipeline mixer, 682
power calculation, 625
reaction injection-molding systems, 650
reactive mixing, 639–653
    equipment types and design guidelines, 649–651
    fundamental concepts, 640–642
    guidelines for small-scale experimentation, 652–653
    ideal flow patterns in reactors and residence-time distribution, 643
    macro-mixing and micro-mixing, 642
    micro-mixing implications, 647
    micro-mixing and segregation, 644
    micro-mixing and selectivity, 644–647
    reactive mixing in multiphase systems, 647–648
    scale-up, 651–652
Reynolds number, 689
Rushton impellers, 675
Rushton turbines, 664
Sauter mean drop size, 689
scale of segregation, 644
series reactions, 641
shear, 671
solid-liquid mixing, 653–660
    equipment, 659–660
    floating solids, 657
    just-suspended conditions, 656
    particle suspension in stirred vessels, 655
    power requirements, 659
    settling solids, 653–655
    solids suspension by jet mixing, 660
    uniform solids concentrations, 657–659
spiral-baffled jacket, 700
static in-line mixers, 682–694
    component flow and viscosity ratios, 686–687
    dispersed-phase size distribution in liquid-liquid and gas-liquid systems, 688–689
    heat-transfer enhancement in laminar flow applications, 691
    hydrodynamics and other characteristics, 683–684
    injection considerations and designs, 693
    interfacial areas for gas-liquid systems, 690
    mass-transfer coefficients for gas-liquid dispersions, 690–691
    mixing efficiency, 692–693
    pressure drop and power requirements, 691–692
    pump selection and flow control, 693–694
    Reynolds number and flow regime, 685–686
    selection and design issues, 684–685
    variation coefficient in blending applications, 687–688
stirred tanks, 623–629
    baffles, 626
    bottom clearance, 629
    flow discharge, 626
    impellers and their characteristics, 623
    mechanical considerations, 628–629
    multiple impellers, 625
    power, 624–625
    vendor's role, 629
    vessel shape, 628
Stokes's law, 654
symbols, dimensionless groups, and terms, 619–623
    common dimensionless groups, 620
    common symbols, 619
    significance of mixing terms, 620–623
turbulent flow conditions, 703
vessel Reynolds number, 620, 637
Weber number, 674, 689
Instrument Society of America (ISA), 1201
Integral equations, 131–143
    Fredholm integral equations, 136
    methods of solution for Fredholm equations, 136–142
    methods of solution for Volterra equations, 132–136

processes governed by, 50–52
    boundary-value problems with mixed derivative boundary condition, 51–52
    determination of pore size distribution in porous media, 50–51
    particle size distribution in continuous comminution process, 50
    Volterra integral equations, 131–132
Integral transforms, 155–163
    Fourier transform, 157–160, 186–187, 188, 189
        application of Fourier cosine transform, 160
        application of Fourier sine transform, 159–160
        application of Fourier transform, 158
        convolution property, 157
    Hankel transform, 162–163, 190
        application, 162–163
        property, 162
    Laplace transform, 156–157, 185–186
        application, 156–157
        convolution property, 156
    Mellin transform, 160–161, 190
        application, 161
        convolution property, 160
    tables of, 185–191
        Fourier cosine transforms, 189
        Fourier sine transforms, 188
        Fourier transforms, 186–187
        Hankel transforms, 191
        Laplace transforms, 185–186
        Mellin transforms, 190
Integrodifferential equations, processes governed by, 52
Intellectual property, *see* Patents and intellectual property
Interaction by exchange with the mean (IEM) model, 647
Internal rate of return (IRR), 1292
International Association for the Properties of Water and Steam (IAPWS), 3
International Solvent Extraction Conferences (ISECs), 712
International Union of Pure and Applied Chemistry (IUPAC), 21
Inventory-sales ratio, 1291
IRR, *see* Internal rate of return
ISA, *see* Instrument Society of America
ISECs, *see* International Solvent Extraction Conferences
Isochores, 286
Isothermal flash, 985
Îto stochastic equation, 53
Îto stochastic integral, 168–170
    application, 169–170
    one-dimensional Îto formula, 169
IUPAC, *see* International Union of Pure and Applied Chemistry

J

Jayaraman–Kulkarni–Doraiswamy model, 891
Jet-loop reactors, 847
Jet mixing, 694–697
    correlations, 695–697
    equipment, 694–695
    principles, 694

K

Kirchhoff's laws, 575
K–L model, *see* Kunii–Levenspiel model
Kremser–Brown relationship, 1088
Kremser equation, 725–726
Kronecker delta, 120, 395
Kunii–Levenspiel (K–L) model, 887

L

Lagrange equation, 102
Langmuir–Hinshelwood–Hougen–Watson (LHHW) model, 758
Langmuir isotherm, 756
Laplace equation, 128
Laplace transform, 156–157
    application, 156–157
    convolution property, 156
Laurent series, 149
LCUs, *see* Local control units
Least squares regression, 233–245
    analysis of variance, 243–245
    generalized multiple linear regression, 239–243
        basic algorithm, 240–241
        estimation of uncertainty, 241–243
    nonlinear regression, 245
    simple linear least squares regression, 234–239
        basic algorithm, 234–237
        estimation of uncertainty, 237–239
Lewis fugacity rule, 355
LHHW kinetics, 763
LHHW model, *see* Langmuir–Hinshelwood–Hougen–Watson model
Lignin extraction, 714
Lime, carbonation of, 925–934
Linear equations, processes governed by, 44–45
    steady-state continuous countercurrent staged extraction, 44–45
    steady-state first-order reactions in stirred tank reactor, 45
Linear partial differential equation, 115
Linear program (LP), 1247
Linear stability analysis, 179
Liquid-liquid equilibrium, 11
Liquid-liquid extraction (LLE), 709–735
    design of extraction systems, 720–726
        countercurrent extractors, 723–725
        Kremser equation, 725–726
        phase diagrams, 721–723
    dispersion, mass transfer, and coalescence, 712–713
    distribution coefficients, 716–720
        empirical distribution models, 720
        nonreactive systems, 716–717
        reactive systems, 718–720
        thermodynamic models, 716
    economic analysis for vertical contactors, 729–732
    entrainer, 712
    industrial extraction equipment, 726–728
        mixer settlers, 726

Index

reciprocating plate columns, 726–728
Internet sites, 729
list of symbols, 731–732
Marshall & Swift index, 730
mix point, 722
Nernst's law, 717, 725
PUREX process, 711, 714
reactive systems, 713–716
  lignin extraction, 714
  simplified TBP reaction models, 714–715
  TBP solvent cleanup, 715–716
Liquid-liquid reactions, industrially important, 787
Liquids, superficial velocities of, 1357
LLE, see Liquid-liquid extraction
Local control units (LCUs), 1184
Lockhart–Martinelli correlation, 913
LP, see Linear program
Lyapunov's function, 180–181

M

MAC, see Mechanically agitated contactor
MACRS, see Modified Accelerated Cost Recovery System
Major plant items (MPI), 1301
Manipulated variable (MV), 1247
Marshall & Swift (M&S) index, 730
Martensite, 1546–1547
Mass transfer, 591–614
  diffusion coefficient, 592
  equilibrium distribution coefficient, 610
  Fick's first law, 591
  heavy phase, 609
  Higbie model, 602
  light phase, 609
  mass transfer across phase boundary, 604–612
    evaporation of spills, 611–612
    interfacial area, 607
    two-film model, 604–606
    volumetric coefficients (gas-liquid), 607–608
    volumetric coefficients (liquid-liquid), 608–611
  mass transfer at phase boundary, 601–604
    penetration model, 602–604
    stagnant-film model, 602
    surface renewal model, 604
  mass transfer coefficient, 602
  molecular diffusion coefficients, 592–601
    diffusion through porous solids, 598–599
    gases (binary mixtures), 592–594
    gases (diffusion through membranes), 600–601
    gases (multicomponent mixtures), 597
    liquids (binary mixtures), 594–597
    liquids (multicomponent mixtures), 598
  nomenclature, 612–613
  slab equation, 598
  summary, 612
  tortuosity factor, 599
  two-resistance theory, 604
MAT, see Minimum approach temperature

Materials of construction (steels and other metals), selection of, 1539–1596
  corrosion, 1561–1567
    anodes, 1562
    basics, 1561–1562
    cathodes, 1562–1580
    corrosion control, 1562–1564
    effect on materials of construction, 1567
    microbiologically influenced corrosion, 1566–1567
    mitigation methods, 1567
    stress corrosion cracking, 1564–1566
  failure modes, 1567–1580
    carbon and low-alloy steel, 1569–1572
    embrittlement phenomena, 1567–1569
    high alloys, 1572
    high-temperature effects, 1572–1574
    hydriding, 1572
    hydrogen gas, 1577–1578
    low-carbon stainless steel, 1574–1577
    nitriding, 1578–1579
    oxidation, 1579
    sulfidation and sulfidic corrosion, 1579–1580
  materials selection process, 1581–1594
    corrosion, 1591
    criteria, 1588
    diagram, 1593–1594
    fabricated equipment, 1589–1590
    grouping process regions, 1590
    high temperature effects, 1590
    low temperature effects, 1590
    piping, 1589
    procedure, 1590–1593
    procedure exceptions, 1589–1590
    procedure organization, 1589
    product contamination, 1588
    pumps, 1589
    reliability, 1588
    template design, 1581–1587
    upset conditions, 1592–1593
  materials selection process, overview, 1540–1542
    governing criteria, 1541
    mandatory requirements, 1541
    materials selection criteria, 1541
    materials selection template, 1542
    special requirements, 1542
  metallurgy, basic, 1542–1561
    alloy designations, 1549
    alloy steel, 1553–1554
    annealing, 1542–1543
    austenite, 1546
    carbon steel, 1552–1553
    cast iron, 1550–1552
    common alloys and metals, 1558–1561
    ferrite, 1546
    heat treatments, 1542–1546
    manufacturing of metals and alloys, 1549–1550
    martensite, 1546–1547
    metallurgical terms, 1547–1549
    metals and alloys, 1550–1561
    microstructural terms, 1546–1547

normalizing, 1543
pearlite, 1547
preheat, 1543
quench and temper, 1545–1546
stainless steel, 1554–1558
stress relief/postweld heat treatment, 1544–1545
Mathematics in chemical engineering, 35–197
   activities related to mathematics, 41
   Adams-Bashforth methods, 99
   Adams-Moulton rule, 100
   algebraic equations, 40, 81–92
      nonlinear equations, 85–92
      system of linear equations, 81–85
   arithmatic-geometric means inequality, 59
   asymptotic approximations and expansions, 170–173
   boundary-value problem, 41, 108, 113, 152, 154
   branch cut, 147
   breakage kernel, 50
   Buckingham's -theorem, 78
   Budan's rule of signs, 86
   calculus of variations, 163–166
      application, 165
      Euler-Lagrange differential equation, 163–164
      Euler-Lagrange equations for functional involving n-order derivative, 165
      Euler-Lagrange equations for functional of n-dependent variables, 164–165
      Euler-Lagrange equations for functional of two independent variables, 165
   Cardano's formula, 87
   Cauchy integral formula, 148
   Cauchy-Riemann equations, 146
   Cauchy-Schwartz inequality, 58
   chaotic problems, 40
   Chebyshev's inequality, 59
   complex variables, 143–155
      analytic functions, 144–149
      argument principle and Rouché theorem, 151–152
      conformal mapping, 152–155
      properties of complex numbers, 143–144
      residue theorem, 149–151
   computational fluid dynamics, 131
   Cramer's rule, 84
   d'Alembert's solution, 124
   degenerate kernels, 136
   Descartes's rule of signs, 86
   diagonal matrix, 82
   difference equations, 92–100
      method of solution for homogeneous equations, 92
      method of solution for inhomogeneous equations, 92–94
      numerical solutions to ordinary differential equations, 94–100
   difference kernel, 133
   differential and integral calculus, 60–66
      derivative, 61–62
      functions, limits, and continuity, 60
      implicit function theorem, 64
      integrals, 64–66
      L'Hôspital's rule, 63–64
      mean value theorem, 62–63

dimensional analysis, 78–81
   applications, 79–81
   theory, 78
dimensional matrix, 78
dimensions of commonly used physical quantities, 79
Dirac delta function, 184
Dirichlet problem, 118
drag coefficient, 80
Duhamel's principle, 120, 121
elliptic equations, 118
empiricism, 39
entire function, 146
equations, 40–53
   difference equations, 42–44
   integral equations, 50–52
   integrodifferential equations, 52
   linear equations, 44–45
   nonlinear equations, 45–47
   ordinary differential equations, 47–48
   partial differential equations, 48–50
   stochastic differential equations, 52–53
Euler-Lagrange equation, 163, 164
evolution equations, 118
Fokker-Planck equation, 167, 168
Fourier transforms, 122
Fredholm equations, 42
Fredholm integral equations, 136
Fredholm resolvent kernels, 137
Galerkin finite element equation, 114
gauge functions, 171
Green's function, 108, 110, 130, 136
harmonic function, 146
Hölder inequality, 59
identity matrix, 82
imaginary number, 143
inflection point, 61
initial-value problems, 41
integral equations, 42, 131–143
   Fredholm integral equations, 136
   methods of solution for Fredholm equations, 136–142
   methods of solution for Volterra equations, 132–136
   Volterra integral equations, 131–132
integral transforms, 155–163
   Fourier transform, 157–160
   Hankel transform, 162–163
   Laplace transform, 156–157
   Mellin transform, 160–161
integrodifferential equations, 42
iterated kernels, 138
Îto stochastic equation, 53
Jacobian matrix, 90
kernels for different transforms and integration limits, 155
Kronecker delta, 120
Laplace equation, 128
Laurent series, 149
linear partial differential equation, 115
linear stability analysis, 179
Lyapunov's function, 179, 180, 181
Markov process, 167

matching strategy, 172
mathematical software, 182–183
MatLab, 81
Minkowski's inequality, 59
modulus of complex number, 143
modulus inequality, 59
multiplicity, 178
natural boundary conditions, 164
Neumann problem, 118
Newton-Raphson iteration, 91
number system, 53–60
    algebraic inequalities, 58–59
    binomial theorem, 58
    comparison test, 56–57
    integral inequalities, 59–60
    integral test, 55–56
    limit comparison test, 57
    ratio test, 56
    real number system, 53–54
    root test, 56
    sequences and series, 54–55
    Taylor series, 57
    tests for convergence of sequence and series, 55–57
ordinary differential equations, 47, 101–115
    linear first-order differential equation, 101
    linear higher-order differential equations, 111–115
    nonlinear first-order differential equation, 101–105
    second-order differential equations, 106–111
overlap region, 172
overrelaxation method, 84
partial differential equations, 115–131
    classification of second-order equations, 118–119
    computational fluid mechanics, 131
    elliptic equations, 128–131
    first-order partial differential equations, 115–118
    hyperbolic equations, 124–128
    parabolic equations, 119–124
Poisson equation, 128, 130
population balance equations, 52
quasilinear partial differential equation, 115
regular asymptotic expansion, 172
regularization, 140, 142
resolvent kernels, 134
Reynolds number, 80
Robin boundary condition, 129
Robin problem, 118
Routh-Hurwitz criterion, 85
rules of differentiation for real functions, 145–146
Runge-Kutta method, 96
scalar difference equations, 42
Simpson's rule, 99
singular expansions, 172
singularity theory, 176, 177
square matrix, 82
steady-state multiplicity and stability, 173–181
    analysis of multiplicity by singularity theory, 176–179
    stability of steady-state solution, 179–181
stochastic differential equations, 42, 166–170
    connection between Fokker-Planck equation and, 167–168
    differential Chapman-Kolmogorov equation, 167
    Îto stochastic integral, 168–170
stochastic integral, 168
symmetric kernels, 139
tables of integral transforms, 185–191
    Fourier cosine transforms, 189
    Fourier sine transforms, 188
    Fourier transforms, 186–187
    Hankel transforms, 191
    Laplace transforms, 185–186
    Mellin transforms, 190
Taylor series expansion, 147
trapezoidal rule, 99
vector analysis, 66–78
    gradients of sum and product, 77–78
    orthogonal curvilinear coordinate systems, 69–76
    vector algebra, 66–68
    vector calculus, 68
    vector integral theorems, 76–77
vector difference equations, 43
Volterra equations, 42, 132, 136
weak formulation of governing differential equation, 114
white noise, 53
Wicke-Kallenbach experiment, 50
Maxwell–Stefan diffusional equations, 1054
McCabe–Thiele diagrams, 986–987
Mechanically agitated contactor (MAC), 938
Mellin transform, 160–161
    application, 161
    convolution property, 160
Metallurgy, basic, 1542–1561
    alloy designations, 1549
    heat treatments, 1542–1546
        annealing, 1542–1543
        normalizing, 1543
        preheat, 1543
        quench and temper, 1545–1546
        stress relief/postweld heat treatment, 1544–1545
    manufacturing of metals and alloys, 1549–1550
    metallurgical terms, 1547–1549
    metals and alloys, 1550–1561
        alloy steel, 1553–1554
        carbon steel, 1552–1553
        cast iron, 1550–1552
        common alloys and metals, 1558–1561
        stainless steel, 1554–1558
    microstructural terms, 1546–1547
        austenite, 1546
        ferrite, 1546
        martensite, 1546–1547
        pearlite, 1547
Metal passivation, 1811
Methylchlorosilanes, manufacture of, 943–954
MFR, *see* Mixed-flow reactor
Microbial cells, 1504–1506
    bacteria, 1505
    culture media, 1506
    fungi, 1505–1506
    microbial nomenclature, 1505
Mie scattering theory, 586

MIMO process, *see* Multiple-input/multiple-output process
MIMO process control, 1242–1246
    SISO controllers and (c, y) pairings, 1242–1245
    tuning decentralized controllers, 1245–1246
Minimum approach temperature (MAT), 1343
Mixed-flow model, 885
Mixed-flow reactor (MFR), 751
Mixer settlers, 726
Mix point, 722
Mixtures, phase behavior of, 291–294
    gas-gas equilibrium, 294
    gas-liquid equilibrium, 291–293
    liquid-liquid equilibrium, 293–294
Miyauchi model, 889, 890
Model(s)
    activity-coefficient, 329–343
        complete local-composition equation, 339–341
        Flory-Huggins equation, 334–336
        nonrandom two-liquids equation, 338–339
        Redlich–Kister equation, 330
        regular solutions, 332–334
        UNIQUAC equation, 341–343
        van Laar equation, 330–332
        Wilson's local-composition equation, 336–338
    AIChE efficiency, 1050
    Bingham plastic, 426
    Bolles–Fair, 1056
    capillary, 1804
    Carreau-Yashuda, 402
    cell, 815
    chain-of-spheres, 302
    Davidson, 885, 888
    distillation, 972–975
        contacting stage, 973–975
        multiple stages, 975
        phase equilibrium, 972–973
    dumbbell rotator, 302
    electrolyte-NRTL, 18
    Eley–Rideal, 758
    Ellis, 401
    engulfment, deformation, and diffusion, 647
    EOS, 8
    FOPDT, 1181
    Fryer–Potter, 891
    Garcia–Fair, 1053
    gas-solid equilibrium, 372–374
    Gibbs energy models of liquid solutions, 347–350
    Hang–Chao–Hilson complete local-composition, 371
    Higbie, 602
    interaction by exchange with the mean, 647
    Jayaraman–Kulkarni–Doraiswamy, 891
    Kunii-Levenspiel, 887
    Langmuir–Hinshelwood–Hougen–Watson, 758
    Lee–Kesler, 1342
    liquid-liquid equilibrium, 367
    mixed-flow, 885
    Miyauchi, 889, 890
    noncompetitive adsorption, 759
    predictive control (MPC), 1246, 1247
    random pore, 773, 783
    SAFT, 11

Separations Research Program, 1061
sharp interface model, 770
shrinking core, 774
Sisko, 401
SUPERTRAPP, 17
TBP reaction, 714
thermodynamic, LLE and, 716
Thiele–Geddes, 993
two-zone, 778
UNIFAC, 717, 981–983
UNIQUAC, 1342
Vacancy Solution, 1141
volume reaction, 776
Modified Accelerated Cost Recovery System (MACRS), 1289
Monochromatic radiation, 568
Moving-bed reactors, 837
MPC, *see* Model predictive control
MPI, *see* Major plant items
M&S index, *see* Marshall & Swift index
Multicomponent vapor, condensation of, 530
Multiphase reactors, 741
Multiple-input/multiple-output (MIMO) process, 1242
Multitubular reactor, 860
Multivariable control (MVC), 1247
Multivariable controller
    general method for designing, 1249–1261
        commissioning of controller, 1260–1261
        conducting of plant test and collection of data, 1254–1255
        design of plant test, 1251–1254
        post audit, 1261
        structuring of controller and analysis of data, 1255–1259
        tuning of controller, 1259–1260
        understanding of process, 1249–1251
    troubleshooting, 1261–1263
        checklist, 1262–1265
        commonsense approach, 1262
        controller still making money, 1264
Murphree point efficiency, 975
MV, *see* Manipulated variable
MVC, *see* Multivariable control

N

National Chemical Laboratory (NCL), 945
National Institute of Standards and Technology (NIST), 28
Natural convection, single-phase heat transfer in, 520–523
    horizontal cylinder, 523
    horizontal plates, 522–523
    other geometries, 523
    two horizontal parallel plates, 523
    vertical plane surface, 521–522
Navier-Stokes equations, 618, 848
NCL, *see* National Chemical Laboratory
Nernst's law, 717, 725
Net positive suction head (NPSH), 1325
Newtonian fluids, 419–422
    all flow regimes, 420–422

Index

laminar flow, 419
rough pipe, 420
turbulent flow, 420
water in Sch 40 pipe, 422
NINA-FBR, *see* Nonisothermal nonadiabatic fixed-bed reactor
NINA reactor, gas-solid reaction in, 878–883
NIST, *see* National Institute of Standards and Technology
NIST, Thermodynamics Research Center, 28
Nonisothermal nonadiabatic fixed-bed reactor (NINA-FBR), 813
Nonlinear equations, 85–92
   numerical solutions, 88–92
      convergence condition, 90–91
      Newton-Raphson iteration, 91–92
      successive substitution, 88–90
   polynomial equations, 85–88
      bounds on real roots, 86
      Budan's rule of signs, 86
      Cardano's formula, 87
      Descartes's rule of signs, 86
      quadratic formula, 86
      quartic formula, 88
      Routh-Hurwitz criterion, 85
   processes governed by, 45–47
      concentration of species in chemical reaction at equilibrium, 45–46
      steady state of continuous stirred-tank reactor, 47
      vapor-liquid equilibria, 46–47
Nonrandom two-liquids (NRTL) equation, 338–339
NPSH, *see* Net positive suction head
NRTL equation, *see* Nonrandom two-liquids equation
Nuclear power generation, 1496–1497
   Chernobyl, 1496–1497
   hazardous waste disposal, 1496
   Three-Mile Island, 1496–1497
   VHTR reactors, 1497
Nucleate boiling, 532
Number system, 53–60
   algebraic inequalities, 58–59
   binomial theorem, 58
   comparison test, 56–57
   integral inequalities, 59–60
   integral test, 55–56
   limit comparison test, 57
   ratio test, 56
   real number system, 53–54
   root test, 56
   sequences and series, 54–55
   Taylor series, 57
   tests for convergence of sequence and series, 55–57
Nusselt number, 505, 764

O

Ohm's law, 1755, 1798
Ordinary differential equations, 47, 101–115
   linear first-order differential equation, 101
   linear higher-order differential equations, 111–115
      application of higher-order equations, 112–114
      finite element method, 114–115
      homogeneous equation, 111–112
      inhomogeneous equation, 112
   nonlinear first-order differential equation, 101–105
      autonomous nonlinear equation, 101
      Bernoulli equation, 103–104
      exact differential equations, 104
      homogeneous equations, 104–105
      implicit equation, 101–102
      Lagrange equation, 102–103
      Ricatti equations, 105
      separable equation, 103
   numerical solutions to, 94–100
      Adams-Bashforth methods, 99
      Adams-Moulton methods, 100
      explicit methods, 96–99
      implicit methods, 99–100
      Runge-Kutta methods, 96–98
      trapezoidal, Simpson, and Runge-Kutta, 99–100
   processes governed by, 47–48
      dynamics of continuous stirred tank reactor, 47
      steady state of tubular reactor, 48
   second-order differential equations, 106–111
      Green's function, 108–110
      Green's function by eigenfunction (Mercer's) expansions, 110–111
      homogeneous linear equations with constant coefficients, 106
      inhomogeneous linear differential equations with constant coefficients, 107–108
Organic compound, hydrogenation of, 934–943
Organic electrochemical reactions, examples of, 1779–1780
Orifice meter, 455–462
   applications, 462
   compressible flow, 460–461
   incompressible flow, 455–460
   loss coefficient, 462
Orthogonal curvilinear coordinate systems, 69–76
   bispherical coordinates, 76
   circular cylindrical coordinates, 71
   conical coordinates, 74
   differential operators in curvilinear coordinate system, 70–71
   ellipsoidal coordinates, 75
   elliptic cylindrical coordinates, 72
   oblate spheroidal coordinates, 73
   parabolic coordinates, 74
   parabolic cylinder coordinates, 73–74
   paraboloidal coordinates, 75–76
   prolate spheroidal coordinates, 72–73
   scale factors and metric tensors, 69–70
   spherical coordinates, 71–72
   toroidal coordinates, 76
OSHA listed chemicals, 1470, 1471–1477
OSHA listed hazardous materials, fire-hazard properties of, 1478
Overpotentials, 1764
Oxydesulfurization of coal, 919
Ozone, oxidation of NO using, 869–870

P

Partial differential equations, 115–131
    classification of second-order equations, 118–119
    computational fluid mechanics, 131
    elliptic equations, 128–131
        Green's function, 130–131
        Poisson integral formula, 129–130
    first-order partial differential equations, 115–118
    hyperbolic equations, 124–128
        d'Alembert's solution, 124–126
        separation of variables, 126–128
    parabolic equations, 119–124
        inhomogeneous boundary conditions, 122
        inhomogeneous equation, Duhamel's principle, 120–121
        inhomogeneous equation in infinite domain, 123–124
        separation of variables, 119–120
        similarity solutions, 122–123
    processes governed by, 48–50
        dynamics of chromatography, 49–50
        dynamics of tubular reactor, 48–49
Patents and intellectual property, 1831–1839
    intellectual property between employer and employee, 1838–1839
        patents, 1838
        rights and obligations created or eliminated by express agreement, 1839
        trade secrets, 1838–1839
    patents outside United States, 1834
    preliminary steps for inventor, 1835–1837
        minimizing chances for invalidation, 1837
        patents as prior art, 1836
        prior art searching, 1835–1836
        record keeping, 1836
        representation before USPTO, 1835
    requirements for patentability, 1834–1835
        nonobviousness, 1835
        novelty, 1834–1835
        utility, 1835
    trade secrets, 1837–1838
    United States patents and related rights and documents, 1832–1834
        copyrights, 1833–1834
        design patent, 1832
        provisional patent application, 1832–1833
        trademarks, 1833
        utility patent, 1832
PBT, see Profiled bottom tank
PC, see Pipeline contactor
PCOR equation, see Polymer chain-of-rotator equation
PCS, see Principle of corresponding states
Pearlite, 1547
Peclet number, 506–507, 855
Peng–Robinson (PR) equation, 11, 299
Peng–Robinson–Stryjek–Vera (PRSV) eos, 349
Perturbed hard-chain theory (PHCT), 301
PFDs, see Process flow drawings
PFR, see Plug-flow reactor
PFTR, see Plug flow tubular reactor
PHCT, see Perturbed hard-chain theory
Physical and chemical properties, 1–34
    apparatus calibration, 21
    aqueous electrolyte solutions, 17–20
        density and enthalpy, 18–19
        transport properties, 19–20
        vapor-liquid equilibria and activity coefficients, 17–18
    binary interaction parameter, 11
    Burnett method, 22
    calorimetry, 22
    combining rule, 11
    corresponding-states methods, 7
    COSMO, 12
    differential ebulliometry, 24
    ebulliometry, 23
    electrolyte-NRTL model, 18
    equation of state, 11
    experience, 21
    flow calorimetry, 23
    fugacity coefficient, 12
    gas saturation method, 24
    Gibbs energy, 12
    group-contribution methods, 7, 20
    Hagen-Poiseuille relationship, 25
    Helgeson–Kirkham–Flowers correlation, 19
    Helmholtz energy, 12
    Henry's law, 13
    ideal-gas law, 9
    ideal mixture volume, 9
    ionic strength, 19
    isochors, 22
    Knudsen method, 24
    liquid-liquid equilibrium , 11
    major data sources, 27–30
        Beilstein, 29–30
        DDB, 29
        DECHEMA, 29
        DIPPR, 28
        Gmelin, 30
        Landolt-Börnstein, 29
        NEL, 29
        NIST, 28
        process simulation software, 30
    measurement of fluid thermophysical properties, 20–27
        density, 22
        electrolyte solutions, 27
        general considerations, 21–22
        heat capacity and caloric properties, 22–23
        liquid–liquid equilibria, 25
        mixture vapor-liquid equilibria, 24–25
        pure-component vapor pressure, 23–24
        thermal conductivity, 26–27
        viscosity, 25–26
        when experiments are necessary, 20–21
    model-substance approach, 18
    oscillating-cup method, 26
    Peng–Robinson equation, 11
    phase equilibria for mixtures, 10–14
        activity-coefficient methods, 12–13
        equation-of-state methods, 11–12

Index

method choice, 13–14
sources of data, 14
types of phase-equilibrium calculations, 10–11
potentiometry, 27
Poynting factor, 12
properties for chemical reaction equilibria, 20
pycnometers, 22
Raoult's law, 13
safety, 21
second virial coefficient, 9
Soave–Redlich–Kwong equation, 11
standard-state volume, 18
static calorimetry, 22
statistical associating fluid theory, 11
SUPCRT92, 19
SUPERTRAPP model, 17
thermodynamic properties of pure fluids, 3–8
    critical constants and acentric factors for pure fluids, 8
    ideal-gas properties, 8
    importance of pure-fluid properties, 3
    pure fluids with little or no data, 7–8
    pure fluids with moderate amounts of data, 5–7
    pure fluids with reference-quality data, 5
    relative importance of different properties, 3
    water and steam, 3–5
thermodynamic properties of saturated water and steam, 4–5
thermodynamic properties of single-phase mixtures, 8–10
    caloric properties, 10
    density, 8–10
transient hot-wire method, 26
transport properties, 14–17
    diffusivity, 17
    kinetic theory, 14–15
    thermal conductivity, 16–17
    viscosity, 15–16
uncertainty, 2
UNIFAC, 12
vapor-liquid equilibrium, 10
vibrating-tube densimeters, 22
Vogel–Tammann–Fulcher equation, 15
VTPR, 12
Wagner equations, 6
Physical properties, measurement of, 1531–1537
  flowmeters, 1535–1536
  level measurements, 1537
  pressure measurements, 1536
  temperature-measuring devices, 1531–1535
    bi-metallic thermometers, 1535
    filled-bulb and glass-stem thermometers, 1534
    pyrometric cones, 1535
    radiation and infrared pyrometers, 1533–1534
    resistance thermometers and thermistors, 1532–1533
    thermcouples, 1532
PID control, advanced, 1227–1246
  antiwindup strategies, 1240–1241
  bumpless transfer, 1241–1242
  cascade control, 1227–1228
  computed manipulated variable control, 1239–1240
  feedforward control, 1230–12333
  inferential control, 1233–1236
  MIMO process control, 1242–1246
  override/select control, 1238–1239
  ratio control, 1229–1230
  scheduling controller tuning, 1236–1238
  split-range flow control, 1242
PID controllers, 1201–1213
  algorithms, 1201–1206
  analysis of P, I, and D action, 1206–1207
    derivative action, 1207
    integral action, 1206–1207
    proportional action, 1206
  analysis of typical control loops, 1209–1213
    composition control loop, 1212–1213
    flow control loop, 1209–1210
    level control loop, 1210–1211
    pressure control loop, 1211
    temperature control loop, 1211–1212
  controller design issues, 1208–1209
    PI control, 1208
    PID control, 1208–1209
    P-only control, 1208
PID tuning, 1213–1227
  controller reliability, 1218–1219
  effect of tuning parameters on dynamic behavior, 1214–1218
    control interval, 1217–1218
    PI control, 1214–1215
    PID control, 1216–1217
  fast-response loops, 1221–1222
  level controller tuning, 1225–1227
  recommended approach to controller tuning, 1218
  selection of tuning criterion, 1219–1220
  slow-response processes, 1222–1225
  tuning criteria and performance assessment, 1213–1214
  tuning of filter on sensor readings, 1220–1221
Pipe flow, 419–443
  analysis, 431–433
  Bingham plastic fluids, 426–428
  compressible flows, 438–443
  economical diameter, 433–435
  fitting losses, 428–431
  flow regimes, 419
  Newtonian fluids, 419–422
  noncircular conduits, 435–437
  power law fluids, 422–426
  pressure-flow relations, 419
  turbulent drag reduction, 437–438
Pipeline contactor (PC), 910
Pitch blade turbines, 623
PLCs, see Programmable logic controllers
Plug-flow reactor (PFR), 751, 954
Plug flow tubular reactor (PFTR), 1410
  bubble reactor, 1415–1416
  empty tube, 1411–1412
  fire tube, 1412
  fixed bed catalyst in tube or vessel, 1412–1413
  fixed bed with radial flow, 1414
  melting cyclone burner, 1420

monolithic, 1418–1418
motionless mixer in tube, 1415
multibed adiabatic with interbed quench, 1414
multiple hearth, 1419
multitube fixed bed catalyst, 1414
packing, 1417–1418
rotary kiln, 1420
shaft furnace, 1420
spray reactor and jet nozzle reactor, 1416
thin film, 1419
transported or slurry, transfer line, 1414–1415
traveling grate, 1419
trays, 1417
trickle bed, 1418
via multistage CSTR, 1420–1421
Poisson equation, 128, 130
Polymer chain-of-rotator (PCOR) equation, 306
Polymers, 1132
Polynomial equations, 85–88
    bounds on real roots, 86
    Budan's rule of signs, 86
    Cardano's formula, 87
    Descartes's rule of signs, 86
    quadratic formula, 86
    quartic formula, 88
    Routh-Hurwitz criterion, 85
Population balance equations, 52
Potentiometry, 27
Pourbaix diagrams, 1807
Powder materials, see Granular and powder materials, dry screening of
Power law fluids, 422–426
    all flow regimes, 426
    laminar flow, 422
Poynting factor, 12, 325
Prandtl number, 506
PR equation, see Peng–Robinson equation
Pressure swing adsorption (PSA), 1122
Prigogine–Flory–Patterson theory of polymer liquids, 302
Principle of corresponding states (PCS), 287
Probability distributions, 201–202, 203–212
    binomial distribution (discrete variable), 204
    characteristic parameters of, 202–203
    chi-square distribution for sample variance (continuous variable), 210–211
    continuous probability density distributions, 201–202
    discrete probability distributions, 201
    F distribution for ratio of sample variances (continuous variable), 211
    geometric distribution (discrete variable), 205
    normal or Gaussian distribution (continuous variable), 206–207
    Poisson distribution (discrete variable), 205–206
    t distribution of sample means (continuous variable), 207–210
Process control, 1173–1265
    adaptive control techniques, 1238
    adjustable-speed pumps, 1191
    advanced PID control, 1227–1246
        antiwindup strategies, 1240–1241
        bumpless transfer, 1241–1242
        cascade control, 1227–1228
        computed manipulated variable control, 1239–1240
        feedforward control, 1230–12333
        inferential control, 1233–1236
        MIMO process control, 1242–1246
        override/select control, 1238–1239
        ratio control, 1229–1230
        scheduling controller tuning, 1236–1238
        split-range flow control, 1242
    ATV tests, 1223, 1246
    boiler drum level control, 1230
    booster relays, 1191
    cavitation, 1190
    closed-loop dynamic behavior, 1181–1182
    comparison of feedback and feedforward control, 1232
    controller cycle time, 1186
    controller/DCS system problems, 1199
    control loop hardware and troubleshooting, 1182–1200
        actuator systems (final control elements), 1187–1191
        distributed control system, 1184–1187
        sensor systems, 1191–1194
        troubleshooting control loops, 1194–1200
    coupled process, 1242
    decentralized control, 1242–1243
    definition of symbols for control diagrams, 1176
    derivative kick, 1202
    direct-acting controller, 1201
    direct-acting final control element, 1190
    distillation column, bottoms composition control of, 1228
    error from setpoint, 1177
    finite control element, common problems with, 1196
    first-order plus deadtime model, 1179–1181
    FOPDT model, 1181
    general dynamic behavior, 1177–1179
    generalized feedback system, 1177
    integral windup, 1240
    interactive PID controller, 1205
    internal reflux control, 1240
    model predictive control, 1246–1264
        description, 1246–1247
        general method for designing multivariable controller, 1249–1261
        history, 1248
        multivariable controller troubleshooting, 1261–1264
        when to use, 1248–1249
    move suppression, 1260
    multiple-input/multiple-output process, 1242
    nonstationary process, 1237
    PID algorithm, 1205
    PID controllers, 1201–1213
        algorithms, 1201–1206
        analysis of P, I, and D action, 1206–1207
        analysis of typical control loops, 1209–1213
        controller design issues, 1208–1209
    PID tuning, 1213–1227
        controller reliability, 1218–1219
        effect of tuning parameters on dynamic behavior, 1214–1218

Index

fast-response loops, 1221–1222
level controller tuning, 1225–1227
recommended approach to controller tuning, 1218
selection of tuning criterion, 1219–1220
slow-response processes, 1222–1225
tuning criteria and performance assessment, 1213–1214
tuning of filter on sensor readings, 1220–1221
quarter-amplitude damping, 1214
rate before reset controller, 1205
reference trajectory, 1260
relay feedback experiment, 1223
reset windup, 1240
reverse-acting controller, 1201
reverse-acting final control element, 1190
ringing, 1181
saturated control valve, 1240
self-regulating process, 1179
self-tuning controllers, 1238
sensor system, common problems, 1198
state of the plant, 1250
stationary process, 1237
underdampened response, 1179
valve deadband, 1190
wastewater neutralization process, 1229
watchdog timer, 1260
Zeigler–Nichols tuning, 1246
Process design, *see* Conceptual process design, process improvement, and troubleshooting
Process flow drawings (PFDs), 1249
Process Safety Management (PSM), 1439
Profiled bottom tank, 843
Programmable logic controllers (PLCs), 1186, 1246
PRSV eos, *see* Peng-Robinson-Stryjek-Vera eos
PSA, *see* Pressure swing adsorption
Pseudo-steady-state (PSS) assumption, 774
PSM, *see* Process Safety Management
PSS assumption, *see* Pseudo-steady-state assumption

Q

QAD, *see* Quarter-amplitude damping
Quarter-amplitude damping (QAD), 1214

R

Radial flow fixed-bed reactor (RF-FBR), 813
Radial-flow turbines (RFT), 623
Radiation heat transfer, 567–589
  bidirectional reflectance distribution function, 574
  blackbody band fraction, 570
  dependent scattering, 586
  diffuse emitter, 569
  discrete ordinates method, 584
  gas and particle radiation, 583–586
    radiative properties of gases and particles, 585–586
    radiative transfer equation, 583–585
  Kirchhoff's laws, 575
  Maxwell's electromagnetic wave equations, 583
  Mie scattering theory, 586
  monochromatic radiation, 568
  radiation thermometry, 586–588
  radiative energy exchange between surfaces, 575–582
    net radiation method, 579–580
    network representation, 580–582
    radiation exchange between blackbody surfaces, 576–578
    radiosity, 575–576
    view factor, 576
  radiative properties of solids, 570–575
    absorptance, reflectance, and transmittance, 573–574
    emissivity, 570–573
    Kirchhoff's law, 574–575
  radiative transitions, 585
  Rayleigh-Jeans formula, 570
  reradiating surface, 581
  scattering phase function, 583
  single scattering albedo, 583
  Stefan-Boltzmann constant, 569
  surface emission, 568–570
    blackbody, 569–570
    intensity, emissive power, and irradiation, 568–569
  third radiation constant, 570
  vibration-rotation bands, 585
  Wien's displacement law, 570
  zonal method, 584
Radiative transfer equation (RTE), 583
Random-pore models, 773, 783
Raoult's law, 13, 325, 975, 977
Rayleigh-Jeans formula, 570
RBC, *see* Rotating biological contactors
RCI, *see* Retreat-curve impellers
RCRA, *see* Resource Conservation and Recovery Act
Reaction analysis, 739
Reaction injection-molding systems (RIM), 650
Reactive distillation, 1005–1006
Reactor, *see also* Plug flow tubular reactor
  adiabatic fixed-bed reactor, 813
  air-lift, 808, 811
  analysis, 739
  Berty, 770
  bubbling-bed reactor, 826
  catalytic wire-gauze, 820
  Choudhary–Doraiswamy, 770
  continuous, design of, 949
  continuous stirred tank reactor, 41, 751
  dead-end, 840
  electrochemical, 1766–1774
    Akzo fluidized bed, 1773
    batch tank reactors, 1771
    characterization, 1766–1769
    choice, 1769–1774
    design experiments, 1766, 1770
    electrode materials, 1768
  external-loop air-lift reactor, 907
  fluid-fluid-reactions, 788
  gradientless, 739
  helical coil, 859
  ideal, 741

jet-loop, 847
kinetic energy (stirred tank reactors), 839–848
    blending, 841, 845
    blending in gas-liquid systems, 844–845
    dead end systems, 846–848
    gas dispersion, 841–843
    heat transfer, 844
    heat transfer in gas-liquid systems, 846
    solid suspension, 843–844
    suspension in gas-liquid-solid systems, 845–846
mixed-flow, 751
moving-bed, 837
multiphase, 741
multitubular, 860
NINA, 879
nonisothermal nonadiabatic fixed-bed reactor, 813
pressure energy, 800–839
    gas-liquid reactors, 800–812
    gas-solid catalytic fixed-bed reactors, 813
    gas-solid catalytic fluidized-bed reactors, 821–835
    gas-solid noncatalytic reactors, 835–839
    liquid-liquid reactors, 812
    solid-liquid reactors, 812–813, 821
radial-flow, 819
radial flow fixed-bed reactor, 813
simulated moving-bed, 740
single-pellet, 770
SPE oil hydrogenation, 1786, 1787
staged, 828
steady-state adiabatic, 386
stirred-tank, 45
three-phase catalytic reactor, 936
three-phase sparged, 921
tubular
    dynamics of, 48
    steady-state multiplicity, 174
turbulent-bed reactor, 826
Real number system, 53–54
    logarithm, 54
    powers and roots, 53–54
Redlich–Kister equation, 330
Redlich and Kwong (RK) equation, 298, 1342
Regularization, method of, 142
Relay feedback experiment, 1223
Resistance temperature detectors (RTD), 1192, 1351
Resource Conservation and Recovery Act (RCRA), 1327, 1499
Retreat-curve impellers (RCI), 623
Retrograde condensation, 292
Reynolds number, 80, 504–505, 518, 689, 764, 864, 912
RF-FBR, see Radial flow fixed-bed reactor
RFT, see Radial-flow turbines
Ricatti equations, 105
RIM, see Reaction injection-molding systems
RK equation, see Redlich and Kwong equation
Robin boundary condition, 129
Robin problem, 118
Rotating biological contactors (RBC), 848
Rouché theorem, 151–152
RTD, see Resistance temperature detectors
RTE, see Radiative transfer equation

S

Safety, see Chemical process safety
Safety interlock system (SIS), 1351
SAFT, see Statistical associating fluid theory
Sales, administration, research, and engineering (SARE) expenses, 1287
SAP catalyst, see Supported aqueous phase catalyst
SARE expenses, see Sales, administration, research, and engineering expenses
SBC, see Sectionalized bubble column
Scalar difference equations, 42
Schmidt number, 1761
SCM, see Shrinking core model
Sectionalized bubble column (SBC), 905
Sensor systems, 1191–1194
    chemical composition analyzers, 1193
    flow measurements, 1193
    level measurements, 1193
    pressure measurements, 1193
    repeatability, accuracy, and dynamic response, 1192
    sampling system, 1193
    temperature measurements, 1192
    thermowells, 1192
    transmitters, 1194
Separations Research Program (SRP) model, 1061
Sharp interface model (SIM), 770
Sherwood number, chemical reaction engineering, 764
Shock wave velocity, 1146
Shrinking core model (SCM), 770, 774
Sieder-tate term, 507
Silicas, 1130–1131
SIM, see Sharp interface model
Simulated moving-bed reactors, 740
Simulated moving-bed systems, 1122
Single-pellet reactor, 770
SIS, see Safety interlock system
Sisko model, 401
SLE, see Solid-liquid extraction
SLS, see Solid/liquid separation
Slugging, 833
Slurry reactor, oxydesulfurization of coal in, 919–925
Soave–Redlich–Kwong (SRK) equation, 11
Software
    CFD, 618, 632
    design, process improvement, and troubleshooting, 1334–1335
    mathematical, 182–183
    MPC modeling software, 1257
    security, process safety management, 1467
    solvent extraction, 729
Solid(s)
    bulk, conveying of, 1729–1736
        air-activated gravity conveyor, 1735
        belt conveyors, 1730–1733
        bucket elevators, 1734
        en-masse conveyors, 1734–1735
        materials characterization, 1729–1730
        pneumatic conveyors, 1735
        screw conveyors, 1733
        vibrating conveyors, 1734

floating, slurries of, 657
-liquid extraction (SLE), 709
-liquid mass transfer coefficient, 930
-liquid reactions, examples, 797
radiative properties, 570–575
    absorptance, 573–574
    emissivity, 570–573
    Kirchhoff's law, 574–575
    reflectance, 573–574
    transmittance, 573–574
Solid-liquid mixing, 653–660
    equipment, 659–660
    floating solids, 657
    just-suspended conditions, 656
    particle suspension in stirred vessels, 655
    power requirements, 659
    settling solids, 653–655
    solids suspension by jet mixing, 660
    uniform solids concentrations, 657–659
Solid/liquid separation (SLS), 1597–1665
    cake washing, 1648–1650
        displaced washing and repulping, 1648
        terms and definitions for displacement washing, 1648–1649
        washing curves, 1649–1650
    classification of operations by particle size, 1599
    equipment and operation, 1629–1648
        bag filters, 1633–1634
        batch pressure filters, 1630–1636
        candle filter, 1632–1633
        cartridge filters, 1634
        centrifugation, 1645–1646
        continuous filters, 1636–1640
        cross-flow filters, 1641
        deep-bed filter, 1640–1641
        expression equipment, 1647–1648
        filter presses, 1634–1636
        horizontal belt filter, 1639
        hydrocyclone, 1646–1647
        indexing belt filter, 1639–1640
        leaf filter, 1631–1632
        membrane filters, 1641–1642
        pressure nutsches, 1630–1631
        rotary drum and disc filters, 1636–1639
        thickening and clarification, 1643–1644
    filtration fundamentals, 1613–1629
        critical pressure drop, 1628–1629
        data analysis, 1616–1620
        deviations from parabolic theory, 1620–1623
        filtration of super-compactible materials, 1623–1629
        material balance, 1614–1616
        strange behavior of super-compactible materials, 1627–1628
        theory, 1613–1614
            volume vs. time, 1616
    laboratory test, 1650–1654
        adverse effect of sedimentation on filtration experiments, 1653
        problems, 1653
        samples, 1653
        test numbers, 1653–1654
        wall friction in C-P cell and small-scale cells, 1653
    operations, 1599–1602
        cake filtration, 1600
        centrifugation, 1601
        cross-flow filtration, 1600–1601
        deep-bed filtration, 1601
        deliquoring, 1601
        flotation, 1601–1602
        hydrocyclone separation, 1601
        membrane filtration, 1601
        sedimentation, 1599
        straining, 1600
        washing, 1601
    overview, 1599–1603
    particles, 1599
    pretreatment (coagulation and flocculation), 1603–1608
        coagulation and flocculation, 1604–1607
        interparticle forces and zeta potential, 1604
        laboratory tests, 1607–1608
    pretreatment (filter aids), 1608–1613
        diatomaceous earth, 1611–1612
        other filter aids, 1613
        perlite, 1612
    stages, 1602–1603
    system design, 1654–1664
        cycle analysis in selection of filter aids, 1661–1663
        decision network for design for cake filtration systems, 1656–1658
        equipment selection, 1654–1565
        filter media selection, 1658–1660
        pump selection in filtration operation, 1660–1661
        scale up, 1663–1664
Solvent extraction services, Internet sites, 729
SOP catalyst, *see* Supported organic phase catalyst
Souders–Brown capacity parameter, 1012
SPCC Act, *see* Spill Prevention Control and Countermeasures Act
Spill Prevention Control and Countermeasures (SPCC) Act, 1499
SRK equation, *see* Soav–Redlich–Kwong equation
SRP model, *see* Separations Research Program model
Staged reactors, 828
Stanton number, 506
Statistical associating fluid theory (SAFT), 11, 308
Steady-state solution, stability of, 179–181
    linear stability analysis, 179–180
    method of Lyapunov's function, 180–181
Steam distillation, 1002
Steam stripping, 1107–1111
Stirred-tank reactor, 45
    kinetic energy, 839–848
        blending, 841
        blending in gas-liquid systems, 844–845
        blending in solid-liquid systems, 845
        dead end systems, 846–848
        gas dispersion, 841–843
        heat transfer, 844, 846
        solid suspension, 843–844
        suspension in gas-liquid-solid systems, 845–846
    linear equations, 45

Stochastic differential equations, 42, 166–170
    connection between Fokker-Planck equation and, 167–168
    differential Chapman-Kolmogorov equation, 167
    Îto stochastic integral, 168–170
        application, 169–170
        one-dimensional Îto formula, 169
    processes governed by, 52–53
Stochastic processes, 166
Stokes's law, 654
Stripping operating line, 986
Sublimation pressure, 284, 372
Sublimation pressure curve, 296
SUPCRT92, 19
Super-compactible materials, filtration of, 1623–1629
    Darcy's law, 1625–1626
    effective pressure, 1623–1625
    empirical constitutive equations, 1626
    integration of Darcy equation, 1626–1627
Superheated vapor, condensation of, 530–531
SUPERTRAPP model, 17
Supported aqueous phase (SAP) catalyst, 754
Supported organic phase (SOP) catalyst, 756

T

Tafel equation, 1751
TBP, *see* Tri-*n*-butyl phosphate
TCs, *see* Thermocouples
Temperature-measuring devices, 1531–1535
    bi-metallic thermometers, 1535
    filled-bulb and glass-stem thermometers, 1534
    pyrometric cones, 1535
    radiation and infrared pyrometers, 1533–1534
    resistance thermometers and thermistors, 1532–1533
    thermcouples, 1532
Temperature swing adsorption (TSA), 1122, 1162–1163
Tetrachloroethane chlorinates, 871
The Natural Step (TNS), 1322
Thermal energy
    direct contact gas-liquid condensers, 1366
    direct contact gas-liquid cooling towers, 1366
    direct contact gas-liquid quenchers, 1366
    direct contact gas-solid fluidized beds, 1365
    direct contact gas-solid kilns, 1363
    direct contact gas-solid multiple-hearth furnaces, 1365
    direct contact liquid-liquid immiscible liquids, 1364
    fluid heat exchangers, 1361
    fluidized bed, 1363
    furnaces, 1360
    gas-gas thermal wheels, 1367
    gas-solid drying of solids, 1365
    heat loss to atmosphere, 1367
    motionless mixers, 1364
    refrigeration, 1367
    solidify liquids, 1367
    steam generation, 1368
Thermocouples (TCs), 1192
Thermodynamics, fluid phase and chemical equilibria, 255–392
    absolute temperature, 267

acentric factor, 288
adiabatic system, 257
analytical solution of groups method, 345
athermal solutions, 334
azeotrope, 292
Boublik–Alder–Chen–Kreglewski eos, 301
bubble-point curve, 291
Carnot cycle, 265
chain-of-rotators equation, 302
chemical reaction equilibria, 375–391
    chemical reaction equilibrium, 375–376
    equilibrium constants, 376–384
    open systems with reaction, 385–388
    phase rule for chemically reacting species, 384–385
    stoichiometric formulation, 388–391
Clapeyron equation, 352
compressed liquid states, 355
degree of freedom, 290
dew-point curve, 291
dimer theory, 306
energy functions, 276
enthalpy balance, 261
entropy, 264
equilibrium ratio, 355
extensive property of system, 257
first law of open systems, 261
fluid-phase equilibria, 351–375
    gas-solid equilibrium models, 372–374
    liquid-liquid equilibrium models, 367–372
    vapor-liquid equilibrium of ideal mixtures, 355–358
    vapor-liquid equilibrium by ϕ-ϕ models, 364–367
    vapor-liquid equilibrium by -ϕ models, 358–364
    vapor-liquid equilibrium in single-component fluid, 351–355
free-energy-matching mixing rules, 349
free volume, 296
fugacity, 279
fugacity coefficient, 322
function of temperature, 377
gas-liquid critical state, 285
Gauss-Jordan elimination, 390
Gibbs-Duhem equation, 281, 283
Gibbs phase rule, 385
Hang–Chao–Hilson complete local-composition model, 371
heat exchangers, 262
Helmholtz free energy, 273
ideal gas definition, 313
ideal-gas equation, 258
ideal-solution law, 325
intensive property of system, 257
isobaric process, 257
isochores, 286
Lewis fugacity rule, 355
light component, 355
liquid solutions, 325–351
    activity coefficient, 363
    activity-coefficient models, 329–343
    Gibbs energy models of liquid solutions, 347–350
    group contribution methods, 343–346
    ideal and excess solution properties, 328–329

Index

ideal and real solutions, 325–328
mechanical instability, 274
model dumbbell rotator, 302
nonstoichiometric formulation, 375
normal fluids, 288
packing fraction, 300, 306
perturbed hard-chain theory, 301
polymer chain-of-rotator equation, 306
Poynting correction factor, 325
Prigogine–Flory–Patterson theory of polymer liquids, 302
principles of thermodynamics, 256–284
    equilibrium energy functions, 268–270
    first law of thermodynamics and internal energy, 258–264
    open system and chemical potential, 277–280
    partial molar quantities and Gibbs-Duhem relation, 280–284
    second law of thermodynamics and entropy, 264–268
    temperature and ideal gas, 257–258
process, 257
Raoult's law, 325
reduced dipole moment, 311
residual functions, 258
retrograde condensation, 292
saturated liquid state, 355
state of equilibrium phase, 257
statistical-associated fluid theory, 308
stoichiometric formulation, 376
sublimation equilibrium, 372
sublimation pressure, 284, 372
sublimation pressure curve, 296
supercritical extraction, 372, 374
surroundings, 256
system, 256
thermodynamic temperature, 265
throttle, 262
UNIQUAC equation, 343, 369
upper critical solution temperature, 294
van der Waals one-fluid mixing rules, 297
vapor pressure curve, 286
volumetric and thermodynamic properties, 284–324
    energy functions of ideal gases and mixtures, 313–316
    equations of state, 295–313
    fugacity, 321–324
    phase behavior of mixtures, 291–294
    phase rule, 290–291
    pressure–volume–temperature relationship, 284–287
    principle of corresponding states, 287–290
    residual functions and energy functions from equations of state, 317–321
work expansion of steam, 272
Thiele–Geddes model, 993
Three-Mile Island, 1496–1497
TNS, *see* The Natural Step
Tortuosity factor, 599
Toxic Substances Control Act (TSCA), 1498
Transition boiling regime, 532

Tri-n-butyl phosphate (TBP), 714, 718
    reaction models, 714
    solvent cleanup, 715
TSA, *see* Temperature swing adsorption
TSCA, *see* Toxic Substances Control Act
Turbulent-bed reactor, 826
Two-zone model, 778

U

UCST, *see* Upper critical solution temperature
Uncertainty, estimation of, 241
UNIFAC model, 717, 981
UNIQUAC equation, 343, 369
UNIQUAC model, 1342
Upper critical solution temperature (UCST), 294
Uranium, distribution coeffieicnt, 715

V

Vacancy Solution model, 1140
Valve trays, 1021
van der Waals (vdW) equation, 295, 330
van der Waals one-fluid mixing rules, 297
van Laar equation, 330–332
Vaporization
    natural and forced convection, 533
    special cases in, 535–536
        boiling outside tube bundles, 535
        enhanced surfaces in boiling, 536
        subcooled boiling, 536
Vapor-liquid equilibrium (VLE), 10, 358, 364
Vapor lock, 447
Vapor pressure curve, 286
VDUs, *see* Video display units
vdW equation, *see* van der Waals equation
Vector analysis, 66–78
    gradients of sum and product, 77–78
    orthogonal curvilinear coordinate systems, 69–76
    vector algebra, 66–68
    vector calculus, 68
    vector integral theorems, 76–77
Vector difference equations, 43
Video display units (VDUs), 1184
View factor, radiation heat transfer and, 576
VLE, *see* Vapor-liquid equilibrium
VOCs, *see* Volatile organic compounds
Vogel–Tammann–Fulcher equation, 15
Volatile organic compounds (VOCs), 1076
Volterra equation, 42, 132, 136
Volterra equations, methods of solution for, 132–136
    degenerate (finite rank) kernels, 133
    difference kernels, 133–134
    generalized Abel integral equation, 135
    kernel as function of dependent variable, 132–133
    method of resolvent kernels, 134
    numerical solution of Volterra integral equations of second kind, 135–136
    separable kernels, 133

Volterra integral equations, 131–132
Volume reaction model, 776
VTPR, 12

W

Weber number, 674, 689
Wegstein acceleration, 1339
Wet corrosion, 1805
Weymouth equation, 440

Wheeler–Robell equation, 1146
White noise, 53
Wien's displacement law, 570
Wilson's local-composition equation, 336–338

Z

Zeigler–Nichols tuning, 1246
Zeolites, 1131